SYSTEMIC LUPUS ERYTHEMATOSUS

FIFTH EDITION

Editor

ROBERT G. LAHITA MD, PhD

*Professor of Medicine, Adjunct Professor
of Biochemistry and Molecular Biology
UMDNJ, New Jersey Medical School
Vice President and Chairman of Medicine
Newark Beth Israel Medical Center
Newark, NJ, USA*

Associate Editors

GEORGE TSOKOS MD

*Department of Medicine
Beth Israel Deaconess Medical Center
Boston, MA, USA*

JILL P. BUYON MD

*NYU Medical Center
Tisch Hospital
New York, NY, USA*

TAKAO KOIKE MD, PhD

*Department of Medicine II
Hokkaido University School of Medicine
Sapporo, Japan*

ELSEVIER

AMSTERDAM • BOSTON • HEIDELBERG • LONDON • NEW YORK • OXFORD
PARIS • SAN DIEGO • SAN FRANCISCO • SINGAPORE • SYDNEY • TOKYO

Academic Press is an imprint of Elsevier

Academic Press is an imprint of Elsevier
32 Jamestown Road, London NW1 7BY, UK
30 Corporate Drive, Suite 400, Burlington, MA 01803, USA
525 B Street, Suite 1800, San Diego, CA 92101-4495, USA

First Edition 1987
Second Edition 1992
Third Edition 1998
Fourth Edition 2004
Fifth Edition 2011

Notice
No responsibility is assumed by the publisher for any injury and/or damage to persons or property as a matter of products liability, negligence or otherwise, or from any use or operation of any methods, products, instructions or ideas contained in the material herein. Because of rapid advances in the medical sciences, in particular, independent verification of diagnoses and drug dosages should be made

Medicine is an ever-changing field. Standard safety precautions must be followed, but as new research and clinical experience broaden our knowledge, changes in treatment and drug therapy may become necessary or appropriate. Readers are advised to check the most current product information provided by the manufacturer of each drug to be administered to verify the recommended dose, the method and duration of administrations, and contraindications. It is the responsibility of the treating physician, relying on experience and knowledge of the patient, to determine dosages and the best treatment for each individual patient. Neither the publisher nor the authors assume any liability for any injury and/or damage to persons or property arising from this publication.

British Library Cataloguing-in-Publication Data
A catalogue record for this book is available from the British Library

Library of Congress Cataloging-in-Publication Data
A catalog record for this book is available from the Library of Congress

ISBN: 978-0-12-374994-9

For information on all Academic Press publications visit our website at www.elsevierdirect.com

Typeset by TNQ Books and Journals

Printed and bound in China

10 11 12 13 10 9 8 7 6 5 4 3 2 1

Working together to grow
libraries in developing countries

www.elsevier.com | www.bookaid.org | www.sabre.org

ELSEVIER   BOOK AID International   Sabre Foundation

To All Lupus Patients and Their Families.

Table of Contents

I

BASIS OF DISEASE PATHOGENESIS

A. GENETICS

B. CELLULAR PATHOGENESIS

C. HUMORAL PATHOGENESIS

D. ENVIRONMENTAL ASPECTS OF PATHOGENESIS

E. MECHANISMS OF TISSUE INJURY

List of Contributors

Joseph M. Ahearn MD Lupus Center of Excellence, University of Pittsburgh Schools of the Health Sciences, Pittsburgh, PA, USA

Olga Amengual MD, PhD Department of Medicine II, Hokkaido University Graduate School of Medicine, Sapporo, Japan

Zahir Amoura MD Service de Médecine, Interne 2 CHU Pitié-Salpêtrière, Paris, France

Cynthia Aranow MD The Center for Autoimmune and Musculoskeletal Disease, The Feinstein Institute for Medical Research, North Shore—LIJ Health System, Manhasset, NY, USA

John P. Atkinson MD Department of Medicine/Division of Rheumatology, Washington University School of Medicine, St. Louis, MO, USA

Tatsuya Atsumi MD, PhD Department of Medicine II, Hokkaido University Graduate School of Medicine, Sapporo, Japan

Ingrid Avalos MD Beth Israel Deaconess Medical Center, Harvard Medical School, Boston, MA, USA

Dimitrios Balomenos MD Department of Immunology and Oncology, Centro Nacional de Biotecnología/CSIC, Cantoblanco, Spain

Jacques Banchereau PhD Baylor Institute for Immunology Research, Dallas, TX, USA

George Bertsias MD Department of Medicine, University of Crete School of Medicine, Heraklion, Greece

Markus Böhm MD Department of Dermatology, University of Münster, Germany

Dimitrios T. Boumpas MD, FACP Clinical Immunology and Allergy, University of Crete School of Medicine, Heraklion, Greece

D. Ware Branch MD Department of Obstetrics and Gynecology H.A., University of Utah Health Sciences Center, Women and Newborns Clinical Program, Intermountain Healthcare, Salt Lake City, UT, USA

Hermine I. Brunner MD, MSc, FAAP, FACR Lupus Center Scientific, Cincinnati Children's Hospital Medical Center, Division of Rheumatology, Cincinnati, OH, USA

Jill P. Buyon MD New York University Medical Center, Tisch Hospital, New York, NY, USA

Livia Casciola-Rosen PhD Johns Hopkins University School of Medicine, Department of Medicine, Baltimore, MD, USA

Ricard Cervera MD, PhD, FRCP Department of Autoimmune Diseases Hospital Clínic, Barcelona, Catalonia, Spain

George Chamilos MD Department of Medicine, Infectious Diseases Unit, University of Crete School of Medicine, Heraklion, Greece

Edward K.L. Chan PhD University of Florida Health Science Center, Gainesville, FL, USA

Robert M. Clancy PhD New York University School of Medicine, New York, NY, USA

Christine A. Clark PhD Mount Sinai Hospital, LifeQuest Centre for Reproductive Medicine, Toronto, ON, Canada

Nathalie Costedoat-Chalumeau MD AP-HP, Service de Médecine Interne, Centre de référence national pour le Lupus et le syndrome des Antiphospholipides, Centre Hospitalier Universitaire Pitié-Salpêtrière, France

Maura Couto MD Rheumatology Department, Coimbra University Hospital, Praceta, Mota Pinto, Portugal

José C. Crispín MD Department of Medicine, Beth Israel Deaconess Medical Center, Harvard Medical School, Boston, MA, USA

Mary K. Crow MD Weill Cornell Medical College, Benjamin M. Rosen Chair in Autoimmunity and Inflammation Research, Hospital for Special Surgery, New York, NY, USA

Michael J. Day BSc, BVMS(Hons), PhD, DSc, DiplECVP, FASM, FRCPath, FRCVS School of Veterinary Science, University of Bristol, Langford, UK

Betty Diamond MD The Center for Autoimmune and Musculoskeletal Disease, The Feinstein Institute for Medical Research, North Shore—LIJ Health System, Manhasset, NY, USA

Iris Dotan MD Department of Gastroenterology and Liver Diseases, Tel Aviv Sourasky Medical Center, Tel Aviv, Israel

Roland M. du Bois MD National Jewish Health, Denver, CO, USA

Yong Du MD, PhD Departments of Internal Medicine (Rheumatology), University of Texas, Southwestern Medical School, Dallas, TX, USA

Catia Duarte MD Rheumatology Department, Coimbra Hospital University, Praceta, Mota Pinto, Portugal

Yun Deng MD, MS Division of Rheumatology, Department of Medicine, David Geffen School of Medicine, University of California, Los Angeles, CA, USA

Jan P. Dutz MD Department of Dermatology and Skin Science, University of British Columbia, Vancouver, Canada

Olga Dvorkina MD SUNY Downstate Medical Center, Brooklyn, New York, NY, USA

Thomas Ernandez MD Vascular Research Division, Department of Pathology, Brigham and Women's Hospital and Harvard Medical School, Boston, MA, USA

Gerard Espinosa MD, PhD Department of Autoimmune Diseases Hospital Clínic, Barcelona, Catalonia, Spain

A. Darise Farris PhD Oklahoma Medical Research Foundation, University of Oklahoma Health Sciences Center, Oklahoma City, OK, USA

Michelle M.A. Fernando MD Clinical Scientist, Faculty of Medicine, Section of Rheumatology, Imperial College London, London, UK

Barri J. Fessler MD, MSPH Division of Clinical Immunology and Rheumatology, University of Alabama, Birmingham, AL, USA

Aryeh Fischer MD Division of Rheumatology and Interstitial Lung Disease Program, National Jewish Health, Denver, CO, USA

Deborah M. Friedman MD, FAAP, FACC New York Medical College, NY, USA

Marvin J. Fritzler PhD, MD Department of Medicine, University of Calgary, Alberta, Canada

Richard Furie MD Albert Einstein College of Medicine, Division of Rheumatology and Allergy-Clinical Immunology, North Shore–Long Island Jewish Health System, Manhasset, NY, USA

Gary S. Gilkeson MD Medical University of South Carolina, Ralph H. Johnson VAMC, Charleston, SC, USA

Ellen M. Ginzler MD Department of Medicine, State University of New York Downstate Medical Center, Brooklyn, New York, NY, USA

John G. Hanly MD Capital District Health Authority and Dalhousie University, Canada

Evelyn V. Hess MD, MACP, MACR Division of Immunology and Allergy, Department of Internal Medicine, College of Medicine University of Cincinnati, OH, USA

Gary S. Hoffman MD, MS Harold C. Schott Professor of Rheumatic and Immunologic Diseases, Center for Vasculitis Care and Research, Lerner College of Medicine, Cleveland Clinic Foundation, Cleveland, OH, USA

Diane Horowitz MD Division of Rheumatology and Allergy-Clinical Immunology, North Shore–Long Island Jewish Health System, Manhasset, NY, USA

Luis Ines MD Rheumatology Department Coimbra University Hospital, Praceta, Mota Pinto, Portugal

Yiannis Ioannou MD Centre for Rheumatology, University College, London, UK

Judith A. James MD, PhD Oklahoma Medical Research Foundation, Oklahoma City, OK, USA

Caroline A. Jefferies MD Molecular and Cellular Therapeutics, Royal College of Surgeons in Ireland, Dublin, Ireland

Kenneth C. Kalunian MD UCSD School of Medicine, Department of Medicine, Division of Rheumatology, Allergy and Immunology, La Jolla, CA, USA

Mariana J. Kaplan MD Division of Rheumatology, University of Michigan, MI, USA

Grainne Kearns MD Department of Rheumatology, Beaumont Hospital & Royal College of Surgeons in Ireland, Dublin, Ireland

Mary Keogan MD Beaumont Hospital & Royal College of Surgeons in Ireland, Dublin, Ireland

Cristián Vera Kellet MD Department of Dermatology, Pontificia Universidad Católica de Chile, Santiago, Chile

Munther A. Khamashta MD, FRCP, PhD The Rayne Institute, St Thomas' Hospital, London, UK

Takao Koike PhD Department of Medicine II, Hokkaido, University Graduate School of Medicine, Sapporo, Japan

Dwight H. Kono MD Department of Immunology and Microbiology Science, The Scripps Research Institute, La Jolla, CA, USA

Steven A. Krilis MB BS UNSW, PhD Syd, FRACP Department of Immunology, Allergy and Infectious Disease, St. George Hospital, University of New South Wales, Australia

Annegret Kuhn MD Department of Dermatology, University of Münster, Germany

Vasileios C. Kyttaris MD Harvard Medical School, Division of Rheumatology, Boston, MA, USA

Robert G. Lahita MD, PhD, FACP, FACR, FRCP UMDNJ, New Jersey Medical School, Newark Beth Israel Medical Center Newark, NJ, USA

Carl A. Laskin MD, FRCPC Departments of Medicine (Rheumatology), Obstetrics and Gynecology and Immunology, University of Toronto, Co-Medical Director, LifeQuest Centre for Reproductive Medicine, Toronto, ON Canada

Gaëlle Leroux MD AP-HP, Service de Médecine Interne, Centre de référence national pour le Lupus et le syndrome des Antiphospholipides, Centre Hospitalier Universitaire Pitié-Salpêtrière, France

Yi Li MD Division of Rheumatology and Clinical Immunology, University of Florida, Gainesville, FL, USA

Matthew H. Liang MD, MPH Harvard Medical School, Boston, Harvard School of Public Health Study, Massachusetts Veterans Epidemiology Research and Information Center, USA

Chau-Ching Liu MD, PhD Lupus Center of Excellence, University of Pittsburgh Schools of Health Sciences, Pittsburgh, PA, USA

Thomas Luger MD Department of Dermatology, University of Münster, Germany

Ian R. Mackay MD Department of Biochemistry and Molecular Biology and Department of Immunology, Monash University, Melbourne, Australia

Meggan Mackay MD, MS The Center for Autoimmune and Musculoskeletal Disease, The Feinstein Institute for Medical Research, North Shore–LIJ Health System, Manhasset, NY, USA

Kathleen Maksimowicz-McKinnon DO University of Pittsburgh, Schools of the Health Sciences, Pittsburgh, PA, USA

Mark J. Mamula PhD Yale University School of Medicine, Section of Rheumatology, New Haven, CT, USA

Susan Manzi MD MPH University of Pittsburgh Schools of the Health Sciences, Pittsburgh, PA, USA

Galina Marder MD Albert Einstein College of Medicine, Division of Rheumatology and Allergy-Clinical Immunology, North Shore–Long Island Jewish Health System, Manhasset, NY, USA

Ahmad K. Mashmoushi Department of Medicine, Division of Rheumatology, Medical University of South Carolina, Charleston, SC, USA

T.N. Mayadas MD Vascular Research Division, Department of Pathology, Brigham and Women's Hospital and Harvard Medical School, Boston, MA, USA

Lloyd Mayer MD Mount Sinai School of Medicine, New York, NY, USA

Terry K. Means PhD Harvard Medical School, Boston, Broad Institute of MIT and Harvard, USA

Joan T. Merrill MD Clinical Pharmacology Research Program, Oklahoma Medical Research Foundation, Oklahoma City, OK, USA

Kristina E. Milan MD Department of Obstetrics and Gynecology, Maternal-Fetal Medicine, University of Utah, Salt Lake City, UT, USA

Rina Mina Dr Cincinnati Children's Hospital Medical Center, University of Cincinnati, Division of Rheumatology, Cincinnati, OH, USA

Chandra Mohan MD, PhD Simmons Arthritis Research Center, Center for Immunology, University of Texas Southwestern Medical Center, Dallas, USA

Anne-Barbara Mongey MD University College Dublin, Dublin, Ireland, Consultant Rheumatologist, St. Vincent's University Hospital, Dublin, Ireland

Seetha U. Monrad MD Division of Rheumatology, University of Michigan, Ann Arbor, MI, USA

Johannes C. Nossent MD Department of Rheumatology, University of Tromso, Norway

Jim C. Oates MD Medical University of South Carolina, Ralph H. Johnson VAMC, USA

Gerlinde Obermoser MD Baylor Institute for Immunology Research, Dallas, TX, USA. Department of Dermatology, University Hospital Innsbruck, Innsbruck Medical University, Austria

Karolina Palucka MD, PhD Baylor Institute for Immunology Research, Dallas, TX, USA

Eva D. Papadimitraki MD Department of Medicine, University of Crete School of Medicine, Heraklion, Greece

Virginia Pascual MD Baylor Institute for Immunology Research, Dallas, TX, USA

Andras Perl MD, PhD Program State University of New York, Upstate Medical University, College of Medicine, New York, USA

Jean-Charles Piette MD AP-HP, Service de Médecine Interne, Centre de référence national pour le Lupus et le syndrome des Antiphospholipides, Centre Hospitalier Universitaire Pitié-Salpêtrière, France

Shiv Pillai MD, PhD Massachusetts General Hospital and Harvard Medical School, Boston, MA, USA

Westley H. Reeves MD Marcia Whitney Schott Professor of Medicine, Division of Rheumatology and Clinical Immunology University of Florida, Gainesville, FL, USA

Bruce C. Richardson MD, PhD Dept of Medicine, University of Michigan and the Ann Arbor VA Medical Center, Ann Arbor, MI, USA

Anthony Rosen MD Division of Rheumatology, Johns Hopkins University School of Medicine, Department of Medicine, Baltimore, MD, USA

Brad H. Rovin MD, FACP, FASN Division of Nephrology, The Ohio State University College of Medicine, Columbus, OH, USA

Minoru Satoh MD, PhD Division of Rheumatology and Clinical Immunology, University of Florida, Gainesville, FL, USA

Peter Schur MD Lupus Center, Brigham and Women's Hospital, Harvard Medical School, Boston, MA, USA

Karen A. Spitzer MSc Mount Sinai Hospital, LifeQuest Centre for Reproductive Medicine, Toronto, ON Canada

C. Michael Stein MD Dan May Professor of Medicine and Pharmacology, Divisions of Rheumatology and Clinical Pharmacology, Nashville, TN, USA

Isaac E. Stillman MD Department of Pathology and Renal Division, Department of Medicine, Beth Israel Deaconess Medical Center, and Harvard Medical School, Boston, MA, USA

Tom J.G. Swaak MD Department of Rheumatology, Ikazia Hospital, Rotterdam, The Netherlands

Kendra N. Taylor MD, PhD Massachusetts General Hospital and Harvard Medical School, Boston, MA, USA

Argyrios N. Theofilopoulos MD Department of Immunology and Microbiology Science, The Scripps Research Institute, La Jolla, CA, USA

Betty P. Tsao PhD Division of Rheumatology, Department of Medicine, University of California, Los Angeles, CA, USA

George Tsokos MD Department of Medicine, Beth Israel Deaconess Medical Center, Harvard Medical School, Boston, MA, USA

Hideki Ueno MD, PhD Baylor Institute for Immunology Research, Dallas, TX, USA

Aziz M. Uluğ PhD   Center for Neurosciences, The Feinstein Institute for Medical Research, North Shore—LIJ Health System, Manhasset, NY, USA

Evan S. Vista MD   Oklahoma Medical Research Foundation, Oklahoma City, OK, USA

Bruce T. Volpe MD   Department of Neurology and Neuroscience, Weill Medical College of Cornell University, Burke Medical Research Institute, White Plains, NY, USA

Tim J. Vyse MD, PhD   Faculty of Medicine, Section of Rheumatology, Imperial College London, London, UK

Mark H. Wener MD   Immunology Division, Department of Laboratory Medicine, Rheumatology Division, Department of Medicine University of Washington Seattle, WA, USA

Victoria P. Werth MD   University of Pennsylvania and Philadelphia, VAMC, PA, USA

Yuan Xu MD   Division of Rheumatology and Clinical Immunology, University of Florida Gainesville, FL, USA

Lijun Yang MD   and Laboratory Medicine, University of Florida, Gainesville, FL, USA

C. Yung Yu MD   Center for Molecular and Human Genetics, The Research Institute at Nationwide Children's Hospital and Department of Pediatrics, The Ohio State University, Columbus, OH, USA

Raymond L. Yung MD   Department of Medicine, University of Michigan and the Ann Arbor VA Medical Center, Ann Arbor, MI, USA

Dun Zhou   The Feinstein Institute for Medical Research, Manhasset, NY, USA

Haoyang Zhuang MD   Division of Rheumatology and Clinical Immunology, University of Florida, Gainesville, FL, USA

Foreword

SYSTEMIC LUPUS ERYTHEMATOSUS — FIFTH EDITION

Systemic Lupus Erythematosus (SLE) is now an international disease. There is probably not a single country which does not have patients with SLE or discoid lupus. There are differences in how the disease expresses itself in these various countries but it is no longer an unknown disorder. Every single country has lupus patients. The recognition of this disorder has resulted in a huge investment in both investigation and research. In recent years, the availability of the many treatments has resulted in greatly improved management of the disorder and a major decrease in mortality from the disease. Despite these advances, there are still areas of the world that lack all of the needed diagnostic and treatment skills required to control this disease.

The importance of various T cell subsets is now well accepted and treatments are aimed at controlling both these cells and B lymphocytes in the hope that they will give insight into the pathogenesis of SLE.. No doubt, in the previous four editions of this wonderful textbook, early diagnosis and more efficient treatments were being developed and because of this, there is a huge increase in the number of survivors of active lupus in many countries. The majority of lupus patients today can live normal lives.

In recent years, there has been a great increase in disease management; not only by Rheumatologists but also by Dermatologists and by many other medical sub-specialties. Earlier diagnosis is easier because of increased education of physicians throughout the world. The availability of these excellent treatments now enable the physician working with patients to be updated and positive, with a disease under better control. Even patients who previously had to have kidneys removed and those with other failed systems are living a markedly improved life span.

Nevertheless, there is continued need for active research into the cause(s) of this disease and for the availability of improved treatments. Perhaps by the next edition of this textbook, we will have "defeated" SLE.

Evelyn Hess MD, MACR, MACP
Cincinnati, OH 2010

Preface

Systemic lupus erythematosus (SLE) remains one of the more remarkable diseases of the last hundred years. Only recently have we heard of a new and approved drug to treat this complex illness. In this book for the first time, we present ways to approach, examine and quantify clinical facts from lupus patients. All are aware of the difficulty facing individual investigators, Pharma, and government regulators with regard to lupus. The organ selectivity and varying severity and chronicity frighten even the most accomplished clinician. Many have said that lupus is more than one disease. They say this because no single illness could incorporate all of the variables we see with SLE. It is with this backdrop of clinical bewilderment that we dedicate ourselves to a comprehensive exploration of this remarkable illness. Science changes quickly and many terms and concepts found in the basic section of this book will be new and unfamiliar to previous editions. Like previous editions we direct this book to the student, researcher, and clinician. We have paid particular attention in this 5th edition to include key opinion leaders from across the globe; a feat that would not be possible without the internet. To each author who has written contributions with dedication, we are grateful. Unlike previous editions, the 5th edition includes the work of three distinguished associate editors. George Tsokos is the editor for the basic science sections, Dr. Jill Buyon for the clinical sections, and R. Takao Koike for antiphospholipid syndrome.

This book was born at the Rockefeller University in 1979 and published in 1983 in the laboratory of Professor Henry G. Kunkel. Now, as then, it serves as a tribute to his pioneering work in immunology. He wrote the first Foreword and he was followed by Eng Tan, Graham Hughes, Murray Urowitz and now Evelyn Hess. Many of the senior authors of previous editions have passed on and the key opinion leaders in immunology and rheumatology today are hybrids of medicine and science. Going from bench to bedside is the translational medicine of today which will give us the innovative ideas and therapies of tomorrow.

Let me borrow a paragraph from the Preface of the last edition:

> Lupus is a complex illness that often requires a physician's considerable time and effort. It is not an easy disease to diagnose or manage. This book is written as an aid to help the physician understand the subtle nuances of this enigma and use knowledge herein to manage the patient. It is written with the idea that new knowledge is urgently needed to diagnose and treat those patients for whom this illness is nothing short of a major catastrophe—a life emergency that deserves all of our efforts and attention.

I thank Elsevier for persisting over many years with previous editions and for this one. Specific thanks go to Megan Wickline and Mara Conner, our developmental editors from Elsevier. The brilliant editing of Jill Buyon, George Tsokos and Takao Koike made this edition possible and particularly good. Additional thanks goes to Carlo Mainardi and Joan Kowalec of our own Division of Rheumatology who helped get us through many pages of manuscript and provided invaluable advice.

Robert G. Lahita
Newark, NJ 2010

Introduction

OVERVIEW

Lupus is a chronic disease of unknown etiology that is very difficult to diagnose and treat, sometimes over-diagnosed and more often than not diagnosed with great delay. Since the last edition of this book, major advances have been noted in the treatment of many autoimmune illnesses using biological therapy, but nothing exists that is able to completely reverse the ravages of an autoimmune disease like lupus. Although there are some promising drugs on the horizon, the absence of significant progress in understanding and treating this disease over a 40-year period is not due to modest research funding or lack of attention, but may be the result of the complexity of the disease and a misunderstanding of the criteria for diagnosis, biomarkers of severity, and instruments to measure clinical activity. The establishment of clear indices that allow regulatory agencies to see some consistency of response is critical to the development of new agents to treat lupus. More important, however, both pharmaceutical and government regulatory agencies have to have a clearer understanding of the disease, its multisystem nature and the nuances of the immune system. This is not a one-cause one-cure illness.

The question of whether lupus is one illness or many may be one of the difficulties. This protean illness presents in a variety of ways and its clinical manifestations depend on the organs chosen by the immune system for attack. Three simple things can be said about SLE; the disease favors women, has a genetic association, and has an elusive trigger. In this 5th edition, these simple facts are explored in lucid detail. In this edition we have revised every chapter, eliminated some and have added many new chapters. The book is organized in sections: the pathogenesis of disease, the clinical aspects of the illness with a special section on the antiphospholipid syndrome, and finally a section on treatment.

The section on the pathogenesis of disease deals with genetics, cellular, humoral, and environmental aspects. We have also included a subsection on tissue injury because all of these aspects of pathogenesis have such importance to a basic understanding of disease. All of the chapters in this section are new with many new authors and provide a view into the scientific thinking of this disease which is not available anywhere else. Clinical aspects of the disease are divided into the clinical presentation and then a focus on organ systems that might be affected by lupus. In this section we have paid special attention to research and clinical trials and have added chapters on design of trials and measurement of disease activity. An additional section entitled "antiphospholipid syndrome" parallels the clinical SLE section and is divided into basic pathogenesis and clinical aspects of the disease as well. At the end of the book is the treatment section which relates the latest drugs approved and development for the disease.

Over 90% of the authors for this edition are new and contributing new topics. The book is enhanced in all areas to provide the reader with the latest information about SLE, its science, presentation and treatment. Lupus gives us a handle on molecular biology, insight into common ailments like atherosclerosis and diseases like multiple sclerosis and Sjögren syndrome.

THE HISTORY OF SYSTEMIC LUPUS ERYTHEMATOSUS

The history of SLE is long and interesting. Lupus is the Latin word for wolf and the Romans used the term loosely to mean "eating and devouring diseases", much as the Greeks referred to the term Herpes which means "to creep." Many terms including leprosy, which means "to scale" in Greek, have come down through the centuries and are used today for specific syndromes. Hippocrates had names for several ulcerating diseases which he called *herpes esthiomenos*. Most historians believe that he included lupus in that group. In antiquity a variety of illnesses, many relevant to the face and disfiguring, were called *"nole me tangere"* (touch me not) and were different than boils of the lower extremities that were called lupus lesions [1]. After 1500 AD, physicians tried to use lupus for lesions of the face, but it was not until the early part of the 19th century when Robert Willon (1757—1812) and his students presented a classification scheme for diseases and lesions of the skin. Vesicular skin lesions were under the heading of Herpes, and destructive or ulcerative skin lesions of the face and nose were called lupus. These classifications were further defined by a student of Willon's, Thomas Bateman. One of Bateman's students, Laurent Biett (1781—1840) (the Parisian school), pursued and enlarged the classification of skin lesions and his student, Cazenave

(1802—1877), was the first physician to use the term lupus erythematosus [2, 3]. This new name evolved over many years. Biett divided Willon's skin lesions called lupus into three categories as it appeared in the fourth edition of their famous dermatology textbook entitled *Abrege Practique Des Maladies de la Peau* [2]. There were the first, *lupus qui detruit en surface* (lupus which destroys on the surface); the second, *lupus qui detruit en profondeur* (lupus which destroys at a depth); and third, *lupus avec hypertrophic* (lupus with hypertrophy). Cazenave was responsible for producing the fourth edition of the *Abrege* textbook in 1851 and in that edition he identified a fourth category of the skin classification noting atrophy, telangiectasias, fixed erythema, adherent scaling and follicular plugging. He called this fourth category *lupus erythemateux*. In 1846, Ferdinand von Hebra (1816—1880), an accomplished Viennese physician, used "butterfly rash" or "bat wing rash" to describe the familiar malar rash of the disease. The butterfly rash was more accepted and is a term used today. In his 1856 book, Von Hebra published the first illustrations of the disease in the *Atlas of Skin Diseases*. In 1872, Moretz Kaposi described the visceral forms of the illness and physicians began to suspect that this illness was more generalized than the skin and used the term *acute disseminated* in their descriptions [4]. Kaposi proposed two types of lupus: disseminated and discoid. In his writings, he supposed that the disseminated form consisted of subcutaneous nodules, arthritis, lymphadenopathy, fever, weight loss, anemia, and central nervous system involvement. In 1894, Payne at Saint Thomas Hospital in London suggested that there was "a vascular disturbance causing hyperemia" that could be influenced by quinine. In 1904, William Osler described two women who developed renal failure within ten months of developing facial erythema, which in retrospect was the facial rash described by Von Hebra [5]. Osler described a number of other illnesses at the time, among them *Henoch Schoenlein purpura* and disseminated *gonococcemia* which could be confused with the lesions in the two women with lupus. In Vienna, at the same time, Jadassohn [6] described similar syndromes in a few patients. Both he and Osler established SLE as a distinct entity by the turn of the century, even though many general practitioners still thought of SLE as a form of skin tuberculosis.

In 1902, Sequira and Balean of the London Hospital published 71 cases of lupus erythematosus: 60 had discoid and 11 had disseminated disease. The new description of acroasphyxia (Raynaud phenomenon) was a common feature.

Typical cases of SLE were reported under a variety of names during this period and not until the 1920s and 1930s were the cases well defined [7]. This was mainly because the pathologists who studied morbid anatomic pathology prior to the 1930s concerned themselves only with the pathologic changes and not the clinical condition of the patients. The atypical bacterial endocarditis of Emmanuel Libman and Benjamin Sacks, reported in 1924, were carefully described and were explained some 60 years later as a consequence of phospholipid antibodies [8]. Following that description, George Baehr published a series of 23 autopsied cases of the renal *wire loop* lesions of lupus nephritis and described the solar sensitivity so well known in this disease. Knowledge of lupus pathology was evolving in the early twentieth century. In 1936 Freiberg, Gross and Wallach autopsied a young woman with lupus and no skin lesions showing that the disease was not restricted to include skin lesions and not associated or caused by tuberculosis. Klempner Pollack and Baehr, in 1941, suggested that collagen was a part of the disease because of the many instances of fibrinoid necrosis that they found with the disease. Thus evolved the name *collagen disease* which was used for many years to describe all of the diseases involving connective tissue.

Hack and Reinhart were the first to describe the false-positive syphilis test in SLE and, in 1940, Keil similarly reported ten cases of SLE with false-positive syphilis tests. Haserick and Lang wrote about an additional series of cases where the presence of the false-positive serology predated the clinical lupus by up to 8 years. In all of these cases, the false-positive syphilis tests probably resulted from the presence of antiphospholipid antibodies, the discovery of which was to take an additional 30 years. In 1955, Moore studied another 148 patients who were positive for syphilis and found that some 7% developed lupus over time, whereas 30% had symptoms relegated to *collagen vascular disease* [9]. In 1949, Phillip Hench discovered cortisone and the future of the connective tissue diseases changed forever — rheumatoid arthritis (RA) patients and those with lupus were manageable and "cures" were reported [10].

In 1948, Hargraves, Richmond, and Morton described the LE cell in the bone marrow of SLE patients [11]. This test was later adapted to peripheral blood. This discovery laid the foundation for our confirmation of the disease lupus as an autoimmune disease. These were phagocytes eating cells coated with autoantibody. Until it was described in other illnesses like RA, it was thought to be pathognomonic of lupus. Dubois, in 1953, and Harvey, in 1954 [12], sought to set the record straight by stressing the chronicity of SLE and recognized the diagnostic importance of the LE test. The presence of this finding in only 50—70% of lupus patients and the rapid development of more specific tests of autoimmune disease made its use limited. Other serum abnormalities like hypergammaglobulinemia detected in newer techniques like immunoelectrophoresis suggested that gamma globulins were abundant in the sera of lupus patients and these were in reality the antibodies that reacted with normal tissues. These were later called autoantibodies. George Friou, in 1957 [13], applied the indirect immuno-fluorescent test of Coons to the study of these autoantibodies. The fluorescent antinuclear antibody test (FANA) is positive in some 95–98% of SLE cases. At about the time of the Friou discovery, Deicher, Holman, and Kunkel at Rockefeller University [14], along with three other groups, described the antibodies to DNA, both double and single stranded. Thus, began one of the most influential and prolific laboratories for the study of lupus in the entire world. In 1956, Eng Tan and Kunkel described an antibody to a glycoprotein called SM after the first initials of one of the many patients followed in the research clinic [15]. The patient's name was Smith and she was one of many patients from whom new autoantibodies were described. Following that autoantibody came the discovery of antibodies to ribonucleoprotein and other cellular components which later helped to dissect cell function and open the doors to molecular biology.

As more and more antibodies were described, clinical conditions like the overlap syndrome of mixed connective tissue disease of Gordon Sharp followed.

Dixon and colleagues [16] assisted with the development of many murine strains of the disease driven by genetics of the mouse strain and sharing many of the features of the human disease. In 1954, Leonhardt described the familial nature of lupus, later studied by Schulman and Arnett at Hopkins [17] and subsequently by Hahn, Harley, and Behrens [18]. Studies provided discussions of multiplex families and twins as well as gender-specific inheritance and the presence of constitutive MHC II and MHC III relationships. All the while, our understanding of the cellular components of the innate and adaptive immune system as well as the fine nuances of the various antibodies in understanding cell functions led to today's revolution in our knowledge of this important illness.

Finally, the work of Hughes [19] and others opened the door to our understanding of antiphospholipid antibody syndrome as an important aspect of lupus. The frequency of this syndrome within the disease lupus is a great curiosity and we have given an entire section to this syndrome within the book.

No history of lupus gives credit to the countless investigators who have provided data to enhance our knowledge of the disease. The history of lupus is not static and there will be many more years of clinical and scientific advances until we find the cause of this disease.

Robert G. Lahita
Newark, NJ 2010

References

[1] C.D. Smith, M. Cyr, The History of Lupus from Hippocrates to Osler, in: J.H. Klippel (Ed.), Rheumatic Disease Clinics of North America, Philadelphia, W.B., Saunders, Inc., 1988, pp. 1–14.

[2] P. Cazenave, H. Schedel, Abrege pratique des maladies de la peau, Labe, Paris, 1847.

[3] P.L.A. Cazanave, Lupus erythemateaux (erytheme centrifuge), Annales de Maladies de la peau et de la syphilis 3 (1850) 297–299.

[4] M.K. Kaposi, Neue Beitrage zur Keantiss des lupus erythematosus, Arch Dermatol Syphilol 4 (1872) 36.

[5] W. Osler, On the visceral manifestations of the erythema group of skin diseases, Am J Med Sci 127 (1904) 1.

[6] J. Jadassohn, Lupus erythematodes. in: F. Mracek (Ed). Hanbuch der Hautkrankheiten. Alfred Holder; 1904, pp. 298–404.

[7] E.C. Reifenstein, Jr.E.C. Refifenstein, G.H. Reifenstein, Variable symptom complex of undetermined etiology with fatal termination, Arch Int Med 63 (1939) 553.

[8] E. Libmann, B. Sacks, A hitherto undescribed form of valvular and murla endocarditis, Arch Int Med 33 (1924) 701.

[9] J.E. Moore, W.B. Lutz, The natural history of systemic lupus erythematosus: an approach to the study through chronic biological false positive reactions, J Chron Dis 2 (1955) 297.

[10] P.S. Hench, The reversibility of certain rheumatic and non-rheumatic conditions by the use of cortisone or of the pituitary adrenocorticotrophic hormone, Ann Int Med 36 (1952) 1.

[11] M.M. Hargraves, H. Richmond, R. Morton, Presentation of two bone marrow elements: The tart cell and the LE cell, Proc Staff Meet Mayo Clinic 23 (1948) 25.

[12] E.L. Dubois, The effect of the LE cell test on the clinical picture of systemic lupus erythematosus, Ann Int Med 38 (1953) 6.

[13] G.J. Friou, Clinical applicatoin of lupus serum nucleoprotein reaction using fluorescent antibody technique, J Clin Invest 36 (1957) 390.

[14] H.R. Deicher, H.R. Holman, H.G. Kunkel, The precipitin reaction between DNA and a serum factorin SLE, J Exp Med 109 (1959) 97.

[15] E.M. Tan, H.G. Kunkel, Characteristics of a soluble nuclear antigen precipitating with sera from patients with systemic lupus erythematosus, J Immunol 96 (1966) 464.

[16] A.N. Theofilopoulos, F.J. Dixon, Murine models of systemic lupus erythematosus, Adv Immunol 37 (1985) 269–390.

[17] F.C. Arnett, L.E. Shulman, Studies in familial systemic lupus erythematosus, Medicine 55 (1976) 313.

[18] J.B. Harley, A.L. Sestak, L.G. Willis, S.M. Fu, J.A. Hansen, M. Reichlin, A model for disease heterogeneity in systemic lupus erythematosus: relationship between histocompatibility antigens, autoantibodies, and lymphopenia or renal disease, Arthritis Rheum 32 (1989) 826–836.

[19] E.N. Harris, A.E. Gharavi, G.R.V. Hughes, Antiphospholipid antibodies, in: R. Williams (Ed.), Clinics in Rheumatic Disease, W.B. Saunders, Philadelphia, 1985, pp. 591–609.

PART I

BASIS OF DISEASE PATHOGENESIS

A. Genetics
(Chapters 1–4)

B. Cellular Pathogenesis
(Chapters 5–12)

C. Humoral Pathogenesis
(Chapters 13–21)

D. Environmental Aspects of Pathogenesis
(Chapters 22–25)

E. Mechanisms of Tissue Injury
(Chapters 26–29)

1

Major Histocompatibility Complex Class II

Michelle M A Fernando, Tim J Vyse

THE MHC REGION

Overview

The major histocompatibility complex (MHC), located on the short arm of chromosome 6, is one of the most extensively studied regions in the human genome because variants at this locus have long been associated with most autoimmune diseases, many infectious and inflammatory diseases as well as transplant compatibility. In particular, human leukocyte antigen (HLA) class I and class II molecules are critical in mediating host defense responses through antigen presentation, and immune tolerance by means of self/non-self recognition.

In 1936, whilst studying the genetic determinants of transplant outcome, Peter Gorer discovered the murine MHC locus, which he named *antigen II* [1—3]. The locus was further characterized and subsequently named *H2*, for its role in histocompatibility 12 years later by George Snell [4]. Shortly afterwards Jean Dausset recognized the human MHC or human leukocyte antigen (HLA) region: so-called because Dausset originally demonstrated MHC antigens on the surface of white blood cells [5]. Subsequently, Baruj Benacerraf described the importance of these antigens in the immune response [6]. Snell, Dausset and Benacerraf's seminal work garnered them the 1980 Nobel Prize for Medicine.

The classical MHC encompasses approximately 3.6 megabase-pairs (Mb) on 6p21.3 and is divided into three subregions: the telomeric class I, class III and the centromeric class II regions (see Figure 1.1). Of the 224 genes within this region, 57% (128 genes) are expressed and 40% of these have putative immunoregulatory function. The concept of the extended MHC (xMHC), spanning about 7.6 Mb of the genome, has been recently established given that LD and MHC-related genes exist outside the classically defined locus. The five subregions of the xMHC comprise the extended class I subregion

(*HIST1H2AA* to *MOG*; 3.9 Mb), classical class I (*C6orf40* to *MICB*; 1.9 Mb), classical class III (*PPIP9* to *NOTCH4*; 0.7 Mb), classical class II (*C6orf10* to *HCG24*; 0.9 Mb), and the extended class II subregions (*COL11A2* to *RPL12P1*; 0.2 Mb) [7]. The clustering of immune and non-immune genes is a particular feature of the xMHC. Indeed, the MHC harbors the two largest gene clusters in the human genome: the histone and tRNA genes [7]. Possible explanations for this phenomenon include the advantage derived from the co-evolution of functionally related (immune) genes and the co-regulation of gene expression through gene-conversion and duplication events and selective forces. In addition, the MHC contains regions of copy number polymorphism: the *RCCX* module in class III (spanning complement *C4*) and the *HLA-DRB* genes in class II.

Subregions of the MHC

The class I and class II regions of the MHC encode the classical HLA genes (*HLA-A, -B, -C, -DR, -DQ* and *-DP*) involved in antigen presentation to T cells and transplant compatibility. The MHC class I and class II molecules belong to the immunoglobulin gene family, and share a similar structure in that they comprise heterodimers with variable extracellular peptide-binding domains and relatively constant transmembrane and intracytoplasmic domains. Their diversity is a consequence of germline polymorphism rather than somatic mutation which governs T-cell receptor and immunoglobulin repertoires. The class III region is the most gene-dense region in the genome encoding a variety of molecules including the early complement components, C2, fFactor B (*CFB*), and C4 as well as the *TNF* cluster, the heat shock protein cluster, and proteins involved in growth and development. The class II region generally encodes molecules involved in the immune response, including proteasome and antigen transporter

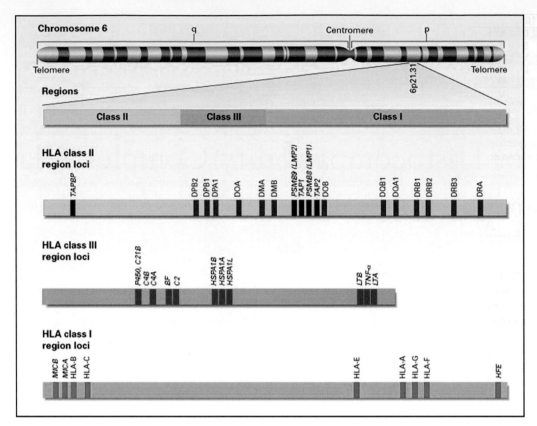

FIGURE 1.1   Location and organization of the MHC on chromosome 6. The MHC is classically divided into three regions: class I, class II and class III. Each region contains several genes but not all are shown. The classical class I and class II HLA genes encoding antigen-presenting molecules are illustrated, as are three gene clusters in class III, from left to right: the complement cluster *C4A, C4B, CFB, C2*, the heat shock protein cluster and the TNF cluster, LTB, TNF and LTA. [92]. *Copyright © 2000 Massachusetts Medical Society. All rights reserved.*

molecules (*PSMB8/LMP7*; proteasome subunit, beta-type, 8/large multifunctional protease 7 and *PSMB9/ LMP2*; proteasome subunit, beta-type, 9/large multifunctional protease 2; *TAP1* and *TAP2*; transporter, ATP-binding cassette, major histocompatibility complex, 1 and 2), *TAPBP* involved in antigen processing, and the antigen-presenting molecules, *HLA-DR*, *HLA-DQ*, and *HLA-DP*. On the other hand, the class I region contains genes that are not of obvious immunological relevance, such as the olfactory receptor gene family.

The primary importance of the classical HLA molecules relates to their role in host defense against attack from foreign pathogens by initiating the adaptive immune response through the presentation of antigen to T cells. Recent studies demonstrate that a variety of other proteins, cytokines and accessory molecules involved in this process are also encoded within the MHC (for example, TNF, complement pathway components *C2*, *C4* and *BF*, *TAP*, *HLA-DM*).

Classical HLA molecules, expressed on the cell surface, display both self- and non-self-antigens in the normal circumstance. Central and peripheral tolerance mechanisms generally ensure that self-antigens do not provoke an immune response. However, when these mechanisms fail, autoimmunity may ensue.

HLA POLYMORPHISM

The extreme polymorphism shown by HLA antigens is thought to be driven by selection against foreign pathogens, thus increasing the diversity of epitopes that can be presented to T cells [8]. The selective pressures thought to act at the MHC include positive natural selection, purifying selection, neutral selection, and balancing selection in the form of frequency-dependent selection and heterozygote advantage.

HLA polymorphism is largely generated by variation within exons 2 and 3 for class I molecules and exon 2 for class II molecules which encode the α1 and α2 domains and α1 and β1 domains of the respective peptide-binding grooves. Interestingly, some bases and hence codons are invariant, whilst others exhibit full nucleotide diversity (Figure 1.2). These exons are approximately 250 bp in length and so are easily amenable to PCR amplification for molecular characterization − the current basis of HLA typing.

	10		20		30		40		50
R F L W Q	L K F E C H F F N G	T E R V R L L E R C	I Y N Q E E S V R F	D S D V G E Y R A V
E E	S T S K Y	Q Y P D S Y	F H H R	N A H Y	R D F W V A
K Y	V M H Q		F H L V	Y L	L
	L G P		G	F	Q
	D Y		R	D	
	V G		H	L	
	P R				

FIGURE 1.2 *HLA-DRB1* gene polymorphism. Codons 6–50 of exon 2 of the *HLA-DRB1* gene are shown. The capital letters represent amino acid codes. Approximately half of the amino acid positions are invariant, while the others show polymorphism with some codons encoding up to seven different amino acids. The top line represents the sequence for *HLA-DRB1*0101*. *Adapted from [10].*

EXTENDED LINKAGE DISEQUILIBRIUM AT THE MHC

Non-random association (or linkage disequilibrium (LD)) in the inheritance of alleles at multiple loci within the MHC was demonstrated as early as 1968 [9]. For example, the recombination rate between *HLA-A* and *HLA-B* is approximately 0.8%; while that between *HLA-DRB1* and *HLA-DQB1* is virtually zero [10]. Upon determination of the physical size of the region, it appeared that LD extended more than 2 Mb in some cases, but not all. This differs from the LD pattern reported for other regions in the genome where strong LD exists in segments of approximately 22 kb [11]. However, closer inspection reveals that the "micro"-structure of LD is similar for the MHC [12, 13]. What appears to be different about a subset of MHC haplotypes is that there is a higher amount of LD observed *between* segments of strong LD. Such tight segment-to-segment LD can pose an important obstacle in MHC research: if one identifies a disease association with a variant in the region, it may not be possible to determine whether the variant is causal or whether its association simply reflects LD with the true causal variation. Most association studies of the MHC region to date have suffered from this caveat.

CONSERVED EXTENDED HAPLOTYPES/ ANCESTRAL HAPLOTYPES

Cepellini first described the existence of MHC haplotypes in 1967 [14]. Since then studies have shown remarkable conservation of particular haplotypes at the MHC extending from *HLA-B* to *HLA-DR/-DQ*. Moreover, such haplotypes account for approximately one-third of total European MHC haplotypes, with half of the population possessing at least one so-called "ancestral haplotype"; a further 35% of MHC haplotypes were found to be simple recombinants of ancestral haplotypes [14–17]. In one Caucasoid population, these extended haplotypes, or simple recombinants thereof, accounted for 73% of unselected MHC haplotypes [15]. These

extended haplotypes are large conserved stretches of MHC DNA traditionally defined and fixed at *HLA-B*, the complotype: *C2*, *CFB*, *C4A*, *C4B* and *HLA-DR*. Such haplotypes are often designated after the *HLA-B* allele they bear. For example the so-called "autoimmune" haplotype, 8.1, is allelically composed of *HLA-B8-SC01-DRB1*0301* (Figure 1.3). Some MHC haplotypes are now assigned based on known genotypes of homozygous cell lines (http://www.ebi.ac.uk/imgt/hla/) [18–20].

THE SEQUENCE OF THE MHC

Given the importance and complexity of the MHC in health and disease, significant effort has been placed into creating an accurate gene map of the region. The first such map of the human MHC was published in 1993 [21]. A sequence-based map of the MHC followed in 1999 [22] — this initial sequence was a "mosaic" or composite and was derived from a number of heterozygous individuals of various or unknown HLA type. The first draft of the Human Genome Project and consequent sequencing of chromosome 6 in its entirety was completed in 2001 [23, 24]). A more comprehensive map of the extended MHC region was published in 2003 [25]. Furthermore, in 2004, 4.6 Mb of the 1999 sequence was replaced with that derived from the homozygous cell line, PGF (*HLA-A*0301*, *-B*0702*, *-C*0702*, *-DRB1*1501*, *-DQA1*0102*, *-DQB1*0602*), and an integrated gene map of the extended MHC was published [7, 26] (Figure 1.4). Given the heterogeneous nature of the MHC sequence at that time, The MHC Haplotype Project was established by The MHC Haplotype Consortium in 2000 [27]. The purpose of the study was to provide a comprehensively annotated reference sequence of a single HLA-homozygous MHC haplotype (PGF) to be used as a base against which variations from seven other homozygous cell lines could be assessed. The eight homozygous consanguineous cell lines of known HLA haplotype were chosen for disease predisposition or protection in type 1 diabetes and multiple sclerosis and for commonality in European populations.

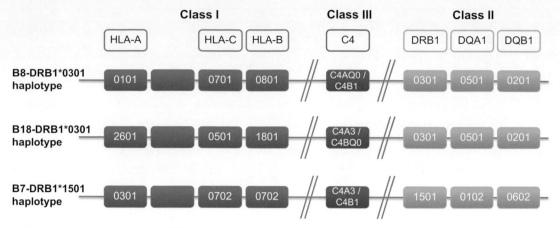

FIGURE 1.3   Extended haplotypes at the MHC. Selected HLA class I, class III and class II alleles are shown to illustrate the concept of extended LD at the MHC for three common European haplotypes. Across the MHC, alleles at distant loci exhibit strong linkage disequilibrium and thus exist as long unbroken haplotypes, sometimes extending over 2 Mb. Such strong LD can hamper the identification of causal variants. The lozenges represent MHC genes, and the alleles specific to each haplotype are shown within.

This task has been recently completed and the data are available through the VEGA/HAVANA database [26–29]. This effort has led to the discovery of haplotype-specific differences in gene annotation and has identified more than 300 loci of which over 160 genes are protein-coding. In addition, more than 44,000 variations were found and submitted to the dbSNP database. Similar projects have been undertaken in the United States and Japan [30–31].

NOMENCLATURE OF HLA AND NON-HLA ALLELES

The nomenclature of HLA alleles can be extremely confusing. Original serological typing defined the class I antigens: *HLA-A*, *-B*, and *-C*. Serological typing of the class II locus followed later. The designation of *HLA-A* and *HLA-B* specificities is exclusive, that is, while *HLA-A1* exists, *HLA-B1* does not, reflecting the original premise that these specificities were derived from a single locus. *HLA-C* antigens were termed "*Cw*" to avoid confusion over complement protein nomenclature (for further details see [32, 33]).

Molecular typing methods have now largely superseded serological and MLC (mixed lymphocyte culture) HLA typing, resulting in a further change in nomenclature in 1987 [32]. Thus, class I and class II alleles are designated by the locus and gene name separated by a hyphen, followed by an asterisk (*HLA-DRB1**). The first two digits after the asterisk define the allele of the gene and this number frequently but not always matches the serological type (*HLA-DRB1*03*); the next two digits define the subtype of the allele (*HLA-DRB1*0301* or *HLA-DRB1*0302*). A further four digits may follow specifying synonymous or non-synonymous amino acid changes within or outside coding sequences (http://www.anthonynolan.org.uk/HIG/lists/nomenlist.html). The presence of a null (not expressed) allele is specified by an "*N*" at the end of an allele sequence (*HLA-DRB4*0103102N*) [32]. Null alleles of the complement genes are defined "*Q0*" (quantity zero), *C4A*Q0* for instance [34].

In order to accommodate the ever-increasing number of HLA alleles identified, the WHO Nomenclature Committee for Factors of the HLA System will implement a further change in HLA nomenclature, planned to take effect from April 2010 (http://hla.alleles.org/nomenclature/nomenclature_2009.html).

HLA typing using so-called next-generation sequencing technology, which allows rapid generation of large quantities of sequence data, is on the horizon. Such platforms have emerged out of the need for cheaper, high-throughput DNA sequencing beyond that possible with current dye-terminator strategies. This technology should allow sequencing of HLA genes to allelic level from single strands of DNA thus improving typing accuracy, eliminating "phase ambiguity" (where heterozygous DNA sequences result in the inability to differentiate one allele from another), and providing greater efficiency in terms of time, cost and quantity of DNA required.

HLA CLASS I MOLECULES

The classical class I molecules (*HLA-A*, *-B*, and *-C*) are responsible for the presentation of endogenous antigen to CD8+ T cells (Figure 1.5). Together with the non-classical class I genes, *HLA-E* and *-G*, they are also involved in innate immunity by mediating natural killer cell responses. Unlike class II genes, there are many class

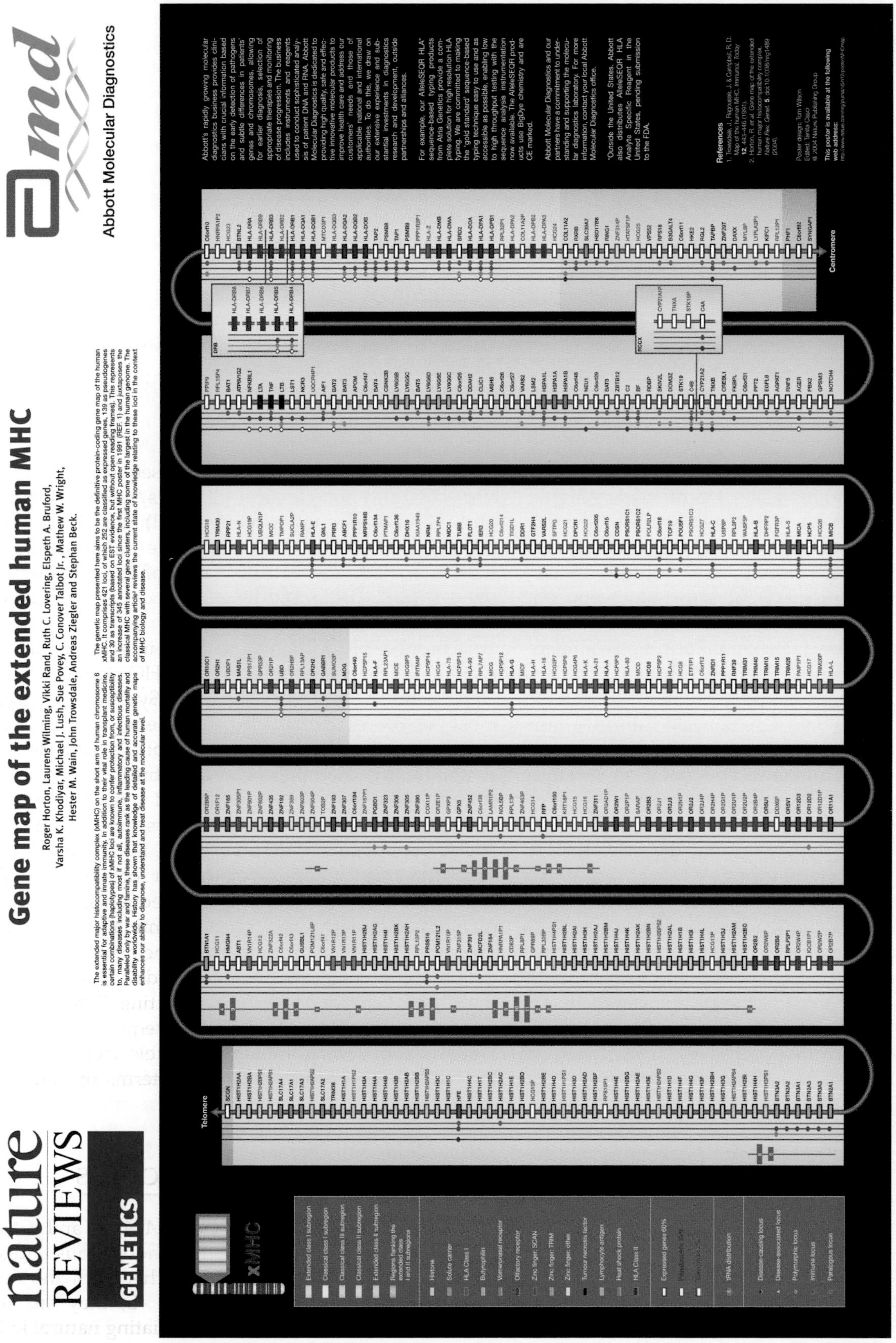

FIGURE 1.4 Gene map of the extended MHC. The gene map is shown from telomere (left) to centromere (right) on the short arm of chromosome 6. The five color-coded subregions of the extended MHC span about 7.6 Mb: extended class I (green block: *HIST1H2AA* to *MOG*; 3.9 Mb), classical class I (yellow block: *C6orf40* to *MICB*; 1.9 Mb), classical class III (orange block: *PPIP9* to *NOTCH4*; 0.7 Mb), classical class II (blue block: *C6orf10* to *HCG24*; 0.9 Mb), extended class II (pink block: *COL11A2* to *RPL12P1*; 0.2 Mb). The regions flanking the xMHC are shown as grey blocks. The insets denote the hypervariable RCCX and DRB regions. Numbers and positions of tRNA genes are represented as indigo bars, the length of which is proportional to the gene number between other loci. Vertical lines connect the two main groupings of tRNA genes of 1.6 Mb and 0.5 Mb of the sequence. Circles to the left of each locus indicate disease status, polymorphism, immune status and paralogy. *Reprinted by permission from Macmillan Publishers Ltd: Nature Reviews Genetics,* [7]*, copyright 2004.*

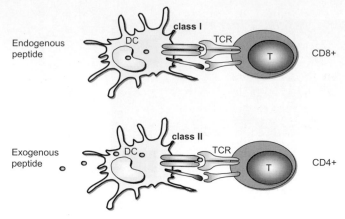

FIGURE 1.5 The "immunological synapse". The immunological synapse comprises the combination of peptide-bound MHC class I (top panel) or class II (lower panel) molecules interacting with the T cell receptor (TCR) in conjunction with costimulatory molecules. DC, dendritic cell; T, T cell.

I-like genes in the genome. Class I proteins are highly polymorphic; in fact *HLA-B* is the most polymorphic gene known in the human genome with 1115 alleles characterized at the present time (http://www.anthonynolan.org.uk/research/hlainformaticsgroup/). The classical (*HLA-A, -B, -C*) and non-classical (*HLA-E, -F, -G and HFE*) class I proteins are heterodimers comprising a polymorphic transmembrane α heavy chain — a glycoprotein of approximately 45 kDa — encoded within MHC class I region, and an invariant water-soluble light chain common to all isoforms — the 12 kDa protein, β_2-microglobulin encoded on chromosome 15. The two chains are held together by non-covalent bonds. Classical class I molecules are widely expressed on all nucleated cells with the exception of a few cell types including neurons.

The heavy α chains of class I genes consist of eight exons: exon 1 encodes the leader peptide; exons 2, 3 and 4 encode the three extracellular domains (α_1, α_2, and α_3) respectively; exon 5 encodes the transmembrane domain; exons 6 and 7 encode the cytoplasmic tail and exon 8 encodes the 3' UTR. The α_1, and α_2 domains form the peptide-binding groove (PBG) which comprises a β-pleated sheet floor with flanking walls that adopt an α-helical conformation. The α_3 domain binds β_2-microglobulin and CD8. Peptide-binding occurs when peptide interacts with pockets within the PBG. The class I PBG is closed and therefore can only bind short peptides that are 8—10 amino acids in length. The majority of the polymorphism exhibited by class I molecules resides within exons 2 and 3 which encode the α_1, and α_2 domains of the PBG.

A single class I or class II molecule can bind different peptides within its PBG. However, peptide binding is limited by specific peptide-binding motifs within the PBG which restrict binding to peptides containing certain amino acid residues. The PBG can be divided into "pockets" which accommodate specific peptide side-chains. The majority of HLA polymorphism occurs at these sites, hence determining which peptides are bound by specific HLA alleles. The PBG of class I molecules comprise six pockets (A—F); while class II molecules comprise nine pockets (P1—P9).

HLA CLASS II MOLECULES

In contrast to class I molecules, the MHC class II proteins comprise polymorphic α (33—35 kDa) and β (26—28 kDa) chains. The chains are largely encoded in tandem in the class II region for each of the five isotypes: *HLA-DM, -DO, -DP, -DQ*, and *-DR*, with the exception of *HLA-DO* α and β chains which are separated by a number of genes. As with class I, the class II genes can be divided into classical (*HLA-DR, -DQ* and *-DP*), and non-classical (*HLA-DM* and *-DO*) types. The *HLA-DRB* genes exhibit copy number variation, existing as several functional genes (*HLA-DRB1, -DRB3, -DRB4*, and *-DRB5*), as well as pseudogenes (*HLA-DRB2, -DRB6, -DRB7, -DRB8*, and *-DRB9*), thus adding a further layer of complexity to this region. There are a number of additional pseudogenes encoded in this region: *HLA-DQA2, HLA-DQB2, HLA-DQB3, HLA-DPA2*, and *HLA-DPB2*. The *HLA-DRB1* gene is the most polymorphic of all class II loci.

The class II genes encoding α and β chains are termed A and B (for example, *HLA-DQA1, HLA-DQB1*) and comprise five and six exons respectively. Both chains show similar genetic organization: exon 1 encodes the leader peptide, exons 2 and 3 encode the two extracellular domains (α_1 and α_2 for the α chain; β_1 and β_2 for the β chain). For the β chain, exon 4 encodes the transmembrane domain, exon 5 the cytoplasmic domain and exon 6 the 3'UTR. In the α chain, exon 4 encodes the transmembrane domain and the cytoplasmic domain, while the 3'UTR is encoded by exon 5. All class II α and β chains are polymorphic with the exception of *HLA-DRA* which exhibits very limited diversity with only three known alleles encoding two proteins (http://www.anthonynolan.org.uk/research/hlainformaticsgroup/). Hence genetic polymorphism for *HLA-DRB* largely derives from exon 2 of the β chain.

HLA-DRB GENE HAPLOTYPES

A variable number of expressed and non-expressed (pseudo) *HLA-DRB* genes map to different "arrangements" or haplotypes of chromosome 6. Thus far five such haplotypes have been identified and their

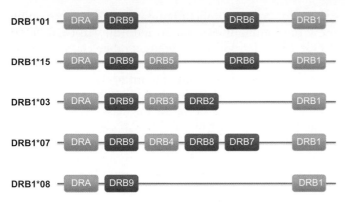

FIGURE 1.6 *HLA-DRB* gene haplotypes. The five different *HLA-DRB* gene haplotypes are illustrated. The green lozenges represent expressed genes; the blue lozenges represent pseudogenes (see also Table 1.1). *Adapted from [32].*

haplotypic structure can be determined given a specific *HLA-DRB1* genotype (Figure 1.6 and Table 1.1). All haplotypes have certain features in common such that the telomeric and centromeric borders are defined by the genes *HLA-DRA* and *HLA-DRB1* respectively. *HLA-DRB9* is also common to all haplotypes and lies adjacent to *HLA-DRA*. There may be zero to three genes or pseudogenes in the interval between *HLA-DRB9* and *HLA-DRB1*; no haplotype contains more than one other additional expressed gene: *HLA-DRB3*, *-DRB4*, or *-DRB5*.

Class II molecules show limited tissue expression (class II restriction) to thymic epithelial cells, and peripheral antigen-presenting cells such as B cells, macrophages and dendritic cells, but can also be induced on capillary endothelial cells and epithelial cells by IFN-γ. Extracellular proteins, which bind MHC class II molecules, enter cells through the process of endocytosis or phagocytosis. The peptide-bound MHC class II complex is then transported from the cytoplasm to the plasma membrane where it can interact with CD4+ T

cells (Figure 1.5). Unlike class I molecules, the PBG of class II polypeptides is "open" and can therefore accommodate longer peptide residues, typically 13—25 amino acids in length. The PBG of class II molecules is similar to that of class I molecules consisting of a β-pleated sheet floor and α-helical walls formed by the α_1 and β_1 domains of the heterodimer.

CD4+ and CD8+ T cells are activated via interactions with peptide-bound MHC molecules in conjunction with co-stimulatory molecules resulting in cell division, differentiation, cytokine release and thus cellular and humoral-mediated adaptive immunity.

THE MHC REGION AND SLE

Interestingly, the MHC has only been significantly linked to SLE in one of the 12 genome-wide linkage scans [35, 36] with supporting evidence from three further studies [37—39]. However, a meta-analysis of linkage studies in lupus has demonstrated evidence of significant linkage at *6p21* [40]. Moreover, recent genome-wide association scans (GWAS) clearly demonstrate that, thus far, variants within the MHC region constitutes the greatest genetic risk for SLE susceptibility in populations of European ancestry, with odds ratios (OR) of 2.0—2.4 [41—43]. The population-attributable risk (PAR) for the MHC in SLE has recently been calculated to be in the region of 13% [44].

The human MHC was first shown to be associated with SLE in 1971 when lupus probands were found to be enriched for the class I alleles, *HL-A8* (now known as *HLA-B8*) and *HLA-W15* (now known as *HLA-B15*), when compared with healthy controls using serological typing methods [45, 46]. These studies in lupus and other autoimmune diseases were consequent upon the known role of MHC alleles in self/non-self recognition. Associations between early complement component deficiencies and lupus were observed in

TABLE 1.1 *HLA-DRB* gene haplotypes. The five human *HLA-DRB* haplotypes are shown (also see Figure 1.6). *HLA-DRB1*08* haplotypes are the least complex and only additionally carry the pseudogene, *HLA-DRB9*, which is common to all *HLA-DRB* haplotypes. *HLA-DRB1*01* and *HLA-DRB1*10* haplotypes are phylogenetically similar and cluster in one group ([94])

Expressed gene			No other expressed gene	
DRB3 (DR52)*	**DRB4 (DR53)***	**DRB5 (DR51)***	**DRB6 (pseudogene)**	**–**
DRB1*03	DRB1*04	DRB1*15	DRB1*01	DRB1*08
DRB1*11	DRB1*07	DRB1*16	DRB1*10	
DRB1*12	DRB1*09			
DRB1*13				
DRB1*14				

* Denotes serological specificity.

the 1980s. Approximately one hundred case-control studies demonstrating association for a number of MHC variants have subsequently followed these initial observations providing clear evidence of the importance of the MHC in SLE susceptibility. However, the majority of these studies failed to encompass the entire locus and were undertaken in small, ethnically diverse cohorts. In addition, most of these studies did not account for LD between associated variants and tested only a handful of the hundreds of genes across the locus: essentially the classical HLA class I and class II genes (for their role in antigen presentation to T cells) as well as complement *C4* alleles given that hereditary and acquired deficiencies of this early classical complement component, leads to a lupus-like syndrome.

The most consistent HLA associations with SLE reside with the class II alleles, *HLA-DR3* (*DRB1*0301*) and to a lesser extent *HLA-DR2* (*DRB1*1501*) and their respective haplotypes in predominantly European-derived populations [47]. In particular, alleles residing within the "autoimmune" haplotype 8.1 (*HLA-A1/B8/Cw*07/ -308A TNFA/C4AQ*0/C4B1/C2*C/B*fS/DRB1*0301/DQA1* 0501/DQB1*0201*) show strong association with disease. Studies in non-white populations are inconsistent. For instance, investigation of the *HLA-DR* locus in the LUMINA study revealed increased frequency of *DRB1* 0301* in Caucasians and Hispanics from Texas [48]. *DRB1*0801* was increased in Hispanics from Texas but not Puerto Rico. African-Americans in the same study were found to have higher frequencies of *DRB1*1503* (*DR2*) compared to matched controls. These populations demonstrated no association with *DRB1*1501*. Other studies in African-Americans have revealed no association of *DRB1* alleles with SLE [49, 50]. An association with *HLA-DR4* was shown in the only study in North Indians [51]. A number of other studies in SLE have demonstrated associations among Mexican [52], Tunisian [53], Korean [54] and Thai [55] populations with *DRB1*15*; Mexican and Tunisian patients with *DRB1*0301* and *DRB3*01/03* in Jamaicans [55].

The MHC class I specificities, *A1* and *B8*, have been linked with SLE. However these alleles reside on the disease-associated *B8-DR3* haplotype and this effect most likely results from LD.

One might expect a close association between class II alleles and autoantibody subsets in SLE if these are indeed the causal variants, given their role in antigen presentation to T cells and subsequent stimulation of B cells to produce autoantibody. A variety of *HLA-DR* and -*DQ* alleles have been associated with autoantibody subsets in ethnically diverse lupus populations. The strongest associations have been demonstrated between anti-Ro and anti-La antibodies and *HLA-DR3* and *HLA-DQ2* (*DQB1*0201*), which are in strong LD [56—60]. Studies of individuals with the antiphospholipid

syndrome and antiphospholipid antibodies in SLE show predominant association with the *HLA-DR4/DQ8* (*DQB1*0302*) haplotype as well as other class II alleles [61—63].

Interestingly, apart from autoantibody subsets, there have been no consistent reports of HLA allele associations with the ACR criteria for the classification of SLE [64, 65].

Despite the fact that the class III region is the most gene-dense in the genome, only complement *C4* and *TNF* polymorphisms have been studied in any detail in SLE. To date the extensive LD across the MHC has hampered attempts to definitively establish a causal role for *C4* deficiency and SLE. MHC class III associations in SLE are further discussed in Chapter 2.

In 2002, Graham *et al.* used a different strategy, that of family-based association to screen the MHC region in a collection of 334, predominantly white, SLE families [66]. This study employed microsatellites as surrogate markers for *HLA-DRB1-DQB1*-containing haplotypes and identified three risk haplotypes: *DRB1*1501/ DQB1*0602*, *DRB1*0301/DQB1*0201*, and *DRB1*0801/ DQB1*0402*. Further analysis of ancestral recombinants in the *DRB1*1501* and *DRB1*0801* haplotypes narrowed the disease-associated region to an area of approximately 500 kb containing *DRB1* and *DQB1* as well as *C6orf10*, *BTNL2*, *HLA-DRA*, *HLA-DRB5* and *HLA-DQA1*. Extensive LD at the *DRB1*0301* haplotype compared to the *DRB1*1501* and *DRB1*0801* haplotypes resulted in fewer ancestral recombinants. Consequently, the disease-risk region on the *DRB1*0301* haplotype could only be narrowed to about 1 Mb of the genome which contains most of the class II and class III regions.

For the reasons outlined above, most published studies of MHC disease association to date have been restricted to small cohorts, each testing a limited number of variants using a variety of typing methodologies. This has resulted in a literature base that can be complex and at times conflicting. A recent pooled analysis of case-control association studies across the region over the last 30 years was undertaken in order to consolidate and evaluate the current literature base in SLE and other autoimmune/inflammatory diseases [33]. This study demonstrated predominant association with variants linked to *HLA-DR3* and *HLA-DR2*-bearing haplotypes in European lupus populations as expected (Figure 1.7). The strongest associations were found with the *HLA-DR3* haplotypes, *HLA-B8-DRB1*0301* and *HLA-B18-DRB1*0301* (odds ratios (OR) 1.5—2.5); while the *HLA-DR2/DRB1*1501* haplotype exhibited OR of approximately 1.7. Some variants in this review showed association that was limited to a particular population: *HLA-DRB1*1503*, a *HLA-DR2* subtype, was seen in African-Americans only; *HLA-DRB1*1602* was observed in Mexican Mestizo, Thai, and Bulgarian populations;

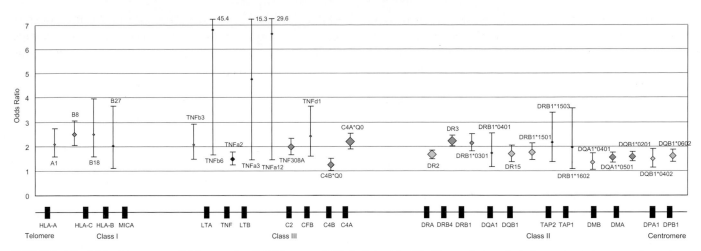

FIGURE 1.7 Major histocompatibility complex (MHC) susceptibility* alleles identified by pooled analysis for systemic lupus erythematosus (*defined as lower confidence interval (CI) greater than 1.0). Odds ratios with 95% CI are shown by the diamonds and whiskers respectively. Beneath is a schematic representation of MHC class I, class III, and class II genes in genomic order but not to scale. Diamond size reflects total number of cases included in pooled analysis for each allele. Diamond colour reflects different disease-relevant ancestral haplotypes. [33].

and *HLA-DRB1*0401* was seen largely in Mexican Mestizo and Hispanic cohorts in whom this allele is uncommon (frequency ~1%). Two further associated class II alleles, *HLA-DQA1*0401* and *HLA-DQB1*0402*, reside on a *HLA-DR8* haplotype which is infrequent in Caucasian populations (frequency ~2%).

Hence, in summary, until the mid-2000s nearly all disease association studies of the MHC had been limited to a subset of ~20 genes in small cohorts of predominant European origin. These genes included the classical HLA loci (*HLA-A, -B, -C, -DRB, -DQA, -DQB, -DPA, -DPB*), *TNF, LTA, LTB*, the *TAP* genes, *MICA, MICB* and the complement loci (*C2, C4, CFB*). Moreover, most of these studies individually investigated only a small proportion of this limited subset of genes, thus impeding the ability to compare the strength of association across study groups. Given that there are 421 genetic loci currently annotated to the xMHC, approximately 252 (60%) of which are thought to be expressed it became necessary that a more comprehensive approach to the study of the MHC in disease be undertaken in conjunction with conditional analyses that address the issue of independent susceptibility loci within the region. In order to differentiate the effects of tightly linked loci one needs a dense map of variation and large cohorts of ethnically and thus haplotypically diverse populations so that rare, distinguishing recombination events can be identified. Conditional analyses can then be applied to separate allelic from haplotypic association. Recent technological and bioinformatic advances are now able to tackle these obstacles, as highlighted by the success of MHC SNP studies in complex diseases such as multiple sclerosis [67], type 1 diabetes [68] and SLE [69, 70].

SNP-BASED MHC ANALYSES IN SLE

The completion of the human genome project in 2001 yielded an immense amount of information including the cataloguing of common sequence variation across the entire genome and dissemination of this information in the form of publicly available databases [23, 24]. This is important because common sequence variation is thought to underpin much of the genetic component of complex disease traits such as SLE, asthma and type 1 diabetes mellitus. Single nucleotide polymorphisms (SNPs) are a form of such sequence variation, and constitute the most common form of polymorphism in the human genome (at least 10 million, or one every 300 base pairs). SNPs were thought to account for approximately 90% of human genetic variation prior to the recognition of the importance of copy number variation (CNV) in the human genome. SNPs are used in the investigation of the genetic component of complex human diseases for a number of reasons. They are numerous and less mutable than other types of polymorphism. SNPs are highly amenable to high-throughput genotyping and population differences in SNP frequencies are advantageous in population-based genetic studies. Further examination of the genetic architecture of the human genome led to the discovery that neighboring SNPs or combinations of SNPs, known as haplotypes, were correlated, or in linkage disequilibrium, with each other [11, 71]. A haplotype is a combination of alleles that lie close together in the genome and are inherited together as a consequence of minimal genetic recombination. These regions of correlation were found to be separated by areas of gene conversion and recombination or "recombination hot-spots", thus giving rise to haplotype blocks which characterize the structure of the majority of the genome. Linkage disequilibrium (LD) or allelic association occurs when alleles at two or more neighboring loci occur together in frequencies significantly different from those expected from the individual allele frequencies; also defined as the non-random association of alleles at linked loci [72]. Maps of these patterns of correlation can be generated using SNP data from cohort studies using publicly available software programs [73]. Haplotype blocks have been found to contain sequences of limited diversity [11, 74]. Commonly, only four to six different haplotypes will account for more than 95% of sequence variation within a given block. Further analysis of haplotype block structure has revealed that a subset of SNPs within each block can be used to identify the common haplotypes therein. Such subsets of SNPs are known as haplotype-tagging SNPs (tag SNPs). These tag SNPs act as surrogate markers for common haplotypes and other sequence variation within a haplotype, thus obviating the need to type every SNP within a genetic region in order to identify sequence variation. Hence, the utility of tag SNPs lies in the fact that a relatively small number of SNPs can be genotyped within a given chromosomal region to allow the assessment of common genetic variation within that region in a given population, making genetic studies more efficient and powerful.

The recent advent of genome-wide association screens (GWAS) in complex human diseases allows the interrogation of the majority of the genome negating the need for *a priori* hypotheses in disease susceptibility. To date there have been four GWA scans published in lupus [41−43, 75]. It is important to note that these studies have been undertaken in individuals of predominant northern European ancestry. The current GWA scans have been very informative not only in that they have confirmed association at known susceptibility loci (alleles at the MHC, *IRF5*, *PTPN22* and *STAT4* for example) but importantly have also shown replicated association with genes not previously known to be involved in lupus pathogenesis (*ITGAM/ITGAX*, *BANK1*). Three of the four GWA scans in lupus demonstrated that the MHC region contributes the greatest genetic risk in SLE cohorts predominantly comprising women of European descent [41−43]. The effect of the MHC region was not reported in the fourth study [75].

Specific studies have focused on examining the association of SNPs within the MHC region and SLE. In a family-based SNP association study of the MHC in SLE, 68 SNPs and *HLA-DRB1* were genotyped across 2.4 Mb of the MHC, encompassing class III and class II, in a cohort of 314 UK Caucasian SLE trios. The SNPs were chosen to tag HLA and non-HLA variation across this region using data from previous high-density SNP maps of the MHC [12, 13, 76]. Transmission disequilibrium testing of the data and conditional analyses demonstrated two distinct and independent signals at the MHC region in SLE: one tagged by *HLA-DRB1*0301* in class II and the other tagged by the T allele of SNP *rs419788* in the class III gene *SKIV2L*.

Examination of haplotype bifurcation plots for transmitted (T) and untransmitted (UT) *HLA-DRB1*0301*-containing haplotypes enabled delineation of the class II association signal to a 180 kb region encompassing three genetic variants, *HLA-DRB1*0301-HLA-DQA1*0501-HLA-DQB1*0201* (Figure 1.8). All three allelic variants represent attractive functional candidates in lupus susceptibility for their role in antigen presentation and stimulation of the adaptive immune response. They may confer disease risk individually or in combination given the strong LD exhibited in this region.

Furthermore, examination of LD structure around the class III signal, *rs419788-T* suggested that this signal could also be delimited to a relatively narrow genomic interval including the genes *CFB*, *RDBP*, *SKIV2L*, *DOM3Z*, and *STK19*. *CFB* is a vital component of the alternate complement pathway and dysregulation may clearly affect the inflammatory response [77]. RD and Skiv2l are proteins potentially involved in RNA processing. Skiv2l is a DEAD box protein with possible function as an RNA helicase. The function of Dom3z is currently unknown, although the homologous yeast protein binds nuclear exoribonuclease. STK19 is a protein kinase of unknown function with primary nuclear localization and forms part of the modular RCCX complex containing complement C4 [78]. Interestingly, *RDBP* and *SKIV2L* are found to be highly expressed in T lymphocytes, B lymphocytes, and dendritic cells [69].

In addition, by constructing haplotype bifurcation plots centred on *HLA-DRB1*0301*, this study illustrated preservation of the common *HLA-DRB1*0301* haplotype in HapMap CEPH and UK SLE populations, while that seen in the HapMap Yoruban cohort was significantly different (Figure 1.9). The class II regions of all three populations are essentially identical across the chosen SNPs; the main differences lie in class III. The difference in African populations in the class III region is one possible explanation for the lack of evidence for an association between *HLA-DRB1*0301* and SLE in African or African-American populations, if this association is driven by class III variants. However, *HLA-DRB1*0301* has a lower frequency (~7%−10%) in African populations compared with Europeans (~13%), and the number of HLA association studies conducted in African populations is very limited.

A number of studies have demonstrated conflicting evidence for and against association with various *TNF*

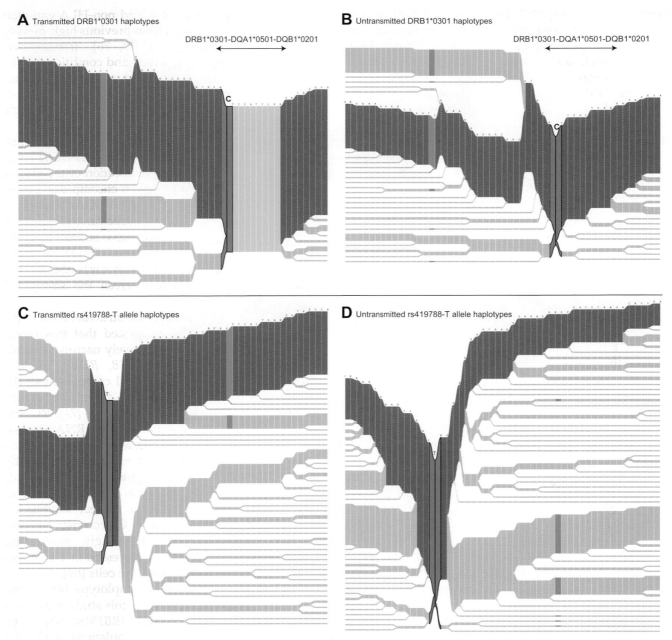

FIGURE 1.8 Structure of transmitted (T) and untransmitted (UT) *HLA-DRB1*0301* and *rs419788-T* allele haplotypes. Haplotype bifurcation plots* were constructed using 120 randomly selected parental chromosomes from each of the four datasets for comparative purposes except (*B*),where there were only 90 chromosomes in the entire dataset: (*A*) T *HLA-DRB1*0301* haplotypes, (*B*) UT *HLA-DRB1*0301* haplotypes, (*C*) T *rs419788-T* allele haplotypes, and (*D*) UT *rs419788-T* allele haplotypes. The allelic composition of the most common haplotype in each subset is shown: the core allele is represented as a dark blue double bar indicating haplotypes to the right and to the left of the core; otherwise, the common haplotype is depicted by dark gray bars. In parts (*A*) and (*B*), the *rs419788-T* allele in class III that shows association independent of *HLA-DRB1*0301* in our cohort is indicated in green, while in parts (*C*) and (*D*), the allele *HLA-DRB1*0301* is shown in green. The key difference between *HLA-DRB1*0301* T and UT haplotypes lies within the class II region of the MHC. All *HLA-DRB1*0301* T haplotypes are identical across a 180 kb region defined by eight SNPs (light blue), whereas the corresponding region within UT *HLA-DRB1*0301* haplotypes exhibits significant recombination. This conserved class II interval encompasses only three expressed genes: *HLA-DRB1*, *HLA-DQA1*, and *HLA-DQB1*. Given the strong LD exhibited by *HLA-DRB1*0301* haplotypes, the allelic composition of this risk region is known to be *HLA-DRB1*0301-HLA-DQA1*0501-HLA-DQB1*0201*. Both T and UT *rs419788-T* allele haplotypes show similar structure overall. The *rs419788-T* allele is clearly present on *HLA-DRB1*0301* and non-*HLA-DRB1*0301*-containing haplotypes, lending credence to our observation that *rs419788-T* or another variant in LD with it constitutes an association signal independent of *HLA-DRB1*0301* in our UK SLE cohort. (*A*) 120 from a total of 176 parental chromosomes, (*B*) 90 from a total of 90 parental chromosomes, (*C*) 120 from a total of 284 parental chromosomes, and (*D*) 120 from a total of 182 parental chromosomes. [69]* The breakdown of LD on core haplotypes is presented using haplotype bifurcation diagrams [93] (also explained in [76]).

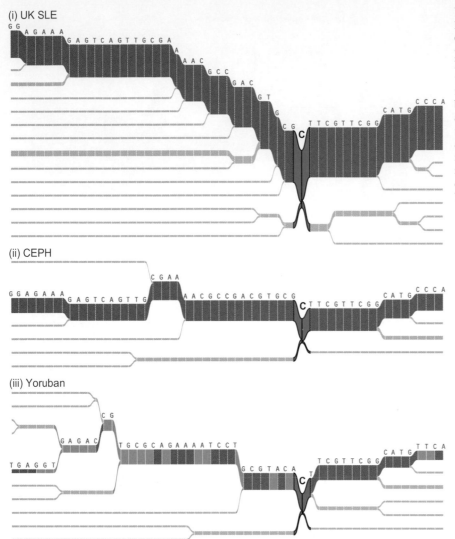

FIGURE 1.9 Comparison of *HLA-DRB1*0301* haplotype bifurcation plots for (i) UK SLE, (ii) CEPH, and (iii) Yoruban populations. We show preservation of the common *HLA-DRB1*0301* haplotype in CEPH and UK SLE, while that seen in the Yorubans is significantly different (differences indicated in yellow, core allele shown in dark blue). The class II regions of all three populations are essentially identical across our chosen SNPs; the main differences lie in class III. One hundred and twenty (of 1,256) randomly selected transmitted and untransmitted parental UK SLE chromosomes used for haplotype bifurcation plot for comparison with 120 total CEPH and 120 total Yoruban chromosomes [69].

locus polymorphisms in SLE [79]. A meta-analysis of the *TNF-308G/A* promoter polymorphism in SLE revealed evidence of association for the minor allele (A) in European populations; however, this study did not account for LD with class II alleles [79]. On conditioning for *HLA-DRB1*0301*, the UK SLE family study demonstrated that the *TNF-308A* promoter signal is lost, suggesting that this association is not independent and is due to LD with *HLA-DRB1*0301* (or another variant in LD with *HLA-DRB1*0301*).

The IMAGEN (International MHC and Autoimmunity GEnetics Network) consortium, a collaborative research effort funded by the NIH, was created with the aim of further characterizing MHC association signals in multiple autoimmune and inflammatory diseases. The initial IMAGEN study examined the association of 1472 tag SNPs, chosen to capture common genetic variation across 3.44 Mb of the classical MHC, in 10,576 DNA samples derived from patients with SLE, Crohn's disease, ulcerative colitis, rheumatoid arthritis, myasthenia gravis, selective IgA deficiency, multiple sclerosis and appropriate control samples. These data demonstrate that MHC associations with these autoimmune diseases result from complex, multilocus effects that span the entire region [70]. In the SLE cohort, comprising 1126 cases of European ancestry (643 UK, 483 US) and 1895 controls (746 UK, 1049 US), the most highly associated SNP (*rs1269852*) was found to be located in the class III region between the genes, *TNXB* and *CREBL1* (OR = 2.4, p = 5.63 × 10^{-29}). This SNP is strongly though not absolutely correlated with the *HLA-DRB1*0301* allele (r^2 = 0.78) − the top HLA allele in this dataset (*HLA-DRB1*0301* tag SNP, *rs2187668*, OR = 2.2, p = 5.62 × 10^{-26}). Conditioning on the top signal (*rs1269852*) identified a number of secondary signals, the best of which were variants in LD with, and including, *HLA-DRB1*1501* (Figure 1.10). Other signals potentially independent of *rs1269852* are seen in class I (between *RNF39* and *TRIM31*), class III (*NOTCH4* and *SKIV2L*) and class II (*HLA-DQB1-DQA2*).

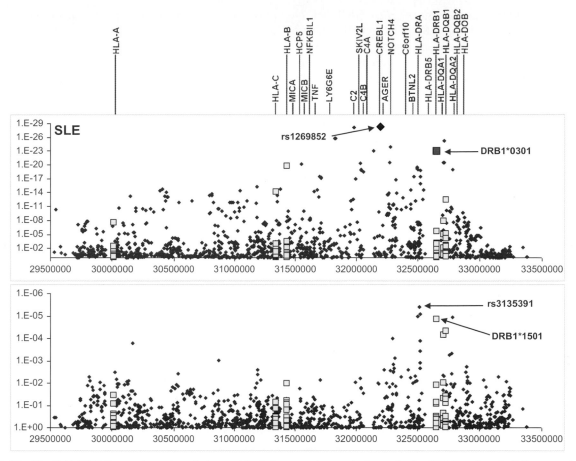

FIGURE 1.10 The IMAGEN Consortium: Association and logistic regression analysis results for SLE. Results of allelic tests of association (top panel) for SNPs (black lozenges) and imputed HLA alleles (yellow boxes). All association results are represented as the −log10 of the p-values (y-axis). The most associated SNP and HLA are highlighted in blue and red, respectively. The dataset was conditioned on the top SNP, rs1269852, in order to identify secondary association signals (lower panel) [70].

Hence, there appear to be at least three separate signals in the class III region of the MHC tagged by variants in and around the *SKIV2L* gene, TNXB-CREBL1 (*rs1269852*), and *NOTCH4*. Together, these data indicate the presence of multiple SLE-risk alleles located across the class I, class II and class III regions. The influence of copy number variation at the complement *C4/RCCX* locus in relation to the association signals demonstrated in these studies remains to be established. This high-density SNP analysis confirms that the predominant signals in SLE map to the class II and class III regions of the MHC. Also of note, there appear to be at least two class III associations, with peaks on either side of the RCCX module, in addition to a further signal centered around *NOTCH4*, a gene involved in development and cell fate.

These SNP studies of the MHC in lupus confirm the association of anti-Ro and anti-La antibody subsets with *HLA-DRB1*0301*. Indeed, data from the IMAGEN study demonstrate that 37% of Ro-positive and 50% of La-positive UK probands possess at least one copy of the *HLA-DRB1*0301* allele (OR = 2.02 and 3.87 respectively).

There was no association between *HLA-DRB1*1501* and *HLA-DRB1*0801* with SLE in the UK family SNP study. These findings are consistent with previous data from the UK [80], Spain [81], The Netherlands [82], Sweden [83], Mexico [84], and the US [48], but conflict with that of other US groups [66, 85]. The IMAGEN study was able to show a significant and independent effect of *HLA-DRB1*1501* in SLE, but not *HLA-DRB1*0801*. These discrepancies likely reflect differences between the various study populations, disease heterogeneity and issues related to power. Interestingly, there are reports of a trend for under-transmission of *HLA-DRB1*0701* — a result observed in UK, US and Canadian lupus studies [69, 70, 80, 86]. Moreover, a protective effect of *HLA-DRB1*0701* has been reported in other autoimmune diseases including Graves disease [87, 88], type 1 diabetes [89], and rheumatoid arthritis [90].

Previous studies have demonstrated increased risk for lupus in individuals carrying particular combinations of microsatellite-inferred *HLA-DRB1-HLA-DQB1* haplotypes [64, 66]. The highest risk genotype was found to be the compound heterozygote *HLA-DRB1*0301-DQB1*0201/HLA-DRB1*1501-DQB1*0602*, while *HLA-DRB1*0301-DQB1*0201*-containing genotypes demonstrated a dose-dependent effect in increasing lupus susceptibility [64, 66]. Analysis of *HLA-DRB1* data from the IMAGEN study show that *HLA-DRB1*0301* homozygotes comprise the greatest HLA risk genotype for lupus susceptibility in this cohort (combined UK and US OR = 3.16, UK OR = 4.68, US OR = 2.11). Individuals heterozygous for *HLA-DRB1*0301/HLA-DRB1*1501* and homozygous for *HLA-DRB1*1501* demonstrated weaker effects (combined UK and US OR = 2.34, UK OR = 2.44, US OR = 2.29 and combined UK and US OR = 1.62, UK OR = 1.42, US OR = 1.84, respectively). The differences between UK and US SLE cohorts may reflect disease, ethnic, and haplotypic heterogeneity. A recent high-density SNP study undertaken in European SLE families has also shown independent association of multiple variants across the MHC region [91].

THE FUTURE: DEFINING CAUSAL VARIATION

In the past, the strong LD observed at the MHC had made it difficult to determine whether single or multiple independent loci were responsible for the MHC associations observed in lupus. It is now clear that the latter is true and unravelling these complex interactions and associations will provide the next challenge in dissecting the MHC contribution to lupus susceptibility.

High-density SNP studies can be powerful in detecting disease association due to common genetic variation; however, their ability to capture variation due to copy number polymorphism is limited or unknown in many instances. Therefore, at the present time, the impact of CNV in the genetic susceptibility to complex diseases cannot be quantified with accuracy using SNPs. Alternative methods such as quantitative PCR, array CGH, Multiplex Ligation-dependent Probe Amplification (MLPA), pyrosequencing and paralog ratio tests are therefore being employed to more accurately determine the influence of such structural variation in disease susceptibility. As the MHC contains two CNV loci — the RCCX module and the *HLA-DRB* genes — the issue is of particular importance in this region. Integration of such CNV data with high-density SNP typing across the MHC will hopefully establish whether complement *C4* associations with lupus are primary or secondary to LD. Moreover, in order to further refine MHC association signals in lupus, higher-density SNP mapping studies in large transethnic cohorts coupled with sequencing of associated intervals/haplotypes will be necessary

and are currently in progress. Lupus subphenotypes including renal disease and antibody subsets, provide more homogeneous cohorts for analysis and should also help in delimiting disease-associated variants.

It seems likely that genetic association studies may not be able to delimit all MHC association signals to specific genes or genetic variants. Indeed, it has been proposed that disease-associated haplotypes may harbor numerous susceptibility alleles which may not be immediately apparent from genetic analyses. For example, many functionally relevant alleles that are in LD with an independent association signal may act in concert with it in predisposing to disease, while having a negligible effect alone. Thus, further refinement of MHC association signals will include gene expression studies of this region in SLE. Future work will also need to include epigenetic studies at the MHC and evaluation of peptide-binding pockets within associated HLA alleles. The possibility that the major genetic influence at the MHC in lupus may arise from the class III region and not classical HLA class II alleles requires confirmation. A meta-analysis of high-density MHC SNP studies in SLE is currently underway as are transethnic subphenotybe analyses. The results of these studies should also help to inform further research efforts in refining MHC association signals in lupus.

References

[1] P.A. Gorer, The detection of a hereditary antigenic difference in the blood of mice by means of human group A serum, J Genet 32 (1936) 17−31.

[2] P.A. Gorer, The detection of antigenic differences in mouse erythrocytes bt the employment of immune sera, Br J Exp Pathol 17 (1936) 42−50.

[3] J. Klein, Seeds of time: fifty years ago Peter A. Gorer discovered the H-2 complex, Immunogenetics 24 (1986) 331−338.

[4] G.D. Snell, Methods for the study of histocompatibility genes, J Genet (1948) 87−108.

[5] J. Dausset, [Iso-leuko-antibodies.], Acta. Haematol 20 (1958) 156−166.

[6] B. Benacerraf, Role of MHC gene products in immune regulation, Science 212 (1981) 1229−1238.

[7] R. Horton, L. Wilming, V. Rand, R.C. Lovering, E.A. Bruford, V.K. Khodiyar, et al., Gene map of the extended human MHC, Nat Rev Genet 5 (2004) 889−899.

[8] D. Zernich, A.W. Purcell, W.A. Macdonald, L. Kjer-Nielsen, L.K. Ely, N. Laham, et al., Natural HLA class I polymorphism controls the pathway of antigen presentation and susceptibility to viral evasion, J Exp Med 200 (2004) 13−24.

[9] B. Amos, F.E. Ward, C.M. Zmijewski, B.G. Hattler, H.F. Seigler, Graft donor selection based upon single locus (haplotype) analysis within families, Transplantation 6 (1968) 524−534.

[10] T.M. Williams, Human leukocyte antigen gene polymorphism and the histocompatibility laboratory, J Mol Diagn 3 (2001) 98−104.

[11] S.B. Gabriel, S.F. Schaffner, H. Nguyen, J.M. Moore, J. Roy, B. Blumenstiel, et al., The structure of haplotype blocks in the human genome, Science 296 (2002) 2225−2229.

[12] E.C. Walsh, K.A. Mather, S.F. Schaffner, L. Farwell, M.J. Daly, N. Patterson, et al., An integrated haplotype map of the human major histocompatibility complex, Am J Hum Genet 73 (2003) 580–590.

[13] M.M. Miretti, E.C. Walsh, X. Ke, M. Delgado, M. Griffiths, S. Hunt, et al., A high-resolution linkage-disequilibrium map of the human major histocompatibility complex and first generation of tag single-nucleotide polymorphisms, Am J Hum Genet 76 (2005) 634–646.

[14] E.J. Yunis, C.E. Larsen, M. Fernandez-Vina, Z.L. Awdeh, T. Romero, J.A. Hansen, et al., Inheritable variable sizes of DNA stretches in the human MHC: conserved extended haplotypes and their fragments or blocks, Tissue Antigens 62 (2003) 1–20.

[15] M.A. Degli-Esposti, A.L. Leaver, F.T. Christiansen, C.S. Witt, L.J. Abraham, R.L. Dawkins, Ancestral haplotypes: conserved population MHC haplotypes, Hum Immunol 34 (1992) 242–252.

[16] M.A. Degli-Esposti, C. Leelayuwat, L.N. Daly, C. Carcassi, L. Contu, L.F. Versluis, et al., Updated characterization of ancestral haplotypes using the Fourth Asia-Oceania Histocompatibility Workshop panel, Hum Immunol 44 (1995) 12–18.

[17] Z.L. Awdeh, D. Raum, E.J. Yunis, C.A. Alper, Extended HLA/complement allele haplotypes: evidence for T/t-like complex in man, Proc Natl Acad Sci U S A 80 (1983) 259–263.

[18] M.T. Dorak, W. Shao, H.K. Machulla, E.S. Lobashevsky, J. Tang, M.H. Park, et al., Conserved extended haplotypes of the major histocompatibility complex: further characterization, Genes Immun 7 (2006) 450–467.

[19] C.V. Jongeneel, L. Briant, I.A. Udalova, A. Sevin, S.A. Nedospasov, A. Cambon-Thomsen, Extensive genetic polymorphism in the human tumor necrosis factor region and relation to extended HLA haplotypes, Proc Natl Acad Sci U S A 88 (1991) 9717–9721.

[20] G. Sturfelt, G. Hellmer, L. Truedsson, TNF microsatellites in systemic lupus erythematosus-a high frequency of the TNFabc 2-3-1 haplotype in multicase SLE families, Lupus 5 (1996) 618–622.

[21] R.D. Campbell, J. Trowsdale, Map of the human MHC, Immunol Today 14 (1993) 349–352.

[22] The MHC sequencing consortium, Complete sequence and gene map of a human major histocompatibility complex. The MHC sequencing consortium, Nature 401 (1999) 921–923.

[23] E.S. Lander, L.M. Linton, B. Birren, C. Nusbaum, M.C. Zody, J. Baldwin, et al., Initial sequencing and analysis of the human genome, Nature 409 (2001) 860–921.

[24] J.C. Venter, M.D. Adams, E.W. Myers, P.W. Li, R.J. Mural, G.G. Sutton, et al., The sequence of the human genome, Science 291 (2001) 1304–1351.

[25] A.J. Mungall, S.A. Palmer, S.K. Sims, C.A. Edwards, J.L. Ashurst, L. Wilming, et al., The DNA sequence and analysis of human chromosome 6, Nature 425 (2003) 805–811.

[26] C.A. Stewart, R. Horton, R.J. Allcock, J.L. Ashurst, A.M. Atrazhev, P. Coggill, et al., Complete MHC haplotype sequencing for common disease gene mapping, Genome Res 14 (2004) 1176–1187.

[27] R. Horton, R. Gibson, P. Coggill, M. Miretti, R.J. Allcock, J. Almeida, et al., Variation analysis and gene annotation of eight MHC haplotypes: the MHC Haplotype Project, Immunogenetics 60 (2008) 1–18.

[28] R.J. Allcock, A.M. Atrazhev, S. Beck, P.J. de Jong, J.F. Elliott, S. Forbes, et al., The MHC haplotype project: a resource for HLA-linked association studies, Tissue Antigens 59 (2002) 520–521.

[29] J.A. Traherne, R. Horton, A.N. Roberts, M.M. Miretti, M.E. Hurles, C.A. Stewart, et al., Genetic analysis of completely sequenced disease-associated MHC haplotypes identifies shuffling of segments in recent human history, PLoS Genet 2 (2006) e9.

[30] W.P. Smith, Q. Vu, S.S. Li, J.A. Hansen, L.P. Zhao, D.E. Geraghty, Toward understanding MHC disease associations: partial resequencing of 46 distinct HLA haplotypes, Genomics 87 (2006) 561–571.

[31] T. Shiina, M. Ota, S. Shimizu, Y. Katsuyama, N. Hashimoto, M. Takasu, et al., Rapid evolution of major histocompatibility complex class I genes in primates generates new disease alleles in humans via hitchhiking diversity, Genetics 173 (2006) 1555–1570.

[32] S.G.E. Marsh, P. Parham, L.D. Barber, The HLA FactsBook, Academic Press, 2000.

[33] M.M. Fernando, C.R. Stevens, E.C. Walsh, P.L. De Jager, P. Goyette, R.M. Plenge, et al., Defining the role of the MHC in autoimmunity: a review and pooled analysis, PLoS Genet 4 (2008) e1000024.

[34] M.C. Pickering, M.J. Walport, Links between complement abnormalities and systemic lupus erythematosus, Rheumatology (Oxford) 39 (2000) 133–141.

[35] P.M. Gaffney, G.M. Kearns, K.B. Shark, W.A. Ortmann, S.A. Selby, M.L. Malmgren, et al., A genome-wide search for susceptibility genes in human systemic lupus erythematosus sib-pair families, Proc Natl Acad Sci U S A 95 (1998) 14875–14879.

[36] P.M. Gaffney, W.A. Ortmann, S.A. Selby, K.B. Shark, T.C. Ockenden, K.E. Rohlf, et al., Genome screening in human systemic lupus erythematosus: results from a second Minnesota cohort and combined analyses of 187 sib-pair families, Am J Hum Genet 66 (2000) 547–556.

[37] R. Shai, F.P. Quismorio Jr., L. Li, O.J. Kwon, J. Morrison, D.J. Wallace, et al., Genome-wide screen for systemic lupus erythematosus susceptibility genes in multiplex families, Hum Mol Genet 8 (1999) 639–644.

[38] A.K. Lindqvist, K. Steinsson, B. Johanneson, H. Kristjansdottir, A. Arnasson, G. Grondal, et al., A susceptibility locus for human systemic lupus erythematosus (hSLE1) on chromosome 2q, J Autoimmun 14 (2000) 169–178.

[39] C. Gray-McGuire, K.L. Moser, P.M. Gaffney, J. Kelly, H. Yu, J.M. Olson, et al., Genome scan of human systemic lupus erythematosus by regression modeling: evidence of linkage and epistasis at 4p16-15.2, Am J Hum Genet 67 (2000) 1460–1469.

[40] P. Forabosco, J.D. Gorman, C. Cleveland, J.A. Kelly, S.A. Fisher, W.A. Ortmann, et al., Meta-analysis of genome-wide linkage studies of systemic lupus erythematosus, Genes Immun 7 (2006) 609–614.

[41] J.B. Harley, M.E. Alarcon-Riquelme, L.A. Criswell, C.O. Jacob, R.P. Kimberly, K.L. Moser, et al., Genome-wide association scan in women with systemic lupus erythematosus identifies susceptibility variants in ITGAM, PXK, KIAA1542 and other loci, Nat. Genet 40 (2008) 204–210.

[42] G. Hom, R.R. Graham, B. Modrek, K.E. Taylor, W. Ortmann, S. Garnier, et al., Association of systemic lupus erythematosus with C8orf13-BLK and ITGAM-ITGAX, N Engl J Med 358 (2008) 900–909.

[43] R.R. Graham, C. Cotsapas, L. Davies, R. Hackett, C.J. Lessard, J.M. Leon, et al., Genetic variants near TNFAIP3 on 6q23 are associated with systemic lupus erythematosus, Nat Genet (2008).

[44] A. Zhernakova, C.C. van Diemen, C. Wijmenga, Detecting shared pathogenesis from the shared genetics of immune-related diseases, Nat Rev Genet 10 (2009) 43–55.

[45] F.C. Grumet, A. Coukell, J.G. Bodmer, W.F. Bodmer, H.O. McDevitt, Histocompatibility (HL-A) antigens associated with systemic lupus erythematosus. A possible genetic predisposition to disease, N Engl J. Med 285 (1971) 193–196.

[46] H. Waters, P. Konrad, R.L. Walford, The distribution of HL-A histocompatibility factors and genes in patients with systemic lupus erythematosus, Tissue Antigens 1 (1971) 68–73.

[47] B.P. Tsao, Update on human systemic lupus erythematosus genetics, Curr Opin Rheumatol 16 (2004) 513–521.

[48] A.G. Uribe, G. McGwin Jr., J.D. Reveille, G.S. Alarcon, What have we learned from a 10-year experience with the LUMINA (Lupus in Minorities; Nature vs. nurture) cohort? Where are we heading? Autoimmun Rev 3 (2004) 321–329.

[49] J.D. Reveille, R.E. Schrohenloher, R.T. Acton, B.O. Barger, DNA analysis of HLA-DR and DQ genes in American blacks with systemic lupus erythematosus, Arthritis Rheum 32 (1989) 1243–1251.

[50] P.F. Howard, M.C. Hochberg, W.B. Bias, F.C. Arnett Jr., R.H. McLean, Relationship between C4 null genes, HLA-D region antigens, and genetic susceptibility to systemic lupus erythematosus in Caucasian and black Americans, Am J Med 81 (1986) 187–193.

[51] N.K. Mehra, I. Pande, V. Taneja, S.S. Uppal, S.P. Saxena, A. Kumar, et al., Major histocompatibility complex genes and susceptibility to systemic lupus erythematosus in northern India, Lupus 2 (1993) 313–314.

[52] L.M. Cortes, L.M. Baltazar, M.G. Lopez-Cardona, N. Olivares, C. Ramos, M. Salazar, et al., HLA class II haplotypes in Mexican systemic lupus erythematosus patients, Hum Immunol 65 (2004) 1469–1476.

[53] K. Ayed, Y. Gorgi, S. Ayed-Jendoubi, R. Bardi, The involvement of HLA -DRB1*, DQA1*, DQB1* and complement C4A loci in diagnosing systemic lupus erythematosus among Tunisians, Ann Saudi Med 24 (2004) 31–35.

[54] H.S. Lee, Y.H. Chung, T.G. Kim, T.H. Kim, J.B. Jun, S. Jung, et al., Independent association of HLA-DR and FCgamma receptor polymorphisms in Korean patients with systemic lupus erythematosus, Rheumatology (Oxford) 42 (2003) 1501–1507.

[55] M. Smikle, N. Christian, K. DeCeulaer, E. Barton, K. Roye-Green, G. Dowe, et al., HLA-DRB alleles and systemic lupus erythematosus in Jamaicans, South Med J 95 (2002) 717–719.

[56] P.H. Schur, Genetics of systemic lupus erythematosus, Lupus 4 (1995) 425–437.

[57] D. Logar, B. Vidan-Jeras, V. Dolzan, B. Bozic, T. Kveder, The contribution of HLA-DQB1 coding and QBP promoter alleles to anti-Ro alone autoantibody response in systemic lupus erythematosus, Rheumatology (Oxford) 41 (2002) 305–311.

[58] M.R. Azizah, S.S. Ainoi, S.H. Kuak, N.C. Kong, Y. Normaznah, M.N. Rahim, The association of the HLA class II antigens with clinical and autoantibody expression in Malaysian Chinese patients with systemic lupus erythematosus, Asian Pac J Allergy Immunol 19 (2001) 93–100.

[59] S. Miyagawa, K. Shinohara, M. Nakajima, K. Kidoguchi, T. Fujita, T. Fukumoto, et al., Polymorphisms of HLA class II genes and autoimmune responses to Ro/SS-A-La/SS-B among Japanese subjects, Arthritis Rheum 41 (1998) 927–934.

[60] M. Galeazzi, G.D. Sebastiani, G. Morozzi, C. Carcassi, G.B. Ferrara, R. Scorza, et al., HLA class II DNA typing in a large series of European patients with systemic lupus erythematosus: correlations with clinical and autoantibody subsets, Medicine (Baltimore) 81 (2002) 169–178.

[61] F.C. Arnett, M.L. Olsen, K.L. Anderson, J.D. Reveille, Molecular analysis of major histocompatibility complex alleles associated with the lupus anticoagulant, J Clin Invest 87 (1991) 1490–1495.

[62] F.C. Arnett, P. Thiagarajan, C. Ahn, J.D. Reveille, Associations of anti-beta2-glycoprotein I autoantibodies with HLA class II alleles in three ethnic groups, Arthritis Rheum 42 (1999) 268–274.

[63] M. Galeazzi, G.D. Sebastiani, A. Tincani, J.C. Piette, F. Allegri, G. Morozzi, et al., HLA class II alleles associations of anti-cardiolipin and anti-beta2GPI antibodies in a large series of European patients with systemic lupus erythematosus, Lupus 9 (2000) 47–55.

[64] R.R. Graham, W. Ortmann, P. Rodine, K. Espe, C. Langefeld, E. Lange, et al., Specific combinations of HLA-DR2 and DR3 class II haplotypes contribute graded risk for disease susceptibility and autoantibodies in human SLE, Eur J Hum Genet 15 (2007) 823–830.

[65] J.B. Harley, K.L. Moser, P.M. Gaffney, T.W. Behrens, The genetics of human systemic lupus erythematosus, Curr Opin Immunol 10 (1998) 690–696.

[66] R.R. Graham, W.A. Ortmann, C.D. Langefeld, D. Jawaheer, S.A. Selby, P.R. Rodine, et al., Visualizing human leukocyte antigen class II risk haplotypes in human systemic lupus erythematosus, Am J Hum Genet 71 (2002) 543–553.

[67] T.W. Yeo, P.L. De Jager, S.G. Gregory, L.F. Barcellos, A. Walton, A. Goris, et al., A second major histocompatibility complex susceptibility locus for multiple sclerosis, Ann Neurol 61 (2007) 228–236.

[68] S. Nejentsev, J.M. Howson, N.M. Walker, J. Szeszko, S.F. Field, H.E. Stevens, et al., Localization of type 1 diabetes susceptibility to the MHC class I genes HLA-B and HLA-A, Nature 450 (2007) 887–892.

[69] M.M. Fernando, C.R. Stevens, P.C. Sabeti, E.C. Walsh, A.J. McWhinnie, A. Shah, et al., Identification of Two Independent Risk Factors for Lupus within the MHC in United Kingdom Families, PLoS Genet 3 (2007) e192.

[70] J.D. Rioux, P. Goyette, T.J. Vyse, L. Hammarstrom, M.M. Fernando, T. Green, et al., Mapping of multiple susceptibility variants within the MHC region for 7 immune-mediated diseases, Proc Natl Acad Sci USA (2009).

[71] L.J. Palmer, L.R. Cardon, Shaking the tree: mapping complex disease genes with linkage disequilibrium, The Lancet. 366 1223–1234.

[72] T. Strachan, A.P. Read, Human Molecular Genetics 2, BIOS Scientific Publishers Ltd, 1999.

[73] J.C. Barrett, B. Fry, J. Maller, M.J. Daly, Haploview: analysis and visualization of LD and haplotype maps, Bioinformatics 21 (2005) 263–265.

[74] D.A. Hafler, P.L. De Jager, Applying a new generation of genetic maps to understand human inflammatory disease, Nature reviews 5 (2005) 83–91.

[75] S.V. Kozyrev, A.K. Abelson, J. Wojcik, A. Zaghlool, M.V. Linga Reddy, E. Sanchez, et al., Functional variants in the B-cell gene BANK1 are associated with systemic lupus erythematosus, Nat Genet 40 (2008) 211–216.

[76] P.I. de Bakker, G. McVean, P.C. Sabeti, M.M. Miretti, T. Green, J. Marchini, et al., A high-resolution HLA and SNP haplotype map for disease association studies in the extended human MHC, Nat Genet 38 (2006) 1166–1172.

[77] P.R. Taylor, J.T. Nash, E. Theodoridis, A.E. Bygrave, M.J. Walport, M. Botto, A targeted disruption of the murine complement factor B gene resulting in loss of expression of three genes in close proximity, factor B, C2, and D17H6S45, J Biol Chem 273 (1998) 1699–1704.

[78] B. Lehner, J.I. Semple, S.E. Brown, D. Counsell, R.D. Campbell, C.M. Sanderson, Analysis of a high-throughput yeast two-hybrid system and its use to predict the function of intracellular proteins encoded within the human MHC class III region, Genomics 83 (2004) 153–167.

[79] Y.H. Lee, J.B. Harley, S.K. Nath, Meta-analysis of TNF-alpha promoter -308 A/G polymorphism and SLE susceptibility, Eur J Hum Genet 14 (2006) 364–371.

[80] N.J. McHugh, P. Owen, B. Cox, J. Dunphy, K. Welsh, MHC class II, tumour necrosis factor alpha, and lymphotoxin alpha gene haplotype associations with serological subsets of systemic lupus erythematosus, Ann Rheum Dis 65 (2006) 488—494.

[81] E. Sanchez, B. Torres, J.R. Vilches, M.A. Lopez-Nevot, N. Ortego-Centeno, J. Jimenez-Alonso, et al., No primary association of MICA polymorphism with systemic lupus erythematosus, Rheumatology (Oxford) 45 (2006) 1096—1100.

[82] M.J. Rood, M.V. van Krugten, E. Zanelli, M.W. van der Linden, V. Keijsers, G.M. Schreuder, et al., TNF-308A and HLA-DR3 alleles contribute independently to susceptibility to systemic lupus erythematosus, Arthritis Rheum 43 (2000) 129—134.

[83] A. Jonsen, A.A. Bengtsson, G. Sturfelt, L. Truedsson, Analysis of HLA DR, HLA DQ, C4A, FcgammaRIIa, FcgammaRIIIa, MBL, and IL-1Ra allelic variants in Caucasian systemic lupus erythematosus patients suggests an effect of the combined FcgammaRIIa R/R and IL-1Ra 2/2 genotypes on disease susceptibility, Arthritis Res Ther 6 (2004) R557—562.

[84] C. Bekker-Mendez, J.K. Yamamoto-Furusho, G. Vargas-Alarcon, D. Ize-Ludlow, J. Alcocer-Varela, J. Granados, Haplotype distribution of class II MHC genes in Mexican patients with systemic lupus erythematosus, Scand J Rheumatol 27 (1998) 373—376.

[85] N. Tsuchiya, A. Kawasaki, B.P. Tsao, T. Komata, J.M. Grossman, K. Tokunaga, Analysis of the association of HLA-DRB1, TNFalpha promoter and TNFR2 (TNFRSF1B) polymorphisms with SLE using transmission disequilibrium test, Genes Immun 2 (2001) 317—322.

[86] D.D. Gladman, M.B. Urowitz, G.A. Darlington, Disease expression and class II HLA antigens in systemic lupus erythematosus, Lupus 8 (1999) 466—470.

[87] Q.Y. Chen, W. Huang, J.X. She, F. Baxter, R. Volpe, N.K. Maclaren, HLA-DRB1*08, DRB1*03/DRB3*0101, and DRB3*0202 are susceptibility genes for Graves' disease in North American Caucasians, whereas DRB1*07 is protective, J Clin Endocrinol Metab 84 (1999) 3182—3186.

[88] M.J. Simmonds, J.M. Howson, J.M. Heward, H.J. Cordell, H. Foxall, J. Carr-Smith, et al., Regression mapping of association between the human leukocyte antigen region and Graves disease, Am J Hum Genet 76 (2005) 157—163.

[89] D.A. Cavan, K.H. Jacobs, M.A. Penny, M.A. Kelly, C. Mijovic, D. Jenkins, et al., Both DQA1 and DQB1 genes are implicated in HLA-associated protection from type 1 (insulin-dependent) diabetes mellitus in a British Caucasian population, Diabetologia 36 (1993) 252—257.

[90] C.M. Weyand, T.G. McCarthy, J.J. Goronzy, Correlation between disease phenotype and genetic heterogeneity in rheumatoid arthritis, J Clin Invest 95 (1995) 2120—2126.

[91] J. Klein, A. Sato, The HLA system. First of two parts, N Engl J Med 343 (2000) 702—709.

[92] Fry, B. (2005) Computational Information Design. Thesis, Massachussetts Institute of Technology

[93] G. Andersson, Evolution of the human HLA-DR region, Front Biosci 3 (1998) d739—745.

I. BASIS OF DISEASE PATHOGENESIS

2

Genetic Susceptibility and Class III Complement Genes

John P. Atkinson, C. Yung Yu

INTRODUCTION

The great interest in this genetic region of chromosome 6 in man relates to the several decades old observation that individuals with complete deficiency of C4 or C2 come to medical attention with the clinical syndrome of SLE. Thus, these two components, along with their friends (C1q, C1r, C1s) in the early part of the classical pathway (CP), are the only known single protein deficiency states that develop SLE or lupus-like syndromes. The question that has been so elusive to solve relates to how these deficiency states lead to SLE. One focus in this chapter is to address this question. After providing background data on the structure and function of the class III complement genes and their protein products, we will review the older data relative to complement's long-appreciated role in immune complex (IC) handling and then move into the more recent results on complement's key role in instructing adaptive immunity and in processing of cellular debris. The new tools in this field, namely knockout mice, have been helpful in establishing an important role for complement in innate and adaptive immunity as well as in preventing autoimmunity.

The class III, or complement (C), region of the human major histocompatibility complex (MHC) is located between the class I (A, C, B) and class II (DR, DQ, DP) regions on the short arm of chromosome 6 (Figure 2.1).

Complement genes in this 130 kb cluster are C4A, C4B, factor B (FB) and C2. C4A and C4B are duplicated (>99% identity) genes. The C2 and factor B genes encode serine proteases which are approximately 40% homologous, indicating a more ancient duplication. C4 and C2 are required for the CP and lectin (LP) pathway activation and factor B is required for alternative pathway (AP) activation (Figure 2.2). Upon engagement of the CP by IC, C4 is cleaved by the serine protease C1s, a subcomponent of the first component of complement (C1). For the LP, the mannan (or mannose)-associated serine proteases (MASPs), which are homologous to C1s, cleave C4. Upon activation, the larger proteolytic fragment of C4, C4b, has the transient capacity to attach covalently to the target antigen (Ag) and to the activating antibody (Ab) (Figure 2.3). Bound C4b has two roles: an opsonic one in which it is a ligand for complement receptors and an enzymatic one in which it serves as a non-catalytic domain of the CP C3 and C5 convertases. C2 and factor B are serine protease precursors (zymogens) and play identical roles in their respective activation pathways. After proteolytic cleavage by C1s or MASPs, C2a combines with C4b to form the classical/lectin pathway C3 convertase (C4bC2a). Factor B binds C3b and then it is cleaved by factor D to form the alternative pathway C3 convertase (C3bBb). If C3b associates with these two C3 convertases, a C5 convertase is formed. Thus, C2a and factor Bb are the catalytic

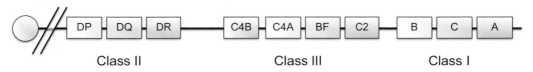

FIGURE 2.1 The human major histocompatibility complex on the short arm of chromosome 6. The class III or complement gene cluster covers a distance of 130 kb and includes complement activation components C4 (two highly homologous genes termed C4A and C4B), factor B (also noted as FB or BF), and C2. *Adapted from Liszewski et al. [170] with permission.*

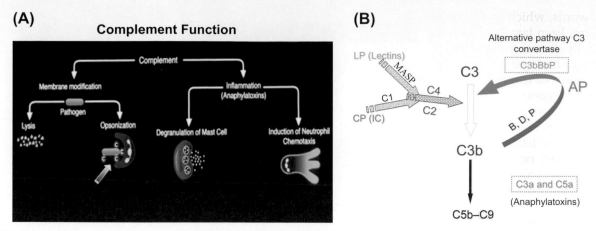

FIGURE 2.2   Complement pathways. (*A*) Function of the complement system. (*B*) Complement activation pathways. Note that C4 and C2 serve the same role in the classical and lectin pathways. C1 cleaves C4 and C2 in the classical pathway and mannan-associated serine protease (MASP) cleaves C4 and C2 in the lectin pathway. Cleavage fragments C4b and C2a then get together to form the C3 convertase. In the alternative pathway, C3b binds FB and the FB is then cleaved by factor D. Properdin stabilizes the alternative pathway C3 convertase.

(serine protease) domains of the CP/LP or AP C3 and C5 convertases, respectively.

Attention was initially drawn to this gene cluster because of its location within the MHC and later because inherited deficiencies of C2 (C2-def) and C4 (C4-def) predispose to SLE. Importantly, a deficiency of any one of the early components of the CP is the only single protein deficiency state that causes SLE in man. Initially, this striking disease association was attributed to the class I or class II markers inherited with the C deficiency. While there may be an independent contribution to disease susceptibility from genetic markers that are in linkage disequilibrium such as the HLA class II genes, C deficiency itself is causative as will be summarized below:

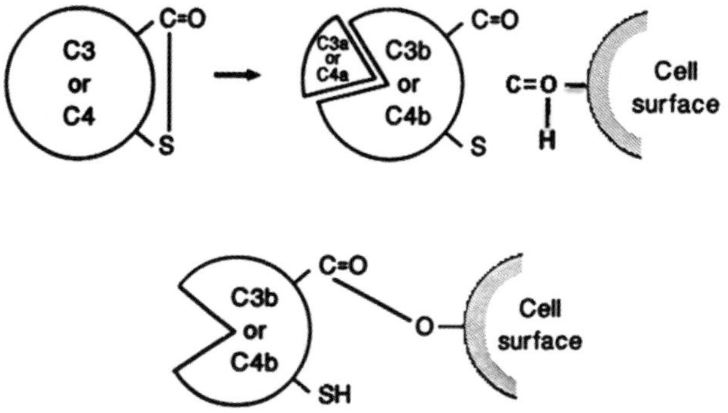

1. The first point is that C4-def and C2-def are inherited on distinct genetic (MHC) backgrounds. Moreover, C4-def is inherited on different MHC backgrounds in, for example, Africans or Asians compared to Caucasians, however the SLE association still holds. Therefore, it is not primarily the company it keeps (i.e. other MHC-linked genes) that produces the disease association but the C deficiency itself.

2. C1q-def and C1r/C1s-def also predispose to SLE [1, 2]. The associations of SLE with C1q and C1r/C1s deficiency are comparable to or even a little stronger (>90% of C1q-def individuals have SLE) than that for C4-def (~80%) and SLE [1]. These subunits of C1 are, of course, immediately upstream of C4 and C2 in the CP, strongly implying that it is activation of the cascade itself through at least C4 that prevents SLE from developing. Further and in support of the first point, the structural genes for these C1 subunits are at distinct genetic loci, outside the MHC [1–4].

3. Low copy number of *C4* genes, especially of a homozygous or heterozygous deficiency of C4A, predisposes an individual to SLE in some populations [1, 3, 4]. In the Caucasian SLE population, C4A-def is observed about two-thirds of the time on the DR3-B8 haplotype while the remaining one-third are associated with a *diverse* array of haplotypes. Moreover, in East-Asian populations, C4A-def appears to be a risk factor for SLE as well, yet the DR3-B8 haplotype is essentially absent in Chinese, Japanese and Korean populations. Thus, a problem in interpretation related to linkage disequilibrium (in

FIGURE 2.3   The thioester bond of C4 and C3. The activation of the thioester bond in C4 and C3 occurs upon their proteolytic cleavage to C4b and C3b and liberation of the C4a and C3a fragments. Condensation with a hydroxyl group (formation of an ester linkage) is shown. Amino groups on the target surface are also bound (formation of an amide linkage). Note that C4A and C4B (with capital letters) refer to the two forms of C4 while C4b and C4a (with lower case letters and which may be derived from either C4A or C4B) refer to cleavage fragments. *Adapted from Hughes-Jones [171], with permission.*

other words, which genes are the disease-promoting ones) has been largely resolved by studies in ethnically diverse populations [1, 3, 4].

4. The major biologic function of these early components of the classical C pathway, C1, C4, and C2, and the central C3 component, is to bind to antigens and thereby facilitate antigen processing (the immune response) and immune complex (IC) formation and handling [5—11]. SLE is the prototypic spontaneously occurring human disease in which many of the clinical manifestations are mediated by the deposition of IC in undesirable locations (e.g., kidney).

5. Patients with hereditary angioedema have an acquired C4-def and C2-def and there is an increased frequently of SLE in this patient population.

Taken together, these data establish deficiency of an early component of the CP as one cause of SLE. Next, we will discuss several caveats related to this statement and deal with several apparent discrepancies.

1. The major function of the early components of the CP is to activate C3 so its C3b fragment can deposit on IC. The expectation might be therefore that C3-def would be an even greater predisposing deficiency state for SLE. However, the rare patients (numbers are small, however) with C3-def present with recurrent pyogenic infections and occasionally glomerulonephritis but usually not with a SLE-like picture. Several explanations are possible. One is that C4b deposition on antigens (structural and functional homolog of C3b) is sufficient such that autoimmunity is avoided. Along this line of reasoning, Michael Carroll and colleagues have proposed a theory that C4b binds to immunocompetent autoreactive bone marrow B cells and thereby inhibits their activation (a negative feedback to prevent autoimmunity) [12]. While this is an attractive hypothesis [1], there is as yet no clear idea as to how this takes place and C1q-def mice do not demonstrate a defect in B cell tolerance [13].

2. Another theory, championed by Mark Walport and colleagues, is the so-called "garbage or waste disposal" hypothesis [1, 14]. It centers on a role for the CP in the clearance of cell debris including apoptotic cells. If cellular debris is not efficiently removed (analogous to IC clearing foreign antigens), then structural modifications may arise in DNA/protein complexes, RNA/protein complexes and other cellular constituents rendering them immunogenic. This line of reasoning would lead one to expect C3 deficiency to be a cause of SLE which, as discussed, is not the case. One possibility is that C1q or C4b binding to the "garbage" is sufficient to prevent autoimmunity.

3. Two other issues that require an explanation in SLE are the striking female predominance and antibody specificity for nuclear material. The waste disposal hypothesis does provide a potential explanation for the latter (see below) [1, 14] while the gender issue continues to defy a plausible explanation. Of interest, in C2-def most of the individuals who develop SLE are females. This further suggests that C2-def is not as predisposing to SLE as C1q- or C4-def.

4. A point emphasized also by the Walport and Botto group is that C1q-def has a higher frequency of SLE (90%), than C4-def (80—90%) or C2-def (10—30%) [1, 15]. This hierarchical relationship suggests that it is the binding of C1q to apoptotic material which is most critical in preventing SLE. These data also imply an interaction with C1q receptors, a field that is in flux as to the identification and function of such proteins. However, C1r/C1s-def (8/14 cases described to date have SLE) [1, 14] and SLE in C4-def and C2-def takes place in the setting of "normal C1q".

5. Therefore, a more plausible hypothesis would seem to be that CP activation through at least C4 and C2 is the critical issue in prevention of SLE. Two experimental approaches have been taken to understand how these CP genes and proteins associate and contribute to lupus disease. The initial and ongoing approach has been to analyze the genetics, protein structure and function of C4 and C2 and the receptors and regulators that interact with these two components. This strategy led to many pioneering discoveries including the cloning of the genes and determination of the primary protein structure as well as their post-translational modifications. The genomic analysis identified common interindividual and heritable gene copy number variations that were unanticipated and inconsistent with the classical human genetic concept of a relatively rigid genome size. The second approach has been to develop mouse models including gene targeting (knockouts) of C1q, C4 or C2 in order to understand disease mechanisms and identify disease pathways. Both types of studies have contributed much to our understanding of complement activation and its association with SLE.

FOURTH COMPONENT OF COMPLEMENT (C4)

Structure

C4 is an abundant (plasma concentration among healthy subjects is 15—45 mg/dl) 200 000 Da, three-chain, disulfide-linked, glycoprotein (Figure 2.4) [16—18]. It is synthesized by the liver and the blood C4

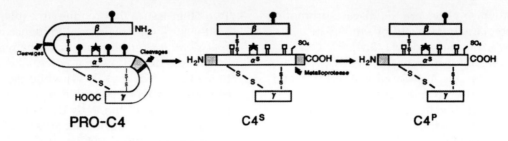

PRO-C4   **C4^S**   **C4^P**

- ● - High mannose type oligosaccharide
- ⬗ - Biantennary complex type oligosaccharide
- ★ - Thiolester bond
- **SO₄** - Sulfate

FIGURE 2.4 Schematic diagram of the structural features of human C4. Pro-C4 is the intracellular, single-chain precursor of C4. Following a series of post-translational modifications that include sulfation, glycosylation, and two proteolytic cleavages, a three-chain protein is secreted (C4^s). The major form of C4 in plasma, however, is C4^P. This form differs from C4^s by removal of a 22-amino-acid peptide from the carboxyl-terminus of the α-chain. The shaded area at the NH₂-terminus of the α-chain represents the C4a fragment. *Modified from Chan and Atkinson [24], with permission.*

is almost entirely derived from this source. Local biosynthesis also occurs in many cell types including peripheral blood monocytes and skin fibroblasts [19–21]. The human precursor C4 protein consists of 1744 amino acids with a leader peptide of 19 amino acids [22, 23]. Prior to secretion, the precursor protein (pro-C4) undergoes proteolytic cleavages at two sites to give rise to the three-chain protein structure linked by disulfide bonds: β, 70 000; α, 95 000; and γ, 35 000. The secreted protein possesses three N-linked complex sugar residues on the α-chain and one N-linked high mannose type carbohydrate on the β-chain [24]. The COOH-terminus of the α-chain undergoes two remarkable post-translational modifications. The first is sulfation of two or three tyrosine residues [25], a modification that increases several-fold the efficiency of its cleavage by C1s [26]. The second modification, following secretion into plasma, involves removal of a 22-amino-acid peptide from the carboxyl-terminus [27, 28]. This cleavage exposes a highly charged, acidic stretch of amino acids, including the sulfation site that

likely is a recognition motif for C1s and MASPs. Thus, truncation and sulfation are designed to improve the efficiency of C4 cleavage by C1s and presumably by the MASPs as well. The complex glycosylation patterns and the incomplete processing at the β–α and α–γ chain junctions lead to the presence of multiple protein products (from each *C4* gene) circulating at low levels in the plasma [17, 29].

Function

C4 is the first component of the classical C pathway to become directly bound to antigen in IC (C1 is bound through C1q subcomponent to the Fc fragment of IgM or IgG). Upon cleavage of the amino-terminus of the α-chain to liberate C4a by C1s or MASPs, C4b undergoes a molecular rearrangement. In so doing, a highly reactive thioester bond in the central portion of the α-chain becomes exposed (residues 991–994; Figures 2.3–2.5). This thioester is formed over a four-amino-acid peptide between a γ-carboxyl group of glutamine-994 and

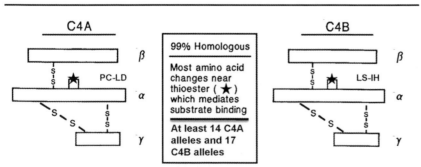

A comparison of C4A and C4B

C4A		C4B
β	**99% Homologous**	β
α PC-LD	Most amino acid changes near thioester (★) which mediates substrate binding	α LS-IH
γ	At least 14 C4A alleles and 17 C4B alleles	γ

Prefers amino substrates
Forms amide linkages

Prefers hydroxyl substrates
Forms ester linkages

FIGURE 2.5 Although highly homologous, C4A (acidic) and C4B (basic) differ in their binding specificities for substrates. *Adapted from Liszewski and Atkinson [172], with permission.*

a sulfhydryl group of cysteine-991 [30, 31]. A reactive carbonyl species is formed that is capable of interacting with a hydroxyl group to form an ester linkage or with an amino group to form an amide bond. The result is the covalent attachment of multiple, clustered C4b molecules to the target and nearby host cell membranes and molecules including immunoglobulins [32–34]. On foreign targets, complement activation by the CP commonly proceeds to the formation of membrane attack complex. When C4b is deposited on healthy autologous cell membranes, the activation process is actively aborted by dissociating the activated (catalytic) domain of the C3 and C5 convertases, and by proteolytic degradation of C4b. The latter leaves behind a remnant (immunologic scar), C4d, covalently attached to surfaces of the cell. Thus, cell-bound C4d is a useful biomarker of past complement activations via the classical and MBL pathways in lupus [35–37], the allogenic immune response in blood transfusions [38, 39] and transplant rejection in kidney allografts [40, 41].

As outlined above, C4b serves two roles in the classical C pathway. First, C4b is a ligand (opsonic role) for complement receptor type one (CR1; CD35; C3b/C4b or immune adherence receptor) of peripheral blood cells (erythrocytes/B-lymphocytes/monocytes/granulocytes) and B-lymphocytes and follicular-dendritic cells in immune organs [42–44]. Second, C4b is the anchor domain of the classical pathway C3 and C5 convertases (Figure 2.2). C4b, covalently bound to the target antigen, captures the C2a cleavage fragment of C2. This bimolecular enzyme complex, C4bC2a, is the classical pathway C3 convertase. C2a is a serine protease with strict substrate specificity; i.e., it cleaves C3 but only if C2a is associated with C4b. If a C3b becomes covalently bound to an acceptor site on C4b, the enzyme complex changes its specificity from C3 to C5 (C4b2a3b is the classical pathway C5 convertase). Thus, the serine protease catalytic domain of C2 cleaves two proteins, C3 and C5.

C4A and C4B Proteins

Plasma C4 is composed of two highly homologous (>99%) isotypes, the acidic C4A and the basic C4B (Figure 2.5). Mature C4A and C4B are each comprised of 1725 amino acids [18, 45, 46]. Eight amino acid differences in positions near the thioester residues (CGEQ 991–994) are responsible for the electrophoretic (fast vs. slow), serologic (Chido and Rodgers blood group antigens), and hemolytic (C4B is several-fold more active than C4A in a hemolytic overlay assay) differences [39, 47]. The isotypic residues at positions 1101–1106 are PCPVLD for human C4A, and LSPVIH for human C4B [39, 46]. Besides their isotypic variations, C4A and C4B each demonstrate additional polymorphisms, defined initially by electrophoresis, giving rise to at least 14 C4A and 17 C4B alleles [48, 49]. Those alleles that have been sequenced result from single amino acid differences in the region containing the thioester (also known as the C4d fragment) [22], but additional amino acid exchanges also exist outside this region (Table 2.1), such as the R458W on the β-chain that disrupts the C5 binding site for the hemolytically inactive C4A6 allele [50, 51].

Genetics of C4

Our concept of *C4* genetics has evolved substantially over the past three decades. Immunologists have been intrigued by the remarkable heterogeneity including the many polymorphisms of the C4 proteins. These are detectable by a variety of experimental methods including electrophoretic mobility in non-denatured agarose gels, serologic properties (particularly allogenic antibodies against C4 proteins produced by recipients of blood transfusion), and different hemolytic activities and chemical reactivities.

At least five structural glycoprotein variants are detectable in EDTA-plasma for each of the *C4* genes. Treatment of plasma samples with neuraminidase reduces the number of glycoprotein variants (as detected in an immunofixation gel) generated from each *C4* gene to three [52, 53]. Additional treatment with carboxyl peptidase B (CPB) eliminates the heterogeneous arginine residues at carboxyl terminus of the β and amino terminus of α chains. These treatments reduce the complexity of protein product from a *C4* gene to one [54], as shown in Figure 2.6. Thus, immunofixation of EDTA-plasma samples pretreated with neuraminidase and CPB is the preferred technique for C4 allotyping [55].

The C4 phenotyping experiments led to the identification of at least 35 variants or allotypes [48, 49]. These can be categorized into two classes or isotypes: C4A is more acidic in nature, has a faster mobility in high-voltage agarose gel electrophoresis (HVAGE), lower activity in hemolytic overlay experiments, a preferential binding affinity for substrates with amino groups (or protein antigens) to form amide bonds, and associates with the Rodgers blood group antigens. C4B is more basic in nature, has a slower migration in HVAGE, greater activity in hemolytic overlay experiments, greater binding affinity to substrates with hydroxyl groups (or carbohydrate antigens) to form a covalent ester linkage, and associates with the Chido blood group antigens [34, 38, 39, 46, 56–58]. Besides the qualitative protein polymorphisms, there are also quantitative variations in the plasma levels of the C4A and C4B proteins.

Figure 2.6B shows the results of a C4 allotyping gel and demonstrates the variations of C4A and C4B plasma

TABLE 2.1 A list of polymorphic sites in human C4A and C4B proteins

			AA Positions[a]		
	Exon no.	Polymorphism	SC=1	NT=1	Remarks/source
β - CHAIN (1 − 656)					
	6	A→V	217	198	AAI71786.1
	9	Y→S	347	328	
	11	V→A	418	399	C4A4
	12	R→W	477	458	C4A6
	12	P→L	478	459	C4B1*hi
	15	A→S	643	624	Source: AAH63289.1
α - CHAIN (661 − 1427); C4A: 661−737; C4D: 938 − 1327					
C4a	17	L→P	726	707	Cos 3A3
C4a	17	R→W	729	710	AAH63289.1
	21	A→T	907	888	short B1, mono-S
	25	D→G	1073	1054	Ch−5/Ch5
	26	P→L	1120	**1101**	**C4A/C4B**
	26	C→S	1121	**1102**	**C4A/C4B**
	26	L→I	1124	**1105**	**C4A/C4B**
	26	D→H	1125	**1106**	**C4A/C4B**
	28	N→S	1176	1157	Ch−6/Ch6
	28	I→K	1182	1163	
	28	T→S	1201	1182	
	28	V→A	1207	1188	Rg1/Ch1
	28	L→R	1210	1191	Rg1/Ch1
	29	P→Q	1245	1226	AAH63289.1
	29	A→S	1286	1267	
	30	I→F	1317	1298	
	33	△ EDY	1400−2	1419−22	C4A4
γ - CHAIN (1432 − 1725)					
	34	D→Y	1497	1478	
	39	R→S	1637	1618	AAI46850.1
	40	E→G	1678	1659	AAI71786.1
POLYMORPHIC RESIDUES DETECTED BY PROTEIN SEQUENCING OF POOLED SERA					
β - chain	15	C→S	635	616	
α - chain	17	D→N	727	708	
α - chain	26	S→I	1109	1090	
α - chain	29	R→V	1300	1281	
α - chain	30	T→G	1305	1286	
α - chain	30	V→G	1306	1287	

[a]Amino acid (AA) numbering based on UniProtKB entry: P0C0L4.1, GI:81175238; SC = 1, the start codon is designated as number 1, with sequence of the leader peptide; NT = 1, the N-terminus of secreted C4 protein is designated as number 1 (adopted in most publications);
C4B1*hi, a C4B1 allotype that is hemolytically inactive [173];
△ EDY, deletion of EDY in C4A4 [45].
References for C4 protein, cDNA and genomic sequences: [18, 22, 23, 45, 46, 73].

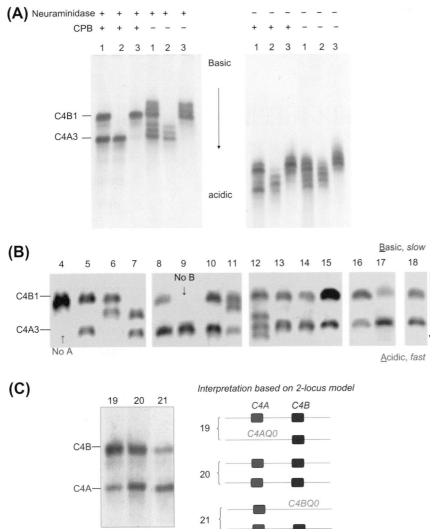

(A)

(B)

(C)

Interpretation based on 2-locus model

The *C4A-C4B* / two-locus model can explain ~50% of the cases. Family and molecular genetic studies reveal high frequencies of haplotypes with:

- single locus coding for either C4A or C4B,

- two loci both coding for C4A or for C4B, i.e., C4A–C4A or C4B–C4B, and

- three or four loci coding for combinations of C4A and C4B proteins.

FIGURE 2.6 Protein structural polymorphisms and the two-locus model for the human complement *C4A* and *C4B* genes. (*A*) Immunofixation experiments to demonstrate the structural protein variants present in EDTA-plasma of three human subjects. C4 proteins were resolved by high-voltage agarose gel electrophoresis (HVAGE), with and without prior neuramindase and/or carboxyl peptidase B (CPB) treatment(s), fixed with a poly-clonal antibody to C4, blotted and washed to remove diffusible proteins, and stained. The C4 structural variants are separated based on differences in electric charge. The acidic C4A proteins migrates faster (farther into the gel) than the more basic C4B proteins. The most common C4A protein is C4A3 and the most common C4B protein is C4B1. Subject number 1 has both C4A and C4B proteins; subject number 2 has C4A only; subject number 3 has C4B protein only. Note that at least five structural variants were detectable in subjects number 2 or 3, when the EDTA-plasma samples were *not* treated with neuraminidase and CPB (right panel). The neuramindase and CPB treatments reduce the heterogeneity caused by glycosylation and incomplete processing at the carboxyl-terminal cleavage sites of the β-chain and the α-chain. (*B*) Polymorphisms of C4A and C4B in 15 human subjects. EDTA plasma from subjects number 4 to 18 were digested with neuraminidase and CPB and processed as described in panel A. Note the absence of C4A in subject number 4 and the absence of C4B in subject number 9; the presence of "more" C4A than C4B in subjects number 8 and 17; the presence of more C4B than C4A in subjects number 5, 11, 15 and 16; and the presence of polymorphic variants other than C4A3 and C4B1 in subjects 6, 7, 11 and 12. (*C*) The two-locus model to explain the C4A and C4B phenotypes. The limitations of the two-locus model are listed. *Modified from [55, 113].*

proteins from 18 human subjects [59]. In subject no. 4, only the slow migrating C4B protein (C4B1) is present. In no. 9, only the fast migrating C4A protein (C4A3) is present. In no. 5 and 15, there are more C4B1 proteins than C4A3 proteins. In no. 8 and 17, there is more C4A3 than C4B1. Polymorphic variants of C4A or C4B are discernible in no. 6, 7, 11 and 12. To explain the phenotypic variations of C4, initially a monogenic model with co-dominant alleles for C4A and C4B (or C4-fast and C4-slow) was proposed [60, 61], but it did not explain the presence of three or four different proteins as observed in subjects No. 11 and 12. A two-locus model on each MHC haplotype, one for *C4A* and one for *C4B*, was then proposed [52, 53, 62–64] and

widely adopted. A "partial deficiency for C4A" was assigned when the intensity of C4A protein appeared lower than that of C4B; a "partial deficiency for C4B" was proposed when the intensity of C4B was lower than that of C4A. A "null" allele was assigned in those cases with a partial C4A or C4B deficiency (panel C, Figure 2.6). However, the *C4A−C4B* (two-locus) model could only explain ~50% of the C4 protein phenotypes in human populations. Family segregation and molecular genetic studies revealed unusually high frequencies of MHC haplotypes with a single *C4* gene coding for either C4A or C4B, haplotypes with two *C4* loci coding for C4A only (*C4A−C4A*) [59, 65−68] or for C4B only (*C4B−C4B*) [59, 67] and three-locus and four-locus

(A) Long *C4* gene, 20.6 kb

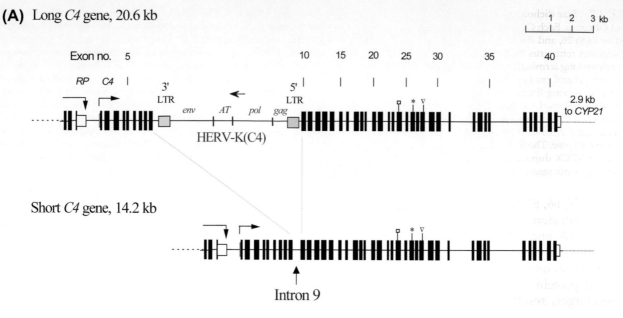

Short *C4* gene, 14.2 kb

Intron 9

(B)

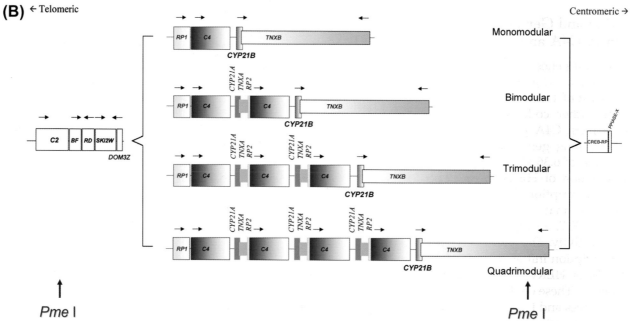

RCCX duplication is discretely modular; each module adds:

- One *CYP21A* (pseudogene) or one *CYP21B* (functional)
- A gene fragment for tenascin *TNXA* (4.0 kb)
- A gene fragment for Ser/Thr kinase *RP2* (0.9 kb)
- One functional *C4* gene: *C4A* or *C4B*; long or short
- Each duplicated module is either 32.7 kb or 26.3 kb in size

◀ **FIGURE 2.7**   Size dichotomy and copy number variation of human *C4* genes and the RCCX modules. (*A*) Exon—intron structure of the long and short *C4* genes. Each *C4* gene consists of 41 exons and 40 introns. The thioester is encoded by exon 24, the C4A and C4B isotypic residues are encoded by exon 26, and the Rodgers 1 and Chido 1 blood group antigens are encoded by exon 28. The long gene is 20.6 kb in size and contains the endogenous retrovirus HERV-K(C4) in its ninth intron. HERV-K(C4) is 6.36 kb in size, consists of sequences similar to *gag*, *pol* and *env*, is flanked by two long terminal repeats (LTRs), and is organized in opposite orientation with respect to the transcription of human *C4*. The reading frames of *gag*, *pol* and *env* in HERV-K(C4) are all closed with nonsense mutations, but its 3' LTR maintains promoter activity. The short *C4* gene is 14.2 kb in size. Among European-Americans, 76% of the *C4* genes belong to the long form and 24% belong to the short form. (*B*) Copy number variation of *C4* and modular duplication of RCCX. Mono-, bi-, tri- and quadric-modular structures of RCCX are frequently present in each copy of chromosome 6 in human populations. The duplications of RCCX are discretely modular as follows: each additional RCCX module carries an intact *CYP21* gene of 3.3 kb that is mostly a mutant *CYP21A* gene, a gene fragment *TNXA* of 4.04 kb, another gene fragment *RP2* of 0.9 kb, and *an intact C4A or C4B gene*. The RCCX module is 32.7 kb if the *C4* gene is long, or 26.3 kb if the *C4* gene is short. Two *Pme*I restriction sites are present outside of the RCCX duplication breakpoints. Therefore, the RCCX modular variations can be shown by size difference of the genomic *Pme*I restriction fragments resolved by pulsed field gel electrophoresis and processed by Southern blot analysis. *Modified from [23, 73, 80].*

haplotypes [59, 66, 67, 69—72]. In addition and a source of further confusion, there were long and short genes coding for C4A and C4B allotypes. Such phenomena were initially bewildering and could not be easily explained by simple genetic models. However, these genetic and protein "problems" relative to human C4 have been largely resolved as outlined below.

Gene Size and Gene Copy Number Variations of Human C4A and C4B

Human *C4* genes exist in two forms: a long gene of 20.6 kb and a short gene of 14.2 kb. Both long and short genes consist of 41 exons [18, 23, 73]. A long or short *C4* gene can either code for a C4A or a C4B protein, although most C4A proteins are encoded by long *C4* genes. The long gene has an endogenous retrovirus, HERV-K(C4) of 6.36 kb, integrated into its ninth intron. The orientation of HERV-K(C4) is opposite to that of *C4* gene transcription [73, 74]. Molecular biologic analyses showed that one of the long terminal repeats (LTR) in HERV-K(C4) maintained gene expression promoter activity. The transcription of a long *C4* gene, or a transcription initiated from regulatory elements in the 3' LTR of HERV-K(C4), would generate antisense RNA species. These could hybridize to exogenous retroviral sequences and trigger the RNA degradation mechanism to protect the host from related retroviral infections [18, 73—77] (Figure 2.7). The size dichotomy of *C4* genes can be traced to non-human primates such as the rhesus macaque [73].

It is not clear why both long and short *C4* genes should coexist in the human populations. Among Europeans and Asians, about one-quarter (24—26%) of *C4* genes belong to the short form and three-quarters belong to the long form [59, 67, 72, 78, 79]. The frequency of short *C4* genes is significantly higher among subjects of African ancestry (~40%) [3, 79]. Population studies reveal that the copy number of short *C4* genes positively associates with higher plasma C4 protein levels, particularly C4B, implying that the presence of the endogenous retrovirus reduces the production of C4 proteins [72, 78].

Characterization of the genomic DNA sequences and organization of *C4A* and *C4B* and their neighboring genes from multiple human subjects enabled a rigorous analysis of gene haplotypes in the class III region of the MHC [80, 81]. In a study of >500 healthy subjects of Northern-European ancestry from the mid-western United States, about 15% of the MHC haplotypes consist of a monomodular RCCX structure characterized by an intact *RP1* gene (also known as *STK19* for serine/threonine kinase 19), an intact gene *C4* that encodes for either C4A or C4B, an intact gene *CYP21B* (*or CYP21A2*) that encodes for cytochrome P450 21-hydroxylase, and an intact gene *TNXB* that encodes for extracellular matrix protein tenascin-X, i.e., *RP1-C4-CYP21B-TNXB* (Figures 2.7, 2.8). In the remaining (~85%) of MHC haplotypes, there can be an additional 1, 2 or 3 duplications of a genetic unit between *C4* and *CYP21B* (Figures 2.7, 2.8). Each duplication unit contains a *CYP21* gene that is usually (but not always) a mutant gene (*CYP21A* or *CYP21A1P*), a 4.04-kb DNA fragment (*TNXA*) corresponding to intron 32 to exon 45 of *TNXB* that is fused to a 0.91-kb DNA fragment (*RP2*) corresponding to part of exon 7 to exon 9 of *RP1*, and an intact *C4* gene that can be either *C4A* or *C4B* [81—84]. This genetic duplication unit is termed RCCX module [80]. Each duplicated RCCX module can be 32.7 kb or 26.3 kb in size, depending on the length of the *C4* gene present. The size variation of *C4* genes and the copy number variation of RCCX modules lead to a repertoire of physical length variants for the MHC class III region (Figure 2.8) that probably plays a role in promoting gene conversion or recombination events during meiosis between RCCX length variants [18, 59, 72, 80, 85].

The copy number of total *C4* among different individuals mainly varies from 2—6 (Figure 2.9), although examples of seven or eight copies have been documented [59, 67, 71, 72, 78, 80]. Among European-Americans, while a majority of healthy subjects (60.8%) have four copies of *C4* genes in a diploid genome, about one-quarter (25.8%) have three copies, and one-tenth have five copies (10.1%). Less than 4% of healthy subjects have either two copies (1.4%) or six copies (2.0%). The distribution of *C4* gene copy number is

(A)

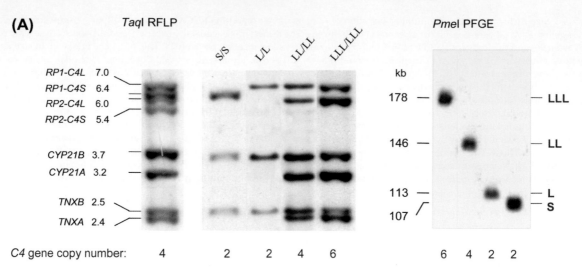

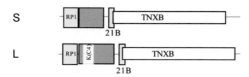

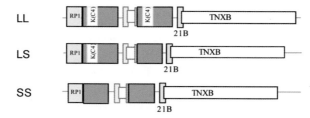

(B)

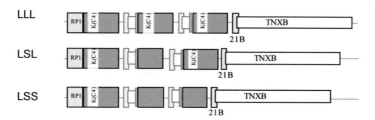

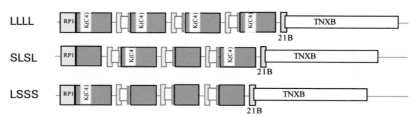

◄ **FIGURE 2.8** Length variants of *RP-C4-CYP21-TNX* (RCCX) haplotypes. (*A*) Upper panels: demonstration of polymorphisms and copy number variations in the constituent genes of the RCCX modules by *Taq*I restriction fragment length polymorphisms (RFLPs) (left panel) and *Pme*I-Pulsed field gel electrophoresis (*Pme*I-PFGE) (right panel). *Left:* Genomic DNA from five different individuals was digested with *Taq*I and probed for genomic DNA specific for *RP-C4*, *CYP21* and *TNX*. The first lane shows the result from a subject with monomodular S and trimodular LLS. The second to fifth lanes show the results from subjects with homozygous S/S, L/L, LL/LL and LLL/LLL, respectively. The copy number of *C4* genes in the diploid genome for each subject is indicated. Note the presence of a 6.4 or 7.0 kb fragment for *RP1-C4S* and *RP1-C4L* in the subjects with monomodular S/S and L/L, respectively. In each of these two subjects, there is an absence of the 3.2 kb fragment for *CYP21A* and the 2.4 kb fragment for *TNXA*. In the subject with bimodular LL/LL, there are equal intensities (numbers) of DNA fragments for *RP1-C4L* and *RP2-C4L*, for *CYP21B* and *CYP21A*, and for *TNXB* and *TNXA*. In the subject with trimodular LLL/LLL, the fragment for *RP2-C4L* is two-fold more intense than that for *RP1-C4L*, the *CYP21A* is two-fold more intense than *CYP21B*, and the *TNXA* fragment is two-fold more intense than *TNXB*. These results illustrate the duplications of the genomic DNA segment(s) spanning *RP2-C4L-CYP21A-TNXA* in bimodular LL and trimodular LLL. *Right:* Relatively large and distinct *Pme*I restriction fragments for trimodular LLL/LLL, bimodular LL/LL, monomodular L/L and monomodular S/S are demonstrated using genomic DNA processed by PFGE and Southern blot analysis. (*B*) *Lower panel*: Length variants of monomodular, bimodular, trimodular and quadrimodular RCCX haplotypes observed in human populations. *Modified from [55, 81, 114].*

skewed negatively, towards the lower copy side. In the healthy population, 27.2% of human subjects have <4 copies of *C4* genes, and those with >4 copies have a frequency of 12.0% [67].

The copy number of *C4A* genes in a diploid genome varies from 0 to 5. The most common gene copy number group of *C4A* is two per diploid genome (56.0%). The *C4A* gene copy number distribution is relatively normal but slightly skewed towards the high copy number side (<2, 18.1%; >2, 25.9%). For *C4B*, the copy number varies from 0 to 4. The most common copy number group of *C4B* is also two per diploid genome (63.6%) but the distribution is skewed towards the low copy number side (<2, 29.9%; >2, 6.6%) [67].

For long *C4* genes, the copy number varies from 0 to 6. The most common gene copy number group is three (36.9%) per diploid genome and an even distribution pattern is present (<3, 31.4%; >3, 31.8%). For short *C4* genes, the copy number varies from 0 to 4. The short gene distribution pattern is skewed significantly to the low copy number side: 35.1% of healthy subjects have no short *C4* genes, 44.4% have one copy and 17.4% have two copies per diploid genome. The frequency of subjects with more than two copies of short *C4* genes is only 3.0% [67].

Similar patterns of *C4* copy number variations with associated polymorphisms were established in a Hungarian population [78] and an Asian-Indian population in the US [72]. It is likely that *C4* copy number variation is present in all human races but there are specific differences in the distribution patterns for each ethnic group. For example, close to one-third (32.8%) of the Asian-Indians had 5, 6 or 7 copies of *C4* genes, which is almost 3-fold more frequent than that in European-Americans. By contrast, only 14.3% of Asian-Indian subjects had 2 or 3 copies of *C4* genes, which is about half of that in European-Americans. It is postulated that copy number variation of *C4* genes creates a diversity of intrinsic strengths in the immune response. Such inherent structural variations possibly empower individuals in a population the capability of responding more efficiently to a variety of environmental challenges, although they could also predispose some subjects to autoimmune diseases. With the application of comparative genomic hybridization in microarray experiments, it has become clear that gene copy number variations, like those present in the case of *C4*, are observed in numerous genomic loci, particularly those involved in host—environment interactions [86]. Besides complement *C4*, other established copy number

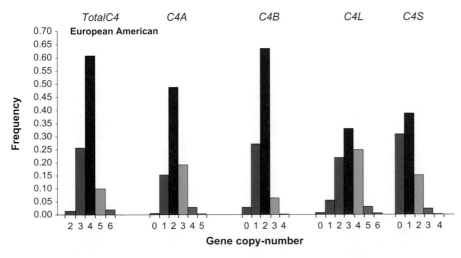

FIGURE 2.9 The inter-individual variations in copy numbers of total *C4*, *C4A*, *C4B*, long *C4* and short *C4* genes among healthy European-Americans. Copy number variations of *C4* genes and associated polymorphisms such as *C4A* and *C4B* isotypes and long (*C4L*) and short (*C4S*) genes were determined and cross-confirmed rigorously in 527 healthy European-American subjects residing in the Midwest (USA). The median for each copy number group is shown in blue, those below the medians are shown in purple and red, and those above the median are shown in shades of green. Note the continuous variations in gene copy numbers of total *C4*, the *C4A* and *C4B*, and the long and short *C4* genes in the population. *Modified from [86].*

variations-loci involved in immune response include the complement factor H-related genes *CFHR3–CFHR1* [87], the HLA class II gene *DRB* [88, 89], the beta-defensin genes [90], the chemokine gene *CCL3L1* [91], the immunoglobulin Fc-receptor gene *FCGR3B* [92], and the killer cell receptor gene complex *KIR* [93].

C2 AND FACTOR B

The *C2* and factor *B* (*BF*) genes are only 421 base pairs apart and reside about 30 kb from the *C4A* locus (Figure 2.1). In structure and function, C2 and factor B are homologous. The *C2* and factor *B* genes encode for mature proteins of 732 and 739 amino acids, respectively [94]. Factor B and C2 consist of three distinct domains: three complement control protein repeating motifs; von Willebrand type A and serine protease (Figure 2.10). Two common polymorphic protein variants of C2 as well as a null allele (C2Q0) have been described. Most individuals (>95%) are homozygous for the common (C2C) variant [95]. A restriction fragment length polymorphism subdivides this *C2C* allele [96]. Below we will discuss the association between C2-def and SLE. There are two common factor B alleles, *F* (fast) and *S* (slow), with gene frequencies of 0.27 and 0.71, respectively, two less common alleles, *F1* and *S07*, and 14 rare variants [95]. However, there is no association between deficiencies of the alternative pathway components and SLE. A patient with homozygous deficiency of FB has not yet been reported. In sum, C2 and FB are serine protease precursors, that following activation by proteolytic cleavage, become the catalytic domain of the C3 and C5 convertases (splitting enzymes) of the CP and AP, respectively.

On clinical aspects of the association between C4-def and C2-def, several points to keep in mind for the discussion to follow are that C4-def and C2-def are inherited on distinct MHC haplotypes and that they are part of an activation cascade. Animals and humans lacking an early component of the classical C pathway, up to and including C3, have markedly abnormal (reduced) clearance of antibody-coated erythrocytes [97] which is representative of a general problem with the handling of IC or apoptotic cell and debris, as well as other processes requiring CP activation.

C4 DEFICIENCY AND GENE COPY NUMBER VARIATION IN SLE

Complete Deficiency of C4 in Human SLE

Complete genetic deficiency of C4A and C4B has been established in 28 individuals [98–100]. They are derived from 19 families with 16 different HLA haplotypes. Of these C4-def subjects, 15 were diagnosed with SLE, 7 had a lupus-like disease and 4 others had renal disease including glomerulonephritis. Among the C4-def subjects who developed SLE, early onset, severe photosensitive skin lesions, the presence of anti-Ro/SSA, and high titers of antinuclear antibodies were clinical features. The molecular defects leading to C4-def have been elucidated in 15 subjects (Figure 2.11), who have the RCCX structures as follows:

1. Monomodular-long (mono-L) with a single *C4A* gene from four different HLA haplotypes with a mini-deletion, mini-insertion or point mutation at exon 13, 20 or 29, leading to a nonsense mutation [98, 101, 102].
2. Bimodular LS with one long *C4A* gene and one short *C4B* gene from three different HLA haplotypes with a nonsense mutation at exons 13, 29 or 36. In two haplotypes, the *C4A* and *C4B* genes have the identical mutation [98, 103], and, in the third haplotype (HLA A2 B12 DR6), two different nonsense mutations were present [104].
3. Bimodular SS with two *C4B* genes that have the identical mutation at the donor splice site of intron 28 [101].

All nonsense mutations except one in these C4-def patients appear to be private or restrictive mutations that were only detectable in a patient's family. The exception is the 2-bp insertion at codon 1213 of exon 29 that has an allelic frequency of 1–4% among Europeans [67, 105, 106].

Copy Number Variation of C4 and SLE

In view of the association between complete complement deficiency and SLE, investigators naturally assessed this relationship in partial C4 deficiency. In the early 1980s, two groups, one in England [107] and one in Australia [108, 109], reported an association between partial C4 deficiency and susceptibility to SLE. Indeed, low C4 protein levels and/or apparent phenotypic deficiencies of C4A or C4B are observed in a majority of SLE patients but the underlying genetics were initially

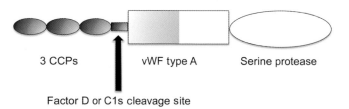

3 CCPs   vWF type A   Serine protease

Factor D or C1s cleavage site

FIGURE 2.10 Domain structure of Factor B and C2. CCP, complement control protein repeat (~60 amino acids each with two disulfide bonds); VWF type A; von Willibrand factor repeat domain.

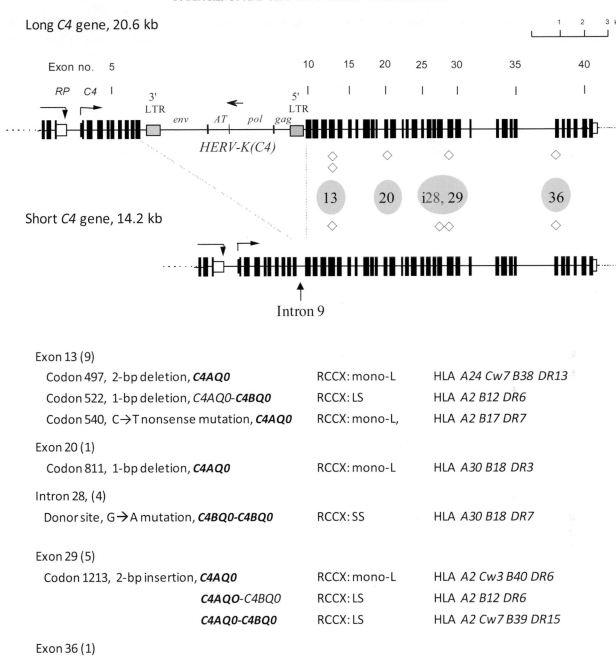

FIGURE 2.11 Molecular basis of complete C4 deficiency. A total of 28 cases of complete C4A and C4B deficiency have been reported. Among them, the molecular basis for the absence of C4 protein production from 15 subjects with eight different HLA haplotypes has been determined. The molecular defects include nonsense mutations caused by point mutations, mini-insertions or mini-deletions in the coding sequence and point mutations at an intron splice site. Exon 13 and exon 29 appear to be hot-spots for mutations. *Modified from [98, 101].*

a subject of controversy. There are three major reasons for such disputes. The first was the lack of awareness of the common *C4* gene copy number and gene-size variations. Most early studies were based on a two-locus model to interpret the C4A and C4B protein phenotypes that was subjected to misinterpretation, particularly if family data were not available. An apparent "partial C4A

deficiency" can be caused by (i) heterozygous deficiency of C4A, (ii) higher copy numbers of *C4B* than *C4A without* a genuine genetic deficiency (e.g., three *C4B* genes and two *C4A* genes), or (iii) higher expression of C4B proteins due to the fact that more *C4B* genes are short genes and more *C4A* genes are long genes. Another reason is the absence of surrogate markers such as microsatellites or

SNPs to tag specific *C4* gene copy number groups. Also, these same issues were some of the reasons why the more recent genome-wide association studies failed to detect an association of *C4* with SLE [110–112] and thus incorrectly dismissed C4's role in disease risk. A last reason is the strong linkage disequilibrium between *C4A* deficiency and HLA-*DR3* among European subjects.

A comprehensive study has been performed to rigorously investigate *C4* copy number variations and associated polymorphisms in SLE. The primary study population included 1241 European-Americans with 233 SLE patients and 356 first-degree relatives, plus 517 unrelated healthy controls [67]. The *C4* and RCCX genotypes were determined by genomic Southern blot analyses. Large genomic DNA fragments derived from *Pme*I digested genomic DNA were resolved by pulsed-field gel electrophoresis to determine the length variation of the RCCX modules. Also, regular genomic restriction fragment length polymorphisms of *Taq*I, *Psh*AI/*Pvu*II digested genomic DNA were utilized to elucidate the details of RCCX constituent genes. The C4 protein phenotypes were determined by immunofixation of neuramindase and CPB digested EDTA-plasma, immunoblot analyses using monoclonal Abs to determine the Rodgers and Chido blood group antigens, and radial immunodiffusion to assay the plasma C4 protein concentrations [55, 113, 114].

The gene copy numbers of total *C4* varied from 0 to 6, *C4A* from 0 to 5, and *C4B* from 0 to 4 and their medians were, as expected, 4, 2 and 2, respectively. In comparison

to healthy controls, SLE patients had their total *C4* and *C4A* but not *C4B* gene copy number groups shifted to the lower copy number side (Figure 2.12). Among the female SLE patients, 9.3% had two copies of *C4* and 6.5% had a homozygous deficiency of *C4A*, compared to 1.5% and 1.3%, respectively, in healthy controls. These data translated to an odds ratio of 6.5 for SLE disease risk among subjects with two copies of total *C4* genes, and an odds ratio of 5.7 for subjects with zero copy (or homozygous deficiency) of *C4A*. The frequency of subjects with a single copy of *C4A* (i.e., heterozygous deficiency, 26.4% in patients, 18.2% in controls; OR = 1.6) or low copy number of total *C4* (32.9% in patients, 27.0% in controls; OR = 1.3) were also increased, although their risks were comparatively weaker. By contrast, human subjects with high copy numbers of total *C4* or *C4A* were protected against SLE disease risk (total *C4*: ≥5 copies; 6.0% in patients, 12.0% in controls; OR = 0.47; *C4A*: ≥3 copies, 15.3% in patients, 23.8% in controls; OR = 0.57). The distribution patterns of total *C4* and *C4A* gene copy number of first-degree relatives were intermediate between those present in SLE and unrelated healthy controls [67].

Among the RCCX structures, a monomodular haplotype with a single short *C4B* gene (mono-S) had a frequency of 0.169 in SLE patients, compared to 0.113 in controls. Family-based transmission tests revealed that monomodular RCCX haplotypes, particularly the mono-S with *C4A* deficiency, were more likely to be transmitted to SLE patients. This mono-S haplotype is

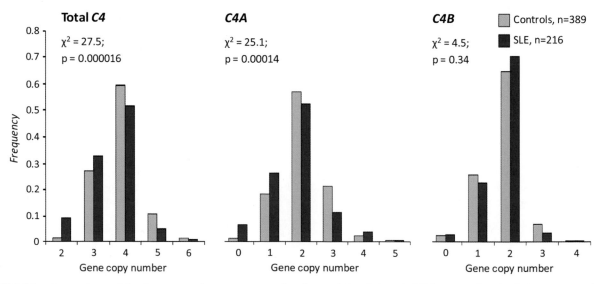

FIGURE 2.12 Comparison of the frequency of gene copy number for total *C4*, *C4A* and *C4B* between female SLE patients and matched healthy controls of European ancestry. There are highly significant differences in the distribution of total *C4* and *C4A* but not *C4B*; namely SLE patients have an increased frequency of subjects with low copy numbers of total *C4* (i.e., 2 or 3 copies) and *C4A* (i.e., 0 or 1 copy), but a reduced frequency of subjects with high copy numbers of total *C4* and *C4A*. Overall, 42.2% of SLE patients have low copy numbers of total *C4*, and close to one-third of SLE patients have a homozygous or heterozygous deficiency of *C4A*. *Modified from* [67].

strongly associated with HLA-*DR3* and the −308A polymorphism of the tumor necrosis factor *TNFA* gene. Subgroup analyses revealed that the effect of *TNFA* −308A was secondary to mono-S of the RCCX [67]. Ongoing studies in this area include deciphering *C4A* and *C4B* copy number variations in SLE patients and matched controls from different racial groups, and dissecting the contribution of different susceptibility genes including HLA-*DR3* and *DR2* in SLE disease risk.

In previous editions of this chapter, there was a table of analyses of C4null (also listed as CAQO or C4BQO which stands for quantity zero) and HLA class II associations in multiple populations (American Caucasoid, British Caucasoid, French Caucasoid, German Caucasoid, Mexican Mestizo, African-American, South-African Black, East-Asians) [115–120]. These were based on the two-gene model and had incomplete *C4* genetic analysis relative to current standards. Most were also underpowered relative to the number of subjects analyzed. For the analysis of Caucasoids, the interested reader should use the results outlined in this chapter and reference [67]. Likewise, the comments herein for Asians are accurate. Also, as noted before, these are as yet inadequate studies characterizing the African-Black and African-American populations. Nevertheless, the results overall are consistent with the hypothesis that it is the deficiency of the C4 protein itself, particularly C4A that predisposes to SLE.

Serum C4 Concentration

Several parameters affect the serum or plasma C4 protein concentration. The *C4* gene copy number (Figure 2.13), the copy number of short *C4* genes and the body-mass index are positively associated with increased C4 protein concentration [72, 78]. Many SLE patients experience low or very low serum or plasma protein levels of complement C4 and C3, persistently or intermittently [121, 122]. In some but not all such patients, these complement levels are biomarkers for lupus disease activity — low serum levels of C3 and/or C4 correlate with a flare and normal C3/C4 levels tend to correspond to a disease remission.

It has been observed recently that lupus patients are characterized by high levels of degradation product C4d attached to red blood cells, reticulocytes and platelets [35–37]. This phenomenon is monitored by flow cytometry using antibodies to C4. It is suggested that levels of RBC-C4d reflect complement activation in the past 60 days; the levels of reticulocyte-C4d reflect during the past 2–3 days, and platelet-C4d levels reflect complement activation during the past 5–10 days.

Longitudinal studies of serum complement protein levels in SLE patients reveal different expression profiles [123] (Figure 2.14). In one group of patients, there were persistently low C4 and C3 protein levels and many of these patients had a low copy number of *C4* genes. The second group of patients was marked by frequent fluctuations of C4 and C3 protein levels and SLE disease activity. The third group of patients was characterized by occasionally low C4 and C3 levels, particularly at the time of disease diagnosis, but their levels often returned to normal. Most patients in the second and third group had the median or higher copy numbers of total *C4* or *C4A*, and the C4 and C3 protein concentrations roughly parallel the SLE disease activities.

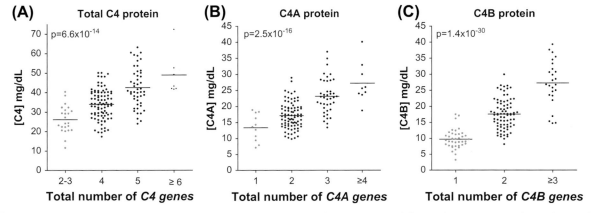

FIGURE 2.13 Quantitative variations of C4 protein among healthy subjects bearing different gene copy number. Plasma C4 protein concentrations from 168 healthy Asian-Indian Americans were determined by radial immunodiffusion assays. C4A and C4B protein polymorphisms of plasma C4 were determined by immunofixation and high-voltage argarose-gel electrophoresis (HVAGE). The quantitative difference between the C4A and C4B protein isotypes in each subject was quantified by densitometry. The total *C4*, *C4A* and *C4B* gene copy numbers were determined by genomic Southern blot analyses of *Taq*I RFLP to elucidate the copy number of total *C4* and RCCX, and by *Psh*AI/*Pvu*II RFLP to determine the ratio of *C4A* and *C4B* genes. There are direct correlations of mean C4, C4A and C4B protein concentrations with their corresponding gene copy numbers. However, other parameters also modulate C4 protein levels and include the size of *C4A* and *C4B* genes, body mass index and physiologic state of the blood donor. *Adapted from [72].*

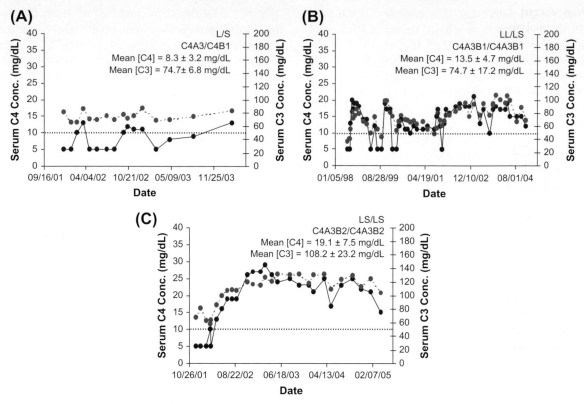

FIGURE 2.14   Typical serial C4 and C3 complement protein profiles in SLE patients. There is a correlation between serum protein concentrations of complement C3 and C4 in most SLE patients. In the first group of patients (Group A), the levels of C4 and C3 are chronically low. Often, even if the C3 level rises into the normal range, C4 levels remain low. Many of these patients have a low copy number of *C4*. In the second group (Group B), frequent fluctuations of C4 and C3 levels are seen. These patients have active disease and low levels of C3 and C4 roughly correlate with disease activity. In Group C, C3 and C4 protein levels stay normal most of the time except at the time of diagnosis or a major relapse. This group of patients tends to have relatively inactive disease. Patients in Groups B and C mostly have normal number of *C4* genes or a heterozygous deficiency of C4A. *Adapted from [123].*

C2 DEFICIENCY AND SUSCEPTIBILITY TO SLE

C2 deficiency occurs with an estimated prevalence of 1/20,000 in individuals of European descent. There are two types of C2 deficiency [99, 124, 125]. Type 1 C2 deficiency is caused by nonsense mutations leading to the absence of protein biosynthesis. The predominant form of type 1 (no protein produced) C2 deficiency is present in the HLA *A10 (A25), B18, DRB1∗15 (DR2), BFS, C2Q0, C4A4, Bf2*, haplotype. The defect is caused by a 28-bp deletion that removes 9-bp from the 3′ end of exon 6 and 19-bp from the 5′ end of intron 6, leading to a skipping of exon 6 in the C2 transcript and generation of premature stop codon [126]. The second form of type 1 C2 deficiency is due to a 2-bp deletion in exon 2 that leads to a nonsense mutation, as seen in HLA *A3, B35, DR4, BFF, C2Q0, C4A3A2* [127]. About 10% of C2 deficiency belongs to the type 2 deficiency, in which the C2 protein is synthesized but not secreted. The

molecular defects were found to be missense mutations C111Y, S189F and G444R [128, 129], although it is not clear how these mutations block secretion of C2.

Homozygous C2 deficiency is associated with SLE (a recent comprehensive review of this deficiency and rheumatologic manifestation in the Swedish population) [130]. Of interest, the female predominance in C2 deficiency approaches that seen in lupus in general. Also, the clinical picture may be suggestive of the deficient state. In several series these patients tended to have early childhood onset of a milder disease process with prominent photosensitive dermatologic manifestations, speckled ANAs (which on further testing demonstrate that the antibody specificity is common for the Ro (SSA) antigen), and a family history of SLE. Anti-DNA antibody tests are usually negative. Severe kidney disease is rare. Although an informative subset, complete C2D accounts for less than 1% of all SLE patients. Heterozygous C2 deficiency may be slightly increased in the lupus patients compared with controls, although a recent comprehensive review concluded that there is probably no association [1]. The review noted above from Sweden found an increased prevalence of anticardiolipin antibodies and anti-C1q antibodies as well as cardiovascular disease. Also, the severity of disease was not that distinct from the overall lupus population.

ACQUIRED C4 AND/OR C2 DEFICIENCY AND SUSCEPTIBILITY TO SYSTEMIC LUPUS ERYTHEMATOSUS

Inherited or acquired deficiency of the C1 inhibitor is associated with an increased frequency of autoimmunity, especially SLE ([125]). With a deficiency of the C1 inhibitor, C1 cleaves, in an unchecked fashion, its natural substrates, C4 and C2, causing a chronic marked reduction in their blood levels. In most patients with C1 inhibitor deficiency, C4 and C2 concentrations are continuously reduced such that it simulates a C4/C2 deficient state. A similar situation exists with autoantibodies that stabilize the C3 convertases (nephritic factors). Such individuals may have chronically low levels and usually present with either lupus (if the classical pathway convertase is stabilized) or glomerulonephritis (if the alternative pathway convertase is stabilized) [10]. Evidence has also been presented that the mechanism whereby hydralazine and isoniazid induce SLE is through inhibition of C4 function [131]. The concentrations of hydralazine and isoniazid (but not procainamide) that inhibited the activity of C4 by 50% were in the range of those obtained during therapy. Perhaps related to this observation was the finding of an increased frequency of C4A null alleles in hydralazine-induced SLE patients [131].

Although the association of complete and partial C deficiency and SLE is accepted by most investigators, there are dissenting opinions. For example, relative to complete C-def, up to 70% of C2-def individuals do not have SLE or a related rheumatic syndrome. Also, the frequency of SLE in C2-def family members or probands with SLE is less than 30%, suggesting a possible ascertainment bias, as individuals with SLE have their C values determined more often than healthy persons or patients with other autoimmune diseases. While the frequency of C2-def in SLE could be somewhat inflated [1], the association with C1q, C1r, C1s or C4 deficiency cannot be accounted for by such an explanation. In a population survey, Japanese investigators examined complement activity in over 150,000 individuals [132, 133]. Total deficiency of early components was not observed, indicating that an ascertainment bias cannot explain the association [1, 10]. Also, C2-def

may not be as severe a defect because there is a means to bypass this component in the cascade [134, 135]. Another explanation is that IC fixation by activated C4b alone leads to binding to CR1 which may be sufficient to solubilize and process some immune complexes.

IMMUNE COMPLEX PROCESSING AND THE COMPLEMENT SYSTEM

The primary goal of the humoral immune response is to bind (recognition function) and process (effector function) foreign antigens. Immune complex formation with C activation is a beneficial and an essential part of this process [5–7, 9–11, 136–139]. A related and newer concept in this arena is that this activity may also be important for the handling of host cellular debris [1, 14, 15, 140–142].

C4b and C3b deposit on the IC and thereby maintain the solubility of the IC [11]. The interaction of Ag with Ab may produce aggregates of increasing size in a network of cross-linked complexes. As the size of the IC increases, so does the possibility of precipitation and cryoglobulin activity. C3 serves a critical role in IC formation by limiting their size and thereby preventing precipitation. This "maintenance of IC solubility" requires the intercalation of C3b into the IC. A precipitated IC, especially in the vascular bed of organs such as the kidney, induces a local inflammatory reaction (glomerulonephritis, vasculitis). A deficiency of one or more CP components, up to and including C3, predisposes to inappropriate formation and handling of IC. Sera from patients deficient in C1q, C1r/C1s, C4, C2, or C3 suboptimally solubilize preformed IC and do not efficiently inhibit the precipitation of IC [1, 5–7, 9–11, 136–138, 143, 144]. Also, particularly with certain types of antigens, C4A is more efficient in preventing IC precipitation than C4B [145, 146]. A corollary here is that the processing of cellular debris including apoptotic cells may be a special case of this IC processing [1, 140]. Increasing evidence points to the CP playing a role in the safe disposal of damaged, degenerative, effete, or apoptotic cells.

C4b and C3b are ligands for C receptors [42–44]. This interaction provides a means to capture IC that form in tissue at sites of inflammation and in the circulation [139]. In tissues, such IC are ingested by dendritic cells, macrophages and monocytes and granulocytes. The antigen-presenting cells migrate to a regional lymph node and then initiate the immune responses through their antigen-processing activities. In the vascular system, IC become bound to cells (Figure 2.15), primarily erythrocytes (a phenomenon known as *immune adherence*), and are then transported to the liver and spleen. In the liver, the IC is transferred to Kupffer cells and the antigen is destroyed. In the spleen, the IC is transferred to follicular-dendritic cells and mononuclear phagocytes for processing. The erythrocyte returns to the circulation, deficient in some of its C3b receptors, but also ready to "shuttle" additional IC. This mechanism probably evolved to facilitate the clearance of IC that arises in the normal course of a humoral immune response to infectious organisms. Analysis of the steps in this pathway indicate that "pathologic" IC may arise from (1) excessive IC formation that overwhelms clearance pathways, (2) inefficient CP activation because of deficient antibody response, e.g., a weak IgG or an IgA response (rather than IgG), (3) C1, C4, C2 or C3 deficiency, or (4) C receptor anomalies. Again, this same set of problems would play out if the CP was involved in the clearance of cellular debris [1, 14, 141, 142, 147, 148].

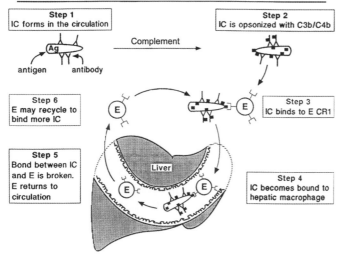

FIGURE 2.15 Erythrocyte processing of immune complexes (IC) or the "EPIC Phenomenon". This diagram delineates a pathway for processing (IC). Complement-coated IC adhere to the erythrocyte (E) C3b/C4b receptor, also known as *complement receptor type 1* (CR1; CD35). This binding of IC through C3b and its receptor to E in primates and to platelets in non-primates is known as the *immune adherence phenomenon*. The erythrocyte transports, like a taxi, shuttle or ferry, the IC to the liver and spleen. In the liver, the IC are transferred to macrophages and destroyed. In the spleen, the IC are transferred to B-lymphocytes, macrophages and follicular-dendritic cells for antigen processing. The E returns to circulation, often stripped of some of its receptors, but available to again bind complement coated IC. This model was initially defined in baboons by Hebert and colleagues; the essential elements of the scheme have now been described in humans, other primates, guinea pigs and rabbits. Several different antigen-antibody systems, including tetanus-anti-tetanus, DNA-anti-DNA and BSA-anti-BSA, have been analyzed. *Adapted from Hebert and Cosio [139], with permission.*

TARGETED GENE KNOCKOUTS

Pertinent to the preceding discussion, C1q, C4, factor B, and C2/factor B deficient mice have been produced [1, 149, 150]. The major lessons gained from these knockouts have been as follows:

1. On most genetic backgrounds, C1q-def or C4-def mice do not develop spontaneous autoimmunity (namely, autoantibodies and a SLE phenotype). When an SLE-like illness did develop spontaneously, it was on a background of autoimmunity. The message is that the genetic background is extremely important and that C1q or C4 *alone* is unable to cause autoimmunity in most strains of mice (in contrast to the outbred human population).

2. In lupus mouse models, C4-def or C1q-def enhanced the lupus phenotype. Disease onset was generally earlier and ANA titers were higher, with a greater frequency of DNA antibodies. There was also more pronounced GN.

3. Factor B and C2/factor B knockouts did not develop spontaneous autoimmunity (on any genetic background tested). Effect of these combined deficiencies on lupus mouse models was mixed. In several cases, there was no disease amelioration [1, 150] while in another there was lessoning in the glomerulonephritis and vasculitis [151].

4. C3-def mouse did not develop spontaneous lupus and, surprisingly, it had little or no effect on the lupus mouse models. This is a surprising result since the dogma would be that C3 activation would mediate some of the tissue damage.

In the following section, more details are given relative to these complement component deficient mice.

C4

High titers of ANA spontaneously developed in C4 KO mice [12, 152] on an autoimmune permissive background. Further, in the Chen study [152], by 10 months of age all female and most male mice had a high titer ANA and IC deposition in the glomerulus. In these two studies, C4 KO mice were also crossed with complement receptor 1 and 2 KO mice and these mice had a disease phenotype like C4 KO alone. The authors concluded that C4 deficiency causes lupus-like autoimmunity through a mechanism independent of CR1/CR2. In other words, C4b did not need to bind to complement receptors to mediate its protective effect against the development of SLE. The investigators favor a direct regulatory effect on autoreactive B cells. Two other points were made in these studies that are worth noting. In the Prodeus experiments,

CR1/CR2 deficiency markedly worsened the disease phenotype (like C4-def) while C3 def had no effect (by itself or in conjunction with C4-def) [12].

On another genetic background, C4 KO mice did not spontaneously develop a SLE phenotype [12, 153]. However, if the C4-def was bred into a strain that develops SLE, a more *severe* SLE-like illness developed [153]. The authors of this report also believe this may be related to defects in central tolerance of B cells. Of note, in all of these C4 def animals, C1q levels are normal and therefore its presence does not prevent the autoimmune phenotype from evolving.

In lupus-prone mice, animals null for both C3 and C4 were studied [153]. Interestingly, the disease process was similar in C4-def and the combined C3/C4 deficient animals. Thus, severe SLE developed in the absence of C3. These data have two straight-forward implications. One is that C3 activation is not required to cause SLE in the mouse and the second is that C4 contributes much more to SLE susceptibility than C3. It should be remembered that complement is involved in susceptibility to as well as in mediating tissue damage in SLE. Further, human and mouse SLE may have distinct causes and pathogenesis. Nevertheless, in both man and mouse, C4 and C1q provide an important protective role against the development of SLE.

Factor B/C2

A combined factor B and C2 [154] or a factor B alone [155] targeted gene deletion mouse did not develop spontaneous autoimmunity. Of interest, factor B/C2 deficient mice crossed into C1q-def did not prevent SLE from developing [156]. Since these animals cannot activate C3, these data again show that C3 is not required for development of lupus phenotype. In a different model system, factor B deficiency was crossed into lupus prone mice. The disease phenotype was milder [151]. Since autoantibody levels were similar, these results suggest that factor B plays a role primarily on the effector side of the equation, presumably via alternative pathway amplification of complement activation.

C1q

C1q is not part of complement gene cluster on chromosome 6 but like C4 and C2 is required for CP activation. About 50% of mice on a mixed genetic but autoimmune prone background developed high titers of ANA [13, 157, 158]. By 8 months of age, 25% of the mice had glomerulonephritis with immune deposits. The authors also demonstrated larger numbers of apoptotic bodies in diseased glomeruli. Their hypothesis is that increased numbers of apoptotic cells are present because of a C-dependent clearance defect for processing autoantigens. It is difficult to accept, however, that C1q-def is the whole story in view of the data with C4-def mice (in the setting of normal C1q). Perhaps it is hierarchy, as pointed out in another publication [13] by these authors, in which C1q > C4 >>> C2 is leading to SLE. To buttress this argument, they point out that C1q could have two roles. One is to serve as a ligand for C1q receptors on phagocytes and the other being to activate the CP thereby providing additional ligands for disposal of apoptotic cells.

In a more recent study of C1q-def, the effect of the genetic background on disease expression was further employed [157]. In two strains, the autoimmune phenotype was not altered but in a third strain the disease process accelerated with production of more autoantibodies and a more severe and earlier onset of glomerulonephritis.

FURTHER SPECULATION ON THE ASSOCIATION BETWEEN SYSTEMIC LUPUS ERYTHEMATOSUS AND COMPLEMENT DEFICIENCY

As discussed, C is activated by IgG or IgM, and C activation through C3 is required to efficiently eliminate many infectious agents. This cannot be readily accomplished in a C-deficient host and therefore would predispose such an individual to more prolonged exposure to infectious agents or damaged self materials.

A problem faced by a C-deficient host revolves around clearance of foreign antigens and damaged self. As discussed in the preceding section, the lack of C3b deposition on an IC leads to impaired antigen localization such as to follicular areas of the spleen. This appears to be a critical factor in the poor antibody response of such individuals [159–161]. For example, if C4A, C2, or C3 def guinea pigs or humans are immunized intravenously with bacteriophage φX174, a relatively low-titer IgM response is produced. Moreover, a secondary IgG response is not made, even with repetitive intravenous immunizations. A failure to localize the antigen to the follicular area of the spleen probably accounts for the defective immune response. The primary and secondary immune response can be rescued in C4-def guinea pigs by adding purified human C4A but not C4B or an adjuvant with the primary immunization [162]. This presumed failure of antigen presentation is more simply explained by C4A being required for efficient opsonization of this antigen. A plausible scenario then for deficient antibody production in C deficiency is defective antigen localization. CR1 (the C3b/C4b receptor) and CR2 (C3d receptor) are expressed by B lymphocytes and follicular-dendritic cells. On both cell types, they facilitate localization of C-coated IC to B cells and enhance B-cell activation [43, 44, 163]. In these model systems, immunization with increased amounts of antigen leads to a normal immune response.

In a C-deficient host there are several avenues to a vasculitic state. Perhaps the simplest is the failure to efficiently eliminate an infectious material. Prolonged antigenemia could result (with or without significant tissue damage to the organ which harbors the infection) such as in hepatitis B and C infections [164]. Chronic antigenemia is the setting for impaired IC formation, such as at Ab-Ag equivalence or slight Ag excess. In experimental serum sickness models, such IC are prone to deposit in glomeruli and other tissues. Although we do not understand the reason for the poor antibody response and failure to eliminate the virus, the consequences are clear: chronic antigenemia, inappropriate IC formation and IC deposition, resulting in glomerulonephritis or a systemic vasculitic syndrome. Hepatitis B and C are surely not the only such infections of mankind that produce this type of a syndrome. More likely, hepatitis is the one that we recognize because of its means of transmission and because of sensitive serologic and biochemical markers for these infections. Complement deficiency would contribute to and potentiate other defects in the immune response. In this context, it is interesting to note that C4A deficiency also appears to play a major role in nonresponders to HBsAg vaccination. It has been demonstrated that this nonresponse is significantly associated with C4A null alleles, independent from the presence of the DRB1*0301 allele [165, 166]. This line of reasoning also suggests that some forms of SLE actually represent immune deficiency syndromes. Lupus associated with C deficiency and hypogammaglobulinemia are two such examples.

In a C-deficient host, even normal amounts of IC (such as those that form because of passive absorption of bacterial antigens from epithelial surfaces) or in association with infections or associated with physiologic tissue turnover and cell death would have a greater likelihood of being inappropriately processed. The prevention of IC precipitation in vascular structures, their adherence to erythrocytes or leukocytes, and their transport to and deposition in the liver and spleen would all likely be impaired. Thus, if such individuals develop a condition leading to chronic antigenemia, they would be at greater risk for mishandling IC than would a normal host. Lachmann and Walport have noted that autoantibodies arise transiently in association with infections like infectious mononucleosis and following tissue necrosis like in myocardial or pulmonary infarctions [10]. They suggest that deficiency of effector mechanisms, especially immune complex processing, induce pathogenic inflammatory reactions at extrahepatic sites. A vicious cycle then ensues as intracellular antigens are exposed and become autoantigens, possibly modified by the inflammatory response with cell injury or death; i.e., the normal decay of an immune response does not occur.

DETECTION OF COMPLEMENT DEFICIENCY, PHENOTYPIC AND GENOTYPIC VARIATIONS

C4 and C3 serum concentrations and the hemolytic complement titer (CH_{50}) are common immunologic parameters monitored in SLE patients. A complete deficiency of a C activating component is usually first suggested by the finding of a very low or zero whole complement (THC or CH_{50}) hemolytic titer [167]. Confirmation of complete deficiency is required and the specific component identified [168, 169]. C2 deficiency is the most common of the complete deficiencies leading to SLE. C4 and C3 protein levels are readily detectable by radial immunodiffusion or nephelometry. Low C4 and C3 serum concentrations are common immunologic features of SLE, particularly during active disease or disease flares [35, 36].

In the case of C4, serum concentration is the cumulative result of gene copy number and gene size for C4, the body mass index, and the physiologic or disease state [72, 78, 123]. C4 protein polymorphisms and homozygous isotype deficiency can be readily detected by immunofixation of neuraminidase and CPB-treated EDTA-plasma that has been stored at $-70°C$ or below. Gene copy number and gene size variations can be efficiently elucidated by quantitative, TaqMan-based real-time PCR [71]. Elucidations of RCCX length variants and constituent gene organizations require more tedious experiments involving Southern blot analyses of PmeI-digested genomic DNA resolved by pulsed field gel electrophoresis, and TaqI, PshAI/PvuII digested genomic DNA resolved by regular agarose gel electrophoresis [55, 65, 77, 95, 114].

References

[1] M.C. Pickering, M. Botto, P.R. Taylor, P.J. Lachmann, M.J. Walport, Systemic lupus erythematosus, complement deficiency and apoptosis, Adv Immunol 76 (2001) 227–324.

[2] A.G. Sjoholm, Inherited complement deficiency states and disease, Complement Inflamm 8 (1991) 341–346.

[3] F.C. Arnett, J.M. Moulds, HLA class III molecules and autoimmune rheumatic diseases, Clin Exp Rheumatol 9 (1991) 289–296.

[4] F.T. Christiansen, W.J. Zhang, M. Griffiths, S.A. Mallal, R.L. Dawkins, Major histocompatibility complex (MHC) complement deficiency, ancestral haplotypes and systemic lupus erythematosus (SLE): C4 deficiency explains some but not all of the influences of the MHC, J Rheumatol 18 (1991) 1350–1358.

[5] J.A. Schifferli, D.K. Peters, Complement, the immune-complex lattice, and the pathophysiology of complement-deficiency syndromes, Lancet 2 (1983) 957–959.

[6] J.P. Atkinson, Complement activation and complement receptors in systemic lupus erythematosus, Springer Semin Immunopathol 9 (1986) 179−194.

[7] J.P. Atkinson, Complement deficiency. Predisposing factor to autoimmune syndromes, Am J Med 85 (1988) 45−47.

[8] J.P. Atkinson, Immune complexes and the role of complement, in: E.C. LeRoy (Ed.), Systemic Vasculitis, Dekker, New York, 1992, pp. 525−546.

[9] J.A. Schifferli, R.P. Taylor, Physiological and pathological aspects of circulating immune complexes, Kidney Int 35 (1989) 993−1003.

[10] P.J. Lachmann, M.J. Walport, Deficiency of the effector mechanisms of the immune response and autoimmunity, in: D. Evered, J. Whelan (Eds.), Autoimmunity and Autoimmune Disease, Wiley & Sons, New York, 1987, p. 129.

[11] M.E. Medof, Complement-dependent maintenance of immune complex solubility, in: K. Rother, G.O. Till (Eds.), "The Complement System", Spring-Verlag, Berlin, 1988, pp. 418−443.

[12] A.P. Prodeus, S. Goerg, L.M. Shen, O.O. Pozdnyakova, L. Chu, E.M. Alicot, et al., A critical role for complement in maintenance of self-tolerance, Immunity 9 (1998) 721−731.

[13] A.J. Cutler, R.J. Cornall, H. Ferry, A.P. Manderson, M. Botto, M.J. Walport, Intact B cell tolerance in the absence of the first component of the classical complement pathway, Eur J Immunol 31 (2001) 2087−2093.

[14] J.S. Navratil, J.M. Ahearn, Apoptosis and autoimmunity: complement deficiency and systemic lupus erythematosus revisited, Curr Rheumatol Rep 2 (2000) 32−38.

[15] K.E. Sullivan, Complement deficiency and autoimmunity, Curr Opin Pediatr 10 (1998) 600−606.

[16] J.P. Atkinson, The Complement System, in: H. Eisen (Ed.), General Immunology, J.B. Lippincott Co., Philadelphia, 1989, pp. 147−166.

[17] A.C. Chan, D.R. Karp, D.C. Shreffler, J.P. Atkinson, The 20 faces of the fourth component of complement, Immunol. Today 5 (1984) 200−203.

[18] C.Y. Yu, E.K. Chung, Y. Yang, C.A. Blanchong, N. Jacobsen, K. Saxena, et al., Dancing with complement C4 and the RP-C4-CYP21-TNX (RCCX) modules of the major histocompatibility complex, Progr Nucl Acid Res Mol Biol 75 (2003) 217−292.

[19] H.E. Feucht, J. Zwirner, D. Bevec, M. Lang, E. Felber, G. Riethmuller, et al., Biosynthesis of complement C4 messenger RNA in normal human kidney, Nephron 53 (1989) 338−342.

[20] H. Tsukamoto, K. Nagasawa, S. Yoshizawa, Y. Tada, A. Ueda, Y. Ueda, et al., Synthesis and regulation of the fourth component of complement (C4) in the human momocytic cell line U937: comparison with that of the third component of complement, Immunology 75 (1992) 565−569.

[21] T.R. Welch, L.S. Beischel, D.P. Witte, Differential expression of complement C3 and C4 in the human kidney, J Clin Invest 92 (1993) 1451−1458.

[22] K.T. Belt, C.Y. Yu, M.C. Carroll, R.R. Porter, Polymorphism of human complement component C4, Immunogenetics 21 (1985) 173−180.

[23] C.Y. Yu, The complete exon-intron structure of a human complement component C4A gene: DNA sequences, polymorphism, and linkage to the 21-hydroxylase gene, J Immunol 146 (1991) 1057−1066.

[24] A.C. Chan, J.P. Atkinson, Oligosaccharide structure of human C4, J Immunol 134 (1985) 1790−1798.

[25] G. Hortin, A.C. Chan, K.F. Rok, A.W. Strause, J.P. Atkinson, Sequence analysis of the COOH terminus of the a-chain of the fourth component of human complement, J Biol Chem 261 (1986) 9065−9069.

[26] G.L. Hortin, T.C. Farries, J.P. Graham, J.P. Atkinson, Sulfation of tyrosinbe residues increase activity of the fourth component of complement, Proc Natl Acad Sci USA 86 (1989) 1338−1342.

[27] A.C. Chan, K.R. Mitchell, T. Munns, D.R. Karp, J.P. Atkinson, Identification and partial characterization of the secreted form of the fourth component of human complement: evidence that it is different from major plasma form, Proc Natl Acad Sci USA 80 (1983) 268−272.

[28] D.M. Lublin, M.K. Liszewski, T.W. Post, M.A. Arce, M.M. Le Beau, M.B. Rebentisch, et al., Molecular cloning and chromosomal localization of human membrane cofactor protein (MCP). Evidence for inclusion in the multigene family of complement-regulatory proteins, J Exp Med 168 (1988) 181−194.

[29] A.C. Chan, J.P. Atkinson, Identification and structural characterization of two incompletely processed forms of the fourth component of human complement, J Clin Invest 72 (1983) 1639−1649.

[30] B.F. Tack, The beta-Cys-gamma-Glu thiolester bond in human C3, C4, and alpha 2-macroglobulin, Springer Semin Immunopathol 6 (1983) 259−282.

[31] S.K. Law, Non-enzymic activation of the covalent binding reaction of the complement protein C3, Biochem J 211 (1983) 381−389.

[32] R.D. Campbell, A.W. Dodds, R.R. Porter, The binding of human complement component C4 to antibody-antigen aggregates, Biochem J 189 (1980) 67−80.

[33] S.K. Law, The covalent binding reaction of C3 and C4, Ann NY Acad Sci 421 (1983) 246−258.

[34] D.E. Isenman, J.R. Young, The molecular basis for the differences in immune hemolysis activity of the Chido and Rodgers isotypes of human complement component C4, J Immunol 132 (1984) 3019−3027.

[35] S. Manzi, J.S. Navratil, M.J. Ruffing, C.C. Liu, N. Danchenko, S.E. Nilson, et-al, Measurement of erythrocyte C4d and complement receptor 1 in systemic lupus erythematosus, Arthritis Rheum 50 (2004) 3596−3604.

[36] C.C. Liu, S. Manzi, A.H. Kao, J.S. Navratil, M.J. Ruffing, J.M. Ahearn, Reticulocytes bearing C4d as biomarkers of disease activity for systemic lupus erythematosus, Arthritis Rheum 52 (2005) 3087−3099.

[37] J.S. Navratil, S. Manzi, A.H. Kao, S. Krishnaswami, C.C. Liu, M.J. Ruffing, Platelet C4d is highly specific for systemic lupus erythematosus, Arthritis Rheum 54 (2006) 670−674.

[38] C.M. Giles, B. Uring-Lambert, J. Goetz, G. Hauptmann, A.H.L. Fielder, W. Ollier, Antigenic determinants expressed by human C4 allotypes: a study of 325 families provides evidence for the structural antigenic model, Immunogenetics 27 (1988) 442−448.

[39] C.Y. Yu, R.D. Campbell, R.R. Porter, A structural model for the location of the Rodgers and the Chido antigenic determinants and their correlation with the human complement C4A/C4B isotypes, Immunogenetics 27 (1988) 399−405.

[40] H.E. Feucht, Complement C4d in graft capillaries − the missing link in the recognition of humoral alloreactivity, Am J Transplant 3 (2003) 646−652.

[41] G.A. Bohmig, G. Bartel, M. Wahrmann, Antibodies, isotypes and complement in allograft rejection, Curr Opin Organ Transplant 13 (2008) 411−418.

[42] G.D. Ross, J.A. Cain, P.J. Lachmann, Membrane complement receptor type three (CR3) has lectin-like properties analogous to bovine conglutinin as functions as a receptor for zymosan and rabbit erythrocytes as well as a receptor for iC3b, J Immunol 134 (1985) 3307−3315.

[43] J.M. Ahearn, D.T. Fearon, Structure and function of the complement receptors, CR1 (CD35) and CR2 (CD21), Adv Immunol 46 (1989) 183–219.

[44] D. Hourcade, V.M. Holers, J.P. Atkinson, The regulators of complement activation (RCA) gene cluster, Adv Immunol 45 (1989) 381–416.

[45] K.T. Belt, M.C. Caroll, R.R. Porter, The structural basis of the multiple forms of human complement component C4, Cell 36 (1984) 907–914.

[46] C.Y. Yu, K.T. Belt, C.M. Giles, R.D. Campbell, R.R. Porter, Structural basis of the polymorphism of human complement component C4A and C4B: gene size, reactivity and antigenicity, EMBO J. 5 (1986) 2873–2881.

[47] M.C. Carroll, D.M. Fathallah, L. Bergamaschini, E.M. Alicot, D.E. Isenman, Substitution of a single amino acid (aspartic acid for histidine) converts the functional activity of human complement C4B to C4A, Proc Natl Acad Sci USA 87 (1990) 6868–6872.

[48] G. Mauff, B. Luther, P.M. Schneider, C. Rittner, B. Strandmann-Bellinghausen, R. Dawkins, et al., Reference typing report for complement component C4, Exp Clin Immunogenet 15 (1998) 249–260.

[49] G. Mauff, C.A. Alper, R. Dawkins, G. Doxiadis, C.M. Giles, G. Hauptmann, C4 nomenclature statement (1990), Complement Inflamm 7 (1990) 261–268.

[50] M.J. Anderson, C.M. Milner, R.G. Cotton, R.D. Campbell, The coding sequence of the hemolytically inactive C4A6 allotype of human complement component C4 reveals that a single arginine to tryptophan substitution at beta-chain residue 458 is the likely cause of the defect, J Immunol 148 (1992) 2795–2802.

[51] A.W. Dodds, S.K. Law, R.R. Porter, The origin of the very variable haemolytic activities of the common human complement component C4 allotypes including C4-A6, EMBO J 4 (1985) 2239–2244.

[52] Z.L. Awdeh, D. Raum, C.A. Alper, Genetic polymorphism of human complement C4 and detection of heterozygotes, Nature 282 (1979) 205–208.

[53] Z.L. Awdeh, C.A. Alper, Inherited structural polymorphism of the fourth component of human complement, Proc Natl Acad Sci USA 77 (1980) 3576–3580.

[54] E. Sim, S. Cross, Phenotyping of human complement component C4, a class III HLA antigen, Biochem J 239 (1986) 763–767.

[55] E.K. Chung, Y.L. Wu, Y. Yang, B. Zhou, C.Y. Yu, Human complement components C4A and C4B genetic diversities: complex genotypes and phenotypes, Curr Protoc Immunol (2005) 13.8.1–13.8.36.

[56] D.E. Isenman, J.R. Young, Covalent binding properties of the CR1 and C4B isotypes of the fourth component of human complement on several C1-bearing cell surfaces, J. Immunol. 136 (1986) 2542–2550.

[57] S.K.A. Law, A.W. Dodds, R.R. Porter, A comparison of the properties of two classes, C4A and C4B, of the human complement component C4, EMBO J 3 (1984) 1819–1823.

[58] A.W. Dodds, X.-D. Ren, A.C. Willis, S.K.A. Law, The reaction mechanism of the internal thioester in the human complement component C4, Nature 379 (1996) 177–179.

[59] C.A. Blanchong, B. Zhou, K.L. Rupert, E.K. Chung, K.N. Jones, J.F. Sotos, et-al, Deficiencies of human complement component C4A and C4B and heterozygosity in length variants of RP-C4-CYP21-TNX (RCCX) modules in Caucasians: the load of RCCX genetic diversity on MHC-associated disease, J Exp Med 191 (2000) 2183–2196.

[60] P. Teisberg, I. Akesson, B. Olaisen, T. Gedde-Dahl Jr., E. Thorsby, Genetic polymorphism of C4 in man and localisation of a structural C4 locus to the HLA gene complex of chromosome 6, Nature 264 (1976) 253–254.

[61] P. Teisberg, B. Olaisen, R. Jonassen, T. Gedde-Dahl Jr., E. Thorsby, The genetic polymorphism of the fourth component of human complement: methodological aspects and a presentation of linkage and association data relevant to its localization in the HLA region, J Exp Med 146 (1977) 1380–1389.

[62] G.J. O'Neill, S.Y. Yang, B. DuPont, Two HLA-linked loci controlling the fourth component of human complement, Proc Natl Acad Sci USA 75 (1978) 5165–5169.

[63] Z.L. Awdeh, H.D. Ochs, C.A. Alper, Genetic analysis of C4 deficiency, J Clin Invest 67 (1981) 260–263.

[64] M.H. Roos, E. Mollenhauer, P. Demant, C. Rittner, A molecular basis for the two locus model of human complement component C4, Nature 298 (1982) 854–855.

[65] C.Y. Yu, R.D. Campbell, Definitive RFLPs to distinguish between the human complement C4A/C4B isotypes and the major Rodgers/Chido determinants: application to the study of C4 null alleles, Immunogenetics 25 (1987) 383–390.

[66] E.K. Chung, Y. Yang, R.M. Rennebohm, M.L. Lokki, G.C. Higgins, K.N. Jones, et al., Genetic sophistication of human complement C4A and C4B and RP-C4-CYP21-TNX (RCCX) modules in the major histocompatibility complex (MHC), Am J Hum Genet 71 (2002) 823–837.

[67] Y. Yang, E.K. Chung, Y.L. Wu, S.L. Savelli, H.N. Nagaraja, B. Zhou, et-al, Gene copy number variation and associated polymorphisms of complement component C4 in human systemic erythematosus (SLE): low copy number is a risk factor for and high copy number is a protective factor against European American SLE disease susceptibility, Am J Hum Genet 80 (2007) 1037–1054.

[68] M.C. Carroll, A. Palsdottir, K.T. Belt, R.R. Porter, Deletion of complement C4 and steroid 21-hydroxylase genes in the HLA class III region, EMBO J 4 (1985) 2547–2552.

[69] M.C. Carroll, K.T. Belt, A. Palsdottir, Y. Yu, Molecular genetics of the fourth component of human complement and the steroid 21-hydroxylase, Immunol Rev 87 (1985) 39–60.

[70] G. Hauptmann, J. Goetz, B. Uring-Lambert, E. Grosshans, Component deficiencies, Prog Allergy 39 (1986) 232–249.

[71] Y.L. Wu, S.L. Savelli, Y. Yang, B. Zhou, B.H. Rovin, D.J. Birmingham, et al., Sensitive and specific real-time PCR Assays to accurately determine copy-number variations (CNVs) of human complement C4A, C4B, C4-Long, C4-Short and RCCX modules: Elucidation of C4 CNVs in 50 consanguineous subjects with defined HLA genotypes, J Immunol 179 (2007) 3012–3025.

[72] K. Saxena, K.J. Kitzmiller, Y.L. Wu, B. Zhou, N. Esack, L. Hiremath, et al., Great genotypic and phenotypic diversities associated with copy-number variations of complement C4 and RP-C4-CYP21-TNX (RCCX) modules: A comparison of Asian-Indian and European American populations, Mol Immunol 46 (2009) 1289–1303.

[73] A.W. Dangel, A.R. Mendoza, B.J. Baker, C.M. Daniel, M.C. Carroll, L.-C. Wu, The dichotomous size variation of human complement C4 gene is mediated by a novel family of endogenous retroviruses which also establishes species-specific genomic patterns among Old World primates, Immunogenetics 40 (1994) 425–436.

[74] X. Chu, C. Rittner, P.M. Schneider, Length polymorphism of the human complement component C4 gene is due to an ancient retroviral integration, Exp Clin Immunogenet 12 (1995) 74–81.

[75] P.M. Schneider, K. Witzel-Schlomp, C. Rittner, L. Zhang, The endogenous retroviral insertion in the human complement C4 gene modulates the expression of homologous genes by antisense inhibition, Immunogenetics 53 (2001) 1–9.

[76] M. Mack, K. Bender, P.M. Schneider, Detection of retroviral antisense transcripts and promoter activity of the HERV-K(C4) insertion in the MHC class III region, Immunogenetics 56 (2004) 321–332.

[77] P.M. Schneider, M.C. Carroll, C.A. Alper, C. Rittner, A.S. Whitehead, E.J. Yunis, et al., Polymorphism of human complement C4 and steroid 21-hydroxylase genes. Restriction fragment length polymorphisms revealing structural deletions, homoduplications, and size variants, J Clin Invest 78 (1986) 650–657.

[78] Y. Yang, E.K. Chung, B. Zhou, C.A. Blanchong, C.Y. Yu, G. Füst, et al., Diversity in intrinsic strengths of the human complement system: serum C4 protein concentrations correlate with C4 gene size and polygenic variations, hemolytic activities and body mass index, J Immunol 171 (2003) 2734–2745.

[79] C.Y. Yu, C.C. Whitacre, Sex, MHC and complement C4 in autoimmune diseases, Trends Immunol 25 (2004) 694–699.

[80] Z. Yang, A.R. Mendoza, T.R. Welch, W.B. Zipf, C.Y. Yu, Modular variations of HLA class III genes for serine/threonine kinase RP, complement C4, steroid 21-hydroxylase CYP21 and tenascin TNX (RCCX): a mechanism for gene deletions and disease associations, J Biol Chem 274 (1999) 12147–12156.

[81] L.M. Shen, L.C. Wu, S. Sanlioglu, R. Chen, A.R. Mendoza, A. Dangel, et al., Structure and genetics of the partially duplicated gene RP located immediately upstream of the complement C4A and C4B genes in the HLA class III region: Molecular cloning, exon-intron structure, composite retroposon and breakpoint of gene duplication, J Biol Chem 269 (1994) 8466–8476.

[82] S.E. Gitelman, J. Bristow, W.L. Miller, Mechanism and consequences of the duplication of the human C4/P450c21/gene X locus, Mol Cell Biol 12 (1992) 2124–2134.

[83] P.C. White, D. Grossberger, B.J. Onufer, M.I. New, B. DuPont, J. Strominger, Two genes encoding steroid 21-hydroxylase are located near the genes encoding the fourth component of complement in man, Proc Natl Acad Sci USA 82 (1985) 1089–1093.

[84] Y. Higashi, H. Yoshioka, M. Yamane, O. Gotoh, Y. Fujii-Kuriyama, Complete nucleotide sequence of two steroid 21-hydroxylase genes tandemly arranged in human chromosome: A pseudogene and a genuine gene, Proc Natl Acad Sci USA 83 (1986) 2841–2845.

[85] R. Dawkins, C. Leelayuwat, S. Gaudieri, G. Tay, J. Hui, S. Cattley, et al., Genomics of the major histocompatibility complex: haplotypes, duplication, retroviruses and disease, Immunol Rev 167 (1999) 275–304.

[86] Y.L. Wu, Y. Yang, E.K. Chung, B. Zhou, K.J. Kitzmiller, S.L. Savelli, et al., Phenotypes, genotypes and disease susceptibility associated with gene copy number variations: complement C4 CNVs in European American healthy subjects and those with systemic lupus erythematosus, Cytogenet Genome Res 123 (2008) 131–141.

[87] A.E. Hughes, N. Orr, H. Esfandiary, M. az-Torres, T. Goodship, U. Chakravarthy, A common CFH haplotype, with deletion of CFHR1 and CFHR3, is associated with lower risk of age-related macular degeneration, Nat Genet 38 (2006) 1173–1177.

[88] R. Horton, L. Wilming, V. Rand, R.C. Lovering, E.A. Bruford, V.K. Khodiyar, M.J. Lush, et al., Gene map of the extended human MHC, Nat Rev Genet 5 (2004) 889–899.

[89] T. Shiina, K. Hosomichi, H. Inoko, J.K. Kulski, The HLA genomic loci map: expression, interaction, diversity and disease, J Hum Genet 54 (2009) 15–39.

[90] P.M. Aldred, E.J. Hollox, J.A. Armour, Copy number polymorphism and expression level variation of the human alpha-defensin genes DEFA1 and DEFA3, Hum Mol Genet 14 (2005) 2045–2052.

[91] E. Gonzalez, H. Kulkarni, H. Bolivar, A. Mangano, R. Sanchez, G. Catano, et al., The influence of CCL3L1 gene-containing segmental duplications on HIV-1/AIDS susceptibility, Science 307 (2005) 1434–1440.

[92] T.J. Aitman, R. Dong, T.J. Vyse, P.J. Norsworthy, M.D. Johnson, J. Smith, et al., Copy number polymorphism in Fcgr3 predisposes to glomerulonephritis in rats and humans, Nature 439 (2006) 851–855.

[93] A.A. Bashirova, M.P. Martin, D.W. McVicar, M. Carrington, The Killer Immunoglobulin-Like Receptor Gene Cluster: Tuning the Genome for Defense (*), Annu Rev Genomics Hum Genet 7 (2006) 277–300.

[94] L.C. Wu, B.J. Morley, R.D. Campbell, Cell-specific expression of the human complement protein factor B gene: evidence for the role of two distinct 5'-flanking elements, Cell 48 (1987) 331–342.

[95] P.M. Schneider, C. Rittner, Complement genetics, in: A.W. Dodds, R.B. Sim (Eds.), Complement: a Practical Approach, IRL Press at Oxford University Press, Oxford, 1997, pp. 165–198.

[96] S.J. Cross, J.H. Edwards, D.R. Bentley, R.D. Campbell, DNA polymorphism of the C2 and factor B genes. Detection of a restriction fragment length polymorphism which subdivides haplotypes carrying the C2C and factor B F alleles, Immunogenetics 21 (1985) 39–48.

[97] C.J. Jaffe, J.P. Atkinson, M.M. Frank, The role of complement in the clearance of cold agglutinin-sensitized erythrocytes in man, J Clin Invest 58 (1976) 942–949.

[98] Y.L. Wu, G. Hauptmann, M. Viguier, C.Y. Yu, Molecular basis of complete complement C4 deficiency in two North-African families with systemic lupus erythematosus, Genes Immun. 10 (2009) 433–445.

[99] C.Y. Yu, G. Hauptmann, Y. Yang, Y.L. Wu, D.J. Birmingham, B.H. Rovin, Complement deficiencies in human systemic lupus erythematosus (SLE) and SLE nephritis: epidemiology and pathogenesis, in: G.C. Tsokos, C. Gordon, J.S. Smolen (Eds.), Systemic Lupus Erythematosus: A Companion to Rheumatology, Elsevier, Philadelphia, 2007, pp. 203–213.

[100] Y. Yang, E.K. Chung, B. Zhou, K. Lhotta, L.A. Hebert, D.J. Birmingham, The intricate role of complement C4 in human systemic lupus erythematosus, Curr. Direct. Autoimmun 7 (2004) 98–132.

[101] Y. Yang, K. Lhotta, E.K. Chung, P. Eder, F. Neumair, C.Y. Yu, Complete complement components C4A and C4B deficiencies in human kidney diseases and systemic lupus erythematosus, J Immunol 173 (2004) 2803–2814.

[102] G.N. Fredrikson, B. Gullstrand, P.M. Schneider, K. Witzel-Schlomp, A.G. Sjoholm, C.A. Alper, et al., Characterization of non-expressed C4 genes in a case of complete C4 deficiency: identification of a novel point mutation leading to a premature stop codon, Hum Immunol 59 (1998) 713–719.

[103] M.-L. Lokki, A. Circolo, P. Ahokas, K.L. Rupert, C.Y. Yu, H.R. Colten, Deficiency of human complement protein C4 due to identical frameshift mutations in the C4A and C4B genes, J Immunol 162 (1999) 3687–3693.

[104] K.L. Rupert, J.M. Moulds, Y. Yang, F.C. Arnett, R.W. Warren, J.D. Reveille, et al., The molecular basis of complete C4A and C4B deficiencies in a systemic lupus erythematosus (SLE) patient with homozygous C4A and C4B mutant genes, J Immunol 169 (2002) 1570–1578.

[105] G. Barba, C. Rittner, P.M. Schneider, Genetic basis of human complement C4A deficiency. Detection of a point mutation leading to nonexpression, J Clin Invest 91 (1993) 1681–1686.

[106] K.E. Sullivan, N.A. Kim, D. Goldman, M.A. Petri, C4A deficiency due to a 2 bp insertion is increased in patients with systemic lupus erythematosus, J Rheumatol 26 (1999) 2144–2147.

[107] A.H.L. Fielder, M.J. Walport, J.R. Batchelor, R.I. Rynes, C.M. Black, I.A. Dodi, et al., Family study of the major histocompatibility complex in patients with systemic lupus erythematosus: importance of null alleles of C4A and C4B in determining disease susceptibility, Br Med J 286 (1983) 425–428.

[108] F.T. Christiansen, G. Uko, R.L. Dawkins, Complement allotyping in systemic lupus erythematosus. Association with C4A null, in: R.L. Dawking (Ed.), Immunogenetics in Rheumatology, Excerta Medica, Amsterdam, 1982, pp. 229–234.

[109] F.T. Christiansen, R.L. Dawkins, G. Uko, Complement allotyping in SLE: Association with C4A null, Aust. NZ J Med 13 (1983) 483.

[110] R.R. Graham, W.A. Ortmann, C.D. Langefeld, D. Jawaheer, S.A. Selby, P.R. Rodine, Visualizing human leukocyte antigen class II risk haplotypes in human systemic lupus erythematosus, Am J Hum Genet 71 (2002) 543–553.

[111] M.M. Fernando, C.R. Stevens, P.C. Sabeti, E.C. Walsh, A.J. McWhinnie, A. Shah, et al., Identification of two independent risk factors for lupus within the MHC in United Kingdom families, PLoS Genet 3 (2007) e192.

[112] J.B. Harley, M.E. arcon-Riquelme, L.A. Criswell, C.O. Jacob, R.P. Kimberly, K.L. Moser, et al., Genome-wide association scan in women with systemic lupus erythematosus identifies susceptibility variants in ITGAM, PXK, KIAA1542 and other loci, Nat Genet 40 (2008) 204–210.

[113] C.Y. Yu, C.A. Blanchong, E.K. Chung, K.L. Rupert, Y. Yang, Z. Yang, Molecular genetic analyses of human complement components C4A and C4B, in: N.R. Rose, R.G. Hamilton, B. Detrick (Eds.), Manuals of Clinical Laboratory Immunology, 6th ed.). ASM Press, Washington, D.C, 2002, pp. 117–131.

[114] E.K. Chung, Y. Yang, K.L. Rupert, K.N. Jones, R.M. Rennebohm, C.A. Blanchong, et al., Determining the one, two, three or four long and short loci of human complement C4 in a major histocompatibility complex haplotype encoding for C4A or C4B proteins, Am J Hum Genet 71 (2002) 810–822.

[115] P.F. Howard, M.C. Hochberg, W.B. Bias, F.C.J. Arnett, R.H. McLean, Relationship between C4 null genes, HLA-D region antigens, and genetic susceptibility to systemic lupus erythematosus in Caucasian and black Americans, Am J Med 81 (1986) 187–193.

[116] H. Dunckley, P.A. Gatenby, B. Hawkins, S. Naito, S.W. Serjeantson, Deficiency of C4A is a genetic determinant of systemic lupus erythematosus in three ethnic groups, J Immunogenet 14 (1987) 209–218.

[117] P.H. Schur, D. Marcus-Bagley, Z. Awdeh, E.J. Yunis, C.A. Alper, The effect of ethnicity on major histocompatibility complex complement allotypes and extended haplotypes in patients with systemic lupus erythematosus, Arthritis Rheum 33 (1990) 985–992.

[118] J. Granados, G. Vargas-Alarcon, F. Andrade, H. Melin-Aldana, J. Cocer-Varela, D. arcon-Segovia, The role of HLA-DR alleles and complotypes through the ethnic barrier in systemic lupus erythematosus in Mexicans, Lupus 5 (1996) 184–189.

[119] W.A. Wilson, M.C. Perez, P.E. Armatis, Partial C4A deficiency is associated with susceptibility to systemic lupus erythematosus in black Americans, Arthritis Rheum 31 (1988) 1171–1175.

[120] M.L. Olsen, R. Goldstein, F.C. Arnett, M. Duvic, M. Pollack, J.D. Reveille, C4A gene deletion and HLA associations on black Americans with systemic lupus erythematosus, Immunogenetics 30 (1989) 27–33.

[121] J.A. Elliot, D.R. Mathieson, Complement in disseminated (systemic) lupus erythematosus, A.M.A. Arch Dermatol Syphilol 68 (1953) 119–128.

[122] E.J. Lewis, C.B. Carpenter, P.H. Schur, Serum complement component levels in human glomerulonephritis, Ann Intern Med 75 (1971) 555–560.

[123] Y.L. Wu, G.C. Higgins, R.M. Rennebohm, E.K. Chung, Y. Yang, B. Zhou, et al., Three distinct profiles of serum complement C4 proteins in pediatric systemic lupus erythematosus (SLE) patients: Tight associations of complement C4 and C3 protein levels in SLE but not in healthy subjects, Adv Exp Med Biol 586 (2006) 227–247.

[124] C.A. Johnson, P. Densen, R. Wetsel, F.S. Cole, N.E. Goeken, H.R. Colten, Molecular heterogeneity of C2 deficiency, New Engl J Med 326 (1992) 874.

[125] V. Agnello, Lupus diseases associated with hereditary and acquired deficiencies of complement, Springer Semin Immunopathol 9 (1986) 161–178.

[126] C.A. Johnson, P. Densen, R.K. Hurford Jr., H.R. Colten, R.A. Wetsel, Type I human complement C2 deficiency. A 28-base pair gene deletion causes skipping of exon 6 during RNA splicing, J Biol Chem 267 (1992) 9347–9353.

[127] X. Wang, A. Circolo, M.L. Lokki, P.G. Shackelford, R.A. Wetsel, H.R. Colten, Molecular heterogeneity in deficiency of complement protein C2 type I, Immunology 93 (1998) 184–191.

[128] R.A. Wetsel, J. Kulics, M.L. Lokki, P. Kiepiela, H. Akama, C.A. Johnson, et al., Type II human complement C2 deficiency. Allele-specific amino acid substitutions (Ser189 −> Phe; Gly444 −> Arg) cause impaired C2 secretion, J Biol Chem 271 (1996) 5824–5831.

[129] Z.B. Zhu, T.P. Atkinson, J.E. Volanakis, A novel type II complement C2 deficiency allele in an African-American family, J Immunol 161 (1998) 578–584.

[130] G. Jonsson, A.G. Sjoholm, L. Truedsson, A.A. Bengtsson, J.H. Braconier, G. Sturfelt, Rheumatological manifestations, organ damage and autoimmunity in hereditary C2 deficiency, Rheumatology (Oxford) 46 (2007) 1133–1139.

[131] E. Sim, E.W. Gill, R.B. Sim, Drugs that induce systemic lupus erythematosus inhibit complement component C4, Lancet ii (1984) 422–424.

[132] Y. Fukumori, K. Yoshimura, S. Ohnoki, H. Yamaguchi, Y. Akagaki, S. Inai, A high incidence of C9 deficiency among healthy blood donors in Osaka, Japan, Int Immunol 1 (1989) 85–89.

[133] S. Inai, Y. Akagaki, T. Moriyama, Y. Fukumori, K. Yoshimura, S. Ohnoki, et al., Inherited deficiencies of the late-acting complement components other than C9 found among healthy blood donors, Int Arch Allergy Appl Immunol 90 (1989) 274–279.

[134] M. Steuer, G. Mauff, C. Adam, M.P. Baur, K. Bender, J. Goetz, An estimate on the frequency of duplicated haplotypes and silent alleles of human C4 protein polymorphism. I. Investigations in healthy Caucasoid families, Tissue Antigens 33 (1989) 501–510.

[135] T.C. Farries, K.L. Steuer, J.P. Atkinson, Evolutionary implications of a new bypass activation pathway of the complement system, Immunol Today 11 (1990) 78–80.

[136] M.J. Walport, P.J. Lachmann, Erythrocyte complement receptor type 1, immune complexes, and rheumatic diseases, Arthritis Rheum 31 (1988) 153–158.

[137] L.E. Kahl, J.P. Atkinson, Autoimmune aspects of complement deficiency, Clinical Aspects of Autoimmunity 2 (1988) 8–20.

[138] J.A. Schifferli, Y.C. Ng, J.P. Paccaud, M.J. Walport, The role of hypocomplementaemia and low erythrocyte complement

receptor type 1 numbers in determining abnormal immune complex clearance in humans, Clin Exp Immunol 75 (1989) 329—335.

[139] L.A. Hebert, G. Cosio, The erythrocyte-immune complex-glomerulonephritis connection in man, Kidney Int 31 (1987) 877—885.

[140] P.R. Taylor, A. Carugati, V.A. Fadok, H.T. Cook, M. Andrews, M.C. Carroll, et al., A hierarchical role for classical pathway complement proteins in the clearance of apoptotic cells in vivo, J Exp Med 192 (2000) 359—366.

[141] J.S. Navratil, L.C. Korb, J.M. Ahearn, Systemic lupus erythematosus and complement deficiency: clues to a novel role for the classical complement pathway in the maintenance of immune tolerance, Immunopharmacology 42 (1999) 47—52.

[142] M. Botto, Links between complement deficiency and apoptosis, Arthritis Res. 3 (2001) 207—210.

[143] J.P. Atkinson, Genetic susceptibility and class III complement genes., chap. 5, in: R.G. Lahita (Ed.), Systemic lupus erythematosus, 2 ed.). Churchill Livingstone, New York, 1992, pp. 87—102.

[144] J.A. Schifferli, G. Steiger, G. Hauptmann, P.J. Spaeth, A.G. Sjoholm, Formation of soluble immune complexes by complement in sera of patients with various hypocomplementemic states. Difference between inhibition of immune precipitation and solubilization, J Clin Invest 76 (1985) 2127—2133.

[145] J.A. Schifferli, G. Hauptmann, J. Pierre-Paccaud, Complement-mediated adherence of immune complexes to human erythrocytes. Difference in the requirements for C4A and C4B, FEBS Lett 213 (1987) 415—418.

[146] L. Paul, V.M. Skanes, J. Mayden, R.P. Levine, C4-mediated inhibition of immune precipitation and differences in inhibitory action of genetic variants, C4A3 and C4B1, Complement 5 (1988) 110—119.

[147] G. Sturfelt, A. Bengtsson, C. Klint, O. Nived, A. Sjoholm, L. Truedsson, Novel roles of complement in systemic lupus erythematosus—hypothesis for a pathogenetic vicious circle, J. Rheumatol. 27 (2000) 661—663.

[148] M. Salmon, C. Gordon, The role of apoptosis in systemic lupus erythematosus, Rheumatology 38 (1999) 1177—1183.

[149] C.M. Reilly, G.S. Gilkeson, Use of genetic knockouts to modulate disease expression in a murine model of lupus, MRL/lpr mice, Immunol Res 25 (2002) 143—153.

[150] V.M. Holers, Phenotypes of complement knockouts, Immunopharmacology 49 (2000) 125—131.

[151] H. Watanabe, G. Garnier, A. Circolo, R.A. Wetsel, P. Ruiz, V.M. Holers, Modulation of renal disease in MRL/lpr mice genetically deficient in the alternative complement pathway factor, B J Immunol 164 (2000) 786—794.

[152] Z. Chen, S.B. Koralov, G. Kelsoe, Complement C4 inhibits systemic autoimmunity through a mechanism independent of complement receptors CR1 and CR2, J Exp Med 192 (2000) 1339—1351.

[153] S. Einav, O.O. Pozdnyakova, H. Ma, M.C. Carroll, Complement C4 is protective for lupus disease independent of C3, J Immunol 168 (2002) 1036—1041.

[154] P.R. Taylor, J.T. Nash, E. Theodoridis, A.E. Bygrave, M.J. Walport, M. Botto, A targeted disruption of the murine complement factor B gene resulting in loss of expression of three genes in close proximity, factor B, C2 and D17H6S45, J Biol Chem 273 (1998) 1699—1704.

[155] M. Matsumoto, W. Fukuda, A. Circolo, J. Goellner, J. Strauss-Schoenberger, X. Wang, Abrogation of the alternative complement pathway by targeted deletion of murine factor B, Proc Natl Acad Sci USA 94 (1997) 8720—8725.

[156] D.A. Mitchell, P.R. Taylor, H.T. Cook, J. Moss, A.E. Bygrave, M.J. Walport, et al., Cutting edge: C1q protects against the development of glomerulonephritis independently of C3 activation, J Immunol 162 (1999) 5676—5679.

[157] D.A. Mitchell, M.C. Pickering, J. Warren, L. Fossati-Jimack, J. Cortes-Hernandez, H.T. Cook, et al., C1q deficiency and autoimmunity: the effects of genetic background on disease expression, J Immunol 168 (2002) 2538—2543.

[158] M. Botto, C. Dell'Agnola, A.E. Bygrave, E.M. Thompson, H.T. Cook, F. Petry, et al., Homozygous C1q deficiency causes glomerulonephritis associated with multiple apoptotic bodies, Nat Genet 19 (1998) 56—59.

[159] L. Ellman, I. Green, F. Judge, M.M. Frank, In vivo studies in C4-deficient guinea pigs, J Exp Med 134 (1971) 162—175.

[160] E.C. Bottger, T. Hoffmann, U. Hadding, D. Bitter-Suermann, Influence of genetically inherited complement deficiencies on humoral immune response in guinea pigs, J Immunol 135 (1985) 4100—4107.

[161] H.D. Ochs, R.J. Wedgwood, M.M. Frank, S.R. Heller, S.W. Hosea, The roles of complement in the induction of antibody responses, Clin Exp Immunol 53 (1983) 208—216.

[162] O. Finco, S. Li, M. Cuccia, F.S. Rosen, M.C. Carroll, Structural differences between the two human complement C4 isotypes affect the humoral immune response, J Exp Med 175 (1992) 537—543.

[163] G.D. Ross, W.J. Yount, M.J. Walport, J.B. Winfield, C.J. Parker, C.R. Fuller, et al., Disease-associated loss of erythrocyte complement receptors (CR1, C3b receptors) in patients with systemic lupus erythematosus and other diseases involving autoantibodies and/or complement activation, J Immunol 135 (1985) 2005—2014.

[164] J. Duffy, M.D. Lidsky, J.T. Sharp, J.S. Davis, D.A. Person, F.B. Hollinger, Polyarthritis, polyarteritis and hepatitis B, Medicine (Baltimore) 55 (1976) 19—37.

[165] D.R. Milich, Influence of C4A deficiency on nonresponse to HBsAg vacination: a new immune response gene, J Hepatol 37 (2002) 396—399.

[166] T. Hohler, B. Stradmann-Bellinghausen, r. Sanger, A. Victor, C. Rittner, P.M. Schneider, C4A deficiency and nonresponse to hepatitis B vaccination, J Hepatol 37 (2002) 387—392.

[167] L. Varga, G. Fust, Assays for complement proteins encoded in the class III region of human MHC, Curr Protoc Immunol (2005) 13.7.1—13.7.20.

[168] S.C. Ross, P. Densen, Complement deficiency states and infection: epidemiology, pathogenesis and consequences of neisserial and other infections in an immune deficiency, Medicine (Baltimore) 63 (1984) 243—273.

[169] J.P. Atkinson, J.L. Kaine, V.M. Holers, A.C. Chan, Complement and the rheumatic diseases, in: G.D. Ross (Ed.), Immunobiology of the Complement System: an Introduction for Research And Clinical Medicine, Academic Press, Orlando, 1986.

[170] M.K. Liszewski, L.E. Kahl, J.P. Atkinson, The functional role of complement genes in SLE and Sjogren's syndrome, Curr Opin Rheumatol 1 (1989) 347—352.

[171] N.C. Hughes-Jones, The classical pathway, in: G.D. Ross (Ed.), Immunobiology of the complement system, Academic Press, San Diego, 1986, p. 21.

[172] M.K. Liszewski, J.P. Atkinson, P. Bigazzi, The role of complement in autoimmunity, in: M.R. Reichlin (Ed.), Systemic Autoimmunity, Marcel Dekker, Inc, New York, 1991, pp. 13—37.

[173] R.H. McLean, G. Niblack, B. Julian, T. Wang, R. Wyatt, J.A. Phillips, et-al, Hemolytically inactive C4B complement allotype caused by a proline to leucine mutation in the C5-binding site, J Biol Chem 269 (1994) 27727—27731.

3

Constitutive Genes and Lupus

Betty P. Tsao, Yun Deng

INTRODUCTION

SLE is a complex and heterogeneous autoimmune disease with a strong genetic component that is modified by environmental exposures. The high heritability (>66%), a higher concordance rate for SLE in monozygotic twins (24—56%) relative to dizygotic twins or siblings (2—5%), and the high sibling risk ratio of SLE patients (8—29-fold of the general population) all support the genetic contribution to the development of SLE [1, 2]. Identifying novel genetic factors for SLE (or SLE subsets) can reveal new paradigms for the pathogenesis of the disease, and may provide new therapeutic targets for disease management.

Approaches to Identify Genetic Variants Predisposing to SLE

The three basic approaches to explore genetic risk factors for diseases are: (1) candidate gene association studies, (2) linkage analysis in multiplex families, and (3) genome-wide association studies (GWAS). Historically, the candidate gene case-control approach is the first methodology used to identify established genetic risk factors for SLE including MHC class II (DR2 and DR3), class III (complement C2, and C4), C1q deficiency, and FcαR variants. Genetic linkage analysis localizes chromosomal regions with shared genetic risk within families containing multiple members affected with disease. Although linkage analysis is a powerful tool in identifying highly penetrant disease genes in Mendelian disorders, a total of 12 genome scans of SLE families have identified a number of putative susceptibility loci and have contributed occasionally to the discovery of new risk genes within the linked interval (reviewed in [3]). Recent advances in tools for genotyping hundreds of thousands of single nucleotide polymorphisms (SNPs) in a single individual have facilitated the genome-wide association approach in mapping

complex disease loci. A total of six GWAS in SLE (four using European-derived populations and two using Asian populations) have increased dramatically the number of convincingly established genetic associations with SLE during the past few years [4—9]. Each study uses thousands of cases and thousands of controls, which has led to the identification of 18 non-MHC loci reaching genome-wide significance ($p < 5 \times 10^{-8}$), 12 of which have been replicated independently. In addition, candidate genes have been selected based on known functions that are likely to play a role in the pathogenesis of SLE for association studies, yielding 14 loci that have been independently confirmed. These 32 non-MHC SLE risk loci are grouped into three biologic pathways involved in SLE (Table 3.1). Here we describe these genetic associations in each pathway.

PATHWAY 1: CLEARANCE AND PROCESSING OF IMMUNE COMPLEXES

Defects in apoptotic cell clearance, processing and presentation to lymphocytes have been identified as a key pathway leading to the development of SLE (reviewed in Ref [10]). Identification of genetic variants associated with SLE at loci related to immune complex (IC) processing provides support for the importance of this theme in SLE pathogenesis.

FCGR2A, FCGR3A, FCGR3B and FCGR2B (1q23)

FCGR2A, FCGR3A, FCGR3B, and *FCGR2B*, a cluster of four functional genes at 1q23, encode low-affinity Fc receptors for immunoglobulin G (FcγR) containing ICs, which are involved in the IC removal and other antibody-dependent responses. Various SNPs in these genes have been identified as risk factors for SLE and

TABLE 3.1 Non-HLA SLE risk loci grouped by biological pathways

Pathway	Gene	Study design[#]	Marker	OR[*]	Population[$]	Ref
Immune complex processing	FCGR2A	Candidate, Linkage, GWAS	rs1801274(H131R)	1.4	E, AA, Koreans	5,12-14
	FCGR3A	Candidate, Linkage	rs396991(F158V)	1.6	E	20
	FCGR3B	Candidate	low copy number	2.2	E	38
	FCGR2B	Candidate	rs1050501I187T)	~2.5	Asian	26-28
	C1q	Candidate	Complete deficiency Some SNPs	~1.0 ~2.1	E E, AA, HA	40 43
	CRP	Candidate	rs3091244 rs3093061	1.4 1.3	Koreans AA, E	51 53
	ITGAM	Candidate, Linkage, GWAS	rs11574637, rs9888739 rs1143679(H77R)	1.3 1.6	E E, EA, AA, HA	5,6 56,57
Toll-like receptor/type I interferon signaling	IRF5	Candidate, GWAS	Four functional variants	1.5	E, AA, Korean, Japanese, Chinese	64,68-72
	STAT4	Candidate, GWAS	rs7574865	1.5	E, EA, Japanese	5-6,81-84
	IRAK1	Candidate	5 SNPs	1.5	EA, AA, HA, AsA	94
	TREX1	Candidate	Rare variant	25	E	89
	TNFAIP3	GWAS	rs5029939, rs2230926	2.3	E	7
	TNIP1	GWAS	rs10036748	1.2	Chinese	8
Immune signal transduction	PDCD1	Candidate, Linkage	PD1.3G/A	1.2	E, Mexican, EA	101,103,105
	PTPN22	Candidate	rs2476601(R620W)	1.3	E	110-113
	IL10	Candidate	rs3024505	1.3	E, Mexican, Asian	117-119
	TNFSF4	Candidate, GWAS	rs2205960	1.4	E, Chinese	8,124
	BLK	GWAS	rs2736340	1.3-1.5	E, Chinese	6,9
	BANK1	GWAS	Three functional variants	1.4	E	4
	LYN	GWAS	rs6983130	1.3	E	5
	ETS1	GWAS	rs1128334	1.3	Chinese	8,9
Other without obvious pathways	KIAA1542	GWAS	rs4963128	1.3	E	5
	PXK	GWAS	rs6445975	1.3	E	5
	XKR6	GWAS	5 SNPs	1.2	E	5
	SCUBE1	GWAS	rs2071725	1.3	E	5
	PRD M1-ATG5	GWAS, Candidate	rs6568431	1.2	E	6,91
	JAZF1	GWAS, Candidate	rs849142	1.2	E	6,91
	UHRF 1BP1	GWAS, Candidate	rs11755393	1.2	E	6,91
	IKZF1	GWAS	rs4917014	1.4	Chinese	8
	RASGRP3	GWAS	rs13385731	1.6	Chinese	8
	SLC15A4	GWAS	rs10847697, rs1385374	1.3	Chinese	8
	WDFY4	GWAS	rs7097397	1.3	Chinese	9

[#]*Candidate: Candidate gene association study; GWAS: Genome-wide association study; Linkage: Linkage study*
[*]*We have converted all ORs to a risk ratio regrardless of how they are presented in the original papers*
[$]*AA: African Americans; AsA: Asian Americans; EA: European Americans; HA: Hispanic Americans; E: European*

may lead to functional changes of their encoded proteins. A non-synonymous SNP (rs1801274) of *FCGR2A* converting the histidine to an arginine at codon 131 (H131R), results in either a low (R131) or a high (H131) affinity for IgG2 opsonized particles and might delay clearance of ICs [11]. The associations between this SNP and the susceptibility to SLE and/or lupus nephritis are established in several independent studies from various ethnic groups with modest sample sizes, including Europeans [12], African-Americans [13] and Koreans [14], whereas inconsistent results are shown in similar studies [15–18]. Ethnic differences, disease heterogeneity, genotyping error (due to extensive sequence homology among FcγR genes), and/or random fluctuations in small samples may be used to explain the inconsistent association results. A meta-analysis of 17 studies of modest sample sizes has concluded that the R131 allele confers a 1.3-fold risk for developing SLE, but not for lupus nephritis [19]. Recently, a GWAS conducted by SLEGEN has confirmed this associated locus with SLE [5].

A non-synonymous SNP (rs396991) in *FCGR3A* (a phenylalanine (F) to valine (V) substitution at amino acid 158 or 176 if the leader sequence is included) alters the binding affinities for IgG1-, IgG3-, and IgG4-containing ICs. Of interest, the low-affinity F allele that clears ICs less efficiently was associated with SLE susceptibility [20], but in SLE patients with renal involvement the high-affinity V allele was associated with progression to end-stage renal disease [21]. Since IgG2 and IgG3 are major subclasses of ICs deposited in renal biopsies of patients with lupus nephritis [22], the relative importance of rs1801274 (*FCGR2A*-H/R131) and rs396991 (*FCGR3A*-V/F158) might depend on the IgG subclass of pathogenic autoantibodies in each patient. They may not be independent risk alleles and often are inherited together as a single-risk haplotype for SLE [23], and the presence of multiple-risk alleles affecting ICs clearance may interact to enhance risk for SLE [24].

A non-synonymous SNP in *FCGR2B* (an isoleucine (I) to threonine (T) substitution at position 187 in the transmembrane domain), which may alter the inhibitory function of FcγRIIb on B cells [25], has been associated with SLE in Asian populations including Japanese [26], Thais [27] and Chinese [28]. The FcγRIIb encoded by the SLE-associated 187T allele is excluded from lipid rafts, resulting in impaired inhibition of B-cell activation promoting autoimmunity [29]. Compared to Asians, the 187T is present at low frequency in European-derived populations, which may explain the lack of association in European-American, Swedish, or African-American populations [25, 30, 31]. In the promoter region of *FCRG2B*, a haplotype (−386G/−120T) that confers stronger promoter activity is associated with SLE in European-Americans [32], and the same −386G/C SNP associated with SLE susceptibility is confirmed in another study of European-Americans [33].

There are six SNPs in *FCRG3B* underlying three different allotypic variants of FcγRIIIb, (NA1, NA2 and SH), including five non-synonymous SNPs and one synonymous SNP. NA1/NA2, are different in four amino acid positions including two potential glycosylation sites [34]. Homozygous NA2 individuals have lower capacity to mediate phagocytosis compared with individuals who possess NA1 [34]. Although a study reports association of NA2 with SLE in a Japanese population, this observation has not been replicated [35], suggesting the association between SLE and this genomic region might be influenced by other genetic variations, for example copy number variations (CNVs). A deficiency of *FCGR3B*, defined as a lack of detection of the NA1, NA2 and SH alleles, was first observed in two healthy individuals [36], and duplication of *FCGR3B* was also reported in normal individuals [37], suggesting CNVs in general populations. Using a real-time PCR assay, association of *FCGR3B* CNVs with SLE has been reported that individuals with fewer than two copies of *FCGR3B* have higher risk for SLE (with and without nephritis) [38]. Functional studies of *FCGR3B* CNVs demonstrate that low copy numbers are associated with decreased protein expression, immune complex uptake, and neutrophil adhesion to immune complexes in a family with *FCGR3B*-deficiency and normal individuals with various CNs [39]. In addition to *FCGR3B* CNVs, *FCGR3A* also has CNVs. An integrated approach to assess simultaneously CNVs, allotypic variants, and SNPs in large-scale case-control, trans-racial studies will help dissect relative contribution of various variants in this complex *FCGR* locus.

C1Q (1p36)

The complement system, through opsonization, assists with clearing apoptotic debris and cellular fragments that may release nuclear antigens that are targets for lupus-associated autoantibodies. The relationship between complement and SLE pathogenesis has long been noted since levels of complement are lower in SLE patients and complement components are found in glomerular biopsies of SLE patients with renal manifestations.

Complement component 1q (C1q, encoded by three genes A, B and C, located at 1p36) is the first component of the classical pathway of complement activation and, together with the enzymatically active components C1r and C1s, forms the C1 complex. Complete deficiency of C1q, though rare, is a powerful disease risk factor since more than 90% individuals with this deficiency develop SLE and have severe disease manifestations and glomerulonephritis [40]. In a mouse model, a significant portion developed glomerulone-phritis in C1q-deficient background, with high numbers of glomerular apoptotic bodies [41]. In addition to the deficiency, a synonymous SNP (rs172378) in the *C1qA* gene that is linked to the decreased levels of serum C1q was associated with subacute cutaneous lupus (SCLE) [42]. Other SNPs in the *C1Q* genes were associated with subphenotypes of SLE (for example, nephritis and photosensitivity) in African-American and Hispanic populations [43]. The pathogenic mechanism in such cases is considered to be defective IC clearance. However, recent studies have found that C1q has a regulatory effect on Toll-like receptor-induced cytokine production [44], as well as IC-induced IFN-α production [45], providing an additional explanation for the elevated disease risk in C1q deficiency.

CRP (1q23.2)

C-reactive protein (CRP), is an important immune modulator that facilitates the clearance and handling of cellular debris and apoptotic bodies [46, 47]. Much of the variance in basal CRP levels is attributed to genetic variation, which makes *CRP* (located on 1q23.2) a SLE candidate locus. An intronic polymorphic GT-repeat has been associated with serum CRP levels, but not with SLE susceptibility [48]. The minor alleles at two SNPs (rs1800947 located within codon 188 and rs1205 in the 3′ UTR) were associated with low basal serum CRP levels, and the minor allele of the latter SNP was also associated with SLE in UK simplex families [49]. Two *CRP* promoter SNP haplotypes (rs3093062 located at −409 and rs3091244 at −390), which could modulate transcription factor binding and the promoter activity, were associated with basal serum CRP levels in healthy controls independent of age, gender, or race [50]. Recently the latter SNP has been identified to contribute to SLE susceptibility in Korean population ($p = 3 \times 10^{-2}$), but not in European-derived populations [51, 52]. Another promoter SNP rs3093061 (located at −707), with a high LD between the haplotypes as described above, appeared to have strong SLE disease association in African-Americans ($p = 6.4 \times 10^{-7}$), and European Americans ($p = 2.1 \times 10^{-6}$) [53]. CRP is a potential protector for SLE that administration of CRP delays disease onset, reverses nephritis and prolongs survival in murine lupus [54]. Relative deficiency of CRP may predispose to the generation of autoantibodies, leading to the development of SLE. Additional genetic factors involved in regulating CRP expression are yet to be discovered.

ITGAM (16p11.2)

Integrin alpha M (*ITGAM*, located on 16p11.2), also known as CD11b or complement receptor 3 (CR3), which encodes the α-chain of the αMβ2 integrin, is an integrin adhesion molecule that binds not only the complement cleavage fragment of C3b, but also a myriad of other possible ligands that are relevant to SLE [55]. Newly discovered association between SLE susceptibility and *ITGAM* or *ITGAM-ITGAX* region was found independently in two GWAS [5, 6]. Supporting these results is a fine-mapping study of chromosome 16p11 locus linked to SLE using European-American and African-American populations that show a non-synonymous variant, rs1143679, at exon 3 of *ITGAM* contributing to SLE risk (meta-analysis OR = 1.74, $p = 6.9 \times 10^{-22}$) [56]. Strong and robust association of this SNP is confirmed in a meta-analysis of multiple cohorts, including Hispanic-Americans ($p = 2.9 \times 10^{-5}$), Europeans ($p = 6.2 \times 10^{-8}$), Colombian ($p = 3.6 \times 10^{-7}$), and Mexican ($p = 2.0 \times 10^{-3}$). The low frequency of non-risk allele in Asians may contribute to the lack of association in Korean and Japanese populations [57], but has been confirmed in Hong Kong Chinese and Thais with a larger sample size [58]. Subphenotype stratification analyses show a strong association of this SNP allele with renal involvement of SLE in European, Hong Kong Chinese and Thai populations [58, 59]. This polymorphism results in converting the arginine at amino acid position 77 to a histidine (R77H substitution), influencing the conformation of the α1 domain (the ligand-binding domain for ICAM-1), with subsequent consequences to αMβ2 ligand binding and altering the structural and functional changes of ITGAM [56]. In addition to rs1143679, another non-synonymous SNP (rs1143683, A858V) may also contribute to SLE risk in Hong Kong Chinese and Thais [58].

PATHWAY 2: TOLL-LIKE RECEPTOR/ TYPE I INTERFERON SIGNALING

The type I interferon pathway bridges the innate and adaptive immune responses and plays an important role in the etiology of autoimmune diseases. Serum levels of IFN-α, as well as the 'interferon (IFN) signature' composed of increased levels of IFN-inducible genes were detected in SLE patients and correlated with more severe disease manifestations [60, 61]. The identification of SLE-associated genes involved in the type I interferon pathway will enhance our understanding of genetic contribution to this dysregulated pathway predisposing to SLE.

IRF5 (7q32)

Interferon regulatory factor 5 (*IRF5*, mapped to the chromosome 7q32), a transcription factor in the type I IFN pathway, has been one of the most consistently associated genes with SLE outside the MHC [62, 63]. Four functional polymorphisms associated with SLE in *IRF5* have been identified in European-derived populations [64, 65] and have been broadly replicated in European-derived, Asian, and admixed African-American, Mexican, and South American populations [66–73] (Figure 3.1). One was a CGGGG insertion/deletion (indel) located 64 bp upstream of the first untranslated exon of *IRF5*, strengthening the binding site for the Sp1 transcription factor [65]. The second was rs2004640 that created a donor splice site in intron 1 resulting in expression of the alternative untranslated exon 1B. Two copies of rs2004640 T allele showed a higher risk for SLE than one copy (odds ratio 2.01 vs 1.27), suggesting a dosage effect of this allele [66]. The third was a 30 bp insertion/deletion (indel) of exon 6, which leads to 10 amino acid inframe indel altering a proline-, glutamate acid-, serine- and threonine-rich (PEST) domain region [74]. The fourth was rs10954213 that alters the polyadenylation site of *IRF5* resulting in a shorter transcript with increased stability [64]. In addition, several linked SNPs, for example rs3807306 located in the first intron of *IRF5*, rs2070197 located in the 3'UTR of *IRF5*, rs10488631 and rs12539741 located downstream of *IRF5* that gave very strong signals of association with SLE were identified [64]. However, no functional role for these SNPs has been identified as yet, suggesting they may serve as markers for causal variants in linkage disequilibrium.

Genetic association studies examining SNPs in *IRF5* have defined distinct haplotypes that confer either susceptibility to, or protection from, SLE in persons with a different genetic background. Four-marker haplotypes composed of rs2004640-indel of exon 6-rs2070197-rs10954213 (Figure 3.1B), or rs2004640-rs3807306-rs10954213-rs2280714 (Figure 3.1C) are associated with increased, decreased, or neutral levels of risk for SLE in European-derived populations [64, 71]. Recently, these data were replicated by studies in African-American, Chinese, Korean and Japanese populations [68–71]. Another four-marker risk haplotype (rs2004640-rs10954213- rs10488631-the CGGGG indel) identified in a Swedish population is associated with enhanced *IRF5*

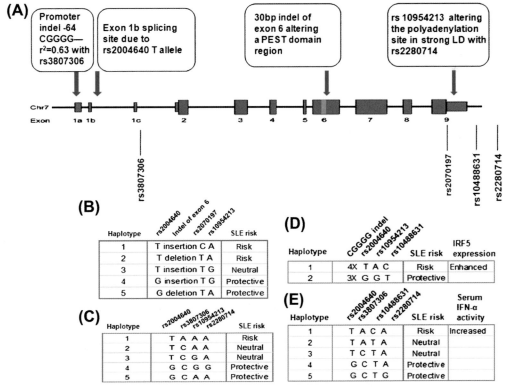

FIGURE 3.1 Summary of a genetic model for IRF5. (*A*) The IRF5 gene diagram is marked to show the location of previously described functional elements (top, white boxes), alternately spliced first exons (1a,1b,1c), subsequent exons 2–9 (blue boxes), the 3' UTR region (gray box), and SNPs associated with SLE disease risk. (*B,C*) Frequencies of the major haplotypes in European-derived SLE patients are shown, categorized as risk, neutral and protective as they relate to SLE susceptibility. Risk haplotypes associated with increased *IRF5* expression levels and serum IFN-α activity are also shown in *D* and *E*.

transcript and protein expression levels in SLE patients (Figure 4.1D) [75]. Since high levels of serum IFN-α activity are a heritable risk factor for SLE [76] and several haplotypes of *IRF5* are identified to associate with SLE risk, recent studies have focused on whether *IRF5* SLE-risk haplotypes predispose to high IFN-α activity. As shown in Figure 3.1E, an *IRF5* risk haplotype (rs2004640-rs3807306-rs10488631-rs2280714) is associated with increased serum IFN-α activity in SLE patients of European and Hispanic ancestry [77]. Normal individuals carrying a *IRF5* risk haplotype may exhibit elevated IRF5, IFN-α, and interferon-inducible chemokine expression as shown in lymphoblastoid cell lines (LCLs) from individuals of European background [78]. A summary of the genetic model for *IRF5* is showed in Figure 3.1. To explore the biological role of IRF5 in lupus pathogenesis, IRF5-deficient or -sufficient $FcγRIIB^{-/-}Yaa$ mice were used to show the absolute requirement of IRF5 for disease development, mediated through pathways beyond that of type I IFN production [79].

STAT4 (2q33)

The signal transducer and activator of transcription 4 (*STAT4*), located at 2q33, is established as a genetic risk factor for SLE. STAT4 transmits signals from the receptors for type I IFN, IL-12 and IL-23, and can contribute to autoimmune responses by affecting the functions of several innate and adaptive immune cells [80]. The minor T allele of rs7574865, located in the third intron of *STAT4*, is first identified for association with SLE in three North-American SLE cohorts of European ancestry [81]. Subsequently, this initial result is replicated and extended in multiple SLE collections with odds ratios of 1.65 in European-Americans [82], 1.62 in Columbians [83], and 1.71 in Japanese [84]. This association is further confirmed in two GWAS of European-derived populations [5, 6] and two GWAS in Asians [8, 9]. Of interest, several studies find association of this rs7574865 risk variant with a more severe SLE phenotype characterized by younger age at disease onset, higher frequency of nephritis and presence of anti-double-stranded DNA antibodies [82, 84, 85], or with an increased sensitivity to IFN-α signaling in peripheral blood mononuclear cells from SLE patients [86]. Efforts to identify causal variants of *STAT4* have led to discovery of several markers independently associated with SLE and/or with differential expression level of *STAT4* [85, 87, 88], and a risk haplotype (spanning 73-kb from the third intron to exon17 of *STAT4*) common to European-Americans, Korean/Asians, and Hispanic-Americans [88]. Located within 25-kb from *STAT4*, *STAT1* exhibits orders of magnitude of weaker association signals, suggesting *STAT4* variants predispose to susceptibility to SLE.

TREX1 (3P21)

DNA repair exonuclease 1 (*TREX1*, located on 3p21.3-p21.2) metabolizes reverse-transcribed single-stranded DNA of endogenous retroelements as a function of cell-intrinsic antiviral surveillance, resulting in a potent type I IFN response [9]. A study of combined UK, Germany and Finland populations has reported mono-allelic frameshift or missense mutations and a single 3' UTR variant of *TREX1* present in SLE patients but absent in controls ($p = 4.1 \times 10^{-7}$) [89]. Although rare, genetic association of *TREX1* with SLE implicates a potential role of this gene in SLE pathogenesis.

TNFAIP3 (6q23) and TNIP1 (5q33)

The tumor necrosis factor-α-induced-protein 3 (*TNFAIP3*, also known as *A20*, located on 6q23.3), and its interacting protein named as TNFAIP3 interacting protein 1 (*TNIP1*, located on 5q33.1) are both key regulators of NF-κB signaling pathway, modulating cell activation, cytokine signaling and apoptosis [90]. Association with SLE of several SNPs spanning the *TNFAIP3*, was recently identified through a GWAS of European-derived populations [22], and subsequently confirmed in both GWA studies of Asian populations [8, 9]. Among these SNPs, the strongest association was observed at rs5029939 (OR = 2.29; meta-analysis $p = 2.89 \times 10^{-12}$). Another SLE-associated SNP rs2230926 (located in the exon 3) identified in European descent has been replicated in Japanese populations [30]. *TNIP1* is identified as a new SLE susceptibility locus in the Chinese GWA study of SLE [23], and replicated in European-derived population [91]. These findings highlight how important the regulation of the NF-κB signaling pathway is in the pathogenesis of SLE.

IRAK1 - MECP2 (Xq28)

Interleukin-1 receptor-associated kinase 1 (*IRAK1*) located on chromosome Xq28 is a serine/threonine protein kinase and involved in Toll/interluekin-1 receptor (TIR) family signaling [92]. It links several immune receptor complexes to the central adapter/activator protein, TRAF6 (TNF receptor-associated factor 6), and regulates multiple pathways in both innate and adaptive responses [93]. A study by Jacob and colleagues offered important insight into *Irak1* function in murine SLE models, with a role in induction of IFN α/γ, regulation of nuclear factor-κB (NF-κB) in T cells and TLR activation [94]. The subsequent study of ~5,000 SLE subjects and healthy controls observes association of five SNPs spanning the *IRAK1* gene with disease (p values reaching 10^{-10}, odds ratio >1.5) in both adult- and childhood-onset SLE from four ethnic groups [94]. Located in the region of linkage disequilibrium with *IRAK1* is another X chromosome gene associated with SLE, Methyl-CpG-binding protein 2 (*MECP2*), which plays a critical role in the transcriptional suppression of methylation-sensitive genes [95]. This SLE-associated region (*IRAK1-MECP2*) has been confirmed in European-derived populations [91]. The demonstration of X chromosome genes as disease susceptibility factors in human SLE raises the possibility that the gender difference in SLE may in part be attributed to sex chromosome genes.

Interaction between Genetic Variants Involved in IFN-α pathway

High levels of serum IFN-α activity are a heritable trait in families containing SLE patients [76]. Genetic factors contributing to this trait are complex, and may require combinations of multiple factors. Another study evaluated interaction between *IRF5* and *STAT4*. The SLE-associated SNP rs7582694 in *STAT4* and the two independent *IRF5* variants (CGGGG indel and rs10488631) displayed a 1.82-fold increased risk for SLE with each additional risk allele, and an OR of 5.8 for those 18% of SLE patients carrying four or more risk alleles [96]. A Finnish group suggests genetic interaction between rs10954213 in *IRF5* and rs2304256 in tyrosine kinase 2 (*TYK2*) indicate that patients homozygous for the *TYK2* risk allele and the *IRF5* risk allele have increased OR than patients homozygous or heterozygous for no-risk alleles (OR = 2.73 vs. 1 or 1.43) [97]. IRF5, STAT4 and TYK2 are all members of the type I IFN pathway and their genetic interaction might increase the risk for SLE, strengthening the role of the type I IFN pathway in the pathogenesis of SLE.

PATHWAY 3: IMMUNE SIGNAL TRANSDUCTION

SLE is characterized by a loss of T- and B-cell tolerance, accounting for the formation of autoantibodies. Multiple genes contained in T- and B-cell signal transduction pathway have been shown to contribute to SLE susceptibility, which illustrate the important role of the lymphocyte signaling pathway involved in SLE pathogenesis.

PDCD1 (2q37)

The programmed cell death 1 (PD-1 or PDCD1), a member of the B7/CD28 family of co-inhibitory receptors, is inducibly expressed on activated T- and B-cells, regulating peripheral tolerance [98]. Studies in animal models indicated that mice deficient in *Pdcd1* develop glomerulonephritis and arthritis in the C57BL/6 background [99], and autoimmune dilated cardiomyopathy in the BALB/c background [100]. These findings make *PDCD1* an excellent positional candidate within 2q37 linked to SLE, which leads to the identification of a *PDCD1* intronic SNP (PD1.3A) associated with SLE in a large collection of European and Mexican patients [101]. This PD1.3A risk allele disrupts the binding site for the runt-related transcription factor 1 (RUNX1/AML1), correlates with decreased expression of PD-1 on CD4+CD25+ T-cells in SLE patients, and suggests the importance of PD-1 pathway in the development of SLE [102]. The association of PD1.3A with SLE and/or SLE with renal involvement has been replicated in several subsequent studies [103, 104], but not in other similar studies [105, 106]. The inconsistent results suggest at least some population heterogeneity that is supported by the geographically decreasing frequency of PD1.3A from Northern to Southern Europe [107].

PTPN22 (1p13)

The gene protein tyrosine phosphatase nonreceptor type 22 (*PTPN22*, located on 1p31) encodes a lymphoid-specific phosphatase (Lyp), which is an important inhibitor mediating T-cell activation [108]. A nonsynonymous SNP rs2476601 (R620W polymorphism of *PTPN22*) is a functional variant that 620W allele encoding a more active phosphatase is associated with risk for SLE and many other autoimmune diseases (reviewed in [109]). Common risk variants to several, but not all, autoimmune diseases will help elucidate both shared and unique pathways underlying these complex disorders. The association of 620W variant is more robust with multiple autoimmune phenotype [110, 111] and familial SLE rather than sporadic SLE [112]. The 620W variant reduces the binding between PTPN22 and the intracellular kinase Csk, which phosphorylates Lck and inhibits activity of Lck. Reduced thresholds for TCR signaling associated with 620W variant have been proposed to potentiate autoimmunity. The 620W allele frequency varies among populations, and is more common in northern Europeans (8−15%) than southern (2−10%) or western Europeans (7−10%), but nearly absent in Asian and African populations [109]. Although *PTPN22* is not a member involved in IFN pathway, a recent study has shown association of the 620W allele with higher serum IFN-α activity and lower TNF-α levels in SLE patients [113]. Another *PTPN22* variant, R263Q in the catalytic domain, has been associated with SLE in European-derived populations [114]. The SLE-associated 263Q variant exhibits reduced phosphatase activity, which is consistent with the proposed model that elevated TCR signaling protects against SLE [114].

IL10 (1q32)

Interleukin-10 (IL-10) is an important cytokine that possesses both immunosuppressive and immunostimulatory properties. An increased IL-10 production by the peripheral blood B-cells and monocytes from SLE patients is shown to correlate with disease activity [115], demonstrating an important role of IL-10 in the disease pathogenesis. The exploration of molecular mechanisms underlying increased IL-10 production in SLE patients leads to the identification of association of *IL10* haplotypes (defined by three SNPs in the promoter) with levels of IL-10 secretion [116]. Association between these *IL10* SNPs and SLE susceptibility has been reported in Europeans [117], Mexican-Americans [118] and Asians [119]. A meta-analysis combining 15 studies showed a significant association of SLE with the *IL10* G11 allele (23 CA repeat at -1.1kb) in multiple ethnic groups, and with a promoter SNP in Asians [120]. A recent large-scale replication study of European-derived populations has confirmed *IL10* as a SLE susceptible locus [91].

TNFSF4 (1q25)

Tumor necrosis factor (ligand) superfamily member 4 (TNFSF4, also known as OX40L) is primarily expressed on antigen-presenting cells (APC). Its receptor, TNFRSF4 (OX40), is expressed on activated T-cells. Interaction between TNFSF4 and TNFRSF4 provides co-stimulatory signals to activate T-cells [121]. TNFSF4-mediated signal inhibits the generation and function of interleukin-10-producing $CD4^+$ type 1 regulatory T-cell (Tr1), but induces B-cell activation and differentiation, as well as interleukin-17 production [122, 123]. A study using both case-control and family-based European-derived samples has identified a SLE risk haplotype marked by a series of tagging SNPs located at the upstream region of *TNFSF4*, which correlates with increased expression of TNFSF4 [124]. Increased TNFSF4 levels are thought to predispose to SLE either by augmenting the interaction between T- and APC-cells, or by influencing the functional consequences of T-cell activation via TNFRSF4 [124]. Associations between some of these *TNFSF4* tagging SNPs and risk for SLE have been confirmed in two recent studies of Chinese population [8, 125], and replicated in four independent SLE datasets from Germany, Italy, Spain and Argentina [126]. Further studies are needed to localize causal variants and to understand how these polymorphisms affect the pathogenesis of SLE.

BLK (8p23)

Recent GWAS in SLE have identified novel associated genes involved in B-cell signaling, which highlights its importance in the disease pathogenesis. B-lymphoid tyrosine kinase (BLK), a member of the Src-family, mediates intracellular signaling and influences proliferation, differentiation and tolerance of B-cells [127]. A GWAS in SLE of European-derived populations identifies association of rs13277113 located within the intergenic region between *C8orf13* and *BLK* (located on 8p23.1), and the SLE-risk allele is associated with reduced expression of BLK and increased expression of C8orf13 [6]. Significant association of a *BLK* intron 1 SNP (rs2248932 located 43 kb downstream of rs13277113) with SLE implicates *BLK* as a risk factor for SLE [5]. Subsequently, these two SNPs are associated with SLE in Chinese [9, 128] and Japanese [129].

BANK1 (4q24)

B-cell scaffold protein with ankyrin repeats 1 (BANK1), a B-cell adapter protein, mediates an interaction between the tyrosine kinase LYN (v-yes-1 Yamaguchi sarcoma viral related oncogene homolog) and the calcium channel IP3R, and facilitates the release of intracellular calcium, altering B-cell activation threshold [130].

In a GWAS from a Swedish population, three functional variants in the *BANK1* (located on 4q24), a non-synonymous substitution (rs10516487, R61H), a branch point-site SNP (rs17266594), and a SNP in the ankyrin domain (rs3733197, A383T), are associated with susceptibility to SLE [4]. These SLE-associated variants may contribute to sustained signaling of B-cell receptor and subsequent B-cell hyperactivity that are commonly seen in SLE [4]. Subsequently, these identified *BANK1* variants are replicated for association with SLE in case-control studies from a European-derived population ($p = 2.0 \times 10^{-5}$, 2.6×10^{-5} for rs17266594 and rs10516487 respectively) [131] and a Chinese Hong Kong population ($p = 4.7 \times 10^{-9}$, 2.0×10^{-2} for rs17266594 and rs3733197 respectively) [125].

LYN (8q13)

Through a GWAS, two SNPs of *LYN* (located on 8q13, a binding partner of BANK1) are significantly associated with SLE in European population [5], but are not consistently replicated in cohorts including European-Americans, African-Americans and Koreans [132]. Of note, another 5'UTR SNP (rs6983130) of *LYN* is associated with SLE in female-only subsets of European-Americans, but not in African-Americans and Koreans [132], suggesting an association of *LYN* with European-derived SLE individuals.

ETS1 (11q24.3)

V-ets erythroblastosis virus E26 oncogene homolog 1 (avian) (ETS-1) is a member of the ETS family transcriptional factors, known to be a negative regulator of differentiation of B-cell and Th17 cell [133, 134]. Ets-1-deficient mice develop lupus-like disease characterized by high-titer autoantibodies, and local activation of complement, implicating a potential role of Ets-1 in SLE pathogenesis [135]. The first evidence for *ETS1* (11q24.3) as a new SLE susceptibility loci came from a GWAS in a Chinese Han population [8]. Recently, another Asian GWAS confirmed the association between *ETS1* and SLE [9]. The identified *ETS1* variant is functional that the 3'UTR rs1128334 risk allele is associated with decreased *ETS-1* expression levels in peripheral blood mononuclear cells from normal controls [9].

OTHER LOCI ASSOCIATED WITH SLE

Recent GWAS have identified many novel loci, some of which are not well characterized as yet, or have no obvious connection to known pathways involved in the pathogenesis of SLE. Two GWAS of European ancestry reported SLE-associated markers in the genomic region containing *PHRF1* (PHD and ring finger domains 1, also known as *KIAA1542*, 11p15.5), *PXK* (PX domain containing serine/threonine kinase, 3p14.3), *XKR6* (XK, Kell blood group complex-related family member 6, 8p23.1), *SCUBE1* (signal peptide-CUB domain-EGF-related 1, 22q13), *PRDM1* (PR-domain zinc finger protein 1, 6q21-22), *JAZF1* (juxtaposed with another zinc finger gene 1, 7p15), and *UHRF1BP1* (UHRF1 binding protein 1, 6p21) [5;91]. Of interest, two GWAS of Asian populations revealed different findings, including new susceptibility loci such as *IKZF1* (IKAROS family zinc finger 1, 7p13-p11.1), *RASGRP3* (RAS guanyl releasing protein 3, 2p25.1-24.1), *SLC15A4* (solute carrier family 15 member 4, 12q24), and *WDFY4* (WDFY family member 4, 10q11) [8, 9], which may help explain the genetic basis of SLE

in various populations. Although the functional roles of these genes and their underlying mechanisms to increase disease risk are poorly understood, some are considered as potential regulators for differentiation, activation, or function of immune cells, for example, *SCUBE1*, *PRDM1*, *IKZF1* and *RASGRP3*. Some have been identified to be associated with other autoimmune diseases, for example, association of *PRDM1* with RA (rheumatoid arthritis) [136], and association of *JAZF1* and *SLC15A4* with type 2 diabetes [137, 138]. In the case of *KIAA1542*, the genetic association with SLE may be due to its close proximity with interferon regulatory factor 7 (*IRF7*), a gene that is also involved in type I IFN signaling [5]. A recent case-control study of European population has replicated two of these loci, *PXK* and *KIAA1542* [139]. Other novel loci await further replication experiments. Identifying new susceptibility loci provides unbiased clues to uncover molecular mechanisms leading to the development of manifestations of SLE.

Shared Loci with Other Autoimmune Disease

The technological advances in cost-effective, high-throughput genotyping have expedited the success in mapping SLE risk loci, which is paralleled by genetic associations identified in other autoimmune diseases, including RA (rheumatoid arthritis), SSc (systemic sclerosis), SS (primary Sjogren's syndrome), APS (primary antiphospholipid syndrome), T1D (type 1 diabetes), IBD (inflammatory bowel disease), CD (Crohn's disease), UC (ulcerative colitis), GD (Graves' disease) and PSO (psoriasis). Some SLE-associated loci also confer risk to at least one additional autoimmune disease, providing the possibility that multiple autoimmune diseases share common pathways during the disease pathogenesis. Table 3.2 provides a list of

TABLE 3.2   Shared loci with autoimmune diseases

Gene	Locations	Function	Risk for disease	Selected Ref*
FCGR2A	1q23	Involved in the IC clearance	SLE, T1D, UC	140, 141
IRF5	7q32	Regulates type 1 IFN pathway	SLE, RA IBD, SSc	142,144-146
STAT4	2q33	Regulates type 1 IFN pathway	SLE, RA, T1D,PSO	81,147-148
TNFAIP3	6q23	Regulators of NF-kB pathway	SLE, RA, T1D, PSO	143,150-151
TNIP1	5q33	Regulators of NF-kB pathway	SLE, PSO	151
PTPN22	1p13	T-cell activation	SLE, RA, T1D,SSc,GD,CD[#]	152-156
BANK1	4q24	B-cell activation	SLE, SSc, RA	157,159
BLK	8p23	B-cell activation	SLE, SSc, APS	158,160
TNFSF4	1q25	T-cell costimulation	SLE, SSc	161

* reference to autoimmune disease not including SLE
[#]PTPN22 620W allele is protective in CD

identified loci shared between multiple autoimmune diseases.

An important SLE-associated locus in the first pathway of IC clearance, *FCGR2A*, also confers an increased risk for T1D and UC from Dutch and Japanese populations, respectively [140, 141]. The SLE risk loci in the IFN and NFκB pathways that are associated with other autoimmune disease include *IRF5*, *STAT4*, *TNFAIP3* and *TNIP1*. Among these four loci, *IRF5*, *STAT4* and *TNFAIP3* are associated with RA [81, 142, 143], suggesting shared aspects of these signaling pathways between RA and SLE. The observed association of *IRF5* with IBD [144] and SSc [145, 146], and association of *STAT4* with SS and SSc [147, 148] suggest the type I IFN pathway plays a role in developing IBD, SSc, and SS. TNFAIP3 (A20) is a ubiquitin-editing enzyme that modifies adaptor proteins RIP and TRAF6 respectively to terminate NFκB-mediated proinflammatory responses downstream of signal transductions through TNFα and TLRs. TNIP1 known as A20 binding and inhibitor of NFκB is a ubiquitin sensor that regulates TNFR signaling, cell activation, and apoptosis [149]. Genetic associations of *TNFAIP3* with T1D [150] and of *TNFAIP3* and *TNIP1* with psoriasis [151] show the pivotal role of the NFκB pathway in these disorders. A number of loci in the immune signal transduction pathway have been associated with autoimmune disorders. The most well-characterized association is with PTPN22, in which 620W allele predisposes to a broad spectrum of autoimmune diseases (SLE, RA, T1D, SSc, or GD) [152−155], but the same allele is protective in another autoimmune disease (e.g. CD) [156]. Other loci like *BANK1* and *BLK* are both associated with SSc [157, 158], or independently with RA [159] and APS [160] in some populations. *TNFSF4* provides co-stimulation to the T-cell activation, and is also associated with SSc [161]. These intriguing observations need to be confirmed in large-scale association studies. Towards this end, the "ImmunoChip Consortium" has been established to build a custom array containing >200,000 SNPs representing genetic associations identified from SLE, RA, TID, CD, UC, psoriasis, primary biliary cirrhosis, autoimmune thyroid disease, multiple sclerosis, celiac disease, IgA deficiency, and ankylosing spondylitis. The availability of this new platform will accelerate the identification of shared variants between multiple autoimmune diseases, and loci that promote disease-specific phenotypes.

CONCLUSION

This chapter has summarized recent advances in the identification of non-HLA genetic variations predisposed to SLE that are generally common SNPs with their genetic associations established in European-derived populations. Similar genetic studies in multiple ethnic groups may identify new genetic associations and help localize causal variants for association with SLE and/ or specific manifestations of SLE, which will refine our current understanding and provide new clues to explain molecular pathways leading to the heterogeneous disease phenotypes. In addition to common SNPs, genetic contribution of rare variants and copy number variations to SLE will be explored in the coming years. Results from these studies will determine the unique and overlapping genetic variants in SLE and other related autoimmune diseases that may reveal underlying disease-specific and common pathways. The information gained from these genetic studies is likely to improve our capacity for diagnosis, prognosis, and treatment of SLE in the near future.

Acknowledgement

This work has been supported by the NIH AR043814. The authors thank Drs. Ornella J Rullo, Wenfeng Tan, and Jian Zhao for helpful comments and suggestions.

References

[1] D. Deapen, A. Escalante, L. Weinrib, D. Horwitz, B. Bachman, P. Roy-Burman, et al., A revised estimate of twin concordance in SLE, Arthritis Rheum 35 (1992) 311.

[2] D. Alarcon-Segovia, M.E. Alarcon-Riquelme, M.H. Cardiel, F. Caeiro, L. Massardo, A.R. Villa, et al., Familial aggregation of systemic lupus erythematosus, rheumatoid arthritis, and other autoimmune diseases in 1,177 lupus patients from the GLADEL cohort, Arthritis Rheum 52 (4) (2004) 1138−1147.

[3] B.P. Tsao, Update on human systemic lupus erythematosus genetics, Curr Opin Rheumatol 16 (5) (2004) 513−521.

[4] S.V. Kozyrev, A.K. Abelson, J. Wojcik, A. Zaghlool, M.V. Linga Reddy, E. Sanchez, et al., Functional variants in the B-cell gene BANK1 are associated with systemic lupus erythematosus, Nat Genet 40 (2) (2008) 211−216.

[5] J.B. Harley, M.E. Alarcon-Riquelme, L.A. Criswell, C.O. Jacob, R.P. Kimberly, K.L. Moser, et al., Genome-wide association scan in women with systemic lupus erythematosus identifies susceptibility variants in ITGAM, PXK, KIAA1542 and other loci, Nat Genet 40 (2) (2008) 204−210.

[6] G. Hom, R.R. Graham, B. Modrek, K.E. Taylor, W. Ortmann, S. Garnier, et al., Association of Systemic Lupus Erythematosus with C8orf13-BLK and ITGAM-ITGAX, N Engl J Med (2008). NEJMoa0707865.

[7] R.R. Graham, C. Cotsapas, L. Davies, R. Hackett, C.J. Lessard, J.M. Leon, et al., Genetic variants near TNFAIP3 on 6q23 are associated with systemic lupus erythematosus, Nat. Genet. 40 (9) (2008) 1059−1061.

[8] J.W. Han, H.F. Zheng, Y. Cui, L.D. Sun, D.Q. Ye, Z. Hu, et al., Genome-wide association study in a Chinese Han population identifies nine new susceptibility loci for systemic lupus erythematosus, Nat Genet 41 (11) (2009) 1234−1237.

[9] W. Yang, N. Shen, D.Q. Ye, Q. Liu, Y. Zhang, X.X. Qian, et al., Genome-Wide Association Study in Asian Populations Identifies Variants in ETS1 and WDFY4 Associated with Systemic Lupus Erythematosus, PLoS Genet 6 (2) (2010) e1000841.

[10] M. Herrmann, R.E. Voll, J.R. Kalden, Etiopathogenesis of systemic lupus erythematosus, Immunol Today 21 (9) (2000) 424—426.

[11] R.G. Bredius, C.E. de Vries, A. Troelstra, A.L. van, R.S. Weening, J.G. van de Winkel, et al., Phagocytosis of Staphylococcus aureus and Haemophilus influenzae type B opsonized with polyclonal human IgG1 and IgG2 antibodies. Functional hFc gamma RIIa polymorphism to IgG2, J Immunol 151 (3) (1993) 1463—1472.

[12] A.J. Duits, H. Bootsma, R.H.W.M. Derksen, P.E. Spronk, L. Kater, C.G.M. Kallenberg, et al., Skewed distribution of IgG Fc receptor IIa (CD32) polymorphism is associated with renal disease in systemic lupus erythematosus patients, Arthritis Rheum 39 (1995) 1832—1836.

[13] J.E. Salmon, S. Millard, L.A. Schachter, F.C. Arnett, E.M. Ginzler, M.F. Gourley, et al., Fc gamma RIIA alleles are heritable risk factors for lupus nephritis in African Americans, J Clin Invest 97 (5) (1996) 1348—1354.

[14] Y.W. Song, C.W. Han, S.W. Kang, H.J. Baek, E.B. Lee, C.H. Shin, et al., Abnormal distribution of Fc gamma receptor type IIa polymorphisms in Korean patients with systemic lupus erythematosus, Arthritis Rheum 41 (3) (1998) 421—426.

[15] M. Botto, E. Theodoridis, E.M. Thompson, H.L. Beynon, D. Briggs, D.A. Isenberg, et al., Fc gamma RIIa polymorphism in systemic lupus erythematosus (SLE): no association with disease, Clin Exp Immunol 104 (2) (1996) 264—268.

[16] K. Manger, R. Repp, B.M. Spriewald, A. Rascu, A. Geiger, R. Wassmuth, et al., Fcγ receptor IIa polymorphism in Caucasian patients with systemic lupus erythematosus: association with clinical symptoms, Arthritis Rheum 41 (7) (1998) 1181—1189.

[17] S.N. Yap, M.E. Phipps, M. Manivasagar, S.Y. Tan, J.J. Bosco, Human Fc gamma receptor IIA (FcgammaRIIA) genotyping and association with systemic lupus erythematosus (SLE) in Chinese and Malays in Malaysia, Lupus 8 (4) (1999) 305—310.

[18] J.Y. Chen, C.M. Wang, K.C. Tsao, Y.H. Chow, J.M. Wu, C.L. Li, et al., Fcgamma receptor IIa, IIIa, and IIIb polymorphisms of systemic lupus erythematosus in Taiwan, Ann Rheum Dis 63 (7) (2004) 877—880.

[19] F.B. Karassa, T.A. Trikalinos, J.P. Ioannidis, Role of the Fcgamma receptor IIa polymorphism in susceptibility to systemic lupus erythematosus and lupus nephritis: a meta-analysis, Arthritis Rheum 46 (6) (2002) 1563—1571.

[20] H.R. Koene, M. Kleijer, A.J. Swaak, K.E. Sullivan, M. Bijl, M.A. Petri, et al., The Fc gammaRIIIA-158F allele is a risk factor for systemic lupus erythematosus, Arthritis Rheum 41 (10) (1998) 1813—1818.

[21] G.S. Alarcon, G. McGwin Jr., M. Petri, R. Ramsey-Goldman, B.J. Fessler, L.M. Vila, et al., Time to renal disease and end-stage renal disease in PROFILE: a multiethnic lupus cohort, PLoS Med 3 (10) (2004) e396.

[22] R. Zuniga, G.S. Markowitz, T. Arkachaisri, E.A. Imperatore, V.D. D'Agati, J.E. Salmon, Identification of IgG subclasses and C-reactive protein in lupus nephritis: the relationship between the composition of immune deposits and FCgamma receptor type IIA alleles, Arthritis Rheum 48 (2) (2003) 460—470.

[23] V. Magnusson, B. Johanneson, G. Lima, J. Odeberg, D. Alarcon-Segovia, M.E. Alarcon-Riquelme, Both risk alleles for FcgammaRIIA and FcgammaRIIIA are susceptibility factors for SLE: a unifying hypothesis, Genes Immun 5 (2) (2004) 130—137.

[24] K.E. Sullivan, A.F. Jawad, L.M. Piliero, N. Kim, X. Luan, D. Goldman, et al., Analysis of polymorphisms affecting immune complex handling in systemic lupus erythematosus, Rheumatology (Oxford) 42 (3) (2003) 446—452.

[25] X. Li, J. Wu, R.H. Carter, J.C. Edberg, K. Su, G.S. Cooper, et al., A novel polymorphism in the Fcgamma receptor IIB (CD32B) transmembrane region alters receptor signaling, Arthritis Rheum 48 (11) (2003) 3242—3252.

[26] C. Kyogoku, H.M. Dijstelbloem, N. Tsuchiya, Y. Hatta, H. Kato, A. Yamaguchi, et al., Fcgamma receptor gene polymorphisms in Japanese patients with systemic lupus erythematosus: contribution of FCGR2B to genetic susceptibility, Arthritis Rheum 46 (5) (2002) 1242—1254.

[27] U. Siriboonrit, N. Tsuchiya, M. Sirikong, C. Kyogoku, S. Bejrachandra, P. Suthipintharm, et al., Association of Fcgamma receptor IIb and IIIb polymorphisms with susceptibility to systemic lupus erythematosus in Thais, Tissue Antigens 61 (5) (2003) 374—383.

[28] Z.T. Chu, N. Tsuchiya, C. Kyogoku, J. Ohashi, Y.P. Qian, S.B. Xu, et al., Association of Fcgamma receptor IIb polymorphism with susceptibility to systemic lupus erythematosus in Chinese: a common susceptibility gene in the Asian populations, Tissue Antigens 63 (1) (2004) 21—27.

[29] R.A. Floto, M.R. Clatworthy, K.R. Heilbronn, D.R. Rosner, P.A. MacAry, A. Rankin, et al., Loss of function of a lupus-associated FcgammaRIIb polymorphism through exclusion from lipid rafts, Nat Med 11 (10) (2005) 1056—1058.

[30] C. Kyogoku, N. Tsuchiya, H. Wu, B.P. Tsao, K. Tokunaga, Association of Fcgamma receptor IIA, but not IIB and IIIA, polymorphisms with systemic lupus erythematosus: A family-based association study in Caucasians, Arthritis Rheum 50 (2) (2004) 671—673.

[31] V. Magnusson, R. Zunec, J. Odeberg, G. Sturfelt, L. Truedsson, I. Gunnarsson, et al., Polymorphisms of the Fc gamma receptor type IIB gene are not associated with systemic lupus erythematosus in the Swedish population, Arthritis Rheum 50 (4) (2004) 1348—1350.

[32] K. Su, J. Wu, J.C. Edberg, X. Li, P. Ferguson, G.S. Cooper, et al., A promoter haplotype of the immunoreceptor tyrosine-based inhibitory motif-bearing FcgammaRIIb alters receptor expression and associates with autoimmunity. I. Regulatory FCGR2B polymorphisms and their association with systemic lupus erythematosus, J Immunol 172 (11) (2004) 7186—7191.

[33] M.C. Blank, R.N. Stefanescu, E. Masuda, F. Marti, P.D. King, P.B. Redecha, et al., Decreased transcription of the human FCGR2B gene mediated by the -343 G/C promoter polymorphism and association with systemic lupus erythematosus, Hum Genet 117 (2-3) (2005) 220—227.

[34] J.E. Salmon, J.C. Edberg, R.P. Kimberly, Fc gamma receptor III on human neutrophils. Allelic variants have functionally distinct capacities, J Clin Invest 85 (4) (1990) 1287—1295.

[35] Y. Hatta, N. Tsuchiya, J. Ohashi, M. Matsushita, K. Fujiwara, K. Hagiwara, et al., Association of Fc gamma Receptor III B polymorphism with SLE, Genes & Immunity 1 (1999) 53—60.

[36] M.R. Clark, L. Liu, S.B. Clarkson, P.A. Ory, I.M. Goldstein, An abnormality of the gene that encodes neutrophil Fc receptor III in a patient with systemic lupus erythematosus, J Clin Invest 86 (1) (1990) 341—346.

[37] H.R. Koene, M. Kleijer, D. Roos, H.M. de, A.E. Von dem Borne, Fc gamma RIIIB gene duplication: evidence for presence and expression of three distinct Fc gamma RIIIB genes in NA(1+,2+)SH(+) individuals, Blood 91 (2) (1998) 673—679.

[38] T.J. Aitman, R. Dong, T.J. Vyse, P.J. Norsworthy, M.D. Johnson, J. Smith, et al., Copy number polymorphism in Fcgr3 predisposes to glomerulonephritis in rats and humans, Nature 439 (7078) (2004) 851—855.

[39] L.C. Willcocks, P.A. Lyons, M.R. Clatworthy, J.I. Robinson, W. Yang, S.A. Newland, et al., Copy number of FCGR3B, which is associated with systemic lupus erythematosus, correlates

with protein expression and immune complex uptake, J Exp Med (2008). jem.

[40] M.J. Walport, K.A. Davies, M. Botto, C1q and systemic lupus erythematosus, Immunobiology 199 (2) (1998) 265–285.

[41] M. Botto, C. Dell'Agnola, A.E. Bygrave, E.M. Thompson, T. Cook, F. Petry, et al., Homozygous C1q deficiency causes glomerulonephritis associated with multiple apoptotic bodies, Nat Genet 19 (1998) 56–59.

[42] D.M. Racila, C.J. Sontheimer, A. Sheffield, J.J. Wisnieski, E. Racila, R.D. Sontheimer, Homozygous single nucleotide polymorphism of the complement C1QA gene is associated with decreased levels of C1q in patients with subacute cutaneous lupus erythematosus, Lupus 12 (2) (2003) 124–132.

[43] B. Namjou, C. Gray-McGuire, A.L. Sestak, G.S. Gilkeson, C.O. Jacob, J.T. Merrill, et al., Evaluation of C1q genomic region in minority racial groups of lupus, Genes Immun 10 (5) (2009) 517–524.

[44] M. Yamada, K. Oritani, T. Kaisho, J. Ishikawa, H. Yoshida, I. Takahashi, et al., Complement C1q regulates LPS-induced cytokine production in bone marrow-derived dendritic cells, Eur J Immunol 34 (1) (2004) 221–230.

[45] C. Lood, B. Gullstrand, L. Truedsson, A.I. Olin, G.V. Alm, L. Ronnblom, et al., C1q inhibits immune complex-induced interferon-alpha production in plasmacytoid dendritic cells: a novel link between C1q deficiency and systemic lupus erythematosus pathogenesis, Arthritis Rheum 60 (10) (2009) 3081–3090.

[46] T.W. Du Clos, C-reactive protein as a regulator of autoimmunity and inflammation, Arthritis Rheum 48 (6) (2003) 1475–1477.

[47] M.C. Carroll, The complement system in regulation of adaptive immunity, Nature Immunol 5 (2004) 981–986.

[48] A.J. Szalai, M.A. McCrory, G.S. Cooper, J. Wu, R.P. Kimberly, Association between baseline levels of C-reactive protein (CRP) and a dinucleotide repeat polymorphism in the intron of the CRP gene, Genes Immun 3 (1) (2002) 14–19.

[49] A.I. Russell, D.S. Cunninghame Graham, C. Shepherd, C.A. Roberton, J. Whittaker, J. Meeks, et al., Polymorphism at the C-reactive protein locus influences gene expression and predisposes to systemic lupus erythematosus, Hum Mol Genet 13 (1) (2004) 137–147.

[50] A.J. Szalai, J. Wu, E.M. Lange, M.A. McCrory, C.D. Langefeld, A. Williams, et al., Single-nucleotide polymorphisms in the C-reactive protein (CRP) gene promoter that affect transcription factor binding, alter transcriptional activity, and associate with differences in baseline serum CRP level, J Mol Med 83 (6) (2005) 440–447.

[51] H.A. Kim, H.Y. Chun, S.H. Kim, H.S. Park, C.H. Suh, C-reactive protein gene polymorphisms in disease susceptibility and clinical manifestations of Korean systemic lupus erythematosus, J Rheumatol 36 (10) (2009) 2238–2243.

[52] P.B. Shih, S. Manzi, P. Shaw, M. Kenney, A.H. Kao, F. Bontempo, et al., Genetic variation in C-reactive protein (CRP) gene may be associated with risk of systemic lupus erythematosus and CRP concentrations, J Rheumatol 35 (11) (2008) 2171–2178.

[53] J.C. Edberg, J. Wu, C.D. Langefeld, E.E. Brown, M.C. Marion, G. McGwin Jr., et al., Genetic variation in the CRP promoter: association with systemic lupus erythematosus, Hum Mol Genet 17 (8) (2008) 1147–1155.

[54] W. Rodriguez, C. Mold, L.L. Marnell, J. Hutt, G.J. Silverman, D. Tran, et al., Prevention and reversal of nephritis in MRL/lpr mice with a single injection of C-reactive protein, Arthritis Rheum 54 (1) (2006) 325–335.

[55] B.H. Luo, C.V. Carman, T.A. Springer, Structural basis of integrin regulation and signaling, Annu Rev Immunol 25 (2007) 619–647.

[56] S.K. Nath, S. Han, X. Kim-Howard, J.A. Kelly, P. Viswanathan, G.S. Gilkeson, et al., A nonsynonymous functional variant in integrin-alpha(M) (encoded by ITGAM) is associated with systemic lupus erythematosus, Nat Genet 40 (2) (2008) 152–154.

[57] S. Han, X. Kim-Howard, H. Deshmukh, Y. Kamatani, P. Viswanathan, J.M. Guthridge, et al., Evaluation of imputation-based association in and around the integrin-alpha-M (ITGAM) gene and replication of robust association between a non-synonymous functional variant within ITGAM and systemic lupus erythematosus (SLE), Hum Mol Genet 18 (6) (2009) 1171–1180.

[58] W. Yang, M. Zhao, N. Hirankarn, C.S. Lau, C.C. Mok, T.M. Chan, et al., ITGAM is associated with disease susceptibility and renal nephritis of systemic lupus erythematosus in Hong Kong Chinese and Thai, Hum Mol Genet 18 (11) (2009) 2063–2070.

[59] X. Kim-Howard, A.K. Maiti, J.M. Anaya, G.R. Bruner, E. Brown, J.T. Merrill, et al., ITGAM coding variant (rs1143679) influences the risk of renal disease, discoid rash, and immunologic manifestations in lupus patients with European ancestry, Ann Rheum Dis (2009).

[60] Y. Kanayama, T. Kim, H. Inariba, N. Negoro, M. Okamura, T. Takeda, et al., Possible involvement of interferon alfa in the pathogenesis of fever in systemic lupus erythematosus, Ann Rheum Dis 48 (10) (1989) 861–863.

[61] A.A. Bengtsson, G. Sturfelt, L. Truedsson, J. Blomberg, G. Alm, H. Vallin, et al., Activation of type I interferon system in systemic lupus erythematosus correlates with disease activity but not with antiretroviral antibodies, Lupus 9 (9) (2000) 664–671.

[62] B.J. Barnes, P.A. Moore, P.M. Pitha, Virus-specific activation of a novel interferon regulatory factor, IRF-5, results in the induction of distinct interferon alpha genes, J Biol Chem 276 (26) (2001) 23382–23390.

[63] S. Sigurdsson, G. Nordmark, H.H. Goring, K. Lindroos, A.C. Wiman, G. Sturfelt, et al., Polymorphisms in the tyrosine kinase 2 and interferon regulatory factor 5 genes are associated with systemic lupus erythematosus, Am J Hum Genet 76 (3) (2005) 528–537.

[64] R.R. Graham, C. Kyogoku, S. Sigurdsson, I.A. Vlasova, L.R. Davies, E.C. Baechler, et al., Three functional variants of IFN regulatory factor 5 (IRF5) define risk and protective haplotypes for human lupus, Proc Natl Acad Sci USA 104 (16) (2007) 6758–6763.

[65] S. Sigurdsson, H.H.H. Goring, G. Kristjansdottir, L. Milani, G. Nordmark, J. Sandling, et al., Comprehensive Evaluation of the Genetic Variants of Interferon Regulatory Factor 5 Reveals a Novel 5bp Length Polymorphism as Strong Risk Factor for Systemic Lupus Erythematosus, Human Molecular Genetics (2007) 359.

[66] R.R. Graham, S.V. Kozyrev, E.C. Baechler, M.V. Reddy, R.M. Plenge, J.W. Bauer, et al., A common haplotype of interferon regulatory factor 5 (IRF5) regulates splicing and expression and is associated with increased risk of systemic lupus erythematosus, Nat Genet 38 (5) (2006) 550–555.

[67] F.Y. Demirci, S. Manzi, R. Ramsey-Goldman, R.L. Minster, M. Kenney, P.S. Shaw, et al., Association of a Common Interferon Regulatory Factor 5 (IRF5) Variant with Increased Risk of Systemic Lupus Erythematosus (SLE), Ann Hum Genet (2006).

[68] H.D. Shin, Y.K. Sung, C.B. Choi, S.O. Lee, H.W. Lee, S.C. Bae, Replication of the genetic effects of IFN regulatory factor 5 (IRF5) on systemic lupus erythematosus in a Korean population, Arthritis Res Ther 9 (2) (2007) R32.

[69] A. Kawasaki, C. Kyogoku, J. Ohashi, R. Miyashita, K. Hikami, M. Kusaoi, et al., Association of IRF5 polymorphisms with systemic lupus erythematosus in a Japanese population: support for a crucial role of intron 1 polymorphisms, Arthritis Rheum 58 (3) (2008) 826–834.

[70] H.O. Siu, W. Yang, C.S. Lau, T.M. Chan, R.W. Wong, W.H. Wong, et al., Association of a haplotype of IRF5 gene with systemic lupus erythematosus in Chinese, J Rheumatol 35 (2) (2008) 360–362.

[71] J.A. Kelly, J.M. Kelley, K.M. Kaufman, J. Kilpatrick, G.R. Bruner, J.T. Merrill, et al., Interferon regulatory factor-5 is genetically associated with systemic lupus erythematosus in African Americans, Genes Immun 9 (3) (2008) 187–194.

[72] M.V. Reddy, R. Velazquez-Cruz, V. Baca, G. Lima, J. Granados, L. Orozco, et al., Genetic association of IRF5 with SLE in Mexicans: higher frequency of the risk haplotype and its homozygosity than Europeans, Hum Genet 121 (6) (2007) 721–727.

[73] S. Lofgren, H. Yin, A.M. Delgado-Vega, E. Sanchez, S. Lewen, B.A. Pons-Estel, et al., Promoter Insertion/Deletion in the IRF5 Gene Is Highly Associated with Susceptibility to Systemic Lupus Erythematosus in Distinct Populations, But Exerts a Modest Effect on Gene Expression in Peripheral Blood Mononuclear Cells, J Rheumatol (2010).

[74] S.V. Kozyrev, S. Lewen, P.M. Reddy, B. Pons-Estel, T. Witte, P. Junker, et al., Structural insertion/deletion variation in IRF5 is associated with a risk haplotype and defines the precise IRF5 isoforms expressed in systemic lupus erythematosus, Arthritis Rheum 56 (4) (2007) 1234–1241.

[75] D. Feng, R.C. Stone, M.L. Eloranta, N. Sangster-Guity, G. Nordmark, S. Sigurdsson, et al., Genetic variants and disease-associated factors contribute to enhanced interferon regulatory factor 5 expression in blood cells of patients with systemic lupus erythematosus, Arthritis Rheum 62 (2) (2010) 562–573.

[76] T.B. Niewold, J. Hua, T.J. Lehman, J.B. Harley, M.K. Crow, High serum IFN-alpha activity is a heritable risk factor for systemic lupus erythematosus, Genes Immun (2007).

[77] T.B. Niewold, J.A. Kelly, M.H. Flesch, L.R. Espinoza, J.B. Harley, M.K. Crow, Association of the IRF5 risk haplotype with high serum interferon-alpha activity in systemic lupus erythematosus patients, Arthritis Rheum 58 (8) (2008) 2481–2487.

[78] O.J. Rullo, J.M. Woo, H. Wu, A.D. Hoftman, P. Maranian, B.A. Brahn, et al., Association of IRF5 polymorphisms with activation of the Interferon-alpha pathway, Ann Rheum Dis (2009).

[79] C. Richez, K. Yasuda, R.G. Bonegio, A.A. Watkins, T. Aprahamian, P. Busto, et al., IFN regulatory factor 5 is required for disease development in the FcgammaRIIB-/-Yaa and FcgammaRIIB-/- mouse models of systemic lupus erythematosus, J Immunol 184 (2) (2010) 796–806.

[80] M.H. Kaplan, STAT4: a critical regulator of inflammation in vivo, Immunol Res 31 (3) (2005) 231–242.

[81] E.F. Remmers, R.M. Plenge, A.T. Lee, R.R. Graham, G. Hom, T.W. Behrens, et al., STAT4 and the risk of rheumatoid arthritis and systemic lupus erythematosus, N Engl J Med 357 (10) (2007) 977–986.

[82] K.E. Taylor, E.F. Remmers, A.T. Lee, W.A. Ortmann, R.M. Plenge, C. Tian, et al., Specificity of the STAT4 genetic association for severe disease manifestations of systemic lupus erythematosus, PLoS Genet 4 (5) (2008) e1000084.

[83] R.J. Palomino-Morales, A. Rojas-Villarraga, C.I. Gonzalez, G. Ramirez, J.M. Anaya, J. Martin, STAT4 but not TRAF1/C5 variants influence the risk of developing rheumatoid arthritis and systemic lupus erythematosus in Colombians, Genes Immun 9 (4) (2008) 379–382.

[84] A. Kawasaki, I. Ito, K. Hikami, J. Ohashi, T. Hayashi, D. Goto, et al., Role of STAT4 polymorphisms in systemic lupus erythematosus in a Japanese population: a case-control association study of the STAT1-STAT4 region, Arthritis Res Ther 10 (5) (2008) R113.

[85] S. Sigurdsson, G. Nordmark, S. Garnier, E. Grundberg, T. Kwan, O. Nilsson, et al., A risk haplotype of STAT4 for systemic lupus erythematosus is over-expressed, correlates with anti-dsDNA and shows additive effects with two risk alleles of IRF5, Hum. Mol Genet 17 (18) (2008) 2868–2876.

[86] S.N. Kariuki, K.A. Kirou, E.J. MacDermott, L. Barillas-Arias, M.K. Crow, T.B. Niewold, Cutting edge: autoimmune disease risk variant of STAT4 confers increased sensitivity to IFN-alpha in lupus patients in vivo, J Immunol 182 (1) (2009) 34–38.

[87] A.K. Abelson, A.M. Delgado-Vega, S.V. Kozyrev, E. Sanchez, R. Velazquez-Cruz, N. Eriksson, et al., STAT4 associates with systemic lupus erythematosus through two independent effects that correlate with gene expression and act additively with IRF5 to increase risk, Ann Rheum Dis 68 (11) (2009) 1746–1753.

[88] B. Namjou, A.L. Sestak, D.L. Armstrong, R. Zidovetzki, J.A. Kelly, N. Jacob, et al., High-density genotyping of STAT4 reveals multiple haplotypic associations with systemic lupus erythematosus in different racial groups, Arthritis Rheum 60 (4) (2009) 1085–1095.

[89] M.A. Lee-Kirsch, M. Gong, D. Chowdhury, L. Senenko, K. Engel, Y.A. Lee, et al., Mutations in the gene encoding the 3'-5' DNA exonuclease TREX1 are associated with systemic lupus erythematosus, Nat Genet 39 (9) (2007) 1065–1067.

[90] R. Beyaert, K. Heyninck, H.S. Van, A20 and A20-binding proteins as cellular inhibitors of nuclear factor-kappa B-dependent gene expression and apoptosis, Biochem Pharmacol 60 (8) (2000) 1143–1151.

[91] V. Gateva, J.K. Sandling, G. Hom, K.E. Taylor, S.A. Chung, X. Sun, et al., A large-scale replication study identifies TNIP1, PRDM1, JAZF1, UHRF1BP1 and IL10 as risk loci for systemic lupus erythematosus, Nat Genet (2009). advance online publication.

[92] M.U. Martin, H. Wesche, Summary and comparison of the signaling mechanisms of the Toll/interleukin-1 receptor family, Biochim Biophys Acta 1592 (3) (2002) 265–280.

[93] C. Kollewe, A.C. Mackensen, D. Neumann, J. Knop, P. Cao, S. Li, et al., Sequential autophosphorylation steps in the interleukin-1 receptor-associated kinase-1 regulate its availability as an adapter in interleukin-1 signaling, J Biol Chem 279 (7) (2004) 5227–5236.

[94] C.O. Jacob, J. Zhu, D.L. Armstrong, M. Yan, J. Han, X.J. Zhou, et al., Identification of IRAK1 as a risk gene with critical role in the pathogenesis of systemic lupus erythematosus, Proc Natl Acad Sci USA 106 (15) (2009) 6256–6261.

[95] A.H. Sawalha, R. Webb, S. Han, J.A. Kelly, K.M. Kaufman, R.P. Kimberly, et al., Common variants within MECP2 confer risk of systemic lupus erythematosus, PLoS ONE 3 (3) (2008) e1727.

[96] S. Sigurdsson, G. Nordmark, S. Garnier, E. Grundberg, T. Kwan, O. Nilsson, et al., A risk haplotype of STAT4 for systemic lupus erythematosus is over-expressed, correlates with anti-dsDNA and shows additive effects with two risk alleles of IRF5, Hum Mol Genet 17 (18) (2008) 2868–2876.

[97] A. Hellquist, T.M. Jarvinen, S. Koskenmies, M. Zucchelli, C. Orsmark-Pietras, L. Berglind, et al., Evidence for genetic association and interaction between the TYK2 and IRF5 genes in systemic lupus erythematosus, J Rheumatol 36 (8) (2009) 1631–1638.

[98] H. Nishimura, T. Honjo, PD-1: an inhibitory immunoreceptor involved in peripheral tolerance, Trends Immunol. 22 (5) (2001) 265–268.

[99] H. Nishimura, M. Nose, H. Hiai, N. Minato, T. Honjo, Development of lupus-like autoimmune diseases by disruption of the PD-1 gene encoding an ITIM motif-carrying immunoreceptor, Immunity 11 (2) (1999) 141–151.

[100] H. Nishimura, T. Okazaki, Y. Tanaka, K. Nakatani, M. Hara, A. Matsumori, et al., Autoimmune dilated cardiomyopathy in PD-1 receptor-deficient mice, Science 291 (5502) (2001) 319–322.

[101] L. Prokunina, C. Castillejo-Lopez, F. Oberg, I. Gunnarsson, L. Berg, V. Magnusson, et al., A regulatory polymorphism in PDCD1 is associated with susceptibility to systemic lupus erythematosus in humans, Nat Genet 32 (4) (2002) 666–669.

[102] H. Kristjansdottir, K. Steinsson, I. Gunnarsson, G. Grondal, K. Erlendsson, M.E. Alarcon-Riquelme, Lower expression levels of the PD1 receptor on CD4+CD25+ T-cells in SLE patients and correlation to the PD-1.3A genotype, Arthritis Rheum (2010).

[103] L. Prokunina, I. Gunnarsson, G. Sturfelt, L. Truedsson, V.A. Seligman, J.L. Olson, et al., The systemic lupus erythematosus-associated PDCD1 polymorphism PD1.3A in lupus nephritis, Arthritis Rheum 50 (1) (2004) 327–328.

[104] M. Johansson, L. Arlestig, B. Moller, S. Rantapaa-Dahlqvist, Association of a PDCD1 polymorphism with renal manifestations in systemic lupus erythematosus, Arthritis Rheum 52 (6) (2005) 1665–1669.

[105] I. Ferreiros-Vidal, J.J. Gomez-Reino, F. Barros, A. Carracedo, P. Carreira, F. Gonzalez-Escribano, et al., Association of PDCD1 with susceptibility to systemic lupus erythematosus: evidence of population-specific effects, Arthritis Rheum 50 (8) (2004) 2590–2597.

[106] D.K. Sanghera, S. Manzi, F. Bontempo, C. Nestlerode, M.I. Kamboh, Role of an intronic polymorphism in the PDCD1 gene with the risk of sporadic systemic lupus erythematosus and the occurrence of antiphospholipid antibodies, Hum Genet 115 (5) (2004) 393–398.

[107] I. Ferreiros-Vidal, S. D'Alfonso, C. Papasteriades, F.N. Skopouli, M. Marchini, R. Scorza, et al., Bias in association studies of systemic lupus erythematosus susceptibility due to geographical variation in the frequency of a programmed cell death 1 polymorphism across Europe, Genes Immun 8 (2) (2007) 138–146.

[108] S. Cohen, H. Dadi, E. Shaoul, N. Sharfe, C.M. Roifman, Cloning and characterization of a lymphoid-specific, inducible human protein tyrosine phosphatase, Lyp. Blood 93 (6) (1999) 2013–2024.

[109] P.K. Gregersen, L.M. Olsson, Recent advances in the genetics of autoimmune disease, Annu Rev Immunol 27 (2009) 363–391.

[110] L.A. Criswell, K.A. Pfeiffer, R.F. Lum, B. Gonzales, J. Novitzke, M. Kern, et al., Analysis of Families in the Multiple Autoimmune Disease Genetics Consortium (MADGC) Collection: the PTPN22 620W Allele Associates with Multiple Autoimmune Phenotypes, Am J Hum Genet 76 (4) (2005) 561–571.

[111] H. Wu, R.M. Cantor, D.S. Graham, C.M. Lingren, L. Farwell, P.L. Jager, et al., Association analysis of the R620W polymorphism of protein tyrosine phosphatase PTPN22 in systemic lupus erythematosus families: Increased t allele frequency in systemic lupus erythematosus patients with autoimmune thyroid disease, Arthritis Rheum 52 (8) (2005) 2396–2402.

[112] K.M. Kaufman, J.A. Kelly, B.J. Herring, A.J. Adler, S.B. Glenn, B. Namjou, et al., Evaluation of the genetic association of the PTPN22 R620W polymorphism in familial and sporadic systemic lupus erythematosus, Arthritis Rheum. 54 (8) (2006) 2533–2540.

[113] S.N. Kariuki, M.K. Crow, T.B. Niewold, The PTPN22 C1858T polymorphism is associated with skewing of cytokine profiles toward high interferon-alpha activity and low tumor necrosis factor alpha levels in patients with lupus, Arthritis Rheum 58 (9) (2008) 2818–2823.

[114] V. Orru, S.J. Tsai, B. Rueda, E. Fiorillo, S.M. Stanford, J. Dasgupta, et al., A loss-of-function variant of PTPN22 is associated with reduced risk of systemic lupus erythematosus, Human Molecular Genetics 18 (3) (2009) 569–679.

[115] E. Hagiwara, M.F. Gourley, S. Lee, D.K. Klinman, Disease severity in patients with systemic lupus erythematosus correlates with an increased ratio of interleukin-10:interferon-gamma- secreting cells in the peripheral blood, Arthritis Rheum 39 (3) (1996) 379–385.

[116] J. Eskdale, G. Gallagher, C.L. Verweij, V. Keijsers, R.G. Westendorp, T.W. Huizinga, Interleukin 10 secretion in relation to human IL-10 locus haplotypes, Proc Natl Acad Sci USA 95 (16) (1998) 9465–9470.

[117] J. Eskdale, P. Wordsworth, S. Bowman, M. Field, G. Gallagher, Association between polymorphisms at the human IL-10 locus and systemic lupus erythematosus [published erratum appears in Tissue Antigens 1997 Dec;50(6):699], Tissue Antigens 49 (6) (1997) 635–639.

[118] R. Mehrian, F.P.J. Quismorio, Strassmann, M.M. Stimmler, D.M. Horowitz, R.C. Kitridou, et al., Synergistic effect between IL-10 and bcl-2 genotypes in determining susceptibility to SLE, Arthritis Rheum 41 (1998) 596–602.

[119] W.P. Chong, W.K. Ip, W.H. Wong, C.S. Lau, T.M. Chan, Y.L. Lau, Association of interleukin-10 promoter polymorphisms with systemic lupus erythematosus, Genes Immun 5 (6) (2004) 484–492.

[120] S.K. Nath, J.B. Harley, Y.H. Lee, Polymorphisms of complement receptor 1 and interleukin-10 genes and systemic lupus erythematosus: a meta-analysis, Hum Genet (2005) 1–10.

[121] E. Stuber, W. Strober, The T cell-B cell interaction via OX40-OX40L is necessary for the T cell-dependent humoral immune response, J Exp Med 183 (3) (1996) 979–989.

[122] T. Ito, OX40 ligand shuts down IL-10-producing regulatory T cells, Proc Natl Acad Sci USA 103 (2006) 13138–13143.

[123] E. Stuber, M. Neurath, D. Calderhead, H.P. Fell, W. Strober, Cross-linking of OX40 ligand, a member of the TNF/NGF cytokine family, induces proliferation and differentiation in murine splenic B cells, Immunity 2 (1995) 507–521.

[124] D.S.C. Graham, R.R. Graham, H. Manku, A.K. Wong, J.C. Whittaker, P.M. Gaffney, et al., Polymorphism at the TNF superfamily gene TNFSF4 confers susceptibility to systemic lupus erythematosus, Nat Genet 40 (1) (2008) 83–89.

[125] Y.K. Chang, W. Yang, M. Zhao, C.C. Mok, T.M. Chan, R.W. Wong, et al., Association of BANK1 and TNFSF4 with systemic lupus erythematosus in Hong Kong Chinese, Genes Immun (2009).

[126] A.M. Delgado-Vega, A.K. Abelson, E. Sanchez, T. Witte, S. D'Alfonso, M. Galeazzi, et al., Replication of the TNFSF4 (OX40L) promoter region association with systemic lupus erythematosus, Genes Immun (2008).

[127] M. Reth, J. Wienands, Initiation and processing of signals from the B cell antigen receptor, Annu Rev Immunol 15 (1997) 453–479.

[128] Z. Zhang, K.J. Zhu, Q. Xu, X.J. Zhang, L.D. Sun, H.F. Zheng, et al., The association of the BLK gene with SLE was replicated in Chinese Han, Arch Dermatol Res (2010).

[129] I. Ito, A. Kawasaki, S. Ito, T. Hayashi, D. Goto, I. Matsumoto, et al., Replication of the association between the C8orf13-BLK region and systemic lupus erythematosus in a Japanese population, Arthritis Rheum 60 (2) (2009) 553–558.

[130] K. Yokoyama, I.H. Su Ih, T. Tezuka, T. Yasuda, K. Mikoshiba, A. Tarakhovsky, et al., BANK regulates BCR-induced calcium

mobilization by promoting tyrosine phosphorylation of IP(3) receptor, EMBO J 21 (1-2) (2002) 83–92.

[131] L. Guo, H. Deshmukh, R. Lu, G.S. Vidal, J.A. Kelly, K.M. Kaufman, et al., Replication of the BANK1 genetic association with systemic lupus erythematosus in a European-derived population, Genes Immun (2009).

[132] R. Lu, G.S. Vidal, J.A. Kelly, A.M. Delgado-Vega, X.K. Howard, S.R. Macwana, et al., Genetic associations of LYN with systemic lupus erythematosus, Genes Immun (2009).

[133] H. Maier, J. Colbert, D. Fitzsimmons, D.R. Clark, J. Hagman, Activation of the early B-cell-specific mb-1 (Ig-alpha) gene by Pax-5 is dependent on an unmethylated Ets binding site, Mol Cell Biol 23 (6) (2003) 1946–1960.

[134] J. Moisan, R. Grenningloh, E. Bettelli, M. Oukka, I.C. Ho, Ets-1 is a negative regulator of Th17 differentiation, J Exp Med 204 (12) (2007) 2825–2835.

[135] D. Wang, S.A. John, J.L. Clements, D.H. Percy, K.P. Barton, L.A. Garrett-Sinha, Ets-1 deficiency leads to altered B cell differentiation, hyperresponsiveness to TLR9 and autoimmune disease, Int Immunol 17 (9) (2005) 1179–1191.

[136] S. Raychaudhuri, B.P. Thomson, E.F. Remmers, S. Eyre, A. Hinks, C. Guiducci, et al., Genetic variants at CD28, PRDM1 and CD2/CD58 are associated with rheumatoid arthritis risk, Nat Genet 41 (12) (2009) 1313–1318.

[137] E. Zeggini, L.J. Scott, R. Saxena, B.F. Voight, J.L. Marchini, T. Hu, et al., Meta-analysis of genome-wide association data and large-scale replication identifies additional susceptibility loci for type 2 diabetes, Nat Genet 40 (5) (2008) 638–645.

[138] F. Takeuchi, Y. Ochiai, M. Serizawa, K. Yanai, N. Kuzuya, H. Kajio, et al., Search for type 2 diabetes susceptibility genes on chromosomes 1q, 3q and 12q, J Hum Genet 53 (4) (2008) 314–324.

[139] M. Suarez-Gestal, M. Calaza, E. Endreffy, R. Pullmann, J. Ordi-Ros, S.G. Domenico, et al., Replication of recently identified systemic lupus erythematosus genetic associations: a case-control study, Arthritis Res Ther 11 (3) (2009). R69.

[140] B.Z. Alizadeh, G. Valdigem, M.J. Coenen, A. Zhernakova, B. Franke, A. Monsuur, et al., Association analysis of functional variants of the FcgRIIa and FcgRIIIa genes with type 1 diabetes, celiac disease and rheumatoid arthritis, Hum Mol Genet 16 (21) (2007) 2552–2559.

[141] K. Asano, T. Matsushita, J. Umeno, N. Hosono, A. Takahashi, T. Kawaguchi, et al., A genome-wide association study identifies three new susceptibility loci for ulcerative colitis in the Japanese population, Nat Genet 41 (12) (2009) 1325–1329.

[142] K. Shimane, Y. Kochi, R. Yamada, Y. Okada, A. Suzuki, A. Miyatake, et al., A single nucleotide polymorphism in the IRF5 promoter region is associated with susceptibility to rheumatoid arthritis in the Japanese patients, Ann Rheum Dis (2008).

[143] R.M. Plenge, C. Cotsapas, L. Davies, A.L. Price, P.I. De Bakker, J. Maller, et al., Two independent alleles at 6q23 associated with risk of rheumatoid arthritis, Nat Genet 39 (12) (2007) 1477–1482.

[144] V. Dideberg, G. Kristjansdottir, L. Milani, C. Libioulle, S. Sigurdsson, E. Louis, et al., An insertion-deletion polymorphism in the Interferon Regulatory Factor 5 (IRF5) gene confers risk of inflammatory bowel diseases, Hum Mol Genet (2007).

[145] P. Dieude, M. Guedj, J. Wipff, J. Avouac, I. Fajardy, E. Diot, et al., Association between the IRF5 rs2004640 functional polymorphism and systemic sclerosis: a new perspective for pulmonary fibrosis, Arthritis Rheum 60 (1) (2009) 225–233.

[146] I. Ito, Y. Kawaguchi, A. Kawasaki, M. Hasegawa, J. Ohashi, K. Hikami, et al., Association of a functional polymorphism in the IRF5 region with systemic sclerosis in a Japanese population, Arthritis Rheum 60 (6) (2009) 1845–1850.

[147] B.D. Korman, M.I. Alba, J.M. Le, I. Alevizos, J.A. Smith, N.P. Nikolov, et al., Variant form of STAT4 is associated with primary Sjogren's syndrome, Genes Immun 9 (3) (2008) 267–270.

[148] B. Rueda, J. Broen, C. Simeon, R. Hesselstrand, B. Diaz, H. Suarez, et al., The STAT4 gene influences the genetic predisposition to systemic sclerosis phenotype, Hum Mol Genet 18 (11) (2009) 2071–2077.

[149] S. Oshima, E.E. Turer, J.A. Callahan, S. Chai, R. Advincula, J. Barrera, et al., ABIN-1 is a ubiquitin sensor that restricts cell death and sustains embryonic development, Nature 457 (7231) (2009) 906–909.

[150] E.Y. Fung, D.J. Smyth, J.M. Howson, J.D. Cooper, N.M. Walker, H. Stevens, et al., Analysis of 17 autoimmune disease-associated variants in type 1 diabetes identifies 6q23/ TNFAIP3 as a susceptibility locus, Genes Immun 10 (2) (2009) 188–191.

[151] R.P. Nair, K.C. Duffin, C. Helms, J. Ding, P.E. Stuart, D. Goldgar, et al., Genome-wide scan reveals association of psoriasis with IL-23 and NF-kappaB pathways, Nat Genet 41 (2) (2009) 199–204.

[152] A.B. Begovich, V.E. Carlton, L.A. Honigberg, S.J. Schrodi, A.P. Chokkalingam, H.C. Alexander, et al., A missense single-nucleotide polymorphism in a gene encoding a protein tyrosine phosphatase (PTPN22) is associated with rheumatoid arthritis, Am J Hum Genet 75 (2) (2004) 330–337.

[153] N. Bottini, L. Musumeci, A. Alonso, S. Rahmouni, K. Nika, M. Rostamkhani, et al., A functional variant of lymphoid tyrosine phosphatase is associated with type I diabetes, Nat Genet 36 (4) (2004) 337–338.

[154] P. Dieude, M. Guedj, J. Wipff, J. Avouac, E. Hachulla, E. Diot, et al., The PTPN22 620W allele confers susceptibility to systemic sclerosis: findings of a large case-control study of European Caucasians and a meta-analysis, Arthritis Rheum 58 (7) (2008) 2183–2188.

[155] M.R. Velaga, V. Wilson, C.E. Jennings, C.J. Owen, S. Herington, P.T. Donaldson, et al., The codon 620 tryptophan allele of the lymphoid tyrosine phosphatase (LYP) gene is a major determinant of Graves' disease, J Clin Endocrinol Metab 89 (11) (2004) 5862–5865.

[156] J.C. Barrett, S. Hansoul, D.L. Nicolae, J.H. Cho, R.H. Duerr, J.D. Rioux, et al., Genome-wide association defines more than 30 distinct susceptibility loci for Crohn's disease, Nat Genet 40 (8) (2008) 955–962.

[157] B. Rueda, P. Gourh, J. Broen, S.K. Agarwal, C.P. Simeon, N. Ortego-Centeno, et al., BANK1 functional variants are associated with susceptibility to diffuse systemic sclerosis in Caucasians, Ann Rheum Dis (2009).

[158] P. Gourh, S.K. Agarwal, E. Martin, D. Divecha, B. Rueda, H. Bunting, et al., Association of the C8orf13-BLK region with systemic sclerosis in North-American and European populations, J Autoimmun 34 (2) (2010) 155–162.

[159] G. Orozco, A.K. Abelson, M.A. Gonzalez-Gay, A. Balsa, D. Pascual-Salcedo, A. Garcia, et al., Study of functional variants of the BANK1 gene in rheumatoid arthritis, Arthritis Rheum 60 (2) (2009) 372–379.

[160] H. Yin, M.O. Borghi, A.M. Delgado-Vega, A. Tincani, P.L. Meroni, M.E. Alarcon-Riquelme, Association of STAT4 and BLK, but not BANK1 or IRF5, with primary antiphospholipid syndrome, Arthritis Rheum 60 (8) (2009) 2468–2471.

[161] P. Gourh, F.C. Arnett, F.K. Tan, S. Assassi, D. Divecha, G. Paz, et al., Association of TNFSF4 (OX40L) polymorphisms with susceptibility to Systemic Sclerosis, Ann Rheum Dis (2009).

Genetics of Lupus in Mice

Dwight H. Kono, Argyrios N. Theofilopoulos

Just over 50 years ago a report of spontaneous auto-immune hemolytic anemia in the NZB/B1 lupus strain [1] provided the first example of a lupus-related manifestation in mice and supported the possibility of this autoimmune disease being an inherited trait. Since then, the importance of genetic predisposition in SLE has been firmly established and recent linkage and association studies have identified many chromosomal regions linked to lupus traits as well as likely candidate genes [2]. Mice, which appear highly susceptible to lupus compared to other autoimmune diseases, have continued to serve an important role in genetic studies of SLE by facilitating the definition of specific genes that promote or inhibit lupus, the characterization of gene-related immunopathogenic mechanisms, and the delineation of gene interactions. This has substantially expanded the scope of genetic variations implicated in the development, severity, and types of lupus pheno-types as well as significantly deepened understanding of the basic underlying etiopathologic processes. Concomitant with this has been a growing constellation of mouse models from which have resulted a greater appreciation for the complexity of genetic susceptibility in SLE and a reconsideration of how knowledge of the genetics of SLE might impact the clinical arena.

MOUSE MODELS OF LUPUS

The availability of a broad range of lupus-prone strains, both spontaneous and induced, has made it possible to study the genetics of lupus in more detail than any other autoimmune disease (Table 4.1). The most commonly studied spontaneous models are the MRL-Fas^{lpr}, (NZBxNZW)F1 (BWF1) hybrid, and BXSB mice, strains that share characteristics such as hyper-gammaglobulinemia, antinuclear antibodies and glomerulonephritis (GN), and also possess unique features, such as arthritis and expanded CD4⁻CD8⁻

(double-negative, DN) T cells in MRL-Fas^{lpr} mice, hemolytic anemia in NZB mice, and monocytosis in BXSB mice. Details of the clinical manifestations and immuno-pathology for these and other mouse models of lupus models have been previously reviewed [3, 4].

Several recombinant inbred (RI) lines, derived from crosses of lupus-prone and non-autoimmune strains have also been generated (Table 4.1). These manifest a spectrum of phenotypes consistent with polygenic inheritance of traits. RI lines, derived from the NZB and NZW strains (NZM/Aeg2410 and NZM/Aeg2328), have been useful for studying recessive susceptibility genes. The BXD2 RI line derived from the non-autoimmune B6 and DBA/2 strains was recently discovered to develop lupus-like manifestations as well as erosive inflammatory arthritis [5]. This and other models have shown that significant autoimmune predisposing alleles are present in non-autoimmune strains. Interval-specific congenic strains with introgressed genomic regions encompassing susceptibility loci have also provided another important resource for genetic studies. Two such notable examples are the B6.Sle set congenics and the B6-$Fcgr2a^{-/-}$ Yaa strain, which are discussed below.

In addition, gene knockout/knockin, transgenic, or mutagenic manipulation of non-autoimmune back-ground strains has generated a large number of novel autoimmune mouse models with manifestations similar to spontaneous SLE (see below and Table 4.4). These strains provide insights into the potential contribution of individual genes in SLE and, in combination, a frame-work from which to build models of both immunopath-ologic processes and overall disease pathogenesis.

DISSECTING THE GENETICS OF LUPUS IN MICE

Genetic alterations that modulate lupus susceptibility in mice represent a continuum ranging from those that

TABLE 4.1 Spontaneous and induced mouse models of lupus

SPONTANEOUS DISEASE MODELS

NZ and related strains

    NZB

    NZW

    (NZBxNZW)F$_1$

    (NZBxSWR)F$_1$

    (NZBxNZW) recombinant inbred (RI) lines "NZM/Aeg" lines [316]

    (NZBxSM)RI lines "(NXSM)RI"

    (NZBxC58)RI lines "(NX8)RI"

MRL (*Fas^{lpr}* and wild-type) and related strains

    MRL-*Fas^{lpr}.ll* (long-lived substrain) [317]

MRL-*Fas^{lpr},Yaa* [318]

    SCG/Kj-*Fas^{lpr}* (BXSBxMRL-*lpr*)RI [319]

BXSB and related strains

    BXSB-*ll* (long-lived; separate B6xSB/Le RI line) [320, 321]

    (NZWxBXSB)F$_1$

    (NZBxBXSB)F$_1$

BXD2 [5]

(SJLxSWR)F$_1$ [322]

Palmerston North [323]

Motheaten strains [195, 196, 324]

INDUCED DISEASE MODELS

Heavy metal-induced autoimmunity [325]

Drug-induced lupus [326]

Pristane (TMPD)-induced [327]

Anti-idiotypic [328]

Graft-versus-host disease

BCG-injected NOD [329, 330]

Bovine thrombin-exposed galactose-alpha1-3-galactose-deficient mice [331]

can alone promote the development of severe lupus in otherwise normal strains to others that completely suppress disease development, with most residing between these extremes. Studies to identify such genetic changes have utilized both forward (phenotype → gene) and reverse (gene → phenotype) approaches. The forward approach, which finds genes based solely on their chromosomal location, has been used to identify predisposing loci and genes in both spontaneous and induced models, including conventional, mercury-induced, and ENU mutagenesis-derived. While the reverse approach, which tests the effects of specific gene mutations on the development of lupus in normal or lupus-prone strains, has been used to identify genes with potential to predispose and/or suppress disease.

PREDISPOSING LOCI AND GENES IN SPONTANEOUS LUPUS MOUSE MODELS

Identification of genes predisposing to quantitative traits, such as those associated with lupus, typically involves four major steps: (1) trait mapping performed by genome-wide scans; (2) generation and analysis of interval-specific congenic strains to confirm mapping results and to identify the major intermediate phenotypes; (3) generation of smaller interval congenics to finely map the location of the susceptibility gene or genes; and (4) identification of candidate gene variations within the fragment selected on the basis of expression, structure, function or other characteristics.

Genome-wide scans to map lupus-associated manifestations have been performed in crosses of the four major lupus-prone strains and several induced or genetically manipulated models. Over 85 named and additional unnamed loci linked to one or more lupus traits distributed over all 19 of the mouse autosomal chromosomes have been identified (Table 4.2). Some loci, identified by different groups, appear likely to represent the same variant, whereas most others are unique loci. Susceptibility in these strains seems to be caused by different sets of a few major loci rather than a large number of common ones.

Several loci have been confirmed with interval congenic mice. These included *Sle1*, *Cgnz1*, *Nba2*, *BXSB1-4*, *Sle16*, and *Mag* (on chromosome 1); *Sle18* (chromosome 3); *Sle2*, *Adnz1*, *Lbw2*, and *Lmb1* (chromosome 4); *Lmb2* (chromosome 5); *Sle3*, *Sle5*, *Nba5*, *Lmb3* (chromosome 7); *Lmb4* (chromosome 10); *Ssb2* (chromsome 12); *Sgp3* and a NZB locus (chromosome 13); and *Sles1* (chromosome 17) [6−28]. These congenics, by isolating single chromosomal regions on new stable backgrounds, permit detailed examination of the effects of a single locus on immune and autoimmune responses. This is currently the most definitive and sensitive method for confirming quantitative trait loci (QTL) and provides the initial basis for precise mapping and gene identification. Characterization of these interval congenic mice has identified specific cellular, developmental, functional, and/or autoimmune phenotypes associated with the specific introgressed QTL intervals.

Studies of interval congenics have provided important insights about the genetic transmission of lupus beyond those obtained from QTL mapping studies that are likely also applicable to human SLE. (1) Phenotypes induced by QTLs in congenic mice do not always correlate with initial mapping studies and may range from no

TABLE 4.2 Susceptibility loci predisposing to lupus-related traits

Name	Chr	Mb	Best assoc Marker	Cross	Phenotype	Parental allele	Ref.
Bxs4	1	20	D1Mit3	B10x(B10xBXSB)F1	LN	BXSB	[332]
Bxs1	1	64	D1Mit5	BXSBx(B10xBXSB)F1	GN/ANA/spleen	BXSB	[333]
—	1	90	D1Mit48	(WxBa)F1xW	IgM ssDNA/IgM histone	BALB/c	[334]
Bxs2	1	63 cM	D1Mit12	BXSBx(B10xBXSB)F1	GN/ANA/spleen	BXSB	[333]
—	1	129	D1Mit494	MRL-lprx(MRL-lprxC3H-lpr)F1	sialadenitis	MRL	[335]
Bana3	1	155	D1Mit396	(NODxBa)xNODBC	ANA (M.bovis)	NOD	[330]
Swrl1	1	170	D1Mit15	Bx(SWRxB)F1	dsDNA/histone	SWR	[336]
Sle1	1	170	D1Mit15	(NZMxB6)xNZM	GN	NZM (NZW)	[337]
	1	170		(NZMxB6)F2	dsDNA/GN/spleen	NZM(NZW)	[338]
Hmr1	1	170	D1Mit15	(SJLxDBA/2)F2	glom. dep. (HgIA resistance)	DBA/2	[339]
Sle16	1	170	D1Mit15	(129xB6)F2	ANA	129	[29]
	1	171	D1Mit36	(BxDBA/2)F2	glom. dep. (HgIA resistance)	DBA/2	[339]
Cgnz1	1	171	D1Mit36	(NZM2328xC57L)F1xNZM2328	chronic GN	NZM2328 (NZW)	[340]
Lbw7	1	171	D1Mit36	BWF2	chr/spleen	NZB	[341]
Nba2	1	171	D1Mit111	(BxSM)xW	GN	NZB	[342]
		173	D1Mit148	(BxSM)xW / $(B6.H2^z \times B) \times B$	ANA/gp70IC/GN	NZB	[343]
		175	Crp/Sap	$((B6.H2^z\ \&\ Ba.H2^z) \times B)F1 \times B$	GN	NZB	[344]
Bxs3	1	178	D1Mit403	BXSBx(B10xBXSB)F1	dsDNA	BXSB	[333]
Agnz1	1	184	D1Mit37	(NZM2328xC57L)F1xNZM2328	acute GN	NZM2328 (NZW)	[340]
—	1	191	D1Mit17	(WxBa)F1xW	ssDNA	NZW	[334]
Swrl5	1	191	D1Mit17	(SWRxB)F2	hyperIgG	SWR	[345]
Mag	1	82–100 cM		B6.MRLc1(82-100) congenic	spl/dsDNA/GN	MRL	[26]
—	2	103	D2Mit12	(MRL-lprxBa)F2	ssDNA/dsDNA	MRL +/+,lpr/+	[346]
Rends (Wbw1)	2	153	D2Mit285	(WxPL)F1xB BXD RI	mortality/GN DNA	NZW DBA/2	[347] [5]
Sles2	3	79	D3Mit137	(B6.NZMc1xW)F1xW	dsDNA/GN (resistance)	NZW	[37]

(Continued)

TABLE 4.2 Susceptibility loci predisposing to lupus-related traits—cont'd

Name	Chr	Mb	Best assoc Marker	Cross	Phenotype	Parental allele	Ref.
Bxs5	3	87	D3Mit40	B10x(B10xBXSB)F1	ANA/IgG3	BXSB	[332]
Sle18	3	125	D3Mit13	(129xB6)F2	ANA	129	[29]
Lprm2	3	137	D3Mit16	MRL-lprx(MRL-lprxC3H-lpr)F1	vasculitis (resistance)	MRL	[348]
Nbwa2	4	35	D4Mit11	(BxBa.H2^z)F2	GN/dsDNA	NZB	[349]
Arvm1	4	46	D4Mit89	(MRL-lprxC3H-lpr)BC & F_2	vasculitis	MRL	[40]
Agnm1	4	56	D4Mit241	(MRL-lprxC3H-lpr)F_2	GN	MRL	[350]
Lprm1	4	64	D4Mit82	MRL-lprx(MRL-lprxC3H-lpr)F1	vasculitis	MRL	[348]
Acla2	4	83	D4Mit79	Wx(WxBXSB)F1	CL	BXSB	[351]
Sle2	4	95	D4Mit9	(NZMxB6)xNZM	GN	NZM (NZW)	[337]
Spm1	4	99	D4Mit58	(B6xNZB)F$_1$xNZB	spleen	NZB	[352]
Adaz1	4	50 cM	D1Mit36	(NZM2328xC57L)F1xNZM2328	dsDNA	NZM2328	[340]
Agnm2	4	100	D4Mit187	(MRL-lprxC3H-lpr)F_2	GN	MRL	[350]
Asm2	4	112	D4Mit199	MRL-lprx(MRL-lprxC3H-lpr)F1	sialadenitis	MRL female	[335]
Lmb1	4	124	D4Mit12	(B6-lprxMRL-lpr)F2	Lprn/dsDNA	B6	[353]
Lbw2	4	124	D4Nds2	BWF2	mortality/GN/spleen	NZB	[341]
Sles2	4	124	D4Mit12	(B6.NZMc1xW)F1xW	dsDNA/GN (resistance)	NZW	[37]
—	4	134	D4Mit70	(BxSM)xW	GN	NZB	[342]
Arvm2	4	125	D4Mit147	(MRL-lprxC3H-lpr)BC & F_2	vasculitis	MRL	[40]
nba1	4	131	Epb4.1(elp-1)	BWF1xW	GN	NZB	[354]
Imh1/Mott	4	141	D4Mit48	BWF1xW	hyper IgM/GN/dsDNA	NZB	[355, 356]
Sle6	5	36	D5Mit4	(B6.NZMc1xNZW)F1xNZW	GN	NZW	[37]
Nba5	5	41	D5Mit353	(SWRxB)F2	GN	NZB	[345]
Lmb2	5	74	D5Mit356	(B6-lprxMRL-lpr)F2	Lprn/dsDNA	MRL	[353]
Lprm4	5	54 cM	D5Mit23	MRL-lprx(MRL-lprxC3H-lpr)F1	spleen	MRL	[348]
Agnm3	5	105	D4Mit187	(MRL-lprxC3H-lpr)F_2	GN	MRL	[350]
Lbw3	5	142	D5Mit101	BWF2	mortality	NZW	[341]
—	6	84	D6Mit8	MRL-lprx(MRL-lprxC3H-lpr)F1	GN (resistance)	MRL	[357]

Locus	Chr	cM	Marker	Cross	Phenotype	Strain	Ref
Lbw4	6	65 cM	D6Mit25	BWF2	mortality	NZB	[341]
—	6	134	D6Mit374	(NZMxB6)F2	dsDNA	B6	[338]
Sle5	7	3	D7Mit178	(NZMxB6)F2	dsDNA	NZM(NZW)	[338]
Lrdm1	7	26	Pou2f2(Otf-2)	(MRL-lprxCAST)F1xMRL-lpr	GN	MRL	[358]
Sle3	7	38	D7Mit25	(NZMxB6)F2	GN	NZM(NZW)	[338]
Lbw5	7	51	D7Nds5(Ngfg)	BWF2	mortality	NZW	[341]
Lmb3	7	28 cM	D7Mit211	(B6-lprxMRL-lpr)F2	Lprn/dsDNA	MRL	[353]
Sle3	7	63	p	(NZMxB6)xNZM	GN	NZM (NZW)	[337]
Aem2	7	89	D7Mit30	(B6xB)F1xB	RBC	NZB	[352]
—	7	116	D7Mit17	(BxSM)xW	GN	NZB	[342]
—	7	126	D7Mit7	(BxW)F1xW	dsDNA	NZB	[359]
Myo1	7	150	D7Mit14	Wx(WxBXSB)F1	MI	BXSB	[351]
Pbat2	8	31	D8Mit96	Wx(WxBXSB)F1	platelet	BXSB	[351]
Asbb1	9	37	D9Mit67	(B6xBa)F2-FcγRIIb$^{-/-}$	spleen (FcγRIIb ko)	BALB/c	[360]
Baa1	9	49	D9Mit22	(WxBa)F1xW	IgM ssDNA/IgM histone	BALB/c	[334]
Gp1	9	105	D9Mit53	BXSBx(B10xBXSB)F1	gp70IC	BXSB	[361]
Baa2	9	114	D9Mit81	(BxBa.H2^z)F2	Tubulointerstitial damage	BALB/c	[349]
Bana2	10	20	D10Mit213	(NODxBa)xNODBC	ANA (M.bovis)	BALB/c	[330]
Asm1	10	70/72	D10Mit115/259	MRL-lprx(MRL-lprxC3H-lpr)F1	sialadenitis	MRL	[335]
Aem3	10	82	D10Mit42	(B6xB)F1xB	RBC	NZB	[352]
Lmb4	10	92	D10Mit11	(B6-lprxMRL-lpr)F2	Lprn/GN	MRL	[353]
—	10	122	D10Mit35	(NZMxB6)F2	GN	NZM & B6#	[338]
—	10	125	D10Mit297	(B10.A^{z}xB)F1xB	chr	B10	[362]
—	11	12/34	D11Mit2/84	(Ba.H2zx B)F1xB	GN	NZB	[344]
—	11	45	D11Mit20	(NZMxB6)F2	GN/dsDNA	NZM	[338]

(Continued)

TABLE 4.2　Susceptibility loci predisposing to lupus-related traits—cont'd

Name	Chr	Mb	Best assoc Marker	Cross	Phenotype	Parental allele	Ref.
Lbw8	11	53	IL4	BWF2	chr	NZB	[341]
—	11	55	D11Mit207	(WxBa)F1xW	ssDNA	NZW	[334]
—	11	94	D11Mit70	(MRL-lprxBa)F2	dsDNA/ssDNA/CL	MRL	[346]
Nbwa1	12		D12Mit291	(BxBa.H2^z)F2	ANA/GN/gp70IC/RBC	NZB	[349]
Asbb2	12	25	D12Mit12	(B6xBa)F2-FcγRIIb$^{-/-}$	ANA (FcγRIIb ko)	B6	[360]
Lrdm2	12	27 cM	D12Nyu3	(MRL-lprxCAST)F1xMRL-lpr	GN	MRL	[358]
Sta-1	12	30	D12Mit85	(NZBxNZW)F1xNZW	CD4 T cell activation	NZB	[363]
Bxs6	13	64	D13Mit253	BXSBx(B10xBXSB)F1 and B10x(B10xBXSB)F1	gp70/gp70IC	BXSB	[361]
Spg3 (Yaa1)	13	56+12	D13Mit250	B6x(WxB6-Yaa)F1	gp70IC	NZW	[364]
Nba6	13	92	D5Mit353	(SWRxB)F2	dsDNA	NZB	[345]
—	13	104	D13Mit226	(B10.A^{z}xB)F1xB	gp70IC/GN	B10	[362]
—	13	115	D13Mit31	(NZMxB6)xNZM	dsDNA	NZM	[337]
—	13	115	D13Mit150	(BxSM)xW	GN	NZB	[342]
—	14	19.5 cM	D14Nds4	(WxBa)F1xW	histone	NZW	[334]
Swrl2	14	63	D14Mit37	Bx(SWRxB)F1	GN/dsDNA	SWR	[336]
—	14	72	D14Mit34	((B6.H2z & Ba.H2z)xB)F1xB	GN	NZB	[344]
Myo2	14	73	D14Mit68	Wx(WxBXSB)F1	MI	BXSB	[351]
Lprm3	14	88	D14Mit195	MRL-lprx(MRL-lprxC3H-lpr)F1	GN (resistance)	MRL	[348]
Paam1	15	32	D15Mit111	MRL-lprx(MRL-lprxC3H-lpr)F1	arthritis in males	MRL	[365]
Lprm5	16	29	D16Mit3	MRL-lprx(MRL-lprxC3H-lpr)F1	dsDNA	MRL	[348]
Bah2	16	32	D16Mit58	(NODxBa)xNODBC	RBC (M.bovis)	BALB/c	[330]

Locus	Chr	cM	Marker	Cross	Phenotype	Strain	Ref
nwa1	16	38 cM	D16Mit5	(WxBa)F1xW (BxW)F1xW	Histone GN/dsDNA	NZW	[334] [359]
Asbb3	17	28	D17Mit198	(B6xBa)F2-FcγRIIb⁻/⁻	ANA/spleen (FcγRIIb ko)	BALB/c	[360]
Acla1	17	34	D17Mit16	Wx(WxBXSB)F1	CL	NZW/BXSB	[351]
Sles1	17	35	H2/D17Mit34	(B6.NZMc1xW)F1xW	GN/dsDNA (resistance)	NZW	[37]
Bana1/Bah1	17	38	D17Mit24	(NODxBa)xNODBC	ANA/RBC (M.bovis)	NOD	[330]
Pbat1	17	19 cM	D17Nds2	Wx(WxBXSB)F1	platelet	NZW/BXSB	[351]
Wbw2	17	49	D17Mit177	(WxPL)F1xB	mortality/GN	NZW	[347]
Agnz2	17	88	D17Mit130	(NZM2328xC57L)F1xNZM2328	acute GN	C57L	[340]
Swrl3	18	40	D18Mit17	Bx(SWRxB)F1	dsDNA/histone	SWR	[336]
—	18	41	D18Mit227	MRL-lprx(MRL-lprxC3H-lpr)F1	sialadenitis	MRL	[335]
Lbw6	18	75	D18Mit8	BWF2	mortality/GN	NZW	[341]
nwa2	19	42	D19Mit11	(WxBa)F1xW	ssDNA	NZW	[334]
—	19	50	D19Mit3	(NZMxB6)xNZM	dsDNA	NZM	[337]

#complex inheritance: either parental strain promotes GN, but heterozygosity protects.

This table includes only named loci with linkages $p < 0.01$ or lod > 1.9. Loci are listed by their approximate chromosomal locations based on the marker with the highest association. Chr = chromosome. Mb and cM distances are based on the Mouse Genome Informatics (Jackson Laboratory). Abbreviations for mouse strains (Cross column): B = NZB, B6 = C57BL/6, B10 = C57BL/10, Ba = BALB/c, CAST = CAST/Ei, lpr = Fas^{lpr}, NOD = NOD/Lt, NZM = NZM/Aeg2410, PL = PL/J, W = NZW, (MRL-lprxC3H-lpr)BC & F2=both MRL-lprx(MRL-lprxC3H-lpr)F1 & (MRL-lprxC3H-lpr)F2 crosses, (NODxBa)xNODBC = NOD backcrossed to (NODxBALB/c)F1 in all four combinations. Original phenotypes that mapped to loci are shown: chr = anti-chromatin autoantibody, dsDNA = anti-dsDNA autoantibody, glom. dep. = glomerular IgG deposits, GN = glomerulonephritis, gp70IC = gp70 immune complexes, histone = anti-histone autoantibody, LN = lymphadenopathy, Lprn = lymphoproliferation, MI = myocardial infarct, platelet = antiplatelet autoAb and thrombocytopenia, RBC = antiRBC autoAb, spleen = splenomegaly. Autoantibodies are IgG unless otherwise specified. Table does not include gp70 loci that were not linked to autoimmunity [366]. Induced or genetically modified models are indicated under phenotype in parentheses: M. bovis = M. bovis i.v., HgIA = mercury-induced autoimmunity.
* formerly named elp-1.

I. BASIS OF DISEASE PATHOGENESIS

discernable phenotype to the presence of multiple additional effects. Thus, the contribution of variants is often highly dependent on other genes and on some backgrounds may be difficult to detect. (2) The association of QTL to lupus traits is complicated by additive and epistatic interactions. This has been shown using double and triple congenic mice as well as by backcrossing QTL onto more than one background. (3) A single locus when defined in more detail often consists of a cluster of loci. This has been documented for most of the QTL examined with subcongenics, including *Sle1*, *Nba2*, *Sle2*, *Lbw2*, *Sle3*, *Lmb3*, and *Lbw5*. (4) QTLs can have highly specific and unexpected effects on disease pathogenesis. For example, NZM2328 congenic mice (NZM2328.C57Lc4) that have the non-susceptible C57L interval replacing the NZM2328 *Adnz1* locus on chromosome 4, develop selective loss of anti-DNA antibodies, but no reduction in the severity of glomerulonephritis [17]. (5) In terms of identifying the responsible genes, several highly attractive candidates have been subsequently excluded or reduced in significance once subcongenic mice were examined. This demonstrates the limitation of identifying potential genes by candidate or expression screening alone and the necessity of using smaller-interval congenics.

Although a detailed description of the various QTL and related congenics is beyond the scope of this chapter, *Sle16*, a 129-derived chromosome 1 locus, is notable because of its implication for genetically engineered mice that develop lupus-like disease [29] (Table 4.2). The reason for generating this congenic line came from the finding that lupus-like manifestations associated with homozygous SAP knockout alleles (on chromosome 1) appeared to depend on the mixed 129xB6 background and not SAP deficiency [29, 30]. Strikingly, mapping studies showed that lupus-like disease including autoantibodies and immune complex GN developed in (129xB6) hybrids, and autoantibody-predisposing loci were associated with chromosomes 1 (*Sle16*, 129-derived), 3 (*Sle18*, B6-derived), and 4 (B6-derived) and this was confirmed in B6.*Sle16* and 129.*Sle18* congenic mice [24, 28, 31, 32]. *Sle16* overlaps with *Sle1*, *Nba2*, *Bxs3*, and *Cgnz1*, and may be caused by some of the same genetic variant(s). Thus, because gene knockout mice are commonly analyzed in 129xB6 mixed lines, lupus manifestations might be completely or partially due to this background (Table 4.4).

Several susceptibility or disease-suppressing genes have been identified through mapping and cloning of variants within lupus-related chromosomal intervals (Table 4.3). These include genes involved in apoptosis (*Fas*, *Fasl*, *Ifi202*), T- and/or B-cell activation (*Ly108*, *Fcgr2b*, *Cr2*), actin dynamics (*Coro1a*), TLR-mediated cell activation (*TLR7*, *Yaa*), and antigen presentation (*H-2*). Within the H-2 complex there are several other potential candidate genes, including TNF [33, 34], complement components C2 and C4 [35], IEX-1 [36] and a recessive NZW locus (*Sles1*) that appears to suppress autoimmunity in NZW mice [37]. Several other possible candidates have also been identified. Among these are CD22 [38], C1q [39], other SLAM family genes [15], CD72 [40], P2X$_7$ receptor [41], *Baff* [42] and *Marco* [43].

SYSTEMIC AUTOIMMUNITY IN NORMAL BACKGROUND GENE KNOCKOUT/MUTATED AND TRANSGENIC MICE

The role of specific genes in the immune system and possible mechanisms of systemic autoimmunity are being defined through genetic manipulation of non-autoimmune background strains. Deletion or over-expression of single genes results in remarkable examples of tolerance loss and systemic autoimmunity that have yielded valuable new models to investigate SLE immunopathogenesis. Thus far, there are nearly 100 lupus-modifying genes reported in the literature, but some of these may be due, as noted previously, to the use of mixed background (B6x129). Nevertheless, the differing autoimmune phenotypes produced by the specific genetic changes and the finding of a high frequency of autoimmunity in ENU mutagenized mice wherein a remarkably high 1 in 7 pedigrees develop ANA positivity [44] suggests the plausibility of the large number of potential susceptibility genes. Although the relevance of many of these models to spontaneous disease is uncertain, they have been particularly informative in dissecting molecular studies of potential mechanisms of autoimmunity. Thus far, gene defects have been shown to enhance B- or T-cell activation, expand DCs, inhibit certain apoptotic pathways, alter antigen presentation, reduce clearance of apoptotic bodies or soluble self-antigens, modify cytokine milieu, alter cell signal transduction, reduce glycosylation, and enhance cell cycling, as well as other mechanisms (Tables 4.4 and 4.5). Common mechanisms derived independently from different molecular defects have also emerged. Individual genes organized by the most likely mechanism will be discussed briefly below with additional details provided in Table 4.4.

B-cell Activation Genes

The fate of B cells following antigen receptor (BCR) engagement is a complex process that involves direct or indirect interaction of the BCR with numerous molecules that can promote or inhibit cell activation. Among these are several tyrosine kinases (lyn, fyn, Btk, Blk, Syk), phosphatases (CD45, SHP-1, SHP-2, and SHIP)

TABLE 4.3   Spontaneous lupus susceptibility or resistant genes

Gene	Chr	Mb	Locus or Allele	Strain*	Type of change	Gene function	Trait In congenic mice	Ref.
Ifi202	1	95.3	Nba2	NZB/B1	Incr. expression	Unknown; anti-apoptotic, cytoplasmic DNA sensor; proliferation	anti-nuclear Ab	[367, 368]
Fasl	1	163.7	gld	C3H/He	Loss-of-function	Pro-apoptotic	lymphoproliferation, DN T cells, autoAbs, GN	[369]
Ly108 (Slamf6)	1	173.8	Sle1b	NZW/Lac	Enhanced function	Co-stimulation	anti-nuclear Ab, impaired B cell tolerance	[370]
Fcgr2b	1	172.9		NZB/B1, NZW/Lac	Loss-of-function	Inhibitory signal: B cells, DC, macrophage	autoAb	[371, 372]
Cr2	1	197.0	Sle1c	NZW/Lac	Loss-of-function	C3 fragment binding; activation B cells	anti-nuclear Ab	[373]
Coro1a	7	133.8	Lmb3	B6-Fas^{lpr}/Scr substrain	Loss-of-function	Actin dynamics	Impaired migration, TCR-mediated activation, and survival of T cells	[374]
H-2	17	34-36	H2	NZW, BXSB	Variant	Antigen presentation	modifies severity of autoimmunity	
Fas	19	34.4	Lpr, Lpr^{cg}	MRL/Mp- Fas^{lpr}	Loss-of-function	Pro-apoptotic	lymphoproliferation, DN T cells, autoAbs, GN	[375]
Tlr7	Y	X-chr dupl.	Yaa	BXSB; SB/Le	Gain-of-function	Activation (ssRNA) B cells, DC	accelerated autoimmunity, enhanced Ab responses to foreign and self-antigens	[376, 377]

* Original strain.

I. BASIS OF DISEASE PATHOGENESIS

TABLE 4.4 Genes associated with lupus-like manifestations in knockout/mutated and transgenic normal background mice

Name	Gene	Chr*	Mb	Major autoimmune manifestations	Ref.
KNOCKOUT/MUTATED					
CTLA-4	*Cd152*	1	60.9	multiorgan lymphoproliferative disease, myocarditis, pancreatitis (mixed 129xB6; 129xBALB/c)	[78-80]
PD-1 (programmed cell death 1)	*Pdcd1*	1	95.9	proliferative arthritis, GN, glomerular IgG3 deposits. (129xB6N11)	[85]
CD45 (protein tyrosine phosphatase, receptor type C	*Ptprc*	1	140.0	lymphoproliferation, dsDNA, splenomegaly, GN (mixed 129xB6) no autoimmunity (B6)	[378]
Ro, SS-A (TROVE domain family, member 2)	*Trove2*	1	145.6	anti-ribosome and anti-chromatin autoAb, GN (129xB6)	[174]
Fasl (spontaneous)	*Fasl*	1	163.7	lymphoproliferation, DN T cells, autoAbs, GN (gld mutation)	[379]
serum amyloid P component	*Apcs*	1	174.8	anti-chromatin Ab, GN, female predominance (129xB6) no autoimmunity (B6)	[177]
mannoside acetyl glucosaminyltransferase 5	*Mgat5*	1	129.1	proliferative GN, enhanced EAE (129)	[114]
roquin (RING CCCH (C3H) domains 1)	*Rc3h1*	1	162.8	autoAbs, lupus-like disease, incr. follicular T helper cells and GCs. M199R mutation. (B6)	[112]
FcγRIIb (Fc receptor, IgG, low affinity IIb)	*Fcgr2b*	1	172.9	exacerbates autoimmunity in B6-*Fas*[lpr] mice, autoAb, GN, arthritis	[380]. Does this fit here
IL-2Rα	*Il2ra*	2	11.6	lymphoproliferation, hyperIgG, autoAb, anti-RBC Ab (129xB6)	[138]
Nrf2 (nuclear, factor, erythroid derived 2, like 2)	*Nfe2l2*	2	75.5	hyperIgG, anti-dsDNA Ab, GN, splenomegaly (129/B6/ICR); (no disease in MRL-lpr Nfe2l2[-/-], see below)	[207]
Ras GRP1 (RAS guanyl releasing protein 1)	*Rasgrp1*	2	117.1	spont. recessive mutation prevents translation of Ras GRP1 protein, CD4[+] T cells resistant to AICD, lymphoprolif., autoAb (129xB6)	[150]
TYRO3 protein tyrosine kinase 3 (Tyro 3 family)	*Tyro3*	2	119.6	triple knockout (*Tyro3*, *Axl*, *Mer*): lymphoproliferation, increased activated T and B cells, autoAb, GN (129xB6)	[168]
c-mer proto-oncogene (Tyro 3 family)	*Mertk*	2	128.5	autoAb (*Mer* knockout alone), (also see TYRO3 above) (129xB6N10)	[170]
Bim (Bcl2-like 11)	*Bcl2l11*	2	128.0	lymphoid/myeloid cell accumulation, autoAb, GN, vasculitis (129xB6)	[129]
IL-2	*Il2*	3	37.0	lymphoproliferation, hyperIgG, autoAb, anti-RBC Ab (129xB6)	[137]
TSAd (SH2 domain protein 2A)	*Sh2d2a*	3	87.7	hyperIgG, autoAbs, GN (129xB6N9)	[381]
Shc1 p66 isoform (p66, ShcA)	*Shc1*	3	89.2	lymphoid hyperplasia, low penetrance autoAb and GN (129)	[110]

TABLE 4.4 Genes associated with lupus-like manifestations in knockout/mutated and transgenic normal background mice—cont'd

Name	Gene	Chr*	Mb	Major autoimmune manifestations	Ref.
lyn	*Lyn*	4	3.6	enhanced B cell activation, splenomegaly, hyperIgM, autoAb, GN (129xB6)	[45, 46]
	Lyn^{up/up}			gain of function (Y508F) mutation: autoAb, GN, reduced survival (B6x129 mixed background)	[47]
CD72	*Cd72*	4	43.5	autoAb, mild GN (129xB6>N7; by genome scan no residual 129 beyond D4Mit53)	[51]
Zinc finger CCCH type containing 12A	*Zc3h12a*	4	124.8	Lymphoid hyperplasia, early mortality, autoAb, anemia (prob. autoimmune), (mixed 129xB6)	[211]
E2F transcription factor 2	*E2f2*	4	135.7	enhanced T cell activation, autoAb, GN, widespread inflammatory infiltrates (129xB6)	[107]
C1q α, β, γ polypeptides (different genes)	*C1qa* *C1qb* *C1qc*	4 4 4	136.5 136.5 136.5	autoAb, GN (129xB6)	[153, 382, 383]
	C1q			Allele − down regulates C1q levels in NZB mice	[39]
GADD45 (growth arrest and DNA-damage-inducible 45 alpha)	*Gadd45a*	6	67.0	autoAb, GN, mortality (129xB6)	[121]
Docking protein 1	*Dok1*	6	83.0	*Dok1/Dok2* dko: autoAb, GN (see chr. 14) (129xB6>N8)	[111]
C-type lectin domain famly 4, member a2 (Dcir)	*Clec4a2*	6	123.1	auto Ab, inflammatory arthritis (129xB6N8)	[180]
SHP-1 (spontaneous)	*Ptpn6*	6	124.7	autoAb (*me* and *me^v* mutations)	[195, 196]
TGFβ1	*Tgfb1*	7	26..5	multiorgan lymphocytic and monocytic infiltrates (129xB6)	[77]
AXL receptor tyrosine kinase (Tyro3 family)	*Axl*	7	26.5	(see TYRO3 above) (129xB6)	
Zfp-36 (tristetraprolin)	*Zfp36*	7	29.2	complex systemic disease: cachexia, dermatitis, arthritis (129xB6)	[187, 188, 384]
CD22	*Cd22*	7	31.6	enhanced B cell activation, autoAb (129xB6)	[385, 386]
MFG-E8 (milk fat globule-EGF factor 8)	*Mfge8*	7	86.3	splenomegaly, incr. GCs, autoAbs, GN. reduced engulfment of apoptotic cells (129xB6)	[160]
LAT (linker for activation of T cells)	*Lat*	7	133.5	mutation inhibits T cell development, but induces Th2 cell lymphoproliferation and polyclonal B cell activation (129xB6)	[98]
PLCγ2	*Plcg2*	8	120.0	D993G (ENU) gain-of-function; arthritis (C3H); autoAb and GN (C3HxB6N2-3; wt littermates were normal)	[66]
Cbl ko (B cell only)	*Cbl*	9	44.0	(*Cbl/Cblb* dko; B cell conditional ko) autoAb, GN (mixed 129xB6)	[103]

(Continued)

TABLE 4.4   Genes associated with lupus-like manifestations in knockout/mutated and transgenic normal background mice—cont'd

Name	Gene	Chr*	Mb	Major autoimmune manifestations	Ref.
Three prime repair exonuclease 1	Trex1	9	109.0	Aicardi-Goutieres syndrome (AGS) and chilblain lupus; mice:	[171, 172]
PCMT (protein-L-isoaspartate (D-aspartate) O-methyltransferase 1	Pcmt1	10	7.3	wild-type mice reconstituted with PCMT(-/-) BM develop high titer anti-DNA Ab and GN (129xB6)	[118]
fyn (+lyn)	Fyn	10	39.1	synergizes with the Lyn ko to accelerate disease (129xB6)	[387]
Traf3 interacting protein 2 (Act1, CIKS)	Traf3ip2	10	39.3	AutoAb, GN, early mortality, Sjogren's syndrome (129xBALB/c); in another study no B cell hyperactivity or lupus-like disease (129xB6>N10; 129xBALB/cN2)	[60, 388, 389] [61]
TACI (tumor necrosis factor receptor superfamily, member 13b)	Tnfrsf13b	11	60.9	fatal lymphoproliferation, autoAb, GN (129xB6)	[390]
Gadd45β	Gadd45b	10	80.4	splenomegaly, glomerular immune complexes, no autoAbs; synergizes with Gadd45β$^{-/-}$ to produce marked splenomegaly, dsDNA/histone Abs, immune complex GN. (129xB6)	[122]
Aiolos (IKAROS family zinc finger 3)	Ikzf3	11	98.3	activated B cells, increased IgG, autoAb (129xB6)	[58]
PECAM-1/CD31	Pecam1	11	106.5	enhanced B cell activation, autoAb, GN (129xB6)	[73]
Stra13 (stimulated by retinoic acid 13)	Stra13	11	120.6	lymphoid organ hyperplasia, autoAb, IC GN (B6x129)	[151]
G protein-coupled receptor 132	Gpr132	12	114.1	lymphoid hyperplasia, hyperIgG, autoAb, GN (129xBALB/cN3-6)	[104]
Src homolgy 2 domain-containing transforming protein C3 (Rai)	Shc3	13	51.5	autoAb, glom. IC dep, mild GN (129xB6N12; 129)	[71]
CD100 (sema domain, immunoglobulin domain, transmembrane domain and short cytoplasmic domain 4D)	Sema4d	13	51.8	suppressed BCR signals, incr MZ B cells, autoAb, GN (129xB6 backcrossed >8 generations with B6)	[52]
Gadd45γ	Gadd45g	13	51.9	synergizes with Gadd45β knockout (chr 10). (129xB6)	[122]
Protein kinase Cδ	Prkcd	14	31.4	splenomegaly, lymphadenopathy, hyperIgM/IgG1/IgG2a, dsDNA, GN (129xB6)	[64]
Docking protein 2	Dok2	14	71.2	Dok1/Dok2 dko: autoAb, GN (see chr 6) (129xB6>N8)	[111]
IL-2Rβ (CD122)	Il2rb	15	78.3	lymphoproliferation, hyperIgG, autoAb, anti-RBC Ab (129xB6)	[139]
Dnase1	Dnase1	16	4.0	ANA, immune complex GN (129xB6)	[156]
SOCS-1 (suppressor of cytokine signaling 1)	Socs1	16	10.8	(+/ko), or ko with expr. in lymphocytes: hyperact. lymphocytes, autoAb, GN (129xB6 backcrossed >7 generations to B6)	[193]

TABLE 4.4 Genes associated with lupus-like manifestations in knockout/mutated and transgenic normal background mice—cont'd

Name	Gene	Chr*	Mb	Major autoimmune manifestations	Ref.
Surrogate light chain: (Vpreb1-λ5 and Vpreb2)	*Vpreb1*	16	16.9	Triple ko: anti-nuclear and cardiolipid autoAb (129xB6, backcrossed to B6 >10 generations, genotyped B6 at Chr 1)	[63]
	Igll1	16	16.9		
	Vpreb2	16	18.0		
Interferon (alpha and beta) receptor 1	*Ifnar1*	16	91.5	exacerbated disease in MRL-*lpr* mice: lymphoproliferation, autoAb, GN	[260]
Cbl-b ko Cbl-b (B cell only)	*Cblb*	16	52.0	multiorgan lymphoid infiltrates, anti-dsDNA Ab (*Cbl/Cblb* dko; B cell conditional ko) autoAb, GN (mixed 129xB6)	[102] [103]
IRF-4 binding protein (differentially expressed in FDCP6, IBP, SLAT)	*Def6*	17	28.3	hyperIgG, autoAb (mixed B6x129); not confirmed	[124, 125]
p21 cyclin-dependent kinase inhibitor 1A	*Cdkn1a*	17	29.2	anti-chromatin Ab, GN, female predominance minimal to no disease (129xB6)	[126] [127, 128]
Complement component 4	*C4*	17	34.9	impaired immune complex clearance, ANA, GN, female predominance (129xB6)	[155]
α-mannosidase II	*Man2a1*	17	65.0	hyperIgG, autoAb, GN (129xB6)	[202]
Signal-induced proliferation associated gene 1 (SPA-1)	*Sipa1*	19	5.7	autoAb, GN (129xB6>N10)	[72]
Emk (ELKL motif kinase, Par-1, MAP/microtubule affinity-regulating kinase)	*Mark2*	19	7.3	growth retardation, hypofertility, splenomegaly, lymphoid infiltrates, immune complex GN (129xB6)	[203]
Metastasis-associated gene family, member 2	*Mta2*	19	9.0	autoAb, GN (mixed B6/129)	[210]
Pten (+/- mice)	*Pten*	19	32.8	lymphadenopathy, autoAb, GN, decreased survival, female predominance	[146]
Fas (spontaneous)	*Fas*	19	34.4	lymphoproliferation, DN T cells, autoAbs, GN, arthritis (*lpr* and *lpr^cg* mutations)	[375]

TRANSGENIC

Name	Gene	Chr*	Mb	Major autoimmune manifestations	Ref.
Bcl-2 (B cell promoter)	*Bcl2*	1	59.8	lymphoid hyperplasia, hyperIgG, autoAb, GN	[131]
CD19	*Cd19*	7	133.6	increased B cell activation, B1 cell population, IgG, autoAb	[49]
BAFF (α1anti-trypsin or β-actin promoter, BLyS, TALL-1, THANK)	*Tnfsf13b*	8	10.0	autoAb (RF, CIC, dsDNA Ab), GN	[54, 55]
Fli-1 (class I promoter)	*Fli1*	9	32.2	lymphoid hyperplasia, autoAb, GN	[68]
IFN-γ (keratin promoter)	*Ifng*	10	117.9	autoAb, GN, female predominance	[179]
IL-4 (class I promoter)	*Il4*	11	53.4	hyperIgG1/IgE, autoAb, GN	[181]
miR-17-92 (CAG romoter/Rosa26 knockin, huCD2-iCre-dependent expression: T and B cells)	*Mir17-92*	14	115.4	Lymphoproliferation, less AICD, autoAb, GN	[74]
IEX-1 (Ig μ enhancer; immediate early response 3)	*Ier3*	17	36.0	lymphoproliferation, autoAb, GN, skin lesions, arthritis	[36]

(Continued)

TABLE 4.4   Genes associated with lupus-like manifestations in knockout/mutated and transgenic normal background mice—cont'd

Name	Gene	Chr*	Mb	Major autoimmune manifestations	Ref.
LIGHT (prox. lck promoter, hu CD2 locus control)	*Tnfsf14*	17	57.3	T cell hyperact., lymphoproliferation, autoAb, GN, lymphoid infiltrattion in peripheral organs	[94]
CD154/CD40L (B cell- or epidermis-specific promoter)	*Cd40lg*	X	54.5	B cell promoter: late onset autoAb, GN epidermis promoter: dermatitis, lymphadenopathy, hyperIgG, dsDNA, GN	[59] [178]
MISC Tg					
6-19 IgG3 anti-IgG2a RF cryoglobulin	*Igh*	12	NA	chronic lethal GN, necrotizing arteritis	[391]

* *ND = not determined.*

Genes are listed in order of their chromosomal locations. Gene names and chromosomal locations are from the Mouse Genome Informatics. Background strain is included in the Major Autoimmune Manifestations column (N = number of backcrosses to the indicated strain).

Abbreviations for autoimmune manifestations: Ab = antibodies, dsDNA = anti-dsDNA autoAb.

and accessory molecules (CD19, CD22, FcRγIIb, CD72). As might be expected, genetic manipulation of many of these B-cell regulatory molecules has resulted in systemic autoimmunity. These include Lyn [45—47], CD22 [48], SHP-1 [48], CD19 [49], FcγRIIb [50], CD72 [51] and its ligand CD100 [52], and CD45 [53].

Several genes affecting B-cell survival and maturation have also been implicated. Overexpression of *Tnfsf13b* (BAFF, BlyS) [54, 55] or deficiency of its inhibitory receptor, *Tnfrsf13b* (TACI), leads to the development of lupus-like disease. Moreover, elevations in *Tnfsf13b* are present in both BWF$_1$ and MRL-*Fas*^{lpr} mice, and blocking *Tnfsf13b* function with a soluble TACI-IgGFc fusion protein can inhibit proteinuria and prolong survival [56]. Importantly, *Tnfrsf13b* inhibition has shown promise in human SLE [413]. BAFF Tg-induced lupus-like disease does not require T cells, but needs MyD88, suggesting surprisingly that Baff-induced activation of B cells in this model is TLR, but not T-cell-dependent [57]. Another gene, *Aiolos*, a zinc finger transcription factor, is highly expressed in mature B cells and, to a lesser extent, in developing bone marrow B cells and thymocytes. Aiolos$^{-/-}$ mice develop defects primarily in the B-cell compartment, with hyperreponsiveness to BCR and CD40 stimulation and increased numbers of conventional B cells, a marked reduction in B1 cells, increased proportion of B cells with activated phenotype, hypergammaglobulinemia (particularly of IgE and IgG1), a three-fold reduction in IgM, and positive ANAs [58]. Likewise, ectopic transgenic expression of CD40L on B cells (normally expressed on T cells) results in enhanced polyclonal IgG, anti-dsDNA and, in about half of mice, the development of immune complex GN [59]. Mice deficient in Act1 (or CISK), a NF-κB-related adaptor protein that negatively regulates BAFF and CD40L, were also reported to develop B-cell

hyperactivity, autoantibody production, immune complex GN, Sjogren syndrome-like disease, and early mortality [60]. Another independent Act1 knockout, however, did not demonstrate B-cell hyperactivity or autoimmunity [61] indicating a significant role for background genetic factors.

Deletion of the surrogate light-chain (SLC, encoded by the VpreB1/2 and λ5 genes), which is thought to play an important role in the maturation and selection of pre-B cells, results in reduced B cell numbers and serum Ig levels [62]. SLC$^{-/-}$ mice, however, have elevated levels of ANA and higher proportion of potentially autoreactive Abs [63]. Thus, it was proposed that the pre-BCR mediates negative selection of self-reactive B cells during early stages of B-cell development.

Protein kinase Cδ (PKC-δ) is highly expressed in developing pro- and pre-B cells, and mediates BCR signaling. PKC-δ$^{-/-}$ mice develop splenomegaly, lymphadenopathy, increased numbers of B2 cells, germinal center formation and IL-6 production, mild increases of IgM, IgG1 and Ig2a, anti-nuclear and IgG1 anti-DNA autoantibodies, as well as GN [64, 65]. BCR-mediated B-cell activation *in vitro* and *in vivo* was not affected, and PKC-δ was postulated to play a role in tolerogenic, but not immunogenic, B-cell responses.

Phospholipase (PL) Cγ2 is a member of the PLC family primarily expressed in B cells that catalyzes phosphoinositides to diacylglycerol and inositol phosphates. A gain-of-function PLCγ2 mutant (D993G, Ali5 mutation) generated by ENU mutagenesis developed lymphocyte-independent inflammatory arthritis in the C3H strain (ENU-treated strain), but anti-DNA and GN in B6xC3H mice [66]. Lupus-like disease occurred in both homozygous and heterozygous mutants, but not in littermates, which strongly suggests that the PLCγ2 mutation and not background genes are responsible.

TABLE 4.5 Mechanisms for induction of systemic autoimmunity

ENHANCED B-CELL ACTIVATION*

Lyn, CD22 or SHP-1 knockout

FcγRIIb knockout

Aiolos knockout

PKCδ knockout (tolerance pathway)

CD19 transgenic

Tnfsf13b (BAFF, BlyS, TALL-1, or THANK) transgenic

Fli-1 transgenic

CD40L transgenic (B cell-specific expression)

PECAM-1/CD31

TACI knockout

CD45 E613R knockin mutation

PLCγ2 D993G (ENU) gain-of-function B cells and innate inflammatory cells

Shc2 (Rai) knockout (also enhanced T-cell activation and increased survival) PI3K/AKT pathway

CD72 knockout (also enhances survival)

Shc1 knockout, also enhances survival and affects T cells; (AgR signaling adapter; also localizes to mitochondria-unknown function)

miR-17-92 transgenic (B and T cells)

REDUCED B-CELL ACTIVATION

CD100 (incr. MZ B cells)

DEFECTIVE NEGATIVE SELECTION OF B CELLS (CENTRAL TOLERANCE DEFECT)

Surrogate light chain (Vpreb1-λ5, Vpreb2) double (3 gene) knockout

ENHANCED B CELL SURVIVAL

BAFF transgenic (ADD)

Act1 knockout (negative regulator of BAFF, CD40L)

ENCHANCED T-CELL ACTIVATION

CTLA-4 knockout

IL-2 or IL-2R knockout

CD45 E613R knockin mutation (B cells as well)

PD-1 knockout (probably B cells as well)

TGF-β deficiency (knockout/dominant negative)

Cbl-b knockout

Gadd45a knockout

Gadd45β knockout (with and without Gadd45γ knockout)

LAT Y136F knockin mutation (Th2 cell lymphoproliferation; triggers polyclonal B cell activation)

G Protein-Coupled Receptor G2A knockout

LIGHT (*Tnfrsf14*) transgenic (T cell-specific promoter)

E2F2 knockout

p21$^{cip1/waf1}$ knockout (pancyclin kinase inhibitor, primarily T cells)

Mgat5 knockout (enhanced T-cell activation)

Roquin M199R mutation

PCMT knockout

DEFECTIVE APOPTOSIS

Fas or FasL mutations (lpr and gld mice, ALPS in humans; also caspase 10)

Bim knockout (not related to Fas)

Bcl-2 transgenic

IEX-1 transgenic

TSAd knockout (primarily T cells)

IL-2 or IL-2R knockout (primarily T cells)

Pten$^{+/-}$ knockout

RasGRP1 mutation (loss-of-function RNA splicing defect)

Stra13 knockout

IRF-4 binding protein (IBP)

DEFECTIVE CLEARANCE OF PROINFLAMMATORY/ IMMUNOSTIMULATORY APOPTOTIC MATERIAL

C1q knockout

C4 knockout

DNase I knockout

Tyro3 family (Tyro3, Axl, Mer) triple-gene knockout

Ro (Trove2) knockout

MFG-E8 knockout

Trex1 knockout

ENHANCED ANTIGEN PRESENTATION

CD40L transgenic (keratin-14 promoter)

SAP knockout (MAYBE)

PLCγ2 D993G (ENU) gain-of-function B cells and innate inflammatory cells (also listed for B cells)

EXPANDED DC POPULATION

Dcir (*Clec4a2*) knockout

CYTOKINE-MEDIATED ACTIVATION

IL-4 transgenic

IFN-γ transgenic

TTP (Zfp-36) deficiency (excessive TNFα)

TNFα transgenic

STAT-4 knockout

Zc3h12a knokcout (RNase) incr. IL-6, IL-12p40, but likely also other proinflammatory agents

DEFECTIVE SIGNAL TRANSDUCTION

Cblb knockout (enhanced activation of T cells, possibly B cells)

Cbl/Cblb double knockout

$Pten^{+/-}$ knockout (defective Fas from increased PIP-3 elevating Akt levels)

SOCS-1 knockout (CD4 T cell dependent; excess cytokines, impaired Treg)

OTHER MECHANISMS

α-mannosidase II knockout

Emk knockout

Nrf2 knockout (anti-oxidant)

Mta2 knockout (expression in non-T cells important) part of NuRD (nucleosome remodeling histone deacetylase) complex

* Genes are categorized according to the most likely or predominant mechanism.

The D993G mutation was associated with increased and sustained IgM-induced Ca^{2+} flux in B cells, enhanced B-cell proliferation, and an increased proportion of marginal zone B cells.

Fli-1 is an Ets transcription factor family member that binds the consensus GGA(A/T) motif [67]. Overexpression of a Fli-1 resulted in lymphoid hyperplasia, hyper-gammaglobulinemia, elevated antinuclear antibodies, and severe immune complex GN associated with hyper-responsive and apoptosis-resistant B cells [68]. Fli-1 expression is also increased in murine SLE and heterozygous $Fli-1^{+/-}$ deletion in MRL-Fas^{lpr} mice (homozygous Fli-1 is lethal) showing reduced serum IgG, autoantibodies, and GN as well as improved survival [69, 70], which further supports the possible importance of Fli-1 in lupus.

Rai (Shc3), an adaptor protein, enhances survival by activating the PI3K/Akt pathway and, in B and T cells, negatively regulates antigen-receptor signaling and cell activation. $Rai^{-/-}$ mice develop lymphoid organ hyperplasia, anti-dsDNA, and mild immune complex GN [71].

SPA-1 is a Rap1 GTPase-activating protein, which inhibits the activity of Rap1 in controlling cell adhesion and MAP kinases. Deletion of SPA-1 leads to inefficient receptor editing in B cells, expansion of the B1a population, and the development of anti-dsDNA, anti-RBC, IgM and C3 glomerular deposits, and GN [72].

Platelet endothelial cell adhesion molecule-1 (PECAM-1/CD31), an Ig superfamily member with two ITIM domains, is expressed on endothelial and hematopoietic cells. $PECAM-1^{-/-}$ mice developed ANA by 9 months of age and immune complex GN by 17 months associated with enhanced B-cell activation, reduced B2, but increased B1a cells, and a block in immature to mature B-cell transition [73].

MicroRNAs (miRNA) are short single-stranded regulatory RNA molecules (~22 nucleotides) that typically pair with a large number of gene-coding mRNAs leading to their degradation or translational repression.

The miR-17-92 cluster, the precursor to seven miRNA molecules, is amplified and overexpressed in certain cancers including lymphomas. Of considerable interest, targeted overexpression of this cluster in lymphocytes resulted in lymphoproliferation, lymphoid organ hyperplasia, elevated Ig, anti-dsDNA, and immune complex GN [74]. Enhanced proliferation and reduced activation-induced apoptosis were observed in transgenic B and T cells, which was attributed at least in part to transgene-mediated reduction in PTEN and Bim expression. Thus, these and other findings suggest that alterations in miRNA expression might affect prediposition to lupus [414].

T-cell Activation Genes

Similar to B-cells, alteration in genes that primarily affect T-cell function have also resulted in systemic autoimmunity (Tables 4.4 and 4.5). These include genes that affect activation, proliferation, and survival. TGF-β1 knockout mice rapidly develop massive necrotizing lymphocytic and monocytic infiltrates in multiple organs and die by 3 weeks of age [75, 76]. Serum IgG autoantibodies to nuclear antigens as well as Ig glomerular deposits are detectable, but they appear to play a minor role in overall disease severity [77].

CTLA-4, a surface glycoprotein expressed exclusively on T cells, acts as an inhibitor of the CD28-B7.1/B7.2 costimulatory pathway in part by binding with higher affinity to B7.1 and B7.2. Deficient mice develop a multiorgan lymphoproliferative disease associated with increased frequency of activated B and T cells, hypergammaglobulinemia and early mortality at 3–4 weeks of age, accompanied by severe myocarditis and pancreatitis [78–80].

PD-1, a 55-kDa ITIM-containing membrane glycoprotein expressed on activated T and B lymphocytes and monocytic cells, is necessary for maintaining homeostasis of lymphocytes and myeloid cells following activation [81, 82]. Engagement of PD-1 by its ligands, PD-L1 and PD-L2, inhibits TCR-mediated lymphocyte proliferation and cytokine secretion [83]. Mice deficient for PD-1 develop moderate hyperplasia of lymphoid and myeloid cells, increases in several Ig isotypes (particularly IgG3), enhanced B-cell responses, and alterations in peritoneal B1 cells [84]. Older B6-PD-1$^{-/-}$ mice spontaneously develop mild GN and proliferative arthritis, but not elevated anti-dsDNA antibodies or rheumatoid factor [85]. Acceleration of GN and arthritis occurs when the PD-1 deletion is combined with the Fas^{lpr} mutation. By contrast, BALB/c mice-PD-1$^{-/-}$ mice develop anti-cardiac troponin I antibodies and cardiomyopathy [86]. SLE has been linked in one population group to a regulatory polymorphism in the PD-1 gene (*PDCD1*) [87] and in other studies to other autoimmune diseases [88–91]. Deficiency of its ligand, PD-L1, also promotes autoimmunity or inflammation depending on the strain background [92, 93].

LIGHT (*Tnfsf14*), a TNF superfamily member expressed transiently on activated T cells and on immature dendritic cells, is a co-stimulatory molecule for T-cell activation [94]. Overexpression of LIGHT on T cells resulted in peripheral lymphoid organ hyperplasia, activation and expansion of mature T cells, increased T-cell cytokine production, including IFN-α and IL-4, as well as anti-DNA antibodies, RF, immune complex GN, and inflammation of the intestines and skin [94]. In contrast, another LIGHT transgenic driven by the CD2 promoter had some of the above manifestations, but no lupus-like manifestations [95], possibly because of the different promoters [96] or background differences.

LAT, a transmembrane scaffolding protein, is tyrosine-phosphorylated following TCR engagement and then recruits multiple signaling molecules critical for T-cell activation [97]. Mutation of the Tyr^{136} position to phenylalanine (Y136F) leads to a severe, but incomplete, block in T-cell development, but paradoxically by 4 weeks of age, lymphoproliferation with activated CD4$^+$ T cells, B cells, macrophages and eosinophils, lymphocytic infiltrates in various organs, high production of IL-4, hypergammaglobulinemia, and autoantibodies to nuclear antigens [98, 99]. This was attributed to defective PLCγ1-mediated calcium signaling in early T-cell development, leading to inefficient negative selection of self-reactive T cells and their exportation to the periphery wherein activation of these cells may not depend on LAT or PLC-γ1 [98].

The cbl-b and cbl adaptor proteins are E3 ubiquitin ligases that inhibit receptor and non-receptor tyrosine kinases by promoting ubiquitination [100]. Cbl-b-deficient T cells exhibit enhanced proliferation to antigen receptor signaling [101, 102]. Cbl-b$^{-/-}$ mice have increased susceptibility to experimental autoimmune encephalomyelitis [101] and develop a generalized autoimmune disease characterized by multiorgan lymphoaccumulation with parenchymal damage, increased plasma cells, and anti-dsDNA by 6 months of age [102]. Spontaneous autoimmunity, however, occurred in only one [102] of two Cbl-b knockout studies, suggesting that background strain, environment or other factors are also important factors. Mice with conditional *Cbl*/ *Cblb* double knockout (dko) in B cells have impaired BCR downmodulation and anergy to self-antigen, and develop spontaneous lupus-like disease with anti-dsDNA, ANA, massive leukocytic infiltrates in multiple organs, and immune complex GN [103].

G2A is a G protein-coupled receptor expressed in various tissues that appears to negatively regulate proliferation and to integrate extracellular signals with cytoskeletal reorganization [104]. G2A$^{-/-}$ mice showed a normal pattern of T and B lineage differentiation, but with age, develop enlarged secondary lymphoid organs, T- and B-cell expansion, enhanced T-cell proliferation responses and, when over 1 year of age, a progressive wasting syndrome, lymphocytic infiltration into various tissues, hypergammaglobulinemia, immune complex GN, and anti-nuclear autoantibodies [104]. A lower threshold for activation coupled with defective Fas-mediated apoptosis has also been considered to be a major contributor to the MRL-*lpr* lymphadenopathy and spontaneous lupus-like disease [105].

E2F2 is a member of the E2F family of DNA-binding heterodimers sequestered by Rb and released after Rb phosphorylation by activated cyclin/CDK kinases [106]. Mice lacking E2F2 developed late-onset autoimmunity consisting of perivascular inflammatory infiltrates in multiple organs, splenomegaly, skin lesions, anti-dsDNA autoAbs, and GN associated with increases in effector/memory T cells [107]. The mechanism appeared to be lowering of the TCR activation threshold and more rapid entry of activated T cells into S phase.

The docking protein Shc1 (ShcA) encodes three isoforms p46, p52, and p66 that have multiple functions in diverse tissues, including tissue morphogenesis and oxidative stress response [108, 109]. Mice deficient for the p66 isoform, although living longer, developed with age spontaneous activation of T and B cells, enlarged spleens, elevated IgG, anti-dsDNA, and mild immune complex GN [110]. Evidence suggests that p66 may function in suppressing antigen-receptor signaling in both T and B lymphocytes. The adaptor proteins, Dok-1 and Dok-2, expressed in hematopoietic cells, negatively regulate the Ras-Erk pathway, and suppress TCR signaling by inhibiting ZAP-70. Dok-1/Dok-1 dko mice develop anti-dsDNA and diffuse proliferative GN with age [111].

Roquin (*Rc3h1*) is a 1130-residue ubiquitously expressed member of the E3 ubiquitin ligase family based on the presence of a highly conserved amino terminal RING-1 zinc finger domain [112]. A loss-of-function M199R mutation within the conserved ROQ domain of Roquin was generated by ENU mutagenesis (*sanroque* mouse strain) that resulted in ANA, anti-dsDNA autoantibody, GN, necrotizing hepatitis, anemia, and immune thrombocytopenia [112]. Lymphoid organ hyperplasia, polyclonal hypergammaglobulinemia, increased germinal centers, and expansion of memory/ effector CD4$^+$ T cells, particularly follicular helper T (T$_{FH}$) cells, were observed, along with overexpression of ICOS and IL-21 [415]. Further studies indicated that systemic autoimmunity was in large part mediated by the *sanroque*-associated accumulation of T$_{FH}$ cells and the resulting enhancement of GC formation [113].

Expansion of circulatory T_{FH}-like cells was also detected in human SLE and associated with severe disease. The presence of a CCCH zinc finger common to several RNA-binding proteins and localization of Roquin to cytosolic RNA granules suggested that Roquin regulates mRNA translation and stability, and likely acts as a repressor of ICOS [112].

Mice deficient for Mgat5 (β1,6 N-acetylglucosaminyl-transferase V), an enzyme in the N-glycosylation pathway, develop proliferative GN suggestive of an autoimmune-mediated disease, however, autoantibodies were not examined [114]. Absence of this enzyme lowered the T-cell activation threshold by enhancing TCR clustering. This was thought to be related to the known function of Mgat5 in initiating GlcNAc β1,6 branching on N-glycans and increasing N-acetylactosamine, the ligand for galectins, which are known to modulate T-cell proliferation and apoptosis [114–116]. The findings indicate that a galectin-glycoprotein lattice strengthened by Mgat5-modified glycans restricts TCR recruitment to the site of antigen presentation, and therefore, dysregulation of Mgat5 increases T-cell activation and susceptibility to autoimmunity.

Protein carboxyl methyltransferase (PCMT) is a highly conserved enzyme that repairs isomerized and racemized derivatives of L-asparaginyl and L-aspartyl residues produced in cells by spontaneous degradation [117]. $PCMT^{-/-}$ mice accumulate significant amounts of aspartyl derivatives in brain, heart, liver and RBCs, have growth retardation, and develop fatal seizures around 42 days of age [117]. T cells from PCMT knockout mice are hyper-responsive to stimulation, while B-cell responses are unaffected [118]. Mice reconstituted with PCMT-deficient bone marrow develop dsDNA autoantibodies 7–9 months after transfer, but no apparent immune complex GN.

The Gadd45 (growth arrest and DNA damage-inducible gene) family is composed of three members, α, β and γ, that play pivotal roles in replication, growth arrest, and apoptosis [119, 120]. Gadd45α is a negative regulator of T-cell proliferation induced by antigen receptor-mediated activation, and deletion leads to the development of a lupus-like syndrome, particularly when coupled with deletion of the p21 cyclin-kinase inhibitor [121]. More recently, it was found that Gadd45β deficiency is also associated with enhanced T-cell proliferation and resistance to activation-induced cell death (ACID), splenomegay, and mild immune complex glomerular deposits [122]. Gadd45$\beta^{-/-}$ mice are also more susceptible to EAE. The related Gadd45γ also regulates proliferation and death of $CD4^+$ T cells, however, Gadd45γ-deficiency is not associated with autoimmunity [122]. The addition of the Gadd45γ knockout mutation to Gadd45$\beta^{-/-}$ mice, however, results in significant worsening of all lupus-like disease parameters suggesting that Gadd45γ deficiency may promote autoimmunity in certain susceptible backgrounds.

IRF-4-binding protein (*Def6*) is a Cdc42/Rac-1 guanine nucleotide exchange factor predominantly expressed in T cells where, following TCR engagement, it is recruited to the immune synapse and acts to promote optimal Ca^{2+} signaling and activation [123]. *Def6*-deficient mice on a mixed 129xB6 background were reported to develop lupus-like disease with increases in effector/memory T cells and memory B cells, hypergammaglobulinemia, and autoantibodies [124]. This, however, was not confirmed in B6-*Def6*$^{-/-}$ mice, suggesting that the development of lupus may be related to the 129xB6 background and not *Def6* deficiency [125]. Similarly, gene knockout of the cyclin inhibitor p21$^{cip1/waf1}$ in mixed background C57BL/6 x 129/Sv mice was also reported to result in systemic autoimmunity characterized by lymphoid hyperplasia, elevated IgG1, IgG antinuclear antibodies, GN and early mortality [126], however, other studies of mixed B6/129 or BXSB female p21-deficient mice were not able to confirm the development of lupus suggesting that background effects may have been responsible for the initial findings [127, 128].

Apoptosis Genes

Altered expression of Bcl2 family members is associated with the development of lupus. Deficiency of Bim, a proapoptotic BH3 member, results in an incomplete-penetrant embryonic-lethal phenotype, but in the surviving offspring, however, alterations in the homeostasis of multiple hematopoietic cell lineages develop with lymphoid hyperplasia, thymus defects, granulocytosis and monocytosis, and, with age, systemic autoimmunity manifested a marked expansion of plasma cells, hyper IgM, IgG and IgA, antinuclear antibodies, immune complex GN, vasculitis and reduced survival [129]. Combined deficiency of Bim(Bcl2l11) and Fas results in accelerated lupus associated with enhanced activation of antigen-presenting cells [130]. Similarly, transgenic expression of the bcl-2 gene in B cells resulted in similar lymphoid hyperplasia, hypergammaglobulinemia, high titers of antinuclear antibodies, and immune complex GN [131]. Constitutive bcl-2 expression may promote autoimmunity by blocking apoptosis of autoantibody-producing B cells that normally arise spontaneously in germinal centers during the primary response to foreign antigens [132–134].

Deficiencies of IL-2 [135–137], IL-2Rα [138], or IL-2Rβ [139] result in late immunosuppression with defective antibody and CTL responses as well as lymphoproliferation, expansion of memory/effector phenotype T cells, polyclonal hypergammaglobulinemia, autoantibodies and immune-mediated hemolytic anemia. Mice lacking

IL-2 [140] or IL-2Rα [138], but not IL-2Rβ (88) also develop an inflammatory bowel disease resembling ulcerative colitis. Autoimmunity has been attributed to resistance of IL-2-deficient T cells to activation-induced cell death and a reduction in T_{reg} cells because of their dependence on IL-2 [141].

The T-cell-specific adaptor protein (TSAd) is encoded by the *Sh2d2a* gene and expressed in thymocytes and activated T cells [142, 143]. TSAd$^{-/-}$ mice develop elevated levels of IgG and IgM, anti-dsDNA, anti-cardiolipin, and IgG (RF), GN, and lymphocytic infiltrates, particularly in the lung [94]. TSAd$^{-/-}$ mice are also more susceptible to TMPD-induced autoimmunity. TSAd-deficient T cells are more resistant to superantigen-induced cell death *in vivo* (similar to Bim knockout mice), which may be caused by reduced IL-2 synthesis [142].

The proto-oncogene PTEN that dephosphorylates phosphatidylinositol (3,4,5)-triphosphate (PIP-3) is associated with a wide range of malignancies as well as the autosomal dominant disorders Cowden disease, Lhermitte-Duclos syndrome and Bannayan-Zonana syndrome [144, 145]. PTEN$^{-/-}$ is embryonic-lethal, but heterozygous mice develop an autoimmune disorder characterized by severe polyclonal lymphadenopathy, diffuse inflammatory cell infiltrates of most organs, hypergammaglobulinemia, anti-DNA, GN and decreased survival [146]. Defective Fas-mediated activation-induced cell death of T and B lymphocytes from increases in PIP3 and the survival factor, Akt [146] and a requirement for reduced PIP3 for induction and maintenance of B-cell anergy [147] have been suggested as possible mechanisms.

IEX-1 (also IER3) is an early response gene that regulates cell growth and apoptosis in response to a variety of external stimuli in part by targeting the mitochondrial F1Fo-ATPase inhibitor for degradation [148]. Mice transgenic for IEX-1 (H-2k^b promoter and Ig heavy chain (μ) enhancer for specific expression in lymphocytes) exhibited decreased apoptosis of activated T cells, increased duration of an immune response effector phase, splenomegaly, lymphadenopathy, accumulation of activated T cells, increased polyclonal IgG2a and anti-dsDNA autoantibodies, alopecia of the skin, arthritis, and immune complex GN [36]. *IEX-1* gene maps within the MHC locus of humans and mice, a region with strong linkage to SLE.

RasGRP1 is a Ras guanine nucleotide exchange factor critical for the transition of the double positive to the single positive thymocyte stage [149]. Despite this, mice with a spontaneous function-impairing RasGRP1 mutation (designated *lag* for lymphoproliferation-autoimmunity-glomerulonephritis), which impaired normal joining of exon 3 to exon 4 resulting in undetectable RasGRP1 protein levels, developed severe systemic autoimmunity by 5—8 months of age [150]. Major manifestations included enlarged spleens and LNs, hyperplastic germinal centers,

hypergammaglobulinemia, ANAs, anti-dsDNA autoantibodies, diffuse proliferative GN with IgG and C3 deposits, and early mortality. *Lag* CD4$^+$ T cells were resistant to activation induced cell death (AICD), while B cells were expanded, but exhibited normal proliferation and apoptosis.

Stra13, a member of the basic helix-loop-helix family of transcriptional repressors, is expressed in lymphoid cells [151, 152]. Although Stra13-deficient CD4+ T cells had reduced proliferation, due in part to impaired IL-2 secretion, Stra13 knockout mice developed systemic autoimmunity by 4—5 months with lymphoid organ hyperplasia, increased numbers and activation of T and B cells, germinal center expansion, antinuclear antibodies, and GN [151]. Gradual accumulation of T and B cells appeared due to greater resistance to AICD because of impaired differentiation of CD4$^+$ T cells into effector cells and reduced expression of FasL following T-cell activation. Consistent with a negative regulatory role, transgenic mice expressing Stra13 (IgH promotor and enhancer) in B- and T-cells had reduced numbers of T- and B-cells, hyporesponsive B cells, and impaired T-dependent humoral response [152].

Defective Clearance of Proinflammatory/ Immunostimulatory Self-antigens

Deficiencies of early complement components (C1q-s, C2, or C4) in humans are known to result in predisposition to SLE, indicating an important regulatory role in suppressing autoimmunity. Homozygous C1q-deficient mice in some backgrounds recapitulated the human disorder with the development of mild, but typical, features of lupus [153]. Notably, an atypical accumulation of apoptotic bodies in the glomeruli of C1q-deficient mice was observed indicating that C1q plays an essential role in the clearance of apoptosis byproducts. As noted above a C1q polymorphism in NZB mice within the *Nba1/Lbw2/Imh1/Mott* interval on chromosome 4 that downregulates C1q levels has been identified [39]. In C4 deficiency, lupus was evident only on a *Fas^{lpr}* background [154], indicating that lack of C4 alone is not sufficient to induce autoimmunity. Nevertheless, early complement components may be vital for maintaining self-tolerance by virtue of their role in presenting tolerizing antigens to B cells [155].

DNase1 is a 32—38-kDa protein that is the major nuclease present in the blood, urine and secretions. Knockout of the DNase1 gene in non-autoimmune background mice was reported to increase the incidence of SLE manifestations, including positive ANA, anti-DNA, and immune complex GN [156]. Reduced DNase1 activity, as observed in sera of lupus patients [157], may contribute to overall SLE susceptibility. Interestingly, an identical heterozygous nonsense mutation in *DNASE1* was detected in two SLE patients [157].

Milk fat globule-EGF factor 8 (MFG-E8) protein functions as a bridging protein between phosphatidylserine on apoptotic cells to αvβ3 or αvβ5 integrins on phagocytic cells, thereby promoting apoptotic cell engulfment [158, 159]. In the spleen and LN, MFG-E8 is primarily expressed on tingible body macrophages (CD68$^+$) within germinal centers [160]. MFG-E8-null mice develop splenomegaly with abundant germinal centers and tingible macrophages with increased numbers of unengulfed apoptotic cells. By 40 weeks, MFG-E8$^{-/-}$ mice have ANA and anti-dsDNA, large glomerular deposits of IgG, and hypercellular glomeruli. Taken together, these findings suggest inefficient phagocytosis of apoptotic B cells within germinal centers promotes autoimmunity. This is consistent with recent studies demonstrating that DNA and RNA in apoptotic material can activate B cells and dendritic cells through TLR9, TLR7, and TLR8 [161–166].

TAM (Tyro3, Axl and Mer) receptors, expressed on APCs, promote clearance of apoptotic cells and inhibit TLR-induced cytokine cascades [167]. Mutant triple TAM knockout mice develop severe systemic autoimmunity characterized by splenomegaly, lymphadenopathy, increases in activated T and B cells, autoantibodies to phospholipids associated with thromboses and hemorrhage, autoantibodies to collagen and dsDNA, and deposition of immune complexes in tissues [168]. Furthermore, mice with a cytoplasmic truncation of *mer* alone are deficient in the clearance of apoptotic thymocytes and develop a mild form of lupus-like disease with antibodies to chromatin [169, 170].

Trex1, the most abundant cytoplasmic 3′ to 5′ ssDNA exonuclease, degrades intracellular nucleic acid and negatively regulates the intracellular IFN-stimulatory DNA response [171, 172]. Deficiency of Trex1 in humans causes Aicardi-Goutieres syndrome and chilblain lupus, but in mice an inflammatory myocarditis and antibodies to heart tissue, not lupus-like disease. Interestingly, Trex1 deficiency was associated with the accumulation of endogenous retroelement-derived DNA. The development of lupus traits in Trex1-deficient humans, but not in Trex1-null mice may be related to differences in background genes.

Ro, a conserved RNA-binding protein encoded by the *Trove2* gene, is a common target for autoantibodies in SLE, Sjogren syndrome, subacute cutaneous LE, neonatal lupus and primary biliary cirrhosis [173]. Mice deficient in Ro were reported to develop ANAs and immune complex GN [174]. Disease severity, however, lessened as the knockout mutation was backcrossed to the B6 background, suggesting the possibility of 129xB6 background effects. More recently, knockout of the autoantigen Ro52 (*Trim21*) also resulted in lupus-like systemic autoimmunity [175]. Despite this null mutation being generated directly in the B6, however, autoimmunity was not observed in Ro52$^{-/-}$ mice generated by another group suggesting the possibility of bystander effects related to the method of gene disruption [176].

Deletion of serum amyloid P component (SAP), a highly conserved plasma protein, was also reported to be associated with lupus [177], however, as described above, subsequent studies, have shown that the association of SAP deficiency with lupus-like disease is mostly or even entirely due to the mixed 129xB6 background used in the initial study [29, 30].

Enhanced Antigen Presentation

Targeted expression of CD40L to the basal keratinocytes of the epidermis of mice (keratin-14 promoter) leads to activation of resident tissue APCs (Langerhans cells) and dermatitis, lymphadenopathy, hypergammaglobulinemia, anti-dsDNA, and immune complex GN [178]. These findings are similar to mice transgenic for IFN-γ under the involucrin promoter [179]. Overall, they indicate that *in situ* activation of APCs in the skin can lead not only to local, but also systemic, autoimmune and inflammatory responses, presumably due to the migration of activated APC to the secondary lymphoid organs, and the activation of self-reactive T cells.

Mice deficient in Dcir (*Clec4a2*), an ITIM-containing C-type lectin receptor highly expressed in DC, develop spontaneous enthesitis, salivary gland infiltrates, and ANA [180]. Evidence suggests this is caused by the expansion of DCs leading to enhanced antigen presentation.

Cytokine Ligand and Receptor Genes

Systemic autoimmunity can develop in mice transgenic for the major Th1 and Th2 cytokines, IFN-γ and IL-4, respectively, in certain circumstances. As noted above, expression of IFN-γ in the suprabasal layer of the epidermis resulted not only in a severe inflammatory skin disorder, but also anti-dsDNA and -histone, and an immune complex proliferative GN [179]. Similarly, C3H mice transgenic for the IL-4 gene (MHC class I promoter) also developed systemic autoimmunity characterized by hyper IgG1 and IgE, anemia, ANAs, and GN from direct IL-4-induced polyclonal activation of B cells [181]. The role of IL-4 in promoting lupus, however, is more complicated, since autoimmunity was not observed in other transgenics expressing IL-4 in B or T cells [182–185] and, in a spontaneous model of lupus, expression of an IL-4 transgene did not exacerbate disease, but was instead protective [186].

Tristetraprolin (TTP or Zfp-36) is a widely expressed zinc-binding protein that binds to an AU-rich element in the TNF-α mRNA destabilizing the mRNA [187].

TTP-null mice develop a complex syndrome of cachexia, patchy alopecia, dermatitis, conjunctivitis, erosive arthritis, myeloid hyperplasia, glomerular mesangial thickening and ANAs [188] due to excessive TNFα production [188, 189]. In contrast, physiological levels of TNF-α may suppress systemic autoimmunity. TNFR1 (p55)-null C57BL/6-Fas^{lpr} mice exhibit accelerated lymphoproliferation and autoimmune disease [190] and deletion of both TNF receptors enhances lupus-like disease in NZM2328 mice [191].

Defective Signal Transduction

Suppressor of cytokine signaling (SOCS)-1 is a negative regulator of Janus kinases that acts as a feedback inhibitor of cytokine signaling [192]. Although complete deficiency of SOCS-1 results in fatty liver degeneration, lymphopenia, macrophage infiltration, and mortality before 3 weeks, partial deficiency was reported to promote lupus-like systemic autoimmunity with increased Ig levels, autoantibodies, glomerular immune complex deposits, and mesangial proliferation, as well as perivascular infiltrates in multiple organs and cutaneous alterations, including eczema and small ulcers [193].

Other Mechanisms

SHP-1 is a protein tyrosine phosphatase expressed in hematopoietic lineage cells and inhibits cell activation following its recruitment by negative regulatory molecules containing ITIMs [194]. Two spontaneous recessive mutations of the SHP-1-encoding $P_{tpn}6$ gene, the motheaten (*me*) and motheaten viable (*me^v*) both lead to similar early lethal phenotypes, which differ slightly in severity because of more complete gene deletion in the *me* variant. Although increased Ig levels and autoantibodies are detected, the major disease manifestations are not similar to spontaneous lupus, and do not require the adaptive immune response. However mice lacking SHP-1 only in B cells develop both autoimmunity and inflammatory disease [417].

The α-mannosidase II enzyme resides in the Golgi apparatus and is critical for the glycosylation of cell surface proteins [200, 201]. Mice deficient in α-mannosidase II develop a lupus-like syndrome characterized by anti-dsDNA, anti-Sm and anti-histone, hyperIg, and GN [202]. No clear explanation for appearance of systemic autoimmunity was evident, but a possibility is that alterations in N-glycan branching among some glycoproteins and tissues may lead to the formation of neoepitopes.

Emk (ELKL motif kinase, MARK2) is a serine/threonine kinase expressed in thymus and mature T and B cells that regulates cell polarity, cell cycle progression, and microtubule dynamics [203]. Emk^{-/-} mice exhibit growth retardation and hypofertility [203, 204], but

with age, develop splenomegaly, lymphadenopathy, activated phenotype T cells, lymphoid infiltrates in various tissues, and membranoproliferative GN [203].

The basic leucine zipper transcription factor Nrf2 (NF-E2-related factor 2) regulates a number of genes encoding detoxifying and anti-oxidant enzymes [205, 206]. Nrf-2-deficient female mice over 5 months of age develop severe immune complex GN along with elevated serum IgG, anti-dsDNA antibody, and slight splenomegaly [207]. Mice deficient in heme-oxygenase-1 (HO-1), a gene potentially regulated by Nrf2, also develop GN resembling that of the Nrf2-deficient mice [208]. In contrast, Nrf2 deficiency in MRL-Fas^{lpr} mice results in increased sensitivity to TNF-mediated apoptosis and disease suppression [209]. These findings suggest that the effect of Nrf2 on lupus depends on the underlying disease pathogenesis or that increased lupus susceptibility in the Nrf2 knockout may be due to background.

Mice deficient for Mta2, a component of the NuRD complex that acts in nucleosome remodeling and histone deacetylation, have greater than 50% embryonic/perinatal lethality, multiple developmental abnormalities, small body size and female infertility, but also develop with age autoantibodies, GN, skin lesions, and liver inflammation [210]. T cells exhibited enhanced proliferation and production of IL-4 and IFN-γ. Findings suggest that defects in chromatin remodeling and histone modification can predispose to lupus, however, the use of mixed 129xB6 mice suggests some caution in interpretation of these results.

Zc3h12a is an Rnase activated by TLR signaling that promotes the degradation of mRNA. Zc3h12a^{-/-} mice have early mortality (50% ~ 6 weeks) associated with severe hemolytic anemia, lymphoproliferation, and ANAs associated with expansions of activated and memory B cells, activated T cells, and plasma cells, as well as enhanced cytokine production [211].

GENE KNOCKOUT AND TRANSGENIC LUPUS BACKGROUND MICE

Direct examination of the role of deleted or overexpressed immune-related genes in lupus-predisposed mice have provided important information on the crucial molecules and pathways, as well as a deeper understanding of the molecular basis for the diverse manifestations. (Table 4.6, listed by chromosome and chromosomal location).

B-cell-related Genes

Studies in MRL-Fas^{lpr} and MRL-+/+ mice with deletion of the Jh locus (no B cells) [212, 213] have clearly

TABLE 4.6 Effects of genetic manipulation of lupus-prone mice

Name and alteration	Gene	Chr#	Mb	Strain*	Result	Ref.
CD28 ko	Cd28	1	60.8	MRL-lpr	reduced GN	[248]
STAT-4	Stat4	1	52.1	NZM2410 B6-TC	exacerbates GN, reduces IgG anti-dsDNA reduces autoAb, GN	[275] [274]
STAT-1	Stat1	1	52.2	BALB/c-TMPD-induced	reduced autoAb	[276]
ICOS	Icos	1	61.0	MRL-lpr	reduced autoAb, no effects on GN or skin	[233, 234]
CD55 (Daf1)	Cd55	1	132.3	MRL-lpr	exacerbates autoimmunity	[291]
IL-10 ko	Il10	1	132.9	MRL-lpr	enhanced lymphoproliferation, IgG2a autoAb, GN, skin lesions, mortality	[277]
CD45 +/- ko	Ptprc	1	140.0	C3H/HeJ-gld	reduced IgG, autoAb, DN T cells	[284]
P-selectin ko	Selp	1	166.0 Mb	MRL-lpr	enhanced GN, mortality, incr. CCL2 (kidney)	[305]
FcR γ-chain ko	Fcer1g	1	173.2	BWF1	same autoAb, glom. dep., but reduced GN, mortality	[299]
				MRL-lpr	no effect on GN	[392]
CD21/CD35 ko	Cr2	1	197.0	lpr	accelerated disease	[35, 154, 393]
soluble Crry Tg (complement receptor related protein)	Crry	1	197.0	MRL-lpr	reduced GN, no effect on glom. dep., autoAb, lymphoproliferation	[290]
Nrf2, Nuclear factor, erythroid derived 2, like 2	Nfe2l2	2	75.5	MRL-lpr	reduced lymphoproliferation, autoAb, GN, mortality; enhanced TNF-mediated apoptosis	[209]
beta-2 microglobulin ko (MHC class I ko)	B2m	2	122.0	MRL-lpr	reduced lymphoproliferation	[220, 221]
				C3H-lpr & -gld	reduced lymphoproliferation	[219]
				NZB	reduced anti-RBC Ab inhibits	[394]
				MRL-lpr	inhibits nephritis, accelerates skin disease	[223]
CD1 ko	Cd1d1	3	86.8	MRL-lpr	no effect on disease reduced skin disease	[223]
				MRL-lpr	exacerbates nephritis	[224]
				TMPD-induced BWF1	exacerbates nephritis	[225, 226]
TNFR2 (p75)	Tnfrsf1b	4	144.8	NZM2328	accelerated autoAb, GN (only with dko: Tnfrsf1a/Tnfrsf1b)	[191]
DNA fragmentation factor, beta subunit (caspase-activated DNase, 40kDa)	Dffb	4	153.3	TMPD-induced	impaired nuclear autoAb, less effect on cytoplasmic and cell surface self Ag, no effect on mild GN	[245]

Description	Gene	Chr	Position	Strain/model	Phenotype	Ref
cappuccino (single nucleotide deletion)	Cno	5	37.1	EOD (MRL-lpr×BXSB)RI	reduced glom. crescent, plat. dysfunction, no effect on autoAb, IC deposits	[304]
osteopontin ko (Eta-1, secreted phosphoprotein 1 ko)	Spp1	5	104.9	B6-lpr	reduced polyclonal and autoAb, delayed lymphoproliferation and kidney disease	[279]
P-selectin ligand ko	Selplg	5	114.3	MRL-lpr	enhanced GN, mortality, incr. CCL2 (kidney)	[305]
CD8 ko	Cd8a / Cd8b	6 / 6	71.3 / 71.3	MRL-lpr	reduced lymphoproliferation	[218]
Arachidonate 5-lipoxygenase ko	Alox5	6	116.4	MRL-lpr	slight accelerated mortality (males only)	[395]
Activation-induced cytidine deaminase (AID) ko	Aicda	6	122.5	MRL-lpr	reduced autoAb, GN, early mortality (incr. IgM autoAb)	[216]
CD4 ko	Cd4	6	124.8	MRL-lpr	increased lymphoproliferation; decreased autoAb, GN, mortality	[218, 396]
TNFR1 (p55)	Tnfrsf1a	6	125.3	B6-lpr	enhanced lymphoproliferation, autoAb, GN	[190]
				NZM2328	accelerated autoAb, GN (only with dko: Tnfrsf1a/Tnfrsf1b)	[191]
apolipoprotein E ko	Apoe	7	20.3	B6-lpr	exacerabated autoAb, GN	[312]
kallikrein related-peptidase family	Klk	7	51.0-51.4	multiple strains	severity of nephritis (nephrotoxic serum model)	[303]
IL-21R ko	Il21r	7	132.7	BXSB	reduced monocytosis, IgG, autoAb, GN, early mortality	[282]
Integrin-αL (CD11a, LFA-1) ko	Itgal	7	134.4	MRL-lpr	reduced autoAb, GN, mortality	[254]
Single immunoglobulin and toll-interleukin 1 receptor (TIR) domain ko	Sigirr	7	148.3	B6-lpr	enhanced lung disease, GN (129×B6, backcrossed 6 generations to B6)	[244]
ICAM-1 ko	Icam1	9	20.8	MRL-lpr	reduced mortality, autoAb, GN, vasculitis	[296]
				MRL-lpr	reduced pulmonary inflammation and mortality, but same autoAb, lymphoproliferation and GN	[295]
Pou domain, class 2, associating factor (OBF-1)	Pou2af1	9	51.0	Aiolos knockout	Prevents development of lupus-like phenotypes	[397]

(Continued)

TABLE 4.6 Effects of genetic manipulation of lupus-prone mice—cont'd

Name and alteration	Gene	Chr#	Mb	Strain*	Result	Ref.
TLR9	Tlr9	9	106.1	MRL-lpr	reduced disease exacerbation of disease	[398]
Myeloid differentiation primary response gene 88	MyD88	9	119.2	MRL-lpr	reduced autoAb, GN, mortality; autoAb & GN induced by poly(I:C)	[240, 399]
ERα (estrogen receptor 1)	Esr1	10	5.3	BWF1 NZM2410, MRL-lpr	females less autoAb, GN, mortality females less autoAb, GN	[294] [293]
IFN-γR ko	Ifngr Ifngr2	10 16	19.3 91.5	MRL-lpr BWF1	reduced GN reduced GN, mortality	[263, 400] [265]
fyn ko	Fyn	10	39.1	MRL-lpr	reduced DN T cells, autoAb, GN	[283, 401]
perforin ko	Prf1	10	60.8	MRL-lpr	accelerated disease	[232]
MIF	Mif	10	75.3	MRL-lpr	reduced GN, early mortality; no other major change	[302]
Integrin-β2 (CD18) ko	Itgb2	10	77.0	MRL-lpr	reduced autoAb, GN, mortality	[254]
complement factor D ko	Cfd	10	79.4	MRL-lpr	reduced C3 dep., kidney pathology; no effect on autoAb, IgG dep. proteinuria, mortality	[287]
IFN-γ ko	Ifng	10	117.9	(MRL-lpr)F2 MRL-lpr HgIA	reduced GN, mortality, autoAb reduced GN, mortality, autoAb reduced autoAb, GN	[262] [264] [266]
STAT-6	Stat6	10	127.1	NZM2410 B6-TC	reduced GN, not IgG anti-dsDNA reduced ANA, but not GN	[275] [274]
IL-12b (p40) Tg (antagonist)	Il12b	11	44.2	MRL-lpr-cg	reduced autoAb, slightly reduced GN and survival	[272]
IL-4 ko	Il4	11	53.4	MRL-lprF2 BXSB	reduced GN no effect	[262] [273]
IRF-1 (interferon regulatory factor 1)	Irf1	11	53.6	MRL-lpr	reduced autoAb, GN, lymphoproliferation, early mortality	[270]

Description	Gene			Strain	Phenotype	References
Nitric oxid synthetase 2 ko	Nos2	11	78.7	MRL-lpr (N4)	same autoAb, GN, arthritis reduced vasculitis	[298] [402]
MCP-1 ko (chemokine (C-C) motif ligand)	Ccl2	11	81.8	MRL-lpr	reduced GN, mortality; same autoAb and glom. dep.	[301]
T-box 21 Tg (T-bet) (Expressed in T cells)	Tbx21	11	97.0	(B6xBXSB)F1	enhanced autoAb, GN, mortality	[269]
T-box 21 ko (T-bet)	Tbx21	11	97.0	MRL-lpr	reduced morality, GN, autoAb, hyperIgG	[268]
Jh (no B cells)	Igh-J	12	114.5	MRL-lpr	no disease	[212] [213]
Jh, mIg Tg (B cells with mIg, no Abs)	Igh-J	12	114.5	MRL-lpr	infiltrate, no GN	[214]
angiotensin II receptor, type 1a (AT1a)	Agtr1a	13	30.4	MRL-lpr	more severe GN and mortality, BM cell independent	[306]
TCRαδ ko (no $\alpha\beta^+$ or $\gamma\delta^+$ T cells)	Tcra/Tcrd	14	54.4		no IgG autoAb, GN	[403]
TCRδ ko (no $\gamma\delta^+$ cells)	Tcrd	14	54.4		disease acceleration	[404]
TCRα ko (no $\alpha\beta^+$ cells)	Tcra	14	54.4	MRL-lpr BXSB	major disease reduction marked disease reduction	[229, 405] [406]
B7.2 ko	Cd86	16	36.6	MRL-lpr	reduced GN	[249]
B7.1 ko	Cd80	16	38.5	MRL-lpr	more severe, distinct GN	[249]
B7.1/B7.2 double ko	Cd80/86	16	37-38	MRL-lpr	marked disease reduction	[407]
Interferon (alpha and beta) receptor 1	Ifnar1	16	91.5	NZB B6/129-lpr	reduced autoAb, GN, anti-RBC reduced autoAb, GN	[255, 256]
MHC class II ko	H2-Aa H2-Ea	17	34.4	MRL-lpr	reduced autoAb, GN, same lymphoproliferation	[217]
human DR Tg	H-2	17	34.4	NZM/Aeg2410	autoAb repertoire alterations	[408]
complement C4 ko	C4	17	34.9	MRL-lpr	accelerated disease	[154, 289]
complement factor B ko	Cfb	17	35.0	MRL-lpr	reduced autoAb, GN, vasculitis	[285]
Tnf +/- ko	Tnf	17	35.3	NZBxB6.129$^{\text{Tnf}/o}$)F1	enhanced disease	[409]

(Continued)

TABLE 4.6 Effects of genetic manipulation of lupus-prone mice—cont'd

Name and alteration	Gene	Chr#	Mb	Strain*	Result	Ref.
C3 ko	C3	17	57.3	MRL-lpr	worse proteinuria and glom. dep., but same GN score, autoAb and CIC	[288]
				129xMRL-lpr/C4 ko	no difference vs. 129xMRL-lpr/C4 ko	[289]
TdT ko	Dntt	19	41.1	BWF1	same autoAb, reduced GN and mortality	[410] [310]
				MRL-lpr	reduced hyperIgG autoAb, GN, mortality, vasculitis	[310]
CD40L ko	CD40lg	X	54.5	MRL-lpr	reduced autoAb, GN	[247]
SH2 domain protein 1A (SAP)	Sh2d1a	X	39.9	129/pristane-induced MRL-lpr	reduced IgG, autoAb, GN spont. mutation; markedly reduced disease (check paper)	[252] [253]
TLR7 ko	Tlr7	X	163.7	MRL-lpr BXSB TMPD	reduced RNA-associated Ab, GN	[237] [411] [238]
protein kinase CK2 alpha ko (casein kinase II) (transgenic)	Csnk2a1 Csnk2a2 Csnk2b	2 8 17	152.0 98.0 35.3	MRL-lpr	accelerated lymphoproliferation, autoAb, GN	[412]

*MRL-lprF₂ = mixed background derived from (MRL-Fas^lpr x(B6x129)F₁)F₂, lpr = Fas^lpr, gld = Fas^gld.
#ND = not determined.
Genes are listed by their approximate chromosomal locations. Genes are deficiencies by homologous recombinant knockout unless otherwise stated.
Abbreviations: antibody = Ab, glom. CIC = circulating immune complexes, glom. dep. = glomerular IgG deposits.

shown the crucial role of B cells in lupus disease manifestations, confirming the central role of autoantibodies in pathogenesis. Furthermore, genetic manipulation of MRL-Fas^{lpr} mice resulting in B-cell expression of surface, but not secreted, immunoglobulin, demonstrated that B cells alone could promote local cellular infiltration and inflammation, but not GN [214]. Indeed, Fas deficiency in germinal center B cells is sufficient for the severe lpr-associated lymphoproliferation [215].

Activation-induced cytidine deaminase (AID) is required for class-switch recombination and somatic hypermutation of immunoglobulins. As expected, IgG autoantibodies are absent in MRL-Fas^{lpr} AID$^{-/-}$ mice, but IgM autoantibodies are increased [216]. These mice have reduced GN and survival is prolonged.

T-cell-related Genes

A wide diversity of intercross of congenic mice rendered defective for MHC, CD4, CD8 or T-cell receptor genes have been used to assess the role of helper and cytotoxic T cells, and $\alpha\beta$- and $\gamma\delta$-T-cell subsets in lupus. In MRL-Fas^{lpr} mice, deletion of MHC class II [217] or CD4 [218] reduced autoantibodies and GN, but had no effect on lymphadenopathy. In contrast, β2m- (class I) [219-222] or CD8-deficient [218] MRL-Fas^{lpr} or C3H-Fas^{lpr} mice showed reduced lymphoproliferation and expansion of DN B220^{+} cells, but only partial diminution in autoantibody levels and GN.

MRL-Fas^{lpr} mice deficient in β2m are discordant in skin and kidney disease, with the former accelerated and the latter ameliorated [223]. The skin disease was not, however, accelerated in CD1-deficient MRL-Fas^{lpr} mice, suggesting that the effect of β2m deletion (affects all MHC class I including CD1) in this manifestation was not mediated by NK T cells that depend on CD1 or by CD1 expression on skin Langerhans or B cells. CD1 deficiency did not affect disease severity, including nephritis and vasculitis. In contrast, another group reported that CD1-deficiency exacerbated skin disease in MRL-Faslpr [224], TMPD-induced lupus nephritis [225], and GN in (NZBxNZW)F1 females [226] consistent with a regulatory role of CD1d and iNKT cells. Findings suggest that local conditions in target organs can affect disease manifestations. In NZB mice, CD4 or β2m deficiency delayed onset and reduced incidence of anti-RBC antibodies [227].

Although the combined results clearly demonstrate the importance of TCR$\alpha\beta$ cells in spontaneous lupus, TCRα gene deletion in MRL-Fas^{lpr} mice only partially inhibited disease [228, 229]. Based on the fact that TCR$\alpha\beta$/TCR$\gamma\delta$ double-knockout MRL-Fas^{lpr} mice fail to generate class-switched autoantibodies and immune complex GN [230, 231], this suggests that TCR$\gamma\delta$ cells can also help drive the autoimmune B cells. Paradoxically, however, MRL-Fas^{lpr} TCR$\gamma\delta$-deficient mice showed disease exacerbation. These results suggest that TCR$\alpha\beta$ cells are the major provider of B-cell help in intact MRL-Fas^{lpr} mice, whereas TCR$\gamma\delta$ cells suppress this process, but can concomitantly provide lesser degrees of non-MHC-restricted polyclonal B-cell help. In contrast, TCR$\alpha\beta$-deficient BXSB mice are resistant to lupus [406]. In addition, knockout of the perforin gene in MRL-Fas^{lpr} mice resulted in disease exacerbation, suggesting that cytolytic cells may be involved in suppressing autoreactivity [232].

ICOS (CD278) is a member of the CD28 family of co-stimulatory molecules on T cells that plays an important role in the formation and function of follicular helper T cells (T$_{FH}$). MRL-Fas^{lpr} ICOS$^{-/-}$ mice had reduced lymphadenopathy, anti-dsDNA, renal vasculitis, but the amount of immune complex deposits in the kidney, severity of GN and cutaneous lupus were unchanged [233, 234].

Genes Related to Nucleic Acid Recognition and Degradation

Studies have shown that certain arms of the innate immune system, particularly the endosomal TLR pathway, are critical for the development of lupus-like disease in mouse models [235]. Specifically, TLR7, Un93b1, and MyD88 have been clearly shown to be required for optimal autoantibody production and end-organ disease in several lupus-prone strains [236–241]. This supports the current paradigm that uptake of nucleic acid complexes into endosomes by antigen receptors on B cells or Fc receptors on DC leads to activation of endosomal TLRs (primarily TLR7), which promotes loss of tolerance to nucleic acids and nucleic-acid-containing material, essential for generating high titers of anti-nucleic acid Abs [235, 242]. In contrast, the role of TLR9 in lupus is less clearly defined. Deletion of TLR9 exacerbated disease in most models, but reductions in anti-DNA Abs and in overall disease severity have also been reported by others [235].

Sigirr (single immunoglobulin and toll-IL1 receptor domain, Tir8) is a member of the IL-1R-like family that is thought to inhibit signaling by IL-1 family members (including IL-1, -18, -33) and TLR-mediated activation [243]. Deletion of Sigirr in B6-Fas^{lpr} mice exacerbates lymphoproliferation and results in severe autoimmune lung and kidney disease [244]. This was attributed to a reduction in CD4+CD25+ T cells and its function as an inhibitor of RNA- and DNA-mediated DC- and B-cell activation.

Caspase-activated DNase (CAD, Dffa) is a major mediator of DNA fragmentation in apoptosis [218].

Mice lacking Dffa have impaired intranuclear chromatin fragmentation, nuclear degradation, and apoptotic membrane blebbing, but are able to phagocytize apoptotic material and have a functional immune system [245]. In the TMPD-model, B6-$Dffa^{-/-}$ mice have markedly reduced autoantibodies to nuclear antigens, but are able to produce antibodies to intracytoplasmic antigens and develop similar mild GN as wt mice.

Co-Stimulatory Molecules

Studies in MRL-Fas^{lpr} mice deficient for CD40L [246, 247] or CD28 [248] demonstrate the prerequisite for co-stimulation in lupus development. MRL-Fas^{lpr} mice deficient for either B7.1 or B7.2 were used to further dissect the CD28-B7 axis [249]. Although neither of these deletions affected autoantibody levels compared to wild-type MRL-Fas^{lpr} mice, GN was substantially worse in B7.1-deleted mice, but less severe in B7.2 knockouts, consistent with findings in BWF1 mice showing antibodies to B7.2, but not B7.1, suppressed disease [250].

SH2D1A (SLAM-associated protein or SAP) is a signal transduction adaptor protein for several SLAM family members and is mostly expressed in T cells [251]. Deficiency of SH2D1A is responsible for the X-linked lymphoproliferative syndrome (XLP) in which patients manifest severe infectious mononucleosis, B-cell lymphomas, and/or dys-gammaglobulinemia because of impaired CTL, NK T cell, and Ab responses. SH2D1A deficiency is associated with reduced lupus-like disease in the TMPD-induced model [252] and in MRL-Fas^{lpr} mice with a spontaneous adenyl nucleotide insertion at codon 21 of the 1st exon that impairs SH2D1A protein function and expression [253]. This was attributed to defective germinal center formation and a reduction in the T-dependent humoral immune response [252].

Leukocyte Adhesion Genes

The leukocyte β-integrins LFA-1 (CD11a/CD18), Mac-1 (CD11b/CD18), and p150/95 (CD11c/CD18) are ligands for ICAM-1 that mediate diverse immune and inflammatory responses. Deficiency of either CD11a (Itgal) or CD18 (Itgb2) in MRL-Fas^{lpr} mice was associated with reduced lymphoproliferation, autoAb, GN and dermatitis [254]. Survival was prolonged in CD11a$^{-/-}$ mice, but not examined in CD18-deficient mice because of occasional fatal infections. In contrast, CD11b-deficiency had no ameliorating effect.

Cytokine Genes

The requirement for IFN-α in lupus has been shown for several lupus-prone strains [255–257] and treatment with exogenous type I IFN has in most instances significantly exacerbated disease [258], consistent with the IFN signature observed in human SLE [259]. In contrast, MRL-Fas^{lpr} mice exhibit the opposite response to type I IFN, with disease worsened by deletion of the IFN-α receptor (Ifnar1) [260] and lessened by IFN-β treatment [261]. This suggests a significant difference in the pathogenesis of MRL-Fas^{lpr} mice and the other mouse models thus far examined for type I IFN-dependence.

Similar to type I IFN, IFN-γ is required for the development of systemic autoimmunity in all models of lupus examined [262–266]. IFN-γ is important in at least two steps of lupus pathogenesis: promotion of the response to self-antigens, perhaps because they are of low antigenicity [264, 266]; and acceleration of local inflammatory responses [264].

A variety of proto-oncogenes, kinases and transcription factors have been implicated in the Th1 versus Th2 polarization process, including interferon regulatory factor 1 (IRF-1) and T-box expressed in T-cell (T-bet, T-box 21) proteins for Th1 cells, and, for Th2 cells, the c-Maf proto-oncogene and the GATA3 zinc finger protein [267]. Deficiency of T-bet in lupus-prone mixed-background MRL-Fas^{lpr} mice resulted in reductions in mortality, GN, autoantibody production and hypergammaglobulinemia [268]. Overexpression of T-bet, on the other hand, reportedly accelerated GN and mortality [269]. MRL-Fas^{lpr} mice deficient for IRF-1 had reduced autoantibodies, less kidney and skin disease, and prolonged survival, although splenomegaly was unaffected [270]. IL-12 is an important regulator of Th1 cell differentiation and IFN-γ production [271]. Transgenic expression of a potent IL-12 antagonist, IL-12p40 in lupus-susceptible MRL-Fas^{lprcg} mice, suppressed the production of Th1 cells, IFN-γ and anti-dsDNA antibodies, but did not significantly improve clinical manifestations of disease, including lymphoproliferation, GN, and mortality [272].

In contrast to the dramatic effects seen with IFN-γ, deletion of IL-4 results in only partial or no reduction in disease, depending on the lupus-prone strain. IL-4$^{-/-}$ MRL-Fas^{lpr} mice produced less serum IgG1 and IgE (Th2-dependent subclasses), but maintained comparable levels of IgG2a, IgG2b, and autoantibodies [262], although reduced lymphoadenopathy and end-organ disease were also observed. In contrast, IL-4-deficient male BXSB mice had similar autoantibody levels, GN severity and mortality as their wild-type counterparts [273], suggesting that IL-4 plays little role in the immunopathogenesis of disease in this strain.

Other studies have examined the role of signaling pathways downstream of cytokine receptors on lupus susceptibility. Deficiency of Stat4, which transduces signals from IL-12, IL-23, and type 1 IFN cytokines, significantly reduced ANAs and GN in B6.TC (sle1-3 triple congenic B6) mice [274], while NZM2410 mice had exacerbation of GN, despite reduction in anti-dsDNA [275]. Lack of Stat6, which primarily mediates IL-4 and IL-13 signals, reduced anti-nuclear antibodies, but not GN in B6.TC mice [274], while NZM2410 mice had reduced GN, but no change in anti-dsDNA [275]. Stat1-deficient BALB/c mice have lower levels of autoantibodies following TMPD treatment [276].

The IL-10-null mutation in MRL-*Fas^lpr* mice exacerbated skin lesions, lymphadenopathy, IFN-γ levels, IgG2a anti-DNA, GN, and mortality, while administration of IL-10 was protective [277]. The protective effect of IL-10 in the MRL-*Fas^lpr* mice was attributed to inhibition of the T_H1 responses that predominate in this mouse.

Eta-1 (osteopontin)-deficient normal background mice were shown to have reduced type 1 immunity for viral and bacterial infection attributed to diminished IL-12 and IFNγ and increased IL-10 production [278]. C57BL/6-*Fas^lpr* mice rendered Eta-1-deficient were reported to display delayed onset of polyclonal B-cell activation and somewhat delayed lymphoaccumulation and kidney disease [279]. Earlier studies have shown increased expression of Eta-1 in MRL-*Fas^lpr* mice [280].

EBV-induced gene 2 (EBI3, IL-17B) is a subunit of IL-27 produced by antigen-presenting cells. Deletion of this gene in MRL-*Fas^lpr* mice altered renal pathology from diffuse proliferative to membranous GN with increased deposits of IgG1 attributed to reduced IFN-γ, but enhanced IL-4 expression [281].

IL-21, a cytokine produced by CD4 T cells, has diverse effects on lymphocytes and NK cells. Delection of the IL-21R in BXSB significantly reduces Ig, autoantibodies, monnocytosis, renal disease and early mortality [282]. IL-21 in BXSB mice is produced by ICOS-expressing CD4 splenic T cells. Interestingly, a polymorphism within IL-21R confers risk for SLE [419]. IL-23 is an IL-21 family member that promotes the stabilization and expansion of the Th17 subset of CD4 T cells. B6-*Fas^lpr* mice deficient for its receptor (IL-23R) had reduced lymphoid hyperplasia, anti-DNA Abs and GN [420].

Cell-signaling Molecules

Fyn is a tyrosine-protein kinase involved in proximal TCR signal transduction. MRL-*Fas^lpr* fyn-deficient mice have reduced frequency of DN B220$^+$ T cells along with decreased autoantibody levels and immunopathology, indicating an essential role for fyn in signal transduction and expansion of DN B220$^+$ T cells [283]. Of interest, *Fasl^gld* mice with only one functional CD45 allele (CD45$^{+/-}$) were reported to have a tenfold reduction in DN T-cell population and decreased Ig and anti-DNA autoantibodies [284]. The finding

suggests that CD45 plays an essential role in the activation of the DN cells, which then accumulate due to the FasL defect.

Complement Components

MRL-*Fas^lpr* mice deficient for the complement factor B (Bf) were utilized to investigate the role of the alternative pathway of complement activation in lupus pathogenesis [285]. Bf is an acute phase reactant required to activate this pathway and local production of Bf can be detected in the kidneys of lupus-prone mice [286]. Deletion of Bf resulted in significant reduction in GN severity and incidence of vasculitis, along with lower levels of anti-dsDNA, IgG3 anti-IgG2a RF, and IgG3 [285]. These findings support a significant role for Bf and the alternative pathway in the immune complex autoimmune pathology of these mice. Factor D deficiency in MRL-*Fas^lpr* mice also mildly reduced glomerular disease with decreases in serum creatinine, glomeruli C3 deposits, and kidney pathology, however, IgG deposits, proteinuria, autoantibodies, and mortality were unaffected [287].

In contrast, MRL-*Fas^lpr* deficient for complement component C3 had the opposite effect with slightly accelerated proteinuria and greater glomerular IgG deposition [288]. Thus, C3, instead of promoting the development of GN, may have a mild beneficial role possibly in the clearance of immune complexes. Moreover, disease severity was the same in mixed 129xMRL-*Fas^lpr* C4 knockout mice with or without the C3 null mutation suggesting little to no role for C3 [289].

MRL-*Fas^lpr* mice overexpressing a soluble CR1-related gene Y (*Crry*) transgene, showed significant inhibition of complement activation and reduced mortality, kidney disease, and C3 deposition despite no effects on lymphadenopathy, autoantibodies, or IgG kidney deposits [290].

Decay-accelerating factor 1 (Daf1, CD55) is a GPI-anchored membrane protein that inhibits complement activation on autologous cells. MRL-*Fas^lpr* Daf1$^{-/-}$ mice have exacerbated disease, consistent with an autoimmune inhibitory role for Daf1 [291]. Suppression of dermatitis, but not lymphoproliferation and anti-chromatin antibodies by Daf1 was dependent on C3 [292].

Sex Hormones

Estrogen receptor 1 (ERα) deficiency in female, but not male, NZM2410 and MRL-*Fas^lpr* mice had reduced GN and survival despite increased or the same autoantibody levels [293]. In (NZBxNZW)F1 mice, absence of ERα resulted in similar results except for a reduction in anti-dsDNA [294].

Local Immune and Inflammatory Response Regulation

Several genes required primarily for the development of local end-organ inflammation have been identified. MRL-*Fas^lpr* mice deficient for ICAM-1 showed improved survival, more as a result of reduced cutaneous vasculitis than changes in autoantibody levels and GN [295, 296] consistent with ICAM-1 being critical for leukocyte adhesion in the skin [297]. Similarly, deletion of the *NOS2* gene in MRL-*Fas^lpr* mice resulted in partial disease abrogation, with significant reduction in vasculitis and IgG rheumatoid factor, but anti-DNA antibody levels and GN were equivalent [298]. In contrast, deletion of the FcR γ-chain in (NZB × NZW)F$_1$ mice resulted in decoupling of the immune and inflammatory processes, resulting in similar autoantibody levels and glomerular deposits of immunoglobulin or complement, but significantly less glomerular destruction and mortality [299]. A similar result was demonstrated in the anti-GBM antibody model [300]. Similar decoupling was observed in MRL-*Faslpr* mice with deletion of MCP-1 (macrophage chemoattractant protein-1, CCL2), a chemokine that recruits macrophages and T cells to tissues [301]. Likewise, MRL-*Fas^lpr* mice with deletion of macrophage migration inhibitory factor (MIF) have reduced GN and skin manifestations and prolonged survival despite no change in autoantibody levels [302]. More recently, kallikrein genes have also been implicated in susceptibility to nephritis in the lupus and GBM antibody-induced nephritis models [303].

The cappuccino (*Cno*) gene encodes the component of the biogenesis of lysosome-related organelle complex 1 (BLOC-1) that functions in organelle trafficking. A spontaneous function-impairing single nucleotide deletion in Cno in a recombinant BXSBxMRL-*Fas^lpr* strain (named EOD) that develops severe crescentic GN was found to significantly reduce crescentic GN, but not to affect autoAb or immune complex deposition [304]. This amelioratory effect of the Cno mutation was attributed to platelet dysfunction caused by the loss of cappuccino function.

P-selectin (*Selp*) and its ligand P-selectin glycoprotein ligand-1 (*Selplg*) promote leukocyte rolling on endothelial cells and subsequent emigration into tissues. Counterintuitively, deficiency of either *Selp* or *Selplg* in MRL-*Fas^lpr* mice resulted in enhanced GN and dermatitis possibly because of an observed enhanced expression of MCP-1 (CCL2) [305]. Thus, P-selectin and its ligand appear to play an unanticipated, but significant role in down-regulating inflammation in lupus.

The renin-angiotensin system is critical for blood pressure homeostasis and vascular reactivity, and may also play roles in immune responses and inflammation. Unexpectedly, MRL-*Fas^lpr* mice with deletion of one of the major angiotensin II receptors AT1a developed accelerated GN and mortality despite similar lymphoproliferation, autoantibodies, and disease in extrarenal tissues [306]. This was due to compensatory increased expression and stimulation of the other angiotensin II receptor, AT1b, which is primarily expressed in podocytes, leading to podocyte injury, and inflammation.

Miscellaneous genes

Terminal deoxynucleotidyl transferase (TdT) is essential for adding nucleotides to the N-regions at the V-(D)-J junctions during B- and T-cell antigen receptor rearrangement, thereby enhancing repertoire diversity [307]. TdT-deficiency in BWF1, B6-*Fas^lpr*, and MRL-*Fas^lpr* mice were all associated with varying degrees of reduced autoimmune disease indicating that a highly diverse repertoire plays a signifcant role in the development of lupus [308—310].

B6 mice deficient in ApoE, a lipoprotein transport protein, have increased anti-oxLDL and anti-cardiolipin autoantibodies, while autoimmunity in GVHD model was exacerbated [311]. B6-*Fas^lpr* mice deficient in ApoE exhibit severe exacerbation with greater autoantibodies and severe proteinuria and enlarged glomerular tufts [312].

Protein kinase CK2 (casein kinase II) is a ubiquitous heterotetrameric serine-theonine kinase composed of lupus and catalytic and regulatory subunits [313]. It can phosphorylate a large range of protein substrates, including those involved in nucleic acid synthesis, nuclear transcription, signal transduction, protein synthesis and the cytoskeleton. CK2 is active in proliferating cells, and high levels are found in certain human cancers. Overexpression of CK2 in MRL-*Fas^lpr* did not change lymphoma incidence, but markedly accelerated lymphoproliferation and autoimmunity, with higher IgG2a and earlier ANA positivity and proliferative GN [314]. CK2-mediated enhancement of T-cell proliferative responses was postulated to be responsible.

Leukotrienes are potent proinflammatory lipid mediators produced through the 5-lipoxygenase pathway of arachidonic acid metabolism. Arachidonic acid metabolites, including the leukotrienes, have been suggested to promote autoimmune pathogenesis in MRL-*Fas^lpr* mice [315], but 5-lipoxygenase deficiency in these mice resulted in a modest acceleration of mortality and a slight increase in the prevalence of arthritis in males only.

IMPLICATIONS FOR HUMAN SLE

Based on mouse studies, a number of inferences can be made about the genetics of human SLE: (1) There is potentially a large number of lupus-predisposing, -suppressing, and -modifying genes that could possibly include over 100 that could, as single mutations, induce ANAs. (2) Both ancestral and more recent genetic alterations are likely to affect the development of SLE, many of which are likely to be recessively inherited. This indicates that current efforts to define SLE-affecting genes by linkage analysis and SNP association studies will identify only a fraction of these genes. (3) Forward genetics approaches in mice have clearly documented the difficulty in identifying susceptibility genes when there is strong linkage dysequilibrium in the region of interest and in confirming even highly appealing candidate genes. These potential impediments will be even more difficult in human SLE and mouse models may provide an important and even necessary part in verification. (4) Based on the high degree of genetic heterogeneity in mouse studies as well as the fact that most of the allelic variants associated with SLE are in unaffected individuals because of the relatively low frequency of SLE (~1:2000) and marker frequency of ~1%, it is unlikely that identification of susceptibility genes will have significant diagnostic or prognostic utility except in possibly highly selected subpopulations. (5) Despite these limitations, identification of SLE-affecting genes will provide important and novel insights into disease pathogenesis and should reveal new targets for intervention.

CONCLUSION

With advancing technologies, it will be feasible in the foreseeable future to consider identification and characterization of most, if not all, of the major genes with potential to predispose and/or suppress lupus-like disease in mice. This should yield vital information about the specific pathways and mechanisms involved in both the maintenance and loss of immunological tolerance. Of particular importance will be studies to identify and characterize genes required for the development of lupus-like disease as these have relevance not only for basic understanding of disease processes, but also for therapy. Definition of the genetics of SLE in mouse models has advanced tremendously over the past decade and studies using these models should continue to generate new insights not only for lupus and autoimmunity, but also for the normal function of the immune system, as well as to provide a valuable resource for investigating human candidate genes.

Acknowledgments

This is Publication Number 20813-IMM from the Department of Immunology, The Scripps Research Institute, 10550 North Torrey Pines Road, La Jolla, CA 92037. The work of the authors reported herein was supported by National Institutes of Health grants AR31203, AR39555, AG15061 and ES08666. The authors thank M. Kat Occhinpinti for editorial assistance.

References

[1] M. Bielschowsky, B.J. Helyer, J.B. Howie, Spontaneous anemia in mice of the NZB/B1 strain, Proc. Unit. Otago Med. School. 37 (1959).

[2] K.L. Moser, J.A. Kelly, C.J. Lessard, J.B. Harley, Recent insights into the genetic basis of systemic lupus erythematosus, Genes Immun 10 (2009) 373—379.

[3] A.N. Theofilopoulos, F.J. Dixon, Murine models of systemic lupus erythematosus, Adv Immunol 37 (1985) 269—390.

[4] A.N. Theofilopoulos, D.H. Kono, Murine lupus models: gene-specific and genome-wide studies, in: R.G. Lahita (Ed.), Systemic Lupus Erythematosus, Academic Press, San Diego, 1999, pp. 145—181.

[5] J.D. Mountz, P. Yang, Q. Wu, J. Zhou, A. Tousson, A. Fitzgerald, et al., Genetic segregation of spontaneous erosive arthritis and generalized autoimmune disease in the BXD2 recombinant inbred strain of mice, Scand J Immunol 61 (2005) 128—138.

[6] L. Morel, C. Mohan, Y. Yu, B.P. Croker, N. Tian, A. Deng, et al., Functional dissection of systemic lupus erythematosus using congenic mouse strains, J Immunol 158 (1997) 6019—6028.

[7] C. Mohan, L. Morel, P. Yang, E.K. Wakeland, Genetic dissection of systemic lupus erythematosus pathogenesis: Sle2 on murine chromosome 4 leads to B cell hyperactivity, J Immunol 159 (1997) 454—465.

[8] C. Mohan, E. Alas, L. Morel, P. Yang, E.K. Wakeland, Genetic dissection of SLE pathogenesis. Sle1 on murine chromosome 1 leads to a selective loss of tolerance to H2A/H2B/DNA subnucleosomes, J Clin Invest 101 (1998) 1362—1372.

[9] C. Mohan, Y. Yu, L. Morel, P. Yang, E.K. Wakeland, Genetic dissection of Sle pathogenesis: Sle3 on murine chromosome 7 impacts T cell activation, differentiation, and cell death, J Immunol 162 (1999) 6492—6502.

[10] L. Morel, K.R. Blenman, B.P. Croker, E.K. Wakeland, The major murine systemic lupus erythematosus susceptibility locus, Sle1, is a cluster of functionally related genes, Proc Natl Acad Sci USA 98 (2001) 1787—1792.

[11] S.J. Rozzo, J.D. Allard, D. Choubey, T.J. Vyse, S. Izui, G. Peltz, et al., Evidence for an interferon-inducible gene, Ifi202, in the susceptibility to systemic lupus, Immunity 15 (2001) 435—443.

[12] J.E. Wither, G. Lajoie, S. Heinrichs, Y.C. Cai, N. Chang, A. Ciofani, et al., Functional dissection of lupus susceptibility loci on the New Zealand black mouse chromosome 1: evidence for independent genetic loci affecting T and B cell activation, J Immunol 171 (2003) 1697—1706.

[13] C. Laporte, B. Ballester, C. Mary, S. Izui, L. Reininger, The Sgp3 locus on mouse chromosome 13 regulates nephritogenic gp70 autoantigen expression and predisposes to autoimmunity, J Immunol 171 (2003) 3872—3877.

[14] P.L. Kong, L. Morel, B.P. Croker, J. Craft, The centromeric region of chromosome 7 from MRL mice (Lmb3) is an epistatic modifier of Fas for autoimmune disease expression, J Immunol 172 (2004) 2785—2794.

[15] A.E. Wandstrat, C. Nguyen, N. Limaye, A.Y. Chan, S. Subramanian, X.H. Tian, et al., Association of extensive polymorphisms in the SLAM/CD2 gene cluster with murine lupus, Immunity 21 (2004) 769—780.

[16] M.E. Haywood, N.J. Rogers, S.J. Rose, J. Boyle, A. McDermott, J.M. Rankin, et al., Dissection of BXSB lupus phenotype using

mice congenic for chromosome 1 demonstrates that separate intervals direct different aspects of disease, J Immunol 173 (2004) 4277–4285.

[17] S.T. Waters, M. McDuffie, H. Bagavant, U.S. Deshmukh, F. Gaskin, C. Jiang, et al., Breaking tolerance to double stranded DNA, nucleosome, and other nuclear antigens is not required for the pathogenesis of lupus glomerulonephritis, J Exp Med 199 (2004) 255–264.

[18] M.K. Haraldsson, N.G. dela Paz, J.G. Kuan, G.S. Gilkeson, A.N. Theofilopoulos, D.H. Kono, Autoimmune alterations induced by the New Zealand Black Lbw2 locus in BWF1 mice, J Immunol 174 (2005) 5065–5073.

[19] J.E. Wither, C. Loh, G. Lajoie, S. Heinrichs, Y.C. Cai, G. Bonventi, et al., Colocalization of expansion of the splenic marginal zone population with abnormal B cell activation and autoantibody production in B6 mice with an introgressed New Zealand Black chromosome 13 interval, J Immunol 175 (2005) 4309–4319.

[20] S. Subramanian, Y.S. Yim, K. Liu, K. Tus, X.J. Zhou, E.K. Wakeland, Epistatic suppression of systemic lupus erythematosus: fine mapping of Sles1 to less than 1 mb, J Immunol 175 (2005) 1062–1072.

[21] S. Kikuchi, L. Fossati-Jimack, T. Moll, H. Amano, E. Amano, A. Ida, et al., Differential role of three major New Zealand Black-derived loci linked with Yaa-induced murine lupus nephritis, J Immunol 174 (2005) 1111–1117.

[22] M.-L. Santiago-Raber, M.K. Haraldsson, A.N. Theofilopoulos, D.H. Kono, Characterization of reciprocal Lmb1-4 interval MRL-Fas^lpr and B6-Fas^lpr congenic mice reveals significant effects from Lmb3, J Immunol 178 (2007) 8195–8202.

[23] K. Liu, Q.Z. Li, Y. Yu, C. Liang, S. Subramanian, Z. Zeng, et al., Sle3 and Sle5 can independently couple with Sle1 to mediate severe lupus nephritis, Genes Immun 8 (2007) 634–645.

[24] F. Carlucci, J. Cortes-Hernandez, L. Fossati-Jimack, A.E. Bygrave, M.J. Walport, T.J. Vyse, et al., Genetic dissection of spontaneous autoimmunity driven by 129-derived chromosome 1 Loci when expressed on C57BL/6 mice, J Immunol 178 (2007) 2352–2360.

[25] C. Loh, Y.C. Cai, G. Bonventi, G. Lajoie, R. Macleod, J.E. Wither, Dissociation of the genetic loci leading to b1a and NKT cell expansions from autoantibody production and renal disease in b6 mice with an introgressed new zealand black chromosome 4 interval, J Immunol 178 (2007) 1608–1617.

[26] O. Ichii, A. Konno, N. Sasaki, D. Endoh, Y. Hashimoto, Y. Kon, Autoimmune glomerulonephritis induced in congenic mouse strain carrying telomeric region of chromosome 1 derived from MRL/MpJ, Histol Histopathol 23 (2008) 411–422.

[27] T. Tarasenko, H.K. Kole, S. Bolland, A lupus-suppressor BALB/c locus restricts IgG2 autoantibodies without altering intrinsic B cell-tolerance mechanisms, J Immunol 180 (2008) 3807–3814.

[28] Y. Heidari, L. Fossati-Jimack, F. Carlucci, M.J. Walport, H.T. Cook, M. Botto, A lupus-susceptibility C57BL/6 locus on chromosome 3 (Sle18) contributes to autoantibody production in 129 mice, Genes Immun 10 (2009) 47–55.

[29] A.E. Bygrave, K.L. Rose, J. Cortes-Hernandez, J. Warren, R.J. Rigby, H.T. Cook, et al., Spontaneous autoimmunity in 129 and C57BL/6 mice-implications for autoimmunity described in gene-targeted mice, PLoS Biol. 2 (2004) E243.

[30] J.D. Gillmore, W.L. Hutchinson, J. Herbert, A. Bybee, D.A. Mitchell, R.P. Hasserjian, et al., Autoimmunity and glomerulonephritis in mice with targeted deletion of the serum amyloid P component gene: SAP deficiency or strain combination? Immunology 112 (2004) 255–264.

[31] Y. Heidari, A.E. Bygrave, R.J. Rigby, K.L. Rose, M.J. Walport, H.T. Cook, et al., Identification of chromosome intervals from 129 and C57BL/6 mouse strains linked to the development of systemic lupus erythematosus, Genes Immun 7 (2006) 592–599.

[32] L. Fossati-Jimack, J. Cortes-Hernandez, P.J. Norsworthy, H.T. Cook, M.J. Walport, M. Botto, Regulation of B cell tolerance by 129-derived chromosome 1 loci in C57BL/6 mice, Arthritis Rheum 58 (2008) 2131–2141.

[33] T. Fujimura, S. Hirose, Y. Jiang, S. Kodera, H. Ohmuro, D. Zhang, et al., Dissection of the effects of tumor necrosis factor-alpha and class II gene polymorphisms within the MHC on murine systemic lupus erythematosus (SLE), Int Immunol 10 (1998) 1467–1472.

[34] C.O. Jacob, S.K. Lee, G. Strassmann, Mutational analysis of TNF-alpha gene reveals a regulatory role for the 3′-untranslated region in the genetic predisposition to lupus-like autoimmune disease, J Immunol 156 (1996) 3043–3050.

[35] M.C. Carroll, The role of complement in B cell activation and tolerance, Adv Immunol 74 (2000) 61–88.

[36] Y. Zhang, S.F. Schlossman, R.A. Edwards, C.N. Ou, J. Gu, M.X. Wu, Impaired apoptosis, extended duration of immune responses, and a lupus-like autoimmune disease in IEX-1-transgenic mice, Proc Natl Acad Sci USA 99 (2002) 878–883.

[37] L. Morel, X.H. Tian, B.P. Croker, E.K. Wakeland, Epistatic modifiers of autoimmunity in a murine model of lupus nephritis, Immunity 11 (1999) 131–139.

[38] C. Mary, C. Laporte, D. Parzy, M.L. Santiago, F. Stefani, F. Lajaunias, et al., Dysregulated expression of the Cd22 gene as a result of a short interspersed nucleotide element insertion in Cd22a lupus-prone mice, J Immunol 165 (2000) 2987–2996.

[39] Y. Miura-Shimura, K. Nakamura, M. Ohtsuji, H. Tomita, Y. Jiang, M. Abe, et al., C1q regulatory region polymorphism down-regulating murine c1q protein levels with linkage to lupus nephritis, J Immunol 169 (2002) 1334–1339.

[40] W. Qu, T. Miyazaki, M. Terada, L. Lu, M. Nishihara, A. Yamada, et al., Genetic dissection of vasculitis in MRL/lpr lupus mice: a novel susceptibility locus involving the CD72^c allele, Eur J Immunol 30 (2000) 2027–2037.

[41] J.I. Elliott, J.H. McVey, C.F. Higgins, The P2X7 receptor is a candidate product of murine and human lupus susceptibility loci: a hypothesis and comparison of murine allelic products, Arthritis Res Ther 7 (2005) R468–475.

[42] Y. Jiang, M. Ohtsuji, M. Abe, N. Li, Y. Xiu, S. Wen, et al., Polymorphism and chromosomal mapping of the mouse gene for B-cell activating factor belonging to the tumor necrosis factor family (Baff) and association with the autoimmune phenotype, Immunogenetics 53 (2001) 810–813.

[43] N.J. Rogers, M.J. Lees, L. Gabriel, E. Maniati, S.J. Rose, P.K. Potter, et al., A defect in Marco expression contributes to systemic lupus erythematosus development via failure to clear apoptotic cells, J Immunol 182 (2009) 1982–1990.

[44] C.G. Vinuesa, C.C. Goodnow, Illuminating autoimmune regulators through controlled variation of the mouse genome sequence, Immunity 20 (2004) 669–679.

[45] M.L. Hibbs, D.M. Tarlinton, J. Armes, D. Grail, G. Hodgson, R. Maglitto, et al., Multiple defects in the immune system of Lyn-deficient mice, culminating in autoimmune disease, Cell 83 (1995) 301–311.

[46] H. Nishizumi, I. Taniuchi, Y. Yamanashi, D. Kitamura, D. Ilic, S. Mori, et al., Impaired proliferation of peripheral B cells and indication of autoimmune disease in lyn-deficient mice, Immunity 3 (1995) 549–560.

[47] M.L. Hibbs, K.W. Harder, J. Armes, N. Kountouri, C. Quilici, F. Casagranda, et al., Sustained activation of Lyn tyrosine

kinase in vivo leads to autoimmunity, J Exp Med 196 (2002) 1593—1604.

[48] R.J. Cornall, J.G. Cyster, M.L. Hibbs, A.R. Dunn, K.L. Otipoby, E.A. Clark, et al., Polygenic autoimmune traits: Lyn, CD22, and SHP-1 are limiting elements of a biochemical pathway regulating BCR signaling and selection, Immunity 8 (1998) 497—508.

[49] T.F. Tedder, M. Inaoki, S. Sato, The CD19-CD21 complex regulates signal transduction thresholds governing humoral immunity and autoimmunity, Immunity 6 (1997) 107—118.

[50] S. Bolland, J.V. Ravetch, Spontaneous autoimmune disease in Fc(gamma)RIIB-deficient mice results from strain-specific epistasis, Immunity 13 (2000) 277—285.

[51] D.H. Li, M.M. Winslow, T.M. Cao, A.H. Chen, C.R. Davis, E.D. Mellins, et al., Modulation of peripheral B cell tolerance by CD72 in a murine model, Arthritis Rheum 58 (2008) 3192—3204.

[52] A. Kumanogoh, T. Shikina, C. Watanabe, N. Takegahara, K. Suzuki, M. Yamamoto, et al., Requirement for CD100-CD72 interactions in fine-tuning of B-cell antigen receptor signaling and homeostatic maintenance of the B-cell compartment, Int Immunol 17 (2005) 1277—1282.

[53] M.L. Hermiston, A.L. Tan, V.A. Gupta, R. Majeti, A. Weiss, The juxtamembrane wedge negatively regulates CD45 function in B cells, Immunity 23 (2005) 635—647.

[54] F. Mackay, S.A. Woodcock, P. Lawton, C. Ambrose, M. Baetscher, P. Schneider, et al., Mice transgenic for BAFF develop lymphocytic disorders along with autoimmune manifestations, J Exp Med 190 (1999) 1697—1710.

[55] S.D. Khare, I. Sarosi, X.Z. Xia, S. McCabe, K. Miner, I. Solovyev, et al., Severe B cell hyperplasia and autoimmune disease in TALL-1 transgenic mice, Proc Natl Acad Sci USA 97 (2000) 3370—3375.

[56] J.A. Gross, J. Johnston, S. Mudri, R. Enselman, S.R. Dillon, K. Madden, et al., TACI and BCMA are receptors for a TNF homologue implicated in B-cell autoimmune disease, Nature 404 (2000) 995—999.

[57] J.R. Groom, C.A. Fletcher, S.N. Walters, S.T. Grey, S.V. Watt, M.J. Sweet, BAFF and MyD88 signals promote a lupuslike disease independent of T cells, J Exp Med 204 (2007) 1959—1971.

[58] J.H. Wang, N. Avitahl, A. Cariappa, C. Friedrich, T. Ikeda, A. Renold, et al., Aiolos regulates B cell activation and maturation to effector state, Immunity 9 (1998) 543—553.

[59] T. Higuchi, Y. Aiba, T. Nomura, J. Matsuda, K. Mochida, M. Suzuki, et al., Cutting Edge: Ectopic expression of CD40 ligand on B cells induces lupus-like autoimmune disease, J Immunol 168 (2002) 9—12.

[60] Y. Qian, J. Qin, G. Cui, M. Naramura, E.C. Snow, C.F. Ware, et al., Act1, a negative regulator in CD40- and BAFF-mediated B cell survival, Immunity 21 (2004) 575—587.

[61] E. Claudio, S.U. Sonder, S. Saret, G. Carvalho, T.R. Ramalingam, T.A. Wynn, et al., The adaptor protein CIKS/Act1 is essential for IL-25-mediated allergic airway inflammation, J Immunol 182 (2009) 1617—1630.

[62] T. Shimizu, C. Mundt, S. Licence, F. Melchers, I.L. Martensson, VpreB1/VpreB2/lambda 5 triple-deficient mice show impaired B cell development but functional allelic exclusion of the IgH locus, J Immunol 168 (2002) 6286—6293.

[63] R.A. Keenan, A. De Riva, B. Corleis, L. Hepburn, S. Licence, T.H. Winkler, et al., Censoring of autoreactive B cell development by the pre-B cell receptor, Science 321 (2008) 696—699.

[64] I. Mecklenbrauker, K. Saijo, N.Y. Zheng, M. Leitges, A. Tarakhovsky, Protein kinase Cdelta controls self-antigen-induced B-cell tolerance, Nature 416 (2002) 860—865.

[65] A. Miyamoto, K. Nakayama, H. Imaki, S. Hirose, Y. Jiang, M. Abe, et al., Increased proliferation of B cells and autoimmunity in mice lacking protein kinase Cdelta, Nature 416 (2002) 865—869.

[66] P. Yu, R. Constien, N. Dear, M. Katan, P. Hanke, T.D. Bunney, et al., Autoimmunity and inflammation due to a gain-of-function mutation in phospholipase C gamma 2 that specifically increases external Ca2+ entry, Immunity 22 (2005) 451—465.

[67] A. Seth, R. Ascione, R.J. Fisher, G.J. Mavrothalassitis, N.K. Bhat, T.S. Papas, The ets gene family, Cell Growth Differ 3 (1992) 327—334.

[68] L. Zhang, A. Eddy, Y.T. Teng, M. Fritzler, M. Kluppel, F. Melet, et al., An immunological renal disease in transgenic mice that overexpress Fli-1, a member of the ets family of transcription factor genes, Mol Cell Biol. 15 (1995) 6961—6970.

[69] X.K. Zhang, S. Gallant, I. Molano, O.M. Moussa, P. Ruiz, D.D. Spyropoulos, et al., Decreased expression of the Ets family transcription factor Fli-1 markedly prolongs survival and significantly reduces renal disease in MRL/lpr mice, J Immunol 173 (2004) 6481—6489.

[70] S. Bradshaw, W.J. Zheng, L.C. Tsoi, G. Gilkeson, X.K. Zhang, A role for Fli-1 in B cell proliferation: implications for SLE pathogenesis, Clin Immunol 129 (2008) 19—30.

[71] M.T. Savino, B. Ortensi, M. Ferro, C. Ulivieri, D. Fanigliulo, E. Paccagnini, et al., Rai acts as a negative regulator of autoimmunity by inhibiting antigen receptor signaling and lymphocyte activation, J Immunol 182 (2009) 301—308.

[72] D. Ishida, L. Su, A. Tamura, Y. Katayama, Y. Kawai, S.F. Wang, et al., Rap1 signal controls B cell receptor repertoire and generation of self-reactive B1a cells, Immunity 24 (2006) 417—427.

[73] R. Wilkinson, A.B. Lyons, D. Roberts, M.X. Wong, P.A. Bartley, D.E. Jackson, Platelet endothelial cell adhesion molecule-1 (PECAM-1/CD31) acts as a regulator of B-cell development, B-cell antigen receptor (BCR)-mediated activation, and autoimmune disease, Blood 100 (2002) 184—193.

[74] C. Xiao, L. Srinivasan, D.P. Calado, H.C. Patterson, B. Zhang, J. Wang, et al., Lymphoproliferative disease and autoimmunity in mice with increased miR-17-92 expression in lymphocytes, Nat Immunol (2008).

[75] M.M. Shull, I. Ormsby, A.B. Kier, S. Pawlowski, R.J. Diebold, M. Yin, et al., Trageted disruption of the mouse transforming growth factor-1 gene results in multifocal inflammatory disease, Nature 359 (1992) 693—699.

[76] A.B. Kulkarni, S. Karlsson, Transforming growth factor-beta 1 knockout mice. A mutation in one cytokine gene causes a dramatic inflammatory disease, Am J Pathol 143 (1993) 3—9.

[77] H. Dang, A.G. Geiser, J.J. Letterio, T. Nakabayashi, L. Kong, G. Fernandes, et al., SLE-like autoantibodies and Sjögren's syndrome-like lymphoproliferation in TGF- knockout mice, J Immunol 155 (1995) 3205—3212.

[78] P. Waterhouse, J.M. Penninger, E. Timms, A. Wakeham, A. Shahinian, K.P. Lee, et al., Lymphoproliferative disorders with early lethality in mice deficient in Ctla-4, Science 270 (1995) 985—988.

[79] E.A. Tivol, F. Borriello, A.N. Schweitzer, W.P. Lynch, J.A. Bluestone, A.H. Sharpe, Loss of CTLA-4 leads to massive lymphoproliferation and fatal multiorgan tissue destruction, revealing a critical negative regulatory role of CTLA-4, Immunity 3 (1995) 541—547.

[80] E.A. Tivol, S.D. Boyd, S. McKeon, F. Borriello, P. Nickerson, T.B. Strom, et al., CTLA4Ig prevents lymphoproliferation and fatal multiorgan tissue destruction in CTLA-4-deficient mice, J Immunol 158 (1997) 5091—5094.

[81] Y. Ishida, Y. Agata, K. Shibahara, T. Honjo, Induced expression of PD-1, a novel member of the immunoglobulin gene superfamily, upon programmed cell death, Embo J 11 (1992) 3887–3895.

[82] R.J. Greenwald, G.J. Freeman, A.H. Sharpe, The B7 family revisited, Annu Rev Immunol 23 (2005) 515–548.

[83] G.J. Freeman, A.J. Long, Y. Iwai, K. Bourque, T. Chernova, H. Nishimura, et al., Engagement of the PD-1 immunoinhibitory receptor by a novel B7 family member leads to negative regulation of lymphocyte activation, J Exp Med 192 (2000) 1027–1034.

[84] H. Nishimura, N. Minato, T. Nakano, T. Honjo, Immunological studies on PD-1 deficient mice: implication of PD-1 as a negative regulator for B cell responses, Int Immunol 10 (1998) 1563–1572.

[85] H. Nishimura, M. Nose, H. Hiai, N. Minato, T. Honjo, Development of lupus-like autoimmune diseases by disruption of the PD-1 gene encoding an ITIM motif-carrying immunoreceptor, Immunity 11 (1999) 141–151.

[86] T. Okazaki, Y. Tanaka, R. Nishio, T. Mitsuiye, A. Mizoguchi, J. Wang, et al., Autoantibodies against cardiac troponin I are responsible for dilated cardiomyopathy in PD-1-deficient mice, Nat Med 9 (2003) 1477–1483.

[87] L. Prokunina, C. Castillejo-Lopez, F. Oberg, I. Gunnarsson, L. Berg, V. Magnusson, et al., A regulatory polymorphism in PDCD1 is associated with susceptibility to systemic lupus erythematosus in humans, Nat Genet. 32 (2002) 666–669.

[88] S.C. Lin, J.H. Yen, J.J. Tsai, W.C. Tsai, T.T. Ou, H.W. Liu, et al., Association of a programmed death 1 gene polymorphism with the development of rheumatoid arthritis, but not systemic lupus erythematosus, Arthritis Rheum 50 (2004) 770–775.

[89] L. Prokunina, L. Padyukov, A. Bennet, U. de Faire, B. Wiman, J. Prince, et al., Association of the PD-1.3A allele of the PDCD1 gene in patients with rheumatoid arthritis negative for rheumatoid factor and the shared epitope, Arthritis Rheum 50 (2004) 1770–1773.

[90] E.K. Kong, L. Prokunina-Olsson, W.H. Wong, C.S. Lau, T.M. Chan, M. Alarcon-Riquelme, et al., A new haplotype of PDCD1 is associated with rheumatoid arthritis in Hong Kong Chinese, Arthritis Rheum 52 (2005) 1058–1062.

[91] A. Kroner, M. Mehling, B. Hemmer, P. Rieckmann, K.V. Toyka, M. Maurer, et al., A PD-1 polymorphism is associated with disease progression in multiple sclerosis, Ann Neurol 58 (2005) 50–57.

[92] J.A. Lucas, J. Menke, W.A. Rabacal, F.J. Schoen, A.H. Sharpe, V.R. Kelley, Programmed death ligand 1 regulates a critical checkpoint for autoimmune myocarditis and pneumonitis in MRL mice, J Immunol 181 (2008) 2513–2521.

[93] J. Menke, J.A. Lucas, G.C. Zeller, M.E. Keir, X.R. Huang, N. Tsuboi, et al., Programmed death 1 ligand (PD-L) 1 and PD-L2 limit autoimmune kidney disease: distinct roles, J Immunol 179 (2007) 7466–7477.

[94] J. Wang, J.C. Lo, A. Foster, P. Yu, H.M. Chen, Y. Wang, et al., The regulation of T cell homeostasis and autoimmunity by T cell-derived LIGHT, J Clin Invest 108 (2001) 1771–1780.

[95] R.B. Shaikh, S. Santee, S.W. Granger, K. Butrovich, T. Cheung, M. Kronenberg, et al., Constitutive expression of LIGHT on T cells leads to lymphocyte activation, inflammation, and tissue destruction, J Immunol 167 (2001) 6330–6337.

[96] S.W. Granger, C.F. Ware, Turning on LIGHT, J Clin Invest 108 (2001) 1741–1742.

[97] D. Yablonski, A. Weiss, Mechanisms of signaling by the hematopoietic-specific adaptor proteins, SLP-76 and LAT and their B cell counterpart, BLNK/SLP-65, Adv Immunol 79 (2001) 93–128.

[98] C.L. Sommers, C.S. Park, J. Lee, C. Feng, C.L. Fuller, A. Grinberg, et al., A LAT mutation that inhibits T cell development yet induces lymphoproliferation, Science 296 (2002) 2040–2043.

[99] M.T. Aguado, R.S. Balderas, R.L. Rubin, M.A. Duchosal, R. Kofler, B.K. Birshtein, et al., Specificity and molecular characteristics of monoclonal IgM rheumatoid factors from arthritic and non-arthritic mice, J Immunol 139 (1987) 1080–1087.

[100] L. Duan, A.L. Reddi, A. Ghosh, M. Dimri, H. Band, The Cbl family and other ubiquitin ligases: destructive forces in control of antigen receptor signaling, Immunity 21 (2004) 7–17.

[101] Y.J. Chiang, H.K. Kole, K. Brown, M. Naramura, S. Fukuhara, R.J. Hu, et al., Cbl-b regulates the CD28 dependence of T-cell activation, Nature 403 (2000) 216–220.

[102] K. Bachmaier, C. Krawczyk, I. Kozieradzki, Y.Y. Kong, T. Sasaki, A. Oliveira-dos-Santos, et al., Negative regulation of lymphocyte activation and autoimmunity by the molecular adaptor Cbl-b, Nature 403 (2000) 211–216.

[103] Y. Kitaura, I.K. Jang, Y. Wang, Y.C. Han, T. Inazu, E.J. Cadera, et al., Control of the B cell-intrinsic tolerance programs by ubiquitin ligases Cbl and Cbl-b, Immunity 26 (2007) 567–578.

[104] L.Q. Le, J.H. Kabarowski, Z. Weng, A.B. Satterthwaite, E.T. Harvill, E.R. Jensen, et al., Mice lacking the orphan G protein-coupled receptor G2A develop a late-onset autoimmune syndrome, Immunity 14 (2001) 561–571.

[105] G.S. Vratsanos, S. Jung, Y.M. Park, J. Craft, CD4(+) T cells from lupus-prone mice are hyperresponsive to T cell receptor engagement with low and high affinity peptide antigens: a model to explain spontaneous T cell activation in lupus, J Exp Med 193 (2001) 329–337.

[106] J. DeGregori, The genetics of the E2F family of transcription factors: shared functions and unique roles, Biochim Biophys Acta 1602 (2002) 131–150.

[107] M. Murga, O. Fernandez-Capetillo, S.J. Field, B. Moreno, R.B. L, Y. Fujiwara, D. Balomenos, et al., Mutation of E2F2 in Mice Causes Enhanced T Lymphocyte Proliferation, Leading to the Development of Autoimmunity, Immunity 15 (2001) 959–970.

[108] W.R. Hardy, L. Li, Z. Wang, J. Sedy, J. Fawcett, E. Frank, et al., Combinatorial ShcA docking interactions support diversity in tissue morphogenesis, Science 317 (2007) 251–256.

[109] E. Migliaccio, M. Giorgio, S. Mele, G. Pelicci, P. Reboldi, P.P. Pandolfi, et al., The p66shc adaptor protein controls oxidative stress response and life span in mammals, Nature 402 (1999) 309–313.

[110] F. Finetti, M. Pellegrini, C. Ulivieri, M.T. Savino, E. Paccagnini, C. Ginanneschi, et al., The proapoptotic and antimitogenic protein p66SHC acts as a negative regulator of lymphocyte activation and autoimmunity, Blood 111 (2008) 5017–5027.

[111] T. Yasuda, K. Bundo, A. Hino, K. Honda, A. Inoue, M. Shirakata, et al., Dok-1 and Dok-2 are negative regulators of T cell receptor signaling, Int Immunol 19 (2007) 487–495.

[112] C.G. Vinuesa, M.C. Cook, C. Angelucci, V. Athanasopoulos, L. Rui, K.M. Hill, et al., A RING-type ubiquitin ligase family member required to repress follicular helper T cells and autoimmunity, Nature 435 (2005) 452–458.

[113] M.A. Linterman, R.J. Rigby, R.K. Wong, D. Yu, R. Brink, J.L. Cannons, et al., Follicular helper T cells are required for systemic autoimmunity, J Exp Med 206 (2009) 561–576.

[114] M. Demetriou, M. Granovsky, S. Quaggin, J.W. Dennis, Negative regulation of T-cell activation and autoimmunity by Mgat5 N-glycosylation, Nature 409 (2001) 733–739.

[115] N.L. Perillo, K.E. Pace, J.J. Seilhamer, L.G. Baum, Apoptosis of T cells mediated by galectin-1, Nature 378 (1995) 736–739.

[116] G.N. Vespa, L.A. Lewis, K.R. Kozak, M. Moran, J.T. Nguyen, L.G. Baum, et al., Galectin-1 specifically modulates TCR signals to enhance TCR apoptosis but inhibit IL-2 production and proliferation, J Immunol 162 (1999) 799–806.

[117] E. Kim, J.D. Lowenson, D.C. MacLaren, S. Clarke, S.G. Young, Deficiency of a protein-repair enzyme results in the accumulation

of altered proteins, retardation of growth, and fatal seizures in mice, Proc Natl Acad Sci U S A 94 (1997) 6132−6137.

[118] H.A. Doyle, R.J. Gee, M.J. Mamula, A failure to repair self-proteins leads to T cell hyperproliferation and autoantibody production, J Immunol 171 (2003) 2840−2847.

[119] W. Fan, G. Richter, A. Cereseto, C. Beadling, K.A. Smith, Cytokine response gene 6 induces p21 and regulates both cell growth and arrest, Oncogene 18 (1999) 6573−6582.

[120] M. Vairapandi, A.G. Balliet, A.J. Fornace Jr., B. Hoffman, D.A. Liebermann, The differentiation primary response gene MyD118, related to GADD45, encodes for a nuclear protein which interacts with PCNA and p21WAF1/CIP1, Oncogene 12 (1996) 2579−2594.

[121] J.M. Salvador, M.C. Hollander, A.T. Nguyen, J.B. Kopp, L. Barisoni, J.K. Moore, et al., Mice lacking the p53-effector gene Gadd45a develop a lupus-like syndrome, Immunity 16 (2002) 499−508.

[122] L. Liu, E. Tran, Y. Zhao, Y. Huang, R. Flavell, B. Lu, Gadd45 beta and Gadd45 gamma are critical for regulating autoimmunity, J Exp Med 202 (2005) 1341−1347.

[123] S. Becart, A. Altman, SWAP-70-like adapter of T cells: a novel Lck-regulated guanine nucleotide exchange factor coordinating actin cytoskeleton reorganization and Ca2+ signaling in T cells, Immunol Rev. 232 (2009) 319−333.

[124] J.C. Fanzo, W. Yang, S.Y. Jang, S. Gupta, Q. Chen, A. Siddiq, et al., Loss of IRF-4-binding protein leads to the spontaneous development of systemic autoimmunity, J Clin Invest 116 (2006) 703−714.

[125] A. Altman, S. Becart, Does Def6 deficiency cause autoimmunity? Immunity 31 (2009) 1−2. author reply 2−3.

[126] D. Balomenos, J. Martin-Caballero, M.I. Garcia, I. Prieto, J.M. Flores, M. Serrano, et al., The cell cycle inhibitor p21 controls T-cell proliferation and sex-linked lupus development, Nat Med 6 (2000) 171−176.

[127] M.L. Santiago-Raber, B.R. Lawson, W. Dummer, M. Barnhouse, S. Koundouris, C.B. Wilson, et al., The role of cyclin kinase inhibitor p21 in systemic autoimmunity, J Immunol 167 (2001) 4067−4074.

[128] B.R. Lawson, D.H. Kono, A.N. Theofilopoulos, Deletion of p21 (WAF-1/Cip1) does not induce systemic autoimmunity in female BXSB mice, J Immunol 168 (2002) 5928−5932.

[129] P. Bouillet, D. Metcalf, D.C. Huang, D.M. Tarlinton, T.W. Kay, F. Kontgen, et al., Proapoptotic Bcl-2 relative Bim required for certain apoptotic responses, leukocyte homeostasis, and to preclude autoimmunity, Science 286 (1999) 1735−1738.

[130] J. Hutcheson, J.C. Scatizzi, A.M. Siddiqui, G.K. Haines 3rd, T. Wu, Q.Z. Li, et al., Combined deficiency of proapoptotic regulators Bim and Fas results in the early onset of systemic autoimmunity, Immunity 28 (2008) 206−217.

[131] A. Strasser, S. Whittingham, D.L. Vaux, M.L. Bath, J.M. Adams, S. Cory, et al., Enforced bcl-2 expression in B-lymphoid cells prolongs antibody responses and elicits autoimmune disease, Proc Natl Acad Sci USA 88 (1991) 8661−8665.

[132] S.K. Ray, C. Putterman, B. Diamond, Pathogenic autoantibodies are routinely generated during the response to foreign antigen: a paradigm for autoimmune disease, Proc Natl Acad Sci U S A 93 (1996) 2019−2024.

[133] P. Kuo, M. Bynoe, B. Diamond, Crossreactive B cells are present during a primary but not secondary response in BALB/c mice expressing a bcl-2 transgene, Mol Immunol 36 (1999) 471−479.

[134] L. Mandik-Nayak, S. Nayak, C. Sokol, A. Eaton-Bassiri, M.P. Madaio, A.J. Caton, et al., The origin of anti-nuclear antibodies in bcl-2 transgenic mice, Int Immunol 12 (2000) 353−364.

[135] H. Schorle, T. Holtschke, T. Hunig, A. Schimpl, I. Horak, Development and function of T cells in mice rendered interleukin-2 deficient by gene targeting, Nature 352 (1991) 621−624.

[136] T.M. Kundig, H. Schorle, M.F. Bachmann, H. Hengartner, R.M. Zinkernagel, I. Horak, Immune responses in interleukin-2-deficient mice, Science 262 (1993) 1059−1061.

[137] B. Sadlack, J. Lohler, H. Schorle, G. Klebb, H. Haber, E. Sickel, et al., Generalized autoimmune disease in interleukin-2-deficient mice is triggered by an uncontrolled activation and proliferation of CD4+ T cells, Eur J Immunol 25 (1995) 3053−3059.

[138] D.M. Willerford, J. Chen, J.A. Ferry, L. Davidson, A. Ma, F.W. Alt, Interleukin-2 receptor alpha chain regulates the size and content of the peripheral lymphoid compartment, Immunity 3 (1995) 521−530.

[139] H. Suzuki, T.M. Kundig, C. Furlonger, A. Wakeham, E. Timms, T. Matsuyama, et al., Deregulated T cell activation and autoimmunity in mice lacking interleukin-2 receptor beta, Science 268 (1995) 1472−1476.

[140] B. Sadlack, H. Merz, H. Schorle, A. Schimpl, A.C. Feller, I. Horak, Ulcerative colitis-like disease in mice with a disrupted interleukin-2 gene, Cell 75 (1993) 253−261.

[141] Y. Rochman, R. Spolski, W.J. Leonard, New insights into the regulation of T cells by gamma(c) family cytokines, Nat Rev Immunol 9 (2009) 480−490.

[142] F. Marti, P.E. Lapinski, P.D. King, The emerging role of the T cell-specific adaptor (TSAd) protein as an autoimmune disease-regulator in mouse and man, Immunol Lett. 97 (2005) 165−170.

[143] V. Sundvold-Gjerstad, S. Granum, T. Mustelin, T.C. Andersen, T. Berge, M.J. Shapiro, et al., The C terminus of T cell-specific adapter protein (TSAd) is necessary for TSAd-mediated inhibition of Lck activity, Eur J Immunol 35 (2005) 1612−1620.

[144] L.C. Cantley, B.G. Neel, New insights into tumor suppression: PTEN suppresses tumor formation by restraining the phosphoinositide 3-kinase/AKT pathway, Proc Natl Acad Sci U S A 96 (1999) 4240−4245.

[145] A. Di Cristofano, P.P. Pandolfi, The multiple roles of PTEN in tumor suppression, Cell 100 (2000) 387−390.

[146] A. Di Cristofano, P. Kotsi, Y.F. Peng, C. Cordon-Cardo, K.B. Elkon, P.P. Pandolfi, Impaired Fas response and autoimmunity in Pten+/- mice, Science 285 (1999) 2122−2125.

[147] C.D. Browne, C.J. Del Nagro, M.H. Cato, H.S. Dengler, R.C. Rickert, Suppression of phosphatidylinositol 3,4,5-trisphosphate production is a key determinant of B cell anergy, Immunity 31 (2009) 749−760.

[148] L. Shen, L. Zhi, W. Hu, M.X. Wu, IEX-1 targets mitochondrial F1Fo-ATPase inhibitor for degradation, Cell Death Differ 16 (2009) 603−612.

[149] N.A. Dower, S.L. Stang, D.A. Bottorff, J.O. Ebinu, P. Dickie, H.L. Ostergaard, J.C. Stone, RasGRP is essential for mouse thymocyte differentiation and TCR signaling, Nat Immunol 1 (2000) 317−321.

[150] K. Layer, G. Lin, A. Nencioni, W. Hu, A. Schmucker, A.N. Antov, et al., Autoimmunity as the consequence of a spontaneous mutation in Rasgrp1, Immunity 19 (2003) 243−255.

[151] H. Sun, B. Lu, R.Q. Li, R.A. Flavell, R. Taneja, Defective T cell activation and autoimmune disorder in Stra13-deficient mice, Nat Immunol 2 (2001) 1040−1047.

[152] M. Seimiya, A. Wada, K. Kawamura, A. Sakamoto, Y. Ohkubo, S. Okada, et al., Impaired lymphocyte development and function in Clast5/Stra13/DEC1-transgenic mice, Eur J Immunol 34 (2004) 1322−1332.

[153] M. Botto, C. Dell'Agnola, A.E. Bygrave, E.M. Thompson, H.T. Cook, F. Petry, et al., Homozygous C1q deficiency causes glomerulonephritis associated with multiple apoptotic bodies, Nat Genet. 19 (1998) 56–59.

[154] A.P. Prodeus, S. Goerg, L.M. Shen, O.O. Pozdnyakova, L. Chu, E.M. Alicot, et al., A critical role for complement in maintenance of self-tolerance, Immunity 9 (1998) 721–731.

[155] Z. Chen, S.B. Koralov, G. Kelsoe, Complement C4 inhibits systemic autoimmunity through a mechanism independent of complement receptors CR1 and CR2, J Exp Med 192 (2000) 1339–1352.

[156] M. Napirei, H. Karsunky, B. Zevnik, H. Stephan, H.G. Mannherz, T. Moroy, Features of systemic lupus erythematosus in Dnase1-deficient mice, Nat Genet. 25 (2000) 177–181.

[157] K. Yasutomo, T. Horiuchi, S. Kagami, H. Tsukamoto, C. Hashimura, M. Urushihara, et al., Mutation of DNASE1 in people with systemic lupus erythematosus, Nat Genet. 28 (2001) 313–314.

[158] R. Hanayama, M. Tanaka, K. Miwa, A. Shinohara, A. Iwamatsu, S. Nagata, Identification of a factor that links apoptotic cells to phagocytes, Nature 417 (2002) 182–187.

[159] S. Akakura, S. Singh, M. Spataro, R. Akakura, J.I. Kim, M.L. Albert, et al., The opsonin MFG-E8 is a ligand for the alphavbeta5 integrin and triggers DOCK180-dependent Rac1 activation for the phagocytosis of apoptotic cells, Exp Cell Res. 292 (2004) 403–416.

[160] R. Hanayama, M. Tanaka, K. Miyasaka, K. Aozasa, M. Koike, Y. Uchiyama, et al., Autoimmune Disease and Impaired Uptake of Apoptotic Cells in MFG-E8-Deficient Mice, Science 304 (2004) 1147–1150.

[161] E.A. Leadbetter, I.R. Rifkin, A.M. Hohlbaum, B.C. Beaudette, M.J. Shlomchik, A. Marshak-Rothstein, Chromatin-IgG complexes activate B cells by dual engagement of IgM and Toll-like receptors, Nature 416 (2002) 603–607.

[162] M.W. Boule, C. Broughton, F. Mackay, S. Akira, A. Marshak-Rothstein, I.R. Rifkin, Toll-like receptor 9-dependent and -independent dendritic cell activation by chromatin-immunoglobulin G complexes, J Exp Med 199 (2004) 1631–1640.

[163] F.J. Barrat, T. Meeker, J. Gregorio, J.H. Chan, S. Uematsu, S. Akira, et al., Nucleic acids of mammalian origin can act as endogenous ligands for Toll-like receptors and may promote systemic lupus erythematosus, J Exp Med 202 (2005) 1131–1139.

[164] C.M. Lau, C. Broughton, A.S. Tabor, S. Akira, R.A. Flavell, M.J. Mamula, et al., RNA-associated autoantigens activate B cells by combined B cell antigen receptor/Toll-like receptor 7 engagement, J Exp Med 202 (2005) 1171–1177.

[165] J. Vollmer, S. Tluk, C. Schmitz, S. Hamm, M. Jurk, A. Forsbach, et al., Immune stimulation mediated by autoantigen binding sites within small nuclear RNAs involves Toll-like receptors 7 and 8, J Exp Med 202 (2005) 1575–1585.

[166] E. Savarese, O.W. Chae, S. Trowitzsch, G. Weber, B. Kastner, S. Akira, et al., U1 small nuclear ribonucleoprotein immune complexes induce type I interferon in plasmacytoid dendritic cells through TLR7, Blood 107 (2006) 3229–3234.

[167] G. Lemke, C.V. Rothlin, Immunobiology of the TAM receptors, Nat Rev Immunol 8 (2008) 327–336.

[168] Q. Lu, G. Lemke, Homeostatic regulation of the immune system by receptor tyrosine kinases of the Tyro 3 family, Science 293 (2001) 306–311.

[169] R.S. Scott, E.J. McMahon, S.M. Pop, E.A. Reap, R. Caricchio, P.L. Cohen, et al., Phagocytosis and clearance of apoptotic cells is mediated by MER, Nature 411 (2001) 207–211.

[170] P.L. Cohen, R. Caricchio, V. Abraham, T.D. Camenisch, J.C. Jennette, R.A. Roubey, et al., Delayed apoptotic cell clearance and lupus-like autoimmunity in mice lacking the c-mer membrane tyrosine kinase, J Exp Med 196 (2002) 135–140.

[171] D.B. Stetson, J.S. Ko, T. Heidmann, R. Medzhitov, Trex1 prevents cell-intrinsic initiation of autoimmunity, Cell 134 (2008) 587–598.

[172] V.G. Bhoj, Z.J. Chen, Linking retroelements to autoimmunity, Cell 134 (2008) 569–571.

[173] F. Franceschini, I. Cavazzana, Anti-Ro/SSA and La/SSB antibodies, Autoimmunity 38 (2005) 55–63.

[174] D. Xue, H. Shi, J.D. Smith, X. Chen, D.A. Noe, T. Cedervall, et al., A lupus-like syndrome develops in mice lacking the Ro 60-kDa protein, a major lupus autoantigen, Proc Natl Acad Sci U S A 100 (2003) 7503–7508.

[175] A. Espinosa, V. Dardalhon, S. Brauner, A. Ambrosi, R. Higgs, F.J. Quintana, et al., Loss of the lupus autoantigen Ro52/Trim21 induces tissue inflammation and systemic autoimmunity by disregulating the IL-23-Th17 pathway, J Exp Med 206 (2009) 1661–1671.

[176] K. Ozato, R. Yoshimi, T.H. Chang, H. Wang, T. Atsumi, H.C. Morse 3rd, Comment on "Gene disruption study reveals a nonredundant role for TRIM21/Ro52 in NF-kappa B-dependent cytokine expression in fibroblasts", J Immunol 183 (2009) 7619. author reply 7720–7611.

[177] M.C. Bickerstaff, M. Botto, W.L. Hutchinson, J. Herbert, G.A. Tennent, A. Bybee, et al., Serum amyloid P component controls chromatin degradation and prevents antinuclear autoimmunity, Nat Med 5 (1999) 694–697.

[178] A. Mehling, K. Loser, G. Varga, D. Metze, T.A. Luger, T. Schwarz, et al., Overexpression of CD40 ligand in murine epidermis results in chronic skin inflammation and systemic autoimmunity, J Exp Med 194 (2001) 615–628.

[179] J.P. Seery, J.M. Carroll, V. Cattell, F.M. Watt, Antinuclear autoantibodies and lupus nephritis in transgenic mice expressing interferon gamma in the epidermis, J Exp Med 186 (1997) 1451–1459.

[180] N. Fujikado, S. Saijo, T. Yonezawa, K. Shimamori, A. Ishii, S. Sugai, et al., Dcir deficiency causes development of autoimmune diseases in mice due to excess expansion of dendritic cells, Nat Med 14 (2008) 176–180.

[181] K.J. Erb, B. Ruger, M. von Brevern, B. Ryffel, A. Schimpl, K. Rivett, Constitutive expression of interleukin (IL)-4 in vivo causes autoimmune-type disorders in mice, J Exp Med 185 (1997) 329–339.

[182] W. Muller, R. Kuhn, K. Rajewsky, Major histocompatibility complex class II hyperexpression on B cells in interleukin 4-transgenic mice does not lead to B cell proliferation and hypergammaglobulinemia, Eur J Immunol 21 (1991) 921–925.

[183] H.J. Burstein, R.I. Tepper, P. Leder, A.K. Abbas, Humoral immune functions in IL-4 transgenic mice, J Immunol 147 (1991) 2950–2956.

[184] R.I. Tepper, D.A. Levinson, B.Z. Stanger, J. Campos-Torres, A.K. Abbas, P. Leder, IL-4 induces allergic-like inflammatory disease and alters T cell development in transgenic mice, Cell 62 (1990) 457–467.

[185] D.B. Lewis, C.C. Yu, K.A. Forbush, J. Carpenter, T.A. Sato, A. Grossman, et al., Interleukin 4 expressed in situ selectively alters thymocyte development, J Exp Med 173 (1991) 89–100.

[186] M. Santiago, L. Fossati, C. Jacquet, W. Muller, S. Izui, L. Reininger, Interleukin-4 protects against a genetically linked lupus-like autoimmune syndrome, J Exp Med 185 (1997) 65–70.

[187] E. Carballo, W.S. Lai, P.J. Blackshear, Feedback inhibition of macrophage tumor necrosis factor-alpha production by tristetraprolin, Science 281 (1998) 1001–1005.

[188] G.A. Taylor, E. Carballo, D.M. Lee, W.S. Lai, M.J. Thompson, D.D. Patel, et al., A pathogenetic role for TNF alpha in the

syndrome of cachexia, arthritis, and autoimmunity resulting from tristetraprolin (TTP) deficiency, Immunity 4 (1996) 445−454.

[189] E. Carballo, G.S. Gilkeson, P.J. Blackshear, Bone marrow transplantation reproduces the tristetraprolin-deficiency syndrome in recombination activating gene-2 (-/-) mice. Evidence that monocyte/macrophage progenitors may be responsible for TNFalpha overproduction, J Clin Invest 100 (1997) 986−995.

[190] T. Zhou, C.K. Edwards, P. Yang, Z. Wang, H. Bluethmann, J.D. Mountz, Greatly accelerated lymphadenopathy and autoimmune disease in lpr mice lacking tumor necrosis factor receptor I, J Immunol 156 (1996) 2661−2665.

[191] N. Jacob, H. Yang, L. Pricop, Y. Liu, X. Gao, S.G. Zheng, et al., Accelerated pathological and clinical nephritis in systemic lupus erythematosus-prone New Zealand Mixed 2328 mice doubly deficient in TNF receptor 1 and TNF receptor 2 via a Th17-associated pathway, J Immunol 182 (2009) 2532−2541.

[192] S.L. Cassel, P.B. Rothman, Chapter 3: Role of SOCS in allergic and innate immune responses, Adv Immunol 103 (2009) 49−76.

[193] M. Fujimoto, H. Tsutsui, O. Xinshou, M. Tokumoto, D. Watanabe, Y. Shima, et al., Inadequate induction of suppressor of cytokine signaling-1 causes systemic autoimmune diseases, Int Immunol 16 (2004) 303−314.

[194] M.L. Thomas, Of ITAMs and ITIMs:turning on and off the B cell antigen receptor, J Exp Med 181 (1995) 1953−1956.

[195] L.D. Shultz, P.A. Schweitzer, T.V. Rajan, T. Yi, J.N. Ihle, R.J. Matthews, et al., Mutations at the murine motheaten locus are within the hematopoietic cell protein-tyrosine phosphatase (Hcph) gene, Cell 73 (1993) 1445−1454.

[196] H.W. Tsui, K.A. Siminovitch, L. deSouza, F.W.L. Tsui, *Motheaten* and *viable motheaten* mice have mutations in the haematopoietic cell phosphatase gene, Nature Genet. 4 (1993) 124−129.

[197] G. Van Zant, L.D. Shultz, Hematologic abnormalities of the immunodeficient mouse mutant, viable motheaten (*me*^{v}), Experimental hematology 17 (1989) 81−87.

[198] C.L. Scribner, C.T. Hansen, D.M. Klinman, A.D. Steinberg, The interaction of the xid and me genes, J Immunol 138 (1987) 3611−3617.

[199] C.C. Yu, H.W. Tsui, B.Y. Ngan, M.J. Shulman, G.E. Wu, F.W. Tsui, B and T cells are not required for the viable motheaten phenotype, J Exp Med 183 (1996) 371−380.

[200] D. Chui, M. Oh-Eda, Y.F. Liao, K. Panneerselvam, A. Lal, K.W. Marek, et al., Alpha-mannosidase-II deficiency results in dyserythropoiesis and unveils an alternate pathway in oligosaccharide biosynthesis, Cell 90 (1997) 157−167.

[201] K.W. Moremen, R.B. Trimble, A. Herscovics, Glycosidases of the asparagine-linked oligosaccharide processing pathway, Glycobiology 4 (1994) 113−125.

[202] D. Chui, G. Sellakumar, R. Green, M. Sutton-Smith, T. McQuistan, K. Marek, et al., Genetic remodeling of protein glycosylation in vivo induces autoimmune disease, Proc Natl Acad Sci U S A 98 (2001) 1142−1147.

[203] J.B. Hurov, T.S. Stappenbeck, C.M. Zmasek, L.S. White, S.H. Ranganath, J.H. Russell, et al., Immune system dysfunction and autoimmune disease in mice lacking Emk (Par-1) protein kinase, Mol Cell Biol. 21 (2001) 3206−3219.

[204] S. Bessone, F. Vidal, Y. Le Bouc, J. Epelbaum, M.T. Bluet-Pajot, M. Darmon, EMK protein kinase-null mice: dwarfism and hypofertility associated with alterations in the somatotrope and prolactin pathways, Dev Biol. 214 (1999) 87−101.

[205] L.M. Zipper, R.T. Mulcahy, The Keap1 BTB/POZ dimerization function is required to sequester Nrf2 in cytoplasm, J Biol Chem. 277 (2002) 36544−36552.

[206] T. Ishii, K. Itoh, M. Yamamoto, Roles of Nrf2 in activation of antioxidant enzyme genes via antioxidant responsive elements, Methods Enzymol 348 (2002) 182−190.

[207] K. Yoh, K. Itoh, A. Enomoto, A. Hirayama, N. Yamaguchi, M. Kobayashi, et al., Nrf2-deficient female mice develop lupus-like autoimmune nephritis, Kidney Int 60 (2001) 1343−1353.

[208] K.D. Poss, S. Tonegawa, Heme oxygenase 1 is required for mammalian iron reutilization, Proc Natl Acad Sci U S A 94 (1997) 10919−10924.

[209] N. Morito, K. Yoh, A. Hirayama, K. Itoh, M. Nose, A. Koyama, et al., Nrf2 deficiency improves autoimmune nephritis caused by the fas mutation lpr, Kidney Int 65 (2004) 1703−1713.

[210] X. Lu, G.I. Kovalev, H. Chang, E. Kallin, G. Knudsen, L. Xia, et al., Inactivation of NuRD component Mta2 causes abnormal T cell activation and lupus-like autoimmune disease in mice, J Biol Chem. 283 (2008) 13825−13833.

[211] K. Matsushita, O. Takeuchi, D.M. Standley, Y. Kumagai, T. Kawagoe, T. Miyake, et al., Zc3h12a is an RNase essential for controlling immune responses by regulating mRNA decay, Nature 458 (2009) 1185−1190.

[212] M.J. Shlomchik, M.P. Madaio, D. Ni, M. Trounstein, D. Huszar, The role of B cells in lpr/lpr-induced autoimmunity, J Exp Med 180 (1994) 1295−1306.

[213] O.T. Chan, M.P. Madaio, M.J. Shlomchik, B cells are required for lupus nephritis in the polygenic, Fas-intact MRL model of systemic autoimmunity, J Immunol 163 (1999) 3592−3596.

[214] O.T. Chan, L.G. Hannum, A.M. Haberman, M.P. Madaio, M.J. Shlomchik, A novel mouse with B cells but lacking serum antibody reveals an antibody-independent role for B cells in murine lupus, J Exp Med 189 (1999) 1639−1648.

[215] Z. Hao, G.S. Duncan, J. Seagal, Y.W. Su, C. Hong, J. Haight, et al., Fas receptor expression in germinal-center B cells is essential for T and B lymphocyte homeostasis, Immunity 29 (2008) 615−627.

[216] C. Jiang, J. Foley, N. Clayton, G. Kissling, M. Jokinen, R. Herbert, et al., Abrogation of lupus nephritis in activation-induced deaminase-deficient MRL/lpr mice, J Immunol 178 (2007) 7422−7431.

[217] A.M. Jevnikar, M.J. Grusby, L.H. Glimcher, Prevention of nephritis in major histocompatibility complex class II-deficient MRL-lpr mice, J Exp Med 179 (1994) 1137−1143.

[218] D.R. Koh, A. Ho, A. Rahemtulla, W.P. Fung-Leung, H. Griesser, T.W. Mak, Murine lupus in MRL/lpr mice lacking CD4 or CD8 T cells, Eur J Immunol 25 (1995) 2558−2562.

[219] T. Giese, W.F. Davidson, CD8+ T cell-deficient lpr/lpr mice, CD4+B220+ and CD4+B220- T cells replace B220+ double-negative T cells as the predominant populations in enlarged lymph nodes, J Immunol 154 (1995) 4986−4995.

[220] T. Ohteki, M. Iwamoto, S. Izui, H.R. MacDonald, Reduced development of CD4-8-B220+ T cells but normal autoantibody production in lpr/lpr mice lacking major histocompatibility complex class I molecules, Eur J Immunol 25 (1995) 37−41.

[221] M.A. Maldonado, R.A. Eisenberg, E. Roper, P.L. Cohen, B.L. Kotzin, Greatly reduced lymphoproliferation in lpr mice lacking major histocompatibility complex class I, J Exp Med 181 (1995) 641−648.

[222] G.J. Christianson, R.L. Blankenburg, T.M. Duffy, D. Panka, J.B. Roths, A. Marshak-Rothstein, et al., beta2-microglobulin dependence of the lupus-like autoimmune syndrome of MRL-lpr mice, J Immunol 156 (1996) 4932−4939.

[223] O.T. Chan, V. Paliwal, J.M. McNiff, S.H. Park, A. Bendelac, M.J. Shlomchik, Deficiency in beta(2)-microglobulin, but not CD1, accelerates spontaneous lupus skin disease while

inhibiting nephritis in MRL-Fas(lpr) nice: an example of disease regulation at the organ level, J Immunol 167 (2001) 2985–2990.

[224] J.Q. Yang, T. Chun, H. Liu, S. Hong, H. Bui, L. Van Kaer, et al., CD1d deficiency exacerbates inflammatory dermatitis in MRL-lpr/lpr mice, Eur J Immunol 34 (2004) 1723–1732.

[225] J.Q. Yang, A.K. Singh, M.T. Wilson, M. Satoh, A.K. Stanic, J.J. Park, et al., Immunoregulatory role of CD1d in the hydrocarbon oil-induced model of lupus nephritis, J Immunol 171 (2003) 2142–2153.

[226] J.Q. Yang, X. Wen, H. Liu, G. Folayan, X. Dong, M. Zhou, et al., Examining the role of CD1d and natural killer T cells in the development of nephritis in a genetically susceptible lupus model, Arthritis Rheum 56 (2007) 1219–1233.

[227] S.Y. Chen, Y. Takeoka, A.A. Ansari, R. Boyd, D.M. Klinman, M.E. Gershwin, The natural history of disease expression in CD4 and CD8 gene-deleted New Zealand black (NZB) mice, J Immunol 157 (1996) 2676–2684.

[228] S.L. Peng, J. Cappadona, J.M. McNiff, M.P. Madaio, M.J. Owen, A.C. Hayday, et al., Pathogenesis of autoimmunity in alphabeta T cell-deficient lupus-prone mice, Clin Exp Immunol 111 (1998) 107–116.

[229] S.L. Peng, M.P. Madaio, D.P. Hughes, I.N. Crispe, M.J. Owen, A.C. Wen, Hayday, et al., Murine lupus in the absence of alpha beta T cells, J Immunol 156 (1996) 4041–4049.

[230] S.L. Peng, J. Craft, The regulation of murine lupus, Ann N Y Acad Sci. 815 (1997) 128–138.

[231] J. Craft, S. Peng, T. Fujii, M. Okada, S. Fatenejad, Autoreactive T cells in murine lupus: origins and roles in autoantibody production, Immunol Res. 19 (1999) 245–257.

[232] S.L. Peng, J. Moslehi, M.E. Robert, J. Craft, Perforin protects against autoimmunity in lupus-prone mice, J Immunol 160 (1998) 652–660.

[233] G.C. Zeller, J. Hirahashi, A. Schwarting, A.H. Sharpe, V.R. Kelley, Inducible co-stimulator null MRL-Faslpr mice: uncoupling of autoantibodies and T cell responses in lupus, J Am Soc Nephrol 17 (2006) 122–130.

[234] Y. Tada, S. Koarada, Y. Tomiyoshi, F. Morito, M. Mitamura, Y. Haruta, et al., Role of inducible costimulator in the development of lupus in MRL/lpr mice, Clin Immunol 120 (2006) 179–188.

[235] A.N. Theofilopoulos, R. Gonzalez-Quintial, B.R. Lawson, Y.T. Koh, M.E. Stern, D.H. Kono, et al., Sensors of the innate immune system: their link to rheumatic diseases, Nat Rev Rheumatol 6 (3) (2010) 146–156.

[236] K.M. Nickerson, S.R. Christensen, J. Shupe, M. Kashgarian, D. Kim, K. Elkon, et al., TLR9 regulates TLR7- and MyD88-dependent autoantibody production and disease in a murine model of lupus, J Immunol 184 (2010) 1840–1848.

[237] S.R. Christensen, J. Shupe, K. Nickerson, M. Kashgarian, R.A. Flavell, M.J. Shlomchik, Toll-like Receptor 7 and TLR9 Dictate Autoantibody Specificity and Have Opposing Inflammatory and Regulatory Roles in a Murine Model of Lupus, Immunity 25 (2006) 417–428.

[238] P.Y. Lee, Y. Kumagai, Y. Li, O. Takeuchi, H. Yoshida, J. Weinstein, et al., TLR7-dependent and FcgammaR-independent production of type I interferon in experimental mouse lupus, J Exp Med 205 (2008) 2995–3006.

[239] D.H. Kono, M.K. Haraldsson, B.R. Lawson, K.M. Pollard, Y.T. Koh, X. Du, et al., Endosomal TLR signaling is required for anti-nucleic acid and rheumatoid factor autoantibodies in lupus, Proc Natl Acad Sci USA 106 (29) (2009) 12061–12066.

[240] A. Sadanaga, H. Nakashima, M. Akahoshi, K. Masutani, K. Miyake, T. Igawa, et al., Protection against autoimmune

nephritis in MyD88-deficient MRL/lpr mice, Arthritis Rheum 56 (2007) 1618–1628.

[241] R.D. Pawar, A. Ramanjaneyulu, O.P. Kulkarni, M. Lech, S. Segerer, H.J. Anders, Inhibition of Toll-like receptor-7 (TLR-7) or TLR-7 plus TLR-9 attenuates glomerulonephritis and lung injury in experimental lupus, J Am Soc Nephrol 18 (2007) 1721–1731.

[242] A. Marshak-Rothstein, I.R. Rifkin, Immunologically active autoantigens: the role of toll-like receptors in the development of chronic inflammatory disease, Annu Rev Immunol 25 (2007) 419–441.

[243] C. Garlanda, H.J. Anders, A. Mantovani, TIR8/SIGIRR: an IL-1R/TLR family member with regulatory functions in inflammation and T cell polarization, Trends Immunol 30 (2009) 439–446.

[244] M. Lech, O.P. Kulkarni, S. Pfeiffer, E. Savarese, A. Krug, C. Garlanda, et al., Tir8/Sigirr prevents murine lupus by suppressing the immunostimulatory effects of lupus autoantigens, J Exp Med 205 (2008) 1879–1888.

[245] L. Frisoni, L. McPhie, S.A. Kang, M. Monestier, M. Madaio, M. Satoh, et al., Lack of chromatin and nuclear fragmentation in vivo impairs the production of lupus anti-nuclear antibodies, J Immunol 179 (2007) 7959–7966.

[246] J. Ma, J. Xu, M.P. Madaio, Q. Peng, J. Zhang, I.S. Grewal, et al., Autoimmune lpr/lpr mice deficient in CD40 ligand: spontaneous Ig class switching with dichotomy of autoantibody responses, J Immunol 157 (1996) 417–426.

[247] S.L. Peng, J.M. McNiff, M.P. Madaio, J. Ma, M.J. Owen, Ra. Flavell, et al., alpha-beta T cell regulation and CD40 ligand dependence in murine systemic autoimmunity, J Immunol 158 (1997) 2464–2470.

[248] Y. Tada, K. Nagasawa, A. Ho, F. Morito, S. Koarada, O. Ushiyama, et al., Role of the costimulatory molecule CD28 in the development of lupus in MRL/lpr mice, J Immunol 163 (1999) 3153–3159.

[249] B. Liang, R.J. Gee, M.J. Kashgarian, A.H. Sharpe, M.J. Mamula, B7 costimulation in the development of lupus: autoimmunity arises either in the absence of B7.1/B7.2 or in the presence of anti-B7.1/B7.2 blocking antibodies, J Immunol 163 (1999) 2322–2329.

[250] A. Nakajima, M. Azuma, S. Kodera, S. Nuriya, A. Terashi, M. Abe, et al., Preferential dependence of autoantibody production in murine lupus on CD86 costimulatory molecule, Eur J Immunol 25 (1995) 3060–3069.

[251] C. Detre, M. Keszei, X. Romero, G.C. Tsokos, C. Terhorst, SLAM family receptors and the SLAM-associated protein (SAP) modulate T cell functions, Semin Immunopathol 32 (2) (2010) 157–171.

[252] J.D. Hron, L. Caplan, A.J. Gerth, P.L. Schwartzberg, S.L. Peng, SH2D1A regulates T-dependent humoral autoimmunity, J Exp Med 200 (2004) 261–266.

[253] H. Komori, H. Furukawa, S. Mori, M.R. Ito, M. Terada, M.C. Zhang, et al., A signal adaptor SLAM-associated protein regulates spontaneous autoimmunity and Fas-dependent lymphoproliferation in MRL-Faslpr lupus mice, J Immunol 176 (2006) 395–400.

[254] C.G. Kevil, M.J. Hicks, X. He, J. Zhang, C.M. Ballantyne, C. Raman, et al., Loss of LFA-1, but not Mac-1, protects MRL/MpJ-Fas(lpr) mice from autoimmune disease, Am J Pathol 165 (2004) 609–616.

[255] M. Santiago-Raber, R. Baccala, M.K. Haraldsson, D. Choubey, T.A. Stewart, D.H. Kono, et al., Type-I interferon receptor deficiency reduces lupus-like disease in NZB mice, J Exp Med 197 (2003) 777–788.

[256] D. Braun, P. Geraldes, J. Demengeot, Type I Interferon controls the onset and severity of autoimmune manifestations in lpr mice, J Autoimmun 20 (2003) 15–25.

[257] D.C. Nacionales, K.M. Kelly-Scumpia, P.Y. Lee, J.S. Weinstein, R. Lyons, E. Sobel, et al., Deficiency of the type I interferon receptor protects mice from experimental lupus, Arthritis Rheum 56 (2007) 3770–3783.

[258] A. Mathian, A. Weinberg, M. Gallegos, J. Banchereau, S. Koutouzov, IFN-alpha induces early lethal lupus in preautoimmune (New Zealand Black x New Zealand White) F1 but not in BALB/c mice, J Immunol 174 (2005) 2499–2506.

[259] A.N. Theofilopoulos, R. Baccala, B. Beutler, D.H. Kono, Type I Interferons (alpha/beta) in immunity and autoimmunity, Annu Rev Immunol 23 (2005) 307–335.

[260] J.D. Hron, S.L. Peng, Type I IFN protects against murine lupus, J Immunol 173 (2004) 2134–2142.

[261] A. Schwarting, K. Paul, S. Tschirner, J. Menke, T. Hansen, W. Brenner, et al., Interferon-beta: a therapeutic for autoimmune lupus in MRL-Faslpr mice, J Am Soc Nephrol 16 (2005) 3264–3272.

[262] S.L. Peng, J. Moslehi, J. Craft, Roles of interferon-gamma and interleukin-4 in murine lupus, J Clin Invest 99 (1997) 1936–1946.

[263] C. Haas, B. Ryffel, H.M. Le, IFN-gamma is essential for the development of autoimmune glomerulonephritis in MRL/Ipr mice, J Immunol 158 (1997) 5484–5491.

[264] D. Balomenos, R. Rumold, A.N. Theofilopoulos, Interferon-gamma is required for lupus-like disease and lymphoaccumulation in MRL-lpr mice, J Clin Invest 101 (1998) 364–371.

[265] C. Haas, B. Ryffel, M. Le Hir, IFN-gamma receptor deletion prevents autoantibody production and glomerulonephritis in lupus-prone (NZB x NZW)F1 mice, J Immunol 160 (1998) 3713–3718.

[266] D.H. Kono, D. Balomenos, D.L. Pearson, M.S. Park, B. Hildebrandt, P. Hultman, et al., The prototypic Th2 autoimmunity induced by mercury is dependent on IFN- gamma and not Th1/Th2 imbalance, J Immunol 161 (1998) 234–240.

[267] I.C. Ho, L.H. Glimcher, Transcription: tantalizing times for T cells, Cell. 109 Suppl (2002) S109–120.

[268] S.L. Peng, S.J. Szabo, L.H. Glimcher, T-bet regulates IgG class switching and pathogenic autoantibody production, Proc Natl Acad Sci U S A 99 (2002) 5545–5550.

[269] H. Shimohata, A. Yamada, K. Yoh, K. Ishizaki, N. Morito, K. Yamagata, et al., Overexpression of T-bet in T cells accelerates autoimmune glomerulonephritis in mice with a dominant Th1 background, J Nephrol 22 (2009) 123–129.

[270] C.M. Reilly, S. Olgun, D. Goodwin, R.M. Gogal Jr., A. Santo, J.W. Romesburg, et al., Interferon regulatory factor-1 gene deletion decreases glomerulonephritis in MRL/lpr mice, Eur J Immunol 36 (2006) 1296–1308.

[271] D.S. Robinson, A. O'Garra, Further checkpoints in Th1 development, Immunity 16 (2002) 755–758.

[272] T. Yasuda, T. Yoshimoto, A. Tsubura, A. Matsuzawa, Clear suppression of Th1 responses but marginal amelioration of autoimmune manifestations by IL-12p40 transgene in MRL-FAS(lprcg)/FAS(lprcg) mice, Cell Immunol 210 (2001) 77–86.

[273] D.H. Kono, D. Balomenos, M.S. Park, A.N. Theofilopoulos, Development of lupus in BXSB mice is independent of IL-4, J Immunol 164 (2000) 38–42.

[274] Z. Xu, B. Duan, B.P. Croker, L. Morel, STAT4 deficiency reduces autoantibody production and glomerulonephritis in a mouse model of lupus, Clin Immunol 120 (2006) 189–198.

[275] R.R. Singh, V. Saxena, S. Zang, L. Li, F.D. Finkelman, D.P. Witte, et al., Differential contribution of IL-4 and STAT6 vs STAT4 to the development of lupus nephritis, J Immunol 170 (2003) 4818–4825.

[276] D.L. Thibault, A.D. Chu, K.L. Graham, I. Balboni, L.Y. Lee, C. Kohlmoos, et al., IRF9 and STAT1 are required for IgG autoantibody production and B cell expression of TLR7 in mice, J Clin Invest 118 (2008) 1417–1426.

[277] Z. Yin, G. Bahtiyar, N. Zhang, L. Liu, P. Zhu, M.E. Robert, et al., IL-10 regulates murine lupus, J Immunol 169 (2002) 2148–2155.

[278] S. Ashkar, G.F. Weber, V. Panoutsakopoulou, M.E. Sanchirico, M. Jansson, S. Zawaideh, et al., Eta-1 (osteopontin): an early component of type-1 (cell-mediated) immunity, Science 287 (2000) 860–864.

[279] G.F. Weber, H. Cantor, Differential roles of osteopontin/Eta-1 in early and late lpr disease, Clin Exp Immunol 126 (2001) 578–583.

[280] R. Patarca, F.Y. Wei, P. Singh, M.I. Morasso, H. Cantor, Dysregulated expression of the T cell cytokine Eta-1 in CD4-8-lymphocytes during the development of murine autoimmune disease, J Exp Med 172 (1990) 1177–1183.

[281] T. Igawa, H. Nakashima, A. Sadanaga, K. Masutani, K. Miyake, S. Shimizu, et al., Deficiency in EBV-induced gene 3 (EBI3) in MRL/lpr mice results in pathological alteration of autoimmune glomerulonephritis and sialadenitis, Mod Rheumatol 19 (2009) 33–41.

[282] J.A. Bubier, T.J. Sproule, O. Foreman, R. Spolski, D.J. Shaffer, H.C. Morse 3rd, et al., A critical role for IL-21 receptor signaling in the pathogenesis of systemic lupus erythematosus in BXSB-Yaa mice, Proc Natl Acad Sci U S A 106 (2009) 1518–1523.

[283] D. Balomenos, R. Rumold, A.N. Theofilopoulos, The proliferative in vivo activities of lpr double-negative T cells and the primary role of p59fyn in their activation and expansion, J Immunol 159 (1997) 2265–2273.

[284] W.P. Brooks, M.A. Lynes, Effects of hemizygous CD45 expression in the autoimmune Fasl(gld/gld) syndrome, Cell Immunol 212 (2001) 24–34.

[285] H. Watanabe, G. Garnier, A. Circolo, R.A. Wetsel, P. Ruiz, V.M. Holers, et al., Modulation of renal disease in MRL/lpr mice genetically deficient in the alternative complement pathway factor B, J Immunol 164 (2000) 786–794.

[286] J. Passwell, G.F. Schreiner, M. Nonaka, H.U. Beuscher, H.R. Colten, Local extrahepatic expression of complement genes C3, factor B, C2, and C4 is increased in murine lupus nephritis, J Clin Invest 82 (1988) 1676–1684.

[287] M.K. Elliott, T. Jarmi, P. Ruiz, Y. Xu, V.M. Holers, G.S. Gilkeson, Effects of complement factor D deficiency on the renal disease of MRL/lpr mice, Kidney Int 65 (2004) 129–138.

[288] H. Sekine, C.M. Reilly, I.D. Molano, G. Garnier, A. Circolo, P. Ruiz, et al., Complement component C3 is not required for full expression of immune complex glomerulonephritis in MRL/lpr mice, J Immunol 166 (2001) 6444–6451.

[289] S. Einav, O.O. Pozdnyakova, M. Ma, M.C. Carroll, Complement C4 is protective for lupus disease independent of C3, J Immunol 168 (2002) 1036–1041.

[290] L. Bao, M. Haas, S.A. Boackle, D.M. Kraus, P.N. Cunningham, P. Park, et al., Transgenic expression of a soluble complement inhibitor protects against renal disease and promotes survival in MRL/lpr mice, J Immunol 168 (2002) 3601–3607.

[291] T. Miwa, M.A. Maldonado, L. Zhou, X. Sun, H.Y. Luo, D. Cai, et al., Deletion of decay-accelerating factor (CD55) exacerbates autoimmune disease development in MRL/lpr mice, Am J Pathol 161 (2002) 1077–1086.

[292] T. Miwa, M.A. Maldonado, L. Zhou, K. Yamada, G.S. Gilkeson, R.A. Eisenberg, et al., Decay-accelerating factor ameliorates systemic autoimmune disease in MRL/lpr mice via both complement-dependent and -independent mechanisms, Am J Pathol 170 (2007) 1258–1266.

[293] J.L. Svenson, J. EuDaly, P. Ruiz, K.S. Korach, G.S. Gilkeson, Impact of estrogen receptor deficiency on disease expression in the NZM2410 lupus prone mouse, Clin Immunol 128 (2008) 259–268.

[294] K.K. Bynote, J.M. Hackenberg, K.S. Korach, D.B. Lubahn, P.H. Lane, K.A. Gould, Estrogen receptor-alpha deficiency attenuates autoimmune disease in (NZB x NZW)F1 mice, Genes Immun 9 (2008) 137–152.

[295] C.M. Lloyd, J.A. Gonzalo, D.J. Salant, J. Just, J.C. Gutierrez-Ramos, Intercellular adhesion molecule-1 deficiency prolongs survival and protects against the development of pulmonary inflammation during murine lupus, J Clin Invest 100 (1997) 963–971.

[296] D.C. Bullard, P.D. King, M.J. Hicks, B. Dupont, A.L. Beaudet, K.B. Elkon, Intercellular adhesion molecule-1 deficiency protects MRL/MpJ-Fas(lpr) mice from early lethality, J Immunol 159 (1997) 2058–2067.

[297] M.U. Norman, W.G. James, M.J. Hickey, Differential roles of ICAM-1 and VCAM-1 in leukocyte-endothelial cell interactions in skin and brain of MRL/faslpr mice, J Leukoc Biol. 84 (2008) 68–76.

[298] G.S. Gilkeson, J.S. Mudgett, M.F. Seldin, P. Ruiz, A.A. Alexander, M.A. Misukonis, et al., Clinical and serologic manifestations of autoimmune disease in MRL-lpr/lpr mice lacking nitric oxide synthase type 2, J Exp Med 186 (1997) 365–373.

[299] R. Clynes, C. Dumitru, J.V. Ravetch, Uncoupling of immune complex formation and kidney damage in autoimmune glomerulonephritis, Science 279 (1998) 1052–1054.

[300] S.Y. Park, S. Ueda, H. Ohno, Y. Hamano, M. Tanaka, T. Shiratori, et al., Resistance of Fc receptor-deficient mice to fatal glomerulonephritis, J Clin Invest 102 (1998) 1229–1238.

[301] G.H. Tesch, S. Maifert, A. Schwarting, B.J. Rollins, V.R. Kelley, Monocyte Chemoattractant Protein 1-dependent Leukocytic Infiltrates Are Responsible for Autoimmune Disease in MRL-Fas(lpr) Mice, J Exp Med 190 (1999) 1813–1824.

[302] A.Y. Hoi, M.J. Hickey, P. Hall, J. Yamana, K.M. O'Sullivan, L.L. Santos, et al., Macrophage migration inhibitory factor deficiency attenuates macrophage recruitment, glomerulonephritis, and lethality in MRL/lpr mice, J Immunol 177 (2006) 5687–5696.

[303] K. Liu, Q.Z. Li, A.M. Delgado-Vega, A.K. Abelson, E. Sanchez, J.A. Kelly, et al., Kallikrein genes are associated with lupus and glomerular basement membrane-specific antibody-induced nephritis in mice and humans, J Clin Invest 119 (2009) 911–923.

[304] M. Yoshida, K. Saiga, T. Hato, S. Iwaki, T. Niiya, N. Arita, et al., Cappuccino mutation in an autoimmune-prone strain of mice suggests a role of platelet function in the progression of immune complex crescentic glomerulonephritis, Arthritis Rheum 54 (2006) 2934–2943.

[305] X. He, T.R. Schoeb, A. Panoskaltsis-Mortari, K.R. Zinn, R.A. Kesterson, J. Zhang, et al., Deficiency of P-selectin or P-selectin glycoprotein ligand-1 leads to accelerated development of glomerulonephritis and increased expression of CC chemokine ligand 2 in lupus-prone mice, J Immunol 177 (2006) 8748–8756.

[306] S.D. Crowley, M.P. Vasievich, P. Ruiz, S.K. Gould, K.K. Parsons, A.K. Pazmino, et al., Glomerular type 1 angiotensin receptors augment kidney injury and inflammation in murine autoimmune nephritis, J Clin Invest 119 (2009) 943–953.

[307] T. Komori, L. Pricop, A. Hatakeyama, C.A. Bona, F.W. Alt, Repertoires of antigen receptors in Tdt congenitally deficient mice, Int Rev Immunol 13 (1996) 317–325.

[308] S. Weller, C. Conde, A.M. Knapp, H. Levallois, S. Gilfillan, J.L. Pasquali, et al., Autoantibodies in mice lacking terminal deoxynucleotidyl transferase: evidence for a role of N region addition in the polyreactivity and in the affinities of anti-DNA antibodies, J Immunol 159 (1997) 3890–3898.

[309] I.D. Molano, M.K. Wloch, A.A. Alexander, H. Watanabe, G.S. Gilkeson, Effect of a genetic deficiency of terminal deoxynucleotidyl transferase on autoantibody production by C57BL6 fas(lpr) mice, Clin Immunol 94 (2000) 24–32.

[310] A.J. Feeney, B.R. Lawson, D.H. Kono, A.N. Theofilopoulos, Terminal deoxynucleotidyl transferase deficiency decreases autoimmune disease in MRL- Fas^{lpr} mice, J Immunol 167 (2001) 3486–3493.

[311] Z. Ma, A. Choudhury, S.A. Kang, M. Monestier, P.L. Cohen, R.A. Eisenberg, Accelerated atherosclerosis in ApoE deficient lupus mouse models, Clin Immunol 127 (2008) 168–175.

[312] X. Feng, H. Li, A.A. Rumbin, X. Wang, A. La Cava, K. Brechtelsbauer, et al., ApoE-/-Fas-/- C57BL/6 mice: a novel murine model simultaneously exhibits lupus nephritis, atherosclerosis, and osteopenia, Journal of lipid research 48 (2007) 794–805.

[313] C.C. Allende, J.E. Allende, Promiscuous subunit interactions: a possible mechanism for the regulation of protein kinase CK2, J Cell Biochem Suppl. 31 (1998) 129–136.

[314] X. Xu, E. Landesman-Bollag, P.L. Channavajhala, D.C. Seldin, Murine protein kinase CK2: gene and oncogene, Mol Cell Biochem 191 (1999) 65–74.

[315] R.F. Spurney, P. Ruiz, D.S. Pisetsky, T.M. Coffman, Enhanced renal leukotriene production in murine lupus: role of lipoxygenase metabolites, Kidney Int 39 (1991) 95–102.

[316] U.H. Rudofsky, B.D. Evans, S.L. Balaban, V.D. Mottironi, A.E. Gabrielsen, Differences in expression of lupus nephritis in New Zealand mixed H-2z homozygous inbred strains of mice derived from New Zealand black and New Zealand white mice. Origins and initial characterization, Lab Invest 68 (1993) 419–426.

[317] L. Fossati, S. Takahashi, R. Merino, M. Iwamoto, J.P. Aubry, M. Nose, et al., An MRL/MpJ-lpr/lpr substrain with a limited expansion of lpr double-negative T cells and a reduced autoimmune syndrome, Int Immunol 5 (1993) 525–532.

[318] H. Suzuka, H. Yoshifusa, Y. Nakamura, S. Miyawaki, Y. Shibata, Morphological analysis of autoimmune disease in MRL-lpr, Yaa male mice with rapidly progressive systemic lupus erythematosus, Autoimmunity 14 (1993) 275–282.

[319] K. Kinjoh, M. Kyogoku, R.A. Good, Genetic selection for crescent formation yields mouse strain with rapidly progressive glomerulonephritis and small vessel vasculitis, Proc Natl Acad Sci USA 90 (1993) 3413–3417.

[320] R. Kofler, P.J. McConahey, M.A. Duchosal, R.S. Balderas, A.N. Theofilopoulos, F.J. Dixon, An autosomal recessive gene that delays expression of lupus in BXSB mice, J Immunol 146 (1991) 1375–1379.

[321] M.E. Haywood, L. Gabriel, S.J. Rose, N.J. Rogers, S. Izui, B.J. Morley, BXSB/long-lived is a recombinant inbred strain containing powerful disease suppressor loci, J Immunol 179 (2007) 2428–2434.

[322] S. Vidal, C. Gelpi, J.L. Rodriguez-Sanchez, (SWR x SJL)F1 mice: a new model of lupus-like disease, J Exp Med 179 (1994) 1429–1435.

[323] S.E. Walker, R.H. Gray, M. Fulton, R.D. Wigley, B. Schnitzer, Palmerston North mice, a new animal model of systemic lupus erythematosus, J Lab Clin Med 92 (1978) 932–945.

[324] J.S. Bignon, K.A. Siminovitch, Identification of PTP1C mutation as the genetic defect in motheaten and viable motheaten mice: a step toward defining the roles of protein tyrosine phosphatases in the regulation of hemopoietic cell differentiation and function, Clin Immunol Immunopathol 73 (1994) 168–179.

[325] K.M. Pollard, P. Hultman, Effects of mercury on the immune system, Met Ions Biol Syst 34 (1997) 421–440.

[326] R.L. Rubin, Drug-induced lupus, in: D.J. Wallace, B.H. Hahn (Eds.), Dubois' Lupus Erythematosus, Williams and Wilkens, Baltimore, 1997, pp. 871–901.

[327] M. Satoh, A. Kumar, Y.S. Kanwar, W.H. Reeves, Anti-nuclear antibody production and immune-complex glomerulonephritis in BALB/c mice treated with pristane, Proc Natl Acad Sci USA 92 (1995) 10934–10938.

[328] S. Mendlovic, B.H. Fricke, Y. Shoenfeld, R. Bakimer, E. Mozes, The genetic regulation of the induction of experimental SLE, Immunol 69 (1990) 228–236.

[329] P.A. Silveira, A.G. Baxter, The NOD mouse as a model of SLE, Autoimmunity 34 (2001) 53–64.

[330] M.A. Jordan, P.A. Silveira, D.P. Shepherd, C. Chu, S.J. Kinder, J. Chen, et al., Linkage analysis of systemic lupus erythematosus induced in diabetes- prone nonobese diabetic mice by Mycobacterium bovis, J Immunol 165 (2000) 1673–1684.

[331] J.G. Schoenecker, R.K. Johnson, A.P. Lesher, J.D. Day, S.D. Love, M.R. Hoffman, et al., Exposure of mice to topical bovine thrombin induces systemic autoimmunity, Am J Pathol 159 (2001) 1957–1969.

[332] M.E. Haywood, M.B. Hogarth, J.H. Slingsby, S.J. Rose, P.J. Allen, E.M. Thompson, et al., Identification of intervals on chromosomes 1, 3, and 13 linked to the development of lupus in BXSB mice, Arthritis Rheum 43 (2000) 349–355.

[333] M.B. Hogarth, J.H. Slingsby, P.J. Allen, E.M. Thompson, P. Chandler, K.A. Davies, et al., Multiple lupus susceptibility loci map to chromosome 1 in BXSB mice, J Immunol 161 (1998) 2753–2761.

[334] T.J. Vyse, L. Morel, F.J. Tanner, E.K. Wakeland, B.L. Kotzin, Backcross analysis of genes linked to autoantibody production in New Zealand White mice, J Immunol 157 (1996) 2719–2727.

[335] M. Nishihara, M. Terada, J. Kamogawa, Y. Ohashi, S. Mori, S. Nakatsuru, et al., Genetic basis of autoimmune sialadenitis in MRL/lpr lupus-prone mice: additive and hierarchical properties of polygenic inheritance, Arthritis Rheum 42 (1999) 2616–2623.

[336] S. Xie, S. Chang, P. Yang, C. Jacob, A. Kaliyaperumal, S.K. Datta, et al., Genetic contributions of nonautoimmune SWR mice toward lupus nephritis, J Immunol 167 (2001) 7141–7149.

[337] L. Morel, U.H. Rudofsky, J.A. Longmate, J. Schiffenbauer, E.K. Wakeland, Polygenic control of susceptibility to murine systemic lupus erythematosus, Immunity 1 (1994) 219–229.

[338] L. Morel, C. Mohan, Y. Yu, J. Schiffenbauer, U.H. Rudofsky, N. Tian, et al., Multiplex inheritance of component phenotypes in a murine model of lupus, Mamm Genome 10 (1999) 176–181.

[339] D.H. Kono, M.S. Park, A. Szydlik, K.M. Haraldsson, J.D. Kuan, D.L. Pearson, et al., Resistance to xenobiotic-induced autoimmunity maps to chromosome 1, J Immunol 167 (2001) 2396–2403.

[340] S.T. Waters, S.M. Fu, F. Gaskin, U.S. Deshmukh, S.S. Sung, C.C. Kannapell, et al., NZM2328: a new mouse model of systemic lupus erythematosus with unique genetic susceptibility loci, Clin Immunol 100 (2001) 372–383.

[341] D.H. Kono, R.W. Burlingame, D.G. Owens, A. Kuramochi, R.S. Balderas, D. Balomenos, et al., Lupus susceptibility loci in New Zealand mice, Proc Natl Acad Sci USA 91 (1994) 10168–10172.

[342] C.G. Drake, S.J. Rozzo, H.F. Hirschfeld, N.P. Smarnworawong, E. Palmer, B.L. Kotzin, Analysis of the New Zealand Black contribution to lupus-like renal disease. Multiple genes that operate in a threshold manner, J Immunol 154 (1995) 2441–2447.

[343] T.J. Vyse, S.J. Rozzo, C.G. Drake, S. Izui, B.L. Kotzin, Control of multiple autoantibodies linked with a lupus nephritis susceptibility locus in New Zealand black mice, J Immunol 158 (1997) 5566–5574.

[344] S.J. Rozzo, T.J. Vyse, C.G. Drake, B.L. Kotzin, Effect of genetic background on the contribution of New Zealand black loci to autoimmune lupus nephritis, Proc Natl Acad Sci USA 93 (1996) 15164–15168.

[345] S. Xie, L. Li, S. Chang, R. Sharma, A. Kaliyaperumal, S.K. Datta, et al., Genetic origin of lupus in NZB/SWR hybrids: lessons from an intercross study, Arthritis Rheum 52 (2005) 659–667.

[346] L. Gu, A. Weinreb, X.P. Wang, D.J. Zack, J.H. Qiao, R. Weisbart, et al., Genetic determinants of autoimmune disease and coronary vasculitis in the MRL-lpr/lpr mouse model of systemic lupus erythematosus, J Immunol 161 (1998) 6999–7006.

[347] Z.S. Rahman, S.K. Tin, P.N. Buenaventura, C.H. Ho, E.P. Yap, R.Y. Yong, et al., A novel susceptibility locus on chromosome 2 in the (New Zealand Black x New Zealand White)F1 hybrid mouse model of systemic lupus erythematosus, J Immunol 168 (2002) 3042–3049.

[348] Y. Wang, M. Nose, T. Kamoto, M. Nishimura, H. Hiai, Host modifier genes affect mouse autoimmunity induced by the lpr gene, Am J Pathol 151 (1997) 1791–1798.

[349] R.J. Rigby, S.J. Rozzo, J.J. Boyle, M. Lewis, B.L. Kotzin, T.J. Vyse, New loci from New Zealand Black and New Zealand White mice on chromosomes 4 and 12 contribute to lupus-like disease in the context of BALB/c, J Immunol 172 (2004) 4609–4617.

[350] T. Miyazaki, M. Ono, W.M. Qu, M.C. Zhang, S. Mori, S. Nakatsuru, et al., Implication of allelic polymorphism of osteopontin in the development of lupus nephritis in MRL/lpr mice, Eur J Immunol 35 (2005) 1510–1520.

[351] A. Ida, S. Hirose, Y. Hamano, S. Kodera, Y. Jiang, M. Abe, et al., Multigenic control of lupus-associated antiphospholipid syndrome in a model of (NZW x BXSB) F1 mice, Eur J Immunol 28 (1998) 2694–2703.

[352] K. Ochiai, S. Ozaki, A. Tanino, S. Watanabe, T. Ueno, K. Mitsui, et al., Genetic regulation of anti-erythrocyte autoantibodies and splenomegaly in autoimmune hemolytic anemia-prone new zealand black mice, Int Immunol 12 (2000) 1–8.

[353] S. Vidal, D.H. Kono, A.N. Theofilopoulos, Loci predisposing to autoimmunity in MRL- Fas^{lpr} and C57BL/6- Fas^{lpr} mice, J Clin Invest 101 (1998) 696–702.

[354] C.G. Drake, S.K. Babcock, E. Palmer, B.L. Kotzin, Genetic analysis of the NZB contribution to lupus-like autoimmune disease in (NZB x NZW)F1 mice, Proc Natl Acad Sci USA 91 (1994) 4062–4066.

[355] S. Hirose, H. Tsurui, H. Nishimura, Y. Jiang, T. Shirai, Mapping of a gene for hypergammaglobulinemia to the distal region on chromosome 4 in NZB mice and its contribution to systemic lupus erythematosus in (NZB x NZW)F1 mice, Int Immunol 6 (1994) 1857–1864.

[356] Y. Jiang, S. Hirose, Y. Hamano, S. Kodera, H. Tsurui, M. Abe, et al., Mapping of a gene for the increased susceptibility of B1 cells to Mott cell formation in murine autoimmune disease, J Immunol 158 (1997) 992–997.

[357] S. Nakatsuru, M. Terada, M. Nishihara, J. Kamogawa, T. Miyazaki, W.M. Qu, et al., Genetic dissection of the complex pathological manifestations of collagen disease in MRL/lpr mice, Pathology international 49 (1999) 974–982.

[358] M.L. Watson, J.K. Rao, G.S. Gilkeson, P. Ruiz, E.M. Eicher, D.S. Pisetsky, et al., Genetic analysis of MRL-lpr mice: relationship of the Fas apoptosis gene to disease manifestations and renal disease-modifying loci, J Exp Med 176 (1992) 1645–1656.

[359] T.J. Vyse, C.G. Drake, S.J. Rozzo, E. Roper, S. Izui, B.L. Kotzin, Genetic linkage of IgG autoantibody production in relation to lupus nephritis in New Zealand hybrid mice, J Clin Invest 98 (1996) 1762–1772.

[360] S. Bolland, Y.S. Yim, K. Tus, E.K. Wakeland, J.V. Ravetch, Genetic modifiers of systemic lupus erythematosus in FcgammaRIIB(-/-) mice, J Exp Med 195 (2002) 1167–1174.

[361] M.E. Haywood, T.J. Vyse, A. McDermott, E.M. Thompson, A. Ida, M.J. Walport, et al., Autoantigen glycoprotein 70 expression is regulated by a single locus, which acts as a checkpoint for pathogenic anti-glycoprotein 70 autoantibody production and hence for the corresponding development of severe nephritis, in lupus-prone PXSB mice, J Immunol 167 (2001) 1728–1733.

[362] S.J. Rozzo, T.J. Vyse, K. Menze, S. Izui, B.L. Kotzin, Enhanced susceptibility to lupus contributed from the nonautoimmune C57BL/10, but not C57BL/6, genome, J Immunol 164 (2000) 5515–5521.

[363] T. Fujii, Y. Iida, M. Yomogida, K. Ikeda, T. Haga, Y. Jikumaru, et al., Genetic control of the spontaneous activation of CD4(+) Th cells in systemic lupus erythematosus-prone (NZB x NZW) F(1) mice, Genes Immun 7 (2006) 647–654.

[364] M.L. Santiago, C. Mary, D. Parzy, C. Jacquet, X. Montagutelli, R.M. Parkhouse, et al., Linkage of a major quantitative trait locus to Yaa gene-induced lupus- like nephritis in (NZW x C57BL/6)F1 mice, Eur J Immunol 28 (1998) 4257–4267.

[365] J. Kamogawa, M. Terada, S. Mizuki, M. Nishihara, H. Yamamoto, S. Mori, et al., Arthritis in MRL/lpr mice is under the control of multiple gene loci with an allelic combination derived from the original inbred strains, Arthritis Rheum 46 (2002) 1067–1074.

[366] R.M. Tucker, T.J. Vyse, S. Rozzo, C.L. Roark, S. Izui, B.L. Kotzin, Genetic control of glycoprotein 70 autoantigen production and its influence on immune complex levels and nephritis in murine lupus, J Immunol 165 (2000) 1665–1672.

[367] D. Choubey, X. Duan, E. Dickerson, L. Ponomareva, R. Panchanathan, H. Shen, et al., Interferon-Inducible p200-Family Proteins as Novel Sensors of Cytoplasmic DNA: Role in Inflammation and Autoimmunity, J Interferon Cytokine Res 30 (6) (2010) 371–380.

[368] Y.H. Cheung, N.H. Chang, Y.C. Cai, G. Bonventi, R. MacLeod, J.E. Wither, Functional interplay between intrinsic B and T cell defects leads to amplification of autoimmune disease in New Zealand black chromosome 1 congenic mice, J Immunol 175 (2005) 8154–8164.

[369] T. Suda, T. Takahashi, P. Golstein, S. Nagata, Molecular cloning and expression of the Fas ligand, a novel member of the tumor necrosis factor family, Cell 75 (1993) 1169–1178.

[370] K.R. Kumar, L. Li, M. Yan, M. Bhaskarabhatla, A.B. Mobley, C. Nguyen, et al., Regulation of B cell tolerance by the lupus susceptibility gene Ly108, Science 312 (2006) 1665–1669.

[371] Z.S. Rahman, T. Manser, Failed up-regulation of the inhibitory IgG Fc receptor Fc gamma RIIB on germinal center B cells in autoimmune-prone mice is not associated with deletion polymorphisms in the promoter region of the Fc gamma RIIB gene, J Immunol 175 (2005) 1440–1449.

[372] Y. Jiang, S. Hirose, R. Sanokawa-Akakura, M. Abe, X. Mi, N. Li, et al., Genetically determined aberrant down-regulation of FcgammaRIIB1 in germinal center B cells associated with hyper-IgG and IgG autoantibodies in murine systemic lupus erythematosus, Int Immunol 11 (1999) 1685–1691.

[373] S.A. Boackle, V.M. Holers, X. Chen, G. Szakonyi, D.R. Karp, E.K. Wakeland, et al., Cr2, a candidate gene in the murine Sle1c lupus susceptibility locus, encodes a dysfunctional protein, Immunity 15 (2001) 775–785.

[374] M.K. Haraldsson, C.A. Louis-Dit-Sully, B.R. Lawson, G. Sternik, M.L. Santiago-Raber, N.R. Gascoigne, et al., The lupus-related Lmb3 locus contains a disease-suppressing Coronin-1A gene mutation, Immunity 28 (2008) 40–51.

[375] R. Watanabe-Fukunaga, C.I. Brannan, N.G. Copeland, N.A. Jenkins, S. Nagata, Lymphoproliferative disorder in mice explained by defects in Fas antigen that mediates apoptosis, Nature 356 (1992) 314–317.

[376] S. Subramanian, K. Tus, Q.Z. Li, A. Wang, X.H. Tian, J. Zhou, et al., A Tlr7 translocation accelerates systemic autoimmunity in murine lupus, Proc Natl Acad Sci U S A 103 (2006) 9970–9975.

[377] P. Pisitkun, J.A. Deane, M.J. Difilippantonio, T. Tarasenko, A.B. Satterthwaite, S. Bolland, Autoreactive B Cell Responses to RNA-Related Antigens Due to TLR7 Gene Duplication, Science 312 (2006) 1669–1672.

[378] R. Majeti, Z. Xu, T.G. Parslow, J.L. Olson, D.I. Daikh, N. Killeen, et al., An inactivating point mutation in the inhibitory wedge of CD45 causes lymphoproliferation and autoimmunity, Cell 103 (2000) 1059–1070.

[379] T. Takahashi, M. Tanaka, C.I. Brannan, N.A. Jenkins, N.G. Copeland, T. Suda, et al., Generalized lymphoproliferative disease in mice, caused by a point mutation in the Fas ligand, Cell 76 (1994) 969–976.

[380] K. Yajima, A. Nakamura, A. Sugahara, T. Takai, FcgammaRIIB deficiency with Fas mutation is sufficient for the development of systemic autoimmune disease, Eur J Immunol 33 (2003) 1020–1029.

[381] J. Drappa, L.A. Kamen, E. Chan, M. Georgiev, D. Ashany, F. Marti, et al., Impaired T cell death and lupus-like autoimmunity in T cell-specific adapter protein-deficient mice, J Exp Med 198 (2003) 809–821.

[382] M.J. Walport, K.A. Davies, M. Botto, C1q and systemic lupus erythematosus, Immunobiology 199 (1998) 265–285.

[383] D.A. Mitchell, M.C. Pickering, J. Warren, L. Fossati-Jimack, J. Cortes-Hernandez, H.T. Cook, et al., C1q deficiency and autoimmunity: the effects of genetic background on disease expression, J Immunol 168 (2002) 2538–2543.

[384] W.S. Lai, E. Carballo, J.R. Strum, E.A. Kennington, R.S. Phillips, P.J. Blackshear, Evidence that tristetraprolin binds to AU-rich elements and promotes the deadenylation and destabilization of tumor necrosis factor alpha mRNA, Mol Cell Biol 19 (1999) 4311–4323.

[385] T.L. O'Keefe, G.T. Williams, S.L. Davies, M.S. Neuberger, Hyperresponsive B cells in CD22-deficient mice, Science 274 (1996) 798–801.

[386] T.L. O'Keefe, G.T. Williams, F.D. Batista, M.S. Neuberger, Deficiency in CD22, a B cell-specific inhibitory receptor, is sufficient to predispose to development of high affinity autoantibodies, J Exp Med 189 (1999) 1307–1313.

[387] C.C. Yu, T.S. Yen, C.A. Lowell, A.L. DeFranco, Lupus-like kidney disease in mice deficient in the Src family tyrosine kinases Lyn and Fyn, Curr Biol 11 (2001) 34–38.

[388] Y. Qian, C. Liu, J. Hartupee, C.Z. Altuntas, M.F. Gulen, D. Jane-Wit, et al., The adaptor Act1 is required for interleukin 17-dependent signaling associated with autoimmune and inflammatory disease, Nat Immunol 8 (2007) 247–256.

[389] Y. Qian, N. Giltiay, J. Xiao, Y. Wang, J. Tian, S. Han, et al., Deficiency of Act1, a critical modulator of B cell function, leads to development of Sjogren's syndrome, Eur J Immunol 38 (2008) 2219–2228.

[390] D. Seshasayee, P. Valdez, M. Yan, V.M. Dixit, D. Tumas, I.S. Grewal, Loss of TACI causes fatal lymphoproliferation and autoimmunity, establishing TACI as an inhibitory BLyS receptor, Immunity 18 (2003) 279–288.

[391] S. Kikuchi, Y. Pastore, L. Fossati-Jimack, A. Kuroki, H. Yoshida, T. Fulpius, et al., A transgenic mouse model of autoimmune glomerulonephritis and necrotizing arteritis associated with cryoglobulinemia, J Immunol 169 (2002) 4644–4650.

[392] K. Matsumoto, N. Watanabe, B. Akikusa, K. Kurasawa, R. Matsumura, Y. Saito, et al., Fc receptor-independent development of autoimmune glomerulonephritis in lupus-prone MRL/lpr mice, Arthritis Rheum 48 (2003) 486—494.

[393] X. Wu, N. Jiang, C. Deppong, J. Singh, G. Dolecki, D. Mao, et al., A role for the Cr2 gene in modifying autoantibody production in systemic lupus erythematosus, J Immunol 169 (2002) 1587—1592.

[394] S.Y. Chen, Y. Takeoka, L. Pike-Nobile, A.A. Ansari, R. Boyd, M.E. Gershwin, Autoantibody production and cytokine profiles of MHC class I (beta2- microglobulin) gene deleted New Zealand black (NZB) mice, Clin Immunol Immunopathol 84 (1997) 318—327.

[395] J.L. Goulet, R.C. Griffiths, P. Ruiz, R.F. Spurney, D.S. Pisetsky, B.H. Koller, et al., Deficiency of 5-lipoxygenase abolishes sex-related survival differences in MRL-lpr/lpr mice, J Immunol 163 (1999) 359—366.

[396] M.S. Chesnutt, B.K. Finck, N. Killeen, M.K. Connolly, H. Goodman, D. Wofsy, Enhanced lymphoproliferation and diminished autoimmunity in CD4- deficient MRL/lpr mice, Clin Immunol Immunopathol 87 (1998) 23—32.

[397] J. Sun, G. Matthias, M.J. Mihatsch, K. Georgopoulos, P. Matthias, Lack of the transcriptional coactivator OBF-1 prevents the development of systemic lupus erythematosus-like phenotypes in Aiolos mutant mice, J Immunol 170 (2003) 1699—1706.

[398] S.R. Christensen, M. Kashgarian, L. Alexopoulou, R.A. Flavell, S. Akira, M.J. Shlomchik, Toll-like receptor 9 controls anti-DNA autoantibody production in murine lupus, J Exp Med 202 (2005) 321—331.

[399] H. Hemmi, T. Kaisho, O. Takeuchi, S. Sato, H. Sanjo, K. Hoshino, et al., Small anti-viral compounds activate immune cells via the TLR7 MyD88-dependent signaling pathway, Nat Immunol 3 (2002) 196—200.

[400] A. Schwarting, T. Wada, K. Kinoshita, G. Tesch, V.R. Kelley, IFN-gamma receptor signaling is essential for the initiation, acceleration, and destruction of autoimmune kidney disease in MRL- Fas(lpr) mice, J Immunol 161 (1998) 494—503.

[401] T. Takahashi, T. Yagi, A. Kakinuma, T. Okada, K. Takatsu, et al., Suppression of autoimmune disease and of massive lymphadenopathy in MRL/Mp-lpr/lpr mice lacking tyrosine kinase Fyn (p59fyn), J Immunol 159 (1997) 2532—2541.

[402] V. Cattell, Nitric oxide and glomerulonephritis, Semin Nephrol 19 (1999) 277—287.

[403] S.L. Peng, J. Craft, T cells in murine lupus: propagation and regulation of disease, Mol Biol Rep 23 (1996) 247—251.

[404] S.L. Peng, M.P. Madaio, A.C. Hayday, J. Craft, Propagation and regulation of systemic autoimmunity by gammadelta T cells, J Immunol 157 (1996) 5689—5698.

[405] L. Wen, W. Pao, F.S. Wong, Q. Peng, J. Craft, B. Zheng, et al., Germinal center formation, immunoglobulin class switching, and autoantibody production driven by "non alpha/beta" T cells, J Exp Med 183 (1996) 2271—2282.

[406] B.R. Lawson, S.I. Koundouris, M. Barnhouse, W. Dummer, R. Baccala, D.H. Kono, et al., The role of alpha beta+ T cells and homeostatic T cell proliferation in Y-chromosome-associated murine lupus, J Immunol 167 (2001) 2354—2360.

[407] K. Kinoshita, G. Tesch, A. Schwarting, R. Maron, A.H. Sharpe, V.R. Kelley, Costimulation by B7-1 and B7-2 is required for autoimmune disease in MRL- *fas*^{lpr} mice, J Immunol 164 (2000) 6046—6056.

[408] T. Paisansinsup, A.N. Vallejo, H. Luthra, C.S. David, HLA-DR modulates autoantibody repertoire, but not mortality, in a humanized mouse model of systemic lupus erythematosus, J Immunol 167 (2001) 4083—4090.

[409] D. Kontoyiannis, G. Kollias, Accelerated autoimmunity and lupus nephritis in NZB mice with an engineered heterozygous deficiency in tumor necrosis factor, Eur J Immunol 30 (2000) 2038—2047.

[410] C. Conde, S. Weller, S. Gilfillan, L. Marcellin, T. Martin, J.L. Pasquali, Terminal deoxynucleotidyl transferase deficiency reduces the incidence of autoimmune nephritis in (New Zealand Black x New Zealand White)F1 mice, J Immunol 161 (1998) 7023—7030.

[411] M.L. Santiago-Raber, S. Kikuchi, P. Borel, S. Uematsu, S. Akira, B.L. Kotzin, et al., Evidence for genes in addition to Tlr7 in the Yaa translocation linked with acceleration of systemic lupus erythematosus, J Immunol 181 (2008) 1556—1562.

[412] I.R. Rifkin, P.L. Channavajhala, H.L. Kiefer, A.J. Carmack, E. Landesman-Bollag, B.C. Beaudette, et al., Acceleration of lpr lymphoproliferative and autoimmune disease by transgenic protein kinase CK2 alpha, J Immunol 161 (1998) 5164—5170.

[413] I. Sanz, F.E. Lee, B cells as therapeutic targets in SLE, Nat Rev Rheumatol 6 (2010) 326—337.

[414] C.G. Vinuesa, R.J. Rigby, D. Yu, Logic and extent of miRNA-mediated control of autoimmune gene expression, Int Rev Immunol 28 (2009) 112—138.

[415] D. Yu, H.A. Tan, X. Hu, V. Athanasopoulos, N. Simpson, D.G. Silva, et al., Roquin suppresses autoimmunity by limiting inducible T-cell co-stimulator messenger RNA, Nature 450 (2007) 299—303.

[416] N. Simpson, P.A. Gatenby, A. Wilson, S. Malik, D.A. Fulcher, S.G. Tangye, et al., Expansion of circulating T cells resembling follicular helper T cells is a fixed phenotype that identifies a subset of severe systemic lupus erythematosus, Arthritis Rheum 62 (2010) 234—344.

[417] L.I. Pao, K.P. Lam, J.M. Henderson, J.L. Kutok, M. Alimzhanov, L. Nitschke, et al., B cell-specific deletion of protein-tyrosine phosphatase Shp1 promotes B-1a cell development and causes systemic autoimmunity, Immunity 27 (2007) 35—48.

[418] S. Nagata, H. Nagase, K. Kawane, N. Mukae, H. Fukuyama, Degradation of chromosomal DNA during apoptosis, Cell Death Differ 10 (2003) 108—116.

[419] R. Webb, J.T. Merrill, J.A. Kelly, A. Sestak, K.M. Kaufman, C.D. Langefeld, et al., A polymorphism within IL21R confers risk for systemic lupus erythematosus, Arthritis Rheum 60 (2009) 2402—2407.

[420] V.C. Kyttaris, Z. Zhang, V.K. Kuchroo, M. Oukka, G.C. Tsokos, Cutting edge: IL-23 receptor deficiency prevents the development of lupus nephritis in C57BL/6-lpr/lpr mice, J Immunol 184 (2010) 4605—4609.

5

B-Cell Tolerance and Autoimmunity

Shiv Pillai, Kendra N. Taylor

A system of host defense made up of immune cells with specific receptors for antigens was first conceived of in the late 19th century by Paul Ehrlich, who also presciently predicted the existence of autoimmunity, a phenomenon that he called "horror autotoxicus" [1]. Over half a century later, the development of the clonal selection hypothesis put forward by Burnet and Talmadge [2] led to studies on clonal deletion and to the experiments by Nossal on clonal anergy as a mechanism for self-tolerance [3]. Tolerance was difficult to study initially because in unmanipulated vertebrates antigen-specific clones constitute very rare populations that cannot be easily identified. Our understanding of tolerance began to grow from the early 1980s when immunoglobulin heavy- and light-chain transgenic mice expressing a receptor specific for a model self antigen were first generated. This was followed a few years later by similar studies using T-cell receptor transgenic mice.

The immune system employs a number of sophisticated mechanisms to prevent the generation of immune responses to self-molecules. These may broadly be divided into two categories. Mechanisms that occur in generative or central lymphoid organs (the thymus for T-cells and the bone marrow for B-cells) are referred to as mechanisms of central tolerance. Tolerance mechanisms that come into effect after lymphocytes mature, either in peripheral lymphoid organs or in peripheral tissues, are collectively referred to as peripheral tolerance. The ability of the immune system to educate or eliminate self-reactive cells is important for the prevention of autoimmunity. Autoimmune diseases are multi-gene disorders that develop in part as a result of triggered immune responses to self-molecules because of a failure to eliminate or rein in autoreactive immune cells either during cellular development in the bone marrow and thymus or in the periphery. In lupus, central and peripheral defects in both T- and B-cell tolerance may contribute to disease pathogenesis. In this review we will examine mechanisms of central and peripheral tolerance in B-cells. We will then attempt to integrate this information in order to connect current information on genetic susceptibility to lupus with potential defects in B-cell tolerance.

B-CELL TOLERANCE

Subjects who develop lupus may have inherited or acquired defects in both T-cell and B-cell tolerance. In lupus the development of somatically mutated IgG autoantibodies against chromatin components is a central feature of the disease. The pathogenesis of lupus can best be explained on the basis of defects in tolerance in both B- and T-lymphocytes, but we will discuss scenarios in which a break in B-cell tolerance assumes particular importance. We will initially discuss central tolerance mechanisms in B-cells and then consider mechanisms of peripheral tolerance that are relevant from a B-cell standpoint.

EARLY B-CELL DEVELOPMENT AND CENTRAL TOLERANCE MECHANISMS

There are at least four different mechanisms that can contribute to central tolerance in the B lineage. These mechanisms include pre-BCR censoring, receptor editing, clonal deletion, and anergy (Figures 5.1 and 5.2). Of these, anergy may be initiated in central lymphoid organs and completed in the periphery, and clonal deletion may occur separately both in central lymphoid organs and in the periphery.

Common lymphoid progenitors receive signals to express the E2A, EBF, and Pax-5 transcription factors and commit to the B lineage. These transcriptional regulators induce the expression of the *Rag1* and *Rag2* genes, that encode the recombinase that drives V(D)J recombination in both T- and B-cells, but also induce the λ5 and

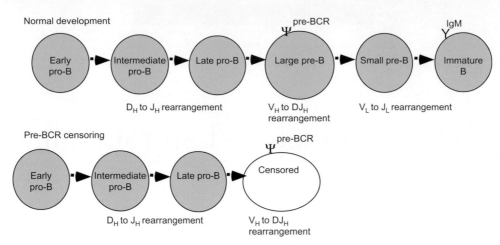

FIGURE 5.1 Pre-B receptor signaling and censoring. Most in-frame rearrangements at the IgH locus lead to assembly of the pre-B cell receptor (pre-BCR) and the positive selection of pro-B cells at the pro-B to pre-B transition, but some self-reactive rearrangements contribute to pre-BCR mediated censoring.

vPreB surrogate light-chain genes, as well as the genes encoding the Igα and Igβ proteins that are an integral part of the B-cell receptor (BCR) and the pre-BCR and contain cytosolic tyrosine-based activating motifs called ITAMs. Committed B lineage cells initially initiate rearrangement of their Ig heavy-chain genes, initially making a D to J_H rearrangement and subsequently a V to DJ_H rearrangement. Since nucleotides are added and removed at junctions during the rearrangement process, and approximately one in three developing B-cells will make a successful productive rearrangement on any one Ig heavy-chain gene chromosome (since two-thirds of the rearrangements may involve junctions made up of a total number of nucleotides that are not a multiple of three). A large fraction of developing B-cells generate two out-of-frame IgH rearrangements and fail to receive signals for survival. Cells in which one productive V-DJ_H rearrangement is made will synthesize a heavy-chain protein, assemble it with surrogate light chains and the Igα and Igβ proteins to form the pre-B cell receptor (pre-BCR) which constitutively provides signals for survival and proliferation [4–9]. Pre-BCR signaling also shuts off rearrangement on the other heavy-chain allele by rendering that chromosome inaccessible at the chromatin level. This process of allelic exclusion — of ensuring that a single productive IgH rearrangement is made in any given B-cell, ensures that clonal specificity is maintained in the immune system. While the pre-BCR is crucial for B-cell development, clonal specificity also guards against autoimmunity by ensuring that anti-self and anti-non-self receptors do not ever decorate a single B-cell clone (this is distinct from the censoring function of the pre-BCR discussed below). Signaling at the pre-BCR checkpoint results in a large expansion of pre-B cells each expressing a single Ig heavy chain, and these signals

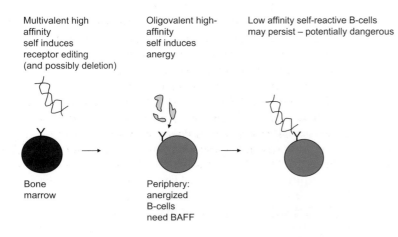

Multivalent high affinity self induces receptor editing (and possibly deletion)

Oligovalent high-affinity self induces anergy

Low affinity self-reactive B-cells may persist – potentially dangerous

Bone marrow

Periphery: anergized B-cells need BAFF

FIGURE 5.2 Mechanisms of tolerance at the immature B-cell stage. These include receptor editing, clonal deletion and anergy. However weakly self-reactive B-cells may still be potentially dangerous.

also contribute to the silencing of surrogate light-chain gene expression and the induction of rearrangement at the Igκ locus. The initial selection of the κ locus and the simple one-step nature of the rearrangement process at this locus both evolved to help mediate self tolerance as will be discussed below.

A role for the pre-BCR in self-tolerance has been suggested from both mouse and human studies. In the absence of surrogate light chains, mice develop antinuclear antibodies and antibodies to double-stranded DNA. The inability to censor certain auto-reactive Ig heavy chains in the absence of a pre-BCR has been demonstrated [10]. This phenomenon also appears to be of relevance in human pre-B cells [11]. The pre-BCR appears therefore not only to contribute to clonal specificity via the positive selection of pre-B cells with in-frame rearrangements of the IgH heavy chain, but a subset of pre-B cells that express certain IgH heavy chains that recognize self-structures are censored − presumably by deletion in a pre-BCR and possibly self-antigen-dependent manner. It may be that signaling via the pre-BCR in the absence of self-antigen recognition contributes to positive selection of pre-B cells with in-frame IgH rearrangements, but superposed recognition of antigen by the pre-BCR takes signaling beyond a survival threshold and induces cell death. Why exactly this censoring mechanism does not occur later in development in the absence of surrogate light chains is not altogether clear.

Once an Igκ light-chain gene is productively rearranged, a developing B-cell in the bone marrow expresses surface IgM for the first time and is called an immature B cell. Immature B-cells that assemble a complete B-cell receptor receive tonic (antigen-independent) signals through the B-cell receptor that mediate cell survival and also result in the shut-off of Rag gene expression. A very large fraction of these immature B-cells are strongly self-reactive and apparently recognize multivalent self-structures such as chromatin and probably also membrane proteins on cells in the bone marrow with high affinity. Strongly self-reactive B-cells are triggered by self antigens and undergo a form of developmental regression, where they revert to the late pre-B stage and reactivate a number of genes including those encoding surrogate light chains, Rag-1 and Rag-2, initiating a process known as receptor editing [12−15]. As a result of this re-induction of Rag gene expression these B-cells initially induce further rearrangement at the Igκ locus. Rearrangement is probably attempted initially on the unmethylated and accessible chromosome that contains a rearranged self-reactive κ light chain, and the rearrangement process utilizes an available upstream Vκ and a downstream Jκ segment (that were not deleted during the first "self-reactive" rearrangement event). Such a secondary κ gene rearrangement event may (if the rearrangement is in-frame)

generate a new κ light chain but also typically deletes the existing self-reactive VJκ exon. If a secondary rearrangement is non-productive, another rearrangement can be attempted if unrearranged upstream Vκ and downstream Jκ segments remain available. Subsequently rearrangements can occur on the other κ light chain bearing chromosome whether or not a previous non-productive VJκ rearrangement had occurred on that chromosome. If no non-self-reactive κ light chain is generated, rearrangement of the Igλ locus can then be attempted. The trans-acting factors that influence λ locus accessibility at the chromatin level also mediate the opening up of a κ deleting element downstream of the Igκ locus [16, 17]. This κ-deleting element contains a version of the recombination signal sequence that is recognized by the Rag1/2 recombinase and is cleaved by Rag2 during V(D)J recombination. Rearrangements of λ light chains occur only in B-cells that are undergoing receptor editing, and are invariably accompanied by the deletion of both Igκ loci mediated by a recombination event involving an upstream Vκ segment and the κ-deleting element at the 3' end of the Igκ locus.

Although we may infer that any B-cell expressing a λ light chain represents a lymphocyte clone that was originally quite strongly self-reactive and has undergone receptor editing, more direct evidence exists to suggest that receptor editing is of clinical significance. Cells that are undergoing receptor editing would be expected to transiently express both surrogate and conventional light chains and such cells have been described in human peripheral blood [18]. One objective measure of receptor editing is the quantification in immature B-cells of DNA circles containing the κ-deleting element. Using this approach a defect in receptor editing has been demonstrated in a proportion of patients with type I diabetes and in a fraction of patients with lupus [19]. The experimental analysis of the autoreactive B-cell repertoire at different stages of human B-cell development has revealed that initially up to 75% of B-cells in the bone marrow are autoreactive, and at an early bone marrow checkpoint this level of autoreactivity is reduced to 7%, presumably largely because of receptor editing [20]. A second peripheral checkpoint, possibly involving the loss of transitional B-cells (see below) results in a more modest reduction of the proportion of autoreactive B-cells [20].

Receptor editing was first described in studies using transgenic BCRs that were self-reactive [12, 13]. In some earlier studies, such as those in which BCR transgenic mice expressing a single transgenic BCR against a model antigen such as hen egg lysozyme were crossed with mice expressing a membrane-bound version of hen egg lysozyme, clonal deletion of B-cells was observed [21]. In these earlier studies the rearranged Ig heavy- and light-chain transgenes were randomly inserted in the genome. Even if receptor editing occurred in these models the original transgenic self-reactive light chain could not be deleted because it was not located in the endogenous κ locus. However, studies using knockin mice, in which rearranged heavy- and light-chain genes are inserted into their actual loci, have revealed that receptor editing is the primary mechanism of central tolerance in B-cells [14, 15]. Clonal deletion probably occurs when the Ig heavy chain in a B-cell clone contributes in a major manner to recognition of a self antigen. It may also be important in transitional cells that have left the bone marrow and have access to peripheral antigens (see next section). From a mechanistic standpoint both receptor editing and deletion depend on effective BCR signaling, and defective BCR signaling (as seen in patients with X-lined agammaglobulinemia, who harbor a defective Btk tyrosine kinase), results in defective receptor editing [21]. A mouse mutation in a SLAM family protein, Ly108, results in a lupus-like syndrome presumably because of defective receptor editing but how Ly108 contributes to receptor editing is unclear [22]. Somewhat counterintuitively, chromatin antigens apparently mediate tolerance by inducing signaling via both the BCR and TLR activation. Although TLR activation may be a part of the disease pathogenesis process in mature B-cells [23], a role for IRAK4 and MyD88 (the former a kinase and the latter an adaptor, both required for TLR signaling) has been demonstrated in tolerance induction in human bone marrow B-cells, and a role for IRAK4, MyD88, and the UNC93B protein (that facilitates activation of TLRs 3, 8, and 9) in a peripheral human tolerance checkpoint has also been postulated [24].

Apart from pre-BCR censoring, receptor editing and clonal deletion, the induction of B-cell anergy may be initiated in the bone marrow but the process depends on the inactivation of B-cells in the periphery. Anergy will be discussed below in the context of peripheral B-cell tolerance.

PERIPHERAL TOLERANCE IN B-CELLS

Peripheral tolerance in B-cells may be maintained in at least four different ways. These mechanisms include anergy, clonal deletion, inactivation by inhibitory receptors, and the elimination of potentially autoreactive B-cells by the FasL/Fas pathway.

Our understanding of anergy in B-cells is derived largely from the study of anti-hen egg lysozyme (HEL) BCR transgenic mice [25]. B-cells that are repeatedly triggered by an antigen that is not membrane bound and is not intrinsically multivalent become unresponsive. Soluble HEL is the best-studied model antigen in this context. Constant stimulation makes these B-cells extremely dependent on BAFF (B-cell-activating factor)

for survival but they are unable to compete with normal follicular B-cells for the BAFF that is available in B-cell follicles [26]. High BAFF levels have been reported in patients with lupus but it is still not clear if this is a consequence of ongoing inflammation or whether high BAFF levels contribute to the onset of disease. Since there are no data at present to indicate that increased BAFF levels are genetically determined in lupus patients it remains unclear if a break in the process of B-cell anergy is of relevance in lupus patients.

Another mechanism of peripheral tolerance that may be of relevance is the clonal deletion of B-cells specific for a multivalent peripheral antigen that is presumably not available in the bone marrow to mediate clonal deletion earlier in development [27]. BCR triggering may induce clonal deletion of transitional T1 cells that are IgM$^+$IgD$^-$ CD23$^-$AA4.1$^+$ B-cells that have recently emigrated from the bone marrow to the spleen and have yet to enter the splenic follicular niche, or of transitional T2 cells that have also recently emigrated from the bone marrow but have entered the follicle. These latter cells are IgM$^+$IgD$^+$ CD23$^+$AA4.1$^+$ B-cells that have recently obtained the ability to recirculate and thus can access antigens that are available in the spleen, lymph nodes, or Peyer's patches. In humans transitional B-cells broadly fall into the peripheral IgM$^+$CD10$^+$CD27$^-$ pool of recent emigrants. It is presumed that transitional B-cells, like immature B-cells in the bone marrow, are more susceptible to apoptosis than mature recirculating B-cells.

While receptor editing, clonal deletion, and anergy all require robust BCR signaling, a distinct mechanism of B-cell tolerance is mediated by inhibitory signaling via sialic acid Ig domain containing lectins or Siglecs, the best-studied inhibitory Siglec in B-cells being CD22. This lectin contains extracellular Ig domains and cytosolic inhibitory tyrosine (ITIM) residues, CD22 recognizes N-glycans with terminal α2, 6, linked sialic acid containing moieties, but only if the 9-OH position of sialic acid is not acetylated. A 9-O sialic acid acetylesterase, SIAE, removes 9-O acetyl moieties from sialated N-glycans, making them accessible to CD22. When CD22 binds to its ligands it is phosphorylated on its ITIM tyrosines by Lyn and then recruits the inhibitory tyrosine phosphatase, SHP-1. CD22 knockout mice exhibit enhanced BCR signaling and develop anti-DNA antibodies, while *Lyn* knockout mice, mice in which SHP-1 in B-cells is conditionally deleted, and *Siae* mutant mice all present with enhanced BCR signaling and develop a lupus-like syndrome [28—31]. While receptor editing, deletion and anergy tolerize B-cells that recognize self structures with high avidity, weakly self-reactive B-cells mature and migrate to the periphery. It is important to prevent these weakly self-reactive B-cells from promiscuously being activated by T-cells, given that T-dependent activation can lead to germinal center formation and somatic mutation. CD22 attenuates BCR signaling, basically setting a threshold for B-cell activation. The likely function of this SIAE/CD22/Lyn/SHP-1 pathway is to prevent the activation of mature B-cells by weak self antigens and to thus reduce the likelihood that these B-cells will endocytose self-antigen-containing complexes and induce the expression of CCR7, a chemokine receptor that draws activated B-cells towards the T-cell zone [32] (Figure 5.3).

During T-dependent immune responses to a protein antigen helper T-cells are initially triggered by dendritic cells that present antigenic peptides on MHC class II molecules and which have been previously triggered by microbial structures via Toll-like receptors in order to express B7 co-stimulatory ligands. The activation of these T-cells via the TCR and CD28 results in the expansion of helper T-cells that recognize a peptide derived from the protein antigen, and these cells reduce the expression of CCR7 and express higher levels of CXCR5 and migrate towards the follicle or B-cell zone. Activated B-cells express CCR7 and migrate towards the T-cell zone. When T-cells and B-cells meet at the T−B interface, some activated B-cells expressing the ICOS ligand will activate ICOS on T helper cells, which in the presence of IL-21 and IL-6 can differentiate into T follicular helper cells (T_{FH} cells) which secrete IL-21 and express high levels of ICOS (inducible co-stimulator) which functions as a co-stimulatory receptor and the CXCR5 chemokine receptor. These T_{FH} cells and activated B-cells enter the follicle and the T_{FH} cells induce the formation of germinal centers (GCs). Somatic mutation occurs in the dark zone of GCs and the highest affinity mutated B-cells are selected by antigen presented on follicular dendritic cells. T_{FH} cells contribute to B-cell selection in a number of ways. IL-21 causes germinal center B-cells to express relatively high levels of the pro-apoptotic BH3 only protein called Bim, setting B-cells up to apoptose unless they are rescued by antigen. High-affinity B-cells that can be triggered by antigen on FDCs may also receive CD40 signals because they most efficiently internalize antigen and thus foster the presentation of peptides that result in prolonged B-cell−T_{FH}-cell interactions.

Mutations in *Fas* and *FasL* lead to a lupus-like syndrome in *lpr* and *gld* mice and mutations in these genes are found in the rare human genetic disorder, the autoimmune lymphoproliferative syndrome or ALPS [33, 34]. The major reason for the breakdown in tolerance in the absence of the Fas pathway appears to be the failure to eliminate self-reactive B-cells generated during T-dependent immune responses [35, 36]. B-cells activated via the BCR induce the expression of c-FLIP (cellular Flice-like protein), an inhibitor of Fas activation, and are protected from apoptosis when FasL on activated T-cells interacts with Fas on activated B-cells. FasL−Fas signaling may activate apoptosis in low-affinity or bystander B-cells activated non-specifically by CD40L on activated T helper cells at the T−B interface; Fas−FasL interactions may also contribute to the elimination of low-affinity self-reactive B-cells in germinal centers when they interact with T_{FH} cells in the light zone during affinity maturation (Figure 5.4). Spontaneous germinal center formation can also lead to a lupus-like syndrome in mutant mice. Mice with a mutation in *Roquin*, a ubiquitin E3 ligase that targets ICOS (inducible costimulator) for degradation, have high ICOS levels, spontaneously form germinal centers and develop a lupus-like syndrome [37]. IL-17 is spontaneously secreted by BXD2 mice. This overproduced cytokine leads to spontaneous germinal center development and BXD2 mice develop a lupus-like phenotype [38]. Germinal center abnormalities and defective selection of activated B-cells deserves further analysis in the context of SLE.

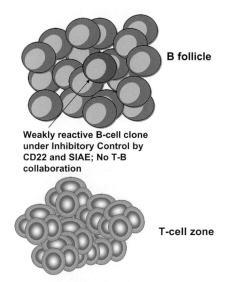

B follicle

Weakly reactive B-cell clone under Inhibitory Control by CD22 and SIAE; No T-B collaboration

T-cell zone

FIGURE 5.3 Weakly self-reactive B-cells may be tolerized by the SIAE pathway. SIAE (sialic acid acetyl esterase) and inhibitory receptors such as CD22 set a threshold for B-cell activation and prevent weakly self-reactive B-cells from receiving T-cell help and entering the germinal center reaction.

LINKS BETWEEN GENETIC SUSCEPTIBILITY TO LUPUS AND DEFECTS IN TOLERANCE

Systemic lupus like most human autoimmune disorders are widely accepted as being non-Mendelian multigene diseases and it is presumed that there is an inherited basis for the loss of B-cell tolerance that represents a crucial event in the pathogenesis of many of these diseases. A number of susceptibility loci for human autoimmune disorders have been uncovered by genome-wide association studies, and the odds ratios for these associations are generally modest.

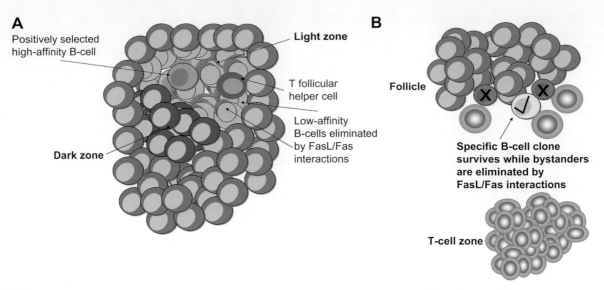

FIGURE 5.4   Maintenance of peripheral B-cell tolerance by the FasL/Fas pathway. (*A*) During a T-dependent B-cell response, Fas activation may contribute to the elimination of somatically mutated B-cells that recognize the immunogen weakly or are self-reactive. (*B*) Fas may be involved in the deletion of promiscuously activated (via CD40) B-cells that are not specific for the immunogen and have not received BCR signals.

An inherited defect in BCR signaling would likely result in a defect in receptor editing, deletion, or anergy and could predispose certain individuals to autoimmunity. Three genes that have a potential role in T- and/ or B-cell signaling have been positively linked in genome-wide association studies to lupus and other autoimmune disorders. BANK1 is an adaptor that participates in B-cell signaling, BLK is a Src-family-kinase expressed in B-cells, and PTPN22 (also known as Lyp) is a tyrosine phosphatase in T and B-cells that contributes to the dephosphorylation of Src- and Syk-family tyrosine kinases and thus attenuates BCR and TCR signaling [39]. The PTPN22 R620W gain-of-function variant is found more frequently in autoimmune patients, and potentially could contribute to attenuated BCR and TCR signaling in response to self-antigens and thus contribute to autoimmunity.

A polymorphic variant of the IRF5 gene seen in SLE patients is associated with high serum levels of type I interferons. It has been suggested that enhanced expression of type I interferons may activate myeloid dendritic cells and thus contribute to a break in tolerance in SLE [40].

It is likely that genome-wide association studies have only begun to scratch the surface in terms of understanding genetic susceptibility to lupus. The analysis of rare variants will involve major sequencing efforts and will require the development of functional assays but will likely contribute in a major way making the link between genetic susceptibility and defects in tolerance.

Functionally defective rare variants of SIAE, a negative regulator of BCR signaling, have been shown to contribute to susceptibility to autoimmune disorders including Lupus [41].

Acknowledgments

This work was supported by grants AI 064930 and AI 076505 from the NIH and a TIL grant from the Alliance for Lupus Research to SP. KNT was supported by a Faculty Development Supplement Award to Promote Diversity in Health Related Research from the NIH.

References

[1] P. Ehrlich, [The partial function of cells. (Nobel Prize address given on 11 December 1908 at Stockholm).], Int Arch Allergy Appl Immunol 5 (1954) 67—86.

[2] M. Burnet, A Darwinian Approach to Immunity, Nature 203 (1964) 451—454.

[3] G.J. Nossal, The cellular and molecular basis of immunological tolerance, Essays Fundam Immunol 0 (1973) 28—43.

[4] A. Kudo, N. Sakaguchi, F. Melchers, Organization of the murine Ig-related lambda 5 gene transcribed selectively in pre-B lymphocytes, Embo J 6 (1987) 103—107.

[5] S. Pillai, D. Baltimore, Formation of disulphide-linked mu 2 omega 2 tetramers in pre-B cells by the 18K omega-immunoglobulin light chain, Nature 329 (1987) 172—174.

[6] M. Reth, Antigen receptors on B lymphocytes, Annu Rev Immunol 10 (1992) 97—121.

[7] Y. Aoki, K.J. Isselbacher, S. Pillai, Bruton tyrosine kinase is tyrosine phosphorylated and activated in pre-B lymphocytes and receptor-ligated B-cells, Proc Natl Acad Sci U S A 91 (1994) 10606—10609.

[8] K. Rajewsky, Clonal selection and learning in the antibody system, Nature 381 (1996) 751—758.

[9] S. Herzog, M. Reth, H. Jumaa, Regulation of B-cell proliferation and differentiation by pre-B cell receptor signalling, Nat Rev Immunol 9 (2009) 195—205.

[10] R.A. Keenan, A. De Riva, B. Corleis, L. Hepburn, S. Licence, T.H. Winkler, I.L. Martensson, Censoring of autoreactive B-cell development by the pre-B cell receptor, Science 321 (2008) 696—699.

[11] Y. Minegishi, M.E. Conley, Negative selection at the pre-BCR checkpoint elicited by human mu heavy chains with unusual CDR3 regions, Immunity 14 (2001) 631—641.

[12] S.L. Tiegs, D.M. Russell, D. Nemazee, Receptor editing in self-reactive bone-marrow B-cells, J.Exp. Med. 177 (1993) 1009—1020.

[13] D. Gay, T. Saunders, S. Camper, M. Weigert, Receptor editing: an approach by autoreactive B-cells to escape tolerance, J Exp Med 177 (1993) 999—1008.

[14] R. Pelanda, S. Schwers, E. Sonoda, R.M. Torres, D. Nemazee, K. Rajewsky, Receptor editing in a transgenic mouse model: site, efficiency, and role in B-cell tolerance and antibody diversification, Immunity 7 (1997) 765—775.

[15] K.L. Hippen, B.R. Schram, L.E. Tze, K.A. Pape, M.K. Jenkins, T.W. Behrens, In vivo assessment of the relative contributions of deletion, anergy, and editing to B-cell self-tolerance, J Immunol 175 (2005) 909—916.

[16] J. Durdik, M.W. Moore, E. Selsing, Novel kappa light-chain gene rearrangements in mouse lambda light chain-producing B lymphocytes, Nature 307 (1984) 749—752.

[17] W.B. Graninger, P.L. Goldman, C.C. Morton, S.J. O'Brien, S.J. Korsmeyer, The kappa-deleting element. Germline and rearranged, duplicated and dispersed forms, J Exp Med 167 (1988) 488—501.

[18] E. Meffre, E. Davis, C. Schiff, C. Cunningham-Rundles, L.B. Ivashkiv, L.M. Staudt, et al., Circulating human B-cells that express surrogate light chains and edited receptors, Nat Immunol 1 (2000) 207—213.

[19] A.K. Panigrahi, N.G. Goodman, R.A. Eisenberg, M.R. Rickels, A. Naji, E.T. Luning Prak, RS rearrangement frequency as a marker of receptor editing in lupus and type 1 diabetes, J Exp Med 205 (2008) 2985—2994.

[20] H. Wardemann, S. Yurasov, A. Schaefer, J.W. Young, E. Meffre, M.C. Nussenzweig, Predominant autoantibody production by early human B-cell precursors, Science 301 (2003) 1374—1377.

[21] Y.S. Ng, H. Wardemann, J. Chelnis, C. Cunningham-Rundles, E. Meffre, Bruton's tyrosine kinase is essential for human B-cell tolerance, J Exp Med 200 (2004) 927—934.

[22] K.R. Kumar, L. Li, M. Yan, M. Bhaskarabhatla, A.B. Mobley, C. Nguyen, et al., Regulation of B-cell tolerance by the lupus susceptibility gene Ly108, Science 312 (2006) 1665—1669.

[23] E.A. Leadbetter, I.R. Rifkin, A.M. Hohlbaum, B.C. Beaudette, M.J. Shlomchik, A. Marshak-Rothstein, Chromatin-IgG complexes activate B-cells by dual engagement of IgM and Toll-like receptors, Nature 416 (2002) 603—607.

[24] I. Isnardi, Y.S. Ng, I. Srdanovic, R. Motaghedi, S. Rudchenko, H. von Bernuth, et al., IRAK-4- and MyD88-dependent pathways are essential for the removal of developing autoreactive B-cells in humans, Immunity 29 (2008) 746—757.

[25] C.C. Goodnow, J. Crosbie, S. Adelstein, T.B. Lavoie, S.J. Smith-Gill, R.A. Brink, et al., Altered immunoglobulin expression and functional silencing of self-reactive B lymphocytes in transgenic mice, Nature 334 (1988) 676—682.

[26] E.H. Ekland, R. Forster, M. Lipp, J.G. Cyster, Requirements for follicular exclusion and competitive elimination of autoantigen-binding B-cells, J Immunol 172 (2004) 4700—4708.

[27] M. Okamoto, M. Murakami, A. Shimizu, S. Ozaki, T. Tsubata, S. Kumagai, T. Honjo, A transgenic model of autoimmune hemolytic anemia, J Exp Med 175 (1992) 71—79.

[28] T.L. O'Keefe, G.T. Williams, F.D. Batista, M.S. Neuberger, Deficiency in CD22, a B-cell-specific inhibitory receptor, is sufficient to predispose to development of high affinity autoantibodies, J Exp Med 189 (1999) 1307—1313.

[29] M.L. Hibbs, D.M. Tarlinton, J. Armes, D. Grail, G. Hodgson, R. Maglitto, et al., Multiple defects in the immune system of Lyn-deficient mice, culminating in autoimmune disease, Cell 83 (1995) 301—311.

[30] L.I. Pao, K.P. Lam, J.M. Henderson, J.L. Kutok, M. Alimzhanov, L. Nitschke, et al., B-cell-specific deletion of protein-tyrosine phosphatase Shp1 promotes B-1a cell development and causes systemic autoimmunity, Immunity 27 (2007) 35—48.

[31] A. Cariappa, H. Takematsu, H. Liu, S. Diaz, K. Haider, C. Boboila, et al., B-cell antigen receptor signal strength and peripheral B-cell development are regulated by a 9-O-acetyl sialic acid esterase, J Exp Med 206 (2009) 125—138.

[32] S. Pillai, A. Cariappa, S. Pirnie, Esterases and autoimmunity: the Sialic acid acetylesterase pathway and the regulation of peripheral B-cell tolerance. Trends In Immunology (in press).

[33] G.H. Fisher, F.J. Rosenberg, S.E. Straus, J.K. Dale, L.A. Middleton, A.Y. Lin, et al., Dominant interfering Fas gene mutations impair apoptosis in a human autoimmune lymphoproliferative syndrome, Cell 81 (1995) 935—946.

[34] M. Del-Rey, J. Ruiz-Contreras, A. Bosque, S. Calleja, J. Gomez-Rial, E. Roldan, et al., A homozygous Fas ligand gene mutation in a patient causes a new type of autoimmune lymphoproliferative syndrome, Blood 108 (2006) 1306—1312.

[35] T.L. Rothstein, J.K. Wang, D.J. Panka, L.C. Foote, Z. Wang, B. Stanger, et al., Protection against Fas-dependent Th1-mediated apoptosis by antigen receptor engagement in B-cells, Nature 374 (1995) 163—165.

[36] Z. Hao, G.S. Duncan, J. Seagal, Y.W. Su, C. Hong, J. Haight, et al., Fas receptor expression in germinal-center B-cells is essential for T and B lymphocyte homeostasis, Immunity 29 (2008) 615—627.

[37] C.G. Vinuesa, I. Sanz, M.C. Cook, Dysregulation of germinal centres in autoimmune disease, Nat Rev Immunol 9 (2009) 845—857.

[38] H.C. Hsu, P. Yang, J. Wang, Q. Wu, R. Myers, J. Chen, et al., Interleukin 17-producing T helper cells and interleukin 17 orchestrate autoreactive germinal center development in autoimmune BXD2 mice, Nat Immunol 9 (2008) 166—175.

[39] A.F. Arechiga, T. Habib, Y. He, X. Zhang, Z.Y. Zhang, A. Funk, et al., Cutting edge: The PTPN22 allelic variant associated with autoimmunity impairs B-cell signaling, J Immunol 182 (2009) 3343—3347.

[40] J. Banchereau, V. Pascual, Type I interferon in systemic lupus erythematosus and other autoimmune diseases, Immunity 25 (2006) 383—392.

[41] I. Surolia, S.P. Pirnie, V. Chellappa, K.N. Taylor, A. Cariappa, J. Moya, et al., Functionally defective germline variants of sialic acid acetylesterase in autoimmunity, Nature (epub 2010).

6

Dendritic Cells in SLE

Gerlinde Obermoser, Karolina Palucka, Hideki Ueno, Jacques Banchereau, Virginia Pascual

INTRODUCTION

The immune system is endowed with the ability to recognize a universe of antigens (Ag), and to generate specific responses to them. Immune responses involve both Ag-non-specific innate immunity and Ag-specific adaptive immunity. The innate and the adaptive immune systems act in concert to eradicate pathogens through cells such as macrophages, granulocytes, dendritic cells (DCs) and lymphocytes, and through effector proteins such as cytokines, antimicrobial peptides, complement, and antibodies [1–4]. Lymphocytes (T, B, NK, and NKT cells) and their products are under the control of DCs [1, 3, 5–6]. DCs reside in peripheral tissues where they are poised to capture antigens. Thus, antigen-loaded DCs migrate from tissues through the afferent lymphatics into the draining lymph nodes. There, they present processed protein and lipid Ags to T cells via both classical (MHC class I and class II) and non-classical (CD1 family) antigen presenting molecules [5] (Figure 6.1). Soluble antigens also reach the draining lymph nodes through lymphatics and conduits where they are captured, processed, and presented by lymph-node resident DCs [7]. In the steady state, non-activated (immature) DCs present self-antigens to T cells [8–10] leading to tolerance. Once activated (mature), antigen-loaded DCs launch antigen-specific immunity [11] leading to T-cell proliferation and differentiation into helper and effector cells with unique functions and cytokine profiles. DCs are also important in launching humoral immunity, partly due to their capacity to directly interact with B cells [12, 13]. DCs can route antigens into non-degradative recycling compartments, which enables the presentation of unprocessed antigens to B cells [14, 15]. DCs appear to be essential for both central tolerance in the thymus and peripheral tolerance [16]. DCs induce immune tolerance partly through T-cell deletion and partly through activation of regulatory T cells (Tregs). Thus, it is not surprising that alterations in DC biology are involved in autoimmune diseases, such as systemic lupus erythematosus (SLE) [17] and in malignant diseases, such as myeloma and breast cancer [18-20].

DCs are composed of multiple subsets with distinct functions. The two major subsets are myeloid DCs (mDCs) and plasmacytoid DCs (pDCs). We will next focus on recently described phenotypical and functional differences among human DC subsets [5, 6, 21].

HUMAN DC SUBSETS

Myeloid DCs

In vivo mDCs exist in at least three compartments: (1) peripheral tissue DCs, (2) secondary lymphoid organ-resident DCs, and (3) circulating blood mDCs.

The best-studied human DC subsets are those from skin, where three subsets can be identified. Langerhans cells (LCs) populate the epidermis, while the dermis contains two DC subsets, CD1a$^+$ DCs and CD14$^+$ DCs, as well as macrophages [22]. Dermal CD14$^+$ DCs express a large number of surface C-type lectins including DC-SIGN, DEC-205, LOX-1, CLEC-6, Dectin-1 and DCIR, while LCs express Langerin and DCIR. Dermal CD14$^+$ DCs express multiple Toll-like receptors (TLRs) recognizing bacterial pathogen-associated molecular patterns (PAMPs), such as TLR2, 4, 5, 6, 8, and 10 [23, 24], while LCs have been reported to express TLR1, 2, 3, 6, and 10 [23, 25].

LCs and CD14$^+$ DCs also differ in their cytokine profiles. While CD14$^+$ DCs produce a large set of soluble factors including IL-1β, IL-6, IL-8, IL-10, IL-12, GM-CSF, MCP and TGF-β in response to stimulation via CD40,

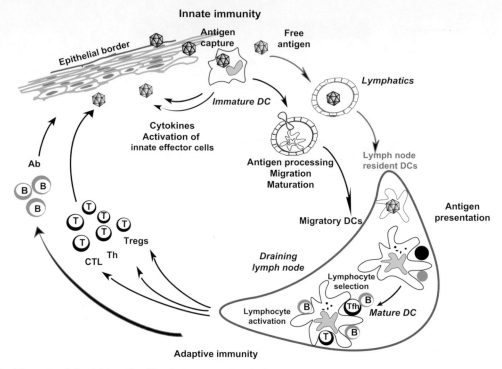

FIGURE 6.1   The life cycle of dendritic cells. Circulating precursor DC enter tissues as immature DC. They can encounter pathogens (e.g.: viruses) directly, which induce secretion of cytokines (e.g.: IFN-α); or indirectly through pathogen effect on stromal cells. Cytokines secreted by DCs in turn activate effector cells of innate immunity such as eosinophils, macrophages, and NK cells. Microbe activation triggers DCs migration towards secondary lymphoid organs and simultaneous activation (maturation). These activated migratory DCs that enter lymphoid organs display pMHC complexes, which allow selection of rare circulating antigen-specific T lymphocytes. Activated T cells help DCs for their terminal maturation, which allows lymphocyte expansion and differentiation. Activated T lymphocytes traverse inflamed epithelia and reach the injured tissue, where they eliminate microbe and/or microbe-infected cells. B cells, activated by DCs and T cells, migrate into various areas where they mature into plasma cells that produce antibodies that neutralize the initial pathogen. Antigen can also reach draining lymph nodes without involvement of peripheral tissue DCs and be captured by lymph node resident DCs [182].

LCs produce only a few cytokines, including IL-15 [26]. As we discuss hereunder, this different cytokine secretion pattern might contribute to the unique biological functions of these DC subsets.

Dermal CD14$^+$ DCs: potent inducers of antibody responses

CD14$^+$ DCs derived from CD34$^+$ hematopoietic progenitor cells (HPCs) induce CD40-activated naïve B cells to differentiate into IgM-producing plasma cells through the secretion of IL-6 and IL-12 [27]. Furthermore, CD14$^+$ DCs induce naïve CD4$^+$ T cells to differentiate into effectors sharing properties with T follicular helper cells (Tfh) [26], a CD4$^+$ T cell subset specialized in B cell help [28, 29]. There, CD4$^+$ T cells primed by CD14$^+$ DCs help naïve B cells produce large amounts of IgM and switch isotypes towards IgG and IgA. This ability to regulate B-cell differentiation appears unique to CD14$^+$ DCs, as LCs are unable to do so.

Acquisition of Tfh phenotype and function by human CD4$^+$ T cells depends on IL-12p70 secreted by DCs as neutralizing antibodies inhibit their development [30]. IL-12 endows activated CD4$^+$ T cells with the capacity

to help the differentiation of antibody-secreting cells (ASCs) via IL-21 [30], a pleotropic cytokine that promotes B-cell growth, differentiation, and class-switch recombination [31]. IL-12 induces human-naïve CD4$^+$ T cells to differentiate into two different types of IL-21-producing T cells: (i) IL-21$^+$ IFN-γ$^+$ Th1 cells expressing T-bet, and (ii) IL-21$^+$ IFN-γ$^-$ non-Th1 cells. The development of both T cell types is dependent on STAT4. Whether both types of IL-21-producing CD4$^+$ T cells display an equivalent capacity to help B cells is currently under investigation.

Thus, IL-12 appears to contribute to humoral immunity in humans through two different paths: a direct path in DC—B interaction, and an indirect path through DC—T-cell interaction and induction of Tfh cells (Figure 6.2). These two paths might act simultaneously *in vivo*, through the interaction of antigen-presenting DCs with antigen-specific T cells and B cells at extrafollicular sites, as recently illustrated through *in vivo* imaging in mice [13, 32].

Mouse studies have demonstrated that IL-12, when administered as vaccine adjuvant, enhances the development of tumor-specific CTL and Th1 responses *in vivo* [33, 34]. In humans, systemically administered

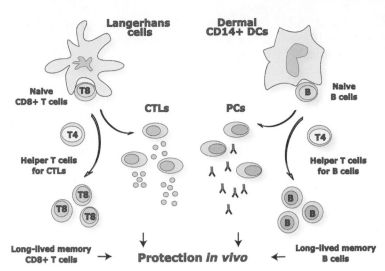

FIGURE 6.2 Human myeloid dendritic cell subsets are highly specialized and activate different arms of the immune system. Distinct mDC subsets result in the generation of different types of immune responses. LCs are potent activators of cellular immunity while Int DCs prime B cells and B helper T cells (Tfh). CTL: cytotoxic T lymphocyte; PC: plasma cell.

IL-12 has thus far shown only very modest clinical efficacy [35, 36]. In contrast, the injection of IL-12 into tumor sites of head and neck cancer patients resulted in the activation of B cells in the draining lymph nodes, which was associated with their infiltration into tumor sites and tumor regression [37]. Thus, adjuvants that promote the secretion of IL-12 might improve vaccines aimed at induction of neutralizing antibodies in humans. On the contrary, blocking of IL-12 might be beneficial to prevent the development of autoreactive B cells in human autoimmune diseases.

Epidermal Langerhans cells: potent activators of CD8$^+$ T cells

LCs are remarkably efficient at inducing CTL responses (Figure 6.2). Thus, LCs, either generated *in vitro* from CD34$^+$ HPCs or isolated from human skin (epidermis), induce a robust proliferation of naïve allogeneic CD8$^+$ T cells when compared to CD14$^+$ DCs [26]. When pulsed with MHC class I peptides derived from tumor or viral antigens, LCs are far more efficient than CD14$^+$ DCs in the priming of antigen-specific CD8$^+$ T cells. LCs are also efficient in cross-presenting peptides from protein antigens to CD8$^+$ T cells. When compared to those induced by CD14$^+$ dermal DCs, the CD8$^+$ T cells primed by LCs show high avidity in tetramer-binding assays and express higher levels of cytotoxic molecules, such as granzymes and perforin. Accordingly, they are remarkably more efficient in killing target cells; in particular tumor cell lines which express low levels of peptide/HLA complexes [26].

While CD14$^+$ DCs educate naïve CD4$^+$ T cells to become IL-21-producing Tfh-like cells, LCs polarize naïve CD4$^+$ T cells into cells secreting type 2 cytokines such as IL-4, IL-5, and IL-13. This is consistent with mouse studies showing the preferential induction of

Th2 responses upon delivery of an antigen to the LC-rich epidermis [38]. Interestingly, IFN-γ-secreting CD4$^+$ T cells are induced at a similar level by human LCs and CD14$^+$DCs. Further studies are necessary to determine whether these IFN-γ-secreting CD4$^+$ T cells share similar biological functions.

The human dermis contains another DC subset, dermal CD1a$^+$ DCs, which seems to share functional properties of both LCs and dermal CD14$^+$ DCs. The biological role of this DC population remains to be addressed. In mice, Langerin$^+$ DCs found in dermis [39–41] share many properties of lymphoid-resident CD8$^+$ DCs, a subset efficient at the induction of CTL responses [42].

The heterogeneity of skin mDCs is mirrored by the existence of at least three blood mDC subsets. In human blood, CD1c$^+$ (BDCA-1$^+$) DCs represent a major population of Lin^{neg}HLA-DR$^+$CD11c$^+$ mDCs. BDCA3$^+$ Lin^{neg}HLA-DR$^+$CD11c$^+$ mDCs [43] represent a minute population that uniquely expresses CLEC9A, a C-type lectin with ITAM-like motif [44]. The third mDC subset expresses 6-Sulfo LacNAc, a modified PSGL-1, and CD16 [45]. The physiological role of these distinct blood mDC subsets remains to be determined.

DC subsets and the control of humoral and cellular immunity

The results summarized above support that the two different arms of adaptive immunity, i.e., humoral and cellular arms, are differentially regulated by the two skin mDC subsets. Thus, *in vitro* studies suggest that humoral immunity is preferentially initiated by CD14$^+$ dermal DCs, while cellular immunity is preferentially regulated by LCs (Figure 6.2). While the formal demonstration of this hypothesis will require *in vivo* studies in humans or non-human primates, this proposed dichotomy is supported by certain mouse studies. In particular, one set of studies indicated that dermal DCs migrate into the lymphoid organ outer paracortex just beneath the B-cell follicles, whereas LCs migrate into the T-cell-rich inner paracortex [46, 47]. Another set of murine studies demonstrate that LCs cross-present skin-derived exogenous antigen to CD8$^+$ T cells and induce effector functions, such as cytokine production and cytotoxicity [47]. Finally, targeting distinct DC subsets *in vivo* with specific antibodies leads to generation of distinct immune responses [48]. While targeting CD8$^+$ DCs with the conjugates of DEC205 and an antigen (OVA) preferentially induced CD8$^+$ T cell response, targeting CD8$^-$ DCs with the conjugates of DCIR2 and an antigen preferentially induced CD4$^+$ T cell response [48]. This study clearly demonstrated the functional specialization of mDC subsets *in vivo*.

Monocytes exposed to different cytokines generate DCs with distinct properties

Upon microbial invasion, monocytes are induced to migrate into inflammatory sites and differentiate into DCs [49]. *In vitro* studies indicate that different cytokines skew the differentiation of monocytes into DCs with different phenotypes and functions. When activated monocytes, for example by GM-CSF, encounter IL-4 (secreted for example by mast cells), they differentiate into IL4-DCs [50]. By contrast, after encounter with IFN-α/β or IL-15 (secreted by pDCs and keratinocytes, respectively) activated monocytes will differentiate into IFN-DCs or IL15-DCs, respectively [51]. This spectrum of DCs represents immune-stimulatory DCs, which generate different types of immune responses. For example, peptide-pulsed IL15-DCs are more efficient than IL4-DCs in the induction of antigen-specific CTL differentiation *in vitro* [52]. This is consistent with our earlier findings on LCs, inasmuch as these DCs might be induced when monocytes encounter GM-CSF and IL15 secreted by keratinocytes [53]. *In vitro*, DCs that exhibit immune-regulatory functions can be generated for example by culturing monocytes in the presence of GM-CSF and IL-10. These DCs can render T cells anergic and allow the expansion of suppressor T cells [54, 55]. The challenge will be to link these distinct DC phenotypes *in vitro* with a specific type of immune response and immune pathology *in vivo*, as exemplified by TNF and IFN-α in autoimmunity.

Plasmacytoid DCs

pDCs circulate in the blood and enter lymphoid organs through high endothelial venules (HEV) rather than afferent lymphatics [56]. These lin^{neg}HLA-DR$^+$ cells express high levels of IL-3Rα chain (CD123), as well as some specific markers such as BDCA-2 [57] and ILT7 [58]. Compared to mDCs, they express a different set of TLRs [59], including TLR7-9 through which they recognize viral components and secrete large amounts of type I IFN [56].

pDCs can be activated by (i) viruses [56, 60]; (ii) IL-3 and CD40 ligand (IL-3/CD40-L) [61, 62], possibly originating from mast cell triggered for example by parasites; (iii) microbial components in the form of CpG DNA [63, 64]; (iv) autoantigens and nucleic-acid-containing immune complexes (ICs) [65–67]. Similar to mDCs, pDCs appear to display a remarkable functional plasticity. There is evidence in mice that resting plasmacytoid DCs are involved in the induction of tolerance [68–70]. However, pDCs exposed to viruses, such as live influenza virus, are able to launch memory responses by inducing the expansion and differentiation of antigen-specific memory B and T lymphocytes into plasma cells [71], and CTLs [72, 73], respectively. On the contrary, pDCs activated with CpG or IL-3/CD40L induce *in vitro* IL-10-secreting regulatory CD4$^+$ T cells [74] as well as suppressor CD8$^+$ T cells through the expression of ICOS ligand [75].

Human pDCs, in fact, are composed of two subsets, distinguished by the expression of CD2 [76]. Both subsets secrete IFN-α and express the cytotoxic molecules Granzyme B and TRAIL. However, the CD2^{high} pDCs are more potent than the CD2^{low} pDCs to induce allogeneic T-cell proliferation. These different functional properties of CD2^{high} pDCs and CD2^{low} pDCs are associated to distinct transcription profiles, differential secretion of IL12 p40 and with differential expression of co-stimulatory molecule CD80 on activation. The differential expression of lymphoid-related genes (RAG1 and Ig rearrangement products) [77], CD4 [78], and Ly49Q [79] also indicate the existence of murine pDC subsets. Similar to our results in the human, murine pDC subsets differ in their capacity to trigger allogeneic T-cell proliferation and to secrete IL12p70 and IFN-γ [77]. Additional studies will be necessary to understand the biological role of these two pDC subsets.

In humans, pDCs can be mobilized *in vivo* with cytokines such as Flt3 ligand and G-CSF [80–82], while mDCs are mobilized by Flt3L only. The differential mobilization of distinct DC subsets or DC precursors by these cytokines offers a novel strategy to manipulate immune responses in humans [21, 80].

DENDRITIC CELLS AND THE CONTROL OF PERIPHERAL TOLERANCE

The role of DCs in the establishment of peripheral tolerance represents a recent area of research in DC biology [83]. It is generally accepted that immature

DCs are tolerogenic while mature DCs are immunogenic. However, mature DCs also play a role in the induction of tolerance. We will thus review the mechanisms that DCs utilize to implement peripheral tolerance.

The thymus steadily produces thymocytes which express newly assembled TCR, some of which may be reactive with components of self. It is accepted that high-affinity autoreactive thymocytes are promptly eliminated upon encountering MHC-self peptide complexes, a phenomenon called central tolerance. Thymic DCs contribute to the deletion of autoreactive thymocytes, but circulating autoreactive T cells are found in the periphery, associated or not to autoimmune diseases, thus indicating that central tolerance mechanisms are insufficient. The failure to purge newly made T cells from their autoreactive elements may depend on the inability of the thymus to display all self-antigens, or on low-affinity TCRs, which are unable to transduce proapoptotic signals.

The first evidence that DCs induce tolerance was obtained by injecting mice with a rat monoclonal antibody specific for an undefined DC antigen (33D1). The mice T cells were subsequently found to be unresponsive to rat IgG [84]. Furthermore, peripheral tolerance to ovalbumin and influenza hemagglutinin expressed in pancreatic islets was found to depend on bone-marrow-derived APCs [84]. Targeting antigens to DC class-II or class-I MHC compartments with antibodies against DEC 205, a C-type lectin expressed by DCs, leads to a short burst of proliferation and subsequent deletion of antigen-specific CD4+ and CD8+ T cells. The subsequent lack of antigen-specific response of the treated animals indicates establishment of tolerance. Co-administration of DEC 205-targeted antigen with activating signals (agonistic anti-CD40 antibody) leads to sustained immunity [85, 86]. Similar results were obtained by using an inducible antigen expression system controlling antigen presentation in resting versus activated DC *in vivo* to endogenous naïve T cells [87].

Apoptotic cells represent an abundant source of self antigens as they are constantly produced in tissues. Accordingly, DCs loaded with apoptotic material are found in the steady state within lymph nodes draining tissues such as the intestinal epithelium [88], the airway, and the skin [89, 90]. The mechanism responsible for developing tolerance to apoptotic cells is different from that implemented by antigen-presenting cells in the human and mouse thymus and lymphoid organs. *In vitro* studies have indeed shown that immature DCs efficiently phagocytose apoptotic cells and, upon maturation, can induce immunity to apoptotic cell-derived antigens [9, 91, 92]. *In vivo* induction of tolerance to apoptotic cell components includes a burst of T-cell proliferation followed by deletion of these cells. The

capture and processing of cell-associated antigens in the periphery has been documented with the proton pump ATPase of the gastric parietal cells [93]. There, gastric ATPase was found in the vesicular compartments of only a few DCs and only in the draining gastric lymph node [93]. During autoimmune gastritis induced by thymectomy, both the number of antigen-containing DCs and their presenting function increased in the draining node. Selective deletion of DCs was found to abrogate the depletion of antigen-specific T cells by antigen-loaded dying splenocytes. Such tolerance mechanisms might permit prevention of or a reduction in the development of autoimmunity when dying cells are generated and processed at the time of infection [94].

DCs can also induce the expansion of regulatory CD4+ T cells. The mouse spleen and lymph nodes display for example a population of "semi-mature" CD45RB^{high} CD11c^{low} DCs which, after activation with LPS or CpG oligonucleotides, secrete IL-10 but do not upregulate MHC class II or co-stimulatory molecules. These DCs are highly potent at inducing tolerance that is mediated through the differentiation of T$_R$ cells *in vivo* [95, 96]. Murine pulmonary DCs induce the development of Tregs in an ICOS-ICOS-L dependent fashion which leads to the production of IL-10 by DCs [97]. The complexity of the lineage and/or the subpopulations of DCs that may be responsible for tolerance induction is further illustrated by unconventional DCs that can induce protection against virally induced type-1 diabetes and display properties of both natural killer (NK) and dendritic cells (DC) [98].

Several strategies have been designed to enhance the ability of immature DCs to induce tolerance, including their culture with IL-10 and/or TGFβ. Actually, immunomodulators such as vitamin D3, steroids, and rapamycin, classically thought to act on T cells, have been found to affect DC maturation [99, 100]. These findings suggest that these agents may at least partially work through DC-mediated tolerance rather than through direct depletion of effector T cells.

In addition to myeloid DCs, allogeneic CD40L-activated pDCs have been shown to prime naïve CD8+T cells into IL-10-producing T cells that inhibit the proliferation of alloreactive naïve CD8 T cells [101].

DCs IN HUMAN AUTOIMMUNE DISEASES

Peripheral, non-lymphoid tissues have been described to display DC/lymphocyte aggregates in a variety of human autoimmune diseases [102–108]. Mature DCs represent a large proportion of cells within rheumatoid synovial fluid and are found to colonize the perivascular infiltrates of rheumatoid synovium [103].

Immature DCs, however, predominate at the lining layer of the rheumatoid synovium [109, 110]. Mature DCs are also observed together with lymphocyte infiltrates in muscle tissues in autoimmune dermatomyositis and polymyositis [111]."Trapping" of misdirected mature DCs may also explain the inflammation of giant cell arteritis [108]. In psoriasis, IL-23, but not IL-12, is expressed by mature DCs and might contribute to the autoimmune reaction [112], as IL-23 plays a critical role in the development of EAE and collagen-induced arthritis [113, 114]. Furthermore, blocking the IL12/Il23 receptor has proven an effective therapy for patients with psoriasis [115].

mDCs in SLE

SLE might be the autoimmune disease that has received the highest attention with regard to dendritic cells, including both myeloid and plasmacytoid DCs. The blood of SLE patients contains unusual CD14+ "monocytes" with properties of DCs [17]. Accordingly, upon exposure to SLE serum, healthy monocytes aggregate and differentiate into CD14+DCs. The generation of these CD14+DCs depends on the presence of type I IFN in SLE serum [17]. Indeed, IFN was pointed out as a factor in SLE pathogenesis more than 20 years ago, though it was nearly forgotten as no link to SLE pathogenesis was established. The finding that type I IFN activates DCs [116], and the identification of the critical role

of immature DCs in the establishment of peripheral tolerance provided a new twist in the study of SLE [117].

As reviewed above, under steady-state conditions immature mDC capture antigen and migrate — without maturing — to draining lymph nodes. There, they present self-peptide MHC complexes in the absence of co-stimulatory signals to self-reactive T cells, which leads to anergy and deletion [5]. Furthermore, immature DC may contribute to peripheral tolerance through the induction and maintenance of regulatory CD4+ T cells. Thus, autoreactive T cells escaping negative selection in the thymus are kept in check by peripheral immune tolerance. IFN-α contributes to disrupting peripheral immune tolerance by promoting mDC maturation [17, 116]. Thus, upregulation of co-stimulatory molecules such as CD80 and CD86 on mature mDCs contributes to the survival, expansion and differentiation of self-reactive T cells. Autoreactive CD4+ T cells provide help to autoreactive CD8+ T cells. Furthermore IFN-α enhances a cytotoxic program of CD8+ T-cell maturation characterized by an increased expression of perforin and granzymes [118]. These autoreactive CD8+ T cells contribute to tissue damage and to the generation of novel granzyme B-dependent autoantigens [119], further fueling immune-complex-driven production of IFN-α (Figure 6.3).

IFN-α was found to be elevated in lupus sera [120], most notably during disease lares, and anecdotal reports described the induction of a lupus-like syndrome following treatment with IFN-α for melanoma or

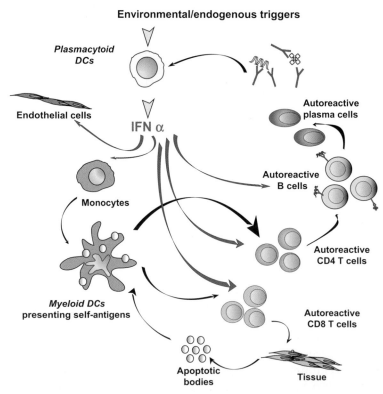

FIGURE 6.3   The central role of type I IFN in SLE. Under the steady state, immature myeloid dendritic cells (DCs) capture apoptotic bodies and present their autoantigens, without costimulatory molecules, to autoreactive T lymphocytes. This results in either their deletion or in the expansion of regulatory T cells. Upon exposure to environmental (i.e. viruses) and/or endogenous (i.e. nucleic acid-containing immune complexes) triggers, pDCs from SLE patients produce IFNα in a sustained fashion. IFN-α activates myeloid DCs, with express co-stimulatory molecules and trigger the expansion and differentiation of autoreactive CD4+ and CD8+ T cells, and possibly mature B cells, into autoreactive effectors. Cytotoxic T cells kill tissue targets thereby generating nucleosomes and granzyme B-dependent autoantigen fragments, which further feed the autoimmune process. B cell tolerance check points are defective in SLE patients, leading to the expansion of anti-nuclear antibody expressing B cells. Type I IFNs, together with other products of activated pDCs such as IL-6, drive these autoreactive B cells to differentiate into plasma cells that secrete autoantibodies. Immune complexes derived from DNA and RNA-containing antibodies can further activate pDCs to release IFN-α, amplifying this pathogenic loop. IFN alpha also directly promotes abnormal vasculogenesis, which might contribute to the development of premature atherosclerosis in SLE.

hepatitis C [121]. This form of therapy may result in induction of anti-nuclear antibodies, though it rarely (0.15—.7%) results in overt autoimmunity such as thyroiditis, diabetes, or SLE (reviewed in [122]). The ability of DNA-containing immune complexes from lupus sera to induce IFN-α production by a novel cell type [123–126] later identified as the plasmacytoid dendritic cell (pDC) [56], was described in the late 1990s. Furthermore, IFN-α-regulated gene transcripts were shown to be significantly upregulated in peripheral blood of pediatric and adult SLE patients upon gene expression profiling [127–129]. Recently, a genetic association with pathways related to type I IFN transcription and/or signaling has been confirmed in SLE (for example IRF5, STAT4, TNFAIP3 or TREX1) [130–142]. Indeed, an excessive production and/or response to type I IFN explains many of the immune alterations observed in the disease [143, 144]. Consequently, IFN-α is a logical therapeutic target [129], a hypothesis that is being tested in clinical trials [145]. Moreover, the IFN-α signature is being assessed as a new biomarker for SLE disease activity [145, 146].

pDCs in SLE

Many cells can produce IFN-α in response to viral infection in small quantities. pDC, previously called "naturally interferon producing cells", are unique in their capacity to produce vast amounts of this family of cytokines [56, 147–149]. A single pDC, for example, can synthesize up to one billion IFN-α molecules within 12 hours (or 3–10 pg per cell), which is 200–1000 times more than the amount produced by any other cell type [56, 150]. This unique ability of pDC can in part be explained by the constitutive expression of TLR7/8, TLR9 and IRF7 [151]. Upon pDC maturation, IFN-α secretion is switched off, the cell acquires a dendritic morphology and assumes "professional" antigen-presenting function [150, 152].

pDC constitute less than 1% of circulating white blood cells [57, 147, 150] and their frequency is decreased up to 100-fold in the peripheral blood of SLE patients [17, 56, 153] due to increased exodus into inflamed tissues like the kidney and skin [154–156]. Interestingly, phototesting can reproduce clinically and histologically photosensitive variants of cutaneous lupus erythematosus, including an influx of pDC following UV irradiation [157].

The capacity of pDCs to produce large amounts of type I IFN in response to pathogens can be traced down to the expression of nucleic acid-sensing receptors such as TLR7–9. To avoid activation and/or cytokine release in response to endogenous nucleic acids, however, most of these receptors are located in endosomal compartments. Additionally, endogenous DNA and RNA sequences are rapidly degraded in the extracellular space by ubiquitous DNAses and RNAses. There is experimental evidence that all of these protective mechanisms can be circumvented in SLE: DNAse1 knock-out mice develop a lupus-like disease [158]; moreover DNAse1 deficiency has been observed in some lupus patients [159]. Molecules like the anti-microbial peptide LL37, produced by neutrophils or damaged keratinocytes [160], or the ubiquitous HMGB1 protect self nucleic acids from extracellular degradation and are also involved in their uptake through lipid-rafts and/or receptors such as RAGE into endosomal TLR7/9 compartments [161]. Moreover, immune complexes can bind to FcgRIIa (CD32) on pDCs and thereby gain access to endosomes by receptor-mediated endocytosis [162, 163]. Hypomethylated DNA rich in linear cytosine-guanosin (CpG) sequences and uracilrich small nuclear RNA (particularly U1 snRNA tightly bound to Sm and other small ribonucloproteins) are infrequent in the human genome; interestingly such unusual sequences are enriched in immune complexes in SLE [164] and are preferentially recognized by endosomal TLRs [67, 165–168].

Nucleic acid ligation of TLR7–TLR9 in the early endosome leads to signaling through a TRAF3–IRAK1–OPN complex, leading to IRF-7 nuclear translocation and induction of type I IFN gene transcription. Linear (monomeric) DNA structures (or their synthetic correlate type B ODN) traffic through early endosomes into more acidic late endolysosomes where TLR7/9 activation recruits a different set of signaling molecules (i.e. NF-κB and IRF5) leading to transcription of inflammatory cytokines such as IL-6 and TNF-α, DC maturation and diminished IFN-α secretion [169]. Synthetic inhibitory compounds that specifically target TLR7–TLR9 to prevent immune-complex-stimulated IFN-α production in pDC promise new therapeutic avenues in the treatment of SLE [170, 171]. In fact, chloroquine, an "old" drug in the treatment of SLE, directly inhibits CpG-driven activation of TLR9, which might explain one of the actions of this antimalarial at the molecular level in autoimmune diseases [172, 173].

Type I IFN at the center of SLE pathogenesis

The multiple effects of type I IFN on cells of the immune system can be explained in part based on the complex transcriptional program induced by this family of cytokines. Thus, upon exposure to type I IFN, hundreds of genes are induced or repressed (reviewed in [174]). These include classical anti-viral/proliferative transcripts (i.e. OAS, MX1, etc.), interferon regulatory factors (IRF5 and IRF7), pro-apoptotic molecules (i.e. FAS and TRAIL), B-cell differentiation factors (i.e. BLyS/BAFF), chemokines and chemokine receptors

(MCP1 and IP10) and, interestingly, even lupus autoantigens such as Ro/SSA [175]. Many IFN-inducible proteins might indeed play a significant role in disease pathogenesis. Fas and TRAIL can for example contribute to nucleosome overload and therefore to increased nuclear antigen presentation. BLyS/BAFF induces the proliferation and differentiation of mature B cells into antibody-secreting cells. Indeed, a neutralizing anti-BAFF/BLyS antibody being tested in clinical trials appears to reduce SLE disease activity and prevent flares [176]. Treating SLE patients with high-dose IV steroids, which are used to control disease flares, results in the silencing of the IFN signature [175]. This is associated with pDC depletion [177], although both the blood pDCs and the IFN signature return in less than 1 week after IV steroid administration (our own unpublished data). Altogether, these observations support the central role of pDCs as producers of type I IFN in lupus pathogenesis.

IFN-α has myelosuppressive effects explaining in part the B-cell and T-cell lymphopenia observed in the peripheral blood of lupus patients. In addition to upregulating BLyS/BAFF, which contributes to the survival of mature, peripheral B cells, IFN-α promotes the differentiation of activated B cells into plasmablasts and, together with IL-6, permits plasmablasts to develop into antibody-secreting plasma cells [71]. Autoreactive B cells generate autoantibodies and thereby provide — through the formation of nucleic-acid-containing immune complexes — a positive feedback loop for enhanced TLR7/9 activation and even more increased IFN-α production. Since B cells express TLR7 and TLR9 similarly to pDC, they too can be targeted by nucleic-acid-containing immune complexes [71, 143, 178] further closing this pathogenic loop (Figure 6.3).

CONCLUSION

This review has highlighted recent developments in the field of DC biology and in the induction and maintenance of peripheral tolerance. Some of these studies are the basis for the development of drugs targeting cytokines such as IFN-α, which alters the balance between immature and mature DCs. Indeed, monoclonal anti-IFN-α antibodies are being tested in clinical trials in SLE patients [179]. Targeting IFN-inducible proteins such as BAFF/BlyS has already shown beneficial effects in these patients [180]. A number of new compounds that inhibit nucleic acid recognizing TLR7 and TLR9 show promising results in animal studies [181] and will be soon tested in humans. Understanding human DCs and their complex interplays with other immune effectors will continue to provide insight into the pathogenesis of autoimmune diseases and help us discover new, better targets for therapeutic intervention.

Acknowledgments

This chapter is dedicated to all the patients and volunteers who participated in our studies and clinical trials. We thank former and current members of the Institute for their contributions to our progresses. The authors are partially supported by U19-A1057234, RO1-CA078846, AR054083-01 (J.B.) and Baylor Health Care foundation (J.B). J.B. holds the W.W. Caruth, Jr. Chair for Transplantation Immunology Research. R01 AR050770-01, ARO54083-01CORT, NIH ARO55503-01CORT, U19- AI082715-01, Alliance for Lupus Research and the Mary Kirkland Foundation to VP.

References

[1] J. Bancherau, R.M. Steinman, Dendritic cells and the control of immunity, Nature 392 (1998) 245–252.

[2] J. Bancherau, F. Briere, C. Caux, J. Davoust, S. Lebecque, Y. Liu, et al., Immunobiology of dendritic cells, Ann Rev Immunol 18 (2000) 767–812.

[3] R.M. Steinman, The dendritic cell system and its role in immunogenicity, Annu Rev Immunol 9 (1991) 271–296.

[4] R. Medzhitov, C.A. Janeway Jr., Innate immunity: the virtues of a nonclonal system of recognition, Cell. 91 (1997) 295–298.

[5] J. Bancherau, F. Briere, C. Caux, J. Davoust, S. Lebecque, Y.J. Liu, et al., Immunobiology of dendritic cells, Annu Rev Immunol 18 (2000) 767–811.

[6] K. Shortman, Y.-J. Liu, Mouse and human dendritic cell subtypes, Nat Rev Immunol 2 (2002) 151–161.

[7] A.A. Itano, M.K. Jenkins, Antigen presentation to naive CD4 T cells in the lymph node, Nat Immunol 4 (2003) 733–739.

[8] M.L. Albert, N. Bhardwaj, Resurrecting the dead: DCs cross-present antigen derived from apoptotic cells on MHC I, The Immunologist 6 (1998) 194–198.

[9] M.L. Albert, S.F. Pearce, L.M. Francisco, B. Sauter, P. Roy, R.L. Silverstein, et al., Immature dendritic cells phagocytose apoptotic cells via alphavbeta5 and CD36, and cross-present antigens to cytotoxic T lymphocytes, J Exp Med 188 (1998) 1359–1368.

[10] W.R. Heath, F.R. Carbone, Cross-presentation, dendritic cells, tolerance and immunity, Annu Rev Immunol 19 (2001) 47–64.

[11] F.D. Finkelman, A. Lees, R. Birnbaum, W.C. Gause, S.C. Morris, Dendritic cells can present antigen in vivo in a tolerogenic or immunogenic fashion, J Immunol 157 (1996) 1406–1414.

[12] G. Jego, V. Pascual, A.K. Palucka, J. Bancherau, Dendritic cells control B cell growth and differentiation, Curr Dir Autoimmun 8 (2005) 124–139.

[13] H. Qi, J.G. Egen, A.Y. Huang, R.N. Germain, Extrafollicular activation of lymph node B cells by antigen-bearing dendritic cells, Science 312 (2006) 1672–1676.

[14] F.D. Batista, N.E. Harwood, The who, how and where of antigen presentation to B cells, Nat Rev Immunol 9 (2009) 15–27.

[15] A. Bergtold, D.D. Desai, A. Gavhane, R. Clynes, Cell surface recycling of internalized antigen permits dendritic cell priming of B cells, Immunity 23 (2005) 503–514.

[16] Y.J. Liu, V. Soumelis, N. Watanabe, T. Ito, Y.H. Wang, W. Malefyt Rde, et al., TSLP: an epithelial cell cytokine that regulates T cell differentiation by conditioning dendritic cell maturation, Annu Rev Immunol 25 (2007) 193–219.

[17] P. Blanco, A.K. Palucka, M. Gill, V. Pascual, J. Bancherau, Induction of dendritic cell differentiation by IFN-alpha in systemic lupus erythematosus, Science 294 (2001) 1540–1543.

[18] C. Aspord, A. Pedroza-Gonzalez, M. Gallegos, S. Tindle, E.C. Burton, D. Su, et al., Breast cancer instructs dendritic cells

to prime interleukin 13-secreting CD4+ T cells that facilitate tumor development, J Exp Med 204 (2007) 1037–1047.

[19] W. Cao, L. Bover, M. Cho, X. Wen, S. Hanabuchi, M. Bao, et al., Regulation of TLR7/9 responses in plasmacytoid dendritic cells by BST2 and ILT7 receptor interaction, J Exp Med 206 (2009) 1603–1614.

[20] M. Gobert, I. Treilleux, N. Bendriss-Vermare, T. Bachelot, S. Goddard-Leon, V. Arfi, et al., Regulatory T cells recruited through CCL22/CCR4 are selectively activated in lymphoid infiltrates surrounding primary breast tumors and lead to an adverse clinical outcome, Cancer Res. 69 (2009) 2000–2009.

[21] B. Pulendran, K. Palucka, J. Banchereau, Sensing pathogens and tuning immune responses, Science 293 (2001) 253–256.

[22] J. Valladeau, S. Saeland, Cutaneous dendritic cells, Semin Immunol 17 (2005) 273–283.

[23] A.M. van der Aar, R.M. Sylva-Steenland, J.D. Bos, M.L. Kapsenberg, E.C. de Jong, M.B. Teunissen, Loss of TLR2, TLR4, and TLR5 on Langerhans cells abolishes bacterial recognition, J Immunol 178 (2007) 1986–1990.

[24] E. Klechevsky, M. Liu, R. Morita, R. Banchereau, L. Thompson-Snipes, A.K. Palucka, et al., Understanding human myeloid dendritic cell subsets for the rational design of novel vaccines, Hum Immunol 70 (2009) 281–288.

[25] V. Flacher, M. Bouschbacher, E. Verronese, C. Massacrier, V. Sisirak, O. Berthier-Vergnes, et al., Human Langerhans cells express a specific TLR profile and differentially respond to viruses and Gram-positive bacteria, J Immunol 177 (2006) 7959–7967.

[26] E. Klechevsky, R. Morita, M. Liu, Y. Cao, S. Coquery, L. Thompson-Snipes, et al., Functional specializations of human epidermal Langerhans cells and CD14+ dermal dendritic cells, Immunity 29 (2008) 497–510.

[27] C. Caux, C. Massacrier, B. Vanbervliet, B. Dubois, I. Durand, M. Cella, et al., CD34+ hematopoietic progenitors from human cord blood differentiate along two independent dendritic cell pathways in response to granulocyte-macrophage colony-stimulating factor plus tumor necrosis factor alpha: II. Functional analysis, Blood 90 (1997) 1458–1470.

[28] N. Fazilleau, L. Mark, L.J. McHeyzer-Williams, M.G. McHeyzer-Williams, Follicular helper T cells: lineage and location, Immunity 30 (2009) 324–335.

[29] C. King, S.G. Tangye, C.R. Mackay, T Follicular Helper (TFH) Cells in Normal and Dysregulated Immune Responses, Annu Rev Immunol 26 (2008) 741–766.

[30] N. Schmitt, R. Morita, L. Bourdery, S.E. Bentebibel, S.M. Zurawski, J. Banchereau, et al., Human dendritic cells induce the differentiation of interleukin-21-producing T follicular helper-like cells through interleukin-12, Immunity 31 (2009) 158–169.

[31] R. Spolski, W.J. Leonard, Interleukin-21: basic biology and implications for cancer and autoimmunity, Annu Rev Immunol 26 (2008) 57–79.

[32] R.N. Germain, M. Bajenoff, F. Castellino, M. Chieppa, J.G. Egen, A.Y. Huang, et al., Making friends in out-of-the-way places: how cells of the immune system get together and how they conduct their business as revealed by intravital imaging, Immunol Rev. 221 (2008) 163–181.

[33] Y. Noguchi, E.C. Richards, Y.T. Chen, L.J. Old, Influence of interleukin 12 on p53 peptide vaccination against established Meth A sarcoma, Proc Natl Acad Sci USA 92 (1995) 2219–2223.

[34] H. Tahara, L. Zitvogel, W.J. Storkus, H.J. Zeh 3rd, T.G. McKinney, R.D. Schreiber, et al., Effective eradication of established murine tumors with IL-12 gene therapy using a polycistronic retroviral vector, J Immunol 154 (1995) 6466–6474.

[35] R.J. Motzer, A. Rakhit, J.A. Thompson, J. Nemunaitis, B.A. Murphy, J. Ellerhorst, et al., Randomized multicenter phase II trial of subcutaneous recombinant human interleukin-12 versus interferon-alpha 2a for patients with advanced renal cell carcinoma, J Interferon Cytokine Res. 21 (2001) 257–263.

[36] M.A. Cheever, Twelve immunotherapy drugs that could cure cancers, Immunol Rev. 222 (2008) 357–368.

[37] C.M. van Herpen, R. van der Voort, J.A. van der Laak, I.S. Klasen, A.O. de Graaf, L.C. van Kempen, et al., Intra-tumoral rhIL-12 administration in head and neck squamous cell carcinoma patients induces B cell activation, Int J Cancer 123 (2008) 2354–2361.

[38] D. Alvarez, G. Harder, R. Fattouh, J. Sun, S. Goncharova, M.R. Stampfli, et al., Cutaneous antigen priming via gene gun leads to skin-selective Th2 immune-inflammatory responses, J Immunol 174 (2005) 1664–1674.

[39] L.S. Bursch, L. Wang, B. Igyarto, A. Kissenpfennig, B. Malissen, D.H. Kaplan, et al., Identification of a novel population of Langerin+ dendritic cells, J Exp Med 204 (2007) 3147–3156.

[40] F. Ginhoux, M.P. Collin, M. Bogunovic, M. Abel, M. Leboeuf, J. Helft, et al., Blood-derived dermal langerin+ dendritic cells survey the skin in the steady state, J Exp Med 204 (2007) 3133–3146.

[41] L.F. Poulin, S. Henri, B. de Bovis, E. Devilard, A. Kissenpfennig, B. Malissen, The dermis contains langerin+ dendritic cells that develop and function independently of epidermal Langerhans cells, J Exp Med 204 (2007) 3119–3131.

[42] S. Bedoui, P.G. Whitney, J. Waithman, L. Eidsmo, L. Wakim, I. Caminschi, et al., Cross-presentation of viral and self antigens by skin-derived CD103+ dendritic cells, Nat Immunol 10 (2009) 488–495.

[43] A. Dzionek, A. Fuchs, P. Schmidt, S. Cremer, M. Zysk, S. Miltenyi, et al., BDCA-2, BDCA-3, and BDCA-4: three markers for distinct subsets of dendritic cells in human peripheral blood, J Immunol 165 (2000) 6037–6046.

[44] C. Huysamen, J.A. Willment, K.M. Dennehy, G.D. Brown, CLEC9A is a novel activation C-type lectin-like receptor expressed on BDCA3+ dendritic cells and a subset of monocytes, J Biol Chem. 283 (2008) 16693–16701.

[45] K. Schakel, R. Kannagi, B. Kniep, Y. Goto, C.J.Z. Mitsuoka, A. Soruri, et al., 6-Sulfo LacNAc, a novel carbohydrate modification of PSGL-1, defines an inflammatory type of human dendritic cells, Immunity 17 (2002) 289–301.

[46] A. Kissenpfennig, S. Henri, B. Dubois, C. Laplace-Builhe, P. Perrin, N. Romani, et al., Dynamics and Function of Langerhans Cells In Vivo: Dermal Dendritic Cells Colonize Lymph Node AreasDistinct from Slower Migrating Langerhans Cells, Immunity 22 (2005) 643.

[47] P. Stoitzner, C.H. Tripp, A. Eberhart, K.M. Price, J.Y. Jung, L. Bursch, et al., Langerhans cells cross-present antigen derived from skin, Proc Natl Acad Sci U S A 103 (2006) 7783–7788.

[48] D. Dudziak, A.O. Kamphorst, G.F. Heidkamp, V. Buchholz, C. Trumpfheller, S. Yamazaki, et al., Differential antigen processing by dendritic cell subsets in vivo, Science 315 (2007) 107–111.

[49] G.J. Randolph, C. Jakubzick, C. Qu, Antigen presentation by monocytes and monocyte-derived cells, Curr Opin Immunol 20 (2008) 52–60.

[50] N. Romani, S. Gruner, D. Brang, E. Kampgen, A. Lenz, B. Trockenbacher, et al., Proliferating dendritic cell progenitors in human blood, J Exp Med 180 (1994) 83–93.

I. BASIS OF DISEASE PATHOGENESIS

[51] J. Banchereau, A.K. Palucka, Dendritic cells as therapeutic vaccines against cancer, Nat Rev Immunol 5 (2005) 296—306.

[52] P. Dubsky, H. Saito, M. Leogier, C. Dantin, J.E. Connolly, J. Banchereau, et al., IL-15-induced human DC efficiently prime melanoma-specific naive CD8(+) T cells to differentiate into CTL, Eur J Immunol 37 (2007) 1678—1690.

[53] M. Mohamadzadeh, F. Berard, G. Essert, C. Chalouni, B. Pulendran, J. Davoust, et al., Interleukin 15 skews monocyte differentiation into dendritic cells with features of Langerhans cells, J Exp Med 194 (2001) 1013—1020.

[54] K. Steinbrink, M. Wolfl, H. Jonuleit, J. Knop, A.H. Enk, Induction of tolerance by IL-10-treated dendritic cells, J Immunol 159 (1997) 4772—4780.

[55] K. Sato, N. Yamashita, M. Baba, T. Matsuyama, Modified myeloid dendritic cells act as regulatory dendritic cells to induce anergic and regulatory T cells, Blood 101 (2003) 3581—3589.

[56] F.P. Siegal, N. Kadowaki, M. Shodell, P.A. Fitzgerald-Bocarsly, K. Shah, S. Ho, et al., The nature of the principal type 1 interferon-producing cells in human blood, Science 284 (1999) 1835—1837.

[57] A. Dzionek, Y. Sohma, J. Nagafune, M. Cella, M. Colonna, F. Facchetti, et al., BDCA-2, a novel plasmacytoid dendritic cell-specific type II C-type lectin, mediates antigen capture and is a potent inhibitor of interferon alpha/beta induction, J Exp Med 194 (2001) 1823—1834.

[58] W. Cao, D.B. Rosen, T. Ito, L. Bover, M. Bao, G. Watanabe, et al., Plasmacytoid dendritic cell-specific receptor ILT7-FcεRIγ inhibits Toll-like receptor-induced interferon production, J Exp Med 203 (2006) 1399—1405.

[59] N. Kadowaki, S. Ho, S. Antonenko, R. de Waal Malefyt, R.A. Kastelein, F. Bazan, et al., Subsets of human dendritic cell precursors express different toll-like receptors and respond to different microbial antigens, J Exp Med 194 (2001) 863—870.

[60] M. Cella, D. Jarrossay, F. Facchetti, O. Alebardi, H. Nakajima, A. Lanzavecchia, et al., Plasmacytoid monocytes migrate to inflamed lymph nodes and produce high levels of type I IFN, Nature Medicine (1999) 919—923.

[61] G. Grouard, M.C. Rissoan, L. Filgueira, I. Durand, J. Banchereau, Y.J. Liu, The enigmatic plasmacytoid T cells develop into dendritic cells with interleukin (IL)-3 and CD40-ligand, J Exp Med 185 (1997) 1101—1111.

[62] M.C. Rissoan, V. Soumelis, N. Kadowaki, G. Grouard, F. Briere, R. de Waal Malefyt, et al., Reciprocal control of T helper cell and dendritic cell differentiation, Science 283 (1999) 1183—1186.

[63] M. Bauer, V. Redecke, J.W. Ellwart, B. Scherer, J.P. Kremer, H. Wagner, et al., Bacterial CpG-DNA triggers activation and maturation of human CD11c-, CD123+ dendritic cells, J Immunol 166 (2001) 5000—5007.

[64] A. Krug, S. Rothenfusser, V. Hornung, B. Jahrsdorfer, S. Blackwell, Z.K. Ballas, et al., Identification of CpG oligonucleotide sequences with high induction of IFN-alpha/beta in plasmacytoid dendritic cells, Eur J Immunol 31 (2001) 2154—2163.

[65] U. Bave, G.V. Alm, L. Ronnblom, The combination of apoptotic U937 cells and lupus IgG is a potent IFN-alpha inducer, J Immunol 165 (2000) 3519—3526.

[66] D. Ganguly, G. Chamilos, R. Lande, J. Gregorio, S. Meller, V. Facchinetti, et al., Self-RNA-antimicrobial peptide complexes activate human dendritic cells through TLR7 and TLR8, J Exp Med 206 (2009) 1983—1994.

[67] F.J. Barrat, T. Meeker, J. Gregorio, J.H. Chan, S. Uematsu, S. Akira, et al., Nucleic acids of mammalian origin can act as endogenous ligands for Toll-like receptors and may promote systemic lupus erythematosus, J Exp Med 202 (2005) 1131—1139.

[68] A. Goubier, B. Dubois, H. Gheit, G. Joubert, F. Villard-Truc, C. Asselin-Paturel, et al., Plasmacytoid dendritic cells mediate oral tolerance, Immunity 29 (2008) 464—475.

[69] H.J. de Heer, H. Hammad, T. Soullie, D. Hijdra, N. Vos, M.A. Willart, et al., Essential role of lung plasmacytoid dendritic cells in preventing asthmatic reactions to harmless inhaled antigen, J Exp Med 200 (2004) 89—98.

[70] J.C. Ochando, C. Homma, Y. Yang, A. Hidalgo, A. Garin, F. Tacke, et al., Alloantigen-presenting plasmacytoid dendritic cells mediate tolerance to vascularized grafts, Nat. Immunol 7 (2006) 652—662.

[71] G. Jego, A.K. Palucka, J.P. Blanck, C. Chalouni, V. Pascual, J. Banchereau, Plasmacytoid dendritic cells induce plasma cell differentiation through type I interferon and interleukin 6, Immunity 19 (2003) 225—234.

[72] J.F. Fonteneau, M. Gilliet, M. Larsson, I. Dasilva, C. Munz, Y.J. Liu, et al., Activation of influenza virus-specific CD4+ and CD8+ T cells: a new role for plasmacytoid dendritic cells in adaptive immunity, Blood 101 (2003) 3520—3526.

[73] T. Di Pucchio, B. Chatterjee, A. Smed-Sorensen, S. Clayton, A. Palazzo, M. Montes, et al., Direct proteasome-independent cross-presentation of viral antigen by plasmacytoid dendritic cells on major histocompatibility complex class I, Nat Immunol 9 (2008) 551—557.

[74] T. Ito, M. Yang, Y.H. Wang, R. Lande, J. Gregorio, O.A. Perng, et al., Plasmacytoid dendritic cells prime IL-10-producing T regulatory cells by inducible costimulator ligand, J Exp Med 204 (2007) 105—115.

[75] M. Gilliet, Y.-J. Liu, Generation of human CD8 T regulatory cells by CD40 ligand-activated plasmacytoid dendritic cells, J Exp Med 195 (2002) 695—704.

[76] T. Matsui, J.E. Connolly, M. Michnevitz, D. Chaussabel, C.I. Yu, C. Glaser, et al., CD2 distinguishes two subsets of human plasmacytoid dendritic cells with distinct phenotype and functions, J Immunol 182 (2009) 6815—6823.

[77] R. Pelayo, J. Hirose, J. Huang, K.P. Garrett, A. Delogu, M. Busslinger, et al., Derivation of 2 categories of plasmacytoid dendritic cells in murine bone marrow, Blood 105 (2005) 4407—4415.

[78] G.X. Yang, Z.X. Lian, K. Kikuchi, Y.J. Liu, A.A. Ansari, S. Ikehara, et al., CD4- plasmacytoid dendritic cells (pDCs) migrate in lymph nodes by CpG inoculation and represent a potent functional subset of pDCs, J Immunol 174 (2005) 3197—3203.

[79] Y. Kamogawa-Schifter, J. Ohkawa, S. Namiki, N. Arai, K. Arai, Y. Liu, Ly49Q defines 2 pDC subsets in mice, Blood 105 (2005) 2787—2792.

[80] B. Pulendran, J. Banchereau, S. Burkeholder, E. Kraus, E. Guinet, C. Chalouni, et al., Flt3-ligand and granulocyte colony-stimulating factor mobilize distinct human dendritic cell subsets in vivo, J Immunol 165 (2000) 566—572.

[81] E. Maraskovsky, K. Brasel, M. Teepe, E.R. Roux, S.D. Lyman, K. Shortman, et al., Dramatic increase in the numbers of functionally mature dendritic cells in Flt3 ligand-treated mice: multiple dendritic cell subpopulations identified, J Exp Med 184 (1996) 1953—1962.

[82] M. Arpinati, C.L. Green, S. Heimfeld, J.E. Heuser, C. Anasetti, Granulocyte-colony stimulating factor mobilizes T helper 2-inducing dendritic cells, Blood 95 (2000) 2484—2490.

[83] R.M. Steinman, D. Hawiger, M.C. Nussenzweig, Tolerogenic dendritic cells, Annu Rev Immunol 21 (2003) 685—711.

[84] C. Kurts, H. Kosaka, F.R. Carbone, J.F. Miller, W.R. Heath, Class I-restricted cross-presentation of exogenous self-antigens leads to deletion of autoreactive CD8(+) T cells, J Exp Med 186 (1997) 239—245.

[85] D. Hawiger, K. Inaba, Y. Dorsett, M. Guo, K. Mahnke, M. Rivera, et al., Dendritic cells induce peripheral T cell unresponsiveness under steady state conditions in vivo, J Exp Med 194 (2001) 769–779.

[86] L. Bonifaz, D. Bonnyay, K. Mahnke, M. Rivera, M.C. Nussenzweig, R.M. Steinman, Efficient targeting of protein antigen to the dendritic cell receptor DEC-205 in the steady state leads to antigen presentation on major histocompatibility complex class I products and peripheral CD8+ T cell tolerance, J Exp Med 196 (2002) 1627–1638.

[87] H.C. Probst, J. Lagnel, G. Kollias, M. van den Broek, Inducible transgenic mice reveal resting dendritic cells as potent inducers of CD8+ T cell tolerance, Immunity 18 (2003) 713–720.

[88] F.P. Huang, N. Platt, M. Wykes, J.R. Major, T.J. Powell, C.D. Jenkins, et al., A discrete subpopulation of dendritic cells transports apoptotic intestinal epithelial cells to T cell areas of mesenteric lymph nodes, J Exp Med 191 (2000) 435–444.

[89] R.M. Steinman, S. Turley, I. Mellman, K. Inaba, The induction of tolerance by dendritic cells that have captured apoptotic cells, J Exp Med 191 (2000) 411–416.

[90] K.Y. Vermaelen, I. Carro-Muino, B.N. Lambrecht, R.A. Pauwels, Specific migratory dendritic cells rapidly transport antigen from the airways to the thoracic lymph nodes, J Exp Med 193 (2001) 51–60.

[91] B. Sauter, M.L. Albert, L. Francisco, M. Larsson, S. Somersan, N. Bhardwaj, Consequences of cell death: exposure to necrotic tumor cells, but not primary tissue cells or apoptotic cells, induces the maturation of immunostimulatory dendritic cells, J Exp Med 191 (2000) 423–434.

[92] F. Berard, P. Blanco, J. Davoust, E.M. Neidhart-Berard, M. Nouri-Shirazi, N. Taquet, et al., Cross-Priming of Naive CD8 T Cells against Melanoma Antigens Using Dendritic Cells Loaded with Killed Allogeneic Melanoma Cells, J Exp Med 192 (2000) 1535–1544.

[93] C. Scheinecker, R. McHugh, E.M. Shevach, R.N. Germain, Constitutive presentation of a natural tissue autoantigen exclusively by dendritic cells in the draining lymph node, J Exp Med 196 (2002) 1079–1090.

[94] K. Liu, T. Iyoda, M. Saternus, Y. Kimura, K. Inaba, R.M. Steinman, Immune tolerance after delivery of dying cells to dendritic cells in situ, J Exp Med 196 (2002) 1091–1097.

[95] A. Wakkach, N. Fournier, V. Brun, J.P. Breittmayer, F. Cottrez, H. Groux, Characterization of dendritic cells that induce tolerance and T regulatory 1 cell differentiation in vivo, Immunity 18 (2003) 605–617.

[96] O. Akbari, R.H. DeKruyff, D.T. Umetsu, Pulmonary dendritic cells producing IL-10 mediate tolerance induced by respiratory exposure to antigen, Nat Immunol 2 (2001) 725–731.

[97] O. Akbari, G.J. Freeman, E.H. Meyer, E.A. Greenfield, T.T. Chang, A.H. Sharpe, et al., Antigen-specific regulatory T cells develop via the ICOS-ICOS-ligand pathway and inhibit allergen-induced airway hyperreactivity, Nat Med 8 (2002) 1024–1032.

[98] D. Homann, A. Jahreis, T. Wolfe, A. Hughes, B. Coon, M.J. van Stipdonk, et al., CD40L blockade prevents autoimmune diabetes by induction of bitypic NK/DC regulatory cells, Immunity 16 (2002) 403–415.

[99] G. Penna, L. Adorini, 1 Alpha,25-dihydroxyvitamin D3 inhibits differentiation, maturation, activation, and survival of dendritic cells leading to impaired alloreactive T cell activation, J Immunol 164 (2000) 2405–2411.

[100] H. Matsue, C. Yang, K. Matsue, D. Edelbaum, M. Mummert, A. Takashima, Contrasting impacts of immunosuppressive agents (rapamycin, FK506, cyclosporin A, and dexamethasone) on bidirectional dendritic cell-T cell interaction during antigen presentation, J Immunol 169 (2002) 3555–3564.

[101] M. Gilliet, Y.J. Liu, Generation of human CD8 T regulatory cells by CD40 ligand-activated plasmacytoid dendritic cells, J Exp Med 195 (2002) 695–704.

[102] A. Jansen, F. Homo-Delarche, H. Hooijkaas, P.J. Leenen, M. Dardenne, H.A. Drexhage, Immunohistochemical characterization of monocytes-macrophages and dendritic cells involved in the initiation of the insulitis and beta-cell destruction in NOD mice, Diabetes 43 (1994) 667–675.

[103] R. Thomas, L.S. Davis, P.E. Lipsky, Rheumatoid synovium is enriched in mature antigen-presenting dendritic cells, J Immunol 152 (1994) 2613–2623.

[104] A. Demidem, J.R. Taylor, S.F. Grammer, J.W. Streilein, T-lymphocyte-activating properties of epidermal antigen-presenting cells from normal and psoriatic skin: evidence that psoriatic epidermal antigen-presenting cells resemble cultured normal Langerhans cells, J Invest Dermatol 97 (1991) 454–460.

[105] B. Serafini, S. Columba-Cabezas, F. Di Rosa, F. Aloisi, Intracerebral recruitment and maturation of dendritic cells in the onset and progression of experimental autoimmune encephalomyelitis, Am J Pathol 157 (2000) 1991–2002.

[106] H.A. Voorby, P.J. Kabel, M. de Haan, P.H. Jeucken, R.D. van der Gaag, M.H. de Baets, et al., Dendritic cells and class II MHC expression on thyrocytes during the autoimmune thyroid disease of the BB rat, Clin Immunol Immunopathol 55 (1990) 9–22.

[107] L. Farkas, K. Beiske, F. Lund-Johansen, P. Brandtzaeg, F.L. Jahnsen, Plasmacytoid Dendritic Cells (Natural Interferon-alpha/beta-Producing Cells) Accumulate in Cutaneous Lupus Erythematosus Lesions, Am J Pathol 159 (2001) 237–243.

[108] W.M. Krupa, M. Dewan, M.S. Jeon, P.J. Kurtin, B.R. Younge, J.J. Goronzy, et al., Trapping of misdirected dendritic cells in the granulomatous lesions of giant cell arteritis, Am J Pathol 161 (2002) 1815–1823.

[109] G. Page, S. Lebecque, P. Miossec, Anatomic localization of immature and mature dendritic cells in an ectopic lymphoid organ: correlation with selective chemokine expression in rheumatoid synovium, J Immunol 168 (2002) 5333–5341.

[110] F. Santiago-Schwarz, P. Anand, S. Liu, S.E. Carsons, Dendritic cells (DCs) in rheumatoid arthritis (RA): progenitor cells and soluble factors contained in RA synovial fluid yield a subset of myeloid DCs that preferentially activate Th1 inflammatory-type responses, J Immunol 167 (2001) 1758–1768.

[111] G. Page, G. Chevrel, P. Miossec, Anatomic localization of immature and mature dendritic cell subsets in dermatomyositis and polymyositis: Interaction with chemokines and Th1 cytokine-producing cells, Arthritis Rheum 50 (2004) 199–208.

[112] E. Lee, W.L. Trepicchio, J.L. Oestreicher, D. Pittman, F. Wang, F. Chamian, et al., Increased Expression of Interleukin 23 p19 and p40 in Lesional Skin of Patients with Psoriasis Vulgaris, J Exp Med 199 (2004) 125–130.

[113] D.J. Cua, J. Sherlock, Y. Chen, C.A. Murphy, B. Joyce, B. Seymour, et al., Interleukin-23 rather than interleukin-12 is the critical cytokine for autoimmune inflammation of the brain, Nature 421 (2003) 744–748.

[114] C.A. Murphy, C.L. Langrish, Y. Chen, W. Blumenschein, T. McClanahan, R.A. Kastelein, et al., Divergent pro- and antiinflammatory roles for IL-23 and IL-12 in joint autoimmune inflammation, J Exp Med 198 (2003) 1951–1957.

[115] K.A. Papp, R.G. Langley, M. Lebwohl, G.G. Krueger, P. Szapary, N. Yeilding, et al., Efficacy and safety of ustekinumab, a human interleukin-12/23 monoclonal antibody, in patients with psoriasis: 52-week results from a randomised, double-blind,

placebo-controlled trial (PHOENIX 2), Lancet 371 (2008) 1675–1684.

[116] S.M. Santini, C. Lapenta, M. Logozzi, S. Parlato, M. Spada, T. Di Pucchio, et al., Type I interferon as a powerful adjuvant for monocyte-derived dendritic cell development and activity in vitro and in Hu-PBL-SCID mice, J Exp Med 191 (2000) 1777–1788.

[117] J. Banchereau, V. Pascual, A.K. Palucka, Autoimmunity through Cytokine-Induced Dendritic Cell Activation, Immunity 20 (2004) 539–550.

[118] P. Blanco, V. Pitard, J.F. Viallard, J.L. Taupin, J.L. Pellegrin, J.F. Moreau, Increase in activated CD8+ T lymphocytes expressing perforin and granzyme B correlates with disease activity in patients with systemic lupus erythematosus, Arthritis Rheum 52 (2005) 201–211.

[119] L. Casciola-Rosen, Cleavage by Granzyme B Is Strongly Predictive of Autoantigen Status: Implications for Initiation of Autoimmunity, J Exp Med 190 (1999) 815–826.

[120] O.T. Preble, R.J. Black, R.M. Friedman, J.H. Klippel, J. Vilcek, Systemic lupus erythematosus: presence in human serum of an unusual acid-labile leukocyte interferon, Science 216 (1982) 429–431.

[121] L.E. Ronnblom, G.V. Alm, K.E. Oberg, Possible induction of systemic lupus erythematosus by interferon-alpha treatment in a patient with a malignant carcinoid tumour, J Intern Med 227 (1990) 207–210.

[122] T.A. Stewart, Neutralizing interferon alpha as a therapeutic approach to autoimmune diseases, Cytokine Growth Factor Rev. 14 (2003) 139–154.

[123] H. Vallin, S. Blomberg, G.V. Alm, B. Cederblad, L. Ronnblom, Patients with systemic lupus erythematosus (SLE) have a circulating inducer of interferon-alpha (IFN-alpha) production acting on leucocytes resembling immature dendritic cells, Clin Exp Immunol 115 (1999) 196–202.

[124] H. Vallin, A. Perers, G.V. Alm, L. Ronnblom, Anti-double-stranded DNA antibodies and immunostimulatory plasmid DNA in combination mimic the endogenous IFN-alpha inducer in systemic lupus erythematosus, J Immunol 163 (1999) 6306–6313.

[125] L. Ronnblom, G.V. Alm, A pivotal role for the natural interferon alpha-producing cells (plasmacytoid dendritic cells) in the pathogenesis of lupus, J Exp Med 194 (2001) F59–63.

[126] T. Lovgren, M.L. Eloranta, U. Bave, G.V. Alm, L. Ronnblom, Induction of interferon-alpha production in plasmacytoid dendritic cells by immune complexes containing nucleic acid released by necrotic or late apoptotic cells and lupus IgG, Arthritis Rheum 50 (2004) 1861–1872.

[127] L. Bennett, K. Palucka, a., E. Arce, V. Cantrell, J. Borvak, J. Banchereau, et al., Interferon and granulopoiesis signatures in systemic lupus erythematosus blood, J Exp Med 197 (2003) 711–723.

[128] E.C. Baechler, F.M. Batliwalla, G. Karypis, P.M. Gaffney, W.A. Ortmann, K.J. Espe, et al., Interferon-inducible gene expression signature in peripheral blood cells of patients with severe lupus, Proc Natl Acad Sci USA 100 (2003) 2610–2615.

[129] M.K. Crow, J. Wohlgemuth, Microarray analysis of gene expression in lupus, Arthritis Res Ther 5 (2003) 279–287.

[130] I.T.W. Harley, K.M. Kaufman, C.D. Langefeld, J.B. Harley, J.a. Kelly, Genetic susceptibility to SLE: new insights from fine mapping and genome-wide association studies. Nature reviews, Genetics 10 (2009) 285–290.

[131] J.B. Harley, M.E. Alarcön-Riquelme, L.a Criswell, C.O. Jacob, R.P. Kimberly, K.L. Moser, et al., Genome-wide association scan in women with systemic lupus erythematosus identifies susceptibility variants in ITGAM, PXK, KIAA1542 and other loci, Nature genetics 40 (2008) 204–210.

[132] G. Hom, R.R. Graham, B. Modrek, K.E. Taylor, W. Ortmann, S. Garnier, et al., Association of systemic lupus erythematosus with C8orf13-BLK and ITGAM-ITGAX, The New England journal of medicine 358 (2008) 900–909.

[133] M.A. Lee-Kirsch, M. Gong, D. Chowdhury, L. Senenko, K. Engel, Y.A. Lee, et al., Mutations in the gene encoding the 3′-5′ DNA exonuclease TREX1 are associated with systemic lupus erythematosus, Nat Genet. 39 (2007) 1065–1067.

[134] D.B. Stetson, J.S. Ko, T. Heidmann, R. Medzhitov, Trex1 prevents cell-intrinsic initiation of autoimmunity, Cell. 134 (2008) 587–598.

[135] D. Feng, R.C. Stone, M.-L. Eloranta, N. Sangster-Guity, G. Nordmark, S. Sigurdsson, et al., Genetic variants and disease-associated factors contribute to enhanced interferon regulatory factor 5 expression in blood cells of patients with systemic lupus erythematosus, Arthritis and rheumatism 62 (2010) 562–573.

[136] V. Gateva, J.K. Sandling, G. Hom, K.E. Taylor, S.a Chung, X. Sun, et al., A large-scale replication study identifies TNIP1, PRDM1, JAZF1, UHRF1BP1 and IL10 as risk loci for systemic lupus erythematosus, Nature genetics 41 (2009) 1–8.

[137] R.R. Graham, C. Kyogoku, S. Sigurdsson, I.A. Vlasova, L.R. Davies, E.C. Baechler, et al., Three functional variants of IFN regulatory factor 5 (IRF5) define risk and protective haplotypes for human lupus, Proc Natl Acad Sci USA 104 (2007) 6758–6763.

[138] K.L. Moser, J.a Kelly, C.J. Lessard, J.B. Harley, Recent insights into the genetic basis of systemic lupus erythematosus, Genes and immunity 10 (2009) 373–379.

[139] S.K. Nath, S. Han, X. Kim-Howard, J.A. Kelly, P. Viswanathan, G.S. Gilkeson, et al., A nonsynonymous functional variant in integrin-alpha(M) (encoded by ITGAM) is associated with systemic lupus erythematosus, Nat Genet 40 (2008) 152–154.

[140] S.V. Kozyrev, A.K. Abelson, J. Wojcik, A. Zaghlool, M.V. Linga Reddy, E. Sanchez, et al., Functional variants in the B-cell gene BANK1 are associated with systemic lupus erythematosus, Nat Genet 40 (2008) 211–216.

[141] D.S.C. Graham, R.R. Graham, H. Manku, A.K. Wong, J.C. Whittaker, P.M. Gaffney, et al., Polymorphism at the TNF superfamily gene TNFSF4 confers susceptibility to systemic lupus erythematosus, Nature Genetics 40 (2008) 83–89.

[142] C.O. Jacob, J. Zhu, D.L. Armstrong, M. Yan, J. Han, X.J. Zhou, et al., Identification of IRAK1 as a risk gene with critical role in the pathogenesis of systemic lupus erythematosus, Proc Natl Acad Sci USA 106 (2009) 6256–6261.

[143] J. Banchereau, V. Pascual, Type I Interferon in Systemic Lupus Erythematosus and Other Autoimmune Diseases, Immunity (2006) 383–392.

[144] A.K. Palucka, J. Banchereau, P. Blanco, V. Pascual, The interplay of dendritic cell subsets in systemic lupus erythematosus, Immunol Cell Biol 80 (2002) 484–488.

[145] Y. Yao, L. Richman, B.W. Higgs, C.a. Morehouse, M. de Los Reyes, P. Brohawn, et al., Neutralization of interferon-alpha/beta-inducible genes and downstream effect in a phase I trial of an anti-interferon-alpha monoclonal antibody in systemic lupus erythematosus, Arthritis and rheumatism 60 (2009) 1785–1796.

[146] L. Ronnblom, G.V. Alm, M.L. Eloranta, Type I interferon and lupus, Curr Opin Rheumatol 21 (2009) 471–477.

[147] M. Cella, D. Jarrossay, F. Facchetti, O. Alebardi, H. Nakajima, A. Lanzavecchia, et al., Plasmacytoid monocytes migrate to

inflamed lymph nodes and produce large amounts of type I interferon, Nat Med 5 (1999) 919—923.

[148] T. Decker, M. Müller, S. Stockinger, The yin and yang of type I interferon activity in bacterial infection, Nature Reviews Immunology (2005) 1—13.

[149] T. Decker, M. Muller, S. Stockinger, The yin and yang of type I interferon activity in bacterial infection, Nat Rev Immunol 5 (2005) 675—687.

[150] Y.-J. Liu, IPC: professional type 1 interferon-producing cells and plasmacytoid dendritic cell precursors, Annual review of immunology 23 (2005) 275—306.

[151] L. Ronnblom, V. Pascual, The innate immune system in SLE: type I interferons and dendritic cells, Lupus 17 (2008) 394—399.

[152] G. Grouard, The Enigmatic Plasmacytoid T Cells Develop into Dendritic Cells with Interleukin (IL)-3 and CD40-Ligand, J Exp Med 185 (1997) 1101—1112.

[153] B. Cederblad, S. Blomberg, H. Vallin, A. Perers, G.V. Alm, L. Ronnblom, Patients with systemic lupus erythematosus have reduced numbers of circulating natural interferon-alpha-producing cells, J Autoimmun 11 (1998) 465—470.

[154] L. Farkas, K. Beiske, F. Lund-Johansen, P. Brandtzaeg, F.L. Jahnsen, Plasmacytoid dendritic cells (natural interferon-alpha/beta-producing cells) accumulate in cutaneous lupus erythematosus lesions, Am J Pathol 159 (2001) 237—243.

[155] S. Blomberg, M.L. Eloranta, B. Cederblad, K. Nordlin, G.V. Alm, L. Ronnblom, Presence of cutaneous interferon-alpha producing cells in patients with systemic lupus erythematosus, Lupus 10 (2001) 484—490.

[156] M. Tucci, C. Quatraro, L. Lombardi, C. Pellegrino, Glomerular Accumulation of Plasmacytoid Dendritic Cells in Active Lupus Nephritis, Arthritis and Rheumatism 58 (2008) 251—262.

[157] G. Obermoser, P. Schwingshackl, F. Weber, G. Stanarevic, B. Zelger, N. Romani, et al., Recruitment of plasmacytoid dendritic cells in ultraviolet irradiation-induced lupus erythematosus tumidus, Br J Dermatol 160 (2009) 197—200.

[158] M. Napirei, H. Karsunky, B. Zevnik, H. Stephan, H.G. Mannherz, T. Moroy, Features of systemic lupus erythematosus in Dnase1-deficient mice, Nat Genet. 25 (2000) 177—181.

[159] M. Botto, C. Dell'Agnola, A.E. Bygrave, E.M. Thompson, H.T. Cook, F. Petry, et al., Homozygous C1q deficiency causes glomerulonephritis associated with multiple apoptotic bodies, Nat Genet. 19 (1998) 56—59.

[160] R. Lande, J. Gregorio, V. Facchinetti, B. Chatterjee, Y.-H. Wang, B. Homey, et al., Plasmacytoid dendritic cells sense self-DNA coupled with antimicrobial peptide, Nature 449 (2007) 564—569.

[161] J. Tian, A.M. Avalos, S.-Y. Mao, B. Chen, K. Senthil, H. Wu, et al., Toll-like receptor 9-dependent activation by DNA-containing immune complexes is mediated by HMGB1 and RAGE, Nature immunology 8 (2007) 487—496.

[162] T.K. Means, E. Latz, F. Hayashi, M.R. Murali, D.T. Golenbock, A.D. Luster, Human lupus autoantibody-DNA complexes activate DCs through cooperation of CD32 and TLR9, J Clin Invest 115 (2005) 407—417.

[163] U. Bave, M. Magnusson, M.L. Eloranta, A. Perers, G.V. Alm, L. Ronnblom, Fc gamma RIIa is expressed on natural IFN-alpha-producing cells (plasmacytoid dendritic cells) and is required for the IFN-alpha production induced by apoptotic cells combined with lupus IgG, J Immunol 171 (2003) 3296—3302.

[164] H. Sano, C. Morimoto, Dna isolated from DNA/anti-DNA antibody immune complexes in systemic lupus erythematosus is rich in guanine-cytosine content, J Immunol 128 (1982) 1341—1345.

[165] J. Vollmer, S. Tluk, C. Schmitz, S. Hamm, M. Jurk, A. Forsbach, et al., Immune stimulation mediated by autoantigen binding sites within small nuclear RNAs involves Toll-like receptors 7 and 8, J Exp Med 202 (2005) 1575—1585.

[166] E. Savarese, O.W. Chae, S. Trowitzsch, G. Weber, B. Kastner, S. Akira, et al., U1 small nuclear ribonucleoprotein immune complexes induce type I interferon in plasmacytoid dendritic cells through TLR7, Blood 107 (2006) 3229—3234.

[167] A. Marshak-Rothstein, Toll-like receptors in systemic autoimmune disease, Nat Rev Immunol 6 (2006) 823—835.

[168] A.M. Krieg, J. Vollmer, Toll-like receptors 7, 8, and 9: linking innate immunity to autoimmunity, Immunol Rev 220 (2007) 251—269.

[169] M. Gilliet, W. Cao, Y.-J. Liu, Plasmacytoid dendritic cells: sensing nucleic acids in viral infection and autoimmune diseases. Nature reviews, Immunology 8 (2008) 594—606.

[170] J. Vollmer, A.M. Krieg, Immunotherapeutic applications of CpG oligodeoxynucleotide TLR9 agonists, Adv Drug Deliv Rev. 61 (2009) 195—204.

[171] F.J. Barrat, R.L. Coffman, Development of TLR inhibitors for the treatment of autoimmune diseases, Immunol Rev 223 (2008) 271—283.

[172] M. Rutz, J. Metzger, T. Gellert, P. Luppa, G.B. Lipford, H. Wagner, et al., Toll-like receptor 9 binds single-stranded CpG-DNA in a sequence- and pH-dependent manner, Eur J Immunol 34 (2004) 2541—2550.

[173] R. Lafyatis, M. York, A. Marshak-Rothstein, Antimalarial agents: closing the gate on Toll-like receptors? Arthritis Rheum 54 (2006) 3068—3070.

[174] M.J. de Veer, M. Holko, M. Frevel, E. Walker, S. Der, J.M. Paranjape, et al., Functional classification of interferon-stimulated genes identified using microarrays, J Leukoc Biol 69 (2001) 912—920.

[175] L. Bennett, A.K. Palucka, E. Arce, V. Cantrell, J. Borvak, J. Banchereau, et al., Interferon and granulopoiesis signatures in systemic lupus erythematosus blood, J Exp Med 197 (2003) 711—723.

[176] C. Ding, Belimumab, an anti-BLyS human monoclonal antibody for potential treatment of inflammatory autoimmune diseases, Expert Opin Biol Ther 8 (2008) 1805—1814.

[177] M. Shodell, K. Shah, F.P. Siegal, Circulating human plasmacytoid dendritic cells are highly sensitive to corticosteroid administration, Lupus 12 (2003) 222—230.

[178] E.A. Leadbetter, I.R. Rifkin, A.M. Hohlbaum, B.C. Beaudette, M.J. Shlomchik, A. Marshak-Rothstein, Chromatin-IgG complexes activate B cells by dual engagement of IgM and Toll-like receptors, Nature 416 (2002) 603—607.

[179] E. Sousa, D. Isenberg, Treating lupus: from serendipity to sense, the rise of the new biologicals and other emerging therapies. Best practice & research, Clinical rheumatology 23 (2009) 563—574.

[180] D.J. Wallace, W. Stohl, R.A. Furie, J.R. Lisse, J.D. McKay, J.T. Merrill, et al., A phase II, randomized, double-blind, placebo-controlled, dose-ranging study of belimumab in patients with active systemic lupus erythematosus, Arthritis and rheumatism 61 (2009) 1168—1178.

[181] F.J. Barrat, R.L. Coffman, Development of TLR inhibitors for the treatment of autoimmune diseases, Immunological Reviews 223 (2008) 271—283.

[182] A.A. Itano, S.J. McSorley, R.L. Reinhardt, B.D. Ehst, E. Ingulli, A.Y. Rudensky, et al., Distinct dendritic cell populations sequentially present antigen to CD4 T cells and stimulate different aspects of cell-mediated immunity, Immunity 19 (2003) 47—57.

T-Cells and Systemic Lupus Erythematosus

José C. Crispín, George C. Tsokos

INTRODUCTION

Although the ultimate cause of systemic lupus erythematosus (SLE) — and other autoimmune diseases — is unknown, there is enough evidence to state that it arises from an unchecked immune response directed against self constituents. Tissue damage in patients with SLE is inflicted by elements of a pathological immune system aroused in the absence of any obvious noxious stimulus. Suppression of the immune system diminishes and in some cases halts organ injury, further supporting the causal role immune mediators have in this disease.

Multiple studies have been performed to identify the changes that allow the immune system to develop a full-blown and sustained response against self tissues. When compared to healthy individuals, multiple abnormalities are found at all levels of the immune system of patients with SLE. In this chapter, we review and discuss the T-cell biochemical and functional aberrations that have been associated with SLE. In some cases these alterations are genetically determined, but in most cases they probably result from faulty adaptations to environmental stimuli. The relative weight of each alteration probably varies among different individuals with SLE and their differential distribution among patients most certainly determines the clinical characteristics as well as the disease severity and response to treatment. The knowledge of the immune defects that lead to SLE will allow us to design better treatments and to more accurately detect disease activity and predict prognosis.

CENTRAL TOLERANCE

T cells originate from bone marrow precursors that migrate to the thymus where they complete their development into mature T cells. Initially devoid of a T-cell receptor, they generate one during their early dwelling in the thymus. The process that yields a T-cell receptor (TCR) is complex and involves DNA recombination; gene segments that code for distinct parts of the receptor are joined in random fashion. The stochastic nature of the process ensures diversity but entails the creation of a large number of flawed receptors. The enormous variety implies that the TCR will be able to recognize a vast number of molecules, including self-antigens. To avoid autoreactive T cells from becoming mature cells, T cells undergo a strict selection. After they complete the genetic recombination process and express a TCR on their surface, they become susceptible to death by apoptosis. In order to complete their maturation, they must receive certain survival signals. Such signals depend on the capacity of the TCR to bind to self MHC molecules (positive selection) with an intermediate affinity to self antigens presented in the thymus. Cells whose TCR display high avidity for antigens are killed (negative selection). Thus, only the T cells whose TCR are able to recognize antigens associated with self MHC molecules but lack strong affinity for the presented self molecules will avoid deletion and exit the thymus as mature naïve T cells.

Central tolerance has been studied in mice with lupus-like diseases [1, 2] and, to a certain extent, indirectly evaluated in patients with SLE by analyzing the frequency of autoreactive T cells in peripheral blood [3, 4]. Studies in murine models have quantified the efficiency of the negative selection process of T cells that bear a transgenic TCR specific for an endogenously expressed protein. In these mice, the TCR recombination process is bypassed because the transgenic TCR is expressed in early stages. Thus, virtually all the T cells express the same TCR. When their cognate antigen is presented in the thymus most of the T cells undergo apoptosis. This process occurs normally in mice with SLE-like diseases [1, 2, 5]. In patients with SLE, histone-reactive T cells have been identified in a similar frequency to healthy controls [3, 4]. Thus, central tolerance is not assumed to be affected in lupus patients.

This notion, however, has limitations. Each TCR is able to recognize several structurally related antigens and it is too simplistic to assume that self antigens and non-self antigens can be clearly separated [6]. One of the most consistent associations reported in genetic studies in patients with SLE is with the MHC locus [7] whose products determine the T-cell repertoire and the assortment of peptides to be presented in each immune response. The composition of the resulting repertoire, along with the assortment of class I and class II MHC molecules present in each individual, determines the array of antigens against which the immune system will respond. Thus even if no gross defects have been found in the central tolerance process of patients with SLE, subtle biases in the T-cell repertoire that are created during the process could contribute to the proclivity of affected individuals to develop self-aimed responses when challenged by infectious agents.

T-CELL ACTIVATION AND SIGNAL TRANSDUCTION

Naïve T cells travel through secondary lymphoid organs and interact with antigen-presenting cells (APC) via the TCR. Most cell-to-cell contacts are short-lived, but when the T cell recognizes an antigen a longer-lasting contact is established between it and the APC and as a consequence the T cell becomes activated. T-cell activation is altered in T cells from SLE

patients. Increases in intracellular calcium concentration and phosphorylation of molecules involved in the early signaling response, two early events that follow antigen recognition by T cells are abnormal; the rise in intracellular calcium and the phosphorylation of tyrosine residues of signaling molecules happen faster and are of a higher magnitude in lupus T cells than in normal cells (Figure 7.1) [8, 9].

Increased Calcium Flux: the Role of CD3ζ and FcRγ

The TCR is associated with a complex of transmembrane proteins collectively called CD3. The CD3 complex is responsible for the transduction of the TCR signal into the cell. One of its components, CD3ζ, a molecule centrally involved in the initiation of the TCR signal, is significantly decreased in T cells from SLE patients [9]. However, its absence is associated with the augmented calcium influx and tyrosine phosphorylation. The explanation for this apparent paradox is that the place of CD3ζ is taken by a closely related molecule normally not expressed in T cells, the common γ chain of the immunoglobulin receptors (FcRγ) [10]. The substitution has profound functional consequences in the manner the TCR-derived signal is transduced. When FcRγ is present, the signal relays through spleen tyrosine kinase (Syk) and not through the canonical ζ-associated protein (ZAP-70) [11]. The consequence is that the intensity of the response is altered and an

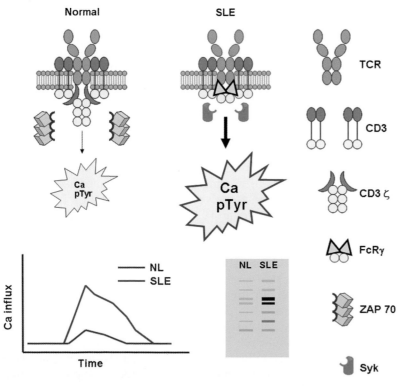

FIGURE 7.1　Early signaling events are altered during the activation process in SLE T cells. Decreased levels of the signaling molecule CD3ζ along with aberrant expression of FcRγ modify the signaling route in SLE T cells. Signal transduction through FcRγ relies on spleen tyrosine kinase (Syk) instead of the canonical ζ-associated protein (ZAP 70). The consequence is a magnified signal that results in a faster and higher calcium influx and a significantly enhanced phosphorylation of tyrosine residues.

abnormally high calcium influx follows TCR engagement (Figure 7.1) [12]. Multiple molecular mechanisms have been described in SLE T cells that contribute to the diminished expression of CD3ζ. These include decreased transcription [13–15], abnormal mRNA splicing [16], decreased mRNA stability [17], and increased caspase 3-mediated protein degradation [18].

Lipid Rafts

A second characteristic of SLE T cells that contributes to the enhanced T-cell activation process is an abnormal reorganization of lipid rafts. Proteins involved in signal transduction are located in lipid rafts — rich cholesterol zones of the cell membrane that associate to the cell cytoskeleton and harbor a high density of signaling molecules. In resting cells, lipid rafts are distributed throughout the cell surface. After cell activation, they coalesce on the pole of the cell that contacts the APC. This allows signal transduction to occur effectively because all the necessary elements are rapidly drawn together, in the area of the membrane where antigen is being presented. In T cells from patients with SLE, lipid rafts are clustered even in the absence of any obvious stimulus [19, 20]. This phenomenon also contributes to the augmented signal transduction that follows stimulation of SLE T cells through the TCR [21, 22]. Evidence for the *in vivo* relevance of these findings was provided by a study which showed that the administration of an agent that favors lipid raft clustering accelerated disease onset in a murine model of lupus (MRl/*lpr*), whereas injection of a drug that disrupts lipid raft clustering had the opposite effect [23].

MAP Kinases

Mitogen-activated protein (MAP) kinases are a group of enzymes that transduce signals from the surface into the cell by forming a phosphorylation cascade. They are involved in several cellular processes such as activation and proliferation, gene expression, and apoptosis induction. In T cells from lupus patients, the MAP kinase response is affected at various levels [24, 25]. Although, MAP kinases are important mediators in the T-cell activation process, they have been linked to tolerance maintenance [26]. In fact, mice deficient in RasGRP1 [27] or in protein kinase C (PKC) δ (a MAP kinase activator) [28] develop spontaneous autoimmune diseases.

Defects in other signaling pathways have been observed in SLE T cells. Cyclic AMP-dependent protein phosphorylation has been reported to be impaired [29], probably due to reduced levels of the protein kinase A [30]. The activities of PKC and Lck have also been reported to be low in SLE T cells [31, 32]. In contrast,

TABLE 7.1 Signaling defects reported in SLE T cells

Molecular defect	References
Decreased activity of PKC	[31]
Decreased PKA levels and activity	[30]
Increased calcium response	[8]
Deficient CD45 phosphatase activity	[144]
Increased phosphorylation of tyrosine residues	[9]
Decreased expression of CD3ζ	[9]
Decreased activity of Lck	[32]
Increased expression of FcRγ	[10]
Decreased MAP kinase activity	[24, 25]
Increased activity of PI3K	[34]
Increased lipid raft clustering	[19, 20]
Increased expression and activity of Syk	[11]

activity of the protein kinase PKR (which is involved in the phosphorylation of translation initiation factors) is increased in SLE T cells [33]. Likewise, activity of phosphatidylinositol-3 kinase (PI3K), the enzyme that produces the second messengers PIP_2 and PIP_3, is increased in mice with a lupus-like disease induced by alloreactivity [34]. The importance of this pathway was further supported by studies which proved that pharmacologic inhibition of class I_B PI3K can ameliorate disease in MRL/lpr mice [35, 36].

In summary, several alterations in the activation process have been found in T cells from patients with SLE (Table 7.1). Prematurely clustered lipid rafts facilitate activation and the abnormally configured transduction machinery causes the calcium influx to be disproportionally high and unbalanced to other signaling pathways such as MAP kinases. The correct transduction of the activation signal is essential for all the functions of T cells. The abnormal activation process observed in SLE T cells results in a T cell with an abnormally low activation threshold susceptible to be activated in the wrong situation, a T cell that is unable to tune an adequate response to external stimuli.

TRANSCRIPTION FACTORS AND GENE EXPRESSION

A consequence of the alterations in SLE T-cell activation and cell signaling is unbalanced activation of transcription factors that results in abnormal gene expression [37]. The abnormal gene transcription profile observed in lupus T cells is complex. Whereas some

TABLE 7.2   Transcription factor abnormalities reported in SLE T cells

Transcription factor defect	References
Decreased NF-κB nuclear activity	[145]
Increased phosphorylated CREM binding to the IL-2 promoter	[47]
Decreased phosphorylated CREB binding to the IL-2 promoter	[47]
Decreased nuclear translocation and DNA binding of AP-1	[24, 48]
Decreased levels of the 98 kDa functional isoform of Elf-1	[13]
Increased levels and nuclear translocation of NFATc2	[39]

genes are not transcribed at sufficient levels, others are overexpressed (Table 7.2). This altered pattern produces a characteristic phenotype that in some aspects resembles that of activated T cells and in others shares characteristics with anergic cells [38].

NFAT

Nuclear factor of activated T cells (NFAT) is a transcription factor regulated by calcium influx. When phosphorylated, it is confined to the cell cytoplasm where it is inactive. After T-cell activation, the ensuing calcium influx activates the phosphatase calcineurin that activates NFAT by dephosphorylating it. The immunosuppressive drugs cyclosporin A and tacrolimus inhibit this process. A consequence of the altered calcium response observed in SLE T cells is increased nuclear translocation of NFATc2 (the main NFAT isoform in T cells) [39]. As a result, the expression of genes regulated by NFAT is altered. The *CD40LG* gene (CD40L), an important co-stimulatory molecule used by T cells to stimulate antibody production and dendritic cell activation, is overexpressed in SLE T cells as a result of the increased NFAT activity [39, 40]. The complexity of the gene regulation defect of SLE T cells is illustrated by the fact that although NFAT also promotes *IL2* transcription, the production of this cytokine is unexpectedly decreased (see below) [41]. This apparently paradoxical phenomenon occurs because in some instances, as is the case with the IL-2 promoter, NFAT must couple to other transcription factors (e.g. AP-1 in the case of IL-2) to exert its effect on transcription [42].

CREM and CREB Balance

A region ~300 bp upstream of the transcription start site of the *IL2* gene defines a dense concentration of transcription factor-binding sites [43]. The transcription factors that bind to this promoter region regulate most of the activity of the *IL2* gene and include NFAT, AP-1, NF-κB, Egr-1, Oct-1, BOB.1/OBF.1, BCL11B, and CREB [42]. At position −180 a cyclic AMP response element (CRE) — a DNA region able to bind to the transcription factors CRE-binding protein (CREB) and CRE-modulator (CREM) — has been shown to be of significant pathogenic importance in SLE [44]. Although CREM and CREB bind to the same site, their effects on the transcription of *IL2* are antagonistic. Binding of CREB favors transcription whereas CREM exerts a repressor effect. In resting T cells the site is occupied by CREB. Following T-cell activation, increased levels of phosphorylated CREB are recruited to the −180 site favoring IL-2 transcription. Afterwards, phosphorylated CREM replaces CREB, silencing the gene [45]. In anergic cells (functionally inactivated T cells that fail to produce IL-2) the site is occupied by CREM [38, 46]. Thus, in normal T cells, the presence of CREM is associated with decreased levels of IL-2 production. In SLE T cells, the balance between CREB and CREM is altered; lower CREB and higher CREM levels contribute to skewed gene expression [47]. Genes that are known to be affected by the disturbed CREB:CREM ratio in SLE T cells include *CD247* (CD3ζ) [15], *FOS* [48], and *CD86* [49]. Since Fos is also a transcription factor, the transcriptional effects of decreased CREB and increased CREM levels extend to genes regulated by Fos, one of the components of AP-1 [48].

Several mechanisms give rise to the decrease in the CREB:CREM ratio of SLE T cells. Activity of the enzyme CaMKIV (calcium/calmodulin-dependent kinase IV) is abnormally increased in SLE T cells [50]. Anti-T-cell antibodies, commonly present in the sera of SLE patients, have been shown to induce the activation and nuclear migration of this enzyme. CaMKIV phosphorylates CREM and activates it increasing its negative effects on gene transcription [50]. This phenomenon illustrates how autoantibodies can alter T-cell function.

Cellular levels and activity of protein phosphatase 2A (PP2A), a highly conserved serine/threonine phosphatase, are increased in T cells from lupus patients [51]. PP2A is the main phosphatase responsible for dephosphorylating — and thus inactivating — CREB [52]. In SLE T cells, heightened activity of PP2A leads to CREB dephosphorylation and hampered IL-2 production. Knockdown of PP2A levels restores the binding of phosphorylated CREB to the *IL2* promoter and IL-2 production [51]. Transcription of the gene that codes for the enzymatic subunit of PP2A (*PPP2CA*) is increased in SLE T cells [51, 53]. This phenomenon occurs at least partially as a consequence of DNA hypomethylation, a disorder in epigenetic regulation common in SLE T cells (see below) [53, 54].

Elf-1

The transcription factor Elf-1 is another molecule whose activity is affected by the increased levels of PP2A in T cells from SLE patients [14]. The inactive form of Elf-1 is an 80-kDa protein that lacks DNA-binding activity and is confined to the cytoplasm of the cell. Phosphorylation and *O*-linked glycosylation increase the molecular weight of Elf-1 to 98 kDa, the active form; 98 kDa Elf-1 binds to the promoter of the gene that codes for CD3ζ inducing its transcription. Interestingly, it has a negative effect on the expression of the FcRγ gene [13]. Thus, in normal T cells 98-kDa Elf-1 maintains the expression of CD3ζ and simultaneously represses the production of FcRγ. Overactivity of PP2A in SLE T cells is associated with decreased expression of the active form of Elf-1 and consequently contributes to the decreased CD3ζ expression [14].

Abnormal Epigenetic Regulation of Gene Expression

DNA methylation and histone modifications (mainly acetylation and methylation) alter the structure of chromatin and the accessibility of transcription factors to DNA thus regulating gene expression at a different level than transcription factors do. These changes are known as epigenetic regulation and represent a level of control that is maintained through cell division. DNA methylation has potent suppressive effects over gene expression. Compared to T cells from healthy individuals, T cells from SLE patients have abnormally low levels of DNA methylation [55]. This defect has been shown to underlie the overexpression of several genes in SLE T cells (Table 7.3). Importantly, drugs that have been associated to drug-induced lupus inhibit DNA methylation (i.e. procainamide and hydralazine) [56, 57] and a lupus-like disease is caused in mice by injection of syngeneic CD4[+] T cells treated with procainamide or 5-azacytidine, an inhibitor of DNA methyltransferases (DNMT) [58].

TABLE 7.3 DNA hypomethylation in SLE T cells has been implicated in the overexpression of certain genes

Gene	References
LFA-1 (CD11a/CD18).	[146]
Perforin	[147]
CD70	[148]
CD40L	[149]
PP2A (catalytic subunit)	[53]
Killer cell Ig-like receptor genes (KIR)	[78]

It is not clear why SLE T cells exhibit DNA hypomethylation. However, some of the signaling alterations mentioned in the preceding section have been associated to the reduced DNA methylation characteristic of SLE T cells. Activity of DNMT1 is regulated by the MAP kinase pathway [59, 60] whose activity is abnormally low in SLE T cells. Defective activity of PKCδ, also reported in SLE T cells, has been associated with hampered DNMT1 activity through a pathway that depends on the faulty activity of MAP kinases [28]. Elevated levels of IL-6, a pro-inflammatory cytokine produced in excess by mononuclear cells from lupus patients, have been reported to abrogate DNA methylation in B cells [61].

Histone acetylation, another epigenetic regulatory mechanism, has been proposed to alter gene expression in SLE T cells. Some of the effects of CREM, particularly its effect on the *IL2* promoter depend on its capacity to recruit histone deacetylase (HDAC) 1 [62]. PP2A also regulates the activity of HDAC and some of its effects are known to be mediated through histone acetylation [63]. Treatment of SLE T cells with trichostatin A, a HDAC inhibitor, diminished the expression of CD40L and the production of IL-10, suggesting that histone acetylation plays a role in the overexpression of these molecules in lupus [64].

In summary, T cells from SLE patients have a grossly distorted pattern of gene expression. This defect, which is in part a consequence of the alterations in cell activation and signaling, affects the phenotype and function of the cell creating a vicious circle in which altered signaling skews gene expression that further alters cell signaling and activation.

CD4 HELPER T CELLS AND CYTOKINE PRODUCTION

Immune responses depend on the coordinated actions of a large variety of cells that respond to potentially harmful agents. To a large extent these actions are controlled by T cells through the production of cytokines. According to signals provided during antigen presentation, naïve CD4 T cells develop into effector subsets defined by the expression of distinct transcription factors that induce a particular phenotype and cytokine production profile (Figure 7.2 and Table 7.4).

After antigen priming, naïve CD4 T cells differentiate into helper cells able to produce distinct cytokines. Each helper subset relies on certain cytokines to activate different effector cell types and steer the system onto a response with particular characteristics that will allow a better clearance of the pathogen that triggered the response. T$_H$1 cells activate the cellular immune response, particularly macrophages, through the production of high levels of IFN-γ and favor the

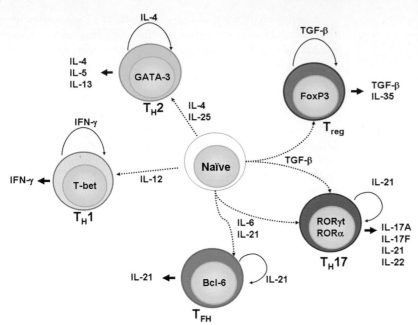

FIGURE 7.2   CD4 T-cell helper differentiation. Naïve CD4 T cells undergo a process of phenotypic differentiation following antigen priming. Cytokines produced locally by antigen-presenting cells as well as by neighboring cells such as NK cells can stimulate the differentiation into effector subsets that are defined by the expression of specific transcription factors and a distinct cytokine production profile. In patients with SLE, increased amounts of cytokines that favor T_{FH} and T_H17 differentiation and hamper T_{reg} generation (i.e. low IL-2; high IL-6 and IL-21) contribute to a skewed pro-inflammatory CD4 helper repertoire.

production of IgG2a antibodies. Type 1 responses are crucial in the defense against intracellular pathogens such as *Listeria monocytogenes*. T_H2 cells produce cytokines that amplify the humoral response, mainly by inducing the production of IgG1 and IgE and recruiting eosinophils. T_H2 responses are induced by extracellular pathogens, particularly helminthes. T_H17 cells produce the pro-inflammatory cytokine IL-17 that acts as a potent neutrophil attractant and activator [65].

Each CD4 helper subset has a mechanism to amplify itself and block the development of the other subsets. Thus, IL-4 produced by T_H2 cells induces T_H2 cell differentiation and blocks that of T_H1 and T_H17 cells. Analogously, IFN-γ amplifies T_H1 differentiation and blocks T_H2 and T_H17. The presence of defined combinations of cytokines induces the development of each subset by stimulating the production of subset-specific transcription factors (Figure 7.2).

TABLE 7.4   Helper CD4 T cell subsets

Subset	Transcription factors	Produced cytokines
T_H1	T-bet	IFN-γ
T_H2	GATA-3	IL-4, IL-5, IL-13
T_H17	ROR-γt, RORα	IL-17A, IL17F, IL-21, IL-22
T_{FH}	Bcl6	IL-21
T_{reg}	FoxP3	TGF-β, IL-35

IL-2

Cytokine production and therefore the effector capacities of T cells from lupus patients are compromised. The production of IL-2, a cytokine centrally involved in the process of T-cell activation and proliferation, is defective in T cells from patients with SLE [41, 66]. This defect, one of the earliest functional abnormalities described in SLE T cells, has broad consequences in the overall function of the immune system. Mice deficient in IL-2 or in components of the IL-2 receptor develop a spontaneous and fatal autoimmune disorder [67]. Autoimmunity caused by the absence of IL-2 has been attributed to the lack of regulatory T cells (T_{reg}) observed in these mice [68]. In the context of SLE, absence of IL-2 is not absolute and T_{reg} are present. However, the IL-2 production defect has broad functional repercussions. Activation-induced cell death (AICD), a process by which clones of T cells that have been activated and expanded are down-regulated, is defective in SLE T cells [69, 70]. IL-2 is indispensable for this process and its decreased production could underlie the AICD defect of SLE T cells [71]. IL-2 is also necessary for the development of CD8 T-cell effector functions and is involved in the differentiation of helper CD4 subsets [72].

IFN-γ

Polymorphisms of the genes that code for IFN-γ [73] and its receptors [74] have been associated with lupus. However, IFN-γ has not been assigned an unambiguous role in the pathogenesis of SLE. Although some studies report that production of IFN-γ by T cells from patients with SLE is decreased [75, 76], others have observed that T cells from SLE patients secrete increased amounts of IFN-γ [77, 78]. Perhaps more significant is the finding of IFN-γ-positive cells and IFN-γ RNA transcripts in glomeruli from kidneys affected by lupus nephritis [79–81]. Mechanistically, increased IFN-γ production could contribute to the pathogenesis of SLE, by stimulating the inflammatory response through macrophages and dendritic cells [77]. In the murine lupus models NZB/NZW and MRL/*lpr*, IFN-γ has an undisputed pathogenic role and its blockage leads to disease improvement [82–84].

IL-10

Single nucleotide polymorphisms (SNP) of the promoter region of the IL-10 gene known to determine the levels of IL-10 production in humans, are associated with SLE in African-American individuals [85]. IL-10 production is increased in patients with SLE and in their healthy relatives compared to non-related healthy individuals [86], further suggesting that a genetic component underlies this defect. Although IL-10 is usually regarded as an anti-inflammatory cytokine, it has been shown to exert stimulatory effects on B cells [87]. Also, it has been shown to promote immunoglobulin production by mononuclear cells of patients with SLE in *in vitro* as well as in *in vivo* models [88]. Interestingly, IL-10 production in patients with SLE was shown to derive mostly from B cells and monocytes [89]. Blockade of IL-10 delayed the disease onset in NZB/NZW mice [90]. Administration of anti-IL-10 to patients with SLE led to joint and skin disease improvement in a small clinical trial [91].

IL-17

IL-17 (IL-17A), the prototypic member of a family of recently described cytokines, and IL-17F, a closely related molecule, are produced mainly by activated T cells [65]. They play an important role in the immune response against certain bacteria and fungi. IL-17 is the signature cytokine of the CD4 T helper subset known as T_H17 (Table 7.4). IL-17-producing cells have been implicated in the pathogenesis of several autoimmune diseases, including multiple sclerosis and SLE [92, 93]. Sera of patients with lupus contains abnormally high levels of IL-17 [94]. Accordingly, an abnormally high fraction of T cells in peripheral blood of patients with SLE produces IL-17 [95, 96]. In these patients CD4 T cells are not the only relevant IL-17-producing T cells, but also $CD4^-CD8^-$ (double negative) T cells [95]. The heightened production of IL-17

in patients with SLE correlates with disease activity [94, 96]. Further, IL-17-producing T cells have been found within kidney infiltrates from patients with lupus nephritis [95]. Release of IL-17 amplifies the inflammatory response by recruiting effector cells to target organs. In addition, when combined with B-cell-activating factor, IL-17 increases the survival and proliferation of B cells and induces their transformation into antibody-secreting cells [94].

In murine models of lupus, IL-17 production is also enhanced. Spleen cells from SNF_1 mice produce increased amounts of IL-17 when cultured in the presence of nucleosomes [97]. Further, cells expressing IL-17 are detected in affected kidneys of mice with lupus-like diseases [97, 98] and clinical improvement achieved by tolerance-inducing treatments is accompanied by a reduction in the percentage of IL-17-producing cells [97, 99]. In MRL/*lpr* mice, as in patients with SLE, DN T cells are robust producers of IL-17 [98]. Mice deficient in the ubiquitin ligase Ro52 develop an autoimmune disease that shares certain characteristics with lupus. Autoimmunity in these mice is driven by the increased production of IL-23 and IL-17 further suggesting a mechanistic role for these cytokines in the inflammatory response of SLE [100].

Differentiation of T cells into the T_H17 subset is induced when priming occurs in the presence of TGF-β and certain inflammatory cytokines such as IL-1β, IL-6 or IL-21 [101]. Interestingly, the differentiation of naïve cells into this pro-inflammatory subset has been proposed to occur in a reciprocal fashion with the development of regulatory T cells (T_{reg}). TGF-β is required for the generation of the antagonist subsets T_H17 and T_{reg} and the presence of an inflammatory signal seems to be the factor that determines if pro-inflammatory or suppressive cells are generated [102]. Increased levels of the T_H17-determining cytokines IL-6 [103] and IL-21 (Juang YT, personal communication) are present in patients with SLE. In the case of IL-6, the increased levels may be genetically determined because polymorphisms that contribute to increased IL-6 transcription are more prevalent in individuals with SLE [104]. Also, increased phosphorylation of STAT-3, a molecule involved in IL-6 and IL-23 signaling has been observed in T cells from SLE patients [105]. On the other hand, IL-2 production — which is required for Treg development and inhibits T_H17 cell generation — is decreased in SLE patients [41, 66]. Thus, although it has not been experimentally proven, several factors present in the cytokine environment of SLE patients favor the development of IL-17-producing cells and hinder that of suppressive T_{reg} (Table 7.5). An analogous phenomenon has been observed in mice with a lupus-like syndrome where differentiation of T_H17 cells is enhanced [106].

TABLE 7.5 Established cytokine alterations in patients with SLE

Cytokine	Defect	Consequences	References
IL-2	Decreased T cell production	Altered T cell effector function Altered T_{reg} development	[41, 66]
IL-4	Increased T cell and NK cell production	B cell stimulation	[139]
IL-6	Increased mononuclear cell production	B cell stimulation Promotion of T_H17 cell differentiation	[103]
IL-10	Increased mononuclear cell production	B cell stimulation	[86, 89]
IL-12	Decreased mononuclear cell production	Promotion of T_H17 cell differentiation	[150]
IL-17	Increased T cell production	Amplified inflammatory response	[95, 96]
IL-21	Increased T cell production	B cell stimulation Promotion of T_H17 cell differentiation	
IL-23	Increased mononuclear cell production	Promotion of T_H17 cell expansion	

T FOLLICULAR HELPER CELLS

Follicular helper T cells (T_{FH}) comprise a recently described CD4 helper subset [107]. As T_H17 cells, their differentiation is induced by IL-6 and IL-21 and depends on the co-stimulation molecule ICOS [108]. However in their case TGF-β must be absent, because it blocks the production of Bcl-6, the lineage-specific transcription factor (Table 7.4) [109, 110]. T_{FH} localize in the B-cell zones of lymph nodes and produce IL-21 and express CD40L. Their main function is to provide B cells with the necessary signals for immunoglobulin production and isotype switch and hypersomatic mutation. The behavior of T_{FH} has not been studied in human SLE. In murine models of lupus, IL-21 and T_{FH} seem to be a necessary element for disease development [111, 112]. Deficiency of ICOS protected MRL/*lpr* mice from developing lupus. This effect was accomplished by the absence of a particular CD4 T cell type that is very similar to T_{FH} and has a similar function in extrafollicular compartments [111].

Further support for the involvement of IL-21 and IL-17 in the pathogenesis of SLE comes from mice deficient in IBP (IFN regulatory factor 4-binding protein). This is a protein that limits IL-17 and IL-21 production [113]. Mice that lack IBP develop an IL-17- and IL-21-driven systemic autoimmune disease that resembles human SLE [114].

REGULATORY T CELLS

Regulatory T cells (T_{reg}) are specialized CD4$^+$ suppressor cells that express the transcription factor FoxP3 and the high-affinity IL-2 receptor (CD25). T_{reg} may be selected in the thymus during the process of central tolerance or may be generated in the periphery from conventional CD4$^+$ T cells that acquire a T_{reg} phenotype under certain activation circumstances [115]. T_{reg} maintain tolerance and limit immune responses and their absence causes a severe autoimmune disorder in mice and humans (IPEX) [116]. Quantitative and functional defects in T_{reg} have been linked to several autoimmune disorders including SLE [117]. Most studies claim that reduced T_{reg} numbers are observed in the peripheral blood of patients with SLE, particularly during active disease periods [118–120]. The suppressive function of SLE-derived T_{reg} has also been studied, but the results are conflicting. Some reports claim they are unable to efficiently suppress proliferation and cytokine production [121, 122]. However, others suggest that their function is conserved and the suboptimal T-cell suppression observed in *in vitro* assays is the consequence of SLE T cells being abnormally resistant to T_{reg}-induced suppression [123, 124].

These defects in T_{reg} number and function may result from the altered cytokine milieu characteristic of SLE patients. Increased levels of the pro-inflammatory cytokines IL-6 and IL-21 combined with the decreased amounts of available IL-2 may decrease conversion of CD4$^+$ T cells into T_{reg} and inactivate their suppressive capacity, as suggested by studies in lupus murine models [125]. T_{reg} defects may contribute to SLE pathology by failing to control T cell responses and autoantibody production [126].

CD8 T CELLS AND CYTOTOXIC RESPONSES

Little is known about CD8 T cells and the autoimmune response in patients with SLE [127]. Some evidence indicates that CD8 T cells are overactive in the peripheral blood of patients with lupus, especially in patients with active disease [128]. A higher fraction of perforin and granzyme-B-positive CD8 T cells have been reported in SLE patients [129] and CD8 T cells have been observed in cellular infiltrates in kidney biopsies from SLE patients, particularly in the interstitial and periglomerular areas [130]. On the other hand, deficient cytotoxic capacity has been observed in CD8 T cells from SLE patients [131]. An intriguing hypothesis suggests that this defect predisposes to the development of SLE because it hampers the capacity of CD8 cells to lyse autoreactive B cells [132]. This phenomenon has been proposed to underlie the fact that perforin-deficient mice have accelerated humoral autoimmunity and lupus-like disease in the MRL background [133]. Altered CD8 T-cell activation and effector differentiation probably explains the expansion of double-negative T cells in patients with SLE (see below) [134].

Activity of natural killer (NK) cells has been reported to be low in patients with SLE [135]. This phenomenon has been linked to the presence of autoantibodies and faulty IL-2 production [136]. Although it is unknown if this defect is involved in the pathogenesis of SLE, it most probably contributes to the increased susceptibility to infections observed in patients with SLE.

DOUBLE-NEGATIVE T CELLS

T lymphocytes that lack the CD4 and CD8 co-receptors are called double-negative (DN) T cells. These cells comprise a scarce population in normal individuals (<5% of T lymphocytes), but are significantly expanded in patients with SLE [95, 137]. Early reports implicated DN T cells in the pathogenesis of SLE by linking their expansion to disease activity in mice with lupus-like diseases [138]. Also, a role in the stimulation of autoantibody production was attributed to them as they were shown to produce IL-4 [139] and induce immunoglobulin and anti-DNA antibody production when co-cultured with B cells [137, 140]. Recent work has shown that they also secrete other chemokines and cytokines such as IL-1, IL-17, and IFN-γ, and are found within cellular infiltrates in kidney biopsies of patients with lupus nephritis [95]. DN cells represent a fraction of CD8$^+$ T cells that upon activation acquire a distinct gene expression profile that induces the loss of CD8 expression and the capacity to produce pro-inflammatory cytokines [134]. A major question that still remains is why the DN T cell population is expanded in patients with SLE.

ADHESION MOLECULES

A fundamental ability of immune cells is their mobility and capacity to patrol the diverse organs and tissues of the body. In order to achieve that, they rely on the expression of surface molecules that bind to receptors on other immune and non-immune cells. T cells depend on adhesion molecules to guide them out of the vascular space into lymphoid organs or peripheral tissues. The expression of the adhesion molecule CD44 is abnormally increased on T cells from patients with SLE [20]. Moreover, the affinity of CD44 for its ligands is enhanced in SLE T cells due to the increased activation state of the cell. This grants the cell an increased capacity to migrate into inflamed organs [20, 141]. The CD44 gene can give rise to several variant isoforms produced by alternative splicing and post-translational modifications. The expression of two of these variants, CD44v3 and v6, is also increased in T cells from lupus patients. This increase correlates with disease activity and presence of renal disease and anti-double-stranded DNA antibodies[142]. The importance of these findings is illustrated by the fact that T cells that infiltrate kidneys from SLE patients express CD44v3 and CD44v6 [143].

CONCLUDING REMARKS

Several intrinsic and extrinsic factors alter the behavior and functional differentiation of T cells in patients with SLE. The consequences of these abnormalities are vast and influence the pathogenesis of the disease as well as the protective immune response. Pro-inflammatory signals originating in SLE T cells stimulate the autoimmune behavior in other immune cells that alter even further the function of T cells. T cells play a central role in this vicious circle and the study and identification of their molecular defects will allow the development of therapies aimed to break the circle and restore the balance of the system. Although patients with SLE have a propensity to autoimmune disease hardwired into their immune system by genetic factors and by incorrect adaptations to previous immune challenges, the understanding of the mechanisms that underlie tissue injury and T-cell activation will be central to devising strategies to decrease the triggers that steer the immune response onto noxious inflammatory responses and away from its normal protective behavior in these patients.

References

[1] B.L. Kotzin, J.W. Kappler, P.C. Marrack, L.R. Herron, T cell tolerance to self antigens in New Zealand hybrid mice with lupus-like disease, J Immunol 143 (1989) 89—94.
[2] L.R. Herron, R.A. Eisenberg, E. Roper, V.N. Kakkanaiah, P.L. Cohen, B.L. Kotzin, Selection of the T cell receptor repertoire in Lpr mice, J Immunol 151 (1993) 3450—3459.

[3] K. Andreassen, S. Bendiksen, E. Kjeldsen, G.M. Van, U. Moens, E. Arnesen, et al., T cell autoimmunity to histones and nucleosomes is a latent property of the normal immune system, Arthritis Rheum 46 (2002) 1270–1281.

[4] R.E. Voll, E.A. Roth, I. Girkontaite, H. Fehr, M. Herrmann, H.M. Lorenz, et al., Histone-specific Th0 and Th1 clones derived from systemic lupus erythematosus patients induce double-stranded DNA antibody production, Arthritis Rheum 40 (1997) 2162–2171.

[5] S. Fatenejad, S.L. Peng, O. Disorbo, J. Craft, Central T cell tolerance in lupus-prone mice: influence of autoimmune background and the lpr mutation, J Immunol 161 (1998) 6427–6432.

[6] K.W. Wucherpfennig, P.M. Allen, F. Celada, I.R. Cohen, B.R. De, K.C. Garcia, et al., Polyspecificity of T cell and B cell receptor recognition, Semin Immunol 19 (2007) 216–224.

[7] I.T. Harley, K.M. Kaufman, C.D. Langefeld, J.B. Harley, J.A. Kelly, Genetic susceptibility to SLE: new insights from fine mapping and genome-wide association studies, Nat Rev Genet 10 (2009) 285–290.

[8] D. Vassilopoulos, B. Kovacs, G.C. Tsokos, TCR/CD3 complex-mediated signal transduction pathway in T cells and T cell lines from patients with systemic lupus erythematosus, J Immunol 155 (1995) 2269–2281.

[9] S.N. Liossis, X.Z. Ding, G.J. Dennis, G.C. Tsokos, Altered pattern of TCR/CD3-mediated protein-tyrosyl phosphorylation in T cells from patients with systemic lupus erythematosus. Deficient expression of the T cell receptor zeta chain, J Clin Invest 101 (1998) 1448–1457.

[10] E.J. Enyedy, M.P. Nambiar, S.N. Liossis, G. Dennis, G.M. Kammer, G.C. Tsokos, Fc epsilon receptor type I gamma chain replaces the deficient T cell receptor zeta chain in T cells of patients with systemic lupus erythematosus, Arthritis Rheum 44 (2001) 1114–1121.

[11] S. Krishnan, Y.T. Juang, B. Chowdhury, A. Magilavy, C.U. Fisher, H. Nguyen, et al., Differential expression and molecular associations of Syk in systemic lupus erythematosus T cells, J Immunol 181 (2008) 8145–8152.

[12] G.C. Tsokos, M.P. Nambiar, K. Tenbrock, Y.T. Juang, Rewiring the T-cell: signaling defects and novel prospects for the treatment of SLE, Trends Immunol 24 (2003) 259–263.

[13] Y.T. Juang, K. Tenbrock, M.P. Nambiar, M.F. Gourley, G.C. Tsokos, Defective production of functional 98-kDa form of Elf-1 is responsible for the decreased expression of TCR zeta-chain in patients with systemic lupus erythematosus, J Immunol 169 (2002) 6048–6055.

[14] Y.T. Juang, Y. Wang, G. Jiang, H.B. Peng, S. Ergin, M. Finnell, et al., PP2A dephosphorylates Elf-1 and determines the expression of CD3zeta and FcRgamma in human systemic lupus erythematosus T cells, J Immunol 181 (2008) 3658–3664.

[15] K. Tenbrock, V.C. Kyttaris, M. Ahlmann, J.M. Ehrchen, M. Tolnay, H. Melkonyan, et al., The cyclic AMP response element modulator regulates transcription of the TCR zeta-chain, J Immunol 175 (2005) 5975–5980.

[16] V.R. Moulton, V.C. Kyttaris, Y.T. Juang, B. Chowdhury, G.C. Tsokos, The RNA-stabilizing protein HuR regulates the expression of zeta chain of the human T cell receptor-associated CD3 complex, J Biol Chem 283 (2008) 20037–20044.

[17] B. Chowdhury, C.G. Tsokos, S. Krishnan, J. Robertson, C.U. Fisher, R.G. Warke, et al., Decreased stability and translation of T cell receptor zeta mRNA with an alternatively spliced 3'-untranslated region contribute to zeta chain downregulation in patients with systemic lupus erythematosus, J Biol Chem 280 (2005) 18959–18966.

[18] S. Krishnan, J.G. Kiang, C.U. Fisher, M.P. Nambiar, H.T. Nguyen, V.C. Kyttaris, et al., Increased caspase-3 expression and activity contribute to reduced CD3zeta expression in systemic lupus erythematosus T cells, J Immunol 175 (2005) 3417–3423.

[19] E.C. Jury, P.S. Kabouridis, F. Flores-Borja, R.A. Mageed, D.A. Isenberg, Altered lipid raft-associated signaling and ganglioside expression in T lymphocytes from patients with systemic lupus erythematosus, J Clin Invest 113 (2004) 1176–1187.

[20] Y. Li, T. Harada, Y.T. Juang, V.C. Kyttaris, Y. Wang, M. Zidanic, et al., Phosphorylated ERM is responsible for increased T cell polarization, adhesion, and migration in patients with systemic lupus erythematosus, J Immunol 178 (2007) 1938–1947.

[21] S. Krishnan, M.P. Nambiar, V.G. Warke, C.U. Fisher, J. Mitchell, N. Delaney, et al., Alterations in lipid raft composition and dynamics contribute to abnormal T cell responses in systemic lupus erythematosus, J Immunol 172 (2004) 7821–7831.

[22] E.C. Jury, D.A. Isenberg, C. Mauri, M.R. Ehrenstein, Atorvastatin restores Lck expression and lipid raft-associated signaling in T cells from patients with systemic lupus erythematosus, J Immunol 177 (2006) 7416–7422.

[23] G.M. Deng, G.C. Tsokos, Cholera toxin B accelerates disease progression in lupus-prone mice by promoting lipid raft aggregation, J Immunol 181 (2008) 4019–4026.

[24] S. Cedeno, D.F. Cifarelli, A.M. Blasini, M. Paris, F. Placeres, G. Alonso, et al., Defective activity of ERK-1 and ERK-2 mitogen-activated protein kinases in peripheral blood T lymphocytes from patients with systemic lupus erythematosus: potential role of altered coupling of Ras guanine nucleotide exchange factor hSos to adapter protein Grb2 in lupus T cells, Clin Immunol 106 (2003) 41–49.

[25] A. Mor, M.R. Philips, M.H. Pillinger, The role of Ras signaling in lupus T lymphocytes: biology and pathogenesis, Clin Immunol 125 (2007) 215–223.

[26] L. Rui, C.G. Vinuesa, J. Blasioli, C.C. Goodnow, Resistance to CpG DNA-induced autoimmunity through tolerogenic B cell antigen receptor ERK signaling, Nat Immunol 4 (2003) 594–600.

[27] K. Layer, G. Lin, A. Nencioni, W. Hu, A. Schmucker, A.N. Antov, et al., Autoimmunity as the consequence of a spontaneous mutation in Rasgrp1, Immunity 19 (2003) 243–255.

[28] G. Gorelik, J.Y. Fang, A. Wu, A.H. Sawalha, B. Richardson, Impaired T cell protein kinase C delta activation decreases ERK pathway signaling in idiopathic and hydralazine-induced lupus, J Immunol 179 (2007) 5553–5563.

[29] R. Mandler, R.E. Birch, S.H. Polmar, G.M. Kammer, S.A. Rudolph, Abnormal adenosine-induced immunosuppression and cAMP metabolism in T lymphocytes of patients with systemic lupus erythematosus, Proc Natl Acad Sci USA 79 (1982) 7542–7546.

[30] G.M. Kammer, I.U. Khan, C.J. Malemud, Deficient type I protein kinase A isozyme activity in systemic lupus erythematosus T lymphocytes, J Clin Invest 94 (1994) 422–430.

[31] Y. Tada, K. Nagasawa, Y. Yamauchi, H. Tsukamoto, Y. Niho, A defect in the protein kinase C system in T cells from patients with systemic lupus erythematosus, Clin Immunol Immunopathol 60 (1991) 220–231.

[32] C. Matache, M. Stefanescu, A. Onu, S. Tanaseanu, I. Matei, et al., p56lck activity and expression in peripheral blood lymphocytes from patients with systemic lupus erythematosus, Autoimmunity 29 (1999) 111–120.

[33] A. Grolleau, M.J. Kaplan, S.M. Hanash, L. Beretta, B. Richardson, Impaired translational response and increased protein kinase PKR expression in T cells from lupus patients, J Clin Invest 106 (2000) 1561–1568.

[34] F. Niculescu, P. Nguyen, T. Niculescu, H. Rus, V. Rus, C.S. Via, Pathogenic T cells in murine lupus exhibit spontaneous

signaling activity through phosphatidylinositol 3-kinase and mitogen-activated protein kinase pathways, Arthritis Rheum 48 (2003) 1071–1079.

[35] D.F. Barber, A. Bartolome, C. Hernandez, J.M. Flores, C. Redondo, C. Fernandez-Arias, et al., PI3Kgamma inhibition blocks glomerulonephritis and extends lifespan in a mouse model of systemic lupus, Nat Med 11 (2005) 933–935.

[36] D.F. Barber, A. Bartolome, C. Hernandez, J.M. Flores, C. Fernandez-Arias, L. Rodriguez-Borlado, et al., Class IB-phosphatidylinositol 3-kinase (PI3K) deficiency ameliorates IA-PI3K-induced systemic lupus but not T cell invasion, J Immunol 176 (2006) 589–593.

[37] J.C. Crispin, V.C. Kyttaris, Y.T. Juang, G.C. Tsokos, How signaling and gene transcription aberrations dictate the systemic lupus erythematosus T cell phenotype, Trends Immunol 29 (2008) 110–115.

[38] R.H. Schwartz, T cell anergy, Annu. Rev Immunol 21 (2003) 305–334.

[39] V.C. Kyttaris, Y. Wang, Y.T. Juang, A. Weinstein, G.C. Tsokos, Increased levels of NF-ATc2 differentially regulate CD154 and IL-2 genes in T cells from patients with systemic lupus erythematosus, J Immunol 178 (2007) 1960–1966.

[40] Y. Yi, M. McNerney, S.K. Datta, Regulatory defects in Cbl and mitogen-activated protein kinase (extracellular signal-related kinase) pathways cause persistent hyperexpression of CD40 ligand in human lupus T cells, J Immunol 165 (2000) 6627–6634.

[41] J. Alcocer-Varela, D. Alarcon-Segovia, Decreased production of and response to interleukin-2 by cultured lymphocytes from patients with systemic lupus erythematosus, J Clin Invest 69 (1982) 1388–1392.

[42] J.C. Crispin, G.C. Tsokos, Transcriptional regulation of IL-2 in health and autoimmunity, Autoimmun Rev 8 (2009) 190–195.

[43] H.P. Kim, J. Imbert, W.J. Leonard, Both integrated and differential regulation of components of the IL-2/IL-2 receptor system, Cytokine Growth Factor Rev 17 (2006) 349–366.

[44] K. Tenbrock, G.C. Tsokos, Transcriptional regulation of interleukin 2 in SLE T cells, Int Rev Immunol 23 (2004) 333–345.

[45] K. Tenbrock, Y.T. Juang, M. Tolnay, G.C. Tsokos, The cyclic adenosine 5'-monophosphate response element modulator suppresses IL-2 production in stimulated T cells by a chromatin-dependent mechanism, J Immunol 170 (2003) 2971–2976.

[46] J.D. Powell, C.G. Lerner, G.R. Ewoldt, R.H. Schwartz, The -180 site of the IL-2 promoter is the target of CREB/CREM binding in T cell anergy, J Immunol 163 (1999) 6631–6639.

[47] E.E. Solomou, Y.T. Juang, M.F. Gourley, G.M. Kammer, G.C. Tsokos, Molecular basis of deficient IL-2 production in T cells from patients with systemic lupus erythematosus, J Immunol 166 (2001) 4216–4222.

[48] V.C. Kyttaris, Y.T. Juang, K. Tenbrock, A. Weinstein, G.C. Tsokos, Cyclic adenosine 5'-monophosphate response element modulator is responsible for the decreased expression of c-fos and activator protein-1 binding in T cells from patients with systemic lupus erythematosus, J Immunol 173 (2004) 3557–3563.

[49] M. Ahlmann, G. Varga, K. Sturm, R. Lippe, K. Benedyk, D. Viemann, et al., The cyclic AMP response element modulator {alpha} suppresses CD86 expression and APC function, J Immunol 182 (2009) 4167–4174.

[50] Y.T. Juang, Y. Wang, E.E. Solomou, Y. Li, C. Mawrin, K. Tenbrock, et al., Systemic lupus erythematosus serum IgG increases CREM binding to the IL-2 promoter and suppresses IL-2 production through CaMKIV, J Clin Invest 115 (2005) 996–1005.

[51] C.G. Katsiari, V.C. Kyttaris, Y.T. Juang, G.C. Tsokos, Protein phosphatase 2A is a negative regulator of IL-2 production in patients with systemic lupus erythematosus, J Clin Invest 115 (2005) 3193–3204.

[52] B.E. Wadzinski, W.H. Wheat, S. Jaspers, L.F. Peruski Jr., R.L. Lickteig, G.L. Johnson, et al., Nuclear protein phosphatase 2A dephosphorylates protein kinase A-phosphorylated CREB and regulates CREB transcriptional stimulation, Mol. Cell Biol 13 (1993) 2822–2834.

[53] K. Sunahori, Y.T. Juang, G.C. Tsokos, Methylation status of CpG islands flanking a cAMP response element motif on the protein phosphatase 2Ac alpha promoter determines CREB binding and activity, J Immunol 182 (2009) 1500–1508.

[54] E. Ballestar, M. Esteller, B.C. Richardson, The epigenetic face of systemic lupus erythematosus, J Immunol 176 (2006) 7143–7147.

[55] B. Richardson, L. Scheinbart, J. Strahler, L. Gross, S. Hanash, M. Johnson, Evidence for impaired T cell DNA methylation in systemic lupus erythematosus and rheumatoid arthritis, Arthritis Rheum 33 (1990) 1665–1673.

[56] L.S. Scheinbart, M.A. Johnson, L.A. Gross, S.R. Edelstein, B.C. Richardson, Procainamide inhibits DNA methyltransferase in a human T cell line, J Rheumatol 18 (1991) 530–534.

[57] C. Deng, Q. Lu, Z. Zhang, T. Rao, J. Attwood, R. Yung, et al., Hydralazine may induce autoimmunity by inhibiting extracellular signal-regulated kinase pathway signaling, Arthritis Rheum 48 (2003) 746–756.

[58] J. Quddus, K.J. Johnson, J. Gavalchin, E.P. Amento, C.E. Chrisp, R.L. Yung, et al., Treating activated CD4+ T cells with either of two distinct DNA methyltransferase inhibitors, 5-azacytidine or procainamide, is sufficient to cause a lupus-like disease in syngeneic mice, J Clin Invest 92 (1993) 38–53.

[59] C. Deng, M.J. Kaplan, J. Yang, D. Ray, Z. Zhang, W.J. McCune, et al., Decreased Ras-mitogen-activated protein kinase signaling may cause DNA hypomethylation in T lymphocytes from lupus patients, Arthritis Rheum 44 (2001) 397–407.

[60] A.H. Sawalha, M. Jeffries, R. Webb, Q. Lu, G. Gorelik, D. Ray, et al., Defective T-cell ERK signaling induces interferon-regulated gene expression and overexpression of methylation-sensitive genes similar to lupus patients, Genes Immun 9 (2008) 368–378.

[61] S. Garaud, D.C. Le, S. Jousse-Joulin, C. Hanrotel-Saliou, A. Saraux, R.A. Mageed, et al., IL-6 modulates CD5 expression in B cells from patients with lupus by regulating DNA methylation, J Immunol 182 (2009) 5623–5632.

[62] K. Tenbrock, Y.T. Juang, N. Leukert, J. Roth, G.C. Tsokos, The transcriptional repressor cAMP response element modulator alpha interacts with histone deacetylase 1 to repress promoter activity, J Immunol 177 (2006) 6159–6164.

[63] M. Martin, M. Potente, V. Janssens, D. Vertommen, J.C. Twizere, M.H. Rider, et al., Protein phosphatase 2A controls the activity of histone deacetylase 7 during T cell apoptosis and angiogenesis, Proc Natl Acad Sci USA 105 (2008) 4727–4732.

[64] N. Mishra, D.R. Brown, I.M. Olorenshaw, G.M. Kammer, Trichostatin A reverses skewed expression of CD154, interleukin-10, and interferon-gamma gene and protein expression in lupus T cells, Proc Natl Acad Sci USA 98 (2001) 2628–2633.

[65] C.T. Weaver, R.D. Hatton, P.R. Mangan, L.E. Harrington, IL-17 family cytokines and the expanding diversity of effector T cell lineages, Annu Rev Immunol 25 (2007) 821–852.

[66] M. Linker-Israeli, A.C. Bakke, R.C. Kitridou, S. Gendler, S. Gillis, D.A. Horwitz, Defective production of interleukin 1 and interleukin 2 in patients with systemic lupus erythematosus (SLE), J Immunol 130 (1983) 2651–2655.

[67] A. Ma, R. Koka, P. Burkett, Diverse functions of IL-2, IL-15, and IL-7 in lymphoid homeostasis, Annu. Rev Immunol 24 (2006) 657–679.

[68] R. Setoguchi, S. Hori, T. Takahashi, S. Sakaguchi, Homeostatic maintenance of natural Foxp3(+) CD25(+) CD4(+) regulatory T cells by interleukin (IL)-2 and induction of autoimmune disease by IL-2 neutralization, J Exp Med 201 (2005) 723–735.

[69] B. Kovacs, D. Vassilopoulos, S.A. Vogelgesang, G.C. Tsokos, Defective CD3-mediated cell death in activated T cells from patients with systemic lupus erythematosus: role of decreased intracellular TNF-alpha, Clin Immunol Immunopathol. 81 (1996) 293–302.

[70] L. Xu, L. Zhang, Y. Yi, H.K. Kang, S.K. Datta, Human lupus T cells resist inactivation and escape death by upregulating COX-2, Nat Med 10 (2004) 411–415.

[71] K.K. Hoyer, H. Dooms, L. Barron, A.K. Abbas, Interleukin-2 in the development and control of inflammatory disease, Immunol Rev 226 (2008) 19–28.

[72] B. Stockinger, Good for Goose, but not for Gander: IL-2 interferes with Th17 differentiation, Immunity 26 (2007) 278–279.

[73] K. Miyake, H. Nakashima, M. Akahoshi, Y. Inoue, S. Nagano, Y. Tanaka, et al., Genetically determined interferon-gamma production influences the histological phenotype of lupus nephritis, Rheumatology (Oxford) 41 (2002) 518–524.

[74] H. Nakashima, H. Inoue, M. Akahoshi, Y. Tanaka, K. Yamaoka, E. Ogami, et al., The combination of polymorphisms within interferon-gamma receptor 1 and receptor 2 associated with the risk of systemic lupus erythematosus, FEBS Lett 453 (1999) 187–190.

[75] G.C. Tsokos, D.T. Boumpas, P.L. Smith, J.Y. Djeu, J.E. Balow, A.H. Rook, Deficient gamma-interferon production in patients with systemic lupus erythematosus, Arthritis Rheum 29 (1986) 1210–1215.

[76] P.A. Neighbour, A.I. Grayzel, Interferon production of vitro by leucocytes from patients with systemic lupus erythematosus and rheumatoid arthritis, Clin Exp Immunol 45 (1981) 576–582.

[77] M. Harigai, M. Kawamoto, M. Hara, T. Kubota, N. Kamatani, N. Miyasaka, Excessive production of IFN-gamma in patients with systemic lupus erythematosus and its contribution to induction of B lymphocyte stimulator/B cell-activating factor/TNF ligand superfamily-13B, J Immunol 181 (2008) 2211–2219.

[78] D. Basu, Y. Liu, A. Wu, S. Yarlagadda, G.J. Gorelik, M.J. Kaplan, et al., Stimulatory and inhibitory killer Ig-like receptor molecules are expressed and functional on lupus T cells, J Immunol 183 (2009) 3481–3487.

[79] K. Masutani, M. Akahoshi, K. Tsuruya, M. Tokumoto, T. Ninomiya, T. Kohsaka, et al., Predominance of Th1 immune response in diffuse proliferative lupus nephritis, Arthritis Rheum 44 (2001) 2097–2106.

[80] W.S. Uhm, K. Na, G.W. Song, S.S. Jung, T. Lee, M.H. Park, et al., Cytokine balance in kidney tissue from lupus nephritis patients, Rheumatology (Oxford) 42 (2003) 935–938.

[81] R.W. Chan, F.M. Lai, E.K. Li, L.S. Tam, K.M. Chow, K.B. Lai, et al., Intrarenal cytokine gene expression in lupus nephritis, Ann Rheum Dis 66 (2007) 886–892.

[82] C. Haas, B. Ryffel, H.M. Le, IFN-gamma receptor deletion prevents autoantibody production and glomerulonephritis in lupus-prone (NZB x NZW)F1 mice, J Immunol 160 (1998) 3713–3718.

[83] A. Schwarting, T. Wada, K. Kinoshita, G. Tesch, V.R. Kelley, IFN-gamma receptor signaling is essential for the initiation, acceleration, and destruction of autoimmune kidney disease in MRL-Fas(lpr) mice, J Immunol 161 (1998) 494–503.

[84] D. Balomenos, R. Rumold, A.N. Theofilopoulos, Interferon-gamma is required for lupus-like disease and lymphoaccumulation in MRL-lpr mice, J Clin Invest 101 (1998) 364–371.

[85] A.W. Gibson, J.C. Edberg, J. Wu, R.G. Westendorp, T.W. Huizinga, R.P. Kimberly, Novel single nucleotide polymorphisms in the distal IL-10 promoter affect IL-10 production and enhance the risk of systemic lupus erythematosus, J Immunol 166 (2001) 3915–3922.

[86] L. Llorente, Y. Richaud-Patin, J. Couderc, D. arcon-Segovia, R. Ruiz-Soto, N. cocer-Castillejos, et al., Dysregulation of interleukin-10 production in relatives of patients with systemic lupus erythematosus, Arthritis Rheum 40 (1997) 1429–1435.

[87] F. Briere, C. Servet-Delprat, J.M. Bridon, J.M. Saint-Remy, J. Banchereau, Human interleukin 10 induces naive surface immunoglobulin D+ (sIgD+) B cells to secrete IgG1 and IgG3, J Exp Med 179 (1994) 757–762.

[88] L. Llorente, W. Zou, Y. Levy, Y. Richaud-Patin, J. Wijdenes, J. cocer-Varela, et al., Role of interleukin 10 in the B lymphocyte hyperactivity and autoantibody production of human systemic lupus erythematosus, J Exp Med 181 (1995) 839–844.

[89] L. Llorente, Y. Richaud-Patin, J. Wijdenes, J. cocer-Varela, M.C. Maillot, I. Durand-Gasselin, et al., Spontaneous production of interleukin-10 by B lymphocytes and monocytes in systemic lupus erythematosus, Eur Cytokine Netw 4 (1993) 421–427.

[90] H. Ishida, T. Muchamuel, S. Sakaguchi, S. Andrade, S. Menon, M. Howard, Continuous administration of anti-interleukin 10 antibodies delays onset of autoimmunity in NZB/W F1 mice, J Exp Med 179 (1994) 305–310.

[91] L. Llorente, Y. Richaud-Patin, C. Garcia-Padilla, E. Claret, J. Jakez-Ocampo, M.H. Cardiel, et al., Clinical and biologic effects of anti-interleukin-10 monoclonal antibody administration in systemic lupus erythematosus, Arthritis Rheum 43 (2000) 1790–1800.

[92] T. Korn, E. Bettelli, M. Oukka, V.K. Kuchroo, IL-17 and Th17 Cells. Annu, Rev Immunol 27 (2009) 485–517.

[93] A. Nalbandian, J.C. Crispin, G.C. Tsokos, Interleukin-17 and systemic lupus erythematosus: current concepts, Clin Exp Immunol 157 (2009) 209–215.

[94] A. Doreau, A. Belot, J. Bastid, B. Riche, M.C. Trescol-Biemont, B. Ranchin, et al., Interleukin 17 acts in synergy with B cell-activating factor to influence B cell biology and the pathophysiology of systemic lupus erythematosus, Nat Immunol 10 (2009) 778–785.

[95] J.C. Crispin, M. Oukka, G. Bayliss, R.A. Cohen, C.A. Van Beek, I.E. Stillman, et al., Expanded double negative T cells in patients with systemic lupus erythematosus produce IL-17 and infiltrate the kidneys, J Immunol 181 (2008) 8761–8766.

[96] J. Yang, Y. Chu, X. Yang, D. Gao, L. Zhu, X. Yang, et al., Th17 and natural Treg cell population dynamics in systemic lupus erythematosus, Arthritis Rheum 60 (2009) 1472–1483.

[97] H.K. Kang, M. Liu, S.K. Datta, Low-dose peptide tolerance therapy of lupus generates plasmacytoid dendritic cells that cause expansion of autoantigen-specific regulatory T cells and contraction of inflammatory Th17 cells, J Immunol 178 (2007) 7849–7858.

[98] Z. Zhang, V.C. Kyttaris, G.C. Tsokos, The role of IL-23/IL-17 axis in lupus nephritis, J Immunol 183 (2009) 3160–3169.

[99] H.Y. Wu, F.J. Quintana, H.L. Weiner, Nasal anti-CD3 antibody ameliorates lupus by inducing an IL-10-secreting CD4+, J Immunol 181 (2008) 6038–6050.

[100] A. Espinosa, V. Dardalhon, S. Brauner, A. Ambrosi, R. Higgs, F.J. Quintana, et al., Loss of the lupus autoantigen Ro52/Trim21 induces tissue inflammation and systemic autoimmunity by disregulating the IL-23-Th17 pathway, J Exp Med 206 (2009) 1661–1671.

[101] L. Yang, D.E. Anderson, C. Baecher-Allan, W.D. Hastings, E. Bettelli, M. Oukka, et al., IL-21 and TGF-beta are required for differentiation of human T(H)17 cells, Nature 454 (2008) 350–352.

[102] E. Bettelli, Y. Carrier, W. Gao, T. Korn, T.B. Strom, M. Oukka, et al., Reciprocal developmental pathways for the generation of pathogenic effector TH17 and regulatory T cells, Nature 441 (2006) 235−238.

[103] M. Linker-Israeli, R.J. Deans, D.J. Wallace, J. Prehn, T. Ozeri-Chen, J.R. Klinenberg, Elevated levels of endogenous IL-6 in systemic lupus erythematosus. A putative role in pathogenesis, J Immunol 147 (1991) 117−123.

[104] M. Linker-Israeli, D.J. Wallace, J. Prehn, D. Michael, M. Honda, K.D. Taylor, et al., Association of IL-6 gene alleles with systemic lupus erythematosus (SLE) and with elevated IL-6 expression, Genes Immun 1 (1999) 45−52.

[105] T. Harada, V. Kyttaris, Y. Li, Y.T. Juang, Y. Wang, G.C. Tsokos, Increased expression of STAT3 in SLE T cells contributes to enhanced chemokine-mediated cell migration, Autoimmunity 40 (2007) 1−8.

[106] D. Wang, S.A. John, J.L. Clements, D.H. Percy, K.P. Barton, L.A. Garrett-Sinha, Ets-1 deficiency leads to altered B cell differentiation, hyperresponsiveness to TLR9 and autoimmune disease, Int Immunol 17 (2005) 1179−1191.

[107] A. Awasthi, V.K. Kuchroo, Immunology. The yin and yang of follicular helper T cells, Science 325 (2009) 953−955.

[108] R.I. Nurieva, Y. Chung, D. Hwang, X.O. Yang, H.S. Kang, L. Ma, et al., Generation of T follicular helper cells is mediated by interleukin-21 but independent of T helper 1, 2, or 17 cell lineages, Immunity 29 (2008) 138−149.

[109] R.J. Johnston, A.C. Poholek, D. DiToro, I. Yusuf, D. Eto, B. Barnett, et al., Bcl6 and Blimp-1 are reciprocal and antagonistic regulators of T follicular helper cell differentiation, Science 325 (2009) 1006−1010.

[110] R.I. Nurieva, Y. Chung, G.J. Martinez, X.O. Yang, S. Tanaka, T.D. Matskevitch, et al., Bcl6 mediates the development of T follicular helper cells, Science 325 (2009) 1001−1005.

[111] J.M. Odegard, B.R. Marks, L.D. DiPlacido, A.C. Poholek, D.H. Kono, C. Dong, et al., ICOS-dependent extrafollicular helper T cells elicit IgG production via IL-21 in systemic autoimmunity, J Exp Med 205 (2008) 2873−2886.

[112] J.A. Bubier, T.J. Sproule, O. Foreman, R. Spolski, D.J. Shaffer, H.C. Morse III, et al., A critical role for IL-21 receptor signaling in the pathogenesis of systemic lupus erythematosus in BXSB-Yaa mice, Proc Natl Acad Sci USA 106 (2009) 1518−1523.

[113] Q. Chen, W. Yang, S. Gupta, P. Biswas, P. Smith, G. Bhagat, et al., IRF-4-binding protein inhibits interleukin-17 and interleukin-21 production by controlling the activity of IRF-4 transcription factor, Immunity 29 (2008) 899−911.

[114] J.C. Fanzo, W. Yang, S.Y. Jang, S. Gupta, Q. Chen, A. Siddiq, et al., Loss of IRF-4-binding protein leads to the spontaneous development of systemic autoimmunity, J Clin Invest 116 (2006) 703−714.

[115] M. Feuerer, J.A. Hill, D. Mathis, C. Benoist, Foxp3+ regulatory T cells: differentiation, specification, subphenotypes, Nat Immunol 10 (2009) 689−695.

[116] B.S. Le, R.S. Geha, IPEX and the role of Foxp3 in the development and function of human Tregs, J Clin Invest 116 (2006) 1473−1475.

[117] J.C. Crispin, M.I. Vargas, J. Alcocer-Varela, Immunoregulatory T cells in autoimmunity, Autoimmun. Rev 3 (2004) 45−51.

[118] J.C. Crispin, A. Martinez, J. Alcocer-Varela, Quantification of regulatory T cells in patients with systemic lupus erythematosus, J Autoimmun 21 (2003) 273−276.

[119] M. Miyara, Z. Amoura, C. Parizot, C. Badoual, K. Dorgham, S. Trad, et al., Global natural regulatory T cell depletion in active systemic lupus erythematosus, J Immunol 175 (2005) 8392−8400.

[120] J.H. Lee, L.C. Wang, Y.T. Lin, Y.H. Yang, D.T. Lin, B.L. Chiang, Inverse correlation between CD4+ regulatory T-cell population

and autoantibody levels in paediatric patients with systemic lupus erythematosus, Immunology 117 (2006) 280−286.

[121] X. Valencia, C. Yarboro, G. Illei, P.E. Lipsky, Deficient CD4+CD25high T regulatory cell function in patients with active systemic lupus erythematosus, J Immunol 178 (2007) 2579−2588.

[122] M. Bonelli, A. Savitskaya, D.K. von, C.W. Steiner, D. Aletaha, J.S. Smolen, C. Scheinecker, Quantitative and qualitative deficiencies of regulatory T cells in patients with systemic lupus erythematosus (SLE), Int Immunol 20 (2008) 861−868.

[123] M.I. Vargas-Rojas, J.C. Crispin, Y. Richaud-Patin, J. cocer-Varela, Quantitative and qualitative normal regulatory T cells are not capable of inducing suppression in SLE patients due to T-cell resistance, Lupus 17 (2008) 289−294.

[124] C.R. Monk, M. Spachidou, F. Rovis, E. Leung, M. Botto, R.I. Lechler, et al., MRL/Mp CD4+CD25- T cells show reduced sensitivity to suppression by CD4+CD25+ regulatory T cells in vitro: a novel defect of T cell regulation in systemic lupus erythematosus, Arthritis Rheum 52 (2005) 1180−1184.

[125] S. Wan, C. Xia, L. Morel, IL-6 produced by dendritic cells from lupus-prone mice inhibits CD4+CD25+ T cell regulatory functions, J Immunol 178 (2007) 271−279.

[126] N. Iikuni, E.V. Lourenco, B.H. Hahn, C.A. La, Cutting edge: Regulatory T cells directly suppress B cells in systemic lupus erythematosus, J Immunol 183 (2009) 1518−1522.

[127] P. Blanco, J.F. Viallard, J.L. Pellegrin, J.F. Moreau, Cytotoxic T lymphocytes and autoimmunity, Curr. Opin. Rheumatol. 17 (2005) 731−734.

[128] J.F. Viallard, C. Bloch-Michel, M. Neau-Cransac, J.L. Taupin, S. Garrigue, V. Miossec, et al., HLA-DR expression on lymphocyte subsets as a marker of disease activity in patients with systemic lupus erythematosus, Clin Exp Immunol 125 (2001) 485−491.

[129] P. Blanco, V. Pitard, J.F. Viallard, J.L. Taupin, J.L. Pellegrin, J.F. Moreau, Increase in activated CD8+ T lymphocytes expressing perforin and granzyme B correlates with disease activity in patients with systemic lupus erythematosus, Arthritis Rheum 52 (2005) 201−211.

[130] L. Couzi, P. Merville, C. Deminiere, J.F. Moreau, C. Combe, J.L. Pellegrin, et al., Predominance of CD8+ T lymphocytes among periglomerular infiltrating cells and link to the prognosis of class III and class IV lupus nephritis, Arthritis Rheum 56 (2007) 2362−2370.

[131] W. Stohl, Impaired polyclonal T cell cytolytic activity. A possible risk factor for systemic lupus erythematosus, Arthritis Rheum 38 (1995) 506−516.

[132] I. Puliaeva, R. Puliaev, C.S. Via, Therapeutic potential of CD8+ cytotoxic T lymphocytes in SLE, Autoimmun. Rev 8 (2009) 219−223.

[133] S.L. Peng, J. Moslehi, M.E. Robert, J. Craft, Perforin protects against autoimmunity in lupus-prone mice, J Immunol 160 (1998) 652−660.

[134] J.C. Crispin, G.C. Tsokos, Human TCR-ab+ CD4- CD8- T cells can derive from CD8+ T cells and display an inflammatory effector phenotype, J Immunol 183 (2009) 4675−4681.

[135] Y.W. Park, S.J. Kee, Y.N. Cho, E.H. Lee, H.Y. Lee, E.M. Kim, et al., Impaired differentiation and cytotoxicity of natural killer cells in systemic lupus erythematosus, Arthritis Rheum 60 (2009) 1753−1763.

[136] G.C. Tsokos, Overview of Cellular Immune Function in Systemic Lupus Erythematosus, in: R.G. Lahita (Ed.), Systemic Lupus Erythematosus, fourth ed., Elsevier, Amsterdam, 2004, pp. 29−92.

[137] S. Shivakumar, G.C. Tsokos, S.K. Datta, T cell receptor alpha/beta expressing double-negative (CD4-/CD8-) and CD4+ T

helper cells in humans augment the production of pathogenic anti-DNA autoantibodies associated with lupus nephritis, J Immunol 143 (1989) 103−112.

[138] T. Masuda, T. Ohteki, T. Abo, S. Seki, S. Nose, H. Nagura, et al., Expansion of the population of double negative CD4-8- T alpha beta-cells in the liver is a common feature of autoimmune mice, J Immunol 147 (1991) 2907−2912.

[139] G.S. Dean, A. Anand, A. Blofeld, D.A. Isenberg, P.M. Lydyard, Characterization of CD3+ CD4- CD8- (double negative) T cells in patients with systemic lupus erythematosus: production of IL-4, Lupus 11 (2002) 501−507.

[140] P.A. Sieling, S.A. Porcelli, B.T. Duong, F. Spada, B.R. Bloom, B. Diamond, et al., Human double-negative T cells in systemic lupus erythematosus provide help for IgG and are restricted by CD1c, J Immunol 165 (2000) 5338−5344.

[141] P. Estess, H.C. DeGrendele, V. Pascual, M.H. Siegelman, Functional activation of lymphocyte CD44 in peripheral blood is a marker of autoimmune disease activity, J Clin Invest 102 (1998) 1173−1182.

[142] J.C. Crispin, B. Keenan, M.D. Finnell, B.L. Bermas, P. Schur, E. Massarotti, et al., Expression of CD44v3 and CD44v6 isoforms is increased on T cells from patients with systemic lupus erythematosus and correlates with disease activity, Arthritis Rheum 62 (2010) 1431−1437.

[143] R.A. Cohen, G. Bayliss, J.C. Crispin, G.F. Kane-Wanger, C.A. Van Beek, V.C. Kyttaris, et al., T cells and in situ cryoglobulin deposition in the pathogenesis of lupus nephritis, Clin Immunol 128 (2008) 1−7.

[144] T. Takeuchi, M. Pang, K. Amano, J. Koide, T. Abe, Reduced protein tyrosine phosphatase (PTPase) activity of CD45 on peripheral blood lymphocytes in patients with systemic lupus erythematosus (SLE), Clin Exp Immunol 109 (1997) 20−26.

[145] H.K. Wong, G.M. Kammer, G. Dennis, G.C. Tsokos, Abnormal NF-kappa B activity in T lymphocytes from patients with systemic lupus erythematosus is associated with decreased p65-RelA protein expression, J Immunol 163 (1999) 1682−1689.

[146] Q. Lu, M. Kaplan, D. Ray, D. Ray, S. Zacharek, D. Gutsch, et al., Demethylation of ITGAL (CD11a) regulatory sequences in systemic lupus erythematosus, Arthritis Rheum 46 (2002) 1282−1291.

[147] M.J. Kaplan, Q. Lu, A. Wu, J. Attwood, B. Richardson, Demethylation of promoter regulatory elements contributes to perforin overexpression in CD4+ lupus T cells, J Immunol 172 (2004) 3652−3661.

[148] Q. Lu, A. Wu, B.C. Richardson, Demethylation of the same promoter sequence increases CD70 expression in lupus T cells and T cells treated with lupus-inducing drugs, J Immunol 174 (2005) 6212−6219.

[149] Q. Lu, A. Wu, L. Tesmer, D. Ray, N. Yousif, B. Richardson, Demethylation of CD40LG on the inactive X in T cells from women with lupus, J Immunol 179 (2007) 6352−6358.

[150] D.A. Horwitz, J.D. Gray, S.C. Behrendsen, M. Kubin, M. Rengaraju, K. Ohtsuka, et al., Decreased production of interleukin-12 and other Th1-type cytokines in patients with recent-onset systemic lupus erythematosus, Arthritis Rheum 41 (1998) 838−844.

8

B-Lymphocyte Biology in SLE

Mark J. Mamula

INTRODUCTION

The diagnostic hallmark of systemic lupus erythematosus (SLE) is the presence of autoantibody responses to a variety of intracellular macromolecules. High-titer, IgG class autoantibodies are the product of both B- and T-lymphocyte interactions that lead to the myriad of tissue pathologies observed in clinical SLE. This chapter will focus on the fundamental biology of B-lymphocyte subsets and their functions, including key events in the selection and tolerance of B cells. We will review the role of B cells as antigen-presenting cells (APCs) in epitope spreading, and their communications with other immune-derived cells, in particular, T cells and dendritic cells. The important features of lupus autoantigens, including post-translational protein modifications and Toll-like receptor (TLR) induction, will be discussed in the context of triggering autoreactive B-cell responses. Finally, murine models of SLE have provided important paradigms for the development of B-cell-directed immunotherapies currently in use for human SLE. We will provide an update of the successes and failures of B-cell-specific therapies in SLE.

FUNDAMENTAL FEATURES OF B-CELL IMMUNE RESPONSES

Clearly, the immune system evolved to protect the host from foreign pathogens. Lupus autoimmunity is the product of both normal intracellular pathways as well as abnormal features that may not be found in the course of immunity to foreign pathogens. B cells have a variety of functions beyond that of antibody secretion (Table 8.1), some of which will be described herein or in other chapters. As with other immune cells, the potential for B-cell autoimmunity resides in early precursor cells in the bone marrow. Again, similar to T cells, the immature repertoire is replete with autoreactive B cells that

may be programmed to die, controlled by suppressor mechanisms, exist without further development, or develop into autoantibody-secreting plasma cells and autoreactive memory cells.

The selection of B cells, their pathway to development or death, is a product of the strength of signal to the B-cell receptor (BCR) and co-stimulatory receptors, its response to BCR signals (death or development), and the presence of cytokines that influence these pathways. Among many potential immune aberrations, lupus is marked by uncontrolled B-cell activation that may be a result of genetic predisposition, the presence of aberrant help from CD4 T cells, or altered cytokine milieu (Table 8.2). Human SLE is marked by abnormal homeostasis of circulating B cells, B-cell lymphopenia and increases in plasma cells (CD19+, CD20-, CD21+, CD38+) and plasmablasts, coincident with the severity of disease (SLEDAI scores) and increased autoantibody titers [1–4]. Two other subsets of B cells, known as transitional B cells and CD38+ pre-germinal B cells are similarly expanded in human SLE (Table 8.2).

TABLE 8.1 General B-lymphocyte functions

Antibody (autoantibody) synthesis dependent or independent of T cell help
Germinal center formation
Antigen presentation to T cells
Activation and/or polarization of T-cell subsets
Antigen transport to dendritic cells and macrophages
Epitope spreading to pathogens and autoantigens
Secretion of antigen/MHC-containing exosomes
Cytokine secretion (IFN-γ, LTα, LTβ, BAFF/BLyS, IL-1β, IL-2, IL-6, IL-10, IL-12, IL-23, IL-27)
Chemokine secretion (MIP1α, MIP1β)

TABLE 8.2   B-cell abnormalities in human SLE or in murine models

Aberrant receptor editing/deletion functions (increased autoreactive B-cell precursors)

Aberrant intracellular signaling

Altered cytokine milieu (increased BAFF/BLyS, IL-6, IFNα)

Increased T-cell helper functions for B cells

Altered B-cell subsets (increased B-1 B cells, plasmablasts, and plasma cells)

Altered B-cell surface proteins (increased CD80/86 co-stimulation, increased CD40L, decreased CR2)

TABLE 8.3   Biological properties of B-1 and B-2 B cells

	B-1 B cells	B-2 B cells
Origins/development	Fetal tissue/fetal liver	Bone marrow/periphery
Variable region diversity	Restricted	Highly diverse
Location	Peritoneum/pleural cavity	Lymph nodes/spleen
T cell help requirement	None	High
Isotype of Ig secretion	IgM; little IgG	IgG in mature cells
Response to carbohydrates	High	Low
Response to protein Ag	Low	High
Degree of somatic mutation	Little/none	Extensive
Memory cell development	Little/none	Extensive

As discussed below, various soluble factors such as B-lymphocyte stimulator (BLyS) and type 1 interferons (IFN), both of which cause expansion of B-cell subsets, are increased in human SLE [5–14]. IFN-α is known to induce CD38 expression on B cells that inhabit the germinal center while also causing naïve B-cell lymphopenia due to its suppressive effects on bone marrow generation of pre-B cells [5]. These B-cell abnormalities are linked to the duration of SLE flares. Existing non-specific immunosuppressive therapy, such as prednisone, does lower many of the abnormal B-cell subset numbers. However, plasma cell subsets are frequently unaltered over the course of any therapies in SLE. What is clear is that the autoimmune repertoire of B cells has become unregulated such that milligram per milliliter quantities of autoantibody may arise in the serum of individual patients. These levels of isotype switched, affinity mature autoantibodies represent a major activation of the B-cell repertoire even beyond that which is found in immunity to foreign antigens.

B lymphocytes have been subdivided into phenotypic groups based on biological activities, activation thresholds of surface receptors, their physiologic location, and their unique surface proteins (Table 8.3). B-cell subsets are generalized as 'fetal' versus 'adult' lineage B cells, B-1 versus B-2 cells, respectively. It is not yet clear whether these B-cell subsets originate from common precursor bone marrow cells. B-1 B cells, so-called CD5+ B cells in humans, are a unique subset of B-cell lineage that predominates immunity in fetal tissues. B-1 cells originate in the fetal liver and populate unique physiologic sites such as the peritoneum and the pleural cavity of mice [6–13]. They constitute less than 20% of the circulating B cell pool in adults and express primarily surface IgM as receptors, with little surface IgD. B-1 cells are self-perpetuating, long-lived cells which secrete highly cross-reactive and low-affinity IgM isotype antibody, so-called "natural antibodies". They contribute little to the adaptive (B-2 cell) immune response. IgM from B-1 B cells binds carbohydrate antigens as well as microbial surface antigens that are often cross-reactive with self.

What is unique about the B-1 subset of cells is in their low BCR signaling threshold and responses compared to other B-cell subsets. Signaling events that dictate positive selection and the induction of tolerance are altered compared to the adult B-2 B-cell population. B-1 cells appear to require a greater threshold of BCR activation for the induction of tolerance in the periphery. In the B-1 B-cell subset, low-affinity cross-linking of BCR by large multivalent antigens provides activation signals in the absence of conventional CD4 T cell help [14, 15]. Indeed, it has been postulated that these polyreactive, low-affinity IgM antibodies frequently bind self proteins in extracellular spaces and protect them from recognition from B-2 B cells. In this manner, it has been postulated that B-1 B-cell responses may regulate or suppress autoimmunity in the B-2 B-cell compartment.

B-1 B cells likely have altered thresholds for triggering receptor editing, as compared to B-2 cells (discussed below). Where low-affinity BCR engagement may normally promote receptor editing in a B-2 cell, this same antigen binding to a low-affinity BCR of B-1 cells may instead promote survival and IgM secretion [14, 16]. Increased numbers of B-1 B cells have been found in some studies of murine SLE and in human autoimmune disease [8–10]. Both B-1 and B-2 cells synthesize DNA-binding autoantibodies typical of SLE. Elimination of B-1 cells will reduce autoantibody levels in some murine models, suggesting a role of B-1 cells at some level of SLE development. As discussed below, B cells have several functions independent of simple antibody production that regulate immune and autoimmune responses. Among these critical B-cell functions is their ability to bind specific autoantigens and activate naïve T cells, their ability to deliver antigen to other professional APCs, such as macrophages and dendritic cells, and their ability to secrete inflammatory cytokines (Table 8.1).

B-CELL TOLERANCE AND RECEPTOR EDITING

The specificity of B lymphocytes arises from a semi-random rearrangement of genes, frequently leading to self-reactive or even non-functional B-cell receptors. Neither of these latter outcomes is useful to the host and mechanisms have evolved to eliminate or repair autoreactive B cells. B-cell tolerance to self proteins is maintained by either the induction of cellular anergy (inactivation of B cells), the active deletion of cells, or the editing of B-cell receptors. In the latter process, the high frequency of autoreactive immature B cells is corrected at the level of DNA gene rearrangements in a process known as "receptor editing". This process occurs when the specificity of the BCR changes from autoreactive to non-autoreactive without eliminating the cell itself. Receptor editing is the most important tolerance mechanism of developing immature B cells in the bone marrow and has been thought to occur in as many as 30–60% of developing B cells [17–19].

As reviewed elsewhere [20], the BCR is composed of rearranged heavy-chain (IgH) and light-chain (IgL) polypeptides with singular specificity unique to each B cell. BCRs that engage self proteins are developmentally blocked and genes for IgL and possibly IgH in the same cell are rearranged to confer new antigenic specificity. Receptor editing processes were first understood from studies of Ig transgenic mice with lupus-like anti-DNA specificity and from B cells specific for MHC class I proteins [21–23]. The latter model allowed analysis of B-cell editing in the presence or absence of the autoantigen, H2b/k or H2d respectively. In the presence of the autoantigen, B cells continually upregulated Rag1 and Rag2 proteins and rearranged IgL chains until autoimmune specificity disappeared. A single IgL rearrangement paired with the original IgH removed autoantigen specificity. The repertoire consisted of B cells with BCR rescued from previously autoreactive immature cells. In the absence of antigen, secondary IgL rearrangements were inhibited resulting in virtually all H2d-specific B cells in the mature repertoire. Once in the peripheral lymphoid compartments, B cells are largely regulated by programmed cell death (apoptosis) upon encounter with self antigen. However,

autoantigens that may exist in low copy or where BCR may not be cross-linked or not bind with sufficient affinity, will not provide sufficient signals for editing to occur. Autoreactive B cells that fail to edit are subject to control in the periphery such as anergy. Receptor editing, however, is not foolproof in that allelic exclusion (the exclusion of the other light-chain allele) may not occur resulting in B cells that express two L chains, one of which may be autoreactive. In murine models of SLE, there is some evidence that receptor editing may, in fact, rearrange to enhance autoimmune receptors [24–26].

TOLL-LIKE RECEPTOR (TLR) CONTROL OF B-CELL ACTIVATION IN LUPUS

Unlike most other autoimmune syndromes, B cells in SLE target a variety of nucleic-acid-containing macromolecules. This feature of lupus autoantigens has remained curious since it is not understood how BCRs and/or TCRs bind to nucleic acid structures. The potential for Toll-like receptor (TLR)-mediated signaling of lupus autoimmune immune cells became appealing as the biology of these receptors has evolved over the past decade [27–31]. In brief, TLRs evolved as cell-associated receptors for the recognition of foreign pathogens. Originally defined as pattern recognition receptors (PRRs), these proteins are found on cells of both the innate and adaptive immune systems. Relevant to the present topic, TLRs are highly expressed on human B-cell subsets (Table 8.4). TLRs exist as early recognition receptors that signal the presence of microbial and viral infection for the rapid activation of immune responses. Unfortunately, these same pathways designed to recognize pathogens have also been found to respond to components of "self". In particular, DNA and RNA of foreign invaders, both bacterial and viral, trigger strong

TABLE 8.4   TLR Expression in Human B cells

TLR	Presence on B cells	TLR ligand
TLR1	yes +++ (surface)	Lipopeptides (with TLR2)
TLR2	yes ++ (surface)	Lipopeptides
TLR3	yes + (endosome)	dsRNA
TLR4	some +/- (surface)	LPS
TLR5	yes + (surface)	Flagellin
TLR6	yes ++ (surface)	Lipopeptides (with TLR2)
TLR7	yes + (endosome)	ssRNA
TLR8	some +/- (endosome)	ssRNA
TLR9	yes ++ (endosome)	CpG motif DNA

signals to TLR 7, 8, and 9 [32—34]. Recent *in vitro* and *in vivo* studies have found that endogenous (host) nucleic acids will also trigger these same receptors and constitute an important pathway in lupus autoimmunity. The source of host nucleic acids in activating TLRs is not clearly understood, although chromatin or ribonucleoproteins from surface blebs of apoptotic cells, apoptotic debris, tissue or cell pathology, may be among the more obvious candidates. As discussed later, tissue inflammation and cell death caused by lupus autoimmune responses may serve to feed the cycle of TLR-based activation, leading to multiple rounds of immune cell activation. In reality, lupus autoimmunity can be considered as having a relatively restricted B- and T-cell autoimmune response in that nucleic-acid-containing autoantigens are the principal targets of immune cells.

A second appealing hypothesis is in the possibility that TLR-based activation may indeed arise in response to nucleic acids of foreign infections that trigger anti-self response arising from the evolutionary conservation of nucleic acid and chromatin components. Studies from our laboratory and others have defined the promiscuity, or cross-reactivity, of lupus autoantibodies in binding ribonucleoprotein particles of both prokaryotic and eukaryotic origin [35]. Other studies demonstrate evidence for the immunologic cross-reactivity between the Epstein-Barr virus EBNA-1 and ribonucleoprotein lupus autoantigens [36—40]. In this scenario, B cells are activated against the EBNA-1 protein. BCRs for this EBNA-1 response may then bind, process and present the snRNPs and/or Ro/SSA ribonucleoproteins in the synthesis of autoantibodies, B-cell epitope spreading, and in the priming of autoreactive T cells. As a theme of this review, this mechanism defines a continuing amplification loop in the spreading of autoimmunity to other ribonucleoprotein determinants. While molecular mimicry is an attractive theoretical mechanism, it does not fully explain the heterogeneity of human SLE autoimmune responses, nor does it explain the selectivity of autoimmunity to nucleic acid macromolecules found in spontaneous murine models of SLE.

TLRs that have received the most attention in lupus autoimmunity are TLR9 and TLR7. In contrast to other TLR family proteins, TLR9 and TLR7, along with TLR3 and TLR8, are found in cytoplasmic components of the endoplasmic reticulum and endosomal/lysosomal vesicles in the cells in which they reside. Interestingly, lupus therapeutic drugs that prevent acidification of intracellular organelles, such as chloroquine, also interfere with signaling through TLR9 and TLR7. Other TLR members are found on the external side of the plasma membrane.

Both bacterial and viral DNA will signal TLR9 [41, 42]. Similarly, TLR7 ligands include both foreign and self RNA, and viral and self ssRNA sequences [43—45]. Important to the etiology of SLE, TLR9 and TLR7 are expressed in B cells and plasmacytoid dendritic cells (Table 8.4). Recent seminal studies have revealed the importance of TLRs in the development of lupus autoantibodies and tissue pathology [34, 45, 46]. These studies were performed by the examination of immune complexes of antichromatin autoantibody with host cell chromatin, acquired from dead or dying cells. A clever immunoglobulin transgenic model was employed, termed 'AM14'. In this model Ig transgenes were derived from lupus-prone MRL mice in which B-cell receptors were of rheumatoid factor specificity, binding IgG2a of the 'a' allotype. These B-cell Ig transgenics remain quiescent on non-autoimmune strains of mice while these same B cells on autoimmune-prone background mice will develop into plasma cells with autoantibody production [47, 48]. *In vitro* studies demonstrated that autoantigen immune complexes would stimulate B cells while foreign protein ICs would fail to do so [46]. Treatment of chromatin ICs with DNAse eliminated this response and failed to stimulate B cells into proliferation. Finally, it was demonstrated that chromatin ICs required signaling through MyD88, a pathway downstream of TLR9-mediated triggering. These studies were the first to indicate that autoantigens require stimulation through pattern recognition proteins on B lymphocytes that led to autoantibody synthesis. The fact that TLR9 resides inside the B-cell compartment showed that BCR engagement as well as internalization and binding of TLR9 is important in the activation of autoantibody-producing B cells. Later studies demonstrated that B cells with BCRs that directly bind chromatin could similarly activate autoantibody-producing cells [34].

In a similar manner as that described above, ribonucleoprotein immune complexes bound to Ig were found to signal through the RNA sentinel, TLR7 [49]. Studies demonstrated that IFN-α, a cytokine that enhances TLR7 expression, also increased signaling and activation of B cells, while RNAse treatment of the IC dampened B-cell activation. In both models of TLR9- and TLR7-mediated B-cell activation, oligodeoxynucleotides (ODNs) that bind and block these TLRs similarly blocked activation of the B cells. As noted above, chloroquine, an agent that interferes with intracellular acidification of endosomal compartments, also inhibited TLR9 and TLR7 signaling by either DNA or RNA immune complexes.

Together, these studies illustrate a novel mechanism by which B cells are activated in lupus autoimmunity. In essence, B-cell surface Ig receptor is engaged and crosslinked by chromatin or ribonucleoprotein particles, or immune complexes of these particles (Figure 8.1). The macromolecular complexes are brought inside of the cell where TLR9 or TLR7 is engaged, sending an amplifying

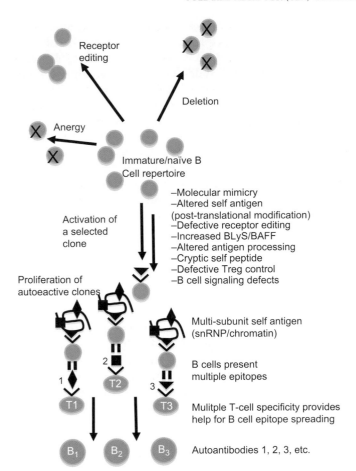

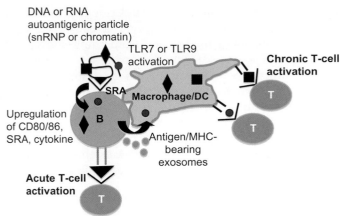

FIGURE 8.2 Interactions of antigen-presenting cell subsets in autoimmunity. As defined in Figure 8.1 and in the text, B cells are important primary APCs in the induction of autoreactive T cells. B cells rapidly bind autoantigens, if present, and activate an acute T-cell response. Indeed, the uptake of nucleic-acid-containing autoantigens, snRNPs or chromatin, activate cells via TLR7 or TLR9, respectively. TLR activation along with BCR cross-linking upregulates a number of processes in the B cell including proliferation, autoantibody synthesis, the production of exosomes, and the expression of cytokines and cell surface molecules including CD80/86. B-cell contact or proximity with other APCs including both human macrophages or DCs, allows the transfer of autoantigen to these latter APCs. TLR activation increases the expression of scavenger receptor A (SRA) and enhances the adhesion and transfer of autoantigens between APCs. Animal models support a role for transferred autoantigen in maintaining chronic activation of autoreactive T-cell subsets.

FIGURE 8.1 B-cell antigen-presenting functions in epitope spreading. Immature B cells in the bone marrow or naïve B cells in the periphery undergo several mechanisms of purging that typically eliminate self reactivity. Encounter with self protein provokes deletion of B cells or conversion to anergy, leading to non-responsiveness to most autoantigens. Alternatively, receptor editing will reprogram B-cell receptor specificity away from binding self proteins (as described in detail in the text). These mechanisms of deleting self-reactive B cells are not perfect and some autoreactive B-cell clones are always found in the periphery. A number of conditions may allow the activation of individual B-cell clones, including encounter with a molecular mimic or post-translationally modified self protein, increased B-cell growth factors (BLyS), or even defects in B-cell signaling functions. These factors lead to a restricted clonal activation of B cells and autoantibody synthesis. Once these B-cell clones have expanded and come in contact with some source of self autoantigen, such as debris from apoptotic or necrotic cells, T-cell activation occurs. Multisubunit autoantigens, such as snRNPs or chromatin are processed by B cells in presenting multiple epitopes of the autoantigen in activating multiple specificities of T cells. T cells, in turn, provide helper functions to successive rounds of B-cell activation (B_1, B_2, B_3, etc.) in the production of autoantibodies to multiple successive determinants. The chronic autoimmune response likely cycles through many rounds of amplification and epitope spreading in this mechanism.

signal to the nucleus of the B cell, whereby plasma cell differentiation occurs leading to antibody synthesis and secretion. As described below, the B cell may also then act in an antigen-processing and presentation fashion, in now presenting multiple peptide

determinants of histones or ribonucleoproteins in amplifying the lupus autoimmune response and in epitope spreading (Figures 8.1 and 8.2). The fact that the normal immune repertoire appears to be replete with autoreactive B cells specific for chromatin presents potential problems in the precarious balance of B-cell regulatory mechanisms versus B-cell autoimmunity. Secondly, nucleic-acid-containing ICs are highly stimulatory for the production of IFN-α from plasmacytoid DCs, cells that possess TLR9 and TLR7 [50–54]. The IFN-α signature in human SLE has been well-characterized and the TLR-based signaling pathway may additionally contribute to this signature. ICs can bind to low-affinity Fc receptors (FcγRIIa; CD32) on pDCs, triggering IFN-α and BLyS/BAFF production from these cells, both of which promote B-cell growth and differentiation. Inhibitors or blocking antibodies for FcγRIIa will inhibit IFN-α production. Importantly, this pathway may activate B-cell responses in a T-cell-independent manner. It is becoming clear that lupus autoimmunity develops with B-cell responses that may not require conventional T-cell helper functions while other stages of B-cell responses may require, or be amplified by CD4 T cells.

Several features of this pathway have yet to be fully understood. For example, the source of autoantigen in lupus B- and T-cell autoimmunity, either chromatin or ribonucleoproteins, have yet to be defined. Debris from cell death or surface blebs of apoptotic cells are candidates for sources of nucleic-acid-containing complexes although it is unclear whether patients with lupus have abnormal levels or numbers of dying cells that predispose them to autoimmunity [46]. Moreover, the fact that immune complexes trigger B-cell activation implies that autoantibodies have been already made by other existing autoreactive B cells (plasma cells) in order to form immune complexes. Certainly the question of original antigenic sin in triggering lupus autoimmunity is far from resolved.

Studies of the roles of TLR in lupus autoimmunity have been supported in murine models of SLE in which these TLRs, 7 or 9, or their signaling pathways, Myd88 have been genetically eliminated. Recent studies utilized TLR9-deficient animals crossed to the MRL/lpr spontaneous model of SLE [34, 55, 56]. MRL mice deficient in TLR9 were found to lack antinuclear antibodies with homogeneous staining patterns consistent with the lack of anti-DNA autoantibodies. However, a significant fraction of the TLR9-deficient mice still generate speckled nuclear autoantibodies as well as antibodies that bind components in the cytoplasm of HEp-2 cells. Speckled nuclear ANA patterns are consistent with autoantibody production to snRNP and/or other ribonucleoprotein particles that are generated in these mice by binding TLR7 remaining in the cells. Interestingly, although TLR9-/- mice have reduced anti-DNA and antichromatin autoantibodies, mice still exhibit significant tissue pathology compared to wild-type lupus-prone animals. MRL/lpr TLR9-/- mice have increased serum Ig levels, enhanced lymphadenopathy and splenomegaly, lymphocyte activation, and increased kidney disease, all leading to significantly reduced survival.

In nearly an opposite manner to TLR9-/- MRL/lpr animals, TLR7 deficiency on the same background still exhibit anti-DNA/chromatin autoantibodies but fail to generate antiribonucleoprotein autoantibodies, anti-Sm/RNP [55, 56]. Some isotypes of Ig are reduced, IgG2a and IgG3, and kidney pathology is somewhat reduced. Overall, studies suggest that TLR9 is critical to antichromatin/anti-DNA production by B cells while TLR7 is required for anti-Sm/RNP autoantibody synthesis [45, 55, 56].

The paradox of how TLR9 deficiency may alter the development of SLE remains under study. It has been hypothesized that the protective role of TLR9 deficiency may be related to the presence of IgM class antichromatin autoantibodies that serve to clear apoptotic chromatin debris, preventing its processing and presentation to the immune system. Alternatively, it has been posited that TLR9 engagement may also initiate anti-inflammatory cytokine production by cells, thereby dampening the pathology associated with SLE.

One recent study, by Nickerson et al., has attempted to resolve this paradox of TLR-mediated models of SLE by developing TLR9/TLR7 double-deficient MRL/lpr animals as well as Myd88-deficient MRL/lpr animals in parallel comparative studies [56]. The authors concluded that TLR9 acts as an upstream suppressor of TLR7-dependent autoantibody synthesis (anti-Sm/RNP autoantibodies). Interestingly, the exacerbated lupus autoimmunity associated with TLR9 deficiency disappeared when TLR7 was also absent in the mice.

Taken together, these observations suggest that TLR9 and TLR7 interact in a co-regulatory manner that controls the production of both antichromatin and antiribonucleoprotein antibodies. The data also demonstrated that no other TLR family proteins were implicated in lupus autoantibody production by B cells although some aspects of murine lupus are independent of any TLR-based signaling. TLR9/7 double-deficient mice and Myd88-/- mice still exhibit dermatitis as well as splenomegaly and lymphadenopathy relative to non-autoimmune-prone mice. What is to be learned from these observations? It is apparent that some elements of B-cell responses in SLE are independent of traditional T-cell helper functions and relies on innate TLR receptor engagement. However, B cells that produce lupus autoantibodies also are subject to conventional T-cell help at some stage, or microenvironment with lymphoid tissues. Further studies will be required to resolve the interactive roles of TLR9 and TLR7 in activating B cells for autoantibody production in the presence of both chromatin and ribonucleoprotein complexes from dead or dying cells.

In contrast, there may also be a role for TLR engagement for the elimination of self-reactive B cells [57]. These provocative studies were performed with B cells from human patients deficient in either IRAK-4 (IL-1R associated kinase), MyD88 (myeloid differentiation factor), or UNC-93B, all downstream signaling complexes of TLRs. In general, these patients all experienced defective central and peripheral B-cell tolerance mechanisms, leading to expanded populations of autoreactive mature B cells in the blood. Single B-cell clones from these patients were examined for autoreactivity to nuclear autoantigens, indicating expanded lupus autoantibodies arise from defective signaling through these pathways. These studies implied that TLR7 and/or TLR9 may play a role in the establishment of B-cell tolerance.

B CELLS IN ANTIGEN PRESENTATION AND EPITOPE SPREADING IN AUTOIMMUNE RESPONSES

A number of studies indicate that the highly diverse autoimmune response in SLE can originate from a single protein or even a single cryptic self peptide of intracellular autoantigens without the need for foreign pathogens or molecular mimics [58–63]. As would be the logical prediction of a normally functioning immune system, mice are unresponsive at both the B- and T-cell levels to immunization with lupus autoantigens, including chromatin and self snRNPs as well as to immunization with most other self proteins. However, mice immunized with either of two cryptic self peptides of the D protein of murine snRNPs elicited strong autoreactive B cells and T cells and autoantibodies to the D protein. If the immune system is then exposed to the self snRNP particle after priming with cryptic self peptides, autoantibody responses arise in a reproducible progression through multiple determinants on other self proteins (the 70K and A proteins). The autoantibodies produced in this artificial system reveal a speckled immunofluorescent pattern (ANAs) resembling patterns found in spontaneous human and murine lupus. This autoimmune response is thus not a collection of independent responses to the various snRNP component proteins but is instead a highly organized development to successive determinants on the target antigen. The mechanisms that may lead to immunologic diversity have been aided by an understanding of novel ways in which self antigens can be processed and presented to the immune system (Figure 8.1). A practical example of these murine studies has been performed in studies of serum autoantibodies long before the onset of clinical disease. Such studies demonstrated the presence of relatively restricted subsets of autoantibody specificities to chromatin and the Ro/SSA and La/SSB ribonucleoproteins years before the onset of active disease. Epitope spreading to other autoantibody specificities appears over the course of active disease in animal models and, presumably, in human disease [60–63].

As noted earlier, antigen-specific B cells elicited with either cryptic self peptides or with molecular mimics can present snRNP autoantigens in a manner that breaks T-cell tolerance in normal mice [58, 59, 64–69]. The overall mechanism of this immune response is illustrated in Figure 8.1. One of at least two pathways for triggering early autoimmune responses may occur, one that relies on antigen-presenting B cells and a second that may include other APC subsets such as DCs or macrophages. As discussed below, there is now substantial evidence that APC subsets may interact in the exchange of antigen (or autoantigens) in fostering a productive autoimmune response (Figure 8.2). In one pathway, professional APCs (dendritic cells or macrophages) present either foreign, cross-reactive antigens or perhaps cryptic self determinants in an initial priming of T cells. Both types of antigen are perceived as foreign by the immune system. Co-stimulation is required for this first step of T-cell activation which, in turn, provides conventional

cognate helper functions to B cells. This first step amplifies B cells bearing surface immunoglobulin receptors that allow these cells to bind and process the native autoantigen. In subsequent steps, the B cells present novel self determinants for a second tier of T-cell priming. The unfortunate product of this mechanism is a diverse and clonally expanded repertoire of autoreactive T cells. What was initially a restricted immune response has evolved to incorporate multiple determinants of the autoantigen.

The central role of B cells in the genesis of lupus autoimmunity was furthered by Chan et al. in having constructed MRL/lpr mice whose B cells are unable to secrete Ig, thus having essentially no circulating autoantibodies (or any soluble immunoglobulin) [70–73]. In brief, these mice developed spontaneous activation of T cells, identical to that found in wild-type MRL/lpr autoimmune mice bearing soluble autoantibody. These studies showed that soluble antibody is not required for T-cell activation. Taken together, it appears clear that B-cell APC function is required for some element of T-cell activation in SLE [70, 74].

The mechanisms described above are dependent on the presence and/or availability of the autoantigen to the B cell as an APC. If there is no endogenous source of self antigen supplied to the B-cell receptor or if T-cell tolerance exists, the expansion and amplification loop of the autoimmune response will not occur as illustrated in Figure 8.1.

Recent studies have now demonstrated synergy in interactions between antigen-presenting cell subsets. Our data demonstrate that antigen-specific B cells capture and process self antigen at very early stages of the autoimmune response [64, 65, 75–77]. Over time *in vivo*, B cells lose their ability to activate responding T cells. This observation may be due to either the turnover of self antigen from the B cells or possibly by the induction of receptor-mediated apoptosis of antigen-presenting B cells. Alternatively, B cells that acquire antigen may differentiate into plasmablasts, a subset of B cells unable to act as APCs. The end product is that once B cells initiate T-cell priming for autoreactivity, other APCs may be important in the ongoing or chronic nature of the lupus autoimmune response (Figure 8.2).

Indeed, we have demonstrated that other professional antigen-presenting cells activate autoreactive T cells later in the immune response. We find that DCs and/or macrophages acquire self antigen over longer time periods *in vivo* (beginning at day 3 or 4 after exposure to self antigen) and are able to activate primed autoreactive T cells [64, 65]. At least two explanations may account for these observations. First, professional APCs, dendritic cells and/or macrophages, may naturally need to mature and phagocytose snRNP antigen over time after it is introduced *in vivo*. This is likely since receptor-mediated uptake via BCR, FcR, CD19 or other receptors, typically occurs more rapidly than passive phagocytic processes [78—80]. This could account for the relatively delayed ability of DCs to activate responding autoreactive T cells. A second mechanism that could explain our observations is by the direct transfer of antigen between B cells and dendritic cells (see Figure 8.2, and discussed below).

Alternatively, antigen (or autoantigen) can also be captured by dendritic cells and preserved in its native state for the activation of antigen-specific B cells. Bergtold et al. have recently shown that dendritic cells can prime B cells by direct presentation of native antigen through the recycling of internalized immune complexes [81]. Taken together, these processes may maintain the antigen in a state required to perpetuate an ongoing immune response in both the B- and T-cell compartments. The B-cell response can be triggered by native antigen preserved in intracellular APC compartments and T-cell activation by antigen that is transferred from B cells to other professional APCs.

The latter mechanism has been supported by prior studies illustrating several mechanisms by which the transfer of antigen may occur between APC subsets [82—85]. Antigenic peptides have been known to be exchanged between APCs via cell-to-cell contact or via gap junctions between cells [82]. Finally, antigen-containing exosomes may also be transferred from B cells to other APC subsets [84, 85]. Exosomes are small vesicles made and released by a number of bone-marrow-derived cells including mature B cells. Exosomes from B cells are known to contain MHC class I and class II proteins as well as antigen that may be acquired by B cells as well as other B-cell surface macromolecules (Figure 8.2).

Studies from our laboratory support a mechanism whereby cell contact between antigen-presenting B cells and DCs or macrophages transfers antigenic protein and/or peptides. In a manner similar to that observed in murine models, human B cells can either bind antigen via BCR and directly activate T cells, or the B cells can transfer antigen to either macrophages or dendritic cells in a manner that requires surface protein, scavenger receptor A (SR-A) in ongoing autoimmunity [64, 65]. SR-A is found on a number of cell types including macrophages and dendritic cells. SR-A mediates adhesion of macrophages to B cells as well and other connective tissues and assists antigen processing in all of these cell subsets through gp96 (endoplasmic reticulum chaperone protein)-associated peptides. Interestingly, activation of TLRs on macrophages and dendritic cells also upregulates SR-A expression and very likely promotes the transfer of antigen from B cells to macrophages and APCs [86]. Thus, RNA- or DNA-associated lupus autoantigens may drive antigen processing and

presentation by activation of TLR7 and TLR9, leading to upregulation of SR-A, and the eventual amplification of the presentation of autoantigens by all of these APC subsets (see Figure 8.2).

The cooperation of B lymphocytes and dendritic cells leads to a more productive immune response as compared to either of these cells functioning as APCs independently [87]. We have demonstrated previously that B cells can initiate an antigen-specific immune response sooner than other APCs; however, persistent immunity to the same antigen requires dendritic cells and/or macrophages to activate T cells as well [64, 75]. In a manner similar to that found in murine systems, we found that human B cells can transfer antigen to either human macrophages or DC subsets [64, 75].

Antigen transfer has been previously characterized for various immune and non-immune cell types [88]. The most relevant to lupus autoimmunity includes transfer between B cells and other APCs such as macrophages and dendritic cells [64, 89—91]. The type of antigen transferred as well as the mechanism employed varies depending on the cell types involved. For instance, B cells have been shown to acquire antigen from subcapsular macrophages in the form of ICs, potentially autoantibody—chromatin complexes, by a complement receptor-dependent mechanism. DCs can transfer non-degraded ICs to B cells in an Fc receptor-mediated process [81, 90, 91]. As for B lymphocytes, these APCs have been found to transfer antigen to DCs either as apoptotic bodies that are endocytosed by the DC or by the formation of cellular exosomes as described above [92, 93].

A common feature associated with antigen transfer from B cells is the requirement for direct contact between the antigen donor and recipient cells [94—97]. Previous studies have demonstrated the direct interaction of B lymphocytes with DCs, and that this interaction is required for dendritic cells to transfer unprocessed antigen to B lymphocytes [97—99]. Confocal microscopy of live cell cultures confirmed that direct contact between B cells and antigen-bearing DCs is a prerequisite for antigen transfer.

Our findings have broad implications in understanding the immunological events leading to a focused immune response to a select group of antigens, whether foreign or self. First, antigen-specific B lymphocytes can capture and present antigen via surface Ig at faster kinetics compared to other APCs in activating T cells [78]. Secondly, primary myeloid dendritic cells can acquire antigen more efficiently by antigen transfer from B cells than by endocytosis of free antigen. Taken together, these findings suggest that antigen transfer from antigen-specific B cells to dendritic cells would lead to a synergistic effort in presenting the same antigen to T cells whereby a more targeted and specific immunologic response would ensue. Antigen transfer requires direct contact between the two APCs followed by the capture of B-cell-derived components by the recipient DC. Moreover, SR-A appears to be a key surface receptor on DCs that mediates the acquisition of antigen from B cells (see Figure 8.2).

The development of autoimmunity may be driven by the collaborative processes between B cells and DCs. For example, systemic lupus erythematosus (SLE) is characterized by B- and T-cell responses that are targeted against a specific subset of intracellular components including nucleosomes and/or ribonucleoproteins [100]. These specific responses arise amid the hundreds of other intracellular proteins that are also potential targets of autoimmunity. The ability of BCR to capture and acquire specific proteins, such as the snRNP autoantigens in SLE, provides a mechanism that first focuses autoimmunity against specific intracellular macromolecules. Thereafter, the transfer of specific intracellular snRNP proteins to other professional APCs would serve to perpetuate the ongoing autoimmune response. As discussed earlier, it is now well demonstrated that B cells are essential to the spontaneous autoimmunity observed in murine models of SLE and in epitope spreading of this syndrome [68].

UNIQUE ANTIGEN-PRESENTING FEATURES OF B CELLS

The mechanism described above was built upon a wealth of information that B cells are highly efficient APCs for the target antigen. Antigen-specific B cells can present peptide to T cells at up to 10,000-fold lower antigen concentrations as compared to non-antigen-specific B cells [78, 79, 101–110]. Moreover, binding of antigen to the B cell surface can induce CD80 and CD86 (B7.1 and B7.2) co-stimulators, as shown with Ig transgenic mice. The surface Ig can bind and concentrate antigen from a complex sea of self antigens, a property that may be relevant in presenting autoantigens that may be in low concentration outside of cells *in vivo*. While small resting B cells can process and present antigen to previously primed T cells, this subclass of B cells lacks the appropriate cell surface co-stimulatory molecules necessary for priming naïve T cells. The isotype of surface Ig (μ, δ, or γ) does not affect the ability or efficiency of the antigen-presenting function of APCs [79]. However, the cytoplasmic domain of the Ig receptor is critical for the internalization of captured antigens and for directing them to the intracellular lysosomal processing compartments [111, 112]. It is likely that the avidity of surface Ig is directly proportional to the efficiency of antigen presentation. In addition, the B cell must undergo activation from sIg as well as

engagement of another surface molecule, CD40, whose ligand, gp39, is found on the surface of CD4 T cells [113]. Antigen activation of B cells has been shown to enhance the surface expression of co-stimulatory molecules, principally CD86 and is required for the ability of B cells to break T-cell tolerance [114, 115].

The fine specificity of surface Ig can directly influence the particular peptide transported to the B-cell surface [104, 105, 116]. Antibodies have been found to exert strong effects on the stability and conformation of proteins and to also protect the sites bound in the antibody cleft from proteolytic degradation. As such, Davidson, Watts, and colleagues have demonstrated that the specificity of surface Ig that captures antigen controls the pattern of antigen fragmentation and the subsequent pattern of T-cell responses elicited [104, 105, 116]. With relevance to the diversification of autoimmune responses described above, it is likely that the determinant bound by surface Ig is protected from (or even selected against) presentation at the B-cell surface. That is, sites located elsewhere on the protein (or ribonucleoprotein particle) are preferentially presented by the autoantigen-specific B-cell APC. Such a mechanism could potentially account for the intramolecular diversification (epitope spreading) of T cell responses over time in the murine model of multiple sclerosis, diabetes, or in the spreading of autoantibody specificity to multiple snRNP determinants in lupus autoimmunity.

The fact that the B-cell repertoire is enriched in particular self-reactive specificities should be reflected in the generation of autoreactive T cells. That is, if B cells are important APCs in lupus autoimmunity, autoreactive T cells should be easily found in the same repertoire. Yan and colleagues have supported this premise in studies that identify snRNP-specific T-cell response in mice transgenic for an immunoglobulin heavy chain that confers B-cell specificity [76, 77]. These studies demonstrated that anti-snRNP B cells trigger snRNP-specific T-cell autoimmunity in lupus-prone MRL mice. In contrast, the same anti-snRNP B cells fail to activate T cells in normal strains of mice and, indeed, trigger tolerance of these autoreactive T cells. In the context of lupus autoimmunity, the role of B cells in presenting autoantigen may be relevant in that T cells from both human SLE and in mouse models have altered TCR signaling pathways and appear more easily activated by antigen [117–121]. In this scenario, autoreactive B cells bearing autoantigens may serve as probes for the characterization of autoreactive T cells.

Defining B-cell–T-cell interactions will be central to understanding the early events in lupus autoimmunity. Activated autoreactive B cells as APCs for lupus autoantigens, chromatin or snRNPs, are capable of breaking peripheral T-cell tolerance by priming ignorant or even anergic autoreactive T cells [68, 69, 122]. Moreover,

B-cell APCs could further activate autoreactive T cells that had already been primed by dendritic cells. As described above, CD4 T cells that provide help for auto-antibody synthesis in SLE may not be regulated or activated by the same threshold of antigen stimulation or even the same requirement for co-stimulation as typically required for immune responses to foreign antigen. An analysis of lupus-prone MRL/lpr mice deficient in either CD80 or CD86 demonstrated essentially normal levels of autoantibody production as well as spontaneous activation of T cells, and nephritis that resembles wild-type autoimmune mice [123, 124]. Given the importance of co-stimulation in B-cell–T-cell interactions, it will be important to better investigate the roles of other co-stimulatory molecules, such as CTLA4, ICOS-B7-RP1, and B7h and its ligand, as well as PD-1 and PD-1L in the context of SLE, as described below. These other surface molecules may be important in sustaining or regulating the chronic phases of lupus autoimmunity.

As illustrated in Figure 8.1, these T-cell–B-cell–dendritic cell interactions participate in an ongoing and self-perpetuating cycle that enhances autoimmune diversification. The key will be in identifying the very first event(s) in the cycle and, thereafter, the process may become difficult to regulate.

CO-STIMULATORY PROCESSES IN B-CELL IMMUNITY

The optimal activation of T lymphocytes requires not only the binding of the T-cell receptor to its appropriate peptide/MHC complex but also the binding of other accessory, or co-stimulatory molecules on its surface [125–127]. T-cell anergy or unresponsiveness results when T-cell receptors find their cognate peptide/MHC in the absence of co-stimulatory signals. This latter mechanism is a second important factor in maintaining T-cell tolerance to many self antigens that fail to drive negative selection in the thymus.

At least four key surface macromolecules are involved in T-cell co-signaling, two on the antigen-presenting cell and two on the T cell. The best-characterized APC surface molecules, CD80 and CD86 (also still known as B7-1 and B7-2, respectively), are members of the Ig supergene family and share only 25% amino acid sequence homology. CD80 and CD86 are independently regulated on the surface of APCs suggesting distinct functional roles for each in the co-stimulation of T cells. CD80 is found primarily on dendritic APCs and in low or marginally detectable levels on B lymphocytes. CD86 is constitutively expressed on dendritic cells and in low levels on B and T cells, and NK cells. CD86 is rapidly upregulated within 24 hours on activated B cells while CD80 levels increase much later in time (after 48 hours) and rapidly decline thereafter. Several modes of B-cell activation including surface Ig cross-linking, LPS mitogen stimulation, and CD40-CD40L interactions all upregulate CD80 and CD86 surface expression which is a critical component in the role of B cells as autoantigen-presenting cells. The rapid and early upregulation of CD86 suggests that this molecule may first influence the decision that a T cell makes in whether to become anergic or activated.

With a demonstrated importance of co-stimulation in conventional immune responses, the natural transition was to elucidate the role of accessory molecules in the development of autoimmunity and/or the maintenance of T cell tolerance. First attempts were performed with antibodies to CTLA4, CD28, CD80 and CD86 administered in vivo. It is likely that CD80 and CD86 co-stimulation has distinct roles in the regulation of autoimmune disease as indicated by the administration of antibodies to either of these molecules. Murine EAE, a Th1-mediated autoimmune syndrome, is inhibited by the sustained treatment with anti-CD80 antibodies while anti-CD86 antibody treatment enhances the severity of disease [127, 128]. Lupus autoimmunity, however, may be more complicated to explain at the level of B7 interactions in eliciting autoreactive T cells with subsequent effects on pathogenic autoantibody responses. As indicated in studies by Nakajima et al. [129], murine SLE in the NZB/NZW F1 mouse can be ameliorated by treatment with a cocktail of anti-CD80 and CD86 antibodies. However, similar studies in our laboratory indicate that MRL lupus autoimmune responses may be more impervious to B7 blockade [123, 124]. Long-term treatment of MRL lpr/lpr mice with anti-B7-1 or anti-B7-2 antibodies, either individually or with both blocking antibodies, did not lessen the anti-snRNP or anti-DNA responses and did not affect long-term survival or kidney pathology. These studies are supported by the analysis of CD80 and CD86 knockout strains of MRL lpr mice in our laboratory. While these mechanisms await further examination, it is possible that MRL autoimmunity bypasses the conventional CD80- and CD86-mediated T-cell activation pathway.

Several studies have demonstrated an importance for CTLA4 signaling in the induction of T-cell tolerance to self antigens. Transgenic mice were developed expressing a soluble fusion protein, CTLA4Ig, which binds endogenous B7 thereby preventing the natural association of B7 with CTLA4 or CD28 [126]. These mice fail to establish CD4 T-cell tolerance to an endogenous superantigen indicating that a B7/CTLA4-dependent signal is required for maintaining unresponsiveness to self constituents. More dramatic evidence for this fact has been provided by studies of CTLA4 knockout mice [130, 131]. The total absence of CTLA4 causes a massive lymphoproliferative disease, hypergammaglobuline-mia, myocarditis, and T-cell infiltration in many organs leading to death by 4 weeks of age. Taken together, it is possible that CTLA4 is critical in deleting autoreactive T cells during thymic selection or in signaling peripheral T cells for tolerance to self constituents.

CTLA4Ig has also been administered to several models of murine autoimmunity to assess the effects of blocking B7 co-stimulatory signals in ongoing disease [126, 132–136]. CTLA4Ig treatment effectively inhibits the induction and perpetuation of EAE, murine dia-betes, and murine SLE. In some examples, inhibition of B7 co-stimulation was capable of ameliorating ongoing disease, an important observation given that immuno-logic insults in human SLE occur long before the patient seeks clinical care.

The surface co-stimulatory proteins described above have served as therapeutic targets for various auto-immune syndromes, including SLE. Indeed, many therapeutic targets along the extended pathway of autoimmunity have been, or are presently, undergoing clinical trials in various rheumatic diseases (Figure 8.3).

In particular, two compounds, abatacept and belata-cept, have been developed that interfere with B7–CD28 interactions that occur between B cells and T cells (or even DCs and T cells). Abatacept is a fusion protein of CTLA4 combined with a portion of human Ig and binds B7 (both CD80 and CD86) with higher affinity than the other natural ligand on T cells, CD28. At the time of this writing, clinical trials of abatacept and cyclophosphamide combination therapy for the treatment of lupus nephritis are just being completed. Non-peer-reviewed reports in the press and preliminary studies presented at the American College of Rheuma-tology in 2009 indicate that abatacept has not improved either primary or secondary endpoints in the treatment of human SLE. This short, 12-month study was per-formed in 175 patients with SLE and active polyarthritis, discoid lesions, and/or serositis. Patients were randomized to abatacept and prednisone (tapered after 30 days). No differences in lupus flares in primary or secondary endpoints were observed in the treatment versus control groups. However, other analyses indi-cated other "quality of life" parameters, such as general physical well-being, fatigue, and sleeping disorders, appeared markedly improved in the abatacept group. Some concern and/or criticism has been noted in several of these early therapeutic trials, both of abatacept and of rituximab, regarding the inadequate length of trials and dosing regimen [137].

BAFF/BLyS IN SLE

B-cell growth factors have been found to be elevated in human and in murine models of SLE (MRL and NZB x NZW models) as well as in rheumatoid arthritis and Sjogren syndrome [5, 150–155]. Of note, two TNF-family members, B-lymphocyte stimulator (BLyS; also referred to as BAFF) and APRIL (A PRoliferation-Inducing Ligand) are important to B-cell development and germinal center formation. BLyS stimulates growth and development of transitional (T2) and germinal center B cells, B-cell subsets shown to be abnormal in SLE (Figure 8.4) [138–141]. Murine models genetically manipulated to overexpress BLyS develop lupus-like autoimmunity, pathology, antinuclear antibodies (with Ig deposits in the kidney) and rheumatoid factors (Table 8.5) [139, 142]. In addition, these mice develop elevated numbers of B-1 B-cell subsets, mature B cells and plasma cells as well as spontaneous germinal center formation. Elevated BLyS/BAFF levels have been reported in SLE patients with correlation to titers of anti-DNA autoantibodies. However, BLyS levels were not clearly shown to correlate with pathology or severity of disease and some patients may show no abnormali-ties in BLyS levels or activity [143, 144].

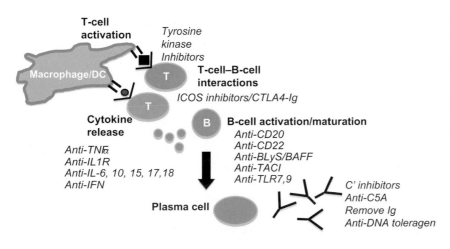

FIGURE 8.3 Therapeutic targets in rheumatic disease. A variety of immunologic targets have been exploited for drug development in various rheumatic diseases. These targets span the spec-trum of immune responses, ranging from early antigen-presenting functions of APCs, signaling inhibitors of responding T cells, cytokine inhibition, B-cell depletion or B-cell development inhibitors, and the removal of circulating autoantibodies.

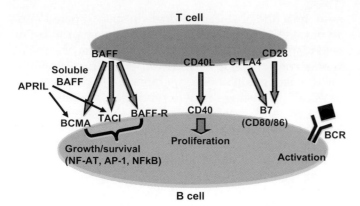

FIGURE 8.4   B-cell development and activation after interactions with T cells. Several surface-mediated signaling events lead to the development and activation of naïve B cells (see text for details). Among the key pathways in lupus autoimmunity are those mediated by BAFF/BLyS and APRIL binding to receptors on the B cell, including BCMA, TACI, and BAFF receptor. Other well-described B-cell developmental triggers include those signaling through CD40 and cross-linking of B cell receptors.

BLyS/BAFF binds and signals B cells through BAFF receptor, BCMA, and TACI and likely mediates most of the B-cell growth and developmental effects [145–148]. APRIL binds B cells in a more restricted manner, only to BCMA and TACI [154]. TACI on B cells triggers signaling through CAML and leads to the activation of transcription factors NF-AT, AP-1, and NFκB (Figure 8.4). TACI and BCMA exhibit significant amino acid sequence morphology. Biologically active BAFF exists either as a membrane-bound protein or as a secreted, soluble protein. BAFF typically forms homotrimers or heterotrimers along with APRIL, both of which are elevated in patients with SLE and RA. Regarding APRIL levels alone, some controversy exists

TABLE 8.5   BLyS/BAFF in animal models of lupus

Mouse strain	Phenotype
BLyS/BAFF-/-	B cell developmental arrest (T1)
TACI-/-	Increased B cells, decreased T-independent responses
BCMA-/-	No overt phenotype
BAFF transgenic	B cell hyperplasia, increased Ig, splenomegaly
	Lymphadenopathy, anti-DNA, kidney disease
TACI-Ig transgenic	Reduced mature B cell numbers
BCMA transgenic	No overt phenotype
TACI-Ig treated mice	Reduced mature B cell numbers
BCMA-Ig treated mice	Reduced mature B cell numbers

as to its association with SLE. Studies have been reported that increased serum levels of APRIL are linked to elevated anti-DNA autoantibody levels, while another report indicated that APRIL was inversely correlated with active disease in SLE [149, 150]. Finally, genetic polymorphisms in APRIL have also been associated with the presence of autoimmunity [151].

BAFF and APRIL macromolecules are critical for B-cell homeostasis once cells have left the bone marrow (reviewed in [152]). Interfering with BAFF binding to BAFF-R on B cells leads to loss of follicular and marginal zone B cells, emphasizing this system as a potentially important therapeutic target in SLE. However, activated T cells also express BAFF-R and TACI, suggesting BAFF may also have a direct role on T-cell immunity or autoimmunity. It has also been found that APRIL can also amplify B-cell antigen-presenting functions after binding to BCMA, a potentially important biological parameter in SLE.

As below, B-cell growth factors have become key therapeutic targets of SLE and related autoimmune syndromes (Table 8.6). Recombinant fusion proteins of receptors, BCMA and TACI linked to the Fc fragment of IgG, inhibits T-cell-mediated immunity in mouse models [147, 153–155]. These agents act via adsorbing soluble BLyS, thereby making it unavailable to B cells as a growth factor. Mice made genetically deficient in BLyS demonstrate B-cell lymphopenia and lowered Ig responses to T-cell-dependent and -independent antigens [155, 156]. In contrast, animal models given exogenous recombinant BLyS experience hypergammaglobulinemia and expansion of virtually all B-cell subsets. Animal models of BLyS biology have provided support for adapting this B-cell growth pathway in the treatment of SLE (Table 8.5). The classical models of SLE, the MRL and NZB x NZW F1 mice, experience elevated levels of BLyS. BLyS was measured both before the onset of murine disease and later in the development of pathology. Nearly one-fifth of the mice demonstrate a three-fold increase in serum BLyS early in disease (14 weeks of age). By 24 weeks of age, all animals experience increases in BLyS up to six-fold over baseline. Remarkably, increases in serum BLyS paralleled similar dramatic increases in kidney disease (proteinuria) and serum autoantibodies. High titers of anti-DNA autoantibodies are correlated with high serum BLyS. These lupus models have been treated with TACI-Ig immunotherapeutics [157]. TACI-Ig prevents the binding of BLyS to BCMA and TACI on receptor bearing B cells. Treatment of NZB x NZW mice in this manner resulted in improved survival, lessened kidney disease and proteinuria.

As noted above, SLE patients have been noted to have elevated levels of serum BLyS. In the treatment of human SLE, belimumab (human anti-BLyS) has

TABLE 8.6   B-cell therapeutic targets

B-cell depleting	Anti-CD20
Co-stimulation inhibition	CTLA4-Ig, anti-ICOSL, anti-CD40L
B-cell inhibition	Anti-FcγRIIb, anti-CD22
B-cell growth/differentiation	Anti-BLyS, anti-IL10, anti-IL6
Inhibitors	Anti-Ifnα, anti-TNFα

been demonstrated to lower levels of activated B cells (CD69+ B cells) [158]. Efficacy of belimumab in SLE therapy has yet to be fully resolved and phase II and III trials are still underway at the present time. Interestingly, increases in BLyS have also been noted in patients treated with rituximab, thought to contribute to the survival and biological functions of memory B-cell subsets. BLyS has been demonstrated to rescue B cells from programmed cell death (apoptosis) and may be responsible for the outgrowth of B cells in the months that follow B-cell-depleting rituximab therapy. Alternatively, BLyS may be one factor responsible for maintaining B-cell populations in those individuals refractory to rituximab therapy. Phase II studies of belimumab in SLE did not fully satisfy primary clinical endpoints although some criticisms of the studies suggest that the length of study may not have been ideal for outcome analyses. In extended studies of ANA-positive individuals, nearly half had significant clinical improvement one year after treatment [159, 160]. Three years post treatment, authors noted a statistic reduction in SLE flares and improvement in disease activity [161]. Among clinical responders, titers of anti-DNA autoantibodies were reduced, as were total Ig levels, while complement levels increased.

A recent 2009 report has reviewed outcomes of a phase II, blinded and placebo-controlled study of belimumab along with standard of care in patients with active SLE [162]. In this study of 449 patients with SELENA scores of ≥4, patients were randomly assigned to dosage groups of belimumab or placebo over a 52-week trial period. Endpoints and SELENA-SLEDAI scores were recorded at week 24 as well as timepoints to the first SLE flare. Although overall primary endpoint differences were not achieved and no dose response to belimumab was observed, some notable outcomes were recorded. In particular, the median time to the first SLE flare in weeks 24–52 was significantly longer in patients treated with belimumab as compared to placebo controls (154 days versus 108 days; $p = 0.03$). When examining a subgroup of sero-active patients (ANA positive ≥1:80; anti-DNA positive; 71% of the total patient group), belimumab treatment led to significantly better outcomes and reduced SELENA-SLEDAI score. Overall, while endpoint outcomes were not

achieved, these promising results in sero-active subsets of SLE patients merit further investigation and support a premise of B-cell immunotherapies.

The basic functions of several other co-stimulatory molecules have been reviewed earlier. In particular, some immunotherapeutics have been developed with the purpose of inhibiting B-cell—T-cell interactions. Indeed, while some T-cell-independent functions of B cells may be critical to control, a role for conventional T cell help in promoting B-cell autoimmunity in lupus remains important as well. CTLA4Ig (abatacept) is one such therapeutic that blocks the co-stimulatory interactions between CD80/86 on B cells and other APCs (DCs and macrophages) to the ligand CD28 on T cells. Clinical trials of abatacept on now large numbers of patients with rheumatoid arthritis, have shown marked improvement in symptoms and a slowing of radiologic joint pathology [163, 164]. Evidence for the treatment of SLE patients with CTLA4Ig has arisen from the successful amelioration of spontaneous murine models treated with this drug. In treatment of the NZB x NZW F1 model of SLE, anti-DNA was reduced, as was proteinuria and kidney pathology, with an increase in survival of the animals. Two separate inhibitors of CD28—CD80/86 interactions have been developed for use in humans, abatacept and belatacept. Based on animal studies, an open-label pilot trial of CTLA4IgG4m in combination with cyclophosphamide in humans with lupus nephritis is underway [165, 166].

While abatacept and rituximab have been approved for use in rheumatoid arthritis [167], their efficacy in SLE has yet to be established. Limited trials in both adults and children with SLE have indicated that rituximab in combination with other immunosuppressive therapeutics improved various clinical parameters including arthritis, nephritis, skin rash, alopecia, hemolytic anemia, and thrombocytopenia [168—173]. Rituximab deletes B cells by its binding to surface CD20, a molecule expressed on pre-B cells through development to memory B cells (Figure 8.5). Rituximab deletion of B cells is via direct, complement-mediated cyotoxicity, or by antibody-dependent cellular cytotoxicity (ADCC). Clearly, B cells at various stages express a variety of unique cell-surface macromolecules, some of which are being exploited as therapeutic targets in autoimmunity (Figure 8.5). However, CD20 is not expressed on human plasma cells, allowing them to escape deletion by rituximab.

Two phase III studies await full analysis, although some early reports suggested no overall differences in clinical outcomes between patients receiving rituximab as compared to placebo controls (EXPLORER trial). However, particular patient subsets, including African-Americans and Hispanics with SLE, may appreciate some clinical amelioration of disease compared to controls [174]. Moreover, the serologic responses appear

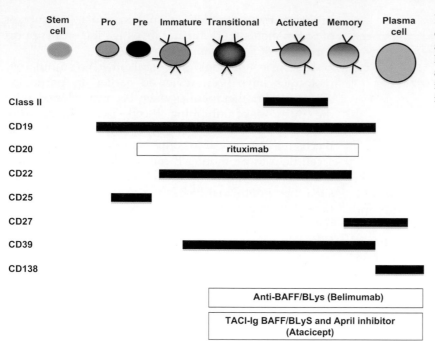

FIGURE 8.5 B-cell surface markers through development. Several key surface macromolecules are unique to various stages of B-cell development. Moreover, molecules targeted by therapeutics, such as rituximab and BAFF/BLyS inhibitors, selectively interfere or eliminate selected subsets of B cells as illustrated.

to improve in the treatment group in that lower titers of ANAs were found ($p < 0.05$). Some concern has been noted that the trial enrolled patients primarily with active SLE and thus treated with moderate- to high-dose steroids. As such, the trial may not have lasted long enough to incur clinical efficacy. Moreover, several studies indicate that rituximab depletion treatment differs between subsets and physiologic location of B cells [175]. Studies in mice demonstrate that depletion of peripheral circulating B cells can occur rapidly, within minutes of therapy, while depletion kinetics is slower in secondary lymphoid organs, the spleen and lymph nodes. B cells most resistant to therapy include peritoneal B cells and marginal zone B cells, the latter subset are only clear to approximately half their basal levels [176–178]. Depletion is not due to differences in CD20 expression, as marginal zone, germinal center, peritoneal B-1 and B-2 cells are virtually identical in expression. Clearly, the exact benefits of B-cell depletion therapies have yet to be fully examined or understood in human SLE and further study will be required.

CONCLUSION

While human SLE is a syndrome marked by genetic predisposition, stochastic events, and multiple cellular abnormalities, it is clear that B cells remain at the center of etiologic and pathologic processes. After all, B cells provide autoantibodies, the diagnostic hallmark of SLE. B-cell functional abnormalities exist in the context of this autoimmune syndrome although the complicated intracellular interactions between B cells and a plethora of other immune and non-immune cells make the mechanisms of disease difficult to tract. As noted here, B-cell autoimmunity does not function in a complete vacuum, as interactions with T cells and other APCs promote disease. B-cell cytokine secretion may alter the pathways of T-cell autoimmunity and DC-cell functions. B cells likely trigger epitope spreading mechanisms that dictate the severity of tissue pathology. Early therapeutic studies that target B-cell populations and B-cell growth factors have yet to be proven fully efficacious in human SLE, although it can be concluded that study parameters may have biased or clouded the interpretations of these approaches. Studies in recent years have rapidly advanced the understanding of B-cell biology in human SLE. Notably, the critical role of TLRs in B-cell activation has provided some explanation to the selectivity of lupus autoimmunity towards nucleic-acid-containing autoantigens. Without question, we await the potential applications of these advances to the therapeutic intervention of this complicated autoimmune syndrome.

Acknowledgments

The author would like to express gratitude to those members of his lab, past and present, who have contributed to understanding B-cell biology in lupus autoimmunity. Their tireless efforts have provided meaningful advances to understanding the etiology of this complicated autoimmune disease. Work from the author is supported by the National Institutes of Health (AR41032 and AI48120) and the Alliance for Lupus Research.

References

[1] M. Odendahl, A. Jacobi, A. Hansen, E. Feist, F. Hiepe, G.R. Burmester, et al., Disturbed peripheral B lymphocyte homeostasis in systemic lupus erythematosus, Journal of Immunology 165 (2000) 5970–5979.

[2] A. Cappione, A. Pugh-Bernard, I. Sanz, Lupus VH4.34-encoded antibodies bind to a B220-specific glycoform of CD45 on the surface of human B lymphocytes, Arthritis Rheum 46 (2002).

[3] J.H. Anolik, J. Barnard, A. Cappione, A.E. Pugh-Bernard, R.E. Felgar, R.J. Looney, et al., Rituximab improves peripheral B cell abnormalities in human systemic lupus erythematosus, Arthritis and Rheumatism 50 (2004) 3580–3590.

[4] Y. Harada, M.M. Kawano, N. Huang, M.S. Mahmoud, I.A. Lisukov, K. Mihara, et al., Identification of early plasma cells in peripheral blood and their clinical significance, British Journal of Haematology 92 (1996) 184–191.

[5] Q. Lin, C. Dong, M.D. Cooper, Impairment of T and B cell development by treatment with a type I interferon, Journal of Experimental Medicine 187 (1998) 79–87.

[6] P. Casali, A.L. Notkins, CD5+ B lymphocytes, polyreactive antibodies and the human B-cell repertoire, Immunol Today 10 (1989) 364–368.

[7] K. Hayakawa, R.R. Hardy, D.R. Parks, L.A. Herzenberg, The "Ly-1 B" cell subpopulation in normal immunodefective, and autoimmune mice, J Exp Med 157 (1983) 202–218.

[8] S.E. Burastero, P. Casali, R.L. Wilder, A.L. Notkins, Monoreactive high affinity and polyreactive low affinity rheumatoid factors are produced by CD5+ B cells from patients with rheumatoid arthritis, Journal of Experimental Medicine 168 (1988) 1979–1992.

[9] M. Dauphinee, Z. Tovar, N. Talal, B cells expressing CD5 are increased in Sjogren's syndrome, Arthritis and Rheumatism 31 (1988) 642–647.

[10] K. Hayakawa, R.R. Hardy, L.A. Herzenberg, Peritoneal Ly-1 B cells: Genetic control, autoantibody production, increased lambda light chain expression, European Journal of Immunology 16 (1986) 450–456.

[11] M. Murakami, H. Yoshioka, T. Shirai, T. Tsubata, T. Honjo, Prevention of autoimmune symptoms in autoimmune-prone mice by elimination of B-1 cells, International Immunology 7 (1995) 877–882.

[12] T.L. Rothstein, Cutting edge commentary: Two B-1 or not to be one, Journal of Immunology 168 (2002) 4257–4261.

[13] A.M. Stall, S.M. Wells, K.P. Lam, B-1 cells: Unique origins and functions, Seminars in Immunology 8 (1996) 45–59.

[14] K. Hayakawa, M. Asano, S.A. Shinton, M. Gui, D. Allman, C.L. Stewart, et al., Positive selection of natural autoreactive B cells, Science 285 (1999) 113–116.

[15] M.J. Chumley, J.M. Dal Porto, S. Kawaguchi, J.C. Cambier, D. Nemazee, R.R. Hardy, A V(H)11V(Œ∫)9 B cell antigen receptor drives generation of CD5+ B cells both in vivo and in vitro, Journal of Immunology 164 (2000) 4586–4593.

[16] K. Hayakawa, M. Asano, S.A. Shinton, M. Gui, L.J. Wen, J. Dashoff, et al., Positive selection of anti-thy-1 autoreactive B-1 cells and natural serum autoantibody production independent from bone marrow B cell development, Journal of Experimental Medicine 197 (2003) 87–99.

[17] R.J. De Boer, A.S. Perelson, How diverse should the immune system be? Proceedings of the Royal Society B: Biological Sciences 252 (1993) 171–175.

[18] Y. Louzoun, E. Luning Prak, T. Friedman, S. Litwin, M. Weigert, Comment on Langman and Cohn, Seminars in Immunology 14 (2002) 231–232.

[19] D. Nemazee, Antigen receptor 'capacity' and the sensitivity of self-tolerance, Immunology Today 17 (1996) 25–29.

[20] D. Nemazee, Receptor editing in lymphocyte development and central tolerance, Nat Rev Immunol 6 (2006) 728–740.

[21] D. Gay, T. Saunders, S. Camper, M. Weigert, Receptor editing: An approach by autoreactive B cells to escape tolerance, Journal of Experimental Medicine 177 (1993) 999–1008.

[22] M.Z. Radic, J. Erikson, S. Litwin, M. Weigert, B lymphocytes may escape tolerance by revising their antigen receptors, Journal of Experimental Medicine 177 (1993) 1165–1173.

[23] S.L. Tiegs, D.M. Russell, D. Nemazee, Receptor editing in self-reactive bone marrow B cells, Journal of Experimental Medicine 177 (1993) 1009–1020.

[24] D.R. Sekiguchi, L. Yunk, D. Gary, D. Charan, B. Srivastava, D. Allman, et al., Development and selection of edited B cells in B6.56R mice, Journal of Immunology 176 (2006) 6879–6887.

[25] E.J. Witsch, H. Cao, H. Fukuyama, M. Weigert, Light chain editing generates polyreactive antibodies in chronic graft-versus-host reaction, Journal of Experimental Medicine 203 (2006) 1761–1772.

[26] C.M. Doyle, J. Han, M.G. Weigert, E.T.L. Prak, Consequences of receptor editing at the Œ^a locus: Multireactivity and light chain secretion, Proceedings of the National Academy of Sciences of the United States of America 103 (2006) 11264–11269.

[27] L. Busconi, C.M. Lau, A.S. Tabor, M.B. Uccellini, Z. Ruhe, S. Akira, et al., DNA and RNA autoantigens as autoadjuvants, J Endotoxin Res 12 (2006) 379–384.

[28] A. Marshak-Rothstein, Toll-like receptors in systemic autoimmune disease, Nat Rev Immunol 6 (2006) 823–835.

[29] A. Marshak-Rothstein, Tolling for autoimmunity-prime time for 7, Immunity 25 (2006) 397–399.

[30] I.R. Rifkin, E.A. Leadbetter, L. Busconi, G. Viglianti, A. Marshak-Rothstein, Toll-like receptors, endogenous ligands, and systemic autoimmune disease, Immunol Rev 204 (2005) 27–42.

[31] A. Marshak-Rothstein, L. Busconi, C.M. Lau, A.S. Tabor, E.A. Leadbetter, S. Akira, et al., Comparison of CpG s-ODNs, chromatin immune complexes, and dsDNA fragment immune complexes in the TLR9-dependent activation of rheumatoid factor B cells, J Endotoxin Res 10 (2004) 247–251.

[32] A. Marshak-Rothstein, I.R. Rifkin, Immunologically active autoantigens: the role of toll-like receptors in the development of chronic inflammatory disease, Annu Rev Immunol 25 (2007) 419–441.

[33] R. Lafyatis, A. Marshak-Rothstein, Toll-like receptors and innate immune responses in systemic lupus erythematosus, Arthritis Res Ther 9 (2007) 222.

[34] M.J. Shlomchik, Activating systemic autoimmunity: B's, T's, and tolls, Curr Opin Immunol 21 (2009) 626–633.

[35] M.J. Mamula, M. Baer, J. Craft, S. Altman, An immunological determinant of RNase P protein is conserved between Escherichia coli and humans, Proc Natl Acad Sci USA 86 (1989) 8717–8721.

[36] M.T. McClain, L.D. Heinlen, G.J. Dennis, J. Roebuck, J.B. Harley, J.A. James, Early events in lupus humoral autoimmunity suggest initiation through molecular mimicry, Nat Med 11 (2005) 85–89.

[37] M.R. Arbuckle, M.T. McClain, M.V. Rubertone, R. Hal Scofield, G.J. Dennis, J.A. James, et al., Development of autoantibodies before the clinical onset of systemic lupus erythematosus, New England Journal of Medicine 349 (2003) 1526–1533.

[38] J.B. Harley, J.A. James, Epstein-Barr virus infection may be an environmental risk factor for systemic lupus erythematosus in children and teenagers, Arthritis Rheum 42 (1999) 1782–1783.

[39] J.A. James, K.M. Kaufman, A.D. Farris, E. Taylor-Albert, T.J. Lehman, J.B. Harley, An increased prevalence of Epstein-Barr virus infection in young patients suggests a possible etiology for systemic lupus erythematosus, J Clin Invest 100 (1997) 3019–3026.

[40] B.D. Poole, R.H. Scofield, J.B. Harley, J.A. James, Epstein-Barr virus and molecular mimicry in systemic lupus erythematosus, Autoimmunity 39 (2006) 63–70.

[41] H. Hemmi, O. Takeuchi, T. Kawai, T. Kaisho, S. Sato, H. Sanjo, et al., A Toll-like receptor recognizes bacterial DNA, Nature 408 (2000) 740–745.

[42] J. Lund, A. Sato, S. Akira, R. Medzhitov, A. Iwasaki, Toll-like receptor 9-mediated recognition of Herpes simplex virus-2 by plasmacytoid dendritic cells, Journal of Experimental Medicine 198 (2003) 513–520.

[43] F. Heil, H. Hemmi, H. Hochrein, F. Ampenberger, C. Kirschning, S. Akira, et al., Species-Specific Recognition of Single-Stranded RNA via Till-like Receptor 7 and 8, Science 303 (2004) 1526–1529.

[44] S.S. Diebold, T. Kaisho, H. Hemmi, S. Akira, C. Reis E Sousa, Innate Antiviral Responses by Means of TLR7-Mediated Recognition of Single-Stranded RNA, Science 303 (2004) 1529–1531.

[45] E.A. Leadbetter, I.R. Rifkin, A.M. Hohlbaum, B.C. Beaudette, M.J. Shlomchik, A. Marshak-Rothstein, Chromatin-IgG complexes activate B cells by dual engagement of IgM and Toll-like receptors, Nature 416 (2002) 603–607.

[46] I.R. Rifkin, E.A. Leadbetter, B.C. Beaudette, C. Kiani, M. Monestier, M.J. Shlomchik, et al., Immune complexes present in the sera of autoimmune mice activate rheumatoid factor B cells, J Immunol 165 (2000) 1626–1633.

[47] H. Wang, M.J. Shlomchik, Autoantigen-specific B cell activation in Fas-deficient rheumatoid factor immunoglobulin transgenic mice, J Exp Med 190 (1999) 639–649.

[48] J. William, C. Euler, E. Leadbetter, A. Marshak-Rothstein, M.J. Shlomchik, Visualizing the onset and evolution of an autoantibody response in systemic autoimmunity, J Immunol 174 (2005) 6872–6878.

[49] C.M. Lau, C. Broughton, A.S. Tabor, S. Akira, R.A. Flavell, M.J. Mamula, et al., RNA-associated autoantigens activate B cells by combined B cell antigen receptor/Toll-like receptor 7 engagement, Journal of Experimental Medicine 202 (2005) 1171–1177.

[50] J. Vollmer, S. Tluk, C. Schmitz, S. Hamm, M. Jurk, A. Forsbach, et al., Immune stimulation mediated by autoantigen binding sites within small nuclear RNAs involves Toll-like receptors 7 and 8, J Exp Med 202 (2005) 1575–1585.

[51] T. Lovgren, M.L. Eloranta, U. Bave, G.V. Alm, L. Ronnblom, Induction of interferon-alpha production in plasmacytoid dendritic cells by immune complexes containing nucleic acid released by necrotic or late apoptotic cells and lupus IgG, Arthritis Rheum 50 (2004) 1861–1872.

[52] M. Magnusson, S. Magnusson, H. Vallin, L. Ronnblom, G.V. Alm, Importance of CpG dinucleotides in activation of natural IFN-alpha-producing cells by a lupus-related oligodeoxynucleotide, Scand J Immunol 54 (2001) 543–550.

[53] T. Lovgren, M.L. Eloranta, B. Kastner, M. Wahren-Herlenius, G.V. Alm, L. Ronnblom, Induction of interferon-alpha by immune complexes or liposomes containing systemic lupus erythematosus autoantigen- and Sjogren's syndrome autoantigen-associated RNA, Arthritis Rheum 54 (2006) 1917–1927.

[54] F.J. Barrat, T. Meeker, J. Gregorio, J.H. Chan, S. Uematsu, S. Akira, et al., Nucleic acids of mammalian origin can act as endogenous ligands for Toll-like receptors and may promote systemic lupus erythematosus, J Exp Med 202 (2005) 1131–1139.

[55] S.R. Christensen, J. Shupe, K. Nickerson, M. Kashgarian, R.A. Flavell, M.J. Shlomchik, Toll-like receptor 7 and TLR9 dictate autoantibody specificity and have opposing inflammatory and regulatory roles in a murine model of lupus, Immunity 25 (2006) 417–428.

[56] K.M. Nickerson, S.R. Christensen, J. Shupe, M. Kashgarian, D. Kim, K. Elkon, et al., TLR9 regulates TLR7- and MyD88-dependent autoantibody production and disease in a murine model of lupus, J Immunol 184 (2010) 1840–1848.

[57] I. Isnardi, Y.S. Ng, I. Srdanovic, R. Motaghedi, S. Rudchenko, H. von Bernuth, et al., IRAK-4- and MyD88-dependent pathways are essential for the removal of developing autoreactive B cells in humans, Immunity 29 (2008) 746–757.

[58] L.K. Bockenstedt, R.J. Gee, M.J. Mamula, Self-peptides in the initiation of lupus autoimmunity, J Immunol 154 (1995) 3516–3524.

[59] R. Roth, R.J. Gee, M.J. Mamula, B lymphocytes as autoantigen-presenting cells in the amplification of autoimmunity, Ann N Y Acad Sci. 815 (1997) 88–104.

[60] J.A. James, T. Gross, R.H. Scofield, J.B. Harley, Immunoglobulin epitope spreading and autoimmune disease after peptide immunization: Sm B/B'-derived PPPGMRPP and PPGIRGP induce spliceosome autoimmunity, J Exp Med 181 (1995) 453–461.

[61] J.A. James, J.B. Harley, Linear epitope mapping of an Sm B/B' polypeptide, J Immunol 148 (1992) 2074–2079.

[62] J.A. James, J.B. Harley, A model of peptide-induced lupus autoimmune B cell epitope spreading is strain specific and is not H-2 restricted in mice, J Immunol 160 (1998) 502–508.

[63] J.A. James, R.H. Scofield, J.B. Harley, Lupus humoral autoimmunity after short peptide immunization, Ann. NY Acad. Sci. 815 (1997) 124–127.

[64] B.P. Harvey, R.J. Gee, A.M. Haberman, M.J. Shlomchik, M.J. Mamula, Antigen presentation and transfer between B cells and macrophages, Eur J Immunol 37 (2007) 1739–1751.

[65] B.P. Harvey, T.E. Quan, B.J. Rudenga, R.M. Roman, J. Craft, M.J. Mamula, Editing antigen presentation: antigen transfer between human B lymphocytes and macrophages mediated by class A scavenger receptors, J Immunol 181 (2008) 4043–4051.

[66] R.H. Lin, M.J. Mamula, J.A. Hardin, C.A. Janeway Jr., Induction of autoreactive B cells allows priming of autoreactive T cells, Journal of Experimental Medicine 173 (1991) 1433–1439.

[67] M.J. Mamula, Epitope spreading: The role of self peptides and autoantigen processing by B lymphocytes, Immunological Reviews 164 (1998) 231–239.

[68] M.J. Mamula, S. Fatenejad, J. Craft, B cells process and present lupus autoantigens that initiate autoimmune T cell responses, J Immunol 152 (1994) 1453–1461.

[69] M.J. Mamula, R.H. Lin, C.A. Janeway Jr., J.A. Hardin, Breaking T cell tolerance with foreign and self co-immunogens. A study of autoimmune B and T cell epitopes of cytochrome c, J Immunol 149 (1992) 789–795.

[70] O.T.M. Chan, L.G. Hannum, A.H. Haberman, M.P. Madaio, M.J. Schlomchik, A novel mouse with B cells but lacking serum antibody reveals an antibody-independent role for B cells in murine lupus, J Exp Med 189 (1999) 1639–1647.

[71] O.T.M. Chan, M.P. Madaio, M.J. Shlomchik, B cells are required for lupus nephritis in the polygenic, Fas-intact MRL model of systemic autoimmunity, Journal of Immunology 163 (1999) 3592–3596.

[72] O.T.M. Chan, M.P. Madaio, M.J. Shlomchik, The central and multiple roles of B cells in lupus pathogenesis, Immunological Reviews 169 (1999) 107–121.

[73] O.T.M. Chan, M.J. Shlomchik, Cutting edge: B cells promote CD8$^+$ T cell activation in MRL-Fas(lpr) mice independently of MHC class I antigen presentation, Journal of Immunology 164 (2000) 1658−1662.

[74] E.S. Sobel, C. Mohan, L. Morel, J. Schiffenbauer, E.K. Wakeland, Genetic dissection of SLE pathogenesis: Adpotive transfer of *Sle1* mediates the loss of tolerance by bone marrow-derived B cells, J Immunol 162 (1999) 2415−2421.

[75] J. Yan, B.P. Harvey, R.J. Gee, M.J. Shlomchik, M.J. Mamula, B cells drive early T cell autoimmunity in vivo prior to dendritic cell-mediated autoantigen presentation, Journal of Immunology 177 (2006) 4481−4487.

[76] J. Yan, M.J. Mamula, B and T cell tolerance and autoimmunity in autoantibody transgenic mice, Int Immunol 14 (2002) 963−971.

[77] J. Yan, M.J. Mamula, Autoreactive T cells revealed in the normal repertoire: escape from negative selection and peripheral tolerance, J Immunol 168 (2002) 3188−3194.

[78] A. Lanzavecchia, Receptor-mediated antigen uptake and its effect on antigen presentation to class II-restricted T lymphocytes, Annu Rev Immunol 8 (1990) 773−793.

[79] S.K. Pierce, J.F. Morris, M.J. Grusby, P. Kaumaya, A. van Buskirk, M. Srinivasan, et al., Antigen-presenting function of B lymphocytes, Immunol Rev 106 (1988) 149−180.

[80] J. Yan, M.J. Wolff, J. Unternaehrer, I. Mellman, M.J. Mamula, Targeting antigen to CD19 on B cells efficiently activates T cells, Int Immunol 17 (2005) 869−877.

[81] A. Bergtold, D.D. Desai, A. Gavhane, R. Clynes, Cell surface recycling of internalized antigen permits dendritic cell priming of B cells, Immunity 23 (2005) 503−514.

[82] J. Neijssen, C. Herberts, J.W. Drijfhout, E. Reits, L. Janssen, J. Neefjes, Cross-presentation by intercellular peptide transfer through gap junctions, Nature 434 (2005) 83−88.

[83] D.M. Patel, R.W. Dudek, M.D. Mannie, Intercellular exchange of class II MHC complexes: ultrastructural localization and functional presentation of adsorbed I-A/peptide complexes, Cell Immunol 214 (2001) 21−34.

[84] G. Raposo, H.W. Nijman, W. Stoorvogel, R. Leijendekker, C.V. Harding, C.J.M. Melief, et al., B lymphocytes secrete antigen-presenting vesicles, Journal of Experimental Medicine 183 (1996) 1161−1172.

[85] R. Wubbolts, R.S. Leckie, P.T.M. Veenhuizen, G. Schwarzmann, W. Möbius, J. Hoernschemeyer, et al., Proteomic and biochemical analyses of human B cell-derived exosomes, Journal of Biological Chemistry 278 (2003) 10963−10972.

[86] Chen, Y., Wermeling, F., Sundqvist, J., Jonsson, A.-B., Tryggvason, K., Pikkarainen, T., et al. A regulatory role for macrophage class A scavenger receptors in TLR4-mediated LPS responses. European Journal of Immunology 40 (2010) 1451−1460

[87] P. Kleindienst, T. Brocker, Concerted antigen presentation by dendritic cells and B cells is necessary for optimal CD4 T-cell immunity in vivo, Immunology 115 (2005) 556−564.

[88] A. Girvan, F.E. Aldwell, G.S. Buchan, L. Faulkner, M.A. Baird, Transfer of macrophage-derived mycobacterial antigens to dendritic cells can induce naive T-cell activation, Scand J Immunol 57 (2003) 107−114.

[89] T.G. Phan, I. Grigorova, T. Okada, J.G. Cyster, Subcapsular encounter and complement-dependent transport of immune complexes by lymph node B cells, Nat Immunol 8 (2007) 992−1000.

[90] Y.R. Carrasco, F.D. Batista, B cells acquire particulate antigen in a macrophage-rich area at the boundary between the follicle and the subcapsular sinus of the lymph node, Immunity 27 (2007) 160−171.

[91] M. Wykes, A. Pombo, C. Jenkins, G.G. MacPherson, Dendritic cells interact directly with naive B lymphocytes to transfer antigen and initiate class switching in a primary T-dependent response, J Immunol 161 (1998) 1313−1319.

[92] R.M. Steinman, S. Turley, I. Mellman, K. Inaba, The induction of tolerance by dendritic cells that have captured apoptotic cells, J Exp Med 191 (2000) 411−416.

[93] K. Inaba, S. Turley, F. Yamaide, T. Iyoda, K. Mahnke, M. Inaba, et al., Efficient presentation of phagocytosed cellular fragments on the major histocompatibility complex class II products of dendritic cells, J Exp Med 188 (1998) 2163−2173.

[94] L.A. Harshyne, M.I. Zimmer, S.C. Watkins, S.M. Barratt-Boyes, A role for class A scavenger receptor in dendritic cell nibbling from live cells, J Immunol 170 (2003) 2302−2309.

[95] K. Denzer, M.v Eijk, M.J. Kleijmeer, E. Jakobson, C.d Groot, H.J. Geuze, Follicular dendritic cells carry MHC class II-expressing microvesicles at their surface, J Immunol 165 (2000) 1259−1265.

[96] T.G. Phan, I. Grigorova, T. Okada, J.G. Cyster, Subcapsular encounter and complement-dependent transport of immune complexes by lymph node B cells, Nat Immunol 8 (2007) 992−1000.

[97] M. Wykes, A. Pombo, C. Jenkins, G.G. MacPherson, Dendritic cells interact directly with naive B lymphocytes to transfer antigen and initiate class switching in a primary T-dependent response, J Immunol 161 (1998) 1313−1319.

[98] H. Qi, J.G. Egen, A.Y.C. Huang, R.N. Germain, Extrafollicular activation of lymph node B cells by antigen-bearing dendritic cells, Science 312 (2006) 1672−1676.

[99] N.-N. Huang, S.B. Han, I.Y. Hwang, J.H. Kehrl, B cells productively engage soluble antigen-pulsed dendritic cells: visualization of live-cell dynamics of B cell-dendritic cell interactions, J Immunol 175 (2005) 7125−7134.

[100] J.A. Hardin, The lupus autoantigens and the pathogenesis of systemic lupus erythematosus, Arthritis and Rheumatism 29 (1986) 457−460.

[101] A. Lanzavecchia, Antigen-specific interaction between T and B cells, Nature 314 (1985) 537−539.

[102] A. Lanzavecchia, Mechanisms of antigen uptake for presentation, Current Opinion in Immunology 8 (1996) 348−354.

[103] E. Roosnek, A. Lanzavecchia, Efficient and selective presentation of antigen-antibody complexes by rheumatoid factor B cells, J Exp Med 173 (1991) 487−489.

[104] C. Watts, A. Lanzavecchia, Suppressive effect of antibody on processing of T cell epitopes, J Exp Med 178 (1993) 1459−1463.

[105] H.W. Davidson, C. Watts, Epitope-directed processing of specific antigen by B lymphocytes, J Cell Biol 109 (1989) 85−92.

[106] H.W. Davidson, M.A. West, C. Watts, Endocytosis, intracellular trafficking, and processing of membrane IgG and monovalent antigen/membrane IgG complexes in B lymphocytes, J Immunol 144 (1990) 4101−4109.

[107] C. Watts, Capture and processing of exogenous antigens for presentation on MHC molecules, Ann Rev Immunol 15 (1997) 821−850.

[108] C. Watts, H.W. Davidson, Endocytosis and recycling of specific antigen by human B cell lines, Embo J 7 (1988) 1937−1945.

[109] C. Watts, P.A. Reid, M.A. West, H.W. Davidson, The antigen processing pathway in B lymphocytes, Semin Immunol 2 (1990) 247−253.

[110] C. Watts, M.A. West, P.A. Reid, H.W. Davidson, Processing of immunoglobulin-associated antigen in B lymphocytes. Cold Spring Harb Symp Quant Biol. 54 Pt 1 (1989) 345−352.

[111] P. Weiser, R. Muller, U. Braun, M. Reth, Endosomal targeting by the cytoplasmic tail of the membrane immunoglobulin, Science 276 (1997) 407−409.

[112] D. Tarlinton, Antigen presentation by memory B cells: the sting is in the tail, Science 276 (1997) 374–375.

[113] R. Noelle, J.A. Ledbetter, A. Aruffo, CD40 and its ligand, an essential ligand-receptor pair for thymus dependent B cell activation, Immunol Today 13 (1992) 431–433.

[114] D.J. Lenschow, A.I. Sperling, M.P. Cooke, G. Freeman, L. Rhee, D.C. Decker, et al., Differential up regulation of the B7-1 and B7-2 co-stimulatory molecules after Ig receptor engagement by antigen, J Immunol 153 (1994) 1990–1997.

[115] W.Y. Ho, M.P. Cooke, C.C. Goodnow, M.M. Davis, Resting and anergic B cells are defective in CD28 dependent co-stimulation of naive CD4+ T cells, J Exp Med 179 (1994) 1539–1549.

[116] S. Ozaki, J.A. Berzofsky, Antibody conjugates mimic specific B cell presentation of antigen: Relationship between T and B cell specificity, J Immunol 138 (1987) 4133–4142.

[117] K. Tenbrock, Y.T. Juang, V.C. Kyttaris, G.C. Tsokos, Altered signal transduction in SLE T cells, Rheumatology (Oxford) 46 (2007) 1525–1530.

[118] K. Tenbrock, G.C. Tsokos, Transcriptional regulation of interleukin 2 in SLE T cells, Int Rev Immunol 23 (2004) 333–345.

[119] M.P. Nambiar, S. Krishnan, G.C. Tsokos, T-cell signaling abnormalities in human systemic lupus erythematosus, Methods Mol Med 102 (2004) 31–47.

[120] G.C. Tsokos, J.P. Mitchell, Y.T. Juang, T cell abnormalities in human and mouse lupus: intrinsic and extrinsic, Curr Opin Rheumatol 15 (2003) 542–547.

[121] G.S. Vratsanos, S. Jung, Y.M. Park, J. Craft, CD4(+) T cells from lupus-prone mice are hyperresponsive to T cell receptor engagement with low and high affinity peptide antigens: a model to explain spontaneous T cell activation in lupus, J Exp Med 193 (2001) 329–337.

[122] F. Monneaux, J.P. Briand, S. Muller, B and T cell immune response to small nuclear ribonucleoprotein particles in lupus mice: Autoreactive CD4+ T cells recognize a T cell epitope located within the RNP80 motif of the 70K protein, European Journal of Immunology 30 (2000) 2191–2200.

[123] B. Liang, R.J. Gee, M.J. Kashgarian, A.H. Sharpe, M.J. Mamula, B7 co-stimulation in the development of lupus: autoimmunity arises either in the absence of B7.1/B7.2 or in the presence of anti-b7.1/B7.2 blocking antibodies, J Immunol 163 (1999) 2322–2329.

[124] B. Liang, M.J. Kashgarian, A.H. Sharpe, M.J. Mamula, Autoantibody responses and pathology regulated by B7-1 and B7-2 co-stimulation in MRL/lpr lupus, J Immunol 165 (2000) 3436–3443.

[125] X. Yu, S. Fournier, J.P. Allison, A.H. Sharpe, R.J. Hodes, The role of B7 co-stimulation in CD4/CD8 T cell homeostasis, J Immunol 164 (2000) 3543–3553.

[126] E.A. Tivol, F. Borriello, A.N. Schweitzer, W.P. Lynch, J.A. Bluestone, A.H. Sharpe, Loss of CTLA-4 leads to massive lymphoproliferation and fatal multiorgan tissue destruction, revealing a critical negative regulatory role of CTLA-4, Immunity 3 (1995) 541–547.

[127] M.K. Racke, D.E. Scott, L. Quigley, G.S. Gray, R. Abe, C.H. June, et al., Distinct roles for B7-1 (CD80) and B7-2 (CD86) in the initiation of experimental allergic encephalomyelitis, J Clin Invest 1995 (1995) 2195–2203.

[128] V.K. Kuchroo, M.P. Das, J.A. Brown, A.M. Ranger, S.S. Zamvil, R.A. Sobel, et al., B7-1 and B7-2 co-stimulatory molecules activate differentially the Th1/Th2 developmetnal pathways: application to autoimmune disease therapy, Cell. 1995 (1995) 707–718.

[129] A. Nakajima, M. Azuma, S. Kodera, S. Nuriya, A. Terashi, M. Abe, et al., Preferential dependence of autoantibody production in murine lupus on CD86 co-stimulatory molecule, Eur J Immunol 25 (1995) 3060–3069.

[130] P. Waterhouse, J.M. Penninger, E. Timms, A. Wakeham, A. Shahinian, K.P. Lee, et al., Lymphoproliferative disorders with early lethality in mice deficient in Ctla-4, Science 270 (1995) 985–988.

[131] A.H. Cross, T.J. Girard, K.S. Giacoletto, R.J. Evans, R.M. Keeling, R.F. Lin, et al., Long-term inhibition of murine experimental autoimmune encephalomyelitis using CTLA-4-Fc supports a key role for CD28 co-stimulation, J Clin Invest 95 (1995) 2783–2789.

[132] P.J. Perrin, D. Scott, L. Quigley, P.S. Albert, O. Feder, G.S. Gray, et al., Role of B7:CD28/CTLA-4 in the induction of chronic relapsing experimental allergic encephalomyelitis, J Immunol 154 (1995) 1481–1490.

[133] P.J. Perrin, D. Scott, C.H. June, M.K. Racke, B7-mediated co-stimulation can either provoke or prevent clinical manifestations of experimental allergic encephalomyelitis, Immunol Res 14 (1995) 189–199.

[134] S.J. Khoury, E. Akalin, A. Chandraker, L.A. Turka, P.S. Linsley, M.H. Sayegh, et al., CD28-B7 co-stimulatory blockade by CTLA4Ig prevents actively induced experimental autoimmune encephalomyelitis and inhibits Th1 but spares Th2 cytokines in the central nervous system, J Immunol 155 (1995) 4521–4524.

[135] B.K. Finck, P.S. Linsley, D. Wofsy, Treatment of murine lupus with CTLA4Ig, Science 265 (1994) 1225–1227.

[136] D.I. Daikh, B.K. Finck, P.S. Linsley, D. Hollenbaugh, D. Wolfy, Long-term inibition of murine lupus by brief simultaneous blockade of the B7/CD28 and CD40/gp39 co-stimulation pathways, J Immunol 159 (1997) 3104–3108.

[137] Bruce, I.N., Gordon, C., Merrill, J.T. and Isenberg, D. Clinical trials in lupus: what have we learned so far? Understanding the gap between reality and expectation. Rheumatology (Oxford) 49 (2010) 1025–1027.

[138] F. Mackay, J.L. Browning, BAFF: A fundamental survival factor for B cells, Nature Reviews Immunology 2 (2002) 465–475.

[139] F. Mackay, S.A. Woodcock, P. Lawton, C. Ambrose, M. Baetscher, P. Schneider, et al., Mice transgenic for BAFF develop lymphocytic disorders along with autoimmune manifestations, Journal of Experimental Medicine 190 (1999) 1697–1710.

[140] S.L. Kalled, The role of BAFF in immune function and implications for autoimmunity, Immunological Reviews 204 (2005) 43–54.

[141] L.G. Ng, A.P.R. Sutherland, R. Newton, F. Qian, T.G. Cachero, M.L. Scott, et al., B cell-activating factor belonging to the TNF family (BAFF)-R is the principal BAFF receptor facilitating BAFF co-stimulation of circulating T and B cells, Journal of Immunology 173 (2004) 807–817.

[142] J. Groom, S.L. Kalled, A.H. Cutler, C. Olson, S.A. Woodcock, P. Schneider, et al., Association of BAFF/BLyS overexpression and altered B cell differentiation with Sjögren's syndrome, Journal of Clinical Investigation 109 (2002) 59–68.

[143] G.S. Cheema, V. Roschke, D.M. Hilbert, W. Stohl, Elevated serum B lymphocyte stimulator levels in patients with systemic immune-based rheumatic diseases, Arthritis and Rheumatism 44 (2001) 1313–1319.

[144] J. Zhang, V. Roschke, K.P. Baker, Z. Wang, G.S. Alarcon, B.J. Fessler, et al., Cutting edge: A role for B lymphocyte stimulator in systemic lupus erythematosus, Journal of Immunology 166 (2001) 6–10.

[145] U. Salzer, S. Jennings, B. Grimbacher, To switch or not to switch - The opposing roles of TACI in terminal B cell differentiation, European Journal of Immunology 37 (2007) 17–20.

[146] J.S. Thompson, S.A. Bixler, F. Qian, K. Vora, M.L. Scott, T.G. Cachero, et al., BAFF-R, a newly identified TNF receptor that specifically interacts with BAFF, Science 293 (2001) 2108–2111.

[147] M. Yan, J.R. Brady, B. Chan, W.P. Lee, B. Hsu, S. Harless, et al., Identification of a novel receptor for B lymphocyte stimulator that is mutated in a mouse strain with severe B cell deficiency, Curr Biol 11 (2001) 1547–1552.

[148] N. Kayagaki, M. Yan, D. Seshasayee, H. Wang, W. Lee, D.M. French, et al., BAFF/BLyS receptor 3 binds the B cell survival factor BAFF ligand through a discrete surface loop and promotes processing of NF-kappaB2, Immunity 17 (2002) 515–524.

[149] W. Stohl, S. Metyas, S.M. Tan, G.S. Cheema, B. Oamar, V. Roschke, et al., Inverse association between circulating APRIL levels and serological and clinical disease activity in patients with systemic lupus erythematosus, Ann Rheum Dis 63 (2004) 1096–1103.

[150] T. Koyama, H. Tsukamoto, Y. Miyagi, D. Himeji, J. Otsuka, H. Miyagawa, et al., Raised serum APRIL levels in patients with systemic lupus erythematosus, Ann Rheum Dis 64 (2005) 1065–1067.

[151] T. Koyama, H. Tsukamoto, K. Masumoto, D. Himeji, K. Hayashi, M. Harada, et al., A novel polymorphism of the human APRIL gene is associated with systemic lupus erythematosus, Rheumatology (Oxford) 42 (2003) 980–985.

[152] M.C. Ryan, I.S. Grewal, Targeting of BAFF and APRIL for Autoimmunity and Oncology, Adv Exp Med Biol 647 (2009) 52–63.

[153] G. Yu, T. Boone, J. Delaney, N. Hawkins, M. Kelley, M. Ramakrishnan, et al., APRIL and TALL-I and receptors BCMA and TACI: system for regulating humoral immunity, Nat Immunol 1 (2000) 252–256.

[154] S.K. Yoshinaga, M. Zhang, J. Pistillo, T. Horan, S.D. Khare, K. Miner, et al., Characterization of a new human B7-related protein: B7RP-1 is the ligand to the co-stimulatory protein ICOS, Int Immunol 12 (2000) 1439–1447.

[155] X.Z. Xia, J. Treanor, G. Senaldi, S.D. Khare, T. Boone, M. Kelley, et al., TACI is a TRAF-interacting receptor for TALL-1, a tumor necrosis factor family member involved in B cell regulation, J Exp Med 192 (2000) 137–143.

[156] J.A. Gross, S.R. Dillon, S. Mudri, J. Johnston, A. Littau, R. Roque, et al., TACI-Ig neutralizes molecules critical for B cell development and autoimmune disease: Impaired B cell maturation in mice lacking BLyS, Immunity 15 (2001) 289–302.

[157] J.A. Gross, J. Johnston, S. Mudri, R. Enselman, S.R. Dillon, K. Madden, et al., TACI and BCMA are receptors for a TNF homologue implicated in B-cell autoimmune disease, Nature 404 (2000) 995–999.

[158] W. Stohl, R.J. Looney, B cell depletion therapy in systemic rheumatic diseases: Different strokes for different folks? Clinical Immunology 121 (2006) 1–12.

[159] R. Furie, M. Petri, M.H. Weisman, Belimumab (fully human monoclonal antibody to BlyS) improved or stabilized systemic lupus erythematosus (SLE) disease activity and reduced flare during 3 years of therapy, Ann Rheum Dis 67 (2008) 53.

[160] E. Ginzler, R. Furie, D. Wallace, Novel combined response endpoint shows that belimumab (fully human monoclonal antibody to B-lymphocyte stimulator [BLYS]) improves or stabilizes SLE disease activity in a phase 2 trial, Ann Rheum Dis 66 (2007) 56.

[161] W. Chatham, C. Aranow, R. Furie, Progressive normalization of autoantibody, immunoglobulin, and complement levels over 3 years of belimumab (fully human monoclonal antibody to BlyS) therapy in systemic lupus erythematosus (SLE) patients, Ann Rheum Dis 67 (2008) 217.

[162] D.J. Wallace, W. Stohl, R.A. Furie, J.R. Lisse, J.D. McKay, J.T. Merrill, et al., A phase II, randomized, double-blind, placebo-controlled, dose-ranging study of belimumab in patients with active systemic lupus erythematosus, Arthritis Rheum 61 (2009) 1168–1178.

[163] E.M. Ruderman, R.M. Pope, Drug insight: Abatacept for the treatment of rheumatoid arthritis, Nature Clinical Practice Rheumatology 2 (2006) 654–660.

[164] F. Vincenti, M. Luggen, T cell co-stimulation: A rational target in the therapeutic armamentarium for autoimmune diseases and transplantation, Annual Review of Medicine 58 (2007) 347–358.

[165] B. Diamond, J. Bluestone, D. Wofsy, The immune tolerance network and rheumatic disease: Immune tolerance comes to the clinic, Arthritis and Rheumatism 44 (2001) 1730–1735.

[166] A. Davidson, B. Diamond, D. Wofsy, D. Daikh, Block and tackle: CTLA4Ig takes on lupus, Lupus 14 (2005) 197–203.

[167] C. Kneitz, M. Wilhelm, H.P. Tony, Effective B cell depletion with rituximab in the treatment of autoimmune diseases, Immunobiology 206 (2002) 519–527.

[168] R. Eisenberg, R.J. Looney, The therapeutic potential of anti-CD20: What do B-cells do? Clinical Immunology 117 (2005) 207–213.

[169] R.J. Looney, B cells as a therapeutic target in autoimmune diseases other than rheumatoid arthritis, Rheumatology 44 (2005) ii13–ii17.

[170] R.J. Looney, J.H. Anolik, D. Campbell, R.E. Felgar, F. Young, L.J. Arend, et al., B cell depletion as a novel treatment for systemic lupus erythematosus: A phase I/II dose-escalation trial of rituximab, Arthritis and Rheumatism 50 (2004) 2580–2589.

[171] M.J. Leandro, G. Cambridge, J.C. Edwards, M.R. Ehrenstein, D.A. Isenberg, B-cell depletion in the treatment of patients with systemic lupus erythematosus: A longitudinal analysis of 24 patients, Rheumatology 44 (2005) 1542–1545.

[172] S.D. Marks, S. Patey, P.A. Brogan, N. Hasson, C. Pilkington, P. Woo, et al., B lymphocyte depletion therapy in children with refractory systemic lupus erythematosus, Arthritis and Rheumatism 52 (2005) 3168–3174.

[173] K.G.C. Smith, R.B. Jones, S.M. Burns, D.R.W. Jayne, Long-term comparison of rituximab treatment for refractory systemic lupus erythematosus and vasculitis: Remission, relapse, and re-treatment, Arthritis and Rheumatism 54 (2006) 2970–2982.

[174] Merrill, J.T., Neuwelt, C.M., Wallace, D.J., Shanahan, J.C., Latinis, K.M., Oates, J.C., et al. Efficacy and safety of rituximab in moderately-to-severely active systemic lupus erythematosus: the randomized, double-blind, phase II/III systemic lupus erythematosus evaluation of rituximab trial. Arthritis Rheum 62 (2010) 222–233.

[175] F. Martin, A.C. Chan, B cell immunobiology in disease: Evolving concepts from the clinic, Annual Review of Immunology 24 (2006) 467–496.

[176] Q. Gong, Q. Ou, S. Ye, W.P. Lee, J. Cornelius, L. Diehl, Y.L. Wei, et al., Importance of cellular microenvironment and circulatory dynamics in B cell immunotherapy, Journal of Immunology 174 (2005) 817–826.

[177] Y. Hamaguchi, J. Uchida, D.W. Cain, G.M. Venturi, J.C. Poe, K.M. Haas, et al., The peritoneal cavity provides a protective niche for B1 and conventional B lymphocytes during anti-CD20 immunotherapy in mice, Journal of Immunology 174 (2005) 4389–4399.

[178] J. Uchida, Y. Hamaguchi, J.A. Oliver, J.V. Ravetch, J.C. Poe, K.M. Haas, et al., The innate mononuclear phagocyte network depletes B lymphocytes through Fc receptor-dependent mechanisms during anti-CD20 antibody immunotherapy, Journal of Experimental Medicine 199 (2004) 1659–1669.

Polymorphonuclear and Endothelial Cells

Robert Clancy

INTRODUCTION

Systemic lupus erythematosus (SLE) is a disease state posing several challenges to the clinician, including heterogeneity of presentation, undulating course, and an extraordinary risk for vascular injury [1, 2] including premature atherosclerosis as well as endothelial injury related to renal disease. Central to this concept is a focus largely on the endothelium, since it provides the physiologic boundary which limits extravasation and diapedesis of inflammatory cells. In this section, we provide a clinical overview and then outline the putative pathogenic events which occur in SLE autoimmune-associated vasculopathy for atherosclerosis and renal disease.

With regard to premature cardiovascular disease, over three decades ago it was noted that the majority of deaths in SLE patients with longer disease duration were attributed to atherosclerosis [1]. Indeed, the rate of myocardial infarction in women aged 35—44 years is 50 times greater than expected [2]. Patients with SLE have an increased atherosclerotic risk despite adjustment for traditional Framingham risk factors [2, 3]. Risk factors among SLE patients are somewhat controversial but may include longer duration of disease and lower likelihood of treatment with prednisone, cyclophosphamide, or hydroxychloroquine [4]. Thus, inflammation related to underlying disease is likely to be contributory. McMahon and coworkers have recently shown that plasma from SLE patients with premature atherosclerosis is enriched in proinflammatory HDL [5]. However, the inflammation may be clinically subtle since detectable cardiovascular events have been unexpectedly reported in SLE patients with extended periods of quiescence [6] and subclinical atherosclerosis has not correlated with disease activity index scores [4].

Functional impairment of the endothelium is reflected by the pattern of "proinjury" mediators such as circulating endothelial cells (CECs), apoptotic circulating endothelial cells and soluble E-selectin. In atherogenic disease, endothelial protection may be subverted because of a loss of boundary function via detachment of endothelial cells into the circulation and/or a change in the endothelial cell phenotype. Increased levels of CECs have been observed in patients with active disease [7] and apoptotic CECs have been reported in SLE patients with diminished flow-mediated dilatation (FMD) [8]. Soluble E-selectin (sE-selectin), likely shed from an abnormally activated endothelium, has been recently associated with atherosclerosis as defined by an abnormal coronary artery calcium assessed by electron beam computed tomography EBCT [9]. Blood vessel homeostasis involves a complex interplay between inflammatory signals, coagulation signals, and other mediators. T-cell recruitment of monocytes into artery walls may be a critical step in the effective handling of cholesterol. Cholesterol in the bloodstream is scavenged by a low-density lipoprotein molecule and deposited in the arterial wall, where monocytes, which have been recruited by activated T cells, are a part of normal homeostasis (Figure 9.1). The monocyte should optimally serve as a temporary depot for fats, differentiating into an efficient cholesterol-metabolizing macrophage until the excess lipid can be picked up by high-density lipoprotein. Contributory factors including upregulation of IFN-γ, IL12 and Th1 cells are found within atherosclerotic lesions. For example, lesions contain T cells, primarily of the Th1 subtype, and IL-12 ([10], a stimulus for Th1 differentiation). Exposure of monocytes to oxidized LDL results in IL-12 production [11]. A pathologic role of this pathway is supported by the finding that atherosclerosis in mice is attenuated by treatment using a IL12-neutralizing antibodies [12]. Complement activation, T-cell cytokine interferon-γ and/or T-cell-stimulated immune complexes promote the activation of endothelial cells to express surface metalloproteinases; concomitantly, there is an impaired lipid handling

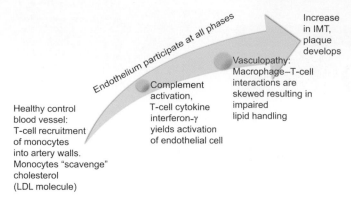

FIGURE 9.1 Macrophage—T-cell interactions in atherosclerosis. Blood vessel homeostasis involves a complex interplay between inflammatory signals, coagulation signals, and other mediators. T-cell recruitment of monocytes into artery walls may be a critical step in the effective handling of cholesterol. Cholesterol in the bloodstream is scavenged by a low-density lipoprotein molecule and deposited in the arterial wall, where monocytes, which have been recruited by activated T cells, are a part of normal homeostasis. Complement activation, T-cell cytokine interferon-γ and/or T-cell-stimulated immune complexes yield the activation of endothelial cell to express surface metalloproteinases and concomitantly, there is an impaired lipid handling with a sequential narrowing of an artery. Once this system becomes overwhelmed, cholesterol plaques develop, which induce both inflammatory and fibrotic reactions. *(Adapted from Xenox et al., 2007.)*

with sequential arterial narrowing. Once this system becomes overwhelmed, cholesterol plaques develop, which induce both inflammatory and fibrotic reactions.

Lupus nephritis (LN) predominantly affects women of African-American and Hispanic minority ethnic/racial populations within the reproductive age group. A higher incidence of progression to end-stage renal disease (ESRD) in these ethnic subsets has been demonstrated, compared to other predominantly Caucasian populations [13]. Although the overall 10-year survival has improved to more than 90% in SLE patients [14], the incidence of LN patients progressing to ESRD has remained constant over the years [13]. A review of the literature has demonstrated a difference in response to treatment modalities for LN based on ethnic/racial differences. For patients with proliferative forms of LN, renal survival is worse with progression to ESRD in African-American and Hispanic patients despite treatment with intravenous cyclophosphamide (IVC) even after controlling for hypertension, initial renal functional impairment, and quantity of corticosteroid therapy [15, 16]. Socioeconomic features (income, educational level, access to health care) may contribute to the poorer prognosis in these populations. In one study, however, the relative risk of progression to ESRD remained higher in the Hispanic population [16]. Several recent studies have found that African-American and Hispanic populations had a better response to mycophenolate mofetil (MMF) than to IVC [17, 18].

No new medications have been FDA-approved for the treatment of SLE or LN specifically, in over 50 years. Most drugs used in the treatment regimen are commercially available and prescribed off-label, or available only as part of an investigational protocol. Recent clinical trials for newer agents in the treatment of LN may not have achieved predetermined endpoints either due to an actual absence of efficacy or because of shortcomings in the design of the study protocol. Further, pharmacogenomics may have a role in determining the efficacy and safety of a medication. It is, therefore, important to know what effect a medication may have in improving the renal survival and overall outcome in select subgroups of patients or even in a specific individual patient before subjecting them to aggressive immunosuppression and its potential side effects.

Nephritis, a life-threatening manifestation of SLE, is strongly influenced by blood vessels due, in part, to a central role by the vasculature to support homeostasis. The contribution of the vascular endothelium to the pathogenesis of renal injury has not been emphasized in lupus nephritis. Despite potential biologic insights and treatment strategies to be gained by studying the endothelium in LN, neither historic WHO classification, NIH chronicity (CI) and activity (AI) indices [19], nor recent ISN/RPS 2003 pathologic classifications of LN [20] specifically address the state of the microvasculature in their definitions. However, recent murine data based on microarray analysis suggest that endothelial activation is a feature shared by progressive glomerulosclerosis compared to non-progressive glomerulosclerosis [21].

For one scenario of putative pathogenic events, the exposure of healthy endothelial cells to such potential stimuli as circulating IFN-α, TNF-α, or immune complexes present in patients with active SLE, results in the expression of NOS2, generation of NO and adhesion molecules. As shown in Figure 9.2, this activated endothelium has now lost its function to serve as a "physiological brake" which normally prevents the infiltration of inflammatory cells that produce Th1 cytokines including IL-12 and IL-18, a potent chemoattractant for pDCs [22]. Endothelial cells may also be activated by IL-18 [23]. pDCs release IFNs which have a paracrine effect on other cell types to express NOS2. In addition to the local inflammatory consequences of activation, endothelial cells are shed into the circulation and membrane EPCR is lost such that EPCR circulates as a soluble form (sEPCR), with a procoagulant effect. Prothrombin F1+2 (a marker of thrombin generation) are generated, a consequence of thrombin which is indirectly responsible for release of sEPCR. The increased release of sEPCR coupled with higher thrombin generation suggests that less membrane-bound EPCR will be available in these individuals for efficient protein C activation.

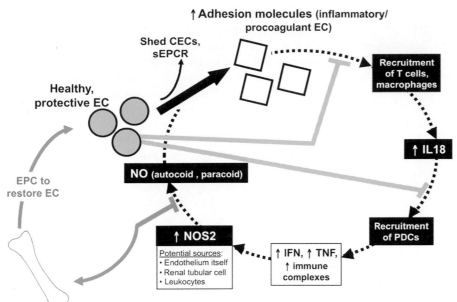

↑ **Adhesion molecules (inflammatory/procoagulant EC)**

Shed CECs, sEPCR

Healthy, protective EC

EPC to restore EC

NO (autocoid, paracoid)

↑ **NOS2**

Potential sources:
• Endothelium itself
• Renal tubular cell
• Leukocytes

↑ IFN, ↑ TNF, ↑ immune complexes

Recruitment of PDCs

↑ IL18

Recruitment of T cells, macrophages

FIGURE 9.2   Amplification of renal injury in SLE. In the initial putative pathogenic events, the exposure of healthy endothelial cells to such potential stimuli as circulating IFN-α, TNF-α, or immune complexes present in patients with active SLE, results in the expression of NOS2, generation of NO and adhesion molecules. This activated endothelium has now lost its function to serve as a "physiological brake", which normally prevents the infiltration of inflammatory cells that produce IL-18, a potent chemoattractant for pDCs [19]. Endothelial cells may also be activated by IL-18. pDCs release IFNs which have a paracrine effect on other cell types to express NOS2. In addition to the local inflammatory consequences of activation, endothelial cells are shed into the circulation and membrane EPCR is lost such that EPCR is now circulating as a soluble form (sEPCR), a procoagulant effect. Prothrombin F1+2 (a marker of thrombin generation) are generated, a consequency of thrombin which is indirectly responsible for release of sEPCR. The increased release of sEPCR coupled with higher thrombin generation suggests that less membrane-bound EPCR will be available in these individuals for efficient protein C activation.

Taken together, adipocyte-derived protein, adiponectin and membrane endothelial protein C receptor (mEPCR) may serve as biomarkers of morbidities involving premature atherosclerosis and vascular injury in nephritis, respectively. In this review, the focus will be on the structure of adiponectin and membrane EPCR and their expression in the context of the proinflammatory CR3-dependent signaling pathways.

PROFILE OF ADIPONECTIN IN HEALTH AND DISEASE

Adiponectin (also known as 30-kDa adipocyte complement-related protein: Acrp30) is a secreted protein which is constitutively produced by adipocytes. Adiponectin, a trimer in serum, is a 30-kDa protein comprised of four domains including signal peptide at the N-terminus, a variable domain, a collagenous domain and a C-terminal globular domain homologous to C1q. The protein is well-characterized regarding its capacity to improve insulin sensitivity. During the early 1990s the best-characterized biological property was enhancing glycogen accumulation and fatty acid oxidation in C2C12 myotubes. The *in vivo* action to regulate serum levels of fatty acids and glucose was linked to adiponectin's effect on the liver to suppress glucose output while acting on muscle to increase glucose uptake and fatty acid oxidation.

However, adiponectin's actions are not restricted to controlling glucose and lipid metabolism Properties involving anti-inflammatory and anti-atherosclerotic functions have also been reported. Adiponectin accumulates in the sub-endothelium of injured human arteries where it inhibits monocyte adhesion to endothelial cells and ultimately inhibits the migration and proliferation of vascular smooth muscle, which contribute to the atherosclerotic process [24] through a mechanism which in part involves the downregulation of adhesion molecules by attenuating the nuclear factor κB pathway [25, 26]. In addition, adiponectin has been found in blood vessel walls after experimental endothelial injury [27], and is strongly expressed around infarcted but not normal myocardium [28], supporting a role in vascular and endothelial remodeling. Specifically, adiponectin was recruited to the affected area of an arterial injury site in balloon-injured rat carotid arteries. Adiponectin knockout (KO) mice were employed to further study the relationship between adiponectin and the properties of the vasculature. The mice displayed an impaired endothelium-dependent vasodilation. A series of experiments were performed which suggested that adiponectin serves a role as a proangiogenic regulator. Angiogenic repair of ischemic hindlimbs was impaired in adiponectin KO mice compared to wild-type. These data suggested that the exogenous supplementation of adiponectin could be beneficial treatment for obesity-related vascular disorders.

mEPCR: PROTECTION OF THE ENDOTHELIUM

Endothelial cell protein C receptor (EPCR), which has been cloned in mice and human tissues [29], is constitutively expressed by endothelial cells, particularly in large blood vessels as well as monocyte/macrophages. EPCR is a 46-kDa type 1 transmembrane glycoprotein with structural features consistent with an antigen-presenting groove analogous to MHC class 1 and CD1 family of proteins. Its structure is comprised of a large extracellular domain (221 amino acids), a transmembrane domain (25 amino acids) and a short highly conserved cytoplasmic sequence (3 amino acids). The 5′ flanking region of the murine endothelial protein C receptor gene was recently examined. It was shown to contain elements which reflect involvement in cell growth and development as well as a thrombin response element. Interestingly, endotoxin and thrombin elevate rodent EPCR mRNA levels and increase receptor shedding *in vivo* [30]. A phospholipid is tightly bound in the position of the antigen-presenting groove, which suggests that EPCR recycles to membrane endosomes and that it participates in antigen presentation. A well-characterized biological property of membrane EPCR is its role as an accessory factor to thrombomodulin–thrombin complexes which results in a dramatic augmentation of the formation of active protein C (APC) [31, 32]. For example, previous studies showed that APC generation was dependent on EPCR and that blocking protein C-EPCR interaction decreased APC formation about 10-fold [33]. In baboons, EPCR-blocking antibodies were found to decrease protein C activation and increase susceptibility to bacterial sepsis. These studies have advanced the notion that high levels of mEPCR are a beneficial property and conversely, lower levels of mEPCR are deleterious. Support was obtained in several studies which have focused on genetic manipulation of EPCR in mice. Over-expression of EPCR under the control of an endothelium-specific Tie2 promoter was shown [34] to dramatically alter patterns of endothelial EPCR expression; yet the mice did not exhibit any gross hemorrhagic abnormalities. They did however, exhibit an eight-fold increase in APC generation in response to infusion of thrombin and were partially resistant to a lethal dose of bacterial LPS. These findings confirm and extend the results of previous studies, which used blocking antibodies (see above).

Cleavage of EPCR from the cell surface by matrix metalloproteinases has been demonstrated [35]. This action is initiated when endothelial cells are treated with LPS, inflammatory cytokines and/or thrombin. A single nucleotide polymorphism (SNP) in exon 4 of the EPCR gene at 20q11, which converts serine 219 to glycine (219 Gly) in a region of the molecule close to the plasma membrane, is associated with increased basal and stimulated shedding of EPCR from endothelial cells[36]. The absence of membrane EPCR resulted in an attenuation of the thrombin/thrombomodulin-dependent protein C activation. Since sEPCR also binds to protein C and active protein C [29], functionally it

results in a loss of a "brake" to thrombosis and inflammation. In a murine model which employs heterozygotes of EPCR knockout mice, administration of endotoxin-induced disseminated intravascular coagulation, which was aggravated in heterozygous protein-C-deficient mice compared to wild-type. It is tempting to speculate that low levels of membrane EPCR secondary to high shedding of EPCR may also have deleterious consequences to coagulation and inflammation. However, studies in humans, which initially focused on deep vein thrombosis have not given uniform results of the risk for venous thrombosis for the genotype in separate cohorts [36, 37].

mEPCR may also have direct effects on endothelial cell phenotype. For example, the shedding of EPCR is associated with the activation of protease-activated receptors (PARs), a novel family of G-protein-coupled receptors that are constitutively expressed on endothelial cells and are involved in the early recruitment of leukocytes [38].

THE CD11B CHAIN SERVES AS A SWITCH HITTER OF THE CR3-DEPENDENT INFLUENCE ON THE CELL-MEDIATED IMMUNE SYSTEM

The complement receptor 3 or CR3 (also known as Mac1, αMβ2, gp165/gp95) is a type I integral membrane heterodimer with expression which is restricted to leukocytes. Major ligands include iC3b, ICAM1, ICAM2 and fibrinogen. Interestingly, CR3 has tremendous flexibility regarding its interaction with varied ligands. For example, it has high-affinity interactions with proteins that possess little sequence homology (<1% identity). There is a strong precedent for CR3 to mediate pro-inflammation. Structural requirements of this signal transducer that mediate inflammation were shown to be restricted to the CD11b chain by a narrow stretch of amino acids contained within the "headpiece" domain (which is comprised of three domains including the beta-propeller domain, I domain and I-like domain). Unexpectedly, it was also shown that CD11b chain mediates anti-injury properties as well.

In this subsection, structural discoveries are presented including CR3's X-ray structure and its analysis after mutation/deletion. The major findings include that the I domain of the integrin faces down toward the membrane in the inactive conformation and it opens or extends upward in a "switchblade"-like opening motion (initially proposed by Shimaoka and coworkers [39]) upon activation (resulting in the high-affinity form). A focal point of ligand binding is a metal-containing site of the CD11b molecule [40, 41], which in its closed conformation, contains a metal ion which interacts with aspartic acid 242 (D242) and serine 144 (S144). In contrast, during the high affinity, there is a shift off at this site with the metal domain interacting with an acidic residue of a ligand with support by threonine 209 (T209), serine 142 (S142) and serine 149 (S149). Of interest, the CD11b chain was initially characterized based on its crystallized presentation in the absence or presence of varied metal ions (Mn^{2+} (closed), Mg^{2+} (open)). For the former, the extracellular domain was characterized by a linear "closed knife-like" structure with the "headpiece" which faces the membrane. These domains are connected in series to "stalk or handle" regions as well as membrane and cytoplasmic domains. The pivotal role by this cluster of amino acids at the metal-binding site has a striking structural homology to small GTPases [42]. For example, the GTP-bound form, which is active in binding to and stimulating effector molecules, and the inactive GDP-bound form have Mg^2 ion coordinations that are analogous to those of the open and closed I domains, respectively.

Other structural features of CD11b include the mutation at Ile-316 involving I316A favors the open configuration [43]. The X-ray studies offer a molecular explanation. Ile-316 contributes to a closed configuration in part due to its capacity to wedge into a hydrophobic socket (a Velcro-like action keeping I domain closed) [41]. Other factors which favor a closed configuration include the capacity of CD18 to constrain its partner alphaM I domain. For structure of CD18, mutations near the junction between the transmembrane and cytoplasmic domains, as well as deletion of the cytoplasmic domains, have been repeatedly shown to result in integrin activation [44]. This finding is consistent with prior studies CD18 which highlight that phosphorylation events regulate ligand binding by integrins. Specifically, there is a phosphorylation of the cytoplasmic tail by protein kinase C of serine (amino acid 756) and threonine (amino acid 758−760 residues) [45, 46]. Dephosphorylation by phosphatases is associated with a return to low affinity. Tyrosine kinase posphorylation also occurs. The precise mechanism relating phosphorylation events to a "loss" of a brake to CR3 is not known. It could be an electrostatic action resulting in the separation of the two chains or the binding of intracellular proteins (due to serine, threonine and tyrosine modifications). Interestingly, monoclonal antibodies to CD18 (60.3, IB4) shared this capacity [47]. Antibodies directed against CD18 (i.e. 60.3 and IB4) were reported to directly stimulate superoxide anion production [47]. Perhaps 60.3 and IB4 attenuate the influence by CD18 on CD11b? In addition, the CD18-mediated transition from its "closed" to "open" configuration results in a separation of alpha and beta chains, an event which results in a display of a cryptic site at EGF-like modules within the stem (mAb 24, a LIBS, a class of monoclonal antibodies which highlight "active" or high-affinity β2 integrin) [48].

CR3 mediates binding of C3bi-coated particles. Regarding the capacity of CR3 to contribute to proinflammatory features of CR3 including superoxide anion production, degranulation and adhesive functions (aggregation, motility and adhesion), it was shown that if an antibody could block the binding of iC3b to CD11b, it would serve as an antagonist to block the proinflammatory functions of CR3. This property was demonstrated for monoclonal antibodies to CD11b (60.1, MN41). For example, the binding of CD11b and homotypic aggregation were blocked by 60.1 [49], MN41 [50] and CBRM1/5 [51] (each binds at the I domain of CD11b). However, it is important to offer a caution concerning the I domain and its ligand binding. Deletion of the I domain of the integrin αM abolishes binding to some ligands and diminishes binding to others [52]. The role of the propeller domain in this residual binding suggests that in some cases both the propeller domain and I domain can directly contribute to ligand binding.

In the early 1990s the function of CR3 was restricted to a select group of proinflammatory functions (e.g. superoxide anion production, degranulation and adhesion functions) and phagocytosis. The latter was mediated by CR3 and its interaction and binding of C3bi-coated particles [53]. A plethora of monoclonal antibodies which recognized CD11b were employed to explore ligand−receptor interactions. If an antibody could block the binding of iC3b to CD11b, it would serve as an antagonist to block the proinflammatory functions of CR3 as well. For example, the binding of CD11b and homotypic aggregation were blocked by 60.1 [49], MN41 [50], ICRF44 (formerly clone 44; [54]) and CBRM1/5 [51] (each binds at the I domain of CD11b).

Surprisingly, several antibodies against CD11b were found to protect and to tolerize. The action of anti-CR3 antibodies (clone 44) included an induction of a calcium flux by a dendritic cell line which was exposed to anti-CD11b [55]. It was demonstrated that crosslinking of CD11b by monoclonal antibody 44 resulted in a CR3 signaling event in human dendritic cells to obviate maturation [56]. CR3 ligation was accompanied by an attenuation of interferon-stimulated secretion of the Th1-promoting cytokine IL12 in addition to IL1b, TNFα and IL-6. These data support that clearance of apoptotic cells (coated with iC3b) preserves dendritic cells in the immature state, a key checkpoint to autoimmunity. Marth and Kelasall observed protection in a sepsis model (against injury) which was provided by prior infusion of iC3b red blood cells [57]. Anti-CR3 antibodies may block the activation of a wide range of inflammatory cells, such as PMNs and macrophages and their trafficking into tissues. This result suggests that therapy provides a direct signal, rather than simply acting to block signaling to a competing CR3 ligand.

ENDOTHELIAL DYSFUNCTION AND PROGRESSION OF ATHEROSCLEROSIS AND RENAL DISEASE

To date, a mechanism to explain endothelial dysfunction and progression of atherosclerosis and renal disease, has fallen short of the mark. The goals of this section are to review pathogenesis in the context of "inflammation" hypothesis and Schwartzman phenomenon [58] and to discuss the implications of a putative risk profile of vascular injury that includes a genetic factor (SNP within an exon of the extracellular domain of the alpha chain of CR3) and environmental factors which are related to pathogenic consequences of Th1 cells and cytokines. Levels of adiponectin and mEPCR may reflect unchecked Th1 immunity.

Many have speculated that inflammation, the central feature of SLE pathogenesis and clinical flares, is linked to the increased atherosclerotic risk among these patients in addition to higher rates of traditional risk factors. However, the absence of association of carotid plaque with overt disease activity in this study and others runs counter to the "inflammation" hypothesis. One simple explanation is that the SELENA-SLEDAI instrument may underestimate disease activity. However, this instrument does capture current inflammation in the cutaneous, renal, serosal, hematologic, neurologic, and serologic systems. In fact, it is generally in these clinical settings of activity that the traditional markers of endothelial (luminal) injury have been previously identified. However, the scope of the vasculopathy may include "normal-appearing" blood vessels. For example, a skin biopsy was obtained from non-lesional, non-sun-exposed skin of the buttocks. Levels of adhesion molecules (VCAM1, ICAM1 and E selectin) were found by immunohistochemistry to be significantly elevated in patients versus normal controls. Since these phenotypic changes occurred in overtly normal-appearing tissue, the "widespread" endothelial activation was portrayed in the context of the Schwartzman phenomenon. It is noteworthy that the expression of the anti-injury molecules (e.g. adiponectin, mEPCR) was not evaluated.

In a cohort of 131 patients and 73 race/ethnicity-matched healthy controls, carotid plaque was observed in over twice the proportion of lupus patients compared to age- and sex-matched controls (43% vs 17%, $p = 0.0002$) [59]. Excess prevalence of plaque was seen beginning in the fifth decade of life. On multivariate analysis, age, SLE disease duration, sE-selectin, and adiponectin were the only independent predictors of plaque. Among lupus patients with plaque, elevations in these biomarkers were persistent over more than one visit in those with multiple measurements. Elevation of sE-selectin was anticipated, because this biomarker reflects activation of the endothelium and has been previously associated with atherosclerosis and cardiovascular risk in both SLE and non-SLE cohorts [9, 60, 61]. The association with elevated adiponectin was unexpected since adiponectin is generally considered to be vasoprotective (see above). In fact, our hypothesis was that adiponecin would be decreased.

This led us to speculate that perhaps the elevated adiponectin represents a continued but unsuccessful attempt at vascular repair. Table 9.1 reports the percentage of plaque for different groups defined according to adiponectin and E-selectin levels. Given the study design, these percentages can be directly interpreted as predictive values or more accurately, predicted probabilities of plaque. For example, the probability of plaque is predicted to be 75% if an individual has high levels of both adiponectin and E-selectin, whereas the probability of plaque is 19% for individuals with low levels of both biomarkers (Table 9.1). It is important to consider the two biomarkers simultaneously when evaluating predictive values since each was found to be independently associated with plaque. As such, the predicted probability of plaque for adiponectin depends on whether the E-selectin level is high or low and vice versa.

sE-selectin and adiponectin have been reported in patients with active disease [62–64]. The elevation of sE-selectin in patients with plaque despite clinical quiescence and the stability of that elevation over time likely indicates that a low level of inflammation or atherogenic injury independent of inflammation does contribute to atherosclerosis in lupus [65]. Further precedent in SLE for the demonstration of endothelial activation in the absence of an inflammatory infiltrate has been reported for the kidney [66] and the skin [67], in which increased

TABLE 9.1 Patients with biomarkers in group sE selectin + adiponectin, high, associated with the development of atherosclerosis

Group	Plasma biomarker	Plaque %
Pts, total	High + low	43
sE selectin	High	52
Adiponectin	High	55
sE selectin + Adiponectin	High	75
sE selectin	Low	33
Adiponectin	Low	32
sE selectin + Adiponectin	Low	19[a]

[a]sE selectin + adiponectin high vs sE selectin + adiponectin low, p = 0.002.

expression of adhesion molecules including E-selectin was observed in the microvasculature of non-lesional areas. Molecules including E-selectin were observed in the microvasculature of non-lesional areas.

Adiponectin has pleiotropic biological activities including improving insulin sensitivity [68, 69]. Beyond these metabolic actions, there is also considerable preclinical evidence that adiponectin exerts direct anti-atherosclerotic and cardioprotective effects [70]. Epidemiological studies evaluating the association of adiponectin with cardiovascular events have reported somewhat conflicting results and have been limited largely to men, especially older men [71−75]. Based on a recent biomarker study from Vanderbilt, performed in a retrospective cross-sectional analysis of 65 patients with SLE and 69 controls, there was no association between adiponectin levels and coronary calcium scores [76−78]. While a definitive explanation for the disparate findings between this study and the one presented herein await larger longitudinal cohorts, several differences are noteworthy. The Vanderbilt cohort comprised a greater percentage of Caucasians, the adiponectin levels were higher in controls than most reported series, subset analysis of patients with low SLEDAI scores was not performed, and assessment was limited to one sampling. In the New York cohort, evaluation across visits in 60% of patients also demonstrated a significant difference between sustained high levels in patients with plaque compared to those without plaque. This is relevant since cross-sectional evaluation of a biomarker for a condition that represents accrued insult has limitations. Most interesting, in both cohorts the adiponectin levels were significantly higher overall in the SLE patients compared to healthy controls, yet these patients are at risk for premature atherosclerosis. This alone distinguishes SLE from most reports of men in which lower levels of adiponectin are associated both with atherosclerosis and risk for progression [79]. We hypothesize that for as yet unknown reasons the endothelial dysfunction characteristic of SLE (reflected by elevated sE-selectin in both studies) drives a higher adiponectin level which is nevertheless not effective in protection. Therefore, increased adiponectin concentration could represent a compensatory mechanism to existing vascular damage. It is notable that children with diabetes (another high-risk group for premature atherosclerosis) were recently reported to have a higher concentration of adiponectin than healthy matched controls [80].

Other biomarkers which have been recently linked to endothelial damage *per se* or inflammatory luminal injury include CECs as well as sEPCR [7, 81]. Neither tracked with plaque, suggesting that inflammation may not be the sole explanation.

Limitations of our study include the challenges in precise quantification of the activity, severity, and disease treatment over the lifetime of the patient with SLE. Atherosclerosis develops over years. Because this study was limited to one year, it is acknowledged that activity at the time of evaluation is not necessarily reflective of the total burden of undulating activity which might lead to plaque. Reliable indices of the cumulative burden of inflammatory disease activity in SLE do not exist. The SLE Damage Index is the closest approximation but that instrument assesses damage accrued since diagnosis, some of which may not be directly related to the disease process *per se* such as avascular necrosis or premature ovarian failure [82]. Although cardiovascular event rates in SLE are markedly higher than matched healthy controls, the rates are too low to permit correlation with biomarkers, thus carotid ultrasound studies are used as proxies of actual events. While blockade of IL-12 function attenuates atherosclerosis (murine models, [12]), the relationship between IL12 and its capacity to increase adiponectin levels is unknown.

As stated above, there is precedence of skin biomarkers which are present despite clinical quiescence. Diverse studies in patients with SLE have confirmed the permissive role of vascular adhesion molecules in the pathogenesis of vasculitis and glomerulonephritis [83−85]. Widespread activation of the endothelium has been suggested by the observation that even in non-lesional, non-sun-exposed (buttock) skin from patients with active SLE, endothelial expression of adhesion molecules as well as inducible nitric oxide synthase (iNOS, NOS2) is upregulated [85, 86]. These findings support the notion that, in SLE, the vascular endothelium in general is "primed" for injury by activated leukocytes and yet there is no overt injury. When another factor is superimposed on widespread priming, vascular lesions develop, contributing to specific organ injury. For example, the deposition of immune complexes in renal tissue initiates a sequence that ultimately involves macrophages, which are recruited to the "primed" endothelium where they secrete inflammatory cytokines such as interferon (IFN) γ [87]. Interestingly, adiponectin promotes an anti-inflammatory phenotype in human monocytes and monocyte-derived dendritic cells by attenuating IL12 [88].

mEPCR regulates the conversion of protein C to activated protein C by presenting it to the thrombin−thrombomodulin complex [81, 89]. mEPCR is shed in a pathologic state to a soluble form, sEPCR, increased levels of which have been reported in two lupus cohorts [81, 89]. Patients with LN had significantly higher levels of sEPCR than those without nephritis [81]. It has been reported that mEPCR expression in the glomerular and interstitial microvasculature was increased in rodents with experimentally induced sepsis [90]. In this model, sepsis was associated with a depletion of activated protein C, the consequence of which may be a compensatory increase in mEPCR as demonstrated in the diseased renal parenchyma. Similarly, increased mEPCR was also seen in renal biopsies from patients with acute kidney injury [90]. A recent retrospective study with chart review by our group reported the analysis of mEPCR (anti-injury molecule) by immunohistochemistry in 59 biopsies from 49 patients with LN [66]. The focus was to score the expression of biomarker using a standard index which is widely used by pathologists during an evaluation of normal-appearing cortical PTCs. mEPCR was expressed in the medulla, arterial endothelium, and cortical peritubular capillaries (PTCs) in all biopsies with LN but not in the cortical PTCs of normal kidney. Positive mEPCR staining in >25% of PTCs was observed in 16/59 biopsies and associated with poor response to therapy. Eleven (84.6%) of 13 patients with positive staining for mEPCR in >25% of PTCs and follow-up at 6 months did not respond to therapy, compared to 8/28 (28.6%) with mEPCR staining in <25% PTCs, $p = 0.0018$. Renal response was defined as per the FDA-sponsored trial led by Dr. Ginzler in which mycophenolate mofetil (MMF) was compared to intravenous cyclophosphamide (IVC) [91]. At 1 year, 10 (83.3%) of 12 patients with positive mEPCR staining in >25% of PTCs did not respond to therapy (with two progressing to end-stage renal disease) compared to 8/24 (33.3%) with positive staining in <25% of PTCs, $p = 0.0116$. Although tubular-interstitial damage (TID) was always accompanied by mEPCR, this endothelial marker was extensively expressed in the absence of TID suggesting that poor response could not be attributed solely to increased TID. mEPCR expression was independent of ISN/RPS class, activity, and chronicity indices.

The relationship between mEPCR expression and renal disease progression is intriguing, since mEPCR is generally considered a protective molecule based on its role in both inflammation and coagulation. mEPCR binds protein C, presenting it to the thrombin/thrombomodulin complex, thus regulating its conversion to activated protein C [32]. Conditions that reduce surface expression of mEPCR on endothelial cells attenuate the efficiency of protein C activation [92]. The decrease in mEPCR is due to a metalloproteinase-dependent cleavage which splits the molecule into a soluble form, sEPCR [93].

Given the prediction that shed mEPCR impairs the integrity of the endothelium and places the net balance of this protective protein in biologic "arrears," a decrease in mEPCR expression was the predicted result in patients with progressive renal injury. Thus, finding increased mEPCR expression in the endothelium of PTCs, compared to controls, was unexpected.

The seemingly paradoxical finding of increased mEPCR in LN is provocative but not without precedent. It has been reported that mEPCR expression in the glomerular and interstitial microvasculature was increased in rodents with experimentally induced sepsis [90]. In this model, sepsis was associated with a depletion of activated protein C, the consequence of which may be a compensatory increase in mEPCR as demonstrated in the diseased renal parenchyma. Similarly, increased mEPCR was also seen in renal biopsies from patients with acute kidney injury [90]. One explanation for the findings herein is that increased mEPCR may represent a thwarted attempt at endothelial defense. An alternative hypothesis is that in LN circulating and/or deposited immune complexes activate the classical complement pathway generating C4b-binding protein, which in turn complexes with protein S, impairing its ability to generate activated protein C [94]. Against this explanation is the absence of a correlation between mEPCR expression and local deposits of immune complexes or complement. However, it remains possible that circulating immune complexes, mimicking the effects of sepsis, may activate the endothelium of the microvasculature. Finally, as in sepsis, LN may induce a low protein C state.

Biomarker studies have corroborated the involvement of the endothelium in murine and human LN. In a murine study, severe LN was associated with high vascular cell adhesion molecule 1 levels in urine as well as increased expression in the tubules and vascular endothelium in LN [95]. In humans, urine levels of adiponectin were significantly elevated prior to and during renal flares [62]. In addition, adiponectin was expressed on endothelial surfaces in the renal microvasculature in patients with LN [62]. Urine levels of the Monocyte Chemoattractant Protein-1, produced by endothelial cells, have been shown to increase significantly during renal flares with decreases paralleling response [96, 97]. Increased nitric oxide production, generated in part by vascular endothelial cells, has been associated with renal damage and poor response to therapy [98].

Several shortcomings are acknowledged with regard to interpretation of the overall data. This study was largely retrospective and the number of patients with biopsy tissue and data for clinical evaluation were limited. In addition, compliance with oral medication was difficult to address and patient numbers were too small to gain insight into the potential benefit of a specific therapy. The use of various treatments, each with a different response rate and time of response, further dictates caution in interpreting these findings. While the progression of nephritis is also linked to Th1 cytokines such as IL12 and IL18 [99, 100], the specific contribution of IL12 and IL18 to expression of mEPCR is unknown.

In summary, positive mEPCR staining >25% in cortical PTCs is associated with a poor renal response to standard therapy. While these results suggest a contribution of the endothelium of the renal parenchyma to the pathophysiology of more progressive lupus nephritis, further studies are needed to distinguish whether the endothelium plays an "active" or "reactive" role. Larger prospective studies are needed to affirm the significance of mEPCR expression as a novel biomarker of unanticipated renal progression and to address the utility of longitudinally measuring sEPCR in the serum and urine.

Considering pathogenic mechanisms of organ-specific vascular disease (i.e. co-morbidity with athero-sclerosis, nephritis), there is an extensive literature which highlights autoantibody-elicited acute vascular injury. For example, autoantibodies can trigger endothelial-cell expression of ICAM-1 and neutrophil expression of αMβ2 integrin. Buyon and coworkers reported that SLE- PMNs expressed higher levels of CR3 than values obtained using PMNs from healthy controls [101]. Recruitment of CR3 from intracellular stores represents an important biomarker which reflects CR3 signals that are pro-inflammatory. Specifically, for resting PMNs, there is a constitutive level of CR3. However, in neutrophils and lymphocytes, there is an intracellular pool of CR3. For PMNs, the pool is derived from the specific granule fraction which is present in the low-affinity, latent form. Stimulation with FMLP, C5a or PMA results in the recruitment to the cell surface (a phenomenon called "inside-out signaling"). Recruitment and activation are distinct processes [102]. The latter was reported to solely occur at the plasma membrane compartment. The second wave of CR3 may be related to special functions of neutrophils including homotypic aggregation and phagocytosis. Recruitment is not restricted to CR3 since there is also an intracellular pool of IL-10 receptors in specific granules of human neutrophils with a differential mobilization by proinflammatory mediators. In subsequent studies, antibody-independent mechanisms of vascular insult in SLE have been proposed by Buyon et al. [67] and Belmont and Abramson [85] and involve the same adhesive interactions.

Harley and coworkers performed a GWA in 3818 SLE individuals of European descent. Their findings have implications to a link between chronic injury in SLE and CR3. Specifically, they reported a novel non-synonymous variant, rs1143679, at exon-3 of the CD11b or ITGAM gene was reported for its association with SLE patients. rs1143679, a SNP, changes the amino acid (R77H) in the vicinity of alpha chain's I domain and may overall modulate the numerous ligand-binding activities of ITGAM in monocytes, PMNs, and dendritic cells. Nath et al. predicted that the effect of the amino acid 77 polymorphism would have an impact on the ICAM-1-binding site. This draws attention to integrin-mediated interactions between leukocytes and endothelial cells and their role in the vasculopathy and vasculitis of SLE. This variant was significantly associated with renal criteria ($p = 0.0003$, OR = 1.39), discoid rash ($p = 0.02$, OR = 1.27), and immunologic manifestations ($p = 0.04$, OR = 1.30) compared to SLE cases without these clinical manifestations [103].

CR3 does not seem to have a classical binding site for its ligand—receptor interactions. Deletion of the I domain of the integrin αM abolishes binding to some ligands and diminishes binding to others [52]. From this result, the role of the propeller domain in this residual binding suggests that in some cases both the propeller domain and I domain can directly contribute to ligand binding. In addition, it would seem that CR3 may represent a fail-safe elimination pathway of iC3b-coated particles. What is the consequence of CR3 containing the variant form of CR3 to its pluripotent role? In one scerario, the variant will have restriction to CR3's function and the varied functions including protective or anti-inflammatory pathways may be attenuated.

Sustained CR3 proinflammatory signals result in recruitment of Th1 cells and Th1 cytokines such as IL12 and IL18 (Figure 9.3). We have identified patient subgroups, including those with high adiponectin reflecting atherogenic injury independent of recognized lupus activity and a patient subgroup where nephritis is

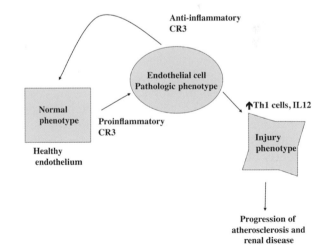

FIGURE 9.3 Conceptual model of the relationship between the vasculature and end-organ damage in SLE. One scenario to explain pathogenesis is an inflammatory role of complement receptor 3 (CR3) and endothelial injury by neutrophils and inflammatory mediators. Paradoxically, CR3 also may mediate a possible negative feedback system to attenuate vascular injury in dendritic cells, which highlight the role of CR3 to protect and to promote tolerance. If a protective form of CR3 is in "arrears", vascular injury will exacerbate the progression towards morbid end-organ damage.

associated with high levels of mEPCR. Patients within these subgroups might strongly benefit from therapy designed to stimulate the anti-inflammatory signaling by CR3. Also, it would be fascinating to evaluate the representation of CR3's variant in these patient populations.

CONCLUSION

Vascular manifestations associated with systemic lupus erythematosus (SLE) span a broad range including vasculopathy. The focus directed at understanding the development and persistence of a vasculopathy in the context of CR3 is relevant to endothelial injury by PMNs and inflammatory mediators. A risk profile of vascular injury would include a genetic factor (SNP within an exon of the extracellular domain of alpha chain of CR3) and environmental factors which are related to pathogenic consequences of Th1 cells and cytokines such as IL12. Biomarkers, which reflect unchecked Th1 immunity including adiponectin and mEPCR, may identify an over-representation of proinflammatory CR3 in vascular injury with co-morbidities involving premature atherosclerosis and SLE nephritis. For example, adiponectin was found to serve as an independent predictor of carotid plaque and its elevations were persistent over more than one visit. Unexpectedly, this biomarker was present despite clinical quiescence. In SLE nephritis, the persistent expression of membrane EPCR occurs at peritubular capillaries. Alternatively, a protective factor is in "arrears". A candidate protective factor is a form of CR3, itself. Paradoxically, CR3 may mediate a possible negative feedback system to control inflammatory responses, which highlight the role of CR3 to protect and to promote tolerance. If a protective form of CR3 is in "arrears" resulting in vascular injury, CR3 may be the cause of — and solution to — all vascular morbidities in SLE. The coming years will no doubt bring further exciting novel insights into the role of CR3 in these mechanisms and its implications to important therapeutic interventions to reverse a vasculopathy in SLE.

References

[1] M.B. Urowitz, A.A. Bookman, B.E. Koehler, D.A. Gordon, H.A. Smythe, M.A. Ogryzlo, The bimodal mortality pattern of systemic lupus erythematosus, Am J Med 60 (1976) 221–225.

[2] S. Manzi, E.N. Meilahn, J.E. Rairie, C.G. Conte, T.A. Medsger Jr., L. Jansen-McWilliams, et al., Age-specific incidence rates of myocardial infarction and angina in women with systemic lupus erythematosus: comparison with the Framingham Study, Am J Epidemiol 145 (1997) 408–415.

[3] J.M. Esdaile, M. Abrahamowicz, T. Grodzicky, Y. Li, C. Panaritis, R. du Berger, et al., Traditional Framingham risk factors fail to fully account for accelerated atherosclerosis in systemic lupus erythematosus, Arthritis Rheum 44 (2001) 2331–2337.

[4] M.J. Roman, B.A. Shanker, A. Davis, M.D. Lockshin, L. Sammaritano, R. Simantov, et al., Prevalence and correlates of accelerated atherosclerosis in systemic lupus erythematosus, N Engl J Med 349 (2003) 2399–2406.

[5] M. McMahon, J. Grossman, J. FitzGerald, E. Dahlin-Lee, D.J. Wallace, B.Y. Thong, et al., Proinflammatory high-density lipoprotein as a biomarker for atherosclerosis in patients with systemic lupus erythematosus and rheumatoid arthritis, Arthritis Rheum 54 (2006) 2541–2549.

[6] N.E. Doherty, R.J. Siegel, Cardiovascular manifestations of systemic lupus erythematosus, Am Heart J 110 (1985) 1257–1265.

[7] R. Clancy, G. Marder, V. Martin, H.M. Belmont, S.B. Abramson, J. Buyon, Circulating activated endothelial cells in systemic lupus erythematosus: further evidence for diffuse vasculopathy, Arthritis Rheum 44 (2001) 1203–1208.

[8] S. Rajagopalan, E.C. Somers, R.D. Brook, C. Kehrer, D. Pfenninger, E. Lewis, et al., Endothelial cell apoptosis in systemic lupus erythematosus: a common pathway for abnormal vascular function and thrombosis propensity, Blood 103 (2004) 3677–3683.

[9] Y.H. Rho, C.P. Chung, A. Oeser, J. Solus, P. Raggi, T. Gebretsadik, et al., Novel cardiovascular risk factors in premature coronary atherosclerosis associated with systemic lupus erythematosus, J Rheumatol 35 (2008) 1789–1794.

[10] J. Frostegard, A.K. Ulfgren, P. Nyberg, U. Hedin, J. Swedenborg, U. Andersson, et al., Cytokine expression in advanced human atherosclerotic plaques: dominance of proinflammatory (Th1) and macrophage-stimulating cytokines, Atherosclerosis 145 (1999) 33–43.

[11] K. Uyemura, L.L. Demer, S.C. Castle, D. Jullien, J.A. Berliner, M.K. Gately, et al., Cross-regulatory roles of interleukin (IL)-12 and IL-10 in atherosclerosis, J Clin Invest 97 (1996) 2130–2138.

[12] A.D. Hauer, C. Uyttenhove, P. de Vos, V. Stroobant, J.C. Renauld, T.J. van Berkel, et al., Blockade of interleukin-12 function by protein vaccination attenuates atherosclerosis, Circulation 112 (2005) 1054–1062.

[13] M.M. Ward, Changes in the incidence of endstage renal disease due to lupus nephritis in the United States, 1996-2004, J Rheumatol 36 (2009) 63–67.

[14] D. Isenberg, P. Lesavre, Lupus nephritis: assessing the evidence, considering the future, Lupus 16 (2007) 210–211.

[15] M.A. Dooley, S. Hogan, C. Jennette, R. Falk, Cyclophosphamide therapy for lupus nephritis: poor renal survival in black Americans. Glomerular Disease Collaborative Network, Kidney Int 51 (1997) 1188–1195.

[16] R.G. Barr, S. Seliger, G.B. Appel, R. Zuniga, V. D'Agati, J. Salmon, et al., Prognosis in proliferative lupus nephritis: the role of socio-economic status and race/ethnicity, Nephrol Dial Transplant 18 (2003) 2039–2046.

[17] D. Isenberg, G.B. Appel, G. Contreras, M.A. Dooley, E.M. Ginzler, D. Jayne, et al., Influence of race/ethnicity on response to lupus nephritis treatment: the ALMS Study. Ann Rheum Dis 20 (2009) 1103–1112.

[18] T.L. Rivera, H.M. Belmont, S. Malani, M. Latorre, L. Benton, J. Weisstuch, et al., Current therapies for lupus nephritis in an ethnically heterogeneous cohort, J Rheumatol 36 (2009) 298–305.

[19] H.A. Austin 3rd, L.R. Muenz, K.M. Joyce, T.A. Antonovych, M.E. Kullick, J.H. Klippel, et al., Prognostic factors in lupus

nephritis. Contribution of renal histologic data, Am J Med 75 (1983) 382–391.

[20] J.J. Weening, V.D. D'Agati, M.M. Schwartz, S.V. Seshan, C.E. Alpers, G.B. Appel, et al., The classification of glomerulonephritis in systemic lupus erythematosus revisited, J Am Soc Nephrol 15 (2004) 241–250.

[21] C. Berthier, R. Bethunaickan, E. Bottinger, al., e, Proliferative SLE nephritis and prgressive non-inflammatory glomerulosclerosis share key gene expression profiles, Arthritis Rheum 58 (Suppl.9) (2008) 902–903.

[22] A. Kaser, S. Kaser, N.C. Kaneider, B. Enrich, C.J. Wiedermann, H. Tilg, Interleukin-18 attracts plasmacytoid dendritic cells (DC2s) and promotes Th1 induction by DC2s through IL-18 receptor expression, Blood 103 (2004) 648–655.

[23] H. Yamagami, K. Kitagawa, T. Hoshi, S. Furukado, H. Hougaku, Y. Nagai, et al., Associations of serum IL-18 levels with carotid intima-media thickness, Arterioscler Thromb Vasc Biol 25 (2005) 1458–1462.

[24] M. Kumada, S. Kihara, S. Sumitsuji, T. Kawamoto, S. Matsumoto, N. Ouchi, et al., Association of hypoadiponectinemia with coronary artery disease in men, Arterioscler Thromb Vasc Biol 23 (2003) 85–89.

[25] N. Ouchi, S. Kihara, Y. Arita, Y. Okamoto, K. Maeda, H. Kuriyama, et al., Adiponectin, an adipocyte-derived plasma protein, inhibits endothelial NF-kappaB signaling through a cAMP-dependent pathway, Circulation 102 (2000) 1296–1301.

[26] N. Ouchi, S. Kihara, Y. Arita, K. Maeda, H. Kuriyama, Y. Okamoto, et al., Novel modulator for endothelial adhesion molecules: adipocyte-derived plasma protein adiponectin, Circulation 100 (1999) 2473–2476.

[27] Y. Okamoto, Y. Arita, M. Nishida, M. Muraguchi, N. Ouchi, M. Takahashi, et al., An adipocyte-derived plasma protein, adiponectin, adheres to injured vascular walls, Horm Metab Res 32 (2000) 47–50.

[28] Y. Ishikawa, Y. Akasaka, T. Ishii, M. Yoda-Murakami, N.H. Choi-Miura, M. Tomita, et al., Changes in the distribution pattern of gelatin-binding protein of 28 kDa (adiponectin) in myocardial remodelling after ischaemic injury, Histopathology 42 (2003) 43–52.

[29] K. Fukudome, S. Kurosawa, D.J. Stearns-Kurosawa, X. He, A.R. Rezaie, C.T. Esmon, The endothelial cell protein C receptor. Cell surface expression and direct ligand binding by the soluble receptor, J Biol Chem 271 (1996) 17491–17498.

[30] J.M. Gu, Y. Katsuura, G.L. Ferrell, P. Grammas, C.T. Esmon, Endotoxin and thrombin elevate rodent endothelial cell protein C receptor mRNA levels and increase receptor shedding in vivo, Blood 95 (2000) 1687–1693.

[31] K. Fukudome, C.T. Esmon, Molecular cloning and expression of murine and bovine endothelial cell protein C/activated protein C receptor (EPCR). The structural and functional conservation in human, bovine, and murine EPCR, J Biol Chem 270 (1995) 5571–5577.

[32] D.J. Stearns-Kurosawa, S. Kurosawa, J.S. Mollica, G.L. Ferrell, C.T. Esmon, The endothelial cell protein C receptor augments protein C activation by the thrombin-thrombomodulin complex, Proc Natl Acad Sci U S A 93 (1996) 10212–10216.

[33] F.B. Taylor Jr., G.T. Peer, M.S. Lockhart, G. Ferrell, C.T. Esmon, Endothelial cell protein C receptor plays an important role in protein C activation in vivo, Blood 97 (2001) 1685–1688.

[34] W. Li, X. Zheng, J. Gu, J. Hunter, G.L. Ferrell, F. Lupu, et al., Overexpressing endothelial cell protein C receptor alters the hemostatic balance and protects mice from endotoxin, J Thromb Haemost 3 (2005) 1351–1359.

[35] J. Xu, D. Qu, N.L. Esmon, C.T. Esmon, Metalloproteolytic release of endothelial cell protein C receptor, J Biol Chem 275 (2000) 6038–6044.

[36] B. Saposnik, J.L. Reny, P. Gaussem, J. Emmerich, M. Aiach, S. Gandrille, A haplotype of the EPCR gene is associated with increased plasma levels of sEPCR and is a candidate risk factor for thrombosis, Blood 103 (2004) 1311–1318.

[37] P. Medina, S. Navarro, A. Estelles, A. Vaya, B. Woodhams, Y. Mira, P. Villa, et al., Contribution of polymorphisms in the endothelial protein C receptor gene to soluble endothelial protein C receptor and circulating activated protein C levels, and thrombotic risk, Thromb Haemost 91 (2004) 905–911.

[38] J.H. Erlich, E.M. Boyle, J. Labriola, J.C. Kovacich, R.A. Santucci, C. Fearns, et al., Inhibition of the tissue factor-thrombin pathway limits infarct size after myocardial ischemia-reperfusion injury by reducing inflammation, Am J Pathol 157 (2000) 1849–1862.

[39] M. Shimaoka, J. Takagi, T.A. Springer, Conformational regulation of integrin structure and function, Annu Rev Biophys Biomol Struct 31 (2002) 485–516.

[40] J.O. Lee, L.A. Bankston, M.A. Arnaout, R.C. Liddington, Two conformations of the integrin A-domain (I-domain): a pathway for activation? Structure 3 (1995) 1333–1340.

[41] J.O. Lee, P. Rieu, M.A. Arnaout, R. Liddington, Crystal structure of the A domain from the alpha subunit of integrin CR3 (CD11b/CD18), Cell 80 (1995) 631–638.

[42] M.V. Milburn, L. Tong, A.M. deVos, A. Brunger, Z. Yamaizumi, S. Nishimura, et al., Molecular switch for signal transduction: structural differences between active and inactive forms of protooncogenic ras proteins, Science 247 (1990) 939–945.

[43] J.P. Xiong, R. Li, M. Essafi, T. Stehle, M.A. Arnaout, An isoleucine-based allosteric switch controls affinity and shape shifting in integrin CD11b A-domain, J Biol Chem 275 (2000) 38762–38767.

[44] J.T. Merrill, S.G. Slade, G. Weissmann, R. Winchester, J.P. Buyon, Two pathways of CD11b/CD18-mediated neutrophil aggregation with different involvement of protein kinase C-dependent phosphorylation, J Immunol 145 (1990) 2608–2615.

[45] S. Fagerholm, N. Morrice, C.G. Gahmberg, P. Cohen, Phosphorylation of the cytoplasmic domain of the integrin CD18 chain by protein kinase C isoforms in leukocytes, J Biol Chem 277 (2002) 1728–1738.

[46] T.J. Hilden, L. Valmu, S. Karkkainen, C.G. Gahmberg, Threonine phosphorylation sites in the beta 2 and beta 7 leukocyte integrin polypeptides, J Immunol 170 (2003) 4170–4177.

[47] G. Berton, C. Laudanna, C. Sorio, F. Rossi, Generation of signals activating neutrophil functions by leukocyte integrins: LFA-1 and gp150/95, but not CR3, are able to stimulate the respiratory burst of human neutrophils, J Cell Biol 116 (1992) 1007–1017.

[48] I. Dransfield, N. Hogg, Regulated expression of Mg2+ binding epitope on leukocyte integrin alpha subunits, EMBO J 8 (1989) 3759–3765.

[49] W.J. Wallis, D.D. Hickstein, B.R. Schwartz, C.H. June, H.D. Ochs, P.G. Beatty, et al., Monoclonal antibody-defined functional epitopes on the adhesion-promoting glycoprotein complex (CDw18) of human neutrophils, Blood 67 (1986) 1007–1013.

[50] J.P. Buyon, S.B. Abramson, M.R. Philips, S.G. Slade, G.D. Ross, G. Weissmann, et al., Dissociation between increased surface expression of gp165/95 and homotypic neutrophil aggregation, J Immunol 140 (1988) 3156–3160.

[51] M.S. Diamond, T.A. Springer, The dynamic regulation of integrin adhesiveness, Curr Biol 4 (1994) 506–517.

[52] P. Yalamanchili, C. Lu, C. Oxvig, T.A. Springer, Folding and function of I domain-deleted Mac-1 and lymphocyte function-associated antigen-1, J Biol Chem 275 (2000) 21877–21882.

[53] M.S. Diamond, J. Garcia-Aguilar, J.K. Bickford, A.L. Corbi, T.A. Springer, The I domain is a major recognition site on the leukocyte integrin Mac-1 (CD11b/CD18) for four distinct adhesion ligands, J Cell Biol 120 (1993) 1031–1043.

[54] T. Ueda, P. Rieu, J. Brayer, M.A. Arnaout, Identification of the complement iC3b binding site in the beta 2 integrin CR3 (CD11b/CD18), Proc Natl Acad Sci U S A 91 (1994) 10680–10684.

[55] F. Leon, N. Contractor, I. Fuss, T. Marth, E. Lahey, S. Iwaki, et al., Antibodies to complement receptor 3 treat established inflammation in murine models of colitis and a novel model of psoriasiform dermatitis, J Immunol 177 (2006) 6974–6982.

[56] M. Skoberne, S. Somersan, W. Almodovar, T. Truong, K. Petrova, P.M. Henson, et al., The apoptotic-cell receptor CR3, but not alphavbeta5, is a regulator of human dendritic-cell immunostimulatory function, Blood 108 (2006) 947–955.

[57] T. Marth, B.L. Kelsall, Regulation of interleukin-12 by complement receptor 3 signaling, J Exp Med 185 (1997) 1987–1995.

[58] L. Thomas, F.W. Denny Jr., J. Floyd, Studies on the generalized Shwartzman reaction. III. Lesions of the myocardium and coronary arteries accompanying the reaction in rabbits prepared by infection with group A streptococci, J Exp Med 97 (1953) 751–766.

[59] H.R. Reynolds, J. Buyon, M. Kim, T. Rivera, P. Izmirly, P. Tunick, et al., Association of plasma soluble E-selectin and adiponectin with carotid plaque in patients with systemic lupus erythematosus, Atherosclerosis (2009).

[60] S.J. Hwang, C.M. Ballantyne, A.R. Sharrett, L.C. Smith, C.E. Davis, A.M. Gotto Jr., et al., Circulating adhesion molecules VCAM-1, ICAM-1, and E-selectin in carotid atherosclerosis and incident coronary heart disease cases: the Atherosclerosis Risk In Communities (ARIC) study, Circulation 96 (1997) 4219–4225.

[61] L.E. Rohde, R.T. Lee, J. Rivero, M. Jamacochian, L.H. Arroyo, W. Briggs, et al., Circulating cell adhesion molecules are correlated with ultrasound-based assessment of carotid atherosclerosis, Arterioscler Thromb Vasc Biol 18 (1998) 1765–1770.

[62] B.H. Rovin, H. Song, L.A. Hebert, T. Nadasdy, G. Nadasdy, D.J. Birmingham, et al., Plasma, urine, and renal expression of adiponectin in human systemic lupus erythematosus, Kidney Int 68 (2005) 1825–1833.

[63] K. Egerer, E. Feist, U. Rohr, A. Pruss, G.R. Burmester, T. Dorner, Increased serum soluble CD14, ICAM-1 and E-selectin correlate with disease activity and prognosis in systemic lupus erythematosus, Lupus 9 (2000) 614–621.

[64] J. Panes, M. Perry, D.N. Granger, Leukocyte-endothelial cell adhesion: avenues for therapeutic intervention, Br J Pharmacol 126 (1999) 537–550.

[65] V. Roldan, F. Marin, G.Y. Lip, A.D. Blann, Soluble E-selectin in cardiovascular disease and its risk factors. A review of the literature, Thromb Haemost 90 (2003) 1007–1020.

[66] P.M. Izmirly, L. Barisoni, J.P. Buyon, M.Y. Kim, T.L. Rivera, J.S. Schwartzman, et al., Expression of endothelial protein C receptor in cortical peritubular capillaries associates with a poor clinical response in lupus nephritis, Rheumatology (Oxford) 48 (2009) 513–519.

[67] H.M. Belmont, J. Buyon, R. Giorno, S. Abramson, Up-regulation of endothelial cell adhesion molecules characterizes disease activity in systemic lupus erythematosus. The Shwartzman phenomenon revisited, Arthritis Rheum 37 (1994) 376–383.

[68] Y. Okamoto, S. Kihara, N. Ouchi, M. Nishida, Y. Arita, M. Kumada, et al., Adiponectin reduces atherosclerosis in apolipoprotein E-deficient mice, Circulation 106 (2002) 2767–2770.

[69] R. Shibata, N. Ouchi, S. Kihara, K. Sato, T. Funahashi, K. Walsh, Adiponectin stimulates angiogenesis in response to tissue ischemia through stimulation of amp-activated protein kinase signaling, J Biol Chem 279 (2004) 28670–28674.

[70] T.A. Hopkins, N. Ouchi, R. Shibata, K. Walsh, Adiponectin actions in the cardiovascular system, Cardiovasc Res 74 (2007) 11–18.

[71] T. Pischon, C.J. Girman, G.S. Hotamisligil, N. Rifai, F.B. Hu, E.B. Rimm, Plasma adiponectin levels and risk of myocardial infarction in men, JAMA 291 (2004) 1730–1737.

[72] J. Frystyk, C. Berne, L. Berglund, K. Jensevik, A. Flyvbjerg, B. Zethelius, Serum adiponectin is a predictor of coronary heart disease: a population-based 10-year follow-up study in elderly men, J Clin Endocrinol Metab 92 (2007) 571–576.

[73] N. Sattar, G. Wannamethee, N. Sarwar, J. Tchernova, L. Cherry, A.M. Wallace, et al., Adiponectin and coronary heart disease: a prospective study and meta-analysis, Circulation 114 (2006) 623–629.

[74] A.M. Kanaya, C. Wassel Fyr, E. Vittinghoff, P.J. Havel, M. Cesari, B. Nicklas, et al., Serum adiponectin and coronary heart disease risk in older Black and White Americans, J Clin Endocrinol Metab 91 (2006) 5044–5050.

[75] J.R. Kizer, J.I. Barzilay, L.H. Kuller, J.S. Gottdiener, Adiponectin and risk of coronary heart disease in older men and women, J Clin Endocrinol Metab 93 (2008) 3357–3364.

[76] C.P. Chung, A.G. Long, J.F. Solus, Y.H. Rho, A. Oeser, P. Raggi, et al., Adipocytokines in systemic lupus erythematosus: relationship to inflammation, insulin resistance and coronary atherosclerosis, Lupus 18 (2009) 799–806.

[77] Y. Asanuma, A. Oeser, A.K. Shintani, E. Turner, N. Olsen, S. Fazio, et al., Premature coronary-artery atherosclerosis in systemic lupus erythematosus, N Engl J Med 349 (2003) 2407–2415.

[78] C.P. Chung, A. Oeser, P. Raggi, T. Gebretsadik, A.K. Shintani, T. Sokka, et al., Increased coronary-artery atherosclerosis in rheumatoid arthritis: relationship to disease duration and cardiovascular risk factors, Arthritis Rheum 52 (2005) 3045–3053.

[79] D.M. Maahs, L.G. Ogden, G.L. Kinney, P. Wadwa, J.K. Snell-Bergeon, D. Dabelea, et al., Low plasma adiponectin levels predict progression of coronary artery calcification, Circulation 111 (2005) 747–753.

[80] K. Heilman, M. Zilmer, K. Zilmer, P. Kool, V. Tillmann, Elevated plasma adiponectin and decreased plasma homocysteine and asymmetric dimethylarginine in children with type 1 diabetes, Scand J Clin Lab Invest 69 (2009) 85–91.

[81] C.A. Sesin, X. Yin, C.T. Esmon, J.P. Buyon, R.M. Clancy, Shedding of endothelial protein C receptor contributes to vasculopathy and renal injury in lupus: In vivo and in vitro evidence, Kidney Int 68 (2005) 110–120.

[82] D. Gladman, E. Ginzler, C. Goldsmith, P. Fortin, M. Liang, M. Urowitz, et al., The development and initial validation of the Systemic Lupus International Collaborating Clinics/American College of Rheumatology damage index for systemic lupus erythematosus, Arthritis Rheum 39 (1996) 363–369.

[83] C.R. Robertson, R.M. McCallum, Changing concepts in pathophysiology of the vasculitides, Curr Opin Rheumatol 6 (1994) 3–10.

[84] R.F. van Vollenhoven, Adhesion molecules, sex steroids, and the pathogenesis of vasculitis syndromes, Curr Opin Rheumatol 7 (1995) 4–10.

[85] H.M. Belmont, D. Levartovsky, A. Goel, A. Amin, R. Giorno, J. Rediske, et al., Increased nitric oxide production accompanied by the up-regulation of inducible nitric oxide synthase in vascular endothelium from patients with systemic lupus erythematosus, Arthritis Rheum 40 (1997) 1810—1816.

[86] J.B. Weinberg, D.L. Granger, D.S. Pisetsky, M.F. Seldin, M.A. Misukonis, S.N. Mason, et al., The role of nitric oxide in the pathogenesis of spontaneous murine autoimmune disease: increased nitric oxide production and nitric oxide synthase expression in MRL-lpr/lpr mice, and reduction of spontaneous glomerulonephritis and arthritis by orally administered NG-monomethyl-L-arginine, J Exp Med 179 (1994) 651—660.

[87] K. Masutani, M. Akahoshi, K. Tsuruya, M. Tokumoto, T. Ninomiya, T. Kohsaka, et al., Predominance of Th1 immune response in diffuse proliferative lupus nephritis, Arthritis Rheum 44 (2001) 2097—2106.

[88] A.M. Wolf, D. Wolf, H. Rumpold, B. Enrich, H. Tilg, Adiponectin induces the anti-inflammatory cytokines IL-10 and IL-1RA in human leukocytes, Biochem Biophys Res Commun 323 (2004) 630—635.

[89] S. Kurosawa, D.J. Stearns-Kurosawa, C.W. Carson, A. D'Angelo, P. Della Valle, C.T. Esmon, Plasma levels of endothelial cell protein C receptor are elevated in patients with sepsis and systemic lupus erythematosus: lack of correlation with thrombomodulin suggests involvement of different pathological processes, Blood 91 (1998) 725—727.

[90] A. Gupta, D.T. Berg, B. Gerlitz, G.R. Sharma, S. Syed, M.A. Richardson, et al., Role of protein C in renal dysfunction after polymicrobial sepsis, J Am Soc Nephrol 18 (2007) 860—867.

[91] E.M. Ginzler, M.A. Dooley, C. Aranow, M.Y. Kim, J. Buyon, J.T. Merrill, et al., Mycophenolate mofetil or intravenous cyclophosphamide for lupus nephritis, N Engl J Med 353 (2005) 2219—2228.

[92] X. Ye, K. Fukudome, N. Tsuneyoshi, T. Satoh, O. Tokunaga, K. Sugawara, et al., The endothelial cell protein C receptor (EPCR) functions as a primary receptor for protein C activation on endothelial cells in arteries, veins, and capillaries, Biochem Biophys Res Commun 259 (1999) 671—677.

[93] D. Qu, Y. Wang, N.L. Esmon, C.T. Esmon, Regulated endothelial protein C receptor shedding is mediated by tumor necrosis factor-alpha converting enzyme/ADAM17, J Thromb Haemost 5 (2007) 395—402.

[94] S.M. Rezende, R.E. Simmonds, D.A. Lane, Coagulation, inflammation, and apoptosis: different roles for protein S and the protein S-C4b binding protein complex, Blood 103 (2004) 1192—1201.

[95] T. Wu, C. Xie, M. Bhaskarabhatla, M. Yan, A. Leone, S.S. Chen, et al., Excreted urinary mediators in an animal model of experimental immune nephritis with potential pathogenic significance, Arthritis Rheum 56 (2007) 949—959.

[96] A. Sica, J.M. Wang, F. Colotta, E. Dejana, A. Mantovani, J.J. Oppenheim, et al., Monocyte chemotactic and activating factor gene expression induced in endothelial cells by IL-1 and tumor necrosis factor, J Immunol 144 (1990) 3034—3038.

[97] B.H. Rovin, H. Song, D.J. Birmingham, L.A. Hebert, C.Y. Yu, H.N. Nagaraja, Urine chemokines as biomarkers of human systemic lupus erythematosus activity, J Am Soc Nephrol 16 (2005) 467—473.

[98] J.C. Oates, S.R. Shaftman, S.E. Self, G.S. Gilkeson, Association of serum nitrate and nitrite levels with longitudinal assessments of disease activity and damage in systemic lupus erythematosus and lupus nephritis, Arthritis Rheum 58 (2008) 263—272.

[99] N. Calvani, M. Tucci, H.B. Richards, P. Tartaglia, F. Silvestris, Th1 cytokines in the pathogenesis of lupus nephritis: the role of IL-18, Autoimmun Rev 4 (2005) 542—548.

[100] A.R. Kitching, A.L. Turner, G.R. Wilson, T. Semple, D. Odobasic, J.R. Timoshanko, et al., IL-12p40 and IL-18 in crescentic glomerulonephritis: IL-12p40 is the key Th1-defining cytokine chain, whereas IL-18 promotes local inflammation and leukocyte recruitment, J Am Soc Nephrol 16 (2005) 2023—2033.

[101] J.P. Buyon, N. Shadick, R. Berkman, P. Hopkins, J. Dalton, G. Weissmann, et al., Surface expression of Gp 165/95, the complement receptor CR3, as a marker of disease activity in systemic Lupus erythematosus, Clin Immunol Immunopathol 46 (1988) 141—149.

[102] J.P. Buyon, M.R. Philips, J.T. Merrill, S.G. Slade, J. Leszczynska-Piziak, S.B. Abramson, Differential phosphorylation of the beta2 integrin CD11b/CD18 in the plasma and specific granule membranes of neutrophils, J Leukoc Biol 61 (1997) 313—321.

[103] X. Kim-Howard, A.K. Maiti, J.M. Anaya, G.R. Bruner, E. Brown, J.T. Merrill, et al., ITGAM coding variant (rs1143679) influences the risk of renal disease, discoid rash, and immunologic manifestations in lupus patients with European ancestry, Ann Rheum Dis 69 (2010) 1329—1332.

10

Mechanisms and Consequences of Mitochondrial Dysfunction and Oxidative Stress in T-Cells of Patients with SLE

Andras Perl

INTRODUCTION

Systemic lupus erythematosus (SLE) is a chronic inflammatory disease characterized by T- and B-cell dysfunction and production of antinuclear antibodies. Potentially autoreactive T and B lymphocytes during development [1] and after completion of an immune response are removed by apoptosis [2, 3]. Paradoxically, lupus T cells exhibit both enhanced spontaneous apoptosis [4] and defective activation-induced cell death [5–9]. Increased spontaneous apoptosis of PBL has been linked to chronic lymphopenia [4] and compartmentalized release of nuclear autoantigens in patients with SLE [10]. By contrast, defective CD3-mediated cell death may be responsible for persistence of autoreactive cells [5].

Both cell proliferation and apoptosis are energy-dependent processes. Energy in the form of ATP is provided through glycolysis and oxidative phosphorylation [11]. The synthesis of ATP is driven by an electrochemical gradient across the inner mitochondrial membrane maintained by an electron transport chain and the membrane potential ($\Delta\psi_m$, negative inside and positive outside). A small fraction of electrons react directly with oxygen and form ROI. Disruption of $\Delta\psi_m$ has been proposed as the point of no return in apoptotic signaling [12–14]. Mitochondrial membrane permeability is subject to regulation by an oxidation-reduction equilibrium of ROI, pyridine nucleotides (NADH/NAD + NADPH/NADP) and GSH levels [15]. Regeneration of GSH by glutathione reductase from its oxidized form, GSSG, depends on NADPH produced by the pentose phosphate pathway (PPP) [16–18]. Metabolic fluxes between glycolysis and the PPP are particularly relevant for balancing cellular requirements for energy and ROI production [19]. While ROI have been considered as toxic by-products of aerobic existence, evidence is now accumulating that controlled levels of ROI modulate various aspects of cellular function and are necessary for signal-transduction pathways, including those mediating T-cell activation and apoptosis [19].

REDOX CONTROL OF T-CELL SIGNAL PROCESSING

Programmed cell death (PCD) or apoptosis is a physiological mechanism for elimination of autoreactive lymphocytes during development. Signaling through the Fas or structurally related TNF family of cell surface death receptors has emerged as a major pathway in elimination of unwanted cells under physiological and disease conditions [20]. Fas and TNF receptors mediate cell death via cytoplasmic death domains (DD) shared by both receptors [21]. They trigger sequential activation of caspases resulting in cleavage of cellular substrates and the characteristic morphologic and biochemical changes of apoptosis [22].

Disruption of the mitochondrial membrane potential ($\Delta\psi_m$) has been proposed as the point of no return in apoptotic signaling that leads to caspase activation and disassembly of the cell [13]. Interestingly, MHP and ROI production precede disruption of $\Delta\psi_m$, activation of caspases, and phosphatidylserine (PS) externalization in Fas- [18], TNFα [23] and H_2O_2-induced apoptosis of Jurkat human leukemia T cells and normal human peripheral blood lymphocytes [24]. Elevation of $\Delta\psi_m$ is independent from activation of caspases and represents an early event in apoptosis [18]. Pretreatment with caspase inhibitors completely abrogated Fas-induced PS externalization, indicating that activation of caspase-3, caspase-8, and related cysteine proteases were absolutely required for cell death [18]. ROI levels were partially inhibited in caspase inhibitor-treated Jurkat cells, suggesting that caspase-3 activation, perhaps through damage of mitochondrial membrane integrity, contributes to ROI production and serves as a positive feedback loop at later stages of the apoptotic process. Cleavage of cytosolic Bid by caspase-8 generates a p15 carboxyterminal fragment that translocates to mitochondria. This may increase mitochondrial membrane permeability and lead to secondary elevation of ROI levels in the Fas and TNF pathway [25]. Apically, trimerization of cell surface Fas receptors through formation of disulfide bonds is required for activation of the death-inducing signaling complex and cleavage of caspase 8 [26]. Alternatively, GSH is required for activity of caspases and its depletion can prevent Fas-dependent apoptosis [27]. ATP is required for apoptosis and its deficiency predisposes to necrosis [28]. Autophagy mediates the bulk degradation of cytoplasmic contents, proteins and organelles including mitochondria, in lysosomes [29]. Rapamycin, a lipophilic, macrolide antibiotic, induces autophagy by inactivating the protein mammalian target of rapamycin (mTOR) [30]. Thus, NO-dependent activation of mTOR may inhibit autophagy in lupus T cells [31].

MITOCHONDRIAL CHECKPOINTS OF CELL DEATH PATHWAY SELECTION: $\Delta \psi_M$, ATP SYNTHESIS, ROI AND NO PRODUCTION, CA^{2+} FLUXING AND REDUCING CAPACITY PROVIDED BY GSH AND NADPH

MHP appears to be the earliest change associated with several apoptosis pathways [9, 19]. Elevation of $\Delta \psi_m$ is also triggered by activation of the CD3−CD28 complex [6] or stimulation with Con A [18], IL-10, IL-3, IFN-γ, or TGFβ [7]. Therefore, MHP represents an early but reversible switch not exclusively associated with apoptosis. MHP is a likely cause of increased ROI production [11] and may be ultimately responsible for increased susceptibility to apoptosis following T-cell activation [6].

MHP in T lymphocytes is associated with a dramatic increase, more than 6-fold, of NO production lasting 24 h after CD3−CD28 co-stimulation. Molecular ordering of T-cell activation-induced NO production revealed critical roles for ROI production and cytoplasmic and mitochondrial Ca^{2+} influx [32]. CD3−CD28 co-stimulation-induced ROI production, similar to H_2O_2, enhances expression of NOS isoforms eNOS and nNOS, which require elevated Ca^{2+} levels for enzymatic activity. These results suggest that T-cell activation-induced ROI and Ca^{2+} signals contribute to NO production with the latter representing a final and dominant step in MHP [32].

MECHANISMS OF MHP AND ATP DEPLETION AND PREDISPOSITION TO NECROSIS IN LUPUS T CELLS

In response to treatment with exogenous H_2O_2, a precursor of ROI, lupus T cells failed to undergo apoptosis and cell death preferentially occurred via necrosis [6]. Endogenous H_2O_2 is generated by superoxide dismutase from ROIs, O_2^- or OH^-, in mitochondria [33]. In turn, H_2O_2 is scavenged by catalase and glutathione peroxidase [16]. While H_2O_2 is freely diffusible, it has no unpaired electrons and, by itself, is not a ROI [33]. Induction of apoptosis by H_2O_2, requires mitochondrial transformation into an ROI, e.g. OH^-, through the Fenton reaction [11, 33]. As previously noted [24], H_2O_2 triggered a rapid increase of $\Delta \psi_m$ and ROI production that was followed by apoptosis of PBL in healthy subjects. By contrast, H_2O_2 failed to elevate $\Delta \psi_m$, ROI production and apoptosis but rather elicited necrosis of lupus T cells. Both CD3−CD28-induced H_2O_2 production and H_2O_2-induced apoptosis require mitochondrial ROI production. Therefore, diminished CD3−CD28-induced H_2O_2 production and H_2O_2-induced apoptosis together with deficient elevation of $\Delta \psi_m$ and ROI levels reveal deviations of key biochemical checkpoints in mitochondria of patients with SLE.

The $\Delta \psi_m$ is dependent upon the electron transport chain transferring electrons from NADH to molecular oxygen and proton transport mediated by the F_0F_1−ATPase complex [11]. During oxidative phosphorylation, the F_0F_1−ATPase converts ADP to ATP utilizing the energy stored in the electrochemical gradient. Alternatively, using the energy of ATP hydrolysis, F_0F_1−ATPase can pump protons out of the mitochondrial matrix into the intermembrane space, causing $\Delta \psi_m$ elevation. Thus, MHP may occur several ways. First, deficiency of cellular ADP could cause diminished utilization of the electrochemical gradient, ATP depletion, and MHP. However, ADP levels were not diminished but slightly elevated in lupus PBL [6]. This suggested that ATP depletion and $\Delta \psi_m$ hyperpolarization were not due to a lack of ADP in patients with SLE. Second, MHP may occur through calcium-activated dephosphorylation of cytochrome c oxidase [34]. Phosphorylation of cytochrome c oxidase is mediated by protein kinase A (PKA), thus, deficiency of PKA may also contribute to MHP in SLE [35]. Third, inhibition of the enzymatic activity of F_0F_1–ATPase would decrease utilization of the electrochemical gradient and cause $\Delta\psi_m$ hyperpolarization, ATP depletion and ADP accumulation. Since blocking of F_0F_1–ATPase by oligomycin led to $\Delta\psi_m$ hyperpolarization and elevated ROI production, prevented H_2O_2- or CD3–CD28-induced elevation of $\Delta\psi_m$ in normal PBL, and sensitized to H_2O_2-induced necrosis, a similar mechanism may also be operational in patients with SLE [6].

NO causes transient MHP via reversible inhibition of complex IV/cytochrome c oxidase [36, 37] and may cause persistent MHP with GSH depletion via S-nitrosylation of complex I in the ETC of the inner mitochondrial membrane [38]. With $\Delta\psi_m$ hyperpolarization and extrusion of H+ ions from the mitochondrial matrix, the cytochromes within the electron transport chain become more reduced which favors generation of ROI [11]. Thus, MHP is a likely cause of increased ROI production and may be ultimately responsible for increased spontaneous apoptosis in patients with SLE. A 28–32% increase of the –200 mV $\Delta\psi_m$ may have a tremendous impact on mitochondrial energy coupling and ATP synthesis [11]. Both T-cell activation and apoptosis require the energy provided by ATP [39]. Intracellular ATP concentration is a key switch in the cell's decision to die via apoptosis or necrosis [28] and, therefore, depletion of ATP may be responsible for defective H_2O_2-induced apoptosis and a shift to necrosis in patients with SLE. Apoptosis is a physiological process that results in nuclear condensation and break-up of the cell into membrane-enclosed apoptotic bodies suitable for phagocytosis by macrophages thus preventing inflammation. By contrast, necrosis is a pathological process that results in cellular swelling, followed by lysis and release of proteases, oxidizing molecules, and other proinflammatory and chemotactic factors resulting in inflammation and tissue damage [39]. Indeed, lymphocyte necrosis occurs in the bone marrow [40] and lymph nodes of lupus patients and may significantly contribute to the inflammatory process [41].

INCREASED NECROSIS PROMOTES PRO-INFLAMMATORY STATE IN PATIENTS WITH SLE

Swollen lymph nodes of patients with SLE harbor increased numbers of necrotic T lymphocytes and dendritic cells (DC) [42]. Necrotic, but not apoptotic, cell death generates inflammatory signals necessary for the activation and maturation of DCs, the most potent antigen-presenting cells [43–45]. High-mobility group B1 (HMGB1) protein, an abundant DNA-binding protein, remains immobilized on chromatin of apoptotic bodies, however, it is released from necrotic cells [46].

HMGB1 stimulates human monocytes to release TNFα, IL-1α, IL-1β, IL-1RA, IL-6, IL-8, macrophage inflammatory protein (MIP)-1α, and MIP-1β, but not IL-10 or IL-12 [47] and induces arthritis [48]. Necrotic but not apoptotic cells also release heat shock proteins (HSPs), HSPgp96, hsp90, hsp70, and calreticulin. HSPs stimulate macrophages to secrete cytokines, and induce expression of antigen-presenting and co-stimulatory molecules on the DC [44]. Mature DCs express high levels of the DC-restricted markers, CD83 and lysosome-associated membrane glycoprotein (DC-LAMP) and the co-stimulatory molecules CD40 and CD86 [45], which may contribute to altered intercellular signaling in SLE [9]. CD14+ monocytes isolated from the blood of lupus patients, but not those from healthy individuals, act as DCs [49]. Their activation is driven by circulating interferon-α (IFN-α) that may come from one of the DC subsets, i.e., plasmacytoid dendritic cells (pDCs) that infiltrate lupus skin lesions. Tissue lesions [50, 51] and blood of patients with SLE harbor activated pDCs which may be responsible for increased production of IFN-α in SLE [49, 52]. Indeed, DCs exposed to necrotic, but not apoptotic, cells induce lupus-like disease in MRL mice and accelerate the disease of MRL/lpr mice [53].

DETECTION OF THE MITOCHONDRIAL GENE EXPRESSION SIGNATURE IN LUPUS T CELLS

The signaling network underlying T-cell dysfunction was investigated by a series of microarray analysis of gene expression and parallel flow cytometry of mitochondrial function in 15 female lupus patients relative to 17 female controls [31]. Among 10 of 15 lupus patients, who have not been exposed to prednisone or cytotoxic drugs, expression of 117 genes was altered. Eighty-two of these genes had identified functions. Although many genes participate in multiple signaling networks, 22 were involved in metabolic control of $\Delta\psi_m$, 19 in calcium signaling, 12 in cytokine/interferon pathway [54–56], 6 in programmed cell death, 9 in small GTPase-mediated intracellular traffic, 8 were transcription factors and 5 were surface receptors/adaptor proteins [31]. While lupus T cells exhibit persistent MHP, exposure of normal human T cells to CD3/CD28 co-stimulation or NO elicit transient MHP [32]. To distinguish the impact of SLE from that of T-cell activation, the influence of CD3/CD28 co-stimulation and NO on gene expression was assessed in negatively isolated "untouched" T cells of four healthy female donors. Direction and extent of CD3/CD28 or NO-induced changes in expression of 48 genes correlated with differences observed in patients with SLE. In contrast, altered expression of 34 genes could not be attributed to CD3/CD28 or NO and thus,

TABLE 10.1   Mitochondrial signaling abnormalities in T cells of patients with SLE

Signal	Effect	References
$\Delta\psi_m$ ↑	ROI ↑, ATP ↓	[6]
ROI ↑	Spontaneous apoptosis ↑, IL-10 production ↑	[6, 7]
GSH ↓	ROI ↑, Spontaneous apoptosis ↑	[6, 17]
Spontaneous apoptosis ↑	Compartmentalized autoantigen release, disease activity ↑	[4, 6, 10]
H_2O_2	Apoptosis ↓, necrosis ↑	[6]
ATP ↓	Necrosis ↑	[6, 81]
Necrosis ↑	Inflammation ↑	[6, 92]
AICD ↓	Persistence of autoreactive cells	[5, 7]
NO ↑	Mitochondrial biogenesis, altered Ca^{2+} fluxing, mTOR ↑	[57, 84, 106]

[a] ↑, increase; ↓, decrease.

they may reflect the disease process in SLE. Expression of 25/34 'lupus-specific' genes were correlated with one or more parameters of mitochondrial dysfunction in T cells; while 13/48 genes coordinately regulated by CD3/CD28 stimulation were correlated with mitochondrial dysfunction. This analysis suggested that lupus-specific changes in gene expression, rather than genes coordinately regulated by CD3/CD28, are more closely linked to mitochondrial dysfunction. Among the 34 genes significantly altered in negatively isolated T cells, 1.8-fold or more, voltage-dependent anion channel 1 (VDAC1), superoxide dismutase type 2 (SOD2), and Rab5A were found to be elevated in negatively isolated lupus T cells and predictive of SLE in each lupus patient using k-nearest-neighbor algorithm (Tables 10.1–10.4).

The functional relevance of changes in gene expression was further investigated on the protein level by western blot analysis [31]. Expression of mitochondrial proteins SOD2 and VDAC1 was elevated in negatively isolated lupus T cells. Increased SOD2 and VDAC1 protein levels are consistent with increased mitochondrial mass and oxidative stress in lupus T cells [57]. Moreover, the 12-kDa FK506-binding protein (FKBP12), the cellular receptor of rapamycin [58], and transaldolase (TAL), a controller of intracellular NADPH, GSH, and $\Delta\psi_m$ [17, 18, 59] were also elevated in lupus T cells [31]. Expression of NOSIP, which inhibits activity of eNOS, was reduced in lupus T cells. Diminished NOSIP expression was associated with elevated production of NO, a key factor eliciting mitochondrial dysfunction in lupus T cells [31, 57].

TABLE 10.2   Biomarkers of abnormal T-cell death in patients with SLE

Signal	Effect	References
Spontaneous apoptosis ↑	Compartmentalized autoantigen release, disease activity ↑	[6, 10]
AICD ↓	Persistence of autoreactive cells	[5, 7]
FasL ↑	Spontaneous apoptosis ↑	[56]
$\Delta\psi_m$ ↑	ROI ↑, ATP ↓	[6]
GSH ↓	ROI ↑, Spontaneous apoptosis ↑	[6, 17]
ATP ↓	AICD ↓, Predisposes for necrosis ↑	[6, 81]
NO ↑	$\Delta\psi_m$ ↑, mitochondrial biogenesis ↑	[27, 106]
ROI ↑	Spontaneous apoptosis ↑	[6, 7]
IL-10 ↑	Spontaneous apoptosis ↓, ROI ↓	[7, 56]
IL-12 ↓	Spontaneous apoptosis ↓, ROI ↓	[7]

↑, increase; ↓, decrease.

TABLE 10.3 Redox-controlled biomarkers of abnormal T-cell activation in patients with SLE

Signal	Effect	References
CD3-induced Ca^{2+} flux ↑	Altered T-cell activation	[127]
Baseline $[Ca^{2+}]_c$ ↑	Ca^{2+} influx ↑	[106, 112]
Baseline $[Ca^{2+}]_m$ ↑	Ca^{2+} influx ↑	[106, 112]
TCR/CD3ζ ↓	Ca^{2+} influx ↑, IL-2 poduction ↓	[146]
FcεRIγ ↑	Ca^{2+} influx ↑	[151]
PP2A ↑	Elf-1 ↓	[66]
Lck ↓	Ca^{2+} influx ↑	[129, 151]
Syk ↑	Ca^{2+} influx ↑	[129, 152]
Lipid raft formation ↑	TCRζ ↓; FcεRIγ ↑; Syk ↑	[128, 129, 152]
Receptor recycling ↑	CD4 ↓, TCRζ ↓	[84]
Rab5 ↑, HRES-1/Rab4 ↑	Receptor recycling ↑	[84]
mTOR ↑	Ca^{2+} influx ↑	[112]
mTOR ↑	Rab5 ↑, HRES-1/Rab4 ↑	[84]
$\Delta\psi_m$ ↑	ROI ↑, ATP ↓	[6]
NO ↑	$\Delta\psi_m$ ↑, mTOR ↑; Ca^{2+} flux ↑	[27, 84, 106]
ROI ↑	IL-10 ↑	[6, 7]
ROI ↑	DNMT1 ↓	[68]
ROI ↑	TCRζ ↓	[48, 153]
DNMT19	PP2A ↑; Elf-1 ↓; TCRζ ↓; FcεRIγ ↑	[67]
PKCδ ↓	ERK ↓, DNMT1 ↓	[154]
IBP ↓	Rho GTPase ↓, ERK ↓	[155]
Stat3 ↑	T-cell migration ↓	[70]
CD4-/CD8-/Th17+ ↑	nephritis	[73, 74]

↑, increase; ↓, decrease.

ASSOCIATION OF NON-CONSERVED CODING SEQUENCE POLYMORPHISMS IN MITOCHONDRIAL DNA WITH SLE

Out of ~1500 mitochondrial proteins, the mitochondrial DNA (mtDNA) encodes 13 polypeptides, seven of the ~45 subunits of complex I, one of 11 members of complex III, three of 13 members of complex IV, two of ~15 members of complex V [60]. The mtDNA of 97 Caucasian patients with SLE, 523 Caucasian patients with MS, and 466 Caucasian healthy controls was recently analyzed using the Sequenom MassAR-RAYTM System [61]. Nine markers within mtDNA (SNPs at nt1719, nt4216, nt4529, nt4917, nt9055, nt10398, nt13708, 14798 and nt16391) and seven markers within the nuclear DNA encoding genes of complex I (two SNPs in NDUFS5 — 1p34.2-p33; three

SNPs in NDUFS7 — 19p13; two SNPs in NDUFA7 — 19p13) were tested. Individual marker associations were detected at nt9055 of the ATP6 gene (subunit of F0F1—ATPase or complex V) and at nt4917 of the NADH dehydrogenase 2 (ND2 gene, subunit of complex I) with both SLE and MS [61]. The nt4917 SNP involved A6G transition, causing an asparagine to aspartic amino acid substitution in the ND2. This protein controls the production of reactive oxygen intermediates by complex I [62]. The A6G transition at nt9055 changes alanine to threonine in the ATP6 subunit of complex V or F0F1—ATPase. Since inhibition of F0F1—ATPase by oligomycin elicits MHP and ATP depletion [6], and thus, mimics the changes persistently observed in lupus T cells [6, 7, 57, 63], it is conceivable that the alanine—threonine change in this enzyme could influence the pathophysiology of SLE.

TABLE 10.4   Mitochondrial gene expression signature of negatively isolated lupus T cells

Affymetrix probe	Gene	FC
214132_at	ATP synthase, H+ transporting, mitochondrial F1 complex, gamma polypeptide	−1.52
213758_at	Cytochrome c oxidase subunit IV isoform 1	−2.72
221550_at	COX15 homolog, cytochrome c oxidase assembly protein	+1.88
218136_s_at	Mitochondrial solute carrier protein	+2.34
217977_at	Methionine-R-sulfoxide reductase (Selenoprotein X 1)	+2.74
225606_at	BCL-2-like 11 (apoptosis facilitator)	−2.35
205681_at	BCL-2-related protein A1	+2.34
229686_at	Purinergic receptor P2Y, G-protein coupled, ↑	−3.42
220005_at	Purinergic receptor P2Y, G-protein coupled, 13	+2.85
204491_at	Phosphodiesterase 4D, cAMP-specific	−2.62
202207_at	ADP-ribosylation factor-like 7	−2.03
218918_at	Mannosidase	−2.68
229553_at	Phosphoglyceromutase 2-like 1	−2.64
208918_s_at	NAD kinase	+1.68
201463_s_at	Transaldolase	+2.04
214183_s_at	Transketolase	−3.02
206177_s_at	Arginase 1	+3.74
217950_at	Nitric oxide synthase interacting protein (NOSIP)	−1.9
201364_s_at	Ornithine decarboxylase antizyme 2	+1.9
215078_at	SOD2	+4.86
212038_s_at	VDAC1	+1.94
204171_at	Ribosomal protein S6 kinase 70 kD (RPS6KB1	+1.98
206857_s_at	FK506-binding protein FKBP1B, 12 kD	+2.04

FC, fold change; +, increase; -, decrease.

ACTIVATION OF THE MAMMALIAN TARGET OF RAPAMYCIN (mTOR) IN LUPUS T CELLS

Rapamycin normalized T-cell mitogen-stimulated splenocyte proliferation and IL-2 production, prevented the typical rise in anti-double-stranded DNA antibody and urinary albumin levels and glomerulonephritis (GN), and prolonged survival of lupus-prone MRL/lpr mice [64]. mTOR is associated with the outer mitochondrial membrane and senses mitochondrial dysfunction and changes in the $\Delta\psi_m$ of T cells [58]. With a focus on mitochondrial dysfunction, we began to utilize rapamycin for treatment of SLE patients resistant or intolerant to conventional medications. Rapamycin markedly improved disease activity [63]. This occurred with the normalization of baseline Ca^{2+} levels in the cytosol and mitochondria and of CD3/CD28-induced Ca^{2+} fluxing while MHP persisted, indicating that increased Ca^{2+} fluxing is downstream or independent of MHP in the pathogenesis of T-cell dysfunction in SLE [63].

The immunosuppressive properties of rapamycin have been attributed to the blocking of TOR complex 1 (TORC1) that is required for transducing T-cell activation initiated by cytokines [65]. TORC2 is required for organization of the actin cytoskeleton and it is rapamycin-insensitive. While mTOR expression was similar in lupus and control T cells, phosphorylation levels of two key mTOR substrates S6K1 and 4E-BP1 were dramatically increased in lupus T cells and such changes were reversed in rapamycin-treated patients [31]. These findings suggest that mTOR kinase activity is increased in lupus T cells and it is reversed by rapamycin treatment. Overexpression of the cellular receptor of rapamycin, FKBP12, was identified as a component of the mitochondrial gene expression signature in lupus T cells

(Table 10.4). In accordance with the persistence of increased mitochondrial mass and MHP in rapamycin-treated patients [63], elevated expression of FKBP12 was unaffected by rapamycin treatment [31], suggesting that mTOR activation may be a consequence of genetic factors underlying MHP and increased mitochondrial biogenesis in lupus T cells [66]. Such genetic factors may also operate in lupus B cells [67, 68]. Autophagy mediates the bulk degradation of cytoplasmic contents, proteins, and organelles including mitochondria in T cells [69], in lysosomes; this process is induced by rapamycin through inactivating mTOR [30]. Interestingly, rapamycin treatment *in vivo* did not affect MHP or mitochondrial mass of lupus T cells [63]. However, rapamycin treatment normalized baseline and T-cell activation-induced Ca^{2+} fluxing. Such outcome may result from modulating the expression of small GTPases Rab5 and HRES-1/Rab4 that control endocytic traffic

(Figure 10.1) and degradation of key molecules of T-cell signal transduction including CD4 [70] and TCR/CD3 [31].

Akt, a kinase that phosphorylates multiple targets in T cells, may be a key link in the chain of events that activate mTOR [66]. Upstream, phosphatidylinositol accumulation by phosphoinositide 3-kinases (PI3K) induces the localization of 3-phosphatidylinositide-dependent protein kinase 1 (PDK1) to the plasma membrane that in turn phosphorylates and activates Akt [71]. Activation of class IA-PI3K in T cells extends CD4+ memory cell survival, triggering an invasive lymphoproliferative disorder and systemic lupus [72]. In turn, the lipid phosphatase PTEN (phosphatase and tensin homolog deleted in chromosome ten) counteracts phosphatidylinositol accumulation by PI3K and mice with PTEN haploinsufficiency also develop systemic autoimmunity [73]. Importantly, the rictor—mTOR complex directly

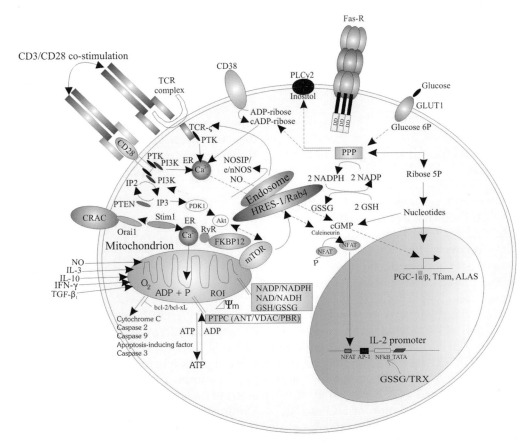

FIGURE 10.1 Schematic diagram of the metabolic pathways controlling (PPP, GSH, NO) and sensing (mTOR, HRES-1/Rab4) mitochondrial hyperpolarization (MHP) in lupus T cells. In normal T cells, MHP and mitochondrial biogenesis are mediated via production of NO by eNOS or nNOS (32) and up-regulation of transcription factors PGC-1α, Tfam, and ALAS [117]. NO production by eNOS may be compartmentalized to the T cell synapse [97]. The PPP regulates the $\Delta\psi_m$ by producing (1) NADPH that serves as a reducing equivalent for GSH regeneration from its oxidized form GSSG and for production of NO by NOS and (2) ribose 5-phosphate for biosynthesis of nucleotides, ADP, ATP, NAD, NAADP, (c) ADP-ribose, and cGMP, the latter is a second messenger of NO. NAADP and (c) ADP-ribose induce Ca2+ release from the endoplasmic reticulum (ER) via ryanodine receptors (RyR). Glucose and dehydroascorbate enter T cells via GLUT1 and fuel NADPH production via the PPP [24]. mTOR senses $\Delta\psi_m$ and regulates IP3R-mediated Ca^{2+} release [84]. The intracellular rapamycin receptor FKBP12 directly binds the RyR. The Bcl-2 family proteins control permeability of the outer mitochondrial membrane and release of apoptosis-inducing factors.

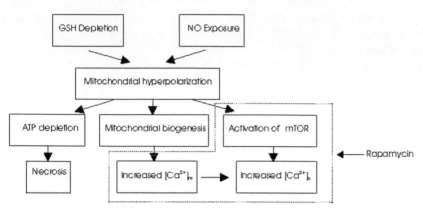

FIGURE 10.2   Schematic cascade of signaling pathways facilitating MHP (NO production and GSH depletion) and sensing MHP (mTOR). The broken line demarcates checkpoints affected by rapamycin.

phosphorylates Akt/PKB on Ser473 *in vitro* and facilitates Thr308 phosphorylation by PDK1 [74]. Thus, activation of mTOR may also account for elevated Akt activity and provide a positive feed-back loop of T-cell activation in SLE. mTOR controls the expression of Foxp3 and development of regulatory T cells [75, 76], which are deficient in patients with SLE [77, 78]. The effectiveness of rapamycin in murine and human SLE suggests that mTOR is a key mediator of autoimmunity in SLE. Therefore, understanding the mechanisms of persistent MHP that lead to mTOR activation and enhanced Ca^{2+} fluxing may be fundamental to the pathogenesis of lupus (Figure 10.2).

MITOCHONDRIAL CHECKPOINTS OF CA^{2+} FLUXING

The Ca^{2+} signal is dysregulated in SLE T cells in a number of ways. SLE T cells have elevated intracellular and mitochondrial Ca^{2+} at baseline [57], while the amount of Ca^{2+} present in the ER is normal [79]. Although the T-cell activation-induced Ca^{2+} flux is elevated initially, the plateau phase is reduced relative to activated control T cells [57]. There are at least three potentially interrelated mechanisms that account for altered Ca^{2+} flux in SLE: (1) abnormal formation of the immunological synapse in T cells [80, 81] and possibly also in B cells [82, 83]; (2) increased mitochondrial biogenesis and Ca^{2+} storage in mitochondria [57]; and (3) mTOR activation enhances IP3R-mediated Ca^{2+} release from the ER [84]. Dysregulation of NO production may play a role in all three mechanisms since normal T cells pre-treated with NO donors recapitulate the Ca^{2+} fluxing abnormalities observed in SLE T cells [57]. NO-induced MHP has recently been confirmed as a checkpoint of Ca^{2+} fluxing via store-operated Ca^{2+} entry [85]. NO induces (1) the expression of HRES-1/Rab4 that regulates receptor recycling and targets CD4 [70] and TCR for lysosomal degradation [31] and (2) promotes mitochondrial biogenesis and mitochondrial

storage of Ca^{2+} [57] and enhances mTOR activation [31]. Considering that Ca^{2+} is required for activation of PKC, calcineurin, NFAT, and production of IL-2, the altered Ca^{2+} homeostasis may account for the inappropriate activation of T cells in SLE [86].

The importance of T-cell synapse formation and Ca^{2+} homeostasis in lupus has been highlighted by the recent discovery that a nonsense mutation of the filamentous actin-inhibiting Coronin-1A gene inhibits T-cell synapse formation and T-cell activation-induced Ca^{2+} fluxing [87]. This mutation suppressed the development of lupus on the MRL/lpr backround which was associated with functional alterations in T cells, reduced T-dependent humoral responses, and no detectable intrinsic B-cell defects. By transfer of T cells it was shown that suppression of autoimmunity could be accounted for by the presence of the Coronin-1A mutation in T cells [87].

TARGETING OF MITOCHONDRIAL CHECKPOINTS IN T-CELL ACTIVATION FOR TREATMENT OF SLE

GSH Depletion

Reduced glutathione (GSH) levels are profoundly diminished in lymphocytes SLE patients [6]. Low GSH in T cells over-expressing TAL predispose to MHP [18]. GSH depletion is robust trigger of MHP via S-nitrosylation of complex I upon exposure to NO [38]. Thus, the effect of NO on MHP is tightly related to GSH depletion. Diminished production of GSH in face of MHP and increased ROI production is suggestive of a metabolic defect in its *de novo* synthesis or maintenance of the reduced state due to deficiency of NADPH [19]. Recent studies showed diminished GSH/GSSG ratios in the kidneys of 8-month-old versus 4-month-old (NZB x NZW) F1 mice; treatment with N-acetylcysteine (NAC), a precursor of GSH and stimulator of its *de novo* biosynthesis, prevented decline of GSH/GSSG ratios, reduced autoantibody production and development of

GN and prolonged survival of (NZB x NZW) F1 mice [88]. Stimulation of GSH levels by conjugated linoleic acid was also associated with reduced anti-DNA antibody production in MRL/lpr mice [89]. GSH levels were found to be >4-fold reduced in B and T cells of the spleen and lymph nodes of lupus-prone MRL/lpr mice relative to control MRL mice [90]. Interestingly, clinical improvement induced by arsenic oxide was associated with elevation of intracellular GSH [90]. Depletion of GSH contributes to autoimmune diseases other than SLE [91]. Oral NAC has been used to treat idiopathic pulmonary fibrosis (IPF) [92]. In a 1-year study of IPF patients treated with prednisone and azathioprine, addition of NAC (3×600 mg/d) improved vital capacity and reduced myelotoxicity in comparison to placebo. Seventy-one per cent (57/80) of NAC-treated patients and 68% (51/75) of placebo-treated patients completed the study. NAC is also useful for inhibiting muscle fatigue [93–95], which is reported to be the most disabling symptom in 53% of patients with SLE [96]. Prospective clinical studies are currently ongoing to assess whether NAC treatment would reduce GSH depletion, correct T-cell-signaling defects and provide clinical benefit to patients with SLE (www.clinicaltrials.gov IND 101320).

NO Production

Nitric oxide (NO) is a particularly interesting molecule in this context because it affects multiple aspects of T cells, providing a potential link between seemingly dissociated features of T-cell activation and mitochondrial function. NO induces MHP and mitochondrial biogenesis, increases Ca^{2+} in the cytosol and mitochondria, and, after 24 pretreatment, recapitulates the enhanced CD3/CD28-induced Ca^{2+} fluxing of lupus T cells [57]. NO is produced by nitric oxide synthases (NOS) that require Ca^{2+} to function and use NADPH and arginine as substrates. There have been three isoforms described: endothelial NOS (eNOS), neuronal NOS (nNOS), and inducible NOS (iNOS), of which T cells express the former two [32]. eNOS is recruited to the site of T-cell receptor engagement, locally increasing NO at the immunological synapse in a calcium- and PI3K-dependent manner, resulting in reduced IL-2 production [97] which is characteristic of SLE [98].

NO is required for the development of GN in the MRL/lpr lupus mouse model [99]. Inactivation of iNOS does not block the development of lupus [100], suggesting a role for eNOS and nNOS isoforms expressed in T cells. However, given the widespread functions of these isoforms in vascular smooth muscle of multiple organ systems and brain, it will be necessary to develop T-cell-specific approaches to avoid potentially deleterious side effects.

Rapamycin

The mammalian target of rapamycin (mTOR) is associated with the outer mitochondrial membrane and senses mitochondrial dysfunction and changes in the $\Delta\psi_m$ of T cells [58]. Rapamycin normalized T-cell mitogen-stimulated splenocyte proliferation and IL-2 production, inhibited the production of anti-double-stranded DNA antibody, the development of albuminuria and GN, and prolonged the survival of lupus-prone MRL/lpr mice [64]. Rapamycin improved the disease activity and reduced prednisone dependence in lupus patients [63]. This occurred with the normalization of baseline Ca^{2+} levels in the cytosol and mitochondria and of CD3/CD28-induced Ca^{2+} fluxing while MHP persisted [63]. This indicated that enhanced Ca^{2+} fluxing is downstream or independent of MHP in the pathogenesis of T-cell dysfunction in SLE. The effectiveness of rapamycin in murine and human SLE is consistent with recent findings that mTOR is a potential sensor and downstream effector of MHP and mediator of T-cell dysfunction and autoimmunity in SLE [31]. Rapamycin can also selectively expand CD4+/CD25+/Foxp3+ regulatory T cells (Tregs) [101], which appear to be deficient in patients with autoimmune diseases, including SLE [77, 102]. Therefore, understanding the mechanism of persistent MHP that leads to mTOR activation and enhanced Ca^{2+} fluxing is critical for the pathogenesis of T-cell dysfunction and the treatment of SLE.

Acknowledgments

This work was supported in part by grants RO1 AI 48079 and AI 072648 from the National Institutes of Health, the Alliance for Lupus Research, and the Central New York Community Foundation.

References

[1] J.J. Cohen, R.C. Duke, V.A. Fadok, K.S. Sellins, Apoptosis and programmed cell death in immunity, Ann Rev Immunol 10 (1992) 267–293.

[2] C.B. Thompson, Apoptosis in the pathogenesis and treatment of disease, Science 267 (1995) 1456–1462.

[3] A. Perl, K. Banki, Molecular mimicry, altered apoptosis, and immunomodulation as mechanisms of viral pathogenesis in systemic lupus erythematosus, in: G.M. Kammer, G.C. Tsokos (Eds.), Lupus: Molecular and Cellular Pathogenesis, Humana Press, Totowa, NJ, 1999, pp. 43–64.

[4] W. Emlen, J.A. Niebur, R. Kadera, Accelerated in vitro apoptosis of lymphocytes from patients with systemic lupus erythematosus, J Immunol 152 (1994) 3685–3692.

[5] B. Kovacs, D. Vassilopoulos, S.A. Vogelgesang, G.C. Tsokos, Defective CD3-mediated cell death in activated T cells from patients with systemic lupus erythematosus: role of decreased intracellular TNF-alpha, Clin Immunol Immunopathol 81 (1996) 293–302.

[6] P.J. Gergely, C. Grossman, B. Niland, F. Puskas, H. Neupane, F. Allam, et al., Mitochondrial hyperpolarization and ATP

depletion in patients with systemic lupus erythematosus, Arth Rheum 46 (2002) 175–190.

[7] P.J. Gergely, B. Niland, N. Gonchoroff, R. Pullmann Jr., P.E. Phillips, A. Perl, Persistent mitochondrial hyperpolarization, increased reactive oxygen intermediate production, and cytoplasmic alkalinization characterize altered IL-10 signaling in patients with systemic lupus erythematosus, J Immunol 169 (2002) 1092–1101.

[8] G.C. Tsokos, N.C. Liossis S-, Immune cell signaling defects in lupus: activation, anergy and death, Immunol Today 20 (1999) 119–124.

[9] A. Perl, P. Gergely Jr., G. Nagy, A. Koncz, K. Banki, Mitochondrial hyperpolarization: a checkpoint of T cell life, death, and autoimmunity, Trends Immunol 25 (2004) 360–367.

[10] L.A. Casciola-Rosen, G. Anhalt, A. Rosen, Autoantigens targeted in systemic lupus erythematosus are clustered in two populations of surface structures on apoptotic keratinocytes, J Exp Med 179 (1994) 1317–1330.

[11] V.P. Skulachev, Mitochondrial physiology and pathology; concepts of programmed death of organelles, cells and organisms, Mol Asp Med 20 (1999) 139–140.

[12] J. Xiang, D.T. Chao, S.J. Korsmeyer, BAX-induced cell death may not require interleukin 1β-converting enzyme-like proteases, Proc Natl Acad Sci USA 93 (1996) 14559–14563.

[13] S.A. Susin, N. Zamzami, M. Castedo, E. Daugas, H.-G. Wang, S. Geley, et al., The central executioner of apoptosis: multiple connections between protease activation and mitochondria in Fas/Apo-1/CD95- and ceramide-induced apoptosis, J Exp Med 186 (1997) 25–37.

[14] M. Vander Heiden, N.S. Chandel, E.K. Williamson, P.T. Schumaker, C.B. Thompson, Bcl-X$_L$ regulates the membrane potential and volume homeostasis of mitochondria, Cell 91 (1997) 627–637.

[15] P. Constantini, B.V. Chernyak, V. Petronilli, P. Bernardi, Modulation of the mitochondrial permeability transition pore by pyridine nucleotides and dithiol oxidation at two separate sites, J Biol Chem 271 (1996) 6746–6751.

[16] P.A. Mayes, The pentose phosphate pathway & other pathways of hexose metabolism, in: R.K. Murray, D.K. Granner, P.A. Mayes, V.W. Rodwell (Eds.), Harper's Biochemistry, Appleton & Lange, Norwalk, CT, 1993, pp. 201–211.

[17] K. Banki, E. Hutter, E. Colombo, N.J. Gonchoroff, A. Perl, Glutathione Levels and Sensitivity to Apoptosis Are Regulated by changes in Transaldolase expression, J Biol Chem 271 (1996) 32994–33001.

[18] K. Banki, E. Hutter, N. Gonchoroff, A. Perl, Elevation of mitochondrial transmembrane potential and reactive oxygen intermediate levels are early events and occur independently from activation of caspases in Fas signaling, J Immunol 162 (1999) 1466–1479.

[19] A. Perl, P. Gergely Jr., F. Puskas, K. Banki, Metabolic switches of T-cell activation and apoptosis, Antiox Redox Signal 4 (5) (2002) 427–443.

[20] S. Nagata, Apoptosis by death factor, Cell 88 (1997) 355–365.

[21] N. Itoh, S. Nagata, A novel protein domain required for apoptosis: Mutational analysis of human Fas antigen, J Biol Chem 268 (1993) 10932–10937.

[22] S.J. Martin, D.R. Green, Protease activation during apoptosis: death by a thousand cuts, Cell 82 (1995) 349–352.

[23] E. Gottlieb, M.G. Vander Heiden, C.B. Thompson, Bcl-X$_L$ prevents the initial decrease in mitochondrial membrane potential and subsequent reactive oxygen species production during tumor necrosis factor alpha-induced apoptosis, Mol Cell Biol 20 (2000) 5680–5689.

[24] F. Puskas, P. Gergely, K. Banki, A. Perl, Stimulation of the pentose phosphate pathway and glutathione levels by dehydroascorbate, the oxidized form of vitamin C, FASEB J 14 (2000) 1352–1361.

[25] A. Gross, J.M. McDonnell, S.J. Korsmeyer, BCL-2 family members and the mitochondria in apoptosis. [Review] [122 refs], Genes Dev 13 (1999) 1899–1911.

[26] L. Casciola-Rosen, D.W. Nicholson, T. Chong, K.R. Rowan, N.A. Thornberry, D.K. Miller, et al., Apopain/CPP32 cleaves proteins that are essential for cellular repair: A fundamental principle of apoptotic cell death, J Exp Med 183 (1996) 1957–1964.

[27] H. Hentze, G. Kunstle, C. Volbracht, W. Ertel, A. Wendel, CD95-Mediated murine hepatic apoptosis requires an intact glutathione status, Hepatology 30 (1999) 177–185.

[28] M. Leist, B. Single, A.F. Castoldi, S. Kuhnle, P. Nicotera, Intracellular Adenosine Triphosphate (ATP) Concentration: A Switch in the Decision Between Apoptosis and Necrosis, J Exp Med 185 (1997) 1481–1486.

[29] J.J. Lemasters, Selective mitochondrial autophagy, or mitophagy, as a targeted defense against oxidative stress, mitochondrial dysfunction, and aging, Rejuvenation Research 8 (1) (2005) 3–5.

[30] M. Kundu, C.B. Thompson, Macroautophagy versus mitochondrial autophagy: a question of fate? [Review] [58 refs], Cell Death Differ 12 (Suppl.2) (2005) 1484–1489.

[31] D.R. Fernandez, T. Telarico, E. Bonilla, Q. Li, S. Banerjee, F.A. Middleton, et al., Activation of mTOR controls the loss of TCR. in lupus T cells through HRES-1/Rab4-regulated lysosomal degradation, J Immunol 182 (2009) 2063–2073.

[32] G. Nagy, A. Koncz, A. Perl, T cell activation-induced mitochondrial hyperpolarization is mediated by Ca^{2+}- and redox-dependent production of nitric oxide, J Immunol 171 (2003) 5188–5197.

[33] B. Halliwell, J.M. Gutteridge, Role of free radicals and catalytic metal ions in human disease: an overview, Meth Enzymol 186 (1990) 1–85.

[34] I. Lee, E. Bender, B. Kadenbach, Control of mitochondrial membrane potential and ROS formation by reversible phosphorylation of cytochrome c oxidase, Molecular & Cellular Biochemistry 234-235 (1-2) (2002) 63–70.

[35] G.M. Kammer, A. Perl, B.C. Richardson, G.C. Tsokos, Abnormal T Cell Signal Transduction in Systemic Lupus Erythematosus, Arth Rheum 46 (5) (2002) 1139–1154.

[36] A. Almeida, J. Almeida, J.P. Bolanos, S. Moncada, Different responses of astrocytes and neurons to nitric oxide: the role of glycolytically generated ATP in astrocyte protection, Proc Natl Acad Sci USA 2001 98 (26) (2001 Dec 18) 15294–15299.

[37] E. Nisoli, E. Clementi, S. Moncada, M.O. Carruba, Mitochondrial biogenesis as a cellular signaling framework. [Review] [153 refs], Biochem Pharmacol 67 (1) (2004) 1–15.

[38] E. Clementi, G.C. Brown, M. Feelisch, S. Moncada, Persistent inhibition of cell respiration by nitric oxide: crucial role of S-nitrosylation of mitochondrial complex I and protective action of glutathione, Proc Natl Acad Sci USA 95 (13) (1998) 7631–7636.

[39] W. Fiers, R. Beyaert, Declercq w, Vandenabeele P. More than one way to die: apoptosis, necrosis and reactive oxygen damage, Oncogene 18 (1999) 7719–7730.

[40] I. Lorand-Metze, M.A. Carvalho, L.T. Costallat, [Morphology of bone marrow in systemic lupus erythematosus]. [German], Pathologie 15 (1994) 292–296.

[41] M.D. Eisner, J. Amory, B. Mullaney, L. Tierney Jr., W.S. Browner, Necrotizing lymphadenitis associated with systemic lupus erythematosus, Sem Arth Rheum 26 (1996) 477–482.

[42] M. Kojima, S. Nakamura, Y. Morishita, H. Itoh, K. Yoshida, Y. Ohno, et al., Reactive follicular hyperplasia in the lymph node lesions from systemic lupus erythematosus patients: a clinicopathological and immunohistological study of 21 cases, Pathol Int 50 (4) (2000) 304–312.

[43] S. Gallucci, M. Lolkema, P. Matzinger, Natural adjuvants: endogenous activators of dendritic cells, Nature Medicine 5 (11) (1999) 1249–1255.

[44] S. Basu, R.J. Binder, R. Suto, K.M. Anderson, P.K. Srivastava, Necrotic but not apoptotic cell death releases heat shock proteins, which deliver a partial maturation signal to dendritic cells and activate the NF-{kappa}B pathway, Int Immunol 12 (11) (2000) 1539–1546.

[45] B. Sauter, M.L. Albert, L. Francisco, M. Larsson, S. Somersan, N. Bhardwaj, Consequences of Cell Death: Exposure to Necrotic Tumor Cells, but Not Primary Tissue Cells or Apoptotic Cells, Induces the Maturation of Immunostimulatory Dendritic Cells, J Exp Med 191 (3) (2000) 423–434.

[46] Bustin M. At the crossroads of necrosis and apoptosis: signaling to multiple cellular targets by HMGB1. [Review] [29 refs]. Science's Stke [Electronic Resource]: Signal Transduction Knowledge Environment 2002; 2002(151)E39.

[47] U. Andersson, H. Wang, K. Palmblad, A.C. Aveberger, O. Bloom, H. Erlandsson-Harris, et al., High mobility group 1 protein (HMG-1) stimulates proinflammatory cytokine synthesis in human monocytes, Journal of Experimental Medicine 192 (4) (2000) 565–570.

[48] R. Pullerits, I.M. Jonsson, M. Verdrengh, M. Bokarewa, U. Andersson, H. Erlandsson-Harris, et al., High mobility group box chromosomal protein 1, a DNA binding cytokine, induces arthritis, Arthritis & Rheumatism 48 (6) (2003) 1693–1700.

[49] V. Pascual, J. Banchereau, A.K. Palucka, The central role of dendritic cells and interferon-alpha in SLE. [Review] [79 refs], Curr Opin Rheumatol 15 (5) (2003) 548–556.

[50] M. Mori, N. Pimpinelli, P. Romagnoli, E. Bernacchi, P. Fabbri, B. Giannotti, Dendritic cells in cutaneous lupus erythematosus: a clue to the pathogenesis of lesions, Histopathology 24 (4) (1994) 311–321.

[51] L. Farkas, K. Beiske, F. Lund-Johansen, P. Brandtzaeg, F.L. Jahnsen, Plasmacytoid dendritic cells (natural interferon-alpha/beta-producing cells) accumulate in cutaneous lupus erythematosus lesions, Am J Pathol 159 (1) (2001) 237–243.

[52] L. Ronnblom, G.V. Alm, The natural interferon-alpha producing cells in systemic lupus erythematosus. [Review] [105 refs], Hum Immunol 63 (12) (2002) 1181–1193.

[53] L. Ma, K.W. Chan, N.J. Trendell-Smith, A. Wu, L. Tian, A.C. Lam, et al., Systemic autoimmune disease induced by dendritic cells that have captured necrotic but not apoptotic cells in susceptible mouse strains, Eur J Immunol 35 (2005) 3364–3375.

[54] L. Bennett, A.K. Palucka, E. Arce, V. Cantrell, J. Borvak, J. Banchereau, et al., Interferon and granulopoiesis signatures in systemic lupus erythematosus blood.[comment], J Exp Med 197 (6) (2003) 711–723.

[55] E.C. Baechler, F.M. Batliwalla, G. Karypis, P.M. Gaffney, W.A. Ortmann, K.J. Espe, et al., Interferon-inducible gene expression signature in peripheral blood cells of patients with severe lupus, Proc Natl Acad Sci USA 100 (5) (2003) 2610–2615.

[56] K.A. Kirou, C. Lee, S. George, K. Louca, I.G. Papagiannis, M.G. Peterson, et al., Coordinate overexpression of interferon-alpha-induced genes in systemic lupus erythematosus, Arth Rheum 50 (2004) 3958–3967.

[57] G. Nagy, M. Barcza, N. Gonchoroff, P.E. Phillips, A. Perl, Nitric Oxide-Dependent Mitochondrial Biogenesis Generates Ca2+

[58] B.N. Desai, B.R. Myers, S.L. Schreiber, FKBP12-rapamycin-associated protein associates with mitochondria and senses osmotic stress via mitochondrial dysfunction, Proc Natl Acad Sci USA 99 (2002) 4319–4324.

Signaling Profile of Lupus T Cells, J Immunol 173 (6) (2004) 3676–3683.

[59] Y. Qian, S. Banerjee, C.E. Grossman, W. Amidon, G. Nagy, M. Barcza, et al., Transaldolase deficiency influences the pentose phosphate pathway, mitochondrial homoeostasis and apoptosis signal processing, Biochem J 415 (1) (2008) 123–134.

[60] D.C. Wallace, Why Do We Still Have a Maternally Inherited Mitochondrial DNA? Insights from Evolutionary Medicine, Ann Rev Biochem 76 (1) (2007) 781–821.

[61] T. Vyshkina, A. Sylvester, S. Sadiq, E. Bonilla, J.A. Canter, A. Perl, et al., Association of common mitochondrial DNA variants with multiple sclerosis and systemic lupus erythematosus, Clin Immunol 129 (2008) 31–35.

[62] A.M. Gusdon, T.V. Votyakova, I.J. Reynolds, C.E. Mathews, Nuclear and Mitochondrial Interaction Involving mt-Nd2 Leads to Increased Mitochondrial Reactive Oxygen Species Production, J Biol Chem 282 (8) (2007) 5171–5179.

[63] D. Fernandez, E. Bonilla, N. Mirza, A. Perl, Rapamycin reduces disease activity and normalizes T-cell activation-induced calcium fluxing in patients with systemic lupus erythematosus, Arth Rheum 54 (9) (2006) 2983–2988.

[64] L.M. Warner, L.M. Adams, S.N. Sehgal, Rapamycin prolongs survival and arrests pathophysiologic changes in murine systemic lupus erythematosus, Arth Rheum 37 (1994) 289–297.

[65] R.A. Kirken, Y.L. Wang, Molecular actions of sirolimus: sirolimus and mTor. [Review] [32 refs], Transplant Proc 35 (3) (2003) 227S–230S.

[66] D. Fernandez, A. Perl, Metabolic control of T cell activation and death in SLE, Autoimmun Rev 8 (2009) 184–189.

[67] T. Wu, X. Qin, Z. Kurepa, K.R. Kumar, K. Liu, H. Kanta, et al., Shared signaling networks active in B cells isolated from genetically distinct mouse models of lupus. [Article], J Clin Invest 117 (8) (2007) 2186–2196.

[68] K. Liu, C. Mohan, Altered B-cell signaling in lupus, Autoimmun Rev 8 (3) (2009) 214–218.

[69] H.H. Pua, J. Guo, M. Komatsu, Y.W. He, Autophagy Is Essential for Mitochondrial Clearance in Mature T Lymphocytes, J Immunol 182 (7) (2009) 4046–4055.

[70] G. Nagy, J. Ward, D.D. Mosser, A. Koncz, P. Gergely, C. Stancato, et al., Regulation of CD4 Expression via Recycling by HRES-1/RAB4 Controls Susceptibility to HIV Infection, J Biol Chem 281 (2006) 34574–34591.

[71] M.M. Juntilla, G.A. Koretzky, Critical roles of the PI3K/Akt signaling pathway in T cell development, Immunol Lett 116 (2) (2008) 104–110.

[72] D.F. Barber, A. Bartolome, C. Hernandez, J.M. Flores, C. Fernandez-Arias, L. Rodriguez-Borlado, et al., Class IB-Phosphatidylinositol 3-Kinase (PI3K) Deficiency Ameliorates IA-PI3K-Induced Systemic Lupus but Not T Cell Invasion, J Immunol 176 (1) (2006) 589–593.

[73] A. Di Cristofano, P. Kotsi, Y.F. Peng, C. Cordon-Cardo, K.B. Elkon, P.P. Pandolfi, Impaired Fas Response and Autoimmunity in Pten+/ Mice, Science 285 (5436) (1999) 2122–2125.

[74] D.D. Sarbassov, D.A. Guertin, S.M. Ali, D.M. Sabatini, Phosphorylation and regulation of Akt/PKB by the rictor-mTOR complex, Science 307 (2005) 1098–1101.

[75] S. Haxhinasto, D. Mathis, C. Benoist, The AKT-mTOR axis regulates de novo differentiation of CD4+Foxp3+ cells, J Exp Med 205 (3) (2008) 565–574.

[76] S. Sauer, L. Bruno, A. Hertweck, D. Finlay, M. Leleu, M. Spivakov, et al., T cell receptor signaling controls Foxp3

expression via PI3K, Akt, and mTOR, Proc Natl Acad Sci USA 105 (22) (2008) 7797–7802.

[77] X. Valencia, C. Yarboro, G. Illei, P.E. Lipsky, Deficient CD4+CD25high T Regulatory Cell Function in Patients with Active Systemic Lupus Erythematosus, J Immunol 178 (4) (2007) 2579–2588.

[78] B.H. Hahn, M. Anderson, E. Le, A. La Cava, Anti-DNA Ig peptides promote Treg cell activity in systemic lupus erythematosus patients, Arth Rheum 58 (8) (2008) 2488–2497.

[79] D. Vassilopoulos, B. Kovacs, G.C. Tsokos, TCR/CD3 complex-mediated signal transduction pathway in T cells and T cell lines from patients with systemic lupus erythematosus, J Immunol 155 (4) (1995) 2269–2281.

[80] E.C. Jury, P.S. Kabouridis, F. Flores-Borja, R.A. Mageed, D.A. Isenberg, Altered lipid raft-associated signaling and ganglioside expression in T lymphocytes from patients with systemic lupus erythematosus, J Clin Invest 113 (2004) 1176–1187.

[81] E.C. Jury, P.S. Kabouridis, A. Abba, R.A. Mageed, D.A. Isenberg, Increased ubiquitination and reduced expression of LCK in T lymphocytes from patients with systemic lupus erythematosus, Arth Rheum 48 (2003) 1343–1354.

[82] F. Flores-Borja, P.S. Kabouridis, E.C. Jury, D.A. Isenberg, R.A. Mageed, Altered lipid raft-associated proximal signaling and translocation of CD45 tyrosine phosphatase in B lymphocytes from patients with systemic lupus erythematosus. [Article], Arth Rheum 56 (1) (2007) 291–302.

[83] F.P. Flores-Borja, P.S.P. Kabouridis, E.C.P. Jury, D.A.M. Isenberg, R.A.P. Mageed, Decreased Lyn Expression and Translocation to Lipid Raft Signaling Domains in B Lymphocytes From Patients With Systemic Lupus Erythematosus. [Article], Arth Rheum 52 (12) (2005) 3955–3965.

[84] D. MacMillan, S. Currie, K.N. Bradley, T.C. Muir, J.G. McCarron, In smooth muscle, FK506-binding protein modulates IP3 receptor-evoked Ca2+ release by mTOR and calcineurin, J Cell Sci 118 (2005) 5443–5451.

[85] R.A. Valero, L. Senovilla, NAαez L, Villalobos C. The role of mitochondrial potential in control of calcium signals involved in cell proliferation, Cell Calcium 44 (3) (2008) 259–269.

[86] V.C. Kyttaris, Y.T. Juang, G.C. Tsokos, Immune cells and cytokines in systemic lupus erythematosus: an update.[see comment]. [Review] [29 refs], Curr Opin Rheumatol 17 (2005) 518–522.

[87] M.K. Haraldsson, C.A. Louis-Dit-Sully, B.R. Lawson, G. Sternik, M.L. Santiago-Raber, N.R.J. Gascoigne, et al., The Lupus-Related Lmb3 Locus Contains a Disease-Suppressing Coronin-1A Gene Mutation, Immunity 28 (1) (2008) 40–51.

[88] S. Suwannaroj, A. Lagoo, D. Keisler, R.W. McMurray, Antioxidants suppress mortality in the female NZB x NZW F1 mouse model of systemic lupus erythematosus (SLE), Lupus 10 (4) (2001) 258–265.

[89] P. Bergamo, D. Luongo, F. Maurano, G. Mazzarella, R. Stefanile, M. Rossi, Conjugated linoleic acid enhances glutathione synthesis and attenuates pathological signs in MRL/lpr mice, J Lipid Res 47 (2006) 2382–2391.

[90] P. Bobe, D. Bonardelle, K. Benihoud, P. Opolon, M.K. Chelbi-Alix, Arsenic trioxide, a novel promising therapeutic agent for lymphoproliferative and autoimmune syndromes in MRL/lpr mice, Blood (2006). blood-2006.

[91] C. Perricone, C. De Carolis, R. Perricone, Glutathione: A key player in autoimmunity, Autoimmun Rev 8 (8) (2009) 697–701.

[92] M. Demedts, J. Behr, R. Buhl, U. Costabel, R. Dekhuijzen, H.M. Jansen, et al., High-dose acetylcysteine in idiopathic pulmonary fibrosis.[see comment], N Engl J Med 353 (2005) 2229–2242.

[93] Y. Matuszczak, M. Farid, J. Jones, S. Lansdowne, M.A. Smith, A.A. Taylor, et al., Effects of N-acetylcysteine on glutathione oxidation and fatigue during handgrip exercise, Muscle Nerve 32 (2005) 633–638.

[94] J.M. Travaline, S. Sudarshan, B.G. Roy, F. Cordova, V. Leyenson, G.J. Criner, Effect of N-acetylcysteine on human diaphragm strength and fatigability, Am J Resp Crit Care Med 156 (1997) 1567–1571.

[95] C. Koechlin, A. Couillard, D. Simar, J.P. Cristol, H. Bellet, M. Hayot, et al., Does oxidative stress alter quadriceps endurance in chronic obstructive pulmonary disease? Am J Resp Crit Care Med 169 (2004) 1022–1027.

[96] L.B. Krupp, N.G. LaRocca, J. Muir, A.D. Steinberg, A study of fatigue in systemic lupus erythematosus, J Rheumatol 17 (1990) 1450–1452.

[97] S. Ibiza, V.M. Victor, I. Bosca, A. Ortega, A. Urzainqui, J.E. O'Connor, et al., Endothelial Nitric Oxide Synthase Regulates T Cell Receptor Signaling at the Immunological Synapse, Immunity 24 (6) (2006) 753–765.

[98] M.P. Nambiar, C.U. Fisher, V.G. Warke, S. Krishnan, J.P. Mitchell, N. Delaney, et al., Reconstitution of deficient T cell receptor zeta chain restores T cell signaling and augments T cell receptor/CD3-induced interleukin-2 production in patients with systemic lupus erythematosus, Arth Rheum 48 (2003) 1948–1955.

[99] J.B. Weinberg, D.L. Granger, D.S. Pisetsky, M.F. Seldin, M.A. Misukonis, S.N. Mason, et al., The role of nitric oxide in the pathogenesis of spontaneous murine autoimmune disease: Increased nitric oxide production and nitric oxide synthase expression in MRL-lpr/lpr mice, and reduction of spontaneous glomerulonephritis and arthritis by orally administered N^G-monomethyl-L-arginine, J Exp Med 179 (1994) 651–660.

[100] G.S. Gilkeson, J.S. Mudgett, M.F. Seldin, P. Ruiz, A.A. Alexander, M.A. Misukonis, et al., Clinical and serologic manifestations of autoimmune disease in MRL-lpr/lpr mice lacking nitric oxide synthase type 2, J Exp Med 186 (1997) 365–373.

[101] M. Battaglia, A. Stabilini, B. Migliavacca, J. Horejs-Hoeck, T. Kaupper, M.G. Roncarolo, Rapamycin Promotes Expansion of Functional CD4+CD25+FOXP3+ Regulatory T Cells of Both Healthy Subjects and Type 1 Diabetic Patients, J Immunol 177 (12) (2006) 8338–8347.

[102] J.C. Crispin, A. Martinez, J. Alcocer-Varela, Quantification of regulatory T cells in patients with systemic lupus erythematosus, J Autoimmun 21 (3) (2003) 273–276.

[103] J.C. Oates, E.F. Christensen, C.M. Reilly, S.E. Self, G.S. Gilkeson, Prospective measure of serum 3-nitrotyrosine levels in systemic lupus erythematosus: correlation with disease activity, Proc Assoc Am Phys 111 (1999) 611–621.

[104] L. Georgescu, R.K. Vakkalanka, K.B. Elkon, M.K. Crow, Interleukin-10 promotes activation-induced cell death of SLE lymphocytes mediated by Fas ligand, J Clin Invest 100 (1997) 2622–2633.

[105] E.J. Enyedy, M.P. Nambiar, S.N. Liossis, G. Dennis, G.M. Kammer, G.C. Tsokos, Fc epsilon receptor type I gamma chain replaces the deficient T cell receptor zeta chain in T cells of patients with systemic lupus erythematosus, Arth Rheum 44 (2001) 1114–1121.

[106] Y.T. Juang, Y. Wang, G. Jiang, H.B. Peng, S. Ergin, M. Finnell, et al., PP2A Dephosphorylates Elf-1 and Determines the Expression of CD3{zeta} and FcR{gamma} in Human Systemic Lupus Erythematosus T Cells, J Immunol 181 (5) (2008) 3658–3664.

[107] S. Krishnan, M.P. Nambiar, V.G. Warke, C.U. Fisher, J. Mitchell, N. Delaney, et al., Alterations in lipid raft composition and

dynamics contribute to abnormal T cell responses in systemic lupus erythematosus, J Immunol 172 (2004) 7821–7831.

[108] R. Franco, O. Schoneveld, A.G. Georgakilas, M.I. Panayiotidis, Oxidative stress, DNA methylation and carcinogenesis, Cancer Lett 266 (1) (2008) 6–11.

[109] M. Otsuji, Y. Kimura, T. Aoe, Y. Okamoto, T. Saito, Oxidative stress by tumor-derived macrophages suppresses the expression of CD3 zeta chain of T-cell receptor complex and antigen-specific T-cell responses, Proc Natl Acad Sci USA 1996 93 (1996 Nov 12) 13119–13124.

[110] M.P. Nambiar, C.U. Fisher, E.J. Enyedy, V.G. Warke, A. Kumar, G.C. Tsokos, Oxidative stress is involved in the heat stress-induced downregulation of TCR zeta chain expression and TCR/CD3-mediated [Ca(2+)](i) response in human T-lymphocytes, Cell Immunol 215 (2) (2002) 151–161.

[111] K. Sunahori, Y.T. Juang, G.C. Tsokos, Methylation Status of CpG Islands Flanking a cAMP Response Element Motif on the Protein Phosphatase 2Ac{alpha} Promoter Determines CREB Binding and Activity, J Immunol 182 (3) (2009) 1500–1508.

[112] G. Gorelik, J.Y. Fang, A. Wu, A.H. Sawalha, B. Richardson, Impaired T Cell Protein Kinase C{delta} Activation Decreases ERK Pathway Signaling in Idiopathic and Hydralazine-Induced Lupus, J Immunol 179 (8) (2007) 5553–5563.

[113] A.B. Pernis, Rho GTPase-mediated pathways in mature CD4+ T cells, Autoimmun Rev 8 (3) (2009) 199–203.

[114] T. Harada, V. Kyttaris, Y. Li, Y.T. Juang, Y. Wang, G.C. Tsokos, Increased expression of STAT3 in SLE T cells contributes to enhanced chemokine-mediated cell migration, Autoimmunity 40 (2007) 1–8.

[115] N. Jacob, H. Yang, L. Pricop, Y. Liu, X. Gao, S.G. Zheng, et al., Accelerated Pathological and Clinical Nephritis in Systemic Lupus Erythematosus-Prone New Zealand Mixed 2328 Mice Doubly Deficient in TNF Receptor 1 and TNF Receptor 2 via a Th17-Associated Pathway, J Immunol 182 (4) (2009) 2532–2541.

[116] J.C. Crispin, M. Oukka, G. Bayliss, R.A. Cohen, C.A. Van Beek, I.E. Stillman, et al., Expanded Double Negative T Cells in Patients with Systemic Lupus Erythematosus Produce IL-17 and Infiltrate the Kidneys, J Immunol 181 (12) (2008) 8761–8766.

[117] E. Nisoli, M.O. Carruba, Nitric oxide and mitochondrial biogenesis, J Cell Sci 119 (14) (2006) 2855–2862.

11

Cell Cycle Regulation and Systemic Lupus Erythematosus

Dimitrios Balomenos

THE CELL CYCLE MACHINERY AND ITS ASSOCIATION WITH T-CELL FUNCTION

Immune cell proliferation is mandatory for the development and function of the immune system. In bone marrow and thymus, immune cell precursors cycle vigorously to expand and differentiate into the diverse constituents of the immune system. T cells proliferate extensively in the thymus, where they differentiate into mature CD4+ or CD8+ cells. Mature T-cell proliferation and expansion in the periphery is required to fight microbial infections.

Cell division is a complex process and its control is fundamental for all cell types. Cell cycle progression depends on signals that affect the cell cycle machinery and are translated into proliferation, cell cycle arrest, or differentiation. In general, cell cycle progression takes place in four phases after stimulation of quiescent G0 cells (Figure 11.1): G1, the first gap phase; S, the DNA replication phase; G2, the second gap phase and M, mitosis, in which genetic information is distributed between two daughter cells. Cell cycle progression is

dependent on cyclins, which associate with catalytically active cyclin-dependent kinases (Cdk) to form active complexes. Cdk inhibitors act as negative regulators of these complexes to control cell cycle progression. This progression is characterized by the cyclin expression, Cdk activation, and phosphorylation of appropriate substrates. A key substrate for cell cycle activation is the retinoblastoma (Rb) gene. Phosphorylation of Rb uncouples its binding to E2F-type transcription factors, resulting in activation of S phase gene transcription [1–3].

The sequence of events required for cell cycle progression is displayed in Figure 11.1. During G1/S phase progression, cyclins D (D1–D3) act in mid-G1, followed by cyclin E and cyclin A at the G1/S boundary; cyclins A and B operate during S and G2/M phases. The Cdk require association with cyclins, as well as phosphorylation for activity. Cdk4 and Cdk6 associate with cyclins D, whereas cyclins A and E assemble with Cdk2. Nevertheless, Cdk2 may not be needed for progression to mitosis, since Cdk2 deficiency is not lethal in mice [4] and Cdk1 (also known as cdc2) can

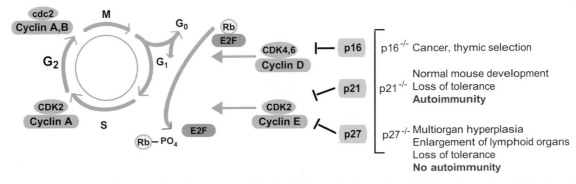

FIGURE 11.1 Immunological effects of cell cycle regulation. The figure shows the four cell cycle phases, the cyclin–Cdk interactions and the cell cycle inhibition points of p16, p21 and p27, as well as their involvement in immunity.

replace Cdk2 function [5]. The activity of cyclin:Cdk complexes is repressed by Cdk inhibitors, which constrain entry into S phase. Two classes of inhibitors have been defined, on the basis of their structural characteristics and Cdk targets. p15, p16, p18 and p19 are defined as INK4 (inhibitors of Cdk4), and associate solely with Cdk4 and Cdk6. The other inhibitor group includes p21, p27 and p57 (the Cip/Kip family), which interfere with cycling by binding to both cyclin and Cdk subunits and inhibit all Cdk involved in G1/S transition [6].

Cell cycle regulation is associated with the function of the immune system, as deficiencies in certain cell cycle inhibitors perturb immune functions (Figure 11.1). p16 deficiency can lead to an overall increase in lymphoid organ size, and affects thymic selection, but is not linked to loss of tolerance (Figure 11.1; [7]). Genetic deletion of p27 leads to multi-organ hyperplasia and enlargement of lymphoid organs [8–10]. Recent findings point to p27 as a molecule involved in peripheral tolerance [11, 12]. Alternatively, p21$^{-/-}$ mice develop normally, progressively lose tolerance to DNA and develop autoimmunity with age [13, 14]. Loss of tolerance is associated with a cyclin:Cdk complex that is targeted by both p21 and p27. It nonetheless appears that p21 is an immune-system-specific cell cycle inhibitor, whereas p27, in addition to its role in immunity, is a broader cell cycle regulator.

FROM T-CELL EXPOSURE TO ANTIGEN TO LOSS OF TOLERANCE

Following microbial invasion, mature T lymphocytes are exposed to antigen, become activated, and undergo intense, prolonged and repeated proliferation to establish a rapid immune response and generate immunological memory. However, this expansion is destined to contract in order to maintain the T-cell population in a stable cellular proportion within the immune system. This condition is known as homeostasis, in which total lymphocyte number, the proportion of naïve and memory T cells, and the extent of T-cell memory expansion are regulated. The basis of homeostasis is the maintenance of a balance between proliferation/expansion and apoptotic death of T cells [15]. Figure 11.2 illustrates the three main events in T cell response such as activation, proliferation/expansion, and apoptosis/death. Deregulation of each of these functions leads to loss of tolerance and autoimmunity development, and will be discussed in more detail below.

Since the discovery of apoptosis in T cells via Fas/FasL and for several years afterward, apoptosis has been considered the principal defect leading to loss of tolerance and development of autoreactivity. Fas has been considered the prototype apoptosis-inducing molecule, that controls T-cell homeostasis and tolerance induction [16]. It was therefore generally accepted that defective apoptosis could lead to break of homeostasis, loss of tolerance and autoimmunity. This view was reinforced by the phenotype of Fas-deficient (*lpr*) mice that develop early lethal glomerulonephritis and lymphadenopathy especially on the MRL background.

More recently several non-apoptotic molecules have been shown to be essential for the establishment and maintenance of tolerance. We will refer to different aspects of the overall T-cell response that contribute to T-cell tolerance establishment, such as T-cell activation, proliferation, and apoptosis (Figure 11.2). Emphasis

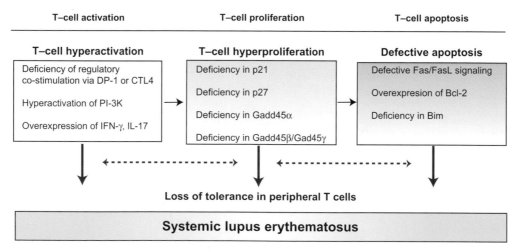

FIGURE 11.2 Loss of tolerance during the response of T cells can be caused by T-cell hyperactivation, hyperproliferation and defective apoptosis. After antigenic presentation to T cells loss of tolerance can occur at different levels: (1) During the T-cell activation process due to T-cell overactivation; (2) Throughout T-cell proliferation due to deregulation of the cell cycle machinery; (3) Upon apoptosis induction due to defects in apoptosis pathways. Examples of molecular defects are presented for each case. Loss of T-cell tolerance may give rise to SLE. The dotted lines suggest the possibility of interaction among the different defective processes of the T-cell response.

will be given on how cell cycle deregulation may lead to tolerance loss.

T-Cell Activation and Loss of Tolerance

Antigen presentation by antigen-presenting cells (APC) to the T-cell receptor (TCR) leads to T-cell activation. Further interaction of APC and T cells via receptor/ligand binding leads to positive co-stimulatory signals that enhance T-cell activation, or to inhibitory co-stimulatory signals that inhibit T-cell hyperactivation. T-cell activation initiates an intricate signaling pathway, causing ERK phosphorylation and NF-κB activation that drive T-cell proliferation.

Uncontrolled T-cell activation could lead to loss of tolerance toward self-antigens (Figure 11.2). Indeed, mice deficient in PD-1, which inhibits T-cell hyperactivation through its binding to PDL1 and PDL2, expressed by APC, lose tolerance and develop autoimmune manifestations [17, 18]. C57BL/6-PD-1-deficient mice develop lupus-like disease [17] (Figure 11.2) and PD-1 is associated with human lupus [19]. CTL4 is another regulator of T-cell activation delivering a negative co-stimulatory signal to T-cell activation. CTL4-deficient mice developed early-SLE-like disease and massive lymphoproliferation [20]. In another setting, increased activation of T cells by hyperactive phosphoinositide 3-kinase (PI-3K) induces autoimmunity [21], while inhibition of hyperactivated PI-3K reduces lupus incidence [22, 23].

The above examples of T-cell hyperactivation clearly associate aberrant T-cell hyperproliferation with loss of tolerance and autoimmunity (Figure 11.2). Another aspect of T-cell activation is the generation of helper T cells, Th1 (IFN-gamma), Th2 (IL-4) and Th-17 (IL-17) that influence autoimmune disease development. IFN-gamma is required for loss of tolerance and anti-DNA autoantibody production [24, 25], while Th2 T cells do not affect loss of tolerance [26]. Recently IL-17 has been implicated in inflammation and autoantibody production in murine models of lupus and in SLE patients [27].

Cell Cycle Deregulation and Loss of Tolerance

Recent findings, including data from our laboratory provide evidence that T-cell homeostasis, tolerance and SLE-like disease are highly dependent on cell cycle regulation of effector /memory T cells. Thus, cell cycle regulation controls T-cell tolerance and homeostasis. Aberrant T-cell proliferation is associated with autoimmunity development [13, 14, 20, 28–30]. More specifically, p21-deficient mice lose tolerance to DNA and develop SLE-like disease, since the cell cycle inhibitor p21 controls memory T-cell homeostasis and tolerance [13, 14, 28]. Another cell cycle inhibitor, p27, is associated with primary T-cell proliferation and anergy induction [11, 12]. The view that cell cycle deregulation can lead to break of tolerance and autoimmune disease [28] has received further support, since mice deficient in other cell-cycle-associated molecules such as the p53 effector gene Gadd45a [31] and E2F2 [32] also show autoimmune manifestations, as do Gadd45β/Gadd45γ double-deficient mice [33]. T-cell proliferation anomalies are also associated with diabetes development [29, 30], suggesting a broader relationship between cell cycle and autoimmunity.

A role for cell cycle regulation in human SLE development is suggested by recent findings showing hyperproliferation of CD3$^+$ CD4$^-$ CD8$^-$ double negative (DN) T cells from SLE patients compared to CD4$^+$ T cells, following anti-CD3 stimulation [34]. Previous studies have established that TCRαβ$^+$ and TCRγδ$^+$ DN T cells are expanded in the peripheral blood of SLE patients and induce pathogenic anti-DNA autoantibodies [35, 36]. Interestingly, in the more recent study the expanded TCRαβ$^+$ DN T cells from SLE patients were found to produce increased IL-17 levels compared to controls [34]. These findings suggest that hyperproliferating DN cells expand in SLE patients and produce increased amounts of pro-inflammatory IL-17. The presence of DN T cells in the kidney associates these cells and IL-17 to lupus nephritis development [34].

The above-mentioned studies establish that in addition to the activation of T-cell receptor, cell cycle regulation has a fundamental role in the establishment of homeostasis and tolerance, which may be relevant to human SLE. Finally, although it is evident that hyperactivation of T cells may lead to hyperproliferation, it would be of interest to examine whether defective cell cycle regulation and hyperproliferation of T cells may influence their activation potential by a feedback mechanism (see dotted lines in Figure 11.2).

Defective Apoptosis and Loss of Tolerance

Deregulation of apoptosis in lymphocytes could provoke autoimmunity since apoptosis defects lead to defective homeostasis and T-cell accumulation, causing breakage of tolerance [15, 37]. This is the case for mice with defects in tumor necrosis factor (TNF) family apoptosis-related molecules such as the Fas/FasL signaling system [38], mice overexpressing Bcl-2 [39] or Bim-deficient mice (40] (Figure 11.2). Fas is the prototype apoptosis-inducing molecule and is essential for T-cell homeostasis and tolerance induction [41]. Upon secondary stimulation, interaction of Fas with its ligand induces apoptosis by recruiting an adaptor protein, the Fas-associated death domain (FADD), which activates caspase-8 and initiates the apoptosis cascade [42].

lpr mice develop lymphadenopathy with increased CD4⁻CD8⁻TCR$^+$ T cells (double-negative, DN). Although, deficient Fas-dependent apoptosis was initially considered the basis of lymphadenopathy development in *lpr* mice, at present its etiology remains elusive [42, 43]. Clear evidence that defective apoptosis alone cannot account for T-cell accumulation was derived from the analysis of transgenic mice overexpressing the caspase-8 inhibitor CrmA in T cells [44, 45]. After secondary stimulation, apoptosis was inhibited *in vitro*, but lymphadenopathy did not develop in the CrmA transgenic mice. This suggests that, in addition to its role in apoptosis, Fas might have another function in homeostasis control [44, 45].

Another unexplained characteristic of *lpr* mice is that accumulating CD4+, CD8+ and DN T cells hyperproliferate *in vivo* compared to their counterparts from normal mice [46, 47]. This T-cell hyperproliferation is also observed in autoimmune lymphoproliferative syndrome (ALPS) patients with mutations in the Fas death domain that present T-cell accumulation and lymphadenopathy [48, 49]. Thus, Fas may exhibit a proliferation-regulating activity in addition to its pro-apoptotic function [50], a point that needs to be formally examined (see dotted line in Figure 11.2).

Overall we have pointed out three different functions that form part of the T-cell response, and relate to T-cell tolerance. We consider it of interest to examine the possibility of an interrelationship of these functions in the establishment of T-cell tolerance.

T-CELL CYCLING REGULATION AND TOLERANCE: THE p21- AND p27-DEFICIENT MOUSE MODELS

Our studies with p21-deficient mice and other reports on p27-deficient mice have established that p21 and p27 are both essential for *in vivo* tolerance induction. To dissect the precise role of p21 in the immune response independently of the influence of the mixed background (129/Sv x C57BL/6), used in many studies of cell-cycle-associated deficiencies, we generated and analyzed C57BL/6 p21$^{-/-}$ mice. Our studies were based on a classical model of memory T-cell generation, in which partial homeostasis is induced by apoptosis [51] (Figure 11.3). After primary mitogenic stimulation, CD4$^+$ T cells were expanded with IL-2; at the end of the second expansion period T cells had acquired effector or memory phenotype (Figure 11.3). Cells were then submitted to secondary stimulation; while the majority of T cells died through Fas-mediated apoptosis, a proportion of memory phenotype T cells survived. Further stimulation of memory T cells was also performed for several rounds.

Our data from the repeated T-cell activation experiments revealed an extraordinary property of p21; p21 controls the proliferation of activated/memory T cells but not of naïve T cells [13] (Figure 11.3). Moreover, absence of p21 does not affect the Fas/FaL-depending apoptosis after secondary stimulation [13] (Figure 11.3). However, apoptosis-resistant T cells require p21 to control their proliferation, revealing a homeostatic mechanism that controls memory T-cell expansion through p21. Effector/memory T cells are generated after extensive IL-2 treatment following primary stimulation and also require p21 in order to control their expansion (Figure 11.3). Since apoptosis is minimal during IL-2-dependent T-cell expansion, cycling regulation by p21 appears to be the major homeostatic mechanism during the effector/memory expansion stage. Overall, it appears that control of proliferation by p21 leads to homeostasis, which limits accumulation of effector/memory T cells during IL-2 expansion and after secondary stimulation. We conclude that certain forms of effector memory T-cell homeostasis are apoptosis-independent and are regulated by cell cycle control through p21 (Figure 11.3).

Our data point to a role for p21 in the regulation of effector/memory but not naïve T-cell proliferation. Regulation of the cell cycle machinery apparently differs in the two types of T cells. One explanation for the distinct role of p21 in naïve and memory T-cell proliferation could be based on epigenetic modifications that occur during the transition from naïve to memory T cells [52]. Epistatically modified memory T cells, which are programmed to respond rapidly after antigen encounter [52], may require p21 to moderate a strong proliferative response.

By regulating homeostasis of memory T cells, p21 assures an adequate proportion of memory cells in the total T-cell population. To explain anti-DNA antibody production and autoimmunity in p21$^{-/-}$ mice, we suggest that p21 deficiency leads to memory T-cell hyperproliferation. Persistent encounter of memory p21$^{-/-}$ T cells with autoantigens leads to deregulated homeostasis and loss of tolerance (Figure 11.3).

Several groups have addressed the role of the cell cycle regulator p27 in primary T-cell proliferation and tolerance [53]. The main conclusion is that p27 has a significant role in anergy induction, at least for CD4$^+$ T cells. p27-deficient T cells escape anergy when activated without CD28 co-stimulation, or in certain other conditions [54–56]. It has thus been generally assumed that p27 is linked to CD4$^+$ anergy induction (Figure 11.3; [57, 58]). Moreover, as is the case for the p21-deficient mice [13], the *in vivo* CD4$^+$ T-cell response to a model antigen is also increased in the absence of p27 [54, 55]. Mice deficient for p27

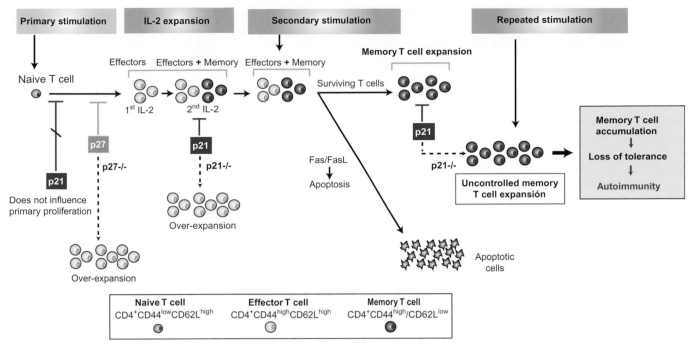

FIGURE 11.3 A model showing proliferation control by p21 of effector/memory T cells after extended stimulation. p21 control checkpoints are represented at various times, including primary stimulation. The red bar indicates that p21 does not affect primary stimulation. p21 deficiency leads to effector/memory T cell hyperproliferation during the IL-2 dependent expansion and after secondary stimulation. Finally, memory p21$^{-/-}$ T cells overproliferate after subsequent stimulations. The overproliferation of memory p21$^{-/-}$ T cells after repeated stimulation leads to loss of tolerance and autoimmunity. Proliferation control by p27 is indicated at the initial steps of the stimulation.

do not develop spontaneous autoimmunity, as is the case for p21$^{-/-}$ mice, nevertheless, it is established that p27$^{-/-}$ T cells are also prone to break tolerance *in vivo* [11, 12]. In one case, blockade of CD28 and CD40 co-stimulation, which leads to long-term allograft survival in wild-type mice, resulted in acute rejection of an MHC-mismatched cardiac allograft in p27$^{-/-}$ mice. This rejection was associated with massive lymphocyte expansion, and increased CD4$^+$ T-cell infiltration and proliferation in the cardiac grafts [12]. In another study, p27 function was examined during tolerance induction in naïve TCR-transgenic T cells. p27$^{-/-}$ T cells resisted tolerance and remained responsive after adoptive transfer into wild-type recipient mice that was followed by a tolerizing stimulus *in vivo* [11].

In conclusion, *in vivo* tolerance induction requires p27 expression in naive T cells or p21 expression in memory T cells. The major difference between the tolerance-inducing effects of these two cell cycle regulators is that p21 is necessary to maintain tolerance under physiological conditions, while the need for p27 is manifested when the tolerance machinery is challenged. Break of tolerance thus occurs spontaneously in p21$^{-/-}$ mice that develop anti-DNA autoantibodies and autoimmune disease, but not in p27$^{-/-}$ mice, which remain free of autoimmune manifestations.

THE ROLE OF CELL CYCLE REGULATION IN AUTOIMMUNITY DEVELOPMENT

Mixed background 129/Sv x C57BL/6 p21$^{-/-}$ female mice develop acute disease that leads to death. It appears that genetic elements of the mixed background [59] and the female hormone environment [60, 61], in conjunction with the lack of p21, control disease severity [14, 31].

In order to avoid the caveats that could be introduced by the mixed 129/Sv x C57BL/6 background, we introduced the p21 deficiency onto the autoimmunity-resistant C57BL/6 background. We found that lack of p21 conferred a moderate predisposition to autoimmunity [13]. This result was anticipated [14, 62] on the basis of the multifactorial nature of lupus and the fact that the genetic background is a determining factor for lupus development. Thus, whereas lack of Fas leads to full-blown lupus in combination with the MRL background, it causes mild disease in C57BL/6 mice [63]. p21 deficiency thus emerges as an autoimmunity accelerator that is background-dependent, as are other major autoimmunity accelerators such as the *lpr* gene.

Deciphering the precise effect of p21 in the overall immune response will help to evaluate the potential of p21-related therapeutic approaches to disease. Indeed, we recently reported that p21 negatively regulates the activation of macrophages by LPS [64] a fact that may result in increased inflammation and enhance autoimmunity in p21-/- mice. Our finding demonstrating that p21 is a potent lupus-autoimmunity suppressor is further supported by data showing that p21 analogs had a therapeutic effect on lupus development in (NZB X NZW)F$_1$ mice [65]. Moreover, T cells from lupus patients which hyperproliferate *in vivo* [20], express significantly reduced levels of p21 and p27 [66].

Both p21 and p27 control tolerance, as discussed above, but only p21$^{-/-}$ mice develop spontaneous autoimmunity. This may be due to the normal development of p21$^{-/-}$ as contrasted with the generalized organomegaly and predisposition to cancer of p27$^{-/-}$ mice [67—69] that may suppress a putative predisposition to autoimmunity. Alternatively, the property of p21 in affecting memory T-cell expansion that are in constant contact with autoantigens, may drive autoimmunity development in p21$^{-/-}$ mice.

In conclusion, molecules classically considered to pertain to the cell cycle machinery are pivotal for tolerance induction, maintenance of T-cell homeostasis, and suppression of autoimmune disease. Therefore, deregulation of cell cycle control has fundamental consequences in the immune response. Since p21 controls T-cell tolerance and is an autoimmunity suppressor, it is evident that therapeutic approaches based on cell cycle inhibition can be considered for targeting autoimmune disease.

PERSPECTIVES: POSSIBLE EFFECT OF CELL CYCLE CONTROL IN AUTOIMMUNE DISEASE TREATMENT

The data showing that p21 and p27 control T-cell tolerance suggest novel approaches for intervention in autoimmunity. Further research could establish the exact pathways through which these two cell cycle inhibitors control T-cell proliferation and tolerance. It appears that since p21$^{-/-}$ mice develop normally, treatments could target the increased T memory cycling, without producing adverse effects on the overall immune response at a systemic level. Indeed, recent research has changed the paradigm of universal cell cycle regulation indicating that different Cdk (other than Cdk1) may be required for different specialized cells [70]. This means that each cell type may require a different type of cell cycle inhibition and thus specialized cell cycle progression inhibitors are designed for different tumors [70]. Based on these findings, it does not appear unusual that p21 is required only for the regulated expansion and tolerance of memory T cells and not for the overall development of mice. It appears, therefore, feasible that specifically designed inhibitors

could regulate the cell cycle of memory T cells and not that of other cells and tissues.

Acknowledgments

This work was supported by grants from the MEC (SAF2005-05264), MICINN (PI081835) and the Community of Madrid (S-BIO-0189-2006). The Department of Immunology and Oncology was founded and is supported by the Spanish National Research Council (CSIC) and by Pfizer.

References

[1] A. Vidal, A. Koff, Cell-cycle inhibitors: three families united by a common cause, Gene 247 (2000) 1—15.

[2] A.D. Wells, Cyclin-dependent kinases: Molecular switches controlling anergy and potential therapeutic targets for tolerance, Semin Immunol 19 (2007) 173—179.

[3] E.A. Rowell, A.D. Wells, The role of cyclin-dependent kinases in T-cell development, proliferation, and function, Crit Rev Immunol 26 (2006) 189—212.

[4] A. Martin, J. Odajima, S.L. Hunt, P. Dubus, S. Ortega, M. Malumbres, et al., Cdk2 is dispensable for cell cycle inhibition and tumor suppression mediated by p27(Kip1) and p21 (Cip1), Cancer Cell 7 (2005) 591—598.

[5] D. Santamaria, C. Barriere, A. Cerqueira, S. Hunt, C. Tardy, K. Newton, et al., Cdk1 is sufficient to drive the mammalian cell cycle, Nature 448 (2007) 811—815.

[6] M. Malumbres, M. Barbacid, Cell cycle kinases in cancer, Curr Opin Genet Dev 17 (2007) 60—65.

[7] T. Bianchi, N. Rufer, H.R. MacDonald, M. Migliaccio, The tumor suppressor p16Ink4a regulates T lymphocyte survival, Oncogene 25 (2006) 4110—4115.

[8] M.L. Fero, M. Rivkin, M. Tasch, P. Porter, C.E. Carow, E. Firpo, et al., A syndrome of multiorgan hyperplasia with features of gigantism, tumorigenesis, and female sterility in p27(Kip1)-deficient mice, Cell 85 (1996) 733—744.

[9] H. Kiyokawa, R.D. Kineman, K.O. Manova-Todorova, V.C. Soares, E.S. Hoffman, M. Ono, et al., Enhanced growth of mice lacking the cyclin-dependent kinase inhibitor function of p27(Kip1), Cell 85 (1996) 721—732.

[10] K. Nakayama, N. Ishida, M. Shirane, A. Inomata, T. Inoue, N. Shishido, et al., Mice lacking p27(Kip1) display increased body size, multiple organ hyperplasia, retinal dysplasia, and pituitary tumors, Cell 85 (1996) 707—720.

[11] L. Li, Y. Iwamoto, A. Berezovskaya, V.A. Boussiotis, A pathway regulated by cell cycle inhibitor p27Kip1 and checkpoint inhibitor Smad3 is involved in the induction of T cell tolerance, Nat Immunol 7 (2006) 1157—1165.

[12] E.A. Rowell, L. Wang, W.W. Hancock, A.D. Wells, The cyclin-dependent kinase inhibitor p27kip1 is required for transplantation tolerance induced by costimulatory blockade, J Immunol 177 (2006) 5169—5176.

[13] C.F. Arias, A. Ballesteros-Tato, M.I. García, J. Martin-Caballero, J.M. Flores, C. Martínez-A, p21CIP1/WAF1 controls proliferation of activated/memory T cells and affects homeostasis and memory T cell responses, J Immunol 178 (2007) 2296—2306.

[14] D. Balomenos, J. Martin-Caballero, M.I. García, I. Prieto, J.M. Flores, M. Serrano, et al., The cell cycle inhibitor p21 controls T-cell proliferation and sex-linked lupus development, Nat Med 6 (2000) 171—176.

[15] A.A. Freitas, B. Rocha, Population biology of lymphocytes: the flight for survival, Annu Rev Immunol 18 (2000) 83—111.

[16] S. Nagata, P. Golstein, The Fas death factor, Science 267 (1995) 1449–1456.

[17] T. Okazaki, T. Honjo, The PD-1-PD-L pathway in immunological tolerance, Trends Immunol. 27 (2006) 195–201.

[18] A.H. Sharpe, E.J. Wherry, R. Ahmed, G.J. Freeman, The function of programmed cell death 1 and its ligands in regulating auto-immunity and infection, Nat Immunol 8 (2007) 239–245.

[19] G.K. Bertsias, M. Nakou, C. Choulaki, A. Raptopoulou, E. Papadimitraki, G. Goulielmos, et al., Genetic, immunologic, and immunohistochemical analysis of the programmed death 1/programmed death ligand 1 pathway in human systemic lupus erythematosus, Arthritis Rheum 60 (2009) 207–218.

[20] E.C. Jury, F. Flores-Borja, H.S. Kalsi, M. Lazarus, D.A. Isenberg, C. Mauri, et al., Abnormal CTLA-4 function in T cells from patients with systemic lupus erythematosus, Eur J Immunol 40 (2009) 569–578.

[21] L.R. Borlado, C. Redondo, B. Alvarez, C. Jimenez, L.M. Criado, J. Flores, et al., Increased phosphoinositide 3-kinase activity induces a lymphoproliferative disorder and contributes to tumor generation in vivo, FASEB J 14 (2000) 895–903.

[22] M. Wymann, D. Balomenos, A.C. Carrera, Class IB-PI3K deficiency ameliorates IA-PI3K-induced systemic lupus but not T cell invasion, J Immunol 176 (2006) 589–593.

[23] D.F. Barber, A. Bartolomé, C. Hernández, J.M. Flores, C. Redondo, C. Fernández Arias, et al., PI3K inhibition blocks glomerulonephritis and extends lifespan in murine models of systemic lupus, Nat Med 11 (2005) 933–935.

[24] D. Balomenos, R. Rumold, A.N. Theofilopoulos, Interferon-gamma is required for lupus-like disease and lymphoaccumulation in MRL-lpr mice. J, Clinical Investigation 101 (1998) 364–371.

[25] D.H. Kono, D. Balomenos, D.L. Pearson, M.S. Park, B. Hildebrandt, P. Hultman, et al., The prototypic Th2 autoimmunity induced by mercury is dependent on IFN-gamma and not Th1/Th2 imbalance, J. Immunol. 161 (1998) 234–240.

[26] D.H. Kono, D. Balomenos, M.S. Park, A.N. Theofilopoulos, Development of lupus in BXSB mice is independent of IL-4, J. Immunol. 167 (2000) 38–42.

[27] A. Nalbandian, J.C. Crispin, G.C. Tsokos, Interlukin-17 and systemic lupus erythematosus: current concepts, Clin Exp Immunol. 157 (2009) 209–215.

[28] D. Balomenos, C. Martínez-A, Cell-cycle regulation in immunity, tolerance and autoimmunity, Immunol Today 21 (2000) 551–555.

[29] C. Carvalho-Pinto, M.I. Garcia, L. Gómez, A. Ballesteros, A. Zaballos, J.M. Flores, et al., Leukocyte attraction through the CCR5 receptor controls progress from insulitis to diabetes in non-obese diabetic mice, Eur J Immunol 34 (2004) 548–557.

[30] C. King, A. Ilic, K. Koelsch, N. Sarvetnick, Homeostatic expansion of T cells during immune insufficiency generates autoimmunity, Cell 117 (2004) 265–277.

[31] J.M. Salvador, M.C. Hollander, A.T. Nguyen, J.B. Kopp, L. Barisoni, J.K. Moore, et al., Mice lacking the p53-effector gene Gadd45a develop a lupus-like syndrome, Immunity 16 (2002) 499–508.

[32] M. Murga, O. Fernandez-Capetillo, S.J. Field, B. Moreno, L.R. Borlado, Y. Fujiwara, et al., Mutation of E2F2 in mice causes enhanced T lymphocyte proliferation, leading to the development of autoimmunity, Immunity 15 (2001) 959–970.

[33] L. Liu, E. Tran, Y. Zhao, Y. Huang, R. Flavell, B. Lu, Gadd45 beta and Gadd45 gamma are critical for regulating autoimmunity, J Exp Med 202 (2005) 1341–1347.

[34] J.C. Crispín, M. Oukka, G. Bayliss, R.A. Cohen, C.A. Van Beek, I.E. Stillman, et al., Expanded double negative T cells in patients with systemic lupus erythematosus produce IL-17 and infiltrate the kidneys, J Immunol 181 (2008) 8761–8766.

[35] S. Shivakumar, G.C. Tsokos, S.K. Datta, T cell receptor alpha/beta expressing double-negative (CD4-/CD8-) and CD4+ T helper cells in humans augment the production of pathogenic anti-DNA autoantibodies associated with lupus nephritis, J Immunol 143 (1989) 103–112.

[36] S. Rajagopalan, T. Zordan, G.C. Tsokos, S.K. Datta, Pathogenic anti-DNA autoantibody-inducing T helper cell lines from patients with active lupus nephritis: isolation of CD4-8- T helper cell lines that express the gamma delta T-cell antigen receptor, Proc Natl Acad Sci U S A 87 (1990) 7020–7024.

[37] Z. Grossman, W.E. Paul, Self-tolerance: context dependent tuning of T cell antigen recognition, Semin Immunol 12 (2000) 197–203.

[38] J. Wang, M.J. Lenardo, Molecules involved in cell death and peripheral tolerance, Curr Opin Immunol 9 (1997) 818–825.

[39] A. Strasser, et al., Enforced BCL2 expression in B-lymphoid cells prolongs antibody responses and elicits autoimmune disease, Proc Natl Acad Sci USA 88 (1991) 8661–8665.

[40] P. Bouillet, D. Metcalf, D.C. Huang, D.M. Tarlinton, T.W. Kay, F. Kontgen, et al., Proapoptotic Bcl-2 relative Bim required for certain apoptotic responses, leukocyte homeostasis, and to preclude autoimmunity, Science 286 (1999) 1735–1738.

[41] S. Nagata, P. Golstein, The Fas death factor, Science 267 (1995) 1449–1456.

[42] N. Bidere, H.C. Su, M.J. Lenardo, Genetic disorders of programmed cell death in the immune system, Annu Rev Immunol 24 (2006) 321–352.

[43] S.E. Straus, M. Sneller, M.J. Lenardo, J.M. Puck, W. Strober, An inherited disorder of lymphocyte apoptosis: the autoimmune lymphoproliferative syndrome, Ann Intern Med 130 (1999) 591–601.

[44] K.G. Smith, A. Strasser, D.L. Vaux, CrmA expression in T lymphocytes of transgenic mice inhibits CD95 (Fas/APO-1)-transduced apoptosis, but does not cause lymphadenopathy or autoimmune disease, EMBO J 15 (1996) 5167–5176.

[45] K. Newton, A.W. Harris, M.L. Bath, K.G. Smith, A. Strasser, A dominant interfering mutant of FADD/MORT1 enhances deletion of autoreactive thymocytes and inhibits proliferation of mature T lymphocytes, EMBO J 17 (1998) 706–718.

[46] D. Balomenos, R. Rumold, A.N. Theofilopoulos, The proliferative in vivo activities of lpr double-negative T cells and the primary role of p59fyn in their activation and expansion, J Immunol 159 (1997) 2265–2273.

[47] K.A. Fortner, R.C. Budd, The death receptor Fas (CD95/APO-1) mediates the deletion of T lymphocytes undergoing homeostatic proliferation, J Immunol 175 (2005) 4374–4382.

[48] F. Le Deist, J.F. Emile, F. Rieux-Laucat, M. Benkerrou, I. Roberts, N. Brousse, et al., Clinical, immunological, and pathological consequences of Fas-deficient conditions, Lancet 348 (1996) 719–723.

[49] M.S. Lim, S.E. Straus, J.K. Dale, T.A. Fleisher, M. Stetler-Stevenson, W. Strober, et al., Pathological findings in human autoimmune lymphoproliferative syndrome, Am J Pathol 153 (1998) 1541–1550.

[50] J.C. Rathmell, C.B. Thompson, Pathways of apoptosis in lymphocyte development, homeostasis, and disease. Cell 109, Suppl (2002) S97–107.

[51] A.K. Abbas, The control of T cell activation vs. tolerance, Autoimmun Rev 2 (2003) 115–118.

[52] E.L. Pearce, H. Shen, Making sense of inflammation, epigenetics, and memory CD8+ T-cell differentiation in the context of infection, Immunol Rev 211 (2006) 197–202.

[53] A.D. Wells, Cyclin-dependent kinases: Molecular switches controlling anergy and potential therapeutic targets for tolerance, Semin Immunol 19 (2007) 173–179.

[54] E.A. Rowell, M.C. Walsh, A.D. Wells, Opposing roles for the cyclin-dependent kinase inhibitor p27kip1 in the control of CD4+ T cell proliferation and effector function, J Immunol 174 (2005) 3359—3368.

[55] L.J. Appleman, A.A. van Puijenbroek, K.M. Shu, L.M. Nadler, V.A. Boussiotis, CD28 costimulation mediates down-regulation of p27kip1 and cell cycle progression by activation of the PI3K/PKB signaling pathway in primary human T cells, J Immunol 168 (2002) 2729—2736.

[56] S. Zhang, V.A. Lawless, M.H. Kaplan, Cytokine-stimulated T lymphocyte proliferation is regulated by p27Kip1, J Immunol 165 (2000) 6270—6277.

[57] C.E. Rudd, Cell cycle 'check points' T cell anergy, Nat Immunol 7 (2006) 1130—1132.

[58] R. Shen, M.H. Kaplan, The homeostasis but not the differentiation of T cells is regulated by p27(Kip1), J Immunol 169 (2002) 714—721.

[59] A.E. Bygrave, K.L. Rose, J. Cortes-Hernandez, J. Warren, R.J. Rigby, H.T. Cook, et al., Spontaneous autoimmunity in 129 and C57BL/6 mice-implications for autoimmunity described in gene-targeted mice, PLoS Biol 2 (2004) E243.

[60] Z.M. Sthoeger, H. Zinger, E. Mozes, Beneficial effects of the anti-oestrogen tamoxifen on systemic lupus erythematosus of (NZBxNZW) F1 female mice are associated with specific reduction of IgG3 autoantibodies, Ann Rheum Dis 62 (2003) 341—346.

[61] R.W. McMurray, Estrogen, prolactin, and autoimmunity: actions and interactions, Int Immunopharmacol 1 (2001) 995—1008.

[62] M.L. Santiago-Raber, B.R. Lawson, W. Dummer, M. Barnhouse, S. Koundouris, C.B. Wilson, D.H. Kono, A.N. Theofilopoulos, Role of cyclin kinase inhibitor p21 in systemic autoimmunity, J Immunol 167 (2001) 4067—4074.

[63] S. Vidal, D.H. Kono, A.N. Theofilopoulos, Loci predisposing to autoimmunity in MRL-Fas lpr and C57BL/6-Faslpr mice, J Clin Invest 101 (1998) 696—702.

[64] M. Trakala, C.F. Arias, M.I. García, M.C. Moreno-Ortic, P.J. Fernández, M. Mellado, et al., Regulation of macrophage activation via p21(WAF1/CIP1), J Immunol 39 (2009) 810—819.

[65] C. Goulvestre, C. Chereau, C. Nicco, L. Mouthon, B. Weill, F. Batteux, A mimic of p21WAF1/CIP1 ameliorates murine lupus, J Immunol 175 (2005) 6959—6967.

[66] H. Tang, G. Tan, Q. Guo, R. Pang, F. Zeng, Abnormal activation of the Akt-GSK3beta signaling pathway in peripheral blood T cells from patients with systemic lupus erythematosus, Cell Cycle 17 (2009) 2789—2793.

[67] M.L. Fero, M. Rivkin, M. Tasch, P. Porter, C.E. Carow, E. Firpo, et al., A syndrome of multiorgan hyperplasia with features of gigantism, tumorigenesis, and female sterility in p27(Kip1)-deficient mice, Cell 85 (1996) 733—744.

[68] H. Kiyokawa, R.D. Kineman, K.O. Manova-Todorova, V.C. Soares, E.S. Hoffman, M. Ono, et al., Enhanced growth of mice lacking the cyclin-dependent kinase inhibitor function of p27(Kip1), Cell 85 (1996) 721—732.

[69] K. Nakayama, N. Ishida, M. Shirane, A. Inomata, T. Inoue, N. Shishido, et al., Mice lacking p27(Kip1) display increased body size, multiple organ hyperplasia, retinal dysplasia, and pituitary tumors, Cell 85 (1996) 707—720.

[70] M. Malumbres, M. Barbacid, Cell cycle, CDKs and cancer: a changing paradigm, Nat. Rev. Cancer. 9 (2009) 153—166.

The Role of Reactive Nitrogen and Oxygen Intermediates in Systemic Lupus Erythematosus

Ahmad K. Mashmoushi, Gary S. Gilkeson, Jim C. Oates

INTRODUCTION

Systemic lupus erythematosus (SLE) is a complex autoimmune multiorgan disease. A unifying feature of lupus is the production of autoantibodies, a function primarily of the acquired immune system, leading to immune complex deposition and consequent inflammatory injury in target organs. However, innate immunity is implicated both in the initiation and pathogenic consequences of autoantibody production in SLE [1]. Reactive intermediates (RIs), important mediators of the innate immune system, are under tight physiologic control. In SLE the disruption of reactive intermediate homeostasis may lead to a break in immune tolerance, increased tissue damage and altered enzyme function [2]. This chapter reviews the latest evidence from both animal and human studies on the role of reactive intermediates in the pathogenesis of SLE.

BIOLOGY OF REACTIVE INTERMEDIATES

Reactive intermediates are short-lived molecules produced by normal cellular metabolism and aid in a multitude of physiological and pathological processes. Nitrogen-based reactive intermediates are known as reactive nitrogen intermediates (RNI) (Table 12.1), while those that are oxygen-based are known as reactive oxygen intermediates (ROI) (Table 12.2). They were first regarded as toxic byproducts [3], but later they were shown to act as signaling molecules that are under tight physiologic control [4]. Reactive intermediates are either produced by reactions involving enzymes such as nicotinamide adenine dinucleotide phosphate *oxidase (NAD (P)H) oxidase [5]*, nitric oxide synthase [6], or by non-enzymatic reactions through the mitochondrial electron transport chain [7], and reduced transition metals [8]. On the other hand, the flux of reactive intermediates is limited by a number of processes and properties including: the short half-life of reactive intermediates [9], enzymatic antioxidants such as superoxide dismutase (SOD) , catalase, glutathione peroxidase (GPx), and non-enzymatic antioxidants such as glutathione, ascorbate (vitamin C), α-tocopherol [10], and the negative-feedback mechanisms that regulate the enzymatic production of reactive intermediates (e.g. the nitric oxide negative feedback inhibition of NOS [11]). A balance should be maintained between reactive intermediate production and clearance; any perturbation in this balance leads to pathology and cellular injury [12] (Figure 12.1). In most cases low to moderate levels of reactive intermediates are physiologic and aid in normal cellular functions such as proliferation [13] and muscle relaxation [14]. However, high levels of these molecules cause direct cellular injury by reacting with lipids, amino acids, and DNA [15–17] or indirect cellular injury by skewing the normal redox pathways that govern a number of cellular processes [12]. In systemic lupus erythematosus [18], overproduction of reactive intermediates in the absence of infection may augment autoreactivity, disrupting normal enzyme function and amplifying tissue damage [2]. The study of reactive intermediate biology is hindered by a number of obstacles that are related to the mere nature of reactive intermediates as well as the lack of sensitive scientific tools that can detect and modulate these molecules. For instance, reactive intermediates can interact with each other and modulate the function of one another. Thus, the outcome of reducing the level of a given reactive species (A) can be due to the primary effect of the reduced level of this species (A), or the secondary effect of another species (B) that is modulated by the first

TABLE 12.1   List of important reactive nitrogen intermediates and their sources, half-lifes, and modes of action

	Source	Half life ($t_{1/2}$) *	Modes of action/effects
Nitric oxide (NO)	Nitric oxide synthase (6)	3–30 s [35]	Nitrosylation of metal centers Redox signaling [70]
Peroxynitrite (ONOO-)	Reaction between NO and SO (70)	1 s [35]	Nitration Oxidation Oxidative stress [70]
Nitrogen dioxide radical (NO_2)	Reaction between NO and O_2 (70)	7×10^{-6} s [a]	Nitration Oxidation Oxidative stress [70]
Dinitrogen trioxide (N_2O_3)	Reaction between NO and NO_2 (70)	7×10^{-4} s [b]	Nitrosation Nitrosative stress [70]

* Note that halflife values are approximate and may vary.
[a]Ford, E., M.N. Hughes, and P. Wardman. (2002) Kinetics of the reactions of nitrogen dioxide with glutathione, cysteine, and uric acid at physiological pH. Free Radic Biol Med. **32**: 1314–1323.
[b]Wink, D.A. and P.C. Ford. (1995) Nitric Oxide Reactions Important to Biological Systems: A Survey of Some Kinetics Investigations. Methods 7(1): p. 14–20.

species (A). For example, when nitric oxide (NO) and superoxide (SO) are produced in a close proximity, peroxynitrite (ONOO-), a stronger oxidant, is produced while SO and NO are consumed [19, 20]. If NO levels are reduced in the system, the functional outcome could either be due to reduced effects of NO, increased availability of SO, or reduced levels of ONOO-.

Reactive Oxygen Intermediates (ROI)

Reactive oxygen intermediates (ROI) encompass all oxygen-based radical (superoxide, hydroxyl radical) and non-radical (hydrogen peroxide, hypochlorous acid) reactive species. High ROI production is associated with SLE as evident by oxidative protein modifications,

TABLE 12.2   List of important reactive oxygen intermediates and their sources, half-lifes, and modes of action

	Source	Half life ($t_{1/2}$) *	Modes of action/effects
Super oxide (SO)	NADPH oxidase [5] Xanthine oxidase [26] Complex I/complex III (mitochondria) [25] 5-lipoxygenase [27] Cyclooxygenase [28] Uncoupled nitric oxide synthase [29]	10^{-9}s [35]	Oxidative stress [24] Redox signaling
Hydrogen peroxide (H_2O_2)	Peroxisomes [a] Super oxide dismutase [38]	10^{-3} s [b]	Redox signaling [38] Oxidative stress
Hydroxyl radical OH	Fenton reaction [39] Haber-Weiss reaction [41]	10^{-9} s [40]	Oxidative stress [40]
Hypochlorous acid (HOCL)	Myeloperoxidase (MPO) [49]	30 s [c]	Chlorination [50] Oxidative stress [50]

* Note that half life values are approximate and may vary.
[a]Valko, M., Izakovic, M., Mazur, M., Rhodes, C. J. and Telser, J. (2004) Role of oxygen radicals in DNA damage and cancer incidence. Mol Cell Biochem. **266**, 37–56.
[b]Vranova, E., Inze, D. and Van Breusegem, F. (2002) Signal transduction during oxidative stress. J. Exp. Bot. **53**, 1227–1236.
[c]Pullar, J. M., Winterbourn, C. C. and Vissers, M. C. (1999) Loss of GSH and thiol enzymes in endothelial cells exposed to sublethal concentrations of hypochlorous acid. Am J Physiol. **277**, 1505–1512.

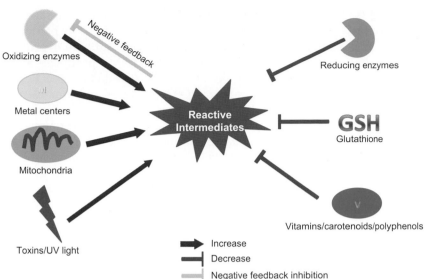

FIGURE 12.1 Production and clearance of reactive intermediates. Major sources of reactive intermediate production include a variety of metallo-enzymes (such as NAD(P)H oxidase and nitric oxide synthase), mitochondrial complexes I and III, transitional metals (Fe^{2+} and Cu^{2+}) and exogenous sources such as toxins and UV light. Reactive intermediate clearance is achieved by a number of mechanisms involving enzymes (catalases and peroxidases) and non-enzymatic antioxidants including glutathione (GSH), vitamins, and carotenoids. Pathology arises when the physiologic balance between reactive intermediate production and clearance is disturbed.

lipid peroxidation [21], and lipoprotein oxidation [22]. Superoxide anion (SO) stands at the center of major physiologic and pathologic processes that take place in a wide variety of biological systems [23, 24]. The mitochondrion is a major source of endogenous SO production. It is estimated that 1–3% of electrons passing through the mitochondrial electron transport chain "leak" and bind with molecular oxygen to form the superoxide anion. Complexes I and III are the major sites of mitochondrial SO production [25]. SO is also produced by a number of enzymes including: nicotine adenine dinucleotide phosphate (NAD(P)H) oxidases [5], xanthine oxidase (XO) [26], 5-lipoxygenase [27], cyclooxygenase [28], and uncoupled NOS [29]. NAD(P)H oxidases produce significant amounts of SO in a wide variety of cells in addition to inflammatory cells. In phagocytes, NAD(P)H oxidase produces high levels of SO that create an oxidative antimicrobial environment known as the respiratory burst [30]. This burst activates redox-sensitive signaling pathways such as nuclear factor-kappa B (NF-κB) and activator protein-1 (AP-1) that in turn regulates the transcription of proinflammatory proteins such as cytokines [31]. In non-phagocytic cells NAD(P)H oxidase produces less SO and may act as an oxygen sensor [32] and a vascular tone regulator [33]. The neutrophil NAD(P)H oxidase is heme-containing enzyme complex made up of six subunits: two membrane-bound subunits (gp91^{phox} and p22^{phox}) that constitute the catalytic cytochrome b558, and four cytosolic subunits (p47^{phox}, p67^{phox}, p40^{phox}, and a small G coupled protein (rac1 or rac2)). Upon activation, the cytosolic subunits are phosphorylated and translocated to cytochrome b558, which binds flavin adenine dinucleotide to form an active enzyme [5, 34].

SO is a very reactive molecule with a very short half life ($t_{1/2} = 10^{-9}$ s [35]). It can be converted into hydrogen peroxide (H_2O_2) either enzymatically (major pathway) under action of superoxide dismutases (SOD) [36] or non-enzymatically (minor pathway) [37]. H_2O_2, in turn, can be converted into water under the action of catalases and peroxidases [38] or hydroxyl radical (OH radical) via either the Fenton [39] or Haber-Weis reaction [40, 41]. H_2O_2 is a very important signaling molecule that is involved in many cellular processes such as proliferation [42], differentiation [43], and immune cell activation [44]. It acts on a number of key proteins and enzymes such as transcription factors, kinases, and phosphatases by oxidizing susceptible cysteine residues in these proteins [38]. A typical example of an H_2O_2 interaction with its target molecule is the oxidation of a cysteine residue located in the catalytic site of protein tyrosine phosphatase 1B (PTP-1B), rendering it inactive [45]. The formation of OH radical by the Fenton and Haber-Weis reactions is accelerated in the presence of iron or copper cations. OH radical can be very detrimental to the cell because of its rapid reactivity with macromolecules. Reactions with DNA can induce mutations [46], while reactions with lipid can initiate cycles of lipid peroxidation and membrane damage [47]. Another important oxygen-derived reactive intermediate is hypochlorous acid (HOCl), a potent bactericide and the active ingredient in bleach [48]. HOCl is produced by the action of myeloperoxidase (MPO), which catalyzes the oxidation of chloride (Cl$^-$) by H_2O_2 [49]. Stimulated phagocytes release their MPO-rich azurophilic granules into phagosomes as well as sites of inflammation to generate HOCl that chlorinates and oxidizes bacterial components and inflicts injury to pathogens [50]. However, growing evidence from the last decade has shown an important role of HOCl in the modulation redox signaling and in the pathogenesis of a number of diseases such as atherosclerosis [51, 52]. Examples of modulation of redox signaling by HOCl are the depletion of NO by oxidizing available NO [53], the uncoupling of endothelial NOS [54], the chlorination of NOS arginine [55] substrate, and the reduction of endothelial NOS transcript stability by a chlorinated form of high-density lipoprotein (HDL) [56]. (Figure 12.2).

Reactive Nitrogen Intermediates (RNI)

Reactive nitrogen intermediates [57] encompass all nitrogen-based reactive species including NO and its nitrogen oxide derivatives (nitrogen dioxide (NO_2), dinitrogen trioxide (N_2O_3), and peroxynitrite (ONOO-)). NO, one of the most widely studied RNI, is overproduced in the setting of lupus activity. NO is a membrane-permeable free radical molecule synthesized by nitric oxide synthase [58] using arginine and oxygen as substrates. Three isoforms of NOS are transcribed from three separate genes. All isoforms dimerize in the presence of cofactors to become active. Each monomer contains a reductase and oxygenase domain. The reductase domain catalyzes the transfer of two electrons to heme iron in the oxygenase domain. Calmodulin, nicotinamide adenine dinucleotide phosphate (NADPH), flavin adenine dinucleotide (FAD), and flavin mononucleotide (FMN) are required cofactors for the reductase domain. Electrons from the reductase domain are transferred to the oxygenase domain of the adjacent monomer, where heme and tetrahydrobiopterin (BH_4) act as cofactors. Here, a reaction between O_2 and L-arginine is catalyzed, resulting in formation of NO and citrulline (Figure 12.3).

Two isoforms (endothelial or eNOS and neuronal or nNOS) are generally constitutively expressed and are dependent on sufficient concentrations of calcium for activity. In the vascular system, NO produced by eNOS is a potent vasodilator and regulator of vascular tone in response to shear stress. Nitroglycerin mimics the activity of eNOS by acting as a donor of NO [59]. The beneficial effect of NO produced by the constitutively expressed NOS isoforms is blunted when NO is produced in an environment high in ROI, as discussed later.

A third NOS gene (NOS2) produces an inducible isoform (iNOS) [58] that is primarily expressed in immune cells, most notably macrophages and macrophage-like cells. INOS is expressed in response to inflammatory stimuli that are well-characterized in murine cells. Among these stimuli are several cytokines and toll-like receptor ligands such as lipopolysaccharide, interleukin-6 (IL6), interferon-γ (IFNγ), IL1β, and tumor necrosis factor-α (TNFα). In human cells, complex mixtures of cytokines are necessary for induction. In most cells, signaling pathways converge on the janus kinase/signal transducer and activator of transcription (JAK/STAT) and/or the nuclear factor kappaB (NF-κB)-pathways [60]. Nuclear hormone receptors may play a role in regulation of iNOS induction. There is evidence to support a role for estrogen as an inducer [61] and PPARγ ligands as inhibitors [62] of iNOS induction in response to IFNγ or IFNγ + LPS stimulation respectively in murine cells. INOS is expressed during pathologic states in human endothelial cells, synovial fibroblasts, polymorphonuclear cells, lymphocytes, and natural killer cells [63]. In normal human tissue, expression is strong in myocytes, skeletal muscle, and Purkinje cells [64].

INOS produces log-fold higher amounts of NO than the constitutively expressed isoforms. In a low arginine environment, iNOS cannot transfer nitrogen to molecular oxygen, and electrons from the reductase domain combine with oxygen to produce SO [65] (Figure 12.3). NO, when combined with SO, forms peroxynitrite (ONOO-), a more reactive and toxic molecule than NO itself. ONOO- produced by immune cells is capable of killing intracellular pathogens and tumor cells. Glutathione peroxidase, catalase, superoxide dismutase, heme oxygenase, and antioxidants serve to protect host cells during inflammatory states by reducing the total ROI burden that can contribute to ONOO- production [66, 67]. Thus, reactive intermediate production depends on catalytic enzyme activity and substrate availability as well as the amount of and activity of detoxifying enzymes.

Despite its simple diatomic structure, NO has the potential to induce different and often contradictory effects [68, 69], a dichotomy that pervades the literature. This is attributed to the unique chemical properties of NO, the variability of the biological milieu where NO signaling takes place, and the wide range of NO responsive targets [70]. The interaction between NO and its targets can be regarded as direct or indirect (Figure 12.4A). Direct interactions, dominant at low

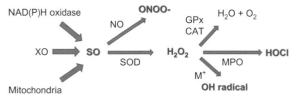

FIGURE 12.2 Fate of superoxide (SO) and its derivatives. SO is produced by a number of enzymes including NAD(P)H oxidase and xanthine oxidase (XO) in addition to the mitochondria. SO is converted in to hydrogen peroxide (H_2O_2) under the action of superoxide dismutase (SOD). NO, reacts with SO to form peroxynitrite (ONOO-). H_2O_2 can be converted into water under the action of glutathione peroxidase (GPx) and/or catalase (CAT), or it can yield OH radical in the presence of metal cations such as Fe^{2+} Cu^{2+}. In phagocytes, H_2O_2 is used to oxidize chloride (Cl-) by myeloperoxidase (MPO) to generate hypochlorous acid (HOCl), a potent bactericide.

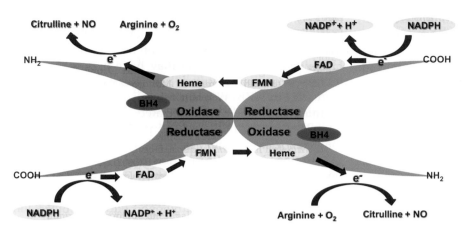

FIGURE 12.3 Structure of nitric oxide synthase and mechanism of NO synthesis. NOS is a homodimer; each monomer is made up of a COOH reductase domain and NH₂ oxidase domain. Two electrons (e), provided by NAD(P)H, are transferred successively to FAD and FMN electron carriers in the reductase domain of one monomer. The electrons then reach the heme group in the oxidase domain of the second monomer. If L-Arg and O₂ substrates are available, the oxidase domain catalyzes two reaction steps, the formation of NN-hydroxy-L-arginine (NHA) followed by conversion of NHA to NO and citrulline. The role of tetrahydrobiopterin (BH4) in this process is unclear, but it may assist in the coupling of NADPH oxidation and NO formation, thus preventing SO formation. In the absence of L-Arg or BH4 the reduced heme group of the oxidase domain transfers electrons to oxygen molecules generating superoxide (SO); this is known as the uncoupling of NOS.

levels of NO (<400 nM), are mediated by reactions between NO and targets such as metal centers and organic free radicals to form nitrosyl adducts. A typical example is the nitrosylation of heme iron in soluble guanylate cyclase [71] and the reaction with alkoxyl radicals limiting lipid peroxidation [72]. Indirect effects, dominant at higher levels of NO (>400 nM), are mediated by reactions between nitrogen oxide derivatives of NO and its target [70, 73]. NO can react with oxygen and SO to produce NO₂ and ONOO- respectively; both species are strong oxidants and mediators of oxidative stress [68]. NO₂ can further react with NO to give N₂O₃, a strong nitrosating agent and mediator of nitrosative stress [68]. Thus, the relative flux of NO with respect to SO in a given biological system determines the nature of indirect NO interactions. When NO level is higher than that of SO, N₂O₃ and nitrosative stress dominates, manifested by nitrosation of thiol, lysine, and zinc fingers [74]. When SO and NO levels are equimolar, ONOO- and oxidative stress dominate, manifested by nitration and oxidation of macromolecules (Figure 12.4B) [75].

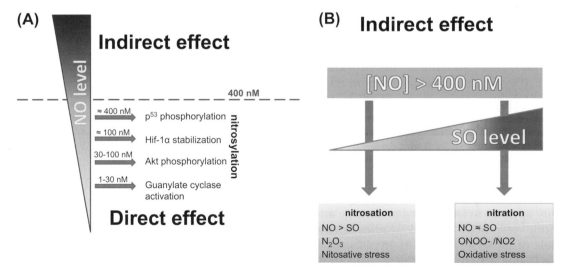

FIGURE 12.4 Modulation of NO effect by local NO concentration and redox state. The effect of NO can be divided in to direct and indirect depending on the local concentration of NO in a given biological system. (*A*) At [NO] < 400 nM, direct effects are dominant; NO reacts directly with targets nitrosylating metal containing groups such as heme groups. NO nitrosylation is concentration dependent where different targets are nitrosylated at different concentrations of NO. For example, guanylate cyclase activation at 1-30 nM NO. Akt phosphorylation at 30-100 nM NO, Hif 1α at 100 nM NO, and p53 at 400 nM NO. (*B*) at [NO] > 400 nM, indirect effects are dominant. Indirect effects are mediated by reactions between nitrogen oxide derivatives of NO and its targets. The nature of the nitrogen-oxide derivative of NO dominant depends on the redox state of the local environment. If [NO] > [SO], nitrogen trioxide (N₂O₃) is dominant and hence nitrosation and nitrosative stress. If [NO] ≈ [SO], nitrogen dioxide (NO2) and peroxynitrite (ONOO-) are dominant and hence nitration/oxidation and oxidative stress.

REACTIVE INTERMEDIATES IN HUMAN LUPUS

Markers of NO Production are Increased in Human Lupus

No study to date has used selective iNOS inhibitors to explore the pathogenic potential of iNOS in humans with lupus. However, several studies have demonstrated elevated markers of NO production in lupus patients compared to controls and a significant correlation between markers of systemic NO production and lupus disease activity [103]. One study demonstrated more prominent increases in markers of NO production among African-Americans with lupus disease activity [104]. A predisposition to produce increased NO in response to inflammatory stimuli may be inherited. Two NOS2 promoter polymorphisms were more prevalent in female African-American SLE subjects than race-matched controls. Increased markers of systemic NO production and improved malaria survival were observed in some African populations with these polymorphisms [105]. These combined observations engender the hypothesis that polymorphisms that are protective in malaria-endemic areas may increase NO production during lupus disease activity.

iNOS Expression in the Skin is Associated with UV Exposure and Disease Activity

The skin often mirrors or heralds disease activity in SLE. iNOS expression in this organ appears to be induced by a known trigger of SLE flares, UV light exposure. Immunostaining for iNOS protein and levels of NOS2 mRNA were elevated in 33% of epidermal tissue from cutaneous lupus subjects before exposure to ultraviolet B (UVB) irradiation, but expression was increased to 100% of skin tissue biopsied after UVB exposure [106]. SLE patients with photosensitivity are often Ro positive. The Ro protein Ro52 is normally expressed in the cytoplasm. In cutaneous lupus patients, Ro52 localizes to the nucleus in the presence of iNOS. In cultured primary keratinocytes *in vitro*, exogenous NO induces such a translocation from the cytoplasm to the nucleus, a phenomenon often seen in stressed cells [58]. These combined observations suggest that UV-induced iNOS expression in cutaneous lupus patients has a functional consequence on Ro52.

iNOS Expression in Proliferative Nephritis is Associated with Mediators of Apoptosis and Inflammation

More recent longitudinal observational studies demonstrate increased markers of systemic NO production (serum NO_X) in lupus patients with proliferative lupus nephritis when compared to those with non-proliferative renal disease or those without nephritis. In nephritis patients, those who did not achieve renal response to therapy had significantly higher serum NO_X levels in the first 3 months of therapy than those who achieved a renal response [107]. How sustained RNI production may mediate renal damage in LN is not completely understood, but several studies shed light on possible mechanisms.

Several groups have described increased glomerular iNOS expression in proliferative lupus nephritis [104, 108, 109]. In such proliferative lesions, citrulline staining increased in parallel with iNOS staining, suggesting that iNOS catalyzed the conversion of arginine to citrulline [110]. Glomerular iNOS staining also co-localized with markers of apoptosis and staining for p53, a redox-sensitive transcription factor that can signal for apoptosis in the presence of damaged DNA [109]. This is consistent with the known effect of higher levels of NO on p53 phosphorylation and activation [76]. Renal biopsies from patients with proliferative LN expressed increased glomerular, tubular, and interstitial cell iNOS levels. Tubulointerstitial iNOS levels correlated with the extent of proteinuria and reduced creatinine clearance at the time of biopsy. INOS and NF-κB co-localized with apoptotic cells in the glomerulus [111]. These data suggest that two mechanisms for iNOS-mediated glomerular damage in proliferative nephritis are through increased signaling for apoptosis via increased p53 activity and for inflammation through activation of NF-κB.

Potential Mechanisms of Reactive Nitrogen and Oxygen-Mediated Vascular Disease in SLE

As described in the background, the ultimate effects of NOS activity are determined by the location and concentration of NO production combined with the proximity of this production to reactive oxygen. Several studies suggest that in SLE patients, NO production is increased in combination with reduced antioxidant capacity. So-called pro-inflammatory HDL (piHDL) cannot prevent oxidation of LDL and aid in cholesterol reverse transport as can normal HDL. The presence of piHDL was significantly associated with carotid plaque in lupus patients. Apolipoprotein A-I (Apo A1) is an HDL protein with both antioxidant and cholesterol efflux properties. SLE patients have increased levels of antibodies to Apo A1 that interfere with function. The presence of these antibodies was associated with elevations in markers of systemic NO production and reductions in total antioxidant capacity [112]. This interference with the antioxidant and anti-inflammatory properties of HDL may have resulted in a more conducive environment for atherosclerosis progression.

ONOO- can also modify lipids. Peroxidation of arachidonate by ONOO- can lead to formation of isoprostanes [113], known to stimulate monocyte adhesion to endothelial cells [114] and induce vasoconstriction in smooth muscles [115]. ONOO- can also oxidize LDL. Oxidized, but not native LDL, complexes with b2-glycroprotein I. This antibody/antigen complex enhances influx of oxidized LDL into foamy macrophages, providing a plausible mechanism for accelerated atherosclerosis in SLE [116]. Some phospholipids within oxidized LDL have platelet-activating factor-like activity and can stimulate growth of smooth muscle cells [82]. Nitrogen dioxide derived from myeloperoxidase and not iNOS can lead to nitration of tryptophan 166 in apo A-I within HDL. Nitration of this amino acid leads to a loss of cholesterol efflux capacity of HDL [117], and this modification along with chlorination of this site confers a 6- to 16-fold risk for cardiovascular disease in non-lupus patients that is independent of Framingham risk factors [118]. The extent to which this phenomenon occurs in SLE subjects is unclear.

Lupus patients often have reduced endothelial function that manifested by reduced endothelium-dependent vasodilation [119]. The mechanism driving this defect is unclear, but increased levels of circulating endothelial cells observed in lupus subjects may be a maker of ongoing direct endothelial injury. This endothelial damage appears to be related to complement-mediated disease activity and is associated with iNOS activity. The level of circulating endothelial cells in lupus subjects correlated positively with disease activity and inversely with complement split product levels. These circulating cells stained for nitrotyrosine, a marker of iNOS or MPO activity [120]. This observation, combined with the observation that endothelial cells stain for iNOS even in non-lesional skin [121], suggests an immune complex-mediated production of ONOO- or NO$_2$ by iNOS or MPO in endothelial cells [120, 121]. Thus, reactive nitrogen and oxygen stress in SLE patients may lead to combined reductions in the protective effects of endothelial and HDL function.

Studies of Human SLE Subjects Suggesting Pathogenic Mechanisms of RNI Production

In SLE, one mechanism through which NO can be pathogenic is through ONOO- or NO$_2$-mediated nitration of self-antigens. This nitration can form neoantigens and serve to break tolerance to self-antigens. This is supported by evidence that SLE serum bound more avidly to nitrated poly-L-tyrosine than native poly-L-tyrosine. Cross reactivity between ONOO- or NO$_2$ modified species and autoantibodies occurs in SLE. Binding of SLE serum with dsDNA antibodies was inhibited by pre-incubation with nitrated poly-L-tyrosine, nitrated

BSA, nitrated DNA, and nitrated chromatin much more effectively than native forms of these antigens. In addition, poly-L-tyrosine immunization of experimental animals induced antibodies that bound avidly to dsDNA and chromatin [122]. Similarly, peroxynitrite-treated DNA is more immunogenic than native DNA as the antigen for dsDNA antibody testing of serum from patients with SLE [123, 124]. These combined studies suggest that ONOO- or NO$_2$ modifications of self-antigens can create neoepitopes with increased binding affinity over native antigens and induce humoral autoimmunity.

NO production can also lead to T-cell dysfunction. Normal T cells expressed eNOS and nNOS but not iNOS with CD3/CD28 co-stimulation. The resulting NO production induced an increase of mitochondrial hyperpolarization (MHP) in normal human T cells [125]. In contrast, T lymphocytes of SLE patients exhibited persistent MHP and mitochondrial mass, accounting for increased production of ROI. These data suggest that mitochondrial dysfunction leading to ATP depletion is ultimately responsible for diminished activation-induced apoptosis and sensitizes lupus T cells to necrosis [126−128]. NO-induced MHP in SLE T cells also leads to activation of mTOR, a sensor of mitochondrial potential and target of the drug rapamycin [129].

CONCLUSION

Observational studies showed a strong association between SLE and reactive oxygen and nitrogen intermediate production in animal models and humans. Moreover, expression of iNOS was shown to be increased in the context of lupus and correlated with disease severity. Conflicting results from murine studies using pharmacologic or genetic manipulation of NOS highlight the complex biology of reactive intermediates in lupus. Further studies should be directed toward investigating potential sources of RIs other than iNOS, defining the key reactive species that are dominant in lupus, and addressing how the interplay between RNIs and ROIs may lead to a break in tolerance and enhanced aggressiveness of lupus disease activity.

ACKNOWLEDGEMENTS

This work was supported by grants from the NIH, VA Merit Review and REAP programs, the Lupus Foundation of America, and the Alliance for Lupus Research.

Some sections of this chapter were previously published in: Oates, J. C. and Gilkeson, G. S. (2006) The biology of nitric oxide and other reactive intermediates in systemic lupus erythematosus. Clin Immunol. **121**,

243–250; Gilkeson, G. S. and Oates, J. C. (2009) Reactive intermediates, inflammation and epigenetics in lupus. In The Epigenetics of Autoimmune Diseases (Zouali, M., ed.). pp. 151–160, John Wiley & Sons Ltd, Chichester.

References

[1] R. Lafyatis, A. Marshak-Rothstein, Toll-like receptors and innate immune responses in systemic lupus erythematosus, Arthritis Res Ther 9 (2007) 222–222.

[2] J.C. Oates, G.S. Gilkeson, The biology of nitric oxide and other reactive intermediates in systemic lupus erythematosus, Clin Immunol 121 (2006) 243–250.

[3] D. Harman, Aging: a theory based on free radical and radiation chemistry, J Gerontol 11 (1956) 298–300.

[4] M. Valko, D. Leibfritz, J. Moncol, M.T.D. Cronin, M. Mazur, J. Telser, Free radicals and antioxidants in normal physiological functions and human disease, Int J Biochem Cell Biol 39 (2007) 44–84.

[5] P.V. Vignais, The superoxide-generating NADPH oxidase: structural aspects and activation mechanism, Cell Mol Life Sci 59 (2002) 1428–1459.

[6] C.-C. Wei, Z.-Q. Wang, D. Durra, C. Hemann, R. Hille, E.D. Garcin, et al., The three nitric-oxide synthases differ in their kinetics of tetrahydrobiopterin radical formation, heme-dioxy reduction, and arginine hydroxylation, J Biol Chem 280 (2005) 8929–8935.

[7] J.F. Turrens, A. Boveris, Generation of superoxide anion by the NADH dehydrogenase of bovine heart mitochondria, Biochem J 191 (1980) 421–427.

[8] B. Chance, H. Sies, A. Boveris, Hydroperoxide metabolism in mammalian organs, Physiol Rev 59 (1979) 527–605.

[9] G.L. Squadrito, W.A. Pryor, Oxidative chemistry of nitric oxide: the roles of superoxide, peroxynitrite, and carbon dioxide, Free Radic Biol Med 25 (1998) 392–403.

[10] H. Sies, Oxidative stress: oxidants and antioxidants, Experimental Physiology 82 (1997) 291–295.

[11] W. Droge, Free radicals in the physiological control of cell function, Physiol Rev 82 (2002) 47–95.

[12] B. D'Autreaux, M.B. Toledano, ROS as signalling molecules: mechanisms that generate specificity in ROS homeostasis, Nat Rev Mol Cell Biol 8 (2007) 813–824.

[13] H. Sauer, M. Wartenberg, J. Hescheler, Reactive oxygen species as intracellular messengers during cell growth and differentiation, Cell Physiol Biochem 11 (2001) 173–186.

[14] L.L. Ji, M.-C. Gomez-Cabrera, J. Vina, Role of free radicals and antioxidant signaling in skeletal muscle health and pathology, Infect Disord Drug Targets 9 (2009) 428–444.

[15] M. Valko, C.J. Rhodes, J. Moncol, M. Izakovic, M. Mazur, Free radicals, metals and antioxidants in oxidative stress-induced cancer, Chem Biol Interact 160 (2006) 1–40.

[16] A.C.E. Campos, F. Molognoni, F.H.M. Melo, L.C. Galdieri, C.R.W. Carneiro, V. D'Almeida, et al., Oxidative stress modulates DNA methylation during melanocyte anchorage blockade associated with malignant transformation, Neoplasia 9 (2007) 1111–1121.

[17] S. Kavak, L. Ayaz, M. Emre, T. Inal, L. Tamer, I. Günay, The effects of rosiglitazone on oxidative stress and lipid profile in left ventricular muscles of diabetic rats, Cell Biochemistry and Function 26 (2008) 478–485.

[18] S.E. Gomez-Mejiba, Z. Zhai, H. Akram, L.J. Deterding, K. Hensley, N. Smith, et al., Immuno-spin trapping of protein and DNA radicals: "tagging" free radicals to locate and understand the redox process, Free Radic Biol Med 46 (2009) 853–865.

[19] A. Trostchansky, V.B. O'Donnell, D.C. Goodwin, L.M. Landino, L.J. Marnett, R. Radi, et al., Interactions between nitric oxide and peroxynitrite during prostaglandin endoperoxide H synthase-1 catalysis: a free radical mechanism of inactivation, Free Radic Biol Med 42 (2007) 1029–1038.

[20] P. Pacher, P. Mukhopadhyay, M. Rajesh, S. Batkai, G. Hasko, C. Szabo, Interplay of superoxide, nitric oxide and peroxynitrite in doxorubicin-induced cell death, FASEB J 22 (2008) 970.912-.

[21] B.T. Kurien, R.H. Scofield, Free radical mediated peroxidative damage in systemic lupus erythematosus, Life Sci 73 (2003) 1655–1666.

[22] W.Y. Craig, Autoantibodies against oxidized low density lipoprotein: a review of clinical findings and assay methodology, J Clin Lab Anal 9 (1995) 70–74.

[23] L. Kopkan, A. Castillo, L.G. Navar, D.S.A. Majid, Enhanced superoxide generation modulates renal function in ANG II-induced hypertensive rats, Am J Physiol Renal Physiol 290 (2006) F80–F86.

[24] M.D. Brand, C. Affourtit, T.C. Esteves, K. Green, A.J. Lambert, S. Miwa, et al., Mitochondrial superoxide: production, biological effects, and activation of uncoupling proteins, Free Radic Biol Med 37 (2004) 755–767.

[25] R.O. Poyton, K.A. Ball, P.R. Castello, Mitochondrial generation of free radicals and hypoxic signaling, Trends Endocrinol Metab 20 (2009) 332–340.

[26] C. Vorbach, R. Harrison, M.R. Capecchi, Xanthine oxidoreductase is central to the evolution and function of the innate immune system, Trends Immunol 24 (2003) 512–517.

[27] G. Bonizzi, J. Piette, M.P. Merville, V. Bours, Cell type-specific role for reactive oxygen species in nuclear factor-kappaB activation by interleukin-1, Biochem Pharmacol 59 (2000) 7–11.

[28] L. Feng, Y. Xia, G.E. Garcia, D. Hwang, C.B. Wilson, Involvement of reactive oxygen intermediates in cyclooxygenase-2 expression induced by interleukin-1, tumor necrosis factor-alpha, and lipopolysaccharide, J Clin Invest 95 (1995) 1669–1675.

[29] M. Satoh, S. Fujimoto, Y. Haruna, S. Arakawa, H. Horike, N. Komai, et al., NAD(P)H oxidase and uncoupled nitric oxide synthase are major sources of glomerular superoxide in rats with experimental diabetic nephropathy, Am J Physiol Renal Physiol 288 (2005) F1144–F1152.

[30] Y. Keisari, L. Braun, E. Flescher, The oxidative burst and related phenomena in mouse macrophages elicited by different sterile inflammatory stimuli, Immunobiology 165 (1983) 78–89.

[31] I. Rahman, S.-R. Yang, S.K. Biswas, Current concepts of redox signaling in the lungs, Antioxid Redox Signal 8 (2006) 681–689.

[32] H. Zhu, H.F. Bunn, Oxygen sensing and signaling: impact on the regulation of physiologically important genes, Respir Physiol 115 (1999) 239–247.

[33] M.S. Wolin, T.M. Burke-Wolin, K.M. Mohazzab-H, Roles for NAD(P)H oxidases and reactive oxygen species in vascular oxygen sensing mechanisms, Respir Physiol 115 (1999) 229–238.

[34] T.E. Decoursey, E. Ligeti, Regulation and termination of NADPH oxidase activity, Cell Mol Life Sci 62 (2005) 2173–2193.

[35] T.C. Zahrt, V. Deretic, Reactive nitrogen and oxygen intermediates and bacterial defenses: unusual adaptations in Mycobacterium tuberculosis, Antioxid Redox Signal 4 (2002) 141–159.

[36] C. Deby, R. Goutier, New perspectives on the biochemistry of superoxide anion and the efficiency of superoxide dismutases, Biochem Pharmacol 39 (1990) 399–405.

[37] M.J. Steinbeck, A.U. Khan, M.J. Karnovsky, Extracellular production of singlet oxygen by stimulated macrophages quantified using 9,10-diphenylanthracene and perylene in a polystyrene film, J Biol Chem 268 (1993) 15649–15654.

[38] E.A. Veal, A.M. Day, B.A. Morgan, Hydrogen peroxide sensing and signaling, Mol Cell 26 (2007) 1–14.

[39] M. Valko, H. Morris, M.T.D. Cronin, Metals, toxicity and oxidative stress, Curr Med Chem 12 (2005) 1161–1208.

[40] N. Pastor, H. Weinstein, E. Jamison, M. Brenowitz, A detailed interpretation of OH radical footprints in a TBP-DNA complex reveals the role of dynamics in the mechanism of sequence-specific binding, J Mol Biol 304 (2000) 55–68.

[41] S.I. Liochev, I. Fridovich, The Haber-Weiss cycle – 70 years later: an alternative view, Redox Rep 7 (2002) 55–57.

[42] J. Foreman, V. Demidchik, J.H.F. Bothwell, P. Mylona, H. Miedema, M.A. Torres, et al., Reactive oxygen species produced by NADPH oxidase regulate plant cell growth, Nature 422 (2003) 442–446.

[43] J. Li, M. Stouffs, L. Serrander, B. Banfi, E. Bettiol, Y. Charnay, et al., The NADPH oxidase NOX4 drives cardiac differentiation: Role in regulating cardiac transcription factors and MAP kinase activation, Mol Biol Cell 17 (2006) 3978–3988.

[44] M. Geiszt, T.L. Leto, The Nox family of NAD(P)H oxidases: host defense and beyond, J Biol Chem 279 (2004) 51715–51718.

[45] A. Salmeen, J.N. Andersen, M.P. Myers, T.-C. Meng, J.A. Hinks, N.K. Tonks, et al., Redox regulation of protein tyrosine phosphatase 1B involves a sulphenyl-amide intermediate, Nature 423 (2003) 769–773.

[46] M. Brandon, P. Baldi, D.C. Wallace, Mitochondrial mutations in cancer, Oncogene 25 (2006) 4647–4662.

[47] K. Stadler, M.G. Bonini, S. Dallas, J. Jiang, R. Radi, R.P. Mason, et al., Involvement of inducible nitric oxide synthase in hydroxyl radical-mediated lipid peroxidation in streptozotocin-induced diabetes, Free Radical Biology and Medicine 45 (2008) 866–874.

[48] S. Mütze, U. Hebling, W. Stremmel, J. Wang, J. Arnhold, K. Pantopoulos, et al., Myeloperoxidase-derived Hypochlorous Acid Antagonizes the Oxidative Stress-mediated Activation of Iron Regulatory Protein 1, Journal of Biological Chemistry 278 (2003) 40542–40549.

[49] C.C. Winterbourn, M.C. Vissers, A.J. Kettle, Myeloperoxidase, Curr Opin Hematol 7 (2000) 53–58.

[50] M.B. Hampton, A.J. Kettle, C.C. Winterbourn, Inside the Neutrophil Phagosome: Oxidants, Myeloperoxidase, and Bacterial Killing, Blood 92 (1998) 3007–3017.

[51] D. Lau, S. Baldus, Myeloperoxidase and its contributory role in inflammatory vascular disease, Pharmacol Ther 111 (2006) 16–26.

[52] S.J. Nicholls, S.L. Hazen, Myeloperoxidase and Cardiovascular Disease, Arterioscler Thromb Vasc Biol 25 (2005) 1102–1111.

[53] J.P. Eiserich, S. Baldus, M.-L. Brennan, W. Ma, C. Zhang, A. Tousson, et al., Myeloperoxidase, a leukocyte-derived vascular NO oxidase, Science 296 (2002) 2391–2394.

[54] J. Xu, Z. Xie, R. Reece, D. Pimental, M.-H. Zou, Uncoupling of Endothelial Nitric Oxidase Synthase by Hypochlorous Acid: Role of NAD(P)H Oxidase-Derived Superoxide and Peroxynitrite, Arterioscler Thromb Vasc Biol 26 (2006) 2688–2695.

[55] C. Zhang, J. Yang, J.D. Jacobs, L.K. Jennings, Interaction of myeloperoxidase with vascular NAD(P)H oxidase-derived reactive oxygen species in vasculature: implications for vascular diseases, Am J Physiol Heart Circ Physiol 285 (2003) 2563–2572.

[56] G. Marsche, R. Heller, G. Fauler, A. Kovacevic, A. Nuszkowski, W. Graier, et al., 2-chlorohexadecanal derived from hypochlorite-modified high-density lipoprotein-associated plasmalogen is a natural inhibitor of endothelial nitric oxide biosynthesis, Arterioscler Thromb Vasc Biol 24 (2004) 2302–2306.

[57] P. Migliorini, F. Pratesi, F. Bongiorni, S. Moscato, M. Scavuzzo, S. Bombardieri, The targets of nephritogenic antibodies in systemic autoimmune disorders, Autoimmun Rev 1 (2002) 168–173.

[58] A. Espinosa, V. Oke, A. Elfving, F. Nyberg, R. Covacu, M. Wahren-Herlenius, The autoantigen Ro52 is an E3 ligase resident in the cytoplasm but enters the nucleus upon cellular exposure to nitric oxide, Exp Cell Res 314 (2008) 3605–3613.

[59] W.K. Alderton, C.E. Cooper, R.G. Knowles, Nitric oxide synthases: structure, function and inhibition, Biochem J 357 (2001) 593–615.

[60] H. Kleinert, A. Pautz, K. Linker, P.M. Schwarz, Regulation of the expression of inducible nitric oxide synthase, Eur. J. Pharmacol 500 (2004) 255–266.

[61] E. Karpuzoglu, S.A. Ahmed, Estrogen regulation of nitric oxide and inducible nitric oxide synthase (iNOS) in immune cells: Implications for immunity, autoimmune diseases, and apoptosis, Nitric Oxide 15 (2006) 177–186.

[62] C.M. Reilly, J.C. Oates, J. Sudian, M.B. Crosby, P.V. Halushka, G.S. Gilkeson, Prostaglandin J(2) inhibition of mesangial cell iNOS expression, Clin. Immunol 98 (2001) 337–345.

[63] J. Lincoln, C.H.V. Hoyle, G. Burnstock, Nitric Oxide in Health and Disease, Cambridge University Press, New York, NY, 1997.

[64] NOS2 expression profiles. Human Protein Atlas 2005,[cited october 2009]; http://www.proteinatlas.org/tissue_profile.php?antibody_id=2014.

[65] Y. Xia, J.L. Zweier, Superoxide and peroxynitrite generation from inducible nitric oxide synthase in macrophages. Proc. Natl. Acad. Sci. U.S. A 94 (1997) 6954–6958.

[66] S.A. Gaertner, U. Janssen, T. Ostendorf, K.M. Koch, J. Floege, W. Gwinner, Glomerular oxidative and antioxidative systems in experimental mesangioproliferative glomerulonephritis, J. Am. Soc. Nephrol 13 (2002) 2930–2937.

[67] N.G. Abraham, A. Kappas, Heme oxygenase and the cardiovascular-renal system, Free Radic. Biol. Med 39 (2005) 1–25.

[68] D.A. Wink, M. Feelisch, J. Fukuto, D. Chistodoulou, D. Jourd'heuil, et al., The cytotoxicity of nitroxyl: possible implications for the pathophysiological role of NO, Arch Biochem Biophys 351 (1998) 66–74.

[69] D.A. Wink, J.B. Mitchell, Chemical biology of nitric oxide: Insights into regulatory, cytotoxic, and cytoprotective mechanisms of nitric oxide, Free Radic Biol Med 25 (1998) 434–456.

[70] D.D. Thomas, L.A. Ridnour, J.S. Isenberg, W. Flores-Santana, C.H. Switzer, S. Donzelli, et al., The chemical biology of nitric oxide: Implications in cellular signaling, Free Radical Biology and Medicine 45 (2008) 18–31.

[71] V.G. Kharitonov, V.S. Sharma, D. Magde, D. Koesling, Kinetics of nitric oxide dissociation from five- and six-coordinate nitrosyl hemes and heme proteins, including soluble guanylate cyclase, Biochemistry 36 (1997) 6814–6818.

[72] S. Padmaja, R.E. Huie, The reaction of nitric oxide with organic peroxyl radicals, Biochem Biophys Res Commun 195 (1993) 539–544.

[73] D.D. Thomas, L.A. Ridnour, M.G. Espey, S. Donzelli, S. Ambs, S.P. Hussain, et al., Superoxide Fluxes Limit Nitric Oxide-induced Signaling, J. Biol. Chem 281 (2006) 25984–25993.

[74] L.A. Ridnour, D.D. Thomas, D. Mancardi, M.G. Espey, K.M. Miranda, N. Paolocci, et al., The chemistry of nitrosative stress induced by nitric oxide and reactive nitrogen oxide species. Putting perspective on stressful biological situations, Biological Chemistry 385 (2005) 1–10.

[75] D. Salvemini, T.M. Doyle, S. Cuzzocrea, Superoxide, peroxynitrite and oxidative/nitrative stress in inflammation, Biochem. Soc. Trans 34 (2006) 965–970.

[76] D.D. Thomas, L.A. Ridnour, J.S. Isenberg, W. Flores-Santana, C.H. Switzer, S. Donzelli, et al., The chemical biology of nitric oxide: implications in cellular signaling, Free Radic. Biol. Med 45 (2008) 18−31.

[77] L. Leon, J.F. Jeannin, A. Bettaieb, Post-translational modifications induced by nitric oxide (NO): implication in cancer cells apoptosis, Nitric Oxide 19 (2008) 77−83.

[78] J. Delgado Alves, L.J. Mason, P.R. Ames, P.P. Chen, J. Rauch, J.S. Levine, et al., Antiphospholipid antibodies are associated with enhanced oxidative stress, decreased plasma nitric oxide and paraoxonase activity in an experimental mouse model, Rheumatology (Oxford) 44 (2005) 1238−1244.

[79] M.H. Zou, A. Yesilkaya, V. Ullrich, Peroxynitrite inactivates prostacyclin synthase by heme-thiolate-catalyzed tyrosine nitration, Drug Metabolism Reviews 31 (1999) 343−349.

[80] D.C. Goodwin, L.M. Landino, L.J. Marnett, Reactions of prostaglandin endoperoxide synthase with nitric oxide and peroxynitrite, Drug Metab. Rev 31 (1999) 273−294.

[81] J.B. Weinberg, D.L. Granger, D.S. Pisetsky, M.F. Seldin, M.A. Misukonis, S.N. Mason, et al., The role of nitric oxide in the pathogenesis of spontaneous murine autoimmune disease: increased nitric oxide production and nitric oxide synthase expression in MRL-lpr/lpr mice, and reduction of spontaneous glomerulonephritis and arthritis by orally administered NG-monomethyl-L- arginine, J. Exp. Med 179 (1994) 651−660.

[82] J.M. Heery, M. Kozak, D.M. Stafforini, D.A. Jones, G.A. Zimmerman, T.M. McIntyre, et al., Oxidatively modified LDL contains phospholipids with platelet-activating factor-like activity and stimulates the growth of smooth muscle cells, Journal of Clinical Investigation 96 (1995) 2322−2330.

[83] T. Keng, C.T. Privalle, G.S. Gilkeson, J.B. Weinberg, Peroxynitrite formation and decreased catalase activity in autoimmune MRL-lpr/lpr mice, Mol. Med 6 (2000) 779−792.

[84] H.P. Wang, T.C. Hsu, G.J. Hsu, S.L. Li, B.S. Tzang, Cystamine attenuates the expressions of NOS- and TLR-associated molecules in the brain of NZB/W F1 mice, Eur. J. Pharmacol 607 (2009) 102−106.

[85] N. Mishra, C.M. Reilly, D.R. Brown, P. Ruiz, G.S. Gilkeson, Histone deacetylase inhibitors modulate renal disease in the MRL-lpr/lpr mouse, J Clin Invest 111 (2003) 539−552.

[86] C.M. Reilly, S. Olgun, D. Goodwin, R.M. Gogal Jr., A. Santo, et al., Interferon regulatory factor-1 gene deletion decreases glomerulonephritis in MRL/lpr mice, Eur J Immunol 36 (2006) 1296−1308.

[87] H.H. Li, H.H. Cheng, K.H. Sun, C.C. Wei, C.F. Li, W.C. Chen, et al., Interleukin-20 targets renal mesangial cells and is associated with lupus nephritis, Clin. Immunol 129 (2008) 277−285.

[88] Y. Takeda, M. Takeno, M. Iwasaki, H. Kobayashi, Y. Kirino, A. Ueda, et al., Chemical induction of HO-1 suppresses lupus nephritis by reducing local iNOS expression and synthesis of anti-dsDNA antibody, Clin Exp Immunol 138 (2004) 237−244.

[89] S.L. Lui, R. Tsang, D. Wong, K.W. Chan, T.M. Chan, P.C. Fung, et al., Effect of mycophenolate mofetil on severity of nephritis and nitric oxide production in lupus-prone MRL/lpr mice, Lupus 11 (2002) 411−418.

[90] C.C. Yu, C.W. Yang, M.S. Wu, Y.C. Ko, C.T. Huang, J.J. Hong, C.C. Huang, Mycophenolate mofetil reduces renal cortical inducible nitric oxide synthase mRNA expression and diminishes glomerulosclerosis in MRL/lpr mice, J Lab Clin Med 138 (2001) 69−77.

[91] C.M. Reilly, L.W. Farrelly, D. Viti, S.T. Redmond, F. Hutchison, P. Ruiz, et al., Modulation of renal disease in MRL/lpr mice by pharmacologic inhibition of inducible nitric oxide synthase, Kidney Int 61 (2002) 839−846.

[92] J.C. Oates, P. Ruiz, A. Alexander, A.M. Pippen, G.S. Gilkeson, Effect of late modulation of nitric oxide production on murine lupus, Clin. Immunol. Immunopathol 83 (1997) 86−92.

[93] G.S. Gilkeson, J.S. Mudgett, M.F. Seldin, P. Ruiz, A.A. Alexander, M.A. Misukonis, et al., Clinical and serologic manifestations of autoimmune disease in MRL-lpr/lpr mice lacking nitric oxide synthase type 2, J Exp Med 186 (1997) 365−373.

[94] C. Njoku, S.E. Self, P. Ruiz, A.F. Hofbauer, G.S. Gilkeson, J.C. Oates, Inducible nitric oxide synthase inhibitor SD-3651 reduces proteinuria in MRL/lpr mice deficient in the NOS2 gene, J Investig Med 56 (2008) 911−919.

[95] C.J. Njoku, K.S. Patrick, P. Ruiz Jr., J.C. Oates, Inducible nitric oxide synthase inhibitors reduce urinary markers of systemic oxidant stress in murine proliferative lupus nephritis, J. Investig. Med 53 (2005) 347−352.

[96] M.H. Zou, R. Cohen, V. Ullrich, Peroxynitrite and vascular endothelial dysfunction in diabetes mellitus, Endothelium 11 (2004) 89−97.

[97] M.H. Zou, C. Shi, R.A. Cohen, Oxidation of the zinc-thiolate complex and uncoupling of endothelial nitric oxide synthase by peroxynitrite, J. Clin. Invest 109 (2002) 817−826.

[98] L.A. Casciola-Rosen, G. Anhalt, A. Rosen, Autoantigens targeted in systemic lupus erythematosus are clustered in two populations of surface structures on apoptotic keratinocytes, J. Exp. Med 179 (1994) 1317−1330.

[99] C.S. Boyd, E. Cadenas, Nitric oxide and cell signaling pathways in mitochondrial-dependent apoptosis, Biol. Chem 383 (2002) 411−423.

[100] J.C. Oates, G.S. Gilkeson, Nitric oxide induces apoptosis in spleen lymphocytes from MRL/lpr mice, J. Investig. Med 52 (2004) 62−71.

[101] A.K. Singh, Lupus in the Fas lane? J.R. Coll, Physicians Lond 29 (1995) 475−478.

[102] H. Ohmori, H. Oka, Y. Nishikawa, H. Shigemitsu, M. Takeuchi, M. Magari, et al., Immunogenicity of autologous IgG bearing the inflammation-associated marker 3-nitrotyrosine, Immunol. Lett 96 (2005) 47−54.

[103] J.C. Oates, G.S. Gilkeson, The biology of nitric oxide and other reactive intermediates in systemic lupus erythematosus, Clin Immunol 121 (2006) 243−250.

[104] J.C. Oates, E.F. Christensen, C.M. Reilly, S.E. Self, G.S. Gilkeson, Prospective measure of serum 3-nitrotyrosine levels in systemic lupus erythematosus: Correlation with disease activity, Proceedings of the Association of American Physicians 111 (1999) 611−621.

[105] J.C. Oates, M.C. Levesque, M.R. Hobbs, E.G. Smith, I.D. Molano, G.P. Page, et al., Nitric oxide synthase 2 promoter polymorphisms and systemic lupus erythematosus in African-Americans, J Rheumatol 30 (2003) 60−67.

[106] A. Kuhn, K. Fehsel, P. Lehmann, J. Krutmann, T. Ruzicka, V. Kolbbachofen, Aberrant timing in epidermal expression of inducible nitric oxide synthase after UV irradiation in cutaneous lupus erythematosus, Journal of Investigative Dermatology 111 (1998) 149−153.

[107] J.C. Oates, S.R. Shaftman, S.E. Self, G.S. Gilkeson, Association of serum nitrate and nitrite levels with longitudinal assessments of disease activity and damage in systemic lupus erythematosus and lupus nephritis, Arthritis Rheum 58 (2008) 263−272.

[108] A. Furusu, M. Miyazaki, K. Abe, S. Tsukasaki, K. Shioshita, O. Sasaki, et al., Expression of endothelial and inducible nitric oxide synthase in human glomerulonephritis, Kidney International 53 (1998) 1760−1768.

[109] J.S. Wang, H.H. Tseng, D.F. Shih, H.S. Jou, L.P. Ger, Expression of inducible nitric oxide synthase and apoptosis in human lupus nephritis, Nephron 77 (1997) 404–411.

[110] Y.G.J.J. Bollain, R. Ramirez-Sandoval, L. Daza, E. Esparza, O. Barbosa, D. Ramirez, et al., Widespread expression of inducible NOS and citrulline in lupus nephritis tissues, Inflammation Research 58 (2009) 61–66.

[111] L. Zheng, R. Sinniah, Renal cell apoptosis and proliferation may be linked to nuclear factor-kappaB activation and expression of inducible nitric oxide synthase in patients with lupus nephritis. S, I. H. H. Hum Pathol 37 (2006) 637–647

[112] J.R. Batuca, P.R. Ames, M. Amaral, C. Favas, D.A. Isenberg, J. Delgado Alves, Anti-atherogenic and anti-inflammatory properties of high-density lipoprotein are affected by specific antibodies in systemic lupus erythematosus, Rheumatology (Oxford) 48 (2009) 26–31.

[113] K.P. Moore, V. Darley-Usmar, J. Morrow, L.J. Roberts 2nd, Formation of F2-isoprostanes during oxidation of human low-density lipoprotein and plasma by peroxynitrite, Circulation Research 77 (1995) 335–341.

[114] J. Huber, V.N. Bochkov, B.R. Binder, N. Leitinger, The isoprostane 8-iso-PGE2 stimulates endothelial cells to bind monocytes via cyclic AMP- and p38 MAP kinase-dependent signaling pathways, Antioxid Redox Signal 5 (2003) 163–169.

[115] M. Fukunaga, N. Makita, L.J.d. Roberts, J.D. Morrow, K. Takahashi, K.F. Badr, Evidence for the existence of F2-isoprostane receptors on rat vascular smooth muscle cells, American Journal of Physiology 264 (1993) C1619–C1624.

[116] L.R. Lopez, K. Kobayashi, Y. Matsunami, E. Matsuura, Immunogenic Oxidized Low-density Lipoprotein/beta2-glycoprotein I Complexes in the Diagnostic Management of Atherosclerosis, Clin. Rev. Allergy Immunol 37 (2009) 12–19.

[117] D.Q. Peng, G. Brubaker, Z. Wu, L. Zheng, B. Willard, M. Kinter, et al., Apolipoprotein A-I tryptophan substitution leads to resistance to myeloperoxidase-mediated loss of function, Arterioscler. Thromb. Vasc. Biol 28 (2008) 2063–2070.

[118] S.J. Nicholls, S.L. Hazen, Myeloperoxidase, modified lipoproteins, and atherogenesis, J. Lipid Res 50 Suppl (2009) S346–S351.

[119] M. El-Magadmi, H. Bodill, Y. Ahmad, P.N. Durrington, M. Mackness, M. Walker, et al., Systemic lupus erythematosus: an independent risk factor for endothelial dysfunction in women, Circulation 110 (2004) 399–404.

[120] R. Clancy, G. Marder, V. Martin, H.M. Belmont, S.B. Abramson, J. Buyon, Circulating activated endothelial cells in systemic lupus erythematosus: further evidence for diffuse vasculopathy, Arthritis Rheum 44 (2001) 1203–1208.

[121] H.M. Belmont, D. Levartovsky, A. Goel, A. Amin, R. Giorno, J. Rediske, et al., Increased nitric oxide production accompanied by the up-regulation of inducible nitric oxide synthase in vascular endothelium from patients with systemic lupus erythematosus, Arthritis Rheum 40 (1997) 1810–1816.

[122] F. Khan, R. Ali, Antibodies against nitric oxide damaged poly L-tyrosine and 3-nitrotyrosine levels in systemic lupus erythematosus, J Biochem Mol Biol 39 (2006) 189–196.

[123] K. Dixit, R. Ali, Role of nitric oxide modified DNA in the etiopathogenesis of systemic lupus erythematosus, Lupus 13 (2004) 95–100.

[124] S. Habib, Moinuddin, R. Ali, Peroxynitrite-modified DNA: a better antigen for systemic lupus erythematosus anti-DNA autoantibodies, Biotechnol Appl Biochem 43 (2006) 65–70.

[125] G. Nagy, A. Koncz, A. Perl, T cell activation-induced mitochondrial hyperpolarization is mediated by Ca2+- and redox-dependent production of nitric oxide, J Immunol 171 (2003) 5188–5197.

[126] P. Gergely Jr., B. Niland, N. Gonchoroff, R. Pullmann Jr., P.E. Phillips, A. Perl, Persistent mitochondrial hyperpolarization, increased reactive oxygen intermediate production, and cytoplasmic alkalinization characterize altered IL-10 signaling in patients with systemic lupus erythematosus, J Immunol 169 (2002) 1092–1101.

[127] G. Nagy, M. Barcza, N. Gonchoroff, P.E. Phillips, A. Perl, Nitric oxide-dependent mitochondrial biogenesis generates Ca2+ signaling profile of lupus T cells, J Immunol 173 (2004) 3676–3683.

[128] P. Gergely Jr., C. Grossman, B. Niland, F. Puskas, H. Neupane, F. Allam, et al., Mitochondrial hyperpolarization and ATP depletion in patients with systemic lupus erythematosus, Arthritis Rheum 46 (2002) 175–190.

[129] D.R. Fernandez, T. Telarico, E. Bonilla, Q. Li, S. Banerjee, A. Middleton, et al., Activation of mammalian target of rapamycin controls the loss of TCRzeta in lupus T cells through HRES-1/Rab4-regulated lysosomal degradation, J Immunol 182 (2009) 2063–2073.

Origins of Antinuclear Antibodies

Westley H. Reeves, Yuan Xu, Haoyang Zhuang, Yi Li, Lijun Yang

INTRODUCTION

The production of antinuclear antibodies (ANAs) is one of the defining features of SLE. Description of the LE cell phenomenon by Hargraves et al. in 1948 was the first evidence for the existence of these autoantibodies [1]. Close examination of the bone marrow from SLE patients reveals not only the classic LE cell phenomenon (phagocytosis of intact nuclei by mature polymorphonuclear leukocytes in the bone marrow) but also adherence of nuclei to and phagocytosis by less mature myeloid cells (promyelocytes, metamyelocytes, and myelocytes) (Figure 13.1A−C). The description of LE cells was followed by the identification of antinuclear and anti-DNA antibodies in 1957 [2, 3]. The discovery of anti-Sm antibodies by Tan and Kunkel in 1966 provided definitive evidence that autoantibodies in lupus recognize nuclear structures other than

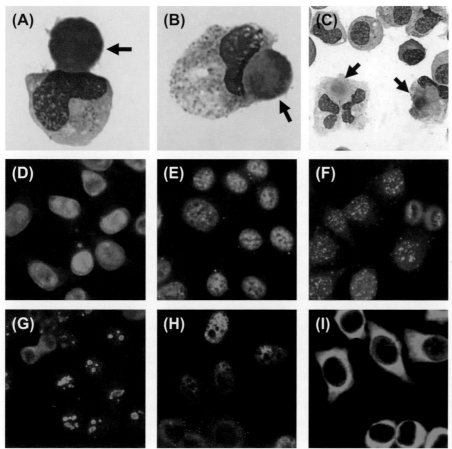

FIGURE 13.1 Detection of antinuclear antibodies. (*A−C*) LE cell phenomenon. Wright-Giemsa staining of a cytospin from a bone marrow aspirate of a 7-year-old girl with lupus nephritis and pancytopenia illustrating granulocytic phagocytosis of astructural nuclear material (arrows) in the bone marrow. (*A, B*) Adherence of nuclei to and phagocytosis of nuclei by immature myeloid cells (*A*, myelocyte and *B*, metamyelocyte, original magnification X1000). (*C*) The "classic" LE cell phenomenon (phagocytosis of nuclei by mature neutrophils) (X1000). (*D−I*) Fluorescent antinuclear antibody test. HeLa cells (human cervical carcinoma cell line) were fixed with methanol and stained with human autoimmune serum at a dilution of 1:40. The binding of antinuclear antibodies to the fixed and permeabilized cells was detected with fluorescein isothiocyanate (FITC)-conjugated goat anti-human IgG antibodies. (*D*) Diffuse (homogeneous) pattern produced by anti-double-stranded DNA antibodies; (*E*) fine speckled pattern sparing the nucleolus produced by anti-nRNP antibodies; (*F*) centromere staining exhibited by serum from a patient with CREST syndrome producing anti-CENP-B autoantibodies; (*G*) nucleolar pattern produced by anti-RNA polymerase I autoantibodies from a patient with scleroderma; (*H*) variable nuclear staining pattern produced by autoantibodies against PCNA; (*I*) cytoplasmic pattern produced by serum containing anti-ribosomal P autoantibodies.

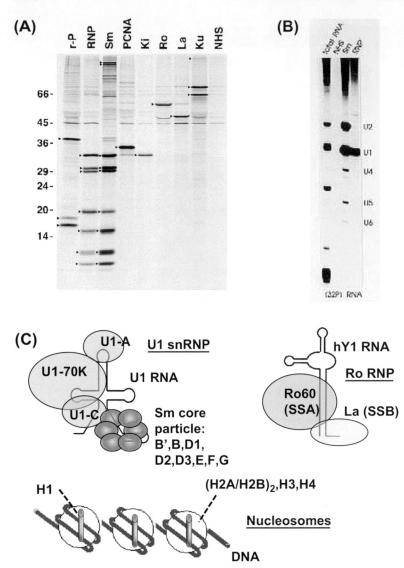

FIGURE 13.2 Antigenic targets of lupus autoantibodies consist of nucleic acid-protein complexes. (*A*) Immunoprecipitation and analysis of [^{35}S]-labeled proteins from K562 cells by SDS-polyacrylamide gel electrophoresis. Immunoprecipitation was performed using prototype sera containing anti- ribosomal P (r-P), nRNP (RNP), Sm, proliferating cell nuclear antigen (PCNA), Ki, Ro (SS-A), La (SS-B), or Ku autoantibodies. Immunoprecipitation with normal human serum (NHS) is shown as a control. Positions of the characteristic protein bands are indicated with arrowheads. Positions of molecular weight standards in kilodaltons are indicated on the left. (*B*) K562 (human erythroleukemia) cells were labeled with [^{32}P] orthophosphate, an extract was immunoprecipitated with anti-Sm or anti-nRNP serum or with normal human serum (NHS). Immunoprecipitates were subjected to phenol extraction, and the radiolabeled RNA was analyzed by gel electrophoresis followed by autoradiography. Positions of U2, U1, U4, U5, and U6 small RNAs are indicated. (*C*) Structure of the U1 small nuclear ribonucleoprotein (U1 snRNP) and Ro ribonucleoprotein (Ro RNP), and nucleosomes. The U1 snRNP is a complex of several proteins (designated 70K, A, B'/B, C, D, E, F, and G) bound to a single molecule of U1 small nuclear RNA. The 70K, A, and C proteins (dark shading) are recognized by anti-nRNP antibodies. The B'/B, D, E, F, and G proteins form a 6S particle termed the Sm core particle. This particle contains the major antigenic determinants recognized by anti-Sm antibodies. The mature hY1 ribonucleoprotein (one of the Ro RNP particles) consists of a single Ro60 (SSA) protein bound to the human Y1 RNA. Newly synthesized hY1 is bound to a second protein, the La (SSB) antigen, which is subsequently cleaved away during maturation of the ribonucleoprotein. Nucleosomes consist of histones H2A/H2B (two copies each) plus H3 and H4 around which DNA is coiled. The DNA between individual nucleosomes is bound to histone H1.

nucleosomes [4], marking the beginning of a 30-year period during which the major autoantibody—autoantigen systems associated with systemic autoimmune diseases were identified and characterized. Over the past 10 years, the pathogenesis of these autoantibodies has been partially elucidated with the discovery that the nucleic acid components of lupus autoantigens are immunostimulatory. This chapter reviews what has been learned about antinuclear antibodies and their molecular targets and pathogenesis.

DIAGNOSTIC IMPORTANCE OF ANAs

The autoantibodies produced in SLE are directed primarily against nuclear antigens, most of which are associated with nucleic acids (DNA or RNA). This is the basis for the fluorescent ANA assay, a screening test for SLE that is positive in > 95% of patients [5]. However, the specificity of a positive ANA for SLE is relatively low [6]. ANAs stain cells in a variety of different patterns, corresponding to reactivity with different subsets of nuclear (or cytoplasmic) antigens (Figures 13.1D—I, 13.2A). Some specificities are uniquely associated with SLE (Table 13.1). Anti-dsDNA antibodies, for example, are found in ~70% of SLE patients at some point during their disease, and are 95% specific for the diagnosis [5]. Anti-Sm antibodies are found in ~10—25% of lupus patients' sera depending on ethnicity [7], and also are virtually pathognomonic of SLE [8]. Antibodies to the ribosomal P0, P1, and P2 antigens [9], proliferating cell nuclear antigen (PCNA) [10], and RNA helicase A [11] are highly specific, but less sensitive, markers of the disease (Table 13.1). These "marker" autoantibodies are highly unusual in drug-induced lupus. In contrast, anti-single-stranded (ss) DNA,

TABLE 13.1 Prevalence of autoantibodies in SLE[a]

Autoantibody	Race/ethnicity			
	White (n = 789)	Black (n = 388)	Latin (n = 62)	Asian (n = 22)
dsDNA [b, c]	21	41	54	39
Sm[d]	11	41	26	12
RNP[d]	12	50	26	18
Ro60 (SS-A)[d]	19	32	37	41
La (SS-B)[d]	6	9	13	5
Su (Ago2)[d]	3	14	19	9
Ribosomal P0, P1, P2[d]	1	4	0	5
PCNA[d]	0.3	0.8	0	0
Ku[d]	0.6	7	3	0
RNA helicase A[d]	6	3	12	17

[a]*University of Florida Center for Autoimmune Disease; patients meeting ACR criteria for the classification of SLE.*
[b]*Crithidia luciliae kinetoplast staining assay. Note that anti-dsDNA antibodies often are present only transiently, but the data are from single serum samples. Estimates in the literature suggest that about 70% of SLE patients will produce anti-dsDNA at some time during their disease.*
[c]***Bold type** indicates autoantibodies that are highly specific for the diagnosis of SLE.*
[d]*Immunoprecipitation assay.*

chromatin/histone, nRNP, Ro-60 (SS-A), La (SS-B), Ro52, Ku, and Su [7], are associated with SLE as well as other systemic autoimmune diseases, but are uncommon in healthy individuals. Remarkably, most of the same antibodies are associated with murine lupus: anti-dsDNA antibodies are produced by (NZB X NZW) F1 mice, anti-dsDNA, Sm, and ribosomal P by MRL mice, and anti-Sm, dsDNA, ribosomal P, and RNA helicase A by mice with pristane-induced lupus [12, 13].

STRUCTURE OF LUPUS AUTOANTIGENS AND IMPLICATIONS FOR AUTOANTIBODY PRODUCTION

Although there are thousands of nuclear proteins, only a few are autoantigens in SLE (Table 13.1, Figure 13.2A). These are mainly RNA—protein or DNA—protein complexes comprised of multiple proteins physically associated with nucleic acid (Table 13.2, Figure 13.2). Importantly, the nucleic acid constituents of these antigens are ligands for innate immune system receptors called "Toll-like receptors" (TLRs) 3, 7, 8, and 9 localized within endosomal compartments [14]. Innate immune responses mediated by TLR7 (a receptor for ssRNA) and TLR9 (a receptor for unmethylated CpG motif in dsDNA) are receiving increasing attention as mediators of autoimmune inflammatory disorders, such as SLE (see below and Chapter 17). TLR7 recognition of U1 RNA, a component of the U1 small nuclear ribonucleoprotein (snRNP, recognized by anti-Sm/RNP autoantibodies), and TLR9 recognition

of DNA, a component of chromatin (recognized by antihistone/DNA autoantibodies) is linked to the production of type I interferon (IFN-I), which is increased in about 60% of lupus patients (see below and Table 13.2). The structures of some of the major nucleic-acid-containing autoantigens in lupus have been defined by immunoprecipitation studies followed by the analysis of the [^{35}S] methionine/cysteine-labeled proteins and [^{32}P]-labeled nucleic acid components of the complexes recognized by lupus autoantibodies (Figure 13.2A and B, respectively).

Anti-Sm and RNP Autoantibodies Recognize the U1 Ribonucleoprotein (see also Chapter 16)

The U1 snRNP, an RNA—protein autoantigen, illustrates the general principles applying to many other autoantigen/autoantibody systems (Figure 13.2). It is a macromolecular complex consisting of a group of proteins designated U1-70K, A, B'/B, C, D1/2/3, E, F, and G associated with U1 small nuclear RNA [15] (Table 13.2, Figure 13.2A, B). The U1 snRNP and the U2, U4—U6, and U5 snRNPs play critical roles in RNA splicing from heterogeneous nuclear RNA. The proteins B'/B, D, E, F, and G assemble into a stable 6S particle (the Sm core particle) reactive with anti-Sm, but not anti-RNP, antibodies. Autoantibodies to the Sm core particle are unique to SLE [4, 8]. In contrast, antibodies to the proteins A, C, and 70K, which carry RNP antigenic determinants, may be found in scleroderma, polymyositis, and other subsets of systemic autoimmune disease, as well as SLE. High levels of anti-nRNP antibodies,

TABLE 13.2   Protein and nucleic acid components of major lupus-associated autoantigens

Autoantibody	Protein component(s)	Nucleic acid component(s)	TLR recognition
Sm (U1, U2, U4-6, U5 snRNPs)	Sm core particle[a] U1: A, C, 70K[b] U2: A', B''[b] U4-U6: 150K[b] U5: eight proteins including 200 kDa doublet[b]	 U1 RNA U2 RNA U4-U6 RNA U5 RNA	TLR7
RNP	Sm core particle[a] U1: A, C, 70K[b]	U1 RNA	TLR7
Ro60 (SS-A)	60K Ro, transiently associates with 45K La	hY1, hY3, hY4, and hY5 RNA	TLR7
La (SS-B)	45K La, transiently associates with 60K Ro	RNA polymerase III precursor transcripts	TLR7
Su (Ago2)	Argonaute 2	Micro RNAs	TLR7
Ribosomal P	P0, P1, P2	Ribosomal RNA	?
dsDNA	H2A/H2B, H3, H4	Genomic DNA	TLR9
PCNA	DNA polymerase δ	Genomic DNA	TLR9
Ku/DNA-PK	Ku70, Ku80, DNA-PK$_{cs}$	Genomic DNA	TLR9
RNA helicase A	150K	Genomic DNA	TLR9

[a]Proteins B'/B, D1/D2/D3, E, F, and G form the 6S Sm core particle (recognized by anti-Sm antibodies), which is shared by all of the U snRNPs listed.
[b]Proteins unique to individual U snRNPs; the U1 snRNP contains three unique proteins, U1-A, U1-C, and U1-70K, which are recognized by anti-RNP antibodies.

without anti-Sm, are seen in mixed connective tissue disease. The U1-A and U1-70K proteins interact directly with U1 RNA via RNA recognition motifs [16]. In addition to the U1 snRNP, other snRNPs, each with a unique uridine-rich (U) RNA species, carry the Sm core particle (Table 13.2). These include the U2, U4/U6, and U5 snRNPs, as well as a number of other less abundant U snRNPs [17]. The U3 ribonucleoprotein is involved in processing of ribosomal RNA and does not carry the Sm/RNP antigenic determinants. The U1, U2, U5, and U4/6 snRNPs are present at levels ranging from 10^6 copies (U1 and U2) to 2×10^5 copies (U5 and U4/U6) per mammalian cell [17]. Along with the Sm core particle, each of these major snRNP particles carries unique protein components (Table 13.2).

Anti-Ro60 and La Autoantibodies Recognize Cytoplasmic Ribonucleoproteins (see also Chapter 15)

Anti-Ro60 (SS-A), and anti-La (SS-B) autoantibodies are associated with sicca (dry eyes and dry mouth) syndrome, and are not lupus-specific. Anti-Ro60 antibodies are seen in 60—80% of patients with primary Sjogren syndrome and in 10—50% of SLE patients (Table 13.1). Anti-La autoantibodies, which are strongly associated with anti-Ro60, are less frequent. Autoantibodies to the Ro60 (SS-A) antigen bind to cytoplasmic ribonucleoproteins containing a 60-kDa protein associated with a single molecule of human Y1, Y3, Y4, or Y5 RNA

(Figure 13.2A, C; Table 13.2). The 47-kDa La (SS-B) antigen is a termination factor for RNA polymerase III that binds transiently to precursors of the Y RNAs. The Ro60 antigen plays an important role in promoting the refolding and/or degradation of damaged (misfolded) U2 and 5S ribosomal RNA molecules [18].

Anti-DNA and Nucleosome Autoantibodies Recognize Histone—DNA Complexes (see also Chapters 14 and 16)

Genomic DNA is packaged in histone and nonhistone proteins. The DNA is wound around nucleosomes, consisting of two molecules each of H2A/H2B, H3, and H4 plus 145 base pairs of DNA (Figure 13.2C). Between the nucleosomes is "linker" DNA, which is packaged in histone H1. More complex folding leads to the formation of chromatin fibers. Autoantibodies against chromatin and nucleosomes in SLE are diverse, recognizing the H2A—H2B—DNA complex, as well as individual histones, ssDNA, and dsDNA. The LE phenomenon is thought to be mediated primarily by autoantibodies against histone H1, which are exposed when cells undergo necrotic cell death (Figure 13.1A—C) [19]. Of the multiple specificities of antichromatin and nucleosome antibodies, only anti-dsDNA antibodies are specific for the diagnosis of SLE [20]. About 70% of SLE patients have anti-dsDNA antibodies at some time in their disease course, and these autoantibodies are thought to play a role in the pathogenesis of immune-complex-mediated glomerulonephritis in SLE [21]. The recent discovery that mammalian RNA and DNA interact with TLR7/8 and TLR9, respectively, raises the possibility that immunostimulatory nucleic acids in immune complexes play a role in the pathogenesis of inflammation in SLE by triggering cytokine production.

Multicomponent Autoantigens and Epitope Spreading

Lupus autoantibodies frequently occur together as groups of interrelated specificities. Mattioli and Reichlin [22] first reported the strong association of anti-Sm antibodies with anti-RNP. Indeed, sera containing anti-Sm antibodies nearly always contain autoantibodies to the RNP (U1-70K, U1A, and U1C) antigens. Many sera contain autoantibodies to multiple polypeptides, and in one study, only one of 29 sera containing anti-nRNP or Sm antibodies recognized a single protein component of the U1 snRNP, and the majority recognized three or more [23]. Anti-RNP antibodies also are associated with autoantibodies to the U1 RNA component of U1 small ribonucleoproteins in ~40% of sera [24].

Similarly, anti-Ro (SS-A) and La (SS-B) antibodies are associated with one another [25] and with antibodies to the Y5 small RNA molecule [26], with which both antigens associate, and autoantibodies to DNA and histones (chromatin) are associated with one another. Thus, the macromolecular complexes illustrated in Figure 13.2C appear to be seen by the immune system as units. This is analogous to the immune responses to viral particles, in which T-cell responses against one protein subunit can provide help for antibody responses to other subunits. In the case of influenza, T cells specific for the nucleoprotein help B cells specific for the hemagglutinin [27]. Since B cells can act as antigen-presenting cells, a B cell specific for one component of an autoantigen may internalize an entire complex and present peptides to T cells specific for any of the constituents, potentially explaining the strong associations between autoantibodies specific for the various components of U1 ribonucleoproteins [28]. Consistent with this model of "epitope spreading", mice immunized with recombinant murine La (SS-A) develop anti-La autoantibodies, as well as anti-Ro60. Conversely, immunization with Ro60 causes the production of anti-La as well as anti-Ro antibodies. Epitope spreading also has been reported in antichromatin/nucleosome responses [29].

In summary, lupus autoantigens typically are multicomponent complexes consisting of both proteins and nucleic acids. Once an immune response is initiated, the physical association of multiple protein and nucleic acid components promotes the spreading of immunity to the other associated components, a phenomenon termed "epitope spreading". As discussed below, the nucleic acid components are immunostimulatory and can signal through TLR7 (RNA) and TLR9 (DNA), perhaps contributing to the selection of autoantigenic targets (see below).

NUCLEIC ACID COMPONENTS OF LUPUS AUTOANTIGENS STIMULATE TYPE I INTERFERON PRODUCTION

Type I Interferon and the Pathogenesis of Lupus Autoantibodies (see also Chapter 18)

Interferons, cytokines with antiviral and antiproliferative effects as well as important effects on the activation of immune effector cells, are likely to be involved in the pathogenesis of SLE. They are classified into type I and type II IFNs based on sequence homology, receptor usage, and the cellular origin. IFN-γ, the sole type II interferon, is produced primarily by T cells, NKT cells, and NK cells and interacts with a dimeric receptor (IFNGR1/IFNGR2).

The type I interferons (IFN-I), which include multiple IFN-α species, IFN-β, and others are produced by leukocytes and fibroblasts and at high levels by plasmacytoid dendritic cells (pDCs) [30]. They all signal via the type I IFN receptor (IFNAR), consisting of two chains, IFNAR1 and IFNAR2, leading to activation of the Janus kinase Jak1 and TYK2 [31]. Jak1/TYK2 then tyrosine-phosphorylate transcription factors of the signal transducer and activator of transcription (STAT) family, including STAT1 and STAT2.

Binding of IFN-I to the IFNAR increases the expression of a group of interferon-stimulated genes (ISGs) regulated by the binding of STAT1 and/or other STATs to a *cis*-acting consensus sequence termed the interferon-stimulated response element (ISRE), which is found in the promoters of all IFN-I inducible genes [31]. ISGs play a key role in the antiviral response and apoptosis. As many as 100 genes are regulated by IFN-I, and the expression of more than 20 of these genes ("interferon signature") is increased in peripheral blood mononuclear cells (PBMCs) from SLE patients, closely reflecting IFN-I levels [32, 33]. Elevated IFN-I expression is associated with severe disease, renal involvement, and autoantibodies against dsDNA and RNA-associated antigens such as Sm/nRNP, SSA/Ro, and SSB/La [34, 35]. Increased IFN-I expression also is seen in lupus skin lesions [36]. IFN-α treatment in hepatitis C infection and certain neoplastic diseases can be complicated by autoimmune phenomena, including the induction of ANA and anti-dsDNA antibodies, and there are case reports of overt SLE [37]. A positive ANA test is seen in up to 22% of patients treated with IFN-α. In addition, humans with partial trisomy of chromosome 9, which contains the type I interferon gene cluster, over-produce IFN-I and develop anti-RNP and anti-Ro60 autoantibodies [38].

IFN-I may play a pathogenic role in murine lupus models. In NZB mice, autoimmune hemolytic anemia is milder in the absence of IFN-I signaling [39] and lupus in (NZB X NZW)F1 (NZB/W) mice is accelerated by IFN-α [40]. Experimental lupus induced by the hydro-carbon 2,6,10,14-tetramethylpentadecane (TMPD, pristane) is associated with the interferon signature and requires signaling through the IFNAR [41]. Thus, over-production of IFN-I may be central to the pathogenesis of lupus and autoantibodies characteristic of SLE.

Cellular Sources of IFN-I in SLE

Although most, if not all, nucleated cells can produce IFN-I, the existence of a minor population of cells in the peripheral blood that produces large amounts of IFN-I was recognized 20 years ago and characterized more recently [30]. These cells are termed "plasmacytoid dendritic cells" (pDC) in view of their eccentrically located nucleus and prominent rough endoplasmic reticulum. In humans, they express CD123 (IL-3 receptor), and HLA-DR, as well as BDCA-2 and BDCA-4. They are CD11c[-] in humans, but CD11c[+] in mice. Increased production of IFN-I in SLE is attributed to pDCs, a view consistent with the ability of DNA and RNA containing immune complexes to stimulate IFN-I production by pDCs *in vitro* [42, 43]. TLR7 and TLR9 are expressed at high levels on human pDCs and B cells, but not conventional (myeloid) dendritic cells, monocytes, or macrophages. Paradoxically, in contrast to healthy controls pDCs are nearly absent from the peripheral blood of SLE patients [34], possibly due to the activation and migration of pDCs to tissues and/or lymphoid organs [36]. However, there is as yet no direct evidence that the pDCs found in the tissues are responsible for the interferon signature. Moreover, *in vitro* depletion of blood pDCs reduces IFN-I production by only ~40%, suggesting that other cell types are involved [44]. A subset of immature Ly6C[hi] monocytes is a major source of IFN-I in murine TMPD-lupus [45].

IFN-I Production in Murine Lupus Results from TLR Signaling (see also Chapter 17)

Several pathways mediate IFN-I production in mammalian cells [14]. TLR3, a sensor for viral dsRNA, and TLR4, a receptor for lipopolysaccharide, both stim-ulate IFN-I secretion through the adaptor protein TRIF. In contrast, TLR7/8 and TLR9 mediate IFN-I production via MyD88 in response to single-stranded (ss) RNA and unmethylated CpG DNA, respectively. Cytoplasmic receptors that recognize intracellular nucleic acids and induce IFN-I also have been described. Retinoic acid inducible gene-I (RIG-I) and melanoma differentiation associated gene-5 (MDA-5) recognize cytoplasmic RNA and trigger IFN-I by activating IPS-1 and IRF-3, whereas cytoplasmic DNA binds to an intracellular sensor triggering IFN-I production via a TBK-1/IRF-3-dependent pathway [14]. The role of each of these path-ways in the pathogenesis of experimental lupus has been examined in the TMPD-lupus model [46]. Autoan-tibody production and appearance of the interferon signature were unaffected by deficiency of TRIF, IPS-1 or TBK-1, but were abolished by MyD88 deficiency. In addition, anti-Sm/RNP/Su autoantibodies and IFN-I over-expression were unaffected by deficiency of TLR9, but abolished in mice deficient in TLR7 [46], strongly implicating endosomal TLRs in the pathogen-esis of lupus autoantibodies.

There is limited evidence linking abnormal IFN-I production in lupus to exogenous triggers, such as microbial infections, and the bulk of evidence points to endogenous triggers, such as antigens released from dying cells (Figure 13.3). DNA-containing immune complexes can stimulate IFN-I production by normal peripheral blood mononuclear cells (PBMC) [42], and in the presence FcγRIIa (CD32), internalized DNA-containing immune complexes acti-vate human dendritic cells via TLR9 [43]. In mice, chromatin–antichromatin immune complexes (which contain DNA) stimulate dendritic cell activation by engaging TLR9 and FcγRIII and stimulate autoanti-body production by antigen-specific B cells by engaging surface immunoglobulin and TLR9 [47, 48]. Similarly, the RNA components of the Sm/RNP antigen (U RNAs) and Ro/SSA antigen (Y RNAs) are TLR7 ligands and stimulate IFN-I production via an endosomal, MyD88-dependent pathway [49–51]. It is likely that endogenous immunostimulatory DNA and RNA molecules originate from apoptotic or necrotic cells [52]. Although Fc receptors are involved in the induction of IFN-I responses *in vitro*, they are dispens-able for IFN-I production *in vivo*, at least in the TMPD-lupus model [46]. Interestingly, the clearance of apoptotic cells is impaired in SLE patients (see below), providing a potential source of endogenous nucleic acids. Thus, the interaction of endogenous RNA/DNA ligands with TLRs may be responsible for the "interferon signature" in SLE, although this remains to be verified experimentally.

Signaling from Endosomal TLRs

Mammalian TLRs sense pathogen-associated molec-ular patterns (PAMPS). Whereas TLRs 1, 2, 4, and 6 are located on the cell surface, TLRs that recognize

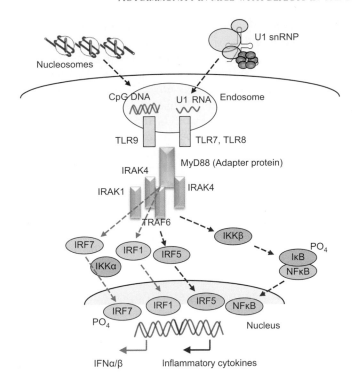

FIGURE 13.3 Nucleic acid components of lupus antigens stimulate IFNα/ß production. Nucleosomal DNA and the small RNA components of ribonucleoproteins, such as the U1 small nuclear ribonucleoprotein (snRNP) are recognized inside of endosomes by TLR9 or TLR7/TLR8, respectively. This leads to the activation of IFN-α/ß gene expression via a pathway involving the adapter protein MyD88, several kinases (IRAK1, IRAK4, TRAF6), and the transcription factors interferon regulatory factor (IRF) 7, IRF1, and IRF5. NFκB also is activated. The transcription factors IRF7/IRF1/IRF5 and NFκB activate expression of IFNα/ß and proinflammatory cytokines, such as TNFα, IL-1, and IL-6, respectively.

foreign nucleic acids (TLRs 3, 7, 8, and 9) are located mainly within the endoplasmic reticulum (ER) and/or endosomes [14] (Figure 13.3). TLR3, 7 and 9 are anchored in the endosomal membrane and detect intralumenal nucleic acids after acidification. Subsequently, MyD88 (TLR7, TLR9) or TRIF (TLR3) are recruited, initiating the signaling cascade. Microbial DNA and RNA ligands for TLR7, TLR8, and TLR9 must be released from the organism before interaction with TLRs is possible, and this requires hydrolase enzymes derived from prelysosomes or lysosomes [53]. The steps involved in activation by TLR9 are known in some detail and may be similar for TLR7 [54]. TLR9 is located in the endoplasmic reticulum (ER) of resting dendritic cells, macrophages, and other cell types and is rapidly recruited to early endosomes and subsequently to lysosomes, where it detects unmethylated CpG motifs in DNA [54, 55]. Activation of TLR9 by CpG DNA (and TLR7 by U1 RNA) requires the acidification of late endosomes/lysosomes, and signaling is abolished by inhibitors

of endosomal acidification, such as chloroquine (used to treat SLE) or bafilomycin A1 [49].

An intrinsic ER protein, UNC-93B, is required for innate responses to nucleic acids mediated by TLR3, 7, and 9 and also is involved in exogenous antigen presentation, providing further evidence that these processes depend on a direct or indirect communication between the ER and endosomes [56]. Interestingly, UNC-93B biases endosomal TLR responses toward DNA and against RNA, and UNC93B deficient DCs are hyperresponsive to TLR7 ligands, but hyporesponsive to TLR9 ligands with no change in TLR3 responses [57].

AUTOIMMUNITY IN MICE WITH DEFECTS IN THE DEGRADATION OF OR RESPONSE TO NUCLEIC ACIDS

The degradation of nucleic acid—protein autoantigens is influenced by normal pathways of programmed cell death (apoptosis), the uptake of apoptotic cells by phagocytic cells, and the degradation of cellular debris by endolysosomal proteases and nucleases. Thus, there is the potential for improperly degraded cellular nucleic acids to engage endosomal TLRs 3, 7/8, and 9.

Autoimmunity with Defects in the Fas-Fas Ligand (FasL) Pathway

Apoptosis is mediated by two major signaling pathways, one utilizing death receptors of the TNF receptor family, such as Fas, which involves cysteine proteases (caspases) that degrade proteins and DNA within the dying cell, and a mitochondrial pathway regulated by the anti-apoptotic protein Bcl-2 [58]. Mutations or deficiency of Fas (CD95), its ligand (FasL) and proteins downstream, such as caspase 10, lead to abnormal programmed cell death and are associated with autoimmunity [58]. In mice, deficiency of Fas (*lpr* mutation) or FasL (*gld* mutation) is associated with massive lymphadenopathy, expansion of double-negative (CD4⁻CD8⁻) T cells, and the production of anti-ssDNA and antichromatin autoantibodies in C57BL/6 *lpr* mice. In MRL/*lpr* or MRL/*gld* mice, there is a striking acceleration of lupus-like autoimmune disease, including anti-dsDNA autoantibody production and immune-complex-mediated glomerulonephritis [58]. Unexpectedly, in humans, Fas, FasL, or caspase 10 deficiency also promotes autoimmunity, but usually not lupus. These individuals develop lymphadenopathy, expansion of double-negative T cells, and autoimmune cytopenias (Coombs positive autoimmune hemolytic anemia, autoimmune thrombocytopenia), but rarely develop antinuclear antibodies or clinical manifestations of lupus. Consistent with defects in which the degradation of DNA is impaired (see below), anti-DNA/chromatin autoantibody production is a prominent feature in *lpr* and *gld* mice (both C57BL/6 and MRL background). It is possible that the failure of caspase 3/7 activation downstream of Fas results in the inability to activate caspase-activated DNase (CAD), a key nuclease involved in internucleosomal DNA cleavage [59], leading to engagement of TLR9 by CpG DNA sequences following the phagocytosis of apoptotic cells.

Autoimmunity with Defects in Endosomal DNA Degradation

Along with proteases, acid nucleases (DNases and RNases) are present within phagolysosomes, and complete the process of degrading nucleic acids initiated by CAD. Mice with a mutation in the lysosomal nuclease DNase II, which degrades DNA from apoptotic cells, die *in utero* due to over-production of IFN-I [60]. In contrast, conditional knockouts of DNase II develop rheumatoid-arthritis-like inflammatory arthritis and anti-DNA auto-antibodies [61]. Although suggestive of the possibility that the incompletely degraded DNA might stimulate IFN-I production via a TLR, cytokine production is TLR9-independent and the mechanism is unknown.

Autoimmunity with Defects in Degradation of Misfolded RNA

Although lysosomal ribonucleases exist, they are poorly studied and knockout mice have not been generated. However, Ro60 knockout mice are deficient in the recognition of misfolded intracellular RNAs, and develop a lupus-like syndrome (antinucleosomal and antiribosomal autoantibodies and glomerulonephritis), suggesting that by promoting the degradation of misfolded and potentially immunostimulatory RNA, Ro60 may protect against the induction of autoimmunity [62]. It is not known whether these mice over-produce IFN-I.

Autoimmunity with Defects in the Clearance of Apoptotic Cells by Phagocytes

Phagocytes express receptors mediating the uptake of apoptotic cells, including complement receptors, CD14, CD36, and scavenger receptor A. Generally, uptake of apoptotic cellular debris by phagocytes is non-inflammatory, whereas the uptake of cells undergoing necrotic death is pro-inflammatory [63]. The C1q and amyloid P molecules are involved in the clearing apoptotic cells and mice deficient in either protein develop lupus-like disease [63]. C57BL/6 mice deficient in the tyrosine kinase MER have a selective defect in the phagocytosis and clearance of apoptotic cells and develop anti-DNA and antichromatin autoantibodies as well as rheumatoid factor [64]. These mice develop only mild renal mesangial changes and proteinuria, and have a normal lifespan, but on a 129Sv background renal disease is more severe. The MER protein associated with GAS6, which can interact with phosphatidylserine exposed on the membrane of cells undergoing apoptosis. TYRO3, AXL, and MER constitute a family tyrosine kinases (TAM receptors) involved in the recognition of apoptotic cells and the suppression of inflammatory responses [65]. Like MER-deficient mice, TYRO3 and AXL knockout mice also develop lupus-like autoimmune disease [63, 65] but anti-dsDNA autoantibody production is most impressive in triple (TYRO3/AXL/MER) knockout mice [66]. Nevertheless, delayed clearance of apoptotic cells, by itself, is insufficient to induce autoantibodies, since CD14 and mannose-binding lectin-deficient mice exhibit defective clearance but do not develop autoimmunity [63].

Autoimmunity with Over-Expression of TLR7 (see also Chapter 17)

The ability of TMPD to induce lupus-like autoantibodies and disease is abolished in TLR7-deficient mice [46] and in MRL mice, both TLR7 and TLR9 have been implicated in autoantibody production [67]. Conversely, over-expression of TLR7 promotes lupus. The Yaa (Y-linked autoimmune accelerator) mutation, first identified in male offspring of B6 X SB/Le cross (BXSB mice), is a 4-megabase duplication of the region on chromosome X containing TLR7, which has been translocated to the Y chromosome [68]. Consequently, male BXSB mice have two transcriptionally active copies of the TLR7 gene, whereas females have only one (due to random X-inactivation). The Yaa mutation does not induce lupus on a normal (C57BL/6) background, but greatly accelerates it on the BXSB background as well as in other autoimmune-prone strains, such as NZB/W and B6 FcγRIIb-/- mice, an effect mediated by TLR7 [69]. Male BXSB mice have high levels of anti-dsDNA autoantibodies and an aggressive form of lupus nephritis, abnormalities not seen in BXSB females. TLR7 also is required for the production of autoantibodies against nucleic acids in C57BL/6 mice transgenic for the immunoglobulin heavy and light chains encoding an antibody reactive with RNA, ssDNA, and nucleosomes [70].

Taken together, these animal models suggest that the production of autoantibodies characteristic of SLE, such as anti-Sm/RNP and anti-DNA, can result from the inability of phagocytes to properly dispose of nucleic-acid–protein complexes or the enhanced expression of endosomal TLRs, notably TLR7, capable of recognizing cellular nucleic acid.

ROLE OF T CELLS IN AUTOANTIBODY PRODUCTION (SEE ALSO CHAPTER 7)

The production of autoantibodies can be dependent or independent of help from CD4$^+$ T cells, and there is evidence for both mechanisms in SLE. T-cell-independent activation of antibody production is stimulated by antigens with repeating epitopes, such as pneumococcal polysaccharide, by TLR ligands, and by B-cell-activating factor of the tumor necrosis factor family (BAFF). T-cell-independent responses generally lead to rapid antibody production by short-lived plasma cells that develop extrafollicularly and produce low-affinity IgM antibodies, although switched isotypes (e.g. IgG) also are produced.

In contrast, T-cell-dependent autoantibody production implies the involvement of secondary lymphoid organs, such as the spleen or lymph nodes, which provide an optimal milieu for interactions between T cells, B cells, and antigen-presenting cells (APCs). Antigen-activated B cells express the chemokine receptor CCR7 and are attracted to the T-cell zones of secondary lymphoid tissues by the chemokine CCL21 (BLC) [71]. Here they receive help from CD4$^+$ T cells after which they may enter two pathways: (1) migration into follicles with the formation of germinal centers, memory B cells, and long-lived antibody-secreting plasma cells ("follicular pathway") and (2) migration to splenic bridging channels or medullary cords with formation of extrafollicular foci of short-lived plasma cells ("extrafollicular pathway"). In the follicular pathway, B cells (centroblasts) proliferate within the germinal centers and then develop into centrocytes, which upon contacting antigen associated with follicular dendritic cells, can activate CD4$^+$ follicular helper cells (T$_{FH}$ cells). This critical T-cell subset provides signals for the development of memory B cells and long-lived plasma cells, a hallmark of the germinal center reaction. Other features of the follicular (germinal center) pathway include the requirement for CD40–CD40L interactions and the induction of somatic hypermutation of immunoglobulin hypervariable regions and class switch recombination from IgM to IgG, IgA, or IgE [72]. In addition to T$_{FH}$, other subsets of CD4$^+$ T cells may promote antibody production, including the T$_{H}$1, T$_{H}$2, and T$_{H}$17 subsets.

Although low-affinity, polyreactive anti-ssDNA autoantibodies produced by B-1 cells bear germline Ig variable region sequences and may be relatively independent of T cell help, much autoantibody production in SLE may be T-cell-dependent [73]. Lupus autoantibodies are skewed toward the T-cell (IFN-γ/ IFN-α/β)-dependent isotypes IgG2a in murine lupus and IgG1 in humans [74]; autoantibody production is decreased in MRL/*lpr* or NZB/W mice treated with anti-CD4 antibodies or CTLA4Ig [75, 76]; and the induction of lupus autoantibodies by TMPD is abolished in T-cell-deficient mice [77]. Serological memory is maintained, at least in part, by long-lived plasma cells [78] thought to be derived mainly from post-germinal center B cells. Additional evidence is the large number of somatic mutations in anti-DNA autoantibody V-regions [79]. Many of the anti-DNA antibodies from MRL/*lpr* mice are members of the same expanded clones [79]. The high frequency of replacement versus silent mutations and their non-random distribution argues that anti-DNA antibodies are selected on the basis of receptor specificity. Thus, the analysis of autoantibody V regions strongly suggests that T cells are involved.

Helper T-Cell Subsets in Autoimmunity

Upon activation by antigen and an appropriate APC-derived co-stimulatory signal, naïve T cells can differentiate along several pathways (Figure 13.4). In 1986, two subsets of CD4$^+$ T cells, designated T$_{H}$1 and T$_{H}$2, were defined in mice [80]. Subsequently, other lineages were identified, including the T$_{H}$17, T$_{FH}$, and regulatory T cell (T$_{reg}$) subsets [81], most of which have been implicated in autoimmunity. Transcriptional programs responsible for establishing these subsets are not stable and the different subsets can interconvert [81].

T$_{H}$1 cells

T$_{H}$1 cell differentiation is promoted by IL-12, which induces expression of the transcription factor T-bet, resulting in IFN-γ and tumor necrosis factor (TNF) β production (Figure 13.4). Although IFN-γ has a major role in macrophage activation, it (along with IFN-α/β) also promotes immunoglobulin isotype switching to IgG1 (in humans) and IgG2a (in mice). As most autoantibodies in human lupus are IgG1 (IgG2a in mice), it is thought that T$_{H}$1 cells play a significant role in generating lupus autoantibodies. Consistent with that possibility, deficiency of IFN-γ decreases autoantibody production (especially of the IgG2a subclass) in mouse models [82–84] and both T$_{H}$1 predominance and activation of IFN-γ signaling have been reported in human SLE [85, 86].

T$_{H}$2 cells

T$_{H}$2 cell differentiation is promoted by IL-4, which induces expression of the transcription factors GATA-3 and STAT6, resulting in IL-4, IL-5, and IL-13 production (Figure 13.4). These cytokines are involved in immune responses to helminths and are instrumental in

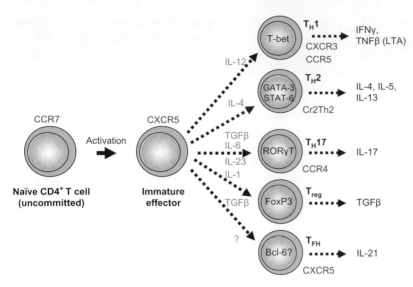

FIGURE 13.4 Development of CD4+ helper T cells implicated in autoantibody production. Upon activation, naïve CD4+ T cells can develop along several pathways in response to polarizing cytokines that induce the expression of specific transcription factors: T-bet in T_H1 cells, GATA-3/STAT-6 in T_H2 cells, RORγT in T_H17 cells, FoxP3 in regulatory T cells (T_{reg}), and Bcl-6 in follicular helper cells (T_{FH}). The individual subsets of mature helper T cells each produce characteristic cytokines that may influence the development of autoimmunity and express chemokine receptors that facilitate homing to specific locations.

promoting antibody responses, especially IgE and also IgG1 (in mice). Despite its importance for immunoglobulin production, there is only limited evidence that IL-4 plays a major role in the pathogenesis of lupus autoantibodies and in the TMPD and BXSB lupus models, IL-4 deficiency actually enhances autoantibody production, presumably by increasing the production of IFNγ [84, 87].

T_H17 cells

T_H17 cell differentiation is promoted by TGFβ, IL-6, and IL-23, which induce expression of the transcription factor RORγT, resulting in IL-17 production (Figure 13.4). T_H17 cells may be important mediators of a number of autoimmune diseases, including rheumatoid arthritis in humans and experimental encephalomyelitis in mice. However, their role in the pathogenesis of lupus is just beginning to come into focus [88]. Elevated IL-17 levels in lupus patients correlate with disease activity and levels of anti-DNA antibodies [89, 90]. In the BXD2 mouse model, development of T_H17 cells is enhanced and deficiency of the IL-17 receptor causes defective generation of autoantibodies against DNA and histone [91]. In NZM2328 mice doubly deficient in tumor necrosis factor receptor (TNFR) 1 and 2, anti-dsDNA autoantibody production is enhanced in association with the production of large numbers of T_H17 memory T cells [92]. IL-17 may act synergistically with BAFF to promote the survival, proliferation, and terminal differentiation of B cells in SLE patients [90].

T_{FH} cells

The cytokines responsible for T_{FH} cell differentiation and the key transcription factor(s) responsible for IL-21 production have not been identified (Figure 13.4). However, the transcription factor Bcl-6, expressed primarily in germinal center B and T_{FH} cells, may be

involved [71]. *Sanroque* mutant mice develop lupus-like autoimmunity with anti-DNA autoantibodies, autoimmune thrombocytopenia, lymphoid hyperplasia, and glomerulonephritis [93]. These mice are deficient in a RING-type ubiquitin ligase (Roquin), causing spontaneous germinal center formation and dysregulation of self-reactive T_{FH} cells expressing high levels of ICOS and IL-21 [93, 94]. Roquin deficiency leads to the production of anti-DNA antibodies reactive in the *Crithidia luciliae* kinetoplast staining assay, but atypical in the absence of IgG2a anti-DNA. Roquin may keep extrafollicular T_{FH} cell development in check, preventing their accumulation. An over-abundance of these cells may provide inappropriate help to autoreactive B cells arising in germinal centers.

Regulatory T cells

Regulatory CD4+ T cell (T_{reg}) differentiation is promoted by TGFβ (Figure 13.4). T_{reg} cells express the surface marker CD25 (IL-2Rα) and the transcription factor Foxp3 [95], which controls the development and function of CD25+CD4+ T_{reg}. CD4+ T_{reg} also are diminished in number and function in SLE, rheumatoid arthritis, and multiple sclerosis [96, 97]. *FoxP3* mutations result in deficiency of CD25+CD4+ T_{reg}, causing autoimmune/inflammatory disease in mice and humans [95]. Scurfy mice with a FoxP3 mutation develop a fatal multi-organ inflammatory disorder affecting skin, eyes, small intestine, pancreas, and other organs [98]. In humans, FoxP3 mutations result in IPEX (immunodysregulation, polyendocrinopathy, enteropathy, X-linked) syndrome, characterized by skin lesions, autoimmune enteropathy, type I diabetes, thyroiditis, chronic inflammation with cytokine production, and autoimmune cytopenias, frequently resulting in death by age 2. However, although autoimmunity is associated with

FoxP3 mutations in both humans and mice, antinuclear antibodies have not been reported. Nevertheless, T_{reg} cells suppress anti-DNA autoantibody production in the NZB/NZW model *in vitro* [99].

Autoantibody Production is Affected by Peripheral T-cell Activation

Autoreactive T cells are censored through clonal deletion and peripheral inactivation (anergy). In the case of T-cell-dependent immune responses, T-cell tolerance is often a more important mechanism than B-cell tolerance for preventing autoantibody production. The main mechanism for removing autoreactive T cells with high-affinity antigen receptors is clonal deletion in the thymus, whereas tolerance to antigens not expressed in the thymus is maintained by peripheral T-cell inactivation.

Importance of co-stimulatory signals

Ligation of the T-cell antigen receptor (signal 1) is insufficient, by itself, to activate naïve T cells to proliferate or differentiate and a second co-stimulatory signal (signal 2) delivered by APCs is required [100]. In the absence of a co-stimulation, T-cell receptor occupancy by antigen results in unresponsiveness (anergy) or apoptosis. The best-characterized co-stimulatory molecules are B7-1 (CD80) and B7-2 (CD86), which are expressed on the surface of macrophages, dendritic cells, activated B cells, and other APCs [100]. Delivery of signal 2 along with signal 1 to a naïve T cell alters gene expression, increasing expression of the IL-2 receptor α-chain (CD25), IL-2, and CD40 ligand (CD154). In naïve T cells, the constitutively expressed CD28 molecule is the only receptor for CD80/CD86, but after cross-linking CD28, expression of a higher-affinity receptor, CTLA-4 (CD152), is induced. CTLA-4 inhibits T-cell activation following T-cell receptor ligation [101]. Due to its 20-fold higher affinity for CD80/CD86, CTLA-4 inhibits IL-2 production, limiting the proliferation of activated T cells. It plays a critical role in peripheral tolerance and is expressed on the $CD4^+CD25^+FoxP3^+$ T_{reg} subset [101].

Additional CD28-like proteins have been identified, including the inducible co-stimulator ICOS, which is a poor inducer of IL-2 but plays a key role in affinity maturation, CD40-mediated class switching, and memory B-cell responses (see T_{FH} cells, above) [102]. Little ICOS is expressed on naïve T cells, whereas it is highly up-regulated on effector/memory T cells, especially T_{FH} cells. ICOS binds to a ligand (ICOSL) expressed on a variety of APCs as well as non-hematopoietic cell types, such as endothelial cells. Humans and mice deficient in ICOS or ICOSL have greatly diminished germinal center responses and are unable to form memory B cells [102].

Therapy directed at co-stimulation

The regulation of co-stimulatory activity influences susceptibility to autoimmune diseases as illustrated by the development of lupus in Roquin mice, which have increased ICOS expression on their T cells (see above) [94].

The B7-CD28 co-stimulatory axis has been exploited therapeutically for treating autoimmune diseases, including lupus. CTLA4-Ig, a fusion of the CTLA4 molecule to the Fc portion of immunoglobulin, is a potent inhibitor of co-stimulation and thus T-cell activation. In NZB/W mice, treatment with CTLA4-Ig decreases IgG (but not IgM) anti-dsDNA antibody production and delays the onset of nephritis while prolonging survival [76, 103]. Similarly, in MRL/*lpr* mice, CTLA4-Ig treatment dramatically improves survival, decreases the severity of nephritis, and abolishes IgG anti-dsDNA autoantibody production [104]. In male BXSB mice, treatment with CTLA4-Ig alone does not significantly reduce anti-dsDNA antibody production, but in combination with anti-CD134 (OX40L) monoclonal antibodies, which recognize a surface marker on $CD4^+$ T cells, a reduction of anti-dsDNA antibody production is seen *in vitro* [105]. However, this is associated with a substantial reduction of IL-6 production, raising the possibility that the effect of CTLA4-Ig in this *in vitro* system may be mediated primarily via effects of IL-6 on plasma cell development (see below).

Activation-induced T-cell death

Most T cells activated in response to foreign antigens ultimately undergo cell death mediated by one of at least two pathways: (1) death receptor-driven apoptosis involving Fas and the TNF receptors, and (2) a Fas-independent mitochondrial pathway involving the Bcl-2 family of proteins [106]. Immunological adjuvants block apoptosis mediated by a third, poorly defined, pathway. Type I interferons, which mediate the adjuvant effect, can rescue T cells from activation-induced death [107, 108], and SLE T cells are resistant to apoptosis [109]. In addition, in mice deficient in IRF-4 binding protein (IBP), T cells are resistant to apoptosis with an accumulation of effector/memory T cells and development of lupus-like autoimmunity [110]. However, although IBP-deficient mice develop anti-DNA antibodies, the overwhelming predominance of IgG1 instead of IgG2a is atypical for lupus.

Summary

There is strong evidence that T cells are involved in autoantibody production, including indirect evidence from characteristics of the autoantibody response (e.g. switched isotypes, evidence of affinity maturation), the inability of TMPD to induce autoantibodies in T-cell-deficient mice, the likely involvement of T_H17 and T_{FH} cells in the induction of lupus autoantibodies, and the diminished autoantibody production following CTLA4-Ig treatment. However, there also is evidence that autoantibodies can arise independently of T cells via extrafollicular activation of autoreactive B cells [111].

ROLE OF B LYMPHOCYTES AND PLASMA CELLS IN AUTOANTIBODY PRODUCTION (SEE ALSO CHAPTER 8)

The two major subsets of B cells, B-1 and B-2, are defined by their surface phenotypes and anatomical distribution [112]. The B-1 subset is enriched in the peritoneal and pleural cavities, and has an IgM^{high}, IgD^{low}, $B220^{high}$, $CD23^{neg}$ phenotype. B-1a cells express moderate levels of CD5, whereas B1-b cells are $CD5^{neg}$. In contrast, the B-2 (conventional) subset is IgM^{low}, IgD^{high}, $B220^{high}$, $CD23^{high}$, $CD5^{neg}$. Most studies suggest that B-1 cells produce polyreactive antibodies, exhibit only limited somatic mutation, develop independently of T cells, and are prone to make low-affinity autoantibodies against repetitive epitopes such as pneumococcal polysaccharide (TI-2 antigens) [112]. In contrast, conventional (B-2) B cells require cognate T cell help and produce high-affinity, somatically mutated antibodies.

B-Cell Activation and Tolerance

Although B-cell responses to repetitive epitopes are often T-cell-independent, the activation of B cells recognizing low-valency antigens generally requires two signals, one from the antigen receptor and another co-stimulatory signal from the interaction of CD40 on the B cell with CD40 ligand (CD40L) on an activated T cell.

It has been estimated that as many as half of immature B cells exhibit autoreactivity [113], emphasizing the importance of mechanisms for eliminating or rendering them anergic. Autoreactive B cells are censored by a variety of mechanisms that depend on the presence or absence of T-cell help, the form of the antigen, and the type of B cell. There are differences in the susceptibility of mature vs. immature B cells as well as between B cells responding to T-cell-independent antigens (B-1 and marginal zone subsets) vs. T-cell-dependent antigens (B-2 subset). Deletion (apoptosis), anergy, receptor editing, and immunological ignorance all play a role in determining whether B cells become activated or are tolerized [114].

Extensive studies of B-cell tolerance to DNA have been carried out in transgenic mouse models [115]. High-affinity anti-dsDNA antibodies are deleted at the pre-B to immature B transitional stage [115] due to the loss of critical homing molecules, such as CD62L, which are important for entry into secondary lymphoid tissues, decreased expression of BAFF, which promotes B-cell survival, and persistent expression of recombination activating genes (RAG) 1 and 2 [114]. Continued expression of RAG1/RAG2 allows the autoreactive B-cell receptors to be edited by replacement with a different immunoglobulin light chain. Autoreactive B cells that are neither deleted nor modified by receptor editing generally die, but can be rescued by engagement of TLRs on the B-cell surface or by increased BAFF expression [116, 117].

The germinal center reaction leads to further immunoglobulin diversification through somatic hypermutation (SHM). This can be particularly dangerous, as the potential to generate high-affinity self-reactive immunoglobulins exists and the B cells producing them enter the long-lived plasma cell and memory compartments. The mechanisms by which these cells are regulated may involve deletion due to the lack of T-cell help, competition for BAFF, CD40L, IL-21, ICOS or other co-factors, or inhibitory $Fc\gamma$ receptor ($Fc\gamma RIIb$) engagement, which inhibits the accumulation of IgG^+ autoreactive plasma cells [114, 118].

The result is that in most transgenic mouse models, anti-DNA antibody-producing B cells are efficiently censored on non-autoimmune backgrounds [115]. However, in BALB/c mice, anti-DNA B cells that are not deleted can be induced to undergo differentiation into autoantibody-secreting plasma cells when provided with help from $CD4^+$ helper cells or in some cases by TLR7/9 signaling [70]. Conversely, the activation of anti-DNA B cells can be modulated by $CD4^+CD25^+$ regulatory T cells [119].

In contrast to the relatively tight regulation of autoreactive B cells in BALB/c mice, B cells bearing anti-DNA autoantibody transgenes are not effectively censored on autoimmune backgrounds [115]. For example, follicular exclusion appears to be an important tolerization mechanism for anti-DNA B cells in MRL/lpr mice [120]. Recent data suggest that numerous checkpoints involved in limiting autoantibody production can be defective in autoimmune-prone mice.

Maturation of Autoreactive B-Cells

B cell development can be broadly divided into antigen-independent (bone marrow) and antigen-dependent phases. After high-avidity autoreactive clones are negatively selected in the bone marrow, the surviving B cells are exported to the periphery and there is a divergence of the B-1 and B-2 (conventional) B-cell lineages. Autoreactive B cells can develop in both pathways. However, resting autoreactive B cells do not secrete immunoglobulin, making the regulation of B-cell differentiation into antibody-secreting plasma cells a key determinant of autoantibody production.

Autoantibody production by B-1 cells

B-1 cells develop without T-cell help into plasma cells. The protein tyrosine phosphatase SHP-1 regulates cell numbers in the B-1a compartment. SHP-1-deficient (motheaten viable) mice have a massive expansion of $CD5^+$ B cells and selective IgM hypergammaglobulinemia. In addition to autoantibodies and glomerular immune complex deposition, motheaten mice have a variety of hematological disorders including the accumulation of macrophages and granulocytes in the lungs, causing death at ~9 weeks of age [121] . Their autoimmune syndrome differs significantly from SLE however. Although anti-DNA antibodies reactive on solid phase assays are produced, they do not give nuclear staining on immunofluorescence [121]. Their cytoplasmic staining pattern suggests that they differ in key respects from the typical anti-dsDNA antibodies of SLE. Most evidence suggests that the prototypical lupus autoantibodies, e.g. IgG anti-Sm, are not likely to be derived from B-1 cells.

Autoantibody production by conventional (B-2) cells

Upon entering the spleen, conventional B cells mature through the transitional 1 and 2 (T1 and T2) stages (Figure 13.5A) [122]. Some T2 cells home to the splenic marginal zone to become marginal zone B cells, a population characterized by germline immunoglobulin receptors, predominantly IgM isotype, and specificity for TI-2 antigens. However, most T2 cells become recirculating naïve follicular B cells, which undergo further maturation upon receiving cognate T-cell help. Marginal zone B cells, follicular B cells, germinal center B cells, and memory B cells all can develop into plasma cells. In the case of B cells responding to T-cell-dependent antigens, there are two rounds of plasma cell differentiation and antibody production, which take place in different locations within secondary lymphoid tissues (Figure 13.5A). B cells enter the spleen via central arterioles, which are surrounded by T cells (the "periarteriolar

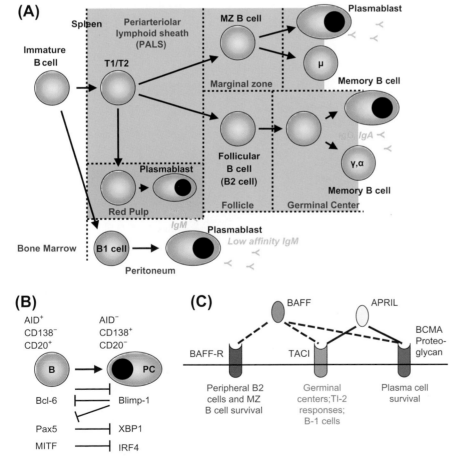

FIGURE 13.5 B-cell development. (*A*) Immature B cells can develop into plasmablasts/plasma cells via several pathways. B-1 cells colonize the peritoneal cavity and can develop there into plasmablasts secreting primarily low-affinity IgM. In contrast, conventional (B-2) B cells can develop into marginal zone B cells that secrete IgM antibodies reactive with T-cell-independent antigens. B-2 cells entering the T-cell zone (the periarteriolar lymphoid sheath) undergo extrafollicular differentiation into short-lived IgG/IgM-secreting plasma cells in the red pulp and also can enter the B-cell follicles, where they may receive T-cell help and develop into proliferating germinal center B cells (centroblasts). The latter can undergo somatic hypermutation and class switching, followed by differentiation into memory B cells and plasma cells (some of them long-lived) that secrete class-switched isotypes such as IgG and IgA. (*B*) Key transcription factors in plasma cell development. The germinal center B-cell transcriptional program is maintained by transcription factors Bxl-6, Pax5, and MITF, which inhibit plasma cell development. Plasma cells develop following the expression of transcription factors Blimp-1, XBP1, and IRF4, which are negatively regulated by Bcl-6, Pax5, and MITF. Conversely, Blimp-1 negatively regulates transcription factors that maintain a B-cell phenotype. The differential expression of characteristic surface markers (CD138, CD20) and intracellular proteins (activation-induced cytidine deaminase, AID) by B cells vs. plasma cells is shown. (*C*) BAFF/APRIL and their receptors. The TNF-like ligands BAFF

and APRIL interact with three receptors, BAFF-R, TACI, and BCMA, as indicated. These interactions regulate the survival of various subsets of B cells, as shown below.

lymphoid sheath"). Antigen-specific B cells receive help from CD4$^+$ T helper cells at the interface of the B- and T-cell zones and after 3–4 days, the activated B cells form extrafollicular foci within the red pulp, leading to the development of short-lived plasma cells (lifespan ~2 weeks). These cells secrete mainly IgM but also can undergo isotype switching. However, their immunoglobulin receptors are unmutated. Subsequently, some of the activated B cells enter primary B-cell follicles and form germinal centers, where they proliferate and undergo somatic hypermutation of their immunoglobulin genes and isotype switching. In general, high-affinity B cells or B cells reactive with repetitive epitopes develop extrafollicularly into plasma cells, whereas low-affinity B cells are directed to germinal centers where they undergo affinity maturation [123]. The germinal center reaction leads to the production of isotype-switched memory B cells and plasma cells (Figure 13.5A), many of which return home to the bone marrow and persist for many years as long-lived plasma cells [78].

Regulation of Plasma Cell Differentiation

The gene expression programs of B cells and plasma cells are distinct, and the decision to remain a B cell or undergo terminal differentiation is regulated by several key transcription factors (Figure 13.5B) [122]. The germinal center phenotype is maintained by PAX5, Bcl-6 and other factors whereas Blimp-1, XBP1, and IRF4 expression characterize plasma cell development. Bcl-6 represses Blimp-1 expression and vice versa [122]. Importantly, IL-21 (the product of T$_{FH}$ cells) strongly induces Blimp-1 and is expressed at high levels in male BXSB mice. Cell surface markers characteristic of B cells, such as CD20, are lost as terminal differentiation progresses, whereas new surface markers characteristic of plasma cells, such as CD138, are expressed (Figure 13.5B).

As the serum half-life of IgG is 1–2 months, the maintenance of antibody levels following immunization must be maintained by continuous antibody secretion. The differentiation of memory B cells into short-lived plasma cells and the generation of long-lived plasma cells are responsible for this "serological memory" [122]. In NZB/W mice, 60% of anti-DNA autoantibody-secreting cells are short-lived and 40% long-lived [124]. Long-lived plasma cells, which can maintain antibody secretion for many years, cannot be eliminated by therapy directed at B-cell-specific markers, such as CD20 (see below and Figure 13.6). The signals regulating whether a plasma cell is

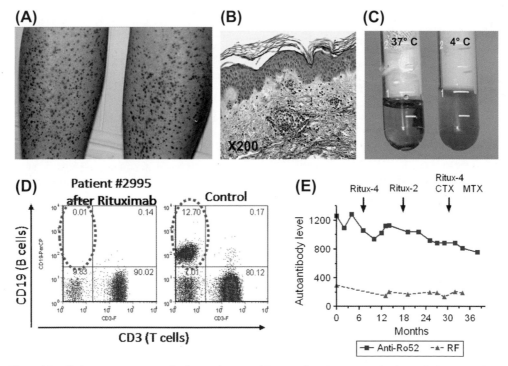

FIGURE 13.6 Effect of B-cell therapy on autoantibody production. (*A*) Vasculitic lesions on the legs of a 21-year-old woman with Sjogren syndrome complicated by cryoglobulinemic vasculitis. (*B*) Biopsy of one of the skin lesions showing small-vessel vasculitis. (*C*) Serum from the patient stored at 37°C and 4°C, illustrating the formation of cryoprecipitate. (*D*) Depletion of the patient's CD19$^+$ B cells following rituximab (anti-CD20) monoclonal antibody therapy. (*E*) Effect of treatment (Ritux, rituximab; CTX, cyclophosphamide; MTX, methotrexate) on the levels of anti-Ro52 and rheumatoid factor (RF) over a period of 36 months.

short-lived or enters the long-lived compartment are just beginning to come into focus [125]. One factor is B-cell receptor affinity: high-affinity B cells tend to develop into short-lived plasma cells, whereas lower-affinity B cells enter germinal centers and are more likely to become long-lived plasma cells and memory cells. The maintenance of long-lived plasma cells requires continuous expression of Blimp-1 as well as factors provided by so-called "survival niches", especially IL-6 and BAFF produced by stromal cells [126, 127]. In the bone marrow, interactions of BAFF with the BCMA receptor (Figure 13.5C) promote the survival of long-lived plasma cells [128].

The persistence of long-lived plasma cells, including anti-DNA plasma cells [118], is regulated, in part, by apoptosis mediated by cross-linking of the inhibitory Fc receptor $Fc\gamma RIIb$, which is expressed on plasma cells [129]. Plasma cells accumulate in the bone marrow of $Fc\gamma RIIb$-deficient mice [129], which also develop auto-antibodies and glomerulonephritis [130]. Interestingly, plasma cells in NZB and MRL mice do not express significant $Fc\gamma RIIb$ and therefore are resistant to apoptosis.

The two mechanisms for maintaining serological memory (differentiation of memory B cells into short-lived plasma cells and bone marrow long-lived plasma cells) may have important implications for maintaining autoantibody production. Anti-dsDNA autoantibody production is often transient and may be associated with disease activity, consistent with production by short-lived plasma cells. Other autoantibodies, particularly anti-Sm, RNP, Ro, and La, tend to persist at relatively stable levels for many years consistent with their production by long-lived plasma cells.

Memory B-Cells

Besides plasma cells, the germinal center reaction generates long-lived memory B cells, which bear the surface marker CD27 in humans and appear to be maintained without further antigenic stimulation. The B-cell signaling threshold helps regulate the choice between extrafollicular plasma cell and germinal center development as well as whether post-germinal center B cells will become plasma cells or memory B cells [123, 125]. In the absence of complement receptors (CD21/CD35), signaling through the B-cell receptor is impaired, decreasing the generation of plasma cells while having little effect on the generation of memory B cells [131]. Signaling through CD40 also is important, as increased CD40 signaling promotes the development of short-lived plasma cells instead of long-lived plasma cells and memory B cells [125]. Although controversial, memory B cells may play a key role in serological

memory due to polyclonal activation by TLR7/9 ligands and the development of short-lived plasma cells [132].

Extrafollicular Generation of Autoantibodies

Some autoantibodies, such as anti-DNA and rheumatoid factors, can develop extrafollicularly without T cells [111, 133]. Despite their T-cell independence, they can exhibit significant somatic hypermutation and class switching, features generally thought to typify germinal center reactions [111]. The extrafollicular activation and differentiation of autoreactive cells is MyD88- and TLR7/9-dependent, and is thought to generate primarily short-lived plasma cells [111, 133]. It is unclear whether or not extrafollicular B-cell responses generate memory B cells or long-lived plasma cells and there remains controversy regarding the relative importance of the extrafollicular vs. germinal center pathway in generating lupus autoantibodies.

Effect of B-Cell Therapy on Autoantibody Levels (see also Chapter 59)

Evidence that both long- and short-lived plasma cells are involved in autoantibody production is provided by the experience using pan-B-cell monoclonal antibodies, such as anti-CD20 (rituximab), to treat lupus [134, 135]. In some patients, anti-CD20 therapy dramatically reduces autoantibody levels, but in others there is little effect, probably reflecting the fact that plasma cells are CD20$^-$. Figure 13.6 shows the effect of B-cell depletion therapy in a 21-year-old woman with cryoglobulinemic vasculitis due to Sjogren syndrome, who exhibited purpuric lesions of the legs (Figure 13.6A), a skin biopsy revealing small vessel vasculitis (Figure 13.6B), and the presence of serum cryoglobulins (Figure 13.6C). Treatment with rituximab profoundly depleted her CD19$^+$ B cells, while having no effect on CD3$^+$ T cells (Figure 13.6D). Concomitantly, the vasculitic lesions disappeared, but they recurred approximately a year later and again responded to rituximab. Lesions recurred again after another year, and the patient again was treated successfully with rituximab plus cyclophosphamide; methotrexate was added to prevent further recurrences. Despite the dramatic response of the patient's skin lesions and complete depletion of circulating B cells on three occasions, B-cell therapy had little effect on serum levels of anti-Ro52 autoantibodies or rheumatoid factor (Figure 13.6E), consistent with the possibility that these autoantibodies were produced mainly by long-lived plasma cells. Although therapy directed at pan-B-cell surface antigens is a potentially promising new therapy for autoimmune disease, autoantibody production (Figure 13.6) and clinical activity are not necessarily improved.

Role of BAFF/APRIL in B-Cell Maturation and Autoantibody Production

B-cell-activating factor of the TNF family (BAFF, also known as BLyS) and the proliferation-inducing ligand APRIL play an important role in the survival and activation of B cells and plasma cells [136]. These two TNF-family ligands interact differentially with three receptors: BAFF-R (also known as BR-3), TACI, and BCMA (Figure 13.5C). BAFF can bind all three receptors (affinity for BCMA is lower), whereas APRIL binds TACI and BCMA (and heparan sulfate proteoglycans), but not BAFF-R. BAFF and APRIL are expressed by neutrophils and antigen-presenting cells, and are inducible by IFN-I as well as other cytokines [136]. The receptors are found on B-cell subsets: BAFF-R on maturing B cells, TACI on all peripheral B cells and at high levels on B1 and marginal zone B cells, and BCMA on plasma cells. TACI expression is induced by TLR7/9 ligands [137]. BAFF/BAFF-R interactions regulate the differentiation and survival of B-2 and MZ B cells, APRIL regulates CD40-independent class switching and plasma cell survival, and TACI regulates TI-2 responses. TACI-deficient mice develop a lupus-like syndrome characterized by proteinuria, reduced survival, and IgG1 anti-dsDNA autoantibodies [138], but it is unclear whether the more characteristic IgG2a anti-DNA antibodies develop. In contrast, patients with TACI deficiency develop common variable immunodeficiency, but no clinical features of lupus [139].

Lupus-like disease in BAFF transgenic mice

BAFF-BCMA interactions are required for the survival of long-lived plasma cells [128], whereas B-cell memory is independent of BAFF/APRIL signaling. BAFF transgenic mice develop hypergammaglobulinemia, autoantibodies, glomerulonephritis, and destruction of the salivary glands [137, 140]. Like lupus patients [141], NZB/W and MRL mice have high levels of circulating soluble BAFF in association with IgG2c (the C57BL/6 equivalent of IgG2a), IgG2b, and IgG3 anti-dsDNA antibodies, rheumatoid factor, and renal disease [137]. Unexpectedly, however, these switched autoantibodies are derived from marginal zone and B-1 cells, and are T-cell-independent. Autoantibody production in BAFF-transgenic mice requires MyD88 and BAFF promotes TLR9-induced isotype switching from IgM to IgG [137]. The production of anti-DNA antibodies is enhanced by TLR4, TLR7, or TLR9 ligand stimulation.

BAFF/APRIL antagonism in the therapy of lupus (see also Chapter 59)

In view of the elevated BAFF levels in SLE and the development of lupus-like disease in BAFF transgenic mice, a human monoclonal anti-BAFF antibody (belimumab) that binds to soluble BAFF and inhibits its interaction with BAFF-R, TACI, and BCMA is in clinical trials, and other agents directed against BAFF, APRIL, and BAFF-R are under development [142]. The relative independence of memory B-cell survival from BAFF/APRIL suggests that therapy might not adversely affect recall responses, e.g. to vaccines. Also, long-lived plasma cells are BAFF/APRIL-dependent and therefore should be depleted. To the extent that autoantibody production is maintained by memory B cells, therapy may not abolish the serological abnormalities.

In a recent phase II clinical trial, belimumab treatment did not significantly reduce disease activity or lupus flares overall [143]. However, in a subset of serologically active patients (defined as ANA titer $\geq$ 1:80 and/or positive anti-dsDNA $\geq$ 30 IU/ml), there was a modest reduction in disease activity. Interestingly, there was a reduction in IgG anti-dsDNA antibodies of 29% vs. 9% for placebo, and a reversion to normal in 15% vs. 3% of placebo-treated controls at 52 weeks. In a preliminary study, anti-Sm and RNP autoantibody levels also appeared to be reduced over a 160-week period. Progressive normalization of autoantibody, immunoglobulin, and complement levels over 3 years of belimumab therapy in systemic lupus erythematosus patients, Poster presentation, PANLAR Congress, Guatemala City, 2008). The relatively low rate of reversion in anti-DNA$^+$ patients treated with belimumab (15% at 1 year, 30% at 3 years) raises the possibility that much of the anti-DNA response is maintained by memory B cells, which are insensitive to BAFF antagonism.

CD40−CD40 Ligand (CD40L) Signaling and Autoantibody Production

Strength of the co-stimulatory signal delivered to the B cell via CD40 engagement of CD40L on the surface of activated T cells contributes to autoimmunity. CD40L expression is abnormally regulated in both human and murine lupus [144]. CD40−CD40L interactions are important for generating high-affinity autoantibodies. Neutralizing antibodies to CD40L have been used by themselves or in combination with CTLA-4Ig to treat murine lupus, improving renal disease and decreasing autoantibody production [145, 146]. In SNF$_1$ lupus mice, anti-CD40 treatment reduces anti-dsDNA and anti-histone/DNA autoantibodies while having no effect on anti-ssDNA or antihistone antibodies and may block nephritis [147]. In limited clinical trials, anti-CD40L treatment reduced anti-dsDNA antibodies (Farr assay) in one study in which all but one subject was anti-DNA$^+$ [148]; however in a second study in which only 17/85 subjects were anti-dsDNA$^+$, a significant reduction was not noted [149]. Unfortunately, trials in human autoimmune disease were stopped due to thrombotic complications [150].

FUTURE DIRECTIONS

Since the previous edition of this book, there has been tremendous progress in our understanding of the origins of lupus autoantibodies. The basis for the auto-antigen selectivity in SLE, which is largely restricted to nucleic-acid-binding proteins, now appears to be related to the fact that the nucleic acid components of these antigens are "endogenous adjuvants" capable of interacting with TLR7/8 or TLR9 and stimulating IFN-I production. The importance of IFN-I production stimulated by TLR7/8/9 signaling is suggested in human studies and confirmed in animal models. In addition, the anatomy of autoantibody production in the B-cell follicles, extrafollicular regions, and marginal zones is coming into clearer focus.

Despite substantial progress, the relative importance of T-cell-dependent vs. -independent B-cell activation in SLE remains unclear. In mouse models, there is evidence for both pathways. Further studies are needed to define to what degree autoantibody production in SLE patients results from cognate T—B interactions and post-germinal center memory/plasma cells and vs. extrafollicular, T-cell-independent (TLR-mediated) responses, a question that may be highly significant for the therapy of SLE. The evidence so far suggests that disruption of T—B interactions (e.g. anti-CD40L, CTLA4-Ig) can partially, but not completely, reduce autoantibody production. Inhibitors of TLR7/8/9 signaling and signaling through the IFN-I receptor are now in clinical trials, and the effects of B-cell depletion therapy and BAFF inhibition are being studied.

Further studies also are needed to determine how IFN-I enhances autoantibody production: is the main effect an enhancement of activated T-cell survival, upregulation of TLR7/9 expression on anergic/ignorant B cells, an increase in BAFF expression and the survival of plasma cells, or some other mechanism? Finally, it will be of interest to know if autoantibody-producing cells are regulated mainly at the level of tolerance (i.e. the generation and survival of autoreactive B cells) or at the level of plasma cell differentiation — i.e. are there are circulating autoreactive B cells just waiting to undergo differentiation into plasma cells?

Acknowledgment

This authors were supported by a research grant AR44731 from NIH/NIAMS.

References

[1] M.M. Hargraves, H. Richmond, R. Morton, Presentation of two bone-marrow elements: the Tart cell and LE cell, Proc. Staff Meetings Mayo Clin. 23 (1948) 25—28.

[2] E.J. Holborow, D.M. Weir, G.D. Johnson, A serum factor in lupus erythematosus with affinity for tissue nuclei, Br. Med. J. 2 (1957) 732—734.

[3] W.C. Robbins, H.R. Holman, H. Deicher, H.G. Kunkel, Complement fixation with cell nuclei and DNA in lupus erythematosus, Proc. Soc. Exp. Biol. Med. 96 (1957) 575—579.

[4] E.M. Tan, H.G. Kunkel, Characteristics of a soluble nuclear antigen precipitating with sera of patients with systemic lupus erythematosus, J. Immunol. 96 (1966) 464—471.

[5] S.M. Edworthy, E. Zatarain, D.J. McShane, D.A. Bloch, Analysis of the 1982 ARA lupus criteria data set by recursive partitioning methodology: new insights into the relative merit of individual criteria, J. Rheumatol. 15 (1988) 1493—1498.

[6] W.C. Shiel, M. Jason, The diagnostic associations of patients with antinuclear antibodies referred to a community rheumatologist, J. Rheumatol. 16 (1989) 782—785.

[7] J. Wang, M. Satoh, F. Kabir, M. Shaw, M.A. Domingo, R. Mansoor, et al., Increased prevalence of autoantibodies to Ku antigen in African-Americans versus Caucasians with systemic lupus erythematosus, Arthritis Rheum. 44 (2001) 2367—2370.

[8] N. Kurata, E.M. Tan, Identification of antibodies to nuclear acidic antigens by counterimmunoelectrophoresis, Arthritis Rheum. 19 (1976) 574—580.

[9] K.B. Elkon, E. Bonfa, N. Brot, Antiribosomal antibodies in systemic lupus erythematosus, Rheum. Dis. Clin. North Am. 18 (1992) 377—390.

[10] Y. Takasaki, D. Fishwild, E.M. Tan, Chrararcterization of proliferating cell nuclear antigen recognized by autoantibodies in lupus sera, J. Exp. Med. 159 (1984) 981—992.

[11] Y. Yamasaki, S. Narain, H. Yoshida, L. Hernandez, T. Barker, P.C. Hahn, et al., Autoantibodies to RNA helicase A: a new serologic marker of early lupus, Arthritis Rheum. 56 (2007) 596—604.

[12] A.N. Theofilopoulos, F.J. Dixon, Murine models of systemic lupus erythematosus, Adv. Immunol. 37 (1985) 269—390.

[13] W.H. Reeves, P.Y. Lee, J.S. Weinstein, M. Satoh, L. Lu, Induction of autoimmunity by pristane and other naturally occurring hydrocarbons, Trends Immunol. 30 (2009) 455—464.

[14] T. Kawai, S. Akira, Innate immune recognition of viral infection, Nat. Immunol. 7 (2006) 131—137.

[15] M.R. Lerner, J.A. Steitz, Antibodies to small nuclear RNAs complexed with proteins are produced by patients with systemic lupus erythematosus, Proc. Natl. Acad. Sci. USA 76 (1979) 5495—5499.

[16] C.C. Query, R.C. Bentley, J.D. Keene, A common RNA recognition motif identified within a defined U1 RNA binding domain of the 70K U1 snRNP protein, Cell 57 (1989) 89—101.

[17] J. Craft, Antibodies to snRNPs in systemic lupus erythematosus, Rheum. Dis. Clin. North Am. 18 (1992) 311—335.

[18] A.J. Stein, G. Fuchs, C. Fu, S.L. Wolin, K.M. Reinisch, Structural insights into RNA quality control: the Ro autoantigen binds misfolded RNAs via its central cavity, Cell 121 (2005) 529—539.

[19] E. Feierl, J.S. Smolen, T. Karonitsch, G.H. Stummvoll, H. Ekhart, C.W. Steiner, et al., Engulfed cell remnants, and not

cells undergoing apoptosis, constitute the LE-cell phenomenon, Autoimmunity 40 (2007) 315–321.

[20] A. Weinstein, B. Bordwell, B. Stone, C. Tibbetts, N.F. Rothfield, Antibodies to native DNA and serum complement (C3) levels: application to diagnosis and classification of systemic lupus erythematosus, Am. J. Med. 74 (1983) 206–216.

[21] D. Koffler, V. Agnello, R. Thoburn, H.G. Kunkel, Systemic lupus erythematosus: prototype of immune complex disease in man, J. Exp. Med. 134 (1971) 169–179.

[22] M. Mattioli, M. Reichlin, Physical association of two nuclear antigens and mutual occurrence of their antibodies: the relationship of the Sm and RNA protein (Mo) systems in SLE sera, J. Immunol. 110 (1973) 1318–1324.

[23] I.M. Pettersson, M. Hinterberger, T. Mimori, E. Gottlieb, J.A. Steitz, The structure of mammalian small nuclear ribonucleoproteins: identification of multiple protein components reactive with anti-(U1)RNP and anti-Sm antibodies, J. Biol. Chem. 259 (1984) 5907–5914.

[24] R.M. Hoet, P. de Weerd, J.K. Gunnewiek, I. Koornneef, W.J. van Venrooij, Epitope regions of U1 small nuclear RNA recognized by anti-U1RNA-specific autoantibodies, J. Clin. Invest. 90 (1992) 1753–1762.

[25] M. Mattioli, M. Reichlin, Heterogeneity of RNA protein antigens reactive with sera of patients with systemic lupus erythematosus: description of a cytoplasmic nonribosomal antigen, Arthritis Rheum. 17 (1974) 421–429.

[26] C. Boulanger, B. Chabot, H. Menard, G. Boire, Autoantibodies in human anti-Ro sera specifically recognize deproteinized hY5 Ro RNA, Clin. Exp. Immunol. 99 (1995) 29–36.

[27] S.M. Russell, F.Y. Liew, T cells primed by influenza virion internal components can cooperate in antibody response to haemagglutinin, Nature (Lond.) 280 (1979) 147–148.

[28] S. Fatenejad, M.J. Mamula, J. Craft, Role of intermolecular/intrastructural B- and T-cell determinants in the diversification of autoantibodies to ribonucleoprotein particles, Proc. Natl. Acad. Sci. USA 90 (1993) 12010–12014.

[29] C. Mohan, S. Adams, V. Stanik, S.K. Datta, Nucleosome: a major immunogen for pathogenic autoantibody-inducing T cells of lupus, J. Exp. Med. 177 (1993) 1367–1381.

[30] P. Fitzgerald-Bocarsly, J. Dai, S. Singh, Plasmacytoid dendritic cells and type I IFN: 50 years of convergent history, Cytokine Growth Factor Rev. 19 (2008) 3–19.

[31] P.Y. Lee, W.H. Reeves, Type I interferon as a target of treatment in SLE, Endocr. Metab Immune. Disord. Drug Targets. 6 (2006) 323–330.

[32] E.C. Baechler, F.M. Batliwalla, G. Karypis, P.M. Gaffney, W.A. Ortmann, K.J. Espe, et al., Interferon-inducible gene expression signature in peripheral blood cells of patients with severe lupus, Proc. Natl. Acad. Sci. USA 100 (2003) 2610–2615.

[33] L. Bennett, A.K. Palucka, E. Arce, V. Cantrell, J. Borvak, J. Banchereau, et al., Interferon and granulopoiesis signatures in systemic lupus erythematosus blood, J. Exp. Med. 197 (2003) 711–723.

[34] H. Zhuang, S. Narain, E. Sobel, P.Y. Lee, D.C. Nacionales, K.M. Kelly, et al., Association of anti-nucleoprotein autoantibodies with upregulation of Type I interferon-inducible gene transcripts and dendritic cell maturation in systemic lupus erythematosus, Clin. Immunol. 117 (2005) 238–250.

[35] K.A. Kirou, C. Lee, S. George, K. Louca, M.G. Peterson, M.K. Crow, Activation of the interferon-alpha pathway identifies a subgroup of systemic lupus erythematosus patients with distinct serologic features and active disease, Arthritis Rheum. 52 (2005) 1491–1503.

[36] L. Farkas, K. Beiske, F. Lund-Johansen, P. Brandtzaeg, F.L. Jahnsen, Plasmacytoid dendritic cells (natural interferon-alpha/beta-producing cells) accumulate in cutaneous lupus erythematosus lesions, Am. J. Pathol. 159 (2001) 237–243.

[37] L.E. Ronnblom, G.V. Alm, K.E. Oberg, Autoimmunity after alpha-interferon therapy for malignant carcinoid tumors, Ann. Intern. Med. 115 (1991) 178–183.

[38] H. Zhuang, M. Kosboth, P. Lee, A. Rice, D.J. Driscoll, R. Zori, et al., Lupus-like disease and high interferon levels with trisomy of the Type I interferon cluster on chromosome 9p, Arthritis Rheum. 54 (2006) 1573–1579.

[39] M.L. Santiago-Raber, R. Baccala, K.M. Haraldsson, D. Choubey, T.A. Stewart, D.H. Kono, et al., Type-I interferon receptor deficiency reduces lupus-like disease in NZB mice, J. Exp. Med. 197 (2003) 777–788.

[40] A. Mathian, A. Weinberg, M. Gallegos, J. Banchereau, S. Koutouzov, IFN-alpha induces early lethal lupus in pre-autoimmune (New Zealand Black x New Zealand White) F1 but not in BALB/c mice, J. Immunol. 174 (2005) 2499–2506.

[41] D.C. Nacionales, K.M. Kelly-Scumpia, P.Y. Lee, J.S. Weinstein, E. Sobel, M. Satoh, et al., Deficiency of the Type I interferon receptor protects mice from experimental lupus, Arthritis Rheum. 56 (2007) 3770–3783.

[42] H. Vallin, A. Perers, G.V. Alm, L. Ronnblom, Anti-double-stranded DNA antibodies and immunostimulatory plasmid DNA in combination mimic the endogenous IFN-alpha inducer in systemic lupus erythematosus, J. Immunol. 163 (1999) 6306–6313.

[43] T.K. Means, E. Latz, F. Hayashi, M.R. Murali, D.T. Golenbock, A.D. Luster, Human lupus autoantibody-DNA complexes activate DCs through cooperation of CD32 and TLR9, J. Clin. Invest 115 (2005) 407–417.

[44] P. Blanco, A.K. Palucka, M. Gill, V. Pascual, J. Banchereau, Induction of dendritic cell differentiation by IFN-alpha in systemic lupus erythematosus, Science (Wash. D.C.) 294 (2001) 1540–1543.

[45] P.Y. Lee, J.S. Weinstein, D.C. Nacionales, P.O. Scumpia, Y. Li, E. Butfiloski, et al., A Novel Type I IFN-Producing Cell Subset in Murine Lupus, J. Immunol. 180 (2008) 5101–5108.

[46] P.Y. Lee, Y. Kumagai, Y. Li, O. Takeuchi, H. Yoshida, J. Weinstein, et al., TLR7-dependent and FcgammaR-independent production of type I interferon in experimental mouse lupus, J. Exp. Med. 205 (2008) 2995–3006.

[47] E.A. Leadbetter, I.R. Rifkin, A.M. Hohlbaum, B.C. Beaudette, M.J. Shlomchik, A. Marshak-Rothstein, Chromatin-IgG complexes activate B cells by dual engagement of IgM and Toll-like receptors, Nature (Lond.) 416 (2002) 603–607.

[48] M.W. Boule, C. Broughton, F. Mackay, S. Akira, A. Marshak-Rothstein, I.R. Rifkin, Toll-like receptor 9-dependent and -independent dendritic cell activation by chromatin-immunoglobulin G complexes, J. Exp. Med. 199 (2004) 1631–1640.

[49] K.M. Kelly, H. Zhuang, D.C. Nacionales, P.O. Scumpia, R. Lyons, J. Akaogi, et al., "Endogenous adjuvant" activity of the RNA components of lupus autoantigens Sm/RNP and Ro 60, Arthritis Rheum. 54 (2006) 1557–1567.

[50] F.J. Barrat, T. Meeker, J. Gregorio, J.H. Chan, S. Uematsu, S. Akira, et al., Nucleic acids of mammalian origin can act as endogenous ligands for Toll-like receptors and may promote systemic lupus erythematosus, J. Exp. Med. 202 (2005) 1131–1139.

[51] E. Savarese, O.W. Chae, S. Trowitzsch, G. Weber, B. Kastner, S. Akira, et al., U1 small nuclear ribonucleoprotein immune complexes induce type I interferon in plasmacytoid dendritic cells through TLR7, Blood 107 (2006) 3229–3234.

[52] A. Marshak-Rothstein, Toll-like receptors in systemic autoimmune disease, Nat. Rev. Immunol. 6 (2006) 823–835.

[53] C. Watts, Immunology. The bell tolls for phagosome maturation, Science (Wash. D.C.) 304 (2004) 976–977.

[54] E. Latz, A. Schoenemeyer, A. Visintin, K.A. Fitzgerald, B.G. Monks, C.F. Knetter, et al., TLR9 signals after translocating from the ER to CpG DNA in the lysosome, Nat. Immunol. 5 (2004) 190–198.

[55] C.A. Leifer, M.N. Kennedy, A. Mazzoni, C. Lee, M.J. Kruhlak, D.M. Segal, TLR9 is localized in the endoplasmic reticulum prior to stimulation, J. Immunol. 173 (2004) 1179–1183.

[56] K. Tabeta, K. Hoebe, E.M. Janssen, X. Du, P. Georgel, K. Crozat, et al., The Unc93b1 mutation 3d disrupts exogenous antigen presentation and signaling via Toll-like receptors 3, 7 and 9, Nat. Immunol. 7 (2006) 156–164.

[57] R. Fukui, S. Saitoh, F. Matsumoto, H. Kozuka-Hata, M. Oyama, K. Tabeta, et al., Unc93B1 biases Toll-like receptor responses to nucleic acid in dendritic cells toward DNA- but against RNA-sensing, J. Exp. Med. 206 (2009) 1339–1350.

[58] A. Strasser, P.J. Jost, S. Nagata, The many roles of FAS receptor signaling in the immune system, Immunity 30 (2009) 180–192.

[59] S. Nagata, DNA degradation in development and programmed cell death, Annu. Rev. Immunol. 23 (2005) 853–875.

[60] Y. Okabe, K. Kawane, S. Akira, T. Taniguchi, S. Nagata, Toll-like receptor-independent gene induction program activated by mammalian DNA escaped from apoptotic DNA degradation, J. Exp. Med. 202 (2005) 1333–1339.

[61] K. Kawane, M. Ohtani, K. Miwa, T. Kizawa, Y. Kanbara, Y. Yoshioka, et al., Chronic polyarthritis caused by mammalian DNA that escapes from degradation in macrophages, Nature (Lond.) 443 (2006) 998–1002.

[62] D. Xue, H. Shi, J.D. Smith, X. Chen, D.A. Noe, T. Cedervall, et al., A lupus-like syndrome develops in mice lacking the Ro 60-kDa protein, a major lupus autoantigen, Proc. Natl. Acad. Sci. USA 100 (2003) 7503–7508.

[63] L.P. Erwig, P.M. Henson, Immunological consequences of apoptotic cell phagocytosis, Am. J. Pathol. 171 (2007) 2–8.

[64] P.L. Cohen, R. Caricchio, V. Abraham, T.D. Camenisch, J.C. Jennette, R.A. Roubey, et al., Delayed apoptotic cell clearance and lupus-like autoimmunity in mice lacking the c-mer membrane tyrosine kinase, J. Exp. Med. 196 (2002) 135–140.

[65] G. Lemke, C.V. Rothlin, Immunobiology of the TAM receptors, Nat. Rev. Immunol. 8 (2008) 327–336.

[66] Q. Lu, G. Lemke, Homeostatic regulation of the immune system by receptor tyrosine kinases of the Tyro 3 family, Science (Wash. D.C.) 293 (2001) 306–311.

[67] S.R. Christensen, J. Shupe, K. Nickerson, M. Kashgarian, R.A. Flavell, M.J. Shlomchik, Toll-like receptor 7 and TLR9 dictate autoantibody specificity and have opposing inflammatory and regulatory roles in a murine model of lupus, Immunity 25 (2006) 417–428.

[68] P. Pisitkun, J.A. Deane, M.J. Difilippantonio, T. Tarasenko, A.B. Satterthwaite, S. Bolland, Autoreactive B cell responses to RNA-related antigens due to TLR7 gene duplication, Science (Wash. D.C.) 312 (2006) 1669–1672.

[69] J.A. Deane, P. Pisitkun, R.S. Barrett, L. Feigenbaum, T. Town, J.M. Ward, et al., Control of toll-like receptor 7 expression is essential to restrict autoimmunity and dendritic cell proliferation, Immunity 27 (2007) 801–810.

[70] R. Berland, L. Fernandez, E. Kari, J.H. Han, I. Lomakin, S. Akira, et al., Toll-like receptor 7-dependent loss of B cell tolerance in pathogenic autoantibody knockin mice, Immunity 25 (2006) 429–440.

[71] C.G. Vinuesa, S.G. Tangye, B. Moser, C.R. Mackay, Follicular B helper T cells in antibody responses and autoimmunity, Nat. Rev. Immunol. 5 (2005) 853–865.

[72] C.D. Allen, T. Okada, J.G. Cyster, Germinal-center organization and cellular dynamics, Immunity 27 (2007) 190–202.

[73] E.S. Sobel, V.N. Kakkanaiah, M. Kakkanaiah, R.L. Cheek, P.L. Cohen, R.A. Eisenberg, T-B collaboration for autoantibody production in lpr mice is cognate and MHC-restricted, J. Immunol. 152 (1994) 6011–6016.

[74] R.A. Eisenberg, S.Y. Craven, P.L. Cohen, Isotype progression and clonality of anti-Sm autoantibodies in MRL/Mp-lpr/lpr mice, J. Immunol. 139 (1987) 728–733.

[75] D. Wofsy, W.E. Seaman, Successful treatment of autoimmunity in NZB/NZW F1 mice with monoclonal antibody to L3T4, J. Exp. Med. 161 (1985) 378–391.

[76] B.K. Finck, P.S. Linsley, D. Wofsy, Treatment of murine lupus with CTLA4Ig, Science (Wash. D.C.) 265 (1994) 1225–1227.

[77] H.B. Richards, M. Satoh, J.C. Jennette, T. Okano, Y.S. Kanwar, W.H. Reeves, Disparate T cell requirements of two subsets of lupus-specific autoantibodies in pristane-treated mice, Clin. Exp. Immunol. 115 (1999) 547–553.

[78] A. Radbruch, G. Muehlinghaus, E.O. Luger, A. Inamine, K.G. Smith, T. Dorner, et al., Competence and competition: the challenge of becoming a long-lived plasma cell, Nat. Rev. Immunol. 6 (2006) 741–750.

[79] M. Shlomchik, M. Mascelli, H. Shan, M.Z. Radic, D. Pisetsky, A. Marshak-Rothstein, et al., Anti-DNA antibodies from autoimmune mice arise by clonal expansion and somatic mutation, J. Exp. Med. 171 (1990) 265–298.

[80] T.R. Mosmann, H. Cherwinski, M.W. Bond, M.A. Geidlin, R.L. Coffman, Two types of murine helper T cell clone. I. Definition according to profiles of lymphokine activities and secreted proteins, J. Immunol. 136 (1986) 2348–2357.

[81] L. Zhou, M.M. Chong, D.R. Littman, Plasticity of CD4+ T cell lineage differentiation, Immunity 30 (2009) 646–655.

[82] C. Haas, M. Le Hir, IFN-gamma is essential for the development of autoimmune glomerulonephritis in MRL/lpr mice, J. Immunol. 158 (1997) 5484–5491.

[83] D. Balomenos, R. Rumold, A.N. Theofilopoulos, Interferon-gamma is required for lupus-like disease and lymphoaccumulation in MRL-lpr mice, J. Clin. Invest. 101 (1998) 364–371.

[84] H.B. Richards, M. Satoh, J.C. Jennette, B.P. Croker, W.H. Reeves, Interferon-gamma promotes lupus nephritis in mice treated with the hydrocarbon oil pristane, Kidney Int. 60 (2001) 2173–2180.

[85] M. Akahoshi, H. Nakashima, Y. Tanaka, T. Kohsaka, S. Nagano, E. Ohgami, et al., Th1/Th2 balance of peripheral T helper cells in systemic lupus erythematosus, Arthritis Rheum. 42 (1999) 1644–1648.

[86] T. Karonitsch, E. Feierl, C.W. Steiner, K. Dalwigk, A. Korb, N. Binder, et al., Activation of the interferon-gamma signaling pathway in systemic lupus erythematosus peripheral blood mononuclear cells, Arthritis Rheum. 60 (2009) 1463–1471.

[87] M.L. Santiago, L. Fossati, C. Jacquet, W. Muller, S. Izui, L. Reininger, Interleukin-4 protects against a genetically linked lupus-like autoimmune syndrome, J. Exp. Med. 185 (1997) 65–70.

[88] L.A. Garrett-Sinha, S. John, S.L. Gaffen, IL-17 and the Th17 lineage in systemic lupus erythematosus, Curr. Opin. Rheumatol. 20 (2008) 519–525.

[89] C.K. Wong, L.C. Lit, L.S. Tam, E.K. Li, P.T. Wong, C.W. Lam, Hyperproduction of IL-23 and IL-17 in patients with systemic lupus erythematosus: implications for Th17-mediated inflammation in auto-immunity, Clin. Immunol. 127 (2008) 385–393.

[90] A. Doreau, A. Belot, J. Bastid, B. Riche, M.C. Trescol-Biemont, B. Ranchin, et al., Interleukin 17 acts in synergy with B cell-activating factor to influence B cell biology and the pathophysiology of systemic lupus erythematosus, Nat. Immunol. 10 (2009) 778–785.

[91] H.C. Hsu, P. Yang, J. Wang, Q. Wu, R. Myers, J. Chen, et al., Interleukin 17-producing T helper cells and interleukin 17 orchestrate autoreactive germinal center development in autoimmune BXD2 mice, Nat. Immunol. 9 (2008) 166–175.

[92] N. Jacob, H. Yang, L. Pricop, Y. Liu, X. Gao, S.G. Zheng, et al., Accelerated pathological and clinical nephritis in systemic lupus erythematosus-prone New Zealand Mixed 2328 mice doubly deficient in TNF receptor 1 and TNF receptor 2 via a Th17-associated pathway, J. Immunol. 182 (2009) 2532–2541.

[93] C.G. Vinuesa, M.C. Cook, C. Angelucci, V. Athanasopoulos, L. Rui, K.M. Hill, et al., A RING-type ubiquitin ligase family member required to repress follicular helper T cells and autoimmunity, Nature (Lond.) 435 (2005) 452–458.

[94] M.A. Linterman, R.J. Rigby, R.K. Wong, D. Yu, R. Brink, J.L. Cannons, et al., Follicular helper T cells are required for systemic autoimmunity, J. Exp. Med. 206 (2009) 561–576.

[95] M. Miyara, S. Sakaguchi, Natural regulatory T cells: mechanisms of suppression, Trends Mol. Med. 13 (2007) 108–116.

[96] M. Miyara, Z. Amoura, C. Parizot, C. Badoual, K. Dorgham, S. Trad, et al., Global natural regulatory T cell depletion in active systemic lupus erythematosus, J. Immunol. 175 (2005) 8392–8400.

[97] X. Valencia, C. Yarboro, G. Illei, P.E. Lipsky, Deficient CD4+CD25high T regulatory cell function in patients with active systemic lupus erythematosus, J. Immunol. 178 (2007) 2579–2588.

[98] H.J. van der Vliet, E.E. Nieuwenhuis, IPEX as a result of mutations in FOXP3, Clin. Dev. Immunol. 2007 (2007) 89017.

[99] N. Iikuni, E.V. Lourenco, B.H. Hahn, C.A. La, Cutting edge: Regulatory T cells directly suppress B cells in systemic lupus erythematosus, J. Immunol. 183 (2009) 1518–1522.

[100] A.J. McAdam, A.N. Schweitzer, A.H. Sharpe, The role of B7 costimulation in activation and differentiation of CD4+ and CD8+ T cells, Immunol. Rev. 165 (1998) 231–247.

[101] R.J. Greenwald, V.A. Boussiotis, R.B. Lorsbach, A.K. Abbas, A.H. Sharpe, CTLA-4 regulates induction of anergy in vivo, Immunity 14 (2001) 145–155.

[102] M.A. Linterman, R.J. Rigby, R. Wong, D. Silva, D. Withers, G. Anderson, et al., Roquin differentiates the specialized functions of duplicated T cell costimulatory receptor genes CD28 and ICOS, Immunity 30 (2009) 228–241.

[103] M. Mihara, I. Tan, Y. Chuzhin, B. Reddy, L. Budhai, A. Holzer, et al., CTLA4Ig inhibits T cell-dependent B-cell maturation in murine systemic lupus erythematosus, J. Clin. Invest. 106 (2000) 91–101.

[104] M. Takiguchi, M. Murakami, I. Nakagawa, I. Saito, A. Hashimoto, T. Uede, CTLA4IgG gene delivery prevents autoantibody production and lupus nephritis in MRL/lpr mice, Life Sci. 66 (2000) 991–1001.

[105] Y.B. Zhou, R.G. Ye, Y.J. Li, C.M. Xie, Y.H. Wu, Effect of anti-CD134L mAb and CTLA4Ig on ConA-induced proliferation, Th cytokine secretion, and anti-dsDNA antibody production in spleen cells from lupus-prone BXSB mice, Autoimmunity 41 (2008) 395–404.

[106] D.A. Hildeman, Y. Zhu, T.C. Mitchell, J. Kappler, P. Marrack, Molecular mechanisms of activated T cell death in vivo, Curr. Opin. Immunol. 14 (2002) 354–359.

[107] A. Le Bon, G. Schiavoni, G. D'Agostino, I. Gresser, F. Belardelli, D.F. Tough, Type I interferons potently enhance humoral immunity and can promote isotype switching by stimulating dendritic cells in vivo, Immunity 14 (2001) 461–470.

[108] P. Marrack, J. Kappler, T. Mitchell, Type I interferons keep activated T cells alive, J. Exp. Med. 189 (1999) 521–529.

[109] L. Xu, L. Zhang, Y. Yi, H.K. Kang, S.K. Datta, Human lupus T cells resist inactivation and escape death by upregulating COX-2, Nat. Med. 10 (2004) 411–415.

[110] J.C. Fanzo, W. Yang, S.Y. Jang, S. Gupta, Q. Chen, A. Siddiq, et al., Loss of IRF-4-binding protein leads to the spontaneous development of systemic autoimmunity, J. Clin. Invest. 116 (2006) 703–714.

[111] M.J. Shlomchik, Sites and stages of autoreactive B cell activation and regulation, Immunity 28 (2008) 18–28.

[112] K. Hayakawa, R.R. Hardy, Development and function of B-1 cells, Curr. Opin. Immunol. 12 (2000) 346–353.

[113] H. Wardemann, S. Yurasov, A. Schaefer, J.W. Young, E. Meffre, M.C. Nussenzweig, Predominant autoantibody production by early human B cell precursors, Science (Wash. D.C.) 301 (2003) 1374–1377.

[114] C.C. Goodnow, J. Sprent, G.B. Fazekas de St, C.G. Vinuesa, Cellular and genetic mechanisms of self tolerance and autoimmunity, Nature (Lond.) 435 (2005) 590–597.

[115] M.L. Fields, J. Erikson, The regulation of lupus-associated autoantibodies: immunoglobulin transgenic models, Curr. Opin. Immunol. 15 (2003) 709–717.

[116] L. Rui, C.G. Vinuesa, J. Blasioli, C.C. Goodnow, Resistance to CpG DNA-induced autoimmunity through tolerogenic B cell antigen receptor ERK signaling, Nat. Immunol. 4 (2003) 594–600.

[117] M. Thien, T.G. Phan, S. Gardam, M. Amesbury, A. Basten, F. Mackay, et al., Excess BAFF rescues self-reactive B cells from peripheral deletion and allows them to enter forbidden follicular and marginal zone niches, Immunity 20 (2004) 785–798.

[118] H. Fukuyama, F. Nimmerjahn, J.V. Ravetch, The inhibitory Fcgamma receptor modulates autoimmunity by limiting the accumulation of immunoglobulin G+ anti-DNA plasma cells, Nat. Immunol. 6 (2005) 99–106.

[119] S.J. Seo, M.L. Fields, J.L. Buckler, A.J. Reed, L. Mandik-Nayak, S.A. Nish, et al., The impact of T helper and T regulatory cells on the regulation of anti-double-stranded DNA B cells, Immunity 16 (2002) 535–546.

[120] L. Mandik-Nayak, S. Seo, C. Sokol, K.M. Potts, A. Bui, J. Erikson, MRL- lpr/lpr mice exhibit a defect in maintaining developmental arrest and follicular exclusion of anti-double-stranded DNA B cells, J. Exp. Med. 189 (1999) 1799–1814.

[121] C.M. Westhoff, A. Whittier, S. Kathol, J. McHugh, C. Zajicek, L.D. Shultz, et al., DNA-binding antibodies from viable motheaten mutant mice. Implications for B cell tolerance, J. Immunol. 159 (1997) 3024–3033.

[122] M. Shapiro-Shelef, K. Calame, Regulation of plasma-cell development, Nat. Rev. Immunol. 5 (2005) 230–242.

[123] D. Paus, T.G. Phan, T.D. Chan, S. Gardam, A. Basten, R. Brink, Antigen recognition strength regulates the choice between extrafollicular plasma cell and germinal center B cell differentiation, J. Exp. Med. 203 (2006) 1081–1091.

[124] B.F. Hoyer, K. Moser, A.E. Hauser, A. Peddinghaus, C. Voigt, D. Eilat, et al., Short-lived plasmablasts and long-lived plasma cells contribute to chronic humoral autoimmunity in NZB/W mice, J. Exp. Med. 199 (2004) 1577–1584.

[125] M.J. Benson, L.D. Erickson, M.W. Gleeson, R.J. Noelle, Affinity of antigen encounter and other early B-cell signals determine B-cell fate, Curr. Opin. Immunol. 19 (2007) 275–280.

[126] M. Shapiro-Shelef, K.I. Lin, L.J. McHeyzer-Williams, J. Liao, M.G. McHeyzer-Williams, K. Calame, Blimp-1 is required for

the formation of immunoglobulin secreting plasma cells and pre-plasma memory B cells, Immunity 19 (2003) 607−620.

[127] K. Moser, K. Tokoyoda, A. Radbruch, I. MacLennan, R.A. Manz, Stromal niches, plasma cell differentiation and survival, Curr. Opin. Immunol. 18 (2006) 265−270.

[128] B.P. O'Connor, V.S. Raman, L.D. Erickson, W.J. Cook, L.K. Weaver, C. Ahonen, et al., BCMA is essential for the survival of long-lived bone marrow plasma cells, J. Exp. Med. 199 (2004) 91−98.

[129] Z. Xiang, A.J. Cutler, R.J. Brownlie, K. Fairfax, K.E. Lawlor, E. Severinson, et al., FcgammaRIIb controls bone marrow plasma cell persistence and apoptosis, Nat. Immunol. 8 (2007) 419−429.

[130] S. Bolland, J.V. Ravetch, Spontaneous autoimmune disease in Fc(gamma)RIIB-deficient mice results from strain-specific epistasis, Immunity 13 (2000) 277−285.

[131] D. Gatto, T. Pfister, A. Jegerlehner, S.W. Martin, M. Kopf, M.F. Bachmann, Complement receptors regulate differentiation of bone marrow plasma cell precursors expressing transcription factors Blimp-1 and XBP-1, J. Exp. Med. 201 (2005) 993−1005.

[132] N.L. Bernasconi, E. Traggiai, A. Lanzavecchia, Maintenance of serological memory by polyclonal activation of human memory B cells, Science (Wash. D.C.) 298 (2002) 2199−2202.

[133] R.A. Herlands, S.R. Christensen, R.A. Sweet, U. Hershberg, M.J. Shlomchik, T Cell-Independent and Toll-like Receptor-Dependent Antigen-Driven Activation of Autoreactive B Cells, Immunity 29 (2008) 249−260.

[134] D.J. DiLillo, Y. Hamaguchi, Y. Ueda, K. Yang, J. Uchida, K.M. Haas, et al., Maintenance of long-lived plasma cells and serological memory despite mature and memory B cell depletion during CD20 immunotherapy in mice, J. Immunol. 180 (2008) 361−371.

[135] J.H. Anolik, J. Barnard, A. Cappione, A.E. Pugh-Bernard, R.E. Felgar, R.J. Looney, et al., Rituximab improves peripheral B cell abnormalities in human systemic lupus erythematosus, Arthritis Rheum. 50 (2004) 3580−3590.

[136] F. Mackay, P. Schneider, Cracking the BAFF code, Nat. Rev. Immunol. 9 (2009) 491−502.

[137] J.R. Groom, C.A. Fletcher, S.N. Walters, S.T. Grey, S.V. Watt, M.J. Sweet, et al., BAFF and MyD88 signals promote a lupus-like disease independent of T cells, J. Exp. Med. 204 (2007) 1959−1971.

[138] D. Seshasayee, P. Valdez, M. Yan, V.M. Dixit, D. Tumas, I.S. Grewal, Loss of TACI causes fatal lymphoproliferation and autoimmunity, establishing TACI as an inhibitory BLyS receptor, Immunity 18 (2003) 279−288.

[139] U. Salzer, B. Grimbacher, TACItly changing tunes: farewell to a yin and yang of BAFF receptor and TACI in humoral immunity? New genetic defects in common variable immunodeficiency, Curr. Opin. Allergy Clin. Immunol. 5 (2005) 496−503.

[140] F. Mackay, S.A. Woodcock, P. Lawton, C. Ambrose, M. Baetscher, P. Schneider, et al., Mice transgenic for BAFF develop lymphocytic disorders along with autoimmune manifestations, J. Exp. Med. 190 (1999) 1697−1710.

[141] J. Zhang, V. Roschke, K.P. Baker, Z. Wang, G.S. Alarcon, B.J. Fessler, et al., Cutting edge: a role for B lymphocyte stimulator in systemic lupus erythematosus, J. Immunol. 166 (2001) 6−10.

[142] M.P. Cancro, D.P. D'Cruz, M.A. Khamashta, The role of B lymphocyte stimulator (BLyS) in systemic lupus erythematosus, J. Clin. Invest 119 (2009) 1066−1073.

[143] D.J. Wallace, W. Stohl, R.A. Furie, J.R. Lisse, J.D. McKay, J.T. Merrill, et al., A phase II, randomized, double-blind, placebo-controlled, dose-ranging study of belimumab in patients with active systemic lupus erythematosus, Arthritis Rheum. 61 (2009) 1168−1178.

[144] A. Desai-Mehta, L. Lu, R. Ramsey-Goldman, S.K. Datta, Hyperexpression of CD40 ligand by B and T cells in human lupus and its role in pathogenic autoantibody production, J. Clin. Invest. 97 (1996) 2063−2073.

[145] G.S. Early, W. Zhao, C.M. Burns, Anti-CD40 ligand antibody treatment prevents the development of lupus-like nephritis in a subset of New Zealand black x New Zealand white mice, J. Immunol. 157 (1996) 3159−3164.

[146] D.I. Daikh, B.K. Finck, P.S. Linsley, D. Hollenbaugh, D. Wofsy, Long-term inhibition of murine lupus by brief simultaneous blockade of the B7/CD28 and CD40/gp39 costimulation pathways, J. Immunol. 159 (1997) 3104−3108.

[147] C. Mohan, Y. Shi, J.D. Laman, S.K. Datta, Interaction between CD40 and its ligand gp39 in the development of murine lupus nephritis, J. Immunol. 154 (1995) 1470−1480.

[148] D.T. Boumpas, R. Furie, S. Manzi, G.G. Illei, D.J. Wallace, J.E. Balow, et al., A short course of BG9588 (anti-CD40 ligand antibody) improves serologic activity and decreases hematuria in patients with proliferative lupus glomerulonephritis, Arthritis Rheum. 48 (2003) 719−727.

[149] K.C. Kalunian, J.C. Davis Jr., J.T. Merrill, M.C. Totoritis, D. Wofsy, Treatment of systemic lupus erythematosus by inhibition of T cell costimulation with anti-CD154: a randomized, double-blind, placebo-controlled trial, Arthritis Rheum. 46 (2002) 3251−3258.

[150] T. Kawai, D. Andrews, R.B. Colvin, D.H. Sachs, A.B. Cosimi, Thromboembolic complications after treatment with monoclonal antibody against CD40 ligand, Nat. Med. 6 (2000) 114.

Anti-DNA Antibodies: Structure, Regulation and Pathogenicity

Cynthia Aranow, Dun Zhou, Betty Diamond

NORMAL IMMUNOLOGY OF ANTIBODIES AND B CELLS

Antibodies: Structure and Assembly

Antibodies are glycoproteins produced by B lymphocytes. Initially, antibodies serve as antigen receptors on the membrane of resting B lymphocytes. Each immunoglobulin (Ig) is composed of two heavy chains and two light chains which combine to form an antibody molecule with two functional regions: a constant region which determines effector functions and a variable region which is involved in antigen binding. Light chains contribute solely to antigen binding, whereas heavy-chain constant regions determine antibody isotype (IgM, IgD, IgG, IgA, or IgE) [1]. Naïve B cells produce IgM antibodies. After antigen exposure and T-cell help, class switching occurs and other isotypes are produced; IgG is the predominant isotype of the secondary (also called memory) immune response.

An antibody contains two identical antigen-binding sites composed of the variable regions of the heavy and light chains. Complementarity-determining regions (CDRs) within the variable region are highly polymorphic and are contact sites for antigen binding [2]. Framework regions (FRs) are more conserved portions of the variable region. Each variable region is composed of four FRs and three CDRs. The variable region of an antibody contains antigenic determinants known as idiotypes. Anti-idiotypic antibodies bind to specific determinants in the CDRs or FRs of another antibody [3]; they can bind to membrane immunoglobulin and, like antigen, can either tolerize or activate the B cell.

The genes encoding Ig light- and heavy-chain variable regions are formed by rearrangement of multiple gene segments. In humans, the heavy-chain segments, V (variable), D (diversity), and J (joining) segments are grouped in tandomly arranged gene clusters on chromosome 14. Light-chain gene segments (κ and λ) are on chromosomes 2 and 22 respectively. Antibody gene rearrangement requires excision of DNA between various gene segments followed by their ligation. V, D, and J segments are brought together to form a heavy-chain variable region gene [1] and V and J segments combine to form a light-chain variable region gene [1]. There are approximately 50 functional V segment genes, 30 functional D gene segments and six J gene segments for the human immunoglobulin heavy chain [4]. Each V, D, and J gene segment is flanked by a conserved heptamer—spacer—nanomer sequence, known as the recombination signal sequence (RSS), which is essential for Ig rearrangement.

Assembly of the heavy-chain locus begins with the joining of a D segment to a J segment. A V gene segment is then added to the assembled DJ unit, forming a complete VDJ variable region. Variable region recombination requires a complex of enzymes called V(D)J recombinase which are present in most cells; however, recombination activating genes, RAG-1 and RAG-2, are found only in lymphocytes [5]. RAGs start the process of VDJ recombination by generating double-stranded DNA breaks. Joining of the coding segments is mediated by several enzymes (Ku 70, Ku 80, DNA-PKs, XRCC4, ligase, Mre 11) [6, 7]. The formation and stabilization of the complex formed of these molecules are regulated by HMG1 and HMG2 from the high-mobility group family of proteins [8]. During rearrangement, imprecise VDJ joining may result in junctions of variable length. Nontemplate-derived nucleotides (N-nucleotides) may be inserted at VD or DJ junctions by the enzyme terminal deoxynucleotidyl transferase (TdT) [9], generating a CDR 3 of variable length and sequence. If the single-stranded DNA that is present at the junction can form

235

a hairpin loop, the resulting double-stranded (P, palindromic) sequences are added at the junction. Alternatively, N-nucleotides or nontemplate-encoded nucleotides are inserted randomly at the VD and DJ junctions. Initially all VDJ rearrangements associate with a μ constant region. After B-cell activation a VDJ can associate with downstream constant region genes [10]. This process, termed class switch recombination, is mediated by switch sequences located upstream of each constant region gene.

Light-chain gene segment rearrangement begins after formation of a functional heavy chain. Rearrangement starts in the κ locus, but will proceed to the λ locus if both rearranged κ alleles are not functional. The ratio between κ and λ light chains is 3:2 in humans and is 20:1 in mice. The light-chain variable region is composed of two gene segments: V and J. The κ light-chain gene locus contains approximately 40 functional V gene segments that are grouped into seven families and five J segments [11]. The λ light-chain locus also contains at least seven families with up to 70 members [12]. Similar to heavy-chain gene rearrangement, V and J elements of the light-chain loci rearrange by recombination at heptamer/nonamer consensus sites. Since TdT is no longer expressed in B cells at the time of light-chain rearrangement, N sequences are rarely inserted at the VJ light-chain junction [13].

Although there are two possible heavy-chain loci and four light-chain loci from which a functional gene can be derived, only one rearranged heavy-chain gene and one rearranged light-chain gene are expressed by a B cell. A productive rearrangement on one heavy- and one light-chain locus inhibits rearrangements on other loci; this process is called allelic exclusion. If a functional antibody binds to self-antigen, additional light-chain rearrangements may occur and change the antibody specificity, preventing autoreactivity. These additional light-chain rearrangements are termed receptor editing and occur frequently in developing B cells [14]. Receptor editing can also occur in mature B cells [15], demonstrated by increased RAG protein expression in germinal centers and extrafollicular B cells.

Somatic Mutation

The immune system can generate enormous immunoglobulin diversity resulting from random combinations of V, D, and J gene segments and V and J segments into heavy- and light-chain genes, respectively; (1) junctional diversity produced by N or P sequence insertion and/or imprecise joining of gene segments; (2) random pairing of heavy and light chains; and (3) somatic mutation of rearranged genes. The first three mechanisms occur before exposure to antigen and are potentially capable of producing a repertoire

of 10^{9-11} antibodies [1], whereas somatic mutation occurs after antigen encounter. Somatic point mutations are single-nucleotide substitutions that occur throughout the heavy- and light-chain variable region genes, and are most numerous in certain hot-spot motifs [16]. They represent site-specific, differentiation stage-specific and lineage-specific events that are important in the generation of high-affinity antibodies. Mutated antibodies can also acquire novel antigenic specificities including the acquisition of autospecificity [17]. Somatic mutation of Ig variable region genes occurs at a much higher rate than mutation of other genes (1 bp per 10^3 bp per cell division) [16]. The DNA mismatch repair system, which generally functions to correct point mutations in DNA, has been implicated in Ig gene mutation. Deficiency in mismatch repair enzymes causes altered nucleotide targeting for mutations. Activation-induced cytidine deaminase (AID), an enzyme discovered in germinal center B cells [18], is also critical for somatic mutation [19] and class switch recombination.

Mutation frequency is higher in IgG than in IgM antibodies. However, not every point mutation results in a change at the amino acid level. Silent (S) mutations are point mutations that do not result in a change in the translated amino acid; replacement (R) mutations cause a change in the amino acid and may result in an antibody with decreased or increased affinity for the triggering antigen. It is possible to indirectly analyze antigen selection during the germinal center response by calculating the ratio of replacement (R) to silent (S) mutations in rearranged antibody genes. Affinity maturation refers to the preferential selection and amplification of B cells producing high-affinity antibodies [20]. B cells producing antibodies with low affinity rarely progress to plasma or memory cells, whereas B cells producing antibodies of higher affinity will continue to be expanded. This selection for increased affinity may be reflected in a greater than random R/S ratio in the CDRs and a random R/S ratio in the FRs of the selected antibodies.

In summary, multiple mechanisms exist to generate a diverse repertoire of antibody molecules and ensure the production of antibodies that are protective against microbial invasion. However, because mechanisms of diversity in both the naïve B-cell repertoire and the antigen-activated B-cell repertoire proceed randomly in the absence of instruction from external antigens, they permit the production of autoantibodies, which must then be regulated.

B-Cell Tolerance

It is now clear that as immunoglobulin genes undergo rearrangement during B-cell development in the bone marrow, many autoreactive B-cell receptors are generated. The mechanisms to purge autoreactive immature B cell in the bone marrow and autoreactive transitional cells in the spleen include receptor editing (the generation of a new light chain which together with the heavy chain might no longer be autoreactive), apoptosis or anergy induction following engagement of the B-cell receptor. These mechanisms eliminate a high percentage of autoreactive B cells, but not all. After maturation to immunocompetence, naïve autoreactive B cells may normally be embargoed from the germinal center response by mechanisms not well understood. There are additional mechanisms after antigen activation that prevent autoreactive B cells from entering the memory B-cell compartment or from becoming long-lived plasma cells. It appears that in lupus patients, many, if not all, of these checkpoints are breached.

There are multiple probable explanations for the failure of normal tolerance mechanisms. Genetic factors that reduce the strength of B-cell receptor signaling clearly contribute to the survival of autoreactive B cells at early stages of B-cell development. High levels of B-cell activating factor (BAFF), a B-cell survival factor, also contribute to the survival of autoreactive B cells [21, 22]. Compensatory elevations in BAFF concentration may be related to the B-cell lymphopenia often seen in SLE patients, perhaps due to cytotoxic medications, or damage to B cells or to B-cell progenitors from disease. Moreover, elevated BAFF levels may be a consequence of nucleic acid containing immune complex (anti-DNA/DNA or anti-RNA/RNA) mediated activation of dendritic cells [23]. High levels of IL-17 may also be involved in the escape and activation of autoreactive B cells in SLE patients [24]. SLE patients overexpress CD40L on B cells; excessive co-stimulation is a mechanism for survival of autoreactive B cells. It is increasingly clear, that once inflammation begins, an amplification process is established that further depresses the efficacy of normal B-cell tolerance checkpoints.

ANTI-DNA ANTIBODIES; ORIGINS, STRUCTURE, IDIOTYPES, ANTIGENIC CROSS-REACTIVITY AND PATHOGENICITY

Anti-DNA antibodies that are disease specific and linked to pathogenicity are IgG isotypes, whereas IgM anti-DNA autoantibodies are members of the "natural autoantibodies" that are postulated to play a role in maintenance of peripheral tolerance through clearance of apoptotic debris. These low-affinity, germline-encoded IgM antibodies react preferentially with ssDNA, but are often polyreactive [25]. They are found in the sera of normal mice and humans and are believed to be non-pathogenic. In their absence, pathogenic IgG anti-DNA antibodies may be made.

B1 cells comprise less than 10% of total B cells and are preferentially localized to the peritoneal and pleural cavity in mice, in contrast to conventional B2 cells, which populate peripheral lymphoid organs [26]. Although both B1 cells and conventional B2 cells can produce anti-DNA antibodies [27, 28], the "natural" anti-DNA antibodies are produced mainly by B1 cells. The physiologic function of B1 cells is still controversial, but it seems increasingly likely that the IgM antibodies routinely participate in the elimination of apoptotic debris. In situations of systemic inflammation, however, these antibodies may undergo class switch recombination and the resulting germline-encoded IgG antibodies may be pathogenic due to IgG-associated effector functions [29]. The autoimmune NZB mouse strain displays an increased number of B1 cells. Elimination of these cells from the peritoneal cavity of NZB/W F1 lupus-prone mice leads to diminished anti-DNA serum titers with subsequent amelioration of kidney disease [30]. Although an increased number of B1 B cells have been reported in some human autoimmune diseases, their association with SLE is not firm. Based on their membrane expression of the 67-kDa molecule CD5, B1 cells can be divided into two subsets: B1a and B1b. The functional significance of the CD5 surface marker has not yet been determined; it may be a negative regulator of B-cell receptor signaling in B1 cells [31]. There is a long-standing notion that B1 cells represent a separate, self-replenishing B-cell population. Mice lacking or with functionally inactive BAFF have a block in B-cell maturation in the periphery with a dramatic decrease in marginal and follicular B2 B cells, but a normal number of peritoneal B1 B cells [32].

Anti-DNA antibodies in humans that are correlated with disease manifestations of SLE, specifically renal pathology, are encoded predominantly by somatically mutated Ig genes of the IgG isotype [33] and have specificity for dsDNA although they may also exhibit reactivity with ssDNA and other autoantigens (see Cross-reactivity section). In the NZB/W lupus-prone mouse, both marginal zone and follicular B cells contribute to anti-DNA antibody production. Follicular B cells are T-cell-dependent, and their involvement in the production of pathogenic autoantibodies is supported by demonstrations that CD4$^+$ T-cell depletion or blockade of B-cell/T-cell co-stimulatory pathways leads to decreased anti-DNA titers in NZB/W mice [34, 35]. NZB/W mice also demonstrate an expansion of marginal zone B cells. Most marginal zone B cells express germline-encoded Ig VH genes [36] and do not require cognate T-cell help for immunoglobulin production. They can provide help to T-cell-dependent responses by acting as antigen-presenting cells (APCs) [37] or by enhancing antigen presentation by follicular DCs [38]. Several studies have implicated marginal zone B cells in murine SLE. The increased number of autoreactive marginal zone B cells in NZB/W mice is linked to a chromosomal region that overlaps with Nba2, a genetic locus associated with lupus nephritis [39]. In a model of estrogen-induced lupus, the majority of anti-dsDNA antibody-secreting B cells are also marginal zone B cells [40].

It becomes increasingly clear that SLE is a heterogeneous disease and that different B-cell populations may be responsible for the majority of autoantibody production in different patients.

Structure of Anti-DNA Antibodies

The anti-DNA antibody response resembles a normal antibody response in its shift from IgM to IgG [41] and in its pattern of somatic mutation [42]. IgG anti-DNA antibodies have features suggesting selection in response to a specific antigen (i.e., a greater than random ratio of replacement to silent mutations in CDRs) [43]. Sequence analysis of some but not all antibodies has demonstrated a pattern of somatic mutation consistent with DNA, either alone or in a complex, as the selecting antigen. Interestingly, however, most of the IgG anti-DNA antibodies have no detectable affinity for DNA when back mutated to their germline configuration. Thus, the triggering antigen for many anti-DNA antibodies is not DNA.

Extensive genetic characterization of anti-DNA antibodies from both lupus patients and mouse models of lupus has been undertaken to understand the molecular basis for DNA binding. In general, the heavy- or light-chain V genes used to produce anti-DNA antibodies are used to encode protective, non-autoreactive antibodies [44]. Restriction fragment length polymorphism analysis comparing the Ig loci of lupus patients with those of non-autoimmune individuals shows that there are no apparent differences, such as gene loss or gene duplication, that distinguish the two groups [45]. Similar conclusions have been reached in mice [44] and it is clear that Ig genes from non-autoimmune strains can encode pathogenic anti-DNA antibodies. Genetic susceptibility to human lupus has not been mapped to the Ig locus.

A large number of heavy- and light-chain variable genes can potentially encode anti-DNA antibodies, but some appear to dominate the response [42]. Although IgM anti-DNA antibodies in humans are encoded by members of all seven V_H families [46] pathogenic IgG anti-DNA antibodies are biased toward members of the V_H3 and V_H4 families [46]. Although the overexpression of V_H3 family members may reflect the large size of this V_H family, preference for the small V_H4 family is intriguing [46].

Light-chain V gene use in both murine and human anti-DNA antibodies is more diverse than heavy-chain V gene use [47] with a preference for light chains with long CDR1 loops containing basic amino acids. The wider repertoire of light chains used to encode anti-DNA antibodies may reflect the proposed heavy-chain dominance in determining DNA reactivity [48] and may explain the polyreactivity of anti-DNA antibodies [49]. However, pairing murine anti-DNA heavy chain with different light chains demonstrates that light-chain identity can greatly influence the fine specificity of dsDNA binding.

Examination of the protein sequence of murine and human anti-DNA antibodies has been fruitful in delineating features that are specific to these antibodies. Several basic amino acids, namely arginine, histidine, and lysine, and some polar amino acids, such as glutamine, tyrosine, and asparagine, contribute to DNA binding [50]. Anti-DNA antibodies appear to be particularly enriched for arginine residues. Arginine is the most versatile amino acid for DNA binding. It is a frequently used residue in protein—nucleic acid interactions due to its potential to form electrostatic or hydrogen bonds with the DNA backbone and to its ability to interact with aromatic DNA bases. Available crystal structures of anti-DNA antibodies suggest that anti-ssDNA and anti-dsDNA antibodies may have distinct DNA-binding sites antibodies to ssDNA contain a deep cleft, whereas antibodies to dsDNA usually have a planar binding site [51].

Idiotypes of Anti-DNA Antibodies

Idiotypes are antigenic determinants located in the variable regions of antibodies (see Antibody structure). Idiotypes shared by multiple antibodies are called cross-reactive idiotypes and may reflect the use of common germline genes used to encode these antibodies. Private idiotypes are unique to specific antibodies. Anti-idiotypic antibodies arise after antibodies are produced in response to antigen and the anti-idiotype network is a complex cascade of antibodies thought to modulate expression of autoantibodies and have been postulated to have therapeutic potential [52]. These anti-idiotypes can bind to all antibodies bearing the specific idiotypic determinant, independent of their antigenic specificity [53].

Anti-DNA antibodies found in sera of lupus patients and lupus-prone mice express cross-reactive idiotypes [54]. These idiotypes are also present on serum Igs of non-autoimmune hosts, but in lupus are present at increased titer. Some are directly correlated with human disease activity and some have been identified in immune deposits in nephritic kidneys [54].

Anti-DNA-associated idiotypes are also present on antibodies that bind microbial antigens. This observation has led to the suggestion that anti-DNA antibodies might arise as a response to bacterial antigen, either by somatic mutation of the antibacterial antibody or through an idiotypic network.

Immunization of mice with an anti-idiotype to an anti-DNA antibody can induce lupus-like disease [55]. This observation, along with multiple murine studies showing that anti-idiotypic antibodies regulate the expression of idiotype-bearing antibodies, led to the hypothesis that the idiotypic network plays a role in triggering SLE or regulating disease activity and to the suggestion that manipulation of idiotype—anti-idiotype interactions is a potential therapeutic approach. One study identified the presence of anti-idiotypic antibodies to anti-DNA antibodies in an SLE patient in remission which were undetectable during disease flares [56]. Similarly, NZB/W lupus-prone mice develop disease at the time that they no longer produce anti-idiotypes to anti-DNA antibodies and passive administration of anti-idiotypic antibodies decreases anti-DNA antibody production and attenuates disease [57].

Administration of peptides derived from idiotypic regions of an anti-DNA antibody can modulate clinical and serologic disease. In mice, SLE-like disease activity is reduced after treatment with idiotypic peptides from the CDR-1 or CDR-3 of an anti-DNA antibody [58]. T regulatory cells appear to play a critical role in these models as diminution in disease activity is dependent on their presence. Furthermore, culture of SLE and healthy peripheral blood mononuclear cells with anti-DNA Ig-derived peptides results in expansion of cells with the phenotype of T regulatory cells in the SLE PBMCs [59]. While these studies are provocative, the role of the idiotype network in SLE remains uncertain.

Antigenic Cross-Reactivity of Anti-DNA Antibodies

The diversity of the autoreactivities found in lupus sera differentiates SLE from many other autoimmune disorders. Although such diversity has been attributed to a polyclonal activation of B cells, studies show that the diversity of antigenic specificities may be due, at least partially, to the cross-reactivity of anti-DNA antibodies with a wide variety of molecules (Table 14.1). IgG anti-DNA-secreting B cells derived from SLE patients are much more cross-reactive than IgG anti-DNA-secreting B cells from non-autoimmune individuals which generally bind single-stranded DNA, and rarely, double-stranded DNA. Moreover, they fail to show a normal decrease in cross-reactivity compared to IgM anti-DNA B cells [60] suggesting that the cross-reactivity reflects more than increased avidity.

The diverse antigenic specificities of IgG anti-DNA antibodies allow them to cross-react with structurally distinct antigens including polynucleotides, phospholipids, polysaccharides, proteoglycans and cell surface and intracellular proteins with variable effects on cell function. Immunization of autoimmune MRL mice with β2 glycoprotein I induces anti-DNA antibodies. This DNA reactivity is inhibited with cardiolipin micelles, demonstrating that these antibodies bind both DNA and cardiolipin [61]. Moreover, monoclonal anti-DNA antibodies from lupus-prone mice bind cardiolipin, as well as other phosphorylated molecules [62]. Analysis of the specificities of these murine and human anti-DNA antibodies suggests that phosphodiester groups common to both DNA and phospholipids constitute the shared epitope recognized by these cross-reactive antibodies. Phospholipids lacking regularly spaced phosphate groups are not bound by anti-DNA antibodies.

Anti-DNA antibodies derived from lupus-prone mice can bind to bacterial surfaces [63]. This binding, which is unaffected by DNase treatment and is competitively inhibited by DNA, is due in part, to the phosphodiester groups of bacterial cell walls. A human monoclonal IgM anti-DNA antibody binds to human commensal bacteria such as *L. acidophyllus, B. bifidum*, and *L. plantarum*; the cross-reactive epitope is a phosphodiester group [64]. Structural similarities between the sugar-phosphate backbone of DNA and lipid A, a common constituent of Gram-negative bacteria [65] provide a possible explanation for the cross-reactivity of human IgM anti-DNA antibodies to these antigens. Antibodies binding both DNA and pneumococcal cell wall and capsular polysaccharide have been isolated from mice [66] and humans [67]. Finally, a human monoclonal antibody that reacts with bacterial capsular polysaccharides of the human pathogens, group B *Meningococcus*, and *Escherichia coli Kl*, also recognizes DNA [68]. Although many cross-reactivities of anti-DNA antibodies can be explained by their recognition of phosphodiester groups, the basis for other cross-reactivities is less apparent. For example, monoclonal anti-DNA antibodies derived from mice and humans with SLE bind to the glycolipid components of the mycobacterial cell.

Two explanations may account for the cross-reactivity observed between DNA and carbohydrate antigens. Either the binding site of anti-DNA antibodies may be able to accommodate a variety of antigenic structures or an as yet unidentified tertiary structure of these molecules may be the common epitope recognized by anti-DNA antibodies. Studies suggesting that negative charge distribution may be the key component responsible for the cross-reactivity between DNA and bacterial capsular polysaccharide support the latter possibility.

TABLE 14.1   Assays for the detection of anti-DNA antibodies

Assay	High-affinity anti-DNA antibodies detected	Low-affinity anti-DNA antibodies detected
Farr assay	+	−
Millipore filter method	+	−
ELISA	+	+
Crithidia lucilia	+	+

Anti-DNA antibodies cross-react with a variety of extracellular matrix components, including proteoglycans. These molecules are rich in repeating, negatively charged units deriving from their glycosaminoglycan constituents hyaluronic acid, chondroitin sulfate, and heparan sulfate. Analysis of the interaction of several murine monoclonal anti-DNA antibodies with extracellular structures within renal glomeruli and tubules reveals cross-reactivity with laminin, a glycoprotein constituent of basement membranes [69]. This binding is inhibited by DNA and cannot be abolished by DNase treatment.

Anti-DNA antibodies have also been shown to interact with cell surface molecules expressed on several human and mouse cell types, including Raji cells [70], B cells [71], T cells, erythrocytes [72], neurons [73], epithelial cell cultures from human kidney and skin [71], and embryonal cells [74]. In general, pretreatment of target cells with proteases abolishes anti-DNA antibody binding whereas pretreatment with enzymes that degrade sugar moieties has no effect on binding [71] suggesting that the reactive cell surface components are proteins, and not the sugar moieties of glycoproteins and proteoglycans. Both murine and human anti-DNA antibodies cross-react with the NR2 subunit of *N*-methyl-D-asparate (NMDA) receptor on neurons [73]. These antibodies can be found in cerebrospinal fluid of lupus patients and in brain tissue. In mouse models, they can induce manifestations of neuropsychiatric SLE [75].

Anti-DNA antibodies cross-react with intracellular proteins as well. Monoclonal human and mouse anti-DNA antibodies react with the intermediate filament vimentin, and other cytoskeletal proteins, such as α-actinin [76] and tubulin [77]. Anti-dsDNA antibodies bind the ribosomal proteins S1 [78], P0 [79], and P1 [80] and the snRNP polypeptides, A and D [81, 82]. Phenylalanine in the C-terminal hydrophobic region of ribosomal protein P1 appears crucial for cross-reactivity with anti-dsDNA antibodies, which may be due to structural similarities between the aromatic ring of phenylalanine and the cyclic forms of a DNA molecule, such as nucleotide bases or pentoses in the backbone.

The search for cross-reactive antigens continues as these antigens may yield clues to the origins and pathogenesis of anti-DNA antibodies.

Pathogenic Potential of Anti-DNA Antibodies

The presence of anti-dsDNA antibodies is virtually diagnostic of SLE and rarely occurs in other conditions. It is clear that anti-DNA antibodies have pathogenic potential and can cause specific organ toxicity or systemic manifestations. Clinically, anti-dsDNA antibody titers correlate with disease activity, nephritis in particular. Anti-DNA antibodies have been eluted from glomeruli from patients with active lupus nephritis [83]. Cross-reactive anti-DNA/anti-NMDAR antibodies have also been eluted from the brains of patients with neuropsychiatric symptoms [75] and when these antibodies are injected into mouse brain, cause neuronal apoptosis. Additionally, levels of these cross-reactive anti-DNA/anti-NMDAR antibodies in cerebral spinal fluid correlate with the presence of active neuropsychiatric lupus [84, 85]. To date, however, much more is known about the association between anti-DNA antibodies and lupus nephritis.

Ex vivo, perfusion of rat kidney with monoclonal mouse and polyclonal human IgG anti-DNA antibodies leads to proteinuria and decreased renal function and administration of anti-DNA antibodies into non-autoimmune mice produces nephritis [86, 87]. Moreover, transgenic mice expressing only the secreted form of an anti-DNA antibody develop lupus nephritis [88]. However, not all anti-dsDNA antibodies are pathogenic. Although approximately two-thirds of lupus patients demonstrate serum anti-DNA antibodies, many do not develop clinical manifestations related to these autoantibodies. The pathogenic potential of anti-DNA antibodies depends on their structural and molecular properties [87–90]. Nephritogenic anti-DNA antibodies are predominantly high-avidity, somatically mutated, IgG, cationic antibodies that fix complement and show preferential reactivity to dsDNA [91]. However, germline-encoded anti-dsDNA antibodies that have undergone T-cell-independent class switching may also bind to the brain and kidney with subsequent tissue damage [29].

Anti-DNA antibody deposition occurs in multiple renal sites of SLE patients and lupus-prone mice including subendothelial and subepithelial spaces, mesangium, and along the basement membrane and tubules, resulting in variable clinical presentations of lupus nephritis [91]. Two mechanisms are hypothesized to explain anti-DNA antibody deposition in the kidney. The first model suggests that direct binding of cross-reactive anti-DNA antibodies to glomerular antigens is the leading mechanism of immune deposit formation. The second is the "planted" antigen hypothesis that suggests that apoptotic debris from dying cells is released into the circulation and binds to sites within the glomerulus. Anti-DNA antibodies bind to these renal-bound determinants. It is no longer assumed that preformed circulating immune complexes become sequested in glomeruli.

Hypothesis 1: Cross-reactivity with renal antigen

A comparison of circulating anti-DNA antibodies derived from normal individuals, patients with active lupus, or renal eluates of patients with active lupus nephritis shows that anti-DNA antibodies eluted from the kidney are extensively cross-reactive, binding to polynucleotides, phospholipids, and the Sm/RNP complex [92]. Similar observations have also been made in a lupus mouse model. The observation that anti-DNA antibodies are cross-reactive for cardiolipin and other negatively charged phospholipids suggests that anti-DNA antibodies may be pathogenetic by reacting with renal antigens that share common epitopes with DNA [62]. This hypothesis is supported by studies showing that an exogenously administered murine monoclonal anti-DNA antibody forms immune deposits by binding directly to the GBM antigens and that this binding is not inhibited by DNase [93]. Moreover, observations on mutated anti-DNA antibodies suggest that binding to DNA is not the sole determinant of renal pathogenicity. Mutants of anti-DNA antibodies with decreased binding to DNA are still pathogenic [94] while mutations that increase DNA binding can be associated with a decrease in glomerular binding.

Clues to the identity of the glomerular antigens bound by anti-DNA antibodies originate from studies demonstrating cross-reactivity between anti-DNA antibodies and constituents of the glomerular basement membrane (GBM) and the mesangium, such as hyaluronic acid, chondroitin sulfate, heparan sulfate [95], laminin [69], and α-actinin [76, 96]. Heparan sulfate is the major glycosaminoglycan constituent of GBM and plays a leading role in the maintenance of the charge and size-selective barrier of the GBM [97]. Masking of glycosaminoglycan polyanionic sites occurs following the deposition of cross-reactive anti-DNA antibodies, resulting in proteinuria and influencing the intraglomerular handling of circulating antigens, antibodies, and immune complexes. This observation presents two possible mechanisms by which the cross-reaction of anti-DNA antibodies with glomerular antigens might

lead to lupus nephritis. Binding of anti-DNA antibodies to renal antigens may (1) disrupt the normal physiologic function of the kidney, or (2) may trigger local inflammation and tissue damage, or both.

In addition to reacting with extracellular components of the kidney, anti-DNA antibodies may bind directly to cell membranes. For example, human and mouse anti-DNA antibodies cross-reacting with A and D snRNP polypeptides exert their deleterious effects by binding directly to intracellular structures following receptor-mediated entry into kidney cells [81, 98]. Whether they interfere with intracellular processes *in vivo* in SLE patients is not known.

Hypothesis 2: Planted antigen

In contrast to the cross-reactivity theory, the planted antigen hypothesis states that anti-DNA antibodies form immune deposits in the kidney by binding to auto-antigens, either DNA or nucleosomes, that have deposited previously [99]. Both DNA and nucleosomes have an affinity for the GBM and glomeruli [100]. Specifically, DNA interacts with collagen [101], fibronectin [102], and laminin [69], whereas nucleosomes have an affinity for heparan sulfate [103].

DNA bound to collagen is recognized by anti-DNA antibodies that readily form immune complexes *in situ*, whereas preformed DNA−anti-DNA complexes do not bind collagen to the same extent. Further support of the planted antigen hypothesis comes from *ex vivo* perfusion studies showing that anti-DNA antibodies do not bind directly to renal antigens, but that binding to the GBM is mediated through histones and DNA. Furthermore, some IgG from patients with SLE nephritis bind human glomerular extracts only in the presence of either DNA or nucleosomes [99].

Although the presence of data supporting each of these two hypotheses may at first be disconcerting, these seemingly contradictory results may simply reflect the diversity of anti-DNA antibodies. These models may not be mutually exclusive. Rather, different pathogenic mechanisms may operate simultaneously or successively in the same lupus patient; alternatively, different mechanisms may be at work in different patients. This may explain observations that the glomerular lesions of SLE represent several distinct forms of nephritis that often differ from patient to patient.

Systemic Effects of Anti-DNA Antibodies

Anti-DNA antibodies complexed with DNA can also activate dendritic cells through engagement of toll-like receptor 9 (TLR 9) with the ensuing secretion of BAFF and pro-inflammatory cytokines [23, 104]. TLR 9, a critical component of the innate immune response, is located in an intracellular endosomal compartment and is expressed in plasmacytoid dendritic cells and B cells. It binds specific DNA sequences, typically of microbial origin, but can also bind mammalian DNA. TLR 9 activation transforms tolerogenic dendritic cells into immunogenic dendritic cells and induces production of interferon (IFN) α [105, 106]. Immunogenic dendritic cells engage the adaptive immune response with T-cell activation and B-cell antibody production.

Overexpression of IFN-α inducible genes in peripheral blood mononuclear cells, the IFN-α signature, is a well-recognized feature of SLE and correlates with disease activity [107, 108]. IFN-α promotes differentiation of monocytes into dendritic cells which subsequently prime naïve T cells, and promote activation and expansion of autoreactive T and B cells [109]. This together with the increased BAFF produced by dendritic cells after TLR 9 activation sustains autoantibody production and leads to an amplication loop with BAFF-induced switching of IgM to IgG antibodies. DNA—anti-DNA immune complexes may also be internalized by the B-cell receptor on DNA-reactive B cells and activate TLR 9 [23], causing heavy-chain class switching and Ig secretion.

Activation of Anti-dsDNA B-Cells

Understanding the process of anti-DNA antibody production necessitates a brief review of B-cell activation. T-cell-dependent B-cell activation occurs following B-cell encounter with antigen that cross-links membrane Ig. Antigen is internalized and complexed to class II MHC molecules within the endoplasmic reticulum and subsequently transported back to the cell surface. Activated T helper cells recognizing the antigen/MHC complex may provide co-stimulatory signals through a variety of cell surface molecules. These co-stimulatory pathways include the interaction of CD40 on the B cell with the CD40 ligand on the T cell, B7 or B7h on the B cell, and CD28, CTLA-4 or ICOS on the T cell, as well as the interaction of BAFF and its B-cell receptors, TACI, BAFF-R, and BCMA (Figure 14.3).

Recent studies have now made it apparent that naive B cells can be activated, undergo class switch recombination and plasma cell differentiation by antigen and various inflammatory mediators such as TLR ligands, BAFF and IL-17 in a fashion independent of cognate T-cell help [110]. Thus, in a permissible pro-inflammatory cytokine milieu, secretion of pathogenic IgG anti-DNA antibodies may occur in the absence of T-cell help.

Identification of the antigen(s) that activate DNA-reactive B cells remains elusive. Some investigators have suggested that self-antigens presented in an immunogenic manner induce SLE, whereas others believe that the precipitating event is a response to foreign antigen. Thus, multiple models exist to explain anti-DNA

antibody production: (1) a novel association between DNA and a foreign antigen such as a viral protein may establish a hapten—carrier complex that triggers a T-cell-dependent anti-DNA response; (2) an antibody elicited in response to a microbial antigen may cross-react with DNA because of shared epitopes between the microbial antigen and DNA (molecular mimicry); (3) somatic mutation of an antimicrobial antibody may generate an autoantibody; (4) apoptotic material including enzymatically degraded DNA and nucleosomal complexes may be internalized by a DNA-specific B cell and activate a cell that would otherwise be tolerized; (5) apoptotic debris may activate dendritic cells and create an immunogenic milieu permissive to the activation of autoreactive cells, and (6) polyclonal activation of either T or B cells may induce anti-DNA antibody production. The first three models propose a role for foreign antigen. If the response to foreign antigen activates T or B cells that are cross-reactive with self-antigen, an immunologic cascade, termed epitope spreading can ensue, leading to autoantibodies against multiple self-antigens. This secondary amplification of target antigens is discussed below.

Immunogenicity of DNA

Hypothesis 1: DNA as a hapten

DNA alone is a poor immunogen. Early attempts to generate anti-dsDNA antibodies by immunizing mice with native eukaryotic bacterial DNA (B-DNA) were unsuccessful Immunization with Z-DNA induces anti-dsDNA antibodies, but these antibodies are specific for Z-DNA and do not cross-react with native B-DNA [111]. Bacterial DNA complexed with bovine serum albumin (BSA) is more immunogenic than eukaryotic DNA. When NZB/W mice are immunized with *Escherichia* (*E.*) *coli* DNA complexed with methylated BSA, they develop antibodies binding both *E. coli* and mammalian DNA [86]; however, immunization of non-autoimmune mice with *E. coli* DNA results in an antibody response that is restricted to bacterial DNA and is rarely cross-reactive with eukaryotic DNA [112]. These data indicate that bacterial DNA can elicit lupus-like antibodies in a host lacking intact mechanisms of self-tolerance, but will only induce antibodies specific for bacterial DNA in a host maintaining self-tolerance.

Based on the poor immunogenicity of DNA, it has been hypothesized that the immunogen in lupus is not naked DNA but consists of DNA complexed to DNA-binding proteins [113]. To determine whether DNA in association with protein might induce anti-dsDNA antibodies, non-autoimmune mice were immunized with a highly immunogenic peptide, Fus 1 (derived from the protozoan *Trypanosoma cruzi*), complexed with calf thymus DNA [113]. In this model it is helpful to consider DNA as a hapten recognized by B cells and the peptide as a carrier recognized by T cells. Immunized mice develop low titers of antibodies to calf thymus DNA, suggesting that DNA can be immunogenic when presented in a complex with a highly immunogenic carrier. DNA coupled to viral proteins can also induce anti-DNA antibodies. Non-autoimmune mice hyperimmunized with polyoma BK virus T antigen develop antibodies to both mammalian and viral DNA. The anti-dsDNA response depends on the DNA-binding property of the T antigen, which serves as the carrier protein for the DNA hapten [114]. By establishing hapten−carrier complexes with DNA, viruses may initiate the production of anti-dsDNA antibodies *in vivo* and persistent viral infection may lead to autoimmunity.

Hypothesis 2: Apoptotic material

Recent studies have provided mounting evidence that apoptotic debris, if not cleared by a variety of mechanisms, including opsonization by natural IgM autoantibodies, can contribute to the development of an SLE-like serology. Deficiencies in C1q, C4, DNase I, and natural IgM all lead to the production of anti-DNA antibodies in mouse models [115−118]. A small number of lupus patients have a deficiency in DNase I, and genetic deficiencies of C1q and C4 predispose to SLE. One recent study has suggested that a polymorphism in Trex 1,

a gene responsible for encoding an exonuclease, predisposes to SLE [119]. These data provide compelling support that the failure to clear apoptotic debris through normal clearance pathways renders that debris immunogenic. How often a failure to clear apoptotic debris initiates human lupus is unclear, but some histologic studies of lymph nodes of lupus patients suggest that poor clearance may be part of the pathogenesis of established disease.

Hypothesis 3: Molecular mimicry

Environmental pathogens have long been suspected to be inducers of autoimmune disease. Autoimmune NZB mice raised in a germ-free environment develop reduced titers of autoantibodies; conversely, elevated anti-DNA antibody titers can be found in patients with microbial infections, suggesting that microbial stimulation of the immune system helps stimulate autoantibody induction [120, 121]. Thus, bacterial exposure plays a role in activating the autoreactive B cells although its role may merely be to establish a level of immune activation. Studies of human and murine monoclonal anti-DNA antibodies demonstrate extensive cross-reactivity with bacterial antigens (discussed earlier), suggesting that bacterial antigens may elicit these antibodies. This cross-reactivity, known as molecular mimicry, offers a potential explanation for the development of autoimmunity (Figure 14.1). Alternatively, during an antimicrobial response, T cells which cross-react with both the

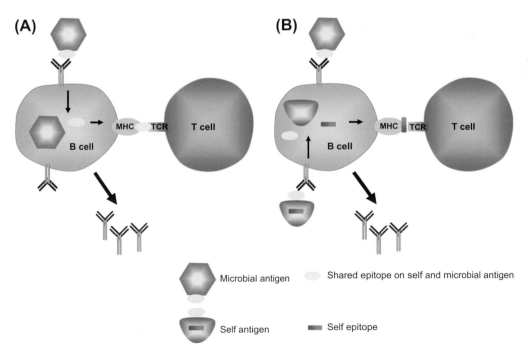

FIGURE 14.1   Molecular mimicry and presentation of self-epitopes can perpetuate an autoimmune response. (*A*) A B cell specific for a determinant shared by a microbial and self-antigen is triggered by a T cell specific for a unique determinant on the foreign antigen. (*B*) The activated B cell internalizes, processes, and presents epitopes of the self-antigen. A T cell that has never seen this epitope before, and has therefore not been tolerized to it, is now activated and continues to provide help to the autoreactive B cell and to other B cells with this autospecificity.

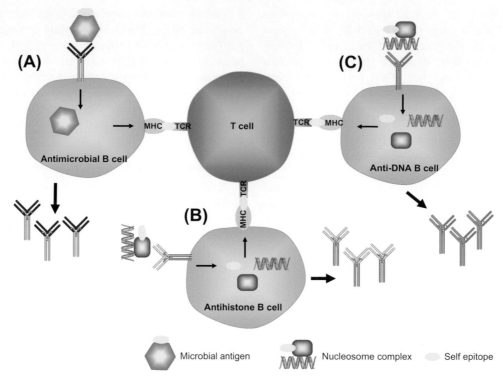

FIGURE 14.2   Epitope spreading leads to the production of anti-dsDNA antibodies. (*A*) A B cell specific for a microbial antigen internalizes, processes, and presents epitopes from the foreign antigen to T cells. (*B*) A T cell recognizing an epitope on a microbial antigen cross-reacts with a self-epitope and therefore activates both antimicrobial- and anti self (histone)-specific B cells, causing them to secrete antibodies. (*C*) An anti-dsDNA B cell recognizes DNA in the nucleosome complex and presents the same histone epitope that has been presented by the antihistone B cells to activated T cells. These T cells can now activate anti-dsDNA B cells, causing them to secrete anti-dsDNA antibodies.

microbial and self-antigens may be activated and provide help to B cells specific for the self-antigen (Figure 14.2).

Hypothesis 4: B-cell somatic mutation

Many pathogenic anti-dsDNA antibodies in human and murine lupus are encoded by somatically mutated Ig genes. The importance of somatic mutation in generating pathogenic anti-DNA antibodies is illustrated by experiments showing that transgenes coding for anti-ssDNA antibody gain specificity for dsDNA by somatic mutation. Also, treatment with a co-stimulatory blockade in NZB/W mice leads to a decreased frequency of somatic mutations in the VHBW-16 gene that encodes pathogenic anti-DNA antibodies, contributing to decreased anti-dsDNA antibody titer and prolonged survival [34].

The observation that many anti-DNA antibodies cross-react with microbial antigens suggests that antibodies to dsDNA arise from antimicrobial antibodies undergoing somatic mutation. Support for this hypothesis is obtained in an *in vitro* culture system in which spontaneous mutation of an antibody to phosphorylcholine (PC) results in an antibody with specificity for dsDNA [17]. To determine if anti-dsDNA antibodies

routinely derive from mutated antimicrobial antibodies, efforts were made to generate anti-dsDNA hybridomas from non-autoimmune mice immunized with PC. Anti-dsDNA antibodies crossreacting with PC occur rarely in serum [66], however many hybridomas generated from splenocytes fused with a bcl-2 transfected fusion partner, cross-react with PC and dsDNA [122]. Bcl-2 is an anti-apoptotic gene [122] which promotes B-cell survival and is upregulated in B cells destined to become plasma cells or memory B cells. These hybridomas display somatically mutated genes, suggesting that autoreactive B-cell clones arise routinely following exposure to a bacterial antigen, but are deleted in normal mice.

Perhaps the most convincing data suggesting that anti-dsDNA antibodies can arise by somatic mutation following a response to bacterial antigen come from studies of human antibodies in which reversion of anti-DNA antibodies to their germ-line sequence leads to a loss of DNA reactivity [123].

Hypothesis 5: Epitope spreading

The production of a diverse repertoire of autoantibodies can begin with the activation of a single autoreactive lymphocyte, either a B cell (Figure 14.1) or a T cell

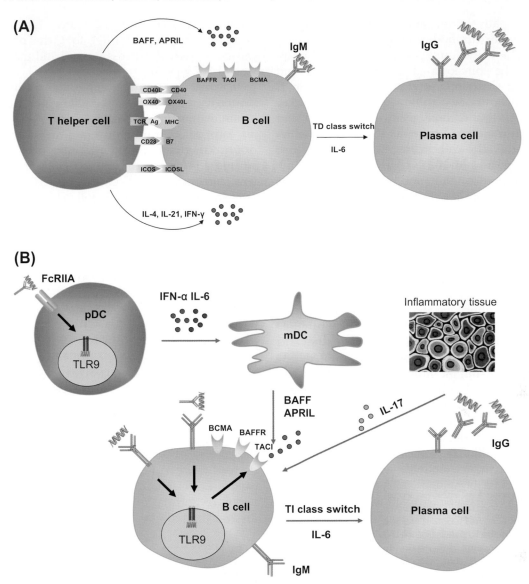

FIGURE 14.3 (*A*) Production of T-dependent pathogenic anti-dsDNA IgG antibodies. Antigen binding to a BCR triggers activation of the B cell. The B cell processes antigen and presents it in the context of MHCII to a T helper cell through an interaction with the TCR. Co-stimulatory molecules enhance the cognate B-cell—T-cell interaction. The activated B cell undergoes somatic mutation to acquire DNA reactivity and differentiates into a plasma cell and produces class switched anti-dsDNA IgG antibodies. (*B*) Production of T-independent pathogenic anti-dsDNA IgG antibodies. DNA-containing immune complexes activate TLR9 within a pDC and B cell. The activated pDC produces IFN-α and IL-6 which triggers mDCs to express high levels of BAFF and APRIL. These molecules together with IL-17 produced by inflammatory tissue promote T-cell-independent class switch recombination. The activated B cell secretes IL-6 and differentiates into plasma cells which produce germline encoded anti-dsDNA IgG antibodies.

(Figure 14.2), followed by the diversification of the initial immune response. Although the initial response may target one epitope of a protein or multimolecular complex, the immune response spreads or diversifies to include other epitopes of the initiating antigen. This process is termed epitope spreading. Native DNA is a component of many multimolecular complexes whose subunits are often targets for autoantibody production in SLE. DNA associates with histone proteins to form chromatin. DNA also associates with replication/repair

proteins, the Ku protein and transcription factors; antibodies to several of these DNA-binding proteins are often present in SLE patients [124]. Cryptic or novel epitopes derived from the protein subunits in DNA protein complexes may be presented by B cells and activate autoreactive T cells, and so, perpetuate and expand an autoantibody response (Figure 14.1).

Evidence for epitope spreading exists in human and mouse SLE [125, 126]. Antichromatin antibodies are detected before anti-DNA antibodies in most lupus patients. Additionally, lupus patients with anti-Sm antibodies often develop antibodies to RNP and DNA over time [82, 126]. This is intriguing as both RNP and Sm are components of the spliceosomal complex. Rabbits and mice immunized with peptides derived from the Sm B protein develop antibodies to Sm as well as to other components of the splicesome, including U RNAs and nRNPs, suggesting that epitope spreading occurs between Sm and RNP [126]. In fact, some animals also developed antibodies to dsDNA, exemplifying how anti-dsDNA antibodies may arise even when DNA is not the eliciting autoantigen if there is physical linkage between the immunizing protein and nucleic acid.

T-Cells Involved in Anti-DNA Antibody Production

While it is clear that pathogenic IgG anti-DNA antibodies can be germline encoded and that the inflammatory milieu present in lupus patients can promote T-cell-independent class switching, many pathogenic anti-dsDNA antibodies are somatically mutated IgG antibodies. Because somatic mutation is usually mediated by T-cell help, it is presumed that cognate T-cell help contributes to the production of autoantibodies [127, 128]. T cells play an important role in autoimmune mouse models; as interfering with T-cell–B-cell interactions ameliorates disease activity [129, 130].

Helper T cells

T cells from SLE patients display activation defects secondary to altered expression of signaling molecules. For example, some lupus T cells have an impaired protein kinase C response to integrin-mediated activation, diminished proliferation in response to anti-CD3 and anti-integrin antibody, and diminished adhesion to fibronectin [131, 132]. Studies have also demonstrated decreased expression of the TCR ζ, chain, a crucial signaling molecule, caused by an increased frequency of mutations and polymorphisms in the TCR ζ chain untranslated region [133]. Upregulated FcRγ chains substitute for the downregulated ζ chain in the T-cell receptor complex resulting in increased T-cell activation with phosphorylation of downstream signaling molecules and increased free intracellular calcium concentrations [133]. The Syk tyrosine kinase is significantly upregulated in lupus T cells and associates with the FcRγ. Signaling through FcRγ-Syk is 100 times more potent than through the ZAP-70 ζ chain association which is predominant in normal T cells [133]. Increased intracellular calcium concentrations cause the translocation of NF-ATc2 into the nucleus resulting in increased expression of CD40L and decreased production of IL-2 following activation [134]. It has been suggested that aberrantly increased protein phosphatase 2a (PP2a) in lupus T cells dephosphorylates Elf-1, resulting in decreased binding of Elf-1 to ζ the promoter which causes decreased expression of ζ and increased expression of FcRγ [135].

In recent years, attention has focused on the antigenic specificity of T cells involved in murine and human lupus. Autoreactive T cells appear to recognize either self-peptides that derive from a nucleoprotein complex or peptides derived from the autoantibodies themselves. An analysis of 15 autoreactive T-cell clones derived from five patients with SLE showed that T cells support the polyclonal activation of B cells [136]. T cells engage in cognate interactions with autoreactive B cells as autoantibody secretion does not occur when antigen presentation is blocked with anti-HLA class II antibodies. Histone-specific T-cell clones generated from autoimmune mice augment the production of anti-DNA antibodies *in vitro* and induce anti-dsDNA production when transferred into autoimmune hosts prior to disease onset [125]. As stated above, because DNA and histones are components of the same nuclear complex, a histone-specific T cell can activate anti-DNA B cells. Similarly, in a peptide-induced model of SLE, activation of class II-restricted CD4$^+$ T cells is necessary for the breakdown of tolerance and autoantibody production [137].

B cells can process their own immunoglobulin molecules and present immunoglobulin-derived peptides in the context of MHC class II molecules [138]. Peptides derived from anti-dsDNA antibodies can prime autoreactive T cells in autoimmune NZB/W mice, lead to activation of anti-dsDNA B cells, and hasten the onset of disease [139].

Treg cells

Regulatory T cells (Treg) are a heterogeneous population of CD4+, CD8+ and NK T cells, and can be classified by their surface markers, cytokine secretion profile and Foxp3 expression [140]. CD4+CD25+Foxp3+ T cells appear to play a significant role in preventing autoimmunity and maintenance of immune homeostasis. Depleting these cells in mice results in severe organ-specific autoimmunity, expansion of autoreactive T cells and accelerated anti-dsDNA autoantibody production in lupus-prone mice [141]. Treg cells inhibit the proliferation of T helper cells *in vitro* and *in vivo*. A Treg cell deficit, therefore, may augment the helper function provided by conventional CD4+ T cells to dsDNA-reactive B cells. Treg cells can also indirectly or directly suppress autoreactive B cells by cell-to-cell contact, cytotoxic factors or TGF-β. The mechanism of direct suppression of B cells by Treg cells is not clear, although the induction of apoptosis of the autoreactive B cells has been suggested [142].

Lupus-prone mice have a significantly lower number of Treg cells than non-autoimmune strains, and the number of Treg cells is negatively correlated with the production of dsDNA autoantibodies [143]. Studies of several lupus-prone strains demonstrate a decrease in the number of Treg cells just before development of high titers of dsDNA autoantibodies [144]. Moreover, adoptive transfer of Treg cells can delay the onset of lupus, or, in mice already expressing high titers of anti-DNA antibodies, transfer of Tregs can decrease their titer, suppress proteinuria and increase survival [145]. Two hypotheses exist to explain the role of Treg cells in the pathogenesis of murine lupus: one is that the number of Treg cells is reduced; the other, that Treg cell function is impaired, due to resistance of target cells to their suppressive effect [146].

Studies of CD4+CD25+Foxp3+ Treg cells in SLE patients are controversial. Most studies show a significantly lower frequency of Treg cells in SLE patients than healthy controls. One study examining samples from untreated patients shows that Treg cells inversely correlate with clinical activity and titers of dsDNA autoantibodies [147]. After corticosteroids or immunosuppressive therapy, the number of Treg cells significantly increase [148]. Some, but not all, studies describe reduced suppressive capacity of Treg cells in SLE. Type I IFN, known to be increased in SLE, was found to contribute to a decreased suppressive capacity of Treg cells [149]. The reasons for the discrepancies in these studies could be heterogeneous patient populations, the effects of treatment on the subjects studied, and technical differences in the assays.

REGULATION OF ANTI-dsDNA B cells

Tolerance Induction

The advent of transgenic technology has enabled investigators to follow the fate of autoreactive B cells and to learn more about the maintenance of tolerance. A number of transgenic mouse models have been generated to study regulation of anti-DNA B cells [150, 151]. These studies have led to the development of a model for tolerance induction in the immature B-cell population. When a B cell encounters self-antigen in the bone marrow, RAG genes are re-expressed and receptor editing occurs. If this process does not lead to successful production of a new non-autoreactive antibody, deletion or anergy will ensue. Deletion occurs under conditions of extensive cross-linking of membrane Ig while anergy results when cross-linking is more modest. Some autoreactive B cells will have an affinity for antigen below a threshold for tolerance induction or will encounter too little antigen to trigger tolerance induction and will remain activatable in the periphery [152].

B-cell numbers are often low in lupus patients. This may reflect the effects of cytotoxic medications or disease-specific factors such as the presence of lymphotoxic antibodies. The relative B-cell lymphopenia leads to elevated levels of serum BAFF. High BAFF levels impair normal B-cell tolerance in mice [22]. Studies of lupus patients show inadequate function of B-cell tolerance mechanisms. Thus, the B-cell lymphopenia may contribute to the survival and maturation of autoreactive B cells. There are also increasing data that at least in some individuals, estrogen may contribute to the survival of autoreactive B cells through increased BAFF production and diminished BCR signaling [153].

Once B cells mature to immunocompetence, they can be activated by antigen and a variety of co-stimulatory signals including, but not limited to, cognate T-cell help. As naïve cells become activated and mature to memory cells and plasma cells, expression of FcRIIb is progressively increased [154]. FcRIIb is an inhibitory Fc receptor and the only Fc receptor expressed on B cells. This maturation program ensures a relatively simple mechanism for inhibiting memory B-cell activation as colligation of the B-cell receptor and FcRIIb by immune complexes delivers a negative signal to the cell. Furthermore, engagement of FcRIIb on plasma cells delivers a death signal, terminating further production of antibody. Mice deficient in FcRIIb develop anti-DNA antibodies [155]; moreover, many lupus-prone mice exhibit low levels of FcRIIb expression and correction of this deficiency leads to reduced titers of anti-DNA antibodies [155]. Many SLE patients fail to upregulate FcRIIb on memory and plasma cells, thereby releasing antigen-activated B cells from normal regulatory pathways [155]. It is also clear that there are as yet undefined mechanisms that prevent B cells from entering a germinal center response or from exiting the germinal center to become a memory B cell or a plasma cell. Analysis of circulating autoreactive, anti-DNA producing B cells from patients with moderately active disease compared to patients with quiescent disease shows more autoreactive antigen-naïve and antigen-experienced B cells in patients with active disease suggesting impaired peripheral tolerance checkpoints in SLE [156].

ANTI-DNA ANTIBODY TRANSGENIC MICE

Transgenic mouse models expressing either the heavy chain of an anti-DNA antibody or both the heavy and the light chain have been analyzed in some detail. In one transgenic system, non-autoimmune BALB/c mice were made transgenic for the heavy chain of an IgM anti-DNA antibody from a MRL/lpr lupus-prone mouse. Pairing of the heavy chain with different endogenous light chains results in either antibodies recognizing ssDNA only or those that also recognize dsDNA. Analysis of anti-DNA B cells suggested that anti-ssDNA B cells are anergized, whereas B cells with an affinity for dsDNA are deleted [150]. Subsequent studies using mice transgenic for IgM heavy chain, as well as particular transgenic light chains, have confirmed these results [157]. Transgenic anti-ssDNA B cells have a normal life span compared to non-transgenic B cells but display functional defects, such as decreased total surface Ig and suboptimal response to T-dependent and T-independent stimuli. Further analysis reveals that receptor editing is responsible for a shift in specificity of the B-cell receptor from dsDNA binding to non-dsDNA binding [157]. SLE patients appear to have diminished receptor editing. B cells from several patients with SLE were unable to edit a V_κ gene that is frequently associated with a cationic anti-DNA antibody whereas normal B cells are capable of editing this light chain [158].

In another transgenic system, both non-autoimmune BALB/c mice and autoimmune NZB/W mice transgenic for the heavy chain of an IgG anti-dsDNA antibody were analyzed [49, 151]. In these mice, the transgenic heavy-chain pairs with a spectrum of endogenous light chains to produce B cells with DNA-binding and non-DNA-binding specificities. Although non-autoimmune mice display negligible titers of transgene-encoded anti-dsDNA antibody due to the induction of tolerance, autoimmune mice secrete elevated titers of transgene-encoded anti-dsDNA antibody. Analysis of autoimmune and non-autoimmune mice reveals the presence of two populations of anti-dsDNA B cells, each of which is subject to a different mechanism of regulation in non-autoimmune hosts. The B-cell population capable of secreting antibodies with high affinity for dsDNA is deleted whereas the population with low-affinity anti-dsDNA antibodies escapes regulation and is termed "indifferent". The immunoglobulin genes from some members of the low-affinity subset of anti-dsDNA B cells differ from the high-affinity subsets by a single base substitution, resulting in a single amino acid replacement and suggest that high-affinity pathogenic anti-DNA B cells can arise from low-affinity, anti-dsDNA B cells that have undergone somatic mutation.

GENES THAT REGULATE B-CELL TOLERANCE

Mouse Models

Several molecules are involved in setting thresholds for the survival, activation, or tolerance of autoreactive B cells. Defects in apoptosis-related genes lead to lymphoproliferation and survival of anti-DNA B cells in some mouse strains. The MRL/lpr mouse model of lupus has a mutation in the Fas gene that inhibits activation-induced T- and B-cell apoptosis, and results in severe lymphoproliferation. As mentioned, MRL/lpr mice transgenic for an IgM anti-DNA antibody fail to display tolerance and do not delete anti-dsDNA B cells [159]. The bcl-2 gene product promotes B-cell survival. Overexpression of bcl-2 in a non-autoimmune mouse, transgenic for the heavy chain of a pathogenic anti-DNA antibody leads to survival of anti-dsDNA B cells that would be deleted in a normal mouse [160]. Furthermore, immunization of these mice with pneumococcus (PC) results in the "rescue" of cross-reactive anti-DNA/anti- (PC) B cells in non-autoimmune mice [161]. The hormones estrogen and prolactin promote upregulation of the bcl-2 gene and the survival of anti-dsDNA B cells [162, 163].

Several regulatory molecules are involved in signaling B cells to proliferate and secrete antibody. These molecules include cell surface receptors and their intracellular components. Motheaten mice spontaneously develop an autoimmune syndrome characterized by the production of autoantibodies including anti-dsDNA [164]. These mice have a mutation that ablates the activity of SHP-1 phosphatase, which helps to regulate B-cell activation. CD22 is a transmembrane molecule that associates with SHP-1 phosphatase and can negatively regulate antigen receptor signaling [165]. Mice deficient in CD22 expression also display elevated titers of anti-DNA antibodies [166]. Lyn is a tyrosine kinase associated with membrane Ig that modulates B-cell responses. Lyn deficiency results in increased anti-DNA titers and immune complex deposits in the glomeruli [167]. CD45 is a transmembrane tyrosine phosphatase involved in B-cell activation [168]. Mice deficient in CD45 are hyper-responsive to antigen receptor cross-linking and display diminished signaling in response to an autoantigen. This results in impaired deletion of immature autoreactive B cells as these cells require more autoantigen to be present to signal their elimination. CD45-deficient mice, have an increased number of autoreactive B cells and an enhanced incidence of autoimmunity [169]. The cell surface molecule CD19 has also been observed to alter signaling thresholds, and mice with an increase in CD19 expression have elevated titers of anti-dsDNA antibodies [170]. As stated above, deficiencies in the B-cell receptor inhibitory co-receptor FcRIIb can lead to a lupus-like phenotype.

Abnormal expression of B-cell co-stimulatory molecules may also influence the induction of anti-DNA antibodies. MRL/lpr mice deficient in CD40L have lower titers of IgG anti-DNA antibodies [171] suggesting that co-stimulatory molecules are critical in disease expression. Indeed, there is an increased expression of CD40 ligand on both B and T lymphocytes in SLE patients [172]. Roquin mice overexpress ICOS on follicular T helper cells, resulting in an anti-DNA response derived from germinal center maturized B cells [173].

Genes involved in innate immunity are also associated with loss of B-cell tolerance and autoantibody production. The lupus susceptibility locus Sle1c includes genes encoding complement receptors CD21/CD35. A single nucleotide polymorphism of CD21 that diminishes C3d binding results in a lupus-like phenotype [174]. Additionally, decreased expression of both CD21 and CD35 has been observed in MRL/lpr mice [175]. The importance of complement in the maintenance of tolerance is further supported by the association of lupus and deficiency of the early complement components. C1q deficiency accelerates disease progression in MRL/lpr mice [117], whereas C4-deleted MRL/lpr mice and mice lacking complement receptors 1 and 2 develop high titers of antinuclear and anti-DNA antibodies and severe glomerulonephritis [176].

Human SLE

Genetic studies of lupus patients have identified several genes that are involved in breaking tolerance of autoreactive B cells. B-lymphoid tyrosine kinase (BLK) is a member of the src-tyrosine kinase family and influences B-cell proliferation, differentiation, and tolerance [177]. A BLK allele is associated with increased risk for SLE. The risk allele of BLK is associated with reduced expression of BLK mRNA and a reduced B-cell receptor signal strength, which could lead to enhanced autoreactivity in the naïve B-cell repertoire [178]. Likewise, an SLE susceptibility allele PTPN22, a phosphatase that inactivates lyn, decreases the strength of the B-cell receptor signaling pathway [179]. B-cell scaffold protein with ankyrin repeats (BANK1) is an adapter protein which is expressed primarily in B cells. An SLE-associated polymorphism of BANKl causes differential expression of BANK1 splicing variants that is expected to lead to increased binding of BANK1 to downstream effector proteins, such as Lyn and IP3R. This may lead to a steady state marked by B-cell hyper-responsiveness or deregulated B-cell activation [180]. There is an association of a Lyn allele with SLE in European patients, especially with anti-dsDNA production. Lyn plays both a positive and negative regulatory role in B-cell receptor signaling. The inhibitory function of Lyn is mediated by its ability to phosphorylate CD22 and FcRIIb [181]. Expression of lyn is significantly decreased in B cells from most SLE patients [182]. The co-stimulatory molecule OX40L is a membrane-bound protein expressed on the surface of antigen-presenting cells and B cells [183]. It provides a strong activating signal when bound to its ligand, OX40, on the surface of CD4+ T cells. An allele of OX40L correlating with increased expression of cell-surface OX40L associates with SLE susceptibility. Another co-stimulatory molecule, PDCD-1, an immunoreceptor of the CD28 receptor family, is expressed on B cells and helps to regulate tolerance. A PDCD-1 allele that increases binding to runt-related transcription factor 1 (RUNX1) which induces B-cell hyperactivity allele is associated with SLE susceptibility in Europeans and Mexicans [184]. FcRIIb (discussed previously) delivers an inhibitory signal which provides critical checkpoints on B-cell activation and plasma cell survival, thus limiting the accumulation of pathogenic autoantibody-producing memory and plasma cells. The Thr187 FcRIIb allele is less effectively distributed to lipid rafts the Ile187 FcRIIb allele, thereby affecting its efficiency in modulating BCR function. The Thr187 allele has been reported as a genetic risk factor of SLE in Asian populations [185]. Polymorphisms identified in the promoter region of FcRIIb, which influence FcRIIb expression levels, also associate with lupus susceptibility [186]. IFN regulatory factor 5 (IRF5) regulates type I IFN-responsive genes and multiple susceptibility alleles have been identified in Asian and African-American cohorts [187]. These studies have led to a model in which TLR signaling, triggered by immune complexes containing apoptotic or necrotic cell debris, results in high levels of type I IFN produced by dendritic cells, mediated by IRF5. Type I IFN promotes the differentiation of mature B cells into plasma cells. Thus, the effects of IFN α on B cells may help explain the breakdown of tolerance to dsDNA in some patients with SLE. Homozygous deficiencies in complement genes are very rare, but the prevalence in developing SLE is estimated to be 93% for C1q, 75% for C4, and 10–25% for C2. Complement components are important in resisting microbial infection and in clearing apoptotic fragments of cell debris that may become autoantigens in SLE patients.

Defects in expression or function of more than one of these molecules are required to sustain a lupus-like autoimmune response. The risk of lupus appears to be related to the number of susceptibility alleles in a given individual.

DETECTION OF ANTI-DNA ANTIBODIES

Anti-DNA antibodies can be detected by several assays. Some of these assays are specific for high-affinity antibodies, whereas others detect low-affinity antibodies as well (Table 14.1). The most frequently utilized assays are the Farr assay, the Millipore filter assay, the enzyme-linked immunosorbent assay (ELISA), and the *Crithidia luciliae* assay [188].

In the Farr assay, radiolabeled single-stranded (ss) or double-stranded (ds) DNA is added to the sample to be tested for anti-DNA antibodies. The immune complexes can then be precipitated by ammonium sulfate and the results are expressed as a percentage of radioactivity that is present in the precipitate. This test can detect both IgG and IgM anti-DNA antibodies but cannot distinguish between the two classes. The Farr assay is an excellent assay for the detection of high-affinity antibodies.

The Millipore filter assay also utilizes radiolabeled dsDNA that is added to the test serum and passed through a filter. Because the filters bind protein, but not dsDNA, DNA—antibody complexes are retained on the filter while free dsDNA passes through. The amount of radioactivity retained on the filter is proportional to the amount of anti-dsDNA antibody in the test sample. This assay cannot be used to detect antibodies to ssDNA because ssDNA, unlike dsDNA, will bind directly to the filter. Neither the Farr assay nor the Millipore assay can distinguish between antibody—DNA complexes and complexes of other serum proteins with DNA.

The ELISA is a rapid, simple, quantitative, and reproducible assay that is the most commonly used clinical assay [189]. Microtiter plates are coated with ssDNA or dsDNA and serum is applied to the coated plates. An enzyme linked anti-Ig antibody is added next. The presence of bound anti-DNA antibody can then be detected by the addition of a substrate that changes color in the presence of the enzyme. The color change can be read in a spectrophotometer. The advantage of this assay is the ease with which large numbers of samples can be screened. ELISAs are highly sensitive and have the advantage of being able to determine antibody isotype, light-chain usage, and idiotypic specificity, depending on the specificity of the secondary antibody. It is currently the most popular assay in clinical use.

The most specific test for anti-dsDNA antibodies utilizes the flagellate *C. luciliae* as a substrate. This organism has a kinetoplast containing circular dsDNA that is not associated with histone proteins. Test samples are incubated with the organism on a glass slide and a fluoresceinated anti-Ig antibody is added. Detection of the fluorescent kinetoplast using a fluorescence microscope indicates the presence of anti-DNA antibodies. Although not as sensitive as the ELISA, this assay is highly specific for dsDNA binding and, like the ELISA, can distinguish structurally distinct subsets of anti-DNA antibodies, depending on the specificity of the secondary fluoresceinated antibody.

Accurate detection of high titers of anti-dsDNA antibodies is useful to diagnose SLE, and an increasing titer is generally correlated with disease flares. Given the heterogeneity of the anti-DNA antibody detection assays available and their specific limitations, the utility of anti-dsDNA antibody detection in the diagnosis and treatment of SLE is enhanced in clinical decision making by a consideration of the sensitivity and specificity of the assay used.

THERAPY

A major goal of understanding the structure, origin, and pathogenicity of anti-DNA antibodies is to develop therapies that will be less harmful than the current non-selective immunosuppressive regimens that are used to treat lupus patients. Several novel approaches to therapy of SLE have been proposed based on our growing understanding of B-cell activation. This section briefly mentions only some of the novel therapeutic approaches to down-regulation of autoantibodies.

The association of anti-DNA antibodies with pathologic events in SLE, especially with lupus nephritis, suggests that removal of anti-DNA antibodies, or blocking their production or activity, may be beneficial. Many anti-DNA antibodies in SLE arise from T-cell-dependent responses in which co-stimulation between B and T cells is mediated through the CD40-CD40L and B7-CD28 pathways (Figures 14.3 and 14.4). Blocking the co-stimulatory pathways, independently or collectively, prevents the generation of a second signal, which is necessary for T-cell-dependent B-cell proliferation, differentiation, and antibody production. Monoclonal antibody to the CD40 ligand leads to improved survival in the NZB/W and SNF mouse models of SLE [190]. Inhibition of the B7-CD28 co-stimulatory pathway by CTLA-4 Ig, a fusion molecule constructed of CTLA-4 and the Fc portion of an immunoglobulin, also blocks anti-DNA antibody production and prolongs life in NZB/W mice [191]. Simultaneous treatment with anti-CD40 and CTLA-4 Ig or cyclophophosphamide and CTLA-4 Ig provides greater benefit than either intervention alone in murine lupus [34, 192]. However, clinical trials with anti-CD40 ligand in patients were hampered by thrombophilic toxicities. A recent phase II clinical trial of CTLA4-Ig for non-renal, non-CNS active lupus failed to demonstrate efficacy compared to placebo [193]. It remains a question whether blockade of the CD40-CD40L or B7-CD28 co-stimulatory pathways or other co-stimulatory pathways with or without concomitant immunosuppressive regimens will be useful therapeutic modalities for SLE, additional clinical trials are underway.

B-cell depletion in murine models of lupus results in amelioration of clinical features of autoimmune disease and lower levels of serum autoantibodies [194]. In

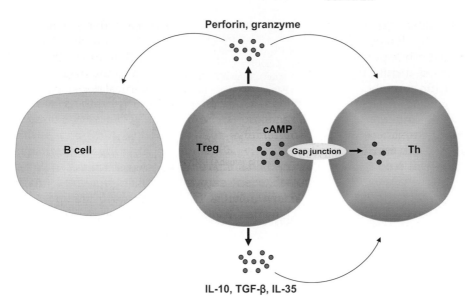

Perforin, granzyme

B cell

Treg

cAMP

Gap junction

Th

IL-10, TGF-β, IL-35

FIGURE 14.4 Suppression of autoreactive B cells by Treg cells. Treg cells produce inhibitory cytokines including IL-10, IL-35, and TGF-β which suppress T helper cells. cAMP in Treg cell traverses the membrane through gap junctions between Tregs and T helper cells and inhibits proliferation and IL-2 synthesis by the T helper cell. Perforin and granzyme produced by Treg cells cause apoptosis of T helper cells. Suppression of T helper cells by Treg cells indirectly suppresses autoreactive B cells. Perforin and granzyme can also kill autoreactive B cells, which might be one of mechanism of direct suppression of B cells by Treg cells.

human lupus, administration of anti-CD20 antibody effectively depletes circulating B cells. Open-label trials with rituximab alone or in combination with mycophenolate mofetil or cyclophosphamide suggested that B-cell depletion might be a promising approach to treat SLE in humans. However, two placebo-controlled studies have failed to show a benefit with this treatment in lupus nephritis and non-renal, non-CNS active SLE and the role of B-cell depletion as a therapeutic approach in human lupus is unclear [193, 195]. Perhaps the lesson is that long-lived plasma cells must be targeted or that B-cell reconstitution is skewed to autoreactivity due to elevated levels of BAFF, diminishing the long-term efficacy of this approach.

Administration of anti-CD22 antibody results in decreased B-cell numbers but not full B-cell depletion. Preliminary results from a study of anti-CD22 in patients with SLE appear promising and suggest a therapeutic effect independent of B-cell depletion (press release).

B-cell survival may also be modulated by BAFF. BAFF inhibition may result in partial B-cell depletion [196] and a skewing away from an autoantibody repertoire [21, 22]. BAFF blockade is beneficial in many but not all models of murine lupus, but is most effective when given before the onset of clinical disease. In one model of spontaneous lupus, eliminating BAFF did not prevent the development of proteinuria, but did improve overall renal survival and death [197]. Belimumab, a monoclonal antibody directed against BAFF has been studied in phase II and III trials in non-renal SLE. Anti-DNA antibodies decreased approximately 10% in the phase II study [198]; this study did not meet its primary endpoint although a *post-hoc* analysis of a subset of patients did show clinical efficacy. Preliminary results of two pivotal

phase 3 studies suggest modest efficacy (press release and [199]). Together, these two trials highlight the role of BAFF in disease maintenance.

The explosion of interest in the contribution of abnormalities of the innate system in the pathogenesis of SLE has led to the design of potential therapeutic agents to inhibit these pathways. Clinical trials of anti-IFN-α antibodies, anti-IFN antibodies and inhibitors of TLR signaling pathways are planned or have been initiated.

SUMMARY

All individuals have the potential to produce high-affinity potentially pathogenic anti-DNA antibodies. Anti-DNA antibodies are encoded by genes that are present in non-autoimmune individuals and that are used routinely to encode protective antibodies. The mechanisms of somatic diversification of the immunoglobulin repertoire appear to be normal in lupus patients. It is clear that anti-DNA antibodies arise in non-autoimmune hosts during microbial infection, but that tolerance is restored when microbial antigens are eliminated. Therefore, it appears that the critical defect in SLE is not in the formation of the B-cell repertoire, but in the regulation of autoreactive B cells.

The antigenic stimulus for anti-DNA antibody production is not known, nor is it clear whether a self or foreign antigen initiates the process. There are probably several triggers for anti-DNA antibody production. Just as the antigenic trigger for anti-DNA antibody production may not always be DNA, the antigenic target may also not be DNA. It is apparent that at least some anti-DNA antibodies can cross-react directly with tissue antigens and so mediate tissue damage. In the near future, transgenic models and human genetic studies in combination should elucidate the defects in B-cell repertoire selection and tolerance induction that can lead to SLE. Major questions need to be addressed in future studies: (1) What abnormalities of B-cell function alter tolerance induction in SLE? (2) What genetic defects underlie these abnormalities? (3) How do anti-DNA antibodies cause tissue injury? (4) What differences in target tissue regulate susceptibility to anti-DNA-mediated tissue damage? Such studies in turn will suggest new and specific therapies.

References

[1] A.K. Abbas, A.H. Lichtman, J.S. Pober, Cellular and Molecular Immunology, third ed., Saunders, Philadelphia, (1997).

[2] E.A. Kabat, T.T. Wu, Attempts to locate complementarity-determining residues in the variable positions of light and heavy chains, Annals of the New York Academy of Sciences 190 (1971) 382–393.

[3] J. Oudin, P.A. Cazenave, Similar idiotypic specificities in immunoglobulin fractions with different antibody functions or even without detectable antibody function, Proceedings of the National Academy of Sciences of the United States of America 68 (1971) 2616–2620.

[4] G.P. Cook, I.M. Tomlinson, The human immunoglobulin VH repertoire, Immunology Today 16 (1995) 237–242.

[5] S.D. Fugmann, A.I. Lee, P.E. Shockett, I.J. Villey, D.G. Schatz, The RAG proteins and V(D)J recombination: complexes, ends, and transposition, Annual Review of Immunology 18 (2000) 495–527.

[6] P.A. Jeggo, G.E. Taccioli, S.P. Jackson, Menage a trois: double strand break repair, V(D)J recombination and DNA-PK, Bioessays 17 (1995) 949–957.

[7] T.T. Paull, M. Gellert, The 3′ to 5′ exonuclease activity of Mre 11 facilitates repair of DNA double-strand breaks, Molecular Cell 1 (1998) 969–979.

[8] D.C. van Gent, K. Hiom, T.T. Paull, M. Gellert, Stimulation of V (D)J cleavage by high mobility group proteins, The EMBO Journal 16 (1997) 2665–2670.

[9] S.V. Desiderio, G.D. Yancopoulos, M. Paskind, E. Thomas, M.A. Boss, N. Landau, et al., Insertion of N regions into heavy-chain genes is correlated with expression of terminal deoxytransferase in B cells, Nature 311 (1984) 752–755.

[10] A.a.H.T. Shimuzu, Immunoglobulin class switching, Cell 36 (1984) 801–803.

[11] H.R. Jaenichen, M. Pech, W. Lindenmaier, N. Wildgruber, H.G. Zachau, Composite human VK genes and a model of their evolution, Nucleic Acids Research 12 (1984) 5249–5263.

[12] J.P. Frippiat, M.P. Lefranc, Genomic organisation of 34 kb of the human immunoglobulin lambda locus (IGLV): restriction map and sequences of new V lambda III genes, Molecular Immunology 31 (1994) 657–670.

[13] M. Heller, J.D. Owens, J.F. Mushinski, S. Rudikoff, Amino acids at the site of V kappa-J kappa recombination not encoded by germline sequences, The Journal of Experimental Medicine 166 (1987) 637–646.

[14] M.W. Retter, D. Nemazee, Receptor editing occurs frequently during normal B cell development, The Journal of Experimental Medicine 188 (1998) 1231–1238.

[15] M. Hikida, M. Mori, T. Takai, K. Tomochika, K. Hamatani, H. Ohmori, Reexpression of RAG-1 and RAG-2 genes in activated mature mouse B cells, Science New York, NY 274 (1996) 2092–2094.

[16] J.U. Peled, F.L. Kuang, M.D. Iglesias-Ussel, S. Roa, S.L. Kalis, M.F. Goodman, et al., The biochemistry of somatic hypermutation, Annual Review of Immunology 26 (2008) 481–511.

[17] B. Diamond, M.D. Scharff, Somatic mutation of the T15 heavy chain gives rise to an antibody with autoantibody specificity, Proceedings of the National Academy of Sciences of the United States of America 81 (1984) 5841–5844.

[18] M. Muramatsu, V.S. Sankaranand, S. Anant, M. Sugai, K. Kinoshita, N.O. Davidson, et al., Specific expression of activation-induced cytidine deaminase (AID), a novel member of the RNA-editing deaminase family in germinal center B cells, The Journal of Biological Chemistry 274 (1999) 18470–18476.

[19] K. Kinoshita, T. Honjo, Linking class-switch recombination with somatic hypermutation, Nature Reviews 2 (2001) 493–503.

[20] C. Kocks, K. Rajewsky, Stepwise intraclonal maturation of antibody affinity through somatic hypermutation, Proceedings of the National Academy of Sciences of the United States of America 85 (1988) 8206–8210.

[21] R. Lesley, Y. Xu, S.L. Kalled, D.M. Hess, S.R. Schwab, H.B. Shu, et al., Reduced competitiveness of autoantigen-engaged B cells due to increased dependence on BAFF, Immunity 20 (2004) 441–453.

[22] M. Thien, T.G. Phan, S. Gardam, M. Amesbury, A. Basten, F. Mackay, et al., Excess BAFF rescues self-reactive B cells from peripheral deletion and allows them to enter forbidden follicular and marginal zone niches, Immunity 20 (2004) 785–798.

[23] M.W. Boule, C. Broughton, F. Mackay, S. Akira, A. Marshak-Rothstein, I.R. Rifkin, Toll-like receptor 9-dependent and -independent dendritic cell activation by chromatin-immunoglobulin G complexes, The Journal of Experimental Medicine 199 (2004) 1631–1640.

[24] D. Tarlinton, IL-17 drives germinal center B cells? Nature Immunology 9 (2008) 124–126.

[25] T.a.A.S. Ternyck, Murine natural monoclonal autoantibodies: A study of their polyspecificities and their affinitites, Immunol. Rev 94 (1986).

[26] K. Hayakawa, R.R. Hardy, Development and function of B-1 cells, Current Opinion in Immunology 12 (2000) 346–353.

[27] P. Casali, A.L. Notkins, Probing the human B-cell repertoire with EBV: polyreactive antibodies and CD5+ B lymphocytes, Annual Review of Immunology 7 (1989) 513–535.

[28] N. Suzuki, T. Sakane, E.G. Engleman, Anti-DNA antibody production by CD5+ and CD5- B cells of patients with systemic lupus erythematosus, The Journal of Clinical Investigation 85 (1990) 238–247.

[29] J. Zhang, A.M. Jacobi, T. Wang, R. Berlin, B.T. Volpe, B. Diamond, Polyreactive autoantibodies in systemic lupus erythematosus have pathogenic potential, Journal of Autoimmunity 33 (2009) 270–274.

[30] M. Murakami, H. Yoshioka, T. Shirai, T. Tsubata, T. Honjo, Prevention of autoimmune symptoms in autoimmune-prone mice by elimination of B-1 cells, International Immunology 7 (1995) 877–882.

[31] G. Bikah, J. Carey, J.R. Ciallella, A. Tarakhovsky, S. Bondada, CD5-mediated negative regulation of antigen receptor-induced growth signals in B-1 B cells, Science New York, NY 274 (1996) 1906–1909.

[32] P. Schneider, H. Takatsuka, A. Wilson, F. Mackay, A. Tardivel, S. Lens, et al., Maturation of marginal zone and follicular B cells requires B cell activating factor of the tumor necrosis factor family and is independent of B cell maturation antigen, The Journal of Experimental Medicine 194 (2001) 1691–1697.

[33] M. Zouali, The structure of human lupus anti-DNA antibodies, Methods San Diego, Calif 11 (1997) 27—35.

[34] X. Wang, W. Huang, M. Mihara, J. Sinha, A. Davidson, Mechanism of action of combined short-term CTLA4Ig and anti-CD40 ligand in murine systemic lupus erythematosus, J Immunol 168 (2002) 2046—2053.

[35] D. Wofsy, W.E. Seaman, Successful treatment of autoimmunity in NZB/NZW F1 mice with monoclonal antibody to L3T4, The Journal of Experimental Medicine 161 (1985) 378—391.

[36] P.M. Dammers, A. Visser, E.R. Popa, P. Nieuwenhuis, F.G. Kroese, Most marginal zone B cells in rat express germline encoded Ig VH genes and are ligand selected, J Immunol 165 (2000) 6156—6169.

[37] J.H. Roark, S.H. Park, J. Jayawardena, U. Kavita, M. Shannon, A. Bendelac, CD1.1 expression by mouse antigen-presenting cells and marginal zone B cells, J Immunol 160 (1998) 3121—3127.

[38] N. Baumgarth, A two-phase model of B-cell activation, Immunological Reviews 176 (2000) 171—180.

[39] J.E. Wither, A.D. Paterson, B. Vukusic, Genetic dissection of B cell traits in New Zealand black mice. The expanded population of B cells expressing up-regulated costimulatory molecules shows linkage to Nba2, European Journal of Immunology 30 (2000) 356—365.

[40] C.M. Grimaldi, D.J. Michael, B. Diamond, Cutting edge: expansion and activation of a population of autoreactive marginal zone B cells in a model of estrogen-induced lupus, J Immunol 167 (2001) 1886—1890.

[41] M.W. Steward, F.C. Hay, Changes in immunoglobulin class and subclass of anti-DNA antibodies with increasing age in N/ZBW F1 hybrid mice, Clinical and Experimental Immunology 26 (1976) 363—370.

[42] M.Z. Radic, M. Weigert, Genetic and structural evidence for antigen selection of anti-DNA antibodies, Annual Review of Immunology 12 (1994) 487—520.

[43] D.M. Tillman, N.T. Jou, R.J. Hill, T.N. Marion, Both IgM and IgG anti-DNA antibodies are the products of clonally selective B cell stimulation in (NZB x NZW)F1 mice, The Journal of Experimental Medicine 176 (1992) 761—779.

[44] R. Kofler, M.A. Duchosal, M.E. Johnson, M.T. Aguado, R. Strohal, G. Kromer, et al., The genetic origin of murine lupus-associated autoantibodies, Immunology Letters 16 (1987) 265—271.

[45] M. Zouali, M.P. Madaio, R.T. Canoso, B.D. Stollar, Restriction fragment length polymorphism analysis of the V kappa locus in human lupus, European Journal of Immunology 19 (1989) 1757—1760.

[46] D.A. Isenberg, M.R. Ehrenstein, C. Longhurst, J.K. Kalsi, The origin, sequence, structure, and consequences of developing anti-DNA antibodies. A human perspective, Arthritis and Rheumatism 37 (1994) 169—180.

[47] R. Kofler, R. Strohal, R.S. Balderas, M.E. Johnson, D.J. Noonan, M.A. Duchosal, et al., Immunoglobulin kappa light chain variable region gene complex organization and immunoglobulin genes encoding anti-DNA autoantibodies in lupus mice, The Journal of Clinical Investigation 82 (1988) 852—860.

[48] M.Z. Radic, M.A. Mascelli, J. Erikson, H. Shan, M. Weigert, Ig H and L chain contributions to autoimmune specificities, J Immunol 146 (1991) 176—182.

[49] L. Spatz, V. Saenko, A. Iliev, L. Jones, L. Geskin, B. Diamond, Light chain usage in anti-double-stranded DNA B cell subsets: role in cell fate determination, The Journal of Experimental Medicine 185 (1997) 1317—1326.

[50] E.A. Padlan, On the nature of antibody combining sites: unusual structural features that may confer on these sites an enhanced capacity for binding ligands, Proteins 7 (1990) 112—124.

[51] J.K. Kalsi, A.C. Martin, Y. Hirabayashi, M. Ehrenstein, C.M. Longhurst, C. Ravirajan, et al., Functional and modelling studies of the binding of human monoclonal anti-DNA antibodies to DNA, Molecular Immunology 33 (1996) 471—483.

[52] W. Zhang, M.B. Frank, M. Reichlin, Production and characterization of human monoclonal anti-idiotype antibodies to anti-dsDNA antibodies, Lupus 11 (2002) 362—369.

[53] N.K. Jerne, Idiotypic networks and other preconceived ideas, Immunological Reviews 79 (1984) 5—24.

[54] E. Paul, A. Manheimer-Lory, A. Livneh, A. Solomon, C. Aranow, C. Ghossein, et al., Pathogenic anti-DNA antibodies in SLE: idiotypic families and genetic origins, International Reviews of Immunology 5 (1990) 295—313.

[55] S. Mendlovic, S. Brocke, Y. Shoenfeld, M. Ben-Bassat, A. Meshorer, R. Bakimer, et al., Induction of a systemic lupus erythematosus-like disease in mice by a common human anti-DNA idiotype, Proceedings of the National Academy of Sciences of the United States of America 85 (1988) 2260—2264.

[56] M. Zouali, A. Eyquem, Idiotypic/antiidiotypic interactions in systemic lupus erythematosus: demonstration of oscillating levels of anti-DNA autoantibodies and reciprocal antiidiotypic activity in a single patient, Annales d'Immunologie 134C (1983) 377—391.

[57] B.H. Hahn, R.R. Singh, W.K. Wong, B.P. Tsao, K. Bulpitt, F.M. Ebling, Treatment with a consensus peptide based on amino acid sequences in autoantibodies prevents T cell activation by autoantigens and delays disease onset in murine lupus, Arthritis and Rheumatism 44 (2001) 432—441.

[58] M.L. Stoll, K.D. Price, C.J. Silvin, F. Jiang, J. Gavalchin, Immunization with peptides derived from the idiotypic region of lupus-associated autoantibodies delays the development of lupus nephritis in the (SWR x NZB)F1 murine model, Journal of Autoimmunity 29 (2007) 30—37.

[59] B.H. Hahn, M. Anderson, E. Le, A. La Cava, Anti-DNA Ig peptides promote Treg cell activity in systemic lupus erythematosus patients, Arthritis and Rheumatism 58 (2008) 2488—2497.

[60] D.M. Klinman, A. Shirai, J. Conover, A.D. Steinberg, Cross-reactivity of IgG anti-DNA-secreting B cells in patients with systemic lupus erythematosus, European Journal of Immunology 24 (1994) 53—58.

[61] A.L. Aron, M.L. Cuellar, R.L. Brey, S. McKeown, L.R. Espinoza, Y. Shoenfeld, A.E. Gharavi, Early onset of autoimmunity in MRL/++ mice following immunization with beta 2 glycoprotein I, Clinical and Experimental Immunology 101 (1995) 78—81.

[62] E.M. Lafer, J. Rauch, C. Andrzejewski Jr., D. Mudd, B. Furie, B. Furie, et al., Polyspecific monoclonal lupus autoantibodies reactive with both polynucleotides and phospholipids, The Journal of Experimental Medicine 153 (1981) 897—909.

[63] P. Carroll, D. Stafford, R.S. Schwartz, B.D. Stollar, Murine monoclonal anti-DNA autoantibodies bind to endogenous bacteria, J Immunol 135 (1985) 1086—1090.

[64] L.A. Dimitrijevic, M.I. Radulovic, B.P. Ciric, M.M. Petricevic, A.B. Inic, D.N. Nikolic, et al., Human monoclonal IgM DJ binds to ssDNA and human commensal bacteria, Human Antibodies 9 (1999) 37—45.

[65] M.B. Spellerberg, C.J. Chapman, C.I. Mockridge, D.A. Isenberg, F.K. Stevenson, Dual recognition of lipid A and DNA by human antibodies encoded by the VH4-21 gene: a possible link between infection and lupus, Human Antibodies and Hybridomas 6 (1995) 52—56.

[66] W. Limpanasithikul, S. Ray, B. Diamond, Cross-reactive antibodies have both protective and pathogenic potential, J Immunol 155 (1995) 967—973.

[67] A. Sharma, D.A. Isenberg, B. Diamond, Crossreactivity of human anti-dsDNA antibodies to phosphorylcholine: clues to their origin, Journal of Autoimmunity 16 (2001) 479—484.

[68] E.A. Kabat, K.G. Nickerson, J. Liao, L. Grossbard, E.F. Osserman, E. Glickman, et al., A human monoclonal macroglobulin with specificity for alpha(2—-8)-linked poly-N-acetyl neuraminic acid, the capsular polysaccharide of group B meningococci and Escherichia coli K1, which crossreacts with polynucleotides and with denatured DNA, The Journal of Experimental Medicine 164 (1986) 642—654.

[69] J. Sabbaga, S.R. Line, P. Potocnjak, M.P. Madaio, A murine nephritogenic monoclonal anti-DNA autoantibody binds directly to mouse laminin, the major non-collagenous protein component of the glomerular basement membrane, European Journal of Immunology 19 (1989) 137—143.

[70] F. Tron, L. Jacob, J.F. Bach, Binding of a murine monoclonal anti-DNA antibody to Raji cells. Implications for the interpretation of the Raji cell assay for immune complexes, European Journal of Immunology 14 (1984) 283—286.

[71] E. Raz, H. Ben-Bassat, T. Davidi, Z. Shlomai, D. Eilat, Cross-reactions of anti-DNA autoantibodies with cell surface proteins, European Journal of Immunology 23 (1993) 383—390.

[72] L. Jacob, F. Tron, J.F. Bach, D. Louvard, A monoclonal anti-DNA antibody also binds to cell-surface protein(s), Proceedings of the National Academy of Sciences of the United States of America 81 (1984) 3843—3845.

[73] L.A. DeGiorgio, K.N. Konstantinov, S.C. Lee, J.A. Hardin, B.T. Volpe, B. Diamond, A subset of lupus anti-DNA antibodies cross-reacts with the NR2 glutamate receptor in systemic lupus erythematosus, Nature Medicine 7 (2001) 1189—1193.

[74] C. Putterman, R. Ulmansky, L. Rasooly, B. Tadmor, H. Ben-Bassat, Y. Naparstek, Down-regulation of surface antigens recognized by systemic lupus erythematosus antibodies on embryonic cells following differentiation and exposure to corticosteroids, European Journal of Immunology 28 (1998) 1656—1662.

[75] C. Kowal, L.A. Degiorgio, J.Y. Lee, M.A. Edgar, P.T. Huerta, B.T. Volpe, et al., Human lupus autoantibodies against NMDA receptors mediate cognitive impairment, Proceedings of the National Academy of Sciences of the United States of America 103 (2006) 19854—19859.

[76] G. Mostoslavsky, R. Fischel, N. Yachimovich, Y. Yarkoni, E. Rosenmann, M. Monestier, et al., Lupus anti-DNA autoantibodies cross-react with a glomerular structural protein: a case for tissue injury by molecular mimicry, European Journal of Immunology 31 (2001) 1221—1227.

[77] J. Andre-Schwartz, S.K. Datta, Y. Shoenfeld, D.A. Isenberg, B.D. Stollar, R.S. Schwartz, Binding of cytoskeletal proteins by monoclonal anti-DNA lupus autoantibodies, Clinical Immunology and Immunopathology 31 (1984) 261—271.

[78] K. Tsuzaka, A.K. Leu, M.B. Frank, B.F. Movafagh, M. Koscec, T.H. Winkler, et al., Lupus autoantibodies to double-stranded DNA cross-react with ribosomal protein S1, J Immunol 156 (1996) 1668—1675.

[79] S. Singh, S. Chatterjee, R. Sohoni, S. Badakere, S. Sharma, Sera from lupus patients inhibit growth of P. falciparum in culture, Autoimmunity 33 (2001) 253—263.

[80] K.H. Sun, C.C. Hong, S.J. Tang, G.H. Sun, W.T. Liu, S.H. Han, et al., Anti-dsDNA autoantibody cross-reacts with the C-terminal hydrophobic cluster region containing phenylalanines in the acidic ribosomal phosphoprotein P1 to exert a cytostatic effect on the cells, Biochemical and Biophysical Research Communications 263 (1999) 334—339.

[81] E. Koren, M. Koscec, M. Wolfson-Reichlin, F.M. Ebling, B. Tsao, B.H. Hahn, et al., Murine and human antibodies to native DNA that cross-react with the A and D SnRNP polypeptides cause direct injury of cultured kidney cells, J Immunol 154 (1995) 4857—4864.

[82] M. Reichlin, A. Martin, E. Taylor-Albert, K. Tsuzaka, W. Zhang, M.W. Reichlin, et al., Lupus autoantibodies to native DNA cross-react with the A and D SnRNP polypeptides, The Journal of Clinical Investigation 93 (1994) 443—449.

[83] D.S. Pisetsky, Autoantibodies and their significance, Current Opinion in Rheumatology 5 (1993) 549—556.

[84] Y. Arinuma, T. Yanagida, S. Hirohata, Association of cerebro-spinal fluid anti-NR2 glutamate receptor antibodies with diffuse neuropsychiatric systemic lupus erythematosus, Arthritis and Rheumatism 58 (2008) 1130—1135.

[85] H. Fragoso-Loyo, J. Cabiedes, A. Orozco-Narvaez, L. Davila-Maldonado, Y. Atisha-Fregoso, B. Diamond, et al., Serum and cerebrospinal fluid autoantibodies in patients with neuropsychiatric lupus erythematosus. Implications for diagnosis and pathogenesis, PloS One 3 (2008) e3347.

[86] G.S. Gilkeson, K. Bernstein, A.M. Pippen, S.H. Clarke, T. Marion, D.S. Pisetsky, et al., The influence of variable-region somatic mutations on the specificity and pathogenicity of murine monoclonal anti-DNA antibodies, Clinical Immunology and Immunopathology 76 (1995) 59—67.

[87] D.V. Vlahakos, M.H. Foster, S. Adams, M. Katz, A.A. Ucci, K.J. Barrett, et al., Anti-DNA antibodies form immune deposits at distinct glomerular and vascular sites, Kidney International 41 (1992) 1690—1700.

[88] B.P. Tsao, K. Ohnishi, H. Cheroutre, B. Mitchell, M. Teitell, P. Mixter, et al., Failed self-tolerance and autoimmunity in IgG anti-DNA transgenic mice, J Immunol 149 (1992) 350—358.

[89] F. Ebling, B.H. Hahn, Restricted subpopulations of DNA antibodies in kidneys of mice with systemic lupus. Comparison of antibodies in serum and renal eluates, Arthritis and Rheumatism 23 (1980) 392—403.

[90] G.S. Gilkeson, A.M. Pippen, D.S. Pisetsky, Induction of cross-reactive anti-dsDNA antibodies in preautoimmune NZB/NZW mice by immunization with bacterial DNA, The Journal of Clinical Investigation 95 (1995) 1398—1402.

[91] M.H. Foster, B. Cizman, M.P. Madaio, Nephritogenic autoantibodies in systemic lupus erythematosus: immunochemical properties, mechanisms of immune deposition, and genetic origins, Laboratory Investigation; a Journal of Technical Methods and Pathology 69 (1993) 494—507.

[92] J. Sabbaga, O.G. Pankewycz, V. Lufft, R.S. Schwartz, M.P. Madaio, Cross-reactivity distinguishes serum and nephritogenic anti-DNA antibodies in human lupus from their natural counterparts in normal serum, Journal of Autoimmunity 3 (1990) 215—235.

[93] M.P. Madaio, J. Carlson, J. Cataldo, A. Ucci, P. Migliorini, O. Pankewycz, Murine monoclonal anti-DNA antibodies bind directly to glomerular antigens and form immune deposits, J Immunol 138 (1987) 2883—2889.

[94] C. Putterman, W. Limpanasithikul, M. Edelman, B. Diamond, The double edged sword of the immune response: mutational analysis of a murine anti-pneumococcal, anti-DNA antibody, The Journal of Clinical Investigation 97 (1996) 2251—2259.

[95] P. Faaber, T.P. Rijke, L.B. van de Putte, P.J. Capel, J.H. Berden, Cross-reactivity of human and murine anti-DNA antibodies with heparan sulfate. The major glycosaminoglycan in glomerular basement membranes, The Journal of Clinical Investigation 77 (1986) 1824—1830.

[96] B. Deocharan, X. Qing, J. Lichauco, C. Putterman, Alpha-actinin is a cross-reactive renal target for pathogenic anti-DNA antibodies, J Immunol 168 (2002) 3072–3078.

[97] J.L. Barnes, R.A. Radnik, E.P. Gilchrist, M.A. Venkatachalam, Size and charge selective permeability defects induced in glomerular basement membrane by a polycation, Kidney International 25 (1984) 11–19.

[98] D. Vlahakos, M.H. Foster, A.A. Ucci, K.J. Barrett, S.K. Datta, M.P. Madaio, Murine monoclonal anti-DNA antibodies penetrate cells, bind to nuclei, and induce glomerular proliferation and proteinuria in vivo, J Am Soc Nephrol 2 (1992) 1345–1354.

[99] J.B. Lefkowith, M. Kiehl, J. Rubenstein, R. DiValerio, K. Bernstein, L. Kahl, et al., Heterogeneity and clinical significance of glomerular-binding antibodies in systemic lupus erythematosus, The Journal of Clinical Investigation 98 (1996) 1373–1380.

[100] G.N. Coritsidis, P.C. Beers, P.M. Rumore, Glomerular uptake of nucleosomes: evidence for receptor-mediated mesangial cell binding, Kidney International 47 (1995) 1258–1265.

[101] S. Gay, M.J. Losman, W.J. Koopman, E.J. Miller, Interaction of DNA with connective tissue matrix proteins reveals preferential binding to type V collagen, J Immunol 135 (1985) 1097–1100.

[102] R.A. Lake, A. Morgan, B. Henderson, N.A. Staines, A key role for fibronectin in the sequential binding of native dsDNA and monoclonal anti-DNA antibodies to components of the extracellular matrix: its possible significance in glomerulonephritis, Immunology 54 (1985) 389–395.

[103] R.M. Termaat, K. Brinkman, F. van Gompel, L.P. van den Heuvel, J.H. Veerkamp, R.J. Smeenk, et al., Cross-reactivity of monoclonal anti-DNA antibodies with heparan sulfate is mediated via bound DNA/histone complexes, Journal of Autoimmunity 3 (1990) 531–545.

[104] V. Pascual, L. Farkas, J. Banchereau, Systemic lupus erythematosus: all roads lead to type I interferons, Current Opinion in Immunology 18 (2006) 676–682.

[105] T. Lovgren, M.L. Eloranta, U. Bave, G.V. Alm, L. Ronnblom, Induction of interferon-alpha production in plasmacytoid dendritic cells by immune complexes containing nucleic acid released by necrotic or late apoptotic cells and lupus IgG, Arthritis and Rheumatism 50 (2004) 1861–1872.

[106] F.P. Siegal, N. Kadowaki, M. Shodell, P.A. Fitzgerald-Bocarsly, K. Shah, S. Ho, et al., The nature of the principal type 1 interferon-producing cells in human blood, Science (New York, N.Y 284 (1999) 1835–1837.

[107] E.C. Baechler, F.M. Batliwalla, G. Karypis, P.M. Gaffney, W.A. Ortmann, K.J. Espe, et al., Interferon-inducible gene expression signature in peripheral blood cells of patients with severe lupus, Proceedings of the National Academy of Sciences of the United States of America 100 (2003) 2610–2615.

[108] L. Bennett, A.K. Palucka, E. Arce, V. Cantrell, J. Borvak, J. Banchereau, et al., Interferon and granulopoiesis signatures in systemic lupus erythematosus blood, The Journal of Experimental Medicine 197 (2003) 711–723.

[109] A.N. Theofilopoulos, R. Baccala, B. Beutler, D.H. Kono, Type I interferons (alpha/beta) in immunity and autoimmunity, Annual Review of Immunology 23 (2005) 307–336.

[110] A. Doreau, A. Belot, J. Bastid, B. Riche, M.C. Trescol-Biemont, B. Ranchin, et al., Interleukin 17 acts in synergy with B cell-activating factor to influence B cell biology and the pathophysiology of systemic lupus erythematosus, Nature Immunology 10 (2009) 778–785.

[111] M.P. Madaio, S. Hodder, R.S. Schwartz, B.D. Stollar, Responsiveness of autoimmune and normal mice to nucleic acid antigens, J Immunol 132 (1984) 872–876.

[112] G.S. Gilkeson, J.P. Grudier, D.G. Karounos, D.S. Pisetsky, Induction of anti-double stranded DNA antibodies in normal mice by immunization with bacterial DNA, J Immunol 142 (1989) 1482–1486.

[113] D.D. Desai, M.R. Krishnan, J.T. Swindle, T.N. Marion, Antigen-specific induction of antibodies against native mammalian DNA in nonautoimmune mice, J Immunol 151 (1993) 1614–1626.

[114] G. Bredholt, E. Olaussen, U. Moens, O.P. Rekvig, Linked production of antibodies to mammalian DNA and to human polyomavirus large T antigen: footprints of a common molecular and cellular process? Arthritis and Rheumatism 42 (1999) 2583–2592.

[115] M.R. Ehrenstein, H.T. Cook, M.S. Neuberger, Deficiency in serum immunoglobulin (Ig)M predisposes to development of IgG autoantibodies, The Journal of Experimental Medicine 191 (2000) 1253–1258.

[116] P. Kouki, J.E. Marsh, S.H. Sacks, N.S. Sheerin, Autoimmune renal injury in C3- and C4-deficient mice: a histological and functional study, Nephron 96 (2004) e14–22.

[117] D.A. Mitchell, M.C. Pickering, J. Warren, L. Fossati-Jimack, J. Cortes-Hernandez, H.T. Cook, et al., C1q deficiency and autoimmunity: the effects of genetic background on disease expression, J Immunol 168 (2002) 2538–2543.

[118] M. Napirei, H. Karsunky, B. Zevnik, H. Stephan, H.G. Mannherz, T. Moroy, Features of systemic lupus erythematosus in Dnase1-deficient mice, Nature Genetics 25 (2000) 177–181.

[119] m O'Driscoll, TRX1 DNA exonuclease deficiency, accumulation of single stranded DNA and complex human gnetic disorders, DNA Repair (Amst) 7 (2008) 997–1003.

[120] A. el-Roiey, O. Sela, D.A. Isenberg, R. Feldman, B.C. Colaco, R.C. Kennedy, et al., The sera of patients with Klebsiella infections contain a common anti-DNA idiotype (16/6) Id and anti-polynucleotide activity, Clinical and Experimental Immunology 67 (1987) 507–515.

[121] K.K. Unni, K.E. Holley, F.C. McDuffie, J.L. Titus, Comparative study of NZB mice under germfree and conventional conditions, The Journal of Rheumatology 2 (1975) 36–44.

[122] S.K. Ray, C. Putterman, B. Diamond, Pathogenic autoantibodies are routinely generated during the response to foreign antigen: a paradigm for autoimmune disease, Proceedings of the National Academy of Sciences of the United States of America 93 (1996) 2019–2024.

[123] U. Wellmann, M. Letz, M. Herrmann, S. Angermuller, J.R. Kalden, T.H. Winkler, The evolution of human anti-double-stranded DNA autoantibodies, Proceedings of the National Academy of Sciences of the United States of America 102 (2005) 9258–9263.

[124] W.H. Reeves, M. Satoh, J. Wang, C.H. Chou, A.K. Ajmani, Systemic lupus erythematosus. Antibodies to DNA, DNA-binding proteins, and histones, Rheumatic Diseases Clinics of North America 20 (1994) 1–28.

[125] S. Fatenejad, M.J. Mamula, J. Craft, Role of intermoleuclar/intrastructural B- and T-cell determinants in the diversification of autoantibodies to ribonu-cleoprotein particles, Proceedings of the National Academy of Sciences of the United States of America 90 (1994) 12010–12040.

[126] J.A. James, T. Gross, R.H. Scofield, J.B. Harley, Immunoglobulin epitope spreading and autoimmune disease after peptide immunization: Sm B/B'-derived PPPGMRPP and PPPGIRGP induce spliceosome autoimmunity, The Journal of Experimental Medicine 181 (1995) 453–461.

[127] D.G. Ando, E.E. Sercarz, B.H. Hahn, Mechanisms of T and B cell collaboration in the in vitro production of anti-DNA antibodies in the NZB/NZW F1 murine SLE model, J Immunol 138 (1987) 3185–3190.

[128] I. Sekigawa, Y. Ishida, S. Hirose, H. Sato, T. Shirai, Cellular basis of in vitro anti-DNA antibody production: evidence for T cell dependence of IgG-class anti-DNA antibody synthesis in the (NZB X NZW)F1 hybrid, J Immunol 136 (1986) 1247–1252.

[129] B.K. Finck, P.S. Linsley, D. Wofsy, Treatment of murine lupus with CTLA4Ig, Science (New York, N.Y 265 (1994) 1225–1227.

[130] C. Mohan, Y. Shi, J.D. Laman, S.K. Datta, Interaction between CD40 and its ligand gp39 in the development of murine lupus nephritis, J Immunol 154 (1995) 1470–1480.

[131] T.T. Ng, I.E. Collins, S.B. Kanner, M.J. Humphries, N. Amft, R.G. Wickremasinghe, et al., Integrin signalling defects in T-lymphocytes in systemic lupus erythematosus, Lupus 8 (1999) 39–51.

[132] Y. Tada, K. Nagasawa, Y. Yamauchi, H. Tsukamoto, Y. Niho, A defect in the protein kinase C system in T cells from patients with systemic lupus erythematosus, Clinical Immunology and Immunopathology 60 (1991) 220–231.

[133] J.C. Crispin, G.C. Tsokos, Novel molecular targets in the treatment of systemic lupus erythematosus, Autoimmunity reviews 7 (2008) 256–261.

[134] V.C. Kyttaris, Y. Wang, Y.T. Juang, A. Weinstein, G.C. Tsokos, Increased levels of NF-ATc2 differentially regulate CD154 and IL-2 genes in T cells from patients with systemic lupus erythematosus, J Immunol 178 (2007) 1960–1966.

[135] Y.T. Juang, Y. Wang, G. Jiang, H.B. Peng, S. Ergin, M. Finnell, et al., PP2A dephosphorylates Elf-1 and determines the expression of CD3zeta and FcRgamma in human systemic lupus erythematosus T cells, J Immunol 181 (2008) 3658–3664.

[136] M. Takeno, H. Nagafuchi, S. Kaneko, S. Wakisaka, K. Oneda, Y. Takeba, et al., Autoreactive T cell clones from patients with systemic lupus erythematosus support polyclonal autoantibody production, J Immunol 158 (1997) 3529–3538.

[137] M. Khalil, K. Inaba, R. Steinman, J. Ravetch, B. Diamond, T cell studies in a peptide-induced model of systemic lupus erythematosus, J Immunol 166 (2001) 1667–1674.

[138] H.W. Davidson, M.A. West, C. Watts, Endocytosis, intracellular trafficking, and processing of membrane IgG and monovalent antigen/membrane IgG complexes in B lymphocytes, J Immunol 144 (1990) 4101–4109.

[139] B.H. Hahn, F.M. Ebling, Suppression of NZB/NZW murine nephritis by administration of a syngeneic monoclonal antibody to DNA. Possible role of anti-idiotypic antibodies, The Journal of Clinical Investigation 71 (1983) 1728–1736.

[140] R.Y. Lan, A.A. Ansari, Z.X. Lian, M.E. Gershwin, Regulatory T cells: development, function and role in autoimmunity, Autoimmunity Reviews 4 (2005) 351–363.

[141] J.M. Kim, J.P. Rasmussen, A.Y. Rudensky, Regulatory T cells prevent catastrophic autoimmunity throughout the lifespan of mice, Nature Immunology 8 (2007) 191–197.

[142] N. Iikuni, E.V. Lourenco, B.H. Hahn, A. La Cava, Cutting edge: Regulatory T cells directly suppress B cells in systemic lupus erythematosus, J Immunol 183 (2009) 1518–1522.

[143] C.M. Cuda, S. Wan, E.S. Sobel, B.P. Croker, L. Morel, Murine lupus susceptibility locus Sle1a controls regulatory T cell number and function through multiple mechanisms, J Immunol 179 (2007) 7439–7447.

[144] W.T. Hsu, J.L. Suen, B.L. Chiang, The role of CD4CD25 T cells in autoantibody production in murine lupus, Clinical and Experimental Immunology 145 (2006) 513–519.

[145] K.J. Scalapino, Q. Tang, J.A. Bluestone, M.L. Bonyhadi, D.I. Daikh, Suppression of disease in New Zealand Black/New Zealand White lupus-prone mice by adoptive transfer of ex vivo expanded regulatory T cells, J Immunol 177 (2006) 1451–1459.

[146] C.R. Monk, M. Spachidou, F. Rovis, E. Leung, M. Botto, R.I. Lechler, et al., MRL/Mp CD4+, CD25- T cells show reduced sensitivity to suppression by CD4+, CD25+ regulatory T cells in vitro: a novel defect of T cell regulation in systemic lupus erythematosus, Arthritis and Rheumatism 52 (2005) 1180–1184.

[147] H.Y. Lee, Y.K. Hong, H.J. Yun, Y.M. Kim, J.R. Kim, W.H. Yoo, Altered frequency and migration capacity of CD4+CD25+ regulatory T cells in systemic lupus erythematosus, Rheumatology (Oxford, England) 47 (2008) 789–794.

[148] N.A. Azab, I.H. Bassyouni, Y. Emad, G.A. Abd El-Wahab, G. Hamdy, M.A. Mashahit, CD4+CD25+ regulatory T cells (TREG) in systemic lupus erythematosus (SLE) patients: the possible influence of treatment with corticosteroids, Clinical Immunology (Orlando, Fla 127 (2008) 151–157.

[149] B. Yan, S. Ye, G. Chen, M. Kuang, N. Shen, S. Chen, Dysfunctional CD4+, CD25+ regulatory T cells in untreated active systemic lupus erythematosus secondary to interferon-alpha-producing antigen-presenting cells, Arthritis and Rheumatism 58 (2008) 801–812.

[150] J. Erikson, M.Z. Radic, S.A. Camper, R.R. Hardy, C. Carmack, M. Weigert, Expression of anti-DNA immunoglobulin transgenes in non-autoimmune mice, Nature 349 (1991) 331–334.

[151] D. Offen, L. Spatz, H. Escowitz, S. Factor, B. Diamond, Induction of tolerance to an IgG autoantibody, Proceedings of the National Academy of Sciences of the United States of America 89 (1992) 8332–8336.

[152] D.A. Fulcher, A. Basten, Whither the anergic B-cell? Autoimmunity 19 (1994) 135–140.

[153] J. Venkatesh, E. Peeva, X. Xu, B. Diamond, Cutting Edge: Hormonal milieu, not antigenic specificity, determines the mature phenotype of autoreactive B cells, J Immunol 176 (2006) 3311–3314.

[154] M. Mackay, A. Stanevsky, T. Wang, C. Aranow, M. Li, S. Koenig, et al., Selective dysregulation of the FcgammaIIB receptor on memory B cells in SLE, The Journal of Experimental Medicine 203 (2006) 2157–2164.

[155] T.L. McGaha, M.C. Karlsson, J.V. Ravetch, FcgammaRIIB deficiency leads to autoimmunity and a defective response to apoptosis in Mrl-MpJ mice, J Immunol 180 (2008) 5670–5679.

[156] A.M. Jacobi, J. Zhang, M. Mackay, C. Aranow, B. Diamond, Phenotypic characterization of autoreactive B cells—checkpoints of B cell tolerance in patients with systemic lupus erythematosus, PloS one 4 (2009) e5776.

[157] C. Chen, Z. Nagy, E.L. Prak, M. Weigert, Immunoglobulin heavy chain gene replacement: a mechanism of receptor editing, Immunity 3 (1995) 747–755.

[158] N. Suzuki, S. Mihara, T. Sakane, Development of pathogenic anti-DNA antibodies in patients with systemic lupus erythematosus, Faseb J 11 (1997) 1033–1038.

[159] J.H. Roark, C.L. Kuntz, K.A. Nguyen, A.J. Caton, J. Erikson, Breakdown of B cell tolerance in a mouse model of systemic lupus erythematosus, The Journal of Experimental Medicine 181 (1995) 1157–1167.

[160] A. Strasser, S. Whittingham, D.L. Vaux, M.L. Bath, J.M. Adams, S. Cory, et al., Enforced BCL2 expression in B-lymphoid cells prolongs antibody responses and elicits autoimmune disease, Proceedings of the National Academy of Sciences of the United States of America 88 (1991) 8661–8665.

[161]. P. Kuo, M.S. Bynoe, C. Wang, B. Diamond, Bcl-2 leads to expression of anti-DNA B cells but no nephritis: a model for a clinical subset, European Journal of Immunology 29 (1999) 3168–3178.

[162] M.S. Bynoe, C.M. Grimaldi, B. Diamond, Estrogen up-regulates Bcl-2 and blocks tolerance induction of naive B cells, Proceedings of the National Academy of Sciences of the United States of America 97 (2000) 2703–2708.

[163] E. Peeva, C. Grimaldi, L. Spatz, B. Diamond, Bromocriptine restores tolerance in estrogen-treated mice, The Journal of Clinical Investigation 106 (2000) 1373—1379.

[164] C.M. Westhoff, A. Whittier, S. Kathol, J. McHugh, C. Zajicek, L.D. Shultz, et al., DNA-binding antibodies from viable motheaten mutant mice: implications for B cell tolerance, J Immunol 159 (1997) 3024—3033.

[165] J.G. Cyster, C.C. Goodnow, Tuning antigen receptor signaling by CD22: integrating cues from antigens and the microenvironment, Immunity 6 (1997) 509—517.

[166] L. Nitschke, R. Carsetti, B. Ocker, G. Kohler, M.C. Lamers, CD22 is a negative regulator of B-cell receptor signalling, Curr Biol 7 (1997) 133—143.

[167] V.W. Chan, F. Meng, P. Soriano, A.L. DeFranco, C.A. Lowell, Characterization of the B lymphocyte populations in Lyn-deficient mice and the role of Lyn in signal initiation and down-regulation, Immunity 7 (1997) 69—81.

[168] L.B. Justement, K.S. Campbell, N.C. Chien, J.C. Cambier, Regulation of B cell antigen receptor signal transduction and phosphorylation by CD45, Science 252 (1991) 1839—1842.

[169] J.G. Cyster, J.I. Healy, K. Kishihara, T.W. Mak, M.L. Thomas, C.C. Goodnow, Regulation of B-lymphocyte negative and positive selection by tyrosine phosphatase CD45, Nature 381 (1996) 325—328.

[170] S. Sato, M. Hasegawa, M. Fujimoto, T.F. Tedder, K. Takehara, Quantitative genetic variation in CD19 expression correlates with autoimmunity, J Immunol 165 (2000) 6635—6643.

[171] J. Ma, J. Xu, M.P. Madaio, Q. Peng, J. Zhang, I.S. Grewal, et al., Autoimmune lpr/lpr mice deficient in CD40 ligand: spontaneous Ig class switching with dichotomy of autoantibody responses, J Immunol 157 (1996) 417—426.

[172] A. Desai-Mehta, L. Lu, R. Ramsey-Goldman, S.K. Datta, Hyperexpression of CD40 ligand by B and T cells in human lupus and its role in pathogenic autoantibody production, The Journal of Clinical Investigation 97 (1996) 2063—2073.

[173] M.A. Linterman, R.J. Rigby, R.K. Wong, D. Yu, R. Brink, J.L. Cannons, et al., Follicular helper T cells are required for systemic autoimmunity, The Journal of experimental medicine 206 (2009) 561—576.

[174] S.A. Boackle, V.M. Holers, X. Chen, G. Szakonyi, D.R. Karp, E.K. Wakeland, et al., Cr2, a candidate gene in the murine Sle1c lupus susceptibility locus, encodes a dysfunctional protein, Immunity 15 (2001) 775—785.

[175] K. Takahashi, Y. Kozono, T.J. Waldschmidt, D. Berthiaume, R.J. Quigg, A. Baron, et al., Mouse complement receptors type 1 (CR1;CD35) and type 2 (CR2;CD21): expression on normal B cell subpopulations and decreased levels during the development of autoimmunity in MRL/lpr mice, J Immunol 159 (1997) 1557—1569.

[176] Z. Chen, S.B. Koralov, G. Kelsoe, Complement C4 inhibits systemic autoimmunity through a mechanism independent of complement receptors CR1 and CR2, The Journal of Experimental Medicine 192 (2000) 1339—1352.

[177] G. Texido, I.H. Su, I. Mecklenbrauker, K. Saijo, S.N. Malek, S. Desiderio, et al., The B-cell-specific Src-family kinase Blk is dispensable for B-cell development and activation, Molecular and Cellular Biology 20 (2000) 1227—1233.

[178] G. Hom, R.R. Graham, B. Modrek, K.E. Taylor, W. Ortmann, S. Garnier, et al., Association of systemic lupus erythematosus with C8orf13-BLK and ITGAM-ITGAX, The New England Journal of Medicine 358 (2008) 900—909.

[179] Y.H. Lee, Y.H. Rho, S.J. Choi, J.D. Ji, G.G. Song, S.K. Nath, et al., The PTPN22 C1858T functional polymorphism and autoimmune diseases—a meta-analysis, Rheumatology (Oxford, England) 46 (2007) 49—56.

[180] S.V. Kozyrev, A.K. Abelson, J. Wojcik, A. Zaghlool, M.V. Linga Reddy, E. Sanchez, et al., Functional variants in the B-cell gene BANK1 are associated with systemic lupus erythematosus, Nature genetics 40 (2008) 211—216.

[181] Y. Xu, K.W. Harder, N.D. Huntington, M.L. Hibbs, D.M. Tarlinton, Lyn tyrosine kinase: accentuating the positive and the negative, Immunity 22 (2005) 9—18.

[182] S.N. Liossis, E.E. Solomou, M.A. Dimopoulos, P. Panayiotidis, M.M. Mavrikakis, P.P. Sfikakis, B-cell kinase lyn deficiency in patients with systemic lupus erythematosus, J Investig Med 49 (2001) 157—165.

[183] E. Stuber, W. Strober, The T cell-B cell interaction via OX40-OX40L is necessary for the T cell-dependent humoral immune response, The Journal of Experimental Medicine 183 (1996) 979—989.

[184] L. Prokunina, C. Castillejo-Lopez, F. Oberg, I. Gunnarsson, L. Berg, V. Magnusson, et al., A regulatory polymorphism in PDCD1 is associated with susceptibility to systemic lupus erythematosus in humans, Nature Genetics 32 (2002) 666—669.

[185] J.Y. Chen, C.M. Wang, C.C. Ma, S.F. Luo, J.C. Edberg, R.P. Kimberly, et al., Association of a transmembrane polymorphism of Fcgamma receptor IIb (FCGR2B) with systemic lupus erythematosus in Taiwanese patients, Arthritis and Rheumatism 54 (2006) 3908—3917.

[186] K. Su, H. Yang, X. Li, X. Li, A.W. Gibson, J.M. Cafardi, et al., Expression profile of FcgammaRIIb on leukocytes and its dysregulation in systemic lupus erythematosus, J Immunol 178 (2007) 3272—3280.

[187] J.A. Kelly, J.M. Kelley, K.M. Kaufman, J. Kilpatrick, G.R. Bruner, J.T. Merrill, et al., Interferon regulatory factor-5 is genetically associated with systemic lupus erythematosus in African Americans, Genes and Immunity 9 (2008) 187—194.

[188] R.J. Smeenk, H.G. van den Brink, K. Brinkman, R.M. Termaat, J.H. Berden, A.J. Swaak, Anti-dsDNA: choice of assay in relation to clinical value, Rheumatology International 11 (1991) 101—107.

[189] Y. Takeuchi, O. Ishikawa, Y. Miyachi, The comparative study of anti-double stranded DNA antibody levels measured by radioimmunoassay and enzyme-linked immunosorbent assay in systemic lupus erythematosus, The Journal of Dermatology 24 (1997) 297—300.

[190] S.L. Kalled, A.H. Cutler, S.K. Datta, D.W. Thomas, Anti-CD40 ligand antibody treatment of SNF1 mice with established nephritis: preservation of kidney function, J Immunol 160 (1998) 2158—2165.

[191] M. Mihara, I. Tan, Y. Chuzhin, B. Reddy, L. Budhai, A. Holzer, et al., CTLA4Ig inhibits T cell-dependent B-cell maturation in murine systemic lupus erythematosus, The Journal of Clinical Investigation 106 (2000) 91—101.

[192] D.I. Daikh, B.K. Finck, P.S. Linsley, D. Hollenbaugh, D. Wofsy, Long-term inhibition of murine lupus by brief simultaneous blockade of the B7/CD28 and CD40/gp39 costimulation pathways, J Immunol 159 (1997) 3104—3108.

[193] J.T. Merrill, J.C. Shanahan, K.M. Latinis, J.C. Oates, T.O. Utset, C. Gordon, et al., Efficacy and safety of Rituximab in patients with moderately to severely active SLE: Results from the randomized, double-blind phase II/III study EXPLORER, Arthritis and Rheumatism (2008). LB12.

[194] Y. Li, F. Chen, M. Putt, Y.K. Koo, M. Madaio, J.C. Cambier, et al., B cell depletion with anti-CD79 mAbs ameliorates autoimmune disease in MRL/lpr mice, J Immunol 181 (2008) 2961—2972.

[195] R. Furie, R.J. Looney, J.M. Rovin, G. Latinish, J. Appel, F.C. Sanchez-Guerrero, et al., Efficacy and safety of Rituximab in subjects with active proliferative Lupous Nephritis (LN): Results from the randomized, Double-Blind Phase III LUNAR study, Arthritis and Rheumatism S1149 (2009).

[196] M. Ramanujam, X. Wang, W. Huang, Z. Liu, L. Schiffer, H. Tao, et al., Similarities and differences between selective and nonselective BAFF blockade in murine SLE, The Journal of Clinical Investigation 116 (2006) 724–734.

[197] C.O. Jacob, L. Pricop, C. Putterman, M.N. Koss, Y. Liu, M. Kollaros, et al., Paucity of clinical disease despite serological autoimmunity and kidney pathology in lupus-prone New Zealand mixed 2328 mice deficient in BAFF, J Immunol 177 (2006) 2671–2680.

[198] D.J. Wallace, W. Stohl, R.A. Furie, J.R. Lisse, J.D. McKay, J.T. Merrill, et al., A phase II, randomized, double-blind, placebo-controlled, dose-ranging study of belimumab in patients with active systemic lupus erythematosus, Arthritis and Rheumatism 61 (2009) 1168–1178.

[199] S. Navarra, R. Guzman, A. Gallacher, R.A. Levy, E.K. Li, M. Thomas, et al., Belimumab, a BLyS-Specific inhibitor, reduced disease activity, flares and prednisone use in patients with active SLE: Efficacy and safety results from the Phase 3 BLISS-52 study, Arthritis and Rheumatism 60 (2009) 3859.

Extractable Nuclear Antigens and SLE: Specificity and Role in Disease Pathogenesis

Mary Keogan, Grainne Kearns, Caroline A. Jefferies

INTRODUCTION

Autoantibodies directed against specific nuclear antigens (proteins and DNA) are central to the pathogenesis of systemic lupus erythematosus (SLE). The detection of antinuclear antibodies (ANA) is considered one of the diagnostic criteria of a number of autoimmune diseases such as SLE, Sjogren's syndrome, and systemic sclerosis. The original identification of the antigens that are recognized by ANA involved salt extraction of the proteins from nucleic acids, and they were therefore dubbed extractable nuclear antigens (ENA) [1]. Anti-ENA antibodies often arise in two grouped sets targeted at specific ribonucleoprotein (RNP) complexes — the Ro/La particle and the U1-RNP complex. The U1-RNP complex is a major component of the spliceosome which catalyzes pre-mRNA splicing into mature mRNA. Autoantigens inherent in this complex include the RNP proteins themselves (RNP70, RNP A, RNP B and RNP C) and the Smith antigen — named after the patient it was first identified in [2]. The Ro/La complex is composed of four small, uridine-rich, cytoplasmic RNA species (hYRNA) and several associated proteins — Ro60, Ro52, and La/SSB. Both Ro60 and La/SSB possess RNA binding motifs and have been shown to bind RNA directly, whereas the association of Ro52 with this complex has become recently unsure [3].

Anti-Sm, anti-RNP, and anti-Ro/La are all associated with systemic lupus erythematosus (SLE), with a prevalence of 20–30%, 30–40%, and 40%, respectively. Anti Ro/SSA and La/SSB are often known as Sjogren's antibodies because of their strong association with this syndrome (anti-Ro: 60–80% and anti-La: 40–60%), whereas mixed connective tissue disease is a clinical syndrome described in the 1970s associated with very high-titer anti-RNP antibodies. Despite clear clinical associations with ENAs it appears that they are most helpful in the diagnosis of autoimmune disease rather than as a tool for measuring disease activity or damage in SLE [4]. Anti-ENA antibodies are therefore useful serological markers for SLE but whether of specific autoantibodies directly correlate with disease progression or severity seems uncertain. Research has clearly shown however that development of ENA autoantibodies precedes the development of clinical manifestations of SLE by years [5–7], with anti-Ro complex autoantibodies being amongst the first observed and anti-Sm and anti-RNP antibodies practically coinciding with onset of clinical symptoms [8]. Interestingly, in addition to their role in immune complex formation and activation of inflammatory responses, there is growing evidence that ENA autoantigens themselves may have a biological role in driving the autoimmune disease either directly or indirectly by contributing to inflammatory responses as discussed in this chapter.

AUTOANTIGEN SPECIFICITY

Ro/SSA Autoantigen Complex — Ro52, Ro60 and La/SSB

Ro/SSA autoantibodies are among the most frequent anti-ENA and are traditionally associated with SLE and Sjögren's syndrome (SS). Interestingly, anti-Ro/SSA antibodies are detectable on average 3.4 years before the diagnosis of SLE [5, 6], which diagnostically makes them very useful. The double name of Ro/SSA derives from the fact that two different research groups identified the target antigen independently — the first described it as a cytoplasmic antigen detected in cells of a SLE patient ("Ro") [9] and the second described it as a nuclear antigen associated with Sjogren's syndrome ("SS") [10]. Originally thought to be autoantibodies directed against a single entity (Ro) that was complexed with or bound to small cytoplasmic RNAs known as hY-RNA [11], we now know that there are in fact two separate proteins that comprise Ro/SSA — the 52-kDa protein Ro52 and 60-kDa protein Ro60 [12]. Whilst Ro60 is indeed an RNA-binding protein, Ro52 does not directly bind RNAs and is in fact not a stable component of the Ro—RNP complex. Although originally thought to be closely related proteins, a direct interaction of Ro52 and Ro60 could never be demonstrated. Both are encoded by separate genes and would appear to have very different functions in the cell. Up-regulated expression of both Ro60 and Ro52 has been reportedly linked with the development of SLE-like disease, nephritis, and UV-induced cutaneous inflammation [13—15].

Ro60 autoantigen

Anti Ro-60 antibodies are among the most prevalent autoantibodies identified in SLE. But what if any functional role do they have that might contribute to the pathogenesis of this disease? Interestingly Ro60-deficient mice develop a lupus-like syndrome, suggesting that the cellular role of these proteins is important in protecting against disease [13]. And this indeed, on closer inspection turns out to be the case. Ro60 binds to misfolded, non-coding RNAs in mammalian cells and likely functions as a molecular chaperone for misfolded RNAs. UV-irradiated cells have been shown to accumulate Ro60 in the nucleus where it binds to SNPs important in slicing and mediates their removal if damaged. Indeed, cells lacking Ro60 have a shorter half-life than do those that have a functioning Ro60. Thus Ro60 appears to function as part of RNA quality control, in a pathway that recognizes and tags misfolded RNAs for degradation [16].

Ro52 autoantigen

Antibodies against Ro52 are associated with systemic autoimmune disease, including SS and SLE. It is encoded on the short arm of chromosome 11 in humans (11p) [17] and whilst it was once thought to act as an RNA-binding protein, more recently this has been shown not to be the case. Ro52 belongs to a family of E3 ubiquitin ligases that play important roles in antiviral defense, cell cycle regulation and cellular activation [18]. The Tri Partite-Motif (TRIM) family members (more than 60 identified to date) all share a common N-terminal RING domain which, through interaction with E2 enzymes, is responsible for the ubiquitin transfer activity of the TRIM members. In addition they all possess a central coiled-coiled domain which is important for protein—protein interactions. The C terminal domain is important for substrate interactions and is different amongst subclasses of the TRIM family. Ro52 or TRIM21 contains a SPRY/PRY domain with which it has been shown to interact with its substrates [19]. Like Ro60, the function of Ro52 has recently been elucidated and it has been shown to play an important role in regulating the production of inflammatory cytokines [20] and type I IFNs [14, 20, 21] in response to activation of innate immune receptors, the Toll-like receptors (TLR). Importantly mice deficient in Ro52 develop a spontaneous autoimmune dermatitis, and show elevated cytokine responses when cells are challenged ex vivo with TLR ligands [14]. Therefore like Ro60, Ro52, in addition to being a target for autoantibody production also plays a critical role in regulating autoimmune responses. One has to ask the question therefore — are these two facts linked? It remains to be seen if autoantibodies against Ro60 and Ro52 functionally inhibit these proteins and thus contribute to autoimmunity or if they simply represent by-products of disease pathogenesis.

La/SSB autoantigen

Anti-La/SSB antibodies are more closely associated with Sjögren's syndrome than with SLE and when present are associated with a less severe form of SLE. Clinically, anti-La/SSB and anti-Ro52 antibodies are also associated with congenital complete heart block in infants of mothers with anti-Ro/anti-La antibodies. SLE patients with Anti-La antibodies seem to have a less severe form of lupus. A review of 1100 SLE patients by Malik et al. showed anti-La antibodies appear to have a protective effect against renal lupus and decreased seizure risk in SLE patients [22].

The human La autoantigen is a 48-kDa, highly abundant protein found associated with all newly synthesized RNA polymerase III products [3]. La/SSB has been shown to be important both at the initiation of and termination of RNA polymerase III transcription and binds a range of RNAs. It binds via the RNA Pol III termination signal at the 3′ end of the transcripts, a polypyrimidine tract (UUU-OH). One of the reported functions of La is to stabilize these products which aids their processing into mature forms. In addition to RNA species involved in regulation of RNA pol III activity, La/SSB also is found complexed to the small hY-RNAs present in the Ro60-containing RNP, albeit at a different site from Ro60 (Figure 15.1). There are two distinct post-translationally modified pools of La — a highly phosphorylated pool found exclusively in the nucleus (pLa) and a non-phosphorylated form (npLa) found in both the cytoplasm and in the nucleolus [23]. Its ability to detect and interact with RNA transcripts in the cytoplasm involves the non-phosphorylated form of La/SSB [24]. Two kinases have been identified

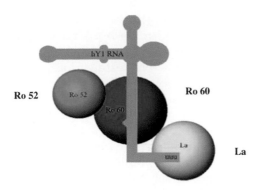

FIGURE 15.1 Ro/La ribonucleoprotein complex. *Reprinted with kind permission by Franceschini et al (2005). Autoimmunity 38, 55—63.*

that phosphorylate La/SSB. Casein kinase 2 (CK2) has been shown to phosphorylate La/SSB and in doing so regulate translation [25]. More recently, AKT phosphorylation on threonine 301 has been shown to regulate nuclear trafficking of La/SSB [26].

RNA pol III has recently been shown to play a critical role in aiding the detection of viral RNA by intracellular pattern recognition receptors such as RIG-I and MDA-5. Both RIG-I and MDA-5 recognize double-stranded RNA viruses. It now appears that RNA polymerase III is responsible for transcribing AT-rich double-stranded DNA into double-stranded RNA which could then be detected by the cytosolic pathogen receptor RIG-I. This system comprises a novel method for detecting viral dsDNA in the cytoplasm and is entirely dependent on RNA polymerase III [27—29]. Interesting both groups demonstrate that RNA pol III was important in sensing small RNAs encoded by Epstein-Barr virus, a virus implicated by many groups to be important in the induction of autoantibody production. As the role for RNA polymerase III expands from regulating tRNA transcription to now being critical for viral RNA detection, the role of La/SSB in this context may be redefined. In support of a potential regulatory role for La/SSB in this system, several reports have established that La/SSB protein directly binds to viral RNAs in the cytoplasm, shielding them from detection by the immune system and enhancing their expression and viral replication [30—32]. Certain viruses may therefore have evolved mechanisms to harness the ability of La/SSB to bind RNA and terminate RNA pol III activity in order to prevent detection of viral nucleic acids by RIG-I. As phosphorylation appears to regulate La/SSB function then this may provide a switch mechanism between potential pro- and antiviral functions.

Human nuclear ribonucleoprotein 1 (hnRNP1) is also part of La/SSA complex that recognizes endogenous RNA pol III (and incidently RNA pol II) transcripts and viral RNA. Again like La, it is an autoantigen, with autoantibodies to hnRNP1 detected in both SLE and rheumatoid arthritis patients. Examining the relative effects of

RNP1 and La on HCV replication using siRNA against both proteins, Domitrovitch *et al.* demonstrated knockdown of La inhibited viral replication (and translation), in keeping with it being a positive regulator of viral replication whereas RNP1 knockdown enhanced viral replication, suggesting that RNP1 functioned to inhibit viral replication [32]. Its ability to do so was through interactions with non-structural proteins (NS) 3 and 5B. The potential for La/SSB and hnRNP1 playing opposing roles in regulating RNA pol III-dependent sensing of DNA viruses remains to be seen.

Smith (Sm) Antigen and Anti-RNP Antigens

The small nuclear ribonucleoparticles (snRNP) are RNA—protein complexes that together with additional proteins form the spliceasome. The snRNP complexes (U1—U5 snRNP) are present in all eukaryotic cells and function to process pre-mRNA to mature mRNA. They are, in essence, RNA particles with two sets of proteins—those that constitute the Smith antigens and those that constitute the RNP autoantigen [2]. Anti-Smith antibodies are directed against seven proteins that constitute the common core of U1, U2, U4, and U5 snRNP complexes. The structure of U1 snRNP is shown in Figure 15.2, in which the protein components (B/B', D, E, F, and G) are clearly marked. These are the autoreactive entities in this complex and anti-Sm antibodies are detected against B/B', D1 and D3, primarily. In the majority of cases anti-Sm antibodies are detected together with anti-RNP. Anti-RNP antibodies are directed against proteins associated with U1 RNA (70kd, A, C), to form U1snRNPs (also clearly marked in Figure 15.2). Anti-RNP antibodies have a prevalence of 25—47% in SLE patients and high titers of anti-RNP are associated with mixed connective tissue disease (MCTD).

There is significant ethnic variation in the prevalence of anti-Sm antibodies. Anti-Sm is observed in approximately 10% of Caucasian patients, whereas it occurs

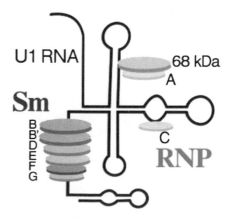

FIGURE 15.2 Sm/RNP ribonucleoprotein complex. *Reprinted with kind permission by Migliorini et al (2005). Autoimmunity 38, 47—54.*

at over twice the frequency (25%) in African-American and Asian sufferers of SLE [3]. Because of the highly selective association of anti-Sm antibodies with SLE, these are very useful predictors of disease. The anti-Sm autoantibody profile seems to develop from a single antigenic epitope on SmB' that has a similar structure in all patients and by intramolecular epitope spreading develops into the mature anti-Sm autoantibody response [34]. As described in detail below antibodies against a peptide in an Epstein-Barr virus-encoded protein, EBNA1, cross-react specifically with SmB autoantigen and in doing so contribute to the generation of an autoantibody response. Due to the close proximity and strong association between Sm antigens and RNP autoantigens in the U1 snRNP, recognition of the U1 snRNP by cross-reactive anti-EBNA1 antibodies results in presentation of RNP-specific peptides and consequent autoantibody production in a process known as intermolecular epitope spreading. Thus in the majority of cases, development of anti-Sm autoantibodies is accompanied by development of anti-RNP antibodies [35].

As yet, unlike their Ro/La counterparts, Sm proteins and U1 70-kDa proteins have not yet been shown to have more diverse roles in regulating RNA processing and hence cellular activities such as immune function.

Genetic Associations with Autoantibody Production

Systemic lupus erythematosus is a complex genetic disease and it is thought that genetic factors may also contribute to autoantibody production. There is significant ethnic variation in expression of anti-ENAs with increased prevalence of anti-Sm in African-Americans and Asians, and increased Anti-Ro in Saudi Arabians along with Puerto Rican Hispanics compared to Texan Hispanics [33, 36, 37]. A strong familial aggregation of autoantibodies (ANA, dsDNA, Ro, La, Sm and RNP) has been observed across families enriched for SLE, supporting the hypothesis that autoantibody production is a genetically complex trait [38]. In addition an association between various HLA haplotypes and anti-ENA prevalence has been observed. HLA-DR2 haplotypes were associated with Sm autoantibodies whereas HLA-DR3 haplotypes were associated with Ro and La autoantibodies both in family members with and without SLE, suggesting that HLA-DR2 and -DR3 haplotypes play a key role in autoantibody and disease susceptibility in SLE [39].

Clinical Detection of Anti-ENA Autoantibodies

Patients with suspected connective tissue diseases are usually screened for antinuclear antibodies (ANA), with confirmatory testing for disease-specific antibodies performed in those found to be ANA positive. There are many different techniques in use, both for ANA screening and follow-on testing for antibodies to dsDNA, and extractable nuclear antigens (Ro, La, Sm, and RNP). Each assay has different sensitivity, specificity, positive and negative predictive values as well as coefficients of variation. It is therefore important that clinicians understand the performance characteristics of methods in use locally, and are aware of any planned changes in laboratory procedures that might affect antibody identification and hence diagnosis.

Antinuclear antibodies (ANAs)

ANAs are usually detected by indirect immunofluorescence (IIF), although in recent years, solid phase assays (ELISA, ELiA, and multiplex bead assays) have also become available. A significant factor affecting the performance of immunofluorescence assays is that it relies on subjective interpretation of the resultant staining pattern. Solid-phase assay performance in contrast is non-subjective, quantitative but can be affected by the quality of the antigen (antigen purity, changes in conformation due to recombinant production methods or denaturation during purification, glycosylation differences). Solid-phase assays may be prone to interference by rheumatoid factors and heterophile antibodies, depending on the assay configuration. Generally, IIF detects moderate- and high-affinity antibodies, while ELISA additionally detects low-affinity antibodies (Egner 2000). American College of Rheumatology (ACR) criteria for the diagnosis of lupus include the detection of an "abnormal titer" ANA at any time point by IIF or equivalent assay [40]. Unfortunately an abnormal titer is not defined, and despite availability of detailed guidelines, standardization of ANA testing between labs is poor [41].

Detection of ANAs by IIF using HEp2 cells allows the pattern of staining to be delineated. Typically the presence of anti-Ro and anti-La antibodies is associated with a fine speckled pattern, while anti-Sm and RNP are associated with a coarse speckled pattern (Figure 15.3A and C). Sm and RNP antigens are located on the U1 small nuclear ribonuclear protein (U1 snRNP) [42], and hence it is not surprising that the staining patterns are indistinguishable. While the ANA pattern gives a guide to the underlying specificity, a common pitfall is the masking of a weak ANA by a stronger ANA with an overlapping pattern in the same sample [43]. The pattern of the ANA staining cannot therefore be considered diagnostic [44] and is not sufficiently reliable to select follow-on tests. ANA screening performed on HEp2 cells is unlikely to miss a clinically significant anti-Sm or anti-RNP antibody. However detection of anti-Ro is more challenging as Ro may be destroyed or

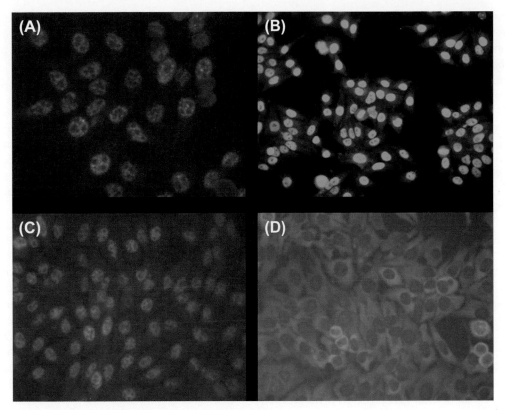

FIGURE 15.3 Staining pattern for antinuclear antibodies (ANA). (*A*) Coarse speckled pattern with anti-RNP positive serum. (*B*) Homogenous pattern for anti-dsDNA antibodies. (*C*) Fine speckled pattern with anti-Ro positive serum. (*D*) Absence of nuclear staining with strong cytoplasmic staining with anti-Ro positive serum.

leach into the cytoplasm during fixation of HEp2 cells (Figure 15.3, compare C and D). It has been suggested that detection of anti-Ro is improved by use of HEp2000 cells [45] although this has been disputed [46].

SPECIFIC ASSAYS FOR THE DETECTION OF AUTOANTIBODIES TO RO, LA, RNP, AND SM

Autoantibodies to classical ENA antigens were originally detected by Ouchterlony double immunodiffusion methods, and many of the studies determining the clinical significance of these antibodies were based on results obtained using this method [41]. As these methods are slow and not suited to processing large numbers of samples, countercurrent immunoelectrophoresis was introduced, with ELISA, immunoblotting and more recently multiplex bead assays following. ELISA is considerably more sensitive than immunodiffusion, with multiplex beads more sensitive again in some studies [47]. However when the same technique is performed with different reagents or assay systems, considerable differences in assay performance are readily identified [48]. To maximize the sensitivity and

specificity of ENA testing, it has been suggested that two methods be used [49], with the results being combined into a single report for the clinician [43].

Detection of antibodies to anti-Ro and anti-La are of particular importance in women of child-bearing age given the association with congenital heart block. Immunoblotting appears to have a poorer sensitivity for detection of anti-Ro [44], compared to other ENA specificities, possibly because of the complex conformation of the antigen. Many ELISA assays do not contain Ro-52, while others use a combined Ro-52 and Ro-60 antigen mix, which may result in failure to detect antibodies to single Ro specificities [12]. Both Ro-52 and Ro-60 and anti-La can all be associated with this complication [50]. The risk of congenital heart block in a woman who is anti-Ro positive is approximately 2%, rising to 17–19% if a previous pregnancy has been affected. No laboratory test or feature (antibody specificity, strength, etc.) can predict the outcome of the pregnancy. It is likely that factors other than autoantibodies play a critical role, as monozygotic twins discordant for congenital heart block have been reported [50].

There is no evidence that quantitative measurement of ENAs is helpful at the time of diagnosis, or that monitoring the level of antibodies is clinically useful [44]. In general repeat testing for ENA is discouraged unless the clinical picture changes. However in pregnant women who were previously anti-Ro negative, it may be prudent to recheck ENA status, as development of an anti-Ro antibody would alter monitoring of the pregnancy.

GENERATION OF AUTOANTIBODIES

The current hypothesis regarding the generation of autoantibodies against ENA is that SLE presents a unique setting in which an increased concentration of products associated with cell death, and a decreased ability to remove them, results in activation of autoreactive B cells in genetically predisposed individuals. The pathogenesis associated with SLE is as a direct result of the anti-ENA or ANA recognizing RNA− or DNA−protein complexes, forming immune complexes that are either recognized by TLRs and trigger type I IFN and pro-inflammatory cytokine production or by macrophages and natural killer cells thus triggering antibody-dependent cell-mediated cytotoxicity (ADCC). This translates into an increase in both local and systemic inflammation, enhanced B-cell and NK-cell activity and thus gives rise to a dangerous cycle of inflammation-induced cell death resulting in further release of autoantigens that trigger exacerbation of the response [51].

What contributes to the production of autoantibodies − what are the triggers? Two major routes behind their generation have been proposed: (1) the concentration of nuclear antigens in apoptotic bodies and (2) viral infection. UV irradiation of cells and the consequent induction of cell damage results in the abnormal distribution of ENA antigens on cell surface apoptotic blebs [52]. In addition viral infection and consequent killing by antigen-specific cytotoxic T cells will also induce apoptosis and cell death. Reduced clearance of apoptotic bodies ensures that these antigens, normally sequestered away from the immune system, are now immunogenic. An interesting feature of this (and other autoimmune pathologies) is the concept of epitope spreading [53]. Innoculation of autoimmune prone mice with a specific recombinant La peptide will induce the activation of antigen-specific T cells and ultimately the activation of specific B cells and production of antibodies to individual non-over-lapping epitopes on La in an example of intramolecular spreading. Moreover the same antigenic stimulus can also give rise to the production of antisera capable of detecting Ro60 − an example of intermolecular spreading [54]. Thus, the availability of the Ro/La complex for the immune recognition could ultimately generate the production of autoantibodies against a range of autoantigens and thus might explain the heterogeneity in autoantibody profiles measured in SLE patients (discussed below).

Relocation of ENA Autoantigens into Apoptotic Bodies

The evidence that apoptosis and the relocation of autoantigens to apoptotic bodies plays a key role in autoantigen recognition can be traced back to initial observations that UV irradiation of keratinocytes derived from patients with SLE results in the cell surface expression of these autoantigens so that they become available for recognition [55]. All of the autoantigens discussed above are intracellular antigens and thus hidden from the immune response. However La, Ro52, Ro60, Sm-proteins and U1-70kDa protein have each been demonstrated to relocate to apoptotic blebs during cell injury − thus allowing them to serve as immunogens for the generation of autoantibodies [56, 57]. With respect to UV-induced apoptosis in keratinocytes, relocation of autoantigens to apoptotic bodies coincides with rapid accumulation of reactive oxygen species (ROS), suggesting that ROS accumulation is critical for these events [58, 59]. Interestingly, the availability of Ro60 and Ro52 epitopes in cells undergoing apoptosis is temporally regulated, with Ro60 available in early apoptotic bodies and available throughout the process, whereas Ro52 is only observed in late-stage apoptotic bodies [60]. Under normal circumstances apoptotic cells are removed without being exposed to the immune system. Critically, impaired phagocytosis and clearance of apoptotic cells has been observed in SLE patients, thus resulting in increased exposure of self-antigens to the immune system [61].

Whilst the mechanism behind translocation of these autoantigens to apoptotic bodies is not understood, one study has demonstrated that La is proteolytically cleaved during apoptosis and may therefore serve as a neo-epitope for the production of anti-ENA antibodies [62]. Redistributed Ro52 and Ro60 may be similarly cleaved or may undergo post-translational modifications (such as phosphorylation or oxidation) in order to further render them immunogenic − in Ro60 the major B-cell epitope between residues 169−190 can be both phosphorylated and citrunillated, thus altering autoantibody affinity and specificity [63]. Our unpublished observations have demonstrated that Ro52 can be inducibly phosphorylated, suggesting that in addition to regulating the activity of this protein, phosphorylation or additional modifications may alter the antigenicity of this autoantigen. Indeed in late-stage apoptosis, the generation of neo-epitopes of La through cleavage is required for a loss of tolerance in experimental autoimmunity [64]. Similar changes in location and post-translational modifications have also been demonstrated for the U1-70kDa protein of the U1 snRNP complex. Early work demonstrated that it was cleaved during apoptosis [65, 66]. Interestingly the U1-70kDa snRNP protein has recently been shown to be phosphorylated at serine 140 during apoptosis and relocated to apoptotic bodies. The changes in phosphorylation associated with apoptosis may explain why this region of U1-70kDa protein becomes a dominant autoantigen during SLE [67].

Role of Viral Infection and Molecular Mimicry in Anti-ENA Induction

The concept that host immune responses to pathogenic organisms are responsible for initiating autoimmunity is not new [68] and has gained credence by recent evidence linking Epstein-Barr virus (EBV) infection, molecular mimicry, and the development of SLE [69]. EBV is very well suited to engendering an immune response. It infects B cells, prevents apoptosis of infected cells and causes dysregulated polyclonal activation and autoantibody production [70–74]. Most importantly, anti-ENA autoantibodies are a hallmark of classic acute EBV-induced infectious mononucleosis [74, 75]. These anti-ENA antibodies arise as a result of molecular mimicry between the highly antigenic EBNA1 protein encoded by EBV and both the Sm and Ro60 autoantigens. An epitope on EBNA1 (PPPGRRP) is specifically cross-reactive with a peptide in the SmB' antigen. This the epitope that is recognized by the first anti-Sm autoantibodies that appear in sera of SLE patients [76]. Injection of this Sm epitope or the EBNA1 peptide induces an SLE-like illness in mice and the autoantibody response expands through epitope spreading to include other autoantibodies against other ribonucleoprotein complex components [35, 77, 78].

In addition to Sm cross-reactivity, early autoantibodies that develop against Ro60 also recognize a single epitope on EBNA1 defined by amino acids 58–72 (GGSGSGPRHRDGVRR) and again, through epitope spreading, inoculation with either single epitope can induce autoantibody production against additional epitopes on Ro60 other ENA antigens — namely anti-La and anti-Ro52 [79]. As with Sm-cross-reactive EBNA1 antigen, immunization of New Zealand White rabbits with either cross-reactive peptide (either EBNA1$_{158-72}$ or Ro60$_{159-180}$) results in autoimmune epitope spreading, recognizing additional epitopes in both Ro60 and other autoantigenic ENA components such as SmB', nRNP and dsDNA, in addition to developing clinical symptoms associated with SLE such as leukopenia, mild thrombocytopenia, and renal dysfunction [79].

Interestingly, although 93% of SLE-negative EBV-positive individuals tested in the Nat Med study transiently produced antibodies that recognized the EBNA$_{158-72}$ peptide, only 18% produced autoantibodies that cross reacted with the original Ro60 epitope (aa 169–180). This epitope in Ro60 has previously been shown to be both citrullinated and phosphorylated thus again raising the possibility that ENA autoantigens may be differentially post-translationally modified in persons genetically susceptible to developing SLE, hence accounting for the differences in their ability to mount cross-reactive antibody responses [63]. In addition, pediatric SLE patients recognize unique epitopes in both the N and C terminal regions of EBNA1 that are not recognized by EBV-infected normal controls to any significant degree [80]. How can the disparate immune and antigenic responses between SLE patients and EBV-infected controls be explained? HLA screening does not appear to explain the differences [80] although perhaps more exhaustive studies into this area might be necessary to definitively rule this possibility out. One explanation might be the ability of EBV to modulate antiviral immunity via interfering with viral nucleic acid sensing and the production and effects of type I IFNs [81–85]. EBV is known to manipulate TLR-mediated responses through the upregulation of a specific isoform of IRF5, Variant 12 [81]. IRF5 is a transcription factor that is activated in response to TLR stimulation of cells and is a known genetic risk factor for SLE [86, 87]. Analysis predicts that Variant 12 would encode a DNA-binding domain of IRF5 without the transactivation domain and would thus predict that this variant would act as a negative regulator of other functional IRF5 isoforms. Interestingly, Ro52 has been shown to interact with IRF5 and target it for degradation, thus limiting proinflammatory cytokine and type I IFN production downstream of TLR7 and 9 [14].

Role of Danger-Associated Molecular Patterns (DAMPs) in Autoantibody Induction

Whereas most individuals develop tolerogenic mechanisms to avoid the sustained production of autoantibodies (such as the activation of NKT cells and T suppressor cells), in genetically susceptible individuals the exposure of intracellular antigens in conjunction with adjuvants results in the induction of a strong autoimmune reaction and generation of autoantibodies. In addition to complexing with RNA, the ability of autoantigens to complex to protein adjuvants may have a large bearing on whether the immune system responds or is tolerized to the antigen in question. Danger-associated molecular chaperones (DAMPs) is a term coined to explain the ability of intracellular proteins such as heat shock proteins (HSPs), molecular chaperones, calreticulin and high-mobility group B1 (HMGB1) protein to function as activators of inflammation once released from stressed or dying cells [88, 89]. DAMPs proteins play important roles as adjuvants in sterile and septic inflammation [90], being able to activate macrophages, dendritic cells, and T and B cells through activation of Toll-like receptors and other cell surface receptors. They can act as peptide carriers thus enhancing immunogenicity of the antigens and also directly bind and activate cell surface receptors such as the TLRs. As potent stimulators of innate and adaptive immune responses it is therefore conceivable that they might also participate in positive feedback loops that perpetuate inflammation thus resulting in autoimmune phenotypes.

In this context, complexes between Ro proteins (Ro52 and Ro60) and molecular chaperones have been implicated as immune triggers [91, 92]. The chaperone protein Grp78 has been shown to co-localize with Ro52 and Ro60 in apoptotic blebs and in doing so may aid activation of the immune response [92] — possibly by acting as a co-stimulator of cells of the innate immune response such as macrophages and dendrtic cells via potentially binding to innate immune receptors such as the TLRs [93]. Another DAMP, Calreticulin, is a potent activator of danger signals, translocates to apoptotic tumour cells, and functions as an "eat-me" signal for dendritic cells thus promoting antitumor immunity [94, 95]. In the context of autoimmunity, calreticulin has long been known to associate and co-localize with Ro52 [91, 96], implying it functions in a similar way in autoimmunity. Interestingly, antibodies against Ro52 and Calreticulin in SLE patients tend to segregate, suggesting a general co-association between these two autoantigens and underlining the possibility of co-presentation to the immune system [97].

Thus the ability of HMGB1 and other DAMPS to complex to self antigens such as La and Ro52 as well as bacterial products and thus activate TLRs promotes the breaking of tolerance and suggests that they may be potential targets for the development of therapeutic strategies in the treatment of autoimmune disorders.

PATHOPHYSIOLOGY OF ANTI-ENA ANTIBODIES — ROLES IN SLE

The direct role for anti-ENA antibodies in pathophysiology of lupus has been difficult to establish, although immunization of mice with antigenic Sm or Ro-60 peptides as discussed above results in kidney disease. In addition, direct administration of an RNP peptide to non-disease-prone mice resulted in the production of autoantibodies and end-organ injury (renal and pulmonary disease) [98]. However, observations performed on autopsy have demonstrated that a heterogeneous group of autoantibodies can be detected in patient glomeruli that react with multiple antigens including anti-Sm, anti-Ro, and anti-La [99]. This suggests that whilst these autoantigens may play a role in initiating the disease, the very nature of the autoimmune response and eptitope spreading results in diverse autoantigens being recognized.

Many studies into autoantibody profiles and disease association have been conducted in SLE patient cohorts over the years. The known clinical associations of anti-ENA antibodies, along with HLA specificity (if known) are summarized in Table 15.1. Studies into antibody clustering and disease association have been exhaustive, with mixed results. Renal disease appears to be associated with anti-dsDNA, anti-Sm, and anti-Ro whereas the presence of anti-RNP appears to predict milder

TABLE 15.1 Clinical association of anti-ENA antibodies

ENA (genetic linkage)	HLA linkage	Clinical association
Anti-Ro (linkage 4q34-35)	HLA DR3	Subacute cutaneous SLE Photosensitivity Discoid lupus Chilblain lupus Cutaneous vasculitis Congenital heart block Neonatal SLE Thrombocytopenia Lymphopenia Neutropenia Liver disease Primary biliary sclerosis Valvular heart disease Atrioventricular block Cardiac abnormalities in childhood SLE C1q, C4 deficiency Common in late onset SLE Secondary Sjogren neuropsychiatric damage Polyautoimmunity
Anti-La (linkage 3q21 and 4q34-35)	HLA DR3	Congenital heart block Protection for renal disease and seizures Secondary Sjögren's
Anti-Sm (linkage 3q27)	HLA DR2	Renal disease Specific for SLE
Anti-RNP (linkage 3q27)	Unknown	Mixed connective tissue disease Raynaud's Polymyositis Scleroderma Arthritis

disease [100]. The presence of other autoimmune diseases such as thyroid, liver, and multiple sclerosis in SLE patients is well documented. Recent studies have shown that female gender, anti-Ro positivity (both Ro52 and Ro60), ethnic origin and family history predicted polyautoimmunity in SLE patients. Again, the presence of anti-RNP antibodies appeared to be protective [101]. Studies in SLE patients over the age of 50 show that disease appears to have a different presentation and appears to be more benign – perhaps correlating with reduced immune responses associated with aging. Interestingly these patients tend to have higher anti-Ro, anti-La, and rheumatoid factor, compared with a younger patient cohort and clinically they are more likely to present with hematologic, lung, and liver involvement and associated Sjögren's syndrome [102]. So a pattern is emerging regarding co-association between antibodies present, gender, and ethnic background that may help in predicting SLE outcome [103].

As autoantibodies of varying specificities are a hallmark of SLE, it is little wonder that understanding the role of B cells and the production of autoantibodies has been the focus of SLE research. But what is the relevance of specific autoantibodies and in particular anti-ENA antibodies to disease pathogenesis? Clearly anti-ENA autoantibodies isolated from SLE patients act as disease markers but thus far only anti-Ro and anti-SM antibodies have been shown to directly participate in tissue damage.

Anti-Ro/SSA and Anti-La/SSB Antibodies

The presence of anti-Ro and anti-La in SLE patients would appear to overlap between SLE and Sjögren's syndrome (SS). Interestingly however a recent review of primary and secondary Sjögren's demonstrated that the occurrence of anti-Ro60 was significantly higher in the SLE/SS group than in patients with primary SS [104]. The clinical associations and relevance of anti-Ro60 and anti-Ro52 have been the most clearly documented of the ENA group of antibodies (see Table 15.1). Unlike the other ENAs, there has also been evidence of direct immunopathogenic effect with antibodies against Ro60 or Ro52. Clinical features described include subacute cutaneous SLE (SCLE), photosensitivity, discoid lupus, and chilblain lupus. More recently, cutaneous vasculitis has been associated with anti-Ro responses in Brazilian patients [105].

Evidence of anti-Ro60 and anti-Ro52 deposition in the skin in response to UV-induced exposure of autoantigens in apoptotic bodies is one of the most extensively described clinical associations with these autoantibodies [55, 106]. It has been suggested that the antibody-bound autoantigens are then recognized via Fc receptors on macrophages or dendritic cells, phagocytosed and presented to the adaptive arm of the immune system, thereby stimulating the autoimmune response [107–110]. Cross-talk between immune complexes and TLR-induced IFN responses play an important role in the development of SCLE. Both RNA- and DNA-containing immune complexes are important in activating plasmacytoid dendritic cells in SCLE and inducing the production of type I IFNs and inflammatory cytokines [110, 111]. Interestingly, loss of the lupus autoantigen Ro52 results in injury-induced autoimmune dermatitis in Ro52-knockout mice, increased autoantibody production and enhanced TLR7 and TLR9 responses. These responses are consistent with its role as a negative regulator of type I IFN production. Loss of Ro60 promotes UV-induced skin damage, although the involvement of innate immune recognition receptors has not yet been evaluated in these mice [13]. These studies suggest that autoantigens such as Ro52 and Ro60 are not just targets for autoantibodies in SLE but actively are involved in controlling disease development.

Neonatal SLE and congenital heart block (CHB) are well-described syndromes associated with anti-Ro52 and anti-La, with evidence for a direct involvement of both autoantibodies in the pathogenesis of this syndrome [112, 113]. The anti-Ro52 IgG antibodies originate through transplacental passage of maternal autoantibodies into the fetal circulation from the 16th to the 20th week of pregnancy and have been detected in the conducting system of the newborn [114, 115]. This syndrome often appears when the mother is completely asymptomatic. An assessment of the true prevalence of CHB was conducted in infants whose mothers were anti-Ro positive with a known connective tissue disease and found it to be 2%. They also found additional electrocardiographic abnormalities such as bradycardia and prolonged QT interval (measuring conductance abnormalities) may be helpful in diagnosis [116]. Specifically, antibodies to a specific region in Ro52 (amino acids 200–239) have been found to directly bind cardiomyocytes and mediate cardiac abnormalities associated with CHB [117]. Confocal studies on cardiac myocytes derived from fetuses determined to have CHB, demonstrated that Ro52 and La/SSB are located at the cell membrane in apoptotic myocytes, indicating that availability of the antigens increases during apoptosis [118]. In addition, anti-Ro52 and anti-La/SSB antibody binding to cardiac myocytes has also been demonstrated to induce the pro-inflammatory cytokine TNF-α thus promoting inflammation [119]. Neonatal lupus is characterized by a transient skin rash similar to that observed in subacute cutaneous lupus, and liver abnormalities which are usually transient in a neonate whose mother is anti-Ro52 positive and clears as the newborn eliminates the mother's IgG. The tissue injury is related to the expression of Ro (Ro60 and Ro52) and La antigens in fetal cardiac tissue and their recognition by autoantibodies mediating ADCC [120].

The role of anti-Ro antibodies with cardiac involvement in adult SLE patients has been of interest because of the known association with congenital conduction abnormalities where the mother is Anti-Ro positive. The association in adults has not been well documented but a recent study from Cairo suggested a correlation with valvular heart disease, where 36% of SLE patients with valvular disease were positive for anti-Ro, whereas only 9% were positive in those without valvular involvement. Serum levels of anti-Ro and Anti-La were also significantly higher in the patients with valve disease [121]. There are also several case reports of SLE presenting with complete atrioventricular block and high-titer Anti-Ro [122]. A study of children with SLE showed 42% had evidence of cardiac involvement and that this occurrence correlated significantly with Anti-Ro and anti-La antibodies [123]. Interestingly, anti-Ro52 can cross-react with, bind, and inhibit the function of the 5-HT4 serotonin receptor, thus providing an additional direct effect of anti-Ro52 autoantibodies on cardiac cells and helping to explain the electrophysiological abnormalities observed in neonatal lupus [124].

Hematological involvement, including lymphopenia, neutropenia, and idiopathic thrombocytopenia (ITP), has been associated with anti-Ro60 antibodies and may predate SLE diagnosis [125]. An anti-Ro60 Fab fragment has been demonstrated to bind to the surface of granulocytes thus fixing complement to the cell membrane and promoting complement-mediated cell death. This can be reversed by recombinant Ro60, demonstrating a direct immunopathogenic role of anti-Ro60 in neutropenia [97, 126]. Hepatic involvement tends to occur in SLE patients who are anti-Ro positive, with antibodies specifically against Ro52 being detected in patients with autoimmune liver disease [127, 128]. Interestingly a review of patients with primary biliary cirrhosis (PBC) showed that 28% were anti-Ro positive and this group had more advanced histology and present with higher serum bilirubin levels at diagnosis. The authors of the study therefore concluded that anti-Ro may helpful in the diagnosis of PBC when antimitochondrial antibodies are not detected [129].

Anti-Sm and Anti-RNP Antibodies and Clinical Manifestations

The clinical significance of anti-Sm and anti-RNP antibodies is a matter of debate. Studies assessing correlation between disease activity/severity and antibody titer have produced varying results [100]. The presence of anti-Sm is helpful in the diagnosis of SLE and its detection along with anti-dsDNA, anti-Ro, and anti-La may be particularly useful when combined with other antibodies in identifying patients more at risk of renal lupus [103]. However, it would appear that fluctuations of anti-Sm antibody levels are not helpful in monitoring disease activity. However, studies would suggest that repeated testing for this antibody is advised when it is initially negative is necessary as patients initially diagnosed as anti-Sm negative may become positive up to 8 years after diagnosis [8]. The appearance of anti-Sm later in the course of SLE may be predictive of developing renal disease.

High levels of anti-RNP antibodies are associated with multiple connective tissue disorder (MCTD). MCTD was the first rheumatic disease to be defined by an autoantibody test, specifically demonstrating high titers of anti-RNP antibodies. The characteristic clinical features of MCTD are Raynaud's syndrome, sclerodactyly, arthritis, polymyositis, and interstitial lung disease. It was initially felt to be a benign disease but studies have shown that it may evolve into SLE or scleroderma, a progression that is associated with decreasing titers of anti-RNP levels [130].

Anti-Sm antibodies have been associated with the severity and presence of renal disease, particularly in patients of African-American descent. This association appears to be stronger when anti-Sm is accompanied with anti-dsDNA antibodies [100]. A recent study showed that the combination of anti-Sm, anti-dsDNA, and lupus anticoagulant (associated with antiphospholipid syndrome) in female patients increased both risk and severity of renal disease [131, 132]. In contrast anti-RNP antibodies are associated with milder renal involvement [100, 131, 133]. In contrast a Canadian study showed the presence and levels of anti-Sm and dsDNA did not predict damage in SLE [134]. In agreement with this, mouse models of experimental lupus would suggest that the anti-Sm autoantibodies are not directly involved in the pathology of lupus nephritis, with autoreactive B cells being identified in nephritic kidneys, producing antibodies against dsDNA and ssDNA but not anti-Sm [135]. When administered systemically these autoantibodies became progressively deposited on the glomerular membrane, mimicking immune-complex deposition and inflammation associated with disease. Although there seems to be controversy surrounding the issue of involvement of anti-Sm antibodies in the pathology of renal lupus, it appears that, regardless of socioeconomic group or lupus-related organ damage, the presence of anti-Sm is a strong predictor of increased mortality in lupus [136].

Whether or not specific autoantibodies are required in development of renal lupus, glomerular immune complex deposition is one of the first signs of renal involvement in SLE. In autoimmune prone mice, immune complex deposition is followed by inflammatory cell migration, activation and resultant tissue destruction as the activated macrophages attempt to clear the immune complexes. However the absolute requirement of immune complex formation in the initiation of nephritis has recently been questioned [137]. Adoptive transfer of CD4$^+$ T lymphocytes from mice immunized with an RNP peptide into syngenic naïve mice resulted in establishment of persistent lupus-like nephritis. Importantly, co-transfer of TLR3-induced myeloid DCs resulted in inhibition of the development of nephritis and instead resulted in the development of mixed connective tissue disease (MCTD)-like lung disease, indicating that different DC populations can help with tissue targeting of anti-RNP autoimmunity [66]. These and other recent studies serve to highlight the importance of TLRs as adjuvants in both the development and pathophysiology of SLE. In particular, TLRs have been strongly implicated in the immune-complex-mediated nephritis [138]. For example, RNA-containing anti-ENA autoantibody-antigen immune complexes contribute to pathology via activation of TLR7-expressing plasmacytoid DCs and consequent IFN-α production. TLR7 ligation also helps activate autoreactive B cells present in renal tissue, resulting in production of more autoantibodies and also the production of inflammatory cytokines and chemokines. Chronic viral infections can not only trigger lupus nephritis but also induce immune complex nephritis in the absence of autoimmunity, highlighting the importance of antiviral mechanisms, and particularly production of type I IFNs, in development of renal disease. Whether resident pDCs or glomerular mesangial cells are responsible for IFN-α induction in renal disease is unclear, although mesangial cells express TLR3 and can respond to polyI:C stimulation by the production of proinflammatory cytokines and type I IFNs. In addition mesangial cells express both RIG-I and MDA-5 and the latter has recently been shown to contribute to polyI:C-mediated production of IFN-α by these cells [139]. Recent reports also implicate bacterial TLRs, TLR2, and TLR4, as being important in autoantibody production and glomerulonephritis [140].

Overall, the evidence suggests that the Sm autoantigen is not directly involved in immune complex recognition via recognition by pathogenic anti-Sm antibodies. However, the ability of Sm or RNP peptides administered to mice to induce a lupus-like nephritis suggests that the autoantigens themselves, either alone or in complex with RNA, can act as danger signals or alarmins. As such they would activate an inappropriate immune response involving dendritic cells, inflammatory macrophages, T and B lymphocytes, and a cycle of inflammatory-mediated tissue damage and the subsequent release of chromatin and nucleosomes from the nucleus — targets for anti-dsDNA antibodies that have been indeed shown to be pathogenic in lupus nephritis [141].

CONCLUSIONS

Anti-ENA antibodies are detected to varying degrees in SLE patients. Our knowledge regarding the genetic susceptibility towards developing SLE and the etiology of autoantibody development has greatly increased in recent years and it is now widely accepted that defective clearance of apoptotic bodies, viral infection, and molecular mimicry all contribute to the activation of autoreactive B cells and autoantibody production. Anti-ENA antibodies develop a number of years prior to disease onset and, with the exception of Anti-Ro and anti-Sm, are not directly implicated in tissue pathogenesis. The role of the IFN system in the pathophysiology of SLE is well recognized. The anti-ENA immune complexes, by virtue of their being bound to RNA, act as potent stimulators of TLR7 and help drive IFN-α production by plasmacytoid dendritic cells and promote the autoimmune cycle. An interesting twist in SLE research in the last few years is the finding that a subset of ENA autoantigens, associated with the Ro/La autoantigen complex, are themselves potentially involved in regulating IFN responses and innate immune detection mechanisms. Are these autoantigens therefore inactivated in SLE-susceptible individuals thus enhancing the autoimmune response? Is there a possibility that autoantibody generation, not only mediates inflammation associated with the recognition of autoantigens, but might also result in the inappropriate sequestering of Ro52 and Ro60 so they cannot function appropriately in dampening immune responses? Whatever the answers, rather than viewing ENA antigens as mere diagnostic criteria for SLE, a critical look at their involvement in the pathogenesis of disease and their potential as therapeutic targets in their own right is warranted.

Acknowledgments

We would like to thank Prof. Jim Johnston for critical comments on this work. We would also like to acknowledge financial support from the Health Research Board and Science Foundation Ireland. The authors would like to thank Ms Aileen Conville for assistance with production of photomicrographs.

References

[1] H.R. Holman, Partial purification and characterization of an extractible nuclear antigen which reacts with SLE sera, Ann N Y Acad Sci 124 (1965) 800—806.

[2] P. Migliorini, C. Baldini, V. Rocchi, S. Bombardieri, Anti-Sm and anti-RNP antibodies, Autoimmunity 38 (2005) 47—54.

[3] F. Franceschini, I. Cavazzana, Anti-Ro/SSA and La/SSB antibodies, Autoimmunity 38 (2005) 55—63.

[4] S. Agarwal, J. Harper, P.D. Kiely, Concentration of antibodies to extractable nuclear antigens and disease activity in systemic lupus erythematosus, Lupus 18 (2009) 407—412.

[5] M.R. Arbuckle, M.T. McClain, M.V. Rubertone, R.H. Scofield, G.J. Dennis, J.A. James, et al., Development of autoantibodies before the clinical onset of systemic lupus erythematosus, N Engl J Med 349 (2003) 1526–1533.

[6] M.R. Arbuckle, J.A. James, K.F. Kohlhase, M.V. Rubertone, G.J. Dennis, J.B. Harley, Development of anti-dsDNA autoantibodies prior to clinical diagnosis of systemic lupus erythematosus, Scand J Immunol 54 (2001) 211–219.

[7] M.T. McClain, C.S. Lutz, K.M. Kaufman, O.Z. Faig, T.F. Gross, J.A. James, Structural availability influences the capacity of autoantigenic epitopes to induce a widespread lupus-like autoimmune response, Proc Natl Acad Sci USA 101 (2004) 3551–3556.

[8] A.C. Faria, K.S. Barcellos, L.E. Andrade, Longitudinal fluctuation of antibodies to extractable nuclear antigens in systemic lupus erythematosus, J Rheumatol 32 (2005) 1267–1272.

[9] G. Clark, M. Reichlin, T.B. Tomasi Jr., Characterization of a soluble cytoplasmic antigen reactive with sera from patients with systemic lupus erythmatosus, J Immunol 102 (1969) 117–122.

[10] M.A. Alspaugh, E.M. Tan, Antibodies to cellular antigens in Sjogren's syndrome, J Clin Invest 55 (1975) 1067–1073.

[11] M.R. Lerner, J.A. Boyle, J.A. Hardin, J.A. Steitz, Two novel classes of small ribonucleoproteins detected by antibodies associated with lupus erythematosus, Science 211 (1981) 400–402.

[12] J. Schulte-Pelkum, M. Fritzler, M. Mahler, Latest update on the Ro/SS-A autoantibody system, Autoimmun Rev 8 (2009) 632–637.

[13] D. Xue, H. Shi, J.D. Smith, X. Chen, D.A. Noe, T. Cedervall, et al., A lupus-like syndrome develops in mice lacking the Ro 60-kDa protein, a major lupus autoantigen, Proc Natl Acad Sci USA 100 (2003) 7503–7508.

[14] A. Espinosa, V. Dardalhon, S. Brauner, A. Ambrosi, R. Higgs, F.J. Quintana, et al., Loss of the lupus autoantigen Ro52/Trim21 induces tissue inflammation and systemic autoimmunity by disregulating the IL-23-Th17 pathway, J Exp Med 206 (2009) 1661–1671.

[15] V. Oke, I. Vassilaki, A. Espinosa, L. Strandberg, V.K. Kuchroo, F. Nyberg, et al., High Ro52 expression in spontaneous and UV-induced cutaneous inflammation, J Invest Dermatol 129 (2009) 2000–2010.

[16] S.L. Wolin, K.M. Reinisch, The Ro 60 kDa autoantigen comes into focus: interpreting epitope mapping experiments on the basis of structure, Autoimmun Rev 5 (2006) 367–372.

[17] Y.C. Kim, Y. Cao, D.M. Pitterle, K.C. O'Briant, G. Bepler, SSA/RO52gene and expressed sequence tags in an 85 kb region of chromosome segment 11p15.5, Int J Cancer 87 (2000) 61–67.

[18] K. Ozato, D.M. Shin, T.H. Chang, H.C. Morse 3rd, TRIM family proteins and their emerging roles in innate immunity, Nat Rev Immunol 8 (2008) 849–860.

[19] D.A. Rhodes, J. Trowsdale, TRIM21 is a trimeric protein that binds IgG Fc via the B30.2 domain, Mol Immunol 44 (2007) 2406–2414.

[20] H.J. Kong, D.E. Anderson, C.H. Lee, M.K. Jang, T. Tamura, P. Tailor, et al., Cutting edge: autoantigen Ro52 is an interferon inducible E3 ligase that ubiquitinates IRF-8 and enhances cytokine expression in macrophages, J Immunol 179 (2007) 26–30.

[21] R. Higgs, J. Ni Gabhann, N. Ben Larbi, E.P. Breen, K.A. Fitzgerald, C.A. Jefferies, The E3 ubiquitin ligase Ro52 negatively regulates IFN-beta production post-pathogen recognition by polyubiquitin-mediated degradation of IRF3, J Immunol 181 (2008) 1780–1786.

[22] S. Malik, G.R. Bruner, C. Williams-Weese, L. Feo, R.H. Scofield, M. Reichlin, et al., Presence of anti-La autoantibody is associated with a lower risk of nephritis and seizures in lupus patients, Lupus 16 (2007) 863–866.

[23] N. Coudevylle, D. Rokas, M. Sakarellos-Daitsiotis, D. Krikorian, E. Panou-Pomonis, C. Sakarellos, et al., Phosphorylated and nonphosphorylated epitopes of the La/SSB autoantigen: comparison of their antigenic and conformational characteristics, Biopolymers 84 (2006) 368–382.

[24] J.A. Fairley, T. Kantidakis, N.S. Kenneth, R.V. Intine, R.J. Maraia, R.J. White, Human La is found at RNA polymerase III-transcribed genes in vivo, Proc Natl Acad Sci USA 102 (2005) 18350–18355.

[25] E.I. Schwartz, R.V. Intine, R.J. Maraia, CK2 is responsible for phosphorylation of human La protein serine-366 and can modulate rpL37 5′-terminal oligopyrimidine mRNA metabolism, Mol Cell Biol 24 (2004) 9580–9591.

[26] F. Brenet, N.D. Socci, N. Sonenberg, E.C. Holland, Akt phosphorylation of La regulates specific mRNA translation in glial progenitors, Oncogene 28 (2009) 128–139.

[27] A. Ablasser, F. Bauernfeind, G. Hartmann, E. Latz, K.A. Fitzgerald, V. Hornung, RIG-I-dependent sensing of poly (dA:dT) through the induction of an RNA polymerase III-transcribed RNA intermediate, Nat Immunol 10 (2009) 1065–1072.

[28] X. Cao, New DNA-sensing pathway feeds RIG-I with RNA, Nat Immunol 10 (2009) 1049–1051.

[29] Y.H. Chiu, J.B. Macmillan, Z.J. Chen, RNA polymerase III detects cytosolic DNA and induces type I interferons through the RIG-I pathway, Cell 138 (2009) 576–591.

[30] V. Bitko, A. Musiyenko, M.A. Bayfield, R.J. Maraia, S. Barik, Cellular La protein shields nonsegmented negative-strand RNA viral leader RNA from RIG-I and enhances virus growth by diverse mechanisms, J Virol 82 (2008) 7977–7987.

[31] M. Costa-Mattioli, Y. Svitkin, N. Sonenberg, La autoantigen is necessary for optimal function of the poliovirus and hepatitis C virus internal ribosome entry site in vivo and in vitro, Mol Cell Biol 24 (2004) 6861–6870.

[32] A.M. Domitrovich, K.W. Diebel, N. Ali, S. Sarker, A. Siddiqui, Role of La autoantigen and polypyrimidine tract-binding protein in HCV replication, Virology 335 (2005) 72–86.

[33] F.C. Arnett, R.G. Hamilton, M.G. Roebber, J.B. Harley, M. Reichlin, Increased frequencies of Sm and nRNP autoantibodies in American blacks compared to whites with systemic lupus erythematosus, J Rheumatol 15 (1988) 1773–1776.

[34] A.H. Sawalha, J.B. Harley, Antinuclear autoantibodies in systemic lupus erythematosus, Curr Opin Rheumatol 16 (2004) 534–540.

[35] J.A. James, T. Gross, R.H. Scofield, J.B. Harley, Immunoglobulin epitope spreading and autoimmune disease after peptide immunization: Sm B/B′-derived PPPGMRPP and PPPGIRGP induce spliceosome autoimmunity, J Exp Med 181 (1995) 453–461.

[36] F.C. Arnett, R.G. Hamilton, J.D. Reveille, W.B. Bias, J.B. Harley, M. Reichlin, Genetic studies of Ro (SS-A) and La (SS-B) autoantibodies in families with systemic lupus erythematosus and primary Sjogren's syndrome, Arthritis Rheum 32 (1989) 413–419.

[37] L.M. Vila, G.S. Alarcon, G. McGwin Jr., A.W. Friedman, B.A. Baethge, H.M. Bastian, et al., Early clinical manifestations, disease activity and damage of systemic lupus erythematosus among two distinct US Hispanic subpopulations, Rheumatology (Oxford) 43 (2004) 358–363.

[38] P.S. Ramos, J.A. Kelly, C. Gray-McGuire, G.R. Bruner, A.N. Leiran, C.M. Meyer, et al., Familial aggregation and linkage analysis of autoantibody traits in pedigrees multiplex for systemic lupus erythematosus, Genes Immun 7 (2006) 417–432.

[39] R.R. Graham, W. Ortmann, P. Rodine, K. Espe, C. Langefeld, E. Lange, et al., Specific combinations of HLA-DR2 and DR3 class II haplotypes contribute graded risk for disease susceptibility and autoantibodies in human SLE, Eur J Hum Genet 15 (2007) 823–830.

[40] M.C. Hochberg, Updating the American College of Rheumatology revised criteria for the classification of systemic lupus erythematosus, Arthritis Rheum 40 (1997) 1725.

[41] A. Kavanaugh, R. Tomar, J. Reveille, D.H. Solomon, H.A. Homburger, Guidelines for clinical use of the antinuclear antibody test and tests for specific autoantibodies to nuclear antigens. American College of Pathologists, Arch Pathol Lab Med 124 (2000) 71–81.

[42] J. Craft, Antibodies to snRNPs in systemic lupus erythematosus, Rheum Dis Clin North Am 18 (1992) 311–335.

[43] J.G. Damoiseaux, J.W. Tervaert, From ANA to ENA: how to proceed? Autoimmun Rev 5 (2006) 10–17.

[44] W. Egner, The use of laboratory tests in the diagnosis of SLE, J Clin Pathol 53 (2000) 424–432.

[45] K. Kidd, K. Cusi, R. Mueller, M. Goodner, B. Boyes, E. Hoy, Detection and identification of significant ANAs in previously determined ANA negative samples, Clin Lab 51 (2005) 517–521.

[46] X. Bossuyt, J. Frans, A. Hendrickx, G. Godefridis, R. Westhovens, G. Marien, Detection of anti-SSA antibodies by indirect immunofluorescence, Clin Chem 50 (2004) 2361–2369.

[47] J.G. Hanly, K. Thompson, G. McCurdy, L. Fougere, C. Theriault, K. Wilton, Measurement of autoantibodies using multiplex methodology in patients with systemic lupus erythematosus, J Immunol Methods (2009).

[48] S.S. Copple, T.B. Martins, C. Masterson, E. Joly, H.R. Hill, Comparison of three multiplex immunoassays for detection of antibodies to extractable nuclear antibodies using clinically defined sera, Ann N Y Acad Sci 1109 (2007) 464–472.

[49] W.J. van Venrooij, P. Charles, R.N. Maini, The consensus workshops for the detection of autoantibodies to intracellular antigens in rheumatic diseases, J Immunol Methods 140 (1991) 181–189.

[50] P.A. Gordon, Congenital heart block: clinical features and therapeutic approaches, Lupus 16 (2007) 642–646.

[51] C. Janko, C. Schorn, G.E. Grossmayer, B. Frey, M. Herrmann, U.S. Gaipl, et al., Inflammatory clearance of apoptotic remnants in systemic lupus erythematosus (SLE), Autoimmun Rev 8 (2008) 9–12.

[52] J.H. Reed, M.W. Jackson, T.P. Gordon, B cell apotopes of the 60-kDa Ro/SSA and La/SSB autoantigens, J Autoimmun 31 (2008) 263–267.

[53] C.L. Vanderlugt, S.D. Miller, Epitope spreading in immune-mediated diseases: implications for immunotherapy, Nat Rev Immunol 2 (2002) 85–95.

[54] F. Topfer, T. Gordon, J. McCluskey, Intra- and intermolecular spreading of autoimmunity involving the nuclear self-antigens La (SS-B) and Ro (SS-A), Proc Natl Acad Sci USA 92 (1995) 875–879.

[55] W.P. LeFeber, D.A. Norris, S.R. Ryan, J.C. Huff, L.A. Lee, M. Kubo, et al., Ultraviolet light induces binding of antibodies to selected nuclear antigens on cultured human keratinocytes, J Clin Invest 74 (1984) 1545–1551.

[56] L.A. Casciola-Rosen, G. Anhalt, A. Rosen, Autoantigens targeted in systemic lupus erythematosus are clustered in two populations of surface structures on apoptotic keratinocytes, J Exp Med 179 (1994) 1317–1330.

[57] M. Ohlsson, R. Jonsson, K.A. Brokstad, Subcellular redistribution and surface exposure of the Ro52, Ro60 and La48 autoantigens during apoptosis in human ductal epithelial cells:

a possible mechanism in the pathogenesis of Sjogren's syndrome, Scand J Immunol 56 (2002) 456–469.

[58] W. Lawley, A. Doherty, S. Denniss, D. Chauhan, G. Pruijn, W.J. van Venrooij, et al., Rapid lupus autoantigen relocalization and reactive oxygen species accumulation following ultraviolet irradiation of human keratinocytes, Rheumatology (Oxford) 39 (2000) 253–261.

[59] J. Saegusa, S. Kawano, M. Koshiba, N. Hayashi, H. Kosaka, Y. Funasaka, et al., Oxidative stress mediates cell surface expression of SS-A/Ro antigen on keratinocytes, Free Radic Biol Med 32 (2002) 1006–1016.

[60] J.H. Reed, P.J. Neufing, M.W. Jackson, R.M. Clancy, P.J. Macardle, J.P. Buyon, et al., Different temporal expression of immunodominant Ro60/60 kDa-SSA and La/SSB apotopes, Clin Exp Immunol 148 (2007) 153–160.

[61] M. Herrmann, R.E. Voll, O.M. Zoller, M. Hagenhofer, B.B. Ponner, J.R. Kalden, Impaired phagocytosis of apoptotic cell material by monocyte-derived macrophages from patients with systemic lupus erythematosus, Arthritis Rheum 41 (1998) 1241–1250.

[62] K. Ayukawa, S. Taniguchi, J. Masumoto, S. Hashimoto, H. Sarvotham, A. Hara, et al., La autoantigen is cleaved in the COOH terminus and loses the nuclear localization signal during apoptosis, J Biol Chem 275 (2000) 34465–34470.

[63] A.G. Terzoglou, J.G. Routsias, H.M. Moutsopoulos, A.G. Tzioufas, Post-translational modifications of the major linear epitope 169-190aa of Ro60 kDa autoantigen alter the autoantibody binding, Clin Exp Immunol 146 (2006) 60–65.

[64] Z.J. Pan, K. Davis, S. Maier, M.P. Bachmann, X.R. Kim-Howard, C. Keech, et al., Neo-epitopes are required for immunogenicity of the La/SS-B nuclear antigen in the context of late apoptotic cells, Clin Exp Immunol 143 (2006) 237–248.

[65] L.A. Casciola-Rosen, D.K. Miller, G.J. Anhalt, A. Rosen, Specific cleavage of the 70-kDa protein component of the U1 small nuclear ribonucleoprotein is a characteristic biochemical feature of apoptotic cell death, J Biol Chem 269 (1994) 30757–30760.

[66] E.L. Greidinger, L. Casciola-Rosen, S.M. Morris, R.W. Hoffman, A. Rosen, Autoantibody recognition of distinctly modified forms of the U1-70-kd antigen is associated with different clinical disease manifestations, Arthritis Rheum 43 (2000) 881–888.

[67] J. Dieker, B. Cisterna, F. Monneaux, M. Decossas, J. van der Vlag, M. Biggiogera, et al., Apoptosis-linked changes in the phosphorylation status and subcellular localization of the spliceosomal autoantigen U1-70K, Cell Death Differ 15 (2008) 793–804.

[68] A. Doria, M. Canova, M. Tonon, M. Zen, E. Rampudda, N. Bassi, et al., Infections as triggers and complications of systemic lupus erythematosus, Autoimmun Rev 8 (2008) 24–28.

[69] J.B. Harley, I.T. Harley, J.M. Guthridge, J.A. James, The curiously suspicious: a role for Epstein-Barr virus in lupus, Lupus 15 (2006) 768–777.

[70] T. Portis, R. Longnecker, Epstein-Barr virus (EBV) LMP2A mediates B-lymphocyte survival through constitutive activation of the Ras/PI3K/Akt pathway, Oncogene 23 (2004) 8619–8628.

[71] Q.Y. Yao, P. Ogan, M. Rowe, M. Wood, A.B. Rickinson, Epstein-Barr virus-infected B cells persist in the circulation of acyclovir-treated virus carriers, Int J Cancer 43 (1989) 67–71.

[72] Q.Y. Yao, P. Ogan, M. Rowe, M. Wood, A.B. Rickinson, The Epstein-Barr virus:host balance in acute infectious mononucleosis patients receiving acyclovir anti-viral therapy, Int J Cancer 43 (1989) 61–66.

[73] S. Henderson, D. Huen, M. Rowe, C. Dawson, G. Johnson, A. Rickinson, et al., Epstein-Barr virus-coded BHRF1 protein, a viral homologue of Bcl-2, protects human B cells from programmed cell death, Proc Natl Acad Sci USA 90 (1993) 8479–8483.

[74] M.T. McClain, E.C. Rapp, J.B. Harley, J.A. James, Infectious mononucleosis patients temporarily recognize a unique, cross-reactive epitope of Epstein-Barr virus nuclear antigen-1, J Med Virol 70 (2003) 253–257.

[75] J.H. Vaughan, M.D. Nguyen, J.R. Valbracht, K. Patrick, G.H. Rhodes, Epstein-Barr virus-induced autoimmune responses. II. Immunoglobulin G autoantibodies to mimicking and nonmimicking epitopes. Presence in autoimmune disease, J Clin Invest 95 (1995) 1316–1327.

[76] J.A. James, J.B. Harley, Linear epitope mapping of an Sm B/B' polypeptide, J Immunol 148 (1992) 2074–2079.

[77] K. Sundar, S. Jacques, P. Gottlieb, R. Villars, M.E. Benito, D.K. Taylor, et al., Expression of the Epstein-Barr virus nuclear antigen-1 (EBNA-1) in the mouse can elicit the production of anti-dsDNA and anti-Sm antibodies, J Autoimmun 23 (2004) 127–140.

[78] B.D. Poole, T. Gross, S. Maier, J.B. Harley, J.A. James, Lupus-like autoantibody development in rabbits and mice after immunization with EBNA-1 fragments, J Autoimmun 31 (2008) 362–371.

[79] M.T. McClain, L.D. Heinlen, G.J. Dennis, J. Roebuck, J.B. Harley, J.A. James, Early events in lupus humoral autoimmunity suggest initiation through molecular mimicry, Nat Med 11 (2005) 85–89.

[80] M.T. McClain, B.D. Poole, B.F. Bruner, K.M. Kaufman, J.B. Harley, J.A. James, An altered immune response to Epstein-Barr nuclear antigen 1 in pediatric systemic lupus erythematosus, Arthritis Rheum 54 (2006) 360–368.

[81] H.J. Martin, J.M. Lee, D. Walls, S.D. Hayward, Manipulation of the toll-like receptor 7 signaling pathway by Epstein-Barr virus, J Virol 81 (2007) 9748–9758.

[82] T.R. Geiger, J.M. Martin, The Epstein-Barr virus-encoded LMP-1 oncoprotein negatively affects Tyk2 phosphorylation and interferon signaling in human B cells, J Virol 80 (2006) 11638–11650.

[83] A. Nanbo, K. Inoue, K. Adachi-Takasawa, K. Takada, Epstein-Barr virus RNA confers resistance to interferon-alpha-induced apoptosis in Burkitt's lymphoma, EMBO J 21 (2002) 954–965.

[84] S. Ning, A.M. Hahn, L.E. Huye, J.S. Pagano, Interferon regulatory factor 7 regulates expression of Epstein-Barr virus latent membrane protein 1: a regulatory circuit, J Virol 77 (2003) 9359–9368.

[85] S. Ning, L.E. Huye, J.S. Pagano, Regulation of the transcriptional activity of the IRF7 promoter by a pathway independent of interferon signaling, J Biol Chem 280 (2005) 12262–12270.

[86] R.R. Graham, S.V. Kozyrev, E.C. Baechler, M.V. Reddy, R.M. Plenge, J.W. Bauer, et al., A common haplotype of interferon regulatory factor 5 (IRF5) regulates splicing and expression and is associated with increased risk of systemic lupus erythematosus, Nat Genet 38 (2006) 550–555.

[87] R.R. Graham, C. Kyogoku, S. Sigurdsson, I.A. Vlasova, L.R. Davies, E.C. Baechler, et al., Three functional variants of IFN regulatory factor 5 (IRF5) define risk and protective haplotypes for human lupus, Proc Natl Acad Sci USA 104 (2007) 6758–6763.

[88] P. Matzinger, The danger model: a renewed sense of self, Science 296 (2002) 301–305.

[89] M.E. Bianchi, DAMPs, PAMPs and alarmins: all we need to know about danger, J Leukoc Biol 81 (2007) 1–5.

[90] W. Van Eden, G. Wick, S. Albani, I. Cohen, Stress, heat shock proteins, and autoimmunity: how immune responses to heat shock proteins are to be used for the control of chronic inflammatory diseases, Ann N Y Acad Sci 1113 (2007) 217–237.

[91] G. Kinoshita, C.L. Keech, R.D. Sontheimer, A. Purcell, J. McCluskey, T.P. Gordon, Spreading of the immune response from 52 kDaRo and 60 kDaRo to calreticulin in experimental autoimmunity, Lupus 7 (1998) 7–11.

[92] G. Kinoshita, A.W. Purcell, C.L. Keech, A.D. Farris, J. McCluskey, T.P. Gordon, Molecular chaperones are targets of autoimmunity in Ro(SS-A) immune mice, Clin Exp Immunol 115 (1999) 268–274.

[93] A.W. Purcell, A. Todd, G. Kinoshita, T.A. Lynch, C.L. Keech, M.J. Gething, et al., Association of stress proteins with auto-antigens: a possible mechanism for triggering autoimmunity? Clin Exp Immunol 132 (2003) 193–200.

[94] N. Chaput, S. De Botton, M. Obeid, L. Apetoh, F. Ghiringhelli, T. Panaretakis, et al., Molecular determinants of immunogenic cell death: surface exposure of calreticulin makes the difference, J Mol Med 85 (2007) 1069–1076.

[95] M. Obeid, A. Tesniere, F. Ghiringhelli, G.M. Fimia, L. Apetoh, J.L. Perfettini, et al., Calreticulin exposure dictates the immunogenicity of cancer cell death, Nat Med 13 (2007) 54–61.

[96] S.T. Cheng, T.Q. Nguyen, Y.S. Yang, J.D. Capra, R.D. Sontheimer, Calreticulin binds hYRNA and the 52-kDa polypeptide component of the Ro/SS-A ribonucleoprotein autoantigen, J Immunol 156 (1996) 4484–4491.

[97] R.H. Scofield, D.M. Racila, T.P. Gordon, B.T. Kurien, R.D. Sontheimer, Anti-calreticulin segregates anti-Ro sera in systemic lupus erythematosus: anti-calreticulin is present in sera with anti-Ro alone but not in anti-Ro sera with anti-La or anti-ribonucleoprotein, J Rheumatol 27 (2000) 128–134.

[98] E.L. Greidinger, Y. Zang, K. Jaimes, S. Hogenmiller, M. Nassiri, P. Bejarano, et al., A murine model of mixed connective tissue disease induced with U1 small nuclear RNP autoantigen, Arthritis Rheum 54 (2006) 661–669.

[99] M. Mannik, C.E. Merrill, L.D. Stamps, M.H. Wener, Multiple autoantibodies form the glomerular immune deposits in patients with systemic lupus erythematosus, J Rheumatol 30 (2003) 1495–1504.

[100] S. Janwityanuchit, O. Verasertniyom, M. Vanichapuntu, M. Vatanasuk, Anti-Sm: its predictive value in systemic lupus erythematosus, Clin Rheumatol 12 (1993) 350–353.

[101] A. Rojas-Villarraga, C.E. Toro, G. Espinosa, Y. Rodriguez-Velosa, C. Duarte-Rey, R.D. Mantilla, et al., Factors influencing polyautoimmunity in systemic lupus erythematosus, Autoimmun Rev (2009).

[102] J. Rovensky, A. Tuchynova, Systemic lupus erythematosus in the elderly, Autoimmun Rev 7 (2008) 235–239.

[103] C.H. To, M. Petri, Is antibody clustering predictive of clinical subsets and damage in systemic lupus erythematosus? Arthritis Rheum 52 (2005) 4003–4010.

[104] I. Gal, G. Lakos, M. Zeher, Comparison of the anti-Ro/SSA autoantibody profile between patients with primary and secondary Sjogren's syndrome, Autoimmunity 32 (2000) 89–92.

[105] F.J. Lopez-Longo, I. Monteagudo, C.M. Gonzalez, R. Grau, L. Carreno, Systemic lupus erythematosus: clinical expression and anti-Ro/SS–a response in patients with and without lesions of subacute cutaneous lupus erythematosus, Lupus 6 (1997) 32–39.

[106] T.D. Golan, K.B. Elkon, A.E. Gharavi, J.G. Krueger, Enhanced membrane binding of autoantibodies to cultured keratinocytes of systemic lupus erythematosus patients after ultraviolet B/ultraviolet A irradiation, J Clin Invest 90 (1992) 1067–1076.

[107] L. Casciola-Rosen, A. Rosen, Ultraviolet light-induced keratinocyte apoptosis: a potential mechanism for the induction of

skin lesions and autoantibody production in LE, Lupus 6 (1997) 175—180.

[108] L. Casciola-Rosen, F. Wigley, A. Rosen, Scleroderma autoantigens are uniquely fragmented by metal-catalyzed oxidation reactions: implications for pathogenesis, J Exp Med 185 (1997) 71—79.

[109] F. Furukawa, T. Itoh, H. Wakita, H. Yagi, Y. Tokura, D.A. Norris, et al., Keratinocytes from patients with lupus erythematosus show enhanced cytotoxicity to ultraviolet radiation and to antibody-mediated cytotoxicity, Clin Exp Immunol 118 (1999) 164—170.

[110] S. Meller, F. Winterberg, M. Gilliet, A. Muller, I. Lauceviciute, J. Rieker, et al., Ultraviolet radiation-induced injury, chemokines, and leukocyte recruitment: An amplification cycle triggering cutaneous lupus erythematosus, Arthritis Rheum 52 (2005) 1504—1516.

[111] L. Farkas, K. Beiske, F. Lund-Johansen, P. Brandtzaeg, F.L. Jahnsen, Plasmacytoid dendritic cells (natural interferon-alpha/beta-producing cells) accumulate in cutaneous lupus erythematosus lesions, Am J Pathol 159 (2001) 237—243.

[112] F. Franceschini, A. Brucato, M. Quinzanini, P.G. Calzavara-Pinton, E. Coluccio, I. Cavazzana, et al., Anti-Ro/SSA autoantibodies over time in mothers of children with congenital complete heart block, Clin Exp Rheumatol 17 (1999) 634—635.

[113] L. Strandberg, S. Salomonsson, K. Bremme, S. Sonesson, M. Wahren-Herlenius, Ro52, Ro60 and La IgG autoantibody levels and Ro52 IgG subclass profiles longitudinally throughout pregnancy in congenital heart block risk pregnancies, Lupus 15 (2006) 346—353.

[114] C. Fritsch, J. Hoebeke, H. Dali, V. Ricchiuti, D.A. Isenberg, O. Meyer, et al., 52-kDa Ro/SSA epitopes preferentially recognized by antibodies from mothers of children with neonatal lupus and congenital heart block, Arthritis Res Ther 8 (2006) R4.

[115] G. Colombo, A. Brucato, E. Coluccio, S. Compasso, C. Luzzana, F. Franceschini, et al., DNA typing of maternal HLA in congenital complete heart block: comparison with systemic lupus erythematosus and primary Sjogren's syndrome, Arthritis Rheum 42 (1999) 1757—1764.

[116] A. Brucato, M. Frassi, F. Franceschini, R. Cimaz, D. Faden, M.P. Pisoni, et al., Risk of congenital complete heart block in newborns of mothers with anti-Ro/SSA antibodies detected by counterimmunoelectrophoresis: a prospective study of 100 women, Arthritis Rheum 44 (2001) 1832—1835.

[117] L. Ottosson, S. Salomonsson, J. Hennig, S.E. Sonesson, T. Dorner, J. Raats, et al., Structurally derived mutations define congenital heart block-related epitopes within the 200-239 amino acid stretch of the Ro52 protein, Scand J Immunol 61 (2005) 109—118.

[118] M.E. Miranda, C.E. Tseng, W. Rashbaum, R.L. Ochs, C.A. Casiano, F. Di Donato, et al., Accessibility of SSA/Ro and SSB/La antigens to maternal autoantibodies in apoptotic human fetal cardiac myocytes, J Immunol 161 (1998) 5061—5069.

[119] M.E. Miranda-Carus, A.D. Askanase, R.M. Clancy, F. Di Donato, T.M. Chou, M.R. Libera, et al., Anti-SSA/Ro and anti-SSB/La autoantibodies bind the surface of apoptotic fetal cardiocytes and promote secretion of TNF-alpha by macrophages, J Immunol 165 (2000) 5345—5351.

[120] A. Brucato, F. Franceschini, A. Doria, R. Cimaz, P.L. Meroni, A. Tincani, et al., Pathogenetic associations of maternal anti-Ro/SSA antibodies, Lupus 11 (2002) 650.

[121] A.C. Oshiro, S.J. Derbes, A.R. Stopa, A. Gedalia, Anti-Ro/SS-A and anti-La/SS-B antibodies associated with cardiac

involvement in childhood systemic lupus erythematosus, Ann Rheum Dis 56 (1997) 272—274.

[122] C.A. Arce-Salinas, M.A. Carmona-Escamilla, F. Rodriguez-Garcia, Complete atrioventricular block as initial manifestation of systemic lupus erythematosus, Clin Exp Rheumatol 27 (2009) 344—346.

[123] A.A. Shahin, H.A. Shahin, M.A. Hamid, M.A. Amin, Cardiac involvement in patients with systemic lupus erythematosus and correlation of valvular lesions with anti-Ro/SS-A and anti-La/SS-B antibody levels, Mod Rheumatol 14 (2004) 117—122.

[124] P. Eftekhari, L. Salle, F. Lezoualc'h, J. Mialet, M. Gastineau, J.P. Briand, et al., Anti-SSA/Ro52 autoantibodies blocking the cardiac 5-HT4 serotoninergic receptor could explain neonatal lupus congenital heart block, Eur J Immunol 30 (2000) 2782—2790.

[125] L.M. Vila, G.S. Alarcon, G. McGwin Jr., H.M. Bastian, B.J. Fessler, J.D. Reveille, Systemic lupus erythematosus in a multiethnic US cohort, XXXVII: association of lymphopenia with clinical manifestations, serologic abnormalities, disease activity, and damage accrual, Arthritis Rheum 55 (2006) 799—806.

[126] B.T. Kurien, J. Newland, C. Paczkowski, K.L. Moore, R.H. Scofield, Association of neutropenia in systemic lupus erythematosus (SLE) with anti-Ro and binding of an immunologically cross-reactive neutrophil membrane antigen, Clin Exp Immunol 120 (2000) 209—217.

[127] A. Granito, P. Muratori, L. Muratori, G. Pappas, F. Cassani, J. Worthington, et al., Antibodies to SS-A/Ro-52kD and centromere in autoimmune liver disease: a clue to diagnosis and prognosis of primary biliary cirrhosis, Aliment Pharmacol Ther 26 (2007) 831—838.

[128] P. Muratori, A. Granito, G. Pappas, L. Muratori, M. Lenzi, F.B. Bianchi, et al., Autoimmune liver disease 2007, Mol Aspects Med 29 (2008) 96—102.

[129] M.C. Lu, K.J. Li, S.C. Hsieh, C.H. Wu, C.L. Yu, Lupus-related advanced liver involvement as the initial presentation of systemic lupus erythematosus, J Microbiol Immunol Infect 39 (2006) 471—475.

[130] P.J. Venables, Mixed connective tissue disease, Lupus 15 (2006) 132—137.

[131] P. Alba, L. Bento, M.J. Cuadrado, Y. Karim, M.F. Tungekar, I. Abbs, et al., Anti-dsDNA, anti-Sm antibodies, and the lupus anticoagulant: significant factors associated with lupus nephritis, Ann Rheum Dis 62 (2003) 556—560.

[132] G.A. McCarty, J.B. Harley, M. Reichlin, A distinctive autoantibody profile in black female patients with lupus nephritis, Arthritis Rheum 36 (1993) 1560—1565.

[133] F.J. Tapanes, M. Vasquez, R. Ramirez, C. Matheus, M.A. Rodriguez, N. Bianco, Cluster analysis of antinuclear autoantibodies in the prognosis of SLE nephropathy: are anti-extractable nuclear antibodies protective? Lupus 9 (2000) 437—444.

[134] R. Prasad, D. Ibanez, D. Gladman, M. Urowitz, Anti-dsDNA and anti-Sm antibodies do not predict damage in systemic lupus erythematosus, Lupus 15 (2006) 285—291.

[135] H. Sekine, H. Watanabe, G.S. Gilkeson, Enrichment of anti-glomerular antigen antibody-producing cells in the kidneys of MRL/MpJ-Fas(lpr) mice, J Immunol 172 (2004) 3913—3921.

[136] C.A. Hitchon, C.A. Peschken, Sm antibodies increase risk of death in systemic lupus erythematosus, Lupus 16 (2007) 186—194.

[137] H. Bagavant, S.M. Fu, Pathogenesis of kidney disease in systemic lupus erythematosus, Curr Opin Rheumatol 21 (2009) 489—494.

I. BASIS OF DISEASE PATHOGENESIS

[138] M.G. Robson, Toll-like receptors and renal disease, Nephron Exp Nephrol 113 (2009) e1–e7.

[139] K. Flur, R. Allam, D. Zecher, O.P. Kulkarni, J. Lichtnekert, M. Schwarz, et al., Viral RNA induces type I interferon-dependent cytokine release and cell death in mesangial cells via melanoma-differentiation-associated gene-5: Implications for viral infection-associated glomerulonephritis, Am J Pathol 175 (2009) 2014–2022.

[140] A. Lartigue, N. Colliou, S. Calbo, A. Francois, S. Jacquot, C. Arnoult, et al., Critical role of TLR2 and TLR4 in autoantibody production and glomerulonephritis in lpr mutation-induced mouse lupus, J Immunol 183 (2009) 6207–6216.

[141] E.S. Mortensen, K.A. Fenton, O.P. Rekvig, Lupus nephritis: the central role of nucleosomes revealed, Am J Pathol 172 (2008) 275–283.

Antihistone and Antispliceosomal Antibodies

Minoru Satoh, Marvin J. Fritzler, Edward K.L. Chan

Autoantibodies directed against intracellular antigens are characteristic features of SLE and other systemic autoimmune diseases [1, 2]. Studies over the past several decades have provided strong evidence that autoantibodies are produced by antigen-driven responses and they can be reporters from the immune system revealing the identity of antigens involved in the disease pathogenesis [3]. Some of these autoantibodies serve as disease-specific markers and are directed against intracellular macromolecular complexes or particles such as nucleosomes which are DNA—histone complexes, small nuclear ribonucleoproteins (snRNPs) which are key components of spliceosomes, and Ro and La cytoplasmic RNPs [1—3]. This chapter will discuss the finer specificity of these autoantibodies in relation to SLE.

ORIGIN OF ANTIHISTONE ANTIBODIES

The identification of the lupus erythematosus (LE) cell phenomenon by Hargraves *et al.* [4] led to the recognition of autoimmune reaction as a major component in SLE. The LE cell phenomenon was based on the observation that cellular components released during cell death (particularly nuclei) can be phagocytosed by neutrophils in the mileu of certain plasma factors. This became the basis for the LE cell test, a widely accepted seological test for SLE that was at one time considered a criterion for classification of SLE [5]. Subsequent studies showed that histones were a key requirement for the LE cell phenomenon [6—9]. Thus, antihistone antibodies hold a distinguished position in the recognition of the autoimmune nature of SLE that lead to the discovery of many other antinuclear antibodies (ANA) including antispliceosomal antibodies. It was generally concluded from these early studies that antibodies that bind deoxyribonucleoprotein are antibodies to histone—DNA complexes, and this specificity was directly related to the LE-cell factor. More recent studies have also supported the notion that antihistone antibodies are responsible for the LE cell phenomenon [10—13].

HISTONES ARE KEY PROTEIN COMPONENT OF CELLULAR CHROMATIN

Histones are normally found in eukaryotic cells in association with genomic DNA and organized into a macromolecular complex referred to as native chromatin. DNA and histone constitute 80% of the chromatin mass, although chromatin contains non-histone mass, chromatin contains non-histone proteins, and other proteins and macromolecules. In addition to histones, many other constituents of chromatin are also targets of autoantibodies in SLE — DNA, RNA, and various proteins such as the centromere proteins, nucleolin, cell-cycle-dependent and high-mobility group proteins [14]. Histones and DNA constitute the repeat subunit of chromatin called the nucleosome which consists of two molecules of each of the "core" histones, H2A, H2B, H3 and H4, forming an octamer, along with a histone H1 molecule and of approximately 200 base pairs of DNA. The stability of the nucleosome is accounted for by multiple interactions between DNA and the octamer. Histone H1 appears to bind close to the exposed face of one of the two H2A molecules, creating asymmetry in the nucleosome providing a ramp where DNA exits and is directed to the adjacent nucleosome [15]. Digestion of chromatin with micrococcal nuclease preferentially cleaves the less protected internucleosomal, linker DNA; H1 tends to dissociate and is removed by treatment with 0.5 M NaCl or weak acids. Unlike native chromatin, H1-stripped chromatin is relatively soluble in physiological solutions. Partial digestion of chromatin with nucleases produces a mixture of core particles, nucleosomes, and oligonucleosomes, resulting in a "ladder" representing

DNA of various sizes. Altering the salt concentration can affect the dissociation of the component H3-H4 tetramer from the two H2A-H2B dimers [16]. An artificial complex between the H2A-H2B dimer and DNA can be formed by *in vitro* reconstitution at physiological conditions and this (H2A-H2B)—DNA complex is an important antigenic target in patients with SLE and drug-induced lupus (DIL).

ASSAYS FOR ANTIHISTONE ANTIBODIES

In clinical laboratories, the indirect immunofluorescence (IIF) assay remains the preferred screening test for ANA [17]. In SLE, most autoantibodies to the histone—DNA complexes can be detected as a homogeneous or diffuse IIF staining pattern (Figure 16.1A), although the staining pattern can be variable depending in part on the cell substrate and the autoantibody titer. For example, Centers for Disease Control (CDC) reference sera have been characterized and evaluated by the Autoantibody Standardization Committee [18]. Figure 16.1B shows the staining of the CDC standard IS2072 (ANA#1) that is expected to produce a homogeneous staining pattern. Although both Figures 16.1A and 1B show staining of condensed mitotic chromatin, there are considerable differences in how "homogeneous" nuclear staining is observed. It should be noted that the homogeneous or diffuse pattern is also seen with other autoantibodies, notably those directed against

dsDNA [19] and it is suggested that these reactivities can be difficult to differentiate from anti-DFS70 (LEDGF) antibodies [20]. It is also important to appreciate that sera with antibodies to certain histone classes (e.g., H1, H3, H4) or hidden determinants (cryptotopes) on native or on denatured histones may show a weak or even a negative ANA [21—24]. Therefore, the identification of histone antibodies is challenging using only IIF staining pattern criteria and more specific assays employing purified analytes are preferable.

Solid-phase Enzyme-Linked Immunosorbent Assays (ELISAs) for Antihistone Antibodies

ELISAs are widely used to detect specific antihistone and antichromatin antibodies and this platform can be adapted to measure reactivity to individual histones, macromolecular histone complexes, and chromatin adsorbed to polystyrene. Using this approach, antibody binding can be quantified with a class-specific, enzyme-labeled secondary antihuman Ig. Most immunoassays for the detection of antihistone antibodies in the past two decades have relied on purified histones. Subnucleosome structures have also been adapted to ELISA formats [25], which allow measurement of autoantibodies requiring these higher-ordered structures. ELISA for chromatin and histone are now available commercially from a number of manufacturers (i.e. INOVA Diagnostics, Inc/Werfen Group, San Diego, CA; Euroimmun, Lubeck, Germany).

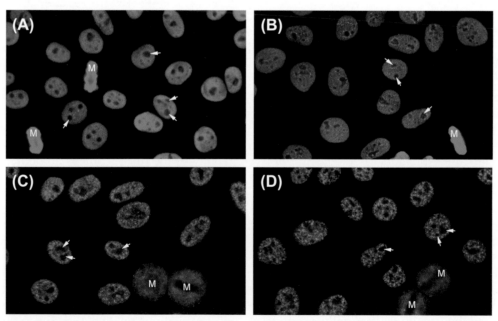

FIGURE 16.1   Indirect immunofluorescence of antihistone and antispliceosomal antibodies on human HEp-2 cells. (A) Prototype anti-DNA/histone serum E4 originally derived from a SLE patient produces a nuclear homogeneous IIF staining pattern. (B—D) Standard sera from the Centers for Disease Control and Prevention (CDC) IS2072 for homogeneous pattern, IS2075 anti-U1RNP, and IS2076 anti-Sm, respectively. HEp-2 substrate was obtained from INOVA Diagnostics, Inc., San Diego. Sera were used at 1:100 dilutions M, mitotic cell; arrows, nucleoli. Original magnification 400x.

Problems and Discrepancies in Measuring Antihistone Antibodies

There are many possible explanations for discrepancies in the literature on the prevalence and fine specificity of antihistone antibodies. The quality of histones used as antigens can be highly variable, and commercial sources were often degraded or contaminated with nonhistone proteins. Inhibition of endogenous and exogenous proteolysis during histone isolation is necessary to obtain intact histones. In addition, particularly in salt extraction protocols, it is challenging to remove all traces of DNA, a serious problem when assessing sera that may also contain anti-DNA antibodies. In addition, the propensity for histones to bind DNA in biological fluids can result in artifacts such as false-positive reactions with anti-DNA antibodies.

Special problems arise when measuring antihistone antibodies in complex biological fluids. Three features of this antigen–antibody system contribute to ambiguous results: (1) histones have a net positive charge and readily bind soluble polyanions such as DNA, RNA or phospholipids in physiological medium or sulfated macromolecules on membranes; (2) DNA and/or histones released from cells are common contaminants of antibody-containing fluids or of other components of the assay; and (3) anti-DNA and antihistone activities commonly co-exist in SLE sera. The presence of soluble DNA in human serum can form a macromolecular 'bridge' between anti-DNA antibodies and histone bound to the solid phase and thus lead to a false-positive antihistone antibody signal. These interactions can result in anti-DNA antibody factitiously binding to histones, a phenomenon that is generally indistinguishable from *bona fide* antihistone antibody reaction [26]. This phenomenon can be observed when harvesting tissue culture supernatants from a hybridoma that secretes anti-DNA antibodies [27]. DNase treatment of the antibody preparation removes the antihistone activity, a finding that is consistent with the involvement of DNA in generation of artifactual antihistone activity.

Similar artifacts can occur in serum. Subiza *et al.* [28] demonstrated that DNase pre-treatment of SLE sera resulted in a significant decrease in antihistone antibody activity in 64% of sera. Affected sera invariably had anti-DNA antibodies (whose activity increased after DNase digestion) and the histone-binding activity could be re-generated by addition of DNA. These results suggest that a significant portion of the antihistone activity in SLE may be attributed to DNA–anti-DNA immune complexes. By comparison, Suzuki *et al.* [29] observed a DNase effect on the antihistone activity in only 27% of SLE sera; this decrease was relatively small and uniform across all histone classes. Therefore, it appears that measurement of antihistone activity in carefully prepared samples of SLE sera, where release of cellular contents is minimized, gives largely valid results even in the presence of anti-DNA antibodies.

Indirect evidence strongly suggests that DNA exists in serum in the form of mono- and oligonucleosomes which can be immunoprecipitated with antihistone antibodies. Therefore, blocking media from these natural biological fluids should be avoided when measuring antihistone antibodies. This type of phenomenon may also have pathologic significance as suggested by the report of Schmiedke *et al.* [30] that circulating nucleohistone binding to the negatively charged residues on heparin sulfate in the glomerular basement membrane may mediate the binding of DNA and anti-DNA antibodies to the glomerulus.

Antibodies to denatured purified histones detected by most assays are common in systemic rheumatic diseases and appear to have limited value as specific diagnostic markers [26]. Other assays which use histone–DNA complexes, including LE cells, fixed cells, deoxyribonucleoprotein, chromatin, soluble (H1-stripped) chromatin, (poly)nucleosomes and (H2A-H2B)–DNA complexes would more likely be detecting reactivity to native autoepitopes. It is possible that in some immunoassays there is overlap of available epitopes in denatured histones and native (DNA-bound) histones; however, for the most part antibodies to histones and antibodies to nucleosome-related antigens should be considered distinct and one cannot be substituted for the other. In the context of autoimmunity, the terms chromatin, nucleosome, and polynucleosome tend to be used interchangeably [31].

PREVALENCE AND DISEASE ASSOCIATION OF ANTIHISTONE AND ANTINUCLEOSOME ANTIBODIES

Reports of antihistone antibodies in various diseases are listed in Table 16.1. Although antihistone antibodies have been observed in various rheumatic diseases, most studies have focused on SLE or DIL. Reported prevalences ranged from 17% to 95% in SLE (average = 51%) and 67% to 100% in DIL (average = 92%) [26]. Antihistone antibodies have also been consistently observed in rheumatoid arthritis (RA) (average prevalence = 11%) and juvenile chronic arthritis (JCA) (average prevalence = 51%). RA patients with ANA were 2–7 times more likely to have antihistone antibodies than ANA-negative RA patients [26]. There are also reports of antihistone antibodies in a substantial proportion of patients with scleroderma-related disorders. Other limited studies on antihistone antibodies in various other syndromes are summarized in

TABLE 16.1   Prevalence of antihistone/antinucleosome antibody in human diseases

Disease/syndrome[a]	Prevalence[b]
RHEUMATIC DISEASES	
SLE	17—95%
Drug-induced lupus	67—100%
Drug-induced ANA	22—95%
Rheumatoid arthritis	0—80%
Vasculitis	31—75%
Felty syndrome	79%
Juvenile chronic arthritis	42—75%
Mixed connective tissue disease	45—90%
Sjögren syndrome	8—67%
Progressive systemic sclerosis disorders	25—50%
Polymyositis/dermatomyositis	17%
OTHER DISEASES	
Primary biliary cirrhosis	50—81%
Hepatic cirrhosis/autoimmune hepatitis	35—50%
Inflammatory bowel disease/ulcerative colitis	13—15%
Neoplastic diseases	14—79%

[a]Modified from Fritzler and Rubin [26].
[b]Prevalence of elevated IgG and/or IgM antibody to total histone or to at least one histone class.

Table 16.1. In some cases a remarkably high occurrence of antihistone antibodies was observed, especially in primary biliary cirrhosis [32, 33], autoimmune hepatitis [33], and ANA-positive neoplastic diseases [34].

In addition to the occurrence of antibodies reactive with isolated histones, patients with lupus-like disorders commonly have autoantibodies to chromatin (nucleosomes). The first report found that ~75% of untreated SLE patients from Singapore had antinucleosomal antibodies [35]; a somewhat lower prevalence and titer was observed in American SLE patients that were treated with medications [35]. In this context, it is important to appreciate that antinucleosome and antinative DNA responses appear to be particularly sensitive to corticosteroid therapy [36]. The bulk of this reactivity was due to binding to the (H2A-H2B)—DNA complex in the higher-ordered structure of chromatin. Other studies from various geographic regions reported that the sensitivity of antinucleosome antibodies in SLE was 38—86%. Most patients with lupus induced by procainamide, penicillamine, isoniazid, acebutolol, methyldopa, timolol and sulfasalazine also have antinucleosome antibodies, predominately reactive with the (H2A-H2B)—DNA complex, although it should be recognized that the majority of these patients had procainamide-

induced lupus. Approximately half the patients with DIL related to quinidine and hydralazine treatment had anti-[(H2A-H2B)—DNA]. Several groups reported antinucleosome antibodies in almost half of the patients with autoimmune hepatitis [37, 38]. Antinucleosome antibodies have also been reported in 25% [39] to almost 50% [40, 41] of patients with SSc (systemic sclerosis, scleroderma), but the specificity of these antibodies may be different from *bona fide* antichromatin antibodies as discussed below. A high prevalence of antinucleosome antibodies have generally not been reported in other rheumatic diseases and are remarkably lower in healthy cohorts, resulting in an overall sensitivity for SLE of 63% and a specificity for SLE compared to other rheumatic diseases of 95% [31].

ANTIHISTONE IN SYSTEMIC LUPUS ERYTHEMATOSUS

Studies on the association of antihistone antibodies with disease activity or severity, with the predominant organ system involvement or with specific clinical features have been inconsistent. Some reports showed no association between the presence or levels of antihistone antibodies and measures of disease activity [26]. In recent years, studies that used nucleosomal antigens have shown strong correlation with symptomatic SLE and a clinical specificity in such patients that was 95—99% [42—44]. Antichromatin and anti-[(H2A-H2B)—DNA] antibodies was significantly correlated with glomerulonephritis [35, 45] and were more specific for this feature than anti-DNA. Antinucleosome antibody of the IgG3 subclass was present at high levels in patients with active SLE, associated with flares and significantly correlated with lupus nephritis [40] or SLE-DAI score [46]. Other groups also observed a higher prevalence and/or amount of antichromatin antibodies in SLE patients with kidney disease [47—49], progression to renal failure and kidney transplant [50] or overall disease activity score [44, 47, 51], although association with disease activity was not seen in all studies [42]. These associations have been supported by more recent reports demonstrating that a systemic and local B-cell response targeting the N-terminal region of histone H2B has a major relationship to lupus pathology in NZB/W mice [52, 53] and that antinucleosome antibodies are a reliable indicator of lupus nephritis [54]. Other studies have shown that nucleosomes can be found in the circulation due to aberrations during apoptosis and/or an ineffective clearance in SLE. During apoptosis, histones can be modified, through acetylation and possibly other modifications, thereby making them more immunogenic (reviewed in [53]). While more definitive comparative studies are still required, current studies indicate that native nucleosomes are more reliable analytes for monitoring SLE than assays using purified histones, and many studies have concluded that the cognitive antibodies are a better marker for SLE than anti-DNA.

ANTIHISTONE IN DRUG-INDUCED LUPUS

Histone-reactive antibodies have been reported in 50—100% of patients with DIL (Table 16.1), depending on the drug and the assay employed. The drugs most commonly associated with DIL include procainamide, hydralazine, quinidine, and isoniazid, although a variety of other drugs have been implicated as well. However, most patients who are treated with procainamide and other lupus-inducing drugs eventually develop antihistone antibodies even though symptomatic disease occurs in only 10—20% of patients [55, 56]. Thus, most antihistone antibodies in this clinical setting are apparently relatively benign, consistent with the seemingly innocuous occurrence of histone-reactive antibodies in many rheumatic and non-rheumatic diseases (Table 16.1). However, examination of their class and fine specificity has revealed that antihistone antibodies in asymptomatic patients are predominantly IgM and display broad reactivity with all the individual histones [57, 58], or are inhibited in binding histones by DNA. In contrast, patients with symptomatic DIL induced by procainamide and many other drugs including penicillamine, isoniazid, acebutolol, methyldopa, timolol, sulfasalazine, and quinidine predominantly develop IgG antihistone antibodies that display pronounced reactivity with the H2A—H2B complex [55], especially when bound to DNA [59]. In fact, anti-(H2A-H2B) has been observed to precede overt clinical symptoms and therefore may have predictive as well as diagnostic value [60, 61]. Anti-(H2A-H2B) has a sensitivity of close to 100% and specificity of >90% for symptomatic procainamide-induced lupus compared to asymptomatic procainamide-treated patients [55] and an even higher specificity when an (H2A-H2B)—DNA complex is used as the screening antigen [59].

A convincing argument that chromatin drives the bulk of the histone-reactive antibody response in SLE and most DIL can be made from the data that compare the antigenicity of various forms of histones. Antibodies from patients with SLE and DIL as well as antibodies from murine lupus bound prominently to a structural epitope in the (H2A-H2B)—DNA complex. Some patients with SLE also bound native DNA and (H3-H4)$_2$—DNA, but reactivity with individual histones was much lower [35, 36, 45]. In murine lupus, antibodies to (H2A-H2B)—DNA were found early in disease, before antibodies to

native DNA and (H3-H4)$_2$—DNA arose [62]. Absorption with chromatin removed most of the antibody reactivity to subnucleosome structures, indicating that regions buried in chromatin were not antigenic in SLE, DIL, or murine lupus. Thus, antihistone antibodies in SLE can be most readily explained by autoimmunization with native chromatin accompanied by sequential loss of tolerance first to the (H2A-H2B)—DNA region and then to (H3-H4)$_2$—DNA and native DNA. Loss of immune tolerance to epitopes on DNA-free histones, which can be considered "denatured histones", and to "denatured DNA" (and to other nuclear antigens) may accompany the immune dysregulation associated with lupus-related disorders, but, because of the complexity of these epitopes and the heterogeneity of this immune response, only with nucleosome-reactive antibodies can a strong case be made for the putative *in vivo* existence of a chromatin-like immunogen [26].

ANTISPLICEOSOMAL ANTIBODIES

The major autoantigens in spliceosome are snRNP and autoantibodies to snRNPs have been a focus of research and clinical immunological studies related to autoantibodies in rheumatic diseases for over three decades [1]. Anti-Sm antibodies, identified over 40 years ago, were one of the first described to target nonhistone proteins in systemic autoimmune rheumatic diseases [63]. Antibodies to the Sm antigen have become known as highly specific (>90%) markers for SLE, although they are less sensitive, being present in 5—30% of unselected SLE populations [1, 64]. The identification of Sm antigen as well-defined proteins bound to U-rich small nuclear RNAs (snRNAs) has been considered a significant advance in biology and these autoantibodies have served as useful probes to help investigate the important components of the spliceosome, which is the multiprotein complex responsible for pre-mRNA splicing of heterogeneous nuclear RNA to mature messenger RNA (mRNA) [65]. In this section, we discuss the major classes of autoantibodies to small nuclear ribonucleoproteins (snRNP) including anti-Sm (Smith), anti-U1RNP (also known as anti-nRNP), anti-U1/U2RNP, U2RNP, and briefly review minor autoantibodies to other classes of U RNPs, such as LSm (Like Sm) proteins.

CELLULAR LOCALIZATION AND FUNCTION OF snRNP

Components of snRNPs

snRNPs are classified by association with specific U-rich snRNAs including the most abundant U1, U2,

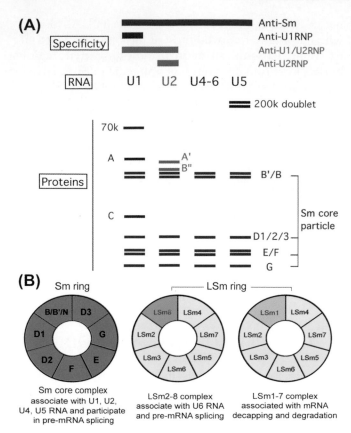

FIGURE 16.2 Structure and components of snRNPs. (*A*) RNA and protein components of snRNPs. Specificity of anti-Sm (red), -U1RNP (blue), -U1/U2RNP, and -U2RNP (green), and RNA and protein components of each snRNP are shown. (*B*) Components of Sm ring and LSm ring. Seven-member Sm and LSm rings have structural similarity in these doughnut-shaped structures. The center of the doughnut-like structure is involved in binding to single-stranded RNA.

U4, U5, and U6 RNAs (Figure 16.2A). U4RNA is always associated with U6RNA and thus, often is designed as U4/U6snRNP. These snRNPs associate with precursor mRNA in a sequential manner to assemble the spliceosome into a functional complex that can catalyze the splicing reaction. U6snRNP uses one of the LSm rings, which are structurally similar to the Sm ring, as its core complex (Figure 16.2B). Other snRNPs, such as U3, U8 and U13, are primarily localized to the nucleolus and are not part of the spliceosomal complex. U11 and U12snRNPs occur in relatively lower abundance and are required for U2snRNP-independent splicing. Other components of the spliceosome include SR proteins, which are serine-arginine-rich proteins involved in the mRNA splicing reaction.

The RNA and protein components of the major snRNPs are illustrated in Figure 16.2A. Each snRNP is an RNA-protein macromolecule of corresponding UsnRNA complexed with several proteins. Common anti-snRNPs autoantibodies are classified into two

major categories: anti-U1RNP antibodies that recognize U1snRNPs; and anti-Sm antibodies that recognize U1, U2, U4-6, and U5 snRNPs (Figure 16.2A, see specificity). The Sm core proteins B or B' (27/28 kDa), D1/D2/D3 (14 kDa), E (12 kDa), F (11 kDa), and G (9 kDa), which are organized as seven-member ring structures (Figure 16.2B, Sm ring, Sm core particle) are shared by U1, U2, U4/U6, and U5 snRNPs. Since these shared Sm core proteins are recognized by anti-Sm antibodies, U1, U2, U4/U6, and U5 snRNAs are immunoprecipitated by anti-Sm antibodies vs. only U1RNA immunoprecipitated by anti-U1RNP antibodies (Figure 16.2A). Crystal structures of Sm protein complexes suggested that the seven Sm core proteins form a closed ring structure and the snRNAs may be bound in the positively charged central hole. Sm B and B' are products of alternative splicing from a single gene whereas Sm N shares 93% amino acid homology with B' and is derived from a different gene with its expression restricted to certain cell types and stages of development.

In addition to the Sm core particle, each snRNP is associated with several unique proteins. U1snRNPs (U1RNP) has U1snRNP specific proteins U1-70k (68/70 kDa), A (33 kDa), and C (22 kDa). U2snRNP has two unique proteins, U2-A' and B'' [66, 67]. U4/U6 has a unique 120−140-kDa protein and U5snRNP has eight unique proteins including 100, 102 kDa, and 200 kDa doublet in addition to the Sm core particle [64]; these are not included in Figure 16.2A except the U5-200 kDa doublet. The U5-specific 200 kDa doublet protein is easy to identify and has been used as supporting evidence of immunoprecipitation of U5snRNPs [68]. Since antibodies directed against U5snRNPs are rare [69], this can be used to differentiate sera with anti-U1RNP alone vs. anti-U1RNP plus anti-Sm in general [68, 70].

REACTIVITY OF ANTI-snRNPs AUTOANTIBODIES

Conventional anti-snRNPs Antibodies

Following immunoprecipitation using sera with anti-snRNP autoantibodies, RNA components and protein components can be analyzed (Figure 16.3). RNA components can be visualized either by autoradiography using ^{32}P- labeled cell extract or silver staining of RNAs from unlabeled cell extract (Figure 16.3A). U1RNA is seen in anti-U1RNP immunoprecipitates (lane U1RNP) while U1, U2, U4, U5, and U6 RNAs are detectable in immunoprecipitates with anti-Sm serum (lane Sm). U1 and U2RNAs are seen with less common anti-U1/U2RNP antibodies and a rare anti-U2RNP antibodies immunoprecipitate only U2RNA.

(A) RNA analysis

(B) Protein analysis

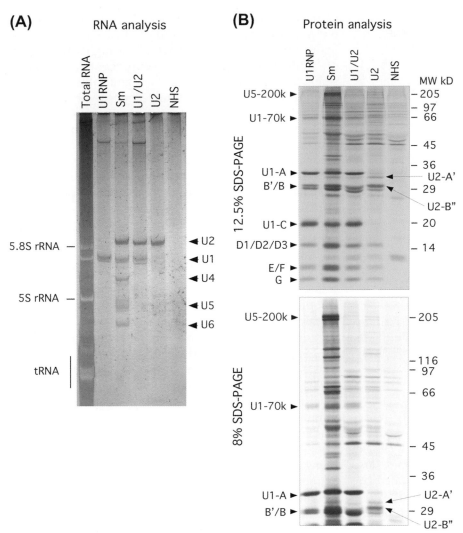

FIGURE 16.3 Immunoprecipitation using anti-snRNPs antibodies. (*A*) RNA analysis of snRNPs immunoprecipitated by human auto-immune sera. RNA components immunoprecipitated from cell extract by human sera were extracted, run on urea-polyacrylamide gel, and identified by silver staining. (*B*) Protein analysis of snRNPs immunoprecipitated by human autoimmune sera. snRNPs associated proteins metabolically radiolabeled with ^{35}S-methionine were immunoprecipitated from cell extract of K562 cells by human autoimmune sera. Proteins were fractionated by 12.5% or 8% SDS-polyacrylamide gel electrophoresis (SDS-PAGE) and visualized by autoradiography.

Protein components are usually analyzed by immunoprecipitation using ^{35}S-methionine-labeled cell extract (Figure 16.3B). Protein components of each snRNP as illustrated in Figure 16.2A are seen. Several unique proteins seen in the anti-Sm lane reflect components of U4/U6RNP and U5RNP [64], however, their structure is complicated and only U5-200kD doublet is indicated. U2-B″ co-migrates with Sm-B′ and cannot be separated clearly in standard high-percentage gel. Radio-immunoprecipitation is the best method to characterize which snRNPs are immunoprecipitated by each serum, however, since snRNPs are relatively stable RNA–protein complexes, all components are immunoprecipitated together regardless the target of antibodies in the serum; any protein component or even RNA in theory.

To identify which proteins are directly recognized, western blot using affinity-purified snRNPs is the standard method (Figure 16.4). Anti-U1RNP sera frequently react with U1-70k, A, B′/B, and less frequently with C,

in various combinations of reactivity [64, 68]. Anti-Sm antibodies react with Sm-B′/B and D1/D2/D3 proteins. Reactivity with E, F, and G of the Sm core particle is rarely reported. Since the vast majority of human anti-Sm sera also have anti-U1RNP antibodies, reactivity with U1 unique proteins are also seen in anti-Sm-positive human autoimmune sera. Anti-U1/U2RNP antibodies primarily recognize the homologous part of U1-A and U2-B″. Crossreactivity of autoantibodies to these two proteins is reported using antibodies eluted from western blot. Anti-U1/U2 sera usually have anti-U1RNP antibodies that react with U1 specific components as well [67]. Sera with predominant anti-U2RNP antibodies produce a precipitin line that is distinct from anti-U1RNP or Sm and strongly react with U2-A′ in immunoblots. Anti-U2RNP antibodies without anti-U1RNP or Sm is rarely reported [66]. It is possible that antibodies that recognize a U2-B″ unique epitope (not cross-reactive with U1-A) can show U2RNP specific reactivity.

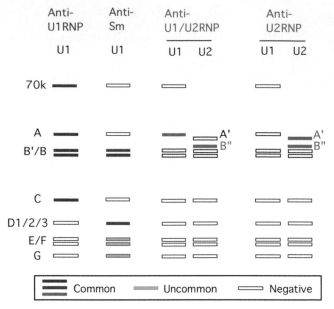

FIGURE 16.4 Western blot reactivity of anti-snRNPs antibodies. Reactivity of antibodies to U1RNP (blue), Sm (red), U1/U2 and U2RNP (green) to components of U1RNP and U2RNP is indicated.

Anti-snRNPs Autoantibodies Recognizing Conformational Structure

In contrast to short linear epitopes for T cells, classic characteristics of autoimmune B-cell epitopes are discontinuous conformational epitopes as indicated by epitope mapping studies [71] and poor reactivity of certain human autoantibodies with denatured proteins in western blot [72]. Studies showing conformational epitopes within each polypeptide of the snRNPs are consistent with this characteristic. In addition, anti-snRNP autoantibodies that recognize the conformational structure of the multiprotein complexes, possibly quaternary structure, have been described. Sm core particle E, F, G proteins are seldom recognized by anti-Sm sera in western blot, however, one study showed the recognition of one or more of these immunoprecipitated proteins by all anti-Sm-positive sera [73]. Furthermore, all anti-Sm-positive sera efficiently immunoprecipitated E-F-G complex formed in vitro, suggesting that many anti-Sm sera recognize unique conformational epitopes in the EFG complex [73].

It was noted that nearly all anti-U1RNP-positive sera react with the native U1-C protein in IP using a condition that dissociates U1-C from the rest of U1snRNPs, in contrast to low frequency of recognition of the SDS-denatured U1-C protein by western blot [68]. Some studies showed that all anti-U1RNP sera had autoantibodies that stabilize the U1-C and the Sm core proteins (some also stabilize U1-A-Sm), possibly by recognizing the quaternary structure [70, 74]. In contrast, the Sm core particle itself was recognized by anti-Sm antibodies but not by most of anti-U1RNP sera. The titers of stabilizing antibodies were comparable or higher than those of antibodies to individual components, suggesting stabilizing autoantibodies are a significant component in anti-U1 RNP autoimmune responses [74]. Another interesting report described the reactivity of a unique monoclonal antibody F78 to a native, heat-labile particle composed of Sm core proteins [75]; however, F78 did not recognize individual Sm components. Two autoimmune sera that had reactivity similar to F78 were described. In addition F78-like reactivity was commonly detected in human anti-Sm positive sera after absorption of antibodies to individual Sm polypeptides [75].

Antibodies to LSm4 and LSm Complex

Recent studies have identified macromolecular complexes composed of Sm-like proteins that form distinct complexes in the form of a Sm ring (Figure 16.2B). They have been named like-Sm proteins, or LSm proteins (LSm1 to LSm8), since they each bear sequence homology with one of the seven Sm core proteins [76]. Autoantibodies to LSm4 were detected in ~80% of all anti-Sm sera analyzed when immunoprecipitation and in vitro translated LSm4 was used [77]. A small fraction (7.2%, 28/391) of the same group of anti-Sm sera immunoprecipitated LSm4 together with the LSm complex from HeLa cell extracts. These reports indicate that IgG autoantibodies to LSm4 are detected in a large overlapping subset of anti-Sm-positive SLE. LSm4 was immunoprecipitated from a few SLE sera that contained little or no co-precipitating antibodies to the Sm core proteins [77]. The clinical significance of anti-LSm antibodies remains to be determined.

HISTORY OF DETECTION OF AUTOANTIBODIES TO snRNPs AND POTENTIAL PROBLEMS

"What is the difference in reactivity of anti-U1RNP vs. anti-Sm?" is probably the most important and basic question that remains unresolved among researchers. This issue has been a subject of controversies and confusion for many years. In addition to the fact that we do not have a clear answer for this question, changes in technology is another factor that further complicates the issue. Changing the standard methods to detect antibodies and the source of antigens without complete understanding of the nature of the immune reaction or how they correspond to each other creates significant challenges. Assay technologies started with Ouchterlony double immunodiffusion (DID) followed by passive hemagglutination (PHA) [78], immunoprecipitation (IP) detection of UsnRNA components and protein components [65], western blot detection of reactivity to individual protein components, eventually evolving to line immunoassays (LIA), ELISA, addressable laser bead immunoassays (ALBIA) and other multiplexed immunoassays (reviewed in [79]). An additional complicating factor is that the source of antigens (analytes) used in various immunoassays; antigens have spanned the spectrum from calf or rabbit thymus extract and extractable nuclear antigen (ENA) to human cell lines and recombinant proteins.

Historically, DID was one of the first assays used to detect anti-Sm reactivity [63]. Anti-U1RNP was originally reported as anti-Mo that recognizes soluble nuclear ribonucleoprotein (nRNP) and makes a distinctive precipitin line from anti-Sm in Ouchterlony DID [80]. However, it was noted that anti-nRNP precipitin lines partially fused with anti-Sm lines, suggesting that they react with common antigenic component(s). Evidence indicating that nRNP antigens are RNA—protein complexes was based on observations that the reactivity of anti-nRNP antibodies was lost by subjecting saline-soluble cell extracts to RNase digestion. Anti-U1RNP antibodies were originally called anti-RNP antibodies based on the knowledge that the target antigen was composed of RNA and proteins. Since other ribosomal RNP antigens recognized by SLE sera localized to nucleoli and the cytoplasm, it was suggested that the appropriate nomenclature should be antinuclear RNP (nRNP).

At about the same time, the same group of specificities (anti-snRNPs) were reported as anti-ENA [78] that was based on a classification into RNase-sensitive anti-ENA (correspond to anti-U1RNP) and RNase-resistant anti-ENA (correspond to anti-Sm) [81]. Saline extractable antigens contain various nuclear antigens, however, somehow only snRNPs appeared to bind to tannic acid-treated sheep red blood cells. Thus, anti-ENA antibodies detected by the passive hemagglutination assay (PHA) were regarded as specific for anti-snRNPs (anti-U1RNP plus anti-Sm). RNase treatment of the ENA-coated RBC eliminated anti-U1RNP reactivity (RNase-sensitive anti-ENA) while anti-Sm was not affected (RNase-resistant anti-ENA antibodies).

After this period, the source of antigens used for immunoassays had gradually shifted from calf or rabbit thymus tissues to human culture cell lines and recombinant human proteins. This was paralleled by a switch in ANA substrates from cryopreserved animal tissues such as rat liver or kidney sections to human cell lines such as HEp-2 (laryngeal cancer) or HeLa (cervical cancer). Characterization of U1RNP, Sm, and other snRNPs antigens by radioimmunoprecipitation was a major breakthrough in the field of autoantibody detection as well

as in cell biology [65], however, this technique has been largely relegated to a research laboratory setting.

DETECTION OF ANTIBODIES TO snRNPs IN CLINICAL PRACTICE

Antibodies to U1RNP and anti-Sm are typical autoantibody specificities that show a nuclear speckled pattern in ANA screening by IIF on HEp-2 substrates (Figure 16.1C and D). Anti-U1RNP, anti-Sm, anti-U1/U2RNP nuclear speckled IIF patterns are indistinguishable. There are many other autoantibody specificities that show a similar nuclear speckled pattern including anti-RNA polymerase III, Ku, La, and others. Thus, IIF alone cannot conclusively identify anti-snRNP antibodies but certainly, anti-U1RNP and Sm are common specificities that should be considered when the nuclear speckled pattern is observed. The titer of ANA is often very high, 1:1280 or even higher, particularly in patients with overlapping features of SLE, SSc, PM/DM in a condition referred to as mixed connective tissue diseases (MCTD) [81], as well as in SLE or undifferentiated connective tissue disease (UCTD) [82, 83].

To confirm anti-U1RNP, -Sm specificities, a commonly used assay in the diagnostic laboratory is ELISA using recombinant or affinity purified antigens. Although ELISA will identify 70—80% of true positives using immunoprecipitation as gold standard, it may be troubled by a significant percentage of false positives, in particular anti-Sm [84]. Improved sensitivity of the anti-U1RNP ELISA using recombinant U1-70k, A, and C proteins was accomplished by adding U1RNA, via the formation of new epitopes resulted from interaction of U1RNA with the proteins antigens [85]. Another improvement in ELISA was the use of symmetric dimethylarginine modified Sm-D-derived peptide to detect anti-Sm antibodies [86] based on earlier studies indicating the importance of this post-translational modification in anti-Sm antibody reactivity [87].

As noted above, DID was the original method to detect and define the anti-U1RNP and Sm specifics [63, 80] and it was used extensively during 1970s to 1980s but less commonly after 1990. Nevertheless, it is a reliable and inexpensive method that is still used in some clinical laboratories. However, with recent emergence of large, high-throughput laboratories the emphasis has shifted to assays, such as ELISA and ALBIA that can be automated, effecting cost savings and rapid turnaround times [88]. In certain European countries and Australia, counter immunoelectrophoresis has been used to detect anti-U1RNP, Sm, and other specificities. Western blot and immunoprecipitation analyses, which have been used in research laboratories for over 30 years, however, never were widely adopted by high throughput clinical diagnostic laboratories. More recently, LIA, which is similar to western blot or dot blot immunoassay but contains multiple autoantigens individually printed on membrane strips facilitates the simultaneous detection of several autoantibodies in a single assay, has been adopted by some laboratories. Other new types of assay include ALBIA [89] in which individual antigens are covalently bound to beads of different fluorochrome composition [89]. Advantages of these assays include the detection of multiple autoantibodies in a single assay and the requirement for only small amounts (typically 10ul) of serum sample. It is clear that the field of diagnostic autoimmunity is rapidly advancing with the application of nanotechnology and the advent of other proprietary technologies.

CLINICAL SIGNIFICANCE OF ANTIBODIES TO snRNPs

Distribution and Coexistence of Anti-U1RNP vs Anti-Sm Antibodies

Antibodies to U1RNP and Sm recognize the U1snRNPs and U1, U2, U4/U6, and U5 snRNPs, respectively. Although anti-U1RNP and Sm often coexist, there are two major differences in clinical significance of these two specificities. First, anti-Sm antibodies are highly specific for the diagnosis of SLE whereas anti-U1RNP can be found in patients with various diagnoses and is, accordingly, not specific for SLE (Figure 16.5) [1]. Second, anti-U1RNP is common in patients with UCTD, primary Raynaud phenomenon, and other unclassified conditions, whereas anti-Sm is rare in undiagnosed or unclassified patients [83, 90].

One of the unique features of anti-U1RNP and anti-Sm antibodies is their apparent inter-relationship; virtually all anti-Sm-positive patients also have anti-U1RNP and anti-Sm alone is rarely found, while only a fraction of patients with anti-U1RNP have anti-Sm antibodies (Figure 16.6) [1, 70]. The relationship of anti-U1RNP vs. anti-Sm is very similar to the relationship between anti-Ro/SS-A and anti-La/SS-B antibodies in Sjögren syndrome, SLE, and other systemic rheumatic diseases in which virtually all anti-La positive sera also contain anti-Ro but only ~25–30% of anti-Ro60-positive sera are positive for anti-La [84]. The reason for this striking 'polarized' B-cell response is intriguing but completely unexplained. In SLE, usually 25–40% of anti-U1RNP-positive patients have anti-Sm, however, this percentage and the number of patients with anti-Sm without anti-U1RNP may vary widely in different population of patients [64]. This ratio (anti-Sm/anti-U1RNP) becomes much lower when all anti-U1RNP-positive cases are analyzed because anti-U1RNP can be seen in diseases

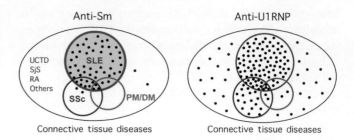

FIGURE 16.5 Distribution pattern of anti-Sm vs. U1RNP in patients with connective tissue diseases. Typical distribution pattern of anti-Sm vs. anti-U1RNP in SLE, SSc, PM/DM, and other connective tissue diseases is illustrated. Data are only to illustrate the trend and not based on actual numbers from two reports [83, 84].

other than SLE such as SSc, PM/DM, Sjögren syndrome and unclassified patients (such as UCTD) although, as noted above, anti-Sm is rarely positive in these conditions (Figure 16.5). The prevalence of anti-U1RNP in UCTD is 6–22% while anti-Sm is only 1–3% [83, 90–92].

Clinical Association of Anti-Sm and U1RNP Antibodies

Anti-Sm is highly specific for the diagnosis of SLE and is one of the most established and widely utilized disease marker antibodies [1]. It is one of the serological criteria, along with anti-dsDNA and antiphospholipid antibodies, embedded as an "immunologic disorder" in the ACR Classification Criteria for SLE [5]. Many anti-Sm-positive patients have typical SLE, however, it should be noted that anti-Sm is often seen in patients with SLE-overlap syndromes, such as those that also have features of SSc and/or PM/DM (Figure 16.5). Some of these patients have been reported to have more than one disease marker antibody as evidenced by the co-existence of anti-Sm and anti-topoisomerase I (Scl-70), a marker of SSc [93]. Anti-Sm has also been occasionally reported in systemic rheumatic diseases without clinical features of SLE, UCTD, or primary Raynaud syndrome

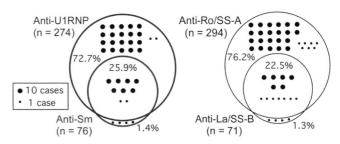

FIGURE 16.6 Relationship between anti-U1RNP vs. anti-Sm antibodies. Anti-Sm antibodies are positive in a subset of anti-U1RNP positive patients and virtually all anti-Sm positive sera also have anti-U1RNP. This is similar to the relationship between anti-Ro vs. anti-La [84].

[83]. Since the production of specific autoantibodies usually precedes the development of typical clinical manifestation [94], a small number of patients who do not meet the classification criteria for definite SLE may also be seen. Unexpected detection of anti-Sm antibodies may be followed in time by overt clinical features of SLE [95]. In contrast, although anti-U1RNP antibodies are common in SLE patients, they are not specific for that disease, rather, they are associated with certain clinical manifestations regardless of the diagnosis. Although by definition virtually all patients with MCTD are positive for anti-U1RNP antibodies, they are not a specific marker for MCTD or overlap syndrome. However, anti-U1RNP-positive patients frequently have anti-U1RNP-associated clinical features such as Raynaud phenomenon, swollen hand, sausage-like finger, leukopenia, etc., regardless of the diagnosis. Why different clinical diagnoses and features are associated with anti-U1RNP vs. anti-Sm, both recognizing closely related UsnRNPs, is not known.

The clinical association and significance of anti-U1RNP and anti-Sm described above appear to be consistent regardless of ethnicity and certain other demographic variables, however, studies on detailed clinical association are more inconsistent. There are many reasons for this inconsistency. One of the reasons may be the different immunoassays used. Consideration of the levels of antibodies may be important since patients with high levels of anti-U1RNP antibodies tend to have more typical features of classical MCTD [81]. Differences in the genetic and environmental background of patients, focus and design of study, and methods in clinical analysis, as well as treatment and follow-up, can also be factors.

The fact that most anti-Sm-positive sera are also positive for anti-U1RNP makes anti-U1RNP antibodies a major confounding factor. The only practical option to analyze the clinical associations of anti-Sm may be an analysis of anti-Sm+U1RNP vs. anti-U1RNP cohorts although this analysis may be the equivalent of comparing SLE vs. MCTD. Similarly, clinical association [96] and immunological cross-reactivity of anti-Sm with antiribosomal P [97] may also bias the data on clinical association of anti-Sm. Association of anti-Sm and anti-dsDNA antibodies [98, 99] could result in a similar bias. This idea is consistent with the observation that clinical manifestations associated with anti-Sm were also associated with either anti-U1RNP antibodies or anti-dsDNA antibodies [100]. Some reports suggested that the titers of anti-Sm antibodies correlate with the disease activity, milder renal and central nervous involvement or late-onset renal disease [100], but such conclusions are controversial [64].

A clinical concept of MCTD characterized by overlapping symptoms and signs of SLE, SSc, and PM/DM was proposed [78, 81]. The main clinical features of MCTD are a high prevalence of Raynaud phenomenon, edema of the fingers, arthritis/arthralgias, myositis, serositis, favorable response to steroid treatment, a relative absence of renal disease and a good prognosis. When anti-U1RNP antibodies are present alone and in high titer, they are often associated with MCTD. Anti-U1RNP antibodies are also detected, usually at a lower titer, in other systemic rheumatic diseases including 30–40% of SLE, and in much lower frequency in RA, SSc, Sjögren syndrome, PM/DM, discoid lupus, and rarely in DIL [1, 64]. Because of the clinical relevance of the titers observed in systemic rheumatic diseases, it is important to detect and quantitate the level of anti-U1RNP antibodies in patient's sera.

It is important to appreciate that some SLE and MCTD sera bind most of the snRNP polypeptides whereas others bind to little, if any, U1-70k or C polypeptides. In one study, only 8% SLE sera containing U1RNP antibodies bound to the U1-70k antigen, but 76% of MCTD sera bound this protein [101]. The observed higher frequency of anti-U1-70k antibodies in MCTD has been supported in several studies and reported to be as high as 95% in MCTD while the range of reactivity in SLE is 20–50%. The frequency of anti-U1RNP is highly dependent on the assay employed because when recombinant U1-70k protein was used in an ELISA, up to 85% of SLE patients showed elevated antibody levels. It has been suggested that the presence of anti-U1-70k is correlated with the presence of Raynaud phenomenon, esophageal dysmotility and myositis, and is a negative indicator for, or perhaps even a protective factor against, the presence or development of renal disease [101–103]. Taken together, the data suggest that antibodies to the U1-70k protein are primarily associated with classic features of MCTD. These may be important observations for the clinician who attempts to identify patients at high risk of developing end organ disease when they present with only a few features (e.g., Raynaud phenomenon, myositis) of other systemic rheumatic diseases. Antibodies to the A and C proteins are found in approximately 25% of unselected SLE cohorts and in 75% of SLE patients preselected for antibodies to U1RNP [101, 102, 104]. Although antibodies to the U1-70k, A, or C protein quantitatively vary during the disease course, there is little evidence that they correlate with disease activity or that they are involved in disease pathogenesis [102, 103].

Other Anti-snRNPs Antibodies

Anti-U1/U2RNP antibodies were reported in two cases of SLE, and in MCTD, RA-Sjögren syndrome [105], and in two patients with psoriasis and Raynaud phenomenon [106]. Eight patients with anti-U1/U2RNP, including four predominantly or strongly reactive with U2RNP and four that had more anti-U1RNP reactivity, were also reported [67]. All these patients had overlapping features of SLE, SSc, and/or PM/DM. Thus, antibodies to U1/U2RNP appear to be associated with patients with overlapping features of MCTD or SLE.

Isolated anti-U4/U6RNP antibodies were reported in SSc [107] and primary Sjögren syndrome [108] and anti-U5RNP antibodies were reported in a single case of SSc-PM overlap syndrome [69]. Whether some anti-Sm-positive sera also have antibodies directed to U4/U6RNP or U5RNP unique proteins was not reported.

Antibodies to the trimethylguanosine cap structure (TMG) of UsnRNA that recognize U1-U5RNAs were reported in four SSc sera [109, 110]; all from Caucasian and three of four had limited cutaneous SSc. Since anti-U3RNP (fibrillarin) antibodies are associated with diffuse SSc, particularly those of African-American descent [111], the clinical association of anti-TMG that immunoprecipitated U1-U5RNP including U3RNP may be different from anti-U3RNP antibodies that recognize fibrillarin.

Antibodies to UsnRNPs were originally thought to target only protein components, however, later studies showed that patients with autoantibodies to snRNPs often have antibodies directed to U1RNA. Anti-U1RNA antibodies were found in 35–38% of anti-snRNP-positive sera [112, 113]. Correlation of levels of anti-U1RNA antibodies and disease activity in nine patients with SLE-overlap syndrome has been reported [113].

IMMUNOLOGIC CHARACTERISTICS OF ANTI-U1RNP AND SM ANTIBODIES

Anti-U1RNP and Sm antibodies have been associated with hypergammaglobulinemia in some patients. In addition to polyclonal B-cell activation, there is evidence that anti-U1RNP and Sm autoantibodies themselves contribute to hypergammaglobulinemia by representing a significant portion of the total IgG. One study reported that 20–30% (up to 8.6 mg/ml) of total IgG is composed of anti-U1RNP and Sm activity in patients with high levels of these autoantibodies [114]. This is in striking contrast to relatively low concentrations of anti-dsDNA antibodies, 1–3 µg/ml [115] and may be one reason why anti-dsDNA fluctuations may reflect disease activity and response to treatment while many patients with anti-snRNPs, for the most part, are unaltered despite disease activity or therapeutic intervention. This also is consistent with the observations that, by comparison to many other autoantibodies, the titers of anti-U1RNP and Sm antibodies tend to be very high in various immunoassays [81].

It has been emphasized in early studies that the titers of anti-U1RNP do not fluctuate significantly over time. While this is true for many patients with MCTD, the levels of anti-U1RNP antibodies may not be as stable as originally believed in particular in patients with SLE [116]. Other factors that may contribute to these initial concepts may be the immunoassays used. In DID or PHA, using serial 1:2 dilutions of sera was the standard method to obtain endpoint titer; two-fold change makes one dilution difference, four-fold change makes two dilution difference. These values may give a different impression than the methods such as ELISA, which give continuous measured values. The issue of stable production of certain autoantibodies was revisited recently because of the new concept of antibody production by short-lived plasmablasts in lymphoid tissue vs. long-lived plasma cells in bone marrow [117] and of B-cell-depleting therapy using Rituximab in patients with systemic rheumatic diseases [118, 119]. The observation was that B-cell depletion dramatically reduced the levels of anti-dsDNA antibodies in patients with SLE but did not affect the levels of anti-U1RNP, Sm, Ro, and La, or total immunoglobulin levels [118, 119]. These data are consistent with the interpretation that anti-U1RNP and Sm antibodies are mainly produced by long-lived plasma cells and may explain why their production is relatively stable after treatment.

PROBLEMS AND CONFUSIONS IN ANTI-SNRNPS ANTIBODY TESTING

Many Anti-U1RNP Positive Sera Show Reactivity with Sm B′/B but Immunoprecipitate Only U1RNP

Confusion in the field of anti-snRNP responses is evidenced by rather simplistic observations that prompt some authors to state that anti-U1RNP antibodies react with U1RNP-specific proteins U1-70k, A, and C whereas anti-Sm antibodies react with B′/B and D1/D2/D3 included in the U1, U2, U4-6, and U5snRNPs. This facile understanding of the immunoprecipitation of only U1RNP by anti-U1RNP sera while anti-Sm sera immunoprecipitate all UsnRNPs that contain Sm core particle, is apparently incorrect. While sera specific for anti-U1RNP do not appear to react with D1/D2/D3, E, F, or G in western blot, binding to B′/B by anti-U1RNP sera is quite common [68, 101]. This is also consistent with the observation that ELISA or other immunoassays that use recombinant B′/B are not specific for anti-Sm-positive sera [102]. The nature of this discrepancy of immunoprecipitation and western blot or ELISAs that employ recombinant B′/B is not completely understood. If anti-U1RNP antibodies recognize the native form of Sm-B'/B in immunoprecipitation, anti-U1RNP sera should immunoprecipitate U2, U4/U6, and U5snRNPs in addition to U1snRNPs. However, they react only with U1RNP in immunoprecipitation. One possible explanation is that anti-B'/B antibodies present in anti-U1RNP sera recognize only the denatured epitopes but not the native B'/B. However, this does not appear to be the case based on one study in which immunoprecipitation using affinity-purified anti-B'/B antibodies was performed [120]. In another study, cross-reactivity of affinity-purified anti-B'/B antibodies from anti-Sm-positive sera with Sm-D was shown while anti-B'/B antibodies from anti-U1RNP positive sera did not cross-react with D in a western blot [121]. Related to this puzzling observation, it is of interest that human anti-U1RNP antibodies react with quaternary structures composed of the Sm core particle and U1-C or U1-A protein. All anti-U1RNP antibody-positive human autoimmune sera stabilize the interaction of the Sm core particle and U1-C or U1-A [70, 74].

Anti-snRNPs Antibodies Recognize Proteins but RNase Treatment Eliminates Reactivity of Anti-U1RNP Antibodies by DID and PHA

As noted above, U1RNP antigens were shown to be sensitive to RNase and trypsin digestion and that was why the antigen was considered as proteins associated with RNA [80]. Similarly, the nuclear speckled pattern of immunofluorescence was also sensitive to RNase digestion. Similarly, RNase treatment of ENA-coated RBCs eliminated the reactivity of anti-U1RNP antibodies by PHA assay [81]. In contrast, immunoprecipitation of RNase digested human cell extracts by anti-U1RNP positive sera showed reactivity with U1-70k, A, and C proteins [70]. Furthermore, anti-U1RNP antibodies recognized U1RNP proteins without U1RNA in western blot or ELISA. These observations are not consistent with the disappearance of anti-U1RNP reactivity by RNase treatment in DID or PHA. Cell extracts become turbid when RNase is added, resulting in an insoluble pellet that is enriched with U1RNP proteins after centrifugation. Thus, it appears that RNase treatment results in an insoluble U1RNP precipitate and the anti-U1RNP precipitin lines do not form. These effects on ENA antigens coated on the surface of red blood cells in the PHA cannot be explained by insolubility or aggregates, however, it may be possible that the bound RNase block the reactivity of anti-U1RNP antibodies.

Reactivity with the Peptide

Other confounding observations are based on studies that utilize synthetic snRNP peptides. For example, the PPPGMRPP sequence has been identified as a major epitope in autoimmune B-cell response to snRNPs; it is frequently recognized and an early target in immune response [122]. However, the original researchers reported that the reactivity with this peptide does not correlate with the presence of conventional anti-Sm antibodies [123, 124]. Thus, although this peptide may be interesting and useful to examine epitope spreading, it does not appear to be particularly useful for anti-Sm antibody testing. The clinical significance of the reactivity of antibodies that bind to the Sm-derived peptide but do not correlate with conventional anti-Sm antibodies is confusing, if not questionable. Reactivity of antibodies with dimethylarginine modified Sm peptide [86] appears specific for SLE and correlates with the presence of anti-Sm, thus differentiating from those employing the PPPGMRPP peptide.

MECHANISM OF PRODUCTION

snRNPs are one of the most common targets of autoimmune response in SLE and it is assumed that snRNPs are somehow more immunogenic than many other intracellular proteins. Despite various hypotheses and observations, the mechanisms of production of autoantibodies in SLE, in particular the mechanism of selection of the target antigens is incompletely understood. Different mechanisms may play a role in breaking tolerance, spreading epitopes, and sustaining autoantibody production. Molecular mimicry may be a trigger while intermolecular-intrastructural help mechanisms may be responsible for the phenomenon of epitope spreading. Apoptosis, microbody production, and TLR stimulation are also likely key factors in these processes. Thus, the actual mechanism is likely multifactorial and, to a certain extent, specific to individual patients based on genotype, environment, and other epigenetic factors.

Molecular mimicry of microbial products such as viral proteins that cross-react with self antigens has been one of the hypotheses proposed that link infections to autoimmunity [122]. Homology and cross-reactivity of snRNP antigens and various viral products has been described. Molony murine leukemia virus p30gag protein and U1-70k protein has sequence homology [125]. In addition to immunological cross-reactivity of human lupus autoantibodies, immunization of animals with p30gag peptide can induce antinuclear antibodies that recognize U1-70k protein [125]. The increased prevalence of Epstein-Barr virus (EBV) infection in patients with SLE and homology and cross-reactivity of EBNA antigen and snRNPs proteins have been reported [122]. Cytomegalovirus (CMV) gB/UL55 has a homology with the U1-70k protein and can induce anti-U1-70k antibodies in a mouse model [126].

UsnRNPs antigens were detected on surface of apoptotic blebs and cleavage of U1-70k protein during apoptosis was described. In one study, reactivity of lupus autoantibodies that react with the apoptotic fragment but not with the U1-70k protein was reported [127]. These data suggest a role of apoptosis in production of anti-snRNPs antibodies, however, significance of this phenomenon in the pathogenesis of autoantibody production will need to be further examined.

Production of anti-snRNPs antibodies in murine models of SLE appears to be dependent on the presence of IFN-γ and IL-12, and type I IFN (I-IFN) [128−130]. Recent studies showed that the anti-snRNPs antibodies in pristane-induced lupus was completely abolished in IFN receptor-deficient mice [131]. However, in humans, the role of cytokines in the production of anti-snRNPs autoantibodies is not well characterized. Levels of I-IFN inducible genes Mx1 and others were higher in patients with autoantibodies to U1RNP/Sm vs. patients without antibodies to RNA−protein complex [132, 133], suggesting the association between I-IFN production and anti-U1RNP/Sm antibodies. However, it is not known whether high I-IFN in these patients is responsible for anti-snRNPs antibody production or this is a result of I-IFN induction by anti-snRNPs immune complex. In mice, stimulation of I-IFN by UsnRNAs was shown to be Fc receptor and TLR7 dependent [134].

There have been controversies for many years about the possibility that anti-U1 RNP antibodies have a unique property of "penetrating" into living cells [135]. One main problem was technical difficulties in clearly differentiating "true" *in vivo* phenomena from *in vitro* phenomena or artifacts. Recent studies on stimulation of intracellular TLR7 by anti-U1RNP antibodies [134, 136] indicate that immune complexes can be internalized via Fc receptor and released U1RNA stimulate TLR7. Thus, antibodies are internalized inside of living cells at least in this condition. When immune response starts from one epitope to other epitopes on the same molecule, or one polypeptide of snRNPs to others, antigen-presenting cells that carry peptide derived from multiprotein complexes may facilitate this process (epitope progression, intermolecular-intrastructural help) [137]. This also has been interpreted as evidence supporting an antigen-driven B-cell response.

References

[1] E.M. Tan, Antinuclear antibodies: diagnostic markers for autoimmune diseases and probes for cell biology, Adv Immunol 44 (1989) 93−151.

[2] J.A. Hardin, The lupus autoantigens and the pathogenesis of systemic lupus erythematosus, Arthritis Rheum 29 (1986) 457−460.

[3] E.M. Tan, E.K.L. Chan, K.F. Sullivan, R.L. Rubin, Antinuclear antibodies (ANAs): diagnostically specific immune markers and clues toward the understanding of systemic autoimmunity, Clin Immunol Immunopathol 47 (1988) 121−141.

[4] M.M. Hargraves, H. Richmond, R. Morton, Presentation of two bone marrow elements: the "Tart" cell and the "L.E." cell. Proc. Staff Mtg, Mayo Clinic 23 (1948) 25−28.

[5] E.M. Tan, A.S. Cohen, J.F. Fries, A.T. Masi, D.J. McShane, N.F. Rothfield, J.G. Schaller, N. Talal, R.J. Winchester, The 1982 revised criteria for the classification of systemic lupus erythematosus, Arthritis Rheum 25 (1982) 1271−1277.

[6] H. Holman, H.R. Deicher, The reaction of the lupus erythematosus (L.E.) cell factor with deoxyribonucleoprotein of the cell nucleus, J Clin Invest 38 (1959) 2059−2072.

[7] H.R. Holman, H.R.G. Deicher, H.G. Kunkel, The L.E. cell and the L.E. serum factors, Bull N Y Acad Med 35 (1959) 409−418.

[8] E.J. Holborow, D.M. Weir, Histone: An essential component for the lupus erythematosus antinuclear reaction, Lancet 1 (1959) 809.

[9] A.C. Aisenberg, Studies on the mechanism of the lupus erythematosus (L.E.) phenomenon, J Clin Invest 38 (1959) 325.

[10] K. Hannestad, O.P. Rekvig, A. Husebekk, Cross-reacting rheumatoid factors and lupus erythematosus (LE)-factors, Springer Semin Immunopathol 4 (1981) 133.

[11] O.P. Rekvig, K. Hannestad, Lupus erythematosus (LE) factors recognize both nucleosomes and viable human leukocytes, Scand J Immunol 13 (1981) 597.

[12] G. Schett, G. Steiner, J.S. Smolen, Nuclear antigen histone H1 is primarily involved in lupus erythematosus cell formation, Arthritis Rheum 41 (1998) 1446−1455.

[13] G. Schett, R.L. Rubin, G. Steiner, H. Hiesberger, S. Muller, J. Smolen, The lupus erythematosus cell phenomenon: comparative analysis of antichromatin antibody specificity in lupus erythematosus cell-positive and -negative sera, Arthritis Rheum 43 (2000) 420−428.

[14] M.J. Fritzler, E.K.L. Chan, R.G. Lahita, Antibodies to nonhistone antigens in systemic lupus erythematosus, in: Systemic Lupus Erythematosus, Elsevier Academic Press, San Diego, 2004, pp. 348−376.

[15] D. Pruss, B. Bartholomew, J. Persinger, J. Hayes, G. Arents, E.N. Moudrianakis, A.P. Wolffe, An asymmetric model for the nucleosome: a binding site for linker histones inside the DNA gyres, Science 274 (1996) 614−617.

[16] J.E. Godfrey, A.D. Baxevanis, E.N. Moudrianakis, Spectropolarimetric analysis of the core histone octamer and its subunits, Biochemistry 29 (1990) 965−972.

[17] M.J. Fritzler, N.R. Rose, E.C. de Macario, J.D. Folds, Immunofluorescent antinuclear antibody test, in: Manual of Clinical Laboratory Immunology, American Society for Microbiology, Washington, D.C, 1997, pp. 920−927.

[18] E.K.L. Chan, M.J. Fritzler, A. Wiik, L.E. Andrade, W.H. Reeves, A. Tincani, P.L. Meroni, AutoAbSC.Org - Autoantibody Standardization Committee in 2006, Autoimmun Rev 6 (2007) 577−580.

[19] C.A. von Muhlen, E.M. Tan, Autoantibodies in the diagnosis of systemic rheumatic diseases, Semin. Arthritis Rheum 24 (1995) 323−358.

[20] R.L. Ochs, Y. Muro, Y. Si, H. Ge, E.K.L. Chan, E.M. Tan, Autoantibodies to DFS70/transcriptional coactivator p75 in atopic dermatitis and other conditions, J Allergy Clin Immunol 105 (2000) 1211−1220.

[21] A. Caturla, J.A. Colome, A. Bustos, M.J. Chamorro, M.A. Figueredo, J.L. Subiza, E.G. de la Concha, Occurrence of antibodies to protease treated histones in a patient with vasculitis, Clin Immunol Immunopathol 60 (1991) 65−71.

[22] J.P. Portanova, R.L. Rubin, F.G. Joslin, V.D. Agnello, E.M. Tan, Reactivity of anti-histone antibodies induced by procainamide and hydralazine, Clin Immunol Immunopathol 25 (1982) 67–79.

[23] D.P. Molden, G.L. Klipple, C.L. Peebles, R.L. Rubin, R.M. Nakamura, E.M. Tan, IgM anti-histone H-3 antibody associated with undifferentiated rheumatic disease syndromes, Arthritis Rheum 29 (1986) 39–43.

[24] J.D. Pauls, E. Silverman, R.M. Laxer, M.J. Fritzler, Antibodies to histones H1 and H5 in sera of patients with juvenile rheumatoid arthritis, Arthritis Rheum 32 (7) (1989) 877–883.

[25] R.W. Burlingame, R.L. Rubin, Subnucleosome structures as substrates in enzyme-linked immunosorbent assays, J Immunol Methods 134 (1990) 187–199.

[26] M.J. Fritzler, R.L. Rubin, Antibodies to histones and nucleosome-related antigens, in: D.J. Wallace, B.H. Hahn (Eds.), Dubois' Lupus Erythematosus, Lippincott Williams & Wilkins, Philadelphia, PA, 2007, pp. 464–486.

[27] R.L. Rubin, A.N. Theofilopoulos, Monoclonal autoantibodies reacting with multiple structurally related and unrelated macromolecules, Intern Rev Immunol 3 (1987) 71–95.

[28] J.L. Subiza, A. Caturla, D. Pascual-salcedo, M.J. Chamorro, E. Gazapo, M.A. Figueredo, et al., DNA-anti-DNA complexes account for part of the antihistone activity found in patients with systemic lupus erythematosus, Arthritis Rheum 32 (4) (1989) 406–412.

[29] T. Suzuki, R.W. Burlingame, C.A. Casiano, M.L. Boey, R.L. Rubin, Antihistone antibodies in systemic lupus erythematosus: Assay dependency and effects of ubiquitination and serum DNA, J Rheumatol 21 (1994) 1081–1091.

[30] T.M.J. Schmiedeke, F.W. Stockl, R. Weber, Y. Sugisaki, S.R. Batsford, A. Vogt, Histones Have High-Affinity for the Glomerular Basement-Membrane - Relevance for Immune-Complex Formation in Lupus Nephritis, J Exp Med 169 (1989) 1879–1894.

[31] R.W. Burlingame, Recent advances in understanding the clinical utility and underlying cause of antinucleosome (antichromatin) autoantibodies, Clin Appl Immunology Reviews 4 (2004) 351–366.

[32] A.J. Czaja, C. Ming, M. Shirai, M. Nishioka, Frequency and significance of antibodies to histones in autoimmune hepatitis, J Hepatol 23 (1995) 32–38.

[33] M. Chen, M. Shirai, A.J. Czaja, K. Kurokohchi, T. Arichi, K. Arima, et al., Characterization of anti-histone antibodies in patients with type 1 autoimmune hepatitis, J Gastroenterol Hepatol 13 (1998) 483–489.

[34] A. Klajman, B. Kafri, T. Shohat, I. Drucker, T. Moalem, A. Jaretzky, The prevalence of antibodies to histones induced by procainamide in old people, in cancer patients, and in rheumatoid-like diseases, Clin Immunol Immunopathol 27 (1983) 1–8.

[35] R.W. Burlingame, M.L. Boey, G. Starkebaum, R.L. Rubin, The central role of chromatin in autoimmune responses to histones and DNA in systemic lupus erythematosus, J Clin Invest 94 (1994) 184–192.

[36] M. Massa, F. De Benedetti, P. Pignatti, S. Albani, G. Di Fuccia, M. Monestier, et al., Anti-double stranded DNA, anti-histone, and anti-nucleosome IgG reactivities in children with systemic lupus erythematosus, Clin Exp Rheumatol 12 (1994) 219–225.

[37] L. Li, M. Chen, D.Y. Huang, M. Nishioka, Frequency and significance of antibodies to chromatin in autoimmune hepatitis type I, J Gastroenterol Hepatol 15 (2000) 1176–1182.

[38] A.J. Czaja, Z. Shums, W.L. Binder, S.J. Lewis, V.J. Nelson, G.L. Norman, Frequency and significance of antibodies to chromatin in autoimmune hepatitis, Dig Dis Sci 48 (2003) 1658–1664.

[39] S. Sato, M. Kodera, M. Hasegawa, M. Fujimoto, K. Takehara, Antinucleosome antibody is a major autoantibody in localized scleroderma, Br J Dermatol 151 (2004) 1182–1188.

[40] Z. Amoura, S. Koutouzov, H. Chabre, P. Cacoub, I. Amoura, L. Musset, et al., Presence of antinucleosome autoantibodies in a restricted set of connective tissue diseases: antinucleosome antibodies of the IgG3 subclass are markers of renal pathogenicity in systemic lupus erythematosus, Arthritis Rheum 43 (2000) 76–84.

[41] D.J. Wallace, H.C. Lin, G.Q. Shen, J.B. Peter, Antibodies to histone (H2A-H2B)-DNA complexes in the absence of antibodies to double-stranded DNA or to (H2A-H2B) complexes are more sensitive and specific for scleroderma-related disorders than for lupus, Arthritis Rheum 37 (1994) 1795–1797.

[42] A. Ghirardello, A. Doria, S. Zampieri, E. Tarricone, R. Tozzoli, D. Villalta, et al., Antinucleosome antibodies in SLE: a two-year follow-up study of 101 patients, J Autoimmun 22 (2004) 235–240.

[43] A.P. Cairns, S.A. McMillan, A.D. Crockard, G.K. Meenagh, E.M. Duffy, D.J. Armstrong, et al., Antinucleosome antibodies in the diagnosis of systemic lupus erythematosus, Ann Rheum Dis 62 (2003) 272–273.

[44] J.A. Simon, J. Cabiedes, E. Ortiz, J. Alcocer-Varela, J. Sanchez-Guerrero, Anti-nucleosome antibodies in patients with systemic lupus erythematosus of recent onset. Potential utility as a diagnostic tool and disease activity marker. Rheumatology, (Oxford) 43 (2004) 220–224.

[45] H. Chabre, Z. Amoura, J.C. Piette, P. Godeau, J.F. Bach, S. Koutouzov, Presence of nucleosome-restricted antibodies in patients with systemic lupus erythematosus, Arthritis Rheum 38 (1995) 1485–1491.

[46] Z. Amoura, J.C. Piette, H. Chabre, P. Cacoub, T. Papo, B. Wechsler, et al., Circulating plasma levels of nucleosomes in patients with systemic lupus erythematosus: correlation with serum antinucleosome antibody titers and absence of clear association with disease activity, Arthritis Rheum 40 (1997) 2217–2225.

[47] R. Cervera, O. Vinas, M. Ramos-Casals, J. Font, M. Garcia-Carrasco, A. Siso, et al., Anti-chromatin antibodies in systemic lupus erythematosus: a useful marker for lupus nephropathy, Ann Rheum Dis 62 (2003) 431–434.

[48] A. Bruns, S. Blass, G. Hausdorf, G.R. Burmester, F. Hiepe, Nucleosomes are major T and B cell autoantigens in systemic lupus erythematosus, Arthritis Rheum 43 (2000) 2307–2315.

[49] D.J. Min, S.J. Kim, S.H. Park, Y.I. Seo, H.J. Kang, W.U. Kim, et al., Anti-nucleosome antibody: significance in lupus patients lacking anti-double-stranded DNA antibody, Clin Exp Rheumatol 20 (2002) 13–18.

[50] L.M. Stinton, S.G. Barr, L.A. Tibbles, A. Sar, H. Benedikttson, M.J. Fritzler, Autoantibodies in lupus nephritis patients requiring renal transplantation, Lupus 16 (2007) 394–400.

[51] M. Benucci, F.L. Gobbi, A. Del Rosso, S. Cesaretti, L. Niccoli, F. Cantini, Disease activity and antinucleosome antibodies in systemic lupus erythematosus, Scand J Rheumatol 32 (2003) 42–45.

[52] S. Lacotte, H. Dumortier, M. Decossas, J.P. Briand, S. Muller, Identification of New Pathogenic Players in Lupus: Autoantibody-Secreting Cells Are Present in Nephritic Kidneys of (NZBxNZW)F1 Mice, J Immunol (2010).

[53] C.C. van Bavel, J. Dieker, S. Muller, J.P. Briand, M. Monestier, J.H. Berden, et al., Apoptosis-associated acetylation on histone H2B is an epitope for lupus autoantibodies, Mol Immunol 47 (2009) 511–516.

I. BASIS OF DISEASE PATHOGENESIS

[54] E. Kiss, G. Lakos, G. Szegedi, G. Poor, P. Szodoray, Anti-nuscleosome antibody, a reliable indicator for lupus nephritis, Autoimmunity 42 (2009) 393–398.

[55] M.C. Totoritis, E.M. Tan, E.M. McNally, R.L. Rubin, Association of antibody to histone complex H2A-H2B with symptomatic procainamide-induced lupus, N Engl J Med 318 (1988) 1431–1436.

[56] A.B. Mongey, R. Donovan-Brand, T.J. Thomas, L.E. Adams, E.V. Hess, Serologic evaluation of patients receiving procainamide, Arthritis Rheum 35 (1992) 219–223.

[57] R.L. Rubin, E.M. McNally, S.R. Nusinow, C.A. Robinson, E.M. Tan, IgG antibodies to the histone complex H2A-H2B characterize procainamide-induced lupus, Clin Immunol Immunopathol 36 (1985) 49–59.

[58] R.N. Hobbs, A.L. Clayton, R.M. Bernstein, Antibodies to the five histones and poly(adenosine diphosphate-ribose) in drug induced lupus: Implications for pathogenesis, Ann Rheum Dis 46 (1987) 408–416.

[59] R.W. Burlingame, R.L. Rubin, Drug-induced anti-histone autoantibodies display two patterns of reactivity with substructures of chromatin, J Clin Invest 88 (1991) 680–690.

[60] R.L. Rubin, S.R. Nusinow, A.D. Johnson, D.S. Rubenson, J.G. Curd, E.M. Tan, Serological changes during induction of lupus-like disease by procainamide, Am J Med 80 (1986) 999–1002.

[61] R.L. Rubin, R.W. Burlingame, J.E. Arnott, M.C. Totoritis, E.M. McNally, A.D. Johnson, IgG but not other classes of anti-[(H2A-H2B)-DNA] is an early sign of procainamide-induced lupus, J Immunol 154 (1995) 2483–2493.

[62] R.W. Burlingame, R.L. Rubin, R.S. Balderas, A.N. Theofilopoulos, Genesis and evolution of antichromatin autoantibodies in murine lupus implicates T-dependent immunization with self antigen, J Clin Invest 91 (1993) 1687–1696.

[63] E.M. Tan, H.G. Kunkel, Characteristics of a soluble nuclear antigen precipitating with sera of patients with systemic lupus erythematosus, J Immunol 96 (1966) 464–471.

[64] J. Craft, Antibodies to snRNPs in systemic lupus erythematosus, Rheum Dis Clin North Am. 18 (1992) 311–335.

[65] M.R. Lerner, J.A. Steitz, Antibodies to small nuclear RNAs complexed with proteins are produced by patients with systemic lupus erythematosus, Proc Natl Acad Sci USA 76 (1979) 5495–5499.

[66] T. Mimori, M. Hinterberger, I. Pettersson, J.A. Steitz, Autoantibodies to the U2 small nuclear ribonucleoprotein in a patient with scleroderma-polymyositis overlap syndrome, J Biol Chem. 259 (1984) 560–565.

[67] J. Craft, T. Mimori, T.L. Olsen, J.A. Hardin, The U2 small nuclear ribonucleoprotein particle as an autoantigen, J Clin Invest 81 (1988) 1716–1724.

[68] M. Satoh, J.J. Langdon, K.J. Hamilton, H.B. Richards, D. Panka, R.A. Eisenberg, W.H. Reeves, Distinctive immune response patterns of human and murine autoimmune sera to U1 small nuclear ribonucleoprotein C protein, J Clin Invest 97 (1996) 2619–2626.

[69] Y. Okano, I.N. Targoff, C.V. Oddis, T. Fujii, M. Kuwana, K. Tsuzaka, M. Hirakata, T. Mimori, J. Craft, J.T.A. Medsger, Anti-U5 small nuclear ribonucleoprotein(snRNP) antibodies: A rare anti-U snRNP specificity, Clin Immunol Immunopathol 81 (1996) 41–47.

[70] M. Satoh, H.B. Richards, K.J. Hamilton, W.H. Reeves, Human anti-nuclear ribonucleoprotein antigen autoimmune sera contain a novel subset of autoantibodies that stabilizes the molecular interaction of U1RNP-C protein with the Sm core proteins, J Immunol 158 (1997) 5017–5025.

[71] M. Mahler, M.J. Fritzler, Epitope specificity and significance in systemic autoimmune diseases, Ann N Y Acad Sci 1183 (2010) 267–287.

[72] Y. Yamasaki, S. Narain, H. Yoshida, L. Hernandez, T. Barker, P.C. Hahn, et al., Autoantibodies to RNA helicase A: A new serological marker of early lupus, Arthritis Rheum 56 (2007) 596–604.

[73] H. Brahms, V.A. Raker, W.J. van Venrooij, R. Luhrmann, A major, novel systemic lupus erythematosus autoantibody class recognizes the E, F, and G Sm snRNP proteins as an E-F-G complex but not in their denatured states, Arthritis Rheum 40 (1997) 672–682.

[74] M. Satoh, J. Akaogi, Y. Kuroda, D.C. Nacionales, H. Yoshida, Y. Yamasaki, et al., Autoantibodies that stabilize U1snRNPs are a significant component of human autoantibodies to snRNPs and delay proteolysis of the Sm antigens, J Rheumatol 31 (2004) 2382–2389.

[75] Y. Takeda, K.S. Wise, G. Wang, G. Grady, E.V. Hess, G.C. Sharp, et al., Human autoantibodies recognizing a native, macromolecular structure composed of Sm core proteins in U small nuclear RNP particles, Arthritis Rheum 41 (1998) 2059–2067.

[76] J. Salgado-Garrido, E. Bragado-Nilsson, S. Kandels-Lewis, B. Seraphin, Sm and Sm-like proteins assemble in two related complexes of deep evolutionary origin, EMBO J 18 (1999) 3451–3462.

[77] T. Eystathioy, C.L. Peebles, J.C. Hamel, J.H. Vaughn, E.K. Chan, Autoantibody to hLSm4 and the heptameric LSm complex in anti-Sm sera, Arthritis Rheum 46 (2002) 726–734.

[78] G.C. Sharp, W.S. Irvin, R.L. LaRoque, C. Velez, V. Daly, A.D. Kaiser, et al., Association of autoantibodies to different nuclear antigens with clinical patterns of rheumatic disease and responsiveness to therapy, J Clin Invest 50 (1971) 350–359.

[79] M.J. Fritzler, Advances and applications of multiplexed diagnostic technologies in autoimmune diseases, Lupus 15 (2006) 422–427.

[80] M. Mattioli, M. Reichlin, Characterization of a soluble nuclear ribonucleoprotein antigen reactive with SLE sera, J Immunol 107 (1971) 1281–1290.

[81] G.C. Sharp, W.S. Irvin, E.M. Tan, R.G. Gould, H.R. Holman, Mixed connective tissue disease-An apparently distinct rheumatic disease syndrome associated with a specific antibody to an extractable nuclear antigen (ENA), Am J Med 52 (1972) 148–159.

[82] M. Mosca, C. Baldini, S. Bombardieri, Undifferentiated connective tissue diseases in 2004, Clin Exp Rheumatol 22 (2004) S14–18.

[83] M. Mosca, A. Tavoni, R. Neri, W. Bencivelli, S. Bombardieri, Undifferentiated connective tissue diseases: the clinical and serological profiles of 91 patients followed for at least 1 year, Lupus 7 (1998) 95–100.

[84] M. Satoh, E.K.L. Chan, E.S. Sobel, D.L. Kimpel, Y. Yamasaki, S. Narain, R. Mansoor, W.H. Reeves, Clinical implication of autoantibodies in patients with systemic rheumatic diseases, Expert Rev Clin Immunol 3 (2007) 721–738.

[85] A. Murakami, K. Kojima, K. Ohya, K. Imamura, Y. Takasaki, A new conformational epitope generated by the binding of recombinant 70-kd protein and U1 RNA to anti-U1 RNP autoantibodies in sera from patients with mixed connective tissue disease, Arthritis Rheum 46 (2002) 3273–3282.

[86] M. Mahler, M.J. Fritzler, M. Bluthner, Identification of a SmD3 epitope with a single symmetrical dimethylation of an arginine residue as a specific target of a subpopulation of anti-Sm antibodies, Arthritis Res Ther 7 (2005) R19–29.

[87] H. Brahms, J. Raymackers, A. Union, F. de Keyser, L. Meheus, R. Luhrmann, The C-terminal RG dipeptide repeats of the spliceosomal Sm proteins D1 and D3 contain symmetrical dimethylarginines, which form a major B-cell epitope for anti-Sm autoantibodies, J Biol Chem 275 (2000) 17122–17129.

[88] M.J. Fritzler, M.L. Fritzler, Microbead-based technologies in diagnostic autoantibody detection, Expert Opin Med Diag 3 (2009) 81–89.

[89] M.J. Fritzler, M.L. Fritzler, The emergence of multiplexed technologies as diagnostic platforms in systemic autoimmune diseases, Curr Med Chem 13 (2006) 2503–2512.

[90] D.O. Clegg, H.J. Williams, J.Z. Singer, V.D. Steen, S. Schlegel, C. Ziminski, et al., Early undifferentiated connective tissue disease. II. The frequency of circulating antinuclear antibodies in patients with early rheumatic diseases, J Rheumatol 18 (1991) 1340–1343.

[91] C.C. Vaz, M. Couto, D. Medeiros, L. Miranda, J. Costa, P. Nero, et al., Undifferentiated connective tissue disease: a seven-center cross-sectional study of 184 patients, Clin Rheumatol 28 (2009) 915–921.

[92] M.G. Danieli, P. Fraticelli, F. Franceschini, R. Cattaneo, A. Farsi, A. Passaleva, M. Pietrogrande, F. Invernizzi, M. Vanoli, R. Scorza, M.G. Sabbadini, R. Gerli, A. Corvetta, G. Farina, F. Salsano, R. Priori, G. Valesini, G. Danieli, Five-year follow-up of 165 Italian patients with undifferentiated connective tissue diseases, Clin Exp Rheumatol 17 (1999) 585–591.

[93] H. Kameda, M. Kuwana, N. Hama, J. Kaburaki, M. Homma, Coexistence of serum anti-DNA topoisomerase I and anti-Sm antibodies: report of 3 cases, J Rheumatol 24 (1997) 400–403.

[94] M.R. Arbuckle, M.T. McClain, M.V. Rubertone, R.H. Scofield, G.J. Dennis, J.A. James, J.B. Harley, Development of autoantibodies before the clinical onset of systemic lupus erythematosus, N Engl J Med 349 (2003) 1526–1533.

[95] M. Satoh, H. Yamagata, F. Watanabe, Y. Matsushita, K. Sakamoto, K. Yoshida, S. Nakayama, M. Akizuki, A case of long-standing classical rheumatoid arthritis complicated by serological and clinical characteristics of systemic lupus erythematosus, Scand J Rheumatol 22 (1993) 138–140.

[96] K.B. Elkon, E. Bonfa, R. Llovet, R.A. Eisenberg, Association between anti-Sm and anti-ribosomal P protein autoantibodies in human systemic lupus erythematosus and MRL/lpr mice, J Immunol 143 (1989) 1549–1554.

[97] L. Caponi, S. Bombardieri, P. Migliorini, Anti-ribosomal antibodies bind the Sm proteins D and B/B', Clin Exp Immunol 112 (1998) 139–143.

[98] M. Reichlin, A. Martin, E. Taylor-Albert, K. Tsuzaka, W. Zhang, M.W. Reichlin, E. Koren, F.M. Ebling, B. Tsao, B.H. Hahn, Lupus autoantibodies to native DNA cross-react with the A and D snRNP polypeptides, J Clin Invest 93 (1994) 443–449.

[99] G.C. Sharp, W.S. Irvin, C.M. May, H.R. Holman, F.C. McDuffie, E.V. Hess, F.R. Schmid, Association of autoantibodies to ribonucleoprotein and Sm antigens with mixed connective-tissue disease, systemic lupus erythematosus and other rheumatic diseases, N Engl J Med 295 (1976) 1149–1154.

[100] M. Homma, T. Mimori, Y. Takeda, H. Akama, T. Yoshida, T. Ogasawara, M. Akizuki, Autoantibodies to the Sm antigen: Immunological approach to clinical aspects of systemic lupus erythematosus, J Rheumatol 14 (suppl) (1987) 188–193.

[101] I. Pettersson, G. Wang, E.I. Smith, H. Wigzell, E. Hedfors, J. Horn, G.C. Sharp, The use of immunoblotting and immunoprecipitation of (U) small nuclear ribonucleoproteins in the analysis of sera of patients with mixed connective tissue disease and systemic lupus erythematosus, Arthritis Rheum 29 (1986) 986–996.

[102] Y. Takeda, G.S. Wang, R.J. Wang, S.K. Anderson, I. Petterson, S. Amaki, G.C. Sharp, Enzyme-linked immunosorbent assay using isolated(U) small nuclear ribonucleoprotein poly-peptides as antigens to investigate the clinical significance of autoantibodies to these polypeptides, Clin Immunol Immunopathol 50 (1989) 213–220.

[103] E.W. St Clair, C.C. Query, R. Bentley, J.D. Keene, R.P. Polisson, N.B. Allen, D.S. Caldwell, J.R. Rice, C. Cox, D.S. Pisetsky, Expression of autoantibodies to recombinant (U1) RNP-associated 70K antigen in systemic lupus erythematosus, Clin Immunol Immunopathol 54 (1990) 266–280.

[104] W.J. Habets, M.H. Hoet, W.J. van Venrooij, Epitope patterns of anti-RNP antibodies in rheumatic diseases. Evidence for an antigen-driven autoimmune response, Arthritis Rheum 33 (1990) 834–841.

[105] W. Habets, M. Hoet, P. Bringmann, R. Luhrmann, W. van Venrooij, Autoantibodies to ribonucleoprotein particles containing U2 small nuclear RNA, EMBO J 4 (1985) 1545–1550.

[106] W.H. Reeves, D.E. Fisher, R. Wisniewolski, A.B. Gottlieb, N. Chiorazzi, Psoriasis and Raynaud's phenomenon associated with autoantibodies to U1 and U2 small nuclear ribonucleoproteins, N Engl J Med 315 (1986) 105–111.

[107] Y. Okano, T.A.J. Medsger, Newly identified U4/U6 snRNP-binding proteins by serum autoantibodies from a patient with systemic sclerosis, J Immunol 146 (1991) 535–542.

[108] T. Fujii, T. Mimori, N. Hama, A. Suwa, M. Akizuki, T. Tojo, M. Homma, Characterization of autoantibodies that recognize U4/U6 small ribonucleoprotein particles in serum from a patient with primary Sjogren's syndrome, J Biol Chem. 267 (1992) 16412–16416.

[109] Y. Okano, T.A.J. Medsger, Novel human autoantibodies reactive with 5'-terminal trimethylguanosine cap structures of U small nuclear RNA, J Immunol 149 (1992) 1093–1098.

[110] A.C. Gilliam, J.A. Steitz, Rare scleroderma autoantibodies to the U11 small nuclear ribonucleoprotein and to the trimethylguanosine cap of U small nuclear RNAs, Proc Natl Acad Sci USA 90 (1993) 6781–6785.

[111] Y. Okano, V.D. Steen, T.A.J. Medsger, Autoantibody to U3 nucleolar ribonucleoprotein(fibrillarin) in patients with systemic sclerosis, Arthritis Rheum 35 (1992) 95–100.

[112] J. Wilusz, J.D. Keene, Autoantibodies specific for U1 RNA and initiator methionine tRNA, J Biol Chem. 261 (1986) 5467–5472.

[113] R.M. Hoet, I. Koornneef, D.J. de Rooij, L.B. van de Putte, W.J. van Venrooij, Changes in anti-U1 RNA antibody levels correlate with disease activity in patients with systemic lupus erythematosus overlap syndrome, Arthritis Rheum 35 (1992) 1202–1210.

[114] P.J. Maddison, M. Reichlin, Quantitation of precipitating antibodies to certain soluble nuclear antigens in SLE, Arthritis Rheum 20 (1977) 819–824.

[115] R.P. Taylor, C. Horgan, Quantitative determination of anti-dsDNA antibodies and antibody/dsDNA stoichiometries in prepared, soluble complement-fixing antibody/dsDNA immune complexes, Mol Immunol 21 (1984) 853–862.

[116] M. Nishikai, Y. Okano, Y. Mukohda, A. Sato, M. Ito, Serial estimation of anti-RNP antibody titers in systemic lupus erythematosus, mixed connective tissue disease and rheumatoid arthritis, J Clin Lab Immunol 13 (1984) 15–19.

[117] R.A. Manz, A.E. Hauser, F. Hiepe, A. Radbruch, Maintenance of serum antibody levels, Annu Rev Immunol 23 (2005) 367–386.

[118] G. Cambridge, M.J. Leandro, M. Teodorescu, J. Manson, A. Rahman, D.A. Isenberg, J.C. Edwards, B cell depletion therapy in systemic lupus erythematosus: effect on autoantibody and antimicrobial antibody profiles, Arthritis Rheum 54 (2006) 3612–3622.

[119] T. Vallerskog, I. Gunnarsson, M. Widhe, A. Risselada, L. Klareskog, R. van Vollenhoven, V. Malmstrom, C. Trollmo, Treatment with rituximab affects both the cellular and the humoral arm of the immune system in patients with SLE, Clin Immunol 122 (2007) 62–74.

[120] Y. Ohosone, T. Mimori, T. Fujii, M. Akizuki, Y. Matsuoka, S. Irimajiri, J.A. Hardin, J. Craft, M. Homma, Autoantigenic epitopes of the B polypeptide of Sm small nuclear RNP particles. Identification of regions accessible only within the U1 small nuclear RNP, Arthritis Rheum 35 (1992) 960–966.

[121] T. Mimori, T. Ogasawara, The structure and antigenic epitopes of U1 small nuclear ribonucleoprotein, in: R. Kasukawa, G.C. Sharp (Eds.), Mixed Connective Tissue Disease and Antinuclear Antibodies, Elsevier, Amsterdam, 1987, pp. 169–174.

[122] J.A. James, J.B. Harley, R.H. Scofield, Epstein-Barr virus and systemic lupus erythematosus, Curr Opin Rheumatol 18 (2006) 462–467.

[123] J.A. James, M.J. Mamula, J.B. Harley, Sequential autoantigenic determinants of the small nuclear ribonucleoprotein Sm D shared by human lupus autoantibodies and MRL lpr/lpr antibodies, Clin Exp Immunol 98 (1994) 419–426.

[124] J.B. Harley, J.A. James, Autoepitopes in lupus, J Lab Clin Med 126 (1995) 509–516.

[125] C.C. Query, J.D. Keene, A human autoimmune protein associated with U1 RNA contains a region of homology that is crossreactive with retroviral p30gag antigen, Cell 51 (1987) 211–220.

[126] J. Lipes, E. Skamene, M.M. Newkirk, The genotype of mice influences the autoimmune response to spliceosome proteins induced by cytomegalovirus gB immunization, Clin Exp Immunol 129 (2002) 19–26.

[127] E.L. Greidinger, M.F. Foecking, S. Ranatunga, R.W. Hoffman, Apoptotic U1-70 kd is antigenically distinct from the intact form of the U1-70-kd molecule, Arthritis Rheum 46 (2002) 1264–1269.

[128] S.L. Peng, J. Moslchi, J. Craft, Roles of interferon-gamma and interleukin-4 in murine lupus, J Clin Invest 99 (1997) 1936–1946.

[129] H.B. Richards, M. Satoh, J.C. Jennete, B.P. Croker, H. Yoshida, W.H. Reeves, Interferon g is required for lupus nephritis in mice treated with the hydrocarbon oil pristane, Kidney Int 60 (2001) 2173–2180.

[130] N. Calvani, M. Satoh, B.P. Croker, W.H. Reeves, H.B. Richards, Nephritogenic autoantibodies but absence of nephritis in Il-12p35-deficient mice with pristane-induced lupus, Kidney Int 64 (2003) 897–905.

[131] D.C. Nacionales, K.M. Kelly-Scumpia, P.Y. Lee, J.S. Weinstein, R. Lyons, E. Sobel, M. Satoh, W.H. Reeves, Deficiency of the type I interferon receptor protects mice from experimental lupus, Arthritis Rheum 56 (2007) 3770–3783.

[132] K.A. Kirou, C. Lee, S. George, K. Louca, M.G. Peterson, M.K. Crow, Activation of the interferon-alpha pathway identifies a subgroup of systemic lupus erythematosus patients with distinct serologic features and active disease, Arthritis Rheum 52 (2005) 1491–1503.

[133] H. Zhuang, S. Narain, E. Sobel, P.Y. Lee, D.C. Nacionales, K.M. Kelly, H.B. Richards, M. Segal, C. Stewart, M. Satoh, W.H. Reeves, Association of anti-nucleoprotein autoantibodies with upregulation of Type I interferon-inducible gene transcripts and dendritic cell maturation in systemic lupus erythematosus, Clin Immunol 117 (2005) 238–250.

[134] P.Y. Lee, Y. Kumagai, Y. Li, O. Takeuchi, H. Yoshida, J. Weinstein, E.S. Kellner, D. Nacionales, T. Barker, K. Kelly-Scumpia, N. van Rooijen, H. Kumar, T. Kawai, M. Satoh, S. Akira, W.H. Reeves, TLR7-dependent and FcγR-independent production of type I interferon in experimental mouse lupus, J Exp Med 205 (2008) 2995–3006.

[135] A. Alarcon-Segovia, A. Ruiz-Arguelles, L. Llorence, Broken dogma: penetration of autoantibodies into living cells, Immunol Today 17 (1996) 163–164.

[136] J. Vollmer, S. Tluk, C. Schmitz, S. Hamm, M. Jurk, A. Forsbach, S. Akira, K.M. Kelly, W.H. Reeves, S. Bauer, A.M. Krieg, Immune stimulation mediated by autoantigen binding sites within small nuclear RNAs involves Toll-like receptors 7 and 8, J Exp Med 202 (2005) 1575–1585.

[137] S. Fatenejad, M. Mamula, J. Craft, Role of intermolecular/intrastructural B and T-cell determinants in the diversification of autoantibodies to ribonucleoprotein particles, Proc Natl Acad Sci USA 90 (1993) 12010–12014.

17

Toll-Like Receptors in SLE

Terry K. Means

INTRODUCTION

SLE is a multisystem autoimmune inflammatory disease characterized by the presence of immune complexes (ICs) composed of autoantibodies to nucleic acids and nuclear proteins. The multiorgan inflammation characteristic of SLE is driven by an inflammatory response induced by IC deposition in tissues that activate complement and stimulate resident tissue cells to secrete inflammatory cytokines and chemokines (Figure 17.1). These secreted inflammatory mediators recruit innate and adaptive immune effector cells to the sites of IC deposition, thereby amplifying the inflammatory response and contributing to immunopathology and end organ damage (Figure 17.1).

Recent efforts have elucidated key molecular mechanisms whereby these autoantibodies incite and drive destructive inflammation in SLE. Members of the Toll-like receptor (TLR) family have been implicated in the pathogenesis of several autoimmune diseases, including SLE. Furthermore, nucleic-acid-containing autoantibody ICs isolated from the peripheral blood of SLE patients activate TLR7 and TLR9, resulting in the production of interferon-alpha (IFN-α) and other inflammatory cytokines and chemokines. Based on these observations, it is believed that cellular activation induced by the engagement of the nucleic-acid-sensing TLRs (TLR7 and TLR9) is central to the genesis of inflammation in human lupus and is responsible for the "interferon signature" that characterizes gene activation in peripheral blood leukocytes in SLE. In support of this hypothesis, recent studies using mouse models of SLE have demonstrated a critical role for TLR7 and TLR9 in disease pathogenesis. This chapter will describe the role of the TLR family in the susceptibility, initiation, progression, and exacerbation of SLE disease.

TOLL-LIKE RECEPTOR FAMILY

It has been nearly 25 years since the first Toll receptor was identified in the fruit fly *Drosophila* [1]. Since that

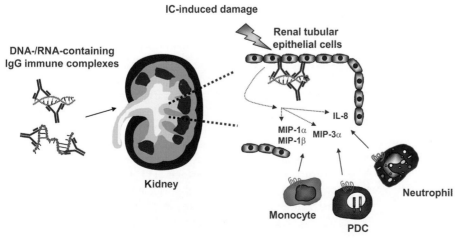

FIGURE 17.1 TLRs in lupus nephritis. Lupus nephritis develops from immune complex deposition in the kidneys that activates renal cells to produce proinflammatory cytokines and chemokines. Subsequent recruitment of monocytes, neutrophils, and pDCs contributes to glomerular injury through additional production of pro-inflammatory mediators. Kidney cells express many TLRs and TLR expression is increased in the kidneys during nephritis. Nucleic-acid-containing immune complexes infiltrating the kidney likely contribute to this inflammatory disorder by activating cells through TLR7 and TLR9. Finally, bacterial and viral infections may exacerbate lupus nephritis by stimulating TLRs expressed on kidneys cells of lupus patients.

time Toll-mutant *Drosophila* were demonstrated to have a defect in dorsal ventral patterning and antifungal immunity [2, 3]. Ten years later the first human ortholog of the *Drosophila* Toll protein was described, a protein later to be designated Toll-like receptor 4 (TLR4) [4]. In humans and mice combined, 13 TLRs have been discovered, however their expression differs between species. Humans, but not mice, express TLR10, and only mice express TLR11, TLR12, and TLR13. TLRs are an evolutionarily conserved family of type I transmembrane receptors that have an extracellular domain comprising leucine-rich repeats and a cytoplasmic domain that shares significant homology with the mammalian type I IL-1 receptor [5]. This family of germ-line-encoded receptors recognizes a myriad of conserved pathogen-associated molecular patterns (PAMPs) found in many different microbes, such as bacteria, fungi, viruses, and parasites. TLR recognition of these PAMPs leads to the initiation of intracellular signaling pathways that elicit the expression of inflammatory genes, such as cytokines essential for host defense. In contrast, it is believed that the inappropriate activation of these receptors by endogenous self ligands can contribute to chronic inflammatory syndromes and autoimmunity. The microbial and endogenous ligands for TLRs are shown in Table 17.1.

The TLR family can be divided into two subgroups, TLRs that are found on the cell surface and TLRs that reside in intracellular vesicles. TLR1, TLR2, TLR4, TLR5, TLR6, TLR10, and TLR11 are expressed on the cell surface, with TLR2 forming heterodimers with TLR1, TLR6, or TLR10. The intracellular TLRs include TLR3, TLR7, TLR8, and TLR9 and they are found localized to intracellular compartments, such as the endoplasmic reticulum, Golgi apparatus, endosomes, and lysosomes (Figure 17.2). Upon ligand binding all TLRs trigger a common signal transduction pathway that starts with the recruitment of intracellular TLR adaptor molecules. The TLR adaptor family includes four members: myeloid differentiation factor 88 (MyD88), MyD88 adaptor-like (MAL), also known as Toll/IL-1R domain-containing adaptor protein (TIRAP), Toll/IL-1R domain-containing adaptor inducing IFN-β (TRIF), and TRIF-related adaptor molecule (TRAM) (Figure 17.2). MyD88 was the first TLR adaptor protein described and its expression is critical for all TLR-mediated responses except TLR3. The other TLR adaptor proteins are used only by specific TLRs. TLR3 responses are mediated exclusively through TRIF, while TRIF, TRAM, MAL/TIRAP mediate the effects of TLR4. MAL/TIRAP also mediates the effects of TLR2 responses. Subsequent signaling events include the activation of IL-1R activated kinases (IRAK1-4), TRAF6, and IKKα/β, which ultimately causes the translocation of NF-κB to the nucleus and the initiation of inflammatory gene transcription (Figure 17.2).

TABLE 17.1 Microbial and endogenous ligands reported to activate cells through human and murine TLRs

	Microbial ligands	Endogenous ligands
TLR1[a]	triacylated lipopeptides, lipoarabinomannan *Borrelia burgdorferi* (OspA)	
TLR2	zymosan, peptidoglycan, *M. tuberculosis*	HSP60, HSP70, HMGB1, gp96, MSU
	Leptospiral LPS	
TLR3	dsRNA	mRNA
TLR4	LPS from Gram-negative bacteria	HSP60, HSP70, gp96, fibrinogen
	GPI from *Trypanosoma cruz*	hyaluron, heparin sulfate, HMGB1
proteins from RSV, MMTV, and *M. tuberculosis*		MSU, β-defensin 2
TLR5	flagellin	
TLR6[a]	diacylated lipopeptides, lipoteichoic acid, zymosan	
TLR7	ssRNA and ssRNA viruses (VSV, NDV, parechovirus)	ssRNA\protein complexes
TLR8	ssRNA , poly T/imidazoquinolines complexes	ssRNA\protein complexes
TLR9	CpG DNA	chromatin, DNA\protein complexes
TLR10	unknown	
TLR11	profilin	
TLR12	unknown	

[a]*TLR1 and TLR6 function only as a heterodimer with TLR2. Murine TLR8 recognizes synthetic RNA analogue complexes (poly T/imidazoquinoline), but not other human TLR8 agonists. TLR10 is present only in humans and TLR11 and TLR12 are present only in mice.*

TLR2 AND ITS TLR CO-RECEPTOR PARTNERS TLR1, TLR6, AND TLR10

TLR2 Biology

TLR2 expression is critical for the recognition of many diverse microbial structures. TLR2 responds to fungal cell walls, lipoproteins from Gram-negative and Gram-positive bacteria, peptidoglycan, mycobacterial lipoarabinomannan, and GPI anchors from parasitic trypanosomes [6–10]. One of the reasons TLR2 can recognize so many diverse ligands is that it can associate with TLR1 and TLR6 to form a heterodimeric receptor complex on the cell surface [11]. For example, TLR2/1 and TLR2/6 recognize distinct bacterial products such as triacylated lipopeptides and diacylated lipopeptides, respectively [12, 13] (Figure 17.2). TLR2 has also been

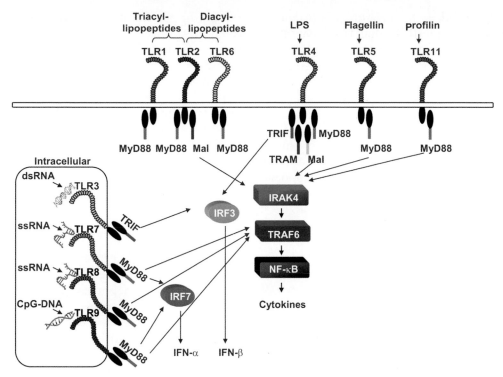

FIGURE 17.2   TLR signaling pathways. Shown is a representation of some of the known signaling intermediates in signaling pathways triggered by TLRs. TLR1, TLR2, TLR4, TLR5, TLR6, and TLR11 are expressed on the cell surface. TLR1 and TLR6 function only as a heterodimer with TLR2. TLR3, TLR7, TLR8, and TLR9 are expressed in intracellular vesicular compartments. All TLRs except TLR3 signal via MyD88-dependent pathways. In addition, TLR3 and TLR4 signal via MyD88-independent pathways.

shown to heterodimerize with TLR10, however to date no defined ligands have been identified for TLR10 [14].

Besides microbial structures, several endogenous TLR2 ligands have been reported and are discussed below.

TLR2 Expression in SLE

PBMCs isolated from SLE patients have been reported to express significantly higher levels of TLR2 mRNA than those from healthy controls [15]. In contrast, surface expression of TLR2 protein on monocytes isolated from SLE patients was reduced compared to healthy subjects [16]. In MRL$^{\text{lpr/lpr}}$ mice, which spontaneously develop a lupus-like syndrome, TLR1, TLR2, and TLR6 expression was shown to be increased in the glomeruli of nephritic kidneys compared to healthy MRL$^{\text{lpr/lpr}}$ mice [17]. Cultured mouse renal tubular cells and glomerular mesangial cells constitutively express TLR1, TLR2, and TLR6 mRNA. Stimulation of these cells with the pro-inflammatory cytokines TNF or IFN-γ induced the expression of TLR1, TLR2, and TLR6. Furthermore *in vitro* stimulation of both types of cells with microbial ligands for TLR2/1 or TLR2/6 induced the production of several inflammatory cytokines and chemokines. While more studies are needed it has been hypothesized that cell-type-specific expression

and regulation of TLR1/TLR2/TLR6 proteins may be involved in infection-associated SLE exacerbations.

In addition to microbial ligands several endogenous TLR2 ligands have been identified. Necrotic but not apoptotic cells activate macrophages via TLR2 [18]. Versican, an extracellular matrix proteoglycan, which is upregulated in many human tumors, activates macrophages through TLR2 and its co-receptor TLR6 [19]. The intracellular heat shock protein 70 has also been reported to induce IL-12 production through TLR2 [20]. It should be noted that endogenous ligand activation via TLR2 is still controversial, as some investigators have found that the recombinant sources of HSPs are contaminated with other TLR ligands, such as LPS, lipoproteins, and flagellin [21–23]. Nevertheless it has been hypothesized that TLR2 and its co-receptors may have a role in the recognition of molecules from cells undergoing cellular injury. While it is unlikely that TLR2 and its co-receptors play a critical role in the initiation of SLE-induced kidney cell injury, they may amplify the inflammatory response through their recognition of ligands released by necrotic kidney cells and the subsequent release of cytokines and chemokines (Figure 17.1). Indeed, TLR2 activation by bacterial lipoproteins on tubular epithelial cells leads to the production of cytokines and chemokines, indicating a role for TLR2 in cell influx and immune cell activation. Thus, while more studies are needed to provide confirmatory evidence, the present data indicate that TLR2 may drive SLE-related inflammation through the recognition of microbial organisms and/or endogenous host ligands released from necrotic cells.

TLR2 Polymorphisms in SLE

TLR2 mediates the innate recognition of many different types of bacteria. Mutant or defective TLR2 genes have been found to suppress macrophage responses to bacteria in animals and humans. Two single nucleotide polymorphisms (SNP) (R677W and R753Q) in TLR2 gene have been identified that impair macrophage responses to mycobacteria in humans. In a small case-control study of 122 SLE patients and 199 healthy individuals, no individuals were found to carry the TLR2-R677W polymorphism whereas the frequency of the TLR2-R753Q polymorphism was too low (only 1%) in the population of subjects to draw any conclusions [24]. Thus, there is currently no evidence for an association of these TLR2 gene polymorphisms with regard to SLE disease predisposition or clinical parameters.

TLR3

TLR3 Biology

TLR3 is highly expressed on human myeloid DCs and mediates the recognition of viral double-stranded RNA [25, 26]. Unlike the other TLRs, TLR3 exclusively uses the TLR adaptor TRIF to mediate its responses [27]. Besides viral RNA, TLR3 has also been shown to induce a potent immune response to mRNA released from necrotic cells [28]. Thus, mRNA escaping from dying cells could serve as an endogenous ligand for TLR3 and induce the production of unwanted inflammatory mediators that could initiate or exacerbate autoimmune disease.

TLR3 Expression in SLE

There is a species-specific difference in TLR3 expression on leukocyte subsets between humans and mice that may have important implications for the study of murine models of human SLE. In humans TLR3 is restricted to myeloid dendritic cells, but in mice TLR3 is also expressed on macrophages [25, 29]. TLR3 has been shown to be expressed in the kidney and on glomerular mesangial cells [17, 30, 31]. In lupus-prone MRL$^{lpr/lpr}$ mice TLR3 expression was shown to be increased in the glomeruli of nephritic kidneys compared to healthy MRL$^{lpr/lpr}$ mice [17]. This increase in TLR3 expression was found on both mesangial cells and macrophages

infiltrating the kidney. Cultured glomerular mesangial cells constitutively express TLR3 mRNA, which is induced upon stimulation with TNF, IFN-γ, or microbial ligands. Furthermore in vitro stimulation of both types of cells with double-stranded RNA, the ligand for TLR3, induced the production of several inflammatory cytokines and chemokines, including IL-6 and CCL2 [17]. In addition, viral dsRNA stimulates TLR3 in human DCs to secrete IFN and other cytokines that are known to be associated with disease activity in SLE. Thus, these data provide a possible explanation for how viral infections trigger or progress autoimmunity.

TLR3 in Murine Lupus

Intraperitoneal injection of synthetic dsRNA (poly I:C) every other day between 16 and 18 weeks of age aggravated lupus nephritis in dsRNA-treated MRL$^{lpr/lpr}$ mice [30]. An increase in proteinuria, glomerular necrosis, chemokine production, and inflammatory cell infiltration was found in the kidney of dsRNA-treated MRL$^{lpr/lpr}$ mice. Interestingly, dsRNA injection in MRL$^{lpr/lpr}$ mice failed to alter serum DNA autoantibody titers or elicit B-cell responses [30]. In addition, ablation of TLR3 in the MRL$^{lpr/lpr}$ background did not inhibit the formation of DNA or RNA autoantibodies or reduce glomerulonephritis in these mice [32]. It appears from these results that TLR3 has no role in the generation of autoantibodies to DNA or RNA. These observations can be explained in several ways. First, TLR3 expression is tightly restricted and is not expressed on B cells. Second, the prototypical RNA-containing immune complexes of lupus do not contain dsRNA, but single-stranded RNA (ssRNA). Thus, TLR3 is most likely not an initiator of SLE however, dsRNA viruses recognized by TLR3 on DCs, macrophages and kidney cells may trigger disease flares and aggravate SLE-related inflammation, providing a mechanism for how viral infections induce disease flares in SLE (Figure 17.1).

TLR4

TLR4 Biology

While Drosophila Toll was originally identified nearly 25 years ago, the springboard for TLR research began shortly after the genetic mystery of the C3H/HeJ mouse was resolved in 1998 [33]. During the mid-1960s a spontaneous mutation occurred in the mouse strain C3H/HeJ that rendered macrophages from these mice hyporesponsive to Gram-negative LPS stimulation. While the genetic defect in these mice was known to arise from a single locus (lps^d), the gene responsible for this deficiency remained unknown. In 1998, positional cloning and sequencing revealed a single missense mutation within the TLR4 coding sequence that resulted in a proline to histidine change at amino acid 712 (P712H) [33]. The P712H mutation is in the cytoplasmic signaling domain of TLR4 and renders it non-functional. Later, TLR4 knockout mice generated by gene targeting would confirm that TLR4 was the sole receptor for LPS isolated from Gram-negative bacteria [34].

While TLR4 is the major component of the LPS receptor complex, it also recognizes other microbial pathogens, including the fusion protein of respiratory syncytial virus, envelope proteins of mouse mammary tumor virus, and glycoinositolphospholipids from *Trypanosoma cruzi* [35−37]. Several endogenous host ligands have been shown to mediate TLR4 responses, including HSP60, HSP70, fibrinogen, gp96, oligosaccharides of hyaluronic acid and heparin sulfate, and β-defensin 2 [38−44]. Similar to the disclaimers for endogenous host ligands of TLR2, whether these endogenous host ligands induce TLR4-specific responses or are related to LPS or lipoprotein contamination remains controversial. Regardless, it has been hypothesized that TLR4 may recognize host endogenous ligands during cellular injury, which may cause additional tissue injury independent of microbial invasion.

TLR4 Expression in SLE

Surface expression levels of TLR4 protein on monocytes isolated from SLE patients were not significantly different from healthy individuals [16]. In 20-week-old nephritic MRL$^{lpr/lpr}$ mice TLR4 mRNA was elevated in the kidney as compared to healthy 5-week-old MRL$^{lpr/lpr}$ mice [17]. In addition, glomerular mesangial cells and renal tubular epithelial cells constitutively express TLR4 mRNA. Furthermore, stimulation of these cells with the TLR4 ligand bacterial LPS induced the production of IL-6, MCP-1, and RANTES [31]. Thus TLR4 expressed by kidney cells may be involved in cytokine and chemokine production following exposure to bacteria and bacterial components, which provides a possible explanation for how bacterial infections could precipitate flares in SLE.

TLR4 signaling is tightly controlled by negative and positive regulators to ensure a balanced immune response. Interestingly, when TLR4 expression is upregulated at the protein or gene level in mice it leads to lupus-like autoimmune disease [45]. Thus increased TLR4 signaling alone even in the absence of other endogenous or exogenous factors can break immunological tolerance. These studies suggest that TLR4 dysregulation, which leads to TLR4 hyper-responsiveness, can cause lupus-like renal disease, at least in mice.

TLR4 Polymorphisms in SLE

TLR4 mediates bacterial LPS-induced cell activation. Mutant or defective TLR4 genes have been found to suppress macrophage responses to LPS [33]. Two SNPs (D299G and T399I) in the TLR4 gene have been identified [46]. These two SNPs are located in exon 4 and encode a portion of the extracellular domain. Macrophages isolated from humans with these SNPs are hyporesponsive to LPS stimulation [46]. In the human population these polymorphic forms of TLR4 have been shown to be associated with an increased incidence of Gram-negative bacterial infections and sepsis. In contrast individuals who express these TLR4 SNPs have been found to be at decreased risk of inflammatory diseases including atherosclerosis and rheumatoid arthritis [24]. Thus, the ability to signal through TLR4 may represent an immunologic balance between being able to elicit a sufficient inflammatory response to control infection versus too vigorous an inflammatory response, resulting in autoimmunity.

Humans with these SNPs were examined in a small case-controlled study consisting of 122 SLE patients and 199 healthy individuals however, no statistically significant evidence was found for an association of TLR4-D299G or TLR4-T399I gene polymorphisms with susceptibility to SLE disease [24].

TLR5

TLR5 Biology

TLR5 recognizes bacterial flagellin from virtually all motile bacteria [47]. To date, this is the only ligand identified for TLR5. Flagellin stimulation of TLR5 induces human myeloid DC maturation, but not murine DCs, which lack TLR5 expression [48].

TLR5 Expression in SLE

Murine glomerular mesangial and renal tubular epithelial cells do not express TLR5 and failed to produce cytokines in response to bacterial flagellin in culture [17, 31].

TLR5 Polymorphisms in SLE

TLR5 recognizes bacterial flagellin, a potent inflammatory stimulus present in the flagellar structure of many bacteria. A stop codon polymorphism in the ligand-binding domain of TLR5 (TLR5-392STOP) was demonstrated to be unable to mediate flagellin signaling and was associated with increased susceptibility to pneumonia caused by *Legionella pneumophila* [49].

The wild-type TLR5, but not the TLR5 stop codon SNP was found to be preferentially transmitted to patients with SLE. In addition, the TLR5 stop polymorphism was preferentially transmitted to unaffected offspring of SLE patients. Together, these results indicate that the TLR5 stop codon polymorphism is associated with protection from the development of SLE [50]. This association was found using transmission disequilibrium testing in a Caucasian cohort. In contrast, two similar studies on Caucasian-American and Caucasian-Spanish SLE patient cohorts failed to replicate this observation [51, 52].

TLR7

TLR7 Biology

TLR7 is expressed in intracellular compartments and mediates the recognition of guanosine- and uridine-rich ssRNA and ssRNA viruses, including influenza, Newcastle disease virus, parechovirus 1, and vesicular stomatitis virus [53–56]. TLR7 also recognizes synthetic antiviral nucleoside analogs such as imiquimod (R848), gardiquimod, and loxoribine [57, 58]. TLR7 is highly expressed in human plasmacytoid dendritic cells (pDCs), which secrete high amounts of IFN-α upon infection with ssRNA viruses.

TLR7 Ligands in SLE

Serologically, the hallmark of SLE is a high level of antinuclear antibodies (ANA) present in nearly all affected individuals (95–99%) [59], with known target antigen specificities of these ANA toward DNA itself and/or nuclear proteins known to form complexes with DNA (i.e., histones) or RNA (i.e., Ro, La, Sm, RNP, others) [60]. Anti-Sm reactivity was first discovered over 40 years ago in a young woman, Stephanie Smith, diagnosed with SLE [61]. The Smith antigen is a complex of uridine-rich small nuclear RNA (snRNA) molecules and several proteins. Since that time several studies have demonstrated a role for RNA-containing autoantibody complexes in the pathogenesis of SLE [62]. In addition, a correlation between the levels of RNA-containing autoantibody complexes in serum and disease severity has been noted [63]. Furthermore, snRNA-containing immune complexes (ICs) isolated from SLE patient serum are taken up through Fc receptors and delivered to intracellular lysosomes where they stimulate TLR7 [64, 65] (Figure 17.3). snRNA-ICs can stimulate pDCs to produce inflammatory, immunoregulatory, and chemotatic cytokines. The specificity of TLR7 activation by snRNA-ICs was confirmed in DCs isolated from TLR7-deficent mice and by the demonstration that

oligonucleotide-based inhibitors of TLR7 blocked snRNA-IC-induced IFN-α production by pDCs [66, 67]. Together these data suggest that snRNA molecules contained in circulating ICs found in the serum of SLE patients, act as endogenous self ligand for TLR7 and may initiate and/or exacerbate disease by inappropriately and chronically stimulating innate immune cells.

Selective recognition of ssRNA by TLR7 and TLR8 is quite remarkable. Total RNA from necrotic cells, but not apoptotic cells, induces DC activation [68]. Furthermore, bacterial RNA is significantly more potent for inducing cell activation than mammalian DNA. Why is self RNA stimulatory? One possible explanation is that the RNA is modified in some way. Indeed, it has been demonstrated that RNA containing modified nucleosides is significantly less stimulatory than unmodified RNA [68]. Most mammalian RNA is modified and less stimulatory than unmodified bacterial RNA. Thus, mammals appear to have developed a system to modify their RNA to suppress its stimulatory potential, thereby providing a mechanism for the innate immune system to distinguish between self and pathogenic sources of RNA. Still some mammalian RNAs, particularly mitochondrial RNA, have very few modifications and are therefore a potential source for the stimulatory self RNA contained in RNA-associated ICs that are found in SLE patients. These data lead to the hypothesis that the RNA contained in anti-RNA ICs is less modified, which would make it a potent TLR7/8 stimulus. More studies analyzing RNA-ICs from human SLE patients are needed to confirm and extend these findings.

TLR7 Expression in SLE

PBMCs isolated from SLE patients have been reported to express significantly higher levels of TLR7 mRNA than those from healthy controls [15]. In addition the levels of TLR7 expression on SLE patient PBMCs correlated with IFN-α mRNA. TLR7 expression is equivalent in cells isolated from female versus male individuals; however TLR7 stimulation induced significantly more IFN-α production from female than male cells [69]. While this observation was found in cells isolated from healthy subjects and needs to be repeated in cells isolated from SLE patients, it may provide an explanation for the nearly 10 times higher prevalence of SLE in females than males.

TLR7 in Murine Lupus

In vitro experiments demonstrate that autoreactive B cells can bind self-RNA through the B-cell antigen receptor (BCR), which delivers it to intracellular vesicles, most likely endosomes or lysosomes, where it sequentially engages TLR7 [70] (Figure 17.3).

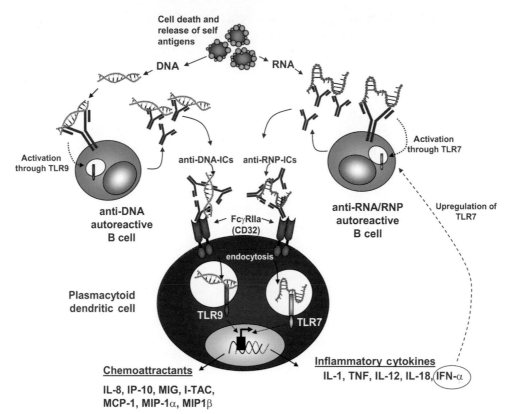

FIGURE 17.3 TLR7 and TLR9 activation by endogenous host RNA and DNA. This schematic represents the innate model of lupus pathogenesis. RNA and DNA released from dying or damaged cells can be recognized by the B-cell receptor on autoreactive B cells. The BCR transports RNA and DNA to intracellular lysosomes where they engage TLR7 and TLR9, respectively. Activation of TLR7 and TLR9 in B cells induces proliferation, cytokine production (IFN-β), and differentiation into plasma cells that secrete autoantibodies that can bind RNA, DNA, or RNA/DNA-associated proteins to form immune complexes. FcγRIIa (CD32), expressed by plasmacytoid dendritic cells, binds nucleic-acid-containing immune complexes and transports them to intracellular lysosomes containing TLR7 and TLR9. Activation of TLR7 and TLR9 in pDCs induces proinflammatory cytokine and chemokine production, which can result in cell recruitment and a vicious cytokine storm that can lead to tissue pathogenesis. In addition, IFN-α secretion by pDCs enhances B-cell activation by nucleic-acid ICs by upregulating TLR7 expression. This may provide a positive-feedback loop by further enhancing autoantibody production and B cell proliferation leading to the exacerbation of autoimmunity.

Engagement of TLR7 then leads to B-cell activation, proliferation, and autoantibody production. In this model, self-RNA is an autoantigen that acts as an adjuvant to trigger in this case an unwanted immune response (Figure 17.3). Moreover, TLR7 expression on B cells can be induced by engagement of the IFN-α/β receptor on B cells [71] (Figure 17.3). The source of IFN-α/β can be autocrine (B cells) or paracrine (pDCs). These *in vitro* experiments highlight the potential involvement of TLR7 in autoimmunity.

To test the involvement of TLR7 in the development of SLE *in vivo* the lupus-prone MRL[lpr/lpr] mice were crossed to TLR7-deficent mice. These mice failed to generate autoantibodies to RNA-associated antigens (Sm), but produced levels of anti-DNA autoantibodies typical for the MRL[lpr/lpr] model [72]. In addition MRL[lpr/lpr] mice deficient in TLR7 had less activated B cells and pDCs in circulation and significantly less renal disease. In a separate study, mice engineered to express

an autoreactive immunoglobulin that binds to the autoantigens ssDNA, ssRNA, and nucleosomes were created on a non-autoimmune background (C57Bl/6) called 546Igi mice [73]. The B cells in 564Igi mice escape anergy and tolerance and produce pathogenic class-switched autoantibodies that result in kidney disease. Notably, anti-RNA autoantibodies are absent in 564Igi mice deficient in TLR7. Together these observations confirm the model that B-cell activation and regulation of autoantibody production against self-RNA-associated antigens is TLR7-dependent.

Since TLR7 serves an important role in murine lupus this receptor has been proposed as a potential therapeutic target for human SLE. Experimentally, oligonucleotide antagonists that specifically inhibit TLR7 signaling significantly decrease anti-RNA and anti-DNA autoantibody levels in the serum and reduce kidney disease when injected into lupus-prone MRL[lpr/lpr] mice [74]. Future clinical trials using TLR7 antagonists in human SLE patients will provide answers to the specificity of the TLR7 pathway in human SLE disease.

TLR7 Polymorphisms and Copy Number in SLE

In 1979 the mouse strain BXSB/MP was demonstrated to develop a spontaneous lupus-like syndrome which was restricted to male mice [75]. The susceptibility locus was mapped to the Y chromosome and called the Y-linked autoimmune accelerator (YAA) locus. Nearly 30 years later it has been shown that YAA is not a mutation but rather a duplication of X chromosomal DNA that was transposed to the Y chromosome. This duplicated 4-Mb gene segment contains TLR7 and 16 other genes [76]. The published data support a role for TLR7 as the main gene responsible for the accelerated lupus-like disease, because reduction of TLR7 copy number abrogated the YAA phenotype [77]. Furthermore, experiments in transgenic mice demonstrated that increasing TLR7 gene dosage directly correlated with the production of autoantibodies directed against RNA autoantigens and the severity of the autoimmune disease [77]. These data suggest that strict regulation of TLR7 expression and function is critical (at least in mice) to prevent spontaneous autoimmunity.

TLR7 may not be the sole gene responsible for the YAA phenotype, because ablation of TLR7 in mice bearing the YAA duplication did not completely protect them against lupus-like disease, which suggests that other genes in the YAA locus may contribute to the autoimmune phenotype [78]. Studies examining the copy number of the TLR7 gene in human SLE patients and healthy controls have not revealed a significant association with the SLE phenotype [79]. Another candidate gene in the YAA locus is TLR8 which is located contiguous to TLR7 on chromosome X. Like TLR7, human TLR8 is activated by specific sequences of guanosine- and uridine-rich ssRNA. In contrast, murine TLR8 is not activated by ssRNA or other TLR8 agonist compounds, which lead to the belief that TLR8 is non-functional in mice. More recently it has been demonstrated that a combination of synthetic imidazoquinline compounds complexed with polyT oligonucleotides specifically activates murine TLR8 [80]. While the natural ligands for murine TLR8 are still unknown it remains plausible that TLR8 contributes to the YAA phenotype. Further studies in YAA mice deficient in TLR7 and TLR8 should help resolve this matter.

Several polymorphisms have been found in human TLR7, including a non-synonymous single nucleotide polymorphism (SNP) in TLR7 that changes a glutamine to a leucine at amino acid 11 in the N-terminus.

This SNP was shown not to confer a role in susceptibility or severity to SLE in a Caucasian-Spanish population [52].

TLR8

TLR8 Biology

Similar to TLR7, TLR8 is expressed in intracellular compartments and detects guanosine- and uridine-rich ssRNA and ssRNA viruses. The TLR8 gene also lies contiguous to TLR7 on the X chromosome. While TLR7 is expressed mainly on pDCS and B cells, TLR8 is more broadly expressed on myeloid DCs, monocytes, differentiated macrophages, and Treg cells [81, 82]. While TLR7 and TLR8 both recognize similar molecules, some functional differences have been noted. For example, TLR8 agonists are more effective than TLR7 agonists at inducing proinflammatory cytokines (TNF, IL-12, MIP-1α), but less effective at inducing IFN-α and IFN-regulated genes (IP-10) [81]. Thus, TLR8 differs from TLR7 in its cell type expression and cytokine induction profile.

TLR8 in SLE

In mice TLR8 expression has not been extensively examined primarily because murine TLR8 was thought to be non-functional because it did not react to human TLR8 ligands. As discussed above the natural ligands for murine TLR8 have not been described, but synthetic imidazoquinoline compounds complexed with polyT react specifically with murine TLR8. The role of TLR8 in human and murine lupus is less clear especially since TLR8 it is not expressed on B cells or pDCs, but rather on myeloid cells. It remains plausible that IgG immune complexes containing unmodified forms of self-RNA may initiate or exacerbate SLE by stimulating TLR8 on myeloid DCs through the induction of IFN-α and other inflammatory cytokines.

Another interesting aspect of TLR8 is its expression and regulation of CD4+ regulatory T cells (Treg) [82]. TLR8 signaling induced by ssRNA can reverse the suppressive function of Treg cells. This effect was shown to be independent of DCs, but dependent on TLR8 and MyD88 signaling. These data suggest that TLR8 signaling in Treg cells could play an important role in controlling immune responses. It is interesting to speculate that the reason why RNA-associated ICs in serum correlates with high levels of IFN-α and renal disease in SLE, is that TLR8 signaling induced by circulating RNA-associated ICs abrogates or reverses the suppressive function of Treg cells in patients with SLE.

TLR9

TLR9 Biology

In 1984 it was demonstrated that bacterial genomic DNA was immunostimulatory and promoted antitumor responses [83]. Other studies demonstrated that unlike mammalian DNA, bacterial and viral DNA induced B-cell proliferation [84]. Later, the immunostimulatory effect of bacterial DNA on B cells was shown to be mediated by the presence of unmethylated 'CpG motifs' [85]. In contrast to bacterial and viral DNA, CpG motifs in mammalian DNA are suppressed and mostly methylated. Thus, most mammalian DNA does not have immunostimulatory activity. TLR9 was shown to be the receptor that mediates cellular responses to CpG-DNA [86]. DCs isolated from TLR9-deficent mice did not show any response to CpG-DNA, including maturation or cytokine production. Thus TLR9 appears to have evolved to specifically recognize microbial DNA from self DNA.

Other cytosolic DNA sensors have been identified, including the HIN-200 (hematopoietic interferon-inducible nuclear proteins with a 200-amino-acid repeat) family member AIM2 (absent in melanoma 2) [87–89]. This protein binds cytosolic DNA through the HIN domain, recruits the adaptor ASC (apoptosis-associated speck-like protein containing a CARD) through its pyrin domain, and forms an inflammasome complex containing caspase-1 that functions in the generation of IL-1β. The role of these proteins in autoimmunity *in vivo* will have to wait for the generation of mouse strains deficient in these proteins.

TLR9 Expression in SLE

Anti-DNA reactivity was first discovered in the sera of patients with SLE a half century ago and it is still used clinically to aid the diagnosis of SLE because of its high specificity for SLE compared with other connective tissue or autoimmune diseases [90, 91]. Since that time several studies have demonstrated a role for DNA-specific antibodies in the pathogenesis of SLE [62]. In addition, a correlation between the levels of anti-DNA reactivity in serum and disease severity has been noted, with DNA-specific immunoglobulin (Ig) being isolated from diseased kidneys of SLE patients in some cases [92]. Interestingly, some, but not all, anti-DNA Ig are pathogenic [93]. This raises the question whether the pathogenicity of DNA-reactive antibodies is related to their affinity for DNA, specificity to DNA sequence/structure, or some combination of both. Notably, SLE serum containing anti-DNA antibodies activates PBMCs to secrete IFN-α [94, 95]. Furthermore, DNA complexed with anti-DNA autoantibodies

purified from SLE patient sera also stimulates PBMCs to produce IFN-α [96]. One possible explanation for this activity is that the DNA contained in DNA–anti-DNA ICs purified from SLE patient serum is stimulatory. Interestingly, DNA sequencing demonstrated that the DNA contained in some anti-DNA ICs includes CpG dinucleotides at a five times higher frequency than expected for random CpG distribution in the genome [97, 98]. While these stimulatory dinucleotide sequences are common in bacterial and viral genomes, they are by comparison relatively suppressed in vertebrate DNA. Synthetic oligonucleotides carrying these CpG motifs are immunostimulatory and induce activation of B cells, pDCs, and monocytes exclusively through TLR9, as cells from TLR9-deficent mice lack these responses [85, 86]. Furthermore, *in vitro* studies demonstrated that anti-DNA ICs isolated from SLE patient serum induce cell activation in a TLR9-dependent manner [99] (Figure 17.3). In quiescent cells TLR9 mainly resides in the endoplasmic reticulum, however upon uptake of DNA-ICs into cells, TLR9 trafficks from the ER to lyosomes [99, 100]. The interaction of TLR9 with DNA-ICs is rapid and specific, as other non-nucleic-acid-containing autoantibody complexes isolated from other rheumatic disease patients fails to activate TLR9 [63, 99]. In pDCs, internalization of DNA-ICs is mediated through FcγRIIa (CD32), which delivers the complexes to intracellular lysosomes containing TLR9, thereby initiating signaling and the production of inflammatory cytokines such as IFN-α [99] (Figure 17.3).

In addition to the two distinct mechanisms described above for DNA-IC recognition by B cells (BCR/TLR9) and pDCs (CD32/TLR9), a common third mechanism that uses high-mobility group protein B1 (HMGB1) and receptor for advanced glycation endproducts (RAGE) for DNA-IC recognition by B cells and pDCs has been demonstrated [101]. HMGB1 is a known ligand for RAGE. Interestingly, HMGB1 has been shown to bind to DNA with high affinity. Moreover, HMGB1 is found in DNA-ICs isolated from SLE patients. Remarkably, RAGE and HMGB1 expression were demonstrated to be required for optimal activation of pDCs and B cells by DNA-ICs [101]. Thus, TLR9 in cooperation with HMGB1/RAGE has the capacity to recognize endogenous host DNA, triggering a signaling cascade that induces inflammatory mediators that may initiate and/or exacerbate SLE disease.

TLR9 expression is higher on B cells and monocytes, but not pDCs isolated from SLE patients as compared to healthy controls [102]. This increase in TLR9 expression also correlated with the presence of circulating anti-DNA antibodies in these patients and treatment of B cells isolated from healthy controls with SLE serum induced TLR9 expression [102].

TLR9 in Murine Lupus

The first paper describing a role for TLR9 in autoimmunity was performed by Marshak-Rothstein and colleagues [103]. They demonstrated that DNA- and RNA-containing IgG complexes activated autoreactive B cells through the sequential engagement of the BCR, which delivers the complexes to intracellular vesicles containing TLR9 or TLR7, respectively (Figure 17.3). The proliferation of autoreactive B cells induced by nucleic acid autoantigens is TLR9- and TLR7-dependent, because autoreactive B cells deficient for TLR9 or TLR7 failed to proliferate in response to both DNA- and RNA-containing complexes, respectively.

Increased TLR9 expression was demonstrated in the kidneys of nephritic $MRL^{lpr/lpr}$ mice and was shown to be expressed on infiltrating macrophages. $MRL^{lpr/lpr}$ mice deficient in TLR9 made less anti-DNA antibodies, but had high levels of RNA-containing autoantibodies (anti-Sm/RNP) [32]. Remarkably, there was an increase in B-cell, T-cell, and pDC activation, increased renal disease, and a higher mortality rate in TLR9-deficent $MRL^{lpr/lpr}$ compared to $MRL^{lpr/lpr}$ mice [104]. This increased cell activation seems to be in direct conflict with the *in vitro* data discussed above and it is still unresolved why, despite the decrease in anti-DNA autoantibodies, immune cell activation, renal disease, and mortality are all increased in TLR9-defcient mice. This apparent paradox might be explained if RNA-containing autoantibodies are more pathogenic or inflammatory, especially in mice, or if TLR9 plays a role in the clearance of apoptotic and necrotic cellular debris. Another explanation may be that Treg cell development or suppressive function is abrogated in TLR9-deficient mice, but so far the current experimental data do not support this concept. Clearly the role of TLR9 in murine SLE disease requires further study.

TLR9 Polymorphisms and Copy Number in SLE

The TLR9 gene (chromosome 3p21.3) is located in one of the defined susceptibility regions for SLE. Several groups have investigated whether genetic variations of TLR9, which include (-1486 T→C, -1237 C→T, $+1174$ A→G, and $+2848$ G→A), are involved in susceptibility to SLE in different populations around the world. The presence of the nucleotide G at position $+1174$ in intron 1 of TLR9 was demonstrated to be significantly associated with an increased risk of SLE in a Japanese cohort. Interestingly, the A→G $+1174$ SNP downregulates TLR9 expression in reporter assays. These data would fit with the current model (discussed above) that TLR9 expression is protective against SLE.

In contrast, other studies performed on American, UK, Korean, and Chinese SLE cohorts did not detect a statistically significant association of any TLR9 gene variations with SLE susceptibility.

RP105

RP105 Biology

RP105 (radio protective) was first discovered as a surface receptor expressed on B cells and ligation of RP105 by anti-RP105 antibodies induces B-cell proliferation, co-stimulatory molecule expression (CD86), and resistance to radiation-induced apoptosis [105]. Similar to the other TLR family members, RP105 is a type I transmembrane receptor with extracellular leucine-rich repeats [106]. In contrast, RP105 does not contain a TIR domain and has a very short 11-amino-acid cytoplasmic tail. The signaling molecules associated with the cytoplasmic tail have not yet been identified, but RP105 ligation triggers Lyn, Bruton tyrosine kinase (BTK), mitogen-activated kinase (MAPK), and NF-κB activation. Immunoprecipitation and transfection experiments demonstrate that RP105 associates with MD-1 to form a heterodimer on the surface of B cells. Moreover MD-1 is essential for RP-105 surface expression, similar to MD-2 for TLR4. B lymphocytes from RP105- and MD-1-deficient mice exhibit reduced activation and proliferation in response to LPS stimulation [107].

RP105 Expression in SLE

As in mice, virtually all human B cells express RP105. Interestingly, SLE patients have a significant population of circulating RP105-negative B cells compared to healthy controls [108]. Moreover, SLE disease and serum ANA levels correlated with the percentage of RP105-negative B cells present in these patients. RP105-negative, but not RP105-positive B cells isolated from SLE patients are highly activated and spontaneously produced anti-DNA autoantibodies [109]. These data suggest RP105 may function as a negative regulator of B-cell activation. Indeed, RP105 has been shown to function as a negative regulator of TLR4 responses [110]. Perhaps, RP105 also regulates TLR7 and TLR9 responses and functions to limit autoreactive B-cell activation.

RP105 in Murine Lupus

$MRL^{lpr/lpr}$ mice deficient in RP105 had slower disease progression, increased renal function, and longer survival as compared to $MRL^{lpr/lpr}$ mice [111]. Surprisingly, there was no change in glomerulonephritis by histology or anti-DNA/chromatin autoantibody production. This study did not report measures of RNA-associated autoantibodies from these animals, so it is possible that their levels were decreased, which might account for the decreases in disease severity.

CONCLUSIONS

It has been over 40 years since a drug has been approved by the FDA for the treatment of SLE, which highlights the vital need for the development of novel therapies that inhibit early steps in the pathogenesis of this refractory disease. Studies on the role of TLRs in SLE will likely lead to the development of novel therapies for SLE and other autoimmune diseases. Murine models of lupus clearly indicate that TLR7 and TLR9 are required for the generation of anti-DNA and anti-RNA autoantibodies, respectively. Thus, these receptors are potential candidates for pharmacological inhibition. Indeed, synthetic oligonucleotide-based inhibitors of TLR7 and TLR9 have already been shown to reduce SLE disease progression in mice. Hopefully, research on TLRs in SLE will continue to move forward at a rapid pace, which will help move new TLR-based therapies into the clinic.

References

[1] K.V. Anderson, G. Jurgens, C. Nusslein-Volhard, Establishment of dorsal-ventral polarity in the Drosophila embryo: genetic studies on the role of the Toll gene product, Cell 42 (1985) 779–789.

[2] C. Hashimoto, K.L. Hudson, K.V. Anderson, The Toll gene of Drosophila, required for dorsal-ventral embryonic polarity, appears to encode a transmembrane protein, Cell 52 (1988) 269–279.

[3] B. Lemaitre, E. Nicolas, L. Michaut, J.M. Reichhart, J.A. Hoffmann, The dorsoventral regulatory gene cassette spatzle/Toll/cactus controls the potent antifungal response in Drosophila adults, Cell 86 (1996) 973–983.

[4] R. Medzhitov, P. Preston-Hurlburt, C.A. Janeway Jr., A human homologue of the Drosophila Toll protein signals activation of adaptive immunity, Nature 388 (1997) 394–397.

[5] N.J. Gay, F.J. Keith, Drosophila Toll and IL-1 receptor, Nature 351 (1991) 355–356.

[6] D.M. Underhill, A. Ozinsky, K.D. Smith, A. Aderem, Toll-like receptor-2 mediates mycobacteria-induced proinflammatory signaling in macrophages, Proc Natl Acad Sci USA 96 (1999) 14459–14463.

[7] D.M. Underhill, A. Ozinsky, A.M. Hajjar, A. Stevens, C.B. Wilson, M. Bassetti, et al., The Toll-like receptor 2 is recruited to macrophage phagosomes and discriminates between pathogens, Nature 401 (1999) 811–815.

[8] T.K. Means, E. Lien, A. Yoshimura, S. Wang, D.T. Golenbock, M.J. Fenton, The CD14 ligands lipoarabinomannan and lipopolysaccharide differ in their requirement for Toll-like receptors, J Immunol 163 (1999) 6748–6755.

[9] A. Ouaissi, E. Guilvard, Y. Delneste, G. Caron, G. Magistrelli, N. Herbault, et al., The Trypanosoma cruzi Tc52-released protein induces human dendritic cell maturation, signals via Toll-like receptor 2, and confers protection against lethal infection, J Immunol 168 (2002) 6366–6374.

[10] E. Lien, T.J. Sellati, A. Yoshimura, T.H. Flo, G. Rawadi, R.W. Finberg, et al., Toll-like receptor 2 functions as a pattern recognition receptor for diverse bacterial products, J Biol Chem 274 (1999) 33419–33425.

[11] A. Ozinsky, D.M. Underhill, J.D. Fontenot, A.M. Hajjar, K.D. Smith, C.B. Wilson, et al., The repertoire for pattern recognition of pathogens by the innate immune system is defined by cooperation between toll-like receptors, Proc Natl Acad Sci USA 97 (2000) 13766–13771.

[12] O. Takeuchi, T. Kawai, P.F. Muhlradt, M. Morr, J.D. Radolf, A. Zychlinsky, et al., Discrimination of bacterial lipoproteins by Toll-like receptor 6, Int Immunol 13 (2001) 933–940.

[13] O. Takeuchi, S. Sato, T. Horiuchi, K. Hoshino, K. Takeda, Z. Dong, et al., Cutting edge: role of Toll-like receptor 1 in mediating immune response to microbial lipoproteins, J Immunol 169 (2002) 10–14.

[14] U. Hasan, C. Chaffois, C. Gaillard, V. Saulnier, E. Merck, S. Tancredi, et al., Human TLR10 is a functional receptor, expressed by B cells and plasmacytoid dendritic cells, which activates gene transcription through MyD88, J Immunol 174 (2005) 2942–2950.

[15] A. Komatsuda, H. Wakui, K. Iwamoto, M. Ozawa, M. Togashi, R. Masai, et al., Up-regulated expression of Toll-like receptors mRNAs in peripheral blood mononuclear cells from patients with systemic lupus erythematosus, Clin Exp Immunol 152 (2008) 482–487.

[16] K. Migita, T. Miyashita, Y. Maeda, M. Nakamura, H. Yatsuhashi, H. Kimura, et al., Toll-like receptor expression in lupus peripheral blood mononuclear cells, J Rheumatol 34 (2007) 493–500.

[17] P.S. Patole, R.D. Pawar, M. Lech, D. Zecher, H. Schmidt, S. Segerer, et al., Expression and regulation of Toll-like receptors in lupus-like immune complex glomerulonephritis of MRL-Fas(lpr) mice, Nephrol Dial Transplant 21 (2006) 3062–3073.

[18] M. Li, D.F. Carpio, Y. Zheng, P. Bruzzo, V. Singh, F. Ouaaz, et al., An essential role of the NF-kappa B/Toll-like receptor pathway in induction of inflammatory and tissue-repair gene expression by necrotic cells, J Immunol 166 (2001) 7128–7135.

[19] S. Kim, H. Takahashi, W.W. Lin, P. Descargues, S. Grivennikov, Y. Kim, et al., Carcinoma-produced factors activate myeloid cells through TLR2 to stimulate metastasis, Nature 457 (2009) 102–106.

[20] R.M. Vabulas, P. Ahmad-Nejad, S. Ghose, C.J. Kirschning, R.D. Issels, H. Wagner, HSP70 as endogenous stimulus of the Toll/interleukin-1 receptor signal pathway, J Biol Chem 277 (2002) 15107–15112.

[21] B. Gao, M.F. Tsan, Recombinant human heat shock protein 60 does not induce the release of tumor necrosis factor alpha from murine macrophages, J Biol Chem 278 (2003) 22523–22529.

[22] B. Gao, M.F. Tsan, Endotoxin contamination in recombinant human heat shock protein 70 (Hsp70) preparation is responsible for the induction of tumor necrosis factor alpha release by murine macrophages, J Biol Chem 278 (2003) 174–179.

[23] Z. Ye, Y.H. Gan, Flagellin contamination of recombinant heat shock protein 70 is responsible for its activity on T cells, J Biol Chem 282 (2007) 4479–4484.

[24] E. Sanchez, G. Orozco, M.A. Lopez-Nevot, J. Jimenez-Alonso, J. Martin, Polymorphisms of toll-like receptor 2 and 4 genes in rheumatoid arthritis and systemic lupus erythematosus, Tissue Antigens 63 (2004) 54–57.

[25] M. Muzio, D. Bosisio, N. Polentarutti, G. D'Amico, A. Stoppacciaro, R. Mancinelli, et al., Differential expression and regulation of toll-like receptors (TLR) in human

leukocytes: selective expression of TLR3 in dendritic cells, J Immunol 164 (2000) 5998—6004.

[26] L. Alexopoulou, A.C. Holt, R. Medzhitov, R.A. Flavell, Recognition of double-stranded RNA and activation of NF-kappaB by Toll-like receptor 3, Nature 413 (2001) 732—738.

[27] M. Yamamoto, S. Sato, K. Mori, K. Hoshino, O. Takeuchi, K. Takeda, et al., Cutting edge: a novel Toll/IL-1 receptor domain-containing adapter that preferentially activates the IFN-beta promoter in the Toll-like receptor signaling, J Immunol 169 (2002) 6668—6672.

[28] K. Kariko, H. Ni, J. Capodici, M. Lamphier, D. Weissman, mRNA is an endogenous ligand for Toll-like receptor 3, J Biol Chem 279 (2004) 12542—12550.

[29] S.E. Applequist, R.P. Wallin, H.G. Ljunggren, Variable expression of Toll-like receptor in murine innate and adaptive immune cell lines, Int Immunol 14 (2002) 1065—1074.

[30] P.S. Patole, H.J. Grone, S. Segerer, R. Ciubar, E. Belemezova, A. Henger, et al., Viral double-stranded RNA aggravates lupus nephritis through Toll-like receptor 3 on glomerular mesangial cells and antigen-presenting cells, J Am Soc Nephrol 16 (2005) 1326—1338.

[31] N. Tsuboi, Y. Yoshikai, S. Matsuo, T. Kikuchi, K. Iwami, Y. Nagai, et al., Roles of toll-like receptors in C-C chemokine production by renal tubular epithelial cells, J Immunol 169 (2002) 2026—2033.

[32] S.R. Christensen, M. Kashgarian, L. Alexopoulou, R.A. Flavell, S. Akira, M.J. Shlomchik, Toll-like receptor 9 controls anti-DNA autoantibody production in murine lupus, J Exp Med 202 (2005) 321—331.

[33] A. Poltorak, X. He, I. Smirnova, M.Y. Liu, C. Van Huffel, X. Du, et al., Defective LPS signaling in C3H/HeJ and C57BL/10ScCr mice: mutations in Tlr4 gene, Science 282 (1998) 2085—2088.

[34] K. Hoshino, O. Takeuchi, T. Kawai, H. Sanjo, T. Ogawa, Y. Takeda, et al., Cutting edge: Toll-like receptor 4 (TLR4)-deficient mice are hyporesponsive to lipopolysaccharide: evidence for TLR4 as the Lps gene product, J Immunol 162 (1999) 3749—3752.

[35] E.A. Kurt-Jones, L. Popova, L. Kwinn, L.M. Haynes, L.P. Jones, R.A. Tripp, et al., Pattern recognition receptors TLR4 and CD14 mediate response to respiratory syncytial virus, Nat Immunol 1 (2000) 398—401.

[36] J.C. Rassa, J.L. Meyers, Y. Zhang, R. Kudaravalli, S.R. Ross, Murine retroviruses activate B cells via interaction with toll-like receptor 4, Proc Natl Acad Sci USA 99 (2002) 2281—2286.

[37] A.C. Oliveira, J.R. Peixoto, L.B. de Arruda, M.A. Campos, R.T. Gazzinelli, D.T. Golenbock, et al., Expression of functional TLR4 confers proinflammatory responsiveness to Trypanosoma cruzi glycoinositolphospholipids and higher resistance to infection with T. cruzi, J Immunol 173 (2004) 5688—5696.

[38] K. Ohashi, V. Burkart, S. Flohe, H. Kolb, Cutting edge: heat shock protein 60 is a putative endogenous ligand of the toll-like receptor-4 complex, J Immunol 164 (2000) 558—561.

[39] A. Asea, M. Rehli, E. Kabingu, J.A. Boch, O. Bare, P.E. Auron, et al., Novel signal transduction pathway utilized by extracellular HSP70: role of toll-like receptor (TLR) 2 and TLR4, J Biol Chem 277 (2002) 15028—15034.

[40] S.T. Smiley, J.A. King, W.W. Hancock, Fibrinogen stimulates macrophage chemokine secretion through toll-like receptor 4, J Immunol 167 (2001) 2887—2894.

[41] T. Warger, N. Hilf, G. Rechtsteiner, P. Haselmayer, D.M. Carrick, H. Jonuleit, et al., Interaction of TLR2 and TLR4 ligands with the N-terminal domain of Gp96 amplifies innate and adaptive immune responses, J Biol Chem 281 (2006) 22545—22553.

[42] C. Termeer, F. Benedix, J. Sleeman, C. Fieber, U. Voith, T. Ahrens, et al., Oligosaccharides of Hyaluronan activate dendritic cells via toll-like receptor 4, J Exp Med 195 (2002) 99—111.

[43] A.H. Tang, G.J. Brunn, M. Cascalho, J.L. Platt, Pivotal advance: endogenous pathway to SIRS, sepsis, and related conditions, J Leukoc Biol 82 (2007) 282—285.

[44] A. Biragyn, P.A. Ruffini, C.A. Leifer, E. Klyushnenkova, A. Shakhov, O. Chertov, et al., Toll-like receptor 4-dependent activation of dendritic cells by beta-defensin 2, Science 298 (2002) 1025—1029.

[45] B. Liu, Y. Yang, J. Dai, R. Medzhitov, M.A. Freudenberg, P.L. Zhang, et al., TLR4 up-regulation at protein or gene level is pathogenic for lupus-like autoimmune disease, J Immunol 177 (2006) 6880—6888.

[46] N.C. Arbour, E. Lorenz, B.C. Schutte, J. Zabner, J.N. Kline, M. Jones, et al., TLR4 mutations are associated with endotoxin hyporesponsiveness in humans, Nat Genet 25 (2000) 187—191.

[47] F. Hayashi, K.D. Smith, A. Ozinsky, T.R. Hawn, E.C. Yi, D.R. Goodlett, et al., The innate immune response to bacterial flagellin is mediated by Toll-like receptor 5, Nature 410 (2001) 1099—1103.

[48] T.K. Means, F. Hayashi, K.D. Smith, A. Aderem, A.D. Luster, The Toll-like receptor 5 stimulus bacterial flagellin induces maturation and chemokine production in human dendritic cells, J Immunol 170 (2003) 5165—5175.

[49] T.R. Hawn, A. Verbon, K.D. Lettinga, L.P. Zhao, S.S. Li, R.J. Laws, et al., A common dominant TLR5 stop codon polymorphism abolishes flagellin signaling and is associated with susceptibility to legionnaires' disease, J Exp Med 198 (2003) 1563—1572.

[50] T.R. Hawn, H. Wu, J.M. Grossman, B.H. Hahn, B.P. Tsao, A. Aderem, A stop codon polymorphism of Toll-like receptor 5 is associated with resistance to systemic lupus erythematosus, Proc Natl Acad Sci USA 102 (2005) 10593—10597.

[51] F.Y. Demirci, S. Manzi, R. Ramsey-Goldman, M. Kenney, P.S. Shaw, C.M. Dunlop-Thomas, et al., Association study of Toll-like receptor 5 (TLR5) and Toll-like receptor 9 (TLR9) polymorphisms in systemic lupus erythematosus, J Rheumatol 34 (2007) 1708—1711.

[52] E. Sanchez, J.L. Callejas-Rubio, J.M. Sabio, M.A. Gonzalez-Gay, J. Jimenez-Alonso, L. Mico, et al., Investigation of TLR5 and TLR7 as candidate genes for susceptibility to systemic lupus erythematosus, Clin Exp Rheumatol 27 (2009) 267—271.

[53] F. Heil, H. Hemmi, H. Hochrein, F. Ampenberger, C. Kirschning, S. Akira, et al., Species-specific recognition of single-stranded RNA via toll-like receptor 7 and 8, Science 303 (2004) 1526—1529.

[54] S.S. Diebold, T. Kaisho, H. Hemmi, S. Akira, C. Reis e Sousa, Innate antiviral responses by means of TLR7-mediated recognition of single-stranded RNA, Science 303 (2004) 1529—1531.

[55] K. Triantafilou, E. Vakakis, G. Orthopoulos, M.A. Ahmed, C. Schumann, P.M. Lepper, et al., TLR8 and TLR7 are involved in the host's immune response to human parechovirus 1, Eur J Immunol 35 (2005) 2416—2423.

[56] J.M. Lund, L. Alexopoulou, A. Sato, M. Karow, N.C. Adams, N.W. Gale, et al., Recognition of single-stranded RNA viruses by Toll-like receptor 7, Proc Natl Acad Sci USA 101 (2004) 5598—5603.

[57] F. Heil, P. Ahmad-Nejad, H. Hemmi, H. Hochrein, F. Ampenberger, T. Gellert, et al., The Toll-like receptor 7 (TLR7)-specific stimulus loxoribine uncovers a strong relationship within the TLR7, 8 and 9 subfamily, Eur J Immunol 33 (2003) 2987—2997.

[58] H. Hemmi, T. Kaisho, O. Takeuchi, S. Sato, H. Sanjo, K. Hoshino, et al., Small anti-viral compounds activate immune cells via the TLR7 MyD88-dependent signaling pathway, Nat Immunol 3 (2002) 196—200.

[59] R.C. Lawrence, C.G. Helmick, F.C. Arnett, R.A. Deyo, D.T. Felson, E.H. Giannini, et al., Estimates of the prevalence of arthritis and selected musculoskeletal disorders in the United States, Arthritis Rheum 41 (1998) 778—799.

[60] E.M. Tan, Antinuclear antibodies: diagnostic markers for autoimmune diseases and probes for cell biology, Adv Immunol 44 (1989) 93—151.

[61] E.M. Tan, H.G. Kunkel, Characteristics of a soluble nuclear antigen precipitating with sera of patients with systemic lupus erythematosus, J Immunol 96 (1966) 464—471.

[62] B.H. Hahn, Antibodies to DNA, N Engl J Med 338 (1998) 1359—1368.

[63] J. Hua, K. Kirou, C. Lee, M.K. Crow, Functional assay of type I interferon in systemic lupus erythematosus plasma and association with anti-RNA binding protein autoantibodies, Arthritis Rheum 54 (2006) 1906—1916.

[64] E. Savarese, O.W. Chae, S. Trowitzsch, G. Weber, B. Kastner, S. Akira, et al., U1 small nuclear ribonucleoprotein immune complexes induce type I interferon in plasmacytoid dendritic cells through TLR7, Blood 107 (2006) 3229—3234.

[65] J. Vollmer, S. Tluk, C. Schmitz, S. Hamm, M. Jurk, A. Forsbach, et al., Immune stimulation mediated by autoantigen binding sites within small nuclear RNAs involves Toll-like receptors 7 and 8, J Exp Med 202 (2005) 1575—1585.

[66] F.J. Barrat, T. Meeker, J. Gregorio, J.H. Chan, S. Uematsu, S. Akira, et al., Nucleic acids of mammalian origin can act as endogenous ligands for Toll-like receptors and may promote systemic lupus erythematosus, J Exp Med 202 (2005) 1131—1139.

[67] K. Yasuda, C. Richez, J.W. Maciaszek, N. Agrawal, S. Akira, A. Marshak-Rothstein, et al., Murine dendritic cell type I IFN production induced by human IgG-RNA immune complexes is IFN regulatory factor (IRF)5 and IRF7 dependent and is required for IL-6 production, J Immunol 178 (2007) 6876—6885.

[68] K. Kariko, M. Buckstein, H. Ni, D. Weissman, Suppression of RNA recognition by Toll-like receptors: the impact of nucleoside modification and the evolutionary origin of RNA, Immunity 23 (2005) 165—175.

[69] B. Berghofer, T. Frommer, G. Haley, L. Fink, G. Bein, H. Hackstein, TLR7 ligands induce higher IFN-alpha production in females, J Immunol 177 (2006) 2088—2096.

[70] C.M. Lau, C. Broughton, A.S. Tabor, S. Akira, R.A. Flavell, M.J. Mamula, et al., RNA-associated autoantigens activate B cells by combined B cell antigen receptor/Toll-like receptor 7 engagement, J Exp Med 202 (2005) 1171—1177.

[71] N.M. Green, A. Laws, K. Kiefer, L. Busconi, Y.M. Kim, M.M. Brinkmann, et al., Murine B cell response to TLR7 ligands depends on an IFN-beta feedback loop, J Immunol 183 (2009) 1569—1576.

[72] S.R. Christensen, J. Shupe, K. Nickerson, M. Kashgarian, R.A. Flavell, M.J. Shlomchik, Toll-like receptor 7 and TLR9 dictate autoantibody specificity and have opposing inflammatory and regulatory roles in a murine model of lupus, Immunity 25 (2006) 417—428.

[73] R. Berland, L. Fernandez, E. Kari, J.H. Han, I. Lomakin, S. Akira, et al., Toll-like receptor 7-dependent loss of B cell tolerance in pathogenic autoantibody knockin mice, Immunity 25 (2006) 429—440.

[74] R.D. Pawar, A. Ramanjaneyulu, O.P. Kulkarni, M. Lech, S. Segerer, H.J. Anders, Inhibition of Toll-like receptor-7 (TLR-7) or TLR-7 plus TLR-9 attenuates glomerulonephritis and lung injury in experimental lupus, J Am Soc Nephrol 18 (2007) 1721—1731.

[75] E.D. Murphy, J.B. Roths, A Y chromosome associated factor in strain BXSB producing accelerated autoimmunity and lymphoproliferation, Arthritis Rheum 22 (1979) 1188—1194.

[76] P. Pisitkun, J.A. Deane, M.J. Difilippantonio, T. Tarasenko, A.B. Satterthwaite, S. Bolland, Autoreactive B cell responses to RNA-related antigens due to TLR7 gene duplication, Science 312 (2006) 1669—1672.

[77] J.A. Deane, P. Pisitkun, R.S. Barrett, L. Feigenbaum, T. Town, J.M. Ward, et al., Control of toll-like receptor 7 expression is essential to restrict autoimmunity and dendritic cell proliferation, Immunity 27 (2007) 801—810.

[78] M.L. Santiago-Raber, S. Kikuchi, P. Borel, S. Uematsu, S. Akira, B.L. Kotzin, et al., Evidence for genes in addition to Tlr7 in the Yaa translocation linked with acceleration of systemic lupus erythematosus, J Immunol 181 (2008) 1556—1562.

[79] J. Kelley, M.R. Johnson, G.S. Alarcon, R.P. Kimberly, J.C. Edberg, Variation in the relative copy number of the TLR7 gene in patients with systemic lupus erythematosus and healthy control subjects, Arthritis Rheum 56 (2007) 3375—3378.

[80] K.K. Gorden, X.X. Qiu, C.C. Binsfeld, J.P. Vasilakos, S.S. Alkan, Cutting edge: activation of murine TLR8 by a combination of imidazoquinoline immune response modifiers and polyT oligodeoxynucleotides, J Immunol 177 (2006) 6584—6587.

[81] K.B. Gorden, K.S. Gorski, S.J. Gibson, R.M. Kedl, W.C. Kieper, X. Qiu, et al., Synthetic TLR agonists reveal functional differences between human TLR7 and TLR8, J Immunol 174 (2005) 1259—1268.

[82] G. Peng, Z. Guo, Y. Kiniwa, K.S. Voo, W. Peng, T. Fu, et al., Toll-like receptor 8-mediated reversal of CD4+ regulatory T cell function, Science 309 (2005) 1380—1384.

[83] T. Tokunaga, H. Yamamoto, S. Shimada, H. Abe, T. Fukuda, Y. Fujisawa, et al., Antitumor activity of deoxyribonucleic acid fraction from Mycobacterium bovis BCG.I. Isolation, physicochemical characterization, and antitumor activity, J Natl Cancer Inst 72 (1984) 955—962.

[84] J.P. Messina, G.S. Gilkeson, D.S. Pisetsky, Stimulation of in vitro murine lymphocyte proliferation by bacterial DNA, J Immunol 147 (1991) 1759—1764.

[85] A.M. Krieg, A.K. Yi, S. Matson, T.J. Waldschmidt, G.A. Bishop, R. Teasdale, et al., CpG motifs in bacterial DNA trigger direct B-cell activation, Nature 374 (1995) 546—549.

[86] H. Hemmi, O. Takeuchi, T. Kawai, T. Kaisho, S. Sato, H. Sanjo, et al., A Toll-like receptor recognizes bacterial DNA, Nature 408 (2000) 740—745.

[87] T.L. Roberts, A. Idris, J.A. Dunn, G.M. Kelly, C.M. Burnton, S. Hodgson, et al., HIN-200 proteins regulate caspase activation in response to foreign cytoplasmic DNA, Science 323 (2009) 1057—1060.

[88] V. Hornung, A. Ablasser, M. Charrel-Dennis, F. Bauernfeind, G. Horvath, D.R. Caffrey, et al., AIM2 recognizes cytosolic dsDNA and forms a caspase-1-activating inflammasome with ASC, Nature 458 (2009) 514—518.

[89] T. Fernandes-Alnemri, J.W. Yu, P. Datta, J. Wu, E.S. Alnemri, AIM2 activates the inflammasome and cell death in response to cytoplasmic DNA, Nature 458 (2009) 509—513.

[90] R. Ceppellini, E. Polli, F. Celada, A DNA-reacting factor in serum of a patient with lupus erythematosus diffusus, Proc Soc Exp Biol Med 96 (1957) 572—574.

[91] E.J. Holborow, D.M. Weir, G.D. Johnson, A serum factor in lupus erythematosus with affinity for tissue nuclei, Br Med J 2 (1957) 732—734.

[92] E.J. ter Borg, G. Horst, E.J. Hummel, P.C. Limburg, C.G. Kallenberg, Measurement of increases in anti-double-stranded DNA antibody levels as a predictor of disease exacerbation in systemic lupus erythematosus. A long-term, prospective study, Arthritis Rheum 33 (1990) 634—643.

[93] F. Ebling, B.H. Hahn, Restricted subpopulations of DNA antibodies in kidneys of mice with systemic lupus. Comparison of antibodies in serum and renal eluates, Arthritis Rheum 23 (1980) 392—403.

[94] H. Vallin, S. Blomberg, G.V. Alm, B. Cederblad, L. Ronnblom, Patients with systemic lupus erythematosus (SLE) have a circulating inducer of interferon-alpha (IFN-alpha) production acting on leucocytes resembling immature dendritic cells, Clin Exp Immunol 115 (1999) 196—202.

[95] H. Vallin, A. Perers, G.V. Alm, L. Ronnblom, Anti-double-stranded DNA antibodies and immunostimulatory plasmid DNA in combination mimic the endogenous IFN-alpha inducer in systemic lupus erythematosus, J Immunol 163 (1999) 6306—6313.

[96] K. Yasuda, C. Richez, M.B. Uccellini, R.J. Richards, R.G. Bonegio, S. Akira, et al., Requirement for DNA CpG content in TLR9-dependent dendritic cell activation induced by DNA-containing immune complexes, J Immunol 183 (2009) 3109—3117.

[97] Y. Sato, M. Miyata, T. Nishimaki, H. Kochi, R. Kasukawa, CpG motif-containing DNA fragments from sera of patients with systemic lupus erythematosus proliferate mononuclear cells in vitro, J Rheumatol 26 (1999) 294—301.

[98] H. Sano, O. Takai, N. Harata, K. Yoshinaga, I. Kodama-Kamada, T. Sasaki, Binding properties of human anti-DNA antibodies to cloned human DNA fragments, Scand J Immunol 30 (1989) 51—63.

[99] T.K. Means, E. Latz, F. Hayashi, M.R. Murali, D.T. Golenbock, A.D. Luster, Human lupus autoantibody-DNA complexes activate DCs through cooperation of CD32 and TLR9, J Clin Invest 115 (2005) 407—417.

[100] E. Latz, A. Schoenemeyer, A. Visintin, K.A. Fitzgerald, B.G. Monks, C.F. Knetter, et al., TLR9 signals after translocating from the ER to CpG DNA in the lysosome, Nat Immunol 5 (2004) 190—198.

[101] J. Tian, A.M. Avalos, S.Y. Mao, B. Chen, K. Senthil, H. Wu, et al., Toll-like receptor 9-dependent activation by DNA-containing immune complexes is mediated by HMGB1 and RAGE, Nat Immunol 8 (2007) 487—496.

[102] E.D. Papadimitraki, C. Choulaki, E. Koutala, G. Bertsias, C. Tsatsanis, I. Gergianaki, et al., Expansion of toll-like receptor 9-expressing B cells in active systemic lupus erythematosus: implications for the induction and maintenance of the autoimmune process, Arthritis Rheum 54 (2006) 3601—3611.

[103] E.A. Leadbetter, I.R. Rifkin, A.M. Hohlbaum, B.C. Beaudette, M.J. Shlomchik, A. Marshak-Rothstein, Chromatin-IgG complexes activate B cells by dual engagement of IgM and Toll-like receptors, Nature 416 (2002) 603—607.

[104] X. Wu, S.L. Peng, Toll-like receptor 9 signaling protects against murine lupus, Arthritis Rheum 54 (2006) 336—342.

[105] K. Miyake, Y. Yamashita, Y. Hitoshi, K. Takatsu, M. Kimoto, Murine B cell proliferation and protection from apoptosis with an antibody against a 105-kD molecule: unresponsiveness of X-linked immunodeficient B cells, J Exp Med 180 (1994) 1217—1224.

[106] K. Miyake, Y. Yamashita, M. Ogata, T. Sudo, M. Kimoto, RP105, a novel B cell surface molecule implicated in B cell activation, is a member of the leucine-rich repeat protein family, J Immunol 154 (1995) 3333—3340.

[107] M. Kimoto, K. Nagasawa, K. Miyake, Role of TLR4/MD-2 and RP105/MD-1 in innate recognition of lipopolysaccharide, Scand J Infect Dis 35 (2003) 568—572.

[108] S. Koarada, Y. Tada, O. Ushiyama, F. Morito, N. Suzuki, A. Ohta, et al., B cells lacking RP105, a novel B cell antigen, in systemic lupus erythematosus, Arthritis Rheum 42 (1999) 2593—2600.

[109] Y. Kikuchi, S. Koarada, Y. Tada, O. Ushiyama, F. Morito, N. Suzuki, et al., RP105-lacking B cells from lupus patients are responsible for the production of immunoglobulins and autoantibodies, Arthritis Rheum 46 (2002) 3259—3265.

[110] S. Divanovic, A. Trompette, S.F. Atabani, R. Madan, D.T. Golenbock, A. Visintin, et al., Negative regulation of Toll-like receptor 4 signaling by the Toll-like receptor homolog RP105, Nat Immunol 6 (2005) 571—578.

[111] T. Kobayashi, K. Takahashi, Y. Nagai, T. Shibata, M. Otani, S. Izui, et al., Tonic B cell activation by Radioprotective105/MD-1 promotes disease progression in MRL/lpr mice, Int Immunol 20 (2008) 881—891.

18

Interferon-Alpha in Systemic Lupus Erythematosus

Mary K. Crow

INTRODUCTION

The idea that interferon-alpha (IFN-α) might play an important role in the pathogenesis of SLE initially emerged when it was observed that serum from patients with active SLE had high levels of that cytokine, based on increased capacity to inhibit virus-induced cell death [1]. Additional data, published in the late 1970s and early 1980s, described an association between IFN activity and the standard serologic indicators of lupus disease activity, anti-DNA antibody titer, and low complement levels [1–3]. A direct pathogenic link between IFN-α and SLE was suggested by clinical observations from patients who had received recombinant IFN-α for viral infection or certain malignancies and developed lupus-associated autoantibodies and occasionally clinical lupus [4–7]. While a *bona fide* lupus syndrome rarely occurred in that setting, the induction of autoantibodies by IFN-α indicated the potential for IFN-α to promote autoimmunity in some individuals. With the advent of microarray technology, it readily became apparent that a characteristic gene expression signature in peripheral blood mononuclear cells (PBMCs) could be observed in many patients with SLE [8–11], and comparison of that "IFN signature" with the gene expression profile of PBMC cultured *in vitro* with IFN-α led to the recognition that type I IFN, or a stimulus that activated identical target genes as IFN-α, was present *in vivo* in patients [12]. The IFN signature appeared to represent a dominant molecular pathway in that disease. A growing body of data relating IFN pathway activation to studies of genetic associations, clinical characterization of patients, *in vitro* studies and results from therapeutic interventions support an important, and possibly central, role for IFN-α in SLE [13, 14].

TYPE I IFNs IN IMMUNE RESPONSES

Type I IFNs

The type I IFN locus on chromosome 9p21 comprises genes encoding 13 IFN-α subtypes, as well as IFN-beta (IFN-β), IFN-omega, IFN-kappa, and IFN-epsilon, the latter mostly restricted to trophoblast cells and produced early in pregnancy [15, 16]. The IFN-α gene complex is likely to have been generated by repeated gene duplications and recombinations [17]. While the need for and function of each of the IFN-α genes is not clear, specific virus infections are associated with induction of one or another IFN-α subtypes [18, 19]. IFN-β is encoded by a single gene and initiates most type I IFN responses. In addition to the type I IFNs, a smaller family of related cytokines, now referred to as type III IFNs, is encoded on a distinct locus [20]. IFN-lambdas (IFN-λ), also called IL-28 and IL-29, have only moderate sequence similarity to IFN-α, bind to a distinct receptor, yet induce genes similar to those induced by IFN-α and IFN-β [21]. The relative functional roles of IFN-λ and the chromosome 9p-encoded IFNs continue to be investigated [22].

Type I IFN-producing Cells

Many cells produce small amounts of type I IFN after infection with DNA or RNA viruses, either through intracellular nucleic acid receptors or after engagement of a Toll-like receptor (TLR) that recognizes nucleic acids, such as TLR3 for double-stranded RNA, TLR7 or 8 for single-stranded RNA, or TLR9 for demethylated CpG-rich DNA [23]. TLR-independent pathways also contribute to type I IFN production in response to small RNAs or intracellular DNA. While IFN-α can be synthesized by many cells, plasmacytoid dendritic cells (pDCs) represent the cell type most capable of high-level type I IFN production [24–28]. pDCs represent a hematopoietic lineage cell type that is distinct from lymphocyte and myeloid cells, and most likely derives from a common dendritic cell progenitor cell that differentiates toward the pDC phenotype when exposed to FLT3 ligand [29, 30]. The balance between the generation of conventional DCs (cDCs) vs. pDCs can be modulated by various growth factors and cytokines, and the gene expression signatures of the two cell types can be distinguished [31]. Leukocyte immunoglobulin-like receptor, subfamily A, member 4 LILRA4 (ILT7), interleukin 3 receptor alpha (IL3RA; CD123), and C-type lectin domain family 4, member C (CLEC4C;BDCA2) are among the gene products that are characteristically expressed on pDCs and can serve to differentiate those cells from lymphocytes, myeloid cells, and other DC populations. Additional genes preferentially expressed in pDCs compared with other DC types include Toll-like receptor 7 (TLR7), interferon-regulatory factor 7 (IRF7), protein kinase C and casein kinase substrate in neurons 1 (PACSIN 1), Src-like adaptor 2 (SLA2), B-cell linker protein (BLNK), transcription factor 4 (TCF4), RNase1, and SPIB [31]. The gene expression profile of pDCs, particularly the high expression of TLR7, an intracellular TLR that senses single-stranded RNA, and IRF7, a transcription factor activated by TLR7 ligands, points to the important role of RNA as an innate immune stimulus for those cells. This gene expression pattern is also consistent with the high-level production of IFN-α by pDCs, a downstream gene target of IRF7.

Many of the genes preferentially expressed in pDCs include those that are predicted to sense the internal and extracellular environment and regulate intracellular trafficking of endocytosed molecules, in contrast to those preferentially expressed by conventional DCs (cDCs) that are associated with responses to pathogens and induction of T-cell activation [31]. The pDC is primed for rapid and high-level production and even when it is not the first responder to microbial stimuli, it likely serves to amplify the level of IFN-α by responding to stimuli that activate TLRs. Those stimuli could include viruses but also comprise cell debris, cathelicidin antimicrobial peptides produced by granulocytes and nucleic-acid-containing immune complexes formed secondary to tissue damage or in the setting of autoimmune reactions [32–35]. The stimuli relevant to lupus patients will be discussed in more detail below. In view of the capacity of pDCs to produce high levels of IFN-α, it is not surprising that those cells have a regulatory system to modulate IFN production. Recent data indicate that pDCs express a B-cell receptor-like ITAM-bearing activation pathway that involves several kinases also involved in B-cell receptor signaling, including Lyn, Syk, Btk, and Blnk [36, 37].

IFN-responsive Cells

Type I IFN production represents the first line of defense in response to viral infection [38, 39]. Following invasion of the host by a virus, IFN-α is secreted by pDC, along with other immune system cells, and binds its receptor, IFNAR, on many target cells, resulting in engagement of intracellular signaling molecules and induction of a gene transcription program [8, 10, 40]. The IFNs, the first cytokines described, have been used as a model system to study cytokine-mediated cell signaling [41–44]. Binding of IFN-α to its cell surface receptor activates Jak-1 and then signal transducer and activator of transcription (STAT1). Subsequently, it was shown that Tyk-2, also a Jak kinase, is constitutively associated with the α subunit of the type I IFN receptor (IFNAR), while Jak-1 is associated with the β subunit of the receptor. Cytokine binding leads to activation of Tyk-2 and Jak-1 and phosphorylation of the α receptor subunit and part of the β subunit. Subsequent events include activation of additional STATs, with formation of STAT1:STAT1 and STAT1:STAT2 dimers that bind to the palindromic IFN response element (pIRE) and IFN-stimulated growth factor 3 (ISGF3), including STAT1, STAT2, and a third protein, p48 (IRF9), binds the IFN-stimulated response element (ISRE) [43, 44], inducing transcription of hundreds of genes that mediate the antiviral effect of IFN-α [40].

pDCs express high levels of one of the chains of the type I IFN receptor, resulting in priming of those cells to be particularly sensitive to RNA-containing stimuli, as both TLR7 and IRF7 are IFN-inducible genes. BDCA1+ DC, a cDC subset that mediates pro-inflammatory responses and MHC class I antigen presentation, express high levels of Jak2, a signaling kinase downstream of the IFNAR, as well as many IFN-inducible genes, including OAS2, OAS3, IFITM1, IFITM2, IFITM3, suggesting that BDCA1+ cDCs are important targets of type I IFN [31]. Their phenotype, with high-level expression of co-stimulatory molecules such as CD86, is also consistent with previous data suggesting that IFN-α present in serum of some lupus patients can promote activation of allogeneic T cells and presumably self-reactive autologous T cells, thereby inducing autoimmunity. However, IFNAR is broadly expressed and IFN-inducible genes are documented in many cell types after *in vitro* culture of PBMC with type I IFN. We have isolated naïve and memory (CD27+) B cells from peripheral blood of lupus patients and see increased expression of IFN-inducible genes in both B-cell subsets compared to expression in B cells from healthy controls (Olferiev, M., and Crow, M.K., unpublished observations).

Functional Effects of IFN-α

The broad expression of the IFNAR on many cell types contributes to the diverse functional responses induced by this cytokine family. IFN-α is an innate immune system product that orchestrates the immune system's initial response to viral infection prior to the activation of T cells. Among its effects on innate immune cells are maturation of dendritic cells by inducing ICAM-1, CD86, MHC class I, and IL-12p70 expression [45—47]. It activates natural killer and NK T cells, induces IFN-γ, and in some cases stimulates apoptosis (it induces transcription of TRAIL, FASL and perforin 1, all with apoptosis-inducing activity) resulting in release of cell debris, including potentially stimulatory self antigens [31, 48, 49]. Type I IFN induces some classic lupus autoantigens, particularly Ro52, perhaps representing an additional mechanism that drives autoimmunity [50]. Type I IFN, along with other inflammatory stimuli, promotes both innate and adaptive immune activation by inducing transcription of chemokine genes, including chemokine, CC motif, ligand 3 (CCL3; MIP1A), CCL5 (RANTES), CXCL10, CXCL11 ,CCL8 (MCP2) and cytokines such as IL-15, but can also contribute to control of inflammation by induction of interleukin 1 receptor antagonist (IL1RN) [31]. Type I IFN also amplifies effective signaling downstream of IFNAR by inducing transcription of STAT1 and STAT2 and might facilitate IFN-γ signaling by increasing JAK2.

Regarding the adaptive immune response IFN-α also induces expression of some T-cell activation molecules (CD69) and it preferentially promotes Th1 responses, by decreasing IL-4 and increasing IFN-γ secretion [51, 52]. IFN-α leads to increased NK- and T-cell-mediated cytotoxicity [48, 49, 53, 54]. This effect on CTL function has been exploited in the treatment of several malignancies with IFN-α in order to augment tumor lysis. IFN-α has antiproliferative effects on T cells, and it is generally described as a suppressor of T-cell immune activity. When CD4$^+$ T cells are cultured with anti-CD3 and anti-CD28 monoclonal antibodies, IFN-α augments IL-10 production, generally considered an anti-inflammatory cytokine [55—58]. IFN-γ does not have these effects and in fact inhibits IL-10 production. Taken together, studies of the impact of type I IFN on T-cell function demonstrate a complex pattern of immune regulation.

Regarding B-cell functions, IFN-α has been shown to promote immunoglobulin (Ig) class switching [59, 60]. At least some of this effect might be attributable to the increased IL-10 induced by IFN-α which can augment B-cell proliferation and differentiation [61, 62] or increased T-cell production of IL-21, a cytokine that contributes to B-cell differentiation [63]. In addition, IFN-α induces expression of B-lymphocyte stimulator (BLyS or BAFF), a mediator that promotes B-cell survival and Ig class switching and represents an additional mechanism that might account for amplification of pathogenic antibody production by IFN-α [64]. Recent demonstrations of enrichment of autoreactive B cells among bone marrow emigrants suggest that effects of IFN-α on central B-cell tolerance might contribute to autoimmunity [65].

In summary, IFN-α helps to initiate an adaptive immune response by promoting maturation of antigen-presenting cells (APCs), increases cytotoxic T- and NK-cell activity, increases antibody production, but decreases T-cell proliferation. IFN-α also contributes to amplification of inflammatory responses. Many of these immune system effects are reminiscent of those observed in patients with SLE.

Beyond its important impact on immune responses, IFN has effects on angiogenesis as it inhibits transcription of VEGF and endoglin, and on vascular integrity as it is associated with reduced numbers of endothelial progenitor cells [31, 66, 67]. Finally, type I IFN increases transcription of CD164 (sialomucin), an observation that is of interest in view of the frequently observed increased presence of mucin in pathologic tissue specimens from lupus patients [68]. Overall, it is striking how many features of the immunologic and clinical profile of lupus patients are consistent with the documented functional effects of type I IFN.

IFN IN SLE

Early Studies of IFN in SLE

Papers published as early as 1979 described increased serum levels of IFN in patients with SLE, particularly those with active disease [1—3, 69, 70]. At that time, the distinct type I and type II IFNs had not yet been documented, but within several years, IFN-α was cloned and it became clear that IFN-α was present in particularly high levels in SLE blood. It was observed that tubuloreticular-like structures in the renal endothelial cells of SLE patients and in murine lupus models were associated with IFN-α and that *in vitro* culture of cell line cells with IFN-α induced similar intracellular structures [71]. These observations suggested that IFN-α was not only increased in concentration in SLE blood but also that it might have a functional impact on cells and perhaps contribute to disease. Another key observation was first reported in 1990 and has been noted many times subsequently. Therapeutic administration of IFN-α to patients with viral infection or malignancy occasionally results in induction of typical lupus autoantibodies and, in some cases, clinical lupus [4—7, 72], suggesting that given the appropriate genetic background SLE could be induced by IFN-α. Twenty to eighty per cent of patients treated with IFN-α have been noted to develop autoantibodies specific for thyroid or nuclear antigens, including anti-DNA autoantibodies [7]. Clinical syndromes consistent with autoimmune thyroiditis, inflammatory arthritis, and SLE have been observed.

Gene Expression Analysis of SLE PBMCs

Microarray studies of PBMC from lupus patients have been highly informative and served to draw attention to the central role of type I IFN in lupus pathogenesis. The observations of an IFN signature, attributed to type I IFNs, that initially came out of several labs have been replicated many times, with both microarray analysis and real-time quantitative polymerase chain reaction (PCR) confirming increased expression of IFN-inducible genes in many patients with SLE [8–12, 73]. Among the overexpressed mRNAs were some that are well known as targets of type I IFN, such as *MX1*, the *OAS* family, *IFIT1* and others. We determined the relative roles of type I and type II IFN in the "IFN signature" and clearly demonstrated the predominant picture of increased levels of type I IFN-induced genes in lupus PBMC [12]. Moreover, the level of expression of those gene products across a population of lupus patients showed a high level of statistically significant correlation of each type I IFN-induced transcript with the others. This pattern strongly suggested that type I IFN present *in vivo* in many lupus patients was driving a broad gene expression program, very similar to what has been seen in patients who have received either recombinant IFN-α or IFN-β for hepatitis C or multiple sclerosis [74]. Some lupus patients also demonstrated increased expression of genes preferentially regulated by IFN-γ, such as *CXCL9* (monokine induced by gamma interferon; MIG), but they were less frequent than those who demonstrated activation of the type I IFN-induced genes [12].

As noted, the type I IFN family includes multiple IFN-α subtypes, but also includes products of related genes, including IFN-β. To determine which of these type I IFNs was most responsible for expression of the IFN-inducible genes, a functional assay of type I IFN activity in plasma or serum was developed and preferential inhibition of that activity in SLE plasmas by neutralizing antibodies to IFN-α was observed [75]. In contrast, only modest inhibition of type I IFN activity was seen when antibodies to IFN-β or IFN-θ were included in the cultures. The data lead us to suggest that IFN-α represents the major type I IFN active *in vivo* in SLE patients, but it is likely that other isoforms contribute a small fraction of the type I IFN activity that alters immune system function in lupus patients.

The proportion of lupus patients demonstrating the IFN signature has varied from one report to another. In some studies of unselected adult patients less than 50% show this gene expression pattern while a study of pediatric lupus patients, most of whom had recently been diagnosed and many of whom had not yet been treated aggressively, saw the IFN signature in nearly all patients [9, 12]. An association of IFN pathway activation with several clinical features of lupus, particularly a history of renal disease and anemia, has been demonstrated in several cohorts, and a relative under-representation of IFN pathway activation has been seen in patients with antiphospholipid antibodies [8, 12, 73, 76]. In view of the acknowledged diversity of disease manifestations in patients with lupus, along with the fluctuating course of disease, it is not surprising that there are differences in prevalence of the IFN signature in cross-sectional studies of lupus patients.

The demonstration of IFN pathway activation in nearly all pediatric lupus patients, with fewer adult patients showing this pattern, raises a question of whether the production or response to IFN-α is a function of age. In that regard, a study characterizing plasma type I IFN activity in SLE patients and healthy first-degree relatives based on age of the subjects showed similar patterns in female and male patients but distinct levels of activity based on age [77]. Interestingly, the age at which plasma IFN activity was greatest corresponded to the peak reproductive years, with females between 12 and 22 showing higher levels than those younger than 12 or older than 22. Female lupus patients and their first-degree relatives showed the lowest levels of type I IFN activity after age 50. Males showed a similar pattern, but with the peak age range several years older (16–29) than the females. IFN levels were not significantly different between females and males in either the patients or relatives. Taken together these data suggest that age of the pediatric lupus cohort, in and of itself, likely contributed to the higher prevalence of IFN signature among those patients compared to studies of adult patients. The molecular basis of this interesting age-related pattern of IFN pathway activation is not known.

Association of IFN-α with Disease Activity in SLE

The first studies of IFN-α in SLE from the 1970s indicated that the circulating levels of the cytokine were associated with serologic activity of SLE [1]. Consistent with those early observations, microarray and quantitative real-time PCR measurement of IFN-inducible gene expression in recent years has clearly shown a relationship to disease activity [8, 12, 73, 76]. What has been less certain is the degree of fluctuation of IFN pathway activation over time in individual patients. The question remains, do patients who do not demonstrate IFN-inducible gene expression in their PBMC or increased plasma type I IFN activity represent a distinct subset of lupus or might a previously active IFN pathway have evolved toward inactivity as the disease became more chronic. Answering this question will require collection of longitudinal biologic samples on patients with a variety of clinical features, along with carefully documented disease activity data. Characterizing the changes in gene expression, serologic activity or immune function that bridge a discrete increase in IFN pathway activation and a flare in disease activity could provide invaluable novel insights into lupus pathogenesis.

Recent Advances Regarding Mechanisms of IFN Pathway Activation in SLE

Genetic factors

The major histocompatibility complex (MHC) provides the most significant contribution to the genetic variations that result in increased risk of developing SLE [78, 79]. Alleles of MHC-encoded class II molecules are likely involved in the capacity to generate autoantigen-specific immune responses and production of autoantibodies. Additional lupus-associated gene variants have been identified in recent years, and large-scale genome-wide association studies (GWAS) and their follow-up investigations, particularly two seminal collaborative studies published in 2008, have made an important contribution to the characterization of the genes or gene loci that contribute to lupus susceptibility [80—82]. The SNPs identified in the published GWAS datasets represent common variants that are widespread in the population and confer a very modest increased risk of SLE. Additionally, several rare genetic variants that are associated with much greater risk of lupus-like disease have been found [83, 84]. Together, the growing body of data on common variants conferring low risk and rare variants conferring higher risk of lupus are pointing to the most important molecular pathways that are involved in lupus pathogenesis.

When the list of lupus-associated gene variants is considered in the context of their known biologic function, the data collected so far support the essential role of the immune system in disease pathogenesis [82, 85]. The next step facing investigators is to characterize the precise functions that are altered by the DNA variants enriched among lupus patients. This work will require the use of the tools of genetics, molecular biology, and cell biology to gain new understanding of lupus disease mechanisms and identify new therapeutic targets. The data generated so far can be interpreted to indicate several essential components of the disease, all of which reflect genetic factors which might augment likelihood of disease development. They include increased generation or impaired clearance of self-antigens,

particularly nucleic acids, and capacity to activate an autoantigen-specific immune response, including T-cell and B-cell activation and differentiation to plasma cells [13, 14, 82, 85]. The recent data identify a growing number of lupus-associated genetic variants, both common and rare, that impact type I IFN production or response.

A role for genetic variation in the increased production of type I IFN seen in many SLE patients was first suggested by a family study in which plasma type I IFN activity was quantified in SLE patients, their first-degree relatives and unrelated individuals [86]. A significant increase in IFN level was documented in healthy first-degree relatives of patients compared with unrelated subjects. Moreover, high IFN levels tended to cluster in families. The conclusion that increased plasma type I IFN was a heritable trait led to follow-up studies to relate lupus-associated genetic variants to activation of the IFN pathway. The contributions of specific gene variants to activation of that pathway have been defined, aided by subphenotyping of lupus patients based on their autoantibody profile [87—90].

Strong support has been presented for the association of a complex set of SNPs in the interferon regulatory factor 5 (IRF5) gene with SLE [91—94]. Dissection of that association points to a role for particular autoantibody specificities, such as anti-Ro and anti-DNA, in the association with the IRF5 risk haplotype [87]. Moreover, that risk haplotype shows increased association with SLE and with increased type I IFN activity in plasma of lupus patients who express those autoantibodies. Together those data link autoantibodies that target nucleic acids or nucleic acid-binding proteins, IRF5 and type I IFN production. With the knowledge that IRF5 is a signaling molecule downstream from several of the intracellular TLRs, the data support the concept that TLRs activated by DNA and RNA signal through IRF5 to induce type I IFN. Additional lupus-associated gene variants that might modulate the TLR pathway and IFN production include IRF7 and TNFAIP3, encoding A20, an inhibitor of the TLR pathway [95, 96]. An association between the lupus risk variants of PTPN22, a lymphocyte phosphatase, and secreted phosphoprotein 1 (SSP1; osteopontin) and plasma type I IFN activity have also been reported although the exact mechanisms by which those variants impact IFN production have not been elucidated [88, 90]. A relationship between polymorphisms in FCGRIIA, one of the first lupus-associated genes to be identified, and IFN production has not been investigated, although that issue might be a productive research direction in view of the role of that Fc receptor in internalization of immune complexes that induce IFN through TLRs [97—99].

The response to type I IFN depends on sequential interaction of the cytokine with the two chains of

IFNAR, the type I IFN receptor, and activation of a series of kinases, including members of the signal transducer and activator of transcription (STAT) and Janus kinase (Jak) families. Although STAT1 has been most often implicated in signaling downstream of IFNAR, the STAT4 gene has been associated with SLE in several GWAS. A study of SLE patients with the risk allele of STAT4 showed normal or even increased levels of plasma type I IFN activity but a significant association with increased IFN-inducible gene expression in the PBMC of those patients [89]. That is, for a given amount of type I IFN those patients with the STAT4 risk allele appeared to have augmented transcription of genes that are regulated by type I IFN.

In any given lupus-susceptible individual the contribution of genetic variants favoring autoantibody production may dominate over those favoring IFN pathway activation [79]. A study of serum from mothers of babies with the neonatal lupus syndrome in whom anti-Ro antibodies were universally present in high titer showed that those antibodies were accompanied by high IFN activity only in those with clinical features of SLE or Sjogren syndrome [100]. These clinical data further support the concept that there are several prerequisites for development of clinical lupus. Some individuals have a genetic load for activation of the IFN pathway and others have increased capacity to form autoantibodies. It is those individuals who engage both arms of the immune system, the production of IFN by the innate immune response and the production of autoantibodies by the adaptive immune response, who are most likely to develop clinical disease [82].

It is possible that a third important prerequisite, generation or impaired clearance of cell-derived self-antigens, is required to complete the requirements for disease pathogenesis. In that regard, while the lupus-associated genetic variants described above represent common variants that result in a modest increased risk of SLE and it is likely that at least several of those risk variants must act together to achieve the threshold needed to initiate disease, several recently described rare variants are associated with a greater risk of disease and appear to generate increased self-antigen. One of these genes, TREX1, encodes a DNAse and another encodes an RNase [83, 84, 101]. Normal function of those gene products is required to dispose of endogenous nucleic acids that might otherwise stimulate an innate immune response. That concept is supported by studies in mice deficient in TREX1 which show increased production of type I IFN [102].

In summary, the data accumulated so far implicate gene variants involved in the intracellular nucleic acid response TLR pathways, as well as at least one gene that might contribute to signaling through the type I IFN receptor pathway, in susceptibility to SLE. A role for

genetic variation in components of the TLR-independent cellular pathways in induction of type I IFN production is currently under investigation. In addition, rare variants are being identified that could result in generation of stimuli for immune responses that result in increased type I IFN. Additional research will be required to determine whether analysis of lupus-risk allelic variants and/or type I IFN activity or the IFN signature will prove practically useful in predicting increased susceptibility to lupus in individuals otherwise at risk, e.g. sisters of lupus patients.

Inducers of IFN pathway activation in SLE

Ronnblom, Alm and colleagues have made important contributions to characterization of the stimuli for induction of IFN-α in lupus patients through their studies of circulating immune complexes [97, 98, 103–107]. They showed that apoptotic or necrotic cell debris, when associated with SLE serum, could induce IFN. That response was inhibited by blockade of Fc receptors as well as chloroquine. At least one proposed mechanism of chloroquine's effects in vitro is its modification of the acidification of intracellular vesicles, a mechanism that is likely to impact signaling downstream of TLRs engaged by nucleic acid ligands. A similar mechanism is likely to be operative in patients treated with hydroxychloroquine.

In view of the apparent contribution of nucleic acids to these stimulatory complexes as well as the role of DNA- and RNA-binding proteins such as histones or Ro as autoantigens in SLE, TLRs triggered by DNA or RNA became prime candidates for the cell receptors mediating the induction of type I IFN gene transcription and synthesis. TLR9, the receptor for demethylated CpG-rich DNA, was an initial top candidate given the well-documented association between anti-DNA antibodies and lupus disease activity. In addition, a publication from Means et al. showed that DNase treatment of immune complexes isolated from SLE sera ablated the capacity of those complexes to induce downstream gene activation by pDCs [108]. That group did not systematically study the impact of RNase treatment on unfractionated SLE serum or total isolated immune complexes. With the assignment of pDCs as the major source of IFN-α, lupus immune complexes were shown to be active inducers of IFN-α by those cells, while additional recent data implicate TLR9 and FcγRIIa in the induction of IFN-α by some of those complexes [26–28, 106, 107, 109].

The suggested role for the ssRNA-responsive TLR7 pathway in induction of type I IFN and SLE pathogenesis was predicted by human studies which demonstrated the RNase sensitivity of stimulatory immune complexes that induce IFN-α production in vivo, an association between the presence of autoantibodies reactive with RNA-binding proteins and type I IFN inducible gene expression in peripheral blood mononuclear cells, as well as a significant correlation between anti-RNP autoantibody titers and plasma type I IFN functional activity [12, 75, 76, 86]. Moreover, genetic studies of a murine lupus model provide additional strong support for an essential role for the TLR7 pathway in development of autoimmunity and disease in some lupus mice [110, 111]. The Y-linked autoimmune accelerator (Yaa) locus of the male-predominant BXSB murine strain, a model characterized by expansion of the monocyte and dendritic cell populations as well as autoimmunity, has been defined as the translocation of a 4-megabase segment of the pseudoautosomal region of the X chromosome, including the TLR7 and several other less-characterized genes, onto the Y chromosome. The effect of this duplication is increased expression of TLR7 mRNA and protein, at a level approximately twice that observed in non-autoimmune mice, along with a shift in the specificity of the autoantibodies toward a nucleolar, RNA-associated, pattern.

In summary, current data support the TLR7 pathway, triggered typically by ssRNA, as the most important innate immune system molecular pathway responsible for induction of excessive type I IFN production in lupus. TLR7 itself is a target of transcriptional regulation by IFN-α, providing a likely positive amplification loop for innate immune system activation. Our own data indicate that TLR7 expression is increased at sites of organ involvement (Kirou, K., Mavragani, C., and Crow, M.K., unpublished observations). Taken together, recent data from both mouse and human systems implicate TLR7 as an important and possibly central innate immune response receptor and pathway that drives IFN-α production in SLE.

Studies to further elucidate the components of the immunostimulatory lupus immune complexes should provide new information regarding the most potent TLR stimuli. Newly constituted immune complexes including either U1RNA (typically associated with Sm or RNP proteins in the spliceosome particle) or hYRNAs (typically associated with Ro or La proteins) have been shown to activate pDCs [32–34, 112, 113]. In addition to these well-characterized RNA components of particles relevant to SLE, an interesting class of possible ligands might include some siRNAs, or miRNAs that can potentially activate signaling pathways through TLR7. The mechanisms that contribute to the described shift in the specificities of autoantigens targeted over time, from Ro and La early in the pre-disease course, to DNA, to Sm and RNP concurrent with onset of clinical manifestations, have not yet been elucidated [114]. Understanding the basis of this shift in targeting of the immune response may come with further characterization of the details of TLR pathway activation.

The induction of IFN by immune complexes cannot fully account for the increased production of type I IFN seen in SLE, as increased plasma IFN is observed in lupus relatives who do not express lupus autoantibodies, and some lupus patients without measurable anti-RBP or anti-DNA antibodies do have an IFN signature [86]. Yet the documented capacity of chloroquine to inhibit immune-complex-mediated IFN production in vitro and the convincing data showing reduced frequency and severity of lupus flares in patients maintained on hydroxychloroquine support an important contribution of TLR signaling to clinical disease activity [115–117]. Our current view is that induction of IFN-α by nucleic acid containing immune complexes represents an important mechanism of augmenting type I IFN production that is influenced by genetic factors. But it does not fully account for type I IFN produced early in the course of preclinical disease, prior to the development of autoantibodies. Additional studies addressing a role for endogenous or environmental triggers, including virus infection, that act on a susceptible genetic substrate should provide more detailed understanding of lupus pathogenesis.

IFN-α and its Targets as Lupus Biomarkers

The body of data supporting IFN-α as a central pathogenic mediator in lupus has contributed to interest in investigating IFN-α or expression of its gene targets as candidate biomarkers of severe disease, active disease, predictors of disease flare, or for identification of patients likely to respond in clinical trials of agents that might modify the IFN pathway. The rationale for these studies is strong, but the practical considerations involved in quantitatively assessing IFN protein, IFN activity, or IFN-inducible gene expression have been challenging. Moreover, the limited number of registries of lupus patients who have been longitudinally followed with collection of disease activity data using validated instruments and with paired biologic samples stored has provided a hurdle to biomarker discovery in SLE that has not yet been fully overcome, in spite of excellent biomarker candidates for further study.

Based on our experience, real-time PCR quantification of a small panel of IFN-inducible genes in RNA isolated from PBMC lysates provides the most sensitive and specific measure of IFN pathway activation. Data being developed by Kyriakos Kirou and colleagues have demonstrated fluctuations in IFN-inducible gene expression over up to 3 years of follow-up of individual lupus patients preselected to enrich for those with IFN pathway activation based on the presence of anti-RBP autoantibodies [118]. Approximately half of those patients show a parallel pattern of fluctuations in disease activity with changes in IFN score, consistent with prior cross-sectional data. Another group of patients shows an intriguing pattern of IFN-inducible gene expression that peaks months prior to clinical disease flare. Additional data will be required to determine whether the relationship of these two events — IFN score and disease activity score — over time has functional significance or is an arbitrary concurrence of an immunologic response and a clinical presentation. We have utilized an assay of type I IFN functional activity in plasma or serum, based on induction of type I IFN-inducible gene expression in the WISH epithelial cell line that is highly responsive to IFN, to identify those patients with increased production of IFN [75]. This assay has proved highly useful in relating levels of IFN activity to various lupus-associated gene alleles and to monitor IFN production over time. Because this assay measures IFN present in the circulation but does not reflect the contribution of expression of IFN receptors or the efficiency of signaling and new gene transcription downstream of IFNAR in patient cells, it does not fully reflect the many determinants of IFN pathway activation and may be less useful as a biomarker of disease activity than the IFN score, which measures expression of IFN-inducible genes in PBMC. However, any conclusion regarding the optimal experimental approach to assessing IFN pathway activation will await a comprehensive comparison of IFN activity and PBMC IFN score in a well-characterized lupus cohort followed regularly for at least two or three years, allowing for sufficient examples of disease flare to assess the relationship of the candidate biomarkers to disease activity.

With the challenges of accurately measuring IFN pathway activation, some investigators have addressed the hypothesis that some of the chemokines that are induced by type I IFN and other stimuli might serve as a more reliable and useful biomarker of lupus disease activity or future flare [119–121]. Several chemokines are highly induced by type I IFN but many of those are also induced by IFN-γ, other cytokines or by microbial or endogenous stimuli of TLRs. Regardless of issues regarding specificity of chemokines as biomarkers of disease activity, several studies suggest that measuring serum chemokines, or alternatively PBMC chemokine transcripts by PCR, might reflect IFN pathway activation, along with other inflammatory triggers, in a manner helpful in predicting generalized lupus flares or occurrence of lupus nephritis. Most promising is a recent study showing the capacity of a chemokine score to predict future flares of lupus nephritis [121]. Whether an assay is based on IFN activity, its specific gene targets or a broader array of chemokine targets, additional studies based on collaborations among several centers and investigators will be required to arrive at a practically useful biomarker that can aid patient management.

IFN-α as a Therapeutic Target

The case for a central pathogenic role for IFN-α is compelling, based on all of the data and observations that were reviewed. With documentation of increased levels of plasma IFN-α as a heritable risk factor, a correlate of disease activity and a global immune response modifier central to the pathogenesis of SLE (and some other systemic autoimmune diseases), this cytokine is an appropriate target for therapeutic modulation [13, 122]. Current efforts are directed at understanding the impact of currently available therapies on IFN pathway activation and development of new agents to inhibit the pathway, directly or indirectly.

A highly effective approach to inhibiting production of IFN-α is administration of intravenous high-dose methylprednisolone. This treatment virtually ablates the IFN signature based on microarray or real-time PCR data from patient PBMC before and after pulse steroid treatment [9]. The presumed basis of this effect is death of the major producers of IFN-α, pDCs, by the high-dose steroids. Recent data suggest that additional mechanisms that modulate the capacity of IRFs to regulate gene transcription might also contribute to reduced IFN pathway activation by high-dose steroids [123]. Although the mechanism by which other frequently used therapeutic agents, such as mycophenolate mofetil (MMF), might inhibit IFN production are only recently coming under study, our preliminary data suggest that MMF treatment is associated with reduction of the IFN score derived from lupus patient PBMC [124]. A recent study implicates a possible effect of MMF on autophagy and the TLR-independent innate immune system pathway, but additional investigation will be required to pursue that suggestion [125].

Hydroxychloroquine inhibits acidification of intracellular vesicles, and *in vitro* studies clearly document the inhibition of IFN production induced by nucleic-acid-containing immune complexes by chloroquine. It is likely that additional mechanisms account for the positive impact of hydroxychloroquine therapy on reduction of lupus flares, but the strong rationale for its use based on the recent IFN pathway data has suggested that additional approaches to inhibition of TLR activation might be even more productive in SLE. Among the approaches under investigation are inhibition of nucleic-acid-mediated TLR activation by oligonucleotide inhibitors of TLR7, TLR8, and TLR9. This approach is quite attractive if the oligonucleotide inhibitors can be modified to assure adequate delivery to target cells.

The most active area of clinical development of therapeutics targeting the IFN pathway involves current clinical trials of monoclonal antibodies specific for numerous IFN-α subtypes. At least three of these monoclonal agents are in clinical development, each presumably slightly different from the others in the range of subtypes targeted. Very promising pharmacodynamic data have demonstrated inhibition of the IFN signature in PBMC and in skin biopsies from at least some lupus patients treated with one of those antibodies [126]. At this time the blockade of interaction of IFN-α with IFNAR by monoclonal anti-IFN-α antibodies appears to be the most feasible and likely to be effective approach to controlling this important innate immune system pathway. Additional antibodies are available that block the IFN receptor. As the receptor not only binds IFN-α but is also activated by the other type I IFNs, including IFN-β, IFN-θ, and others, it would seem that receptor blockade might produce a more complete blockade of downstream gene expression than the anti-IFN-α antibodies, but might present a greater risk of complications.

The essential role of type I IFN in host defense against virus infection is clearly evident from the obvious effort expended by the collective human genome over evolutionary time to generate a variety of similar but nonidentical type I IFN subtypes. The high impact of this system on generating effective and comprehensive immune responses triggered by virus infection is emphasized when considering the numerous approaches used by viruses to hijack the normal host response. Blockade of any system that holds such responsibility for maintaining the intactness of the host in the setting of a viral assault should only be modulated with great care. Development of any of the therapeutic approaches described will be accompanied by careful monitoring for viral infection. Considering the different options, it would seem that blockade of IFNAR might be most risky, while TLR blockade or inhibition of selective type I IFNs (such as inhibition of IFN-α with monoclonal antibodies) would allow some other routes for production of IFN or response to other isoforms to be available. Although the role of type I IFN in viral host defense has garnered the most extensive investigation, IFN-α is also active in modulating certain hematologic malignancies such as hairy cell leukemia, and the role of the type I IFNs in regulating myeloid differentiation is not fully understood, suggesting a further need for caution as therapeutic trials move forward.

SUMMARY: A MODEL FOR TYPE I IFN EXPRESSION AND ITS PATHOGENIC ROLE IN SLE

Production of IFN-α and overexpression of its gene targets are central features of the altered immune system regulation that characterizes SLE, and a role for nucleic-acid-containing immune complexes in the activation and amplification of the IFN pathway is strongly supported. A more fundamental issue is the potential role of IFN-α as a primary etiologic factor in SLE. Based on the family data demonstrating high plasma IFN levels in first-degree relatives of lupus patients, along with the growing body of data identifying lupus-associated genetic polymorphisms that impact IFN-α production or response, there is a strong case to be made for altered regulation of the IFN pathway as one of the earliest events that ultimately result in clinical lupus disease. We propose that IFN-α likely acts at two points in lupus pathogenesis. First, based on genetic factors, elevated constitutive expression of type I IFN primes the immune system to become more readily activated by either endogenous or environmental innate immune system triggers. Exposure of the primed immune system either to endogenous stimuli, including particles containing self-antigens and associated nucleic acid present in apoptotic debris or endogenous nucleic acids that are not adequately cleared by nucleases, or to exogenous stimuli such as viruses or toxins would then promote low-level secretion of self-directed autoantibodies that form nucleic-acid-containing immune complexes. Second, those immune complexes would amplify the production of IFN-α through activation of TLR pathways. Effective immune system activation by immunostimulatory immune complexes would not only activate the IFN pathway but also generate transcription and production of pro-inflammatory and immunomodulatory gene products that are responsible for promoting inflammation and tissue damage.

Once IFN-α production is sufficiently established such that immunostimulatory autoantibody-containing complexes are available to the immune system, a spectrum of pathogenic mechanisms come into play, some mediated by IFN-α itself and others mediated by distinct gene products induced by the immunostimulatory complexes. It is likely that IFN-α is responsible for many of the altered immune functions that have been described in SLE patients. These include altered antigen-presenting cell capacity, increased Ig class switching, possibly altered central B-cell tolerance resulting in a pro-autoimmune repertoire, inhibition of T-cell proliferation, and increased production of pro-inflammatory cytokines and chemokines. Determinant spreading of the autoimmune response to include the classic SLE autoantibody specificities and complement activation might be promoted by excessive IFN-α. IFN-α might also promote tissue damage, as has been recently demonstrated in a murine lupus model where induction of type I IFN was associated with recruitment of a macrophage cell population to the kidney and a sclerotic pathologic process.

IFN-α might also contribute to clinically important disease manifestations that are not obviously related to immune system function. We propose that IFN-α might alter the metabolism of cells in the central nervous system and contribute to cognitive dysfunction or depression, as has been observed in some patients with hepatitis C infection who have been treated with recombinant IFN-α [127]. The IFN pathway might also contribute to the development of premature atherosclerosis [66, 67, 128]. Our data document local expression of IFN-α transcripts in renal tissue from patients with class IV glomerulonephritis, and recent data from the murine system suggest that type I IFN can promote macrophage infiltration into kidneys and crescent formation [129, 130]. In addition to the direct contribution of IFN-α to disease, additional downstream targets of immune-complex-mediated cell activation will include products triggered through other signaling pathways, including reactive oxygen intermediates in addition to cytokines. It will be exciting to test the validity of the proposed scenarios in the context of the ongoing clinical development programs testing anti-IFN-α monoclonal antibodies in patients with SLE.

References

[1] J.J. Hooks, H.M. Moutsopoulos, S.A. Geis, N.I. Stahl, J.L. Decker, A.L. Notkins, Immune interferon in the circulation of patients with autoimmune disease, N Engl J Med 301 (1979) 5—8.

[2] J.J. Hooks, G.W. Jordan, T. Cupps, H.M. Moutsopoulos, A.S. Fauci, A.L. Notkins, Multiple interferons in the circulation of patients with systemic lupus erythematosus and vasculitis, Arthritis Rheum 25 (1982) 396—400.

[3] O.T. Preble, R.J. Black, R.M. Friedman, J.H. Klippel, J. Vilcek, Systemic lupus erythematosus: presence in human serum of an unusual acid-labile leukocyte interferon, Science 216 (1982) 429—431.

[4] L.E. Ronnblom, G.V. Alm, K.E. Oberg, Possible induction of systemic lupus erythematosus by interferon-alpha treatment in a patient with a malignant carcinoid tumour, J Intern Med 227 (1990) 207—210.

[5] P.J. Schilling, R. Kurzrock, H. Kantarjian, J.U. Gutterman, M. Talpaz, Development of systemic lupus erythematosus after interferon therapy for chronic myelogenous leukemia, Cancer 68 (1991) 1536—1537.

[6] E. Pittau, A. Bogliolo, A. Tinti, Q. Mela, G. Ibba, G. Salis, et al., Development of arthritis and hypothyroidism during alph-interferon therapy for chronic hepatitis C, Clin Exp Rheumatol 15 (1997) 415—419.

[7] C. Gota, L. Calabrese, Induction of clinical autoimmune disease by therapeutic interferon-alpha, Autoimmunity 36 (2003) 511—518.

[8] E.C. Baechler, F.M. Batliwalla, G. Karypis, P.M. Gaffney, W.A. Ortmann, K.J. Espe, et al., Interferon-inducible gene expression signature in peripheral blood cells of patients with severe lupus, Proc Natl Acad Sci USA 100 (2003) 2610—2615.

[9] L. Bennett, A.K. Palucka, E. Arce, V. Cantrell, J. Borvak, J. Banchereau, et al., Interferon and granulopoiesis signatures in systemic lupus erythematosus blood, J Exp Med 197 (2003) 711—723.

[10] M.K. Crow, K.A. Kirou, J. Wohlgemuth, Microarray analysis of interferon-regulated genes in SLE, Autoimmunity 36 (2003) 481—490.

[11] G.M. Han, S.L. Chen, N. Shen, S. Ye, C.D. Bao, Y.Y. Gu, Analysis of gene expression profiles in human systemic lupus erythematosus using oligonucleotide microarray, Genes Immun 4 (2003) 177—186.

[12] K.A. Kirou, C. Lee, S. George, K. Louca, I.G. Papagiannis, M.G. Peterson, et al., Coordinate overexpression of interferon-alpha-induced genes in systemic lupus erythematosus, Arthritis Rheum 520 (2004) 3958—3967.

[13] M.K. Crow, Interferon-α. A new target for therapy in systemic lupus erythematosus? Arthritis Rheum 48 (2003) 2396—2401.

[14] M.K. Crow, Type I interferon in systemic lupus erythematosus, Curr Top Microbiol Immunol 316 (2007) 359—386.

[15] J.W. Fountain, M. Karayiorgou, D. Taruscio, S.L. Graw, A.J. Buckler, D.C. Ward, et al., Genetic and physical map of the interferon region on chromosome 9p, Genomics 14 (1992) 105—112.

[16] J.L. Martal, N.M. Chêne, L.P. Huynh, R.M. L'Haridon, P.B. Reinaud, M.W. Guillomot, et al., IFN-tau: A novel subtype I IFN[1] Structural characteristics, non-ubiquitous expression, structure-function relationships, a pregnancy hormonal embryonic signal and cross-species therapeutic potentialities, Biochimie 80 (1998) 755—777.

[17] C.H. Woelk, S.D. Frost, D.D. Richman, P.E. Higley, S.L. Kosakovsky Pond, Evolution of the interferon alpha gene family in eutherian mammals, Gene 397 (2007) 38—50.

[18] B.J. Barnes, P.A. Moore, P.M. Pitha, Virus-specific activation of a novel interferon regulatory factor 5, results in the induction of distinct interferon alpha genes, J Biol Chem 276 (2001) 23382—23390.

[19] R. Lin, P. Génin, Y. Mamane, J. Hiscott, Selective DNA binding and association with the CREB binding protein coactivator contribute to differential activation of alpha interferon genes by interferon regulatory factors 3 and [7] Mol Cell Biol 20 (2000) 6342—6354.

[20] P. Sheppard, W. Kindsvogel, W. Xu, K. Henderson, S. Schlutsmeyer, T.E. Whitmore, et al., IL-28, IL-29 and their class II cytokine receptor IL-28R, Nat Immunol 4 (2003) 63—68.

[21] S.V. Kotenko, G. Gallagher, V.V. Baurin, A. Lewis-Antes, M. Shen, N.K. Shah, et al., IFN-lambdas mediate antiviral protection through a distinct class II cytokine receptor complex, Nat Immunol 4 (2003) 69—77.

[22] E.M. Coccia, M. Severa, E. Giacomini, D. Monneron, M.E. Remoli, I. Julkunen, et al., Viral infection and Toll-like receptor agonists induce a differential expressión of type I and lambda interferons in human plasmacytoid and monocyte-derived dendritic cells, Eur J Immunol 34 (2004) 796—805.

[23] H. Kumar, T. Kawai, S. Akira, Pathogen recognition in the innate immune response, Biochem J 420 (2009) 1—16.

[24] H. Svensson, A. Johannisson, T. Nikkilä, G.V. Alm, B. Cederblad, The cell surface phenotype of human natural interferon-α producing cells as determined by flow cytometry, Scand J Immunol 44 (1996) 164—172.

[25] F.P. Siegal, N. Kadowaki, M. Shodell, P.A. Fitzgerald-Bocarsly, K. Shah, S. Ho, et al., The nature of the principal type I interferon-producing cells in human blood, Science 284 (1999) 1835—1837.

[26] L. Ronnblom, G.V. Alm, A pivotal role for the natural interferon α-producing cells (plasmacytoid dendritic cells) in the pathogenesis of lupus, J Exp Med 194 (2001) 59—63.

[27] L. Farkas, K. Beiske, F. Lund-Johansen, P. Brandtzaeg, F.L. Jahnsen, Plasmacytoid dendritic cells (natural interferon-alpha/beta—producing cells) accumulate in cutaneous lupus erythematosus lesions, Am J Pathol 159 (2001) 237—243.

[28] C. Richez, K. Yasuda, A.A. Watkins, S. Akira, R. Lafyatis, J.M. van Seventer, et al., TLR4 ligands induce IFN-alpha production by mouse conventional dendritic cells and human

monocytes after IFN-beta priming, J Immunol 182 (2009) 820–828.

[29] N. Onai, A. Obata-Onai, M.A. Schmid, T. Ohteki, D. Jarrossay, M.G. Manz, Identification of clonogenic common Flt3+M-CSFR + plasmacytoid and conventional dendritic cell progenitors in mouse bone marrow, Nat Immunol 8 (2007) 1207–1216.

[30] S.H. Naik, P. Sathe, H.Y. Park, D. Metcalf, A.I. Proietto, A. Dakic, et al., Development of plasmacytoid and conventional dendritic cell subtypes from single precursor cells derived in vitro and in vivo, Nat Immunol 8 (2007) 1217–1226.

[31] S.H. Robbins, T. Walzer, D. Dembélé, C. Thibault, A. Defays, G. Bessou, et al., Novel insights into the relationships between dendritic cell subsets in human and mouse revealed by genome-wide expression profiling, Genome Biology 9 (2008) R17.

[32] F.J. Barrat, T. Meeker, J. Gregorio, J.H. Chan, S. Uematsu, S. Akira, et al., Nucleic acids of mammalian origin can act as endogenous ligands for Toll-like receptors and may promote systemic lupus erythematosus, J Exp Med 202 (2005) 1131–1139.

[33] J. Vollmer, S. Tluk, C. Schmitz, S. Hamm, M. Jurk, A. Forsbach, et al., Immune stimulation mediated by autoantigen binding sites within small nuclear RNAs involves Toll-like receptors 7 and [8] J Exp Med 202 (2005) 1575–1585.

[34] K.M. Kelly, H. Zhuang, D.C. Nacionales, P.O. Scumpia, R. Lyons, J. Akaogi, et al., "Endogenous adjuvant" activity of the RNA components of lupus autoantigens Sm/RNP and Ro [60] Arthritis Rheum 54 (2006) 1557–1567.

[35] R. Lande, J. Gregorio, V. Facchinetti, B. Chatterjee, Y.H. Wang, B. Homey, et al., Plasmacytoid dendritic cells sense self-DNA coupled with antimicrobial peptide, Nature 449 (2007) 564–569.

[36] J. Röck, E. Schneider, J.R. Grün, A. Grützkau, R. Küppers, J. Schmitz, et al., CD303 (BDCA-2) signals in plasmacytoid dendritic cells via a BCR-like signalosome involving Syk, Slp65 and PLCgamma[2] Eur J Immunol 37 (2007) 3564–3575.

[37] W. Cao, L. Zhang, D.B. Rosen, L. Bover, G. Watanabe, M. Bao, et al., BDCA2/Fc epsilon RI gamma complex signals through a novel BCR-like pathway in human plasmacytoid dendritic cells, PLoS Biol 5 (2007) e248.

[38] A. Issacs, J. Lindenmann, Virus interference. [1] The interferon, Proc R Soc Lond B Biol Sci 147 (1957) 258–267.

[39] J. Vilcek, Fifty years of interferon research: aiming at a moving target, Immunity 25 (2006) 343–348.

[40] S.D. Der, A. Zhou, B.R. Williams, R.H. Silverman, Identification of genes differentially regulated by interferon alpha, beta, or gamma using oligonucleotide arrays, Proc Natl Acad Sci USA 95 (1998) 15623–15628.

[41] J.E. Darnell Jr., I.M. Kerr, G.R. Stark, Jak-STAT pathways and transcriptional activation in response to IFNs and other extracellular signaling proteins, Science 264 (1994) 1415–1421.

[42] N.C. Reich, J.E. Darnell Jr., Differential binding of interferon-induced factors to an oligonucleotide that mediates transcriptional activation, Nucleic Acids Res 17 (1989) 3415–3424.

[43] S.A. Veals, C. Schindler, D. Leonard, X.Y. Fu, R. Aebersold, J.E. Darnell Jr., et al., Subunit of an alpha-interferon-responsive transcription factor is related to interferon regulatory factor and Myb families of DNA-binding proteins, Mol Cell Biol 12 (1992) 3315–3324.

[44] C. Daly, N.C. Reich, Characterization of specific DNA-binding factors activated by double-stranded RNA as positive regulators of interferon alpha/beta-stimulated genes, J Biol Chem 279 (1995) 23739–23746.

[45] T. Luft, K.C. Pang, E. Thomas, P. Hertzog, D.N. Hart, J. Trapani, et al., Type I IFNs enhance the terminal differentiation of dendritic cells, J Immunol 161 (1998) 1947–1953.

[46] L.G. Radvanyi, A. Banerjee, M. Weir, H. Messner, Low levels of interferon-alpha induce CD86 (B7.2) expression and accelerates dendritic cell maturation from human peripheral blood mononuclear cells, Scand J Immunol 50 (1999) 499–509.

[47] P. Blanco, A.K. Palucka, M. Gill, V. Pascual, J. Banchereau, Induction of dendritic cell differentiation by IFN-alpha in systemic lupus erythematosus, Science 294 (2001) 1540–1543.

[48] N. Kayagaki, N. Yamaguchi, M. Nakayama, H. Eto, K. Okumura, H. Yagita, Type I interferons (IFNs) regulate tumor necrosis factor-related apoptosis-inducing ligand (TRAIL) expression on human T cells: A novel mechanism for the antitumor effects of type I IFNs, J Exp Med 189 (1999) 1451–1460.

[49] K.A. Kirou, R.K. Vakkalanka, M.J. Butler, M.K. Crow, Induction of Fas ligand-mediated apoptosis by interferon-alpha, Clin Immunol 95 (2000) 218–226.

[50] L. Strandberg, A. Ambrosi, A. Espinosa, L. Ottosson, M.L. Eloranta, W. Zhou, et al., Interferon-alpha induces up-regulation and nuclear translocation of the Ro52 autoantigen as detected by a panel of novel Ro52-specific monoclonal antibodies, J Clin Immunol 28 (2008) 220–231.

[51] V. Brinkmann, T. Geiger, S. Alkan, C.H. Heusser, Interferon alpha increases the frequency of interferon gamma-producing human CD4+ T cells, J Exp Med 178 (1993) 1655–1663.

[52] D. Chakrabarti, B. Hultgren, T.A. Stewart, Ifn-alpha induces autoimmune T cells through induction of intracellular adhesion molecule-1 and B7.[2] J Immunol 157 (1996) 522–528.

[53] J.Y. Djeu, N. Stocks, K. Zoon, G.J. Stanton, T. Timonen, R.B. Herberman, Positive self regulation of cytotoxicity in human natural killer cells by production of interferon upon exposure to influenza and herpes viruses, J Exp Med 156 (1982) 1222–1234.

[54] G. Trinchieri, D. Santoli, Antiviral activity induced by culturing lymphocytes with tumor-derived or virus-transformed cells. Enhancement of natural killer cell activity by interferon and antagonistic inhibition of susceptibility of target cell to lysis, J Exp Med 147 (1978) 1314–1333.

[55] M.J. Aman, T. Tretter, I. Eisenbeis, G. Bug, T. Decker, W.E. Aulitzky, et al., Interferon-alpha stimulates production of interleukin-10 in activated CD4+ T cells and monocytes, Blood 87 (1996) 4731–4736.

[56] L. Ding, E.M. Shevach, IL-10 inhibits mitogen-induced T cell proliferation by selectively inhibiting macrophage costimulatory function, J Immunol 148 (1992) 3133–3139.

[57] R. de Waal Malefyt, H. Yssel, J.E. de Vries, Direct effects of IL-10 on subsets of human CD4+ T cell clones and resting T cells. Specific inhibition of IL-2 production and proliferation, J Immunol 150 (1993) 4754–4765.

[58] K. Taga, H. Mostowski, G. Tosato, Human interleukin-10 can directly inhibit T cell growth, Blood 81 (1993) 2964–2971.

[59] A. Le Bon, G. Schiavoni, G. D'Agostino, I. Gresser, F. Belardelli, D.F. Tough, Type I interferons potently enhance humoral immunity and can promote isotype switching by stimulating dendritic cells in vivo, Immunity 14 (2001) 461–470.

[60] G. Jego, A.K. Palucka, J.P. Blanck, C. Chalouni, V. Pascual, J. Banchereau, Plasmacytoid dendritic cells induce plasma cell differentiation through type I interferon and interleukin [6] Immunity 19 (2003) 225–234.

[61] K. Itoh, S. Hirohata, The role of IL-10 in human B cell activation, proliferation, and differentiation, J Immunol 154 (1995) 4341–4350.

[62] F. Malisan, F. Brière, J.M. Bridon, N. Harindranath, F.C. Mills, E.E. Max, et al., Interleukin-10 induces immunoglobulin G isotype switch recombination in human CD40-activated naive B lymphocytes, J Exp Med 183 (1996) 937–947.

[63] M. Strengell, I. Julkunen, S. Matikainen, IFN-alpha regulates IL-21 and IL-21R expression in human NK and T cells, J Leukoc Biol. 76 (2004) 416–422.

[64] M. Ittah, C. Miceli-Richard, J. Eric Gottenberg, F. Lavie, T. Lazure, N. Ba, et al., B cell-activating factor of the tumor necrosis factor family (BAFF) is expressed under stimulation by interferon in salivary gland epithelial cells in primary Sjogren's syndrome, Arthritis Res Ther 8 (2006) R51.

[65] S. Yurasov, H. Wardemann, J. Hammersen, M. Tsuiji, E. Meffre, V. Pascual, et al., Defective B cell tolerance checkpoints in systemic lupus erythematosus, J Exp Med 201 (2005) 702–711.

[66] M.F. Denny, S. Thacker, H. Mehta, E.C. Somers, T. Dodick, F.J. Barrat, et al., Interferon-alpha promotes abnormal vasculogenesis in lupus: a potential pathway for premature atherosclerosis, Blood 110 (2007) 2907–2915.

[67] M.F. Denny, S. Yalavarthi, W. Zhao, S.G. Thacker, M. Anderson, A.R. Sandy, et al., A Distinct Subset of Proinflammatory Neutrophils Isolated from Patients with Systemic Lupus Erythematosus Induces Vascular Damage and Synthesizes Type I IFNs, J Immunol (2010). Feb [17] [Epub ahead of print].

[68] I. Arrue, A. Saiz, P.L. Ortiz-Romero, J.L. Rodríguez-Peralto, Lupus-like reaction to interferon at the injection site: report of five cases, J Cutan Pathol 34 (2007) 18–21.

[69] S.N. Shi, S.F. Feng, Y.M. Wen, L.F. He, Y.X. Huang, Serum interferon in systemic lupus erythematosus, Br J Dermatol 117 (1987) 155–159.

[70] A.M. Yee, Y.K. Yip, H.D. Fischer, J.P. Buyon, Serum activity that confers acid lability to alpha-interferon in systemic lupus erythematosus: its association with disease activity and its independence from circulating alpha-interferon, Arthritis Rheum 33 (1990) 563–568.

[71] S.A. Rich, Human lupus inclusions and interferon, Science 213 (1981) 772–775.

[72] U.B. Wandl, M. Nagel-Hiemke, D. May, E. Kreuzfelder, O. Kloke, M. Kranzhoff, et al., Lupus-like autoimmune disease induced by interferon therapy for myeloproliferative disorders, Clin Immunol Immunopathol 65 (1992) 70–74.

[73] X. Feng, H. Wu, J.M. Grossman, P. Hanvivadhanakul, J.D. FitzGerald, G.S. Park, et al., Association of increased interferon-inducible gene expression with disease activity and lupus nephritis in patients with systemic lupus erythematosus, Arthritis Rheum 54 (2006) 2951–2962.

[74] M.K. Singh, T.F. Scott, W.A. LaFramboise, F.Z. Hu, J.C. Post, G.D. Ehrlich, Gene expression changes in peripheral blood mononuclear cells from multiple sclerosis patients undergoing beta-interferon therapy, J Neurol Sci 258 (2007) 52–59.

[75] J. Hua, K. Kirou, C. Lee, M.K. Crow, Functional assay of type I interferon in systemic lupus erythematosus plasma and association with anti-RNA binding protein autoantibodies, Arthritis Rheum 54 (2006) 1906–1916.

[76] K.A. Kirou, C. Lee, S. George, K. Louca, M.G. Peterson, M.K. Crow, Interferon-alpha pathway activation identifies a subgroup of systemic lupus erythematosus patients with distinct serologic features and active disease, Arthritis Rheum 52 (2005) 1491–1503.

[77] T.B. Niewold, J.E. Adler, S.B. Glenn, T.J. Lehman, J.B. Harley, M.K. Crow, Age- and sex-related patterns of serum interferon-alpha activity in lupus families, Arthritis Rheum 58 (2008) 2113–2119.

[78] I.T. Harley, K.M. Kaufman, C.D. Langefeld, J.B. Harley, J.A. Kelly, Genetic susceptibility to SLE: new insights from fine mapping and genome-wide association studies, Nat Rev Genet 10 (2009) 285–290.

[79] P.S. Ramos, J.A. Kelly, C. Gray-McGuire, G.R. Bruner, A.N. Leiran, C.M. Meyer, et al., Familial aggregation and linkage analysis of autoantibody traits in pedigrees multiplex for systemic lupus erythematosus, Genes Immun 7 (2006) 417–432.

[80] J.B. Harley, M.E. Alarcón-Riquelme, L.A. Criswell, C.O. Jacob, R.P. Kimberly, K.L. Moser, et al., International Consortium for Systemic Lupus Erythematosus Genetics (SLEGEN), Genome-wide association scan in women with systemic lupus erythematosus identifies susceptibility variants in ITGAM, PXK, KIAA1542 and other loci, Nat Genet 40 (2008) 204–210.

[81] G. Hom, R.R. Graham, B. Modrek, K.E. Taylor, W. Ortmann, S. Garnier, et al., Association of systemic lupus erythematosus with C8orf13-BLK and ITGAM-ITGAX, N Engl J Med 358 (2008) 900–909.

[82] M.K. Crow, Developments in the clinical understanding of lupus, Arth Res Ther 11 (2009) 245.

[83] G. Rice, W.G. Newman, J. Dean, T. Patrick, R. Parmar, K. Flintoff, et al., Heterozygous mutations in TREX1 cause familial chilblain lupus and dominant Aicardi-Goutieres syndrome, Am J Hum Genet 80 (2007) 811–815.

[84] Y.J. Crow, J. Rehwinkel, Aicardi-Goutieres syndrome and related phenotypes: linking nucleic acid metabolism with autoimmunity, Hum Mol Genet 18 (2009) R130–136.

[85] M.K. Crow, Collaboration, genetic associations, and lupus erythematosus, New Engl J Med 358 (2008) 956–961.

[86] T.B. Niewold, J. Hua, T.J. Lehman, J.B. Harley, M.K. Crow, High serum IFN-alpha activity is a heritable risk factor for systemic lupus erythematosus, Genes Immun 8 (2007) 492–502.

[87] T.B. Niewold, J.A. Kelly, M.H. Flesch, L.R. Espinoza, J.B. Harley, M.K. Crow, Association of the IRF5 risk haplotype with high serum interferon-alpha activity in systemic lupus erythematosus patients, Arthritis Rheum 58 (2008) 2481–2487.

[88] S.N. Kariuki, M.K. Crow, T.B. Niewold, The PTPN22 C1858T polymorphism is associated with skewing of cytokine profiles toward high IFN-alpha activity and low tumor necrosis factor-alpha levels in patients with lupus, Arthritis Rheum 58 (2008) 2818–2823.

[89] S.N. Kariuki, K.A. Kirou, E.J. MacDermott, L. Barillas-Arias, M.K. Crow, T.B. Niewold, Cutting edge: Autoimmune disease risk variant of STAT4 confers increased sensitivity to IFN-alpha in lupus patients in vivo, J Immunol 182 (2009) 34–38.

[90] S.N. Kariuki, J.G. Moore, K.A. Kirou, M.K. Crow, T.O. Utset, T.B. Niewold, Age- and gender-specific modulation of serum osteopontin and interferon-α by osteopontin genotype in systemic lupus erythematosus, Genes and Immunity 10 (2009) 487–494.

[91] R.R. Graham, S.V. Kozyrev, E.C. Baechler, M.V. Reddy, R.M. Plenge, J.W. Bauer, et al., A common haplotype of interferon regulatory factor 5 (IRF5) regulates splicing and expression and is associated with increased risk of systemic lupus erythematosus, Nat Genet 38 (2006) 550–555.

[92] R.R. Graham, C. Kyogoku, S. Sigurdsson, I.A. Vlasova, L.R. Davies, E.C. Baechler, et al., Three functional variants of IFN regulatory factor 5 (IRF5) define risk and protective haplotypes for human lupus, Proc Natl Acad Sci USA 104 (2007) 6758–6763.

[93] S. Sigurdsson, H.H. Göring, G. Kristjansdottir, L. Milani, G. Nordmark, J.K. Sandling, et al., Comprehensive evaluation of the genetic variants of interferon regulatory factor 5 (IRF5) reveals a novel 5 bp length polymorphism as strong risk factor for systemic lupus erythematosus, Hum Mol Genet 17 (2008) 872–881.

[94] D. Feng, R.C. Stone, M.L. Eloranta, N. Sangster-Guity, G. Nordmark, S. Sigurdsson, et al., Genetic variants and disease-associated factors contribute to enhanced interferon regulatory factor 5 expression in blood cells of patients with systemic lupus erythematosus, Arthritis Rheum 62 (2010) 562–573.

[95] R. Salloum, B.S. Franek, S.N. Kariuki, L. Rhee, R.A. Mikolaitis, M. Jolly, et al., Genetic variation at the IRF7/PHRF1 locus is associated with autoantibody profile and serum interferon-alpha activity in lupus patients, Arthritis Rheum 62 (2010) 553–561.

[96] S.L. Musone, K.E. Taylor, T.T. Lu, J. Nititham, R.C. Ferreira, W. Ortmann, et al., Multiple polymorphisms in the TNFAIP3 region are independently associated with systemic lupus erythematosus, Nat Genet 40 (2008) 1062–1064.

[97] T. Lövgren, M.L. Eloranta, U. Båve, G.V. Alm, L. Rönnblom, Induction of interferon-alpha production in plasmacytoid dendritic cells by immune complexes containing nucleic acid releases by necrotic or late apoptotic cells and lupus IgG, Arthritis Rheum 50 (2004) 1861–1872.

[98] M. Magnusson, S. Magnusson, H. Vallin, L. Rönnblom, G.V. Alm, Importance of CpG dinucleotides in activation of natural IFN-alpha-producing cells by a lupus-related oligo-deoxynucleotide, Scand J Immunol 54 (2001) 543–550.

[99] J.E. Salmon, S. Millard, L.A. Schachter, F.C. Arnett, E.M. Ginzler, M.F. Gourley, et al., Fc gamma RIIA alleles are heritable risk factors for lupus nephritis in African Americans, J Clin Invest 97 (1996) 1348–1354.

[100] T.B. Niewold, T.L. Rivera, J.P. Buyon, M.K. Crow, Serum type I interferon activity is dependent on maternal diagnosis in anti-SSA/Ro-positive mothers of children with neonatal lupus, Arthritis Rheum 58 (2008) 541–546.

[101] F.W. Perrino, S. Harvey, N.M. Shaban, T. Hollis, RNaseH2 mutants that cause Aicardi-Goutieres syndrome are active nucleases, J Mol Med 87 (2009) 25–30.

[102] D.B. Stetson, J.S. Ko, T. Heidmann, R. Medzhitov, Trex1 prevents cell-intrinsic initiation of autoimmunity, Cell 134 (2008) 569–571.

[103] H. Vallin, S. Blomberg, G.V. Alm, B. Cederblad, L. Rönnblom, Patients with systemic lupus erythematosus (SLE) have a circulating inducer of interferon-alpha (IFN-α) production acting on leukocytes resembling immature dendritic cells, Clin Exp Immunol 115 (1999) 196–202.

[104] H. Vallin, A. Perers, G.V. Alm, L. Rönnblom, Anti-double-stranded DNA antibodies and immunostimulatory plasmid DNA in combination mimic the endogenous IFN-alpha inducer in systemic lupus erythematosus, J Immunol 163 (1999) 6306–6313.

[105] U. Båve, G.V. Alm, L. Rönnblom, The combination of apoptotic U937 cells and lupus IgG is a potent IFN-alpha inducer, J Immunol 165 (2000) 3519–3526.

[106] U. Båve, H. Vallin, G.V. Alm, L. Rönnblom, Activation of natural interferon-alpha producing cells by apoptotic U937 cells combined with lupus IgG and its regulation by cytokines, J Autoimmun 17 (2001) 71–80.

[107] L. Rönnblom, G.V. Alm, A pivotal role for the natural interferon alpha-producing cells (plasmacytoid dendritic cells) in the pathogenesis of lupus, J Exp Med 194 (2001) F59–63.

[108] T.K. Means, E. Latz, F. Hayashi, M.R. Murali, D.T. Golenbock, A.D. Luster, Human lupus autoantibody-DNA complexes activate DCs through cooperation of CD32 and TLR[9] J Clin Invest 115 (2005) 407–417.

[109] S. Blomberg, M.L. Eloranta, M. Magnusson, G.V. Alm, L. Rönnblom, Expression of the markers BDCA-2 and BDCA-4 and production of interferon-alpha by plasmacytoid dendritic cells in systemic lupus erythematosus, Arthritis Rheum 48 (2003) 2524–2532.

[110] P. Pisitkun, J.A. Deane, M.J. Difilippantonio, T. Tarasenko, A.B. Satterthwaite, S. Bolland, Autoreactive B cell responses to RNA-related antigens due to TLR7 gene duplication, Science 312 (2006) 1669–1672.

[111] S. Subramanian, K. Tus, Q.Z. Li, A. Wang, X.H. Tian, J. Zhou, et al., A Tlr7 translocation accelerates systemic autoimmunity in murine lupus, Proc Natl Acad Sci USA 103 (2006) 9970–9975.

[112] R.W. Hoffman, T. Gazitt, M.F. Foecking, R.A. Ortmann, M. Misfeldt, R. Jorgenson, et al., U1 RNA induces innate immunity signaling, Arthritis Rheum 50 (2004) 2891–2899.

[113] E. Savarese, O.W. Chae, S. Trowitzsch, G. Weber, B. Kastner, S. Akira, et al., U1 small nuclear ribonucleoprotein immune complexes induce type I interferon in plasmacytoid dendritic cells through TLR[7] Blood 107 (2006) 3229–3234.

[114] M.R. Arbuckle, M.T. McClain, M.V. Rubertone, R.H. Scofield, G.J. Dennis, J.A. James, et al., Development of autoantibodies before the clinical onset of systemic lupus erythematosus, N Engl J Med 349 (2003) 1526–1533.

[115] The Canadian Hydroxychloroquine Study Group, A random-ized study of the effect of withdrawing hydroxychloroquine sulfate in systemic lupus erythematosus, N Engl J Med 324 (1991) 150–154.

[116] E. Tsakonas, L. Joseph, J.M. Esdaile, D. Choquette, J.L. Senécal, A. Cividino, et al., A long-term study of hydroxychloroquine withdrawal on exacerbations in systemic lupus erythematosus. The Canadian Hydroxychloroquine Study Group, Lupus 7 (1998) 80–85.

[117] I.M. Meinão, E.I. Sato, L.E. Andrade, M.B. Ferraz, E. Atra, Controlled trial with chloroquine diphosphate in systemic lupus erythematosus, Lupus 5 (1996) 237–241.

[118] L. Barillas-Arias, E.J. MacDermott, R. Duculan, A.G. Santiago, J. Gordon, P. Cole, et al., Longitudinal prospective study of Type I interferon pathway activation as a biomarker of disease activity in patients with systemic lupus erythematosus(SLE) - Interim analysis, Arthritis Rheum 56 (2007) 4245.

[119] J.W. Bauer, E.C. Baechler, M. Petri, F.M. Batliwalla, D. Crawford, W.A. Ortmann, et al., Elevated serum levels of interferon-regulated chemokines are biomarkers for active human systemic lupus erythematosus, PLoS Med 3 (2006) e491.

[120] Q. Fu, X. Chen, H. Cui, Y. Guo, J. Chen, N. Shen, et al., Asso-ciation of elevated transcript levels of interferon-inducible chemokines with disease activity and organ damage in systemic lupus erythematosus patients, Arthritis Res Ther 10 (2008) R112.

[121] J.W. Bauer, M. Petri, F.M. Batliwalla, T. Koeuth, J. Wilson, C. Slattery, et al., Interferon-regulated chemokines as biomarkers of systemic lupus erythematosus disease activity: a validation study, Arthritis Rheum 60 (2009) 3098–3107.

[122] T.A. Stewart, Neutralizing interferon alpha as a therapeutic approach to autoimmune diseases, Cytokine Growth Factor Rev 14 (2003) 139–154.

[123] Y. Chinenov, I. Rogatsky, Glucocorticoids and the innate immune system: crosstalk with the toll-like receptor signaling network, Mol Cell Endocrinol 275 (2007) 30–42.

[124] S. Gold, J. Cherian, N. Blank, A. Santiago, M.K. Crow, K.A. Kirou, Type I interferon pathway activation parallels therapeutic response in patients with SLE, Arthritis Rheum 60 (2009) S338–339.

[125] B. Chaigne-Delalande, G. Guidicelli, L. Couzi, P. Merville, W. Mahfouf, S. Bouchet, et al., The immunosuppressor myco-phenolic acid kills activated lymphocytes by inducing a nonclassical actin-dependent necrotic signal, J Immunol 181 (2008) 7630–7638.

[126] Y. Yao, L. Richman, B.W. Higgs, C.A. Morehouse, M. de los Reyes, P. Brohawn, et al., Neutralization of interferon-alpha/beta-inducible genes and downstream effect in a phase I trial of an anti-interferon-alpha monoclonal antibody in systemic lupus erythematosus, Arthritis Rheum 60 (2009) 1785–1796.

[127] A. Reichenberg, J.M. Gorman, D.T. Dieterich, Interferon-induced depression and cognitive impairment in hepatitis C virus patients: a 72 week prospective study, AIDS 19 (2005) S174–178.

[128] M.J. Roman, B.A. Shanker, A. Davis, M.D. Lockshin, L. Sammaritano, R. Simantov, et al., Prevalence and correlates of accelerated atherosclerosis in systemic lupus erythematosus, N Engl J Med 349 (2003) 2399–2406.

[129] A.M. Fairhurst, C. Xie, Y. Fu, A. Wang, C. Boudreaux, X.J. Zhou, et al., Type I interferons produced by resident renal cells may promote end-organ disease in autoantibody-mediated glomerulonephritis, J Immunol 183 (2009) 6831–6838.

[130] A. Triantafyllopoulou, C.W. Franzke, S.V. Seshan, G. Perino, G.D. Kalliolias, M. Ramanujam, et al., Proliferative lesions and metalloproteinase activity in murine lupus nephritis mediated by type I interferons and macrophages, Proc Natl Acad Sci U S A 107 (2010) 3012–3017.

Immune Complexes in Systemic Lupus Erythematosus

Mark H. Wener

INTRODUCTION

Systemic lupus erythematosus (SLE) is the prototype of autoimmune diseases that are principally mediated by immune complexes. The evidence for tissue injury by immune complexes in SLE is marshaled from many observations. The recognition of a number of autoantibodies in this disease served as an important initial finding. The presence of immunoglobulin and complement deposits in target organs and the characteristics of pathologic changes allowed the comparison of SLE to experimental models of serum sickness [1]. Further evidence of involvement of immune complexes in the genesis of tissue lesions was provided by the identification of specific antibodies in the glomeruli of patients with SLE [2].

Injury to specific target organs in immune-complex-mediated diseases results from the presence of antigen—antibody complexes in tissue. The immune complex deposits activate complement and interact with cell receptors leading to release of cytokines from cells and culminating in tissue damage and organ dysfunction. Immune complexes containing nucleic acids and other antigens that can activate the innate immune system may play a central role in upregulating the inflammatory response and autoimmunity, in part by augmenting production of type I interferon by dendritic cells. The antigen—antibody deposits in tissue may arise from deposition in tissues of circulating immune complexes, from formation of immune complexes at the site of their existence in tissues, or a combination of both. Deposition of ICs within tissues is responsible for glomerulonephritis and vasculitis and probably also for arthritis and some forms of cutaneous lupus.

CHARACTERISTICS OF IMMUNE COMPLEXES

Immune complexes are composed of antigens and antibodies. The chemical features and physical characteristics of antigens can be quite variable, ranging from simple molecules to complex macromolecules or tissue components. Antibodies in immune complexes may belong to any of the classes of immunoglobulins and endow the complexes with biologic properties unique to the particular class of immunoglobulin. In addition, the nature of the bond between the antigen and antibody influences the biologic properties of the resultant immune complexes.

Antigens

Antigens may be proteins, polysaccharides, nucleic acids, lipoproteins, phospholipids, or other chemicals, and can be simple or complex. The number of antigenic determinants on a given molecule defines its valence for the interaction with specific antibodies. Small hapten antigens typically contain a single antigenic determinant. Proteins or other substances of larger molecular size may possess multiple epitopes. The valence of antigens profoundly alters the nature of immune complexes formed upon interaction with antibodies.

The chemical features of an antigen may significantly influence the biologic properties of immune complexes, independent of the properties provided by antibodies in the complexes. For example, activation of Toll-like receptors that recognize DNA and RNA plays a central role in activation of the immune response, and cell activation by nucleic acids is enhanced when they are in immune complexes [3]. Highly cationic (positively charged) antigens can interact with fixed negative charges in glomeruli and thereby influence the localization of immune complexes in these structures. The clearance of immune complexes from the circulation can be altered by the nature of antigens present in immune complexes. For example, under conditions in which circulating immune complexes containing the mucoprotein orosomucoid remain in the circulation, similar immune complexes prepared with asialo-orosomucoid (with exposed galactose) are removed rapidly from the circulation by the interaction of galactose with galactose receptors on hepatocytes [4]. In a similar manner, large-molecular-weight DNA is quickly removed from the circulation by the liver because of the presence of receptors for DNA on non-parenchymal cells [5]. As a consequence, immune complexes containing such DNA are also removed faster from circulation than immune aggregates with comparable numbers of antibody molecules and protein antigens [6]. These animal studies emphasize the potential role of antigens on the biologic properties of immune complexes.

Whereas in the classic serum sickness immune complex model, the antigen in the immune complex bears little relevance to the resultant pathology, in SLE and other human immune complex diseases the antigen constituents within the immune complex could influence the pattern of clinical sequelae. For example, immune complexes containing lipids could augment the risk for cardiovascular disease. Uptake of lipids and formation of foam cells are enhanced if the lipids are bound to IgG, i.e., as an immune complex [7]. This uptake is mediated, at least in part, by Fcγ receptors (FcγR) [8]. The FcγR-mediated uptake of lipids can lead to formation and activation of the lipid-laden macrophage [9], including promotion of an atherogenic, oxidative phenotype of the macrophage [10]. Serum from patients with SLE has been reported to contain LDL-bearing immune complexes which enhanced uptake of lipids into macrophages [11].

Some antiphospholipid antibodies bind to other families of lipids, including oxidized low-density lipoproteins (LDL) [12]. Immune complexes containing antibodies to lipoproteins known to be associated with atherogenesis could play a role in development of coronary artery disease. Hasunama *et al.* found that the anti-cardiolipin cofactor β2-glycoprotein I (β2-GPI) bound preferentially to oxidized plasma lipoproteins, i.e., oxidized (ox)VLDL, oxLDL, and oxHDL, in comparison with the native forms of the lipoproteins [13]. Antibodies to β2-GPI bound to the β2-glycoprotein I—oxLDL complex. Whereas binding of β2-GPI to oxLDL inhibited the uptake of oxLDL by macrophages, the uptake was enhanced in the presence of immune complexes containing antiβ2-GPI and β2-GPI-oxLDL. Uptake of oxLDL by macrophages predisposes to the formation of foam cells leading to intimal disease and atherosclerosis, thus the enhanced uptake caused by lipoprotein-containing immune complexes could contribute to accelerated atherosclerosis [14] as well as immune complex disease.

Evidence has accumulated that immune complexes containing nucleosomes are important in the pathogenesis of lupus nephritis, and the antigen within those complexes may play a critical role leading to localization of immune complexes in the glomerular basement membrane. Renal biopsies from patients with diffuse proliferative lupus nephritis and membranous lupus nephritis contain nucleosomes and histones [15]. The DNA found in the blood of patients with SLE is probably nucleosomal [16]. Furthermore, immune complexes containing nucleosomes are found circulating in the MRL/lpr lupus model mice [17]. It has been suggested that nucleosomes also interact with heparan sulfate proteoglycans in the kidney, promoting glomerular deposition of circulating nucleosome—antinucleosome immune complexes, or *in situ* formation of immune complexes after initial deposition of nucleosomes in the glomerulus [18]. Heparin treatment of MRL/lpr lupus mice led to less severe nephritis, compared with untreated MRL mice, perhaps by competing with glomerular basement membrane structural heparan sulfate for binding to nucleosomes [19]. Collectively, these data suggest that nucleosomes within immune complexes of lupus patients could influence the deposition of those immune complexes, and again demonstrate that the antigen within an immune complex influences the pathogenicity of an immune complex, in contrast to the minimal role of antigen in immune complexes in classical serum sickness.

Antibodies

IgG and monomeric IgA antibody molecules have two combining sites and hence a valence of two. IgM molecules exhibit a valence of 10 for small antigenic molecules or a valence of 5 for large antigenic molecules. The distance between the two antibody-combining sites can also alter the nature of the formed complexes. In a given antibody molecule, this distance is somewhat variable, depending on the flexibility of the two Fab arms at the hinge region. The flexibility of the hinge region depends substantially on the isotype of the immunoglobulin, with human IgG3 having a long, flexible hinge region, and IgG2 having a shorter, more constrained hinge region. The effector functions of the antibody molecules are localized in the constant regions of the two heavy chains. The important effector functions in relation to immune complexes are complement activation and interaction with specific cell receptors, and depend on the isotype of the antibody in the immune complex.

Nature of the Antigen–Antibody Complex

The interaction of antigens and antibodies leads to the formation of immune complexes. A variety of immune complexes can be formed, ranging from the union of one antigen molecule and one antibody molecule to interactions of many molecules of each of the reactants. The *lattice* of immune complexes is defined as the number of antigen and the number of antibody molecules in a given immune complex. The lattice of immune complexes influences the biologic properties of the complexes.

The valence of antibodies and the valence of antigens, among other variables, dictate the nature of the formed lattice. A monovalent antigen can only form Ag_2Ab_1 or Ag_1Ab_1 complexes, and larger lattices and immune precipitates cannot be formed. Multivalent antigens may form increasingly larger complexes with high degrees of lattice formation, and immune precipitates may form. When an antigen is polyvalent with an appropriately spaced, repeating antigenic determinant, both antibody-combining sites of a bivalent antibody molecule preferentially bind to the same antigenic molecule because of the formation of two antigen–antibody bonds between the reactants. This type of reaction is termed *monogamous bivalent binding* and does not favor the formation of immune complexes with many antigen and many antibody molecules. Antibodies to DNA, for example, may bind with monogamous bivalent interactions to DNA of appropriate size. The strand of DNA bound with high-avidity antibodies to DNA by monogamous bivalent binding is protected from degradation by DNase. The protected length of DNA corresponded to the distance between the two combining sites of an IgG molecule [20]. In contrast, low-avidity anti-DNA that bound to a DNA molecule with only a single antigen-binding site allowed complete degradation of DNA by DNases. Resistance of DNA to DNase digestion mediated by high-avidity anti-DNA antibodies provides a mechanism for persistence and pathogenicity of DNA-containing immune complexes.

Another variable in the nature of antigen–antibody interaction is the molar ratio of the reactants; this is well illustrated by the classical precipitin curves (Figure 19.1). When an increasing amount of antigen is added to a constant amount of antibody, an increasing amount of precipitate is formed under conditions of antibody excess. The maximum amount of precipitate is formed at the point of equivalence. At this ratio of the interactants, free antigen and free antibody are not detectable in the supernatant. The addition of antigen beyond the point of equivalence leads to the formation of soluble immune complexes, and the amount of formed precipitate decreases. At low degrees of antigen excess the soluble complexes have relatively large lattices ($>Ag_2Ab_2$); these large-latticed immune

complexes can activate complement, interact with $Fc\gamma R$, and deposit in tissues (discussed further below). With the addition of higher degrees of antigen excess, the soluble immune complexes become smaller. With very high degrees of antigen excess, only small-latticed immune complexes (e.g. Ag_1Ab_1, Ag_2Ab_2, or Ag_2Ab_1) are formed, depending on the characteristics of the antigen. These complexes are inefficient in complement activation and interactions with $Fc\gamma R$ and do not deposit in tissues. Furthermore, the absolute concentration of antigen and antibody influences lattice formation independent of the antigen–antibody molar ratio. Very low concentrations of antigens and antibodies tend to form more small-latticed complexes than are formed at higher antigen and antibody concentrations at the same antigen–antibody molar ratio. Finally, the association constant between the antigen and antibody influences the lattice formed by these interactants, with low-affinity antibodies forming smaller lattices than high-affinity antibodies. With increasing lattice formation, biologic properties such as

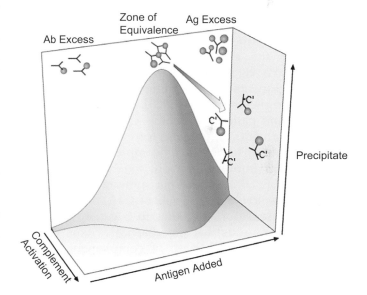

FIGURE 19.1 Schematic representation of immune precipitation and the effect of complement on this process. The classical precipitin curve demonstrates the effect of adding increasing amounts of antigen to a constant amount of antibody. With increasing amounts of antigen, the amount of precipitate increases until the zone of equivalence. Upon the addition of even more antigen, the amount of precipitate decreases, owing to the formation of soluble immune complexes. This relationship is shown by the two-dimensional precipitin curve shown as the 'back' of this schematic. This three-dimensional schematic also illustrates the effect of complement on the amount of immune precipitate. Increasing amounts of added complement prevent formation of immune precipitates and large immune complexes via the classical pathway, and the alternate pathway solubilizes pre-formed immune precipitates. The complement protein fragment C3b interferes with formation of the immune complex lattice, or disrupts the lattice that has been formed. The large arrow represents solubilization of immune complexes by increasing amounts of added complement.

complement activation and interaction with cell receptors increase. Interactions between other parts of antibody molecules, such as the Fc regions, are thought to contribute to non-specific aggregation, cryoprecipitability, and to formation of precipitates of the immune complexes. These events are slower than the initial antigen—antibody interaction.

Complement components have significant influence on antigen—antibody complexes. First, when small amounts of immune complexes at equivalence are formed in the presence of complement-replete serum, precipitation is prevented and only soluble complexes are generated (Figure 19.1). If complement components are inactivated, then immune precipitates are formed under the same conditions. In this prevention of precipitation, the classical complement pathway plays a significant role [21]. Second, small quantities of already-formed immune precipitates can be converted to soluble immune complexes by activation of the alternative pathway of the complement system [22]. The solubilized immune complexes contain covalently bound C3b and no longer are able to form immune precipitates. These observations in conjunction with the observed increased prevalence of immune-complex-mediated diseases in patients with deficiencies of the early components of the classical pathway have suggested an important role for complement components in retarding the deposition of immune complexes in tissues [23]. Hypocomplementemic sera from patients with SLE, in comparison with normal sera, fail to prevent formation of immune precipitates. This defective complement-dependent prevention of immune precipitation is seen in early cases of SLE. Prevention of immune precipitation correlates positively with levels of C4A, and inversely with the presence of antibodies to C1q [24, 25]. Thus, the possibility exists that the absence of complement components or the decrease of complement components due to activation by immune complexes may contribute to formation of immune complexes with increased pathogenic potential in tissues or in the circulation. On the other hand, inflammatory cells may be recruited to the sites of immune deposits less efficiently in the presence of complement deficiency. These possibilities and the association between genetic abnormalities of the complement system and SLE are considered further in Chapters 2 and 20.

BIOLOGIC PROPERTIES OF IMMUNE COMPLEXES

Among the many properties of immune complexes, activation of the complement systems, interaction with cell receptors, and deposition in tissues are most relevant to SLE pathogenesis. Interactions with cells are mediated largely through Fc and complement receptors present on a variety of cell types. The degree of these biologic activities depends in part on the class or subclass of antibodies in the immune complexes and on the degree of lattice formation in the complexes. Immune complexes, especially those containing DNA or other nucleic acid antigens, play a central role as initiators and amplifiers of the abnormal inflammatory and immune response characteristic of SLE.

Complement Activation by Immune Complexes

Activation of complement by immune complexes may proceed through the classical or the alternative pathways. Antibodies of the IgG and IgM class activate complement by the classical pathway, if sufficient lattice formation is present. The IgG1, IgG2, and IgG3 subclasses are effective in complement activation, and the IgG4 subclass is ineffective. Activation of complement requires that the IgG-containing immune complexes form a multivalent ligand for interaction with C1. In general, immune complexes become increasingly more effective in binding C1 and activation of the complement system with increasing numbers of IgG molecules [26].

Complement activation by immune complexes has several consequences in relation to disease processes. For example, the generation of chemotactic factors leads to an influx of phagocytic cells to tissues that contain deposits of immune complexes. The phagocytic cells ingest the immune deposits and during this process release enzymes and other factors that contribute to the damaging effects of inflammation. In addition, the Fc receptor-mediated interaction of IgG-containing immune complexes with phagocytic cells is enhanced by the presence of complement components on immune complexes and complement receptors on these cells. The binding of complement components to the immune deposits contributes to tissue damage and organ malfunction. For instance, complement components of the membrane-attack complex have been identified in skin and in glomeruli of patients with SLE [27, 28]. Studies in experimental animals have shown that the binding of complement components to the subepithelial deposits in glomeruli initiates the proteinuria associated with this lesion [29]. There is, however, a critical role of FcγRs in the mediation of experimental murine immune complex nephritis, as will be discussed below and in Chapter 21.

The potential role of complement in mediating immune complex disease has been challenged by experiments involving animals without functional complement. In mice lacking complement proteins due to experimental manipulation, immune-mediated renal disease may be altered. The acute renal disease caused by administration of antibodies to glomerular basement membrane (GBM) antigens is less severe in mice deficient in C3 or C4 [30], whereas the chronic immune complex phase, mediated by antibodies to the administered anti-GBM, is little different or possibly slightly more severe in mice with complement deficiency. In MRL/lpr mice with C3 deficiency, a larger number of glomerular immune deposits were observed but the renal disease was no different histopathologically than found in MRL/lpr mice with normal C3 genes [31]. Together, these data indicate that an intact complement system is not necessary for many forms of immune complex tissue damage. Nevertheless, other evidence indicates that complement deficiency can ameliorate immune complex disease and decrease tissue damage (see Chapter 20).

Interaction of Immune Complexes with Cell Receptors: Binding and Functional Consequences

Fcγ receptors (FcγR) reacting with IgG molecules exist in three major forms, termed FcγRI (CD64), FcγRII (CD32), and FcγRIII (CD16) (reviewed in [32] and in Chapter 21). Subtypes and splice variants of several of the FcγRs add to their complexity. The FcγRs on Kupffer cells of the liver may have a role in removal of circulating IgG-containing immune complexes, and FcγRs on liver endothelial cells also may have a role in immune complex binding [33]. FcγRII are expressed on mononuclear phagocytes and preferentially bind IgG2 subclass. The FcγRII polymorphism variant associated with decreased uptake of IgG2 is also associated with an increased risk of SLE, and particularly with increased serum concentrations of IgG2-containing antiC1q and immune complexes in SLE patients [34]. A lower copy number of the FcγR III gene is associated with SLE, possibly because it leads to decreased expression of the FcγR III low-affinity receptor on neutrophils, and therefore with diminished immune complex clearance [35]. Expression of FcRs on neutrophils alone is sufficient to result in immune complex nephritis and the cutaneous Arthus reactions in experimental models in which other cells lack FcRs [36].

The lattice of immune complexes significantly influences the interaction of complexes with FcγR. Monomeric IgG molecules have a weak interaction with FcγRII and FcγRIII and these interactions do not trigger phagocytosis. The attachment of immune complexes with sufficient lattice, however, results in phagocytosis of the complexes. Investigations have suggested that the interaction of large-latticed complexes with Fcγ receptors results in increased local concentration of immune complexes that leads to further condensation or rearrangement of the complexes to even larger lattices before interiorization [37].

Immune complexes can alter the immune response and lymphocyte functions. For example, immune complexes are 100−1000-fold more effective in inducing proliferation of primed T cells than antigen alone [38]. FcγRs may have an important role in modulating T-cell-mediated responses by targeting immune complexes to antigen-presenting dendritic cells, leading to internalization of antigen, maturation of dendritic cells, activating or inhibiting dendritic cell function, and antigen presentation to T cells [39].

By this means, immune complexes play a central role in the development of SLE by harnessing the specificity of the acquired immune system (high-affinity anti-DNA autoantibodies) to augment the potent but less-specific inflammatory response of the innate immune response (Figure 19.2). Dysfunctional overactivity of the type I interferon system is considered to be a central factor in SLE. The cell that is the most potent producer of type I interferon is the plasmacytoid dendritic cell (pDC), and the trigger for type I interferon production by pDCs is typically viral DNA and/or RNA in response to infection. Immune complexes composed of nucleic acids and IgG from SLE patients are potent inducers of type I interferon production by pDCs, which can lead to activation of monocytoid dendritic cells and thence

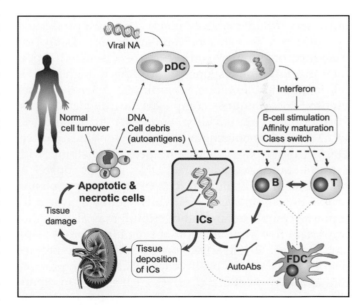

FIGURE 19.2 Central immunoregulatory and tissue-damaging roles of immune complexes in pathogenesis of SLE. DNA is delivered to plasmacytoid dendritic cells in the form of immune complexes, with internalization into the cell augmented by cellular receptors for IgG. DNA entry and binding to Toll-like receptors leads to activation and secretion of interferon. This in turn leads to T- and B-cell stimulation and development of higher avidity autoantibodies. Immune complexes also deposit in tissues, causing tissue damage, which also leads to an increase in dying cells and more circulating DNA and cell debris autoantigens. *(From Wener, MH, 'Immune Complexes', in Tsokos, et al. Systemic Lupus Erythematosus − A Companion to Rheumatology, 2007. Used by permission of Mosby−Elsevier).*

to T-cell and B-cell activation [40]. The nucleic acids responsible for interferon production also can be released from apoptotic or necrotic cells. Efficient stimulation of type I interferon production by SLE IgG requires Fcγ receptors and intact IgG, whereas F(ab) or F(ab')$_2$ fragments of IgG are not sufficient [41]. Thus, a key mechanism leading to sustained production of interferon in SLE is thought to be immune complexes containing nucleic acids, including most prominently DNA and anti-DNA, and also RNA and antibodies to RNA—protein complexes such as the traditional lupus autoantigen U1-RNP [42]. Internalization of the immune complexes is mediated by Fcγ receptors, and the internalized nucleic acid then binds to toll-like receptors (TLRs) including TLR 9. Binding of characteristic CpG motifs to TLR9 induces cellular signaling pathways leading to enhanced interferon mRNA transcription and protein release. In addition, ICs binding to germinal center follicular dendritic cells (FDC) facilitates antigen presentation by FDCs and thereby promotes the ability of FDCs to interact with B cells and cause affinity maturation and class switching. It has been proposed that immune-complex-bearing FDCs may be required for development of high-affinity IgG antibodies, including anti-DNA [43a]. Thus, nucleic-acid-containing immune complexes augment the autoimmune response in SLE by leading to sustained overproduction of type I interferon and affinity maturation and continued production of anti-DNA and related autoantibodies. Depressed clearance of immune complexes, as discussed above, may magnify this response by leading to higher levels of circulating immune complexes and/or higher binding of immune complexes to dendritic cells. Furthermore, high-avidity anti-DNA that binds to DNA with monogamous bivalent binding protects DNA from DNase digestion, possibly helping with intracellular delivery of intact DNA strands capable of binding and activating TLRs. An unusual population of low-density granulocytes which can be activated by immune complexes and other stimuli may also play an important role in production of type I interferons [43b].

Polyclonal B-cell activation by immune complexes also could serve as a positive feedback loop to promote autoantibody-mediated diseases such as SLE. In experimental animals, immune complexes have been demonstrated to stimulate proliferation of B cells and synthesis of antibodies that are not related to the constituents of immune complexes [44]. Monocyte induction of the B-cell-stimulating cytokines IL-10 and IL-6 is enhanced by immune complexes from lupus sera, via a mechanism dependent on FcγRII on the monocytes [45]. Insoluble immune complexes lead to IL10 release by human monocytes which could be further enhanced by activators of Toll-like receptors, which in turn could augment B-cell stimulation and production of

antibodies [46]. While immune aggregates increased production of IL-10, they also down-regulated the production of IL-12 in the presence of complement, whereas the down-regulation of IL-12 was lessened in the absence of the complement [47]. Thus the effect of immune complexes and aggregates may depend in part on the complement milieu: in the presence of complement, a Th2 phenotype dominates with up-regulation of IL10 and down-regulation of IL-12, whereas the up-regulation of IL-12 production in the absence of complement would tend to lead to a Th1 skewing.

Binding of immune complexes to Fc receptors leads to aggregation of those receptors, and triggers intracellular signaling pathways [48]. Immune complexes binding via the FcγRIIB receptor have been shown to exert a down-regulatory immunomodulatory effect, counterbalancing the activation induced by immune complexes binding to cells via the FcγRI and FcγRIII receptors [48]. Soluble and insoluble immune complexes appear to activate neutrophils by different mechanisms and via somewhat different FcγR involvement. Activation by soluble complexes is mediated by binding to both FcγRII and FcγRIIIb and leads to the release of secreted inflammatory mediators from cytokine-primed neutrophils, whereas insoluble complex activation is mediated by FcγRIIIb without FcγRII and leads to the production of intracellular reactive oxygen species and release of granule enzymes from neutrophils [49].

FcγRs are necessary for mediating immune complex disease in several situations. Mice that lack the transmembrane, signal-transducing gamma chain, which is found on IgG FcγRI and FcγRIII, are resistant to many forms of immune complex disease. In cutaneous Arthus reaction models in those mice, as well as experiments with autoimmune mice cross-bred with the Fc receptor knock-out mice, neutrophil infiltration and organ dysfunction required intact Fc receptors despite the presence of immune complexes in tissues [50]. Furthermore, inflammation was not altered by complement deficiency. The susceptibility to murine lupus nephritis [51] and to collagen-induced arthritis was also found to be altered in FcγR-deficient mice [52], indicating that similar mechanisms were important in these diseases. In a murine model of immune complex peritonitis, neutrophil migration was attenuated after complement depletion, but totally abolished in mice lacking the FcR gamma chain [53]. It has been proposed that local microenvironments within different tissues could influence expression of FcR on macrophages at different sites, thus modulating the local inflammation and other tissue effects of circulating or deposited immune complexes [54].

As already pointed out, human erythrocytes possess the C3b receptor, complement receptor type 1 (CR1). The average number of CR1 molecules per erythrocyte varies considerably among normal persons and also among erythrocytes of an individual. In patients with SLE, the number of CR1 molecules per erythrocyte is decreased [55]. The CR1 molecules exist in clusters, rendering them highly effective in interacting with immune complexes that contain multiple molecules of C3b or C4b [56, 57]. Therefore, because of the large total red cell mass, these receptors effectively bind the very large, complement-containing immune complexes and thus prevent them from depositing in tissues. *In vitro* experiments have shown that the immune complexes with complement components transfer effectively from erythrocytes to monocytes with FcγR and CR1 receptors. This reaction is much faster than the uptake of the same complexes from a fluid phase [58]. Experiments in monkeys suggest that binding of very large, complement-containing immune complexes to erythrocytes may diminish immune deposits in glomeruli [59]. This system is applicable only to large immune complexes that have bound complement components.

Complement receptors have also been described in tissues. Of particular interest has been the presence of C3b receptors in glomerular epithelial cells [60]. These receptors have been found to be decreased in glomeruli of patients with proliferative but not membranous lupus glomerulonephritis, in whom the deposits are adjacent to the epithelial cells. The role of the C3b receptors in localization of immune complexes in glomeruli has not been well established. In experimental animals immune deposits form in the same location in species that do not have demonstrable C3b receptors in glomeruli. These observations suggest that the presence of the C3b receptors in glomeruli do not play a significant role in the deposition of immune complexes.

Fate of Circulating Immune Complexes

Studies in experimental animals have indicated that with increasing loads of circulating immune complexes, deposition in tissues is enhanced. Protective mechanisms, however, exist to remove complexes from circulation, thereby decreasing the possibility of deposition in tissues. The principles involved in removal of circulating immune complexes have been examined in experimental animals. Once immune complexes are formed in circulation or gain access to circulation, the removal by protective mechanisms is influenced by the lattice of immune complexes, the status of the mononuclear phagocyte system, the nature of antibodies in the complexes, and to some degree the nature of the antigen molecules in the complexes.

Large-latticed IgG immune complexes are removed from the circulation by the mononuclear phagocyte system. The interaction of complexes with the FcγR of Kupffer cells in the liver mediated this removal from circulation. In contrast, small-latticed complexes persisted longer in circulation but are removed faster than antibodies alone. The spleen accounts for only a relatively small proportion of the large-latticed immune complexes removed from circulation [61]. Furthermore, complement components are not involved in this rapid uptake of large-latticed immune complexes in mice and rabbits. The rapid removal of immune complexes by Kupffer cells is aided by the fact that these cells in hepatic sinusoids are not covered by endothelial cells and thus are directly exposed to circulating materials.

The mononuclear phagocyte system in the liver is involved in removal of a variety of substances from circulation that do not depend on the FcγR. This nonspecific uptake is a saturable process. The FcγR-specific uptake of large-latticed circulating immune complexes is also saturable, as established by injecting increasing doses of immune complexes. The saturation of the hepatic uptake of immune complexes causes prolonged circulation of the large-latticed immune complexes and increased deposition in glomeruli [61].

Several studies of soluble immune complex clearance have been conducted in humans. Large immune complexes were formed with tetanus toxoid and antibodies to tetanus toxoid [62]. The injected large complexes rapidly incorporated complement components. In normal persons, these complexes were cleared with a single exponential component, ranging from 9.9% to 18.7% of the injected complexes per minute. In 11 of 15 patients with SLE or other immune complex diseases, an initial very rapid component of removal was present. This was thought to result from trapping in tissues. The second phase of clearance in patients ranged from 8.6% to 32.2% per minute. The binding of injected complexes to erythrocytes during the first minute correlated with the average number of CR1 molecules per erythrocyte. Furthermore, when immune complexes with complement components were bound to autologous erythrocytes and then injected, a large fraction of the complexes was dissociated from erythrocytes during the first minute. This release was inversely proportional to the average number of CR1 molecules per erythrocyte. Thus, these studies demonstrate that patients with SLE have neither excessive persistence of immune complexes in circulation nor a decreased hepatic uptake of immune complexes. This is in contrast to the previously described decreased splenic uptake and prolonged circulation of IgG-coated red blood cells in patients with SLE [63]. The splenic uptake of cellular probes is highly dependent on splenic blood flow [64].

Another study examined the clearance from circulation and uptake by the liver and spleen of immune complexes prepared with hepatitis B surface antigen (HBsAg) and antibodies to this protein, in normal persons (n = 12) and in patients with SLE (n = 10) [65]. The polymeric HBsAg had a molecular mass of about 3000 kDa and the formed immune complexes were quite large and fixed complement effectively. Upon intravenous injection some of the very large soluble immune complexes, which had been prepared in antibody excess, bound to erythrocytes to a higher percentage in normal persons than in patients with SLE. The clearance from blood, however, was faster in patients with SLE than in normal persons. The median half-life in patients was 2.2 min (range 1.3—6.6) and in normal persons 5.2 min (range 3.6—14). In patients with SLE the very largest immune complexes were not removed selectively from plasma. In contrast, the same analysis of plasma from normal persons showed that the very largest immune complexes had been selectively removed. External counting of the ^{123}I on HBsAg in immune complexes demonstrated that the radioactive probe was taken up more rapidly in the liver in SLE patients than in normal persons. In the SLE patients a high correlation was found between the half-life of immune complexes and the C4 concentration. These findings clearly indicate the role of the complement system in the clearance from circulation of the very large immune complexes. Once complement components are bound to the complexes, they become attached to the erythrocytes via the CR1, are no longer in plasma and are then delivered mainly to the spleen. The large complexes that are less effective in activating complement are mainly removed by the liver. In patients with SLE and with decreased complement levels the binding of the very large immune complexes to erythrocytes is decreased and these complexes are diverted to the liver. This accounts for the observed initial rapid removal from blood, the increased hepatic uptake and the decreased splenic uptake. This interpretation is also consistent with the already discussed decreased hepatic uptake by the spleen of IgG-coated erythrocytes [63]. In addition, this interpretation was confirmed by a study in a patient with hereditary homozygous C2 deficiency [66]. The initial clearance of the complexes was rapid without binding to erythrocytes, the hepatic uptake was rapid, and no uptake was observed in the spleen. After administration of fresh frozen plasma containing complement the initial rapid removal was decreased, immune complexes bound to erythrocytes, hepatic uptake was decreased, and splenic uptake was enhanced.

Radiolabeled, aggregated human IgG has been used in normal persons and small numbers of patients with SLE to determine its clearance from circulation and uptake by the liver. The material used, however, was a mixture of aggregated and monomeric IgG, thereby complicating the analysis of the collected data. The study involved 22 patients with SLE, and the kinetics of the removal of the aggregates were analyzed by a model assuming homogeneity of the material and a distribution phase faster than the elimination phase. The authors concluded that the initial phase of removal of the IgG aggregates was faster in patients with SLE than in normal persons and that the amount of aggregates bound to red cells was higher in normal persons [67].

Collectively, these human studies indicate that very large immune complexes in circulation that have already bound complement components bind effectively to erythrocytes. Thus, these complexes are no longer available for deposition in tissues. A significant proportion of the erythrocyte-bound immune complexes are delivered to the spleen. As pointed out above, in patients with SLE and decreased complement levels or in patients with absent early complement components, the delivery of immune complexes to the spleen is defective. This defect in patients with SLE was also detected by IgG-coated erythrocytes that are primarily removed by the spleen [63]. Small immune complexes that still contain several IgG molecules do not bind to erythrocytes and are cleared by the liver, mainly by Kupffer cells via FcγR. When the binding of very large immune complexes to erythrocytes is defective, these complexes are also rapidly cleared by the liver. Smaller immune complexes, containing one or two IgG molecules, persist in circulation and are not deposited in tissues (see Figure 19.3). In C1q-deficient mice, clearance of experimental immune complexes in the spleen was shown to be decreased, whereas the initial phase of immune complex clearance in the liver was enhanced [68]. Since mice lack the CR1-dependent immune complex binding to erythrocytes, these experiments indicate that activation of classical pathway of complement may have additional roles in clearance of immune complexes.

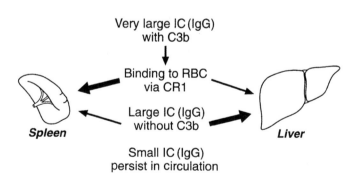

FIGURE 19.3 Schematic representation of the fate of circulating immune complexes (IC). Very large IC formed in antibody excess activate complement effectively, bind C3b, and become attached to red blood cells (RBC) via the complement receptor one (CR1). The IC bound to RBC do not deposit in tissues and are delivered principally to the spleen. The splenic uptake of IC bound to RBC is decreased in patients with SLE and hypocomplementemia and the IC with C3b are diverted to the liver. Large IC without bound C3b are taken up principally by the liver, both in normal persons and patients with SLE. Small IC (e.g., Ag_1Ab_1, Ag_2Ab_1, or Ag_2Ab_2) are not rapidly removed and do not deposit in tissues.

As pointed out under the discussion of antigens, the nature of antigens in the immune complexes can alter the fate of immune complexes. If the antigen alone is rapidly removed from circulation, then the presence of such antigens in the immune complexes can enhance the removal of complexes from circulation. In experimental animals the clearance of immune complexes containing glycosylated antigens is governed in part by specific carbohydrate receptors on hepatocytes [4]. A serum carbohydrate binding protein, mannose-binding lectin (MBL, also known as mannan- or mannose-binding protein), may have an important role in clearing immune complexes containing antigens with selected carbohydrate residues [69]. A member of the collagen motif-containing collectin family of proteins, MBL binds terminal mannose, fucose, glucose, or N-acetylglucosamine residues, can serve as an opsonin, activates the lectin pathway of complement, and activates macrophages via the C1q receptor [70]. Genetic polymorphisms responsible for depressed function and serum levels of MBL are associated with SLE in African-Americans [71] and some other ethnic and national groups [72, 73]. Furthermore, certain ribonucleoprotein autoantigens, including the U1-specific 68 kDa and A proteins and the U2-specific B'' protein, are glycoproteins, with mannose, glucose, and N-acetylglucosamine detected on the 68-kDa protein [74]. Thus, it is conceivable that the clearance of glycoprotein antigens or immune complexes containing such antigens, including the U1-RNP particle, could be influenced by MBL polymorphisms. These considerations suggest that MBL polymorphisms could participate in the pathogenesis of SLE by influencing immune complex clearance, analogous to the role of polymorphisms in complement components and FcR. MBL contributes to immune complex clearance from the circulation, particularly in patients with defects in complement-mediated immune complex clearance, as demonstrated by testing individuals with C2 deficiency in whom the classical pathway of complement activation is defective [75]. Polymorphisms in genes that influence immune complex clearance (e.g., Fcγ receptors, classical and MBL pathways of complement activation, complement receptors) that individually incur a small risk for SLE may act synergistically to increase the risk of SLE substantially [76].

TISSUE DEPOSITION OF IMMUNE COMPLEXES

In patients with SLE, deposits of immune complexes have been identified in several organs, including glomeruli, renal peritubular capillaries, small and medium-sized blood vessels in several organs, dermal–epidermal junction, and choroid plexus. In most of these sites the presence of immune complexes has been inferred from the presence of immunoglobulins and complement components. In the glomeruli and blood vessels, specific antibodies have been identified. The presence of immune complexes at various sites may arise from deposition of circulating immune complexes or from local formation of antigen–antibody complexes in situ. The local formation of immune complexes can arise from selective deposition or presence of an antigen at a given location, followed by binding of specific antibodies to the localized antigens. At this time, information is not available to distinguish immune complexes that have arisen in the given location by deposition from circulation or by local formation.

Localization of Immune Complexes in Glomeruli

Several unique structural features of glomeruli may well contribute to immune complex deposition. First, large volumes of plasma are filtered through the capillary wall with transient retention of macromolecules, including immune complexes. Second, the glomerular capillary wall contains a fenestrated endothelium, allowing larger molecules to traverse the glomerulus than other capillary beds. Third, fixed negative charges are present in the glomerular basement membrane in the subendothelial subepithelial areas.

In lupus nephritis, in other forms of glomerulonephritides, and in chronic serum sickness models in experimental animals the renal lesions may contain mesangial, subendothelial, and subepithelial immune deposits (Figure 19.4). Experiments with injection of preformed immune complexes into unimmunized animals have indicated that the complexes in circulation deposit principally in the subendothelial and mesangial areas. In a number of studies of this nature the immune complexes injected into circulation caused no deposits in the subepithelial area [61, 77].

Several lines of evidence have indicated that the lattice of circulating immune complexes is highly important to their deposition in glomeruli. First, when mixtures of large-latticed and small-latticed immune complexes in antigen excess were injected into mice, glomerular deposition of complexes progressed only while the large-latticed complexes remained in circulation [61]. Second, the injection of only small-latticed immune complexes into the circulation of mice caused no deposits in the glomeruli [78]. Third, when large-latticed immune complexes were deposited in glomeruli, the injection of a large excess of antigen resulted in the complete removal of the antigen–antibody deposits within hours, presumably by conversion of immune deposits into small-latticed immune complexes [79]. All these studies have indicated that the lattice of immune

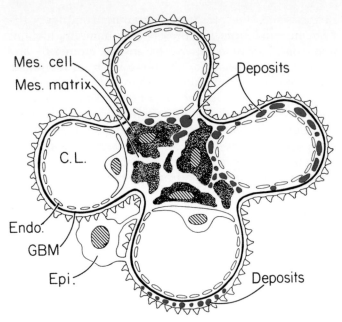

FIGURE 19.4 Schematic representation of the location of immune complexes in glomeruli. The diagram represents glomerular capillaries cut in cross-section. The capillary lumen (CL) is lined with fenestrated endothelium (Endo). The glomerular basement membrane (GBM) is a size and charge barrier to circulating macromolecules. Negative charges are present in the subendothelial area (lamina rara interna) and in the subepithelial area (lamina rara externa). Beyond the GBM are the foot processes of the epithelial cells (Epi). The area between the capillary loops is the mesangium, consisting of mesangial matrix (mes. matrix) and mesangial cells (mes. cell). Immune complexes, recognized as electron-dense deposits by electron microscopy, may be only in the mesangium. Immune complexes may exist in the subendothelial area, but these are accompanied by immune complexes in the mesangium. Immune complexes may be present in the subepithelial area, as seen in membranous glomerulonephritis.

complexes has a pivotal role in their deposition in the glomeruli.

The immune deposits in the subendothelial or mesangial area, as visualized by electron microscopy, are considerably larger than immune complexes that have been injected into experimental animals. This suggests that after the initial deposition in glomeruli, the immune complexes must undergo further condensation or rearrangement into even larger deposits. The need for further condensation or rearrangement of immune complexes to persist in glomeruli and to become visible as electron-dense deposits was demonstrated by the use of covalently cross-linked immune complexes. The covalent cross-linking established a fixed lattice in the immune complexes so that they could not rearrange into large lattices or precipitate. When such complexes were injected into mice, only transient deposits in glomeruli were identified by immunofluorescence microscopy, and large electron-dense deposits did not evolve. In contrast, when similar complexes were prepared without covalent bonds and administered to

mice, then deposits evolved that persisted and became visible as electron-dense deposits [80]. These studies suggest that large-latticed immune complexes become locally concentrated in glomeruli, either as a result of filtration of plasma or because of interactions with glomerular structures. As a consequence of this increased local concentration, rearrangement and condensation occur, leading to larger lattices. Thus, non-precipitating antigen—antibody systems would not form persisting immune deposits in glomeruli and would not lead to the formation of electron-dense deposits.

Electrostatic interactions constitute one possible mechanism for attachment to glomeruli and local increase in concentration of immune complexes. When chemically cationized (positively charged) antibodies were used to prepare soluble immune complexes, the injection of these complexes caused extensive subendothelial and mesangial deposits in glomeruli. In contrast to immune complexes prepared with unaltered antibodies, the complexes with cationic antibodies caused extensive subendothelial deposits that persisted in this location for several days, even when injected in small doses. The lattice of the complexes was still highly important with the cationized antibodies. When only small-latticed immune complexes were prepared with these antibodies and injected into mice, the deposits persisted in glomeruli by immunofluorescence microscopy comparable to antibodies alone [81]. These results suggest that, when large-latticed immune complexes with more than two antibody molecules attach to the fixed negative charges in the glomerular basement membrane their local concentration is increased and leads to rearrangement and formation of larger deposits that become visible as electron-dense deposits in the subendothelial area. Small-latticed immune complexes would initially attach in a comparable manner, but they would not condense into larger deposits and therefore remained in glomeruli only for a few hours. The initial interaction of cationic immune complexes with the fixed negative sites on the glomerular basement membrane is a charge—charge interaction. Once the condensation to larger deposits occurred, then the deposits become more permanent [82]. The mechanism for removal of immune complexes from the basement membrane and other organs is not well understood. Absence of C3 in experimental immune-mediated renal disease may lead to alteration in the distribution of immune deposits within the glomerulus, suggesting that complement may play a role in immune complex rearrangement within the kidney [83]. The complement system, particularly C3, probably plays a role not only in solubilization of immune precipitates, but also in removal and trafficking of immune deposits within the basement membrane of the kidney [84].

The formation of mesangial deposits occurs with large-latticed immune complexes containing cationized antibodies. Similar deposits, however, occurred also with immune complexes prepared with anionic (negatively charged) antibodies and anionic antigens, suggesting that charge—charge interactions were not important in the formation of mesangial deposits [85]. It is of interest that FcγR were not expressed by unstimulated human mesangial cells in culture. Upon stimulation with IL-1 and lipopolysaccharide, however, these cells contained mRNA of FcγRIII [86]. The expression of receptors on mesangial cells may enhance localization and retention of immune complexes in the glomerular mesangium.

Several experimental models exist for local immune complex formation in glomeruli due to antigens that become attached or planted in glomeruli [77]. Intravenously injected aggregated IgG or aggregated albumin becomes entrapped in the mesangial matrix. When this event is followed by the injection of specific antibodies, immune deposits form and cause an inflammatory response in the mesangial area of glomeruli [87]. Considerable evidence indicates that the subepithelial immune deposits, as seen in membranous lupus glomerulonephritis, are locally formed and not deposited from circulation as immune complexes [88]. The high density of fixed negative charges in the subepithelial area of the glomerular basement membrane serves as for planting positively charged antigens in glomeruli. The administration of cationized antigens followed by administration of specific antibodies leads to persisting immune deposits in the subepithelial area and the formation of electron-dense deposits [89]. Similar results have been achieved with chronic administration of cationic antigens, leading to endogenous immune response and to development of extensive membranous glomerulonephritis with subepithelial deposits [90]. The formation of persisting immune deposits in the subepithelial area and the formation of electron-dense deposits also required precipitating antigen—antibody systems. With non-precipitating antigen—antibody systems, only transient deposits formed in the subepithelial area as detected by immunofluorescence microscopy, and electron-dense deposits were not formed [91].

These studies in experimental animals suggest that the formation of subepithelial immune deposits in human disorders, including SLE, involves cationic antigens or cationic antibodies that become localized in this area due to electrostatic interactions. The subsequent access to this area of glomeruli by specific antibodies or antigens, respectively, would then lead to the formation of immune deposits. Studies in experimental animals have shown that cationic histones bind to glomeruli by charge—charge interactions, followed by DNA that bind to the planted histones, and finally antibodies to histones and to DNA bind to these antigens in glomeruli [92]. In support of this concept was the identification of histones in the majority of glomeruli in renal biopsies of patients with SLE whereas in the renal biopsies of other forms of glomerular disease, these molecules were rarely present [93].

In patients with lupus the presence of a number of serum autoantibodies has been associated with renal disease [94]. Serum antibodies to dsDNA have a strong association with renal disease. In addition, antibodies to histones and to chromatin share this association. Extraction of antibodies from glomeruli of patients with SLE has demonstrated the presence of antibodies to dsDNA, ssDNA, SS-A/Ro, and nucleoproteins [2, 95]. In patients with proliferative lupus glomerulonephritis, serum autoantibodies to the collagen-like region of C1q (antiC1q-CLR) are found enriched in the glomerular basement membrane, typically together with other autoantibodies [96]. The presence of antiC1q-CLR serves as a better biomarker predictor of developing proliferative glomerulonephritis than a rise in the level of antibodies to dsDNA [97]. The antibodies to C1q-CLR do not react with circulating C1 and do not bind well to C1q in fluid phase. The anti C1q-CLR, however, bind well to C1q attached to immune complexes in vitro or in vivo [98]. Therefore, it is likely that the anti C1q-CLR contribute to the glomerular inflammation by binding to already deposited immune complexes with bound C1q (Figure 19.5). Alternatively, these antibodies may bind to C1q attached to DNA, because DNA in glomeruli may bind C1q directly without the presence of antibodies [99]. Further studies have shown that multiple autoantibody specificities, including anti-dsDNA, anti-C1q-CLR, anti-SSA, anti-Sm, anti-SSB, antihistone, and antichromatin, are frequently and concurrently present and enriched in the glomerular basement membranes from kidneys of patients with lupus [96]. In contrast, control antibodies such as antibodies to tetanus and to EB-virus antigens are present in concentrations that reflected only the serum concentration of those antibodies, and were not enriched in the kidneys. Furthermore, elution of the antibodies from lupus kidneys required strong denaturants and even disruption of covalent bonds, indicating that in chronic human immune complex disease the immune complexes form covalent bonds with components in the kidney. Murine models have demonstrated that C1q-containing immune complexes need to be present in the glomerular basement membrane in order for antiC1q to be pathogenic in causing immune complex nephritis [100]. Thus, antiC1q serves as a mechanism for magnifying and enhancing immune complex nephritis. This is particularly relevant in SLE, since multiple immune complex systems with different antigens and antibodies contribute to lupus nephritis. Diverse small immune complexes from different antigen—antibody systems would not necessarily condense to form large-

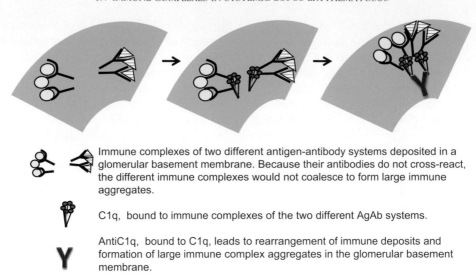

Immune complexes of two different antigen-antibody systems deposited in a glomerular basement membrane. Because their antibodies do not cross-react, the different immune complexes would not coalesce to form large immune aggregates.

C1q, bound to immune complexes of the two different AgAb systems.

AntiC1q, bound to C1q, leads to rearrangement of immune deposits and formation of large immune complex aggregates in the glomerular basement membrane.

FIGURE 19.5 Role of antibodies to C1q to augment and enlarge immune complex deposits in glomerular basement membrane. Antigen–antibody systems containing antibodies that do not cross-react do not form large deposits even if they are in close proximity, because they cannot rearrange their lattice structure to form large aggregates (left panel). As long as the immune complexes contain at least two IgG molecules ($\geq$AgAb2), they still would be able to bind C1q (middle panel). Adjacent C1q-bound immune complexes could coalesce into large immune aggregates when cross-linked by antiC1q (right panel). This model suggests a mechanism by which smaller amounts of immune complexes of multiple different antigen–antibody systems, as might be present in SLE, may coalesce to form larger immune deposits.

lattice, electron dense deposits with pathogenic potential. However, if they bind complement and then antiC1q, they can coalesce to form larger and more pathogenic glomerular immune deposits (Figure 19.5).

Localization of Immune Complexes in Other Organs

The localization of immune complexes in other organs has not been examined in the detail described for glomeruli. The existence of immunoglobulins and complement components at the dermal–epidermal junction of patients with SLE is well established. Antibodies to nuclear antigens and antibodies to DNA and chromatin have been recovered from these deposits, suggesting that the detected deposits exist as immune complexes at this location [101]. Some evidence indicates that keratinocytes at the basal layer of skin contain the Ro (SS-A) antigen. As this antigen is extruded, immune complexes form at the dermal–epidermal junction with antibodies to Ro [102].

Studies with preformed immune complexes have suggested that charge–charge interactions contribute to the formation of immune complexes at this site. First, cationic antibodies alone bound to the dermal–epidermal junction of mice, as detected by immunofluorescence microscopy. This finding raises the possibility that *in situ* immune complex formation may occur with the initial deposition of, for example, cationic antibodies to DNA and subsequent immune complex formation with DNA. Second, large-latticed circulating

immune complexes, prepared with cationic antibodies, deposited at this location in the skin. The deposited immune complexes persisted in this location longer than the cationic antibodies alone [103]. Third, the chronic injection of a cationized antigen caused immune deposits at the dermal–epidermal junction, whereas the injection of the unaltered antigen according to the same schedule and dose did not cause these deposits [104].

In patients with SLE, immune complexes have been described in the choroid plexus, between the basement membrane and the base of the epithelial cells. Similar lesions have been observed in NZB/NZW mice and in patients with chronic serum sickness. The injection of very large doses of preformed immune complexes has resulted in some localization of the injected immune complexes in the choroid plexus [105]. The injection of cationic antigens has resulted in deposits of the antigen in the choroid plexus. When antibodies were injected later, immune complexes were formed in the choroid plexus and persisted in this area. Of interest was the observation that a cationic antigen persisted for many days in the choroid plexus, whereas the bulk of cationized antigens lasted in glomeruli for less than a day. These findings suggest differences in the nature or the turnover of the fixed negative charges in glomeruli compared with the choroid plexus [106]. Immune complexes may also form within the cerebrospinal fluid, when antigens released into the CSF from the brain after damage caused by neurocytotoxic autoantibodies or other causes react with autoantibodies in the CSF. Those autoantibodies are able to stimulate type I interferon release when incubated with plasmacytoid dendritic cells, suggesting a mechanism for localized cytokine production which could bathe the brain and neural tissue of SLE patients [107].

The immune deposits in blood vessel walls clearly are associated with vasculitis in patients with SLE. The mechanisms for formation of these deposits are not known. In experimental animals serum sickness is associated with vascular deposits of immune complexes; increased permeability of vessels facilitates the development of these deposits [108]. The injection of a single bolus of preformed immune complexes into mice also caused vascular deposits, as recognized by immunofluorescence microscopy. These deposits, however, tended to persist for a much shorter period in the dermal, myocardial, and peritubular renal small vessels than in glomeruli [103].

MEASUREMENT OF IMMUNE COMPLEXES AND THE USE OF THESE TESTS IN PATIENTS WITH SLE

Since the clinical manifestations of SLE and other systemic immune complex diseases are related at least in part to circulating immune complexes, it seems desirable to quantify these materials in serum. A number of methods have been developed to detect and to quantify immune complexes in the serum or other body fluids. The methods for detection of immune complexes rely on either the physical or the biologic properties of immune complexes. As already pointed out, both of these categories of properties of immune complexes are highly variable because of the variations in antigens, antibodies, and numbers of reactants that are present in a given complex. The physical methods rely largely on the fact that immune complexes differ from antibodies alone by size, solubility, or cryoprecipitability. The tests based on biologic properties recognize immune complexes because antibodies have acquired properties different from free antibodies or because complement components have been attached to the immune complexes. Unfortunately, assays for detection and quantification of immune complexes in general are not specific for any given antigen or disease and therefore provide little assistance in reaching a specific diagnosis.

The physical methods for detection of immune complexes are based on the fact that complexes are larger than antibodies alone or have different solubility. Ultracentrifugation and chromatography methods are time-consuming, and it can be difficult to detect small immune complexes, especially in the setting of hypergammaglobulinemia and therefore these methods are not used by clinical laboratories but remain tools for research laboratories. The addition to serum specimens of low concentrations of polyethylene glycol precipitates some immune complexes, however, this approach has not provided a reliable tool for quantitation of circulating immune complexes.

Several assays for measurement of immune complexes are based on recognizing antibody molecules that have acquired biological properties that distinguish them from free antibody molecules. This category of assays includes complement activation, C1q-binding assays, and assays that depend on the presence of complement components. When the sizes of immune complexes in patients with SLE were examined, those with proliferative lupus glomerulonephritis had IgG-containing complexes but also had monomeric IgG that bound to C1q in the solid-phase assay [109]. This reactivity was shown to result from the presence of autoantibodies to C1q, directed to the collagen-like region of these molecules [110]. Therefore, when the solid-phase C1q assay is used as an assay for immune complexes, the presence of immune complexes and autoantibodies to the collagen-like region of C1q are both potentially detected. The two types of IgG reactants with solid-phase C1q can be optimally distinguished by use of purified collagen-like regions of C1q for identification of the autoantibodies [111]. Binding of IgG to C1q in the presence of high salt concentrations is a reasonable surrogate for measurement of autoantibodies to C1q, however the two experimental approaches are not identical [112]. Autoantibodies to C1q bind to solid-phase but not soluble C1q. Autoantibodies to C1q also display restricted binding to cell-bound C1q, in that these autoantibodies bind specifically to early apoptotic cells, but not to other cell-associated C1q [113].

Other tests for immune complexes depend on the fact that complement components have bound to immune complexes *in vivo*. The prototypic historical assay in this category is the Raji cell assay which detects immune complexes that have C3b and other complement components bound to them [114].

Measuring IgG bound to soluble complement receptor 1 (CR1), which is the erythrocyte and leukocyte receptor for C3b- and C4b-bearing immune complexes, also has been reported as a means to detect circulating immune complexes in SLE sera [115]. Immune complexes in SLE sera have been characterized as containing soluble complement membrane attack complex in conjunction with the complement regulatory proteins clusterin (also known as apolipoprotein J) and vitronectin (S40) [116] . High levels of clusterin and vitronectin were associated with lupus nephritis, and detection of complexes with these constituents has been suggested as a biomarker for SLE.

The relationships between clinical activity of SLE and the concentrations of immune complexes by one or more tests have been examined by several laboratories [117].

Positive correlations have been found between clinical activity of disease in patients with SLE and immune complexes detected by various assays, including the solid-phase C1q assay and Raji cell assay. Other observers have, however, reached the conclusion that a definite relationship does not exist between disease activity or renal involvement with one form or another of lupus nephritis and the measured levels of immune complexes [118]. Clinicians familiar with the interpretation of these tests may use the tests for measurement of immune complexes as an adjunct with other tests or in place of other tests (e.g., total hemolytic complement or complement component levels) along with clinical information to make therapeutic decisions. [119].

DEVELOPMENT OF THERAPIES BASED ON THE IMMUNE COMPLEX MODEL

The immune complex model for the cause of tissue damage in SLE has been the dominant paradigm for several decades, and it remains so. Therapeutic approaches based on this paradigm, however, have been relatively disappointing. For example, whereas plasmapheresis for the treatment of SLE originally met with great enthusiasm, a controlled clinical trial of plasmapheresis in patients with lupus nephritis was unsuccessful [120].

Affinity columns containing silica-bound staphylococcal protein A (SPA) have been approved for use in patients with severe, refractory rheumatoid arthritis [121]. The rationale for use of SPA immunoabsorbants arises from the observation that IgG within immune complexes binds preferentially to SPA, compared with monomeric non-complexed IgG. It has been suggested that other mechanisms besides immune complex removal may play a role in the improvement [122], nevertheless this successful therapy was developed because of the immune complex disease model. C1q immunoadsorption has been employed in the treatment of patients with SLE [123, 124].

Evidence has been presented that small soluble immune complexes could down-regulate autoimmune diseases, which could play a role in immune thrombocytopenia and other autoimmune diseases. Blocking or saturation of Fc receptors by soluble immune complexes has been proposed as a mechanism explaining some of the efficacy of intravenous gamma globulin therapy of autoimmune diseases such as immune-mediated thrombocytopenia [125]. An alternative approach has been the administration of soluble Fc receptors, in hopes of binding immune complexes before they target tissues or cells that would lead to inflammation. Administration of soluble forms of the high-affinity receptors FcgRI and FcgR1A (CD64A) has been used successfully to treat the passive cutaneous Arthus reaction and collagen-induced arthritis in experimental animals [126–128]. Administration of the low-affinity soluble receptor FcγRIII (CD16) prevented neutrophil emigration to a site of immune complex deposition and prevented inflammation [129]. Soluble Fcγ receptors also inhibit immune complex precipitation, complement activation, and activation of cellular receptors by soluble immune complexes. Thus, therapeutic approaches targeting the antigens in immune complexes as well as approaches that are not antigen-specific could prove helpful in treatment of SLE. Immune complexes play a pivotal role in the pathogenesis of lupus nephritis as well as other manifestations of SLE, and understanding the pathophysiology could provide a useful basis for further therapies.

References

[1] F.J. Dixon, The role of antigen-antibody complexes in disease, The Harvey Lectures 58 (1963) 21–52.

[2] D. Koffler, P.H. Schur, H.G. Kunkel, Immunological studies concerning the nephritis of systemic lupus erythematosus, J Exp Med 126 (1967) 607–624.

[3] R. Lafyatis, A. Marshak-Rothstein, Toll-like receptors and innate immune responses in systemic lupus erythematosus, Arthritis Res Ther 9 (2007) 222. doi:10.1186/ar2321.

[4] D.S. Finbloom, D.B. Magilavy, J.B. Hartford, A. Rifai, P.H. Plotz, The influence of antigen on immune complex behavior in mice, J Clin Invest 68 (1981) 214–224.

[5] W. Emlen, A. Rifai, D. Magilavy, M. Mannik, Hepatic binding of DNA is mediated by a receptor on non-parenchymal cells, Am J Pathol 133 (1988) 54–60.

[6] W. Emlen, M. Mannik, Clearance of circulating DNA-antiDNA immune complexes in mice, J Exp Med 155 (1982) 1210–1215.

[7] J.C. Khoo, E. Miller, F. Pio, D. Steinberg, J.L. Witztum, Monoclonal antibodies against LDL further enhance macrophage uptake of LDL aggregates, Arterioscler Thromb 12 (1992) 1258-166.

[8] M. Lopes-Virella, N. Binzafar, S. Rackley, A. Takei, M. La Via, G. Virella, The uptake of LDL-IC by human macrophages: Predominant involvement of the fcg RI receptor, Atherosclerosis 135 (1997) 161–170.

[9] Y. Huang, A. Jaffa, S. Koskinen, A. Takei, M. Lopes-Virella, Oxidized LDL-containing immune complexes induce fc gamma receptor I-mediated mitogen-activated protein kinase activation in THP-1 macrophages, Arterioscler Thromb Vasc Biol 19 (1999) 1600-167.

[10] P.A. Kiener, B.M. Rankin, P.M. Davis, S.A. Yocum, G.A. Warr, R.I. Grove, Immune complexes of LDL induce atherogenic responses in human monocytic cells, Arterioscler Thromb Vasc Biol 15 (1995) 990–999.

[11] A.E. Kabakov, V.V. Tertov, V.A. Saenko, A.M. Poverenny, A.N. Orekhov, The atherogenic effect of lupus sera: Systemic lupus erythematosus-derived immune complexes stimulate the accumulation of cholesterol in cultured smooth muscle cells from human aorta. Clin Immunol Immunopathol 63 (1992) 214–220.

[12] O. Vaarala, G. Alfthan, M. Jauhiainen, M. Leirisalo-Repo, K. Aho, T. Palosuo, Crossreaction between antibodies to oxidized low-density lipoprotein and to cardiolipin in systemic lupus erythematosus, Lancet 341 (1993) 923–925.

[13] Y. Hasunuma, E. Matsuura, Z. Makita, T. Katahira, S. Nishi, T. Koike, Involvement of b2-glycoprotein I and anticardiolipin antibodies in oxidatively modified low-density lipoprotein uptake by macrophages, Clin Exp Immunol 107 (1997) 569—573.

[14] M. Puurunen, M. Manttari, V. Manninen, al, e. Antibodies against oxidized low density lipoprotein predicting myocardial infarction, Arch Intern Med 154 (1994) 2605—2609.

[15] M.C. van Bruggen, C. Kramers, B. Walgreen, J.D. Elema, C.G. Kallenberg, d.B. van, et al., Nucleosomes and histones are present in glomerular deposits in human lupus nephritis, Nephrol Dial Transplant 12 (1997) 57—66.

[16] P.M. Rumore, C.R. Steinman, Endogenous circulating DNA in systemic lupus erythematosus. occurrence as multimeric complexes bound to histone, J Clin Invest 86 (1990) 69—74.

[17] R. Licht, M.C. van Bruggen, B. Oppers-Walgreen, T.P. Rijke, J.H. Berden, Plasma levels of nucleosomes and nucleosome-autoantibody complexes in murine lupus: Effects of disease progression and lipopolyssacharide administration, Arthritis Rheum 44 (2001) 1320-130.

[18] C. Kramers, M.N. Hylkema, M.C. van Bruggen, d. L. van, H.B. Dijkman, K.J. Assmann, et al., Anti-nucleosome antibodies complexed to nucleosomal antigens show anti-DNA reactivity and bind to rat glomerular basement membrane in vivo, J Clin Invest 94 (1994) 568—577

[19] M.C. van Bruggen, B. Walgreen, T.P. Rijke, M.J. Corsius, K.J. Assmann, R.J. Smeenk, et al., Heparin and heparinoids prevent the binding of immune complexes containing nucleosomal antigens to the GBM and delay nephritis in MRL/lpr mice, Kidney Int 50 (1996) 1555-164.

[20] W. Emlen, R. Ansari, G. Burdick, DNA-antiDNA immune complexes: Antibody protection of a discrete DNA fragment from DNase digestion in vitro, J Clin Invest 74 (1984) 185—190.

[21] J.A. Schifferli, P. Woo, D.K. Peters, Complement-mediated inhibition of immune precipitation. I. role of the classical and alternative pathways, Clin Exp Immunol 47 (1982) 555—562.

[22] M. Takahashi, J. Czop, A. Ferreira, V. Nussenzweig, Mechanism of solubilization of immune aggregates by complement. implications for immunopathology, Transplant Review 32 (1976) 121—139.

[23] J.A. Schifferli, D.K. Peters, Complement, the immune complex lattice, and the pathophysiology of complement-deficiency syndromes, Lancet 2 (1983) 957—959.

[24] G.J. Arason, R. Kolka, A.B. Hreidarsson, H. Gudjonsson, P.M. Schneider, L. Fry, et al., Defective prevention of immune precipitation in autoimmune diseases is independent of C4A*Q0, Clin. Exp. Immunol 140 (2005) 572—579. doi:10.1111/j.1365-2249.2005.02794.x.

[25] G.J. Arason, K. Steinsson, R. Kolka, T. Vikingsdottir, M.S. D'Ambrogio, H. Valdimarsson, Patients with systemic lupus erythematosus are deficient in complement-dependent prevention of immune precipitation, Rheumatology (Oxford) 43 (2004) 783—789. doi:10.1093/rheumatology/keh183.

[26] Y.M.L. Valim, P.J. Lachmann, The effect of antibody isotype and antigenic epitope density on the complement-fixing activity of immune complexes: A systematic study using chimaeric anti-NIP antibodies with human fc regions, Clin Exp Immunol 84 (1991) 1—8.

[27] G. Biesecker, L. Lavin, M. Ziskind, D. Koffler, Cutaneous localization of the membrane attack complex in discoid and systemic lupus erythematosus, N Engl J Med 306 (1982) 264—270.

[28] R.G. Biesecker, S. Katz, D. Koffler, Renal localization of the membrane attack complex (MAC) in SLE nephritis, J Exp Med 154 (1981) 1779—1794.

[29] D.J. Salant, S. Belok, M.P. Madaio, W.G. Couser, A new role for complement in experimental membranous nephropathy in rats, J Clin Invest 66 (1980) 1339—1350.

[30] N.S. Sheerin, T. Springall, M.C. Carroll, B. Hartley, S.H. Sacks, Protection against anti-glomerular basement membrane (GBM)-mediated nephritis in C3- and C4-deficient mice, Clin Exp Immunol 110 (1997) 403.

[31] H. Sekine, C.M. Reilly, I.D. Molano, G. Garnier, A. Circolo, P. Ruiz, et al., Complement component C3 is not required for full expression of immune complex glomerulonephritis in MRL/lpr mice, J Immunol 166 (2001) 6444—6651.

[32] E.E. Brown, J.C. Edberg, R.P. Kimberly, Fc receptor genes and the systemic lupus erythematosus diathesis, Autoimmunity 40 (2007) 567—581. doi:10.1080/08916930701763710.

[33] T. Lovdal, E. Andersen, A. Brech, T. Berg, Fc receptor mediated endocytosis of small soluble immunoglobulin G immune complexes in kupffer and endothelial cells from rat liver, J Cell Sci 113 (2000) 3255—3366.

[34] L.A. Haseley, J.J. Wisnieski, M.R. Denburg, A.R. Michael-Grossman, E.M. Ginzler, M.F. Gourley, et al., Antibodies to C1q in systemic lupus erythematosus: Characteristics and relation to fc gamma RIIA alleles, Kidney Int 52 (1997) 1375—1380.

[35] L.C. Willcocks, P.A. Lyons, M.R. Clatworthy, J.I. Robinson, W. Yang, S.A. Newland, et al., Copy number of FCGR3B, which is associated with systemic lupus erythematosus., correlates with protein expression and immune complex uptake, J Exp Med 205 (2008) 1573—1582. doi:10.1084/jem.20072413.

[36] N. Tsuboi, K. Asano, M. Lauterbach, T.N. Mayadas, Human neutrophil fcgamma receptors initiate and play specialized nonredundant roles in antibody-mediated inflammatory diseases, Immunity 28 (2008) 833—846. doi:10.1016/j.immuni.2008.04.013.

[37] S.K. Dower, C. Delisi, J.A. Titus, D.M. Segal, Mechanism of binding of multivalent immune complexes to fc receptors. 1. equilibrium binding, Biochemistry 20 (1981) 6326—6334.

[38] S. Marusic-Galesic, K. Pavelic, B. Pokric, Cellular immune response to antigen administered as an immune complex, Immunology 72 (1991) 526—531.

[39] S. Amigorena, Fcg receptors and cross-presentation in dendritic cells, J Exp Med 195 (2002) F1—F3.

[40] L. Ronnblom, V. Pascual, The innate immune system in SLE: Type I interferons and dendritic cells, Lupus 17 (2008) 394—399. doi:10.1177/0961203308090020.

[41] U. Bave, M. Magnusson, M.L. Eloranta, A. Perers, G.V. Alm, L. Ronnblom, Fc gamma RIIa is expressed on natural IFN-alpha-producing cells (plasmacytoid dendritic cells) and is required for the IFN-alpha production induced by apoptotic cells combined with lupus IgG, J Immunol 171 (2003) 3296—3302.

[42] M.L. Eloranta, T. Lovgren, D. Finke, L. Mathsson, J. Ronnelid, B. Kastner, et al., Regulation of the interferon-alpha production induced by RNA-containing immune.complexes in plasmacytoid dendritic cells, Arthritis Rheum 60 (2009) 2418—2427. doi:10.1002/art.24686.

[43a] Y. Wu, S. Sukumar, M.E. El Shikh, A.M. Best, A.K. Szakal, J.G. Tew, Immune complex-bearing follicular dendritic cells deliver a late antigenic signal that promotes somatic hypermutation, J Immunol 180 (2008) 281—290.

[43b] M.F. Denny, S. Yalavarthi, W. Zhao, S.G. Thacker, M. Anderson, A.R. Sandy, W.J. McCune, M.J. Kaplan, A distinct subset of proinflammatory neutrophils isolated from patients with systemic lupus erythematosus induces vascular damage and synthesizes type I IFNs, J Immunol 184 (6) (March 15, 2010) 3284—3297.

[44] E.L. Morgan, W.O. Weigle, Polyclonal activation of murine B lymphocytes by immune complexes, J Immunol 130 (1983) 1066—1070.

[45] J. Ronnelid, A. Tejde, L. Mathsson, K. Nilsson-Ekdahl, B. Nilsson, Immune complexes from SLE sera induce IL10 production from normal peripheral blood mononuclear cells by an FcgammaRII dependent mechanism: Implications for a possible vicious cycle maintaining B cell hyperactivity in SLE, Ann Rheum Dis 62 (2003) 37–42.

[46] S.J. DiMartino, W. Yuan, P. Redecha, L.B. Ivashkiv, J.E. Salmon, Insoluble immune complexes are most effective at triggering IL-10 production in.human monocytes and synergize with TLR ligands and C5a, Clin Immunol 127 (2008) 56–65. doi:10.1016/j.clim.2007.11.014.

[47] A. Tejde, L. Mathsson, K.N. Ekdahl, B. Nilsson, J. Ronnelid, Immune complex-stimulated production of interleukin-12 in peripheral blood mononuclear cells is regulated by the complement system, Clin Exp Immunol 137 (2004) 521–528. doi:10.1111/j.1365-2249.2004.02569.x.

[48] J.V. Ravetch, S. Bolland, IgG fc receptors, Annu Rev Immunol 19 (2001) 275–290.

[49] G. Fossati, R.C. Bucknall, S.W. Edwards, Insoluble and soluble immune complexes activate neutrophils by distinct activation mechanisms: Changes in functional responses induced by priming with cytokines, Ann Rheum Dis 61 (2002) 13–19.

[50] D. Sylvestre, R. Clynes, M. Ma, H. Warren, M.C. Carroll, J.V. Ravetch, Immunoglobulin G-mediated inflammatory responses develop normally in complement-deficient mice, J Exp Med 184 (1996) 2385–2392.

[51] R. Clynes, C. Dumitru, J.V. Ravetch, Uncoupling of immune complex formation and kidney damage in autoimmune glomerulonephritis, Science 279 (1998) 1052–1054.

[52] T. Yuasa, S. Kubo, T. Yoshino, A. Ujike, K. Matsumura, M. Ono, et al., Deletion of fcgamma receptor IIB renders H-2(b) mice susceptible to collagen-induced arthritis, J Exp Med 189 (1999) 187–194.

[53] T. Heller, J.E. Gessner, R.E. Schmidt, A. Klos, W. Bautsch, J. Kohl, Cutting edge: Fc receptor type I for IgG on macrophages and complement mediate the inflammatory response in immune complex peritonitis, J Immunol 162 (1999) 5657–5661.

[54] A. Bhatia, S. Blades, G. Cambridge, J.C. Edwards, Differential distribution of fc gamma RIIIa in normal human tissues and co-localization with DAF and fibrillin-1: Implications for immunological microenvironments, Immunology 94 (1998) 56–63.

[55] M.J. Walport, P.J. Lachmann, Erythrocyte complement receptor type 1, immune complexes, the rheumatic diseases, Arthritis Rheum 31 (1988) 153–158.

[56] J. Chevalier, M.D. Kazatchkine, Distribution in clusters of complement receptor type one (CR1) on human erythrocytes, J Immunol 142 (1989) 2031–2036.

[57] J.P. Paccaud, J.L. Carpentier, J.A. Schifferli, Direct evidence for the clustered nature of complement receptor type 1 on the erythrocyte membrane, J Immunol 141 (1988) 3889–3894.

[58] W. Emlen, G. Burdick, V. Carl, P.J. Lachmann, Binding of model immune complexes to erythrocyte CR1 facilitates immune complex uptake by U937 cells, J Immunol 142 (1989) 4366–4371.

[59] L.A. Hebert, F.G. Cosio, D.J. Birmingham, J.D. Mahan, H.M. Sharma, W.L. Smead, et al., Experimental immune complex-mediated glomerulonephritis in the nonhuman primate, Kidney Int 39 (1991) 44–56.

[60] M.D. Kazatchkine, D.T. Fearon, M.D. Appay, C. Mandet, J. Bariety, Immunohistochemical study of the human glomerular C3b receptor in normal kidney and seventy-five cases of renal diseases. loss of C3b receptor antigen in focal hyalinosis and proliferative nephritis of systemic lupus erythematosus, J Clin Invest 69 (1982) 900–912.

[61] M. Mannik, Pathophysiology of circulating immune complexes, Arthritis Rheum 25 (1982) 783–787.

[62] J.A. Schifferli, Y.C. Ng, J.P. Paccaud, M.J. Walport, The role of hypocomplementemia and low erythrocyte complement receptor type 1 numbers in determining abnormal immune complex clearance in humans. Clin Exp Immunol 75 (1989) 329–335.

[63] R.P. Kimberly, R. Ralph, Endocytosis by the mononuclear phagocyte system and autoimmune disease, Am J Med 74 (1983) 481–493.

[64] C. Halma, M.R. Daha, L.A. van Es, In vivo clearance by the mononuclear phagocyte system in humans: An overview of methods and their interpretation, Clin Exp Immunol 89 (1992) 1–7.

[65] K.A. Davies, A.M. Peters, H.L.C. Beynon, M.J. Walport, Immune complex processing in patients with systemic lupus erythematosus. in vivo imaging and clearance sutdies, J Clin Invest 90 (1992) 2075–2083.

[66] K.A. Davies, K. Erlendsson, H.L.C. Beynon, A.M. Peters, K. Steinsson, H. Vladimarsson, et al., Splenic uptake of immune complexes in man is complement-dependent, J Immunol 151 (1993) 3866–3873.

[67] C. Halma, F.C. Breedveld, M.R. Daha, D. Blok, J. Evers-Schouten, J. Hermans, et al., Elimination of soluble 123I-labeled aggregates of IgG in patients with systemic lupus erythematosus. effect of serum IgG and number of erythrocyte complement receptor type 1, Arthritis Rheum 34 (1991) 442–452.

[68] J.T. Nash, P.R. Taylor, M. Botto, P.J. Norsworthy, K.A. Davies, M.J. Walport, Immune complex processing in C1q-deficient mice, Clin Exp Immunol 123 (2001) 196–202.

[69] M.W. Turner, R.M. Hamvas, Mannose-binding lectin: Structure, function, genetics and disease associations, Rev Immunogenet 2 (2000) 305–322.

[70] A. Tenner, S. Robinson, R. Ezekowitz, Mannose binding protein enhances mononuclear phagocytic function via a receptor that contains the 126,000 mr component of the C1q receptor, Immunity 3 (1995) 485–493.

[71] K.E. Sullivan, C. Wooten, D. Goldman, M. Petri, Mannose-binding protein genetic polymorphisms in black patients with systemic lupus erythematosus, Arthritis Rheum 39 (1996) 2046–2251.

[72] A. Tsutsumi, K. Sasaki, N. Wakamiya, K. Ichikawa, T. Atsumi, K. Ohtani, et al., Mannose-binding lectin gene: Polymorphisms in japanese patients with systemic lupus erythematosus, rheumatoid arthritis and sjogren's syndrome, Genes Immun 2 (2001) 99–104.

[73] J. Villarreal, D. Crosdale, W. Ollier, A. Hajeer, W. Thomson, J. Ordi, et al., Mannose binding lectin and FcgammaRIIa (CD32) polymorphism in spanish systemic lupus erythematosus patients, Rheumatology (Oxford) 40 (2001) 1009–1112.

[74] J.A.P. Chen, Small nuclear ribonucleoprotein particles contain glycoproteins recognized by rheumatic disease-associated autoantibodies, Lupus 1 (1992) 119–124.

[75] S. Saevarsdottir, K. Steinsson, B.R. Ludviksson, G. Grondal, H. Valdimarsson, Mannan-binding lectin may facilitate the clearance of circulating immune.complexes—implications from a study on C2-deficient individuals, Clin Exp Immunol 148 (2007) 248–253. doi:10.1111/j.1365-2249.2007.03349.x.

[76] K.E. Sullivan, A.F. Jawad, L.M. Piliero, N. Kim, X. Luan, D. Goldman, et al., Analysis of polymorphisms affecting immune complex handling in systemic lupus erythematosus, Rheumatology (Oxford) 42 (2003) 446–452.

[77] W.G. Couser, D.J. Salant, In situ immune complex formation and glomerular injury, Kidney Int 17 (1980) 1–13.

[78] A.O. Haakenstad, G.E. Striker, M. Mannik, The disappearance kinetics and glomerular deposition of small-latticed soluble immune complexes, Immunology 47 (1982) 407–414.

[79] M. Mannik, G.E. Striker, Removal of glomerular deposits of immune complexes in mice by administration of excess antigen, Lab Invest 42 (1980) 483–489.

[80] M. Mannik, L.Y.C. Agodoa, K.A. David, Rearrangement of immune complexes in glomeruli leads to persistence and development of electron dense deposits, J Exp Med 157 (1983) 1516–1528.

[81] V.J. Gauthier, M. Mannik, G.E. Striker, Effect of cationized antibodies in preformed immune complexes on deposition and persistence in renal glomeruli, J Exp Med 156 (1982) 766–777.

[82] V.J. Gauthier, M. Mannik, Only the initial binding of cationic immune complexes to glomerular anionic sites is mediated by charge-charge interactions, J Immunol 136 (1986) 3266–3271.

[83] N.S. Sheerin, T. Springall, M. Carroll, S.H. Sacks, Altered distribution of intraglomerular immune complexes in C3-deficient mice, Immunology 97 (1999) 393-39.

[84] Y. Fujigaki, S.R. Batsford, D. Bitter-Suermann, A. Vogt, Complement system promotes transfer of immune complex across glomerular filtration barrier, Lab Invest 72 (1995) 25–33.

[85] V.J. Gauthier, G.E. Striker, M. Mannik, Glomerular localization of preformed immune complexes prepared with anionic antibodies or with cationic antigens, Lab Invest 50 (1984) 636–644.

[86] H.H. Radeke, J.E. Gessner, P. Uciechowski, H.J. Mägert, R.E. Schmidt, K. Resh, Intrinsic human glomerular mesangial cells can express receptors for IgG complexes (hFc gamma RIII-A) and the associated fc epsilon RI gamma-chain, J Immunol 153 (1994) 1281–1292.

[87] A.F. Michael, W.F. Keane, L. Raij, R.L. Vernier, S.M. Mauer, The glomerular mesangium, Kidney Int 17 (1980) 141–154.

[88] M. Nangaku, W.G. Couser, Mechanisms of immune-deposit formation and the mediation of immune renal injury, Clin. Exp. Nephrol 9 (2005) 183–191. doi:10.1007/s10157-005-0357-8.

[89] T. Oite, S.R. Batsford, M.J. Mihatsch, H. Takamiya, A. Vogt, Quantitative studies of in situ immune complex glomerulo-nephritis in the rat induced by planted, cationized antigen, J Exp Med 155 (1982) 460–474.

[90] W.A. Border, H.J. Ward, E.S. Kamil, A.H. Cohen, Induction of membranous nephropathy in rabbits by administration of an exogenous cationic antigen: Demonstration of a pathogenic role for electrical charge, J Clin Invest 69 (1982) 451–461.

[91] L.Y.C. Agodoa, V.J. Gauthier, M. Mannik, Precipitating antigen-antibody systems are required for the formation of sub-epithelial electron dense immune deposits in rat glomeruli, J Exp Med 153 (1983) 1259–1271.

[92] T.M.J. Schmiedeke, F.W. Stšckl, R. Weber, Y. Sugisaki, S.R. Batsford, A. Vogt, Histones have high affinity for the glomerular basement membrane. effect of serum IgG and number of erythrocyte complement receptor type 1, J Exp Med 169 (1989) 1879–1894.

[93] F.W. Stšckl, S. Muller, S. Batsford, T. Schmiedeke, R. Waldherr, K. Andrassy, et al., A role of histones and ubiquitin in lupus nephritis? Clinical Nephrol 41 (1994) 10–17.

[94] J.B. Lefkowith, G.S. Gilkeson, Nephritogenic autoantibodies in lupus. Current concepts and continuing controversies, Arthritis Rheum 39 (1996) 894–903.

[95] D. Koffler, V. Agnello, R. Thoburn, H.G. Kunkel, Systemic lupus erythematosus: Prototype of immune complex nephritis in man, J Exp Med 134 (1971) 169s–179s.

[96] M. Mannik, C.E. Merrill, L.D. Stamps, M.H. Wener, Multiple autoantibodies form the glomerular immune deposits in patients with systemic lupus erythematosus, J Rheumatol 30 (2003) 1495–1504.

[97] J.E.M. Coremans, P.E. Spronk, H. Bootsma, M.R. Daha, d. V. van, L. Kater, et al., Changes in antibodies to C1q predict renal relapse in systemic lupus erythematosus, Am J Kidney Dis 26 (1995) 595–601.

[98] S. Uwatoku, M.J. Gauthier, M. Mannik, Autoantibodies to the collagen-like region of C1q deposit in glomeruli via C1q in immune deposits, Clin Immunol Immunopathol 61 (1991) 268–273.

[99] S. Uwatoku, M. Mannik, The location of binding sites on C1q for DNA, J Immunol 144 (1990) 3484–3488.

[100] L.A. Trouw, T.W. Groeneveld, M.A. Seelen, J.M. Duijs, I.M. Bajema, F.A. Prins, et al., Anti-C1q autoantibodies deposit in glomeruli but are only pathogenic in combination with glomerular C1q-containing immune complexes, J Clin Invest 114 (2004) 679–688. doi:10.1172/JCI21075.

[101] S. Fismen, A. Hedberg, K.A. Fenton, S. Jacobsen, E. Krarup, A.L. Kamper, et al., Circulating chromatin-anti-chromatin antibody complexes bind with high affinity.to dermo-epidermal structures in murine and human lupus nephritis, Lupus 18 (2009) 597–607. doi:10.1177/0961203308100512.

[102] L.A. Lee, W.L. Weston, G.G. Krueger, M. Emam, M. Reichlin, J.O. Stevens, et al., An animal model of antibody binding in cutaneous lupus, Arthritis Rheum 29 (1986) 782–788.

[103] S.A. Joselow, M. Mannik, Localization of preformed, circu-lating immune complexes in murine skin, J Invest Dermatol 82 (1984) 335–340.

[104] S.A. Joselow, A. Gown, M. Mannik, Cutaneous deposition of immune complexes in chronic serum sickness of mice induced with cationized or unaltered antigen, J Invest Dermatol 85 (1985) 559–563.

[105] N.S. Peress, F. Miller, W. Palu, The choroid plexus in passive serum sickness, J Neuropathol Exp Neurol 36 (1977) 561–566.

[106] J.T. Huang, M. Mannik, J. Gleisner, In situ formation of immune complexes in the choroid plexus of rats by sequential injection of a cationized antigen and unaltered antibodies, J Neuropathol Exp Neurol 43 (1984) 489–499.

[107] D.M. Santer, T. Yoshio, S. Minota, T. Moller, K.B. Elkon, Potent induction of IFN-alpha and chemokines by autoantibodies in the.cerebrospinal fluid of patients with neuropsychiatric lupus, J Immunol 182 (2009) 1192–1201.

[108] C.G. Cochrane, Mechanisms involved in the deposition of immune complexes in tissues, J Exp Med 134 (1971) 75S–89S.

[109] M.H. Wener, M. Mannik, M.M. Schwartz, E.J. Lewis, Rela-tionship between renal pathology and the size of circulating immune complexes in patients with systemic lupus eryth-ematosus, Medicine 66 (1987) 85–97.

[110] S. Uwatoko, M. Mannik, Low molecular weight C1q-binding IgG in patients with systemic lupus erythematosus consists of autoantibodies to the collagen-like region of C1q, J Clin Invest 82 (1988) 816–824.

[111] M.H. Wener, S. Uwatoko, M. Mannik, Antibodies to the collagen-like region of C1q in sera of patients with autoim-mune rheumatic diseases, Arthritis Rheum 32 (1989) 544–551.

[112] J. Kohro-Kawata, M.H. Wener, M. Mannik, The effect of high salt concentration on detection of serum immune complexes and autoantibodies to C1q in patients with systemic lupus erythematosus, J Rheumatol 29 (2002) 84–89.

[113] C. Bigler, M. Schaller, I. Perahud, M. Osthoff, M. Trendelenburg, Autoantibodies against complement C1q specifically target C1q bound on early.apoptotic cells, J Immunol (2009). doi:10.4049/jimmunol.0803573.

[114] A.N. Theofilopoulos, F.J. Dixon, The biology and detection of immune complexes. Adv, Immunol 28 (1979) 89–220.

[115] I. Csipo, E. Kiss, E. Bako, G. Szegedi, M. Kavai, Soluble complement receptor 1 (CD35) bound to immune complexes in sera of patients with systemic lupus erythematosus, Arthritis Rheum 52 (2005) 2950–2951. doi:10.1002/art.21245.

[116] A.K. Chauhan, T.L. Moore, Presence of plasma complement regulatory proteins clusterin (apo J) and vitronectin (S40) on circulating immune complexes (CIC), Clin Exp Immunol 145 (2006) 398–406. doi:10.1111/j.1365-2249.2006.03135.x.

[117] R.D. Inman, Immune complexes in SLE, Clin Rheum Dis 8 (1982) 49–62.

[118] W. Lloyd, P.H. Schur, Immune complexes, complement, and anti-DNA in excerbations of systemic lupus erythematosus (SLE), Medicine 60 (1981) 208–217.

[119] Z. Bentwich, N. Bianco, L. Jager, V. Houba, P. Lambert, W. Knapp, et al., Use and abuse of laboratory tests in clnical immunology: Critical considerations of eight widely used diagnostic procedures, Clin Exp Immunol 46 (1981) 662–674.

[120] E.J. Lewis, L.G. Hunsicker, S.P. Lan, R.D. Rohde, J.M. Lachin, A controlled trial of plasmapheresis therapy in severe lupus nephritis. the lupus nephritis collaborative study group. N Engl J Med 326 (1992) 1373–1379.

[121] D.T. Felson, M.P. LaValley, A.R. Baldassare, J.A. Block, J.R. Caldwell, G.W. Cannon, et al., The prosorba column for treatment of refractory rheumatoid arthritis: A randomized, double-blind, sham-controlled trial, Arthritis Rheum 42 (1999) 2153–2219.

[122] E.H. Sasso, C. Merrill, T.E. Furst, Immunoglobulin binding properties of the prosorba immunadsorption column in treatment of rheumatoid arthritis, Ther Apher 5 (2001) 84–91.

[123] B. Pfueller, K. Wolbart, A. Bruns, G.R. Burmester, F. Hiepe, Successful treatment of patients with systemic lupus erythematosus by immunoadsorption with a C1q column: A pilot study, Arthritis Rheum 44 (2001) 1962-193.

[124] B. Berner, A.K. Scheel, V. Schettler, K.M. Hummel, M. Reuss-Borst, G.A. Muller, et al., Rapid improvement of SLE-specific cutaneous lesions by C1q immunoadsorption, Ann Rheum Dis 60 (2001) 898-89.

[125] R. Clynes, Immune complexes as therapy for autoimmunity, J Clin Invest 115 (2005) 25–27. doi:10.1172/JCI23994.

[126] J.L. Ellsworth, M. Maurer, B. Harder, N. Hamacher, M. Lantry, K.B. Lewis, et al., Targeting immune complex-mediated hypersensitivity with recombinant soluble human Fcgamma-RIA (CD64A), J Immunol 180 (2008) 580–589.

[127] J.L. Ellsworth, N. Hamacher, B. Harder, K. Bannink, T.R. Bukowski, K. Byrnes-Blake, et al., Recombinant soluble human FcgammaR1A (CD64A) reduces inflammation in murine collagen-induced arthritis, J Immunol 182 (2009) 7272–7279. doi:10.4049/jimmunol.0803497.

[128] F.L. Ierino, M.S. Powell, I.F. McKenzie, P.M. Hogarth, Recombinant soluble human fc gamma RII: Production, characterization, and inhibition of the arthus reaction, J Exp Med 178 (1993) 1617–1628.

[129] R. Shashidharamurthy, R.A. Hennigar, S. Fuchs, P. Palaniswami, M. Sherman, P. Selvaraj, Extravasations and emigration of neutrophils to the inflammatory site depend on the interaction of immune-complex with fcgamma receptors and can be effectively blocked by decoy fcgamma receptors, Blood 111 (2008) 894–904.

Complement and Tissue Injury in SLE

Chau-Ching Liu, Joseph M. Ahearn

HISTORICAL OVERVIEW

Although its name may imply an ancillary role in immunity, recent studies have demonstrated that the complement system is not only a vital component of host defense, through participation in innate immune response and adaptive immunity, but also an "accidental" culprit in the pathogenesis of immune-inflammatory diseases.

Investigation of complement originated in the late 19th century, when a heat-labile serum component with nonspecific activity (now known to be complement) was found capable of facilitating the killing of bacteria by a heat-stable serum component with antigen specificity (now known to be antibodies) [1]. Subsequently, similar heat-labile and heat-stable factors were demonstrated to mediate lysis of erythrocytes sensitized by immune sera. In 1899, the term "complement" was introduced by Paul Ehrlich to emphasize that the heat-labile factors present in fresh serum "complemented" the heat-stable specific factors mediating immune bacteriolysis and hemolysis [2]. At that time, there was substantial debate over whether complement was a group of factors or a uniform substance. Protein biochemistry studies conducted in the early 1900s soon began to provide some answers. Since then, laborious studies by many investigators have established that complement is not a single substance but consists of multiple proteins (reviewed in [3]). By the late 1960s and throughout the1970s and 1980s, the biochemical nature of complement was nearly resolved and research in complement genes was rapidly progressing. Consequently, significant investigative efforts were geared toward delineating the biological functions and regulatory mechanisms of complement. As a result, an unexpectedly large number of regulatory proteins and receptors have been identified and characterized [4–6]. These studies have laid a solid foundation for our understanding of complement in health and disease.

The complement system is arguably linked more intimately to systemic lupus erythematosus (SLE) than to any other human disease. This association has been recognized for decades and, until recently, was viewed as inexplicably paradoxical. Two seemingly irreconcilable early observations formed the foundation for this conundrum. First, since 1951, Vaughan [7] and subsequently other investigators have independently reported that decreased complement levels in the sera are characteristic of patients with SLE. Prominent among those earlier studies, Schur and colleagues suggested that complement levels were of particular value in following and evaluating patients with SLE, especially those with nephritis [8]. Those seminal observations were followed by a large body of work from many laboratories demonstrating that complement activation, reflected by depressed serum levels of C3, C4, and CH_{50}, plays a major role in the tissue inflammation and organ damage that result from lupus pathogenesis.

Seemingly paradoxical to these findings was a second set of observations that demonstrated a strong association between hereditary homozygous deficiency of the classical pathway components and development of SLE (reviewed in [9]). In fact, inherited complement deficiency is still recognized as conferring the greatest known risk for development of SLE. Thus, for decades this paradox has been pondered. How is it that complement deficiency may be causative in SLE, yet activation of this same inflammatory cascade is detrimental in patients who already have the disease?

Equipped with a wealth of information accumulated over the past one and a half centuries and an armory of sophisticated techniques, investigators are now poised to elucidate this perplexing link between complement and SLE and identify potential strategies and opportunities for mining the complement system for lupus genes, biomarkers, and therapeutics.

BIOLOGY OF THE COMPLEMENT SYSTEM

The complement system comprises more than 30 plasma and membrane-bound proteins that form three distinct pathways (classical, alternative, and lectin) designed to protect against invading pathogens (Figure 20.1). Many of the complement proteins exist in plasma as functionally inactive pro-proteins until appropriate events trigger their activation. Once activated, the proteins within each pathway undergo a cascade of sequential serine protease-mediated cleavage events, release biologically active fragments, and self-assemble into multimolecular complexes. This activation process is a series of enzymatic reactions in which each enzyme (complement) molecule generated at one step can generate multiple enzyme (complement) molecules at the next step, thus allowing for tremendous amplification of activated molecules. Mechanistically, the activation of the complement system can be viewed as a two-phase process. The first phase, unique to each of the three activation pathways, involves the early complement components that lead to the formation of the so-called C3 convertases. The second phase, common to all three pathways once they converge, results in the formation of a lytic complex consisting of the terminal complement components (Table 20.1).

Complement Activation Pathways

Classical pathway

The classical pathway of complement activation, which is responsible for executing a major effector mechanism of antibody-mediated immune responses, is thought to play an important role in SLE pathogenesis. Five proteins, C1q, C1r, C1s, C4, and C2, are specific to activation of the classical pathway (Figure 20.1). Activation of this pathway begins when C1q binds to the Fc portion of IgM or IgG (particularly IgG1 and IgG3) molecules that are bound to an antigen. The binding of C1q to an antigen—antibody complex (immune complex) leads to activation of C1r (a serine protease), which, in turn, leads to activation of C1s (also a serine protease). C1s enzymatically cleaves the other two classical pathway proteins, C4 and C2, generating and releasing two small soluble polypeptides, C4a and C2b. At the same time, this proteolytic cleavage leads to the formation of a surface-bound bimolecular complex, C4b2a, which functions as an enzyme and is referred to as the classical pathway C3 convertase. Recent studies have shown that in addition to immune complexes, a variety of other molecules, which are generally products of inflammatory reactions, can also initiate activation of the classical pathway by interacting with C1q. These include C-reactive protein [10], amyloid P component [11], β-amyloid protein [12], and DNA [13].

Alternative pathway

In parallel with the much-noticed studies on antibody-dependent activation of complement, Pillemer and colleagues made a series of observations in the 1940s and 1950s that eventually led to the discovery of the alternative pathway of complement activation [14, 15]. Unlike activation of the classical pathway, activation of the alternative pathway is not dependent on antibodies. Three plasma proteins, Factor B, Factor D, and properdin, are unique to the alternative pathway (Figure 20.1). Normally, complement C3 undergoes a so-called C3 tickover process, a spontaneous, low-rate hydrolysis that generates $C3(H_2O)$ (also referred to as C3i*) and subsequently C3b fragments [16]. A fraction of these spontaneously generated $C3(H_2O)$ and C3b may covalently attach to the surface of microbial pathogens and host cells via thioester bonds, which in turn are capable of binding Factor B. Once bound, Factor B is

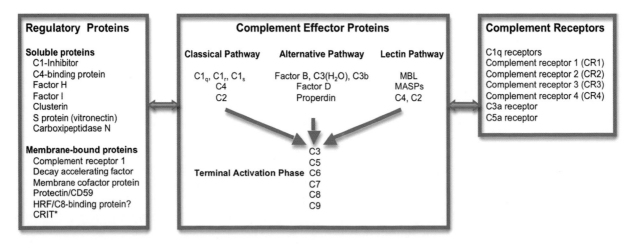

FIGURE 20.1   The complement system.

TABLE 20.1 Effector Components of the Human Complement System

Effector proteins	Pathway involved/ function	M_r (KD)
C1q	Classical/recognition, binding	450 (a six-subunit bundle)
C1r	Classical/serine protease	85
C1s	Classical/serine protease	85
C4[a]	Classical/serine protease (C4b); anaphylatoxin (C4a)	205 (a 3-chain, $\alpha\beta\gamma$, complex)
C2	Classical/serine protease (C2a); small fragment with kinin-activity (C2b)	102
C3[b]	Terminal/membrane binding, opsonization (C3b); anaphylatoxin (C3a)	190 (a 2-chain, $\alpha\beta$, complex)
C5	Terminal/MAC component (C5b), anaphylatoxin (C5a)	190 (a 2-chain, $\alpha\beta$, complex)
C6	Terminal/MAC component	110
C7	Terminal/MAC component	100
C8	Terminal/MAC component	150 (a 3-chain, $\alpha\beta\gamma$, complex)
C9	Terminal/MAC component	70
Factor B	Alternative/serine protease	90
Factor D	Alternative/serine protease	24
Properdin	Alternative/stabilizing C3bBb complexes	55 (monomers); 110, 165, 220, or higher (oligomers)
MBL	Lectin/recognition, binding	200-400 (2–4 subunits with three 32 KD chains each)
MASP-1	Lectin/serine protease	100
MASP-2	Lectin/serine protease	76

[a]*Serum concentration range considered normal: 20—50 mg/dL.*
[b]*Serum concentration range considered normal: 55—120 mg/dL.*

normally degraded by several regulatory proteins, thereby preventing self-damage of the host cells and tissue. However, the C3bBb complexes associated with microbial pathogens, which do not express these regulatory proteins, will remain intact and can be further stabilized by the binding of properdin.

Lectin pathway

A third complement activation pathway was discovered in the late 1980s [17]. The lectin pathway shares several components with the classical pathway (Figure 20.1). Initiation of the lectin pathway is mediated through the binding of mannose-binding lectin (MBL) to a variety of repetitive carbohydrate moieties such as mannose, N-acetyl-D-glucosamine, and N-acetyl-mannosamine, which are abundantly present on a variety of microorganisms [18, 19]. MBL, structurally similar to C1q, forms complexes in the plasma with mannose-binding protein-associated serine proteases (MASPs, such as MASP-1, MASP-2, and MASP-3) [20, 21]. Under physiological conditions, MBL does not bind to mammalian cells, probably because these cells lack mannose residues on their surface. Once bound to microbial pathogens, MASPs, particularly MASP-2 within the MBL complex, can cleave C4 and initiate the complement cascade. At this point, the lectin pathway intersects with the classical pathway and a C3 convertase, i.e., the C4b2a complex, is eventually generated.

Terminal activation phase

C3 convertases, generated during the first stage of complement activation, cleave C3, the central and most abundant component of the complement system. This proteolytic cleavage gives rise to a smaller C3a fragment and a larger C3b fragment. Similar to C4a, C3a is a soluble polypeptide that diffuses away. C3b molecules, if not hydrolyzed and inactivated in the fluid phase, can bind covalently to the surface of microbial pathogens or to immune complexes initially responsible for activating the system. The C3b molecules associate with C4bC2a or C3bBb complexes to form the C5 convertase. The C5 convertase cleaves C5 and initiates activation and assembly of the terminal components, C5, C6, C7, C8, and C9, into the C_{5b-9} membrane attack complex (MAC; also known as terminal complement complex [TCC]) on the surface of foreign pathogens [22, 23].

Regulators of Complement Activation and Complement Receptors

Regulators of complement activation

In humans and other mammals, the complement system is controlled by many regulatory proteins to ensure that this effective machinery does not become

cleaved into Ba and Bb fragments by Factor D (a serine protease). While the small, soluble Ba fragment diffuses away from the activation site, the Bb fragment remains associated with C3b. Similar to the C4b2a complex in the classical pathway, the surface-bound C3bBb complex serves as the alternative pathway C3 convertase. The C3bBb complexes, if bound to mammalian cells, are overactive and inflict undesirable inflammatory reactions on the host (Table 20.2) (see [5, 6] for recent reviews).

In humans, six membrane-anchored or transmembrane regulatory proteins have been identified: decay accelerating factor (DAF; also known as CD55; a glycophosphatidylinositol [GPI]-anchored protein), membrane cofactor protein (MCP; CD46; a transmembrane protein), complement receptor 1 (CR1; CD35; a transmembrane protein), protectin (CD59; a GPI-anchored protein), homologous restriction factor (HRF; a transmembrane protein), and C2 receptor inhibitor trispanning (CRIT; a transmembrane protein). DAF, MCP, and CR1 share structural similarities with two soluble regulatory proteins, Factor H and C4-binding protein (C4BP). Collectively, they belong to the so-called regulators of complement (RCA) family, whose members contain the characteristic 60–70 amino acids-long domains called short consensus repeats (SCRs) [24]. Genes encoding these regulatory proteins are closely located in the RCA cluster on chromosome 1. Recently, CRIT, a novel transmembrane regulatory protein, was identified and shown to be capable of blocking C2 cleavage and thus preventing the formation of C4bC2a complexes [25]. Other soluble regulatory proteins, although not belonging to the RCA family proteins, exhibit their respective structures and execute unique regulatory functions [5].

To control the potent consequence of complement activation, soluble or cell-membrane regulatory proteins need to act at multiple steps of the activation pathways using different mechanisms, functioning as proteolytic enzymes (Factor I, carboxypeptidase), cofactors for proteolytic enzymes (Factor H, CR1, MCP), protease inhibitors (C1-INH), physical dissociators of multimolecular convertases (C4BP, DAF, Factor H, CRIT), or inhibitors of MAC formation and membrane insertion (CD59, HRF) (Figure 20.2).

Receptors for complement components

Receptors for proteolytic fragments of complement proteins (e.g., C3a, C3b, C4b, iC3b, C3d, and C5a) and, in some circumstances, for complement proteins with altered conformation (e.g., C1q) are expressed by a wide

TABLE 20.2   Regulatory Proteins of the Human Complement System

Soluble regulatory protein	Function	M_r (KD)
C1-inhibitor (C1-INH)	Removing activated C1r and C1s from the C1 complex	105
C4-binding protein (C4BP)	Binding to C4b and displacing C2a in the C4bC2a complex; accelerating decay of C3 convertase; cofactor for Factor I	570 (an 8-subunit complex)
Factor H	Displacing Bb in the C3bBb complex; cofactor for Factor I	160
Factor I	Serine protease capable of cleaving C3b and C4b	88
Clusterin	Preventing insertion of soluble C5b-7 complexes into cell membranes	70
S protein (Vitronectin)	Preventing insertion of soluble C5b-7 complexes into cell membranes	84
Carboxypeptidase N	Inactivating anaphylatoxins	280 (a multisubunit complex)

Membrane-bound regulatory protein	Function	M_r (KD)
CR1[a] (CD35)	Binding C3b and C4b; co-factor for Factor I	160–250 (4 isoforms[g])
MCP[b] (CD46)	Promoting C3b and C4b inactivation by Factor I	45–70 (different glycosylation forms)
DAF[c] (CD55)	Accelerating decay of the C3bBb and C4b2a complexes	70
Protectin (CD59)[d]	Preventing C9 incorporation into the MAC in a homologous restriction manner	18–20
HRF[e]	Binding to C8; preventing MAC formation/membrane insertion in a homologous restriction manner	65
CRIT[f]	Binding C2 and preventing formation of C3 convertase	31–32

[a]Complement receptor 1.
[b]Membrane cofactor protein.
[c]Decay accelerating factor.
[d]Also known as H19; to be differentiated from HRF.
[e]Homologous restriction factor; also known as C8-binding protein.
[f]C2 receptor inhibitor traspanning.
[g]Four isoforms with different numbers of SCR and displaying distinct M_r under reducing condition: CR1-A (220 kD), CR1-B (250 kD), CR1-C (190 kD), and CR1-D (280 kD).

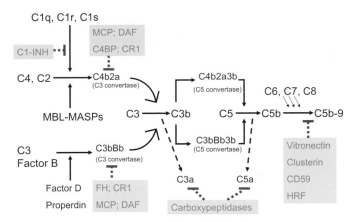

FIGURE 20.2 Regulation of complement activation.

spectrum of cells and serve pivotal roles in executing many of the effector functions of complement (Table 20.3). Receptors for anaphylatoxins C3a and C5a have been identified and cloned. The C3a receptor (C3aR) and C5a receptor (C5aR; CD88) belong to the rhodopsin family of the G-protein-coupled 7-transmembrane-domain receptors, and are expressed on leukocytes, endothelial cells, podocytes, and proximal tubular epithelial cells in the kidney [26, 27]. Interactions of C3aR and C5aR with their respective ligands (C3a, C3adesArg, C5a, and C5adesArg) are essential for the intracellular signaling processes that lead to leukocyte degranulation, production of cytokines, release of vasoactive substances, and other anaphylaxis and inflammatory responses.

Recent studies have led to the identification of at least four potential receptors for C1q cC1qR (calreticulin; a collectin receptor), gC1qR, C1qRp, and CR1 (CD35) [28–31]. Binding of C1q-opsonized immune complexes to endothelial cells via C1q receptors has been reported to induce expression of adhesion molecules on endothelial cells and thus enhance leukocyte binding/extravasation [32]. On other cell types, C1q binding, presumably via distinct receptors, has been shown to enhance phagocytosis, increase generation of reactive oxygen intermediates, and activate platelets [33–35].

CR1 (CD35) and CR2 (CD21), two major receptors for C3- and C4-derived fragments, belong to the RCA family [4, 24]. CR1 is widely expressed by erythrocytes, neutrophils, monocytes/macrophages, B lymphocytes, some T lymphocytes, and glomerular podocytes [36, 37]. CR1 binds primarily C3b and C4b. One important function of CR1 expressed on erythrocytes is to bind and clear immune complexes [38]. CR1 also plays a role in regulation of complement activation by serving as a co-factor for Factor I, which is responsible for cleaving C3b and C4b to iC3b and iC4b [39]. CR2 is expressed mainly on B lymphocytes, activated T lymphocytes, and follicular dendritic cells, and binds

primarily iC3b, C3dg, and C3d [40, 41]. CR2, together with its cognate complement ligands, is a critical link between the innate and adaptive immune systems (see further discussion in "Complement: An Important Bridge Between Innate Immunity and Adaptive Immunity" below).

CR3 and CR4 belong to the β_2 integrin family and are composed of two subunits: a common β chain (CD18) and a specific α chain (CD11b in CR3 and CD11c in CR4). These receptors are expressed on phagocytic cells (e.g., neutrophils, monocytes, and macrophages), antigen-presenting cells (e.g., dendritic cells), and follicular dendritic cells [42]. CR3 and CR4 not only play important roles in phagocytic removal of C3-opsonized pathogens, but also participate in mediating adhesion of phagocytes to endothelial cells [43, 44].

Effector Functions of Complement

The complement system is traditionally thought to have four biological functions:

1. Opsonization;
2. Activation of inflammation;
3. Clearance of immune complexes;
4. Osmotic lysis of invading microorganisms [45].

These functions are mediated by the soluble or surface-bound fragments of activated complement proteins (generally referred to as complement split products or complement activation products), which in turn interact with specific membrane receptors expressed on various cell types.

The soluble proteolytic fragments, C3a, C4a, and C5a, are highly potent pro-inflammatory molecules. They attract and activate leukocytes by binding to specific receptors (e.g., C5a receptor) expressed on those cells. The larger fragments, C3b, C4b, and their derivatives (e.g., iC3b and iC4b), can remain bound to the surface of microbial pathogens (opsonization) and facilitate recognition and uptake of the opsonized pathogens by phagocytic cells. This function is mediated through the binding of these complement-derived fragments to CR1 (for C3b and C4b), CR3 (for iC3b), and perhaps CR4 (for iC3b) expressed on phagocytes.

The binding of C4b and C3b to immune complexes prevents their aggregation into insoluble complexes and enhances their clearance. The clearance of C3b/C4b-opsonized immune complexes is believed to be mediated by erythrocytes that express CR1 and are capable of transporting immune complexes to macrophages of the reticuloendothelial system in the spleen and liver [38]. Finally, the C_{5b-9} MACs may perturb the osmotic equilibrium and/or disrupt the integrity of the surface membrane of target cells, thereby causing lysis of these

TABLE 20.3　Receptors of the Human Complement System

Complement receptor	Complement ligand(s)[a]	M_r (KD)
CR1 (CD35)	C3b; C4b; iC3b; C1q	190–280[b] (single chain)
CR2 (CD21)	C3dg/C3d; iC3b;	140–145 (single chain)
CR3 (CD11b/CD18)	iC3b	170/95 (2-chain, $\alpha\beta$)
CR4 (CD11c/CD18)	iC3b	150/95 (2-chain, $\alpha\beta$)
cC1qR (calreticulin)	C1q (collagenous tail); MBL	60 (single chain)
gC1qR	C1q (globular head)	33 (tetramer of 33 KD subunits)
C1qRP	C1q (collagenous tail)	126 (single chain)
C3a receptor	C3a	50? (single chain)
C5a receptor (CD88)	C5a	50 (single chain)

[a]Noncomplement ligands (e.g., Epstein-Barr virus for CR2 and fibrinogen for CR3 and CR4) not listed.
[b]Four isoforms with different numbers of SCR and displaying distinct M_r under reducing condition: CR1-A (220 kD), CR1-B (250 kD), CR1-C (190 kD), and CR1-D (280 kD).

cells. Interestingly, on nucleated cells, sublytic levels of MACs can instead stimulate cellular activities including Ca^{2+} influx, activation of phospholipases and protein kinases, and production of pro-inflammatory cytokines and arachidonic acid-derived mediators [46–48].

Complement: An Important Bridge Between Innate Immunity and Adaptive Immunity

For several decades, the role of the complement system was thought to be limited to these four effector functions. However, there has been a recent explosion in discovery of additional roles for complement. An increasing number of studies have shown that innate immunity and adaptive immunity, the two arms of the immune system, collaborate in an intricate way to elicit efficient immune responses against infectious agents [49, 50]. The complement system, particularly C3, its derivative fragments, and their cognate receptors, plays an important role in this collaboration. First, antigens (and immune complexes) decorated with C3d, the end cleavage product of C3 and a major ligand for CR2, are capable of cross-linking the B cell receptors to the CR2/CD19/TAPA-1 co-receptor complexes and thus facilitating B cell activation and enhancing humoral immune responses [49]. Second, antigens (and immune complexes) opsonized by C3 can be retained in the germinal centers of secondary lymphoid follicles via binding to CR2-expressing follicular dendritic cells [41]; the retained antigens provide essential signals for survival and affinity maturation of B cells as well as for generation of memory B cells [51]. Third, opsonization of pathogens by complement components facilitates their uptake by phagocytes and antigen-presenting cells, and thus may enhance presentation of antigens and initiation of specific immune responses. Fourth, complement activation products generated at sites of infection

can recruit inflammatory cells and immune effector cells to help eliminate pathogenic antigens. In addition, complement appears to play an important role in opsonizing apoptotic cells and facilitating their clearance (see further discussion in "Possible Mechanisms Underlying the Complement Deficiency–SLE Association").

COMPLEMENT AND INFLAMMATION

Inflammation is a complex pathophysiologic process that can be initiated by any stimuli causing cell injury and involve numerous cellular and soluble mediators. The ultimate goal of the body's inflammatory response is to eliminate invading microorganisms or other injurious stimuli and to repair tissue damage caused by the inciting agents. Because the complement system is actively involved in various steps during the execution and regulation of an inflammatory response, tissue damage mediated by complement almost always occurs when the complement system is faithfully executing its physiologic function. Many of the classical clinical signs of inflammation — *tumor* (swelling), *rubor* (redness), *calor* (warmth), and *dolor* (pain) — are mediated directly or indirectly by the products of complement activation (reviewed in [52]).

Initial tissue insult by invading agents or harmful stimuli triggers an orchestrated cascade of responses including activation of the complement system. C3a and C5a, anaphylatoxins generated as a result of complement activation, are well-established chemotactic factors for leukocytes and are also capable of upregulating the expression of adhesion molecules on endothelial cells [53]. Thus, complement anaphylatoxins, particularly C5a, play a major role in recruiting neutrophils, macrophages, and other inflammatory cells to the infected site, where they release large bursts of cytokines, chemokines, and inflammatory mediators that can alter the vasculature tone/permeability, activate endothelial cells, recruit additional inflammatory cells, and inadvertently damage surrounding tissues. For example, during reperfusion of ischemic tissue, complement may be activated by natural antibodies that bind to ischemia-injured cells and inflict significant damage on surrounding cells and tissue. Profound activation of the complement system during bacterial sepsis undoubtedly also enhances secretion of cytokines (e.g., TNFα, IL-1β, and IL-6) and other injurious mediators (eicosanoids, tissue metalloprotease, etc.) that in turn intensify local and systemic tissue damage [54—56].

Another major function of complement is to facilitate removal of immune (antigen—antibody) complexes and dead cells. Immune complexes, if not solubilized and cleared properly, can form insoluble aggregates that deposit in various tissues and may bind C1q and trigger complement activation. Apoptotic and necrotic cells, if remaining in the tissue for a sufficient time, may also activate the complement system. Once activated, the products of the complement cascade can initiate and perpetuate the above-mentioned pathophysiologic responses, culminating in damage of the surrounding tissue. The kidneys, skin, and vessels appear to be most susceptible to such pathogenesis. The resulting glomerulonephritis, skin rashes, and vasculitis commonly occur in patients with immune complex diseases such as SLE (see further discussion in the next section).

COMPLEMENT AND SLE

The participation of complement in the etiopathogenesis of SLE has been scrutinized over the past several decades. Suffice it to say that the role of complement in SLE is both complex and paradoxically intriguing. On the one hand, activation of the complement system is thought to play an important role in tissue inflammation/damage in SLE as a consequence of tissue deposition of immune complexes formed by autoantigens and autoantibodies [57—59]. On the other hand, a hereditary deficiency of a component of the classical pathway (C1, C2, or C4) has been associated with the development of SLE [9, 60]. These seemingly discordant roles for complement may be reconciled by studies conducted over the past several years. These studies have demonstrated that, while the complement system plays a role in tissue-destructive inflammatory processes once SLE is established in a patient [58, 59], it also participates in maintaining immune tolerance to prevent the development of SLE [57, 61, 62]. This section will be devoted to the detrimental role of complement, while the protective role of complement will be discussed in the next section.

Immune Complex Abnormalities and Complement Activation

SLE has long been viewed as the prototype of autoimmune complex diseases. Considerable evidence has indicated that many of the clinical and pathologic manifestations of SLE can be attributed to decreased solubility and impaired disposal of immune complexes, consequent deposition of immune complexes in various tissues, and activation of the complement system. Detailed mechanistic factors influencing immune complex formation and deposition are covered elsewhere in this volume. Briefly, in patients with SLE, decreased serum levels of C3 and C4 (due to genetic and/or acquired factors) may not permit sufficient binding of C3 and C4 fragments to the antigen—antibody lattice and thereby prevent the formation of small, soluble immune complexes [63]. Furthermore, reduced levels of CR1 on erythrocytes, frequently detected in patients with SLE, may lead to impaired binding, processing, and transporting of immune complexes to phagocytes of the reticuloendothelial system [64, 65]. Consequently, abnormally large quantities of immune complexes are likely to circulate for prolonged periods of time and form insoluble aggregates that may deposit in various tissues. Alternatively, insoluble immune complexes may form *in situ* as a result of the "planting" of autoantigens and subsequent binding of autoantibodies at specific loci (see the next section). Deposited immune complexes do not seem to cause tissue damage directly but provide ample binding sites for complement components. The ensuing activation of the complement system causes the release of various mediators, promotes cellular infiltration and interaction, and culminates in tissue damage.

Mechanisms of Immune Complex/Complement-mediated Tissue Injury

In patients with SLE, immune complexes have been identified in the circulation and in various tissues/organs, such as glomeruli, small and medium-sized blood vessels, and at the dermal—epidermal junction. The susceptibility of certain anatomical sites to immune complex deposition and resultant inflammatory reactions provides a molecular basis underlying diverse clinical manifestations of SLE. As a case study, the role of complement in renal injury in SLE patients and animal models will be outlined here. Detailed discussion of the histopathologic findings and pathogenesis of other organ-specific clinical manifestations in SLE can be found elsewhere in this volume.

Lupus nephritis remains the foremost cause of morbidity and mortality in SLE. The diverse and complex histopathologic features of lupus nephritis, involving glomeruli, tubulointerstitia, and vasculature, form the basis of various classification schemes of lupus nephritis [66, 67]. The characteristic pathology of lupus nephritis includes mesangial hypercellularity, focal (defined as <50% of glomeruli involved) or diffuse involvement, infiltration of inflammatory cells, and deposition of immune complexes. The locations of immune complex deposition distinguish various subtypes of lupus nephritis. For example, mesangial, subendothelial, and subepithelial immune deposits are characteristic features of mesangial proliferative lupus nephritis (WHO class II), proliferative glomerulonephritis (WHO classes III and IV), and lupus membranous nephropathy (WHO class V), respectively [67]. Immuno-detection of different immunoglobulin isotypes including IgG, IgM, and IgA, in combination with C3, C4, and C1q, often referred to as the "full house" picture, has been found in about one-quarter of patients with SLE but almost never in patients with non-lupus renal disease [68]. Decreased C3 and increased anti-dsDNA levels in the serum are common in patients with lupus nephritis [69]. Moreover, multiple autoantibodies, including anti-dsDNA, antihistone, antichromatin, anti-Sm, anti-C1q, anti-SSA, and anti-SSB, have been eluted from kidneys affected by lupus nephritis [70, 71]. These findings provide supportive, albeit mostly circumstantial, evidence for the involvement of immune complexes/complement in the pathogenesis of lupus nephritis.

The nature of the "nephritogenic" autoantibodies/immune complexes has been a subject of intense debate [72]. The autoantibodies most closely associated with lupus nephritis are anti-dsDNA antibodies and antinucleosome antibodies [73–75]. Several theories have been put forth to explain the nephritogenicity, i.e., the ability to bind to the kidney and provoke tissue damage, of these autoantibodies. First, earlier studies showed that multiple subtypes of anti-dsDNA antibodies exist and that immune complexes deposited in the kidneys are predominantly those consisting of high-affinity IgG antibodies [73, 76–78]. Interestingly, it has been shown that proliferative glomerulonephritis is primarily associated with IgG1 and IgG3 immune complexes, whereas membranous nephropathy is mainly associated with the IgG4 subclass [79–81]. Second, physicochemical characters of autoantibodies/immune complexes, e.g., cationic charge and size, also appear to be important in dictating their potential for depositing in the kidney and inducing nephritis. For example, the cationic charge of nucleosomes (most likely due to histones) and nucleosome-containing immune complexes may make them bind preferentially to the anionic components of glomerular basement membranes (e.g., heparan sulfate) [82–86]. Third, the propensity of certain anti-DNA or antinucleosome autoantibodies to bind to renal tissues is thought to originate from their polyreactivity to not only nuclear autoantigens, but also various glomerular constituents. For example, several investigators have shown that subsets of lupus autoantibodies are capable of binding to human mesangial cells, endothelial cells, and glomerular basement membrane [87–90]. One of the candidate cross-reactive antigens has recently been identified as α-actinin [91]. Based on currently available data, it is reasonable to conclude that: (1) Nephritogenic autoantibodies/immune complexes encompass a broad spectrum of specificities; (2) Distinct nephritogenic auto-antibodies/immune complexes may play unique roles in any given patients with lupus nephritis.

Where do the kidney-destined immune complexes, with the aforementioned dissimilar physicochemical characteristics and antigenic specificities, originate? First, it is conventionally thought that excessive levels of preformed immune complexes in the circulation lead to their extravasation and passive trapping in the mesangium and subendothelial space within the glomeruli. This entrapping process is probably aided by the unique physiological (e.g., large volume of plasma constantly filtered through the glomeruli) and anatomical (e.g., fenestrated endothelium of glomerular capillaries) features of renal glomeruli [80]. However, the actual concentrations of circulating immune complexes in patients with SLE are often relatively low and the presence of anti-dsDNA in these complexes has not been convincingly demonstrated [75, 92], rendering this mechanism of immune complex deposition less plausible. Because of the lack of solid support for the circulating immune complex hypothesis, a second hypothetical *in situ* formation mechanism has been proposed [93]. The *in situ* model postulates that immune complexes may develop on site when antibodies bind (specifically or cross-reactively) to intrinsic glomerular constituents or to extrinsic antigens "preplanted" in the glomeruli via cationic/anion charge interactions. According to the *in situ* model, immune complexes can form locally at mesangial, subendothelial, or subepithelial sites, correlating with the typical histopathologic changes of mesangial lupus nephritis, proliferative glomerulonephritis, and lupus membranous nephropathy, respectively. This alternative source of renal immune complex deposits has received considerable support from observational studies in humans and experimental studies in animals (reviewed in [80]).

Once immune complexes are trapped or form *in situ* in the glomeruli, they are ready to ignite the inflammatory arsenal and inflict tissue injury. The complement system is clearly the principal weapon. Upon binding to the immunoglobulin components of the immune complexes, C1q initiates the activation of the classical pathway, leading to local generation of anaphylatoxins (C3a and C5a) and formation of C_{5b-9} MACs. If generated at a site accessible to the circulation (e.g., subendothelial space or mesangium), C3a and C5a not only can act as chemoattractants to recruit leukocytes from the circulation, but also are capable of stimulating infiltrating inflammatory cells and endogenous glomerular cells (e.g., mesangial cells, endothelial cells, and epithelial cells) via receptors expressed on these cells. In turn, activated inflammatory cells (e.g., neutrophils, macrophages, and mast cells) and endothelial cells upregulate expression of adhesion molecules, secrete vasoactive substances, chemokines, cytokines, proteases, and growth factors, and may also activate the coagulation system. Activated mesangial cells may overproliferate and produce growth factors, cytokines, and extracellular matrix. A recent study showed increased expression of $C3aR$ in the kidneys affected by lupus nephritis but not other diseases [94], supporting a role for C3a in mediating renal injury in SLE.

The final products of complement activation, the C_{5b-9} MACs, may damage renal cells via a direct cytolytic mechanism. In addition, these multimolecular complexes, if inserted in sublytic quantities into the membranes of glomerular cells and infiltrating inflammatory cells, may activate these cells and further contribute to the inflammatory response [46−48, 95, 96]. Such a concerted and interconnected cascade of reactions leads to tremendous tissue injury and results in histopathologic changes characteristic of proliferative glomerulonephritis − mesangial hypercellularity, fibrinoid necrosis, and microthrombi.

In comparison with subendothelial immune deposits, immune complexes deposited in the subepithelial space, as occurs in lupus membranous nephropathy, may trigger complement-mediated injury of podocytes but generally do not induce apparent inflammatory reactions [80]. This is probably due to the lack of direct access of circulating inflammatory cells to the subepithelial space that is separated by the glomerular basement membrane [80]. In this case, the diverse cellular effects of glomerular cells triggered by membrane insertion of sublytic amounts of the C_{5b-9} MACs, including release of cytokines and oxidants, activation of phospholipases, and activation of the apoptotic machinery, are likely to be responsible for the renal pathophysiologic changes.

COMPLEMENT DEFICIENCY AND SLE (AND OTHER DISEASES)

Hereditary and Acquired Deficiency of Complement Effector Proteins

Hereditary complement deficiency in humans has been reported for almost every component of the complement system, and a deficiency of any complement component is significantly associated with specific human diseases including SLE. These intriguing associations are discussed in Chapter 2, "Genetic Susceptibility and Class III Complement Genes."

Complement deficiency may also occur as an acquired phenomenon. When complement is increasingly consumed during heightened active states of an underlying disease, acquired deficiency can occur and usually involves multiple components simultaneously. For example, C1 inhibitor (C1-INH) is capable of inhibiting not only the complement system but also the clotting and kinin-kallikrein systems. C1-INH deficiency causes unregulated activation of the complement classical and lectin pathways (C1r and C1s directly and C4 and C2 consequently). As a result, patients with C1-INH deficiency have reduced serum levels of C4 and C2 and typically suffer from hereditary angioneurotic edema. However, development of SLE in some C1-INH-deficient patients has been reported [97, 98]. These reports not only echo the association between hereditary complement deficiency and SLE described above, but also underscore the notion that early components of the complement classical pathway play a protective role against development of SLE.

In patients with SLE, autoantibodies against complement components have also been found to cause acquired complement deficiency. For example, C3 nephritic factor, an autoantibody capable of stabilizing the alternative pathway C3 convertase BbC3b, can cause consumption of complement proteins via unregulated activation of the alternative and terminal pathways [99]. Another autoantibody reactive with the first complement component, C1q, has been detected at increased frequencies in patients with SLE [100, 101]. A significant portion of patients with SLE also develop functional C1q deficiency secondary to the presence of anti-C1q antibodies [9]. Although the pathophysiologic role of anti-C1q in SLE is largely unknown, its presence in patients with SLE has been associated with lupus nephritis [101−103].

Deficiency and Dysfunction of Complement Regulatory Proteins

Deficiency or functional abnormalities of complement regulatory proteins have also been associated with human diseases, in which the disease manifestations and pathogenesis appear attributable to dysregulated complement-mediated injury of target cells and tissues (Table 20.4). The most extensively studied association is that of C1-INH deficiency with hereditary angioneurotic edema, as mentioned in the preceding section. As for other regulatory proteins, deficiencies of DAF and CD59 on bone marrow-derived cells have been identified as a result of somatic mutations of a gene essential for the synthesis of the GPI anchor. Erythrocytes deficient in DAF and CD59 are highly susceptible to spontaneous complement-mediated lysis, as seen during episodes of hemolysis in patients with paroxysmal nocturnal hematuria (PNH) [104]. Germline mutation of the *Daf* gene leading to a global DAF deficiency has been identified in a small number of individuals [105]. Erythrocytes of these individuals are deficient in the Cromer blood group antigen associated with DAF (*Inab* phenotype) and were shown to be sensitive to hemolysis *in vitro* [106]. However, no episodes of spontaneous hemolysis *in vivo* have been observed in individuals with the *Inab* phenotype. Inherited complete deficiency of CD59 has been reported in one patient who developed PNH [107]. Although no evidence is currently available for direct links between genetic deficiency of DAF or CD59 and SLE, it has been reported that the levels of CD55 and CD59 on erythrocytes were downregulated in some patients with SLE [108].

Evidence accumulated in recent years has linked genetic polymorphism-induced dysfunction of Factor H to age-related macular degeneration (AMD) [109]. The retinal deposits of AMD contain several complement components such as C3a, C5a, and C_{5b-9} and other inflammatory constituents, suggesting impaired regulation of alternative pathway activation as an underlying pathogenic mechanism. Interestingly, genetic polymorphisms leading to gain-of-function of Factor B, the initiating component of the alternative pathway, have also been associated with AMD [110]. Genetic mutations with Factor H, Factor B, and another complement regulator, MCP, have also been associated with atypical hemolytic uremic syndrome (aHUS) [111–113]. These associations indicate that dysregulated complement activation, particularly of the alternative pathway, may play an important role in the pathogenesis of aHUS. Because Factor H is the major inhibitor of the alternative pathway, its deficiency leads to uncontrolled activation/consumption of C3. Thus, it is not surprising that Factor H deficiency has also been associated with recurrent infections, SLE, and membranoproliferative glomerulonephritis in both human and experimental animal studies [114, 115].

Reduced levels of CR1 expressed on the erythrocytes of patients with SLE have been reported by several investigators [116–121]. Studies conducted in the1980s led some investigators to conclude that erythrocyte CR1 deficiency is inherited [116, 118], while later studies using SLE patients of various ethnic backgrounds suggested that erythrocyte CR1 deficiency is acquired [119–121]. Some studies showed that reduced E-CR1 levels correlated with disease activity [119, 120]. Different experimental methods and ethnic populations may account for these conflicting results. Based on these conflicting studies, it is most reasonable to conclude that both genetic and acquired factors are likely to contribute to the observed deficiency of erythrocyte CR1 in SLE patients. However, the precise nature of these factors and the pathogenic implication of reduced CR1 in SLE remain to be elucidated.

Collectively, the aforementioned studies provide convincing evidence supporting that uncontrolled complement activation, due to genetic deficiencies or polymorphisms of complementary regulatory proteins, plays a pivotal role in mediating tissue injury occurring in various inflammatory and autoimmune diseases.

TABLE 20.4   Complement Deficiencies/Dysfunction and Associated Diseases

Deficient/dysfunctional effector component	Associated diseases
C1	SLE[a,b]; bacterial infections
C2	SLE[a,b]; bacterial infections
C4	SLE[a,b]; bacterial infections
C3	Bacterial infections[c]; SLE[d]
C5	Bacterial infections[c]; SLE[d]
C6	Bacterial infections[c]; SLE[d]
C7	Bacterial infections[c]; SLE[d]
C8	Bacterial infections[c]; SLE[d]
C9	Bacterial infections[c]; SLE[d]
Properdin	Bacterial infections[c]; SLE[d]
Factor B	AMD[e]; aHUS[f]
MBL	Bacterial and viral infections; SLE

Deficient/dysfunctional regulatory protein	Associated diseases
C1-inhibitor	Hereditary angioneurotic edema; SLE
Factor H	AMD[e]; aHUS[f]; bacterial infections
MCP	AMD[e]; aHUS[f]
CD55/CD59	PNH[g]

[a]Predominant phenotype.
[b]Risk hierarchy for developing SLE : C1 deficiency (~90%) > C4 deficiency (~75%) > C2 deficiency (~40%).
[c]Most frequently infections with encapsulated bacteria, especially Neisseria meningitidis
[d]Case reports only.
[e]Age-related macular degeneration.
[f]Atypical hemolytic uremic syndrome.
[g]Paroxysmal nocturnal hematuria.

Animal Studies of the Roles of Complement in Tissue Injury in SLE

Despite observational and descriptive studies in humans that have lent strong support for the involvement of complement in tissue injury *in vivo*, they are limited in the scope and possibility of experimental exploration. In contrast, animal studies allow for systematic dissection and manipulation of individual pathogenic mediators in specific diseases. Indeed, insightful information has continuously been obtained using gene knockout mice, transgenic mice, or experimental complement inhibitors over the past two decades. These studies will be briefly reviewed here.

Depletion or inhibition of complement using experimental agents

Early studies often utilized cobra venom factor to deplete complement in experimental animals. Cobra venom factor is capable of massively activating complement at the C3 step, leading to depletion of C3 and consequently absence of downstream complement activity. The results of these studies suggested a role for complement in a number of renal disease models, including nephrotoxic serum nephritis (a mouse model of autoantibody-mediated glomerulonephritis, induced using antibodies reactive with glomerular basement membrane), mesangial proliferative glomerulonephritis, and membranous nephropathy [122−124]. Subsequently, investigators used soluble inhibitors, such as recombinant soluble CR1 (sCR1) and chimeric molecules of CR1-related gene/protein y (Crry-Ig), to examine whether inhibition of the complement system would modify clinical measures and pathology in different models of renal diseases [125−128]. Both sCR1 and Crry-Ig target the steps of C3 and C5 convertases, making them efficient inhibitors of all pathways of complement activation. Particularly relevant to SLE are the studies involving MRL/*lpr* mice (a lupus mouse strain). MRL/*lpr* mice treated with Crry-Ig, compared to the untreated control mice, showed reduction in renal injury, as measured by proteinuria and renal failure, and glomerular scarring. Transgenic expression of Crry systemically or locally in the kidney was also shown to reduce renal disease in MRL/*lpr* mice [127]. These observations indicate direct involvement of complement in the pathogenesis of lupus nephritis.

More recently, inhibitors targeting the terminal activation phase of complement activation have been developed and used in lupus-prone mice. Significantly, treatment of the F1 mice resulted from a cross between New Zealand White and New Zealand Black mice (NZW/B F1; another mouse lupus model) with an inhibitory anti-C5 monoclonal antibody markedly prevented the development of glomerulonephritis and improved the survival of these mice [129]. These findings suggest that C5a and/or C_{5b-9} MAC are important mediators of lupus nephritis.

Knockout mouse studies of genes encoding complement effector proteins and receptors

Clinical studies of patients with SLE and deficiency of complement C1, C4, and C2 strongly suggest that early components of the classical pathway play a protective role against the development of SLE. The observed association between early complement component deficiency and susceptibility of SLE in humans has been corroborated by recent animal studies. Several laboratories have generated mice deficient in individual complement proteins and complement receptors using gene-targeting techniques. C1q-knockout ($C1q^{-/-}$) and C4-knockout ($C4^{-/-}$) mice on the 129/C57BL6 hybrid genetic background were shown to spontaneously develop higher levels of antinuclear autoantibodies (ANA) than strain-matched wild-type mice [130−133]. Moreover, $C1q^{-/-}$ mice backcrossed onto the MRL/*lpr* background developed accelerated, severe autoimmune disease and showed histopathologic changes resembling glomerulonephritis [130, 132]. In contrast to $C1q^{-/-}$ and $C4^{-/-}$ mice, C3-knockout ($C3^{-/-}$) mice on the 129/B6 hybrid background were not found to develop autoimmune phenotypes [133-135]. In a recent study, MRL/*lpr* mice deficient in C4, C3, or both were generated by cross-breeding and were examined for the development of autoimmune phenotype/disease [134]. Autoimmune disease, characterized by marked splenomegaly, lymphadenopathy, high titers of ANA and anti-dsDNA, and glomerulonephritis, were noted to develop in $C4^{-/-}/lpr$ as severely as in $C4^{-/-}/C3^{-/-}/lpr$ mice, but the disease did not occur in $C3^{-/-}/lpr$ mice. These findings suggest that deficiency of C4, but not C3, predisposes to the development of SLE in mice.

Given the established role of C3 and its activation products as important mediators in inflammatory responses, it was surprising when another study showed that C3 deficiency affected neither the progression nor the severity of lupus nephritis in MRL/*lpr* mice [136]. Detailed serological studies showed that $C3^{-/-}$ MRL/*lpr* mice developed autoantibody responses similar to those of C3-intact control mice, but had increased deposit of immune complexes in the glomeruli. One plausible explanation for these unexpected findings is that the roles played by C3 in the kidney are both beneficial (solubilizing/clearing immune complex deposits) and detrimental (mediating the inflammatory reaction in the kidney); deficiency of C3 may equally affect and hence negate these opposing effects, resulting in an unremarkable disease phenotype.

Apart from the intriguing role of the early components of the classical pathway in protection against the development of SLE and SLE-like disease, other components of the complement cascade are thought to mediate autoantibody/immune complex-induced tissue injury.

This notion is supported by mostly descriptive findings of immune complexes and complement in human kidney biopsies. To prove the "evildoer" role of complement in SLE, especially lupus nephritis, investigators have again turned to the gene knockout mouse approach. Targeted deletion of Factor B [137] or Factor D [138] was shown to protect MRL/*lpr* mice from developing renal disease. These results may appear surprising at first glance since the classical pathway is generally thought to be the culprit in immune complex diseases such as lupus nephritis. However, it is possible that the actual role the classical pathway plays in this disease setting is to generate a threshold amount of C3b sufficient to initiate and amplify the alternative pathway, which in turn generates the actual mediators of tissue injury (e.g., C5a and C_{5b-9} MAC).

More unexpected results indicative of an extremely intricate balancing act of complement in inflammatory tissue injury came from recent studies by Quigg and colleagues as well as Braun and colleagues. The Quigg group recently reported that expression levels of the C3a receptor (C3aR) in glomerular tissues were significantly increased in prediseased and diseased kidneys of MRL/*lpr* mice [139]. Moreover, treatment with a C3aR antagonist protected MRL/*lpr* mice from developing renal disease, reduced inflammatory cell infiltration into the kidneys, and prolonged their survival. These results, combined, suggest that C3a/C3aR interactions play a relevant role in mouse lupus nephritis and may be targeted for therapeutic intervention in human lupus nephritis. Independently, as a means to better understand the roles of C3a and C5a in inflammatory nephritis, the Braun group knocked out the gene encoding the receptor for C3aR or C5aR in MRL/*lpr* mice [140, 141]. While $C5aR^{-/-}$ MRL/*lpr* mice had attenuated renal disease [141], $C3aR^{-/-}$ MRL/*lpr* mice, compared to control MRL/*lpr* mice, had elevated autoantibody titers, earlier onset of renal disease, increased expression of chemokines and chemokine receptors in the kidney, and accelerated interstitial fibrosis [140]. These findings not only show the opposing functional roles of C3aR and C5aR in the development and progression of renal disease in MRL/*lpr* mice, but also suggest that C3aR plays a protective, rather than proinflammatory, role in mouse lupus nephritis. It is unclear why C3aR appeared to play a detrimental role in the study by the Quigg group but a protective role in the Braun study. These conflicting findings, nevertheless, imply a complement-mediated mechanism far more complex than originally depicted.

Knockout mouse studies of genes encoding complement regulatory proteins

To harness the extent and dynamics of complement activation, complement regulatory proteins are indispensable. Without proper regulation, unchecked activation of the complement system can inflict considerable injury of virtually any cell and tissue. Mice with targeted deletion of specific complement regulatory proteins have proved to be powerful tools for dissecting the role of complement in tissue injury in immune-inflammatory diseases [142–145]. There are, however, some limitations and caveats in these studies. For example, the complement regulatory protein system of the mouse is quite different from that of the human. Two forms of DAF coded by different genes (*Daf-1* and *Daf-2*) exist in the mouse. *Daf-1* encodes a GPI-anchored protein that is considered equivalent to human DAF, whereas *Daf-2* encodes a transmembrane protein expressed only in the testis. Similarly, two CD59 genes (*CD59a* and *CD59b*) exist in mouse and *CD59a* is considered the mouse ortholog of human CD59. In mouse, MCP is expressed only in the testis and the CR1-related gene/protein y (Crry) is considered a functional homolog of human MCP [145]. Therefore, the observations made in mice deficient in complement regulatory proteins should be interpreted and related to human diseases with caution.

DAF-knockout ($DAF-1^{-/-}$) mice in the C57BL/6 genetic background have been generated and appeared grossly normal. However, these mice were found to be more susceptible to nephrotoxic serum-induced nephritis [142]. Compared with wild-type control mice, DAF-knockout mice suffered from exacerbated renal injury that appeared to correlate with enhanced C3 deposition in the glomeruli. These findings suggest that a complement-dependent mechanism is responsible for disease exacerbation. To further investigate the relevance of DAF in autoimmune diseases, DAF-knockout mice were backcrossed onto the MRL/*lpr* genetic background to create $DAF^{-/-}$ MRL/*lpr* mice [143]. Interestingly, these mice developed severe dermatitis, more so in female mice than in male mice. Analysis of early skin lesions of $DAF^{-/-}$MRL/*lpr* mice revealed inflammatory cell infiltration and C3 deposition along the dermal–epidermal junction. These results suggest that DAF is a critical complement regulator in the skin and, in the absence of DAF, uncontrolled complement activation propels local inflammatory responses and ultimately leads to severe dermatitis. In a subsequent study, $DAF^{-/-}$MRL/*lpr* mice were crossbred with $C3^{-/-}$MRL/*lpr* mice to generate $DAF^{-/-}C3^{-/-}$MRL/*lpr* mice and $DAF^{-/-}$$C3^{+/+}$MRL/*lpr* littermate controls. Compared to $DAF^{-/-}C3^{+/+}$MRL/*lpr* mice, fewer $DAF^{-/-}C3^{-/-}$MRL/*lpr* mice developed less severe skin disease [143]. This latter study confirmed directly that DAF is responsible for regulating a C3-dependent pathogenic mechanism to prevent cutaneous inflammatory disease in SLE.

CD59a-knockout mice were originally generated in a mixed (129 × C57BL/6) genetic background and studied in the experimental allergic encephalomyelitis model [146]. Disease incidence and severity as well as the extent of pathologic changes (inflammation, demyelination, and axonal injury in the spinal cord) were significantly increased in $CD59a^{-/-}$ mice. Moreover, deposits of C_{5b-9} MACs were found in the demyelinated lesions. These results support a pathogenic role of complement in autoimmune injury of the central nervous system. In renal disease models, deletion of CD59a was shown to further exacerbate nephrotoxic serum nephritis in $DAF^{-/-}$ mice [144], suggesting that DAF (regulating the early step of complement activation) may serve as the major barrier to complement-mediated injury with CD59 providing the second line of defense. To investigate more definitely the role of CD59 in murine SLE, $CD59a^{-/-}$ mice were backcrossed onto the MRL/*lpr* background. Similar to $DAF^{-/-}$MRL/*lpr* mice, $CD59a^{-/-}$MRL/*lpr* developed severe inflammatory skin disease [147]. This finding points again to significant involvement of complement in mediating cutaneous inflammatory disease in murine, and possibly human, SLE.

The role of complement in mediating renal injury in SLE has been questioned due to a recent study demonstrating that deficiency of C4BP, a major regulator of the classical pathway, had no significant impact on serologic changes (e.g., autoantibody titers), disease severity (measured by proteinuria and serum creatinine), renal histopathology (hypercellularity, sclerosis, Ig/C3/C4/C1q staining), or survival of the MRL/*lpr* model of lupus nephritis [148]. Although seemingly surprising, these findings may in fact serve as a reminder that the inflammatory mechanisms mediating tissue injury are extremely complex and our understanding of these mechanisms is still evolving. In this context, it should be noted that immune complexes can also exert their pathogenic effects through interactions with Fc receptors expressed on infiltrating inflammatory cells and endogenous glomerular cells. Such a complement-independent mechanism was unraveled by a seminal study of Fc receptor-knockout mice [149]. Additional mechanisms of tissue injury in SLE and other immune-inflammatory diseases may exist and still await discovery. It is highly likely that different mechanisms interact in a co-regulated and delicately balanced manner. Numerous factors, such as the nature of inciting agents, surrounding tissue environment, accessibility to the circulation, timing of the injurious stimulus, and other co-morbidities of the underlying disease may determine the dynamics, quantity, and quality of a given inflammatory response and ensuing disease manifestations. These issues are undoubtedly important for a better understanding of SLE pathogenesis and treatment opportunities. Interested readers should refer to pertinent chapters elsewhere in this volume for detailed information.

Alternative animal studies of complement-mediated tissue injury

The complement system is inevitably activated in almost all clinical conditions encompassing inflammatory responses. Complement activation may cause alterations in vascular tone and blood flow in the inflamed tissue, resulting in functional/physical vascular ischemia that in turn further injures the tissue. The ensuing reperfusion of the ischemic tissue often leads to a second round of inflammatory reaction further aggravating tissue injury. A clinical scenario of insidious ischemia and reperfusion is conceivable in SLE, wherein vasculopathy, vasculitis, and antiphospholipid antibodies may precipitate endothelial dysfunction, thrombosis, and reduced blood flow. Furthermore, recent studies have clearly shown that patients with SLE have increased incidence of myocardial infarction and stroke [150, 151], where acute ischemia and reperfusion contribute to substantial morbidity and mortality [152, 153]. In this context, a review of the mechanisms underlying ischemia/reperfusion (I/R) injury may shed light on the understanding of tissue injury in SLE.

Although the pathogenic mechanisms of I/R injury are undoubtedly complex and may be tissue/organ-dependent, recent studies have suggested a key role of the complement system and natural antibodies in mediating I/R-provoked injury in organs such as the heart, intestine, skeletal muscle, and brain (see [154] for a recent review). Evidence supporting a critical role of complement in I/R injury has been provided by animal studies. Blockage of complement activation at the point of C3 activation, using inhibitors for all complement activation pathways (e.g., soluble CR1 and Crry and C3/CVF hybrid proteins), has been shown to prevent or attenuate intestinal, myocardial, and skeletal muscle I/R injury in mice and rats [155–157]. Likewise, inhibition of complement activation at the point of C5 activation, using C5a receptor antagonists or anti-C5 antibodies, has also been demonstrated to attenuate I/R injury rats [158–160].

Additional lines of evidence elucidating the mechanisms of complement activation in I/R injury have derived from animal studies using immunodeficient or gene knockout mice. Immunodeficient RAG-1 deficient ($RAG-1^{-/-}$) mice (deficient in immunoglobulins and lymphocytes) are resistant to intestinal I/R injury, but infusion of normal mouse IgM reinstates the I/R-induced tissue damage [161]. Similarly, CR2-knockout ($Cr2^{-/-}$) mice, which have been shown to produce decreased levels of natural antibodies, are also resistant to intestinal I/R injury and infusion of IgM and IgG reconstitutes the inflammatory response triggered by I/R [162]. Moreover, C4-deficient ($C4^{-/-}$) or C3-deficient ($C3^{-/-}$) mice (with normal levels of immunoglobulin) displayed significantly reduced tissue injury in an experimental intestinal I/R model as compared to wild-type mice [161]. Taken together, these results suggest that activation of the classical pathway of complement by natural antibodies plays a pivotal role in initiating I/R injury.

Natural antibodies, spontaneously produced polyreactive antibodies with low affinity to multiple autoantigens, have been found in mice and humans [163]. The self-reactivity of natural antibodies is reminiscent of autoantibodies present in patients with SLE and mice with SLE-like diseases. This autoantigenic feature has led Fleming and colleagues to hypothesize that autoantibodies present in the sera of SLE-prone mice should be capable of mediating I/R injury, as do natural antibodies, and thus autoimmune-prone mice (e.g., B6.MRL/*lpr* mice) should exhibit aggravated or accelerated I/R-induced tissue injury [164]. Indeed, these investigators found that, when subjected to experimental intestinal I/R, 5-month-old B6.MRL/*lpr* mice (autoimmunity developed) displayed tissue damage at a time point earlier than did 2-month-old B6.MRL/*lpr* mice (autoimmunity not yet developed) or normal 5-month-old B6 mice. Injection of serum IgG derived from 5-month-old, but not 2-month-old, B6.MRL/*lpr* mice into I/R jury-resistant RAG-1$^{-/-}$ mice restored their response to intestinal I/R. Furthermore, murine anti-DNA or anti-histone monoclonal antibody was also able to reconstitute tissue injury in RAG-1$^{-/-}$ mice subjected to experimental intestinal I/R. In a subsequent study, these investigators showed that administration of either murine monoclonal or human polyclonal antibodies specific for phospholipids or β2-glycoprotein I into Cr2$^{-/-}$ mice (previously shown to resist I/R injury) can restore tissue damage following experimental intestinal I/R [165].

Independently, several investigators have also demonstrated that natural IgM antibodies, of both mouse and human origins, play a substantial role in triggering activation of the complement classical pathway and mediating tissue damage in various I/R injury models [166, 167]. The most plausible explanation for these observations is as follows. The initial ischemic insult may damage cells and unravel neo-(auto)antigens. Natural antibodies (or autoantibodies) bind to such neo-(auto) antigens expressed on the surface of injured cells and activate the classical pathway of complement [168]. The identities of ischemia-induced neo-(auto)antigens are being investigated in several laboratories [169, 170].

Although the complement classical pathway is strongly implicated in I/R injury, experimental data also suggest involvement of the alternative and lectin pathways. For example, mice treated with selective inhibitors of the alternative pathway (e.g., anti-Factor B antibody, chimeric CRIg-Fc or CR2-factor H molecules) appear to be protected from tissue injury in various I/R models [171–174]. Furthermore, mice deficient in MBL are protected from tissue damage in myocardial and gastrointestinal I/R models [175, 176]. Collectively, these studies not only support antibody/complement-dependent mechanisms of I/R-provoked tissue injury, but also suggest that similar pathogenic mechanisms may be involved in tissue injury in SLE.

Possible Mechanisms Underlying the Complement Deficiency-SLE Association

Currently, there are three nonmutually exclusive hypotheses explaining the intriguing clinical association between complement deficiency and SLE. The first hypothetical mechanism envisions that impaired clearance of immune complexes in the absence of early complement components may trigger or augment the development of SLE. It is interesting to note that of the two isotypes of human C4, C4A has predominantly been implicated in the binding and solubilization of immune complexes [177]. Consequently, it is not unexpected that the prevalence of C4A deficiency is reportedly higher in SLE patients than in the general population [9, 178]. Several studies have demonstrated abnormal processing of immune complexes in SLE patients [179, 180], supporting the concept that impaired clearance of immune complexes may contribute to the development of SLE in the context of complement deficiency.

The second, "waste disposal," hypothetical mechanism originated from the discovery that C1q can bind directly to apoptotic keratinocytes [181]. Further observations leading to this hypothesis include (1) endothelial cells and monocytes that are undergoing apoptosis also bind C1q [182] and (2) this binding can subsequently trigger activation and deposition of C4 and C3 on these apoptotic cells [183, 184].

Thus, apoptotic cells and blebs become opsonized and can be effectively taken up by phagocytic cells via a complement receptor-mediated mechanism [183, 184]. During apoptosis, normally hidden intracellular constituents are often biochemically modified and redistributed/segregated into surface blebs of dying cells [185]. Impaired clearance of apoptotic cells due to complement deficiency may lead to persistence of such altered-self constituents, which may be recognized by the immune system, breach immune tolerance, and trigger autoimmune responses [186]. Taken together, these studies suggest that complement is involved in facilitating the clearance of autoantigen-containing apoptotic bodies and therefore complement deficiency may result in accumulation of autoantigens and breakdown of immune tolerance [9, 57, 62].

Using a mouse model, Botto *et al.* were first to report accumulation of apoptotic cells in the kidneys and spontaneous development of autoimmune responses to nuclear autoantigens and glomerulonephritis in the absence of C1q [130]. Subsequently, Taylor *et al.* reported similar, but less severe, defects in the clearance of apoptotic cells and spontaneous autoantibody production in C4-deficient mice [131]. Results from these animal studies suggest that clearance of apoptotic cells is impaired or delayed in the absence of C1q, and less so in the absence of C4. In humans, reduced phagocytic activity of neutrophils, monocytes, and macrophages of SLE patients has previously been observed in *in vitro* studies [187, 188]. Baumann *et al.* reported an abnormal accumulation of apoptotic cells, accompanied by a significantly decreased number of tangible body macrophages (cells responsible for removing apoptotic nuclei), in the germinal centers of lymph nodes in a small subset of SLE patients [189]. Collectively, data from both animal and human studies not only substantiate the observed hierarchical correlations between the deficiency of C1, C4, or C2 and the risks for developing SLE, but also provide a strong mechanistic basis linking complement deficiency and SLE.

The third hypothetical mechanism relates to the capacity of complement to determine activation thresholds of B and T lymphocytes, suggesting that complement deficiency may alter the normal mechanism of negative selection of self-reactive lymphocytes [50, 61]. Since co-ligation of CR2 and BCR augments B cell activation by decreasing the threshold to antigenic stimulation [49, 61], it has been postulated that self-antigens not opsonized by C4b or C3b (due to complement deficiency) are unlikely to trigger sufficient activation of self-reactive B cells, and, as a result, these cells may escape negative selection. The escaped cells may become activated once encountering relevant autoantigens in the periphery, and thus breach self-tolerance.

COMPLEMENT AS BIOMARKERS FOR SLE DIAGNOSIS AND MONITORING

During flares of SLE, complement proteins would presumably be consumed at a rate proportional to activity of the disease. Measuring complement activation may therefore be useful for diagnosing disease, assessing its activity, and determining response to therapy.

Measurement of Complement

Conventionally, the complement system is measured by one of two types of assays, functional assays and immunochemical assays. Complement-mediated hemolytic activity such as CH_{50} (indicative of the activity of the classical pathway) and APH_{50} (indicative of the activity of the alternative pathway) are measured by functional assays. It is common to check CH_{50} in patients with SLE, since complement activation in SLE is triggered predominantly by immune complexes that activate the classical pathway. Immunochemical assays measure serum concentrations of complement split products or activation products, and individual complement components.

Measuring serum complement levels

Measurement of serum levels of individual complement components is commonly used to identify deficiencies of specific complement proteins and to diagnose and assess disease activity in SLE. Selecting a proper method depends on several factors: the level of sensitivity required, the availability of specific antibody, and the number and types of samples. Nephelometry is routinely used to measure complement components that are present at relatively high concentrations in the serum (e.g., C3 and C4). Radial immunodiffusion (RID) or enzyme-linked immunosorbent assay (ELISA) can be used to measure other components that are usually present at low concentrations (e.g., C1, C2, C5–C9, Factor B, Factor D, properdin, etc.) When C3 and C4 concentrations are too low to be measured accurately by nephelometry, i.e., less than 20 mg/dL and less than 10 mg/dL, respectively, RID is the alternative method of choice. ELISA is the most practical method to use for other body fluids or cell culture supernatants, in which the levels of complement components may be very low.

Measuring complement split products

Complement components constantly undergo synthesis, activation, and catabolism, so measurement of their serum concentrations is essentially a static appraisal of an extremely dynamic process. Because complement split products are generated only when complement activation occurs, direct determination of these products is thought to reflect more precisely the activation process of complement *in vivo* and hence the disease activity. Many clinical immunology laboratories currently measure complement split products in the plasma, yielded from activation of the classical pathway (C1rs-C1 inhibitor complex, C4a, and C4d), alternative pathway (Bb and C3bBbP), lectin pathway (C4a and C4d), and terminal activation phase (C3a, iC3b, C3d, C5a, and sC_{5b-9}).

Generally, plasma is used to measure complement split products. Using EDTA-anticoagulated plasma will avoid generating complement split products *in vitro*. ELISA and EIA are the most practical methods for the measurement of split products, since only low levels may be present in the circulation, even after significantly increased complement activation. Complement split products have commonly short, but different, half-lives in the plasma. Some split products that have very short half-lives, such as C3a, C4a, and C5a, are quickly converted to more stable, less active forms, such as C3a-desArg, C4a-desArg, and C5a-desArg. In contrast, some complement products, like those that form multimolecular complexes, usually have a relatively long half-life in the circulation. Examples of these include products of classical pathway activation, such as C1rs-C1 inhibitor complexes, products of alternative pathway, like C3bBbP complexes, and sC_{5b-9}. sC_{5b-9}, the soluble form of MAC, is constituted of protein S and C_{5b-9} MAC.

Problems associated with complement measurement

Several factors that particularly confound measurement of serum C3 and C4 exist and should be taken into consideration during data interpretation. First, traditional concentration measurements reflect the presence of C3 and C4 protein entities irrespective of their functional integrity. Second, there is wide variation in serum C3 and C4 levels among healthy individuals; this range overlaps with that observed in patients with different diseases. Third, genetic variations such as partial deficiency of C4, which is commonly present in the general population and in patients with autoimmune diseases [9, 178], may result in lower than normal serum C4 levels in some patients because of decreased synthesis rather than increased complement consumption during disease flares. Fourth, acute-phase responses during inflammation may lead to an increase in C4 and C3 synthesis [190], which can counterbalance the consumption of these proteins during activation. Fifth, enhanced catabolism [191] as well as altered synthesis of C3 and C4 [192] have been reported to occur in patients with SLE, which clearly can interfere with static measures of serum C3 and C4 levels. Sixth, tissue deposition of immune complexes may result in complement activation at local sites in patients with certain diseases; therefore, the levels of complement products in the systemic circulation may not faithfully reflect such local activity. Additional concerns should be raised regarding the measurements of complement split products. Notably, complement activation can easily occur *in vitro* after blood sampling, and many of the split products have undefined, most likely short, half-lives both *in vivo* and *in vitro*. These factors, in combination, may hamper accurate measures of soluble activation products that are derived from complement activation occurring in patients.

Complement as Biomarkers for SLE Diagnosis and Disease Activity Monitoring

In clinical practice, patients with SLE are often monitored by measures of serum C3, C4, and CH_{50}. Studies to evaluate the utility of these assays in the diagnosis and monitoring of SLE, with their noteworthy data and precautions, are succinctly reviewed here.

First, there is no consensus regarding the actual value of complement measures in SLE monitoring, in spite of the conventional notion that decreased levels of complement components reflect activation of the classical and/or alternative pathway and correlate with clinical disease activity. Some investigators have found CH_{50} as well as serum C4 and C3 levels to be valuable as markers of SLE activity, for the following reasons:

1. Patients with increased SLE disease activity, manifested by active nephritis and extrarenal involvement, have demonstrated significantly decreased levels of CH_{50} and serum C3 and C4 [8, 69, 193, 194].
2. Clinical exacerbation has been preceded by a decrease in serum C4 levels [69, 195].
3. Remission/relapse of lupus nephritis has coincided significantly with an increase/decrease in serum C3 levels [195].
4. An impending flare of SLE may be heralded by a progressive fall of serum C3 or C4 levels.
5. On resolution of disease flares, serum C3 and C4 levels have normalized [69].

On the contrary, the following observations argue against the usefulness of conventional complement measurement:

1. The extent of changes in serum C3 and C4 levels do not correlate quantitatively with disease severity [196].
2. During disease flares, serum C4 and C3 levels have been found to remain normal in some patients.
3. In SLE patients with inactive disease, persistently low C4 levels have been detected [197].
4. Increases in the split products (e.g., C3a, C3d, and C4d) have not always accompanied decreases in C3 and C4.

Second, controversy regarding the utility of complement split product measures still remains. Studies have generally shown that plasma concentrations of complement split products, including C1-C1-INH complex, C3a, C4a, C5a, C3d, C4d, C_{5b-9}, Ba, and Bb, increased before or during clinical exacerbation, and in some cases, the plasma levels correlated strongly with SLE disease activity [198–204]. These studies would support the direct measurement of split complement products. On the contrary, elevated levels of split products such as C1-INH and C3d have been reported not only in almost all clinically ill patients, but also in a significant fraction of patients with quiescent disease, suggesting that plasma C1-INH and C3d levels bear little relationship with clinical activity, and arguing against their use.

Furthermore, inconsistent results have been reported for the utility of plasma levels of a given complement split product in distinguishing patients with different levels of disease activity and severity.

Given the numerous confounding factors outlined here, it is not surprising that irreconcilable results regarding complement measurement have prevailed in the research arena of complement and SLE disease activity. However, if complement measurements are performed chronologically in the same patient and interpretation is based on the specific genetic and clinical characteristics of the patient, they may still be informative.

Complement as Biomarkers for Organ-specific Involvement in SLE

Measurement of complement in body fluid other than the plasma may also be valuable in assessing specific organ involvement or predicting outcome under particular conditions such as pregnancy. Lupus nephritis is one of the most serious clinical manifestations of SLE and nephritic flare has been shown to be a predictor of a poor long-term outcome in SLE patients [205]. Measurements of complement components and activation products in the plasma or in the urine may be useful for evaluating the extent of active inflammation in the kidneys. SLE patients with renal involvement were found to have markedly reduced serum levels of C3 and C4 more frequently than patients with only extrarenal involvement [69, 193]. Correspondingly, SLE patients with normal C3 and C4 levels were rarely found to have active nephritis [69, 193]. Therefore, the absence of a low C3 or C4 level in a patient with SLE may serve to exclude the possibility of ongoing renal disease. Because low C3 levels have been reported to be predictive of persistently active glomerular disease [206] and associated with end-stage renal disease [207], low C3 and C4 levels may also be helpful in predicting long-term outcome in patients with SLE. In addition to low C3 and C4, very low levels of serum C1q were detected in SLE patients with active renal disease [69], but not in those without it. Persistently low C1q levels, before and after intense treatment for lupus nephritis, have been shown to be indicative of continuously progressive damage to the kidneys and are thus associated with a poor outcome [208].

Measurement of C3d in the urine has been explored as a test for specific and accurate estimation of inflammation in the kidney, because it is likely that C3d generated in the kidney at sites of immune complex deposition would pass into the urine. Kelly et al. [209] and Manzi et al. [202] have reported the detection of C3d in the urine of SLE patients with acute nephritis, but also in that of patients without evidence of renal involvement. Thus, urinary C3d may also come from

nonrenal origins and therefore may not be viewed as a specific marker of acute nephritis or a prognostic indicator of renal disease. Nevertheless, in the same study by Manzi et al. [202] urinary C3d was better than serum C3, plasma C4d, Bb, and C_{5b-9} in distinguishing patients with acute lupus nephritis from those without such disease activity. In addition, Negi et al. recently reported that C3d levels were elevated in the urine of patients with active disease. Patients with active lupus nephritis had significantly higher levels of urinary C3d than patients with extrarenal disease or with inactive lupus nephritis [210].

In addition to renal disease, hematologic manifestations have also been associated with low serum C3 and C4 levels. Ho and colleagues reported that decreases in C3 were correlated with concurrent decreases in platelet and white blood cell counts, as well as in hematocrit [211]. However, these investigators did not find that decreased complement levels were consistently associated with SLE flares. Investigations have also explored the relationship between CNS disease in SLE and complement levels. One study showed that plasma C3a levels increased in SLE patients and were particularly high in five patients who had acute CNS dysfunction; additionally, four of these five patients had significantly elevated plasma C5a levels as well [199]. Similarly, Rother et al. found significantly higher levels of plasma C3d in SLE patients with CNS involvement than in those without such involvement [203].

Measurement of Cell-bound Complement Activation Products

Recent studies have explored the hypothesis that cell-bound complement activation products (CB-CAPs) may serve as biomarkers for SLE diagnosis and monitoring [212–214]. This hypothesis was based upon the following rationale. First, as discussed in the preceding section, serum C3 and C4 levels have limited diagnostic and monitoring utility. Second, although measurement of soluble CAPs does have utility in certain clinical situations, these assays have yet to replace measurement of serum C3 and C4 in clinical practice. Third, receptors for complement activation products, present on all circulating cells, may confound accurate and reliable measurement of the soluble activation products. Fourth, C3- and C4-derived CAPs are capable of covalent attachment to cell surfaces, and this property may increase longevity in the circulation, making CB-CAPs more likely candidates for use in clinical practice than soluble CAPs. Fifth, C4-derived CAPs are known to be present on surfaces of normal erythrocytes, although the physiologic significance of this phenomenon is unknown [215]. Sixth, CB-CAPs on specific cell types with unique cellular properties, such as erythrocytes and reticulocytes, might provide additional clues to disease diagnosis, activity, and pathogenesis.

The first report of CB-CAPs was a cross-sectional flow cytometric study examining erythrocyte-bound C4d (E-C4d) levels in patients with SLE, patients with other inflammatory and immune-mediated diseases, and healthy controls [212]. In light of the previous reported association of low E-CR1 levels in SLE, erythrocyte-CR1 (E-CR1) was determined simultaneously. This study demonstrated that: (1) patients with SLE have significantly higher levels of E-C4d than those with other diseases and healthy individuals and (2) there is a pattern of abnormally high E-C4d levels that, in conjunction with abnormally low E-CR1 levels, has high diagnostic sensitivity (81%) and specificity (91%) for SLE, as compared with healthy individuals, and with patients with other inflammatory diseases (72% sensitive; 79% specific).

A subsequent study took advantage of the knowledge that erythrocytes develop from hematopoietic stem cells in the bone marrow and emerge as reticulocytes, which then maintain distinct phenotypic features for 1–2 days before fully maturing into erythrocytes. Reticulocytes, if released into the peripheral circulation during an active disease state, may immediately be exposed to and bind C4-derived fragments generated from activation of the complement system. Therefore, it was hypothesized that the levels of C4d bound on reticulocytes (R-C4d) may effectively and precisely reflect the current disease activity in a given SLE patient at a specific point in time. The results of a cross-sectional study showed that: (1) R-C4d levels of SLE patients were significantly higher than those of patients with other diseases or healthy controls and (2) R-C4d levels fluctuated over time in patients with SLE and correlated with clinical disease activity as measured by the Systemic Lupus Erythematosus Disease Activity Index (SLEDAI) and Systemic Lupus Activity Measurement (SLAM) indices [213].

These findings suggest that C4d-bearing reticulocytes may provide clues to current and perhaps impending disease flares, thereby serving as biomarkers for SLE disease activity.

Subsequent studies explored the potential of platelet-bound C4d (P-C4d) as a source of complement-based biomarkers [214]. Cross-sectional determination of P-C4d in SLE patients, patients with other diseases, and healthy controls demonstrated significant levels of C4d on platelets from 18% of SLE patients, 1.7% of patients with other diseases, and 0% of healthy controls. Accordingly, detection of C4d on platelet surfaces is 100% specific in distinguishing SLE patients from healthy controls, and 98% specific in distinguishing SLE patients from patients with other diseases. These results demonstrate great potential for P-C4d measurement as a diagnostic assay for SLE. Moreover, the presence of C4d on platelets from SLE patients correlated significantly with a history of a neurologic event and the presence of antiphospholipid antibodies. These data suggest that detection of P-C4d may identify SLE patients who are at increased risk for cerebrovascular disease and perhaps other vascular complications of SLE.

ANTICOMPLEMENT THERAPEUTICS FOR SLE

The fundamental role of complement activation in SLE pathogenesis has led naturally to exploration of the complement system as a target for therapeutic intervention. To date, a variety of reagents that inhibit or modulate complement activation at different steps of the cascade have been developed. These reagents can be classified into two broad categories: (1) inhibitors of the early steps of complement activation and (2) inhibitors of the terminal pathway that do not interfere with early activation events [216, 217] (Figure 20.3).

Examples of the first group include soluble CR1 (sCR1; capable of regulating the generation of C3/C4 fragments and C3 convertases) [125, 218], heparin (a polyanionic glycosamine; capable of binding/inhibiting C1,

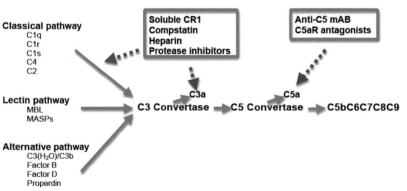

FIGURE 20.3   Overview of potential anticomplement therapeutics.

inhibiting C1q binding to immune complexes, blocking C3 convertase formation, and interfering with MAC assembly) [219], compstatin (a synthetic peptide capable of binding C3 and preventing its proteolytic cleavage), and protease inhibitors [220]. Prominent among the second group are anti-C5 monoclonal antibodies (mAbs) that can bind C5, block its cleavage and formation of C5a, and abrogate MAC assembly [129, 221, 222]. Synthetic antagonists of C5a receptors also belong to the second group and have been exploited to block the anaphylactic and chemotactic effects of C5a [223, 224]. Considering that C3b opsonization of pathogens and immune complexes is crucial for host defense and for prevention of immune complex-associated adverse reactions, it is reasonable to postulate that inhibitors of complement activation at a downstream step, such as C5 cleavage, will have therapeutic effects for patients with inflammatory diseases but will less likely increase the risk for infection in these patients. Recently, a phase I clinical trial of Eculizumab in patients with SLE concluded that Eculizumab was safe and well tolerated, without significant adverse effects [225]. Moreover, heparin has recently been demonstrated to prevent anti-phospholipid antibody/complement-induced fetal loss in a murine model [226]. This seminal observation suggests that heparin at subtherapeutic (nonanticoagulating) doses may be beneficial in pathological situations where excess complement activation is unfavorable, such as ischemia or reperfusion injury, antiphospholipid antibody syndrome, and lupus nephritis.

ACKNOWLEDGEMENTS

We thank our colleagues in the Lupus Center of Excellence and Division of Rheumatology and Clinical Immunology for providing clinical samples, helpful discussion, and skilled technical as well as administrative support. Investigations in the authors' laboratory were supported by grants from the National Institutes of Health (RO1HL074335 and RO1AI077995), the Alliance for Lupus Research, and the Lupus Research Institute.

References

[1] J. Bordet, O. Gengou, Sur l'existence de substance sensibilisatrices dans la plupart des sérums antimicrobiens, Ann Inst Pasteur 15 (1901) 289–302.

[2] P. Ehrlich, J. Morgenroth, Uber Hamolysine. Berlin Klin, Wochenschr 36 (1899) 481–486.

[3] H.J. Muller-Eberhard, Molecular organization and function of the complement system, Annu Rev Biochem 57 (1988) 321–347.

[4] G.D. Ross, J.P. Atkinson, Complement receptor structure and function, Immunol. Today 6 (1985) 115–119.

[5] M.K. Liszewski, T.C. Farries, D.M. Lublin, I.A. Rooney, J.P. Atkinson, Control of the complement system, Adv Immunol 61 (1996) 201–283.

[6] D.D. Kim, W.C. Song, Membrane complement regulatory proteins, Clin Immunol 118 (2006) 127–136.

[7] J.H. Vaughan, T.B. Bayles, C.B. Favour, The response of serum gamma globulin level and complement titer to adrenocorticotropic hormone (ACTH) therapy in lupus erythematosus disseminatus, J Lab Clin Med 37 (1951) 698–702.

[8] P.H. Schur, J. Sandson, Immunological factors and clinical activity in systemic lupus erythematosus, N Engl J Med 278 (1968) 533–538.

[9] M.C. Pickering, M. Botto, P.R. Taylor, P.J. Lachmann, M.J. Walport, Systemic lupus erythematosus, complement deficiency, and apoptosis, Adv Immunol 76 (2000) 227–324.

[10] H.X. Jiang, J.N. Siegel, H. Gewurz, Binding and complement activation by C-reactive protein via the collagen-like region of C1q and inhibition of these reactions by monoclonal antibodies to C-reactive protein and C1q, J Immunol 146 (1991) 2324–2330.

[11] S.C. Ying, A.T. Gewurz, H. Jiang, H. Gewurz, Human serum amyloid P component oligomers bind and activate the classical complement pathway via residues 14-26 and 76-92 of the A chain collagen-like region of C1q, J Immunol 150 (1993) 169–176.

[12] J. Rogers, N.R. Cooper, S. Webster, et al., Complement activation by beta-amyloid in Alzheimer disease, Proc Natl Acad Sci USA 89 (1992) 10016–10020.

[13] H. Jiang, B. Cooper, F.A. Robey, H. Gewurz, DNA binds and activates complement via residues 14-26 of the human C1q A chain, J Biol Chem 267 (1992) 25597–25601.

[14] L. Pillemer, I.H. Lepow, L. Blum, The requirement for a hydrozine-sensitive serum factor and heat-labile serum factors in the inactivation of human C'3 by zymosan, J Immunol 71 (1953) 339–345.

[15] L. Pillemer, L. Blum, I.H. Lepow, O.A. Ross, E.W. Todd, A.C. Wardlaw, The properdin system and immunity. I. Demonstration and isolation of a new serum protein, properdin, and its role in immune phenomena, Science 120 (1954) 279–285.

[16] M.K. Pangburn, H.J. Muller-Eberhard, Initiation of the alternative complement pathway due to spontaneous hydrolysis of the thioester of C3, Annals NY Acad Sci 421 (1983) 291–298.

[17] K. Ikeda, T. Sannoh, N. Kawasaki, T. Kawasaki, I. Yamashina, Serum lectin with known structure activates complement through the classical pathway, J Biol Chem 262 (1987) 7451–7454.

[18] M.W. Turner, R.M. Hamvas, Mannose-binding lectin: structure, function, genetics and disease associations, Reviews in Immunogenetics 2 (2000) 305–322.

[19] R. Wallis, Structural and functional aspects of complement activation by mannose-binding protein, Immunobiol 205 (2002) 433–445.

[20] S. Thiel, T. Vorup-Jensen, c. M. Stover, W. schwaeble, S.B. Laursen, K. Poulsen, et al., A second serine protease associated with mannan-binding lectin that activates complement, Nature 386 (1997) 506–510.

[21] M.R. Dahl, S. Thiel, M.C. Matsushita, T. Fujita, a. C. willis, T. Christensen, et al., A new mannan-binding lectin associated serine protease, MASP-3, and its association with distinct complexes of the MBL complement activation pathway, Immunity 15 (2001) 1–10.

[22] M.M. Mayer, Mechanism of cytolysis by complement, Proc Natl Acad Sci USA 69 (1972) 2954–2958.

[23] H.J. Muller-Eberhard, The membrane attack complex of complement, Annu Rev Immunol 4 (1986) 503–528.

[24] D. Hourcade, V.M. Holers, J.P. Atkinson, The regulators of complement activation (RCA) gene cluster, Adv Immunol 45 (1989) 381–416.

[25] J.M. Inal, K.M. Hui, S. Miot, S. Lange, M.I. Ramirez, B. Schneider, et al., Complement C2 receptor inhibitor trispanning: a novel human complement inhibitory receptor, J Immunol 174 (2005) 356–366.

[26] C. Gerard, N.P. Gerard, C5A anaphylatoxin and its seven transmembrane-segment receptor, Ann Rev Immunol 12 (1994) 775–808.

[27] R.A. Wetsel, Structure, function and cellular expression of complement anaphylatoxin receptors, Curr Opin Immunol 7 (1995) 48–53.

[28] R. Malhotra, A.C. Willis, J.C. Jensenius, J. Jackson, R.B. Sim, Structure and homology of human C1q receptor (collectin receptor), Immunology 78 (1993) 341–348.

[29] B. Ghebrehiwet, B.L. Lim, R. Kumar, X. Feng, E.I. Peerschke, gC1q-R/p33, a member of a new class of multifunctional and multicompartmental cellular proteins, is involved in inflammation and infection, Immunol Rev 180 (2001) 65–77.

[30] R.R. Nepomuceno, A.J. Tenner, C1qRP, the C1q receptor that enhances phagocytosis, is detected specifically in human cells of myeloid lineage, endothelial cells, and platelets, J Immunol 160 (1998) 1929–1935.

[31] L.B. Klickstein, S.F. Barbashov, T. Liu, R.M. Jack, A. Nicholson-Weller, Complement receptor type 1 (CR1, CD35) is a receptor for C1q, Immunity 7 (1997) 345–355.

[32] C. Lozada, R.I. Levin, M. Huie, R. Hirschhorn, D. Naime, M. Whitlow, et al., Identification of C1q as the heat-labile serum cofactor required for immune complexes to stimulate endothelial expression of the adhesion molecules E-selectin and intercellular and vascular cell adhesion molecules 1, Proc Nat Acad Sci USA 92 (1995) 8378–8382.

[33] P. Eggleton, B. Ghebrehiwet, J.P. Coburn, K.N. Sastry, K.S. Zaner, A.I. Tauber, Characterization of the human neutrophil C1q receptor and functional effects of free ligand on activated neutrophils, Blood 84 (1994) 1640–1649.

[34] E.I. Peerschke, B. Ghebrehiwet, C1q augments platelet activation in response to aggregated Ig, J Immunol 159 (1997) 5594–5598.

[35] E. Guan, S.L. Robinson, E.B. Goodman, A.J. Tenner, Cell-surface protein identified on phagocytic cells modulates the C1q-mediated enhancement of phagocytosis, J Immunol 152 (1994) 4005–4016.

[36] M.C. Gelfand, M.M. Frank, I. Green, A receptor for the third component of complement in the human renal glomerulus, J Exp Med 142 (1975) 1029–1034.

[37] D.T. Fearon, Identification of the membrane glycoprotein that is the C3b receptor of the human erythrocyte, polymorphonuclear leukocyte, B lymphocyte, and monocyte, J Exp Med 152 (1980) 20–30.

[38] J.A. Schifferli, The role of complement and its receptor in the elimination of immune complexes, N Engl J Med 315 (1986) 488–495.

[39] M.E. Medof, K. Iida, C. Mold, V. Nussenzweig, Unique role of the complement receptor CR1 in the degradation of C3b associated with immune complexes, J Exp Med 156 (1982) 1739–1754.

[40] J.J. Weis, T.F. Tedder, D.T. Fearon, Identification of a 145,000 Mr membrane protein as the C3d receptor (CR2) of human B lymphocytes, Proc Nat Acad Sci USA 81 (1984) 881–885.

[41] Y. Fang, C. Xu, Y.X. Fu, V.M. Holers, H. Molina, Expression of complement receptors 1 and 2 on follicular dendritic cells is necessary for the generation of a strong antigen-specific IgG response, J Immunol 160 (1998) 5273–5279.

[42] L.J. Miller, R. Schwarting, T.A. Springer, Regulated expression of the Mac-1, LFA-1, p150,95 glycoprotein family during leukocyte differentiation, J Immunol 137 (1986) 2891–2900.

[43] S.K. Lo, G.A. Van Seventer, S.M. Levin, S.D. Wright, Two leukocyte receptors (CD11a/CD18 and CD11b/CD18) mediate transient adhesion to endothelium by binding to different ligands, J Immunol 143 (1989) 3325–3329.

[44] C.W. Smith, S.D. Marlin, R. Rothlein, C. Toman, D.C. Anderson, Cooperative interactions of LFA-1 and Mac-1 with intercellular adhesion molecule-1 in facilitating adherence and transendothelial migration of human neutrophils in vitro, J Clin Invest 83 (1989) 2008–2017.

[45] M.J. Walport, Complement. First of two parts, N Engl J Med 344 (2001) 1058–1066.

[46] B.P. Morgan, Effects of the membrane attack complex of complement on nucleated cells, Curr Top Microbiol Immunol 178 (1992) 115–140.

[47] D.K. Imagawa, N.E. Osifchin, W.A. Paznekas, et al., Consequences of cell membrane attack by complement: release of arachidonate and formation of inflammatory derivatives. Proc Natl Acad Sci USA 80 (1983) 6647–6651.

[48] M. Schonermark, R. Deppisch, G. Riedasch, K. Rother, G.M. Hansch, Induction of mediator release from human glomerular mesangial cells by the terminal complement components C5b-9, Int Arch Allergy Appl Immunol 96 (1991) 331–337.

[49] D.T. Fearon, R.H. Carter, The CD19/CR2/TAPA-1 complex of B lymphocytes: linking natural to acquired immunity, Annu Rev Immunol 13 (1995) 127–149.

[50] M.C. Carroll, The role of complement and complement receptors in induction and regulation of immunity, Annu Rev Immunol 16 (1998) 545–568.

[51] M.B. Fischer, S. Goerg, L. Shen, A.P. Prodeus, C.C. Goodnow, G. Kelsoe, et al., Dependence of germinal center B cells on expression of CD21/CD35 for survival, Science 280 (1998) 582–585.

[52] M.M. Markiewski, J.D. Lambris, The role of complement in inflammatory diseases from behind the scenes into the spotlight, Am J Pathol 171 (2007) 715–727.

[53] E.A. Albrecht, A.M. Chinnaiyan, S. Varambally, C. Kumar-Sinha, T.R. Barrette, J.V. Sarma, et al., C5a-induced gene expression in human umbilical vein endothelial cells, Am J Pathol 164 (2004) 849–859.

[54] T. Takabayashi, E. Vannier, B.D. Clark, N.H. Margolis, C.A. Dinarello, J.F. Burke, et al., A new biologic role for C3a and C3a desArg: regulation of TNF-alpha and IL-1 beta synthesis, J Immunol 156 (1996) 3455–3460.

[55] R.M. Clancy, C.A. Dahinden, T.E. Hugli, Complement-mediated arachidonate metabolism, Prog Biochem Pharmacol 20 (1985) 120–131.

[56] W.H. Fischer, M.A. Jagels, T.E. Hugli, Regulation of IL-6 synthesis in human peripheral blood mononuclear cells by C3a and C3a(desArg), J Immunol 162 (1999) 453–459.

[57] A.P. Manderson, M. Botto, M.J. Walport, The role of complement in the development of systemic lupus erythematosus, Ann Rev Immunol 22 (2004) 431–456.

[58] P.H. Schur, Complement and lupus erythematosus, Arthrit Rheum 25 (1982) 793–798.

[59] H.T. Cook, M. Botto, Mechanisms of Disease: the complement system and the pathogenesis of systemic lupus erythematosus, Nature Clinical Practice Rheumatology 2 (2006) 330–337.

[60] M.L. Barilla-LaBarca, J.P. Atkinson, Rheumatic syndromes associated with complement deficiency, Curr Opin Rheumatol 15 (2003) 55–60.

[61] M.C. Carroll, The role of complement in B cell activation and tolerance, Adv Immunol 74 (2000) 61–88.

[62] C.-C. Liu, J.S. Navratil, J.M. Sabatine, J.M. Ahearn, Apoptosis, Complement, and SLE: A mechanistic view, Current Directions of Autoimmunity 7 (2004) 49–86.

[63] J.A. Schifferli, D.K. Peters, Complement, the immune-complex lattice, and the pathophysiology of complement-deficiency syndromes, Lancet 2 (1983) 957–959.

[64] M.J. Walport, P.J. Lachmann, Erythrocyte complement receptor type 1, immune complexes, and the rheumatic diseases, Arthritis & Rheumatism 31 (1988) 153–158.

[65] J.A. Schifferli, Y.C. Ng, J.P. Paccaud, M.J. Walport, The role of hypocomplementaemia and low erythrocyte complement receptor type 1 numbers in determining abnormal immune complex clearance in humans, Clin Exp Immunol 75 (1989) 329–335.

[66] J.J. Weening, V.D. D'Agati, M.M. Schwartz, S.V. Seshan, C.E. Alpers, G.B. Appel, et al., The classification of glomerulonephritis in systemic lupus erythematosus revisited, Kidney Int 65 (2004) 521–530.

[67] E.J. Lewis, M.M. Schwartz, Pathology of lupus nephritis, Lupus 14 (2005) 31–38.

[68] S. Tang, S.L. Lui, K.N. Lai, Pathogenesis of lupus nephritis: an update, Nephrology (Carlton) 10 (2005) 174–179.

[69] W. Lloyd, P.H. Schur, Immune complexes, complement, and anti-DNA in exacerbations of systemic lupus erythematosus (SLE), Medicine 60 (1981) 208–217.

[70] D. Koffler, P.H. Schur, H.G. Kunkel, Immunological studies concerning the nephritis of systemic lupus erythematosus, J Exp Med 126 (1967) 607–624.

[71] M. Mannik, C.E. Merrill, L.D. Stamps, M.H. Wener, Multiple autoantibodies form the glomerular immune deposits in patients with systemic lupus erythematosus, J Rheumatol 30 (2003) 1495–1504.

[72] J.B. Lefkowith, G.S. Gilkeson, Nephritogenic autoantibodies in lupus: current concepts and continuing controversies, Arthritis Rheum 39 (1996) 894–903.

[73] B.H. Hahn, Antibodies to DNA, N Engl J Med 338 (1998) 1359–1368.

[74] Z. Amoura, J.C. Piette, J.F. Bach, S. Koutouzov, The key role of nucleosomes in lupus, Arthritis Rheum 42 (1999) 833–843.

[75] D.A. Isenberg, J.J. Manson, M.R. Ehrenstein, A. Rahman, Fifty years of anti-ds DNA antibodies: are we approaching journey's end? Rheumatology (Oxford) 46 (2007) 1052–1056.

[76] D.V. Vlahakos, M.H. Foster, S. Adams, M. Katz, A.A. Ucci, K.J. Barrett, et al., Anti-DNA antibodies form immune deposits at distinct glomerular and vascular sites, Kidney Int 41 (1992) 1690–1700.

[77] O.P. Rekvig, M. Kalaaji, H. Nossent, Anti-DNA antibody subpopulations and lupus nephritis, Autoimmun Rev 3 (2004) 1–6.

[78] J.B. Winfield, I. Faiferman, D. Koffler, Avidity of anti-DNA antibodies in serum and IgG glomerular eluates from patients with systemic lupus erythematosus. Association of high avidity antinative DNA antibody with glomerulonephritis, J Clin Invest 59 (1977) 90–96.

[79] J.L. Roberts, R.J. Wyatt, M.M. Schwartz, E.J. Lewis, Differential characteristics of immune-bound antibodies in diffuse proliferative and membranous forms of lupus glomerulonephritis, Clin Immunol Immunopathol 29 (1983) 223–241.

[80] M. Nangaku, W.G. Couser, Mechanisms of immune-deposit formation and the mediation of immune renal injury, Clin Exp Nephrol 9 (2005) 183–191.

[81] H. Liapis, G.C. Tsokos, Pathology and immunology of lupus glomerulonephritis: can we bridge the two? Int Urol Nephrol 39 (2007) 223–231.

[82] G.R. Gallo, T. Caulin-Glaser, M.E. Lamm, Charge of circulating immune complexes as a factor in glomerular basement membrane localization in mice, J Clin Invest 67 (1981) 1305–1313.

[83] V.J. Gauthier, M. Mannik, G.E. Striker, Effect of cationized antibodies in performed immune complexes on deposition and persistence in renal glomeruli, J Exp Med 156 (1982) 766–777.

[84] T.M. Schmiedeke, F.W. Stockl, R. Weber, Y. Sugisaki, S.R. Batsford, A. Vogt, Histones have high affinity for the glomerular basement membrane. Relevance for immune complex formation in lupus nephritis, J Exp Med 169 (1989) 1879–1894.

[85] C. Kramers, M.N. Hylkema, M.C. van Bruggen, R. van de Lagemaat, H.B. Dijkman, K.J. Assmann, et al., Anti-nucleosome antibodies complexed to nucleosomal antigens show anti-DNA reactivity and bind to rat glomerular basement membrane in vivo, J Clin Invest 94 (1994) 568–577.

[86] M. Kalaaji, E. Mortensen, L. Jorgensen, R. Olsen, O.P. Rekvig, Nephritogenic lupus antibodies recognize glomerular basement membrane-associated chromatin fragments released from apoptotic intraglomerular cells, Am J Pathol 168 (2006) 1779–1792.

[87] P. Faaber, T.P. Rijke, L.B. van de Putte, P.J. Capel, J.H. Berden, Cross-reactivity of human and murine anti-DNA antibodies with heparan sulfate. The major glycosaminoglycan in glomerular basement membranes, J Clin Invest 77 (1986) 1824–1830.

[88] J. Sabbaga, O.G. Pankewycz, V. Lufft, R.S. Schwartz, M.P. Madaio, Cross-reactivity distinguishes serum and nephritogenic anti-DNA antibodies in human lupus from their natural counterparts in normal serum, J Autoimmun 3 (1990) 215–235.

[89] D.M. D'Andrea, B. Coupaye-Gerard, T.R. Kleyman, M.H. Foster, M.P. Madaio, Lupus autoantibodies interact directly with distinct glomerular and vascular cell surface antigens, Kidney Int 49 (1996) 1214–1221.

[90] T.M. Chan, J.K. Leung, S.K. Ho, S. Yung, Mesangial cell-binding anti-DNA antibodies in patients with systemic lupus erythematosus, J Am Soc Nephrol 13 (2002) 1219–1229.

[91] Z. Zhao, E. Weinstein, M. Tuzova, A. Davidson, P. Mundel, P. Marambio, et al., Cross-reactivity of human lupus anti-DNA antibodies with alpha-actinin and nephritogenic potential, Arthritis Rheum 52 (2005) 522–530.

[92] S. Izui, P.H. Lambert, P.A. Miescher, Failure to detect circulating DNA–anti-DNA complexes by four radio-immunological methods in patients with systemic lupus erythematosus, Clin Exp Immunol 30 (1977) 384–392.

[93] W.G. Couser, D.J. Salant, In situ immune complex formation and glomerular injury, Kidney Int 17 (1980) 1–13.

[94] M. Mizuno, S. Blanchin, P. Gasque, K. Nishikawa, S. Matsuo, High levels of complement C3a receptor in the glomeruli in lupus nephritis, Am J Kidney Dis 49 (2007) 598–606.

[95] A.V. Cybulsky, J.C. Monge, J. Papillon, A.J. McTavish, Complement C5b-9 activates cytosolic phospholipase A2 in glomerular epithelial cells, Am J Physiol 269 (1995) F739–749.

[96] D.H. Lovett, G.M. Haensch, M. Goppelt, K. Resch, D. Gemsa, Activation of glomerular mesangial cells by the terminal membrane attack complex of complement, J Immunol 138 (1987) 2473–2480.

[97] V. Agnello, Association of systemic lupus erythematosus and SLE-like syndromes with hereditary and acquired complement deficiency states, Arthritis & Rheumatism 21 (1978) S146–152.

[98] M. Koide, S. Shirahama, Y. Tokura, M. Takigawa, M. Hayakawa, F. Furukawa, Lupus erythematosus associated with C1 inhibitor deficiency, J Dermatol 29 (2002) 503–507.

[99] M.J. Walport, K.A. Davies, M. Botto, M.A. Naughton, D.A. Isenberg, D. Biasi, et al., C3 nephritic factor and SLE: report of four cases and review of the literature, QJM 87 (1994) 609–615.

[100] L. Horvath, L. Czirjak, B. Fekete, L. Jakab, T. Pozsonyi, L. Kalabay, et al., High levels of antibodies against C1q are associated with disease activity and nephritis but not with other organ manifestations in SLE patients, Clinical & Experimental Rheumatology 19 (2001) 667–672.

[101] M. Trendelenburg, M. Lopez-Trascasa, E. Potlukova, S. Moll, S. Regenass, V. Fremeaux-Bacchi, et al., High prevalence of anti-C1q antibodies in biopsy-proven active lupus nephritis, Nephrology Dialysis Transplantation 21 (2006) 3115–3121.

[102] G. Moroni, M. Trendelenburg, N.D. Papa, S. Quaglini, E. Raschi, P. Panzeri, et al., Anti-C1q antibodies may help in diagnosing a renal flare in lupus nephritis, Am J Kidney Dis 37 (2001) 490–498.

[103] E. Potlukova, P. Kralikova, Complement component c1q and anti-c1q antibodies in theory and in clinical practice, Scandinavian Journal of Immunology 67 (2008) 423–430.

[104] A. Nicholson-Weller, J.P. March, S.I. Rosenfeld, K.F. Austen, Affected erythrocytes of patients with paroxysmal nocturnal hemoglobinuria are deficient in the complement regulatory protein, decay accelerating factor, Proc Natl Acad Sci USA 80 (1983) 5066–5070.

[105] M.J. Telen, S.E. Hall, A.M. Green, J.J. Moulds, W.F. Rosse, Identification of human erythrocyte blood group antigens on decay-accelerating factor (DAF) and an erythrocyte phenotype negative for DAF, J Exp Med 167 (1988) 1993–1998.

[106] L. Wang, M. Uchikawa, H. Tsuneyama, K. Tokunaga, K. Tadokoro, T. Juji, Molecular cloning and characterization of decay-accelerating factor deficiency in Cromer blood group Inab phenotype, Blood 91 (1998) 680–684.

[107] M. Yamashina, E. Ueda, T. Kinoshita, T. Takami, A. Ojima, H. Ono, et al., Inherited complete deficiency of 20-kilodalton homologous restriction factor (CD59) as a cause of paroxysmal nocturnal hemoglobinuria, N Engl J Med 323 (1990) 1184–1189.

[108] Y. Richaud-Patin, B. Perez-Romano, E. Carrillo-Maravilla, A.B. Rodriguez, A.J. Simon, J. Cabiedes, et al., Deficiency of red cell bound CD55 and CD59 in patients with systemic lupus erythematosus, Immunol Lett 88 (2003) 95–99.

[109] A. Thakkinstian, P. Han, M. McEvoy, W. Smith, J. Hoh, K. Magnusson, et al., Systematic review and meta-analysis of the association between complement factor H Y402H polymorphisms and age-related macular degeneration, Hum Mol Genet 15 (2006) 2784–2790.

[110] S.R. Montezuma, L. Sobrin, J.M. Seddon, Review of genetics in age-related macular degeneration, Semin Ophthalmol 22 (2007) 229–240.

[111] N. Rougier, M.D. Kazatchkine, J.P. Rougier, V. Fremeaux-Bacchi, J. Blouin, G. Deschenes, et al., Human complement factor H deficiency associated with hemolytic uremic syndrome, J Am Soc Nephrol 9 (1998) 2318–2326.

[112] M. Noris, S. Brioschi, J. Caprioli, M. Todeschini, E. Bresin, F. Porrati, et al., Familial haemolytic uraemic syndrome and an MCP mutation, Lancet 362 (2003) 1542–1547.

[113] E. Goicoechea de Jorge, C.L. Harris, J. Esparza-Gordillo, L. Carreras, E.A. Arranz, C.A. Garrido, et al., Gain-of-function mutations in complement factor B are associated with atypical hemolytic uremic syndrome, Proc Natl Acad Sci USA 104 (2007) 240–245.

[114] M.C. Pickering, H.T. Cook, J. Warren, A.E. Bygrave, J. Moss, M.J. Walport, et al., Uncontrolled C3 activation causes membranoproliferative glomerulonephritis in mice deficient in complement factor H, Nat Genet 31 (2002) 424–428.

[115] M.C. Pickering, H.T. Cook, Translational mini-review series on complement factor H: renal diseases associated with complement factor H: novel insights from humans and animals, Clin Exp Immunol 151 (2008) 210–230.

[116] Y. Miyakawa, A. Yamada, K. Kosaka, F. Tsuda, E. Kosugi, M. Mayumi, Defective immune-adherence (C3b) receptor on erythrocytes from patients with systemic lupus erythematosus, Lancet 2 (1981) 493–497.

[117] K. Iida, R. Mornaghi, V. Nussenzweig, Complement receptor (CR1) deficiency in erythrocytes from patients with systemic lupus erythematosus, J Exp Med 155 (1982) 1427–1438.

[118] J.G. Wilson, W.W. Wong, P.H. Schur, D.T. Fearon, Mode of inheritance of decreased C3b receptors on erythrocytes of patients with systemic lupus erythematosus, N Engl J Med 307 (1982) 981–986.

[119] G. Uko, R.L. Dawkins, P. Kay, F.T. Christiansen, P.N. Hollingsworth, CR1 deficiency in SLE: acquired or genetic? Clinical & Experimental Immunology 62 (1985) 329–336.

[120] G.D. Ross, W.J. Yount, M.J. Walport, J.B. Winfield, C.J. Parker, C.R. Fuller, et al., Disease-associated loss of erythrocyte complement receptors (CR1, C3b receptors) in patients with systemic lupus erythematosus and other diseases involving autoantibodies and/or complement activation, J Immunol 135 (1985) 2005–2014.

[121] J.H. Cohen, H.U. Lutz, J.L. Pennaforte, A. Bouchard, M.D. Kazatchkine, Peripheral catabolism of CR1 (the C3b receptor, CD35) on erythrocytes from healthy individuals and patients with systemic lupus erythematosus (SLE), Clinical & Experimental Immunology 87 (1992) 422–428.

[122] E. Unanue, F.J. Dixon, Experimental Glomerulonephritis. IV. Participation of Complement in Nephrotoxic Nephritis, J Exp Med 119 (1964) 965–982.

[123] T. Yamamoto, C.B. Wilson, Complement dependence of antibody-induced mesangial cell injury in the rat, J Immunol 138 (1987) 3758–3765.

[124] D.J. Salant, S. Belok, M.P. Madaio, W.G. Couser, A new role for complement in experimental membranous nephropathy in rats, J Clin Invest 66 (1980) 1339–1350.

[125] W.G. Couser, R.J. Johnson, B.A. Young, C.G. Yeh, C.A. Toth, A.R. Rudolph, The effects of soluble recombinant complement receptor 1 on complement-mediated experimental glomerulonephritis, J Am Soc Nephrol 5 (1995) 1888–1894.

[126] R.J. Quigg, Y. Kozono, D. Berthiaume, A. Lim, D.J. Salant, A. Weinfeld, et al., Blockade of antibody-induced glomerulonephritis with Crry-Ig, a soluble murine complement inhibitor, J Immunol 160 (1998) 4553–4560.

[127] L. Bao, M. Haas, D.M. Kraus, B.K. Hack, J.K. Rakstang, V.M. Holers, et al., Administration of a soluble recombinant complement C3 inhibitor protects against renal disease in MRL/lpr mice, J Am Soc Nephrol 14 (2003) 670–679.

[128] C. Atkinson, F. Qiao, H. Song, G.S. Gilkeson, S. Tomlinson, Low-dose targeted complement inhibition protects against renal disease and other manifestations of autoimmune disease in MRL/lpr mice, J Immunol 180 (2008) 1231–1238.

[129] Y. Wang, Q. Hu, J.A. Madri, S.A. Rollins, A. Chodera, L.A. Matis, Amelioration of lupus-like autoimmune disease in NZB/WF1 mice after treatment with a blocking monoclonal antibody specific for complement component C5, Proc Natl Acad Sci USA 93 (1996) 8563–8568.

[130] M. Botto, C. Dell'Agnola, A.E. Bygrave, E.M. Thompson, H.T. Cook, F. Petry, et al., Homozygous C1q deficiency causes glomerulonephritis associated with multiple apoptotic bodies.[see comment], Nature Genetics 19 (1998) 56–59.

[131] P.R. Taylor, A. Carugati, V.A. Fadok, H.T. Cook, M. Andrews, M.C. Carroll, et al., A hierarchical role for classical pathway complement proteins in the clearance of apoptotic cells in vivo, J Exp Med 192 (2000) 359—366.

[132] D.A. Mitchell, M.C. Pickering, J. Warren, L. Fossati-Jimack, J. Cortes-Hernandez, H.T. Cook, et al., C1q deficiency and autoimmunity: the effects of genetic background on disease expression, J Immunol 168 (2002) 2538—2543.

[133] E. Paul, O.O. Pozdnyakova, E. Mitchell, M.C. Carroll, Anti-DNA autoreactivity in C4-deficient mice, Eur J Immunol 32 (2002) 2672—2679.

[134] S. Einav, O.O. Pozdnyakova, M. Ma, M.C. Carroll, Complement C4 is protective for lupus disease independent of C3, J Immunol 168 (2002) 1036—1041.

[135] R.J. Quigg, A. Lim, M. Haas, J.J. Alexander, C. He, M.C. Carroll, Immune complex glomerulonephritis in C4- and C3-deficient mice, Kidney Int 53 (1998) 320—330.

[136] H. Sekine, C.M. Reilly, I.D. Molano, G. Garnier, A. Circolo, P. Ruiz, et al., Complement component C3 is not required for full expression of immune complex glomerulonephritis in MRL/lpr mice, J Immunol 166 (2001) 6444—6451.

[137] H. Watanabe, G. Garnier, A. Circolo, R.A. Wetsel, P. Ruiz, V.M. Holers, et al., Modulation of renal disease in MRL/lpr mice genetically deficient in the alternative complement pathway factor B, J Immunol 164 (2000) 786—794.

[138] M.K. Elliott, T. Jarmi, P. Ruiz, Y. Xu, V.M. Holers, G.S. Gilkeson, Effects of complement factor D deficiency on the renal disease of MRL/lpr mice, Kidney Int 65 (2004) 129—138.

[139] L. Bao, I. Osawe, M. Haas, R.J. Quigg, Signaling through up-regulated C3a receptor is key to the development of experimental lupus nephritis, J Immunol 175 (2005) 1947—1955.

[140] S.E. Wenderfer, H. Wang, B. Ke, R.A. Wetsel, M.C. Braun, C3a receptor deficiency accelerates the onset of renal injury in the MRL/lpr mouse, Mol Immunol 46 (2009) 1397—1404.

[141] S.E. Wenderfer, B. Ke, T.J. Hollmann, R.A. Wetsel, H.Y. Lan, M.C. Braun, C5a receptor deficiency attenuates T cell function and renal disease in MRLlpr mice, J Am Soc Nephrol 16 (2005) 3572—3582.

[142] H. Sogabe, M. Nangaku, Y. Ishibashi, T. Wada, T. Fujita, X. Sun, et al., Increased susceptibility of decay-accelerating factor deficient mice to anti-glomerular basement membrane glomerulonephritis, J Immunol 167 (2001) 2791—2797.

[143] T. Miwa, M.A. Maldonado, L. Zhou, X. Sun, H.Y. Luo, D. Cai, et al., Deletion of decay-accelerating factor (CD55) exacerbates autoimmune disease development in MRL/lpr mice, Am J Pathol 161 (2002) 1077—1086.

[144] D. Turnberg, M. Botto, J. Warren, B.P. Morgan, M.J. Walport, H.T. Cook, CD59a deficiency exacerbates accelerated nephrotoxic nephritis in mice, J Am Soc Nephrol 14 (2003) 2271—2279.

[145] W.C. Song, Complement regulatory proteins and autoimmunity, Autoimmunity 39 (2006) 403—410.

[146] R.J. Mead, J.W. Neal, M.R. Griffiths, C. Linington, M. Botto, H. Lassmann, et al., Deficiency of the complement regulator CD59a enhances disease severity, demyelination and axonal injury in murine acute experimental allergic encephalomyelitis, Lab Invest 84 (2004) 21—28.

[147] t. Miwa, M.A. Maldonado, L. Zhou, R.A. Eisenberg, W.C. Song, Deletion of CD59a exacerbates autoimmune disease in MRL/lpr mice, Mol Immunol 41 (2004) 278—279.

[148] S.E. Wenderfer, K. Soimo, R.A. Wetsel, M.C. Braun, Analysis of C4 and the C4 binding protein in the MRL/lpr mouse, Arthritis Res Ther 9 (2007). R114.

[149] R. Clynes, C. Dumitru, J.V. Ravetch, Uncoupling of immune complex formation and kidney damage in autoimmune glomerulonephritis, Science 279 (1998) 1052—1054.

[150] S. Manzi, E.N. Meilahn, J.E. Rairie, C.G. Conte, T.A.J. Medsger, L. Jansen-McWilliams, et al., Age-specific incidence rates of myocardial infarction and angina in women with systemic lupus erythematosus: comparison with the Farmingham study, Am J Epidemiol 145 (1997) 408—415.

[151] S. Eustace, M. Hutchinson, B. Bresnihan, Acute cerebrovascular episodes in systemic lupus erythematosus, Quart J Med 80 (1991) 739—750.

[152] D.M. Yellon, D.J. Hausenloy, Myocardial reperfusion injury, N Engl J Med 357 (2007) 1121—1135.

[153] R.J. Komotar, G.H. Kim, M.L. Otten, B. Hassid, J. Mocco, M.E. Sughrue, et al., The role of complement in stroke therapy, Adv Exp Med Biol 632 (2008) 23—33.

[154] G.M. Diepenhorst, T.M. van Gulik, C.E. Hack, Complement-mediated ischemia-reperfusion injury: lessons learned from animal and clinical studies, Ann Surg 249 (2009) 889—899.

[155] H.F. Weisman, T. Bartow, M.K. Leppo, H.C. Marsh Jr., G.R. Carson, M.F. Concino, et al., Soluble human complement receptor type 1: in vivo inhibitor of complement suppressing post-ischemic myocardial inflammation and necrosis, Science 249 (1990) 146—151.

[156] M. Pemberton, G. Anderson, V. Vetvicka, D.E. Justus, G.D. Ross, Microvascular effects of complement blockade with soluble recombinant CR1 on ischemia/reperfusion injury of skeletal muscle, J Immunol 150 (1993) 5104—5113.

[157] W.B. Gorsuch, B.J. Guikema, D.C. Fritzinger, C.W. Vogel, G.L. Stahl, Humanized cobra venom factor decreases myocardial ischemia-reperfusion injury, Mol Immunol 47 (2009) 506—510.

[158] B. de Vries, J. Kohl, W.K. Leclercq, T.G. Wolfs, A.A. van Bijnen, P. Heeringa, et al., Complement factor C5a mediates renal ischemia-reperfusion injury independent from neutrophils, J Immunol 170 (2003) 3883—3889.

[159] C. Costa, L. Zhao, Y. Shen, X. Su, L. Hao, S.P. Colgan, et al., Role of complement component C5 in cerebral ischemia/reperfusion injury, Brain Res 1100 (2006) 142—151.

[160] G.H. Kim, J. Mocco, D.K. Hahn, C.P. Kellner, R.J. Komotar, A.F. Ducruet, et al., Protective effect of C5a receptor inhibition after murine reperfused stroke, Neurosurgery 63 (2008) 122—125. discussion 125—126.

[161] J.P. Williams, T.T. Pechet, M.R. Weiser, R. Reid, L. Kobzik, F.D. Moore Jr., et al., Intestinal reperfusion injury is mediated by IgM and complement, J Appl Physiol 86 (1999) 938—942.

[162] S.D. Fleming, T. Shea-Donohue, J.M. Guthridge, L. Kulik, T.J. Waldschmidt, M.G. Gipson, et al., Mice deficient in complement receptors 1 and 2 lack a tissue injury-inducing subset of the natural antibody repertoire, J Immunol 169 (2002) 2126—2133.

[163] R.R. Hardy, K. Hayakawa, B cell development pathways, Annu Rev Immunol 19 (2001) 595—621.

[164] S.D. Fleming, M. Monestier, G.C. Tsokos, Accelerated ischemia/reperfusion-induced injury in autoimmunity-prone mice, J Immunol 173 (2004) 4230—4235.

[165] S.D. Fleming, R.P. Egan, C. Chai, G. Girardi, V.M. Holers, J. Salmon, et al., Anti-phospholipid antibodies restore mesenteric ischemia/reperfusion-induced injury in complement receptor 2/complement receptor 1-deficient mice, J Immunol 173 (2004) 7055—7061.

[166] M.R. Weiser, J.P. Williams, F.D. Moore Jr., L. Kobzik, M. Ma, H.B. Hechtman, et al., Reperfusion injury of ischemic skeletal muscle is mediated by natural antibody and complement, J Exp Med 183 (1996) 2343—2348.

[167] M. Zhang, E.M. Alicot, M.C. Carroll, Human natural IgM can induce ischemia/reperfusion injury in a murine intestinal model, Mol Immunol 45 (2008) 4036—4039.

[168] G.C. Tsokos, S.D. Fleming, Autoimmunity, complement activation, tissue injury and reciprocal effects, Curr Dir Autoimmun 7 (2004) 149–164.

[169] M. Zhang, E.M. Alicot, I. Chiu, J. Li, N. Verna, T. Vorup-Jensen, et al., Identification of the target self-antigens in reperfusion injury, J Exp Med 203 (2006) 141–152.

[170] L. Kulik, S.D. Fleming, C. Moratz, J.W. Reuter, A. Novikov, K. Chen, et al., Pathogenic natural antibodies recognizing annexin IV are required to develop intestinal ischemia-reperfusion injury, J Immunol 182 (2009) 5363–5373.

[171] S. Rehrig, S.D. Fleming, J. Anderson, J.M. Guthridge, J. Rakstang, C.E. McQueen, et al., Complement inhibitor, complement receptor 1-related gene/protein y-Ig attenuates intestinal damage after the onset of mesenteric ischemia/reperfusion injury in mice, J Immunol 167 (2001) 5921–5927.

[172] J. Chen, J.C. Crispin, J. Dalle Lucca, G.C. Tsokos, A Novel Inhibitor of the Alternative Pathway of Complement Attenuates Intestinal Ischemia/Reperfusion-Induced Injury, J Surg Res [Epub ahead of print] (2009).

[173] J.M. Thurman, P.A. Royer, D. Ljubanovic, B. Dursun, A.M. Lenderink, C.L. Edelstein, et al., Treatment with an inhibitory monoclonal antibody to mouse factor B protects mice from induction of apoptosis and renal ischemia/reperfusion injury, J Am Soc Nephrol 17 (2006) 707–715.

[174] Y. Huang, F. Qiao, C. Atkinson, V.M. Holers, S. Tomlinson, A novel targeted inhibitor of the alternative pathway of complement and its therapeutic application in ischemia/reperfusion injury, J Immunol 181 (2008) 8068–8076.

[175] M.C. Walsh, T. Bourcier, K. Takahashi, L. Shi, M.N. Busche, R.P. Rother, et al., Mannose-binding lectin is a regulator of inflammation that accompanies myocardial ischemia and reperfusion injury, J Immunol 175 (2005) 541–546.

[176] M.L. Hart, K.A. Ceonzo, L.A. Shaffer, K. Takahashi, R.P. Rother, W.R. Reenstra, et al., Gastrointestinal ischemia-reperfusion injury is lectin complement pathway dependent without involving C1q, J Immunol 174 (2005) 6373–6380.

[177] J.A. Schifferli, G. Steiger, J.P. Paccaud, A.G. Sjoholm, G. Hauptmann, Difference in the biological properties of the two forms of the fourth component of human complement (C4), Clinical & Experimental Immunology 63 (1986) 473–477.

[178] Y. Yang, E.K. Chung, B. Zhou, K. Lhotta, L.A. Hebert, D.J. Birmingham, et al., The intricate role of complement component C4 in human systemic lupus erythematosus, Current Directions in Autoimmunity 7 (2004) 98–132.

[179] R.P. Taylor, C. Horgan, R. Buschbacher, C.M. Brunner, C.E. Hess, W.M. O'Brien, et al., Decreased complement mediated binding of antibody/3H-dsDNA immune complexes to the red blood cells of patients with systemic lupus erythematosus, rheumatoid arthritis, and hematologic malignancies, Arthritis & Rheumatism 26 (1983) 736–744.

[180] K.A. Davies, A.M. Peters, H.L. Beynon, M.J. Walport, Immune complex processing in patients with systemic lupus erythematosus. In vivo imaging and clearance studies, J Clin Invest 90 (1992) 2075–2083.

[181] L.C. Korb, J.M. Ahearn, C1q binds directly and specifically to surface blebs of apoptotic human keratinocytes. Complement deficiency and systemic lupus erythematosus revisited, J Immunol 158 (1997) 4525–4528.

[182] J.S. Navratil, S.C. Watkins, J.J. Wisnieski, J.M. Ahearn, The globular heads of C1q specifically recognize surface blebs of apoptotic vascular endothelial cells, J Immunol 166 (2001) 3231–3239.

[183] D. Mevorach, J.O. Mascarenhas, D. Gershov, K.B. Elkon, Complement-dependent clearance of apoptotic cells by human macrophages, J Exp Med 188 (1998) 2313–2320.

[184] A.J. Nauta, L.A. Trouw, M.R. Daha, O. Tijsma, R. Nieuwland, W.J. Schwaeble, et al., Direct binding of C1q to apoptotic cells and cell blebs induces complement activation, Eur J Immunol 32 (2002) 1726–1736.

[185] L.A. Casciola-Rosen, G. Anhalt, A. Rosen, Autoantigens targeted in systemic lupus erythematosus are clustered in two populations of surface structures on apoptotic keratinocytes, J Exp Med 179 (1994) 1317–1330.

[186] A. Rosen, L.A. Casciola-Rosen, Autoantigens as substrates for apoptotic proteases: implications for the pathogenesis of systemic autoimmune disease, Cell Death and Differentiation 6 (1999) 6–12.

[187] L. Brandt, H. Hedberg, Impaired phagocytosis by peripheral blood granulocytes in systemic lupus erythematosus, Scand J Haematol 6 (1969) 348–353.

[188] M. Herrmann, R.E. Voll, O.M. Zoller, M. Hagenhofer, B.B. Ponner, J.R. Kalden, Impaired phagocytosis of apoptotic cell meterial by monocyte-derived macrophages from patients with systmic lupus erythematosus, Arthrit Rheum 41 (1998) 1241–1250.

[189] I. Baumann, W. Kolowos, R.E. Voll, B. Manger, U. Gaipl, W.L. Neuhuber, et al., Impaired uptake of apoptotic cells into tingible body macrophages in germinal centers of patients with systemic lupus erythematosus, Arthritis Rheum 46 (2002) 191–201.

[190] G. Sturfelt, A.G. Sjoholm, Complement components, complement activation, and acute phase response in systemic lupus erythematosus, Int Arch Allergy Appl Immunol 75 (1984) 75–83.

[191] J.A. Charlesworth, P.W. Peake, J. Golding, J.D. Mackie, Hypercatabolism of C3 and C4 in active and inactive systemic lupus erythematosus, Ann Rheum Dis 48 (1989) 153–159.

[192] H. Tsukamoto, A. Ueda, K. Nagasawa, Y. Tada, Y. Niho, Increased production of the third component of complement (C3) by monocytes from patients with systemic lupus erythematosus, Clin Exp Immunol 82 (1990) 257–261.

[193] R.M. Valentijn, H. van Overhagen, I.I.M. Hazevoet, J. Hermans, A. Cats, M.R. Daha, et al., The value of complement and immune complex determinations in monitoring disease activity in patients with systemic lupus erythematosus, Arthrit Rheum 28 (1985) 904–913.

[194] D.M. Ricker, L.A. Hebert, R. Rohde, D.D. Sedmak, E.J. Lewis, J.D. Clough, et al., Serum C3 levels are diagnostically more sensitive and specific for systemic lupus erythematosus activity than are serum C4 slevels, Am J Kidney Dis 19 (1991) 678–685.

[195] A.J. Swaak, J. Groenwold, A. Hannema, C.E. Hack, Correlation of disease activity with circulating immune complexes (C1qbA) and complement breakdown products (C3d) in patients with systemic lupus erythematosus, Rheumatology International 5 (1985) 215–220.

[196] C.K. Abrass, K.M. Nies, J.S. Louie, W.A. Border, R.J. Glassock, Correlation and predictive accuracy of circulating immune complexes with disease activity in patients with systemic lupus erythematosus, Arthrit Rheum 23 (1980) 273–282.

[197] B.A. Walz LeBlanc, D.D. Gladman, M.B. Urowitz, Serologically active clinically quiescent systemic lupus erythematosus-predictors of clinical flares, J Rheumatol 21 (1994) 2239–2241.

[198] L.D. Kerr, B.R. Adelsberg, P. Schulman, H. Spiera, Factor B activation products in patients with systemic lupus erythematosus, Arthrit Rheum 32 (1989) 1406–1413.

[199] P. Hopkins, H.M. Belmont, J.P. Buyon, M. Philips, G. Weissmann, S.B. Abramson, Increased levels of plasma anaphylatoxins in systemic lupus erythematosus predict flares

of the disease and may elicit vascular injury in lupus cerebritis, Arthrit Rheum 31 (1988) 632–641.

[200] J.M. Porcel, J. Ordi, A. Castro-Salomo, M. Vilardell, M.J. Rodrigo, T. Gene, et al., The value of complement activation products in the assessment of systmeic lupus erythematosus flares, Clin Immunol Immunopath 74 (1995) 283–288.

[201] J.P. Buyon, J. Tamerius, H.M. Belmont, S.B. Abramson, Assessment of disease activity and impending flare in patients with systemic lupus erythematosus: comparison of the use of complement split products and conventional measurements of complement, Arthrit Rheum 35 (1992) 1028–1037.

[202] S. Manzi, J.E. Rairie, A.B. Carpenter, R.H. Kelly, S.P. Jagarlapudi, S.M. Sereika, et al., Sensitivity and specificity of plasma and urine complement split products as indicators of lupus disease activity, Arthrit Rheum 39 (1996) 1178–1188.

[203] E. Rother, B. Lang, R. Coldewey, K. Hartung, H.H. Peter, Complement split product C3d as an indicator of disease activity in systemic lupus erythematosus, Clinical Rheumatology 12 (1993) 31–35.

[204] R.J. Falk, M.D. Agustin, P. Dalmasso, Y. Kim, S. Lam, A. Michael, Radioimmunoassay of the attack complex of complement in serum from patients with systemic lupus erythematosus, N Engl J Med 312 (1985) 1594–1599.

[205] C. Ponticelli, G. Moroni, Lupus nephritis, Journal of Nephrology 13 (2000) 385–399.

[206] S.R. Pillemer, H.A.I. Austin, G.C. Tsokos, J.E. Balow, Lupus nephritis: Association between serology and renal biopsy measures, J Rheumatol 15 (1988) 284–288.

[207] K. Manger, B. Manger, R. Repp, M. Geisselbrecht, A. Geiger, A. Pfahlberg, et al., Definition of risk factors for death, end stage renal disease, and thromboembolic events in a monocentric cohort of 338 patients with systemci lupus erythematosus, Ann Rheum Dis 61 (2002) 1065–1070.

[208] I. Gunnarsson, B. Sundelin, M. Heimburger, J. Forslid, R. van Vollenhoven, I. Lundberg, et al., Repeated renal biopsy in proliferative lupus nephritis - Predictive role of serum C1q and albuminuria, J Rheumatol 29 (2002) 693–699.

[209] R.H. Kelly, A.B. Carpenter, K.S. Sudol, S.P. Jugariapudi, S. Manzi, Complement C3 fragments in urine detection in systemic lupus erythematosus patients by Western Blotting, Applied and Theoretical Electrophoresis 3 (1993) 265–269.

[210] V.S. Negi, A. Aggarwal, R. Dayal, S. Naik, R. Misra, Complement degradation product C3d in urine: marker of lupus nephritis, J Rheumatol 27 (2000) 380–383.

[211] A. Ho, S.G. Barr, L.S. Magder, M. Petri, A decrease in complement is associated with increased renal and hematologic activity in patients with systemic lupus erythematosus, Arthrit Rheum 44 (2001) 2350–2357.

[212] S. Manzi, J.S. Navratil, M.J. Ruffing, C.-C. Liu, N. Danchenko, S.E. Nilson, et al., Measurement of erythrocyte C4d and complement receptor 1 in the diagnosis of systemic lupus erythematosus, Arthrit Rheum 50 (2004) 3596–3604.

[213] C.C. Liu, S. Manzi, A.H. Kao, M.J. Ruffing, J.S. Navratil, J.M. Ahearn, Reticulocytes bearing C4d as biomarkers of disease activity for systemic lupus erythematosus, Arthritis & Rheumatism 52 (2005) 3087–3099.

[214] J.S. Navratil, S. Manzi, A.H. Kao, S. Krishnaswami, C.C. Liu, M.J. Ruffing, et al., Platelet C4d is highly specific for systemic lupus erythematosus, Arthritis & Rheumatism 54 (2006) 670–674.

[215] C.A. Tieley, D.G. Romans, M.C. Crookston, Localization of Chido and Rodgers determinants to the C4d fragment of human C4, Nature 276 (1978) 713–715.

[216] B.P. Morgan, C.L. Harris, Complement therapeutics; history and current progress, Mol.Immunol 40 (2003) 159–170.

[217] D. Ricklin, J.D. Lambris, Complement-targeted therapeutics, Nat Biotechnol 25 (2007) 1265–1275.

[218] H.F. Weisman, T. Bartow, M.K. Leppo, H.C. Marsh Jr., G.R. Carson, M.F. Concino, et al., Soluble human complement receptor type 1: in vivo inhibitor of complement suppressing post-ischemic myocardial inflammation and necrosis, Science 249 (1990) 146–151.

[219] J.M. Weiler, R.E. Edens, R.J. Linhardt, D.P. Kapelanski, Heparin and modified heparin inhibit complement activation in vivo, J Immunol 148 (1992) 3210–3215.

[220] D. Ricklin, J.D. Lambris, Compstatin: a complement inhibitor on its way to clinical application, Adv Exp Med Biol 632 (2008) 273–292.

[221] P.A. Whiss, Pexelizumab Alexion, Current Opinion in Investigational Drugs 3 (2002) 870–877.

[222] M. Kaplan, Eculizumab (Alexion), Current Opinion in Investigational Drugs 3 (2002) 1017–1023.

[223] T.V. Arumugam, I.A. Shiels, A.J. Strachan, G. Abbenante, D.P. Fairlie, S.M. Taylor, A small molecule C5a receptor antagonist protects kidneys from ischemia/reperfusion injury in rats, Kidney Internat 63 (2003) 134–142.

[224] H. Sumichika, C5a receptor antagonists for the treatment of inflammation, Current Opinion in Investigational Drugs 5 (2004) 505–510.

[225] R.P. Rother, C.F. Mojcik, E.W. McCroskery, Inhibition of terminal complement: a novel therapeutic approach for the treatment of systemic lupus erythematosus, Lupus 13 (2004) 328–334.

[226] G. Girardi, P. Redecha, J.E. Salmon, Heparin prevents antiphospholipid antibody-induced fetal loss by inhibiting complement activation, Nature Medicine 10 (2004) 1222–1226.

Humoral Pathogenesis
Fcγ Receptors in Autoimmunity and End-Organ Damage

T. Ernandez, T.N. Mayadas*

INTRODUCTION

Antibodies are potent inducers of inflammation. A consistent feature of systemic lupus erythematosus (SLE) is the presence of autoantibodies and complement-fixing immune complexes resulting in inflammatory lesions in multiple organ systems. The receptors for antibodies are the Fc receptors, which are a group of transmembrane glycoproteins central to host defense against pathogens but not surprisingly can also inflict significant tissue damage when antibodies are directed against host antigens as in SLE. Fc receptors are largely expressed in hematopoietic cells and mediate a wide array of immune functions such as the recruitment and the activation of inflammatory cells, degranulation, antibody-dependent cell-mediated cytotoxicity (ADCC), phagocytosis, enhancement of antigen presentation, regulation of B cell antibody production, and immune-complex (IC) clearance. Although Fc receptors exist for all the different immunoglobulin classes (IgA, IgD, IgE, IgG, IgM), our discussion will focus on the Fcγ receptors (FcγR), the receptors for IgG as studies in mouse models suggest primary roles for FcγRs in development of SLE and functional polymorphisms in FcγRs are associated with disease susceptibility in humans.

FCγRS: STRUCTURE AND FUNCTIONS

The extracellular region of FcγRs contains two or three extracellular immunoglobulin-like domains, which is a common feature of the immunoglobulin superfamily. In humans, three different classes of FcγRs, FcγRI, FcγRII, and FcγRIII exist that are defined by their structural differences, signaling capacity, and variable affinity for the different IgG subclasses [1, 2]. The FcγRI (CD64) and FcγRIII (CD16) are expressed as oligomeric complexes with γ (monocytes and macrophages) or ζ (uniquely for human FcγRIIIA in NK cells) chain. The adaptor chains contain the cytoplasmic immunoreceptor tyrosine-based activation motif (ITAM), which links FcγRs to protein tyrosine-based intracellular signaling. Partnering with the γ chain, through amino acid interactions between charged residues present in the transmembrane domains of the adaptor and FcγR, is also required for the surface expression of both FcγRI and FcγRIII [3]. On the other hand, members of the FcγRII class (CD32) exist as single polypeptide transmembrane receptors containing either an activating (FcγRIIA) or inhibitory (FcγRIIB) signaling motif within their own cytoplasmic tail. Genetic deletion of FcγRs in mice has revealed primary roles for these receptors in a number of autoimmune diseases including SLE. However, the repertoire of FcγRs between human and mice differ in some important aspects, which should be considered when extrapolating data from mice models to human biology. Both human and mouse share a structurally common FcγRI, FcγRIIB, and FcγRIII. However, mice express the species-specific FcγRIV and lack the uniquely human FcγRIIIB and FcγRIIA (Figure 21.1). The mouse FcγRIV and FcγRIII are considered orthologs of the human FcγRIIIA and FcγRIIA, respectively, based on sequence similarity in their extracellular domain [4]. The repertoire of FcγRs expressed on specific cell types can also differ between humans and mice (Figure 21.1). For example, human monocytes/macrophages and neutrophils express FcγRIIA, while these cell types in mice express the species-specific FcγRIV [5]. Moreover, human platelets express FcγRIIA, while mouse platelets do not express any FcγRs [6].

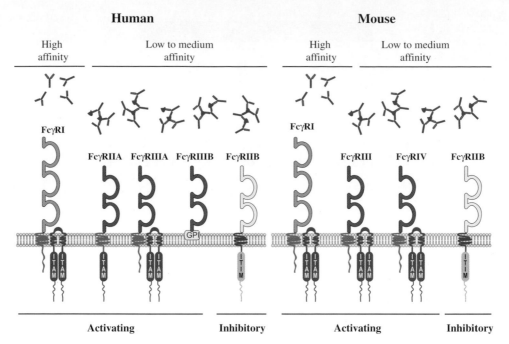

FIGURE 21.1 Fcγ receptor repertoire in humans and mice Human FcγRs include the high affinity receptor FcγRI and the low- and medium-affinity receptors FcγRIIA, FcγRIIB, FcγRIIIA, and FcγRIIIB. Mice do not express FcγRIIA and FcγRIIIB, but express a FcγRIII and a species-specific FcγRIV. The activating FcγRs are coupled to an Fc γ chain containing two immunoreceptor tyrosine-based activation motifs (ITAM). The human FcγRIIA contains an ITAM motif included in its own cytoplasmic tail. In contrast, the inhibitory FcγRIIB of both humans and mice signals through a distinct immunoreceptor tyrosine-based inhibitory motif (ITIM) present in its cytoplasmic tail.

FcγRI

The FcγRI is the only receptor able to efficiently bind monomeric IgG and maybe also aggregated IgG [7] and is therefore referred to as a high-affinity receptor. It differs structurally from the other FcγRs as it contains three extracellular immunoglobulin-like domains [8]. In contrast to the low-affinity FcγRII and FcγRIII, FcγRI is saturated with circulating monomeric IgG and triggers intracellular signals only after cross-linking with antigens, which cluster the FcγR and thus enable signal transduction [4]. FcγRI preferentially binds IgG1 and IgG3 and to a lesser extent IgG4 and IgG2. It is constitutively expressed on monocytes, macrophages, and dendritic cells and can be upregulated by IFN-γ on neutrophils (also G-CSF), eosinophils, and glomerular mesangial cells (also IL-10) [1, 9, 10]. Induction of FcγRI during inflammation by cytokines such as IFN-γ or IL-10 may trigger critical immune effector-cell functions such as phagocytosis, ADCC, and enhanced antigen presentation [1].

FcγRII

The FcγRII class encompasses two functionally very distinct receptors, the FcγRIIA and the FcγRIIB. The FcγRIIA is an activating receptor, which signals through an ITAM motif similar to that of FcγRI and FcγRIIIA. FcγRIIB is an inhibitory FcγR, which signals through its immunoreceptor tyrosine-based inhibitory motif (ITIM) to inhibit signals generated by ITAM-containing receptors [11, 12]. Thus FcγRIIB sets the threshold of activation and determines the level of cellular responses. Both FcγRII receptors are called low-affinity receptors as they do not bind monomeric IgG, but only aggregated IgG such as ICs. Both receptors exhibit specificity for IgG1, IgG2, and IgG3 but do not bind IgG4 immune complexes. FcγRIIA and IIB are widely expressed on leukocytes (Table 21.1). FcγRIIA is present on neutrophils, eosinophils, basophils, platelets [13, 14], dendritic cells [15–17], monocytes, macrophages, Langerhans cells, and transiently in immature T cells [18, 19]. FcγRIIA on neutrophils and macrophages promotes phagocytosis, NADPH oxidase triggered oxidative burst, leukotriene release, ADCC, and cellular activation [1]. On neutrophils FcγRIIA supports their recruitment to IC-coated endothelial cells [20]. In dendritic cells, FcγRIIA promotes phagocytosis and facilitates the antigen-presentation required for an efficient adaptive immune response [15, 16]. FcγRIIA is transiently expressed in immature T cells before their TCR rearrangement and may be important in thymocyte maturation [18]. FcγRIIA on human platelets appears to play an important role in αIIbβ3 integrin outside-in signaling and thereby participates in platelet activation and spreading after

TABLE 21.1 Cellular Distribution and Functions of FcγRs in Human Leukocytes

	FcγR expression	FcγR mediated functions
Neutrophils	Activating IIA, IIIB Inhibitory IIB	• Phagocytosis • Oxidative burst • Degranulation and release of proteolytic enzymes • Release of chemoattractants • ADCC
Mast cells, basophils	Activating IIA, IIIA Inhibitory IIB	• Release of vasoactive amines • Release of chemoattractants • Release of cytokines (TNFα) and prostaglandins
Macrophages	Activating I, IIA, IIIA Inhibitory IIB	• Phagocytosis • Oxidative burst • ADCC • Release of cytokines (TNFα, IL-1, IL-6)
NK cells	Activating IIIA No inhibitory FcγRs	• ADCC • Release of cytokines (TNFα, INFγ)
Dendritic cells	Activating I, IIA No inhibitory FcγRs	• Phagocytosis • Enhancement of antigen presentation • Maturation • Release of cytokines (IL-12, INFγ)
B cells	No activating FcγRs Inhibitory IIB	• Modulation of B cell activation • Inhibition of antibody production • Apoptosis (plasma cells)

binding of αIIbβ3 to fibrinogen [13]. In platelets, this receptor may also facilitate the clearance of circulating ICs [21]. FcγRIIB is predominantly expressed on B cells but also on neutrophils and monocytes at a lower level [22–24]. FcγRIIB is also present on dendritic cells but its expression is variable and may be related to the stage of cell maturation [16, 22]. The inhibitory function of FcγRIIB has been extensively studied and appears to play a critical role in maintaining peripheral tolerance (detailed in the section titled "Maintenance of Peripheral Tolerance") [11, 12] as even a small decrease in its expression leads to a breakdown of tolerance, autoantibody production, and autoimmune disease [25]. FcγRIIB expression in other cell types also restrains the adaptive immune response, as recently demonstrated for instance for dendritic cells. Indeed, FcγRIIB engagement in dendritic cells limits the phagocytosis of ICs as well as TNFα production, therefore impairing the antigen presentation to T cells and their subsequent activation and proliferation [16, 17, 26].

FcγRIII

The FcγRIII class includes FcγRIIIA and FcγRIIIB, which are both low-affinity receptors. They preferentially bind IgG1 and IgG3 and exhibit virtually no binding of IgG2 and IgG4. Despite a 98% identity of their extracellular domains, the two receptors are structurally distinct: while the FcγRIIIA is expressed as a transmembrane protein and requires interaction with the common γ-chain for both expression and signaling [1, 19, 27], the FcγRIIIB attaches to the membrane through a glycosylphosphatidylinositol (GPI) anchor and lacks direct signaling capacity [4, 28, 29]. The FcγRIIIA is expressed on macrophages (Table 21.1), NK, γδ T cells [30–32] and in some reports also on a small subset of terminally differentiated αβ T cells [33, 34]. FcγRIIIA is the only FcγR expressed on NK cells and it principally mediates ADCC and proinflammatory cytokine production [35]. On γδ T cells, FcγRIIIA also mediates cytolysis via ADCC [36]. FcγRIIIA is expressed on glomerular mesangial cells after IFN-γ treatment [1, 37], where it induces the production of IL-6 and may be a key factor in the progression of glomerulonephritis [3]. In contrast to the relatively broad expression of FcγRIIIA, the GPI-linked FcγRIIIB is only expressed on granulocytes (i.e., neutrophils and eosinophils) [38, 39] and has recently been described in basophils [40]. It is the most abundant FcγR in neutrophils with a surface expression that is four- to fivefold greater than FcγRIIA [41]. On neutrophils, it partners with other receptors such as FcγRIIA and the complement receptor CR3 to induce neutrophil activation [42–44]. The role of two structurally distinct activating receptors in neutrophils begs the question of whether each has evolved distinct functions. FcγRIIIB cross-linking induces calcium mobilization, triggers degranulation [45, 46], leukotriene release, oxidative burst [43], and activation of integrins [47]. Recent work demonstrates that FcγRIIIB-specific engagement triggers a robust increase in nuclear-restricted phosphorylation of ERK and the transcription factor, Elk-1 [48]. In vitro, human FcγRIIIB tethers neutrophils to immobilized ICs under physiological flow conditions, suggesting a role for this receptor in neutrophil recruitment [20]. Studies in transgenic mice expressing human FcγRIIIB and/or FcγRIIA on neutrophils of mice lacking their endogenous FcγRs has provided in vivo evidence that both receptors promote neutrophil recruitment that is context-dependent [49].

It is noteworthy that although FcγRIIIB is a true low-affinity receptor, the FcγRIIIA in tissue macrophages is categorized as having intermediate affinity for monomeric IgG [50]. The affinity of FcγRIIIA is modulated by its glycosylation state and different glycoforms of this receptor with variable affinity pattern have been identified on monocytes/macrophages and NK cells [51]. FcγRIIIA with enhanced affinity is particularly well suited for the capture and the processing of ICs by macrophages resident in the spleen and liver, which may be critical for IC clearance from the circulation [51]. Additional complexity in the Fcγ receptor system is introduced by the presence of different allelic variants and copy number variations. These have functional consequences and may in fact dictate one aspect of the genetically predetermined susceptibility to autoimmune diseases such as SLE, as detailed in the section titled "FcγR Polymorphisms and Copy Number Variation in Lupus" [51–54].

Mirroring the complex organization of FcγRs, IgG are also subdivided into four classes that exhibit different patterns of affinity for their receptors. This topic was recently revisited in a study where the affinity of the different IgG isotypes for all four FcγRs and known allelic variants were compared in parallel *in vitro* [7]. The distinct affinity pattern of the different IgG subclasses for FcγRs may account for their variable pathogenicity both in mice and humans. For example, in mouse models of autoimmune anemia and tumor cell cytotoxicity, higher pathogenicity is attributed to IgG2a and IgG2b compared to other isotypes [55, 56], likely because of their preferential affinity for FcγRIV, which is highly proinflammatory in several models of disease in mice [57]. Lupus-prone mice strains also express predominantly IgG2a and IgG2b: IgG2a is the dominant subclass in glomerular eluates from NZB/W mice [58, 59], whereas IgG2b is prevalent in BXSB mice [59]. IgG2a and IgG2b are equally common in MRL/lpr mice [59]. Furthermore, C57BL/6 mice genetically engineered with a deficiency in FcγRIIb develop spontaneous SLE (see the section titled "Maintenance of Peripheral Tolerance") [60] that correlates with high serum titers and glomerular deposits of IgG2a and IgG2b [61]. FcγRIV may be predicted to be the predominant FcγR responsible for tissue injury in lupus models in mice. In human SLE, IgG3 and IgG1 are the dominant isotypes involved in the anti-DNA response and preferentially engage FcγRIIA (ortholog of FcγRIV) and FcγRIIIA [62]. Anti-dsDNA IgG3 antibody appears to be more specifically involved in SLE among other connective tissue diseases [63] and in human lupus nephritis, IgG3 and IgG1 renal deposition is the most commonly observed [64–66]. However, IgG2 deposition is also frequently observed in lupus nephritis and interestingly intense IgG2 deposition in the kidney was associated with the FcγRIIA *R131* allele [66], a polymorphism shown to be associated with susceptibility to SLE. The FcγRIIA R131 polymorphism exhibits a decrease in affinity for IgG2 that may impair clearance of IgG2 containing IC clearance, thus resulting in an increase in IC deposition, and lupus nephritis (see the section titled "FcγR Polymorphisms and Copy Number Variation in Lupus") [54, 67].

As discussed, the FcγRs differ in their expression pattern, their surface levels and affinity for IgG subclasses, and their activating ratio (i.e., activating versus inhibitory receptors) on the cell surface [57]. All these variables are crucial in determining the immune effector response and therefore the nature of IC mediated diseases. Thus it is not surprising that inflammatory mediators such as cytokines play a key role in modulating these parameters. IFN-γ and C5a upregulate the expression of FcγRI and FcγRIII and downregulate the inhibitory FcγRIIB [9, 10, 68, 69]. On the other hand, Th-2 cytokines such as IL-4, IL-10, or TGF-β upregulate the inhibitory FcγRIIB and downregulate the activating FcγRs [70–72]. Post-transcriptional effects of cytokines on FcγRs have also been described. For instance, inflammatory mediators promote FcγRIIIB shedding from the surface, but can also induce FcγRIIIB translocation from intracellular stores, thus increasing surface levels [73, 74]. On the other hand, the surface levels of FcγRIIA do not change, but its activity may be regulated at the ligand-binding stage. Its binding activity for IgG-ICs transiently increases on neutrophils following treatment with the bacterial chemotactic peptide fmlp [44], and on eosinophils following GM-CSF, IL-3, and IL-5 [75]. In addition to their effects on FcγRs, cytokines regulate IgG isotype switching. The Th-1 cytokine IFN-γ preferentially induces the more pathogenic IgG2a subclass, whereas the Th-2 cytokines IL-4 induces switching to more indolent murine IgG1 [2, 76].

FcγRs and Complement

The complement system is a major effector mechanism in innate immune responses. Complement can be activated via the classical, alternative, or mannose-binding lectin pathways, each of which converges on C3. Deficiency in components of the complement cascade and in particular the classical pathway has been strongly associated with autoimmunity and SLE [77]. In the classical complement pathway, the first component of the cascade, C1q, binds to and is activated by the Fc-portion of immunoglobulins in immune complexes [78]. C1q has protective and pro-inflammatory roles that have been shown to potentially contribute to the development of SLE, as discussed in detail elsewhere [79]. In particular, C1q deficiency has been linked to defects in binding and subsequent clearance of apoptotic cells, but this is likely through non-FcγR related mechanisms [80]. On the other hand, several of C1q's pro-inflammatory effects are manifested directly or indirectly through FcγRs. C1q may aid in the deposition of immune complexes within the vasculature [81], which in turn triggers FcγR-mediated immune cell recruitment. Complement C1q triggers the production of C3b, which subsequently catalyzes C5 to its active form C5a [82] and these components have been shown to directly modulate FcγR functions [24]. For instance, C5a transcriptionally upregulates FcγRIII and downregulates FcγRIIB in alveolar macrophages in a murine model of acute pulmonary ICs hypersensitivity [68]. Similarly, C5a induces the expression of FcγRI and FcγRIII in a model of autoimmune hemolytic anemia [83]. In turn, FcγRs induce the production of C5a, as shown *in vitro* following cross-linking on macrophages [84, 85] and *in vivo* in a model of autoimmune hemolytic anemia [83]. C5a also appears to increase FcγRIIA binding affinity to ICs through a mechanism that remains to be elucidated (N. Tsuboi and T. Mayadas, unpublished data). Unexpectedly, although C5aR deficient mice exhibit defects in FcγR-mediated inflammatory responses in the lung, peritoneum and kidney of mice [86–88], C3-deficient mice are not protected. These data suggest a complex interaction between C5a and FcγR that is potentially independent of C3b production and thereby bypasses the upstream classical and alternative complement cascades [86, 89]. In support of this, C5aR blockade significantly attenuated neutrophil accumulation and edema in C3-deficient mice subjected to the reverse passive Arthur's reaction (RPA) [89]. In models that require C5aR but not complement C3, C5a may also be generated by a proteolytic pathway independent of C4b. For example, C5a can be generated by thrombin, a serine protease in the coagulation pathway [90]. Thus FcγRs and C5aRs may play co-dominant roles in IC-induced inflammatory responses. Clarification of the mechanisms underlying their collaboration may thereby uncover critical steps involved in autoimmune mediated end-organ damage.

Activating and Inhibitory FcγR Signaling

Activating and inhibitory FcγRs are classified by their signaling properties through ITAM- and ITIM-based motifs, respectively. The activating FcγRI, FcγRIIA, FcγRIIIA as well as the murine FcγRIII and FcγRIV initiate signaling through ITAM motifs. FcγRI, FcγRIII, and FcγRIV need to interact with the Fcγ chain bearing two ITAM motifs, while the human FcγRIIA is a single polypeptide chain containing an ITAM motif in its own cytoplasmic domain. FcγRIIIB does not directly interact with an ITAM-bearing adaptor. It may instead transduce signals by interacting with other membrane proteins (such as CD18-integrins or FcγRIIA) [91] and accumulate in lipid rafts to trigger an activating signal through src-family tyrosine kinases, such as hck [92,93]. FcγRIIB, the only inhibitory receptor,

downmodulates activating signals through an ITIM motif included in its cytoplasmic domain (Figure 21.2).

After binding of ICs, FcγR cluster in lipid rafts at the cell surface, which induces the phosphorylation of the ITAM motif by an src protein kinase family member. The syk kinase is subsequently recruited by the phosphorylated ITAM through its src-homology 2 (SH2) domain leading to the formation of a signaling complex at the membrane, in which Syk phosphorylates and activates several other substrates (Figure 21.2) [94]. As a result, the phosphatidylinositol 3-kinase (PI3K) pathway is activated and its lipid products (PIP3) mediate the membrane localization and activation of several signaling proteins including Akt and the phospholipase Cγ (PLCγ). PLCγ induces the production of inositol-3-phosphate (IP3), mediating a sustained intracellular increase in calcium (Ca^{++}) as a result of mobilization of stores from the endoplasmic reticulum and the opening of plasma membrane Ca^{++} channels. Syk also activates the exchange factor Sos, leading to the activation of the ras-raf-MAPK pathway [95]. The Rac pathway is also activated by Syk through the Rho/Rac GTPase guanine exchange factor Vav critically involved in neutrophil and macrophage FcγR-mediated oxidative burst and cytoskeleton reorganization [96, 97]. These signaling events can lead to cellular activation, degranulation,

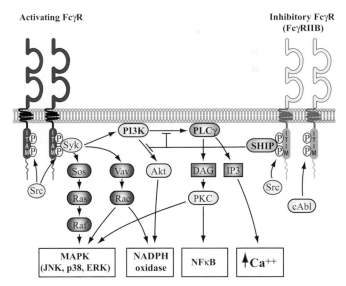

FIGURE 21.2 Activating and inhibitory Fcγ receptor signaling IgG-immune complexes cluster the FcγRs at the membrane surface, which results in the phosphorylation of the ITAM motif (activating FcγRs) or the ITIM motif (inhibitory FcγRIIB) by src family kinases (Src). As depicted, this triggers opposing signaling events. ITAM signals promote proinflammatory cellular functions such as cell activation, degranulation, phagocytosis, oxidative burst, transcription, and cytokine release. ITIM signals directly counteract ITAM signals initiated by FcγR and other ITAM linked receptors such as BCR, through activation of SHIP. ITIM can also induce plasma cell apoptosis through a cAbl-dependent pathway.

phagocytosis, oxidative burst, ADCC, proinflammatory genes expression, and release of cytokines [4].

FcγRIIB, the only inhibitory FcγR, downmodulates activating signals through an ITIM motif included in its cytoplasmic domain (Figure 21.2). FcγRIIB triggers signaling only if co-engaged with either BCR in B cells, FcγRIII or FcεRI in mast cells, or FcγRIII or FcγRIA in macrophages [98]. The co-aggregation of FcγRIIB induces the phosphorylation of the ITIM motif by the src-family kinase Lyn, triggering the recruitment of SH2-domain-containing phosphatases such as SHP1, SHP2, and the inositol polyphosphate 5′ phosphatase (SHIP) [99, 100]. As the primary substrates of SHIP are the PI3K lipid products (PIP3), FcγRIIB inhibits ITAM, activating signaling primarily by blocking the production of the second messengers inositol-3-phosphate (IP3) and diacylglycerol (DAG) generated by PI-3k/PLCγ. This in turn suppresses Ca^{2+} mobilization from intracellular stores [12]. An ITIM- and SHIP-independent pathway engaged by FcγRIIB has also been identified in B cells. Cross-linking of FcγRIIB specifically mediates plasma cell apoptosis via an Abl-family kinase pathway. This may play an important role in eliminating autoantibody producing B-cells and thereby maintaining tolerance [101].

The importance of counterbalancing ITAM and ITIM signals in titrating effector responses emerges as a theme for other immune receptor systems as well, including TCR, BCR, and NK activating receptor [102–105]. Albeit it is important to recognize that ITAM-containing receptors are not always activating. For example, γ-chain associated with IgA receptor (FcαR) inhibits IgG-mediated functions in monocytes and mast cells and the ITAM-containing adaptor DAP12 limits TLR-induced cytotoxic production [3]. However, these are instances of heterologous receptors with distinct ligand interactions. In the case of FcγRIIB and IIA, which bind the same ligands with equal affinity, the view that ITAM is activating and ITIM is inhibitory appears to hold true with the net signaling effect depending on the ratio of activating and inhibitory signal. This ratio may be influenced by the IgG subclasses that offer differential affinity for activating and inhibitory receptors, as well as the local cytokine milieu, which directly modulates the expression of activating and inhibitory receptors at the cell surface.

ROLES OF FCγR IN SLE

The roles of FcγRs in autoimmune disease can be divided into three distinct steps [106]:

1. Defect of FcγRIIB in controlling autoreactive B cell activation leading to the breakdown of peripheral tolerance;

2. Deficiency in clearance of circulating autoantibody IgG-containing ICs by FcγRs on macrophages in liver and spleen;

3. Tissue IC deposition and activation of FcγR bearing immune effector cells (e.g., macrophages and neutrophils) that directly promote end-organ injury, and clear autoantibody-opsonized circulating cells (erythrocytes, platelets), thus leading to anemia/thrombocytopenia.

The available data supporting such mechanisms in SLE will be detailed in this section.

Maintenance of Peripheral Tolerance

Fcγ receptors are centrally involved in the regulation and the specificity of immunoglobulins produced and are therefore an essential checkpoint in the adaptive immune response [107]. FcγIIB plays a critical role in maintaining immune homeostasis under physiological conditions and an increasing amount of data suggests that dysfunction of this receptor causes autoimmune disorders. The importance of FcγRIIB in tolerance maintenance has been largely elucidated following the generation of the FcγRIIB-deficient mice [108] and the discovery that a deficiency in this receptor in C57Bl/6 mice is associated with production of autoantibodies and lethal autoimmune glomerulonephritis [60] (Table 21.3). However, this appears to be strain-specific as mice on the BALB/c background did not develop these features of SLE, suggesting epistatic modifications by other background genes [60]. As FcγRIIB is the only Fcγ receptor expressed on B cells, its deficiency may be a critical determinant in development of autoreactive B cells and autoantibody production in SLE [4, 106].

Indeed, the inhibitory FcγRIIB on B cells regulates the adaptive humoral immune response at two different levels, B cell activation and plasmocyte survival [12, 109, 110]. First, FcγRIIB controls the magnitude and persistence of response to antigen by modulating the B cell receptor signal. Second, this receptor excludes low-affinity or self-reactive B cells. That is, during the affinity maturation of B cells, high-affinity B cells receive a signal form both BCR and FcγRIIB when interacting with follicular dendritic cells in peripheral immune organs, while low-affinity or self-reactive lymphocytes receive a signal only through FcγRIIB and thus undergo apoptosis [4]. FcγRIIB also directly modulates plasma cell survival. In response to immunization, FcγRIIB-deficient mice have a greater number of plasma cells due to their greater production [11] and a longer lifespan [110]. With regards to the latter, cross-linking of FcγRIIB on plasma cells from wild-type mice induces apoptosis independently of B cell receptor activation, possibly through the cAbl kinase pathway, and therefore controls their persistence in the bone marrow [100, 101, 110, 111]. Finally, cross-linking of FcγRIIB expressed on human plasmablasts or on the human myeloma cell line (EJM) also induces apoptosis, suggesting that this mechanism is relevant in humans [110]. Thus, a dysregulation of FcγRIIB expression in B cells could enhance autoantibody production and thereby autoimmunity. This is supported by a study showing in active SLE patients a higher percentage of plasma B cells and lower levels of FcγRIIB on memory B lymphocytes (CD19$^+$CD27$^+$) and plasma B cells (CD19^{low}CD27^{high}) compared to normal donors [22].

Murine models of autoimmune disorders also support the thesis that FcγRIIB regulates humoral immunity. Mouse strains susceptible to autoimmunity including SLE (NZB, MRL, BXBB, SB/Le, 129, and NOD mice) all contain a deletion in the promoter of FcγRIIB that correlates with a lower cell surface expression of this receptor on mature B cells and macrophages compared to control mouse strains (BALB/c, C57BL/6, or DBA2) [112]. Furthermore, partial restoration of expression of FcγRIIB in B cells, but also in immature thymocytes and macrophages in autoimmune-prone mouse strains (NZM 2410, BXSB, B6. FcγRIIB$^{-/-}$), limits autoantibody generation and greatly improves the survival of these mice [113]. The role of FcγRIIB specifically on B cells in the development of autoimmunity is also demonstrated by the finding that transgenic expression of FcγRIIB specifically in the B cell compartment of FcγRIIB-deficient mice reduces SLE, leads to early resolution of collagen-induced arthritis, and suppresses T-cell-dependent IgG immune responses. On the other hand, FcγRIIB expression in macrophages has no effect on these parameters but decreases survival after *Streptococcus pneumoniae* infection [114]. Furthermore, Rag1- or IgH-deficient mice transplanted with bone marrow from FcγRIIB-deficient mice develop autoantibodies and glomerulonephritis in spite of FcγRIIB expression in the monocytic compartment, but not in B cells [60]. A more general immunomodulatory role for FcγRIIB is also apparent in inflammatory mouse models that don't rely on B cell responses. FcγRIIB-deficient mice exhibit an increase in collagen-induced arthritis [115, 116] and Goodpasture-like syndrome induced by anti-collagen IV antibody [117].

Immune Complex Clearance

Effective processing of ICs diminishes their deposition in tissues and in the context of host defense may help in the delivery of antigens to specific sites where antigen presentation may occur. Defects in IC clearance are observed in many autoimmune disorders such as SLE [118, 119]. However, these are complex disorders and clear evidence demonstrating a cause and effect in

patients still remains elusive. Clearance of circulating ICs is accomplished by the mononuclear phagocyte system predominantly in the liver and spleen. In addition erythrocytes bind ICs, thus preventing their interaction with the endothelium and/or extravasation into the tissues [120, 121]. Sinusoidal phagocytes subsequently remove ICs bound to erythrocytes when erythrocytes circulate through the liver and spleen [122].

The two primary molecular mediators of IC clearance are the complement proteins and their receptors, and FcγRs. Complement prevents IC precipitation by disrupting the stability and decreasing the size of the ICs, and it coats the ICs with C4b and C3b for their safe transport within the circulation via erythrocytes bearing the C3b/C4b receptor CR1 (CD35) [123]. It also mediates uptake and disposal of ICs via complement receptors present on phagocytes of the liver and spleen. Impaired processing and uptake of ICs in SLE patients is attributed to deficiencies in early complement components, decreased CR1 expression, and impaired FcγR function. A deficiency in early classical complement proteins including C1q and C4 is the strongest genetic susceptibility factor for SLE [77]. C3 deficiency and a deficiency in the complement receptor for C3, Mac-1 (Complement receptor 3, CR3) is also associated with SLE [124−126]. FcγRIIA on macrophages have been implicated in IC clearance from the circulation and patients with expression of the low IgG binding variant of FcγRIIA R131 show reduced IC endocytosis by macrophages [127, 128].

In addition to clearance of circulating ICs, removal of tissue-deposited ICs is also required. Endothelial cells of certain vascular beds may promote endocytosis of intravascular ICs through FcγRs [129−132] and epithelial cells in close contact with the basement membrane in the glomerulus may remove trapped ICs via an active transport system involving FcRn, an IgG transport receptor [133]. Circulating phagocytes (neutrophils, monocytes) may play a role in clearing ICs deposited within the vasculature, which may be particularly important in the glomerulus, a frequent site of IC trapping [134]. Indeed, glomerular ICs can trigger transient accumulation of neutrophils that is associated with a complete clearance of the immune deposits and restoration of the structural integrity of the glomerulus within 24 h of IC deposition [135]. There is indirect evidence to suggest that FcγRIIIB plays a role in this process. FcγRIIIB promotes tethering of neutrophils to ICs under flow conditions [20] and may endocytose preformed ICs [136, 137]. The presence of FcγRIIIB on microvilli [138], its predicted fast mobility in the membrane bilayer [41], coupled with its weak signaling capacity, may suit FcγRIIIB for efficient capture and internalization of ICs deposited within the vasculature, without the potential injurious consequences of overt neutrophil activation. By extension, a potential attenuation in clearance of tissue-deposited IC as a result of impaired FcγRIIIB function may explain the observed increase in the risk of lupus nephritis in patients with a functional polymorphism in FcγRIIIB, (see the section titled "Fcγ Receptor Polymorphisms and Copy Number Variations in Lupus").

Pathogenesis of Lupus Nephritis

Significant strides have been made in identifying pathogenic processes in vivo that are dependent on activating FcγRs. Using mouse models, several studies have shown that FcγRs are required for leukocyte influx and injury induced by autoimmune responses in several organs including the skin, lung, kidney, and joints. In SLE, murine activating FcγRs play an important role in the development of lupus nephritis, as disease is significantly attenuated in Fcγ chain-deficient mice bred to NZB/NZW [139] or BXSB mice [140, 141] despite normal IC deposition and complement activation (Table 21.3). Further confidence in these results is provided by a careful analysis of congenic strains with Fcγ chain deficiency, which has ruled out the possibility that the protective effect seen in these animals is due to the restoration of the altered inhibitory FcγRIIB function [141]. It is noteworthy that Fcγ chain-deficient BXSB mice had significantly less renal damage and mortality compared to wild-type BXSB mice despite comparable levels of anti-DNA autoantibodies and renal IC deposits. On the other hand, BXSB mouse strains bred to C57Bl/B6 animals to restore FcγRIIB expression levels were also protected but in this case correlated with a suppression of anti-DNA antibody production and IC deposition [141]. This set of experiments definitively showed that the protection provided by Fcγ chain deficiency in lupus models is likely related to the impaired activation of effector immune cells independently of FcγRIIB function, production of auto-antibodies, and IC deposition. On the other hand, Fcγ chain deficiency in the MRL/lpr was not protective [142]. The difference may be explained by the fact that among several lupus-prone mice, including NZB/NZW and BXSB, MRL-Fas^lpr mice produce the largest amount of IgG3 cryoglobulins (IgG that form insoluble aggregates at low temperature), which activate complement but largely fail to engage FcγRs [143, 144]. This is corroborated by findings that the level of IgG3 autoantibodies correlates with lupus nephritis in the MRL-Fas^lpr strain [145]. A recent study also suggests that activating FcγRs are required for the generation of a hyper-reactive Gr-1⁻ monocyte subset in BXSB mice that are characterized by high expression of activating FcγRIV and low levels of FcγRIIB. This subset, which contributes to the generation of tissue macrophages and dendritic cells, may be involved in tissue damage in this lupus-prone mouse strain. Indeed, BXSB mice lacking the Fcγ chain are protected against renal damage, as reported earlier, and do not exhibit the usual monocytosis observed in the BXSB mouse strain [146].

There remains limited information on the specific contribution of each Fc-bearing cell type in the effector responses in lupus as well as other autoimmune models. With respect specifically to SLE, Fcγ chain on circulating bone marrow-derived cells has been shown to play a major role in tissue injury [140]. FcγR may also play an important role once induced on the surface of tissue-resident cells. Mesangial cells express both FcγRIA and FcγRIIIA after stimulation with IFN-γ. Engagement of FcγRIIIA on these cells triggers the production of IL-6, CCL2 (MCP-1), and CSF-1 [37, 147], which can promote leukocyte recruitment in glomerulonephritis and thereby renal tissue damage.

The paucity of data at present examining the role of specific FcγRs directly in lupus nephritis is likely related to the complex genetic background of lupus-prone mouse strains, and the fact that a detailed and comprehensive analysis of genes of interest in lupus is both time- and labor-intensive. Lupus nephritis is the most common and is often a major determinant in prognosis in SLE patients. The phenotype of knock-out mice subjected to anti-GBM nephritis (also referred to as nephrotoxic serum, induced by the administration of serum raised against the glomerular basement membrane), may predict the contribution of the genes of interest in lupus nephritis [148]. The anti-GBM model may thus represent a valuable tool to study the involvement of specific genes in lupus nephritis. By this argument, FcγRI and FcγRIII shown to participate in the pathogenesis of anti-GBM nephritis may also be involved in SLE [138, 149—151]. The anti-GBM model may also be useful to identify which cell type bearing the activating FcγRs may be specifically involved in the lupus nephritis pathogenesis. For instance, the specific re-expression of the γ-chain and therefore the activating FcγRs in macrophages of mice restores susceptibility to anti-GBM nephritis. However, the restoration is partial [140], suggesting that other Fc-bearing leukocyte subsets are more important. Indeed, the specific knock-in of human FcγRIIA or FcγRIIIB in neutrophils of mice lacking their endogenous FcγRs is sufficient to restore susceptibility to nephrotoxic nephritis, which might suggest overall a secondary role of FcγR expression in mesangial cells and macrophages in determining glomerulonephritis pathogenesis [49]. As neutrophils are consistently present in human renal biopsies from patients with membranoproliferative, crescentic and lupus glomerulonephritis [152—154], activating FcγRs on neutrophils, may be key players in kidney injury in human SLE. Further studies are required, however, to clarify the role of FcγRs in the effector phase and end-organ injury in animal models of SLE.

Fcγ RECEPTOR POLYMORPHISMS AND COPY NUMBER VARIATION IN LUPUS

Besides the well-known class II major histocompatibility complex polymorphisms [155, 156], genome-wide linkage analysis studies have linked the chromosomal region 1q21-1q23 to SLE susceptibility [52]. This chromosomal region contains, among others, the FcγRs genes; further candidate gene analysis largely confirmed the association between polymorphisms of FcγRIIA, FcγRIIB, FcγRIIIA, FcγRIIIB and SLE, or specifically with lupus nephritis susceptibility [51]. However, the numerous influencing factors and the overall small effect of these polymorphisms in SLE susceptibility, if taken in isolation, require large cohorts or meta-analysis to achieve a sufficiently high statistical power. The observed contrasting effects of FcγR polymorphisms in different ethnic backgrounds are also suggestive of a more complex multigenic background dictating SLE susceptibility. It is noteworthy that although some FcγR polymorphisms such as FcγRIIIA-F/F158, FcγRIIA-R/R131, and FcγRIIIB-NA2/NA2 were not always associated with an increased risk of SLE in different cohorts, they still appeared to influence the course of the disease and the occurrence of specific manifestations such as nephritis [157, 158], hematological manifestations [159, 160], serositis, and arthritis [160]. However, these data remain isolated and larger cohorts are required to definitively demonstrate these links. In this section, we summarize the most frequently reported FcγRs polymorphisms associated with SLE or lupus nephritis (Table 21.2). In conjunction with functional explanations, these data may aid in the understanding of the underlying mechanisms of autoimmunity in SLE.

FcγIIB T/I 232 Polymorphism

As the only inhibitory FcγR, FcγRIIB appears to play a critical role in immune tolerance in mouse models. Thus, any gain or loss of function of this receptor by a genetic variant may be expected to influence the susceptibility to SLE in humans. Indeed, a recently identified polymorphism of FcγRIIB, T/I 232 (also named T/I 187 if the signal peptide is omitted), has been associated with SLE [161–164], although this has not been confirmed in other reports [165, 166]. A recent meta-analysis confirms an association between the polymorphism FcγRIIB T/I 232 and an increased susceptibility for SLE in Asian-descent subjects (OR, 1.332 for the T allele) but not in Europeans [167]. Functionally, the isoleucine (I) to threonine (T) change at amino acid position 232 (transmembrane domain) may exclude the receptor from lipid rafts and thus prevent its association with BCR. The loss of its inhibitory function in B cells may lead to the subsequent emergence of autoreactive B cells [168, 169]. However, this polymorphism may be interacting epistatically with a more complex susceptible genetic background and therefore is biologically significant only in certain ethnicities such as Asians. Nonetheless, these data support a regulatory role for FcγRIIB in the maintenance of tolerance with downmodulation of its function resulting in autoimmunity in SLE.

FcγIIB Promoter Polymorphism

The -386C-120A promoter polymorphism of FcγRIIB has also been associated with an increase in susceptibility to SLE in a Caucasian population (OR, 1.6) [170]. Although a second study confirms this association in another European cohort [25], this link with SLE has to be cautiously interpreted because of the relatively small sample size and the absence of a convincing mechanistic

TABLE 21.2 FcγR Polymorphisms and Copy Number Variants Associated with SLE

Alleles	Effect	Suspected mechanisms
FcγRIIA R/H 131	R131: increased SLE susceptibility in Europeans and Africans	Altered affinity for IgG2 potentially resulting in impaired IgG2-based IC clearance
FcγRIIIA F/V 158	Increased susceptibility for lupus nephritis in Europeans, possible trend in Asians and Africans	Altered affinity for IgG1 and IgG3 potentially resulting in impaired IgG1- and IgG3-based IC clearance
FcγRIIIB NA1/NA2	NA2: suspected increase in SLE susceptibility and lupus nephritis in Asians and Europeans	Impaired neutrophil-mediated phagocytosis
FcγRIIIB copy number variant	Reduced FcγRIIIB gene copy number: increase in susceptibility for lupus nephritis in Europeans	Reduced surface expression of FcγRIIIB potentially resulting in impaired IC clearance
FcγRIIB T/I 232	T232: increase in SLE susceptibility in Asians	Exclusion of FcγRIIB from lipid rafts, resulting in altered signaling capacity and inhibitory functions
FcγRIIB -386C-120A	Increase in SLE susceptibility in Europeans	Unclear mechanism. Possibly altered expression of FcγRIIB

TABLE 21.3 FcγR Functions in Animal Models of SLE

Mouse strain	Phenotype	FcγR functions involved
FcγRIIB knock out [60]		
Sv129/C57BL/6 mixed background	Normal	
BALB/c background	Normal	
C57BL/6 background	Systemic chronic inflammatory syndrome with spontaneous progressive glomerulonephritis resulting in high level of proteinuria and increased mortality. High titer of antinuclear antibodies at 4–5 months.	Impaired inhibitory signal in B cells resulting in an activated phenotype of B cells with increased number of mature IgG producing B cells. Possibilities include decreased plasma cell apoptosis [110] and/or impaired inhibitory signal in antigen presenting cells resulting in an increase in the CD4$^+$ lymphocytes subset [26].
Wild-type C57BL/6 FcγRIIB restoration in BXSB background [141]	Absence of lupus syndrome. Absence of autoantibody production and glomerulonephritis. Normal lifespan.	Restoration of functional level of FcγRIIB expression in genetically impaired BXSB background (promoter polymorphism). Therefore restoration of FcγRIIB inhibitory functions in immune cells and prevention of tolerance breakdown in BXSB mice.
Fcγ chain knock out		
NZB/NZW background [139, 140]	Delayed onset and reduced proteinuria in Fcγ$^{-/-}$ mice despite IC as well as C3 glomerular deposition. Prolonged survival compared to Fcγ$^{+/-}$ littermate.	Disruption of the FcγR-mediated response of immune effector cells.
BXSB [141]	Absence of glomerulonephritis and prolonged survival despite comparable autoantibody titers and similar IC and C3 glomerular deposition compared to BXSB Fcγ$^{+/+}$.	Disruption of the FcγR-mediated response of immune effector cells. Impaired development of hyperreactive Gr-1$^-$ monocytes subset [146].
MRL-Fas^{lpr} [142]	Compared to Fcγ$^{+/+}$ background, similar incidence of glomerulonephritis and level of proteinuria. Comparable survival. Similar levels of anti-DNA antibodies and circulating ICs.	Inflammatory syndrome and tissue damage observed in MRL-Fas^{lpr} mice appear to be FcγR independent possibly mediated by IgG3 autoantibody cryoglobulins, which do not bind FcγRs. In this mouse strain, tissue damage may be primarily mediated by complement activation [143].

explanation. Indeed, the two studies present conflicting results on the activity of this promoter haplotype. While the first study observed increased -386C-120A promoter activity *ex vivo* as well as in transfected cell lines [170, 171], an opposite effect was observed in the second study [25]. Therefore, further studies are required to clarify the significance of this polymorphism.

FcγIIA R/H 131

FcγRIIA R/H 131 (also named FCGR2A G458A), a polymorphism resulting in a single amino-acid substitution at position 131 (extracellular domain) has been associated with SLE in many reports. In a meta-analysis of 17 reports, the FcγRIIA R/H 131 polymorphism is linked with SLE development. The RR131 as well as

the RH131 genotypes were significantly associated with an increase in susceptibility to SLE in European- and African-descent populations, whereas only a (nonsignificant) trend was observed in Asian cohorts [172]. Although the risk appears to be modest (OR, 1.3 for RR/RH versus HH), it still may have a significant effect at a population level as this polymorphism is co-dominantly expressed and the *R131* is a common allele. Mechanistically, the enhanced susceptibility may be best explained by the altered affinity for IgG2 of the low-binding *R131* allele compared to the high-binding H131 [127], which may result in impaired handling of IgG2-based ICs by FcγRIIA R131. Indeed, isolated neutrophils from FcγRIIA R131 patients exhibit dramatically impaired phagocytosis of erythrocytes coated with human IgG2, while no differences were observed with erythrocytes coated with IgG1, IgG3, and IgG4 [173]. The *R131* allele might therefore be associated with decreased IC clearance capacity and consequently with enhanced end-organ tissue damage as a result of IC deposition. However, no clear association between the FcγRIIA *R131* allele and lupus nephritis was observed. Since IgG1 and IgG3 are predominantly involved in proliferative glomerulonephritis [63, 174, 175], it was anticipated that the R131 polymorphism with altered binding to IgG2 may have mild or no effect on lupus nephritis susceptibility. Interestingly, one study links FcγRIIA R131 with SLE severity and in particular with renal involvement in IgG2 anti-C1q autoantibody-positive patients. In this case, the R131 genotype may indeed lead to a specific defect in IgG2-based ICs clearance [67].

FcγRIIIA F/V158 Polymorphisms

A meta-analysis of 13 reports has evaluated the link between the FcγRIIIA F/V158 polymorphism (also named FCGR3A T559G) and SLE or lupus nephritis. These co-dominant alleles differ at amino acid position 158 coding for either a phenylalanine (F) or a valine (V) in the extracellular domain of the receptor. The F158 genotype appears to be significantly associated with lupus nephritis in people of European descent (OR, 1.21) and a strong trend (but not significant) is observed in Asian and African populations [176]. However, no significant association was observed between overall SLE susceptibility and the *F158* allele, despite a clear trend in European and African cohorts. Nevertheless, the FcγRIIIA *F158* allele enhances the risk of renal involvement and may thereby negatively influence the prognosis of SLE patients. Functionally, the FcγRIIIA *F158* allele has a reduced affinity for IgG1 and IgG3 [177, 178]. Similarly to FcγRIIA R131 polymorphism, this may influence the ability of phagocytic cells to clear ICs [51]. Thus this polymorphism may be particularly relevant in renal involvement as IgG1 and IgG3 ICs are predominantly implicated in the pathogenesis of proliferative glomerulonephritis [63, 174, 175].

FcγRIIIB NA1/NA2

Two main co-dominant allelic forms of FcγRIIIB receptor exist, *NA1* and *NA2*. *NA2* has been functionally associated with impaired phagocytosis function in neutrophils [179]. However, despite an initial report suggesting an increased risk of SLE associated with the *NA2* allele [180], a recent meta-analysis has shown no association between the *NA2* allele and SLE or lupus nephritis in either Asian or European populations [167]. Interestingly, the apparent initial positive association between the *NA2* allele and SLE susceptibility may

have resulted from a strong linkage disequilibrium with FcγRIIB, which is located in close proximity to FcγRIIIB [162]. Although the *NA2* genotype is not definitively associated with SLE incidence, *NA2* does appear to be associated with increased risk of lupus nephritis and hematological manifestations [157, 159].

Copy Number Variant of FcγRIIIB

In addition to single-nucleotide polymorphisms, copy number variants (CNVs) are another type of genetic variation that can effect FcγR function. CNVs are structural variations defined as DNA segments at least 1 kb in size present at variable number of copies in comparison with a reference genome. These genetic variations are being increasingly recognized as a source of gene expression modulation and thereby phenotypic variation [181]. FcγRIIIB is the first gene CNV implicated in autoimmunity. Comparing 60 lupus nephritis patients to 109 unrelated seronegative controls, reduced numbers of copies of FcγRIIIB was significantly associated with lupus nephritis. Moreover, this effect was independent of the FcγRIIA R/H131 and FcγRIIIA F/V158 polymorphism in a logistic regression analysis model [182]. Another study recently confirmed this association in a European cohort, but found no significant link in Chinese patients [136]. A correlation between FcγRIIIB CNV and surface expression on human neutrophils has been recently reported [136]. Moreover, low CNV was also associated with impaired neutrophil adhesion to immunoglobulin-coated surface under flow conditions and reduced IC uptake [136]. As FcγRIIIB tethers neutrophils to immobilized ICs *in vitro* [138] as well as intravascular ICs *in vivo* [49] and may be able to phagocytose ICs [136, 137], a reduced number of copies of the FcγRIIIB gene associated with a lower expression of this receptor may impair glomerular clearance of ICs and thereby increase susceptibility to renal injury in SLE and possibly to other autoimmune disorders.

Other Polymorphisms that Potentially Influence FcγR Functions

Recently, a polymorphism of the CD18 receptor family and complement receptor Mac1 (ITGAM, CD11b/CD18, CR3) has been strongly associated with SLE (OR, 1.78) in European, African, and American populations [124–126]. This nonsynonymous polymorphism is predicted to alter the ligand binding capacity of this integrin. As Mac1 is known to interact with FcγRs at the cell surface, in particular with FcγRIIIB and FcγRIIA [91, 183], and thereby modulate their function, this polymorphism may indirectly impact FcγR-mediated IC clearance. However, further functional studies are required to understand the potential role of this Mac1 genetic variant in modulating FcγR functions.

FCγR TARGETED TREATMENTS

To date, IVIG is the only commonly used immunosuppressive therapy that appears to target FcγR functions. Pooled normal polyspecific IgG obtained from healthy donors (IVIG) have been efficiently used for many years in the treatment of several autoimmune and systemic inflammatory diseases such as central nervous inflammatory diseases (multiple sclerosis, Guillain-Barré syndrome, myasthenia gravis), vasculitis, and idiopathic thrombocytopenic purpura. Off-label prescription of IVIG also includes SLE [184, 185] and antiphospholipid syndrome [186], and shows clear benefits. However, controlled randomized clinical trials are still required to thoroughly assess the risks and benefits of such treatment in lupus.

Since IVIG is an important therapeutic option in many autoimmune diseases, clarification of its anti-inflammatory function might help to elucidate key events in autoimmunity, including SLE. Furthermore, given that IVIG is available in limited supply, is a human derived product and required in multiple doses to achieve efficacy, alternative approaches are needed. F(ab)2-dependent mechanisms have been speculated and specific so-called natural antibodies against molecules implicated in inflammation have been identified in IVIG preparation (for instance, anti-integrins, anti-Fas, anticytokines, or anticytokine receptors) [187–189]. However, strong functional *in vivo* data are missing and the aforementioned effects might be anecdotal. On the other hand, the antibody Fc fragments of the pooled IVIG appear to play a dominant anti-inflammatory role. This is supported by one clinical trial with idiopathic thrombocytopenic purpura that demonstrated a similar anti-inflammatory effect of both complete IVIG and isolated Fc fragments [190]. In animal models, Fc fragments have also been reported to play a significant protective role in immune thrombocytopenia, arthritis, and proliferative nephritis models [191–193].

Numerous appealing explanations have been provided for these Fc-dependent anti-inflammatory activities such as saturation of neonatal Fc receptors (critically involved in maintaining immunoglobulin level in the serum), binding and activation of C3b and C4b leading to less complement deposition and associated tissue damage, and blockade of the activating FcγR and subsequent inhibition of pro-inflammatory cytokine release [193, 194]. The most convincing *in vivo* data support the idea that IVIG works principally by upregulating the expression of the inhibitory FcγRIIB and thus counteracting the FcγR activation by IC. Indeed the anti-inflammatory activity of IVIG is lost in FcγRIIB-deficient mice subjected to immune thrombocytopenia, arthritis, or proliferative nephritis models [191, 193, 195]. Another promising explanation is that IVIG is enriched in IgG glycosylated on their asparagine 297 residue, which might preferentially interact with a subtype of regulatory macrophages[194, 195]. Indeed, various IgG glycovariants have been identified in humans and deglycosylated IVIG preparations have been shown to lose their anti-inflammatory effect. Moreover, enrichment of Fc fragments for terminal sialic acid residues enhanced the IVIG activity, suggesting a specific anti-inflammatory activity of this subset of salicylated antibodies [192]. Therefore, overall, IVIG may activate a specific subset of anti-inflammatory regulatory macrophages, while at the same time inhibit pro-inflammatory effector macrophages by engaging FcγRIIB [194]. Seemingly non-FcγR-related effects of IVIG have also been described as reviewed in detail elsewhere [194]. Interestingly, in recent study, IVIG was shown to prevent leukocyte rolling on the endothelium through effects on members of the selectin and integrin family of adhesion molecules in a model of ischemia and reperfusion [196]. Thus, IVIG, which is a complex mixture, likely engages multiple targets to reduce inflammation.

Recently, a specific inhibitor of the syk kinase (R788/R406, fostamatinib) has been developed. Although the syk inhibitor blocks ITAM-activating signals downstream of FcγRs, its targets are broader because syk is downstream of many ITAM-containing receptors as well as non-ITAM receptors such as integrins in cells of both hematopoietic and nonhematopoietic lineage [197]. It has been shown to efficiently reduce arthritis severity in the collagen-induced arthritis mouse model [198] and has demonstrated a clear therapeutic benefit in an extended randomized double-blind placebo-controlled clinical trial in 189 rheumatoid patients [199]. This new medication has also shown encouraging results in a mouse model of immune thrombocytopenia as well as in a phase 2 clinical trial in 16 patients with chronic refractory immune thrombocytopenic purpura [200]. Fostamatinib was also efficient in reducing the severity of nephritis in the lupus-prone NZB/NZW F1 mice either preemptively or as a treatment of established nephritis [201]. To date, no data are available on fostamatinib in human SLE.

An isolated report underscores the potential of FcγR-targeted treatment in autoimmune diseases and SLE. A soluble humanized FcγRII substantially reduced the severity of the lupus-like disease once administered prophylactically in lupus prone NZB/NZW F1 mice [202]. Similarly, blocking activating FcγRI or FcγRIII in patients with refractory immune thrombocytopenic purpura has also produced encouraging results [203]. In summary, FcγRs represent potential therapeutic targets for treatment of SLE: Approaches may include blocking the activating receptors, modulating the expression of the inhibitory FcγRIIB, or altering the activating signaling pathway of the ITAM-bearing FcγRs.

FUTURE DIRECTIONS

Both inhibitory and activating FcγRs have emerged as key regulators of the autoimmune response and end-organ damage in a wide range of autoimmune diseases, based largely on the study of genetically engineered mice in models of autoimmunity. Despite these advances, given the complexity of the FcγR family, the differences between the repertoire of murine and human receptors, and the involvement of FcγRs in several stages of autoimmune disorders from the immune response to end-stage autoimmunity-mediated inflammatory responses, it is not surprising that much more work needs to be done. There is a paucity of studies directly addressing the relative importance of FcγRs in mouse models of SLE and the epistatic environmental and genetic factors that dictate the FcγRIIB dependence of B cell autoreactivity. The potential role of FcγRs in human disease has come from associations between polymorphisms in several different FcγRs and SLE disease susceptibility or disease course. However, definitive evidence that these polymorphisms have functional consequences for development of SLE is needed. Advances in this area require a greater understanding of the mechanisms of FcγR function, as well as the relevant cell types bearing activating FcγRs that promote end-organ damage in SLE. Other areas that deserve attention are determining whether the pro-inflammatory environment existing in SLE patients modulates FcγR expression and function. As SLE is more prevalent in women, another intriguing question is how gender can influence FcγRs function or expression. For example, gender may modulate the penetrance of certain FcγRs polymorphisms, as illustrated for instance from the significant enrichment of FcγRIIB T232 (which reduces FcγRIIB's signaling capacity) in male lupus patients compared to females [161]. These areas of study are just a small sampling of the opportunities in FcγR research that could provide insights into the mechanisms by which FcγRs promote susceptibility to autoimmunity and its manifestations. This in turn could result in rational FcγRs-targeting drug design that potentially blocks the initiation of the humoral response to autoantigens, and/or attenuates IC-induced end-organ damage.

References

[1] J.E. Gessner, H. Heiken, A. Tamm, R.E. Schmidt, The IgG Fc receptor family, Ann Hematol 76 (1998) 231–248.

[2] F. Nimmerjahn, J.V. Ravetch, Fcgamma receptors: old friends and new family members, Immunity 24 (2006) 19–28.

[3] J.A. Hamerman, L.L. Lanier, Inhibition of immune responses by ITAM-bearing receptors, Sci STKE 2006, re1 (2006) 19–28.

[4] F. Nimmerjahn, J.V. Ravetch, Fcgamma receptors as regulators of immune responses, Nat Rev Immunol 8 (2008) 34–47.

[5] F. Nimmerjahn, P. Bruhns, K. Horiuchi, J.V. Ravetch, FcgammaRIV: a novel FcR with distinct IgG subclass specificity, Immunity 23 (2005) 41–51.

[6] M.E. McKenzie, P.A. Gurbel, D.J. Levine, V.L. Serebruany, Clinical utility of available methods for determining platelet function, Cardiology 92 (1999) 240–247.

[7] P. Bruhns, B. Iannascoli, P. England, D.A. Mancardi, N. Fernandez, S. Jorieux, et al., Specificity and affinity of human Fcgamma receptors and their polymorphic variants for human IgG subclasses, Blood 113 (2009) 3716–3725.

[8] J.M. Allen, B. Seed, Isolation and expression of functional high-affinity Fc receptor complementary DNAs, Science 243 (1989) 378–381.

[9] P. Uciechowski, M. Schwarz, J.E. Gessner, R.E. Schmidt, K. Resch, H.H. Radeke, IFN-gamma induces the high-affinity Fc receptor I for IgG (CD64) on human glomerular mesangial cells, Eur J Immunol 28 (1998) 2928–2935.

[10] F. Hoffmeyer, K. Witte, R.E. Schmidt, The high-affinity Fc gamma RI on PMN: regulation of expression and signal transduction, Immunology 92 (1997) 544–552.

[11] H. Fukuyama, F. Nimmerjahn, J.V. Ravetch, The inhibitory Fcgamma receptor modulates autoimmunity by limiting the accumulation of immunoglobulin G+ anti-DNA plasma cells, Nat Immunol 6 (2005) 99–106.

[12] S. Bolland, J.V. Ravetch, Inhibitory pathways triggered by ITIM-containing receptors, Adv Immunol 72 (1999) 149–177.

[13] B. Boylan, C. Gao, V. Rathore, J.C. Gill, D.K. Newman, P.J. Newman, Identification of FcgammaRIIa as the ITAM-bearing receptor mediating alphaIIbbeta3 outside-in integrin signaling in human platelets, Blood 112 (2008) 2780–2786.

[14] D.J. Schneider, H.S. Taatjes-Sommer, Augmentation of megakaryocyte expression of FcgammaRIIa by interferon gamma, Arterioscler Thromb Vasc Biol 29 (2009) 1138–1143.

[15] D. Benitez-Ribas, G.J. Adema, G. Winkels, I.S. Klasen, C.J. Punt, C.G. Figdor, et al., Plasmacytoid dendritic cells of melanoma patients present exogenous proteins to CD4+ T cells after Fc gamma RII-mediated uptake, J Exp Med 203 (2006) 1629–1635.

[16] Y. Liu, X. Gao, E. Masuda, P.B. Redecha, M.C. Blank, L. Pricop, Regulated expression of FcgammaR in human dendritic cells controls cross-presentation of antigen-antibody complexes, J Immunol 177 (2006) 8440–8447.

[17] A.M. Boruchov, G. Heller, M.C. Veri, E. Bonvini, J.V. Ravetch, J.W. Young, Activating and inhibitory IgG Fc receptors on human DCs mediate opposing functions, J Clin Invest 115 (2005) 2914–2923.

[18] R.G. Lynch, Rous-Whipple Award lecture. The biology and pathology of lymphocyte Fc receptors, Am J Pathol 152 (1998) 631–639.

[19] J.V. Ravetch, S. Bolland, IgG Fc receptors, Annu Rev Immunol 19 (2001) 275–290.

[20] F.W. Luscinskas, T. Mayadas, Fc{gamma}Rs join in the cascade, Blood 109 (2007) 3615–3616.

[21] R.G. Worth, C.D. Chien, P. Chien, M.P. Reilly, S.E. McKenzie, A.D. Schreiber, Platelet FcgammaRIIA binds and internalizes IgG-containing complexes, Exp Hematol 34 (2006) 1490–1495.

[22] K. Su, H. Yang, X. Li, A.W. Gibson, J.M. Cafardi, T. Zhou, et al., Expression profile of FcgammaRIIb on leukocytes and its dysregulation in systemic lupus erythematosus, J Immunol 178 (2007) 3272–3280.

[23] M.D. Hulett, P.M. Hogarth, Molecular basis of Fc receptor function, Adv Immunol 57 (1994) 1–127.

[24] R.E. Schmidt, J.E. Gessner, Fc receptors and their interaction with complement in autoimmunity, Immunol Lett 100 (2005) 56–67.

[25] M.C. Blank, R.N. Stefanescu, E. Masuda, F. Marti, P.D. King, P.B. Redecha, et al., Decreased transcription of the human FCGR2B gene mediated by the -343 G/C promoter polymorphism and association with systemic lupus erythematosus, Hum Genet 117 (2005) 220–227.

[26] D.D. Desai, S.O. Harbers, M. Flores, L. Colonna, M.P. Downie, A. Bergtold, et al., Fc gamma receptor IIB on dendritic cells enforces peripheral tolerance by inhibiting effector T cell responses, J Immunol 178 (2007) 6217–6226.

[27] T. Takai, M. Li, D. Sylvestre, R. Clynes, J.V. Ravetch, FcR gamma chain deletion results in pleiotrophic effector cell defects, Cell 76 (1994) 519–529.

[28] J.V. Ravetch, B. Perussia, Alternative membrane forms of Fc gamma RIII(CD16) on human natural killer cells and neutrophils. Cell type-specific expression of two genes that differ in single nucleotide substitutions, J Exp Med 170 (1989) 481–497.

[29] D. Simmons, B. Seed, The Fc gamma receptor of natural killer cells is a phospholipid-linked membrane protein, Nature 333 (1988) 568–570.

[30] V. Groh, S. Porcelli, M. Fabbi, L.L. Lanier, L.J. Picker, T. Anderson, et al., Human lymphocytes bearing T cell receptor gamma/delta are phenotypically diverse and evenly distributed throughout the lymphoid system, J Exp Med 169 (1989) 1277–1294.

[31] D.F. Angelini, G. Borsellino, M. Poupot, A. Diamantini, R. Poupot, G. Bernardi, et al., FcgammaRIII discriminates between 2 subsets of Vgamma9Vdelta2 effector cells with different responses and activation pathways, Blood 104 (2004) 1801–1807.

[32] V. Lafont, J. Liautard, J.P. Liautard, J. Favero, Production of TNF-alpha by human V gamma 9V delta 2 T cells via engagement of Fc gamma RIIIA, the low affinity type 3 receptor for the Fc portion of IgG, expressed upon TCR activation by nonpeptidic antigen, J Immunol 166 (2001) 7190–7199.

[33] N.K. Bjorkstrom, V.D. Gonzalez, K.J. Malmberg, K. Falconer, A. Alaeus, G. Nowak, et al., Elevated numbers of Fc gamma RIIIA+ (CD16+) effector CD8 T cells with NK cell-like function in chronic hepatitis C virus infection, J Immunol 181 (2008) 4219–4228.

[34] K. Oshimi, Y. Oshimi, O. Yamada, M. Wada, T. Hara, H. Mizoguchi, Cytotoxic T lymphocyte triggering via CD16 is regulated by CD3 and CD8 antigens. Studies with T cell receptor (TCR)-alpha beta+/CD3+16+ and TCR-gamma delta+/CD3+16+ granular lymphocytes, J Immunol 144 (1990) 3312–3317.

[35] P.J. Leibson, Signal transduction during natural killer cell activation: inside the mind of a killer, Immunity 6 (1997) 655–661.

[36] Z. Chen, M.S. Freedman, CD16+ gammadelta T cells mediate antibody dependent cellular cytotoxicity: potential mechanism in the pathogenesis of multiple sclerosis, Clin Immunol 128 (2008) 219–227.

[37] H.H. Radeke, J.E. Gessner, P. Uciechowski, H.J. Magert, R.E. Schmidt, K. Resch, Intrinsic human glomerular mesangial cells can express receptors for IgG complexes (hFc gamma RIII-A) and the associated Fc epsilon RI gamma-chain, J Immunol 153 (1994) 1281–1292.

[38] B. Perussia, J.V. Ravetch, Fc gamma RIII (CD16) on human macrophages is a functional product of the Fc gamma RIII-2 gene, Eur J Immunol 21 (1991) 425–429.

[39] M. Li, U. Wirthmueller, J.V. Ravetch, Reconstitution of human Fc gamma RIII cell type specificity in transgenic mice, J Exp Med 183 (1996) 1259–1263.

[40] N. Meknache, F. Jonsson, J. Laurent, M.T. Guinnepain, M. Daeron, Human basophils express the glycosylphosphatidylinositol-anchored low-affinity IgG receptor FcgammaRIIIB (CD16B), J Immunol 182 (2009) 2542–2550.

[41] P. Selvaraj, W.F. Rosse, R. Silber, T.A. Springer, The major Fc receptor in blood has a phosphatidylinositol anchor and is deficient in paroxysmal nocturnal haemoglobinuria, Nature 333 (1988) 565–567.

[42] J.C. Unkeless, Z. Shen, C.W. Lin, E. DeBeus, Function of human Fc gamma RIIA and Fc gamma RIIIB, Semin Immunol 7 (1995) 37–44.

[43] M. Hundt, R.E. Schmidt, The glycosylphosphatidylinositol-linked Fc gamma receptor III represents the dominant receptor structure for immune complex activation of neutrophils, Eur J Immunol 22 (1992) 811–816.

[44] P. Selvaraj, N. Fifadara, S. Nagarajan, A. Cimino, G. Wang, Functional regulation of human neutrophil Fc gamma receptors, Immunol Res 29 (2004) 219–230.

[45] T.W. Huizinga, K.M. Dolman, N.J. van der Linden, M. Kleijer, J.H. Nuijens, A.E. von dem Borne, et al., Phosphatidylinositol-linked FcRIII mediates exocytosis of neutrophil granule proteins, but does not mediate initiation of the respiratory burst, J Immunol 144 (1990) 1432–1437.

[46] P. Boros, J.A. Odin, T. Muryoi, S.K. Masur, C. Bona, J.C. Unkeless, IgM anti-Fc gamma R autoantibodies trigger neutrophil degranulation, J Exp Med 173 (1991) 1473–1482.

[47] A. Ortiz-Stern, C. Rosales, Fc gammaRIIIB stimulation promotes beta1 integrin activation in human neutrophils, J Leukoc Biol 77 (2005) 787–799.

[48] E. Garcia-Garcia, G. Nieto-Castaneda, M. Ruiz-Saldana, N. Mora, C. Rosales, FcgammaRIIA and FcgammaRIIIB mediate nuclear factor activation through separate signaling pathways in human neutrophils, J Immunol 182 (2009) 4547–4556.

[49] N. Tsuboi, K. Asano, M. Lauterbach, T.N. Mayadas, Human neutrophil Fcgamma receptors initiate and play specialized nonredundant roles in antibody-mediated inflammatory diseases, Immunity 28 (2008) 833–846.

[50] K.L. Miller, A.M. Duchemin, C.L. Anderson, A novel role for the Fc receptor gamma subunit: enhancement of Fc gamma R ligand affinity, J Exp Med 183 (1996) 2227–2233.

[51] X. Li, T.S. Ptacek, E.E. Brown, J.C. Edberg, Fcgamma receptors: structure, function and role as genetic risk factors in SLE, Genes Immun 10 (2009) 380–389.

[52] J. Castro, E. Balada, J. Ordi-Ros, M. Vilardell-Tarres, The complex immunogenetic basis of systemic lupus erythematosus, Autoimmun Rev 7 (2008) 345–351.

[53] S.K. Nath, J. Kilpatrick, J.B. Harley, Genetics of human systemic lupus erythematosus: the emerging picture, Curr Opin Immunol 16 (2004) 794–800.

[54] B.P. Tsao, The genetics of human systemic lupus erythematosus, Trends Immunol 24 (2003) 595–602.

[55] L. Fossati-Jimack, A. Ioan-Facsinay, L. Reininger, Y. Chicheportiche, N. Watanabe, T. Saito, et al., Markedly different pathogenicity of four immunoglobulin G isotype-switch variants of an antierythrocyte autoantibody is based on their capacity to interact in vivo with the low-affinity Fcgamma receptor III. J Exp Med 191 (2000) 1293–1302.

[56] T.J. Kipps, P. Parham, J. Punt, L.A. Herzenberg, Importance of immunoglobulin isotype in human antibody-dependent,

cell-mediated cytotoxicity directed by murine monoclonal antibodies, J Exp Med 161 (1985) 1—17.

[57] F. Nimmerjahn, J.V. Ravetch, Divergent immunoglobulin g subclass activity through selective Fc receptor binding, Science 310 (2005) 1510—1512.

[58] P.H. Lambert, F.J. Dixon, Pathogenesis of the glomerulonephritis of NZB/W mice, J Exp Med 127 (1968) 507—522.

[59] J.H. Slack, L. Hang, J. Barkley, R.J. Fulton, L. D'Hoostelaere, A. Robinson, et al., Isotypes of spontaneous and mitogen-induced autoantibodies in SLE-prone mice, J Immunol 132 (1984) 1271—1275.

[60] S. Bolland, J.V. Ravetch, Spontaneous autoimmune disease in Fc(gamma)RIIB-deficient mice results from strain-specific epistasis, Immunity 13 (2000) 277—285.

[61] M. Ehlers, H. Fukuyama, T.L. McGaha, A. Aderem, J.V. Ravetch, TLR9/MyD88 signaling is required for class switching to pathogenic IgG2a and 2b autoantibodies in SLE, J Exp Med 203 (2006) 553—561.

[62] P.J. Maddison, Autoantibodies in SLE. Disease associations, Adv Exp Med Biol 455 (1999) 141—145.

[63] Z. Amoura, S. Koutouzov, H. Chabre, P. Cacoub, I. Amoura, L. Musset, et al., Presence of antinucleosome autoantibodies in a restricted set of connective tissue diseases: antinucleosome antibodies of the IgG3 subclass are markers of renal pathogenicity in systemic lupus erythematosus, Arthritis Rheum 43 (2000) 76—84.

[64] M. Haas, IgG subclass deposits in glomeruli of lupus and nonlupus membranous nephropathies, Am J Kidney Dis 23 (1994) 358—364.

[65] H. Imai, K. Hamai, A. Komatsuda, H. Ohtani, A.B. Miura, IgG subclasses in patients with membranoproliferative glomerulonephritis, membranous nephropathy, and lupus nephritis, Kidney Int 51 (1997) 270—276.

[66] R. Zuniga, G.S. Markowitz, T. Arkachaisri, E.A. Imperatore, V.D. D'Agati, J.E. Salmon, Identification of IgG subclasses and C-reactive protein in lupus nephritis: the relationship between the composition of immune deposits and FCgamma receptor type IIA alleles, Arthritis Rheum 48 (2003) 460—470.

[67] L.A. Haseley, J.J. Wisnieski, M.R. Denburg, A.R. Michael-Grossman, E.M. Ginzler, M.F. Gourley, et al., Antibodies to C1q in systemic lupus erythematosus: characteristics and relation to Fc gamma RIIA alleles, Kidney Int 52 (1997) 1375—1380.

[68] N. Shushakova, J. Skokowa, J. Schulman, U. Baumann, J. Zwirner, R.E. Schmidt, et al., C5a anaphylatoxin is a major regulator of activating versus inhibitory FcgammaRs in immune complex-induced lung disease, J Clin Invest 110 (2002) 1823—1830.

[69] Y. Okayama, A.S. Kirshenbaum, D.D. Metcalfe, Expression of a functional high-affinity IgG receptor, Fc gamma RI, on human mast cells: Up-regulation by IFN-gamma, J Immunol 164 (2000) 4332—4339.

[70] L. Pricop, P. Redecha, J.L. Teillaud, J. Frey, W.H. Fridman, C. Sautes-Fridman, et al., Differential modulation of stimulatory and inhibitory Fc gamma receptors on human monocytes by Th1 and Th2 cytokines, J Immunol 166 (2001) 531—537.

[71] H.H. Radeke, I. Janssen-Graalfs, E.N. Sowa, N. Chouchakova, J. Skokowa, F. Loscher, et al., Opposite regulation of type II and III receptors for immunoglobulin G in mouse glomerular mesangial cells and in the induction of anti-glomerular basement membrane (GBM) nephritis, J Biol Chem 277 (2002) 27535—27544.

[72] S. Tridandapani, R. Wardrop, C.P. Baran, Y. Wang, J.M. Opalek, M.A. Caligiuri, et al., TGF-beta 1 suppresses [correction of supresses] myeloid Fc gamma receptor function by regulating the expression and function of the common gamma-subunit, J Immunol 170 (2003) 4572—4577.

[73] T.W. Huizinga, C.E. van der Schoot, C. Jost, R. Klaassen, M. Kleijer, A.E. von dem Borne, et al., The PI-linked receptor FcRIII is released on stimulation of neutrophils, Nature 333 (1988) 667—669.

[74] P.J. Middelhoven, J.D. Van Buul, P.L. Hordijk, D. Roos, Different proteolytic mechanisms involved in Fc gamma RIIIb shedding from human neutrophils, Clin Exp Immunol 125 (2001) 169—175.

[75] L. Koenderman, S.W. Hermans, P.J. Capel, J.G. van de Winkel, Granulocyte-macrophage colony-stimulating factor induces sequential activation and deactivation of binding via a low-affinity IgG Fc receptor, hFc gamma RII, on human eosinophils, Blood 81 (1993) 2413—2419.

[76] F.D. Finkelman, J. Holmes, I.M. Katona, J.F. Urban Jr., M.P. Beckmann, L.S. Park, et al., Lymphokine control of in vivo immunoglobulin isotype selection, Annu Rev Immunol 8 (1990) 303—333.

[77] D.R. Karp, Complement and systemic lupus erythematosus, Curr Opin Rheumatol 17 (2005) 538—542.

[78] R.B. Sim, K.B. Reid, C1: molecular interactions with activating systems, Immunol Today 12 (1991) 307—311.

[79] M.J. Lewis, M. Botto, Complement deficiencies in humans and animals: links to autoimmunity, Autoimmunity 39 (2006) 367—378.

[80] C.A. Ogden, A. deCathelineau, P.R. Hoffmann, D. Bratton, B. Ghebrehiwet, V.A. Fadok, et al., C1q and mannose binding lectin engagement of cell surface calreticulin and CD91 initiates macropinocytosis and uptake of apoptotic cells, J Exp Med 194 (2001) 781—795.

[81] T. Stokol, P. O'Donnell, L. Xiao, S. Knight, G. Stavrakis, M. Botto, et al., C1q governs deposition of circulating immune complexes and leukocyte Fcgamma receptors mediate subsequent neutrophil recruitment, J Exp Med 200 (2004) 835—846.

[82] P. Gasque, Complement: a unique innate immune sensor for danger signals, Mol Immunol 41 (2004) 1089—1098.

[83] V. Kumar, S.R. Ali, S. Konrad, J. Zwirner, J.S. Verbeek, R.E. Schmidt, et al., Cell-derived anaphylatoxins as key mediators of antibody-dependent type II autoimmunity in mice, J Clin Invest 116 (2006) 512—520.

[84] J. Godau, T. Heller, H. Hawlisch, M. Trappe, E. Howells, J. Best, et al., C5a initiates the inflammatory cascade in immune complex peritonitis, J Immunol 173 (2004) 3437—3445.

[85] J. Skokowa, S.R. Ali, O. Felda, V. Kumar, S. Konrad, N. Shushakova, et al., Macrophages induce the inflammatory response in the pulmonary Arthus reaction through G alpha i2 activation that controls C5aR and Fc receptor cooperation, J Immunol 174 (2005) 3041—3050.

[86] U. Baumann, J. Kohl, T. Tschernig, K. Schwerter-Strumpf, J.S. Verbeek, R.E. Schmidt, et al., A codominant role of Fc gamma RI/III and C5aR in the reverse Arthus reaction, J Immunol 164 (2000) 1065—1070.

[87] T.R. Welch, M. Frenzke, D. Witte, A.E. Davis, C5a is important in the tubulointerstitial component of experimental immune complex glomerulonephritis, Clin Exp Immunol 130 (2002) 43—48.

[88] L. Bao, I. Osawe, T. Puri, J.D. Lambris, M. Haas, R.J. Quigg, C5a promotes development of experimental lupus nephritis which can be blocked with a specific receptor antagonist, Eur J Immunol 35 (2005) 2496—2506.

[89] U. Baumann, N. Chouchakova, B. Gewecke, J. Kohl, M.C. Carroll, R.E. Schmidt, et al., Distinct tissue site-specific requirements of mast cells and complement components

C3/C5a receptor in IgG immune complex-induced injury of skin and lung, J Immunol 167 (2001) 1022–1027.

[90] M. Huber-Lang, J.V. Sarma, F.S. Zetoune, D. Rittirsch, T.A. Neff, S.R. McGuire, et al., Generation of C5a in the absence of C3: a new complement activation pathway, Nat Med 12 (2006) 682–687.

[91] H.R. Petty, R.G. Worth, R.F. Todd 3rd, Interactions of integrins with their partner proteins in leukocyte membranes, Immunol Res 25 (2002) 75–95.

[92] I. Stefanova, V. Horejsi, I.J. Ansotegui, W. Knapp, H. Stockinger, GPI-anchored cell-surface molecules complexed to protein tyrosine kinases, Science 254 (1991) 1016–1019.

[93] F. Barabe, C. Gilbert, N. Liao, S.G. Bourgoin, P.H. Naccache, Crystal-induced neutrophil activation VI. Involvment of FcgammaRIIIB (CD16) and CD11b in response to inflammatory microcrystals, FASEB J 12 (1998) 209–220.

[94] G. Berton, A. Mocsai, C.A. Lowell, Src and Syk kinases: key regulators of phagocytic cell activation, Trends Immunol 26 (2005) 208–214.

[95] M. Daeron, Fc receptor biology, Annu Rev Immunol 15 (1997) 203–234.

[96] A. Utomo, X. Cullere, M. Glogauer, W. Swat, T.N. Mayadas, Vav proteins in neutrophils are required for FcgammaR-mediated signaling to Rac GTPases and nicotinamide adenine dinucleotide phosphate oxidase component p40(phox), J Immunol 177 (2006) 6388–6397.

[97] C.M. Wells, P.J. Bhavsar, I.R. Evans, E. Vigorito, M. Turner, V. Tybulewicz, et al., Vav1 and Vav2 play different roles in macrophage migration and cytoskeletal organization, Exp Cell Res 310 (2005) 303–310.

[98] M. Daeron, S. Latour, O. Malbec, E. Espinosa, P. Pina, S. Pasmans, et al., The same tyrosine-based inhibition motif, in the intracytoplasmic domain of Fc gamma RIIB, regulates negatively BCR-, TCR-, and FcR-dependent cell activation, Immunity 3 (1995) 635–646.

[99] M. Ono, S. Bolland, P. Tempst, J.V. Ravetch, Role of the inositol phosphatase SHIP in negative regulation of the immune system by the receptor Fc(gamma)RIIB, Nature 383 (1996) 263–266.

[100] M. Ono, H. Okada, S. Bolland, S. Yanagi, T. Kurosaki, J.V. Ravetch, Deletion of SHIP or SHP-1 reveals two distinct pathways for inhibitory signaling, Cell 90 (1997) 293–301.

[101] S.J. Tzeng, S. Bolland, K. Inabe, T. Kurosaki, S.K. Pierce, The B cell inhibitory Fc receptor triggers apoptosis by a novel c-Abl family kinase-dependent pathway, J Biol Chem 280 (2005) 35247–35254.

[102] J. Ruland, T.W. Mak, From antigen to activation: specific signal transduction pathways linking antigen receptors to NF-kappaB, Semin Immunol 15 (2003) 177–183.

[103] T. Dorner, P.E. Lipsky, Signalling pathways in B cells: implications for autoimmunity, Curr Top Microbiol Immunol 305 (2006) 213–240.

[104] J.W. Chung, S.R. Yoon, I. Choi, The regulation of NK cell function and development, Front Biosci 13 (2008) 6432–6442.

[105] L.L. Lanier, Up on the tightrope: natural killer cell activation and inhibition, Nat Immunol 9 (2008) 495–502.

[106] T. Takai, Roles of Fc receptors in autoimmunity, Nat Rev Immunol 2 (2002) 580–592.

[107] C.C. Goodnow, J. Sprent, B. Fazekas de St Groth, C.G. Vinuesa, Cellular and genetic mechanisms of self tolerance and autoimmunity, Nature 435 (2005) 590–597.

[108] T. Takai, M. Ono, M. Hikida, H. Ohmori, J.V. Ravetch, Augmented humoral and anaphylactic responses in Fc gamma RII-deficient mice, Nature 379 (1996) 346–349.

[109] S. Amigorena, C. Bonnerot, J.R. Drake, D. Choquet, W. Hunziker, J.G. Guillet, et al., Cytoplasmic domain heterogeneity and functions of IgG Fc receptors in B lymphocytes, Science 256 (1992) 1808–1812.

[110] Z. Xiang, A.J. Cutler, R.J. Brownlie, K. Fairfax, K.E. Lawlor, E. Severinson, et al., FcgammaRIIb controls bone marrow plasma cell persistence and apoptosis, Nat Immunol 8 (2007) 419–429.

[111] R.N. Pearse, T. Kawabe, S. Bolland, R. Guinamard, T. Kurosaki, J.V. Ravetch, SHIP recruitment attenuates Fc gamma RIIB-induced B cell apoptosis, Immunity 10 (1999) 753–760.

[112] N.R. Pritchard, A.J. Cutler, S. Uribe, S.J. Chadban, B.J. Morley, K.G. Smith, Autoimmune-prone mice share a promoter haplotype associated with reduced expression and function of the Fc receptor FcgammaRII, Curr Biol 10 (2000) 227–230.

[113] T.L. McGaha, B. Sorrentino, J.V. Ravetch, Restoration of tolerance in lupus by targeted inhibitory receptor expression, Science 307 (2005) 590–593.

[114] R.J. Brownlie, K.E. Lawlor, H.A. Niederer, A.J. Cutler, Z. Xiang, M.R. Clatworthy, et al., Distinct cell-specific control of autoimmunity and infection by FcgammaRIIb, J Exp Med 205 (2008) 883–895.

[115] T. Yuasa, S. Kubo, T. Yoshino, A. Ujike, K. Matsumura, M. Ono, et al., Deletion of fcgamma receptor IIB renders H-2(b) mice susceptible to collagen-induced arthritis, J Exp Med 189 (1999) 187–194.

[116] S. Kleinau, P. Martinsson, B. Heyman, Induction and suppression of collagen-induced arthritis is dependent on distinct fcgamma receptors, J Exp Med 191 (2000) 1611–1616.

[117] A. Nakamura, T. Yuasa, A. Ujike, M. Ono, T. Nukiwa, J.V. Ravetch, et al., Fcgamma receptor IIB-deficient mice develop Goodpasture's syndrome upon immunization with type IV collagen: a novel murine model for autoimmune glomerular basement membrane disease, J Exp Med 191 (2000) 899–906.

[118] M. Herrmann, R.E. Voll, O.M. Zoller, M. Hagenhofer, B.B. Ponner, J.R. Kalden, Impaired phagocytosis of apoptotic cell material by monocyte-derived macrophages from patients with systemic lupus erythematosus, Arthritis Rheum 41 (1998) 1241–1250.

[119] M. Kavai, G. Szegedi, Immune complex clearance by monocytes and macrophages in systemic lupus erythematosus, Autoimmun Rev 6 (2007) 497–502.

[120] W.M. Bogers, R.K. Stad, L.A. Van Es, M.R. Daha, Both Kupffer cells and liver endothelial cells play an important role in the clearance of IgA and IgG immune complexes, Res Immunol 143 (1992) 219–224.

[121] J.B. Cornacoff, L.A. Hebert, W.L. Smead, M.E. VanAman, D.J. Birmingham, F.J. Waxman, Primate erythrocyte-immune complex-clearing mechanism, J Clin Invest 71 (1983) 236–247.

[122] A.L. Henderson, M.A. Lindorfer, A.D. Kennedy, P.L. Foley, R.P. Taylor, Concerted clearance of immune complexes bound to the human erythrocyte complement receptor: development of a heterologous mouse model, J Immunol Methods 270 (2002) 183–197.

[123] J.Q. He, C. Wiesmann, M. van Lookeren Campagne, A role of macrophage complement receptor CRIg in immune clearance and inflammation, Mol Immunol 45 (2008) 4041–4047.

[124] G. Hom, R.R. Graham, B. Modrek, K.E. Taylor, W. Ortmann, S. Garnier, et al., Association of systemic lupus erythematosus with C8orf13-BLK and ITGAM-ITGAX, N Engl J Med 358 (2008) 900–909.

[125] J.B. Harley, M.E. Alarcon-Riquelme, L.A. Criswell, C.O. Jacob, R.P. Kimberly, K.L. Moser, et al., Genome-wide association scan in women with systemic lupus erythematosus identifies susceptibility variants in ITGAM, PXK, KIAA1542 and other loci, Nat Genet 40 (2008) 204–210.

[126] S.K. Nath, S. Han, X. Kim-Howard, J.A. Kelly, P. Viswanathan, G.S. Gilkeson, et al., A nonsynonymous functional variant in

integrin-alpha(M) (encoded by ITGAM) is associated with systemic lupus erythematosus, Nat Genet 40 (2008) 152–154.

[127] J.E. Salmon, S. Millard, L.A. Schachter, F.C. Arnett, E.M. Ginzler, M.F. Gourley, et al., Fc gamma RIIA alleles are heritable risk factors for lupus nephritis in African Americans, J Clin Invest 97 (1996) 1348–1354.

[128] G. Fossati, R.C. Bucknall, S.W. Edwards, Fcgamma receptors in autoimmune diseases, Eur J Clin Invest 31 (2001) 821–831.

[129] D.D. Sedmak, D.H. Davis, U. Singh, J.G. van de Winkel, C.L. Anderson, Expression of IgG Fc receptor antigens in placenta and on endothelial cells in humans. An immunohistochemical study, Am J Pathol 138 (1991) 175–181.

[130] M. Groger, G.F. Fischer, K. Wolff, P. Petzelbauer, Immune complexes from vasculitis patients bind to endothelial Fc receptors independent of the allelic polymorphism of FcgammaRIIa, J Invest Dermatol 113 (1999) 56–60.

[131] S. Vielma, G. Virella, A. Gorod, M. Lopes-Virella, Chlamydophila pneumoniae infection of human aortic endothelial cells induces the expression of FC gamma receptor II (FcgammaRII), Clin Immunol 104 (2002) 265–273.

[132] S.A. Mousavi, M. Sporstol, C. Fladeby, R. Kjeken, N. Barois, T. Berg, Receptor-mediated endocytosis of immune complexes in rat liver sinusoidal endothelial cells is mediated by FcgammaRIIb2, Hepatology 46 (2007) 871–884.

[133] S. Akilesh, T.B. Huber, H. Wu, G. Wang, B. Hartleben, J.B. Kopp, et al., Podocytes use FcRn to clear IgG from the glomerular basement membrane, Proc Natl Acad Sci U S A 105 (2008) 967–972.

[134] M. Nangaku, W.G. Couser, Mechanisms of immune-deposit formation and the mediation of immune renal injury, Clin Exp Nephrol 9 (2005) 183–191.

[135] J.W. Fries, D.L. Mendrick, H.G. Rennke, Determinants of immune complex-mediated glomerulonephritis, Kidney Int 34 (1988) 333–345.

[136] L.C. Willcocks, P.A. Lyons, M.R. Clatworthy, J.I. Robinson, W. Yang, S.A. Newland, et al., Copy number of FCGR3B, which is associated with systemic lupus erythematosus, correlates with protein expression and immune complex uptake, J Exp Med 205 (2008) 1573–1582.

[137] S. Nagarajan, K. Venkiteswaran, M. Anderson, U. Sayed, C. Zhu, P. Selvaraj, Cell-specific, activation-dependent regulation of neutrophil CD32A ligand-binding function, Blood 95 (2000) 1069–1077.

[138] A. Coxon, X. Cullere, S. Knight, S. Sethi, M.W. Wakelin, G. Stavrakis, et al., Fc gamma RIII mediates neutrophil recruitment to immune complexes. a mechanism for neutrophil accumulation in immune-mediated inflammation, Immunity 14 (2001) 693–704.

[139] R. Clynes, C. Dumitru, J.V. Ravetch, Uncoupling of immune complex formation and kidney damage in autoimmune glomerulonephritis, Science 279 (1998) 1052–1054.

[140] A. Bergtold, A. Gavhane, V. D'Agati, M. Madaio, R. Clynes, FcR-bearing myeloid cells are responsible for triggering murine lupus nephritis, J Immunol 177 (2006) 7287–7295.

[141] Q. Lin, Y. Xiu, Y. Jiang, H. Tsurui, K. Nakamura, S. Kodera, et al., Genetic dissection of the effects of stimulatory and inhibitory IgG Fc receptors on murine lupus, J Immunol 177 (2006) 1646–1654.

[142] K. Matsumoto, N. Watanabe, B. Akikusa, K. Kurasawa, R. Matsumura, Y. Saito, et al., Fc receptor-independent development of autoimmune glomerulonephritis in lupus-prone MRL/lpr mice, Arthritis Rheum 48 (2003) 486–494.

[143] C. Atkinson, F. Qiao, H. Song, G.S. Gilkeson, S. Tomlinson, Low-dose targeted complement inhibition protects against renal disease and other manifestations of autoimmune disease in MRL/lpr mice, J Immunol 180 (2008) 1231–1238.

[144] H. Sekine, C.M. Reilly, I.D. Molano, G. Garnier, A. Circolo, P. Ruiz, et al., Complement component C3 is not required for full expression of immune complex glomerulonephritis in MRL/lpr mice, J Immunol 166 (2001) 6444–6451.

[145] L. Baudino, F. Nimmerjahn, S. Azeredo da Silveira, E. Martinez-Soria, T. Saito, M. Carroll, et al., Differential contribution of three activating IgG Fc receptors (FcgammaRI, FcgammaRIII, and FcgammaRIV) to IgG2a- and IgG2b-induced autoimmune hemolytic anemia in mice, J Immunol 180 (2008) 1948–1953.

[146] M.L. Santiago-Raber, H. Amano, E. Amano, L. Baudino, M. Otani, Q. Lin, et al., Fcgamma receptor-dependent expansion of a hyperactive monocyte subset in lupus-prone mice, Arthritis Rheum 60 (2009) 2408–2417.

[147] K. Hora, J.A. Satriano, A. Santiago, T. Mori, E.R. Stanley, Z. Shan, et al., Receptors for IgG complexes activate synthesis of monocyte chemoattractant peptide 1 and colony-stimulating factor 1, Proc Natl Acad Sci U S A 89 (1992) 1745–1749.

[148] Y. Fu, Y. Du, C. Mohan, Experimental anti-GBM disease as a tool for studying spontaneous lupus nephritis, Clin Immunol 124 (2007) 109–118.

[149] Y. Suzuki, I. Shirato, K. Okumura, J.V. Ravetch, T. Takai, Y. Tomino, et al., Distinct contribution of Fc receptors and angiotensin II-dependent pathways in anti-GBM glomerulonephritis, Kidney Int 54 (1998) 1166–1174.

[150] R.M. Tarzi, K.A. Davies, J.W. Claassens, J.S. Verbeek, M.J. Walport, H.T. Cook, Both Fcgamma receptor I and Fcgamma receptor III mediate disease in accelerated nephrotoxic nephritis, Am J Pathol 162 (2003) 1677–1683.

[151] S.Y. Park, S. Ueda, H. Ohno, Y. Hamano, M. Tanaka, T. Shiratori, et al., Resistance of Fc receptor- deficient mice to fatal glomerulonephritis, J Clin Invest 102 (1998) 1229–1238.

[152] S. Segerer, A. Henger, H. Schmid, M. Kretzler, D. Draganovici, U. Brandt, et al., Expression of the chemokine receptor CXCR1 in human glomerular diseases, Kidney Int 69 (2006) 1765–1773.

[153] G. Camussi, F.C. Cappio, M. Messina, R. Coppo, P. Stratta, A. Vercellone, The polymorphonuclear neutrophil (PMN) immunohistological technique: detection of immune complexes bound to the PMN membrane in acute post-streptococcal and lupus nephritis, Clin Nephrol 14 (1980) 280–287.

[154] D.H. Hooke, D.C. Gee, R.C. Atkins, Leukocyte analysis using monoclonal antibodies in human glomerulonephritis, Kidney Int 31 (1987) 964–972.

[155] J.A. Croker, R.P. Kimberly, Genetics of susceptibility and severity in systemic lupus erythematosus, Curr Opin Rheumatol 17 (2005) 529–537.

[156] A. Davidson, B. Diamond, Autoimmune diseases, N Engl J Med 345 (2001) 340–350.

[157] K. Manger, R. Repp, M. Jansen, M. Geisselbrecht, R. Wassmuth, N.A. Westerdaal, et al., Fcgamma receptor IIa, IIIa, and IIIb polymorphisms in German patients with systemic lupus erythematosus: association with clinical symptoms, Ann Rheum Dis 61 (2002) 786–792.

[158] A.J. Duits, H. Bootsma, R.H. Derksen, P.E. Spronk, L. Kater, C.G. Kallenberg, et al., Skewed distribution of IgG Fc receptor IIa (CD32) polymorphism is associated with renal disease in systemic lupus erythematosus patients, Arthritis Rheum 38 (1995) 1832–1836.

[159] C.H. Hong, J.S. Lee, H.S. Lee, S.C. Bae, D.H. Yoo, The association between fcgammaRIIIB polymorphisms and systemic lupus erythematosus in Korea, Lupus 14 (2005) 346–350.

I. BASIS OF DISEASE PATHOGENESIS

[160] H.M. Dijstelbloem, M. Bijl, R. Fijnheer, R.H. Scheepers, W.W. Oost, M.D. Jansen, et al., Fcgamma receptor polymorphisms in systemic lupus erythematosus: association with disease and in vivo clearance of immune complexes, Arthritis Rheum 43 (2000) 2793–2800.

[161] J.Y. Chen, C.M. Wang, C.C. Ma, S.F. Luo, J.C. Edberg, R.P. Kimberly, et al., Association of a transmembrane polymorphism of Fcgamma receptor IIb (FCGR2B) with systemic lupus erythematosus in Taiwanese patients, Arthritis Rheum 54 (2006) 3908–3917.

[162] C. Kyogoku, H.M. Dijstelbloem, N. Tsuchiya, Y. Hatta, H. Kato, A. Yamaguchi, et al., Fcgamma receptor gene polymorphisms in Japanese patients with systemic lupus erythematosus: contribution of FCGR2B to genetic susceptibility, Arthritis Rheum 46 (2002) 1242–1254.

[163] U. Siriboonrit, N. Tsuchiya, M. Sirikong, C. Kyogoku, S. Bejrachandra, P. Suthipinittharm, et al., Association of Fcgamma receptor IIb and IIIb polymorphisms with susceptibility to systemic lupus erythematosus in Thais, Tissue Antigens 61 (2003) 374–383.

[164] Z.T. Chu, N. Tsuchiya, C. Kyogoku, J. Ohashi, Y.P. Qian, S.B. Xu, et al., Association of Fcgamma receptor IIb polymorphism with susceptibility to systemic lupus erythematosus in Chinese: a common susceptibility gene in the Asian populations, Tissue Antigens 63 (2004) 21–27.

[165] V. Magnusson, R. Zunec, J. Odeberg, G. Sturfelt, L. Truedsson, I. Gunnarsson, et al., Polymorphisms of the Fc gamma receptor type IIB gene are not associated with systemic lupus erythematosus in the Swedish population, Arthritis Rheum 50 (2004) 1348–1350.

[166] X. Li, J. Wu, R.H. Carter, J.C. Edberg, K. Su, G.S. Cooper, et al., A novel polymorphism in the Fcgamma receptor IIB (CD32B) transmembrane region alters receptor signaling, Arthritis Rheum 48 (2003) 3242–3252.

[167] Y. Lee, J. Ji, G. Song, Fc{gamma} receptor IIB and IIIB polymorphisms and susceptibility to systemic lupus erythematosus and lupus nephritis: a meta-analysis, Lupus 18 (2009) 727–734.

[168] R.A. Floto, M.R. Clatworthy, K.R. Heilbronn, D.R. Rosner, P.A. MacAry, A. Rankin, et al., Loss of function of a lupus-associated FcgammaRIIb polymorphism through exclusion from lipid rafts, Nat Med 11 (2005) 1056–1058.

[169] H. Kono, C. Kyogoku, T. Suzuki, N. Tsuchiya, H. Honda, K. Yamamoto, et al., FcgammaRIIB Ile232Thr transmembrane polymorphism associated with human systemic lupus erythematosus decreases affinity to lipid rafts and attenuates inhibitory effects on B cell receptor signaling, Hum Mol Genet 14 (2005) 2881–2892.

[170] K. Su, J. Wu, J.C. Edberg, X. Li, P. Ferguson, G.S. Cooper, et al., A promoter haplotype of the immunoreceptor tyrosine-based inhibitory motif-bearing FcgammaRIIb alters receptor expression and associates with autoimmunity. I. Regulatory FCGR2B polymorphisms and their association with systemic lupus erythematosus, J Immunol 172 (2004) 7186–7191.

[171] K. Su, X. Li, J.C. Edberg, J. Wu, P. Ferguson, R.P. Kimberly, A promoter haplotype of the immunoreceptor tyrosine-based inhibitory motif-bearing FcgammaRIIb alters receptor expression and associates with autoimmunity. II. Differential binding of GATA4 and Yin-Yang1 transcription factors and correlated receptor expression and function, J Immunol 172 (2004) 7192–7199.

[172] F.B. Karassa, T.A. Trikalinos, J.P. Ioannidis, Role of the Fcgamma receptor IIa polymorphism in susceptibility to systemic lupus erythematosus and lupus nephritis: a meta-analysis, Arthritis Rheum 46 (2002) 1563–1571.

[173] J.E. Salmon, J.C. Edberg, N.L. Brogle, R.P. Kimberly, Allelic polymorphisms of human Fc gamma receptor IIA and Fc gamma receptor IIIB. Independent mechanisms for differences in human phagocyte function, J Clin Invest 89 (1992) 1274–1281.

[174] T. Doi, M. Mayumi, K. Kanatsu, F. Suehiro, Y. Hamashima, Distribution of IgG subclasses in membranous nephropathy, Clin Exp Immunol 58 (1984) 57–62.

[175] J.L. Roberts, R.J. Wyatt, M.M. Schwartz, E.J. Lewis, Differential characteristics of immune-bound antibodies in diffuse proliferative and membranous forms of lupus glomerulonephritis, Clin Immunol Immunopathol 29 (1983) 223–241.

[176] F.B. Karassa, T.A. Trikalinos, J.P. Ioannidis, The Fc gamma RIIIA-F158 allele is a risk factor for the development of lupus nephritis: a meta-analysis, Kidney Int 63 (2003) 1475–1482.

[177] J. Wu, J.C. Edberg, P.B. Redecha, V. Bansal, P.M. Guyre, K. Coleman, et al., A novel polymorphism of FcgammaRIIIa (CD16) alters receptor function and predisposes to autoimmune disease, J Clin Invest 100 (1997) 1059–1070.

[178] H.R. Koene, M. Kleijer, J. Algra, D. Roos, A.E. von dem Borne, M. de Haas, Fc gammaRIIIa-158V/F polymorphism influences the binding of IgG by natural killer cell Fc gammaRIIIa, independently of the Fc gammaRIIIa-48L/R/H phenotype, Blood 90 (1997) 1109–1114.

[179] J.E. Salmon, J.C. Edberg, R.P. Kimberly, Fc gamma receptor III on human neutrophils. Allelic variants have functionally distinct capacities, J Clin Invest 85 (1990) 1287–1295.

[180] Y. Hatta, N. Tsuchiya, J. Ohashi, M. Matsushita, K. Fujiwara, K. Hagiwara, et al., Association of Fc gamma receptor IIIB, but not of Fc gamma receptor IIA and IIIA polymorphisms with systemic lupus erythematosus in Japanese, Genes Immun 1 (1999) 53–60.

[181] B.E. Stranger, M.S. Forrest, M. Dunning, C.E. Ingle, C. Beazley, N. Thorne, et al., Relative impact of nucleotide and copy number variation on gene expression phenotypes, Science 315 (2007) 848–853.

[182] T.J. Aitman, R. Dong, T.J. Vyse, P.J. Norsworthy, M.D. Johnson, J. Smith, et al., Copy number polymorphism in Fcgr3 predisposes to glomerulonephritis in rats and humans, Nature 439 (2006) 851–855.

[183] J. Galon, J.F. Gauchat, N. Mazieres, R. Spagnoli, W. Storkus, M. Lotze, et al., Soluble Fcgamma receptor type III (FcgammaRIII, CD16) triggers cell activation through interaction with complement receptors, J Immunol 157 (1996) 1184–1192.

[184] G. Zandman-Goddard, M. Blank, Y. Shoenfeld, Intravenous immunoglobulins in systemic lupus erythematosus: from the bench to the bedside, Lupus 18 (2009) 884–888.

[185] M.Y. Karim, C.N. Pisoni, M.A. Khamashta, Update on immunotherapy for systemic lupus erythematosus—what's hot and what's not!, Rheumatology (Oxford) 48 (2009) 332–341.

[186] S. Bucciarelli, D. Erkan, G. Espinosa, R. Cervera, Catastrophic antiphospholipid syndrome: treatment, prognosis, and the risk of relapse, Clin Rev Allergy Immunol 36 (2009) 80–84.

[187] M. Basta, F. Van Goor, S. Luccioli, E.M. Billings, A.O. Vortmeyer, L. Baranyi, et al., F(ab)'2-mediated neutralization of C3a and C5a anaphylatoxins: a novel effector function of immunoglobulins, Nat Med 9 (2003) 431–438.

[188] T.L. Vassilev, M.D. Kazatchkine, J.P. Van Huyen, M. Mekrache, E. Bonnin, J.C. Mani, et al., Inhibition of cell adhesion by antibodies to Arg-Gly-Asp (RGD) in normal immunoglobulin for therapeutic use (intravenous immunoglobulin, IVIg), Blood 93 (1999) 3624–3631.

[189] V.S. Negi, S. Elluru, S. Siberil, S. Graff-Dubois, L. Mouthon, M.D. Kazatchkine, et al., Intravenous immunoglobulin: an update on the clinical use and mechanisms of action, J Clin Immunol 27 (2007) 233–245.

[190] M. Debre, M.C. Bonnet, W.H. Fridman, E. Carosella, N. Philippe, P. Reinert, et al., Infusion of Fc gamma fragments for treatment of children with acute immune thrombocytopenic purpura, Lancet 342 (1993) 945—949.

[191] Y. Kaneko, F. Nimmerjahn, M.P. Madaio, J.V. Ravetch, Pathology and protection in nephrotoxic nephritis is determined by selective engagement of specific Fc receptors, J Exp Med 203 (2006) 789—797.

[192] Y. Kaneko, F. Nimmerjahn, J.V. Ravetch, Anti-inflammatory activity of immunoglobulin G resulting from Fc sialylation, Science 313 (2006) 670—673.

[193] A. Samuelsson, T.L. Towers, J.V. Ravetch, Anti-inflammatory activity of IVIG mediated through the inhibitory Fc receptor, Science 291 (2001) 484—486.

[194] F. Nimmerjahn, J.V. Ravetch, Anti-inflammatory actions of intravenous immunoglobulin, Annu Rev Immunol 26 (2008) 513—533.

[195] P. Bruhns, A. Samuelsson, J.W. Pollard, J.V. Ravetch, Colony-stimulating factor-1-dependent macrophages are responsible for IVIG protection in antibody-induced autoimmune disease, Immunity 18 (2003) 573—581.

[196] V. Gill, C. Doig, D. Knight, E. Love, P. Kubes, Targeting adhesion molecules as a potential mechanism of action for intravenous immunoglobulin, Circulation 112 (2005) 2031—2039.

[197] F. Duta, M. Ulanova, D. Seidel, L. Puttagunta, S. Musat-Marcu, K.S. Harrod, et al., Differential expression of spleen tyrosine kinase Syk isoforms in tissues: Effects of the microbial flora, Histochem Cell Biol 126 (2006) 495—505.

[198] P.R. Pine, B. Chang, N. Schoettler, M.L. Banquerigo, S. Wang, A. Lau, et al., Inflammation and bone erosion are suppressed in models of rheumatoid arthritis following treatment with a novel Syk inhibitor, Clin Immunol 124 (2007) 244—257.

[199] M.E. Weinblatt, A. Kavanaugh, R. Burgos-Vargas, A.H. Dikranian, G. Medrano-Ramirez, J.L. Morales-Torres, et al., Treatment of rheumatoid arthritis with a Syk kinase inhibitor: a twelve-week, randomized, placebo-controlled trial, Arthritis Rheum 58 (2008) 3309—3318.

[200] A. Podolanczuk, A.H. Lazarus, A.R. Crow, E. Grossbard, J.B. Bussel, Of mice and men: an open-label pilot study for treatment of immune thrombocytopenic purpura by an inhibitor of Syk, Blood 113 (2009) 3154—3160.

[201] F.R. Bahjat, P.R. Pine, A. Reitsma, G. Cassafer, M. Baluom, S. Grillo, et al., An orally bioavailable spleen tyrosine kinase inhibitor delays disease progression and prolongs survival in murine lupus, Arthritis Rheum 58 (2008) 1433—1444.

[202] S. Werwitzke, D. Trick, P. Sondermann, K. Kamino, B. Schlegelberger, K. Kniesch, et al., Treatment of lupus-prone NZB/NZW F1 mice with recombinant soluble Fc gamma receptor II (CD32), Ann Rheum Dis 67 (2008) 154—161.

[203] J.B. Bussel, Fc receptor blockade and immune thrombocytopenic purpura, Semin Hematol 37 (2000) 261—266.

Drug-Induced Lupus Mechanisms

Raymond L. Yung, Bruce C. Richardson

INTRODUCTION

Genes and the environment play important roles in the development of lupus-like autoimmunity, and recent advances have identified some of the genes contributing to lupus. However, how the environment interacts with the immune system to cause lupus in genetically predisposed people remains obscure. Drug-induced lupus (DIL) provides a clear example of environmental agents directly causing a systemic lupus-like disease, and reports of iatrogenic lupus continue to increase. The observation that specific drugs can induce autoimmunity in susceptible individuals provides an opportunity to study early steps in the pathogenesis of lupus, and the overlap in signs, symptoms, and serologic findings in idiopathic systemic lupus erythematosus (SLE) and DIL suggests that DIL is a valid model for the study of idiopathic lupus.

Despite the growing number of drugs implicated in DIL, most investigations into the mechanisms of DIL continue to focus on procainamide and hydralazine. It is worth pointing out that drug-induced autoimmunity needs to be distinguished from classic drug hypersensitivity reactions. The immune response in DIL develops much more gradually and is not directed to the offending medication. Interestingly, different drugs can cause diverse clinical manifestations with different autoantibody specificities, suggesting heterogeneity in the environmental agents triggering idiopathic lupus. It is now clear that more than one mechanism is involved in causing DIL. Advances, including the development of murine models of DIL, have helped clarify the pathophysiology of this disease and its relationship to idiopathic lupus. We will review the current understanding of the pathogenic mechanisms contributing to the development of DIL, with an emphasis on *in vivo* models.

STRUCTURES OF DRUGS CAUSING DRUG-INDUCED LUPUS

The drugs implicated in DIL have diverse chemical structures and pharmacologic properties. Those most commonly associated with DIL contain aromatic amines, hydrazines, sulfhydryl groups, or a phenol ring (Figure 22.1). The considerable overlap between clinical features of DIL caused by drugs with these diverse chemical structures suggests that no single configuration is responsible for DIL.

ANTIDRUG ANTIBODIES

The formation of antidrug antibodies is a relatively common event, and in some instances correlates with allergic or anaphylactic reactions. However, lupus-inducing drugs are generally poor immunogens, and antibodies raised against them have not been shown to be pathogenic in humans or experimental systems. Indeed, only 6% of patients with procainamide-induced lupus have antibodies against procainamide, compared to 50% of all patients receiving the drug [1]. Interestingly, "natural" antibodies to procainamide have been found in sera from 19—35% of patients with idiopathic lupus and rheumatoid arthritis and in up to 30% of healthy adults never exposed to the drug. These data suggest that development of clinical DIL from procainamide does not depend on antidrug antibodies, and that these antibodies are unlikely to cross-react with autoantigens *in vivo* to cause DIL. However, antibodies to a drug-macromolecule complex may cross-react with unrelated autoantigens to cause clinical disease. In one report, all patients with active hydralazine-induced lupus demonstrated antihydralazine antibodies [2]. Another study reported that 76% of

385

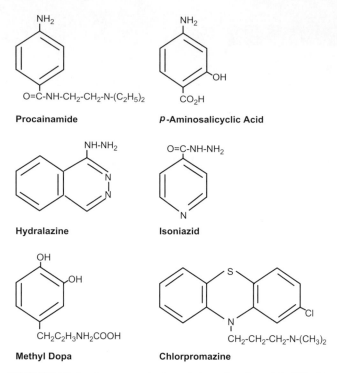

FIGURE 22.1 Structure of selected drugs implicated in DIL. Procainamide is an example of aromatic amines, while hydralazine and isoniazid are example of hydrazines.

patients developed antihydralazine antibodies after 1 year of treatment [3]. The high incidence of antihydralazine antibodies, together with structural similarities between hydralazine and the nucleotide adenosine, led to the hypothesis of molecular mimicry and antibody cross-reactivity as pathogenic mechanisms for hydralazine-induced lupus. In support of this, Yamauchi *et al.* [4] demonstrated that rabbits hyperimmunized with hydralazine conjugated to human serum albumin can produce antihydralazine antibodies that cross-react with DNA.

Treatment with anti-TNF (tumor-necrosis factor) therapies has been associated with the appearance of ANAs, anti-double-stranded (ds) DNA antibodies and occasionally clinical lupus [5]. A recent French study reported a 0.19% incidence of DIL among etanercept and infliximab users [6]. Approximately 13% of patients exposed to infliximab develop infliximab-specific antibodies. Antibody development is lower in recipients on concurrent immunosuppressants such as methotrexate or azathioprine. The presence of antibodies to infliximab increases the likelihood of an infusion reaction but does not predict the development of autoantibodies or DIL. Non-neutralizing antibodies to the TNF receptor portion or other protein components of etanercept are present in up to 16% of rheumatoid arthritis

patient receiving this drug. The majority of the autoantibodies are IgM and there is no correlation between antibody development and clinical response or autoimmunity. Similarly, antiadalimumab antibodies have been described in 17% of rheumatoid arthritis patients after 28 weeks of drug treatment [7]. Interestingly, adalimumab nonresponders are more likely to have antidrug antibodies. Similar to etanercept and infliximab, the use of adalimumab has been linked to lupus-like illness [8]. However, there is no evidence that the presence of antidrug antibodies correlates with the appearance of autoantibodies or the development of drug-induced lupus.

DRUG—DNA INTERACTIONS

Both procainamide and hydralazine have been shown to bind polynucleotides *in vitro*. It has been suggested that the resulting drug—DNA structures may potentially be more immunogenic than the native DNA molecule. The amino and hydrazine groups of procainamide and hydralazine, respectively, can also bind to nucleoproteins [9—11] which may augment their immunogenicity. This raises the possibility that DNA may be modified by these agents in a fashion analogous to haptens, making the DNA more antigenic. These two drugs, and to a lesser extent, isoniazid and D-penicillamine, have also been shown to induce a stable transition from the B (right-handed) to Z (left-handed) configuration of synthetic DNA [12]. DNA in the Z configuration is more immunogenic [13, 14] and the drug—DNA complex may cause the formation of some of the autoantibodies found in DIL. Several studies have demonstrated anti-Z-DNA antibodies in hydralazine-treated patients [15—17] with one reporting that 82% of sera from hypertensive patients receiving hydralazine had evidence of anti-Z-DNA antibodies [16]. In a longitudinal study, investigators were able to show that the antibodies developed after the initiation of hydralazine. In this context, it is significant that increased levels of natural polyamines have been demonstrated in sera from patients with active idiopathic lupus [18] and some of these polyamines are capable of inducing Z-DNA conformation, possibly contributing to the anti-Z-DNA antibodies demonstrated in idiopathic lupus [19—21]. Hydralazine may also stabilize another unusual and potentially immunogenic form of DNA, known as triplex DNA [22]. Up to 72% of sera from hydralazine-treated patients may have antibodies against triplex DNA. At present, the clinical relevance of these observations is unclear.

DRUG–HISTONE INTERACTIONS AND ANTIHISTONE ANTIBODIES

Histones are DNA-binding proteins found in eukaryotic cell nuclei. There are five major histone subclasses: H1 (or H5), H2A, H2B, H3, and H4. Histones are responsible for packaging DNA into chromatin, a highly regulated and tightly compressed complex of histones and DNA. The basic chromatin subunit is the nucleosome, consisting of DNA wrapped twice around a protein core containing two copies each of histones H2A, H2B, H3, and H4. Histone H1 is involved in the formation of higher-order chromatin structures. It has become increasingly clear that chromatin is not a passive structure. Dynamic modifications of the histones, such as acetylation, methylation, phosphorylation, ubiquitination, and ADP-ribosylation have important regulatory effects on gene expression. Although it is tempting to postulate that selected drugs may directly bind chromatin structure to alter gene expression, there is as yet no evidence that this actually occurs.

Antibodies against histone subunits are commonly detected in patients with DIL. However, the precise role of these and other autoantibodies in the pathogenesis of DIL is unclear. Indeed, it would be a mistake to equate the presence of antihistone antibodies to the diagnosis of DIL. There are considerable discrepancies among published studies on the fine specificity of antihistone antibodies in patients with DIL. Part of this may be due to the purity of the substrates used and the different assay conditions employed [23]. In addition, the specificity of antihistone antibodies appears to differ from those found in idiopathic SLE [24, 25]. Hydralazine has been shown to interact with soluble DNA–histone complexes [26]. The resulting structural change makes histones more resistant to proteolytic digestion. The altered structure may also be more immunogenic, suggesting an explanation for the high incidence of antihistone antibodies in these patients. Different drugs also induce antihistone antibodies with different fine specificity. It is worth noting that antihistone antibodies can have higher affinity to native than purified histones. Burlingame et al. [27] examined the reactivity of antihistone antibodies in lupus patients and demonstrated preferential binding to chromatin and selected histone–DNA subnucleosome complexes compared to purified DNA. Consistent with this, Mohan et al. [28] studied autoreactive T-cell clones from lupus-prone mice, and showed that 50% of the clones had specificity against nucleosome antigens and not free DNA or histones. These observations raise the possibility that the immunogenic form of DNA may be the nucleosome, possibly modified by drugs. Interestingly, one study suggested a statistical correlation between the presence of anti-(H2A-H2B), anti-(H2A-H2B)-DNA, antichromatin antibodies, and proteinuria in idiopathic lupus [27].

ABNORMAL IMMUNE REGULATION

Depending on the concentrations tested, procainamide has been reported to enhance or reduce the mitogenic response of human peripheral blood lymphocytes [29–32]. Low concentrations of procainamide hydroxylamine (PAHA), a reactive oxidized metabolite of procainamide, have been shown to enhance lymphocyte mitogenic responses and promote B-cell immunoglobulin (Ig) secretion. Opposite effects, however, have been reported when higher concentrations of the drug were tested [31, 32]. Another report found that sera from patients receiving procainamide will inhibit the Concanavalin-A mitogenic response of peripheral blood mononuclear cells from normal and procainamide-treated patients [33]. Others have reported that T cells from patients with procainamide-induced lupus demonstrate decreased proliferative responses [34]. Most, but not all, researchers report B-lymphocyte activation and increased Ig production in procainamide-treated patients [35–37]. Patients with active procainamide-induced lupus have circulating B cells that spontaneously secrete Ig, and B-cell Ig production in response to pokeweed mitogen is enhanced [38, 39] similar to patients with idiopathic lupus. In one in vitro study, researchers showed that procainamide may inhibit T-cell suppressor activity, resulting in excessive B-cell activation. Yet another report found increased Ig production in procainamide-treated patients without concomitant change in T-cell suppressor function [39]. The reason for the inconsistent reporting is not clear, but may in part be related to conditions under which the assays were performed.

Inconsistent results have also been reported for other lupus-inducing drugs. Peripheral blood lymphocytes of patients with hydralazine-induced lupus have been shown to have a higher transformation index compared with lymphocytes from untreated patients [40]. Litwin et al. [41, 42] examined hydralazine-treated patients and found normal delayed hypersensitivity responses as measured by skin testing. The peripheral blood lymphocytes from the same group of patients had normal mitogenic responses, but had increased proliferation after incubation with hydralazine–albumin conjugates. Decreased suppressor T-cell function, with enhanced B-cell Ig production, has been reported in patients taking methyldopa and in normal T cells treated with methyldopa in vitro. Others, however, found depressed peripheral blood mononuclear cell mitogenic responses after methyldopa treatment [43].

Diphenylhydantoin was found to inhibit lymphocyte proliferation [44] but had no effect on T-cell suppressor function [45].

More recently, a number of cytokine/anticytokine therapies have been reported to cause autoimmunity in humans. Type I interferons including interferon-α (IFN-α) play a critical role in antiviral responses. Between 0.15% and 0.7% of IFN-α-treated patients develop a lupus-like illness [46, 47]. Interestingly, IFN-α levels are increased in idiopathic lupus and may correlate with anti-DNA antibody titers [48]. It has been postulated that IFN-α may modulate autoimmunity through its effects on apoptosis. However, this is uncertain because IFN-α has been shown to upregulate the expression of both the proapoptotic Fas ligand [49] and the antiapoptotic Bcl-2 [50] genes in lymphoid cells. Since IFN-α induces IFN-γ expression, it is also possible that IFN-α-induced lupus occurs in part through IFN-γ-dependant mechanisms. Additional data shows that the cytokine may activate autoreactive T cells by stimulating viral superantigen production [51]. Recent data suggest that IFN-α may induce dendritic cell differentiation in lupus patients that may in turn play a role in the disease pathogenesis [52]. Another report suggests that IFN-α may drive the autoimmune response in lupus by inducing monocyte differentiation into antigen-presenting dendritic cells [53].

The mechanism behind anti-TNF-associated lupus is unclear. Serum levels of TNF and soluble TNF receptors are increased in patients with idiopathic lupus, and may correlate with lupus disease activity [54]. TNF and interleukin-1 (IL-1) β expression are also increased in the kidneys of MRL/lpr mice with autoimmune lupus nephritis [55]. In contrast, NZB/NZW F1 mice have low serum TNF levels. TNF supplementation prior to the onset of nephritis in these mice delays proteinuria and improves survival, but does not affect anti-DNA antibody production [56]. Additionally, TNF-α blockade may suppress production of Th1 cytokines, thereby skewing the overall immune response toward a Th2 response. Another postulated mechanism is that anti-TNF antibodies bind to TNF-α-bearing cells, inducing apoptosis and the release of potentially immunogenic nuclear antigens into the circulation [5]. In support of this, anti-TNF treatment has been reported to induce apoptosis in T cells and monocytes [57, 58]. Infliximab, but not etanercept, was also shown to cause leukocyte apoptosis in the lamina propria of Crohn's disease patients [59]. Reduced leukocyte CD44 expression and impaired apoptotic cell clearance have been described in lupus patients. TNF-α has been shown to upregulate CD44 expression, and inhibiting the TNF pathway may therefore impair the natural clearance of apoptotic cells [60]. Finally, antinucleosome antibodies have been reported to correlate with the development of antinuclear and anti-dsDNA antibodies in rheumatoid arthritis patients receiving TNF-α blocking agents [61]. This provides further support to the hypothesis that TNF-associated apoptosis plays a role in the pathogenesis of autoimmunity in patients receiving anti-TNF agents. Interestingly, newly described members of the TNF/TNF-receptor protein family, including B-cell-activating factor (BAFF, also known as BlyS zTNF4) and APRIL, and their receptors TACI, BCMA, and BAFF-R, have been implicated in the pathogenesis of lupus, and anti-BlyS therapy may become available [62]. A number of biologic agents, including IFNs and anti-TNF agents, have been modified by processes such as PEGylation to extend the half-life in the circulation. Whether this will result in a greater incidence of DIL is unclear [63].

Minocycline has also been reported to induce a variety of immune and autoimmune phenomena such as autoimmune hepatitis, immune thrombocytopenia, and cutaneous lupus. This drug has a number of effects on the immune and inflammatory systems. For example, minocycline inhibits matrix metalloproteinases including collagenase (MMP-1), gelatinase (MMP-2), and stromelysin (MMP-3). Additionally, minocycline selectively downregulates T cell TNF-α production but paradoxically enhances IL-6 and TNF-α release from monocytes [64]. Although untested, it is possible that minocycline may induce autoimmune features *in vivo* by mechanisms similar to that of blockade with anti-TNF agents.

GENETIC FACTORS

Human Leukocyte Antigen and Complement

Many studies have documented an association between certain major histocompatibility complex (MHC) alleles and the development of idiopathic lupus. These include human leukocyte antigen (HLA)-DR2 and HLA-DR3, and class III C4A and C4B null complement alleles. In contrast, there are only limited and sometimes conflicting reports of HLA associations in DIL. Initial studies from England by Batchelor *et al.* [65] showed that 79% of patients with hydralazine-induced lupus expressed the HLA-DR4 allele, compared with 25% of asymptomatic patients receiving hydralazine. Twenty-five of the 26 patients with hydralazine-induced disease were also slow acetylators. No association between the presence of DR4 and acetylator phenotype was found. In 1987, a second English study [66] also reported an association between the HLA-DR4 allele and hydralazine-induced lupus. These researchers found that 70% of their 20 patients were HLA-DR4-positive, compared with 33% in the general population. Of note, six of the 20 patients were also included in the initial study by

Batchelor *et al.* [65]. In contrast, a third study from Australia failed to find an increased incidence of HLA-DR4 in 18 patients with hydralazine-induced disease when compared to untreated controls [67]. It now appears that the putative HLA association in the initial English study may be the result of linkage disequilibrium with the complement C4-null allele. C4 is encoded at two loci, C4A and C4B, located between the class I MHC and HLA-DR genes. Multiple alleles, including null or nonfunctioning alleles, are present within each of these loci. When the researchers from the English study reexamined their patients, they found a much higher incidence of the C4-null allele in the study group (16 of 21 patients, or 76%) compared with controls (35 of 82 patients, or 43%) [68]. The observed HLA-DR4 association is possibly the result of linkage disequilibrium between the C4-null-DR4 haplotype and the C4 null trait. Of note, partial deficiencies of C4, especially the C4A isotype, are risk factors for idiopathic lupus. The occasional patients with total C4 deficiency have a high incidence of the disease.

Another potential pathogenic DIL mechanism also involves the complement system. The classic complement pathway plays an important role in removing circulating immune complexes. Some of the drugs implicated in DIL, including hydralazine, penicillamine, isoniazid, and metabolic products of procainamide, have been shown to inhibit the covalent binding of C4 to C2 [69, 70]. This may inhibit the activation of C3 by the classic pathway, resulting in defective immune-complex clearance. Increased amounts of circulating immune complexes have been reported, but not consistently, in patients with DIL [71].

Very few studies have examined possible HLA associations in other forms of DIL. One report suggested an association between HLA-DR6Y and procainamide-induced lupus [72]. Fifty-three percent of the clinically affected patients were HLA-DR6Y-positive, compared with 25% of procainamide-treated patients who were asymptomatic but ANA-positive, and 17% of controls. However, this has not been confirmed by other investigators [73]. Another report suggested an association between HLA-Bw44 and chlorpromazine-related autoimmunity [74]. Forty-six percent of patients with chlorpromazine-induced autoantibodies were HLA-Bw44 positive, compared with 19% of the normal population. Another study examining Caucasian psychiatric patients on chronic chlorpromazine therapy found that the presence of lupus anticoagulant correlated positively with HLA-DR7 [75]. HLA-DR4 was linked with penicillamine-induced lupus in one study [76]. Another study, however, failed to confirm the association, but found increased frequencies of HLA-A11 and B15 in a group of penicillamine-treated rheumatoid arthritis patients [77]. In summary, there has not been any consistent

report of an HLA association in DIL. Genetic factors predisposing patients to DIL are likely to lie outside the MHC region.

Large-scale genome-wide association studies from around the world have now identified an ever increasing number of risk-associated genes in systemic lupus erythematosus. However, similar studies have not been conducted in patients with DIL. It is therefore unclear if the same genetic risk factors exist for both conditions.

Acetylator Phenotype and DIL

Acetylation by the enzyme *N*-acetyltransferase is the dominant pathway for metabolizing aromatic amines and hydrazines (Figure 22.2). Although the enzyme(s) is found primarily in the liver and gut, acetylation activity has also been detected in human erythrocytes and peripheral blood mononuclear cells. The rate of drug acetylation is influenced by genetic factors. Individuals who are phenotypically slow acetylators are homozygous for the recessive allele, while fast acetylators are either heterozygous or homozygous for a dominant allele of the hepatic acetyltransferase gene. People with the slow acetylator phenotype have decreased or absent *N*-acetyltransferase activities in the liver. Drugs known to be metabolized by this pathway include procainamide, hydralazine, isoniazid, sulfapyridine, sulfadimidine, dapsone, an amine metabolite of nitrazepam, and some carcinogenic aromatic amines. Nongenetic factors may also affect the rate of acetylation. For example, the acetylation rate is decreased in patients with renal insufficiency and increased with ethanol ingestion. Since renal function naturally declines with age, the resulting drop in acetylation may contribute to the higher likelihood of DIL in this age group.

There are approximately equal numbers of people with the slow and fast acetylator phenotypes in North America. Although DIL can occur in patients with either phenotype, the majority of patients who develop procainamide or hydralazine-induced lupus have the slow acetylator phenotype [78–80]. Individuals who are slow acetylators develop autoantibodies after a shorter

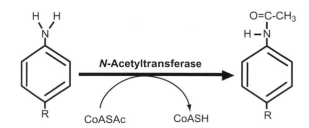

FIGURE 22.2 N-acetyltransferase catalyzes the acetylation of aromatic amines by acetyl-coenzyme A (CoASAc).

exposure to hydralazine than do fast acetylators (Figure 22.3) and are also more prone to developing hydralazine-induced lupus. Perry *et al.* [81] were the first to document the relationship between hepatic acetyltransferase activity, hydralazine, and the development of ANA and DIL. Of the 57 patients receiving hydralazine in their study, 33 were slow acetylators and 24 were fast acetylators. Twelve patients developed hydralazine-induced lupus, and all were slow acetylators. The cumulative dose and the steady-state concentration of the drugs are important factors in determining the onset of DIL. Importantly, the incidence of DIL is about the same in patients with the slow and fast acetylator phenotypes, if the serum concentration of procainamide is maintained at the same level [82]. *N*-Acetylprocainamide (NAPA), a procainamide metabolite, does not have the same lupus-inducing potential as the parent drug, suggesting that the aromatic amino group in procainamide may be important in inducing this disease. This suggests that the primary amino or hydrazine group on the parent drugs, rather then their metabolites, may be important elements in the pathogenesis of procainamide and hydralazine-induced lupus.

Only a small number of drugs implicated in DIL utilize the acetylation metabolic pathway. The contribution of a patient's acetylator phenotype to determining the development of DIL also varies among the different drugs metabolized by this pathway. Although acetylation is the primary determinant of the serum half-life of isoniazid, patients who are slow acetylators and receive the drug have been reported to be both equally likely [83] or more likely [84, 85] to develop ANAs and clinical disease, compared with fast acetylators.

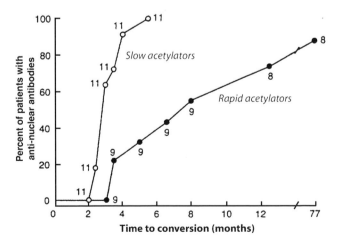

FIGURE 22.3 Rate of development of procainamide-induced ANA in slow acetylators (open circles) and rapid acetylators (closed circles) with time. Reproduced with permission from Massachusetts Medical Society, Woolsey *et al.* [182].

Acetylator Phenotype and Idiopathic SLE

The association between slow acetylator phenotype and some forms of DIL suggests that acetylation inactivates the inciting drug, thereby providing protection. Unless other environmental aromatic amines play a role in the etiology of idiopathic lupus, there should be no relationship between acetylation phenotype and idiopathic lupus. Initial studies by Reidenberg and Martin [86] showed that 71% of patients with idiopathic SLE may be slow acetylators, compared with the 52% expected incidence in the general population. Subsequent studies, however, showed positive associations, statistically insignificant associations, or no association between acetylator status and idiopathic lupus [87, 88]. The association, if present, appears to be small. It is highly doubtful that genetic slow acetylation plays a significant role in the pathogenesis of idiopathic lupus. The varying results in different studies may be in part due to the use of historical controls or variation of other unidentified geographic factors important in idiopathic lupus.

NEUTROPHIL-MEDIATED OXIDATIVE METABOLISM

Because lupus-inducing drugs have diverse chemical structures and pharmacological actions, it has been postulated that these agents may cause clinical autoimmunity via common reactive oxygen metabolites. This model postulates that reactive oxygen species released by neutrophils modify xenobiotics, resulting in toxic metabolites. Immune cells including monocytes and Langerhans cells possess more limited amounts of enzymes such as myeloperoxidase and prostaglandin H synthase but may have some activity. Hepatic metabolism involving the microsomal mono-oxygenase system and esterases also produces some potentially toxic compounds. These highly reactive agents are very labile and the active forms generally do not circulate in the body. A role for these reactive metabolites in immune responses therefore must involve contact with immunocytes in the liver, or locally within areas of inflammation (Figure 22.4). Peripheral blood neutrophils can oxidize procainamide to the toxic metabolite PAHA [89]. Mixing experiments and cell-free studies suggest that myeloperoxidase and reactive oxygen radicals are released by activated neutrophils during this chemical transformation [89–91]. Therapeutic levels of six different lupus-inducing drugs, including hydralazine, were found to be cytotoxic to EL4 cells when cultured in the presence of activated neutrophils [91]. It is possible that PAHA, oxygen radicals, and other metabolic by products contribute to the cytotoxicity [91].

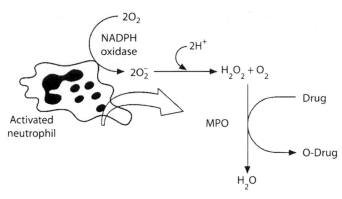

FIGURE 22.4 Mechanism for biotransformation of drugs by activated neutrophils. Reproduced with permission from SAGE Publications, Rubin, R. L. [183].

Kinetics of inhibition studies suggest that the observed cytotoxicity may be related to the ability of these drugs to serve as substrates for the myeloperoxidase enzyme. However, these highly reactive drug metabolites are unstable, and it has been difficult to detect their presence *in vivo*. Of interest but uncertain significance are the observations that PAHA and its nitroso derivatives have also been shown to bind covalently to histones, and autoantibodies to myeloperoxidase have been found in the serum of patients with DIL [92, 93]. These observations are consistent with the notion that myeloperoxidase-mediated metabolism may be involved in the pathogenesis of drug-induced autoimmunity.

Kretz-Rommel and Rubin examined the possibility that local production of PAHA may affect processes involved in T-cell tolerance to self-antigens [94]. Partial activation of T cells through T-cell receptor (TCR) engagement in the absence of co-stimulating signals can induce anergy to peptide antigens. Interference with this process may potentially affect autoimmune tolerance, resulting in the production of autoreactive T cells. The investigators showed that treatment with PAHA, but not procainamide, will block the induction of anergy in cloned T cells by anti-CD3 antibody. Although PAHA is unstable in solution, its effects on T-cell anergy may be up to 8 h long. This model suggests that colocalization of potentially autoreactive T cells, activated neutrophils, and persistent inflammation is required to maintain the tolerant state. However, it is uncertain whether the concentrations of PAHA achieved *in vivo* approach the 1- to 5-μM concentrations necessary to modify tolerance *in vitro*.

EPIGENETIC MECHANISMS

Epigenetic mechanisms may also play a role in the pathogenesis of DIL and idiopathic lupus. In particular, biochemical modifications of chromatin can affect gene expression and may be important in the induction of autoimmune diseases. These modifications include methylation of DNA and the phosphorylation, acetylation, and adenosine diphosphate (ADP)-ribosylation of histones.

ADP-Ribosylation and DIL

Limited studies have been conducted examining the effects of lupus-inducing agents on histone ADP-ribosylation. Poly-(ADP-ribose)-polymerase (PADPRP) is a chromatin-bound enzyme that catalyzes the transfer of ADP-ribose from nicotinamide to histones, including HI and H2B. Autoantibodies reactive to HI and H2B, and to poly-(ADP-ribose) (PADPR) and PADPRP, have been demonstrated in patients with DIL and idiopathic lupus [95–97]. Interestingly, ultraviolet light (UV), which has been implicated in activating lupus, also activates PADPR [98]. These observations prompted further studies. Ayer *et al.* [99] examined the effect of procainamide and hydralazine on PADPRP activity in human T- and B-cell lines. These investigators found that therapeutic procainamide concentrations increased PADPRP activity up to 2.5-fold. Hydralazine and NAPA also increased the enzyme's activity, though to a smaller degree. It is possible that ADP-ribosylation of intracellular molecules, including histone proteins, may alter the structure and increase the immunogenicity of the molecules.

T-Cell DNA Hypomethylation and DIL

Evidence also supports an association between abnormal methylation of T-lymphocyte DNA and some forms of DIL. DNA methylation refers to the post-synthetic methylation of cytosine at the 5 position [100] and is one of the mechanisms regulating gene expression. For genes regulated by DNA methylation, hypomethylation of regulatory sequences correlates with gene expression, whereas methylation results in transcriptional suppression. DNA methylation can affect gene expression by several mechanisms. The methylation of cytosine residues can prevent the binding of some transcription factors if located in their recognition sequences. In addition, a family of specific nuclear proteins recognize and bind the methylated sequences, preventing the binding of transcription factors. Finally, it is now clear that the process of DNA methylation acts in concert with histone deacetylation, a related epigenetic process, in chromatin remodeling and suppression of gene transcription. Some methylcytosine binding proteins such as MeCP2 attract a chromatin inactivation complex containing histone deacetylases, which condenses chromatin into an inactive configuration, inhibiting interactions with transcription factors required for gene expression.

DNA methylation patterns are replicated during mitosis by the enzyme DNA (cytosine-5-)-methyltransferase 1 (Dnmt1). During mitosis, Dnmt1 recognizes hemimethylated CpG dinucleotides in the parent and daughter DNA strands, and catalyzes the transfer of the methyl group from *S*-adenosylmethionine to the cytosine residues in the unmethylated daughter DNA strand, producing symmetrically methylated sites and maintaining methylation patterns (Figure 22.5). The importance of DNA methylation is evidenced by its role in X-chromosome inactivation, and in genomic imprinting. Disruption of the *Dnmt1* gene in embryonal stem cells results in abnormal development and embryonic death, suggesting that DNA methylation is important in ontogeny. Additional DNA methyltransferases have been identified, including Dnmt3a and Dnmt3b. In contrast to Dnmt1, these enzymes will methylate both unmethylated and hemimethylated DNA equally well, and function as *de novo* methyltransferases. Homozygous deletion of these enzymes also causes death during embryonic development or in the early neonatal period.

Certain drugs inhibit the DNA methyltransferases, modifying methylation patterns with sometimes striking effects. For example, 5-azacytidine (5-azaC), an irreversible DNA methyltransferase inhibitor, causes 10T1/2 cells to differentiate into myocytes, adipocytes, and chondrocytes by altering expression of genes determining cellular differentiation [101]. This approach has been used to clone the *Myo D* gene, which regulates myocytic differentiation. The observation that classic DNA hypomethylating agents like 5-azaC can affect the activity of histone deacetylases suggests another potential mechanism whereby lupus-inducing drugs

may alter gene expression and promote autoimmunity. However, the role of histone deacetylation has not been examined in DIL.

The role of DNA methylation in regulating T-cell function and gene expression has more recently been examined. 5-azaC treatment of cloned, antigen-specific CD4$^+$ cells caused them to lose their requirement for nominal antigen and respond to autologous antigen-presenting cells (APCs) alone [102–105], thus becoming autoreactive. This was confirmed using a panel of CD4$^+$ antigen-specific human cloned cells, PHA activated polyclonal CD4$^+$ human T cells, cloned murine antigen-specific CD4$^+$ T cells, alloreactive murine CD4$^+$ T cells, and Concanavalin-A-activated CD4$^+$ murine T cells. This autoreactive response is specific for autologous or syngeneic APCs and requires specific MHC class II and T-cell antigen—MHC receptor interactions, indicating that the autoreactive response uses the TCR to transmit one of the activation signals.

The autoreactivity correlates with an approximately tenfold overexpression of leukocyte function antigen-1 (LFA-1) (CDlla-CD18) on the surface of the treated cells [102–105], and transfecting human or murine T-cell clones with LFA-I causes an identical autoreactivity [106, 107]. In either 5-azaC-treated or LFA-1-transfected cells, small amounts of antibody to CD11a or intercellular adhesion molecule I (ICAM-1) (CD54), an LFA-1 ligand [104, 106, 107], completely inhibits the autoreactive response to APCs without antigen, while only minimally inhibiting the response to the same APCs presenting specific antigen. This suggests that LFA-I overexpression might contribute to the autoreactivity, and inhibiting the effects of the additional LFA-1 molecules reverses the effect. LFA-1 is a member of the β2 integrin family of adhesion molecules, and serves to stabilize the interaction of the TCR with class II MHC molecules on APCs. Others have also shown that LFA-1-ICAM interactions are particularly important when the affinity of the TCR-class II interaction is low. LFA-1 overexpression may cause autoreactivity by overstabilizing the normally low-affinity interaction between the TCR and class II MHC molecules, allowing T-cell activation by the class II determinants alone or with other peptides than those normally recognized by the T cell.

The observation that normal, antigen-specific T cells can be made autoreactive through chemical exposure prompted studies examining the relevance of this phenomenon to autoimmunity. A similar situation, in which CD4$^+$ T cells respond to host MHC determinants, occurs in the murine chronic graft-vs.-host (GVH) disease model. In this system, adoptive transfer of CD4$^+$ T cells from a parent strain into the F1 hybrid of certain histoincompatible parental strains results in the development of an autoimmune disease with

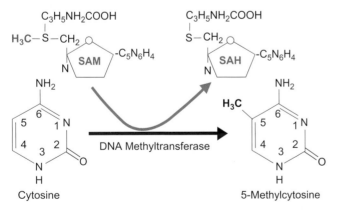

FIGURE 22.5 Formation of 5-methylcytosine. DNA methyltransferase enzyme catalyzes the transfer of the methyl group (CH3) from S-adenosylmethionine (SAM) to (deoxy) cytosin producing 5-(deoxy)methylcytosine and S-adenosylhomocystine (SAH). Reproduced with permission from Springer Verlag, Attwood et al. [100].

features of human lupus [108]. The recipients develop an immune-complex glomerulonephritis, splenomegaly, and arthritis, as well as an autoimmune liver disease. Serologic features include hypergammaglobulinemia, ANAs, and anti-double-stranded DNA (dsDNA) antibodies. Subsequent studies demonstrated that therapeutic concentrations of procainamide or hydralazine inhibit DNA methylation in human and murine T cells [109–111]. Procainamide was subsequently shown to reversibly inhibit human T-cell DNA MTase activity in a dose-dependent fashion. Activated normal human and murine CD4$^+$ T-cell clones treated with therapeutic concentrations of procainamide or hydralazine also became autoreactive, similar to 5-azaC-treated cells [102–105]. This autoreactivity also correlates with increased LFA-1 expression. Conversely, NAPA, a metabolite of procainamide that does not cause lupus, is about 100-fold less potent in its ability to induce T-cell autoreactivity *in vitro*. More recently, it was shown that UV light, implicated in activating lupus, also inhibits T-cell DNA methylation, increases LFA-I expression, and induces autoreactivity at relatively modest doses [106]. It is likely that DNA demethylating drugs also affect other methylation-sensitive genes that may contribute to autoimmunity. Our group examined the effects of 5-azaC on the expression of approximately 7000 T-cell genes using a microarray genome scanning approach [112]. One hundred eighteen genes were identified that have greater than twofold increases in expression following the drug treatment, including IFN-γ and CD70 as B-cell co-stimulating molecules, and perforin as a molecule participating in cytotoxic responses.

Defects in signal transduction have been identified in lupus T cells and may account for the aberrant cellular and humoral immune responses in lupus patients [113]. Evidence has shown that Dnmt1 is upregulated following T-cell stimulation in part by signals transmitted through the MAPKK-catalyzed ERK phosphorylation pathway, and that decreased signaling in the pathway may result in T-cell DNA hypomethylation in lupus patients [114]. Interestingly, although hydralazine has no direct effect on Dnmt1 activity, it may indirectly decrease Dnmt1 expression and function via inhibition of the ERK signaling pathway [115, 116]. It now appears that hydralazine may specifically inhibit protein kinase C (PKC) δ, a signaling protein upstream of ERK, to induce demethylation of selected T cell genes that are important in the pathogenesis of lupus [117]. Together, these studies suggest that certain environmental agents may also induce T-cell autoreactivity by inhibiting T-cell DNA methylation, thereby altering T-cell gene expression. A recent study examined the effect of hydralazine on B cell tolerance, and showed that the drug induces upregulation of the RAG-2 gene and reduces secondary Ig gene rearrangements. Furthermore, the investigators found that a specific MEK1/2 inhibitor suppresses B cell receptor editing in a dose-dependent manner. Adoptive transfer of bone marrow B cells exposed to hydralazine or a MEK inhibitor into syngeneic mice induces autoantibody production [118]. These results complement the T cell work and provide further evidence that selected drugs may induce autoimmunity via their effects on the ERK signaling pathway.

T cells made autoreactive by treatment with DNA methylation inhibitors interact with other cells of the immune system in a way that could contribute to the development of autoimmunity. For example, co-culture of B lymphocytes with CD4$^+$ T cells treated with DNA methylation inhibitors, including 5-azaC, procainamide, and hydralazine, induces differentiation of the B cells into IgG-secreting cells, without the addition of antigen or mitogen [119]. An identical amount of antibody secretion was induced with supernatant from activated treated or untreated cells, suggesting that the autoreactive T cells were activated by MHC determinants on autologous B cells, and the T cells then secreted B-cell differentiation factors, such as IFN-γ, interleukin-4 (IL-4) and/or IL-6, activating the B cells. However, a role for cell–cell communication through surface molecules such as CD40 and the CD40 ligand or CD70 and CD27 is also possible. This mechanism could contribute to the polyclonal B-cell activation observed in patients with procainamide-induced lupus. In addition, the autoreactive cells spontaneously induce apoptosis in autologous or syngeneic macrophages through pathways including Fas, TWEAK, and TRAIL [120]. These studies have led to the proposal that a similar T-cell modification occurring *in vivo* could result in the release of intracellular and nuclear antigens. If the autoreactive T cells traffic to lymphatic tissue such as the lymph nodes or spleen, macrophage lysis in the immediate proximity of B cells and other T cells, together with the release of cytokines such as IL-4 and IL-6, might contribute to the generation of autoantibodies. Macrophage lysis in the liver or spleen may also contribute to impaired immune-complex clearance.

T-Cell DNA Methylation in Idiopathic Lupus

The DNA hypomethylation model suggests that certain chemical agents can inhibit T-cell DNA methylation, modify T-cell LFA-1 expression, and induce autoreactivity. Furthermore, the autoreactive T cells can interact with MHC-identical B cells and macrophages, inducing macrophage apoptosis and B-cell activation, and *in vivo* the same cells can cause a disease with many manifestations of human lupus (see

or with untreated cells plus conalbumin. Chromium and fluorescein labeling studies demonstrated that autoreactive as well as untreated cells homed to the spleen, as predicted by the model and by earlier trafficking studies. Adoptive transfer of the autoreactive D10 cells into AKR mice produced a more severe immune-complex glomerulonephritis, pulmonary alveolitis, and central nervous system lesions resembling human lupus with vascular fibrinoid necrosis, cerebral infarcts, and sterile meningitis. Other pathologic findings included liver lesions with periportal inflammatory infiltration and bile duct proliferation resembling primary biliary cirrhosis. The mice also developed significant titers of anti-ssDNA, anti-dsDNA, and antihistone antibodies. The reason for the difference in disease severity between the DBA/2 and AKR models is not known. Possible explanations include the use of a cloned Th2 line in the AKR model, with less inhibition by Th1 cells, or differences in the genetic makeup of the hosts. However, a more recent study showed that cloned Th1 cells over-express LFA-1 and become autoreactive following 5-azaC treatment, similar to Th2 cells [141]. Adoptive transfer of the autoreactive Th1 cells also induces similar anti-DNA antibodies. Comparison of effector mechanisms suggests that the T-cell cytokine repertoire is probably not crucial in this system. In contrast, macrophage/monocyte killing is common to all three models and may be an important early step providing antigenic nucleosomes and decreasing clearance of apoptotic materials.

The D10 model has also been used to compare relative potency of procainamide, NAPA, hydralazine, and phthalazine. NAPA does not cause DIL, whereas phthalazine is the parent compound of hydralazine but lacks the hydrazine side chain, which has been implicated in autoimmunity. Procainamide and hydralazine are also more potent DNA methylation inhibitors than NAPA. Mice receiving hydralazine and procainamide developed a similar disease. Phthalazine was less potent than hydralazine, and NAPA had no effect in this system. These results lend further credibility to the DNA hypomethylation model, and suggest a reason for the differing potencies of these drugs for inducing DIL.

Gender predisposition to lupus has also been examined in the DNA hypomethylation model. Lupus is predominantly a female disease, and sex hormones have been shown to play an important role in modulating immune responses in humans and murine models of lupus. Consistent with data from humans and auto-immune prone mice, normal female mice receiving pro-cainamide-treated T cells develop a more severe disease and have higher titers of autoantibodies than age-matched male recipients [142]. Oophorectomized female mice develop a much milder disease, with anti-DNA titers similar to orchiectomized and normal male mice receiving drug-treated cells.

Further studies demonstrated gender-specific effects of T-cell trafficking *in vivo*, with up to seven times as many T cells homing to the female spleens compared to male spleens, and this female-specific increase in splenic homing also decreased following oophorectomy. Splenectomized females did not develop autoantibodies or autoimmune pathology, supporting an important pathogenic role for the spleen in this model. These studies also indicated that relatively few cells accumulating in the spleen (a total of approximately 360,000) are sufficient to induce autoimmunity in mice [142]. These observations suggest that female-specific increases in T-cell trafficking to the spleen may cause increased disease severity in this model, presumably due to effects of female sex hormones on T-cell trafficking patterns. Estrogens have been reported to upregulate endothelial cell adhesion molecule expression [143], providing a plausible explanation for the gender-specific differences in T-cell trafficking observed. However, the effect of gender appears to be less prominent in DIL. Since DIL affects primarily older individuals, this may be due to the decline in sex-hormone production with aging. Alternately, gender-specific factors may play a less important role in DIL.

T cells from patients with lupus or treated with hydralazine have defective extracellular signal-regulated kinase (ERK) phosphorylation. This is at least in part due to impaired T cell protein kinase C (PKC) δ activation, as both pharmacologic inhibition of PKCδ or transfection with a dominant negative PKCδ mutant caused demethylation of the CD70 promoter similar to lupus and hydralazine-treated T cells [117]. Dnmt1 levels are regulated in part by the extracellular signal-regulated kinase (ERK)-mitogen-activated protein kinase (MAPK) pathway [144]. Using a transgenic mouse that inducibly expresses a dominant-negative MEK in T cells in the presence of doxycycline, it was shown that decreased ERK pathway signaling results in lower expression of Dnmt1 and overexpression of the methylation-sensitive genes CD11a and CD70, similar to T cells from human patients [145]. Together, these *in vitro* and *in vivo* studies provide a strong support of a link between T cell signaling defects and DNA hypomethylation in idiopathic and drug-induced lupus.

XENOBIOTICS

Environmental exposure to certain toxins, known as xenobiotics, has been linked to the development of lupus-like illnesses and autoimmunity in humans and in several animal models. Selected environmental factors include heavy metals, occupational and industrial contaminants, foods, and dietary supplements. This is a broad field, and it is difficult to discuss every environmental agent linked to autoimmunity. Therefore, only selected agents, including those with more recent associations, will be discussed. Interested readers are referred to some excellent reviews on this subject [146, 147].

Foods and Dietary Supplements

L-canavanine

A lupus-like syndrome associated with the ingestion of the nonessential amino acid L-canavanine has been described in humans and animals. Dietary supplements containing alfalfa were initially implicated in triggering disease exacerbations in lupus patients [148]. In 1981, Malinow et al. [149] reported that ingestion of alfalfa seeds was the cause of a reversible autoimmune hemolytic anemia and pancytopenia in one patient. L-Canavanine was subsequently shown to be the principal ingredient in alfalfa seeds responsible for this disease. Primates fed alfalfa seeds, alfalfa sprouts, or L-canavanine developed a similar disease, with hemolytic anemia, high-titer ANA, and anti-dsDNA antibodies. Withdrawal of alfalfa seeds from the diet results in resolution of the clinical condition, although autoantibodies remain detectable for up to 2 years [150, 151].

L-Canavanine is an amino acid structurally similar to L-arginine and can compete with L-arginine for the charging of transfer RNA (tRNA) Arg. The isoelectric point of canavanine is lower than arginine. The substitution may change the structure and function of the resulting proteins, including arginine-rich histone proteins, altering their immunogenicity [152]. L-Arginine also affects the immune system, including the arginine-nitrous oxide pathway. L-Canavanine may therefore influence the immune system by several mechanisms. At high doses, this nonessential amino acid inhibits the proliferative response of human mononuclear cells to mitogens through its effect on DNA synthesis. L-Canavanine may also affect B-cell function. Under *in vitro* conditions, L-canavanine can alter the surface charge of B cells, but not T cells, from NZB/NZW F1 mice [153]. It is possible that the activated B cells cause increased incorporation of canavanine into surface proteins, modifying the surface charge [154]. Interestingly, L-canavanine can also stabilize DNA in the more antigenic Z configuration. The enhanced antigenicity may contribute to the development of anti-DNA antibodies.

Anilines and L-tryptophan

Two more recent epidemics of environmentally induced autoimmunity highlight the pathologic potential of chemicals the human body may be exposed to

and their clinical sequelae. Ingestion of rapeseed oil denatured with aniline was responsible for reports of toxic oil syndrome (TOS) in Spain in 1981 [155]. Affected patients developed acute pneumonitis, followed by a chronic illness characterized by a scleroderma-like skin disease, neuromyopathy, and sicca syndrome. Eosinophilia, elevated IgE, and occasionally the presence of ANAs and other autoantibodies were also reported. A second outbreak of a similar disease, the eosinophilic myalgia syndrome (EMS), was reported in the late 1980s [155, 156]. Epidemiologic studies identified the dietary supplement L-tryptophan, an essential amino acid, as the causative agent in the outbreak. Although not conclusive, it appears that contaminants or by-products introduced during the manufacturing process by a single manufacturer may be the culprit [156]. The precise mechanism underlying this disease has not been worked out, and only a small percentage of people who ingested the offending agents were affected. Human and animal studies suggest that immunogenetic and metabolic phenotypes are important factors in the expression of both diseases.

Increased frequencies of HLA-DR3 and HLA-DR4, enhanced T-suppressor cell activity, and antibodies to T and B cells have all been reported in patients with TOS. Interestingly, patients with TOS may also have abnormal tryptophan metabolism, including abnormal IFN-γ activation of indoleamine-2, 3-dioxygenase, an enzyme involved in tryptophan breakdown metabolism. A number of studies have been performed using a variety of animal systems. Unfortunately, no animal model of TOS that truly resembles the human disease has been described. Mice fed aniline develop high serum IgE titers and increased IL-1β, IL-6, and IFN-γ [157]. It was suggested that a breakdown in tolerance and polyclonal B-cell activation may be important in this disease. Only one study reported lung toxicity in rats [158], with other researchers reporting uniformly negative lung pathology in a number of animal systems [159—161]. In one long-term study, rats were fed for 200 days with "case oil" obtained from the National Center for Food and Nutrition in Majadahonda [162]. These rats developed a large increase in dermal collagen deposition. Electron microscopic analysis showed increased collagen fibers of uniform diameter arranged in thick bundles. A pathogenic mechanism for TOS akin to GVH has also been proposed [163—165]. Both diseases display a biphasic clinical course. 1-phenyl-5-vinyl-2-imidazolidinethione (IZT), a possible constituent of unrefined rapeseed oils and a possible reaction product of aniline, shares structural similarities with anticonvulsants (such as phenytoin) and antithyroid drugs known to be capable of inducing GVH or lupus-like illnesses in humans. However, only traces of IZT could be found in the case oils. The role of IZT in TOS is currently unclear [166, 167].

Eosinophilic myalgia syndrome has been linked to HLA-Aw33, HLA-B44, and HLA-DR6 [168]. Abnormal tryptophan metabolism, enhanced kynurenine production, and roles for IFN-γ and transforming growth factor β (TGF-β) have all been proposed for this disease. Kynurenine is a metabolic product of L-tryptophan. Of interest, increased urinary excretion of kynurenine and plasma kynurenine levels has been reported in patients with scleroderma, CREST syndrome, and eosinophilic fasciitis [169]. Female Lewis rats fed L-tryptophan have enhanced tryptophan metabolism with increased plasma L-kynurenine levels [170]. Other features of this model include suppression of the hypothalamic-pituitary-adrenal (HPA) axis and pathologic findings including fasciitis and perimyositis. However, several features of the human disease are absent in this model, including pulmonary and skin involvement and peripheral or tissue eosinophilia. The role of eosinophils in this and other eosinophilic rheumatic diseases is unclear. Cytokine-mediated activation of eosinophils, with resulting degranulation and release of toxic products (e.g., major basic protein, eosinophil-derived neurotoxin), may contribute to tissue injury. Other alterations of cellular immunity have also been described, including phenotypic changes in peripheral blood mononuclear cells and activation of peripheral T lymphocytes. A contaminant, termed *peak E*, was isolated from contaminated oil [171]. The compound was determined to be 1, 1'-ethylidenebis(tryptophan), composed of two L-tryptophan molecules joined by an ethylidene bridge. A definitive pathogenic role of this novel amino acid in EMS has not been established. Despite clinical similarities with TOS, a convincing link between EMS and TOS has not been shown.

Silica, Silicon, and Silicones

Silicon is the second most abundant element on the earth. In nature, it is found in combination with oxygen to yield various forms of silicas, silicates, glasses, and sand. Silicone is a synthetic polymer containing a repeating Si—O backbone, and the organic groups attached to the silicon atom via silicon—carbon bonds determine the class of silicone. Exposure to silica dust has been linked to the development of autoimmune diseases, including scleroderma, and to a lesser extent rheumatoid arthritis. A more recent Spanish report examined the incidence of systemic autoimmune diseases in a group of workers at a scouring-powder factory [172]. The scouring powder is composed of 90% silica. Seventy-two percent of the patients were found to be ANA-positive, and 64% of the 50 studied workers were found to have a systemic autoimmune illness. These included six patients with Sjögren's syndrome, five with scleroderma, three with SLE, five with a lupus-scleroderma overlap, and 19 with an undefined collagen disease. The incidence of lupus and other autoimmune diseases in this group is much higher than would be expected in the general population. Although interesting, further studies will be needed to determine if the association truly exists.

Silica particles have been shown to induce IL-1 production by macrophages [173]. Animal studies also suggest a potential role for TNF-α. This proinflammatory cytokine may potentially play a role in inflammatory and fibrotic reactions. Workers exposed to silica may have decreased numbers of circulating T cells and increased numbers of circulating B cells [174]. Hilar lymph nodes in some patients also showed reversal of normal T-helper cells: suppressor-cell ratios, with increased T-helper and decreased T-suppressor cells.

Considerable controversy exists regarding the role of silicone breast implants in the induction of autoimmune diseases, including scleroderma and SLE [175, 176]. The pathologic and immunologic effects of silicone have been studied in detail [177, 178]. However, the clinical significance of these observations is unclear. Local inflammatory changes are common. Pathologic studies often show granulomatous, macrophage-rich chronic inflammatory reactions. Silicone has been found to exhibit adjuvant activity in a rat model, and monocytes and macrophages produce IL-1β when exposed to silicone rubber in culture. Fibroblasts have twice as much rough endoplasmic reticulum when they are grown on silicone surfaces compared with polystyrene, and mice exposed to silicone gel have increased numbers of granulocyte-macrophage colony-forming units in the bone marrow, increased hepatic macrophage uptake of radio-labeled sheep erythrocytes, and reduced responsiveness of natural killer cells against YAC target cells. However, despite anecdotal case reports of an association between silicone breast implants and autoimmunity, large numbers of epidemiologic studies including meta-analysis of up to 24 studies have failed to show a causal relationship with any specific autoimmune disease including lupus [179–181].

CONCLUSIONS

Many mechanisms have been proposed for how various drugs may cause DIL. Of these, the evidence is best for procainamide and hydralazine, in part because these have been the most extensively studied. The more recent development of animal models for autoimmunity induced by these two drugs is likely to greatly accelerate our understanding of the underlying mechanisms involved.

References

[1] A.S. Russell, M. Ziff, Natural antibodies to procainamide, Clin Exp Immunol 3 (1968) 901−909.

[2] B.H. Hahn, G.C. Sharp, W.S. Irvin, et al., Immune response to hydralazine and nuclear antigens in hydralazine-induced lupus erythematosus, Ann Intern Med 76 (1972) 365−374.

[3] J.R. Carpenter, F.C. McDuffie, S.G. Sheps, et al., Prospective study of immune response to hydralazine and development of antideoxyribonucleo-protein in patients receiving hydralazine, Am J Med 69 (1980) 395−400.

[4] Y. Yamauchi, A. Litwin, L.E. Adams, H. Zimmer, E.V. Hess, Induction of antibodies to nuclear antigens in rabbits by immunization with hydralazine-human serum albumin conjugates, J Clin Invest 56 (1975) 958−969.

[5] P.J. Charles, R.J.T. Smeenk, J. De Jong, M. Feldmann, R.N. Maini, Assessment of antibodies to double-stranded DNA induced in rheumatoid arthritis patients following treatment with infliximab, a monoclonal antibody to tumor necrosis factor α, Arthritis Rheum 43 (2000) 2383−2390.

[6] M. De Bandt, J. Sibilia, X. Le Löet, et al., Systemic lupus erythematosus induced by anti-tumor necrosis factor alpha therapy: a French national survey, Arthritis Res Ther 7 (2005) R545−R551.

[7] G.M. Bartelds, C.A. Wijbrandts, M.T. Nurmohamed, et al., Clinical response to adalimumab: relationship to anti-adalimumab antibodies and serum adalimumab concentrations in rheumatoid arthritis, Ann Rheum Dis 66 (2007) 921−926.

[8] M. Manosa, E. Domenech, L. Marin, et al., Adalimumab-induced lupus erythematosus in Crohn's disease patients previously treated with infliximab, Gut 57 (2008) 559.

[9] H.G. Bluestein, D. Redelman, N.J. Zvaifler, Procainamide-lymphocyte reactions. A possible explanation for drug-induced autoimmunity, Arthritis Rheum 24 (1981) 1019−1023.

[10] L.M. Dubroff, R. Reid Jr., M. Papalian, Molecular models for hydralazine-related systemic lupus erythematosus, Arthritis Rheum 24 (1981) 1082−1085.

[11] L.M. Dubroff, R. Reid Jr., Hydralazine-pyrimidine interactions may explain hydralazine-induced lupus erythematosus, Science 208 (1980) 404−406.

[12] T.J. Thomas, R.P. Messner, Effects of lupus-inducing drugs on the B to Z transition of synthetic DNA, Arthritis Rheum 29 (1986) 638−645.

[13] E.M. Lafer, A. Moller, A. Nordheim, B.D. Stoller, A. Rich, Antibodies specific for left-handed Z-DNA, Proc Natl Acad Sci USA 78 (1981) 3546−3550.

[14] U.B. Gunnia, T. Thomas, T.J. Thomas, The effects of polyamines on the immunogenicity of polynucleotides, Immunol Invest 20 (1991) 337−350.

[15] R.L. Rubin, S.A. Bell, R.W. Burlingame, Autoantibodies associated with lupus induced by diverse drugs target a similar epitope in the (H2A-H2B)-DNA complex, J Clin Invest 90 (1992) 165−173.

[16] T.J. Thomas, J.R. Seibold, L.E. Adams, E.V. Hess, Hydralazine induces Z-DNA conformation in a polynucleotide and elicits anti (Z-DNA) antibodies in treated patients, Biochem J 294 (1993) 419−425.

[17] R.W. Burlingame, R.L. Rubin, Drug-induced anti-histone autoantibodies display two patterns of reactivity with substructures of chromatin, J Clin Invest 88 (1991) 680−690.

[18] T.J. Thomas, U.B. Gunnia, J.R. Seibold, T. Thomas, Defective signal-transduction pathways in T-cells from autoimmune MRL-lpr/lpr mice are associated with increased polyamine concentrations, Biochem J 311 (1995) 175−182.

[19] E.M. Lafer, R.-P. Vae, A. Moller, et al., Z-DNA-specific antibodies in human systemic lupus erythematosus, J Clin Invest 71 (1983) 314−321.

[20] H. Sano, C. Morirnoto, DNA isolated from DNA/anti-DNA antibody immune complexes in systemic lupus erythematosus is rich in guanine-cytosine content, J Immunol 128 (1982) 1341−1345.

[21] H. Sano, C. Moirimoto, Isolation of DNA from DNA/anti-DNA antibody immune complexes in systemic lupus erythematosus, J Immunol 126 (1981) 538−539.

[22] T.J. Thomas, J.R. Seibold, L.E. Adams, E.V. Hess, Triplex-DNA stabilization by hydralazine and the presence of anti-(triplex DNA) antibodies in patients treated with hydralazine, Biochem J 311 (1995) 183−188.

[23] M. Monestier, T.M. Fasy, T.M. Debbas, K. Patel, Deoxyribonuclease I treatment of histones for the detection of anti-histone antibodies in solid-phase immunoassays. Effect of protease contamination in commercial deoxyribonuclease I preparations, J Immunol Methods 127 (1990) 289−291.

[24] R.M. Bernstein, R.N. Hobbs, D.J. Lea, D.J. Ward, G.R.V. Hughes, Patterns of antihistone antibody specificity in systemic rheumatic disease 1. Systemic lupus erythematosus, mixed connective tissue disease, primary sicca syndrome, and rheumatoid arthritis with vasculitis, Arthritis Rheum 28 (1985) 285−293.

[25] J. Gohill, P.D. Cary, M. Couppez, M.J. Fritzier, Antibodies from patients with drug-induced and idiopathic lupus erythematosus react with epitopes restricted to the arnino and carboxyl termini of histone, J Immunol 135 (1985) 3116−3121.

[26] J.P. Portanova, R.E. Arndt, E.M. Tan, B.L. Kotzin, Anti-histone antibodies in idiopathic and drug-induced lupus recognize distinct intrahistone regions, J Immunol 138 (1987) 446−451.

[27] R.W. Burlingame, M.L. Boey, G. Starkebaum, R.L. Rubin, The central role of chromatin in autoimmune responses to histones and DNA in systemic lupus erythematosus, J Clin Invest 94 (1994) 184−192.

[28] C. Mohan, S. Adams, V. Stanik, S.K. Datta, Nucleosome: A major immunogen for pathogenic autoantibody-inducing T cells of lupus, J Exp Med 17 (1993) 1367−1381.

[29] H.G. Bluestein, N.J. Zvaifler, M.H. Weisman, R.F. Shapiro, Lymphocyte alteration by procainamide: Relation to drug-induced lupus erythematosus syndrome, Lancet 2 (1979) 816−819.

[30] G. de Boccardo, D. Drayer, A.L. Rubin, et al., Inhibition of pokeweed mitogen-induced B cell differentiation by compounds containing primary amine or hydralazine groups, Clin Exp Immunol 59 (1985) 69−76.

[31] T. Ochi, E.A. Goldings, P.E. Lipsky, M. Ziff, Immunomodulatory effect of procainamide in man, Inhibition of human suppressor T-cell activity in vitro. J Clin Invest 71 (1983) 36−45.

[32] L.E. Adams, C.E. Sanders Jr., R.A. Budinsky, et al., Immunomodulatory effects of procainamide metabolites: Their implications in drug-related lupus, J Lab Clin Med 113 (1989) 482−492.

[33] R.H. Tannen, S. Cunningham-Rundles, Inhibition of Con A mitogenesis by serum from procainamide-treated patients and patients with systemic lupus erythematosus, Immunol Commun 11 (1982) 33−45.

[34] C.L. Yu, M. Ziff, Effects of long-term procainamide therapy on immunoglobulin synthesis, Arthritis Rheum 28 (1985) 276−284.

[35] J. Forrester, J. Golbus, D. Brede, J. Hudson, B. Richardson, B cell activation in patients with active procainamide induced lupus, J Rhumatol 15 (1988) 1384−1388.

[36] A. Klajman, N. Camin-Belsky, A. Kimchi, S. Ben-Efraim, Occurrence, immunoglobulin pattern and specificity of anti-nuclear antibodies in sera of procainamide treated patients, Clin Exp Immunol 7 (1970) 641—649.

[37] C.L. Yu, M. Ziff, Effects of long-term procainamide therapy on immunoglobulin synthesis, Arthritis Rheum 28 (1985) 276—284.

[38] B.J. Green, D.G. Wyse, H.J. Duff, L.B. Mitchell, D.S. Matheson, Procainamide in vivo modulates suppressor T cell activity, Clin Invest Med 11 (1988) 425—429.

[39] K.B. Miller, D. Salem, Immune regulatory abnormalities produced by procainamide in vitro suppressor cell function of IgG secretion, Am J Med 73 (1982) 487—492.

[40] B.H. Hahn, G.C. Sharp, W.S. Irvin, et al., Immune responses to hydralazine and nuclear antigens in hydralazine-induced lupus erythematosus, Ann Intern Med 76 (1972) 365—374.

[41] A. Litwin, L.E. Adams, H. Zimmer, et al., Prospective study of immunologic effects of hydralazine in hypertensive patients, Clin Pharmacol Ther 29 (1981) 447—456.

[42] A. Litwin, L.E. Adams, H. Zimmer, E.V. Hess, Immunologic effects of hydralazine in hypertensive patients, Arthritis Rheum 24 (1981) 1074—1078.

[43] H.H. Kirtland, D.N. Mohler, D.A. Horwitz, Methyldopa inhibition of suppressor-lymphocyte function: A proposed cause of autoimmune hemolytic anemia, N Engl J Med 302 (1980). 825-291.

[44] A.A. MacKinney, H.E. Booker, Diphenyl-hydantoin effects on human lymphocytes in vitro and in vivo. A hypothesis to explain some drug interactions, Arch Intern Med 129 (1972) 988—992.

[45] D. Alarcon-Segovia, R. Palacios, Differences in immunoregulatory T cell circuits between diphenylhydantoin-related and spontaneously occurring systemic lupus erythematosus, Arthritis Rheum 24 (1981) 1086—1092.

[46] T. Okanoue, S. Sakamoto, Y. Itoh, M. Minami, K. Yasui, M. Sakamoto, et al., Side effects of high-dose interferon therapy for chronic hepatitis C, J Hepatol 25 (1996) 283—291.

[47] L.E. Ronnblom, G.V. Alm, K.E. Oberg, Autoimmunity after alpha-interferon therapy for malignant carcinoid tumors, Ann Intern Med 324 (1991) 509—514.

[48] T. Kim, Y. Kanayama, N. Negoro, M. Okamura, T. Takeda, T. Inoue, Serum levels of inteferons in patients with systemic lupus erythematosus, Clin Exp Immunol 70 (1987) 562—569.

[49] K.A. Kirou, R.K. Vakkalanka, M.J. Butler, M.K. Crow, Induction of Fas ligand-mediated apoptosis by interferon-alpha, Clin Immunol 95 (2000) 218—226.

[50] A.P. Jewell, C.P. Worman, P.M. Lydyard, K.L. Yong, F.J. Giles, A.H. Goldstone, Interferon-alpha up-regulates bcl-2 expression and protects B-CLL cells from apoptosis in vitro and in vivo, Br J Haematol 88 (1994) 268—274.

[51] Y. Stauffer, S. Marguerat, F. Meylan, C. Ucla, N. Sutkowski, B. Huber, et al., Interferon-α-induced endogenous super-antigen: A model linking environment and autoimmunity, Immunity 15 (2001) 591—601.

[52] L. Ronnblom, G.V. Alm, A pivotal role for the natural interferon alpha-producing cells (plasmacytoid dendritic cells) on the pathogenesis of lupus, J Exp Med 194 (2001) F59—F63.

[53] P. Blanco, A.K. Palucka, M. Gill, V. Pascual, J. Banchereau, Induction of dendritic cell differentiation by IFN-alpha in systemic lupus erythematosus, Science 294 (2001) 1540—1543.

[54] B. Heilig, C. Fiehn, M. Brockhaus, et al., Evaluation of soluble tumor necrosis factor alpha and its soluble receptors parallel clinical disease and autoimmune activity in systemic lupus erythematosus, Br J Rheumatol 35 (1996) 1067—1074.

[55] J.M. Boswell, M.A. Yui, D.W. Burt, V.E. Kelley, Increased tumor necrosis factor and IL-1 beta gene expression in the kidneys of mice with lupus nephritis, J Immunol 141 (1988) 3050—3054.

[56] C.O. Jacob, H.O. McDevitt, Tumor necrosis factor-α in murine autoimmune "lupus" nephritis, Nature 331 (1988) 356—358.

[57] C. Shen, P. Maerten, G. Assche, K. Geboes, P. Putgeerts, J. Ceuppens, Anti-murine TNF-alpha induces apoptosis of peritoneal macrophages from SCID mice, Gastroenterology 124 (2005) A486.

[58] C. Shen, G.V. Assche, S. Colpaert, P. Maerten, K. Geboes, P. Putgeerts, et al., Adalimumab induces apoptosis of human monocytes: a comparative study with infliximab and etanercept, Alimentary Pharm Therap 21 (2005) 251—258.

[59] T. Ten Hove, C. Van Mtfrans, M.P. Peppelenbosch, S.J. Van Deventer, Infliximab treatment induces apoptosis of lamina propria T lymphocytes in Crohn's disease, Gut 50 (2002) 206—211.

[60] S.P. Hart, G.J. Dougherty, C. Hasle, I. Dransfield, CD44 regulates phagocytosis of apoptotic neutrophil granulocytes, but not apoptotic lymphocytes, by human macrophages, J Immunol 159 (1997) 919—925.

[61] M. Benucci, G. Saviola, P. Baiardi, E. Cammelli, M. Manfredi, Anti-nucleosome antibodies as prediction factor of development of autoantibodies during therapy with three different TNFα blocking agents in rheumatoid arthritis, Clin Rheumatol 27 (2008) 91—95.

[62] T. Dorner, C. Putterman, B cells, BAFF/zTNF4, TACI, and sytemic lupus erythematosus, Arthritis Res 3 (2001) 197—199.

[63] S. Yilmaz, K.A. Cimen, Pegylated interferon alfa-2B induced lupus in a patient with chronic hepatitis B virus infection: case report, Clin Rheumatol 28 (2009) 1241—1243.

[64] M. Kloppenburg, B.M.N. Brinkman, H.H. de Rooij-Dijk, et al., The tetracycline derivative minocycline differentially affects cytokine production by monocytes and T lymphocytes, Antimicrob Agents Chemother 40 (1996) 934—940.

[65] J.R. Batchelor, K.I. Welsh, R.M. Tinoco, et al., Hydralazine-induced systemic lupus erythematosus: Influence of HLA-DR and sex on susceptibility, Lancet 1 (1980) 1107—1109.

[66] G.I. Russell, R.F. Bing, J.A. Jones, H. Thurston, J.D. Swales, Hydralazine sensitivity: Clinical features, autoantibody changes and HLA-DR phenotype, QJ Med 65 (1987) 845—852.

[67] C. Brand, A. Davidson, G. Littlejohn, P. Ryan, Hydralazine-induced lupus: No association with HLA-DR4, Lancet 1 (1984) 462—463.

[68] C. Speirs, A.H. Fielder, U. Chapel, N.J. Davey, J.R. Batchelor, Complement system protein C4 and susceptibility to hydralazine-induced systemic lupus erythematosus, Lancet 1 (1989) 922—924.

[69] E. Sim, A.W. Dodds, A. Goldin, Inhibition of the covalent binding reaction of complement component C4 by penicillamine, an anti-rheumatic agent, Biochem J 259 (1989) 415—419.

[70] E. Sim, E.W. Gill, R.B. Sim, Drugs that induce systemic lupus erythematosus inhibit complement component C4, Lancet 2 (1984) 422—424.

[71] J.A. Mitchell, J.R. Batchelor, H. Chapel, et al., Erythrocyte complement receptor type I (CRI) expression and circulating immune complex (CIC) levels in hydralazine-induced SLE, Clin Exp Immunol 68 (1987) 446—456.

[72] T. Whiteside, L. Mulhern, R. Buckingham, J. Luksick, Procainamide-induced lupus (PIL) is associated with an increased frequency of HLA-DR 6Y (abstract), Arthritis Rheum 25 (1982) S41.

[73] M.C. Totoritis, E.M. Tan, E.M. McNally, R.L. Rubin, Association of antibody to histone complex H2A-H2B with symptomatic

procainamide-induced lupus, N Engl J Med 318 (1988) 1431–1436.

[74] R.T. Canoso, M.E. Lewis, E.J. Yunis, Association of HLA-Bw44 with chlorpromazine-induced autoantibodies, Clin Immunol Immunopathol 25 (1982) 278–282.

[75] G. Vargas-Alarcón, J.K. Yamamoto-Furusho, J. Zuñiga, R. Canoso, J. ans Granados, HLA-DR7 in Association with Chlorpromazine-induced Lupus Anticoagulant, J Autoimmun 10 (1997) 579–583.

[76] G.L. Chin, N.C. Kong, B.C. Lee, I.M. Rose, Penicillamine induced lupus-like syndrome in a patient with classical rheumatoid arthritis (letter), J Rheumatol 18 (1991) 947–948.

[77] A. Chalmers, D. Thompson, H.E. Stein, G. Reid, A.C. Patterson, Systemic lupus erythematosus during penicillamine therapy for rheumatoid arthritis, Ann Intern Med 97 (1982) 659–663.

[78] H.M. Perry Jr., E.M. Tan, S. Cordody, A. Sahamato, Relationship of acetyl transferase activity to antinuclear antibodies and toxic symptoms in hypertensive patients treated with hydralazine, J Lab Clin Med 76 (1970) 114–125.

[79] I. Strandberg, G. Boman, L. Hassler, F. Sjoqvist, Acetylator phenotype in patients with hydralazine-induced lupoid syndrome, Acta Med Scand 200 (1976) 367–371.

[80] R.L. Woosley, D.E. Drayer, M.M. Reidenberg, et al., Effect of acetylator phenotype on the rate at which procainamide induces antinuclear antibodies and the lupus syndrome, N Engl J Med 298 (1978) 1157–1159.

[81] H.M. Perry Jr., E.M. Tan, S. Cordody, A. Sahamato, Relationship of acetyl transferase activity to antinuclear antibodies and toxic symptoms in hypertensive patients treated with hydralazine, J Lab Clin Med 76 (1970) 114–125.

[82] C. Sonnhag, E. Karlsson, J. Hed, Procainamide-induced lupus erythematosus-like syndrome in relation to acetylator phenotype and plasma levels of procainamide, Acta Med Scand 206 (1979) 245–251.

[83] S.L. Lee, I. Rivero, M. Siegel, Activation of systemic lupus erythematosus: A critical review, Semin. Arthritis Rheum 5 (1975) 83–103.

[84] D. Alarcon-Segovia, E. Fishbein, H. Alcala, Isoniazid acetylation rate and development of antinuclear antibodies upon isoniazid treatment, Arthritis Rheum 14 (1971) 748–752.

[85] D.A. Evans, M.F. Bullen, J. Houston, C.A. Hopkins, J.M. Vetters, Antinuclear factor in rapid and slow acetylator patients treated with isoniazid, J Med Genet 9 (1972) 53–56.

[86] N.A.M. Reidenberg, J.H. Martin, The acetylator phenotype of patients with systemic lupus erythematosus, Drug Metab Dispos 2 (1974) 71–73.

[87] M.M. Reidenberg, D.E. Drayer, B. Lorenzo, et al., Acetylation phenotypes and environmental chemical exposure of people with idiopathic systemic lupus erythematosus, Arthritis Rheum 36 (1993) 971–973.

[88] M.L. Ong, T.G. Mant, K. Veerapen, et al., The lack of relationship between acetylator phenotype and idiopathic systemic lupus erythematosus in a Southeast Asian population: A study of Indians, Malays and Malaysian Chinese, Br J Rheumatol 29 (1990) 462–464.

[89] R.L. Rubin, J.T. Curnutte, Metabolism of procainamide to the cytotoxic hydroxylamine by neutrophils activated, in vitro J Clin Invest 83 (1989) 1336–1343.

[90] R.L. Rubin, Autoantibody specificity in drug-induced lupus and neutrophil-mediated metabolism of lupus-inducing drugs, Clin Biochem 25 (1992) 223–234.

[91] X. Jiang, G. Khursigara, R.L. Rubin, Transformation of lupus-inducing drugs to cytotoxic products by activated neutrophils, Science 266 (1994) 810–813.

[92] L. Nassberger, H. Jonsson, A.G. Sjoholm, G. Stuefelt, A. Akesson, Autoantibodies against neutrophil cytoplasmic components in systemic lupus erythematosus and in hydralazine-induced lupus, Clin Exp Immunol 81 (1990) 380–383.

[93] G. Cambridge, U. Wallace, R.M. Bernstein, B. Leaker, Autoantibodies to myeloperoxidase in idiopathic and drug-induced systemic lupus erythematosus and vasculitis, Br J Rheumatol 33 (1994) 109–114.

[94] A. Kretz-Rommel, R.L. Rubin, A metabolite of the lupus-inducing drug procainamide prevents anergy induction in T cell clones, J Immunol 158 (1997) 4465–4470.

[95] Y. Kanai, Y. Kawaminami, M. Miwa, T. Matsushima, T. Sugimura, Naturally-occurring antibodies to poly(ADP-ribose) in patients with systemic lupus erythematosus, Nature 265 (1977) 175–177.

[96] Y. Kanai, T. Sugirnura, Comparative studies on antibodies to poly(ADP-ribose) in rabbits and patients with systemic lupus erythematosus, Immunology 43 (1981) 101–110.

[97] R.N. Hobbs, A.L. Clayton, R.M. Bernstein, Antibodies to the five histones and poly (adenosine diphosphate-ribose) in drug induced lupus: Implications for pathogenesis, Ann Rheum Dis 46 (1987) 408–416.

[98] N.A. Berger, G.W. Sikorski, S.J. Petzold, K.K. Kurohara, Association of poly(adenosine diphosphoribose) synthesis with DNA damage and repair in normal human lymphocytes, J Clin Invest 63 (1979) 1164–1171.

[99] L.M. Ayer, S.M. Edworthy, M.J. Fritzler, Effects of procainamide and hydralazine on poly (ADP-ribosylation) in cell lines, Lupus 2 (1993) 167–172.

[100] J.T. Attwood, R.L. Yung, B.C. Richardson, DNA methylation and the regulation of gene transcription, Cell Mol Life Sci 59 (2002) 241–257.

[101] S.M. Taylor, P.A. Jones, Multiple new phenotypes induced in 10T1/2 and 3T3 cells treated with 5-azacytidine, Cell 17 (1979) 771–779.

[102] B. Richardson, L. Kahn, E.J. Lovett, J. Hudson, Effect of an inhibitor of DNA methylation on T cells. I. 5-Azacytidine induces T4 expression on T8+ T cells, J Immunol 137 (1986) 35–39.

[103] B. Richardson, Effect of an inhibitor of DNA methylation on T cells. II. 5-Azacytidine induces self-reactivity in antigen-specific T4+ cells, Hum Immunol 17 (1986) 456–470.

[104] J. Quddus, K.J. Johnson, J. Gavalchin, et al., Treating activated CD4+ T cells with either of two distinct DNA methyltransferase inhibitors, 5-Azacytidine or procainamide, is sufficient to cause a lupus-like disease in syngeneic mice, J Clin Invest 92 (1993) 38–53.

[105] R.L. Yung, J. Quddus, C.E. Chrisp, K.J. Johnson, B.C. Richardson, Mechanisms of drug induced lupus. I. Cloned Th2 cells modified with DNA methylation inhibitors in vitro cause autoimmunity in vivo, J Immunol 154 (1995) 3025–3035.

[106] B.C. Richardson, D. Powers, F. Hooper, R.L. Yung, K. O'Rourke, Lymphocyte function-associated antigen I overexpression and T cell autoreactivity, Arthritis Rheum 37 (1994) 1363–1372.

[107] R.L. Yung, D. Powers, K. Johnson, et al., Mechanisms of drug-induced lupus. II. T cells overex-pressing LFA-1 cause a lupus-like disease in syngeneic mice, J Clin Invest 97 (1996) 2866–2871.

[108] A.G. Rolink, E. Gleichmann, Allosuppressor- and allohelper-T cells in acute and chronic graft-vs-host (GvH) disease. 111. Different Lyt subsets of donor T cells induce different pathological syndromes, J Exp Med 158 (1983) 546–558.

[109] L.S. Scheinbart, M.A. Johnson, L.A. Gross, S.R. Edelstein, B.C. Richardson, Procainamide inhibits DNA methyltransferase in a human T cell line, J Rheumatol 18 (1991) 530—534.

[110] E. Cornacchia, J. Golbus, J. Maybaum, et al., Hydralazine and procainamide inhibit T cell DNA methylation and induce autoreactivity, J Immunol 140 (1988) 2197—2200.

[111] R. Yung, S. Chang, N. Hemati, K. Johnson, B. Richardson, Mechanisms of drug-induced lupus IV. Comparison of procainamide and hydralazine with analogs in vitro and in vivo, Arthritis Rheum 40 (1997) 1436—1443.

[112] B.C. Richardson, J.T. Attwood, D. Ray, D.K. Richardson, C. Deng, Identification of methylation sensitive T cell genes capable of participating in autoimmunity, Arthritis Rheum 44 (2001) S201.

[113] G.M. Kammer, A. Perl, B.C. Richardson, G.C. Tsokos, Abnormal T cell signal transduction in sytemic lupus erythematosus, Arthritis Rheum 46 (2002) 1139—1154.

[114] C. Deng, M.J. kaplan, J. Yang, D. Ray, Z. Zhang, W.J. McCune, et al., Decreased Ras-mitogen-activated protein kinase signaling may cause DNA hypomethylation in T lymphocytes from lupus patients, Arthritis Rheum 44 (2001) 397—407.

[115] C. Deng, Z.Y. Zhang, T. Rao, J. Attwood, B.C. Richardson, Hydralazine inhibits ERK pathway signaling in human T cells, Arthritis Rheum 48 (2003) 746—756.

[116] C. Deng, J. Yang, J. Scott, S. Hanash, B.C. Richardson, Role of the ras-MAPK signaling pathway in the DNA methyltransferase response to DNA hypomethylation, Biochemistry 379 (1998) 1113—1120.

[117] G. Gorelik, J. Fang, A. Wu, B. Richardson, Impaired T cell PKCδ activation decreases ERK pathway signaling in idiopathic and hydralazine-induced lupus, J Immunol 179 (2007) 5553—5563.

[118] L. Mazari, M. Ouarzane, M. Zouali, Subversion of B lymphocyte tolerance by hydralazine, a potential mechanism for drug-induced lupus, PNAS 104 (2007) 6317—6322.

[119] B.C. Richardson, M.R. Liebling, J.L. Hudson, CD4+ cells treated with DNA methylation inhibitors induce autologous B cell differentiation, Clin Immunol Immunopathol 55 (1990) 368—381.

[120] M.J. Kaplan, D. Ray, R.R. Mo, R.L. Yung, B.C. Richardson, TRAIL (APO2 ligand) and TWEAK (Apo3 ligand) mediate CD4+ T cell killing of antigen-presenting macrophages, J Immunol 164 (2000) 2897—2904.

[121] B.C. Richardson, J.R. Strahler, S. Pivirotto, et al., Phenotypic and functional similarities between 5-azacytidine-treated cells and a T cell subset in patients with active systemic lupus erythematosus, Arthritis Rheum 35 (1992) 647—662.

[122] A. Corvetta, R. Della Bitta, M.M. Luchetti, G. Pomponio, 5-Methylcytosine content of DNA in blood, synovial mononuclear cells and synovial tissue from patients affected by autoimmune rheumatic diseases, J Chromatogr 566 (1991) 481—491.

[123] J. Yang, C. Deng, N. Hemati, S.M. Hanash, B.C. Richardson, Effect of mitogenic stimulation and DNA methylation on human T cell DNA methyl-transferase expression and activity, J Immunol 159 (1997) 1303—1309.

[124] T. Takeuchi, K. Amano, H. Sekine, J. Koide, T. Abe, Upregulated expression and function of integrin adhesive receptors in systemic lupus erythematosus patients with vasculitis, J Clin Invest 92 (1993) 3008—3016.

[125] B.C. Richardson, R.L. Yung, K.J. Johnson, P.E. Rowse, N.D. Lalwani, Monocyte apoptosis in patients with active lupus, Arthritis Rheum 39 (1996) 1432—1434.

[126] M. Linker-Israeli, R.J. Deans, D.J. Wallace, et al., Elevated levels of endogenous IL-6 in systemic lupus erythematosus. A putative role in pathogenesis, J Immunol 147 (1991) 117—123.

[127] P. Comens, Experimental hydralazine disease and its similarity to disseminated lupus erythematosus, J Lab Clin Med 47 (1956) 444—454.

[128] E.L. Dubois, Y.J. Katz, V. Freeman, F. Garbak, Chronic toxicity studies of apresoline in dogs with particular reference to the production of the "hydralazine syndrome", J Lab Clin Med 50 (1957) 119—126.

[129] A.E. Carrera, M.V. Reid, N.B. Kurnick, Differences in susceptibility of polymorphonuclear leucocytes from several species to alteration by S.L.E. serum: Application to a more sensitive L.E. phenomenon, Test Blood 9 (1954) 1165—1171.

[130] I.M. Braverman, A.B. Lerner, Hydralazine disease in the guinea-pig as an experimental model for lupus erythematosus, J Invest Dermatol 39 (1967) 317—327.

[131] A. Leovey, G. Szegedi, J. Bobory, I. Devenyi, Experimental "hydralazine erythematodes" of the guinea-pig, Acta Rheum Scand 13 (1967) 119—136.

[132] L. Ellman, J. Inman, I. Green, Strain difference in the immune response to hydralazine in inbred guinea-pigs, Clin Exp Immunol 9 (1971) 927—937.

[133] A. Cannat, M. Seligmann, Induction by isoniazid and hydralazine of antinuclear factors in mice, Clin Exp Immunol 3 (1968) 99—105.

[134] J.H. Ten Veen, T.E.W. Feltkamp, Studies on drug induced lupus erythematosus in mice. I. Drug induced antinuclear antibodies (ANA), Clin Exp Immunol 11 (1972) 265—276.

[135] R. Thoburn, D. Koffler, U.G. Kunkel, Distribution of antibodies to native DNA, single-stranded DNA, and double-stranded RNA in mouse serums, Proc Soc Exp Biol Med 136 (1971) 711—714.

[136] J.H. Ten Veen, T.M. Feltkamp-Vroom, Studies on drug induced lupus erythematosus in mice. 111. Renal lesions and splenomegaly in drug-induced lupus erythematosus, Clin Exp Immunol 15 (1973) 591—600.

[137] Y. Yamauchi, A. Litwin, L. Adams, H. Zimmer, E.V. Hess, Induction of antibodies to nuclear antigens in rabbits by immunization with hydralazine-human serum albumin conjugates, J Clin Invest 56 (1975) 958—969.

[138] H. Mollerberg, Attempts to produce the "hydralazine syndrome" in the albino rat, Acta Med Scand 161 (1958) 443—445.

[139] F.W. McCoy, W.J. Leach, Experimental attempt to produce the L.E. syndrome (arthritis) in swine with hydralazine, Proc Soc Exp Biol Med 101 (1959) 183.

[140] A. Kretz-Rommel, S.R. Duncan, R.L. Rubin, Autoimmunity caused by disruption of central T cell tolerance. A murine model of drug-induced lupus, J Clin Invest 99 (1997) 1888—1896.

[141] R. Yung, M. Kaplan, D. Ray, K. Schneider, R.R. Mo, K. Johnson, et al., Autoreactive murine Th1 and Th2 cells kill syngeneic macrophages and induce autoantibodies, Lupus 10 (2001) 539—546.

[142] R. Yung, R. Williams, K. Johnson, et al., Mechanisms of drug-induced lupus III. Sex-specific differences in T cell homing may explain increased disease severity in female mice, Arthritis Rheum 140 (1997) 1334—1343.

[143] C. Cid, H.K. Kleinman, D.S. Grant, et al., Estradiol enhances leukocyte binding to tumor necrosis factor (TNF)-stimulated endothelial cells via an increase in TNF-induced adhesion molecules E-selectin, intercellular adhesion molecule type I and vascular cell adhesion molecule type I, J Clin Invest 93 (1994) 17—25.

[144] C. Deng, J. Yang, J. Scott, S. Hanash, B.C. Richardson, Role of the ras-MAPK signaling pathway in the DNA methyltransferase response to DNA hypomethylation, Biol Chem 379 (1998) 1113—1120.

I. BASIS OF DISEASE PATHOGENESIS

[145] A. Sawalha, M. Jeffries, R. Webb, Q. Lu, G. Gorelik, D. Ray, et al., Defective T-cell ERK signaling induces interferon-regulated gene expression and overexpression of methylation sensitive genes, Genes Immunity 9 (2008) 368—378.

[146] S. Yoshida, M.E. Gershwin, Autoimmunity and selected environmental factors of disease induction, Semin Arthritis Rheum 22 (1993) 399—419.

[147] M. Kaplan, B.C. Richardson, Mechanisms of autoimunity in environmentally-induced connective tissue diseases. In "Rheumatic Diseases and the Environment", in: J. Varga, L. Kauffman (Eds.), McGraw-Hill, New York, 1999, pp. 19—31.

[148] J.L. Roberts, J.A. Hayashi, Exacerbation of SLE associated with alfalfa ingestion, N Engl J Med 380 (1983) 1361.

[149] M.R. Malinow, E.J. Bardana Jr., S.H. Goodnight Jr., Pancytopenia during ingestion of alfalfa seeds, Lancet 1 (1981) 615.

[150] E.J. Bardana, M.R. Malinow, D.C. Houghton, et al., Diet-induced systemic lupus erythematosus (SLE) in primates, Am J Kidney Dis 1 (1982) 345—352.

[151] M.R. Malinow, E.J. Bardana Jr., B. Pirofsky, S. Craig, P McLaughlin, Systemic lupus erythematosus like syndrome in monkeys fed alfalfa sprouts: Role of nonprotein amino acid, Science 216 (1982) 415—417.

[152] J. Alcocer-Varela, A. Iglesias, L. Llorente, D. Alarcon-Segovia, Effects of l-canavanine on T cells may explain the induction of systemic lupus erythematosus by alfalfa, Arthritis Rheum 28 (1985) 52—57.

[153] P.E. Prete, Effects of l-canavanine on immune function in normal and autoimmune mice: Disordered B cell function by a dietary amino acid in the irnmuno-regulation of autoimmune disease, Can J Physiol Pharmacol 63 (1985) 843—854.

[154] P.E. Prete, Membrane surface properties of lymphocytes of normal DBA/2 and autoimmune NZB/NZW Fl hybrid mice: Effects of l-canavanine and a proposed mechanism for diet-induced autoimmune disease, Can J Physiol Pharmacol 64 (1986) 1189—1196.

[155] R.M. Philen, M. Posada, Toxic oil syndrome and eosinophilia-myalgia syndrome. 8—10 May 1991, World Health Organization meeting report, Semin Arthritis Rheum 23 (1993) 104—124.

[156] J. Varga, S.A. Jimenez, J. Uitto, l-Tryptophan and the eosinophilia-myalgia syndrome: Current understanding of the etiology and pathogenesis, J Invest Dermatol 100 (1993) 97S—105S.

[157] S.A. Bell, M.V. Hobbs, R.L. Rubin, Isotype-restricted hyperimmunity in a murine model of the toxic oil syndrome, J Immunol 148 (1992) 3369—3376.

[158] G. Tena, Fatty acid anilides and the toxic oil syndrome, Lancet 1 (1982) 98—99.

[159] W.N. Aldridge, T.A. Connors, Toxic oil syndrome in Spain, Food Chem Toxicol 20 (1982) 989—992.

[160] C. Casals, P. Garcia-Barreno, A.M. Municio, Lipogenesis in liver, lung and adipose tissue of rats fed with oleoylanilide, Biochem J 212 (1983) 339—344.

[161] A. Suarez, M.D. Viloria, P. Garcia-Barreno, A.M. Municio, Toxic oil syndrome, Spain: Effect of oleoylanilide on the release of polysaturated fatty acids and lipid peroxidation in rats, Arch Environ Contam Toxicol 14 (1985) 131—139.

[162] W.N. Aldridge, Experimental studies, in: "Toxic Oil Syndrome Current Knowledge and Future Perspectives", WHO Regional Publications, 1992. European Series, No. 42, pp. 67—97.

[163] M.E. Kammuller, N. Bloksma, W. Seinen, Chemical-induced autoimmune reactions and Spanish toxic oil syndrome. Focus on hydantoins and related compounds, J Toxicol Clin Toxicol 26 (1988) 157—174.

[164] M.E. Kammuller, A.H. Penninks, W. Seinen, Spanish toxic oil syndrome and chemically induced graft-versus-host-like reactions (letter), Lancet 2 (1984) 805—806.

[165] H. Gleichmann, E. Gleichmann, GVHD, a model for Spanish toxic oil syndrome? Lancet 1 (1984) 1474.

[166] M.E. Kammuller, H.J. Verhaar, C. Versluis, et al., 1-Phenyl-5-vinyl-2-imidazolidinethione, a proposed causative agent of Spanish toxic oil syndrome: Synthesis, and identification in one of a group of case-associated oil samples, Food Chem Toxicol 26 (1988) 119—127.

[167] H.J. Verhaar, M.E. Kammuller, J.K. Terlouw, L. Brandsma, W. Seinen, Spanish toxic oil syndrome: An isothiocyanate-derived compound cannot be substantiated as a causative agent (letter), Food Chem Toxicol 27 (1989) 205—206.

[168] T. Mizutani, H. Mizutani, K. Hashimoto, et al., Simultaneous development of two cases of eosinophilia-myalgia syndrome with the same lot of l-tryptophan in Japan, J Am Acad Dermatol 25 (1991) 512—517.

[169] R.M. Silver, S.E. Sutherland, P. Carreira, M.P. Heyes, Alterations in tryptophan metabolism in the toxic oil syndrome and in the eosinophilia-myalgia syndrome, J Rheumatol 19 (1992) 69—73.

[170] L.A. Love, J.I. Rader, L.J. Crofford, et al., Pathological and immunological effects of ingesting l-tryptophan and 1,1'-ethylidenebis (l-tryptophan) in Lewis rats, J Clin Invest 91 (1993) 804—811.

[171] A.N. Mayeno, F. Lin, C.S. Foote, et al., Characterization of "peak E", a novel amino acid associated with eosinophilia-myalgia syndrome, Science 250 (1990) 1707—1708.

[172] J. Sanchez-Roman, I. Wichmann, J. Salaberri, J.M. Varela, A. Nunez-Roldan, Multiple clinical and biological autoimmune manifestations in 50 workers after occupational exposure to silica, Ann Rheum Dis 52 (1993) 534—538.

[173] A.C. Allison, Fibrogenic and other biological effects of silica, Curr Top Microbiol Immunol 210 (1996) 147—158.

[174] R.K. Scheule, A. Holian, Immunologic aspects of pneumoconiosis, Exp Lung Res 17 (1991) 661—685.

[175] H.R. Smith, Do silicone breast implants cause autoimmune rheumatic diseases? J Biomater Sci Polym Ed 7 (1995) 115—121.

[176] T.J. Laing, D. Schottenfeld, J.V. Lacey Jr., B.W. Gillespie, D.H. Garabrant, B.C. Cooper, et al., Potential risk factors for undifferentiated connective tissue disease among women: Implanted medical devices, Am J Epidemiol 154 (2001) 610—617.

[177] N. Kossovsky, C.J. Freiman, Silicon breast implant pathology. Clinical data and immunologic consequences, Arch Pathol Lab Med 118 (1994) 686—693.

[178] J.O. Naim, C.J. van Oss, Silicone gels as adjuvants. Effects on humoral and cell-mediated immune responses, Adv Exp Med Biol 383 (1995) 1—6.

[179] J. Sanchez-Guerrero, G.A. Colditz, E.W. Karlson, et al., Silicone breast implants and the risk of connective-tissue diseases and symptoms, N Engl J Med 332 (1995) 1666—1670.

[180] E.C. Janowsky, L.L. Kupper, B.S. Hulka, Meta-analysis of the relationship between silicone breast implants and the risk of connective-tissue disease, N Engl J Med 342 (2000) 781—790.

[181] P. Tugwell, G. Wells, J. Peterson, V. Welch, J. Page, C. Davison, et al., Do silicon breast implants cause rheumatic disorders? A systemic review for a court-appointed national science panel, Arthritis Rheum 44 (2001) 2477—2484.

[182] Woolsey, et al., N Engl J Med 298 (1987) 1157.

[183] R.L. Rubin, Role of xenobiotic oxidative metabolism, Lupus 3 (1994) 479—482.

23

Gender and Age in Lupus

Robert G. Lahita

LATE-ONSET SYSTEMIC LUPUS ERYTHEMATOSUS

Introduction

The onset of systemic lupus erythematosus (SLE) and its clinical course are affected by age in many ways. The relationship between age and the morbidity and mortality of SLE is important and the incidence of disease varies in males and females depending on age. A discussion of SLE in the neonatal and prepubertal child can be found in Chapters 31 and 32, but this chapter limits the discussion of the older patient to people over age 50.

The ratio of males to females is 1:8 in the older patient instead of 1:15 [1, 2]. Patients with drug-induced lupus are distinctly older and are usually male [3, 4] because commonly used drugs such as procainamide and hydralazine are widely used for conditions in people over age 50 and there are more male patients taking drugs such as procainamide. The topic of drug-induced SLE is also covered extensively in Chapter 22.

SLE has no predilection for males who are older; however, in younger male patients who are less than 50 years old, the number of men with SLE is greater. About 25% of the males in one study were over age 50 years [5]. Most investigators report that the ratios of females to males with SLE after age 50 decreases from 10–15:1 to 8:1. The older male presentation of SLE was atypical [6] in one study and in the largest series of SLE patients studied who were 50 years of age or older, 79% were women [7]. There are more older men than older women getting lupus for the first time and the reasons for this may relate to a variety of factors such as sexual senescence, environmental agents, and possibly an increase of prolactin [8].

Clinical Presentation of the Aged Patient with SLE

In almost all studies of aged people with SLE, the majority (>90%) met four or more of the revised American College of Rheumatology (ACR) criteria for SLE. In another study, 81% of older patients met four or more criteria, whereas 19% met three or fewer [9]. Because of late age of onset, the initial diagnosis of SLE was incorrect in 55% of patients. This is not surprising as physicians are reluctant to diagnose SLE and because this age group presents with atypical signs and symptoms. Therefore, the diagnosis of older-age SLE can be delayed for up to 2 years. While there is no question that age may modify the clinical presentation of the disease, it is imperative that physicians learn to include SLE as part of any differential diagnosis in patients of older age presenting with myalgia, arthralgias, unexplained weight loss, fatigue, and low-grade fever.

The presentation of SLE in the older patient (aged 50–80 years) is different. The older SLE patient presents clinically more like a patient with drug-induced SLE than one with idiopathic disease [10]. Older patients have a low to absent incidence of renal disease and primarily pleuropericardial and musculoskeletal symptoms [11]; moreover, they rarely have low complement levels and native DNA antibodies, whereas anti-single-stranded (ss) DNA antibodies are common [12]. Meta analysis indicates that Serositis, interstitial pulmonary disease, anti-La antibodies, and Sjögren's syndrome symptoms (secondary sicca) are commonly associated with the disease of the older age group [13]. Alopecia, Raynaud's phenomenon, fever, lymphadenopathy, hypocomplementemia, and neuropsychiatric disease are not common in this age group. As with all aspects of medicine, generalizations cannot be made. Some older patients do have native DNA antibody and clinical signs of renal disease [14, 15] but they are the exception.

The data suggest that age works in favor of the female gender. In a retrospective study of 102 patients with late-onset SLE, there was loss of the female preponderance, a more insidious onset of presentation, fewer skin manifestations, and more serositis. Table 23.1 shows 13 patients from a study compared to a meta analysis of other studies. No differences in renal, neuropsychiatric, cardiac, pulmonary, or liver involvement were found when the older group was compared to the new [16].

TABLE 23.1   Clinical and Laboratory Features in 13 Late-Onset Patients with SLE Compared to a Meta-analysis of Other Series[a]

	13 Patients, present study (OR)	170 Patients (pooled OR)
CLINICAL		
Fever	0.51	0.21 ↓
Butterfly rash	0.23 ↓	
Discoid rash	0.86	
Photosensitivity	0.83 ↓	1.10
Mouth ulcer	0.85	0.74
Alopecia	0.21 ↓	0.47 ↓
Raynaud's syndrome	0.38	0.48 ↓
Arthritis	0.41	
Pleural	7.32 ↑	
Pericarditis	3.51	
Serositis	5.32 ↑	1.63 ↑
Lung involvement	7.33	3.48 ↑
Cardiac involvement	0.87	
Nephritis	1.71	0.79
CNS involvement	0.71	0.52 ↓
Myositis	1.84	0.68
Lymphadenopathy	0.54	0.39 ↓
LABORATORY		
Leukopenia	0.87	0.67
Thrombocytopenia	1.17	
Anti-dsDNA	0.56	0.76
Anti-Ro antibody	0.66	
Anti-La antibody	3.44	4.24 ↑
Hypocomplementemia	1.18	0.54 ↓

[a] ↑/↓: Increase/decrease incidence of feature compared to patients with early-onset disease.
Reproduced with permission from The Jouranl of Rheumatology, Mak et al. [234].

Race and Age

The ethnic group with the patients whose SLE occurs at an older age are Caucasians [17]. In one study [12], 59% of the younger SLE group were Black patients, whereas only 20% of Black patients were in the older age group (>50 years). In still another study, 87% of the patients in the older group were Caucasian and about 6.5% were Black [18, 19]. The reasons for this unusual racial predilection for older-age SLE patients are unknown and further investigation is needed, but it may be related to a greater severity of the disease in Blacks. In fact, data from the National Institutes of Health suggest that the overall mortality of lupus in the young Black female is five times greater than that found in Caucasians. It is clear that age and race influence both morbidity and mortality. An older study [20, 21] of survival in SLE in which follow-up was 75.6 months revealed that race and socioeconomic status had significant effects on survival. While no diminished lupus-related mortality was associated with age, morbidity in women decreases with age. The authors suggested that race distribution between young and old patients with SLE might confound the results with regard to gender differences.

Course and Therapy in the Older Patient

Older SLE patients had a mean age of 59.5 years at disease onset in one study (range, 50–81 years). In that study the mean age at diagnosis was 62.4 years (range 52–83 years). This means that there was about a 3-year lapse in diagnosis from time of onset. These patients were maintained on lower doses of corticosteroid, and conservative therapy was the rule. Only 42% of these patients required high-dose corticosteroid (>25 mg/day × 1 month), whereas 23% could be maintained on 1–25 mg per day, and 35% never needed steroids at all [9]. Overall, patients with late-onset disease require less medication and lower doses of immunosuppressive agents [22].

Survival in the older age group is good; the 5-year survival rate was 92.3 ± 5.3%, and 83.1 ± 10% had survived 9 years at follow-up. One conclusion might be that older patients rarely die of their disease or the complications of therapy when they are treated conservatively. However, at least one older study suggested that the major causes of death in the older age groups were related to therapy [23]. Unlike younger patients, where the major mortality was atherosclerosis, infections and perforated peptic ulcers remained the most common cause of death in older patients [24]. These observations are probably relevant to today's patients as well.

10 to 20% of cases (ages 50–65 years) occur in older patients. Arthritis, fever, serositis, sicca symptoms,

Raynaud's syndrome, lung disease, and neuropsychiatric symptoms are more common in patients with elderly onset SLE. Moreover, the diagnosis of SLE in this population may be delayed longer than that of younger patients [17].

In older-onset SLE patients, arthritis, malar rash, and nephropathy are less frequently found. In a more recent study, high-dose corticosteroids and immunosuppressive agents are used with less frequency in older patients [25]. However, deaths are more common in this age group with a 10-year survival rate of 71% vs. 95% in the younger age group ($p < 0.01$). But this reduced survival was due to aging and its attendant morbidity, not disease severity.

Comments on the Elderly SLE Patient

There are many factors that hypothetically could predispose the older patient to SLE, including liver and thyroid disease or exposure to certain exogenous agents, such as hormones [26]. However, no studies conclusively implicate these factors at this time. In the aging immune system, cellular phenomena such as programmed cell death may vary, and Fas and Fas ligand (FasL), which are members of the TNF family directly involved in apoptosis, could underlie autoimmune mechanisms. Data suggest that apoptosis resistance is modulated during aging [27] and may be at the root of late-onset disease.

Other aging factors are based on immune senescence. Two phenomena exist in senescence: a decrease of immune function and a decrease of autoantibody production [28]. One reason why the immune system becomes hyperactive at a time when a human ages is that there is a decline of immune surveillance. The appearance of autoantibodies in older people without disease has always suggested that immune tolerance might decline with age. For example, elderly people do not handle infections as effectively as younger people and recovery generally takes longer. Diseases that rely on immune function such as Alzheimer's, type II diabetes, and atherosclerosis are linked to immune function. Changes in expression of the T-cell receptor and its co-receptor are affected by immunosenescence [29].

GENDER AND SLE

The data suggest that the most important contributor to autoimmune disease is gender. Gender and more specifically sex steroids play an important role in the maturation of organ systems that affect animals throughout life and are therefore important to the health of most vertebrates. Even though the most obvious effects occur at puberty in the development of secondary sexual characteristics, hormones affect major developmental changes even before parturition. Diverse biologic functions including behavior, intelligence, sexual preference, physical stature, and the immune system, are likely targets for these hormones.

The effects of sex steroids on the immune system are profound and long-lasting because they control the growth and maturation of various cell systems and may also influence susceptibility to disease through modulation of immune cell populations, alteration of cytokine levels, control of cell populations through processes such as apoptosis, and change of very basic molecular mechanisms. Lupus is one disease that is highly affected by hormones.

Historical Background: Murine Models of Autoimmunity and Sex Hormones

In the history of SLE, animal models have been and continue to be background for the study of gender effects on experimental SLE. Several strains of mice are used as models of autoimmune disease [30]. Like human illness, disease in some murine models predominates in females. Examples are the *New Zealand white/black (NZWZNZB) F1* hybrid and the MRL *lpr* strains in which females die before males [31, 32] of disease. In one specific strain of mice, the disease favors males, the *BXSB* (derived from the *C57Bl/6* and satin beige mouse), and this strain is unusual in that the males die of early SLE-like disease (33). Unlike the female predisposed strains of mice, disease manifestations in the *BXSB* mouse are not mediated hormonally because neither administration of sex steroids nor removal of the gonads improves or changes the disease. One strain of mice bred because the mouse develops polyarteritis nodosa, called the *Palmerston North* (PN) mouse [34, 35], also develops an SLE-like illness that favors females.

In the NZB/W mouse, circulating DNA-containing immune complexes and a variety of autoantibodies like those to native DNA and ssDNA [36–38] also accompanied the clinical disease as in the human. Aberrant B-cell function in NZB mice is reflected by their resistance to the induction and maintenance of tolerance [39]. Historically, this last characteristic was often mentioned as part of the female character of most of the murine SLE strains. Sex hormones also enhance the response of the NZB strain to thymus-dependent antigens, and high-dose testosterone therapy decreases and orchidectomy increases plaque-forming colonies (PFCs) in NZB males if the mice are irradiated sublethally after treatment with sex hormones [40]. Early estrogen therapy accelerated the disease in females, whereas ovariectomy or testosterone therapy prolongs the life of the mouse. In males of this strain, castration caused early death and estrogen produced a mortality rate similar to that of females [41]. Overall, testosterone therapy of female mice prolonged life and decreased overall morbidity[41].

Early on sex steroids were thought to modulate the generation of antigen-specific suppressor cells and true sex differences in the responses to T- and B-cell mitogens were said to exist in this strain. Androgens increased suppressor cell activity in NZB/NZW mice by increasing levels of interleukin 2 and androgens also suppressed the development of thymocytotoxic antibodies [41–44]. Increased prolactin levels accelerate autoimmune diseases in NZB/W F1 mice.

The MRL (lpr/lpr or n/n) strain of mice has a lupus-like disease that is closer serologically to the human disease for various reasons and it produces a greater variety of antibodies to antigens such as DNA, RNP, SM, and rheumatoid factor [45]. Because of the appearance of rheumatoid factor, the strain was a useful model for the study of rheumatoid arthritis [46]. They are also models for Sjögren's disease because of serologic characteristics and the development of clinical signs such as lymphadenopathy commonly found in the human disease. Again in this strain the female gender predominates (death occurs at 17 months for females and 22 months for males). Hormone therapy of these mice using high doses of the androgens testosterone and 5β-dihydrotestosterone delay the occurrence of both lymphadenopathy and autoantibody production.

Finally, another strain, the BXSB, was unique in that spontaneous autoimmune disease was more common in the males (male life span, 5 months; female life span, 15 months). Both B- and T-cell functions were found to be normal in this strain. The accelerated autoimmune disease in these mice was linked to the Y chromosome, and castration did not alter the course of the disease. Transfer of bone marrow cells from diseased males to nondiseased males led to development of autoimmune disease [47, 48].

New Murine Observations about Hormones and SLE

Estrogen-induced receptor binding is observed in thymic tissues from male NZB/NZW mice, and estrogens may act entirely on the thymus gland or at the level of the alpha or beta estrogen receptors on lymphoid cells. The role of sex steroids in the pathogenesis of lupus-like disease in the inbred mouse is based on simple observations that alteration of sex steroid levels or gonadectomy affects the progression of disease. Importantly, the mouse will succumb to lupus whether male or female, but the sex steroids regulate disease activity and, depending on the steroid, accelerate mortality [49].

The most exciting data about B-cell regulation in SLE comes from the work of the Diamond Laboratory. Estrogen and prolactin have a role in the breakdown of B-cell tolerance in a nonautoimmune mouse model, the Balb/c mouse transgenic for anti-double stranded DNA antibody heavy chain called R4A.

Estrogen has been known for many years to be an inhibitor of B cell lymphopoesis. Pregnancy and other times when plasma estrogen levels are high result in decreased B cell lymphopoesis, whereas low estrogen levels lead to increased B cell lymphopoesis [50–52].

Treatment of the mice with testosterone results in a reduction of B cell [53, 54] lymphopoesis as well. Some of this effect may come from interleukin-7, which increases B cell lymphopoesis and is inhibited by sex steroids [55, 56].

In elegant work with the Balb/c-R4A producing strain of mice, Diamond and her colleagues indicated that pathogenic high-affinity anti-DNA antibody could be induced by treatment with estrogen [57]. In addition, the effects of estradiol in this strain increase the percentage of marginal zone B cells within the whole population [58]. As shown in other studies, estradiol can exert a direct effect on B cells as they express both estrogen receptor (ER) α and ERβ. Estrogen upregulates molecules of the apoptotic pathway (Bcl2) and B-cell receptor signaling pathways (SHP1 and CD22) [59]. There is a 20% increase in expression of this BCR pathway. This process may rescue autoreactive B cells from apoptosis and place them in the marginal zone area.

Prolactin enhances pro-B cell generation in mice. In lupus-prone mice, prolactin causes enhanced mortality and accelerated onset of disease. Treatment with bromocriptine reverses this process and it improves survival [60, 60]. Prolactin used in the Diamond Balb/c model resulted in rescue of B cells that secrete anti-DNA antibody and that were normally destined for deletion as autoreactive. The mechanism was the upregulation of CD40 on B cells and CD40L on T cells [61].

HUMAN DISEASE

Sex Chromosomes and SLE

Chromosomal effects may be important to patients with SLE since immune responses to DNA are linked to the presence of the X chromosome in the NZB/NZW mouse. NZB or NZW males mated to BXSB females result in female offspring that die of SLE nephritis at an early age. In early work, this suggested a role for the recessive X chromosome in the male parent in early studies [62]. Recent examination of Klinefelter males and patients with Turner syndrome and SLE suggest that the risk of SLE could be due to a gene dose effect for the X chromosome. Abnormal activation of genes on the X chromosome as demonstrated for CD 40L or genetic polymorphism as demonstrated by Xq28. Because of this there is a lower risk of SLE in patients with Turner's syndrome (45XO) [63]. A gene that escapes inactivation on the X chromosome will have a higher level of expression in persons with 2Xs [64] and could explain the SLE in the Klinefelter patient.

Sex Steroids and Human SLE

In human SLE, there is little consistency of disease manifestation from patient to patient even within the same family. A patient or sibling from one family may present with autoimmune hemolytic anemia, whereas another will present with encephalopathy. Moreover, siblings from the same family who are of different gender might be equally affected.

Therefore, while gender plays a major role in the expression of SLE, its role in the etiology of SLE is unclear. The vast majority of patients who present with SLE after puberty are females. Present evidence suggests that estrogens or those steroids that are exceptionally feminizing exacerbate SLE [65]. Among such compounds are synthetic estrogens such as birth control agents [66, 67]. There are data to suggest that previous estrogen use is a predisposing factor for lupus [68, 69] and data to show that it is not [67, 70].

The role of the ovary in lupus is important and possible relationships go back many years [71]. In many women with SLE, clinical activity varies with the menstrual period [72] and pregnancy [73], both before and after parturition. These specific times may worsen the activity of SLE and depend on the levels of both sex steroids and possibly gonadotrophins. Because of the large numbers of women with lupus, the X chromosome could play a role as mentioned above [64, 74], but there are no convincing data at this time. A major concern regarding women with SLE is the fact that they are 50 times more likely to have a myocardial infarction than men (35- to 44-age groups). In those who had a cardiovascular event, an older age at diagnosis, longer disease duration, more steroid use, high cholesterol levels, and most importantly postmenopausal status were associated with an increase of cardiac disease [75].

Patients with Klinefelter syndrome (XXY) are described with clinical SLE and disordered estrogen metabolism [64, 76]. Study of such males provoked the initial series of experiments designed to look at the patterns of estrogen and androgen metabolism in SLE. These studies revealed that patients with SLE have altered sex steroid metabolism.

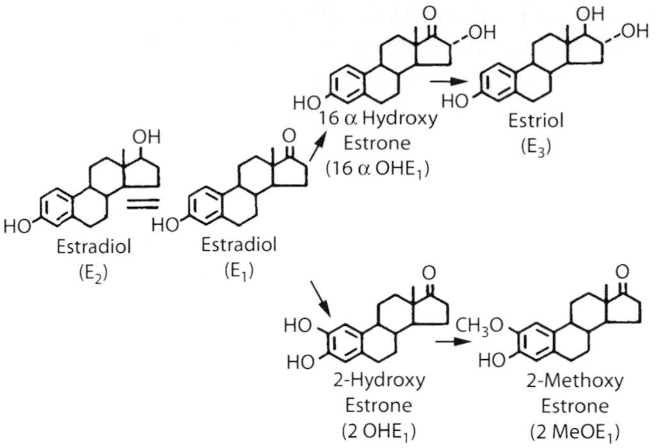

FIGURE 23.1 The metabolism of estradiol. Estradiol is converted to estrone that can go in one of two directions: toward the 16-feminizing metabolites or the catechol estrogens that are the 2-hydroxylated metabolites. In patients with lupus, the preferential direction of estrone metabolism is in the 16-hydroxylated direction.

Estrogen Metabolism

Females are likely to have more aberrations of the immune system than males. Male and female patients with SLE have normal levels of estrogen; however, the overall metabolism of such compounds favors more feminizing compounds such as estriol and 16β-OH estrone [77, 78]. Specifically, the pattern of hydroxylation of estrone (Figure 23.1) favors the 16-hydroxylated compounds over the catechol estrogens [65]. Patients of both sexes have increased levels of 16α-hydroxyestrone (Table 23.2), whereas only SLE females and Klinefelter (XXY) males had elevated levels of estriol. The compound 16α-hydroxyestrone and its elevation in SLE patients could have some special significance, as later studies demonstrated that this hormone had unique qualities [79].

One estrogen, 16α-hydroxyestrone, is feminizing, highly uterotropic, and modestly bound to cytosol receptors and testosterone—estradiol-binding globulin

TABLE 23.2 Rate of Hydroxylation in SLE

| | Number studied | Rate | |
		2-Hydroxylation[a]	16-Hydroxylation[a]
MALES			
Normal	13	21.3 ($\pm$8.7) (7.4—34.2)	7.5 ($\pm$2.1) (3.2—11.1)
SLE	6	21.9 ($\pm$13.4) (8.6—35.0)	12.3 ($\pm$2.7) (8.8—16.8)[b]
FEMALES			
Normal	14	40.4 ($\pm$5.9) (17.7—54.7)	9.5 ($\pm$2.2) (5.9—14.1)[b]
	15	25.6 ($\pm$17.4) (6.0—67.1)	15.7 ($\pm$5.0) (9.0—30.0)

[a]Mean, standard deviation, and range.
[b]$P < 0.01$.

in contradistinction to compounds such as 17β-estradiol [80]. Radioimmunoassay failed to show uniformly elevated levels of 16α-hydroxyestrone in active SLE patients, meaning that this altered metabolic pathway may not be found in all patients [81]. Clinical studies on the steroid 16α-hydroxyestrone showed that it had interesting properties *in vivo* that might explain its possible role in disease, including covalent binding of this steroid to erythrocytes and lymphocytes via a Heyn's rearrangement *in vivo* and the possibility that such covalent binding might occur at the level of the estrogen receptor or the T-cell receptor and result in an alteration of immune function [8283]. Studies of family members of SLE patients indicated that elevated hydroxylation of estradiol was commonly observed in nonaffected first-degree relatives as well as patients [77].

A concerted effort to shift the hydroxylation of estrone toward the two compounds in an attempt to decrease the overall effects of very feminizing estrogens involves the ingestion of cruciferate vegetables, thyroid hormone, and drugs such as naltrexone.

Enhanced binding of 16α-hydroxyestrone to various cell proteins was found in normal women ingesting oral contraceptives. Of particular interest, specific anti-estrogen−protein adduct immunoglobulins were isolated from normal and SLE patients ingesting oral contraceptives, which means that these adducts are common. This finding suggested a common pathway to adduct formation in all women who ingest large amounts of estradiol or for one reason or another have an endogenous source of high estrogen level [79]. Males with SLE were also reported who had hormone protein adduct-specific IgG in their sera. Reversing the 16 hydroxylation of estrone is a goal of modern therapeutics. Using the NZB/W F1 murine model, investigators have shown a role for indole-3-carbinaol, a compound derived from cruciferate vegetables. Investigators were able to show a delay of T-cell maturation and decreased CD4/CD8 T-cell ratio [84].

Androgen Metabolism

There have been several studies of androgen levels in plasma of men and women with SLE and of androgen metabolism in men and women with SLE [85, 86]. Early studies suggested that the reasons for SLE in the male are the result of too little androgen and too much estrogen [87, 88]. However, to date most studies indicate that young men with SLE are hormonally normal and that estrogen:androgen ratios might be minimally elevated (if at all). Furthermore, data from studies of males with SLE do not help explain the large numbers of females who predominate with the disease. Studies of androgen metabolism in SLE females indicate that a difference in the overall metabolism of androgens

can be found (Figure 23.2). The oxidation of testosterone at C17 in SLE females is increased in comparison with males, who have both normal oxidation of testosterone and normal plasma levels of androgenic steroids [89]. Several studies of females with active SLE who never took corticosteroids had decreased plasma levels of androgen [89—91]. This observation was also found in Klinefelter patients, as well as in women with lupus. Low plasma androgens in women with SLE form the basis for androgen replacement therapy in this disease. Clinical studies involving the use of DHEA as a therapy for lupus are a result of this observation.

Prolactin and SLE

There is great interest in the effects of prolactin on immune function in mice and humans [92—97]. There are diverse effects of hyperprolactinemia in both male and female mice, and multiple studies show that prolactin stimulates the appearance of murine disease. Early studies found that the male NZW/NZB mouse autoimmune disease was accelerated and mortality was worsened by prolactin. Murine lupus, as noted earlier, is similar to that found in the human in that prolactin has a role to play in the exacerbation and possibly the initiation of lupus in the human. Estrogens stimulate prolactin and may be the overriding mechanism. Bromocriptine inhibits the development of murine lupus [98—101]. Murine studies even demonstrate that estrogen itself might depend on the stimulation of prolactin for activity [102]. Humans with hyperprolactinemia and SLE related to micro- or macroadenomas are of significant interest and indentified in large cohorts of females or males with SLE.

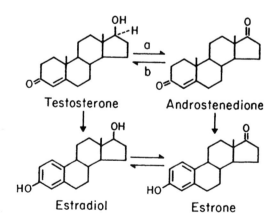

FIGURE 23.2 The oxidation of testosterone. Testosterone has several possible fates in the human. Testosterone can be converted to the weak androgen androstenedione, which can then become estrone or testosterone can be aromatized to estradiol directly. In lupus patients, the oxidation of testosterone at C17 (reaction a) is increased in women with lupus and not in men.

Prolactin is an immunomodulatory pituitary hormone and could be considered a cytokine itself. In the human, prolactin elevations have been observed in SLE juveniles and correlated with both disease activity and central nervous system (CNS) manifestations [103]. This finding is supported by *in vitro* work showing IgG- and IgM-induced anti-DNA antibodies by both normal and SLE lymphocytes in the presence of high levels of prolactin. Lectins did not produce this effect [104, 105]. One study correlates elevated prolactin levels with elevated cortisol levels [106].

Women who are pregnant have higher serum prolactin levels than women who are not pregnant [107]. Investigators have associated the decline in serum testosterone during pregnancy in SLE patients to hyperprolactinemia, but the most significant descriptions of hyperprolactinemia have been in men with SLE [108]. These descriptions of hyperprolactinemia are of particular interest as hyperprolactinemia is treated readily with bromocriptine analogs. Studies using bromocriptine to treat SLE are routine in some clinics around the world.

There are studies that refute the significance of prolactin in human SLE [109, 110]. In a detailed study of a Chinese cohort of patients with SLE, Mok *et al.* [111] found no correlation of clinical activity with levels of prolactin in 72 lupus patients. A look at autoantibodies in patients with lupus found no association of prolactin levels with specific lupus autoantibodies [112]. One particular study suggests that prolactin is complexed with IgG but remains biologically active in the lupus patient [113–115]. This binding of prolactin (a specific 23-kD nonglycosylated form) to IgG is not covalent. Delayed clearance of this complex due to its high molecular weight is the proposed reason for the activity of the prolactin in most patients with lupus.

Many investigators associate prolactin with disease activity [116]. In one study, 61.9% of patients with hyperprolactinemia had active disease. The SLEDAI correlated with the levels of prolactin in these patients as well when prolactin levels were measured by both tests of immunoradioactivity and biological activity. Another series noted an association of hyperprolactinemia in the serum and urine of patients with severe renal disease [117]. The statistical power of most of these prolactin studies could explain the lack of consistency among research laboratories. After review of five studies, two of the studies did not have the statistical power to conclude that there was an association with lupus activity and prolactin levels, whereas three studies did suggest an association [118]. High prolactin levels are found in 20–30% of patients with SLE in some clinics [119]. In addition to the suggestion that prolactin is itself a cytokine that provokes the synthesis of immunoglobulins by lymphocytes [96], German investigators found that using lymphocytes from SLE patients and measuring clinical activity with the European activity measure (ECLAM) that SLE patients have lymphocytes that are sensitive to levels of prolactin and are likely to be activated by physiological concentrations [95]. These data are supported by the findings that T lymphocytes from SLE patients secrete more prolactin than controls, suggesting a difference in the regulation of genes responsible for control of this cytokine. In fact, single nucleotide polymorphisms in the upstream promoter regions of both pituitary and nonpituitary prolactin secretion exist. Such polymorphisms that are specific can affect prolactin transcription and possibly disease association in a cohort of lupus patients. This was the case in a study where patients had an increased frequency of the prolactin-*1149G* allele when compared to control subjects [120]. Whether prolactin is a cytokine or is itself increased by cytokines common in certain active lupus patients remains the subject of major investigation.

Finally, prolactin levels of mothers who were pregnant or breast-feeding and had lupus were studied as reproductive risk factors for the development of disease [121]. Surprisingly, breast-feeding was associated with a decreasing risk of developing SLE. In addition, the numbers of pregnancies or live births with lupus activity showed no relationship to levels of prolactin. These authors found no association of elevated prolactin levels with an increased risk of lupus. Prolactin upregulates antiapoptotic molecules such as BCL2 and downregulates BAX [122]. *In vitro* prolactin stimulates IgM, IgG, antiDNA antibody, and interferon-γ by PBMC of SLE patients [96, 123, 124].

Males with SLE

SLE onset in males is distributed more evenly, as one-fourth are diagnosed after age 50 years [125]. Hormonal metabolic study data suggest that an increase in feminizing 16-hydroxylated estrogenic metabolites is found in SLE males, although there is no phenotypic evidence of hyperestrogenism [126–128]. There is an animal model, the BXSB mouse, which develops SLE-like disease in a nonhormone-dependent fashion. In this strain, the presence of the Y chromosome is most important. A group of human male relatives has been described who resemble the mouse strain BXSB in that male-predominant families exist in which SLE occurs in men in preference to females (Figure 23.3) [129]. In one interesting study of males with lupus, females that have a male relative with lupus were more likely to have renal disease [130].

Males with SLE are in some series reported to be clinically different from females with the disease [131]. While several male studies show no clinical difference

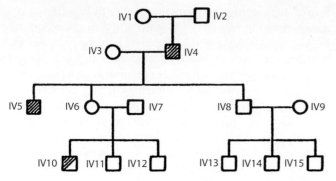

FIGURE 23.3 This pedigree shows one of three families where lupus occurs in males of each family. No females have acquired lupus in any of these families, whereas three generations of males are affected.

TABLE 23.3 Vascular Manifestations in Male Patients with SLE

Feature	%
Digital vasculitis	33
Livedo vasculitis	41
Ulcers	10
Raynaud's phenomenon	29
VASCULAR INVOLVEMENT OF	
Lungs	19
CNS	12
Heart	15
Avascular necrosis of bones	15

Adapted with permission from Terapeutisheskii Arkhiv, Folomeev et al. [140].

in severity of disease between women and men, others have suggested that males have a more severe form of the disease [132, 133]. Pleuropericardial disease and peripheral neuropathy are said to be more common in males and men were found to have more discoid lupus erythematosus and papular nodular mucinosis [134]. In a Spanish series of 261 SLE patients, 11.5% were males and they had less arthritis, more serositis, and a greater propensity for discoid rashes. A database on males from Malta also found more cardiorespiratory problems in males [135], and a Taiwan study suggested that males have a significantly lower $Fc\gamma R$ distribution on monocytes and neutrophils and high prolactin levels that might have a role in the pathogenesis of lupus in this sex [136]. In a recent Israeli study, males presented in middle age with hematological abnormalities, renal involvement, and neurological manifestations, which preceded the onset of the skin and joint complaints [16]. In a Brazilian series, the males had much worse lupus nephritis than their female counterparts (a higher creatinine and renal activity index) [137]. All of these data are collected from small numbers of men and such things as statistical bias might be significant.

Sex hormone profiles indicate that men with lupus have significantly higher levels of gonadotrophins such as follicle-stimulating hormone (FSH) or luteinizing hormone (LH) than controls. A small percentage of patients (14%) in one study had low testosterone and elevated LH levels [138]. Patients with low androgens had more CNS disease and serositis when compared to controls. Finally, the prolactin to testosterone levels correlated with the SLEDAI scores in these men.

Studies of Russian male lupus patients are perhaps the most insightful [139, 140]. Investigators described elevated LH and FSH in SLE males; a lower trochanteric index (1.89 vs. 2.00 for normal men), which is indicative of a lack of androgen effect on bone growth; severe aortic insufficiency and sacroileitis (12% of all men); and overall

a greater incidence of severe vascular diseases, such as Raynaud's phenomenon and digital vasculitis [141].

Russian investigators found more severe disease in men, with 63% dying from end-stage renal disease. Table 23.3 illustrates the vascular manifestations of male patients with SLE. Table 23.4 shows the frequencies of various clinical manifestations in both male and female patients. Note that the only significant increases

TABLE 23.4 Frequency (%) of SLE Manifestations in Male and Female Patients[a]

Symptom	Females (n = 200)	Males (n = 170)	P <
Skin involvement	100	94	—
Malar rash	42.5	65	0.02
Discoid lupus erythematosus	20	25	—
Digital vasculitis	18.2	33.3	0.02
Articular involvement	99.7	92.5	—
Cardiac involvement	87	79.9	0.05
Pneumonitis	22.3	17	0.05
Pleuritis	72	65.2	—
Pericarditis	51	41	—
Nephritis	58.3	75	0.02
CNS involvement	47.3	34.8	0.02
Peripheral nerve system involvement	36.1	8.7	0.01
Raynaud's phenomenon	16.6	29	0.02
Sacroiliitis	n.d.	49	—

Adapted with permission from Terapeutisheskii Arkhiv, Folomeev et al. [140].

in Russian males with SLE are the incidences of nephritis, Raynaud's phenomenon, and malar rash.

Finally, the Russian study also included male SLE patients with profound impotence. The causes of such impotence in young SLE males remain unknown (Table 23.5). Elderly males who present with SLE are also found to have low androgen levels and are hypogonadal [142]. Such males might respond to androgen therapy.

Males with Klinefelter Syndrome

Patients with Klinefelter syndrome can also have a variety of rheumatic diseases such as SLE and scleroderma [143]. Such males commonly have gynecomastia, infertility, a female fat phenotype, and the usual sequelae of hypogonadism. These Klinefelter males have met ARA criteria for SLE both serologically and clinically (Figure 23.4). The incidence of SLE or any other autoimmune disease is not increased in patients with Klinefelter syndrome, even though this is frequently stated. Patients with SLE and Klinefelter syndrome together have the estrogen and androgen metabolism of females with SLE [76], but low levels of both androgens and estrogens and are hypergonadotropic as a result. When SLE does occur in young Klinefelter males, it can be treated with the synthetic androgens. Androgens such as methyl testosterone as a tablet, androgen patch, or gel can alleviate the symptoms of the disease in Klinefelter males but do not alter the amount or type of autoantibody. The male with Klinefelter syndrome and SLE oxidizes testosterone in an exaggerated fashion, like females with SLE. This increased oxidation is not found in the SLE male with a normal XY karyotype. The hypergonadotropic state of men with Klinefelter syndrome and SLE has not been investigated adequately at this time and may have some role in the etiopathogenesis of the disease.

Hyperestrogenism

Hyperestrogenic conditions are associated with autoantibodies in both normal males and females. In the human female, periods of hormonal change such as pregnancy and disturbed menses are associated with changes in immune function. These changes are also observed in mice. Clinical syndromes such as insulin-resistant diabetes mellitus, hirsutism, and cystic ovaries are found in patients with autoantibodies [144]. Polycystic ovary disease is associated with autoimmune disease, and autoantibodies could result from the unopposed actions of plasma estrogen [145]. Autoantibodies have been described in patients with endometriosis, and many such patients have a lupus-like illness [146, 147]. The uterus and the ovaries of humans and animals are a source of cytokines [148] and are probably affected by higher levels of estrogens.

Estrogens and Human Lupus

From work in mice, research on the effects of estrogens and androgens on immune function suggests that they act on immune tolerance. Mice that develop autoimmune diseases have defective tolerance to a variety of antigens, and both estrogen and androgen levels could be responsible [86]. Androgens modulate the T-cell phenotype in transition from marrow to the thymus [149, 150], and the target of androgen action is likely to be the thymocyte. Using 20α-hydroxysteroid dehydrogenase as a T-cell marker, investigators noted a decrease

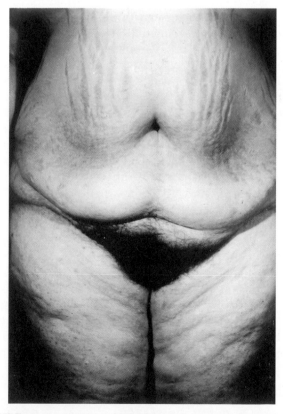

FIGURE 23.4 This is the midsection of a Klinefelter male with SLE. Note the ambiguous genitalia and the female distribution of fat.

TABLE 23.5 Results of Studies of Impotence in Systemic Lupus Erythematosus[a]

1. Thirty-six percent of males with SLE have sexual disturbances

2. Impotence develops before SLE onset

3. Decreased libido and satisfactory effect of androgen therapy suggest an imbalance of sex hormone levels

4. Impotent patients with SLE have a twofold decrease in testosterone levels

[a]*These are male SLE patients with mild and moderate disease activity who did not receive cytotoxic or corticosteroid drugs.*

in thymocytes and a rise in CD8 cells after androgen administration [151]. The hypothesis that androgens act in this manner is attractive because androgens afford a method of altering the absolute numbers of T cells.

Data from mice and humans have shown that unopposed estrogens exacerbate SLE. Data from human studies, however, are largely *in vitro* or are derived from anecdotal studies of patients on oral contraceptives [152]. Most studies involving oral contraceptives and SLE or hormone replacement therapy in postmenopausal women with SLE suggest that the disease worsens with these agents. However, recent prospective studies conclude that the use of hormone replacement therapy (HRT) is safe and does not predispose women to develop SLE [69], although there may be a small risk of mild to moderate flares in women with SLE that take HRT [153]. Results of the contraceptive/hormone replacement trial, a double-blinded control study designed to look at oral contraceptive use and HRT in lupus women indicate that both are safe in the absence of a procoagulant state. Current epidemiological studies do not support the idea that female hormones increase the risk of developing SLE [154]. This is contested by some investigators who have shown that the use of combined oral contraceptives increases the risk of SLE [155]. Interestingly, these authors attribute this risk to the dose of ethinyl estradiol. Sex steroids have been considered for use in inflammatory diseases such as gout and osteoarthritis because estrogens suppress neutrophil function, which is why gout is uncommon in the premenopausal female [156, 157]. The overall gender-related mechanisms for the selectivity of one sex for a disease over another continues to evade explanation and there is debate about this from many quarters [158].

Estrogens may be effective in the treatment of rheumatoid arthritis, and several clinical studies have argued both for [159, 160] and against their use [161]. The mechanism of suppression of inflammation in cases of rheumatoid arthritis may give some significant insight into the pathogenesis of this common disease and is based on two observations: 1. patients improve when pregnant and 2. oral contraceptives may have a protective effect.

Menopause is an interesting time for most lupus patients. Patients who reach menopause are known by most clinicians to have fewer flares. These conclusions were without basis until menopause was conclusively associated with a remission of illness [162]. Most authors conclude that a "modest" decrease of activity occurs in those women reaching menopause [72].

Androgens and Human Lupus

Therapy of SLE with anabolic steroids such as Danazol or cyproterone acetate is not effective therapy for either murine or human SLE [163–167]. However, the use of hormones such as Danazol in idiopathic thrombocytopenic purpura increases platelet numbers and coagulation factors such as factor VIII. This suggests a role for certain patients with SLE. However, early on there was evidence of a worsening of lupus in some males treated with androgen [168]. Other androgens under study, such as 19-nortestosterone, have also had limited use in the treatment of SLE in women. Doses used were 100 mg/ml per week intramuscularly. While the overall condition of female patients improved and patients admitted increased energy, loss of joint pains, and resolution of systemic abnormalities such as skin rashes or anemia (Figure 23.5); no significant change of serologies was noted. Anti-DNA and ANA titers remained elevated. Most curious in this treatment group was an overall worsening of lupus symptoms in male patients [169]. The worsening of males on 19-nortestosterone was associated with lowered endogenous testosterone levels and elevated estradiol levels.

Another steroid, cyproterone acetate, a potent antigonadotrophic agent [170], resulted in resolution of some systemic abnormalities such as oral ulcers after 50 mg daily for a mean of 63 months. As with other androgens, no improvement in serologic features was observed.

Another approach to the treatment of lupus involves the use of gonadotropic hormone-releasing hormone agonists. These agents are used effectively to treat endometriosis and prostate carcinoma. Data about their use in SLE are inconclusive because of the small numbers studied [171, 172].

One androgen, studied extensively in women, is dehyroepiandrosterone (DHEA), the so-called antiaging drug that was of brief interest to rheumatologists [173, 174]. At doses of 200 mg per day in tablet form, this agent produced modest clinical improvement as measured by standard indices and was steroid sparing, i.e., the total dose of prednisone might be lowered with its use [175–178]. Early data showed suppressed cytokine levels such as IL-4, -5, and -6 and increases IL-2 during treatment in both humans and animals [179, 180]. Natural killer cell activity also increases with this agent [181–183]. Despite these early observations, DHEA did not achieve significant clinical benchmarks in order to allow its approval in the US for treatment of SLE.

Sex Hormones, Behavior, and Autoimmune Diseases

Sex steroids are important to the development of various organs, and a major one is the brain. Attention has been directed to cerebral development, cerebral dominance, and the incidence of autoimmune disease [184, 185]. Some studies say that patients with

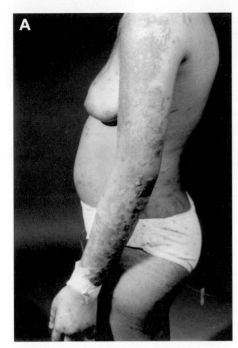 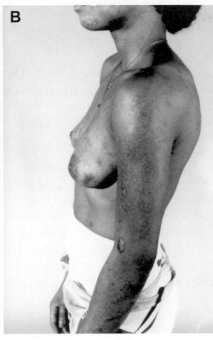

FIGURE 23.5 (*A*) A young woman prior to the institution of 19-nortestosterone therapy. (*B*) The same woman 2 years later after intramuscular injection of 100 mg/week of 19-nortestosterone.

autoimmune disease are predominantly left-handed, indicating dominance of the right cerebral hemisphere. This variation is reported in patients with SLE [186, 187] and other diseases of the immune system. Neuronal migration is under the influence of steroids such as testosterone, and data suggest that SLE mice have aberrant neuronal migration patterns consistent with those observed in humans with learning disabilities such as dyslexia. Handedness in patients with SLE is more directed to the left than in the normal population [188, 189]. The finding of increased learning disabilities such as developmental dyslexia in patients with SLE, their unaffected male offspring, and unaffected male siblings has been confirmed by several groups and is a subject of considerable interest.

The Basic biology of Sex Hormones and Immune Function

Androgens

Androgens have many effects on immune function. They are immunosuppressive because of *in vitro* observations in normal lymphocytes and because of their effects on the disease manifestations of inbred autoimmune mice [190]. Testosterone suppresses anti-DNA antibody production in peripheral blood mononuclear cells from patients with SLE [191] and it may do so through inhibition of B cell lymphopoesis mentioned earlier; the downregulation of IL-6 and the inhibition of B-cell activity [192]. Graft rejection in rodents is delayed by the injection of testosterone. Resistance to certain viral infections is reduced, whereas in some cases resistance to certain infections is enhanced when androgens are given at certain doses [193, 194]. A consistent early effect of androgens in animals was the immunosuppression of chickens through the retardation of the function and development of the bursa of Fabricius [195–197] an effect replicated elegantly by Diamond et al. Androgens also have an effect on the pluripotential stem cells of the bone marrow: namely the accelerated proliferation and differentiation of such stem cells into compartments of cells that number among them lymphoid elements. The important effects of androgens on immune maturation are reflected by the discovery of receptors for estrogen and dihydrotestosterone, a 5α-reduced metabolite of testosterone in the thymus [198; 199]. Androgens induce thymic atrophy and castration produces thymic hypertrophy [200]. In men treated with medical castration there is a reduction in levels of T reg cells and activation of CD8 cells [201]. Androgen receptors are found on lymphocytes, but data are inconsistent. Androgens inhibit B- and T-cell maturation, reduce B-cell synthesis of immunoglobulins, and suppress the phytohemagglutinin-induced blast transformation of lymphocytes. Androgens are implicated as modifiers of regulatory genes that influence the function of structural genes. For example, guinea pig mammary epithelial la antigens are increased in number through the effects of estrogens and prolactin and are decreased by testosterone [202]. The wide range of effects that this steroid displays might be due to variable androgen sensitivity of certain cell groups or the *in vivo* conversion of androgens to estrogens through well-recognized pathways.

Estrogens

It has been known for many years that female mice make more antibody to foreign antigens than males, and this difference is reported for a variety of antigens [203]. Depending on the dose, estrogens have been both immunosuppressors and immunostimulants. Historically, the steroid 17β-estradiol prolonged first and second set skin grafts in mice after X irradiation and inhibited corneal graft acceptance in preimmunized rabbits [204]. Skin allograft rejection was naturally better in females than in males [205]. Estrogens regulate immunity by way of the thymus in rodents [206] and decrease the overall thymic population of small lymphocytes. Thymic atrophy could be induced by estrogen prevention of proliferation of thymic precursor cells [207]. The effect was that of estradiol on the thymic membrane receptor. Estrogen also affects dendritic cells [208]. Estrogen given prior to bone marrow transplantation results in increased graft failure. Estradiol and diethylstilbestrol (in concentrations of 10–50 mg/ml) were known to reduce the phytohemagglutinin and concanavalin A response of lymphocytes in vitro [209]. The mixed lymphocyte reaction was enhanced by estradiol. Fluctuant lymphocyte responses are observed during normal menses, during pregnancy, and during the use of oral contraceptives [72, 210]. Castrated male and certain female mice display hyperplastic spleens and thymuses after challenge with thymic-dependent antigens, indicating an effect on thymic cell activity [150]. Castration of males leads to accelerated allograft rejection. Syngeneic grafts of ovaries in males and grafts of testes in females have no significant effect on allograft rejection [211]. Sustained levels of estrogens in mice led to a marked reduction in natural killer cell activity. In other studies, estrogens have been found to depress cell-mediated immunity, natural killer cell function, and cancer cell immune surveillance. Estrogens also deplete thymic hormones and are known to produce a relative lymphopenia. Estrogen enhances TH1 responses in nonpregnant women, whereas high doses enhance TH2 responses. Estradiol also influences chemokine production by activated splenocytes [212] and influences T cell trafficking. Estradiol can convert CD25$^-$ to CD25$^+$ cells [213].

Estrogens are detrimental to animals with SLE and this refers not only to mice with SLE, but also other animals, such as dogs. In fact, Tamoxifen and the antiestradiol antibody have beneficial effects on experimental SLE. This benefit occurs via cytokine regulation [214]. Men with prostatic cancer who are given diethylstilbestrol have markedly depressed cell-mediated immunity. Using normal lymphocytes, estradiol treatment of pokeweed mitogen-treated B cells shows an increase in plaque-forming cells (in vitro) [215, 216]. Both ERα and ERβ receptors are found on CD8$^+$ T lymphocytes, macrophage antigen-presenting cells, dendritic cells, and PBMCs. [217–219]. Studies have shown that CD4$^+$ T helper cells increase after estrogen therapy. By some studies, estrogens are thought to act by inhibiting CD8$^+$ suppressor T cells. Consequently, helper T cells would be enhanced with the resulting polyclonal B-cell immunoglobulin production. An exciting new direction is the inhibition of apoptosis in vitro using peripheral blood mononuclear cells from women with normal menses. The estradiol decreases tumor necrosis factor-α and apoptosis in SLE PBMCs and increased IL-10, but this was not seen in cells from normal people [220, 221]. Studies of estrogen receptor in SLE have shown that ERβ expression is downregulated in PBMCs, while ERα is upregulated [222, 223].

Effects on T cells

SLE is characterized by exaggerated T-cell signaling and a high degree of T-cell activity. Mitogen-activated protein kinase (MAPK) is involved in the maintenance of T cell tolerance that fails in patients with SLE. Estrogen can stimulate second messengers and affect calcium flux and kinase activation. One area of exploration involves estrogen-mediated phosphorylation of extracellular signal-regulated kinase ½ (ERK1/2). However, this suppression of MAKP by estrogen appears to only involve patients with inactive or mild disease [224].

Effects on B cells

Diamond has shown that B-cell maturation is affected by both prolactin and estrogen permitting these cells to mature to competence in a mouse model of SLE [191, 225]. This was accomplished using anti-DNA antibodies in murine models of SLE. DNA-reactive B cells mature to specific anti-DNA antibody subsets depending on the hormonal environment. Anti-DNA B cells in estrogen-treated mice become marginal cells, while prolactin-treated mice become follicular cells [226]. Diamond and colleagues found that estrogen exacerbated disease and accelerated death in murine models of SLE, but also induced a lupus phenotype in nonspontaneously autoimmune mice. Estrogen rescues naïve autoreactive B cells and makes them mature to a marginal zone phenotype. Tamoxifen, an estrogen receptor-blocking agent, prevented this [227]. Estrogen produces an increase of high affinity anti-DNA B cells, whereas low-affinity B cells are decreased. Estrogen receptor-alpha (ERα) is the key in the B-cell story. Knockout mice for ERα do not show these findings.

It therefore appears that estradiol decreases transitional zone B cells and increases marginal zone B cells. This is done through a decreased calcium flux, phosphorylation, ERK1/2, and apoptosis. Estrogen also increases expression of Bcl2 and CD22 on B cells.

Progestogens

Progesterone is a hormone present in men and women. It is present in women during various times in the menstrual cycle and in high levels during pregnancy. Other sex steroids such as progesterone have been considered effective therapeutic agents in diseases such as SLE. Progesterone is an immunosuppressive agent [228, 229], and levels of this substance rise during pregnancy when the placenta assumes an active role in its synthesis and secretion. At concentrations of 10–15 mg/ml, this steroid reduces lymphocyte responses to phytohemagglutinin and to concanavalin A *in vitro*. Other analogs (20α-hydroxyprogesterone) have similar effects [230].

Progesterone has been known to increase the relative amounts of CD8$^+$ suppressor cells and to decrease them in mice. In addition, progesterone has been invoked to explain many of the suppressive effects found in the sera of pregnant females. This hormone promotes TH2 differentiation *in vitro* [231]. It enhances production of self-protective antibodies [232] or those that are asymmetrically glycosylated so that they fail to trigger immune effector mechanisms. In the lupus mouse, progesterone protected castrated male NZB/W mice against development of lupus [233].

SUMMARY

Sex steroids are potent modulators of immunity in all animal systems. They are very important in patients with SLE and may help explain the fluctuant activity in lupus disease. Their role in the maturation of the immune system and their effect on the development of organs such as the brain could explain some of the abnormalities found in these systems.

References

[1] R.G. Lahita, Sex and Age in systemic lupus erythematosus, in: R.G. Lahita (Ed.), Systemic Lupus Erythematosus, 1 ed)., John Wiley and Sons, New York, 1986, pp. 523–539.

[2] K.H. Costenbader, D. Feskanich, M.J. Stampfer, E.W. Karlson, Reproductive and Menopausal Factors and Risk of Systemic Lupus Erythematosus, Arthritis and Rheumatism 56 (4) (2007) 1251–1262.

[3] J.J. Cush, E.A. Goldings, Drug induced lupus: clinical spectrum and pathogenesis, American Journal of Medical Science 290 (1985) 36–45.

[4] R.L. Yung, B.C. Richardson, Pathophysiology of Drug Induced Lupus, in: R.G. Lahita (Ed.), Systemic Lupus Erythematosus, 3 ed)., Academic Press, San Diego, 1999, pp. 909–928.

[5] D.D. Urowitz, D. Gladman, Evolving spectrum of mortailty and morbidity in SLE, Lupus 8 (1999) 253–255.

[6] M. Maragou, F. Siotsiou, H. Sfondouris, Z. Nicolia, G. Vayopoulos, P. Dantis, Late-onset systemic lupus erythematosus presenting as polymyalgia rheumatica, Clin Rheumatol 8 (1989) 91–97.

[7] R.E. Kellum, J.R. Haserick, Systemic Lupus Erythematosus, a statistical evaluation of mortality based on a conservative series of 299 patients, Arch Int Med 113 (1964) 200.

[8] C.C. Mok, C.S. Lau, K.W. Lee, R.W. Wong, Hyperprolactinemia in males with systemic lupus erythematosus, J Rheumatol 25 (12) (1998) 2357–2363.

[9] S.B. Baker, J.R. Rovira, E.W. Campion, J.A. Mills, Late onset SLE, Am J Med 66 (1979) 727–732.

[10] H.A. Wilson, M.E. Hamilton, D.A. Spyker, Age influences the clinical and serologic expression of SLE, Arth Rheum 24 (1981) 1230–1235.

[11] B.S.I. Foadd, R.P. Sheon, A.B. Kirsner, Systemic Lupus Erythematosus in the elderly, Arch Int Med 130 (1972) 743–746.

[12] S.P. Ballou, M.A. Khan, I. Kushner, Clinical features of SLE: differences related to race and age of onset, Arth Rheum 25 (1982) 55–60.

[13] P.J. Maddison, D.A. Isenberg, N.J. Goulding, J. Leddy, R.P. Skinner, Anti La(SSB) identifies a distinctive subgroup of systemic lupus erythematosus, Br J Rheumatol 27 (1988) 27–31.

[14] D. Adu, D.G. Williams, D. Taube, Late onset lupus-like disease in patients with apparent idiopathic glomerulonephritis, Quart J Med 52 (1983) 471–487.

[15] A. Baer, T. Pincus, Occult systemic lupus erythematosus in elderly men, JAMA 249 (1983) 3350–3352.

[16] N.L. Ambrose, G. Kearns, A. Mohammad, Male lupus: a diagnosis often delayed-a case series and review of the literature, Ir J Med Sci (2009). [Epub ahead of print].

[17] D. Lazaro, Elderly-onset systemic lupus erythematosus: prevalence, clinical course and treatment, Drugs Aging 24 (9) (2007) 701–715.

[18] E. Krishnan, H.B. Hubert, Ethnicity and mortality from systemic lupus erythematosus in the US, Ann Rheum Dis 65 (11) (2006) 1500–1505.

[19] G.S. Alarcon, J. Calvo-Alen, G. McGwin Jr., A.G. Uribe, S.M. Toloza, J.M. Roseman, et al., Systemic lupus erythematosus in a multiethnic cohort: LUMINA XXXV. Predictive factors of high disease activity over time, Ann Rheum Dis 65 (9) (2006) 1168–1174.

[20] S. Studenski, N.B. Allen, D.S. Caldwell, J.R. Rice, R.P. Polisson, Survival in systemic lupus erythematosus. A multivariate analysis of demographic factors, Arthritis Rheum 30 (1987) 1326–1332.

[21] M.M. Ward, S. Studenski, Age associated clinical manifestations of systemic lupus erythematosus: a multivariate regression analysis, J Rheum 17 (1990) 476–481.

[22] R, K Rashkov, K. unev, [The evolution of systemic lupus erythematosus], Vutr Boles 29 (1990) 91–96.

[23] H. Hashimoto, H. Tsuda, T. Hirano, Y. Takasaki, T. Matsumoto, S. Hirose, Differences in clinical and immunological findings of systemic lupus erythematosus related to age, J Rheumatol 14 (1987) 497–501.

[24] M.C. Hochberg, Mortality from systemic lupus erythematosus in England and Wales, 1974-1983, Br J Rheumatol 26 (1987) 437–441.

[25] J. Boddaert, D.L. Huong, Z. Amoura, B. Wechsler, P. Godeau, J.C. Piette, Late-onset systemic lupus erythematosus:

a personal series of 47 patients and pooled analysis of 714 cases in the literature, Medicine (Baltimore) 83 (6) (2004) 348–359.

[26] J. Calvo-Alen, S.M. Toloza, M. Fernandez, H.M. Bastian, B.J. Fessler, J.M. Roseman, et al., Systemic lupus erythematosus in a multiethnic US cohort (LUMINA). XXV. Smoking, older age, disease activity, lupus anticoagulant, and glucocorticoid dose as risk factors for the occurrence of venous thrombosis in lupus patients, Arthritis Rheum 52 (7) (2005) 2060–2068.

[27] M.C. Turi, M. D'Urbano, E. Celletti, C. Alessandri, G. Valesini, R. Paganelli, Serum sFas/sFasL ratio in systemic lupus erythematosus (SLE) is a function of age, Arch Gerontol Geriatr 49 (Suppl 1) (2009) 221–226.

[28] E. Rosato, F. Salsano, Immunity, autoimmunity and autoimmune diseases in older people, J Biol Regul Homeost Agents 22 (4) (2008) 217–224.

[29] A. Larbi, T. Fulop, G. Pawelec, Immune receptor signaling, aging and autoimmunity, Adv Exp Med Biol 640 (2008) 312–324.

[30] E. Raveche, J.H. Tijo, A.D. Steinberg, Genetic studies in NZB mice. IV. The effects of sex hormones on the spontaneous production of anti-T cell antibodies, Arth Rheum 23 (1980) 48.

[31] J.B. Howie, B.J. Helyer, Autoimmune diseases in mice, Ann N Y Acad Sci 124 (1965) 167–177.

[32] J.B. Howie, B.J. Helyer, The immunology and pathology of NZB mice, Adv Immunol 9 (1968) 215–266.

[33] E.D. Murphy, Jb. Roths, A Y chromosome associated factor in strain BXSB producing accelerated autoimmunity and lymphoproliferation, Arth Rheum 22 (1979) 1188–1193.

[34] S.E. Walker, R.H. Gray, M. Fulton, Palmerston north mice, a new animal model of systemic lupus erythematosus, J Lab Clin Med 92 (1978) 932.

[35] S.E. Walker, A.B. Kier, E.C. Siegfied, B.G. Harris, J.S. Schultz, Accelerated autoimmune disease and lymphoreticular neoplasms in F1 hybrid PN/NZB and NZB/PN mice, Clin Immunol Immunopathol 39 (1986) 81–92.

[36] P.H. Lambert, F.J. Dixon, Pathogenesis of the glomerulonephritis of NZB/W mice, J Exp Med 127 (1968) 507–522.

[37] P.O. Teague, G.J. Irion, J.J. Myers, Antinuclear antibodies in mice. I Influence of age and possible genetic factors on spontaneous and induced responses, J Immunol 101 (1968) 791–798.

[38] G. Tonietti, M.B.A. Oldstone, F.J. Dixon, The effect of induced chronic viraql infections on the immunological lesions of New Zealand white mice, J Exp Med 132 (1970) 89–109.

[39] P.J. Staples, A.D. Steinberg, N. Talal, Induction of immunological tolerance in older New Zealnd mice repopulated with young spleen, bone marrow, or thymus, J Exp Med 131 (1980) 1223–1238.

[40] J.I. Morton, D.A. Weyant, B.V. Seigel, B. Golding, Androgen sensitivity and autoimmune disease I. Influence of sex and testosterone on the humoral immune responses of autoimmune and nonautoimmune mouse strains to SRBC, Immunology 44 (1981) 661–669.

[41] J.R. Roubinian, N. Talal, J.S. Greenspan, J.R. Goodman, P.K. Siiteri, Effect of castration and sex hormone treatment on survival, anti-nucleic acid antibodies, and glomerulonephritis in NZB/NZW F1 mice, J Exp Med 147 (1978) 1568–1583.

[42] J.R. Roubinian, R. Papoian, N. Talal, Androgenic hormones modulate autoantibody reponses and improve survival in murine lupus, J Clin Invest 59 (1977) 1066–1070.

[43] J.R. Roubinian, R. Papoian, N. Talal, Effects of neonatal thymectomy and splenectomy on survival and regulation of autoantibody formation in NZB/NZW F1 mice, J Immunol 118 (1977) 1524–1529.

[44] J.R. Roubinian, N. Talal, J.S. Greenspan, Delayed androgen treatment survival in murine lupus, J Clin Invest 63 (1979) 902–911.

[45] A.N. Theofilopoulos, F. Dixon, Experimental murine systemic lupus erythematosus, in: R.G. Lahita (Ed.), Systemic Lupus Erythematosus, 1 ed)., John Wiley and Sons, New York, 1986, pp. 121–202.

[46] L. Hang, A.N. Theofilopoulos, F.J. Dixon, A spontaneous rheumatoid arthritis-like disease in MRL/l mice, J Exp Med 155 (1982) 1690–1701.

[47] R.A. Eisenberg, F.J. Dixon, Effect of castration on male determined acceleration of autoimmune disease in BXSB mice, J Immunol 125 (1980) 1939–1963.

[48] R.A. Eisenberg, S. Izui, P.J. McConahey, Male determined accelerated autoimmune disease in BXSB mice: transfer by bone marrow and spleen cells, J Immunol 125 (1980) 1032–1036.

[49] S. Saha, A. Tieng, K.P. Pepeljugoski, G. Zandamn-Goddard, E. Peeva, Prolactin, Systemic Lupus Erythematosus, and Autoreactive B Cells: Lessons Learnt from Murine Models, Clin Rev Allergy Immunol (2009). [Epub ahead of print].

[50] K.L. Medina, G. Smithson, P.W. Kincade, Suppression of B lymphopoiesis during normal pregnancy, J Exp Med 178 (5) (1993) 1507–1515.

[51] T. Masuzawa, C. Miyaura, Y. Onoe, K. Kusano, H. Ohta, S. Nozawa, et al., Estrogen deficiency stimulates B lymphopoiesis in mouse bone marrow, J Clin Invest 94 (3) (1994) 1090–1097.

[52] K.L. Medina, P.W. Kincade, Pregnancy-related steroids are potential negative regulators of B lymphopoiesis, Proc Natl Acad Sci U S A 91 (12) (1994) 5382–5386.

[53] S.M. Viselli, K.R. Reese, J. Fan, W.J. Kovacs, N.J. Olsen, Androgens alter B cell development in normal male mice, Cell Immunol 182 (2) (1997) 99–104.

[54] T.M. Ellis, M.T. Moser, P.T. Le, R.C. Flanigan, E.D. Kwon, Alterations in peripheral B cells and B cell progenitors following androgen ablation in mice, Int Immunol 13 (4) (2001) 553–558.

[55] C. Miyaura, Y. Onoe, M. Inada, K. Maki, K. Ikuta, M. Ito, et al., Increased B-lymphopoiesis by interleukin 7 induces bone loss in mice with intact ovarian function: similarity to estrogen deficiency, Proc Natl Acad Sci USA 94 (17) (1997) 9360–9365.

[56] G. Smithson, K. Medina, I. Ponting, P.W. Kincade, Estrogen suppresses stromal cell-dependent lymphopoiesis in culture, J Immunol 155 (7) (1995) 3409–3417.

[57] M.S. Bynoe, C.M. Grimaldi, B. Diamond, Estrogen up-regulates Bcl-2 and blocks tolerance induction of naive B cells, Proc Natl Acad Sci USA 97 (6) (2000) 2703–2708.

[58] C.M. Grimaldi, D.J. Michael, B. Diamond, Cutting edge: expansion and activation of a population of autoreactive marginal zone B cells in a model of estrogen-induced lupus, J Immunol 167 (4) (2001) 1886–1890.

[59] S. Hande, E. Notidis, T. Manser, Bcl-2 obstructs negative selection of autoreactive, hypermutated antibody V regions during memory B cell development, Immunity 8 (2) (1998) 189–198.

[60] E. Peeva, J. Venkatesh, D. Michael, B. Diamond, Prolactin as a modulator of B cell function: implications for SLE, Biomed Pharmacother 58 (5) (2004) 310–319.

[61] E. Peeva, C. Grimaldi, L. Spatz, B. Diamond, Bromocriptine restores tolerance in estrogen-treated mice, J Clin Invest 106 (11) (2000) 1373–1379.

[62] E. Mazes, S. Fuchs, Linkage between immune response potential to DNA and X chromosome, Nature 249 (1974) 167–168.

[63] C.M. Cooney, G.R. Bruner, T. Aberle, B. Namjou-Khales, L.K. Myers, L. Feo, et al., 46, X, del(X)(q13) Turner's syndrome women with systemic lupus erythematosus in a pedigree multiplex for SLE, Genes Immun 10 (5) (2009) 478–481.

[64] A.H. Sawalha, J.B. Harley, R. Scofield, Autoimmunity and Klinefelter's syndrome: when men have two X chromosomes, J Autoimmun 33 (1) (2009) 31–34.

[65] R.G. Lahita, The importance of estrogens in systemic lupus erythematosus, Clin Immunol Immunopathol 63 (1992) 17–18.

[66] T.A. Chapel, R.E. Burns, Oral contraceptives and exacerbations of SLE. Am J Obst Gynec 110 (1971) 366–369.

[67] M. Petri, M.Y. Kim, K.C. Kalunian, J. Grossman, B.H. Hahn, L.R. Sammaritano, et al., Combined oral contraceptives in women with systemic lupus erythematosus, N Engl J Med 353 (24) (2005) 2550–2558.

[68] R.A. Asherson, N.E. Harris, A.E. Gharavi, G.R.V. Hughes, Systemic lupus erythematosus, antiphospholipid antibodies, chorea and oral contraceptives, Arth Rheum 29 (1986) 1535–1536.

[69] J. Sanchez-Guerrero, M.H. Liang, E.W. Karlson, D.J. Hunter, G.A. Colditz, Past use of oral contraceptives and the risk of developing systemic lupus erythematosus, Ann Int Med 122 (1995) 430–433.

[70] M. Petri, Exogenous estrogen in systemic lupus erythematosus: oral contraceptives and hormone replacement therapy, Lupus 10 (3) (2001) 222–226.

[71] E. Rose, D.M. Pillsbury, Lupus erythematosus (erythematodes) and ovarian function: Observations on a possible relationship with a report of six cases, Ann Int Med 21 (1944) 1022–1024.

[72] J. Sanchez-Guerrero, A. Villegas, A. Mendoza-Fuentes, J. Romero-Diaz, G. Moreno-Coutino, M.C. Cravioto, Disease activity during the premenopausal and postmenopausal periods in women with systemic lupus erythematosus, Am J Med 111 (6) (2001) 464–468.

[73] A. Doria, A. Ghirardello, L. Iaccarino, S. Zampieri, L. Punzi, E. Tarricone, et al., Pregnancy, cytokines, and disease activity in systemic lupus erythematosus, Arthritis Rheum 51 (6) (2004) 989–995.

[74] R. Scofield, G.R. Bruner, B. Namjou-Khales, R.P. Kimberly, R. Ramsey-Goldman, M. Petri, et al., Klinefelter's syndrome (47, XXY) in male systemic lupus erythematosus patients: support for the notion of a gene-dose effect from the X chromosome, Arth Rheum 58 (8) (2008) 2511–2517.

[75] S. Manzi, E.N. Meilahn, J.E. Rairie, C.G. Conte, T.A. Medsger, L. Jansen-McWilliams, et al., Age-specific incidence rates of myocardial infarction and angina in women with systemic lupus erythematosus: comparison with the Framingham Study, Am J Epidemiol 145 (5) (1997) 408–415.

[76] R.G. Lahita, H.L. Bradlow, Klinefelter's syndrome: hormone metabolism in hypogonadal males with systemic lupus erythematosus, J Rheumatol 14 (Suppl 13) (1987) 154–157.

[77] R.G. Lahita, H.L. Bradlow, J. Fishman, H.G. Kunkel, Estrogen metabolism in systemic lupus erythematosus: patients and family members, Arth Rheum 25 (1982) 843–846.

[78] R.G. Lahita, H.L. Bradlow, H.G. Kunkel, J. Fishman, Increased 16 alpha hydroxylation of estradiol in systemic lupus erythematosus, J Clin Endo Metab 53 (1981) 174–178.

[79] R. Bucala, R.G. Lahita, J. Fishman, A. Cerami, Anti-estrogen antibodies in users of oral contraceptives and in patients with systemic lupus erythematosus, Clinical and Experimental Immunology 67 (1987) 167–175.

[80] J. Fishman, C. Martucci, Biological properties of 16-alpha hydroxyestrone: implicationso in estrogen physiology and pathophysiology, J Clin Endo Metab 51 (1980) 611–615.

[81] S. Ikegawa, R.G. Lahita, J. Fishman, Concentration of 16 alpha hydroxyestrone in human plasma as measured by a specific RIA, J Steroid Biochem 18 (1983) 329–332.

[82] R. Bucala, J. Fishman, A. Cerami, The reaction of 16-a hydroxyestrone with erythrocytes in vitro and in vivo, Eur J Biochem 140 (1984) 593–598.

[83] R. Bucala, J. Fishman, A. Cerami, Formation of co-valent adducts bewteen cortisol and 16-alpha-hydroxyestrone and protein, possible role in pathogenesis of cortisol topxicity and SLE, Proc Nat Acad Sci 79 (1982) 3320–3324.

[84] X.J. Yan, M. Qi, G. Telusma, S. Yancopoulos, M. Madaio, M. Satoh, et al., Indole-3-carbinol improves survival in lupus-prone mice by inducing tandem B- and T-cell differentiation blockades, Clin Immunol 131 (3) (2009) 481–494.

[85] M. Cutolo, A. Sulli, S. Capellino, B. Villaggio, P. Montagna, B. Seriolo, et al., Sex hormones influence on the immune system: basic and clinical aspects in autoimmunity, Lupus 13 (9) (2004) 635–638.

[86] C.C. Chen, C.R. Parker Jr., Adrenal androgens and the immune system, Semin Reprod Med 22 (4) (2004) 369–377.

[87] R.D. Inman, Immunologic Sex differences and the female preponderance in systemic lupus erythematosus, Arth Rheum 21 (1978) 849–852.

[88] R.D. Inman, L. Jovanovic, J.A. Markenson, Systemic lupus erythematosus in men: genetic and endocrine features, Arch Int Med 142 (1982) 1813–1815.

[89] R.G. Lahita, H.L. Bradlow, H.G. Kunkel, J. Fishman, Increased oxidation of testosterone in systemic lupus erythematosus, Arth Rheum 26 (1983) 1517–1521.

[90] P. Jungers, C. Pelissier, J.F. Bach, K. Nahoul, Les androgènes plasmatiques chez les femmes atteintes de lupus érythémateux disseminé (LED), Pathologie Biologie 28 (1980) 391–392.

[91] P. Jungers, K. Nahoul, C. Pelissier, Low plasma androgens in women with active or quiescent SLE, Arth Rheum 25 (1982) 454–457.

[92] C.M. Grimaldi, L. Hill, X. Xu, E. Peeva, B. Diamond, Hormonal modulation of B cell development and repertoire selection, Mol Immunol 42 (7) (2005) 811–820.

[93] C.M. Huang, C.T. Chou, Hyperprolactinemia in systemic lupus erythematosus, Chung Hua I Hsueh Tsa Chih (Taipei) 59 (1) (1997) 37–41.

[94] G.R. Hughes, M.A. Khamashta, Prolactin, the immune response and lupus, Rheumatology (Oxford) 38 (4) (1999) 296–297.

[95] A.M. Jacobi, W. Rohde, M. Ventz, G. Riemekasten, G.R. Burmester, F. Hiepe, Enhanced serum prolactin (PRL) in patients with systemic lupus erythematosus: PRL levels are related to the disease activity, Lupus 10 (8) (2001) 554–561.

[96] A.M. Jacobi, W. Rohde, H.D. Volk, T. Dorner, G.R. Burmester, F. Hiepe, Prolactin enhances the in vitro production of IgG in peripheral blood mononuclear cells from patients with systemic lupus erythematosus but not from healthy controls, Ann Rheum Dis 60 (3) (2001) 242–247.

[97] L.J. Jara, O. Vera-Lastra, J.M. Miranda, M. Alcala, J. Alvarez-Nemegyei, Prolactin in human systemic lupus erythematosus, Lupus 10 (10) (2001) 748–756.

[98] S.E. Walker, Impaired hypothalamic function, prolactinomas, and autoimmune diseases, J Rheumatol 33 (6) (2006) 1036–1037.

[99] S.E. Walker, Bromocriptine treatment of systemic lupus erythematosus, Lupus 10 (10) (2001) 762–768.

[100] S.E. Walker, Modulation of hormones in the treatment of lupus, Am J Manag Care 7 (16 Suppl) (2001) S486–S489.

[101] S.E. Walker, Treatment of systemic lupus erythematosus with bromocriptine, Lupus 10 (3) (2001) 197–202.

[102] K.B. Elbourne, D. Keisler, R.W. McMurray, Differential effects of estrogen and prolactin on autoimmune disease in the NZB/NZW F1 mouse model of systemic lupus erythematosus, Lupus 7 (6) (1998) 420–427.

[103] A. El-Garf, S. Salah, M. Shaarawy, S. Zaki, S. Anwer, Prolactin hormone in juvenile systemic lupus erythematosus: a possible relationship to disease activity and CNS manifestations, J Rheumatol 23 (2) (1996) 374–377.

[104] M.A. Gutierrez, J.F. Molina, L.J. Jara, C. Garcia, S. Gutierrez-Urena, M.L. Cuellar, et al., Prolactin-induced immunoglobulin and autoantibody production by peripheral blood mononuclear cells from systemic lupus erythematosus and normal individuals, Int Arch Allergy Immunol 109 (3) (1996) 229–235.

[105] M.A. Gutierrez, J.F. Molina, L.J. Jara, M.L. Cuellar, C. Garcia, S. Gutierrez-Urena, et al., Prolactin and systemic lupus erythematosus: prolactin secretion by SLE lymphocytes and proliferative (autocrine) activity, Lupus 4 (5) (1995) 348–352.

[106] A.A. Lash, Why so many women? Part 1. Systemic lupus erythematosus, Medsurg Nurs 2 (1993) 259–264.

[107] L. Jara-Quezada, A. Graef, C. Lavalle, Prolactin and gonadal hormones during pregnancy in systemic lupus erythematosus, J Rheum 18 (1991) 349–353.

[108] C. Lavalle, E. Loyo, R. Paniagua, J.A. Bermudez, J. Herrera, A. Graef, et al., Correlation study between prolactin and androgens in male patients with systemic lupus erythematosus, J Rheumatol 14 (1987) 268–272.

[109] C.C. Mok, C.S. Lau, Lack of association between prolactin levels and clinical activity in patients with systemic lupus erythematosus [letter; comment], J Rheumatol 23 (12) (1996) 2185–2186.

[110] B. Ostendorf, R. Fischer, R. Santen, B. Schmitz-Linneweber, C. Specker, M. Schneider, Hyperprolactinemia in systemic lupus erythematosus? Scand J Rheumatol 25 (2) (1996) 97–102.

[111] C.C. Mok, C.S. Lau, S.C. Tam, Prolactin profile in a cohort of Chinese systemic lupus erythematosus patients, Br J Rheumatol 36 (9) (1997) 986–989.

[112] D. Kozakova, J. Rovensky, L. Cebecauer, V. Bosak, E. Jahnova, M. Vigas, Prolactin levels and autoantibodies in female patients with systemic lupus erythematosus, Z Rheumatol 59 (Suppl 2) (2000) II/80–II/84.

[113] A. Leanos-Miranda, D. Pascoe-Lira, K.A. Chavez-Rueda, F. Blanco-Favela, Persistence of macroprolactinemia due to antiprolactin autoantibody before, during, and after pregnancy in a woman with systemic lupus erythematosus, J Clin Endocrinol Metab 86 (6) (2001) 2619–2624.

[114] A. Leanos-Miranda, D. Pascoe-Lira, K.A. Chavez-Rueda, F. Blanco-Favela, Antiprolactin autoantibodies in systemic lupus erythematosus: frequency and correlation with prolactinemia and disease activity, J Rheumatol 28 (7) (2001) 1546–1553.

[115] A. Leanos-Miranda, K.A. Chavez-Rueda, F. Blanco-Favela, Biologic activity and plasma clearance of prolactin-IgG complex in patients with systemic lupus erythematosus, Arthritis Rheum 44 (4) (2001) 866–875.

[116] M. Pacilio, S. Migliaresi, R. Meli, L. Ambrosone, B. Bigliardo, R. Di Carlo, Elevated bioactive prolactin levels in systemic lupus erythematosus–association with disease activity, J Rheumatol 28 (10) (2001) 2216–2221.

[117] J.M. Miranda, R.E. Prieto, R. Paniagua, G. Garcia, D. Amato, L. Barile, et al., Clinical significance of serum and urine prolactin levels in lupus glomerulonephritis, Lupus 7 (6) (1998) 387–391.

[118] F. Blanco-Favela, G. Quintal-Alvarez, A. Leanos-Miranda, Association between prolactin and disease activity in systemic lupus erythematosus. Influence of statistical power, J Rheumatol 26 (1) (1999) 55–59.

[119] L.J. Jara, O. Vera-Lastra, J.M. Miranda, M. Alcala, J. Alvarez-Nemegyei, Prolactin in human systemic lupus erythematosus, Lupus 10 (10) (2001) 748–756.

[120] A. Stevens, D.W. Ray, J. Worthington, J.R. Davis, Polymorphisms of the human prolactin gene–implications for production of lymphocyte prolactin and systemic lupus erythematosus, Lupus 10 (10) (2001) 676–683.

[121] G.S. Cooper, M.A. Dooley, E.L. Treadwell, E.W. St Clair, G.S. Gilkeson, Hormonal and reproductive risk factors for development of systemic lupus erythematosus: results of a population-based, case-control study, Arthritis Rheum 46 (7) (2002) 1830–1839.

[122] J.S. Krumenacker, D.J. Buckley, M.A. Leff, Prolactin regulated apoptosis of Nb2 lymphoma cells:pim-1, bcl-2, and Bax expression, Endocrine 9 (1998) 163–170.

[123] M.A. Gutierez, J.M. Anaya, G.E. Cabrera, O. Vindrola, L.R. Espinoza, [Prolactin, a link between the neuroendocrine and immune systems. Role in the pathogenesis of rheumatic diseases], Rev Rhum Ed Fr 61 (4) (1994) 278–285.

[124] T.C. Cesario, S. Yousefi, G. Carandang, N. Sadati, J. Le, N. Vaziri, Enhanced yields of gamma interferon in prolactin treated human peripheral blood mononuclear cells, Proc Soc Exp Biol Med 205 (1) (1994) 89–95.

[125] N. Stahl, J. Decker, Androgenic status of males with SLE, Arth Rheum 21 (1978) 665–668.

[126] R.G. Lahita, Sex hormones as immunomodulators of disease, Ann N Y Acad Sci 685 (1993) 278–287.

[127] R.G. Lahita, Sex steroids and the rheumatic diseases, Arthritis Rheum 28 (1985) 121–126.

[128] R.G. Lahita, H.L. Bradlow, H.G. Kunkel, J. Fishman, Increased 16 alpha hydroxylation of estradiol in systemic lupus erythematosus, J Clin Endo Metab 53 (1981) 174–178.

[129] R.G. Lahita, N. Chiorazzi, A. Gibofsky, Familial systemic lupus erythematosus in males, Arth Rheum 26 (1983) 39–44.

[130] C.M. Stein, J.M. Olson, C. Gray-McGuire, G.R. Bruner, J.B. Harley, K.L. Moser, Increased prevalence of renal disease in systemic lupus erythematosus families with affected male relatives, Arthritis Rheum 46 (2) (2002) 428–435.

[131] M.H. Miller, M.B. Urowitz, D.D. Gladman, D.W. Killinger, Systemic lupus erythematosus in males, Medicine (Baltimore) 62 (5) (1983) 327–334.

[132] L.D. Kaufman, J.J. Gomez-Reino, M.H. Heinicke, P.D. Gorevic, Male lupus: retrospective analysis of the clinical and laboratory features of 52 patients, with a review of the literature, Semin Arthritis Rheum 18 (1989) 189–197.

[133] J.F. Sequeira, G. Keser, B. Greenstein, M.J. Wheeler, P.C. Duarte, M.A. Khamashta, et al., Systemic lupus erythematosus: sex hormones in male patients, Lupus 2 (1993) 315–317.

[134] H. Anisman, M.G. Baines, I. Berczi, C.N. Bernstein, M.G. Blennerhassett, R.M. Gorczynski, et al., Neuroimmune mechanisms in health and disease: 2, Disease. Can Med Assoc J 155 (8) (1996) 1075–1082.

[135] F. Camilleri, C. Mallia, Male SLE patients in Malta, Adv Exp Med Biol 455 (1999) 173–179.

[136] D.M. Chang, C.C. Chang, S.Y. Kuo, S.J. Chu, M.L. Chang, Hormonal profiles and immunological studies of male lupus in Taiwan, Clin Rheumatol 18 (2) (1999) 158–162.

[137] J.F. de Carvalho, A.P. do Nascimento, L.A. Testagrossa, R.T. Barros, E. Bonfa, Male gender results in more severe lupus nephritis, Rheumatol Int (2009). [Epub ahead of print].

[138] C.C. Mok, C.S. Lau, Profile of sex hormones in male patients with systemic lupus erythematosus, Lupus 9 (4) (2000) 252–257.

[139] Z.S. Alekberova, M.I. Folomeev, [Sexual dimorphism in rheumatic diseases], Revmatologiia (Mosk) (1985) 58–61.

[140] M. Folomeev, Z. Alekberova, J. Polyntsev, The role of estrogen androgen imbalance in rheumatic diseases. (Rus), Terapeuticheskii Arkhiv 62 (5) (1990) 17–21.

[141] Z. Alekberova, G. Kotelnikova, M. Folomeev, Aortic defects in systemic lupus erythematosus. (Rus), Terapeuticheskii Arkhiv 61 (5) (1989) 35–38.

[142] M. Folomeev, Z. Alekberova, Impotence in systemic lupus erythematosus, J Rheum 17 (1990) 117–119.

[143] N. Aoki, Klinefelter's syndrome, autoimmunity, and associated endocrinopathies, Intern Med 38 (11) (1999) 838–839.

[144] R.G. Lahita, The connective tissue diseases and the overall influence of gender, Int J Fertil Menopausal Stud 41 (2) (1996) 156–165.

[145] L.C. Harrison, B. Dean, I. Peluso, S. Clark, G. Ward, Insulin resistance, acanthosis nigricans, and polycystic ovaries associated with a circulating inhibitor of postbinding insulin action, J Clin Endo Metab 60 (1985) 1047–1052.

[146] W.P. Dmowski, H.M. Gebel, R.G. Rawlins, Immunological aspects of endometriosis. Obstetrics and Gynecology clinics of North America, W.B. Saunders Inc, Philadelphia, Penn, 1989. 93–103.

[147] R.W. Steele, W.P. Dmowski, D.J. Marmer, Immunologic aspects of human endometriosis, Am J Reprod Immunol Microbiol 6 (1984) 33.

[148] E.Y. Adashi, Do cytokines play a role in the regulation of ovarian function? PNE 3 (1990) 11–17.

[149] N.J. Olsen, G. Olson, S.M. Viselli, X. Gu, W.J. Kovacs, Androgen Receptors in thymic epithelium modulate thymus size and thymocyte development, Endocrinology 142 (3) (2001) 1278–1283.

[150] N.J. Olsen, M.B. Watson, G.S. Henderson, W.J. Kovacs, Androgen deprivation induces phenotypic and functional changes in the thymus of adult male mice, Endocrinology 129 (1991) 2471–2476.

[151] Y. Weinstein, Y. Isakov, Effects of testosterone metabolites and of anabolic androgens on the bone marrow and thymus in castrated female mice, Immunopharm 5 (1983) 229–237.

[152] A. Gompel, J.C. Piette, Is there a place for postmenopausal hormone therapy use in women with lupus? Panminerva MEdicine 50 (3) (2008) 247–254.

[153] C. Holroyd, C.J. Edwards, The Effects of hormone replacement therapy on autoimmune diseases: rheumatoid arthritis and systemic lupus erythemaotsus, Climacteric 12 (5) (2009) 378–386.

[154] M.H. Liang, E.W. Karlson, Female hormone therapy and the risk of developing or exacerbating systemic lupus erythematosus or rheumatoid arthritis, Proc Assoc Am Physicians 108 (1) (1996) 25–28.

[155] M.O. Bernier, Y. Mikaeloff, M. Hudson, S. Suissa, Combined oral contraceptives and the risk of systemic lupus erythematosus, Arthritis Rheum 61 (4) (2009) 476–481.

[156] J. Buyon, H.M. Korchack, L.E. Rutherford, Female hormones reduce neutrophil responsiveness in vitro, Arth Rheum 27 (1984) 623–630.

[157] P. Bodel, G.M. Dillard, S.S. Kaplan, S.E. Malawista, Antiinflammatory effects of estradiol on human blood leukocytes, J Lab Clin Med 80 (1972) 373–384.

[158] M. Lockshin, Genome and hormones: gender difference in physiology-invited review; sex ratio and rheumatic disease, Journal of Applied Physiology 91 (2001) 2366–2373.

[159] A. Linos, J.W. Worthing, W.M. O'Fallon, L.T. Kurland, The epidemiology of rheumatoid arthritis in Rochester, Minnesota: a study of its incidence, prevalence, and mortality, Am J Epidem 111 (1989) 87–98.

[160] A. Linos, W.M. O'Fallon, J.W. Worthington, L.T. Kurland, Case control study of rheumatoid arthritis and prior use of oral contraceptives, Lancet 1 (1983) 1299–1300.

[161] M. Oka, U. Vainio, Effect of pregnancy on the prognosis and serology of rheumatoid arthritis, Acta Rheum Scand 12 (1966) 47–52.

[162] C.C. Mok, C.S. Lau, C.T. Ho, R.W. Wong, Do flares of systemic lupus erythematosus decline after menopause? Scand J Rheumatol 28 (6) (1999) 357–362.

[163] V. Agnello, K. Pariser, J. Gell, Preliminary observation on Danazol therapy of systemic lupus: effects on DNA antibodies, thrombocytopenia and complement, J Rheumatol 10 (1983) 682–687.

[164] Y.S. Ahn, W.J. Harrington, S.R. Simon, R. Mylvaganam, L.M. Pall, A.G. So, Danazol for the treatment of idiopathic thrombocytopenic purpura, N Eng J Med 308 (1983) 1396–1399.

[165] Y.S. Ahn, R. Mylvaganam, R.O. Garcia, C.I. Kim, D. Palow, W.J. Harrington, Low-dose danazol therapy in idiopathic thrombocytopenic purpura, Ann Intern Med 107 (1987) 177–181.

[166] V.H. Donaldson, E.V. Hess, Effect of danazol on lupus erythematosus like disease in hereditary angioneurotic edema, Lancet 2 (1980) 1143.

[167] M. Dougados, C. Job-Deselandre, B. Amor, J. Menkes, Danazol therapy in systemic lupus erythematosus, Clin Trials J 24 (1987) 191–200.

[168] M. Fretwell, L.O. Altman, Exacerbation of a lupus erythematosus-like syndrome during treatment of non-C1 esterase inhibitor dependent angioedema with danazol, J Allergy Clin Immunol 69 (1982) 306–310.

[169] R.G. Lahita, C.Y. Cheng, C. Monder, C.W. Bardin, Experience with 19-nortestosterone in the therapy of systemic lupus erythematosus: worsened disease after treatment with 19-nortestosterone in men and lack of improvement in women, J Rheumatol 19 (1992) 547–555.

[170] F. Liote, V. Dehaine, C. Pelissier, F. Kuttenn, P. Jungers, Hormonal modulation with cyproterone acetate in systemic lupus erythematosus: a long term follow-up study, Arth Rheum 28 (11) (1991) 1243–1250.

[171] A. El-Roeiy, W.P. Dmowski, N. Gleicher, Effect of danazol and GnRH agonists (GnRH-a) on the immune system in endometriosis, Soc Gynecol Invest 283 (1988).

[172] K. Sudo, K. Shiota, T. Masaki, T. Fujita, Effects of TAP-144-SR, a sustained-release formulation of a potent GnRH agonist, on experimental endometriosis in the rat, Endocrinol Jpn 38 (1991) 39–45.

[173] O. Khorram, L. Vu, S.C.C. Yen, Activation of immune function by dehydroepiandrosterone (DHEA) in age-advanced men, Journal of Gerontology 52A (1) (1997) 1–7.

[174] N.F.L. Spencer, M.E. Poynter, J.D. Hennebold, H.H. Mu, R.A. Daynes, Does DHEAS restore immune competence in aged animals through its capacity to function as a natural modulator of peroxisome activities? in: F.L. Bellino, R.D. Daynes, P.J. Hornsby, D.H. Lavren, J.E. Nestler (Eds.), Dehydroepiandrosterone (DHEA) and Aging, 1 ed)., New York Academy of Sciences, New York, 1996, pp. 200–216.

[175] J.T. Merrill, Dehydroepiandrosterone, a sex steroid metabolite in development for systemic lupus erythematosus, Expert Opin Investig Drugs 12 (6) (2003) 1017–1025.

[176] K.E. Schwartz, Autoimmunity, dehydroepiandrosterone (DHEA), and stress, J Adolesc Health 30 (4 Suppl) (2002) 37–43.

[177] M.A. Petri, R.G. Lahita, R.F. Van Vollenhoven, J.T. Merrill, M. Schiff, E.M. Ginzler, et al., Effects of prasterone on corticosteroid requirements of women with systemic lupus erythematosus: a double-blind, randomized, placebo-controlled trial, Arthritis Rheum 46 (7) (2002) 1820–1829.

[178] M.A. Petri, P.J. Mease, J.T. Merrill, R.G. Lahita, M.J. Iannini, D.E. Yocum, et al., Effects of prasterone on disease activity and symptoms in women with active systemic lupus erythematosus, Arthritis Rheum 50 (9) (2004) 2858–2868.

[179] V.R. Van, E.G. Engleman, J.L. McGuire, Dehydroepiandrosterone in systemic lupus erythematosus. Results of a double-blind, placebo-controlled, randomized clinical trial, Arthritis Rheum 38 (12) (1995) 1826–1831.

[180] V.R. Van, J.L. McGuire, Studies of dehydroepiandrosterone (DHEA) as a therapeutic agent in systemic lupus erythematosus, Ann Med Interne (Paris) 147 (4) (1996) 290–296.

[181] B.A. Araneo, R.A. Daynes, Dehydroepiandrosterone functions as more than an antiglucocorticoid in preserving immunocompetence after thermal injury, Endocrinology 136 (2) (1995) 393–401.

[182] R.A. Daynes, B.A. Araneo, Programming of lymphocyte responses to activation: extrinsic factors, provided micro-environementally, confer flexibility and compartmentalization to T cell function, Chem Immunol 54 (1992) 1–20.

[183] R.A. Daynes, D.J. Dudley, B.A. Araneo, Regulation of murine lymphokine production in vivo II. Dehydroepiandrosterone is a natural enhancer of interleukin 2 synthesis by helper T cells, J Immunol 20 (1990) 793–802.

[184] K. Nandy, L. Harbous, D. Bennet, M. Bennet, Correlation between learning disorder and elevated brain reactive antibodies in aged C57B1/6 and young NZB mice, Life Sciences 33 (1983) 1499–1503.

[185] N. Geschwind, P. Behan, Left handedness: association with immune disease, migraine and developmental learning disorders, Proc Nat Acad Sci 74 (1972) 5097–5100.

[186] R.G. Lahita, Systemic Lupus Erythematosus: learning disability in the male offspring of female patients and relationship to laterality, Psychoneuroendocrinology 13 (1988) 385–396.

[187] Wood Lc, D.S. Cooper, Autoimmune thyroid disease, left-handedness, and developmental dyslexia, Psychoneuroendocrinology 17 (1) (1992) 95–99.

[188] K.B. Boone, R.S. Swerdloff, B.L. Miller, D.H. Geschwind, J. Razani, A. Lee, et al., Neuropsychological profiles of adults with Klinefelter syndrome, J Int Neuropsychol Soc 7 (4) (2001) 446–456.

[189] W.H. James, The sex ratios of probands and of secondary cases in conditions of multifactorial inheritance where liability varies with sex, J Med Genet 28 (1) (1991) 41–43.

[190] S.A. Ahmed, M. Dauphinee, N. Talal, Effects of short term administration of sex hormones on normal and autoimmune mice, J Immunol 134 (1985) 204–210.

[191] J.F. Cohen-Solal, V. Jeganathan, L. Hill, D. Kawabata, D. Rodriguez-Pinto, C. Grimaldi, et al., Hormonal regulation of B-cell function and systemic lupus erythematosus, Lupus 17 (6) (2008) 528–532.

[192] C.M. Grimaldi, R. Hicks, B. Diamond, B cell selection and susceptibility to autoimmunity, J Immunol 174 (4) (2005) 1775–1781.

[193] R.J. Ablin, G.R. Bruns, J. Guinan, I.M. Bush, Antiandrogenic suppression of lymphocytic blastogenesis: in vitro and in vivo observations, Experientia 30 (1974) 1351–1353.

[194] S.A. Huber, J. Kupperman, M.K. Newell, Hormonal regulation of CD4(+) T-cell responses in coxsackievirus B3-induced myocarditis in mice, J Virol 73 (6) (1999) 4689–4695.

[195] Y. Hirota, T. Suzuki, Y. Chazono, Y. Bito, Humoral immune responses characteristics of testosterone propionate treated chickens, Immunology 30 (1976) 341–348.

[196] Y. Hirota, T. Suzuki, Y. Bito, The B-cell development independent of bursa of Fabricius but dependent upon the thymus in chickens treated with testosterone propionate, Immunology 39 (1980) 37–46.

[197] Y. Hirota, T. Suzuki, Y. Bito, The development of unusual B-cell functions in the testosterone-propionate treated chicken, Immunology 39 (1980) 29–36.

[198] C.J. Grossman, P. Nathan, B.B. Taylor, L.J. Sholitan, Rat thymic dihydrotestosterone receptor: preparation, location, and physicochemical properties, Steroids 34 (1979) 539–553.

[199] C.J. Grossman, L.J. Sholitan, G.A. Roselle, Dihydrotestosterone regulation of thymocyte function in the rat: mediation by serum factors, J Steroid Biochem 19 (1983) 1459–1467.

[200] M. Hince, S. Sakkal, K. Vlahos, J. Dudakov, R. Boyd, A. Chidgey, The role of sex steroids and gonadectomy in the control of thymic involution, Cell Immunol 252 (1-2) (2008) 122–138.

[201] S.T. Page, S.R. Plymate, W.J. Bremner, A.M. Matsumoto, D.L. Hess, D.W. Lin, et al., Effect of medical castration on CD4+ CD25+ T cells, CD8+ T cell IFN-gamma expression, and NK cells: a physiological role for testosterone and/or its metabolites, Am J Physiol Endocrinol Metab 290 (5) (2006) E856–E863.

[202] L. Klareskog, U. Forsum, P.A. Peterson, Hormonal regulation of the expression of Ia antigens on mammary gland epithelia, Eur J Immunol 10 (1980) 958–963.

[203] G. Terres, S.L. Morrison, G.L. Habicht, A quantitative difference in the immune response between male and female mice, Proc Exp Biol Med 27 (1968) 664–667.

[204] J.S. Thompson, M.K. Crawford, R. Reilly, C. Stevenson, Estrogenic hormones in immune responses in normal and X irradiated mice, J Immunol 98 (1957) 331–335.

[205] S.R. Waltman, R.M. Brude, J. Benios, Prevention of corneal rejection by estrogens, Transplantation 11 (1971) 194–196.

[206] W.H. Stimson, I.C. Hunter, Oestrogen-induced immunoregulation mediated through the thymus, J Clin Lab immunology 4 (1) (1980) 27–33.

[207] C. Wang, B. Dehghani, Y. Li, L.J. Kaler, T. Proctor, A.A. Vandenbark, et al., Membrane estrogen receptor regulates experimental autoimmune encephalomyelitis through up-regulation of programmed death 1, J Immunol 182 (5) (2009) 3294–3303.

[208] J. Ding, B.T. Zhu, Unique effect of the pregnancy hormone estriol on antigen-induced production of specific antibodies in female BALB/c mice, Steroids 73 (3) (2008) 289–298.

[209] F.A. Wyle, J.R. Kent, Immunosuppression by sex steroid hormones. I. The effect upon PHA- and PPD-stimulated lymphocytes, Clinical and Experimental Immunology 27 (1977) 407–415.

[210] P.S. Satoh, W.E. Fleming, K.A. Johnstone, J.M. Ozmun, Active E rosette formation in women taking oral contraceptives, N Eng J Med 296 (1977) 1–54.

[211] R.J. Graff, M.A. Lappe, G.D. Snell, The influence of the gonads and the adrenal glands on the immune response to skin grafts, Transplantation 7 (1969) 105–111.

[212] A.J. Lengi, R.A. Phillips, E. Karpuzoglu, S.A. Ahmed, 17beta-estradiol downregulates interferon regulatory factor-1 in murine splenocytes, J Mol Endocrinol 37 (3) (2006) 421–432.

[213] P. Tai, J. Wang, H. Jin, X. Song, J. Yan, Y. Kang, et al., Induction of regulatory T cells by physiological level estrogen, J Cell Physiol 214 (2) (2008) 456–464.

[214] D. Buskila, M. Berezin, H. Gur, H.C. Lin, I. Alosachie, J.W. Terryberry, et al., Autoantibody profile in the sera of women with hyperprolactinemia, J Autoimmun 8 (3) (1995) 415–424.

[215] Z. Sthoeger, N. Chiorazzi, R.G. Lahita, Regulation of the immune response by sex steroids, J Immunol 141 (1988) 91–98.

[216] T. Paavonen, H. Aronen, S. Pyrhonen, A. Hajba, L.C. Andersson, The effects of anti-estrogen therapy on

lymphocyte functions in breast cancer patients, APMIS 99 (1991) 163–170.

[217] J.H.M. Cohen, L. Danel, G. Cordier, S. Saez, J. Revillard, Sex steroid receptors in peripheral T cells: absence of androgen receptors and restriction of estrogen receptors to OKt8 positive cells, J Immunology 131 (1983) 2767–2771.

[218] L. Danel, G. Sovweine, J.C. Monier, S. Saez, Specific estrogen binding sites in human lymphoid cells and thymic cells, J Steroid Biochem 18 (1983) 559.

[219] Y.J. Lee, K.S. Shin, S.W. Kang, C.K. Lee, B. Yoo, H.S. Cha, Association of the estrogen receptor alpha gene polymorphisms with disease onset in systemic lupus erythematosus, Annals of Rheumatic Disease 63 (2004) 1244–1249.

[220] N. Kanda, K. Tamaki, Estrogen enhances immunoglobulin production by human PBMCs, J Allergy Clin Immunol 103 (2 Pt 1) (1999) 282–288.

[221] M.J. Evans, S. MacLaughlin, R.D. Marvin, N.I. Abdou, Estrogen decreases in vitro apoptosis of peripheral blood mononuclear cells from women with normal menstrual cycles and decreases TNF-alpha production in SLE but not in normal cultures, Clin Immunol Immunopathol 82 (3) (1997) 258–262.

[222] A. Inui, H. Ogaswara, T. Naito, Estroegn reeceptor expression by peripheral blood mononuclear cells of patients with systemic lupus erythematosus, Clin Rheumatol 26 (2007) 1675–1678.

[223] K.L. Phiel, R.A. Henderson, S.J. Adelman, M.M. Elloso, Differential estrogen receptor gene expression in human peripheral blood mononuclear cell populations, Immunol Lett 97 (2005) 107–113.

[224] S. Gorjestani, V. Rider, B.F. Kimler, C. Greenwell, N.I. Abdou, Extracellular signal-regulated kinase 1/2 signalling in SLE T cells is influenced by estrogen and disease activity, Lupus 17 (6) (2008) 548–554.

[225] J.F. Cohen-Solal, V. Jeganathan, C.M. Grimaldi, E. Peeva, B. Diamond, Sex hormones and SLE: influencing the fate of autoreactive B cells, Curr Top Microbiol Immunol 305 (2006) 67–88.

[226] J. Venkatesh, E. Peeva, X. Xu, B. Diamond, Cutting Edge: Hormonal milieu, not antigenic specificity, determines the mature phenotype of autoreactive B cells, J Immunol 176 (6) (2006) 3311–3314.

[227] E. Peeva, J. Venkatesh, B. Diamond, Tamoxifen blocks estrogen-induced B cell maturation but not survival, J Immunol 175 (3) (2005) 1415–1423.

[228] K.W. Beagley, C.M. Gockel, Regulation of innate and adaptive immunity by the female sex hormones estradiol and progesterone, FEMS Immunol Med Microbiology 38 (2003) 13–22.

[229] J.P. Buyon, M.A. Petri, M.Y. Kim, K.C. Kalunian, J. Grossman, B.H. Hahn, et al., The effect of combined estrogen and progesterone hormone replacement therapy on disease activity in systemic lupus erythematosus: a randomized trial, Ann Intern Med 142 (12 Pt 1) (2005) 953–962.

[230] T. Mori, H. Kobayashi, H. Nishimoto, Inhibitory effect of progesterone and 20-hydroxy-pregn-4-en-3-one on the phytohemagglutinin-induced transformation of human lymphocytes, Am J Obst Gynec 127 (1977) 151–157.

[231] M.P. Piccinni, M.G. Giudizi, R. Biagiotti, L. Beloni, L. Giannarini, S. Sampognaro, et al., Progesterone favors the development of human T helper cells producing Th2-type cytokines and promotes both IL-4 production and membrane CD30 expression in established Th1 cell clones, J Immunol 155 (1) (1995) 128–133.

[232] A. Canellada, S. Blois, T. Gentile, R.A. Margni Idehu, In vitro modulation of protective antibody responses by estrogen, progesterone and interleukin-6, Am J Reprod Immunol 48 (5) (2002) 334–343.

[233] J.R. Roubinian, N. Talal, J.S. Greenspan, J.R. Goodman, P.K. Siiteri, Effect of castration and sex hormone treatment on survival, anti-nucleic acid antibodies, and glomerulonephritis in NZB/NZW F1 mice, J Exp Med 147 (1978) 1568.

[234] S.K. Mak, E.K. Lam, A.K. Wong, Clinical profile of patients with late-onset SLE: not a benign subgroup, Lupus 7, no 1 (1998) 23–28.

Roles for Infections in Systemic Lupus Erythematosus Pathogenesis

Evan S. Vista, A. Darise Farris, Judith A. James

INTRODUCTION

Systemic lupus erythematosus (SLE) remains a complicated, heterogeneous, multifactorial disease and many studies have suggested involvement of specific genetic, hormonal, immunologic, and environmental factors, which broadly include infection (Figure 24.1). Hyper-responsive B cells, dysregulated T cells, increased numbers of plasmablasts/plasmacytes, autoantibody production, and immune complex formation are common immune regulation abnormalities in lupus [1–7]. Defective phagocytosis and subsequent defective clearance of apoptotic cells in SLE permits antigen persistence and the development of unusually modified antigens with the potential to trigger autoimmune cascades [8]. Dysregulated immune systems in SLE, whether intrinsic or induced by immunomodulatory drugs, promote various levels of immunodeficiency, thus predisposing lupus patients to infections that increase morbidity and mortality.

Through complex interactions of genetic predisposition, hormonal influences, dysregulated immune responses after infection, ultraviolet light exposure and likely other factors, the balance can be tipped to initiate SLE in select individuals. In addition, by driving different cytokine pathways, immune cell differentiation/subsets and pathogen-specific responses, infection may also trigger disease flares and/or progression. Recent advances in identifying and confirming genetic associations with SLE are leading to examination of potential interactions between infections and select genetic polymorphisms. This chapter focuses on examining roles for infection in the initiation and pathogenesis of SLE.

INFECTION AND SLE INITATION

SLE is an interesting disorder in that individuals can live for decades without evidence of clinical disease. Although some genetic defects, such as select complement deficiencies, may be sufficient to eventually lead to lupus, in most individuals multiple genetic alleles are required for a clinical phenotype. Indeed, in monozygotic twins where one twin has lupus, the concordance rate for disease is only 20–40% [9–11]. Therefore, environmental factors that might in part trigger lupus in susceptible individuals have been sought. Some ideal features of a pathogenic agent that may promote triggering of autoimmunity are presented in Table 24.1. Individuals who subsequently develop lupus initially have no detectable autoantibodies, then develop limited autoimmunity and progressively accrue autoantibody specificities as clinical disease develops [12]. Based upon these observations, molecular mimicry between an infectious agent and a common autoantigen could trigger autoantibody production and subsequently clinical disease. Various pathogenic mechanisms, such as molecular mimicry, bystander lymphocyte activation, presence of superantigens, polyclonal activation, and epitope spreading have been implicated. We will first present accumulating data for a role of Epstein-Barr virus in lupus initiation, followed by work in other pathogens linked with SLE autoimmunity.

Epstein-Barr Virus as a Model for Lupus Autoimmunity

Epstein-Barr virus (EBV) is human herpesvirus 4 that infects nearly all individuals worldwide by early adulthood. This virus infects and remains latent in memory

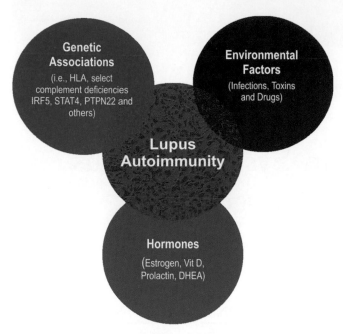

FIGURE 24.1 Schematic representation of the potential interplay of different types of etiologic factors in generation of lupus autoimmunity and clinical disease. Genetic predisposition to SLE has been long supported by familial aggregation and increased disease concordance in monozygotic twins. Recent research has found and confirmed association of SLE with a number of susceptibility genes, of which only a few examples are provided in this figure. Environmental factors including select pathogens, ultraviolet light exposure, smoking, and select vitamin deficiencies are also important. Finally, SLE is increased in females, which has led to examination of hormonal imbalances in lupus autoimmunity. Abbreviations: HLA, human leukocyte antigen; VitD, vitamin D; DHEA, dehydroepiandrosterone.

B cells of most individuals with episodic reactivation. Given the ubiquitous nature of EBV as an exposure risk in the general population and its predilection for B cells, this pathogen has been explored for association with select autoimmune diseases, such as SLE. SLE patients were first used as a control group in a pediatric study which showed association with EBV seroconversion and SLE in 1969 [13]. The hypothesis that arose from this association was that frequent EBV reactivation

TABLE 24.1 Proposed Properties for an Infectious Trigger of Lupus Autoimmunity

Ubiquitous infection

Persistent infection

Infection precedes autoimmunity

Triggers production of pathogenic autoreactive lymphocytes

Inhibits apoptosis of infected cells

Induces Type 1 interferons

Inhibits its own destruction

Breaks tolerance

might lead to chronic immune stimulation, as well as the observation that EBV may promote antibodies that cross-react with self-antigens. Historical debate raged, considering whether these serologic associations were byproducts of intrinsic or extrinsic immune dysregulation within SLE or they had pathogenic association. Please see [13–15] for detailed reviews.

More recently, studies have focused on serologic and deoxyribonucleic acid (DNA) associations of EBV exposure and SLE. Virtually all pediatric SLE patients have serologic evidence of EBV infection, while only 70% of pediatric controls are seropositive (odds ratio [OR], 49.9; 95% confidence interval [95% CI], 9.3–1,025, $p < 0.0001$), a result that is confirmed at the level of recovered EBV DNA [16]. Further studies have also found strong EBV serologic association in antinuclear antibody (ANA)-positive adult SLE patients (seroconversion in 195 of 196 patients and 370 of 392 matched controls: OR, 9.35; 95% CI, 1.45 to infinity, $p = 0.014$) [17]. Various studies from several other groups throughout the world have confirmed and expanded these associations (please see [18] for a recent review). EBV appears well suited to initiate autoimmune disease in that it infects B cells, promotes dysregulated polyclonal B cell activation, and promotes autoantibody production. EBV latency provides lifelong antigenic challenge. EBV also produces viral proteins that prevent apoptosis (including a Bcl-2 homolog) and exhibit interleukin-10-like activity [18].

Approximately 90% of healthy, EBV-infected adults make antibodies directed against the major EBV latency protein, Epstein-Barr nuclear antigen 1 (EBNA-1). The healthy humoral immune response toward this protein is dominated by antibodies binding the repetitive glycine–alanine-rich central region of EBNA-1 [19, 20]. This antigen exhibits a very unusual antigen presentation, since defective, truncated ribosomal products of EBNA-1 seem to be required for presentation of EBNA-1 peptides through the class I pathway, limiting this response [21, 22]. The normal level of expression of EBNA-1 in physiologic conditions also further impairs processing and presentation of EBNA-1 peptides to stimulate EBNA-1-specific CD4-positive T cells [23]. Lupus patients generate higher concentrations of antibodies against the amino and carboxyl regions of EBNA-1, while normal EBV-positive controls actually produced higher levels of antibodies against the glycine–alanine-rich middle domain than did lupus patients [20]. Thus, the immune response to EBV is fundamentally altered in SLE.

Studies from our group and others have previously demonstrated molecular mimicry links between select sequences of EBV and lupus humoral autoimmunity [13–15, 24–37]. We have shown that common human antigenic targets of the lupus autoantigens, Sm B′

(PPPGMRPP) and Sm D1 ((GR)$_x$), cross-react with sequences from EBNA-1 PPPGRRP and ((GR)$_x$), respectively [24, 25, 27, 31, 36]. Other groups have described cross-reactivity between antigenic regions of EBNA-1 or EBNA-2 and the Sm D1 epitope [33—35]. Immunization of rabbits and select strains of mice with cross-reactive EBNA-1 peptides or fragments also induce lupus autoimmunity and select aspects of clinical disease [27, 32, 36, 37].

To further test the potential generality that lupus autoimmunity may arise from anti-EBV heteroimmunity and taking a hint from the PPPGMRPP epitope of anti-Sm, we investigated early events in anti-60-kD Ro autoimmunity, one of the earliest responses measured in preclinical lupus patient sera [12]. SLE patients who developed anti-Ro without anti-La were most likely to target one sequential region of 60 kD Ro, Ro169-180 or TKYKQRNGWSHKD (Ro169). Affinity-purified antibodies against Ro169 also bound specifically to EBNA-1. This cross-reactivity was mapped to the GGSGSGPRHRDGVRR sequence of EBNA-1, which bears no significant primary homology with Ro169. Rabbits immunized with Ro169 or the cross-reactive EBNA-1 sequence developed lupus autoimmunity and select lupus-like clinical features [29]. The data in both the Sm and Ro examples presented above are consistent with a cross-reactive antibody being critically important in the induction of lupus autoimmunity. Once the autoantibody response is initiated, the autoantigen is processed, and B-cell epitope spreading occurs. The affected individual thus develops the autoimmune response that places her or him at increased risk for developing clinical lupus.

Pediatric SLE patients have unique humoral immune responses to EBNA-1, some of which cross-react with self antigens [20]. SLE patients have also been found to have aberrant EBV T-cell responses, with suggestions of increased interferon-producing CD4$^+$ EBV-specific cells and dysfunctional CD8$^+$ T-cell responses [38, 39].

The striking concept that EBV uses mature B cell biology to access memory B cells as a site of persistent infection has also been shown by Thorley-Lawson and colleagues. EBV adapts its gene expression profile to the state of the B cell it resides in and that the level of infection is stable over time [37, 40]. Infection of B lymphocytes *in vitro* invariably leads to their transformation into proliferating, activated lymphoblasts through the expression of nine latent proteins (the growth transcription program or latency III). In contrast, the virus is quiescent *in vivo*, persisting in resting, peripheral memory B cells that express none of the latent proteins (the latency program). Although patients with SLE have frequencies of infected cells comparable to those seen in immunosuppressed membrane patients, in SLE the effect was independent of immunosuppressive therapy. Aberrant expression of viral lytic (BZLF1)

and latency (LMP d1 and 2a) genes was detected in the blood of SLE patients [37].

Interestingly, Kang and colleagues have shown that patients with SLE have defective control of latent EBV infection from probable altered T-cell responses [38]. Virus-specific T-cell responses to EBV viral loads using whole blood recall assays, HLA-A2 tetramers, and real-time quantitative PCR were analyzed. Patients with SLE had an approximately 40-fold increase in EBV viral loads compared with controls, a finding not explained by disease activity or immunosuppressive medications. The frequency of EBV-specific CD69$^+$ CD4$^+$ T cells producing IFN-γ was higher in patients with SLE than in controls. EBV viral loads were inversely correlated with the frequency of EBV-specific CD69$^+$ CD4$^+$ T cells producing IFN-γ and were positively correlated with the frequencies of CD69$^+$ CD8$^+$ T cells producing IFN-γ and with EBV-specific, HLA-A2 tetramer-positive CD8$^+$ T cells. Patients with SLE have increased EBV viral loads in peripheral blood mononuclear cells (PBMCs) that probably stem from inadequate CD8$^+$ T-cell responses against EBV. However, patients with SLE can mount appropriate EBV-specific CD4$^+$ T-cell responses that account for the lack of development of EBV-associated lymphoproliferative diseases. Please see Figure 24.2 for a working model of EBV-triggered lupus autoimmunity.

Role of Other Infection-Derived Antigens

Aside from EBV, other infectious pathogens may also play important roles in the induction of autoimmunity and/or to precipitation of overt clinical disease. Infection-derived antigens (IDAs) found in common and endogenous microbial flora (staphylococci, streptococci, mycoplasma, certain viruses) activate large numbers of T cells regardless of antigen specificity. Acting as superantigens, they bind avidly to Class II MHC molecules at sites distinct from the conventional antigen-binding clefts, allowing for productive interactions between T lymphocytes and antigen-presenting cells that do not require recognition of specific antigens. This permits activation of normally tolerant self-reactive T lymphocytes, with consequent B-cell activation [41].

In a study of 22 female patients with SLE, evaluating the presence and quantity of human cytomegalovirus (CMV) genome, the authors found the viral genome in 100% of the patients, whereas it was present only in 73% of 15 healthy female controls ($p = 0.02$) [42]. Another study showed that immunization with the structural CMV pp65 antigen induced lupus-associated autoantibodies and severe glomerulonephritis. Reports of acute human parvovirus B19 (B19) infection inducing SLE-like symptoms exist, but no definite association has so far been proven [43—46].

A Model for the Development of Systemic Lupus Erythematosus (SLE)

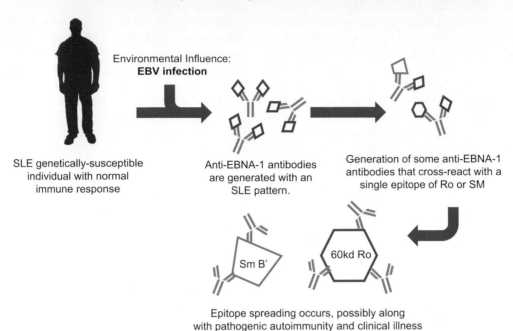

FIGURE 24.2   Proposed model for how Epstein-Barr virus exposure and inappropriate humoral immune response against the major latent EBV protein (EBNA-1) could lead to lupus autoimmunity and clinical disease. Abbreviations: EBNA-1, Epstein Barr nuclear antigen-1; EBV, Epstein-Barr virus.

Retroviruses acting as superantigens and perpetuating the autoimmune process have also been demonstrated [47–50]. In addition, molecular mimicry between human T-cell lymphotropic virus-1-related endogenous sequence (HRES)-1 and snRNP complex epitopes may serve as a trigger for initiation of an autoimmune response leading to clinical lupus. HRES-1 encodes a 28-kDa nuclear autoantigen, HRES-1/p28, antibodies to which are detected in 21–30% of SLE patients [49]. Interestingly, these human endogenous retroviruses are heritable in a stable Mendelian fashion and comprise 0.1–0.6% of human DNA, thus contributing substantially to the architecture of the human genome by integration. Undoubtedly, a subset of these is integrated into sites of immune regulation.

Anti-dsDNA antibodies derived from naïve lupus-prone mice can bind to the cell surfaces of murine endogenous microbial flora, suggesting that bacterial antigens may elicit these antibodies in some settings [51]. Monoclonal anti-dsDNA antibodies derived from mice and humans with SLE were shown to bind the glycolipid components of the mycobacterial cell wall [52]. A possible link between infection with Burkholderia bacteria and SLE based on epitope mimicry was demonstrated by Zhang and Reichlin [53]. Their studies showed that purified anti-dsDNA antibodies specifically bind a 15-mer peptide ASPVTARVLWKASH, and the 15-mer peptide partial sequence ARVLWKASH is

shared with Burkholderia bacterial cytochrome B561 partial sequence ARVLWRATH. Sera with anti-dsDNA antibodies also bind to the RAGTDEGFG peptide, which is shared with a Burkholderia bacterial transcription regulator protein sequence.

An experimental model of protection from lupus immunity by infection has also been reported. Infection of mice with the murine retrovirus BM5 murine leukemia virus ameliorated the severity of experimental lupus induced by administration of an anti-DNA antibody bearing the 16/6 idiotype [54]. This model explores the possible interaction between SLE and retroviral infection, as HIV infection has been cogitated to have a beneficial immunological impact over disease activity of SLE. Recent work has also presented evidence that the two most broadly reactive human immunodeficiency virus (HIV)-1 envelope gp41 human monoclonal antibodies bind to phospholipid cardiolipin, raising the question of whether anticardiolipin responses found in SLE patient sera may also inhibit HIV, thereby in theory somewhat protecting SLE patients from HIV infection [55–57].

Early Pathogen Exposures and Their Influence of SLE Onset

Environmental factors experienced in childhood may program changes in the immune system that lead to the production of autoantibodies and autoimmune disease years later [58–62]. The cumulative effect of repeated infectious exposures alters the level of inflammation in a person in a long-standing fashion. These include B- and T-cell repertoire selection and alterations in the relative levels of pro- and anti-inflammatory cytokines produced upon immune cell reactivation. The Hygiene Hypothesis has been suggested to operate as a consequence not only of exposure to infectious pathogens but also as a consequence of administration of vaccines aimed at preventing certain infections. Aggregate exposure to common organisms including CMV, hepatitis A virus, herpes simplex virus I and II, *Chlamydia pneumonia* and *Helicobacter pylori* results in increased circulating levels of IL-6 and c-reactive protein (CRP). Parasitic infections including schistosomiasis increase production of the anti-inflammatory cytokine IL-10 [63, 64]. Early infections may also decrease the likelihood of ANA in later life [61]. This is the opposite of the situation for Rheumatoid Factor and rheumatoid arthritis, with ANA being associated with increased viral infections in infancy. Previous studies have also shown that exposure to microbial antigens in the form of vaccines may have various influences the likelihood of developing different autoimmune diseases [58].

INFECTIONS EXACERBATING SLE DISEASE

In a previous section we discussed a concise model wherein a predisposed individual could develop lupus autoimmunity through the interplay of immune dysregulation and IDA. Several studies have also investigated reasons for the great variation in SLE clinical expression. Again, multiple factors such as inherited genetic deficiencies, susceptibility genes, other environmental determinants and physical parameters ultimately determine the diverse phenotypic expression of SLE. This section provides a high level overview that dissects direct correlations between certain infectious processes to the gamut of serologic and clinical manifestations seen among SLE patients.

Activators of Innate Immunity

The role of innate immunity is primarily to protect humans from external harm or threat by employing hard-wired recognition of pathogen-specific molecules. One of the initial sets of events is the activation of a repair program. This program involves the direct modification of bystander cells, which secrete growth and survival factors, and the expansion of local precursor cells. It also involves the attraction and recruitment of inflammatory leukocytes and circulating stem cells. Resident and newly recruited cells secrete cytokines and growth factors and form granulation tissue; eventually, restoration also requires the organization of novel extracellular matrix, via deposition by activated fibroblasts and degradation by infiltrating leukocytes. A second set of events is the activation of a defense program. Microbes are likely to exploit disruption of the physical (such as the epithelia) and chemical barriers (such as mucoid/protein secretions) of the host caused by trauma or toxins. Bystander cells and inflammatory leukocytes activate mechanisms of both passive and active resistance to infection, some of which are preventative and occur constitutively even in the absence of infection. These functional outcomes do not substantially differ from those elicited by the direct recognition of the molecular structures shared by pathogens, referred to as pathogen-associated molecular patterns (PAMPs). Recognition by these signaling pathways promotes the production of pro-inflammatory cytokines and chemokines and the expression of costimulatory molecules. These factors are essential for induction and regulation of the adaptive immune response to foreign antigens. The study of sterile inflammation clearly indicates that endogenous triggers are equally effective at activating a reparative and protective response, including efficient activation of the immune system.

A study of the role played by high-mobility group protein (HMGB1) as a signal of tissue damage, an activator of defense programs, and potential inducer of autoimmunity depicts this highly evolved process [65]. HMGB1 has special significance because it was the first molecule convincingly implicated in sterile inflammation, although it can also be involved in pathogen-response inflammation. The state of alert of the immune system, induced by dying cells regardless of the actual presence of invading pathogens, is likely to be evolutionarily advantageous in some sense. However, the process permits immune responses to be initiated against the antigens that local antigen-presenting cells (APCs) are processing at the time of the injury. This event can lead to autoimmunity in the genetically susceptible host because such antigens are self-antigens. A pictorial summary of how pathogens may interact in part with the immune system in autoimmune disease is presented in Figure 24.3.

Interferons and Infection in SLE

Interferons (IFNs) are a family of cytokines that are commonly driven by infectious processes. SLE patients with active disease have elevated levels of IFN-α in their sera, with a predilection for more severe SLE among

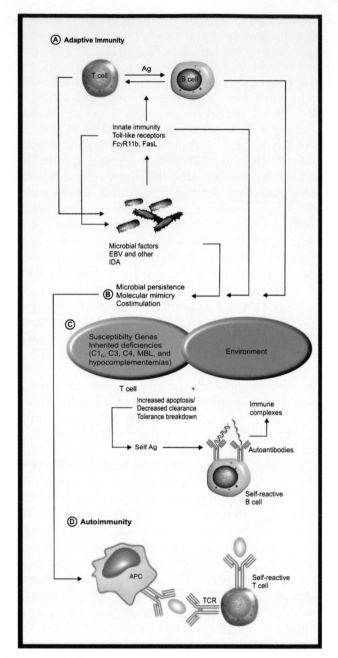

FIGURE 24.3 Schematic drawing showing potential roles for infections in autoimmunity. *A* The adaptive immune response to infection triggered by *B* constant exposure to IDA activates the innate immune response. Those with *C* preexisting genetic predisposition and environmental determinants promotes molecular mimicry, epitope spreading, polyclonal activation, and bystander activation, all of which lead to the development of *D* autoimmunity. Abbreviations: Ag, antigen; EBV, Epstein-Barr Virus; IDA, Iinfection-derived antigens; APC, antigen-presenting cell; TCR, T-cell receptor.

high Type I IFN producers. Studies have also demonstrated that peripheral blood mononuclear cells from SLE patients exhibit a gene expression profile indicative of active IFN-α-signaling [66–68]. Investigation is ongoing to understand the role of infections in driving the interferon-induced gene expression profiles found in lupus. Please see the interferon and SLE chapter for more details.

Roles of Toll-like Receptors

During a state of infection, the human body produces low levels of transient autoantibodies and autoreactive T cells [69]. This implies that when the innate and adaptive immune responses are highly activated, such as is the case during infections, the majority of individuals experience subclinical and potentially transient loss of self-tolerance. The lupus autoantigens, which include double- or single-stranded DNA, histones, chromatin, and RNA binding proteins (including nRNP, Sm, Ro, and La), are composed of or intimately associated with nucleic acids [70]. Nucleic acid recognition is a prime strategy for the innate immune system's detection of microbial pathogens, especially viral pathogens. The role of Toll-like receptors (TLRs) in lupus pathogenesis is more fully discussed in Chapter 17.

FcγRIIb and Infections in SLE

Fc receptors for IgG are essential in regulating innate and adaptive immunity. FcγRIIb is an inhibitory Fc receptor expressed on B lymphocytes and myeloid cells that plays a unique role in regulating immune responses. FcγRIIb is the only inhibitory Fc receptor for IgG and the only Fc receptor expressed on B cells, where it controls the magnitude and persistence of antigenic responses through effects on both plasma cells as well as on mature and memory B cells. In addition to this central role in controlling susceptibility to autoimmunity, FcγRIIb controls survival after bacterial infection, balancing bacterial clearance and the risk of septic shock [71].

FcγRIIb deficiency in mice has been reported to contribute to the development of spontaneous SLE on certain genetic backgrounds and to autoantibody production in certain B-cell subsets [72]. Mice deficient in FcγRIIb develop enhanced inducible immune diseases, such as collagen-induced arthritis (CIA) [73]. In addition, naturally occurring polymorphisms in the promoter of FcγRIIb reduce expression of the receptor in mice and may contribute to spontaneous lupus in other mouse models [74].

Stringent control of FcγRIIb expression is thought to be critical in determining its effect on the immune system. IL-4 enhances FcγRIIb expression and functions on human macrophages, whereas the same cytokine abolishes FcγRIIb function and reduces expression on B cells [75]. Expression of FcγRIIb during dendritic cell (DC) maturation is also modulated by cytokines, as shown by an increase in FcγRIIb on mature human DCs after treatment with IL-10 and IL-13 [76]. Stimulation of B cells and DCs with a physiologically relevant concentration of immune complexes results in their increased expression of FcγRIIb and subsequent apoptosis. Brownlie and colleagues overexpressed FcγRIIb on B cells or macrophages in mice [77]. Overexpression of FcγRIIb on B cells reduced the IgG component of T-dependent immune responses, led to early resolution of CIA, and reduced spontaneous lupus. In contrast, overexpression on macrophages had no effect on immune responses, CIA, or SLE but increased mortality after *Streptococcus pneumoniae* infection.

Mannose Binding Lectin, Complement Deficiencies and Lupus

Abnormal clearance of apoptotic cells due to mannose binding lectin (MBL) and complement (C)1q deficiencies may provide a source of autoantigens in SLE. If we support the postulate that infection is believed to be one of the causes of SLE, the presence of lupus subsets with MBL or C1q deficiency may lead to more frequent infections and thus greatly impact the severity and progression of disease in the affected lupus patient. MBL and C1q can bind to and initiate uptake of apoptotic cells into macrophages, and abnormal clearance of apoptotic cells caused by MBL or C1q deficiency may result in overexpression of autoantigens [79]. This area is more fully discussed in [79–81] and Chapter 20 on complement in SLE.

Hypogammaglobulinemia and Infections in SLE

A major immune pathology observed in SLE is the occurrence of polyclonal hypergammaglobulinemia. However, various forms of antigen-specific hypogammaglobulinemia, both congenital and acquired, can occur in patients with SLE. This functionally immunodeficient state clearly alters the course of SLE, predisposing the patients to developing intractable infections that further increase morbidity within this disease subgroup. Persistent antigen stimulation, recurrent tissue damage, defective clearance of immune complexes and immune dysregulation seen in various primary immunodeficiency states, have all been postulated to contribute toward the progression and exacerbation of SLE [82].

Development of B-cell-specific autoantibodies, early senescence of hyperactive B cells, and B-cell maturation defects have been proposed to be the underlying mechanisms of B-cell dysfunction and antibody deficiency seen among lupus patients [83]. BLyS/BAFF (B-lymphocyte stimulator/B-cell activating factor) is a vital B-cell survival factor. Overexpression of BLyS in mice may lead to SLE-like disease, and BLyS overexpression is common in human SLE [84]. Mutations have been identified in transmembrane activator and calcium-modulator and cyclophilin ligand (CAML) interactor (TACI), a molecule involved in signaling to B cells [85]. Elimination of TACI in mice has been shown to cause lupus-like features, including B-cell hyperplasia, autoantibody formation, and glomerulonephritis [86]. Defects in TACI are thought to account for approximately 10% of patients with common variable immunodeficiency (CVID) or selective IgA deficiency [87].

The occurrence of IgA deficiency in SLE has been reported at a prevalence of 5% in 96 SLE patients in one cohort, suggesting a significant association [88]. Another recent study estimated the prevalence of IgA deficiency at 2.6% in adults ($n = 152$) and 5.2% in children ($n = 77$) with SLE. These particular patients did not have an altered clinical course compared with SLE patients without IgA deficiency [89]. Selective IgM deficiency has been described in patients with SLE and there is a suggestion that it correlates with more severe or long-standing SLE [90].

Reduction in immunoglobulin levels is seen among patients with lymphoproliferative disorders, including myeloma, chronic lymphocytic leukemia, and lymphoma, the latter of which has been associated with SLE [85]. Routine measurement of Ig levels in patients with SLE is a valuable tool for assisting in the identification of those patients at greater risk of infection. Effective therapies exist for treatment of hypogammaglobulinemia, and these can be utilized to improve the care of such patients, once they have been identified.

Association of Infection and Antiphospholipid Antibodies

Antiphospholipid antibodies (aPL) were originally detected in human serum by Wasserman almost 100 years ago, when his complement fixation test was first used for the diagnosis of syphilis and when the Venereal Disease Research Laboratory (VDRL) test was described [91]. A phospholipid termed cardiolipin was the major tissue used in this assay. The VDRL test was subsequently found to not be specific for syphilis but also gave positive results in autoimmune diseases such as SLE. Syphilis was thus the first infection to be recognized as being linked to aPL. The importance of aPL in the pathogenesis of clotting is revealed in antiphospholipid syndrome (APS), which occurs in systemic autoimmune diseases, particularly SLE. Many infections may not only trigger the production of aPL, but also appear to be accompanied by clinical manifestations of APS itself.

In 1990, it was found that the binding of the aPL to phospholipid was enhanced in autoimmune conditions by a co-factor known as β2 glycoprotein I (β2GPI) — a glycoprotein with anticoagulant properties — whereas the nonthrombogenic aPL did not require this cofactor to enhance binding [92, 93]. The two types of aPL were referred to as autoimmune and infectious types. This distinction, however, was subsequently found not to be absolute [94], and it was postulated that infections may be a trigger factor for the induction of pathogenic aPL in certain predisposed individuals.

A hexapeptide (TLRVYK) has been identified by Blank et al. [95] that is specifically recognized by a pathogenic anti-β2GPI monoclonal antibody. An evaluation of the pathogenic potential of a variety of microbial pathogens carrying sequences related to this hexapeptide in mice was carried out by the same group by infusing IgG specific to the peptide intravenously into naive mice. High titers of antipeptide anti-β2GPI antibodies were observed in mice immunized with H. influenza, N. gonorrhoeae, and tetanus toxoid. Significant thrombocytopenia, prolonged activated partial thromboplastin times, and increased percentages of fetal loss were also observed. Zhang et al. recently identified an S. aureus protein (Sbi) that also bound β2GPI and could serve as a target molecule for IgG binding [96]. Gharavi et al. showed that synthetic peptides that share both structural similarity with the putative phospholipid binding region of the β2GPI molecule and a high homology with CMV were able to induce aPL in NIH/Swiss mice [97].

Many viral infections may be accompanied by increases in aPL. Among these, HCV, HIV, and parvovirus B19 infections have been intensively studied [98, 99]. In 1986, Bloom et al. first documented lupus anticoagulant in 44% of AIDS patients and in 43% of asymptomatic HIV-positive individuals (in which they may be transient). The anti-cardiolipin antibodies described in HIV patients are of both the pathogenic (β2GPI co-factor-dependent) and the infectious type (β2GPI-independent). As HIV infection leads to immunosuppression affecting mainly CD4$^+$ cells and macrophages, the pathophysiologic mechanism of APS associated with HIV is likely to be fundamentally different from that in other infections.

Bacterial infections are also associated with increased levels of aPL; however, these autoantibodies are not usually associated with thrombotic events and are not β2GPI-dependent. On the other hand, in patients with leprosy (particularly in the multibacillary type of leprosy) anticardiolipin antibodies may be β2GPI-dependent, as is found in autoimmune diseases [100]. Lucio's phenomenon is a rare manifestation of leprosy in which the histopathologic findings are related to microvasculature thromboses in the absence of inflammatory infiltration of the vessel walls. Levy et al. showed that this type of leprosy was associated with β2GPI dependency of the anticardiolipin antibodies [101]. Streptococcal infections, leptospirosis, and Q fever caused by Coxiella burnetti were also found to be associated with a high frequency of anticardiolipin antibody positivity [91, 98].

Of particular interest is the unusual but potentially fatal subset of catastrophic APS. The latest published analysis showed that no less than 24% of catastrophic APS cases were preceded by infections [102]. These were comprised of respiratory (10%), cutaneous, including infected leg ulcers (4%), urinary tract (4%), gastrointestinal (2%), generalized septic (1%), and other infections (3%) [102].

Metzger et al. developed a biosensor with the ability to distinguish between disease-relevant anti-β2GPI autoantibodies and pathogen-specific β2GPI cross-reactive antibodies that occur transiently during infections [103]. Surface plasmon resonance biosensor analyses demonstrate that covalent attachment of β2GPI through a covalent amide linkage to the self-assembled monolayer (SAM)-coated sensor chip is a suitable method for obtaining highly reproducible APL measurements in patient sera without noteworthy loss of activity after more than 50 measurement cycles.

SUMMARY

Infections may act as major environmental triggers inducing, promoting and/or complicating SLE. This chapter provides a discussion on how environmental exposures, particularly EBV, may influence SLE development in select individuals as well as other processes that exacerbates SLE. Molecular mimicry, epitope spreading, exaggerated local activation of antigen-presenting cells, and priming of large numbers of T cells with broad specificities may all have important roles in lupus autoimmunity. The interplay of infections and select lupus genetic polymorphisms are also beginning to be evaluated. Polyclonal activation of infected B cells resulting in B-cell proliferation, enhanced antibody production, and generation of circulating immune complexes, as well as bystander T cell activation, where enhanced cytokine production induces the expansion of autoreactive T cells, may promote the development of SLE; an illness termed by clinicians and researchers alike as one of the most complex amongst all chronic multisystem diseases.

References

[1] A.C. Grammer, P.E. Lipsky, B cell abnormalities in systemic lupus erythematosus, Arthritis Res Ther 5 (Suppl 4) (2003) S22–S27. Review.
[2] F. Mackay, P.A. Silveira, R. Brink, B cells and the BAFF/APRIL axis: fast-forward on autoimmunity and signaling, Curr Opin Immunol 19 (3) (2007 Jun) 327–336.

[3] A. La Cava, Lupus and T cells, Lupus 18 (3) (2009) 196–201.

[4] R.H. Shmerling, Autoantibodies in systemic lupus erythematosus—there before you know it, N Engl J Med 349 (16) (2003 Oct 16) 1499–1500.

[5] R. Lyons, S. Narain, C. Nichols, M. Satoh, W.H. Reeves, Effective use of autoantibody tests in the diagnosis of systemic autoimmune disease, Ann N Y Acad Sci 1050 (2005 Jun) 217–228. Review.

[6] V.C. Kyttaris, C.G. Katsiari, Y.T. Juang, G.C. Tsokos, New insights into the pathogenesis of systemic lupus erythematosus, Curr Rheumatol Rep 7 (6) (2005 Dec) 469–475. Review.

[7] F. Conti, C. Alessandri, D. Bompane, M. Bombardieri, F.R. Spinelli, A.C. Rusconi, et al., Autoantibody profile in systemic lupus erythematosus with psychiatric manifestations: a role for anti-endothelial-cell antibodies, Arthritis Res Ther 6 (4) (2004) R366–R372.

[8] C. Schulze, L.E. Munoz, S. Franz, K. Sarter, R.A. Chaurio, U.S. Gaipl, et al., Clearance deficiency—a potential link between infections and autoimmunity, Autoimmun Rev. 8 (1) (2008) 5–8.

[9] I.T. Harley, K.M. Kaufman, C.D. Langefeld, J.B. Harley, J.A. Kelly, Genetic susceptibility to SLE: new insights from fine mapping and genome-wide association studies, Nat Rev Genet. 10 (5) (2009 May) 285–290.

[10] D. Alarcón-Segovia, M.E. Alarcón-Riquelme, M.H. Cardiel, F. Caeiro, L. Massardo, A.R. Villa, et al., Grupo Latino americano de Estudio del Lupus Eritematoso (GLADEL). Familial aggregation of systemic lupus erythematosus, rheumatoid arthritis, and other autoimmune diseases in 1,177 lupus patients from the GLADEL cohort, Arthritis Rheum 52 (4) (2005 Apr) 1138–1147.

[11] D. Deapen, A. Escalante, L. Weinrib, D. Horwitz, B. Bachman, P. Roy-Burman, A. Walker, T.M. Mack, A revised estimate of twin concordance in systemic lupus erythematosus, Arthritis Rheum 35 (3) (1992 Mar) 311–318.

[12] M.R. Arbuckle, M.T. McClain, M.V. Rubertone, R.H. Scofield, G.J. Dennis, J.A. James, J.B. Hartley, Development of autoantibodies before the clinical onset of systemic lupus erythematosus, N Engl J Med 349 (16) (2003 Oct 16) 1526–1533.

[13] M.T. McClain, J.B. Harley, J.A. James, The role of Epstein-Barr virus in systemic lupus erythematosus, Front Biosci 6 (2001) e137–e147.

[14] B.D. Poole, R.H. Scofield, J.B. Harley, J.A. James, Epstein-Barr virus and molecular mimicry in systemic lupus erythematosus, Autoimmunity 39 (2006) 63–70.

[15] J.B. Harley, Guthridge, J.A. James, The curiously suspicious: a role for Epstein-Barr virus in lupus, Lupus 15 (2006) 768–777.

[16] J.A. James, K.M. Kaufman, A.D. Farris, E. Taylor-Albert, T.J.A. Lehman, J.B. Harley, An increased prevalence of Epstein-Barr virus infection in young patients suggests a possible etiology for systemic lupus erythematosus, J Clin Invest 100 (1997) 3019–3026.

[17] J.A. James, B.R. Neas, K.L. Moser, T. Hall, G.R. Bruner, A.L. Sestak, et al., Systemic lupus Erythematosus in adults is associated with previous Epstein-Barr virus exposure, Arthritis Rheum 44 (5) (2001 May) 1122–1126.

[18] J.A. James, J.B. Harley, R.H. Scofield, Epstein-Barr virus and systemic lupus erythematosus, Curr Opin Rheumatol 18 (5) (2006 Sep) 462–467. Review.

[19] M.T. McClain, E.C. Rapp, J.B. Harley, J.A. James, Infectious mononucleosis patients temporarily recognize a unique, cross-reactive epitope of Epstein-Barr virus nuclear antigen-1, J Med Virol 70 (2003) 253–257.

[20] M.T. McClain, B.D. Poole, B.F. Bruner, K.M. Kaufman, J.B. Harley, J.A. James, An altered immune response to Epstein-Barr nuclear antigen 1 in pediatric systemic lupus erythematosus, Arthritis Rheum 54 (1) (2006 Jan) 360–368.

[21] J. Levitskaya, A. Sharipo, A. Leonchiks, A. Ciechanover, M.G. Masucci, Inhibition of ubiquitin/proteasome-dependent protein degradation by the Gly-Ala repeat domain of the Epstein-Barr virus nuclear antigen 1, Proc Natl Acad Sci USA 94 (1997) 12616–12621.

[22] Y. Yin, B. Manoury, R. Fahraeus, Self-inhibition of synthesis and antigen presentation by Epstein-Barr virus-encoded EBNA-1, Science 301 (2003) 1371–1374.

[23] J. Mautner, D. Pich, F. Nimmerjahn, S. Milosevic, D. Adhikary, H. Christoph, et al., Epstein-Barr virus nuclear antigen 1 evades direct immune recognition by CD4+ T helper cells, Eur J Immunol 34 (2004) 2500–2509.

[24] K.M. Kaufman, M.Y. Kirby, J.B. Harley, J.A. James, Peptide mimics of a major lupus epitope of SmB/B′, Ann N Y Acad Sci 987 (2003 Apr) 215–229.

[25] J.A. James, J.B. Harley, Linear epitope mapping of a Sm B/B′ polypeptide, J Immunol 148 (1992) 2074–2080.

[26] B.D. Poole, A.K. Templeton, J.M. Guthridge, E.J. Brown, J.B. Harley, J.A. James, Aberrant Epstein-Barr viral infection in systemic lupus erythematosus, Autoimmun Rev. 8 (4) (2009 Feb) 337–342.

[27] B.D. Poole, T. Gross, S. Maier, J.B. Harley, J.A. James, Lupus-like autoantibody development in rabbits and mice after immunization with EBNA-1 fragments, J Autoimmun 31 (4) (2008 Dec) 362–371.

[28] J.B. Harley, J.A. James, Epstein-Barr virus infection induces lupus autoimmunity, Bull NYU Hosp Jt Dis 64 (1-2) (2006) 45–50. Review.

[29] M.T. McClain, L.D. Heinlen, G.J. Dennis, J. Roebuck, J.B. Harley, J.A. James, Early events in lupus humoral autoimmunity suggest initiation through molecular mimicry, Nat Med 11 (1) (2005 Jan) 85–89.

[30] J.B. Harley, J.A. James, Epstein-Barr virus infection may be an environmental risk factor for systemic lupus erythematosus in children and teenagers, Arthritis Rheum 42 (8) (1999 Aug) 1782–1783.

[31] J.A. James, J.B. Harley, B-cell epitope spreading in autoimmunity, Immunol Rev. 164 (1998) 185–200.

[32] J.A. James, R.H. Scofield, J.B. Harley, Lupus humoral autoimmunity after short peptide immunization, Ann N Y Acad Sci 5 (815) (1997 Apr) 124–127.

[33] E. Rivkin, M.J. Vella, R.G. Lahita, A heterogeneous immune response to an SmD-like epitope by SLE patients, J Autoimmun 7 (1994) 119–132.

[34] B. Marchini, M.P. Dolcher, A. Sabbatini, G. Klein, P. Migliorini, Immune response to different sequences of the EBNA-1 molecule in Epstein-Barr virus-related disorders and in autoimmune diseases, J Autoimmun 7 (1994) 179–191.

[35] M. Incaprera, L. Rindi, A. Bazzichi, C. Garzelli, Potential role of the Epstein-Barr virus in systemic lupus erythematosus autoimmunity, Clin Exp Rheumatol 16 (1998) 289–294.

[36] J.A. James, M.J. Mamula, J.B. Harley, Sequential autoantigenic determinants of the small nuclear ribonucleoprotein Sm D shared by human lupus autoantibodies and MRL lpr/lpr antibodies, Clin Exp Immunol 98 (1994) 419–426.

[37] K. Sundar, S. Jacques, P. Gottlieb, R. Villars, M.E. Benito, D.K. Taylor, et al., Expression of the Epstein-Barr virus nuclear antigen-1 (EBNA-1) in the mouse can elicit the production of anti-dsDNA and anti-Sm antibodies, J Autoimmun 23 (2004) 127–140.

I. BASIS OF DISEASE PATHOGENESIS

[38] A.J. Gross, D. Hochberg, W.M. Rand, D.A. Thorley-Lawson, EBV and systemic lupus erythematosus: a new perspective, J Immunol 174 (11) (2005) 6599—6607.

[39] I. Kang, T. Quan, H. Nolasco, S.H. Park, M.S. Hong, J. Crouch, E.G. Pamer, J.G. Howe, J. Craft, Defective control of latent Epstein-Barr virus infection in systemic lupus erythematosus, J Immunol 172 (2) (2004 Jan 15) 1287—1294.

[40] D.A. Thorley-Lawson, K.A. Duca, M. Shapiro, Epstein-Barr virus: a paradigm for persistent infection - for real and in virtual reality, Trends Immunol 29 (4) (2008 Apr) 195—201.

[41] S.V. Navarra, Role of Infectious Agents in Autoimmunity, Future Rheumatol 2 (3) (2007) 1—8.

[42] A. Hrycek, D. Kusmierz, U. Mazurek, T. Wilczok, Human cytomegalovirus in patients with systemic lupus erythematosus, Autoimmunity 38 (2005) 487—491.

[43] M.C. Severin, Shoenfeld. Parvovirus B19 infection and its association with autoimmune disease, in: Y. Shoenfeld, N.R. Rose (Eds.), Infection and autoimmunity, 1st edn, Elsevier, Amsterdam, 2004, pp. 181—188.

[44] A. Bengtsson, A. Widell, S. Elmstahl, G. Sturfelt, No serological indications that systemic lupus erythematosus is linked with exposure to human parvovirus B19, Ann Rheum Dis 59 (2000) 64—66. 2000.

[45] A. Hemauer, K. Beckenlehner, H. Wolf, B. Lang, S. Modrow, Acute parvovirus B19 infection in connection with a flare of systemic lupus erythematodes in a female patient, J Clin Virol 14 (1999) 73—77.

[46] M. Seishima, Z. Oyama, M. Yamamura, Two-year followup study after human parvovirus B19 infection, Dermatology 206 (2003) 192—196.

[47] M.K. Adelman, J.J. Marchalonis, Endogenous retroviruses in systemic lupus erythematosus: candidate lupus viruses, Clin Immunol 102 (2) (2002 Feb) 107—116.

[48] A. Perl, Role of endogenous retroviruses in autoimmune diseases, Rheum Dis Clin North Am. 29 (1) (2003 Feb) 123—143. vii.

[49] A. Perl, E. Colombo, H. Dai, R. Agarwal, K.A. Mark, K. Banki, B.J. Poiesz, P.E. Phillips, S.O. Hock, J.D. Reveille, Antibody reactivity to the HRES-1 endogenous retroviral element identifies a subset of patients with systemic lupus erythematosus and overlap syndromes. Correlation with antinuclear antibodies and HLA class II alleles, Arthritis Rheum 38 (11) (1995 Nov) 1660—1671.

[50] C. Magistrelli, E. Samoilova, R.K. Agarwal, K. Banki, P. Ferrante, A. Vladutiu, et al., Polymorphic genotypes of the HRES-1 human endogenous retrovirus locus correlate with systemic lupus erythematosus and autoreactivity, Immunogenetics 49 (10) (1999 Sep) 829—834.

[51] P. Carroll, D. Stafford, R.S. Schwartz, B.D. Stollar, Murine monoclonal anti-DNA autoantibodies bind to endogenous bacteria, Journal of Immunology 135 (2) (1985) 1086—1090.

[52] Y. Shoenfeld, Y. Vilner, A.R. Coates, Monoclonal antituberculosis antibodies react with DNA, andmonoclonal anti-DNA autoantibodies react with Mycobacterium tuberculosis, Clinical & Experimental Immunology 66 (2) (1986) 255—261.

[53] W. Zhang, M. Reichlin, A possible link between infection with burkholderia bacteria and systemic lupus erythematosus based on epitope mimicry, Clin Dev Immunol 2008 (2008) 683489.

[54] B.B. Mittleman, H.C. Morse 3rd, S.M. Payne, G.M. Shearer, E. Mozes, Amelioration of experimental systemic lupus Erythematosus by retrovirus infection, J Clin Immunol 16 (1996) 230—236.

[55] B.F. Haynes, J. Fleming, E.W. St Clair, H. Katinger, G. Stiegler, R. Kunert, et al., Cardiolipin polyspecific autoreactivity in two broadly neutralizing HIV-1 antibodies, Science 308 (2005) 1906—1908.

[56] S.M. Alam, M. McAdams, D. Boren, M. Rak, R.M. Scearce, F. Gao, et al., The role of antibody polyspecificity and lipid reactivity in binding of broadly neutralizing anti-HIV-1 envelope human monoclonal antibodies 2F5 and 4E10 to glycoprotein 41 membrane proximal envelope epitopes, J Immunol 178 (7) (2007) 4424—4435.

[57] S.M. Alam, R.M. Scearce, R.J. Parks, K. Plonk, S.G. Plonk, L.L. Sutherland, et al., Human immunodeficiency virus type 1 gp41 antibodies that mask membrane proximal region epitopes: antibody binding kinetics, induction, and potential for regulation in acute infection, J Virol 82 (2008) 115—125.

[58] F. Silvestris, M.A. Frassanito, P. Cafforio, D. Potenza, M. Di Loreto, M. Tucci, et al., Antiphosphatidylserine antibodies in human immunodeficiency virus-1 patients with evidence of T-cell apoptosis and mediate antibody-dependent cellular cytotoxicity, Blood 87 (12) (1996 Jun 15) 5185—5195.

[59] C.J. Edwards, C. Cooper, Early environmental exposure and the development of lupus, Lupus 15 (11) (2006) 814—819.

[60] C.J. Edwards, J.A. James, Making lupus: a complex blend of genes and environmental factors is required to cross the disease threshold, Lupus 15 (11) (2006) 713—714.

[61] C.J. Edwards, H. Syddall, R. Goswami, P. Goswami, E.M. Dennison, C. Cooper, Infections in infancy and the presence of antinuclear antibodies in adult life, Lupus 15 (4) (2006) 213—217.

[62] C.J. Edwards, Environmental factors and lupus: are we looking too late? Lupus 14 (6) (2005) 423—425.

[63] N.E. Mangan, R.E. Fallon, P. Smith, R.N. van, A.N. McKenzie, P.G. Fallon, Helminth infection protects mice from anaphylaxis via IL-10-producing B cells, J Immunol 173 (2004) 6346—6356.

[64] A.H. Van den Biggelaar, R.R. van, L.C. Rodrigues, Decreased atopy in children infected with Schistosoma haematobium: a role for parasite induced interleukin-10, Lancet 356 (2000) 1723—1727.

[65] M.F. Bianchi, A.A. Manfredi, High-mobility group box 1 (HMGB1) protein at the crossroads between innate and adaptive immunity, Immunol Rev. 220 (2007 Dec) 35—46.

[66] E.C. Baechler, F.M. Batliwalla, G. Karypis, P.M. Gaffney, W.A. Ortmann, K.J. Espe, K.B. Shark, W.J. Grande, K.M. Hughes, V. Kapur, P.K. Gregersen, T.W. Behrens, et al., Interferon-inducible gene expression signature in peripheral blood cells of patients with severe lupus, Proc Natl Acad Sci USA 100 (2003) 2610—2615.

[67] X. Feng, H. Wu, J.M. Grossman, P. Hanvivadhanakul, J.D. FitzGerald, G.S. Park, et al., Association of increased interferon-inducible gene expression with disease activity and lupus nephritis in patients with systemic lupus erythematosus, Arthritis Rheum 54 (9) (2006 Sep) 2951—2962.

[68] T.B. Niewold, J. Harley, J. Hua, M. Crow, T. Lehman, Over-expression of interferon-α in lupus families: evidence for a complex heritable trait, Clin Immunol 123 (2007) 89.

[69] M.B. Uccellini, A.M. Avalos, A. Marshak-Rothstein, G.A. Viglianti, Toll-like receptor-dependent immune complex activation of B cells and dendritic cells, Methods Mol Biol. 517 (2009) 363—380.

[70] D.S. Pisetsky, The role of innate immunity in the induction of autoimmunity, Autoimmun Rev. 8 (1) (2008 Oct) 69—72.

[71] M. Lech, O. Kulkarni, S. Pfeiffer, E. Savarese, A. Krug, C. Garlanda, A. Mantovani, H.J. Anders, Tir8/Sigirr prevents murine lupus by suppressing the immunostimulatory effects of lupus autoantigens, J Exp Med 205 (8) (2008 Aug 4) 1879—1888.

[72] M.R. Clatworthy, K.G.C. Smith, FcγRIIb balances efficient pathogen clearance and the cytokine-mediated consequences of sepsis, J. Exp. Med 199 (2004) 717–723.

[73] S. Bolland, Y.S. Yim, K. Tus, E.K. Wakeland, J.V. Ravetch, Genetic modifiers of systemic lupus erythematosus in FcγRIIb mice, J. Exp. Med 195 (2002) 1167–1174.

[74] T. Yuasa, S. Kubo, T. Yoshino, A. Ujike, K. Matsumura, M. Ono, et al., Deletion of FcγRIIb render H-2b mice susceptible to collagen-induced arthritis, J. Exp. Med 189 (1999) 187–194.

[75] N.R. Pritchard, A.J. Cutler, S. Uribe, S.J. Chadban, B.J. Morley, K.G.C. Smith, Autoimmune-prone mice share a promoter haplotype associated with reduced expression and function of the Fc receptor FcgammaRII, Curr. Biol 10 (2000) 227–230.

[76] Y. Liu, E. Masuda, M.C. Blank, K.A. Kirou, X. Gao, M.S. Park, L. Pricop, Cytokine-mediated regulation of activating and inhibitory Fc gamma receptors in human monocytes, J. Leukoc. Biol. 2005 (77) (2005) 767–776.

[77] R.J. Brownlie, K.E. Lawlor, H.A. Niederer, A.J. Cutler, Z. Xiang, M.R. Clatworthy, R.A. Floto, D.R. Greaves, P.A. Lyons, K.G. Smith, Distinct cell-specific control of autoimmunity and infection by FcgammaRIIb, J Exp Med 205 (4) (2008 Apr 14) 883–895.

[78] R. Takahashi, A. Tsutsumi, K. Ohtani, Y. Muraki, D. Goto, I. Matsumoto, M. Wakamiya, T. Sumida, Association of mannose binding lectin (MBL) gene polymorphism and serum MBL concentration with characteristics and progression of systemic lupus erythematosus, Ann Rheum Dis 64 (2) (2005 Feb) 311–314.

[79] M.C. Pickering, M. Botto, P.R. Taylor, P.J. Lachmann, M.J. Walport, Systemic lupus erythematosus, complement deficiency, and apoptosis, Adv Immunol 76 (2000) 227–324.

[80] B.P. Morgan, M.J. Walport, Complement deficiency and disease, Immunol Today 12 (1991) 301–306.

[81] P. Garred, A. Voss, H.O. Madsen, P. Junker, Association of mannose-binding lectin gene variation with disease severity and infections in a population-based cohort of systemic lupus erythematosus patients, Genes Immun 2 (8) (2001 Dec) 442–450.

[82] P.F. Yong, L. Aslam, M.Y. Karim, M.A. Khamashta, Management of hypogammaglobulinaemia occurring in patients with systemic lupus erythematosus, Rheumatology (Oxford) 47 (9) (2008 Sep) 1400–1405.

[83] T.K. Tarrant, D.H. Frazer, J.P. Aya-Ay, D.D. Patel, B cell loss leading to remission in severe systemic lupus erythematosus, J Rheumatol 30 (2003) 412–414.

[84] W. Stohl, SLE–systemic lupus erythematosus: a BLySful, yet BAFFling, disorder, Arthritis Res Ther 5 (3) (2003) 136–138.

[85] D. Seshasayee, P. Valdez, M. Yan, V.M. Dixit, D. Tumas, I.S. Grewal, Loss of TACI causes fatal lymphoproliferation and autoimmunity, establishing TACI as an inhibitory BLyS receptor, Immunity 18 (2003) 279–288.

[86] U. Salzer, H.M. Chapel, A.D. Webster, et al., Mutations in TNFRSF13B encoding TACI are associated with common variable immunodeficiency in humans, Nat Genet 37 (2005) 820–828.

[87] E. Castigli, S.A. Wilson, L. Garibyan, R. Rachid, F. Bonilla, L. Schneider, R.S. Geha, TACI is mutant in common variable immunodeficiency and IgA deficiency, Nat Genet 37 (2005) 829–834.

[88] E.C. Rankin, D.A. Isenberg, IgA deficiency and SLE: prevalence in a clinic population and a review of the literature, Lupus 6 (1997) 390–394.

[89] J.T. Cassidy, R.K. Kitson, C.L. Selby, Selective IgA deficiency in children and adults with systemic lupus erythematosus, Lupus 16 (2007) 647–650.

[90] T. Takeuchi, T. Nakagawa, Y. Maeda, S. Hirano, M. Sasaki-Hayashi, S. Makino, A. Functional defect of B lymphocytes in a patient with selective IgM deficiency associated with systemic lupus erythematosus, Autoimmunity 34 (2001) 115–122.

[91] R. Cervera, R.A. Asherson, M.L. Acevedo, J.A. Gómez-Puerta, G. Espinosa, G. De La Red, et al., Antiphospholipid syndrome associated with infections: clinical and microbiological characteristics of 100 patients, Ann Rheum Dis 63 (10) (2004 Oct) 1312–1317. Review.

[92] E. Matsuura, Y. Igarashi, M. Fujimoto, K. Ichikawa, T. Koike, Anticardiolipin cofactor(s) and differential diagnosis of autoimmune disease [letter], Lancet 336 (1990) 177–178.

[93] J.E. Hunt, H.P. McNeil, G.J. Morgan, I.R. Cramer, S.A. Krilis, A phospholipid B-B2- glycoprotein I complex is an antigen for anticardiolipin antibodies occurring in autoimmune disease but not with infection, Lupus 1 (1992) 75–81.

[94] A. Elbeialy, K. Strassburger-Lorna, T. Atsumi, M.L. Bertolaccini, O. Amengual, M. Hanafi, M.A Khamashta, G.R. Hughes, et al., Antiphospholipid antibodies in leprotic patients: a correlation with disease manifestations, Clin Exp Rheumatol 18 (2000) 492–494.

[95] M. Blank, I. Krause, M. Fridkin, N. Keller, J. Kopolovic, I. Goldberg, et al., Bacterial induction of autoantibodies to beta2-glycoprotein-I accounts for the infectious etiology of antiphospholipid syndrome, J Clin Invest 109 (6) (2002 Mar) 797–804.

[96] I. Zhang, K. Jaconsson, K. Strom, M. Londberg, K.L. Frykberg, Staphylococcus aureus expresses a cell surface protein that binds both IgG and B2-glycoprotein 1, Microbiology 145 (1999) 177–183.

[97] A.E. Gharavi, S.S. Pierangeli, R.G. Espinola, X. Liu, M. Coden-Stanfield, E.N. Harris, Antiphospholipid antibodies induced in mice by immunization with a cytomegalovirus-derived peptide cause thrombosis and activation of endothelial cells in vivo, Arthritis Rheum 46 (2002) 545–552.

[98] D. Sène, J.C. Piette, P. Cacoub, Antiphospholipid antibodies, antiphospholipid syndrome and infections, Autoimmun Rev. 7 (4) (2008 Feb) 272–277. Review.

[99] I.W. Uthman, A.E. Gharavi, Viral infections and antiphospholipid antibodies, Semin Arthritis Rheum 31 (4) (2002 Feb) 256–263. Review.

[100] P. Fiallo, C. Travaglino, E. Nunzi, P.P. Cardo, Beta-2 glycoprotein dependence of anticardiolipin antibodies in multibacillary leprosy patients, Lepr Rev 69 (1998) 376–381.

[101] R.A. Levy, S.A. Pierangeli, R.G. Espinola, Antiphospholipid beta-2 glycoprotein 1 dependency assay to determine antibody pathogenicity, Arthritis Rheum 43s (2000) 1476.

[102] R.A. Asherson, R. Cervera, J.C. Piette, Y. Shoenfeld, G. Espinosa, M.A. Petri, et al., Catastrophic antiphospholipid syndrome: clues to the pathogenesis from a series of 80 patients, Medicine (Baltimore) 80 (2001) 355–377.

[103] J. Metzger, P. von Landenberg, M. Kehrel, A. Buhl, K.J. Lackner, P.B. Luppa, Biosensor analysis of beta2-glycoprotein I-reactive autoantibodies: evidence for isotype-specific binding and differentiation of pathogenic from infection-induced antibodies, Clin Chem. 53 (6) (2007 Jun) 1137–1143.

Systemic Lupus Erythematosus in Domestic Animals

Michael J. Day

INTRODUCTION

The study of spontaneously arising diseases of domestic animals is of great value in understanding the pathogenesis of equivalent disorders of humans. Comparative research of this type is an important component of the "one medicine" concept that has gathered momentum over recent years [1]. The dog provides the single most effective model for such research. Dogs develop a spectrum of immune-mediated, neoplastic, infectious, and degenerative disorders that are close clinical and pathological mimics of equivalent human diseases [2]. The dog increasingly shares a domestic environment with humans and is consequently exposed to the same range of environmental triggers of disease. Their size, life-span, and temperament mean that it is feasible to keep this species as an experimental animal and much has been learned from the study of breeding colonies of dogs with specific inherited diseases. The recent sequencing of the canine genome has led to unparalleled opportunities in comparative research. The canine genome is much more homologous to that of humans than is the genome of laboratory rodent species [3]. Moreover, the extreme phenotypic diversity of canine breeds has led to highly conserved genetic subgroups within a single species, and a wide range of inherited breed-related diseases. The genetic basis of canine disease is now being rapidly unraveled. Initial studies demonstrating linkage between alleles and haplotypes of major histocompatibility complex (MHC) genes and diseases such as rheumatoid arthritis [4], diabetes mellitus [5], lymphocytic thyroiditis [6], autoimmune hemolytic anemia [7], and anal furunculosis [8] are now published. More recently, the application of genome-wide association studies using commercial chips containing thousands of single nucleotide polymorphisms (snps) is beginning to pay dividends.

A spectrum of organ-specific autoimmune diseases is now well-defined in the dog and the list of canine disorders with an autoimmune basis continues to grow. The clinical presentation, pathogenesis, and pathology of the autoimmune blood dyscrasias (anemia, thrombocytopenia, and neutropenia), polyarthropathies, myopathies, endocrinopathies (lymphocytic thyroiditis, diabetes mellitus, hypoadrenocorticism), and skin diseases are widely documented [2]. Most are still regarded as primary idiopathic autoimmune events, although it is increasingly recognized that canine immune-mediated disease may occur secondary to infection, neoplasia, or drug and vaccine administration.

Multisystemic autoimmune disease is recognized less commonly in the dog with the most frequent co-presentation being of autoimmune hemolytic anemia and thrombocytopenia (Evans syndrome). Systemic lupus erythematosus (SLE), when defined by strict criteria, is in fact a relatively rare disorder [9]. Autoimmune disease appears far less common in the two other major companion animal species — the cat and horse.

CANINE SYSTEMIC LUPUS ERYTHEMATOSUS

Definition of SLE in Dogs

Canine SLE was first described in 1965 with a series of case studies reported by Lewis *et al.* [10]. Over the years, different approaches to the clinical definition of canine SLE have been adopted. Initially, attempts were made to extrapolate the human criteria defined by the American College of Rheumatology (ACR) [11, 12], but

437

some of these did not occur in the dog or were difficult to evaluate. Moreover, immunodiagnostic testing remains relatively rudimentary in veterinary practice, and although much is now known of the immune basis of canine autoimmune disease, the relevant diagnostic tests are not widely available to practicing veterinarians. For these reasons, the currently proposed criteria for definition of SLE in the dog [9] are relatively simple:

1. The patient should have multisystemic disease involving at least two body systems. The organ systems most commonly involved are the joints, skin, kidney, and blood.
2. The disease affecting each body system should be proven to be immune-mediated in nature by demonstration of relevant autoantibodies in serum or tissue.
3. The patient should have a high-titered serum antinuclear antibody (ANA).
4. There should be no evidence of underlying disease or recognized trigger factors (e.g., drug or vaccine administration, infectious or neoplastic disease).

However, even though these criteria are straightforward, clinical cases that satisfy them are infrequently documented and it is considered that true SLE is an uncommon disease of the dog. Estimates have suggested that the incidence of SLE among the canine population in the United States may be as low as 0.027% [13]. In contrast, there are many more patients who may partially satisfy the criteria and may be considered to have probable SLE or an SLE-overlap syndrome. One well-recognized SLE-overlap syndrome occurs in German shepherd dogs with pyrexia, lethargy, wasting, polyarthritis, and high-titered serum ANA [14]. It is also of note that many dogs presenting with vague clinical illness termed pyrexia of unknown origin are ANA-positive and proposed to have some form of immune-mediated disease [15, 16].

Clinical Signs of SLE in Dogs

Canine SLE is a chronic multisystemic disease that often takes a relapsing and remitting clinical course. A wide range of body systems may be involved and these may be affected concurrently or sequentially. An indication of the relative frequency of clinical signs is given in Table 25.1.

Joint disease is the most common presenting sign and is present in 40 to more than 90% of patients, and is multi-articular, nonerosive, and generally symmetrical and nondeforming [11–13, 17–22]. Synovial fluid aspirates contain significant numbers of neutrophils (10–220 $\times 10^9$/l). Histopathology of the synovial membrane reveals hypertrophy and hyperplasia of the synoviocytes,

TABLE 25.1 Frequency (%) of Clinical Signs in Canine and Feline SLE[a]

Clinical feature	Canine SLE (n = 260) Frequency (%)	Feline SLE (n = 23) Frequency (%)
Polyarthritis	75	43
Mucocutaneous lesions	49	48
Proteinuria/ glomerulopathy	44	48
Anemia, thrombocytopenia, or leukopenia	43	57
Pleuritis or pericarditis	5.7	0
Neurological signs	4.9	29

[a]Table modified from [21].

and a mixed cellular infiltrate with lymphocytes and plasma cells predominating. Vasculitis may be present. Polymyositis may accompany arthritis in some 5% of cases and it may be severe and lead to pronounced muscular atrophy [23].

The cutaneous manifestations of canine SLE are highly variable and range from a trivial, mild, alopecic, scarring lesion to severe and widespread ulceration [11]. Symmetry is a striking feature. Some cases present with a pruritic, generalized seborrheic dermatitis. In others, ulceration of the mucocutaneous junctions is apparent, which may spread to involve other parts of the body. Ulcerative and/or hyperkeratotic footpads may be seen. Focal alopecia is often noted, which may extend to involve large areas. Apparent photosensitization may occur, with a history of lesions involving the hairless areas of the body flaring up following exposure to sunlight. Depigmentation of the nose and eyelids may be apparent, and facial edema is sometimes noted. Some cases proceed to severe nasal ulceration (Figure 25.1).

Although vesiculation with subsequent ulceration is not reported as a feature of the cutaneous signs of canine SLE, one report features a bichon frise with severe erosive lesions of the right elbow, axilla, and lateral thorax [24] (Figure 25.2). Histopathology revealed subepidermal clefting with an absence of dermal inflammation. Immunological studies revealed a circulating antibasement membrane antibody (titer >20,000) and serum antibody bound to the NC1 domain of recombinant human type VII collagen. In addition, the patient suffered from persistent proteinuria, Coombs-positive anemia, thrombocytopenia, pleuritis, and hepatitis. The authors proposed that this represented the canine equivalent of type 1 bullous SLE of humans [25]. Another report details a dog with SLE showing

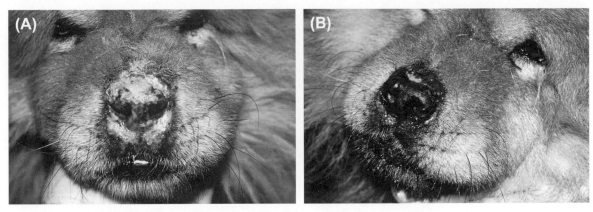

FIGURE 25.1 Depigmentation and ulceration of the nose of a chow (A). The dog also had oral ulcers and arthritis. There was a good response to prednisolone and azathioprine within 8 weeks (B).

Coombs-positive anemia, thrombocytopenia, and polyarthritis with an ANA titer of 10,240 [26]. However, instead of cutaneous signs of SLE, the patient had both clinical and histopathological signs suggestive of pemphigus foliaceus. Immunohistochemistry revealed desmosomal deposition of IgG. The dog subsequently developed a B-cell lymphoma. This appears to represent concomitant SLE and pemphigus — a phenomenon also reported occasionally in humans [25].

Histopathology of the cutaneous changes in SLE is classically characterized by a lichenoid, largely lymphoid infiltration of the dermoepidermal junction (Figure 25.3) [11, 27, 28]. Vacuolar alteration and apoptosis of individual basal cells are major diagnostic criteria. Leukocytoclastic vasculitis may be seen in the dermis, and thickening of the basement membrane with pigmentary incontinence is often a feature. Immunofluorescence or immunohistochemical studies reveal immunoglobulin and/or complement deposits at the dermoepidermal junction (Figure 25.4).

Dogs also suffer from a spectrum of disorders collectively referred to as cutaneous lupus erythematosus (CLE), in which only skin disease is present in the absence of ANA [29]. The most common of this group is nasal planum CLE in which there is initial depigmentation followed by erythema and ulceration (Figure 25.5). Lesions are exacerbated by sunlight. Histopathology is similar to that of skin lesions in SLE, and immunofluorescence studies may reveal immunoglobulin deposits at the dermoepidermal junction.

Vesicular CLE of Shetland sheepdogs and rough collies represents a further variant of cutaneous LE [30, 31]. Affected animals show ulceration of the hairless portion of the ventral abdomen and thorax that is exacerbated in the summer (Figure 25.6). Some also develop buccal ulceration. Other recognized variants of CLE include oral cavity or mucocutaneous CLE, generalized CLE, exfoliative CLE of German shorthaired pointers, and lupoid onychodystrophy [29]. Ulcerative stomatitis occurs in 10–20% of cases.

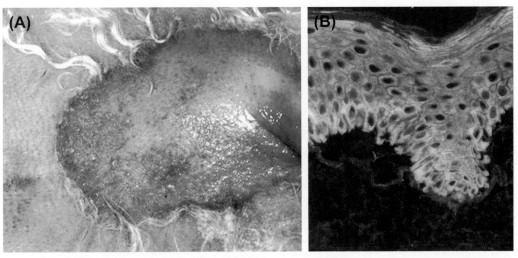

FIGURE 25.2 (A) Ulcerative lesions in a bichon frisé suffering from type 1 bullous SLE. (B) Indirect immunofluorescence in a case of type 1 bullous SLE using canine salt-spit lip. *Reproduced with permission from British Veterinary Association, Olivry et al. [24].*

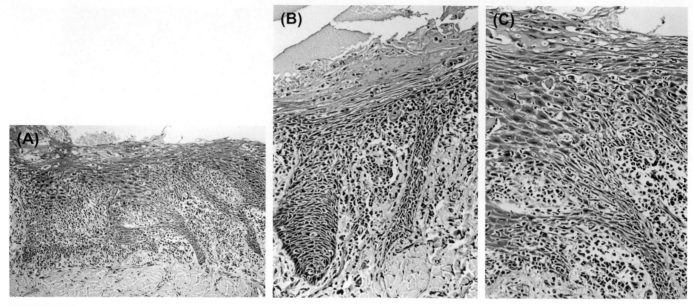

FIGURE 25.3　Histopathology in a case of SLE (×200). There is a striking mononuclear cell infiltrate of the dermoepidermal junction and basal hydropic change with disruption of the epidermis. *Courtesy of Professor. R. Else, University of Edinburgh.*

It may involve buccal mucosal surfaces, the hard or soft palate, and less frequently the tongue. Histopathology of the affected areas reveals similar changes to those described for the skin.

Some 30—60% of cases are presented with anemia. Of these dogs, approximately half are apparently immune-mediated, with evidence of hemolysis with positive direct Coombs tests or autoagglutination. The remaining patients have anemia of chronic disease. Leukopenia and leukocytosis affect 20—30% of cases each. In a proportion of the cases with leukopenia, the change is attributed to marked lymphopenia. In some cases, thrombocytopenia may be the initial cause for presentation. This may be manifested as petechial or ecchymotic

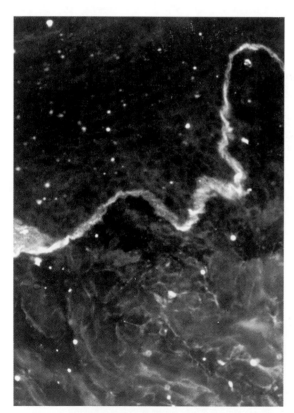

FIGURE 25.4　Immunoglobulin deposition at the dermoepidermal junction in a case of SLE. *Courtesy of Dr. L. Werner, University of California. Reproduced with permission from WB Saunders, Halliwell and Gorman [108].*

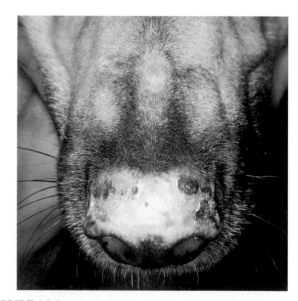

FIGURE 25.5　Nasal planum CLE in a dog. The depigmentation is followed by erythema and inflammation. The disease was exacerbated by sunlight, but ANA was negative and there were no other systemic signs.

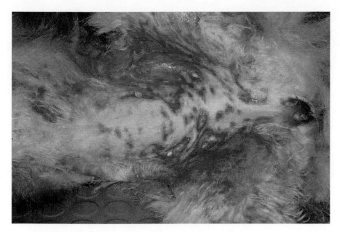

FIGURE 25.6 Vesicular cutaneous LE in a Shetland sheepdog. *Reproduced with permission from Wiley-Blackwell, Jackson and Olivry [30].*

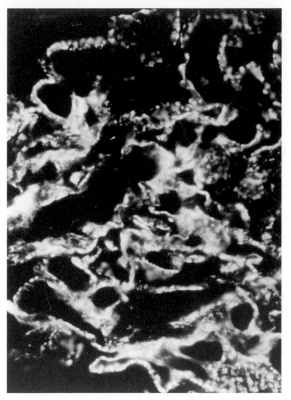

FIGURE 25.7 Immunoglobulin deposits in the renal glomerulus. Glomerulonephritis is associated with a case of SLE.

hemorrhages of mucocutaneous surfaces, intraocular hemorrhage, or excessive bleeding following minor trauma or venepuncture.

Marked proteinuria may be present in some 50% of cases. Azotemia may be present on presentation and is a frequent and often life-threatening complication in severe cases. Biopsies usually reveal a proliferative, membranous glomerulonephritis, and immune complexes are revealed readily using fluorescein-conjugated anti-IgG and anti-C3 (Figure 25.7) or electron microscopy.

Neurological abnormalities are rarely recorded; however, psychological and behavioral abnormalities are occasionally documented. Polyneuropathies characterized by hyperesthesia, convulsions, and/or obvious behavioral changes are noted in some 2–5% of cases.

Lymphadenopathy (lymphadenomegaly) is detectable in some 50% of cases and generally affects, symmetrically, the majority of the palpable nodes. The microscopic appearance of affected nodes is of reactive hyperplasia.

Pleuritis/pericarditis is reported as affecting some 5–10% of cases and chronic pneumonia may be a rare complication [17]. Pyrexia, which may be cyclical, is present in a high proportion of cases [13].

Differential Diagnosis

As described above, canine immune-mediated disease may often be secondary to underlying trigger factors or disease processes. The single most important differential diagnosis for canine multisystemic immune-mediated disease in endemic areas of infection is leishmaniosis. Infection of susceptible dogs with sandfly-transmitted *Leishmania infantum* may lead to multisystemic disease that is an excellent clinical mimic for SLE. Dogs with leishmaniosis may present with combinations of pyrexia, lymphadenopathy, skin disease, polyarthritis, immune-mediated anemia or thrombocytopenia, protein losing nephropathy or uveitis. Affected animals may also have high-titered serum ANA [32]. In geographical areas endemic for arthropod-borne infectious diseases, or in animals that have traveled temporarily to such a location, ruling out infection by combinations of cytology, serology, and real-time polymerase chain reaction (PCR) is essential.

Recent history of drug and vaccine administration should also be recorded. Although rare, there is one case report of a dog with multisystemic immune-mediated disease (defined as Sweet's syndrome) following drug administration [33]. Beagle dogs treated with hydralazine were reported to become ANA-positive but did not show clinical abnormalities [34]. Vaccine-associated immune-mediated disease in the dog is widely recognized [35], but SLE as an adverse event is rare [36].

Immunological Findings in Dogs with SLE

Antinuclear antibody

A diagnosis of SLE in the dog requires the presence of high-titered serum ANA. Most commercial veterinary diagnostic clinical pathology laboratories will offer simple ANA testing, but there is no international standard to which the test should be performed. Substrates employed have included sections of mouse and rat liver, and HeLa, Vero or HEp-2 cells monolayers [37]. Either immunofluorescence or immunoperoxidase methodology is employed for the detection of serum ANA. Patient serum should be fully titrated and the pattern of nuclear immunoreactivity recorded. The reported sensitivity of the test varies between laboratories and with the substrate employed. The individual laboratory should advise on the significance of titer as this may depend upon the test methodology employed.

It is important to emphasize that serum ANA may be present in many dogs with chronic inflammatory, infectious or neoplastic diseases and may be present in dogs with immune-mediated diseases other than SLE [38]. In addition to leishmaniosis (see "Differential Diagnosis" above), dogs with tick-transmitted monocytic ehrlichiosis (*Ehrlichia canis*) or infection with *Bartonella vinsonii* subspp. *B.V. berkhoffi* may be ANA-positive, and the likelihood of ANA positivity increases when co-infections with these pathogens are found [39]. ANA may also be found in sera from normal dogs. In one study, the incidence of ANA was assessed in sera from 100 normal dogs presented for routine heartworm prophylaxis using HeLa cells [40]. Two of these showed a titer of 320 and two of 80. In 19 animals, a titer of 10 was recorded, three had a titer of 20, and one of 40. Forty-seven were negative when assessed undiluted. Another study examined the incidence of ANA in a line-bred colony of English cocker spaniels in which there was a high incidence of cardiomyopathy and in which the existence of at least one case of SLE had been confirmed [41]. Using Vero cells, the incidence in diseased dogs ranged from 47 to 60%. The incidence in 58 clinically normal dogs in the same kennel was 46.5%. Antibody titers ranged from 10 to 2560. In contrast, none of 25 sera from English cocker spaniels from a control kennel yielded positive results.

Most ANA tests aim to detect antibodies of the IgG class. Canine ANA are mostly of the IgG1, IgG3, and IgG4 subclasses, with IgG2 rarely represented. There is no association between IgG subclass and the pattern of nuclear labeling, but there may be a link between IgG subclass profile and the nature of clinical presentation [42].

Antigen specificities of canine ANA

Anti-dsDNA antibodies are rarely encountered in dogs with SLE. In dogs, an acidic β-globulin, which is not antibody, binds to DNA and also to the synthetic double-stranded DNA analog polydeoxyadenylate-deoxythymidylate (dAdT) [43]. This protein leads to false-positive reactions when techniques based on the Farr assay are employed [19, 43, 44]. The nature of this protein is poorly characterized. It is heat labile and destroyed by heating for 30 min at 60°C, but not at 58°C [45]. It does not bind to staphylococcal protein A, would not appear to be a glycoprotein, as evidenced by a failure to bind to lentil lectin, and is partially inhibited by the addition of EDTA and more completely by the addition of dextran sulfate [45]. The interference can also be inhibited by the addition of sodium dodecyl sulfate [19] and by the use of increased ionic strength and pH of the buffers [46].

The detection of anti-dsDNA in canine sera is, therefore, problematic. Other detection systems have included indirect immunofluorescence employing *Crithidia luciliae* or *Trypanosoma brucei* and ELISA techniques employing highly purified dsDNA. In all studies, immunofluorescence techniques show at best a low incidence of weak positive reactions [14]. In two more studies using a combination of indirect immunofluorescence and ELISA, positive reactions were rarely seen [47, 48]. The DNA-binding protein in normal canine sera could, in addition to interfering with the Farr assay, also interfere with the binding of specific anti-dsDNA antibodies in ELISA. However, in a study that addressed this question, none of the canine sera assayed were able to inhibit binding of a murine anti-DNA monoclonal antibody [48]. However, one study employing a commercial ELISA system showed relatively high DNA binding in sera from dogs with SLE, and to a lesser extent from dogs with other arthropathies [49]. The latter authors noted that sera that gave high dsDNA-binding values also bound strongly to single-stranded (ss)DNA, suggesting possible contamination of the dsDNA substrate with ssDNA [49].

Demonstration of the presence of anti-ssDNA is of no value in the diagnosis of SLE in dogs. In one study, sera from 21% of 100 dogs with SLE gave positive results, as did 14% of dogs with positive ANA but less than four criteria for SLE, and 26.8% of 56 dogs with miscellaneous infectious diseases, including leishmaniosis [47].

The presence of antihistone antibodies is highly correlated with a positive diagnosis of SLE [47–51]. In the most extensive study, 71 of 100 sera from patients fulfilling at least four ACR criteria gave positive results, as compared with 6.7% of 120 normal controls [47]. The pattern of recognition, however, differs from that in humans. Using immunoblots, 54% of sera from dogs with SLE recognized H4, with the same number recognizing H3. Eight percent recognized H1, 22% H2A, and 20% H2B [47]. In another study, antibodies to H2B were not detected in any of 43 sera from dogs with SLE [48]. In contrast, H1 and H2B are more prominent autoantigens in man [52, 53]. A further difference is that the histone determinants against which canine antibodies are directed are mostly trypsin-resistant [47]. A recent study reported the development of a flow cytometric bead-based assay for the detection of canine anti-histone antibodies, but there was poor correlation between this assay and the standard immunofluorescence test for ANA [54]. The prevalence of antihistone antibodies has recently been examined in 43 dogs with leishmaniosis [35]. There was a significantly greater prevalence of such antibodies when infected dogs had glomerulonephritis (22/25 animals) compared with infected dogs without renal damage (7/18).

Many of the other antigens that are recognized by sera from human SLE patients are also recognized by canine sera. Some 7% of sera from dogs with SLE precipitate ribonucleoprotein (RNP), and a further 12% detect both RNP and Sm antigen [47]. High-motility group (HMG) proteins were recognized by 20% of sera, with HMG1 detected by 6%, and HMG2 by 18%. No sera recognized HMG14 and HMG17. In the aforementioned study, three sera also contained anti-Sjögren's syndrome A antibody (anti-SSA), but anti-SSB has yet to be detected.

Of great interest are antibodies with a specificity that appears to be unique for sera from canine patients with SLE, which were initially termed anti-type 1 (or T1) and anti-type 2 (or T2) [47, 50]. These are reactive with soluble nuclear extracts, but the T2 antigen is absent from extracted nuclear antigen preparations. Anti-T1 is directed against a major 43-kDa nuclear antigen [47]. Studies have confirmed that this is identical with the 43-kDa glycoprotein recognized by some canine SLE sera in the studies of Soulard et al. [56]. This has now been identified as hnRNP G [57] and epitope mapping studies have demonstrated binding to a central 33 amino acid motif in addition to a second N-terminal region of the molecule [58]. Interestingly, many sera from dogs with SLE also react with a ribosomal antigen of identical molecular weight [59].

Four cases, with clinical signs compatible with SLE, have been reported in which a high-titer antinucleolar antibody was detected. In three of these, polyarthritis was a major presenting sign [60], although one case had concomitant immune-mediated thrombocytopenia, anemia, leukopenia, and skin eruption [61]. The antibody specificity has been examined, and sera detect three polypeptides of 110, 95, and 45 kDa, which are antigenically cross-reactive [60].

Canine ANA patterns include homogenous, speckled, rim, and nucleolar labeling. The homogeneous and speckled patterns are most common. A number of studies have attempted to correlate specific antibody activity with nuclear labeling patterns. Early studies had demonstrated that anti-T1 (hnRNP G) gives a relatively fine speckled pattern, anti-Sm and anti-RNP give a coarser speckled, or reticulonodular pattern, and the antihistone antibody gives a generally homogeneous pattern [47]. More recent studies have employed immunodiffusion [62], ELISA and immunoblots [63].

Sera that exhibit chromosomal reactivity exhibit a homogeneous nuclear fluorescence pattern and fail to precipitate on immunodiffusion analysis using commercially available ENA preparations. Sera that show negative chromosomal reactivity are more likely to exhibit speckled patterns, and a proportion are positive on immunodiffusion. Lines of identity with human reactive sera were shown in the case of anti-RNP and anti-Sm [62]. The autoantigen recognized most frequently in ELISA and immunoblots was hnRNP G, an antigen absent from the commercially available extract made for the examination of human sera. Examination of anti-RNP activity confirmed that canine sera react with the same full major antigenic region of the 70K protein as human sera [63].

A recent study has suggested for the first time that the pattern of nuclear labeling, considered in conjunction with the presence or absence of precipitating antibodies, may have clinical significance. Speckled nuclear labeling in the absence of precipitating antibodies more frequently occurred in dogs with musculoskeletal disease, while dogs with multisystemic immune-mediated disease more often had homogenous nuclear labeling and precipitating antibodies [64].

T-cell subsets in canine SLE

A study of 20 dogs with SLE in differing stages showed striking abnormalities in lymphocytes and in T-cell subsets [65]. First, animals in the active phase showed marked lymphopenia, with mean counts of $1.05 \times 10^9/l$ as compared with $2.13 \times 10^9/l$ in controls. In the former, CD4$^+$ and CD8$^+$ cells comprised a mean of 56.7 and 10.9%, respectively (CD4:CD8 ratio, 5.2). Comparative data for the controls were 40.5 and 18%, which implies a striking deficiency of CD8$^+$ cells. Moreover, the percentage of 2B3$^+$, or activated T cells, increased in the active phases of the disease from a mean of 46.5 to 64.1%. A combination treatment with prednisone and levamisole was effective in inducing long-term remission, and successful therapy was associated with normalization of the CD4:CD8 ratio.

Lupus anticoagulant

The lupus anticoagulant has been investigated in dogs with autoimmune hemolytic anemia with secondary pulmonary thromboembolism [66]. However, this antibody is rarely found and such patients also lack autoantibodies reactive with endothelium [67].

Antithyroglobulin antibodies

Hypothyroidism as a result of lymphocytic thyroiditis is a relatively common disease in dogs, and the presence of antithyroglobulin antibodies is a useful diagnostic aid [42]. However, although thyroiditis has been reported in cases of SLE [41], it appears to be an uncommon manifestation.

Rheumatoid factor

Serum IgM or IgA rheumatoid factor is reported in a significant proportion of dogs with SLE but also occurs in dogs with leishmaniosis, heart worm infection, or pyometra [68, 69].

Genetic and Experimental Studies of Canine SLE

Breeding studies

As for most canine autoimmune diseases, it is suggested that there is a genetic influence on the development of SLE. In fact, the first reports of canine SLE, dating from the 1960s, resulted from breeding studies conducted by Lewis and Schwartz [70]. These authors established a breeding colony by mating two dogs with SLE. The progeny of these animals, through several generations, had high-titered serum ANA but clinical disease (including SLE, rheumatoid arthritis, and autoimmune thyroiditis) did not develop until the dogs reached middle age.

Another colony was established by Monier and colleagues in France, starting with the mating of an affected German shepherd to an affected Belgian shepherd [71]. Both the incidence of positive ANA and clinical signs increased with each inbred generation. In the F1 family, the incidence of positive ANA and of clinical signs was 3/6 and 0/6, respectively. In the F2 it was 2/4 and 1/4, and in the F3 it was 5/6 and 4/6. Lymphocytes from affected animals showed higher sensitivity to concanavalin A (Con A) in lymphocyte blastogenesis assays, reduced Con A-induced suppression, lower levels of thymulin, and resistance to the induction of immunological tolerance using intravenous aggregate-free human IgG. In general, the appearance of ANA preceded that of clinical signs; in one case by as much as 3.8 years.

Hubert et al. [72] examined another colony of German shepherd dogs that were being bred for show and sale. In contrast to the earlier study, the incidence of ANA and of clinical signs was reduced in each generation, which could be ascribed to the use of out-breeding rather than inbreeding once the clinical signs had become evident. It was of interest that of the 34 dogs that showed one or more ACR criteria for SLE, 22 were males and 12 were females. It was also noted that the appearance of the different clinical signs could be spread over some time. For example, in one animal, arthritis developed 16 months prior to the onset of proteinuria and cutaneous manifestations [72].

Conclusions of these three studies involving artificial or spontaneous breeding colonies are that the occurrence of SLE cannot be ascribed solely to genetic influences and that environmental influences must also play a role.

Breed and sex incidence

In most clinical studies, the German shepherd breed appears to be overrepresented, and in a study from Lyon, France, German shepherd dogs comprised 46.7% of 75 cases of SLE fulfilling at least four ACR criteria, as opposed to a 25% incidence of this breed in the base population [73]. In another study of 58 dogs from California, in which the diagnosis was aided by the application of log-linear and logistic regression models in disease prediction, 17.4% of 58 cases were German shepherd dogs, as compared with a base clinic population of 7.8% ($p<0.01$) [74]. SLE occurs as one of a number of autoimmune manifestations in the susceptible English cocker spaniel [41] and old English sheepdog [75] breeds.

The existence of a sex bias is controversial and complicated by the fact that many pet dogs are neutered. The only study suggesting a gender predisposition was that of Drazner [18] in which 23 of 30 affected animals were female. However, two important points relating to the latter study are that some animals diagnosed as suffering from SLE satisfied only two ACR criteria, and it was also noteworthy that only three out of the 23 females were neutered. A male sex bias was suggested in both the French study (51 males versus 24 females) and the Californian study in which intact males were significantly overrepresented ($p<0.025$) [73, 74].

Genetic analysis

Possible associations between SLE and MHC class I and II antigens, as well as associations with C4 allotypes, have been investigated in three breeding colonies in which there is a high incidence of SLE [76]. There was a significant positive association of SLE with DLA-A7 (relative risk [RR], 11.93) and negative associations with both DLA-A1 and DLA-B5 (RR, 0.06 and 0.05). No associations were found with class II antigens, and there were no associations with the C4 allotype. The latter finding contrasts with a previous study on eight dogs with polyarthritis, all of which had positive ANA, but only two of which satisfied the current ACR criteria for SLE [77]. Five (out of five studied) had the same C4 allotype, and three had markedly lowered C4 levels, which rose following glucocorticoid treatment. The same C4 allotype was over-represented in a breeding colony of English cocker spaniel dogs with a high frequency of various autoimmune diseases [41]. The dogs in that colony also often had concurrent IgA deficiency. Current genomic technology has not yet been applied to studies of canine SLE — which likely reflects the relative rarity of the disease and the difficulty in obtaining samples from well-phenotyped populations.

Possible role of transmissible agents

The early investigations of Lewis and colleagues suggested that serological abnormalities could be transmitted by a filterable agent [78]. Cell-free extracts of spleens from the offspring of dogs with SLE were injected into newborn puppies and into mice and rats. The dogs and the mice, but none of the rats, became ANA-positive. None of the animals developed SLE. This led to a search for possible evidence of transmission of serological abnormalities from man to dog, and *vice versa*. The first report concerned two dogs living in a household where a number of family members had SLE [79]. Skin disease and arthritis were noted in the dogs, which were not subjected to detailed veterinary investigations. Both had anti-dsDNA by Farr assay.

The implication that a transmissible agent might cross the species barrier was refuted by later studies [80–82], but the controversy was reawakened by study in which sera from 15 dogs living with patients with SLE were compared with sera from nine dogs with various immune-mediated diseases and control sera from ten laboratory-reared dogs [83]. The control group was not matched for breed, age, or environment. Dogs from both the immune-mediated disease group and the SLE patient group had higher levels of dsDNA by ELISA, but no sera from any group contained anti-dsDNA by indirect immunofluorescence and none of the sera from dogs living with SLE patients were ANA-positive.

The issue, however, still fails to go away, and recent review again draws heavily on anecdotal observations [84]. Two dogs were described that lived in the home of a rheumatologist whose mother-in-law suffered from SLE. One of the dogs had clinical signs compatible with SLE, a positive ANA, and elevated anti-RNP as assessed by ELISA. The other dog had an ANA titer of 25 and what was described as ocular Sjögren syndrome. Canine Sjögren syndrome is relatively rare, and the dog probably was suffering from keratoconjunctivitis sicca, which is a common but distinct immune-mediated disease.

Experimental model

The most innovative recent advance in the study of canine SLE has been the description of an experimental model of the disease [85]. Beagle dogs repeatedly immunized with heparan sulfate developed an SLE-like disease with cutaneous changes, protein-losing nephropathy and serum ANA. The skin changes included alopecia, erythema, crusting, and scaling with histological features of lymphoplasmacytic dermatitis with basement membrane immunoglobulin and complement deposition. The model has been used to study the effect of a novel gene therapy in which dogs were injected intravenously with a peptide-gene construct including the gene encoding the canine CTLA-4 extracellular domain coupled to sequence from the canine immunoglobulin alpha heavy chain CH_2-CH_3 domain. These sequences were inserted between the cytomegalovirus promoter and the polyadenylation sequence of bovine growth hormone and this construct was attached to a leader peptide through a linker DNA sequence. CTLA-4 is normally expressed by regulatory T cells and binds to the CD80/86 molecules expressed on the surface of antigen-presenting cells (APCs). This prevents these APC receptors from being ligated by CD28 expressed by potential effector T cells. Treated dogs had resolution of the clinical and serological abnormalities.

Treatment

Therapy of dogs with SLE generally involves the use of immunosuppressive doses of glucocorticoids, administered in a gradually tapered protocol. Adjunct cytotoxic drugs may be considered depending upon the perceived severity of the clinical presentation. In veterinary medicine, these adjunct agents have traditionally included azathioprine, cyclophosphamide, or chlorambucil [86, 87]. In cases where immune-mediated thrombocytopenia is part of the presentation, the use of vincristine may be employed. The most recent adjunct therapeutics used for other canine autoimmune diseases include ciclosporin and high-dose intravenous human immunoglobulin (IVIG) therapy. There have been no large multicenter trials comparing outcome for different immunosuppressive protocols employed in canine medicine. In general, immunosuppressive therapy is tapered and withdrawn after clinical response. Relapses of disease will require further cycles of therapy and in some individual patients, life-long low-dose maintenance therapy is required to control clinical signs.

Plasmapheresis was used on four dogs with SLE that were refractory to conventional treatment [88]. Three of the four dogs had good responses lasting from 1 to 6 months. The therapy led to an appreciable fall in the ANA titer and to an increase in the levels of C3 and total hemolytic complement. The authors recommend the treatment only for severe cases that are minimally responsive to anti-inflammatory therapy and cytotoxic drugs; however, the equipment required for this technique is not generally available in veterinary hospitals.

Excellent results have been reported with the combined use of prednisone with and the anthelminthic drug levamisole, which is also suggested to have immunomodulatory properties [73, 89]. In a comparative study, one group of dogs received initial treatment with 1–2 mg/kg of prednisone, which was gradually tapered once control was affected. The other received the same dose of prednisone with levamisole orally at a dose of 3–7 mg/kg/day on alternate days. The first group continued to receive minimal doses of glucocorticoids for maintenance, whereas in the case of the second group, prednisone was discontinued, with levamisole alone used when there was a recrudescence. In the group treated with glucocorticoids alone, clinical improvement always occurred, but it was not maintained when prednisone was withdrawn. In the group in which combined therapy was used, prolonged remission was obtained in more than 50% of cases, which was as long as 9 years. When relapses occurred, levamisole alone was effective in achieving a further sustained remission. Successful treatment with levamisole caused a normalization of the lymphopenia, as well as a normalization in the abnormal CD4:CD8 ratio [65]. Despite this, the use of levamisole in dogs can no longer be recommended due to the relatively high incidence of severe side effects associated with this drug.

FELINE SYSTEMIC LUPUS ERYTHEMATOSUS

Introduction

Spontaneously arising SLE in the cat is at best a rare occurrence. There is a relatively sparse literature and this is reviewed in the following sections. However, in addition to the spontaneous disease, cats are known to develop drug-induced lupus. This species suffers relatively commonly from hyperthyroidism, generally related to the presence of a functional thyroid adenoma. Medical management of this condition initially involved the use of propylthiouracil, but this has more recently been superceded by thiamazole (methimazole) and carbimazole. All of these drugs have immune-mediated side effects [90–92]. In experimental studies, 53% of cats given 150 mg propylthiouracil daily became ill within 4–8 weeks, manifesting weight loss and lymphadenopathy. Affected cats also showed Coombs-positive anemia and positive ANA [91]. Serum ANA positivity, Coombs-positive anemia, and thrombocytopenia are also recognized side effects of thiamazole and carbimazole [92].

The potential for arthropod-transmitted infectious agents to trigger SLE-like disease in the cat should also be considered. A recent report describes three cats with infection by an *Ehrlichia canis*-like organism that developed polyarthritis, anemia, thrombocytopenia and serum ANA [93].

Clinical Signs

The first report of an SLE-like disease in the cat appeared in 1971 and was described as a case of immune complex glomerulonephritis, which was accompanied by a positive LE-cell preparation [94]. This was followed by papers detailing a cat with autoimmune hemolytic anemia [95] and one with immune-mediated thrombocytopenia [96], both accompanied by positive LE-cell preparations

Scott *et al.* [97] described two cats whose presenting signs were dermatological with low-titered serum ANA (10 and 40). One case was presented with recurrent fever and lymphadenopathy, paronychia, and a symmetric crusting alopecic rash involving the head and ears. Direct immunofluorescence revealed immunoglobulin deposits at the dermoepidermal junction. The second case had an extensive, ulcerative, crusting skin disease and anemia. Histopathology revealed striking hydropic degeneration of the epidermal basal cells.

A further report discussed results of ANA screening of sera from 107 cats presented to the University of Florida College of Veterinary Medicine over a 4-year period with signs suggestive of immune-mediated disease [98]. Five sera yielded titers over 10. Two of the cats had ulcerative plasma cell pododermatitis accompanied by glomerulonephritis and anemia. Two other cats showed skin lesions that were histologically compatible with pemphigus foliaceus, one of which showed linear deposits of immunoglobulin at the dermoepidermal junction. The final case had a history of progressive, erosive polyarthritis and later developed erosive nasal lesions, periocular erythema, and glomerulonephritis. Three of these cases satisfied the revised ACR criteria for a diagnosis of SLE.

Another report described 11 cats with a wide range of clinical signs, including various neurological and behavioral changes, glomerulonephritis, nonerosive arthritis, anemia, stomatitis, lymphadenopathy, and a range of skin changes [99]. Six of these cats satisfied the revised ACR criteria by showing at least four classical signs. ANA titers ranged from 40 to 1280. Two of these cats were Siamese and two more were Siamese-crosses. A further series of 13 cats with positive ANA (titers 10–160) has been reported [100]. Seven of these cats had immune-mediated polyarthritis (and four of five tested also had positive serum rheumatoid factor), three had immune-mediated skin disease, one had nephrotic syndrome, one only nonspecific pyrexia and lethargy, and only a single cat satisfied criteria for diagnosis of SLE. Three of these cats were Siamese, four were Persian, and five were domestic shorthairs (one not recorded). The cats ranged in age from 2 to 11 years (mean, 4.85 years).

The latest reports feature (i) a cat with symmetrical facial dermatitis (Figure 25.8), thrombocytopenia, and an ANA titer of 160 [101] and (ii) a cat with autoimmune hemolytic anemia, thrombocytopenia, probable

FIGURE 25.8   Facial lesions in a cat with SLE. *Reproduced with permission from Wiley-Blackwell, Vitale et al. [101].*

neurologic disease, and an ANA titer of 4096 [102]. The latter case did not fulfill the criteria for definite SLE. Histopathology of the skin in the former case showed an interface dermatitis and folliculitis with occasional basal cell vacuolation and necrosis and pigmentary incontinence. The cat responded to glucocorticoid therapy, but the skin disease relapsed following cessation of the treatment with, in addition, ulceration involving the hard palate. It was concluded that the cat had now satisfied four of the ACR criteria.

A summary of the incidence of the clinical signs ascribed to feline lupus is detailed in Table 26.1.

Diagnosis and Therapy

The approach to the diagnosis in the cat follows broadly the same principles as outlined for the dog. However, care must be taken to exclude underlying viral diseases, in particular feline leukemia and feline infectious peritonitis, both of which can induce positive ANA tests [98].

High-dose glucocorticoid represents the cornerstone of the therapeutic approach, with the addition of chlorambucil or cyclophosphamide if the former approach alone proves insufficient or where an alternate day regimen for the corticosteroids cannot be obtained. In comparison to dogs, cats are relatively resistant to the side effects of glucocorticoids, which may reflect differences in plasma binding or in tissue receptors. This also means that generally higher levels are required to effect immunosuppression than in dogs [86].

EQUINE SYSTEMIC LUPUS ERYTHEMATOSUS

SLE is rarely reported in horses [103, 104]. The literature contains full descriptions of only two cases, although there are anecdotal reports of additional cases. The reported clinical signs include cutaneous lesions (distal limb edema, panniculitis, mucocutaneous ulceration, focal alopecia, scaling and leukoderma) (Figure 25.9), polyarthritis, anemia, purpura, lymphadenopathy, fever, anorexia, and weight loss.

The first report described a horse presenting with polyarthritis, proteinuria, thrombocytopenia, and a positive ANA test [105]. The second contained a full description, with postmortem findings. The patient was presented with a history of weight loss, Coombs-positive anemia, and bilaterally symmetrical alopecia [106]. Clinical examination also showed an ulcerative glossitis and generalized lymphadenopathy. Histopathological examination of skin biopsies showed a superficial and deep lymphocytic infiltration, with hydropic degeneration and frequent single-cell apoptosis of the follicular and interfollicular basal cells. Immunoperoxidase labeling of paraffin wax-embedded sections revealed linear deposits of immunoglobulin at the dermoepidermal junction, and the ANA test was positive with a titer of 320.

Although the anemia responded to the administration of dexamethasone (0.2 mg/kg IV daily), the animal became recumbent and was humanely destroyed. In addition to the findings noted premortem, necropsy examination showed focal lymphoid infiltration of the liver, fibrinous synovitis, and membranous glomerulonephritis accompanied by immunoglobulin deposits on the basement membrane. The diagnosis of SLE was convincingly made.

FIGURE 25.9   Facial lesions of a horse with cutaneous LE. *Reproduced with permission from Wiley-Blackwell, Stannard [104].*

Serum ANA was not detected in horses with a range of degenerative or traumatic joint diseases [107], but none of these animals had SLE.

Acknowledgments

This chapter in the previous edition of this text was written by Professor R.E.W. Halliwell of the University of Edinburgh. Professor Halliwell has now retired and devolved responsibility for the chapter to the present author. I would like to acknowledge the substantial and excellent review by Professor Halliwell that forms the basis for this revised chapter.

References

[1] M.J. Day, One health and the legacy of John M'Fadyean, Journal of Comparative Pathology 139 (2008) 151—153.

[2] M.J. Day, The basis of immune-mediated disease, in: M.J. Day (Ed.), "Clinical Immunology of the Dog and Cat", Manson Publishing Ltd, London, 2008, pp. 75—93.

[3] E.F. Kirkness, V. Bafna, A.L. Halpern, S. Levy, K. Remington, D.B. Rusch, et al., The dog genome: survey sequencing and comparative analysis, Science 301 (2003) 1898—1903.

[4] W.E.R. Ollier, L.J. Kennedy, W. Thompson, A.N. Barnes, S.C. Bell, D. Bennett, et al., Dog MHC alleles containing the human RA shared epitope confer susceptibility to canine rheumatoid arthritis, Immunogenetics 53 (2001) 669—673.

[5] L.J. Kennedy, L.J. Davison, A. Barnes, A.D. Short, N. Fretwell, C.A. Jones, et al., Identification of susceptibility and protective major histocompatibility complex haplotypes in canine diabetes mellitus, Tissue Antigens 68 (2006) 467—476.

[6] L.J. Kennedy, S. Quarmby, G.M. Happ, A. Barnes, I.K. Ramsey, R.M. Dixon, et al., Association of canine hypothyroidism with a common major histocompatibility complex DLA class II allele, Tissue Antigens 68 (2006) 82—86.

[7] L.J. Kennedy, A. Barnes, W.E.R. Ollier, M.J. Day, Association of a common DLA class II haplotype with canine primary immune-mediated haemolytic anaemia, Tissue Antigens 68 (2006) 502—506.

[8] A. Barnes, T. O'Neill, L.J. Kennedy, A.D. Short, B. Catchpole, A. House, et al., Association of canine anal furunculosis with TNFA is secondary to linkage disequilibrium with DLA-DRB1*, Tissue Antigens 73 (2009) 218—224.

[9] M.J. Day, Multisystem and intercurrent immune-mediated disease, in: M.J. Day (Ed.), "Clinical Immunology of the Dog and Cat", Manson Publishing Ltd., London, 2008, pp. 356—368.

[10] R.M. Lewis, R. Schwartz, W.B. Henry, Canine systemic lupus erythematosus, Blood 25 (1965) 143—160.

[11] D.W. Scott, D.K. Walton, C.A. Manning, Smith, R.M. Lewis, Canine lupus erythematosus. I. Systemic lupus erythematosus, Journal of the American Animal Hospital Association 19 (1983) 461—479.

[12] C.B. Grindem, K.H. Johnson, Systemic lupus erythematosus: literature review and report of 43 new canine cases, Journal of the American Animal Hospital Association 19 (1983) 489—503.

[13] L. Chabanne, C. Fournel, J.-C. Monier, D. Rigal, Canine systemic lupus erythematosus. 1. Clinical and biological aspects, Compendium of Continuing Education for the Practicing Veterinarian 21 (1999) 135—141.

[14] K. Thoren-Tolling, L. Ryden, Serum auto antibodies and clinical/pathological features in German shepherd dogs with a lupuslike syndrome, Acta Veterinaria Scandinavica 32 (1991) 15—26.

[15] K.J. Dunn, J.K. Dunn, Diagnostic investigations in 101 dogs with pyrexia of unknown origin, Journal of Small Animal Practice 39 (1998) 574—580.

[16] J.O. Bohnhorst, I. Hanssen, T. Moen, Immune-mediated fever in the dog. Occurrence of antinuclear antibodies, rheumatoid factor, tumour necrosis factor and interleukin-6 in serum, Acta Veterinaria Scandinavica 43 (2002) 165—171.

[17] N.C. Pedersen, K. Weisner, J.J. Castels, G.V. Ling, G. Weiser, Non-infectious canine arthritis: the inflammatory nonerosive arthritides, Journal of the American Veterinary Medical Association 169 (1976) 304—310.

[18] F.H. Drazner, Systemic lupus erythematosus in the dog, Compendium of Continuing Education for the Practicing Veterinarian 3 (1980) 243—254.

[19] J.C. Monier, M. Dardenne, D. Rigal, O. Costa, C. Fournel, M. Lapras, Clinical and laboratory features of canine lupus syndromes, Arthritis and Rheumatism 23 (1980) 294—301.

[20] D. Bennett, Immune-based non-erosive inflammatory joint disease of the dog. 1. Canine systemic lupus erythematosus, Journal of Small Animal Practice 28 (1987) 871—899.

[21] O.A. Garden, M.J. Day, C.M. Elwood, H.C. Rutgers, Polysystemic immune-mediated disease in the dog and cat. Part I. Pathogenesis and clinicopathological features, Veterinary International 2 (1997) 3—16.

[22] M.J. Day, D. Bennett, Immune-mediated musculoskeletal and neurological disease, in: M.J. Day (Ed.), "Clinical Immunology of the Dog and Cat", Manson Publishing Ltd., London, 2008, pp. 172—200.

[23] S.H. Krum, G.H. Cardinet, B.C. Anderson, T.A. Holliday, Polymyositis and polyarthritis associated with systemic lupus erythematosus in a dog, Journal of the American Veterinary Medical Association 170 (1977) 61—64.

[24] T. Olivry, K.C.M. Savary, K.M. Murphy, S.M. Dunston, M. Chen, Bullous systemic lupus erythematosus (type 1) in a dog, Veterinary Record 145 (1999) 165—169.

[25] R.D. Sontheimer, The lexicon of cutaneous lupus erythematosus—A review and personal perspective on the nomenclature and classification of the cutaneous manifestations of lupus erythematosus, Lupus 6 (1997) 84—95.

[26] A.P. Foster, C.P. Sturgess, D.J. Gould, T. Iwasaki, M.J. Day, Pemphigus foliaceus in association with systemic lupus erythematosus, and subsequent lymphoma in a cocker spaniel, Journal of Small Animal Practice 41 (2000) 266—270.

[27] J.A. Yager, B.P. Wilcock, Systemic lupus erythematosus, in: "Color Atlas and Text of Surgical Pathology of the Dog and Cat," Wolfe, London, 1994, pp. 95—96.

[28] T.L. Gross, P.J. Ihrke, E.J. Walder, V.K. Affolter, Systemic lupus erythematosus, in: "Skin Diseases of the Dog and Cat: Clinical and Histopathologic Diagnosis," Blackwell Science Ltd, Oxford, 2005, pp. 55—57.

[29] M.J. Day, S.E. Shaw, Immune-mediated skin disease, in: M.J. Day (Ed.), "Clinical Immunology of the Dog and Cat", Manson Publishing Ltd., London, 2008, pp. 122—171.

[30] H.A. Jackson, T. Olivry, Ulcerative dermatosis of the Shetland sheepdog and rough collie may represent a novel vesicular variant of cutaneous lupus erythematosus, Veterinary Dermatology 12 (2001) 19—27.

[31] H.A. Jackson, Eleven cases of vesicular cutaneous lupus erythematosus in Shetland sheepdogs and rough collies: clinical management and prognosis, Veterinary Dermatology 15 (2004) 37—41.

[32] G. Baneth, M.J. Day, X. Roura, S.E. Shaw, Leishmaniosis, in: S.E. Shaw, M.J. Day (Eds.), "Arthropod-Borne Infectious Diseases of the Dog and Cat", Manson Publishing Ltd., London, 2005, pp. 89—99.

[33] P.J. Mellor, A.J.A. Roulois, M.J. Day, B.A. Blacklaws, S.J. Knivett, M.E. Herrtage, Neutrophilic dermatitis and immune-mediated haematological disease in a dog: suspected adverse reaction to carprofen, Journal of Small Animal Practice 46 (2005) 237—242.

[34] T. Balazs, C.J.G. Robinson, N. Balter, Hydralazine-induced antinuclear antibodies in beagle dogs, Toxicology and Applied Pharmacology 57 (1981) 452—456.

[35] M.J. Day, Vaccine side effects — fact and fiction, Veterinary Microbiology 117 (2006) 51—58.

[36] L. Ackerman, H. Bargman, Postvaccinal signs suggestive of systemic lupus erythematosus in a dog, Modern Veterinary Practice 66 (1985) 867—870.

[37] H. Hansson, G. Trowald Wigh, A. KarlssonParra, Detection of antinuclear antibodies by indirect immunofluorescence in dog sera: comparison of rat liver tissue and human epithelial-2 cells as antigenic substrate, Journal of Veterinary Internal Medicine 10 (1996) 199—203.

[38] N.M. Smee, K.R. Harkin, M.J. Wilkerson, Measurement of serum antinuclear antibody titer in dogs with and without systemic lupus erythematosus: 120 cases (1997-2005), Journal of the American Veterinary Medical Association 230 (2007) 1180—1183.

[39] B.E. Smith, M.B. Tompkins, E.B. Breitschwerdt, Antinuclear antibodies can be detected in dog sera reactive to *Bartonella vinsonii* subsp. *berkhoffii, Ehrlichia canis,* or *Leishmania infantum* antigens, Journal of Veterinary Internal Medicine 18 (2004) 47—51.

[40] R.E.W. Halliwell, Skin diseases associated with autoimmunity II. The nonbullous autoimmune skin diseases, Compendium of Continuing Education for the Practicing Veterinarian 3 (1981) 156.

[41] M.J. Day, Inheritance of serum autoantibody, reduced serum IgA and autoimmune disease in a canine breeding colony, Veterinary Immunology and Immunopathology 53 (1996) 207—219.

[42] M.J. Day, IgG subclasses of canine anti-erythrocyte, anti-nuclear and anti-thyroglobulin autoantibodies, Research in Veterinary Science 61 (1996) 129—135.

[43] R. Thorburn, A.I. Hurvitz, H.G. Kunkel, A DNA-binding protein in the serum of certain mammalian species, Proceedings of the National Academy of Sciences USA 69 (1972) 3327—3330.

[44] D. Bennett, D. Kirkham, The laboratory identification of serum antinuclear antibody in the dog, Journal of Comparative Pathology 97 (1987) 523—539.

[45] J. Zeromski, K. Thoren-Tolling, R. Bergqvist, V. Stejskal, DNA binding proteins in canine sera: A method for removal of nonspecific DNA binding in the Farr assay, Veterinary Immunology and Immunopathology 7 (1984) 169.

[46] R.M. Shull, H.A. Miller, A.R. Chilina, Investigation of the nature and specificity of antinuclear antibody in dogs, American Journal of Veterinary Research 44 (1983) 2004—2008.

[47] J.C. Monier, J. Ritter, C. Caux, L. Chabanne, C. Fournel, C. Venet, et al., Canine systemic lupus erythematosus. II. Antinuclear antibodies, Lupus 1 (1992) 287—293.

[48] M. Monestier, K.E. Novick, E.T. Karam, L. Chabanne, J.-C. Monier, D. Rigal, Autoantibodies to histone, DNA and nucleosome antigens in canine systemic lupus erythematosus, Clinical and Experimental Immunology 99 (1995) 37—41.

[49] S.C. Bell, D.E. Hughes, D. Bennet, A.S.M. Bari, D.F. Kelly, S.D. Carter, Analysis and significance of anti-nuclear antibodies in dogs, Research in Veterinary Science 62 (1997) 83—84.

[50] O. Costa, C. Fournel, E. Lotchouang, J.C. Monier, M. Fontaine, Specificities of antinuclear antibodies detected in dogs with systemic lupus erythematosus, Veterinary Immunology and Immunopathology 7 (1984) 369—382.

[51] A. Brinet, C. Fournel, J.R. Faure, C. Venet, J.C. Monier, Anti-histone antibodies (ELISA and immunoblot) in canine lupus erythematosus, Clinical and Experimental Immunology 74 (1988) 105—109.

[52] O. Costa, J.C. Monier, Antihistone antibodies detected by ELISA and immunoblotting in systemic lupus erythematosus and rheumatoid arthritis, Journal of Rheumatology 13 (1986) 722—725.

[53] O. Costa, J.C. Tchouatcha-Tchouassom, B. Roux, J.C. Monier, Anti-H1 histone antibodies in systemic lupus erythematosus: Epitope localization after immunoblotting of chymotrypsin digested H1, Clinical and Experimental Immunology 63 (1986) 608—613.

[54] S. Paul, M.J. Wilkerson, W. Shuman, K.R. Harkin, Development and evaluation of a flow cytometry microsphere assay to detect anti-histone antibody in dogs, Veterinary Immunology and Immunopathology 107 (2005) 315—325.

[55] P.J. Ginel, S. Camacho, R. Lucena, Anti-histone antibodies in dogs with leishmaniasis and glomerulonephritis, Research in Veterinary Science 85 (2008) 510—514.

[56] M. Soulard, J.-P. Barque, V. Delia Valle, D. Hernandez Verdun, C. Masson, F. Danon, et al., A novel 43-kDa glycoprotein is detected in the nucleus of mammalian cells by autoantibodies from dogs with autoimmune disorders, Experimental Cell Research 193 (1991) 59—71.

[57] M. Soulard, V. Dellavalle, G. Monod, P. Prelaud, J.C. Lacroix, C.J. Larsen, The 1 protein of the heterogeneous nuclear ribonucleoprotein is a novel dog nuclear autoantigen, Journal of Autoimmunity 9 (1996) 599—608.

[58] M. Soulard, V. Della Valle, C.J. Larsen, Autoimmune antibodies to hnRNPG protein in dogs with systemic lupus erythematosus: epitope mapping of the antigen, Journal of Autoimmunity 18 (2002) 221—229.

[59] M. Absi, J.P. La Vergne, A. Marzouki, F. Giraud, D. Rigal, A.M. Reboud, et al., Heterogeneity of ribosomal autoantibodies from human and canine connective tissue diseases, Immunology Letters 23 (1989) 35—41.

[60] M. Soulard, S. Lagaye, V. Delia Valle, F. Danon, C.J. Larsen, J.P. Barque, Nucleolar proteins identified in human cells as antigens by sera from dogs with autoimmune disorders, Experimental Cell Research 182 (1989) 482—498.

[61] L.L. Werner, M.S. Bloomberg, M.B. Calderwood Mays, N. Ackerman, Progressive polysystemic immune-mediated disease in a dog, Veterinary Immunology and Immunopathology 8 (1985) 183—192.

[62] H. Hansson, A. Karlsson-Parra, Canine antinuclear antibodies: Comparison of immunofluorescence staining patterns and precipitin reactivity, Acta Veterinaria Scandinavica 40 (1999) 205—212.

[63] E.W. Henriksson, H. Hansson, A. Karlsson-Parra, I. Petersson, Autoantibody profiles in canine ANA-positive sera investigated by immunoblot and ELISA, Veterinary Immunology and Immunopathology 61 (1998) 157—170.

[64] H. Hansson-Hamlin, I. Lilliehook, G. Trowald-Wigh, Subgroups of canine antinuclear antibodies in relation to laboratory and clinical findings in immune-mediated disease, Veterinary Clinical Pathology 35 (2006) 397—404.

[65] L. Chabanne, C. Fournel, C. Caux, J. Bernaud, C. Bonnefond, J.C. Monier, et al., Abnormalities of lymphocyte subsets in canine systemic lupus erythematosus, Autoimmunity 22 (1995) 1—8.

[66] M.S. Stone, I.B. Johnstone, M. Brooks, T.K. Bollinger, S.M. Cotter, Lupus-type anticoagulant in a dog with

hemolysis and thrombosis, Journal of Veterinary Internal Medicine 8 (1994) 57–61.

[67] R. Wells, A. Guth, M. Lappin, S. Dow, Anti-endothelial cell antibodies in dogs with immune-mediated hemolytic anemia and other diseases associated with high risk of thromboembolism, Journal of Veterinary Internal Medicine 23 (2009) 295–300.

[68] L. Chabanne, C. Fournel, J.R. Faure, C.M. Veysseyre, D. Rigal, J.P. Bringuier, et al., IgM and IgA rheumatoid factors in canine polyarthritis, Veterinary Immunology and Immunopathology 39 (1993) 365–379.

[69] S.C. Bell, S.D. Carter, C. May, D. Bennett, IgA and IgM rheumatoid factors in canine rheumatoid arthritis, Journal of Small Animal Practice 34 (1993) 259–264.

[70] R.M. Lewis, R.S. Schwartz, Canine systemic lupus erythematosus, genetic analysis of an established breeding colony, Journal of Experimental Medicine 134 (1971) 417–438.

[71] J.C. Monier, C. Fournel, M. Lapras, M. Dardenne, T. Randle, M. Fontaine, Systemic lupus erythematosus in a colony of dogs, American Journal of Veterinary Research 49 (1988) 46–51.

[72] B. Hubert, M. Teichner, C. Fournel, J.C. Monier, Spontaneous familial lupus erythematosus in a canine breeding colony, Journal of Comparative Pathology 98 (1988) 81–89.

[73] C. Fournel, L. Chabanne, C. Caux, J.R. Faure, D. Rigal, J.P. Magnol, et al., Canine systemic lupus erythematosus. I. A study of 75 cases, Lupus 1 (1992) 133–139.

[74] P.H. Kass, T.B. Farver, D.R. Strombeck, A.A. Ardans, Application of the log-linear and logistic regression models in the prediction of systemic lupus erythematosus in the dog, American Journal of Veterinary Research 46 (1985) 2340–2345.

[75] M.J. Day, W.J. Penhale, Immune-mediated disease in the old English sheepdog, Research in Veterinary Science 53 (1992) 87–92.

[76] M. Teichner, K. Krumbacher, I. Doxiadis, G. Doxiadis, C. Fournel, D. Rigal, et al., Systemic lupus erythematosus in dogs: Association to the major histocompatibility complex antigen DLA-A7, Clinical Immunology and Immunopathology 55 (1990) 255–262.

[77] M.J. Day, P.H. Kay, W.T. Clark, S.E. Shaw, W.J. Penhale, R.L. Dawkins, Complement C4 allotype association with and serum C4 concentration in an autoimmune disease in the dog, Clinical Immunology and Immunopathology 35 (1985) 85–91.

[78] R.M. Lewis, A.J. Schwartz, G. Harris, M.S. Hirsch, P.H. Black, R.S. Schwartz, Canine SLE, transmission of serological abnormalities by cell free filtrates, Journal of Clinical Investigation 52 (1973) 1893–1907.

[79] W.R. Beaucher, R.H. Garman, J.J. Condemi, Familial LE: Antibodies to DNA in household dogs, New England Journal of Medicine 296 (1977) 982–984.

[80] D. Clair, R.J. DeHoratius, J. Wolfe., R.E.W. Halliwell, Autoantibodies in human contacts of SLE dogs, Arthritis and Rheumatism 23 (1980) 251–253.

[81] S. Kristensen, A. Flagstad, H. Jansen, G. Bendixen, R. Manthorpe, P. Oxholm, et al., The absence of evidence suggesting that systemic lupus erythematosus is a zoonosis of dogs, Veterinary Record 105 (1979) 422–423.

[82] J.L. Reinertsen, R.A. Kaslow, J.H. Kippel, N.J. Zvaifler, N. Rothfield, R.M. Lewis, et al., Epidemiologic study of households exposed to canine systemic lupus erythematosus, Arthritis and Rheumatism 22 (1979) 649–650.

[83] D.R.E. Jones, N.D. Hopkinson, R.J. Powell, Autoantibodies in pet dogs owned by patients with systemic lupus erythematosus, Lancet 339 (1992) 1378–1380.

[84] R.S. Panush, M.L. Levine, M. Reichlein, Do I need an ANA? Some thoughts about man's best friend and the transmissibility of lupus, Journal of Rheumatology 27 (2000) 287–291.

[85] E.W. Choi, I.S. Shin, H.Y. Youn, D.Y. Kim, H. Lee, Y.J. Chae, et al., Gene therapy using non-viral peptide vector in a canine systemic lupus erythematosus model, Veterinary Immunology and Immunopathology 103 (2005) 223–233.

[86] M.J. Day, Glucocorticoids and antihistamines, in: J.E. Maddison, S.W. Page, D.B. Church (Eds.), "Small Animal Clinical Pharmacology", Second ed., Saunders Elsevier, Philadelphia, 2008, pp. 261–269.

[87] M.J. Day, Immunomodulatory therapy, in: J.E. Maddison, S.W. Page, D.B. Church (Eds.), "Small Animal Clinical Pharmacology", Second ed., Saunders Elsevier, Philadelphia, 2008, pp. 270–286.

[88] R.E. Matus, B.R. Gordon, C.E. Leifer, S. Saal, A.I. Hurvitz, Plasmapheresis in five dogs with systemic immune-mediated disease, Journal of the American Veterinary Medical Association 187 (1985) 595–599.

[89] L. Chabanne, C. Fournel, D. Rigal, J.-C. Monier, Canine systemic lupus erythematosus. II. Diagnosis and treatment, Compendium of Continuing Education for the Practicing Veterinarian 21 (1999) 402–410.

[90] M.E. Petersen, A.I. Hurvitz, M.S. Leib, P.G. Cavanagh, R.E. Dutton, Propylthiouracil-associated hemolytic anemia, thrombocytopenia, and antinuclear antibodies in cats with hyperthyroidism, Journal of the American Veterinary Medical Association 184 (1984) 806–808.

[91] D.P. Aucoin, M.E. Petersen, A.I. Hurvitz, D.E. Drayer, R.G. Lahita, F.W. Quimby, et al., Propylthiouracil-induced immune-mediated disease in the cat, Journal of Pharmacology and Experimental Therapeutics 234 (1985) 13–18.

[92] M.E. Peterson, P.P. Kintzer, A.I. Hurvitz, Methimazole treatment of 262 cats with hyperthryoidism, Journal of Veterinary Internal Medicine 2 (1988) 150–157.

[93] E.B. Breitschwerdt, A.C.G. Abrams-Ogg, M.R. Lappin, D. Bienzle, S.I. Hancock, S.M. Cowan, et al., Molecular evidence supporting Ehrlichia canis-like infection in cats, Journal of Veterinary Internal Medicine 16 (2002) 642–649.

[94] D.O. Slauson, S.W. Russell, R.D. Schechter, Naturally occurring immune complex glomerulonephritis in a cat, American Journal of Pathology 103 (1971) 131.

[95] S.C. Heise, R.S. Smith, O.W. Schalm, Lupus erythematosus with hemolytic anemia in an cat, Feline Practice 3 (1973) 14–19.

[96] N.H. Gabbert, Systemic lupus erythematosus in a cat with thrombocytopenia, Veterinary Medicine: Small Animal Clinician 78 (1983) 77–78.

[97] D.W. Scott, K.H. Haupt, B.F. Knowlton, R.M. Lewis, A glucocorticoid-responsive dermatitis in cats, resembling systemic lupus erythematosus in man, Journal of the American Animal Hospital Association 15 (1979) 157–171.

[98] L.L. Werner, N.T. Gorman, Immune-mediated disorders of cats, Veterinary Clinics of North America Small Animal Practice 14 (1984) 1039–1064.

[99] N.C. Pedersen, J.E. Barlough, Systemic lupus erythematosus in the cat, Feline Practice 19 (1991) 5–13.

[100] M.J. Day, Diagnostic assessment of the feline immune system. Part 2, Feline Practice 24 (1996) 14–25.

[101] C.B. Vitale, P.J. Ihrke, T.L. Gross, L.L. Werner, Systemic lupus erythematosus in a cat: fulfillment of the American Rheumatism Association criteria with supportive skin histopathology, Veterinary Dermatology 8 (1997) 133–138.

[102] D. Lusson, B. Billiemaz, J.L. Chabanne, Circulating lupus anticoagulant and probable systemic lupus erythematosus in

a cat, Journal of Feline Medicine and Surgery 1 (1999) 193–196.

[103] D.W. Scott, W.H. Miller, Immune-mediated disorders, in: "Equine Dermatology", Saunders Elsevier, St Louis, 2003, pp. 475–547.

[104] A. Stannard, Stannard's illustrated equine dermatology notes, Veterinary Dermatology 11 (177) (2000) 163–178.

[105] A. Vrins, B. Feldman, Lupus-erythematosus-like syndrome in a horse, Equine Practice 5 (1983) 18–25.

[106] R.J. Geor, E.G. Clark, D.M. Haines, P.G. Napier, Systemic lupus erythematosus in a filly, Journal of the American Veterinary Medical Association 197 (1990) 1489–1492.

[107] S.D. Carter, A.C. Osborne, S.A. May, D. Bennett, Rheumatoid factor, anti-heat shock protein (65 kDa) antibodies and anti-nuclear antibodies in equine joint diseases, Equine Veterinary Journal 27 (1995) 288–295.

[108] R.E.W. Halliwell, N.T. Gorman, "Veterinary Clinical Immunology", WB Saunders, Philadelphia, 1989.

Pathogenesis of Lupus Nephritis

Yong Du, Chandra Mohan

THE PATHOGENESIS OF LUPUS NEPHRITIS

Renal disease ranks as the foremost cause of morbidity and mortality in systemic lupus erythematosus (SLE). Yet, the pathogenesis of lupus nephritis has not been fully elucidated. Numerous mediators and mechanisms have been suggested as drivers of the disease. These include dysregulation of the innate immune system, pro-inflammatory cytokine overproduction, impaired B- and T-cell tolerance/regulation, as well as increased expression of various chemokines and other regulators. Experiments, mostly based on lupus-prone mouse models, suggest that the steps and events leading to the development of SLE can be parsed into two phases: systemic autoimmunity and target organ injury, as illustrated in Figure 26.1.

The systemic autoimmune response in lupus is characterized by the activation of self-reactive B cells and T cells, largely directed to nuclear antigens. The consequence of this lymphocyte activation is the elaboration of anti-DNA autoantibodies, one of the hallmarks of this disease. In addition to activation of the adaptive arm of the immune system, cells in the innate arm of the immune system, including dendritic cells (DCs) and macrophages, are also activated in this disease. It is now apparent that different genetic loci (and genes) are responsible for activation of the two broad arms of the immune system in the context of lupus, as recently reviewed [1]. The detailed mechanisms underlying the systemic immune response in SLE, and the cellular players involved have been discussed in Chapters 6, 7, 8, and 19.

Although several different end organs can be targeted in SLE, the pathogenic events leading to disease are fairly similar, involving both infiltrating leukocytes (i.e., the product of the systemic autoimmune response) as well as resident nonimmune cells. Briefly, the deposition of autoantibodies (e.g., anti-DNA antibodies) and

immune complexes in the kidneys can initiate tissue injury, followed by recruitment of inflammatory leukocytes into the kidneys. Cytokines, chemokines, and various mediators released by the infiltrating leukocytes as well as resident renal cells can lead to a vicious cycle of inflammation, leading to acute, then chronic tissue injury. These systemic and local steps are captioned in Figure. 26.1, and elaborated further below.

TOOLS FOR STUDYING LUPUS NEPHRITIS PATHOGENESIS

Two distinct classes of animal models have played important roles in deepening our understanding of lupus nephritis. First, several inbred strains have been noted to develop lupus and nephritis spontaneously. Among the spontaneous mouse models, the New Zealand Black (NZB), F1 hybrids of NZB × NEW Zealand White (NZW) (NZB/W F1), MRL/MpJ-*lpr/lpr* (MRL/*lpr*), and BXSB strains are the most widely studied. These strains exhibit many characteristic features of human SLE, such as increased nuclear-antigen targeting autoantibodies, immune complex deposition, B- and T-cell activation/tolerance, as well as kidney disease [2—4]. However, these strains show some differences in genetic origin, gender bias, age of onset and the pattern of end organ involvement.

The NZB mouse is characterized by mild lupus-like symptoms, in addition to red blood cell autoimmunity. However, NZB/W F1 hybrids develop severe humoral (including antinuclear) autoimmunity, T- and B-cell activation, and a high incidence of lupus nephritis, owing to one or more NZW gene(s). Similar to human SLE, NZB/W F1 mice have a strong female bias, with a mortality rate of 50% at 9 months among the females [2, 5]. Compared to other strains, NZB and NZB/W F1 mice also exhibit hemolytic anemia. The BXSB mouse strain is a recombinant inbred strain derived from a hybrid

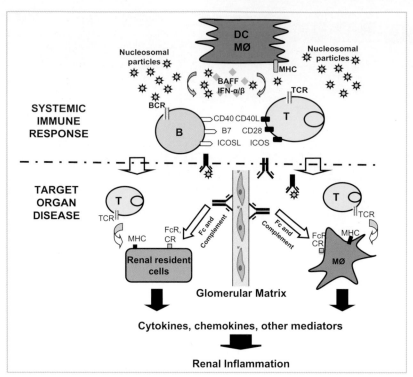

FIGURE 26.1 Two phases in the pathogenesis of lupus nephritis. The systemic events leading to lupus nephritis involve both the innate and adaptive arms of immune system. Hyperactive or pro-inflammatory dendritic cells play pivotal roles in activating T cells and possibly B cells. Activated T cells in turn help autoreactive B cells elaborate potentially pathogenic autoantibodies. Nucleosomal materials derived from apoptotic cells/debris are likely to contribute to the vicious cycle in several ways. Several mediators, notably, type I interferon and BAFF, play facilitatory roles in amplifying the systemic autoimmune response. The consequence of these events is the generation of a cadre of pathogenic effector cells and molecules that can inflict injury in many different target organs. Within the kidneys (and other target organs), autoantibodies and immune complexes deposited in the glomerular matrix initiate local inflammation and recruit myeloid cells, including macrophages and leukocytes, into the kidneys. Immune complexes can then engage FcR and complement receptors on infiltrating leukocytes and resident renal cells, hence activating these effector cells. Pathogenic T cells generated through systemic events infiltrate into the kidneys. A large number of mediators, mainly cytokines, chemokines, and adhesion molecules, expressed or induced by pathogenic T cells, infiltrating myeloid cells, as well as resident renal cells collude to inflict damage. Several molecular intermediates, including CD40:CD40L and B7:CD28, play key roles in facilitating cell−cell contact both systemically and within the kidneys, as diagramed. The various cells and molecules diagramed also constitute attractive therapeutic targets in lupus nephritis.

of the SB/Le and C57B6J strains [4, 6]. The disease of the BXSB mouse strain is facilitated by an accelerating locus located on the Y chromosome, named Y-linked autoimmune acceleration (*Yaa*). Therefore, BSXB male mice develop the disease, with high levels of autoantibodies, and fatal lupus nephritis. It is believed that the *Yaa* contribution is a consequence of a translocation from the telomeric end of the X chromosome onto the Y chromosome [7-8]. *TLR7*, located within the translocated segment of the X chromosome, has been proposed to confer leukocyte activation, and hence systemic lupus [7-9]. More recent findings have indicated that *TLR7* gene duplication may not be sufficient for the *Yaa*-mediated enhancement of disease, suggesting that additional gene(s) present in the same translocated segment may also contribute to the development of lupus [10]. Another strain that develops spontaneous lupus is MRL/*lpr*. Compared to NZB/W F1 and BXSB mice, the MRL/*lpr* strain has a more complex genomic background. It is derived from a complex cross of several inbred strains, including LG/J (75%), AKR/J (12.6%), C3H/HeDi (12.1%), and C57BL/6 (0.3%) [3]. MRL/*lpr* mice carry the *lpr* mutation of *Fas* on the lupus-prone MRL background. As a receptor of the TNF receptor superfamily, *Fas* is an important mediator of apoptosis. Hence, the *lpr* mutation leads to the absence of *Fas*, resulting in impaired lymphocyte apoptosis and subsequent T- and B-cell autoactivation

[2, 4, 11]. However, the deletion in *Fas* alone is not sufficient to trigger lupus, but rather the interaction of MRL background genes with the *Fas* defect accelerates the nephritis. This strain exhibits high titers of ANAs as early as 2−3 months of age, high levels of immune complexes beginning at 4−5 months age, and high mortality (with 100% by 9 months and 50% around 5 months). Additionally, MRL/*lpr* mice exhibit other end-organ phenotypes such as dermatitis, arthritis, and arteritis, which are not prominent in either the NZB/W F1 or BXSB strains [4−5]. Though most of our understanding of murine lupus and lupus nephritis have arisen from these three strains, additional insights have been gained from studies of other strains such as NZM2410 (and the derived congenic strains), NZM2328, and SNF1.

A second category of animal models that has been instructive are mice subjected to antiglomerular basement membrane (GBM) antibody-induced experimental nephritis. Anti-GBM disease is an autoimmune disorder characterized by the presence of anti-GBM antibodies and rapidly progressive crescentic glomerulonephritis. This disease is regarded as the prototype of immune complex-induced nephritis, in which anti-GBM antibodies bind to specific antigenic targets on the glomerular and/or pulmonary basement membranes [12]. It is now widely recognized that the NC1 domain of α3 (IV) collagen is the main autoantigen recognized by the anti-GBM antibody [13-14]. Increasing evidence suggests that other mediators, such as T cells, cytokines, chemokines, as well as resident renal cells, also play important roles in the pathogenesis of this disease. In analyzing the roles of more than 25 different molecules examined in both anti-GBM nephritis and spontaneous lupus nephritis, it has been recognized that these diverse molecules and cells impact both diseases in a concordant fashion, suggesting that these two disease settings may share common downstream pathways and cellular mechanisms leading to tissue damage and renal dysfunction, as detailed elsewhere [15]. Given that anti-GBM nephritis can be studied within a time-frame of 2–3 weeks, unlike spontaneous lupus nephritis, the experimental anti-GBM nephritis model has emerged as a powerful tool for rapidly unraveling the molecular cascades leading to the pathogenesis of spontaneous lupus nephritis. Insights gained from both categories of animal model, supplanted with human studies, have yielded a fairly detailed view of the pathogenic processes leading to lupus nephritis, as summarized in the following section.

ROLE OF ANTI-DNA ANTIBODIES AND IMMUNE COMPLEXES IN LUPUS NEPHRITIS

The presence of autoantibodies directed against several nuclear and cytoplasmic antigens is a hallmark of SLE. Among these spectrum of autoantibodies present in lupus patients, anti-DNA antibodies emerge as a key diagnostic and prognostic marker, being actively involved in the pathogenesis of lupus nephritis owing to their ability to bind to cell surface antigens or components of the glomerular matrix either directly (cross-reactivity) or indirectly (via chromatin-containing bridges).

Its pathogenic role in lupus nephritis was first implicated through its elution from the kidneys of both lupus patients and lupus mouse models in 1957 [16]. Later, it was found that intraperitoneal injection of human or murine anti-DNA antibodies could trigger lupus-like syndromes [17–20]. Likewise, in transgenic mouse models, the engineered hypersecretion of IgG anti-DNA antibodies was noted to elicit many features similar to human lupus nephritis, again underscoring the importance of this autoantibody specificity in lupus [21].

Three mechanisms have been proposed to explain the ability of anti-dsDNA antibodies to localize to the kidney [22, 23]. The first mechanism is based on the formation of immune complexes, where anti-DNA antibodies form complexes with DNA/nucleosomes released from apoptotic cells, and these immune complexes can deposit in the kidney and drive downstream inflammatory cascades. The second mechanism, referred to as the planted antigen hypothesis, suggests that anti-DNA antibodies can directly target the kidney by reacting with DNA/nucleosomes trapped in the glomerular matrix. The trapping of DNA/nucleosomes in the glomerular matrix and the GBM has been attributed to interactions between the positively charged histone DNA and negatively charged glomerular matrix. The third mechanism relies on the cross-reactivity between intrinsic non-DNA glomerular antigens and anti-DNA antibodies. Laminin, heparan sulfate, as well as collagen type IV of GBM, alpha-actinin and ribosomal P protein of mesangial cells, and ribosomal P protein of endothelial cells, are all potential non-DNA autoantigens that can be recognized by anti-DNA antibodies in intrinsic kidney cells [23–24]. Once again, charge-based interaction (between the cationic anti-DNA antibodies and the anionic antigens) may contribute to this cross-reactivity. It appears that all three mechanisms may be playing a role in anti-DNA binding to the glomerular substrate in lupus nephritis.

Once antibodies and immune complexes have been deposited onto the glomerular matrix, they essentially signal to the immune system the need to destroy the target, through two highly conserved molecular adaptors: FcγR-based interaction and complement. In support of this is the observation that NZB/W F1 mice lacking functional FcγR develop autoantibodies and have equivalent immune complex deposition but do not develop glomerulonephritis [25]. Moreover, antibody-mediated glomerular disease can be attenuated by soluble FcR [26–27]. Likewise, the notion that anti-DNA mediated nephritis is complement-dependent is substantiated by gene knockout studies and pharmacological inhibition studies [28–31]. Besides the Fc segment of autoantibody, it appears that the antigen-reactive Fab segments may also have some role in promoting disease, once they mediate targeting to specific glomerular antigens. Thus, for example, it has been documented that anti-DNA antibodies can also directly modulate gene expression in mesangial cells through both FcR-dependent and non-FcR-dependent mechanisms [32]. As a result, cellular functions, including proliferation, the synthesis of extracellular matrix components and inflammatory cytokine production, can be influenced [22, 33].

Not all anti-DNA antibodies, however, have nephritogenic property, and this poses the question: what features make an anti-DNA antibody pathogenic? Previous studies have linked pathogenicity with IgG class, especially IgG1 and IgG3, and their ability to activate complement or engage Fc receptors. Several reports indicate that subtle changes in the amino acid residues at the antigen-binding site of the antibody can determine the antigen-binding fine-specificity. Anti-DNA antibodies with charged amino acid residues such as arginine, asparagine, and lysine within the heavy chain complementarity determining regions (CDRs) have been shown to be nephritogenic [34]. Indeed, the binding effect and pathogenicity can be abrogated by single amino acid substitutions in the CDR and/or framework regions of the heavy chain of murine anti-DNA antibodies [35]. Thus, subtle differences in binding affinity (to nuclear antigens) or fine-specificity and/or cross-reactivity to other renal antigens could also dictate whether anti-DNA antibodies are pathogenic [23], though our understanding of this is far from complete. In this context, another dimension that appears to be important in dictating the degree of pathogenicity of anti-DNA or anti-glomerular antibodies is the degree of concomitant innate stimuli delivered via toll receptors [36–37].

In contrast to IgG anti-DNA antibodies, IgM anti-DNA antibodies show a possible protective effect on lupus nephritis. In murine models of SLE, the lack of secreted IgM was associated with more severe glomerulonephritis [38]. Werwitzke and colleagues treated lupus-prone NZB/W F1 mice with murine IgM anti-dsDNA monoclonal antibodies. The administration of IgM antibodies delayed the onset of proteinuria, reduced the severity of renal pathology, and significantly increased survival [39]. In a study of 202 SLE patients, renal involvement was strongly associated with the absence of IgM anti-DNA antibodies, as SLE patients with a ratio of IgG to IgM less than 0.8 were unlikely to develop renal disease [40]. A similar protective effect for IgM anti-DNA autoantibodies was noted when autoantigen proteome arrays were used to screen autoantibodies in human SLE [41]. The detailed mechanism of protection is still elusive, though it is postulated that IgM antibodies against dsDNA may bind to dsDNA and subsequently block the formation of pathogenic immune complexes bearing IgG anti-DNA antibodies.

Despite the strong association between anti-DNA antibodies and lupus nephritis, and the mechanistic evidence for their pathogenic role in disease, exceptions to this rule have been documented, where there is a clear discordance between anti-DNA antibody titers and disease in lupus patients [42–44]. Indeed, similar findings have also been noted in murine models, with some interesting insights [45–48]. Conversely, mice with high titers of IgG anti-dsDNA antibodies need not always be nephritic. Examples include FcR$^{-/-}$ BWF1 [25], BAFF^{Tg} [49], and B6.Sle1.CD19^{Tg} mice [50]. These examples may allude to the need for certain factors to be present within the kidneys in order for nephritis to ensue or may relate to the fine-specificity and/or cross-reactivity profiles of the anti-DNA antibodies.

What is more interesting is that lupus nephritis can develop in the absence of any anti-DNA antibodies. Waters et al. generated NZM2328Lc4 congenic mice which carried a small interval of chromosome 4 from lupus-resistant C57L/J mice on the NZM2328 background. These congenic mice did not exhibit any anti-DNA antibodies. However, these mice had severe immune complex deposition in their kidneys and eventually developed fatal lupus nephritis [48]. Antibodies eluted from the kidneys of these mice reacted with several renal antigens but not with dsDNA [51]. Similar findings were reported by Christensen et al. employing mice deficient in Toll like receptor 9 (*TLR9*) and by Chan and colleagues, who utilized the mIgM.MRL/*lpr* transgenic mice, where mild nephritis developed in the absence of anti-dsDNA antibodies [52–53]. Together, these data indicate that DNA reactivity is not a prerequisite for nephritis or for the nephritogenic potential of autoantibodies. In view of these findings, one has to entertain the possibility that other, non-DNA-binding autoantibodies may also be pathogenic. It is also conceivable that non-autoantibody-related factors (e.g., other serum-derived factors and intrinsic abnormalities within resident renal cells) could also be contributing to the pathogenesis of nephritis. Both of these clearly warrant further investigation.

THE ROLE OF NUCLEOSOMES IN LUPUS NEPHRITIS

The nucleosome is the fundamental packing unit of DNA, playing a key role in controlling the expression of genetic information by regulating access to certain proteins (e.g., transcription factors). It is believed to be a major autoantigen in SLE. Free nucleosome has been documented in the serum of SLE patients [54]. The prevalence of antinucleosome antibodies in SLE patients ranges from 31 to 100% in different studies, being significantly higher than that of anti-DNA antibodies and antihistone antibodies, with a strong association with the disease activity index, SLEDAI [55]. Likewise, Burlingame and colleagues reported the nucleosome to be among the strongest autoantigenic substrates, 88% of SLE patients being positive with high IgG reactivity to nucleosomes, when compared to reactivity to DNA or histones alone [56]. Importantly, the production of antinucleosome antibodies is considered to be an early event in the disease process and can be detected before anti-dsDNA and antihistone antibodies emerge in both lupus patients and the murine model [57]. Whereas DNA by itself is poorly immunogenic, nucleosomes injected into BALB/c mice can induce lupus-like syndromes, accompanied by autoantibody production and immune-mediated glomerulonephritis [58], while the administration of nucleosomes to young SNF1 lupus-prone mice significantly accelerated nephritis [59], indicating the potential pathogenic role of nucleosomes in driving lupus. Taken together, the above findings suggest that nucleosomes may be a key early driver of autoantibody production in SLE.

The production of autoantibodies in SLE is driven by cognate interactions between select populations of autoactivated Th and B cells, as discussed in Chapters 5, 7, and 8. Anti-DNA B cells and antinucleosome B cells can pick up nucleosome-containing nuclear material at high efficiency, process these complex antigens into peptides, and then present peptides (i.e., derived from the nucleosomes) to specific T cells, resulting in the activation of nucleosome-specific T cells and histone-reactive T cells, which can then reciprocate by activating the antigen-presenting B cells, resulting in ANA production. In support of this model, Datta and colleagues have cloned pathogenic autoantibody-inducing Th cells from both SLE patients and lupus-prone mouse strains [60–62]. In the SNF1 mouse model, the same group reported that nucleosome-specific T cells were detectable long before they produce pathogenic autoantibodies, and that pathogenic autoantibody-inducing Th clones could induce immune-deposit glomerulonephritis when transferred into young preautoimmune mice [59]. This strongly indicates that nucleosomes may be primary immunogens that initiate the cognate interaction between pathogenic Th and B cells in lupus. In support of this is the finding that the administration of apoptotic thymocytes can induce transient anti-DNA and APL-antibody production in normal mice [63]. Hence, it is not a surprise that genetic defects that lead to impaired clearance of apoptotic cells (e.g., defects in C1q, MER, SAP, DNAse1, etc.) lead to lupus-like features, including lupus nephritis, as reviewed elsewhere [64–65].

Of causal importance, antinucleosome and anti-DNA autoantibodies are believed to be involved in the pathogenesis of lupus nephritis. Antinucleosome and anti-DNA autoantibodies can be found in kidney eluates from several lupus-prone strains, including MRL/lpr [57, 66], NZB/W F1 [67], and SNF1 [68], as well as from human lupus nephritis eluates [69]. Nucleosome-specific monoclonal Abs administered into RAG-1-deficient mice can localize within glomeruli [70]. Antinucleosome antibodies can also bind to intrinsic renal cells and matrix through the same mechanisms as anti-DNA antibodies, as already discussed above. In particular, apoptotic nuclear material within the kidneys can serve as targets for circulating antinucleosome antibodies [55]. After deposition within the glomeruli, IgG-bearing immune complexes may then induce tissue injury via FcγR-mediated activation of infiltrating phagocytes and neutrophils or through the activation of complement as discussed above. Finally, it appears that free nucleosomes (e.g., from apoptotic cells) can contribute directly to renal inflammation in yet another way. Coritsidis et al. have reported that mesangial cells stimulated by free nucleosome showed increased total matrix protein, total collagen, and, specifically, collagen type I synthesis [71]. Hence, nucleosome and antinucleosome antibodies can contribute to the pathogenesis of lupus nephritis in a wide variety of ways.

ROLE OF LYMPHOCYTES IN LUPUS NEPHRITIS

A growing body of evidence indicates that B cells can contribute to the pathogenesis of lupus nephritis in both antibody-dependent and antibody-independent ways [72–73], though it was traditionally viewed as simply the producer of antibodies in this disease. Given that B cells can elaborate autoantibodies, it stands to reason that genes and molecules that can breach B-cell tolerance can impact autoantibody formation, and hence antibody-mediated nephritis. By engineering the overexpression or ablation of various genes, researchers have uncovered a growing number of genes that influence B-cell tolerance and by extrapolation, autoimmunity (and lupus nephritis). Examples include FcR, Bim, CD22, Lyn, CD72 as well as programmed cell death gene-1 (PDCD-1), molecules involved in regulating either innate or adaptive immunity [73–77], as elaborated in Chapter 8.

Conversely, researchers have recently shown that a gene that causes spontaneous lupus in mice actually functions by abrogating B-cell tolerance [78]. One of the strongest susceptibility loci for lupus in the commonly studied mouse models, BWF1 or NZM2410, is located on murine chromosome 1. Researchers have identified a key culprit gene within this interval as Ly108 [78–79]. A polymorphism in this gene dictates the ratio of two expression isoforms encoded by this gene: Ly108.1 and Ly108.2. Whereas the normally expressed Ly108.2 isoform was found to enhance the sensitivity of immature B cells to receptor editing and deletion by altering the strength of B cell signaling after BCR cross-linking, the lupus-associated isoform, Ly108.1, failed to delete or edit self-ligated B-cells efficiently, leading to the persistence of self-reactive B cells [78].

Though the primary contribution of B cells may be to elaborate pathogenic autoantibodies, it soon became apparent that B cells could mediate disease in yet other ways, including the presentation of antigens to T cells, and thereby activating them, as demonstrated by the elegant mouse models generated by Shlomchik et al. [53, 80, 81]. Those seminal studies suggested that the glomerular disease in lupus nephritis may be largely antibody-dependent, whereas the tubulointerstitial disease may be antibody-independent and possibly influenced by T cells. Yet another antibody-independent function of B cells in autoimmune disease is cytokine and chemokine release, including IL-6, IL-10, TNF-α, and INF-γ [82–85], as elaborated in Chapters 8 and 17.

Like B cells, T cells also play a key role in the pathogenesis of lupus nephritis. The role that T cells play in driving the systemic autoimmunity in lupus is detailed in Chapter 7. In addition, T cells are among the most conspicuous inflammatory cells to enter the inflamed kidney in both SLE patients and lupus mouse models [86–87]. Importantly, T cells cloned from the cortical interstitium of MRL/*lpr* lupus mice are autoreactive and kidney-specific, can induce tubular epithelial and mesangial cell proliferation [88] and produce cytokines such as IFN-γ [89]. In this same mouse model of lupus nephritis, the pathogenic role of T cells within the kidneys has been demonstrated through the use of renal transplantation in MHC II-deficient MRL/lpr mice [90], MRL/lpr CD4−/− mice [91–92] and treatment of MRL/lpr mice with anti-CD4 antibody [86, 93], where the abrogation of nephritis was related to the loss of autoaggressive MHC II-dependent CD4+ T cells.

In resonance with the above findings, Datta and colleagues have documented that representative pathogenic autoantibody-inducing Th-cell clones derived from the nephritic SNF1 mice rapidly induce lupus nephritis when injected into young prenephritic SNF1 mice [61, 94], while Radeke *et al.* have demonstrated that CD4$^+$ T cells alone were sufficient as initiators and effectors in nephritis, by recognizing specific antigens expressed within the glomeruli in a mouse model [95]. Moreover, in a closely related experimentally induced mouse model, anti-GBM antibody induced glomerulonephritis, it has been reported that a CD4$^+$ T cell-mediated mechanism is sufficient not only to induce glomerular injury, but also to trigger autoantibody responses to diversified glomerular antigens [96].

In terms of effector mechanisms, T cells are likely to contribute to lupus nephritis through cell–cell contact-dependent interactions, resulting in the activation of various effector cells, including B cells, macrophages, as well as renal resident cells. Activated T cells express many costimulatory molecules, which play a role in regulating the function of target cells. Among these molecules, CD40L, a member of TNF family, is arguably the most important [97–107], as discussed in Chapter 7. It is now apparent that the pathogenic contributions of the CD40–CD40L interaction in lupus nephritis arise both systemically as well as in the kidneys. The interaction of T-cell CD40L and CD40 expressed on B cells plays a central role in humoral immune responses, having the capacity to induce clonal expansion, immunoglobulin isotype switch, and differentiation of B cells into antibody-producing plasma cells. In anti-CD40L mAb-treated NZB/W F1 lupus-prone mice, the delayed onset of disease was strongly associated with a decrease in splenic B-cell numbers, suppression of B-cell isotype switching, and up to a tenfold decrease in anti-dsDNA autoantibodies [105].

Apart from B cells, CD40 can be expressed on various effectors cells, such as macrophages, neutrophils, DCs, as well as resident renal cells, suggesting that CD40–CD40L interactions may be important in driving effector functions of other CD40-expressing cells within the kidneys [108–111]. Its expression is markedly upregulated in proliferative lupus nephritis, in parallel with the increased presence of CD40L-bearing T cells in kidneys [53, 112]. Activated T cell–renal tubular epithelial cell co-cultures induce the production of high levels of MCP-1, RANTES, IL-8, and IFN-inducible protein-10 from tubular epithelial cells [113–114]. In these co-cultures, cell contact has been found to be essential, as the blockade of CD40–CD40L dramatically decreased chemokine generation by renal tubular epithelial cells following T-cell contact [114]. It is apparent that the contact between activated T cells and other resident renal cells also enhances adhesion molecule expression and mediator production, with subsequent impact on leukocyte infiltration and egress. Though CD40–CD40L interactions are likely to be critical for lupus nephritis, there is still no definitive evidence that CD40–CD40L interaction within the kidney is absolutely essential for lupus nephritis to ensue.

Besides directing and regulating the function of other cells, T cells can also impact lupus nephritis through the release of various cytokines, as highlighted in Chapters 7 and 17. In human lupus nephritis, a predominance of Th1-type response has been suggested by several studies [115–120]. In lupus-prone mice, administration of IFN-γ to NZB/W F1 mice accelerates disease and IFN-γ deficiency in MRL/*lpr* mice ameliorates nephritis [121–123]. Likewise, inhibition of IL-18, an inducer of IFN-γ and a Th1 cytokine, effectively reduced spontaneous lymphoproliferation, IFN-γ production and glomerulonephritis while improving survival [124]. Similarly, the production of nephritogenic autoantibodies in lupus has also been shown to be Th1-dependent [125–127].

However, there is also some evidence that Th2 cytokines can also have a potential impact on the nature of lupus nephritis. In several lupus-prone mouse models, including NZM 2410 [46], NZB/W F1 [128], (MRL/*lpr* × B6) F2 [122], antibody-mediated neutralization of IL-4 or genetic deficiency of IL-4 or its signaling molecule, STAT6, can protect mice from developing advanced lupus nephritis, without having any effect on anti-DNA antibody production and inflammatory cell infiltration. Consistent with these observations, transgenic overexpression of IL-4 leads to glomerulosclerosis, which has been found to be independent of immunoglobulin deposition [129]. The fibrogenic role of IL-4 has also been implicated by several studies, in which IL-4 exhibits a direct effect on fibroblast proliferation, collagen gene expression, collagen synthesis as well as TGF-β production [130]. Taken together, it appears that IL-4 may directly act upon renal cells to perpetuate glomerulosclerosis and chronic renal fibrosis, partly through its effect on extracellular matrix generation. Finally, one should keep in mind that T cells generate a wide spectrum of additional cytokines, chemokines, and mediators. Whether any of these plays a role in the pathogenesis of lupus nephritis remains an open question.

ROLE OF MYELOID CELLS IN LUPUS NEPHRITIS

The Role of DCs in Lupus Nephritis

DCs constitute the most powerful APCs and are crucial for both innate and adaptive immunity. However, the exact mechanisms by which DCs, particularly renal DCs, contribute to lupus nephritis are poorly understood [131]. Within normal human kidneys, at least two myeloid DC (mDCs) subtypes characterized by BDCA-1$^+$DC-SIGN$^+$ and BDCA-1$^+$DC-SIGN$^-$, and one plasmacytoid DC (pDCs) subtype defined as BDCA-2$^+$ DC-SIGN$^-$, are abundantly located in the tubulointerstitium, but rarely observed within glomeruli [132]. Similar to these findings in humans, others have documented in murine kidneys, CD11c$^+$ cells, mostly CD11b^{lo}, located in the tubulointerstitial spaces and close to, but mostly outside of peritubular capillaries [133], forming an intricate network patrolling the renal tubulointerstitium [134].

In lupus nephritis patients, strong renal infiltrates of BDCA1$^+$, BDCA3$^+$, and BDCA4$^+$ DCs have been reported. Notably, DCs infiltrated both the tubulointerstitium and the glomeruli, and the extent of infiltration was higher in severe renal damage (e.g., class III/IV LN) [135—136]. In other renal diseases such as rejection and IgA nephropathy, DC-SIGN$^+$ DCs have been noted to be significantly increased and intraglomerular DC infiltration has also been observed [132]. These observations suggest that the number of renal DCs and their anatomical distribution might change under pathophysiological conditions [132]. As in normal kidneys, DC infiltrates in diseased human kidneys were mostly immature, marked by the absence of DC-LAMP$^+$ cells [132, 135]. In contrast to the renal DCs, a significant decrease of mDCs and/or pDCs in peripheral blood of activated lupus patients has been documented by

several independent studies [135—138]. It has been suggested that the decreased number of DCs in peripheral blood may be a consequence of their enhanced maturation and accelerated migration into the kidneys [136, 139—140].

Studies in murine models have also yielded some insights into the role of DCs in lupus nephritis. The increased infiltration of DCs into the renal glomeruli and tubulointerstitium has been documented in several murine nephritis models, including NZB/W F1 [141, 142], anti-GBM nephritis [133, 143—144], and chronic glomerulonephritis induced in NOH transgenic mice [145]. In the latter mouse model, the model antigens ovalbumin and hen egg lysozyme were selectively expressed in glomerular podocytes. Co-injection of ovalbumin-specific transgenic CD8$^+$ CTLs and CD4$^+$ Th cells into this model resulted in nephritis resembling human chronic glomerulonephritis. The critical role of renal DCs in progression of renal disease was documented by the fact that selective CD11c$^+$ depletion rapidly resolved established renal immunopathology [145]. Conversely, a renoprotective effect of DCs has also been documented in anti-GBM-induced experimental nephritis, where the depletion of CD11c$^+$ was associated with aggravated pathological damage and worse renal function [144]. Given that DCs from lupus-susceptible genomes may be functionally more active and pro-inflammatory [146], it is conceivable that renal DCs in lupus-prone subjects may be intrinsically more pathogenic.

Despite the above findings, little is known about how renal infiltrating DCs impact the pathogenesis of lupus nephritis. Based on the known function of DCs, some mechanisms can be postulated. First, DCs are known to generate several pro-inflammatory and pro-fibrotic factors, including TNF-α, IL-6, IL-1, and TGF-β [139, 147], particularly on lupus-prone genetic backgrounds [146]. These cytokines can potentially enhance renal inflammation and hence lead to renal fibrosis. Importantly, mDCs are the major producer of IL-18 [148] and pDCs are primary IFN-α-producing cells [149], and both these cytokines are causally related to lupus nephritis. Second, as important APCs, DCs can migrate to local lymph nodes and potentially present renal autoantigens to T (and B) lymphocytes directing an autoimmune response toward self [135, 150—151]. Another possible scenario is exemplified by the NOH^{Tg} mice, where the intrarenal presentation of glomerular antigens by DCs plays an essential role in eliciting cytokine production by specific Th cells [145]. Third, since renal DCs express various co-stimulatory molecules such as CD40L, MHC-II, and chemokine receptors such as CCR1 and CCR5 [136, 152], they can interact with intrinsic renal cells and other infiltrating inflammatory cells through these molecules [153, 154]. Another open question relates to how DCs migrate into the diseased kidneys. Activated DCs can express IL-18R [136, 148] and the blockade of IL-18 by anti-IL-18 monoclonal antibody abrogates migration of pDCs, suggesting that IL-18 is a strong chemoattractant for pDCs [148]. CCR5 also contributes to the process, as CCR5 antagonism also inhibits DC migration and intrarenal accumulation [152]. The relevance of these two players and other chemokines in recruiting DCs (and other leukocytes) in the context of lupus nephritis remains to be examined.

The Role of Macrophages in Lupus Nephritis

Macrophages are key cellular determinants of pathogenesis and progression in lupus nephritis. Their recruitment has been documented in both lupus-prone mice and lupus nephritis patients [87, 142, 155–157]. Recruited macrophages are located in both the glomerular tuft and tubulointerstitium [87, 142, 155–158] and constitute the major cell type in glomerular crescents [158]. Importantly, renal infiltrating macrophages are mainly activated type II macrophages, with high expression of CD11b, Ox40L, CD80, and CD86, being markers of disease onset in lupus nephritis [142]. The critical role of macrophages in the pathogenesis of lupus nephritis has been demonstrated by several depletion and transfer studies conducted in experimental anti-GBM nephritis [159–164], an experimental model that closely resembles lupus nephritis. Consistent with these observations, MCP-1-deficient lupus-prone mice exhibit reduced renal disease and improved survival, accompanied by fewer interstitial macrophages [165].

The glomerular recruitment of macrophages is likely to be initiated by local signals triggered in response to the deposition of immune complexes in the glomerular matrix [166]. These signals include chemotactic factors (e.g., MCP-1 and RANTES), adhesion molecules (e.g., ICAM-1 and VCAM-1), as well as various cytokines (e.g., CSF-1 and MIF). The detailed mechanisms of macrophage recruitment and activation by these released mediators will be discussed below and in Chapter 6

Once recruited, it is believed that activated macrophages play a wide variety of roles in mediating renal injury. Of note, activated glomerular macrophages have the capacity to produce a variety of proinflammatory mediators which in turn can amplify leukocyte recruitment, activate or injure intrinsic renal cells, and amplify local inflammatory injury. Indeed, macrophage-derived products have been implicated in a number of pathological processes leading to glomerulonephritis. For example, macrophage-derived reactive oxygen species and proteolytic enzymes can disrupt the integrity of the GBM, injure other glomerular cells, and induce widespread necrosis within the glomerular tuft [167]. The pro-inflammatory role of TNF, IL-1, and other macrophage-derived cytokines has been demonstrated in experimental models of immune-mediated nephritis through antibody inhibition studies [168, 169] and the use of cytokine-deficient mice [170–172]. Moreover, growth factors generated by macrophages, including PDGF and TGF-β, contribute to mesangial cell proliferation and tissue sclerosis in a variety of experimental and human glomerulonephritis, including lupus nephritis [173–176]. Besides releasing various toxic mediators, there is little evidence indicating that macrophages can serve as APCs within the renal milieu.

The Role of Neutrophils in Lupus Nephritis

Besides macrophage and DCs, another major category of myeloid cells, the neutrophils, are also likely to contribute to the pathogenesis of antibody-meditated nephritis, particularly to acute changes [166, 177], as confirmed by the depletion studies [177, 178] and adoptive transfer experiments [179]. Recruited neutrophils are mostly located within the glomerular tuft in several types of glomerulonephritis, including lupus nephritis [180] and crescentic glomerulonephritis [177, 178,181]. In addition to their role in pro-inflammatory cytokine production [182, 183], neutrophils can release proteases and reactive oxygen intermediates from their granules. These proteases, including cathepsin, elastase, MPO, etc., have been documented to be capable of destroying glomerular structures both in vivo and in vitro [177–181, 184, 185], whereas these effects can be diminished through the use of specific inhibitors [186]. One of the proteases, gelatinase-B-associated lipocalin, has recently been shown to be a good biomarker of lupus nephritis [187–190].

In sum, our understanding of how myeloid cells contribute locally to nephritis in SLE is incomplete. How myeloid cells migrate into the kidneys in lupus nephritis, what mediators they elaborate, and whether their role in disease pathogenesis is obligatory are all open questions.

The Role of Intrinsic Renal Cells in Lupus Nephritis

The major resident cells of the kidney include mesangial cells, endothelial cells, as well as epithelial cells. These intrinsic renal cells turn out to be both the cause and the victim of various types of insults leading to nephritis, including lupus nephritis. The use of gene-deficient mice and in vivo inhibition studies has indicated that intrinsic renal cells contribute to pathogenesis in many ways: (1) the release of pro-inflammatory mediators, (2) functioning as APCs and co-stimulators, (3) cell proliferation, (4) and fibrogenesis [191, 192].

Perhaps the most compelling evidence that intrinsic renal cells play an important role in immune-meditated glomerulonephritis has come from bone-marrow transfer or kidney-transplant studies in mice subjected to anti-GBM nephritis. Studies of this nature have helped outline the role of MHC-II, TNF, and Fn14 on intrinsic renal cells. Using MHC-II chimeric mice, Tipping and colleagues pin-pointed the pivotal role of MHC-II expression by intrinsic renal cells in secondary antigen recognition and directing delayed-type hypersensitivity responses in anti-GBM models, where crescentic glomerulonephritis failed to develop with the deletion of MHC-II on resident kidney cells [193], suggesting that intrinsic renal cells could function as APCs. Using anti-GBM-challenged chimeric mice, they also showed that mice with TNF-deficient leukocytes but intact intrinsic renal cell-derived TNF promoted severe proliferative glomerulonephritis (GN), and that TNFR2 deficiency on intrinsic renal cells, not TNFR1 deficiency, dominated over leukocyte expression of TNFR in the development of anti-GBM GN [172]. These studies clearly implicated intrinsic renal cells as both the cellular source and target for TNF-mediated inflammatory injury in crescentic GN. Yet another example is Fn14: the presence of Fn14 on resident kidney cells alone was sufficient to initiate nephritis in cGVH-induced murine lupus model [194]. Besides the isolated examples above, we know very little about whether other molecules need to be intrinsically expressed within resident renal cells in order for immune-mediated nephritis or lupus nephritis to ensue.

Independent of whether or not certain molecules need to be intrinsically expressed within the kidney in order for lupus nephritis to progress, it is clear that the intrinsic renal cells posses an interesting array of functional capabilities. Compelling evidence indicates that intrinsic renal cells, such as mesangial cells, and tubular epithelial cells can produce a variety of pro-inflammatory mediators *in vitro*, including cytokines, chemokines, nitric oxide, etc. [192, 195, 196]. Mesangial cells stimulated by IFN-γ *in vitro* can express MHC-II and ICAM-1, and present antigen to class II-restricted T-cell hybridomas [197, 198]. Similar to mesangial cells, podoctyes also can express MHC-I, MHC-II, and ICAM-1 upon IFN-γ treatment *in vitro* [199], while increased *in vivo* expression of these molecules on podocytes has been documented in a murine model of necrotizing crescentic glomerulonephritis [200], alluding to the APC function of podocytes under inflammatory conditions. Another co-stimulatory molecule, B7-1, has been found to be expressed only on podocytes, and its increased expression strongly correlates with disease stage and severity of proteinuria in human and murine lupus nephritis, while mice deficient in B7-1 exhibit reduced proteinuria [201]. Moreover, podocytes have been identified in the population of cells comprising glomerular crescents both in lupus nephritis and anti-GBM nephritis, in patients and mouse models [202–204]. Tubular epithelial cells can present antigen for recognition by CD4+ T cells and provide co-stimulation *in vitro* to support both T-cell proliferation and IL-12 production [205]. ICAM-1, B7, and CD40 expression on tubular epithelial cells has also been reported [206–209]. Though a couple of different resident cells can potentially function as APCs, the extent to which any of these cells contribute as APCs during ongoing disease remains unclear.

Besides its potential pro-inflammatory contributions and APC function, another contribution of tubular epithelial cells in the development of GN is referred to as epithelial-to-mesenchymal (EMT) transition, in which epithelial cells can transition into fibroblasts and lead to renal fibrosis. EMT has been demonstrated to occur in both lupus patients and lupus-prone mice [210]. Interestingly, in a recent report, Zeisberg *et al.* found 30–50% of normal renal interstitial cells to co-express the endothelial cell marker CD31 with fibroblast/myofibroblast markers FSP1 and α-SMA, suggesting that endothelial cells might be an additional fibroblast precursor [211]. As with many other features, the extent to which intrinsic renal cells in lupus-prone individuals are prone to EMT remains unknown.

Another question relates to whether resident renal cells from lupus-prone individuals are intrinsically aberrant because of yet-to-be-defined genetic factors. One report indicated that mesangial cells derived form NZB/W F1 lupus-prone mice could elaborate more MCP-1 and OPN [212]. Others have found that MRL/*lpr* mesangial cells, even before disease onset, demonstrated an intrinsic decreased threshold for the production of inflammatory mediators in response to inflammatory stimuli [213, 214]. In addition, mesangial cells derived from MRL/*lpr* lupus-prone mice also over-express α-actinin and have stronger binding capacity to bind pathogenic murine anti-dsDNA/anti-actinin Ab, whereas BALB/c mesangial cells do not, suggesting that the increased expression of α-actinin in MRL/*lpr* mice may enhance their susceptibility to antibody-meditated renal injury [215]. Importantly, novel sequence polymorphisms between MRL/*lpr* and BALB/c in the gene for α-actinin have been identified by the same group, though the functional roles of these mutations are still not clear [216]. In contrast to these clues from murine studies, we have absolutely no idea if intrinsic renal cells may be fundamentally anomalous in human lupus nephritis, owing to poor tissue accessibility.

Collectively, compared to what we know about the role of infiltrating leukocytes in lupus nephritis, our understanding of how intrinsic renal cells contribute to disease is very rudimentary. As is true for the other cell types, it would also be important to ascertain if genes that predispose to lupus can impact the function of intrinsic renal cells.

The Role of Cytokines, Chemokines, and Other Mediators in Lupus Nephritis

Besides the wide array of leukocytes and autoantibodies described above, many released mediators, notably cytokines and chemokines, also contribute to the pathogenesis of lupus nephritis. Cytokines are mediators released by cells of both the adaptive and innate immune system. They function to initiate or sustain the immune response or inflammation. Chemokines and their receptors play crucial roles in mediating leukocyte recruitment to diseased tissues, including the kidney, both during the initiation as well as the progression phases [15, 164]. Although they are primarily produced by T cells and monocytes/macrophages, it is clear that several cell types, including resident renal cells, can produce cytokines and chemokines, and collective interplay of these mediators and the leukocytes that they recruit orchestrate end organ disease in SLE.

Some cytokines that aggravate lupus and lupus nephritis may act predominantly in a systemic fashion, with little evidence for a direct role within the end organs. On such example is BAFF. BAFF, also called B Lymphocyte Stimulator (Blys), is a crucial homeostatic cytokine regulating the selection and survival of B cells. Given the essential and varied role currently assigned to B cells in lupus pathogenesis, it is apparent that BAFF contributes to SLE through a variety of systemic mechanisms [217]. This molecule is discussed in detail in Chapter 8 and 17.

In contrast to BAFF, other cytokines have been shown to have a role in systemic autoimmunity as well as renal disease. Three such examples include IL-17, IFN-α, and TGF-β. IL-17 is a potent proinflammatory cytokine produced by activated T cells, particularly Th17 cells, a CD4$^+$ effector T-cell subset. It is postulated that the cytokine milieu commonly associated with SLE (e.g. lack of IL-2, high levels of IL-6 and IL-21) could promote Th17 cell differentiation and expansion [218]. Increased IL-17-producing T cells were detected in kidneys from both SLE patients and SNF1 lupus-prone mice [219, 220], raising the possibility that this cytokine may contribute to renal damage. Importantly, treatment of SNF1 lupus-prone mice with a histone-derived peptide significantly reduced the disease, accompanied by decreased IL-17 production and IL-17-producing cells, and abrogation of IL-17-positive kidney-infiltrating T cells [220]. Consistent with these observations, the disease relevance of TNF in NZM2328 mice has been found to rely on the presence of CD4$^+$ T cells, which exhibit a Th17 gene profile, again suggesting that that acceleration of nephritis in SLE may be dependent on the IL-17/Th17 pathway [221].

Type I interferons (IFN-α/β) are immunoregulatory cytokines that play a crucial role in DC maturation, antigen presentation, and several other immune functions, as reviewed in Chapter 18. Several independent experiment have found PBMCs from SLE patients to exhibit a prominent type I IFN-inducible gene expression profile, referred to as the IFN signature, supporting the hypothesis that type I IFNs may play a key role in lupus pathogenesis [222—224]. However, in murine lupus models, the role of IFN in SLE pathogenesis appears to be complex, depending on the genetic background of the strain examined. Deliberate administration or upregulation of IFN-α aggravated diseases in the NZB/W F1 [225, 226], NZB [227], and B6.Sle1.Sle2.-Sle3 [228] murine lupus models, but not in MRL/lpr mice [229]. Conversely, the ablation of the IFN-I receptor dampened disease in the NZB [230], NZB/W F1 [231], and TMPD-induced lupus models [232], but not in MRL/lpr mice [233]. Though IFN-I is known to impact systemic immunity in a variety of ways, recent evidence indicates that IFN-I produced by resident renal cells may also contribute to renal inflammation [234].

TGF-β is a potent multifunctional cytokine with a broad spectrum of biological activities. In system immunity, TGF-β plays an anti-inflammatory and immunosuppressive role, whereas it acts as a major inducer of fibrosis locally within the disease kidneys, in almost all types of glomerulonephritis. Recent studies have demonstrated that TGF-β is an essential signaling intermediate for collagen production in response to a variety of injurious stimuli [235]. Studies indicate that the impact of persistent, dysregulated TGF-β production on extracellular matrix provides the basis for progressive renal disease in patients with lupus nephritis [236]. Elevated TGF-β expression has been found in renal tissue from both lupus patients and lupus-prone models, and the renal cortical levels of TGF-β1 mRNA are correlated with histological activity [237—239]. In contrast, disease remission in lupus nephritis is related to decreased renal TGF-β expression [238]. In NZB/W F1 lupus-prone mice, the collective data in the field strongly indicate that reduced TGF-β in immune cells predispose mice to immune dysregulation and autoantibody production, whereas enhanced TGF-β expression in kidneys leads to dysregulated tissue repair, progressive fibrogenesis, and eventual end-organ damage [176]. Consistent with this, in experimentally induced anti-GBM nephritis, overexpression of Smad7, a negative regulator of TGF/Smads signaling, confers protection from crescent formation [240]. Hence, TGF-β is a double-edged sword, subduing systemic immunity, but aggravating chronic nephritis [176].

Other cytokines that have been shown to be important for antibody-mediated renal disease and lupus nephritis include IL-1, IL-6, IL-10, and TNF-α, based mostly on gene-ablation studies [241−244] and anticytokine antibody-mediated therapies [82, 124, 245−247], as reviewed elsewhere [15, 164]. Since all of these cytokines can be elaborated systemically, as well as locally within the kidneys, it remains to be established whether renal expression of any of these molecules is necessary for lupus nephritis.

There is strong evidence that some chemokines may also be important in driving lupus nephritis. MCP-1, which can recruit and activate inflammatory cells, is overexpressed in the kidneys from both lupus-prone mice and patients with lupus nephritis [248−251]. Patients with lupus nephritis have significantly higher levels of MCP-1 in their urine, but not serum, in comparison to healthy controls, alluding to the predominant local production of this chemokine in diseased kidneys [252,253]. Conversely, although MCP-1 knockout mice show normal numbers of circulating leukocytes and resident macrophages, they are unable to recruit monocytes after intraperitoneal thioglycollate administration [254]. MCP-1-deficient MRL/*lpr* mice show prolonged survival and reduced glomerular and interstitial macrophage infiltration [165]. Furthermore, anti-MCP-1 therapy inhibited the progression of nephritis in MRL/*lpr* mice by protecting against renal injury because of reduced leukocyte infiltration [255−257]. RANTES is another chemokine that has been observed to be elevated in the kidneys during lupus nephritis. However, the functional significance of this observation remains unclear. Available data on blocking RANTES and its receptors are inconsistent. One report indicated that both RANTES antagonists, Met-RANTES and AOP-RANTES, aggravated glomerular damage and proteinuria in murine immune-complex glomerulonephritis [258], whereas others showed improved outcome [259, 260].

The pathogenic role of CXCR4/CXCL12 axis in lupus nephritis has garnered much interest recently. CXCR4 has been found significantly upregulated on leukocytes, while its ligand, CXCL12, has been significantly elevated within nephritic kidneys in lupus patients and several murine lupus models including BXSB, MRL/*lpr*, B6.*Sle1.Yaa*, and NZB/W F1 mice [261−263]. Treatment of murine lupus with a peptide antagonist of CXCR4 [262] or anti-CXCR4 mAb [263] ameliorated several lupus phenotypes. Likewise, mice lacking another chemokine receptor CXCR3 had significantly reduced renal T cell infiltrates, accompanied by improved renal function and pathology, when challenged with anti-GBM serum [264]. MRL/*lpr* lupus mice with CXCR3 deficiency also showed amelioration of nephritis with reduced renal T-cell recruitment, reduced renal IFN-γ producing Th1 cells and IL-17 producing Th17 T cells [265]. The intact systemic humoral immune response

noted in both CXCR3$^{-/-}$ anti-GBM nephritis model and MRL/*lpr* CXCR3$^{-/-}$ mice pointed to the crucial role of CXCR3 in the development of nephritis by directing pathogenic T cells into the kidneys [264, 265]. Hence, it is not a surprise that these chemokines and their receptors constitute potential therapeutic targets in lupus nephritis.

Besides cytokines and chemokines, several other mediators with wild ranging biological properties have also been implicated in the pathogenesis of lupus nephritis, including kallikreins (*KLK*), SOD, CSF-1, MIF, NO, etc. *KLK* genes are localized within a spontaneous lupus susceptibility interval on chromosome 7 [266, 267]. Liu and colleagues have reported that the downregulation of genes from the *KLK* family could distinguish the nephritis-sensitive mouse strains from the control strains, through the use of microarray-based transcriptomic analysis [268]. Importantly, blocking the *KLK* pathway was shown to aggravate renal disease, while bradykinin agonists dampened the severity of anti-GBM nephritis [268]. In addition, *KLK1* gene delivery ameliorated anti-GBM nephritis in mice [269]. The genetic association of human *KLK1* gene and *KLK3* promoter polymorphisms with lupus and lupus nephritis has also been confirmed in several independent cohorts of patients [268]. Collectively, these studies suggest that the KLK family of serine proteases play a major role in modulating renal inflammation, echoing their well established renoprotective role in other renal diseases [270−274].

Increased oxidative stress is a significant feature of lupus, with overproduction of nitric oxide (NO), being one of the readouts. In lupus patients and MRL/*lpr* lupus mice, increased expression of inducible NO synthase has been reported in the glomeruli and infiltrating cells [275−277]. Treating mice with NO synthase inhibitor significantly reduced glomerular proliferation and proteinuria [276, 277]. Superoxide dismutase (SOD), on other hand, serves as a defense mechanism by degrading potentially injurious superoxides, and attenuating local inflammation [278]. Elevated levels of SOD have been observed in the blood, urine, and kidneys in both experimental anti-GBM disease and spontaneous lupus nephritis [279−281].

As discussed in "The Role of Macrophages in Lupus Nephritis", macrophages play a central role in mediating lupus nephritis. Therefore, CSF-1, the principal macrophage growth factor, and macrophage migration inhibitory factor (MIF), key proinflammatory cytokines regulating macrophage recruitment have also been documented as central players in lupus nephritis. Renal resident cells, most notably tubular epithelial cells, are the primary source of CSF-1 during renal disease [282, 283]. Increased renal expression of CSF-1 has been found before overt renal pathology and becomes more abundant with advancing lupus nephritis [284]. Studies using gene transfer and CSF-1 knockout mice have demonstrated the obligatory role of CSF-1 in the pathogenesis of lupus nephritis [285, 286]. Upregulation of MIF has also been reported in both human and experimental renal disease, including proliferative lupus nephritis and crescentic glomerulonephritis. This molecule contributes significantly to macrophage and T-cell accumulation and progressive renal injury [287–291]. The functional importance of MIF in renal disease is demonstrated by the finding that mice deficient in MIF are protected from lupus nephritis [290], whereas MIF transgenic mice show podocyte injury and progressive mesangial sclerosis [291]. In addition, the renal injury in experimental anti-GBM nephritis can be prevented or reversed by the administration of a neutralizing anti-MIF antibody [292, 293].

It is likely that the mediators described above represent only a handful of the molecules that mediate lupus nephritis. Ongoing transcriptome-profiling studies and genetic initiatives are likely to expand our current understanding beyond what appears to be the tip of the iceberg. Indeed, recent genome-wide association studies in human SLE have begun to unravel an expanding set of culprit genes, as detailed in Chapters 2 and 3. It remains to be ascertained if any of these genes directly impact the degree of mediator production or function and end-organ disease.

CONCLUSION

Given that nephritis continues to be a leading cause of morbidity and mortality in lupus, it is absolutely vital to understand the molecular basis of this end-organ manifestation. Relative to what we have learned about how the immune system affects lupus development, we know very little about the events that take place within the kidneys. This chapter has captured some of the cells, molecules, and mechanisms that interplay to orchestrate nephritis. With ongoing transcriptomic profiling studies and genome-wide association studies in human and murine lupus, the potential list of culprit molecules is likely to grow exponentially over the next decade. Decoding the intricate tapestry of intrarenal mechanisms and molecules that drive the pathogenesis of lupus nephritis is absolutely key for the discovery of better rationalized therapeutics for this devastating disease.

References

[1] H. Kanta, C. Mohan, Three checkpoints in lupus development: central tolerance in adaptive immunity, peripheral amplification by innate immunity and end-organ inflammation, Genes Immun 10 (5) (2009) 390–396.

[2] B.S. Andrews, R.A. Eisenberg, A.N. Theofilopoulos, S. Izui, C.B. Wilson, P.J. McConahey, et al., Spontaneous murine lupus-like syndromes. Clinical and immunopathological manifestations in several strains, J Exp Med 148 (5) (1978) 1198–1215.

[3] A.N. Theofilopoulos, F.J. Dixon, Murine models of systemic lupus erythematosus, Adv Immunol 37 (1985) 269–390.

[4] A.M. Fairhurst, A.E. Wandstrat, E.K. Wakeland, Systemic lupus erythematosus: multiple immunological phenotypes in a complex genetic disease, Adv Immunol 92 (2006) 1–69.

[5] U.H. Rudofsky, B.D. Evans, S.L. Balaban, V.D. Mottironi, A.E. Gabrielsen, Differences in expression of lupus nephritis in New Zealand mixed H-2z homozygous inbred strains of mice derived from New Zealand black and New Zealand white mice. Origins and initial characterization, Lab Invest 68 (4) (1993) 419–426.

[6] M.L. Santiago-Raber, C. Laporte, L. Reininger, S. Izui, Genetic basis of murine lupus, Autoimmun Rev 3 (1) (2004) 33–39.

[7] P. Pisitkun, J.A. Deane, M.J. Difilippantonio, T. Tarasenko, A.B. Satterthwaite, S. Bolland, Autoreactive B cell responses to RNA-related antigens due to TLR7 gene duplication, Science 312 (5780) (2006) 1669–1672.

[8] S. Subramanian, K. Tus, Q.Z. Li, A. Wang, X.H. Tian, J. Zhou, et al., A Tlr7 translocation accelerates systemic autoimmunity in murine lupus, Proc Natl Acad Sci USA 103 (26) (2006) 9970–9975.

[9] J.A. Deane, P. Pisitkun, R.S. Barrett, L. Feigenbaum, T. Town, J.M. Ward, et al., Control of Toll-like receptor 7 expression is essential to restrict autoimmunity and dendritic cell proliferation, Immunity 27 (5) (2007) 801–810.

[10] M.L. Santiago-Raber, S. Kikuchi, P. Borel, S. Uematsu, S. Akira, B.L. Kotzin, et al., Evidence for genes in addition to Tlr7 in the Yaa translocation linked with cceleration of systemic lupus erythematosus, J Immunol 181 (2) (2008) 1556–1562.

[11] S. Nagata, T. Suda, Fas and Fas ligand: lpr and gld mutations, Immunol Today 16 (1) (1996) 39–43.

[12] C.D. Pusey, Anti-glomerular basement membrane disease, Kidney Int. 64 (4) (2006) 1535–1550.

[13] R.J. Butkowski, J.P. Langeveld, J. Wieslander, J. Hamilton, B.G. Hudson, Localization of the Goodpasture epitope to a novel chain of basement membrane collagen, J Biol Chem 262 (16) (1987) 7874–7877.

[14] J. Saus, J. Wieslander, J.P. Langeveld, S. Quinones, B.G. Hudson, Identification of the Goodpasture antigen as the alpha 3(IV) chain of collagen IV, J Biol Chem 263 (26) (1998) 13374–13380.

[15] Y. Fu, Y. Du, C. Mohan, Experimental anti-GBM disease as a tool for studying spontaneous lupus nephritis, Clin Immunol 124 (2) (2007) 109–118.

[16] R. Ceppellini, E. Polli, F. Celada, A DNA-reacting factor in serum of a patient with lupus erythematosus diffusus, Proc Soc Exp Biol Med 96 (3) (1957) 572–574.

[17] E. Koren, M. Koscec, M. Wolfson-Reichlin, F.M. Ebling, B. Tsao, B.H. Hahn, et al., Murine and human antibodies to native DNA that cross-react with the A and D SnRNP polypeptides cause direct injury of cultured kidney cells, J Immunol 154 (9) (1995) 4857–4864.

[18] M.R. Ehrenstein, D.R. Katz, M.H. Griffiths, M.H. Griffiths, L. Papadaki, T.H. Winkler, et al., Human IgG anti-DNA antibodies deposit in kidneys and induce proteinuria in SCID mice, Kidney Int 48 (3) (1995) 705–711.

[19] L.J. Mason, C.T. Ravirajan, D.S. Latchman, D.A. Isenberg, A human anti-dsDNA monoclonal antibody caused hyaline thrombi formation in kidneys of 'leaky' SCID mice, Clin Exp Immunol 126 (1) (2001) 137–142.

[20] Z. Liang, C. Xie, C. Chen, D. Kreska, K. Hsu, L. Li, et al., Pathogenic profiles and molecular signatures of antinuclear autoantibodies rescued from NZM2410 lupus mice, J Exp Med 199 (3) (2004) 381—398.

[21] B.P. Tsao, K. Ohnishi, H. Cheroutre, B. Mitchell, M. Teitell, P. Mixter, et al., Failed self-tolerance and autoimmunity in IgG anti-DNA transgenic mice, J Immunol 149 (1) (1992) 350—358.

[22] K.A. Fenton, O.P. Rekvig, A central role of nucleosomes in lupus nephritis, Ann N Y Acad Sci 1108 (2007) 104—113.

[23] S. Yung, T.M. Chan, Anti-DNA antibodies in the pathogenesis of lupus nephritis- The emerging mechanisms, Autoimmun Rev 7 (4) (2008) 317—321.

[24] E.S. Mortensen, O.P. Rekvig, Nephritogenic potential of anti-DNA antibodies against necrotic nucleosomes, J Am Soc Nephrol 20 (4) (2009) 696—704.

[25] R. Clynes, C. Dumitru, J.V. Ravetch, Uncoupling of immune complex formation and kidney damage in autoimmune glomerulonephritis, Science 279 (5353) (1998) 1052—1054.

[26] H. Watanabe, D. Sherris, G.S. Gilkeson, Soluble CD16 in the treatment of murine lupus nephritis, Clin Immunol Immunopathol 88 (1) (1998) 91—95.

[27] S. Werwitzke, D. Trick, P. Sondermann, K. Kamino, B. Schlegelberger, K. Kniesch, et al., Treatment of lupus-prone NZB/NZW F1 mice with recombinant soluble Fc gamma receptor II (CD32), Ann Rheum Dis 67 (2) (2008) 154—161.

[28] L. Bao, J. Zhou, V.M. Holers, R.J. Quigg, Excessive matrix accumulation in the kidneys of MRL/lpr lupus mice is dependent on complement activation, J Am Soc Nephrol 14 (10) (2003) 2516—2525.

[29] L. Bao, M. Haas, S.A. Boackle, D.M. Kraus, P.N. Cunningham, P. Park, et al., Transgenic expression of a soluble complement inhibitor protects against renal disease and promotes survival in MRL/lpr mice, J Immunol 168 (7) (2002) 3601—3607.

[30] M.G. Robson, H.T. Cook, M. Botto, P.R. Taylor, N. Busso, R. Salvi, et al., Accelerated nephrotoxic nephritis is exacerbated in C1q-deficient mice, J Immunol 166 (11) (2001) 6820—6828.

[31] M.J. Hébert, T. Takano, A. Papayianni, H.G. Rennke, A. Minto, D.J. Salant, et al., Acute nephrotoxic serum nephritis in complement knockout mice: relative roles of the classical and alternate pathways in neutrophil recruitment and proteinuria, Nephrol Dial Transplant 13 (11) (1998) 2799—2803.

[32] X. Qing, J. Zavadil, M.B. Crosby, M.P. Hogarth, B.H. Hahn, C. Mohan, et al., Nephritogenic anti-DNA antibodies regulate gene expression in MRL/lpr mouse glomerular mesangial cells, Arthritis Rheum 54 (7) (2006) 2198—2210.

[33] U.S. Deshmukh, H. Bagavant, S.M. Fu, Role of anti-DNA antibodies in the pathogenesis of lupus nephritis, Autoimmun Rev 5 (6) (2006) 414—418.

[34] M.Z. Radic, M. Weigert, Genetic and structural evidence for antigen selection of anti-DNA antibodies, Annu Rev Immunol 12 (1994) 487—520.

[35] J.B. Katz, W. Limpanasithikul, B. Diamond, Mutational analysis of an autoantibody: differential binding and pathogenicity, J Exp Med 180 (3) (1994) 925—932.

[36] N. Limaye, C. Mohan, Pathogenicity of anti-DNA and anti-glomerular antibodies: weighing the evidence, Drug Discovery Today: Disease Models 1 (4) (2004) 395—403.

[37] Y. Fu, C. Xie, J. Chen, J. Zhu, H. Zhou, J. Thomas, C. Mohan, et al., Innate stimuli accentuate end-organ damage by nephrotoxic antibodies via Fc receptor and TLR stimulation and IL-1/TNF-alpha production, J Immunol 176 (1) (2006) 632—639.

[38] M. Boes, T. Schmidt, K. Linkemann, B.C. Beaudette, A. Marshak-Rothstein, J. Chen, Accelerated development of IgG autoantibodies and autoimmune disease in the absence of secreted IgM, Proc Natl Acad Sci USA 97 (3) (2000) 1184—1189.

[39] S. Werwitzke, D. Trick, K. Kamino, T. Matthias, K. Kniesch, B. Schlegelberger, et al., Inhibition of lupus disease by anti-double-stranded DNA antibodies of the IgM isotype in the (NZB x NZW) F1 mouse, Arthritis Rheum 52 (11) (2005) 3629—3638.

[40] F. Forger, T. Matthias, M. Oppermann, H. Becker, K. Helmke, Clinical significance of anti-dsDNA antibody isotypes: IgG/IgM ratio of anti-dsDNA antibodies as a prognostic marker for lupus nephritis, Lupus 13 (1) (2004) 36—44.

[41] Q.Z. Li, C. Xie, T. Wu, M. Mackay, C. Aranow, C. Putterman, et al., Identification of autoantibody clusters that best predict lupus disease activity using glomerular proteome arrays, J Clin Invest 115 (12) (2005) 3428—3439.

[42] P. Alba, L. Bento, M.J. Cuadrado, Y. Karim, M.F. Tungekar, I. Abbs, et al., Anti-dsDNA, anti-Sm antibodies, and the lupus anticoagulant: significant factors associated with lupus nephritis, Ann Rheum Dis 62 (6) (2003) 556—560.

[43] D.D. Gladman, M.B. Urowitz, E.C. Keystone, Serologically active clinically quiescent systemic lupus erythematosus: a discordance between clinical and serologic features, Am J Med 66 (2) (1979) 210—215.

[44] D.D. Gladman, N. Hirani, D. Ibañez, M.B. Urowitz, Clinically active serologically quiescent systemic lupus erythematosus, J Rheumatol 30 (9) (2003) 1960—1962.

[45] P.L. Kong, T. Zhu, M.P. Madaio, J. Craft, Role of the H-2 haplotype in Fas-intact lupus-prone MRL mice: association with autoantibodies but not renal disease, Arthritis Rheum 48 (10) (2003) 2992—2995.

[46] R.R. Singh, V. Saxena, S. Zang, L. Li, F.D. Finkelman, D.P. Witte, et al., Differential contribution of IL-4 and STAT6 vs. STAT4 to the development of lupus nephritis, J Immunol 170 (9) (2003) 4818—4825.

[47] C.O. Jacob, S. Zang, L. Li, V. Ciobanu, F. Quismorio, A. Mizutani, et al., Pivotal role of Stat4 and Stat6 in the pathogenesis of the lupus-like disease in the New Zealand mixed 2328 mice, J Immunol 171 (3) (2003) 1564—1571.

[48] S.T. Waters, M. McDuffie, H. Bagavant, U.S. Deshmukh, F. Gaskin, C. Jiang, et al., Breaking tolerance to double stranded DNA, nucleosome, and other nuclear antigens is not required for the pathogenesis of lupus glomerulonephritis, J Exp Med 199 (2) (2004) 255—264.

[49] W. Stohl, D. Xu, K.S. Kim, M.N. Koss, T.N. Jorgensen, B. Deocharan, et al., BAFF overexpression and accelerated glomerular disease in mice with an incomplete genetic predisposition to systemic lupus erythematosus, Arthritis Rheum 52 (7) (2005) 2080—2091.

[50] X. Shi, C. Xie, S. Chang, X.J. Zhou, T. Tedder, C. Mohan, CD19 hyperexpression augments Sle1-induced humoral autoimmunity but not clinical nephritis, Arthritis Rheum 56 (9) (2007) 3057—3069.

[51] S.T. Waters, S.M. Fu, F. Gaskin, U.S. Deshmukh, S.S. Sung, C.C. Kannapell, et al., NZM2328: a new mouse model of systemic lupus erythematosus with unique genetic susceptibility loci, Clin Immunol 100 (3) (2001) 372—383.

[52] S.R. Christensen, M. Kashgarian, L. Alexopoulou, R.A. Flavell, S. Akira, M.J. Shlomchik, Toll-like receptor 9 controls anti-DNA autoantibody production in murine lupus, J Exp Med 202 (2) (2005) 321—331.

[53] O.T. Chan, L.G. Hannum, A.M. Haberman, M.P. Madaio, M.J. Shlomchik, A novel mouse with B cells but lacking serum antibody reveals an antibody-independent role for B cells in murine lupus, J Exp Med 189 (10) (1999) 1639—1648.

[54] P.M. Rumore, C.R. Steinman, Endogenous circulating DNA in systemic lupus erythematosus. Occurrence as multimeric complexes bound to histone, J Clin Invest 86 (1) (1990) 69—74.

[55] P. Decker, Nucleosome autoantibodies, Clin Chim Acta 366 (1-2) (2006) 48–60.

[56] R.W. Burlingame, M.L. Boey, G. Starkebaum, R.L. Rubin, The central role of chromatin in autoimmune responses to histones and DNA in systemic lupus erythematosus, J Clin Invest 94 (1) (1994) 184–192.

[57] Z. Amoura, H. Chabre, S. Koutouzov, C. Lotton, A. Cabrespines, J.F. Bach, et al., Nucleosome-restricted antibodies are detected before anti-dsDNA and/or antihistone antibodies in serum of MRL-Mp lpr/lpr and +/+ mice, and are present in kidney eluates of lupus mice with proteinuria, Arthritis Rheum 37 (11) (1994) 1684–1688.

[58] E.N. Voynova, A.I. Tchorbanov, T.A. Todorov, T.L. Vassilev, Breaking of tolerance to native DNA in nonautoimmune mice by immunization with natural protein/DNA complexes, Lupus 14 (7) (2005) 543–550.

[59] C. Mohan, S. Adams, V. Stanik, S.K. Datta, Nucleosome: a major immunogen for pathogenic autoantibody-inducing T cells of lupus, J Exp Med 177 (5) (1993) 1367–13681.

[60] K. Sainis, S.K. Datta, CD4$^+$ T cell lines with selective patterns of autoreactivity as well as CD4$^-$/CD8$^-$ T helper cell lines augment the production of idiotypes shared by pathogenic anti-DNA autoantibodies in the NZB x SWR model of lupus nephritis, J Immunol 140 (7) (1998) 2215–2224.

[61] S. Adams, P. Leblanc, S.K. Datta, Junctional region sequences of T-cell receptor beta-chain genes expressed by pathogenic anti-DNA autoantibody-inducing helper T cells from lupus mice: possible selection by cationic autoantigens, Proc Natl Acad Sci USA 88 (24) (1991) 11271–11275.

[62] S. Rajagopalan, T. Zordan, G.C. Tsokos, S.K. Datta, Pathogenic anti-DNA autoantibody-inducing T helper cell lines from patients with active lupus nephritis: isolation of CD4-8- T helper cell lines that express the gamma delta T-cell antigen receptor, Proc Natl Acad Sci USA 87 (18) (1990) 7020–7024.

[63] D. Mevorach, J.L. Zhou, X. Song, K.B. Elkon, Systemic exposure to irradiated apoptotic cells induces autoantibody production, J Exp Med 188 (2) (1998) 387–392.

[64] L.E. Munoz, C. van Bavel, S. Franz, J. Berden, M. Herrmann, J. van der Vlag, Apoptosis in the pathogenesis of systemic lupus erythematosus, Lupus 17 (5) (2008) 371–375.

[65] P.L. Cohen, Apoptotic cell death and lupus, Springer Semin Immunopathol 28 (2) (2006) 145–152.

[66] M.C. van Bruggen, C. Kramers, M.N. Hylkema, R.J. Smeenk, J.H. Berden, Significance of anti-nuclear and anti-extracellular matrix autoantibodies for albuminuria in murine lupus nephritis; a longitudinal study on plasma and glomerular eluates in MRL/l mice, Clin Exp Immunol 105 (1) (1996) 132–139.

[67] F. Elouaai, J. Lulé, H. Benoist, S. Appolinaire-Pilipenko, C. Atanassov, S. Muller, et al., Autoimmunity to histones, ubiquitin, and ubiquitinated histone H2A in NZB x NZW and MRL-lpr/lpr mice. Anti-histone antibodies are concentrated in glomerular eluates of lupus mice, Nephrol Dial Transplant 9 (4) (1994) 362–366.

[68] C. Xie, Z. Liang, S. Chang, C. Mohan, Use of a novel elution regimen reveals the dominance of polyreactive antinuclear autoantibodies in lupus kidneys, Arthritis Rheum 48 (8) (2003) 2343–2352.

[69] C. Krishnan, M.H. Kaplan, Immunopathologic studies of systemic lupus erythematosus. II. Antinuclear reaction of gamma-globulin eluted from homogenates and isolated glomeruli of kidneys from patients with lupus nephritis, J Clin Invest 46 (4) (1967) 569–579.

[70] F. Jovelin, G. Mostoslavsky, Z. Amoura, H. Chabre, D. Gilbert, D. Eilat, et al., Early anti-nucleosome autoantibodies from a single MRL+/+ mouse: fine specificity, V gene structure and pathogenicity, Eur J Immunol 28 (11) (1998) 3411–3422.

[71] G.N. Coritsidis, F. Lombardo, P. Rumore, S.F. Kuo, R. Izzo, R. Mir, et al., Nucleosome effects on mesangial cell matrix and proliferation: a possible role in early lupus nephritis, Exp Nephrol 10 (3) (2002) 216–226.

[72] M.J. Shlomchik, M.P. Madaio, The role of antibodies and B cells in the pathogenesis of lupus nephritis, Springer Semin Immunopathol 24 (4) (2003) 363–375.

[73] M.R. Clatworthy, K.G. Smith, B cells in glomerulonephritis: focus on lupus nephritis, Semin Immunopathol 29 (4) (2007) 337–353.

[74] M.J. Shlomchik, Sites and stages of autoreactive B cell activation and regulation, Immunity 28 (1) (2008) 18–28.

[75] M.L. Hibbs, D.M. Tarlinton, J. Armes, D. Grail, G. Hodgson, R. Maglitto, et al., Multiple defects in the immune system of Lyn-deficient mice, culminating in autoimmune disease, Cell 83 (2) (1995) 301–311.

[76] D.H. Li, M.M. Winslow, T.M. Cao, A.H. Chen, C.R. Davis, E.D. Mellins, et al., Modulation of peripheral B cell tolerance by CD72 in a murine model, Arthritis Rheum 58 (10) (2008) 3192–3204.

[77] H. Nishimura, M. Nose, H. Hiai, N. Minato, T. Honjo, Development of lupus-like autoimmune diseases by disruption of the PD-1 gene encoding an ITIM motif-carrying immunoreceptor, Immunity 11 (2) (1999) 141–151.

[78] K.R. Kumar, L. Li, M. Yan, M. Bhaskarabhatla, A.B. Mobley, C. Nguyen, et al., Regulation of B cell tolerance by the lupus susceptibility gene Ly108, Science 312 (5780) (2006) 1665–1669.

[79] A.E. Wandstrat, C. Nguyen, N. Limaye, A.Y. Chan, S. Subramanian, X.H. Tian, et al., Association of extensive polymorphisms in the SLAM/CD2 gene cluster with murine lupus, Immunity 21 (6) (2004) 769–780.

[80] O. Chan, M.J. Shlomchik, A new role for B cells in systemic autoimmunity: B cells promote spontaneous T cell activation in MRL-lpr/lpr mice, J Immunol 160 (1) (1998) 51–59.

[81] S.K. Pierce, J.F. Morris, M.J. Grusby, P. Kaumaya, A. van Buskirk, M. Srinivasan, et al., Antigen-presenting function of B lymphocytes, Immunol Rev 106 (1988) 149–180.

[82] B. Liang, D.B. Gardner, D.E. Griswold, P.J. Bugelski, X.Y. Song, Anti-interleukin-6 monoclonal antibody inhibits autoimmune responses in a murine model of systemic lupus erythematosus, Immunology 119 (3) (2006) 296–305.

[83] H. Ishida, Clinical implication of IL-10 in patients with immune and inflammatory diseases, Rinsho Byori 42 (8) (1994) 843–852.

[84] D.P. Harris, L. Haynes, P.C. Sayles, D.K. Duso, S.M. Eaton, N.M. Lepak, et al., Reciprocal regulation of polarized cytokine production by effector B and T cells, Nat Immunol 1 (6) (2000) 475–482.

[85] P. Bhat, J. Radhakrishnan, B lymphocytes and lupus nephritis: new insights into pathogenesis and targeted therapies, Kidney Int 73 (3) (2008) 261–268.

[86] D.A. Jabs, C.L. Burek, Q. Hu, R.C. Kuppers, B. Lee, R.A. Prendergast, Anti-CD4 monoclonal antibody therapy suppresses autoimmune disease in MRL/Mp-lpr/lpr mice, Cell Immunol 141 (2) (1992) 496–507.

[87] V.D. D'Agati, G.B. Appel, D. Estes, D.M. Knowles 2nd, C.L. Pirani, Monoclonal antibody identification of infiltrating mononuclear leukocytes in lupus nephritis, Kidney Int 30 (4) (1986) 573–581.

[88] C. Díaz Gallo, A.M. Jevnikar, D.C. Brennan, S. Florquin, A. Pacheco-Silva, V.R. Kelley, Autoreactive kidney-infiltrating T-cell clones in murine lupus nephritis, Kidney Int 42 (4) (1992) 851–859.

[89] T. Giese, W.F. Davidson, Evidence for early onset, polyclonal activation of T cell subsets in mice homozygous for lpr, J Immunol 149 (9) (1992) 3097–3106.

[90] R. Mukherjee, Z. Zhang, R. Zhong, Z.Q. Yin, D.C. Roopenian, A.M. Jevnikar, Lupus nephritis in the absence of renal major histocompatibility complex class I and class II molecules, J Am Soc Nephrol 7 (11) (1996) 2445–2452.

[91] D.R. Koh, A. Ho, A. Rahemtulla, W.P. Fung-Leung, H. Griesser, T.W. Mak, Murine lupus in MRL/lpr mice lacking CD4 or CD8 T cells, Eur J Immunol 25 (9) (1995) 2558–2562.

[92] M.S. Chesnutt, B.K. Finck, N. Killeen, M.K. Connolly, H. Goodman, D. Wofsy, Enhanced lymphoproliferation and diminished autoimmunity in CD4-deficient MRL/lpr mice, Clin Immunol Immunopathol 87 (1) (1998) 23–32.

[93] D.A. Jabs, R.C. Kuppers, A.M. Saboori, C.L. Burek, C. Enger, B. Lee, et al., Effects of early and late treatment with anti-CD4 monoclonal antibody on autoimmune disease in MRL/MP-lpr/lpr mice, Cell Immunol 154 (1) (1994) 66–76.

[94] C. Mohan, Y. Shi, J.D. Laman, S.K. Datta, Interaction between CD40 and its ligand gp39 in the development of murine lupus nephritis, J Immunol 154 (3) (1995) 1470–1480.

[95] H.H. Radeke, T. Tschernig, A. Karulin, G. Schumm, S.N. Emancipator, K. Resch, et al., CD4+ T cells recognizing specific antigen deposited in glomeruli cause glomerulonephritis-like kidney injury, Clin Immunol 104 (2) (2002) 161–173.

[96] Y.H. Lou, Anti-GBM glomerulonephritis: a T cell-mediated autoimmune disease? Arch Immunol Ther Exp (Warsz) 52 (2) (2004) 96–103.

[97] A. Desai-Mehta, L. Lu, R. Ramsey-Goldman, S.K. Datta, Hyperexpression of CD40 ligand by B and T cells in human lupus and its role in pathogenic autoantibody production, J Clin Invest 97 (9) (1996) 2063–2073.

[98] M. Koshy, D. Berge, M.K. Crow, Increased expression of CD40 ligand on systemic lupus erythematosus lymphocytes, J Clin Invest 98 (3) (1996) 826–837.

[99] B.S. Devi, S. Van Noordin, T. Krausz, K.A. Davies, Peripheral blood lymphocytes in SLE–hyperexpression of CD154 on T and B lymphocytes and increased number of double negative T cells, J Autoimmun 11 (5) (1998) 471–475.

[100] S.K. Datta, Major peptide autoepitopes for nucleosome-centered T and B cell interaction in human and murine lupus, Ann N Y Acad Sci 987 (2003) 79–90.

[101] Y. Yi, M. McNerney, S.K. Datta, Regulatory defects in Cbl and mitogen-activated protein kinase (extracellular signal-related kinase) pathways cause persistent hyperexpression of CD40 ligand in human lupus T cells, J Immunol 165 (11) (2001) 6627–6634.

[102] A. Mehling, K. Loser, G. Varga, D. Metze, T.A. Luger, T. Schwarz, et al., Overexpression of CD40 ligand in murine epidermis results in chronic skin inflammation and systemic autoimmunity, J Exp Med 194 (5) (2001) 615–628.

[103] S.L. Kalled, A.H. Cutler, S.K. Datta, D.W. Thomas, Anti-CD40 ligand antibody treatment of SNF1 mice with established nephritis: preservation of kidney function, J Immunol 160 (5) (1998) 2158–2165.

[104] X. Wang, W. Huang, L.E. Schiffer, M. Mihara, A. Akkerman, K. Hiromatsu, et al., Effects of anti-CD154 treatment on B cells in murine systemic lupus erythematosus, Arthritis Rheum 48 (2) (2003) 495–506.

[105] G.S. Early, W. Zhao, C.M. Burns, Anti-CD40 ligand antibody treatment prevents the development of lupus-like nephritis in a subset of New Zealand Black x New Zealand White mice. Response correlates with the absence of an anti-antibody response, J Immunol 157 (7) (1996) 3159–3164.

[106] I.S. Grewal, J. Xu, R.A. Flavell, Impairment of antigen-specific T-cell priming in mice lacking CD40 ligand, Nature 378 (6557) (1995) 617–620.

[107] Y. Wu, J. Xu, S. Shinde, I.S. Grewal, T. Henderson, R.A. Flavell, et al., Rapid induction of a novel costimulatory activity on B cells by CD40 ligand, Curr Biol 5 (11) (1995) 1303–1311.

[108] U. Schönbeck, P. Libby, The CD40/CD154 receptor/ligand dyad, Cell Mol Life Sci 58 (1) (2001) 4–43.

[109] M.K. Crow, Modification of accessory molecule signaling, Springer Semin Immunopathol 27 (4) (2006) 409–424.

[110] M.R. Alderson, R.J. Armitage, T.W. Tough, L. Strockbine, W.C. Fanslow, M.K. Spriggs, CD40 expression by human monocytes: regulation by cytokines and activation of monocytes by the ligand for CD40, J Exp Med 178 (2) (1993) 669–674.

[111] C. Caux, C. Massacrier, B. Vanbervliet, B. Dubois, C. van Kooten, I. Durand, et al., Activation of human dendritic cells through CD40 cross-linking, J Exp Med 180 (4) (1994) 1263–1272.

[112] M.J. Yellin, V. D'Agati, G. Parkinson, A.S. Han, A. Szema, D. Baum, et al., Immunohistologic analysis of renal CD40 and CD40L expression in lupus nephritis and other glomerulonephritides, Arthritis Rheum 40 (1) (1997) 124–134.

[113] T. Kuroiwa, R. Schlimgen, G.G. Illei, I.B. McInnes, D.T. Boumpas, Distinct T cell/renal tubular epithelial cell interactions define differential chemokine production: implications for tubulointerstitial injury in chronic glomerulonephritides, J Immunol 164 (6) (2000) 3323–3329.

[114] C. van Kooten, J.S. Gerritsma, M.E. Paape, L.A. van Es, J. Banchereau, M.R. Daha, Possible role for CD40-CD40L in the regulation of interstitial infiltration in the kidney, Kidney Int 51 (3) (1997) 711–721.

[115] W.S. Uhm, K. Na, G.W. Song, S.S. Jung, T. Lee, M.H. Park, et al., Cytokine balance in kidney tissue from lupus nephritis patients, Rheumatology (Oxford) 42 (8) (2003) 935–938.

[116] N. Calvani, H.B. Richards, M. Tucci, G. Pannarale, F. Silvestris, Up-regulation of IL-18 and predominance of a Th1 immune response is a hallmark of lupus nephritis, Clin Exp Immunol 138 (1) (2004) 171–178.

[117] K. Masutani, M. Akahoshi, K. Tsuruya, M. Tokumoto, T. Ninomiya, T. Kohsaka, et al., Predominance of Th1 immune response in diffuse proliferative lupus nephritis, Arthritis Rheum 44 (9) (2001) 2097–2106.

[118] M. Tucci, L. Lombardi, H.B. Richards, F. Dammacco, F. Silvestris, Overexpression of interleukin-12 and T helper 1 predominance in lupus nephritis, Clin Exp Immunol 154 (2) (2008) 247–254.

[119] R.W. Chan, L.S. Tam, E.K. Li, F.M. Lai, K.M. Chow, K.B. Lai, et al., Inflammatory cytokine gene expression in the urinary sediment of patients with lupus nephritis, Arthritis Rheum 48 (5) (2003) 1326–1331.

[120] M. Akahoshi, H. Nakashima, Y. Tanaka, T. Kohsaka, S. Nagano, E. Ohgami, et al., Th1/Th2 balance of peripheral T helper cells in systemic lupus erythematosus, Arthritis Rheum 42 (8) (1999) 1644–1648.

[121] C.O. Jacob, P.H. van der Meide, H.O. McDevitt, In vivo treatment of (NZB x NZW) F1 lupus-like nephritis with monoclonal antibody to gamma interferon, J Exp Med 166 (3) (1987) 798–803.

[122] S.L. Peng, J. Moslehi, J. Craft, Roles of interferon-gamma and interleukin-4 in murine lupus, J Clin Invest 99 (8) (1997) 1936–1946.

[123] D. Balomenos, R. Rumold, A.N. Theoilopoulos, Interferon-γ is required for lupus-like disease and lymphoaccumulation in MRL-lpr mice, J Clin Invest 101 (2) (1998) 364–371.

[124] P. Bossù, D. Neumann, E. Del Giudice, A. Ciaramella, I. Gloaguen, G. Fantuzzi, et al., IL-18 cDNA vaccination protects mice from spontaneous lupus-like autoimmune disease, Proc Natl Acad Sci USA 100 (24) (2003) 14181–14816.

[125] A. Kaliyaperumal, C. Mohan, W. Wu, S.K. Datta, Nucleosomal peptide epitopes for nephritis-inducing T helper cells of murine lupus, J Exp Med 183 (6) (1996) 2459–2469.

[126] M. Khalil, K. Inaba, R. Steinman, J. Ravetch, B. Diamond, T cell studies in a peptide-induced model of systemic lupus erythematosus, J Immunol 166 (3) (2001) 1667–1674.

[127] K. Teramoto, N. Negoro, K. Kitamoto, T. Iwai, H. Iwao, M. Okamura, et al., Microarray analysis of glomerular gene expression in murine lupus nephritis, J Pharmacol Sci 106 (1) (2008) 56–67.

[128] A. Nakajima, S. Hirose, H. Yagita, K. Okumura, Roles of IL-4 and IL-12 in the development of lupus in NZB/W F1 mice, J Immunol 158 (3) (1997) 1466–1472.

[129] B.M. Rüger, K.J. Erb, Y. He, J.M. Lane, P.F. Davis, Q. Hasan, Interleukin-4 transgenic mice develop glomerulosclerosis independent of immunoglobulin deposition, Eur J Immunol 30 (9) (2000) 2698–2703.

[130] R.R. Singh, IL-4 and many roads to lupuslike autoimmunity, Clin Immunol 108 (2) (2003) 73–79.

[131] C. Kurts, Dendritic cells: Not just another cell type in the kidney, but a complex immune sentinel network, Kidney Int 70 (3) (2006) 412–414.

[132] A.M. Woltman, J.W. de Fijter, K. Zuidwijk, A.G. Vlug, I.M. Bajema, S.W. van der Kooij, et al., Quantification of dendritic cell subsets in human renal tissue under normal and pathological conditions, Kidney Int 71 (10) (2007) 1001–1008.

[133] T. Krüger, D. Benke, F. Eitner, A. Lang, M. Wirtz, E.E. Hamilton-Williams, et al., Identification and functional characterization of dendritic cells in the healthy murine kidney and in experimental glomerulonephritis, J Am Soc Nephrol 15 (3) (2004) 613–621.

[134] T.J. Soos, T.N. Sims, L. Barisoni, K. Lin, D.R. Littman, M.L. Dustin, et al., CX(3)CR1(+) interstitial dendritic cells form a contiguous network throughout the entire kidney, Kidney Int 70 (3) (2006) 591–596.

[135] N. Fiore, G. Castellano, A. Blasi, C. Capobianco, A. Loverre, V. Montinaro, et al., Immature myeloid and plasmacytoid dendritic cells infiltrate renal tubulointerstitium in patients with lupus nephritis, Mol Immunol 45 (1) (2008) 259–265.

[136] M. Tucci, C. Quatraro, L. Lombardi, C. Pellegrino, F. Dammacco, F. Silvestris, Glomerular accumulation of plasmacytoid dendritic cells in active lupus nephritis: role of interleukin-18, Arthritis Rheum 58 (1) (2008) 251–262.

[137] B. Cederblad, S. Blomberg, H. Vallin, A. Perers, G.V. Alm, L. Rönnblom, Patients with systemic lupus erythematosus have reduced numbers of circulating natural interferon-alpha-producing cells, J Autoimmun 11 (5) (1998) 465–470.

[138] M.A. Gill, P. Blanco, E. Arce, V. Pascual, J. Banchereau, A.K. Palucka, Blood dendritic cells and DC-poietins in systemic lupus erythematosus, Hum Immunol 63 (12) (2002) 1172–1180.

[139] J. Banchereau, F. Briere, C. Caux, J. Davoust, S. Lebecque, Y.J. Liu, et al., Immunobiology of dendritic cells, Annu Rev Immunol 18 (2000) 767–811.

[140] L. Farkas, K. Beiske, F. Lund-Johansen, P. Brandtzaeg, F.L. Jahnsen, Plasmacytoid dendritic cells (natural interferon-alpha/beta-producing cells) accumulate in cutaneous lupus erythematosus lesions, Am J Pathol 159 (1) (2001) 237–243.

[141] H. Bagavant, U.S. Deshmukh, H. Wang, T. Ly, S.M. Fu, Role for nephritogenic T cells in lupus glomerulonephritis: progression to renal failure is accompanied by T cell activation and expansion in regional lymph nodes, J Immunol 177 (11) (2006) 8258–8265.

[142] L. Schiffer, R. Bethunaickan, M. Ramanujam, W. Huang, M. Schiffer, H. Tao, et al., Activated renal macrophages are markers of disease onset and disease remission in lupus nephritis, J Immunol 180 (3) (2008) 1938–1947.

[143] H. Fujinaka, M. Nameta, P. Kovalenko, A. Matsuki, N. Kato, G. Nishimoto, et al., Periglomerular accumulation of dendritic cells in rat crescentic glomerulonephritis, J Nephrol 20 (3) (2007) 357–363.

[144] J. Scholz, V. Lukacs-Kornek, D.R. Engel, E. Specht, E. Kiss, F. Eitner, et al., Renal dendritic cells stimulate IL-10 production and attenuate nephrotoxic nephritis, J Am Soc Nephrol 19 (3) (2008) 527–537.

[145] F. Heymann, C. Meyer-Schwesinger, E.E. Hamilton-Williams, L. Hammerich, U. Panzer, S. Kaden, et al., Kidney dendritic cell activation is required for progression of renal disease in a mouse model of glomerular injury, J Clin Invest 119 (5) (2009) 1286–1297.

[146] J. Zhu, X. Liu, C. Xie, M. Yan, Y. Yu, E.S. Sobel, et al., T cell hyperactivity in lupus as a consequence of hyperstimulatory antigen-presenting cells, J Clin Invest 115 (7) (2005) 1869–1878.

[147] S. Monrad, M.J. Kaplan, Dendritic cells and the immunopathogenesis of systemic lupus erythematosus, Immunol Res 37 (2) (2007) 135–145.

[148] A. Kaser, S. Kaser, N.C. Kaneider, B. Enrich, C.J. Wiedermann, H. Tilg, Interleukin-18 attracts plasmacytoid dendritic cells (DC2s) and promotes Th1 induction by DC2s through IL-18 receptor expression, Blood 103 (2) (2004) 648–655.

[149] F.P. Siegal, N. Kadowaki, M. Shodell, P.A. Fitzgerald-Bocarsly, K. Shah, S. Ho, et al., The nature of the principal type 1 interferon-producing cells in human blood, Science 284 (5421) (1999) 1835–1837.

[150] J. Banchereau, V. Pascual, A.K. Palucka, Autoimmunity through cytokine-induced dendritic cell activation, Immunity 20 (5) (2004) 539–550.

[151] C. Kurts, W.R. Heath, F.R. Carbone, J. Allison, J.F. Miller, H. Kosaka, Constitutive class I-restricted exogenous presentation of self antigens in vivo, J Exp Med 184 (3) (1996) 923–930.

[152] P.T. Coates, B.L. Colvin, A. Ranganathan, F.J. Duncan, Y.Y. Lan, W.J. Shufesky, et al., CCR and CC chemokine expression in relation to Flt3 ligand-induced renal dendritic cell mobilization, Kidney Int 66 (5) (2004) 1907–1917.

[153] F. Castellino, A.Y. Huang, G. Altan-Bonnet, S. Stoll, C. Scheinecker, et al., Chemokines enhance immunity by guiding naive CD8+ T cells to sites of CD4+ T cell-dendritic cell interaction, Nature 440 (7086) (2006) 890–895.

[154] M. Tucci, N. Calvani, H.B. Richards, C. Quatraro, F. Silvestris, The interplay of chemokines and dendritic cells in the pathogenesis of lupus nephritis, Ann N Y Acad Sci 1051 (2005) 421–432.

[155] D.H. Hooke, D.C. Gee, R.C. Atkins, Leukocyte analysis using monoclonal antibodies in human glomerulonephritis, Kidney Int 31 (4) (1987) 964–972.

[156] E. Alexopoulos, D. Seron, R.B. Hartley, J.S. Cameron, Lupus nephritis: correlation of interstitial cells with glomerular function, Kidney Int 37 (1) (1990) 100–109.

[157] C.J. Kootstra, M. Sutmuller, H.J. Baelde, E. de Heer, J.A. Bruijn, Association between leukocyte infiltration and development of glomerulosclerosis in experimental lupus nephritis, J Pathol 184 (2) (1998) 219–225.

[158] H.Y. Lan, D.J. Nikolic-Paterson, W. Mu, R.C. Atkins, Local macrophage proliferation in the pathogenesis of glomerular crescent formation in rat anti-glomerular basement membrane (GBM) glomerulonephritis, Clin Exp Immunol 110 (2) (1997) 233–240.

[159] X.R. Huang, P.G. Tipping, J. Apostolopoulos, C. Oettinger, M. D'Souza, G. Milton, et al., Mechanisms of T cell-induced glomerular injury in anti-glomerular basement membrane (GBM) glomerulonephritis in rats, Clin Exp Immunol 109 (1) (1997) 134—142.

[160] S.R. Holdsworth, T.J. Neale, C.B. Wilson, Abrogation of macrophage-dependent injury in experimental glomerulonephritis in the rabbit. Use of an antimacrophage serum, J Clin Invest 68 (3) (1981) 686—698.

[161] J.S. Duffield, P.G. Tipping, T. Kipari, J.F. Cailhier, S. Clay, R. Lang, et al., Conditional ablation of macrophages halts progression of crescentic glomerulonephritis, Am J Pathol 167 (5) (2005) 1207—1219.

[162] Y. Ikezumi, L.A. Hurst, T. Masaki, R.C. Atkins, D.J. Nikolic-Paterson, Adoptive transfer studies demonstrate that macrophages can induce proteinuria and mesangial cell proliferation, Kidney Int 63 (1) (2003) 83—95.

[163] S.R. Holdsworth, T.J. Neale, Macrophage-induced glomerular injury. Cell transfer studies in passive autologous anti-glomerular basement membrane antibody-initiated experimental glomerulonephritis, Lab Invest 51 (2) (1984) 172—180.

[164] Y. Du, Y. Fu, C. Mohan, Experimental anti-GBM nephritis as an analytical tool for studying spontaneous lupus nephritis, Arch Immunol Ther Exp (Warsz) 56 (1) (2008) 31—40.

[165] G.H. Tesch, S. Maifert, A. Schwarting, B.J. Rollins, V.R. Kelley, Monocyte chemoattractant protein 1-dependent leukocytic infiltrates are responsible for autoimmune disease in MRL-Fas (lpr) mice, J Exp Med 190 (12) (1999) 1813—1824.

[166] S.R. Holdsworth, P.G. Tipping, Leukocytes in glomerular injury, Semin Immunopathol 29 (4) (2007) 355—374.

[167] N.W. Boyce, P.G. Tipping, S.R. Holdsworth, Glomerular macrophages produce reactive oxygen species in experimental glomerulonephritis, Kidney Int 35 (3) (1989) 778—782.

[168] S.B. Khan, H.T. Cook, G. Bhangal, J. Smith, F.W. Tam, C.D. Pusey, Antibody blockade of TNF-alpha reduces inflammation and scarring in experimental crescentic glomerulonephritis, Kidney Int 67 (5) (2005) 1812—1820.

[169] M.A. Little, G. Bhangal, C.L. Smyth, M.T. Nakada, H.T. Cook, S. Nourshargh, et al., Therapeutic effect of anti-TNF alpha antibodies in an experimental model of anti-neutrophil cytoplasm antibody-associated systemic vasculitis, J Am Soc Nephrol 17 (1) (2006) 160—169.

[170] J.R. Timoshanko, A.R. Kitching, Y. Iwakura, S.R. Holdsworth, P.G. Tipping, Leukocyte derived interleukin-1beta interacts with renal interleukin-1 receptor I to promote renal tumor necrosis factor and glomerular injury in murine crescentic glomerulonephritis, Am J Pathol 164 (6) (2004) 1967—1977.

[171] J.R. Timoshanko, A.R. Kitching, Y. Iwakura, S.R. Holdsworth, P.G. Tipping, Contributions of IL-1beta and IL-1alpha to crescentic glomerulonephritis in mice, J Am Soc Nephrol 15 (4) (2004) 910—918.

[172] J.R. Timoshanko, J.D. Sedgwick, S.R. Holdsworth, P.G. Tipping, Intrinsic renal cells are the major source of tumor necrosis factor contributing to renal injury in murine crescentic glomerulonephritis, J Am Soc Nephrol 14 (7) (2003) 1785—1793.

[173] C. Entani, K. Izumino, M. Takata, A. Futamura, Y. Nakagawa, H. Inoue, et al., Expression of platelet-derived growth factor in lupus nephritis in MRL/MpJ-lpr/lpr mice, Nephron 77 (1) (1997) 100—104.

[174] D.A. Troyer, B. Chandrasekar, J.L. Barnes, G. Fernandes, Calorie restriction decreases platelet-derived growth factor (PDGF)-A and thrombin receptor mRNA expression in autoimmune murine lupus nephritis, Clin Exp Immunol 108 (1) (1997) 58—62.

[175] A. Sadanaga, H. Nakashima, K. Masutani, K. Miyake, S. Shimizu, T. Igawa, et al., Amelioration of autoimmune nephritis by imatinib in MRL/lpr mice, Arthritis Rheum 52 (12) (2005) 3987—3996.

[176] V. Saxena, D.W. Lienesch, M. Zhou, R. Bommireddy, M. Azhar, T. Doetschman, et al., Dual roles of immunoregulatory cytokine TGF-beta in the pathogenesis of autoimmunity-mediated organ damage, J Immunol 180 (3) (2008) 1903—1912.

[177] C.G. Cochrane, E.R. Unanue, F.J. Dixon, A role of polymorphonuclear leukocytes and complement in nephrotoxic nephritis, J Exp Med 122 (1) (1965) 99—116.

[178] P.F. Naish, N.M. Thomson, I.J. Simpson, D.K. Peters, The role of polymorphonuclear leucocytes in the autologous phase of nephrotoxic nephritis, Clin Exp Immunol 22 (1) (1975) 102—111.

[179] P.M. Henson, Pathologic mechanisms in neutrophil-mediated injury, Am J Pathol 68 (3) (1972) 593—612.

[180] G. Camussi, F.C. Cappio, M. Messina, R. Coppo, P. Stratta, A. Vercellone, The polymorphonuclear neutrophil (PMN) immunohistological technique: detection of immune complexes bound to the PMN membrane in acute poststreptococcal and lupus nephritis, Clin Nephrol 14 (6) (1980) 280—287.

[181] T. Oda, O. Hotta, Y. Taguma, H. Kitamura, K. Sudo, I. Horigome, et al., Involvement of neutrophil elastase in crescentic glomerulonephritis, Hum Pathol 28 (6) (1997) 720—728.

[182] P. Scapini, A. Carletto, B. Nardelli, F. Calzetti, V. Roschke, F. Merigo, et al., Proinflammatory mediators elicit secretion of the intracellular B-lymphocyte stimulator pool (BLyS) that is stored in activated neutrophils: implications for inflammatory diseases, Blood 105 (2) (2005) 830—837.

[183] K.P. van Gisbergen, M. Sanchez-Hernandez, T.B. Geijtenbeek, Y. van Kooyk, Neutrophils mediate immune modulation of dendritic cells through glycosylation-dependent interactions between Mac-1 and DC-SIGN, J Exp Med 201 (8) (2005) 1281—1292.

[184] R.J. Johnson, W.G. Couser, C.E. Alpers, M. Vissers, M. Schulze, S.J. Klebanoff, The human neutrophil serine proteinases, elastase and cathepsin G, can mediate glomerular injury in vivo, J Exp Med 168 (3) (1988) 1169—1174.

[185] R.J. Johnson, D. Lovett, R.I. Lehrer, W.G. Couser, S.J. Klebanoff, Role of oxidants and proteases in glomerular injury, Kidney Int 45 (2) (1994) 352—359.

[186] S. Suzuki, F. Gejyo, T. Kuroda, J.J. Kazama, N. Imai, H. Kimura, et al., Effects of a novel elastase inhibitor, ONO-5046, on nephrotoxic serum nephritis in rats, Kidney Int 53 (5) (1998) 1201—1208.

[187] C.H. Hinze, M. Suzuki, M. Klein-Gitelman, M.H. Passo, J. Olson, N.G. Singer, et al., Neutrophil gelatinase-associated lipocalin is a predictor of the course of global and renal childhood-onset systemic lupus erythematosus disease activity, Arthritis Rheum 60 (9) (2009) 2772—2781.

[188] M. Suzuki, K.M. Wiers, M.S. Klein-Gitelman, K.A. Haines, J. Olson, K.B. Onel, et al., Neutrophil gelatinase-associated lipocalin as a biomarker of disease activity in pediatric lupus nephritis, Pediatr Nephrol 23 (3) (2008) 403—412.

[189] N. Schwartz, J.S. Michaelson, C. Putterman, Lipocalin-2, TWEAK, and other cytokines as urinary biomarkers for lupus nephritis, Ann N Y Acad Sci 1109 (2007) 265—274.

[190] S.M. Ka, A. Rifai, J.H. Chen, C.W. Cheng, H.A. Shui, H.S. Lee, et al., Glomerular crescent-related biomarkers in a murine model of chronic graft versus host disease, Nephrol Dial Transplant 21 (2) (2006) 288—298.

[191] P.G. Tipping, J. Timoshanko, Contributions of intrinsic renal cells to crescentic glomerulonephritis, Nephron Exp Nephrol 101 (4) (2005) e173—178.

I. BASIS OF DISEASE PATHOGENESIS

[192] J.R. Timoshanko, P.G. Tipping, Resident kidney cells and their involvement in glomerulonephritis, Curr Drug Targets Inflamm Allergy 4 (3) (2005) 353−362.

[193] S. Li, C. Kurts, F. Kontgen, S.R. Holdsworth, P.G. Tipping, Major histocompatibility complex class II expression by intrinsic renal cells is required for crescentic glomerulonephritis, J Exp Med 188 (3) (1998) 597−602.

[194] A. Molano, P. Lakhani, A. Aran, L.C. Burkly, J.S. Michaelson, C. Putterman, TWEAK stimulation of kidney resident cells in the pathogenesis of graft versus host induced lupus nephritis, Immunol Lett 125 (2) (2009) 119−128.

[195] C. Gómez-Guerrero, P. Hernández-Vargas, O. López-Franco, G. Ortiz-Muñoz, J. Egido, Mesangial cells and glomerular inflammation: from the pathogenesis to novel therapeutic approaches, Curr Drug Targets Inflamm Allergy 4 (3) (2005) 341−351.

[196] H.H. Radeke, K. Resch, The inflammatory function of renal glomerular mesangial cells and their interaction with the cellular immune system, Clin Investig 70 (9) (1992) 825−842.

[197] M. Martin, R. Schwinzer, H. Schellekens, K. Resch, Glomerular mesangial cells in local inflammation. Induction of the expression of MHC class II antigens by IFN-gamma, J Immunol 142 (6) (1989) 1887−1894.

[198] D.C. Brennan, A.M. Jevnikar, F. Takei, V.E. Reubin-Kelley, Mesangial cell accessory functions: mediation by intercellular adhesion molecule-1, Kidney Int 38 (6) (1990) 1039−1046.

[199] C. Baudeau, F. Delarue, C.J. Hé, G. Nguyen, C. Adida, M.N. Peraldi, et al., Induction of MHC class II molecules HLA-DR, -DP and -DQ and ICAM 1 in human podocytes by gamma-interferon, Exp Nephrol 2 (5) (1994) 306−312.

[200] W. Coers, E. Brouwer, J.T. Vos, A. Chand, S. Huitema, P. Heeringa, et al., Podocyte expression of MHC class I and II and intercellular adhesion molecule-1 (ICAM-1) in experimental pauci-immune crescentic glomerulonephritis, Clin Exp Immunol 98 (2) (1994) 279−286.

[201] J. Reiser, G. von Gersdorff, M. Loos, J. Oh, K. Asanuma, L. Giardino, et al., Induction of B7-1 in podocytes is associated with nephrotic syndrome, J Clin Invest 113 (10) (2004) 1390−1397.

[202] J. Bariéty, P. Bruneval, A. Meyrier, C. Mandet, G. Hill, C. Jacquot, Podocyte involvement in human immune crescentic glomerulonephritis, Kidney Int 68 (3) (2005) 1109−1119.

[203] P.S. Thorner, M. Ho, V. Eremina, Y. Sado, S. Quaggin, Podocytes contribute to the formation of glomerular crescents, J Am Soc Nephrol 19 (3) (2008) 495−502.

[204] M.J. Moeller, A. Soofi, I. Hartmann, M. Le Hir, R. Wiggins, W. Kriz, et al., Podocytes populate cellular crescents in a murine model of inflammatory glomerulonephritis, J Am Soc Nephrol 15 (1) (2004) 61−67.

[205] H. Yokoyama, X. Zheng, T.B. Strom, V.R. Kelley, B7(+)-transfectant tubular epithelial cells induce T cell anergy, ignorance or proliferation, Kidney Int 45 (4) (1994) 1105−1112.

[206] H. Ishikura, C. Takahashi, K. Kanagawa, H. Hirata, K. Imai, T. Yoshiki, Cytokine regulation of ICAM-1 expression on human renal tubular epithelial cells in vitro, Transplantation 51 (6) (1991) 1272−1275.

[207] S. Laxmanan, D. Datta, C. Geehan, D.M. Briscoe, S. Pal, CD40: a mediator of pro- and anti-inflammatory signals in renal tubular epithelial cells, J Am Soc Nephrol 16 (9) (2005) 2714−2723.

[208] U. Niemann-Masanek, A. Mueller, B.A. Yard, R. Waldherr, F.J. van der Woude, B7-1 (CD80) and B7-2 (CD 86) expression in human tubular epithelial cells in vivo and in vitro, Nephron 92 (3) (2002) 542−556.

[209] Y. Chen, C. Yang, Z. Xie, L. Zou, Z. Ruan, X. Zhang, et al., Expression of the novel co-stimulatory molecule B7-H4 by renal tubular epithelial cells, Kidney Int 70 (12) (2006) 2092−2099.

[210] A.E. Postlethwaite, H. Shigemitsu, S. Kanangat, Cellular origins of fibroblasts: possible implications for organ fibrosis in systemic sclerosis, Curr Opin Rheumatol 16 (6) (2004) 733−738.

[211] E.M. Zeisberg, S.E. Potenta, H. Sugimoto, M. Zeisberg, R. Kalluri, Fibroblasts in Kidney Fibrosis Emerge via Endothelialto- Mesenchymal Transition, J Am Soc Nephrol 19 (12) (2008) 2282−2287.

[212] S.M. Ka, C.W. Cheng, H.A. Shui, W.M. Wu, D.M. Chang, Y.C. Lin, et al., Mesangial cells of lupus-prone mice are sensitive to chemokine production, Arthritis Res Ther 9 (4) (2007) R67.

[213] C.M. Reilly, J.C. Oates, J.A. Cook, J.D. Morrow, P.V. Halushka, G.S. Gilkeson, Inhibition of mesangial cell nitric oxide in MRL/lpr mice by prostaglandin J2 and proliferator activation receptor-gamma agonists, J Immunol 164 (3) (2000) 1498−1504.

[214] C.M. Reilly, J.C. Oates, J. Sudian, M.B. Crosby, P.V. Halushka, G.S. Gilkeson, Prostaglandin J(2) inhibition of mesangial cell iNOS expression, Clin Immunol 98 (3) (2001) 337−345.

[215] B. Deocharan, X. Qing, J. Lichauco, C. Putterman, Alpha-actinin is a cross-reactive renal target for pathogenic anti-DNAantibodies, J Immunol 168 (6) (2002) 3072−3078.

[216] Z. Zhao, B. Deocharan, P.E. Scherer, L.J. Ozelius, C. Putterman, Differential binding of cross-reactive anti-DNA antibodies to mesangial cells: the role of alpha-actinin, J Immunol 176 (12) (2006) 7704−7714.

[217] M.P. Cancro, D.P. D'Cruz, M.A. Khamashta, The role of B lymphocyte stimulator (BLyS) in systemic lupus erythematosus, J Clin Invest 119 (5) (2009) 1066−1073.

[218] A. Nalbandian, J.C. Crispín, G.C. Tsokos, Interleukin-17 and systemic lupus erythematosus: current concepts, Clin Exp Immunol 157 (2) (2009) 209−215.

[219] J.C. Crispín, M. Oukka, G. Bayliss, R.A. Cohen, C.A. Van Beek, I.E. Stillman, et al., Expanded double negative T cells in patients with systemic lupus erythematosus produce IL-17 and infiltrate the kidneys, J Immunol 181 (12) (2008) 8761−8766.

[220] H.K. Kang, M. Liu, S.K. Datta, Low-dose peptide tolerance therapy of lupus generates plasmacytoid dendritic cells that cause expansion of autoantigen-specific regulatory T cells and contraction of inflammatory Th17 cells, J Immunol 178 (12) (2007) 7849−7858.

[221] N. Jacob, H. Yang, L. Pricop, Y. Liu, X. Gao, S.G. Zheng, et al., Accelerated pathological and clinical nephritis in systemic lupus erythematosus-prone New Zealand Mixed 2328 mice doubly deficient in TNF receptor 1 and TNF receptor 2 via a Th17-associated pathway, J Immunol 182 (4) (2009) 2532−2541.

[222] L. Bennett, A.K. Palucka, E. Arce, V. Cantrell, J. Borvak, J. Banchereau, et al., Interferon and granulopoiesis signatures in systemic lupus erythematosus blood, J Exp Med 197 (6) (2003) 711−723.

[223] E.C. Baechler, F.M. Batliwalla, G. Karypis, P.M. Gaffney, W.A. Ortmann, K.J. Espe, et al., Interferon-inducible gene expression signature in peripheral blood cells of patients with severe lupus, Proc Natl Acad Sci USA 100 (5) (2003) 2610−2615.

[224] K.A. Kirou, C. Lee, S. George, K. Louca, I.G. Papagiannis, M.G. Peterson, et al., Coordinate overexpression of interferon-alpha induced genes in systemic lupus erythematosus, Arthritis Rheum 50 (12) (2004) 3958−3967.

[225] A. Mathian, A. Weinberg, M. Gallegos, J. Banchereau, S. Koutouzov, IFN-alpha induces early lethal lupus in pre-autoimmune (New Zealand Black × New Zealand White) F1 but not in BALB/c mice, J Immunol 174 (5) (2005) 2499−2506.

I. BASIS OF DISEASE PATHOGENESIS

[226] K. Hasegawa, T. Hayashi, K. Maeda, Promotion of lupus in NZB x NZWF1 mice by plasmids encoding interferon (IFN)-gamma but not by those encoding interleukin (IL)-4, J Comp Pathol 127 (1) (2002) 1—6.

[227] D. Sergiescu, I. Cerutti, E. Efthymiou, A. Kahan, C. Chany, Adverse effects of interferon treatment on the life span of NZB mice, Biomedicine 31 (2) (1979) 48—51.

[228] A.M. Fairhurst, A. Mathian, J.E. Connolly, A. Wang, H.F. Gray, T.A. George, et al., Systemic IFN-α drives kidney nephritis in B6. *Sle123* mice, Eur J Immunol 38 (7) (2008) 1948—1960.

[229] A. Schwarting, K. Paul, S. Tschirner, J. Menke, T. Hansen, W. Brenner, et al., Interferon-beta: a therapeutic for autoimmune lupus in MRL-Faslpr mice, J Am Soc Nephrol 16 (11) (2005) 3264—3272.

[230] M.L. Santiago-Raber, R. Baccala, K.M. Haraldsson, D. Choubey, T.A. Stewart, D.H. Kono, et al., Type-I interferon receptor deficiency reduces lupus-like disease in NZB mice, J Exp Med 197 (6) (2003) 777—788.

[231] C. Haas, B. Ryffel, M. Le Hir, IFN-gamma receptor deletion prevents autoantibody production and glomerulonephritis in lupus-prone (NZB x NZW) F1 mice, J Immunol 160 (8) (1998) 3713—3718.

[232] D.C. Nacionales, K.M. Kelly-Scumpia, P.Y. Lee, J.S. Weinstein, R. Lyons, E. Sobel, et al., Deficiency of the type I interferon receptor protects mice from experimental lupus, Arthritis Rheum 56 (11) (2007) 3770—3783.

[233] J.D. Hron, S.L. Peng, Type I IFN protects against murine lupus, J Immunol 173 (3) (2004) 2134—2142.

[234] A.M. Fairhurst, C. Xie, Y.Y. Fu, A. Wang, C.D. Boudreaux, X.J. Zhou, et al., Type I interferons produced by resident renal cells may promote end-organ disease in autoantibody-mediated glomerulonephritis, J Immunol (2009). in press.

[235] J.P. Grande, Mechanisms of progression of renal damage in lupus nephritis: Pathogenesis of renal scarring, Lupus 7 (9) (1998) 604—610.

[236] W.A. Border, E. Ruoslahti, Transforming growth factor beta in disease: the dark side of tissue repair, J Clin Invest 90 (1) (1992) 1—7.

[237] T. Nakamura, I. Ebihara, I. Shirato, Y. Tomino, H. Koide, Increased steady-state levels of mRNA coding for extracellular matrix components in kidneys of NZB/W F1 mice, Am J Pathol 139 (2) (1991) 437—450.

[238] T. Nakamura, I. Ebihara, I. Nagaoka, S. Osada, Y. Tomino, H. Koide, Effect of methylprednisolone on transforming growth factor-beta, insulin-like growth factor-I, and basic fibroblast growth factor gene expression in the kidneys of NZB/W F1 mice, Ren Physiol Biochem 16 (3) (1993) 105—116.

[239] Y. Taniguchi, N. Yorioka, T. Masaki, K. Yamashita, T. Ito, H. Ueda, et al., Role of transforming growth factor-beta 1 in glomerulonephritis, J Int Med Res 25 (2) (1997) 71—80.

[240] S.M. Ka, X.R. Huang, H.Y. Lan, P.Y. Tsai, S.M. Yang, H.A. Shui, et al., Smad7 gene therapya meliorates an autoimmune crescentic glomerulonephritis in mice. J Am Soc Nephrol 18 (6), 1777—1788.

[241] E. Voronov, M. Dayan, H. Zinger, L. Gayvoronsky, J.P. Lin, Y. Iwakura, et al., IL-1 beta-deficient mice are resistant to induction of experimental SLE, Eur Cytokine Netw 17 (2) (2006) 109—116.

[242] H.B. Richards, M. Satoh, M. Shaw, C. Libert, V. Poli, W.H. Reeves, Interleukin 6 dependence of anti-DNA antibody production: evidence for two pathways of autoantibody formation in pristane-induced lupus, J Exp Med 188 (5) (1998) 985—990.

[243] Z. Yin, G. Bahtiyar, N. Zhang, L. Liu, P. Zhu, M.E. Robert, et al., IL-10 regulates murine lupus, J Immunol 169 (4) (2000) 2148—2155.

[244] D. Kontoyiannis, G. Kollias, Accelerated autoimmunity and lupus nephritis in NZB mice with an engineered heterozygous deficiency in tumor necrosis factor, Eur J Immunol 30 (7) (2000) 2038—2047.

[245] H.U. Schorlemmer, E.J. Kanzy, K.D. Langner, R. Kurrle, Immunoregulation of SLE-like disease by the IL-1 receptor: disease modifying activity on BDF1 hybrid mice and MRL autoimmune mice, Agents Actions 39 (1993) C117—120.

[246] H. Ishida, T. Muchamuel, S. Sakaguchi, S. Andrade, S. Menon, M. Howard, Continuous administration of anti-interleukin 10 antibodies delays onset of autoimmunity in NZB/W F1 mice, J Exp Med 179 (1) (1994) 305—310.

[247] J.F. McHale, O.A. Harari, D. Marshall, D.O. Haskard, TNF-alpha and IL-1 sequentially induce endothelial ICAM-1 and VCAM-1 expression in MRL/lpr lupus-prone mice, J Immunol 163 (7) (1999) 3993—4000.

[248] C. Zoja, X.H. Liu, R. Donadelli, M. Abbate, D. Testa, D. Corna, et al., Renal expression of monocyte chemoattractant protein-1 in lupus autoimmune mice, J Am Soc Nephrol 8 (5) (1997) 720—729.

[249] G. Pérez de Lema, H. Maier, E. Nieto, V. Vielhauer, B. Luckow, F. Mampaso, et al., Chemokine expression precedes inflammatory cell infiltration and chemokine receptor and cytokine expression during the initiation of murine lupus nephritis, J Am Soc Nephrol 12 (7) (2001) 1369—1382.

[250] S.D. Marks, S.J. Williams, K. Tullus, N.J. Sebire, Glomerular expression of monocyte chemoattractant protein-1 is predictive of poor renal prognosis in pediatric lupus nephritis, Nephrol Dial Transplant 23 (11) (2008) 3521—3526.

[251] R.W. Chan, F.M. Lai, E.K. Li, L.S. Tam, K.M. Chow, K.B. Lai, et al., Intrarenal cytokine gene expression in lupus nephritis, Ann Rheum Dis 66 (7) (2007) 886—892.

[252] M. Noris, S. Bernasconi, F. Casiraghi, S. Sozzani, E. Gotti, G. Remuzzi, et al., Monocyte chemoattractant protein-1 is excreted in excessive amounts in the urine of patients with lupus nephritis, Lab Invest 73 (6) (1995) 804—809.

[253] T. Wada, H. Yokoyama, S.B. Su, N. Mukaida, M. Iwano, K. Dohi, et al., Monitoring urinary levels of monocyte chemotactic and activating factor reflects disease activity of lupus nephritis, Kidney Int 49 (3) (1996) 761—767.

[254] B. Lu, B.J. Rutledge, L. Gu, J. Fiorillo, N.W. Lukacs, S.L. Kunkel, et al., Abnormalities in monocyte recruitment and cytokine expression in monocyte chemoattractant protein 1-deficient mice, J Exp Med 187 (4) (1998) 601—608.

[255] H. Hasegawa, M. Kohno, M. Sasaki, A. Inoue, M.R. Ito, M. Terada, et al., Antagonist of monocyte chemoattractant protein 1 ameliorates the initiation and progression of lupus nephritis and renal vasculitis in MRL/lpr mice, Arthritis Rheum 48 (9) (2003) 2555—2566.

[256] S. Shimizu, H. Nakashima, K. Masutani, Y. Inoue, K. Miyake, M. Akahoshi, et al., Anti-monocyte chemoattractant protein-1 gene therapy attenuates nephritis in MRL/lpr mice, Rheumatology (Oxford) 43 (9) (2004) 1121—1128.

[257] S. Shimizu, H. Nakashima, K. Karube, K. Ohshima, K. Egashira, Monocyte chemoattractant protein-1 activates a regional Th1 immunoresponse in nephritis of MRL/lpr mice, Clin Exp Rheumatol 23 (2) (2005) 239—242.

[258] H.J. Anders, M. Frink., Y. Linde, B. Banas, M. Wörnle, C.D. Cohen, et al., CC chemokine ligand 5/RANTES chemokine antagonists aggravate glomerulonephritis despite reduction of glomerular leukocyte infiltration, J Immunol 170 (11) (2003) 5658—5666.

[259] C.M. Lloyd, A.W. Minto, M.E. Dorf, A. Proudfoot, T.N. Wells, D.J. Salant, et al., RANTES and monocyte chemoattractant protein-1 (MCP-1) play an important role in the inflammatory phase of crescentic nephritis, but only MCP-1 is involved in crescent formation and interstitial fibrosis, J Exp Med 185 (7) (1997) 1371–1380.

[260] H.J. Anders, E. Belemezova, V. Eis, S. Segerer, V. Vielhauer, G. Perez de Lema, et al., Late onset of treatment with a chemokine receptor CCR1 antagonist prevents progression of lupus nephritis in MRL-Fas(lpr) mice, J Am Soc Nephrol 15 (6) (2004) 1504–1513.

[261] B.F. Chong, C. Mohan, Targeting the CXCR4/CXCL12 axis in systemic lupus erythematosus, Expert Opin Ther Targets 13 (10) (2009) 1147–1153.

[262] A. Wang, A.M. Fairhurst, K. Tus, S. Subramanian, Y. Liu, F. Lin, et al., CXCR4/CXCL12 hyperexpression plays a pivotal role in the pathogenesis of lupus, J Immunol 182 (7) (2009) 4448–4458.

[263] K. Balabanian, J. Couderc, L. Bouchet-Delbos, A. Amara, D. Berrebi, A. Foussat, et al., Role of the chemokine stromal cell-derived factor 1 in autoantibody production and nephritis in murine lupus, J Immunol 170 (6) (2003) 3392–3400.

[264] U. Panzer, O.M. Steinmetz, H.J. Paust, C. Meyer-Schwesinger, A. Peters, J.E. Turner, et al., Chemokine receptor CXCR3 mediates T cell recruitment and tissue injury in nephrotoxic nephritis in mice, J Am Soc Nephrol 18 (7) (2007) 2071–2084.

[265] O.M. Steinmetz, J.E. Turner, H.J. Paust, M. Lindner, A. Peters, K. Heiss, et al., CXCR3 Mediates Renal Th1 and Th17 Immune Response in Murine Lupus Nephritis, J Immunol 1863 (7) (2009) 4693–4704.

[266] C. Mohan, L. Morel, P. Yang, H. Watanabe, B. Croker, G. Gilkeson, E.K. Wakeland, Genetic dissection of lupus pathogenesis: a recipe for nephrophilic autoantibodies, J Clin Invest 103 (12) (1999) 1685–1695.

[267] K. Liu, Q.Z. Li, Y. Yu, C. Liang, S. Subramanian, Z. Zeng, et al., Sle3 and Sle5 can independently couple with Sle1 to mediate severe lupus nephritis, Genes Immun 8 (8) (2007) 634–645.

[268] K. Liu, Q.Z. Li, A.M. Delgado-Vega, A.K. Abelson, E. Sánchez, J.A. Kelly, et al., Kallikrein genes are associated with lupus and glomerular basement membrane-specific antibody-induced nephritis in mice and humans, J Clin Invest 119 (4) (2009) 911–923.

[269] Q.Z. Li, J. Zhou, R. Yang, M. Yan, Q. Ye, K. Liu, et al., The lupus-susceptibility gene kallikrein downmodulates antibody-mediated glomerulonephritis, Genes Immun 10 (5) (2009) 503–508.

[270] J. Chao, J.J. Zhang, K.F. Lin, L. Chao, Adenovirus-mediated kallikrein gene delivery reverses salt-induced renal injury in Dahl salt-sensitive rats, Kidney Int 54 (4) (1998) 1250–1260.

[271] G. Bledsoe, S. Crickman, J. Mao, C.F. Xia, H. Murakami, L. Chao, et al., Kallikrein/kinin protects against gentamicin-induced nephrotoxicity by inhibition of inflammation and apoptosis, Nephrol Dial Transplant 21 (3) (2006) 624–633.

[272] J.P. Schanstra, E. Neau, P. Drogoz, M.A. Arevalo Gomez, J.M. Lopez Novoa, D. Calise, et al., In vivo bradykinin B2 receptor activation reduces renal fibrosis, J Clin Invest 110 (3) (2002) 371–379.

[273] M. Kakoki, N. Takahashi, J.C. Jennette, O. Smithies, Diabetic nephropathy is markedly enhanced in mice lacking the bradykinin B2 receptor, Proc Natl Acad Sci USA 101 (36) (2004) 13302–133305.

[274] J. Chao, G. Bledsoe, H. Yin, L. Chao, The tissue kallikrein-kinin system protects against cardiovascular and renal diseases and ischemic stroke independently of blood pressure reduction, Biol Chem 387 (6) (2006) 665–675.

[275] A. Furusu, M. Miyazaki, K. Abe, S. Tsukasaki, K. Shioshita, O. Sasaki, et al., Expression of endothelial and inducible nitric oxide synthase in human glomerulonephritis, Kidney Int 53 (6) (1998) 1760–1768.

[276] J.B. Weinberg, D.L. Granger, D.S. Pisetsky, M.F. Seldin, M.A. Misukonis, S.N. Mason, et al., The role of nitric oxide in the pathogenesis of spontaneous murine autoimmune disease: increased nitric oxide production and nitric oxide synthase expression in MRL-lpr/lpr mice, and reduction of spontaneous glomerulonephritis and arthritis by orally administered NG-monomethyl-L-arginine, J Exp Med 179 (2) (1994) 651–660.

[277] C.M. Reilly, L.W. Farrelly, D. Viti, S.T. Redmond, F. Hutchison, P. Ruiz, et al., Modulation of renal disease in MRL/lpr mice by pharmacologic inhibition of inducible nitric oxide synthase, Kidney Int 61 (3) (2002) 839–846.

[278] K. Yasui, A. Baba, Therapeutic potential of superoxide dismutase (SOD) for resolution of inflammation, Inflamm Res 55 (9) (2006) 359–363.

[279] T. Wu, Y. Fu, D. Brekken, M. Yan, X.J. Zhou, K. Vanarsa, et al., Urine proteome scans uncover total urinary protease, prostaglandin D synthase, serum amyloid P, and superoxide dismutase as potential markers of lupus nephritis, J Immunol 184 (4) (2010) 2183–2193.

[280] J.S. Wang, L.P. Ger, H.H. Tseng, Expression of glomerular antioxidant enzymes in human glomerulonephritis, Nephron 76 (1) (1997) 32–38.

[281] S. Taysi, M. Gul, R.A. Sari, F. Akcay, N. Bakan, Serum oxidant/antioxidant status of patients with systemic lupus erythematosus, Clin Chem Lab Med 40 (7) (2002) 684–688.

[282] R.D. Bloom, S. Florquin, G.G. Singer, D.C. Brennan, V.R. Kelley, Colony stimulating factor-1 in the induction of lupus nephritis, Kidney Int 143 (5) (1993) 1000–1009.

[283] N.M. Isbel, P.A. Hill, R. Foti, W. Mu, L.A. Hurst, C. Stambe, et al., Tubules are the major site of M-CSF production in experimental kidney disease: correlation with local macrophage proliferation, Kidney Int 60 (2) (2001) 614–625.

[284] M.A. Yui, W.H. Brissette, D.C. Brennan, R.P. Wuthrich, V.E. Rubin-Kelley, Increased macrophage colony-stimulating factor in neonatal and adult autoimmune MRL-lpr mice, Am J Pathol 139 (2) (1991) 255–261.

[285] T. Naito, H. Yokoyama, K.J. Moore, G. Dranoff, R.C. Mulligan, V.R. Kelley, Macrophage growth factors introduced into the kidney initiate renal injury, Mol Med 2 (3) (1996) 297–312.

[286] D.M. Lenda, E.R. Stanley, V.R. Kelley, Negative role of colonystimulating factor-1 in macrophage, T cell, and B cell mediated autoimmune disease in MRL-Fas(lpr) mice, J Immunol 173 (7) (2004) 4744–4754.

[287] H.Y. Lan, N. Yang, D.J. Nikolic-Paterson, X.Q. Yu, W. Mu, N.M. Isbel, et al., Expression of macrophage migration inhibitory factor in human glomerulonephritis, Kidney Int 57 (2) (2000) 499–509.

[288] H.Y. Lan, W. Mu, N. Yang, A. Meinhardt, D.J. Nikolic-Paterson, Y.Y. Ng, et al., De novo renal expression of macrophage migration inhibitory factor during the development of rat crescentic glomerulonephritis, Am J Pathol 149 (4) (1996) 1119–1127.

[289] H. Otukesh, M. Chalian, R. Hoseini, H. Chalian, N. Hooman, A. Bedayat, et al., Urine macrophage migration inhibitory factor in pediatric systemic lupus erythematosus, Clin Rheumatol 26 (12) (2007) 2105–2107.

[290] A.Y. Hoi, M.J. Hickey, P. Hall, J. Yamana, K.M. O'Sullivan, L.L. Santos, et al., Macrophage migration inhibitory factor deficiency attenuates macrophage recruitment, glomerulonephritis, and lethality in MRL/lpr mice, J Immunol 177 (8) (2006) 5687–5696.

[291] S. Sasaki, J. Nishihira, T. Ishibashi, Y. Yamasaki, K. Obikane, M. Echigoya, et al., Transgene of MIF induces podocyte injury and progressive mesangial sclerosis in the mouse kidney, Kidney Int 65 (2) (2004) 469−481.

[292] N. Yang, D.J. Nikolic-Patersonm, Y.Y. Ng, W. Mu, C. Metz, M. Bacher, et al., Reversal of established rat crescentic glomerulonephritis by blockade of macrophage migration inhibitory factor (MIF): potential role of MIF in regulating glucocorticoid production, Mol Med 4 (6) (1998) 413−424.

[293] H.Y. Lan, M. Bacher, N. Yang, W. Mu, D.J. Nikolic-Paterson, C. Metz, et al., The pathogenic role of macrophage migration inhibitory factor in immunologically induced kidney disease in the rat, J Exp Med 185 (8) (1997) 1455−1465.

27

Tissue Injury and the Skin

Annegret Kuhn, Markus Böhm, Thomas A. Luger

INTRODUCTION

The complexity of lupus erythematosus (LE) makes understanding the underlying pathophysiology problematic and simultaneously implies a multifactorial nature of the pathogenic course. It should be noted that correct classification of this autoimmune disease into two major categories, systemic LE (SLE) and cutaneous LE (CLE), still relies on criteria established by the American College of Rheumatology (ACR) in 1971 [1] and revised in 1982 [2] and 1997 [3]. These criteria were originally intended only for the classification of SLE, providing some degree of uniformity to the patient populations of clinical studies. In patients with primarily cutaneous manifestations, however, application of the ACR criteria often results in overestimation of SLE [4]. In particular, photosensitivity is listed as one of the ACR criteria despite being poorly defined as "a result of an unusual reaction to sunlight by patient's history or physician's observation." Meanwhile, it is known that various environmental factors besides ultraviolet (UV) radiation influence the clinical expression of LE and that irritative stimuli may lead to the induction of primarily cutaneous manifestations [5]. Moreover, genetic factors seem to be involved in the development of the different CLE subtypes. For instance, a strong association with HLA-A1, -B8, and -DR3 has been demonstrated in patients with subacute cutaneous LE (SCLE) [6]. In addition, SCLE seems to be significantly associated with a single nucleotide polymorphism in the tumor necrosis factor (TNF)-alpha gene promoter (-308A), which results in high TNF-α expression [7]. Congenital homozygous deficiency of C1q is a further risk factor for photosensitive CLE. Recently, it was shown that the presence or even accumulation of apoptotic cells in UV-irradiated skin of susceptible patients with different subtypes of the disease results in the development of characteristic inflammatory skin lesions, probably due to delayed clearance of these apoptotic cells [8]. Additional factors such as a decreased number of regulatory T cells (T_{reg}) at the site of inflammation and the expression of interferon (IFN)-inducible cytokines, including protein myxovirus protein A (MxA) and CXCL10, have been shown to be involved in the development of CLE [9, 10].

Classification of Skin Lesions in LE

The clinical expression of CLE exhibits great variation and, consequently, has led to the practice of identifying different subtypes of the disease. The development of a unified classification of skin manifestations, however, has proven difficult. The classification system developed by Gilliam in 1977 divided cutaneous lesions into LE-specific and LE-nonspecific manifestations by histological analysis of skin biopsy specimens [11]. Skin lesions such as urticarial vasculitis and livedo reticularis are some of the most common LE-nonspecific cutaneous lesions and are primarily associated with SLE, reflecting potentially serious complications [12]. In contrast, the LE-specific cutaneous findings encompass the various subtypes of CLE, which are subdivided into three different categories defined by constellations of clinical features, histological changes, serological abnormalities, and average duration of skin lesions: acute cutaneous LE (ACLE), SCLE, and chronic cutaneous LE (CCLE), including discoid LE (DLE), Chilblain LE (CHLE), and LE panniculitis (LEP). Since the initial creation of Gilliam's nomenclature and classification system more than two decades ago, several attempts have been made to improve this system and to provide new approaches to the problem of CLE classification. An additional subtype with characteristic clinical, histological, and photobiological features, termed LET, was first described in 1909 and has only recently been defined as a separate entity of CLE [13, 14]. The course and prognosis of patients with LET are generally more favorable than in patients with other subtypes of CLE; therefore, a revised classification system that includes LET as an intermittent subtype of CLE (ICLE) was suggested in 2004 [15].

Histopathology

The different subtypes of CLE often show similar histological features with superficial and deep perivascular and periadnexal lymphocytic infiltrates, in addition to the so-called interface-dermatitis. Vacuolar degeneration of the dermoepidermal junction and scattered dead keratinocytes (Civatte bodies) in the lower epidermal layers are characteristic, followed by thickening of the basement membrane as a late consequence (Table 27.1) [16]. These changes are routinely identified by hematoxylin and eosin staining in formalin-fixed biopsy specimens. In addition, thickening of the basement membrane can be visualized using the periodic acid-Schiff (PAS) stain, while dermal mucin deposition, another consistent feature of virtually all LE-specific lesions, can be visualized by colloidal iron staining or Alcian blue staining [17]. As a novel technique, reflectance confocal microscopy (RCM) was recently evaluated in the context of DLE by taking 4-mm punch biopsies from skin lesions [18]. RCM was able to identify interface changes and epidermal, dermal, and adnexal inflammatory cell infiltration in a high percentage of cases. This technique might be helpful in biopsy site selection; however, further studies are necessary.

TABLE 27.1   Histopathology and direct immunofluorescence in different subtypes of CLE[a]

CLE Subtype	1.3ACLE	SCLE	DLE	LET
Histopathology				
Orthohyperkeratosis	0	(+)	++	0
Interface dermatitis	+	++	++	0
Thickened basement membrane	+	++	+++	0
Lymphocytic infiltrate	+	++	+++	+++
Interstitial mucin deposition	+	++(+)	++	+++
Direct immunofluorescence				
Lesional skin	+++	++	+++	+
Nonlesional, sun-protected skin	++(+)	(+)	0	n.d.

[a]Modified from Kuhn et al. [164].
In this table, specific characteristics have been simplified and graded since percentages given for DIF and antibody serology vary greatly in the literature due to small patient groups.

Epidermal changes

In ACLE, histological findings such as vacuolar degeneration of the basal layer and necrotic keratinocytes are usually less prominent relative to the impressive clinical picture of patients with this subtype. In contrast, various epidermal changes, such as atrophy and compact orthohyperkeratosis, are key features in DLE lesions [17, 19, 20]. Moreover, epidermal changes can typically include adnexal structures (follicular plugging), and in the hypertrophic form of DLE, epidermal hyperkeratosis is even more increased along with parakeratosis and acanthosis. SCLE lesions generally have less hyperkeratosis and follicular plugging compared with DLE but show prominent vacuolar degeneration of the dermoepidermal junction and increased accumulation of inflammatory cells in the basal membrane zone, which in rare cases can proceed to subepidermal blister formation [21]. Basement membrane thickening, an additional pathologic finding of the dermoepidermal junction, is primarily seen in long-standing CLE, which is most apparent in DLE, less apparent in SCLE and often not present in ACLE lesions [17, 19]. In CHLE, which is characterized by painful purple plaques in cold-exposed acral areas (particularly on the toes and fingers), vacuolar degeneration of the dermoepidermal junction as well as epidermal hyperkeratosis and atrophy can be found [22, 23]. Furthermore, atrophy, follicular plugging, vacuolar degeneration of the dermoepidermal junction, and basement membrane thickening may also occur in LEP lesions, particularly when there is overlying DLE [24, 25]. Epidermal changes such as atrophy, follicular plugging, vacuolar degeneration of the dermoepidermal junction, and basement membrane thickening, however, are absent or show minimal and focal alterations in patients with LET [13, 26, 27].

Dermal changes

Another prominent feature of all CLE lesions is the presence of an inflammatory infiltrate composed mostly of lymphocytes, which can vary in its location depending on the disease subtype. Besides its striking perivascular and periadnexal localization, in most lesions the inflammatory infiltrate displays a patchy, sometimes band-like pattern. In ACLE lesions, the lymphocytic infiltrate is typically located in the upper dermis around the blood vessels and may be quite sparse, especially in early lesions [17]. It becomes more prominent in advanced lesions involving the dermoepidermal junction (interface dermatitis), sometimes with extravasation of erythrocytes and deposition of fibrinoid material around blood vessels and between collagen fibers. Some biopsy specimens from patients with SLE also display signs of leukocytoclastic vasculitis, e.g., nuclear dust, fibrinoid necrosis of the vessel wall, neutrophil infiltration, and extravasation of erythrocytes. In SCLE lesions the lymphocytes are largely confined to the upper dermis, which leads to a band-like infiltrate with interface dermatitis and erythrocyte extravasation; dermal fibrin deposition may also be present [17, 19, 20]. A periadnexal inflammatory infiltrate is a prominent feature in DLE [19, 28], and epidermal changes that are particularly found in the hypertrophic form of this subtype (follicular plugging due to hyperkeratosis) may represent a follicular reaction to proinflammatory and proliferative stimuli released by infiltrating lymphocytes (Figure 27.1). The inflammatory infiltrate of CHLE skin lesions is similarly situated in the upper dermis around the blood vesicles and occasionally around the hair follicles [22], while the infiltrates in LEP are mainly present in deeper layers of the skin. In LEP, there is also a lymphocytic infiltrate, sometimes together with eosinophils, of the subcutaneous fat that leads to panniculitis, fat necrosis and hyalinization of adipose lobules [24, 25]. The pattern of the panniculitis is lobular, sometimes paraseptal, and a subcutaneous panniculitis-like T-cell lymphoma is an important differential diagnosis that may be misdiagnosed as LEP [29]. In LET, lymphocytes situated perivascularly and periadnexally are present in the superficial and deep dermis; moreover, abundant mucin deposition between collagen bundles is characteristic of this entity compared to other subtypes of CLE (Figure 27.2) [13, 26, 27]. Nevertheless, mucin deposition alone is not a specific dermatopathologic finding; it may frequently be found in other inflammatory and noninflammatory skin conditions such as reticular erythematous mucinosis.

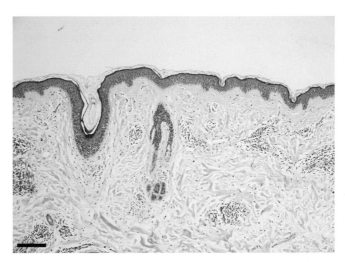

FIGURE 27.2 Histopathology of lupus erythematosus tumidus. Note the lymphocytic infiltrate in the superficial and deep dermis along with mucin deposition between collagen bundles. Scale bar = 200μm.

Direct Immunofluorescence

Direct immunofluorescence (DIF) can be used on skin biopsy specimens of patients with CLE to detect deposits of the immunoglobulins IgG, IgM and, in rare cases, IgA, as well as the complement component C3 at the dermoepidermal junction (Figure 27.3). It is most reliably performed on snap-frozen skin specimens, and the intensity of fluorescence in skin specimens depends on biopsy site, acuity of the lesion, and previous treatment. Sun-damaged skin, especially in the face, may give false-positive results, and those treated with topical corticosteroids and immunomodulators may yield false-negative results. Therefore, a positive DIF must be considered in the context of the whole clinical picture, including immoserological

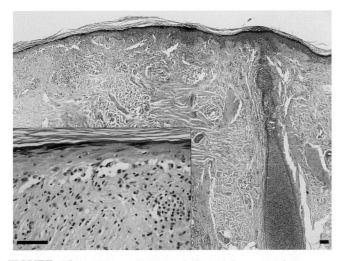

FIGURE 27.1 Histopathology of discoid lupus erythematosus. Note the vacuolar degeneration of the dermoepidermal junction, follicular plugging and periadnexal lymphocytic infiltrate. Scale bar = 100μm.

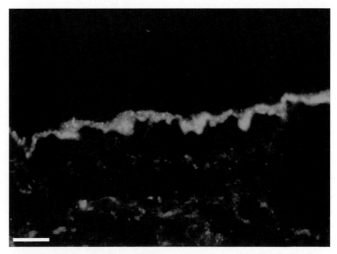

FIGURE 27.3 Direct immunofluorescence. Note the linear immunofluorescence pattern due to deposition of immunoglobulins along the dermoepidermal junction. Scale bar = 100μm.

data of the patient [12]. The strongest clinical association of a positive DIF in nonlesional, sun-protected skin is with SLE. Some groups use the term "lupus band test" (LBT) to refer to lesional and nonlesional DIF findings in SLE, whereas others reserve this designation for reference to immunofluorescence examination of only nonlesional skin [30, 31].

In some cases, DIF studies may be beneficial in discriminating between inflammatory skin disorders with similar histopathologic presentations to CLE and provide some information on the pathogenesis of the disease. Generally, most DIF studies have been undertaken in patients with SLE, DLE and SCLE, while uncommon forms of CLE (LEP and CHLE) have been investigated infrequently or have proved negative. The DIF is positive in 50–90% of nonlesional, sun-protected skin samples from patients with SLE, but only about 25% of the patients with SLE who have DLE lesions have positive nonlesional immunofluorescence [32]. Immune deposits can be found, however, in about 60–95% of lesional skin samples in DLE, where IgG3 and C3 are found most frequently and typically display a linear band-like or sometimes granular fluorescence pattern along the dermoepidermal junction. Generally, the DIF is negative in nonlesional skin of patients with DLE (without systemic organ involvement), although in some cases deposits of C3 and IgM have been detected. A similar pattern and composition of immunoglobulins and complement in the basement membrane zone can be found in lesional skin from patients with SCLE as compared to DLE. The DIF is positive in 60–100% of skin biopsy specimens taken from lesional skin of patients with SCLE and is approximately 25% positive in nonlesional skin [20]. For LET, the results of DIF are rather heterogeneous in the literature [26, 27]. In a study by our group in 2002 [27], biopsy specimens from primary skin lesions demonstrated immunoglobulin deposits (IgG, IgM) along the dermoepidermal junction in 24% of LET patients, but additional immunoglobulin classes (IgA) as well as complement components were not identified in any of the specimens from primary or UV-induced lesions. Moreover, Alexiades-Armenakas *et al.* [26] demonstrated that in five out of ten cases with LET, the DIF analysis was negative. Vieira et al. [33] performed DIF in 15 of 26 patients with LET, and 11 were negative. Two patients showed slight IgG deposition at the basal membrane: one had moderate deposition of IgG and slight deposition of C3, and the other had slight deposition of IgM, C3, and C1q. Choonhakarn et al. [34], however, reported a negative DIF in all 15 patients analyzed with LET.

The exact mechanism of immune deposition at the dermoepidermal junction in the skin of patients with CLE remains incompletely understood. A pathogenetic role for deposited immunoglobulins has been emphasized in one subset of patients with bullous SLE. These patients have circulating antibasement membrane zone antibodies (mostly IgG, less frequently IgA) directed against type VII collagen [35, 36]. Autoantibodies directed against several undefined proteins of 230, 200, 180, 130 and 97 kD from epidermal extracts and 75 kD from dermal extracts were identified using the salt-split skin technique [37]. It has been accepted for some time that the fluorescence intensity of a positive DIF in SLE correlates with disease activity [38] and the serum titer of antinuclear antibodies, which suggests a causal relationship [39]. Both native and single-stranded DNA antibodies exhibit an affinity for collagen in the basement membrane, quite possibly leading to *in vivo* fixation of such antibodies [21].

Inflammatory Cells and Molecules

Inflammatory cells, primarily lymphocytes, are consistently found in dermal infiltrates of CLE lesions as outlined earlier in the section titled "Histopathology." T helper cells have critical immune functions that are dependent on distinct cytokine profiles. Classically, T helper cells are defined as type I (Th1) or Ths, based on the expression profiles that comprise distinct cytokines, including IFNγ and IL4, respectively. A newly discovered subset of $CD4^+$ T cells are identified on the basis of their ability to produce IL-17A, IL-17F, IL-21, and IL-22 [40, 41]. Another cytokine, IL-23, a member of the IL-12 family, also plays a crucial role in the expansion and maintenance of committed Th17 cells [42, 43].

Large numbers of $CD4^+$ and $CD8^+$ lymphocytes are found in lesional skin of patients with CLE by immunophenotyping infiltrating dermal inflammatory cells [44]. The involvement of $CD4^+$ cells in cutaneous inflammation is supported by successful treatment of patients suffering from severe CLE with a chimeric CD4 monoclonal antibody, cM-T412 [45]. The concept of Th17 cells was conceived within the setting of experimental autoimmune diseases and recently, these cells are beginning to be implicated in the pathogenesis of SLE [46, 47]. For example, SLE patients have higher serum levels of IL-17 and IL-23 compared to healthy controls [48, 49]. In addition, the frequency of IL17-producing T cells is increased in peripheral blood of SLE patients [49, 50]. Accordingly, IL-17 production is increased in *in vitro* stimulated lymphocytes from SLE patients when compared to healthy controls and plasma IL-17 levels show a positive correlation with SLE disease activity. Recently, $IL-17A^+$ cells were identified in the kidneys of lupus-prone MRL/lpr mice with active nephritis, a finding that is in accordance with human data showing expression of IL-17A in the kidney of SLE patients [50, 51]. However, it is still unclear whether Th17 cells are involved in the pathogenesis of skin lesions of this disease.

Regulatory T cells

CD4[+]CD25[+] regulatory T cells (T$_{reg}$) play a critical role in suppressing responses of the immune system to self-antigens [52, 53]. Various groups have analyzed the number of T$_{reg}$ cells in the peripheral blood of patients with SLE and have reported a decrease in the proportion of circulating T$_{reg}$ cells isolated from patients with active disease [54–57]. In addition, Miyara *et al.* [58] suggested increased sensitivity of T$_{reg}$ cells to CD95L-mediated apoptosis, termed activation-induced cell death. These results were confirmed by the finding of a significant decrease in T$_{reg}$ cells in pediatric patients with SLE [59]. An inverse correlation was found, however, between the number of these cells and disease activity, as well as autoantibody level. Interestingly, significantly reduced levels of T$_{reg}$ cells have been correlated with autoantibody production and the onset of a lupus-like disease in different murine models [60]. Therefore, the pathogenic mechanisms of SLE might be tied to a defect in the homeostatic control of the T$_{reg}$ subpopulation, although conflicting data on the role of CD4[+]CD25[+] T$_{reg}$ cells in human autoimmune diseases have been reported in the literature [61]. Alternatively, CD4[+]CD25[-] conventional T cells in patients with this disease might have reduced sensitivity to suppression by T$_{reg}$ cells as reported for mice of the MRL/Mp strain [62, 63].

Recently, a decrease in T$_{reg}$ cells was observed at sites of inflammation in skin lesions of patients with different subtypes of CLE [9]. The number of T$_{reg}$ cells was not reduced in other inflammatory skin diseases, however, such as atopic dermatitis, psoriasis, and lichen planus. Thus, a reduction in the number of T$_{reg}$ cells in skin lesions of patients with CLE might be caused by specific factors of the disease. This reduction of T$_{reg}$ cells in the dermal infiltrate was independent of disease subtype; however, patients with CLE did not show a general T$_{reg}$ defect as supported by a normal frequency of circulating T$_{reg}$ cells, their capacity to suppress conventional T cell proliferation and normal sensitivity of T$_{reg}$ cells to CD95L-mediated apoptosis. These data suggest an organ-specific abnormality of T$_{reg}$ cells in CLE skin lesions rather than global dysfunction as reported for patients with systemic manifestations of the disease.

Dendritic cells

In addition to T cell subtypes, there is evidence to suggest the involvement of epidermal dendritic cells, such as Langerhans cells (LCs), in the pathogenesis of CLE. A gradual decrease in epidermal LCs during UV irradiation and reduced numbers of these cells were found in lesional epidermis of CLE patients [64]. Recently, the receptor activator for NF-κB (RANK), a molecule of the TNF superfamily expressed on LCs,

and its ligand, RANKL, were found to play an important role in the regulation of numerous DC functions such as enhancement of T-cell activation capacity [65], increased release of proinflammatory cytokines [66], and prolonged cell survival [67, 68]. Furthermore, RANKL is expressed on inflamed or activated keratinocytes and seems to rewire local LCs in order to regulate the quantity of peripheral CD4[+]CD25[+] T$_{reg}$ cells [69]. Recent data further suggest that RANKL might be important for the development of skin lesions in inflammatory diseases such as CLE [70]. On the other hand, CD36[+] dendritic macrophages were found to be increased in lesional skin of patients with of CLE (similar to UV-irradiated skin), suggesting a stimulatory role of the autoimmune response by activation of CD45RA[+] cells [71]. Plasmacytoid dendritic cells, which naturally produce IFN-α/β, have been found to accumulate in skin lesions of patients with both SLE and DLE, with predominant accumulation along the dermal–epidermal junction, around hair follicles, and perivascularly [72, 73]. The number of these cells in lesional skin is correlated with those cells expressing the IFN-α/β-inducible protein MxA [74]. Additionally, the number of plasmacytoid dendritic cells in lesional skin of CLE patients coincides with the L-selectin ligand peripheral lymph node addressin on dermal endothelial cells. Hence, plasmacytoid dendritic cells may contribute to the pathogenesis of skin lesions in CLE via their ability to stimulate lymphocyte extravasation and activation.

Adhesion molecules

Cellular adhesion molecules (CAM) are key mediators regulating the influx of T cells into the skin of patients with CLE. They are members of the immunoglobulin superfamily and include intercellular CAM-1 (ICAM-1), vascular adhesion molecule-1 (VCAM-1), and E-selectin. All are strongly induced by proinflammatory signals such as interleukin (IL)-1, TNF-α, or UV irradiation. Accordingly, increased *in situ* expression of these CAMs has been detected not only in various subtypes of CLE, but also in other inflammatory skin conditions such as polymorphous light eruption (PLE), scleroderma and lichen planus [75–79]. Efforts have been made to identify distinct patterns of these CAMs in lesional skin of SCLE, DLE, and LET patients in comparison with other inflammatory skin diseases; however, the overall picture is not clear. The most consistent finding appears to be overexpression of ICAM-1 detectable either throughout the epidermis, sometimes linearly, with more focally distributed overexpression in the basal layers of the epidermis in SCLE compared to DLE. VCAM-1 has been found to be overexpressed in endothelial cells of lesional skin in CLE patients and in nonlesional skin of SLE patients, underscoring the role of an activated endothelium in the latter disease subtype. The increased *in situ* expression of these CAMs may reflect the upregulation of these molecules by TNF-α (which is strongly induced by UVB exposure) and/or UVA/B. In addition to overexpression of ICAM-1 and VCAM-1 in skin lesions of patients with CLE, increased serum levels of the corresponding soluble forms of these CAMs have been detected [80]. Interestingly, increased serum VCAM-1 levels have been found to correlate with disease activity of SLE. Likewise, elevated levels of sSELE have been reported in patients with active widespread CLE lesions, suggesting the usefulness of both sCAMs as LE activity markers. While the precise functional roles of the increased serum levels of sCAM-1, sVCAM-1, and sE-selectin in the pathogenesis of skin damage of CLE remain to be defined, the pathogenetic role of ICAM-1 is seen by blockade of ICAM-1 in SLE-prone MRL/lpr mice. Intraperitoneal injection of an anti-ICAM-1 antibody prevents both neurological disease and the development of vasculitic skin lesions in the treated animals [81].

Photobiology

Abnormal reactivity to UV light is an important factor in the pathogenesis of CLE, and photosensitivity shows a strong association with all disease subtypes [82]. More than 100 years ago, it was first suspected that sunlight was involved in the induction of LE, partly because the disease seemed to accumulate in the spring and summer. In the following years, several articles in the literature documented that in all forms of LE, skin lesions were found predominantly on sun-exposed areas such as the face, V area of the neck and extensor surfaces of the arms. In addition, it was noted that exposure to sunlight could induce new skin eruptions, exacerbate existing lesions, cause progression of the disease to non-UV-exposed areas, and induce systemic activity [83].

Photosensitivity

In 1982, photosensitivity was listed as one of the ACR criteria for the classification of SLE, although it was poorly defined as "a result of an unusual reaction to sunlight by patient's history or physician's observation" [2]. This is an extremely broad definition that can be fulfilled by a variety of other diseases such as PLE, photoallergic contact dermatitis, and dermatomyositis. In addition, Albrecht and co-workers [4] recently criticized the malar rash, another ACR criterion for the classification of SLE, as often being indistinguishable from photosensitivity, making both criteria not independent. In the opinion of these authors, a control group is needed to develop new criteria, which should include not only patients with connective tissue diseases but also patients with photodermatoses such as PLE.

A detailed clinical history is important for the diagnosis and assessment of photosensitivity in patients with CLE, including several key components such as the morphology of the rash, duration, distribution, and the relationship to sun exposure and specific symptoms (e.g., pain, pruritus, burning, blistering, and swelling). A negative history of photosensitivity, however, does not necessarily exclude sensitivity to sunlight [82]. This might be due to the fact that the development of UV-induced skin lesions in patients with CLE is characterized by a latency period of up to several weeks. Thus, a relationship between sun exposure and exacerbation of CLE does not seem obvious to the patient, and it might be difficult for some patients to link sun exposure to their disease. In addition, the occurrence of photosensitivity varies among the different subtypes of CLE and often differs between various ethnic groups such as African Blacks [84]. Walchner and co-workers [85] observed that primarily patients younger than 40 years old report photosensitivity, suggesting that age at the onset of disease also plays a role.

Photoprovocation

Due to the clinical evidence demonstrating a clear relationship between sunlight exposure and skin manifestations of CLE, experimental light testing with different wavelengths has been developed to better define UV sensitivity in patients with a photosensitive form of the disease [82]. Results of earlier studies of experimental photoprovocation studies indicated that characteristic skin lesions of CLE could be induced by repeatedly delivering high doses of UVB to the same test site [86, 87]. In 1986, Lehmann and co-workers [88] developed a standardized protocol and demonstrated that the action spectrum of CLE also reaches into the long-wave UVA region. In the following years, a total of 128 patients were studied, and characteristic skin lesions clinically and histologically compatible with CLE were experimentally induced by UVA or UVB irradiation in 43% of patients [89]. A practical consequence of UVA sensitivity is that patients with CLE are not adequately protected by window glass or conventional sunscreens, most of which absorb UVA poorly. Moreover, high-intensity UVA sources in tanning salons might be dangerous for these patients [90, 91].

Phototesting can be carried out by irradiating defined test areas with single doses of 60–100 J/cm^2 UVA and/or 1.5 MED UVB daily for 3 consecutive days using a standardized protocol [92]. Interestingly, there are substantial differences in the clinical subtypes of CLE with regard to different UV wavelength. Patients with LET have been found to be the most photosensitive subtype of CLE, with phototesting revealing characteristic skin lesions in 72% of these patients [93]. In contrast, pathologic skin reactions were induced by UV irradiation in 63% of patients with SCLE, 60% of patients with ACLE, and 45% of patients with DLE. It remains unclear, however, why skin lesions cannot be reproduced under the same conditions several months after the initial phototesting and why UV irradiation does not show positive results in all patients tested. This gives indirect evidence for variant factors in the pathophysiology of CLE. Additionally, it is still unclear why UV-induced skin lesions in patients with this disease are characterized by a latency period of 8.0 ± 4.6 days (range, 1 day to 3 weeks). Therefore, this testing regimen is an optimal procedure to evaluate photosensitivity in CLE patients; the capacity of UVA and UVB irradiation to reproduce skin lesions in this disease also makes it an ideal model for several experimental approaches to allow the study of inflammatory and immunologic events during the course of lesion formation [94].

Accumulation of apoptotic keratinocytes after UV irradiation

As part of their normal program of differentiation, keratinocytes die by apoptosis [95−97]; however, the molecular mechanisms controlling this programmed cell death in the epidermis are complex and still incompletely understood. Basal keratinocytes have been found to be resistant to apoptosis induced by a variety of stimuli [98], but it has long been known that suprabasal keratinocytes are more sensitive to UV-induced apoptotic cell death, earning them the title "sunburn cells," coined by morphologists [99].

Interestingly, a significant increase in apoptotic keratinocytes has been found in genuine and UV-induced skin lesions of patients with CLE compared to healthy controls [100−102]. Furthermore, the number of apoptotic nuclei increased significantly in the majority of patients with CLE between day 1 and day 3 after a single UV exposure [8, 102]. This suggests that late apoptotic cells accumulate in the skin of a large subgroup of patients with different subtypes of the disease. Although detection of an increased number of apoptotic cells in the epidermis of CLE patients may reflect an increase in apoptosis, deficient clearance of apoptotic debris could also lead to an increased number of apoptotic keratinocytes.

A study in patients with SLE demonstrated that the clearance of apoptotic lymphocytes by macrophages is in fact impaired in some patients [103], and impaired uptake of apoptotic cells by macrophages was also noted in germinal centers of these patients [104]. The presence of specialized phagocytes, which are usually referred to as tingible body macrophages (TBMs), is a distinct feature of germinal centers. Under healthy conditions, TBMs remove apoptotic cells very efficiently in the early phase of apoptosis. In a subgroup of patients with SLE, however, apoptotic cells accumulate in the germinal centers of the lymph nodes, which may be due to impaired phagocytic activity or caused by the absence of TBMs [105]. The magnitude of a clearance defect in patients with CLE remains unclear, although a number of cellular signals and receptors for the phagocytosis of apoptotic debris have been identified. Furthermore, it is unknown whether impaired clearance is secondary to a defect in the recognition and binding of apoptotic particles or in macrophage phagocytosis.

Systemic autoimmunity has also been noted in mice deficient for molecules potentially involved in the clearance of apoptotic cells such as serum amyloid P (SAP), c-Mer, C4, IgM and C1q [106]. SAP is a member of a family of proteins termed pentraxins, which bind to apoptotic cells and then directly interact with phagocyte receptors or with C1q. The surface blebs of apoptotic keratinocytes bind C1q, an early component of the complement cascade [107]. This suggests that the clearance of these cells is aided by macrophages expressing a C1q cell surface receptor [108]. A potential role for C1q in the clearance of apoptotic debris in LE may be deduced from two observations: (1) mice with C1q deficiency develop an LE-like disease associated with an accumulation of apoptotic cells in the kidney [109] and (2) patients with complete congenital C1q deficiency frequently develop LE-like photosensitive eruptions and SLE at an early age [110]. The clearance of UV-induced apoptotic keratinocytes, however, was not found to be altered in C1q-deficient mice [111].

Nevertheless, it is still unclear whether the increased number of apoptotic cells in the skin of CLE patients leads to systemic consequences. There is evidence that the biochemical processes of apoptosis generate novel antigens that are uniquely targeted by autoantibodies. Casciola-Rosen and co-workers [112] showed that the translocation of autoantigens such as Ro/SSa and La/SSB to the cell surface of apoptotic blebs may allow circulating autoantibodies to gain access to these autoantigens, which are usually sequestered inside the cells. Furthermore, this group demonstrated that caspases activated during apoptosis cleave intracellular proteins into fragments that are bound by autoantibodies from patients with SLE [113]. In addition, proteins specifically phosphorylated during stress-induced apoptosis are targeted by autoantibodies from SLE patient sera [114, 115]. It has been noted that patients with skin disease have autoantibodies that preferentially recognize apoptosis-modified U1-70-kd RNP antigen compared to patients without cutaneous manifestations [116]. These data provide evidence that immune recognition of modified forms of self-antigens occurs in CLE and suggest that apoptosis-derived antigens may be involved in the pathogenesis of this disease.

UV-induction of chemokines and inducible nitric oxide synthase

It has recently been demonstrated that UVB irradiation of primary human keratinocytes in the presence of pro-inflammatory cytokines such as IL-1 and TNF-α or IFN-γ significantly enhances the expression of the inflammatory chemokines CCL5, CCL20, CCL22, and CXCL [117]. This is relevant in CLE since expression of CCL5 and CXCL8 has been reported to be highly upregulated in skin lesions of patients with this disease. Elevated levels of CCL27, a novel skin-specific chemokine known to recruit memory T cells into the skin, was also found in the dermis of CLE patients after phototesting. Furthermore, the CXCR3 ligands CXCL9, CXCL10, and CXCL11 have been identified as the most abundantly expressed genes in patients with this disease [118]. It has been reported that the CCR4 ligand TARC/CCL17 is strongly expressed in skin lesions and elevated in the serum of patients with CLE. The functional relevance of lymphocytic CCR4 expression was confirmed by TARC/CCL17-specific *in vitro* migration assays, suggesting that CCR4 and TARC/CCL17 plays a role in the pathophysiology of this disease.

Gene array studies have demonstrated that type I IFN-α/β plays a significant role in the pathogenesis of SLE [119]. Interestingly, strong expression of MxA, a protein specifically induced by type I IFNs, has been found in lesional skin of patients with CLE, and large numbers of infiltrating CXCR3$^+$ lymphocytes were closely correlated with lesional MxA expression [74]. Furthermore, natural IFN-α-producing cells, termed plasmacytoid dendritic cells, have been detected in CLE lesions [72, 73] and are also associated with the presence of MxA, as mentioned above [74]. This suggests that local IFN-α production by these cells promotes T-helper 1 (Th1)-biased inflammation. Although the pathophysiologic role of skin-infiltrating lymphocytes is undisputed, their recruitment and activation pathways in inflammatory skin diseases are not entirely clear.

UV exposure has also been shown to modulate local nitric oxide (NO) production in the skin via inducible nitric oxide synthase (iNOS) [120]. The NO liberated following UV irradiation plays a significant role in initiating melanogenesis, erythema, and immunosuppression. The evidence further suggests that it may be involved in protecting the keratinocytes against UV-induced apoptosis by increasing Bcl-2 expression and inhibiting UVA-induced overexpression of Bax protein in endothelial cells *in vitro* [121]. It has further been demonstrated that the presence of nitrite, not nitrate, during the irradiation of endothelial cells exerts a potent and concentration-dependent protection against UVA-induced apoptotic cell death [121]. An antiapoptotic role for NO in keratinocytes following exposure to UVB has been suggested [122]; however, when applied to normal, nonirradiated skin, NO induced the accumulation of CD4$^+$ and CD8$^+$ T cells, expression of ICAM-1 and VCAM-1, and accumulation of p53, followed by keratinocyte apoptosis [123]. Varied expression of this molecule may provide another link between dysregulated keratinocyte apoptosis and inflammation. Moreover, it has been demonstrated that iNOS is expressed in skin lesions of patients with CLE after UVA and UVB irradiation [124]. An iNOS-specific signal appeared at day 3 after UV exposure, in contrast to healthy controls in which iNOS expression was already present at day 1 after irradiation and absent by day 3.

Overall, the data suggest that the kinetics of iNOS induction and the time span of local iNOS expression possibly play an important role in the pathogenesis of CLE. Given the evidence of delayed and prolonged expression of iNOS in the skin of patients with CLE after UV exposure, it should be determined whether NO via chemical donors is a promising target for therapeutic intervention. NO production is also increased in patients with SLE, perhaps due to upregulated iNOS expression in activated endothelial cells and keratinocytes [125]. In addition, serum nitrite is correlated with measures of disease activity and titers of anti-dsDNA antibodies in these patients. Polymorphisms in the iNOS gene promoter, however, do not seem to be relevant to the pathogenesis of patients with SLE [126].

Immunogenetics

Based on initial observations in pairs of sisters and twins and subsequent case—control studies, a genetic basis for CLE has long been considered [127]. Carriers and patients with certain hereditary disorders such as X-linked and autosomal-recessive chronic granulomatosis disease and non-X-linked hyperimmunoglobulin M syndrome have an increased risk of developing CLE [128, 129], indicating that the genes affected in the above disorders could be involved in the pathogenesis of CLE. Furthermore, patients with SCLE and DLE exhibit a high prevalence of additional PLE, suggesting a common genetic background [130, 131]. The methodological approach to studying the genetic base of CLE has relied on association studies, family linkage analysis, and transmission equilibrium testing. All studies collectively indicate that the development of CLE is controlled by multiple genes.

MHC complex

The most important locus associated with genetic susceptibility to SCLE appears to be the MHC locus at 6p21.3, which includes a number of genes that control inflammatory and immune responses. Region I contains the class I human leukocyte antigens (HLA), including HLA-A, -B, and Cw, while region II contains the HLA II antigens (DP, DQ, and DR). Region III contains several complement genes, TNF and heat shock protein (HSP) 70. Accordingly, the HLA A1, DR3, B8, DQ2, DRw52, C4null haplotype has been identified to be highly associated with susceptibility to SCLE in Caucasians with circulating anti-Ro/SSA antibodies [130]. This haplotype is also linked with susceptibility to other autoimmune disorders such as myasthenia gravis, dermatitis herpetiformis Duhring, and insulin-dependent diabetes [132]. Furthermore, the haplotype HLA DQ1 and DQ2 is associated with the highest serum anti-Ro/SSA antibodies in patients with SCLE [133]. These data suggest that the MHC complex controls the tendency of an individual to mount an immune response against the Ro/SSA antigen being exposed on the surface of UV-irradiated keratinocytes. In contrast to SCLE, studies investigating HLA associations with DLE have had conflicting results, with some studies reporting no HLA association and others confirming the A1, B8, DR3 and B7, DR7 haplotypes.

Non-HLA genes of the MHC complex

Region III of the MHC locus contains genes for complement C2, C4A, C4B, and factor B. Inherited C2 and C4 deficiency is strongly associated with circulating anti-Ro/SSA antibodies and the development of SCLE [134–138]. Moreover, LEP is associated with partial deficiency of C4 [139]. Since complement factors are involved in macrophage activation and clearance of antibody—antigen complexes, these findings may indicate defective clearance of apoptotic cells or immune complexes containing antinuclear antibodies, including anti-Ro/SSA antibodies [140]. Alternatively, linkage disequilibrium with the real disease-predisposing locus may exist. In addition to C2 and C4, the proinflammatory cytokine TNF-α has been implicated in the pathogenesis of CLE. TNF-α is strongly induced in epidermal keratinocytes upon UV irradiation and stimulates expression of the Ro/SSA antigen in these cells [141]. The TNF polymorphism -308A has been associated with increased TNF production after UVB irradiation of epidermal keratinocytes and with the development of SCLE [142]. It has been shown that the above polymorphism confers susceptibility to SLE independent of the HLA-DR3 haplotype. Finally, a polymorphism of the locus encoding HSP 70 (HSPA1A, B and L) has been investigated in patients with LE [143, 144]. In some populations, the PstI site containing allele B was found to be significantly increased in SLE patients compared to healthy controls. Although the genetic role of HSP70 in CLE remains undefined, HSP70 expression is known to be induced in keratinocytes upon cellular stress such as UV irradiation. HSP immunostaining in lesional skin of patients with SLE and DLE is diffusely distributed in the whole epidermis, hair follicles, and sweat gland cells. Others have detected HSP70 immunoreactivity along the dermoepidermal junction and around papillary vessels in lesional skin of patients with SLE. Using double fluorescence labeling, the latter authors found that immune deposits of IgM, IgG, and C3 are co-localized with Hsp70, suggesting that Hsp70 could shuttle autoantigens to the dermoepidermal junction [145].

Candidate loci outside the MHC locus

Several loci outside the MHC complex have been found to confer susceptibility to CLE. The respective loci contain genes that encode for various components of the immune system (cytokines, cytokine receptors), oxidant defense system, and apoptosis machinery, with the latter involved in UV-mediated damage of epidermal keratinocytes. The 2q13 locus encoding the IL-1 cluster (IL-1α, -β, and IL-1 receptor antagonist, IL-RA) has been linked to UV photosensitivity and DLE [146]. Single nucleotide polymorphisms (SNPs) have also been found for the IL-10 promoter [147]. The identified SNPs gave reduced *in vitro* production of IL-10 by mononuclear cells [148]. Since IL-10 has potent immunosuppressive functions and is highly expressed by T$_{reg}$, representing one of the main effector molecules produced by such cells, failure to produce this cytokine in sufficient amounts may facilitate the development of an abnormal immune response. Genes encoding the T-cell receptor C (TCR) are also associated with susceptibility to SLE. Restriction fragment length polymorphisms for the TCR genes C$_{\beta}$1 and C$_{\beta}$2 were found in 76% of patients with SLE and circulating anti-Ro/SSA antibodies, while they were found in only 41% of patients without these antibodies. Linkage analysis using genome-wide scans further revealed a susceptibility locus at 1q23 encoding the Fcγ receptor II gene, which mediates binding of immunoglobulins to lymphocytes to generate antibody-dependent cellular cytotoxicity [149]. Like the loci encoding the IL-1 cluster and the cytokine IL-10, it is possible that changes in both of the latter genes predispose to a break in the immune tolerance to the Ro/SSA antigen. Finally, polymorphisms in two genes involved in UV light-induced genotoxicity and apoptosis, glutathione S-transferase (GST) and Fas (CD95), have been found to be associated with SLE [150, 151]. GST is critically involved in detoxifying intracellular reactive oxidative species induced by both UVA and UVB. Fas is a member of the TNF receptor family and is activated both by binding to TNF-α and by ligand-independent receptor aggregation following UVB exposure. A GSTM1 null polymorphism is highly associated with SLE and presence of anti-Ro/SSA antibodies, suggesting that deregulation of oxidative stress is important in the break of tolerance to Ro/SSA antigens. The SNP identified at the -670 nucleotide position of Fas was associated with photosensitive SLE when homozygous for Mval*2, suggesting that correct Fas signaling is involved in controlling the extent of cutaneous UV sensitivity and possibly the tendency of cells to expose antigens such as Ro/SSA to the immune system. Supporting an important role of Fas is the *lpr* mouse, which carries a point mutation in *Fas* that subsequently leads to defective signaling and increased susceptibility to autoimmunity, including LE-like skin changes [152]. Recently, there has been evidence that estrogen receptor α gene polymorphisms could play a role in susceptibility to skin involvement in patients with SLE [153]. Estrogens are considered to be important environmental (and physiologic) LE triggers in patients with SLE and SCLE since they can activate mature peripheral B cells to produce immunoglobulins, including anti-dsDNA antibodies. Pregnancy, oral contraceptives containing estrogens, and hormonal replacement therapy are also well known triggers of disease activity in patients with SLE and SCLE. The Pvull C and the Xbal G alleles of the estrogen receptor α have been associated with a milder form skin involvement in patients with SLE [153].

Trex1 mutation in familial CHLE

Recently, the first monogenic form of CLE, termed familial chilblain lupus (familial CHLE), was described, and the genetic locus was mapped to chromosome 3p [154]. Heterozygous mutations in TREX1 encoding an intracellular 3′-5′ DNA exonuclease [155] have been identified by several groups as the cause of familial CHLE [156, 157], and it has been shown that loss of exonuclease activity leads to impairment of granzyme A-mediated cell death, a caspase-independent form apoptosis initiated by cytotoxic T lymphocytes and natural killer cells [157, 158]. Biallelic mutations in TREX1 cause Aicardi-Goutières syndrome (AGS), a rare infantile encephalopathy resembling congenital viral infection characterized by increased IFN-α in the cerebrospinal fluid and chilblain-like lesions in some affected children [159]. AGS is also caused by mutations in three nonallelic genes encoding the subunits of the RNASEH2 complex [160], suggesting that defects in nucleic acid metabolism could trigger an innate immune response.

The findings in monogenic familial CHLE were extended to SLE, and TREX1 mutations in 2.6% of patients with SLE were identified [157]. Lee-Kirsch et al. proposed that in addition to its role in granzyme A-mediated apoptosis, TREX1 may also function as a cytosolic DNA sensor and participate in TLR9-independent type I-IFN responses [161]. Recently, it was shown that TREX1 deficiency leads to constitutive activation of the ATM-dependent DNA-damage checkpoint, which results in the accumulation of cytosolic ssDNA [162]. Thus, defects in TREX1-related pathways may underlie photosensitivity in response to UV-induced DNA damage. Furthermore, the findings suggest that rare alleles may significantly contribute to the overall genetic risk for autoimmunity and highlight the validity of deep resequencing of candidate genes in

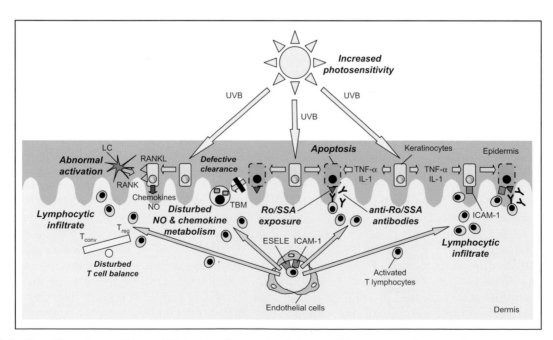

FIGURE 27.4   Simplified scheme of the multifactorial pathogenesis of photosensitive cutaneous lupus erythematosus.

the genetic dissection of complex diseases. In addition, there is a strong history of the analysis of complex traits in the rat. The discovery of SNPs was contributed to the genome sequence, and the first genome-wide study integrating transcriptional profiling and linkage analysis for the identification of disease genes was conducted in the rat [163].

CONCLUSION

Although the exact pathogenesis of specific skin lesions in patients with CLE is still not completely understood, there is evidence that genetics can confer susceptibility to LE-specific skin disease (Figure 27.4). Consequently, genes located in previously identified susceptibility loci for CLE have been implicated as integral players in inflammatory responses. Thus, certain haplotypes may lead to the predisposition to an abnormal immune response as seen in CLE. Furthermore, UV-mediated apoptosis and exposure of autoantigens appear to be central pathogenetic steps. Finally, immune cells, including T cells and dendritic cells, are important executers of the abnormal immune response that leads to skin damage in CLE.

References

[1] A.S. Cohen, W.E. Reynolds, E.C. Franklin, J.P. Kulka, M.W. Ropes, L.E. Shulman, et al., Preliminary criteria for the classification of systemic lupus erythematosus, Bull Rheum Dis 21 (1971) 643—648.

[2] E.M. Tan, A.S. Cohen, J.F. Fries, A.T. Masi, D.J. McShane, N.F. Rothfield, et al., The 1982 revised criteria for the classification of systemic lupus erythematosus, Arthritis Rheum 25 (1982) 1271—1277.

[3] M.C. Hochberg, Updating the American College of Rheumatology revised criteria for the classification of systemic lupus erythematosus, Arthritis Rheum 40 (1997) 1725.

[4] J. Albrecht, J.A. Berlin, I.M. Braverman, J.P. Callen, M.K. Connolly, M.I. Costner, et al., Dermatology position paper on the revision of the 1982 ACR criteria for systemic lupus erythematosus, Lupus 13 (2004) 839—849.

[5] A. Kuhn, E., C., B. Aberer, I. Foeldvari, M. Haust, C. Iking-Konert, P. Kind, et al., Leitlinien Kutaner Lupus Erythematodes (Entwicklungsstufe 1), in: H.C. Korting, R. Callies, M. Reusch, M. Schlaeger, W. Sterry (Eds.), Dermatologische Qualitätssicherung: Leitlinien und Empfehlungen, ABW Wissenschaftsverlag GmbH, Berlin, 2009, pp. 214—257.

[6] R.D. Sontheimer, Subacute cutaneous lupus erythematosus: 25-year evolution of a prototypic subset (subphenotype) of lupus erythematosus defined by characteristic cutaneous, pathological, immunological, and genetic findings, Autoimmun Rev 4 (2005) 253—263.

[7] V.P. Werth, W. Zhang, K. Dortzbach, K. Sullivan, Association of a promoter polymorphism of tumor necrosis factor-alpha with subacute cutaneous lupus erythematosus and distinct photoregulation of transcription, J Invest Dermatol 115 (2000) 726—730.

[8] A. Kuhn, M. Herrmann, S. Kleber, M. Beckmann-Welle, K. Fehsel, A. Martin-Villalba, et al., Accumulation of apoptotic cells in the epidermis of patients with cutaneous lupus erythematosus after ultraviolet irradiation, Arthritis Rheum 54 (2006) 939—950.

[9] B. Franz, B. Fritzsching, A. Riehl, N. Oberle, C.D. Klemke, J. Sykora, et al., Low number of regulatory T cells in skin lesions of patients with cutaneous lupus erythematosus, Arthritis Rheum 56 (2007) 1910—1920.

[10] J. Wenzel, S. Zahn, T. Bieber, T. Tuting, Type I interferon-associated cytotoxic inflammation in cutaneous lupus erythematosus, Arch Dermatol Res 301 (2009) 83—86.

[11] J.N. Gilliam, The cutaneous signs of lupus erythematosus, Cont Educ Fam Phys. 6 (1977) 34—70.

[12] M.I. Costner, R.D. Sontheimer, T.T. Provost, Lupus erythematosus, in: R.D. Sontheimer, T.T. Provost (Eds.), Cutaneous Manifestations of Rheumatic Diseases, Williams & Wilkins, Philadelphia, 2003, pp. 15—64.

[13] V. Schmitt, A.M. Meuth, S. Amler, E. Kuehn, M. Haust, G. Messer, et al., Lupus erythematosus tumidus is a separate subtype of cutaneous lupus erythematosus, Br J Dermatol 162 (2010) 64—73.

[14] A. Kuhn, D. Bein, G. Bonsmann, The 100th anniversary of lupus erythematosus tumidus, Autoimmun Rev. 8 (2009) 441—448.

[15] A. Kuhn, T. Ruzicka, Classification of cutaneous lupus erythematosus, in: A. Kuhn, P. Lehmann, T. Ruzicka (Eds.), Cutaneous Lupus Erythematosus, Springer, Heidelberg, 2004, pp. 53—58.

[16] A.B. Ackermann, Lupus erythematosus, in: Histologic Diagnosis of Inflammatory Skin Diseases, Williams & Wilkins, Baltimore, 1997, pp. 525—546.

[17] M.S. Jerdan, A.F. Hood, G.W. Moore, J.P. Callen, Histopathologic comparison of the subsets of lupus erythematosus, Arch Dermatol 126 (1990) 52—55.

[18] M. Ardigo, I. Maliszewski, C. Cota, A. Scope, G. Sacerdoti, S. Gonzalez, et al., Preliminary evaluation of in vivo reflectance confocal microscopy features of Discoid lupus erythematosus, Br J Dermatol 156 (2007) 1196—1203.

[19] J.L. Bangert, R.G. Freemann, R.D. Sontheimer, J.N. Gilliam, Subacute cutaneous lupus erythematosus and discoid lupus erythematosus. Comparative histopathologic findings, Arch Dermatol 120 (1984) 332—337.

[20] R.D. Sontheimer, J.R. Thomas, J.N. Gilliam, Subacute cutaneous lupus erythematosus: a cutaneous marker for a distinct lupus erythematosus subset, Arch Dermatol 115 (1979) 1409—1415.

[21] S. Vassileva, Bullous systemic lupus erythematosus, Clin Dermatol 22 (2004) 129—138.

[22] M.S. Doutre, C. Beylot, J. Beylot, E. Pompougnac, P. Royer, Chilblain lupus erythematosus: report of 15 cases, Dermatology 184 (1992) 26—28.

[23] L. Pock, P. Petrovska, R. Becvar, V. Mandys, J. Hercogova, Verrucous form of chilblain lupus erythematosus, J Eur Acad Dermatol Venereol 15 (2001) 448—451.

[24] C. Massone, K. Kodama, W. Salmhofer, R. Abe, H. Shimizu, A. Parodi, et al., Lupus erythematosus panniculitis (lupus profundus): clinical, histopathological, and molecular analysis of nine cases, J Cutan Pathol 32 (2005) 396—404.

[25] N.P. Sanchez, M.S. Peters, R.K. Winkelmann, The histopathology of lupus eyrthematosus panniculitis, J Am Acad Dermatol 122 (1981) 576—582.

[26] M.R. Alexiades-Armenakas, M. Baldassano, B. Bince, V. Werth, J.C. Bystryn, H. Kamino, et al., Tumid lupus erythematosus:

criteria for classification with immunohistochemical analysis, Arthritis Rheum 15 (2003) 494–500.

[27] A. Kuhn, M. Sonntag, T. Ruzicka, P. Lehmann, M. Megahed, Histopathologic findings in lupus erythematosus tumidus: review of 80 patients, J Am Acad Dermatol 48 (2003) 901–908.

[28] C.L. Hexsel, S.D. Bangert, A.A. Hebert, H.W. Lim, Current sunscreen issues: 2007 Food and Drug Administration sunscreen labelling recommendations and combination sunscreen/insect repellent products, J Am Acad Dermatol 59 (2008) 316–323.

[29] E.G. Gonzalez, E. Selvi, S. Lorenzini, R. Maggio, S. Mannucci, M. Galeazzi, et al., Subcutaneous panniculitis-like T-cell lymphoma misdiagnosed as lupus erythematosus panniculitis, Clin Rheumatol 26 (2007) 244–246.

[30] C. Cardinali, M. Caproni, P. Fabbri, The utility of the lupus band test on sun-protected non-lesional skin for the diagnosis of systemic lupus erythematosus, Clin Exp Rheumatol 17 (1999) 427–432.

[31] C. Cardinali, M. Caproni, P. Fabbri, The composition of the lupus band test (LBT) on the sun-protected non-lesional (SPNL) skin in patients with cutaneous lupus erythematosus (CLE), Lupus 8 (1999) 755–760.

[32] S.D. Prystowsky, J.H. Herndon, J.N. Gilliam, Chronic cutaneous lupus erythematosus (DLE): a clinical and laboratory investigation of 80 patients, Medicine 55 (1975) 183.

[33] V. Vieira, J. Del Pozo, M.T. Yebra-Pimentel, W. Martinez, E. Fonseca, Lupus erythematosus tumidus: a series of 26 cases, Int J Dermatol 45 (2006) 512–517.

[34] C. Choonhakarn, A. Poonsriaram, J. Chaivoramukul, Lupus erythematosus tumidus, Int J Dermatol 43 (2004) 815–818.

[35] W.R. Gammon, R.A. Briggaman, Bullous SLE: a phenotypically distinctive but immunologically heterogeneous bullous disorder, J Invest Dermatol 100 (1993) 28S–34S.

[36] Y. Nitta, C. Kawamura, T. Hashimoto, Vesiculobullous systemic lupus erythematosus: a case with circulating IgG and IgA autoantibodies to type VII collagen, J Am Acad Dermatol 47 (2002) S283–286.

[37] S. Yang, Y. Gao, Y. Song, J. Liu, C. Yang, Z. Wang, et al., The study of the participation of basement membrane zone antibodies in the formation of the lupus band in systemic lupus erythematosus, Int J Dermatol 43 (2004) 420–427.

[38] P. Halberg, S. Ullman, F. Jorgensen, The lupus band test as a measure of disease activity in systemic lupus erythematosus, Arch Dermatol 118 (1982) 572–576.

[39] R.D. Sontheimer, J.N. Gilliam, A reappraisal of the relationship between subepidermal immunoglobulin deposits and DNA antibodies in systemic lupus erythematosus: a study using the Crithidia luciliae immunofluorescence anti-DNA assay, J Invest Dermatol 72 (1979) 29–32.

[40] L.E. Harrington, R.D. Hatton, P.R. Mangan, H. Turner, T.L. Murphy, K.M. Murphy, et al., Interleukin 17-producing CD4+ effector T cells develop via a lineage distinct from the T helper type 1 and 2 lineages, Nat Immunol 6 (2005) 1123–1132.

[41] H. Park, Z. Li, X.O. Yang, S.H. Chang, R. Nurieva, Y.H. Wang, et al., A distinct lineage of CD4 T cells regulates tissue inflammation by producing interleukin 17, Nat Immunol 6 (2005) 1133–1141.

[42] L. Zhou, Ivanov II, R. Spolski, R. Min, K. Shenderov, T. Egawa, et al., IL-6 programs T(H)-17 cell differentiation by promoting sequential engagement of the IL-21 and IL-23 pathways, Nat Immunol 8 (2007) 967–974.

[43] G.L. Stritesky, N. Yeh, M.H. Kaplan, IL-23 promotes maintenance but not commitment to the Th17 lineage, J Immunol 181 (2008) 5948–5955.

[44] T. Hasan, E. Stephansson, A. Ranki, Distribution of naive and memory T-cells in photoprovoked and spontaneous skin lesions of discoid lupus erythematosus and polymorphous light eruption, Acta Derm Venereol 79 (1999) 437–442.

[45] J.C. Prinz, M. Meurer, C. Reiter, E.P. Rieber, G. Plewig, G. Riethmuller, Treatment of severe cutaneous lupus erythematosus with a chimeric CD4 monoclonal antibody, cM-T412, J Am Acad Dermatol 34 (1996) 244–252.

[46] A.B. Pernis, Th17 cells in rheumatoid arthritis and systemic lupus erythematosus, J Intern Med 265 (2009) 644–652.

[47] A. Nalbandian, J.C. Crispin, G.C. Tsokos, Interleukin-17 and systemic lupus erythematosus: current concepts, Clin Exp Immunol 157 (2009) 209–215.

[48] C.K. Wong, C.Y. Ho, E.K. Li, C.W. Lam, Elevation of proinflammatory cytokine (IL-18, IL-17, IL-12) and Th2 cytokine (IL-4) concentrations in patients with systemic lupus erythematosus, Lupus 9 (2000) 589–593.

[49] C.K. Wong, L.C. Lit, L.S. Tam, E.K. Li, P.T. Wong, C.W. Lam, Hyperproduction of IL-23 and IL-17 in patients with systemic lupus erythematosus: implications for Th17-mediated inflammation in auto-immunity, Clin Immunol 127 (2008) 385–393.

[50] J.C. Crispin, M. Oukka, G. Bayliss, R.A. Cohen, C.A. Van Beek, I.E. Stillman, et al., Expanded double negative T cells in patients with systemic lupus erythematosus produce IL-17 and infiltrate the kidneys, J Immunol 181 (2008) 8761–8766.

[51] Z. Zhang, V.C. Kyttaris, G.C. Tsokos, The role of IL-23/IL-17 axis in lupus nephritis, J Immunol 183 (2009) 3160–3169.

[52] E.M. Shevach, CD4+ CD25+ suppressor T cells: more questions than answers, Nat Rev Immunol 2 (2002) 389–400.

[53] S. Sakaguchi, Naturally arising Foxp3-expressing CD25+CD4+ regulatory T cells in immunological tolerance to self and nonself, Nat Immunol 6 (2005) 345–352.

[54] J.C. Crispin, A. Martinez, J. Alcocer-Varela, Quantification of regulatory T cells in patients with systemic lupus erythematosus, J Autoimmun 21 (2003) 273–276.

[55] S. Mellor-Pita, M.J. Citores, R. Castejon, P. Tutor-Ureta, M. Yebra-Bango, J.L. Andreu, et al., Decrease of regulatory T cells in patients with systemic lupus erythematosus, Ann Rheum Dis 65 (2006) 553–554.

[56] E.Y. Lyssuk, A.V. Torgashina, S.K. Soloviev, E.L. Nassonov, S.N. Bykovskaia, Reduced number and function of CD4+ CD25highFoxP3+ regulatory T cells in patients with systemic lupus erythematosus, Adv Exp Med Biol 601 (2007) 113–119.

[57] X. Valencia, C. Yarboro, G. Illei, P.E. Lipsky, Deficient CD4+CD25high T regulatory cell function in patients with active systemic lupus erythematosus, J Immunol 178 (2007) 2579–2588.

[58] M. Miyara, Z. Amoura, C. Parizot, C. Badoual, K. Dorgham, S. Trad, et al., Global natural regulatory T cell depletion in active systemic lupus erythematosus, J Immunol 175 (2005) 8392–8400.

[59] J.H. Lee, L.C. Wang, Y.T. Lin, Y.H. Yang, D.T. Lin, B.L. Chiang, Inverse correlation between CD4+ regulatory T-cell population and autoantibody levels in paediatric patients with systemic lupus erythematosus, Immunology 117 (2006) 280–286.

[60] H.Y. Wu, N.A. Staines, A deficiency of CD4+CD25+ T cells permits the development of spontaneous lupus-like disease in mice, and can be reversed by induction of mucosal tolerance to histone peptide autoantigen, Lupus 13 (2004) 192–200.

[61] A. Kuhn, S. Beissert, P.H. Krammer, CD4(+)CD25 (+) regulatory T cells in human lupus erythematosus, Arch Dermatol Res. 301 (2009) 71–81.

[62] C.R. Monk, M. Spachidou, F. Rovis, E. Leung, M. Botto, R.I. Lechler, et al., MRL/Mp CD4+, CD25- T cells show reduced sensitivity to suppression by CD4+, CD25+ regulatory T cells in vitro: a novel defect of T cell regulation in systemic lupus erythematosus, Arthritis Rheum 52 (2005) 1180–1184.

[63] R.K. Venigalla, T. Tretter, S. Krienke, R. Max, V. Eckstein, N. Blank, et al., Reduced CD4+, CD25- T cell sensitivity to the suppressive function of CD4+, CD25high, CD127 -/low regulatory T cells in patients with active systemic lupus erythematosus, Arthritis Rheum 58 (2008) 2120−2130.

[64] A.S. Janssens, E.E. Lashley, C.J. Out-Luiting, R. Willemze, S. Pavel, F.R. de Gruijl, UVB-induced leucocyte trafficking in the epidermis of photosensitive lupus erythematosus patients: normal depletion of Langerhans cells, Exp Dermatol 14 (2005) 138−142.

[65] D.M. Anderson, E. Maraskovsky, W.L. Billingsley, W.C. Dougall, M.E. Tometsko, E.R. Roux, et al., A homologue of the TNF receptor and its ligand enhance T-cell growth and dendritic-cell function, Nature 390 (1997) 175−179.

[66] R. Josien, B.R. Wong, H.L. Li, R.M. Steinman, Y. Choi, TRANCE, a TNF family member, is differentially expressed on T cell subsets and induces cytokine production in dendritic cells, J Immunol 162 (1999) 2562−2568.

[67] B.R. Wong, R. Josien, S.Y. Lee, B. Sauter, H.L. Li, R.M. Steinman, et al., TRANCE (tumor necrosis factor [TNF]-related activation-induced cytokine), a new TNF family member predominantly expressed in T cells, is a dendritic cell-specific survival factor, J Exp Med 186 (1997) 2075−2080.

[68] I. Cremer, M.C. Dieu-Nosjean, S. Marechal, C. Dezutter-Dambuyant, S. Goddard, D. Adams, et al., Long-lived immature dendritic cells mediated by TRANCE-RANK interaction, Blood 100 (2002) 3646−3655.

[69] K. Loser, A. Mehling, S. Loeser, J. Apelt, A. Kuhn, S. Grabbe, et al., Epidermal RANKL controls regulatory T-cell numbers via activation of dendritic cells, Nat Med 12 (2006) 1372−1379.

[70] F. Toberer, J. Sykora, D. Göttel, V. Schmitt, K. Loser, S. Beissert, et al., Differential expression of RANKL in infammatory skin diseases, J Invest Derm 129 (2009). Abstract ESDR.

[71] B.S. Andrews, A. Schenk, R. Barr, G. Friou, G. Mirick, P. Ross, Immunopathology of cutaneous human lupus erythematosus defined by murine monoclonal antibodies, J Am Acad Dermatol 15 (1986) 474−481.

[72] L. Farkas, K. Beiske, F. Lund-Johansen, P. Brandtzaeg, F.L. Jahnsen, Plasmacytoid dendritic cells (natural interferon-alpha/beta-producing cells) accumulate in cutaneous lupus erythematosus lesions, Am J Pathol 159 (2001) 237−243.

[73] A. Wollenberg, M. Wagner, S. Gunther, A. Towarowski, E. Tuma, M. Moderer, et al., Plasmacytoid dendritic cells: a new cutaneous dendritic cell subset with distinct role in inflammatory skin diseases, J Invest Dermatol 119 (2002) 1096−1102.

[74] J. Wenzel, E. Worenkamper, S. Freutel, S. Henze, O. Haller, T. Bieber, et al., Enhanced type I interferon signalling promotes Th1-biased inflammation in cutaneous lupus erythematosus, J Pathol 205 (2005) 435−442.

[75] F. Nyberg, T. Hasan, C. Skoglund, E. Stephansson, Early events in ultraviolet light-induced skin lesions in lupus erythematosus: expression patterns of adhesion molecules ICAM-1, VCAM-1 and E-selectin, Acta Derm Venereol 79 (1999) 431−436.

[76] E. Stephansson, A.M. Ros, Expression of intercellular adhesion molecule-1 (ICAM-1) and OKM5 in UVA- and UVB-induced lesions in patients with lupus erythematosus and poly-morphous light eruption, Arch Dermatol Res 285 (1993) 328−333.

[77] S.M. Jones, C.M. Mathew, J. Dixey, C.R. Lovell, N.J. McHugh, VCAM-1 expression on endothelium in lesions from cutaneous lupus erythematosus is increased compared with systemic and localized scleroderma, Br J Dermatol 135 (1996) 678−686.

[78] H.M. Belmont, J. Buyon, R. Giorno, S. Abramson, Up-regulation of endothelial cell adhesion molecules characterizes disease activity in systemic lupus erythematosus. The Shwartzman phenomenon revisited, Arthritis Rheum 37 (1994) 376−383.

[79] A. Kuhn, M. Sonntag, P. Lehmann, M. Megahed, D. Vestweber, T. Ruzicka, Characterization of the inflammatory infiltrate and expression of endothelial cell adhesion molecules in lupus erythematosus tumidus, Arch Dermatol Res 294 (2002) 6−13.

[80] F. Nyberg, F. Acevedo, E. Stephansson, Different patterns of soluble adhesion molecules in systemic and cutaneous lupus erythematosus, Exp Dermatol 6 (1997) 230−235.

[81] R.L. Brey, B. Sakic, H. Szechtman, J.A. Denburg, Animal models for nervous system disease in systemic lupus erythematosus, Ann N Y Acad Sci 823 (1997) 97−106.

[82] A. Kuhn, S. Beissert, Photosensitivity in lupus erythematosus, Autoimmunity 38 (2005) 519−529.

[83] W.A. Pusey, Attacks of lupus erythematosus following exposure to sunlight or other weather factors, Arch Dermatol Syph 33 (1915) 388.

[84] P.G. Sutej, A.J. Gear, R.C. Morrison, M. Tikly, M. de Beer, L. Dos Santos, et al., Photosensitivity and anti-Ro (SS-A) antibodies in black patients with systemic lupus erythematosus (SLE), Br J Dermatol 28 (1989) 321−324.

[85] M. Walchner, G. Messer, P. Kind, Phototesting and photoprotection in LE, Lupus 6 (1997) 167−174.

[86] D.J. Cripps, J. Rankin, Action spectra of lupus erythematosus and experimental immunofluorescence, Arch Dermatol 107 (1973) 563−567.

[87] R.G. Freeman, J.M. Knox, D.W. Owens, Cutaneous lesions of lupus erythematosus induced by monochromatic light, Arch Dermatol 100 (1969) 677−682.

[88] P. Lehmann, E. Hölzle, R. von Kries, G. Plewig, Lichtdiagnostische Verfahren bei Patienten mit Verdacht auf Photodermatosen, Zentralblatt Haut- und Geschlechtskrankheiten 152 (1986) 667−682.

[89] P. Lehmann, E. Hölzle, P. Kind, G. Goerz, G. Plewig, Experimental reproduction of skin lesions in lupus erythematosus by UVA and UVB radiation, J Am Acad Dermatol 22 (1990) 181−187.

[90] R.S. Stern, W. Docken, An exarcerbation of SLE after vistiting a tanning salon, JAMA 255 (1986) 3120.

[91] H. Tronnier, H. Petri, P. Pierchalla, UV-provozierte bullöse Hautveränderungen bei systemischem Lupus erytematodes, Z Hautkr 154 (1988) 616−617.

[92] A. Kuhn, M. Sonntag, D. Richter-Hintz, C. Oslislo, M. Megahed, T. Ruzicka, et al., Phototesting in lupus erythematosus: a 15-year experience, J Am Acad Dermatol 45 (2001) 86−95.

[93] A. Kuhn, M. Sonntag, D. Richter-Hintz, C. Oslislo, M. Megahed, T. Ruzicka, et al., Phototesting in lupus erythematosus tumidus−review of 60 patients, Photochem Photobiol 73 (2001) 532−536.

[94] A. Kuhn, M. Bijl, Pathogenesis of cutaneous lupus erythematosus, Lupus 17 (2008) 389−393.

[95] D. Weedon, G. Strutton, Apoptosis as the mechanism of the involution of hair follicles in catagen transformation, Acta Derm Venereol 61 (1981) 335−339.

[96] P.E. Budtz, I. Spies, Epidermal homeostasis - apoptosis and cell emigration as mechanisms of controlled cell deletion in the epidermis of the toad, Bufo bufo, Cell Tissue Res 256 (1989) 475−486.

[97] C.A. McCall, J.J. Cohen, Programmed cell death in terminally differentiating keratinocytes: role of endogenous endonuclease, J Invest Dermatol 97 (1991) 111−114.

[98] D.A. Norris, K. Whang, K. David-Bajar, S.D. Bennion, The influence of ultraviolet light on immunological cytotoxicity in the skin, Photochem Photobiol 65 (1997) 636−646.

[99] F. Daniels Jr., D. Brophy, W.C. Lobitz Jr., Histochemical responses of human skin following ultraviolet irradiation, J Invest Dermatol 37 (1961) 351–357.

[100] B. Baima, M. Sticherling, Apoptosis in different cutaneous manifestations of lupus erythematosus, Br J Dermatol 144 (2001) 958–966.

[101] J.H. Chung, O.S. Kwon, H.C. Eun, J.I. Youn, Y.W. Song, J.G. Kim, et al., Apoptosis in the pathogenesis of cutaneous lupus erythematosus, Am J Dermatopathol 20 (1998) 233–241.

[102] J.L. Pablos, B. Santiago, M. Galindo, P.E. Carreira, C. Ballestin, J.J. Gomez-Reino, Keratinocyte apoptosis and p53 expression in cutaneous lupus and dermatomyositis, J Pathol 188 (1999) 63–68.

[103] M. Herrmann, R.E. Voll, O.M. Zoller, M. Hagenhofer, B.B. Ponner, J.R. Kalden, Impaired phagocytosis of apoptotic cell material by monocyte-derived macrophages from patients with systemic lupus erythematosus, Arthritis Rheum 41 (1998) 1241–1250.

[104] I. Baumann, W. Kolowos, R.E. Voll, B. Manger, U. Gaipl, W.L. Neuhuber, et al., Impaired uptake of apoptotic cells into tingible body macrophages in germinal centers of patients with systemic lupus erythematosus, Arthritis Rheum 46 (2002) 191–201.

[105] U.S. Gaipl, A. Kuhn, A. Sheriff, L.E. Munoz, S. Franz, R.E. Voll, et al., Clearance of apoptotic cells in human SLE, Curr Dir Autoimmun 9 (2006) 173–187.

[106] A. Roos, W. Xu, G. Castellano, A.J. Nauta, P. Garred, M.R. Daha, et al., Mini-review: A pivotal role for innate immunity in the clearance of apoptotic cells, Eur J Immunol 34 (2004) 921–929.

[107] L.C. Korb, J.M. Ahearn, C1q binds directly and specifically to surface blebs of apoptotic human keratinocytes: complement deficiency and systemic lupus erythematosus revisited, J Immunol 158 (1997) 4525–4528.

[108] R.R. Nepomuceno, A.H. Henschen-Edman, W.H. Burgess, A.J. Tenner, cDNA cloning and primary structure analysis of C1qR(P), the human C1q/MBL/SPA receptor that mediates enhanced phagocytosis in vitro, Immunity 6 (1997) 119–129.

[109] M. Botto, C. Dell'Agnola, A.E. Bygrave, E.M. Thompson, H.T. Cook, F. Petry, et al., Homozygous C1q deficiency causes glomerulonephritis associated with multiple apoptotic bodies, Nat Genet 19 (1998) 56–59.

[110] P. Bowness, K.A. Davies, P.J. Norsworthy, P. Athanassiou, J. Taylor-Wiedeman, L.K. Borysiewicz, et al., Hereditary C1q deficiency and systemic lupus erythematosus, Qjm 87 (1994) 455–464.

[111] M.C. Pickering, S. Fischer, M.R. Lewis, M.J. Walport, M. Botto, H.T. Cook, Ultraviolet-radiation-induced keratinocyte apoptosis in C1q-deficient mice, J Invest Dermatol 117 (2001) 52–58.

[112] L.A. Casciola-Rosen, G. Anhalt, A. Rosen, Autoantigens targeted in systemic lupus erythematosus are clustered in two populations of surface structures on apoptotic keratinocytes, J Exp Med 179 (1994) 1317–1330.

[113] L.A. Casciola-Rosen, G.J. Anhalt, A. Rosen, DNA-dependent protein kinase is one of a subset of autoantigens specifically cleaved early during apoptosis, J Exp Med 182 (1995) 1625–1634.

[114] P.J. Utz, M. Hottelet, P.H. Schur, P. Anderson, Proteins phosphorylated during stress-induced apoptosis are common targets for autoantibody production in patients with systemic lupus erythematosus, J Exp Med 185 (1997) 843–854.

[115] P.J. Utz, P. Anderson, Posttranslational protein modifications, apoptosis, and the bypass of tolerance to autoantigens, Arthritis Rheum 41 (1998) 1152–1160.

[116] E.L. Greidinger, L. Casciola-Rosen, S.M. Morris, R.W. Hoffman, A. Rosen, Autoantibody recognition of distinctly modified forms of the U1-70-kd antigen is associated with different clinical disease manifestations, Arthritis Rheum 43 (2000) 881–888.

[117] S. Meller, F. Winterberg, M. Gilliet, A. Muller, I. Lauceviciute, J. Rieker, et al., Ultraviolet radiation-induced injury, chemokines, and leukocyte recruitment: An amplification cycle triggering cutaneous lupus erythematosus, Arthritis Rheum 52 (2005) 1504–1516.

[118] J. Wenzel, S. Henze, E. Worenkamper, Role of the chemokine receptor CCR4 and its ligand thymus- and activation-regulated chemokine/CCL 17 for lymphocyte recruitment in cutaneous lupus erythematosus, J Invest Dermatol 124 (2005). 1241–1248.

[119] E.C. Baechler, P.K. Gregersen, T.W. Behrens, The emerging role of interferon in human systemic lupus erythematosus, Curr Opin Immunol 16 (2004) 801–807.

[120] M.M. Cals-Grierson, A.D. Ormerod, Nitric oxide function in the skin, Nitric Oxide 10 (2004) 179–193.

[121] C.V. Suschek, V. Krischel, D. Bruch-Gerharz, D. Berendji, J. Krutmann, K.D. Kroncke, et al., Nitric oxide fully protects against UVA-induced apoptosis in tight correlation with Bcl-2 up-regulation, J Biol Chem 274 (1999) 6130–6137.

[122] R. Weller, A. Schwentker, T.R. Billiar, Y. Vodovotz, Autologous nitric oxide protects mouse and human keratinocytes from ultraviolet B radiation-induced apoptosis, Am J Physiol Cell Physiol 284 (2003) C1140–1148.

[123] A.D. Ormerod, P. Copeland, I. Hay, A. Husain, S.W. Ewen, The inflammatory and cytotoxic effects of a nitric oxide releasing cream on normal skin, J Invest Dermatol 113 (1999) 392–397.

[124] A. Kuhn, K. Fehsel, P. Lehmann, J. Krutmann, T. Ruzicka, V. Kolb-Bachofen, Aberrant timing in epidermal expression of inducible nitric oxide synthase after UV irradiation in cutaneous lupus erythematosus, J Invest Dermatol 111 (1998) 149–153.

[125] H.M. Belmont, D. Levartovsky, A. Goel, A. Amin, R. Giorno, J. Rediske, et al., Increased nitric oxide production accompanied by the up regulation of inducible nitric oxide synthase in vascular endothelium from patients with systemic lupus erythematosus, Arthritis Rheum 40 (1997) 1810–1816.

[126] M.A. Lopez-Nevot, L. Ramal, J. Jimenez-Alonso, J. Martin, The inducible nitric oxide synthase promoter polymorphism does not confer susceptibility to systemic lupus erythematosus, Rheumatology (Oxford) 42 (2003) 113–116.

[127] J.S. Lawrence, C.L. Martins, G.L. Drake, A family survey of lupus erythematosus. 1. Heritability, J Rheumatol 14 (1987) 913–921.

[128] G.R. Yeaman, K. Froebel, G. Galea, A. Ormerod, S.J. Urbaniak, Discoid lupus erythematosus in an X-linked cytochrome-positive carrier of chronic granulomatous disease, Br J Dermatol 126 (1992) 60–65.

[129] K.A. Wolpert, A.D. Webster, S.J. Whittaker, Discoid lupus erythematosus associated with a primary immunodeficiency syndrome showing features of non-X-linked hyper-IgM syndrome, Br J Dermatol 138 (1998) 1053–1057.

[130] T.P. Millard, E. Kondeatis, R.W. Vaughan, C.M. Lewis, M.A. Khamashta, G.R. Hughes, et al., Polymorphic light eruption and the HLA DRB1*0301 extended haplotype are independent risk factors for cutaneous lupus erythematosus, Lupus 10 (2001) 473–479.

[131] T.P. Millard, C.M. Lewis, M.A. Khamashta, G.R. Hughes, J.L. Hawk, J.M. McGregor, Familial clustering of polymorphic light eruption in relatives of patients with lupus erythematosus: evidence of a shared pathogenesis, Br J Dermatol 144 (2001) 334–338.

[132] P. Price, C. Witt, R. Allcock, D. Sayer, M. Garlepp, C.C. Kok, et al., The genetic basis for the association of the 8.1 ancestral haplotype (A1, B8, DR3) with multiple immunopathological diseases, Immunol Rev 167 (1999) 257–274.

[133] J.B. Harley, M. Reichlin, F.C. Arnett, E.L. Alexander, W.B. Bias, T.T. Provost, Gene interaction at HLA-DQ enhances autoantibody production in primary Sjogren's syndrome, Science 232 (1986) 1145–1147.

[134] L.R. Braathen, A. Bratlie, P. Teisberg, HLA genotypes in a family with a case of homozygous C2 deficiency and discoid lupus erythematosus, Acta Derm Venereol 66 (1986) 419–422.

[135] T.T. Provost, F.C. Arnett, M. Reichlin, Homozygous C2 deficiency, lupus erythematosus, and anti-Ro (SSA) antibodies, Arthritis Rheum 26 (1983) 1279–1282.

[136] J.P. Callen, S.J. Hodge, K.B. Kulick, G. Stelzer, J.J. Buchino, Subacute cutaneous lupus erythematosus in multiple members of a family with C2 deficiency, Arch Dermatol 123 (1987) 66–70.

[137] S.B. Levy, S.R. Pinnell, L. Meadows, R. Snyderman, F.E. Ward, Hereditary C2 deficiency associated with cutaneous lupus erythematosus: clinical, laboratory, and genetic studies, Arch Dermatol 115 (1979) 57–61.

[138] O. Meyer, G. Hauptmann, G. Tappeiner, H.D. Ochs, F. Mascart-Lemone, Genetic deficiency of C4, C2 or C1q and lupus syndromes. Association with anti-Ro (SS-A) antibodies, Clin Exp Immunol 62 (1985) 678–684.

[139] H.C. Nousari, A. Kimyai-Asadi, T.T. Provost, Generalized lupus erythematosus profundus in a patient with genetic partial deficiency of C4, J Am Acad Dermatol 41 (1999) 362–364.

[140] M. Shapiro, A.C. Sosis, J.M. Junkins-Hopkins, V.P. Werth, Lupus erythematosus induced by medications, ultraviolet radiation, and other exogenous agents: a review, with special focus on the development of subacute cutaneous lupus erythematosus in a genetically predisposed individual, Int J Dermatol 43 (2004) 87–94.

[141] D.A. Norris, Pathomechanisms of photosensitive lupus erythematosus, J Invest Dermatol 100 (1993) 58S–68S.

[142] M.J. Rood, M.V. van Krugten, E. Zanelli, M.W. van der Linden, V. Keijsers, G.M. Schreuder, et al., TNF-308A and HLA-DR3 alleles contribute independently to susceptibility to systemic lupus erythematosus, Arthritis Rheum 43 (2000) 129–134.

[143] J.L. Pablos, P.E. Carreira, J.M. Martin-Villa, G. Montalvo, A. Arnaiz-Villena, J.J. Gomez-Reino, Polymorphism of the heat-shock protein gene HSP70-2 in systemic lupus erythematosus, Br J Rheumatol 34 (1995) 721–723.

[144] M. Ghoreishi, I. Katayama, H. Yokozeki, K. Nishioka, Analysis of 70 KD heat shock protein (HSP70) expression in the lesional skin of lupus erythematosus (LE) and LE related diseases, J Dermatol 20 (1993) 400–405.

[145] R. Villalobos-Hurtado, S.H. Sanchez-Rogriguez, E. Avalos-Diaz, R. Herrera-Esparza, Possible role of Hsp70 in autoantigen shuttling to the dermo-epidermal junction in systemic lupus erythematosus, Reumatismo 55 (2003) 155–158.

[146] H. Suzuki, Y. Matsui, H. Kashiwagi, Interleukin-1 receptor antagonist gene polymorphism in Japanese patients with systemic lupus erythematosus, Arthritis Rheum 40 (1997) 389–390.

[147] J. Eskdale, D. Kube, H. Tesch, G. Gallagher, Mapping of the human IL10 gene and further characterization of the 5' flanking sequence, Immunogenetics 46 (1997) 120–128.

[148] M. Lazarus, A.H. Hajeer, D. Turner, P. Sinnott, J. Worthington, W.E. Ollier, et al., Genetic variation in the interleukin 10 gene promoter and systemic lupus erythematosus, J Rheumatol 24 (1997) 2314–2317.

[149] M.B. Frank, R. McArthur, J.B. Harley, A. Fujisaku, Anti-Ro (SSA) autoantibodies are associated with T cell receptor beta genes in systemic lupus erythematosus patients, J Clin Invest 85 (1990) 33–39.

[150] W. Ollier, E. Davies, N. Snowden, J. Alldersea, A. Fryer, P. Jones, et al., Association of homozygosity for glutathione-S-transferase GSTM1 null alleles with the Ro+/La- autoantibody profile in patients with systemic lupus erythematosus, Arthritis Rheum 39 (1996) 1763–1764.

[151] Q.R. Huang, V. Danis, M. Lassere, J. Edmonds, N. Manolios, Evaluation of a new Apo-1/Fas promoter polymorphism in rheumatoid arthritis and systemic lupus erythematosus patients, Rheumatology (Oxford) 38 (1999) 645–651.

[152] F. Furukawa, H. Kanauchi, H. Wakita, Y. Tokura, T. Tachibana, Y. Horiguchi, et al., Spontaneous autoimmune skin lesions of MRL/n mice: autoimmune disease-prone genetic background in relation to Fas-defect MRL/1pr mice, J Invest Dermatol 107 (1996) 95–100.

[153] M. Johansson, L. Arlestig, B. Moller, T. Smedby, S. Rantapaa-Dahlqvist, Oestrogen receptor {alpha} gene polymorphisms in systemic lupus erythematosus, Ann Rheum Dis 64 (2005) 1611–1617.

[154] M.A. Lee-Kirsch, M. Gong, H. Schulz, F. Ruschendorf, A. Stein, C. Pfeiffer, et al., Familial chilblain lupus, a monogenic form of cutaneous lupus erythematosus, maps to chromosome 3p, Am J Hum Genet 79 (2006) 731–737.

[155] D.J. Mazur, F.W. Perrino, Identification and expression of the TREX1 and TREX2 cDNA sequences encoding mammalian 3'->5' exonucleases, J Biol Chem 274 (1999) 19655–19660.

[156] G. Rice, W.G. Newman, J. Dean, T. Patrick, R. Parmar, K. Flintoff, et al., Heterozygous mutations in TREX1 cause familial chilblain lupus and dominant Aicardi-Goutieres syndrome, Am J Hum Genet 80 (2007) 811–815.

[157] M.A. Lee-Kirsch, D. Chowdhury, S. Harvey, M. Gong, L. Senenko, K. Engel, et al., A mutation in TREX1 that impairs susceptibility to granzyme A-mediated cell death underlies familial chilblain lupus, J Mol Med 85 (2007) 531–537.

[158] D. Chowdhury, P.J. Beresford, P. Zhu, D. Zhang, J.S. Sung, B. Demple, et al., The exonuclease TREX1 is in the SET complex and acts in concert with NM23-H1 to degrade DNA during granzyme A-mediated cell death, Mol Cell 23 (2006) 133–142.

[159] Y.J. Crow, B.E. Hayward, R. Parmar, P. Robins, A. Leitch, M. Ali, et al., Mutations in the gene encoding the 3'-5' DNA exonuclease TREX1 cause Aicardi-Goutieres syndrome at the AGS1 locus, Nat Genet 38 (2006) 917–920.

[160] Y.J. Crow, A. Leitch, B.E. Hayward, A. Garner, R. Parmar, E. Griffith, et al., Mutations in genes encoding ribonuclease H2 subunits cause Aicardi-Goutieres syndrome and mimic congenital viral brain infection, Nat Genet 38 (2006) 910–916.

[161] M.A. Lee-Kirsch, M. Gong, D. Chowdhury, L. Senenko, K. Engel, Y.A. Lee, et al., Mutations in the gene encoding the 3'-5' DNA exonuclease TREX1 are associated with systemic lupus erythematosus, Nat Genet 39 (2007) 1065–1067.

[162] Y.G. Yang, T. Lindahl, D.E. Barnes, Trex1 exonuclease degrades ssDNA to prevent chronic checkpoint activation and autoimmune disease, Cell 131 (2007) 873–886.

[163] N. Hubner, C.A. Wallace, H. Zimdahl, E. Petretto, H. Schulz, F. Maciver, et al., Integrated transcriptional profiling and linkage analysis for identification of genes underlying disease, Nat Genet 37 (2005) 243–253.

[164] A. Kuhn, K. Gensch, S. Stander, G. Bonsmann, Cutaneous lupus erythematosus: Part 2: Diagnostics and therapy, Hautarzt 57 (2006) 345–360.

Neuropsychiatric Systemic Lupus Erythematosus: Mechanisms of Injury

Meggan Mackay, Aziz M. Uluğ, Bruce T. Volpe

INTRODUCTION

Neuropsychiatric SLE (NPSLE), comprised of numerous, complex central and peripheral nervous system symptoms, poses significant challenges at the bedside and even more in the laboratory. While it is expected that some immunopathogenic mechanisms implicated in central nervous system (CNS) disease may also be responsible for peripheral nerve pathology, anatomic differences will strongly determine aspects of immune assault on different areas of the nervous system. For example, the presence of the blood—brain barrier (BBB) has led to the traditional portrayal of the brain as an immune privileged area that is not readily accessible to the inflammatory initiators associated with SLE such as autoantibodies, immune complexes, cytokines, and activated immune cells. Conversely, peripheral nerves are not afforded protection by the BBB and are directly susceptible to immune attack. For this reason, mechanisms that underlie NPSLE symptoms are best considered in terms of anatomic location. A consideration of pathologic mechanisms stratified by anatomic location allows enhanced understanding of diverse symptomatology, kinetics of disease, symptom reversibility and regional brain vulnerability. Mechanisms of injury to brain vasculature, brain parenchyma, and the peripheral nervous system will be presented, with a discussion of the BBB included to inform studies of damage to brain parenchyma. Insults to blood vessels are often mediated by endothelial cell activation. Moreover, the BBB links endothelial or vascular damage to CNS damage. Activated endothelial cells can lead to vasculitis, but more often to alterations in BBB integrity that permit circulating autoantibodies and inflammatory molecules to gain access to vulnerable cellular components of the CNS.

It is critical to state that an understanding of pathogenic mechanisms of NPSLE requires the correct attribution of neuropsychiatric symptoms to SLE and not to secondary causes of peripheral nerve or brain dysfunction that frequently accompany SLE and include infection, metabolic disturbances, hormonal effects, medication effects, and the specific NPSLE mimics, thrombotic thrombocytopenic purpura and posterior reversible encephalopathy.

VASCULAR PATHOLOGY

It is now clear that inflammatory mechanisms can be associated even with thrombosis and thrombotic angiopathy, as well as with vasculitis and accelerated atherosclerosis. Postmortem neuropathologic examination of SLE patients highlights the prevalence of microvascular injury with evidence of widespread microinfarcts, perivascular lymphocytic infiltrates, and endothelial cell proliferation in patients with and without NPSLE [1—3]. Thus, much cerebrovascular disease in SLE is subclinical. True vasculitis with transmural neutrophilic infiltrate and fibrinoid necrosis within the vessel wall is rarely reported; if present, however, it associates with clinical symptoms of NPSLE [1, 2, 4—6].

The common cerebral vasculopathy with arterial intimal hyperplasia that leads to occlusive, bland thrombotic stroke syndromes is often linked to antiphospholipid antibody. The antiphospholipid (APL) syndrome [7, 8] is associated with a family of autoantibodies that recognize epitopes on phospholipids and occasionally phospholipid-binding plasma proteins, most commonly beta(2)-glycoprotein-I (β2GP1). APL antibodies occur in the circulation as anticardiolipin and/or anti-β2GP1 antibodies (IgG, IgM or IgA) and the lupus anticoagulant

[7]. These antibodies activate endothelial cells, either through cross-reactivity to an endothelial cell antigen or through generation of a lattice of phospholipid on the surface of endothelial cells, which activates a clotting cascade. This in turn leads to activation of monocytes, complement, and platelets in vascular beds, inducing intracellular signal transduction and induction of inflammatory and vasoconstrictive mediators [9, 10]. This hypercoaguable state, in combination with a "second hit," accorded either by the inflammation of lupus activity itself or by infection, frequently gives rise to thrombosis. In the brain, impaired blood flow secondary to thrombus or embolus results in infarction and hemorrhage and focal neuron injury; accompanying symptoms depend on the location, duration, and degree of vascular compromise. The association of these antibodies and cerebrovascular disease (stroke, TIA, cognitive decline associated with chronic microvascular ischemia) can be explained by the prothrombotic effects of these antibodies. However, a number of other NPSLE syndromes including headache, seizures, chorea, psychosis, and depression have also been associated with APL antibodies in the absence of stroke or vascular compromise [11–14].

Murine studies have provided more information on potential roles for APL antibodies in CNS pathology that do not involve vessel occlusion. There is growing evidence, as suggested above, that APL antibodies alter brain function through direct binding to brain endothelial cells. This not only activates a clotting cascade, but also activates the endothelial cells and affects permeability of the BBB and induction of neuroinflammatory responses. Exposure of brain endothelial cells to purified APL antibodies in vitro results in brain endothelial cell activation with upregulation of cellular adhesion molecules ICAM-1, E-selectin, and VCAM-1 and increased production of proinflammatory cytokines such as IL-1 and IL-6 [15, 16]. Increased expression of adhesion molecules on brain endothelial cells in the presence of inflammatory molecules can alter the integrity of the BBB through effects on local recruitment of inflammatory cells and subsequent activation of microglial cells (see the section titled "Blood–Brain Barrier"). APL antibodies and β2GP1 antibodies have also been demonstrated to bind in vitro to elements of brain tissue including neurons, myelin, astrocytes, vascular endothelium, choroidal epithelium, and ependymal cells [17–19]. Concentration-dependent, depolarizing, and permeabilizing effects of purified human APL antibodies on synaptosomes in vitro [20] and on glutamate receptor-mediated neurotoxicity [21] suggest that these autoantibodies are also capable of inducing changes in neuronal function. Therefore, once APL antibodies gain access to the CNS, neuronal dysfunction may be mediated in part by the APL antibodies themselves.

For example, nonautoimmune Balb/c mice immunized with β2GP1 develop behavioral abnormalities manifested as increased anxiety, hyperactivity, and exploratory behavior and cognitive deficits several months after immunization [22, 23]. Brain histopathology reveals perivascular infiltrates containing lymphocytes and macrophages and evidence of microglial activation in white matter, cortex, and parahippocampal areas with no micro or macro infarcts [24]. Significantly elevated brain levels of TNF-α and prostaglandin E (PGE) in these mice were attributed to antibody effects on brain endothelial cells and perivascular and meningeal cells. Of note, APL and anti-β2GP1 antibodies exhibit a great diversity with respect to antigen-binding properties, probably responsible for studies that have failed to demonstrate associations between presence of serum APL antibodies and NPSLE.

Atherosclerotic mechanisms contribute to NPSLE through the conventional risk factors (age, sex, family history, hypercholesterolemia, hyperhomocysteinemia, hypertension, diabetes mellitus, smoking, sedentary life style, elevated creatinine, elevated waist–hip ratio, and postmenopausal status), and some novel risk factors such as inflammation and pro-inflammatory HDL [25, 26]. The connection between atherosclerosis and inflammation has been well described (reviewed in [26, 27]) and atherosclerotic risk is clearly accelerated in SLE. Increased circulating levels of IFN-γ and C1q-fixing immune complexes result in decreased cholesterol removal from vessel walls due to effects on cholesterol 27-hydroxylase in endothelial cells and macrophages. IL-1 and TNF-α both alter endothelial cell function resulting in upregulation of cell adhesion molecules and increased endothelial cell shedding. Low-density lipoproteins (LDL) enter the subendothelial spaces created by endothelial cell shedding where they undergo oxidation. Endothelial cells express cellular adhesion molecules, pro-inflammatory cytokines, and chemokines in response to oxidized LDL; macrophages are then recruited to the site where they ingest oxidized LDL and become foam cells that form the core of the plaque. These activated macrophages are highly pro-inflammatory and secrete metalloproteinases in addition to other pro-inflammatory molecules. Increased risk of atherosclerotic disease in SLE is generally attributed to an ongoing, sometimes low-grade, inflammatory process; however, the exact mechanisms responsible are not clear. Recent work with lymphocytes obtained from LDL-deficient animals immunized with β2-glycoprotein-I (β2GPI) and transferred to syngeneic animals demonstrated increased atherosclerosis. These data suggest that autoantibodies may exacerbate atherosclerotic mechanisms in NPSLE [27]. Additionally, increased circulating levels of autoantibodies to oxidized LDL, cardiolipin, dsDNA, apolipoprotein A-1, and lipoprotein lipase have been found in individuals with atherosclerosis, with and without autoimmune disease.

Mechanisms of vascular disruption — thrombosis, vasculitis, atherosclerosis and endothelial cell activation — can all lead to compromised blood flow and tissue ischemia. Equally important however, is the effect of endothelial cell activation on integrity of the BBB that exposes CNS tissue to the damaging effects of inflammatory molecules and autoantibodies.

BLOOD—BRAIN BARRIER

Immune responses in the brain require a break in the BBB such that either cells, soluble molecules, or both, can gain access to brain tissue. The concept of the BBB as an impermeable barrier designed for neural protection has evolved to that of a more dynamic gatekeeper with transport, signaling and secretory functions. While exclusionary tight junctions between brain endothelial cells maintain a physical barrier, specialized transport mechanisms allow the flow of essential nutrients from blood to brain and metabolites and toxins from brain to blood [28]. The BBB has more recently been demonstrated to respond to molecular signals from brain parenchyma. Pro-inflammatory molecules such as cytokines, chemokines, prostaglandins, and others are also produced by cells in the BBB, allowing for autocrine and paracrine mechanisms of BBB disruption [29–31]. It is, therefore, possible to consider that permeability of the BBB is altered by both systemic and CNS mediators and that BBB dysfunction occurs more frequently than previously thought and may play a pivotal role in the pathogenesis of neurodegenerative diseases including NPSLE [32].

Components of the BBB

The BBB is comprised of brain endothelial cells, pericytes, basement membrane, and astrocytes. The cells primarily responsible for BBB integrity are the brain endothelial cells. Tight junctions between brain endothelial cells are regulated by a group of intracellular cytoskeletal and scaffolding proteins and junctional adhesion molecules [33, 34]. Resulting intercellular seals prevent access of macromolecules to the CNS unless they can be transported across the BBB via pinocytosis or active transport mechanisms. Destabilization of the tight junctions results from alterations in structure of the adhesion molecules or cytoskeletal proteins in response to cytokines produced in the brain or the periphery. Intracellular enzymes degrade lipid-soluble molecules that diffuse into brain endothelial cells and a complex series of transport systems on the luminal and abluminal sides of the brain endothelial cells

actively regulate transport of nutrients and other essential molecules to and from the brain. The basement membrane, composed of type 4 collagen, elastin, laminin, fibronectin, and extracellular matrix and signaling proteins, anchors the brain endothelial cells and pericytes and disruption of the basement membrane can indirectly affect integrity of the tight junctions. Pericytes are interspersed between brain endothelial cells in some areas of the brain and they are thought to modulate blood flow through receptor-mediated effects on vascular tone and contractility [35]. End-foot processes of astrocytes coalesce to form another layer of barrier, the glia limitans, which is adjacent to the vascular basement membrane. The perivascular space between the basement membrane and glial limitans can become a niche for immune cells to cluster resulting in perivascular cuffing. Astrocytes are involved in water transport [36] and may secrete soluble factors that regulate tight junctions [28]. Microglia are considered to be resident immune cells of the CNS; they are capable of transforming into phagocytic cells resembling macrophages and producing inflammatory molecules such as cytokines, chemokines, and metalloproteinases (MMPs) [37]. Abnormally activated microglial cells have been implicated in the pathogenesis of several neurodegenerative diseases such as Parkinson disease, Alzheimer dementia, HIV dementia, and multiple sclerosis [38–40]. The origins of parenchymal microglial cells are not well understood; however, perivascular microglial cells have been demonstrated to arise from the bone marrow and, unlike the parenchymal microglial cells, are continually replenished.

Regulation and Compromise of the BBB

There is now increased awareness that bi-directional communication occurs between blood and brain utilizing the BBB as a signaling interface and that communication between peripheral cells and soluble molecules, cells comprising the BBB, and brain parenchyma all influence BBB functions [41]. There are several known examples of blood-to-brain signaling. Bacterial lipopolysaccharide (LPS) has been shown to cause disruption of the BBB [42]. Direct effects of LPS on barrier integrity include destabilization of tight junctions [43] and release of pro-inflammatory mediators by brain endothelial cells. LPS stimulation of brain endothelial cells *in vitro* results in release of IL-6, IL-10, TNF-α, and granulocyte macrophage colony stimulating factor (GMCSF) from brain endothelial cells into the brain parenchyma [44]. Other effects of LPS include enhanced absorptive endocytosis, increased expression of adhesion molecules, and selectins, facilitating increased transmigration of leukocytes into brain tissue and decreased brain-to-blood transport of toxic metabolites (reviewed in [38]). Systemic infections can also cause peripheral release of soluble molecules such as TNF-α, IL-1, and IL-6 that activate receptors on brain endothelial cells resulting in upregulation of cell adhesion molecules and destabilization of structural proteins, thereby compromising integrity of the BBB [38, 45]. Murine models of LPS-induced encephalopathy also demonstrate pericyte damage and increased microglial activation consistent with BBB disruption [46]. Likewise, human studies of sepsis-induced encephalopathy demonstrate evidence for BBB disruption [47].

Complement activation is a hallmark of SLE disease activity. Activation of the complement cascade is associated with poor outcomes in sepsis and anaphylatoxin C5a is a potent mediator of end-organ toxicity including the brain [48, 49]. C5a is associated with BBB dysfunction and blockade of C5a receptor ameliorates disruption of the BBB in mice [50]. Brain endothelial cells exposed to C5a *in vitro* increase expression of iNOS and ROS, leading to cytoskeletal reorganization and altered BBB permeability [51]. Resident brain cells, neurons, astrocytes, and microglia, are all capable of expressing complement components and receptors [40, 52] and so can amplify systemic effects of complement activation. Increased expression of complement receptors and components in brain tissue subjected to trauma, neurodegenerative processes such as Alzheimer disease, hypertension, and ischemic insults suggests an important role for the complement pathway in tissue destruction [52]. Significant upregulation of pituitary expression of TNF and IL-6 mRNA in response to sepsis suggests that the pituitary (one of the circumventricular organs not protected by the BBB) is another source for pro-inflammatory cytokines that can alter BBB function [50].

Barrier impairment in response to stress has been demonstrated in murine models with increased diffusion of drugs and neurotransmitters across the BBB (reviewed in [39]). Peripheral pain models in mice demonstrate increased permeability of the BBB from disrupted tight junctions and increased expression of cellular adhesion molecules [53, 54]. Exposure of brain endothelial cells to tobacco smoke and nicotine *in vitro* results in upregulation of pro-inflammatory cytokines and MMPs as well as alterations in brain endothelial cells' tight junction integrity [55, 56]. Since it is probable that many insults to BBB integrity can act synergistically, it is possible that complement activation and ischemia, which may be direct results of lupus activity, synergize with nicotine or infection in particular individuals to cause episodic or chronic impairments of barrier integrity.

Brain-to-blood signaling occurs by an "inside-out" signaling process that originates with cytokines or other soluble molecules that have accessed the CNS or are synthesized by resident cells in the brain [32]. For example, several *in vitro* and *in vivo* murine studies have demonstrated the ability of perivascular microglial cells to upregulate MHC II for participation in antigen presentation and T-cell activation and to secrete soluble pro-inflammatory and immunoregulatory molecules (TNFα, IL-1-β, nitric oxide (NO), prostaglandin E2 (PGE2), transforming growth factor beta (TGFβ) and nerve growth factor (NGF)) [40]. Cytokines can also enter the brain through certain areas not protected by the BBB; circumventricular organs, comprised of the pineal gland, pituitary, hypothalamus, and choroid plexus have specialized fenestrations that facilitate cytokine signaling to resident brain cells that then communicate with the BBB using cytokine cascades and other soluble molecules such as prostaglandins [38, 57]. There are some limited data to suggest that brain endothelial cells express *N*-methyl-*D*-aspartate receptors (NMDAR) and that glutamate binding results in increased BBB permeability, suggesting that anti-NMDAR antibodies may have direct effects on BBB integrity [58, 59].

Enhanced immune cell entry into the CNS is an important consequence of BBB dysfunction. Normally, diapedesis of monocytes into the CNS is a carefully controlled facet of routine immune surveillance that is orchestrated by brain endothelial cells, monocytes, and chemokine gradients (reviewed in [60]). Increased display of cellular adhesion molecules by activated brain endothelial cells results in increased leukocyte rolling along the vascular lumen. Aided by release of MMPs that digest basement membrane and chemokines or by leaky tight junctions weakened under the influence of cytokines, leukocytes gain access to brain parenchyma (reviewed in [39]). Once inside the brain parenchyma, inflammatory cells may continue to alter BBB functions. Moreover, their function can be altered by the brain environment. For example, TGFβ and GMCSF secreted by the BBB may play a role in the transformation of naïve CD4$^+$ T cells into Th17 effector cells [61].

BBB disruption has been implicated in a variety of chronic neurodegenerative diseases [28, 39]. In NPSLE, appreciation of the BBB must be interpreted in the context of the central importance of serum antibodies capable of binding to brain tissue. Evidence for BBB disruption in SLE is accumulating. Autoantibodies, the hallmark of SLE, have been shown to traverse the BBB in murine and human studies [62-66]. For example, antibodies to the NMDAR (anti-NMDAR) are present in the sera of approximately 50% of lupus patients and these autoantibodies have been shown to bind NMDAR on neurons resulting in temporary neuron dysfunction and apoptotic neuronal death (see the section titled "Central Nervous System"). Importantly, in animal models these autoantibodies do not gain access to the CNS unless the BBB is breached by systemic administration of LPS or norepinephrine. The murine model demonstrates that, once inside the CNS, anti-NMDAR antibodies cause regional neuronal dysfunction associated with specific behavioral changes. In human SLE, CSF titers of anti-NMDAR antibodies correlate with diffuse NPSLE symptoms, whereas serum levels do not [62, 67–69]. NMDAR autoantibodies, as well as other autoantibodies including APL, antiribosomal P, and anti-dsDNA antibodies have been found in the CSF of lupus patients with no evidence of active infection or NPSLE symptoms [62], suggesting the possibility of increased BBB permeability unrelated to clinically apparent disease or active infection. Magnetic resonance spectroscopy (MRS) imaging of lupus patients experiencing systemic flare and no NPSLE symptoms has shown abnormally low NAA:Cr ratios in brain parenchyma [70], also suggesting increased permeability of the BBB allowing for reversible, subclinical neuronal dysfunction. It has recently been demonstrated that depressive symptoms in MRL/lpr mice occur early in the disease process, prior to the onset of other organ involvement, and depressive symptoms correlate with elevated peripheral blood titers of autoantibodies to dsDNA, NMDAR, and cardiolipin [71]. Abnormal function in the hippocampus and related cortical areas in MRS imaging of these mice suggests evidence of altered neuronal function as a consequence of BBB dysfunction and autoantibody infiltration of brain parenchyma [71, 72].

Complement activation is thought to play a critical role in the pathogenesis of lupus and elevated levels of complement split products, C5a and C3a, have been associated with disease activity and poor outcome [73]. As described above, C5a has potent pro-inflammatory effects on the BBB and brain parenchyma and studies of murine lupus have demonstrated that inhibiting the complement activation pathway ameliorates CNS disease [74, 75]. Unopposed complement activation in MRL/lpr mice leads to disruption of the BBB and free autoantibody access to brain parenchyma that is abrogated by treatment with Crry-Ig, a soluble C3/C5 convertase inhibitor [74]. Crry-Ig also blocked transcription of aquaporin 4, a protein associated with formation of brain edema and brain water homeostasis [76]. Systemic blockade of soluble C5a receptor with Crry-Ig in MRL/lpr mice resulted in maintenance of an intact BBB and no parenchymal deposition of autoantibody [51]. This has potentially important implications since blockade of C5a receptor does not affect other protective functions of the complement cascade. Increased expression of complement factor B (a protein required for activation of the alternative pathway) in brains of MRL/lpr mice leads to immune deposits, inflammatory infiltrate, and neuronal apoptosis associated with behavioral alterations that are alleviated

with blockade of the alternative pathway [75]. It was recently suggested that lupus patients with systemic flare associated with low serum complement levels and evidence of BBB disruption (measured by a positive Q albumin) in the absence of overt NPSLE symptoms are at high risk for corticosteroid-induced psychosis [77], perhaps suggesting a synergy between corticosteroids and inflammatory cytokines or autoantibodies.

Regional BBB Dysfunction

Murine studies and human analogs demonstrate regional rather than diffuse alterations in the BBB in response to various stimuli. The reasons for this are likely to be multifactorial. It is possible that focal BBB disruption is guided by regional microglial activation and cytokine bombardment of the BBB [39]. Alternatively, the agent responsible for BBB disruption can produce variable effects; LPS induces relatively increased permeability in dorsal hippocampal regions, whereas norepinephrine induces relatively increased permeability in ventral brain and in the amygdalar regions. It is also possible that receptors for the soluble molecules and antibodies associated with BBB and brain impairment are not uniformly distributed in the BBB. Therefore, location of BBB disruption will depend on the offending agent, resulting in regional loss of BBB integrity, regional brain damage, and corresponding behavioral alterations.

CENTRAL NERVOUS SYSTEM

Neuropathology

The NPSLE syndromes encompass abnormalities at the neuromuscular junction, the peripheral and autonomic nerves, the nerve roots, the spinal cord, and the brain. The nervous system may be affected by focal anatomical injury, as in stroke or transverse myelitis, or nonfocal or diffuse pathophysiology, as in cognitive or mood disorder, psychosis, acute confusional state, or seizures. Neuropathological evidence for a locus and cellular composition of injury and evidence of neuron loss has remained an important standard in establishing the disease mechanism, yet, unlike histopathological findings in the kidney and spleen, no characteristic pathology exists for NPSLE. Vasculitis is exceedingly rare from the neuropathological point of view [3], but ischemic and thrombotic brain infarction or hemorrhage are not [1, 3, 78, 79]. Multiple small infarcts and widespread microvascular injury has also been recorded (see the section titled "Vascular Pathology") [2], but vascular compromise alone does not explain the range of symptoms associated with NPSLE. There is the added complexity of understanding mechanisms of neural tissue damage because neuropathological changes may not be temporally related to symptoms and the confluence of vascular and direct brain injury may be difficult to untangle. For example, two of the three prototype examples of SLE patients with vasculitis in the brain had characteristic inflammatory components throughout the vessel walls that were revealed at postmortem analysis 5 days and 6 weeks after the onset of neurological disease. The third patient had seizures for 6 years as the signature NPSLE symptom before a terminal hemorrhagic brain infarct revealed active vasculitis [3]. Therefore, damaged or lost structure may reflect historical disease rather than acute disturbances.

Cellular elements of the CNS are subject to injury from a variety of inflammatory mechanisms and autoantibodies. These insults affect behavior, cognition, and movement and can alter the wakefulness or the level of consciousness. Clearly, the pathophysiology need not be lethal to neurons as many of the NPSLE syndromes exhibit transient symptomatology. Lupus-prone mice (MRL/lpr) display mononuclear cell infiltrates (primarily $CD3^+$ T cells) infiltrating the choroid plexus, hippocampus, meninges, and cerebellar parenchyma early in their disease. As the mice age, they exhibit increased brain atrophy that correlates with increased mononuclear cell infiltrate and the appearance of $CD19^+$ B cells and $CD138^+$ plasma cells, suggesting the ability for intrathecal antibody production once the BBB has been disrupted [80, 81]. Evidence supportive of cellular damage in human NPSLE is demonstrated by increased levels of soluble molecules such as neurofilaments, Tau (axonal degradation product) and astroglial fibrillary acidic protein in the CSF of patients with NPSLE compared to non-NPSLE patients [82, 83].

Neuronal damage, whether caused by cytokines [84], antibody [7], or cells capable of producing these molecules, requires that these agents are either produced in the brain or cross the BBB. Supporting evidence for intrathecal production of autoantibodies is demonstrated by elevated CSF IgG levels, increased IgG index, oligoclonal bands, and identification of B cells in brain parenchyma in SLE patients. The BBB and its role in NPSLE has been discussed (see the section titled "Blood—Brain Barrier"). Because of the protection provided by the BBB and the mechanisms for altering barrier integrity, circulating neurotoxic molecules will gain access to brain parenchyma only intermittently and CNS involvement may occur independently of disease activity and proceed without associated systemic flares. Therefore, it is expected that serum levels of potentially neurotoxic molecules will routinely fail to correlate directly with NPSLE symptoms [67], whereas CSF levels of neurotoxic molecules will more likely correlate with brain dysfunction [62, 85—87].

Autoantibodies

One of the hallmarks of SLE is the production of a wide variety of antibodies that are directed against intracellular components; absence of autoantibodies effectively rules out SLE. The presence of autoantibodies in systemic disease or neurological disease is variable across studies and syndromes. The assays are most easily performed on serum but assays of CSF are more informative with regard to CNS syndromes of NPSLE for reasons cited above. Antibody-mediated toxicity can involve direct binding effects, complement activation leading to an inflammatory cascade, or antibody-dependent cellular cytotoxicity. The paucity of vasculitis and parenchymal inflammatory infiltrate reported in NPSLE neuropathology suggests that neurotoxic effects of autoantibodies or other soluble substances penetrating the BBB may be more important.

Serum antineuronal antibodies have been recognized for years; however, interest in these autoantibodies waned due to lack of correlation of serum titers with NPSLE symptoms and the absence of functional consequences of binding to neuronal tissue. More recently immunoproteomic assays have been used to identify specific brain antigens targeted by SLE sera [88]. NPSLE patients and non-NPSLE patients were found to have distinctive patterns of IgG reactivity against human brain tissue, raising the possibility of both neuropathogenic and neuroprotective autoantibodies; however, no associations with specific NPSLE syndromes have been identified.

Serum and CSF anti-ribosomal P (anti-P) antibodies have been inconsistently associated with thought and mood disorders [89—91]. Recent work has begun to illustrate a mechanism through which these autoantibodies might affect brain function [92]. It has been demonstrated that the anti-P antibodies recognize an integral membrane protein on the neuronal cell surface, and when added to brain cell cultures, anti-P antibodies caused a rapid and sustained increase in calcium influx in neurons, resulting in apoptotic cell death. This cellular mechanism leaves aside the issue of how neuron apoptosis and death account for symptoms as complex as thought or mood disorders. Another example of complex neurologic function is the sense of smell, which relies, in part, on an intact limbic system. Anti-P antibodies stain neurons in the hippocampus, cingulate cortex and olfactory piriform cortex, and when injected into the brain of living mice lead to depression and hyposmia [63, 93]. Interestingly, measurement of decreased olfactory function in SLE patients has been correlated with disease activity and past history of NPSLE [94].

A subset of anti-dsDNA autoantibodies cross-reacts with the *N*-methyl *D*-aspartate receptor (NMDAR)

[95]. NMDARs are widely distributed across the brain with preferential density in the hippocampus and amygdala. They are localized within the glutamatergic synapses [96]. NMDAR activity underlies excitatory synaptic transmission, through control of externalization of another receptor, the amino-3-hydroxy-5-methyl-isoxazole-4-propionic acid receptors (AMPAR). These receptors regulate intracellular levels of sodium and calcium. It has become increasingly clear that there is a fine balance in receptor activation so that these ion fluxes may lead to survival signals [97, 98] or death cascades and neuronal loss under abnormal conditions [99]. Anti-DNA, anti-NMDAR antibodies bind to the NR2A and NR2B subunits of the NMDAR and synergize with glutamate to cause an excitatory, noninflammatory cell death. Experiments in murine models demonstrate neuron loss in nonautoimmune mice immunized to produce anti-DNA, anti-NMDAR antibodies under the conditions of altering BBB properties. Loss in the hippocampus or the amygdala is correlated with behavioral dysfunction that co-varies across the region of brain damage [64, 65, 100]. Mice with hippocampal neuron loss exhibit memory deficits and mice with amygdala damage fail to respond appropriately to a fear conditioning paradigm. Of particular importance is the observation that regional brain toxicity of the anti-NMDAR antibody is dependent on mode of antibody access to the CNS. Anti-NMDAR antibodies preferentially affect the hippocampus when LPS is used to breach the BBB, whereas they affect the amygdala when epinephrine is used to breach the BBB. Mice with circulating anti-NMDAR antibody and an intact BBB have no clinical or pathologic evidence of neuronal damage. These data demonstrate several important points. First, serum anti-brain antibodies do not correlate with NPSLE disease. Additionally, regional disruption of the BBB is dependent on the agent used to modify the BBB and the same antibody can cause variable behavioral changes that are dependent on the region of brain exposed to antibody. Moreover, higher antibody concentrations result in cellular apoptosis and lower concentrations affect synaptic plasticity; these experimental approaches may provide a basis for understanding both transient and long-lasting cognitive and behavioral changes observed in SLE patients. It should also be noted that polyclonal cross-reactive anti-NMDAR/anti-DNA antibodies may have variable effects on NMDAR activation possibly due to differences in isotype, fine specificity, and affinity.

Clinical correlates in human SLE support the observations from the laboratory. Serum anti-DNA, anti-NMDAR autoantibodies do not correlate consistently with cognitive impairment or depression in cross-sectional studies [67, 68, 101, 102]. However, two studies have demonstrated significant correlations between CSF anti-NMDAR antibody titers and central, diffuse NPSLE syndromes (seizures, acute confusional state, mood and anxiety disorders, psychosis, severe cognitive dysfunction) [87, 103]. A third study takes this observation one step further and additionally demonstrates that presence of the anti-NMDAR antibody in the CSF distinguishes central diffuse NPSLE from peripheral nerve involvement and SLE patients without NPSLE [62]. Additionally, the anti-NMDAR antibodies have been eluted from brain tissue of patients with lupus [95]. Given the exclusionary role of the BBB, it is not surprising that studies of CSF have proven to be more enlightening than serum studies.

Cytokines and Other Soluble Molecules

Cytokines are produced by a wide variety of cells to facilitate cellular communication and regulate immune response and inflammation. In the brain, astrocytes and microglia contribute to the level of intrathecal cytokines. These molecules can cause endothelial dysfunction, the microangiopathy of vasculitis and contribute to the progressive vasculopathy of focal ischemic and hemorrhagic brain disease. A number of pro-inflammatory cytokines have been identified in the CSF of NPSLE patients including fractalkine [104], IL-6 [105, 106], Il-8 [107], IFN-α, IL-1, BAFF, and APRIL [108]. IL-6 is a highly pro-inflammatory cytokine that has been associated with numerous inflammatory conditions, autoimmune diseases, and neurologic conditions including CNS infections, cerebrovascular events, and myelitis [109]. It has been implicated in focal disruption of the BBB in MS [110] and has been identified in CSF from patients with central NPSLE syndromes [111—113]. The IgG index and Q-albumin are traditionally used as measures of intrathecal IgG synthesis and integrity of the BBB, respectively. Evidence of an association between CSF IL-6 and the IgG index [107] suggests the possibility that increased intrathecal IL-6 may increase B-cell activation in the CNS. Recently, CSF levels of BAFF and APRIL were demonstrated to be significantly increased in patients with diffuse NPSLE syndromes compared to SLE patients without CNS symptoms [108]. BAFF and APRIL are known to promote B-cell survival, isotype-switching, and CD-40L-independent antibody production.

The IFN-α amplification loop has recently been ascribed a central role in SLE pathogenesis and this cytokine also has been associated with numerous neuropsychiatric symptoms. Treatment of chronic liver disease or cancer with IFN-α has resulted in seizures in 1—4% of patients, depressive symptoms in one of three patients and occasionally psychosis, confusion, and mania [114]. IFN-α has also been identified in NPSLE CSF [115, 116]. Furthermore, immune complexes formed by CSF from NPSLE patients with diffuse syndromes (mood and anxiety disorders, seizures, psychosis, acute confusional states), combined with antigen, stimulate significant expression of IFN-α by IFN-producing cells [115].

Other molecules identified in CSF of active NPSLE patients include chemokines (fractalkine, IL-8, MCP-1, IP-10, RANTES, MIG) and the metalloprotease, MMP-9 [112, 117]. While these are all potent inflammatory mediators capable of disrupting brain parenchyma, their role in NPSLE needs further study. That MMP9 may play a role in CNS disease is suggested by a study showing that systemic treatment of NZB/W lupus mice with cystamine, which blocks MMP-9, results in decreased neuronal apoptosis [118].

Fetal Brain

Intrauterine and perinatal exposure to maternal immunoglobulins (IgGs) is a normal, physiologic event. During pregnancy, IgG is transported from mother to fetus across the placenta [119]. Maternal IgG transfer begins by the first trimester and may exceed that in the maternal circulation by term. Postnatal acquisition of maternal antibody is essential for the newborn's immunity. While maternal antibodies confer protective immunity, it has also been documented that they can mediate tissue injury. For example, neonatal lupus is characterized by cutaneous lesions and congenital heart block, caused by maternal anti-Ro autoantibodies. A well-defined immune-mediated complication in fetal development that occurs in the offspring of lupus mothers is the development of complete heart block [120–123]. In other experiments, dams that harbor anti-DNA, anti-NMDAR antibodies throughout pregnancy have offspring with cognitive impairment and histological abnormalities in the cortex. Embryological brain development is abnormal throughout gestation. It appears that anti-DNA, anti-NMDAR antibody bind fetal brain tissue, but the mechanism(s) for the behavioral and histological abnormalities in adult offspring are not clear. Whether this animal model will explain the high frequency of children with learning disabilities born to mothers, but not fathers, with SLE remains to be tested [124].

PERIPHERAL AND AUTONOMIC DISEASE

Recent studies of autoimmune peripheral neuropathies suggest that patients with SLE have the widest spectrum of peripheral nervous system (PNS) involvement. Peripheral nervous system derangements in SLE include motor, sensory, or mixed sensorimotor polyneuropathy, mononeuropathy (single and multiplex),

plexopathy, and acute inflammatory demyelinating polyneuropathy (AIDP) (Guillain-Barré type) and autonomic dysfunction [125]. Clinical presentation depends of the distribution of involved nerves and is characterized by hypesthesia, motor dysfunction, and hyporeflexia. Distinction between axonal degeneration and demyelination relies on electrophysiologic testing and nerve conduction studies. Nerve biopsy often shows axonal degeneration and vasculitis of epineural arteries, often accompanied by local vascular compromise [126]. Alternative mechanisms include antibodies to nicotinic ganglionic acetylcholine receptor (alpha3-AChR), which have been associated with SLE [127]. Current work has focused on these autoantibodies as a biomarker for autoimmune disease and cancer rather than a quantitative correlation of antibody and symptom severity [127]. One study correlated peripheral neuropathy in SLE with anti-Sm antibodies [128]. Anti-GM1 and GM3 antibodies can exert neurotoxic effects on peripheral nerves in SLE similar to Guillain-Barré syndrome [129].

Peripheral nerves have no anatomic barrier to impede antibody interaction with the target antigen; therefore, unlike central NPSLE syndromes that might be caused by autoantibody, there should be a direct relationship between the serum antibody titer and clinical symptoms. Because complement components are not limited in the circulation, antibody-mediated damage in the PNS might be more likely to involve complement activation than antibody-mediated damage in the CNS. Peripheral nerve conduction velocities (NCV) in randomly selected lupus patients (without clinical evidence of peripheral neuropathy) demonstrate fluctuations over time, both shortening and lengthening of the NCV [130]. These data are reminiscent of the longitudinal studies of cognition in SLE that also suggest variability and reversible deficits. Response to treatment with immunosuppression and plasmapheresis provides indirect evidence for immune or antibody-mediated disease. APL antibodies have also been implicated in pathogenesis.

Differences in definition and ascertainment may account for the extreme variability in frequency reported (6–93%) for autonomic dysfunction in lupus. Autonomic neuropathies involve the sympathetic and/or parasympathetic neurons; symptoms, if present at all, are nonspecific and testing is infrequently done. Cardiac autonomic tone is measured in heart rate variability, baroreflex sensitivity, and various cardiovascular reflex tests. Most cross-sectional studies of autonomic dysfunction in lupus demonstrate no associations between disease activity, damage, neuropsychiatric disease, or presence of peripheral neuropathy [131–133]. However, depressed vagal activity has been demonstrated in many diseases other than SLE and consistently associates with poor prognosis [134].

Pupillary autonomic neuropathy, presumably a sign of CNS involvement, has been associated [135] and acute autonomic neuropathy has also been postulated as a cause of sudden death in SLE [136]. One study has recently reported a correlation between low vagal tone and serum albumin in SLE, suggesting that low albumin may be a surrogate marker for autonomic dysfunction in SLE [131]. Correlation with APL antibodies has been demonstrated in some studies but not others [133, 137] and autoantibodies to the alpha3-AChR have also been associated with autonomic dysfunction in small numbers of patients [127].

CLINICAL STUDIES OF PATHOGENESIS

NPSLE Syndromes

Neuropathological evidence demonstrates that NPSLE causes tissue damage through vascular occlusion, vascular inflammation, disruption the blood—brain barrier, and direct antibody or cytokine toxicity. It may be instructive to review the potential tissue destructive mechanisms that are implicated in the particular NPSLE syndromes.

The classification of NP syndromes by the ACR Ad Hoc committee resulted in 19 case definitions complete with diagnostic criteria, exclusions, and guidelines for ascertainment, including laboratory and imaging techniques [138]. The intent was that these case definitions would be used as an organizational framework providing the clinical community with a defined set of syndromes that could attributed to NPSLE. It is critical to appreciate that none of the defined syndromes are unique to SLE and therefore it is not surprising that evaluation of the ACR NPSLE criteria for performance characteristics [139] yielded a low specificity of 0.46. It is also certain that many of the NPSLE syndromes can be the end result of different pathogenic insults. Therefore, some have argued that NPSLE may be better conceptualized as anatomically delineated categories of neurologic injury [140] or grouped as diffuse vs. focal disease [141]. The utility of these classification schemes has been effectively demonstrated in a recent report from Fragoso-Loyo et al. that shows that CSF antibodies rather than serum antibodies correlate with NPSLE symptoms in the CNS. Moreover, centrally occurring autoantibodies may become meaningful if patients are grouped according to patterns of neurologic injury rather than combined into an inclusive group of NPSLE [62, 142].

Central, Diffuse Syndromes

Diffuse NPSLE syndromes affecting the CNS include psychosis, cognitive dysfunction, acute confusional state, anxiety and mood disorders, headache, seizure disorders, and aseptic meningitis.

Psychosis

Patients presenting with psychosis tend to have normal CSF and brain MRI at the time of presentation compared to those who develop psychosis later in the course of disease, suggesting that the triggering insult to the CNS may not be an inflammatory one but rather one that alters neuron or neurotransmitter function. Investigators found a striking lack of CSF Ig in NPSLE patients with psychosis compared to those with cognitive deficits only [142]. There is little evidence to suggest an association between APL antibodies and psychosis. Initial associations of peripherally circulating antiribosomal P antibodies with psychosis [89] were contradicted by numerous studies [91]. Further studies focusing on CSF analyses rather than peripheral blood levels may elicit a more definitive role for antiribosomal P antibodies in lupus psychosis. It has recently been demonstrated that increased CSF IL-6 levels greater than 4.3 pg/ml have a sensitivity and specificity for the diagnosis of lupus psychosis of 87.5% and 92.3%. This may be a consequence of the disruptive effects of IL-6 on the BBB [113]. Additionally, patients with acute confusional state, anxiety disorder, mood disorder, and cognitive dysfunction all display elevated IL-6 levels. It is worth noting that patients with psychosis are also at higher risk for development of other diffuse NPSLE manifestations such as cognitive impairment, mood disturbances, intractable headache, seizures, and stroke.

Acute confusional states

The acute confusional states (ACSs) that cause fluctuation in the level of consciousness and encompass agitation, delirium, obtundation, stupor, or coma are among the more severe manifestations of NPSLE. Comparison between SLE patients without NP disease and NPSLE patients with and without ACS failed to demonstrate differences in peripheral serology [107]. Although CSF levels IL-6, IL-8, and IFN-α, BAFF, APRIL, and the autoantibodies anti-DNA, NMDAR, ribosomal P, and cardiolipin are all elevated in patients with ACS, they do not distinguish between ACS and other NPSLE syndromes. It is likely that ACS encompasses differing pathologic insults. For example, coma in SLE has been attributed to transient regional brain hyperperfusion demonstrated on SPECT scan [142], cerebral edema with CSF protein greater than 1600 [143], and diffuse cerebral white and grey matter lesions demonstrated on MRI [144].

Cognitive dysfunction

Cognitive dysfunction is recognized as one of the most prevalent NPSLE syndromes. The pathogenesis of insidious and progressive cognitive dysfunction may depend on a combination of autoantibodies, cytokines, metalloproteinases, neuropeptides, and endocrine factors. Although both antineuronal and anti-NMDAR antibodies have been demonstrated to bind neurons, only anti-NMDAR antibodies have been shown in a mouse model to mediate cognitive dysfunction. Complex attention, executive skills (e.g., planning, organizing, sequencing), memory (e.g., learning, recall), visual-spatial processing, language (e.g., verbal fluency), and psychomotor speed are all part of the fragile network of skills required for normal cognitive function. The variation in reported prevalence of cognitive dysfunction can be attributed to differences in assessment, classification of dysfunction, and populations selected. The complexity of the multitude of test batteries and the insensitivity of combining Z-scores on these batteries to define domains of cognitive impairment, and the lack of age-matched control non-patient performance compounded by the large number of confounding influences on cognition in SLE patients (depression, medications, pain, fatigue) highlight the need for multicenter, longitudinal studies using common methodologies to more precisely define the problem.

Mood and anxiety disorders

Mood and anxiety disorders are also among the more commonly reported manifestations of NPSLE; however, attribution of these diagnoses to SLE is particularly difficult. The ACR case definitions for mood and anxiety disorders are based on the DSM IV classification criteria and include major depressive-like episode; mood disorder with depressive, manic, or mixed features; prominent anxiety; panic disorder; panic attacks; obsessions; or compulsions. Both antiribosomal P (anti-P) and anti-NMDAR antibodies have been shown to mediate direct effects on neuronal cells. Human anti-P injected into brains of naive C3H mice results in depressive symptoms, with no discernible cognitive or motor effects, linked to neuronal apoptosis in hippocampus, amygdala, and olfactory regions (cingulate and piriform) [63]. Gao et al. recently reported that depressive symptoms in MRL/lpr mice occur prior to the onset of other organ involvement (renal, arthritis, skin, lymphadenopathy) and associate with elevated peripheral blood titers of antibodies to dsDNA, NMDAR, chromatin, and cardiolipin [71]. Additionally, these mice demonstrated abnormal function in the hippocampus and related cortical areas on MRS imaging. Anti-NMDAR antibodies have been shown to preferentially target the hippocampus and amygdala in nonimmune mice (depending on mode of entry into the CNS) resulting in abnormal memory and fear-conditioning responses; these mice were not tested for depressive symptoms, although the amygdala is a mediator for anxiety [145, 146]. Increased peripheral blood titers of anti-P and anti-NMDAR antibodies have been linked to depressive symptoms in some studies [101, 102, 147], but others have shown no associations [67–69, 101, 148]. Importantly, neurotoxic and behavioral effects of both antibodies are blocked with immune-targeted therapies, supporting evidence that psychiatric disturbances may represent primary manifestations of NPSLE. This is also supported by clinically isolated case reports of SLE-related catatonia, a severe form of mood disorder, responding to immunosuppression [149].

Lupus headache

Lupus headache (HA) is a controversial marker of NPSLE. The ACR classification scheme for HA disorders includes migraine with and without aura, tension HA, cluster HA, HA secondary to intracranial hypertension, intractable HA, and dural sinus thrombosis [138, 150]. One recent prospective study found a prevalence of 95%, although previously studies report a range from 33 to 78% [151]. A recent meta-analysis of published literature on lupus HA concluded that, with the caveat that all of the studies were retrospective and largely uncontrolled, there is no association between other SLE disease activity and HA, there are no unique characteristics of "lupus HA," and the prevalence of HA appears to be the same in patients and controls [152]. Furthermore, attempts to identify pathogenic mechanisms that would help assign attribution have failed to demonstrate clear associations between serum APL antibodies or Raynaud phenomenon and HA. However, at least one study has shown elevated APL antibodies in the CSF of patients with lupus headache [153]. Imaging studies demonstrate a variety of nonspecific lesions that neither distinguish between lupus HA and controls, nor between HA and other NP disorders such as cognitive dysfunction, depression, anxiety, and confusion [154–156]. Again, exceptions occur, as demonstrated by a recent study that showed structural lesions evolving over time in individuals with lupus HA compared to those with HA that do not meet the criteria for lupus HA. Thus, lupus HA may be associated with some mechanisms of neurotoxicity [154].

Seizure disorders

Seizure disorders have been reported in up to 51% of patients, although a range of 10–20% is more common. In all reported series, generalized tonic-clonic seizures occur more commonly than partial seizures [157–159] and they are the presenting symptom in approximately one-third of SLE patients with seizure disorders. CSF is usually unrevealing with normal protein levels and occasionally a mild pleocytosis. Brain imaging studies reveal evidence of ischemic stroke or hyperintense lesions suggestive of small vessel ischemia 25—45% of the time, highlighting the classically reported association between APL, cerebrovascular events, and seizures in SLE. Increased risk of seizures in SLE is associated with higher baseline disease activity, other manifestations of NPSLE (psychosis, stroke), abnormal MRI, ACL, and anti-Smith antibodies. Risks for recurrent seizures include male gender, prior strokes, partial seizures, onset of seizures within the first year of diagnosis, and APLS [157—159]. Ischemic vascular disease as a result of hypercoagulability has been thought to account for the strong association between APL antibodies and seizures; however, additional studies have suggested a direct neurotoxic role for these autoantibodies through direct binding [20] or inhibition of GABA (γ-aminobutyric acid) receptor-ion channel complex resulting in increased neuronal excitability [160]. Interestingly, autoantibodies directed against glutamate receptors (AMPA receptor GluR3 and NMDA receptor) have been identified in sera and CSF of subpopulations of otherwise nonautoimmune epilepsy patients [161]. This discussion of autoantibodies to NMDAR (particularly to the NR1 subunit) leaves aside the subject of paraneoplastic limbic encephalitis; a paraneoplastic syndrome mediated by anti-NMDAR autoantibodies associated primarily with ovarian tumors.

Aseptic meningitis

The clinical presentation of aseptic meningitis is identical to septic meningitis and it remains a diagnosis of exclusion. One analysis of SLE patients presenting with symptoms of meningeal irritation found that 60% were secondary to CNS infection [162]. Importantly, there were no differences between the septic and aseptic groups in terms of age, disease duration, clinical manifestations, CSF characteristics (leukocytosis and increased protein), and SLEDAI scores (both were associated with active disease at the time of occurrence). A low CSF/serum glucose ratio was not predictive of infection. Risk factors for both groups included peripheral lymphopenia and evidence of active SLE. Aseptic meningitis attributable to SLE has also been associated with presence of anti-U1RNP antibody [163]. Pathologic identification of vasculitic lesions in the leptomeninges is rare [1—3, 78] and the underlying inflammatory mechanisms remain unknown.

Central, Focal Syndromes

Central, focal NPSLE syndromes include stroke, demyelinating syndromes, movement disorders, and cranial neuropathies.

Cerebrovascular disease

Cerebrovascular insults to the nervous system may result in strokes, transient ischemic attacks, chronic cerebrovascular disease, movement disorders, myelopathy, and seizures. The most commonly accepted etiology for cerebrovascular disease in NPSLE is thrombosis secondary to presence of APL antibodies or the lupus anticoagulant (see the section titled "Vascular Pathology"). Although vasculitis is found commonly elsewhere in lupus patients, several autopsy studies have shown that this is distinctly rare in the CNS, occurring in less than 7% of cases [1—3, 78].

Movement disorders

Movement disorders, while rare, may appear as chorea or Parkinsonism. Parkinsonism has been reported in pediatric as well as adult SLE and symptoms include cogwheel rigidity, coarse tremors, poverty of facial expression, akinesia, and shuffling gait [164]. MRI inconsistently demonstrates basal ganglia lesions, whereas SPECT or PET may be reliable and sensitive techniques for detection of specific lesions [165]. Pathophysiology is unclear; vasculitis of the thalamostriate arteries, APL-associated coagulopathy leading to stroke, vasculopathy, direct toxic effects of immune complexes, and cytokines have all been suggested. Parkinsonism associated only with APL antibodies has been reported in patients with primary APLS, suggesting a vulnerability of brain endothelium to these autoantibodies [166—168]. Additionally, antibodies against dopaminergic cells have been demonstrated in SLE-related Parkinsonism [169]. Contrary to hypokinetic parkinsonism, patients with NPSLE may present with hyperkinetic movement disorders that include chorea and athetotis. Focal dystonias are also reported. The rarity of this syndrome (less than 2% of NPSLE) prohibits extensive research; however, several case reports and retrospective cohort reviews [12, 170—173] suggest that most lupus patients with chorea have circulating APL antibodies. The prognosis for parkinsonism or chorea related to SLE appears favorable. Reversibility suggests transient ischemia rather than stroke.

Cranial neuropathies

Cranial neuropathies are rarely reported manifestations of NPSLE. Involvement of the optic nerve (CII) is the most common, although prevalence is around 1%. Although ocular motor abnormalities have been reported in up to 29% of patients [174], true oculomotor palsy (involving cranial nerves III, IV, and VI) is rare [175, 176]. Isolated occurrences of vagal (C10) and hypoglossal (C12) nerve palsies presenting with hoarseness, nasal speech, nasal regurgitation of fluid, and marked dysphagia due to weakness of the palatal, laryngeal, and pharyngeal muscles have been reported in SLE [177—179]. Cranial neuropathies occurring as focal events are generally attributable to APL antibodies; however, improvement with immunosuppressive treatment suggests a more diffuse CNS pathology.

Optic neuritis is associated with neuromyelitis optica and is characterized by seropositivity for IgG antibody directed against aquporin-4, a water channel expressed on astrocytes that controls brain water homeostasis [180]. One recent case report suggests that these antibodies can also occur in SLE [181].

Pseudotumor cerebri typically presents with severe, intractable headache and is further characterized by the presence of papilledema, occasional VIth nerve palsy (abducens), absence of ventricular enlargement or space-occupying lesion on imaging studies, normal CSF, and markedly elevated intracranial pressure (ICP). Increased ICP is caused by overproduction or impaired absorption of CSF and/or increased cerebral blood flow. The pathophysiology of pseudotumor cerebri in SLE has been attributed to APL antibodies and cerebral venous thrombosis, but vasculitis, immune complex deposition, and direct antibody-mediated effects have also been implicated. [182].

Transverse myelitis (TM) is thought to be caused by a vasculitis and ischemic infarction of a segment of the spinal cord but has also been reported in the context of APL antibodies [183].

Neuroimaging Techniques

Brain tissue for pathophysiological studies is rarely available. Neuroimaging has therefore become an extremely important part of the diagnostic work-up for NPSLE. It is hoped that newer imaging techniques for assessment of nervous system damage and functional impairment together with studies of CSF antibody or cytokine profile will further inform our understanding of pathogenic mechanisms. MRI remains the modality of choice for assessment of anatomic structure; ischemic stroke, hemorrhage, and demyelinating lesions are all readily appreciated with MRI, but MRI provides few clues to etiology. Patients with diffuse presentations of NPSLE syndromes will frequently have normal MRI scans [184]. Moreover, up to 50% of SLE patients with no history of NPSLE have abnormal MRI scans and up to 75% of NPSLE patients have evidence of chronic lesions on MRI [185]. Nonetheless, recent postmortem studies demonstrate that premortem new MRI findings accurately represent pathologically confirmed cerebrovascular and brain injury [186].

The proposed pathologic mechanisms, vasculitis, vasculopathy, immune complexes, direct antibody-mediated neurotoxicity, thrombosis, and cytokine-induced inflammation, can lead to perfusion abnormalities, neuronal dysfunction, axonal damage, and microstructural damage that may not be detectable on MRI. Hypoperfusion without infarction may result in some of the reversible syndromes that are seen clinically. Therefore, magnetic resonance spectroscopy (MRS), positron emission tomography (PET), single photon emission computed tomography (SPECT), magnetization transfer imaging (MTI), diffusion-weighted imaging (DWI), and functional MRI (fMRI) have all been used for assessment of brain function to supplement anatomical MRI findings.

Global cerebral atrophy is frequently reported as one of the more common findings in pathologic and radiographic examinations of NPSLE and this has been attributed to both axonal damage with anterograde and retrograde changes [185] and white matter pathology. Associations with corticosteroids, older age, APL antibodies, disease duration and severity, cognitive impairment, depression and anxiety are not consistent across studies [187–190]. However, investigations of regional brain atrophy may prove to be more enlightening. Volumetric measurements of cerebral structures can be obtained from routine brain MRI. Thus far, specific reductions in corpus callosum and hippocampal volumes, in the context of global atrophy, have been shown to correlate both with disease duration, history of NPSLE events, and cognitive dysfunction irrespective of the presence of APL antibodies and corticosteroid doses [70, 187–190]. Additionally, reduction in amygdala volume has been correlated with NPSLE and presence of serum anti-NMDAR antibodies [191]. A recent study reported MRI frequencies of brain atrophy and focal lesions 35% and 8% of the time, respectively, in a cohort of 97 newly diagnosed SLE subjects, again demonstrating that brain pathology progresses in the absence of clinical symptoms.

FDG-PET (fluoro-deoxy-glucose PET) is a form of radionucleotide imaging that measures neuronal glucose uptake as a correlate of brain function. Hypometabolism in specific areas of the brain that appear normal on MRI scan has been documented in several SLE cohorts [192–195] and persistent regional hypoperfusion is predictive of significant clinical deterioration [195]. There is no correlation as yet with particular CSF abnormalities.

Magnetization transfer ratio imaging (MTRI) is an advanced MRI technique sensitive to the presence of macromolecules in brain tissue. The MTR is reflective of the amount of magnetization transferred between liquid and semisolid components of tissue and provides quantitative information on the integrity of brain tissue. Cross-sectional studies of whole brain assessments using MTR histograms have been performed in NPSLE [189, 196–199]. Decreased peak heights of whole-brain MTR histograms have been shown to differentiate NPSLE from healthy controls [198, 200] and distinguish active NPSLE from chronic NPSLE [196], although a more recent study did not report any differences between SLE and NPSLE patients [198]. One longitudinal study found a correlation between the peak height of whole-brain MTR histograms and clinical status of individual NPSLE patients[197] who demonstrated both improvement and deterioration over time. Importantly, no correlations have been observed between disease duration and global brain damage assessed with MTRI [189]. However, all of these studies have included a wide variety of both peripheral and central NPSLE syndromes, including cerebrovascular disease, which makes interpretation of the findings more difficult. A reported selective decrease in MTRI parameters specifically in the gray matter of NPSLE subjects compared with normal controls suggests neuronal targeting [199]. Another study has demonstrated significant associations between lower peak heights of MTR histograms in NPSLE brain parenchyma, white and grey matter, and a lower NAA:creatine ratio on MRS [198], again suggestive of neuronal dysfunction. Additionally, lower MTR parameters in white matter associated significantly with cognitive dysfunction. Decreased MTR parameters were correlated with the presence of serum APL IgM antibodies in patients with NPSLE and normal MRI scans [201], but MTRI has not been investigated in patients with anti-P and anti-NMDAR antibodies.

MRS provides assessment of the biochemical composition of tissue. N-acetyl aspartate:creatine (NAA:Cr) and choline:creatine (Cho:Cr) ratios represent neuronal/axonal integrity and demyelination or inflammation, respectively. Abnormally low NAA:Cr ratios have been demonstrated in areas represented as focal lesions on MRI and also in normal-appearing brain parenchyma. Clinically, these low NAA:Cr ratios have also correlated with diffuse NPSLE syndromes such as seizures, severe cognitive impairment, delirium, stroke, or coma compared to those with headache, minor affective disorders or cognitive disorders [202–205]. Increased Cho:Cre ratios have been associated specifically with cognitive impairment in two studies, suggesting increased destruction or inflammation in white matter [206, 207] as a pathologic mechanism. While there is evidence suggestive of reversible axonal dysfunction in some patients with lower NAA:Cr ratios at the time of active SLE with or without clinical NPSLE that improve with treatment (and vice versa) [70], it has also been shown that increased Cho:Cr ratios may be predictive of future hyperintense lesions on T2-weighted MRI at follow-up [208]. Two studies have combined MRS with SPECT scans (measuring perfusion) in patients with normal MRI and demonstrate that abnormal MRS metabolites combined with regional perfusion abnormalities also predict future development of new MRI lesions [209, 210]. Interestingly, abnormally low NAA:Cr ratios have been reported in SLE patients with active disease without clinical CNS involvement [70], perhaps suggesting permeability of

the BBB during disease flares. Thus, MRS may be a sensitive and useful modality for distinguishing between axonal damage and demyelinating or inflammatory processes, although its usefulness in differentiating between the different NPSLE syndromes remains limited.

SPECT imaging measures cerebral blood flow and thus is a surrogate measure of tissue integrity. The increased sensitivity of SPECT over MRI for detecting abnormal brain function has been well documented, particularly during acute episodes [192, 211–213]. As expected, patients with evidence of infarct display hypoperfusion in corresponding areas, however, virtually all patients with normal MRI and active CNS disease demonstrate perfusion defects, most commonly in the parietal lobes. Abnormal perfusion on SPECT has been correlated with increased severity of NPSLE syndromes [192]. Unfortunately, SPECT has not been successfully used to distinguish between different NPSLE subtypes [192, 211–214]. Longitudinal studies demonstrate improvement or complete reversal of perfusion deficits in some patients but not in others despite clinical improvement [212, 213, 215]. One study has utilized both SPECT and PET scans in patients with normal MRI scans and reported that abnormal PET and SPECT scans associate with serious NPSLE syndromes but normal glucose uptake in the context of decreased perfusion can be seen in SLE patients with or without NP symptoms [192]. The pitfalls of SPECT imaging are highlighted in a more recent study where it was demonstrated that although the combination of abnormal MRI and abnormal SPECT scans were a more specific measure of active NPSLE than either alone, 49% of SLE patients without CNS symptoms had abnormal SPECT, 37% had abnormal MRI, and 27% had abnormalities on both [216].

Diffusion-weighted imaging (DWI) is an MRI technique where the MR image contrast is made sensitive to the random motion of water molecules and therefore microstructural changes in the brain are reflected in increased diffusivity of water molecules within the extracellular space. DWI is used to calculate an apparent diffusion constant (ADC) of the tissue of interest and diffusion tensor imaging (DTI) can be utilized to obtain further structural parameters, such as directional diffusion properties of the water molecules. These combined data yield valuable information about the integrity of brain parenchyma and connectivity through direct measurement of white matter tracts [217, 218]. Regional increased diffusion has been demonstrated in the frontal lobe, splenium of the corpus callosum, the anterior internal capsule, thalamus, and amygdala of SLE patients with and without abnormal MRI, highlighting the increased sensitivity of this modality [191, 219]. Significantly decreased anisotropy, reflective of disrupted white matter tracts, was also seen in the context of normal MRI. Whole-brain tractography showed fewer fibers when compared with normal controls, suggestive of global white matter damage. Unfortunately, there were no serological or clinical correlates to NPSLE syndromes in this study. A smaller study of eight patients with active NPSLE also found significant regional increases in ADC and anisotropy in normal-appearing gray and white matter parenchyma on MRI [220] due to demyelination. Selective amygdala damage is more prevalent in SLE patients than controls. Furthermore, as previously stated, a trend was noted between amygdala damage and presence of the anti-NMDAR antibody, highlighting the potential utility of imaging on ascertainment of pathogenic mechanisms [191].

Functional magnetic resonance Imaging (fMRI) is a noninvasive imaging technology that produces high resolution images of regional brain activation patterns during a specific task performance by measuring oxygenated hemoglobin blood flow. fMRI has been used to demonstrate differences in brain activation patterns in response to a variety of memory, attention, and language processing tasks in pediatric lupus patients without NPSLE compared to healthy controls [221]. All of the SLE patients had increased cerebral activation compared to healthy controls. Six of ten of the pediatric SLE patients had evidence of mild cognitive impairment on formal testing. Another study demonstrated increased regional brain activation patterns in parietal and frontal lobes in response to a working memory task in NPSLE patients (nine SLE patients with central and peripheral, focal, and diffuse NPSLE syndromes) compared to patients with rheumatoid arthritis and healthy controls [222]. Additionally, Rocca et al. have reported increased cortical activation patterns in the frontal and parietal lobes in NPSLE subjects (mixture of cerebrovascular disease, cognitive dysfunction, aseptic meningitis, and mood disorder) during the performance of a simple motor task [223]. Increased cortical activation patterns have been described in other neurodegenerative diseases such as Alzheimer disease, HIV, and traumatic injury [224–227], and in those recovering from stroke [228], investigators hypothesize that disruption of normal neural circuitry leads to increased compensatory neural recruitment. The fMRI patterns to date suggest focal neurologic disruption but have not yet helped to identify pathogenic mechanisms.

In conclusion, the expanding use of PET, SPECT, MRS, MTI, DTI, and fMRI has added sensitivity for the detection of subclinical, regional,and reversible CNS abnormalities. Successful identification of mechanisms is likely to require the use of standardized cohorts of patients selected on the basis of specific NPSLE syndromes rather than combinations of NPSLE patients whose manifestations most likely represent different mechanisms such as acute confusional state or psychosis and stroke. Moreover, identifying CSF abnormalities that associate with particular imaging abnormalities will provide some clues to pathogenesis.

CONCLUSION

NP syndromes in SLE are likely to be mediated by autoantibodies and inflammatory mediators. The exact nature of the syndrome will depend on the anatomic location in the brain of the insult and also on the potential reversibility. Increased understanding of the BBB and its role in regulating access to the CNS has led to an appreciation of its permissive and preventive role in pathologic conditions. Once the BBB has been disrupted, there is potential for direct toxicity to cells of the CNS by autoantibodies and cytokines. New information about autoantibodies and cytokines suggest they contribute to the pathogenesis of the diffuse nervous system disorders through neuronal death and dysfunction. Vasculopathy and clotting abnormalities, also the result of autoantibodies and inflammation, contribute strongly to the focal nervous system disorders. Systemic levels of autoantibodies may not be so predictive of brain injury as cerebrospinal fluid levels. New imaging technology promises better diagnosis for focal brain injury, but more and larger studies with carefully selected subsets of NPSLE patients and correlations to CSF abnormalities are needed if imaging methods are to become a useful tool in studying mechanisms of brain injury.

References

[1] S.G. Ellis, M.A. Verity, Central nervous system involvement in systemic lupus erythematosus: a review of neuropathologic findings in 57 cases, 1955–1977, Semin Arthritis Rheum 8 (1979) 212–221.

[2] J.G. Hanly, N.M. Walsh, V. Sangalang, Brain pathology in systemic lupus erythematosus, J Rheumatol 19 (1992) 732–741.

[3] R.T. Johnson, E.P. Richardson, The neurological manifestations of systemic lupus erythematosus, Medicine (Baltimore) 47 (1968) 337–369.

[4] D. Goel, S.R. Reddy, C. Sundaram, A.K. Prayaga, L. Rajasekhar, G. Narsimulu, Active necrotizing cerebral vasculitis in systemic lupus erythematosus, Neuropathology 27 (2007) 561–565.

[5] T.J. Kleinig, B. Koszyca, P.C. Blumbergs, P. Thompson, Fulminant leucocytoclastic brainstem vasculitis in a patient with otherwise indolent systemic lupus erythematosus, Lupus 18 (2009) 486–490.

[6] T. Rizos, M. Siegelin, S. Hahnel, B. Storch-Hagenlocher, A. Hug, Fulminant onset of cerebral immunocomplex vasculitis as first manifestation of neuropsychiatric systemic lupus erythematosus (NPSLE), Lupus 18 (2009) 361–363.

[7] R.A. Asherson, R. Cervera, J.T. Merrill, D. Erkan, Antiphospholipid antibodies and the antiphospholipid syndrome: clinical significance and treatment, Seminars in thrombosis and hemostasis 34 (2008) 256—266.

[8] J.E. Salmon, P.G. de Groot, Pathogenic role of antiphospholipid antibodies, Lupus 17 (2008) 405—411.

[9] S.S. Pierangeli, P.P. Chen, E.B. Gonzalez, Antiphospholipid antibodies and the antiphospholipid syndrome: an update on treatment and pathogenic mechanisms, Current opinion in hematology 13 (2006) 366—375.

[10] J.F. Roldan, R.L. Brey, Neurologic manifestations of the antiphospholipid syndrome, Curr Rheumatol Rep 9 (2007) 109—115.

[11] A. Afeltra, P. Garzia, A.P. Mitterhofer, M. Vadacca, S. Galluzzo, F. Del Porto, et al., Neuropsychiatric lupus syndromes: relationship with antiphospholipid antibodies, Neurology 61 (2003) 108—110.

[12] T. Avcin, S.M. Benseler, P.N. Tyrrell, S. Cucnik, E.D. Silverman, A followup study of antiphospholipid antibodies and associated neuropsychiatric manifestations in 137 children with systemic lupus erythematosus, Arthritis and rheumatism 59 (2008) 206—213.

[13] L. Harel, C. Sandborg, T. Lee, E. von Scheven, Neuropsychiatric manifestations in pediatric systemic lupus erythematosus and association with antiphospholipid antibodies, J Rheumatol 33 (2006) 1873—1877.

[14] G. Sanna, M.L. Bertolaccini, M.J. Cuadrado, H. Laing, M.A. Khamashta, A. Mathieu, et al., Neuropsychiatric manifestations in systemic lupus erythematosus: prevalence and association with antiphospholipid antibodies, J Rheumatol 30 (2003) 985—992.

[15] H. Zaccagni, J. Fried, J. Cornell, P. Padilla, R.L. Brey, Soluble adhesion molecule levels, neuropsychiatric lupus and lupus-related damage, Front Biosci 9 (2004) 1654—1659.

[16] R. Aronovich, D. Gurwitz, Y. Kloog, J. Chapman, Antiphospholipid antibodies, thrombin and LPS activate brain endothelial cells and Ras-dependent pathways through distinct mechanisms, Immunobiology 210 (2005) 781—788.

[17] B. Caronti, V. Pittoni, G. Palladini, G. Valesini, Anti-beta 2-glycoprotein I antibodies bind to central nervous system, Journal of the neurological sciences 156 (1998) 211—219.

[18] M. Kent, F. Alvarez, E. Vogt, R. Fyffe, A.K. Ng, N. Rote, Monoclonal antiphosphatidylserine antibodies react directly with feline and murine central nervous system, J Rheumatol 24 (1997) 1725—1733.

[19] M.N. Kent, F.J. Alvarez, A.K. Ng, N.S. Rote, Ultrastructural localization of monoclonal antiphospholipid antibody binding to rat brain, Experimental neurology 163 (2000) 173—179.

[20] J. Chapman, M. Cohen-Armon, Y. Shoenfeld, A.D. Korczyn, Antiphospholipid antibodies permeabilize and depolarize brain synaptoneurosomes, Lupus 8 (1999) 127—133.

[21] A. Riccio, C. Andreassi, M.L. Eboli, Antiphospholipid antibodies bind to rat cerebellar granule cells: the role of N-methyl-D-aspartate receptors, Neurosci Lett 257 (1998) 116—118.

[22] A. Katzav, C.G. Pick, A.D. Korczyn, E. Oest, M. Blank, Y. Shoenfeld, et al., Hyperactivity in a mouse model of the antiphospholipid syndrome, Lupus 10 (2001) 496—499.

[23] S. Shrot, A. Katzav, A.D. Korczyn, Y. Litvinju, R. Hershenson, C.G. Pick, et al., Behavioral and cognitive deficits occur only after prolonged exposure of mice to antiphospholipid antibodies, Lupus 11 (2002) 736—743.

[24] D. Tanne, A. Katzav, O. Beilin, N.C. Grigoriadis, M. Blank, C.G. Pick, et al., Interaction of inflammation, thrombosis, aspirin and enoxaparin in CNS experimental antiphospholipid syndrome, Neurobiol Dis 30 (2008) 56—64.

[25] M.J. Roman, B.A. Shanker, A. Davis, M.D. Lockshin, L. Sammaritano, R. Simantov, et al., Prevalence and correlates of accelerated atherosclerosis in systemic lupus erythematosus, N Engl J Med 349 (2003) 2399—2406.

[26] B.H. Hahn, J. Grossman, W. Chen, M. McMahon, The pathogenesis of atherosclerosis in autoimmune rheumatic diseases: roles of inflammation and dyslipidemia, J Autoimmun 28 (2007) 69—75.

[27] Y. Sherer, H. Zinger, Y. Shoenfeld, Atherosclerosis in systemic lupus erythematosus. Autoimmunity 43, 98—102

[28] N.J. Abbott, A.A. Patabendige, D.E. Dolman, S.R. Yusof, D.J. Begley, Structure and function of the blood-brain barrier Neurobiol Dis 37, 13—25

[29] C.D. Breder, C. Hazuka, T. Ghayur, C. Klug, M. Huginin, K. Yasuda, et al., Regional induction of tumor necrosis factor alpha expression in the mouse brain after systemic lipopolysaccharide administration, Proceedings of the National Academy of Sciences of the United States of America 91 (1994) 11393—11397.

[30] H.E. de Vries, M.C. Blom-Roosemalen, M. van Oosten, A.G. de Boer, T.J. van Berkel, D.D. Breimer, et al., The influence of cytokines on the integrity of the blood-brain barrier in vitro, Journal of neuroimmunology 64 (1996) 37—43.

[31] Z. Fabry, K.M. Fitzsimmons, J.A. Herlein, T.O. Moninger, M.B. Dobbs, M.N. Hart, Production of the cytokines interleukin 1 and 6 by murine brain microvessel endothelium and smooth muscle pericytes, Journal of neuroimmunology 47 (1993) 23—34.

[32] B. Diamond, P.T. Huerta, P. Mina-Osorio, C. Kowal, B.T. Volpe, Losing your nerves? Maybe it's the antibodies, Nat Rev Immunol 9 (2009) 449—456.

[33] H. Wolburg, A. Lippoldt, Tight junctions of the blood-brain barrier: development, composition and regulation, Vascular pharmacology 38 (2002) 323—337.

[34] H. Wolburg, S. Noell, A. Mack, K. Wolburg-Buchholz, P. Fallier-Becker, Brain endothelial cells and the glio-vascular complex, Cell and tissue research 335 (2009) 75—96.

[35] C.M. Peppiatt, C. Howarth, P. Mobbs, D. Attwell, Bidirectional control of CNS capillary diameter by pericytes, Nature 443 (2006) 700—704.

[36] J. Satoh, H. Tabunoki, T. Yamamura, K. Arima, H. Konno, Human astrocytes express aquaporin-1 and aquaporin-4 in vitro and in vivo, Neuropathology 27 (2007) 245—256.

[37] G. Stoll, S. Jander, The role of microglia and macrophages in the pathophysiology of the CNS, Progress in neurobiology 58 (1999) 233—247.

[38] W.A. Banks, M.A. Erickson, The blood-brain barrier and immune function and dysfunction. Neurobiol Dis 37, 26—32

[39] P.M. Carvey, B. Hendey, A.J. Monahan, The blood-brain barrier in neurodegenerative disease: a rhetorical perspective, J Neurochem 111 (2009) 291—314.

[40] L. Minghetti, M.A. Ajmone-Cat, M.A. De Berardinis, R. De Simone, Microglial activation in chronic neurodegenerative diseases: roles of apoptotic neurons and chronic stimulation, Brain Res Brain Res Rev 48 (2005) 251—256.

[41] E. Neuwelt, N.J. Abbott, L. Abrey, W.A. Banks, B. Blakley, T. Davis, et al., Strategies to advance translational research into brain barriers, Lancet neurology 7 (2008) 84—96.

[42] B. Wispelwey, A.J. Lesse, E.J. Hansen, W.M. Scheld, Haemophilus influenzae lipopolysaccharide-induced blood brain barrier permeability during experimental meningitis in the rat, J Clin Invest 82 (1988) 1339—1346.

[43] L.B. Jaeger, S. Dohgu, R. Sultana, J.L. Lynch, J.B. Owen, M.A. Erickson, et al., Lipopolysaccharide alters the blood-brain barrier transport of amyloid beta protein: a mechanism for

inflammation in the progression of Alzheimer's disease, Brain Behav Immun 23 (2009) 507–517.

[44] S. Verma, R. Nakaoke, S. Dohgu, W.A. Banks, Release of cytokines by brain endothelial cells: A polarized response to lipopolysaccharide, Brain Behav Immun 20 (2006) 449–455.

[45] Q. Li, Q. Zhang, C. Wang, X. Liu, N. Li, J. Li, Disruption of tight junctions during polymicrobial sepsis in vivo, J Pathol 218 (2009) 210–221.

[46] T. Nishioku, S. Dohgu, F. Takata, T. Eto, N. Ishikawa, K.B. Kodama, et al., Detachment of brain pericytes from the basal lamina is involved in disruption of the blood-brain barrier caused by lipopolysaccharide-induced sepsis in mice, Cell Mol Neurobiol 29 (2009) 309–316.

[47] S.A. Hamed, E.A. Hamed, M.M. Abdella, Septic encephalopathy: relationship to serum and cerebrospinal fluid levels of adhesion molecules, lipid peroxides and S-100. B protein, Neuropediatrics 40 (2009) 66–72.

[48] C. Gerard, Complement C5. a in the sepsis syndrome—too much of a good thing? N Engl J Med 348 (2003) 167–169.

[49] P.A. Ward, The dark side of C5. a in sepsis, Nat Rev Immunol 4 (2004) 133–142.

[50] M.A. Flierl, P.F. Stahel, D. Rittirsch, M. Huber-Lang, A.D. Niederbichler, L.M. Hoesel, et al., Inhibition of complement C5. a prevents breakdown of the blood-brain barrier and pituitary dysfunction in experimental sepsis, Critical care (London, England) 13 (2009) R12.

[51] A. Jacob, B. Hack, E. Chiang, J.G. Garcia, R.J. Quigg, J.J. Alexander, C5 a alters blood-brain barrier integrity in experimental lupus, FASEB J 24 (2010) 1682–1688.

[52] P. Gasque, J.W. Neal, S.K. Singhrao, E.P. McGreal, Y.D. Dean, B.J. Van, et al., Roles of the complement system in human neurodegenerative disorders: pro-inflammatory and tissue remodeling activities, Molecular neurobiology 25 (2002) 1–17.

[53] T.A. Brooks, B.T. Hawkins, J.D. Huber, R.D. Egleton, T.P. Davis, Chronic inflammatory pain leads to increased blood-brain barrier permeability and tight junction protein alterations, American journal of physiology 289 (2005) H738–H743.

[54] G. McCaffrey, M.J. Seelbach, W.D. Staatz, N. Nametz, C. Quigley, C.R. Campos, et al., Occludin oligomeric assembly at tight junctions of the blood-brain barrier is disrupted by peripheral inflammatory hyperalgesia, J Neurochem 106 (2008) 2395–2409.

[55] B.T. Hawkins, T.J. Abbruscato, R.D. Egleton, R.C. Brown, J.D. Huber, C.R. Campos, et al., Nicotine increases in vivo blood-brain barrier permeability and alters cerebral microvascular tight junction protein distribution, Brain Res 1027 (2004) 48–58.

[56] M. Hossain, T. Sathe, V. Fazio, P. Mazzone, B. Weksler, D. Janigro, et al., Tobacco smoke: a critical etiological factor for vascular impairment at the blood-brain barrier, Brain Res 1287 (2009) 192–205.

[57] J. Roth, E.M. Harre, C. Rummel, R. Gerstberger, T. Hubschle, Signaling the brain in systemic inflammation: role of sensory circumventricular organs, Front Biosci 9 (2004) 290–300.

[58] C.R. Kuhlmann, C.M. Zehendner, M. Gerigk, D. Closhen, B. Bender, P. Friedl, et al., MK801 blocks hypoxic blood-brain-barrier disruption and leukocyte adhesion, Neurosci Lett 449 (2009) 168–172.

[59] A. Reijerkerk, G. Kooij, S.M. van der Pol, T. Leyen, K. Lakeman, B. van Het Hof, et al., The NR1 subunit of NMDA receptor regulates monocyte transmigration through the brain endothelial cell barrier, J Neurochem 113 (2010) 447–453.

[60] S. Man, E.E. Ubogu, R.M. Ransohoff, Inflammatory cell migration into the central nervous system: a few new twists on an old tale, Brain pathology (Zurich, Switzerland) 17 (2007) 243–250.

[61] I. Ifergan, H. Kebir, M. Bernard, K. Wosik, A. Dodelet-Devillers, R. Cayrol, et al., The blood-brain barrier induces differentiation of migrating monocytes into Th17-polarizing dendritic cells, Brain 131 (2008) 785–799.

[62] H. Fragoso-Loyo, J. Cabiedes, A. Orozco-Narvaez, L. Davila-Maldonado, Y. Atisha-Fregoso, B. Diamond, et al., Serum and cerebrospinal fluid autoantibodies in patients with neuropsychiatric lupus erythematosus, Implications for diagnosis and pathogenesis (2008). PloS one 3, e3347.

[63] A. Katzav, I. Solodeev, O. Brodsky, J. Chapman, C.G. Pick, M. Blank, et al., Induction of autoimmune depression in mice by anti-ribosomal P antibodies via the limbic system, Arthritis and rheumatism 56 (2007) 938–948.

[64] C. Kowal, L.A. Degiorgio, J.Y. Lee, M.A. Edgar, P.T. Huerta, B.T. Volpe, et al., Human lupus autoantibodies against NMDA receptors mediate cognitive impairment, Proceedings of the National Academy of Sciences of the United States of America 103 (2006) 19854–19859.

[65] C. Kowal, L.A. DeGiorgio, T. Nakaoka, H. Hetherington, P.T. Huerta, B. Diamond, et al., Cognition and immunity; antibody impairs memory, Immunity 21 (2004) 179–188.

[66] M.M. Sidor, B. Sakic, P.M. Malinowski, D.A. Ballok, C.J. Oleschuk, J. Macri, Elevated immunoglobulin levels in the cerebrospinal fluid from lupus-prone mice, Journal of neuroimmunology 165 (2005) 104–113.

[67] J.G. Hanly, J. Robichaud, J.D. Fisk, Anti-NR2 glutamate receptor antibodies and cognitive function in systemic lupus erythematosus, J Rheumatol 33 (2006) 1553–1558.

[68] M.J. Harrison, L.D. Ravdin, M.D. Lockshin, Relationship between serum NR2. a antibodies and cognitive dysfunction in systemic lupus erythematosus, Arthritis and rheumatism 54 (2006) 2515–2522.

[69] G. Steup-Beekman, S. Steens, M. van Buchem, T. Huizinga, Anti-NMDA receptor autoantibodies in patients with systemic lupus erythematosus and their first-degree relatives, Lupus 16 (2007) 329–334.

[70] S. Appenzeller, L.M. Li, L.T. Costallat, F. Cendes, Evidence of reversible axonal dysfunction in systemic lupus erythematosus: a proton MRS study, Brain 128 (2005) 2933–2940.

[71] H.X. Gao, S.R. Campbell, M.H. Cui, P. Zong, J. Hee-Hwang, M. Gulinello, et al., Depression is an early disease manifestation in lupus-prone MRL/lpr mice, Journal of neuroimmunology 207 (2009) 45–56.

[72] J.J. Alexander, C. Zwingmann, R. Quigg, MRL/lpr mice have alterations in brain metabolism as shown with [1. H-13. C] NMR spectroscopy, Neurochemistry international 47 (2005) 143–151.

[73] H.M. Belmont, P. Hopkins, H.S. Edelson, H.B. Kaplan, R. Ludewig, G. Weissmann, et al., Complement activation during systemic lupus erythematosus. C3. a and C5. a anaphylatoxins circulate during exacerbations of disease, Arthritis and rheumatism 29 (1986) 1085–1089.

[74] J.J. Alexander, L. Bao, A. Jacob, D.M. Kraus, V.M. Holers, R.J. Quigg, Administration of the soluble complement inhibitor, Crry-Ig, reduces inflammation and aquaporin 4 expression in lupus cerebritis, Biochimica et biophysica acta 1639 (2003) 169–176.

[75] J.J. Alexander, A. Jacob, P. Vezina, H. Sekine, G.S. Gilkeson, R.J. Quigg, Absence of functional alternative complement pathway alleviates lupus cerebritis, Eur J Immunol 37 (2007) 1691–1701.

[76] J.S. Jung, R.V. Bhat, G.M. Preston, W.B. Guggino, J.M. Baraban, P. Agre, Molecular characterization of an aquaporin cDNA from brain: candidate osmoreceptor and regulator of water balance, Proceedings of the National Academy of Sciences of the United States of America 91 (1994) 13052–13056.

[77] K. Nishimura, M. Harigai, M. Omori, E. Sato, M. Hara, Blood-brain barrier damage as a risk factor for corticosteroid-induced psychiatric disorders in systemic lupus erythematosus, Psychoneuroendocrinology 33 (2008) 395–403.

[78] O. Devinsky, C.K. Petito, D.R. Alonso, Clinical and neuropathological findings in systemic lupus erythematosus: the role of vasculitis, heart emboli, and thrombotic thrombocytopenic purpura, Ann Neurol 23 (1988) 380–384.

[79] N.J. Scolding, F.G. Joseph, The neuropathology and pathogenesis of systemic lupus erythematosus, Neuropathol Appl Neurobiol 28 (2002) 173–189.

[80] X. Ma, J. Foster, B. Sakic, Distribution and prevalence of leukocyte phenotypes in brains of lupus-prone mice, Journal of neuroimmunology 179 (2006) 26–36.

[81] J.G. Sled, S. Spring, M. van Eede, J.P. Lerch, S. Ullal, B. Sakic, Time course and nature of brain atrophy in the MRL mouse model of central nervous system lupus, Arthritis and rheumatism 60 (2009) 1764–1774.

[82] E. Trysberg, K. Blennow, O. Zachrisson, A. Tarkowski, Intrathecal levels of matrix metalloproteinases in systemic lupus erythematosus with central nervous system engagement, Arthritis Res Ther 6 (2004) R551–R556.

[83] E. Trysberg, K. Nylen, L.E. Rosengren, A. Tarkowski, Neuronal and astrocytic damage in systemic lupus erythematosus patients with central nervous system involvement, Arthritis and rheumatism 48 (2003) 2881–2887.

[84] T. Kasama, T. Odai, K. Wakabayashi, N. Yajima, Y. Miwa, Chemokines in systemic lupus erythematosus involving the central nervous system, Front Biosci 13 (2008) 2527–2536.

[85] E.S. Husebye, Z.M. Sthoeger, M. Dayan, H. Zinger, D. Elbirt, M. Levite, et al., Autoantibodies to a NR2. A peptide of the glutamate/NMDA receptor in sera of patients with systemic lupus erythematosus, Annals of the rheumatic diseases 64 (2005) 1210–1213.

[86] T. Yoshio, H. Okamoto, S. Minota, Antibodies to bovine serum albumin do not affect the results of enzyme-linked immunosorbent assays for IgG anti-NR2 glutamate receptor antibodies: reply to the letter by Hirohata et al. Arthritis and rheumatism 56 (2007) 2813–2814.

[87] T. Yoshio, K. Onda, H. Nara, S. Minota, Association of IgG anti-NR2 glutamate receptor antibodies in cerebrospinal fluid with neuropsychiatric systemic lupus erythematosus, Arthritis and rheumatism 54 (2006) 675–678.

[88] D. Lefranc, D. Launay, S. Dubucquoi, J. de Seze, P. Dussart, M. Vermersch, et al., Characterization of discriminant human brain antigenic targets in neuropsychiatric systemic lupus erythematosus using an immunoproteomic approach, Arthritis and rheumatism 56 (2007) 3420–3432.

[89] E. Bonfa, S.J. Golombek, L.D. Kaufman, S. Skelly, H. Weissbach, N. Brot, et al., Association between lupus psychosis and anti-ribosomal P protein antibodies, N Engl J Med 317 (1987) 265–271.

[90] K.B. Elkon, A.P. Parnassa, C.L. Foster, Lupus autoantibodies target ribosomal P proteins, J Exp Med 162 (1985) 459–471.

[91] F.B. Karassa, A. Afeltra, A. Ambrozic, D.M. Chang, F. De Keyser, A. Doria, et al., Accuracy of anti-ribosomal P protein antibody testing for the diagnosis of neuropsychiatric systemic lupus erythematosus: an international meta-analysis, Arthritis and rheumatism 54 (2006) 312–324.

[92] S. Matus, P.V. Burgos, M. Bravo-Zehnder, R. Kraft, O.H. Porras, P. Farias, et al., Antiribosomal-P autoantibodies from psychiatric lupus target a novel neuronal surface protein causing calcium influx and apoptosis, J Exp Med 204 (2007) 3221–3234.

[93] A. Katzav, T. Ben-Ziv, J. Chapman, M. Blank, M. Reichlin, Y. Shoenfeld, Anti-P ribosomal antibodies induce defect in smell capability in a model of CNS -SLE (depression), J Autoimmun 31 (2008) 393–398.

[94] N. Shoenfeld, N. Agmon-Levin, I. Flitman-Katzevman, D. Paran, B.S. Katz, S. Kivity, et al., The sense of smell in systemic lupus erythematosus, Arthritis and rheumatism 60 (2009) 1484–1487.

[95] L.A. DeGiorgio, K.N. Konstantinov, S.C. Lee, J.A. Hardin, B.T. Volpe, B. Diamond, A subset of lupus anti-DNA antibodies cross-reacts with the NR2 glutamate receptor in systemic lupus erythematosus, Nat Med 7 (2001) 1189–1193.

[96] G.L. Collingridge, S.J. Kehl, H. McLennan, Excitatory amino acids in synaptic transmission in the Schaffer collateral-commissural pathway of the rat hippocampus, J Physiol 334 (1983) 33–46.

[97] T.V. Bliss, T. Lomo, Long-lasting potentiation of synaptic transmission in the dentate area of the anaesthetized rabbit following stimulation of the perforant path, J Physiol 232 (1973) 331–356.

[98] P.T. Huerta, B.T. Volpe, Transcranial magnetic stimulation, synaptic plasticity and network oscillations, Journal of neuroengineering and rehabilitation 6 (2009) 7.

[99] G.E. Hardingham, Y. Fukunaga, H. Bading, Extrasynaptic NMDARs oppose synaptic NMDARs by triggering CREB shutoff and cell death pathways, Nat Neurosci 5 (2002) 405–414.

[100] P.T. Huerta, C. Kowal, L.A. DeGiorgio, B.T. Volpe, B. Diamond, Immunity and behavior: antibodies alter emotion, Proceedings of the National Academy of Sciences of the United States of America 103 (2006) 678–683.

[101] L. Lapteva, M. Nowak, C.H. Yarboro, K. Takada, T. Roebuck-Spencer, T. Weickert, et al., Anti-N-methyl-D-aspartate receptor antibodies, cognitive dysfunction, and depression in systemic lupus erythematosus, Arthritis and rheumatism 54 (2006) 2505–2514.

[102] R. Omdal, K. Brokstad, K. Waterloo, W. Koldingsnes, R. Jonsson, S.I. Mellgren, Neuropsychiatric disturbances in SLE are associated with antibodies against NMDA receptors, Eur J Neurol 12 (2005) 392–398.

[103] Y. Arinuma, T. Yanagida, S. Hirohata, Association of cerebrospinal fluid anti-NR2 glutamate receptor antibodies with diffuse neuropsychiatric systemic lupus erythematosus, Arthritis and rheumatism 58 (2008) 1130–1135.

[104] N. Yajima, T. Kasama, T. Isozaki, T. Odai, M. Matsunawa, M. Negishi, et al., Elevated levels of soluble fractalkine in active systemic lupus erythematosus: potential involvement in neuropsychiatric manifestations, Arthritis and rheumatism 52 (2005) 1670–1675.

[105] S. Hirohata, T. Miyamoto, Elevated levels of interleukin-6 in cerebrospinal fluid from patients with systemic lupus erythematosus and central nervous system involvement, Arthritis and rheumatism 33 (1990) 644–649.

[106] E. Trysberg, H. Carlsten, A. Tarkowski, Intrathecal cytokines in systemic lupus erythematosus with central nervous system involvement, Lupus 9 (2000) 498–503.

[107] Y. Katsumata, M. Harigai, Y. Kawaguchi, C. Fukasawa, M. Soejima, K. Takagi, et al., Diagnostic reliability of cerebral spinal fluid tests for acute confusional state (delirium) in patients with systemic lupus erythematosus: interleukin 6 (IL-6), IL-8, interferon-alpha, IgG index, and Q-albumin, J Rheumatol 34 (2007) 2010–2017.

[108] A. George-Chandy, E. Trysberg, K. Eriksson, Raised intrathecal levels of APRIL and BAFF in patients with systemic lupus erythematosus: relationship to neuropsychiatric symptoms, Arthritis Res Ther 10 (2008) R97.

I. BASIS OF DISEASE PATHOGENESIS

[109] A.I. Kaplin, D.M. Deshpande, E. Scott, C. Krishnan, J.S. Carmen, I. Shats, et al., IL-6 induces regionally selective spinal cord injury in patients with the neuroinflammatory disorder transverse myelitis, J Clin Invest 115 (2005) 2731–2741.

[110] A. Quintana, M. Muller, R.F. Frausto, R. Ramos, D.R. Getts, E. Sanz, et al., Site-specific production of IL-6 in the central nervous system retargets and enhances the inflammatory response in experimental autoimmune encephalomyelitis, J Immunol 183 (2009) 2079–2088.

[111] J. Alcocer-Varela, D. Aleman-Hoey, D. Alarcon-Segovia, Interleukin-1 and interleukin-6 activities are increased in the cerebrospinal fluid of patients with CNS lupus erythematosus and correlate with local late T-cell activation markers, Lupus 1 (1992) 111–117.

[112] H. Fragoso-Loyo, Y. Richaud-Patin, A. Orozco-Narvaez, L. Davila-Maldonado, Y. Atisha-Fregoso, L. Llorente, et al., Interleukin-6 and chemokines in the neuropsychiatric manifestations of systemic lupus erythematosus, Arthritis and rheumatism 56 (2007) 1242–1250.

[113] S. Hirohata, Y. Kanai, A. Mitsuo, Y. Tokano, H. Hashimoto, Accuracy of cerebrospinal fluid IL-6 testing for diagnosis of lupus psychosis. A multicenter retrospective study, Clin Rheumatol 28 (2009) 1319–1323.

[114] C.L. Raison, M. Demetrashvili, L. Capuron, A.H. Miller, Neuropsychiatric adverse effects of interferon-alpha: recognition and management, CNS drugs 19 (2005) 105–123.

[115] D.M. Santer, T. Yoshio, S. Minota, T. Moller, K.B. Elkon, Potent induction of IFN-alpha and chemokines by autoantibodies in the cerebrospinal fluid of patients with neuropsychiatric lupus, J Immunol 182 (2009) 1192–1201.

[116] J.B. Winfield, M. Shaw, L.M. Silverman, R.A. Eisenberg, H.A. Wilson 3. rd, D. Koffler, Intrathecal IgG synthesis and blood-brain barrier impairment in patients with systemic lupus erythematosus and central nervous system dysfunction, Am J Med 74 (1983) 837–844.

[117] E. Trysberg, K. Hoglund, E. Svenungsson, K. Blennow, A. Tarkowski, Decreased levels of soluble amyloid beta-protein precursor and beta-amyloid protein in cerebrospinal fluid of patients with systemic lupus erythematosus, Arthritis Res Ther 6 (2004) R129–R136.

[118] T.C. Hsu, Y.C. Chen, W.X. Lai, S.Y. Chiang, C.Y. Huang, B.S. Tzang, Beneficial effects of treatment with cystamine on brain in NZB/W F1 mice, European journal of pharmacology 591 (2008) 307–314.

[119] L.G. Morphis, D. Gitlin, Maturation of the maternofoetal transport system for human gamma-globulin in the mouse, Nature 228 (1970) 573.

[120] J.P. Buyon, S.H. Swersky, H.E. Fox, F.Z. Bierman, R.J. Winchester, Intrauterine therapy for presumptive fetal myocarditis with acquired heart block due to systemic lupus erythematosus. Experience in a mother with a predominance of SS-B (La) antibodies, Arthritis and rheumatism 30 (1987) 44–49.

[121] L. Chameides, R.C. Truex, V. Vetter, W.J. Rashkind, F.M. Galioto Jr., J.A. Noonan, Association of maternal systemic lupus erythematosus with congenital complete heart block, N Engl J Med 297 (1977) 1204–1207.

[122] J.S. Scott, P.J. Maddison, P.V. Taylor, E. Esscher, O. Scott, R.P. Skinner, Connective-tissue disease, antibodies to ribonucleoprotein, and congenital heart block, N Engl J Med 309 (1983) 209–212.

[123] R.B. Winkler, A.H. Nora, J.J. Nora, Familial congenital complete heart block and maternal systemic lupus erythematosis, Circulation 56 (1977) 1103–1107.

[124] J.Y. Lee, P.T. Huerta, J. Zhang, C. Kowal, E. Bertini, B.T. Volpe, et al., Neurotoxic autoantibodies mediate congenital cortical

impairment of offspring in maternal lupus, Nat Med 15 (2009) 91–96.

[125] L. Servioli, C. Perez, S. Consani, A. Suarez, G. Sehabiaga, C. Collazo, et al., Prevalence and characteristics of immuno-mediated neuropathies in a group of patients with autoimmune diseases, Journal of clinical neuromuscular disease 9 (2007) 285–290.

[126] P.A. McCombe, J.G. McLeod, J.D. Pollard, Y.P. Guo, T.J. Ingall, Peripheral sensorimotor and autonomic neuropathy associated with systemic lupus erythematosus. Clinical, pathological and immunological features, Brain 110 (Pt 2) (1987) 533–549.

[127] A. McKeon, V.A. Lennon, D.H. Lachance, R.D. Fealey, S.J. Pittock, Ganglionic acetylcholine receptor autoantibody: oncological, neurological, and serological accompaniments, Arch Neurol 66 (2009) 735–741.

[128] C. Huynh, S.L. Ho, K.Y. Fong, R.T. Cheung, C.C. Mok, C.S. Lau, Peripheral neuropathy in systemic lupus erythematosus, J Clin Neurophysiol 16 (1999) 164–168.

[129] Y. Matsuki, T. Hidaka, M. Matsumoto, K. Fukushima, K. Suzuki, Systemic lupus erythematosus demonstrating serum anti-GM1 antibody, with sudden onset of drop foot as the initial presentation, Intern Med 38 (1999) 729–732.

[130] R. Omdal, S. Loseth, T. Torbergsen, W. Koldingsnes, G. Husby, S.I. Mellgren, Peripheral neuropathy in systemic lupus erythematosus—a longitudinal study, Acta neurologica Scandinavica 103 (2001) 386–391.

[131] S.T. Huang, G.Y. Chen, C.H. Wu, C.D. Kuo, Effect of disease activity and position on autonomic nervous modulation in patients with systemic lupus erythematosus, Clin Rheumatol 27 (2008) 295–300.

[132] R. Shalimar, Handa, K.K. Deepak, M. Bhatia, P. Aggarwal, R.M. Pandey, Autonomic dysfunction in systemic lupus erythematosus, Rheumatology international 26 (2006) 837–840.

[133] L. Stojanovich, B. Milovanovich, S.R. de Luka, D. Popovich-Kuzmanovich, V. Bisenich, B. Djukanovich, et al., Cardiovascular autonomic dysfunction in systemic lupus, rheumatoid arthritis, primary Sjogren syndrome and other autoimmune diseases, Lupus 16 (2007) 181–185.

[134] T.G. Buchman, P.K. Stein, B. Goldstein, Heart rate variability in critical illness and critical care, Current opinion in critical care 8 (2002) 311–315.

[135] R.H. Straub, M. Zeuner, G. Lock, H. Rath, R. Hein, J. Scholmerich, et al., Autonomic and sensorimotor neuropathy in patients with systemic lupus erythematosus and systemic sclerosis, J Rheumatol 23 (1996) 87–92.

[136] C.J. Laversuch, H. Seo, H. Modarres, D.A. Collins, W. McKenna, B.E. Bourke, Reduction in heart rate variability in patients with systemic lupus erythematosus, J Rheumatol 24 (1997) 1540–1544.

[137] M. Magaro, L. Mirone, L. Altomonte, A. Zoli, S. Angelosante, Lack of correlation between anticardiolipin antibodies and peripheral autonomic nerve involvement in systemic lupus erythematosus, Clin Rheumatol 11 (1992) 231–234.

[138] The American College of Rheumatology nomenclature and case definitions for neuropsychiatric lupus syndromes, Arthritis and rheumatism 42 (1999) 599–608.

[139] H. Ainiala, A. Hietaharju, J. Loukkola, J. Peltola, M. Korpela, R. Metsanoja, et al., Validity of the new American College of Rheumatology criteria for neuropsychiatric lupus syndromes: a population-based evaluation, Arthritis and rheumatism 45 (2001) 419–423.

[140] J. Sanchez-Guerrero, C. Aranow, M. Mackay, B. Volpe, B. Diamond, Neuropsychiatric systemic lupus erythematosus reconsidered, Nat Clin Pract Rheumatol 4 (2008) 112–113.

[141] J.G. Hanly, New insights into central nervous system lupus: a clinical perspective, Curr Rheumatol Rep 9 (2007) 116–124.

[142] L. Stojanovich, D. Smiljanich-Miljkovich, R. Omdal, B. Sakic, Neuropsychiatric lupus and association with cerebrospinal fluid immunoglobulins: a pilot study, Isr Med Assoc J 11 (2009) 359–362.

[143] M.T. Bianchi, C. Lavigne, F. Sorond, B. Bermas, Transient life-threatening cerebral edema in a patient with systemic lupus erythematosus, J Clin Rheumatol 15 (2009) 181–184.

[144] M. Shibata, T. Kibe, S. Fujimoto, T. Ishikawa, M. Murakami, T. Ichiki, et al., Diffuse central nervous system lupus involving white matter, basal ganglia, thalami and brainstem, Brain & development 21 (1999) 337–340.

[145] Etkin, A., Egner, T., Peraza, D.M., Kandel, E.R., Hirsch, J. Resolving emotional conflict: a model for amygdalar modulation by the rostral anterior cingulate cortex. Neuron 51 (2006) 1–12.

[146] A. Etkin, K.C. Klemenhagen, J.T. Dudman, M.T. Rogan, R. Hen, E.R. Kandel, et al., Individual differences in trait anxiety predict the response of the basolateral amygdala to unconsciously processed fearful faces, Neuron 44 (2004) 1043–1055.

[147] T. Eber, J. Chapman, Y. Shoenfeld, Anti-ribosomal P-protein and its role in psychiatric manifestations of systemic lupus erythematosus: myth or reality? Lupus 14 (2005) 571–575.

[148] F.G. Nery, E.F. Borba, V.S. Viana, J.P. Hatch, J.C. Soares, E. Bonfa, et al., Prevalence of depressive and anxiety disorders in systemic lupus erythematosus and their association with anti-ribosomal P antibodies, Progress in neuro-psychopharmacology & biological psychiatry 32 (2008) 695–700.

[149] L. Brelinski, O. Cottencin, D. Guardia, J.D. Anguill, V. Queyrel, P.Y. Hatron, et al., Catatonia and systemic lupus erythematosus: a clinical study of three cases, General hospital psychiatry 31 (2009) 90–92.

[150] B.A. Cruz, L.A. Santos, R.P. Damasceno, L.S. Ribeiro, G.A. Xavier Jr., S.V. Nunes, et al., Dural sinus thrombosis in childhood systemic lupus erythematosus, J Rheumatol 28 (2001) 2140–2141.

[151] R. Davey, J. Bamford, P. Emery, The ACR classification criteria for headache disorders in SLE fail to classify certain prevalent headache types, Cephalalgia 28 (2008) 296–299.

[152] D.D. Mitsikostas, P.P. Sfikakis, P.J. Goadsby, A meta-analysis for headache in systemic lupus erythematosus: the evidence and the myth, Brain 127 (2004) 1200–1209.

[153] N.S. Lai, J.L. Lan, Evaluation of cerebrospinal anticardiolipin antibodies in lupus patients with neuropsychiatric manifestations, Lupus 9 (2000) 353–357.

[154] S. Bicakci, S. Ozbek, K. Bicakci, K. Aslan, B. Kara, Y. Sarica, Recurrent headache and MRI findings in systemic lupus erythematosus, Journal of the National Medical Association 100 (2008) 323–326.

[155] F. Nobili, A. Mignone, E. Rossi, S. Morbelli, A. Piccardo, F. Puppo, et al., Migraine during systemic lupus erythematosus: findings from brain single photon emission computed tomography, J Rheumatol 33 (2006) 2184–2191.

[156] C.L. Rozell, W.L. Sibbitt Jr., W.M. Brooks, Structural and neurochemical markers of brain injury in the migraine diathesis of systemic lupus erythematosus, Cephalalgia 18 (1998) 209–215.

[157] S. Appenzeller, F. Cendes, L.T. Costallat, Epileptic seizures in systemic lupus erythematosus, Neurology 63 (2004) 1808–1812.

[158] A. Gonzalez-Duarte, C.G. Cantu-Brito, L. Ruano-Calderon, G. Garcia-Ramos, Clinical description of seizures in patients with systemic lupus erythematosus, European neurology 59 (2008) 320–323.

[159] J. Mikdashi, A. Krumholz, B. Handwerger, Factors at diagnosis predict subsequent occurrence of seizures in systemic lupus erythematosus, Neurology 64 (2005) 2102–2107.

[160] H.H. Liou, C.R. Wang, H.C. Chou, V.L. Arvanov, R.C. Chen, Y.C. Chang, et al., Anticardiolipin antisera from lupus patients with seizures reduce a GABA receptor-mediated chloride current in snail neurons, Life Sci 54 (1994) 1119–1125.

[161] Y. Ganor, H. Goldberg-Stern, T. Lerman-Sagie, V.I. Teichberg, M. Levite, Autoimmune epilepsy: Distinct subpopulations of epilepsy patients harbor serum autoantibodies to either glutamate/AMPA receptor GluR3, glutamate/NMDA receptor subunit NR2. A or double-stranded DNA, Epilepsy Res 65 (2005) 11–22.

[162] J.F. Baizabal-Carvallo, G. Delgadillo-Marquez, B. Estanol, G. Garcia-Ramos, Clinical characteristics and outcomes of the meningitides in systemic lupus erythematosus, European neurology 61 (2009) 143–148.

[163] J. Okada, T. Hamana, H. Kondo, Anti-U1. RNP antibody and aseptic meningitis in connective tissue diseases, Scand J Rheumatol 32 (2003) 247–252.

[164] R.P. Khubchandani, V. Viswanathan, J. Desai, Unusual neurologic manifestations (I): Parkinsonism in juvenile SLE, Lupus 16 (2007) 572–575.

[165] A.C. Felicio, M.C. Shih, C. Godeiro-Junior, L.A. Andrade, R.A. Bressan, H.B. Ferraz, Molecular imaging studies in Parkinson disease: reducing diagnostic uncertainty, The neurologist 15 (2009) 6–16.

[166] W.H. Chen, H.S. Lin, Y.F. Kao, M.Y. Lan, J.S. Liu, Postoperative parkinsonism and lupus anticoagulant: a model of autoantibody-mediated neurotoxicity in stress, Brain Inj 21 (2007) 539–543.

[167] Y.C. Huang, R.K. Lyu, S.T. Chen, Y.C. Chu, Y.R. Wu, Parkinsonism in a patient with antiphospholipid syndrome—case report and literature review, Journal of the neurological sciences 267 (2008) 166–169.

[168] W. Miesbach, A. Gilzinger, B. Gokpinar, D. Claus, I. Scharrer, Prevalence of antiphospholipid antibodies in patients with neurological symptoms, Clinical neurology and neurosurgery 108 (2006) 135–142.

[169] R.C. Kunas, A. McRae, J. Kesselring, P.M. Villiger, Anti-dopaminergic antibodies in a patient with a complex autoimmune disorder and rapidly progressing Parkinson's disease, The Journal of allergy and clinical immunology 96 (1995) 688–690.

[170] R.A. Asherson, R.H. Derksen, E.N. Harris, B.N. Bouma, A.E. Gharavi, L. Kater, et al., Chorea in systemic lupus erythematosus and "lupus-like" disease: association with antiphospholipid antibodies, Semin Arthritis Rheum 16 (1987) 253–259.

[171] M.A. Khamashta, A. Gil, B. Anciones, P. Lavilla, M.E. Valencia, V. Pintado, et al., Chorea in systemic lupus erythematosus: association with antiphospholipid antibodies, Annals of the rheumatic diseases 47 (1988) 681–683.

[172] N.M. Orzechowski, A.P. Wolanskyj, J.E. Ahlskog, N. Kumar, K.G. Moder, Antiphospholipid antibody-associated chorea, J Rheumatol 35 (2008) 2165–2170.

[173] T. Watanabe, H. Onda, Hemichorea with antiphospholipid antibodies in a patient with lupus nephritis, Pediatric nephrology (Berlin, Germany) 19 (2004) 451–453.

[174] J.R. Keane, Eye movement abnormalities in systemic lupus erythematosus, Arch Neurol 52 (1995) 1145–1149.

[175] S. Genevay, G. Hayem, S. Hamza, E. Palazzo, O. Meyer, M.F. Kahn, Oculomotor palsy in six patients with systemic lupus erythematosus. A possible role of antiphospholipid syndrome, Lupus 11 (2002) 313–316.

[176] A.S. Friedman, V. Folkert, G.A. Khan, Recurrence of systemic lupus erythematosus in a hemodialysis patient presenting as a unilateral abducens nerve palsy, Clin Nephrol 44 (1995) 338–339.

[177] C.N. Chan, E. Li, F.M. Lai, J.A. Pang, An unusual case of systemic lupus erythematosus with isolated hypoglossal nerve palsy, fulminant acute pneumonitis, and pulmonary amyloidosis, Annals of the rheumatic diseases 48 (1989) 236–239.

[178] J.H. Vaile, P. Davis, Isolated unilateral vagus nerve palsy in systemic lupus erythematosus, J Rheumatol 25 (1998) 2287–2288.

[179] K.H. Yu, C.H. Yang, C.C. Chu, Swallowing disturbance due to isolated vagus nerve involvement in systemic lupus erythematosus, Lupus 16 (2007) 746–749.

[180] D.M. Wingerchuk, W.F. Hogancamp, P.C. O'Brien, B.G. Weinshenker, The clinical course of neuromyelitis optica (Devic's syndrome), Neurology 53 (1999) 1107–1114.

[181] J. Birnbaum, D. Kerr, Optic neuritis and recurrent myelitis in a woman with systemic lupus erythematosus, Nat Clin Pract Rheumatol 4 (2008) 381–386.

[182] A.Y. Hershko, Y. Berkun, D. Mevorach, A. Rubinow, Y. Naparstek, Increased intracranial pressure related to systemic lupus erythematosus: a 26-year experience, Semin Arthritis Rheum 38 (2008) 110–115.

[183] A. Bhat, S. Naguwa, G. Cheema, M.E. Gershwin, The epidemiology of transverse myelitis, Autoimmun Rev 9 (2010) A395–A399.

[184] W.L. Sibbitt Jr., R.R. Sibbitt, W.M. Brooks, Neuroimaging in neuropsychiatric systemic lupus erythematosus, Arthritis and rheumatism 42 (1999) 2026–2038.

[185] S. Appenzeller, G.B. Pike, A.E. Clarke, Magnetic resonance imaging in the evaluation of central nervous system manifestations in systemic lupus erythematosus, Clin Rev Allergy Immunol 34 (2008) 361–366.

[186] W.L. Sibbitt Jr., W.M. Brooks, M. Kornfeld, B.L. Hart, A.D. Bankhurst, C.A. Roldan, Magnetic Resonance Imaging and Brain Histopathology in Neuropsychiatric Systemic Lupus Erythematosus, Semin Arthritis Rheum [Epub ahead of print] (2009).

[187] H. Ainiala, P. Dastidar, J. Loukkola, T. Lehtimaki, M. Korpela, J. Peltola, et al., Cerebral MRI abnormalities and their association with neuropsychiatric manifestations in SLE: a population-based study, Scand J Rheumatol 34 (2005) 376–382.

[188] S. Appenzeller, A.D. Carnevalle, L.M. Li, L.T. Costallat, F. Cendes, Hippocampal atrophy in systemic lupus erythematosus, Annals of the rheumatic diseases 65 (2006) 1585–1589.

[189] G.P. Bosma, H.A. Middelkoop, M.J. Rood, E.L. Bollen, T.W. Huizinga, M.A. van Buchem, Association of global brain damage and clinical functioning in neuropsychiatric systemic lupus erythematosus, Arthritis and rheumatism 46 (2002) 2665–2672.

[190] A. Cauli, C. Montaldo, M.T. Peltz, P. Nurchis, G. Sanna, P. Garau, et al., Abnormalities of magnetic resonance imaging of the central nervous system in patients with systemic lupus erythematosus correlate with disease severity, Clin Rheumatol 13 (1994) 615–618.

[191] B.J. Emmer, J. van der Grond, G.M. Steup-Beekman, T.W. Huizinga, M.A. van Buchem, Selective involvement of the amygdala in systemic lupus erythematosus, PLoS medicine 3 (2006) e499.

[192] C.H. Kao, Y.J. Ho, J.L. Lan, S.P. Changlai, K.K. Liao, P.U. Chieng, Discrepancy between regional cerebral blood flow and glucose metabolism of the brain in systemic lupus erythematosus patients with normal brain magnetic resonance imaging findings, Arthritis and rheumatism 42 (1999) 61–68.

[193] N. Komatsu, K. Kodama, N. Yamanouchi, S. Okada, S. Noda, Y. Nawata, et al., Decreased regional cerebral metabolic rate for glucose in systemic lupus erythematosus patients with psychiatric symptoms, European neurology 42 (1999) 41–48.

[194] A. Otte, S.M. Weiner, H.H. Peter, J. Mueller-Brand, M. Goetze, E. Moser, et al., Brain glucose utilization in systemic lupus erythematosus with neuropsychiatric symptoms: a controlled positron emission tomography study, European journal of nuclear medicine 24 (1997) 787–791.

[195] S.M. Weiner, A. Otte, M. Schumacher, R. Klein, J. Gutfleisch, I. Brink, et al., Diagnosis and monitoring of central nervous system involvement in systemic lupus erythematosus: value of F-18 fluorodeoxyglucose PET, Annals of the rheumatic diseases 59 (2000) 377–385.

[196] G.P. Bosma, M.J. Rood, T.W. Huizinga, B.A. de Jong, E.L. Bollen, M.A. van Buchem, Detection of cerebral involvement in patients with active neuropsychiatric systemic lupus erythematosus by the use of volumetric magnetization transfer imaging, Arthritis and rheumatism 43 (2000) 2428–2436.

[197] B.J. Emmer, S.C. Steens, G.M. Steup-Beekman, J. van der Grond, F. Admiraal-Behloul, H. Olofsen, et al., Detection of change in CNS involvement in neuropsychiatric SLE: a magnetization transfer study, J Magn Reson Imaging 24 (2006) 812–816.

[198] B.J. Emmer, G.M. Steup-Beekman, S.C. Steens, T.W. Huizinga, M.A. van Buchem, J. van der Grond, Correlation of magnetization transfer ratio histogram parameters with neuropsychiatric systemic lupus erythematosus criteria and proton magnetic resonance spectroscopy: association of magnetization transfer ratio peak height with neuronal and cognitive dysfunction, Arthritis and rheumatism 58 (2008) 1451–1457.

[199] S.C. Steens, F. Admiraal-Behloul, G.P. Bosma, G.M. Steup-Beekman, H. Olofsen, S. Le Cessie, et al., Selective gray matter damage in neuropsychiatric lupus, Arthritis and rheumatism 50 (2004) 2877–2881.

[200] G.P. Bosma, M.J. Rood, A.H. Zwinderman, T.W. Huizinga, M.A. van Buchem, Evidence of central nervous system damage in patients with neuropsychiatric systemic lupus erythematosus, demonstrated by magnetization transfer imaging, Arthritis and rheumatism 43 (2000) 48–54.

[201] S.C. Steens, G.P. Bosma, G.M. Steup-Beekman, S. le Cessie, T.W. Huizinga, M.A. van Buchem, Association between microscopic brain damage as indicated by magnetization transfer imaging and anticardiolipin antibodies in neuropsychiatric lupus, Arthritis Res Ther 8 (2006) R38.

[202] S. Appenzeller, L.T. Costallat, L.M. Li, F. Cendes, Magnetic resonance spectroscopy in the evaluation of central nervous system manifestations of systemic lupus erythematosus, Arthritis and rheumatism 55 (2006) 807–811.

[203] J.S. Axford, F.A. Howe, C. Heron, J.R. Griffiths, Sensitivity of quantitative (1)H magnetic resonance spectroscopy of the brain in detecting early neuronal damage in systemic lupus erythematosus, Annals of the rheumatic diseases 60 (2001) 106–111.

[204] W.L. Sibbitt Jr., L.J. Haseler, R.R. Griffey, S.D. Friedman, W.M. Brooks, Neurometabolism of active neuropsychiatric lupus determined with proton MR spectroscopy, Ajnr 18 (1997) 1271–1277.

[205] M.K. Lim, C.H. Suh, H.J. Kim, Y.K. Cho, S.H. Choi, J.H. Kang, et al., Systemic lupus erythematosus: brain MR imaging and single-voxel hydrogen 1 MR spectroscopy, Radiology 217 (2000) 43–49.

[206] W.M. Brooks, R.E. Jung, C.C. Ford, E.J. Greinel, W.L. Sibbitt Jr., Relationship between neurometabolite derangement and neurocognitive dysfunction in systemic lupus erythematosus, J Rheumatol 26 (1999) 81–85.

[207] E. Kozora, D.B. Arciniegas, C.M. Filley, M.C. Ellison, S.G. West, M.S. Brown, et al., Cognition, MRS Neurometabolites, and MRI Volumetrics in Non-Neuropsychiatric Systemic Lupus Erythematosus: Preliminary Data, Cogn Behav Neurol 18 (2005) 159–162.

[208] S. Appenzeller, L.M. Li, L.T. Costallat, F. Cendes, Neurometabolic changes in normal white matter may predict appearance of hyperintense lesions in systemic lupus erythematosus, Lupus 16 (2007) 963–971.

[209] G. Castellino, M. Govoni, M. Padovan, P. Colamussi, M. Borrelli, F. Trotta, Proton magnetic resonance spectroscopy may predict future brain lesions in SLE patients: a functional multi-imaging approach and follow up, Annals of the rheumatic diseases 64 (2005) 1022–1027.

[210] R. Handa, P. Sahota, M. Kumar, N.R. Jagannathan, C.S. Bal, M. Gulati, et al., In vivo proton magnetic resonance spectroscopy (MRS) and single photon emission computerized tomography (SPECT) in systemic lupus erythematosus (SLE), Magnetic resonance imaging 21 (2003) 1033–1037.

[211] J.J. Chen, R.F. Yen, A. Kao, C.C. Lin, C.C. Lee, Abnormal regional cerebral blood flow found by technetium-99. m ethyl cysteinate dimer brain single photon emission computed tomography in systemic lupus erythematosus patients with normal brain MRI findings, Clin Rheumatol 21 (2002) 516–519.

[212] F. Falcini, M.T. De Cristofaro, M. Ermini, M. Guarnieri, G. Massai, M. Olmastroni, et al., Regional cerebral blood flow in juvenile systemic lupus erythematosus: a prospective SPECT study. Single photon emission computed tomography, J Rheumatol 25 (1998) 583–588.

[213] A. Reiff, J. Miller, B. Shaham, B. Bernstein, I.S. Szer, Childhood central nervous system lupus; longitudinal assessment using single photon emission computed tomography, J Rheumatol 24 (1997) 2461–2465.

[214] S. Appenzeller, B.J. Amorim, C.D. Ramos, P.A. Rio, C.E.E.C. de, E.E. Camargo, et al., Voxel-based morphometry of brain SPECT can detect the presence of active central nervous system involvement in systemic lupus erythematosus, Rheumatology (Oxford) 46 (2007) 467–472.

[215] S.S. Sun, W.S. Huang, J.J. Chen, C.P. Chang, C.H. Kao, J.J. Wang, Evaluation of the effects of methylprednisolone pulse therapy in patients with systemic lupus erythematosus with brain involvement by Tc-99. m HMPAO brain SPECT, European radiology 14 (2004) 1311–1315.

[216] G. Castellino, M. Padovan, A. Bortoluzzi, M. Borrelli, L. Feggi, M.L. Caniatti, et al., Single photon emission computed tomography and magnetic resonance imaging evaluation in SLE patients with and without neuropsychiatric involvement, Rheumatology (Oxford) 47 (2008) 319–323.

[217] C.H. Sotak, Nuclear magnetic resonance (NMR) measurement of the apparent diffusion coefficient (ADC) of tissue water and its relationship to cell volume changes in pathological states, Neurochemistry international 45 (2004) 569–582.

[218] A.M. Ulug, D.F. Moore, A.S. Bojko, R.D. Zimmerman, Clinical use of diffusion-tensor imaging for diseases causing neuronal and axonal damage, Ajnr 20 (1999) 1044–1048.

[219] L. Zhang, M. Harrison, L.A. Heier, R.D. Zimmerman, L. Ravdin, M. Lockshin, et al., Diffusion changes in patients with systemic lupus erythematosus, Magnetic resonance imaging 25 (2007) 399–405.

[220] M. Hughes, P.C. Sundgren, X. Fan, B. Foerster, B. Nan, R.C. Welsh, et al., Diffusion tensor imaging in patients with acute onset of neuropsychiatric systemic lupus erythematosus: a prospective study of apparent diffusion coefficient, fractional anisotropy values, and eigenvalues in different regions of the brain, Acta Radiol 48 (2007) 213–222.

[221] M.W. DiFrancesco, S.K. Holland, M.D. Ris, C.M. Adler, S. Nelson, M.P. DelBello, et al., Functional magnetic resonance imaging assessment of cognitive function in childhood-onset systemic lupus erythematosus: a pilot study, Arthritis and rheumatism 56 (2007) 4151–4163.

[222] B.M. Fitzgibbon, S.L. Fairhall, I.J. Kirk, M. Kalev-Zylinska, K. Pui, N. Dalbeth, et al., Functional MRI in NPSLE patients reveals increased parietal and frontal brain activation during a working memory task compared with controls, Rheumatology (Oxford) 47 (2008) 50–53.

[223] M.A. Rocca, F. Agosta, D.M. Mezzapesa, G. Ciboddo, A. Falini, G. Comi, et al., An fMRI study of the motor system in patients with neuropsychiatric systemic lupus erythematosus, Neuroimage 30 (2006) 478–484.

[224] J.A. Bobholz, S.M. Rao, L. Lobeck, C. Elsinger, A. Gleason, J. Kanz, et al., fMRI study of episodic memory in relapsing-remitting MS: correlation with T2 lesion volume, Neurology 67 (2006) 1640–1645.

[225] F.G. Hillary, Neuroimaging of working memory dysfunction and the dilemma with brain reorganization hypotheses, J Int Neuropsychol Soc 14 (2008) 526–534.

[226] M.A. Rocca, A. Falini, B. Colombo, G. Scotti, G. Comi, M. Filippi, Adaptive functional changes in the cerebral cortex of patients with nondisabling multiple sclerosis correlate with the extent of brain structural damage, Ann Neurol 51 (2002) 330–339.

[227] F.Z. Yetkin, R.N. Rosenberg, M.F. Weiner, P.D. Purdy, C.M. Cullum, FMRI of working memory in patients with mild cognitive impairment and probable Alzheimer's disease, European radiology 16 (2006) 193–206.

[228] C. Calautti, J.C. Baron, Functional neuroimaging studies of motor recovery after stroke in adults: a review, Stroke 34 (2003) 1553–1566.

29

Atherosclerosis and Tissue Injury in Systemic Lupus Erythematosus

Ingrid Avalos, C. Michael Stein

INTRODUCTION

Systemic lupus erythematosus (SLE) can cause tissue damage in many organs. Damage affecting organs such as the kidneys is most clinically obvious and causes substantial morbidity and mortality. As the treatment of renal and other complications of lupus has improved, the mortality associated with acute lupus has decreased and survival has improved. With this increased long-term survival has come a clearer recognition of tissue damage affecting the vasculature, particularly the coronary vasculature, and the impact of coronary heart disease (CHD) on long-term morbidity and mortality in lupus.

RECOGNITION OF INCREASED CORONARY HEART DISEASE RISK IN SLE

A bimodal pattern of mortality in SLE, with early deaths being most likely due to active lupus or infection, and atherosclerotic vascular disease a prominent cause of death in patients with more established disease, was initially reported 1976 [1]. A follow-up study 10 years later confirmed this pattern in 51 patients who had died. Of 21 patients with disease longer than 5 years duration, six died from myocardial infarction or sudden death, causes that were thought to be unrelated to lupus itself [2]. A more recent study evaluated changing patterns in SLE survival and co-morbidities over a 36-year follow-up period. A cohort of 1240 patients was divided into four groups based on when they entered the cohort: 1970–1978, 1979–1987, 1988–1996, and 1997–2005. The standardized mortality ratio decreased over time and was 12.5 in the first, and 3.5 in last group; disease activity also decreased over the decades, likely

attributable to increased use of immunosuppressive and antimalarial medications. However, over time there was an increasing risk of organ damage, osteonecrosis, and CHD,[3].

Studies such as these have led to efforts to: (1) quantify the morbidity and mortality due to CHD in SLE, (2) use noninvasive measures to identify atherosclerosis in SLE, (3) understand the relationship between traditional cardiovascular risk factors and increased CHD, and (4) characterize the factors specific to patients with lupus that increase CHD risk and identify their underlying mechanisms.

Morbidity and Mortality Due to Coronary Heart Disease in SLE

Clinically evident CHD affects approximately 6–10% of patients with SLE [4–6]. In a cohort of 498 women with SLE, the incidence of myocardial infarction was higher in all age groups than in women of similar age in the Framingham cohort. There was a 52-fold increase in the risk of MI in women with SLE who were 35–44 years of age compared to age- and sex-matched controls [5]. In women with SLE older than 44 years, the risk was increased two- to fourfold. Factors that were more common in women who had a cardiovascular event included older age at diagnosis, longer disease duration, and longer duration of corticosteroid treatment [5]. The relative risk was 10 for nonfatal MI, 17 for CHD death, and 8 for stroke [7] in another study of 296 SLE patients compared to the Framingham cohort.

Concordant with the increased rate of cardiovascular events in SLE, cardiovascular morbidity, measured as hospitalization rates, are also increased. In the California Hospital Discharge Database, young women with SLE were 2.3 times more likely to be hospitalized with an MI, and 2.05 times more likely to be hospitalized with a cerebrovascular accident, compared to age-matched women without SLE. In several studies it has been striking that the greatest increase in risk, relative to women without SLE, was in young patients between the ages of 18 and 44 [5, 8], an age group not typically affected by CHD. This increase in CHD risk among young women in epidemiological studies, and the frequency of CHD detected by myocardial perfusion studies [9], coronary angiography [10], and at autopsy [2] have resulted in efforts to identify subclinical atherosclerotic disease noninvasively, so that interventions can be targeted at the individuals at highest risk before clinical events occur.

Noninvasive Measures to Identify Atherosclerosis and Its Risk Factors in SLE

Several noninvasive vascular measurements suitable for performance in large groups of patients predict cardiovascular risk in the general population. These include techniques such as carotid ultrasound and cardiac electron beam or multislice detector computed tomography (CT) that identify clinical and subclinical atherosclerotic structural vascular changes in the carotid artery and coronary vasculature, respectively. Other techniques that measure impairment of functional measures, for example, endothelium-dependant vasodilation, provide a less specific measure of vascular health that is affected by many CHD risk factors.

Carotid Ultrasound

Several studies have used carotid ultrasound to determine if patients with SLE have a higher prevalence of plaque or increased intima-media thickness (IMT), structural measures of preclinical carotid artery atherosclerosis that are associated with the risk of CHD and stroke in the general population.

In a large cohort of 197 patients with SLE and 197 matched controls (94% female; mean age, 44 years) the prevalence of carotid plaque (37.1% vs. 15.2%, $p<0.001$) was higher in patients with lupus compared to controls. This increased prevalence of carotid atherosclerosis was associated with age, longer disease duration, higher damage index, and less use of cyclophosphamide. This suggested a role for disease-related factors and also for aggressive pharmacologic management in order to potentially modify atherosclerotic disease. Interestingly, despite the increased prevalence of plaque, IMT measurements did not differ significantly among patients with lupus and controls [11]. Factors associated with carotid artery plaque in patients with SLE in two other studies

were as follows: older age, systolic blood pressure, LDL cholesterol, prolonged treatment with prednisone, and a previous coronary event ($n=175$) [12]; and age, male gender, hypertension, diabetes, serum creatinine, and systolic blood pressure ($n=605$) [13] .

Several smaller studies have also reported an increased prevalence of plaque, but no increase in IMT in lupus [14–16]. However, in patients with SLE who had cardiovascular disease IMT was significantly greater compared to those who did not [17, 18].

Carotid ultrasound has also been used to follow the progression of atherosclerosis in patients with lupus [19, 20]. Patients with SLE ($n=217$) had more carotid artery plaque at baseline and had accelerated progression, but IMT at baseline, and progression of IMT, did not differ from that of 104 controls [20].

Coronary Artery Calcification

Electron beam or multislice detector CT is a sensitive, noninvasive method used to assess the presence and extent of coronary artery calcification. The amount of coronary calcification detected on CT correlates well with the amount of atherosclerosis present and predicts future CHD risk in the general population [21]. Patients with SLE had a higher prevalence (31%) of coronary calcification than age- and sex-matched control subjects (9%) ($p=0.002$); calcification scores were also significantly higher in SLE, indicating that coronary atherosclerosis was not only more prevalent but more severe [22] (Figure 29.1). Coronary calcification was associated with older age and male sex [22] and was detected in women with SLE who were younger than 40 years of age, a finding that emphasizes the need for early detection and aggressive management of cardiovascular risk factors. Another study confirmed the presence of coronary calcification at an early age; 28% of 75 SLE patients without clinical signs of CHD had coronary calcification by CT [23]. The median age of this cohort was 38.8 years, and the youngest patient with documented coronary artery calcification was less than 30 years old. A multivariable analysis identified factors such as smoking, cumulative prednisone dose greater than 30 g, reduced renal function, and high C3 levels as factors associated with coronary calcification [23]. There was no significant difference in disease activity, measured by the European Consensus Lupus Activity Measurement (ECLAM), between patients with SLE with and without coronary calcification [23].

Additional larger studies have confirmed a higher prevalence of coronary artery calcification in patients with lupus compared to controls [24] and that coronary calcium was common (43%) and associated with several CHD risk factors, but not with measures of lupus disease activity [25].

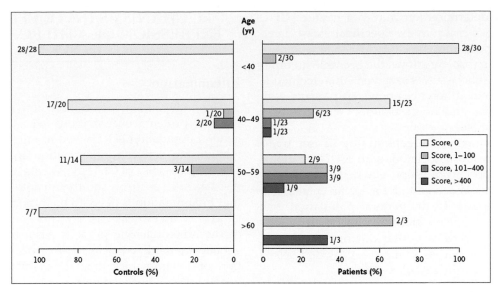

FIGURE 29.1   Frequency of coronary artery calcium scores among patients with lupus and control subjects, according to age. Higher scores indicate more extensive calcification. *Reproduced with permission from the Massachusetts Medical Society, Asanuma et al. [22].*

Endothelial Function

Endothelial dysfunction, most often measured as impairment of endothelium-dependent arterial dilation, occurs early in the atherosclerotic process, and is associated with the presence of several cardiovascular risk factors and early atherosclerosis [26]. Potential causes of endothelial dysfunction that may lead to atherosclerosis or accelerate it include dyslipidemia, smoking, hypertension, diabetes mellitus, and hyperhomocysteinemia [27]. Doppler ultrasound of the brachial artery can be used to measure the dilation of the brachial artery in response to an increase in flow induced by ischemia. This measure of endothelial function is termed flow-mediated dilation (FMD).

Several studies have found that patients with SLE have impaired FMD compared to controls [28–31], even after adjusting for traditional cardiovascular risk factors [28]. In addition, patients with SLE had impairment of FMD similar to that observed in control subjects with established CHD [31]. In keeping with impaired endothelial function, measures of vascular stiffness are also increased in SLE [32, 33].

NONINVASIVE MEASURES OF ATHEROSCLEROSIS TO PREDICT CHD OUTCOMES IN SLE

There are no long-term studies that define the proportion of CHD risk that is predicted by measures of subclinical atherosclerosis in SLE. Nevertheless, given their predictive value for risk stratification in the general population, there is substantial interest in using these techniques for the early identification of those patients with SLE who have the greatest risk of CHD so that preventive strategies can be implemented.

THE RELATIONSHIP BETWEEN TRADITIONAL CARDIOVASCULAR RISK FACTORS AND INCREASED CHD IN SLE

The mechanisms underlying the increased risk of CHD in lupus are not well established. In several studies of subclinical atherosclerosis in SLE, traditional cardiovascular risk factors such as age, smoking, and hypertension have been associated with coronary calcification or carotid plaque and thus appear to be important contributors to atherosclerosis in lupus, as they are in the general population.

The prevalence of several traditional cardiovascular risk factors has been reported to be increased in patients with SLE compared to control subjects in some studies. These include hypercholesterolemia [34, 35], hypertension, and diabetes [36]. Hypercholesterolemia, as is the case in the general population, is associated with increased risk of clinical CHD [4, 5] and stroke [37] in patients with SLE. A longitudinal study found that a persistently elevated total cholesterol level was a predictor of a new cardiovascular event over a 12- to 14-year follow-up period [38]. However, hypercholesterolemia is not generally associated with SLE; more frequent findings are decreased high-density lipoprotein (HDL) cholesterol and increased triglyceride concentrations, changes that are associated with inflammation [39].

Routine lipid determination may not capture CHD risk optimally since risk may be specifically associated with particular lipid subclasses. Nuclear magnetic resonance (NMR) spectroscopy allows the accurate quantification of lipid subclasses. However, this technique revealed only minor differences between patients with lupus and controls, and no association with coronary artery calcification [40].

Thus, just as occurs in the general population, traditional risk factors are associated with CHD in lupus. However, the important question is whether the excess CHD risk in lupus is accounted for by an increase in traditional risk factors. In the general population, risk factors for CHD have been identified and their predictive contribution to cardiovascular outcomes validated in large prospective studies that have provided scores that predict future cardiovascular risk.

The Framingham risk score is one such model used to estimate the risk of cardiovascular events in the general population. It takes into account age, sex, blood pressure, cholesterol concentrations, and the presence of smoking and diabetes. Several studies indicate that traditional cardiovascular risk factors, as measured by the Framingham score, do not account for the excess cardiovascular risk in SLE [7, 41, 42]. For example, the Framingham score did not differ significantly among women with lupus and controls, although coronary artery calcification was more prevalent and severe in those with lupus [41]. Furthermore, 99% of women with lupus were classified as low-risk, with a median Framingham 10-year predicted risk of an event of 1%, despite the presence of coronary calcium in 19% [41]. Findings such as these suggest that additional risk factors not captured by measures such as the Framingham risk score contribute to accelerated atherosclerosis in SLE. Potential risk factors include the inflammatory, immunological, and metabolic changes associated with lupus (Figure 29.2).

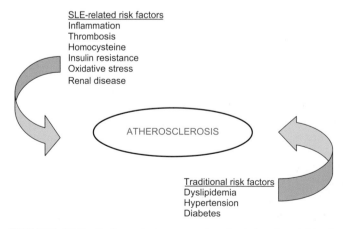

FIGURE 29.2 Pathogenic processes involved in atherosclerosis in SLE.

MECHANISMS INVOLVED IN THE EXPRESSION OF ATHEROSCLEROSIS IN SLE

Inflammation

The independent association between increased C-reactive protein concentrations and CHD risk in the general population [43] has led to intense interest in the relationship between inflammation and atherosclerosis. The increased risk of CHD in patients with inflammatory diseases such as lupus and rheumatoid arthritis suggests that inflammation is likely to be important in the pathogenesis of atherosclerosis, and not merely a byproduct of the atherosclerotic process. Also, in animal models genetically modified to increase or decrease the production of inflammatory mediators, inflammation has in general been associated with increased atherosclerosis.

Infiltrating cells

Inflammation plays a key role in the pathogenesis of atherosclerosis and is present throughout the various stages of the disease. In the early lesion of atherosclerosis, the fatty streak, inflammatory involvement with T lymphocytes and macrophages is prominent [44]. Migration of leukocytes and mast cells into the vascular wall is facilitated by inflammatory mediators such as monocyte chemoattractant protein-1 (MCP-1) [27]. Activated smooth muscle cells produce MCP-1, and its increased expression is upregulated in part by cytokines such as TNF-α and IL-1 [45]. An increase in the gap between endothelial cells allows the entry of monocytes and low-density lipoproteins (LDL) into the subendothelial space. There LDL can be oxidized; this oxidized LDL (OxLDL) can facilitate the formation of proatherogenic antibodies (anti-oxLDL antibodies), activate endothelial cells, and lead to the formation of foam cells [46].

Cytokines and adhesion molecules

Once atherosclerotic plaque is established, inflammation also plays a role and transforms stable plaque into unstable plaque that is more likely to rupture and lead to thrombosis [47, 48]. Activation of immune cells in the plaque leads to production of inflammatory mediators and proteolytic enzymes [49]. One of the inflammatory cytokines produced within the atherosclerotic lesion by activated T cells is interferon-γ (IFN-γ), which in turn is associated with increased production of tumor necrosis factor (TNF) and interleukin-1 [50]. T cells may also play a systemic role in the process of atherosclerotic plaque rupture. Clonal expansion of a particular subset of T lymphocytes, CD4+CD28^{null} cells, is found in the peripheral blood of patients with unstable angina compared to those with stable disease [51]. This subset of T cells is known to produce high levels of IFN-γ

[52], which may further contribute to atherosclerotic plaque vulnerability. Clonal expansion of this subset of $CD4^+$ cells has also been seen in patients with RA and correlates with extra-articular manifestations of RA [53]. There is no information regarding this subset of cells in SLE patients. However, in a murine model of atherosclerosis and SLE, atherosclerotic plaques had an increase in CD3 and CD4 T cells, as well as in lipid-laden macrophages [54].

In patients with lupus, coronary calcification was associated with higher concentrations of the inflammatory cytokine interleukin-6 (IL-6) [55]. Another study in the same cohort found that concentrations of TNF-α and the adhesion molecules, E-selectin, vascular cell adhesion molecule (VCAM), and intercellular adhesion molecule (ICAM), were associated with coronary calcification, independent of Framingham risk score. However, myeloperoxidase (MPO), matrix metalloproteinase-9 (MMP-9), and IL-1α concentrations were not associated with coronary calcification [56], suggesting that specific mediators rather than a generalized inflammatory process are associated with atherosclerosis in SLE. Concordant with these observations, concentrations of TNF-α were higher in a subset of patients with SLE and a history of cardiovascular disease than in lupus patients without cardiovascular disease and control subjects [57].

In addition to its direct effects on atherogenesis, inflammation can also contribute indirectly to the pathogenesis of atherosclerosis through modification of traditional risk factors. For example, inflammation lowers HDL cholesterol and increases triglyceride concentrations. Lower concentrations of HDL cholesterol in SLE were independently associated with IL-6, ESR, and TNF-α concentrations. This finding remained significant even after statistical adjustment for exposure to drugs such as corticosteroids and hydroxychloroquine that may affect the lipid profile [39]. Moreover, TNF-α concentrations correlated with triglyceride and VLDL cholesterol concentrations [57].

Endothelial Dysfunction

An imbalance between vascular damage and repair is one mechanism that may contribute to endothelial dysfunction. Injury to the endothelium is followed by endothelial cell (EC) apoptosis and a subsequent increased number of circulating ECs [58]. EC apoptosis can disrupt the endothelium by several mechanisms, including loss of nitric oxide release [59]. Studies in lupus suggest that there is an increase in circulating activated ECs in active SLE [31, 60] and this correlates with endothelial dysfunction as measured by brachial artery FMD [31]. There is also a decrease in the number of

endothelial progenitor cells (EPCs) [61] and an inability of EPCs and myelomonocytic circulating angiogenic cells (MCACs) to become mature endothelial cells with proangiogenic activity. This inability to mediate vascular repair is triggered by increased expression of IFN-γ, which has been shown to induce EPC and MCAC apoptosis [62].

Inflammation can also induce endothelial dysfunction directly, as suggested by a study in which local infusion of the TNF antagonist infliximab into the brachial artery improved endothelial function in that vessel in patients with rheumatoid arthritis [63].

The Metabolic Syndrome and Insulin Resistance

A sedentary lifestyle [34, 36], obesity [34]' [64], and a constellation of cardiovascular risk factors that cluster together, termed the metabolic syndrome [65, 66], may be more prevalent in patients with SLE than in age-matched controls. The metabolic syndrome encompasses several cardiovascular risk factors including hypertension, abnormal glucose metabolism or insulin resistance, central obesity, and dyslipidemia. In the general population, the metabolic syndrome predicts future cardiovascular risk; women with the metabolic syndrome have a twofold increased risk of a major cardiovascular events, including death [67].

Insulin resistance, as part of the abnormal glucose metabolism found in the metabolic syndrome, occurs more frequently in patients with SLE than in controls [65, 66], and is associated with higher levels of inflammation [65], providing further evidence implicating inflammation as a contributor to increased cardiovascular risk. Corticosteroid use could predispose to many of the features of the metabolic syndrome, but there were no differences in cumulative corticosteroid use in patients with and without the metabolic syndrome [68]. Moreover, in another study, insulin resistance did not correlate with current or recent steroid therapy [66]. Interestingly, there was a significant correlation between insulin resistance and levels of oxidized-LDL, suggesting a common pathway through which metabolic and inflammatory processes may interact [66].

Insulin resistance associated with inflammation may differ in RA and SLE. Both diseases are associated with insulin resistance, but obesity, measured as body mass index, appeared to contribute more in SLE, while IL-6 and TNF-α concentrations contributed more in RA [68].

Homocysteine

Hyperhomocysteinemia is associated with increased cardiovascular risk in the general population [69].

Homocysteine concentrations are modestly elevated in several autoimmune diseases, including SLE [70]. Pathogenic mechanisms resulting in deficiency of vitamin B and folate, co-factors in homocysteine metabolism, have been proposed as possible pathways leading to hyperhomocysteinemia in the setting of autoimmune diseases. The persistent inflammatory activation may result in increased vitamin consumption via increased DNA synthesis in immune cells. Moreover, vitamin deficiency may result from decreased gastrointestinal absorption as well as interference with folate metabolism related to medications such as methotrexate. [70] In SLE, high plasma homocysteine concentrations are associated with a higher risk of CHD [36, 71] and thrombosis [72], and correlate independently with lupus disease duration and increased coronary artery calcification [24]. Although low levels of folate and vitamin B_{12} are associated with hyperhomocysteinemia in SLE [72] and in the general population, supplementation with folic acid and vitamins B_6 and B_{12} in the general population do not reduce the risk of cardiovascular events in individuals with established vascular disease [73]. There is no information about the efficacy or lack of efficacy of such strategies in patients with SLE.

Impaired HDL Cholesterol Function

Oxidized LDL generates oxidation-specific epitopes that are highly immunogenic and induce formation of autoantibodies [74]. The presence of similar epitopes in apoptotic cells undergoing oxidative stress has also been found by binding of the murine monoclonal EO IgM antibodies (specific for epitopes of OxLDL) to apoptotic cells. These antibodies interfere with macrophage phagocytosis of such apoptotic cells. [75]. This is particularly relevant in the setting of SLE, where autoantibodies recognize antigens that are present on the surface of apoptotic cells [76]. Such oxidation-specific epitopes in apoptotic cells may also activate endothelial cells, further contributing to the process of atherosclerosis [77].

HDL is typically considered to be antiatherogenic; it transports cholesterol out of cells, stimulates endothelial nitric oxide production [78], and exerts anti-inflammatory and antioxidant actions, in part by decreasing the formation of OxLDL. However, when the antiatherogenic function of HDL is impaired, it becomes pro-inflammatory and may be a risk factor for atherosclerosis [79]. A cohort of 154 patients with SLE had significantly more pro-inflammatory HDL compared to healthy controls (44.7% of SLE patients vs. 4.1% of controls). Moreover, within the SLE group, out of 14 patients with documented clinical atherosclerosis, 57%

had pro-inflammatory HDL [80]. In a subsequent study, pro-inflammatory HDL was found in 48% of 276 patients with SLE, and in 39 of 45 (87%) of these patients who had carotid plaque on ultrasound [81].

Techniques for the reliable and convenient measurement of pro-inflammatory HDL are not readily available. Quantitative HDL measurements do not suffice, as the pro-inflammatory properties of HDL are not related to HDL cholesterol concentrations measured by standard techniques [82].

Thrombosis

Antiphospholipid antibodies (aPL) may be present in up to two-thirds of patients with SLE, and their presence confers an increased risk for arterial and venous thromboembolism [83]. Thromboembolic clinical manifestations associated with antiphospholipid antibodies include deep venous thrombosis, myocardial infarction, stroke, and obstetric complications.

Antiphospholipid antibodies have been proposed as a potential risk factor for atherosclerosis, although the exact mechanisms are not clear. In animal models, immunization with anticardiolipin antibodies [84] or with β2-glycoprotein [85] accelerated atherosclerosis. Patients with primary APL have functional and structural arterial abnormalities. They have significantly increased carotid IMT, pulse wave velocity, and decreased FMD when compared to controls matched for age, sex, and CV risk factors [86]. In patients with APL antibodies, carotid IMT is independently associated with levels of paraoxonase activity, an enzyme with antioxidant properties that prevents lipid peroxidation [86]. Intima media thickness measurements are also greater in APL patients with a history of documented thrombotic events compared to those patients without. Furthermore, the HDL from patients with APL has impaired inflammatory and antioxidant properties when compared to HDL from controls [86], and therefore lacks the protective antiatherogenic properties usually associated with HDL cholesterol.

Altered endothelial cell function is present in patients with primary and secondary APL syndrome and they have higher levels of soluble ICAM, tissue plasminogen activator, and von Willebrand factor compared to controls [87]. Numbers of circulating endothelial cells were also higher in patients with APL syndrome [87].

Platelet Activation

Platelets have the ability to modulate inflammatory reactions and immune responses. During the process of platelet activation, platelets undergo conformational changes and release a variety of chemokines and lipid mediators that contribute to inflammation and atherosclerosis [88]. Abnormal platelet activation has been documented extensively in SLE, and concentrations of thromboxane are elevated [89]. Platelet-leukocyte aggregates, a sensitive marker of platelet activation, are higher in patients with SLE when compared to controls [90]. However, these small studies have not defined the extent to which activated platelets play a role in atherosclerosis in SLE; nevertheless, it is a potential contributing mechanism that warrants further study.

Oxidative Stress

Increased oxidative stress results from an imbalance between products of oxidation and antioxidant defenses. Free radicals that are generated in this process contribute to tissue damage. Conditions that confer increased cardiovascular risk such as smoking [91], diabetes mellitus [92], obesity [93], hypercholesterolemia [94], and homocysteinemia [95] are associated with increased oxidative stress. There are many methods to quantify oxidative stress and its consequences. Among these are the formation of immunogenic complexes of OxLDL binding $\beta2$ glycoprotein. Circulating immune complexes and IgG antibodies directed against them are increased in patients with SLE when compared to controls [96, 97], and are associated with arterial thromboses in patients with antiphospholipid syndrome [97].

Several studies have reported increased oxidative stress in SLE, but these have generally included patients with impaired renal function, a factor known to increase oxidative stress markedly. In contrast, in a large study the concentrations of F_2-isoprostanes, a robust marker of oxidative stress *in vivo*, did not differ among 95 patients with SLE who had low-to-modest disease activity, and 103 control subjects [98]. Thus, oxidation of LDL or antibodies against OxLDL may be related to increased oxidative stress locally, but it seems unlikely that a global increase in oxidative stress is a major contributor to accelerated atherosclerosis in SLE.

SUMMARY

In summary, there have been significant advances in the diagnosis of early atherosclerosis in SLE and in defining the contribution of both traditional and nontraditional cardiovascular risk factors to accelerated atherosclerosis in lupus. A better understanding of the mechanisms underlying vascular damage, plaque formation and stability, and thrombosis, and an improved ability to accurately identify and intervene in patients at highest cardiovascular risk, will greatly facilitate the long-term care of patients with lupus.

References

[1] M.B. Urowitz, A.A. Bookman, B.E. Koehler, D.A. Gordon, H.A. Smythe, M.A. Ogryzlo, The bimodal mortality pattern of systemic lupus erythematosus, Am J Med 60 (2) (1976 February) 221−225.

[2] L.A. Rubin, M.B. Urowitz, D.D. Gladman, Mortality in systemic lupus erythematosus: the bimodal pattern revisited, Q J Med 55 (216) (1985 April) 87−98.

[3] M.B. Urowitz, D.D. Gladman, B.D. Tom, D. Ibanez, V.T. Farewell, Changing patterns in mortality and disease outcomes for patients with systemic lupus erythematosus, J Rheumatol 35 (11) (2008 November) 2152−2158.

[4] M. Petri, S. Perez-Gutthann, D. Spence, M.C. Hochberg, Risk factors for coronary artery disease in patients with systemic lupus erythematosus, Am J Med 93 (5) (1992 November) 513−519.

[5] S. Manzi, E.N. Meilahn, J.E. Rairie, et al., Age-specific incidence rates of myocardial infarction and angina in women with systemic lupus erythematosus: comparison with the Framingham Study, Am J Epidemiol 145 (5) (1997 March 1) 408−415.

[6] I.N. Bruce, D.D. Gladman, M.B. Urowitz, Premature atherosclerosis in systemic lupus erythematosus, Rheum Dis Clin North Am 26 (2) (2000 May) 257−278.

[7] J.M. Esdaile, M. Abrahamowicz, T. Grodzicky, et al., Traditional Framingham risk factors fail to fully account for accelerated atherosclerosis in systemic lupus erythematosus, Arthritis Rheum 44 (10) (2001 October) 2331−2337.

[8] M.M. Ward, Premature morbidity from cardiovascular and cerebrovascular diseases in women with systemic lupus erythematosus, Arthritis Rheum 42 (2) (1999 February) 338−346.

[9] I.N. Bruce, R.J. Burns, D.D. Gladman, M.B. Urowitz, Single photon emission computed tomography dual isotope myocardial perfusion imaging in women with systemic lupus erythematosus. I. Prevalence and distribution of abnormalities, J Rheumatol 27 (10) (2000 October) 2372−2377.

[10] E.M. Sella, E.I. Sato, A. Barbieri, Coronary artery angiography in systemic lupus erythematosus patients with abnormal myocardial perfusion scintigraphy, Arthritis Rheum 48 (11) (2003 November) 3168−3175.

[11] M.J. Roman, B.A. Shanker, A. Davis, et al., Prevalence and correlates of accelerated atherosclerosis in systemic lupus erythematosus, N Engl J Med 349 (25) (2003 December 18) 2399−2406.

[12] S. Manzi, F. Selzer, K. Sutton-Tyrrell, et al., Prevalence and risk factors of carotid plaque in women with systemic lupus erythematosus, Arthritis Rheum 42 (1) (1999 January) 51−60.

[13] K. Maksimowicz-McKinnon, L.S. Magder, M. Petri, Predictors of carotid atherosclerosis in systemic lupus erythematosus, J Rheumatol 33 (12) (2006 December) 2458−2463.

[14] S. Jimenez, M.A. Garcia-Criado, D. Tassies, et al., Preclinical vascular disease in systemic lupus erythematosus and primary antiphospholipid syndrome, Rheumatology (Oxford) 44 (6) (2005 June) 756−761.

[15] P.G. Vlachoyiannopoulos, P.G. Kanellopoulos, J.P. Ioannidis, M.G. Tektonidou, I. Mastorakou, H.M. Moutsopoulos, Atherosclerosis in premenopausal women with antiphospholipid syndrome and systemic lupus erythematosus: a controlled study, Rheumatology (Oxford) 42 (5) (2003 May) 645−651.

[16] M.J. Roman, R.B. Devereux, J.E. Schwartz, et al., Arterial stiffness in chronic inflammatory diseases, Hypertension 46 (1) (2005 July) 194−199.

[17] E. Svenungsson, K. Jensen-Urstad, M. Heimburger, et al., Risk factors for cardiovascular disease in systemic lupus erythematosus, Circulation 104 (16) (2001 October 16) 1887−1893.

[18] A. Cederholm, E. Svenungsson, D. Stengel, et al., Platelet-activating factor-acetylhydrolase and other novel risk and protective factors for cardiovascular disease in systemic lupus erythematosus, Arthritis Rheum 50 (9) (2004 September) 2869–2876.

[19] M.J. Roman, M.K. Crow, M.D. Lockshin, et al., Rate and determinants of progression of atherosclerosis in systemic lupus erythematosus, Arthritis Rheum 56 (10) (2007 October) 3412–3419.

[20] T. Thompson, K. Sutton-Tyrrell, R.P. Wildman, et al., Progression of carotid intima-media thickness and plaque in women with systemic lupus erythematosus, Arthritis Rheum 58 (3) (2008 March) 835–842.

[21] P.C. Keelan, L.F. Bielak, K. Ashai, et al., Long-term prognostic value of coronary calcification detected by electron-beam computed tomography in patients undergoing coronary angiography, Circulation 104 (4) (2001 July 24) 412–417.

[22] Y. Asanuma, A. Oeser, A.K. Shintani, et al., Premature coronary-artery atherosclerosis in systemic lupus erythematosus, N Engl J Med 349 (25) (2003 December 18) 2407–2415.

[23] K. Manger, M. Kusus, C. Forster, et al., Factors associated with coronary artery calcification in young female patients with SLE, Ann Rheum Dis 62 (9) (2003 September) 846–850.

[24] J.M. Von Feldt, L.V. Scalzi, A.J. Cucchiara, et al., Homocysteine levels and disease duration independently correlate with coronary artery calcification in patients with systemic lupus erythematosus, Arthritis Rheum 54 (7) (2006 July) 2220–2227.

[25] A.N. Kiani, L. Magder, M. Petri, Coronary calcium in systemic lupus erythematosus is associated with traditional cardiovascular risk factors, but not with disease activity, J Rheumatol 35 (7) (2008 July) 1300–1306.

[26] D.S. Celermajer, K.E. Sorensen, C. Bull, J. Robinson, J.E. Deanfield, Endothelium-dependent dilation in the systemic arteries of asymptomatic subjects relates to coronary risk factors and their interaction, J Am Coll Cardiol 24 (6) (1994 November 15) 1468–1474.

[27] R. Ross, Atherosclerosis—an inflammatory disease, N Engl J Med 340 (2) (1999 January 14) 115–126.

[28] M. El-Magadmi, H. Bodill, Y. Ahmad, et al., Systemic lupus erythematosus: an independent risk factor for endothelial dysfunction in women, Circulation 110 (4) (2004 July 27) 399–404.

[29] D.S. Lima, E.I. Sato, V.C. Lima, F. Miranda Jr., F.H. Hatta, Brachial endothelial function is impaired in patients with systemic lupus erythematosus, J Rheumatol 29 (2) (2002 February) 292–297.

[30] S.A. Wright, F.M. O'Prey, D.J. Rea, et al., Microcirculatory hemodynamics and endothelial dysfunction in systemic lupus erythematosus, Arterioscler Thromb Vasc Biol 26 (10) (2006 October) 2281–2287.

[31] S. Rajagopalan, E.C. Somers, R.D. Brook, et al., Endothelial cell apoptosis in systemic lupus erythematosus: a common pathway for abnormal vascular function and thrombosis propensity, Blood 103 (10) (2004 May 15) 3677–3683.

[32] F. Selzer, K. Sutton-Tyrrell, S. Fitzgerald, R. Tracy, L. Kuller, S. Manzi, Vascular stiffness in women with systemic lupus erythematosus, Hypertension 37 (4) (2001 April) 1075–1082.

[33] C.P. Chung, J.F. Solus, A. Oeser, et al., N-terminal pro-brain natriuretic peptide in systemic lupus erythematosus: relationship with inflammation, augmentation index, and coronary calcification, J Rheumatol 35 (7) (2008 July) 1314–1319.

[34] M. Petri, D. Spence, L.R. Bone, M.C. Hochberg, Coronary artery disease risk factors in the Johns Hopkins Lupus Cohort: prevalence, recognition by patients, and preventive practices, Medicine (Baltimore) 71 (5) (1992 September) 291–302.

[35] E.F. Borba, E. Bonfa, Dyslipoproteinemias in systemic lupus erythematosus: influence of disease, activity, and anticardiolipin antibodies, Lupus 6 (6) (1997) 533–539.

[36] I.N. Bruce, M.B. Urowitz, D.D. Gladman, D. Ibanez, G. Steiner, Risk factors for coronary heart disease in women with systemic lupus erythematosus: the Toronto Risk Factor Study, Arthritis Rheum 48 (11) (2003 November) 3159–3167.

[37] P. Rahman, S. Aguero, D.D. Gladman, D. Hallett, M.B. Urowitz, Vascular events in hypertensive patients with systemic lupus erythematosus, Lupus 9 (9) (2000) 672–675.

[38] I.N. Bruce, M.B. Urowitz, D.D. Gladman, D.C. Hallett, Natural history of hypercholesterolemia in systemic lupus erythematosus, J Rheumatol 26 (10) (1999 October) 2137–2143.

[39] C.P. Chung, A. Oeser, J. Solus, et al., Inflammatory mechanisms affecting the lipid profile in patients with systemic lupus erythematosus, J Rheumatol 34 (9) (2007 September) 1849–1854.

[40] C.P. Chung, A. Oeser, P. Raggi, et al., Lipoprotein subclasses and particle size determined by nuclear magnetic resonance spectroscopy in systemic lupus erythematosus, Clin Rheumatol 27 (2008 April 18) 1227–1233.

[41] C.P. Chung, A. Oeser, I. Avalos, P. Raggi, C.M. Stein, Cardiovascular risk scores and the presence of subclinical coronary artery atherosclerosis in women with systemic lupus erythematosus, Lupus 15 (9) (2006) 562–569.

[42] A.B. Lee, T. Godfrey, K.G. Rowley, et al., Traditional risk factor assessment does not capture the extent of cardiovascular risk in systemic lupus erythematosus, Intern Med J 36 (4) (2006 April) 237–243.

[43] G.K. Hansson, Atherosclerosis—an immune disease: The Anitschkov Lecture 2007, Atherosclerosis 202 (1) (2009 January) 2–10.

[44] H.C. Stary, A.B. Chandler, S. Glagov, et al., A definition of initial, fatty streak, and intermediate lesions of atherosclerosis. A report from the Committee on Vascular Lesions of the Council on Arteriosclerosis, American Heart Association, Circulation 89 (5) (1994 May) 2462–2478.

[45] J.M. Wang, A. Sica, G. Peri, et al., Expression of monocyte chemotactic protein and interleukin-8 by cytokine-activated human vascular smooth muscle cells, Arterioscler Thromb 11 (5) (1991 September) 1166–1174.

[46] A.B. Reiss, A.D. Glass, Atherosclerosis: immune and inflammatory aspects, J Investig Med 54 (3) (2006 April) 123–131.

[47] P.T. Kovanen, M. Kaartinen, T. Paavonen, Infiltrates of activated mast cells at the site of coronary atheromatous erosion or rupture in myocardial infarction, Circulation 92 (5) (1995 September 1) 1084–1088.

[48] L. Jonasson, J. Holm, O. Skalli, G. Bondjers, G.K. Hansson, Regional accumulations of T cells, macrophages, and smooth muscle cells in the human atherosclerotic plaque, Arteriosclerosis 6 (2) (1986 March) 131–138.

[49] A.C. van der Wal, A.E. Becker, C.M. van der Loos, P.K. Das, Site of intimal rupture or erosion of thrombosed coronary atherosclerotic plaques is characterized by an inflammatory process irrespective of the dominant plaque morphology, Circulation 89 (1) (1994 January) 36–44.

[50] S.J. Szabo, B.M. Sullivan, S.L. Peng, L.H. Glimcher, Molecular mechanisms regulating Th1 immune responses, Annu Rev Immunol 21 (2003) 713–758.

[51] G. Liuzzo, J.J. Goronzy, H. Yang, et al., Monoclonal T-cell proliferation and plaque instability in acute coronary syndromes, Circulation 101 (25) (2000 June 27) 2883–2888.

[52] W. Park, C.M. Weyand, D. Schmidt, J.J. Goronzy, Co-stimulatory pathways controlling activation and peripheral tolerance of human CD4+, Eur J Immunol 27 (5) (1997 May) 1082–1090.

[53] P.B. Martens, J.J. Goronzy, D. Schaid, C.M. Weyand, Expansion of unusual CD4+ T cells in severe rheumatoid arthritis, Arthritis Rheum 40 (6) (1997 June) 1106–1114.

[54] A.K. Stanic, C.M. Stein, A.C. Morgan, et al., Immune dysregulation accelerates atherosclerosis and modulates plaque composition in systemic lupus erythematosus, Proc Natl Acad Sci U S A 103 (18) (2006 May 2) 7018–7023.

[55] Y. Asanuma, C.P. Chung, A. Oeser, et al., Increased concentration of proatherogenic inflammatory cytokines in systemic lupus erythematosus: relationship to cardiovascular risk factors, J Rheumatol 33 (3) (2006 March) 539–545.

[56] Y.H. Rho, C.P. Chung, A. Oeser, et al., Novel Cardiovascular Risk Factors in Premature Coronary Atherosclerosis Associated with Systemic Lupus Erythematosus, J Rheumatol 35 (2008) 1789–1794.

[57] E. Svenungsson, G.Z. Fei, K. Jensen-Urstad, U. de Faire, A. Hamsten, J. Frostegard, TNF-alpha: a link between hypertriglyceridaemia and inflammation in SLE patients with cardiovascular disease, Lupus 12 (6) (2003) 454–461.

[58] M.J. Kaplan, Premature vascular damage in systemic lupus erythematosus: an imbalance of damage and repair? Transl Res 154 (2) (2009 August) 61–69.

[59] S. Dimmeler, A.M. Zeiher, Endothelial cell apoptosis in angiogenesis and vessel regression, Circ Res 87 (6) (2000 September 15) 434–439.

[60] R. Clancy, G. Marder, V. Martin, H.M. Belmont, S.B. Abramson, J. Buyon, Circulating activated endothelial cells in systemic lupus erythematosus: further evidence for diffuse vasculopathy, Arthritis Rheum 44 (5) (2001 May) 1203–1208.

[61] P.E. Westerweel, M.C. Verhaar, Endothelial progenitor cell dysfunction in rheumatic disease, Nat Rev Rheumatol 5 (6) (2009 June) 332–340.

[62] M.F. Denny, S. Thacker, H. Mehta, et al., Interferon-alpha promotes abnormal vasculogenesis in lupus: a potential pathway for premature atherosclerosis, Blood 110 (8) (2007 October 15) 2907–2915.

[63] C. Cardillo, F. Schinzari, N. Mores, et al., Intravascular tumor necrosis factor alpha blockade reverses endothelial dysfunction in rheumatoid arthritis, Clin Pharmacol Ther 80 (3) (2006 September) 275–281.

[64] A. Oeser, C.P. Chung, Y. Asanuma, I. Avalos, C.M. Stein, Obesity is an independent contributor to functional capacity and inflammation in systemic lupus erythematosus, Arthritis Rheum 52 (11) (2005 November) 3651–3659.

[65] C.P. Chung, I. Avalos, A. Oeser, et al., High prevalence of the metabolic syndrome in patients with systemic lupus erythematosus: association with disease characteristics and cardiovascular risk factors, Ann Rheum Dis 66 (2) (2007 February) 208–214.

[66] M.M. El, Y. Ahmad, W. Turkie, et al., Hyperinsulinemia, insulin resistance, and circulating oxidized low density lipoprotein in women with systemic lupus erythematosus, J Rheumatol 33 (1) (2006 January) 50–56.

[67] K.E. Kip, O.C. Marroquin, D.E. Kelley, et al., Clinical importance of obesity versus the metabolic syndrome in cardiovascular risk in women: a report from the Women's Ischemia Syndrome Evaluation (WISE) study, Circulation 109 (6) (2004 February 17) 706–713.

[68] C.P. Chung, A. Oeser, J.F. Solus, et al., Inflammation-associated insulin resistance: differential effects in rheumatoid arthritis and systemic lupus erythematosus define potential mechanisms, Arthritis Rheum 58 (7) (2008 July) 2105–2112.

[69] D.S. Wald, M. Law, J.K. Morris, Homocysteine and cardiovascular disease: evidence on causality from a meta-analysis, BMJ 325 (7374) (2002 November 23) 1202.

[70] P.E. Lazzerini, P.L. Capecchi, E. Selvi, et al., Hyperhomocysteinemia: a cardiovascular risk factor in autoimmune diseases? Lupus 16 (11) (2007) 852–862.

[71] M. Petri, R. Roubenoff, G.E. Dallal, M.R. Nadeau, J. Selhub, I.H. Rosenberg, Plasma homocysteine as a risk factor for atherothrombotic events in systemic lupus erythematosus, Lancet 348 (9035) (1996 October 26) 1120–1124.

[72] T.M. Refai, I.H. Al-Salem, D. Nkansa-Dwamena, M.H. Al-Salem, Hyperhomocysteinaemia and risk of thrombosis in systemic lupus erythematosus patients, Clin Rheumatol 21 (6) (2002 November) 457–461.

[73] E. Lonn, S. Yusuf, M.J. Arnold, et al., Homocysteine lowering with folic acid and B vitamins in vascular disease, N Engl J Med 354 (15) (2006 April 13) 1567–1577.

[74] S. Horkko, C.J. Binder, P.X. Shaw, et al., Immunological responses to oxidized LDL, Free Radic Biol Med 28 (12) (2000 June 15) 1771–1779.

[75] M.K. Chang, C. Bergmark, A. Laurila, et al., Monoclonal antibodies against oxidized low-density lipoprotein bind to apoptotic cells and inhibit their phagocytosis by elicited macrophages: evidence that oxidation-specific epitopes mediate macrophage recognition, Proc Natl Acad Sci U S A 96 (11) (1999 May 25) 6353–6358.

[76] L.A. Casciola-Rosen, G. Anhalt, A. Rosen, Autoantigens targeted in systemic lupus erythematosus are clustered in two populations of surface structures on apoptotic keratinocytes, J Exp Med 179 (4) (1994 April 1) 1317–1330.

[77] M.K. Chang, C.J. Binder, Y.I. Miller, et al., Apoptotic cells with oxidation-specific epitopes are immunogenic and proinflammatory, J Exp Med 200 (11) (2004 December 6) 1359–1370.

[78] I.S. Yuhanna, Y. Zhu, B.E. Cox, et al., High-density lipoprotein binding to scavenger receptor-BI activates endothelial nitric oxide synthase, Nat Med 7 (7) (2001 July) 853–857.

[79] M. Navab, S.Y. Hama, S.T. Reddy, et al., Oxidized lipids as mediators of coronary heart disease, Curr Opin Lipidol 13 (4) (2002 August) 363–372.

[80] M. McMahon, J. Grossman, J. FitzGerald, et al., Proinflammatory high-density lipoprotein as a biomarker for atherosclerosis in patients with systemic lupus erythematosus and rheumatoid arthritis, Arthritis Rheum 54 (8) (2006 August) 2541–2549.

[81] M. McMahon, J. Grossman, B. Skaggs, et al., Dysfunctional proinflammatory high-density lipoproteins confer increased risk of atherosclerosis in women with systemic lupus erythematosus, Arthritis Rheum 60 (8) (2009 August) 2428–2437.

[82] B.J. Ansell, M. Navab, S. Hama, et al., Inflammatory/antiinflammatory properties of high-density lipoprotein distinguish patients from control subjects better than high-density lipoprotein cholesterol levels and are favorably affected by simvastatin treatment, Circulation 108 (22) (2003 December 2) 2751–2756.

[83] E. Somers, L.S. Magder, M. Petri, Antiphospholipid antibodies and incidence of venous thrombosis in a cohort of patients with systemic lupus erythematosus, J Rheumatol 29 (12) (2002 December) 2531–2536.

[84] J. George, A. Afek, B. Gilburd, et al., Atherosclerosis in LDL-receptor knockout mice is accelerated by immunization with anticardiolipin antibodies, Lupus 6 (9) (1997) 723–729.

[85] J. George, A. Afek, B. Gilburd, et al., Induction of early atherosclerosis in LDL-receptor-deficient mice immunized with beta2-glycoprotein I, Circulation 98 (11) (1998 September 15) 1108–1115.

[86] M. Charakida, C. Besler, J.R. Batuca, et al., Vascular abnormalities, paraoxonase activity, and dysfunctional HDL in primary

I. BASIS OF DISEASE PATHOGENESIS

antiphospholipid syndrome, JAMA 302 (11) (2009 September 16) 1210—1217.

[87] M. Cugno, M.O. Borghi, L.M. Lonati, et al., Patients with anti-phospholipid syndrome display endothelial perturbation, J Autoimmun 34 (2009) 105—110.

[88] H.P. von, C. Weber, Platelets as immune cells: bridging inflammation and cardiovascular disease, Circ Res 100 (1) (2007 January 5) 27—40.

[89] I. Avalos, C.P. Chung, A. Oeser, et al., Aspirin therapy and thromboxane biosynthesis in systemic lupus erythematosus, Lupus 16 (12) (2007) 981—986.

[90] J.E. Joseph, P. Harrison, I.J. Mackie, D.A. Isenberg, S.J. Machin, Increased circulating platelet-leucocyte complexes and platelet activation in patients with antiphospholipid syndrome, systemic lupus erythematosus and rheumatoid arthritis, Br J Haematol 115 (2) (2001 November) 451—459.

[91] J.D. Morrow, B. Frei, A.W. Longmire, et al., Increase in circulating products of lipid peroxidation (F2-isoprostanes) in smokers. Smoking as a cause of oxidative damage, N Engl J Med 332 (18) (1995 May 4) 1198—1203.

[92] G. Davi, G. Ciabattoni, A. Consoli, et al., In vivo formation of 8-iso-prostaglandin f2alpha and platelet activation in diabetes mellitus: effects of improved metabolic control and vitamin E supplementation, Circulation 99 (2) (1999 January 19) 224—229.

[93] G. Davi, M.T. Guagnano, G. Ciabattoni, et al., Platelet activation in obese women: role of inflammation and oxidant stress, JAMA 288 (16) (2002 October 23) 2008—2014.

[94] G. Davi, P. Alessandrini, A. Mezzetti, et al., In vivo formation of 8-Epi-prostaglandin F2 alpha is increased in hypercholesterolemia, Arterioscler Thromb Vasc Biol 17 (11) (1997 November) 3230—3235.

[95] G. Davi, G. Di Minno, A. Coppola, et al., Oxidative stress and platelet activation in homozygous homocystinuria, Circulation 104 (10) (2001 September 4) 1124—1128.

[96] L.R. Lopez, M. Salazar-Paramo, C. Palafox-Sanchez, B.L. Hurley, E. Matsuura, L.T.I. Garcia-De, Oxidized low-density lipoprotein and beta2-glycoprotein I in patients with systemic lupus erythematosus and increased carotid intima-media thickness: implications in autoimmune-mediated atherosclerosis, Lupus 15 (2) (2006) 80—86.

[97] L.R. Lopez, D.F. Simpson, B.L. Hurley, E. Matsuura, OxLDL/beta2GPI complexes and autoantibodies in patients with systemic lupus erythematosus, systemic sclerosis, and anti-phospholipid syndrome: pathogenic implications for vascular involvement, Ann N Y Acad Sci 1051 (2005 June) 313—322.

[98] I. Avalos, C.P. Chung, A. Oeser, et al., Oxidative Stress in Systemic Lupus Erythematosus: Relationship to Disease Activity and Symptoms, Lupus 16 (3) (2007) 195—200.

PART II

CLINICAL ASPECTS OF DISEASE

A. Clinical Presentation
(Chapters 30–39)

B. Organ Systems
(Chapters 40–49)

The Clinical Presentation of Systemic Lupus Erythematosus

Robert G. Lahita

INTRODUCTION

Systemic lupus erythematosus (SLE) is an auto-immune disease that affects many organ systems, is more prevalent in females, and has no known etiology. Moreover, the presentation of patients with the disease can be as diverse as the many systems in the body that it can affect. The systems commonly involved include muscle and joints, brain and peripheral nervous system, lungs, heart, kidneys, skin, serous membranes, and components of the blood. While other systems can be affected, they are with lesser frequency. This disease is complex and often affects one organ system to the exclusion of others. Moreover, the clinical manifestations are protean, overlap with other illnesses, and are often subtle. The pathogenetic factors causing lupus remain enigmatic, and while predominantly immunological, are influenced and modified by multiple systems such as the endocrine or clotting systems. The complex nature of lupus could explain the convergence of seemingly unrelated abnormalities and the varied nature of pathology observed from patient to patient in this very interesting illness.

The predilection of this disease for females after puberty cannot be adequately explained; however, sus-pected pathogenetic factors, such as sex steroid hormones or gonadotrophins such as prolactin, may play a role in the severity of the disease and the different clinical presentations. Environmental factors such as drugs, diet, and toxins have also been implicated in the pathogenesis of SLE.

Although lupus is largely associated with specific immune response genes in the MHC class II or class III regions, there are new and exciting genetic associations that include other loci. Certain genetic MHC II alleles are found more commonly with certain autoantibody groups and patients with inherited complement defi-ciencies also develop variants of lupus with specific clin-ical characteristics.

Although there are a few nonimmunologic laboratory characteristics of SLE, these are not specific. Immuno-logical tests, which are specific to most autoimmune diseases such as lupus, include the presence of specific cytotoxic lymphocytes and a variety of autoantibodies, including a persistently positive antinuclear antibody (ANA). Specific areas and systems are covered in great detail throughout the book.

In this chapter there is much history, and many of the references are classic and hold true today.

GENERAL SYMPTOMS

One major problem diagnosing the patient with lupus is the nature of the general symptoms associated with the disease. The most problematic symptoms that tax the mind of the clinician include fatigue, weight loss, and fever [1, 2]. These symptoms are often the initial complaint and are usually attributed to causes other than lupus at first glance. Fatigue is well recognized and is the most common and often the most debilitating symptom of SLE, as it is commonly overlooked, cannot be quantified except perhaps by exercise tolerance testing, and is often the only symptom to remain after therapy of an acute flare [3]. Patients liken the fatigue of SLE to a bout of influenza. A curious pattern of fatigue is described in SLE when compared to patients with other connective tissue diseases [4]. In SLE, fatigue decreased in the morning and increased in the evening in contrast to other conditions such as scleroderma where the opposite is true. Weight loss is common in patients with lupus and worsen when there is malabsorption due to an overlapping illness such as CREST syndrome (calcinosis, Raynaud, esophageal dysmotility, sclerodactyly, and telangiectasia) or mixed connective tissue disease (MCTD). Although anorexia is common in severely debilitated patients with associated organ disease such as renal failure, it is not an isolated finding in SLE. The fever of lupus is usually low grade and rarely exceeds 102°F. A temperature greater than 102°F should warrant a search for infection [5]. Of course, patients who are taking immunosuppressive drugs and have fever must be handled differently from the untreated patient, as high fevers may be masked or lowered by such agents and infections in immunosuppressed patients are more likely.

CRITERIA FOR LUPUS

The diagnosis of SLE should be made principally on clinical grounds with the support of laboratory tests, not the other way around. While diagnostic criteria have been proposed for the classification of SLE [6], they are not universally applied in practice. Eleven criteria have been designated by the American College of Rheumatology (ACR) (Table 30.1) for classification. The presence of four or more criteria is mandatory for the appropriate classification of SLE. When used, these are of value in clinical practice and are 96% sensitive and specific [7]. In some studies of patients with cutaneous lupus, ACR criteria showed a sensitivity of 88%, a specificity of 79%, a positive predictive value of 56%, and a negative predictive value of 96% [8]. Despite these

TABLE 30.1   Classification Criteria for SLE

Criterion	Definition
1. Malar rash	Fixed erythema, flat or raised, over the malar eminence, tending to spare the nasolabial folds
2. Discoid rash	Erythematous raised patches with adherent keratotic scaling and follicular plugging; atrophic scarring may occur in older lesions
3. Photosensitivity	Skin rash as a result of unusual reaction to sunlight, by patient history or physician observation
4. Oral ulcers	Oral or nasopharyngeal ulceration, usually painless, observed by a physician
5. Arthritis	Nonerosive arthritis involving two or more peripheral joints, characterized by tenderness, swelling, or effusion
6. Serositis	a. Pleuritis: convincing history of pleuritic pain or rub heard by physician or evidence of pleural effusion
	b. Pericarditis: documented by ECG or rub or evidence of pericardial effusion
7. Renal disorder	a. Persistent proteinuria greater than 0.5 g per day or greater than 3+ if quantitation not performed
	b. Cellular casts: may be red cell, hemoglobin, granular, tubular, or mixed
8. Neurologic disorder	a. Seizures: in the absence of offending drugs or known metabolic derangement, e.g., uremia, ketoacidosis, or electrolyte imbalance
	b. Psychosis: in the absence of offending drugs or known metabolic derangement, e.g., uremia, ketoacidosis, or electrolyte imbalance
9. Hematologic disorder	a. Hemolytic anemia: with reticulocytosis
	b. Leukopenia: less than 4000/mm^3 total on two or more occasions
	c. Lymphopenia: less than 1500/mm^3 on two or more occasions
	d. Thrombocytopenia: less than 100,000/mm^3 in the absence of offending drugs
10. Immunologic disorder[b]	a. Anti-DNA: antibody to native DNA in abnormal titer
	b. Anti-SM: presence of antibody to SM nuclear antigen
	c. Positive finding of antiphospholipid antibodies based on (i) an abnormal serum level of IgG or IgM anticardiolipin antibodies, (ii) a positive test result for lupus anticoagulant using a standard method, or (iii) a false-positive serologic test for syphilis known to be positive for at least 6 months and confirmed by *Treponema pallidum* immobilization or fluorescent treponemal antibody absorption test
11. ANA	An abnormal titer of ANA by immunofluorescence or an equivalent assay at any point in time and in the absence of drugs known to be associated with "drug-induced" lupus syndrome

[a] *This classification is based on 11 criteria. For the purpose of identifying patients in clinical studies, a person must have SLE if any four or more of the 11 criteria are present, serially or simultaneously, during any interval of observation.*
[b]*The modifications to criterion number 10 were made in 1996 [6].*

facts, the diagnosis of SLE is given routinely to patients in practice who fail to meet any criteria [9] and it should be discouraged. Patients who fail to meet four of the 11 criteria may have what is called a lupus-like syndrome or incomplete lupus. Such patients may also have a related autoimmune disease. It is important to realize that lupus criteria can be acquired over a period of time. These revised criteria have been reviewed and applied to a number of cohorts [10]. The criteria have also been criticized from a number of groups because of their failure to apply to every lupus-like condition [11].

The Revised ACR criteria for lupus, while the standard for establishing eligibility of subjects for epidemiologic and clinical lupus studies, exclude patients with limited disease. Such exclusions are a problem for researchers and clinicians. In 2002 the Boston weighted criteria for clinical studies were established, and these criteria were compared to the clinical rheumatologists' diagnosis of SLE. Using these criteria, the investigators were able to identify a larger number of cases with SLE. The criteria had a sensitivity of 93% and specificity of 69% compared to the ACR criteria, and 7% more patients with lupus were identified [12].

In 2001, Wilson *et al.* published the criteria for antiphospholipid syndrome. These criteria, which may need revision over time, are very helpful to this important variant of a condition that can be primary or secondary to SLE [13, 14].

GENERAL CONSIDERATIONS REGARDING LUPUS

The initial diagnosis of lupus depends on the manner of clinical presentation. Table 30.2 [15–17], can be so nonspecific as to be confused with other illnesses. In other cases, the signs and symptoms may be ignored. As a result, the mean length of time from onset of symptoms to diagnosis can be as long as 5 years, a lapse that could seriously affect prognosis. There are cases of SLE that elude diagnosis until autopsy. Lupus is not always apparent at the initial visit and continued follow-up with interval histories, clinical examinations, and directed serological evaluations may be necessary in order to finalize the diagnosis. It is not uncommon for patients to present with isolated arthralgias, Raynaud phenomenon, hypercoagulable states, fever of unknown origin, respiratory symptoms such as dyspnea and pleural effusions or overt renal failure. The protean clinical manifestations of this illness can make the diagnosis difficult and if access to specialty care is unavailable the patient may elude diagnosis altogether. In many cases, patients are often misdiagnosed with rheumatoid arthritis (RA) [18], fever of unknown origin [19],

TABLE 30.2 Initial Manifestations of SLE[a]

Manifestation	Data from Estes and Christian [15]	Data from Dubois and Tuiffanelli [17]
Arthritis or arthralgia, cutaneous	53[a]	46
Discoid	9	11
Malar rash	9	8
Other skin manifestations	1	
Nephritis	6	4
Overlap	5	3
Other	17	16
Serositis (pleurisy, pericarditis)	5	
Seizures	3	
Raynaud phenomenon	3	
Anemia	2	
Thrombocytopenia	2	
False-positive syphilis test	1	
Jaundice	1	

[a]Expressed as percentages.

fibromyalgia [20, 21], or asked to seek psychiatric help for a functional disorder [22].

Laboratory data in the form of an isolated false-positive syphilis test, a low platelet count, elevated PTT, or leukopenia may suggest many other diseases that are often considered before SLE.

SLE is really a clinical diagnosis that relies heavily on a careful history and physical examination because auto-antibodies are so frequently associated with nondisease; they alone are supportive but not diagnostic. A history of specific medication is particularly important, as reversible drug-induced lupus is responsible for some 10% of cases [23] and is often seen in older patients, particularly males. Lupus can be related to other illnesses that can occur in the patient's family [24]. There are class II antigens, D locus associations to these diseases as well. These include rheumatoid arthritis, multiple sclerosis, idiopathic thrombocytopenic purpura (ITP), rheumatic fever, overlap syndromes such as Sjögren's, scleroderma, thyroiditis, and the inflammatory diseases of muscle.

Fibromyalgia is an extremely important condition that often occurs with SLE and it is often difficult to differentiate the pain of fibromyalgia from that of the patient with SLE. Physicians should be cautioned not to consider all pain in lupus patients as that of a lupus flare. Many patients are mistakenly treated with immunosuppressives for pain that is caused by fibromyalgia [20]. A positive ANA and muscle pain must be differentiated from SLE-induced arthropathy and myopathy [21].

Many lupus patients are also known to have increased allergies by some early reports [25] to a variety of agents, especially sulfur. Sulfur-containing drugs should be avoided where possible in patients with SLE. The most common drug used to treat lupus patients for urinary tract infections is Bactrim and this should be avoided in most cases. However, clinical judgment is required, because sulfa drugs like Bactrim are often used for *Pneumocystis* prophylaxis in those patients on large doses of immunosuppressive drugs.

The frequencies of some common clinical presentations are given in Table 30.3. Simple laboratory tests can be helpful adjuncts in the diagnosis of SLE but should not be used by themselves as sole criteria. Common laboratory assays can aid the physician in establishing the diagnosis. Most useful is the white blood cell (WBC) count, which often shows leukopenia and lymphopenia [26]. Anemia of chronic disease or, in rare instances autoimmune hemolytic anemia, can be differentiated by the examination of red cell indices, a reticulocyte count, a peripheral smear, and a positive Coombs test. Thrombocytopenia coupled with elevated coagulation tests such as the PTT might suggest the presence of antiphospholipid antibodies and the lupus anticoagulant. A false-positive syphilis test along with an abnormal PTT also suggests phospholipid antibody syndrome. The antibodies are not sufficient criteria for a diagnosis of the antiphospholipid syndrome since the presence of a blood clot or second trimester miscarriage are necessary criteria for the diagnosis. An abnormal urinalysis with the appearance of WBCs, red blood cells, granular casts, and proteinuria can also be helpful and raise a suspicion of lupus nephritis. While blood urea nitrogen (BUN) and creatinine levels are not usually elevated at the outset of the disease, they can be useful rarely as baseline values in a patient who progresses to azotemia. A chest x-ray and electrocardiogram should be obtained initially to rule out pulmonary pathology, to explain an enlarged cardiac silhouette, and differentiate EKG signs of pericarditis, enlarged cardiac chambers, or signs of ischemia. This is of particular importance since accelerated atherosclerosis and early heart and vascular disease is so common to this condition [27].

Depending on the clinician, more complex immunologic laboratory tests may solidify the diagnosis when suspected clinically or contribute to the erroneous diagnosis of SLE in the absence of clinical signs (Table 30.4). The widespread availability of such testing, the lack of standardization, and inappropriate application of such tests contribute to this confusion. The most useful tests for SLE are the fluorescent antinuclear antibody test, antinative DNA assay, and total hemolytic complement (CH50). All of these tests are subject to wide variation, the ANA because of substrate variability, and the

TABLE 30.3   Frequency (%) of Some Common Clinical Manifestations of SLE

Manifestation	Data from Estes and Christian [15]	Data from Hochberg et al. [26]
Musculoskeletal	95	83
Arthritis	95	76
Avascular necrosis	7	24
Myositis	5	5
Cutaneous	88	81
Malar rash	39	61
Alopecia	37	45
Photosensitivity	NA	45
Dermal vasculitis	21	27
Raynaud phenomenon	21	44
Discoid lesions	14	15
Rheumatoid nodules	11	12
Oral ulcers	7	23
Fever	77	NA
Serositis	NA	63
Pleurisy	40	57
Pericarditis	19	23
Peritonitis	NA	8
Neuropsychiatric	59	55
Psychosis	37	16
Neurosis	5	NA
Grand mal seizures	13	26
Peripheral neuropathy	7	21
Cranial nerve palsies	NA	5
Hemiparesis	NA	5
Renal (nephritis)	53	31
Nephrotic syndrome	26	13
Hypertension	46	NA
Pulmonary	NA	NA
Lupus pneumonia	9	NA
Fibrosis	6	NA
Cardiac	NA	NA
Myocarditis	8	NA
Sinus tachycardia	13	NA
Heart failure	11	NA

NA, not available.

TABLE 30.4 Frequency (%) of Common Laboratory Manifestations of SLE[a]

Manifestation	Data from Estes and Christian [15]	Data from Hochberg et al. [26]
Hematologic		
Anemia	73	57
Leukopenia[b]	66	41
Thrombocytopenia[c]	19	45
Positive direct Coombs	27	27
Immunologic		
Antinuclear antibodies	87	94
Hypocomplementemia[d]	NA	59
Rheumatoid factor	21	34
Hyperglobulinemia[e]	77	30
False-positive syphilis test	24	26
LE cells	78	71
Anti-dsDNA	NA	28
Anti-Sm	NA	17
Anti-RNP	NA	34

[a]Defined as hemoglobin <11 g/dl (Estes and Christian) or hematocrit <35% (Hochberg et al.).
[b]Defined as white blood cells <4500/mm^3 (Estes and Christian) or <4000/mm^3 (Hochberg et al.).
[c]Defined as platelets <100,000/mm^3 (Estes and Christian) or <150,000/mm^3 (Hochberg et al.).
[d]Defined as CH50 <26 units.
[e]Defined as globulins >1.5 g/dl (Estes and Christian) or >4 g/dl (Hochberg et al.).
NA, not available.

TABLE 30.5 Some Common Autoantibodies in SLE and Drug-Induced Lupus

Condition	Autoantibody	% Positive	Comment
SLE	Anti-dsDNA	30—70	Associated with nephritis, marker for SLE
	Anti-Sm	20—40	Marker for SLE
	Anti-RNP	40—60	Also seen in MCTD and PSS
	Anti-Ro/SS-A	30-40	Associated with sicca syndrome, also seen in Sjögren syndrome
	Anti-PCNA	5—10	
	Anti-Ku	30—40	Also seen in overlap syndromes
	Antilarnin B	5—10	Also seen in autoimmune liver disease
	Antiribosomal P	5—10	Associated with psychosis
	Antihistone	30	Seen in many disorders
	Anti-ssDNA		Seen in many disorders
Drug-induced lupus	Antihistone	95—100	Seen in many disorders
	Anti-ssDNA		Seen in many disorders

tendency to use ELISA and other nonspecific modalities; the anti-DNA because of single-stranded DNA contamination and cross-reactivity with phospholipid; and finally the CH50 because of the temperature lability of complement components [28, 29].

Many rheumatologists believe that ANA-negative lupus exists [30], but this is probably an artifact of substrate. Occasionally, other reactive antibodies, such as anti-Ro, anti-La, and other ribonucleoprotein antibodies, might be found in the absence of an ANA. Conversely, ANAs can be found in a wide variety of nonlupus individuals, such as the elderly, pregnant females, those ingesting certain medications, and in the setting of certain viral syndromes [7, 31, 32]. In these cases, the ANA are likely to be transient. Antibodies to native DNA [33] and antibodies to Sm antigen, a nuclear ribonucleoprotein, are also quite specific for SLE [34]. Relevant autoantibodies and their associations are given in Table 30.5 [35].

MUSCULOSKELETAL SYSTEM

A common presenting symptom of lupus is arthritis. It is generally nonerosive, nondeforming, symmetric arthropathy. Multiple joints are involved, and 80—95% of them are tender, swollen, and effusive joints. The joints are the most frequently involved organs in SLE. Chapter 49 by Furie, Marder and Horowitz details these important aspects of the disease. The most frequently involved joints are the proximal interphalangeal, metacarpal phalangeal, wrists, and knees. The deforming arthritis of Jaccoud can occur in the lupus patient, and in such patients, swan neck deformity and profound ulnar deviation can be reduced and this form of joint involvement lacks erosive changes. The most frequent musculoskeletal x-ray changes are soft tissue swelling, acral sclerosis, and periarticular demineralization. When rare erosions of bone occur, one must consider a form of overlap syndrome [36]. Joints such as the temporomandibular and the sacroiliac may also be involved (the latter particularly in males) in the SLE patient. However, involvement of these joints suggests other diseases such as RA [37], MCTD, or ankylosing spondylitis. Overlap syndromes such as MCTD resemble RA or scleroderma more than SLE, and erosions are more likely to be found on x-ray in these patients. Rheumatoid nodules can occur in SLE with the presence of a high rheumatoid factor [38, 39] but this is rare. Erosive joint changes can contribute to confusion about making a diagnosis of SLE, particularly in the elderly patient, where SLE is distinctly uncommon [40].

Avascular necrosis is a particular source of joint pain in SLE patients and should be a part of every differential diagnosis. It is a feature found in patients who are ingesting corticosteroids and those with phospholipid antibodies [41]. Avascular necrosis (AVN) is commonly found in the hips, carpal bones of the wrist, and heads of the humeri, and the knees. Less commonly, the shafts of the long bones can be affected. Anywhere from 5 to 10% of patients with SLE can have AVN [42] and these findings are not always associated with steroid use. In many cases, AVN can be asymptomatic and detected by routine x-ray evaluation. In decreasing order of sensitivity, magnetic resonance imaging (MRI), bone scan (Tc99), and plain x-ray are useful in detecting AVN; however, MRI is often positive when all other diagnostic modalities are negative.

Septic arthritis must also be considered as a cause of SLE joint pain [43], particularly when there is swelling and intense warmth of a joint coupled with peripheral leukocytosis. An aspiration of an effusive SLE joint with subsequent culture is mandatory and can be lifesaving.

Overt clinical myositis can present in 3–5% of SLE patients with creatinine phosphokinase (CPK) greater than 1000 [44, 45], but clinical features such as myalgias distinct from fibromyalgia can be found in as many as 50% of patients. The CPK is rarely elevated above 1000, but an electromyogram (EMG) can be very abnormal. Biopsy of the muscle is rarely required for a definitive diagnosis, but if the CPK is exceptionally high or there is diagnostic confusion, one may wish to consider alternative diagnoses and measure aldolase and obtain synthetase antibodies. The lymphocytic, monocytic, and plasma cell infiltration found in primary immune myopathies can be observed in varying degrees with SLE patients. A vacuolar myopathy is rare in SLE but may be found in untreated patients [45] in distal as well as proximal muscles. Myopathy in SLE can also be secondary to corticosteroid therapy [46], and a form of myopathy can be found in patients ingesting antimalarials [47].

CLINICAL INVOLVEMENT OF THE RENAL SYSTEM

Clinical evidence of kidney involvement is found in one-half to two-thirds of patients with SLE [48]. Biopsy evidence of immune complex deposition is found in all kidneys of all patients with SLE [49, 50], regardless of

TABLE 30.6 World Health Organization Classification of Lupus Nephritis[a]

Class	Pattern	Site of immune complex deposition
I	Normal	None
II	Mesangial	Mesangial only
III	Focal and segmental proliferative	Mesangial, subendothelial, ± subepithelial
IV	Diffuse proliferative	Mesangial, subendothelial, ± subepithelial
V	Membranous	Mesangial, subepithelial

[a]Adapted from Appel et al. [48].

urine sediment. Table 30.6 shows that there are various forms of glomerular disease listed by the World Health Organization (WHO) that are helpful in establishing activity and chronicity as well as degrees of severity. Both diffuse proliferative glomerulonephritis and progressive forms of focal proliferative nephritis have poorer prognoses than membranous and mesangial forms of the disease [51, 52]. A major clinical point is that each renal lesion has activity. The greater the activity, the more important the need to treat the patient aggressively with large-dose steroids or immunosuppressive agents. Those patients with inactive lesions (i.e., membranous nephropathy, sclerotic glomeruli, fibrous crescents, tubular atrophy, or interstitial fibrosis) may not require aggressive therapy. The extent of renal pathology can best be supported with a renal biopsy, and then must always include two components: light microscopy and immunofluorescence. Serial renal biopsy has prognostic value and is recommended for the regulation of chemotherapy [53, 54]. A biopsy with immunofluorescent analysis is a good way to determine lupus renal activity. If feasible, electron microscopy should also be done.

In the years since the last edition of this book, the International Society of Nephrology/Renal Pathology Society (ISN/RPS) has re-classified lupus nephritis. This classification eliminates certain WHO categories and has led to higher interobserver reproducibility and a change in the interpretation of certain types of renal pathology such as an increase of type IV and a reciprocal decrease of type III and type V nephritis. This will be detailed in Chapters 26 and 42 [55, 56].

A renal biopsy done under fluoroscopy or by CT guidance is best for the patient with SLE. An adequate number of glomeruli should be obtained for verifiable diagnosis. Hypothetical problems encountered with the renal biopsy include: (1) that the biopsy is static or reflective of one point in time, (2) the efficacy of the activity/chronicity index is questionable in designing treatment, and (3) no single glomerular lesion can reflect the entire renal picture.

The most serious complications of renal biopsies include pericapsular hemorrhage or clot obstruction in those patients with lupus procoagulants.

In those patients with WBC, erythrocyte, hyaline, or granular casts, a BUN and creatinine are helpful in order to assess renal function. For most patients, renal function early in the course of the disease is normal despite abnormal urine sediment. If the activity of the disease progresses unchecked, these parameters can change rapidly. When proteinuria is found qualitatively by urine dipstick (1+ or greater), a 24-h urine protein and a creatinine ratio should be obtained to quantify the amounts. Urine protein is a useful measure of renal lupus activity. An incremental change of 500 mg of protein excretion is significant to renal pathology, just as a decrement can herald clinical improvement.

There can be many other reasons for decreased renal function in SLE. These include concomitant infections, use of aspirin, NSAIDs, or ACE inhibitors, all of which might induce a decrease of renal circulation, as well as obstruction or thrombosis of the renal vein. Sonograms, contrast studies, or renal scans can be helpful in the evaluation of renal function when causes other than lupus nephropathy are suspected. Platt *et al.* [57] suggest that ultrasound might be of utility in the prediction of worsened renal disease.

Less common forms of renal involvement in SLE include interstitial nephritis, which can be the result of immune complexes [58] or the result of drug therapy [59]. This form of renal disease may not present as a glomerulopathy but rather as a disorder of acidification and potassium transport or regulation [60]. Renal pathology could indicate that thrombosis of the glomerulus [61] and intraglomerular thrombi [62] are alternate causes of proteinuria and renal failure in SLE.

Renal transplantation in lupus is as successful as that in the non-SLE population. However, renal disease can occur again in the transplanted kidneys [63]. For unknown reasons, the alleviation of overall disease activity of lupus erythematosus after dialysis for chronic renal disease has been reported.

CLINICAL INVOLVEMENT OF THE CENTRAL NERVOUS SYSTEM

Both cognitive dysfunction and neuropsychiatric lupus are reviewed extensively in Chapters 28 and 40.

Neuropsychiatric manifestations can be found in as many as 66% of patients with SLE [64–67]. The pathophysiology of this clinical manifestation is not widely understood; however, thrombosis and vasculitis are not responsible for the large number of neuropsychiatric manifestations observed. It is possible that components that are both nonimmunologic and immunologic in nature might be involved, such as hormones and a direct effect of antibody. Central nervous system (CNS) manifestations include seizures, psychiatric illness, and disorders of the cranial nerves. The frequency of organic CNS manifestations in SLE has been reported as between 35 and 75% [68]. The peripheral nervous system is involved in as many as 18% of patients [69].

Seizures are found in 15–20% of SLE patients [70, 71]. These can be the result of the disease process, such as lupus vasculitis or acute thrombosis, steroid therapy and its attendant hypertension, or a resulting metabolic problem such as uremia [65] or infection. Grand-mal tonic-clonic seizures are most common, although other seizures, such as Jacksonian, psychomotor, and absence attacks, have all been reported. On rare occasions, patients with SLE can present with status epilepticus. Treatment of seizures in SLE requires anticonvulsants. Corticosteroid therapy when given as pulse therapy, is associated with status epilepticus and should be used with caution [72].

Lupus can cause profound psychiatric disturbances in 50–67% of patients. Overt psychosis can occur in 12% of cases as well as a variety of organic brain syndromes. Severe depression is common to lupus patients and is thought to be a disease manifestation rather than reactive depression from chronic disease. Sleep disturbances are common in lupus and not usually related to depression. Steroid psychosis is common in lupus patients on high-dose steroids for long periods [73]. Antiribosomal P protein antibodies are positive in over 60% of patients with SLE-related psychosis [74] and may help distinguish these patients from those with steroid psychosis, although there is no specific antibody associated with cerebral SLE.

Ten percent of patients can have cranial nerve abnormalities, and they can be the presenting symptom in a small number. Although spinal cord involvement in SLE is rare, three types of cord involvement are seen: transverse myelitis [75], demyelination [76], and spinal cord occlusion because of thrombosis [77]. The latter three spinal manifestations are all commonly associated with antiphospholipid antibodies [78].

In the first 5 years of the disease, the incidence of cerebrovascular accident (CVA) is high (6.6% occurring in the first year alone) [79]. Patients with antiphospholipid antibodies also have an increased risk of stroke [80].

Even though movement disorders are not common in lupus, chorea is common in children with SLE and is found in adults and children with phospholipid antibody [81]. It is virtually indistinguishable from Sydenham chorea. Parkinsonism and cerebellar ataxia are rare [82, 83]. Rarer forms of CNS involvement include pseudotumor cerebri, hypothalamic dysfunction (especially due to thalamic infarcts), aseptic meningitis (particularly related to NSAID use), myasthenia, Eaton-Lambert syndrome, and a TTP-like syndrome.

The presence of microadenomata in the pituitary glands of rare patients causes hyperprolactinemia, which itself causes an SLE-like syndrome [84].

Peripheral nervous system disease is found in 3—18% of patients and is largely a sensory only or combined sensorimotor neuropathy [85]. Guillain-Barré syndrome, mononeuropathy, or mononeuritis multiplex has also been reported [86].

The laboratory diagnosis of CNS disease in SLE is difficult [67]. Spinal fluid pleocytosis and/or high spinal fluid protein levels are the only helpful indicators that CNS disease is present [87]. MRI and position electron tomography (PET) scanning show the most promise for diagnosing disease of the brain. Use of the newer modalities, such as Tc-99-HMAAQ brain SPECT, may have better utility in the diagnosis of CNS SLE [88]. Infarctions and demyelinating lesions of the brain are best found with these modalities [89]. CT scans are good for detecting focal lesions but are often unreliable. Antibodies to ribosomal P protein and antineuronal antibodies, as well as the finding of cytotoxic lymphocytes against myelin in blood, may eventually prove useful, but now are not very specific for cerebritis.

An exciting development in the area of lupus cerebritis is the discovery of antibodies to the glutamate receptors in the brain [90]. Originally based on a murine model of neuropsychiatric lupus and an antibody that cross-reacted with dsDNA and the N-methyl-D-aspartate (NMDA) receptors, the antibody from human SLE patients causes neuronal apoptosis when injected into mouse brains and demonstrable brain pathology.

CLINICAL PRESENTATION OF CARDIOVASCULAR LUPUS

Cardiac involvement is very common in SLE, and some 30—50% of patients suffer from some form of heart disease [91]. Pericarditis, the most common form of acute heart disease occurs in 19—48% [92, 93] of patients. Pleural-pericardial pain can occur at any time. Pericardial tamponade is rare but can be the initial presentation, and most patients present with pain and small pericardial effusions. Echocardiography is the best diagnostic test for this manifestation. Pericarditis in SLE can be managed with NSAIDS and/or low doses of corticosteroids. The differential diagnosis of pericardial effusion in SLE should include tuberculosis, uremia, or bacterial peritonitis if the patient is immunosuppressed, and in instances when pericardiocentesis is necessary, it is essential to send the appropriate cultures. Myocarditis is rare in SLE, involving only 5—10% of patients and usually presents with fever, conduction abnormalities, elevated CPK, skeletal myositis, and serositis [92]. In the SLE patient with hypertensive cardiomyopathy, the differential

diagnosis can be difficult and may require extensive investigation until a final diagnosis is made. Myocardial infarctions [94] and conduction abnormalities can be secondary to the myocardiopathy. A myocardial biopsy would give a definitive answer, but as with most invasive procedures carries a high morbidity in the SLE patient. SLE in the human results in a higher incidence of myocardial infarction because of accelerated atherosclerosis, coronary vasculitis, or coronary emboli [95].

Systolic cardiac murmurs are heard in up to 70% of SLE patients. These may be related to anemia, fever, or hypoxemia and are found with Libman-Sacks endocarditis, a component more frequent with antiphospholipid antibodies. The mitral and aortic valves are involved most commonly [96, 97].

Pulmonary arterial hypertension is common in patients with phospholipid antibodies [98], and a pulmonic murmur or a loud second heart sound in the presence of an elevated PTT are clues to this diagnosis, which should be confirmed by echocardiography or cardiac angiography. Pharmacological dilatation of the pulmonary arterial tree may be required in choosing a proper therapy. Ultimately, those patients who have severe pulmonary hypertension may benefit from lung transplantation. Patients with extremely low ejection fractions as a result of severe valve disease can benefit from valvuloplasty. Those with myocardiopathy might benefit from heart transplant.

The most common cardiac murmur in the SLE patient is likely to be a mitral valve prolapse. Disturbed autonomic function is common in SLE [99] and may be a reason for many of the cardiac symptoms such as tachycardia. Cardiac disease and particularly diastolic impairment is also common in patients that do not have active disease [100].

Vasculitis is common in SLE and may be reflected in the presence of splinter hemorrhages, digital infarcts, or ecthymic skin lesions (Table 30.7). Involvement of small- and medium-sized arteries may mimic polyarteritis nodosa and produce localized signs. This variant can be life-threatening. Vasculitis of the coronary or mesenteric vessels is not rare.

Raynaud phenomenon in SLE is not as common as in scleroderma or primary antiphospholipid syndrome (APLS) and is present in only 20% of patients [101, 102]. It is often associated with pulmonary hypertension and may be a sign of overlap syndrome. Digital vasoconstriction can produce digital ulcers, gangrene, and autoamputation. This phenomenon is often treated with vasodilators, calcium channel blockade, or, in severe cases, laser sympathectomy.

The most common cause of death in SLE is early onset cardiovascular disease. Hyperlipidemia is often implicated as one of the reasons for early arteriosclerotic cardiovascular disease (ASCVD) [103]. Aranow and Ginzler

TABLE 30.7 Vasculitic Involvement in SLE

Manifestation	Vessels involved and site
Leukocytoclastic angiitis (urticaria, palpable purpura)	Postcapillary venules, upper dermis
Atrophie blanche, livedo vasculitis	Small vessels, middle and lower dermis
Subcutaneous nodules	Medium-sized arteries, panniculus, and lower dermis
Livedo reticularis	Medium-sized arteries, panniculus, and lower dermis
Coronary arteritis	Coronary arteries
Mononeuritis multiplex	Vasa nervorum
Cerebral infarcts	Primarily small vessels
Mesenteric arteritis	Small and medium-sized arteries
Retinopathy	Arterioles, venules

[104] reviewed this topic in detail. This is also reviewed by Macsimowicz-McKinnon and Manzi in Chapter 43 on the Heart and by Avalos and Stein in Chapter 29 on Atherosclerosis. Risk factors for accelerated ASCVD are under intense study [105]. The antioxidant capacity of normal high-density lipoproteins (HDL) is lost during inflammation and the dysfunctional HDLs predispose to atherosclerosis. These dysfunctional HDLs are thought to be the single factor that increases the risk of developing subclinical atherosclerosis in SLE [106–108].

PULMONARY DISEASE AND LUPUS

The lungs are commonly affected in lupus patients [109–111] and discussed in detail in Chapter 45 by Fischer and duBois. Over 50% of SLE patients have had some form of pleural disease in their lifetime and pleural effusions, which are mostly exudative (>3 g protein), are less common than the pain and findings associated with simple pleuritis. The pain of pleuritis can be quite severe and must be distinguished from pulmonary embolus, infectious, or other forms of pneumonia. Pericarditis may also occur at the same time as the pleuritis.

Evidence of severe restrictive or obstructive disease in SLE is unusual, although this can be the major manifestation in some variant forms of the disease [112] and is suspected when the carbon monoxide diffusing capacity is very low (DLCO).

Parenchymal lung involvement can present suddenly as acute pneumonitis, dyspnea, or pleuritic pain. Subsequent to such a presentation, the patient might continue with ongoing pulmonary problems. Interstitial pneumonitis and chronic fibrosis can result from early inflammatory disease [113–115]. SLE parenchymal disease of the lung can be treated with high-dose corticosteroids, and improved pulmonary function is the desired result.

Pulmonary capillaritis is also part of parenchymal disease [114]. Hemoptysis and overt pulmonary hemorrhage are emergencies in SLE patients and are either the result of pneumonitis or pulmonary embolus, which are reversible. There is also an association of alveolar hemorrhage with renal microangiopathy [116]. Shrinking lung found on x-ray in some lupus patients is the result of diaphragmatic weakness [117] or paralysis. Pulmonary hypertension, even in patients with minimal lupus activity, can be severe and progressive [118].

HEMATOLOGY OF SLE

Both the cellular elements of the blood and the coagulation pathway can be adversely affected in the SLE patient. The latter is affected as a result of the antiphospholipid syndrome and is covered in a special section titled "Antiphospholipid Syndrome" (Chapters 50–55).

Sixty to 80% of lupus patients have anemia of chronic disease. Other kinds of anemia, such as autoimmune hemolytic anemia, are rare and are found in less than 10% of patients; however, a positive Coombs test can be found in 20–60% of patients [119, 120] and indicates that red cell antigens are prominent.

Leukopenia can be found in over 50% of patients with SLE and is associated with either granulocytopenia or lymphopenia [121]. Antibodies can be directed to either of these cellular elements at any point in their maturation pathways. When directed against stem cells, these antibodies cause aplastic anemia. Most low cell counts in SLE can be reversed with immunosuppressive therapy. Leukopenia is often a good general sign of disease exacerbation but also occurs in response to cytotoxic agents used in lupus therapy. Chapter 48 deals with cellular hematology.

Thrombocytopenia, found in 30–50% of SLE patients [122], is caused by either antiplatelet antibodies [123] or phospholipid antibodies. Either can cause profound thrombocytopenia (<50,000), which usually responds to corticosteroid, immunosuppressive drugs such as azathioprine, intravenous γ-globulin, or Retuximab. Platelet transfusions are contraindicated in most SLE patients except on occasions where platelets reach dangerous levels, because of the possibility that patients will be exposed to new platelet antigens that make them more refractory. Anticlotting factor antibodies have been found in SLE and are often associated with bleeding. Antibodies are directed most commonly to factors II, VIII, IX, XI, or XII. Acquired von Willebrand syndrome is also seen. Lupus anticoagulants are found commonly in patients with SLE and are associated with mild to profoundly elevated partial thromboplastin times. This abnormality is usually associated with hypercoagulation and not with bleeding [124, 125]. Associations have been observed, and the triad of the lupus anticoagulant, recurrent abortions, and the presence of false-positive tests for syphilis is often found in patients. Patients with lupus can be hypercoagulable for a variety of reasons other than procoagulant antibodies and these include hereditary deficiencies of factors C, S, or antithrombin III. One acquired reason for hypercoagulability is the loss of antithrombin III in the urine of patients with nephrotic syndrome.

If in rare instances a patient requires a splenectomy to control thrombocytopenia, laparoscopy is the best way to remove the spleen. Open laparotomy is not advisable in any except the most complicated cases.

CLINICAL PRESENTATION OF LUPUS OF THE SKIN

Ninety percent of lupus patients have some involvement of the skin. This is covered in Chapter 41 by Werth, Vera Kellet and Dutz. Only 40% of patients experience sensitivity to ultraviolet (UV) light and these are mostly Caucasians [126, 127]. Black patients are less sensitive to UV light. The actual percentage prevalence is 57% Caucasian vs 11% African-American. The lupus band test, originally considered by many to be the definitive test for cutaneous lupus, measures immunoglobulin and complement deposition at the dermal epidermal junction in nonlesional skin in greater than 60% of patients [128—130]. However, false-positives are encountered in rosacea, rheumatoid arthritis, MCTD, renal diseases, and many other disorders. Its true value is probably the differentiation of discoid lupus from SLE. In discoid lupus, only lesional skin stains positive, whereas in SLE, both lesional and nonlesional skin stains with immunoglobulin at the dermal—epidermal junction.

Acute cutaneous lupus (30—50%) and subacute cutaneous lupus (10—15%) comprise the vast majority of patients with dermal disease. The butterfly malar rash found in 40% of patients is part of acute cutaneous lupus. This rash is acute at onset and usually heals without scarring. Widespread morbilliform eruptions or bullous lesions can be confused with drug eruptions or erythema multiforme. The bullous lesions usually respond to Dapsone therapy.

Subacute cutaneous lupus (SCLE) is an annular, widespread, nonscarring or papulosquamous/psoriasis-form lesion that is worsened by sun exposure. This form of lupus is associated with HLA-DR3, anti-Ro antibody, and high titers of ANA. SCLE has also been associated with complement component deficiencies of C2 [131], C1q, and C1s.

Chronic forms of lupus skin disease include several forms of discoid lupus and lupus profundus. These discoid lesions are usually localized to the head, scalp, and external ear, but more widespread involvement is possible. Unlike those of subacute cutaneous lupus, these lesions can occur in non-sun-exposed areas. Discoid lesions can be chronic and not associated with the other manifestations of SLE. Patients with isolated discoid lupus have a 2—10% chance of developing systemic disease, whereas 10—20% of SLE patients have discoid lesions. Discoid LE is more common in African-Americans.

Lupus panniculitis, or lupus profundus, is an unusual form of chronic cutaneous lupus that is manifested by indurated subcutaneous nodules on the extremities. It is rare, occurs in 2% of patients, and responds to Dapsone therapy.

Alopecia can occur in all patients with SLE at some time in their disease presentation and can also be the result of therapy or hypothyroidism, which is very common in patients with SLE. In lupus it is usually reversible alopecia. Livido reticularis and livedo racemosa, a more significant form of skin lesion, are found in the antiphospholipid syndrome.

GASTROINTESTINAL TRACT AND LIVER DISEASE IN SLE

The most common manifestation of gastrointestinal lupus is the occurrence of painless ulcers in the nose and mouth of patients. Many patients develop these ulcers at some time during the course of their disease and they indicate a flare of disease [132]. Esophageal ulcerations and dysphagia are rarely found [133]. Thirty percent of patients have nausea and vomiting. Dotan and Mayer covers this topic in Chapter 47 titled Gastrointestinal: Non-Hepatic.

Abdominal pain can result from a variety of causes [134—137]. Pancreatitis [138], ischemic bowel, perforation, or mesenteric vasculitis can suggest a surgical abdomen and mandate a laparotomy. The tumor marker CA125 can be elevated in cases of SLE-induced gastrointestinal ischemia [139]. Serositis and, in some cases, vasculitis may also simulate a surgical abdomen, but a prudent trial of steroids before invasive abdominal surgery may be curative. Abdominal surgery is associated with an extremely high mortality rate in the active SLE patient.

Lupus peritonitis [140] is the result of small vessel involvement in the bowel serosa or retroperitoneum or the result of actual perforation of the bowel. Rebound tenderness, fever, nausea, vomiting, and diarrhea can be found [141]. Imaging with computerized tomography will be helpful in differentiating the severity of the bowel lesion. Bacterial peritonitis is quite common in patients with nephrotic syndrome.

Parenchymal liver disease as a result of lupus is uncommon and more likely represents chronic active hepatitis or cirrhosis. Lupoid hepatitis is a separate entity and is not part of SLE [142]. The entire spectrum of liver disease in lupus is detailed by MacKay in Chapter 46. Liver function studies may be very abnormal in patients with lupus. These abnormalities are usually secondary to drug therapy, aspirin ingestion [143], and rarely with thrombosis (Budd-Chiari syndrome) secondary to antiphospholipid syndrome. When ascites is produced, paracentesis is of diagnostic help.

THE EYE AND SLE

The eye is not commonly involved in SLE [144, 145]. Only 10% or less of patients have episcleritis or conjunctivitis. In a prospective study, retinopathy was detected in 7% of SLE patients. This retinopathy consists of microangiopathic lesions with cotton wool spots and hemorrhages that can be a significant problem in someone with a bleeding diathesis or one who is anticoagulated. Optic neuritis, papilledema, and retinal vein occlusion are also major problems [146]. Lupus retinopathy is common in patients with active SLE (88%) and in those with lupus cerebritis (73%). Patients can also have uveitis, cytoid bodies, and angle-closure glaucoma.

MISCELLANEOUS ORGAN INVOLVEMENT

Less common in the patient with SLE is the sicca syndrome. Patients who are older are particularly affected. Patients with lupus can have asymptomatic parotid gland enlargement with abnormal labial biopsies that suggest the Sjögren syndrome [147]. Many of these patients have a positive anti-Ro antibody and a Sjögren syndrome overlap. Patients with lupus may have vocal cord paralysis or present with hoarseness because of vasculitis of the recurrent laryngeal nerve.

Lupus may be a cause of sensorineural learning loss. The mechanism of ear damage is not known [148, 149].

References

[1] J.I. Stahl, J.H. Klippel, J.L. Decker, Fever in systemic lupus erythematosus, Am J of Med 67 (1979) 933–940.

[2] L.C. Robb-Nicholson, M.H. Liang, L. Daltroy, H. Eaton, V. Gall, J. Schwartz, et al., Effects of aerobic conditioning in lupus fatigue: a pilot study, J Rheum 28 (1989) 500–505.

[3] P.L. Dobkin, D. Da Costa, P.R. Fortin, S. Edworthy, S. Barr, J.M. Esdaile, et al., Living with lupus: a prospective pan-Canadian study, J Rheumatol 28 (11) (2001) 2442–2448.

[4] G.L. Godaert, A. Hartkamp, R. Geenen, A. Garssen, A.A. Kruize, J.W. Bijlsma, et al., Fatigue in daily life in patients with primary Sjögren's syndrome and systemic lupus erythematosus, Ann N Y Acad Sci 966 (2002) 320–326.

[5] B.H. Rovin, Y. Tang, J. Sun, H.N. Nagaraja, K.V. Hackshaw, L. Gray, et al., Clinical significance of fever in the systemic lupus erythematosus patient receiving steroid therapy, Kidney Int 68 (2) (2005) 747–759.

[6] M.C. Hochberg, Updating the American College of Rheumatology revised criteria for the classification of systemic lupus erythematosus, Arthritis Rheum 40 (9) (1997) 1725.

[7] N. Gleicher, Autoantibodies in normal and abnormal pregnancy, Am J Reprod Immunol 28 (1992) 269–273.

[8] A. Parodi, A. Rebora, ARA and EADV criteria for classification of systemic lupus erythematosus in patients with cutaneous lupus erythematosus, Dermatology 194 (3) (1997) 217–220.

[9] R.A. Asherson, R. Cervera, R.G. Lahita, Latent, incomplete or Lupus at all? J Rheum 18 (12) (1991) 1783–1786.

[10] I.M. Gilboe, G. Husby, Application of the 1982 revised criteria for the classification of systemic lupus erythematosus on a cohort of 346 Norwegian patients with connective tissue disease, Scand J Rheumatol 28 (2) (1999) 81–87.

[11] E.L. Smith, R.H. Shmerling, The American College of Rheumatology criteria for the classification of systemic lupus erythematosus: strengths, weaknesses, and opportunities for improvement, Lupus 8 (8) (1999) 586–595.

[12] K.H. Costenbader, E.W. Karlson, L.A. Mandl, Defining lupus cases for clinical studies: the Boston weighted criteria for the classification of systemic lupus erythematosus, J Rheumatol 29 (12) (2002) 2545–2550.

[13] W.A. Wilson, A. Gharavi, T. Koike, M.C. Lockshin, D.W. Branch, J.C. Piette, et al., International Consensus statement on preliminary classification criteria for definite antiphospholipid syndrome: report of an international workshop, Arthritis Rheum 42 (7) (2000) 1309–1311.

[14] W.A. Wilson Classification Crtieria for the Antiphospholipid Syndrome. 3[27], 499–505. 8-1-2001. Philadelphia, W.B. Saunders Co. Rheumatology Clinics of North America. Khamashta, M. A.

[15] D. Estes, C.L. Christian, The natural history of systemic lupus erythematosus by prospective analysis, Medicine 50 (1971) 85–95.

[16] E.L. Dubois, The clinical picture of systemic lupus erythematosus, in: E.L. Dubois (Ed.), Lupus Erythematosus, 2nd ed., University of California Press, Los Angeles, 1974, pp. 232–250.

[17] E.L. Dubois, D.L. Tuiffanelli, Clinical manifestations of systemic lupus erythematosus: computer analysis of 500 cases, JAMA 190 (1964) 104–111.

[18] G.S. Hoffman, Polyarthritis: the differential diagnosis of rheumatoid arthritis, Semin Arthritis Rheum 8 (1978) 115–141.

[19] A.M. Harvey, L.E. Shulman, P.A. Tumulty, Systemic lupus erythematosus: review of the literature and clinical analysis of 138 cases, Medicine 33 (1954) 291–437.

[20] G.S. Alarcon, Arthralgias, myalgias, facial erythema, and a positive ANA: not necessarily SLE, Cleve Clin J Med 64 (7) (1997) 361–364.

[21] D.E. Blumenthal, Tired, aching, ANA-positive: does your patient have lupus or fibromyalgia? Cleve Clin J Med 69 (2) (2002). 143-2.

[22] S. Perry, F. Miller, The psychiatric aspects of systemic lupus erythematosus, in: R.G. Lahita (Ed.), Systemic Lupus Erythematosus, 2nd ed., Churchill Livingstone, New York, 1992, pp. 845–856.

[23] J.J. Cush, E.A. Goldings, Drug induced lupus: clinical spectrum and pathogenesis, Am J of Med Sci 290 (1985) 36–45.

[24] F.C. Arnett, L.E. Shulman, Studies in familial systemic lupus erythematosus, Medicine 55 (1976) 313–322.

[25] M. Honey, Systemic lupus erythematosus presenting with sulfonamide hypersensitivity, Brit Med J 1 (1956) 1272–1275.

[26] M.C. Hochberg, R.E. Boyd, J.M. Ahearn, Systemic lupus erythematosus: a review of clinical laboratory features and genetic markers in 150 patients with emphasis on demographic subsets, Medicine 64 (1985) 285–295.

[27] J.E. Salmon, M.J. Roman, Accelerated atherosclerosis in systemic lupus erythematosus: implications for patient management, Curr Opin Rheumatol 13 (5) (2001) 341–344.

[28] J.P. Atkinson, P.M. Schneider, Genetic Susceptibliity and Class III complement genes, in: R.G. Lahita (Ed.), Systemic Lupus Erythematosus, 3rd ed., Academic Press, San Diego, 1999, pp. 91–104.

[29] L. Cook, V. Agnello. Complement Deficiency, systemic lupus erythematosus, in: R.G. Lahita (Ed.), Systemic Lupus Erythematosus, 3rd ed., Academic Press, San Diego, 1999, pp. 105–128.

[30] M.B. Urowitz, D.D. Gladman, Antinuclear antibody negative lupus, in: R.G. Lahita (Ed.), Systemic Lupus Erythematosus, 2nd ed., Churchill Livingstone, New York, 1992, pp. 561–580.

[31] S.A. Fernandez, A.Z. Lobo, Z.N. Oliveira, L.M. Fukumori, A.M. Pr, E.A. Rivitti, Prevalence of antinuclear autoantibodies in the serum of normal blood dornors, Rev Hosp Clin Fac Med Sao Paulo 58 (6) (2003) 315–319.

[32] D.J. Tobin, N. Orentreich, J.C. Bystryn, Autoantibodies to hair follicles in normal individuals, Arch Dermatol 130 (3) (1994) 395–396.

[33] B.D. Stollar, The antigenic potential and specificity of nucleic acids, nucleoproteins and their modified derivatives, Arthritis and Rheum 24 (1981) 1010–1018.

[34] M. Abu-Shakra, M. Krup, H. Slor, Y. Shoenfeld, Anti-Sm-RNP activity in sera of patients with rheumatic and autoimmune diseases, Clin Rheumatol 9 (1990) 346–355.

[35] E.M. Tan, J.S. Smolen, J.S. McDougal, M.J. Fritzler, T. Gordon, J.A. Hardin, et al., A critical evaluation of enzyme immuno-assay kits for detection of antinuclear autoantibodies of defined specificities. II. Potential for quantitation of antibody content, J Rheumatol 29 (1) (2002) 68–74.

[36] R.M. van Vugt, R.H. Derksen, L. Kater, J.W. Bijlsma, Deforming arthropathy or lupus and rhupus hands in systemic lupus erythematosus, Ann Rheum Dis 57 (9) (1998) 540–544.

[37] R.C. Lawrence, M.C. Hochberg, J.L. Kelsey, F.C. McDuffie, T.A. Medsger Jr., W.R. Felts, et al., Estimates of the prevalence of selected arthritic and musculoskeletal diseases in the United States, J Rheumatol 16 (1989) 427–441.

[38] J.K. Schofield, R. Cerio, K. Grice, Systemic lupus erythematosus presenting with 'rheumatoid nodules', Clin Exp Dermatol 17 (1) (1992) 53–55.

[39] M. Suarez-Gestal, M. Calaza, R. Dieguez-Gonzalez, E. Perez-Pampin, J.L. Pablos, F. Navarro, et al., Rheumatoid arthritis does not share most of the newly identified systemic lupus erythematosus genetic factors, Arthritis Rheum 60 (9) (2009) 2558–2564.

[40] K.A. Egol, L.M. Jazrawi, H. De Wal, E. Su, M.P. Leslie, P.E. Di Cesare, Orthopaedic manifestations of systemic lupus erythematosus, Bull Hosp Jt Dis 60 (1) (2001) 29–34.

[41] D.D. Gladman, M.B. Urowitz, V. Chaudhry-Ahluwalia, D.C. Hallet, R.J. Cook, Predictive factors for symptomatic osteonecrosis in patients with systemic lupus erythematosus, J Rheumatol 28 (4) (2001) 761–765.

[42] G. Murphy Nancy, A. Koolvisoot, H.R. Schumacher, M. Feldt Joan, E. Callegari Peter, Musculoskeletal features in systemic lupus erythematosus and their relationship with disability [record supplied by Aries Systems], J Clin Rheumatol 4 (5) (1998) 238–245.

[43] F.P. Quismorio, E.L. Dubois, Septic arthritis in systemic lupus erythematosus, J Rheum 2 (1975) 73–82.

[44] D. Isenberg, Myositis in other connective tissue diseases, Clinics of Rheumatic Diseases 10 (1984) 151–174.

[45] R.A. Yood, T.W. Smith, Iclusion body myositis and systemic lupus erythematosus, J Rheum 12 (1985) 568570.

[46] A. Askari, P.J. Vignos, R.W. Moskowitz, Steroid myopathy in connective tissue disease, Am J Med 61 (1976) 485–492.

[47] E. Casado, J. Gratacos, C. Tolosa, J.M. Martinez, I. Ojanguren, A. Ariza, et al., Antimalarial myopathy: an underdiagnosed complication? Prospective longitudinal study of 119 patients, Ann Rheum Dis 65 (3) (2006) 385–390.

[48] G.B. Appel, F.G. Silva, C.L. Pirani, Renal involvement in systemic lupus erythematosus (SLE): a study of 56 patients emphasizing histologic classification, Medicine 57 (1978) 371.

[49] D. Koffler, V. Agnello, R.I. Carr, Variable patterns of immuno-globulin and complement deposition in the kidneys of patients with systemic lupus erythematosus, Am J Path 56 (1969) 305–316.

[50] J.R. McLaughlin, C. Bombardier, V.T. Farewell, D.D. Gladman, M.B. Urowitz, Kidney biopsy in systemic lupus erythematosus. III. Survival analysis controlling for clinical and laboratory variables, Arthritis Rheum 37 (1994) 559–567.

[51] D.S. Baldwin, J. Lowenstein, N.F. Rothfield, The clinical course of the proliferative and membranous forms of lupus, Ann Intern Med 73 (1970) 929–942.

[52] D.D. Gladman, D.D. Urowitz, E. Cole, Kidney Biopsy in SLE. I. A clinical-morphologic evaluation, Quart J Med 73 (1989) 1125–1133.

[53] S. Bajaj, L. Albert, D.D. Gladman, M.B. Urowitz, D.C. Hallett, S. Ritchie, Serial renal biopsy in systemic lupus erythematosus, J Rheumatol 27 (12) (2000) 2822–2826.

[54] C.W. Yoo, M.K. Kim, H.S. Lee, Predictors of renal outcome in diffuse proliferative lupus nephropathy: data from repeat renal biopsy, Nephrol Dial Transplant 15 (10) (2000) 1604–1608.

[55] G.S. Markowitz, V.D. D'Agati, Classification of lupus nephritis, Curr Opin Nephrol Hypertens 18 (3) (2009) 220–225.

[56] K.E. Sada, H. Makino, Usefulness of ISN/RPS classification of lupus nephritis, J Korean Med Sci 24 (Suppl) (2009) S7–10.

[57] J.F. Platt, J.M. Rubin, J.H. Ellis, Lupus nephritis: predictive value of conventional and Doppler US and comparison with serologic and biopsy parameters, Radiology 203 (1) (1997) 82–86.

[58] G.B. Appel, J. Radhakrishnan, E.M. Ginzler, Use of mycophe-nolate mofetil in autoimmune and renal diseases, Trans-plantation 80 (Suppl. 2) (2005) S265–S271.

[59] J. Arias, M. Fernandez-Rivas, P. Panadero, Selective fixed drug eruption to amoxycillin, Clin Exp Dermatol 20 (4) (1995) 339–340.

[60] R.A. DeFronzo, C.R. Cooke, M. Goldberg, Impaired renal tubular potassium secretion in systemic lupus erythematosus, Ann Intern Med 86 (1977) 268–271.

[61] M.R. Uwonkunda, J.P. Cosyns, J.P. Devogelaer, F.A. Houssiau, Glomerular thrombosis: an unusual cause of renal failure in systemic lupus erythematosus, Acta Clin Belg 53 (6) (1998) 371–373.

[62] S. Bhandari, P. Harnden, A.M. Brownjohn, J.H. Turney, Asso-ciation of anticardiolipin antibodies with intraglomerular

thrombi and renal dysfunction in lupus nephritis, QJM 91 (6) (1998) 401—409.

[63] W.J.C. Amend, F. Vincenti, N.J. Feduska, Recurrent systemic lupus erythematosus involving renal allografts, Ann Intern Med 94 (1981) 444.

[64] L.V. Calabrese, T.A. Stern, Neuropsychiatric manifestations of systemic lupus erythematosus, Psychosomatics 36 (4) (1995) 344—359.

[65] E.J. Feinglass, F.C. Arnett, C.A. Dorsch, Neuropsychiatric manifestations of systemic lupus erythematosus: diagnosis, clinical spectrum, and relationship to other features of the disease, Medicine 55 (1976) 323—339.

[66] M.J. Harrison, L.D. Ravdin, Cognitive dysfunction in neuro-psychiatric systemic lupus erythematosus, Curr Opin Rheumatol 14 (5) (2002) 510—514.

[67] R. Omdal, Some controversies of neuropsychiatric systemic lupus erythematosus, Scand J Rheumatol 31 (4) (2002) 192—197.

[68] W.J. McCune, J. Golbus, Neuropsychiatric Lupus. Rheum Dis Clin North Am (1988 April) 149—167.

[69] R.T. Johnson, E.P. Richardson, The neurological manifestations of systemic lupus erythematosus, Medicine 47 (1968) 337—369.

[70] E. Ozgencil, C. Gulucu, S. Yalcyn, Z. Alanoglu, N. Unal, M. Oral, et al., Seizures and loss of vision in a patient with systemic lupus erythematosus, Neth J Med 65 (7) (2007) 274.

[71] J.F. Baizabal-Carvallo, C. Cantu-Brito, G. Garcia-Ramos, Acute neurolupus manifested by seizures is associated with high frequency of abnormal cerebral blood flow velocities, Cerebrovasc Dis 25 (4) (2008) 348—354.

[72] R.M. Andrade, G.S. Alarcon, L.A. Gonzalez, M. Fernandez, M. Apte, L.M. Vila, et al., Seizures in patients with systemic lupus erythematosus: data from LUMINA, a multiethnic cohort (LUMINA LIV), Ann Rheum Dis 67 (6) (2008) 829—834.

[73] M. Kohen, R.A. Asherson, A.E. Gharavi, R.G. Lahita, Lupus psychosis: differentiation from the steroid-induced state, Clin Exp Rheumatol 11 (1993) 323—326.

[74] E. Bonfa, S.J. Golombek, L.D. Kaufman, Association between lupus psychosis and anti-ribosomal P protein antibodies, N Eng J Med 317 (1987) 265—271.

[75] A.S. Penn, A.J. Rowan, Myelopathy in systemic lupus erythematosus, Archives of Neurology 18 (1968) 337—349.

[76] K.W.M. Fulford, R.D. Catterall, J.J. Delhanty, D. Doniach, M. Kremer, A collagen disorder of the nervous system presenting as multiple sclerosis, Brain 95 (1972) 373—386.

[77] S.R. Levine, K.M.A. Welch, The spectrum of neurologic disease associated with anti-phospholipid antibodies, Archives of Neurology 44 (1987) 876—883.

[78] C. Lavalle, S. Pizarro, C. Drenkard, Transverse myelitis: a manifestation of systemic lupus erythematosus strongly associated with antiphospholipid antibodies, J Rheum 17 (1990) 34—37.

[79] N. Futrell, C. Millikan, Frequency, etiology, and prevention of stroke in patients with systemic lupus erythematosus, Stroke 20 (1989) 583—591.

[80] K.S. Ginsburg, M.H. Liang, L. Newcomer, S.Z. Golhaber, P.H. Schur, C.H. Hennekens, Anticardiolipin antibodies and the risk for ischemic stroke and venous thrombosis, Ann Intern Med 117 (1992) 997—1002.

[81] J.R. Groothuis, D.R. Groothuis, D. Mukhopadhyay, Lupus associated chorea in childhood, Am J Dis Child 131 (1977) 1131—1134.

[82] E. Lahat, G. Eshel, E. Azizi, Chorea associated with systemic lupus erythematosus in children: a case report, Israel J Med Sci 25 (1989) 568—570.

[83] B. Brechnian, CNS Lupus, Clin Rheum Dis 8 (1982) 183—192.

[84] L.J. Jara, O. Vera-Lastra, J.M. Miranda, M. Alcala, J. Alvarez-Nemegyei, Prolactin in human systemic lupus erythematosus, Lupus 10 (10) (2001) 748—756.

[85] D.B. Hellmann, T.J. Laing, M. Petri, Mononeuritis multiplex: the yield of evaluations for occult rheumatic diseases, Medicine 67 (1988) 145—153.

[86] J.A. Simpson, Peripheral neuropathy: etiological and clinical aspects, Proc R Soc Med 64 (3) (1971) 291—293.

[87] P. Small, M.F. Mass, P.T. Kohler, Central nervous system involvement in systemic lupus erythematosus, Diagnostic profile and clinical features, Arth Rheum 20 (1977) 869—878.

[88] P. Lass, M. Koseda, P. Lyczak, Technetium-99m-HMPAO brain SPECT in systemic lupus erythematosus with central nervous system involvement, J Nucl Med 39 (5) (1998) 930—931.

[89] W.L. Sibbit Jr., R.R. Sibbit, R.H. Griffey, Magnetic resonance and computed tomographic imaging in the evaluation of acute neuropsychiatric disease in SLE, Annals of Rheumatic Disease 48 (1989) 1014—1022.

[90] C. Kowal, L.A. Degiorgio, J.Y. Lee, M.A. Edgar, P.T. Huerta, B.T. Volpe, et al., Human lupus autoantibodies against NMDA receptors mediate cognitive impairment, Proc Natl Acad Sci USA 103 (52) (2006) 19854—19859.

[91] M. Petri, Systemic lupus erythematosus and the heart, in: R.G. Lahita (Ed.), Systemic Lupus Erythematosus, 3rd ed., Academic Press, San Diego, 1999, pp. 687—706.

[92] J. Lemos, L. Santos, I. Martins, L. Nunes, O. Santos, P. Henriques, Systemic lupus erythematosus — a diagnosis in cardiology, Rev Port Cardiol 27 (6) (2008) 841—849.

[93] A.S. Cohen, J.J. Canoso, Pericarditis in the rheumatologic diseases, Cardiovasc Clin 7 (3) (1976) 237—255.

[94] D.C. Knockaert, Cardiac involvement in systemic inflammatory diseases, Eur Heart J 28 (15) (2007) 1797—1804.

[95] R.W. Chang, Cardiac manifestations of SLE, Clin Rheum Dis 8 (1) (1982) 197—206.

[96] L.A. Kalashnikova, E.L. Nasonov, V.V. Borisenko, V.B. Usman, L.Z. Prudnikova, V.U. Kovaljov, et al., Sneddon's syndrome: cardiac pathology and antiphospholipid antibodies, Clin Exp Rheumatol 9 (4) (1991) 357—361.

[97] F.R. Hierro, J.M. Corretger, [Cardiovascular manifestations of children's collagenosis (author's transl)], An Esp Pediatr 15 (1) (1981) 75—81.

[98] Y. Bayraktar, N. Tanaci, T. Egesel, A. Gokoz, F. Balkanci, Antiphospholipid syndrome presenting as portopulmonary hypertension, J Clin Gastroenterol 32 (4) (2001) 359—361.

[99] M.B. Hogarth, L. Judd, C.J. Mathias, J. Ritchie, D. Stephens, R.G. Rees, Cardiovascular autonomic function in systemic lupus erythematosus, Lupus 11 (5) (2002) 308—312.

[100] E. Astorri, P. Fiorina, G.A. Contini, D. Albertini, E. Ridolo, P. Dall'Aglio, Diastolic impairment in asymptomatic systemic lupus erythematosus patients, Clin Rheumatol 16 (3) (1997) 320—321.

[101] A. Kasparian, A. Floros, E. Gialafos, M. Kanakis, S. Tassiopoulos, N. Kafasi, et al., Raynaud's phenomenon is correlated with elevated systolic pulmonary arterial pressure in patients with systemic lupus erythematosus, Lupus 16 (7) (2007) 505—508.

[102] J.E. Pope, The diagnosis and treatment of Raynaud's phenomenon: a practical approach, Drugs 67 (4) (2007) 517—525.

[103] A.S. Wierzbicki, Lipids, cardiovascular disease and atherosclerosis in systemic lupus erythematosus, Lupus 9 (3) (2000) 194—201.

[104] C. Aranow, E.M. Ginzler, Epidemiology of cardiovascular disease in systemic lupus erythematosus, Lupus 9 (3) (2000) 166—169.

[105] E. Svenungsson, K. Jensen-Urstad, M. Heimburger, A. Silveira, A. Hamsten, U. de Faire, et al., Risk factors for cardiovascular disease in systemic lupus erythematosus, Circulation 104 (16) (2001) 1887–1893.

[106] B.H. Hahn, M. McMahon, Atherosclerosis and systemic lupus erythematosus: the role of altered lipids and of autoantibodies, Lupus 17 (5) (2008) 368–370.

[107] M. McMahon, B.H. Hahn, Atherosclerosis and systemic lupus erythematosus: mechanistic basis of the association, Curr Opin Immunol 19 (6) (2007) 633–639.

[108] M. McMahon, J. Grossman, B. Skaggs, J. Fitzgerald, et al., Dysfunctional proinflammatory high-density lipoproteins confer increased risk of atherosclerosis in women with systemic lupus erythematosus, Arth Rheum 60 (8) (2009) 2428–2437.

[109] A. Marchesoni, K. Messina, P. Carrieri, Pulmonary hypertension and systemic lupus erythematosus, Clin Exp Rheum 1 (1983) 24–250.

[110] K. Onomura, H. Nakata, Y. Tanaka, T. Tsuda, Pulmonary hemorrhage in patients with systemic lupus erythematosus, J Thorac Imaging 6 (1991) 57–61.

[111] J.B. Orens, F.J. Martinez, Lynch JP3. Pleuropulmonary manifestations of systemic lupus erythematosus, Rheum Dis Clin North Am 20 (1994) 159–193.

[112] J.L. Callejas-Rubio, L. Lopez-Perez, E. Moreno-Escobar, N. Ortego-Centeno, Raynaud's phenomenon and pulmonary arterial hypertension, Lupus 17 (4) (2008) 355.

[113] P.T. Dalcin, S.S. Barreto, R.D. Cunha, R.M. Xavier, J.C. Brenol, B.J. Marroni, Lung clearance of 99mTc-DTPA in systemic lupus erythematosus, Braz J Med Biol Res 35 (6) (2002) 663–668.

[114] T.J. Franks, M.N. Koss, Pulmonary capillaritis, Curr Opin Pulm Med 6 (5) (2000) 430–435.

[115] G. Zandman-Goddard, M. Ehrenfeld, Y. Levy, S. Tal, Diffuse alveolar hemorrhage in systemic lupus erythematosus, Isr Med Assoc J 4 (6) (2002) 470.

[116] M.D. Hughson, Z. He, J. Henegar, R. McMurray, Alveolar hemorrhage and renal microangiopathy in systemic lupus erythematosus, Arch Pathol Lab Med 125 (4) (2001) 475–483.

[117] P.G. Wilcox, H.B. Stein, S.D. Clarke, Phrenic nerve function in patients with diaphragmatic weakness and systemic lupus erythematosus, Chest 93 (1988) 352–358.

[118] T.O. Cheng, Pulmonary hypertension in systemic lupus erythematosus, Mayo Clin Proc 74 (8) (1999) 845.

[119] S. Shapiro, M. Long, Hematology: coagulation problems, in: R.G. Lahita (Ed.), Systemic Lupus Erythematosus, 3rd ed., Academic Press, San Diego, 1999, pp. 871–886.

[120] R. Simantov, J. Laurence, R. Nachman, The cellular hematology of systemic lupus erythematosus, in: R.G. Lahita (Ed.), Systemic Lupus Erythematosus, 3rd ed., Academic Press, San Diego, 1999, pp. 765–792.

[121] J. Laurence, J.E.L. Wong, R. Nachman, The cellular hematology of systemic lupus erythematosus, in: R.G. Lahita (Ed.), Systemic Lupus Erythematosus, 2nd ed., Churchill Livingstone, New York, 1992, pp. 771–806.

[122] D.B. Cines, H. Liebman, R. Stasi, Pathobiology of secondary immune thrombocytopenia, Semin Hematol 46 (1 Suppl. 2) (2009) S2–14.

[123] D.C. Pierrot-Deseilligny, M. Michel, M. Khellaf, M. Gouault, L. Intrator, P. Bierling, et al., Antiphospholipid antibodies in adults with immune thrombocytopenic purpura, Br J Haematol 142 (4) (2008) 638–643.

[124] D. Erkan, Y. Yazici, M.D. Lockshin, Lupus anticoagulant–hypoprothrombinemia syndrome and pregnancy, Lupus 10 (4) (2001) 311–312.

[125] R.E. Joseph, J. Radhakrishnan, G.B. Appel, Antiphospholipid antibody syndrome and renal disease, Curr Opin Nephrol Hypertens 10 (2) (2001) 175–181.

[126] R.D. Zecevic, D. Vojvodic, B. Ristic, M.D. Pavlovic, D. Stefanovic, D. Karadaglic, Skin lesions — an indicator of disease activity in systemic lupus erythematosus? Lupus 10 (5) (2001) 364–367.

[127] M. Blaszczyk, S. Jablonska, T.P. Chorzelski, M. Jarzabek-Chorzelska, E.H. Beutner, V. Kumar, Clinical relevance of immunologic findings in cutaneous lupus erythematosus, Clin Dermatol 10 (1992) 399–406.

[128] W.R. Gammon, J.D. Fine, M. Forbes, R.A. Briggaman, Immunofluorescence on split skin for the detection and differentiation of basement membrane zone autoantibodies, J Am Acad Dermatol 27 (1) (1992) 79–87.

[129] J. Gilliam, Systemic lupus erythematosus and the skin, in: R.G. Lahita (Ed.), Systemic Lupus Erythematosus, 1st ed., Wiley, New York, 1987, pp. 616–650.

[130] T.D. Golan, A.E. Gharavi, K.B. Elkon, Penetration of autoantibodies into living epithelial cells, J Invest Dermatol 100 (1993) 316–322.

[131] T.T. Provost, F. Arnett, M. Reichlin, C2 deficiency, lupus erythematosus and anticytoplasmic Ro (SSA antibodies), Arth Rheum 26 (1983) 1279–1282.

[132] S.M. Sultan, Y. Ioannou, D.A. Isenberg, A review of gastrointestinal manifestations of systemic lupus erythematosus, Rheumatology (Oxford) 38 (10) (1999) 917–932.

[133] M. Tatelman, M.K. Keech, Esophageal motililty in systemic lupus erythematosus, rheumatoid arthritis and scleroderma, Radiology 86 (1966) 1041–1046.

[134] H.H. Chng, Lupus the great mimic: gastrointestinal manifestations, Singapore Med J 42 (8) (2001) 342–345.

[135] D.S. Hallegua, D.J. Wallace, Gastrointestinal manifestations of systemic lupus erythematosus, Curr Opin Rheumatol 12 (5) (2000) 379–385.

[136] C.K. Lee, M.S. Ahn, E.Y. Lee, J.H. Shin, Y.S. Cho, H.K. Ha, et al., Acute abdominal pain in systemic lupus erythematosus: focus on lupus enteritis (gastrointestinal vasculitis), Ann Rheum Dis 61 (6) (2002) 547–550.

[137] W. Luman, K.B. Chua, W.K. Cheong, H.S. Ng, Gastrointestinal manifestations of systemic lupus erythematosus, Singapore Med J 42 (8) (2001) 380–384.

[138] M. Baron, M.L. Brisson, Pancreatitis in systemic lupus erythematosus, Arth Rheum 25 (1982) 1006–1009.

[139] M.S. Koh, K. Sunil, H.S. Howe, Late onset systemic lupus erythematosus with elevated CA125 and gastrointestinal ischaemia, Intern Med J 32 (3) (2002) 117–118.

[140] V.E. Pollack, W.J. Grove, R.M. Karkin, Systemic lupus erythematosus simulating acute surgical condition of the abdomen, N Eng J Med 259 (1958) 258–266.

[141] P. Salomon, L. Mayer, Non-hepatic gastrointestinal manifestations of systemic lupus erythematosus, in: R.G. Lahita (Ed.), Systemic Lupus Erythematosus, 2nd ed., Churchill Livingstone, New York, 1992, pp. 747–760.

[142] A.G. Bearn, H.G. Kunkel, R.J. Slater, The problem of chronic liver disease in young women, Am J Med 21 (1956) 3–15.

[143] W.E. Seaman, P.H. Plotz, Effect of aspirin on liver tests in patients with RA or SLE and in normal volunteers, Arth Rheum 19 (1976) 155–160.

[144] D.H. Gold, D.A. Morris, P. Henkind, Ocular findings in systemic lupus erythematosus, Brit J Opthalmol 56 (1972) 800–804.

[145] F. Stafford-Brady, M.B. Urowitz, D.D. Gladman, Lupus retinopathy, Arth Rheum 31 (1988) 1105–1110.

[146] K. Wong, A. Everett, J. Verier-Jones, Visual loss as the initial symptom of systemic lupus erythematosus, Am J Opthalmol 92 (1981) 238–244.

[147] J.T. Rosenbaum, R. Wernick, The utility of routine screening of patients with uveitis for systemic lupus erythematosus or tuberculosis. A Bayesian analysis, Archives of Opthalmology 108 (1990) 1291–1293.

[148] S. Harari, G. Paciocco, S. Aramu, Ear and nose involvement in systemic diseases, Monaldi Arch Chest Dis 55 (6) (2000) 466–470.

[149] I. Kastanioudakis, N. Ziavra, P.V. Voulgari, G. Exarchakos, A. Skevas, A.A. Drosos, Ear involvement in systemic lupus erythematosus patients: a comparative study, J Laryngol Otol 116 (2) (2002) 103–107.

31

Neonatal Lupus

Jill P. Buyon, Deborah M. Friedman

INTRODUCTION

In 1901 congenital heart block (CHB) was first reported by Morquio [1] and in 1945 was diagnosed antepartum by Plant and Steven [2]. CHB can occur in association with structural heart diseases such as atrio-ventricular septal defects, left atrial isomerism, and abnormalities of the great arteries [3], with tumors [4], or as an isolated defect. In 1928, Aylward reported the occurrence of CHB in two children whose mother "suffered from Mikulicz's disease" [5]. This curious clinical observation was further solidified by the 1970s with reports of CHB, absent major structural abnormalities, in children whose mothers had autoimmune diseases [6–8] and by the finding that the maternal sera contained antibodies to SSA/Ro ribonucleoproteins [9–12]. It was subsequently noted that many mothers also had antibodies to SSB/La [13–17]. Other abnormalities affecting the skin, liver, and blood elements were reported to be associated with anti-SSA/Ro-SSB/La antibodies in the maternal and fetal circulation and are now grouped under the overall heading of neonatal lupus syndromes (NLS), neonatal lupus erythematosus (NLE), or simply neonatal lupus (NL) [18–21]. NL was so termed because the cutaneous lesions of the neonate resembled those seen in systemic lupus erythematosus (SLE) [20, 21]. The name is misleading and often a cause of undue concern because the neonate does not have SLE and often neither does the mother.

NL, albeit infrequently encountered by those caring for pregnant patients with SLE, presents a unique clinical challenge to the team of rheumatologist, perinatologist, neonatologist, pediatric cardiologist, and dermatologist. For the immunologist, the study of these syndromes may yield important insights into the pathogenesis of autoimmune-mediated tissue damage, while the embryologist may uncover relevant milestones in fetal antigen expression. The biology of human viviparity involves a wide variety of fetal–maternal relationships, several of which may facilitate the occurrence of passively acquired autoimmunity [22]. Tissue injury in the fetus is presumed to be dependent on the transplacental passage of maternal IgG autoantibodies from the mother who has an autoimmune process but who may be clinically asymptomatic. Disease in the offspring parallels the presence of maternal antibodies in the fetal and neonatal circulation and disappears, except for CHB, with the clearance of the maternal antibodies by the 6th to 8th month of postnatal life. The transient hematologic abnormalities and skin disease of the neonate reflect the effect of passively acquired autoantibodies on those organ systems that have the capacity of continual regeneration. In contrast, these regenerative processes apparently do not occur in cardiac tissue; the reversal of complete block has never been sustained. Curiously, the transient manifestations of this passively acquired autoimmune syndrome closely mimic the disease manifestations observed in adolescents or adults with SLE, while heart block is not observed in the adult, despite the presence of identical antibodies in the maternal circulation. In fact, heart block has only once been reported in the mother [23].

The study of NLS exemplifies not only translational research, which inherently draws upon clinical observations and explores them in the laboratory, but "integrational" research, which attempts to fit critical clinical and basic observations together, even those seemingly at odds. When this chapter was initially written in 1991 and further updated in 1998 and again in 2004, many questions based on important clinical observations were raised and are again reiterated as follows: What accounts for the spectrum of neonatal disease, i.e., why do some offspring remain totally healthy despite circulating maternal autoantibodies while others have transient rashes and still others die *in utero* from acute myocarditis? Why is complete heart block irreversible and why is there specificity for the conducting system? Is this an acquired disease that occurs in a previously healthy fetus? Why is the maternal heart unaffected? What is the mechanism of tissue injury? What is the autoantigen—antibody system and how does it relate to the SSA/Ro-SSB/La polypeptides, which are intracytoplasmic cellular components unlikely to be present on the cell surface? Is there a genetic basis for these syndromes? What is the fine specificity of the maternal autoantibody response that is most diagnostically predictive and how can this knowledge guide the approach to therapy? Finally, can we safely tell patients "No antibodies to SSA/Ro or SSB/La, no affected offspring"? While many of these questions remain incompletely answered, the field has advanced considerably since 2004 and exciting new information from both the bench and bedside will be reviewed in this updated chapter.

PASSIVELY ACQUIRED CHB: AN IRREVERSIBLE MANIFESTATION OF NL WHEN COMPLETE BLOCK

Cardiac Histopathology

Histopathologic studies constitute a major basis for formulating hypotheses regarding the pathogenesis of CHB. It appears logical to assume that the time of death relative to initial immune attack may influence the pathologic findings. Evidence of a cellular infiltrate might be present if death occurs close to the time a bradyarrhythmia is first detected, but calcifications may be the sole pathologic finding if death has occurred months later. Two of the early publications on autopsy findings suggested an inflammatory component as evidenced by the demonstration of a mononuclear cell infiltration in the myocardium of a fetus dying *in utero* at 18 weeks of gestation [24] and patchy lymphoid aggregates throughout the myocardium of an infant delivered at 30 weeks and dying in the immediate postnatal period [25]. Moreover, immunofluorescent studies have shown deposition of IgG, complement (including $C1_q$, C4, C3d, C6, and C9), and fibrin [25, 26]. Thus, it might be reasonable to hypothesize that the first cardiac lesion is a global pancarditis with inflammation of the pericardium, myocardium, and endocardium resulting in subsequent fibrosis of the conducting system clinically manifest as permanent heart block. Litsey and colleagues identified IgG deposits in the epicardial, myocardial, and endocardial tissue of the right atrium on postmortem analysis of a neonate with CHB [26]. Although the published literature on serial echocardiograms in mothers at risk of a pregnancy complicated by CHB has significantly expanded (discussed in detail in the section Treatment and Prevention of Cardiac NL), it remains the general experience that the first clinically apparent abnormality

in cardiac function is not myocarditis (e.g., effusions, ventricular dysfunction, valvular abnormalities) but bradycardia. This implies that: (1) early inflammation, which may not even be clinically detectable, is confined to the conduction system and occurs independent of an inflammatory pancarditis, and/or (2) the tools available are not sufficiently sensitive, and/or (3) signs of inflammation beyond the conduction system are missed since they are not expected. Specific vulnerability of the conducting system, most commonly the atrioventricular (AV) node, is unexplained. Ho *et al.* described the histopathology of seven hearts with CHB and associated maternal antibodies to the SSA/Ro polypeptide [27]. In all of these hearts, there was atrial-axis discontinuity, the AV node was replaced by varying degrees of fibrosis or fatty tissue [27]. The distribution of the distal conducting system was normal. In five cases of CHB compiled by Carter and colleagues, there was disruption of the AV conduction system by a process of uncertain cause [4]. In all five instances, the presence of microscopic crystalline structures was associated with the conduction system and with fibrous structures of the heart. These deposits have been designated as products of connective tissue degeneration resulting from an intrauterine inflammatory process. Hogg has reported hematoxylin bodies in the AV node [28]. Further support for an inflammatory process is demonstrated by the findings of calcification along ventricular portions of the conducting system and area of the sinoatrial (SA) node [26, 29].

The origin of diffuse fibroelastosis, which has also been reported in some of these affected babies, is considered to result from dilatation of the cardiac chambers secondary to the compensatory increased stroke volume present in CHB [4]. However, Nield and colleagues [30] have reported 13 CHB patients with endocardial fibroelastosis (EFE), six diagnosed *in utero* and seven in the postnatal period, despite ventricular pacing of all but one infant. EFE is associated with significant mortality and morbidity: nine (70%) of these 13 patients died, and two (15%) required heart transplants. These same authors have also reported anti-SSA/Ro-associated EFE without conduction defects [31].

Given the importance of histologic data to infer pathogenic mechanisms, medical records of all families enrolled in the Research Registry for Neonatal Lupus (RRNL established in September 1994 [32]) were reviewed to determine the incidence and timing of death, with emphasis on the pathologic findings in the affected fetal hearts [33]. Complete autopsy reports were available in 11 cases. The mean time from detection of CHB to autopsy was 11 weeks. Although in three cases there were various lesions of the tricuspid valve, the pathologic descriptions were more suggestive of an imposed injury than a true developmental defect. These included nodularity, dysplasia, hypoplasia and fusion of valve leaflets, and fibrosis. The pulmonary valve was abnormal in two cases, one was described as stenotic dysplastic, and the other nodular and dysplastic. Aortic valve insufficiency and stenosis and hypoplasia of the mitral valve leaflet were observed in one. Endocardial fibroelastosis (EFE) of the right and left ventricles (RV, LV), with or without calcification, was present in seven. Chronic changes in the myocardium were documented in ten, and included biventricular hypertrophy and increased RV and LV walls, thickened but hypoplastic RV, and hyperchromatic nuclei of the myocytes. Abnormalities of the AV node or vicinity were noted in eight with involution, fibrosis, fatty infiltration, or calcification. However, in two the AV node *per se* appeared normal: in one there was calcification in adjacent tissue, and in another there was an atrophic His bundle with replacement by dense focally calcified fibrous tissue and scarring of the left and right bundle branches. Although previously unappreciated, autopsies obtained from the RRNL revealed a high incidence of valvular abnormalities. While there were sufficient changes in the AV node to account for CHB in most cases, clinical conduction abnormalities may have been secondary to a functional exit block in a normal-appearing node. Fibrosis of the SA node was also observed and thus SA nodal disease expands the spectrum of conduction dysfunction (further addressed in the section titled "Arrhythmogenicity of Maternal Autoantibodies and Perturbation of L-Type Calcium Channels", page 552).

These studies leave little doubt that the signature lesion of autoantibody-associated CHB is fibrosis, which can clearly extend beyond the conduction system. As will be discussed in the update of pathogenesis, the cascade leading to fibrosis is a major focus of investigation. Of interest, our own group reported immunohistologic evidence supporting exaggerated apoptosis, IgG deposition, and macrophage/fibroblast cross-talk in four available autopsy specimens from the RRNL [34]. Apoptosis was extensive in fetuses dying early and most pronounced in regions containing conduction tissue. Deposition of IgG was observed in hearts from fetuses with CHB/myocarditis, but not in three control hearts, and was co-localized with apoptotic cells. Giant cells and macrophages (frequently seen in close proximity to IgG and apoptotic cells) were present in septal and thickened fibrous subendocardial regions, most apparent in the youngest fetuses. Septal tissue also revealed extensive areas of fibrosis and microcalcification in which a predominant smooth muscle actin (SMA)-positive infiltrate (myofibroblast scarring phenotype) was observed. In contrast, there were no macrophages or SMA-positive cells (other than those lining blood vessels) in septal tissue from control hearts, although rare macrophages were seen in the working

myocardium. These data support the notion that transdifferentiation of cardiac fibroblasts to a scarring phenotype may be a pathologic process initiated by maternal antibodies, and persistence of this phenotype even after birth may relate to the progression of block seen in some infants postpartum [35]. In two fetal CHB hearts, protein expression of TGF-β, but not TNF-α, was demonstrated in septal regions, extracellularly in the fibrous matrix, and intracellularly in macrophage infiltrates [36]. Age-matched fetal hearts from voluntary terminations expressed neither cytokine.

CHB Occurs in an Anatomically Developed Heart and Coincides with Placental Transport

The presence of maternal immunoglobulins in the fetal circulation is directly related to the normal physiology of antibody traffic across the placenta [37]. FcRn is the receptor which mediates placental transport of maternal IgG to the fetus [38]. Maternal antibodies interact with Fc receptors on the trophoblastic cell surface in a specific transport process. Each Fc receptor has a different ligand specificity and affinity for the IgG subclasses but all receptors bind IgG1 and IgG3 with greater affinity than IgG2 or IgG4 [39]. IgG1, IgG2, and IgG3 are transported relatively early with detectable levels noted at 6–11 weeks of gestation. In contrast, IgG4 is present in the fetal circulation after 19 weeks [40]. The resultant fetal concentrations of total IgG are marginally detectable in the first trimester (<100 mg/dl) and remain low until after 17 weeks, at which time they steadily increase reaching 400 mg/dl by 24 weeks and 800 mg/dl by 32 weeks as placental transfer becomes more efficient [41]. IgM and IgA antibodies do not cross the placenta.

Perhaps CHB occurs in offspring of mothers whose antibodies are of IgG1 and IgG3 subclasses, while those whose antibodies are predominantly IgG2 and IgG4 are protected. However, no significant differences in subclass distribution have been observed between mothers with anti-SSA/Ro and/or anti-SSB/La antibodies who had pregnancies complicated by CHB or those who had normal pregnancies [42]. For both groups of mothers, IgG1 antibodies were significantly increased over the other three subclasses in the anti-52kD and 60kD SSA/Ro responses. IgG1 and IgG3 were the major subclasses represented in the 48kD SSB/La responses. All subclasses, including IgG2 and IgG4, were observed in one-third to one-half of the maternal serum with anti-52-kD and 48-kD responses. In contrast, anti-60-kD antibodies were, with rare exception, confined to IgG1. Accordingly, the IgG subclasses of anti-48-kD SSB/La, 52-kD, or 60-kD SSA/Ro antibodies do not likely account for the susceptibility of one fetus versus another for the development of CHB.

The stage of cardiac ontogeny that coincides with the initiation of transplacental passage of maternal autoantibodies into the fetal circulation may influence the extent of tissue injury and permanent dysfunction. The human heart attains most of its adult characteristics by the 6th to 8th week of gestation [43]. Parasympathetic innervation of the heart occurs very early in fetal development while sympathetic innervation develops much later and is completed some months after birth. The SA node can be recognized in the first trimester and by 10 weeks of fetal age attains its own artery. Landmarks of the three internodal pathways from SA to AV nodes appear in the 2nd month of gestation, although the septal course of these pathways does not become fully developed until the closing of the foramen ovale cordis, which occurs shortly after birth. The AV node arises separately from the bundle of His and is joined to it at 8 weeks. The human His bundle undergoes extensive postnatal remodeling to achieve its adult form. Clearly the fetal conduction system has reached functional maturity before maternal antibodies gain access to the fetal circulation.

To support a pathogenic role of maternal autoantibodies in the development of CHB, the onset of bradycardia would be predicted to coincide with effective placental transport and occur in a previously normal heart. As expected if the hypothesis is correct, Deng et al. reported the lack of immunoglobulin deposition in cardiac tissue from 10-week fetuses of two mothers with anti-SSA/Ro antibodies [44]. There is likely to be a window of vulnerability when maternal antibodies gain access to the fetal circulation and recognize an antigen unique to the developing heart that is absent or otherwise inaccessible in the maternal heart. However, it does need to be acknowledged that even at 20 weeks of human gestation, maternal levels of IgG in the fetal circulation are far lower than at birth, suggesting that it requires very little antibody to achieve a pathogenic effect.

An early review of data obtained from the RRNL revealed that in 71 (82%) of 87 fetuses, bradycardia was identified before 30 weeks of pregnancy [32]. Detection was most frequently clustered between 20 and 24 weeks. Fourteen (16%) cases were first identified in the third trimester, five of which were noted at the time of delivery. The median time of in utero detection was 23 weeks. Of the 85 fetuses diagnosed with CHB during pregnancy, 15 were born between 1970 and 1987, eight of whom were diagnosed before 30 weeks of gestation. Seventy pregnancies occurred between 1988 and 1997, in which 63 were diagnosed with CHB prior to 30 weeks ($p < 0.003$, earlier detection of CHB in pregnancies after 1988 compared to pregnancies before 1988). As enrollment in the RRNL has increased over the years, the timing of bradycardia has remained consistent, with the vulnerable period being greatest between 20 and 24 weeks of gestation.

Spectrum of Cardiac Manifestations of NL

There still exists some confusion regarding the assignment of "congenital" heart block. Webster defines congenital as that "existing at or dating from birth." Data from Hübscher et al. strongly suggest that postnatally "acquired" heart block (even without structural abnormalities) is not associated with maternal antibodies to SSA/Ro or SSB/La [45] and is probably best categorized as a separate entity.

Again, the issue of complete versus incomplete heart block warrants clarification. At the time of writing the initial chapter on NL, heart block was operationally defined as third degree or congenital complete heart block (CCHB), although it was hypothesized that heart block might progress through various stages. Of 187 children in the RRNL with CHB associated with anti-SSA/Ro-SSB/La antibodies in the mother, nine had a prolonged PR interval on EKG at birth, four of whom progressed to more advanced AV block [35]. A child whose younger sibling had third-degree block was diagnosed with first-degree block at age 10 years at the time of surgery for a broken wrist. Two children diagnosed in utero with second-degree block were treated with dexamethasone and reversed to normal sinus rhythm by birth, but ultimately progressed to third-degree block. Four children had second-degree block at birth, two of whom progressed to third-degree block. Accordingly, the general abbreviation CCHB is not precise and CHB has been adopted. To complicate matters, the search for early markers of progressive conduction defects has identified in utero prolongation of the PR interval, the strict definition of which remains somewhat controversial [46–48]. Furthermore, it is unclear whether first-degree block detected in utero progresses to advanced conduction defects since all in utero prolonged PR intervals have either prompted treatment or spontaneously resolved within the 1st month of postnatal life. The implications of this spectrum of conduction abnormalities from the perspectives of pathobiology and treatment are discussed in subsequent sections.

When CHB is associated with major structural abnormalities such as l-transposition of the great vessels, maternal autoantibodies are generally not present. This defect of cardiogenesis occurs prior to the 10th week of gestation and results in the disruption of the development of the AV conducting system. Such cases should be considered as a distinct classification of CHB secondary to abnormal cardiac embryogenesis and not likely the result of passively acquired autoimmunity. Examination of a heart from a child with this type of

CHB, whose maternal serum did not contain autoantibodies, revealed nodoventricular discontinuity, with the AV node well formed and normally situated [27]. Similarly, mass lesions developing in the conducting system, such as mesotheliomas of the AV node, would constitute another distinct type of CHB.

As increased awareness of cardiac manifestations associated with maternal anti-SSA/Ro-SSB/La antibodies continues, the spectrum expands. As discussed, EFE with concomitant evidence of cardiac dysfunction has been reported in isolation and in association with conduction defects [30,31]. Furthermore, heart block can progress to a fatal cardiomyopathy [49].

THE TRANSIENT SKIN RASH

In striking contrast to the typical late second-trimester onset of CHB, the skin lesions generally become manifest several weeks into postnatal life. Less commonly the rash is present at birth. Ultraviolet exposure may be an initiating factor and can exacerbate an existing rash [50, 51]. Thornton *et al.* report that telangiectasia may be a presenting feature and can occur in sun-protected sites independent of lupus dermatitis [52]. Cutaneous activity, inclusive of erythema and the continued appearance of new lesions, is generally present for several weeks with resolution by 6–8 months of age, coincident with the clearance of maternal autoantibodies from the baby's circulation. However, hypopigmentation may persist into the 2nd year of life. Lee and her colleagues have observed two children with persistent telangiectasias and one child with a small but persistent patch of hyperpigmentation [51]; 11 other children, including several with extensive skin disease, have had complete resolution of the rash.

The rash frequently involves the face and scalp with a characteristic predilection for the upper eyelids (Figure 31.1) [53]. In some instances, the rash is present in other locations and can cover virtually the entire body. A review of the corporeal distribution of 57 infants (20 males, 37 females) diagnosed with cutaneous NLE (absent heart disease) enrolled in the RRNL [50] revealed that all had facial involvement (periorbital region most common) followed by the scalp, trunk, extremities, neck, intertriginous areas, and rarely the palms and soles. In most cases, the infant's rash was temporally related to UV exposure; mean age of detection was 6 weeks and duration 22 weeks. The lesions were described as superficial inflammatory plaques resembling subacute cutaneous lupus erythematosus of the adult [54]. They were typically annular or elliptical with erythema and scaling. Hypopigmentation was frequent and may be a prominent feature. The more characteristic lesions of adult discoid lupus such

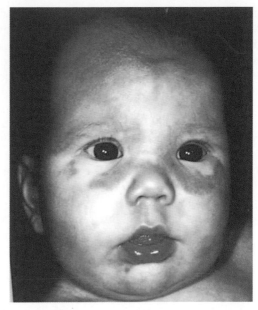

FIGURE 31.1 Typical skin rash with predilection for the periorbital area. This rash resolved completely without scarring. *Reproduced with permission, courtesy of Dr. Susan Manzi.*

as follicular plugging, dermal atrophy, and scarring were generally not observed in the neonatal skin rash. The lesional histology supports the clinical descriptions of subacute cutaneous lupus with basal cell damage in the epidermis and a superficial mononuclear cell infiltrate in the upper dermis [55, 56]. As observed in subacute cutaneous lupus, immunofluorescence is positive with the finding of a particulate pattern of IgG in the epidermis [51].

LESS COMMONLY ENCOUNTERED MANIFESTATIONS OF NL

The clinical spectrum of NL includes hepatic involvement, a manifestation that could well be underestimated since routine neonatal evaluation does not include a liver profile [18, 57–61]. Laxer *et al.* have described three living infants and one perinatal death with NL associated with significant hepatic involvement [18]. The living infants presented with neonatal cholestasis as a major component of their clinical picture. Pathologic changes included giant cell transformation, ductal obstruction, and extramedullary hematopoiesis. The authors speculated that an inflammatory hepatitis proceeding to hepatic fibrosis may ensue, analogous to the mechanism hypothesized to occur in cardiac tissue. Rosh *et al.* reported an infant born with CHB in whom severe neonatal cholestasis developed, requiring surgical exploration to exclude extrahepatic biliary atresia [57]. The clinical picture included an elevation of the serum glutamic-pyruvic transaminase, glutamic oxaloacetic transaminase, alkaline phosphatase, and gamma-GTP. A percutaneous liver biopsy revealed mild fibrosis, bile ductular proliferation, and a mixed inflammatory infiltrate in the portal tracts. Lee *et al.* described three infants with hepatic dysfunction; all had laboratory and histologic evidence of significant cholestasis [58]. One occurred in the setting of intractable congestive heart failure. At autopsy, immunofluorescence revealed widespread deposits of IgG. A second infant had thrombocytopenia and hepatosplenomegaly at birth followed at 3 weeks of age by a cutaneous eruption characteristic of NL. Liver biopsy revealed hepatocellular cholestasis, lobular disarray, and mild pseudoacinar formation. A third neonate developed a typical rash at 2 weeks and transaminitis with jaundice by 8 weeks. Liver biopsy revealed canalicular and hepatocellular cholestasis. Lee *et al.* [59] investigated the incidence of hepatobiliary manifestations among 219 NL patients in the RRNL. Nineteen (9%) had probable or possible hepatobiliary disease, appearing as the sole manifestation of NL in three cases, and in association with cardiac or cutaneous manifestations in 16 cases. Six (including one previously reported by Schoenlebe *et al.* [60]) of the 19 infants died, either during gestation or within the first few weeks of life. Three clinical variants were observed: (1) severe liver failure present during gestation or in the neonatal period, often with the phenotype of neonatal iron storage disease (most uncommon); (2) conjugated hyperbilirubinemia with mild or no elevations of aminotransferases, occurring in the first few weeks of life; and (3) mild elevations of aminotransferases occurring at approximately 2–3 months of life. The prognosis for the children in the last two categories was excellent.

These reports and others [61] suggest that the diagnosis of NL-related liver disease should be considered in situations in which the liver enzymes and bilirubin levels are most consistent with cholestasis in the absence of a major structural abnormality of the biliary tree. Reassuringly, in babies who survive cardiac manifestations, the general observation is that hepatic disease resolves.

While the nervous system has not been regarded as an organ characteristically affected in NL, clinical detection may be a limiting factor. There have been several reports of neurologic sequelae in NL. Specifically, aseptic meningitis occurred in an infant with CHB and circulating maternal anti-SSA/Ro and anti-SSB/La antibodies [62]. There has been one case report of NL and transient hypocalcemia with seizures [63]. Wong and colleagues observed sonographic evidence of infantile lenticulostriate vasculopathy (LSV) in a case of NL [64]. The authors suggest that sonographic LSV is a nonspecific marker of a previous insult to the developing brain, the clinical significance of which is

uncertain. Bourke *et al.* describe an infant with thrombocytopenia and a generalized annular rash with scattered telangiectases at birth in the setting of antibodies to SSA/Ro and SSB/La [65]. At the age of 1 year the child was found to have an abnormal gait and examination revealed mild spastic diplegia of the lower limbs. The authors appropriately point out that the central nervous system abnormalities might have been due to an intracerebral hemorrhage in the neonatal period. This curious observation of late-onset lower limb spasticity has been reported in one other infant with antibodies to SSA/Ro and butterfly rash at birth [66]. In a recent publication from Toronto in which 87 infants exposed to maternal Ro/SSA were studied, seven children were reported to have hydrocephalus [67]. Ten infants had head circumference measurements which were abnormally elevated (most often between 8 and 24 months). The authors concluded that hydrocephalus and macrocephaly are manifestations of NL. However, unlike CHB, the abnormalities tended to resolve spontaneously, with only one of the infants with hydrocephalus requiring surgical intervention. Notably in this Toronto study, the prevalence of NL was close to 50% (15% CHB, not known if this reflects recurrences), which is higher than reported in most other prospective studies of anti-SSA/Ro mothers and their pregnancy outcomes. In sum, the mechanism of these abnormal neurologic findings is entirely elusive and awaits further observation in larger cohorts of neonates born to mothers with anti-SSA/Ro-SSB/La antibodies.

Hematologic abnormalities have been described as a manifestation of NL. Thrombocytopenia has been observed together with other manifestations of NL [19]. Some of these infants have a petechial or purpuric eruption as the initial feature. Thrombocytopenia was present in 10% of the neonates referred to Lee and her colleagues [51]. Gastrointestinal bleeding occurred in one of these infants. In contrast to the cardiac manifestations which do not parallel disease in the mother and the cutaneous manifestations which occasionally occur in the mother but often not in synchrony with her affected offspring, the hematologic manifestations may more closely parallel maternal disease. However, Watson has described thrombocytopenia in offspring of anti-SSA/Ro positive mothers with no apparent history of thrombocytopenia [19]. Despite this, the presence of thrombocytopenia raises some questions as to whether antiplatelet antibodies rather than antibodies to SSA/Ro-SSB/La *per se* are targeting the surface of fetal cells. These manifestations, while secondary to passively acquired autoimmunity (consistent with NL), may be more analogous to the neonatal thrombocytopenia of idiopathic thrombocytopenic purpura (ITP). In addition, the disparate fetal and adult vulnerability appears more pronounced in NL than in ITP.

Kanagasegar *et al.* [68] reported an infant with neutropenia and mildly abnormal liver function, but no cardiac or cutaneous manifestations of NL, born to a mother with anti-SSA/Ro-SSB/La antibodies. The child's neutropenia improved as maternal antibody was metabolized. Sera from this child and mother, as well as sera from two mothers enrolled in the RRNL who had given birth to infants with CHB and neutropenia, were shown to bind the cell surface of intact neutrophils [68]. Binding to neutrophils was inhibited (>80%) by incubating the sera with 60-kD Ro antigen, suggesting that anti-60-kD SSA/Ro is directly involved in the pathogenesis of neutropenia.

Wolach *et al.* extend the hematologic spectrum of NL [69]. They describe a 5-month-old infant with anti-SSA/Ro antibodies and typical cutaneous involvement in the setting of complete marrow aplasia who recovered at 8 months with the disappearance of anti-SSA/Ro antibodies. However, before this complication is added to the official list of manifestations, it is curious that the mother herself was said to have tested negative for anti-SSA/Ro antibodies. The child died at 16 months from Gram-negative sepsis.

CANDIDATE ANTIGEN-ANTIBODY SYSTEMS IN NL

The Target Autoantigens of the SSA/Ro-SSB/La System

Antibodies to SSA/Ro ribonucleoproteins in the maternal sera, often in association with SSB/La, have been almost universally demonstrated when isolated CHB develops in an offspring [9—11]. Anti-SSA/Ro antibodies are characteristically found in the majority of patients with Sjögren's syndrome (SS) [70] and close to half of those with SLE [71] but can also be detected in asymptomatic individuals [72, 73]. Fritzler *et al.* detected a 0.5% frequency of anti-SSA/Ro positivity after testing several thousand healthy female blood donors of childbearing age [72]. A similar frequency was detected by Harmon *et al.* in a study of 800 asymptomatic pregnant women [73]. Reactivity against SSB/La is also strongly associated with the development of CHB and is frequently detected in patients with SS [70, 74]. Its prevalence in SLE is only about 15% [71].

The candidate autoantigens are ubiquitous and present in all cells. The major antigenic component of SSA/Ro was initially described as a polypeptide of 60-kD. The gene has been cloned in several laboratories with accumulating evidence that there are isoforms in T-cell lines with at least two slightly different copies in the haploid genome [75—78]. The zinc finger in human 60-kD SSA/Ro is not conserved across species [79]. A

focus on the role of SSA/Ro 60 as a ribonuclear-binding protein has highlighted an RNA-binding protein consensus motif, which is comprised of the outer surface at a ring of largely alpha helical HEAT repeats and its adjacent hole. The former binds small cytoplasmic hY-RNAs while the latter serves as a docking site of a single-stranded RNA (misfolded pre-5S rRNA) [80, 81]. Regarding the latter, Ro60 serves as part of a novel quality control or discard pathway for 5S rRNA production in Xenopus oocytes [81].

Anti-SSB/La antibodies recognize a 48-kD polypeptide that does not share antigenic determinants with either 52-kD or 60-kD SSA/Ro [82]. The SSB/La polypeptide is comprised of at least two structural domains on the native protein, each of which contains a distinct antigenic binding site [83]. SSB/La facilitates maturation of RNA polymerase III transcripts, directly binds a spectrum of RNAs and associates at least transiently with 60-kD SSA/Ro [84, 85].

In addition to the well-characterized 60-kD SSA/Ro and 48-kD SSB/La antigens, another target of the autoimmune response in mothers whose children have CHB is the 52-kD SSA/Ro protein [86]. The full-length protein, 52α (also referred to as TRIM21), has three distinct domains: an N-terminal region rich in cysteine/histidine motifs containing two distinct zinc fingers known as RING finger and B-box; a central region containing two coiled coils with heptad periodicity, one being a leucine zipper with potential for intramolecular dimerization; and a C-terminal rfp-like domain [87, 88]. The 52-kD SSA/Ro protein has a high degree of homology with the ret finger protein, rfp, which is part of the transforming gene ret; this raises the question of whether the 52-kD SSA/Ro protein might have similar transforming potential [87]. Moreover, the ret protein has the structure of a cell surface receptor. Ro52 has now been identified as an E3 ubiquitin ligase and has been shown to catalyze the ubiquitination of several proteins, including Ro52 itself (89). Most recently, based on the generation of Ro52-null mice, it has been suggested that Ro52 may be a negative regulator of proinflammatory cytokine production (90).

Cloning and sequencing of a novel autoantigen identified by sera containing predominant reactivity with a 60-kD SSA/Ro RNA binding protein was found to be more than 96% identical to rabbit calreticulin [91, 92]. Preliminary work demonstrated that many sera from mothers of children with CHB react with human calreticulin by ELISA [93]. This finding may be of potential relevance since calreticulin is an acidic high-affinity calcium-binding protein of M_r 46,567, common to both sarcoplasmic reticulum membranes in muscle and endoplasmic membranes in nonmuscle tissues. The carboxyl-terminal end of calreticulin contains the tetrapeptide Lys-Asp-Glu-Leu (KDEL) salvage or endoplasmic reticulum retention signal. Future work on the association of CHB and antibodies to proteins of the endoplasmic reticulum may be of interest, but at the time of writing no recent data have emerged.

Eftekhari *et al.* reported that antibodies reactive with the serotoninergic 5-hydroxytryptamine (5-HT)$_{4A}$ receptor, cloned from human adult atrium, also bind 52-kD SSA/Ro [94]. Moreover, affinity-purified 5-HT$_4$ antibodies antagonized the serotonin-induced L-type Ca channel activation in human atrial cells. Two peptides in the C terminus of 52-kD SSA/Ro, aa365-382 and aa380-396, were identified that shared some similarity with the 5-HT$_4$ receptor. The former was recognized by sera from mothers of children with NL and it was this peptide that was reported to be cross-reactive with peptide aa165-185, derived from the second extracellular loop of the 5-HT$_4$ receptor. These findings are of particular importance, since over 75% of serum from mothers whose children have CHB contain antibodies to 52-kD SSA/Ro as detected by ELISA, immunoblot, and immunoprecipitation (further discussed in the section titled "Evaluation of the Fine Specificities of the Maternal SSA/Ro-SSB/La Autoantibody Response") [95–100].

Given the intriguing possibility that antibodies to the 5-HT$_4$ receptor might represent the hitherto elusive reactivity that could directly contribute to CHB, the 5-HT$_4$ receptor was examined as a target of the immune response in these mothers. Initial experiments demonstrated mRNA expression of the 5-HT$_4$ receptor in the human fetal atrium [101]. Electrophysiologic studies established that human fetal atrial cells express functional 5-HT$_4$ receptors. Sera from 116 mothers enrolled in the RRNL, whose children have CHB, were evaluated. Ninety-nine (85%) of these maternal sera contained antibodies to SSA/Ro, 84% of which were reactive with the 52-kD SSA/Ro component by immunoblot. However, none of the 116 sera were reactive with the peptide spanning aa165-185 of the serotoninergic receptor [101]. Rabbit antisera, which recognized this peptide, did not react with 52-kD SSA/Ro. A collaborative follow-up study revealed that of 75 sera from mothers of children with CHB, 12 were reactive with the 5-HT$_4$ peptide [102]. Accordingly, although 5-HT$_4$ receptors are present and functional in the human fetal heart, maternal antibodies to the 5-HT$_4$ receptor are not likely a major factor associated with the development of CHB.

Although it is not known how maternal antibodies influence the development of cardiac versus cutaneous manifestations of NL, to date antibodies to U1RNP in the absence of reactivity to anti-SSA/Ro and/or SSB/La have never been reported in children with CHB (however, transient PR prolongation has been noted in one fetus exposed to maternal anti-u1 RNP. Personal communication, Acherman and Friedman). Sheth and colleagues have added two cases to eight previously reported in which anti-U1RNP antibodies but not anti-SSA/Ro-SSB/La antibodies were present in infants with cutaneous disease alone [103]. Furthermore, Solomon *et al.* describe anti-U1RNP-positive, anti-SSA/Ro-SSB/La-negative dizygotic twins discordant for cutaneous manifestations of NLS (neither twin had cardiac disease) [104]. The segregation of anti-U1RNP antibodies with cutaneous disease may be a useful maternal marker and should guide research efforts in sorting out cardiac vs. cutaneous susceptibility to antibody-mediated injury.

Evaluation of the Fine Specificities of the Maternal SSA/Ro-SSB/La Autoantibody Response

Issues regarding the management of a pregnancy and the outcome of both the mother and her child are important from the clinician's perspective. Can our rapidly expanding knowledge of SSA/Ro and SSB/La autoantibodies and their cognate antigens help define the mother at highest risk for having a pregnancy complicated by NL? While seemingly straightforward, the task is a challenging one. Several laboratories in the United States and abroad have tackled this problem using various techniques including immunodiffusion, immunoblot of various tissues and cell lysates, ELISA employing recombinant and purified proteins, and immunoprecipitation of radiolabeled *in vitro* translation products and cell lysates.

Our laboratory initially used gel separation methods that vary the quantity of acrylamide to bis-acrylamide to obtain a more precise molecular characterization of the relevant antigenic structures identified by the immune response in women whose children have NLS [17, 95, 105]. Specifically, increasing the ratio of monomer to cross-linker from 37.5 (used in a standard Laemmli buffer system [106]) to 172.4 in a 15% acrylamide solution readily separates the 48-kD SSB/La polypeptide from the 52-kD SSA/Ro component. We employed this method of gel separation for immunoblot, as well as ELISA, to evaluate antibody frequency, titer, and fine specificity in the sera from four groups of mothers segregated according to the status of the child: 57 whose children had CHB, 12 whose children had transient dermatologic or hepatic manifestations of NL but no detectable cardiac involvement, 152 with SLE and related autoimmune diseases who gave birth to healthy children, and 30 with autoimmune diseases whose pregnancy resulted in miscarriage, fetal demise, or early postpartum death unrelated to NLS [95]. Anti-SSA/Ro antibodies were identified by ELISA in 100, 91, 47, and 43% of mothers of infants with CHB, transient NL, healthy children, and fetal demise, respectively. High titers of anti-SSA/Ro antibodies were more often present in mothers of children with cardiac or cutaneous manifestations of NL than in either of the other two groups. Maternal antibodies to SSB/La were detected by ELISA in 76% of the CHB group, 73% of the cutaneous/hepatic group, 15% with healthy children, and 7% with fetal demise. On immunoblot, 91% of the heart block group who had antibodies to SSA/Ro but not to SSB/La recognized at least one SSA/Ro antigen, with significantly greater reactivity against the 52-kD component. In contrast, 62% of the anti-SSA/Ro-positive SSB/La-negative responders in the healthy group recognized either the 52-kD and/or 60kD components. Although there was no profile of anti-SSA/Ro response unique to the mothers of children with heart block or cutaneous manifestations of NLS, only 1% of normal infants were born of mothers with antibodies directed to both the 52-kD SSA/Ro and 48kD SSA/La antigens and not to the 60-kD SSA/Ro antigen, compared to 21% with cardiac and 25% with cutaneous manifestations of NL.

Dörner *et al.* [107] investigated quantitative and qualitative differences of anti-SSA/Ro-SSB/La antibodies by ELISA in sera from 16 infants with CHB and their mothers compared to eight healthy anti-SSA/Ro positive infants born to SLE mothers. No serum sample contained IgM autoantibodies. All 16 (100%) CHB infants had anti-52-kD SSA/Ro antibodies, 14 (88%) had anti-SSB/La, and nine (56%) had anti-60-kD SSA/Ro compared to six (75%) anti-52, three (38%) anti-48, and two (25%) anti-60 in the control infants. The anti-52-kD SSA/Ro and anti-SSB/La antibody levels were significantly higher in CHB infants than in the controls. The anti-60-kD SSA/Ro IgG levels of sera from infants and mothers from the CHB and control groups were similar.

Based on analysis by immunoblot and ELISA of sera from 14 mothers of children with CHB and 12 SLE patients with healthy offspring, Meilof *et al.* [108] concluded that the fine specificity of the autoantibody response to SSA/Ro and SSB/La proteins does not predict the occurrence of CHB. These results are not surprising, since it can be predicted from the absence of complete concordance of disease expression in twins and the less than 20% recurrence rate in subsequent pregnancies (see the section titled "Recurrence Rates", page 563) that another factor (probably fetal) is operative and that the discovery of a unique risk profile is unlikely.

In a study of sera from 31 mothers of children with CHB, Julkunen *et al.* [96] demonstrated 97% reactivity with 52-kD SSA/Ro by ELISA, 77% with 60-kD SSA/Ro, and 39% with SSB/La. Compared to 45 mothers with SLE and healthy children, mothers of CHB children had higher titers of antibodies to recombinant 52-kD and 60-kD SSA/Ro proteins. However, compared to 19 mothers with primary SS and healthy offspring, the autoantibody responses were similar. No differences in the titer of anti-SSB/La antibodies were found between mothers within these three groups. In agreement with previous data, these investigators did not find a specific antibody profile unique to CHB.

Silverman and colleagues [109] evaluated the maternal antibody profile in two groups of sera, 41 obtained from mothers whose children had manifestations of NL (21 CHB, 20 cutaneous) and 19 from SLE patients known to have anti-SSA/Ro and/or anti-SSB/La antibodies and healthy children. Significantly higher levels of anti-SSB/La and anti-52-kD SSA/Ro antibodies were demonstrated in the mothers of affected children. Mothers whose children had cutaneous manifestations had higher titers of anti-SSB/La antibodies than mothers whose children had CHB. Fine delineation of the anti-SSB/La responses revealed that a small carboxyl terminus polypeptide of recombinant SSB/La, DD, was recognized by 30% of mothers whose children had NL but none who had healthy children.

Defining the risk of CHB based on a particular antibody profile depends in part on the control group chosen. For example, if the control group is mothers with SLE, as in our study done in 1993 [95] and that of Dörner *et al.* [107], it would not be surprising to find a greater frequency of anti-48-kD SSB/La and 52-kD SSA/Ro in the CHB mothers. It has been previously reported that in SLE, anti-60kD responses predominate over anti-52-kD responses [110]. If the control mothers have SS, the antibody profiles of the CHB mothers and those with healthy offspring might be equivalent as supported by the studies of Julkunen *et al.* [96, 111]. In effect these observations reinforce the findings of most studies that the serologic profile of mothers whose children have CHB closely resembles that of SS. One approach to defining risk would be to subset the mothers of children with CHB and compare SS-CHB to SS-healthy, and SLE-CHB to SLE-healthy. However, because many mothers are asymptomatic and are only identified by the birth of their affected child, an appropriate control group is difficult to assemble.

The fact that SDS-immunoblot favors recognition of denatured epitopes complicates the interpretation of antibody reactivity. It is likely that all anti-SSA/Ro responses would include reactivity with the 60-kD SSA/Ro component if immunoprecipitation of the native protein was the assay employed [112]. Perhaps instead of defining a high-risk profile we should define a low-risk profile. In our experience, mothers who do not have anti-SSB/La antibodies and have anti-SSA/Ro antibodies of low titer that do not recognize either the 60-kD or 52-kD component on SDS-immunoblot appear to be at lower risk [113]. In the last edition of this chapter, we evaluated sera from 150 mothers whose children have CHB. All had antibodies to SSA/Ro or SSB/La by ELISA. An isolated response to the denatured 60-kD component on immunoblot was observed in one (<1%) serum and only two sera (<2%) did not recognize any component of SSA/Ro or SSB/La on immunoblot (unpublished observation, RRNL). However, these findings may also be characteristic of mothers who have healthy children as well.

A note of caution is in order. Reporting of antibody profiles is highly dependent on the assays employed. For example, Neidenbach and Sahn reported an infant with cutaneous manifestations of NL associated with maternal antibodies to SSB/La in the absence of associated antibodies to SSA/Ro [114]. Testing was done by ELISA and immunoblot. In our experience and that of others, immunoprecipitation of the native 60-kD SSA/Ro antigen has always been positive when anti-SSB/La antibodies are present [113], and such may well be the case in this unique infant. Anti-idiotype antibodies to SSB/La may be protective against NL. This was demonstrated in one study that reported a lower risk of NL in women with anti-idiotype antibodies to SSB/La [115].

The most recent data on evaluation of fine specificity of maternal antibody and risk of CHB comes from the group of Wahren-Herlenius [97, 100]. These investigators have confirmed and extended earlier work on epitope mapping of the Ro52 response and risk of CHB [98]. Based on nine CHB-mothers and 26 anti-SSA/Ro+ mothers of healthy children, the Swedish group posited that antibody to aa200-239 of Ro52 (p200) predicted CHB with greater certainty than currently available testing for either Ro60 or Ro52 [97]. Utilizing an extensive serum bank from the RRNL, we found that reactivity to p200 is a dominant but not uniform anti-Ro 52 response in women whose children have CHB and exposure to this antibody specificity was observed with a similar frequency in children without CHB [99]. Most recently, an international exchange of sera revealed higher levels of Ro52 p200 antibodies in mothers of children with CHB (in Sweden and in the United States, but not in Finland), but noted that these same specificities can also occur in women with unaffected children [100].

PROPOSED MECHANISMS OF IMMUNE INJURY

This model of passively acquired autoimmunity offers an exceptional opportunity to examine the effector arm of immunity and define the pathogenicity of an autoantibody in mediating tissue injury. A molecular scenario in which maternal anti-SSA/Ro-SSB/La antibodies convincingly contribute to the pathogenesis

of cardiac scarring has yet to be formulated. Two points, one clinical and the other cellular, are particularly difficult to reconcile. Only 2% of neonates born to mothers with the candidate antibodies have CHB, yet these antibodies are present in over 85% of mothers whose fetuses are identified with conduction abnormalities in a structurally normal heart. Accordingly, the antibodies are necessary but insufficient to cause CHB. The final pathway leading to fibrosis may be variable, kept totally in check in most fetuses (normal sinus rhythm), subclinical in others (first degree block), and fully executed in very few (advanced block). The intracellular location of the target antigens raises fundamental questions regarding accessibility to maternal antibodies.

Accessibility of Fetal Antigen to Maternal Antibody

A mechanism whereby antibodies might interrupt critical intracellular events in fetal cardiomyocytes or specialized Purkinje cells is largely unknown. As previously stated by Tan, "the question is whether autoantibody reacts with intrinsic antigens to perturb the biologic function of normal cells" [116]. For anti-SSA/Ro and -SSB/La antibodies to be causal in the development of NL, three basic requirements should be satisfied. First, the candidate antigens must be present in the target fetal tissues; second, the cognate maternal autoantibodies must be present in the fetal circulation; and third, these antigens must be accessible to the maternal antibodies.

Earlier studies have firmly demonstrated reactivity of anti-SSA/Ro antibodies with fetal cardiac tissues, including the conduction system [17, 117]. The preferential vulnerability of the fetal versus the adult heart has been approached by a study demonstrating that a 23-week fetal heart contained a greater quantity of SSA/Ro per milligram of protein than 18- to 22-week hearts or an adult heart [118]. The presence of SSA/Ro and SSB/La antigens in the fetal heart is well established and therefore satisfies the first requirement. Similarly, the second requirement has been readily fulfilled by studies demonstrating anti-SSA/Ro-SSB/La antibodies in the fetal circulation as assessed by measurements in cord blood [119].

The third requirement, accessibility, has been more difficult to establish. It has been suggested that autoantibodies can penetrate living cells, subsequently alter function, and cause cell death [120], but this notion still remains controversial. Alternatively, if the antibody cannot cross the cell membrane, then is the antigen trafficked to the cell surface? Finally, anti-SSA/Ro-SSB/La antibodies could potentially cross-react with other surface cardiac antigens.

Several lines of evidence have been advanced to support the possibility that otherwise sequestered intracellular autoantigens can be expressed on the cell surface. Baboonian *et al.* have demonstrated the sequential expression of the SSB/La antigen from the nucleus through the cytoplasm and ultimately on the cell surface of Hep 2 cells infected with adenovirus [121]. There is now accumulating data supporting surface expression of SSA/Ro in keratinocytes, after exposure to ultraviolet light [122-124] or following incubation with TNF-α [125]. Potentially relevant to NL, 17β-estradiol at $10^{-7}M$, a concentration reached during the third trimester of pregnancy [126], enhances binding of anti-SSA/Ro antibodies to keratinocytes [127] and induces up to fivefold increases in the expression of 52α and 60Ro mRNA [128]. Biologic significance is strengthened by the reports of estrogen receptors in cardiac tissue [129, 130]. Furthermore, one mismatch consensus estrogen response element was identified upstream of the transcriptional start site in the gene encoding 52-kD SSA/Ro, and an estrogen response element similar to that of c-myc was detected in the human gene encoding 60-kD SSA/Ro [128].

Reichlin *et al.* have bolstered support for the accessibility of the candidate antigens to the respective maternal antibodies by the finding of antibodies to native 60-kD SSA/Ro and denatured 52-kD SSA/Ro in acid eluates of a heart from a fetus with CHB who died at 34 weeks gestation [131]. The enrichment was apparently selective since these antibodies were not detected in eluates from the brain, kidney, or skin. Furthermore, Horsfall *et al.* have demonstrated maternal IgG-bearing anti-SSB/La idiotypes on the surface of fetal myocardial fibers on autopsy of a neonate with CHB [132].

Apoptosis has been traditionally conceptualized from an immunologic point of view as either a means of maintaining B- and T-cell tolerance [133, 134] or as a mechanism for providing accessibility of intracellular antigens to induce an immune response [135]. Casciola-Rosen *et al.* have demonstrated that autoantigens are clustered in two distinct populations of surface blebs on keratinocytes [135]. The larger blebs, so-called apoptotic bodies, derived from the apoptotic nucleus, contain both SSA/Ro and SSB/La proteins with SSB/La detected at the cell surface surrounding large blebs in the later stages of apoptosis. The 52-kD protein was not specifically identified but rather deduced since evaluation was done with a patient serum considered monospecific for 52-kD SSA/Ro antibodies. The smaller blebs, arising from fragmented rough endoplasmic reticulum and ribosomes, contain SSA/Ro presumably of cytoplasmic origin. SSB/La was not contained in these blebs.

Apoptosis may be highly relevant in the pathogenesis of NL. It is a selective process of physiological cell deletion in embryogenesis and normal tissue turnover and plays an important role in shaping morphological and functional maturity [136]. Apoptosis is a process that affects scattered single cells rather than tracts of contiguous cells. In the normal adult myocardium, apoptosis has been observed only rarely [137, 138]. In contrast, apoptosis does occur during the development of the heart. In the 1970s, Pexeider extensively characterized the temporal and spatial distribution of cell death in the hearts of chicken, rat, and human embryos [139]. Major foci included the AV cushions and their zones of fusion, the bulbar cushions and their zones of fusion, and the aortic and pulmonary valves. Albeit much of the cell death was noted in non-myocytes, a focus of myocyte death was apparent in the muscular interventricular septum as it grew toward the AV cushions in mid-gestation. Takeda and colleagues demonstrated apoptosis in mid-gestational rat hearts using terminal deoxynucleotidyl transferase dUTP nick end labeling, an *in situ* technique, which detects DNA strand breaks in tissue sections [140]. Although not coincident with the precise timing of CHB, it has also been suggested that apoptosis contributes to the postnatal morphogenesis of the SA node, AV node, and His bundle [141]. Perhaps a novel view of apoptosis is that it facilitates the placing of cardiac target autoantigens in a location accessible to previously generated maternal autoantibodies. Tissue damage might be a consequence of being in the right place at the wrong time.

Available autopsy specimens from fetuses dying with CHB have been highly informative [34]. The initial analysis included one fetus diagnosed with CHB at 19 weeks who died 2 h after delivery at 40 weeks [34]. This limited us to evaluations of late changes (referring to time of death) since the earliest cellular events accompanying AV nodal fibrosis might not be present unless they were persistent. Autopsy sections from three additional CHB-hearts dying early became available: 18 week/20 week (diagnosis/death), 20 week/22 week, and 34 week/34 week (isolated myocarditis). Apoptosis was increased in septal tissue from each of these three fetuses compared to the late-CHB [34] and 22- to 23-week control hearts. IgG deposition was also greater in the early-CHB fetuses compared to the late-CHB. In contrast, IgG was not found in the septum of the normal 22- and 23-week fetal hearts. Most recently, we have evaluated two additional early-CHB hearts and one unaffected heart and increased apoptosis in the septal region of the affected hearts was confirmed. Our conclusion from these data is that the exaggerated apoptosis supports dysregulated physiologic clearance.

In vitro data generated from cultured human fetal cardiocytes [142, 143] and *in vivo* work in a murine system [144, 145] confirm the binding of anti-SSA/Ro and SSB/La to the surface of apoptotic cells. The consideration of exaggerated apoptosis as the initial link between maternal autoantibodies and tissue injury led to the observation that cardiocytes are capable of phagocytosing autologous apoptotic cardiocytes and that anti-SSA/Ro-SSB/La antibodies inhibit this function [143]. Recognizing that this perturbation of physiologic efferocytosis might divert uptake to professional FcγR-bearing phagocytes fit well with earlier work demonstrating macrophage secretion of pro-inflammatory and fibrosing cytokines when co-incubated with apoptotic cardiocytes bound by anti-SSA/Ro-SSB/La antibodies [142, 146]. That macrophages engage Toll-like receptors (TLRs) via binding to the RNA moiety of the target autoantigen is clearly an area which may provide an additional clue to pathogenesis.

Consistent with the findings on cardiocytes, Reed et al. demonstrated that anti-Ro60 binds specifically to early apoptotic Jurkat cells and remains accessible on the cell surface throughout early and late apoptosis [147]. Anti-SSB/La bound exclusively to late apoptotic cells in experiments controlled for nonspecific membrane leakage of IgG. An immunodominant internal epitope of Ro60 that contains the RNA recognition motif (RRM) and is recognized by the majority of sera from mothers of children with CHB, was also accessible as an apotope on early apoptotic cells. The distinct temporal expression of the immunodominant Ro60 and SSB/La apotopes indicates that these intracellular autoantigens translocate independently to the cell surface and supports a model in which maternal antibody populations against both Ro60 and SSA/La apotopes act in an additive fashion to increase the risk of tissue damage in CHB [145].

A potential contributor to the amplification of the proposed cascade derives from the recent finding that Ro60 functions as a novel receptor for beta(2)GPI on the surface of apoptotic cells. The formation of Ro60-beta(2)GPI complexes may protect against anti-Ro60 autoantibody-mediated tissue injury [148].

Complement Regulatory Proteins

With regard to the mechanism of tissue injury, consideration should be given to the possibility that protective molecules may be diminished on the surface of fetal cells. In adulthood, the organism's cells are protected from complement-mediated damage by membrane-bound proteins such as decay accelerating factor (DAF, CD55), protectin (CD59), and membrane cofactor protein (MCP, CD46) [149–151]. DAF regulates C3/C5 convertases and CD59 regulates assembly of the terminal components of the membrane attack complex. MCP patrols cell membranes, inactivating the C4b and C3b that is inadvertently bound by acting as a cofactor for their factor I-mediated cleavage [152]. MCP is expressed on a wide variety of cells including those of the epithelial, fibroblast, and endothelial cell lineage [153].

To examine fetal characteristics that might influence autoantibody-mediated diseases acquired in utero, such as heart block in NL, the tissue expression of MCP was studied [154]. Immunoblots of organs from six fetuses (aged 19–24 weeks) probed with rabbit anti-MCP antibodies revealed a band at 60 kD in addition to the known 65-kD and 55-kD isoforms that comprise the codominant allelic system of MCP. Five fetuses expressed the most common MCP polymorphism (predominance of the 65-kD isoform, upper band α phenotype) in the kidney, spleen, liver, and lung. In contrast, all hearts from these five fetuses demonstrated a different pattern in which there was a marked decrease in the intensity of the 65-kD band and accentuation of the lower molecular weight bands. In a sixth fetus that expressed the second most common polymorphism (equal expression of the 65-kD and 55-kD MCP isoforms, αβ phenotype), the heart was similar to the other tissues. Preferential expression of the MCP β isoform in five of six fetal hearts, irrespective of the phenotype of other organs, suggests tissue-specific RNA splicing or post-translational modification. The clinical significance of this tissue-specific phenotype is unknown at present but may provide another clue to the susceptibility of the fetal heart to antibody-mediated damage.

Arrhythmogenicity of Maternal Autoantibodies and Perturbation of L-Type Calcium Channels

Two earlier publications, both in animal models, indirectly invoked arrhythmogenic effects of anti-SSA/Ro-SSB/La antibodies. Alexander et al. reported that superfusion of newborn rabbit ventricular papillary muscles with IgG-enriched fractions from sera containing anti-SSA/Ro-SSB/La antibodies specifically reduced the plateau phase of the action potential consistent with an alteration of calcium influx [155]. Garcia et al., using isolated adult rabbit hearts, showed that IgG fractions from women with anti-SSA/Ro-SSB/La antibodies induced conduction abnormalities and reduced the peak slow inward current (I_{Ca}) in patch-clamp experiments of isolated rabbit ventricular myocytes [156]. Boutjdir et al. demonstrated a reduction of I_{Ca} by antisera from CHB mothers in isolated human fetal myocytes [157] and inhibition of expressed cardiac calcium channels by these antisera [158], which further supports the contribution of Ca^{2+} channels to the conduction abnormalities observed in the whole heart. This work has been extended with the report that 17 of 118 (14.4%) sera from mothers with CHB children reacted with the extracellular loop of domain I S5-S6 region (E1) of the L-type Ca channel alpha(1D) subunit [159].

Given the data in the rabbit and human heart, it is tempting to conclude that inhibition of L-type Ca^{2+} channels explains the pathogenicity of anti-SSA/Ro (perhaps anti-SSB/La) antibodies in the development of CHB. Several facts are supportive of this conclusion. AV nodal electrogenesis is dependent on L-type Ca^{2+} currents. Ca^{2+} channel density is lower and sarcoplasmic reticulum less abundant in fetal compared to adult cardiac cells, increasing the dependency on trans-sarcolemmal Ca^{2+} entry [160]. Prolonged exposure of fetal Ca^{2+} channels to maternal anti-SSA/Ro-SSB/La antibodies may lead to internalization and degradation of the channel, cell death, and ultimately fibrosis. Inhibition of ventricular Ca^{2+} channels may result in decreased contraction and congestive failure. Alternatively, antibody binding to the channel may result in opsonization and phagocytosis by macrophages with inflammatory/fibrotic sequelae (similar to the apoptosis theory described above in this section).

SA nodal electrogenesis is also dependent on L-type Ca^{2+} currents [161]. Interestingly, evaluation of autopsies done on children in the RRNL revealed pathology in some cases at the SA node (as described in the section title "Cardiac Histopathology", page 542) [33]. Mazel et al. observed sinus bradycardia in a murine model of passive immunity with anti-SSA/Ro antibodies [162]. Highly relevant to these histologic and functional observations is Brucato's identification of sinus bradycardia in four of 24 EKGs from otherwise healthy newborns born to a cohort of mothers with SSA/Ro antibodies [23]. Of 187 cases in the RRNL [35], atrial rates from postnatal EKGs were available for 40 neonates; the mean rate was 137 bpm ± 20 SD (range, 75—200). The one slow rate of 75 bpm was obtained during sleep and increased to 140 bpm when awake. In an additional child, the records stated sinus bradycardia; however, no EKG was available and subsequent records were not sent to the RRNL. In both our review and the study by Brucato, the sinus bradycardia was not permanent. Although this suggests that the nature of the initial insult and/or the subsequent reparative processes may be different for the SA and AV nodes, normal atrial rates may reflect the functioning of other pacer foci in the atria.

Murine Model of CHB

While clinical data leave little doubt regarding the association of anti-SSA/Ro and/or SSB/La antibodies with the development of CHB, and experimental data are beginning to suggest pathogenicity, efforts to establish an animal model have been not only limited but also disappointing. Kalush et al. reported that offspring of BALB/c mice immunized with the monoclonal anti-DNA idiotype 16/6 had conduction abnormalities [163]. Of 31 pups born to mothers with experimental SLE, eight had first-degree heart block, two had second-degree heart block, two had complete block, ten had bradycardia, and eight demonstrated widening of the QRS complex. None of these disorders could be detected in the 20 offspring of healthy control mice. One of the difficulties in interpreting these findings is that the immunized mothers synthesized a variety of autoantibodies including antibodies reactive with 16/6 Id, ss/dsDNA, Sm, RNP, cardiolipin, SSA/Ro, and SSB/La. Accordingly, it was not possible to segregate out which specific antibody might be responsible for the arrhythmias detected in these pups. The electrocardiographic data are provocative; however, no histologic data were provided to assess the status of the SA or AV node, or the presence of myocarditis.

To further establish an antibody-specific murine model to correlate arrhythmogenic effects of maternal autoantibodies with the in vivo genesis of CHB, Miranda-Carus et al. reported that following immunization of female Balb/C mice with 100 μg of one of the following 6×His human recombinant proteins purified by Ni^{2+} affinity chromatography: 48-kD SSB/La, 60-kD SSA/Ro, 52-kD SSA/Ro (52α full-length), and 52β and mating following demonstration of immune reactivity, there was minimal evidence of conduction abnormalities [164]. Of 54 pups born to six fertile mice immunized with 60-kD SSA/Ro, none had CHB; of 27 pups born to three fertile mice immunized with 48-kD SSB/La, none had CHB. In contrast, of 78 pups born to five fertile mice immunized with 52α and 86 pups born to five fertile mice immunized with 52β, two and five pups, respectively, had advanced AV block. Accordingly, this antibody-specific animal model provided preliminary evidence for a pathogenic role of antibodies reactive with 52-kD SSA/Ro, particularly the 52β form, in the development of CHB. Moreover, analogous to the frequency of 2% given for women with SLE who have anti-SSA/Ro and/or SSB/La antibodies, this model suggests that additional factors promote disease expression.

Further studies revealed that in an in vivo rodent model and in vitro culturing system anti-p200 52Ro antibodies bind neonatal rodent cardiocytes and alter calcium homeostasis [165]. While 19% of rat pups born to mothers actively immunized with p200 had first-degree block, unfortunately none had third-degree.

Lessons from Monozygotic and Dizygotic Twins

The study of twin gestations further emphasizes the complexity of NL. These informative in vivo experiments of nature provide clues to the true relevance of the maternal antibody and host factors. Discordance of disease expression in monozygotic twins would be particularly intriguing since the placenta is shared and the fetal genetics are identical. Twins born to mothers with anti-SSA/Ro-SSB/La antibodies provide a unique opportunity to gauge the effect of a specific antibody profile on disease phenotype. Regardless of whether twins share a common placenta or not, it is likely that even if a mother had three types of antibodies (Type A causing CHB, Type B causing skin rashes, and Type C being nonpathogenic), the proportion of these antibodies should be similar in each fetal circulation. Types A and B by definition must be of the correct subclass and isotype (i.e., IgG, subclasses 1−3 optimal) to be transported early across the trophoblast, while Type C might not be transported at all. One might envision quantitative differences in each twin, but it is difficult to conceive of a mechanism to explain why one placenta would transport antibodies more selectively than another unless Fc receptors were differentially expressed.

As shown in Table 31.1, the number of twins reported in which at least one child had CHB is quite limited and thus conclusions are conditional at best [10, 11, 32, 166−178]. The current numbers for monozygotic twins, inclusive of subjects unreported from the RRNL, total nine, of which three are concordant (33%) (albeit one did not report maternal antibody status). The current numbers for dizygotic twins, inclusive of subjects unreported from the RRNL, total 16, of which one is concordant (6%). This provides evidence (similar to SLE in some series) that the genetics of the affected fetus contribute to the underlying basis of autoimmune CHB. Notably, there were two additional cases of concordant CHB but zygosity was not determined. Further supportive of a genetic risk is the recurrence of CHB in subsequent pregnancies following the birth of a child with CHB, which is 25 of 134 (18.6%) in the RRNL database [179]. Recurrence rates for siblings of affected infants are 3000 times higher than the population prevalence. This apparent λs is exceptionally high for complex genetic diseases. However, the number may be overinflated because the denominator uses the population prevalence for CHB (1:15,000 live births) and the risk of recurrence, which approaches 20% of anti-SSA/Ro exposed pregnancies, i.e., $0.2*15,000 = 3,000$. Thus, it may be more realistic to use the predicted rate of 2% in the denominator since anti-SSA/Ro antibodies are required for disease expression and 2% is the frequency of CHB in these women (λs is $0.2*50 = 10$). Finally, since anti-SSA/Ro antibodies are likely underreported, (see Arbuckle et al. [180]) the λs is difficult to define. In aggregate, it is probably somewhere between 10 and 3,000. The recurrence rate as well as a familial recurrence rate (λmd, between 1.8 and 5.5), serve as a rational basis for the identification of genetic variants that confer risk to developing the clinical phenotype.

GENETICS OF THE NL CHILD AND THE MOTHER

Likely due to the rarity of disease, there are limited genetic studies of NL in the literature. Siren et al. [168] performed HLA typing on a series of 24 children with CHB from Finland, and several suggestive associations were found. Regarding Class I molecules, HLACw3 was found in increased frequency in the affected children (18/23; 75%; $p=0.001$) in comparison to population norms (27%). No Class II allele was significantly enriched in the children. HLA alleles enriched in the mothers were DRB1*03, DQB1*02, DQA1*05, and HLA Cw7 [168], a finding corroborated by Colombo and co-workers [181]. Miyagawa et al. reported that the maternal extended haplotype DRB1*11-DQA1*05-DQB1*03 was associated with the development of cutaneous manifestations of NL, but not CHB [182]. Maternal HLA-DQB1*06 was associated with CHB. Infant HLA type was not correlated with outcome. The association of DQA1*0501 with rash in the absence of CHB in this small Japanese series contrasts with the data presented above by Siren et al., in which DQA1*05 was associated with mothers of children with CHB [168], suggesting there may be racial variation.

Several studies have documented the near universal presence of DR3 alleles in the mothers of affected offspring, frequently associated with the extended haplotype A1-B8 [11, 111, 118, 176, 185, 186]. There have also been reports of an increased frequency of HLA-DR2 in anti-SSA/Ro-positive mothers of normal infants [11]. This may be expected since sera from the majority of mothers whose offspring have CHB contain both anti-SSA/Ro and -SSB/La antibodies and not anti-SSA/Ro alone [14, 17, 95]. It is the former subgroup that is more strongly associated with the linked HLA alleles B8, DR3, DRw52, DQw2 and the latter subgroup with DR2, DQw1 [185].

As has been suggested for other HLA-linked diseases, it is possible that several genes within a particular haplotype may contribute to enhanced susceptibility. Arnaiz-Villena et al. found that Class III antigens BS and C4QOB1 are increased together with the A1;B8;DR3 haplotype in SSA/Ro positive mothers whose offspring have CHB [186]. Moreover, these class II genes were not significantly increased in a group of SSA/Ro positive mothers whose offspring did not have CHB. Of particular interest was the observation that the most common DR3-bearing haplotype found in the Spanish population for adult SLE (A30, B18, DR3, BfF1, C2C, C4A3BQ0) was not increased among the mothers of offspring who had CHB. This study did not find an increased frequency of DR2 in the anti-SSA/Ro-positive mothers of healthy offspring, i.e., $DR2^+ = 3/15$ (20%) vs. 23% in their

TABLE 31.1   Neonatal Lupus in Twins

Study	Maternal serology	Zygosity	Disease manifestations	
			Sib 1	Sib 2
Gawkrodger and Beveridge [166]	SSA/Ro, SSB/La	Monozygotic	CHB/rash/anemia	Rash
Cooley et al. [167]	SSA/Ro	Monozygotic	CHB	Healthy
Cooley et al. 167]	SSA/Ro	Monozygotic	CHB	Healthy
Buyon et al. [32]	SSA/Ro, SSB/La	Monozygotic	CHB	Healthy
Buyon (RRNL unpublished)	SSA/Ro, SSB/La	Monozygotic	CHB	Healthy
Buyon (RRNL, unpublished)	SSA/Ro, SSB/La	Monozygotic	CHB	Healthy
Siren et al. [168]	SSA/Ro, SSB/La	Monozygotic, HLA Cw3	CHB	CHB
Rios et al. [169]	Unknown	Monozygotic	CHB	CHB
Barquero, Genoves et al. [170]	SSA/Ro	Monozygotic	CHB	CHB
Machado et al. [171]	SSA/Ro	Not reported	CHB	CHB
McCue et al. [172] *	+ RF, − ANA,	Not reported	CHB-death *	CHB *
Siren et al. [168]	SSA/Ro, SSB/La	Dizygotic, HLA Cw3	CHB	CHB
Silverman [173]	SSA/Ro, SSB/La	Dizygotic	CHB	Rash
Scott et al. [10]	SSA/Ro	Dizygotic	CHB	Healthy
Harley et al. [174]	SSA/Ro	Dizygotic	CHB	Healthy
Watson et al. [11]	SSA/Ro, SSB/La	Dizygotic, HLA identical	CHB	Healthy
Brucato et al. [175]	SSA/Ro	Dizygotic, HLA identical	CHB	Healthy
Kaaja et al. [176]	SSA/Ro	Dizygotic, HLA identical	CHB	Healthy
Buyon et al. [32]	SSA/Ro	Dizygotic	CHB	Healthy
Buyon et al. [32]	SSA/Ro	Dizygotic	CHB	Healthy
Eronen et al. [177]	not reported	Dizygotic	CHB	Healthy
Lockshin et al. [178]	SSA/Ro	Dizygotic	CHB	Healthy
Buyon (RRNL unpublished)	SSA/Ro, SSB/La	Dizygotic	CHB	Healthy
Buyon (RRNL unpublished)	SSA/Ro, SSB/La	Dizygotic	CHB	Healthy
Buyon (RRNL unpublished)	SSA/Ro, SSB/La	Dizygotic	CHB	Healthy
Buyon (RRNL unpublished)	SSA/Ro, SSB/La	Dizygotic	CHB	Healthy
Buyon (RRNL unpublished)	SSA/Ro, SSB/La	Dizygotic	CHB	Healthy

* Both twins had CHB recognized at birth; twin 2 had associated fibroelastosis and was asymptomatic at 1 year of age, twin 1 had l-transposition, ventricular septal defect, coarctation, and hypoplastic right ventricle, and died at 15 days of age.
Note: Sib 1 & 2 arbitrarily refer to the twin sibling but do not imply sequence of birth.

normal population. In this same study, four of five offspring with CHB were DR3-positive compared to two of five unaffected siblings [186].

Brucato *et al.* reviewed the literature and reported that of 28 mothers whose children have CHB, 50% were A1, 62% B8, and 96% DR3 [184]. In their own cohort of 15 Italian mothers of children with CHB, Brucato found a significantly increased prevalence of DR3 and DQ2 and B8/DR3, DR3/DQ2, and A1/Cw7/B8/DR3/DQ2 haplotypes [184]. In a study of 31 Finnish mothers (all anti-SSA/Ro- and/or SSB/La-positive) of children with CHB, HLA B8, and DR3 were present in 71 and 74% of the affected mothers, respectively [111]. Thus, the genetic background of CHB mothers is similar in Anglo-Saxon populations and populations from southern and northern Europe [111]. The same Finnish study demonstrated that as a group, CHB-mothers were genetically more closely related to primary SS than to SLE [111].

Perhaps fetal HLA, not as an isolated risk factor, but as it relates to maternal HLA, is a fetal factor contributing to injury. A plausible hypothesis might be that tissue damage occurs either when the HLA relationship between the fetus and mother is bi-directionally compatible or uni-directionally compatible from the child's perspective. Relevant to the hypothesis that macrophage phagocytosis of apoptotic cardiocytes opsonized by autoantibodies initiates an inflammatory cascade leading to CHB, HLA class II compatible maternal Th1 cells would be able to provide help to fetal macrophages. In support of this notion, Miyagawa and colleagues reported on a limited study on 13 Japanese families in which children with CHB shared both HLA class II alleles with their mothers significantly more often than children without CHB (4/9 CHB vs. 0/12 healthy siblings, $p < 0.02$) [187]. Furthermore, Stevens *et al.* [188] have demonstrated maternal cells in the AV node, myocardium, and liver of two infants with NL who died shortly after birth, possibly representing alloreactive hematopoietic cells that trigger inflammation leading to tissue destruction. However, the maternal cells expressed markers consistent with cardiac myocytes, suggesting the possibility that maternal cells could contribute to a secondary process of tissue repair. Overall the fact that CHB children may be enriched in genes at 6p21 highlights this region for its potential importance. In the developing heart, these genes may represent a permissive factor for tissue injury, while in the mother they may be permissive for generation of the requisite autoantibody.

A number of excellent candidates exist for further genetic characterization. There is a SNP for TGF-β at 19q3 that is associated with fibrosis (rs1982073; the variant is called Leu10). Based on data generated on families in the RRNL, the Leu[10] polymorphism of the TGF-β gene (potentially conferring increased scarring) occurs more frequently in children with CHB than their unaffected siblings [36]. Specifically, the allelic frequency of Leu[10] was 78% in 40 children with CHB, 64% in 17 children with rash, 56% in 31 unaffected children, and 65% of 74 mothers. The number of studies has been expanded and this polymorphism continues to be more significantly represented in the children with CHB compared to anti-SSA/Ro-exposed siblings without CHB.

A genome-wide association in Caucasian children with CHB enrolled in the RRNL is underway at the time of writing this chapter.

MATERNAL DISEASE AT IDENTIFICATION OF NEONATAL LUPUS AND PROGRESSION

In 1987, McCune and colleagues assessed the health status of 21 mothers and their children with NL [189]. This study suggested that all mothers eventually developed symptoms of a rheumatic disease but larger series have not reached similar conclusions [96, 190–192].

In a study from Finland reported by Julkunen *et al.* [96], 15 (48%) of 31 mothers whose children had CHB were asymptomatic before the index delivery, seven of whom remained asymptomatic after a mean follow-up of 8 years. Two mothers had SLE, one mother had definite primary SS, one had probable SS, one had autoimmune hypothyroidism, and one had Graves disease. Six (19%) gave a self-reported diagnosis of a chronic autoimmune disease antedating the index delivery. Of two mothers who died, one subsequently developed SLE and died of a fatal cardiac arrhythmia 5 years after the birth of the affected child. The other died of alcoholic liver disease and had developed SLE after 11 years. No patient with SLE had nephritis. Importantly, these authors noted that as a group, mothers of CHB children had clinical and immunologic characteristics more closely related to primary SS than SLE.

In a study of 64 CHB-mothers from Toronto, 42 (66%) were asymptomatic, two (3%) had SLE, two (3%) had linear scleroderma, two (3%) had RA, three (5%) had a history of rheumatic fever, one (2%) had SS, and 12 (2%) had undifferentiated autoimmune syndrome (UAS) [190]. Three of the 12 mothers with UAS progressed to SLE and two developed SS. Thirty-six of the 42 initially healthy mothers remained well, one developed SLE, one hyperthyroidism, one ankylosing spondylitis, and three UAS. Unlike the other larger studies, not all mothers included in this cohort were documented to have anti-SSA/Ro-SSB/La antibodies.

Maternal diagnoses and follow-up were recently reported utilizing the database of the RRNL [192]. Of the 321 mothers enrolled, follow-up of at least 6 months was available in 229. Of the 51 mothers who were asymptomatic at the NL child's birth, 26 progressed: 14 developed UAS (12-pauci, 2-poly), seven SS, four SLE, and one SLE/SS. The median time to develop any symptom was 3.15 years. Of the 37 mothers classified as pauci-UAS at the NL child's birth, 16 progressed: five developed poly-UAS, six SS, four SLE, and one SLE/SS. Of the pauci-UAS mothers enrolled within 1 year, the median time to progression was 6.7 years. Four mothers developed lupus nephritis (two asymptomatic, two pauci-UAS). The probability of an asymptomatic mother developing SLE by 10 years was 18.6%, and developing probable/definite SS was 27.9%. NL manifestations did not predict disease progression in an asymptomatic mother. Mothers with both anti-SSA/Ro and SSB/La were nearly twice as likely to develop an autoimmune disease as mothers with anti-SSA/Ro only. These data suggest that continued follow-up of asymptomatic NL mothers is warranted since nearly half progress, albeit few develop life-threatening SLE. While the anti-SSB/La antibodies may be a risk factor for progression, further work is needed to determine reliable biomarkers in otherwise healthy women with anti-SSA/Ro antibodies identified solely on the basis of an NL child.

The prevalence of hypothyroidism may be increased among women with anti-SSA/Ro antibodies, as indicated in data from a prospective study of 87 women with such antibodies, of whom nine either were hypothyroid or had a history of hypothyroidism [193]. In addition, the incidence of CHB in the offspring of the nine women with hypothyroidism was significantly higher than in those without (56% vs. 13%, respectively; odds ratio, 8.63; 95% CI, 1.63—48). In a second study utilizing the RRNL, 23 (33%) of 69 mothers of children with NL had antithyroglobulin antibodies and 15 (21.7%) had antithyroperoxidase antibodies [194]. These numbers are higher than in the general population. Eleven mothers had clinical thyroid disease (nine hypothyroidism, two hyperthyroidism), which was consistent with the earlier study. A mechanistic explanation for an association between NL and antithyroid antibodies is not apparent at this time, but evaluation of thyroid disorders in mothers who might otherwise complain of hair loss and fatigue is warranted.

TREATMENT AND PREVENTION OF CARDIAC NL

The clinical approach to cardiac manifestations of NL includes obstetric and rheumatologic management of: (1) the fetus identified with cardiac NL, and (2) the fetus with a normal heart beat but at high risk of developing cardiac NL.

Our current approach to CHB diagnosed *in utero* is presented in Figure 31.2.

It is imperative to note that these recommendations are considered experimental, based on the authors' experience, and have not been tested in a controlled trial. The side effects of treatment with dexamethasone must be weighed against the absence of established benefit.

Management of Identified Cardiac NL

To address the treatment of identified cardiac NL (herein considered manifestations discussed in the section titled "Spectrum of Cardiac Manifestations of NL", page 544) one needs to know if the presence of the cardiac abnormality ranging from a prolonged PR interval to bradycardia to overt cardiac dysfunction represents a reversible process or irreversible fibrosis and if continued autoimmune tissue injury will cause progressive damage. McCue has reported a neonate described with first-degree heart block at birth that resolved at 6 months [6]. In contrast, Geggel and colleagues reported an infant born with second-degree heart block that progressed to third-degree block by 9 weeks of age [195]. As discussed in preceding sections, data from the RRNL also emphasize that incomplete AV block is not always immutable and the degree of block is variable [35]. There is a spectrum of conduction abnormalities even at the time of initial detection and *in utero* injury can have continued sequelae in some cases despite clearance of maternal antibodies from the neonatal circulation.

The rationale for treatment of identified heart block or cardiomyopathy and prevention of potential heart block is to diminish a generalized inflammatory insult and consequent fibrosing reaction and reduce or eliminate maternal autoantibodies. Accordingly, several intra-uterine therapeutic regimens have been tried including dexamethasone (DEX), which is not metabolized by the placenta and is available to the fetus in an active form, and plasmapheresis. Although, at the time of writing, there is no documentation in the literature regarding the sustained reversal of third-degree heart block (complete fibrosis of the AV node would not be expected to reverse) by maternal treatment with dexamethasone alone, the potential for diminishing an inflammatory fetal response attacking the myocardium is plausible. This would be an effect independent of a decrease in antibody titer. Precedent for this therapeutic rationale is the resolution of associated pleuro-pericardial effusions and ascites reported in separate investigations [13, 196—200]. In the largest retrospective study published to date, it was observed that fluorinated glucocorticoids ameliorated incomplete atrioventricular block and hydropic changes in CHB but did not reverse established third-degree block [197].

GUIDELINES FOR A THERAPEUTIC APPROACH TO CHB DIAGNOSED *IN UTERO**

Situation	Treatment
1. Degree of block at presentation	
3rd degree (>1 wk from detection)	Evaluate by weekly echocardiography and obstetrical sonography; no therapy is initiated.
3rd degree (≤1 wk from detection)	4 mg p.o. dexamethasone (DEX) daily for 1-2 wks. If no change, discontinue. If reversal to 2nd degree or better, consider continuing 4mg DEX for 4-6 wks weighed against side effects, continue weekly echos.
Alternating 2nd/3rd degree 2nd degree	4 mg p.o. dex daily for 1-2 wks. If progression to 3rd degree, discontinue. If reversal to 2nd or lesser degree, consider continuing 4mg DEX for 4-6 wks weighed against side effects, continue weekly echos.
Prolonged mechanical PR interval (1st degree)	Repeat echo within 1-2 days. If still prolonged, 4 mg p.o. DEX and repeat echo in 1 wk. If reversal, might continue for 2 wks; not clear if should continue until delivery. If persistent, may continue for 2 wks but not clear that therapy makes a difference.
2. Block associated with early signs of myocarditis, congestive heart failure, and/or hydropic changes	4 mg p.o. DEX daily until improvement.
3. Severely hydropic fetus	4 mg p.o. DEX daily, with apheresis as a last resort to rapidly remove maternal antibodies, or consider terbutaline to raise fetal heart rate, deliver if lungs mature.
4. Postnatal management of affected newborn (any degree of block or myocarditis)	Pacing is indicated by the clinical situation, as is the use of anticongestive medications. Consider use of prednisone and or/IVIG.

* These guidelines are based on expert opinion and anecdotal reports. They do not represent a sole course of action.

FIGURE 31.2   Guidelines for Therapeutic approach to CHB diagnosed *in utero*. Based on expert opinion and anecdotal reports. They do not represent a sole course of action.

The results of a more recent multicenter, open-label, nonrandomized study involving 30 pregnancies treated with DEX (22 with third-degree block, six with second-degree block, two with first-degree block) and ten untreated (nine with third-degree block, one with first-degree block) have been reported [198]. There was no reversal of third-degree block with therapy or spontaneously. In fetuses treated with DEX, one in six with second-degree block progressed to third-degree block and three remained in second-degree block (postnatally one paced, two progressed to third-degree); two reverted to normal sinus rhythm (NSR; postnatally one progressed to second-degree). DEX reversed the two fetuses with first-degree block to NSR by 7 days with no regression at discontinuation. Absent DEX, the one with first-degree block detected at 38 weeks had NSR at birth (overall stability or improvement in four of eight in the DEX group versus one of one in the non-DEX group). However, there are risks of steroid therapy to both the mother (e.g., infection, hypertension, insulin resistance, and gestational diabetes) and the infant (e.g., oligohydramnios, growth restriction, and the still undetermined potential effect upon neurocognitive development).

Jaeggi *et al.* [199] reported extensive clinical data on their cohorts of prenatally diagnosed complete AV block without structural heart disease. This was a single-institution timed series. Between 1990 and 1997, CHB was diagnosed in 16 pregnancies, of which four were treated, resulting in 80% live births and 47% 1-year survival. In the 1997–2003 time period, 21 additional cases of *in utero* CHB were identified. In these cases, the investigators employed a standardized treatment approach of maternal DEX as well as β-adrenergic stimulation for fetuses with a heart rate under 55 bpm. Eighteen of these 21 pregnancies were treated, resulting in a 95% live birth and 1-year survival rate. Of the fetuses treated with DEX, there was a 90% survival rate while the untreated fetuses had only a 46% survival rate, and four of these nine untreated patients developed myocarditis, hepatitis, or cardiomyopathy resulting in death or transplantation. These authors concluded that maternal DEX as well as β-adrenergic stimulation for bradycardia was an effective treatment program.

Finally, the Meijboom group [200] reported on their experience. They present a case of progression of incomplete AV block to third-degree block despite the use of DEX. Moreover, complications of DEX in this case included intrauterine growth retardation, oligohydramnios, prolonged adrenal suppression, and late learning disabilities. Disturbed by this case experience, these investigators reviewed the literature on the use of steroids for CHB. Ninety-three cases were identified in which CHB was treated with maternal steroids. Importantly, complete block was always irreversible. Only three of the 13 cases of incomplete heart block improved. There were multiple steroid side effects. This European group therefore concluded that maternal DEX therapy as prevention or treatment for CHB is questionable at best and not recommended clinically.

Regarding potential neuropsychological impairment, Brucato *et al.* reported normal IQ testing and absence of learning disabilities or dyslexia in 11 children exposed to prolonged DEX *in utero* for treatment of CHB [201]. This same group has also described normal T-cell development and function in seven children with CHB exposed to DEX *in utero* [202]. Intervention with glucocorticoids might decrease acute inflammation but not necessarily prevent subsequent fibrosis. Finally, apheresis in addition to DEX has not reversed third-degree heart block [13], although titers of maternal anti-SSA/Ro-SSB/La have been profoundly decreased [13, 203].

Available data support serial cardiac monitoring of all fetuses with any conduction defect or signs of cardiomyopathy. Fetal echocardiogram is essential to diagnose and follow the course of disease and may suggest the presence of an associated myocarditis by the finding of decreased contractility in addition to the secondary changes associated with myocarditis such as an increase of cardiac size, pericardial effusions, and tricuspid regurgitation. The obstetric management should be guided by the degree of cardiac failure noted on the ultrasound images. The *in utero* environment is preferred as long as possible to prevent the added complication of prematurity.

It is important to emphasize that because first-degree block is clinically silent and can progress postpartum, an EKG should be obtained on all neonates of mothers with anti-SSA/Ro and/or SSB/La antibodies. For those neonates with NSR but who had any cardiac abnormalities detected *in utero* (including those that resolved with therapy), continued postnatal follow-up is recommended. It is unclear whether postnatal treatment with steroids for prevention of disease progression is needed. If there are no signs of adrenal suppression, postnatal steroids may not be indicated.

Prophylactic Approach

The initiation of dexamethasone or plasmapheresis as a preventative measure has been considered. With regard to prophylactic therapy of even the very highest-risk mother (documentation of high-titer anti-SSA/Ro and SSB/La antibodies and a previous child with NL), administration of prednisone, dexamethasone, or plasmapheresis is not justified at the present time. Maternal prednisone (at least in low and moderate doses) early in pregnancy does not prevent the development of CHB [204]. This might be anticipated since prednisone given to the mother is not active in the fetus [205] and

levels of anti-SSA/Ro and anti-SSB/La antibodies remain relatively constant during steroid therapy. However, Shinohara *et al.* reported uncontrolled data suggesting that prenatal use of maternal glucocorticoids has prophylactic merit [206].

With regard to plasmapheresis as a prophylactic therapy, Barclay and colleagues initiated plasmapheresis during the late second trimester in a woman with anti-SSA/Ro antibodies and a history of four unsuccessful pregnancies, including a 32-week stillbirth with unexplained prenatal bradycardia. The pregnancy resulted in a healthy birth and the titer of anti-SSA/Ro antibodies was decreased by 75%, although detectable antibodies were present in the cord blood [207]. We similarly utilized "prophylactic" plasmapheresis in a pregnant woman with SS and a previous child with CHB [203]. In this case, the fetus, despite having circulating levels of these antibodies detectable at birth, was exposed to only 10% of the potential maternal antibody load during the second and third trimesters. The child, now 20 years of age, never had any manifestations of NL. Because the recurrence rate is less than 20%, these cases are extremely difficult to interpret. Therefore, in the absence of controlled studies, which may never be feasible given the rarity of CHB, plasmapheresis should be considered highly experimental and only reserved for those cases in which the fetus is in a life-threatening situation with hydrops and deteriorating cardiac function.

Prenatal screening for anti-SSA/Ro and anti-SSB/La antibodies is warranted for all women with any manifestation of an autoimmune disease such as SLE, SS, UAS, or rheumatoid arthritis, or NL in a previous pregnancy. However, it is notable that women with anti-SSA/Ro antibodies can be clinically free of any disease. It is therefore unfortunate that current prenatal screening for completely healthy women does not include at least anti-SSA/Ro antibodies (anti-SSB/La antibodies absent anti-SSA/Ro antibodies is extremely rare). The detection of anti-SSA/Ro antibodies and or anti-SSB/La identifies a group of women at increased risk of having a pregnancy complicated by CHB who may be appropriate for frequent fetal echocardiographic testing. The most vulnerable period for the fetus is from 16 to 24 weeks of gestation. New onset of heart block is less likely from the 25th through the 30th weeks, and it rarely develops after 30 weeks of pregnancy.

Three prospective studies have been published since the last writing of this chapter and each is informative.

In a prospective study in the United States, 95 women with anti-Ro/SSA antibodies were studied during 98 pregnancies [46]. In this study, first-degree block was defined as a PR interval greater than 150 ms, 3 SD above normal. Ninety-two fetuses had normal PR intervals. NL developed in ten cases; four were rash only. Three fetuses had third-degree block; none had a preceding abnormal PR interval, although in two fetuses more than 1 week elapsed between echocardiographic evaluations. Tricuspid regurgitation preceded third-degree block in one fetus, and an atrial echodensity preceded block in a second. Two fetuses had PR intervals greater than 150 ms. Both were detected at or before 22 weeks, and each reversed within 1 week with 4 mg DEX. The ECG of one additional newborn revealed a prolonged PR interval persistent at 3 years despite normal intervals throughout gestation. No first-degree block developed after a normal ECG at birth.

Sonesson *et al.* reported the first prospective study in which the mechanical PR interval was used to identify early conduction disease in 24 pregnancies of mothers with anti-SSA/Ro antibodies [47]. In contrast to the low percentage of fetuses affected in the US study, one-third of the fetuses in that study had signs of a prolonged PR interval. While one explanation for this high frequency might have been the inclusion requirement for antibodies reactive against the Ro52 in all patients, resulting in an "enriched" cohort, this seroreactivity was observed in 80% of the mothers in the US study. Since information on previous pregnancies was not provided, no inference can be drawn with regard to an increased incidence of injury expected based on the recurrence rate of 20%. Perhaps most important is the definition of a prolonged PR interval that was set at 135 ms (2 SD > mean, derived from their previous studies in nearly 300 pregnancies). Reevaluation of the US study utilizing this lower threshold revealed a consistency between the two studies, with about one-third of the fetuses in the US study having a prolonged PR interval by the Sonesson criterion. However, all cases in the US study with a PR between 135 and 150 ms spontaneously reversed by the next echocardiogram. In the Sonesson study, only two fetuses had a prolonged PR interval as defined by the US criterion. One fetus had a PR approaching 150 ms at 24 weeks, which decreased to 145 ms by 26 weeks. No information regarding treatment was provided. The other fetus had second-degree block, which reversed to first-degree after treatment, but it was not clear whether there was an initial progression through first-degree prior to second-degree. The one fetus said to progress from a prolonged PR interval to third-degree block within 6 days had a PR of 140 ms. The plasticity of the PR interval prolongation was further supported by the return of all abnormal values obtained on the newborn ECG to normal values several weeks later.

In a prospective study from Israel, tissue velocity-based fetal kinetocardiogram (FKCG) was performed in 70 fetuses of 56 mothers with anti-SSA/Ro and/or SSB/La antibodies [48]. Six fetuses developed first-degree block (AV conduction time > 2 SD above mean) between 21 and 34 gestational weeks. Immediate treatment of the six fetuses with 4 mg DEX resulted in normalization of AV conduction in all within 3—14 days. There were no cases of advanced CHB.

At the time of writing, the clinical significance of isolated first-degree block (detected as defined by the different techniques) is still unknown since progression to more advanced block in the absence of treatment has not been established.

The morbidity and mortality of third-degree block suggests the need for development of a new prophylactic therapy, other than DEX, to be given early in pregnancy before the onset of disease, perhaps targeted to the highest risk pregnancies such as those with a prior affected fetus. Therapy should either be targeted to eliminating the "necessary" factor (no antibody, no disease) or modifying the inflammatory component before it provokes an irreversible scarring phenotype of the fibroblast. IgG pooled from the plasma of healthy donors (intravenous immune globulin therapy, also known as IVIG) is a promising agent that might have an effect at several levels of the proposed pathologic cascade. In a study comprised of eight pregnancies in mothers with anti-SSA/Ro antibodies and a previous child with CHB, treatment with 1 g/kg of IVIG at the 14th and 18th week of gestation prevented CHB in seven cases [208]. Although even one case of CHB is disappointing, it is less than the predicted recurrence rate of one in five. However, the dosing schedule of IVIG may not have been optimal. Arguably, initiation of IVIG at 14 weeks might be too late since it is at least 2 weeks after maternal antibody transfer has become effective and these 2 weeks might be a critical time for prevention. Furthermore, discontinuation at 18 weeks may be premature.

In addition to its efficacy in affording resistance to infections, the intravenous administration of high doses of IVIG has been of benefit in a variety of immune-mediated and inflammatory diseases and is FDA approved for primary immunodeficiency, idiopathic thrombocytopenia purpura, Kawasaki disease, B-cell chronic lymphocytic leukemia with hypogammaglobulinemia, pediatric HIV infection, and for allogenic bone marrow transplant in adults [209]. Many of the proposed effects of IVIG may be entirely nonspecific, mediated through the constant regions of IVIG. These include blocking FcRn, which would enhance autoantibody half-life, and blocking of stimulatory FcγR and stimulating inhibitory FcγR, FcγRIIB [210, 211]. Zhao *et al.* emphasized that the sialic rich IgG fraction of IVIG confers enhanced anti-inflammatory activity [212].

Accordingly, IVIG may be particularly effective in prevention of the passively acquired autoimmune disease of CHB. The rationale considers several potential mechanisms; the first two relate to lowering or even eliminating maternal antibody in the fetal circulation (maternal perspective): (1) increased catabolism of maternal antibody, and (2) decreased placental transport of maternal antibody. By decreasing antibody levels, there would be less antibody available to bind apoptotic cardiocytes. Thus, the initial cascade to injury might be abrogated. The third consideration is an effect of IVIG transported into the fetal circulation where it might act to upregulate surface expression of the inhibitory FcγRIIB receptor on fetal macrophages, thereby decreasing secretion of TNFα and exaggerated TGFβ (fetal perspective). Highly speculative would be an anti-apoptotic effect of IVIG, which would certainly be relevant to the pathogenesis of CHB, in which there is accumulating evidence that apoptosis of cardiocytes provides an essential link between antibody and fibrosis.

The first phase of the PITCH study (Preventive IVIG Therapy for Congenital Heart Block) has been recently completed (ClinicalTrials.gov identifier No. NCT00460928) [213]. This was a multicenter open-label study based on Simon's two-stage optimal design. Enrollment criteria included: maternal anti-SSA/Ro antibody, a previous child with CHB/rash, 20 mg prednisone or less, less than 12 weeks pregnant. IVIG (400 mg/kg) was given every 3 weeks from 12 to 24 weeks of gestation. The primary outcome was the development of second- or third-degree block. Twenty mothers completed the IVIG protocol before reaching the predetermined stopping rule of three cases of advanced CHB. CHB was detected at 19, 20, and 25 weeks; none followed an abnormal PR interval. One of these mothers had two previous children with CHB. One child without CHB developed a transient rash consistent with NL. Sixteen children had no manifestations of NL at birth. No significant changes in maternal antibody titers to SSA/Ro, SSB/La, or Ro52 were detected over the course of therapy or at delivery. There were no safety issues. This study established the safety of IVIG and feasibility of recruiting pregnant women with previously affected CHB children. However, IVIG at low doses consistent with replacement did not prevent the recurrence of CHB or reduce maternal antibody titers.

In parallel with the enrollment of the PITCH study, a European study was initiated in December 2004. The treatment protocol was identical to that utilized in PITCH. That study was terminated after three cases of advanced block were identified following enrollment of 15 mothers who had previously affected children with CHB [214]. Combining data generated from the two studies in which the prior pregnancy was CHB, not rash, there were six cases of recurrent advanced block in 33 mothers (15 UK and 18 US), which is consistent with the recently reported recurrence rate of 17.4% [179] and confirms that IVIG at 400 mg/kg dosed every 3 weeks from 12 gestational weeks is not effective in reducing the incidence of recurrent disease. Whether higher doses of IVIG will achieve a therapeutic effect is unknown.

Breast-Feeding

Mothers often ask about the risks of breast-feeding. Askanase and colleagues [215] have addressed whether human breast milk contains antibodies to components of the SSA/Ro-SSB/La complex and, if so, whether breast-feeding might be associated with the postnatal manifestations of NL. To accomplish these goals, breast milk from nine mothers with anti-SSA/Ro and/or SSB/La antibodies was examined by ELISA and immunoblot for the presence of these same antibodies. Five of these breast-fed infants were healthy without any manifestations of NL, one had an isolated cardiomyopathy and died, two had cutaneous manifestations of NL, and one had both CHB and rash. IgA and IgG antibodies to all components of the SSA/Ro–SSB/La complex were present in breast milk. Not unexpectedly, the antibody profiles of the breast milk paralleled those observed in the serum.

Of 237 mothers enrolled in the RRNL as of September 2000, 129 mothers answered a questionnaire regarding breast-feeding of their 266 children. NL was present in 149 of the children (55 with rash alone, 72 with CHB alone, and 22 with both manifestations) and 117 were unaffected. The frequency of breast-feeding in the mothers enrolled was slightly lower (51%) than the national average (60%) of mothers breast-feeding upon leaving the hospital [216]. This might be explained by the fact that mothers with anti-SSA/Ro-SSB/La antibodies are discouraged from breastfeeding by their physicians or in some cases might be too ill or on other medications. Overall, a total of 136 children (51%) were breast-fed. Of 55 children with rashes, 33 were breast-fed. Of 22 with both rash and CHB, 12 were breast-fed. For the unaffected siblings, 60 of 117 were breast-fed. Fisher's exact test revealed no significant differences between the breast-fed and non-breast-fed children with isolated rash, or those with associated CHB, compared to the unaffected children. Although there was a trend for the children who were breast-fed to have the cutaneous manifestations of NL appear at a later age than those who had rashes and were formula-fed, the difference in mean age of presentation of the rash did not reach statistical significance, 9.69 weeks vs. 7.55 weeks (p = NS). The duration of the rash was not influenced by breast-feeding: 14.7 weeks in the breast-fed group vs. 19 weeks in those not breastfed (p = NS). Not unexpectedly, of 72 children with isolated CHB, 31 were breast-fed, which did not differ significantly from the unaffected children. Of seven infants in whom cardiomyopathy was detected after birth, four were breast-fed. Accordingly, the available data do not suggest that breastfeeding has pathologic consequences. Specifically, children with skin rashes were not breast-fed more frequently than those who remained healthy. Furthermore, prematurity was not a factor contributing to the development of skin rash in breast-fed infants. There was a trend toward later presentation of the rash in the children who were breast-fed but there was no increase in duration compared to children who received formula. Since maternal antibodies transferred to the fetus during gestation would still be present for several months postpartum, the additional contribution of antibodies from breast milk appear to be inconsequential, even if intestinal transport is effective.

Klauninger et al. also addressed breast-feeding and the development of NL [217]. These authors investigated postnatal SSA/Ro and SSB/La IgG, IgA, and IgM antibody levels up to 1 year of age in 32 children born to SSA/Ro-positive mothers. Skin lesions developed independently of breast-feeding. The conclusion paralleled that of the RRNL study [215] and supports the recommendation that refraining from breast-feeding does not protect from skin involvement.

In summary, mothers should be advised that autoantibodies are present in their breast milk but reassured that, at least within the limits of the published literature, breast-feeding is not associated with NL. Given the potential for intestinal transport of the maternal antibodies, in the unusual circumstance of a worsening skin rash or developing cardiomyopathy, consideration should be given to discontinuation of breast-feeding.

Treatment of and Outcome of Cutaneous Manifestations

Infants with CHB should be protected from excessive sun exposure since they are at risk for developing skin lesions until 8–12 months of age. In the absence of precise information regarding specific pathogenicity of the maternal antibody response, it is reasonable to consider all offspring of mothers with antibodies to components of the SSA/Ro-SSB/La ribonucleoproteins at risk for cutaneous disease in the first few months of postnatal life. Topical steroids, preferably those that are nonfluorinated, have been recommended for babies who develop lesions [51]. High-potency topical steroids to the skin of an infant can result in systemic effects. Since the lesions are transient and generally benign, systemic therapies such as antimalarials have not been recommended [52].

Neiman and colleagues [50] reported on the therapy and outcome of 57 infants (20 males, 37 females) diagnosed with cutaneous NL (absent heart disease) between 1981 and 1997. Thirty-four (60%) were treated. Thirty-one were given only low- to medium-potency topical corticosteroid preparations. Three children were initially given topical antifungal agents and then subsequently treated with topical steroid preparations.

None received systemic glucocorticoid therapy. The active rash resolved in all children regardless of treatment. In 51 children for whom reliable follow-up data were available, 37 rashes completely resolved without sequelae, of which 21 were treated and 16 received no therapy except avoidance of sun exposure. However, in 14 there were residual skin abnormalities: ten had telangiectasias, two had hyperpigmentation of the affected areas, and ten had what was described as pitting, scarring, or atrophy after at least 2 years of follow-up. Of these 14 children, ten were treated and four untreated. Although there was no significant difference in outcome between treated and untreated children by Fisher's exact test, firm conclusions are limited by the small number of cases.

Outcome of Cardiac NL

Data published from the RRNL [32] reveal that 22 (19%) of 113 offspring with CHB whose mothers were documented to have anti-SSA/Ro and/or SSB/La antibodies have died (12 boys, ten girls). Six of these deaths occurred in utero. Ten neonates died in the first 3 months of life. Six children died between 3 months and 3 years of age. None of the 67 children between 3 and 10 years old remaining in the cohort have died. Twenty-two children are older than 10 years. However, the mortality is markedly reduced in those children born at later gestational ages. Specifically, only eight (9%) of 86 children born at or after 34 weeks have died [32]. Those infants who survive the neonatal period have an excellent prognosis. The cumulative probability of survival at 3 years is 79%.

With regard to the morbidity of CHB, 67 (63%) of the 107 children born alive have required pacemakers, 35 within the first 9 days of life. Fifteen additional children have been paced in the 1st year, and 17 after 1 year. One infant had a cardiac transplant at 8 months of age because of intractable cardiomyopathy. Moak et al. [49] have underscored the need to recognize late-onset cardiomyopathy as a sequela of CHB with their report of 16 cases that developed despite adequate pacing.

In a study of 15 CHB patients followed by Silverman et al., there were three neonatal deaths, two late deaths due to pacemaker failure, and six who have required pacemaker therapy [14]. Similarly, McCune reported a follow-up of 14 neonates with CHB of whom five required pacemakers [189].

Long-Term Follow-up of Children with Varied Manifestations of NLS

Given the rarity of the disease, little information is available on the general health outcome of children with NL and their unaffected siblings. This is of interest from several perspectives. There is a genetic susceptibility for the development of SLE [218–222], and relatives of SLE patients can have autoantibodies in the absence of clinical disease [223]. Perhaps in addition to autoantibodies, the expression of SLE in the mother increases the risk of autoimmune disease in her offspring independent of whether the child has NL or not. This generates the hypothesis that children with NL, as well as their unaffected siblings, whose mothers have SLE might be at greater risk for the development of subsequent disease than children whose mothers are asymptomatic. Perhaps vulnerability to NL is a marker for susceptibility to, or protection from, the development of actively acquired autoimmunity.

Martin and colleagues [224] reported on the health of children 8 years of age or older who had manifestations of NL and on the health of their unaffected siblings. Questionnaires were sent to mothers (with anti-SSA/Ro-SSB/La antibodies) enrolled in the RRNL, and a control group comprising children of healthy mothers referred by the RRNL enrollees. Responses to the questionnaires were confirmed and expanded by review of medical records. Fifty-five mothers enrolled in the RRNL returned questionnaires on 49 children with NL and their 45 unaffected siblings. Six children were identified with definite rheumatologic/autoimmune diseases: two with juvenile rheumatoid arthritis, one with Hashimoto's thyroiditis, one with psoriasis and iritis, one with diabetes mellitus and psoriasis, and one with congenital hypothyroidism and nephrotic syndrome. All had manifestations of NL and their mothers have manifestations of autoimmune diseases: four SS, one SLE/SS, and one UAS. In four of 55 sera tested, the ANA was positive (two of 33 affected children and two of 22 unaffected children). No serum contained antibodies reactive with SSA/Ro or SSB/La antigens. These data suggest that children with NL require continued follow-up, especially prior to adolescence and if the mother herself has an autoimmune disease. While there was no apparent increased risk of SLE, the development of some form of autoimmune disease (systemic or organ-specific) in early childhood may be of concern. During adolescence and young adulthood, individuals with NL and their unaffected siblings do not appear to have an increased risk of developing systemic rheumatic diseases but clearly longer follow-up is needed.

In one small series, Brucato et al. [175] found that at a mean follow-up of 18 years none of 13 children with CHB developed clinical symptoms or serological abnormalities suggesting immune disease.

Despite these encouraging results, there have been seven cases published in which an autoimmune disease developed in a child with manifestations of NL. All the children were female and in each instance except one, the mother had SLE. Specifically, Esscher and Scott [225] reported a female with CHB who developed SLE at age 15 years. Jackson and Gulliver [226] reported an infant with cutaneous NL who developed SLE at age 13, and Fox et al. [227] described a similar patient who developed SLE at age 19 years. In a patient described by Waterworth [228], CHB was identified at age 6 years and SLE diagnosed at age 13 years; no information was provided on the mother. Lanham et al. [229] reported two children with CHB; one subsequently developed primary SS at age 23 years, and the other arthritis, positive ANA, and antibodies to dsDNA at age 19 years. Hübscher et al. [230] reported the development of scleroderma of the face, puffy hands, and Raynaud phenomenon in a 13-year-old girl with CHB. At age 15, she was found to be seropositive for anti-SSA/Ro and U1 RNP. It should be emphasized that maternal anti-SSA/Ro-SSB/La antibodies were not actually documented for many of these mothers.

Recurrence Rates

The frequency of a second child with cardiac manifestations of NL is critical to understanding the pathogenesis of anti-SSA/Ro-mediated injury, counseling regarding future pregnancies, and powering preventative trials. The recurrence of cardiac NL was recently evaluated in the RRNL [179]. The overall rate of recurrence of cardiac NL in 161 pregnancies of 129 mothers with anti-SSA/Ro antibodies was 17.4% (95% CI, 11.1–23.6%). For evaluation of potential risk factors, data from 129 mothers with a pregnancy immediately following the birth of a cardiac NL child were analyzed. Maternal diagnosis was not associated with outcome: 23% of asymptomatic/UAS mothers had a second child with cardiac NL compared to 14% of mothers with SLE or SS ($p=0.25$). The recurrence rate was not statistically significant in mothers who used steroids compared to no steroids, (16% vs. 21%, respectively; $p=0.78$). Antibody status of the mother did not predict outcome. Death of the first child with cardiac NL did not predict recurrence in the subsequent pregnancy ($p=0.31$). The risk of cardiac NL was similar in male and female children (17.2% vs. 18.3%, respectively; $p=1.0$).

In a second recent analysis of the RRNL, the impact of cutaneous NL on the risk of cardiac disease in future pregnancies was addressed [229]. In 58 families enrolled in the RRNL, a pregnancy occurred subsequent to the child with cutaneous NL. The majority (78%) of the 58 mothers were Caucasian. Of 77 pregnancies following a child with cutaneous NL, the overall recurrence rate for any manifestation of NL was 49% (95%CI, 37–62%); 14 (18%) were complicated by cardiac NL, 23

(30%) by cutaneous NL, and one (1%) by hematologic/hepatic NL. A subset analysis was restricted to the 39 children born prospectively after the initial cutaneous NL child was enrolled in the RRNL. The overall recurrence rate for NL was 36% (95%CI, 20–52%); five (13%) had cardiac NL and nine (23%) had cutaneous NL. There were no significant differences in the following maternal risk factors for having a subsequent child with cardiac or cutaneous NL: age, race/ethnicity, anti-SSB/La status, diagnosis, use of nonfluorinated steroids, or breast-feeding. Fetal gender of the subsequent child did not influence the development of cardiac or cutaneous NL. Based on data from this large cohort, the identification of cutaneous NL in an anti-SSA/Ro-exposed infant is particularly important since it predicts a six- to tenfold risk for a subsequent child with cardiac NL.

CONCLUSIONS

Fetal exposure to maternal antibodies reactive with several components of the SSA/Ro-SSB/La ribonucleoprotein complex is a requisite but not sufficient factor for the development of NL (albeit rarely anti-RNP for cutaneous NL). Although NL is uncommon, its discussion is an integral part of all pregnancy counseling of women with SLE, SS, and UAS. Overall, studies suggest that there is unacceptable morbidity and mortality. A major clue to defining the pathogenesis of antibody-mediated damage is the selective vulnerability of the fetal heart. Vulnerability could relate to a direct or indirect antigen target differentially expressed in the developing human heart. Alternatively, but not mutually exclusively, biologic events operative during fetal life, such as apoptosis, could facilitate accessibility of intracellular antigens to the extracellular environment. Opsonization of apoptotic cardiocytes might promote an inflammatory/fibrosing cascade leading to tissue injury. A reproducible murine model of CHB would be a critical tool to define antibody pathogenicity at the histologic and molecular level with subsequent application for testing prophylactic and therapeutic interventions. The availability of a Research Registry devoted to NL provides an invaluable resource for basic and clinical research. The reporting of recurrence rates, mortality and morbidity, and maternal outcomes in a large number of patients has greatly facilitated family planning. The need for biomarkers of early cardiac injury continues, with some promise for the measurement of fetal PR intervals during pregnancy. The genetic contributions to disease may offer further insights. We have come a long way in our understanding of NL and a review of our translational knowledge to date is summarized in Table 31.2.

TABLE 31.2 Clinical Take Home Points

1. CHB[a] (absent structural abnormalities) diagnosed between 16 and 24 weeks of gestation, predicts anti-SSA/Ro antibodies in nearly all mothers regardless of any past medical history of a rheumatic disease. In general, the titer will be high and will not change. Thus, postponing pregnancy for a decline in titer is futile, as is the use of steroids to lower the titer.

2. Mothers with anti-SSA/Ro antibodies face a 2% risk of having a child with CHB if this is a first pregnancy or if previous babies have all been healthy.

3. A previous child with CHB or a previous child with rash raises the risk of having a subsequent child with CHB tenfold and fivefold, respectively.

4. The maternal heart is not affected.

5. Normal sinus rhythm can progress to complete block in 7 days, thus frequent echocardiographic monitoring of a pregnancy in a mother with anti-SSA/Ro antibodies is appropriate.

6. A mechanical PR interval of greater than 150 ms is consistent with first-degree block and warrants an immediate discussion regarding the use of a fluorinated steroid to potentially reverse the situation.

7. IVIG at 400 mg/kg does not prevent recurrent CHB.

8. Breast feeding is not contraindicated in a mother with anti-SSA/Ro antibodies.

[a]Congenital heart block (CHB) is the term used here since most of these points have been addressed with third-degree block, although as discussed in the text, cardiac NL includes a spectrum of conduction disease and isolated cardiomyopathy.

Acknowledgments

The work reported herein has been supported in part by NIH/NIAMS Contract No. NO1-AR-4-2220 (Research Registry for Neonatal Lupus), NIH Grant No. RO1 AR42455, the S.L.E. Foundation, Inc., the American Heart Association New York Chapter, the March of Dimes, the Kirkland Center at the Hospital for Special Surgery, Alliance for Lupus Research, and the many families of the RRNL that have contributed their time and donations.

References

[1] L. Morquio, Sur une maladie infantile et familiale characterisee par des modifications permanentes du pouls, des attaques syncopales et epileptiformes et al mort subite, Arch Med Inf 4 (1901) 467–475.

[2] R.K. Plant, R.A. Steven, Complete A-V block in a fetus, Am Heart J 30 (1945) 615–618

[3] M.V.L. Machado, M.J. Tynan, P.V.L. Curry, L.D. Allan, Fetal complete heart block, Brit Heart J 60 (1988) 512–515

[4] J.B. Carter, L.C. Blieden, J.E. Edwards, Congenital heart block: Anatomic correlations and review of the literature, Arch Pathol 97 (1974) 51–57

[5] R.D. Aylward, Congenital heart-block, Brit Med J (1928) 943

[6] C.M. McCue, M.E. Mantakas, J.B. Tingelstad, S. Ruddy, Congenital heart block in newborns of mothers with connective tissue disease, Circulation 56 (1977) 82–90

[7] L. Chameides, R.C. Truex, V. Vetter, W.J. Rashkind, F.M. Galioto, J.A. Noonan, Association of maternal systemic lupus erythematosus with congenital complete heart block, N Engl J Med 297 (1977) 1204–1207

[8] R.B. Winkler, A.H. Nora, J.J. Nora, Familial congenital complete heart block and maternal systemic lupus erythematosus, Circulation 56 (1977) 1103–1107.

[9] L.A. Lee, B.R. Reed, C. Harmon, Autoantibodies to SS-A/Ro in congenital heart block, Arthritis Rheum 20 (1983) S24.

[10] J.S. Scott, P.J. Maddison, P.V. Taylor, E. Esscher, O. Scott, R.P. Skinner, Connective-tissue disease, antibodies to ribonucleoprotein, and congenital heart block, New Engl J Med 309 (1983) 209–212.

[11] R.M. Watson, A.T. Lane, N.K. Barnett, W.B. Bias, F.C. Arnett, T.T. Provost, Neonatal lupus erythematosus: a clinical, serological and immunogenetic study with review of the literature, Medicine (Baltimore) 63 (1984) 362–378.

[12] R. Ramsey-Goldman, D. Hom, J.-S. Deng, G.C. Ziegler, L.E. Kahl, V.D. Steen, et al., Anti-SS-A antibodies and fetal outcome in maternal systemic lupus erythematosus, Arthritis Rheum 29 (1986) 1269–1273.

[13] J.P. Buyon, S. Swersky, H. Fox, F. Bierman, R.J. Winchester, Intrauterine therapy for presumptive fetal myocarditis with acquired heart block due to systemic lupus erythematosus: experience in a mother with a predominance of SSB/La antibodies, Arthritis Rheum 30 (1987) 44–49.

[14] E.D. Silverman, M.J. Mamula, J.A. Hardin, R.M. Laxer, The importance of the immune response to the Ro/La particle in the development of complete heart block and neonatal lupus erythematosus, J Rheumatol 18 (1991) 120–124.

[15] P.V. Taylor, K.F. Taylor, A. Norman, S. Griffiths, J.S. Scott, Prevalence of maternal Ro(SS-A) and La(SS-B) autoantibodies in relation to congenital heart block, British J Rheum 27 (1988) 128–132.

[16] J.P. Buyon, S.G. Slade, K. Elkon, E.K.L. Chan, R. Winchester, M. Lockshin, Combined assays for Ro/La identify risk of congenital heart block (CHB), Arthritis Rheum 33 (1990) S28.

[17] J.P. Buyon, E. Ben-Chetrit, S. Karp, R.A.S. Roubey, L. Pompeo, W.H. Reeves, et al., Acquired congenital heart block: pattern of maternal antibody response to biochemically defined antigens of the SSA/Ro-SSB/La system in neonatal lupus, J Clin Inv 84 (1989) 627–634.

[18] R.M. Laxer, E.A. Roberts, K.R. Gross, J.R. Britton, E. Cutz, J. Dimmick, et al., Liver disease in neonatal lupus erythematosus, J Pediatr 116 (1990) 238–242.

[19] R. Watson, J.E. Kang, M. May, M. Hudak, T. Kickler, T.T. Provost, Thrombocytopenia in the neonatal lupus syndrome, Arch Dermatol 124 (1988) 560–563.

[20] C.H. McCuistion, E.P. Schoch, Possible discoid lupus erythematosus in a newborn infant. Report of a case with subsequent development of acute systemic lupus erythematosus in the mother, Arch Dermatol Syph 70 (1954) 782–785.

[21] D.C. Kephardt, A.F. Hood, T.T. Provost, Neonatal lupus erythematosus: new serologic findings, J Invest Derm 77 (1987) 331–333.

[22] J. Buyon, I. Szer, Passively acquired autoimmunity and the maternal fetal dyad in systemic lupus erythematosus, Sem Immunopathol 9 (1986) 283–304.

[23] A. Brucato, M. Frassi, F. Franceschini, R. Cimaz, D. Faden, M.P. Pisoni, et al., Risk of congenital complete heart block in newborns of mothers with Anti-Ro/SSA antibodies detected by counterimmunoelectrophoresis. A prospective study of 100 women, Arthritis Rheum 44 (2001) 1832–1835.

[24] G. Herreman, N. Galezewski, Maternal connective tissue disease and congenital heart block, (letter) New Engl J Med 312 (1985) 1329.

[25] L.A. Lee, S. Coulter, S. Erner, H. Chu, Cardiac immunoglobulin deposition in congenital heart block associated with maternal anti-Ro antibody, Am J Med 83 (1987) 793–796.

[26] S.E. Litsey, J.A. Noonan, W.N. O'Connor, C.M. Cottrill, B. Mitchell, Maternal connective tissue disease and congenital heart block. Demonstration of immunoglobulin in cardiac tissue, New Engl J Med 312 (1985) 98–100.

[27] Y.S. Ho, E. Esscher, R.H. Anderson, M. Michaelsson, Anatomy of congenital complete heart block and relation to maternal anti-Ro antibodies, Am J Cardiol 58 (1986) 291–294.

[28] G.R. Hogg, Congenital acute lupus erythematosus associated with subendocardial fibroelastosis: report of a case, Am J Clin Pathol 28 (1957) 648–654.

[29] B.H. Singsen, J.E. Akhter, M.M. Weinstein, G.S. Sharp, Congenital complete heart block and SSA antibodies: obstetric implications, Am J Obstet Gynecol 152 (1985) 655–658.

[30] L.E. Nield, E.D. Silverman, C.P. Taylor, J.F. Smallhorn, J.B. Mullen, N.H. Silverman, et al., Maternal anti-Ro and anti-La antibody-associated endocardial fibroelastosis, Circulation 105 (2002) 843–848.

[31] L.E. Nield, E.D. Silverman, J.F. Smallhorn, G.P. Taylor, J.B. Mullen, L.N. Benson, et al., Endocardial fibroelastosis associated with maternal anti-Ro and anti-La antibodies in the absence of atrioventricular block, J Am Coll Cardiol 40 (2002) 796–802.

[32] J.P. Buyon, R. Hiebert, J. Copel, J. Craft, D. Friedman, M. Katholi, et al., Autoimmune-associated congenital heart block: demographics, mortality, morbidity and recurrence rates obtained from a national neonatal lupus registry, J Am Coll Cardiol 31 (1998) 1658–1666.

[33] C. Tseng, D. Friedman, J.P. Buyon, Spectrum of cardiac histopathology in cases of autoimmune-associated congenital heart block (CHB) obtained from the Research Registry for Neonatal Lupus [abstract], Arthritis Rheum 40 (Suppl) (1997) S333.

[34] R.M. Clancy, R.P. Kapur, Y. Molad, A.D. Askanase, J.P. Buyon, Immunohistologic evidence supports apoptosis, IgG deposition, and novel macrophage/fibroblast crosstalk in the pathologic cascade leading to congenital heart block, Arthritis Rheum 50 (2004) 173–182.

[35] A.D. Askanase, D.M. Friedman, M.R. Dische, A. Dubin, T. Starc, M.C. Katholi, et al., Spectrum and progression of conduction abnormalities in infants born to mothers with anti-SSA/Ro-SSB/La antibodies, Lupus 11 (2002) 145–151.

[36] R.M. Clancy, C.B. Backer, X. Yin, R.P. Kapur, Y. Molad, J.P. Buyon, Cytokine polymorphisms and histologic expression in autopsy studies: contribution of TNF-alpha and TGF-beta 1 to the pathogenesis of autoimmune-associated congenital heart block, J Immunol 171 (2003) 3253–3261.

[37] N.S. Rote, Maternal fetal immunology, in: J.R. Scott, N.S. Rote (Eds.), Immunology in Obstetrics and Gynecology, Appelton-Crofts, Norwalk, CT, 1985, p. 55.

[38] J.L. Leach, D.D. Sedmak, J.M. Osborne, B. Rahill, M.D. Lairmore, C.L. Anderson, Isolation from human placenta of the IgG transporter, FcRn, and localization to the syncytiotrophoblast: implications for maternal-fetal antibody transport, J Immunol 157 (1996) 3317–3322.

[39] C.T. Lin, Z. Shen, P. Boros, J.C. Unkeless, Fc receptor-mediated signal transduction, J Clin Immunol 14 (1992) 1–13.

[40] M. Wells, H. Fox, Immunology and immunopathology of the maternofetal interface, in: C.B. Coulam, W.P. Faulk, J.A. McIntyre (Eds.), Immunological Obstetrics, W.W. Norton & Company, New York, 1992, pp. 166–176.

[41] E.R. Stiehm, Fetal defense mechanisms, Am J Dis Child 129 (1975) 438–443.

[42] C. Tseng, K. Caldwell, S. Feit, E.K.L. Chan, J.P. Buyon, Subclass distribution of maternal and neonatal anti-SSA/Ro and -SSB/La antibodies in congenital heart block, J Rheum 23 (1996) 925–932.

II. CLINICAL ASPECTS OF DISEASE

[43] T.N. James, Cardiac conduction system: fetal and postnatal development, Am J Cardiol 25 (1970) 213–226.

[44] J.S. Deng, L.W. Bair Jr., S. Shen-Schwartz, R. Ramsey-Goldman, T. Medsger Jr., Localization of Ro(SS-A) antigen in the cardiac conduction system, Arthritis Rheum 30 (1987) 1232–1238.

[45] O. Hübscher, N. Batista, S. Rivero, C. Marletta, M. Arriagada, G. Boire, et al., Clinical and serological identification of 2 forms of complete heart block in children, J Rheum 22 (1995) 1352–1355.

[46] D.M. Friedman, M.Y. Kim, J.A. Copel, C. Davis, C.K. Phoon, J.S. Glickstein, et al., Utility of cardiac monitoring in fetuses at risk for congenital heart block: the PR Interval and Dexamethasone Evaluation (PRIDE) prospective study, Circulation 29 (2008) 485–493.

[47] S.E. Sonesson, S. Salomonsson, L.A. Jacobsson, K. Bremme, M. Wahren-Herlenius, Signs of first-degree heart block occur in one-third of fetuses of pregnant women with anti-SSA/Ro 52-kd antibodies, Arthritis Rheum 50 (2004) 1253–1261.

[48] A.J. Rein, D. Mevorach, Z. Perles, S. Gavri, M. Nadjari, A. Nir, et al., Early diagnosis and treatment of atrioventricular block in the fetus exposed to maternal anti-SSA/Ro-SSB/La antibodies: a prospective, observational, fetal kinetocardiogram-based study, Circulation 14 (2009) 1867–1872.

[49] J.P. Moak, K.S. Barron, T.J. Hougen, H.B. Wiles, S. Balaji, N. Sreeram, et al., Congenital heart block: development of late-onset cardiomyopathy, a previously underappreciated sequela, J Am Coll Cardiol 37 (2001) 238–242.

[50] A.R. Neiman, L.A. Lee, W.L. Weston, J.P. Buyon, Cutaneous manifestations of neonatal lupus without heart block: characteristics of mothers and children enrolled in a national registry, J Pediatr 13 (2000) 674–680.

[51] L.A. Lee, Maternal autoantibodies and pregnancy-II: The neonatal lupus syndrome, in: A.L. Parke (Ed.), Bailliere's Clinical Rheumatology, Pregnancy and the Rheumatic Diseases, Bailliere Tindall, 1990, pp. 69–84.

[52] C.M. Thornton, L.F. Eichenfield, E.A. Shinall, E. Siegfried, L.G. Rabinowitz, N.B. Esterly, et al., Cutaneous telangiectases in neonatal lupus erythematosus, J Am Acad Dermatol 33 (1995) 19–25.

[53] J.P. Buyon, Neonatal lupus, in: J.H. Klippel, P.A. Dieppe (Eds.), Rheumatology, 2nd ed., Mosby-Wolfe Publishers, London, 1998.

[54] L.A. Lee, W.L. Weston, Neonatal lupus erythematosus, Sem Dermatol 7 (1988) 66–72.

[55] J.L. Bangert, R.G. Freeman, R.D. Sontheimer, J.N. Gilliam, Subacute cutaneous lupus erythematosus and discoid lupus erythematosus: comparative histopathologic findings, Arch Dermatol 120 (1984) 332–337.

[56] K.M. David, S.D. Bennion, J.D. DeSpain, L.E. Golitz, L.A. Lee, Immunoreactants in lesions and uninvolved skin in lupus (abstract), Clin Res 37 (1989) 748A.

[57] J.R. Rosh, E.D. Silverman, G. Groisman, S. Dolgin, N.S. LeLeiko, Intrahepatic cholestasis in neonatal lupus erythematosus, J Pediatr Gastroenterol Nutr 17 (1993) 310–312.

[58] L.A. Lee, M. Reichlin, S.Z. Ruyle, W.L. Weston, Neonatal lupus liver disease, Lupus 2 (1993) 333–338.

[59] L.A. Lee, R.J. Sokol, J.P. Buyon, Hepatobiliary disease in neonatal lupus erythematosus: Prevalence and clinical characteristics in cases enrolled in a national registry, Pediatrics 109 (2002) e11.

[60] J. Schoenlebe, J.P. Buyon, B.J. Zitelli, D. Friedman, M.A. Greco, A.S. Knisely, Neonatal hemochromatosis associated with maternal autoantibodies against Ro/SS-A and La/SS-B ribonucleoproteins, Am J Dis Child 147 (1993) 1072–1075.

[61] N. Evans, K. Gaskin, Liver disease in association with neonatal lupus erythematosus, J Paed Child Health 29 (1993) 478–480.

[62] S.A. Wallace, A.M. Aron, I. Taff, Neonatal lupus involving the central nervous system, Ann Neurol 16 (1984) 399.

[63] A. Moudgil, K. Kishore, R.N. Srivastava, Neonatal lupus erythematosus, late onset hypocalcaemia, and recurrent seizures, Arch Dis Childhood 62 (1987) 736–739.

[64] H.S. Wong, M.F. Kuo, T.C. Chang, Sonographic lenticulostriate vasculopathy in infants: some associations and a hypothesis, Am J Neuroradiol 16 (1995) 97–102.

[65] J.F. Bourke, D.A. Burns, Neonatal lupus erythematosus with persistent telangiectasia and spastic paraparesis, Clinical Exp Dermatol 18 (1993) 271–273.

[66] E.M. Kaye, I.J. Butler, S. Conley, Myelopathy in neonatal and infantile lupus erythematosus, J Neurol Neurosurg Psych 50 (1987) 923–926.

[67] C.A. Boros, D. Spence, S. Blaser, E.D. Silverman, Hydrocephalus and macrocephaly: new manifestations of neonatal lupus erythematosus, Arthritis Rheum 15 (2007) 261–266.

[68] S. Kanagasegar, R. Cimaz, B.T. Kurien, A. Brucato, R.H. Scofield, Neonatal lupus manifests as isolated neutropenia and mildly abnormal liver functions, J Rheumatol 29 (2002) 187–191.

[69] B. Wolach, L. Choc, A. Pomeranz, Y. Ben Ari, D. Douer, A. Metzker, Aplastic anemia in neonatal lupus erythematosus, Am J Dis Child 147 (1993) 941–944.

[70] J.B. Harley, Autoantibodies in Sjögren's syndrome, J Autoimmun 2 (1989) 283–394.

[71] M. Reichlin, Antibodies to cytoplasmic antigens, in: R. Lahita (Ed.), Systemic Lupus Erythematosus, 2nd ed., Churchill Livingstone, London, 1992, pp. 237–246.

[72] M.J. Fritzler, J.D. Pauls, T.D. Kinsella, T.J. Bowen, Antinuclear, anticytoplasmic, and anti-Sjögren's syndrome antigen A (SS-A/Ro) antibodies in female blood donors, Clin Immunol Immunopathol 36 (1985) 120–128.

[73] C.E. Harmon, L.A. Lee, J.C. Huff, D.A. Norris, W.L. Weston, The frequency of autoantibodies to the SS-A/Ro antigen in pregnancy sera [abstract], Arthritis Rheum 27 (1984) S20.

[74] C.L. Keech, J. McCluskey, T.P. Gordon, SS-B(La) autoantibodies, in: J.B. Peter, Y. Shoenfeld (Eds.), Autoantibodies, Elsevier, Amsterdam, 1996, pp. 789–797.

[75] E. Ben-Chetrit, B.J. Gandy, E.M. Tan, K.F. Sullivan, Isolation and characterization of a cDNA clone encoding the 60-kD component of the human SS-A/Ro ribonucleoprotein autoantigen, J Clin Invest 83 (1989) 1284–1292.

[76] S.L. Deutscher, J.B. Harley, J.D. Keene, Molecular analysis of the 60kDa human Ro ribonucleoprotein, Proc Natl Acad Sci USA 85 (1988) 9479–9483.

[77] E.K.L. Chan, J.C. Hamel, C.L. Peebles, E.M. Tan, Molecular heterogeneity in SS-A/Ro autoantigens, Arthritis Rheum 33 (1990) S73.

[78] M.R. Lerner, J.A. Boyle, J.A. Hardin, J.A. Steitz, Two novel classes of small ribonucleoproteins detected by antibodies associated with lupus erythematosus, Science 211 (1981) 400–402.

[79] D. Wang, J.P. Buyon, E.K.L. Chan, Cloning and expression of mouse 60kDa ribonucleoprotein SS-A/Ro, Mol. Biol. Reports 23 (1996) 205–210.

[80] G. Fuchs, A.J. Stein, C. Fu, K.M. Reinisch, S.L. Wolin, Structural and biochemical basis for misfolded RNA recognition by the Ro autoantigen, Nat Struct Mol Biol 13 (2006) 1002–1009.

[81] C.A. O'Brien, S.L. Wolin, A possible role for the 60-kD Ro autoantigen in a discard pathway for defective 5s rRNA precursors, Genes and Development 8 (1994) 2891–2903.

[82] J.C. Chambers, D. Kenan, B.J. Martin, J.D. Keene, Genomic structure and amino acid sequence domains of the human La autoantigen, J Biol Chem 263 (1988) 18043–18051.

[83] E.K.L. Chan, A.M. Francour, E.M. Tan, Epitopes, structural domains and asymmetry of amino acid residues in SS-B/La nuclear protein, J Immunol 136 (1986) 3744—3749.

[84] E. Gottlieb, J.A. Steitz, Function of the mammalian La protein: evidence for its action in transcription termination by RNA polymerase III, EMBO J 8 (1989) 851—861.

[85] G. Boire, J. Craft, Human Ro ribonucleoprotein particles: characterization of native structure and stable association with the La polypeptide, J Clin Invest 85 (1990) 1182—1190.

[86] E. Ben-Chetrit, E.K.L. Chan, K.F. Sullivan, E.M. Tan, A 52 kD protein is a novel component of the SS-A/Ro antigenic particle, J Exp Med 162 (1988) 1560—1571.

[87] E.K.L. Chan, J.C. Hamel, J.P. Buyon, E.M. Tan, Molecular definition and sequence motifs of the 52-kD component of human SS-A/Ro autoantigen, J Clin Invest 87 (1991) 68—76.

[88] K. Itoh, Y. Itoh, M.B. Frank, Protein heterogeneity in the human Ro/SSA ribonucleoproteins, J Clin Invest 87 (1991) 177—186.

[89] K. Wada, T. Kamitani, Autoantigen Ro52 is an E3 ubiquitin ligase, Biochem Biophys Res Commun 339 (2006) 415—421.

[90] A. Espinosa, V. Dardalhon, S. Brauner, A. Ambrosi, R. Higgs, F.J. Quintana., et al., Loss of the lupus autoantigen Ro52/Trim21 induces tissue inflammation and systemic autoimmunity by disregulating the IL-23-Th17 pathway, J Exp Med 206 (2009) 1661—1671.

[91] D.P. McCauliffe, A.F. Lux, T. Lieu, I. Sanz, J. Hanke, M.M. Newkirk, et al., Molecular cloning, expression, and chromosome 19 localization of a human Ro/SS-A autoantigen, J Clin Inv 85 (1990) 1379—1391.

[92] L. Fliegel, K. Burns, D.H. MacLennan, R.A.F. Reithmeier, M. Michalak, Molecular cloning of the high affinity calcium-binding protein (calreticulin) of skeletal muscle sarcoplasmic reticulum, J Biol Chem 264 (1989) 21522—21528.

[93] R.D. Sontheimer, T.O. Nguyen, J.P. Buyon, L.A. Lee, R.P. Hall, Y.S. Yang, et al., Clinical correlations of autoantibodies to a recombinant, hYRNA-binding form of human calreticulin. [abstract], Arthritis Rheum 39 (Suppl.) (1996) S38.

[94] P. Eftekhari, L. Salle, F. Lezoualc'h, J. Mialet, M. Gastineau, J.P. Briand, et al., Anti-SSA/Ro52 autoantibodies blocking the cardiac 5-HT$_4$ serotoninergic receptor could explain neonatal lupus congenital heart block, Eur J Immunol 30 (2000) 2782—2790.

[95] J.P. Buyon, R.J. Winchester, S.G. Slade, F. Arnett, J. Copel, D. Friedman, et al., Identification of mothers at risk for congenital heart block and other neonatal lupus syndromes in their children, Arthritis Rheum 36 (1993) 1263—1273.

[96] H. Julkunen, P. Kurki, R. Kaaja, R. Heikkila, I. Ilkka, E.K.L. Chan, et al., Isolated congenital heart block: Long-term outcome of mothers and characterization of the immune response to SS-A/Ro and to SS-B/La, Arthritis Rheum 36 (1993) 1588—1598.

[97] S. Salomonsson, T. Dörner, E. Theander, K. Bremme, P. Larsson, M. Wahren-Herlenius, A serologic marker for fetal risk of congenital heart block, Arthritis Rheum 46 (2002) 1233—1241.

[98] J.P. Buyon, S.G. Slade, J.D. Reveille, J.C. Hamel, E.K. Chan, Autoantibody responses to the "native" 52-kDa SS-A/Ro protein in neonatal lupus syndromes, systemic lupus erythematosus, and Sjögren's syndrome, J Immunol 152 (1994) 3675—3684.

[99] R.M. Clancy, J.P. Buyon, K. Ikeda, K. Nozawa, D.A. Argyle, D.M. Friedman, et al., Maternal antibody responses to the 52-kd SSA/Ro p200 peptide and the development of fetal conduction defects, Arthritis Rheum 52 (2005) 3079—3086.

[100] L. Strandberg, O. Winqvist, S.E. Sonesson, S. Mohseni, S. Salomonsson, K. Bremme, et al., Antibodies to amino acid 200-239 (p200) of Ro52 as serological markers for the risk of developing congenital heart block, Clin Exp Immunol 154 (2008) 30—37.

[101] J.P. Buyon, R. Clancy, F. Di Donato, M.E. Miranda-Carus, A.D. Askanase, J. Garcia, et al., Cardiac 5-HT(4) serotoninergic receptors, 52kD SSA/Ro and autoimmune-associated congenital heart block, J Autoimmun 19 (2002) 79—86.

[102] R. Kamel, P. Eftekhari, R. Clancy, J.P. Buyon, J. Hoebeke, Autoantibodies against the serotoninergic 5-HT4 receptor and congenital heart block: a reassessment, J Autoimmun 25 (2005) 72—76.

[103] A.P. Sheth, N.B. Esterly, S.L. Ratoosh, J.P. Smith, A.A. Hebert, E. Silverman, U₁RNP positive neonatal lupus erythematosus: association with anti-La antibodies? Brit J Dermatol 132 (1995) 520—526.

[104] B.A. Solomon, T.A. Laude, A.R. Shalita, Neonatal lupus erythematosus: Discordant disease expression of U$_1$RNP-positive antibodies in fraternal twins — Is this a subset of neonatal lupus erythematosus or a new distinct syndrome? J Am Acad Dermatol 32 (1995) 858—862.

[105] J.P. Buyon, S.G. Slade, E.K.L. Chan, E.M. Tan, R.J. Winchester, Effective separation of the 52 SSA/Ro polypeptide from the 48kD polypeptide by altering conditions of gel electrophoresis, J Immunologic Methods 129 (1989) 207—210.

[106] U.K. Laemmli, Cleavage of structural proteins during the assembly of the head of bacteriophage T4, Nature 227 (1970) 680—685.

[107] T. Dörner, R. Chaoui, E. Feist, B. Göldner, K. Yamamoto, F. Hiepe, Significantly increased maternal and fetal IgG autoantibody levels to 52 kD Ro(SS-A) and La(SS-B) in complete congenital heart block, J Autoimmunity 8 (1995) 675—684.

[108] J.F. Meilof, I.M.E. Frohn-Mulder, P.A. Stewart, A. Szatmari, J. Hess, C.H.A. Veldhoven, et al., Maternal autoantibodies and congenital heart block: No evidence for the existence of a unique heart block-associated anti-Ro/SS-A autoantibody profile, Lupus 2 (1993) 239—246.

[109] E.D. Silverman, J. Buyon, R.M. Laxer, R. Hamilton, P. Bini, J.L. Chu, et al., Autoantibody response to the Ro/La particle may predict outcome in neonatal lupus erythematosus, Clin Exp Immunol 100 (1995) 499—505.

[110] E. Ben-Chetrit, R.I. Fox, E.M. Tan, Dissociation of immune responses to the SS-A/Ro 52-kD and 60-kD polypeptides in systemic lupus erythematosus and Sjögrens syndrome, Arthritis Rheum 33 (1990) 349—355.

[111] H. Julkunen, M.K. Siren, R. Kaaja, P. Kurki, C. Friman, S. Koskimies, Maternal HLA antigens and antibodies to SS-A/Ro and SS-B/La. Comparison with systemic lupus erythematosus and primary Sjögren's syndrome, Brit J Rheum 34 (1995) 901—907.

[112] Y. Itoh, M. Reichlin, Autoantibodies to the Ro/SSA antigen are conformation dependent. I. Anti-60kD antibodies are mainly directed to the native protein; anti-52kD antibodies are mainly directed to the denatured protein, Autoimmunity 14 (1992) 57—65.

[113] J.P. Buyon, Congenital complete heart block, Lupus 2 (1993) 291—295.

[114] P.J. Neidenbach, E.E. Sahn, La (SS-B)-positive neonatal lupus erythematosus: Report of a case with unusual features, J Am Acad Dermatol 29 (1993) 848—852.

[115] E.A. Stea, J.G. Routsias, R.M. Clancy, J.P. Buyon, H.M. Moutsopoulos, A.G. Tzioufas, Anti-La/SSB antiidiotypic antibodies in maternal serum: a marker of low risk for neonatal lupus in an offspring, Arthritis Rheum 54 (2006) 2228—2234.

[116] E.M. Tan, Do autoantibodies inhibit function of the cognate antigens in vivo? Arthritis Rheum 32 (1989) 924—925.

[117] P.V. Taylor, J.S. Scott, L.M. Gerlis, F.R.C. Path, E. Esscher, O. Scott, Maternal antibodies against fetal cardiac antigens in congenital complete heart block, New Engl J Med 315 (1986) 667–672.

[118] J.B. Harley, J.L. Kaine, O.F. Fox, M. Reichlin, B. Gruber, Ro(SS-A) antibody and antigen in a patient with congenital complete heart block, Arthritis Rheum 28 (1985) 1321–1325.

[119] J.P. Buyon, J. Waltuck, B. Crawford, S. Slade, J. Copel, E.K.L. Chan, Relationship between maternal and neonatal levels of antibodies to 48 kD SSB/La, 52 kD SSA/Ro and 60 kD, SSA/Ro in pregnancies complicated by congenital heart block, J Rheumatol 21 (1994) 1943–1950.

[120] D. Alarcon-Segovia, A. Ruiz-Arguelles, L. Llorente, Broken dogma: penetration of autoantibodies into living cells, Immunology Today 17 (1996) 163–164.

[121] C. Baboonian, P.J.W. Venables, J. Booth, D.G. Williams, L.M. Roffe, R.N. Maini, Virus infection induces redistribution and membrane localization of the nuclear antigen La (SS-B): a possible mechanism for autoimmunity, Clin Exp Immunol 78 (1989) 454–459.

[122] W.P. LeFeber, D.A. Norris, S.B. Ryan, J.C. Huff, L.A. Lee, M. Kubo, et al., Ultraviolet light induces binding of antibodies to selected nuclear antigens on cultured human keratinocytes, J Clin Invest 74 (1984) 1545–1551.

[123] F. Furukawa, M. Kashihara-Sawami, M.B. Lyons, D.A. Norris, Binding of antibodies to the extractable nuclear antigens of SS-A/Ro and SS-B/La is induced on the surface of human keratinocytes by ultraviolet light (UVL): implications for the pathogenesis of photosensitive cutaneous lupus, J Invest Dermatol 94 (1990) 77–85.

[124] J. Zhu, Ultraviolet B irradiation and cytomegalovirus infection synergize to induce the cell surface expression of 52-kD Ro antigen, Clin Exp Immunol 103 (1996) 47–53.

[125] T. Dörner, M. Hucko, W.J. Mayet, U. Trefzer, G. Burmester, F. Hiepe, Enhanced membrane expression of the 52kD Ro (SSA) and La (SSB) antigens by human keratinocytes induced by TNF alpha, Ann Rheum Dis 54 (1996) 904–909.

[126] M. Levitz, B.K. Young, Estrogens in pregnancy, Vitamin Hormone 35 (1977) 109–147.

[127] F. Furukawa, M.B. Lyons, L.A. Lee, S.N. Coulter, D.A. Norris, Estradiol enhances binding to cultured human keratinocytes of antibodies specific for SS-A/Ro and SS-B/La, J Immunol 141 (1988) 1480–1488.

[128] D. Wang, E.K.L. Chan, 17-β-estradiol increases expression of 52-kDa and 60-kDa SS-A/Ro autoantigens in human keratinocytes and breast cancer cell line MCF-7, J Invest Dermatol 107 (1996) 610–614.

[129] D.R. Ciocca, L.M. Vargas Roig, Estrogen receptors in human nontarget tissues: biological and clinical implications, Endocrine Reviews 16 (1995) 35–62.

[130] A.L. Lin, S.A. Shain, Estrogen-mediated cytoplasmic and nuclear distribution of rat cardiovascular estrogen receptors, Arteriosclerosis 5 (1985) 668–677.

[131] M. Reichlin, A. Brucato, M.B. Frank, P.J. Maddison, V.R. McCubbin, M. Wolfson-Reichlin, et al., Concentration of autoantibodies to native 60-kd Ro/SS-A and denatured 52-kd Ro/SS-A in eluates from the heart of a child who died with congenital complete heart block, Arthritis Rheum 37 (1994) 1698–1703.

[132] A.C. Horsfall, P.J.W. Venables, P.V. Taylor, R.N. Maini, Ro and La antigens and maternal autoantibody idiotype on the surface of myocardial fibres in congenital heart block, J Autoimmunity 4 (1991) 165–176.

[133] R. Watanabe-Fukunaga, C.L. Brannan, N.G. Copeland, N.A. Jenkins, S. Nagata, Lymphoproliferation disorder in mice explained by defects in Fas antigen that mediates apoptosis, Nature 356 (1993) 314–317.

[134] P. Bretscher, The two-signal model of lymphocyte activation twenty-one years later, Immunol Today 13 (1992) 74–76.

[135] L.A. Casciola-Rosen, G. Anhalt, A. Rosen, Autoantigens targeted in systemic lupus erythematosus are clustered in two populations of surface structures on apoptotic keratinocytes, J Exp Med 179 (1994) 1317–1330.

[136] D.S. Ucker, Death by suicide: one way to go in mammalian cellular development? New Biol 3 (1991) 103–109.

[137] J. Kajstura, W. Cheng, K. Reiss, W.A. Clark, E.H. Sonneblick, S. Krajewski, et al., Apoptotic and necrotic myocyte cell deaths are independent contributing variables of infarct size in rats, Lab Invest 74 (1996) 86–107.

[138] W. Cheng, B. Li, J. Kajstura, P. Li, M.S. Wolin, E.H. Sonneblick, et al., Stretch-induced programmed myocyte cell death, J Clin Invest 96 (1995) 2247–2259.

[139] T. Pexeider, Cell death in the morphogenesis and teratogenesis of the heart, Adv Anat Embryo Cell Bio 51 (1975) 1–100.

[140] K. Takeda, Z.X. Yu, T. Nishikawa, M. Tanaka, S. Hosoda, V.J. Ferrans, et al., Apoptosis and DNA fragmentation in the bulbus cordis of the developing rat heart, J Mol Cell Cardiol 28 (1996) 209–215.

[141] T.N. James, Normal and abnormal consequences of apoptosis in the human heart: from postnatal morphogenesis to paroxysmal arrhythmias, Circulation 90 (1994) 556–573.

[142] M.E. Miranda-Carús, A. Dinu Askanase, R.M. Clancy, F. Di Donato, T.M. Chou, M.R. Libera, et al., Anti-SSA/Ro and anti-SSB/La autoantibodies bind the surface of apoptotic fetal cardiocytes and promote secretion of tumor necrosis factor α by macrophages, J Immunol 165 (2000) 5345–5351.

[143] R.M. Clancy, P.J. Neufing, P. Zheng, M. O'Mahony, F. Nimmerjahn, T.P. Gordon, et al., Impaired clearance of apoptotic cardiocytes is linked to anti-SSA/Ro and -SSB/La antibodies in the pathogenesis of congenital heart block, J Clin Invest 116 (2006) 2413–2422.

[144] H.B. Tran, M. Ohlsson, D. Beroukas, J. Hiscock, J. Bradley, J.P. Buyon, et al., Subcellular redistribution of La(SS-B) autoantigen during physiologic apoptosis in the fetal mouse heart and conduction system: a clue to the pathogenesis of congenital heart block, Arthritis Rheum 46 (2002) 202–208.

[145] H.B. Tran, P.J. Macardle, J. Hiscock, D. Cavill, J. Bradley, J.P. Buyon, et al., Anti-La/SSB antibodies transported across the placenta bind apoptotic cells in fetal organs targeted in neonatal lupus, Arthritis Rheum 46 (2002) 1572–1579.

[146] R.M. Clancy, A.D. Askanase, R.P. Kapur, E. Chiopelas, N. Azar, M.E. Miranda-Carus, et al., Transdifferentiation of cardiac fibroblasts, a fetal factor in anti-SSA/Ro-SSB/La antibody-mediated congenital heart block, J Immunol 169 (2002) 2156–2163.

[147] J.H. Reed, P.J. Neufing, M.W. Jackson, R.M. Clancy, P.J. Macardle, J.P. Buyon, et al., Different temporal expression of immunodominant Ro60/60 kDa-SSA and La/SSB apotopes, Clin Exp Immunol 148 (2007) 153–160.

[148] J.H. Reed, B. Giannakopoulos, M.W. Jackson, S.A. Krilis, T.P. Gordon, Ro 60 functions as a receptor for beta(2)-glycoprotein I on apoptotic cells, Arthritis Rheum 60 (2009) 860–869.

[149] A. Nicholson-Weller, J. Burge, D.T. Fearon, P.F. Weller, K.F. Austen, Isolation of a human erythrocyte membrane glycoprotein with decay accelerating activity for C3 convertases of the complement system, J Immunol 129 (1982) 184–189.

[150] L.L. Ballard, N.S. Bora, G.H. Yu, J.P. Atkinson, Biochemical characterization of membrane cofactor protein of the complement system, J Immunol 141 (1988) 3923–3929.

[151] D.M. Ojicius, S. Jiang, J.D.E. Young, Restriction factors of homologous complement: a new candidate? Immunol Today 11 (1990) 47–49.

II. CLINICAL ASPECTS OF DISEASE

[152] T. Seya, J.R. Turner, J.P. Atkinson, Purification and character-ization of a membrane protein (gp 45-70) that is a cofactor for cleavage of C3b and C4b, J Exp Med 163 (1986) 837—855.

[153] T. Seya, L. Ballard, N. Bora, T. McNearney, J.P. Atkinson, Membrane cofactor protein (MCP or gp 45-70): a distinct complement regulatory protein with a wide tissue distribution, Complement 4 (1987) 225.

[154] A. Gorelick, T. Oglesby, W. Rashbaum, J. Atkinson, J.P. Buyon, Ontogeny of membrane cofactor protein: phenotypic diver-gence in the fetal heart, Lupus 4 (1995) 293—296.

[155] E. Alexander, J.P. Buyon, T.T. Provost, T. Guarnieri, Anti-Ro/SS-A antibodies in the pathophysiology of congenital heart block in neonatal lupus syndrome, an experimental model: *in vitro* electrophysiologic and immunocytochemical studies, Arthritis Rheum 35 (1992) 176—189.

[156] S. Garcia, J.H.M. Nascimento, E. Bonfa, R. Levy, S.F. Oliveira, A.V. Tavares, et al., Cellular mechanism of the conduction abnormalities induced by serum from anti-Ro/SSA-positive patients in rabbit hearts, J Clin Invest 93 (1994) 718—724.

[157] M. Boutjdir, L. Chen, Z. Zhang, C. Tseng, F. DiDonato, W. Rashbaum, et al., Arrhythmogenicity of IgG and anti-52kD SSA/Ro affinity purified antibodies from mothers of children with congenital heart block, Circ Res 80 (1997) 354—362.

[158] G.Q. Xiao, K. Hu, M. Boutjdir, Direct inhibition of expressed cardiac L- and T-type calcium channels by IgG from mothers whose children have congenital heart block, Circulation 103 (2001) 1599—1604.

[159] E. Karnabi, Y. Qu, R. Wadgaonkar, S. Mancarella, Y. Yue, M. Chahine, et al., Congenital heart block: Identification of autoantibody binding site on the extracellular loop (domain I, S5-S6) of alpha(1D) L-type Ca channel, J Autoimmun 34 (2009) 80—86.

[160] L. Mahony, Development of myocardial structure and function, in: G.C. Emmanouilides, T.A. Riemenschneider, H.D. Allen, H.P. Gutgesell (Eds.), 5th ed., Moss and Adams Heart Disease in Infants, Children, and Adolescents, Including the Fetus and Young Adults, Volume I, Williams and Wilkins, Baltimore, 1995, pp. 17—28.

[161] I.R. Josephson, N. Sperelakis, Initiation and propagation of the cardiac action potential, in: N. Sperelakis, R.O. Banks (Eds.), Physiology, Little, Brown and Co, Boston, 1993, pp. 251—269.

[162] J.A. Mazel, N. El-Sherif, J. Buyon, M. Boutjdir, Electrocardio-graphic abnormalities in a murine model injected with IgG from mothers of children with congenital heart block, Circu-lation 99 (1999) 1914—1918.

[163] F. Kalush, E. Rimon, A. Keller, E. Mozes, Neonatal lupus erythematosus with cardiac involvement in offspring of mothers with experimental systemic lupus erythematosus, J Clin Immunol 14 (1994) 314—322.

[164] M.E. Miranda-Carús, M. Boutjdir, C.E. Tseng, F. DiDonato, E.K.L. Chan, J.P. Buyon, Induction of antibodies reactive with SSA/Ro-SSB/La and development of congenital heart block in a murine model, J Immunol 161 (1998) 5886—5892.

[165] S. Salomonsson, S.E. Sonesson, L. Ottosson, S. Muhallab, T. Olsson, M. Sunnerhagen, et al., Ro/SSA autoantibodies directly bind cardiomyocytes, disturb calcium homeostasis, and mediate congenital heart block, J Exp Med 201 (2005) 11—17.

[166] D.J. Gawkrodger, G.W. Beveridge, Neonatal lupus eryth-ematosus in four successive siblings born to a mother with discoid lupus erythematosus, British J Dermatol 111 (1984) 683—687.

[167] H.M. Cooley, C.L. Keech, B.J. Melny, S. Menahem, G. Morahan, T.W. Kay, Monozygotic twins discordant for congenital complete heart block, Arthritis Rheum 40 (1997) 381—384.

[168] M.K. Siren, H. Julkunen, R. Kaaja, H. Ekblad, S. Koskimies, Role of HLA in congenital heart block: susceptibility alleles in children, Lupus 8 (1999) 60—67.

[169] B. Rios, J. Duff, J.W. Simpson, Endocardial fibroelastosis with congenital complete heart block in identical twins, Am Heart J 107 (1984) 1290—1293.

[170] I.M. Barquero Genoves, J. Figueras Aloy, M. Guerola Serret, R. Jimenez Gonzalez, A complete congenital heart block in twins, An Esp Pediatr 44 (1996) 164—166.

[171] M.V.L. Machado, M.J. Tynan, P.V.L. Curry, L.D. Allan, Fetal complete heart block, Brit Heart J 60 (1988) 512—515.

[172] C.M. McCue, M.E. Mantakas, J.B. Tingelstad, Congenital heart block in newborns of mothers with connective tissue disease, Circulation 56 (1977) 82—90.

[173] E.D. Silverman, Congenital heart block and neonatal lupus erythematosus: prevention is the goal, J Rheumatol 20 (1993) 1101—1104.

[174] J.B. Harley, E.L. Alexander, W.B. Bias, O.F. Fox, T.T. Provost, M. Reichlin, et al., Anti-Ro (SS-A) and anti-La (SS-B) in patients with Sjögren's syndrome, Arthritis Rheum 29 (1986) 196—206.

[175] A. Brucato, M. Gasparini, G. Vignati, S. Riccobono, E. De Juli, M. Quinzanini, et al., Isolated congenital heart block: longterm outcome of children and immunogenetic study, J Rheum 22 (1995) 541—543.

[176] R. Kaaja, H. Julkunen, P. Ammala, P. Kurki, S. Koskimies, Congenital heart block in one of the HLA identical twins, Eur J Obstet Gynecol Reprod Biol 51 (1993) 78—80.

[177] M. Eronen, M.K. Siren, H. Ekblad, T. Tikanoja, H. Julkunen, T. Paavilainen, Short- and long-term outcome of children with congenital complete heart block diagnosed *in utero* or as a newborn, Pediatrics 106 (1 Pt 1) (2000) 86—91.

[178] M.D. Lockshin, E. Bonfa, K. Elkon, M.L. Druzin., Neonatal risk to newborns of mothers with systemic lupus erythematosus, Arthritis Rheum 31 (1988) 697—701.

[179] C. Llanos, P. Izmirly, M. Katholi, R.M. Clancy, D.M. Friedman, M.Y. Kim, et al., Recurrence rates of cardiac manifestations associated with neonatal lupus and maternal/fetal risk factors, Arthritis Rheum 60 (2009) 3091—3097.

[180] M.R. Arbuckle, M.T. McClain, M.V. Rubertone, R.H. Scofield, G.J. Dennis, J.A. James, et al., Development of autoantibodies before the clinical onset of systemic lupus erythematosus, N Engl J Med 349 (2003) 1526—1533.

[181] G. Colombo, A. Brucato, E. Coluccio, S. Compasso, C. Luzzana, F. Franceschini, et al., DNA typing of maternal HLA in congenital complete heart block: comparison with systemic lupus erythematosus and primary Sjogren's syndrome, Arthritis Rheum 42 (1999) 1757—1764.

[182] S. Miyagawa, K. Shinohara, K. Kidoguchi, T. Fujita, T. Fukumoto, Y. Yamashina, et al., Neonatal lupus eryth-ematosus: HLA-DR and -DQ distributions are different among the groups of anti-Ro/SSA-positive mothers with different neonatal outcomes, J Invest Dermatol 108 (6) (1997) 881—885.

[183] L.A. Lee, W.B. Bias, F.C. Arnett, C. Huff, D.A. Norris, C. Harmon, et al., Immunogenetics of the neonatal lupus syndrome, Ann Intern Med 99 (1983) 592—596.

[184] A. Brucato, F. Franceschini, M. Gasparini, E. De Juli, G. Ferraro, M. Quinzanini, et al., Isolated congenital complete heart block: longterm outcome of mothers, maternal antibody specificity and immunogenetic background, J Rheum 22 (1995) 533—540.

[185] R.G. Hamilton, J. Harley, W. Bias, M. Roebber, M. Reichlin, M. Hochberg, et al., Two Ro(SS-A) autoantibody responses in SLE. Correlation of HLA-DR/DQ specificities with quantita-tive expression of Ro(SSA) autoantibody, Arthritis Rheum 31 (1988) 496—505.

[186] A. Arnaiz-Villena, J.J. Vazquez-Rodriguez, J.L. Vicario, P. Lavilla, D. Pascual, F. Moreno, et al., Congenital heart block immnogenetics: Evidence of an additional role of HLA class III antigens and independence of Ro autoantibodies, Arthritis Rheum 32 (1989) 1421—1426.

[187] S. Miyagawa, T. Fukumoto, K. Hashimoto, A. Yoshioka, T. Shirai, K. Shinohara, et al., Neonatal lupus erythematosus: haplotypic analysis of HLA class II alleles in child/mother pairs, Arthritis Rheum 40 (1997) 982—983.

[188] A.M. Stevens, H.M. Hermes, J.C. Rutledge, J.P. Buyon, J.L. Nelson, Myocardial-tissue-specific phenotype of maternal microchimerism in neonatal lupus congenital heart block, Lancet 15 (2003) 1617—1623.

[189] A.B. McCune, W.L. Weston, L.A. Lee, Maternal and fetal outcome in neonatal lupus erythematosus, Ann Intern Med 106 (1987) 518—523.

[190] J. Press, Y. Uziel, R.M. Laxer, L. Luy, R.M. Hamilton, E.D. Silverman, Long-term outcome of mothers of children with complete congenital heart block, Am J Med 100 (1996) 328—332.

[191] S. Lawrence, L. Luy, R. Laxer, B. Krafchik, E. Silverman, The health of mothers with cutaneous neonatal lupus erythematosus differs from that of mothers with congenital heart block, Am J Med 15 (2000) 705—709.

[192] T.L. Rivera, P.M. Izmirly, B.K. Birnbaum, P. Byrne, J.B. Brauth, M. Katholi, et al., Disease progression in mothers of children enrolled in the Research Registry for Neonatal Lupus, Ann Rheum Dis 68 (2009) 828—835.

[193] D. Spence, R.L. Hornberger, R. Hamilton, E.D. Silverman, Increased risk of complete congenital heart block in infants born to women with hypothyroidism and anti-Ro and/or anti-La antibodies, J Rheumatol 33 (2006) 167—170.

[194] A.D. Askanase, I. Iloh, J.P. Buyon, Hypothyroidism and antithyroglobulin and antithyroperoxidase antibodies in the pathogenesis of autoimmune associated congenital heart block, J Rheumatol 33 (2006) 2099.

[195] R.L. Geggel, L. Tucker, I. Szer, Postnatal progression from second- to third-degree heart block in neonatal lupus syndrome, J Pediatr 113 (1988) 1049—1052.

[196] F.Z. Bierman, L. Baxi, I. Jaffe, J. Driscoll, Fetal hydrops and congenital complete heart block: response to maternal steroid, J Pediatr 112 (1988) 646—648.

[197] S. Saleeb, J. Copel, D. Friedman, J.P. Buyon, Comparison of treatment with fluorinated glucocorticoids to natural history of autoantibody-associated congenital heart block: Retrospective review of the Research Registry for Neonatal Lupus, Arthritis Rheum 42 (1999) 2335—2345.

[198] D.M. Friedman, M.Y. Kim, J.A. Copel, C. Llanos, C. Davis, J.P. Buyon, Prospective evaluation of fetuses with autoimmune-associated congenital heart block followed in the PR Interval and Dexamethasone Evaluation (PRIDE) Study, Am J Cardiol 103 (2009) 1102—1106.

[199] E.T. Jaeggi, J.C. Fouron, E.D. Silverman, G. Ryan, J. Smallhorn, L.K. Hornberger, Transplacental fetal treatment improves the outcome of prenatally diagnosed complete atrioventricular block without structural heart disease, Circulation 110 (2004) 1542—1548.

[200] J.M. Breur, G.H. Visser, A.A. Kruize, P. Stoutenbeek, E.J. Meijboom, Treatment of fetal heart block with maternal steroid therapy: case report and review of the literature, Ultrasound Obstet Gynecol 24 (2004) 467—472.

[201] A. Brucato, M.G. Astori, R. Cimaz, P. Villa, M. Li Destri, L. Chimini, et al., Normal neuropsychological development in children with congenital complete heart block who may or may not be exposed to high-dose dexamethasone in utero, Ann Rheum Dis 65 (2006) 1422—1426.

[202] P. Airo', M. Scarsi, A. Brucato, T. Benicchi, F. Malacarne, I. Cavazzana, et al., Characterization of T-cell population in children with prolonged fetal exposure to dexamethasone for anti-Ro/SS-A antibodies associated congenital heart block, Lupus 15 (2006) 553—561.

[203] J. Buyon, R. Roubey, S. Swersky, L. Pompeo, A. Parke, L. Baxi, et al., Complete congenital heart block: risk of occurrence and therapeutic approach to prevention, J Rheum 15 (1988) 1104—1108.

[204] J. Waltuck, J.P. Buyon, Autoantibody-associated congenital heart block: outcome in mothers and children, Ann Intern Med 120 (1994) 544—551.

[205] A.T. Blanford, B.E. Pearson Murphy, In vitro metabolism of prednisolone, dexamethasone, betamethasone, and cortisol by the human placenta, Am J Obstet Gynecol 127 (1977) 264—267.

[206] K. Shinohara, S. Miyagawa, T. Fujita, T. Aono, K. Kidoguchi, Neonatal lupus erythematosus: results of maternal corticosteroid therapy, Obstet Gynecol 93 (1999) 952—957.

[207] C.S. Barclay, M.A.H. French, L.D. Ross, R.J. Sokol, Successful pregnancy following steroid therapy and plasma exchange in a woman with anti-Ro (SS-A) antibodies. Case report, Br J Obstet Gynecol 94 (1987) 369—371.

[208] R. Kaaja, H. Julkunen, Prevention of recurrence of congenital heart block with intravenous immunoglobulin and corticosteroid therapy: comment on the editorial by Buyon et al, Arthritis Rheum 50 (2004) 280—281.

[209] R.J. Looney, J. Huggins, Use of intravenous immunoglobulin (IVIG), Best Practice Res Clin Haematol 18 (2006) 3—25.

[210] N. Li, M. Zhao, J. Hilario-Vargas, P. Prisayanh, S. Warren, L.A. Diaz, et al., Complete FcRn dependence for intravenous Ig therapy in autoimmune skin blistering diseases, J Clin Invest 115 (2005) 3440—3450.

[211] A. Samuelsson, T.L. Towers, J.V. Ravetch, Anti-inflammatory activity of IVIG mediated through the inhibitory Fc receptor, Science 291 (2001) 445—446.

[212] R.M. Anthony, F. Wermeling, M.C. Karlsson, J.V. Ravetch, Identification of a receptor required for the anti-inflammatory activity of IVIG, Proc Natl Acad Sci USA 50 (2008) 19571—19578.

[213] D.M. Friedman, C. Llanos, P.M. Izmirly, B. Brock, J. Byron, J. Copel, et al., Evaluation of fetuses in a study of intravenous immunoglobulin as preventive therapy for congenital heart block, Arthritis Rheum (2010) 1138—1146. 62n.

[214] C.N. Pisoni, A. Brucato, A. Ruffatti, G. Espinosa, R. Cervera, M. Belmonte-Serrano, et al., Failure of intravenous immunoglobulin to prevent congenital heart block: Findings of a multicenter, prospective, observational study, Arthritis Rheum 62 (4) (2010) 1147—1152.

[215] A.D. Askanase, M.E. Miranda-Carus, X. Tang, M. Katholi, J.P. Buyon, The presence of IgG antibodies reactive with components of the SSA/Ro-SSB/La complex in human breast milk: Implications in neonatal lupus, Arthritis Rheum 46 (2002) 269—271.

[216] A.S. Ryan, The resurgence of breastfeeding in the United States, Pediatrics 99 (1997) E12.

[217] R. Klauninger, A. Skog, L. Horvath, O. Winqvist, A. Edner, K. Bremme, et al., Serologic follow-up of children born to mothers with Ro/SSA autoantibodies, Lupus 18 (2009) 792—798.

[218] M. Hochberg, P. Florsheim, J. Scott, F. Arnett, Familial aggregation of systemic lupus erythematosus, Am J Epidemiol 122 (1985) 526—527.

[219] J.S. Lawrence, L. Martins, G. Drake, A family survey of lupus erythematosus: 1, Heritability J Rheumatol 14 (1987) 913—921.

[220] S.R. Block, J.B. Winfield, M.D. Lockshin, W.A. D'Angelo, C.L. Christian, Studies of twins with systemic lupus erythematosus: A review of the literature and presentation of 12 additional sets, Am J Med 59 (1975) 533–552.

[221] F.C. Arnett, L.E. Shulman, Studies in familial systemic lupus erythematosus, Medicine 55 (1976) 313–322.

[222] D. Deapen, A. Escalante, L. Weinrib, D. Horwitz, B. Bachman, P. Roy-Burman, et al., A revised estimate of twin concordance in systemic lupus erythematosus, Arthritis Rheum 35 (1992) 311–318.

[223] Y. Shoenfeld, H. Slor, S. Shafrir, I. Krause, J. Granados, G.M. Villarreal, et al., Diversity and pattern of inheritance of autoantibodies in families with multiple cases of systemic lupus erythematosus, Ann Rheum Dis 51 (1992) 611–618.

[224] V. Martin, L.A. Lee, A.D. Askanase, M. Katholi, J.P. Buyon, Long term follow-up of children with neonatal lupus and their unaffected siblings, Arthritis Rheum 46 (2002) 2377–2383.

[225] E. Esscher, J.S. Scott, Congenital heart block and maternal systemic lupus erythematosus, Brit Med J 1 (1979) 1235–1238.

[226] R. Jackson, M. Gulliver, Neonatal lupus erythematosus progressing into systemic lupus erythematosus. A 15 year follow-up, Br J Dermatol 101 (1979) 81–86.

[227] R.J. Fox, C.H. McCuistion, E.P. Schoch Jr., Systemic lupus erythematosus. Association with previous neonatal lupus erythematosus [abstract], Arch Dermatol 115 (1979) 340.

[228] R.F. Waterworth, Systemic lupus erythematosus occurring with congenital complete heart block, N Z Med J 92 (1980) 311–312.

[229] J.G. Lanham, M.J. Walport, G.R. Hughes, Congenital heart block and familial connective disease, J Rheumatol 10 (1983) 823–825.

[230] O. Hübscher, D. Carrillo, M. Reichlin, Congenital heart block and subsequent connective tissue disorder in adolescence, Lupus 6 (1997) 283–284.

[231] P.M. Izmirly, C. Llanos, L.A. Lee, A. Askanase, M.Y. Kim, J.P. Buyon, Cutaneous manifestations of neonatal lupus and risk for subsequent congenital heart block, Arthritis Rheum 2010 (62) (2010) 1153–1157.

SLE in Children

Rina Mina, Hermine I. Brunner

INTRODUCTION

The terms childhood-onset SLE (cSLE), pediatric SLE, and juvenile-onset SLE are used interchangeably in the medical literature to refer to SLE with diagnosis during childhood or adolescence. In this chapter we will refer to this subset of SLE patients as having cSLE. An estimated 10–20% of patients experience an onset of SLE prior to adulthood. More precise estimates are difficult due to a lack of a clear age limit up to which a patient can be diagnosed with cSLE. Upper limits of age at diagnosis range from 14 to 20 years, with most studies defining cSLE with diagnosis prior to age 17–18 years. This chapter focuses on the clinical presentation, clinical features and their treatment, laboratory features, genetic background, special management issues, and prognosis of cSLE, with emphasis on the differences between cSLE and SLE diagnosed during adulthood (aSLE).

The American College of Rheumatology (ACR) Classification Criteria for SLE were developed to help standardize research in aSLE. Previous studies in cSLE also employed these criteria, supporting the notion that the ACR Criteria can be applied to cSLE. Nonetheless, there are only a few studies that formally tested the validity of the ACR Criteria when used in cSLE. The most reliable data come from a study of 204 Brazilian children with rheumatic diseases [1], where the 1982 ACR Criteria were 96% sensitive and 100% specific for the diagnosis of cSLE. The cSLE group (n=103) fulfilled a median of six classification criteria compared to only one criterion in the control group of children with other rheumatic diseases [1]. cSLE is most commonly diagnosed between the ages of 12 and 14 years and occurs rarely in children younger than 5 years of age, as is supported by infrequent case reports of toddlers with cSLE [2].

The female preponderance with cSLE is less pronounced than that with aSLE. The female to male ratio in cSLE changes from 4:3 in the first decade of life, to 4:1 in the second decade to 9:1 in aSLE and decreases to 5:1 in late onset aSLE with diagnosis at age 50 years or older. A recent study in 135 children with cSLE suggests that the overall female to male ratio in cSLE is about 4–4.5:1 [3].

PRESENTATION AT DIAGNOSIS WITH cSLE

Nonspecific and variable signs and symptoms at the time of the initial clinical presentation contribute to the delay in the diagnosis with cSLE of, on average, 10 months [4]. The most common physical findings reported at the time of diagnosis with cSLE are malar rash, lymphadenopathy, and renal disease [5, 6]. At diagnosis, constitutional symptoms are exceedingly common, with fever present in 70–90%, weight loss in 35%, and fatigue in 50% of the cSLE patients [5, 6]. Fever is reported in 41–86%, and weight loss in 31–51% of the patients in some aSLE cohorts [7]. A small case series comparing clinical features at disease onset between patients with prepubertal versus postpubertal onset of cSLE supports that hematological and renal involvement are more common in the former, while cutaneous and musculoskeletal features are more characteristic for the latter group [2, 8].

About one-third of the children and adolescents with cSLE present with anemia, thrombocytopenia, or leukopenia [9, 10]. Lymphopenia is present in up to 59% of cSLE patients at the time of diagnosis [9, 10], and hepatomegaly is noted in 25–40% of cSLE patients at presentation. Based on the findings of a single study approximately one-third of the cSLE patients test positive for anti-Smith, anti-ribonuclear protein (RNP) and anti-Ro antibodies, but only 13% have elevated levels of anti-La antibodies at the time of diagnosis [6, 11]. For reasons not completely understood, but especially when kidney involvement is present, newly diagnosed cSLE patients exhibit a distinct pattern of dyslipoproteinemia with increased triglyceride and depressed HDL-C levels [12].

There are several studies directly comparing the initial clinical presentation of cSLE with that of aSLE, suggesting that patients with cSLE have a more severe disease onset with a higher frequency of renal, neurologic, and hematologic manifestations than patients with aSLE [13]. More acute disease onset of cSLE compared to aSLE, as measured by the SLEDAI, has been reported in a study comparing an inception cohort of cSLE patients ($n=67$; SLEDAI score mean $\pm$ standard deviation (SD):16.8$\pm$10.1) with an inception cohort of 131 aSLE patients (SLEDAI score mean $\pm$ SD: 9.3$\pm$7.3) [14].

MUCOCUTANEOUS MANIFESTATIONS

cSLE-related skin changes are reported in 50–80% at presentation and in up to 85% of the children during the course of the disease [15, 16]. This compares to reports of 40–60% in aSLE [5, 11, 15]. The typical malar rash is seen in about 60–80% of children with cSLE, which is somewhat more frequent than with aSLE [5, 15, 16]. Some children develop petechial, purpuric, urticarial, or even blistering lesions, suggesting cutaneous vasculitis. Palmar erythema is commonly observed as is livedo reticularis, particularly among patients who test positive for antiphospholipid (aPL) antibodies (Figure 32.1).

Up to 53% of cSLE patients, especially if Caucasian, are photosensitive [15]. This SLE feature is associated with positivity for anti-Ro and anti-La antibodies in both children and adults. Children with complement deficiencies, notably C1q deficiency, often present with treatment-resistant skin rashes.

Isolated discoid skin lesions in the absence of other systemic features of cSLE, i.e., childhood discoid lupus erythematosus (cDLE), are very rare. Children under the age of 10 years and adolescents younger than 15 years account for less than 3% and 5% of all cases of DLE,

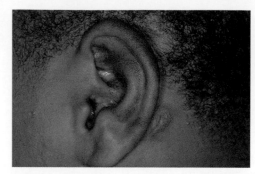

FIGURE 32.2 Discoid lupus on the pinna and posterior auricular area of a child with systemic lupus erythematosus.

respectively [17]. cDLE affects both sexes equally [18], and a positive family history of DLE or SLE is more common with cDLE than DLE diagnosed in adults (11% vs. 4%). Twenty-five to 30% of children with cDLE will progress to cSLE over time, which compares to only 5–10% of adults [18]. Risk factors for the progression of cDLE to cSLE are antinuclear antibody (ANA) positivity and the presence of disseminated (as opposed to localized) discoid lesions [18, 19] (Figure 32.2).

Alopecia occurs in 40–50% of the children with SLE and appears to be similarly common in aSLE [11, 15, 16]. A Korean cohort study of adults and children with SLE ($n=86$) suggests that nonscarring diffuse alopecia is the most common type of hair loss with aSLE and cSLE (65% of all patients with alopecia), followed by nonscarring patchy alopecia (15%) [20]. Mucosal lesions in cSLE are frequently asymmetrically distributed in the oral cavity (palate, buccal, mucosa, tongue) and variably painful (Figure 32.3).

Treatment of mucocutaneous lesions and photosensitivity in cSLE include topical and/or systemic glucocorticoids, antimalarial drugs, dapsone, topical calcineurin inhibitors, azathioprine, and mycophenolate mofetil.

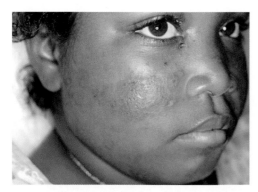

FIGURE 32.1 Malar erythema and skin ulceration at the epicanthus in a child with systemic lupus erythematosus.

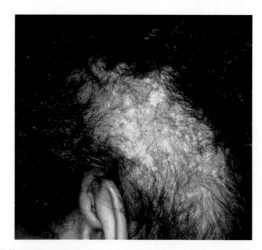

FIGURE 32.3 Alopecia in a child with systemic lupus erythematosus

Photoprotection and avoidance of sun exposure, including tanning beds, are mandatory.

RAYNAUD PHENOMENON

Raynaud phenomenon occurs in 12–39% of cSLE patients, is similarly frequent with aSLE [5, 11, 16], and may precede the diagnosis of cSLE for many years. In a recent study only 3.6% of 250 children with Raynaud phenomenon were eventually diagnosed with cSLE after a median follow-up of 2 years (range, 1–11 years) [21]. The presence of ANA and nailfold capillary abnormalities were both positively associated with the subsequent development of a connective tissue disease [21].

Cold avoidance and warm multilayered clothing are the first-line therapy for patients with Raynaud phenomenon. Vasodilators, usually calcium channel blockers or topical nitroprusside, are prescribed for moderate to severe disease. Iloprost, a prostacyclin analog, and sildenafil, a selective inhibitor of phosphodiesterase type 5, have been used successfully in cSLE patients for the treatment of ischemic ulcers that are part of a severe Raynaud phenomenon [22, 23].

MUSCULOSKELETAL MANIFESTATIONS

Direct comparisons of cSLE and aSLE cohorts suggest arthralgias to be more common in aSLE than cSLE, as are Jaccoud arthropathy and drug-induced myopathy, while arthritis is equally common [5, 11, 15, 24]. Arthritis is usually migratory, symmetric, rarely erosive, and affects both large and small joints. Joint effusions are usually mild to moderate, and arthropathy is associated with various degrees of pain. Nonsteroidal anti-inflammatory drugs (NSAIDs) provide useful control of arthritis and musculoskeletal manifestations.

Myalgia is reported in 20–30% of cSLE patients, while true myositis is rare and often occurs in cSLE patients who also have features of systemic vasculitis. Drug-induced myopathy from statins, antimalarials, and corticosteroids, albeit uncommon in children, must be considered in cSLE patients who have a poor response of myopathy to anti-inflammatory therapy [25], as well as noninflammatory musculoskeletal pain that is part of pain amplification syndrome.

Avascular bone necrosis may occur in as many as 10–40% of the cSLE patients [26], with large weightbearing bones being most commonly affected. At a reported point prevalence of 4–20%, avascular bone necrosis appears to be less common in aSLE [27]. Although often asymptomatic initially, pain often progresses over time and then may necessitate arthroplasty. Efforts are made to postpone joint replacement until final adult height is reached, as joint replacement raises special issues for the growing child. Activity restriction for cSLE patients with avascular bone necrosis is typically self-regulated. Avascular bone necrosis in cSLE has been associated with corticosteroid use and inconsistently with hypercoagulable states, but may also be due to the underlying disease process [28].

The prevalence of osteopenia is at 40% and osteoporosis at 5% among premenopausal women with aSLE, and considerably higher estimates are reported from cohorts that included postmenopausal patients [24, 29, 30]. Osteoporosis, defined as a bone mineral density Z-score of less than −2.5, is reported in up to 20% of cSLE patients [31, 32], and osteoporotic fractures occur in 6–10% of them [29, 31, 33, 34].

Children and adolescents with cSLE often are unable to reach their peak bone mass due to inflammation, corticosteroid use, inadequate calcium and vitamin D supplementation, reduced physical activity, avoidance of sun exposure, and in rare cases, premature ovarian failure [29, 31, 33, 34]. The estimated annual decline of bone mass with cSLE is at 3.4% and compares to a predicted increase in bone mass of 8% among age-matched healthy children [34]. The urinary concentrations of the bone turnover markers, deoxypyridinoline and collagen cross-links, are somewhat higher, while bone formation markers are similar, in cSLE as compared to healthy controls [33]. Based on small studies, the relative bone loss in cSLE is greatest in mid to late adolescence [33–35]. Long disease duration and the cumulative corticosteroid dose are both associated with low bone mass in cSLE [32], while global measures of disease activity appear unrelated [36].

As in adults, hypovitaminosis D appears to be common in cSLE. Severe vitamin D deficiency, defined as 25-hydroxyvitamin D levels under 10 ng/ml by 125 I-labeled radioimmunoassay in one study, is more common in children with cSLE than healthy controls (36.8% vs. 9.2%; $p < 0.001$) [37]. Optimal levels of vitamin D, to avoid osteoporosis in cSLE, have not been established.

Calcium plus vitamin D supplementation is often used in an effort to improve the bone mass in cSLE. Conversely, treatment of children with bisphosphonates remains controversial due to insufficient long-term safety data. Sclerotic metaphyseal bands, paralleling the growth plates, are seen on radiographs of some children treated with bisphosphonates [38], leading to concerns of altered bone remodeling and undertubulation. Animal studies suggest unfavorable effects of bisphosphonates on the fetal skeleton, a concern for female adolescents and children who are expected to become pregnant later in life.

Conversely, no major side effects of bisphosphonate therapy were observed in a controlled study of children with various rheumatic and systemic diseases, including six with cSLE requiring corticosteroid therapy. Alendronate improved the volumetric bone mineral density at the lumbar spine compared to children who did not receive a bisphosphonate [39].

Currently, data are inadequate to justify the use of bisphosphonates in children with asymptomatic reduction of bone mass. Until there is more evidence of the safety of these drugs in children, we suggest bisphosphonate use be restricted to children with recurrent fractures and/or symptomatic vertebral collapse due to insufficient bone mass.

HEMATOLOGIC MANIFESTATIONS

Anemia of chronic disease is the most common cause of anemia in both cSLE and aSLE. There is a higher incidence of anemia in the very young, with 77% of patients in infantile SLE, i.e., SLE with onset prior to age 1 year, as compared to only 35% in children older than 1 year at onset of cSLE [2]. Anemia in cSLE is usually mild to moderate, normochromic and normocytic but microcytic and hypochromic if chronic [40]. Rarely, anemia is the result of autoantibodies against erythrocytes or stem cells.

A positive Coombs test is present in up to 49% of cSLE patients [9], and hemolytic anemia may be more common in cSLE than aSLE [11, 15, 16]. The Latin American Group for the Study of Lupus (GLADEL) cohort reports a frequency of hemolytic anemia in 16% of 230 children with cSLE [15], and an even higher prevalence is reported for other racial cSLE cohorts [11, 15].

As with aSLE, hypersplenism, gastrointestinal or genitourinary blood losses, drug-sensitivity, hemoglobinopathy, and microangiopathy must all be considered in a cSLE patient who develops or presents with anemia. Rarely, hemolytic uremic syndrome or thrombotic thrombocytopenic purpura (TTP) is part of the initial presentation of cSLE [41, 42] ANA positivity and high-grade proteinuria at the time of presentation are both risk factors for children with TTP to subsequently develop cSLE [43]. There are also case reports of co-existent cSLE and sickle cell disease which led to a delay in the diagnosis of cSLE due to overlapping symptoms [44].

Somewhat more frequent than with aSLE, leukopenia is reported in 45—65% and lymphopenia in 59—68% of the cSLE patients during the disease course. Both are thought to be multifactorial, e.g., due to medication side effects, infections, cytokine-induced apoptosis, autoantibody-induced destruction, and suppression of progenitor cell growth. Leukopenia may be less frequent with infantile SLE as compared to cSLE with onset at or after puberty [2].

Neutropenia has been associated with thrombocytopenia and neuropsychiatric manifestations in cSLE and aSLE as well as an increased risk of infections [45]. Lymphopenia is present in up to 59% of cSLE patients at the time of diagnosis [9, 10], changes with disease activity is correlated with anti-double stranded DNA antibody levels and associated with neuropsychiatric and mucocutaneous involvement in both cSLE and aSLE [46].

Thrombocytopenia occurs in about two-thirds of children with infantile SLE and in 25% of cSLE patients with disease onset later in life [2]. Overall, the prevalence of thrombocytopenia appears to be somewhat higher in cSLE than aSLE [11, 15]. Low platelet counts are associated with the presence of antibodies to aPL and platelets. Impaired thrombopoiesis, microangiopathy, hemophagocytic syndrome, and drug side effects must all be considered as possible causes of cSLE-associated thrombocytopenia.

Children with thrombocytopenia may have carried a diagnosis of idiopathic thrombocytopenic purpura (ITP) prior to developing cSLE. Progression to cSLE was rare in a retrospective study of 365 Turkish children with idiopathic ITP. ANA titers of at least 1:80 were present in 9% of the children but none developed cSLE during the mean follow-up of 3.6 years [47]. ANA positivity was associated with a chronic course of ITP [48]. This low progression of ANA-positive children to cSLE in Turkey may not be representative for other parts of the world. In our experience, careful follow-up of any child with ITP who tests high-titer-positive for ANA appears warranted. Corticosteroids, intravenous immunoglobulins, azathioprine, dapsone, vincristine, mycophenolate mofetil, cyclophosphamide, and cyclosporine have all been used to treat cSLE-associated thrombocytopenia [49, 50]. Successful therapy of chronic treatment-resistant hemolytic anemia and thrombocytopenia in cSLE with rituximab is reported from a single center; no serious infections were observed, despite prolonged B-cell depletion [51].

Bone marrow findings in aSLE are variable and reflect the diverse causes of cytopenias. There is very limited literature on bone marrow changes with cSLE. In a case report of a 12-year-old girl whose cSLE presentation included pancytopenia, serositis, neuropsychiatric features, and nephritis, bone marrow biopsy revealed polynuclear erythrocytes and pseudo-Pelger-Huet anomaly in the myeloid lineage [52], i.e., pathological features similar to those with aSLE.

Macrophage activation syndrome, or reactive hemophagocytosis, can be the initial manifestation of cSLE [53] and is better recognized since the description of virus-associated hemophagocytosis in a series of patients which included one with SLE by Risdall et al. [54]. Histological features with cSLE are indistinguishable from those seen with etiologies [53]. Pulse methylprednisolone remains the cornerstone of therapy. For steroid resistant cases, intravenous immunoglobulins and cyclosporine are recommended, although other medications have been used successfully, including cyclophosphamide, tumor necrosis blockers antagonists, and plasma exchange.

NEUROPSYCHIATRIC MANIFESTATIONS

Neuropsychiatric involvement with cSLE (cNPSLE) occurs at least as frequently as with aSLE [55, 56], with a reported point prevalence in various cohort studies of 20–95% [40, 56, 57]. The wide range in the reported frequency of cNPSLE may be explained by differences in study designs and definitions of cNPSLE employed, but intrinsic variation due to race or ethnicity cannot be excluded. In 70% of the children who will develop cNPSLE, symptoms commence within the 1st year after diagnosis [58].

Rare features of cNPSLE include cranial neuropathy, optic neuritis, glaucoma, transverse myelitis, and parkinsonism. Conversely, headache is the most common feature of cNPSLE and may be associated with antiphospholipid (aPL) antibody positivity [59]. The most common mood disorder with cSLE is depression [60, 61] and, if moderate to severe, has a pronounced negative impact on school performance [62]. Besides cSLE itself, reactive depression and depression secondary to corticosteroid therapy must be considered.

Approximately 20% of children with cNPSLE will develop psychosis, which usually presents with visual hallucinations [56, 63]. Cerebrovascular disease is reported in up to 25% of children with cNPSLE, thus more frequently than with aSLE [55]. Cerebral vein thrombosis is reported in 15–25% of cSLE patients, often presenting with severe headache and lupus anticoagulant (LAC) positivity [64, 65].

Chorea occurs more commonly in children than adults with SLE and is present in approximately 5% of children with cNPSLE, virtually all of them test positive for LAC [57]. cNPSLE-associated chorea is often unilateral, single episode, and more common with cSLE than aSLE [57, 66].

Generalized seizures, as opposed to focal, have been reported in 4–20% of the children with cNPSLE [67] and are often observed in conjunction with cerebrovascular disease and cognitive dysfunction [67]. Studies directly comparing cSLE with aSLE suggest that seizures are more common in children [68, 69]. Seizures may be a risk factor for permanent neurological damage. Hypertensive encephalopathy due to severe renal involvement may manifest as seizures and must be considered in a child with cSLE presenting with seizures.

Neurocognitive dysfunction is reported in as many as 30–60% of patients with cSLE [70]. This wide range of numeric estimates reported in the literature is likely due to differences in design and case ascertainment between studies. Nonetheless, neurocognitive dysfunction appears similarly frequent in cSLE as with aSLE [6, 70–72]. The 1999 ACR case definitions of NPSLE have not been validated for use in children and adolescents, and the proposed 1-h ACR battery of standardized tests to assess neuropsychiatric function is not suited for use in pediatrics [73]. More recently an alternative battery for children has been developed [74] (Table 32.1).

Maturity for voluntary control of inappropriate behavior, working memory, and processing speed becomes evident between ages 14 and 19 years [70], while cognitive development is not completed until early adulthood. Studies suggest that cNPSLE is associated with impairment in complex problem solving, working memory, verbal memory, attention, and visuomotor integration. The diagnosis of neurocognitive dysfunction in cSLE should be made by neuropsychological testing. A single-center study suggests that the Pediatric Automated Neuropsychological Assessment Metrics (Ped-ANAM) has concurrent validity with formal neuropsychological testing in cSLE and may be a useful screening tool for cSLE-associated neurocognitive dysfunction [75]. The Ped-ANAM is the pediatric adaptation of a computerized testing battery for adults, does not require prior computer skills of the participant and can be completed in less than 45 min.

Although longer duration of cSLE has been associated with lower cognitive ability, long-term outcome studies of cNPSLE are lacking [70]. Neurocognitive dysfunction in aSLE is associated with the presence of aPL, antineuronal, and anti-N-methyl-D-aspartate (NMDA) receptor (anti-NR2) antibodies as well as cerebrovascular disease [76, 77]. Inconsistently, these associations have been reported in cSLE as well [78].

In the absence of a specific laboratory test, establishing a diagnosis of cNPSLE continues to be a challenge. Infections, metabolic aberration, malignancy, and drug side effects all must be considered in this setting. Cerebrospinal fluid (CSF) evaluation is mandatory to exclude an infectious etiology of neuropsychiatric features. Once infection has been excluded, the presence of oligoclonal bands, increased protein or pleocytosis on CSF analysis, or abnormal opening pressure on lumbar puncture may be supportive of cNPSLE association with various autoantibodies has been reported, but their diagnostic utility remains to be determined [67]. Based on current data, antiribosomal-P antibodies in the blood or CSF are neither sensitive nor specific for cNPSLE-associated psychosis [79]. As with aSLE, neither a normal ESR nor absence of specific autoantibodies can exclude cNPSLE.

TABLE 32.1   Childhood Arthritis and Rheumatology Research Alliance (CARRA) NPSLE Standardized Battery for Children with cSLE

Name of measure	Duration of test	Cognitive domain tested
Wechsler Abbreviated Scales of Intelligence-2 Subtest	20 min	General intelligence
Wechsler Intelligence Scales subtests: Coding and Symbol Search	10 min	Psychomotor speed
Wechsler Intelligence Scales subtests: Digit Span and Letter-Number Sequencing	20 min	Verbal working memory
Conners Continuous Performance Task — II (CPT-II)	15 min	Attention, processing speed
Woodcock-Johnson III Tests of Achievement subtests: Letter-Word Identification, Reading Fluency, Calculation, Math Fluency	30 min	Academic skills mastery
Wide Range Assessment of Memory and Learning-2 (WRAML-2) Screening subtests: Story Memory, Verbal Learning, Picture Memory, Design Memory	30 min	Verbal and visual memory
Delis-Kaplan Executive Function System (D-KEFS) Color-Word Interference Test	10 min	Executive function: attention, cognitive flexibility, response inhibition
Child Behavior Checklist — Parent Form	20 min	Behavior Problem Scales

Unless there is the suspicion of a cerebrovascular event, magnetic resonance imaging (MRI) is considered the imaging modality of choice for the evaluation of cNPSLE. Abnormalities on traditional MRI are reported in as many as 30% of the patients with cNPSLE [62] and may include changes reflecting prior vascular or demyelinating events. The most common MRI abnormality observed is a nonspecific reduction of brain volume, which may represent loss of neurons. Single photon emission computed tomography, fluoro-2-deoxy-D-glucose positron emission tomography, MR spectroscopy, and electroencephalography have all been used in the diagnostic work-up of cNPSLE. To date, none of these modalities provides cNPSLE-specific information that would unequivocally allow the attribution of a given clinical presentation to cSLE as opposed to other etiologies [80, 81].

Children and adolescents with cNPSLE appear to have an excellent response to treatment, and the majority of them will experience a resolution of symptoms. As with NPSLE in adults, high-dose glucocorticoids, often combined with immunosuppressive medications including cyclophosphamide, are the mainstay of cNPSLE treatment [57, 82, 83]. For cNPSLE-associated cerebral vein thrombosis or arterial strokes anticoagulation is added to the anti-inflammatory drug regimen [84]. There are no longitudinal randomized studies addressing target parameters for anticoagulant therapy or antithrombotic prophylaxis in cNPSLE. Academic delays of children and adolescents with cNPSLE have not been quantified but are likely considerable. Educational interventions and psychological support appear important for patients with cNPSLE, based on studies in aSLE.

LUPUS NEPHRITIS

Despite large variations between racial groups, most studies suggest lupus nephritis (LN) to be present in 50—60% of the patients with cSLE [9, 15, 85]. This compares to a reported prevalence of LN in aSLE at about 20—47% [28, 68, 86, 87]. LN is often diagnosed early in the disease course, with most children presenting with proteinuria and/or persistent microscopic hematuria. In this setting, interstitial cystitis with active cSLE must be excluded. Pyuria in the absence of urinary tract infection can be an early manifestation of LN [88]. Acute renal failure as an isolated initial manifestation of cSLE or nephrotic syndrome preceding cSLE is uncommon [89]. Severe proteinuria, pronounced hematuria, and low serum albumin at disease presentation are all associated with more severe renal involvement with cSLE [90]. An acute onset of massive proteinuria in cSLE necessitates a work-up for renal vein thrombosis. Hypertension occurs in 40% of cSLE patients with LN [28, 68, 87, 91], and African-American children, particularly boys, may have a significantly higher risk of hypertension than Caucasians [92].

Irrespective of the mode of presentation, an initial kidney biopsy is necessary to establish the diagnosis and, to a certain extent, the prognosis of LN. Similar to what has been described in aSLE, diffuse proliferative glomerulonephritis (Class IV) is reported in 40—60%, focal proliferative glomerulonephritis (Class III) in 10–20%, and membranous nephritis (Class V) in 3–28% of cSLE patients at presentation, while the other classes make up 3–15%. Overlap between proliferative and membranous changes have been reported in 12% (4% Classes III and V, 8% Classes IV and V) by Marks *et al.* [93].

The value of quantifying features of LN activity and chronicity in cSLE to predict LN long-term outcomes have not been firmly established [85]. Sequential biopsies in children with LN often show progression of Classes II and III to Class IV or even Class V [94], but improvement of active features of LN over time can occur [95]. In one study, nonimprovement on sequential kidney biopsies predicted the risk of future renal flares and progression of LN in cSLE after cyclophosphamide therapy [96].

Clinical monitoring of LN in cSLE warrants a careful examination of the urinary sediment for cellular casts. The traditional urine collection over a 24-h period is often difficult, thus inaccurate for quantifying proteinuria in pediatric populations. Calculation of the protein to creatinine ratio in a random spot urine sample correlates well with the 24-h urine protein excretion and now is commonly used for the assessment of cSLE-associated proteinuria in clinical practice [97]. Random spot urine may, however, not adequately capture pronounced proteinuria as can be present with nephrotic flares.

In recent years, there has been much interest in isolating and validating biomarkers to predict renal flares in cSLE and aSLE. While several novel renal biomarkers have been discovered, none of them are sufficiently validated to allow its use in clinical care. Among the more promising biomarkers are urinary neutrophil gelatinase associate lipocalin (NGAL) that was demonstrated to predict LN flares in cSLE [98]. Other potential biomarkers of LN flares are melanoma-associated antigen gene-B2 (MAGE-B2), chemokine ligands 1, 2, and 5, transferrin, ceruloplasmin, lipocalin-like prostaglandin synthetase-D, and orosomucoid [99–101].

Children with Classes III and IV have more severe renal disease and worse prognosis compared to patients with Classes I, II, or V [94]. As with aSLE, treatment of cSLE patients with LN must include good blood pressure control, as this may improve renal outcome and prevent complications [92]. Combination antihypertensive therapy is often needed and should, whenever possible, include an angiotensin-converting enzyme inhibitor and/or angiotensin II type 1 receptor blocker [102]. Angiotensin-converting enzyme inhibitors were associated with a delay in development of renal involvement and decreased disease activity in the LUMINA cohort [103].

The best induction and maintenance therapy for cSLE-associated LN remains to be determined. In children with severe LN cyclophosphamide use is associated with better renal survival compared to corticosteroid therapy alone [104]. Lehman *et al.* treated 16 patients with Class III or Class IV LN with monthly intravenous pulse cyclophosphamide for 6 months, followed by three monthly infusions for a total of 36 months. A significant reduction of proteinuria, disease activity, and prednisone dosage at 1 year after initial diagnosis was reported as well as improvement of LN on re-biopsy. No significant treatment complications were reported by this group during the 3-year study [95]. The combination therapy of cyclophosphamide and methotrexate for refractory Class IV LN has been suggested but is rarely used in clinical practice, given its potentially severe side effects [105].

Azathioprine is used mostly for the maintenance therapy of LN, but promising results of azathioprine use for induction therapy in cSLE have been reported in two cohort studies [94, 106]. No statistically significant differences in serum creatinine, time to renal flare, overall renal survival, disease activity, disease damage, mean annual corticosteroid dose, and rate of infection were observed in cSLE patients with Classes III and IV LN treated with azathioprine compared to cyclophosphamide during a mean follow-up of about 3 years [106].

At present, mycophenolate mofetil is used primarily for the maintenance therapy of LN in cSLE. The efficacy and safety of mycophenolate mofetil in controlling the LN in children has been reported. In a case series, mycophenolate mofetil was effective in five of 13 patients (38%), partially effective in four (31%), and ineffective in four (31%); no severe side effects were observed [107]. As in aSLE, several questions remain regarding the use of mycophenolate mofetil in cSLE, such as durability of response and optimal dosing. There is initial evidence in cSLE that weight-adjusted dosing does not predict exposure to active drug or the response of LN well [108].

The combination therapy of cyclophosphamide and methotrexate for refractory Class IV therapy in children with SLE has been suggested but is rarely used in pediatric or adult clinical practice, given its potentially severe side effects [105].

The benefits of B-cell depletion with rituximab have been studied in 18 children with Classes III, IV, and V LN [109]. Global disease activity, renal function, and proteinuria improved in virtually all patients during the average follow-up time of 3 years. B-cell depletion persisted for up to 1 year in some children, and one patient died of infectious endocarditis related to severe immunosuppression [109].

Although it may be beneficial for the treatment of Class V LN [110], cyclosporine is used rarely for the therapy of other types of LN in children, as neither the creatinine clearance improves significantly nor is relapse prevented [111]. Therapies proposed for Class V include glucocorticoids combined with azathioprine and calcineurin inhibitors [112]. A recent study in 15 French children with Class V LN reports complete remission in five patients, partial response in nine, and chronic renal insufficiency in one after combination treatment of corticosteroids and immunosuppressive medications [113]. However, the efficacy of mycophenolate mofetil in Class V LN remains to be confirmed, and cyclophosphamide therapy appears not to result in a significant improvement of Class V based on a case series of 26 cSLE patients [114].

Overall kidney survival in cSLE has improved in recent years [115], and end-stage renal disease is less common than in aSLE. Reports on the prognosis of LN in cSLE vary between studies and may reflect differences in patient population and/or adherence to medical regimens. While short-term renal outcomes between Black and Caucasians appear to be similar [113], long-term outcomes are worse among Black cSLE patients [94]. Furthermore, LN is significantly more common in non-Caucasian than in Caucasian cSLE patients (62% vs. 45%; $p = 0.01$) [116]. Other risk factors for poor outcome are the presence of Class IV LN, tubular atrophy, a high degree of chronicity on kidney biopsy, persistent hypertension or anemia, possibly nephrotic syndrome at presentation, and delay in implementing immunosuppressive therapy [117–119].

Some studies suggest that proliferative LN in cSLE progresses in 9–15% of the cases to end-stage renal disease within 5 years [93, 94], which is comparable to what has been reported for some aSLE cohorts [120]. LN accounts for 3% of end-stage renal disease leading to kidney transplantations among children in North America [121], while adults with SLE account for 1.9% of the adult kidney transplant population, based on the US Renal Data System. A review of the North American Renal Transplant Cooperative Study database supports 1-year graft survival rates with cSLE at 91% after living donor and 78% after cadaveric transplants. These rates were not significantly different from kidney graft survivals in other pediatric patients [122]. Overall, cSLE patient survival was 89% at 3 years and again equivalent to that of the control group. Death in the cSLE group was primarily due to infection [122].

CARDIOVASCULAR MANIFESTATIONS

Cardiac involvement with cSLE is reported in 12–54% of the children, similarly frequent as with aSLE [123]. Pericarditis occurs in as many as 25% of cSLE patients who have any cardiac involvement [124] and is the most common cardiac feature with SLE.

Pericarditis may be associated with anti-Ro, anti-La, anti-Sm, and anti-RNP antibodies [123, 125].

Myocarditis is difficult to ascertain in cSLE as it is often asymptomatic [123]. Recent palpitations were reported by four out of five cSLE patients with silent cardiac abnormalities on electrocardiogram or echocardiogram in one study [126]. Global hypokinesia and low left ventricular ejection fraction are echocardiographic findings supportive of myocarditis. Cardiac failure with cSLE is usually the sequelae of co-morbid conditions, involving the kidneys or the lung, rather than due to intrinsic cardiac disease. Cardiac dysfunction with cSLE may be underrecognized, as is supported by a case series of 27 asymptomatic cSLE patients; although exercise responses and echocardiography were normal in all patients, five children (20%) had a marginally low left ventricular ejection fraction at rest by MUGA (multigated acquisition) scan [127]. Differences in ejection fraction, fractional shortening, peak early diastolic filling velocity, and deceleration between healthy children and those with cSLE have been reported, suggesting asymptomatic systolic and diastolic dysfunction with cSLE [128, 129]. The significance of the above findings for cSLE at large will require study in larger longitudinal cohorts [130].

As in aSLE, valvular abnormalities in cSLE include Liebman-Sacks vegetations, stenotic valves, and regurgitant valves. The exact frequency of valvular insufficiency is unknown but a case series reported three of 19 (16%) asymptomatic children with cSLE to have valvular changes [126].

Sinus arrhythmia is the most common arrhythmia in cSLE but atrial premature complexes, supraventricular tachycardia, and right bundle branch block can also occur [124].

Coronary arteritis is uncommon in both aSLE and cSLE. Both symptomatic coronary heart disease and myocardial infarction are rare in cSLE before adulthood but there are several reports of myocardial infarction in young adults with cSLE [131, 132]. Evidence of early subclinical atherosclerosis is present in adolescents with cSLE [126]. In a prospective study of 31 asymptomatic cSLE patients, four children had (16%) coronary perfusion defects as measured by thallium myocardial perfusion scans, radionuclide angiography with MUGA, or echocardiography. All coronary changes were reversible except for one large fixed perfusion defect [127]. Higher carotid intima—media thickness, an indicator of atherosclerosis used in clinical studies, and increased carotid arterial stiffness were observed in cSLE patients when compared to healthy age- and sex-matched controls [133, 134]. Carotid arterial stiffness may be an independent predictor of left ventricular mass in cSLE [128]. Atherosclerosis in aSLE is associated with traditional and nontraditional risk factors.

Increased body mass index, male sex, increasing age, nephrotic range proteinuria, high disease activity, and azathioprine use have all been associated with higher carotid intima–media thickness in cSLE [133, 134]. The exact pro-atherogenic effects of azathioprine are unknown at this time but may include elevated plasma homocysteine levels. Evidence of the pro-atherogenic consequences of corticosteroid use in cSLE is inconsistent. A prospective study of 221 cSLE patients suggests that prednisone at moderate doses (defined as prednisone dosage 0.15–0.40 mg/kg/day) is associated with a lower carotid intima–media thickness than with either higher or lower doses, possibly reflecting inhibition of inflammation that is associated with premature atherosclerosis. An ongoing multicenter randomized controlled trial aims to evaluate the use of statins in preventing premature atherosclerosis in cSLE [135].

ENDOCRINE MANIFESTATIONS

Dyslipidemia occurs in 50–85% of aSLE and cSLE patients [136, 137] and appears to be multifactorial. As with aSLE, children and adolescents with cSLE have higher insulin levels compared to controls which, if persistent, may lead to metabolic syndrome [138–140]. Other risk factors of dyslipidemia to be considered are inflammation, renal disease, diet, inactivity, and drug side effects. Adolescents with cSLE have lower aerobic fitness than age-matched peers [141]. Newly diagnosed cSLE patients have elevated levels of triglycerides and depressed high-density lipoprotein (HDL) levels, especially with high disease activity and when LN is present [12].

Despite ample evidence of dyslipidemia in cSLE, there is no universally accepted therapeutic guideline. Fish oil supplementation and dietary modifications have been studied prospectively in 11 cSLE patients, resulting in a decrease in triglycerides but not of total cholesterol, HDL or LDL [142]. Given the limited evidence of its benefits, fish oil supplementation is not routinely recommended. Hydroxychloroquine use is associated with lower total serum cholesterol and decreased thrombotic risk in aSLE. There are no specific studies on the cholesterol-lowering effects of hydroxychloroquine in cSLE.

Neuroendocrine abnormalities have been recognized in both cSLE and aSLE. Autoimmune thyroid disorders are associated with cSLE and are present in about 15–20% of the patients, a frequency similar to that seen in aSLE [143, 144]. Thyroid antibodies, antimicrosomal and antithyroglobulin antibodies, are present in 14–43% of cSLE patients, often leading to hypothyroidism. Acute suppurative thyroiditis or hyperthyroidism occurs rarely in cSLE.

There is a trend toward higher levels of follicle stimulating hormone (FSH), luteinizing hormone, and prolactin in cSLE patients compared to healthy children [145]. High prolactin levels are described in a subset of cSLE patients with active disease and those with NPSLE [146]. cSLE patients with antiprolactin autoantibodies have significantly higher levels of serum prolactin than those without antiprolactin autoantibodies (41.9 vs. 17.8 ng/ml; $p = 0.01$) [147].

Levels of estrogen of both prepubertal and postpubertal cSLE patients are normal and similar to those of children with juvenile idiopathic arthritis [145]. Transient or permanent amenorrhea is reported in 12% of 298 adolescents with cSLE and is positively correlated with both the presence of disease activity and damage [146]. The risk of overt ovarian failure after cyclophosphamide therapy is considerably lower in cSLE than in aSLE [148]. Based on limited information from case series, the average risk of premature ovarian failure is 11% in females with cSLE who are younger than 21 years old [148, 149]. In a cohort of 77 patients with cSLE, of whom 47% were treated with cyclophosphamide, a reduced ovarian reserve, but not overt ovarian failure, was observed in 31% of the females who were treated with cyclophosphamide [148]. Ovarian protection for cSLE has not been studied, but a randomized trial to study the usefulness of gonadotropin-releasing hormone agonists is ongoing [148].

There is no generally accepted approach to contraception in adolescents with cSLE. In varying reports, worsening SLE disease activity has been reported with oral contraceptive use, but likely, based on adult studies, low estrogen oral contraceptives are safe for use in adolescents with cSLE who test negative for aPL antibodies. Using evidence again from studies in aSLE, adolescents testing positive for aPL antibodies should not receive contraceptives containing estrogen or third-generation progestins. Second-generation progesterone only contraceptives (oral, injection, or intrauterine devices) are preferred by most pediatric rheumatologists but have the potential to further decrease bone mineral density.

While the use of the levonorgestrel intrauterine device (LNG-IUD) is well established in the healthy adult population, and despite common off-label use of this IUD in some countries, there are no data on the safety and efficacy when used in females with SLE. In a nationwide survey of 179 mostly nulliparous adolescents in New Zealand using the LNG-IUD, the spontaneous expulsion rate was 8%. There were few infections or other side effects [150]. None of the adolescents in this study were treated with immunosuppressive therapy but three had a bleeding disorder.

As with aSLE, semen abnormalities, low testicular volumes, and high gonadotropin levels appear to be more frequent in males with cSLE than their healthy peers, especially following cyclophosphamide therapy with initiation after the onset of puberty [151, 152]. Sertoli cell dysfunction occurs more often in cSLE males than in healthy adolescents. Low levels of inhibin B and high levels of FSH have been associated with oligozoospermia in cSLE.

GASTROINTESTINAL MANIFESTATIONS

Gastrointestinal manifestations are present in about 20% of children with cSLE and are reported in 8–40% of aSLE patients [153–155]. The most common gastrointestinal symptom is abdominal pain, reported in 87% of those with cSLE-associated GI involvement [153]. Adults with SLE are often diagnosed with age- or medication-related abdominal pathology [156], while in children symptoms are more frequently due to SLE itself [153]. The frequency of ascites in cSLE is comparable to that in aSLE [153]. Distinguishing surgical from nonsurgical cases of acute abdomen is challenging, especially in children with cSLE. Acute onset of even mild abdominal pain and low-grade fever in an otherwise well-controlled patient on immunosuppressive medications may be due to bacterial peritonitis. Abnormalities on ultrasound are present in about half of the children with cSLE with abdominal pathology, while abnormalities on computer tomography are seen in 80% of the cases [153].

Although it can be a presenting feature [157], cSLE-associated hepatitis is usually mild, and cirrhosis is particularly uncommon. Antiribosomal-P antibodies may be a risk factor for aSLE and cSLE-associated hepatic involvement [158, 159]. Transaminitis has been reported in 20–25% of patients with both cSLE and aSLE. Besides cSLE-associated hepatitis, other causes of transaminitis in a child with cSLE include medication side effects, including steatohepatitis, liver infarction due to vasculitis, and hepatic congestion due to fluid overload.

Overlap of cSLE and autoimmune hepatitis has been reported [159]. In these studies, hepatitis preceded the development of overt cSLE, on average, by 22 months [159]. The absence of significant and persistent liver function abnormalities distinguishes cSLE from chronic active infectious hepatitis. Viral hepatitis is equally common in SLE patients, as in the general population, based on studies that included children [160].

Although it may be a presenting feature of cSLE, pancreatitis is thought to be uncommon and is reported in less than 2% of the patients [153, 161]. Pancreatitis in cSLE may be secondary to the underlying disease or iatrogenic, most commonly due to corticosteroid use [161]. Complications of acute pancreatitis, such as pseudocyst formation and chronic pancreatic insufficiency, have been reported in cSLE [162, 163]. Mesenteric vasculitis, lymphadenitis, ischemic colitis, and protein-losing enteropathy can occur in cSLE. There are case reports of co-existent cSLE and inflammatory bowel disease or celiac disease [164].

PULMONARY MANIFESTATIONS

Symptomatic and asymptomatic pulmonary manifestations occur in up to 60% of the children and adolescents with cSLE. This compares to estimates at 20–90% in aSLE [165, 166].

Clinically important parenchymal lung involvement, primarily diffuse interstitial infiltrates, is seen in only 5% of the patients with cSLE. Nonetheless, one study showed that restrictive lung disease is found in 40% and impaired diffusion capacity in 38% of the cSLE patients undergoing pulmonary function testing. Indeed, abnormal pulmonary function has been reported in as many as 60% of children with cSLE who lack clinical evidence of lung involvement [167]. Longitudinal pulmonary function testing of 15 cSLE patients who were asymptomatic for respiratory disease revealed no progression of abnormalities as per pulmonary function testing over time [167]. Pulmonary function abnormalities with cSLE have been inconsistently associated with global disease activity.

Shrinking lung syndrome, restrictive lung disease combined with diaphragmatic paralysis, is a rare complication of SLE with less than 150 cases in adults and fewer than ten cases reported in pediatrics[168–172].

Another rare but potentially fatal complication of cSLE is pulmonary hemorrhage [173]. As with aSLE, pulmonary hemorrhage presents in cSLE with sudden dyspnea, hypoxemia, anemia, and bilateral lung infiltrates, and early aggressive therapy appears crucial. Early use of high-dose pulse intravenous methylprednisolone is recommended, followed by high-dose oral corticosteroids and an immunosuppressive drug, such as intravenous cyclophosphamide, mycophenolic acid, or azathioprine. Plasmapheresis is rarely used, and its efficacy in cSLE-associated pulmonary hemorrhage has not been verified [173]. There are case reports of the use of rituximab for pulmonary hemorrhage treatment in aSLE but not cSLE. Pulmonary hemorrhage is still linked to high mortality rates in children and adults with SLE [173, 174].

Discriminating cSLE-associated pneumonitis from infectious pneumonia can be difficult. Bronchoalveolar lavage yields a definitive diagnosis of infectious causes in only 70% of immunocompromised pediatric populations, and no data are available regarding the sensitivity and specificity of bronchoalveolar lavage with cSLE in particular [175]. Empiric anti-infectious therapy is strongly considered, especially when cSLE management mandates corticosteroids and immunosuppressive drugs.

IMMUNOLOGICAL MARKERS

As with aSLE, circulating ANA are the hallmark of cSLE, and present in virtually all children. An estimated 30—50% of children referred to pediatric rheumatologists who test ANA positive have a musculoskeletal pain syndrome and not cSLE. In a retrospective study of 110 children who tested positive for ANA, ten children eventually developed cSLE after follow-up for 4 years or less. The median ANA titer of the children who developed cSLE was 1:1080 [176]. Conversely, ANA titers of 1:320 or lower appeared not to confer a sizeable risk for the subsequent development of cSLE [176].

Elevated levels of anti-double-stranded DNA (anti-dsDNA) antibodies are more common in cSLE than aSLE (61% vs. 25%) [11]. An estimated 92% of children with infantile SLE test positive for anti-dsDNA antibodies [2]. As with aSLE, changes in the titer of anti-dsDNA antibodies are used to monitor disease activity in cSLE. There are conflicting reports as to whether anti-dsDNA antibodies and complement levels are predictive of future cSLE flares [177, 178]. Although traditionally viewed to be specific for SLE, anti-dsDNA antibodies can be present with autoimmune hepatitis, Epstein-Barr virus infections, rheumatoid arthritis, and very rarely in healthy people.

Besides anti-dsDNA antibodies, antihistone and antiribosomal P antibodies are all more frequently observed in cSLE than aSLE [11]. Antiribosomal P antibodies are elevated in 25—42% of patients with cSLE [11], compared to only 6—11% of patients with aSLE [11]. This is of concern, as antiribosomal P antibodies have been associated with cNPSLE.

The prevalence of antihistone, anti-Smith (Sm), antiribonucleoprotein (anti-RNP), anti-Ro/SSA, and anti-La/SSB antibodies is unlikely different between cSLE and aSLE [11]. Antihistone antibodies have been reported in 39%, anti-Sm antibodies in up to 58%, anti-RNP antibodies in about 37%, anti-Ro antibodies in 33%, and anti-La antibodies in 13% of the cSLE patients during the course of their disease [6, 179]. Antibody clustering in cSLE may be different from that observed in aSLE [180]. Similar to what has been reported in aSLE, there are ethnic variations in the frequency of autoantibodies with cSLE; anti-U1RNP and anti-Sm antibodies occur more frequently in non-Caucasian patients with cSLE [181]. Antinucleosome antibodies, autoantibodies against intranuclear antigens exposed by apoptosis,

have been implicated in the pathogenesis of LN with cSLE, and global SLE disease activity in general [182]. The sensitivity and specificity of antinucleosome antibodies for the diagnosis of cSLE is about 53% and 98%, respectively [183]. Rheumatoid factor (RF) is positive in 10—54% of cSLE patients, exceeding the frequency at which RF is reported in juvenile idiopathic arthritis [16, 88]. The aPL antibodies are discussed later in the section titled "Antiphospholipid Antibody Syndromes."

Hyperglobulinemia occurs in about 20% of patients with aSLE, thus is as frequent as with cSLE. In particular, persistently low levels of IgG are suggestive of a concomitant immunoglobulin deficiency. There are several case reports of adolescents with both cSLE and co-existent common variable immunodeficiency (CVID) [184]. It remains unclear whether CVID constitutes a risk factor for cSLE, as there is no known common pathway in humans that could explain the co-existence of both disorders.

As with aSLE, serum complement levels presently are used as indicators of disease activity in cSLE. Persistently low complement levels, especially if discordant with other cSLE features of disease activity, may reflect partial or complete deficiency of the complement system components, which is likely secondary to low copy numbers of C4 genes [185].

Similar to aSLE, the ESR rather than C-reactive protein is the preferred measure of systemic inflammation in cSLE. C-reactive protein tends to be normal or only mildly elevated in active cSLE and aSLE, while marked elevations of the C-reactive protein are strongly suggestive of bacterial infections or serositis.

GENETICS

Little is known about the relative risk of children with different races to develop cSLE. However, as with aSLE non-Caucasian children appear to be at a higher risk of developing cSLE. Genetic studies in cSLE are for the most part confirmation studies of genetic variants reported in aSLE. At this time, none of the studies can explain differences in disease presentation, activity, and outcomes between cSLE and aSLE.

The best early evidence of a genetic contribution to SLE is based on twin concordance rates which are at 20—30% for monozygotic twins and 2—3% for dizygotic twins [186, 187]. A case report of monozygotic triplets with cSLE suggests similar clinical features at onset but different disease courses over time in siblings with cSLE [188]. The risk of siblings of a patient with cSLE to develop SLE is about 20-fold higher than that of the general population [189].

Congenital complement deficiencies are seen in about 1% of patients with SLE [190]. SLE-like disease attributed to complement deficiencies usually presents at a younger age with prominent skin lesions and infections but less renal, pulmonary, and pericardial involvement. ANA and anti-dsDNA antibodies are usually negative or present at low titers but anti-Ro antibodies are typically elevated. The closest association between SLE and complement deficiencies is seen with C1q deficiency, where there is a risk of over 90% to develop a lupus-like illness with onset early in life [190]. Deficiency of C1q, a rare condition with only about 50 cases reported in the literature, can lead to decreased clearance of apoptotic keratinocytes, which are potential autoantigens [191]. Other complement deficiencies include homozygous C4 deficiency at a 75% association rate with SLE [192] and C1s or C1r deficiency, which has a 50% association rate with SLE that presents primarily in infancy or childhood. Homozygous C2 deficiency is present in 1/10,000 to 1/30,000 Caucasians, and 10–30% of them may develop SLE [193].

Based on microarray analysis of blood, the innate and the adaptive immune system are involved in the processes resulting in cSLE, and interferon alpha is a key cytokine [194]. Both Class II and Class III, but not Class I major histocompatibility complex (MHC) alleles, confer susceptibility for SLE. Reveille *et al.* report a high frequency of HLA DRW*08 among American Blacks who develop cSLE [195]. In another study, HLA DRB1*15:03, DRB5*01:01, DQA1*01:02, and DQB1*06:02 were found to be more common in American Blacks with SLE, independent of age at disease onset. HLA DRB1*03:01, DQA1*05:01, DQB1*02:01 were all shown to be more common in Caucasians with SLE, again independent of age at disease onset [196]. Each of the triplets mentioned in the preceding paragraph was found to have HLA DRB1*13:03 and DQB1*06:02 [188]. More recently, non-HLA loci have been associated with autoimmunity and SLE. A polymorphism (-28C/G polymorphism) of the RANTES gene, a chemokine that plays a role in recruiting leukocytes into inflammatory sites, has been associated with cSLE in Chinese children [197]. The 1858T single nucleotide polymorphism (SNP) of *PTPN22*, a gene that encodes lymphocyte phosphatase (Lyp), which inhibits T-cell activation, has been associated with sporadic cSLE in Mexican children [198]. A genome-wide scan of patients with cSLE identified both the N673S polymorphism of the P-selectin gene (SELP) and the C203S polymorphism of the interleukin-1 receptor-associated kinase 1 gene (IRAK1) to be associated with cSLE [199]. SELP is a cell adhesion molecule of activated endothelial cells, and IRAK1 is involved in the signaling cascade of the toll/interleukin-1 receptor family [199].

The PD1.3G/A SNP, located within the regulatory area of the programmed cell death 1 (PDCD1) gene, has also been associated with susceptibility to aSLE in Swedes, European Americans, and Mexicans. The same SNP has been associated with cSLE in Mexico [200].

At present, genetic studies in cSLE must all be considered exploratory and preliminary. Confirmation of the above-mentioned genetic variations and polymorphisms is needed in large well-phenotyped cohorts.

OVERLAP WITH OTHER AUTOIMMUNE DISEASE

Overlap features between cSLE and other autoimmune diseases, such as juvenile dermatomyositis, juvenile scleroderma, and Wegener granulomatosis, although uncommon, have been reported in case reports and small case series. Patients with cSLE frequently have signs and symptoms of aPL syndrome, and infrequently, Sjögren syndrome.

ANTIPHOSPHOLIPID ANTIBODY SYNDROME

Primary aPL syndrome needs to be discriminated from secondary aPL syndrome that is associated with cSLE. Compared to those with secondary aPL syndrome, children with primary aPL syndrome tend to be younger and encounter more often arterial thrombotic events [201]. Primary aPL syndrome may herald cSLE. In a study of 14 children with primary aPL syndrome, two eventually developed cSLE during a median follow-up period of 6 years [202].

The reported prevalence of aPL in cSLE varies widely between series, likely reflecting differences in assays, frequency of the antibody testing, and progression of disease over time [203]. The prevalence of anticardiolipin (aCL) antibodies and LAC appears similar in groups of adults and children with SLE [2, 204]. In cSLE, LAC has been reported in 19–62%, anti-beta2 glycoprotein-I antibodies in 29%, and anticardiolipin (aCL) antibodies in 27–87% of the patients, respectively. The prevalence of aCL antibodies and LAC appears to be independent from patient age with cSLE [2]. Only 71% of 115 cSLE patients who initially presented with aPL had persistence of these antibodies over time, suggesting that aPL production is amenable to anti-inflammatory therapy [64]. Titers of aCL IgG correlate with the levels of complement C3, ESR, and disease activity [64]. Hepatomegaly, thrombocytopenia, and hemolytic anemia are all statistically more frequent among cSLE patients who test positive for aPL [205].

Thromboses occur more often in aSLE than cSLE patients who test positive for aPL antibodies, which is likely reflective of underlying co-morbidities and longer disease duration in aSLE. Disease onset during childhood is found to be protective of thrombosis with aPL syndrome (odds ratio, 0.52 for age ≤ 20; $p= 0.001$) [206]. Love and Santoro report that 42% of LAC-positive and 40% of aCL-positive patients with aSLE have a history of thrombosis [207]. The prevalence of cSLE-associated aPL syndrome ranges from 0 to 24% [203]. In cSLE with aPL syndrome, venous thromboses are more common than arterial clots [203]. Thrombotic events may be a presenting feature of cSLE, although the average time interval from the diagnosis of cSLE to the initial thrombotic event is 15 months in children who test positive for aPL antibodies [84]. Particularly LAC positivity rather than cSLE itself, predisposes to thromboses as cSLE patients who test negative for aPL rarely encounter thrombotic events [208]. The sites of thrombosis include the brain, lung, kidney, cardiac, peripheral arteries, and the deep veins of extremities [84]. A large cSLE cohort study found sustained thrombocytopenia and LAC to be risk factors of thromboembolic events [209]. cSLE patients who tested persistently positive for LAC have a 28-fold increased risk of experiencing a thrombotic event compared to cSLE patients without LAC [208]. About one-third of cSLE patients with aPL syndrome will experience recurrent thromboses within 13 months of the initial event, especially if anticoagulation is discontinued or if other predisposing thrombophilic factors are present. The association between venous thrombosis and Factor V Leiden mutation shown in aSLE has not been verified by a cohort study of 149 cSLE patients [84].

SJÖGREN SYNDROME

While 6.5% of all aSLE patients have Sjögren syndrome, it is less common in children with cSLE [210]. Although a case series (n=34) from Japan reports that 41% of the children with SLE have secondary Sjögren syndrome, the prevalence of Sicca syndrome at 3.9% in cSLE vs. 9.3% in aSLE was considerably lower in the much larger GLADEL cohort [15, 211]. Primary, as opposed to secondary Sjögren syndrome, is very rare prior to adulthood. In a cohort study of 180 children with primary pediatric Sjögren syndrome, the mean age at diagnosis was 9.8 years [212]. The common clinical manifestation of pediatric Sjögren syndrome is bilateral parotid swelling, which is present in 70% of the cases, while extraglandular manifestations are reported in 5% of the children in one case series [213]: xerophthalmia occurred in 66% and xerostomia in 43% of children with Sjögren syndrome. Other common extraglandular features are arthritis and arthralgia (75%),

Raynaud phenomenon (50%), and myalgias (45%). ANA positivity is less frequently observed in children than in adults with Sjögren syndrome [214, 215]. Serum anti-α-fodrin antibodies have been described in pediatric Sjögren syndrome, but these antibodies are not specific to pediatric Sjögren syndrome [216, 217].

As in adults, therapy of pediatric Sjögren syndrome is largely supportive, using artificial tears and artificial saliva [218]. Parotitis can be treated effectively with short courses of corticosteroids [213]. Hydroxychloroquine appears not to be beneficial for the management of pediatric Sjögren syndrome [218]. The use of TNF-blockers and rituximab are reported for the treatment of extraglandular symptoms of Sjögren syndrome in adults only [219].

SPECIAL THERAPEUTIC CONSIDERATIONS

The chronicity of illness and uncertainty of prognosis is a source of apprehension for patients with cSLE and their parents alike. Growth, development, long residual lifespan, incomplete brain development, and reproductive potential all need to be considered in the treatment of any pediatric disease, including cSLE. The most commonly used medications including the doses suggested are summarized in Table 32.2. More detailed information on drug dosing and monitoring for cSLE is provided elsewhere [220].

As with aSLE, the therapeutic approaches to cSLE are nonstandardized at present and are influenced by organ involvement, severity of global disease, availability of drug, and concomitant damage. Antimalarials are taken by about two-thirds of all cSLE patients [6], and in some centers virtually all cSLE patients are prescribed hydroxychloroquine. Patients with SLE of all ages older than 6 years are equally likely to be prescribed antimalarials [14, 87]. Adults and children with SLE are comparable in their use of nonsteroidal anti-inflammatory medications [5, 87], except for cyclooxygenase 2 inhibitors, which appear to be less commonly prescribed to children and adolescents [87].

In comparing medication choice in cSLE and aSLE, a larger proportion of children and adolescents are treated with high-dose corticosteroids and immunosuppressive medications than patients with aSLE [14]. Currently azathioprine, methotrexate, mycophenolate mofetil, cyclophosphamide, and leflunomide are most often used in cSLE.

Based on several observational studies, virtually all cSLE patients will be exposed to oral corticosteroids during the course of their disease [87]. In a study of patients followed at Canadian tertiary centers, 97% of the 67 cSLE patients compared to only 70% of the 131 aSLE patients were prescribed oral corticosteroids [14]. Patients with cSLE in this study received high-dose intravenous methylprednisolone therapy more frequently than aSLE patients, 30% vs. 11%, respectively. Similarly, immunosuppressive medications were more often prescribed to children as compared to patients with aSLE (66% vs. 37%), while the choice of specific immunosuppressive medications appears quite similar in both groups [14]. An exception may be the more common use of methotrexate in aSLE than cSLE patients (31% vs. 9%) [14].

The most common indication for cyclophosphamide treatment in cSLE is cNPSLE and LN [6]. How the more common use of mycophenolate mofetil and CD20-depleting therapies in recent years influences the overall medication profiles of cSLE is unknown at present. Infection rates with immunosuppression in cSLE vary considerably, likely reflecting differences in health milieu and local practices. Fatal infections with fungus or Gram-negative bacilli occur more commonly in patients treated with immunosuppressive medications for active LN.

TABLE 32.2   Anti-inflammatory Medications for Children with cSLE

Drug(s)	Suggested doses	Usual maximum dose	Useful for	Remarks
NSAIDS				
Naproxen	10–25 mg/kg/day	1000 mg orally divided in b.i.d.	Mild disease	Musculoskeletal disease; needs monitoring for effect on NPSLE and kidneys
Ibuprofen	20–40 mg/kg/day	2400 ma orally divided in t.i.d.		
Diclofenac	1–3 mg/kg/day	150 mg orally divided in b.i.d.		
CORTICOSTEROIDS				
Prednisone	1–2 mg/kg/day	80 mg orally per day	Rapid control of acute disease symptoms of moderate to severe degree	Rarely exceed 60 mg daily; may be divided in quid dosing, if necessary. Some patients will require low-dose oral corticosteroids for maintenance therapy.
Oral methylprednisolone	1–2 mg/kg/day	60 mg orally per day		Use for patients with liver involvement.
Intravenous methylprednisolone	10–30 mg/dose	1000 mg per dose		Acute manifestations of NPSLE, kidney, and hematological disease.
IMMUNOSUPPRESSIVES				
Azathioprine	0.5–2.5 mg/kg/day	200 mg orally once daily	Moderate or severe disease	Vasculitis, NPSLE, glomerulonephritis Steroid sparing medications; use is associated with improved outcome of NPSLE and kidney disease
Oral cyclophosphamide	0.5–2 mg/kg/day	2000 mg orally divided in b.i.d.		
Intravenous cyclophosphamide	500–1000 mg/m^2	2500 mg per dose		
Mycophenolate mofetil	60 mg/kg/day	3000 mg orally divided in b.i.d.		
OTHERS				
Hydroxychloroquine	5–7 mg/kg/day	400 mg orally once daily		Mucocutaneous disease, arthritis
Dapsone	2 mg/kg/day	100 mg orally once daily		Skin disease and skin vasculitis
Immunoglobulins	1–2 mg/kg/dose	25 mg orally or subcutaneously once every week		Hematological disease
Methotrexate	15–20 mg/m^2/week			Arthritis; not in patients with kidney disease

At this point, there are no data on the life-long risk of malignancy in cSLE patients exposed to immunosuppressive drugs. There are no reports of pediatric malignancy in cSLE patients treated with cyclophosphamide [91].

ADHERENCE

Adolescents' concerns with body image and peer acceptance often make for a challenging treatment process. Nonadherence to visits and medications is a universal challenge in both cSLE and aSLE. In a single center study, 39% of 55 SLE patients were nonadherent (adherence rates less than 80%) to prednisone and 51% to hydroxychloroquine, based on pharmacy refill data [221]. Parental adherence to therapy appears to be key to the child's adherence. Inaccurate perception of disease severity, fear of medication side effects, including changes in physical appearance, all deter adolescents from taking their medications regularly. In cSLE, as with other pediatric rheumatic diseases, the complexity of the medical regimen, delayed onset of improvement, and chronic medication intake are all risk factors of nonadherence [222]. Low adherence likely compounds poor clinical outcomes in patients with SLE. Enhancement of physician—parent/patient communication and education remains the cornerstone of adherence improvement. Initial studies suggest that the use of e-mail and text messaging are promising venues to enhance adherence to cSLE therapies [223].

OUTCOME MEASURES

Measures of disease activity and damage developed for aSLE have all been found suitable for use in cSLE [43, 224]. Indices for disease activity that have been validated for use in cSLE are the Systemic Lupus Erythematosus Disease Activity Index (SLEDAI), Systemic Lupus Activity Measure (SLAM), British Isles Lupus Assessment Group Index (BILAG), and the European Consensus Lupus Activity Measurement (ECLAM) [43, 225]. None of the indices seems convincingly superior in terms of sensitivity or responsiveness to change in cSLE [43]. The Systemic Lupus International Collaborating Clinics/American College of Rheumatology (SLICC/ACR) Damage index (SDI) has face, construct, and content validity when used in cSLE. In a single study, an adaptation of the SDI with added items for growth and pubertal progression, the Pedi-SDI, has been suggested for cSLE [226]. However, both of these items are theoretically reversible, thus may not strictly constitute irreversible organ damage [226].

As with other rheumatic diseases, it is unlikely that improvement of cSLE can be captured by changes of disease activity scores alone [227]. A set of cSLE core variables has been proposed to measure improvement:

1. Physician global assessment of disease activity on a 10-cm length VAS (visual analog scale).
2. Patient/parent's assessment of overall well-being on 10-cm length VAS.
3. 24-h Proteinuria.
4. ECLAM or SLAM or SLEDAI score.
5. Physical summary score of the Child Health Questionnaire [228].

Using the cSLE core set, improvement in cSLE is defined as at least 50% improvement in any two core variables with no more than one of the remaining core variables deteriorating by 30% or more [228]. These criteria have been validated in another cohort, supporting high specificity but only moderate sensitivity to clinically relevant improvement of cSLE [228].

Disease flare in cSLE is described in different ways in the literature [229]. Recently a consensus process led to a definition of flare as "a measurable worsening of cSLE disease activity in at least one organ system, involving new or worse signs of disease that may be accompanied by new or worse SLE symptoms. Depending on the severity of the flare, more intensive therapy may be required" [230].

The Childhood Health Assessment Questionnaire (C-HAQ) has been validated as a measure of physical function and disability in cSLE [231]. The Simple Measure of the Impact of Lupus Erythematosus in Youngsters (SMILEY) is a cSLE-specific tool to assess health-related quality of life (HRQOL) of children with cSLE ages 4–18 years [232]. The SMILEY shows moderate agreement between the patient and their parent when reporting HRQOL [232]. Generic HRQOL measures validated for cSLE include the Pediatric Quality of Life Inventory (PedsQL) Generic Core Scale, the PedsQL Rheumatology Module, the PedsQL Multidimensional Fatigue Scale, and the Child Health Questionnaire. HRQOL is significantly lower in cSLE than in healthy control populations [233]. Higher disease damage and activity are also inversely related to HRQOL [233]. Disease activity in the neurological, musculoskeletal, general, and vascular organ systems in particular have a negative impact on HRQOL in cSLE [233].

COST OF CARE

Cost of care for children with cSLE is considerably higher than that of aSLE [234, 235]. The estimated economic burden of cSLE ranges from $146–650 million annually in the United States [234]. In a study utilizing administrative databases from two tertiary pediatric rheumatology centers in the United States, the annual cost of care of cSLE was $14,944 (cost basis: 2000), excluding outpatient medication expenses. Cost was accrued mostly by hospitalizations (28%), laboratory testing (21%), and outpatient clinic visits (20%), while emergency department visits contributed only 1% to the total cost of care. Renal replacement therapy, although only required for three of the 119 children, constituted 11% of the total cost.

Previous studies examining the cost of health services in aSLE used a slightly different valuation system than that used for the above-mentioned study in children. Nonetheless, using a conservative estimate, the direct cost of care for a child with SLE appears to be roughly three times higher than for an adult. Whether this difference in cost between adults and children is due to differences in healthcare delivery systems, adherence to therapies, or differences in disease severity remains to be determined [236].

PROGNOSIS AND OUTCOMES

Despite improved survival rates in cSLE, there remains substantial morbidity due to disease damage, medication side effects, concomitant illness, as well as interruption in scholastic activity and employment.

About 50–70% of adults with SLE will have accrued some disease damage at 10 years postdiagnosis [237, 238]. In a study of 1015 children with cSLE from 39 countries, 40% acquired some disease damage after average disease durations of only 4 years. This percentage increased to 58% in those with more than 5 years of disease duration [226]. The kidneys were the most common site of damage reported (13%), followed by neuropsychiatric (11%), musculoskeletal (11%), ocular (8%), and cutaneous damage (8%). Growth failure (15%) and delayed puberty (11%) occurred as well [226]. Similar frequencies and distribution of disease damage have been described in other contemporary cohorts [6, 28, 87, 239]. Whether age at cSLE onset is a risk factor of poor disease outcome is still controversial [28, 239, 240]. Statistically significant difference in the mean SDI score was seen in the Canadian cohort (1.7 in cSLE vs. 0.76 in aSLE); ocular and musculoskeletal damage were statistically more common in children than adults who, on the other hand, had more malignancy damage [14]. In the LUMINA cohort (including only adolescent-onset SLE) there was a trend toward more disease damage

as measured by the SDI score at the time of the last follow-up in the adolescent-onset group as compared to the aSLE group (2.3 vs. 1.6) [68], and renal damage was statistically more common in those with disease onset prior to adulthood ($p=0.023$). Similarly, in the LUMINA cohort, there was somewhat more neuropsychiatric, ocular, and musculoskeletal damage with the adolescent-onset SLE than aSLE, but diabetes and peripheral vascular damage were common in those with disease onset during adulthood [68].

In a study comparing SLE in adolescents and adults over a follow-up period of 5.1 years and 4 years, respectively, the former had more disease activity, higher damage accrual, and a twofold higher mortality rate than the latter [234].

In a study conducted in the United States in the 1950s, 26 (70%) of the 37 children died within 5 years of disease onset [241]. Conversely, in a single-center study of cSLE in 1983, the 5-year mortality rate was 8% and has improved since then [242, 243]. With reported mortality rates as high as 38% between 2 and 31 months after disease onset, infantile SLE appears to have worse outcomes than cSLE with disease onset later in life [244]. The primary causes of death or poor outcomes in cSLE are renal disease, severe flares, infections, and cNPSLE [242, 243, 245]. There is some evidence that severe thrombocytopenia is a risk factor for poor prognosis with cSLE [246].

Delay in aSLE diagnosis is associated with a lower likelihood for achieving remission and worse survival [247]. At present, it is unknown whether the same holds true for cSLE but remission is generally considered a very rare event in children, and predictors for achieving remission have not been identified [87].

TRANSITION

Transition is defined by the American Society of Adolescent Medicine as "purposeful, planned movement of adolescents and young adults with chronic physical and medical conditions from child-centered to adult-oriented health care systems" [248].

There are few data on the optimal process by which cSLE patients should be transitioned to adult health care providers [249]. The best timing of transition depends on numerous factors, including social maturation level, disease activity, and preparedness of the parents to relinquish control, available medical resources, and cultural mindset. A multidisciplinary approach of disease-focused transition appears most appropriate and should occur during a period of disease quiescence. This is often not possible as cSLE has often a persistently active disease course. Transition concepts should be introduced early and patient self-management skills fostered during adolescence [249]. Implementation of a transition clinic where pediatric rheumatologists, adult rheumatologists, and interdisciplinary staff work together in the same setting is one way of formalizing the transition process. Further information is provided in a recent review [249].

SUMMARY

Clinical findings, genetic background, and immunological profiles of cSLE are, for the most part, similar to aSLE. Nevertheless, several key studies suggest important differences between aSLE and cSLE. The presence of certain complement deficiencies are risk factors for cSLE. Children generally have more pronounced features at cSLE onset, often more active disease over time, accrue more damage, and require more intensive drug therapy than patients with disease onset later in life. Renal damage especially is more common in cSLE than aSLE. Treatment considerations unique to children include their longer lifespan, growth, and maturation potential. The overall prognosis for cSLE has improved markedly over the last two decades, but challenges in providing the best therapeutic approaches to cSLE remain.

References

[1] M.B. Ferraz, J. Goldenberg, M.O. Hilario, W.A. Bastos, S.K. Oliveira, E.C. Azevedo, et al., Evaluation of the 1982 ARA lupus criteria data set in pediatric patients. Committees of Pediatric Rheumatology of the Brazilian Society of Pediatrics and the Brazilian Society of Rheumatology, Clin Exp Rheumatol 12 (1) (1994) 83—87.

[2] F.R. Pluchinotta, B. Schiavo, F. Vittadello, G. Martini, G. Perilongo, F. Zulian, Distinctive clinical features of pediatric systemic lupus erythematosus in three different age classes, Lupus 16 (8) (2007) 550—555.

[3] J.T. Lo, M.J. Tsai, L.H. Wang, M.T. Huang, Y.H. Yang, Y.T. Lin, et al., Sex differences in pediatric systemic lupus erythematosus: a retrospective analysis of 135 cases, J Microbiol Immunol Infect 32 (3) (1999) 173—178.

[4] A.G. Meislin, N. Rothfield, Systemic lupus erythematosus in childhood. Analysis of 42 cases, with comparative data on 200 adult cases followed concurrently, Pediatrics 42 (1) (1968) 37—49.

[5] J. Font, R. Cervera, G. Espinosa, L. Pallares, M. Ramos-Casals, S. Jimenez, et al., Systemic lupus erythematosus (SLE) in childhood: analysis of clinical and immunological findings in 34 patients and comparison with SLE characteristics in adults, Ann Rheum Dis 57 (8) (1998) 456—459.

[6] L.T. Hiraki, S.M. Benseler, P.N. Tyrrell, D. Hebert, E. Harvey, E.D. Silverman, Clinical and laboratory characteristics and long-term outcome of pediatric systemic lupus erythematosus: a longitudinal study, J Pediatr 152 (4) (2008) 550—556.

[7] B.A. Pons-Estel, L.J. Catoggio, M.H. Cardiel, E.R. Soriano, S. Gentiletti, A.R. Villa, et al., The GLADEL multinational Latin American prospective inception cohort of 1,214 patients with systemic lupus erythematosus: ethnic and disease heterogeneity among "Hispanics", Medicine (Baltimore) 83 (1) (2004) 1—17.

[8] E. Descloux, I. Durieu, P. Cochat, D. Vital-Durand, J. Ninet, N. Fabien, et al., Influence of age at disease onset in the outcome of paediatric systemic lupus erythematosus, Rheumatology (Oxford) 48 (7) (2009) 779—784.

[9] B. Bader-Meunier, J.B. Armengaud, E. Haddad, R. Salomon, G. Deschenes, I. Kone-Paut, et al., Initial presentation of childhood-onset systemic lupus erythematosus: a French multicenter study, J Pediatr 146 (5) (2005) 648—653.

[10] S. Iqbal, M.R. Sher, R.A. Good, G.D. Cawkwell, Diversity in presenting manifestations of systemic lupus erythematosus in children, J Pediatr 135 (4) (1999) 500—505.

[11] I.E. Hoffman, B.R. Lauwerys, F. De Keyser, T.W. Huizinga, D. Isenberg, L. Cebecauer, et al., Juvenile-onset systemic lupus erythematosus: different clinical and serological pattern than adult-onset systemic lupus erythematosus, Ann Rheum Dis 68 (3) (2009) 412—415.

[12] P.N. Tyrrell, J. Beyene, S.M. Benseler, T. Sarkissian, E.D. Silverman, Predictors of lipid abnormalities in children with new-onset systemic lupus erythematosus, J Rheumatol 34 (10) (2007) 2112—2119.

[13] L.B.U.A. Tucker, M. Fernandez, Clinical differences between juvenile and adult onset patients with systemic lupus erythematosus: results from a multiethnic longitudinal cohort, Arthritis Rheum 54 (6) (2006) S162.

[14] H.I. Brunner, D.D. Gladman, D. Ibanez, M.D. Urowitz, E.D. Silverman, Difference in disease features between childhood-onset and adult-onset systemic lupus erythematosus, Arthritis Rheum 58 (2) (2008) 556—562.

[15] L.A. Ramirez Gomez, O. Uribe Uribe, O. Osio Uribe, H. Grisales Romero, M.H. Cardiel, D. Wojdyla, et al., Childhood systemic lupus erythematosus in Latin America. The GLADEL experience in 230 children, Lupus 17 (6) (2008) 596—604.

[16] M.J. Rood, R. ten Cate, L.W. van Suijlekom-Smit, E.J. den Ouden, F.E. Ouwerkerk, F.C. Breedveld, et al., Childhood-onset Systemic Lupus Erythematosus: clinical presentation and prognosis in 31 patients, Scand J Rheumatol 28 (4) (1999) 222—226.

[17] C. Requena, A. Torrelo, I. de Prada, A. Zambrano, Linear childhood cutaneous lupus erythematosus following Blaschko lines, J Eur Acad Dermatol Venereol 16 (6) (2002) 618—620.

[18] M.C. Sampaio, Z.N. de Oliveira, M.C. Machado, V.M. dos Reis, M.A. Vilela, Discoid lupus erythematosus in children—a retrospective study of 34 patients, Pediatr Dermatol 25 (2) (2008) 163—167.

[19] J. Del Boz, T. Martin, E. Samaniego, A. Vera, A. Sanz, V. Crespo, Childhood discoid lupus in identical twins, Pediatr Dermatol 25 (6) (2008) 648—649.

[20] S.J. Yun, J.W. Lee, H.J. Yoon, S.S. Lee, S.Y. Kim, J.B. Lee, et al., Cross-sectional study of hair loss patterns in 122 Korean systemic lupus erythematosus patients: a frequent finding of non-scarring patch alopecia, J Dermatol 34 (7) (2007) 451—455.

[21] S. Pavlov-Dolijanovic, N. Damjanov, P. Ostojic, G. Susic, R. Stojanovic, D. Gacic, et al., The prognostic value of nailfold capillary changes for the development of connective tissue disease in children and adolescents with primary raynaud phenomenon: a follow-up study of 250 patients, Pediatr Dermatol 23 (5) (2006) 437—442.

[22] F. Zulian, F. Corona, V. Gerloni, F. Falcini, A. Buoncompagni, M. Scarazatti, et al., Safety and efficacy of iloprost for the treatment of ischaemic digits in paediatric connective tissue diseases, Rheumatology (Oxford) 43 (2) (2004) 229—233.

[23] M. Garcia-Carrasco, M. Jimenez-Hernandez, R.O. Escarcega, C. Mendoza-Pinto, R. Pardo-Santos, R. Levy, et al., Treatment of Raynaud's phenomenon, Autoimmun Rev 8 (1) (2008) 62—68.

[24] R. Cervera, M.A. Khamashta, J. Font, G.D. Sebastiani, A. Gil, P. Lavilla, et al., Morbidity and mortality in systemic lupus erythematosus during a 10-year period: a comparison of early and late manifestations in a cohort of 1,000 patients, Medicine (Baltimore) 82 (5) (2003) 299—308.

[25] J.E. Nord, P.K. Shah, R.Z. Rinaldi, M.H. Weisman, Hydroxychloroquine cardiotoxicity in systemic lupus erythematosus: a report of 2 cases and review of the literature, Semin Arthritis Rheum 33 (5) (2004) 336—351.

[26] R.M. Hurley, R.H. Steinberg, H. Patriquin, K.N. Drummond, A vascular necrosis of the femoral head in childhood systemic lupus erythematosus, Can Med Assoc J 111 (8) (1974) 781—784.

[27] D. Ibanez, D.D. Gladman, M.B. Urowitz, Adjusted mean Systemic Lupus Erythematosus Disease Activity Index-2K is a predictor of outcome in SLE, J Rheumatol 32 (5) (2005) 824—827.

[28] H.I. Brunner, E.D. Silverman, T. To, C. Bombardier, B.M. Feldman, Risk factors for damage in childhood-onset systemic lupus erythematosus: cumulative disease activity and medication use predict disease damage, Arthritis Rheum 46 (2) (2002) 436—444.

[29] C. Mendoza-Pinto, M. Garcia-Carrasco, H. Sandoval-Cruz, R.O. Escarcega, M. Jimenez-Hernandez, I. Etchegaray-Morales, et al., Risks factors for low bone mineral density in premenopausal Mexican women with systemic lupus erythematosus, Clin Rheumatol 28 (1) (2009) 65—70.

[30] C.A. Pineau, M.B. Urowitz, P.J. Fortin, D. Ibanez, D.D. Gladman, Osteoporosis in systemic lupus erythematosus: factors associated with referral for bone mineral density studies, prevalence of osteoporosis and factors associated with reduced bone density, Lupus 13 (6) (2004) 436—441.

[31] K.A. Alsufyani, O. Ortiz-Alvarez, D.A. Cabral, L.B. Tucker, R.E. Petty, H. Nadel, et al., Bone mineral density in children and adolescents with systemic lupus erythematosus, juvenile dermatomyositis, and systemic vasculitis: relationship to disease duration, cumulative corticosteroid dose, calcium intake, and exercise, J Rheumatol 32 (4) (2005) 729—733.

[32] S. Compeyrot-Lacassagne, P.N. Tyrrell, E. Atenafu, A.S. Doria, D. Stephens, D. Gilday, et al., Prevalence and etiology of low bone mineral density in juvenile systemic lupus erythematosus, Arthritis Rheum 56 (6) (2007) 1966—1973.

[33] V. Lilleby, Bone status in juvenile systemic lupus erythematosus, Lupus 16 (8) (2007) 580—586.

[34] S. Trapani, R. Civinini, M. Ermini, E. Paci, F. Falcini, Osteoporosis in juvenile systemic lupus erythematosus: a longitudinal study on the effect of steroids on bone mineral density, Rheumatol Int 18 (2) (1998) 45—49.

[35] V. Lilleby, G. Lien, K. Frey Froslie, M. Haugen, B. Flato, O. Forre, Frequency of osteopenia in children and young adults with childhood-onset systemic lupus erythematosus, Arthritis Rheum 52 (7) (2005) 2051—2059.

[36] T.C. Castro, M.T. Terreri, V.L. Szejnfeld, C.H. Castro, M. Fisberg, M. Gabay, et al., Bone mineral density in juvenile systemic lupus erythematosus, Braz J Med Biol Res 35 (10) (2002) 1159—1163.

[37] T.B. Wright, J. Shulls, M.B. Leonard, B.S. Zemel, J.M. Burnham, Hypovitaminosis D is associated with greater body mass index and disease activity in pediatric systemic lupus erythematosus, J Pediatr 155 (2) (2009) 260—265. Epub 2009 May 15.

[38] L.E. Grissom, H.T. Harcke, Radiographic features of bisphosphonate therapy in pediatric patients, Pediatr Radiol 33 (4) (2003) 226—229.

[39] S. Rudge, S. Hailwood, A. Horne, J. Lucas, F. Wu, T. Cundy, Effects of once-weekly oral alendronate on bone in children on glucocorticoid treatment, Rheumatology (Oxford) 44 (6) (2005) 813—818.

[40] S.M. Benseler, E.D. Silverman, Neuropsychiatric involvement in pediatric systemic lupus erythematosus, Lupus 16 (8) (2007) 564—571.

[41] H.I. Brunner, M. Freedman, E.D. Silverman, Close relationship between systemic lupus erythematosus and thrombotic thrombocytopenic purpura in childhood, Arthritis Rheum 42 (11) (1999) 2346—2355.

[42] C.Y. Wu, Y.T. Su, J.S. Wang, Y.H. Chiou, Childhood hemolytic uremic syndrome associated with systemic lupus erythematosus, Lupus 16 (12) (2007) 1006—1010.

[43] H.I. Brunner, B.M. Feldman, C. Bombardier, E.D. Silverman, Sensitivity of the Systemic Lupus Erythematosus Disease Activity Index, British Isles Lupus Assessment Group Index, and Systemic Lupus Activity Measure in the evaluation of clinical change in childhood-onset systemic lupus erythematosus, Arthritis Rheum 42 (7) (1999) 1354—1360.

[44] V.R. Saxena, R. Mina, H.J. Moallem, S.P. Rao, S.T. Miller, Systemic lupus erythematosus in children with sickle cell disease, J Pediatr Hematol Oncol 25 (8) (2003) 668—671.

[45] B.W. Kinder, M.M. Freemer, T.E. King Jr., R.F. Lum, J. Nititham, K. Taylor, et al., Clinical and genetic risk factors for pneumonia in systemic lupus erythematosus, Arthritis Rheum 56 (8) (2007) 2679—2686.

[46] H.H. Yu, L.C. Wang, J.H. Lee, C.C. Lee, Y.H. Yang, B.L. Chiang, Lymphopenia is associated with neuropsychiatric manifestations and disease activity in paediatric systemic lupus erythematosus patients, Rheumatology (Oxford) 46 (9) (2007) 1492—1494.

[47] A. Altintas, A. Ozel, N. Okur, T. Cil, S. Pasa, O. Ayyildiz, Prevalence and clinical significance of elevated antinuclear antibody test in children and adult patients with idiopathic thrombocytopenic purpura, J Thromb Thrombolysis 24 (2) (2007) 163—168.

[48] N.L. Kao, P.K. Musto, G.W. Richmond, Refractory thrombocytopenia in a patient with systemic lupus erythematosus and prior immune thrombocytopenia, not responsive to an accessory splenectomy, South Med J 87 (9) (1994) 941—943.

[49] C. Lindholm, K. Borjesson-Asp, K. Zendjanchi, A.C. Sundqvist, A. Tarkowski, M. Bokarewa, Longterm clinical and immunological effects of anti-CD20 treatment in patients with refractory systemic lupus erythematosus, J Rheumatol 35 (5) (2008) 826—833.

[50] L. Quartuccio, S. Sacco, N. Franzolini, A. Perin, G. Ferraccioli, S. De Vita, Efficacy of cyclosporin-A in the long-term management of thrombocytopenia associated with systemic lupus erythematosus, Lupus 15 (2) (2006) 76—79.

[51] S. Kumar, S.M. Benseler, M. Kirby-Allen, E.D. Silverman, B-cell depletion for autoimmune thrombocytopenia and autoimmune hemolytic anemia in pediatric systemic lupus erythematosus, Pediatrics 123 (1) (2009) e159–e163.

[52] Y. Oka, J. Kameoka, Y. Hirabayashi, R. Takahashi, T. Ishii, T. Sasaki, et al., Reversible bone marrow dysplasia in patients with systemic lupus erythematosus, Intern Med 47 (8) (2008) 737–742.

[53] T. Avcin, S.M. Tse, R. Schneider, B. Ngan, E.D. Silverman, Macrophage activation syndrome as the presenting manifestation of rheumatic diseases in childhood, J Pediatr 148 (5) (2006) 683–686.

[54] R.J. Risdall, R.W. McKenna, M.E. Nesbit, W. Krivit, H.H. Balfour Jr., R.L. Simmons, et al., Virus-associated hemophagocytic syndrome: a benign histiocytic proliferation distinct from malignant histiocytosis, Cancer 44 (3) (1979) 993–1002.

[55] R.L. Brey, S.L. Holliday, A.R. Saklad, M.G. Navarrete, D. Hermosillo-Romo, C.L. Stallworth, et al., Neuropsychiatric syndromes in lupus: prevalence using standardized definitions, Neurology 58 (8) (2002) 1214–1220.

[56] W.L. Sibbitt Jr., J.R. Brandt, C.R. Johnson, M.E. Maldonado, S.R. Patel, C.C. Ford, et al., The incidence and prevalence of neuropsychiatric syndromes in pediatric onset systemic lupus erythematosus, J Rheumatol 29 (7) (2002) 1536–1542.

[57] M.O. Olfat, S.M. Al-Mayouf, M.A. Muzaffer, Pattern of neuropsychiatric manifestations and outcome in juvenile systemic lupus erythematosus, Clin Rheumatol 23 (5) (2004) 395–399.

[58] E. Muscal, B.L. Myones, The role of autoantibodies in pediatric neuropsychiatric systemic lupus erythematosus, Autoimmun Rev 6 (4) (2007) 215–217.

[59] H.I. Brunner, O.Y. Jones, D.J. Lovell, A.M. Johnson, P. Alexander, M.S. Klein-Gitelman, Lupus headaches in childhood-onset systemic lupus erythematosus: relationship to disease activity as measured by the systemic lupus erythematosus disease activity index (SLEDAI) and disease damage, Lupus 12 (8) (2003) 600–606.

[60] M.I. Steinlin, S.I. Blaser, D.L. Gilday, A.A. Eddy, W.J. Logan, R.M. Laxer, et al., Neurologic manifestations of pediatric systemic lupus erythematosus, Pediatr Neurol 13 (3) (1995) 191–197.

[61] S.B. Turkel, J.H. Miller, A. Reiff, Case series: neuropsychiatric symptoms with pediatric systemic lupus erythematosus, J Am Acad Child Adolesc Psychiatry 40 (4) (2001) 482–485.

[62] E. Demirkaya, Y. Bilginer, N. Aktay-Ayaz, D. Yalnizoglu, K. Karli-Oguz, V. Isikhan, et al., Neuropsychiatric involvement in juvenile systemic lupus erythematosus, Turk J Pediatr 50 (2) (2008) 126–131.

[63] A. Reiff, J. Miller, B. Shaham, B. Bernstein, I.S. Szer, Childhood central nervous system lupus; longitudinal assessment using single photon emission computed tomography, J Rheumatol 24 (12) (1997) 2461–2465.

[64] T. Avcin, S.M. Benseler, P.N. Tyrrell, S. Cucnik, E.D. Silverman, A followup study of antiphospholipid antibodies and associated neuropsychiatric manifestations in 137 children with systemic lupus erythematosus, Arthritis Rheum 59 (2) (2008) 206–213.

[65] Y. Uziel, R.M. Laxer, S. Blaser, M. Andrew, R. Schneider, E.D. Silverman, Cerebral vein thrombosis in childhood systemic lupus erythematosus, J Pediatr 126 (5 Pt 1) (1995) 722–727.

[66] S. Parikh, K.F. Swaiman, Y. Kim, Neurologic characteristics of childhood lupus erythematosus, Pediatr Neurol 13 (3) (1995) 198–201.

[67] M.J. Spinosa, M. Bandeira, P.B. Liberalesso, S.C. Vieira, L.L. Janz Jr., E.G. Sa, A. Lohr Jr., Clinical, laboratory and neuroimage findings in juvenile systemic lupus erythematosus presenting involvement of the nervous system, Arq Neuropsiquiatr 65 (2B) (2007) 433–439.

[68] L.B. Tucker, A.G. Uribe, M. Fernandez, L.M. Vila, G. McGwin, M. Apte, et al., Adolescent onset of lupus results in more aggressive disease and worse outcomes: results of a nested matched case-control study within LUMINA, a multiethnic US cohort (LUMINA LVII), Lupus 17 (4) (2008) 314–322.

[69] S. Appenzeller, F. Cendes, L.T. Costallat, Epileptic seizures in systemic lupus erythematosus, Neurology 63 (10) (2004) 1808–1812.

[70] D.M. Levy, S.P. Ardoin, L.E. Schanberg, Neurocognitive impairment in children and adolescents with systemic lupus erythematosus, Nat Clin Pract Rheumatol 5 (2) (2009) 106–114.

[71] W.L. Sibbitt Jr., P.J. Schmidt, B.L. Hart, W.M. Brooks, Fluid Attenuated Inversion Recovery (FLAIR) imaging in neuropsychiatric systemic lupus erythematosus, J Rheumatol 30 (9) (2003) 1983–1989.

[72] H.H. Yu, J.H. Lee, L.C. Wang, Y.H. Yang, B.L. Chiang, Neuropsychiatric manifestations in pediatric systemic lupus erythematosus: a 20-year study, Lupus 15 (10) (2006) 651–657.

[73] J.A.E.J. Mikdashi, G.S. Alarcón, L. Crofford, B.J. Fessler, L. Schanberg, H. Brunner, et al., for the Ad Hoc Committee on Lupus Response Criteria: Cognition Sub-committee. Proposed response criteria for neurocognitive impairment in systemic lupus erythematosus clinical trials, Lupus 16 (2007) 418–425.

[74] H.I. Brunner, F. Zelko, L. Schanberg, K. Anthony, G. Ross, M.S. Klein-Gitelman (Eds.), Standardizing the Neuropsychological Evaluation of Children with SLE. 7th European Lupus Congress, Sage, Pub, Amsterdam, 2008.

[75] H.I. Brunner, N.M. Ruth, A. German, S. Nelson, M.H. Passo, T. Roebuck-Spencer, et al., Initial validation of the Pediatric Automated Neuropsychological Assessment Metrics for childhood-onset systemic lupus erythematosus, Arthritis Rheum 57 (7) (2007) 1174–1182.

[76] L. Lapteva, M. Nowak, C.H. Yarboro, K. Takada, T. Roebuck-Spencer, T. Weickert, et al., Anti-N-methyl-D-aspartate receptor antibodies, cognitive dysfunction, and depression in systemic lupus erythematosus, Arthritis Rheum 54 (8) (2006) 2505–2514.

[77] E.Y. McLaurin, S.L. Holliday, P. Williams, R.L. Brey, Predictors of cognitive dysfunction in patients with systemic lupus erythematosus, Neurology 64 (2) (2005) 297–303.

[78] D. Levy, Anti-NR-2 antibodies are prevalent in childhood-onset SLE (cSLE), but do not predict neurocognitive dysfunction [abstract], Arthritis Rheum 55 (2007) S143.

[79] M. Reichlin, T.F. Broyles, O. Hubscher, J. James, T.A. Lehman, R. Palermo, et al., Prevalence of autoantibodies to ribosomal P proteins in juvenile-onset systemic lupus erythematosus compared with the adult disease, Arthritis Rheum 42 (1) (1999) 69–75.

[80] M.W. DiFrancesco, S.K. Holland, M.D. Ris, C.M. Adler, S. Nelson, M.P. DelBello, et al., Functional magnetic resonance imaging assessment of cognitive function in childhood-onset systemic lupus erythematosus: a pilot study, Arthritis Rheum 56 (12) (2007) 4151–4163.

[81] M. Mortilla, M. Ermini, M. Nistri, G. Dal Pozzo, F. Falcini, Brain study using magnetic resonance imaging and proton MR spectroscopy in pediatric onset systemic lupus erythematosus, Clin Exp Rheumatol 21 (1) (2003) 129–135.

[82] C.L. Yancey, R.A. Doughty, B.H. Athreya, Central nervous system involvement in childhood systemic lupus erythematosus, Arthritis Rheum 24 (11) (1981) 1389–1395.

[83] V. Baca, C. Lavalle, R. Garcia, T. Catalan, J.M. Sauceda, G. Sanchez, et al., Favorable response to intravenous methylprednisolone and cyclophosphamide in children with severe neuropsychiatric lupus, J Rheumatol 26 (2) (1999) 432–439.

[84] D.M. Levy, M.P. Massicotte, E. Harvey, D. Hebert, E.D. Silverman, Thromboembolism in paediatric lupus patients, Lupus 12 (10) (2003) 741–746.

[85] M. Zappitelli, C.M. Duffy, C. Bernard, I.R. Gupta, Evaluation of activity, chronicity and tubulointerstitial indices for childhood lupus nephritis, Pediatr Nephrol 23 (1) (2008) 83–91.

[86] A.S. Al Arfaj, N. Khalil, S. Al Saleh, Lupus nephritis among 624 cases of systemic lupus erythematosus in Riyadh, Saudi Arabia, Rheumatol Int 29 (9) (2009) 1057–1067.

[87] A.O. Hersh, E. von Scheven, J. Yazdany, P. Panopalis, L. Trupin, L. Julian, et al., Differences in long-term disease activity and treatment of adult patients with childhood- and adult-onset systemic lupus erythematosus, Arthritis Rheum 61 (1) (2009) 13–20.

[88] L.B. Tucker, S. Menon, J.G. Schaller, D.A. Isenberg, Adult- and childhood-onset systemic lupus erythematosus: a comparison of onset, clinical features, serology, and outcome, Br J Rheumatol 34 (9) (1995) 866–872.

[89] J. Dudley, T. Fenton, J. Unsworth, T. Chambers, A. MacIver, J. Tizard, Systemic lupus erythematosus presenting as congenital nephrotic syndrome, Pediatr Nephrol 10 (6) (1996) 752–755.

[90] M. Zappitelli, C. Duffy, C. Bernard, R. Scuccimarri, K. Watanabe Duffy, R. Kagan, et al., Clinicopathological study of the WHO classification in childhood lupus nephritis, Pediatr Nephrol 19 (5) (2004) 503–510.

[91] F. Perfumo, A. Martini, Lupus nephritis in children, Lupus 14 (1) (2005) 83–88.

[92] B.E. Ostrov, W. Min, A.H. Eichenfield, D.P. Goldsmith, B. Kaplan, B.H. Athreya, Hypertension in children with systemic lupus erythematosus, Semin Arthritis Rheum 19 (2) (1989) 90–98.

[93] S.D. Marks, N.J. Sebire, C. Pilkington, K. Tullus, Clinicopathological correlations of paediatric lupus nephritis, Pediatr Nephrol 22 (1) (2007) 77–83.

[94] S. Hagelberg, Y. Lee, J. Bargman, G. Mah, R. Schneider, C. Laskin, et al., Longterm followup of childhood lupus nephritis, J Rheumatol 29 (12) (2002) 2635–2642.

[95] T.J. Lehman, K. Onel, Intermittent intravenous cyclophosphamide arrests progression of the renal chronicity index in childhood systemic lupus erythematosus, J Pediatr 136 (2) (2000) 243–247.

[96] D. Askenazi, B. Myones, A. Kamdar, R. Warren, M. Perez, M. De Guzman, et al., Outcomes of children with proliferative lupus nephritis: the role of protocol renal biopsy, Pediatr Nephrol 22 (7) (2007) 981–986.

[97] Y.Y. Leung, C.C. Szeto, L.S. Tam, C.W. Lam, E.K. Li, K.C. Wong, et al., Urine protein-to-creatinine ratio in an untimed urine collection is a reliable measure of proteinuria in lupus nephritis, Rheumatology (Oxford) 46 (4) (2007) 649–652.

[98] C.H. Hinze, M. Suzuki, M. Klein-Gitelman, M.H. Passo, J. Olson, N.G. Singer, et al., Neutrophil Gelatinase-associated Lipocalin (NGAL) Predicts the course of global and renal childhood-onset systemic lupus erythematosus (SLE) disease activity, Arthritis Rheum 59 (60) (2009) 2272–2281.

[99] S.D. Marks, S.J. Williams, K. Tullus, N.J. Sebire, Glomerular expression of monocyte chemoattractant protein-1 is predictive of poor renal prognosis in pediatric lupus nephritis, Nephrol Dial Transplant 23 (11) (2008) 3521–3526.

[100] A.D. Hoftman, L.Q. Tai, S. Tze, D. Seligson, R.A. Gatti, D.K. McCurdy, MAGE-B2 autoantibody: a new biomarker for pediatric systemic lupus erythematosus, J Rheumatol 35 (12) (2008) 2430–2438.

[101] M. Suzuki, K. Wires, K.D. Greis, E.B. Brooks, K. Haines, M.S. Klein-Gitelman, J. Olson, et al., Initial validation of a novel protein biomarker panel for active pediatric lupus nephritis, Pediatr Res 65 (5) (2009) 530–536.

[102] J.V. Morales, R. Weber, M.B. Wagner, E.J. Barros, Is morning urinary protein/creatinine ratio a reliable estimator of 24-hour proteinuria in patients with glomerulonephritis and different levels of renal function? J Nephrol 17 (5) (2004) 666–672.

[103] S. Duran-Barragan, G. McGwin Jr., L.M. Vila, J.D. Reveille, G.S. Alarcon, Angiotensin-converting enzyme inhibitors delay the occurrence of renal involvement and are associated with a decreased risk of disease activity in patients with systemic lupus erythematosus—results from LUMINA (LIX): a multiethnic US cohort, Rheumatology (Oxford) 47 (7) (2008) 1093–1096.

[104] G. Barbano, R. Gusmano, B. Damasio, M.G. Alpigiani, A. Buoncompagni, M. Gattorno, et al., Childhood-onset lupus nephritis: a single-center experience of pulse intravenous cyclophosphamide therapy, J Nephrol 15 (2) (2002) 123–129.

[105] T.J. Lehman, B.S. Edelheit, K.B. Onel, Combined intravenous methotrexate and cyclophosphamide for refractory childhood lupus nephritis, Ann Rheum Dis 63 (3) (2004) 321–323.

[106] S.M. Benseler, J. Bargman, B.M. Feldman, P.N. Tyrrell, E. Harvey, D. Hebert, E.D. Silverman, Acute renal failure in paediatric systemic lupus erythematosus: treatment and outcome 48 (2) (2009) 176–182.

[107] F. Falcini, S. Capannini, G. Martini, F. La Torre, A. Vitale, F. Mangiantini, et al., Mycophenolate mofetil for the treatment of juvenile onset SLE: a multicenter study, Lupus 18 (2) (2009) 139–143.

[108] A.C. Sagcal-Gironella, K. Wiers, M. Klein-Gitelman, S. Nelson, D. Blair, S. Cox, et al. (Eds.), Pharmacokinetics (PK) and Pharmacodynamics (PD) of Mycophenolic Acid (MPA) and Their Relation to Response to Therapy of Childhood Systemic Lupus Erythematosus (cSLE) ACR, Wiley, San Francisco, CA, 2008.

[109] O. Nwobi, C.L. Abitbol, J. Chandar, W. Seeherunvong, G. Zilleruelo, Rituximab therapy for juvenile-onset systemic lupus erythematosus, Pediatr Nephrol 23 (3) (2008) 413–419.

[110] A. Segarra, M. Praga, N. Ramos, N. Polanco, I. Cargol E. Gutierrez-Solis M.R. Gomez, et al., Successful treatment of membranous glomerulonephritis with rituximab in calcineurin inhibitor-dependent patients Clin J Am Soc Nephrol. 4 (6) 1083–1088 Epub 2009 May 28 (2009).

[111] V. Baca, T. Catalan, M. Villasis-Keever, G. Ramon, A.M. Morales, F. Rodriguez-Leyva, Effect of low-dose cyclosporine A in the treatment of refractory proteinuria in childhood-onset lupus nephritis, Lupus 15 (8) (2006) 490–495.

[112] K.K. Lau, D.P. Jones, B.H. Ault, Prognosis of lupus membranous nephritis in children, Lupus 16 (1) (2007) 70.

[113] K.K. Lau, D.P. Jones, M.C. Hastings, L.W. Gaber, B.H. Ault, Short-term outcomes of severe lupus nephritis in a cohort of predominantly African-American children, Pediatr Nephrol 21 (5) (2006) 655–662.

[114] C.H. Cramer 2nd, M. Mills, R.P. Valentini, W.E. Smoyer, H. Haftel, P.D. Brophy, Clinical presentation and outcome in a cohort of paediatric patients with membranous lupus nephritis, Nephrol Dial Transplant 22 (12) (2007) 3495–3500.

[115] N. Baqi, S. Moazami, A. Singh, H. Ahmad, S. Balachandra, A. Tejani, Lupus nephritis in children: a longitudinal study of prognostic factors and therapy, J Am Soc Nephrol 7 (6) (1996) 924–929.

[116] L. Hiraki, S. Benseler, P. Tyrrell, E. Harvey, D. Hebert, E. Silverman, Ethnic Differences in Pediatric Systemic Lupus Erythematosus. J Rheumatol published online before print 36 (October 15, 2009) 2539–2546.

[117] T. Apenteng, B. Kaplan, K. Meyers, Renal outcomes in children with lupus and a family history of autoimmune disease, Lupus 15 (2) (2006) 65–70.

[118] S. Emre, I. Bilge, A. Sirin, I. Kilicaslan, A. Nayir, F. Oktem, et al., Lupus nephritis in children: prognostic significance of clinicopathological findings, Nephron 87 (2) (2001) 118–126.

[119] M. Faurschou, H. Starklint, P. Halberg, S. Jacobsen, Prognostic factors in lupus nephritis: diagnostic and therapeutic delay increases the risk of terminal renal failure, J Rheumatol 33 (8) (2006) 1563–1569.

[120] S. Lionaki, P.P. Kapitsinou, A. Iniotaki, A. Kostakis, H.M. Moutsopoulos, J.N. Boletis, Kidney transplantation in lupus patients: a case-control study from a single centre, Lupus 17 (7) (2008) 670–675.

[121] P. Cochat, S. Fargue, G. Mestrallet, T. Jungraithmayr, P. Koch-Nogueira, B. Ranchin, et al., Disease recurrence in paediatric renal transplantation, Pediatr Nephrol 24 (2009) 2097–2108.

[122] S.M. Bartosh, R.N. Fine, E.K. Sullivan, Outcome after transplantation of young patients with systemic lupus erythematosus: a report of the North American pediatric renal transplant cooperative study, Transplantation 72 (5) (2001) 973–978.

[123] M.W. Beresford, A.G. Cleary, J.A. Sills, J. Couriel, J.E. Davidson, Cardio-pulmonary involvement in juvenile systemic lupus erythematosus, Lupus 14 (2) (2005) 152–158.

[124] T.T. Yeh, Y.H. Yang, Y.T. Lin, C.S. Lu, B.L. Chiang, Cardiopulmonary involvement in pediatric systemic lupus erythematosus: a twenty-year retrospective analysis, J Microbiol Immunol Infect 40 (6) (2007) 525–531.

[125] A.C. Oshiro, S.J. Derbes, A.R. Stopa, A. Gedalia, Anti-Ro/SS-A and anti-La/SS-B antibodies associated with cardiac involvement in childhood systemic lupus erythematosus, Ann Rheum Dis 56 (4) (1997) 272–274.

[126] J.P. Guevara, B.J. Clark, B.H. Athreya, Point prevalence of cardiac abnormalities in children with systemic lupus erythematosus, J Rheumatol 28 (4) (2001) 854–859.

[127] M. Gazarian, B.M. Feldman, L.N. Benson, D.L. Gilday, R.M. Laxer, E.D. Silverman, Assessment of myocardial perfusion and function in childhood systemic lupus erythematosus, J Pediatr 132 (1) (1998) 109–116.

[128] P.C. Chow, M.H. Ho, T.L. Lee, Y.L. Lau, Y.F. Cheung, Relation of arterial stiffness to left ventricular structure and function in adolescents and young adults with pediatric-onset systemic lupus erythematosus, J Rheumatol 34 (6) (2007) 1345–1352.

[129] N. Gunal, N. Kara, N. Akkok, N. Cakar, O. Kahramanyol, N. Akalin, Cardiac abnormalities in children with systemic lupus erythematosus, Turk J Pediatr 45 (4) (2003) 301–305.

[130] A.J. Al-Abbad, D.A. Cabral, S. Sanatani, G.G. Sandor, M. Seear, R.E. Petty, et al., Echocardiography and pulmonary function testing in childhood onset systemic lupus erythematosus, Lupus 10 (1) (2001) 32–37.

[131] G.J. Pons-Estel, L.A. Gonzalez, J. Zhang, P.I. Burgos, J.D. Reveille, L.M. Vila, et al., Predictors of cardiovascular damage in patients with systemic lupus erythematosus: data from LUMINA (LXVIII), a multiethnic US cohort, Rheumatology (Oxford) 48 (7) (2009) 817–822.

[132] C. Korkmaz, D.U. Cansu, T. Kasifoglu, Myocardial infarction in young patients (< or =35 years of age) with systemic lupus erythematosus: a case report and clinical analysis of the literature, Lupus 16 (4) (2007) 289–297.

[133] F. Falaschi, A. Ravelli, A. Martignoni, D. Migliavacca, M. Sartori, et al., Nephrotic-range proteinuria, the major risk factor for early atherosclerosis in juvenile-onset systemic lupus erythematosus, Arthritis Rheum 43 (6) (2000) 1405–1409.

[134] L.E. Schanberg, C. Sandborg, H.X. Barnhart, S.P. Ardoin, E. Yow, G.W. Evans, et al., Premature atherosclerosis in pediatric systemic lupus erythematosus: risk factors for increased carotid intima-media thickness in the atherosclerosis prevention in pediatric lupus erythematosus cohort, Arthritis Rheum 60 (5) (2009) 1496–1507.

[135] L.E. Schanberg, C. Sandborg, Dyslipoproteinemia and premature atherosclerosis in pediatric systemic lupus erythematosus, Curr Rheumatol Rep 6 (6) (2004) 425–433.

[136] S.P. Ardoin, C. Sandborg, L.E. Schanberg, Management of dyslipidemia in children and adolescents with systemic lupus erythematosus, Lupus 16 (8) (2007) 618–626.

[137] J.B. Soep, M. Mietus-Snyder, M.J. Malloy, J.L. Witztum, E. von Scheven, Assessment of atherosclerotic risk factors and endothelial function in children and young adults with pediatric-onset systemic lupus erythematosus, Arthritis Rheum 51 (3) (2004) 451–457.

[138] C. Posadas-Romero, M. Torres-Tamayo, J. Zamora-Gonzalez, B.E. Aguilar-Herrera, R. Posadas-Sanchez, G. Cardoso-Saldana, et al., High insulin levels and increased low-density lipoprotein oxidizability in pediatric patients with systemic lupus erythematosus, Arthritis Rheum 50 (1) (2004) 160–165.

[139] M. El Magadmi, Y. Ahmad, W. Turkie, A.P. Yates, N. Sheikh, R.M. Bernstein, et al., Hyperinsulinemia, insulin resistance, and circulating oxidized low density lipoprotein in women with systemic lupus erythematosus, J Rheumatol 33 (1) (2006) 50–56.

[140] C.P. Chung, I. Avalos, A. Oeser, T. Gebretsadik, A. Shintani, P. Raggi, et al., High prevalence of the metabolic syndrome in patients with systemic lupus erythematosus: association with disease characteristics and cardiovascular risk factors, Ann Rheum Dis 66 (2) (2007) 208–214.

[141] K.M. Houghton, L.B. Tucker, J.E. Potts, D.C. McKenzie, Fitness, fatigue, disease activity, and quality of life in pediatric lupus, Arthritis Rheum 59 (4) (2008) 537–545.

[142] N.T. Ilowite, N. Copperman, T. Leicht, T. Kwong, M.S. Jacobson, Effects of dietary modification and fish oil supplementation on dyslipoproteinemia in pediatric systemic lupus erythematosus, J Rheumatol 22 (7) (1995) 1347–1351.

[143] B.A. Eberhard, R.M. Laxer, A.A. Eddy, E.D. Silverman, Presence of thyroid abnormalities in children with systemic lupus erythematosus, J Pediatr 119 (2) (1991) 277–279.

[144] T.C. Robazzi, C. Alves, M. Mendonca, Acute suppurative thyroiditis as the initial presentation of juvenile systemic lupus erythematosus, J Pediatr Endocrinol Metab 22 (4) (2009) 379–383.

[145] B.H. Athreya, J.H. Rafferty, G.S. Sehgal, R.G. Lahita, Adenohypophyseal and sex hormones in pediatric rheumatic diseases, J Rheumatol 20 (4) (1993) 725–730.

[146] C.A. Silva, M.O. Hilario, M.V. Febronio, S.K. Oliveira, M.T. Terreri, S.B. Sacchetti, et al., Risk factors for amenorrhea in juvenile systemic lupus erythematosus (JSLE): a Brazilian multicentre cohort study, Lupus 16 (7) (2007) 531–536.

[147] F. Blanco-Favela, G. Quintal Ma, A.K. Chavez-Rueda, A. Leanos-Miranda, R. Berron-Peres, V. Baca-Ruiz, et al., Anti-prolactin autoantibodies in paediatric systemic lupus erythematosus patients, Lupus 10 (11) (2001) 803–808.

[148] C.A. Silva, H.I. Brunner, Gonadal functioning and preservation of reproductive fitness with juvenile systemic lupus erythematosus, Lupus 16 (8) (2007) 593–599.

[149] H.I. Brunner, A. Bishnoi, A.C. Barron, L.J. Houk, A. Ware, Y. Farhey, et al., Disease outcomes and ovarian function of childhood-onset systemic lupus erythematosus, Lupus 15 (4) (2006) 198–206.

[150] H. Paterson, J. Ashton, M. Harrison-Woolrych, A nationwide cohort study of the use of the levonorgestrel intrauterine device in New Zealand adolescents, Contraception 79 (6) (2009) 433–438.

[151] R.M. Suehiro, E.F. Borba, E. Bonfa, T.S. Okay, M. Cocuzza, P.M. Soares, et al., Testicular Sertoli cell function in male systemic lupus erythematosus, Rheumatology (Oxford) 47 (11) (2008) 1692–1697.

[152] C.A. Silva, J. Hallak, F.F. Pasqualotto, M.F. Barba, M.I. Saito, M.H. Kiss, Gonadal function in male adolescents and young males with juvenile onset systemic lupus erythematosus, J Rheumatol 29 (9) (2002) 2000–2005.

[153] O. Richer, T. Ulinski, I. Lemelle, B. Ranchin, C. Loirat, J.C. Piette, et al., Abdominal manifestations in childhood-onset systemic lupus erythematosus, Ann Rheum Dis 66 (2) (2007) 174–178.

[154] D.S. Hallegua, D.J. Wallace, Gastrointestinal manifestations of systemic lupus erythematosus, Curr Opin Rheumatol 12 (5) (2000) 379–385.

[155] S.M. Sultan, Y. Ioannou, D.A. Isenberg, A review of gastrointestinal manifestations of systemic lupus erythematosus, Rheumatology (Oxford) 38 (10) (1999) 917–932.

[156] C.K. Lee, M.S. Ahn, E.Y. Lee, J.H. Shin, Y.S. Cho, H.K. Ha, et al., Acute abdominal pain in systemic lupus erythematosus: focus on lupus enteritis (gastrointestinal vasculitis), Ann Rheum Dis 61 (6) (2002) 547–550.

[157] R.A. Apak, N. Besbas, S. Ozdemir, H. Ozen, A. Bakkaloglu, U. Saatci, Hepatitis as the presenting symptom of childhood systemic lupus erythematosus, Turk J Pediatr 41 (4) (1999) 541–544.

[158] F.C. Arnett, M. Reichlin, Lupus hepatitis: an under-recognized disease feature associated with autoantibodies to ribosomal P, Am J Med 99 (5) (1995) 465–472.

[159] Y. Usta, F. Gurakan, Z. Akcoren, S. Ozen, An overlap syndrome involving autoimmune hepatitis and systemic lupus erythematosus in childhood, World J Gastroenterol 13 (19) (2007) 2764–2767.

[160] Y. Karakoc, K. Dilek, M. Gullulu, M. Yavuz, A. Ersoy, H. Akalyn, et al., Prevalence of hepatitis C virus antibody in patients with systemic lupus erythematosus, Ann Rheum Dis 56 (9) (1997) 570–571.

[161] L. Perrin, I. Giurgea, V. Baudet-Bonneville, G. Deschenes, A. Bensman, T. Ulinski, Acute pancreatitis in paediatric systemic lupus erythematosus, Acta Paediatr 95 (1) (2006) 121–124.

[162] S.M. Al-Mayouf, M. Majeed, A. Al-Mehaidib, H. Alsuhaibani, Pancreatic pseudocyst in paediatric systemic lupus erythematosus, Clin Rheumatol 21 (3) (2002) 264–266.

[163] J.C. Penalva, J. Martinez, E. Pascual, V.M. Palanca, F. Lluis, F. Peiro, et al., Chronic pancreatitis associated with systemic lupus erythematosus in a young girl, Pancreas 27 (3) (2003) 275–277.

[164] M. Mukamel, Y. Rosenbach, I. Zahavi, M. Mimouni, G. Dinari, Celiac disease associated with systemic lupus erythematosus, Isr J Med Sci 30 (8) (1994) 656–658.

[165] J.J. Swigris, A. Fischer, J. Gillis, R.T. Meehan, K.K. Brown, Pulmonary and thrombotic manifestations of systemic lupus erythematosus, Chest 133 (1) (2008) 271–280.

[166] B. Memet, E.M. Ginzler, Pulmonary manifestations of systemic lupus erythematosus, Semin Respir Crit Care Med 28 (4) (2007) 441–450.

[167] S. Trapani, G. Camiciottoli, M. Ermini, W. Castellani, F. Falcini, Pulmonary involvement in juvenile systemic lupus erythematosus: a study on lung function in patients asymptomatic for respiratory disease, Lupus 7 (8) (1998) 545–550.

[168] C.M. Laroche, D.A. Mulvey, P.N. Hawkins, M.J. Walport, B. Strickland, J. Moxham, et al., Diaphragm strength in the shrinking lung syndrome of systemic lupus erythematosus, Q J Med 71 (265) (1989) 429–439.

[169] P.J. Thompson, D.P. Dhillon, J. Ledingham, M. Turner-Warwick, Shrinking lungs, diaphragmatic dysfunction, and systemic lupus erythematosus, Am Rev Respir Dis 132 (4) (1985) 926–928.

[170] P.J. Ferguson, M. Weinberger, Shrinking lung syndrome in a 14-year-old boy with systemic lupus erythematosus, Pediatr Pulmonol 41 (2) (2006) 194–197.

[171] K.J. Warrington, K.G. Moder, W.M. Brutinel, The shrinking lungs syndrome in systemic lupus erythematosus, Mayo Clin Proc 75 (5) (2000) 467–472.

[172] M.Y. Karim, L.C. Miranda, C.M. Tench, P.A. Gordon, D.P. D'Cruz, M.A. Khamashta, et al., Presentation and prognosis of the shrinking lung syndrome in systemic lupus erythematosus, Semin Arthritis Rheum 31 (5) (2002) 289–298.

[173] A.S. Samad, C.B. Lindsley, Treatment of pulmonary hemorrhage in childhood systemic lupus erythematosus with myco-phenolate mofetil, South Med J 96 (7) (2003) 705–707.

[174] M.F. Liu, J.H. Lee, T.H. Weng, Y.Y. Lee, Clinical experience of 13 cases with severe pulmonary hemorrhage in systemic lupus erythematosus with active nephritis, Scand J Rheumatol 27 (4) (1998) 291–295.

[175] E.N. Pattishall, B.E. Noyes, D.M. Orenstein, Use of bronchoalveolar lavage in immunocompromised children with pneumonia, Pediatr Pulmonol 5 (1) (1988) 1–5.

[176] J.L. McGhee, L.M. Kickingbird, J.N. Jarvis, Clinical utility of antinuclear antibody tests in children, BMC Pediatr 4 (2004) 13.

[177] N.A. Dayal, C. Gordon, L. Tucker, D.A. Isenberg, The SLICC damage index: past, present and future, Lupus 11 (4) (2002) 261–265.

[178] L.B. Tucker, Controversies and advances in the management of systemic lupus erythematosus in children and adolescents, Best Pract Res Clin Rheumatol 16 (3) (2002) 471–480.

[179] L. Carreno, F.J. Lopez-Longo, I. Monteagudo, M. Rodriguez-Mahou, M. Bascones, C.M. Gonzalez, et al., Immunological and clinical differences between juvenile and adult onset of systemic lupus erythematosus, Lupus 8 (4) (1999) 287–292.

[180] C.H. To, M. Petri, Is antibody clustering predictive of clinical subsets and damage in systemic lupus erythematosus? Arthritis Rheum 52 (12) (2005) 4003–4010.

[181] R. Jurencak, M. Fritzler, P. Tyrrell, L. Hiraki, S. Benseler, E. Silverman, Autoantibodies in pediatric systemic lupus erythematosus: ethnic grouping, cluster analysis, and clinical correlations, J Rheumatol 36 (2) (2009) 416–421.

[182] A. Souza, L.M. da Silva, F.R. Oliveira, A.M. Roselino, P. Louzada-Junior, Anti-nucleosome and anti-chromatin antibodies are present in active systemic lupus erythematosus but not in the cutaneous form of the disease, Lupus 18 (3) (2009) 223–229.

[183] L.M. Campos, M.H. Kiss, M.A. Scheinberg, C.L. Mangueira, C.A. Silva, Antinucleosome antibodies in patients with juvenile systemic lupus erythematosus, Lupus 15 (8) (2006) 496–500.

[184] P.F. Yong, L. Aslam, M.Y. Karim, M.A. Khamashta, Management of hypogammaglobulinaemia occurring in patients with systemic lupus erythematosus, Rheumatology (Oxford) 47 (9) (2008) 1400–1405.

[185] Y.L. Wu, G.C. Higgins, R.M. Rennebohm, E.K. Chung, Y. Yang, B. Zhou, et al., Three distinct profiles of serum complement C4 proteins in pediatric systemic lupus erythematosus (SLE) patients: tight associations of complement C4 and C3 protein levels in SLE but not in healthy subjects, Adv Exp Med Biol 586 (2006) 227–247.

[186] S.R. Block, J.B. Winfield, M.D. Lockshin, W.A. D'Angelo, C.L. Christian, Studies of twins with systemic lupus erythematosus. A review of the literature and presentation of 12 additional sets, Am J Med 59 (4) (1975) 533–552.

[187] D. Deapen, A. Escalante, L. Weinrib, D. Horwitz, B. Bachman, P. Roy-Burman, et al., A revised estimate of twin concordance in systemic lupus erythematosus, Arthritis Rheum 35 (3) (1992) 311–318.

[188] S.S. Kamat, P.H. Pepmueller, T.L. Moore, Triplets with systemic lupus erythematosus, Arthritis Rheum 48 (11) (2003) 3176–3180.

[189] T.J. Vyse, J.A. Todd, Genetic analysis of autoimmune disease, Cell 85 (3) (1996) 311–318.

[190] G. Bussone, L. Mouthon, Autoimmune manifestations in primary immune deficiencies, Autoimmun Rev 8 (4) (2009) 332–336.

[191] N.M. Stone, A. Williams, J.D. Wilkinson, G. Bird, Systemic lupus erythematosus with C1q deficiency, Br J Dermatol 142 (3) (2000) 521–524.

[192] Y. Yang, K. Lhotta, E.K. Chung, P. Eder, F. Neumair, C.Y. Yu, Complete complement components C4A and C4B deficiencies in human kidney diseases and systemic lupus erythematosus, J Immunol 173 (4) (2004) 2803–2814.

[193] M.L. Barilla-LaBarca, J.P. Atkinson, Rheumatic syndromes associated with complement deficiency, Curr Opin Rheumatol 15 (1) (2003) 55–60.

[194] L.A. Bennett, A.E. Palucka, E. Arce, V. Cantrell, J. Borvak, J. Banchereau, et al., Interferon and Granulopoiesis Signatures in Systemic Lupus Erythematosus Blood, The Journal of Experimental Medicine, vol. 197 (Number 6) (2003) 711–723.

[195] J.D. Reveille, R.E. Schrohenloher, R.T. Acton, B.O. Barger, DNA analysis of HLA-DR and DQ genes in American blacks with systemic lupus erythematosus, Arthritis Rheum 32 (10) (1989) 1243–1251.

[196] K.S. Barron, E.D. Silverman, J. Gonzales, J.D. Reveille, Clinical, serologic, and immunogenetic studies in childhood-onset systemic lupus erythematosus, Arthritis Rheum 36 (3) (1993) 348–354.

[197] C.H. Liao, T.C. Yao, H.T. Chung, L.C. See, M.L. Kuo, J.L. Huang, Polymorphisms in the promoter region of RANTES and the regulatory region of monocyte chemoattractant protein-1 among Chinese children with systemic lupus erythematosus, J Rheumatol 31 (10) (2004) 2062–2067.

[198] G. Lettre, J.D. Rioux, Autoimmune diseases: insights from genome-wide association studies, Hum Mol Genet 17 (R2) (2008) R116–R121.

[199] C.O. Jacob, A. Reiff, D.L. Armstrong, B.L. Myones, E. Silverman, M. Klein-Gitelman, et al., Identification of novel susceptibility genes in childhood-onset systemic lupus erythematosus using a uniquely designed candidate gene pathway platform, Arthritis Rheum 56 (12) (2007) 4164–4173.

[200] R. Velazquez-Cruz, L. Orozco, F. Espinosa-Rosales, R. Carreno-Manjarrez, E. Solis-Vallejo, N.D. Lopez-Lara, et al., Association of PDCD1 polymorphisms with childhood-onset systemic lupus erythematosus, Eur J Hum Genet 15 (3) (2007) 336–341.

[201] M. Gattorno, F. Falcini, A. Ravelli, F. Zulian, A. Buoncompagni, G. Martini, et al., Outcome of primary antiphospholipid syndrome in childhood, Lupus 12 (6) (2003) 449–453.

[202] M. Gattorno, A.C. Molinari, A. Buoncompagni, M. Acquila, S. Amato, P. Picco, Recurrent antiphospholipid-related deep vein thrombosis as presenting manifestation of systemic lupus erythematosus, Eur J Pediatr 159 (3) (2000) 211–214.

[203] L.M. Campos, M.H. Kiss, E.A. D'Amico, C.A. Silva, Antiphospholipid antibodies in 57 children and adolescents with systemic lupus erythematosus, Rev Hosp Clin Fac Med Sao Paulo 58 (3) (2003) 157–162.

[204] A. Ravelli, A. Martini, G.R. Burgio, Antiphospholipid antibodies in paediatrics, Eur J Pediatr 153 (7) (1994) 472–479.

[205] A. Ravelli, R. Caporali, G. Di Fuccia, L. Zonta, C. Montecucco, A. Martini, Anticardiolipin antibodies in pediatric systemic lupus erythematosus, Arch Pediatr Adolesc Med 148 (4) (1994) 398–402.

[206] R. Kaiser, C.M. Cleveland, L.A. Criswell, Risk and protective factors for thrombosis in systemic lupus erythematosus: results from a large, multi-ethnic cohort, Ann Rheum Dis 68 (2) (2009) 238–241.

[207] P.E. Love, S.A. Santoro, Antiphospholipid antibodies: anticardiolipin and the lupus anticoagulant in systemic lupus erythematosus (SLE) and in non-SLE disorders. Prevalence and clinical significance, Ann Intern Med 112 (9) (1990) 682–698.

[208] C. Berube, L. Mitchell, E. Silverman, M. David, C. Saint Cyr, R. Laxer, et al., The relationship of antiphospholipid antibodies to thromboembolic events in pediatric patients with systemic lupus erythematosus: a cross-sectional study, Pediatr Res 44 (3) (1998) 351–356.

[209] M. Schmugge, S. Revel-Vilk, L. Hiraki, M.L. Rand, V.S. Blanchette, E.D. Silverman, Thrombocytopenia and thromboembolism in pediatric systemic lupus erythematosus, J Pediatr 143 (5) (2003) 666–669.

[210] H.F. Pan, D.Q. Ye, Q. Wang, W.X. Li, N. Zhang, X.P. Li, et al., Clinical and laboratory profiles of systemic lupus erythematosus associated with Sjogren syndrome in China: a study of 542 patients, Clin Rheumatol 27 (3) (2008) 339–343.

[211] N. Iwata, M. Mori, T. Miyamae, S. Ito, T. Imagawa, S. Yokota, [Sjogren's syndrome associated with childhood-onset systemic lupus erythematosus], Nihon Rinsho Meneki Gakkai Kaishi 31 (3) (2008) 166–171.

[212] P.A. Ostuni, A. Ianniello, P. Sfriso, G. Mazzola, M. Andretta, P.F. Gambari, Juvenile onset of primary Sjogren's syndrome: report of 10 cases, Clin Exp Rheumatol 14 (6) (1996) 689–693.

[213] J.M. Anaya, N. Ogawa, N. Talal, Sjogren's syndrome in childhood, J Rheumatol 22 (6) (1995) 1152–1158.

[214] M. Civilibal, N. Canpolat, A. Yurt, S. Kurugoglu, S. Erdamar, O. Bagci, et al., A child with primary Sjogren syndrome and a review of the literature, Clin Pediatr (Phila) 46 (8) (2007) 738–742.

[215] M. Stiller, W. Golder, E. Doring, T. Biedermann, Primary and secondary Sjogren's syndrome in children—a comparative study, Clin Oral Investig 4 (3) (2000) 176—182.

[216] N. Maeno, S. Takei, H. Imanaka, H. Oda, K. Yanagi, Y. Hayashi, et al., Anti-alpha-fodrin antibodies in Sjogren's syndrome in children, J Rheumatol 28 (4) (2001) 860—864.

[217] K. Takahashi, O. Tatsuzawa, K. Yanagi, Y. Hayashi, H. Takahashi, Alpha-fodrin auto-antibody in Sjogren syndrome and other auto-immune diseases in childhood, Eur J Pediatr 160 (8) (2001) 520—521.

[218] N.G. Singer, I. Tomanova-Soltys, R. Lowe, Sjogren's syndrome in childhood, Curr Rheumatol Rep 10 (2) (2008) 147—155.

[219] V. Sankar, M.T. Brennan, M.R. Kok, R.A. Leakan, J.A. Smith, J. Manny, et al., Etanercept in Sjogren's syndrome: a twelve-week randomized, double-blind, placebo-controlled pilot clinical trial, Arthritis Rheum 50 (7) (2004) 2240—2245.

[220] J.T. Cassidy, R.E. Petty, Textbook of pediatric rheumatology, chapter 5, pharmacology and drug therapy, fifth ed., Elsevier Saunders, Philadelphia, PA, 2005, p. S76-131.

[221] S. Koneru, M. Shishov, A. Ware, Y. Farhey, A.B. Mongey, T.B. Graham, et al., Effectively measuring adherence to medications for systemic lupus erythematosus in a clinical setting, Arthritis Rheum 57 (6) (2007) 1000—1006.

[222] M.A. Rapoff, Management of adherence and chronic rheumatic disease in children and adolescents, Best Pract Res Clin Rheumatol 20 (2) (2006) 301—314.

[223] V.K.D. Ting, S. Nelson, J. Huggins, J. Eaton, J. Rammel, D. Drotar, et al., Use of Cellular Text Messaging to Improve Visit Adherence in Adolescents with Childhood-onset Systemic Lupus Erythematosus (cSLE), ACR (2009).

[224] H.I. Brunner, B.M. Feldman, M.B. Urowitz, D.D. Gladman, Item weightings for the Systemic Lupus International Collaborating Clinics/American College of Rheumatology Disease Damage Index using Rasch analysis do not lead to an important improvement, J Rheumatol 30 (2) (2003) 292—297.

[225] H.I. Brunner, E.D. Silverman, C. Bombardier, B.M. Feldman, European Consensus Lupus Activity Measurement is sensitive to change in disease activity in childhood-onset systemic lupus erythematosus, Arthritis Rheum 49 (3) (2003) 335—341.

[226] R. Gutierrez-Suarez, N. Ruperto, R. Gastaldi, A. Pistorio, E. Felici, R. Burgos-Vargas, et al., A proposal for a pediatric version of the Systemic Lupus International Collaborating Clinics/American College of Rheumatology Damage Index based on the analysis of 1,015 patients with juvenile-onset systemic lupus erythematosus, Arthritis Rheum 54 (9) (2006) 2989—2996.

[227] H.I. Brunner, M.S. Klein-Gitelman, M.J. Miller, A. Barron, N. Baldwin, M. Trombley, et al., Minimal clinically important differences of the childhood health assessment questionnaire, J Rheumatol 32 (1) (2005) 150—161.

[228] N. Ruperto, A. Ravelli, S. Oliveira, M. Alessio, D. Mihaylova, S. Pasic, et al., The Pediatric Rheumatology International Trials Organization/American College of Rheumatology provisional criteria for the evaluation of response to therapy in juvenile systemic lupus erythematosus: prospective validation of the definition of improvement, Arthritis Rheum 55 (3) (2006) 355—363.

[229] H.H.G. Brunner, J. Ying, K. Wiers, T. Graham, S. Lapidus, J. Nocton, et al., Sensitivity & Specificity of Adult Lupus Flare Measures When Used in Pediatric Systemic Lupus Erythematosus, Abstract, American College of Rheumatology, Boston MA, (2007) S885.

[230] E.A. Mina R, E. von Scheven, G. Higgins, S. Lapidus, J. Eaton, L. Schanberg, et al., Defining and Measuring Global Flares of Juvenile Systemic Lupus Erythematosus, American College of Rheumatology Abstract, Philadelphia, PA (2009).

[231] S. Meiorin, A. Pistorio, A. Ravelli, S.M. Iusan, G. Filocamo, L. Trail, et al., Validation of the Childhood Health Assessment Questionnaire in active juvenile systemic lupus erythematosus, Arthritis Rheum 59 (8) (2008) 1112—1119.

[232] L.N. Moorthy, M.G. Peterson, M. Baratelli, M.J. Harrison, K.B. Onel, E.C. Chalom, et al., Multicenter validation of a new quality of life measure in pediatric lupus, Arthritis Rheum 57 (7) (2007) 1165—1173.

[233] H.I. Brunner, G.C. Higgins, K. Wiers, S.K. Lapidus, J.C. Olson, K. Onel, M. Punaro, et al., Health-related quality of life and its relationship to patient disease course in childhood-onset systemic lupus erythematosus, J Rheumatol 2009 36 (7) (2009) 1536—1545.

[234] H.I. Brunner, T.M. Sherrard, M.S. Klein-Gitelman, Cost of treatment of childhood-onset systemic lupus erythematosus, Arthritis Rheum 55 (2) (2006) 184—188.

[235] C.S. Lau, A. Mak, The socioeconomic burden of SLE, Nat Rev Rheumatol 2009 5 (7) (2009) 400—404.

[236] N. Sutcliffe, A.E. Clarke, R. Taylor, C. Frost, D.A. Isenberg, Total costs and predictors of costs in patients with systemic lupus erythematosus, Rheumatology (Oxford) 40 (1) (2001) 37—47.

[237] S.A. Chambers, E. Allen, A. Rahman, D. Isenberg, Damage and mortality in a group of British patients with systemic lupus erythematosus followed up for over 10 years, Rheumatology (Oxford) 48 (6) (2009) 673—675.

[238] A. Becker-Merok, H.C. Nossent, Damage accumulation in systemic lupus erythematosus and its relation to disease activity and mortality, J Rheumatol 33 (8) (2006) 1570—1577.

[239] E. Descloux, I. Durieu, P. Cochat, D. Vital Durand, J. Ninet, N. Fabien, et al., Paediatric systemic lupus erythematosus: prognostic impact of antiphospholipid antibodies, Rheumatology (Oxford) 47 (2) (2008) 183—187.

[240] M. Bandeira, S. Buratti, M. Bartoli, C. Gasparini, L. Breda, A. Pistorio, et al., Relationship between damage accrual, disease flares and cumulative drug therapies in juvenile-onset systemic lupus erythematosus, Lupus 15 (8) (2006) 515—520.

[241] C.D. Cook, R.J. Wedgwood, J.M. Craig, J.R. Hartmann, C.A. Janeway, Systemic lupus erythematosus. Description of 37 cases in children and a discussion of endocrine therapy in 32 of the cases, Pediatrics 26 (1960) 570—585.

[242] R.S. Glidden, E.C. Mantzouranis, Y. Borel, Systemic lupus erythematosus in childhood: clinical manifestations and improved survival in fifty-five patients, Clin Immunol Immunopathol 29 (2) (1983) 196—210.

[243] B. Gonzalez, P. Hernandez, H. Olguin, M. Miranda, L. Lira, M. Toso, et al., Changes in the survival of patients with systemic lupus erythematosus in childhood: 30 years experience in Chile, Lupus 14 (11) (2005) 918—923.

[244] F. Zulian, F. Pluchinotta, G. Martini, L. Da Dalt, G. Zacchello, Severe clinical course of systemic lupus erythematosus in the first year of life, Lupus 17 (9) (2008) 780—786.

[245] R. Cervera, M. Abarca-Costalago, D. Abramovicz, F. Allegri, P. Annunziata, A.O. Aydintug, et al., Lessons from the "Euro-Lupus Cohort", Ann Med Interne (Paris) 153 (8) (2002) 530—536.

[246] P.D. Ziakas, U.G. Dafni, S. Giannouli, A.G. Tzioufas, M. Voulgarelis, Thrombocytopaenia in lupus as a marker of adverse outcome—seeking Ariadne's thread, Rheumatology (Oxford) 45 (10) (2006) 1261—1265.

[247] C. Drenkard, A.R. Villa, C. Garcia-Padilla, M.E. Perez-Vazquez, D. Alarcon-Segovia, Remission of systematic lupus erythematosus, Medicine (Baltimore) 75 (2) (1996) 88—98.

[248] R.W. Blum, D. Garell, C.H. Hodgman, T.W. Jorissen, N.A. Okinow, D.P. Orr, et al., Transition from child-centered to adult health-care systems for adolescents with chronic conditions. A position paper of the Society for Adolescent Medicine, J Adolesc Health 14 (7) (1993) 570—576.

[249] L.B. Tucker, D.A. Cabral, Transition of the adolescent patient with rheumatic disease: issues to consider, Rheum Dis Clin North Am 33 (3) (2007) 661—672.

Drug-Induced Disease

Anne-Barbara Mongey, Evelyn V. Hess

INTRODUCTION

Autoimmunity can follow exposure to drugs or other agents. There are increasing numbers of reports and case series of patients who develop autoantibodies and lupus-like syndromes following exposure to certain medications and environmental agents. Thus far, over 80 pharmacologic agents have been implicated in the development of drug-related lupus (DRL) with more recent reports implicating biologic agents, such as tumor necrosis factor (TNF) alpha inhibitors (Table 33.1). The most frequently reported clinical features include constitutional symptoms, arthralgias/arthritis, serositis, and the development of autoantibodies. Typically there is resolution of clinical features on discontinuation of the drug, although autoantibodies may persist for prolonged periods of time. Studies suggest that genetic factors may play a role in determining susceptibility of patients to developing autoimmunity. Physicians need to be aware of possible associations between the development of autoimmune-like features and exposure to certain pharmacologic agents for both diagnostic and therapeutic reasons.

Systemic lupus erythematosus (SLE) is an autoimmune disease for which the etiology(ies) remain undetermined despite intensive research. There is evidence to suggest that genetic factors play a major role in the pathogenesis of this disease. However, results from monozygotic twin studies [1] would indicate that other, most likely environmental, factors are required for the induction of the disease. There have been many reports in the literature of the development of autoimmune phenomenon, such as autoantibodies and autoimmune-like diseases, following exposure to certain environmental agents including chemicals and drugs. The term "drug-related lupus" refers to the development of a lupus-like syndrome following exposure to a certain drug. Typically there is a rapid resolution of the lupus-like clinical features following removal of the offending

agent, although autoantibodies may persist for some time. With the introduction of new pharmacological agents, in particular the biological agents, there has been an increase in the number of reports of drug-related autoantibodies and lupus-like syndromes. Unfortunately, many of these reports are in the form of case reports that provide weak evidence for an association between the ingestion of a particular agent and the development of a lupus-like syndrome. In the absence of good controlled prospective studies on large numbers of patients, it can be difficult to confirm a definitive association between the ingestion of an agent and the development of an autoimmune-like disease. In some cases, exposure to the so-called offending agent may either have been coincidental with, or even possibly have resulted in, the induction of idiopathic SLE. It is important to try to make the distinction between this and the development of drug-related lupus since, in the case of the latter, removal of the offending agent should result in resolution of the lupus-like syndrome. Ideally, one would like to confirm the association by rechallenge with the offending agent; however, this is seldom done in clinical practice. With the ever-increasing availability of new medications, biological therapies, and herbal and other alternative remedies, physicians need to be aware of possible associations between the development of autoimmune-like diseases and the ingestion of these agents.

HISTORY

The first report of an association between ingestion of a drug and a lupus-like syndrome was made in 1945 by Hoffman [2] in which he described the development of fever, rash, myalgia, and nephritis in a 19-year-old man receiving sulfadiazine. Further reports implicating sulfonamides and penicillin followed in the 1950s [3—6] but since then, there has been no further evidence to

TABLE 33.1   Drugs Definitively Associated with Drug-Related Lupus

| Chlorpromazine |
| Hydralazine |
| Isoniazid |
| Methyldopa |
| Minocycline |
| Procainamide |
| Quinidine |

OTHER DRUGS ASSOCIATED WITH DRUG-RELATED LUPUS

Acebutolol	Griseofulvin
Acecainide	Guanoxan
Adalimumab	Ibuprofen
Allopurinol	Infliximab
Aminoglutethimide	Interferon alpha
Amiodarone	Interferon gamma
Amoproxan	Interleukin 2
Anthiomaline	Labetalol
Atenolol	Leuprolide acetate
Atorvastatin	Levadopa
Benoxaprofen	Levomeprazone
Captopril	Lithium Carbonate
Carbamazepine	Lovastatin
Chlorprothixene	Mephenytoin
Chlorthalidone	Mesalazine
Cimetidine	Methimazole
Cinnarazine	Methylsergide[a]
Clonidine	Methylthiouracil[a]
Dasatinib	Metoprolol
Danazol	Metrizamide
Diclofenac	Minoxidil
1−2-dimethyl-3 hydroxy-pyride-4−1	Nalidixic acid
Diphenylhydantoin	Nitrofurantoin
Disopyriamide	Nomifensine[a]
Enalapril	Oxyphenisatin
Estrogens	Oxyprenolol
Etanercept	para-amino salicyclic acid
Ethosuximide	Penicillamine
Ethylphenacemide	Penicillin
Gold salts	Perazine

(Continued)

TABLE 33.1   Drugs Definitively Associated with Drug-Related Lupus—cont'd

Perphenazine	Rifampicin
Phenelzine	Sertraline
Phenopyrazone[a]	Simvastatin
Phenylbutazone[a]	Spironolactone
Phenylethylacetylureaa	Streptomycin
Phenytoin	Sulindac
Practolol[a]	Sulfadimethoxine
Prazosin	Sulfamethoxypyridazine
Primidone	Sulfasalazine
Prindolol	Tetracyclines
Promethazine	Tetrazine
Propafenone	Thionamidea
Prophythiouracil	Thioridazide
Psopranolol	Ticlopidine
Psoralen[a]	Timolol eyedrops
Pyrathiazine	Tolazamide
Pyrithoxine	Tolmetin
Quinine	Trimethadione
Reserpine	

[a]No longer available.

support a definite association. It is now thought that an association with either of these drugs is unlikely and that previous reports may represent coincidental development of idiopathic SLE. It is also possible that the infection for which the antibiotic was prescribed may have played a role in the development of SLE, as suggested by the current increasing speculation on a relationship between viral and bacterial infections and SLE. Hydralazine was first incriminated in 1953 by Morrow et al. [7]. Since that time, there have been numerous other reports confirming this association [8, 9]. Anticonvulsants were first incriminated in 1957 when Lindqvist reported the development of a lupus-like syndrome in two patients administered mesantoin [10]. Procainamide, which is the most common cause of DRL, was first implicated by Ladd in 1962 [11]. A report from France in 1966 suggested an association with isoniazid, which was subsequently supported by others [12, 13]. In 1971, Fabius and Gaulhofer [14] implicated the psychotrophic agents. Since the 1970s, at least 80 pharmacological and other agents have been reported to be associated with DRL (Table 33.1). These include a variety of antiarrhythmic, antihypertensive, antipsychotic, antithyroid, antibiotic, antirheumatic, and various biological agents. With increasing physician awareness and introduction of new medications, it is likely that this number will continue to increase. Of note, it is more common to find reports of an association with serological changes alone than with the development of DRL.

DEFINITION

In contrast to idiopathic SLE for which the American College of Rheumatology has identified specific diagnostic criteria, there are no definitive criteria for DRL [15]. An important feature of DRL is a temporal association between the ingestion of an agent and development of lupus-like features with remission of the features and decrease in autoantibody titers following withdrawal of the offending agent. Recurrence of the lupus-like features following reintroduction of the agent would provide confirmatory evidence, although this is seldom done in clinical practice. Many patients develop autoantibodies while taking certain medications, but only a minority develop lupus-like features. Hence, the development of autoantibodies in the absence of clinical features is insufficient to make the diagnosis or reason to discontinue the medication.

PREVALENCE

The real prevalence of DRL is unknown. It has been estimated that approximately one million people in the United States have idiopathic SLE and that up to 5% of these may be related to the ingestion of a pharmacological agent. Procainamide is considered to be the commonest cause of DRL, with up to 30% of patients receiving this medication developing DRL. However, due to the lack of prospective studies it is difficult to predict the incidence of DRL for other medications.

The prevalence of DRL among certain populations is influenced by the pattern of drug prescription within these groups. With the increasing availability of alternative drugs and therapies, it is likely that the incidence of procainamide- and hydralazine-related lupus will decrease. However, physicians need to be alert to the possibility of new associations, including possible associations with biological agents, herbs, and other such compounds, which are being used more frequently.

DEMOGRAPHIC FEATURES

The demographic features of DRL largely reflect those of the diseases for which the incriminated drugs are being prescribed. Hence, DRL occurs more frequently in the elderly population because of their higher incidence of diseases for which the drugs are prescribed. Exceptions include the anticonvulsant drugs and minocycline. Anticonvulsants are more commonly prescribed for children and, hence, most of the reports of anticonvulsant-related lupus have been described in younger age groups [16, 17]. Minocycline is generally used to treat acne in young females and, therefore, minocycline-related lupus has been most frequently reported in this population. In contrast to idiopathic SLE, which has a male to female ratio of 1:9, DRL occurs only slightly more frequently in females [16, 18]. This reflects the demographic features of the underlying diseases for which the drugs are prescribed. Drug-related lupus in the United States occurs more frequently in Caucasians than in African-Americans, unlike idiopathic SLE.

CLINICAL FEATURES

Typical features of DRL include constitutional symptoms, arthralgias/arthritis, myalgias, serositis, elevated ESR, and a positive ANA, which is generally accepted as a requirement for the diagnosis. Specific clinical features for the diagnosis of DRL have not as yet been defined (Table 33.2). Hence, the ARA revised criteria for the classification of SLE are often used as a guideline. However, as noted, most patients with DRL have fewer than four criteria. A set of ascertainment criteria has been proposed by the Environmental Study Section of the American College of Rheumatology, which include a positive ANA [19]. Features take weeks to months to develop following exposure to the medication. Withdrawal of the offending agent generally leads to significant improvement in symptoms within days and rarely weeks, but autoantibodies tend to persist for months and even years.

In the majority of patients, the symptoms are relatively mild compared with idiopathic SLE and acutely ill patients are rarely seen. DRL patients frequently present with constitutional symptoms such as malaise, low-grade fever, and myalgia, which may occur acutely or insidiously. Articular features such as arthralgias and arthritis occur in over 80%. Arthralgias occur more commonly than frank arthritis. As seen with idiopathic SLE, there is typically symmetrical polyarticular involvement of mild to moderate intensity. Articular involvement most commonly affects the small joints of the hands followed by wrists, elbows, shoulders, ankles, and knees. Typically the synovial fluid is noninflammatory, although inflammatory fluids have also been reported [20]; effusions are an uncommon finding. Pleuropulmonary disease occurs most frequently in procainamide-related lupus. Manifestations include pleurisy, pleural effusion, and pulmonary infiltrates.

TABLE 33.2 Frequency of Clinical and Laboratory Features in Idiopathic SLE and DRL

Features	Idiopathic SLE	DRL
Constitutional	40—85	40—50
Arthralgias/arthritis	75—95	80—95
Myalgias	40—80	35—57
Rash	50—70	0—30
Lymphadenopathy	23—67	<15
Pleurisy	42—60	0—52
Pleural effusion	16—20	0—33
Pulmonary infiltrates	0—10	5—40
Pericarditis	20—30	0—18
Hepatomegaly	10—31	0—25
Splenomegaly	9—46	0—20
Renal involvement	50	0—13 (Hydralazine)
Neurological involvement	25—70	0—2
Anemia	30—90	0—53
Leukopenia	35—66	0—33
Thrombocytopenia	20—50	0—10
Positive Coombs	18—30	0—23
Elevated ESR	50—75	60—93
Antinuclear antibodies	>95	100
Antibodies to histones	50—70	>95
Antibodies to dsDNA	50	<5
Anti-Sm antibodies	25	<5
Hypocomplementemia	40—65	0—25
Rheumatoid factor	25	20—40

Pulmonary parenchymal disease has been reported in up to 40% of these patients unlike idiopathic SLE in which it is relatively uncommon [21]. LE cells may be found in pleural fluid of both DRL and idiopathic SLE patients and may be helpful in ruling out other possible causes for effusions, such as pneumonia and pulmonary infarction, which can occur in the elderly population. Pericarditis is another manifestation of DRL occurring most frequently in procainamide-related lupus [22]. It may be complicated by the development of a pericardial effusion and even tamponade [23, 24]. LE cells may be detected in the pericardial fluid. Constricted pericarditis has been reported, although it is an extremely uncommon manifestation of DRL [25]. Pericarditis has also been reported with DRL associated with the use of mesalazine and TNF-α inhibitors [26—31].

Other clinical features of idiopathic SLE, in particular dermatological, renal, and neurological features, are rare manifestations of the classical form of DRL. There have been reports of discoid lesions occurring with the use of griseofulvin and etanercept [32—34] and two reports of acute neutrophilic dermatosis (Sweet syndrome) related to the use of hydralazine [35, 36]. The classic malar rash is rarely noted. Elkayam et al. [37] reported cutaneous involvement among five of seven patients with minocycline-related lupus, of whom three had livedo reticularis and two had subcutaneous nodules. None of the three patients with livedo had anti-cardiolipin antibodies. Tournigand et al. [38] reported the recurrence of a lupus-like syndrome, characterized by arthralgias, fever, and a maculopapular eruption in a 14-year-old girl who had restarted minocycline after previously developing minocycline-related lupus. A few patients have been reported to have hepatosplenomegaly [39] and minocycline has been implicated in the development of a case of autoimmune hepatitis [40]. In contrast, the lupus-like syndrome related to the use of TNF-α inhibitors is associated with a higher incidence of cutaneous manifestations, including malar rash and discoid lupus [41].

Renal involvement has been noted in patients receiving hydralazine [42, 43], although renal biopsy findings are generally minimal and often difficult to distinguish from hypertensive-related changes. However, there have been some reports of patients developing acute nephritis that has resolved on discontinuation of the hydralazine [44]. Reported histopathological findings included focal, segmental, membranous proliferative, and necrotizing glomerulonephritis, which are indistinguishable from those seen in idiopathic SLE [42, 45, 46]. While improvement was noted in the majority of patients following discontinuation of the drug, some required treatment with steroids and other immunosuppressant agents, while others failed to respond at all. It is unclear whether failure to respond reflected some underlying renal insufficiency secondary to another disease process such as hypertensive-related renal insufficiency. Clinical neurological disease occurs very rarely in DRL, although there are a few reports of peripheral and cranial neuropathy occurring in procainamide and hydralazine DRL [47—49]. Vasculitis has also been reported as a manifestation of DRL [45, 50—55]. Lawson et al. reported the occurrence of cutaneous vasculitis in two of 23 patients with minocycline-induced lupus [54]. Cases of anti-neutrophil cytoplasmic antibody (ANCA)-positive vasculitis, particularly when associated with antimyeloperoxidase antibodies, have been described with exposure to certain medications, such as hydralazine and propylthiouracil [55]. One review reported that 118 cases of vasculitis have occurred in association with the use of TNF-α inhibitors, primarily in rheumatoid arthritis patients [41]. Purpura was the most common presentation with nearly all cases resolving following discontinuation of the TNF inhibitor with recurrence upon rechallenge. Subacute cutaneous lupus erythematosus (SCLE) has been reported to develop in association with the use of certain medications, primarily cardiovascular agents [56].

LABORATORY FINDINGS

Laboratory, including hematological, abnormalities occur less commonly in DRL than idiopathic SLE. One needs to keep in mind that many of the patients will exhibit abnormalities as a result of the underlying disease for which the medication has been prescribed. The erythrocyte sedimentation rate (ESR) is frequently elevated and often reaches moderate to high values; it quickly returns to the normal range with resolution of the clinical features of DRL. C-reactive protein is increased in cases of minocycline-induced lupus. A mild, generally normochromic, normocytic, anemia may be seen, although severe hematological abnormalities are rare [57]. Mild leukopenia and thrombobcytopenia can also occur. A positive Coombs test has been reported in patients receiving methyldopa, procainamide, and chlorpromazine [58–63]. Hemolytic anemia has been reported in some patients receiving methyldopa [64].

Increased immunoglobulin levels may occur but less frequently than with idiopathic SLE. Circulating immune complexes have been described in procainamide- and hydralazine-related lupus [65, 66]. There have been a few reports of decreased C3 and C4 levels in patients with procainamide- and quinidine-related lupus [67, 68]. However, levels tend to return to normal on discontinuation of the drug. Persistently low complement levels should suggest idiopathic SLE. Deposition of immunoglobulins at the dermal-epidermal junction (lupus band test) has been reported in patients with DRL secondary to procainamide and penicillamine [69, 70]. Immunoglobulin and complement deposition has also been noted in the glomeruli of a few patients with hydralazine-related lupus [43, 46, 71].

It is generally accepted that antinuclear antibodies need to be present in order to diagnose DRL, although their presence alone is not sufficient to make the diagnosis. Typically a diffuse/homogeneous staining pattern is seen, although other patterns have occasionally been reported. The autoantibody profile associated with DRL is more restrictive compared with the vast multitude of autoantibodies seen in idiopathic SLE [72,73]. In particular, antibodies to the Sm antigen have rarely been described in DRL and their presence should strongly suggest idiopathic SLE. Antibodies to

double-stranded dsDNA are uncommon findings in drug-related lupus, but they have been reported to develop in patients treated with TNF-α inhibitors, sulphasalazine, and minocycline [74–75]. Studies by Charles et al. indicated that the incidence of anti-dsDNA antibodies among patients receiving infliximab varied according to the assay used [76]. Their data suggested that high-affinity anti-dsDNA antibodies were more predictive of the development of lupus-like syndromes. Antibodies to single-stranded DNA are a relatively common finding but are considered nonspecific. Positive rheumatoid factors have also been reported [61] and considered to be nonspecific, most likely representing cross-reactivity with antihistone antibodies.

The homogeneous/diffuse fluorescent ANA staining pattern seen in DRL represents binding of antibodies to chromatin, which consists of DNA and histones. There are five different histone proteins seen in humans, namely H1, H2A, H2B, H3, and H4. The specificity of the autoantibody for these different histone proteins varies between the different drugs. Antihistone antibodies seen in patients receiving procainamide commonly react with the H2A–2B dimer, its complex with DNA, and also with chromatin itself [77, 78]. It has been suggested that IgG antibodies to (H2A–2B)-DNA complex and to chromatin are specific for drug-related lupus, while IgM antibodies to this complex and antibodies directed against the H2A-2B are most commonly found in asymptomatic patients receiving procainamide [78]. Antibodies against the (H2A–2B)-DNA complex have also been reported in DRL secondary to isoniazid, quinidine, sulfasalazine, penicillamine, and timolol [68, 79–81]. Antibodies from procainamide patients also exhibit reactivity with individual subtractions, most commonly the H1 [82]. Antihistone antibodies from patients with hydralazine-related lupus recognize a broader array of autoantigenic epitopes including the individual histones H1, H3, and H4 and the H2A-2B and H3–H4 complexes [83–85]. Antibodies from procainamide-related lupus patients have also demonstrated reactivity with the H3–H4 complex but to a lesser degree than that to the H2A–2B complex [84]. Antibodies to the H1 and H2B subfractions have been reported in quinidine-related lupus [86]. It has been suggested that a switch from the IgM to IgG isotype precedes the development of lupus-like syndrome in patients receiving procainamide and this may also be true for other drugs [87]. While antihistone antibodies are frequently found in DRL, they are also found in idiopathic SLE and hence are not considered diagnostic for drug-related lupus [72, 82, 88]. In addition, antihistone antibodies are not a common feature of minocycline-related lupus. In contrast, antibodies to double-stranded DNA are highly specific for idiopathic SLE and their presence would tend to rule out a diagnosis of drug-related lupus, although they have been described in patients with minocycline-related lupus and patients receiving TNF-α inhibitors [37, 41, 76].

Antiphospholipid antibodies and lupus anticoagulants have been reported with drug-related lupus. They have been described in patients receiving chlorpromazine, perphenazine, procainamide, quinine, quinidine, hydralazine, sulfasalazine, and minocycline [37, 54, 63, 89—106]. These antibodies are most frequently of the IgM subclass and in prospective studies have not been associated with a significant increase in thrombotic events [89, 106], although there have been individual case reports suggesting an association [92, 93, 98].

Antineutrophilic cytoplasmic antibodies giving a perinuclear pattern (p-ANCA) on indirect immunofluorescence have been associated with the use of some medications. These include minocycline, sulphasalazine, hydralazine, and propylthiouracil [37, 54, 55]. Antineutrophilic cytoplasmic antibodies have been described with hydralazine-related lupus; these antibodies react with myeloperoxidase and elastase [107—109]. Choi *et al.* determined the prevalence of exposure to medications associated with ANCA-positive vasculitis among 30 patients with the highest titers of antimyeloperoxidase antibodies among 250 patients in whom p-ANCA and antimyeloperoxidase antibodies had been assayed [55]. Of the 30 patients, ten (33%) had been exposed to hydralazine, three (10%) to propylthiouracil, two to penicillamine, one to sulphasalazine, and two to allopurinol. All patients had clinical and histological findings typical of ANCA-positive vasculitis. There was a strong association between the presence of antielastase and/or antilactoferrin antibodies and exposure to these medications. Although not all the 30 patients had documented histological findings diagnostic of ANCA-positive vasculitis, all had the typical features of this entity. There are also reports associating antimyeloperoxidase antibodies with exposure to hydralazine and propylthiouracil in the absence of vasculitis [110—112].

It is difficult to estimate the incidence of drug-induced ANCAs because of the lack of prospective studies. Choi *et al.* evaluated sera prospectively collected as part of three separate double-blind controlled trials for ANCA seroconversion [113]. The three prospective studies consisted of a 40-week trial of minocycline and a 37-week trial of sulfasalasine, both for treatment of rheumatoid arthritis, and a 104-week trial of penicillamine for the treatment of early systemic sclerosis. Sera of 12 patients from the combined total of 228 patients contained antimyeloperoxidase antibodies, but no patients seroconverted to a positive ANCA during the duration of the respective studies. Hence, the exact role of drugs in the development of ANCAs remains unclear.

Lymphocytotoxic antibodies have been reported in patients with drug-related lupus secondary to procainamide, hydralazine, and phenytoin [114—117].

Antibodies to hydralazine and procainamide have been reported in patients receiving these medications [118, 119]. Antibodies against the incriminated drug appear to occur more frequently in patients with hydralazine-related lupus compared to those with procainamide-related lupus. The actual significance of these antibodies has yet to be defined.

DIAGNOSIS

There are as yet no established criteria for the diagnosis of DRL. The American College of Rheumatology requires four of 11 specified criteria to be present in order to diagnose idiopathic SLE [15]. However, patients with DRL frequently present with less than four of these features. A set of ascertainment criteria has been proposed by the Environmental Study Section of the American College of Rheumatology, which include a positive ANA [19]. A temporal association between the ingestion of an agent and the development of the lupus-like features is required to make the diagnosis. Unlike drug hypersensitivity, it generally requires weeks to months of exposure to an agent before DRL develops. There should be remission of the clinical features with a decrease in the autoantibody titers following withdrawal of the offending agent. Recurrence of the syndrome following reintroduction of the agent would provide confirmatory evidence for an association, but this is seldom done in clinical practice given the large number of alternative medications available. The development of antinuclear antibodies (ANAs) alone is not sufficient to diagnose DRL. It is now known that many patients will develop antinuclear antibodies while taking certain medications, but the majority do not appear to develop the features of a lupus-like syndrome. Hence, the development of a positive ANA in the absence of other clinical features is not sufficient reason to discontinue the medication.

DIFFERENTIAL DIAGNOSIS

Since many of the features of drug-related lupus are relatively nonspecific, it can be difficult to make a definitive diagnosis. A variety of inflammatory, infectious, and other autoimmune diseases can present with similar features and patients may not give accurate histories of previous medication usage, as noted by Gordon and Porter [75]. In particular, it is often difficult to differentiate DRL from idiopathic SLE. Since renal, cutaneous, neurological, and severe hematological involvement is rarely seen in drug-related lupus, the presence of these features would suggest idiopathic SLE. Demographic features may also be helpful in distinguishing between these two diseases since, with the exception of anticonvulsants and minocycline, DRL generally occurs more frequently in elderly Caucasian males in contrast to idiopathic SLE. The presence of antibodies to double-stranded DNA, Smith, and Ro antigens would favor idiopathic SLE. Other differential diagnoses would include infections, in particular viral syndromes that are a recognized cause of arthropathies. However, in most cases of virus-related arthropathy the symptoms resolve spontaneously within a matter of weeks to months. Osteoarthritis is a frequent finding in the elderly population and the articular symptoms of DRL may be mistaken for this, especially if the symptoms are of an insidious onset. The diagnosis of DRL related to the use of minocycline and TNF-α inhibitors may be particularly problematic since both of these drugs are used to treat rheumatoid arthritis and it may be difficult to distinguish the development of DRL from a flare of rheumatoid arthtitis. Supervised rechallenge with the suspected medication may be helpful in confirming the diagnosis, as suggested by the study by Lawson *et al.* [54].

THERAPY

The definitive treatment is withdrawal of the offending agent. In some cases, this can be difficult since many patients are on multiple medications. Generally, there is significant improvement in the symptoms within days of discontinuation of the medication. Rarely, it may take some weeks for improvement to occur. However, occasionally some patients with persistent symptoms may require further therapy. Nonsteroidal anti-inflammatory agents are useful to treat constitutional symptoms such as low-grade fevers, arthralgias, and myalgias and may also be used to treat pleuritis and pericarditis. In patients with more severe involvement, such as large pleural or pericardial effusions, steroids are required. Generally these can be successfully tapered once the clinical features have resolved. Pericardectomy has been used in the treatment of constrictive pericarditis secondary to procainamide. In the rare instance where a patient requires continuation of the offending medication for medical reasons, symptoms have been treated with either a nonsteroidal anti-inflammatory agent or steroids. However, with the greater availability of alternative medications, in particular the antiarrythmic agents, this is rarely if ever necessary.

Mechanisms

A number of hypotheses have been proposed to explain the pathogenic mechanisms for drug-induced autoimmunity [120]. These include the following:

1. Hapten or altered self-antigen. The drug or one of its metabolites act as a hapten by binding to a self-macromolecule or altering a self-antigen, resulting in stimulation of lymphocytes with the production of antibodies that cross-react with or cause spreading of the immune response to native self-macromolecules.
2. Cytotoxicity. Reactive molecules of some of the drugs associated with DRL can cause cell death via a nonimmune mechanism such as direct cytoxicity or through enhancement of the production of reactive oxygen species by leucocytes leading to apoptosis.
3. Nonspecific activation of lymphocytes. Inhibition of methylation of T-cell DNA by drugs, such as procainamide or hydralazine, can alter gene expression and lymphocyte function, resulting in the generation of autoreactive T lympocytes.
4. Dysruption of central immune tolerance. In murine studies, intrathymic injection with a reactive molecule of procainamide led to the export of chromatin-reactive T lymphocytes to the peripheral circulation and the sustained production of antichromatin antibodies.

Many of the incriminated drugs have: (1) an aryl-amine or hydrazine functional group, (2) a sulfhydryl or thiono sulfur group, or (3) are hydantoins possibly metabolized to phenol. Since not all individuals exposed to a certain drug will develop autoimmunity, this suggests that other factors may be very important. Certain individuals may metabolize drugs and other agents at a faster rate than others. Polymorphisms of genes coding for enzyme systems exist and these may be important in determining exposure time of the host to certain agents or their metabolites. A number of polymorphisms have been reported for the cytochrome P450 enzyme system, which is known to play an important role in the metabolism of drugs. The risk of developing drug-related lupus in patients receiving hydralazine has been linked to the acetylator phenotype of the patient, which is under genetic control [121].

COMMENTS ON CERTAIN DRUGS ASSOCIATED WITH DRUG-RELATED LUPUS

Hydralazine

Hydralazine was introduced as a new antihypertensive agent in 1952 and it is generally accepted to be the first drug definitively associated with DRL [7]. Reports have indicated that between 24 and 54% of patients receiving hydralazine develop antinuclear antibodies; however, only 2−21% of patients develop lupus-like symptoms [9]. The most common symptoms are fever, malaise, arthralgias, and arthritis. Serositis may also be seen but occurs less commonly than with procainamide-related lupus. Renal involvement has also been reported with this drug, although in some cases it is unclear whether the renal manifestations may not in fact be related to the underlying hypertension. The development of hydralazine-related lupus is dose-related. Patients receiving greater than 200 mg per day of hydralazine have a significant increased risk of developing DRL compared with those receiving smaller doses [51]. The risk has also been related to the acetylator phenotype, which is under genetic control [121]. DRL occurs more commonly in patients who are slow acetylators. This is believed to be due to the slower rate of inactivation of the parent compound through acetylation of its hydrazine group by the hepatic acetyl transferase enzymes.

The development of hydralazine-related lupus has been linked to the HLA Class II and III genes. In 1980, Batchelor *et al.* reported a frequency of 79% for the HLA-DR4 antigen in patients who developed hydralazine-related lupus compared to 25% in asymptomatic hydralazine-treated controls [18]. However, a subsequent study by Brand *et al.* of 18 Australian patients with hydralazine-related lupus failed to show an increased frequency of the HLA-DR4 antigen compared with a control group from Melbourne [122]. A study by Russell *et al.* [123], reported a frequency of HLA-DR4 antigen of 70% among 20 patients with hydralazine-related lupus compared with a frequency of 33% in the general population. Six of these 20 patients had been included in the 1980 study by Batchelor *et al.* A further study by the Batchelor group examining a possible role of polymorphisms of the HLA Class III gene noted a significant increase in the frequency of one or more null alleles of the genes coding for C2, C4A, and C4B among 21 cases with hydralazine DRL compared with 82 normal controls [124]. Of interest was the finding of a significantly higher frequency of the HLA-DR4 antigen among the patients compared with the controls. The genes that code for C2, C4A, and C4B are located on chromosome 6 adjacent to the HLA DR regions and their alleles exhibit linkage disequilibrium with HLA DR alleles. In particular, the C4B null allele is in linkage disequilibrium with HLA-DR4 antigen. It has therefore been postulated that it may be the C4 null allele in linkage disequilibrium with the DR4 antigen that is important in determining susceptibility to the development of DRL in patients receiving hydralazine. If no linkage disequilibrium exists between the HLA-DR4

and the C4 null alleles among the Australian population, this could explain the discrepancy between the results of the previously mentioned studies.

Antiphospholipid, lymphocytotoxic, antineutrophilic cytoplasmic antibodies and antibodies to poly A and Z-DNA have been described in patients receiving hydralazine in addition to antinuclear and antihistone antibodies [104, 107−109, 116, 125, 126]. The pathogenic significance of these autoantibodies has yet to be elucidated. The frequency of hydralazine-related lupus appears to have decreased during the more recent years as the dosages and use of the drug have decreased.

Procainamide

Procainamide is a class I antiarrhythmic agent and is believed to be the most frequent cause of DRL in the United States. A positive ANA has been reported to develop in up to 90% of patients [77]. Approximately 30% of these patients will subsequently develop DRL [127]. The duration of therapy prior to the onset of symptoms varies from 1 month to as late as 12 years, with an average of 1 year [61]. Common features include constitutional and musculoskeletal symptoms, with approximately half of the patients developing pleuropulmonary involvement and/or pericarditis. Although there is a weaker relationship with acetylator status compared to hydralazine, it has been reported that patients with a slow acetylator phenotype will develop ANAs and DRL after a shorter duration of therapy compared to those with a fast acetylator phenotype [127]. However, this difference in phenotypes is less obvious if the dose is titrated such that the plasma levels of the drug are equivalent between the two groups [128]. A more recent study did not find an association between slow acetylator status and the development of procainamide-related lupus, although patient numbers were small [129]. However, this suggests that other factors are involved. A positive association of DQβ1 DQw7 (split of DQw3) with antibodies to histone subfractions (H2A−2B), Z-DNA, and poly A has been reported in asymptomatic patients receiving procainamide [130].

As with hydralazine, lymphocytotoxic, antiphospholipid antibodies, lupus anticoagulants, and antibodies to poly A and Z-DNA have been described in patients on procainamide [71, 87−93, 108]. Antibodies to procainamide have also been reported [119].

Isoniazid

Antinuclear antibodies develop in approximately 20−25% of patients receiving isoniazid; however, less than 1% of patients develop a lupus-like syndrome [131]. Common features of isoniazid-related lupus include fever, arthralgias, and arthritis and less commonly serositis.

IgG antinuclear antibodies against the nucleohistone complex (H2A−H2B)−DNA have been reported in a patient with isoniazid-related lupus [132]. Since other antituberculous agents and tuberculosis itself have been reported to be associated with ANAs and rheumatoid factors, it may be difficult for the clinician to determine if isoniazid has a pathogenic role in the development of new lupus-like symptoms. Although isoniazid is also inactivated through acetylation, no association between acetylator phenotype and the development of ANA and/or DRL has been reported [133, 134].

Quinidine

Quinidine is another class I antiarrhythmic agent associated with the development of DRL. The first report by Kendal and Hawkins was published in 1970 [135]. At least 29 further reports of quinidine-related lupus have appeared in the English literature since then [136]. Quinidine-related lupus occurs primarily in older white patients (mean age, 63 years) and equally in men and women. There has been one case report in an 11-year-old girl. Typically, symptoms develop after 1−18 months of therapy, and resolve within 1−4 weeks. Of interest is the report of development of quinidine-related lupus in two patients who had a previous history of procainamide-related lupus [137]. In both cases, there had been resolution of the symptoms following discontinuation of procainamide and prior to the institution of quinidine.

The most common clinical features are those of arthralgias and arthritis, which are characterized by symmetrical involvement of the hands and wrists and less commonly involvement of the shoulders and knees. Hence, it may be mistaken for rheumatoid arthritis. Less frequently reported features include fever, rash, serositis, and hepatic and neurological involvement, primarily manifested by carpal tunnel syndrome and peripheral neuropathy. Thrombocytopenia and leukopenia have been reported in 47 and 24% of patients, respectively. Hypocomplementemia has been reported in a few patients. While the ANA fluorescence pattern associated with DRL is typically homogeneous, both speckled and homogeneous patterns have been seen in patients with quinidine-related lupus. Both IgG and IgM classes of antihistone antibodies have been reported, which may demonstrate reactivity with H1, H2B, and the complexes H2A−2B and H3−H4 [138]. The lupus anticoagulant has also been described.

There have been three cases of development of polyarthropathy alone in patients receiving quinidine that resolved on discontinuation of the drug and recurred on rechallenge [63]. None of these three patients, however, had a positive ANA and the articular symptoms were milder than those usually seen with quinidine-related lupus. The duration of therapy was shorter and the symptoms resolved within 1 week of discontinuation of the drug. It is unclear whether these may represent a variant of quinidine-related lupus.

Quinine, an antimalarial drug analog of quinidine, has also been reported to be associated with the development of autoantibodies and a lupus-like syndrome [103].

Anticonvulsants

Many anticonvulsants have been implicated in the development of DRL, of which the most common are the hydantoins, ethosuximide, trimethadione, and carbamazepine [16, 17, 138−142]. There have also been reports implicating primidone and valproate [143]. The demographic features of anticonvulsant-related lupus differ from the other forms of DRL in that the majority of patients have been children and young females. Lupus-like syndromes have been reported to occur more frequently with ethosuximide than with the other anticonvulsants [144]. Singsen et al. reported the development of lupus-like syndromes, characterized by fever, malaise, rash, arthritis, and serositis, among five children who were receiving ethosuximide; resolution of these symptoms occurred in four of five children following discontinuation of ethosuximide, although three required treatment with prednisone [17]. Lupus-like syndromes with renal and cerebral involvement have been reported in association with the use of ethosuximide [145, 146]. One report described the case of a man who developed fulminant myopericarditis from chronic phenytoin use [147]. Interestingly, there does not appear to be a relationship between phenobarbital and DRL and there is only one case report suggesting a relationship with one of the benzodiazepines. The report described a photosensitive skin eruption resembling discoid lupus, mild leukopenia, and positive ANA and rheumatoid factor in a 7-year-old girl during clobazam therapy [148]. There was resolution of the cutaneous lesions and autoantibody disappearance following discontinuation of this drug.

Some studies have reported a higher frequency of ANAs in children receiving anticonvulsants. Beernink and Miller [16] reported that 11 of 48 children receiving anticonvulsant therapy at a pediatric neurology clinic had antinuclear antibodies, although none had any symptoms suggestive of lupus, in addition to five patients with antinuclear antibodies and lupus-like symptoms. Singsen and colleagues reported that 14 of 70 patients receiving anticonvulsants, which have been associated with DRL had high titers of ANA compared with low titers in five of 23 receiving phenobarbitol and one of 50 controls [17]. Histone antibodies have been described in a patient who developed DRL secondary to ethosuximide [142]. A study by Verrot *et al.* of 163 patients attending a medical center specializing in epilepsy found that 41 patients (25%) had antinuclear antibodies compared with 10% of the controls; no statistical association was found between the presence of antinuclear antibodies and the type of epilepsy, the kind of antiepileptic drug, or the age or sex of the patient [149]. However, the numbers were relatively small and hence may not have allowed for an association to be detected. IgG class anticardiolipin antibodies were found in 31 patients, although no patient had a history of a thrombotic event. Anti-dsDNA antibodies have been reported in association with the development of a lupus-like syndrome in a patient receiving ethosuximide [150].

Problems arise in deciding whether the lupus-like symptoms in patients receiving anticonvulsants represent a form of DRL or are in fact a manifestation of idiopathic lupus of which the seizure disorder is the presenting feature. This may pose particular diagnostic and management problems in adults, as there may be a history of continuing convulsions and possibly psychiatric problems.

Chlorpromazine

Antinuclear antibodies occur in 20–50% of patients treated with chlorpromazine [151], but a lupus-like syndrome develops in less than 1%. Typical features are fever, arthralgias/arthritis, serositis, and cutaneous features [92, 152]. Photosensitivity and malar rashes have been reported that have resolved on discontinuation of the drug. Over 50% of patients have been reported to have splenomegaly. Significant elevation of IgM, prolongation of PTT, and both lupus anticoagulant and anticardiolipin antibodies have been described in patients receiving long-term chlorpromazine therapy [89, 106]. Canoso and de Oliveira [89] noted that 54 of 96 of these patients had IgM lupus anticoagulants and, of these, 31 had IgM and four had IgG anticardiolipin antibodies; five patients had anticardiolipin antibodies alone. Only three patients developed thrombotic episodes over a median follow-up period of 5 years, suggesting a low risk for thrombosis. A study by Zarrabi *et al.* [63] noted that 41 of 47 patients receiving chlorpromazine for greater than 2.5 years had a circulating anticoagulant. It should be noted that many of the patients studied in the reports were hospitalized for psychiatric and psychological disorders, raising the question of whether or not these disorders may have represented a manifestation of SLE. This points out the great importance of follow-up studies for all drugs and environmental agents to observe the consequences of discontinuation of the possible offending agent.

Beta-Blockers

Antinuclear antibodies occur in 10–30% of patients receiving beta-blockers. Of these, acebutolol has been the drug most commonly implicated, followed by labetalol (14–16%), oxprenolol (13%), metoprolol (12%), propranolol (10%), and pindolol (10%) [153–157]. ANA titers tend to rise with duration of therapy and decrease following discontinuation of the drugs. Lupus-like syndromes rarely occur with these agents, although there are reports implicating practolol, acebutolol, labetalol, and pindolol. There is also a case report of a lupus-like syndrome developing in a patient receiving ophthalmic timolol, which was associated with markedly elevated IgG antibodies to the (H2A–2B)–DNA complex [81]. Beta-blockers have also been implicated in the development of subacute cutaneous lupus erythematosus [158].

ACE Inhibitors

Antinuclear antibodies have been reported in patients receiving captopril and enalapril [159–165]. Reidenberg *et al.* [159] noted that ten of 37 patients developed antinuclear antibodies while receiving high doses of captopril. Three of nine patients receiving enalapril developed a positive ANA test (titer greater than 1:160) during a 6-week follow-up study by Schwartz *et al.* [163]. Of interest was the fact that none of these patients were ANA-positive prior to starting the drug. There have been a few case reports implicating captopril as a cause of DRL. Two patients developed serositis and antinuclear antibodies that resolved on discontinuation of the drug [161, 164]. Another patient developed myalgia, arthralgia, photosensitivity, antinuclear antibodies, and IgG antibodies to H2A–2B dimer after taking captopril for 2.5 years; there was significant improvement in the clinical symptoms following withdrawal of the medication [165]. Subacute cutaneous lupus erythematosus and other forms of cutaneous lupus in association with Ro/SS-A antibodies have been described in patients receiving ACE inhibitors [166].

Sulfasalazine

There have been a number of reports implicating sulfasalazine with the development of DRL, although the association is unclear. Sulfasalazine was first implicated in drug-related lupus in 1965 when Alarcon-Segovia *et al.* described the development of a lupus-like syndrome and LE cells in five patients receiving sulfasalazine for treatment of ulcerative colitis [167]. Since then, there have been a number of reports on the development of autoantibodies and lupus-like syndromes in patients receiving this drug for treatment of inflammatory bowel disease and arthropathies [74, 80, 105, 167–176]. In one study, antinuclear antibodies developed in 13 of 77 patients with spondyloarthropathy while receiving sulfasalazine [168]. Antihistone antibodies including IgG antibodies to the H2A–2B dimer and the (H2A–2B)–DNA complex have been noted [80]. Some patients have had anticardiolipin and antineutrophil cytoplasmic antibodies [107, 176]. Mielke *et al.* [74] reported that 20 of 24 patients with inflammatory arthritis treated with sulfasalazine developed double-stranded DNA antibodies, although only one developed symptoms of DRL. The occurrence of these antibodies was transient in the majority of patients.

The most common clinical manifestations of sulfasalazine-related lupus are arthralgias, arthritis, pleuritis, pericarditis, and pleural effusions. There have been two reports of a patient developing cardiac tamponade [170, 174]. Fevers and a variety of cutaneous lesions have also been described. Gunnarsson *et al.* reported that four out of 41 consecutive patients with early rheumatoid arthritis, who were treated with sulfasalazine for a minimum of 6 months, developed DRL [177]. Two patients developed anti-dsDNA antibodies after approximately 12 months of treatment and one of these patients also developed a photosensitive rash. Another patient developed biopsy-proven membranous nephritis, compatible with lupus nephritis, after 26 months of treatment with sulphasalazine. One patient developed a sun rash and increasing titers of ANA after 6 months of therapy, and a fourth patient, who developed anti-dsDNA antibodies, remained asymptomatic. All four patients were ANA-positive. Sixteen (39%) patients had ANAs prior to institution of sulphasalazine and 13 (32%) were ANA-positive following 6 months of therapy. Patients who developed lupus-like features did not have a significantly higher ANA titer. However, there was a correlation between ANA positivity and a speckled pattern prior to initiation of treatment, regardless of titer, and the development of lupus-like features. Three of the four patients who developed lupus-like features had the HLA DR 0301 haplotype compared with four out of 37 patients without lupus-like symptoms. The occurrence of this SLE-related HLA haplotype among those patients who developed a lupus-like syndrome suggests a genetic role for the development of sulphasazine-induced lupus. Increased interleukin (IL)-10 levels were also seen in these patients.

In contrast, in a randomized prospective study of patients comparing the use of sulfasalazine with auranofin for the treatment of rheumatoid arthritis, no patient developed a lupus-like syndrome, although 14 (19%) of 72 patients treated with sulfasalazine who were either ANA-negative or weakly ANA-positive at the start of the study became strongly ANA-positive during the study [178]. This suggests that the occurrence of ANA

positivity either prior to or during treatment with sulphasalazine does not preclude treatment with this medication. However, although the association with sulfasalazine is unclear, one should consider the possibility of DRL in a RA patient who develops apparent arthritis flare-ups or increases in articular symptoms while receiving sulfasalazine.

It has been suggested that the sulfonamide component of sulfasalazine is responsible for the development of the autoimmunity since sulfonamides have previously been associated with DRL. However, there are a number of reports implicating mesalazine as a cause of DRL, suggesting that the presence of sulfapyridine is not necessary. Dent *et al.* described the development of pleuropericarditis, fever, and a positive ANA in a patient receiving mesalazine [26]. Kirkpatrick *et al.* reported that four patients with inflammatory bowel disease, treated with 5-aminosalicylic acid (5-ASA), developed ANA-positive arthralgias and inflammation that resolved rapidly following discontinuation of the 5-ASA compounds [179]. Timsit *et al.* reported the case of a patient with Crohn disease who developed polyarthritis, alopecia, lymphoneutropenia, positive ANA, and antibodies to histones, Smith, and RNP while being treated with mesalazine [180]. There was rapid resolution of the clinical features and a decrease in the levels of autoantibodies with discontinuation of the medication. Vayre *et al.* described the case of a 53-year-old male with Crohn disease, for which he had been treated for approximately 8 years with mesalazine, who developed chest pain, ST-segment elevation, and a small pericardial effusion [27]. Resolution of these features occurred following discontinuation of mesalazine. Molnar *et al.* reported the case of a 29-year-old woman with inflammatory bowel disease, treated with mesalazine for approximately 2 months, who developed pericarditis and a moderate-size pericardial effusion that responded to the institution of steroids, even though she remained on mesalazine [28]. She also had a positive p-ANCA. Sentongo and Piccoli reported the case of a child who developed pericardits, believed to be secondary to treatment of mesalazine [29]. Gunnarsson *et al.* described the case of a 43-year-old female with Crohn disease who developed polyarthritis involving the small joints of her hands, wrists, ankles, and MTP joints [181]. ANA and IgG anticardiolipin antibodies were markedly positive, antihistone antibodies were positive, and antibodies to dsDNA were borderline elevated. Her clinical features resolved within 2–3 weeeks of discontinuation of the medication, although her ANA and IgG anticardiolipin antibodies were still positive 1 year later. She was noted to be DQA1*0501 haplotype-positive. This haplotype has previously been reported to occur more frequently among patients with sulphasalazine-induced lupus syndrome, suggesting a genetic predisposition [182].

Minocycline

Minocycline hydrochloride is a semisynthetic tetracycline used for a variety of infections and for the long-term treatment of acne vulgaris. It was first associated with DRL in 1992 by Matsuura et al. [183]. There have been at least 56 further cases reported in the literature since then [37, 38, 40, 54, 76, 184—190]. Minocycline has also been implicated in the development of autoimmune hepatitis [40].

A review by Schlienger et al. reported 57 published cases of lupus-like syndromes that were suspected to be associated with minocycline use. These cases tended to resolve quickly after stopping minocycline and in general no systemic treatment was required. The mean age of the patients was 21.6 years and the median time of exposure was 19 months, ranging from 3 days to 6 years [191]. There have been 267 cases of possible minocycline-induced lupus reported to the WHO adverse drug reaction databases [192]. In addition to case reports, there have been a number of case series of lupus-like syndromes developing among patients receiving minocycline [184, 190]. In contrast to other forms of DRL, a similar gender profile to that of idiopathic SLE has been observed since the great majority of these cases have occurred in young women who were being treated for acne, although it may occur in men. Of note, autoimmune hepatitis has been reported to occur in the majority of men who develop minocycline-related lupus [37].

The most commonly reported features include constitutional symptoms, rash, polyarthralgias, an inflammatory arthritis that tends to affect the small joints of the hands, wrists, shoulders, and ankles in a symmetrical fashion, and autoimmune hepatitis [38, 188, 189, 193]. Pleuropulmonary involvement, thrombocytopenia, anemia, and cutaneous involvment such as livedo reticularis, subcutaneous nodules, erythema, and urticaria have also been described [37, 40, 187, 188, 193]. Elkayam et al. reported cutaneous involvment among five of seven patients with minocycline-related lupus, of whom three had livedo reticularis and two had subcutaneous nodules [37]. None of the three patients with livedo had anticardiolipin antibodies and only two had positive ANAs. Dunphy et al. reported that two patients treated with minocycline developed erythema on their legs, two an urticarial eruption, one patient thrombocytopenia, and another patient anemia [193]. Autoimmune hepatitis occurs in the majority of men who develop minocycline-related lupus [37]. There are also reports of an allergic type fulminant type hepatitis and of hypersensitivity-like reactions consisting of fever, lymphadenopathy, and skin eruption developing in patients receiving minocycline [188].

Antinuclear antibodies with titers varying from 1:80 to 1:1280 are a common finding, although they may only be transiently present in some patients [37, 54, 193]. Anti-dsDNA and anticardiolipin antibodies occur less often and antihistone antibodies, which are a frequent finding in other forms of drug-related lupus, are not commonly seen in patients with minocycline-induced lupus. Patients often have positive p-ANCAs at titers of 1:160 or greater with activity directed to myeloperoxidase and/or elastase, which may be a marker for the development of DRL in genetically susceptible individuals prescribed minocycline [54, 193].

Duration of minocycline therapy prior to the onset of symptoms varies from a few weeks to up to 10 years. Resolution or significant improvement of the clinical features occurs in most cases within 2—56 weeks of drug withdrawal, although some patients have required steroid therapy [54, 75]. The autoimmune hepatitis tends to resolve more slowly [37]. There is no correlation between duration of treatment with minocycline and the time taken for resolution of the symptoms. Although serological features also improve, ANAs may remain positive in some patients.

Gordon and Porter [75] reported the clinical and serological features of a case series of 20 patients diagnosed with minocycline-related lupus over a 5-year period in the west of Scotland. Minocycline had been prescribed for a mean of 25 months (range, 3—60 months) prior to the onset of symptoms. Fifteen patients were female and five were male, with the mean age of 24 years. Twenty patients had arthritis with some patients complaing of lethargy, myalgia, and fever. Discontinuation of minocycline led to resolution of symptoms in all patients after a mean of 15.7 weeks (range, 2—56 weeks). All patients had a positive ANA, two patients had anticardiolipin antibodies, and one patient had a positive p-ANCA. Eight of 20 (40%) patients had antibodies to dsDNA, of which two were positive using the Crithidia assay. Antibodies to dsDNA had disappeared and ANA titers decreased during the follow-up period. Six of the patients who had initially denied talking minocycline developed a recurrence of symptoms after a mean of 4.5 days (range, 2—7 days) of restarting minocycline. Thus, denial of previous use of minocycline does not rule out the possibility of minocycline-related lupus and supervised rechallenge may be helpful in confirming the diagnosis.

Patients rechallenged with minocycline generally developed a recurrence of their symptoms within a few hours to a few days of restarting the medication. Lawson et al. reported on the series of 23 patients with minocycline-induced lupus [54]. All the patients complained of polyarthralgias, with three patients having synovitis of the MCP and PIP joints and one patient an effusion. Cutaneous vasculitis occurred in two patients.

Five patients complained of impaired concentration and poor memory and one patient had neuropathy. Nineteen patients had positive ANAs with titers varying from 1:40 to greater than 1:16,000. Four patients had antibodies to ds-DNA by ELISA and *Crithidia* assays and ten patients had positive p-ANCAs, with the majority having titers of 1:160 or greater. Of these ten patients, seven had antibodies to myeloperoxidase. Antihistone antibodies were not detected in the nine patients in whom they were sought. Following discontinuation of minocycline, ten patients became asymptomatic within 1 month, nine patients had improved within 1–6 months, and three within 6–12 months; it took 2 years for the symptoms to resolve in one patient. Four patients with severe disease required treatment with corticosteroids. There was no correlation between duration of treatment with minocycline and the time taken for resolution of the symptoms following discontinuation of the drug. Antinuclear antibodies remained strongly positive in four patients but became negative in the remainder of patients within 2 years of withdrawal of minocycline. Recurrence of symptoms occurred within hours of restarting minocycline in five patients. Ten additional patients underwent a supervised rechallenge with minocycline. There was a recurrence of symptoms in seven patients within 12 h and in the remaining three patients within 72 h of ingesting a single 100-mg dose of minocycline. Two patients who were rechallenged required a short course of steroids; all subjects were asymptomatic again within 2 weeks. Elkayam *et al.* rechallenged five patients with a single 50-mg dose of minocycline: all five patients developed arthralgias within 24 h that subsequently resolved spontaneously [37]. Thus, supervised rechallenge may be helpful in confirming the diagnosis in some patients.

A well-designed nested case–control study of 27,688 acne patients based on "prescriptions used" estimated the risk of developing drug-related lupus to have an odds ratio of 8.5 (2.1–35) comparing current users of minocycline with previous and nonusers of tetracycline combined [194]. Current exposure to minocycline and a cumulative dose above 100 defined daily doses increased the risk 16-fold. Among females, there were 17.2 cases of lupus-like syndrome per 100,000 prescriptions for oxytetracycline and 52.8 cases per 100,000 prescriptions for minocycline, suggesting that minocycline is more likely to induce lupus-like syndromes compared with other forms of tetracycline. However, this study was not able to exclude an association with the other tetracyclines. There are early reports implicating tetracycline in DRL, but it is generally considered that the association with tetracyclines is weak and reported cases may in fact represent an exacerbation of SLE.

A retrospective study of 96,694 individuals with a history of acne who were followed for about 520,000 person-years identified 51 patients who had developed lupus-like syndromes based on physician reports [195]. The overall hazard ratio for the association of minocycline to the development of lupus was 2.64 (95%CI, 1.51–4.66) and when adjusted for age and gender was 3.11 (95%CI, 1.77–5.48). No association was noted for doxycycline or the other tetracyclines as compared with nonusers. Women were much more likely to develop lupus-like syndromes than men, with a hazard ratio of 9.2 (95%CI, 3.20–25.6). There was a strong relationship between duration of exposure to minocycline and LE. Some patients required systemic treatment.

Work by McHugh *et al.* suggests that minocycline-induced toxicity may be mediated through an oxidized metabolite and be myeloperoxidase-dependent [196]. The enzymes involved in the metabolism of minocycline may bind to a reactive intermediate acting as a hapten, leading to the formation of a neoantigen with consequent induction of an autoimmune response. Herzog *et al.* suggested that there may be an antibody reaction to a metabolite of minocycline that cross-reacts with microsomal chromosomes in the liver [197]. They reported that serum obtained from a patient with minocycline induced autoimmune hepatitis reacted with a 50-kDa protein from rat liver microsomes. Minocycline is known to have immunomodulating properties and has been demonstrated to affect production of cytokines, including TNF-α [198]. These effects on cytokine production may play a role in the development of minocycline-related autoimmunity.

Reports have also suggested a role for MHC Class II genes in the development of minocycline-related lupus. Dunphy *et al.* reported a case series of 14 patients who developed minocycline-induced lupus after using minocycline for a median of 3.8 years (range, 1–10 years) [193]. Discontinuation of minocycline led to resolution of symptoms in all patients within days to 6 months. Recurrence of symptoms was seen among patients, who were rechallenged with minocycline. Sera from all 14 patients had ANCAs giving a perinuclear pattern on indirect immunofluorescence. Eleven patients had elevated antimyeloperoxidase antibodies and ten had antielastase antibodies that decreased following discontinuation of medication. All 13 patients who were tested were either HLA-DR4 (nine patients) or HLA-DR2- (four patients) positive and all had an HLA-DQB1 allele. The authors suggested that the presence of p-ANCA may be a marker for the development of lupus-like symptoms in genetically susceptible individuals prescribed minocycline. Elkayam *et al.* reported that four out of six patients whom they had HLA typed had HLA DRβ*0104 that compared to an expected frequency of 12% in the general Israeli population [37].

Lipid-Lowering Agents

Lupus-like syndromes have also been reported in association with the 3-hydroxy-3-methylglutaryl coenzyme A (HMG-COA) reductase inhibitors (statins). In a review of the Medline database from 1966 to September 2005, Noel identified 28 cases of autoimmune diseases related to statin therapy, which included lovastatin, simvastatin, pravastatin, and fluvastatin [199]. These included ten cases of lupus-like syndromes, three cases of biopsy-proven subacute cutaneous lupus erythematosus (SCLE), 14 cases of dermatomyositis and polymyositis, and one case of lichen planus pemphigoides. Autoimmune hepatitis occurred in two patients with lupus. The mean time to onset of clinical features was 21.1 months for lupus-like syndromes and 12.7 months for SCLE.

The lupus-like features that have been described in association with the use of statins have included malaise and inflammatory polyarthritis in two patients receiving lovastatin and one patient receiving simvastatin [200, 201]. Other manifestations included the development of serosistis in three patients receiving simvastatin that was manifested by pleural and pericardial effusions in one patient, pleurisy in another, and pleuropericarditis in the third patient [202-204]. Myalgia, polyarthralgias, rash, leukopenia, photosensitivity, and Raynaud phenomenon have also been described [204-206]. There was also a report of a patient who developed focal segmental glomerulosclerosis with fibrotic lung disease while taking fluvastatin [206]. All patients had a positive ANA at a high titer, varying from 1:160 to 1:1280, with both homogeneous and speckled patterns being reported. ANAs remained positive for many months after discontinuation of the statin despite improvement in clinical symptoms. Antihistone antibodies were positive in six patients and negative in four patients; antibodies to dsDNA occurred in three patients. Most patients were treated with prednisone with slow resolution of their symptoms.

One patient developed severe autoimmune hepatitis, rash, polyarthralgias, ANA, antihistone antibodies, and strongly positive anti-dsDNA antibodies while receiving atorvastatin [208]. Biopsy of the rash confirmed the diagnosis of lupus and a liver biopsy revealed highly active chronic hepatitis, which initially did not respond to high-dose steroid therapy but improved when treated with mycophenolate mofetil and tacrolimus. Sridhar and Abdulla reported the development of polyarthritis affecting the hands and right knee, myalgias, and a generalized rash in a 67-year-old woman approximately 1 week after she started fluvastatin [208]; however, symptoms did not resolve in spite of discontinuation of medication for more than 1 month. Initial testing was negative for anti-dsDNA antibodies, but subsequent tests were positive for anti-dsDNA antibodies, although antihistone antibodies were negative. The patient subsequently went on to develop ARDS syndrome and died. Although this patient was reported as a case of fluvastatin-induced lupus, the features are somewhat atypical in that the symptoms developed relatively quickly after starting the medication and did not resolve with discontinuation.

Three cases of biopsy-proven subacute cutaneous lupus erythematosus related to the use of pravastatin and simvastatin were also identified in the Medline database review [199]. All three patients had a positive ANA and interface dermatitis with perivascular and periadnexal lymphocytic infiltrates and responded to topical steroids. The lupus band test was positive for two patients and negative for the third patient. There have been additional reports of statin-induced cutaneous lupus associated with anti-SSA/Ro antibodies [166, 209]. Biopsies were reported as showing interface dermatitis with IgG deposition and perivascular lymphocytic infiltrate. Cutaneous lesions generally resolved with discontinuation of the statin, although some patients were treated with corticosteroids.

The Medline database review study also identified 14 patients with statin-related polymyositis and dermatomyositis of whom 11 had muscle biopsies that revealed perivascular and perifascicular T-cell inflammation [199]. Anti-Jo-1 antibodies were present in three patients. Most patients required treatment with immunosuppressive agents and ANAs remained positive in five patients despite improvement in clinical features.

Statins have been demonstrated to have a number of immunomodulatory activities that may play a role in the pathogenesis of statin-induced lupus [210].

D-Penicillamine

D-Penicillamine-related lupus was first described in patients with rheumatoid arthritis by Harkcom et al. in 1978 [211] but has also been reported in those with other autoimmune diseases such as scleroderma and in patients with Wilson disease and cystinuria. The frequency of D-penicillamine-related lupus has been reported as high as 2% of rheumatoid arthritis [212] patients and up to 7% of patients with Wilson disease [213]. The most common clinical manifestations include polyarthritis, pleurisy, leukopenia, thrombocytopenia, and cutaneous involvement. In addition to antinuclear antibodies, some patients have been reported to develop antibodies to double-stranded DNA and hypocomplementemia [211, 213-215]. Penicillamine therapy has also been associated with the development of other autoimmune phenomena such as myasthenia gravis and polymyositis.

Hormonal Factors

Hormonal factors are believed to play a role in the development of SLE. Bae *et al.* suggested that growth hormone may have precipitated a flare in a 19-year-old male with a 11-year-old history of lupus after he had been treated with a growth hormone for 9 months because of growth retardation [216]. The flare was manifested by a decrease in C3 and C4, elevated anti-dsDNA antibodies, and the development of nephrotic range proteinuria at 1.5 g/24 h. He required treatment with monthly IV bolus doses of cyclophosphamide and high doses of prednisolone; in addition, his growth hormone was discontinued. It has been postulated that growth hormone may play a role in preventing apoptosis of mononuclear cells with consequent increased survival of autoreactive immune cells. A study by Ogueta *et al.* reported that transgenic mice that expressed bovine growth hormone developed arthritis and autoantibodies [217]. Higher basal prolactin levels have been found and hyperprolactinemia has been reported to play a role in the development and progression of lupus [218, 219]. These findings are of interest since both growth hormone and prolactin belong to the same receptor family.

Aromatase Inhibitors

Increasing numbers of patients are receiving aromatase inhibitors for the treatment or prevention of breast cancer. Clinical trials data indicate that the use of these medications have been associated with a higher incidence of arthralgias and inflammatory arthropathies compared with patients treated with placebo or tamoxifen. The ATAC (Arimidex Tamoxifen Alone or in Combination) Trials group reported that 27.8% of patients receiving anastrozole developed musculoskeletal disorders (no specific details were provided), which was statistically significantly more frequent compared with patients treated with tamoxifen or a placebo [220]. In a study of 5187 women who had completed 5 years of tamoxifen therapy for the treatment of breast cancer, 21.3% of patients treated with letrozole developed arthralgias and 5.6% develop arthritis compared with 16.6 and 3.5%, respectively, of patients receiving placebo [221]. The symptoms were generally low grade with few women discontinuing the study because of the side effects. Coombes *et al.* reported that 33% of women receiving exemestane developed "aches or pains" and 5.4% arthralgias compared with 29.4 and 3.6%, respectively, for the tamoxifen group in a double-blind, randomized trial that switched patients to either exemestane or tamoxifen after 2–3 years therapy with tamoxifen [222]. In a study of 77 patients with metastatic breast cancer, 12 (16%) patients developed arthralgias, primarily involving the hands, knees, hips, lower back, and shoulders, within 2 months of starting anastrozole [223]. Of these, four patients (5%) discontinued anastrozole because of the severity of the arthralgias, which subsequently resolved. Bonneterre *et al.* reported that 6.3% of patients developed "bone pain" while receiving anastrozole compared with 6.1% of tamoxifen in their double-blind multicenter randomized study of 668 patients receiving these drugs as first-line treatment for advanced breast cancer [224]. In a study of 551 patients with advanced breast cancer previously treated with antiestrogens, 8.5% of patients receiving letrozole 0.5 mg q.d. and 13.2% of those receiving letrozole 2.5 mg q.d. developed arthralgias compared with 7.9% of those patients receiving megestrol acetate [225]. The higher incidence of arthralgias among the higher dose of letrozole would suggest a possible dose-related effect. With the increase in use of these agents, the clinician should be aware of the possibility of a drug effect in patients presenting with arthralgias or inflammatory arthritis who are receiving aromatase inhibitors.

BIOLOGICAL AGENTS

Biological agents are being used increasingly to treat a variety of diseases including malignancies as well as infectious and autoimmune diseases such as rheumatoid arthritis, systemic lupus erythematosus, systemic sclerosis, multiple sclerosis, and psoriasis. These agents include a number of cytokines, such as interferon-α, -β, -γ and interleukin-2 (IL-2), which have immunomodulating affects. Despite considerable therapeutic success, many of the patients receiving these agents develop autoantibodies and in some cases autoimmune-like diseases [226–231]. Case reports and follow-up studies of large cohorts of patients treated with interferons indicate that immune-mediated abnormalities can occur and affect different organ systems, although they occur infrequently [232]. Interferon (INF) therapy can precipitate the development of these immune-mediated abnormalities either *de novo* or by exacerbation of an existing autoimmune predisposition, as manifested by increasing antibody levels and the development of autoimmune disease in patients with preexisting autoantibodies. A rise in titres of preexisting autoantibodies may occur in over 50% patients with hepatitis C treated with interferon, while the development of overt autoimmune pathologies only occurs in 1–2% of these patients [233]. In addition, the finding of non-organ-specific antibodies at baseline may increase the likelihood of developing autoimmune hepatitis.

Treatment with INF-α and -β and IL-2 has been associated with the development of thyroid autoantibodies, anti-DNA, and antinuclear antibodies, which may develop as early as the 1st month of treatment and generally disappear following discontinuation of treatment [227, 228, 231]. Autoimmune disorders have been reported to develop in 15—30% of patients receiving interferon-α therapy [234]. Ronnblom *et al.* [227] reported that 25 (19%) of their 135 patients with carcinoid tumors developed evidence of autoimmune diseases after treatment with INF-α. Some patients develop autoimmune thyroid disease, autoimmune hemolytic anemia, autoimmune thrombocytopenia, lupus-like syndromes, pernicious anemia, and vasculitis [227, 235, 236]. Reports suggest that autoimmune thyroid disease is more likely to develop in patients with preexisting thyroid antibodies treated with INF-α [227, 231]. Over 60% of patients with thyroid antibodies who received INF-α were reported to develop autoimmune thyroid disease compared with 7% of those who were antibody-negative [228]. Treatment of thyropathy often must be continued because of persistent thyroid dysfunction despite discontinuation of the INF therapy [233]. Anti-islet cell antibodies have also been reported to develop in patients receiving INF therapy.

Obermoser *et al.* implicated INF-α in the development of a lupus-like syndrome in an 80-year-old woman treated with INF-α for chronic myelogenous leukemia who developed fatigue, fever, leukopenia, and a multiforme-like erythema on the arms and legs; skin biopsy revealed an interface dermatitis suggestive of lupus erythematosus [237]. ANA was positive at a 1:320 titer, but antibodies to dsDNA and ENA were negative. INF was discontinued and the patient was treated with steroids with subsequent resolution of her clinical features. A lupus-like syndrome has also been reported in association with the use pegylated-IFN-α-2b for treatment of chronic hepatitis B virus infection [238].

Data suggest that the development of autoimmunity is associated with a favorable antitumor effect in patients receiving type 1 INFs. In one report, the risk of recurrence of melanoma was reduced by a factor of 50 in patients in whom autoimmunity developed during treatment with INF-α-2b [239]. Similarly, the development of autoimmune thyroid disease, antiphospholipid antibody syndrome, and vitiligo has been reported to confer a survival benefit in patients with melanoma treated with IL-2.

INF-α and IL-2 have been associated with the induction or exacerbation of symmetric inflammatory arthropathies [235, 240]. Three patients treated with IL-2 developed a positive rheumatoid factor, antinuclear antibodies, and inflammatory arthritis [240]. It has been postulated that high-dose IL-2 may enhance expression of HLA Class II antigen, resulting in the development of autoimmunity.

Thyroid disease has also been reported with the use of granulocyte-monocyte colony stimulating factor (GMCSF). Reversible hypothyroidism developed in two patients treated with GMCSF, both of whom had antibodies to thyroid peroxidase prior to commencement of therapy [241].

Lupus nephritis has been described with the use of anti-CTLA4 antibody. A 64-year-old male with metastatic melanoma was reported to have developed nephrotic syndrome and anti-dsDNA antibodies after receiving two injections of ipilimumab, a monoclonal antibody against cytotoxic T-lymphocyte-associated antigen 4. Biopsy revealed evidence suggestive of lupus nephritis with deposits of IgG and C3 seen on immunofluorescence. Anti-dsDNA antibodies disappeared and the nephritic syndrome improved following withdrawal of ipilimumab, although the patient also received high-dose steroids [242].

Dasatinib, a tyrosin kinase inhibitor, has been associated with the development of a lupus-like syndrome. A 74-year-old female with CML developed arthralgia, fatigue, fever, pleural and pericardial effusions, hepatosplenomegaly, new onset ANA, and mildly positive anti-dsDNA antibodies approximately 3 months after starting dasatinib. Her clinical features resolved within 1 month, anti-dsDNA antibodies within 2 months, and the ANA within 8 months of discontinuation of dasatinib [243]. DeLavallade *et al.* reported that a prior history of autoimmune disease was a strong independent risk factor ($p = 0.0008$) for the development of pleural effusion in patients receiving dasatinib, suggesting that these effusions may have an immune-mediated pathogenesis [244].

Tumor Necrosis Factor Alpha Inhibitors

Treatment with tumor necrosis factor (TNF)-α inhibitors has been associated with the development of autoantibodies and lupus-like syndromes [245, 246]. The range of clinical features as well as the pattern of autoantibody profiles of drug-induced autoimmunity secondary to TNF inhibitors are broader than that seen in classical DRL. Antinuclear antibodies and antibodies to dsDNA, which are considered specific for systemic lupus erythematous, have been detected frequently in patients treated with TNF-α inhibitors. However, in contrast, relatively few patients have been reported to develop a lupus-like syndrome.

Prevalence

Although DRL is a rare but well-recognized side effect of TNF-α inhibitors, the exact incidence is difficult to estimate. Prospective studies would suggest that DRL secondary to TNF inhibitors is a rare occurrence despite the relatively high incidence of autoantibodies. Reported incidence rates have varied from 0.1—0.6% for RA patients to 1.6% for patients with Crohn disease, although actual numbers are very small [247–251]. Clinical trial data from Centocor indicated that DRL had occurred in less than 1% of the patients who received infliximab and that these lupus-like syndromes resolved on discontinuation of infliximab infusions [252]. Reported rates for adalimumab have varied from 0.06 to 0.13 events/100 patient years [251]. Review of published data suggests that infliximab has been the TNF-α inhibitor most frequently implicated in the development of DRL. A retrospective analysis revealed 92 cases of DRL related to the use of TNF-α inhibitors, of which 40 were related to the use of infliximab, 37 to etanercept, and 15 to adalimumab [253]. The lower incidence of DRL secondary to adalimumab may be related to structural or pharmacokinetic differences or to the fact that relatively fewer patients have been treated with adalimumab.

In contrast, in an online survey 37% of rheumatologists surveyed reported seeing one to three cases of DRL among their RA patients treated with TNF-α inhibitors, although no specific details of the manifestations were provided. Based on the responses, the event risk per patient was estimated to be 0.87% [254]. Some confounding factors in correctly diagnosing drug-related lupus secondary to TNF-α inhibitors may include the fact that most patients treated with the medications have arthritis, which is also a manifestation of drug-related lupus, and many patients are receiving methotrexate, which may be complicated by hematologic abnormalities.

Autoantibodies have been detected in patients with rheumatoid arthritis (RA), Crohn disease, and spondyloarthropathy, suggesting that the development of these antibodies are independent of the disease background. However, the lupus-like syndromes have primarily been reported in RA patients. This may be due to the relatively larger number of RA patients treated with these agents compared with the other diseases or alternatively to an immunogenetic susceptibility of RA patients to this syndrome.

Autoantibodies

The reported incidence of ANAs in patients treated with TNF inhibitors is variable, ranging from 23 to 57%, with most ANAs having activity directed against chromatin [247, 255]. In clinical trials, 50% of patients developed ANAs compared with 20% of placebos; new anti-dsDNA antibodies occurred in 4% of certolizumab pegol patients, 10–12% of adalimumab patients, 15% of etanercept patients, and 20% of infliximab patients [256]. The reported incidence of antibodies to dsDNA varies from 9 to 81%, but antihistone antibodies, although reported, do not appear to be a common feature [257]. Antinuclear antibodies have been reported to develop in 34% and anti-dsDNA antibodies in 9% of

patients after approximately 4–12 weeks of treatment with infliximab for Crohn disease; anti-dsDNA antibodies disappeared when infliximab was discontinued. Both IgG and IgM anticardiolipin antibodies have been detected in up to 25% RA patients receiving TNF-α inhibitors, although thrombotic events have only occurred in about 4% of patients [257, 258].

Data from Centocor indicated that low titers of autoantibodies developed in less than 10% of a total of 777 patients who received infliximab as part of several placebo-controlled, randomized clinical, and open trials [252]. In the ATTRACT study, ANAs were detected in 23% of patients receiving infliximab compared with 6% of placebo recipients [255]. Anti-dsDNA antibodies occurred in 16% of patients receiving infliximab but were not detected in placebo recipients. Antibodies appeared after a mean of 6.5 weeks (range, 4–10 weeks) following initiation of treatment without any correlation with infliximab dosages. They disappeared spontaneously during or after treatment.

In a placebo-controlled trial of 156 patients with rheumatoid arthritis, the incidence of ANAs increased from 30 to 53% and anti-dsDNA antibodies from 0 to 14% in patients treated with infliximab [248]. Assayed using three methods — Crithidia luciliae indirect immunofluorescence assay, a Farr assay using mammalian DNA, and a Farr assay that utilized ^{125}I-labeled circular plasmid DNA [76] — the anti-dsDNA antibodies developed after a mean of 6.3 weeks (range, 4–10 weeks). These antibodies were of the IgM class in all patients except for one patient who also had IgG and IgA antibodies to dsDNA. One patient with anti-dsDNA antibodies detected by all three methodologies developed a lupus-like syndrome characterized by fever, dry cough, rash, chronic chest pain, and a pericardial effusion approximately 11 weeks after the initiation of treatment with infliximab [76]. The clinical features and the anti-dsDNA antibodies disappeared within 8 weeks of discontinuation of infliximab. In contrast, only one patient treated with placebo developed a positive ANA and none developed anti-dsDNA antibodies. In a prospective trial, with a mean follow-up of 20.5 months, eight of 26 RA patients treated with infliximab developed ANAs [259]. In another study, ANAs developed in 19 of 62 RA patients and 25 of 35 spondyloarthropathy patients treated with infliximab; seven of the RA patients and six of the spondyloarthropathy patients developed antibodies to dsDNA [260]. Haraoui and Keystone reported a 12.6% incidence of ANAs developing in patients receiving adalimumab compared with 7.3% of placebo-treated patients using clinical trial data [261]. The incidence of ANAs was reported to be 11% and anti-dsDNA antibodies 15% in patients treated with etanercept compared with 5 and 4%, respectively, of placebo-treated patients using data from randomized control trials [249]. Antihistone antibodies were also present.

In a 2-year prospective study comparing TNF-α inhibitors, ANAs developed in 62% of infliximab-treated compared with 15% of etanercept-treated spondyloarthropathy patients and in 41% of infliximab-treated compared with 10% of etanercept-treated RA patients at year 1 [262]. Anti-dsDNA antibodies developed in 71% of infliximab-treated spondyloarthropathy and 49% of infliximab-treated RA patients compared with 0% among the etanercept-treated patients. Isotyping of the anti-dsDNA antibodies revealed a predominance of IgM antibodies, some IgA antibodies but no IgG antibodies to dsDNA. The IgM antibodies developed after approximately 6 weeks of treatment with infliximab, persisted for up to 2 years without switching to the IgG isotype and resolved rapidly after discontinuation of treatment. In another prospective study of RA patients, the prevalence of ANAs increased from 24 to 77% and IgG and IgM antibodies to dsDNA developed in 64 and 81%, respectively, after 30 weeks of infliximab treatment [257]. None of the etanercept-treated patients developed autoantibodies. These studies suggest that autoantibodies are more likely to develop in patients treated with infliximab compared with etanercept. ANAs have also been reported to occur in higher titers in patients receiving the higher dose of infliximab [263].

The development of autoantibodies among patients receiving TNF-α inhibitors may be related to a general nonspecific B-cell activation, the induction of apoptosis, and/or reduction in clearance of nuclear debris as a result of downregulation of C-reactive protein. There has been speculation that the development of ANAs and anti-dsDNA antibodies among rheumatoid arthritis patients receiving infliximab may result from increased amounts of nucleosomal antigens in the synovium as a result of a rapid decrease in C-reactive protein that occurs during the early phase of treatment with infliximab. C-reactive protein has been postulated to be important in the clearance of this apoptotic material and a sudden decrease in its level may lead to impaired clearance of nucleosomal antigens with subsequent induction of an autoantibody response. Studies have also demonstrated increased numbers of apoptotic lamina propria T cells and blood monocytes from patients with Crohn disease treated with infliximab and increased concentrations of apoptotic monocytes/macrophages in the synovium of RA patients treated with etanercept and infliximab [264]. TNF-inhibitor-induced apoptosis of leukocytes may release DNA and other lupus autoantigens, which may elicit antinuclear antibodies and clinical symptomatology. There is wide fluctuation in the serum concentrations of infliximab in contrast to the relative constant concentrations of etanercept and adalimumab with initial concentrations of infliximab as high as 40-fold greater than steady-state concentrations of etanercept and adalimumab; in studies of RA patients, undetectable trough concentrations were found in 22—30% of patients receiving infliximab 3 mg/kg every 8 weeks [265]. The higher reported ANA response to infliximab may relate to this unique pharmacokinetic profile whereby the high plasma concentrations shortly after infusion may trigger apoptosis of transmembrane TNF (tmTNF)-bearing cells and release nucleosomes.

Clinical Features

There have been several reports of lupus-like features consisting of constitutional symptoms, pericarditis, pleural or pericardial effusions, worsening polyarthritis, and glomerulonephritis developing in patients treated with TNF-α inhibitors [30, 31, 248, 259, 266]. In contrast to other forms of DRL, antibodies to dsDNA and cutaneous features, such as a malar rash, photosensitivity, biopsy-proven discoid and subacute cutaneous lupus lesions, and cutaneous vasculitis have been commonly reported as manifestations of DRL associated with TNF-α inhibitors, with positive lupus band test noted on histopathology. The time to onset and resolution of DRL is variable [249, 250, 267, 268]. In a French survey, DRL developed after a mean of 9 months for infliximab and 4 months for etanercept, with the majority of patients having resolution of their features within 3—16 weeks after discontinuation of the TNF-α inhibitor. In other reports, symptoms resolved within 3 weeks to 12 months.

DeBandt et al. reported the development of lupus-like syndromes in two patients treated with etanercept [34]. One patient developed cutaneous lesions on the face and scalp, and diffuse pain 3 months after starting etanercept in addition to a positive ANA, anti-DNA and anticardiolipin antibodies, and elevated muscle enzymes: cutaneous lesions resolved within 1 month and biological abnormalities returned to normal within 4 months of discontinuation of the etanercept. The second was a 50-year-old woman who developed a diffuse erythematous and purpuric eruption with fine scaling of the hands and fingers approximately 4 months after starting etanercept in addition to lymphopenia, thrombocytopenia, an elevated ESR, a positive ANA at a 1:640 titer. The cutaneous manifestations resolved following discontinuation of the etanercept. Brion et al. reported the development of cutaneous lesions in two patients treated with etanercept [33]. One patient developed a diffuse erythematous rash 4 days after her fourth etanercept injection and a biopsy revealed acute discoid lupus; neither ANA or anti-dsDNA antibodies were detected. The other patient developed a widespread rash after her seventh etanercept injection with purpuric lesion on the lower extremities and buttocks; biopsy revealed a necrotizing vasculitis and anti-dsDNA antibodies were negative. Both patients' rashes resolved within 2 weeks following discontinuation of etanercept.

In data from Centocor three (<0.5%) of approximately 388 patients treated with infliximab as part of clinical trials developed drug-induced lupus: one patient had Crohn disease and the others RA. One of these patients developed serositis, fever, and high levels of both ANA and anti-dsDNA antibodies and the other patient developed a rash with evidence of mild perivascular infiltrate. These features resolved on discontinuation of infliximab and treatment with steroids. Combined verifiable data from the Crohn disease and rheumatoid arthritis trials indicated that four of 987 (0.2%) patients developed a lupus-like syndrome primarily manifested by rash and fevers with one patient developing pleuropericarditis. Of the four patients, three had rheumatoid arthritis and one had Crohn disease. Postmarketing surveillance by Centocor revealed a 0.04% incidence of DRL among 170,000 patients.

According to FDA data, there have been 14 cases of patients with Crohn disease who have developed ANAs and lupus-like syndromes after receiving a median of three infusions. Eight patients developed antibodies to dsDNA. Following discontinuation of the infliximab infusions, there was resolution of the lupus-like syndrome in six patients, but one patient failed to improve. There were no follow-up data available for the other patients. Using the FDA database, Mohan *et al.* identified 16 cases of new-onset SLE that developed after a median of 4 months (range of 2–18 months) after initiation of etanercept [269]. Discoid lupus was reported in eight, photosensitivity in six, and malar rash in four patients; seven patients had biopsies consistent with subacute cutaneous lupus and three with discoid lupus. Resolution of clinical features occurred within 1–4 months after discontinuation of the agent. In another report, DRL, manifested by arthritis, discoid lupus, malar rash and pleuritis, developed in four patients after 6 weeks to 14 months of treatment with etanercept, which resolved within 2–6 weeks of discontinuing etanercept [249].

A retrospective study using a French database of patients treated with TNF-α inhibitors reported that 12 patients developed systemic and 10 additional patients cutaneous manifestations of lupus, which developed after a mean of 9 months (range, 3–16 months) for those treated with infliximab and 4 months (range, 2–5 months) for patients treated with etanercept [250]. All patients had anti-dsDNA antibodies. Eight patients received steroids and the features resolved in all but one patient. The incidence of DRL was estimated to be 0.19% for patients treated with infliximab and 0.18% for those treated with etanercept.

There are reports of patients developing glomerulonephritis and vasculitis while being treated with TNF-α inhibitors [270–272]. Mohan *et al.* reported 35 cases of leukocytoclastic vasculitis (LCV) of which 20 occurred after treatment with etanercept and 15 with infliximab [272]. Seventeen of the patients had biopsy-proven vasculitis and 22 (62.8%) had complete or marked improvement of their cutaneous lesions following discontinuation of the drug. Six patients developed a recurrence of their leukocytoclastic vasculitis following rechallenge with the TNF-α inhibitor, but three patients did not.

In a recent review Ramos-Casals *et al.* identified a total of 379 cases of autoimmune diseases secondary to TNF-α inhibitors: these comprised 105 cases of lupus-like syndromes, 183 of vasculitis, 50 cases of psoriasis, 34 of interstitial lung disease, 32 cases of antiphospholipid-like syndromes, 19 cases of ocular autoimmune diseases, ten cases of sarcoidosis, seven cases of autoimmune hepatitis, and four cases of inflammatory myopathies; 183 of these cases were associated with the use of infliximab, 163 with etanercept, and 42 with adalimumab [41]. Of those who developed the lupus-like syndrome, 43% had been receiving infliximab, 36% etanercept, and 21% adalimumab. The most frequently reported clinical features were malar rash, arthritis, discoid lupus, and serositis. Subacute and chronic cutaneous features occurred more frequently in patients receiving etanercept, whereas serositis occurred more frequently in patients treated with infliximab. Antinuclear antibodies were present in 91% and anti-dsDNA antibodies in 64% of the 89 patients for whom immunologic data was available; in contrast, only 7% of these patients had antihistone antibodies. Of the 118 patients who developed vasculitis, 60 (51%) had been receiving etanercept, 51 (43%) had been receiving infliximab, and five (4%) had been receiving adalimumab; the other two patients had been treated with other agents. Cutaneous vasculitis occurred in 86% of patients and included purpura in 24%, ulcerative lesions, nodules, digital vasculitis, maculopapular rash, chillblain, and other cutaneous lesions. Other areas of vasculitic involvement included the peripheral nervous system in 18 patients, kidney in 16 patients, CNS in five patients, and lung in three patients. The most common histopathologic finding was leukocytoclatic vasculitis which was seen in 69% of skin biopsies. Other histopathologic findings included necrotizing vasculitis, and lymphocytic vasculitis. Serologic studies revealed ANAs in 27 patients, ANCAs in 11, anti-dsDNA antibodies in four, antiphospholipid antibodies in three, and antiRo/La antibodies in one patient and five patients had cryoglobulins. Treatment with the TNF-α inhibitor was discontinued in 94% patients; 26% of patients received corticosteroids and 15% immunosuppressive therapy.

II. CLINICAL ASPECTS OF DISEASE

There was complete resolution of clinical features in 71 patients, partial resolution in 25 patients, and no improvement in eight patients, with one patient dying from PAN and one from rapidly progressive glomerulonephritis. Rechallenge with the TNF inhibitor resulted in recurrence or worsening of features in 75% of reported cases. In some patients treated with etanercept, the cutaneous vasculitic lesions began at the injection site and spread to involve other areas, suggesting a possible direct antigen-mediated hypersensitivity pathogenic mechanism.

Mechanisms

The development of autoantibodies has been associated with the use of all three TNF-α inhibitors. Possible mechanisms by which these antibodies may be induced by TNF-α inhibitors include dysregulation of apoptosis with release of autoimmunogenic plasma nucleosomes from apoptotic cells that trigger formation of autoantibodies against cytoplasmic and nuclear compounds or inhibition of a cytotoxic T-lymphocyte response that normally suppressives autoreactive B cells. Greater concentrations and frequencies of ANAs, anticardiolipin and anti-dsDNA antibodies have been described with infliximab compared with etanercept or adalimumab, suggesting that factors other than TNF blockade may be important [262, 264].

As monotherapy infliximab is considered to be the most immunogenic of the three TNF-α inhibitors, although its immunogenicity decreases with higher doses. The higher immunogenicity of infliximab, which most likely results from infliximab being a chimeric while adalimumab is a fully humanized monoclonal antibody and etanercept a soluble fusion protein, may play a role in the pathogenesis of DRL. As a result of its pharmacokinetic profile, infliximab may also reach higher concentrations in tissue microenvironments than etanercept or adalimumab and, thereby, may have a greater opportunity to bind to tmTNF on cells and induce reverse signaling or cytotoxicity of tmTNF-bearing cells by Fc-dependent mechanisms, such as complement-dependent cytotoxicity and antibody-dependent cellular toxicity. There is evidence for differential induction of cytokine suppression through reverse signaling with some studies showing more complete suppression by infliximab and adalimumab than etanercept [264]. Reverse signaling may also be important in apoptosis and/or other cellular events that may play a role in the pathogenesis of DRL.

It has been proposed that the development of DRL may be related to the observation that removal of TNF may result in an increased activity of T and B cells that react with autoantigens and foreign antigens [264]. However, we reported the case of a patient with Crohn disease who was successfully switched from one TNF-α inhibitor after the development of DRL, suggesting that other mechanisms may be more important [273]. Our patient developed joint pain and swelling, facial rash, photosensitivity, and oral ulcers after 11 months of treatment with infliximab; ANA was positive at a 1:320 titre and antibodies to dsDNA were elevated at 65. Infliximab was discontinued with complete resolution of his symptoms within 3 months without the use of steroids or immunosuppressive agents. Subsequently, adalimumab was instituted for treatment of a flare-up of his Crohn disease without recurrence of any lupus-like features.

Studies comparing infliximab to adalimumab demonstrated that both agents can induce apoptosis in normal blood monocytes and T cells and a human acute monocytic leukemia cell line [274]. A recent study demonstrated concentration dependence of infliximab- and adalimumab-induced apoptosis of normal blood T cells [275]. Concentration dependence of apoptosis may be important to the development of autoimmunity secondary to TNF-α inhibitors and explain the higher incidence seen with infliximab, which has wide peak-trough fluctuations.

Psoriasis

The use of TNF-α inhibitors has been associated with the development and worsening of psoriasis. In a retrospective review using the Medline database between 2005 and 2007, Grinblat and Scheinberg found that 50 patients had developed psoriasis or psoriasiform lesions following treatment with a TNF inhibitor [276]. Of these, 30 patients had been receiving infliximab, 11 were receiving etanercept, and nine patients adalimumab. Different types of presentation were reported with the plaque form being the most frequent. Seven (19%) of 37 patients had an exacerbation of their previous psoriasis. Based on clinical and immunological observations, the authors suggested that a cytokine disequilibrium resulting from anti-TNF therapy may be the pathogenic mechanism for the development of these lesions. Similarly, in their recent review Ramos-Casals et al. identified 50 cases of psoriasis secondary to TNF-α inhibitors [41].

Ko et al. reported a prevalence ranging from 0.6 to 5.3% in their analysis of 127 published cases of psoriasis occurring in patients treated with TNF-α inhibitors [277]. Seventy of these patients were receiving infliximab, 35 were on etanercept, and 22 were on adalimumab. Onset varied from a few days to 4 years. A relatively higher prevalence occurred among patients with ankylosing spondylitis and Behçet disease. Palmoplantar pustular psoriasis was most frequently reported, occurring in 40.5%, with plaque-like psoriasis occurring in 33.1% of patients. Discontinuation of the

TNF-α inhibitor with systemic therapy led to improvement in 64.3% of cases, use of topical steroids resulted in improvement in 26.8% cases, and switching to an alternative TNF-α inhibitor led to resolution in 15.4% cases

SUBACUTE CUTANEOUS LUPUS ERYTHEMATOSUS

There have been at least 71 cases reported of subacute cutaneous lupus erythematosus associated with the use of medications [56]. Medications that have been implicated in the development of subacute cutaneous lupus are listed in Table 33.3. The first cases were reported by Reed *et al.*, who described five patients who developed subacute cutaneous lupus erythematosus lesions while taking hydrochlorothiazide, which resolved following discontinuation of the drug [278]. All the patients had anti-Ro antibodies, which disappeared in one-third of the patients. Srivastava *et al.* found that 15 of 70 patients with anti-RO/SSA-positive cutaneous lupus erythematosus had a potential drug induction [166]. Implicated medications included pravastatin, simvastatin, hydrochlorothiazide, enalapril, lisinopril, diltiazem, nifedipine, INF-α, and INF-β. In their population-based study of 514 patients diagnosed with cutaneous lupus between 1965 and 2005, Durosaro *et al.* found 14 cases of drug-induced cutaneous lupus [279]. A review study identified 71 published cases of drug-induced SCLE as of June 2007 [56]. The mean age of these patients was 59 years of age and the mean duration of medication use varied from weeks to years before the onset of the cutaneous lesions. Those medications that were most frequently implicated were cardiovascular drugs, especially calcium channel blockers, thiazides, angiotensin-converting enzyme (ACE) inhibitors and beta-blockers, and antifungal agents, such as terbinafine. Mean time to clearance of the lesions following discontinuation of the suspected drug was 5.76 weeks with a range of 1–24 weeks. Anti-Ro antibodies that were present in the majority of the patients tended to persist following resolution of the rash. Unlike typical drug-induced lupus, antihistone antibodies were present in less than 50% of patients. Callen has suggested that the prevalence of drug-induced SCLE may be increasing: in their most recent study, Callen and colleagues found three patients with drug-induced SCLE among their 76 patients with SCLE, while their previous studies had not revealed any patients [280]. In addition, the list of drugs associated with the development of SCLE has increased.

Of interest is the finding that most of the implicated drugs have a propensity to produce photosensitivity

TABLE 33.3 Drugs Associated with Drug-Related Subacute Cutaneous Lupus

CARDIOVASCULAR DRUGS
Hydrochlorothiazide[a]
Spironolactone[a]
Diltiazem[a]
Nifedipine[a]
Nitrendipine[a]
Oxprenolol[a]
Captopril
Enalapril
Lisinopril
Cilazapril
Procainamide
ANTIFUNGAL DRUGS
Griseofulvin[a]
Terbinafine[a]
CHEMOTHERAPEUTIC DRUGS
Doxetaxel[a]
COL-3
Capecitabine
Doxorubicin
IMMUNOMODULATING DRUGS
Hydroxychloroquine
Leflunomide
Etanercept
Infliximab
Interferon-α
Interferon-β
D-Penicillamine
LIPID-LOWERING DRUGS
Pravastatin
Simvastatin
OTHERS
Cinnarazine/Triethylperazine[a]
Naproxen
Piroxicam
Omeprazole
Ranitidine
Phenytoin
Glyburide
Insecticides

[a]*Most frequently associated with SCLE.*

and lichenoid drug reactions, suggesting that drug-induced SCLE may represent a photo-induced iso-morphic/Koebner response in an immunogenetically predisposed host.

CONCLUSION

Autoantibodies and autoimmune syndromes can develop with the use of certain medications and biologic agents. Although autoantibodies occur relatively frequently, only a minority of patients develop autoimmune diseases. The development of autoantibodies alone is not sufficient reason to withhold the drug. The spectrum of drug-induced autoantibodies and disease manifestations have broadened, particularly with the use of biologic agents. Generally the clinical manifestations resolve following discontinuation of the drug, although the antibodies may persist. With the ever-increasing number of new drugs, in particular the biologic agents, the incidence of drug-induced autoimmunity is likely to increase. Physicians need to be aware of possible associations between the development of autoimmune-like features and exposure to certain pharmacologic agents for both diagnostic and thera-peutic reasons.

References

[1] D. Deapen, A. Escalante, L. Weinrib, D. Horwitz, B. Bachman, P. Roy-Burman, et al., A revised estimate of twin concordance in systemic lupus erythematosus, Arthritis Rheum 35 (1992) 311–318.

[2] B.J. Hoffman, Sensitivity of sulfadiazine resembling acute disseminated lupus erythematosus, Arch Dermatol Syph 51 (1945) 190–192.

[3] S. Gold, Role of sulphonamides and penicillin in the patho-genesis in systemic lupus erythematosus, Lancet 1 (1951) 268–272.

[4] J.R. Walsh, J.H. Zimmerman, The demonstration of the "LE" phenomenon in patients with penicillin hypersensitivity, Blood 8 (1953) 65–71.

[5] A.M. Paull, Occurrence of the "LE" phenomenon in a patient with a severe penicillin reaction, N Engl J Med 252 (1955) 128–129.

[6] M. Honey, SLE presenting with sulphonamide hypersensi-tivity, Br Med J 1 (1956) 1272–1275.

[7] J.D. Morrow, H.A. Schroeder, H.M. Perry Jr., Studies on the control of hypertension by Hyphex: II. Toxic reactions and side effects, Circulation 8 (1953) 829–839.

[8] H.M. Perry, H.A. Schroeder, Syndrome simulating collagen disease caused by hydralazine (Apresoline), JAMA 154 (1954) 670–673.

[9] S.L. Lee, P.H. Chase, Drug-induced systemic lupus eryth-ematosus: A critical review, Semin. Arthritis Rheum 5 (1975) 83–103.

[10] T. Lindqvist, Lupus erythematosus disseminatus after admin-istration of mesantoin. Report of two cases, Acta Med Scand 158 (1957) 131–138.

[11] A.T. Ladd, Procainamide-induced lupus erythematosus, N Engl J Med 267 (1962) 1357.

[12] A. Cannat, M. Seligmann, Possible induction of antinuclear antibodies by isoniazid, Lancet 1 (1966) 825–827.

[13] N.F. Rothfield, W.G. Bierer, J.W. Garfield, Isoniazid induction of antinuclear antibodies: A prospective study, Ann Intern Med 88 (1978) 650–652.

[14] A.J.M. Fabius, W.K. Gaulhofer, Systemic lupus erythematosus induced by psychotropic drugs, Acta Rheumatol Scand 17 (1971) 137.

[15] E.M. Tan, A.S. Cohen, J.F. Fries, A.T. Masi, D.J. McShane, N.F. Rothfield, et al., The 1982 revised criteria for the classifi-cation of systemic lupus erythematosus, Arthritis Rhuem 25 (1982) 1271–1277.

[16] D.H. Beernink, J.J. Miller, Anticonvulsant-induced antinuclear antibodies and lupus-like disease in children, J Pediatr 82 (1973) 113–117.

[17] B.H. Singsen, L. Fishman, V. Hanson, Antinuclear antibodies and lupus-like syndromes in children receiving anticonvul-sants, Pediatrics 57 (1976) 529–534.

[18] J.R. Batchelor, K.I. Welsh, R.M. Tinoco, et al., Hydralazine-induced systemic lupus erythematosus: Influence of HLA-DR and sex on susceptibility, Lancet 1 (1980) 1102–1109.

[19] F.W. Miller, E.V. Hess, D.J. Clauw, P.A. Hertzman, T. Pincus, R.M. Silver, et al., Approaches for identifying and defining environmentally associated rheumatic disorders, Arthritis Rheum 43 (2000) 243–249.

[20] F.B. Vivino, H.R. Schumacher, Synovial fluid characteristics and the lupus erythematosus cell phenomenon in drug-induced lupus, Arthritis Rheum 32 (1989) 560–568.

[21] J.J. Cush, E.A. Goldings, Drug-induced lupus: Clinical spec-trum and pathogenesis, Am J Med Sci 290 (1985) 36–45.

[22] M.J. Goldberg, M. Hsain, W.J. Wajszczuk, M. Rubenfire, Pro-cainamide-induced lupus erythematosus pericarditis encoun-tered during coronary bypass surgery, Am J Med 69 (1980) 159–162.

[23] A.M. Carey, M. Coleman, A. Feder, Pericardial tamponade: A major presenting manifestation of hydralazine induced lupus syndrome, Am J Med 54 (1973) 84–87.

[24] M.K. Ghose, Pericardial tamponade. A presenting manifesta-tion of procainamide-induced lupus erythematosus, Am J Med 58 (1975) 581–585.

[25] S.K. Sunder, A. Shah, Constrictive pericarditis in procaina-mide-induced lupus erythematosus syndrome, Am. J. Cardiol 36 (1975) 960–962.

[26] M.T. Dent, S. Ganapathy, C.D. Holdsworth, K.C. Channer, Mesalazine induced lupus-like syndrome, Br Med J 305 (1992) 159.

[27] F. Vayre, L. Vayre-Oundjian, J.J. Monsuez, Pericarditis associ-ated with longstanding mesalazine administration in a patient, Int. J. Cardiol 68 (1999) 243–245.

[28] T. Molnar, M. Hogye, F. Nagy, J. Lonovics, Pericarditis associ-ated with inflammatory bowel disease: Case report, Am J Gastroenterol 94 (1999) 1099–1100.

[29] T.A. Sentongo, D.A. Piccoli, Recurrent pericarditis due to mesalazine hypersensitivity: A pediatric case report and review of the literature, J. Pediatric. Gastroenterol. Nutr 27 (1998) 344–347.

[30] I. Porfyridis, I. Kalomenidis, I. Psallidas, G. Stratakos, C. Roussos, T. Vassilakopoulos, et al., Etanercept-induced pleuropericardial lupus-like syndrome, Eur Respir J 33 (2009) 939–941.

[31] J. Abunasser, F.A. Forouhar, M.L. Metersky, Etanercept-induced lupus erythematosus presenting as a unilateral pleural effusion, Chest 134 (2008) 850–853.

[32] S. Alexander, Lupus erythematosus in two patients after griseofulvin treatment of trichophyton rubrum infection, Br J Dermatol 74 (1962) 72–74.

[33] P.H. Brion, A. Mittal-Henke, K.C. Kalunian, Autoimmune skin rashes associated with etanercept for rheumatoid arthritis, Ann Intern Med 131 (1999) 634.

[34] M. DeBandt, V. Descamps, O. Meyer, Etanercept induced systemic lupus erythematosus: Two patients with rheumatoid arthritis, Lupus 10 (2001) S118.

[35] O. Servitje, M. Ribera, X. Juanola, J. Rodriguez-Moreno, Acute neutrophilic dermatosis associated with hydralazine-induced lupus, Arch Dermatol 123 (1987) 1435–1436.

[36] W. Sequeira, R.B. Polisky, D.P. Alrenga, Neutrophilic dermatosis (Sweet's syndrome): Association with hydralazine-induced lupus syndrome, Am J Med 81 (1986) 558–560.

[37] O. Elkayam, D. Levarovosky, C. Brautbar, et al., Clinical and immunological study of 7 patients with minocycline-induced autoimmune phenomena, Am J Med 105 (1998) 484–487.

[38] C. Tournigand, T. Genereau, M. Prudent, M.-C. Dimert, S. Herson, O. Chosidow, Minocycline-induced clinical and biological lupus-like disease, Lupus 8 (1999) 773–774.

[39] R.R. Hope, L.A. Bates, The frequency of procainamide-induced systemic lupus erythematosus, Med J Aust 2 (1972) 298–303.

[40] A. Gough, S. Chapman, K. Wagstaff, P. Emery, E. Elias, Minocycline induced autoimmune hepatitis and systemic lupus erythematosus-like syndrome, Br Med J 312 (1996) 169–172.

[41] M. Ramos-Casals, P. Brito-Zeron, M.J. Soto, et al., Autoimmune diseases induced by TNF-targeted therapies, Best Pract Res Clin Rheumatol 22 (2008) 847–861.

[42] T.K. Sheikh, R.C. Charron, A. Katz, Renal manifestations of drug-induced systemic lupus erythematosus, Am J Clin Pathol 75 (1981) 755–762.

[43] K.S. Shapiro, V.W. Pinn, J.T. Harrington, et al., Immune complex glomerulonephritis in hydralazine-induced SLE, Am J Kidney Dis 3 (1984) 270–274.

[44] H.P. Dustan, R.D. Taylor, A.C. Corcoran, et al., Rheumatic and febrile syndromes during prolonged hydralazine therapy, JAMA 154 (1954) 23–29.

[45] S.G. Sturman, D. Kumararatne, D.G. Beevers, Fatal hydralazine-induced systemic lupus erythematosus [letter], Lancet 2 (1988) 1304.

[46] S. Bjorck, C. Svalander, G. Westberg, Hydralazine-associated glomerulonephritis, Acta Med Scand 218 (1985) 261–269.

[47] D. Alarcon-Segovia, K.G. Wakim, J.W. Worthington, et al., Clinical and experimental studies on the hydralazine syndrome and its relationship to systemic lupus erythematosus, Medicine 46 (1967) 1–33.

[48] Z. Sahenk, J.R. Mendell, J.L. Rossio, et al., Polyradiculoneuropathy accompanying procainamide-induced lupus erythematosus: Evidence for drug-induced enhanced sensitization to peripheral nerve myelin, Ann Neurol 1 (1977) 378–384.

[49] S. Ahmad, Procainamide and peripheral neuropathy, South Med. J 74 (1981) 509–510.

[50] R.M. Bernstein, J. Egerton-Vernon, J. Webster, Hydralazine-induced cutaneous vasculitis, Br. Med. J 280 (1980) 156–157.

[51] H.A. Cameron, L.E. Ramsay, The lupus syndrome induced by hydralazine: A common complication with low dose treatment, Br Med J 289 (1984) 410–412.

[52] J.P. Knox, S.E. Welykyj, R. Grandini, et al., Procainamide-induced urticarial vasculitis, Cutis 42 (1988) 469–472.

[53] J.M. Rosin, Vasculitis following procaineamide therapy, Am J Med 42 (1967) 625–629.

[54] T.M. Lawson, N. Amos, D. Bulgen, B.D. William, Minocycline induced lupus: Clinical features and response to rechallenge, Rheumatology 40 (2001) 329–335.

[55] H.K. Choi, P.A. Merkel, A.M. Walker, J.L. Niles, Drug-associated antineutrophil cytoplasmic antibody-positive vasculitis, Arthritis Rheum 43 (2000) 405–413.

[56] R.D. Sontheimer, C.L. Henderson, R.H. Grau, Drug-induced subacute cutaneous lupus erythematosus: a paradigm for bedside-to-bench patient-orientated translational clinical investigation, Arch Dermatol Res 301 (2009) 65–70.

[57] D. Alarcon-Segovia, Drug-induced lupus syndromes, Mayo Clin Proc 44 (1969) 664–681.

[58] A. Dupont, R. Six, Lupus-like syndrome induced by methyldopa, Br Med J 285 (1982) 693–694.

[59] J.D. Sherman, D.E. Love, J.F. Harrington, Anemia, positive lupus and rheumatoid factors with methyldopa. A report of three cases, Arch Intern Med 120 (1967) 321–326.

[60] M. Harth, L.E. cells and positive direct Coomb's test induced by methyldopa, Can Med Assoc J 99 (1968) 277–280.

[61] S.E. Blomgren, J.J. Condemi, J.H. Vaughan, Procainamide-induced lupus erythematosus. Clinical and laboratory observations, Am J Med 52 (1972) 338–348.

[62] S. Kleinman, R. Nelson, L. Smith, D. Goldfinger, Positive direct antiglobulin tests and immune hemolytic anemia in patients receiving procainamide, N Engl J Med 311 (1984) 809–812.

[63] M.H. Zarrabi, S. Zucker, F. Miller, R.M. Derman, G.S. Romano, J.A. Hartnett, et al., Immunologic and coagulation disorders in chlorpromazine-treated patients, Ann Intern Med 91 (1979) 194–199.

[64] D.M. Nordstrom, S.G. West, R.L. Rubin, Methyldopa-induced systemic lupus erythematosus, Arthritis Rheum 32 (1989) 205–208.

[65] M. Becker, A. Klajman, T. Moalem, et al., Circulating immune complexes in sera from patients receiving procainamide, Clin Immunol Immunopathol 12 (1979) 220–227.

[66] J.A. Mitchell, J.R. Batchelor, H. Chapel, C.N. Spiers, F. Sim., Erythrocyte complement receptor type I (CRI) expression and circulating immune complex (CIC) levels in hydralazine-induced SLE, Clin Exp Immunol 68 (1987) 446–456.

[67] P.D. Utsinger, N.J. Zvaifler, H.G. Bluestein, Hypocomplementemia in procainamide-associated systemic lupus erythematosus, Ann Intern Med 84 (1976) 293.

[68] M.G. Cohen, S. Kevat, M.V. Prowse, et al., Two distinct quinidine-induced rheumatic syndromes, Ann Intern Med 108 (1988) 369–371.

[69] J. Grossman, M.L. Callerame, J.J. Condemi, Skin immunofluorescence studies on lupus erythematosus and other antinuclear-antibody-positive diseases, Ann Intern Med 80 (1974) 496–500.

[70] J.D. Kirby, P.A. Dieppe, E.C. Huskisson, et al., d-Penicillamine and immune complex deposition, Ann Rheum Dis 38 (1979) 344–346.

[71] Y. Naparstek, J. Kopolovic, R. Tur-Kaspa, et al., Focal glomerulonephritis in the course of hydralazine-induced lupus syndrome, Arthritis Rheum 27 (1984) 822–825.

[72] M.J. Fritzler, E.M. Tan, Antibodies to histones in drug-induced and idiopathic lupus erythematosus, J Clin Invest 62 (1978) 560–567.

[73] Y. Sherer, et al., Autoantibody explosion in systemic lupus erythematosus: more than 100 different antibodies found in SLE patients, Semin Arthritis Rheum 34 (2004) 501–537.

[74] H. Mielke, K. Wildhagen, W. Mau, H. Zeidler, Follow-up of patients with double-stranded DNA antibodies induced by

sulfasalazine during the treatment of inflammatory rheumatic diseases, Rheumatology 22 (1993) 229–301.

[75] M.-M. Gordon, D. Porter, Minocycline related lupus: Case series in the West of Scotland, J Rheumatol 28 (2001) 1004–1006.

[76] P.J. Charles, R.J.T. Smeenk, J. DeJonj, M. Feldmann, R.N. Manie, Assessment of antibodies to double-stranded DNA induced in Rheumatoid Arthritis patients following treatment with Infliximab, a monoclonal antibody to tumor necrosis factor α: Findings in open-label and randomized placebo-controlled trials, Arthritis Rheum 43 (2001) 2383–2390.

[77] A.B. Mongey, R. Donovan-Brand, T.J. Thomas, et al., Serologic evaluation of patients receiving procainamide, Arthritis Rheum 35 (1992) 219–223.

[78] R.W. Burlingame, R.L. Rubin, Drug-induced anti-histone autoantibodies display two patterns of reactivity with substructures of chromatin, J Clin Invest 88 (1991) 680–690.

[79] R.L. Rubin, S.A. Bell, A.W. Burlingame, Autoantibodies associated with lupus induced by diverse drugs target a similar epitope in the (H2A–2B)-DNA complex, J Clin Invest 90 (1992) 165–173.

[80] V.J. Bray, S.G. West, K.T. Schultz, D.T. Boumpas, R.L. Rubin, Antihistone autobody profile in sulfasalazine induced lupus, J Rheumatol 21 (1994) 2157–2158.

[81] R. Zamber, H. Martens, R.L. Rubin, G. Starkebaum, Drug induced lupus due to ophthalmic timolol, J Rheumatol 19 (1992) 977–979.

[82] J. Gohill, P.D. Cary, M. Couppez, M.J. Fritzler, Antibodies from patients with drug-induced and idiopathic lupus erythematosus react with epitopes restricted to the amino and carboxyl termini of histone, J Immunol 135 (1985) 3116–3121.

[83] J.E. Craft, J.A. Radding, M.A. Harding, R.M. Bernstein, J.A. Hardin, Autoantigenic histone epitopes: A comparison between procainamide- and hydralazine-induced lupus, Arthritis Rheum 30 (1987) 689–694.

[84] J.P. Portanova, R.L. Rubin, F.G. Joslin, et al., Reactivity of anti-histone antibodies induced by procainamide and hydralazine, Clin Immunol Immunopathol 25 (1982) 67–79.

[85] J.P. Portanova, R.E. Arndt, E.M. Tan, et al., Anti-histone antibodies in idiopathic and drug-induced lupus recognize distinct intrahistone regions, J Immunol 138 (1987) 446–451.

[86] S.G. West, M. McMahon, J.P. Portanova, Quinidine-induced lupus erythematosus, Ann. Intern. Med 100 (1984) 840–842.

[87] M.C. Totoritis, E.M. Tan, E.M. McNally, et al., Association of antibody to histone complex H2A-H2B with symptomatic procainamide-induced lupus [published erratum appears in N. Engl. J. Med. 319, 256, 1988], N Engl J Med 318 (1988) 1431–1436.

[88] Y. Shoenfeld, O. Segol, Anti-histone antibodies in SLE and other autoimmune diseases, Clin Exp Rheumatol 7 (1989) 265–271.

[89] R.T. Canoso, R.M. de Oliveira, Chlorpromazine-induced anti-cardiolipin antibodies and lupus anticoagulant: Absence of thrombosis, Am J Hematol 27 (1988) 272–275.

[90] R.T. Canoso, M.E. Lewis, E.J. Yunis, Association of HLA-Bw44 with chlorpromazine-induced autoantibodies, Clin Immunol Immunopathol 25 (1982) 278–282.

[91] G. Tollefson, K. Rodysill, M. Cusulos, A circulating lupus-like coagulation inhibitor induced by chlorpromazine, J Clin Psych Pharmacol 4 (1984) 49–51.

[92] V.D. Steen, R. Ramsey-Goldman, Phenothiazine-induced systemic lupus erythematosus with superior vena cava syndrome: Case report and review of the literature, Arthritis Rheum 31 (1988) 923–926.

[93] M.T. Weber, W.G. Hocking, Procainamide induced lupus anticoagulant, Wis Med J 87 (1988) 30–32.

[94] R.A. Asherson, J. Zulman, G.R.V. Hughes, Pulmonary thromboembolism associated with procainamide-induced lupus syndrome and anticardiolipin antibodies, Ann Rheum Dis 48 (1989) 232–235.

[95] R. Chokron, A. Robert, L. Rozensztajn, Procainamide-induced lupus with circulating anticoagulant [letter], Nouv Presse Med 11 (1982) 2568.

[96] S. Davis, B.C. Furie, J.H. Griffin, B. Furie, Circulating inhibitors of blood coagulation associated with procainamide-induced lupus erythematosus, Am J Hematol 4 (1978) 401–407.

[97] R.L. Edwards, M.E. Rick, C.J. Wakem, Studies on a circulating anticoagulant in procainamide-induced lupus erythematosus, Arch Intern Med 141 (1981) 1688–1690.

[98] A.F. List, D.C. Doll, Thrombosis associated with procainamide-induced lupus anticoagulant, Acta Haematol 82 (1989) 50–52.

[99] J.T. Merrill, C. Shen, M. Gugnani, R.G. Lahita, A.-B. Mongey, High prevalence of antiphospholipid antibodies in patients taking procainamide, J Rheumatol 24 (1997) 1083–1088.

[100] D.A. Gastineau, F.J. Kazmier, W.L. Nichols, et al., Lupus anticoagulant: An analysis of the clinical and laboratory of 219 cases, Am J Hematol 19 (1985) 265–275.

[101] C.J. Lavie, J. Biundo, R.J. Quinet, et al., Systemic lupus erythematosus (SLE) induced by quinidine, Arch Intern Med 145 (1985) 446–448.

[102] M.R. Bird, A.I. O'Neill, R.R.C. Buchanan, K.M. Ibrahim, J.D. Parkin, Lupus anticoagulant in the elderly may be associated with both quinine and quinidine usage, Pathology 27 (1995) 136–139.

[103] D. Rosare, F. Garcia, J. Gascon, J. Angrill, R. Cervera, Quinine induced lupus-like syndrome and cardiolipin antibodies, Ann Rheum Dis 55 (1996) 559–560.

[104] B. Anderson, M.T. Stillman, False-positive FTA-ABS in hydralazine induced lupus, JAMA 239 (1978) 1392–1393.

[105] T. Vyse, A.K. So, Sulphasalazine induced autoimmune syndrome, Br. J. Rheumatol 31 (1992) 115–116.

[106] R.T. Canoso, H.S. Sise, Chlorpromazine-induced lupus anticoagulant and associated immunologic abnormalities, Am J Hematol 13 (1982) 121–129.

[107] L. Nassberger, A.C. Johansson, S. Bjorck, et al., Antibodies to neutrophil granulocyte myeloperoxidase and elastase: Autoimmune responses in glomerulonephritis due to hydralazine treatment, J Intern Med 229 (1991) 261–265.

[108] L. Nassberger, A.G. Sjoholm, H. Jonsson, et al., Autoantibodies against neutrophil cytoplasm components in systemic lupus erythematosus and in hydralazine-induced lupus, Clin Exp Immunol 81 (1990) 380–383.

[109] G. Cambridge, H. Wallace, R.M. Bernstein, B. Leaker, Autoantibodies to myeloperoxidase in idiopathic and drug-induced systemic lupus erythematosus and vasculitis, Br J Rheumatol 33 (1994) 109–114.

[110] G. Cambridge, H. Wallace, R.M. Bernstein, B. Leaker, Autoantibodies to myeloperoxidase in idiopathic and drug-induced systemic lupus erythematosus and vasculitis, Br J Rheumatol 33 (1994) 109–114.

[111] H. Honda, T. Shibata, H. Hara, Y. Ban, T. Sugisaki, MPO-ANCA in patients with Grave's disease; strong association with propylthiouracil therapy (abstract), Sarcoidosis Vasc Diffuse Lung Dis 13 (1996) 280.

[112] J.W. Cohen Tervaert, W. Tel, O. de Vries, C.A. Links, C.A. Stegman, The occurrence of anti-neutrophil cytoplasmic

antibodies with specificity for proteinase 3, myeloperoxidase, and/or elastase during treatment with antithyroid drugs (abstract), Sarcoidosis Vasc Diffuse Lung Dis 13 (1996) 280.

[113] H.K. Choi, M.C. Slot, G. Pan, C.A. Weissbach, J.L. Niles, P.A. Merkel, Evaluation of anti-neutrophil cytoplasmic antibody seroconversion induced by minocycline, sulfasalazine or penicillamine, Arthritis Rheum 43 (2000) 2488–2492.

[114] H.G. Bluestein, D. Redelman, N.J. Zvaifler, Procainamide-lymphocyte reactions. A possible explanation for drug-induced autoimmunity, Arthritis Rheum 24 (1981) 1019–1023.

[115] G.R. Hughes, R.I. Rynes, A. Gharavi, et al., The heterogeneity of serologic findings and predisposing host factors in drug-induced lupus erythematosus, Arthritis Rheum 24 (1981) 1070–1073.

[116] P.F. Ryan, G.R. Hughes, R. Bernstein, et al., Lymphocytotoxic antibodies in hydralazine-induced lupus erythematosus [letter], Lancet 2 (1979) 1248–1249.

[117] B.S. Ooi, K.S. Kant, I.B. Hanenson, A.J. Pesce, V.E. Pollack, Lymphocytotoxins in epileptic patients receiving phenytoin, Clin Exp Immunol 30 (1977) 56–61.

[118] B.H. Hahn, G.C. Sharp, W.S. Irvin, O.S. Kantor, C.A. Gardner, et al., Immune response to hydralazine and nuclear antigens in hydralazine-induced lupus erythematosus, Ann Intern Med 76 (1972) 365–374.

[119] A.S. Russell, M. Ziff, Natural antibodies to procainamide, Clin Exp Immunol 3 (1968) 901–909.

[120] R.L. Rubin, Drug-induced lupus, Toxicology 209 (2005) 135–147.

[121] H.M. Perry Jr., E.M. Tan, S. Carmody, et al., Relationship of acetyl transferase activity to antinuclear antibodies and toxic symptoms in hypertensive patients treated with hydralazine, J Lab Clin Med 76 (1970) 114–125.

[122] C. Brand, A. Davidson, G. Littlejohn, P. Ryan, Hydralazine-induced lupus: No association with HLA-DR4, Lancet 2 (1984) 462.

[123] G.I. Russell, R.F. Bing, J.A. Jones, H. Thurston, J.D. Swales, Hydralazine sensitivity: Clinical features, autoantibody changes and HLA-DR phenotype, QJ Med 65 (1987) 845–852.

[124] C. Speirs, A.H. Fielder, H. Chapel, et al., Complement system protein C4 and susceptibility to hydralazine-induced systemic lupus erythematosus, Lancet 1 (1989) 922–924.

[125] A. Litwin, L.E. Adams, H. Zimmer, E.V. Hess, Immunologic effects of hydralazine in hypertensive patients, Arthritis Rheum 24 (1981) 1074–1077.

[126] T.J. Thomas, J.R. Seibold, L.E. Adams, E.V. Hess, Hydralazine induces Z-DNA confirmation in a polynucleotide and elicits anti (Z-DNA) antibodies in treated patients, Biochem. J 294 (1993) 419–425.

[127] R.L. Woosley, D.E. Drayer, M.D. Reidenberg, A.S. Nies, K. Carr, et al., Effector of acetylator phenotype on the rate at which procainamide induces antinuclear antibodies and the lupus syndrome, N. Engl. J. Med 298 (1978) 1157–1179.

[128] C. Sonnhag, E. Karlsson, J. Hed, Procainamide-induced lupus erythematosus-like syndrome in relation to acetylator phenotype and plasma levels of procainamide, Acta Med Scand 206 (1979) 245–251.

[129] A.-B. Mongey, E. Sim, A. Risch, E.V. Hess, Acetylation status is associated with serological changes but not clinically significant disease in patients receiving procainamide, J Rheumatol 26 (1999) 1721–1726.

[130] L.E. Adams, K. Balakrishnan, S.M. Roberts, R. Belcher, A.-B. Mongey, T.J. Thomas, et al., Genetic, immunologic and biotransformation studies of patients on procainamide, Lupus 2 (1993) 89–98.

[131] N.F. Rothfield, W.F. Bierer, J.W. Garfield, Isoniazid induction of antinuclear antibodies. A prospective study, Ann Intern Med 88 (1978) 650–652.

[132] M. Salazar-Proamo, R.L. Rubin, I. Garcia-de la Torre, Isoniazid-induced systemic lupus erythematosus, Ann Rheum Dis 51 (1987) 1085–1087.

[133] D. Alarcon-Segovia, E. Fishbein, H. Alcala, Isoniazid acetylation rate and development of antinuclear antibodies upon isoniazid treatment, Arthritis Rheum 14 (1971) 748–752.

[134] D.A. Evans, M.F. Bullen, J. Houston, et al., Antinuclear factor in rapid and slow acetylator patients treated with isoniazid, J Med Genet 9 (1972) 53–56.

[135] M.J. Kendall, C.F. Hawkins, Quinidine-induced systemic lupus erythematosus, Postgrad Med J 46 (1970) 729–731.

[136] J.A. Alloway, M.P. Salata, Quinidine-induced rheumatic syndromes, Semin. Arthritis Rheum 24 (1995) 315–322.

[137] P. Amadio Jr., D.M. Cummings, L. Dashow, Procainamide, quinidine, and lupus erythematosus [letter], Ann Intern Med 102 (1985) 419.

[138] C.M. De Giorgio, A.L. Rabinowicz, R.D. Olivas, Carbamazepine-induced antinuclear antibodies and systemic lupus erythematosus-like syndrome, Epilepsia 32 (1991) 128–129.

[139] S.T. Schmidt, M. Welcker, W. Greil, M. Schattenkirchner, Carbamazepine-induced systemic lupus erythematosus, Br J Psych 161 (1992) 560–561.

[140] S. Alballa, M. Fritzler, P. Davis, A case of drug induced lupus due to carbamazepine, J Rheumatol 14 (1987) 599–600.

[141] V.E. Drory, A.D. Korczyn, Hypersensitivity vasculitis and systemic lupus erythematosus induced by anticonvulsants, Clin Neuropharmacol 16 (1993) 19–29.

[142] B.M. Ansell, Drug-induced systemic lupus erythematosus in a nine-year-old boy, Lupus 2 (1993) 193–194.

[143] J.J. Asconape, K.R. Manning, M.E. Lancman, Systemic lupus erythematosus associated with use of valproate, Epilepsia 35 (1994) 162–163.

[144] S.J. Wallace, A comparative review of the adverse effects of anticonvulsants in children with epilepsy, Drug Saf 15 (6) (1996) 378–393.

[145] K. Casteels, C. VanGeet, K. Wouters, Ethosuximide-associated lupus with cerebral and renal manifestations, Eur J Pediatric 157 (1988) 780.

[146] S. Takeda, F. Koizumi, E. Takazakura, Ethosuximide-induced lupus-like syndrome with renal involvement, Intern Med 35 (7) (1996 Jul) 587–591.

[147] B.D. Atwater, Z. Ai, M.R. Wolff, Fulminant myopericarditis from phenytoin-induced systemic lupus erythematosus, WMJ 107 (2008) 298–300.

[148] P. Caramaschi, D. Biasi, A. Carletto, T. Manzo, L.M. Bambara, Clobazam induced lupus erythematosus [letter], Lupus 3 (1994) 69.

[149] D. Verrot, M. San-Marco, C. Dravet, et al., Prevalence and signification of antinuclear and anticardiolipin antibodies in patients with epilepsy, Am J Med 103 (1997) 33–37.

[150] A. Crespel, R. Velizarova, M. Agullo, P. Gelisse, Ethosuximide-induced de novo systemic lupus erythematous with anti-double-strand DNA antibodies: case report with definite evidence, Epilepsia 2009 (2003) 50.

[151] F.P. Quismorio, D.F. Bjarnason, W.F. Kiely, et al., Antinuclear antibodies in chronic psychotic patients treated with chlorpromazine, Am J Psychiatry 132 (1975) 1204–1206.

[152] G.P. Pavlidakey, K. Hashimoto, G.L. Heller, S. Daneshvar, Chlorpromazine-induced lupus-like disease. Case report and review of the literature, Am J Acad Dermatol 13 (1985) 109–115.

[153] J.D. Wilson, Antinuclear antibodies and cardiovascular drugs, Drugs 19 (1980) 292–305.

[154] N.B. Record Jr., Acebutolol-induced pleuropulmonary lupus syndrome, Ann Intern Med 95 (1981) 326–327.

[155] E.B. Raferty, A.M. Denman, Systemic lupus erythematosus induced by practolol, Br Med J 2 (1973) 452–455.

[156] J. Bensaid, J.C. Aldigier, N. Gaulde, SLE syndrome induced by pindolol, Br Med J 1 (1979) 1603–1604.

[157] R.C. Brown, J. Cooke, M.S. Losowsky, SLE syndrome probably induced by labetalol, Postgrad Med J 57 (1987) 189–190.

[158] Vedove CD Sontheimer, DelGiglio M, Schena D, Girolomoni G. Drug-induced lupus erythematosus, Arch Dermatol Res 301 (2009) 99–105.

[159] M.M. Reidenberg, D.B. Case, D.E. Drayer, S. Reis, B. Lorenzo, Development of antinuclear antibody in patients treated with high doses of captopril, Arthritis Rheum 27 (1984) 579–581.

[160] C.G. Kallenberg, Antibodies during captopril treatment, Arthritis Rheum 28 (1985) 597–598.

[161] C. Sieber, E. Grimm, F. Follath, Captopril and systemic lupus erythematosus, Br Med J 301 (1990) 669.

[162] P.V. Pigott, Captopril and drug-induced lupus, Br Med J 284 (1982) 1786.

[163] D. Schwartz, A. Pines, M. Averbuch, Y. Levo, Enalapril-induced antinuclear antibodies, Lancet 336 (1990) 187.

[164] M. Pelayo, V. Vargas, A. Gonzales, A. Vallano, R. Esteban, J. Guardia, Drug-induced lupus-like reaction and captopril, Ann Pharmacother 27 (1993) 1541–1542.

[165] P. Benin, J. Kamdem, C. Bonnet, M. Arnaud, R. Treves, Captopril induced lupus, Clin. Exp. Rheumatol 11 (1993) 695.

[166] M. Srivastava, A. Rencic, G. Diglio, H. Santana, et al., Drug-induced, Ro/SSA-positive cutaneous lupus erythematosus, Arch Dermatol 139 (2003) 45–49.

[167] D. Alarcon-Segovia, T. Herslovic, W.H. Dearing, L. Bartholomew, J. Cain, R. Shorter, Lupus erythematosus cell phenomenon in patients with chronic ulcerative colitis, Gut 6 (1965) 39–47.

[168] F. Dekeyser, H. Mielants, J. Praet, S. Goemaere, E.M. Veys, Changes in antinuclear serology in patients with spondyloarthropathy under sulphasalazine treatment, Br J Rheumatol 32 (1993) 521.

[169] A.R. Siam, M. Hammoudeh, Sulfasalazine-induced systemic lupus erythematosus in a patient with rheumatoid arthritis [letter], J Rheumatol 20 (1993) 207.

[170] G.L. Clementz, B.J. Dolin, Sulfasalazine-induced lupus erythematosus, Am J Med 84 (1988) 535–538.

[171] C.J. Laversuch, D.A. Collins, P.J. Charles, B.E. Bourke, Sulphasalazine-induced autoimmune abnormalities in patients with rheumatic disease, Br J Rheumatol 34 (1995) 435–439.

[172] E.M. Walker, J.E. Carty, Sulphasalazine-induced systemic lupus erythematosus in a patient with erosive arthritis, Br J Rheumatol 33 (1994) 175–176.

[173] D.J. Veale, M. Ho, K.D. Morley, Sulphasalazine-induced lupus in psoriatic arthritis, Br J Rheumatol 34 (1995) 383–384.

[174] G. Deboever, R. Devogelaere, G. Holvoet, Sulphasalazine-induced lupus-like syndrome with cardiac tamponade in a patient with ulcerative colitis, Am J Gastroenterol 84 (1989) 85–86.

[175] D.L. Carr-Locke, Sulfasalazine-induced lupus syndrome in a patient with Crohn's disease, Am J Gastroenterol 77 (1982) 614–616.

[176] M. Caulier, C. Dromer, V. Andrieu, P. LeGuennec, B. Fourne, Sulfasalazine induced lupus in rheumatoid arthritis, J Rheumatol 21 (1994) 750–751.

[177] I. Gunnarsson, B. Nordmark, A. Hassan Bakri, G. Grondal, P. Larsson, J. Forslid, et al., Development of lupus-related side-effects in patients with early RA during sulphasalazine treatment—the role of IL-10 and HLA, Rheumatology 39 (2000) 886–893.

[178] M.-M. Gordon, D.R. Porter, H.A. Capell, Does sulphasalazine cause drug induced lupus erythematosus? No effect evident in a prospective randomized trial of 200 rheumatoid patients treated with sulphasalazine or auranofin over five years, Ann Rheum Dis 58 (1999) 288–290.

[179] A.W. Kirkpatrick, A.A. Bookman, F. Habal, Lupus-like syndrome caused by 5-aminosalicylic acid in patients with inflammatory bowel like disease, Can J Gastroenterol 13 (1999) 159–162.

[180] M.A. Timsit, D. Anglicheau, F. Liote, P. Marteau, A. Dryll, Mesalazine-induced lupus, Rev Rheum Engl Ed 64 (1997) 586–588.

[181] I. Gunnarsson, J. Forslid, B. Ringertz, Mesalazine-induced lupus syndrome, Lupus 8 (1999) 486–491.

[182] I. Gunnarsson, L. Kanerud, E. Pettersson, et al., Predisposing factors in sulphasalazine-induced systemic erythmatosus, Br J Rheumatol 36 (1997) 1089–1094.

[183] T. Matsuura, Y. Shimizu, H. Fujimoto, T. Miyazaki, S. Kano, Minocycline-related lupus, Lancet 340 (1992) 1553.

[184] P.A.C. Byrne, B.D. Williams, M.H. Pritchard, Minocycline-related lupus, Br. J. Rheumatol 33 (1994) 674–676.

[185] B. Quilty, N. McHugh, Lupus-like syndrome associated with the use of minocycline, Br J Rheumatol 33 (1994) 1197–1198.

[186] D.Y. Bulgen, Minocycline-related lupus, Br J Rheumatol 34 (1995) 398.

[187] O. Elkayam, M. Yaron, D. Caspi, Minocycline induced arthritis associated with fever, livedo reticularis and p ANCA, Ann Rheum Dis 55 (1996) 769–771.

[188] S.R. Knowles, L. Shapiro, N.H. Shear, Serious adverse reactions induced by minocycline, Arch Dermatol 132 (1996) 934–939.

[189] A. Gough, S. Chapman, K. Wagstaff, P. Emery, E. Elias, Minocycline induced autoimmune hepatitis and systemic lupus erythematosus-like syndrome, Br Med J 312 (1996) 169–172.

[190] L.E. Shapiro, S.R. Knowles, N.H. Shear, Comparative safety of tetracycline, minocycline and doxycycline, Arch Dermatol 133 (1997) 1224–1230.

[191] R.G. Schlienger, A.J. Bircher, C.R. Meier, Minocycline-induced lupus. A systematic review, Dermatology 200 (2000) 223–231.

[192] F. Ahmed, P.R. Kelsey, N. Shariff, Lupus syndrome with neutropenia following minocycline therapy — a case report, Int Jnl Lab Hem 30 (2008) 543–545.

[193] J. Dunphy, M. Oliver, A.L. Rands, C.R. Lovell, N.J. McHugh, Antineutrophil cytoplasmic antibodies and HLA class II alleles in minocycline-induced lupus like syndrome, Br J Dermatol 142 (2000) 461–467.

[194] M.C. Sturkenboom, C.R. Meier, H. Jick, B.H. Sticker, Minocycline and lupus-like syndrome in acne patients, Arch Intern Med 159 (1999) 493–497.

[195] D.J. Margolis, O. Hoffstad, W. Bilker, Association or lack of association between tetracycline class antibiotics used for acne vulgaris and lupus erythematosus, Br J Dermatol 157 (2007) 540–546.

[196] N.J. McHugh, J. Dunphy, A. Rands, Antimyeloperoxidase antibodies in minocycline-induced lupus, Arthritis Rheum 39 (Suppl.) (1996) S110.

[197] D. Herzog, O. Hajoui, P. Russo, F. Alvarez, Study of immune reactivity of minocycline-induced chronic active hepatitis, Dig Dis Sci 42 (1997) 1002−1103.

[198] M. Kloppenburg, B.M.N. Binkman, H.H. deRooij-Dijk, et al., The tetracycline derivative minocycline differentially affects cytokine production by monocytes and T-lymphocytes, Antimicrob Agents Chemother 40 (1996) 934−940.

[199] B. Noel, Lupus erythematosus and other autoimmune diseases related to statin therapy a systematic review, J Eur Acad Dermatol Venereol 21 (2007) 17−24.

[200] S. Ahmad, Lovastatin-induced lupus erythematosus, Arch Intern Med 151 (1991) 1667−1668.

[201] B. Bannwarth, G. Miremont, P.-M. Papapietro, Lupus like syndrome associated with simvastatin, Arch Intern Med 152 (1992) 1093.

[202] H. Kaur, D. Singh, G. Bollin, Simvastatin-induced lupus erythematosus, ACP-ASIM Ohio Chapter Scientific Meeting 74 (2001).

[203] A. Ahmad, M.T. Fletcher, T.M. Roy, Simvastatin-induced lupus-like syndrome, Tenn Med 93 (2000) 21−22.

[204] R. Khosla, A.N. Butman, D.F. Hammer, Simvastatin-induced lupus erythematosus, South Med J 91 (1998) 873−874.

[205] J. Hanson, D. Bossingham, Lupus-like syndrome associated with simvastatin, Lancet 352 (1998) 1070.

[206] L. Rudski, M.A. Rabinovitch, D. Danoff, Systemic immune reactions to HMG-CoA reductase inhibitors. Report of 4 cases and review of the literature, Medicine (Baltimore) 77 (1998) 378−383.

[207] G. Obermoserg, I. Graziadei, N. Sepp, W. Bogel, Lupus-like syndrome associated with Atorvastatin, Lupus 10 (2001) S121.

[208] M.K. Sridhar, A. Abdulla, Fatal lupus-like syndrome and ARDS induced by fluvastatin, Lancet 352 (1998) 114.

[209] R. Suchak, K. Benson, V. Swale, Statin-induced Ro/SSa-positive subacute cutaneous lupus erythematosus, Clin Exp Dermatol 32 (2007) 589−591.

[210] L.J. Raggatt, N.C. Partridge, HMG-CoA reductase inhibitors as immunomodulators: potential use in transplant rejection, Drugs 62 (2002) 2185−2191.

[211] T.M. Harkcom, D.L. Conn, K.E. Holley, d-Penicillamine and lupus erythematosus-like syndrome [letter], Ann Intern Med 89 (1978) 1012.

[212] A. Chalmers, D. Thompson, H.E. Stein, G. Reid, A.L. Patterson, Systemic lupus erythematosus during penicillamine therapy for rheumatoid arthritis, Ann Intern Med 97 (1982) 659−663.

[213] J.M. Walshe, Penicillamine and the SLE syndrome, J Rheumatol Suppl 8 (1981) 155−160.

[214] J. Thorvaldsen, Penicillamine-induced lupus-like reaction in rheumatoid arthritis and vasculitis, Dermatologica 162 (1981) 277−280.

[215] J.P. Camus, J.C. Homberg, J. Crouzet, et al., Autoantibody formation in d-penicillamine-treated rheumatoid arthritis, J Rheumatol Suppl 7 (1981) 80−83.

[216] Y.-S. Bae, S.-L. Bae, S.-W. Lee, D.-H. Yoo, T.Y. Kim, S.Y. Kim, Lupus flare associated with growth hormone, Lupus 10 (2001) 448−450.

[217] S. Ogueta, I. Olazabal, I. Santos, E. Delgado-Baeza, J.P. Garcia-Ruiz, Transgenic mice expressing bovine GH develop arthritic disorder and self-antibodies, J Endocrinol 165 (2000) 321−328.

[218] J. Rovensky, S. Blazickova, L. Rauoval, et al., The hypothalamic-pituitary response in SLE. Regulation of prolactin growth hormone and Cortisol release, Lupus 7 (1998) 409−413.

[219] R.W. McMurray, S.H. Allen, A.L. Braun, S. Rodriguez, S.E. Walker, Long-standing hyperpro-clatinemia associated with systemic lupus erythematosus: Possible hormonal stimulation of an autoimmune disease, J Rheumatol 21 (1994) 843−850.

[220] Baum, et al., Anastrozole alone or in combination with tamoxifen for adjuvant treatment of postmenopausal women with early breast cancer; first results of the ATAC randomized trial, Lancet 359 (2002) 2131−2139.

[221] P.E. Goss, et al., A randomized trail of letrozole in postmenopausal women after years of tamoxifen therapy for early-stage breast cancer, N Engl J Med 349 (2003) 1793−1802.

[222] R.C. Coombes, et al., A randomized trial of exemestane after two to three years of tamoxifen therapy in postmenopausal women with primary breast cancer, N Engl J Med 350 (2004) 1081−1092.

[223] P.P. Donnellan, et al., Aromatase inhibitors and arthralgia (letter), J Clin Oncol 19 (2001) 2767.

[224] J. Bonneterre, et al., Anastrozole versus tamoxifen as first-line therapy for advanced breast cancer in 668 postmenopausal women: results of the Tamoxifen or Arimidex Randomized Group Efficacy and Tolerability study, J Clin Oncol 18 (2000) 3748−3757.

[225] P. Dombernowsky, et al., Letrozole, a new oral aromatase inhibitor for advanced breast cancer: double-blind randomized trial showing a dose effect and improved efficacy and tolerability compared with megestrol acetate, J Clin Oncol 16 (1998) 453−461.

[226] U.B. Wandl, M. Nagel-Hiemke, D. May, E. Kreuzfelder, O. Kloke, M. Kranzhoff, et al., Lupus-like autoimmune disease induced by interferon therapy for myeloproliferative disorders, Clin Immunol Immunopathol 65 (1992) 70−74.

[227] L.E. Ronnblom, G.V. Alm, K.E. Oberg, Autoimmunity after alpha-interferon therapy for malignant carcinoid tumors, Ann Intern Med 115 (1991) 178−183.

[228] W.-J. Mayet, G. Hess, G. Gerken, S. Rossel, R. Voth, M. Manns, et al., Treatment of chronic type B hepatitis with recombinant α-interferon induced autoantibodies not specified for autoimmune chronic hepatitis, Hepatology 10 (1989) 24−28.

[229] M.B. Atkins, J.W. Mier, D.R. Parkinson, J.A. Gould, E.M. Berkman, M.M. Kaplan, Hypothryoidism after treatment with interleukin 2 and lymphokine activated killer cells, N Engl J Med 318 (1988) 1557−1563.

[230] M.R. Ehrenstein, E. McSweeney, M. Swana, C.P. Worman, A.H. Goldstone, D.A. Isenberg, Appearance of anti-DNA antibodies in patients treated with interferon-α, Arthritis Rheum 36 (1993) 279−280.

[231] H. Gisslinger, B. Gilly, W. Woloszczuk, W.R. Mayr, L. Havelec, W. Linkesch, Thyroid autoimmunity and hypothyroidism during long-term treatment with recombinant interferon-alpha, Clin Exp Immunol 90 (1992) 363−367.

[232] F.A. Borg, D.A. Isenberg, Syndromes and complications of interferon therapy, Curr Opin Rheumatol 19 (2007) 61−66.

[233] R. Pellicano, A. Smedile, S. Peyre, et al., Autoimmune manifestations during interferon therapy in patients with hepatitis C; the hepatologist's view, Minerva Gastroenterol Dietol 51 (2005) 55−61.

[234] H. Koon, M. Atkins, Autoimmunity and immunotherapy for cancer, N Engl J Med 354 (2006) 758−760.

[235] K.C. Conlon, W.J. Urba, J.W. Smith, R.G. Steis, D.L. Longo, J. Clark, Exacerbation of symptoms of autoimmune disease in patients receiving alpha-interferon therapy, Cancer 65 (1990) 2237−2242.

[236] A. Flores, A. Olive, E. Feliu, X. Tena, Systemic lupus erythematosus following interferon therapy, Br J Rheumatol 33 (1994) 787−792.

[237] G. Obermoser, B. Semenitz, J. Thaler, N. Sepp, Lupus-like syndrome following interferon therapy in a patient with chronic myelogenous leukemia, Lupus 10 (2001) S121.

[238] S. Yilmaz, K.A. Cimen, Pegylated interferon alpha-2B induced lupus in a patient with chronic hepatitis B virus infection: case report, Clin Rheumatol 28 (2009) 1241–1243.

[239] H. Gogas, J. Ioannovich, U. Dafni, et al., Prognostic significance of autoimmunity during treatment of melanoma with interferon, N Engl J Med 354 (7) (2006 Feb 16) 709–718.

[240] E.M. Massarotti, N.Y. Liu, J. Nier, M.B. Atkins, Chronic inflammatory arthritis after treatment with high-dose interleukin-2 for malignancy, Am J Med 92 (1992) 693–697.

[241] K. Hoekman, B.M. von Blomberg-van der Flier, J. Wagstaff, H.A. Drexhage, H.M. Pinedo, Reversible thyroid dysfunction during treatment with GM-CSF, Lancet 2 (1991) 541–542.

[242] F. Fadel, K. El Karoui, B. Knebelmann, Anti-CTLA4 antibody-induced lupus nephritis, N Engl J Med 361 (2) (2009) 211–212.

[243] D. Rea, A. Bergeron, C. Fieschi, et al., Dasatanib-induced lupus, Lancet 30 (372) (2008) 713–714.

[244] H. DeLavallade, S. Punnialingam, D. Milojkovic, et al., Pleural effusions in patients with chronic myeloid leukemia treated with dasatinib may have an immune-mediated pathogenesis

[245] Feldman, M. (1997). TNF blockade in rheumatoid arthritis. Prog. AAI, CIS, AAAI Annual Meeting, February.

[246] J.M. Martin, et al., Adalimumab-induced lupus erythematosus, Lupus 17 (2008) 676–678.

[247] S. Vermeire, et al., Autoimmunity associated with anti-tumor necrosis factor alpha treatment in Crohn's disease: a prospective cohort study, Gastroenterology 125 (2003) 32–39.

[248] P.J. Charles, et al., Assessment of antibodies to double-stranded DNA induced in rheumatoid arthritis patients following treatment with infliximab, a monoclonal antibody to tumor necrosis factor alpha: findings in open-label and randomized placebo-controlled trials, Arthritis Rheum 43 (2000) 2383–2390.

[249] N. Shakoor, M. Michalska, C. Harris, et al., Drug-induced systemic lupus erythematosus associated with etanercept therapy, Lancet 52 (2002) 579–580.

[250] M. DeBandt, J. Sibilia, X. Le Loët, et al., Systemic lupus erythematosus induced by anti-tumor necrosis factor alpha therapy: a French national survey, Arthritis Res Ther 7 (2005) R545–R551.

[251] Schiff, et al., Safety analyses of adalimumab (HUMIRA) in global clinical trials and US postmarketing surveillance of patients with rheumatoid arthritis, Ann Rheum Dis 65 (2006) 889–894.

[252] T.F. Schaible, Long term safety of Infliximab, Can J Gastroenterol 14 (Suppl. C) (2000) 29C–32C.

[253] M. Ramos-Casals, P. Brito-Zeron, S. Munoz, et al., Autoimmune Diseases Induced by TNF-Targeted Therapies: Analysis of 233 Cases, Medicine 86 (2007) 242–251.

[254] J.J. Cush, Biological drug use: US perspectives on indications and monitoring, Ann Rheum Dis 64 (Supplement 4) (2005) 18–23.

[255] A. Markham, H.M. Lamb, Infliximab: A review of its use in the management of rheumatoid arthritis, Drugs 59 (2000) 1341–1359.

[256] J.F. Kerbleski, A.B. Gottlieb, Dermatological complications and safety of anti-TNF treatments, Gut 58 (2009) 1033–1039.

[257] C. Eriksson, S. Engstrand, K.G. Sundqvist, S. Rantapää-Dahlqvist, Autoantibody formation in patients with rheumatoid arthritis treated with anti-TNF alpha, Ann Rheum Dis 64 (3) (2005) 403–407.

[258] T. Jonsdottir, J. Forslid, A. van Vollenhoven, et al., Treatment with tumour necrosis factor alpha antagonists in patients with rheumatoid arthritis induces anticardiolipin antibodies, Ann Rheum Dis 63 (2004) 1075–1078.

[259] O. Elkayam, M. Burke, N. Vardinon, et al., Autoantibodies profile of rheumatoid arthritis patients during treatment with infliximab, Autoimmunity 38 (2) (2005 Mar) 155–160.

[260] L. DeRycke, et al., Antinuclear antibodies following infliximab treatment in patients with rheumatoid arthritis or spondyloarthropathy, Arthritis Rheum 48 (2003) 1015–1023.

[261] B. Haraoui, E. Keystone, Musculoskeletal manifestations and autoimmune diseases related to new biologic agents, Curr Opin Rheumatol 18 (2006) 96–100.

[262] L. De Rycke, D. Baeten, E. Kruithof, et al., Infliximab, but not etanercept, induces IgM anti-double-stranded DNA autoantibodies as main antinuclear reactivity: biologic and clinical implications in autoimmune arthritis, Arthritis Rheum 52 (2005). 2192-20.

[263] M. DeBandt, Lessons for lupus from tumor necrosis factor blockade, Lupus 15 (2006) 762–767.

[264] D. Tracey, L. Klareskog, E.H. Sasso, et al., Tumor necrosis factor antagonist mechanisms of action: A comprehensive review, Pharmacology & Therapeutics 117 (2008) 244–279.

[265] E.W. St. Clair, C.L. Wagner, A.A. Fasanmade, et al., The relationship of serum infliximab concentration to clinical improvements in rheumatoid arthritis: results from ATTRACT, a multicenter, randomized, double-blind, placebo-controlled trial, Arthritis Rheum 46 (2002) 1451–1459.

[266] A.P. Cairns, M.K. Duncan, A.E. Hinder, A.J. Taggart, New onset systemic lupus erythematosus in a patient receiving etanercept for rheumatoid arthritis, Ann Rheum Dis 61 (2002) 1031–1032.

[267] M. Benucci, F. Li Gobbi, F. Fossi, et al., Drug-induced lupus after treatment with infliximab in rheumatoid arthritis, J Clin Rheumatol 11 (2005) 47–49.

[268] E.G. Favalli, L. Sinigaglia, M. Varenna, et al., Drug-induced lupus following treatment with infliximab in rheumatoid arthritis, Lupus 11 (2002) 753–755.

[269] A.K. Mohan, et al., Drug-induced systemic lupus erythematosus and TNF-α blockers, Lancet 360 (2002) 646.

[270] A. Mor, C. Bingham 3rd, L. Barisoni, E. Lydon, H.M. Belmont, Proliferative lupus nephritis and leukocytoclastic vasculitis during treatment with etanercept, J Rheumatol 32 (2005) 740–743.

[271] M.B. Stokes, K. Foster, G.S. Markowitz, F. Ebrahimi, W. Hines, D. Kaufman, et al., Development of glomerulonephritis during anti-TNF-alpha therapy for rheumatoid arthritis, Nephrol Dial Transplant 20 (2005) 1400–1406.

[272] N. Mohan, E.T. Edwards, T.R. Cupps, N. Slifman, J.H. Lee, J.N. Siegel, et al., Leukocytoclastic vasculitis associated with tumor necrosis factor-alpha blocking agents, J Rheumatol 31 (2004) 1955–1958.

[273] L. Kocharla, B. Mongey A-, Is the development of drug-related lupus a contra-indication for switching from one TNF alpha inhibitor to another? Lupus 18 (2) (2009) 169–171.

[274] C. Shen, G.V. Assche, S. Colpaert, et al., Adalimumab induces apoptosis of human monocytes: a comparative study with infliximab and etanercept, Aliment Pharmacol Ther 21 (2005) 251–258.

[275] R. Chaudhary, M. Butler, R.J. Playford, et al., Anti-TNF antibody induced stimulated T lymphocyte apoptosis depends on the concentration of the antibody and etanercept induces apoptosis at rates equivalent to infliximab and adalimumab at 10 micrograms per ml concentration, Gastroenterology (2006) 130. A696.

[276] B. Grinblat, M. Scheinberg, The enigmatic development of psoriasis and psoriasiform lesions during anti-TNF therapy: a review, Semin Arthritis Rheum 37 (2008) 251–255.

[277] J. Ko, A.B. Gottlieb, J.F. Kerbleski, Induction and exacerbation of psoriasis with TNF blockade therapy: A review and analysis of 127 cases, J Dermatol Treatment 20 (2009) 100–108.

[278] B.R. Reed, J.C. Huff, S.K. Jones, D.W. Orton, L.A. Lee, D.A. Norris, Subacute cutaneous lupus erythematosus associated with hydrochlorothiazide therapy, Ann Intern Med 103 (1985) 49–51.

[279] O. Durosaro, M.D.P. Davis, K.B. Reed, A.L. Rohlinger, Incidence of cutaneous lupus erythematosus, 1965-2005: a population-based study, Arch Dermatol 145 (2009) 249–253.

[280] J.P. Callen, Clinically relevant information about cutaneaous lupus erythematosus, Arch Dermatol 145 (2009) 316–319.

Laboratory Evaluation of Patients with Systemic Lupus Erythematosus

Peter H. Schur

INTRODUCTION

The use of laboratory tests has greatly enhanced our ability to diagnosis systemic lupus erythematosus (SLE) and related disorders. Thus tests such as the antinuclear antibody (ANA) test and tests for antiphospholipid antibodies have in fact been incorporated into the diagnosis of SLE. However, it is important to remember when using these tests, as with any other test, in a clinical setting to consider their sensitivity, specificity, and predictive value. In addition, laboratory tests are often used to assess the activity of SLE. In this respect, serum complement levels, anti-double-stranded (ds)DNA, and acute-phase reactants have often proved to be useful. Laboratory tests are also used as in any disease to assess organ involvement and function (e.g., renal function). Finally, laboratory tests are often used to evaluate the patient's response to therapy, and again many tests have been useful in this respect (e.g., complement, anti-dsDNA, CBC, renal function).

ANTINUCLEAR ANTIBODIES

Antinuclear antibodies were initially discovered in the 1940s while investigating the immunology of the LE cell test. LE cell tests are rarely performed and have been replaced by the immunofluorescence assay to detect ANAs.

Performance of the ANA Assay by Immunofluorescence

The titer and specificity of IF antinuclear antibodies vary depending on the antigen substrate used for the immunofluorescence (IF ANA) assay. Most laboratories currently use HEp2 cells (human epithelial cell tumor line), which provide certain advantages over the frozen sections of murine livers or kidneys [1]. HEp2 cells offer a standardized substrate with larger nucleoli. They also provide better sensitivity for antibodies to nuclear antigens present during cell division, such as the centromere antigens. In addition, ANA titers are almost always higher when measured on HEp2 cells than on frozen sections of murine tissue.

The assay is performed by first incubating acetone-fixed HEp2 cells with the patient's serum and then overlaying this combination with fluorescein-tagged antihuman g globulin [1]. When viewed through a fluorescent microscope, antibodies bound to nuclear antigens produce an apple-green nuclear pattern. The pattern of fluorescence and the dilution at which nuclear fluorescence disappears (titer) are subsequently noted. Differences in titer of one tube dilution commonly occur but are without clinical significance. Certain antibody specificities, such as an anticentromere, can be read directly from the ANA.

In the past, the tremendous interlaboratory variability of ANA results led to a great deal of confusion. However, it is the author's opinion that the ANA is presently reliable and reproducible, as most laboratories use commercially available tissue culture cell substrates (e.g., HEp2 cells) and participate in proficiency testing.

Enzyme-Linked Immunosorbent Assay

Over the last few years, investigators and biotech firms have been developing solid-phase immunoassays [2–22] to replace the gold standard, the immunofluorescent ANA test [1]. The rationale behind this attempt relates to performance characteristics of the immunofluorescent technique. This test is very labor-intensive and is subject to variation due to different interpretations by technicians. Also complicating testing is the fading of the image as it is examined in a fluorescent microscope. Furthermore, the immunofluorescent technique uses serial dilutions of patient sera, which will give results that may not be linear. Variations in titer by twofold are common in day-to-day testing on the same sample; fourfold differences are said to be "significant." In contrast, solid-phase immunoassays are automated and highly reproducible. The results are linear, and the technique is less labor intensive and thus cheaper to perform.

Because of these considerations, there should be economic savings in employing a solid-phase immuno-assay for quantitating an ANA. Thus, in an attempt to develop solid phase ANA immunoassays, a number of groups have put onto the solid phase whatever antigens are typically assayed in the more specific ANA immuno-assays (e.g., DNA, Sm, RNP, Ro/SSA, La/SSB, nucleo-protein, cell extracts, etc.). Each kit relies on a different method for preparing and coating the nuclear antigens on the microtiter wells, which may account for the variations noted when several kits are compared [2]. In published reports [2–21], the correlation coefficient between IF ANA titers and these solid-phase assays is quite good. Thus, many commercial firms have switched their IF ANAs to these solid-phase ANA immunoassays. Of concern, however, is the high frequency or percentage of false-negative results in these solid-phase ANA immunoassays in patients with known SLE and related diseases, as well as the continued high frequency of false-positives (e.g., a positive ANA in someone without SLE) in these studies [4, 5, 8, 10, 18, 19]. Further work is needed to improve the sensitivity and especially the specificity of these solid-phase immunoassays to ensure that patients with SLE and related diseases are not missed by these solid-phase immunoassays. In the meantime, some commercial laboratories will still perform a IF ANA if specifically requested.

Diseases Associated with a Positive IF ANA

A positive ANA can be seen with systemic autoimmune diseases, organ-specific autoimmune diseases, and a variety of infections. Their presence does not mandate the presence of illness, as they can also be found in otherwise normal individuals.

Systemic Autoimmune Disease

A positive ANA is an essential component of the definition of some systemic autoimmune disorders, such as systemic lupus erythematosus, but can also be found in association with many autoimmune disorders that are not defined by these antibodies. As a result, the sensitivity of a positive IF ANA for a particular autoimmune disease can vary widely (see Table 34.1) [23]:

- SLE: sensitivity, 93%
- Scleroderma: 85%
- Mixed connective tissue disease: 100%
- Polymyositis/dermatomyositis: 61%
- Rheumatoid arthritis (RA): 41%
- Rheumatoid vasculitis: 33%
- Sjögren syndrome: 48%
- Drug-induced lupus: 100%
- Discoid lupus: 15%
- Juvenile chronic arthritis: 57%
- Raynaud phenomenon: 64%
- Antiphospholipid syndrome: 50%

Specific Organ Autoimmune Disease

Positive ANAs are occasionally seen in patients with autoimmune diseases that are limited to a specific organ,

TABLE 34.1   Summary of Test Performance Characteristics of the ANA for Major Rheumatic Diseases

Disease	Sensitivity Overall	Specificity				Likelihood Ratio	
		Other CTD	Non-CTD Rheumatic	Healthy	Overall	Positive	Negative
SLE	93%	49%	75%	78%	57%	2.2	0.11
SSc	85%	44%	75%	71%	54%	1.86	0.27
PM/DM	61%	52%	91%	82%	63%	1.67	0.61
Sjögren's	48%	44%	91%	71%	52%	0.99	1.01
Raynaud's	36%	52%	92%	85%	59%	0.88	1.08
JCA	57%	na	na	na	39%	0.95	1.08
With uvetis	80%	na	na	na	53%	1.68	0.39
RA	41%	38%	85%	82%	56%	0.93	1.06

such as the thyroid gland, liver, or lung. The following sensitivities have been reported in these disorders:

- Hashimoto thyroiditis: 46% [24]
- Graves disease: 35–50% [24, 25]
- Autoimmune hepatitis: 44–71% [26, 27]
- Primary autoimmune cholangitis: 100% [28]
- Primary pulmonary hypertension: 40% [29]

Others

Other well-recognized disorders associated with a positive ANA titer include chronic infectious diseases, such as mononucleosis [30], subacute bacterial endocarditis, and tuberculosis, and some lymphoproliferative diseases [31, 32]. ANAs have also been identified in up to 50% of patients taking certain drugs; however, most of these patients do not develop drug-induced lupus.

False Positives

False-positive ANAs (i.e., ANAs in the absence of autoimmune disease or known antigenic stimuli) are seen more commonly in women and in elderly patients. They are invariably in low titer. Positive ANAs are commonly found in the normal population. When HEp2 cells are used as a substrate, one study of 125 normal individuals found an ANA titer above 1:40 in 32%, above 1:80 in 13%, and above 1:320 in 3% [33]. No patient had anti-dsDNA antibodies. Antibody titers in healthy individuals usually remain relatively constant over time, a finding that can also be seen in patients with known disease.

It is felt that the very low specificity of a positive ANA in the absence of clinical findings of an autoimmune disorder precludes its use as a screening test for disease in the general healthy population.

False Negatives

Certain antinuclear antibodies can be detected at high titer using certain methods and techniques, but are only found at low titer or may even be absent when assayed using other techniques. These confounding findings result from a number of technical and physical nuances, including the method of substrate fixation, the solubility of the antigen (e.g., Ro, La, PCNA, and Ku), and the localization of the antigen outside the nucleus (i.e., Jo-1 and single-stranded DNA). Due to this limitation, a patient with a negative ANA and strong clinical evidence of a systemic autoimmune disorder may require specific antibody assays to accurately diagnosis a rheumatic disease.

In a review of the literature on ANA, including 65 articles that dealt with ANAs and SLE, the sensitivity of a positive IF ANA was 93%, the specificity vs. other connective tissue diseases (CTDs) was 49%, vs non-CTD rheumatic was 75%, vs. healthy was 78% (the overall specificity was 57%) [23]. A positive test had a 2.2 likelihood ratio (LR) for the diagnosis of SLE, whereas a negative test had a 0.11 likelihood ratio [23] (see Table 34.1).

Serial testing of ANAs in patients with SLE has an unknown value [23].

Usefulness of ANA Titer

Unlike nuclear pattern recognition, determination of the titer of antinuclear antibodies still provides clinically relevant information.

1. The presence of very high concentrations of antibody (titer > 1:640) should arouse suspicion of an autoimmune disorder. However, its presence alone is not diagnostic of disease. If no initial diagnosis can be made, it is our practice to watch the patient carefully over time and to exclude ANA-associated diseases.
2. The combination of very low titers of antibody (<1:80) and no signs or symptoms of disease portend a much less ominous prognosis. As a result, these patients need to be re-evaluated far less frequently than those with extremely high antibody titers.
3. Little if any data suggest a correlation between ANA titer and the activity of SLE [23].

Types and Usefulness of Staining Pattern

Antinuclear antibodies produce a wide range of different staining patterns, reflecting the presence of antibodies to one or a combination of nuclear antigens. The nuclear staining pattern has been recognized to have a relatively low sensitivity and specificity for different autoimmune disorders, although it was commonly used in the past to detect specific antibody and antigen specificity. At present, specific tests have largely supplanted the use of patterns [34, 35]

- The homogeneous or diffuse pattern represents antibodies to the DNA–histone complex, also called deoxyribonucleoprotein or nucleosome. It is believed that these antibodies are responsible for the LE phenomenon.
- The peripheral or rim pattern is produced by antibodies to DNA.
- Antibodies to Sm, RNP, Ro/SSA, La/SSB, Scl-70, centromere, PCNA, and other antigens produce the speckled pattern.
- The nucleolar pattern is produced by antibodies to RNA polymerase I, fibrillarin, and NOR-90.
- Antibodies to centromeres produce the centromeric pattern.

Despite these general observations, it is increasingly clear that accurate interpretation of different nuclear patterns is confounded by the following difficulties.

- The recognition of specific patterns is operator-dependent and does not produce a permanent record. Because the fluorescence fades in 1–2 days, one cannot compare a result with other samples without photographing each test result.
- Different serum dilutions can produce varying nuclear patterns.
- One nuclear pattern may obscure and prevent the recognition of another pattern if several antibodies are present simultaneously.
- Certain specificities are visible only on specific substrates. Anti-Ro antibodies and anticentromere antibodies, for example, are not detected with murine organs, but can be found with HEp2 cells.
- Nuclear patterns are neither sensitive nor specific. As a result, no single pattern denotes a single disease and, conversely, several diseases may produce a particular ANA pattern.

Given these difficulties, specific nuclear pattern recognition is not as useful as previously thought. Pattern recognition has been increasingly supplanted by assays, which detect specific antibodies to an ever-increasing array of nuclear antigens and to antigens not normally found in the nucleus.

TYPES OF ANTINUCLEAR ANTIBODIES

Their target antigen defines the different types of ANAs, including single- and double-stranded DNA, nuclear histones, nonhistone nuclear proteins, and RNA–protein complexes. As will be seen, some of these antibodies are relatively specific for a particular disease or for specific clinical manifestations in patients with lupus.

REFLEX TESTING

Some laboratories have a policy that when an IF ANA test is ordered they will routinely perform a screening IF ANA test. Most laboratories, if the screen is positive, will titer the ANA and report patterns. Some laboratories perform what is called reflex testing. This entails carrying out tests for anti-dsDNA, Sm, RNP, Ro/SSA, and La/SSB by solid-phase immunoassays whenever the ANA is positive. This reflexology has the advantage of helping to discriminate ANAs associated with SLE (e.g., positive in one of these assays) from nonspecific ANAs immediately instead of waiting for the patient and physician to resort to a second blood specimen and an additional visit to the office or laboratory, making reflex ANA testing cost-effective.

Anti-DNA Antibodies

Autoantibodies to DNA were first described in the 1950s. These are the best-recognized specific autoantibodies found in patients with SLE. Antibodies to DNA can be divided primarily into two groups: those reactive with denatured (single-stranded) DNA and those recognizing native (double-stranded) DNA.

Measurement of Anti-dsDNA Antibodies

There are currently three methods commonly used by most clinical laboratories to quantitate anti-dsDNA antibodies. Most of these tests measure both high- and low-avidity antibodies.

The Farr assay is based on the precipitation of radioactively labeled DNA–anti-DNA antibody complexes in high salt concentrations. This assay detects high-affinity antibodies to dsDNA [36]. Approximately 40–100% of all patients with SLE have elevated titers of anti-DNA antibodies measured by this method; the titers appear to correlate closely with disease activity, especially with active proliferative nephritis [36, 37]. Because this method requires the use of a radioactive antigen, its routine use has been limited.

The *Crithidia luciliae* assay is an indirect immunofluorescent assay that makes use of the fact that the kinetoplast of this unicellular flagellate is very rich in double-stranded DNA in the absence of other nuclear antigens [36]. This method, while of comparable sensitivity to the Farr assay, is more cumbersome to quantitate and the antibodies detected correlate less closely with active nephritis [38, 39].

A third method in routine use utilizes the solid phase immunoassay (e.g., the ELISA technique) [36]. Double-stranded DNA adherent to polystyrene microwells, treated to increase their adhesiveness, serves as an antigen to capture antibodies. These antibodies are then quantitated using a second antiserum to human immunoglobulin conjugated to a detector enzyme. This method is positive in approximately 70% of patients with SLE. The IgG antibody titers correlate moderately well with active nephritis and, in the author's experience, there is a good correlation with disease activity in general.

Properties of Anti-dsDNA Antibodies

Anti-dsDNA antibodies can demonstrate different properties based on avidity that affects their usefulness as a diagnostic tool. As an example, high-affinity IgG anti-dsDNA antibodies can be demonstrated in 70–80% of patients with SLE when their disease is active [40]. In contrast, some patients with SLE have predominantly IgM or low-avidity IgG antibodies to dsDNA. These antibodies are less useful diagnostically, as they can be found in association with drug-induced lupus, rheumatoid arthritis, Sjögren syndrome, other connective tissue diseases, chronic infection, chronic liver disease, and normal aging [37]; in these instances, the antibodies have no clinical significance. It is possible that lower-avidity antibodies are actually reacting with ssDNA fragments in the DNA preparations used as antigenic substrates.

A number of properties of anti-dsDNA antibodies other than avidity also affect their pathogenicity, including the isoelectric point, isotype, and idiotype. Anti-DNA antibodies that are IgG, cationic, and bind with high-affinity correlate best with renal activity. These antibody properties may vary from individual to individual, over time in the same individual, or with disease activity, even in the face of a stable antibody titer. Thus, there are some patients with SLE who have persistent anti-dsDNA antibody activity despite an improvement in disease activity. Similarly, a small but significant minority of patients have active nephritis without elevations of anti-dsDNA antibody titer. These findings may be related in part to differences in the anti-DNA antibodies over time. Anti-DNA antibodies present in some patients during periods of active nephritis may differ idiotypically from the anti-DNA antibodies present in the same patient during periods of inactive disease [41].

Clinical Relevance of Anti-dsDNA Antibodies

1. They are relatively specific (97%) for SLE, making them very useful for diagnosis (see Table 34.2) [37, 38]. They have been found occasionally in autoimmune hepatitis and in a few patients receiving minocycline, etanercept, infliximab, and penicillamine.
2. There is a well-recognized association of high titers of IgG anti-dsDNA titers with active glomerulonephritis [36]; there also appear to be highly enriched amounts of anti-dsDNA antibodies in the glomerular deposits of immune complexes found in patients with lupus nephritis. These observations have led many investigators to believe that anti-dsDNA antibodies are of primary importance in the pathogenesis of lupus nephritis [42–44].
3. Titers rise when disease is active and usually fall (generally into the normal range) when the flare subsides [42, 43, 45]. Early studies reported a tight

correlation between high-titer anti-dsDNA antibodies and nephritic activity [18, 26], particularly in the setting of hypocomplementemia [42, 43]. More recent studies, however, have reported exceptions to this correlation of titers and disease activity, e.g., some patients have elevated titers of anti-dsDNA antibodies in the setting of inactive or minimally active lupus [46]. A review of the literature is summarized in Table 34.2 [36].

Current sensitive assays are more likely to detect low-avidity antibodies, and may have been missed by earlier studies/assays. As mentioned earlier, clinical correlations that hold about high-avidity antibodies appear not to be applicable to those of lower avidity, and current assays are unable to distinguish between high- and low-avidity, i.e., pathogenic and less pathogenic, antibodies. Thus, while the correlation between antibody titer and disease activity holds for the majority of patients with SLE, there may be some patients for whom this assay has limited clinical utility.

The association between anti-DNA antibodies and other disease manifestations of SLE is far less clear. As an example, there is no relationship between antibody titer and disease activity for lupus cerebritis [47].

Distinguishing active lupus from infectious complications (e.g., often secondary to treatment with immunosuppressive agents), from toxic effects of drugs, and from unrelated disease is always a challenge. Anti-DNA antibodies may be helpful in some patients in making this distinction. Although there is clear variability among patients, the test becomes very useful once one has demonstrated that a given patient follows the characteristic pattern of rising DNA and falling complement in the setting of a flare. However, once a particular patient shows a disassociation between his or her anti-DNA antibody titer and clinical evidence of nephritis, future changes in anti-DNA antibody activity are unlikely to accurately reflect disease activity. In this setting, therapeutic decisions must be guided by the clinical picture and perhaps by other serological findings such as complement levels.

Anti-ssDNA Antibodies

Antibodies that identify denatured DNA (ssDNA) are probably reacting with the purine and pyrimidine bases that are accessible on single-stranded DNA, but are buried within the b helix of double-stranded DNA. Thus, antidenatured DNA antibodies do not cross-react with native DNA. Antibodies to ssDNA have the following general properties:

1. They have been eluted from the kidneys of patients with proliferative nephritis and may therefore be of pathogenic significance [48].

TABLE 34.2 Anti ds DNA

	Sensitivity	Specificity	Positive LR	Negative LR
SLE vs. normals & other diseases	0.573	0.974	16.4	0.49
SLE: active vs. inactive	0.66	0.66	4.14	0.51
Lupus nephritis:				
present vs. absent	0.65	0.41	1.7	0.76
active vs. inactive	0.86	0.45	1.7	0.3

2. They are much less specific for SLE than antibodies to dsDNA. As an example, anti-ssDNA antibodies have been reported in rheumatoid arthritis, drug-related lupus, healthy relatives of patients with SLE [49], and, less commonly, in other rheumatic diseases. Thus, anti-ssDNA has limited usefulness for the diagnosis of SLE.
3. They do not correlate well with disease activity and are therefore not useful for disease management.

Anti-Smith Antibodies and Anti-RNP Antibodies

The anti-Sm and anti-RNP systems are considered together because they coexist in many patients with SLE and bind to related but distinct antigens.

Anti-Sm Antibodies

The Smith antigen is a nuclear nonhistone protein that was characterized in 1966 and was the first nuclear protein autoantigen to be described in SLE [50]. The antigen to which anti-Sm antibodies bind consists of a series of proteins: B, B′, D, E, F, and G, complexed with small nuclear RNAs: U1, U2, U4–6, and U5. These complexes of nuclear proteins and RNAs are called small nuclear ribonucleoprotein particles (snRNPs); they are important in the splicing of precursor messenger RNA [51], an integral step in the processing of RNA transcribed from DNA.

The anti-Sm immune reaction consists of multiple antibodies binding to multiple protein antigens. Thus, although we speak of the anti-Sm antibody, it is actually better described as an antibody system.

Anti-RNP Antibodies

The anti-RNP system binds to antigens that are different from but related to Sm antigens. These antibodies bind to proteins containing only U1-RNA. The U1-RNP particle is involved in splicing heterogeneous nuclear RNA into messenger RNA. Anti-RNP antibodies are not specific for SLE, but are a defining feature in the related syndrome, mixed connective tissue disease. The antibody is present in lower titers in several other rheumatic diseases, including scleroderma.

Methods of Measurement

While anti-Sm antibodies and anti-RNP antibodies can be detected by immunoprecipitation in agarose gels using radial immunodiffusion or counterimmunoelectrophoresis, these methods are relatively insensitive and difficult to quantitate [52]. Most clinical laboratories now employ solid-phase immunoassays (e.g., ELISA) to detect these antibodies [52]. Both methods are sensitive, although the ELISA method is quantitated more easily.

Anti-Sm antibodies are insensitive (10–41% depending on the assay used) [52], but highly specific for SLE, and generally remain positive when titers of anti-DNA antibodies have fallen into the normal range and the clinical activity of SLE has waned. Thus, the measurement of anti-Sm titers may be useful diagnostically, particularly at a time when DNA antibodies are undetectable.

Usefulness

Anti-Sm antibodies occur more frequently in African-Americans and Asians than in Caucasians with SLE. As an example, one series reported a prevalence rate of 25% in African-Americans and 10% in Caucasians (via immunodiffusion) [53]. The prevalence of anti-Sm antibodies in SLE using the ELISA assay varies depending on whether Sm is measured alone or in combination with RNP. Generally accepted rates of anti-Sm antibody positivity in SLE are in the range of 10–30% in Caucasians and 30–40% in Asians and African-Americans [53].

Antibody titers may fluctuate somewhat over time, although this is a controversial issue.

Many studies have tried to correlate anti-Sm antibody with disease activity in general and with specific disease manifestations. These investigations have often yielded conflicting results.

- Anti-Sm does not predict lupus nephritis, there is no association with lupus nephritis, and therefore is useless for the diagnosis of lupus nephritis [52]. There is some association between anti-Sm and serositis [52].
- There is conflicting data regarding the possible association between anti-Sm and neuropsychiatric manifestations of SLE [52].

These often-contradictory results may be explained by the use of different methods of antibody assay by different investigators, as well as by the application of inappropriate statistical methods.

At this point, there is no evidence that anti-Sm antibodies will be useful for following or predicting disease activity in the way that anti-dsDNA antibodies are used. Antibodies to Sm do function as an important diagnostic marker for SLE. As an example, the presence of antibodies to the Sm antigen system in a patient in whom we suspect the diagnosis of SLE strongly supports the diagnosis. Similarly, in a patient with a less convincing clinical picture, we believe that the presence of these antibodies suggests that even a nonspecific symptom complex is likely to progress to SLE. Given their relatively low prevalence, however, a negative value in no way excludes a presumptive diagnosis of SLE.

Anti-Ro/Anti-SSA and Anti-La/Anti-SSB Antibodies

Anti-Ro/SSA and anti-La/SSB antibodies are ANAs that have been detected with high frequency in patients with SLE and in patients with Sjögren syndrome [54].

Immunodiffusion, ELISA, Western blot, and RNA immunoprecipitation detect antibodies to Ro clinically. Most but not all sera react in all three assays. ELISA assays are ten to 100 times more sensitive than immunodiffusion assays; they are also quantitative [55]. The Ro antigen is extracted from human or bovine spleen, Wil-2 cell lines, or represents recombinant fusion proteins [55].

These antinuclear antibodies recognize cellular proteins with molecular masses of approximately 52 and 60 kDa. The 60-kDa protein is complexed with the hy1−5 species of small nuclear RNAs [56].

Relationship Between Anti-Ro and Anti-SSA Antibodies

In 1969, Clark et al. [57] described the presence of antibodies in the sera of some patients with SLE that reacted with ribonucleoprotein antigens present in saline extracts of rabbit and human spleen. When purified preparations of this antibody were incubated with a human epithelial cell line (HEp2), both the nuclei and the cytoplasm were stained in a speckled pattern. These investigators named the antibody anti-Ro after the original patient in whom the antibodies were identified.

At about the same time, Alspaugh and Tan [58] noted antibody activity in sera from many patients with Sjögren syndrome that gave the same immunofluorescent staining pattern as that reported for anti-Ro antibodies. These workers referred to this antibody as anti-SSA. It was soon shown that these two antibody systems (Ro and SSA) produced a line of identity on immunodiffusion, indicating that they were reacting with the same nuclear RNA protein [59]. Subsequent research found that anti-Ro antibodies in SLE and anti-SSA antibodies in primary Sjögren syndrome actually react with different epitopes on the same 60-kDa particle [60].

Anti-Ro/SSA antibodies recognize at least two proteins: a 52-kDa protein (475 amino acids) and a 60-kDa protein (525 amino acids). Sera of all patients with anti-Ro/SSA activity detected by immunoprecipitation bind the 60-kD antigen-a, whereas sera from only a subset bind the 52-kDa antigen. The additional antigenic reactivity appears to result from a molecular interaction between peptides of the 60- and 52-kDa proteins. No consistent disease association has been noted for any of the fine specificities of these antibodies.

Clinical Significance in SLE

Anti-Ro/SSA antibodies are found by immunodiffusion in approximately 38.6% of patients with SLE; the specificity is 99.986%; and the positive LR is 27571 and the negative LR is 0.4 [55]. Anti-Ro/SSA antibodies are found by ELISA in 29.6% of patients with SLE; the specificity is 97.6%; the positive LR is 12 and the negative LR is 0.7 [55]. Anti-Ro/SSA is found by immunoblotting in 40.6% of SLE patients; specificity 100%; positive LR is 68 and the negative LR is 0.6 [55] (see Table 34.3). Anti-Ro/SSA has been associated with photosensitivity, a rash known as subacute cutaneous lupus, cutaneous vasculitis (palpable purpura), interstitial lung disease, neonatal lupus, and congenital heart block (CCHB) [54, 61−64]. CCHB occurred in 35 of 278 anti-Ro/SSA-positive (immunodiffusion) SLE mothers and in one of 398 anti-Ro-negative mothers (sensitivity 13%; specificity 99.7%; positive LR, 43) [55]. There appears to be no predisposition of mothers with anti-52-kDa vs. 48-kDa Ro/SSA antibodies to develop CCHB [55] (Table 34.4). In addition, anti-Ro/SSA did not appear to predict either spontaneous abortions or stillbirths among SLE patients [55].

Association of Anti-Ro Antibodies with Other Disorders

Anti-Ro/SSA antibodies have been noted in 49−69% of patients with Sjögren syndrome [55] and have also occasionally been noted in patients with rheumatoid arthritis (9.9−20%), progressive systemic sclerosis

TABLE 34. 3 Anti-Ro

	Sensitivity	Specificity	Positive LR	Negative LR
SLE vs. normals by ID	38.6	99.986	27,571	0.4
SLE vs. normals by ELISA	29.6	97.6	12	0.7
SLE vs. normals by IB	40.6	100	>68	0.6
SS vs. normals by ID	49	99.7	163	0.5
SS vs. normals by ELISA	69	97.3	26	0.3
SS vs. normals by IB	65	100	>108	0.4
PSS (vs. normals)	7.9−21	98−100	4/6−11	0.8−0.9
DM/PM (vs. normals)	23−24	98−100	12−14	0.8
RA (vs. normals)	9.9−20	98−100	8.3−11	0.8−0.9

TABLE 34. 4   Anti-Ro & Pregnancy

	Sensitivity	Specificity	Positive LR	Negative LR
Prevalence of anti-Ro in mothers of children with CCHB	85			
Predictive value at anti-Ro testing for NLB/CCHB in offspring of SLE patients	13	99.7	43	0.9
Predictive value of anti-Ro testing for fetal wastage/ prematurity in mothers with SLE	24	65	0.69	1.2

TABLE 34. 5   Anti-Ro

	Sensitivity SLE	Specificity SS	Specificity (+)	Specificity (−)
Anti-52 kd alone	44	92	5.5	0.6
Anti-60 kd alone	3	88	0.3	1.1
Anti-52 + 60	28	75	1.1	1.0
All 52 kd	72	67	2.2	0.4
All 60 kd	31	63	0.8	1.1

(7.9−21%), cutaneous vasculitis, dermatomyositis/ polymyositis (23−24%), undifferentiated connective tissue disease, mixed connective tissue disease, juvenile rheumatoid arthritis, chronic active hepatitis, primary biliary cirrhosis, and homozygous C2 or C4 deficiency [55, 56] (Table 34.3). They are also found in 0.1−0.5% of normal subjects; such individuals may have an enhanced sensitivity to ultraviolet light.

Relationship to ANA-Negative SLE

In the 1970s, there were several reports of patients who met the American College of Rheumatology criteria for SLE, but were persistently negative for IF ANA [65]. Although not recognized at the time, this consistently negative finding occurred because sera were tested using mouse and not human tissue as the substrate [48]. By comparison, anti-Ro/SSA antibodies were found in most of these patients when a human cell line extract was used as the substrate for the Ro antigens.

More recently, the substitution of HEp2 cells (a human cell line) for mouse tissue sections in the IF ANA test has resulted in only a small number of SLE patients who test persistently negative for ANA. Nevertheless, on rare occasions, the anti-Ro/SSA antibody test may be useful in suggesting a diagnosis of systemic autoimmune disease in the face of a negative ANA. In one study of 4025 sera sent for ANA testing, for example, 64 patients were ANA-negative by immunofluorescence but anti-Ro-positive by ELISA [66]. Of these, 12 and five patients were diagnosed with SLE and cutaneous LE, respectively.

Finding anti-60-kDa Ro is somewhat useful in distinguishing SLE from primary Sjögren syndrome (sensitivity, 12%; specificity, 97%), whereas detecting anti-52-kDa Ro is more helpful in distinguishing primary Sjögren syndrome from SLE (sensitivity, 44%; specificity, 92%) [55] (Table 34.5).

ANTI-LA/SSB ANTIBODIES

Approximately 50% of sera from patients with SLE, which have anti-Ro antibody activity, also contain antibody to La, a closely related RNP antigen.

The antigen appears to be a nuclear phosphoprotein with a molecular mass of 48 kDa and is complexed with the Ro particle [67]. In addition to other functions, the La protein serves as a termination factor for RNA polymerase III.

Immunodiffusion, ELISA, Western blot, and RNA immunoprecipitation detect anti-La/SSB antibodies clinically. The La/SSB antigen is derived from rabbit thymus extracts or recombinant fusion proteins. ELISA assays are ten to 100 times more sensitive than immunodiffusion assays; they are also quantitative [55].

Clinical Associations of Anti-La

Anti-La/SSB antibodies are found in the following circumstances:

- It is very unusual to encounter sera that contain anti-La/SSB activity without demonstrable antibodies to Ro/SSA in patients with SLE or Sjögren syndrome.
- Antibodies to the La/SSB by immunodiffusion are present in 13% of patients with SLE (specificity 100% vs. normals; positive LR, >33; negative LR, 0.9). Antibodies to La/SSB by ELISA are found in 24% of SLE patients (specificity 99% vs. normals; positive LR, 20; negative LR, 0.8) [55] (Table 34.6).
- Anti-La/SSB antibody activity has also been detected in 22% (by immunodiffusion) to 71% (by ELISA) of

TABLE 34. 6   Anti-La

	Sensitivity	Specificity	Positive LR	Negative LR
SLE vs. normals ID	13	100	>33	0.9
SLE vs. normals ELISA	24	99	20	0.8
SS vs. normals ID	22	100	>17	0.8
SS vs. normals ELISA	71	93	10	0.3
SS vs. normals IB	50	100	28	0.5

patients with Sjögren and has been seen in some patients with scleroderma (2–37%), dermatomyositis or polymyositis (3%), rheumatoid arthritis (1%), primary biliary cirrhosis, and autoimmune hepatitis [55] (Table 34.6).

INDICATIONS FOR TESTING FOR ANTI-RO/SSA AND ANTI-LA/SSB

In the author's opinion, indications for ordering an anti-Ro/SSA antibody test are as follows: women with SLE who have become pregnant; women who have a history of giving birth to a child with heart block or myocarditis; patients with a history of unexplained photosensitive skin eruptions; patients suspected of having a systemic connective tissue disease in whom the screening ANA test is negative; and patients with symptoms of xerostomia, keratoconjunctivitis sicca, and/or salivary and lacrimal gland enlargement.

Once a patient is either Ro/SSA- and or La/SSB-positive or -negative, this rarely changes. There is also little if any justification for performing serial determinations of anti-Ro/SSA and/or anti-La/SSB [55].

NUCLEOSOME-SPECIFIC ANTIBODIES

The first autoantibody described in systemic lupus erythematosus, the LE factor was a nucleosome-specific autoantibody. These antibodies are now termed nucleosome-restricted antibodies because they bind to the complex of DNA and histones, but not to the individual components.

Pathogenic Role

An original report suggested that nucleosomes might be critical antigens in the eventual generation of antinuclear antibodies in SLE [60]. Subsequent observations have provided some support for this hypothesis [69, 70]. The appearance of these antibodies may precede the emergence of antibodies to DNA or histones.

Methods of Detection

The detection of antinucleosome-specific antibodies remains problematic because the nucleosomal material used as a substrate also binds anti-dsDNA and antihistone antibodies. The traditional method of circumventing this problem, absorption techniques, is cumbersome and difficult to perform. An epitope recognized only by nucleosome-specific antibodies has not yet been identified.

Incidence and Accuracy

Because of the difficulties in detection, there is a paucity of data concerning the incidence and accuracy of antinucleosome-specific autoantibodies. A prevalence of 12% to over 80% has been noted in patients with SLE, particularly those with active disease; they are uncommon in those with cutaneous LE [71–73]. They are usually, but not always, associated with anti-dsDNA and antihistone antibodies [63]. The specificity of these antibodies remains a source of controversy.

Associated Clinical Manifestations

Nucleosome-specific autoantibodies have been linked with lupus nephritis in murine models of SLE. In humans, however, the results are conflicting. Some series have found a weak but significant association [74, 75], whereas others have not [70, 72]. One study of a large cohort of patients with varied connective tissue disorders and control sera from those with hepatitis C virus infection found that IgG antinucleosomal antibodies were found exclusively in patients with SLE, systemic sclerosis, and MCTD, with prevalences of 72, 46, and 45%, respectively [75]. In addition, antinucleosomal antibodies of the IgG3 subclass were observed only in patients with SLE and correlated more closely with active disease and particularly with nephritis than anti-DNA antibodies.

Summary

The clinical utility of antinucleosome-specific autoantibodies remains undefined, as their sensitivity, specificity, and association with particular manifestations of SLE remain undefined. Nevertheless, these antibodies are a source of interest because they may have a role in the pathogenesis of autoantibody production in SLE.

Anti-hnRP Antibodies

The spliceosome refers to a complex of nuclear RNA-binding proteins, which processes pre-mRNA into spliced mature mRNA. Among the subunits of the spliceosome are a group of 30 structurally related proteins known as heterogeneous nuclear ribonuclear proteins, which are called hnRNP [76, 77].

Methods of Detection

Autoantibodies specific for the RNA-binding regions of two of the hnRNP proteins, A2 (also called RA33) and A1, produce a finely speckled pattern of staining in the nucleoplasm on indirect immunofluorescence microscopy. Their presence must be confirmed by either immunoblotting or dot blot assays or ELISA [77, 78].

Incidence and Accuracy

Anti-hnRNP antibodies against the A1 and A2 proteins have been reported with the following frequencies [69, 70]:

- Rheumatoid arthritis: 50% for the A1 protein and 35% for the A2 protein.
- SLE: 22—38% for A1 and 23—60% for A2, especially in active disease [77].
- Mixed connective tissue disease: 40% for A1 and 38% for A2.
- Scleroderma, myositis, Sjögren syndrome: <5% for A2.
- No connective tissue disease: <7% for A1 and A2.

The clinical utility of anti-hnRNP antibodies is currently limited because they cross-react with multiple antigens, and antibodies directed against hnRNP antigens other than A1 and A2 may be found in different autoimmune disorders.

Summary

Anti-hnRNP antibodies have limited utility in the diagnosis of connective tissue diseases as they are found at significant levels in multiple disorders and their presence appears to correlate highly with other specific antibodies (such as anti-Sm antibodies).

Anti-PCNA Antibodies

A number of proteins accumulate in the nucleoplasm of cells during the G1/S phase of the cell cycle, which corresponds to the period of DNA synthesis. Approximately 2—10% of sera from patients with SLE sera contain antibodies, which react with one such protein: proliferating cell nuclear antigen (PCNA) [81].

Methods of Detection

Anti-PCNA antibodies can be detected by indirect immunofluorescence as a finely granular nuclear staining pattern on rapidly dividing cells. An enzyme-linked immunoassay for the detection of these antibodies has been reported. The dominant epitope of PCNA is an area of 14 contiguous amino acids [82].

Accuracy

Anti-PCNA antibodies appear to be specific (>95%) for SLE; they have not been found in sera from patients with RA or other connective tissue diseases or in normal sera [81]. The incidence of this antibody varies, as its production is rapidly inhibited by the administration of corticosteroids and immunosuppressive drugs.

Clinical Association

The presence of anti-PCNA antibodies in patients with SLE is not associated with any particular clinical manifestation other than arthritis.

Summary

Despite its high specificity, the anti-PCNA antibody test has limited clinical utility because of its very low sensitivity.

Antibodies to Ribosomal P Proteins

Antibodies to ribosomal P proteins are detected by either Western immunoblotting or an ELISA with the latter being the test of choice. The ELISA, which is more sensitive using affinity-purified human P protein rather than bovine P protein, is quantitative and more convenient [83].

The targets of these antibodies were three phosphoproteins (P proteins) located on the 60S subunit of ribosomes [84, 85]. The proteins were designated P0, P1, and P2, with molecular masses of 35, 19, and 17 kDa, respectively. They were thought to be involved in protein synthesis; specifically, in the interaction of EF-1a and EF-2 with ribosomes.

Clinical Associations

All published studies agree on the specificity of antiribosomal P protein antibodies for SLE [86—88]. These antibodies have not been found using conventional assays in normal controls [77, 79] or in patients with other autoimmune diseases (e.g., rheumatoid arthritis, scleroderma, and myositis) [90].

The antibodies have been detected in 10—20% of patients with SLE [86, 87, 91, 92]; however, the incidence may approach 50% in Asian populations and children [90, 93, 94].

RIBOSOMAL P PROTEIN ANTIBODIES AND LUPUS CEREBRITIS

The presence of ribosomal P protein antibodies has been strongly associated with neuropsychiatric manifestations of lupus by some [87, 89, 90] but not by others [92, 95, 96].

ANTIBODIES TO RIBOSOMAL P PROTEIN AND NON-CNS LUPUS

The presence of antiribosomal P protein antibodies has been linked with the involvement of other organ systems in SLE, specifically the liver and the kidney.

1. Liver disease: In a case—control study, liver disease was found in seven of 20 patients with SLE who had antiribosomal P antibodies, but in only one of 20 patients without these antibodies [97]. Antibodies to ribosomal P proteins were not found in sera of patients with chronic active hepatitis, primary biliary cirrhosis, or sclerosing cholangitis; these findings suggest that these autoantibodies were not simply a by-product of immune-related hepatic injury. Others have made similar observations [98].
2. Renal disease: Antiribosomal P antibodies have also been associated with lupus nephritis in some studies [97], including fluctuations of antibody levels with the activity of renal disease [89]; however, others have noted no such association [87, 92].

3. An association between antiribosomal P protein antibodies and antibodies to cardiolipin has been noted [100]. It is unknown whether the coexistence of such antibodies is associated with a higher risk of neuropsychiatric manifestations.

VARIABILITY OF ANTIRIBOSOMAL P PROTEIN ANTIBODY RESULTS

Any or all of the following possibilities may distort the results of studies evaluating the frequency of antibodies to ribosomal P proteins and their clinical associations:

1. Differences in methodology for detection of antibody.
2. Variable ethnic composition of populations in different studies.
3. Small sample sizes.
4. Problems establishing a definition.

Perhaps the biggest confounding factor is the problem of definition. Lupus-related psychosis and depression remain exceedingly difficult to diagnose and still harder to quantify. In addition, authors use different criteria to diagnose these conditions. As examples, one study included in their group of lupus patients with depression only those requiring admission for their psychiatric manifestations [87], whereas another included patients with much less severe disease [95]. Results might therefore not be comparable.

UTILITY OF ANTIRIBOSOMAL P ANTIBODIES IN SLE

It is believed that testing for antibodies to ribosomal P proteins may be useful in certain circumstances as they are quite specific, but not sensitive (12—16%), for SLE [86, 87, 101]. A positive antiribosomal P antibody might be used diagnostically in a fashion similar to that of antibodies to Sm. As an example, presence of an anti-Sm antibody suggests a diagnosis of SLE even in a patient with otherwise undifferentiated disease. Positive antibodies to ribosomal P proteins may also favor this diagnosis in this setting. In addition, the presence of these antibodies is independent of antibodies to dsDNA [89, 92]. They may therefore remain elevated when other marker antibodies have returned to normal and might again be helpful in diagnosing SLE even with quiescent disease. The utility of these antibodies in these scenarios, as well as their exact role in NPSLE, liver disease, and renal disease, requires further investigation.

Other Autoantibodies

Anti-C1q Antibodies

See the section titled "Complement."

COOMBS

Overt autoimmune hemolytic anemia (AIHA), characterized by an elevated reticulocyte count, low haptoglobin levels, an increased indirect bilirubin concentration, and a positive direct Coombs test, has been noted in approximately 10% of patients with SLE. Even more patients have a positive Coombs test without evidence of overt hemolysis. The presence of both immunoglobulin and complement on the red cell is usually associated with some degree of hemolysis, whereas the presence of complement alone (e.g., C3 and/or C4) is often not associated with hemolysis. The antibodies are "warm," IgG, and are directed against Rh determinants. IgM-mediated cold agglutinin hemolysis is uncommon. Anemia associated with a positive Coombs may result more from increased splenic sequestration than to direct hemolysis.

RHEUMATOID FACTORS

Rheumatoid factors (RF) represent autoantibodies to the Fc portion of IgG. Most RF are IgM, although RF of all Ig classes have been described. RF, as detected by latex fixation, have been found in 20—60% (mean, 33%) of patients with SLE [92]. While earlier observations suggested a negative correlation between a positive RF test and lupus nephritis, subsequent studies have failed to confirm this. While RF titers fluctuate little in patients with RA, they tend to both vary in titer and come and go in patients with SLE, to some extent in correlation with the activity of SLE. It is the author's opinion that there is no merit in testing for RF in patients with SLE.

ANTIPLATELET ANTIBODIES

Antiplatelet antibodies demonstrable as platelet-bound IgG have been shown to be increased in virtually all SLE patients with thrombocytopenia [102]. While the serum antiplatelet antibodies in chronic ITP are IgG3, in SLE sera they are IgG1—4 [103, 104]. Serum antibody assays for antiplatelet antibodies generally, however, do not correlate well with the presence of thrombocytopenia. In contrast, all IgG subclasses are found bound to platelets (in patients with thrombocytopenia) in both SLE and ITP [105]. Testing for platelet-bound IgG may be helpful in evaluating SLE patients with thrombocytopenia. The presence of the antibody suggests an antibody-mediated thrombocytopenia, whereas an absence of the antibody suggests that thrombocytopenia is not due to an ITP antibody-mediated mechanism. The IgG bound to platelets mediates thrombocytopenia by causing increased phagocytosis of opsonized platelets in the spleen.

Thrombocytopenia in a patient with SLE should also make one suspect the antiphospholipid antibody syndrome (APS) (see the section titled "Antiphospholipid Antibodies").

ANTI-WBC ANTIBODIES

Anti-white blood cell antibodies, especially anti-lymphocyte antibodies, have been noted in many SLE patients. This subject is reviewed in Chapter 48. Detection of these antibodies is performed primarily for research.

ANTIPHOSPHOLIPID ANTIBODIES

The antiphospholipid syndrome (APS) is defined by two major components [106]:

1. Presence in the serum of at least one type of autoantibody known as an antiphospholipid antibody (aPL). APLs are directed against phospholipid-binding plasma proteins.
2. The occurrence of at least one clinical feature from a diverse list of potential disease manifestations, the most common of which are categorized as venous or arterial thromboses, pregnancy morbidity, or thrombocytopenia.

Although the clinical manifestations of APS occur in other disease populations, in the APS they occur by definition in the context of aPL. APL may be detected by:

1. Lupus anticoagulant tests.
2. Anticardiolipin antibody ELISA.
3. Anti-β2 glycoprotein-I ELISA.

The full clinical significance of other autoantibodies, including those directed against prothrombin, annexin V, phosphatidylserine, and phosphatidylinositol, remain unclear.

APS occurs either as a primary condition or in the setting of an underlying disease, particularly systemic lupus erythematosus (SLE).

Lupus Anticoagulant

The lupus anticoagulant phenomenon refers to the ability of aPL to cause prolongation of *in vitro* clotting assays such as the activated partial thromboplastin time (aPTT), the dilute Russell viper venom time (dRVVT), the kaolin clotting time or, infrequently, the prothrombin time. This prolongation is not reversed when the patient's plasma is diluted 1:1 with normal platelet-free plasma. In contrast, such mixing studies do correct the clotting abnormality associated with factor deficiencies [106].

One common effect of aPL detected by routine laboratory testing is the prolongation of the aPTT. However,

only about one-half of patients with LAs have prolongations of the aPTT. Thus, if APS is suspected strongly, additional testing, usually with a dRVVT, is essential.

In 1995, an international committee recommended the following guidelines for the detection of LAs [106]; additional explanation of technical details can be found in three excellent reviews [106–108].

- Both patient and normal plasma should be as platelet-free as possible. For testing on fresh plasma, frozen samples, and pooled normal plasma, the platelet count should be less than 10,000/μl. Platelet filtration, the preferred method for preparing platelet-free plasma, may interfere with testing of other coagulation factors.
- Two or more tests should be used to screen for LAs [106]. The most commonly used assays are the dRVVT and the dilute activated PTT (dAPTT). Other tests include the kaolin clotting time (KCT) and the dilute prothrombin time (dPT) [106]. An LA is present if any of the tests is positive [106].
- Inhibitor activity should be documented by the effect of patient plasma on pooled normal plasma. Confirmatory studies need to be performed to document the phospholipid dependence of the inhibitor.

Laboratory experience is important in LA assays. In one study, one-quarter of all plasma samples diagnosed as having LA activity were found to be false-positive tests upon measurement in a reference laboratory [106].

There is no standardized assay for lupus anticoagulant activity [108]. Thus, confirmatory tests are required, such as an abnormal dilute Russell viper venom test and neutralization of the inhibitor with excess phospholipids, particularly a hexagonal phase phospholipid [109–111]. Multiple tests should be performed when the clinical index of suspicion is high, as a single test will detect only 60–80% of cases [110]. Furthermore, the test chosen, the source of the reagents, and the instruments used in the assay all affect the ability to detect a lupus anticoagulant [112]. As an example, 59 laboratories were surveyed for their ability to detect the presence, in normal pooled plasma, of a murine monoclonal antibody against human b2-glycoprotein 1 with lupus anticoagulant activity [112]. Results of the survey showed that positivity was found by 82 and 37% of the laboratories for the high- and low-potency samples, respectively; overall test responsiveness was greatest for the dilute prothrombin time, dilute Russell viper venom time, and kaolin clotting time; and PTT LA and Innovin reagents showed the greatest responsiveness in activated partial thromboplastin time and dilute prothrombin time, respectively.

The presence of lupus anticoagulant activity can be confirmed by: (1) employing an aPTT assay that is insensitive to the presence of a lupus anticoagulant, or (2) adding an additional source of phospholipid to counteract the antibody. This can be in the form of a hexagonal phospholipid or a phospholipid from formalin-fixed or freeze-dried platelets [112].

These confirmatory tests are consistent with guidelines concerning the detection of lupus anticoagulant activity [133]. Such activity is verified if the following results are obtained.

- Prolonged phospholipid-dependent coagulation should be demonstrated on a screening test, such as activated partial thromboplastin time, kaolin clotting time, and others.
- Mixing with normal platelet-poor plasma should not correct the prolonged coagulation time.
  The prolonged coagulation time is shortened or corrected by the addition of excess phospholipid. Other coagulopathies are excluded.

Anticardiolipin Antibodies

Anticardiolipin antibodies (aCL) react with proteins (e.g., b2-glycoprotein 1, prothrombin, and annexin V) bound to anionic phospholipids, such as cardiolipin and phosphatidylserine. There is an approximate 85% concordance between the presence of a LA and aCL. In many cases, however, the LA is a separate population of antibodies from aCL [113–115]. Thus, testing should be performed for both LA and aCL if APS is clinically suspected. LA positivity incurs a somewhat greater risk for thrombosis than aCL.

Different immunoglobulin isotypes and subclasses are associated with aCL, including IgG, IgA, IgM, and IgG subclasses 1–4. Elevated levels of IgG aCL (particularly IgG2) incur a greater risk of thrombosis than other immunoglobulin isotypes [110].

IgA aCL is the most prevalent isotype in African-Americans [116]. The predominance of this isotype in combination with the low prevalence of significantly elevated levels of aCL may account for the low frequency of APS in these patients.

Anti-b2 Glycoprotein I Antibodies

Anti-b2GP1 antibodies directly bind to b2GP1 as opposed to aCL, which frequently bind to a complex of anionic phospholipids and b2GPl. Antibodies to b2-glycoprotein 1, a phospholipid-binding inhibitor of coagulation, are found in a large percentage of patients with primary or secondary APS [117]. Although antibodies to b2-glycoprotein 1 are found commonly in those with other antiphospholipid antibodies, they are the sole antiphospholipid antibody found in approximately 11% of patients with APS [117].

Antiprothrombin Antibodies

Antiprothrombin antibodies have been described in association with both clotting and pulmonary hemorrhage [118].

Patients with antiphospholipid antibodies may also have antibodies directed against other proteins, including heparin/heparin sulfate, prothrombin, platelet-activating factor, tissue-type plasminogen activator, protein S, annexin IV and V, thromboplastin, oxidized low-density lipoproteins, thrombomodulin, and kininogen [119–129].

False-Positive Serologic Test for Syphilis

Some patients with SLE have a false-positive serologic test for syphilis. When patient sera contain anticardiolipin antibodies, the false-positive STS occurs because the syphilis antigen used in the test is embedded in cardiolipin. As a result, a reaction against this molecule will be interpreted incorrectly as being directed against the treponemal antigen. STS should not be used to screen for APL because it has a low sensitivity and specificity [107].

Prevalence of Antiphospholipid Antibodies

Although antiphospholipid antibodies are associated with a propensity for thrombosis and with various autoimmune disorders, they can sometimes be found in normal asymptomatic individuals, as illustrated by the following observations:

- Normal individuals occasionally have elevated levels of either IgG or IgM aCL [130, 131]. In one study, for example, the prevalence was 5% on a first test but only 2% on retesting [131].
- Increased levels of IgG or IgM aCL have been observed in 12–52% of the elderly [130, 132].
- One study found that 40 of 499 normal blood donors had a positive LA test (frequently young women); however, only three had elevated levels of aCL [133].

Associated Disorders

Antiphospholipid antibodies have been noted in increased frequency in patients with SLE: approximately 31% of patients have a LA, 23–47% have an aCL, and 20% have antibodies to b2-glycoprotein 1 [131, 134–136]. However, roughly 50% of patients with a LA have SLE [111, 115]. Antiphospholipid antibodies also occur with increased frequency (5–10%) in women with greater than three spontaneous recurrent abortions [112].

Both LA and aCL have also been found in patients with a variety of autoimmune and rheumatic diseases, including [111]:

- Hemolytic anemia.
- Idiopathic thrombocytopenic purpura (up to 30%).
- Juvenile arthritis.
- Rheumatoid arthritis (7–50%).
- Psoriatic arthritis (28%).
- Scleroderma (25%), especially with severe disease [137].
- Behçet syndrome (20%).
- Sjögren syndrome (25–42%).
- Mixed connective tissue disease (22%).
- Polymyositis and dermatomyositis.
- Polymyalgia rheumatica (20%) [138].
- Osteoarthritis (less than 14%).
- Occasionally in gout and in multiple sclerosis.
- Chronic discoid LE [139].
- Eosinophilia myalgia and toxic oil syndrome [140].
- Raynaud phenomenon [141].

Antiphospholipid antibodies have also been noted in patients with infections and after the administration of certain drugs. These are usually IgM aCL antibodies, which may occasionally result in thrombotic events [111, 142]. Furthermore, the antibodies do not appear to have anti-b2-glycoprotein 1 antibody activity [143]. The infections that have been associated with these antibodies include the hepatitis A and C virus (see the section titled "Immunoglobulins"), mumps, bacterial septicemia, HIV infection, syphilis, HTLV-I, malaria, *Pneumocystis carinii*, infectious mononucleosis, and rubella. Among the drugs that have been implicated are phenothiazines (chlorpromazine), phenytoin, hydralazine, procainamide, quinidine, quinine, dilantin, interferon-α, amoxicillin, and propranolol [142, 144]. A discussion of the clinical manifestations of antiphospholipid antibodies is found in Chapter 50.

IMMUNOGLOBULINS

SLE is characterized by a polyclonal gammopathy. In fact, increased levels of Ig have been noted in up to 76% of patients [145]. Elevated levels tend to correlate with active disease. Elevations of IgG (including all four IgG subclasses), IgA, IgM, and IgE have been noted. Three series have suggested an increased frequency (3–4%) of IgA deficiency [146]. Acquired hypogammaglobulinemia, especially of IgG subclasses, has rarely been noted, but should be suspected in patients with recurrent infections. The hypogamma IgG may relate to chronic immunosuppressive therapy.

IMMUNE COMPLEXES

SLE is considered to be the prototype for immune complex-mediated diseases. A number of methods have been developed for the detection of immune complexes in serum based on the (highly) variable biological and physical characteristics of these immune complexes. These methods include complement activation, C1q binding, rheumatoid factor binding, the Raji cell assay, conglutinin assay, assay with C3, and cryoprecipitation. This subject is discussed in depth in Chapter 19. While early results suggested good correlation with clinical activity in patients with SLE, especially lupus nephritis, subsequent studies showed highly variable results. These assays are therefore rarely employed nowadays.

COMPLEMENT

Considerable evidence shows that much of the pathology in patients with systemic lupus erythematosus can be attributed to immune complexes [147, 148] and, thereby, complement activation [148, 149]. These immune complexes may either form in the circulation and later deposit in tissues or form *in situ*. Immune complexes may cause tissue inflammation directly or through activation of the complement system. Complement activation causes the release of various mediators, promotes cell interaction, and ultimately results in inflammation. Evaluation of the complement system, which can serve as an indirect measure of the presence of immune complexes, often correlates with clinical aspects of SLE. Monitoring blood levels of complement may be useful in adjusting therapy for the patient.

Measurement of Complement

The complement system can be assessed by:

1. The measurement of total hemolytic complement activity (CH_{50}).
2. The immunochemical or hemolytic measurement of individual components.
3. Measurements of complement fragments, activation-dependent neoepitopes, and complexes that arise from complement activation.
4. The determination of complement metabolism.

The total hemolytic complement level (CH_{50}) represents the sum of all components of the system. By adding diluted serum to antibody-sensitized sheep erythrocytes and quantitating the amount of released hemoglobin, one can measure this level in units that represent the reciprocal of that dilution of serum that causes 50% of cells to lyse (i.e., CH_{50}). A CH_{100} represents that dilution of serum that lyses 100% of sensitized cells. This measurement is not nearly as accurate as a CH_{50}, which reflects actual hemolysis in a more linear fashion than a CH_{100} [150–153].

Separate complement components are generally measured by immunochemical means: radial immuno-diffusion, electroimmunodiffusion, or nephelometry, and recently by flow cytometry [154]. However, such determinations yield no information about the functional biologic integrity or hemolytic potential of the components being measured. Such functional tests of complement components are rarely employed because of their difficulty in performing routine clinical assays.

When measuring CH_{50} or components, it is important to remember that some components are thermolabile. Serum stored at room temperature, even for a few days, is adequate for the immunochemical measurement of individual components as proteins (as, for instance, by immunodiffusion). However, it is essential to store samples (preferably EDTA plasma) as soon as possible at $-70°C$ when measurement of the hemolytic levels of individual components is desired. The CH_{50} will remain relatively stable for a few hours at room temperature or overnight at $-20°C$; however, samples are best stored at $-70°C$ as soon as possible. Heating for 30 min at $56°C$ is also known to inactivate complement components and decrease or abolish CH_{50} activity.

These assays give but a glimpse of the dynamic state of the complement system. Serial values may be helpful in assessing this dynamic state: falling values suggest more catabolism (e.g., via immune complex fixation) than synthesis, whereas rising levels suggest more synthesis than catabolism is taking place.

Measurement of both the native components and the activation products of individual components is a way of assessing the catabolic state of the complement system without necessarily doing serial measurements [155]. Activation products in EDTA plasma generally are measured by ELISA using monoclonal antibodies [155], and by flow analysis [154]. However, the levels of complement activation factors can also be affected by binding to complement (fragment) receptors, degradation by serum proteases, and renal and hepatic clearance [151]. Studies suggest that erythrocyte C4d levels are elevated and erythrocyte Cr1 are decreased in patients with SLE, especially those with active disease [154].

Complement metabolism is a better way of assessing the dynamic state (i.e., synthesis and catabolism) of individual complement components [156]. However, these studies are rarely performed because of the great difficulty both in isolating hemolytically and biologically active purified components and in maintaining their activity after radiolabeling. The liver is the primary source of synthesis of complement components, with the notable exception of C1q, which is probably produced primarily by macrophages [157]. Patients with inherited complement deficiencies, especially homozygous deficiency of either C1q, C4, or C2, are prone to develop SLE. These patients typically have an absent CH_{50}, and the absent complement component; the other complement component levels are normal [158]. A more complete discussion of complement abnormalities is found in Chapter 20.

ACUTE-PHASE REACTANTS

The acute-phase response is a major pathophysiologic phenomenon that accompanies inflammation [159, 160]. With this reaction, normal homeostatic mechanisms are replaced by new set points, which presumably contribute to defensive or adaptive capabilities. Focus on this phenomenon first occurred with the discovery of elevated serum concentrations of C-reactive protein (CRP) during the acute phase of pneumococcal pneumonia [161].

Despite its name, the acute phase response accompanies both acute and chronic inflammatory states. It can occur in association with a wide variety of disorders, including infection, trauma, infarction, inflammatory arthritides, and various neoplasms.

Acute-phase proteins are defined as those proteins whose plasma concentrations increase (positive acute-phase proteins) or decrease (negative acute-phase proteins) by at least 25% during inflammatory states [162]. These changes largely reflect their production by hepatocytes.

Increases in acute-phase proteins may vary from approximately 50% with ceruloplasmin and several complement components to 1000-fold for CRP and serum amyloid A. Other positive acute-phase proteins include fibrinogen, a1-antitrypsin, haptoglobin, and ferritin, whereas negative reactants include albumin, transferrin, and transthyretin.

Clinical Relevance of Acute-Phase Reactants

Despite the lack of diagnostic specificity, the measurement of serum levels of acute-phase proteins is useful because it may reflect the presence and intensity of an inflammatory process. Currently, the most widely used indicators of the acute-phase protein response are the erythrocyte sedimentation rate (ESR) and CRP. The rate at which erythrocytes fall through plasma, the ESR, depends largely on the plasma concentration of fibrinogen [163]. These tests may be useful both diagnostically: (1) in helping to differentiate inflammatory from noninflammatory conditions, and (2) in patient management because they may generally reflect the response to and need for therapeutic intervention.

Comparison of ESR and CRP

Compared to the measurement of CRP, the ESR has the advantages of familiarity, simplicity, and an abundant literature compiled over the past seven decades. However, the ESR has a number of disadvantages compared to the CRP determination.

1. The ESR is only an indirect measurement of plasma acute-phase protein concentrations; it can be greatly influenced by the size, shape, and number of red cells, as well as by other plasma constituents, such as immunoglobulins. Thus, results may be imprecise and sometimes misleading.
2. As a patient's condition worsens or improves, the ESR changes relatively slowly; by comparison, CRP concentrations change rapidly.
3. The range of abnormal values for CRP is greater than for the ESR, with accompanying clinical implications: as an example, among patients with CRP concentrations greater than 10 mg/dl (100 mg/l), 80–85% have bacterial infections [162].
4. Normal values for the ESR are slightly higher in women than men.

Population studies reveal a skewed, rather than normal, distribution of the plasma CRP concentration. Although most normal subjects have CRP concentrations of 0.2 mg/dl (2 mg/l) or less, some individuals have concentrations as high as 1.0 mg/dl (10 mg/l). These higher values have been attributed to a modest stimulation by minimally apparent low-grade processes such as gingivitis or trivial injury. This observation has led to the suggestion that values less than 1.0 mg/dl (10 mg/l) should be regarded as clinically insignificant [162].

Information drawn from the Third National Health and Nutrition Evaluation Survey indicates that the upper reference limits in a representative sample of the population vary with age, gender, and race [164]. The following formulas provide values for CRP levels that closely approximate the 95th percentile for subjects without identified inflammatory conditions:

- Females: For those 25–70 years of age, the upper limit of reference range = (age/65) + 0.7 mg/dl.
- Males: For those 25–70 years of age, the upper limit = (age/65) + 0.1 mg/dl.

For most clinical purposes, CRP values less than 0.1 or 0.2 mg/dl can be regarded as normal and values over 1.0 as indicating clinically significant inflammation.

However, although CRP is a sensitive reflector of inflammation, it is not specific. Values between 0.2 and 1.0 mg/dl may reflect minor degrees of inflammation, but may also reflect obesity, cigarette smoking, diabetes mellitus, or other noninflammatory causes of modest elevations in the CRP.

Rationale for Employing Multiple tests

Although elevations in multiple components of the acute-phase response commonly occur together, not all happen uniformly in all patients. Discordance between concentrations of different acute-phase proteins is common; some may be elevated, whereas others are not. These variations may be explained by differences in the production of specific cytokines or their modulators in different diseases [161].

Discrepancies between ESR and CRP are found with some frequency. An elevated ESR observed together with a normal CRP is often a false-positive value for the ESR; this may reflect the effects of blood constituents that are not related to inflammation but that can influence the ESR. However, this conclusion is not always valid. As an example, the ESR may be markedly elevated in patients with active SLE, while the CRP is normal.

As there undoubtedly are a number of other clinical situations in which similar discrepancies occur, there probably is no single best laboratory test to reflect inflammation. Currently, the optimal use of acute-phase protein measurements may be to obtain several measurements rather than a single test, the results of which are interpreted in light of the clinical context.

ESR can be high with no obvious clinical activity and normal with active disease [165]. SLE represents an exception to the generalization that CRP concentrations correlate with the extent and severity of inflammation in patients with rheumatic disorders. Many patients with active SLE do not have elevated CRP concentrations, although they may have marked increases in CRP concentrations during bacterial infection [166]. This finding can be applied to the differential diagnosis of fever in patients with SLE, but one must remember that CRP concentrations are high in patients with active lupus serositis [167] or chronic synovitis [168].

Serum amyloid A levels also tend to be lower in SLE than in other chronic inflammatory disorders, which may explain why secondary amyloidosis is unusual in this disease [169].

HEMATOLOGICAL ABNORMALITIES

Abnormalities of the formed elements of the blood, and of the clotting, fibrinolytic, and related systems, are very common in SLE [170]. The major clinical manifestations are anemia, leukopoenia, and thrombocytopenia. Thus, whenever the author sees a patient with SLE he tends to order a CBC.

Anemia

Anemia is a frequent occurrence in SLE, affecting most patients at some time in the course of their disease. Multiple mechanisms contribute, including inflammation, renal insufficiency, blood loss, dietary insufficiency, medications, and immune hemolysis [170–177], as well as antibodies to erythropoietin [177].

Chronic Inflammation

The most frequent cause of anemia in SLE is suppressed erythropoiesis from chronic inflammation (anemia of chronic disease) and/or renal insufficiency [174, 175].The anemia is normocytic and normochromic, with a relatively low reticulocyte count. Although serum iron levels may be reduced, bone marrow stores are adequate and the serum ferritin concentration is elevated.

As in other chronic illnesses, serum erythropoietin levels may be inappropriately low for the degree of anemia. However, some of the apparent reduction in serum erythropoietin may be spurious [179].

Blood Loss

Anemia may reflect acute or chronic blood loss from the gastrointestinal tract, usually secondary to medications (nonsteroidal anti-inflammatory drugs or steroids), or may be due to excessive menstrual bleeding. Iron-deficiency anemia is not uncommon, especially among teenagers or young women.

Red Cell Aplasia

Red cell aplasia, probably due to antibodies directed against bone marrow erythroblasts, has been rarely observed [175, 176, 180]. This form of anemia may respond to steroids, although cyclophosphamide and cyclosporine have been employed successfully. Even rarer are isolated case reports of aplastic anemia that is presumably mediated by autoantibodies against bone marrow precursors; immunosuppressive therapy may also be effective in this setting [181–183].

In addition, bone marrow suppression can also be induced by medications, including antimalarials and immunosuppressive drugs.

Autoimmune Hemolytic Anemia

Overt autoimmune hemolytic anemia (AIHA), characterized by an elevated reticulocyte count, low haptoglobin levels, increased indirect bilirubin concentration, and a positive direct Coombs test, has been noted in approximately 10–22% of patients with SLE [170, 172–174, 185, 186]. Other patients have a positive Coombs test without evidence of overt hemolysis. The presence of both immunoglobulin and complement on the red cell is usually associated with some degree of hemolysis, whereas the presence of complement alone (e.g., C3 and/or C4) is often not associated with hemolysis [171–174]. The antibodies are "warm," IgG, and are directed against Rh determinants. IgM-mediated cold agglutinin hemolysis is uncommon.

Lupus has also been associated with a thrombotic microangiopathic hemolytic anemia [187, 188]. Most affected patients also have thrombocytopenia, kidney involvement, fever, and neurologic symptoms, producing a picture of thrombotic thrombocytopenic purpura; some do not have fever or neurologic disease, producing a pattern of hemolytic–uremic syndrome. The pathogenesis of this syndrome is not completely understood. In one report of four patients plus 24 others identified from a literature review, antiphospholipid antibodies were searched for in eight and found in five [187].

Leukopenia

Leukopenia is common in SLE and may reflect disease activity. A white blood cell count of less than 4500/ml has been noted in approximately 50% of patients, especially those with active disease [170, 173, 174], whereas lymphocytopenia occurs in approximately 20% [161]. In comparison, a white blood cell count below 4000/ml (the ARA criteria) occurs in only 15–20% of patients [173, 189].

Leukopenia in patients with SLE can result from immune mechanisms, medications (e.g., cyclophosphamide or azathioprine), bone marrow dysfunction, or hypersplenism [173, 174, 189, 190]. Functional defects of neutrophils have also been noted. They are thought to be induced by immune abnormalities (e.g., immune complexes, inhibition of complement-derived chemotactic factors) and/or medications (e.g., corticosteroids) [191, 192].

Although leukopenia is more common, leukocytosis (mostly granulocytes) can occur in SLE. It is usually due to infection or to the use of corticosteroids (in high doses) [193], but may be seen during acute exacerbations of SLE. A shift of granulocytes to more immature forms (a "left" shift) suggests infection.

White cell constituents other than neutrophils can also be affected in SLE.

1. Lymphocytopenia (less than 1500 cells/ml), especially involving suppressor T cells, has been observed in 20–75% of patients, particularly during active disease [170, 171–173, 190, 194]. This finding is

strongly associated with IgM, cold reactive, complement fixing, and presumably cytotoxic antilymphocyte antibodies. In one study, for example, these antibodies were noted in 26 of 29 patients with SLE; the antibody titer correlated directly with the degree of lymphopenia. Another potential mechanism of lymphopenia is increased apoptosis, as reflected by the increased expression of Fas antigen on T cells [195].

2. Steroid therapy may result in low absolute eosinophil and monocyte counts [196].

3. The number of basophils may also be decreased in SLE, particularly during active disease [197]. Basophil degranulation with release of platelet-activating factor and other mediators may play a role in immune complex deposition and vascular permeability.

Thrombocytopenia

Mild thrombocytopenia (platelet counts between 100,000 and 150,000/ml) have been noted in 25–50% of patients, whereas counts of less than 50,000/ml occur in only 10% [170, 173, 174, 189, 198]. The major mechanism is increased platelet destruction by antiplatelet antibodies, leading to phagocytosis via their Fc receptors in the spleen, as in idiopathic thrombocytopenic purpura (ITP) [199]. Other important mechanisms in selected patients include bone marrow suppression by immunosuppressive drugs (other than corticosteroids) and increased consumption due to a thrombotic microangiopathy [187] or the antiphospholipid antibody syndrome.

ITP may be the first sign of SLE, followed by other symptoms as long as many years later. It has been estimated that 3–15% of patients with apparently isolated ITP go on to develop SLE [200]. Evan syndrome, i.e., both autoimmune thrombocytopenia and autoimmune hemolytic anemia, may also precede the onset of SLE. Patients with thrombocytopenia have a poor prognosis [198].

Splenectomy in ITP was originally thought to predispose to the development of SLE [201]. However, this hypothesis was refuted in subsequent studies [202].

Antibodies to Clotting Factors

Antibodies to a number of clotting factors, including VIII, IX, XI, XII, and XIII, have been noted in patients with SLE [171, 172, 189]. These antibodies may not only cause abnormalities of *in vitro* coagulation tests, but may also cause bleeding. Much more common and more important are antiphospholipid antibodies, the presence of which has been associated with a prolongation of the partial thromboplastin time and an increased risk of arterial and venous thrombosis, thrombocytopenia, and fetal loss.

RENAL FUNCTION

This subject is covered in Chapter 42. The author requests a routine urinanlysis most times when lupus patients are seen. In a patient without a history of renal disease, it should be done at least annually. For those patients with renal disease, routine urinanlysis should be done on a more routine basis, the frequency depending on the severity of the renal disease.

Likewise, a serum creatinine should be done at least on an annual basis. In someone with renal disease, it should be measured more frequently depending on the severity of the disease.

The author tends to do periodic glomerular filtration rate (GFR) (usually as creatinine clearances) in patients with renal disease to monitor their response to therapy as well as progression.

For patients with proteinuria, especially if they are clinically nephritic, the author monitors periodic 24-h urine for protein, although measuring and monitoring serum albumin is easier to do and is just as useful.

SYNOVIAL FLUID

Joint effusions are generally infrequent and small in volume. The synovial fluid viscosity is generally normal and mucin clots are normal [203]. The white blood cell count is usually under 2000 cells/mm^3 and is infrequently over 10,000. Most of the white blood cells tend to be mononuclear. Total protein tends to be elevated, but complement levels are usually depressed markedly, even more so than in blood, suggesting local activation of the complement system (the same is also found in rheumatoid arthritis). Sugar levels tend to be normal, in contrast to rheumatoid arthritis where they are often low. LE cells can be found but are not diagnostic. ANAs reflect diffusion from plasma.

PLEURAL FLUID

Pleural effusion in SLE is a mild exudate characterized by an elevation in pleural fluid LDH but not signs of marked inflammation. The findings differ from those in other conditions, particularly rheumatoid arthritis.

• The total white cell count (with a predominance of either lymphocytes or polymorphonuclear cells) is lower in lupus-related effusions [204, 205]. However, because there is substantial overlap, the white count in an individual patient cannot be used to establish the diagnosis.

- Pleural glucose levels in lupus effusions are slightly lower than serum blood levels; by comparison, pleural glucose levels in rheumatoid effusions are significantly reduced [204].
- Low complement levels characterize lupus and rheumatoid arthritis effusions.
- The protein concentration tends to be lower in lupus effusions than in patients with RA, reflecting their transudative as well as slight inflammatory nature.

Although the presence of rheumatoid factor or ANA in pleural fluid suggests RA and SLE, respectively, these findings provide no additional diagnostic information beyond that obtained from the measurement of these autoantibodies in serum and should therefore not be performed. Additional causes of pleural effusions, such as infection, congestive heart failure, and uremia, must also be excluded.

PERICARDIAL FLUID

The pericardial fluid is a fibrinous exudate or transudate that may contain antinuclear antibodies, LE cells, low complement levels, and immune complexes similar to those seen in lupus pleural effusions [206]. The glucose concentration is normal and the protein concentration is variable, being low with a transudate and elevated with an exudate. The pericardium may reveal foci of inflammatory lesions with immune complexes. There is usually a predominance of mononuclear cells, but scarring may be the primary finding in healed disease.

CEREBROSPINAL FLUID

Routine evaluation of the cerebrospinal fluid (CSF) is usually normal in patients with CNS lupus, except in cases of aseptic meningitis and transverse myelitis. Some reports, however, have noted some pleocytosis (usually mononuclear cells), as well as immunologic abnormalities, including elevated levels of total protein, anti-DNA antibodies, IgG, oligoclonal banding, immune complexes, and interleukin-6, which may roughly correlate with neuropsychiatric disease [207–209]. Other antibrain antibodies have been detected, but these assays should be considered strictly research. While considered useful in one paper, measuring complement levels has been shown not to be of any use clinically. CSF examination is performed primarily to exclude infection and/or bleeding.

Acknowledgment

Much of this chapter has been adapted from the relevant chapters in UPTODATE in Rheumatology.

References

[1] A. Kavanaugh, R. Tomar, J. Reveille, et al., Guidelines for clinical use of the antinuclear antibody test and tests for specific autoantibodies to nuclear antigens: American College of Pathologists, Arch Pathol Lab Med 124 (2000) 71–81.

[2] D. Sinclair, M. Saas, D. Williams, M. Hart, R. Goswami, Can an ELISA replace immunofluorescence for the detection of anti-nuclear antibodies? The routine use of anti-nuclear antibody screening ELISAs, Clin Lab 53 (2007) 183–191.

[3] E. Avaniss-Aghajani, S. Berzon, A. Sarkissian, Clinical value of multiplexed bead-based immunoassays for detection of auto-antibodies to nuclear antigens, Clin Vaccine Immunol 14 (2007) 505–509.

[4] M. Lopez-Hoyos, V. Rodriguez-Valverde, V. Martinez-Taboada, Performance of antinuclear antibody connective tissue disease screen, Ann N Y Acad Sci 1109 (2007) 322–329.

[5] P. Ghillani, A.M. Rouquette, C. Desgruelles, N. Hauguel, C. Le Pendeven, J.C. Piette, et al., Evaluation of the LIAISON ANA screen assay for antinuclear antibody testing in autoimmune diseases, Ann N Y Acad Sci 1109 (2007) 407–413.

[6] S.S. Copple, T.B. Martins, C. Masterson, E. Joly, H.R. Hill, Comparison of three multiplex immunoassays for detection of antibodies to extractable nuclear antibodies using clinically defined sera, Ann N Y Acad Sci 1109 (2007) 464–472.

[7] P. Caramaschi, O. Ruzzenente, S. Pieropan, A. Volpe, A. Carletto, L.M. Bambara, et al., Determination of ANA specificity using multiplexed fluorescent microsphere immunoassay in patients with ANA positivity at high titres after infliximab treatment: preliminary results, Rheumatol Int 27 (2007) 649–654.

[8] E. Bonilla, L. Francis, F. Allam, M. Ogrinc, H. Neupane, P.E. Phillips, et al., Immunofluorescence microscopy is superior to fluorescent beads for detection of antinuclear antibody reactivity in systemic lupus erythematosus patients, Clin Immunol 124 (2007) 18–21.

[9] R.E. Biagini, C.G. Parks, J.P. Smith, D.L. Sammons, S.A. Robertson, Analytical performance of the AtheNA MultiLyte ANA II assay in sera from lupus patients with multiple positive ANAs, Anal Bioanal Chem 388 (2007) 613–618.

[10] A.P. Nifli, G. Notas, M. Mamoulaki, M. Niniraki, V. Ampartzaki, P.A. Theodoropoulos, et al., Comparison of a multiplex, bead-based fluorescent assay and immunofluorescence methods for the detection of ANA and ANCA autoantibodies in human serum, J Immunol Methods 311 (2006) 189–197.

[11] O. Shovman, B. Gilburd, G. Zandman-Goddard, A. Yehiely, P. Langevitz, Y. Shoenfeld, Multiplexed AtheNA multi-lyte immunoassay for ANA screening in autoimmune diseases, Autoimmunity 38 (2005) 105–109.

[12] C. Gonzalez, B. Garcia-Berrocal, M. Perez, J.A. Navajo, O. Herraez, J.M. Gonzalez-Buitrago, Laboratory screening of connective tissue diseases by a new automated ENA screening assay (EliA Symphony) in clinically defined patients, Clin Chim Acta 359 (2005) 109–114.

[13] P. Eissfeller, M. Sticherling, D. Scholz, K. Hennig, T. Luttich, M. Motz, et al., Comparison of different test systems for

simultaneous autoantibody detection in connective tissue diseases, Ann N Y Acad Sci 1050 (2005) 327–339.

[14] J. Smith, D. Onley, C. Garey, S. Crowther, N. Cahir, A. Johanson, et al., Determination of ANA specificity using the UltraPlex platform, Ann N Y Acad Sci 1050 (2005) 286–294.

[15] T.B. Martins, R. Burlingame, C.A. von Muhlen, T.D. Jaskowski, C.M. Litwin, H.R. Hill, Evaluation of multiplexed fluorescent microsphere immunoassay for detection of autoantibodies to nuclear antigens, Clin Diagn Lab Immunol 11 (2004) 1054–1059.

[16] S. Bernardini, M. Infantino, L. Bellincampi, M. Nuccetelli, A. Afeltra, R. Lori, et al., Screening of antinuclear antibodies: comparison between enzyme immunoassay based on nuclear homogenates, purified or recombinant antigens and immunofluorescence assay, Clin Chem Lab Med 42 (2004) 1155–1160.

[17] H. Nossent, O.P. Rekvig, Antinuclear antibody screening in this new millennium: farewell to the microscope? Scand J Rheumatol 30 (2001) 123–126. discussion 127–128.

[18] E. Ulvestad, Performance characteristics and clinical utility of a hybrid ELISA for detection of ANA, Apmis 109 (2001) 217–222.

[19] E. Olaussen, O.P. Rekvig, Screening tests for antinuclear antibodies (ANA): selective use of central nuclear antigens as a rational basis for screening by ELISA, J Autoimmun 13 (1999) 95–102.

[20] H.A. Homburger, Y.D. Cahen, J. Griffiths, G.L. Jacob, Detection of antinuclear antibodies: comparative evaluation of enzyme immunoassay and indirect immunofluorescence methods, Arch Pathol Lab Med 122 (11) (1998 Nov) 993–999.

[21] R.A. Gniewek, D.P. Stites, T.M. McHugh, J.F. Hilton, M. Nakagawa, Comparison of antinuclear antibody testing methods: immunofluorescence assay versus enzyme immunoassay, Clin Diagn Lab Immunol 4 (2) (1997 Mar) 185–188.

[22] G. Shen, N. Lal, M. Patnaik, B. Dardashit, Comparison of EIA screen and IFA for antinuclear antibodies, Arthritis Rheum 43 (2000) S153.

[23] D.H. Solomon, A. Kavanaugh, P. Schur, R. Lahita, J.D. Reveille, Y. Sherrer, Evidence-based guidelines for the use of immunologic tests: Antinuclear antibody testing, Arthritis Rheum 47 (2002) 434–444.

[24] M. Petri, E.W. Karlson., D.S. Cooper, P.W. Ladenson, Autoantibody tests in autoimmune thyroid disease: A case-control study, J Rheumatol 18 (1991) 1529–1531.

[25] M.G. Tektonidou, M. Anapliotou, P. Vlachoyiannopoulos, H.M. Moutsopoulos, Presence of systemic autoimmune disorders in patients with autoimmune thyroid diseases, Ann Rheum Dis 63 (2004) 1159–1161.

[26] A.J. Czaja, M. Nishioka, S.A. Morshed, T. Hachiya, Patterns of nuclear immunofluorescence and reactivities to recombinant nuclear antigens in autoimmune hepatitis, Gastroenterology 107 (1994) 200–207.

[27] A.J. Czaja, F. Cassani, M. Cataleta, et al., Antinuclear antibodies and patterns of nuclear immunofluoresence in type 1 autoimmune hepatitis, Digestive Diseases & Sciences 42 (1997) 1688–1696.

[28] S.L. Taylor, P.J. Dean, C.A. Riely, Primary autoimmune cholangitis: An alternative to antimitochondrial antibody-negative primary biliary cirrhosis, Am J Surg Pathol 18 (1994) 91–99.

[29] S. Rich, K. Kieras, K. Hart, et al., Antinuclear antibodies in primary pulmonary hypertension, J Am Coll Cardiol 8 (1986) 1307–1311.

[30] M. Kaplan, E. Tan, Antinuclear antibodies in infectious mononucleosis, Lancet 1 (1968) 561–563.

[31] T. Burnham, Antinuclear antibodies in patients with malignancies, Lancet 2 (1972) 436.

[32] M. Seiner, E. Klein, G. Klein, Antinuclear reactivity or sera in patients with leukemia and other neoplastic diseases, Clin Immunol Immunopathol 4 (1975) 374–381.

[33] E.M. Tan, T.E. Feltkamp, J.S. Smolen, et al., Range of antinuclear antibodies in "healthy" individuals, Arthritis Rheum 40 (1997) 1601–1611.

[34] J.P. Buyon, E. Ben-Chetrit, S. Karp, et al., Acquired congenital heart block: Pattern of maternal antibody response to biochemically defined antigens of the SSA/Ro-SSB/La system in neonatal lupus, J Clin Invest 84 (1989) 627.

[35] L. Cook, New methods for detection of antinuclear antibodies, Clin Immunol Immunopathol 88 (1998) 211–220.

[36] A. Kavanaugh, D. Solomon, et al., Guidelines for immunological laboratory testing in the Rheumatic Diseases: Anti-DNA antibody tests, Arthritis Rheum 47 (2002) 546–555.

[37] B.H. Hahn, Antibodies to DNA, N Engl J Med 338 (1998) 1359–1368.

[38] R. Smeenk, G. Lily, L. Aarden, Avidity of antibodies to dsDNA: Comparison of IFT on Crithidia luciliae, Farr assay and PEG assay, J Immunol 128 (1982) 73–78.

[39] W. Crowe, I. Kushner, J.D. Clough, Comparison of the Crithidia luciliae, illipore filter, Farr and hemagglutination methods for detection of antibodies to DNA, Arthritis Rheum 21 (1978) 390–391.

[40] R. Cervera, M. Kharmasta, J. Font, et al., Systemic lupus erythematosus: Clinical and immunological patterns of disease expression in a cohort of 1,000 patients, Medicine (Baltimore) 72 (1993) 113–124.

[41] S. Ronsuke, M. Evans, N. Abou, Idiotypic and immunochemical differences of anti-DNA antibodies of lupus patients during active and inactive disease, Clin Immunol Immunopathol 61 (1991) 320–331.

[42] P.H. Schur, J. Sandson, Immunologic factors and clinical activity in systemic lupus erythematosus, N Engl J Med 278 (1968) 533–538.

[43] W. Lloyd, P. Schur, Immune complexes, complement and anti-DNA in exacerbations of systemic lupus erythematosus (SLE), Medicine (Baltimore) 60 (1981) 208–217.

[44] A.J. Swaak, J. Groenwold, W. Bronsveld, Predictive value of complement profiles and anti-dsDNA in systemic lupus erythematosus, Ann Rheum Dis 45 (1986) 359–366.

[45] E.J. Ter Borg, G. Horst, Hummel, et al., Measurement of increases in anti-double-stranded DNA antibody levels as a predictor of disease exacerbation in systemic lupus erythematosus, Arthritis Rheum 33 (1990) 634–643.

[46] D.D. Gladman, M.B. Urowitz, E.C. Keystone, Serologicially active clinically quiescent systemic lupus erythematosus. 31, Am J Med 66 (1979) 210–215.

[47] J. Winfield, C. Brunner, D. Koffler, Serologic studies in patients with systemic lupus erythematosus and central nervous system dysfunction, Arthritis Rheum 21 (1978) 289–294.

[48] D. Koffler, V. Agnello, R. Winschester, H.G. Kunkel, The occurrence of single-stranded DNA in the serum of patients with systemic lupus erythematosus and other diseases, J Clin Invest 52 (1973) 198–204.

[49] S. Miles, D.A. Isenberg, A review of serological abnormalities in the relatives of patients with systemic lupus erythematosus, Lupus 2 (1993) 145–150.

[50] E.M. Tan, H.G. Kunkel, Characteristics of a soluble nuclear antigen precipitating with sera of patients with systemic lupus erythematosus, J Immunol 96 (1966) 464–471.

[51] H. Bush, R. Reddy, L. Ruthblum, Y.C. Choy, SnRNAs, SnRNPs, and RNA processing, Annu Rev Biochem 51 (1982) 617–654.

[52] E. Benito-Garcia, P.H. Schur, R. Lahita, et al., Guidelines for immunologic laboratory testing in the rheumatic diseases: Anti-Sm and Anti-RNP antibody tests, Arthritis Rheum 51 (2004) 1030–1044.

[53] F.C. Arnett, R.G. Hamilton, M.G. Roebber, et al., Increased frequencies of Sm and nRNP auto-antibodies in American blacks compared to whites with systemic lupus erythematosus, J Rheumatol 15 (1988) 1773–1776.

[54] J. Sanchez-Guerro, R.A. Lew, A.H. Fossel, P.H. Schur, .Utility of anti-Sm, anti-RNP, anti-Ro/SS-A, and anti-La/SS-B (extractable nuclear antigens) detected by enzyme-linked immunosorbent assay for the diagnosis of systemic lupus erythematosus, Arthritis Rheum 39 (1996) 1055–1061.

[55] Reveille, J.D., Solomon, D.H., Schur, P., and Kavanaugh, A. (2010). Evidence based guidelines for the use of immunological laboratory tests: Anti-Ro (SS-A) and La (SS-B). in press.

[56] T.T. Provost, R. Watson, E. Simmons-O'Brien, Significance of the anti-Ro (SS-A) antibody in the evaluation of patients with cutaneous manifestations of a connective tissue disease, J Am Acad Dermatol 35 (1996) 147–169.

[57] G. Clark, M. Reichlin, T. Tomeri, Characterization of a soluble cytoplasmic antigen reactive with sera from patients with systemic lupus erythematosus, J Immunol 102 (1969) 117–122.

[58] M.A. Alspaugh, E.M. Tan, Antibodies to cellular antigens in Sjögren's syndrome, J Clin Invest 55 (1975) 1067–1073.

[59] M. Alspaugh, P. Maddison, Resolution of the identity of certain antigen-antibody systems in systemic lupus erythematosus and Sjögren's syndrome: An inter-laboratory collaboration, Arthritis Rheum 22 (1979) 796–798.

[60] S. Barakat, G. Meyer, F. Torterotot, et al., IgG antibodies from patients with primary Sjögren's syndrome and systemic lupus erythematosus recognize different epitopes in 60-Kd-A/Ro protein, Clin Exp Immunol 89 (1992) 38–45.

[61] P. Maddeson, H. Mogavero, T.T. Provost, et al., The clinical significance of autoantibodies to a soluble cytoplasmic antigen in systemic lupus erythematosus and other connective tissue diseases, J Rheumatol 6 (1979) 189–195.

[62] M.D. Lockshin, E. Bonja, K. Elkon, M.L. Druzen, Neonatal lupus risk to newborns of mothers with systemic lupus erythematosus, Arthritis Rheum 31 (1988) 697–701.

[63] J.P. Buyon, E. Ben-Chetrit, S. Karp, et al., Acquired congenital heart block: Pattern of maternal antibody response to biochemically defined antigens of the SSA/Ro-SSB/La system in neonatal lupus, J Clin Invest 84 (1989) 627–634.

[64] S. Garcia, J.H. Nascimento, E. Bonfa, et al., Cellular mechanism of the conduction abnormalities induced by serum from anti-Ro/SSA positive patients in rabbit heart, J Clin Invest 93 (1994) 718–724.

[65] P. Maddison, T. Provost, M. Reichlin, Serological findings in patients with "ANA-negative" systemic lupus erythematosus, Medicine 60 (1981) 87–94.

[66] S. Blomberg, L. Ronnblom, A.C. Wallgren, et al., Anti-SSA/Ro antibody determination by enzyme-linked immunosorbent assay as a supplement to standard immunofluorescence in antinuclear antibody screening, Scand J Immunol 51 (2000) 612–617.

[67] R.H. Scofield, A.D. Farris, A.C. Horsfall, J.B. Harley, Fine specificity of the autoimmune response to the Ro/SSA and La/SSB ribonucleoproteins, Arthritis Rheum 42 (1999) 199–209.

[68] J. Hardin, The lupus autoantigens and the pathogenesis of systemic lupus erythematosus, Arthritis Rheum 29 (1986) 457–460.

[69] R. Burlingame, R. Rubin, R. Balderas, A. Theofilopoulos, Genesis and evolution of antichromatin autoantibodies in murine lupus implicates T-dependent immunization with self antigen, J Clin Invest 91 (1993) 1687–1696.

[70] Z. Amoura, J.-C. Piette, J.-F. Bach, S. Koutouzov, The key role of nucleosomes in lupus, Arthritis Rheum 42 (1999) 833–843.

[71] H. Chabre, Z. Amoura, J. Piette, et al., Presence of nucleosome-restricted antibodies in patients with systemic lupus erythematosus, Arthritis Rheum 38 (1995) 1485–1492.

[72] M. Massa, F. de Benedetti, P. Pignatti, Antidouble-stranded DNA, antihistone and antinucleosome IgG reactivities in children with systemic lupus erythematosus, Clin Exp Rheumatol 12 (1994) 219–225.

[73] A. Souza, L.M. da Silva, F.R. Oliveira, A.M.F. Roslino, P. Louzada-Junior, Anti-nucleosome and anti-chromatin antibodies are present in active systemic lupus eryhthematosus but not in the cutaneous form of the disease, Lupus 18 (2009) 223–229.

[74] M. Tupchong, J. Wither, D. Hallett, D. Gladman, Anti-nucleosome antibodies as markers for renal disease in lupus nephritis (abstract), Arthritis Rheum 40 (1997) S304.

[75] Z. Amoura, S. Koutouzov, H. Chabre, et al., Presence of anti-nucleosome autoantibodies in a restricted set of connective tissue diseases:Antinucleosome antibodies of the IgG3 subclass are markers of renal pathogenicity in systemic lupus erythematosus, Arthritis Rheum 43 (2000) 76–84.

[76] G. Dreyfuss, M.J. Matunis, S. Pinol-Roma, C.G. Burd, hnRNP proteins and the biogenesis of mRNA, Annu Rev Biochem 62 (1993) 289–321.

[77] G. Schett, H. Dumortier, E. Hoefler, et al., B cell epitopes of the heterogeneous nuclear ribonucleoprotein A2: identification of a new specific antibody marker for active lupus disease, Ann Rheum Dis 68 (2009) 729–735.

[78] C. Montecucco, R. Caporali, C. Negri, et al., Antibodies from patients with rheumatoid arthritis and systemic lupus erythematosus recognize different epitopes of a single heterogeneous nuclear RNP core protein, Arthritis Rheum 33 (1990) 180–186.

[79] W. Hassfeld, G. Steiner, A. Sludnicka-Benke, et al., Auto-immune response to the spliceosome: An immunologic link between rheumatoid arthritis, mixed connective tissue disease, and systemic lupus erythematosus, Arthritis Rheum 38 (1995) 777–785.

[80] G. Biamonti, C. Ghigna, R. Caporali, C. Montecucco, Heterogeneous nuclear ribonucleoproteins (hnRNPs): An emerging family of autoantigens in rheumatic diseases, Clin Exp Rheum 16 (1998) 317–326.

[81] M.J. Fritzler, G.A. McCarty, J.P. Ryan, T.D. Kinsella, Clinical features of patients with antibodies directed against proliferating cell nuclear antigen, Arthritis Rheum 26 (1983) 140–145.

[82] G. Roos, G. Landberg, J.P. Huff, et al., Analysis of the epitopes of proliferating cell nuclear antigen recognized by monoclonal antibodies, Lab Invest 68 (1993) 204–210.

[83] K. Rayno, M. Reichlin, Evaluation of assays for the detection of autoantibodies to the ribosomal P proteins, Clin Immunol 95 (2000) 99–103.

[84] K.B. Elkon, A.P. Parnassa, C.L. Foster, Lupus autoantibodies target ribosomal P proteins, J Exp Med 162 (1985) 459–471.

[85] A.M. Francour, C.I. Peebles, K.J. Heckman, et al., Identification of ribosomal protein autoantigens, J Immunol 135 (1985) 2378–2384.

[86] E. Bonfa, K.B. Elkon, Clinical and serologic associations of the antiribosomal P protein antibody, Arthritis Rheum 29 (1986) 981–985.

[87] A.B. Schneebaum, J.D. Singleton, S.G. West, J.K. Blodgett, Association of psychiatric manifestations with antibodies to ribosomal P proteins in systemic lupus erythematosus, Am J Med 90 (1991) 54–62.

[88] D. Koffler, T.E. Miller, R.G. Lahita, Studies on the specificity and clinical correlation of antiribosomal antibodies in systemic lupus erythematosus sera, Arthritis Rheum 22 (1979) 463–470.

[89] E. Bonfa, S. Golombek, L. Kaufman, et al., Association between lupus psychosis and antiribosomal P protein antibodies, N Engl J Med 317 (1987) 265–271.

[90] T. Sato, T. Uchiumi, T. Ozawa, et al., Autoantibodies against ribosomal proteins found with high frequency in patients with systemic lupus erythematosus with active disease, J Rheumatol 18 (1991) 1681–1684.

[91] E. Koren, M.W. Reichlin, M. Koscec, et al., Autoantibodies to the ribosomal P proteins react with a plasma membrane-related target on human cells, J Clin Invest 89 (1992) 1236–1241.

[92] A. Van Dam, H. Nossent, J. de Jong, et al., Diagnostic value of antibodies against ribosomal phospho-proteins: A cross-sectional and longitudinal study, J Rheumatol 18 (1991) 1026–1034.

[93] Y. Nojima, S. Minota, A. Yamada, et al., Correlation of antibodies to ribosomal P protein with psychosis in patients with systemic lupus erythematosus, Ann Rheum Dis 51 (1992) 1053–1055.

[94] M. Reichlin, T.F. Broyles, O. Hubscher, et al., Prevalence of autoantibodies to ribosomal P proteins in juvenile-onset systemic lupus erythematosus compared with the adult disease, Arthritis Rheum 42 (1999) 69–75.

[95] L. Teh, A. Bedwell, D. Isenberg, et al., Antibodies to protein P in systemic lupus erythematosus, Ann Rheum Dis 51 (1992) 489–494.

[96] R.H. Derksen, A.P. van Dam, F.H. Gmelig Meyling, et al., A prospective study on antiribosomal P proteins in two cases of familial lupus and recurrent psychosis, Ann Rheum Dis 49 (1990) 779–782.

[97] M. Hulsey, R. Goldstein, L. Scully, et al., Antiribosomal P antibodies in systemic lupus erythematosus: A case-control study correlating hepatic and renal disease, Clin Immunol Immunopathol 74 (1995) 252–256.

[98] F.C. Arnett, M. Reichlin, Lupus hepatitis: An under-recognized disease feature associated with autoantibodies to ribosomal P, Am J Med 99 (1995) 465–472.

[99] V. Chindalore, B. Neas, M. Reichlin, The association between anti-ribosomal P antibodies and active nephritis in systemic lupus erythematosus, Clin Immunol Immunopathol 87 (1998) 292–296.

[100] A. Ghirardello, A. Doria, S. Zampieri, et al., Anti-ribosomal P protein antibodies detected by immunoblotting in patients with connective tissue diseases: Their specificity for SLE and association with IgG anticardiolipin antibodies, Ann Rheum Dis 59 (2000) 975–981.

[101] K. Miyachi, E. Tan, Antibodies reacting with ribosomal ribonucleoprotein in connective tissue diseases, Arthritis Rheum 22 (1979) 87–93.

[102] F.P. Quismorio, Other serologic abnormalities in SLE, in: D.J. Wallace, B.H. Hahn (Eds.), "Dubois' Lupus Erythematosus", 4th Ed., Lea & Febiger, Philadelphia, 1993, pp. 264–276.

[103] S. Karpatkin, P.H. Schur, N. Strick, G.W. Siskind, Heavy chain subclass of human antiplatelet antibodies, Clin Immunol Immunopathol 2 (1973) 1–8.

[104] R.H. Dixon, W.F. Rosse, Platelet antibody in auto-immune thrombocytopenia, Br J Hematol 31 (1975) 129–134.

[105] K. Hymes, P.H. Schur, S. Karpatkin, Heavy-chain subclass of bound antiplatelet IgG autoimmune hemolytic anemia, Blood 56 (1980) 84–87.

[106] L.R. Sammaritano, A.E. Gharavi, M.D. Lockshin, Antiphospholipid antibody syndrome: Immunologic and clinical aspects, Semin Arthritis Rheum 20 (1990) 81–96.

[107] S.A. Santoro, Antiphospholipid antibodies and thrombotic predisposition: Underlying pathogenetic mechanisms, Blood 83 (1994) 2389–2391.

[108] R.A. Roubey, Autoantibodies to phospholipid-binding plasma proteins: A new view of lupus anticoagulants and other "antiphospholipid" autoantibodies, Blood 84 (1994) 2854–2867.

[109] J. Rauch, Lupus anticoagulant antibodies: Recognition of phospholipid-binding protein complexes, Lupus 7 (Suppl. 2) (1998) S29–S31.

[110] L.R. Sammaritano, S. Ng, R. Sobel, et al., Anticardiolipin IgG subclasses: Association of IgG2 with arterial and/or venous thrombosis, Arthritis Rheum 40 (1997) 1998–2006.

[111] H.P. McNeil, C.N. Chesterman, S.A. Krilis, Immunology and clinical importance of antiphospholipid antibodies, Adv Immunol 49 (1991) 193–280.

[112] A. Melk, G. Mueller Eckhardt, B. Polten, et al., Diagnostic and prognostic significance of anticardiolipin antibodies in patients with recurrent spontaneous abortions, Am J Reprod Immunol 33 (1995) 228–233.

[113] R.A. Roubey, Immunology of the antiphospholipid antibody syndrome, Arthritis Rheum 39 (1996) 1444–1454.

[114] J.S. Ginsberg, P.S. Wells, P. Brill-Edwards, et al., Antiphospholipid antibodies and venous thromboembolism, Blood 86 (1995) 3685–3691.

[115] D.A. Triplett, J.T. Brandt, K.A. Musgrave, Relationship between lupus anticoagulants and antibodies to phospholipids, JAMA 259 (1988) 550–554.

[116] E. Cucurull, A.E. Gharavi, E. Diri, et al., IgA anticardiolipin and anti-beta(2)-glycoprotein I are the most prevalent isotypes in African American patients with systemic lupus erythematosus, Am J Med Sci 318 (1999) 55–60.

[117] H.M. Day, P. Thiagarajan, C. Ahn, et al., Autoantibodies to beta2-glycoprotein I in systemic lupus erythematosus and primary antiphospholipid antibody syndrome: Clinical correlations in comparison with other antiphospholipid antibody tests, J Rheumatol 25 (1998) 667–674.

[118] M. Galli, Should we include anti-prothrombin antibodies in the screening for the antiphospholipid syndrome? J Autoimmun 15 (2000) 101–105.

[119] M. Puurunen, O. Vaarala, H. Julkunen, K. Aho, et al., Antibodies to phospholipid-binding plasma proteins and occurrence of thrombosis in patients with SLE, Clin Immunol Immunopathol 80 (1996) 16–22.

[120] S. Shibata, P.C. Harpel, A. Gharavi, et al., Autoantibodies to heparin from patients with antiphospholipid syndrome inhibit formation of antithrombin III-thrombin complexes, Blood 83 (1994) 2532–2540.

[121] O. Vaarala, Antiphospholipid antibodies and atherosclerosis, Lupus 5 (1996) 442–447.

[122] M. Galli, Non beta 2-glycoprotein cofactors for antiphospholipid antibodies, Lupus 5 (1996) 388–392.

[123] R.H. Wu, S. Nityanand, L. Berglund, et al., Antibodies against cardiolipin and oxidatively modified LDL in 50-year-old men predict myocardial infarction, Arterioscler Thromb Vasc Biol 17 (1997) 3159–3163.

[124] A. Satoh, K. Suzuki, E. Takayama, et al., Detection of anti-annexin IV and V antibodies in patients with antiphospholipid syndrome and systemic lupus erythematosus, J Rheumatol 26 (1999) 1715–1720.

[125] K.S. Song, Y.S. Park, H.K. Kim, Prevalence of anti-protein S antibodies in patients with systemic lupus erythematosus, Arthritis Rheum 43 (2000) 557−560.

[126] F.J. Munoz-Rodriguez, J.C. Reverter, J. Font, et al., Prevalence and clinical significance of antiprothrombin antibodies in patients with systemic lupus erythematosus or with primary antiphospholipid syndrome, Haematologica 85 (2000) 632−637.

[127] M.G. Tektonidou, C.A. Petrovas, J.P.A. Ioannidis, et al., Clinical importance of antibodies against platelet activation factor in antiphospholipid syndrome manifestations, Eur J Clin Invest 30 (2000) 646−652.

[128] M. Cugno, M. Dominguez, M. Cabibbe, et al., Antibodies to tissue-type plasminogen activator in plasma from patients with primary antiphospholipid syndrome, Br J Haematol 108 (2000) 871−875.

[129] E. Gromnica-Ihle, W. Schlosser, Antiphospholipid syndrome, Int Arch Allergy Immunol 123 (2000) 67−76.

[130] R.A. Fields, H. Toubbeh, R.P. Searles, A.D. Bankhurst, The prevalence of anticardiolipin antibodies in a healthy elderly population and its association with antinuclear antibodies, J Rheumatol 16 (1989) 623−625.

[131] P. Vila, M.C. Hernandez, M.F. Lopez-Fernandez, J. Battle, Prevalence, follow-up and clinical significance of the anti-cardiolipin antibodies in normal subjects, Thromb Haemost 72 (1994) 209−213.

[132] M.N. Manoussakis, A.G. Tzioufas., M.P. Silis, et al., High prevalence of anticardiolipin and other autoantibodies in a healthy elderly population, Clin Exp Immunol 68 (1987) 557−565.

[133] W. Shi, S.A. Krilis, B.H. Chong, et al., Prevalence of lupus anticoagulant in a healthy population: Lack of correlation with anticardiolipin antibodies, Aust NZ J Med 20 (1990) 2319236.

[134] P.E. Love, S.A. Santoro, Antiphospholipid antibodies: Anti-cardiolipin and the lupus anticoagulant in systemic lupus erythematosus (SLE) and non-SLE disorders, Ann Intern Med 112 (1990) 682−698.

[135] M. Abu-Shakra, D. Gladman, M.B. Urowitz, V. Farewell, Anticardiolipin antibodies in systemic lupus erythematosus: Clinical and laboratory correlations, Am J Med 99 (1995) 624−626.

[136] G.D. Sebastiani, M. Galeazzi, A. Tincani, et al., Anti-cardiolipin and anti-beta2GPI antibodies in a large series of European patients with systemic lupus erythematosus: Prevalence and clinical associations, Scand J Rheumatol 28 (1999) 344−351.

[137] U. Picillo, S. Migliaresi, M.R. Marcialis, et al., Clinical signifi-cance of anticardiolipin antibodies in patients with systemic sclerosis, Autoimmunity 20 (1995) 1−7.

[138] K. Chakravarty, G. Pountain, P. Merry, et al., A longitudinal study of anticardiolipin antibody in polymyalgia rheumatica and giant cell arteritis, J Rheumatol 22 (1995) 1694−1697.

[139] A. Ruffatti, C. Veller-Fornasa, G.M. Patrassi, et al., Anti-cardiolipin antibodies and antiphospholipid syndrome in chronic discoid lupus erythematosus, Clin Rheumatol 14 (1995) 402−404.

[140] P.E. Carreira, M.G. Montalvo, L.D. Kaufman, et al., Anti-phospholipid antibodies in patients with eosinophilia myalgia and toxic oil syndrome, J Rheumatol 24 (1997) 69−72.

[141] M. Vayssairat, N. Abuaf, N. Baudot, et al., Abnormal IgG cardiolipin antibody titers in patients with Raynaud's phenomenon and/or related disorders: Prevalence and clinical significance, J Am Acad Dermatol 38 (1998) 555−558.

[142] D.A. Triplett, Many faces of lupus anticoagulants, Lupus 7 (Suppl. 2) (1998) S18−S22.

[143] T. McNally, G. Purdy, I.J. Mackie, et al., The use of anti-beta 2-glycoprotein-I assay for discrimination between anticardiolipin antibodies associated with infection and increased risk of thrombosis, Br J Haematol 91 (1995) 471−473.

[144] J.T. Merrill, C. Shen, M. Gugnani, et al., High prevalence of antiphospholipid antibodies in patients taking procainamide, J Rheumatol 24 (1997) 1083−1088.

[145] D.J. Wallace, Serum and plasma protein abnormalities and other clinical laboratory determinations in SLE, in: D.J. Wallace, B.H. Hahn (Eds.), "Dubois' Lupus Eryth-ematosus", 4th Ed., Lea & Febiger, Philadelphia, 1993, pp. 457−460.

[146] J.T. Cassidy, R.K. Kitson, C.L. Selby, Selective IgA deficiency in children and adults with systemic lupus erythematosus, Lupus 16 (2007) 647−650.

[147] P.A. Gatenby, The role of complement in the aetiopathogenesis of systemic lupus erythematosus, Autoimmunity 11 (1991) 61−66.

[148] J.M. Porcel, D. Vergani, Complement and lupus: Old concepts and new directions, Lupus 1 (1992) 343−349.

[149] M.C. Pickering, M.J. Walport, Links between complement abnormalities and systemic lupus erythematosus, Rheuma-tology 39 (2000) 133−141.

[150] P.H. Schur, Complement testing in the diagnosis of immune and autoimmune disease, Am J Clin Pathol 68 (1977) 647−659.

[151] M. Oppermann, U. Hopken, O. Gotze, Assessment of complement activation in vivo, Immunopharmacology 24 (1992) 119−134.

[152] J.M. Porcel, M. Peakman, G. Senaldi, D. Vergani, Methods for assessing complement activation in the clinical immunology laboratory, J Immunol Methods 157 (1993) 1−9.

[153] M.J. Walport, Advances in Immunology: Complement, N Engl J Med 344 (2001) 1056−1066. 1140-1144.

[154] S. Manzi, J.S. Navratil, M.J. Ruffing, et al., Measurement of erythrocyte C4d and complement receptor 1 in systemic lupus erythematosus, Arthritis Rheum 50 (2004) 3596−3604.

[155] G. Sturfelt, L. Truedsson, Complement and its breakdown products in SLE, Rheumatologu 44 (2005) 1227−1232.

[156] S. Ruddy, C.B. Carpenter, K.W. Chin, J.N. Knostman, N.A. Soter, O. Gotze, et al., Human complement metabolism: An analysis of 144 studies, Medicine 54 (1975) 165−178.

[157] H.R. Colten, Biosynthesis of complement, Adv Immunol 22 (1976) 67−118.

[158] L. Truedsson, A.A. Bengtsson, G. Sturfelt, Complement defi-ciencies an systemic lupus erythematosus, Autoimmunity 40 (2007) 560−566.

[159] I. Kushner, The phenomenon of the acute phase response, Ann NY Acad Sci 389 (1982) 39−48.

[160] C. Gabay, I. Kushner, Acute-phase proteins and other systemic responses to inflammation, N Engl J Med 340 (1999) 448−454.

[161] W.S. Tillet, T.J. Francis, Serological reactions in pneumonia with a non-protein somatic fraction of pneumococcus, J Exp Med 52 (1930) 561−571.

[162] J.J. Morley, I. Kushner, Serum C-reactive protein levels in disease, Ann NY Acad Sci 389 (1982) 406−418.

[163] S.E. Bedell, B.T. Bush, Erythrocyte sedimentation rate: From folklore to facts, Am J Med 78 (1985) 1001−1009.

[164] M.H. Wener, P.R. Daum, G.M. McQuillan, The influence of age, sex, and race on the upper reference limit of serum C-reactive protein concentration, J Rheumatol 27 (2000) 2351−2359.

[165] D.J. Wallace, Serum and plasma protein abnormalities and other clinical laboratory determinations in SLE, in: D.J. Wallace, B.H. Hahn (Eds.), "Dubois' Lupus

Erythematosus", 4th Ed., Lea and Febiger, Philadelphia, 1993, pp. 457–460.

[166] M.B. Pepys, J.G. Lanham, F.C. De Beer, C-reactive protein in SLE, Clin Rheum Dis 8 (1982) 91–103.

[167] E.J. ter Borg, G. Horst, P.C. Limburg, et al., C-reactive protein levels during disease exacerbations and infections in systemic lupus erythematosus: A prospective longitudinal study, J Rheumatol 17 (1990) 1642–1648.

[168] H.M. Moutsopoulos, A.K. Mavridis, N.C. Acritidis, P.C. Avgerinos, High C-reactive protein response in lupus polyarthritis, Clin Exp Rheumatol 1 (1983) 53–55.

[169] F.C. De Beer, R.K. Mallya, E.A. Fagan, et al., Serum amyloid-A protein concentration in inflammatory diseases and its relationship to the incidence of reactive systemic amyloidosis, Lancet 320 (1982) 231–234.

[170] E. Beyan, C. Beyan, M. Turan, Hematological presentation in systemic lupus erythematosus and its relationship with disease activity, Hematology 12 (2007) 257–261.

[171] R. Simantov, J. Laurence, R. Nachman, The Cellular hematology of systemic lupus erythematosus, in: R.G. Lahita (Ed.), "Systemic Lupus Erythematosus", 4th Ed., Churchill Livingstone, New York, 2004, pp. 1019–1036.

[172] Y. Shoenfeld, M. Ehrenfeld, Hematologic manifestations. In, in: P.H. Schur (Ed.), "The Clinical Management of Systemic Lupus Erythematosus", 2nd Ed., Lippincott, Philadelphia, 1996, pp. 95–108.

[173] J.C. Nossent, A.J. Swaak, Prevalence and significance of haematological abnormalities in patients with SLE. Q.J, Med 80 (1991) 605–612.

[174] D.M. Keeling, D.A. Isenberg, Haematological manifestations of SLE, Blood Rev 7 (1993) 199–207.

[175] H. Liu, K. Ozaki, Y. Matsuzaki, et al., Suppression of haematopoiesis by IgG autoantibodies from patients with systemic lupus erythematosus (SLE), Clin Exp Immunol 100 (1995) 480–485.

[176] J. Cavalcant, R.K. Shadduck, A. Winkelstein, et al., Red-cell hypoplasia and increased bone marrow reticulin in systemic lupus erythematosus: Reversal with corticosteroid therapy, Am J Hematol 5 (1978) 253–263.

[177] M. Voulgarelis, S.I. Kokori, J.P. Ioannidis, et al., Anaemia in systemic lupus erythematosus: Aetiological profile and the role of erythropoietin, Ann Rheum Dis 59 (2000) 217–222.

[178] A.G. Tzioufas, S.I. Kokori, C.I. Petrovas, H.M. Moutsopoulos, Autoantibodies to human recombinant erythropoietin in patients with systemic lupus erythematosus: correlation with anemia, Arthritis Rheum 40 (1997) 2212–2216.

[179] G. Schett, U. Firbas, W. Fureder, et al., Decreased serum erythropoietin and its relation to antierythropoietin antibodies in anaemia of systemic lupus erythematosus, Rheumatology (Oxford) 40 (2001) 424–431.

[180] G.S. Habib, W.R. Saliba, P. Froom, Pure red cell aplasia and lupus, Seminars in Arthritis Rheumatism 31 (2002) 279–283.

[181] A. Winkler, R.W. Jackson, D.S. Kay, et al., High dose intravenous cyclophosphamide treatment of systemic lupus erythematosus associated aplastic anemia, Arthritis Rheum 31 (1988) 693–694.

[182] B.J. Brooks Jr., H.E. Broxmeyer, C.F. Bryan, S.H. Leech, Serum inhibitor in systemic lupus erythematosus associated with aplastic anemia, Arch Intern Med 144 (1984) 1474–1477.

[183] C. Roffe, M.R. Cahill, A. Samanta, et al., Aplastic anemia in systemic lupus erythematosus: A cellular immune mechanism? Br J Rheumatol 30 (1991) 301–304.

[184] S.I. Kokori, J.P. Ioannidis, M. Voulgarelis, et al., Autoimmune hemolytic anemia in patients with systemic lupus erythematosus, Am J Med 108 (2000) 198–204.

[185] S. Duran, M. Apte, G.S. Alarcon, et al., Features associated with, and the impact of, hemolytic anemia in patients with systemic lupus erythematosus: LX, results from a multiethnic cohort, Arthritis Rheum 59 (2008) 1332–1340.

[186] S.I. Kokori, J.P. Ioannidis, M. Voulgarelis, et al., Autoimmune hemolytic anemia in patients with systemic lupus erythematosus, Am J Med 108 (2000) 198–204.

[187] G. Nesher, V.E. Hanna, T.L. Moore, et al., Thrombotic microangiopathic hemolytic anemia in systemic lupus erythematosus, Semin Arthritis Rheum 24 (1994) 165–172.

[188] A.A. Shah, J.P. Higgins, E.F. Chakravarty, Thrombotic microangiopathic hemolytic anemia in a patients with SLE, Nature Clinical Practice Rheumatology 3 (2007) 357–362.

[189] D.R. Budman, A.D. Steinberg, Hematologic aspects of systemic lupus erythematosus: Current concepts, Ann Intern Med 86 (1977) 220–229.

[190] L.M. Vita, G.S. Alarcon, G. McGwin Jr., et al., Systemic lupus erythematosus in a multiethnic cohort, XXXVII: association of lymphopenia with clinical manifestations, serologic abnormalities, disease activity, and damage accrual, Arthritis Rheum 55 (2006) 799–806.

[191] H.D. Perez, M. Lipton, I.M. Goldstein, A specific inhibitor of complement (C5)-derived chemotactic activity in serum from patients with systemic lupus erythematosus, J Clin Invest 62 (1978) 29–38.

[192] S.B. Abramson, W.P. Gren, H.S. Eddson, G. Weissman, Neutrophil aggregation induced by sera from patients with active systemic lupus erythematosus, Arthritis Rheum 26 (1983) 630–636.

[193] D.T. Boumpas, G.P. Chrousos, R.L. Wilder, et al., Glucocorticoid therapy for immune-mediated diseases: Basic and clinical correlates, Ann Intern Med 119 (1993) 1198–1208.

[194] S.J. Rivero, E. Diaz-Jouanen, D. Alarcon-Segovia, Lymphopenia in systemic lupus erythematosus, Arthritis Rheum 21 (1978) 295–305.

[195] Y. Amasaki, S. Kobayashi, T. Takeda, et al., Up-regulated expression of Fas antigen (CD 95) by peripheral naive and memory T cell subsets in patients with SLE: A possible mechanism for lymphopenia, Clin Exp Immunol 99 (1995) 245–250.

[196] D.A. Isenberg, K.G. Patterson, A. Todd-Pokropek, et al., Haematological aspects of SLE: A reappraisal using automated methods, Acta Haematol 67 (1982) 242–248.

[197] G. Camussi, C. Tetta, R. Coda, J. Benveniste, Release of platelet-activating factor in human pathology. I. Evidence for the occurrence of basophil degranulation and release of platelet-activating factor in systemic lupus erythematosus, Lab Invest 44 (1981) 241–251.

[198] Fernandez, G.S. Alarcon, M. Apte, et al., Systemic lupus erythematosus in a multiethnic US cohort: XLIII. The significance of thrombocytopenia as a prognostic factor, Arthritis Rheum 56 (2007) 614–621.

[199] M. Pujol, A. Ribera, M. Vilardell, et al., High prevalence of platelet autoantibodies in patients with SLE, Br J Haematol 89 (1995) 137–141.

[200] S. Karpatkin, Autoimmune thrombocytopenic purpura, Blood 56 (1980) 329–343.

[201] W. Dameshek, W.H. Reeves, Exacerbation of lupus erythematosus following splenectomy in "idiopathic" thrombocytopenic purpura and autoimmune hemolytic anemia, Am J Med 21 (1956). 560–566.

[202] Y.N. You, A. Tefferi, D.M. Nagorney, Outcome of splenectomy for thrombocytopenia associated with systemic lupus erythematosus, Annals of Surgery 240 (2004) 286–292.

[203] C.M. Ziminski, Musculoskeletal manifestations. in: P.H. Schur (Ed.), "The Clinical Management of Systemic Lupus", 2nd Ed., Lippincott, Philadelphia, 1996, pp. 47–65.

[204] J.M. Pego-Reigosa, D.A. Medeiros, D.A. Isenberg, Respiratory manifestations of systemic lupus erythematosus: old and new concepts, Best Practice & Research in Clinical Rheumatology 23 (2009) 469–480.

[205] B. Memet, E.M. Ginzler, Pulmonary manifestations of systemic lupus erythematosus, Seminars in Respiratory & Critical Care Medicine 28 (2007) 441–450.

[206] G.G. Hunder, B.J. Mullen, F.C. McDuffie, Complement in pericardial fluid of lupus erythematosus, Ann Intern Med 80 (1974) 453–458.

[207] S.G. West, W. Emlen, M.H. Wener, B.L. Kotzin, Neuropsychiatric lupus erythematosus: A 10-year prospective study on the value of diagnostic tests, Am J Med 99 (1995) 153–163.

[208] S. Hirohata, T. Miyamoto, Elevated levels of interleukin-6 in cerebrospinal fluid from patients with systemic lupus erythematosus, Arthritis Rheum 33 (1990) 644–649.

[209] B.N. McLean, D. Miller, E.J. Thompson, Oligoclonal banding of IgG in CSF, blood-brain barrier function, and MRI findings in patients with sarcoidosis, SLE, and Behcet's disease involving the nervous system, J Neurol Neurosurg Psych 58 (1995) 548–554.

Pregnancy and Reproductive Concerns in Systemic Lupus Erythematosus

Carl A. Laskin, Karen A. Spitzer, Christine A. Clark

INTRODUCTION

Systemic lupus erythematosus (SLE) is a multisystem, autoimmune disease predominantly affecting women in their child-bearing years, with a female-to-male ratio of 9:1. It is therefore not surprising that those physiological states associated with hyperestrogenicity or other changes in sex hormone levels may affect disease status. The hormonal changes associated with pregnancy may indeed have a profound influence on lupus disease activity [1, 2]. Management of pregnant women with SLE presents a challenge to all attending physicians whether rheumatologists, general internists, medical subspecialists, or obstetricians. A coordinated management plan is essential for attending to both the medical and obstetrical needs of the patient. Ideally a woman with lupus should be evaluated prior to a pregnancy to ensure that there is no medical contraindication. On too many occasions, a patient may first present in a medical office pregnant either with active disease or on unsafe medications. Critical decisions must be made at these times that may be difficult emotionally and ethically.

In addition to pregnancy, other issues that are relevant to the woman with SLE include contraception and the investigation and management of infertility. This chapter will concentrate on pregnancy issues but will also address these broader issues concerning reproduction.

PHYSIOLOGICAL CHANGES RELEVANT TO WOMEN WITH UNDERLYING SLE

Pregnancy is accompanied by major physiological changes, which may adversely affect the woman with underlying SLE. In addition, such changes will affect numerous clinical and laboratory parameters complicating the assessment of disease activity in the pregnant woman with SLE. Among those systemic changes relevant to the woman with SLE are: cardiovascular with potential cardiac, renal, and hypertensive complications; hematologic including changes in blood volume and control mechanisms of coagulation such as changes in platelet counts, clotting factor levels, and regulatory proteins; immunologic including the regulation of both humoral and cellular immunity, complement kinetics, and the development of immune complexes; and hormonal with changing levels of estrogen, progesterone, prolactin, and protein-hormone binding globulins [3].

Cardiovascular

Associated with pregnancy are dramatic changes in maternal hemodynamics due to increases in blood volume, heart rate, stroke volume, and a decrease in systemic vascular resistance [3, 4]. Cardiac output increases by 30−50% with a 20% increase in heart rate and a 5- to 10-mmHg decrease in mean arterial pressure in pregnancy. The lupus patient with underlying cardiac, renal, or other vascular complications of the disease may be at risk of further major organ deterioration during pregnancy.

Hematologic

The hematologic changes that occur normally in pregnancy compound the changes in the cardiovascular system. Plasma volume increases 30−50% beyond that seen in the nonpregnant state with the red cell mass increasing 20−40% [5, 6, 7]. Again, such changes may not bode well for the lupus patient with cardiovascular or renal compromise. Other alterations include an increase in white blood cell count and a mild decrease in platelet count [8, 9]. Although the WBC increases during pregnancy, the absolute lymphocyte count does not change. In contrast, changes in platelet counts may manifest as asymptomatic thrombocytopenia in the third trimester in many normal pregnancies. These changes observed in the CBC may confound the assessment of disease activity in the pregnant woman with SLE.

It is well appreciated that pregnancy is a prothrombotic state due to increases in the levels of several clotting factors (factors II, VII, VIII, X, and von Willebrand); a decrease in the levels or activity of naturally occurring anticoagulants such as Protein S as well as an increase in activated Protein C resistance; and an increase in the activity of fibrinolytic inhibitors [10–14]. Owing to these changes in pregnancy, there may be an increased theoretical risk of thrombotic events in those with antiphospholipid antibodies. Supporting data regarding such an increased risk is still lacking.

Immunologic

During pregnancy, alterations in immune function occur affecting lymphocyte function, humoral immunity, and the inflammatory response. Awareness of these changes is of paramount importance in the assessment of a patient with known SLE. Human pregnancy is associated with an increase in immune suppressor activity, leading to a decrease in humoral B cell function. In lupus there is dysregulation of the immune response, leading to an impairment of suppressor activity, resulting in polyclonal B cell activation.

Pregnancy-specific proteins such as α-fetoprotein, β_1-glycoprotein, α_2-macroglobulin, and others suppress lymphocyte function, as do increases in endogenous corticosteroids. Although their roles are not completely understood, interleukins-1 and -3 (IL-1, IL-3), tumor necrosis factor α (TNF-α), interferon-γ (IFN-γ), granulocyte–macrophage colony-stimulating factor (GM-CSF), and other cytokines may be critical in sustaining pregnancy [15]. This is relevant in that the production or metabolism of some cytokines is thought to be abnormal in rheumatic illness.

Therefore, the superimposition of pregnancy on the immunoregulatory abnormalities present in a woman with lupus will alter the immune environment. The interaction of these two altered immune states must be appreciated in order to account for changes in disease activity.

Hormonal

Estrogen, progesterone, prolactin, and other hormone levels rise in pregnancy [16]. Estrogens upregulate and androgens downregulate T-cell responses

and immunoglobulin synthesis [17]. IL-1, IL-2, IL-6, and TNF-α can be regulated by sex hormones. Since circulating and local levels of cytokines may be abnormal in SLE, it is likely that these hormonally induced alterations in immune function may affect disease activity in the pregnant woman with SLE.

THE IMMUNOLOGY OF IMPLANTATION, PREGNANCY, AND LABOR AND DELIVERY

The fundamental theory of the establishment of pregnancy has been the Th-1/Th-2 paradigm [1]. Helper T-cell clones can be divided into two phenotypes that secrete a distinct cytokine pattern [18]. During pregnancy, there is an overall suppression of Th-1-mediated cellular immunity and an enhancement of Th-2-mediated humoral immunity. The maintenance of pregnancy requires the downregulation of the pro-inflammatory Th-1 cytokines (TNF-α, IFN-γ, and IL-2) and the upregulation of anti-inflammatory Th-2 cytokines (IL-4, IL-6, and IL-10). During embryo implantation, uterine epithelial cells surrounding the blastocyst undergo apoptosis. These apoptotic cells are then taken up by macrophages found in excess at the implantation site. The uptake of the apoptotic cells into the macrophages promotes the secretion of Th-2 cytokines and suppresses the release of Th-1 cytokines including TNF-α from the macrophages. The Th-2-rich anti-inflammatory environment surrounds and protects the developing embryo. Therefore, pregnancy appears to be a Th-2 phenomenon. Recent data, however, suggest that this paradigm may be overly simplistic. Although TNF-α and IFN-γ have been implicated in failed implantation or early pregnancy loss, it now appears that small quantities of these cytokines may be necessary for the successful implantation of the embryo. Moreover, it may be that Th-1 activity both accompanies and predominates over Th-2-mediated events during the early implantation period, and premature and term labor. Th-1 activity plays an important role in the promotion of the Th-2 response, regulation of the placentation process, defense against infections, and initiation of delivery. The new paradigm should more properly be thought of as "Th-1–Th-2 cooperation" rather than a Th-2-dominant phenomenon.

THE PLACENTA

The placenta is often the target organ culminating in adverse pregnancy outcome in women with SLE. Abnormalities in placentation, uteroplacental vascular insufficiency, and placental infarction may be the underlying cause of intrauterine growth restriction (IUGR), preeclampsia, and preterm delivery often described as complicating factors in lupus pregnancies [19, 20]. Therefore, an understanding of normal placental physiology and morphology is necessary to appreciate the pathological changes that may occur in the placenta during pregnancy.

The placenta brings maternal and fetal blood in close relationship. The placenta is partly of fetal origin (the trophoblast) and partly of maternal origin (transformation of the uterine mucosa). The fetal trophoblast ensures implantation. It is composed of two layers: the inner cellular layer (cytotrophoblast) and the exterior syncytial layer (syncytiotrophoblast). The placenta is well defined by the 3rd month. At term it weighs 500 g and is 20 cm in diameter. The human placenta is villous, hemochorial, and chorioallantoic, meaning that the placental villi are bathed directly in maternal blood and are traversed by vessels coming from the allantoic circulation of the fetus. While maintenance of pregnancy prior to implantation is assured by both ovarian and pituitary hormones, after implantation the placenta assumes an important role in hormone production: chorionic gonadotrophins are discernible several days after nidation, peak at the 60th day, and then fall to low levels until completion of pregnancy [21].

MANAGEMENT OF PREGNANT WOMEN WITH SLE

General Principles

In any woman with an underlying medical problem, pregnancy should be planned in advance to minimize the risks to mother and fetus, thereby maximizing the probabilities of a successful outcome. A prepregnancy medical and obstetrical evaluation represents the ideal management plan. In the case of rheumatic diseases, specifically SLE, a rheumatologist should assess the patient when a pregnancy is being considered. The objective of the assessment is to note the manifestations of the underlying disease, the past history of exacerbations and past and current medications. With this approach, the attending physician will then be able to document: (1) the clinical profile; (2) the laboratory profile; (3) the frequency and pattern of the most recent disease flare; and (4) the current medications.

- Clinical profile. This profile is constructed from the history and physical examination. In addition, the past history is utilized to record how the disease initially presented and the usual manifestations of a flare.
- Laboratory profile. The laboratory tests that best reflect the patient's disease status are used as

indicators of disease activity during a pregnancy. The profile is constructed from the current laboratory tests and any relevant serology; as well as noting which laboratory variables are associated with disease activity.
- Most recent disease flare. It is important to determine and document the most recent exacerbation of the disease as well as the severity. The timing of a pregnancy will determine the safety of such an undertaking. Appropriate counseling of the patient can then be undertaken. This information should be incorporated into the patient's clinical profile.
- Medications. A detailed history of current medications is essential to provide appropriate counseling to the patient regarding the timing of a pregnancy. Many antirheumatic drugs are unsafe in pregnancy, and because many are also long-acting, they may have to be discontinued months prior to conception.

The Prepregnancy Evaluation

Women with underlying SLE should consider a planned pregnancy to ensure their safety as well as the best outcome for themselves and their newborn. Whenever possible, women with SLE contemplating pregnancy should be evaluated prior to conception to determine disease status and activity. The clinician should determine the patient's clinical and laboratory profile, date of most recent flare of the disease, and current medications. Using a methodical approach, the physician may grasp the subtle nuances of the underlying disease, permitting better distinction of disease activity and the physiological or pathophysiological changes associated with pregnancy.

Pregnancy in any woman with an underlying medical problem should be managed by both an internist/subspecialist and obstetrician/perinatologist. In the case of SLE, that internist should be a rheumatologist or at the very least an internist familiar with the management of lupus. The prepregnancy evaluation should form the basis of a recommendation to the patient regarding the medical management plan. Appropriate counseling regarding the potential effect of the disease on the pregnancy and neonate, and similarly the effect of the pregnancy on the disease should be provided.

The prepregnancy evaluation must include a detailed history, noting presenting and usual manifestations of the disease, which will assist in distinguishing a disease flare from other pregnancy-related complications. The timing of the most recent flare as well as documenting a frequency of flare pattern will indicate the safety of undertaking a pregnancy at a particular time. The current medications as well as a recent medication history are essential as many drugs used to treat SLE are relatively or absolutely contraindicated in pregnancy and some require a specific washout period of time prior to conception. The physical examination must be documented, especially any abnormalities to ensure an accurate baseline clinical evaluation prior to pregnancy. Laboratory investigations should include baseline CBC, liver function tests, renal function, and serology. A baseline urine evaluation is essential and should include not only a dipstick with microscopic evaluation of the sediment but also a protein/creatinine ratio on a spot sample or 24-h collection. A recommendation and plan should be discussed with the patient and relayed to all physicians involved in the management. Even if the opinion is not to undertake pregnancy at this time, the physician should present to the patient the reasons for the negative recommendation as well as the goals that need to be achieved before a favorable recommendation can be given. Booking a reassessment visit may prevent the occurrence of an unplanned pregnancy by a woman who simply decides to proceed despite the risks.

Once the decision has been made to proceed with a pregnancy, the management plan should be discussed with the patient. This will involve presenting the risks of potential problems, proper use of medications and their safety, frequency of visits and monitoring specifics, as well as a recommendation for obstetrical management. For individuals with complicated but stable disease, who will undertake a pregnancy in the near future, a pre-pregnancy obstetrical assessment should be considered. The ideal obstetrician is a perinatologist familiar with lupus in pregnancy. The perinatologist should be informed of the medical management plan and the obstetrical and medical prenatal assessments should be coordinated to minimize the inconvenience to the patient and maximize the lines of communication among all involved.

Prenatal Assessments

The first prenatal visit will include a complete history, physical exam, and laboratory assessment, which will be compared to that obtained at the prepregnancy evaluation. This will serve as the baseline for future assessments during the pregnancy, noting any clinical or lab evidence consistent with disease activity. The frequency of visits should be at minimum once per trimester in those cases of full remission with minor organ disease. In those women with a history of major organ disease or any evidence of active lupus, the frequency should be at least once monthly or coincide with obstetrical visits. Where the patient shows any medical instability, visits will be more frequent or the patient may require hospital admission. Regular, frequent medical assessments are the norm to minimize the probability of an unexpected disease exacerbation.

The Effect of Pregnancy on SLE

In all pregnant women with an underlying medical problem, the physician must view the situation as the effect of the pregnancy on the disease and the effect of the disease on the pregnancy. The former scenario is more medical, whereas the latter scenario is related to things obstetrical. This is the exact situation in SLE and all physicians involved in patient management must be cognizant of this "two-way street."

There is some consistency in the response of SLE to pregnancy, but some patients do not "behave" as expected. As early as 1952, it was noted that some women with SLE flare during pregnancy [22]. Other studies have also noted an increase in flares during pregnancy [23–26]. The more recent literature, however, notes that the frequency of disease exacerbation during pregnancy and postpartum is less than that reported earlier [27, 28]. In a case–control prospective study comparing pregnant and nonpregnant women with similar manifestations of SLE, Lockshin et al. found no increase in flares during pregnancy [29, 30]. In contrast, Petri concluded that pregnancy was associated with an increased rate of disease flares in her population with a frequency of 1.63 per person years compared to 0.64–0.65 in a postpartum group or in nonpregnant controls [31]. Ruiz-Irastorza et al. observed findings similar to Petri with a 65% flare rate during pregnancy compared to 42% in the control group [32]. Studies by Cortes-Hernandez et al. found that 33% of their lupus patients flared during pregnancy, with 26% in the second trimester and 51% postpartum [33]. The major predictors of a flare were the discontinuation of antimalarial treatment, a history of more than three flares before the pregnancy, and a SLEDAI (Systemic Lupus Erythematosus Disease Activity Index) score of 5 or more during these flares. In one study of 46 women with SLE who underwent 61 pregnancies, Urowitz et al. observed no increased frequency of lupus flares, using the SLEDAI, during pregnancy compared with controls [34]. Indeed there was a reduced chance of flare during a pregnancy if the patient had inactive disease for 6 months prior to conception. Clearly there is a lack of consensus among these studies. This may be due to dissimilar entry criteria, differing definitions of a flare, distinct patient populations, and differing control groups [31, 32]. Despite the lack of consensus regarding the relationship between SLE flare and pregnancy, it is clear that many patients do well during pregnancy and that vigilance for flare on the part of the treating physician is always necessary. Modifications have been suggested and tested for measuring SLE flare during pregnancy using the SELENA SLEDAI (SLEP-DAI, systemic lupus erythematosus pregnancy disease activity index), the LAI (LAI-P, lupus activity index pregnancy), or the SLAM-R (m-SLAM, modified systemic lupus activity measure) [35, 36].

The incidence of adverse pregnancy outcomes in women with SLE is increased. In a retrospective analysis, 555 women with SLE had an adverse outcome apart from manifestations of SLE compared to a group of 600,000 controls [37]. These outcomes included hypertension, renal disease, preterm delivery, nonelective cesarean section, postpartum hemorrhage, and delivery-related deep venous thrombosis. Clark *et al.* noted 38.9% of women with SLE had a preterm delivery (before 37 weeks gestation) in a group of 72 pregnancies [38]. The observation of preterm delivery was associated with disease activity and the presence of IgG anticardiolipin antibody.

Renal Disease

Renal function may decline during pregnancy in those with renal disease, but it is determined, not surprisingly, by the severity of the underlying renal dysfunction. A permanent decline will occur in up to 10% of women with glomerular filtration rates initially normal or mildly reduced (serum creatinine less than 132 µmol/l). Patients with hypertension are much more likely to have progressive disease regardless of pregnancy status [39–42]. In those with moderate renal insufficiency (serum creatinine, 132–255 µmol/l), the serum creatinine declines over the first half of pregnancy similar to that in those without renal disease, but then may increase significantly higher and above that observed at the onset of pregnancy [43, 44]. This decline in renal function may be irreversible, especially in those with hypertension [45].

Any renal disease, including lupus glomerulonephritis, predisposes patients to preeclampsia. There is some disagreement whether patients with renal SLE do less well than those with other forms of glomerulonephritis. In general, those with preserved renal function and minimal hypertension do well. Even those with severe renal disease often have surprisingly good pregnancy outcomes. Normal urinary protein in pregnancy may be as high as 500 mg/day. Modest increases in urinary protein (particularly in patients with fixed proteinuria), accompanied by proportional increases in glomerular filtration in mid- to late pregnancy and not accompanied by other signs of active lupus, likely do not indicate lupus nephritis and need not be treated. Our experience leads us to expect an increase in proteinuria after 20 weeks in almost all patients who have had any renal involvement. The increase in proteinuria is related to the normal increase in plasma volume seen in pregnancy, which appears to overwhelm the filtering capacity of the previously damaged glomeruli. In patients with marked increases in urinary protein, with or without hypertension, it may be impossible to distinguish between preeclampsia and active lupus nephritis, and may therefore be necessary to treat if delivery is not likely in the immediate future. Isolated hypocomplementemia may suggest increased vigilance but by itself does not warrant intervention.

Renal disease may flare during a pregnancy, which may respond to treating with corticosteroids or increasing the dosage of administered steroids [32]. Studies have been inconsistent in finding deterioration in renal function associated with pregnancy [27, 37, 46–50]. Tozman *et al.* noted no recurrence of renal disease in 11 of 18 patients with similar findings by Jungers *et al.* and Huong *et al.* All noted that the best prognosis was associated with remission of the disease at pregnancy onset. Furthermore, they all noted a higher risk of preeclampsia and premature birth than expected. In contrast, others noted an increase of renal flare in pregnancy [48, 49]. These authors and others also conclude that the only predictor of a favorable maternal outcome in a pregnancy is quiescence of renal disease [34].

Renal Disease and Preeclampsia

Preeclampsia is a syndrome occurring after 20 weeks gestation. The first manifestations are hypertension and proteinuria, but other clinical features may include headache, visual disturbances, epigastric pain, thrombocytopenia, and abnormalities in liver function tests [51]. The underlying pathogenesis is a microangiopathy affecting the brain, liver, kidney, and placenta [52]. The condition often precipitates preterm delivery owing to fetal compromise characterized by IUGR due to placental insufficiency. The inciting event actually occurs early in pregnancy with abnormalities in the development of the placental vasculature leading to underperfusion. The hypoperfusion then leads to the production of antiangiogenic factors, which upon release into the maternal circulation, adversely affect endothelial cell function, resulting in the manifestations noted above.

Preeclampsia is not uncommon in lupus pregnancies occurring in approximately 13% of such patients [53]. In those with underlying renal disease, the risk of developing preeclampsia may be much higher with a reported frequency that may be as high as 66% [54]. Superimposed on the increased risk associated with SLE and renal disease is an even greater risk in those patients with antiphospholipid antibodies (aPL), diabetes mellitus, or preeclampsia having occurred in a prior pregnancy [55].

A major diagnostic challenge to the clinician is to distinguish preeclampsia from a lupus flare during a pregnancy. Defining the clinical and laboratory profile of the patient before pregnancy will often assist in distinguishing these two conditions. If there are clinical features of active lupus and positive serology, this is likely a lupus flare. Depressed complement levels are characteristic of active lupus, whereas elevated complement levels are seen in pregnancy and usually do not change in preeclampsia. Although proteinuria may be seen both in preeclampsia and lupus nephritis, the presence of active sediment is a feature of active nephritis and not preeclampsia. Thus, a simple microscopic evaluation of the urine may be extremely informative, although it is acknowledged that Class V membranous glomerulonephritis can be associated with a totally benign sediment. Elevated liver function tests and uric acid with thrombocytopenia and decreased urinary excretion of calcium are characteristic of preeclampsia, whereas thrombocytopenia alone or, if there is renal insufficiency, elevated serum urate can be seen in active SLE. Knowledge of your patient and how her disease has manifested itself in the past will assist immeasurably in correctly assessing this difficult diagnostic situation.

Hematologic Disease

THROMBOCYTOPENIA

Thrombocytopenia is common during a pregnancy, especially in women with underlying SLE. There are a number of causes of thrombocytopenia, which may be unrelated to the underlying autoimmune disease. As is the case of any pregnant woman with a low platelet count, a methodical approach to the differential diagnosis must be undertaken lest the clinician erroneously attribute the thrombocytopenia to active lupus. Management should be directed to the underlying cause without simply assuming lupus to be the culprit. Platelet counts in normal pregnancies may be toward the lower limit of normal but in general, they remain within the normal range throughout pregnancy [56—59].

Before concluding that thrombocytopenia is indeed present, be certain that the finding is not merely a spurious lab result due to EDTA-induced platelet aggregation. Once ruled out, the differential diagnosis should be undertaken in the pregnant woman with underlying SLE with consideration of the following possibilities:

1. Gestational thrombocytopenia is typically observed late in gestation with counts $70-100 \times 10^9/l$. The condition is benign without fetal/neonatal thrombocytopenia, and resolves after delivery [60—62]. Should this occur in a woman with SLE, close follow-up is necessary to ensure the platelet count stabilizes at a safe level and is not a manifestation of active lupus.

2. Thrombocytopenia occurring after 25 weeks is most often due to preeclampsia. Platelet counts are lower in preeclampsia than that seen in gestational thrombocytopenia, with an incidence of 15% in women who develop preeclampsia. Preeclampsia tends to worsen as pregnancy progresses, is associated with worsening fetal health, and remits after delivery, although not always immediately. It is not associated with other signs or symptoms of active SLE. Biochemical and clinical manifestations of preeclampsia or the HELLP syndrome (hemolysis, elevated liver enzymes, low platelet count) may be present. Treatment directed at early delivery is usually indicated. Since preeclampsia occurs with greater frequency in women with SLE, the clinician can expect thrombocytopenia more often in the pregnant woman with SLE. Distinction between preeclampsia and a lupus flare may be problematic but it is an important distinction to make nonetheless. The clinical and lab assessment as discussed above is necessary to help clarify the diagnosis.

3. Thrombocytopenia associated with an antiphospholipid antibody can occur at any time but is often seen early, usually before 15 weeks. The platelet count will reach a nadir greater than $50 \times 10^9/l$ and remain constant throughout most of pregnancy. The condition typically remits after delivery, whether or not treatment is administered, and tends to recur with subsequent pregnancies. It is unaccompanied by signs of SLE activity in other organ systems; fragmented erythrocytes and increased lactic dehydrogenase (LDH) do not occur. Aspirin may reverse the thrombocytopenia [63].

4. Severe thrombocytopenia (often $<10 \times 10^9/l$) similar to that seen in idiopathic thrombocytopenic purpura may occur in lupus pregnancies but is more likely in those women with a history of thrombocytopenia as a major manifestation of SLE. When it occurs, it is often without other manifestations of active SLE and may occur at any time during the pregnancy. It will usually respond to high doses of prednisone and may require continuation of these higher doses. Intravenous gamma globulin can also be used as a temporizing measure, particularly to avoid splenectomy. The condition usually resolves postpartum but tends to recur in subsequent pregnancies. It is noteworthy that there is a 5—10% risk of neonatal thrombocytopenia (an example of passively acquired autoimmunity) in those born to mothers having had episodes of platelet counts below $50 \times 10^9/L$ during the pregnancy with a greater risk if the mother has had a splenectomy [64—66]. Aspirin is contraindicated in those women with this type of thrombocytopenia.

5. In those women with other manifestations of SLE, thrombocytopenia may occur, usually in the moderate range of $50-120 \times 10^9/l$. These counts are characterized by persistence over time and

responsiveness to prednisone. The low platelet count will be observed in those women where this is a known manifestation of lupus outside of the pregnant state.

No specific test clearly differentiates one type of thrombocytopenia from another in pregnant patients with SLE. Unfortunately, tests for platelet-associated immunoglobulin or for antiplatelet antibodies seldom help differentiate among the types of thrombocytopenia.

ANEMIA

Anemia is a common complication of SLE. The anemia "of chronic disease" is the most frequent attributed cause. A true autoimmune hemolytic anemia is uncommon, but does occur, with the risk being higher in those in whom there is a history of this being a manifestation of lupus [67]. SLE-induced anemia amplifies the normal dilutional anemia of pregnancy and, if severe, may indicate intervention using corticosteroids or blood transfusion.

Cutaneous Disease

Normal pregnancy erythema, particularly of the face and hands, resembles that seen in active SLE. Skin areas where there has been prior lupus eruption frequently become more erythematous as cutaneous blood flow increases in pregnancy. With experience, distinguishing between active SLE cutaneous disease and pregnancy-induced change is not difficult.

Arthritis

Joints previously affected by lupus arthritis often become painful and may even develop noninflammatory effusions during pregnancy, particularly late in gestation as ligamentous laxity occurs. The decision that a given complaint is due to active SLE rather than to physiologic changes of pregnancy depends on the demonstration that the arthritis is inflammatory. If extra-articular manifestations of active SLE are present, it is likely that the articular effusion represents active disease. If warranted, arthrocentesis may be informative with regard to leukocyte count and culture since there is no contraindication to this procedure during pregnancy.

Neurologic SLE

Neurologic events during the course of pregnancy are rare. There are isolated reports but no cogent evidence that SLE neurologic events, including chorea and transverse myelitis, are induced or exacerbated by pregnancy [68–70]. There are also occasional reports of SLE patients suffering stroke postpartum, usually in the setting of coexisting antiphospholipid antibodies.

One of the more difficult diagnostic decisions concerns the occurrence of seizures late in pregnancy, when hypertension and renal failure are also present. The circumstances in which the events occur, the presence of other clinical and serologic evidence of active SLE, and response to therapy may all be required to make the distinction between active neurologic lupus and eclampsia. Treatment for both may be indicated.

The Effect of SLE on Pregnancy

Adverse pregnancy outcome is more common in SLE than in any other rheumatic disease. Appropriate pre-pregnancy evaluation and counseling will maximize the probability of a successful outcome for both the mother and the neonate. Maintaining the viability of the pregnancy requires close collaboration between the obstetrician/perinatologist and the internist/rheumatologist.

The incidence of fetal wastage in SLE pregnancies has been reported in the past to be as high as 50%, including spontaneous abortion (miscarriage), prematurity, and stillbirth [23, 26, 31–34, 71–73]. A more recent analysis of long-term data over the past 40 years noted a decline in the spontaneous abortion rate from 50% to less than 20% [74]. Those risk factors associated with adverse outcomes included antiphospholipid antibodies, hypocomplementemia, and hypertension during pregnancy [33]. In addition, persistent, significant proteinuria (>0.5 g/day) would also likely contribute to adverse pregnancy and maternal outcomes. Although some studies report that infants born to mothers with SLE are small for gestational age, this has not been a consistent observation, even in cases where placental size is reduced.

The increased frequency of fetal loss in SLE may be due to several factors: (1) active lupus resulting in decidual vasculitis, which in turn compromises placental blood flow depriving the fetus; (2) trophoblast-reactive lymphocytotoxic antibodies; (3) anti-Ro/SSA or anti-La/SSB antibodies with their associated compromise of the fetal cardiac conduction system; and (4) antiphospholipid antibodies with resulting placental vascular thrombosis and insufficiency culminating in ischemic pregnancy loss.

Antiphospholipid Antibodies and Pregnancy

Antiphospholipid antibodies (aPL) in low to moderate titer may be seen in 40–60% of lupus patients with active disease. The aPL family of antibodies includes IgG or IgM anticardiolipin antibodies (aCL) and a nonspecific in vitro inhibitor of coagulation often referred to as the lupus anticoagulant (LAC). Although the mere presence of aPL in a woman with lupus may not be associated with any particular manifestation, there may be an increased risk of adverse pregnancy outcome in such individuals. A review of ten studies comprising 554 women with SLE observed fetal loss more frequently in the presence of aPL (39–59%) compared to those without aPL (16–20%) [75]. In addition, aPL have been associated with preeclampsia and placental abruption [76–79]. Women with SLE and aPL and the associated clinical manifestations including thromboembolism and pregnancy wastage, are classified as secondary antiphospholipid syndrome (APS). Those with only aPL and an associated clinical feature are classified as primary APS since they lack any other feature of SLE.

LAC and aCL have often been used interchangeably to indicate the presence of antiphospholipid antibodies. While there is certainly a correlation between these two antibodies, they are not identical. Lockshin *et al.* reported that while both the LAC and aCL were associated with fetal loss, higher levels of aCL appear to be more predictive of fetal distress or fetal death among pregnant women with SLE [79]. However, other observations by Clark *et al.* and Salmon *et al.* have shown that the LAC may have a higher association with pregnancy loss and adverse outcome than aCL [80, 81]. The risk of fetal loss in women with circulating aPL has lately become controversial. Although some studies suggest that the presence of aCL or features of the antiphospholipid syndrome (APS) are associated with increased risk of pregnancy loss in women with SLE, others have not found this to be the case [73, 82–89]. Questions have arisen regarding the association of aPL with early vs. late pregnancy loss; the significance of IgM and IgA isotypes of aCL [90], and whether the presence of LAC or aCL in women with no history of thromboembolism or adverse pregnancy outcome is sufficient indication for intervention.

Treatment of pregnant women with APS, whether primary or secondary, has also become somewhat controversial. Although initial studies supported the use of prednisone and aspirin to promote live birth in a woman with a history of pregnancy loss and aPL, a double-blind, randomized controlled trial failed to show a benefit beyond placebo [91, 92]. Heparin (both unfractionated and more recently, low molecular weight) and aspirin have become the accepted treatment for the prevention of pregnancy loss in women with APS [93–98], although studies supporting the use of such therapy prior to 2000 contrast with those conducted after that date where ASA alone was found to be at least as efficacious as heparin with ASA [99–100].

Fetal and Neonatal Considerations

In general, term births from women with SLE are at no greater risk of congenital anomalies than those born to mothers without SLE. The major exception is those babies born to mothers possessing anti-Ro (SSA) and/or anti-La (SSB) antibodies. Although abnormalities occur in only 1–2% of the neonates [101, 102], the manifestations form collectively the neonatal lupus syndrome. Transient abnormalities include skin lesions, liver disease, and cytopenias. Each generally cleared by 6–8 months of life, coincident with the clearance of the maternal antibodies. However, cardiac anomalies including conduction defects and more rarely myocardial involvement are permanent, with the latter often fatal. The antibody may be found in up to 25–40% of women with SLE but may also be an isolated finding in the general population. Once a woman with anti-Ro/La antibodies has given birth to an infant with congenital heart block, the risk in a subsequent pregnancy is about 17% [103]

Maternal IgG-mediated thrombocytopenia may be transmitted to the fetus, but most infants born of thrombocytopenic mothers with SLE have normal platelet counts. Occasionally the IgG Coombs hemolytic antibody may be transmitted to and cause hemolysis in the fetus and newborn. The anti-dsDNA antibody, also transmitted to the infant, has no apparent pathologic effect. Recent studies on a subset of anti-dsDNA antibodies, anti-NR2, which cross-react with neuronal tissue, destroy such tissue by excitotoxicity and apoptosis [104, 105]. These maternal antibodies may cross the placenta and result in cognitive dysfunction in the offspring of mothers with the autoantibody. Although aPL have been reported to be associated with placental insufficiency, intrauterine growth restriction, and fetal death, they do not usually cause abnormalities in the infant.

Summary of the Management of SLE During Pregnancy

Ideally, lupus should be inactive or under excellent control on minimal therapy for at least 6 months prior to pregnancy. Should the disease flare during pregnancy, treatment must be instituted immediately using the safest, most effective regimen. Prednisone (being non-fluorinated and thereby limiting its ability to cross the placenta owing to inactivation by placental 11-beta dehydrogenase) has few adverse effects on the fetus and should be used if necessary. Other drugs with very good safety profiles in SLE include some nonsteroidal anti-inflammatory drugs (naproxen, ibuprofen), antimalarials, and azathioprine. Indications for the treatment of the pregnant patient with lupus do not differ from that when considering treatment in the nonpregnant woman. Clinical disease flares should be treated with the same dose of corticosteroids regardless of pregnancy status.

Monitoring

Both the obstetrician and internist should concurrently follow pregnant women with SLE. Efforts should be coordinated to avoid duplication of laboratory tests and inappropriate, overly liberal consultation with other medical specialties. Maternal clinical evaluation in addition to appropriate laboratory testing will determine disease activity. If the patient's serology is concordant with disease activity, then rising anti-dsDNA antibody and falling complement levels will be the markers of a disease flare. Be aware that complement levels in pregnancy are typically elevated, so a falling level should be noted, not just a low level. Validated measures of disease activity usually restricted to research protocols can be adapted for use in the clinic setting [106, 107]. Echocardiograms for fetuses of anti-Ro antibody-positive mothers should be initiated between 16 and 18 weeks with frequencies determined by the managing physicians but often are recommended once weekly [101].

Labor and Delivery

The obstetrician should make the decision regarding the mode of delivery. Most patients with SLE will deliver vaginally. Corticosteroid supplementation should be administered at labor or prior to a cesarean section in women currently or recently using such medications. Of note, there is no rationale for prophylactically increasing the dose of corticosteroids to prevent a postpartum flare of the disease [106].

Postpartum

Some women may flare postpartum, but it is usually those who had active disease at conception with continued activity throughout the pregnancy [108]. There is no indication for the routine use of adding or increasing corticosteroids just prior to delivery as a means of preventing an exacerbation of disease. The risk of a postpartum flare of lupus is approximately 20%, but the vast majority of patients remain as they were during the pregnancy. However, it is recommended that all patients be assessed shortly after delivery (4–6 weeks postpartum) and reassessed periodically over the following 3 months. Treatment of those women with active lupus postpartum is the same as in the case of nonpregnant women with SLE. However, if treatment is required and the woman is breast-feeding, it may be necessary for her to discontinue depending upon medications necessary to manage the disease.

Breast-Feeding

The major concern in nursing the neonate is exposure to certain medications that may enter the breast milk. The underlying disease per se is not an issue in terms of the safety of the infant. If the disease is active, caution must be exercised due to some medications with which the mother may be treated. Prednisone (up to 30–40 mg/day), hydroxychloroquine, and short-acting nonsteroidal agents (naproxen, ibuprofen) are considered compatible with breast-feeding.

PHARMACOLOGIC TREATMENT OF SLE IN PREGNANCY

The assessment of any pregnant patient with rheumatic disease or those contemplating pregnancy must include consideration of current medications and, in some cases, previous exposure to specific pharmacologic agents. Although most drugs may be safe while the couple is attempting to conceive, some must be discontinued during pregnancy and yet others require discontinuation some time prior to conception. Under ideal conditions, medications should be discontinued prior to pregnancy but not at the expense of the patient's well-being. An understanding of the woman's medical condition including disease status and any compromised organ function such as renal insufficiency must be taken into consideration when deciding upon the use of specific therapeutic agents. Moreover, the physician must have extensive knowledge of specific drugs or have knowledge of the resources available to obtain such information when recommending implementation or discontinuation of specific drugs.

This section will address those agents commonly used in the treatment of SLE and its various manifestations. As evidence accumulates, recommendations regarding the safe use of particular drugs in pregnancy will change. Therefore, it is incumbent upon the clinician to be current with respect to the safe use of any of these agents in pregnancy.

Nonsteroidal Anti-inflammatory Drugs (NSAIDs)

NSAIDs are commonly used in the treatment of the arthritis manifested in many rheumatic diseases. Many are now available over the counter, making it of paramount importance that the obstetrician and internist be aware of all medications consumed by the patient whether prescribed or self-administered.

Most NSAIDs are safe to use for couples attempting to conceive. However, for women, some agents can inhibit follicular rupture, preventing the release of the oocyte and thereby contributing to subfertility [109, 110]. This is not a common phenomenon, occurring in about 10% of women taking such agents (personal observations). In addition, NSAIDs may inhibit the motility of the fallopian tubes and by extension, the passage of the oocyte down the tubes [111, 112]. Although these issues might arise with the use of any

NSAID due to the inhibition of cyclooxygenase, the most widely studied agent is indomethacin [113, 114].

During pregnancy, naproxen and ibuprofen are the two most commonly used NSAIDs. Indeed, there is too little in the literature regarding any other drug in this class other than the contraindicated indomethacin to have any knowledge regarding their safety profiles. When considering naproxen and ibuprofen in pregnancy, the fetal risk category is B but is reclassified as C when used in high doses. Peripartum there is concern with respect to the neonate regarding intracranial hemorrhage, premature closure of the ductus arteriosus, and impaired renal function leading to a decrease in amniotic fluid volume. The patient should be informed that NSAIDs will start being tapered by week 25 and completely discontinued by week 32 at the latest (6—8 weeks prior to the expected date of delivery).

A population-based study in Denmark indicated that there might be some association with the use of NSAIDs and early pregnancy loss [115]. However, the study fails to indicate the reason for the use of the drugs or if there was an underlying disorder that might predispose to pregnancy loss. It is interesting to note that in women with rheumatoid arthritis, in whom the use of NSAIDs would be quite extensive, there does not appear to be any increased incidence of pregnancy loss. Clearly, further studies need to be undertaken to support or refute the Danish observation.

Aspirin (ASA) has become a commonly used drug in pregnancy [116]. Despite a continuing lack of consensus, it has been variously recommended as a pretreatment for ovulation induction, for the promotion of implantation in *in vitro* fertilization cycles, and for the prevention of pregnancy loss [117—119]. It has been shown to provide moderate but consistent reductions in the relative risk of preeclampsia and preterm delivery [120]. ASA appears to be a safe agent to use in pregnancy with the exception of a reported questionable increased incidence of gastroschisis in the offspring of women taking ASA in the first trimester [121, 122]. Regardless of the very low frequency of this side effect (1/3000 compared to 1/10,000 in the general population), patients should still be counseled regarding the potential association prior to initiating ASA therapy during pregnancy. ASA is also capable of prolonging labor and may cause an increase in antepartum and postpartum bleeding. It is actually for the latter reasons that it has an FDA classification of C/D when used in the third trimester. To avoid the issues surrounding labor and delivery, discontinuation of ASA 4 weeks prior to the expected date of delivery should be implemented.

The American Academy of Pediatrics (AAP) considers ASA, naproxen, and ibuprofen compatible with breast-feeding but as always, the lowest effective dose should be used.

Anticoagulants

The patient with APS and previous thrombosis should receive anticoagulation treatment throughout their pregnancy and during the postpartum period [123]. Data supporting the use of heparin treatment in women with pregnancy loss and APL with no history of thrombosis are inconclusive and evidence-based standards of care both during the peripartum and postpartum periods are lacking [124]. Once the decision to use heparin has been made, low-molecular-weight heparin (LMWH) is perceived to be more desirable than unfractionated heparin (UFH), despite a Cochrane Review recommending the use of unfractionated heparin [125]. LMWH carries less risk of osteoporosis and thrombocytopenia and can be administered once daily [126]. Two small studies comparing UFH and LMWH did not show differences in efficacy but larger prospective studies are needed [127, 128]. LMWH treatment over the duration of a pregnancy can be associated with osteopenia and therefore calcium and vitamin D supplementation should be recommended during pregnancy [129, 130].

Antimalarial Agents

Antimalarials are used extensively in the treatment of SLE. The major side effect is retinal toxicity, which requires ophthalmologic monitoring every 6—12 months. The favored antimalarial is hydroxychloroquine (FDA Category C), which appears to have a lower incidence of retinal toxicity.

Numerous studies have attested to the safety of antimalarials in pregnancy [131—133]. All of these investigators have commented that the risk of a flare of disease far outweighs any risk of fetal toxicity. In a follow-up study of the children born to mothers on hydroxychloroquine during pregnancy, Klinger *et al.* found no evidence of retinal toxicity, prompting these authors to conclude that there appears to be little or no risk of ocular toxicity in children exposed to hydroxychloroquine *in utero* [134]. This observation has been confirmed by Motta *et al.* [135].

Although hydroxychloroquine is eliminated slowly and theoretically could accumulate in neonatal tissues, only 2% of the maternal dose can be detected in breast milk [131]. The AAP have therefore determined that hydroxychloroquine is compatible with breast-feeding.

Corticosteroids

Corticosteroids are commonly used in the treatment of most rheumatic diseases and are associated with a rapid or relatively rapid therapeutic response. In SLE, prednisone doses in treating minor organ disease range from 10 to 40 mg/day and major organ manifestations may be treated with 40–80 mg/day. In all cases, the lowest effective dose of prednisone should be used.

The use of prednisone in pregnancy is associated with few adverse side effects on the fetus. Maternal side effects are dose-related. The commonest side effects are hypertension and gestational diabetes mellitus. In a double-blind, randomized controlled trial, Laskin et al. observed an incidence of gestational diabetes mellitus of 15% and hypertension 13% in the prednisone-treated group compared to 5% in the placebo group for either condition [92]. The cushingoid side effects and osteopenia occur similarly to that seen in the nonpregnant state, the latter necessitating appropriate supplementation with calcium and vitamin D.

Fetal side effects are few and uncommon. Orofacial clefting in the offspring of mothers treated with corticosteroids during the first trimester has been reported [136, 137]. The results in these studies have been supported by the findings in a meta-analysis where the prevalence of orofacial clefting in prednisone-exposed infants was 1 in 400 compared to 1 in 800 in the general population [138]. In spite of the low risk of this potential side effect, any pregnant woman on corticosteroids should be counseled appropriately.

Premature birth has been described in pregnant women treated with corticosteroids. In a randomized trial referred to above, Laskin et al. found premature births before 37 weeks gestation in 62% of the prednisone-treated group compared to 11% in the placebo group [92]. The neonates were all appropriate size for gestational age. Prednisone is FDA classified as D when used in the first trimester. The physician must weigh potential risks vs. benefits when prescribing these agents.

The AAP considers prednisone to be compatible with breast-feeding. There appears to be minimal exposure with maternal doses at 30–40 mg/day.

Azathioprine

Azathioprine (AZA) is used in many rheumatic diseases for its immunosuppressive properties and as a steroid-sparing agent. Among all immunosuppressive agents, AZA appears to be the safest in pregnancy. The placenta reportedly forms a relative barrier to AZA and its metabolites [139]. In one study of 189 women exposed to AZA compared to 230 controls not exposed to any teratogens during pregnancy, outcomes associated with AZA included increased rates of spontaneous abortions, intrauterine growth restriction (IUGR), and prematurity [140], but no increase in the occurrence of major malformations. However, as larger studies are required to confirm these results, AZA continues to be a Class D drug.

With few data available regarding the safety of AZA in breast-feeding, the AAP has recommended that mothers avoid nursing while being treated with this agent.

Methotrexate

Methotrexate (MTX), a folic acid antagonist, has been used with good success in rheumatoid arthritis and is occasionally used in the treatment of SLE. Use in pregnancy has been associated with spontaneous abortions due to embryotoxicity. The drug has definite association with numerous fetal anomalies as well as IUGR [141–143]. MTX is not to be used in pregnancy and has an FDA Category X rating.

Owing to MTX binding to tissues, it is recommended that it be discontinued at least 3 months prior to conception. A similar recommendation applies to men taking MTX, but evidence is lacking to support such a recommendation [142]. However, until the situation is clarified, it is recommended that both men and women avoid pregnancy for at least 3 months after discontinuing MTX [141–143].

MTX is only excreted in breast milk to a very small degree. However, it binds to neonatal tissues and therefore accumulates, leading to toxicity [143]. The AAP categorizes MTX as contraindicated in nursing mothers.

Cyclophosphamide

Cyclophosphamide (CTX) is a cytotoxic, alkylating agent used in the treatment of severe major organ involvement in SLE. The FDA has categorized it as a Class D drug. CTX is embryotoxic and associated with many anomalies upon exposure in the first trimester. However, it does not appear to be associated with abnormalities if used in the second and third trimesters. Regardless, the drug should not be used in pregnancy unless there is a life-threatening problem and even then restricted to use late in the pregnancy. CTX is contraindicated in the nursing mother owing to the risk of neutropenia, immunosuppression, growth disturbances, and potential carcinogenesis in the neonate [143–145].

Cyclosporine

Cyclosporine A (CSA, FDA Classification C) is used to treat certain manifestations of SLE renal disease. Most of the literature surrounding the use of CSA in pregnancy deals with renal transplantation. There appears to be little evidence that CSA crosses the placenta significantly [146–148] and current data indicate that it is probably safe in pregnancy with no specific anomalies described [149–151]. Adverse pregnancy outcomes such as prematurity and IUGR likely have more to do with the underlying disease process than treatment with CSA [152].

CSA is excreted in breast milk and is associated with immunosuppression, neutropenia, and growth disturbances in the neonate. The AAP categorizes CSA as contraindicated in the nursing mother.

Mycophenolate Mofetil

Mycophenolate mofetil (MMF) is a purine biosynthesis inhibitor. Its use in pregnancy is accompanied by several congenital anomalies and spontaneous abortions [149, 153–157]. These findings have been noted not only in animal studies but also in humans. In addition, a possible characteristic phenotype has been described [154, 158]. Although MMF does not appear to impact male fertility, paternal exposure may be associated with congenital anomalies [151]. Recommendations in women regarding discontinuation of MMF prior to conception vary from 3 to 12 weeks. It would appear that avoidance of this drug by females at least 6 weeks prior to conceiving is most appropriate, whereas for males, the recommendation is 12 weeks' avoidance. This is an FDA Class D drug and should be avoided in pregnancy.

Since MMF is excreted into breast milk, it should not be administered to nursing mothers.

Angiotensin-Converting Enzyme Inhibitors (ACE) and Angiotensin Receptor Blockers (ARB)

ACE inhibitors and ARBs are commonly used in patients with SLE. Their use in pregnancy, however, is contraindicated owing to adverse effects on the fetus. Most of the studies have been performed on patients exposed to ACE inhibitors during pregnancy, but the concerns equally apply to the use of ARBs [159, 160].

Angiotensin II appears to play a major role in the regulation of uteroplacental blood flow as well as fetal growth, with particular emphasis on the growth of the fetal kidney [159]. Thus use of these antihypertensives may adversely affect maternal blood flow and fetal development. It is necessary to avoid the use of ACE inhibitors and ARBs even in the first trimester due to the risk of cardiovascular and CNS anomalies [160]. Exposure in the second and third trimesters can result in severe fetal adverse effects owing to the effect on fetal renal hemodynamics [161, 162]. The fetal circulation is dependent upon high circulating levels of angiotensin II, which is necessary to maintain the GFR in the normally low-pressure fetal circulation [163]. Use of these agents at this time results in a rapid fall in angiotensin II levels with subsequent fall in GFR, which in turn

leads to a decrease in fetal urine production, resulting in a decrease in amniotic fluid volume. Low amniotic fluid volume can then result in developmental abnormalities such as limb contractures, lung hypoplasia, and craniofacial deformities owing to abnormal cranial ossification [161].

Although some reports indicate recovery of renal function after birth, some renal insufficiency may remain [164]. Therefore, ACE inhibitors and ARBs should be avoided in any woman contemplating a pregnancy and during pregnancy. Alternatives should be prescribed and their use stabilized prior to conception. In addition, those women taking an ACE inhibitor or ARB for the treatment of proteinuria should be assessed after discontinuing the agent to determine if the return of proteinuria now contraindicates pregnancy.

ACE inhibitors and ARBs are found in breast milk in very low levels. However, owing to the extreme sensitivity of the neonatal kidney to ACE inhibitors in the first few weeks of life, nursing while taking these agents should be avoided until approximately 6 weeks after birth [165].

FAMILY PLANNING AND COUNSELING

As noted above, a prepregnancy evaluation should be undertaken whenever possible. Apart from the medical assessment, this visit represents the most appropriate time to discuss planning a pregnancy, thereby maximizing the safety of the woman and gaining the highest probability for the desired obstetrical outcome. Many studies have consistently demonstrated that planned pregnancies have better outcomes than those that are unplanned [166]. This is especially the case in those women with an underlying medical problem. Optimizing the health of the prospective mother will improve pregnancy outcome. In the case of the woman with SLE, this will not only involve the assessment of disease activity, but address medication issues. Based upon the findings at this visit, a recommendation for timing a pregnancy will be presented. If there are any medication issues, then substituting a pregnancy-safe medication should be undertaken immediately with plans for follow-up over an appropriate time to ensure that the disease is adequately treated with the newly prescribed alternative medication.

The mother with lupus contemplating a pregnancy must be counseled not only with respect to her disease entity and the interaction of lupus with pregnancy, but a detailed risk assessment includes a discussion of issues relevant to any woman planning a pregnancy. The counseling session should involve a discussion of maternal age factors and the impact of advanced maternal age on conception and pregnancy; weight and nutrition; reproductive and family histories; substance use including tobacco, caffeine, and alcohol; environmental exposures; and psychosocial issues. The clinician must ensure that the patient's partner is enrolled in the decision regarding pregnancy and is fully cognizant of any risks to proceeding.

The decision to enter into a pregnancy is one that is quite personal and emotional. The prepregnancy evaluation visit provides a unique opportunity for the clinician to establish a trusting relationship with the patient to maximize the probability that she will undertake a pregnancy informed to ensure her safety and that of her future child.

SLE AND FERTILITY

Fertility in women with SLE has been reported as normal with pregnancy rates of 2–2.5 observed even in the presence of active disease [160, 167]. Although such studies indicate no difference in fertility rates in women with SLE compared to the general population, one must interpret these findings with caution [168]. Any active inflammatory disorder will impact the pituitary–ovarian axis leading to anovulation. In addition, the adverse effects of certain medications such as cyclophosphamide may severely compromise gonadal function. Therefore, under these circumstances, fertility rates will be affected.

There are women with well-controlled SLE who may require fertility treatment often in the form of ovulation induction therapy. This treatment requires the use of high doses of exogenous follicle stimulating hormone (FSH), leading to a hyperestrogenic state. This transient hyperestrogenicity may exacerbate lupus. There are several studies of ovulation induction therapy in women with SLE [169–172]. For the most part, the women have fared well, although some complications associated with disease exacerbation have occurred. Since the studies are small, no conclusions can be drawn. Although fertility therapy is not contraindicated in women with SLE, it is advisable to initiate ovulation induction or superovulation only in those individuals whose disease is under excellent control for an extended period such as 6 months. In addition, the patient must be assessed for disease activity and advisability of entering into a pregnancy.

CONTRACEPTION

At one time, combined oral contraceptives (estrogen-containing agents) were considered to be at least relatively contraindicated in women with SLE [173–175].

This may still be the case in those with aPL as there appears to be increased risk in those patients of thromboembolism [176]. However, in those lacking aPL the controversies have been limited to patients with some degree of disease activity. Two studies clarified this issue [177, 178]. Neither found any association between combined oral contraceptive use and exacerbation of SLE, providing the woman has stable disease.

CONCLUSION

The pregnant woman with SLE represents a unique and potentially complicated management problem for both the rheumatologist/internist and obstetrician/perinatologist. The multisystem nature of the disease and the impact of pregnancy on compromised organ function will challenge the skills of all clinicians involved with the patient. The attending rheumatologist must have detailed knowledge of the clinical disease entity as well as the specific details of the patient's disease history including characteristic clinical manifestations, laboratory markers, and indicators of disease flare. The clinician must know if the patient demonstrates concordancy or discordancy between clinical disease activity and laboratory indicators. Under ideal circumstances, the pregnancy should be planned with the patient and her partner. The woman should be counseled to undertake a pregnancy when the disease is quiescent, after withdrawal of any fetotoxic drugs, and stabilization on medications known to be safe in pregnancy. With improvements in disease management and perinatal monitoring, in addition to collaboration between rheumatologists and perinatologists, there is now a good prognosis for both mother and fetus for the majority of women with systemic lupus erythematosus.

References

[1] A. Doria, L. Iaccarino, S. Arienti, et al., Th2 immune deviation induced by pregnancy: the two faces of autoimmune rheumatic diseases, Reprod Toxicol 22 (2006) 234–241.

[2] C.A. Laskin, Pregnancy and the rheumatic diseases, in: G.N. Burrow, T.P. Duffy, J.A. Copel (Eds.), Medical Complications During Pregnancy, sixth ed., Elsevier Saunders, Philadelphia, PA, 2004, pp. 429–449.

[3] D.W. Branch, Physiologic adaptations of pregnancy, Am J Reprod Immunol. 28 (1992) 120–122.

[4] S.L. Clark, D.B. Cotton, W. Lee, et al., Central hemodynamic assessment of normal term pregnancy, Am J Obstet Gynecol 161 (1989) 1439–1442.

[5] C.J. Lund, J.C. Donovan, Blood volume during pregnancy. Significance of plasma and red cell volumes, Am J Obstet Gynecol 98 (1967) 394–403.

[6] J.A. Pritchard, Changes in blood volume during pregnancy and delivery, Anesthesiology 26 (1965) 393–399.

[7] K. Ueland, Maternal cardiovascular dynamics. VII. Intra-partum blood volume changes, Am J Obstet Gynecol 126 (1976) 671−677.

[8] J.H. Matthews, S. Benjamin, D.S. Gill, et al., Pregnancy associated thrombocytopenia: Definition, incidence and natural history, Acta Haematol 84 (1990) 24−29.

[9] S.F. Kuvin, G. Brecher, Differential neutrophil counts in pregnancy, N Engl J Med 266 (1962) 877−878.

[10] M.J. Paidas, D.H. Ku, Y.S. Arkel, Screening and management of inherited thrombophilias in the setting of adverse pregnancy outcome, Clin Perinatol 31 (2004) 783−805.

[11] P. Lindqvist, B. Dahlback, K. Marsal, Thrombotic risk during pregnancy: a population study, Obstet Gynecol 94 (1999) 595−599.

[12] M. Hellgren, M. Blomback, Studies on blood coagulation and fibrinolysis in pregnancy, during delivery and in the puerperium, Gynecol Obstet Invest 12 (1981) 141−154.

[13] Y. Stirling, L. Woolf, W.R. North, et al., Haemostasis in normal pregnancy, Thromb Haemost 52 (1984) 176−182.

[14] K.A. Bremme, Haemostatic changes in pregnancy, Best Pract Res Clin Haematol 16 (2003) 153−168.

[15] J.A. Hill, Cytokines considered critical in pregnancy, Am J Reprod Immunol 28 (1992) 123.

[16] R.G. Lahita, The effects of sex hormones on the immune system in pregnancy, Am J Reprod Immunol 28 (1992) 136−137.

[17] Z.M. Sthoeger, N. Chiroazzi, R.G. Lahita, Regulation of the immune response by sex hormones. 1. *In vivo* effects of estradiol and testosterone on poekeweed mitogen induced B-cell differentiation, J Immunol 141 (1988) 91−98.

[18] I.M. Roitt, P.J. Delves, Roitt's Essential Immunology, tenth ed., Blackwell Science, London, 2001.

[19] J.C. Kingdom, P. Kaufmann, Oxygen and placental villous development: origins of fetal hypoxia, Placenta 18 (8) (1997) 613−621.

[20] S. Viero, V. Chaddha, F. Alkazaleh, et al., Prognostic value of placental ultrasound in pregnancies complicated by absent end diastolic flow velocity in umbilical arteries, Placenta 25 (2004) 735−741.

[21] H. Tuchman-Duplessis, G. David, P. Haegel, "Illustrated Human Embryology", vol. I, Springer Verlag, New York, 1975. 62−84.

[22] F.A. Ellis, E.S. Bereston, Lupus erythematosus associated with pregnancy and menopause, AMA Arch Dermatol Syphilol 69 (1952) 170−176.

[23] M. Garsenstein, V.E. Pollak, R.M. Kark, Systemic lupus erythematosus and pregnancy, New England Journal of Medicine 267 (1962) 165−169.

[24] A. Mund, J. Simson, N. Rothfield, Effect of pregnancy on the course of systemic lupus erythematosus, JAMA 183 (1963) 917−920.

[25] R.B. Zurier, Systemic lupus erythematosus and pregnancy, Clinics in Rheumatic Diseases 1 (1975) 613−616.

[26] R.R. Grigor, P.C. Shervington, G.R.V. Hughes, D.F. Hawkins, Outcome of pregnancy in systemic lupus erythematosus. Proc R Soc Med 70: 99−100.

[27] E.C. Tozman, M.B. Urowitz, D.D. Gladman, Systemic lupus erythematosus and pregnancy, J Rheumatol 7 (1980) 624−632.

[28] M.I. Zulman, N. Talal, G.S. Hoffman, W.V. Epstein, Problems associated with the management of pregnancies in patients with systemic lupus erythematosus, J Rheumatol 7 (1980) 37−49.

[29] M.D. Lockshin, E. Reinitz, M.L. Druzin, M. Murrman, D. Estes, Lupus pregnancy: case control study demonstrating absence of lupus exacerbations during and after pregnancy, Am J Med 77 (1984) 893−898.

[30] M.D. Lockshin, Pregnancy does not cause systemic lupus erythematosus to worsen, Arthritis Rheum 32 (1989) 665−670.

[31] M. Petri, Hopkins lupus pregnancy center: 1987-1996, Rheum Dis Clin North Am 23 (1997) 1−13.

[32] G. Ruiz-Irastorza, F. Lima, J. Alves, et al., Increased rate of lupus flare during pregnancy and the puerperium: a prospective study of 78 pregnancies, Br J Rheumatol 35 (1996) 133−138.

[33] J. Cortes-Hernandez, J. Ordi-Ros, F. Paredes, M. Casellas, F. Castillo, M. Vilardell-Tarres, Clinical predictors of fetal and maternal outcome in systemic lupus erythematosus: a prospective stud of 103 pregnancies, Rheumatology (Oxford) 41 (2002) 643−650.

[34] M.B. Urowitz, D.D. Gladman, V.T. Farewell, J. Stewart, J. McDonald, Lupus and pregnancy studies, Arthritis Rheum 36 (1993) 1392−1397.

[35] J.P. Buyon, K.C. Kalunian, R. Ramsey-Goldman, M.A. Petri, G. Ruiz-Irastorza, M. Khamashta, Assessing disease activity in SLE patients during pregnancy, Lupus 8 (1999) 677−684.

[36] G. Ruiz-Irastorza, M.A. Khamashta, Evaluation of systemic lupus erythematosus activity during pregnancy, Lupus 13 (2004) 679−682.

[37] S. Yasmeen, E.E. Wilkins, N.T. Field, R.A. Sheikh, W.M. Gilbert, Pregnancy outcomes in women with systemic lupus erythematosus, J Maternal Fetal Med 10 (2001) 91−96.

[38] C.A. Clark, K.A. Spitzer, J.N. Nadler, C.A. Laskin, Preterm deliveries in women with systemic lupus erythematosus, J Rheumatol 30 (2003) 2127−2132.

[39] J.P. Hayslett, Kidney disease and pregnancy, Kidney Int 25 (1984) 579−587.

[40] M. Surian, E. Imbasciati, P. Cosci, Glomerular disease and pregnancy: A study of 123 pregnancies inpatients with primary and secondary glomerular diseases, Nephron 36 (1984) 101−105.

[41] P. Jungers, P. Houillier, D. Forget, et al., Influence of pregnancy on the course of primary chronic glomerulonephritis, Lancet 346 (1995) 1122−1224.

[42] S. Hou, Pregnancy in women with chronic renal disease, N Engl J Med 312 (1985) 836−839.

[43] E. Imbasciati, G. Pardi, P. Capetta, et al., Pregnancy in women with chronic renal failure, Am J Nephrol 6 (1986) 193−198.

[44] E. Imbasciati, G. Gregorini, G. Cabiddu, et al., Pregnancy in CKD stages 3 to5: Fetal and maternal outcomes, Am J Kidney Dis 49 (2007) 753−762.

[45] P. Jungers, P. Houillier, D. Forget, M. Henry-Amar, Specific controversies concerning the natural history of renal disease in pregnancy, Am J Kidney Dis 17 (1991) 116−122.

[46] P. Jungers, M. Dougados, C. Pelissier, et al., Lupus nephropathy and pregnancy, Arch Intern Med 142 (1982) 771−776.

[47] D.L. Huong, B. Wechsler, D. Vauthier-Brouzes, et al., Pregnancy in past or present lupus nephritis: a study of 32 pregnancies from a single centre, Ann Rheum Dis 60 (2001) 599−604.

[48] G. Bobrie, F. Liote, P. Houillier, et al., Pregnancy in lupus nephritis and related disorders, Am J Kidney Dis 9 (1987) 339−343.

[49] G. Moroni, S. Quaglini, G. Banfi, et al., Pregnancy in lupus nephritis, Am J Kidney Dis 40 (2002) 713−720.

[50] A. Tandon, D. Ibaniz, D.D. Gladman, M.B. Urowitz, The effect of pregnancy on lupus nephritis, Arthritis Rheum 50 (2004) 3941−3946.

[51] ACOG Committee on Practice Bulletins − Obstetrics, ACOG practice bulletin. Diagnosis and management of preeclampsia and eclampsia, Obstet Gynecol 99 (2002) 159−167.

[52] K.Y. Lain, J.M. Roberts, Contemporary concepts of the pathogenesis and management of preeclampsia, JAMA 287 (2002) 3183−3186.

[53] M.D. Lockshin, Pregnancy does not cause SLE to worsen, Arthritis Rheum 32 (1989) 665-370.

[54] H.C. Nossent, T.J. Swaak, Systemic lupus erythematosus. VI. Analysis of the interrelationship with pregnancy, J Rheumatol 17 (1990) 771–776.

[55] G. Ruiz-Irastorza, M.A. Khamashta, Lupus and pregnancy: ten questions and some answers, Lupus 17 (2008) 416–420.

[56] J.H. Matthews, S. Benjamin, D.S. Gill, et al., Pregnancy-associated thrombocytopenia: Definition, incidence and natural history, Acta Haematol 84 (1990) 24–29.

[57] V. Fenton, K. Saunders, I. Cavill, The platelet count in pregnancy, J Clin Pathol 30 (1977) 68–69.

[58] P.R. Sill, T. Lind, W. Walker, Platelet values during normal pregnancy, Br J Obstet Gynaecol 92 (1985) 480–483.

[59] A.S. Sejeny, R.D. Eastham, S.R. Baker, Platelet counts during normal pregnancy, J Clin Pathol 28 (1975) 812–813.

[60] R.F. Burrows, J.G. Kelton, Fetal thrombocytopenia and its relation to maternal thrombocytopenia, N Engl J Med 329 (1993) 1463–1466.

[61] J.N. George, S.H. Woolf, G.E. Raskob, et al., Idiopathic thrombocytopenic purpura. A practice guideline developed by explicit methods for the America Society of Hematology, Blood 88 (1996) 3–40.

[62] R.F. Burrows, J.G. Kelton, Platelets and pregnancy, in: R.V. Lee (Ed.), Current Obstetric Medicine, vol. 2, Mosby-Year Book, St. Louis, 1993, pp. 83–106.

[63] D. Alarcon-Segovia, J. Sanchez-Guerrero, Correction of thrombocytopenia with small dose aspirin in the primary antiphospholipid antibody syndrome, J Rheumatol 16 (1989) 1359–1361.

[64] S.D. Payne, R. Resnik, T.R. Moore, et al., Maternal characteristics and risk of severe neonatal thrombocytopenia and intracranial hemorrhage in pregnancies complicated by autoimmune thrombocytopenia, Am J Obstet Gyencol 177 (1997) 149–155.

[65] A.S. Valat, M.T. Caulier, P. Devos, et al., Relationships between severe neonatal thrombocytopenia and maternal characteristics in pregnancies associated with autoimmune thrombocytopenia, Br J Haematol 103 (1998) 397–401.

[66] K. Fujimura, Y. Harada, T. Fujimoto, et al., Nationwide study of idiopathic thrombocytopenic purpura in pregnant women and the clinical influence on neonates, Int J Hematol 75 (2002) 426–433.

[67] J.O. Martinez-Rueda, C.A. Arce-Salinas, A. Kraus, et al., Factors associated with fetal losses in severe systemic lupus erythematosus, Lupus 5 (1996) 113–119.

[68] C.S. Kuzis, M.L. Druzin, R.E. Lambert, Case report: A patient with severe CNS lupus during pregnancy, Ann Med Intern 147 (1996) 274–275.

[69] R.E. Wolf, J.G. McBeath, Chorea gravidarum in systemic lupus erythematosus, J Rheumatol 12 (1985) 992–993.

[70] M. Marabani, A. Zoma, D. Hadley, et al., Transverse myelitis occurring during pregnancy in a patient with systemic lupus erythematosus, Ann Rheum Dis 48 (1989) 160–162.

[71] W.I. Morris, Pregnancy in rheumatoid arthritis and systemic lupus erythematosus, Aust N Z J Obstet Gynecol 9 (1969) 136–144.

[72] E.A. Friedman, J.W. Rutherford, Pregnancy and lupus erythematosus, Obstet Gynecol 8 (1956) 601–610.

[73] F. Lima, N.M.M. Buchanan, M.A. Kamashta, S. Kerslake, G.R. Hughes, Obstetric outcome in systemic lupus erythematosus, Semin Arthritis Rheum 25 (1995) 184–192.

[74] C.A. Clark, K.A. Spitzer, C.A. Laskin, Decrease in pregnancy loss rates in patients with systemic lupus erythematosus over a 40-year period, J Rheumatol 32 (2005) 1709–1712.

[75] H.P. McNeill, C.N. Chesterman, S.A. Krilis, Immunology and clinical importance of antiphospholipid antibodies, Adv Immunol 8 (1991) 425–426.

[76] C.E. Tseng, J.P. Buyon, Neonatal lupus syndrome, Rheumatic Dis Clin North Am 23 (1997) 31–54.

[77] M. Elias, A. Eldor, Thromboembolism in patients with the "lupus" type circulating anticoagulant, Arch Intern Med 144 (1984) 510–515.

[78] J.R. Mueh, K.D. Herbst, S.I. Rapaport, Thrombosis in patients with the lupus anticoagulant, Ann Intern Med 92 (1980) 156–159.

[79] M.D. Lockshin, M.L. Druzin, S. Goei, et al., Antibody to cardiolipin as a predictor of fetal distress or death in pregnant patients with systemic lupus erythematosus, N Engl J Med 313 (1985) 152–156.

[80] C.A. Clark-Soloninka, K.A. Spitzer, J.N. Nadler, C.A. Laskin, Evaluation of screening 590 plasma samples for the lupus anticoagulant using a panel of four tests, Arthritis Rheum 41 (Suppl.) (1998) S168.

[81] J.E. Salmon, M. Kim, M. Guerra, et al., Absence of lupus anticoagulant is a strong predictor of uncomplicated pregnancy in patients with APL antibodies, Arthritis Rheum 58 (Suppl.) (2008) S403.

[82] J.O. Martinez-Rueda, C.A. Arce-Salinas, A. Kraus, J. Alcocer-Varela, D. Alarcon-Segovia, Factors associated with fetal loss in severe systemic lupus erythematosus, Lupus 5 (1996) 113–119.

[83] J.G. Hanly, D.D. Gladman, T.H. Rose, C.A. Laskin, M.B. Urowitz, Lupus pregnancy: A prospective study of placental changes, Arthritis Rheum 31 (1998) 358–366.

[84] C.R. Abramowsky, M.E. Vegas, G. Swinehart, M.T. Gyves, Decidual vasculopathy of the placenta in lupus erythematosus, N Engl J Med 303 (1980) 668–672.

[85] E. Guzman, H. Schulman, L. Bracero, B. Rochelson, G. Farmakides, A. Coury, Uterine-umbilical artery Doppler velocimetry in pregnant women with systemic lupus erythematosus, J Ultrasound Med 11 (1992) 275–281.

[86] L.O. Carreras, G. Defreyn, S.J. Machin, et al., Arterial thrombosis, intrauterine death and "lupus" anticoagulant: Detection of immunoglobulin interfering with prostacyclin formation, Lancet 1 (1981) 244–246.

[87] W. Geis, D.W. Branch, Obstetric implications of antiphospholipid antibodies: pregnancy loss and other complications, Clin Obstet Gynecol 44 (2001) 2–10.

[88] H.J. Out, H.W. Bruinsem, G.C.M.L. Chrustuaebs, et al., A prospective, controlled multicenter study on the obstetric risks of pregnant women with antiphospholipid antibodies, Am J Obstet Gynecol 167 (1992) 26–32.

[89] D. Vinatier, P. Dufour, M. Cosson, J.L. Houpeau, Antiphospholipid syndrome and recurrent miscarriages, Eur J Obstet Gynecol Reprod Biol 96 (2001) 37–50.

[90] C.A. Soloninka, C.A. Laskin, J. Wither, D. Wong, C. Bombardier, J. Raboud, Clinical utility and specificity of anticardiolipin antibodies, J Rheumatol 18 (1991) 1849–1855.

[91] W.F.F. Lubbe, W.S. Butler, S.J. Palmer, G.C. Liggins, Fetal survival after prednisone suppression of maternal lupus anticoagulant, Lancet 2 (1983) 1361–1363.

[92] C.A. Laskin, C. Bombardier, M. Hannah, et al., Prednisone and aspirin in women with autoantibodies and unexplained recurrent fetal loss, N Engl J Med 337 (1997) 148–153.

[93] R. Rai, H. Cohen, M. Dave, L. Regan, Randomized controlled trial of aspirin and aspirin plus heparin in pregnant women with recurrent miscarriage associated with phospholipids antibodies (or antiphospholipid antibodies), Br Med J 314 (1997) 253–257.

[94] D.L. Huong, B. Wechsler, O. Bletry, D. Vauthier-Brouzes, G. Lefebvre, J.C. Piette, A study of 75 pregnancies in patients with antiphospholipid syndrome, J Rheumatol 28 (2001) 2025–2030.

[95] R.G. Farquharson, S. Quenby, M. Greaves, Antiphospholipid syndrome in pregnancy: a randomized, controlled trial of treatment, Obstet Gynecol 100 (2002) 408–413.

[96] H.A. Shehata, C. Nelson-Piercy, M.A. Khamashta, Management of pregnancy in antiphospholipid syndrome, Rheum Dis Clin North Am 27 (2001) 643–659.

[97] C.A. Laskin, J. Ginsberg, M. Crowther, et al., Results from the HepASA trial: low molecular weight heparin + ASA vs ASA only for recurrent pregnancy loss, Arthritis Rheum 54 (Suppl.) (2006) S558.

[98] B. Bresnihan, R.R. Grigor, M. Oliver, et al., Immunological mechanism for spontaneous abortion in systemic lupus erythematosus, Lancet 2 (1997) 1205–1207.

[99] M. Empson, M. Lassere, J.C. Craig, J.R. Scott, Recurrent pregnancy loss with antiphospholipid antibody: a systematic review of therapeutic trials, Obstet Gynecol 100 (2002) 135–144.

[100] F. Meccacci, B. Bianchi, A. Peiralll, et al., Pregnancy outcome in systemic lupus erythematosus complicated by anti-phospholipid antibodies, Rheumatology (Oxford) 48 (2009) 246–249.

[101] D.M. Freidman, M.Y. Kim, J.A. Copel, et al., Utility of cardiac monitoring in fetuses at risk for congenital heart block: the PR Interval and Dexamethasone Evaluation (PRIDE) prospective study, Circulation 117 (2008) 485–493.

[102] A. Brucato, A. Doria, M. Frassi, et al., Pregnancy outcome in 100 women with autoimmune diseases and anti-Ro/SSA antibodies: a prospective controlled study, Lupus 11 (2002) 716–721.

[103] C. Llanos, P.M. Izmirly, M. Katholi, et al., Recurrence rates of cardiac manifestations associated with neonatal lupus and maternal/fetal risk factors, Arthritis Rheum 60 (2009) 3091–3097.

[104] J.Y. Lee, P.T. Huerta, J. Zhang, C. Kowal, E. Bertini, B.T. Volpe, et al., Maternal lupus and congenital cortical impairment, Nat Med 15 (2009) 91–96.

[105] J. Sanchez-Guerrero, C. Aranow, M. Mackay, B. Volpe, B. Diamond, Neuropsychiatric systemic lupus erythematosus reconsidered, Nat Clin Pract Rheumatol 4 (2009) 112–113.

[106] G. Ruiz-Irastorza, F. Lima, J. Alvers, et al., Increased rate of lupus flare during pregnancy and the puerperium: a prospective study of 78 pregnancies, Br J Rheumatol 35 (1996) 133–138.

[107] M.B. Urowitz, D.D. Gladman, Measures of disease activity and damage in SLE, Baillière's Clin Rheumatol 12 (1998) 405–413.

[108] R.M. Andrade, G. McGwin Jr., G.S. Alarcon, et al., Predictors of post-partum damage accrual in systemic lupus erythematosus: data from LUMINA, a multiethnic US cohort (XXXVII), Rheumatology (Oxford) 45 (2006) 1380–1384.

[109] S. Stone, M.A. Khamashta, C. Nelson-Piercy, Nonsteroidal anti-inflammatory drugs and reversible female infertility: is there a link? Drug Safety 25 (2002) 545–551.

[110] S. Killick, M. Elstein, Pharmacologic production of luteinized unruptured follicles by prostaglandin synthetase inhibitors, Fertil Steril 47 (1987) 773–777.

[111] M.G. Elder, L. Myatt, G. Chaudhuri, The role of prostaglandins in the spontaneous motility of lthe fallopian tube, Fertil Steril 28 (1997) 86–90.

[112] A. Laszlo, G.L. Nadasy, E. Monos, B. Zsolnai, Effect of pharmacological agents on the activity of the circular and longitudinal smooth muscle layers of human fallopian tube ampullar segments, Acta Physiol Hungarica 72 (1988) 123–133.

[113] N. Sookvanichsilp, P. Pulbutr, Anti-implantation effects of indomethacin and celecoxib in rats, Contraception 65 (2002) 373–378.

[114] H. Lim, B.C. Paria, S.K. Das, et al., Mulitple female reproductive failures in cyclooxygenase 2-deficient mice, Cell 91 (1997) 197–208.

[115] G.L. Nielsen, H.T. Sorensen, H. Larsen, L. Pedersen, Risk of adverse birth outcome and miscarriage in pregnant users of non-steroidal anti-inflammatory drugs: population based observational study and case-control study, Br Med J 322 (2001) 266–270.

[116] A.H. James, L.R. Brancazio, T. Price, Aspirin and reproductive outcomes, Obstet Gynecol Surv 63 (2008) 49–57.

[117] J.G. Bromer, M.B. Cetinkaya, A. Arici, Tretreatments before the induction of ovulation in assisted reproduction technologies: evidence-based medicine in 2007, Ann N Y Acad Sci 1127 (2008) 31–40.

[118] J.L. Frattarelli, G.D. McWilliams, M.J. Hill, K.A. Miller, R.T. Scott Jr., Low-dose aspirin use does not improve in vitro fertilization outcomes in poor responders, Fertil Steril 89 (2008) 1113–1117.

[119] E. Jauniaux, R.G. Farquharson, O.B. Christiansen, N. Exalto, Evidence-based guidelines for the investigation and medical treatment of recurrent miscarriage, Human Reprod 21 (2006) 2216–2222.

[120] L.M. Askie, L. Duley, D.J. Henderson-Smart, L.A. Stewart, and the PARIS Collaborative Group. Antiplatelet gents for prevention of pre-eclampsia: a meta-analysis of individual patient data, Lancet 369 (2007) 1791–1798.

[121] E. Kozer, S. Nikfar, A. Costei, R. Boskovic, I. Hulman, G. Koren, Aspirin consumption during the first trimester of pregnancy and congenital anomalies: a meta-analysis, Am J Obstet Gynecol 187 (2002) 1623–1630.

[122] B. Norgard, E. Puho, A.E. Czeizel, M.V. Skriver, H.T. Sorensen, Aspirin use during early pregnancy and the risk of congenital abnormalities: a population-based case-control study, Am J Obstet Gynecol 192 (2005) 922–923.

[123] G. Riuz-Irastorza, M.A. Khamashta, Management of thrombosis in antiphospholipid syndrome and systemic lupus erythematosus in pregnancy, Ann N Y Acad Sci 1051 (2005) 606–612.

[124] K.A. Spitzer, D. Murphy, M. Crowther, C.A. Clark, C.A. Laskin, Postpartum management of women at increased risk of thrombosis: results of a Canadian pilot study, J Rheumatol 33 (2006) 2222–2226.

[125] M. Empson, M. Lassere, J. Craig, J. Scott, Prevention of recurrent miscarriage for women with antiphospholipid antibody or lupus anticoagulant, Cochrane Database System Reviews 18 (2005) CDD002859.

[126] I.A. Greer, Anticoagulants in pregnancy, J Thromb Thrombolysis 21 (2006) 57–65.

[127] M.D. Stephenson, P.J. Ballem, P. Tsang, Treatment of antiphospholipid antibody syndrome (APS) in pregnancy: a randomized pilot comparing low molecular weight heparin to unfractionated heparin, Canad J Obstet Gynecol 26 (2004) 729–734.

[128] L.S. Noble, W.H. Kutteh, N. Lashey, R.D. Frandklin, J. Herrad, Antiphospholipid antibodies associated with recurrent pregnancy loss: prospective, multicenter, controlled pilot study comparing treatment with low molecular weight heparin vs unfractionated heparin, Fertil Steril 83 (2005) 684–690.

[129] G. Ruiz-Irastorza, M.A. Khamashta, C. Nelson-Piercy, G.R. Hughes, Lupus pregnancy: is heparin a risk factor for osteoporosis? Lupus 10 (2001) 597–600.

[130] P. Deruelle, C. Coulon, The use of low molecular weight heparins in pregnancy—how safe are they? Curr Opin Obstet Gynecol 19 (2007) 573—577.

[131] A. Al-Herz, M. Schulzer, J.M. Esdaile, Survey of antimalarial use in lupus pregnancy and lactation, J Rheumatol 29 (2002) 700—706.

[132] M.E. Clowse, L. Magder, R. Witter, M. Petri, Hydroxy-chloroquine in lupus pregnancy, Arthritis Rheum 54 (2006) 3640—3647.

[133] M.A. Khamashta, N.M. Buchanan, G.R. Hughes, The use of hydroxychloroquine in lupus pregnancy: the British experience, Lupus 5 (Suppl. 1) (1996) S65—S66.

[134] G. Klinger, Y. Morad, C.A. Westall, et al., Ocular toxicity and antenatal exposure to chloroquine or hydroxychloroquine for rheumatic diseases, Lancet 358 (2001) 813—814.

[135] M. Motta, A. Tincai, D. Faden, et al., Follow-up of infants exposed to hydroxychloroquine given to mothers during pregnancy and lactation, J Perinatol 25 (2005) 86—89.

[136] S.L. Carmichael, G.M. Shaw, C. Ma, et al., and the National Birth Defects Prevention Study, Maternal corticosteroid use and orofacial cleft, Am J Obstet Gynecol 197 (2007) 585—587.

[137] P. Pradat, E. Robert-Gnansia, G.L. Di Tanna, et al., and contributors to the MADRE database, 2003 First trimester exposure to corticosteroids and oral clefts, Birth Defects Res A Clin Mol Teratol 67 (2003) 968—970.

[138] L. Park-Wyllie, P. Mazzotta, A. Pastuszak, et al., Birth defects after maternal exposure to corticosteroids: prospective cohort study and meta analysis of epidemiological studies, Teratology 62 (2000) 385—392.

[139] N.K. de Boer, S.V. Jarbandhan, P. de Graaf, C.J. Mulder, R.M. van Elburg, A.A. van Bodergraven, Azathioprine use during pregnancy: unexpected intrauterine exposure to metabolites, Am J Gastroenterol 101 (2006) 1390—1392.

[140] L.H. Goldstein, G. Dolinsky, R. Greenberg, et al., Birth Defects, Res A Clin Mol Teratol 79 (2007) 696—701.

[141] R. Ramsey-Goldman, E. Schilling, Immunosuppressive drug use during pregnancy, Rheum Dis Clin North Am 23 (1997) 149—167.

[142] A.E. French, G. Koren, and the Motherisk Team. Effect of methotrexate on male fertility, Can Fam Physician 49 (2003) 577—578.

[143] K.K. Temprano, R. Branlamudi, T.L. Moore, Antirheumatic drugs in pregnancy and lactation, Semin Arthritis Rheum 35 (2005) 112—121.

[144] N.M. Janssen, M.S. Genta, The effects of immunosuppressive and anti-inflammatory medication on fertility, pregnancy, and lactation, Arch Intern Med 160 (2000) 610—619.

[145] M. Ostensen, Treatment with immunosuppressive and disease modifying drugs during pregnancy and lactation, Am J Reprod Immunol 28 (1992) 148—152.

[146] M. Nandakumaran, A.S. Eldeen, Transfer of cyclosporine in the perfused human placenta, Dev Pharmacol Ther 15 (1990) 101—105.

[147] R. Venkataramanan, B. Koneru, C.C. Wang, G.J. Burckart, S.N. Caritis, T.E. Starzl, Cyclosporine and its metabolites in mother and baby, Transplantation 46 (1988) 468—469.

[148] S. Di Paolo, R. Monno, G. Stallone, et al., Placental imbalance of vasoactive factors does not affect pregnancy outcome in patients treated with Cyclosporine A after transplantation, Am J Kidney Dis 39 (2002) 776—783.

[149] M. Ostensen, M. Khamashta, M. Lockshin, et al., Anti-inflammatory and immunosuppressive drugs and reproduction, Arthritis Res Ther 8 (2006) 209.

[150] I. Cockburn, P. Krupp, C. Monka, Present experience of San-dimmun in pregnancy, Transplant Proc 21 (1989) 3730—3732.

[151] V.T. Armenti, K.M. Ahlswede, B.A. Ahlswede, et al., National Transplantation Pregnancy Registry: Outcomes of 154 pregnancies in cyclosporine-treated female kidney transplant recipients, Transplantation 57 (1994) 502—506.

[152] B. Bar Oz, R. Hackman, T. Einarson, G. Koren, Pregnancy outcome after cyclosporine therapy during pregnancy: a meta-analysis, Transplantation 71 (2001) 1051—1055.

[153] N.M. Sifontis, L.A. Coscia, S. Constaninescu, A.F. Lavelanet, M.J. Moritz, V.T. Armenti, Pregnancy outcomes in solid organ transplant recipients with exposure to mycophenolate mofetil or sirolimus, Transplantation 82 (2006) 1698—1702.

[154] A. Perez-Aytess, A. Ledo, V. Boso, et al., In utero exposure to mycophenolate mofetil: a characteristic phenotype? Am J Med Genet 146 (2008) 1—7.

[155] S.M. Downs, Induction of meiotic maturation in vivo in the mouse by IMP dehydrogenase inhibitors: effects on the developmental capacity of ova, Mol Reprod Develop 38 (1994) 293—302.

[156] M.A. Parisi, H. Zayed, A.M. Slavotinek, J.C. Rutledge, Congential diaphragmatic hernia and microtia in a newborn with mycophenolate mofetil (MMF) exposure: phenocopy for Fryne syndrome or broad spectrum of teratogenic effects? Am J Med Genet A 149A (2009) 1237—1240.

[157] M.T. Anderka, A.E. Lin, D.N. Abuelo, A.A. Mitchell, S.A. Rasmussen, Reviewing the evidence of mycophenolate mofetil as a new teratogen: case report and review of the literature, Am J Med Genet A 149A (2009) 1241—1248.

[158] P. Merlob, B. Stahl, G. Klinger, Tetrada of possible mycophe-nolate mofetil embyopathy: a review, Reprod Toxicol 28 (2009) 105—108.

[159] A. Shotan, J. Widerhorn, A. Hurst, U. Elkayam, Risks of angiotensin-converting enzyme inhibition during pregnancy: experimental and clinical evidence, potential mechanisms, and recommendations for use, Am J Med 96 (1994) 451—456.

[160] W.O. Cooper, S. Hernandez-Diaz, P.G. Arbogast, J.A. Dudley, S. Dyer, P.S. Gideon, K. Hall, W.A. Ray, Major congenital malformations after first-trimester exposure to ACE inhibitors, N Engl J Med 354 (2006) 2443—2451.

[161] P.G. Pryde, A.B. Sedman, C.E. Nugent, M. Barr, Jr., Angiotensin-converting enzyme inhibitor fetopathy, J Am Soc Nephrol 3 (1993) 1575—1582.

[162] S. Tabacova, R. Little, Y. Tsong, A. Vega, C.A. Kimmel, Adverse pregnancy outcomes associated with maternal enalapril anti-hypertensive treatment, Pharmacoepidemiol Drug Saf 12 (2003) 633—646.

[163] G. Guron, P. Friberg, An intact renin-angiotensin system is a prerequisite for normal renal development, J Hypertens 18 (2000) 123—137.

[164] G.D. Simonetti, T. Baumann, J.M. Pachlopnik, et al., Non-lethal fetal toxicity of the angiotensin receptor blocker candesartan, Pediatr Nephrol 21 (2006) 1329.

[165] American Academy of Pediatrics, Committee on Drugs. The transfer of drugs and other chemicals into human milk, Pediatrics 108 (2001) 776—789.

[166] K. Johnson, S.F. Posner, J. Biermann, et al., Recommendations to improve preconception health and health care, United States A report of the CDC/ATSDR Preconception Care Work Group and the Select Panel on Preconception Care, MMWR Recomm Rep 55 (2006) 1.

[167] C.C. Mok, R.W. Wong, Pregnancy in systemic lupus erythematosus, Postgrad Med J 77 (2001) 157—165.

[168] A. Fraga, G. Mintz, J. Orozco, Sterility and fertility rates, fetal wastage and maternal morbidity in systemic lupus erythematosus, J Rheumatol 1 (1974) 283—288.

[169] J. Sanchez-Guerrero, E.W. Karlson, M.H. Liang, D.J. Hunter, F.E. Speizer, G.A. Colditz, Past use of oral contraceptives and the risk of developing systemic lupus erythematosus, Arthritis Rheum 40 (1997) 804–808.

[170] D.L. Huong, B. Wechsler, J.C. Piette, Risks of ovulation induction therapy in systemic lupus erythematosus, Br J Rheumatol 35 (1996) 1184–1186.

[171] D.L. Huong, B. Wechsler, D. Vauthier-Brouzes, et al., Importance of planning ovulation in systemic lupus erythematosus and antiphospholipid syndrome: a single center retrospective study of 21 cases and 114 cycles, Semin Arthritis Rheum 32 (2002) 174–188.

[172] N. Guballa, L. Sammaritano, S. Schwartzman, J. Buyon, M.D. Lockshin, Ovulation induction and *in vitro* fertilization in systemic lupus erythematosus and antiphospholipid syndrome, Arthritis Rheum 43 (2002) 550–556.

[173] G.S. Cooper, M.A. Dooley, E.L. Treadwell, E.W. St Clair, G.S. Gilkeson, Hormonal and reproductive risk factors for development of systemic lupus erythematosus: results of a population-based, case-control study, Arthritis Rheum 46 (2002) 1830–1839.

[174] D. Gill, Rheumatic complaints of women using antiovulatory drugs, J Chron Dis 21 (1968) 435–444.

[175] E.L. Dubois, L. Strain, M. Ehn, G. Bernstein, G.J. Friou, LE cells after oral contraceptives, Lancet 2 (1968) 679.

[176] T.A. Chapel, R.E. Burns, Oral contraceptives and exacerbation of lupus erythematosus, Am J Obstet Gynecol 110 (1971) 366–369.

[177] M. Petri, M.Y. Kim, K.C. Kalunian, et al., Combined oral contraceptives in women with systemic lupus erythematosus, N Engl J Med 353 (2005) 2550–2558.

[178] J. Sanchez-Guerrero, A.G. Uribe, L. Jimenez-Santana, et al., A trial of contraceptive methods in women with systemic lupus erythematosus, N Engl J Med 353 (2005) 2539–2549.

Epidemiology of Systemic Lupus Erythematosus

Catia Duarte, Maura Couto, Luis Ines, Matthew H. Liang

INTRODUCTION

Epidemiology is the study of the distribution of illnesses and diseases and the factors affecting their incidence and clinical course in the population. As such it serves as the foundation and basis of interventions made in the interest of public health; and in patient care, in evidence-based medicine for identifying risk factors for disease and determining optimal treatment.

Systemic lupus erythematosus (SLE) an autoimmune disease, with a broad spectrum of clinical and immunological manifestations, is a major challenge to study epidemiologically, but research on SLE has been carried out in many parts of the world. These include descriptive studies of incidence and prevalence, observational studies of SLE prognosis, and the identification of potentially preventable causes of morbidity and mortality. Analytic and genetic epidemiologic studies suggest a multifactorial etiology of SLE. Better understanding of the epidemiology of SLE should help us understand its etiology, identify predictors of morbidity and mortality, and improve SLE care and outcomes.

EPIDEMIOLOGIC APPROACHES IN SLE

Epidemiologic studies provide important insights into risk factors for the development of diseases, but all have strengths and limitations. These are highlighted in Table 36.1. Cohort studies, in which potential risk factors are collected in normal individuals before they develop SLE can provide some of the strongest evidence that a factor is important in its etiology. Understandably, this type of data is unusual in SLE research. With more limitations, but a practical alternative in SLE, an uncommon disease, are case—control or cross-sectional studies. In these studies, subjects are selected on the basis of whether they have SLE (cases) or not (controls) and various exposures are compared between the groups. Risk, the association between exposure and disease, is expressed as a relative risk (RR) in cohort studies or odds ratio (OR) in case—control studies. A RR is the ratio between incidence in exposed and in nonexposed persons. In a case—control study, the true incidence of the disease among those exposed and not exposed cannot be calculated. In this case, one compares the frequency of exposure among cases and controls. This is an OR or the odds that a case is exposed divided by the odds that a control is exposed (Table 36.1). Case reports and case series can suggest possible risk factors but need to be corroborated with studies in other populations or with data from experimental studies or cohort studies.

TABLE 36.1 Comparison Between Cohort and Case—Control Studies

	Cohort	Case—control
Strengths	• Particular value when exposure is rare • Examine multiple aspects of a single exposure • Establish temporal relationship • Minimizes bias (if prospective) • Direct measurement of incidence of disease.	• Relatively quick and inexpensive • Diseases with long latency periods • Evaluation of rare diseases • Examine multiple etiologic factors for a single disease.
Limitations	• Inefficient for rare diseases • Expensive and time-consuming • If retrospective, requires availability of adequate records • Validity affected by losses of follow-up.	• Inefficient for rare exposure • Cannot directly compute incidence rates of disease in exposed and nonexposed patients • Difficult to establish temporal relationship • Bias (selection and recall bias).

In epidemiologic research, a completely definitive study is usual and it is usually the weight of accumulating data and replication of findings that allows one to come to a conclusion. In this chapter, we selectively review areas of SLE in which epidemiologic data are available and emphasize the more rigorous studies in a given question.

Classification Criteria for SLE

A case definition for SLE is required for both the conduct of epidemiological research and its interpretation [1]. It follows that the incidence and prevalence of the disease can vary depending on how the syndrome is defined. In 1971, the first criteria for the classification for SLE was published by the Diagnostic and Therapeutic Committee of the American Rheumatism Association (now American College of Rheumatology) and the criteria had a sensitivity of 90% and a specificity of 99% in separating SLE from rheumatoid arthritis and 98% separating SLE from other diseases [2].

Later, the 1982 revised criteria included a fluorescent antinuclear antibody and antibody to native DNA and Sm antigen, and different syndromes within an organ system were aggregated into a single criterion. Raynaud phenomenon and alopecia were not included [3]. These criteria were 96% sensitive and 96% specific when tested with SLE patients and disease controls [4]. Several studies confirmed the improved sensitivity and specificity over the 1971 criteria [5–7].

Although very useful, certain limitations of these criteria have become evident and improvements have been suggested. For example, with these criteria, a patient with a single major organ involved such as the kidney would not be classified as having SLE. Using Bayes' Theorem, Clough *et al.* developed alternative criteria for the SLE, assigning separate weights for each criterion [8]. Costenbader *et al.* [9] used different weights (Table 36.2), with a negative weight in case of negative ANA, in the Boston Weighted Criteria, and compared the new criteria with the 1982 ACR criteria. The new criteria had better psychometric properties with a sensitivity of 93%, specificity 69%, positive predictive value (PPV) 84%, and negative predictive value (NPV) of 85% [9]. These criteria were also evaluated by Sanchez [10]. He found that the Boston Weighted Criteria had higher sensitivity than the 1982 Criteria (90.3% vs. 86.5%), but lower specificity; PPV, NPV, and accuracy were comparable. The Boston Weighted Criteria would capture more patients than the ACR criteria in studies where the widest spectrum of disease is desired and where external validity or generalizability is desired. However, it has not been used widely.

It should be emphasized that the criteria used for classifying patients with SLE were not designed for

TABLE 36.2 Modified Weighted Criteria for the Classification of Systemic Lupus Erythematosus

Criteria	Weight
Malar rash	1.5
Photosensitivity	0.6
Discoid lupus	1.5
Oral ulcers	0.1
Arthritis	0.5
Serositis	0.6
Neuropsychiatric	0.7
Renal	2
• WHO Histopathology Class III–VI	1.5
• Cellular casts	1
• Proteinuria	
Hematologic	1.5
Positive ANA	0.5
Anti-DNA antibodies	0.5
Anti-Sm antibodies	0.5
Antiphospholipid antibodies	0.5
Negative ANA	−1.8

diagnostic purposes and as such lack sensitivity for milder cases, incomplete SLE case, or latent lupus [11]. Depending on the study question, one might make an argument to include or exclude these potential subjects. In assessing population burden, for instance, including all phenotypes irrespective of severity might be justified [12]. Despite the limitations, classification criteria have been useful in identifying homogeneous populations in analytic epidemiological studies [11].

Morbidity Rates: Incidence and Prevalence

In general, the frequency of a disease or an event in the course of disease is expressed as incidence and prevalence. Prevalence is the proportion of people with SLE or a complication at a given point in time. Incidence, on the other hand, is the proportion of people initially free of SLE or a complication who develop SLE or a complication over a defined period of time. Prevalence = incidence × the duration of a condition. Both incidence and prevalence, in this chapter, are expressed as number of SLE per 100,000 of the population per year. Ideally these rates need to be adjusted for age and gender as a basis for comparing rates from different populations. Both prevalence and incidence rates are influenced by the definition for SLE, case ascertainment strategy, sensitivity of the diagnostic method, and changes in the community population (migration, mortality).

Different techniques for ascertaining SLE have strengths and weaknesses and may contribute to variability in results [13]. Early studies from hospital data failed to capture cases of mild SLE not requiring hospitalization [14]. Studies relying solely on self-reported SLE without validation have reported higher prevalence rates [13]. Bernatsky [15] noted that the prevalence and incidence of SLE using billing data or hospitalization were similar, but that only a small number of prevalent cases of SLE were identified in common.

Incidence of SLE

Between 1956 and 1965, Siegel [16—18] conducted studies in the United States in New York and Alabama and reported an overall age-adjusted incidence for both genders of 1.4 among Whites, 2.3 among Puerto Ricans, and 4.6 among Blacks. Since then, studies from the United States report incidence rates between 1.8 and 5.56 [19—23]. In all studies with race- and gender-specific rates, the incidence was higher among women and Blacks. The study by Naleway et al. [22] reported an incidence rate for all races of 5.1. In Canada, Peschken and Esdaile reported a crude annual incidence rates between 2.0 and 7.4 for American Indians and 0.9 and 2.3 for other Canadian populations [24].

Nived et al. [25] conducted a hospital-based study in southern Sweden during 1981—1982 and reported an overall incidence of 4.5. In 1981—1986, Jonsson et al. [26] repeated the study and found an overall incidence of 4. Similar rates were reported by Stahl-Hallengren et al. [27] (1986—1991 = 4.8 and 1981—1986 = 4.5).

In the UK, a hospital and clinic based study [28] showed an overall age-adjusted incidence in Nottingham of 4. A study by Nightingale et al. [29], based on the General Practice Research Database, showed an overall crude incidence of 3.0 and probably gives a truer picture of incidence at the primary care level.

Studies in Iceland [30], Norway [31], Spain [32], Italy [33], Denmark [34], and Greece [35] show incidence rates between 1.04 and 3.3. (Table 36.2) and are generally lower than observed in the United States.

Studies from other countries are scarce. In Australia, a hospital-based study reported an overall crude incidence of 11 among Australian aborigines [36]. In Japan, Iseki et al. [37] reported an overall crude incidence of 0.9 in 1972 and 2.9 in 1991. Overall incidence rates of 8.7 and 3.1 were reported in the Brazil (Natal) [38] and China [39], respectively.

Secular Trends

A temporal trend in incidence among Whites can be inferred from Rochester, Minnesota data. Rates nearly tripled from 1.5 in the 1950—1979 cohort to 5.56 per 100,000 in the 1980—1992 cohort [23]. An increase in incidence was also reported in Iceland (1975—1984) [30], Greece (1982—2001) [35], the arctic region of Norway (1978—1996) [31], and Japan (1972—1991) [37]. However, stable incidence rates over the time were reported in studies in Sweden (1981—1991) [27], UK (1992—1998) [29], Denmark (1995—2002) [34], and China (2000—2006) [39].

Trivial explanations for these observed differences include the case ascertainment mechanism, the case definition, particularly the assay used for detection of ANA. Real differences in incidence might be explained by changes in environmental exposures such as increased exposure to oral contraceptives and estrogen replacement therapy, ultraviolet light because of depletion in the ozone layer because of global warming and pollution [13].

Prevalence

In the United States, Michet [19] studied definite SLE and reported an overall age-adjusted prevalence of 40 in Rochester, Minnesota area between 1950 and 1979. Uramoto [23] repeated the study in 1980—1992 and found a higher prevalence rate of 122. Naleway et al. [22] showed an overall adjusted prevalence in rural Wisconsin of 78.5, which increased to 130 if both incomplete and definite SLE were considered. Using hospitalization discharges in a 2000 database, the estimated prevalence was 107.6 in California and 149.5 in Pennsylvania [40].

In Canada, Peschken and Esdaile showed a crude prevalence rate of 22.1, with higher rates among Indians (crude prevalence rate of 33.4) [24].

Prevalence studies from European countries [25, 27, 31—35, 41—43] are generally lower than in US studies. The lowest rate of 6.5 was reported by Hochberg [42], who used data from general practices in England and Wales. Nived et al. [25] reported an overall prevalence of 36.3 in southern Sweden. Higher rates were reported from Sweden by Stahl-Hallengren [27] in 1986 (42) and in 1991 (68).

In Curacao, Nossent [44] reported a prevalence rate of 47.6. In the region of Darwin, Australia, the overall crude prevalence rate [36] was 52.6. In Australia, Segasothy [45] showed a prevalence of 73.5 among aborigines compared to 19.3 in Caucasians. A prevalence rate of 89.3 was reported in another study among aborigines from Northern Queensland (Australia) [46]. In Martinique, the prevalence rate was 64.2 in 1999 [47] (Table 36.3).

Secular Trends

Increased prevalence of SLE over time has been reported from various parts of the world [25, 27, 34, 37, 48]. In Japan, Iseki et al. [37] reported a prevalence rate of 3.7 in 1972, which increased to 37.7 in 1991. Studies from southern Sweden conducted in 1981—1991 showed increased prevalence rates from 36.3 (1981—1982) to 68 (1987—1991) [25, 27]. The rise in

TABLE 36.3   Incidence and Prevalence of SLE

Author	Country/year	Methods	Incidence	Prevalence
Helve, 1985 [41]	Finland 1976–1978	Definite SLE National Board of Health		28
Nived et al., 1985 [25]	Southern Sweden 1981–1982	Definite SLE (ACR 1982) Hospital and primary care	4.5 F = 7.6 M = 1.3	36.3 F = 62.3 M = 9.1
Hochberg, 1987 [42]	England and Wales 1981–1982	General practice		6.5 F = 12.5
Gudmundsson and Steinsson, 1990 [30]	Iceland 1975–1984	Definite SLE (ACR 1982) Hospitals	3.3 F = 5.9 M = 0.8	
Nossent, 1992 [44]	Curacao 1980–1989	Definite SLE (ACR 1982) Multiple	4.63 F = 7.86 M = 1.13	47.6 F = 83.8 M = 8.5
Hopkinson et al., 1993 [28]	UK (Nottingham) 1989–1990	Definite SLE Multiple	4.4 M = 1.5 F = 6.5	24 F = 45.4 M = 3.7
Uramoto et al., 1999 [23]	US (Rochester, MN) 1980–1992	Definite SLE (ACR 1982) Multiple	5.56 F = 9.4 M = 1.54	122
Stahl-Hallengren et al., 2000 [27]	Southern Sweden 1981–1986 1987–1991	Definite SLE Multiple	4.5 4.8	42 68
Peschken and Esdaile, 2000 [24]	Canada (Manitoba) 1980–1986	Definite SLE (ACR 1982) Multiple	Indian = 0–7.4 Non-Indian = 0.9–2.3	22.1 Indian = 42.3 Non-Indian = 20.6
Nossent, 2001 [31]	Norway 1978–1996	Definite SLE (ACR 1982) Multiple	2.6 M = 0.6 F = 4.6	49.7 F = 89.3 M = 9.7
Segasothy and Phillips, 2001 [45]	Australia 1990–1999	Definite SLE (ACR 1982) Multiple		Aborigines = 73.5 Caucasians = 19.3
Vilar and Sato, 2002 [38]	Brazil (Natal) 2000	Definite SLE (ACR 1982) Multiple	8.7 F = 14.1 M = 2.2	
Deligny et al., 2002 [47]	Martinique 1990–1999	Definite SLE (ACR 1982) Population-based	4.7	64.2
Al-Arfaj et al., 2002 [221]	Saudi Arabia	Definite SLE (ACR 1982)		19.28
Lopez et al., 2003 [32]	Spain (Asturias) 1992–2002	Definite SLE (ACR 1982) Hospital	2.15 F = 3.64 M = 0.54	34.12 F = 57.91 M = 8.33
Alamanos et al., 2003 [35]	Northwest Greece 1982–2001	Definite SLE (ACR 1982) Multiple	1.9 F/M = 7.4	38.12
Naleway et al., 2005 [22]	US (Wisconsin) 1991–2001	Definite SLE (ACR 1982) Marshfield Clinic	5.1 F = 8.2 M = 1.9	78.5 F = 131.5 M = 24.8
Nightingale et al., 2006 [29]	UK 1992–1998	Definite SLE GPRB	3.02 F = 5.3 M = 0.45	
Govoni et al., 2006 [33]	Italy (Ferrera) 1996–2002	Definite SLE (ACR 1982) Hospital	2000: 2.01 2001: 1.15 2002: 2.6	57.9 F = 100.1 M = 12

(Continued)

TABLE 36.3 Incidence and Prevalence of SLE—cont'd

Author	Country/year	Methods	Incidence	Prevalence
Somers *et al.*, 2007 [52]	England 1990–1999	Definite SLE GPRB	4.71 F = 7.89 M = 1.53	
Mok *et al.*, 2008 [39]	Southern China 2000–2006	Definite SLE Hospital	3.1 F = 5.4	
Laustrup *et al.*, 2009 [34]	Denmark (Funen) 1995–2002	Definite SLE and Incomplete SLE	D-SLE: 1.04 I-SLE: 0.36	D-SLE: 28.3 I-SLE: 7.53

F, female rates; M, male rates; D-SLE, Definite SLE; I-SLE, Incomplete SLE; GPRB, General practice research database

prevalence reflects the difference between incidence and mortality rates and probably represents the cumulative effect of earlier diagnosis, improved recognition of the disease, and improved medical management with higher survival rates.

The Influence of Sex, Age, and Race on SLE Incidence and Prevalence

All studies show a marked female predominance in SLE. In large cohorts in Europe, the United States, and Latin America, the subjects are 90% female (90.8, 88, and 90%, respectively) [49–51] and the ratio of female to male subjects varies between 4.3 and 11.7. The incidence of SLE in women is higher than in men in all ages and the difference is greatest in the 15- to 40-year-old group, with fewer differences in children and after 70 years of age [22, 29, 32, 35, 44, 48].

Incidence and prevalence rates of SLE are lower in children and in the elderly in the majority of the studies [21, 31, 32, 42, 52]. In studies from Europe [53, 54], North America [55], and Japan [56], the annual incidence rate of SLE in children is less than 1 per 100,000 persons at risk. Prevalence rates of 6.3 and 8.8 are reported for Taiwanese children [57] and the children of native populations of British Columbia [58], respectively.

The mean age at diagnosis of SLE varies between 35 and 45 years, [21, 24, 32, 33, 35, 38, 45] and in all studies the mean age at diagnosis is lower in females. The peak incidence rate for women is during puberty and during the child-bearing years. In a study conducted by Nossent in Norway, the peak age for women was 30–49 years old [31] and is similar to Greece [35] and Curacao (15–44 years) [44]. However, Nived *et al.* in Sweden [25] and Somers *et al.* [52] in England found the peak age in women to be 45–64 and 40–54, respectively. In men, the peak age incidence and prevalence rates is usually after 60 years [22, 29, 32, 35, 41, 42, 44, 52].

SLE is more common in non-Caucasians, including persons of African descent, Asians, and aboriginal populations from the Americas and Australia. In 1964, Siegel *et al.* reported a higher incidence and prevalence in non-Whites in a study in Manhattan, and Blacks had the highest rate followed by Puerto Ricans [48]. In England, Hopkinson *et al.* found the highest rates in Afro-Caribbeans (age-standardized prevalence and incidence rates of 207 and 31.9) followed by Asians (age-standardized prevalence and incidence rates of 48.8 and 4.1) [59]. The higher incidence and prevalence in Blacks is well documented in studies from the U.S. and Europe, both in males and females, with the highest rates in women of child-bearing age [20, 21, 28, 40, 48, 59, 60]. SLE is believed to be rare in Africa but ascertainment bias is likely. Nevertheless, case series and some population data suggest a gradient of SLE prevalence from West Africa, the genetic root of Afro-Europeans, Afro-Caribbeans, and Afro-Americans, along the route of the Slave trade [61, 62].

The majority of studies reported that SLE is more common among Asians than in the White population [40, 59, 63–65]. In a study conducted in Hawaii [63], Serdula and Rhoads reported an overall age-adjusted prevalence higher in all Asian groups (9.7–20.4) than in Whites (5.8), with the highest rates among the Chinese (24.1). A threefold excess prevalence in Polynesians was found in New Zealand [64]. In England [59], Asians had a higher incidence and prevalence rates than Whites but lower than Afro-Caribbeans. SLE is higher in the Chinese than other Asian groups (92.9 in Chinese, 48.8 in other Asian ethnic groups), but studies conducted in Japan and China report incidence and prevalence rates comparable with rates found in studies conducted in the West [37, 39], which may indicate that there is a gradient of SLE in Asian populations similar to that observed in African groups, raising the possibility that environmental factors may be important. A higher prevalence of SLE is observed in North American Indians [24, 66] and in Australian aborigines [45] than in Caucasians in the same region.

Etiology

SLE is a complex, multifactorial disease, where it is believed that one or more environmental factors induce SLE in a genetically predisposed individual [67] (Figure 36.1). SLE, however, is a genetically complex

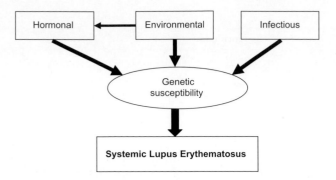

FIGURE 36.1 Multifactorial model and possible interactions between factors potentially involved in etiology of SLE. Infectious, hormonal, and environmental factors could trigger SLE in genetically susceptible subjects. Environmental factors could also cause SLE directly (*viz.*, drug-induced SLE) or act as modulators.

disease, with several disease predisposing genetic loci and non-Mendelian patterns of inheritance [68].

Genetic Factors

The importance of genetic factors in SLE is apparent [69]. A family history of autoimmune disease is a risk factor for developing SLE and the risk increases with the number of first-degree relatives with autoimmune disease [70, 71]. A family history of SLE was identified as a risk factor for developing SLE (OR, 3.3–6.8) in a case–control study from Sweden and the United States with 350 cases and 560 age-matched controls [72, 73]. In studies from European and Asian countries, familial clustering has been demonstrated with 4–11% of SLE patients having an affected first-degree relative [74–77]. The overall prevalence of SLE in first-degree relatives was 1.37 in Taiwan [75].

Differences in the concordance for SLE among monozygotic and dizygotic twins also support the importance of genetic susceptibility. The concordance rate for SLE among monozygotic twins is 24–69% compared to 2–9% concordance for dizygotic pairs [78, 79]. The lack of complete concordance of SLE among monozygotic twins also highlights the importance of nongenetic factors.

The development of powerful genetic probes in genome-wide association (GWA) studies has identified robustly replicated novel loci (e.g., ITGAM, BLK, BANK1, KIAA1542, PXK, and TNFAIP3), confirmed the association at other previously implicated loci (e.g., STAT4-STAT1, PTPN22, HLA-DR2, HALA-DR3), and generated large second-tier loci for further study (e.g., CRP, TREX1, MECP2-IRAK1) [80]. These are discussed in detail in Chapter 2.

Reproductive Factors

The marked predominance of women with SLE, its highest incidence and prevalence found in women of child-bearing age, and the equal prevalence in men and women before puberty and after menopause suggest the important role of endogenous estrogens and androgens in SLE etiology. Experimental data in mouse models of SLE (NZB/NZW, MRL/1pr, and BALB/c), discussed in greater detail in Chapter 4, show the role of sex hormones and its receptors in SLE onset and development. SLE nephritis is associated with estrogen level and androgenic hormones delay onset of SLE in these mice and are associated with milder disease [68, 81–87].

In humans, it has been reported over two decades ago that increased levels of estrogens and low levels of androgens are present in patients with SLE as compared to controls (healthy or individuals with other non-autoimmune disease) [88–92].

Early menarche and late menopause may increase exposure to endogenous estrogens and might theoretically increase risk of SLE. Results from studies are conflicting but the ones least subject to bias that are prospective and/or of larger size support the hypothesis that early menarche is a modest risk factor.

A prospective cohort study of 238,308 women in the Nurses Health Study (NHS) (121,700 women ages 30–55 in 1976 and 116,608 women ages 25–42 in 1989) found that early menarche (<10 years) was associated with a higher risk of SLE (RR, 2.1 [95%CI, 1.2–3.1]) [93]. One case–control study from Japan (282 cases, 292 controls) reported an increased risk associated with late menarche (>15 years) compared with early menarche (OR, 3.92; 95%CI, 1.66–8.81) [71]. In two other case–control studies, one from the Carolina Lupus Study [94] and another of Swedish patients [73], no association between age at menarche and SLE was found.

Available data do not support the hypothesis that late menopause is a risk factor to SLE. An increased risk of SLE in women with early menopause was found in the NHS (RR, 2.4, 95%CI, 1.1–5.1 for menopause <47 years) [93]. The population-based case–control study from South Carolina reported similar results with an increased risk of SLE conferred by natural menopause (HR, 2.4; 95% CI, 1.3–4.4), which is also seen in specific race analysis, with a risk of 3.6 (95% CI, 0.9–13.9) among African-Americans and 2.2 (95% CI, 1.0–4.6) among Whites [94].

During pregnancy, estrogens increase nearly 100-fold over peak menstrual levels [226]. As a consequence of increased estrogen levels, pregnancy could theoretically be associated with SLE onset. However, in two case–control studies (Sweden and Carolina Lupus study) [73, 94] and one cohort study (NHS and NHSII) [93], no association was found between parity, age at first birth, and number of pregnancies and SLE.

Only one epidemiologic cohort, the NHS, has been used to study the role of exogenous estrogens in SLE etiology. The NHS was assembled in 1976 and includes

121,000 female registered nurses aged 30—55 years living in 11 states who are contacted for updated exposures and other data every 2 years and in between for supplementary information when relevant. Analyzing data from this cohort, past users of oral contraceptives (OC) had a 1.4 RR (95% CI, 0.9—.1) adjusted for age and postmenopausal hormones of developing SLE compared with never-users. However, there was no significant increased risk with duration of OC use or time since first or last use of OCs [95]. In this cohort, ever-users of postmenopausal hormones had an age-adjusted RR of 2.1 (95% CI, 1.1—4.0), with higher risk in the current users than in past users (2.5 vs. 1.8) [96]. The risk was similar in a follow-up study of the same cohort by Costenbader *et al.* [93], but there was no association with the duration of postmenopausal hormones or time since last use. OCs were associated with an increased risk of developing SLE, but the risk was highest among women with short duration of OC use, and no association was found with type of hormones or the hormone potency of the OC.

The Carolina Lupus Study came to a different conclusion. This study assembled 240 SLE patients from community-based rheumatologists in South Carolina and compared them to control subjects identified through driver's license records frequency-matched to cases within 5 years of age, sex, and state. They found no correlation between OCs or HRT with SLE [94].

The difference among these studies could be explained by different methodologies, subjects assembled (nurses can provide better information), and eventually different composition and dosages in OCs and HRT treatments.

Prolactin hormone is a modulator of the immune system, and its receptors are expressed in many cells of the immune system, including T and B cells, macrophages, and NK cells. This hormone is involved in lymphocyte survival, activation, and proliferation, increases the production of IgM, IgG, and anti-dsDNA antibodies by peripheral mononuclear cells, and induces increased production of IFN-γ and IL-2 [97].

Consequently, there has been interest in a possible role of prolactin in SLE. Studies in murine lupus (NZB/W mice) have shown clear evidence that prolactin has a deleterious effect on disease activity [98—101]. In humans, some studies report increased prolactin level among SLE patients [102—106]. In a case—control study, the mean PRL level was higher in SLE patients than in controls ($p < 0.01$) [106]. However, in a cross-sectional study of 259 Mexican SLE patients, Leanos-Miranda and Cardenas-Mondragon reported that only some specific isoforms were increased in SLE patients (little Prolactin) [107]. Moreover, studies have not shown an increased risk associated with either the frequency or duration of breast-feeding [93, 94].

Infection

An early clinical observation in SLE was its onset following an infection. The variety of immunologic phenomena in response to infection has similarities to phenomena observed in SLE. These include complement changes, cytokines (such as IL-1, IL-2, IL-6, IL-10, INF), increased immunoglobulin classes, and diminished lymphocytes and granulocytes stimulated by infections, particularly viral infections. Nevertheless, a causal role for any specific infection has been difficult to prove or disprove [67]. To show that a specific pathogen is an etiologic factor in SLE, certain evidence would be required [108]:

1. The pathogen should be demonstrated.
2. The pathogen should be present before the onset of SLE.
3. The pathogen should induce immunological responses that could explain the pathogenesis of SLE and its autoimmune deregulation (such as inhibiting apoptosis, inducing humoral autoimmunity, generating INF).
4. The possibility of active SLE mimicking the infection should be excluded.
5. The syndrome is chronic or persistent.

The infections best studied in SLE have been viral, particularly parvovirus, cytomegalovirus (CMV), and Epstein-Barr Virus (EBV), and the most intriguing is the potential link is to EBV. A higher prevalence of seroreactivity to EBV among adults and children with SLE compared to sex- and age-matched healthy controls has been demonstrated. [109—115]. A higher prevalence of IgA antibody against EBV capsid antigen (EBV CA), which is increased in case of reactivation or reinfection, was found in SLE patients than in controls [113]. The association between EBV-IgA seroprevalence and SLE was stronger in African-Americans (OR, 5.6; 95%CI, 3—10.6) than in Caucasians (OR, 1.6; 95%CI, 10.6—10.4) [109]. Anti-EBNA-1 (Epstein-Barr nuclear antigen 1), a marker of previous infection, which appears 2—4 months after acute infection resolution and persists throughout life, was identified in higher titers in SLE patients than in controls [110, 112, 115].

The association between SLE and CMV, parvovirus, or herpes zoster is unconvincing. Three cases have been reported linking cytomegalovirus infection and onset of SLE [116—118]. Sekigawa *et al.* [119] reviewed three cases in which CMV infection was associated with SLE onset. Rider *et al.* reported a higher prevalence of seropositivity to CMV among SLE patients (OR, 14.53; 95%CI, 6.39—33.04) [120], but three case—control studies failed to confirm this association [111, 114, 115].

Overt parvovirus B19 infection has multisystem symptoms resembling SLE, which makes it very difficult to differentiate if it is a trigger of, or simply mimicking SLE. In a review by Ramos Casals and co-workers [121], 16 new onset SLE cases were linked to parvovirus B19. However, in a study of 99 SLE patients matched for age and sex with control subjects, IgG to parvovirus B19 was not increased in SLE patients [122].

In three case–control studies, which are vulnerable to recall bias, involving a total of 521 SLE subjects and 671 controls, there was a 2.5- to 6.4-fold risk of SLE with herpes zoster based on patient recall [72, 123, 124]. However, in four studies totaling 570 SLE patients and 857 controls, both children and adults, there was no serologic evidence of association between varicella and SLE [109, 111, 114, 115].

More controversial is the association between SLE and retrovirus, mainly human endogenous retrovirus. Studies with lupus-mice and in patients have shown a little smoke but no fire, and only a variable but small percentage of SLE patients have messenger RNA of retroviral origin [227–230].

Immunizations/Vaccinations

Cases of SLE associated with vaccination and immunization (tetanus, typhoid/paratyphoid, hepatitis B, anthrax) have been described and raise the possibility that they could be triggers for SLE in susceptible individuals or those with latent lupus [125–128]. The vaccine against hepatitis B has been the best studied, but the data are inconclusive. Using data from the Vaccine Adverse Events Reporting System, hepatitis B vaccine was associated with an increased risk of SLE compared to tetanus vaccine (OR 9.1; 95%CI, 2.3–7.6) [129]. However, one case–control study found no association between SLE and hepatitis B immunization [72]. For the other vaccinations and immunizations, there are either no data or systematic study of adequate size or with examination of these putative vaccines to prove or disprove their possible role in disease etiology.

Vaccines contain microbial and adjuvant components and it is conceivable that these may play a role in SLE etiology. The microbial components are generally considered responsible for adverse reactions and induced autoimmunity and little importance is given to adjuvant components. Squalene, pristane, and incomplete Freund adjuvant (FIA) are oil adjuvants used in vaccines. Experimental data in animal models show that these components can induce lupus-related autoantibody production, hypergammaglobulinemia, induce lupus-like syndromes, and accelerate lupus in lupus models (NZB/W F1, MRL$^{+/+}$, and BXSB) [130, 131].

A proposed link with autoimmune symptoms and diseases such as SLE and squalene, an adjuvant used in anthrax vaccine to develop more rapid (weeks compared to months) immunity during deployment of armed forces, is supported by animal studies and indirect and anecdotal evidence [130, 131] To date there has been no formal study and the Department of Defense does not permit study of active duty military [132]. An Institute of Medicine panel of experts reviewed the totality of the data and concluded that it was not a factor [240].

Environmental/Occupational Exposures

Many environmental and occupational exposures have been studied in the etiology of SLE. Medications are a category, technically, of environmental exposures, where the connection is strong for certain medications; these are discussed in Chapter 22. The challenge to identifying specific environmental or occupational exposures is having a valid and reliable measure of the exposure, its dose, and a sufficiently large population at risk for lupus, and the potential long latency of some agents. These exposures could be at the work site or home or both. Over time, many environmental factors have been linked to SLE etiology. From epidemiological research, some of them are definite a risk factor but do not explain most cases of SLE; others were refuted and the most of them remain as putative causes (Table 36.4).

SUN EXPOSURE

Sun exposure is a historic and widely recognized risk factor for SLE onset and flare, but its mechanism and epidemiology is unclear. Sunlight and ultraviolet radiation, mainly ultraviolet B (UV-B) radiation exposure, may induce apoptosis and/or the clearance of apoptotic cells induced by UV-B radiation [133, 134]. Werth [241] suggests that increased keratinocyte apoptosis results from increased TNF-α that induces apoptosis or to decreased clearance of apoptotic cells due to polymorphisms associated with decreased levels of collectins

TABLE 36.4 Environmental Risk Factors for SLE

Established	Putative	Refuted
• Ultraviolet exposure	• Anthrax vaccine	• Hair dye
• Medications (procainamide, hydralazine, etc.)	• Vaccines containing squalene	• Silicone breast implants
• Silica	• Organochlorines	
• Postmenopausal estrogen therapy	• Organic solvents	
• Current smoking	• Petroleum by-products	
	• Protein from soy bean, corn, spinach, and carrots which cross-react with anti-Sm antibody	
	• Heavy metals (Hg, Pb, Cd)	
	• Pesticides	
	• Hydrazine	
	• Alfalfa sprouts	

such as C1q and mannose-binding lectin. The role of UV-Bas a risk factor for SLE onset comes mainly from mouse data (MRL/lpr mice, BXSB, and NOD mice) [135]. In contrast to the pathogenic effect of UV-B, other studies show no deleterious effect and even a protective and therapeutic role for UV-A1 in SLE patients [136, 137]. Only one case—control (85 cases and 205 controls) study from Sweden has evaluated sun exposure as a risk factor for SLE onset and reported an increased risk associated with skin type I—II (skin type I, White, very fair, red, or blond hair, blue eyes, freckles, always burns, never tans; skin type II, White, fair, red, or blond hair, blue, hazel, or green eyes, usually burns, tans with difficulty) (OR, 2.9; 95%CI, 1.6—5.1) and with a history of serious sunburn as a youth (OR, 2.2; 95%CI, 1.2—4.1) [73]. Antibodies to SSA/Ro are associated with photosensitivity and this suggests that sun exposure may be more risky in these patients.

Although sunscreen and sun avoidance are standard recommendations in SLE to prevent flares, there are no clinical trials of their efficacy. The likelihood that rigorous trials could be mounted is doubtful since the risk—benefit ratio is small and sun avoidance and screens are recommended for skin cancer prevention in the general population.

SMOKING

Cigarette smoke contains many potentially toxic substances and metabolites. A small increase risk of SLE is associated with smoking, but the underlying mechanism is unclear. It is possible that individual response may be genetically determined and/or that smoke can act synergistically with other unknown toxins.

In a meta-analysis of nine studies (seven case—control and two cohort studies) from the United States, Japan, Sweden, and the UK) [138], current smokers had an increased risk for developing SLE compared with never-smokers (pooled OR estimate, 1.5; 95%CI, 1.09—2.08). Comparing ex-smokers with never smokers, the combined OR of the nine studies was 0.98 (95%CI, 0.75—1.27). After adjustment for, or stratified by age, sex, and race, the risk remains slightly increased (OR, 1.68; 95%CI, 1.12—2.52). Five studies evaluated the dose—response relationship and only one found a dose effect [139]. Two case—control studies from Sweden and Japan, not included in this meta-analysis, found an increased risk of SLE associated with smoking (OR, 1.8 and 2.26, respectively) [73, 140].

ALCOHOL CONSUMPTION

A meta-analysis [141] of five case—control studies (a total of 621 cases and 1439 controls) and one cohort study (including 64,500 African-American women) from populations in the US, Japan, the UK, and Sweden reported a protective effect for SLE for mild consumption of alcohol. Three studies, including the cohort study, found no association between alcohol consumption and SLE onset. Two of the studies, from the UK and Sweden, including 235 patients and 505 controls, reported a protective effect only for high doses and one reported a high risk of SLE associated with higher alcohol consumption. Therefore, "mild" (not reliably defined) alcohol consumption seems to have a protective effect, but more studies are necessary.

HAIR DYES

Drugs that contain aromatic amines, such as hydralazine and procainamide, have been associated with SLE-like syndromes. Hair dyes contain a similar component to these drugs, the aromatic amine hydrazine, which can be absorbed by the scalp and metabolized through acetylation. For this reason, the association between hair dyes and SLE has been the aim of different studies over the time. However, the weight of the evidence suggests no risk. Use of hair dyes and SLE onset was observed in two case—control studies from the US [142, 143], but not in three case—control studies from Sweden, the US, and the UK (500 SLE patients and 860 sex- and age-matched controls) [73, 144, 145] and the NHS cohort [146], in which no association with hair dye use, frequency of use, or duration of use was found.

SILICA

Occupational exposures to silica occur in a variety of industries, including construction, mining, road building, manufacturing, and repair services. Silica can act as an immune stimulant or adjuvant, resulting in the increased production of pro-inflammatory cytokines, including tumor necrosis factor and interleukin-1 [147, 148]. Silica may also result in increased exposure to self-antigens through generation of apoptotic material (silica is toxic to macrophages, resulting in both apoptosis and necrosis). For example, silica exposure exacerbated the development of SLE in the mouse models (NZB/NZW) [148], and autoantibodies from these mice recognized specific epitopes on apoptotic macrophages [149].

Silica exposure is associated with rheumatoid arthritis, SLE, scleroderma, and small vessel vasculitis with renal involvement [147]. Although silica exposure probably explains the minority of SLE cases, the epidemiologic evidence supports its importance. A high prevalence of SLE (93/100,000) was found among heavily silica-exposed uranium miners in Germany [150]. In the case—control Carolina Lupus Study, a strong association between SLE and silica exposure was found and it was dose-dependent (high exposure, OR, 4.6; low exposure, OR, 1.6) [151]. Exposure to silica was also associated with an increased risk of SLE among urban women in the Roxbury Lupus Project, mostly African-Americans from economically disadvantaged communities. In this case—control study, with 95 cases and 191 controls from the same area, exposure to silica was associated with a 4.3-fold increased risk (95%CI, 1.7—11.2) and an exposure response was seen for longer duration of exposures (for exposure to silica for 1—5 years, OR, 4; 95%CI, 1.2—12.9; for exposure for more than 5 years, OR, 4.9; 95%CI, 1.1—21.9) [152].

SILICONE-CONTAINING IMPLANTS

The established association between silica and autoimmune diseases and case reports of connective tissue disease in women who had received silicone breast implants raised the possibility that these and silicone-containing penile implants could cause autoimmune or connective tissue disease.

In a meta-analysis of nine cohort studies, nine case—control studies, and two cross-sectional studies, conducted in the US, Canada, Australia, the UK, and Northern Europe, found no association between silicone breast implants and connective tissue disease. In this meta-analysis, there was no increased risk of SLE among patients with breast implants when compared with patients without breast implants (RR, 0.65; 95%CI, 0.35—1.23) [153].

TOXINS/TOXIC WASTE/POLLUTION

Other environmental agents such as organic solvents, mercury, and pesticides have been of interest as a potential risk factor for SLE, but there has been little evidence for this to date.

In two epidemiologic studies, the Carolina Lupus Study and the Roxbury Lupus Project (total of 360 cases and 466 controls), no association was found between organic solvents and SLE [152, 154]. In the Carolina Lupus Study, exposure to mercury was associated with an increased risk of SLE with an OR of 4.8 (95%CI, 1.8—13) [154]. The association between SLE and pesticides, a diverse group of compounds, was evaluated in the Carolina Lupus Study. In this study, mixing pesticides for agricultural work was associated with an increased risk of SLE with an OR of 7.4 (95%CI, 1.4—40), but no increased risk was found in pesticide appliers [154].

The Roxbury Lupus Project included women predominantly from three African-American neighborhoods in Boston where 416 hazardous waste sites were identified and did not show an independent association between residential proximity to hazardous waste sites and the risk of earlier SLE [155].

MORBIDITY

Advances in the diagnosis of and the improved treatment of SLE have led to improved survival, making lupus a truly chronic ailment in many patients. As patients live longer, new complications have become apparent. In some of these, notably the cardiovascular complications and possibly avascular necrosis, it is not clear whether these are the result of SLE itself, its treatment, or a combination. Examples of morbidity more closely linked to the treatment are cyclophosphamide-associated bladder carcinoma and ovarian failure, steroid-associated cataracts, and osteopenia, and examples of morbidity more attributable to the effects of organ damage from active lupus include endstage lupus nephritis.

FLARE

SLE flare can be defined as exacerbation of overall or organ-specific SLE activity and manifested by inflammatory features. Operationally this might be defined as the change in clinical state prompting closer surveillance and/or necessitating a change in therapy (commonly increasing steroids and/or adding an immunosuppressive agent). In practice, most clinicians agree on when a severe lupus flare occurs but not on mild to moderate flares. Reproducible, valid, semi-quantitative SLE diseases activity measures, designed to be rated by physicians experienced in SLE evaluation, include recent versions of SLEDAI [231], SELENA-SLEDAI [156], BILAG [232], SLAM-R [233], and ECLAM [234]. Flare has been operationally defined for each by the ACR Response Criteria [235].

There is an effort in progress to develop, validate, and establish reliability in flare criteria for BILAG and SLEDAI using a common definition of flare, but as of this writing no full-length publications have appeared. Depending on the definition used, the incidence of overall flares is 0.07—0.6 [157—160] per patient per year. The majority of flares that SLE patients experience are mild — usually involving the skin, joints, constitutional symptoms — and not flares of renal or CNS systems [157].

What triggers flares has been much studied but remains a mystery. This stems in part from the practical and conceptual difficulties in studying the phenomena. The chief limitation may be that for any SLE population, the actual incidence of flare is low and requires a large well-characterized SLE population and appropriate analysis to deal with time-varying confounding variables such as treatment that might influence the number and the severity of the flare.

Based on animal studies and case reports, exogenous female hormones, both oral contraceptives (OC) and hormone replacement therapy (HRT), were once relatively contraindicated in SLE for fear of exacerbating the disease. However, two randomized clinical trials provide firm data to guide practice. The Safety of Estrogen in Lupus Erythematosus National Assessment (SELENA) was a double-blind randomized placebo-controlled equivalence trial of hormone replacement therapy in postmenopausal women [161] and a similar trial of oral contraceptive therapy in premenopausal women [162]. In both studies, the definition of flare and its classification as mild to moderate and severe was the same and based on the Physicians Global Assessment, SLEDAI, and additional items [156]. The SELENA study of oral contraceptives [162] included 183 women with inactive or stable active SLE, who were randomly assigned to receive either OCs (triphasic ethinyl estradiol 35 μg plus norethindrone at a dose of 0.5−1 mg for 12 cycles of 28 days) or placebo. No increase of risk of flare was observed in treated patients compared to the placebo group. For evaluation of HRT, 351 menopausal women, with inactive or mild disease, were randomly assigned to receive either 0.625 mg of conjugated estrogen daily plus 5 mg of medroxyprogesterone for days 1−12 of the month or inert placebo for 12 months. This study showed a slight increase of flare (RR of any flare, 1.39; $p = 0.02$) but only for mild to moderate flare (RR, 1.34; $p = 0.01$) and no increased risk for severe flare [161].

A double-blind, randomized, placebo controlled, 24-month clinical trial of 106 postmenopausal Mexican women reported that HRT in a continuous sequential regimen (0.625 mg of conjugated estrogen daily plus 5 mg of medroxyprogesterone for days 1−10 of the month) did not increase SLE flares (defined as an increase in SLEDAI scores ≥3 and severe flares as an increase ≥12) in lupus patients [163].

Pregnancy can be a trigger for flare of SLE; however, its frequency, particularly in subsets of SLE patients, is disputed [164−166]. This is discussed in detail in Chapter 35.

It is widely believed that ultraviolet light can aggravate SLE, mainly skin symptoms. However, the data on this issue are unclear. One retrospective study, involving 66 patients with SLE from France, reported more SLE flares not involving the skin after than before the summer period (RR, 1.75; $p = 0.006$) [167]. However, these results were not confirmed in a study from Israel, where no seasonal pattern in the overall disease activity was observed, but photosensitivity was increased during the summer months [168]. Similar findings were reported in a study conducted from Norway involving 21 patients [169]. A prospective study from Finland of 33 patients showed an increase of SLE activity in spring and summer compared to the winter period [170]. During the winter period, 12 volunteer patients were submitted to a photo-provocation test. No significant activity increase was observed among the photo-provoked patients compared to the nonprovoked group [170].

Stress is commonly believed by patients and some physicians to be a trigger of SLE exacerbations, but the evidence is limited by the number of subjects studied, by the definition of good stress and bad stress, the completeness of ascertainment of stress, the lack of a biologically informed mechanism, and the failure to address the dose response and probable recall bias. In two prospective studies involving 87 subjects with SLE, daily stress as measured by the Daily Stress Inventory [171] and a 44-item *ad hoc* questionnaire [172] was associated with SLE flares. However, another prospective study on 23 patients reported no increased SLE activity 2 weeks after a large increase in depression or anxiety [173].

MORBIDITY FROM DISEASE AND/OR TREATMENT

Organ Damage

SLE is an autoimmune inflammatory disease that ebbs and flows. During periods of uncontrolled disease activity, target organs may be irreversibly damaged. Cumulative organ damage in SLE implies irreversible damage but may be the result of the disease medication used to its treatment, both, and/or coexisting conditions or co-morbid illnesses. Attribution to a specific cause is commonly difficult. Making the determination that a patient's manifestations are due to damage rather than active inflammation is critical but not always possible with complete confidence. Treating an organ system damaged beyond repair with toxic agents is hazardous but stopping an agent that is controlling active disease is equally so. A formal evaluation used in studies to rate damage is the American College of Rheumatology/Systemic Lupus International Collaborating Clinics Damage Index (ACR-SLICC/DI). It makes no assumption about cause, only that damage exists, and is a valid and reproducible scale that includes 41 items on 12 organ systems and describes cumulative damage since the onset of SLE. ACR-SLICC/DI ranges from 0 to 49, with higher scores indicating greater damage [174].

Studies in Caucasian, Hispanic, Chinese, and African-American SLE patients show that damage increases gradually over time and the mean SDI score ranges from 0.3 at the first evaluation to 2.46 at the last one, with follow-up from 3 to 10 years [175−178]. Cardiovascular, neuropsychiatric, renal, and musculoskeletal systems were the most commonly damaged organs [175−178]. In a multiethnic cohort, Lupus in Minority Populations, Nature vs. Nurture (LUMINA), involving 258 subjects, a larger proportion of Hispanics (61%) than African-Americans (51%) or Caucasians (44%) had accrued any damage after a mean follow-up period of 5 years. ($p = 0.038$), with the greater increase between the first and last evaluation for Hispanics [178]. In this study, non-Caucasians had more renal damage than Caucasians (Hispanics 25%, African-Americans 17%, and Caucasians 7%, $p = 0.0102$). Integument damage (alopecia and skin scarring) was more frequent among African-Americans (Hispanics 10%, African-Americans 22%, and Caucasians 5%; $p = 0.002$) [178]. Ocular damage occurred in 15% of the group over a mean 5-year follow-up and cataracts were the most common manifestation [178]. A high prevalence of cataracts was also reported in a study from Canada (31.5% at 15 years) [177]. In ten European centers with 187 patients, the most frequent types of damage were hypertension (40%), cardiovascular disease (15%), osteoporosis, renal impairment, and central nervous system disease (14%) [179]. In the Euro-lupus cohort of 1000 subjects [49], over a 10-year prospective study, the most common complications were infections, hypertension, and osteoporosis.

In the LUMINA Study, predictors for cumulative damage were: older age, maximum dose of corticosteroid at the beginning, numbers of ACR criterion met, and increased disease activity at the beginning (T0) [178]. Mok *et al.* found that the number of flares and cyclophosphamide use were independent predictors for cumulative damage in Chinese patients [176]. In a Canadian study, the contribution of damage related to corticosteroids was lower in the early phase of the disease but constituted most of the damage at 15 years of disease [177].

Cardiovascular Disease

SLE patients have a five- to sixfold increased risk of cardiovascular disease compared to healthy persons and the risk was especially pronounced in young Caucasian women for whom the excess risk is greater than 50-fold [180]. A high prevalence of the Framingham traditional risk factors (hypertension, hypercholesterolemia, diabetes mellitus, obesity, and sedentary lifestyle) are seen in SLE patients [181]. However, these risk factors for cardiovascular disease do not completely explain the increased risk for cardiovascular disease among SLE patients. In a retrospective study with 263 SLE patients, the risk for any vascular outcome (7.5 for any coronary heart disease and 7.9 for stroke) remains high after controlling for traditional risk factors [182]. Factors such as renal disease, disease activity, antiphospholipid antibodies, corticosteroid dose, and hydroxychloroquine use may be contributing factors or effect modifiers but are difficult to study.

Although there have been strong recommendations to be more vigilant about risk factor detection and to vigorously manage these risk factors, it should be pointed out that there are no controlled studies to support these being efficacious. Nevertheless, studies on practice patterns in elite academic centers suggest a gap between these recommendations and what is practiced.

Infections

Infections are common complications in patients with SLE. Longitudinal studies indicate that 30–40% of SLE patients have at least one serious infection [183–185]. This has been confirmed by a prospective, controlled study of 110 patients followed for 3 years (RR, 1.63) [184]. Urinary tract infections, skin infections, pneumonia, and bacteremia without focus are the most common [183–186]. Active disease [183, 186], immunosuppressive therapy, corticosteroids [184, 185], hypocomplementemia [184, 186] are risk factors for infection.

Neoplasm

Determining whether SLE patients are at higher risk of developing a neoplasm compared to normal individuals or to other types of patients [187–190] requires epidemiologic evidence from populations of sufficient size with appropriate controls having the same likelihood of a cancer diagnosis. The best study in this regard over a calendar period of 9547 SLE patients between 1957 and 2000 (an average follow-up period of 8 years) showed an increased risk of cancer in SLE patients compared to what would be expected depending on age, sex, and geographic area (standardized incidence ratios (SIR), 1.15; 95%CI, 1.05–1.27), especially hematological cancers (SIR, 2.75; 95%CI, 2.13–3.49) (non-Hodgkin lymphoma is the most common with a SIR of 3.64) [190].

Health-Related Quality of Life

With the increased survival of SLE patients, health-related quality of life (HRQoL) and disability become important outcomes. SLE patients, as might be expected, have poorer quality of life compared to healthy individuals and comparable to those with illnesses such as AIDS, rheumatoid arthritis (RA), and Sjögren syndrome (SS). Attempts to understand the determinants of impaired quality of life show that disease activity, organ damage, age, disease duration, fatigue, and psychosocial factors probably affect HRQoL, but their interaction is not understood, nor what might be modifiable [191].

Morbidity Related to Treatment

Active lupus involving major organs is inevitably treated with large doses of steroids and other immunosuppressive agents as well as with many other agents, each with a possibility of untoward acute and chronic effects. Attributing these effects to a single drug exposure is difficult to do with certainty. Corticosteroids, despite their usefulness, are also associated with cataracts, thinning of the skin, increased blood pressure, fluid retention, facial hair, osteoporosis, Cushingoid syndrome, metabolic syndrome, diabetes, and avascular necrosis. Gastrointestinal bleeding is associated with corticosteroid use and the risk potentially increases with concomitant use of nonsteroidal anti-inflammatory drugs.

Antimalarials cause retinopathy in approximately 2.6% of SLE patients on chloroquine (total of 647 patients in four retrospective or cross-sectional studies) and 0.1% of those medicated with hydroxychloroquine (total of 2043 patients, one prospective and five retrospective studies), with a higher risk associated with chloroquine (OR, 25.88; 95%CI, 6.05−232.8; $p < 0.001$) [192].

Cyclophosphamide, an immunosuppressive drug used in SLE patients with major organ involvement, is mainly associated with hematuria, bladder cancer, and ovarian failure. The frequencies of these effects are increased with higher dose, longer duration of exposure, and younger age of the patients.

Mortality

Historically, lupus was once considered a rapidly fatal disease, as treatment options were limited or nonexistent. Although survival and its prognosis has improved remarkably in the last decades [193, 194], it remains a serious condition with greater mortality than the general population. Decreasing standardized mortality rates over the last few decades has been observed in case series and pooled case series from academic centers, but, significantly, when studied in the population with proportionate representation of Whites and minorities, the mortality rate declined between 1968 and 1977 and then increased between 1979 and 1998 (Table 36.5).

In 1955, Merrell and Shulman [195] published the first study of SLE prognosis in a case series from Johns Hopkins. In 99 patients, they found a 51% cumulative survival rate at 4 years after diagnosis. Since then, studies of patients with SLE have shown that survival in SLE patients has improved dramatically in the last five decades. Studies from America, Asia, and Europe show 5- and 10-year survival rates of 95% and 90%, respectively. The 20-year survival rate is around the 70% [196, 197].

In a pooled case series from the US, Canada, the UK, Sweden, Iceland, and South Korea, Bernatsky et al. showed a dramatic 60% decrease over time in the standardized all-cause mortality rates from 1970−1979 (SMR = 4.9) to 1990−2001 (SMR= 2.0) [194]. Two other reports from centers included in the Bernatsky study, Johns Hopkins and the University of Toronto, but over slightly different periods of time [197, 198] and another by Mok et al. in a Southern China population [39], show a similar trend. The increased survival rate is likely attributable to improved classification of the

TABLE 36.5 Five-Year and 10-Year Survival Rates for SLE

Reference	Year	N	Location	5-Year survival (%)	10-Year survival (%)
Merrell and Shulman [195]	1955	99	US	50	—
Urowitz et al. [216]	1976	81	Canada	75	63
Wallace et al. [222]	1981	609	US	88	79
Swaak et al. [223]	1989	110	Holland	—	87
Nossent [44]	1992	94	Curacao	60.1	45.7
Doria et al. [217]	2006	207	Italy	96	93
Kasitanon et al. [224]	2002	349	Thailand	84	74.9
Cervera et al. [49]	2003	908	Europe	95	92
Alamanos et al. [225]	2003	185	Greece	96.2	87.4
Pons-Estel [50]	2004	1214	Latin America	95	—
Mok et al. [207]	2005	285	China	92	83
Kasitanon et al. [197]	2006	1378	US	95	91
Funauchi et al. [196]	2007	306	Japan	94	92
Wadee et al. [204]	2007	226	South Africa	57−72	—

disease, early diagnosis, inclusion of milder cases in many cohorts, and improvement in treating concomitant co-morbidities [199].

When true population data are used, a different picture emerges. Kaslow and Masi [200] identified 3614 deaths attributed to SLE for the 1968−1978 period from national data in the United States. The average annual adjusted mortality rates were about four times higher for females (6.3 per million per person-year) than for males (1.6 per million per person-year). The overall rate for Blacks (8.8 per million per person-year) was 2.6 times that for Whites (3.4 per million per person-year). The highest mortality was observed among Black females: 14.8 per million per person-year.

Walsh and DeChello [201] repeated the study for the US (excluding Alaska and Hawaii) from 1984 to 1993. The crude mortality rate was 4.6 per million per person-year. Sex- and race-specific rates were higher in Blacks and women (White men = 1.5 per million per person-year; White women = 5.2 per million per person-year; Black men = 3.3 per million per person-year; Black women = 20.2 per million per person-year).

Using population data from the U.S. National Center for Health Statistics, Gordon et al. [193] documented a decline in the mortality rates between 1968 and 1977. Comparing mortality rates attributed to SLE in the United States between these two periods, 1968 through mid-1972 and 1972−1976, there was an overall decline in the age-adjusted mortality rates for the younger age group (1−49 years old).

A major study from the U.S. National Center of Chronic Disease Prevention and Health Promotion at the Centers for Disease Control analyzed the mortality from SLE during 1979−1998 and showed the crude death rate increased from 39 to 52 per 10 million population. The mortality rates were more than five times higher among women than men, and were more than three times higher among Blacks than Whites. Importantly, the highest rate was among Black women and increased most among those aged 45−64 (70%) [202]. Later diagnosis, more aggressive disease, problems in access to health care, less effective treatments, or poorer compliance may explain these disparities. [202]. Understanding these might help identify new public health strategies.

In Finland, the average annual mortality rate was 4.7 per million person-year in 1972−1978 [41]. In 1974−1983, a study from England and Wales on Caucasians only showed an overall mortality rate of 2.5 per million person-years, with a higher rate in females than males (3.94 vs. 1.02 per million persons-year) [203].

Bernatsky et al. [194] reported an overall Standardized Mortality Rate (SMR) of 2.9 from 1958 to 2001 in 9547 SLE patients from the academic health centers described above in this section. Using the United States as the referent group, the adjusted SMR point estimates were higher for Canadian and English patients, 1.8 and 1.6, respectively. The lowest SMR point estimates were observed in Sweden (0.8) and South Korea (0.7). The SMR for the US patients was higher in African-Americans than Caucasians (2.6 vs. 1.4).

In developing countries, survival rates at 5 years are significantly lower. In South Africa, a case series of 226 SLE patients the 5-year survival rate was 72% (1986−2003) [204]. In a study from Curacao between 1980 and 1989, 94 patients of African descent had 5-year and 10-year survival rates of 60.1 and 45.7%, respectively [44].

Predictors of Mortality

Despite overall improvement of survival in patients with SLE, patients remain at significantly increased risk of death and patients with SLE still succumb relatively early in the course of their disease. Mortality in SLE is clearly multifactorial and various studies have examined possible predictors of mortality, including sociodemographic and clinical features. The identification of these predictors is crucial and could provide useful information in the development of interventions aimed to further reduce mortality in SLE.

The effect of the gender on mortality has been controversial. Older studies showed a higher mortality in women [200]. Some studies report that males have higher mortality rates than do females with the disease. In a Spanish case series of 306 SLE patients, Blanco et al. found that male gender was associated with higher mortality rate (RR, 3.17; $p = 0.012$) [205]. In the Hopkins Lupus Cohort, a case series from one center, a similar increased risk of mortality was observed with male patients (HR, 2.4; $p = 0.004$) [197]. However, other studies suggest no gender effect in mortality rates [50, 206, 207].

Race plays a role in prognosis of SLE, although it has been difficult to separate out the effects of the race and socioeconomic status. Studies consistently show an increased mortality in persons of African descent in African-Americans, African-Caribbeans, and Hispanics compared with Caucasians [193, 197, 201, 202, 208].

Poverty is a poor prognostic factor in all diseases studied and SLE appears to be no exception. In the LUMINA Study, a significantly higher proportion of patients below the poverty level among the deceased patients than among the surviving patients (64% vs. 31.6%, $p = 0.002$) was found, and after multivariate analysis it remained significant (OR, 4.06; 95%CI, 1.5−11.01) [206]. Two other studies from the United States confirm the increased mortality associated with poverty [197, 209].

Lower education attainment is associated with poorer prognosis in most diseases studied, including SLE. In the GLADEL Multinational cohort from Latin America, a lower education level was a predictor of higher mortality ($\leq$10 years vs. 10 years; OR, 3.2; 95% CI, 1.3–7.6) [50]. Ward, using education level as a measure of socioeconomic status, found that mortality was higher in White men (OR, 2.12) and women (OR, 1.51) with education level less then 12 years. In this same study, lower mortality rates were found in Black patients with a lower education level [210]. In the LUMINA Cohort, education level was not a predictor of mortality [206].

The link between age at diagnosis and mortality remains controversial. In a study including all patients from the Hopkins Lupus Cohort from 1984 through 2004, Kasitanon et al. [197] found that late onset (>50 years of age) was related to lower survival (adjusted HR, 5.9; 95%CI, 2.5–14.4). In a nested case–control study in the LUMINA study, Bertoli et al. reported that patients with late-onset SLE (>50 years of age) have a higher risk of mortality, with an OR of 10.4 (95%CI, 3.07–37.56) [211]. In a cohort of 245 subjects from Spain followed between 1978 and 2001, later age at diagnosis (>60 years) was associated with low survival [212]. In a Southern Chinese population, Mok et al. showed that survival rates were significantly worse in late-onset patients (>50 years old) ($p < 0.0001$) [207]. However, age at onset was not a significant predictor of mortality in the studies conducted by Alarcón et al. in the LUMINA cohort [206] and Blanco et al. in Spanish patients [205], and late-onset SLE was not associated with worst prognosis in these studies.

Clinical Features and Mortality

Renal, neurological, and hematological involvement have been linked to worse prognosis and reduced survival in SLE patients.

Renal impairment at baseline, defined as an increase of creatinine level, was more common among the deceased patients in the LUMINA Study [206]. In the Hopkins Lupus Cohort, renal involvement was more common among patients who did not survive; however, in the adjusted analysis for demographic and other clinical features at baseline, renal involvement was not a predictor of mortality [197]. Bellomio et al., in a study from Argentina including 366 patients from 1990 through to 1998, found that renal involvement was associated with increased risk of mortality, which remained an independent predictor in the multivariate analysis (RR, 2.62; 95%CI, 1.13–6.10) [213]. In a small study of 41 patients in Mexico requiring hospitalization, renal involvement was associated with an increased risk of

mortality (OR, 4.6; 95%CI, 1.0–20.6) [214]. In a study from Japan including 306 patients, renal involvement was more frequent among patients who died [196]. However, renal involvement was not associated with lower survival in other studies [205, 207].

In the study conducted by Funauchi et al. in Japan, neurological involvement was more common among fatal cases of SLE [196]. Bellomio et al. found neurological involvement was associated with higher mortality; however, in the multivariate analysis it was not a predictor [213]. In studies from Spain, the United States, Japan, and Mexico, neurological involvement was not identified as a predictor of mortality [197, 205, 207, 214].

Hematological involvement includes leukopenia and lymphopenia, thrombocytopenia, and hemolytic anemia. However, their relative prognostic importance is not clear. In the studies from Argentina and Mexico, thrombocytopenia was associated with higher mortality, with an OR of 2.18 and 4.0, respectively [213, 214]. In the LUMINA study, patients who died had lower hematocrits compared to patients who survived ($p < 0.0001$) [206]. In the Hopkins Lupus Cohort, hemolytic anemia was associated with poor prognosis (adjusted HR, 2.1; 95%CI, 1.1–3.7) [197].

Low complement, a common serological finding in SLE, was associated with poor prognosis in the Hopkins Lupus Cohort. In this cohort, both low C3 and C4 were associated with increased risk of mortality with a fully adjusted HR of 2.7 and 2.2, respectively [197]. None of the studies reported an association between anti-dsDNA and mortality [206, 207].

Disease activity and damage due to SLE are important to define the health status of these patients and are commonly used in clinical practice. Studies have shown that both are associated with poor prognosis.

In the LUMINA Cohort [206], patients who died had higher disease activity at diagnosis, at enrollment in the study, and across all visits ($p < 0.001$). In the Mexican study, Zonana-Nacach et al. found that disease activity, assessed using SLEDAI, was a predictor of mortality [214]. In a multicenter, international cohort from 12 European countries with 2500 patients, Nossent et al. found that disease activity, assessed at time of death by ECLAM and SLEDAI, was high. When comparing early and late mortality, higher scores were associated with early death [215].

Organ damage accrual assessed by the SLICC/ACR Damage Index was associated with poor prognosis in studies from the LUMINA cohort in Mexico and China and a multinational cohort from Latin America [50, 206, 207, 214]. In this multinational cohort from Latin America, a SLICC/ACR DI one of 1 or higher was associated with an increased risk of mortality with a OR of 2.8 (95%CI, 1.2–6.4) [50]. In the European cohort, Nossent *et al.* found that SLICC/ACR DI was higher among deceased patients, but higher scores were more common in the late mortality group (4 vs. 7.2; $p = 0.034$) [215].

Causes of Death

A bimodal pattern of mortality in SLE was noted by Urowitz *et al.* on the basis of just 11 patients [216]. Based on this, they hypothesized that early mortality in SLE was related to severe disease activity and that late mortality was more likely due to complications of long-standing disease and immunosuppressive therapy as well as coexisting illnesses such as infections or atherosclerosis.

Thirty years after this editorial, Nossent *et al.*, studying some 2500 European patients, observed a low frequency of death within the 1st year and demonstrated that, in fact, a nearly constant mortality rate is observed over time [215].

Prolonged survival of patients with SLE makes it a chronic disease and longer periods of follow-up have demonstrated new morbidity such as accelerated atherosclerosis, osteoporosis, malignancy, and infection. Ongoing studies to parse out their cause continue, but it is likely that lupus itself and/or the treatment of its serious organ manifestations with steroids and immunosuppressive agents are the most important.

Active disease is itself directly responsible for increased mortality and is the cause of death in a large number of patients. The reported variation may be due to how "active disease" is defined, case selection, and differences in treatment in individual studies, which would be interesting to study. In 1985, Helve reported that active disease was the cause of death in 52% of SLE patients in Finland [41]. During the 10-year period of follow-up of the Euro-Lupus Cohort, active disease was the cause of death in 26.5% of the patients [49]. Higher disease activity was responsible for 35% of the deaths in an Italian cohort (35.3%) [217] and the GLADEL cohort (35%) [50] and 63% of European patients [215]. In this prospective study, 70% of the patients had active disease by ECLAM (score >2) and 56% of patients by SLEDAI (score ≥10) at the time of death. Comparing patients who died less than 5 years after onset with those who succumbed later, active disease contributed more often to early death (61.1% vs. 34.5%; $p = 0.018$).

However, SLE is a multiorgan disease and many systems may contribute to mortality, often simultaneously: renal, neurologic, hematological, and cardiopulmonary involvement are the most common causes of death due to active disease [41, 49, 217].

The risk of infection is increased in SLE patients and is associated with immunosuppressive therapy and/or innate impairment of the SLE immune system to protect against infections. Infections are a common cause of death [41, 49, 50]. In a SLE cohort reported by Nossent *et al.*, infection was the cause of death [215] in 30% of patients with a single cause of death and in 23% of subjects with more than one cause of death. The most common infections were pneumonia (36%) and sepsis of unknown origin (30%) [215]. Bernatsky *et al.* reported a SMR of 5 for infection (not including pneumonia) as the cause of death and a SMR of 2.6 for pneumonia as the cause of death [194].

Infections are blamed in 33% of the deaths in South Africa [204], 55% in China [207], and 31% in Mexico [214].

Cardiovascular Disease

Although all-cause mortality in SLE has declined, the risk of death due to cardiovascular disease remains essentially unchanged worldwide [49, 194, 215, 216, 218–220]. The best evidence suggests that atherosclerotic heart disease is some 50 times more common in women with SLE than the control population [236]. Correcting for the traditional risk factors of family history, hypertension, diabetes, cholesterol, and smoking, SLE and/or its treatment still increases that risk some seven- to 17-fold [237–239], suggesting that even controlling known risks factors optimally may not improve outcomes [240].

Malignancy

It is not clear that malignancies occur more frequently in SLE patients compared with the general population, but as a cause of death in SLE this cause varies. In the Euro-Lupus cohort, malignant neoplasms were the cause of death in 5.9% of patients [49]. Nossent *et al.* reported that 8% of deaths were due to malignancies (5% hematological malignancies) [215]. In the largest cohort of 9547 patients, with a long period of observation (1958–2001), Bernatsky *et al.* identified an increased risk of death due to specific malignancies, including hematological malignancies (SMR, 2.1), particularly non-Hodgkin lymphoma (SMR, 2.8), and lung cancer (SMR, 2.3) [194].

CONCLUSIONS

Descriptive and hypothesis testing or analytical epidemiologic studies have provided useful insights into the cause, course, and modifying factors for outcome. Much of the epidemiologic work on SLE is neither truly population-based (but from single centers or assemblies of patients from multiple centers), nor standardized by age, gender, or ethnicity, which is essential for meaningful comparisons. Small numbers lead to unstable estimates and imprecision.

Epidemiologic clues for potential cause(s) remain elusive, not so much from lack of effort, but lack of strong signals. Even the most established risk factors, except for gender, explain the minority of cases. Overall, epidemiologic studies demonstrate a good and possibly improved prognosis since its first description, making SLE a chronic but potentially dangerous illness. Whether this is due to improved diagnosis, improved treatment, diagnosis of milder phenotypes, and/or that SLE has changed (herd immunity) may never be resolved. Studies also indicate considerable variations in outcome and differences between ethnic groups and between industrial and preindustrial countries that are probably increasing, comprising a major research challenge and a possible public health opportunity to correct these inequalities.

Long-term morbidity is appreciated but its study is only just beginning. Future studies will need to look at what is really modifiable and to test interventions rigorously.

The era of molecular epidemiology and genomics is well underway but harvesting its riches for patients will require increased cooperation to establish sufficient numbers of carefully phenotyped patients with all the varied presentations of this heterogeneous disorder and, importantly, over time.

References

[1] O. Nived, G. Sturfelt, Epidemiology of systemic lupus erythematosus, Monogr Allergy 21 (1987) 197–214.

[2] A.S. Cohen, W.E. Reynolds, E.C. Franklin, J.P. Kulka, M.W. Ropes, L.E. Shulman, et al., Preliminary criteria for the classification of systemic lupus erythematosus, Bull Rheum Dis 21 (1971) 643–648.

[3] E.M. Tan, A.S. Cohen, J.F. Fries, A.T. Masi, D.J. McShane, N.F. Rothfield, et al., The 1982 revised criteria for the classification of systemic lupus erythematosus, Arthritis Rheum 25 (1982) 1271–1277.

[4] J.F. Fries, Methodology of validation of criteria for SLE, Scand J Rheumatol Suppl 65 (1987) 25–30.

[5] C.M. Passas, R.L. Wong, M. Peterson, M.A. Testa, N.F. Rothfield, A comparison of the specificity of the 1971 and 1982 American Rheumatism Association criteria for the classification of systemic lupus erythematosus, Arthritis Rheum 28 (1985) 620–623.

[6] R.E. Levin, A. Weinstein, M. Peterson, M.A. Testa, N.F. Rothfield, A comparison of the sensitivity of the 1971 and 1982 American Rheumatism Association criteria for the classification of systemic lupus erythematosus, Arthritis Rheum 27 (1984) 530–538.

[7] R. Yokohari, T. Tsunematsu, Application, to Japanese patients, of the 1982 American Rheumatism Association revised criteria for the classification of systemic lupus erythematosus, Arthritis Rheum 28 (1985) 693–698.

[8] J.D. Clough, M. Elrazak, L.H. Calabrese, R. Valenzuela, W.B. Braun, G.W. Williams, Weighted criteria for the diagnosis of systemic lupus erythematosus, Arch Intern Med 144 (1984) 281–285.

[9] K.H. Costenbader, E.W. Karlson, L.A. Mandl, Defining lupus cases for clinical studies: the Boston weighted criteria for the classification of systemic lupus erythematosus, J Rheumatol 29 (2002) 2545–2550.

[10] M.L. Sanchez, G.S. Alarcon, G. McGwin Jr., B.J. Fessler, R.P. Kimberly, Can the weighted criteria improve our ability to capture a larger number of lupus patients into observational and interventional studies? A comparison with the American College of Rheumatology criteria, Lupus 12 (2003) 468–470.

[11] E.L. Smith, R.H. Shmerling, The American College of Rheumatology criteria for the classification of systemic lupus erythematosus: strengths, weaknesses, and opportunities for improvement, Lupus 8 (1999) 586–595.

[12] R.C. Lawrence, C.G. Helmick, F.C. Arnett, R.A. Deyo, D.T. Felson, E.H. Giannini, et al., Estimates of the prevalence of arthritis and selected musculoskeletal disorders in the United States, Arthritis Rheum 41 (1998) 778–799.

[13] N. Danchenko, J.A. Satia, M.S. Anthony, Epidemiology of systemic lupus erythematosus: a comparison of worldwide disease burden, Lupus 15 (2006) 308–318.

[14] G.J. Pons-Estel, G.S. Alarcon, L. Scofield, L. Reinlib, G.S. Cooper, Understanding the epidemiology and progression of Systemic Lupus Erythematosus, Semin Arthritis Rheum 39 (2010) 257–268.

[15] S. Bernatsky, L. Joseph, C.A. Pineau, R. Tamblyn, D.E. Feldman, A.E. Clarke, A population-based assessment of systemic lupus erythematosus incidence and prevalence—results and implications of using administrative data for epidemiological studies, Rheumatology (Oxford) 46 (2007) 1814–1818.

[16] M. Siegel, S.L. Lee, The epidemiology of systemic lupus erythematosus, Semin Arthritis Rheum 3 (1973) 1–54.

[17] M. Siegel, H.L. Holley, S.L. Lee, Epidemiologic studies on systemic lupus erythematosus. Comparative data for New York City and Jefferson County, Alabama, 1956-1965, Arthritis Rheum 13 (1970) 802–811.

[18] M. Siegel, S.L. Lee, D. Widelock, G.J. Reillyeb, Wise, S.B. Zingale, H.T. Fuerst, The epidemiology of systemic lupus erythematosus: preliminary results in New York City, J Chronic Dis 15 (1962) 131–140.

[19] C.J. Michet Jr., C.H. McKenna, L.R. Elveback, R.A. Kaslow, L.T. Kurland, Epidemiology of systemic lupus erythematosus and other connective tissue diseases in Rochester, Minnesota, 1950 through [1979] Mayo Clin Proc 60 (1985) 105–113.

[20] D.J. McCarty, S. Manzi, T.A. Medsger Jr., R. Ramsey-Goldman, R.E. LaPorte, C.K. Kwoh, Incidence of systemic lupus erythematosus. Race and gender differences, Arthritis Rheum 38 (1995) 1260–1270.

[21] M.C. Hochberg, The incidence of systemic lupus erythematosus in Baltimore, Maryland, 1970-1977, Arthritis Rheum 28 (1985) 80–86.

[22] A.L. Naleway, M.E. Davis, R.T. Greenlee, D.A. Wilson, D.J. McCarty, Epidemiology of systemic lupus erythematosus in rural Wisconsin, Lupus 14 (2005) 862–866.

[23] K.M. Uramoto, C.J. Michet Jr., J. Thumboo, J. Sunku, W.M. O'Fallon, S.E. Gabriel, Trends in the incidence and mortality of systemic lupus erythematosus, 1950-[1992] Arthritis Rheum 42 (1999) 46–50.

[24] C.A. Peschken, J.M. Esdaile, Systemic lupus erythematosus in North American Indians: a population based study, J Rheumatol 27 (2000) 1884–1891.

[25] O. Nived, G. Sturfelt, F. Wollheim, Systemic lupus erythematosus in an adult population in southern Sweden: incidence, prevalence and validity of ARA revised classification criteria, Br J Rheumatol 24 (1985) 147–154.

[26] H. Jonsson, O. Nived, G. Sturfelt, A. Silman, Estimating the incidence of systemic lupus erythematosus in a defined population using multiple sources of retrieval, Br J Rheumatol 29 (1990) 185—188.

[27] C. Stahl-Hallengren, A. Jonsen, O. Nived, G. Sturfelt, Incidence studies of systemic lupus erythematosus in Southern Sweden: increasing age, decreasing frequency of renal manifestations and good prognosis, J Rheumatol 27 (2000) 685—691.

[28] N.D. Hopkinson, M. Doherty, R.J. Powell, The prevalence and incidence of systemic lupus erythematosus in Nottingham, UK, 1989-[1990] Br J Rheumatol 32 (1993) 110—115.

[29] A.L. Nightingale, R.D. Farmer, C.S. de Vries, Incidence of clinically diagnosed systemic lupus erythematosus 1992-1998 using the UK General Practice Research Database, Pharmacoepidemiol Drug Saf 15 (2006) 656—661.

[30] S. Gudmundsson, K. Steinsson, Systemic lupus erythematosus in Iceland 1975 through [1984] A nationwide epidemiological study in an unselected population, J Rheumatol 17 (1990) 1162—1167.

[31] H.C. Nossent, Systemic lupus erythematosus in the Arctic region of Norway, J Rheumatol 28 (2001) 539—546.

[32] P. Lopez, L. Mozo, C. Gutierrez, A. Suarez, Epidemiology of systemic lupus erythematosus in a northern Spanish population: gender and age influence on immunological features, Lupus 12 (2003) 860—865.

[33] M. Govoni, G. Castellino, S. Bosi, N. Napoli, F. Trotta, Incidence and prevalence of systemic lupus erythematosus in a district of north Italy, Lupus 15 (2006) 110—113.

[34] H. Laustrup, A. Voss, A. Green, P. Junker, Occurrence of systemic lupus erythematosus in a Danish community: an 8-year prospective study, Scand J Rheumatol 38 (2009) 128—132.

[35] Y. Alamanos, P.V. Voulgari, C. Siozos, P. Katsimpri, S. Tsintzos, G. Dimou, et al., Epidemiology of systemic lupus erythematosus in northwest Greece 1982-[2001] J Rheumatol 30 (2003) 731—735.

[36] N.M. Anstey, I. Bastian, H. Dunckley, B.J. Currie, Systemic lupus erythematosus (SLE): different prevalences in different populations of Australian aborigines. Aust N Z J Med 25 (1995) 736.

[37] K. Iseki, F. Miyasato, T. Oura, H. Uehara, K. Nishime, K. Fukiyama, An epidemiologic analysis of end-stage lupus nephritis, Am J Kidney Dis 23 (1994) 547—554.

[38] M.J. Vilar, E.I. Sato, Estimating the incidence of systemic lupus erythematosus in a tropical region (Natal, Brazil), Lupus 11 (2002) 528—532.

[39] C.C. Mok, C.H. To, L.Y. Ho, K.L. Yu, Incidence and mortality of systemic lupus erythematosus in a southern Chinese population, 2000-2006, J Rheumatol 35 (2008) 1978—1982.

[40] E.F. Chakravarty, T.M. Bush, S. Manzi, A.E. Clarke, M.M. Ward, Prevalence of adult systemic lupus erythematosus in California and Pennsylvania in 2000: estimates obtained using hospitalization data, Arthritis Rheum 56 (2007) 2092—2094.

[41] T. Helve, Prevalence and mortality rates of systemic lupus erythematosus and causes of death in SLE patients in Finland, Scand J Rheumatol 14 (1985) 43—46.

[42] M.C. Hochberg, Prevalence of systemic lupus erythematosus in England and Wales, 1981-2, Ann Rheum Dis 46 (1987) 664—666.

[43] G.O. Eilertsen, A. Becker-Merok, J.C. Nossent, The influence of the 1997 updated classification criteria for systemic lupus erythematosus: epidemiology, disease presentation, and patient management, J Rheumatol 36 (2009) 552—559.

[44] J.C. Nossent, Systemic lupus erythematosus on the Caribbean island of Curacao: an epidemiological investigation, Ann Rheum Dis 51 (1992) 1197—1201.

[45] M. Segasothy, P.A. Phillips, Systemic lupus erythematosus in Aborigines and Caucasians in central Australia: a comparative study, Lupus 10 (2001) 439—444.

[46] D.M. Grennan, D. Bossingham, Systemic lupus erythematosus (SLE): different prevalences in different populations of Australian aboriginals. Aust N Z J Med 25 (1995) 182—183.

[47] C. Deligny, L. Thomas, F. Dubreuil, C. Theodose, A.M. Garsaud, P. Numeric, et al., Systemic lupus erythematosus in Martinique: an epidemiologic study, Rev Med Interne 23 (2002) 21—29.

[48] M. Siegel, E.B. Reilly, S.L. Lee, H.T. Fuerst, M. Seelenfreund, Epidemiology of systemic lupus erythematosus: Time trend and racial differences, Am J Public Health Nations Health 54 (1964) 33—43.

[49] R. Cervera, M.A. Khamashta, J. Font, G.D. Sebastiani, A. Gil, P. Lavilla, et al., Morbidity and mortality in systemic lupus erythematosus during a 10-year period: a comparison of early and late manifestations in a cohort of 1,000 patients, Medicine (Baltimore) 82 (2003) 299—308.

[50] B.A. Pons-Estel, L.J. Catoggio, M.H. Cardiel, E.R. Soriano, S. Gentiletti, A.R. Villa, et al., The GLADEL multinational Latin American prospective inception cohort of 1,214 patients with systemic lupus erythematosus: ethnic and disease heterogeneity among "Hispanics," Medicine (Baltimore) 83 (2004) 1—17.

[51] E.M. Ginzler, H.S. Diamond, M. Weiner, M. Schlesinger, J.F. Fries, C. Wasner, et al., A multicenter study of outcome in systemic lupus erythematosus. I. Entry variables as predictors of prognosis, Arthritis Rheum 25 (1982) 601—611.

[52] E.C. Somers, S.L. Thomas, L. Smeeth, W.M. Schoonen, A.J. Hall, Incidence of systemic lupus erythematosus in the United Kingdom, 1990-[1999] Arthritis Rheum 57 (2007) 612—618.

[53] C. Huemer, M. Huemer, T. Dorner, J. Falger, H. Schacherl, M. Bernecker, et al., Incidence of pediatric rheumatic diseases in a regional population in Austria, J Rheumatol 28 (2001) 2116—2119.

[54] P.M. Pelkonen, H.J. Jalanko, R.K. Lantto, A.L. Makela, M.A. Pietikainen, H.A. Savolainen, et al., Incidence of systemic connective tissue diseases in children: a nationwide prospective study in Finland, J Rheumatol 21 (1994) 2143—2146.

[55] P.N. Malleson, M.Y. Fung, A.M. Rosenberg, The incidence of pediatric rheumatic diseases: results from the Canadian Pediatric Rheumatology Association Disease Registry, J Rheumatol 23 (1996) 1981—1987.

[56] S. Fujikawa, M. Okuni, A nationwide surveillance study of rheumatic diseases among Japanese children, Acta Paediatr Jpn 39 (1997) 242—244.

[57] J.L. Huang, T.C. Yao, L.C. See, Prevalence of pediatric systemic lupus erythematosus and juvenile chronic arthritis in a Chinese population: a nation-wide prospective population-based study in Taiwan, Clin Exp Rheumatol 22 (2004) 776—780.

[58] K.M. Houghton, J. Page, D.A. Cabral, R.E. Petty, L.B. Tucker, Systemic lupus erythematosus in the pediatric North American Native population of British Columbia, J Rheumatol 33 (2006) 161—163.

[59] N.D. Hopkinson, M. Doherty, R.J. Powell, Clinical features and race-specific incidence/prevalence rates of systemic lupus erythematosus in a geographically complete cohort of patients, Ann Rheum Dis 53 (1994) 675—680.

[60] A.E. Johnson, C. Gordon, R.G. Palmer, P.A. Bacon, The prevalence and incidence of systemic lupus erythematosus in Birmingham, England. Relationship to ethnicity and country of birth, Arthritis Rheum 38 (1995) 551—558.

[61] D.P. Symmons, Frequency of lupus in people of African origin, Lupus 4 (1995) 176—178.

[62] S.C. Bae, P. Fraser, M.H. Liang, The epidemiology of systemic lupus erythematosus in populations of African ancestry: a critical review of the "prevalence gradient hypothesis," Arthritis Rheum 41 (1998) 2091–2099.

[63] M.K. Serdula, G.G. Rhoads, Frequency of systemic lupus erythematosus in different ethnic groups in Hawaii, Arthritis Rheum 22 (1979) 328–333.

[64] H.H. Hart, R.R. Grigor, D.E. Caughey, Ethnic difference in the prevalence of systemic lupus erythematosus, Ann Rheum Dis 42 (1983) 529–532.

[65] A. Samanta, J. Feehally, S. Roy, F.E. Nichol, P.J. Sheldon, J. Walls, High prevalence of systemic disease and mortality in Asian subjects with systemic lupus erythematosus, Ann Rheum Dis 50 (1991) 490–492.

[66] G.S. Boyer, D.W. Templin, A.P. Lanier, Rheumatic diseases in Alaskan Indians of the southeast coast: high prevalence of rheumatoid arthritis and systemic lupus erythematosus, J Rheumatol 18 (1991) 1477–1484.

[67] G.S. Cooper, M.A. Dooley, E.L. Treadwell, E.W. St Clair, C.G. Parks, G.S. Gilkeson, Hormonal, environmental, and infectious risk factors for developing systemic lupus erythematosus, Arthritis Rheum 41 (1998) 1714–1724.

[68] Z.M. Sthoeger, Z. Bentwich, H. Zinger, E. Mozes, The beneficial effect of the estrogen antagonist, tamoxifen, on experimental systemic lupus erythematosus, J Rheumatol 21 (1994) 2231–2238.

[69] L.A. Criswell, The genetic contribution to systemic lupus erythematosus, Bull NYU Hosp Jt Dis 66 (2008) 176–183.

[70] R. Priori, E. Medda, F. Conti, E.A. Cassara, M.G. Danieli, R. Gerli, et al., Familial autoimmunity as a risk factor for systemic lupus erythematosus and vice versa: a case-control study, Lupus 12 (2003) 735–740.

[71] C. Nagata, S. Fujita, H. Iwata, Y. Kurosawa, K. Kobayashi, M. Kobayashi, et al., Systemic lupus erythematosus: a case-control epidemiologic study in Japan, Int J Dermatol 34 (1995) 333–337.

[72] G.S. Cooper, M.A. Dooley, E.L. Treadwell, E.W. St Clair, G.S. Gilkeson, Risk factors for development of systemic lupus erythematosus: allergies, infections, and family history, J Clin Epidemiol 55 (2002) 982–989.

[73] A.A. Bengtsson, L. Rylander, L. Hagmar, O. Nived, G. Sturfelt, Risk factors for developing systemic lupus erythematosus: a case-control study in southern Sweden, Rheumatology (Oxford) 41 (2002) 563–571.

[74] S. Koskenmies, E. Widen, J. Kere, H. Julkunen, Familial systemic lupus erythematosus in Finland, J Rheumatol 28 (2001) 758–760.

[75] H.M. Yeh, J.R. Chen, J.J. Tsai, W.J. Tsai, J.H. Yen, H.W. Liu, Prevalence of familial systemic lupus erythematosus in Taiwan, Gaoxiong Yi Xue Ke Xue Za Zhi 9 (1993) 664–667.

[76] J. Wang, S. Yang, J.J. Chen, S.M. Zhou, S.M. He, Y.H. Liang, et al., Systemic lupus erythematosus: a genetic epidemiology study of 695 patients from China, Arch Dermatol Res 298 (2007) 485–491.

[77] J.S. Lawrence, C.L. Martins, G.L. Drake, A family survey of lupus erythematosus. [1] Heritability, J Rheumatol 14 (1987) 913–921.

[78] D. Deapen, A. Escalante, L. Weinrib, D. Horwitz, B. Bachman, P. Roy-Burman, et al., A revised estimate of twin concordance in systemic lupus erythematosus, Arthritis Rheum 35 (1992) 311–318.

[79] S.R. Block, J.B. Winfield, M.D. Lockshin, W.A. D'Angelo, C.L. Christian, Studies of twins with systemic lupus erythematosus. A review of the literature and presentation of 12 additional sets, Am J Med 59 (1975) 533–552.

[80] I.T. Harley, K.M. Kaufman, C.D. Langefeld, J.B. Harley, J.A. Kelly, Genetic susceptibility to SLE: new insights from fine mapping and genome-wide association studies, Nat Rev Genet 10 (2009) 285–290.

[81] J.R. Roubinian, R. Papoian, N. Talal, Androgenic hormones modulate autoantibody responses and improve survival in murine lupus, J Clin Invest 59 (1977) 1066–1070.

[82] H. Carlsten, A. Tarkowski, R. Holmdahl, L.A. Nilsson, Oestrogen is a potent disease accelerator in SLE-prone MRL lpr/lpr mice, Clin Exp Immunol 80 (1990) 467–473.

[83] G. Kokeny, M. Godo, E. Nagy, M. Kardos, K. Kotsch, P. Casalis, et al., Skin disease is prevented but nephritis is accelerated by multiple pregnancies in autoimmune MRL/LPR mice, Lupus 16 (2007) 465–477.

[84] Y.Y. Dhaher, B. Greenstein, E. de Fougerolles Nunn, M. Khamashta, G.R. Hughes, Strain differences in binding properties of estrogen receptors in immature and adult BALB/c and MRL/MP-lpr/lpr mice, a model of systemic lupus erythematosus, Int J Immunopharmacol 22 (2000) 247–254.

[85] K.K. Bynote, J.M. Hackenberg, K.S. Korach, D.B. Lubahn, P.H. Lane, K.A. Gould, Estrogen receptor-alpha deficiency attenuates autoimmune disease in (NZB x NZW) F1 mice, Genes Immun 9 (2008) 137–152.

[86] B. Greenstein, R. Roa, Y. Dhaher, E. Nunn, A. Greenstein, M. Khamashta, et al., Estrogen and progesterone receptors in murine models of systemic lupus erythematosus, Int Immunopharmacol 1 (2001) 1025–1035.

[87] M. Blank, S. Mendlovic, H. Fricke, E. Mozes, N. Talal, Y. Shoenfeld, Sex hormone involvement in the induction of experimental systemic lupus erythematosus by a pathogenic anti-DNA idiotype in naive mice, J Rheumatol 17 (1990) 311–317.

[88] R.G. Lahita, H.L. Bradlow, E. Ginzler, S. Pang, M. New, Low plasma androgens in women with systemic lupus erythematosus, Arthritis Rheum 30 (1987) 241–248.

[89] R.G. Lahita, H.L. Bradlow, H.G. Kunkel, J. Fishman, Alterations of estrogen metabolism in systemic lupus erythematosus, Arthritis Rheum 22 (1979) 1195–1198.

[90] M. Folomeev, M. Dougados, J. Beaune, J.C. Kouyoumdjian, K. Nahoul, B. Amor, et al., Plasma sex hormones and aromatase activity in tissues of patients with systemic lupus erythematosus, Lupus 1 (1992) 191–195.

[91] P. Jungers, K. Nahoul, C. Pelissier, M. Dougados, N. Athea, F. Tron, et al., Plasma androgens in women with disseminated lupus erythematosus, Presse Med 12 (1983) 685–688.

[92] C. Lavalle, E. Loyo, R. Paniagua, J.A. Bermudez, J. Herrera, A. Graef, et al., Correlation study between prolactin and androgens in male patients with systemic lupus erythematosus, J Rheumatol 14 (1987) 268–272.

[93] K.H. Costenbader, D. Feskanich, M.J. Stampfer, E.W. Karlson, Reproductive and menopausal factors and risk of systemic lupus erythematosus in women, Arthritis Rheum 56 (2007) 1251–1262.

[94] G.S. Cooper, M.A. Dooley, E.L. Treadwell, E.W. St Clair, G.S. Gilkeson, Hormonal and reproductive risk factors for development of systemic lupus erythematosus: results of a population-based, case-control study, Arthritis Rheum 46 (2002) 1830–1839.

[95] J. Sanchez-Guerrero, E.W. Karlson, M.H. Liang, D.J. Hunter, F.E. Speizer, G.A. Colditz, Past use of oral contraceptives and the risk of developing systemic lupus erythematosus, Arthritis Rheum 40 (1997) 804–808.

[96] J. Sanchez-Guerrero, M.H. Liang, E.W. Karlson, D.J. Hunter, G.A. Colditz, Postmenopausal estrogen therapy and the risk for developing systemic lupus erythematosus, Ann Intern Med 122 (1995) 430–433.

[97] E. Peeva, J. Venkatesh, D. Michael, B. Diamond, Prolactin as a modulator of B cell function: implications for SLE, Biomed Pharmacother 58 (2004) 310–319.

[98] R. McMurray, D. Keisler, S. Izui, S.E. Walker, Hyperprolactinemia in male NZB/NZW (B/W) F1 mice: accelerated autoimmune disease with normal circulating testosterone, Clin Immunol Immunopathol 71 (1994) 338–343.

[99] R. McMurray, D. Keisler, K. Kanuckel, S. Izui, S.E. Walker, Prolactin influences autoimmune disease activity in the female B/W mouse, J Immunol 147 (1991) 3780–3787.

[100] R.W. McMurray, D. Keisler, S. Izui, S.E. Walker, Effects of parturition, suckling and pseudopregnancy on variables of disease activity in the B/W mouse model of systemic lupus erythematosus, J Rheumatol 20 (1993) 1143–1151.

[101] E. Peeva, C. Grimaldi, L. Spatz, B. Diamond, Bromocriptine restores tolerance in estrogen-treated mice, J Clin Invest 106 (2000) 1373–1379.

[102] Z. Rezaieyazdi, A. Hesamifard, Correlation between serum prolactin levels and lupus activity, Rheumatol Int 26 (2006) 1036–1039.

[103] M. Pacilio, S. Migliaresi, R. Meli, L. Ambrosone, B. Bigliardo, R. Di Carlo, Elevated bioactive prolactin levels in systemic lupus erythematosus—association with disease activity, J Rheumatol 28 (2001) 2216–2221.

[104] A.M. Jacobi, W. Rohde, M. Ventz, G. Riemekasten, G.R. Burmester, F. Hiepe, Enhanced serum prolactin (PRL) in patients with systemic lupus erythematosus: PRL levels are related to the disease activity, Lupus 10 (2001) 554–561.

[105] C.C. Mok, C.S. Lau, S.C. Tam, Prolactin profile in a cohort of Chinese systemic lupus erythematosus patients, Br J Rheumatol 36 (1997) 986–989.

[106] P. Jimena, M.A. Aguirre, A. Lopez-Curbelo, M. de Andres, C. Garcia-Courtay, M.J. Cuadrado, Prolactin levels in patients with systemic lupus erythematosus: a case controlled study, Lupus 7 (1998) 383–386.

[107] A. Leanos-Miranda, G. Cardenas-Mondragon, Serum free prolactin concentrations in patients with systemic lupus erythematosus are associated with lupus activity, Rheumatology (Oxford) 45 (2006) 97–101.

[108] J.B. Harley, I.T. Harley, J.M. Guthridge, J.A. James, The curiously suspicious: a role for Epstein-Barr virus in lupus, Lupus 15 (2006) 768–777.

[109] C.G. Parks, G.S. Cooper, L.L. Hudson, M.A. Dooley, E.L. Treadwell, E.W. St Clair, et al., Association of Epstein-Barr virus with systemic lupus erythematosus: effect modification by race, age, and cytotoxic T lymphocyte-associated antigen 4 genotype, Arthritis Rheum 52 (2005) 1148–1159.

[110] J.J. Lu, D.Y. Chen, C.W. Hsieh, J.L. Lan, F.J. Lin, S.H. Lin, Association of Epstein-Barr virus infection with systemic lupus erythematosus in Taiwan, Lupus 16 (2007) 168–175.

[111] J.A. James, B.R. Neas, K.L. Moser, T. Hall, G.R. Bruner, A.L. Sestak, et al., Systemic lupus erythematosus in adults is associated with previous Epstein-Barr virus exposure, Arthritis Rheum 44 (2001) 1122–1126.

[112] H. Kitagawa, S. Iho, T. Yokochi, T. Hoshino, Detection of antibodies to the Epstein-Barr virus nuclear antigens in the sera from patients with systemic lupus erythematosus, Immunol Lett 17 (1988) 249–252.

[113] C.J. Chen, K.H. Lin, S.C. Lin, W.C. Tsai, J.H. Yen, S.J. Chang, et al., High prevalence of immunoglobulin A antibody against Epstein-Barr virus capsid antigen in adult patients with lupus with disease flare: case control studies, J Rheumatol 32 (2005) 44–47.

[114] J.A. James, K.M. Kaufman, A.D. Farris, E. Taylor-Albert, T.J. Lehman, J.B. Harley, An increased prevalence of Epstein-Barr virus infection in young patients suggests a possible etiology for systemic lupus erythematosus, J Clin Invest 100 (1997) 3019–3026.

[115] M.T. McClain, B.D. Poole, B.F. Bruner, K.M. Kaufman, J.B. Harley, J.A. James, An altered immune response to Epstein-Barr nuclear antigen 1 in pediatric systemic lupus erythematosus, Arthritis Rheum 54 (2006) 360–368.

[116] H. Bezanahary, R. Inaoui, V. Allot, K. Ly, S. Rogez, E. Liozon, et al., Systemic lupus erythematosus and herpes virus infection: three new observations, Rev Med Interne 23 (2002) 1018–1021.

[117] M. Nawata, N. Seta, M. Yamada, I. Sekigawa, N. Lida, H. Hashimoto, Possible triggering effect of cytomegalovirus infection on systemic lupus erythematosus, Scand J Rheumatol 30 (2001) 360–362.

[118] S. Akagi, H. Ichikawa, J. Suzuki, H. Makino, Systemic lupus erythematosus associated with cytomegalovirus infection, Scand J Rheumatol 33 (2004) 58–59.

[119] I. Sekigawa, M. Nawata, N. Seta, M. Yamada, N. Iida, H. Hashimoto, Cytomegalovirus infection in patients with systemic lupus erythematosus, Clin Exp Rheumatol 20 (2002) 559–564.

[120] J.R. Rider, W.E. Ollier, R.J. Lock, S.T. Brookes, D.H. Pamphilon, Human cytomegalovirus infection and systemic lupus erythematosus, Clin Exp Rheumatol 15 (1997) 405–409.

[121] M. Ramos-Casals, M.J. Cuadrado, P. Alba, G. Sanna, P. Brito-Zeron, L. Bertolaccini, et al., Acute viral infections in patients with systemic lupus erythematosus: description of 23 cases and review of the literature, Medicine (Baltimore) 87 (2008) 311–318.

[122] A. Bengtsson, A. Widell, S. Elmstahl, G. Sturfelt, No serological indications that systemic lupus erythematosus is linked with exposure to human parvovirus B[19], Ann Rheum Dis 59 (2000) 64–66.

[123] B.L. Strom, M.M. Reidenberg, S. West, E.S. Snyder, B. Freundlich, P.D. Stolley, Shingles, allergies, family medical history, oral contraceptives, and other potential risk factors for systemic lupus erythematosus, Am J Epidemiol 140 (1994) 632–642.

[124] J.E. Pope, A. Krizova, J.M. Ouimet, J.L. Goodwin, M. Lankin, Close association of herpes zoster reactivation and systemic lupus erythematosus (SLE) diagnosis: case-control study of patients with SLE or noninflammatory nusculoskeletal disorders, J Rheumatol 31 (2004) 274–279.

[125] L.F. Ayvazian, T.L. Badger, Disseminated lupus erythematosus occurring among student nurses, N Engl J Med 239 (1948) 565–570.

[126] S.A. Older, D.F. Battafarano, R.J. Enzenauer, A.M. Krieg, Can immunization precipitate connective tissue disease? Report of five cases of systemic lupus erythematosus and review of the literature, Semin Arthritis Rheum 29 (1999) 131–139.

[127] A. Choffray, L. Pinquier, H. Bachelez, Exacerbation of lupus panniculitis following anti-hepatitis-B vaccination, Dermatology 215 (2007) 152–154.

[128] D. Santoro, M. Stella, G. Montalto, S. Castellino, Lupus nephritis after hepatitis B vaccination: an uncommon complication, Clin Nephrol 67 (2007) 61–63.

[129] M.R. Geier, D.A. Geier, A case-series of adverse events, positive re-challenge of symptoms, and events in identical twins following hepatitis B vaccination: analysis of the Vaccine Adverse Event Reporting System (VAERS) database and literature review, Clin Exp Rheumatol 22 (2004) 749–755.

[130] Y. Kuroda, J. Akaogi, D.C. Nacionales, S.C. Wasdo, N.J. Szabo, W.H. Reeves, et al., Distinctive patterns of autoimmune response induced by different types of mineral oil, Toxicol Sci 78 (2004) 222–228.

[131] Y. Kuroda, D.C. Nacionales, J. Akaogi, W.H. Reeves, M. Satoh, Autoimmunity induced by adjuvant hydrocarbon oil components of vaccine, Biomed Pharmacother 58 (2004) 325–337.

[132] P.B. Asa, Y. Cao, R.F. Garry, Antibodies to squalene in Gulf War syndrome, Exp Mol Pathol 68 (2000) 55—64.

[133] M.K. Kuechle, K.B. Elkon, Shining light on lupus and UV, Arthritis Res Ther 9 (2007) 101.

[134] E. Reefman, H. Kuiper, M.F. Jonkman, P.C. Limburg, C.G. Kallenberg, M. Bijl, Skin sensitivity to UVB irradiation in systemic lupus erythematosus is not related to the level of apoptosis induction in keratinocytes, Rheumatology (Oxford) 45 (2006) 538—544.

[135] M. Ghoreishi, J.P. Dutz, Murine models of cutaneous involvement in lupus erythematosus, Autoimmun Rev 8 (2009) 484—487.

[136] H. McGrath Jr., Ultraviolet-A1 irradiation decreases clinical disease activity and autoantibodies in patients with systemic lupus erythematosus, Clin Exp Rheumatol 12 (1994) 129—135.

[137] H. McGrath Jr., Ultraviolet A1 (340-400 nm) irradiation and systemic lupus erythematosus, J Investig Dermatol Symp Proc 4 (1999) 79—84.

[138] K.H. Costenbader, D.J. Kim, J. Peerzada, S. Lockman, D. Nobles-Knight, M. Petri, et al., Cigarette smoking and the risk of systemic lupus erythematosus: a meta-analysis, Arthritis Rheum 50 (2004) 849—857.

[139] C.J. Hardy, B.P. Palmer, K.R. Muir, A.J. Sutton, R.J. Powell, Smoking history, alcohol consumption, and systemic lupus erythematosus: a case-control study, Ann Rheum Dis 57 (1998) 451—455.

[140] C. Kiyohara, M. Washio, T. Horiuchi, Y. Tada, T. Asami, S. Ide, et al., Cigarette Smoking, STAT4 and TNFRSF[1] B Polymorphisms, and Systemic Lupus Erythematosus in a Japanese Population, J Rheumatol 18 (2009) 630—638.

[141] J. Wang, H.F. Pan, D.Q. Ye, H. Su, X.P. Li, Moderate alcohol drinking might be protective for systemic lupus erythematosus: a systematic review and meta-analysis, Clin Rheumatol 27 (2008) 1557—1563.

[142] L.W. Freni-Titulaer, D.B. Kelley, A.G. Grow, T.W. McKinley, F.C. Arnett, M.C. Hochberg, Connective tissue disease in southeastern Georgia: a case-control study of etiologic factors, Am J Epidemiol 130 (1989) 404—409.

[143] G.S. Cooper, M.A. Dooley, E.L. Treadwell, E.W. St Clair, G.S. Gilkeson, Smoking and use of hair treatments in relation to risk of developing systemic lupus erythematosus, J Rheumatol 28 (2001) 2653—2656.

[144] C.J. Hardy, B.P. Palmer, K.R. Muir, R.J. Powell, Systemic lupus erythematosus (SLE) and hair treatment: a large community based case-control study, Lupus 8 (1999) 541—544.

[145] M. Petri, J. Allbritton, Hair product use in systemic lupus erythematosus. A case-control study, Arthritis Rheum 35 (1992) 625—629.

[146] J. Sanchez-Guerrero, E.W. Karlson, G.A. Colditz, D.J. Hunter, F.E. Speizer, M.H. Liang, Hair dye use and the risk of developing systemic lupus erythematosus, Arthritis Rheum 39 (1996) 657—662.

[147] C.G. Parks, K. Conrad, G.S. Cooper, Occupational exposure to crystalline silica and autoimmune disease, Environ Health Perspect 5 (107 Suppl) (1999) 793—802.

[148] J.M. Brown, A.J. Archer, J.C. Pfau, A. Holian, Silica accelerated systemic autoimmune disease in lupus-prone New Zealand mixed mice, Clin Exp Immunol 131 (2003) 415—421.

[149] J.C. Pfau, J.M. Brown, A. Holian, Silica-exposed mice generate autoantibodies to apoptotic cells, Toxicology 195 (2004) 167—176.

[150] K. Conrad, J. Mehlhorn, K. Luthke, T. Dorner, K.H. Frank, Systemic lupus erythematosus after heavy exposure to quartz dust in uranium mines: clinical and serological characteristics, Lupus 5 (1996) 62—69.

[151] C.G. Parks, G.S. Cooper, L.A. Nylander-French, W.T. Sanderson, J.M. Dement, P.L. Cohen, et al., Occupational exposure to crystalline silica and risk of systemic lupus erythematosus: a population-based, case-control study in the southeastern United States, Arthritis Rheum 46 (2002) 1840—1850.

[152] A. Finckh, G.S. Cooper, L.B. Chibnik, K.H. Costenbader, J. Watts, H. Pankey, et al., Occupational silica and solvent exposures and risk of systemic lupus erythematosus in urban women, Arthritis Rheum 54 (2006) 3648—3654.

[153] E.C. Janowsky, L.L. Kupper, B.S. Hulka, Meta-analyses of the relation between silicone breast implants and the risk of connective-tissue diseases, N Engl J Med 342 (2000) 781—790.

[154] G.S. Cooper, C.G. Parks, E.L. Treadwell, E.W. St Clair, G.S. Gilkeson, M.A. Dooley, Occupational risk factors for the development of systemic lupus erythematosus, J Rheumatol 31 (2004) 1928—1933.

[155] E.W. Karlson, J. Watts, J. Signorovitch, M. Bonetti, E. Wright, G.S. Cooper, et al., Effect of glutathione S-transferase polymorphisms and proximity to hazardous waste sites on time to systemic lupus erythematosus diagnosis: results from the Roxbury lupus project, Arthritis Rheum 56 (2007) 244—254.

[156] M. Petri, J. Buyon, M. Kim, Classification and definition of major flares in SLE clinical trials, Lupus 8 (1999) 685—691.

[157] M. Petri, M. Genovese, E. Engle, M. Hochberg, Definition, incidence, and clinical description of flare in systemic lupus erythematosus. A prospective cohort study, Arthritis Rheum 34 (1991) 937—944.

[158] H. Jonsson, O. Nived, G. Sturfelt, Outcome in systemic lupus erythematosus: a prospective study of patients from a defined population, Medicine (Baltimore) 68 (1989) 141—150.

[159] J.E. Weiss, C.P. Sison, N.T. Ilowite, B.S. Gottlieb, B.A. Eberhard, Flares in pediatric systemic lupus erythematosus, J Rheumatol 34 (2007) 1341—1344.

[160] P.G. Vlachoyiannopoulos, F.B. Karassa, K.X. Karakostas, A.A. Drosos, H.M. Moutsopoulos, Systemic lupus erythematosus in Greece. Clinical features, evolution and outcome: a descriptive analysis of 292 patients, Lupus 2 (1993) 303—312.

[161] J.P. Buyon, M.A. Petri, M.Y. Kim, K.C. Kalunian, J. Grossman, B.H. Hahn, et al., The effect of combined estrogen and progesterone hormone replacement therapy on disease activity in systemic lupus erythematosus: a randomized trial, Ann Intern Med 142 (2005) 953—962.

[162] M. Petri, M.Y. Kim, K.C. Kalunian, J. Grossman, B.H. Hahn, L.R. Sammaritano, et al., Combined oral contraceptives in women with systemic lupus erythematosus, N Engl J Med 353 (2005) 2550—2558.

[163] J. Sanchez-Guerrero, M. Gonzalez-Perez, M. Durand-Carbajal, P. Lara-Reyes, L. Jimenez-Santana, J. Romero-Diaz, et al., Menopause hormonal therapy in women with systemic lupus erythematosus, Arthritis Rheum 56 (2007) 3070—3079.

[164] J. Cortes-Hernandez, J. Ordi-Ros, F. Paredes, M. Casellas, F. Castillo, M. Vilardell-Tarres, Clinical predictors of fetal and maternal outcome in systemic lupus erythematosus: a prospective study of 103 pregnancies, Rheumatology (Oxford) 41 (2002) 643—650.

[165] M. Petri, D. Howard, J. Repke, Frequency of lupus flare in pregnancy. The Hopkins Lupus Pregnancy Center experience, Arthritis Rheum 34 (1991) 1538—1545.

[166] G. Ruiz-Irastorza, F. Lima, J. Alves, M.A. Khamashta, J. Simpson, G.R. Hughes, et al., Increased rate of lupus flare during pregnancy and the puerperium: a prospective study of 78 pregnancies, Br J Rheumatol 35 (1996) 133—138.

[167] J. Leone, J.L. Pennaforte, V. Delhinger, J. Detour, K. Lefondre, J.P. Eschard, et al., Influence of seasons on risk of flare-up of

systemic lupus: retrospective study of 66 patients, Rev Med Interne 18 (1997) 286–291.

[168] M. Amit, Y. Molad, S. Kiss, A.J. Wysenbeek, Seasonal variations in manifestations and activity of systemic lupus erythematosus, Br J Rheumatol 36 (1997) 449–452.

[169] H.J. Haga, J.G. Brun, O.P. Rekvig, L. Wetterberg, Seasonal variations in activity of systemic lupus erythematosus in a subarctic region, Lupus 8 (1999) 269–273.

[170] T. Hasan, M. Pertovaara, U. Yli-Kerttula, T. Luukkaala, M. Korpela, Seasonal variation of disease activity of systemic lupus erythematosus in Finland: a 1 year follow up study, Ann Rheum Dis 63 (2004) 1498–1500.

[171] M.I. Peralta-Ramirez, J. Jimenez-Alonso, J.F. Godoy-Garcia, M. Perez-Garcia, The effects of daily stress and stressful life events on the clinical symptomatology of patients with lupus erythematosus, Psychosom Med 66 (2004) 788–794.

[172] C.R. Pawlak, T. Witte, H. Heiken, M. Hundt, J. Schubert, B. Wiese, et al., Flares in patients with systemic lupus erythematosus are associated with daily psychological stress, Psychother Psychosom 72 (2003) 159–165.

[173] M.M. Ward, A.S. Marx, N.N. Barry, Psychological distress and changes in the activity of systemic lupus erythematosus, Rheumatology (Oxford) 41 (2002) 184–188.

[174] D. Gladman, E. Ginzler, C. Goldsmith, P. Fortin, M. Liang, M. Urowitz, et al., The development and initial validation of the Systemic Lupus International Collaborating Clinics/American College of Rheumatology damage index for systemic lupus erythematosus, Arthritis Rheum 39 (1996) 363–369.

[175] G. Cassano, S. Roverano, S. Paira, V. Bellomio, E. Lucero, A. Berman, et al., Accrual of organ damage over time in Argentine patients with systemic lupus erythematosus: a multi-centre study, Clin Rheumatol 26 (2007) 2017–2022.

[176] C.C. Mok, C.T. Ho, R.W. Wong, C.S. Lau, Damage accrual in southern Chinese patients with systemic lupus erythematosus, J Rheumatol 30 (2003) 1513–1519.

[177] D.D. Gladman, M.B. Urowitz, P. Rahman, D. Ibanez, L.S. Tam, Accrual of organ damage over time in patients with systemic lupus erythematosus, J Rheumatol 30 (2003) 1955–1959.

[178] G.S. Alarcon, G. McGwin Jr., A.A. Bartolucci, J. Roseman, J. Lisse, B.J. Fessler, et al., Systemic lupus erythematosus in three ethnic groups. IX. Differences in damage accrual, Arthritis Rheum 44 (2001) 2797–2806.

[179] A.J. Swaak, H.G. van den Brink, R.J. Smeenk, K. Manger, J.R. Kalden, S. Tosi, et al., Systemic lupus erythematosus: clinical features in patients with a disease duration of over 10 years, first evaluation, Rheumatology (Oxford) 38 (1999) 953–958.

[180] S. Manzi, E.N. Meilahn, J.E. Rairie, C.G. Conte, T.A. Medsger Jr., L. Jansen-McWilliams, et al., Age-specific incidence rates of myocardial infarction and angina in women with systemic lupus erythematosus: comparison with the Framingham Study, Am J Epidemiol 145 (1997) 408–415.

[181] I.N. Bruce, 'Not only…but also': factors that contribute to accelerated atherosclerosis and premature coronary heart disease in systemic lupus erythematosus, Rheumatology (Oxford) 44 (2005) 1492–1502.

[182] J.M. Esdaile, M. Abrahamowicz, T. Grodzicky, Y. Li, C. Panaritis, R. du Berger, et al., Traditional Framingham risk factors fail to fully account for accelerated atherosclerosis in systemic lupus erythematosus, Arthritis Rheum 44 (2001) 2331–2337.

[183] A. Zonana-Nacach, A. Camargo-Coronel, P. Yanez, L. Sanchez, F.J. Jimenez-Balderas, A. Fraga, Infections in outpatients with systemic lupus erythematosus: a prospective study, Lupus 10 (2001) 505–510.

[184] X. Bosch, A. Guilabert, L. Pallares, R. Cerveral, M. Ramos-Casals, A. Bove, et al., Infections in systemic lupus erythematosus: a prospective and controlled study of 110 patients, Lupus 15 (2006) 584–589.

[185] V. Noel, O. Lortholary, P. Casassus, P. Cohen, T. Genereau, M.H. Andre, et al., Risk factors and prognostic influence of infection in a single cohort of 87 adults with systemic lupus erythematosus, Ann Rheum Dis 60 (2001) 1141–1144.

[186] S.J. Jeong, H. Choi, H.S. Lee, S.H. Han, B.S. Chin, J.H. Baek, et al., Incidence and risk factors of infection in a single cohort of 110 adults with systemic lupus erythematosus, Scand J Infect Dis 41 (2009) 268–274.

[187] O. Nived, A. Bengtsson, A. Jonsen, G. Sturfelt, H. Olsson, Malignancies during follow-up in an epidemiologically defined systemic lupus erythematosus inception cohort in southern Sweden, Lupus 10 (2001) 500–504.

[188] D.M. Sweeney, S. Manzi, J. Janosky, K.J. Selvaggi, W. Ferri, T.A. Medsger Jr., et al., Risk of malignancy in women with systemic lupus erythematosus, J Rheumatol 22 (1995) 1478–1482.

[189] T. Pettersson, E. Pukkala, L. Teppo, C. Friman, Increased risk of cancer in patients with systemic lupus erythematosus, Ann Rheum Dis 51 (1992) 437–439.

[190] S. Bernatsky, J.F. Boivin, L. Joseph, R. Rajan, A. Zoma, S. Manzi, et al., An international cohort study of cancer in systemic lupus erythematosus, Arthritis Rheum 52 (2005) 1481–1490.

[191] K. McElhone, J. Abbott, L.S. Teh, A review of health related quality of life in systemic lupus erythematosus, Lupus 15 (2006) 633–643.

[192] G. Ruiz-Irastorza, M. Ramos-Casals, P. Brito-Zeron, M.A. Khamashta, Clinical efficacy and side effects of antimalarials in systemic lupus erythematosus: a systematic review, Ann Rheum Dis 69 (2009) 20–28.

[193] M.F. Gordon, P.D. Stolley, R. Schinnar, Trends in recent systemic lupus erythematosus mortality rates, Arthritis Rheum 24 (1981) 762–769.

[194] S. Bernatsky, J.F. Boivin, L. Joseph, S. Manzi, E. Ginzler, D.D. Gladman, et al., Mortality in systemic lupus erythematosus, Arthritis Rheum 54 (2006) 2550–2557.

[195] M. Merrell, L.E. Shulman, Determination of prognosis in chronic disease, illustrated by systemic lupus erythematosus, J Chronic Dis 1 (1955) 12–32.

[196] M. Funauchi, H. Shimadzu, C. Tamaki, T. Yamagata, Y. Nozaki, M. Sugiyama, et al., Survival study by organ disorders in 306 Japanese patients with systemic lupus erythematosus: results from a single center, Rheumatol Int 27 (2007) 243–249.

[197] N. Kasitanon, L.S. Magder, M. Petri, Predictors of survival in systemic lupus erythematosus, Medicine (Baltimore) 85 (2006) 147–156.

[198] M.B. Urowitz, D.D. Gladman, B.D. Tom, D. Ibanez, V.T. Farewell, Changing patterns in mortality and disease outcomes for patients with systemic lupus erythematosus, J Rheumatol 35 (2008) 2152–2158.

[199] A. Ippolito, M. Petri, An update on mortality in systemic lupus erythematosus, Clin Exp Rheumatol 26 (2008) S72–S79.

[200] R.A. Kaslow, A.T. Masi, Age, sex, and race effects on mortality from systemic lupus erythematosus in the United States, Arthritis Rheum 21 (1978) 473–479.

[201] S.J. Walsh, L.M. DeChello, Geographical variation in mortality from systemic lupus erythematosus in the United States, Lupus 10 (2001) 637–646.

[202] CDC, Trends in Deaths from Systemic Lupus Erythematosus-United States, 1979-[1998] MMVR 51 (2002) 371–373.

[203] M.C. Hochberg, Mortality from systemic lupus erythematosus in England and Wales, 1974-[1983] Br J Rheumatol 26 (1987) 437–441.

[204] S. Wadee, M. Tikly, M. Hopley, Causes and predictors of death in South Africans with systemic lupus erythematosus, Rheumatology (Oxford) 46 (2007) 1487–1491.

[205] F.J. Blanco, J.J. Gomez-Reino, J. de la Mata, A. Corrales, V. Rodriguez-Valverde, J.C. Rosas, et al., Survival analysis of 306 European Spanish patients with systemic lupus erythematosus, Lupus 7 (1998) 159–163.

[206] G.S. Alarcón, G. McGwin Jr., H.M. Bastian, J. Roseman, J. Lisse, B.J. Fessler, et al., Systemic lupus erythematosus in three ethnic groups. VII [correction of VIII]. Predictors of early mortality in the LUMINA cohort. LUMINA Study Group, Arthritis Rheum 45 (2001) 191–202.

[207] C.C. Mok, A. Mak, W.P. Chu, C.H. To, S.N. Wong, Long-term survival of southern Chinese patients with systemic lupus erythematosus: a prospective study of all age-groups, Medicine (Baltimore) 84 (2005) 218–224.

[208] E. Krishnan, H.B. Hubert, Ethnicity and mortality from systemic lupus erythematosus in the US, Ann Rheum Dis 65 (2006) 1500–1505.

[209] M. Fernandez, G.S. Alarcón, J. Calvo-Alen, R. Andrade, G. McGwin Jr., L.M. Vila, et al., A multiethnic, multicenter cohort of patients with systemic lupus erythematosus (SLE) as a model for the study of ethnic disparities in SLE, Arthritis Rheum 57 (2007) 576–584.

[210] M.M. Ward, Education level and mortality in systemic lupus erythematosus (SLE): evidence of underascertainment of deaths due to SLE in ethnic minorities with low education levels, Arthritis Rheum 51 (2004) 616–624.

[211] A.M. Bertoli, G.S. Alarcón, J. Calvo-Alen, M. Fernandez, L.M. Vila, J.D. Reveille, Systemic lupus erythematosus in a multiethnic US cohort. XXXIII. Clinical [corrected] features, course, and outcome in patients with late-onset disease, Arthritis Rheum 54 (2006) 1580–1587.

[212] S. Bujan, J. Ordi-Ros, J. Paredes, M. Mauri, L. Matas, J. Cortes, et al., Contribution of the initial features of systemic lupus erythematosus to the clinical evolution and survival of a cohort of Mediterranean patients, Ann Rheum Dis 62 (2003) 859–865.

[213] V. Bellomio, A. Spindler, E. Lucero, A. Berman, M. Santana, C. Moreno, et al., Systemic lupus erythematosus: mortality and survival in Argentina. A multicenter study, Lupus 9 (2000) 377–381.

[214] A. Zonana-Nacach, P. Yanez, F.J. Jimenez-Balderas, A. Camargo-Coronel, Disease activity, damage and survival in Mexican patients with acute severe systemic lupus erythematosus, Lupus 16 (2007) 997–1000.

[215] J. Nossent, N. Cikes, E. Kiss, A. Marchesoni, V. Nassonova, M. Mosca, et al., Current causes of death in systemic lupus erythematosus in Europe, 2000–2004: relation to disease activity and damage accrual, Lupus 16 (2007) 309–317.

[216] M.B. Urowitz, A.A. Bookman, B.E. Koehler, D.A. Gordon, H.A. Smythe, M.A. Ogryzlo, The bimodal mortality pattern of systemic lupus erythematosus, Am J Med 60 (1976) 221–225.

[217] A. Doria, L. Iaccarino, A. Ghirardello, S. Zampieri, S. Arienti, P. Sarzi-Puttini, et al., Long-term prognosis and causes of death in systemic lupus erythematosus, Am J Med 119 (2006) 700–706.

[218] L. Bjornadal, L. Yin, F. Granath, L. Klareskog, A. Ekbom, Cardiovascular disease a hazard despite improved prognosis in patients with systemic lupus erythematosus: results from a Swedish population based study 1964-95, J Rheumatol 31 (2004) 713–719.

[219] M.B. Urowitz, D.D. Gladman, M. Abu-Shakra, V.T. Farewell, Mortality studies in systemic lupus erythematosus. Results from a single center. III. Improved survival over 24 years, J Rheumatol 24 (1997) 1061–1065.

[220] M. Abu-Shakra, M.B. Urowitz, D.D. Gladman, J. Gough, Mortality studies in systemic lupus erythematosus. Results from a single center. I. Causes of death, J Rheumatol 22 (1995) 1259–1264.

[221] A.S. Al-Arfaj, S.R. Al-Balla, A.N. Al-Dalaan, S.S. Al-Saleh, S.A. Bahabri, M.M. Mousa, et al., Prevalence of systemic lupus erythematosus in central Saudi Arabia, Saudi Med J 23 (2002) 87–89.

[222] D.J. Wallace, T. Podell, J. Weiner, J.R. Klinenberg, S. Forouzesh, E.L. Dubois, Systemic lupus erythematosus—survival patterns. Experience with 609 patients, JAMA 245 (1981) 934–938.

[223] A.J. Swaak, J.C. Nossent, W. Bronsveld, A. Van Rooyen, E.J. Nieuwenhuys, L. Theuns, et al., Systemic lupus erythematosus. I. Outcome and survival: Dutch experience with 110 patients studied prospectively, Ann Rheum Dis 48 (1989) 447–454.

[224] N. Kasitanon, W. Louthrenoo, W. Sukitawut, R. Vichainun, Causes of death and prognostic factors in Thai patients with systemic lupus erythematosus, Asian Pac J Allergy Immunol 20 (2002) 85–91.

[225] Y. Alamanos, P.V. Voulgari, M. Papassava, K. Tsamandouraki, A.A. Drosos, Survival and mortality rates of systemic lupus erythematosus patients in northwest Greece. Study of a 21-year incidence cohort, Rheumatology (Oxford) 42 (2003) 1122–1123.

[226] D.L. Loriaux, H.J. Ruder, D.R. Knab, M.B. Lipsett, Estrone sulfate, estrone, estradiol and estriol plasma levels in human pregnancy, J Clin Endocrinol Metab 35 (1972) 887–891.

[227] L. Baudino, K. Yoshinobu, N. Morito, S. Kikuchi, L. Fossati-Jimack, B.J. Morley, et al., Dissection of genetic mechanisms governing the expression of serum retroviral gp70 implicated in murine lupus nephritis, J Immunol 181 (2008) 2846–2854.

[228] P.C. Piotrowski, S. Duriagin, P.P. Jagodzinski, Expression of human endogenous retrovirus clone 4-1 may correlate with blood plasma concentration of anti-U1 RNP and anti-Sm nuclear antibodies, Clin Rheumatol 24 (2005) 620–624.

[229] M. Okada, H. Ogasawara, H. Kaneko, T. Hishikawa, I. Sekigawa, H. Hashimoto, et al., Role of DNA methylation in transcription of human endogenous retrovirus in the pathogenesis of systemic lupus erythematosus, J Rheumatol 29 (2002) 1678–1682.

[230] H. Ogasawara, T. Naito, H. Kaneko, T. Hishikawa, I. Sekigawa, H. Hashimoto, et al., Quantitative analyses of messenger RNA of human endogenous retrovirus in patients with systemic lupus erythematosus, J Rheumatol 28 (2001) 533–538.

[231] C. Bombardier, D.D. Gladman, M.B. Urowitz, D. Caron, D.H. Chang, the Committee on Prognosis Studies in SLE. (1992) Derivation of the SLEDAI: a disease activity index for lupus patients, Arthritis Rheum 35 (1992) 630–640.

[232] E.M. Hay, P.A. Bacon, C. Gordon, D.A. Isenberg, P. Maddison, M.L. Snaith, et al., The BILAG index: a reliable and valid instrument for measuring clinical disease activity in systemic lupus erythematosus. QJM 86 (1993) 447–458.

[233] S.C. Bae, H.K. Koh, D.K. Chang, M.H. Kim, J.K. Park, S.Y. Kim, Reliability and validity of Systemic Lupus Activity Measure-Revised (SLAM-R) for measuring clinical disease activity in systemic lupus erythematosus, Lupus 10 (2001) 405–409.

[234] W. Bencivelli, C. Vitali, D.A. Isenberg, J.S. Smolen, M.L. Snaith, M. Sciuto, et al., Disease activity in systemic lupus erythematosus: report of the Consensus Study Group of the European Workshop for Rheumatology Research. III. Development of a computerised clinical chart and its application to the comparison of different indices of disease activity. The European Consensus Study Group for Disease Activity in SLE, Clin Exp Rheumatol 10 (1992) 549–554.

[235] American College of Rheumatology Ad Hoc Committee on Systemic Lupus Erythematosus Response Criteria, The American College of Rheumatology response criteria for systemic lupus erythematosus clinical trials: measures of overall disease activity, Arthritis Rheum 50 (2004) 3418—3426.

[236] S. Manzi, E.N. Meilahn, J.E. Rairie, C.G. Conte, T.A. Medsger Jr., L. Jansen-McWilliams, et al., Age-specific incidence rates of myocardial infarction and angina in women with systemic lupus erythematosus: comparison with the Framingham Study, Am J Epidemiol 145 (1997) 408—415.

[237] J.M. Esdaile, M. Abrahamowicz, T. Grodzicky, Y. Li, C. Panaritis, R. du Berger, et al., Traditional Framingham risk factors fail to fully account for accelerated atherosclerosis in systemic lupus erythematosus, Arthritis Rheum 44 (2001) 2331—2337.

[238] R. Bessant, A. Hingorani, L. Patel, A. MacGregor, D.A. Isenberg, A. Rahman, Risk of coronary heart disease and stroke in a large British cohort of patients with systemic lupus erythematosus, Rheumatology (Oxford) 43 (2004) 924—929.

[239] A.B. Lee, T. Godfrey, K.G. Rowley, C.S. Karschimkus, G. Dragicevic, E. Romas, et al., Traditional risk factor assessment does not capture the extent of cardiovascular risk in systemic lupus erythematosus, Intern Med J 36 (2006) 237—243.

[240] A. Schattner, M.H. Liang, The Cardiovascular Burden of Lupus: A complex challenge, Arch Int Med 163 (2003) 1507—1510.

[240] Committee to Review the CDC Anthrax Vaccine Safety and Efficacy Research Program, An Assessment of the CDC Anthrax Vaccine Safety and Efficacy Research Program, The National Academies Press, Washington, DC, 2002. 1—184.

[241] V.P. Werth, M. Bashir, W. Zhang, Photosensitivity in Rheumatic diseases, J Investig Dermatol Symp Proc 9 (1) (2004) 57—63.

Monitoring Disease Activity

Kenneth C. Kalunian, Joan T. Merrill

Multiple organizations have devised instruments for the purpose of monitoring SLE disease activity, primarily with the goal of following the course of clinical disease expression in epidemiological studies. Several of these indices have been adapted for use in clinical trial outcomes. Instruments differ in their focus, but variously address the assessment of global and organ-specific disease activity, disease flare, fatigue, quality of life, and damage. Validated instruments have been successful at measuring SLE activity. Although they seem to be sensitive at capturing change in disease activity with time, it has become evident that sensitivity to change does not necessarily optimize the assessment of clinically significant responses to an intervention or clinically significant flares, and those are important outcomes for interventional trials. Although not a true assessment of disease activity, the concept of damage caused by disease activity, co-morbid conditions, and consequences of disease-specific therapy have each emerged as important measures for the longitudinal monitoring of patients and each may be useful during long-term follow-up assessments in clinical trials. Fatigue and quality-of-life indices are also fundamental to longitudinal assessments, but may integrate other aspects of the patient's overall health that might be independent of disease activity.

This chapter will focus on several validated instruments of disease activity that have been widely utilized in clinical practice, are accepted outcome indices for clinical trials, and/or have been used as endpoints for the evaluation of many experimental drugs [1]. These include several iterations of the BILAG (British Isles Lupus Assessment Group) Index [2], the Systemic Lupus Erythematosus Disease Activity Index (SLEDAI) (and modified versions) [3] and ways in which these instruments have been combined in composite outcome measures or modified to specifically address the capturing of flares. The basic instruments from which the rest have been derived are the BILAG Index and SLEDAI, both of which have been in wide use since the 1980s. Both of these validated instruments have been demonstrated to reliably assess disease activity and are sensitive to changes in disease activity over time. The BILAG Index links the grading of disease activity to the presence or absence of an intent to change therapy, and is intended to mirror the assessment of clinicians experienced with lupus patient care. The BILAG Index ranks symptoms by organ and by severity. The SLEDAI charts the presence or absence of symptoms but does not have the same level of complexity as the BILAG. Although it is not designed to assess different organs separately, it can be used that way, and it may have an advantage over the BILAG in being able to separate the assessment of multiple features within one organ, especially in the mucocutaneous system, where multiple signs or symptoms are common. Both instruments can be useful in clinical trials, but an understanding of their strengths and weaknesses is very important in designing outcomes for trials. The relevance of defining flare for certain clinical trial designs must also be underscored, and will be discussed further below, since both the BILAG and SLEDAI have had pitfalls in this area. Measuring fatigue and quality of life may be important for the clinical assessment of patients and for clinical trials as these assessments may capture aspects of lupus that are not adequately assessed by other outcome measurements.

MEASURING DISEASE ACTIVITY

Over 40 disease activity instruments have been developed [1]; however, more recent observational and epidemiological studies and most multicenter interventional clinical trials have used the BILAG Index and/or the SLEDAI. These instruments assess disease activity within a confined time, have demonstrated inter- and intrarater reliability, and are sensitive to change and can measure change with time [4, 5]. The duration of time over which the activity is assessed depends on the particular instrument, but each has been modified in some trials. Originally the BILAG instrument assessed disease activity during the month prior to and up until the assessment visit, and the SLEDAI only assessed back to 10 days before the visit. Most studies now use both instruments to assess back 1 month, so that all activity between monthly visits can be captured in a prospective study. The use of an updated version of SLEDAI, the SLEDAI-2K, has been validated for use as assessment of disease activity for the previous month [6]. Some studies have tried to use the BILAG Index or SLEDAI to capture all activity between tri-monthly visits, but this approach has obvious pitfalls and has never been validated. The pitfalls include overly heavy reliance on patient history going too far back for reliability, and how to handle multiple flares in a 3-month time period. One study tried to formalize the capture of the first consecutive flare in that period; however, this made no sense if the second flare was more severe.

The BILAG Index was developed by clinical investigators from four centers in the United Kingdom and Ireland [2]. In its original version, it rated activity of SLE in eight organ systems. Scoring in each organ system was based on the principle of intention to treat using the following ratings:

A "Action" meaning severe disease that requires urgent, disease-modifying therapy
B "Beware" meaning moderate disease that demands close attention and, perhaps, modifications of minor therapy (e.g., addition of low-dose corticosteroids or hydroxychloroquine) with maintenance but not institution of new major modalities
C "Content" meaning mild, improving, or static disease requiring no treatment or only symptomatic therapy (including pain medications and nonsteroidal anti-inflammatory drugs)
D Absence of symptoms or laboratory abnormalities.

Ratings are given for each organ system based on the assessor's evaluation of the patient's clinical condition in the last month prior to evaluation as compared to the previous month. For statistical comparison with other numerically-based indices, the following weights have been given to the four categories: A = 9; B = 3, C = 1, and D = 0. Theoretically possible scores could vary from 0 to 72. Patients with severe disease in more than two or three organ systems are likely to be critically ill. However, scores greater than 30 are rare.

The BILAG instrument has been updated in order to address some pitfalls in the use of the original instrument in clinical trials and to include two new organ systems that can produce important, although relatively rare manifestations of lupus, the gastrointestinal and ophthalmic systems. The resultant instrument, the BILAG-2004 [7] (available at http://rheumatology. oxfordjournals.org/cgi/content/full/kep064/DC1? maxtoshow=&hits=10&RESULTFORMAT=&fulltext= bilag-2004&searchid=1&FIRSTINDEX=0&resourcetype= HWCIT), has been subjected to validation studies [8, 9] and may be a particularly valuable tool for use in clinical trials that assess new therapeutics for SLE. A nominal consensus approach was undertaken in updating the original BILAG by members of the BILAG to improve the original index. The new instrument was evaluated in two real-patient exercises as part of its validation testing. Reliability of scoring in terms of the ability of the instrument to differentiate patients was assessed by calculating intraclass correlation coefficients. These exercises demonstrated good instrument reliability and high levels of physician agreement were achieved in the constitutional, mucocutaneous, neurological, cardiorespiratory, renal, ophthalmic, and hematological systems. In contrast, the musculoskeletal system did not score as well, although it was felt that reliability may be greatly improved by providing a more clear-cut glossary. The BILAG-2004 instrument retains the philosophy of the original BILAG instrument with respect to scoring based on the principle of the physician's intention to treat. The BILAG-2004 instrument continues to include the A−D categories as well as an additional E category that was not present in the original BILAG but added in 1993 [10]. The E category denotes no active disease in a particular organ system at any point in the past or present, and is differentiated from D, which signifies no disease in the past month, but with a history of activity in that organ. The BILAG-2004 instrument has modified or elaborated the definitions for the A−C categories and these definitions apply to each of the nine currently described organ systems (one previous system, vasculitis, was incorporated into the mucocutaneous system when the two new systems were added, so that total systems increased from eight on the classic BILAG index to nine on the BILAG-2004). The weights of the different categories have been changed in BILAG-2004; these changes have been validated and appear to better distinguish change in clinically meaningful ways. In BILAG-2004, A = 12, B = 8, C = 1 and both D and E are equal to 0 [9].

Category A is now defined as severe disease, which would deserve the consideration of any of the following treatments:

1. Systemic high-dose oral glucocorticoids (equivalent to prednisolone greater than 20 mg daily).
2. Intravenous pulse glucocorticoids (equivalent to pulse methylprednisolone greater to or equal to 500 mg).

3. Systemic immunomodulators (including biologicals, immunoglobulins, and/or plasmapheresis).
4. Therapeutic high-dose anticoagulation (e.g., warfarin with target INR 3–4) in the presence of high-dose glucocorticoids or immunomodulators.

Category B is defined as moderate disease activity consistent with any of the following treatments:

1. Systemic low-dose oral glucocorticoids (equivalent to prednisolone of less than or equal to 20 mg daily).
2. Intramuscular or intraarticular or soft tissue glucocorticoid injection (equivalent to methylprednisolone less than 500 mg).
3. Topical glucocorticoids.
4. Topical immunomodulators.
5. Antimalarials or thalidomide or prasterone or acitretin.
6. Symptomatic therapy (e.g., nonsteroidal anti-inflammatory drugs).

It is important to appreciate that the change in treatment is not literally required in order to give a score, since there are many reasons why patients may not have a literal change in treatment (e.g., patient refusal, clinical trial protocol requirements, hope that a treatment started earlier might still provide benefit if more time were given, toxicity, or infection). The intent to treat guidelines are meant to be used not as a literal imperative in the scoring but as a landmark against which the disease activity should be measured. If a symptom is not severe enough so that it could not possibly warrant a BILAG A score (based on what treatment you would prescribe) then it should not be given a BILAG A score.

Within each organ system, there are multiple individual clinical parameters that are captured; scoring for each of the features is rated as not present, improving, same, worse, or new, based on its severity in the month leading up to this assessment compared to the previous month. Only features attributable to SLE are recorded and refer to the last 4 weeks compared with the previous 4 weeks. From these very practical, clinically based assessments, the A, B, C, D, or E score is eventually derived for each organ system, through a set of rules that are also published with the index. The weakness of the BILAG Index, which is not shared by the SLEDAI, is that if there is more than one manifestation within an organ system, only the highest scoring one counts. If there are two high scoring manifestations, only one is credited to the final score. Therefore, cumulative manifestations in an individual organ are "lumped" and one could not differentiate a patient with one B manifestation from a patient with two B- and one C-qualifying manifestations within a single organ.

The BILAG-2004 Index has been shown to be sensitive to change [12], which demonstrates its utility in longitudinal epidemiological outcomes and clinical trial settings. For this exercise, a prospective multicenter study of 1761 assessments from 347 SLE patients were analyzed by determining overall BILAG-2004 Index scores as well as the scores of the individual organ-specific systems. Using multinomial logistic regression, sensitivity to change was assessed by noting the relationship between change in disease activity and therapy changes between consecutive visit assessments. In this observational period, 22.7% of subjects had an increase in therapy between visits, 37.3% had a decrease, and in 40% there was no change. There was an observed relationship between changes in total BILAG-2004 scores and therapy changes, with a greater score change associated with greater predictive power. Whereas there was an association between increases and decreases in overall index scores associating with increases and decreases in therapy, respectively, increases in most system scores were associated with increases in therapy, but there were less consistent associations between decreases in these individual scores and decreases in therapy. It should be stressed that not all A or B scores were associated with actual changes in treatment, even in the hands of these experts; therefore the "intent to treat" guideline cannot be considered proscriptive in scoring, but rather a landmark of the thought process in helping a rater determine severity.

The SLEDAI was developed at the University of Toronto [3]. The importance of 37 variables in defining SLE activity were rated by an expert panel of physicians. Using the highest-rated variables, 39 patient case descriptions were invented and 14 rheumatologists were asked to rank these cases in terms of disease activity. Implied weights of each variable in contributing to the judgment of activity in this group of fictitious patients were derived from multiple regression analyses. Real patients were then used to compare the instrument with the physician's global assessment of activity and significant correlations were found. The resulting index is a simple, one-page form with 24 items. Definitions of items are provided on the form. Items that are present are noted and scoring is calculated by summing predetermined weights for each item. Items that are usually life-threatening have higher weights. Possible scores using the instrument vary theoretically from zero to 105, although it is rare for a real patient with SLE to score higher than 20. Although the SLEDAI avoids the pitfall in the BILAG of lumping all manifestations within an organ to one score that does not increase with addition of similarly ranked signs or symptoms, the original SLEDAI index does not recognize either partial improvement or worsening of symptoms. Furthermore, some disorders such as thrombocytopenia, which is usually a benign condition, receive a low score whenever it occurs, even in the rare cases when it is life-threatening. The instrument appears to be heavily weighted toward neuropsychiatric and renal manifestations of disease rather than other manifestations because the dichotomous nature of the scale makes it difficult to assess whether manifestations are improving, worsening, or remaining stable. Another problem exists with the arthritis criterion: in order for it to be scored, a patient would need to have more than two swollen joints; for example, a patient with markedly swollen bilateral knee arthritis disabling the patient from walking would not be given points for arthritis despite the presence of severe arthritis.

In the classic SLEDAI, the manifestations being scored must have been present in the 10 days preceding evaluation. Two subsequent modifications of the SLEDAI have been developed. The first modification, the Safety of Estrogens in Lupus Erythematosus-National Assessment (SELENA)-SLEDAI index [13] (available at http://content.nejm.org/cgi/content/full/353/24/2550/DC1) was adapted from the SLEDAI for use in the NIH-funded multicenter safety studies of estrogens in women with SLE [13, 14]. It was validated through its prospective use in the two SELENA studies. It differs from the original SLEDAI in the definitions of some descriptors, which were modified for the purposes of clarification and attribution. The definition of the seizure descriptor, for example, was expanded to exclude seizures resulting from past irreversible central nervous system damage. Scleritis and episcleritis were added to the visual disturbance descriptor and vertigo was added to the cranial nerve disorder descriptor. The cerebrovascular accident descriptor excluded hypertensive causes of stroke. The pleurisy and pericarditis descriptors required attribution to lupus with the added qualification of severe pain. Descriptors were changed to ensure that organ system involvement also included ongoing disease activity; for rash, alopecia, and mucosal ulcers, not just new activity. The SELENA-SLEDAI descriptors are counted if there are ongoing manifestations in these areas due to lupus rather than being required to be new or recurrent signs or symptoms as defined in the original SLEDAI. By only recording new or recurrent signs or symptoms in the original SLEDAI, the SELENA-SLEDAI authors felt that some ongoing activity was not being captured. The original SLEDAI descriptor of proteinuria included the new onset or increase of more than 500 mg per 24 h. The SELENA SLEDAI clarified the time increment of the increase, allowing a score each time there was an increase of more than 500 mg even if the increase occurred over more than 1 month.

A subsequent iteration, the SLEDAI-2K [15] (available at www.jrheum.org/content/29/2/288.full.pdf), was modified by the original SLEDAI group. It is similar to the SELENA-SLEDAI, but not identical. Like the SELENA-SLEDAI, descriptors have been modified to ensure that organ system involvement reflects ongoing disease activity. The definition for proteinuria has been dramatically simplified so that all proteinuria over 500 mg per 24 h is scored if attributed to active lupus. The SLEDAI-2K was validated against the original SLEDAI as a predictor for mortality and as a measure of global disease activity in the clinical setting. In order to validate the newer instrument, all visits in the University of Toronto lupus cohort of 960 patients were used to correlate SLEDAI-2K scores against the original SLEDAI, and the whole cohort was used to validate SLEDAI-2K as a predictor of mortality. A subgroup of 212 patients who had had five regular visits, 3–6 months apart, were also included. An uninvolved clinician evaluated each patient record and assigned a clinical activity level. The scores were then calculated according to both the original and modified instruments. The SLEDAI-2K correlated highly ($r = 0.97$) with the original SLEDAI. Both methods for scoring predicted mortality equally ($p = 0.0001$) and described similarly the range of disease activity as recognized by the clinician. It appears from these studies that the SLEDAI-2K is suitable for use in clinical trials and studies of prognoses in SLE.

From a study in 2009, it appears that the SLEDAI-2K can also be used in clinical studies and clinical trials to describe disease activity over the previous 30 days rather than the traditional 10-day period utilized by this instrument [16], which is important because it is practical for most clinical trials to schedule monthly visits, and it would not be practical if the outcome measure could not cover the entire time between visits. This study rated 41 subjects for the previous 10- and 30-day periods at a single center at monthly intervals for 12 months. In all but one patient visit, there was complete agreement between the two uses of the instrument; this suggested that it is unusual to have a manifestation of disease activity that is present at 11–30 day prior to a clinical visit and then have complete resolution of the particular manifestation in the 10 days prior to that visit. However, in an international year-long trial of 800 or 900 patients, this could potentially occur enough to become a problem.

Although there has been demonstrated validity of activity instruments such as the BILAG and the SLEDAI and their derivates for cutaneous manifestations of SLE, the practicality and usefulness of these instruments as used by dermatologists and for skin-specific lupus has been questioned [17, 18]. To address these concerns, the Cutaneous Lupus Erythematosus Disease Activity and Severity Index (CLASI) [17] was developed as a quantitative instrument that assesses both skin disease activity and damage. This measurement (available at http://www.interscience.wiley.com/jpages/0004-3591:1/suppmat/index.html) is designed to quantify disease activity in the skin, scalp, and oral mucosa as well as damage from previous inflammation in these same areas; the total theoretically possible scores for activity and damage are 0–70 and 0–56, respectively, although it would be rare for a patient to have skin rash on each segment of the body with damage at every location, which would be necessary to meet the highest scores. Nevertheless, patients with subacute cutaneous lupus can develop quite high activity scores, and patients with severe discoid lupus can sometimes approach the highest activity and damage scores. In an attempt to emphasize body areas that are more apparent in daily life, the CLASI assigns relative weights of body surface areas with this concept in mind; for example, the malar area of the face and the nose together have the same weight as the back and buttock areas together. The instrument has demonstrated construct and face validity and good inter-rater and intrarater reliability when used by dermatologists considered experts in cutaneous lupus. Additionally, it has been demonstrated to be responsive to change in the setting of treatment implementation. A more recent validation of the CLASI was performed for its use by rheumatologists with additional validation testing for dermatologists as part of this study [19]. The design of this study involved assessing 14 subjects (ten with cutaneous lupus, one with a skin disease that appeared clinically similar to cutaneous lupus, and three with both types of lesions) by five academic dermatologists and five academic rheumatologists who were all considered lupus experts. The results confirmed the reliability of the instrument when utilized by dermatologists and demonstrated reliability when used by rheumatologists as the dermatology intraclass correlation coefficient was 0.92 for activity and 0.82 for damage and the rheumatology intraclass correlation coefficient was 0.83 for activity and 0.86 for damage; intrarater reliability was above 0.9 for both activity and damage for both rheumatologists and dermatologists.

Several alternative approaches have been taken to assess disease activity over time with an emphasis on attempting to define response in the setting of a clinical trial and/or with respect to a clinically meaningful change. The Responder Index for Lupus Erythematosus (RIFLE) [20] was developed from the perspective that it may be more clinically meaningful to measure outcome from a particular therapeutic intervention by utilizing an instrument with predefined specific variables of both worsening and improvements in organ-specific clinical manifestations rather than by comparing global or organ specific disease activity scores by BILAG or SLEDAI assessments. By developing an instrument designed specifically to measure response to treatment, rather than "intent to treat" or change in clinical activity, the hypothesis was that outcome from an intervention may have more clinical relevance. The RIFLE, however, has never been validated.

An American College of Rheumatology Ad Hoc Committee on SLE Response Criteria defined clinically meaningful improvement, no change, or worsening in six clinical measures of disease activity in order to gauge the clinical relevance of any observed change in disease activity and to aid clinical trial design by enabling clinical investigators to estimate sample size determinants based on meaningful effect sizes [21]. Medical records from 310 patients were abstracted into standard formatted vignettes that included both clinical and laboratory data from two to three clinical visits. Either from the time of each visit or through retrospective review, ratings on six instruments (BILAG, SLEDAI, the revised Systemic Lupus Activity Measure, the European Consensus Lupus Activity Measure, the SELENA-SLEDAI, and the RIFLE) were obtained for the patients; from these vignettes, five common vignettes and ten randomly selected vignettes were rated through a secure internet site by 88 international SLE experts who were blinded to the activity measure scores. The raters assessed the patient's clinical condition as worsened, improved, or unchanged relative to the previous visit and these ratings were statistically transformed into performance characteristic curves that related the physicians' agreement on whether a particular patient improved, worsened, or remained the same to changes on a particular activity measure. A committee that was blinded to the actual instrument used determined the level of expert agreement that would be utilized to determine clinically meaningful change. The six activity indices showed separation of clinical conditions and the committee determined that 70% agreement by physicians was the point on the performance characteristic curves at which meaningful change in an instrument's score could be noted, and for each instrument the units of change required to indicate improvement or worsening were determined. A more recent approach to defining response was developed as a *post hoc* assessment measure of outcome after the completion of a phase 2 randomized, controlled clinical trial of belimumab, an investigational human monoclonal antibody against a B-lymphocyte stimulator [22]. This instrument, the SLE Responder Index, was subsequently used as the primary endpoint of two phase 3 randomized, controlled clinical trials of belimumab and demonstrated differences between the active and placebo groups were seen with this instrument. The index, a composite of outcome measurements, is defined by meeting all of the following:

1. A reduction from baseline of at least four points on the SELENA-SLEDAI index.
2. No worsening of disease as measured by the Physician's Global Assessment (worsening defined as an increase of 0.30 points or more from baseline).

3. No new BILAG A organ domain score (which indicates a severe flare of disease activity) and no more than one new BILAG B organ domain score (which indicates a moderate flare of disease activity, but can sometimes indicate a minor, transient change, especially when it is a single B).

It should be emphasized that in some previous trials the development of one new BILAG B would define a nonresponder. Since one new BILAG B could be a minor, transient fluctuation in disease activity, this could potentially overestimate nonresponders to a treatment, at least from the point of view of practical clinical practice.

MEASURING FLARE

As the course of disease in patients with SLE is characterized by unpredictable flares of disease activity, validated indices that capture and describe these events are critical for the purposes of assessing the utility of an experimental drug or other intervention. Carefully constructed indices that assess flares would be particularly helpful as meaningful outcomes in SLE clinical trials. Currently, investigators use instruments that were primarily developed to assess disease activity, such as the BILAG index and the SELENA-SLEDAI instrument to measure flare; however, their use may lead to both over- and under-reporting of flares due to difficulties with operational definitions utilized by these instruments. In order to optimize the definitions of flare and to translate these into validated instruments to assess change in flare, the Lupus Foundation of America (LFA) sponsored two international conferences on lupus flare. At these conferences, academic thought leaders as well as lupus leaders from the National Institutes of Health (NIH), the U.S. Food and Drug Administration (FDA) and the pharmaceutical and biotechnology industry discussed the usefulness of current indices and made recommendations for improving the ability to capture flares. At the first conference in 2006, it was concluded that improvements were necessary in order to improve the sensitivity and accuracy in reporting flares for the purposes of both clinical investigational trials and clinical practice. Participants agreed that this could optimally be accomplished by improving the existing instruments based on the BILAG and SELENA-SLEDAI instruments [23]. Additionally, the participants identified the need for a flare instrument specific for dermatological manifestations of SLE and advocated the further development of the CLASI as an instrument to capture dermatological flare for use in both clinical practice and clinical trials. Committees were developed to formulate improvements in these instruments; one important specific goal was to distinguish between mild and moderate flares as the original

instruments tended to combine these different clinical events and some of these events were felt to be clinically significant and others not. During the second conference in 2008, representatives from academia, the NIH, the FDA, and the pharmaceutical/biotechnology industry representing rheumatology, immunology, dermatology, nephrology, neurology, pulmonary medicine, epidemiology, and biostatistics discussed the efforts that been accomplished by the committees since the initial meeting, finalized a consensus definition of flare, addressed ways to modify the original indices, and proposed validation exercises for modified instruments.

In developing an instrument that assessed flare, experts from the first LFA-sponsored conference felt that it was essential that a definition of flare needed to be established that reflected the clinical and therapeutic implications of varying degrees of flare from a broad community of experts [23]. In order to achieve this goal, the Delphi technique was utilized to develop an international consensus among SLE clinicians on numerous aspects of defining and measuring flare. This effort was accomplished utilizing two internet-based questionnaires. Initially, 120 rheumatologists from around the world were identified as experts in the clinical management of SLE; these individuals received a survey of 13 questions and 69 (57.5%) responded to the invitation. Of the 67 who agreed to participate, 54 of the 69 (78%) participants completed the survey. The responses were analyzed and utilized to develop a second survey in which 12 questions were utilized and for each question, the participants were asked to react to a proposed definition. This survey was completed by 87 of 118 (74%) of the participants with greater than 80% agreement for most of the definitions that had been modified according to the comments from the original survey. The respondents noted the importance of two key concepts in defining flare: (1) there must be a measurable increase in disease activity that can be attributed to SLE, and (2) the treating physician must consider a therapeutic change as a response to the change in the clinical characteristics of the patient. Based on these concepts, and after discussion at the second conference [23], three draft flare definitions were agreed upon. One version included language that included a statement of effects on activities of daily living, another included the clause of "consideration of treatment," and the third did not include this clause. These three proposed definitions were sent to an international group of experts in a third survey. Among the 146 experts polled, 116 responded (79.5%), and an agreement was reached by 71 of 116 (61%) of the respondents for the following definition:

A flare is a measurable increase in disease activity in one or more organ systems involving new or worse clinical signs and symptoms and/or laboratory measurements. It must be considered clinically significant by the assessor, and usually there would be at least consideration of initiation or increase in treatment.

Rather than requiring a treatment change to fulfill the definition of flare, the experts agreed that consideration of treatment should be made. It was agreed that an actual initiation or change of treatment was not a mandate since many factors may prevent a treatment change, such as co-morbid conditions, insurance challenges, patient compliance, and issues related to safety. "Consideration of treatment" implies that a treatment change would be initiated if possible and links the flare definition to an "intent to treat." This is similar to the way the BILAG is scored.

Committees were organized at the second conference to propose preliminary flare index scoring systems derived from the SELENA-SLEDAI, BILAG, and CLASI instruments. The BILAG and the SELENA-SLEDAI instruments were chosen as they have been validated for both disease activity and flare assessment and have been extensively used in pharmaceutical/biotechnology company-sponsored clinical trials. The CLASI instrument was chosen for modification by the conference given its relevance to dermatologic-based assessments of flare. Despite the widespread use and validity of the BILAG and SELENA-SLEDAI instruments, they both are currently unable to adequately distinguish mild from moderate flares and both have inherent problems of inadequately addressing the "threshold effect," whereby a small change in disease activity crosses a threshold from categorical characterization as a mild, moderate, or severe flare despite little real change in disease activity [22]. Each committee addressed the adaptation of one of these three instruments [24].

The current definition of a severe flare using the BILAG-2004 Index is [23]: (1) a new score of A in any organ system that previously was categorized as C, D, or E, or (2) a new score of A in any organ system that previously was categorized as a B if the manifestations were new or worse after improving significantly. The current definition of a moderate flare is: (1) a new score of B in any system that previously was categorized as a D or E, or (2) a new score of B in any organ system that previously was categorized as a C if the manifestations were new or worse or if the manifestations were worse after a period of significant improvement.

Because of scoring rules (which were not designed to measure flares), the use of the BILAG instrument to determine flare was met with unanticipated issues with regard to outcomes. Because of this scoring artifact, patients could mistakenly be labeled with a flare if a patient had no change in disease activity (the phantom flare phenomenon). This situation arose, for example, if a patient who previously fulfilled criteria for moderate activity in a particular organ system (BILAG B) had subsequent partial improvement and was then labeled a C (minimal disease). However, if the partial improvement remained unchanged at the next assessment but fulfilled threshold criteria for moderate disease, then the patient would be mistakenly labeled as a flare (new B) without any real change in disease activity. It is now recognized that all new B scores are not flares unless accompanied by an increase in disease activity. Fortunately, data analysts do have access to that information in clinical trials since the BILAG scoring sheet indicates whether a descriptor is the same, better, or worse. However, this does not completely solve the problem of the threshold effect since a true new B flare could represent a range of disease activity from a relatively transient or mild change to a moderately severe change. Clinical trial design can potentially address this problem by considering a moderate flare to require two B scores or by requiring B flares to either be treated or to persist longer than 1 month.

At the second LFA conference [23], it was thought that the SELENA-SLEDAI Flare Index (SFI), which categorizes flares into mild/moderate or severe flares, might be improved by defining mild and moderate flares separately as mild flares generally do not prompt therapeutic changes and are often clinically less significant. Additionally, the threshold effect was felt to be problematic for this instrument, especially in categorizing patients who reached the threshold score for severe flares of 13 when a previous score was 12. This transition most likely would only rarely be considered a flare but as currently applied using the rules of scoring the SELENA-SLEDAI flare instrument, the patient would be categorized as having had a severe flare.

Through a series of teleconferences, the SELENA-SLEDAI Instrument Analysis Committee then created a new set of rules for the SELENA flare instrument by employing an organ-system approach and defining within each organ system mild flares that rarely required treatment and either moderate or severe ones that did require significant treatment changes [25]. With the new system, it was decided that in some patients, clinical manifestations and treatment choices may not always match, and when they did not, treatment choice would take precedence over clinical definitions. Unfortunately, this becomes problematic when a severe manifestation is treated with a moderate change of therapy or when it becomes unclear how to score a mild rash in a patient being treated for severe nephritis. Other changes made to the instrument include: treatment changes due to intolerance, toxicity, or safety considerations would not constitute a flare; clinical manifestations must be attributed to active lupus activity rather than co-morbidities or organ damage; and the "intent to treat" decision to institute to treatment is noted without regard as to whether it is instituted. The revised SFI differs significantly from the original SELENA-SLEDAI and can now be used independently of that instrument.

Given the validation of the CLASI, which assesses cutaneous lupus disease activity and damage, another outcome of the second LFA conference was the assignment of the CLASI Instrument Analysis Committee to evaluate the utility of the CLASI in defining and measuring lupus-specific skin flares [24]. The CLASI Instrument Analysis Committee has developed an internet-based prospective patient database that notes cutaneous characteristics, CLASI scores, quality-of-life assessments, physician and patient global activity assessments, and pain assessments. These data are intended to correlate CLASI score changes with other flare indicators such as medication changes or descriptive changes in skin lesions. By comparing the clinical meaningfulness of different CLASI score changes over time, the Committee has demonstrated that a CLASI change of 4 or more correlates with clinical indicators of flaring skin disease that prompts therapeutic changes. As both mild/moderate and severe CLASI flare definitions require change, the threshold effect is avoided with this instrument. Further work is required to define a mild cutaneous flare with the CLASI and efforts to correlate cutaneous flares by CLASI and the other generalized disease flare instruments must be undertaken. An important aspect of the CLASI that is fundamentally different than the current use of SLEDAI and BILAG is that most of the score is restricted to mucocutaneous findings seen in the clinic on the day of the visit, and would not reflect disease activity over an entire month.

In order to validate the three flare instruments, both assessments of content validity (to determine whether the instruments measure what is intended) and discriminant validity (to discriminate between no flare and mild flare, mild flare and moderate flare, moderate flare and severe flare) need to be undertaken. The approach that emerged was a two-step plan; the first approach, which has been completed, consisted of a real-patient assessment utilizing lupus experts assessing actual patients, and the second approach that is planned will be a prospective assessment of a large group of patients possibly in the context of an ongoing prospective multi-center randomized clinical trial.

For the initial validation study using actual patients, the SLICC group developed an exercise with the main objective being to determine whether the BILAG-2004 Index flare definitions and the SFI reliably identified flare as compared with a physician assessment of flare. Secondary objectives of this study were to determine whether these instruments can distinguish mild, moderate, and severe flares from each other and from the absence of flare. For this exercise, 16 patients who met four or more of the 1997 ACR revised classification criteria for SLE consented to participate; the patients were recruited from the University College London and St. Thomas' Hospital (London) lupus clinics; most of these patients had been selected because they were experiencing a flare [26]. Sixteen experienced rheumatologists from North America and Europe served as assessors; each patient was seen for up to 1h by four different physicians and each physician assessed four different patients. The order in which patients were seen by the assessors was randomized using four separate four-by-four Latin squares. A panel of five lupus specialists reviewed the patients' historical presentations and examined them collectively in the hour prior to the start of the exercise; each patient was well known to at least one of these panel members and these assessors designated each patient as having either a severe, moderate, mild, or no flare based on their collective clinical assessments. Upon entering each patient's room, the assessing physician was provided with the patient's history that detailed events, signs, and symptoms up to 1 month prior to the exercise date, current laboratory data, and data collection forms that included the BILAG-2004 Index, the SFI, and a four-point physicians' global assessment of flare (indexed for none, mild, moderate, and severe). Detailed glossaries for the indices were available in each room, as was the most recent definition of flare agreed upon by the LFA task force. The assessors had been trained in the use of the indices 1 day prior to the exercise. On exiting the room, the assessors presented their forms to two lupus expert physicians to ensure that all forms had been completed. For each of the four separate groups of four physicians and four patients, a percentage agreement between the physician ratings and the rating by the committee of five who had collectively categorized the patients was calculated for the three clinical assessment scores. A percentage agreement was also calculated between the BILAG flare and SFI scores and an assessment of the internal reliability of the three forms was based on the calculation of intraclass correlation coefficients with 95% confidence intervals. The rate of complete agreement of the four individual examining physicians for any flare versus no flare was 81% (55%; 94%), 75% (49%; 90%), and 75% (49%; 90%) for the BILAG-2004, SFI, and physician global assessment, respectively. Overall agreement of the flare defined by BILAG-2004 and the SFI was 81 and 52% when type of flare was considered. Intraclass correlation coefficients were 0.54 (0.32; 0.78) for BILAG-2004 flare compared to 0.21 (0.08; 0.48) for SFI and 0.18 (0.06; 0.45) for physician global assessment. There appeared to be good agreement between indices for severe flare with less agreement for mild/moderate flare.

References

[1] M.H. Liang, S.A. Socher, W.N. Roberts, J.M. Esdaile, Measurement of systemic lupus erythematosus activity in clinical research, Arthritis Rheum 31 (1988) 817—825.

[2] D.P.M. Symmons, J.S. Coppock, P.A. Bacon, B. Bresnihan, D.A. Isenberg, P. Maddison, et al., Development and assessment of a computerized index of clinical disease activity in systemic lupus erythematosus, Q J Med 69 (1988) 927—937.

[3] C. Bombardier, D.D. Gladman, M.B. Urowitz, D. Caron, D.H. Chang, and the Committee on Prognosis Studies in SLE Derivation of the SLEDAI, Arthritis Rheum 35 (1992) 630—640.

[4] D.D. Gladman, C.H. Goldsmith, M.B. Urowitz, P.A. Bacon, C. Bombardier, D. Isenberg, et al., Crosscultural validation and reliability of three disease activity indices in systemic lupus erythematosus, J Rheum 19 (1992) 608—611.

[5] D.D. Gladman, C.H. Goldsmith, M.B. Urowitz, P.A. Bacon, C. Bombardier, D. Isenberg, et al., Sensitivity to change of three SLE disease activity indices: international validation, J Rheum 21 (1994) 14568—14571.

[6] Z. Touma, M.B. Urowitz, D. Ibanez, D. Gladman, SLEDAI-2K 10 days vs. SLEDAI-2K 30 days in a longitudinal evaluation, Arthritis Rheum 60 (Suppl. 10) (2009) S898.

[7] C.S. Yee, V. Farewell, D.A. Isenberg, A. Prabu, K. Sokoll, L.S. Teh, et al., Revised British Isles Lupus Assessment Group 2004 index: a reliable tool for assessment of systemic lupus erythematosus activity, Arthritis Rheum 54 (2006) 3300—3305.

[8] D.A. Isenberg, A. Rahman, E. Allen, V. Farewell, M. Akil, I.N. Bruce, D. D'Cruz, et al., BILAG 2004. Development and initial validation of an updated version of the British Isles Lupus Assessment Group's disease activity index for patients with systemic lupus erythematosus, Rheumatol (Oxford) 44 (2005) 902—906.

[9] C.S. Yee, V. Farewell, D.A. Isenberg, A. Rahman, L.S. Teh, B. Griffiths, et al., British Isles Lupus Assessment Group 2004 index is valid for assessment of disease activity in systemic lupus erythematosus, Arthritis Rheum 56 (2007) 4113—4119.

[10] E.M. Hay, P.A. Bacon, C. Godron, D.A. Isenberg, P. Maddison, M.L. Snaith, et al., The BILAG index: a reliable and valid instrument for measuring clinical disease activity in systemic lupus erythematosus Q J Med 86 (1993) 447—458.

[11] C.-S. Yee, L. Cresswell, V. Farewell, A. Rahman, L.-S. Teh, B. Griffiths, et al., Numerical scoring for the BILAG-2004 Index, Rheumatol, doi:10.1093/rheumatology/kea026.

[12] C.S. Yee, V. Farewell, D.A. Isenberg, B. Griffiths, L.S. Teh, I.N. Bruce, et al., The BILAG-2004 index is sensitive to change for assessment of SLE disease activity, Rheumatol (Oxford) 48 (2009) 691—695.

[13] J.P. Buyon, M.A. Petri, M.Y. Kim, K.C. Kalunian, J. Grossman, B.H. Hahn, et al., The effect of combined estrogen and progesterone hormone replacement on disease activity in systemic lupus erythematosus: a randomized trial, Ann Intern Med 142 (2005) 953—962.

[14] M. Petri, M.Y. Kim, K.C. Kalunian, J. Grossman, B.H. Hahn, L.R. Sammaritano, et al., Combined oral contraceptives in women with systemic lupus erythematosus, N Eng J Med 15 (353) (2005) 2550—2558.

[15] D.D. Gladman, D. Ibanez, M.B. Urowitz, Systemic lupus erythematosus disease activity index 2000, J Rheumatol 29 (2002) 288—291.

[16] Z. Touma, M.B. Urowitz, D. Ibanez, D. Gladman, SLEDAI-2K 10 days versus SLEDAI-2K 30 days in a longitudinal evaluation (abstract), Arthritis Rheum 60 (2009) S898.

[17] J. Albrecht, L. Taylor, J.A. Berlin, S. Dulay, G. Ang, S. Fakharzadeh, et al., The CLASI (Cutaneous Lupus Erythematosus Disease Area and Severity Index): an outcome instrument for cutaneous lupus erythematosus, J Invest Dermatol 125 (2005) 889—894.

[18] A. Parodi, C. Massone, M. Cacciapuoti, M.G. Aragone, P. Bondavalli, G. Cattarini, et al., Measuring the activity of the disease in patients with cutaneous lupus erythematosus, Br J Dermatol 142 (2000) 457—460.

[19] M.S. Krathen, J. Dunham, E. Gaines, J. Junkins-Hopkins, E. Kim, S.L. Kolasinski, et al., The Cutaneous Lupus Erythematosus Disease Activity and Severity Index: expansion for rheumatology and dermatology, Arthritis Care Res 59 (2008) 338—344.

[20] M. Petri, S.G. Barr, J. Buyon, J. Davis, E. Ginzler, K. Kalunian, J.T. Merrill, RIFLE: Responder Index for Lupus Erythematosus (abstract), Arthritis Rheum 43 (2000) S244.

[21] American College of Rheumatology Ad Hoc Committee on Systemic Lupus Erythematosus Response Criteria. The American College of Rheumatology, Response Criteria for Systemic Lupus Erythematosus Clinical Trials: measures of overall disease activity, Arthritis Rheum 50 (2004) 3418—3426.

[22] R.A. Furie, M.A. Petri, D.J. Wallace, E.M. Ginzler, J.T. Merrill, W. Stoll, et al., Novel evidence-based systemic lupus erythematosus responder index, Arthritis Rheum 15 (61) (2009) 1143—1151.

[23] N. Ruperto, L.M. Hanrahan, G.S. Alarcon, H.M. Belmont, P. Brunetta, J.P. Buyon, et al. International consensus to define flare in lupus (submitted).

[24] Second international lupus flare conference. June 5-6, 2008 Conference summary, Lupus Foundation of America (2008) 1—17.

[25] J. Buyon, K.C. Kalunian, M.B. Urowitz, M.A. Dooley, V. Strand, J. Merrill, et al., Revision of the SELENA Flare Index (abstract), Arthritis Rheum 60 (2009) S339.

[26] D.A. Isenberg, E. Allen, V. Farewell, D. D'Cruz, G.S. Alarcon, C. Aranow, et al. An assessment of disease flare in patients with systemic lupus erythematosus: a comparison of BILAG-2004 and the flare version of SLEDAI (submitted).

Incomplete Lupus Erythematosus

Johannes C. Nossent, Tom J.G. Swaak

DEFINITION

Incomplete lupus erythematosus (ILE) refers to a condition where patients present with signs of systemic autoimmunity and clinical symptoms compatible with systemic lupus erythematosus (SLE), but presently do not fulfill four of the ACR criteria required for classification of SLE [1, 2].

BACKGROUND

The introduction and subsequent updating of the classification criteria has been a major step forward in studying the epidemiology, prognosis, and underlying immune disturbances in SLE. The statistical procedures underlying classification aim for the lowest amount of heterogeneity in cohorts and thus excluded patients with fewer than four typical or other symptoms suggestive of SLE. This stringent classification process has also introduced a clinical dilemma in diagnosing certain patients. Currently, the fulfilment of ACR criteria is often regarded and recommended as the absolute requirement for the diagnosis of SLE in clinical practice, even though the proposed model was only meant to facilitate formal communication in scientific papers and to better ensure uniformity in enrollment for clinical trials. The authors involved in the classification process had not proposed their use as diagnostic criteria [1] and it should be remembered that during the classification process a considerable amount of clinically useful information was excluded to achieve a model that best discriminated between SLE patients and other disease types. Most rheumatologists will readily acknowledge that SLE is a much more diverse clinical syndrome than suggested by the narrowly defined ACR criteria set and recognize situations where SLE is diagnosed clinically and managed based on a understanding of pathogenesis in spite of less than four classification

criteria being present [3]. This is also reflected in the manner by which various scoring systems for disease activity such as SLEDAI, BILAG, and ECLAM were developed; they incorporate a wide range of disease manifestations (e.g., vasculitis, fever) not included in the ACR criteria, because these manifestations are considered important in individual patient management. Finding isolated immune complex-mediated glomerulonephritis in an ANA-positive patient will carry a different weight during patient evaluation than finding isolated UV hypersensitivity and discoid lesions in that ANA-positive patient. The importance of severe manifestations often overrules the number of classification criteria fulfilled in the diagnostic process and subsequent management [4]; this seems of particular relevance for childhood SLE, where presentation is diverse and often severe [5]. Figure 38.1 details the frequency of a variety of other skin, vascular, and serologic manifestations of SLE in comparison to ACR criteria from a cohort study of European patients with severe SLE [6].

To compensate for this lack of flexibility of the ACR classification, several authors have proposed weighting of the various ACR criteria [5, 6]. In this process, different point scores are given to a range of skin, renal, CNS, joint, and serological manifestations. Applying the latest version of the Boston weighted score allowed SLE diagnosis in patients with, for example, leukocytopenia and a positive ANA and increased total SLE cohort size by 20% [7]. However, weighing of ACR criteria has so far not considered alternative SLE-compatible symptoms. Strict use of the ACR criteria set in clinical practice has an additional weakness, as it fails to take into account that several disease manifestations in SLE are interdependent. This is best exemplified by the strong correlation between anti-dsDNA Ab and positive ANA, which leads to a situation where in reality only three classification criteria are fulfilled (as the anti-dsDNA antibodies also cause the positive ANA), but patients

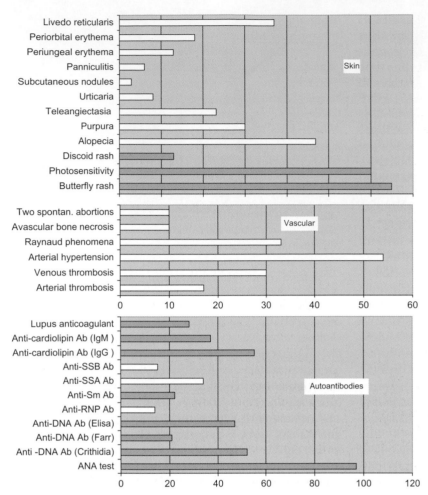

FIGURE 38.1 Frequency distribution of ACR criteria (gray bars) and other skin, cardiovascular, and serological manifestations (white bars) that are not included in this criteria set.

nonetheless are readily classified as SLE. A similar situation exists for the large number of mucocutaneous symptoms in the criteria set, and the resulting skewing of clinical and scientific data (i.e., patients do not have four independent criteria) needs to be recognized [8]. All the above serves in our opinion to strengthen the concept that when ACR criteria are applied as a gold standard in clinical practice, ILE constitutes a genuine clinical entity and is an appropriate acronym for patients presenting with less than four ACR criteria and/or other lupus-related manifestations. ILE is also being referred to as subclinical lupus, variant lupus, latent lupus, or nonclassic lupus [9–11].

Our current assumption that ILE patients are at risk of developing a more severe disease entity in the future provides not only a clinical challenge, but also an opportunity for learning and possibly prevention of classic SLE development. Recognition of ILE is naturally and directly important in patient management, but increased attention on ILE may also lead to a better understanding of the mechanisms by which the immune disturbances occurring in ILE patients do not (yet) translate into a severe clinical phenotype. Scientific studies on early recognition of

ILE and therapeutic measures to prevent the development of more severe disease may thus be one of the most fruitful measures in our efforts to relieve the burden of disease for SLE patients. This can provide important ideas for the whole field of systemic autoimmunity.

ILE vs. Undifferentiated Connective Tissue Disease

SLE is a disease that develops gradually over time and only a minority of patients present with acute multisystem disease, which can be promptly classified according to the presence of four or more ACR criteria. In the LUMINA study cohort, 15% of patients presented in this manner and the average time needed to accrue four ACR criteria was more than 3 years for patients with only one ACR criterion at first presentation and 2.5 and 1 year, respectively, for patients presenting with two or three validated ACR criteria [12]. One can thus say that most SLE patients in the LUMINA study have been going through a period where their disease was best termed ILE, which ultimately became progressive. The longest lag time between ILE onset and fulfilment of the scientific diagnosis of SLE was 328 months, indicating that even very late disease progression may occur in ILE [12]. Similarly, in studies based on SLE patients for whom prediagnosis serum was available from a US military serum repository, 80% of patients had a gradual accrual of serological and clinical disease criteria [13–15]. While these cohort studies excluded patients with nonprogressive ILE, they clearly show that in most SLE patients a prior window of opportunity exists, where early ILE diagnosis and intervention can take place in order to prevent development of more severe disease.

After ANA testing became more widely available between 1980 and 1990, the term "undifferentiated connective tissue diseases" (UCTD) was introduced to designate an increasing number of patients that were seen with sporadic manifestations compatible with systemic autoimmune disease and a positive ANA test. However, despite proposed criteria for its classification, UCTD is not an established entity [16] with the name reserved for those patients with features strongly suggestive of an autoimmune rheumatic disease, but not readily classified as a specific CTD [17]. Published UCTD series contain variable numbers of ILE patients in addition to patients with not yet fully recognised RA or scleroderma [18, 19]. As UCTD and ILE share a considerable number of characteristics, the distinction between the two conditions is not very clear, suggesting to some authors that they should be considered as one and the same condition [9, 20]. Others prefer to group the two conditions together under the heading of undifferentiated autoimmune syndrome (UAS) since both imply autoimmunity and not necessarily or strictly connective tissue involvement [21]. The question will not be solved until our current lack of data on the subject is addressed. For all practical purposes, one might consider ILE to be a subset of UCTD, where presenting features already indicate the potential for SLE development (Figure 38.2).

EPIDEMIOLOGY

There has been only one report on the epidemiologic characteristics of ILE. A long-term disease registry in the Danish county of Funen with a population of nearly 400,000 has, with the use of capture/recapture techniques, estimated the annual ILE incidence to be 0.4/100,000 and ILE point prevalence in 2003 around 7/100,000. It should be noted that these are data from an area with one of the lowest reported SLE incidence (1/10,000) and prevalence (29/100,000) rates worldwide [22, 23]. Given the additional selection bias toward definite SLE, these rates thus represent only minimum estimates of ILE epidemiology. Based on compiled long-term observational data from southern Sweden,

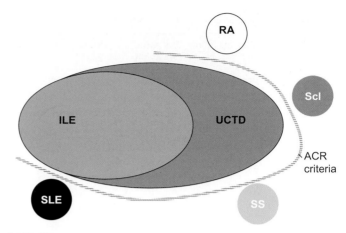

FIGURE 38.2 Graphical representation of the interrelationships between ILE, UCTD, and defined CTS.

an annual ILE incidence of nearly 2/100,000 can be inferred [24, 25]. Other cross-sectional studies indicate that ILE is not uncommon and occurs with similar frequency in Caucasian as in other ethnic populations [9, 10, 26, 27]. Epidemiological data on ILE from areas with a high incidence rate of SLE are clearly needed to address the question "to what extent can ILE be considered a subset of SLE in various populations?"

CLINICAL PRESENTATION

At the time of writing, seven studies have been reporting on dedicated ILE cohorts in several regions [9, 10, 23, 25, 26, 28, 29]. Despite slightly differing inclusion criteria, the demographics of patients included in these studies show that age and gender distribution are largely similar to patient characteristics in SLE cohorts (Table 38.1). The frequencies of the main clinical findings in these studies are depicted in Figure 38.3. Skin and joint affection and cytopenias are the most frequent manifestations. With the exception of the Swedish study, there is a very low frequency of renal involvement with no cases of end-stage renal failure reported. All studies report a virtual absence of neurological ACR criteria, which is not surprising given their low sensitivity in SLE classification. Serological findings (Figure 38.4) show that the presence of ANA is most frequently an isolated finding and when subspecificities are detected, the rate of anti-Ro/La antibodies exceeds the rate for anti-Sm or anti-dsDNA antibodies. Whether this indicates an increased possibility of developing Sjögren syndrome (SS) rather than SLE is not known at present, as the relationship between the two diseases processes is much debated. Rivera *et al.* found a higher tendency to SS development than SLE development in asymptomatic anti-Ro-positive mothers of children with neonatal lupus [21] and the development of autoimmune disease was enhanced by

TABLE 38.1   Characteristics at Presentation of Patients Included in Clinical ILE Studies

	Laustrup et al., 2009 [23]	Al Attia, 2006 [28]	Stahl et al., 2004 [25]	Swaak et al., 2001 [29]	Vila et al., 2000 [26]	Greer and Panush, 1989 [9]	Ganczarczyk et al., 1989 [10]	All
Design	Population	Single center	Population	Multicenter	Single center	Single center	Single center	–
Geography	Denmark	United Arab Emirates	Sweden	Europe	Puerto Rico	US	Canada	–
Inclusion	≥1 ACR <4	≥1 ACR <4	≥1 ACR <4	ANA + 1 organ	≥1 ACR <4	≥2 ACR < 4	≥1 ACR ≤2 + one other symptom	–
No. included	37	12	28	122	87	38	22	346
Female (%)	100	80	93	99	94	84	90	91
Age (years)	39	30	45	40	34	37	38	38
Symptom duration	–	15				38		20

the concomitant presence of anti-La antibodies. SS and SLE may be considered as separate diseases that occasionally have overlapping clinical and serologic manifestations with anti-Ro and anti-La antibodies as common markers for (development of) both diseases.

DISEASE COURSE IN ILE

Clinical Phenotype

The rate of progression to an ACR criteria-based definition of SLE in dedicated ILE studies is summarized in Table 38.2. The highest progression rates were described in studies with the longest follow-up, consistent with the finding that ILE may continue to evolve to classical SLE after prolonged periods. Few clinical features consistently predict progression to SLE. The rate of SLE progression was independent of gender and ethnicity and the timeline was similar to that seen in the US military study on pre-SLE autoantibody development; however, presentation with malar rash was related with more rapid progression in two studies [15, 25]. The presence of anti-dsDNA and antiphospholipid antibodies as well as low complement levels were found to be the most reliable serological predictors of progression to SLE [25, 26].

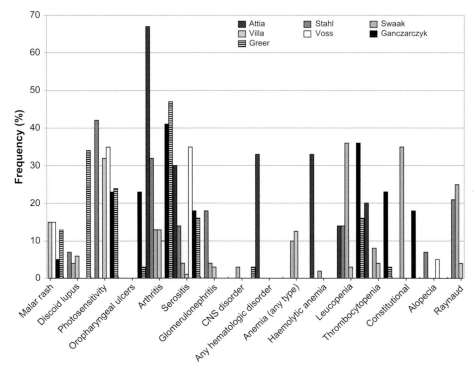

FIGURE 38.3   Frequency of reported disease manifestations in ILE patients.

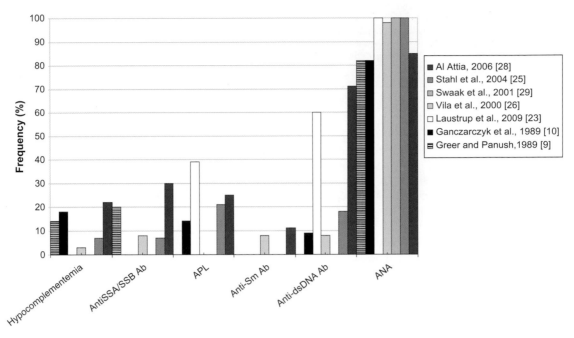

FIGURE 38.4 Frequency of reported autoantibodies in ILE patients.

Thus, while prolonged observation of ILE patients has been indicated for many years, the risk for SLE progression is highest within the first 4 years and may be predicted by the presence of antibody clusters [30, 31]. Several UCTD cohort studies have also reported rates of progression to SLE. In these more broadly defined cohorts, the percentage of patients with presumed ILE that subsequently progressed to SLE varied strongly from 12 to 95% [32–34]. Among asymptomatic or pauci-symptomatic mothers with anti-Ro antibodies in the Research Registry for Neonatal Lupus, progression to SLE occurred in 8% within 5 years and 19% within 10 years in [21].

Disease Activity

Two studies have reported measures of disease activity in ILE cohorts. The European multicenter study with the largest cohort ($n = 122$) followed patients over 3 years [29]. The development of disease activity over time (SLEDAI and ECLAM scores) (Table 38.3) shows that ILE patients had persistent low disease activity

over the 3-year period in contrast to patients progressing to SLE. In these patients, disease activity scores increased sharply, indicating that this progression was characterized by multisystem activity rather than the gradual accrual of one more ACR criterion. However, the validity of these scoring systems in ILE patients has not been assessed, and given their organ-based approach, they may well lead to underestimates of disease activity in ILE. This can be illustrated in a more detailed analysis of the changes in specific disease manifestations. At group levels, there is relatively little change in the frequency of specific organ manifestations over 3 years of follow-up (Figure 38.5), with the exception of an increase in malar rash and nonhemolytic anemia prevalence. More detailed information on the type and number of ILE patients experiencing disease activity can be gathered from Figure 38.6, which captures dynamics at the individual level. Here the annual number of ILE patients experiencing specific clinical features is given as a function of that manifestation in the previous year (represented as the percentage of patients with increasing,

TABLE 38.2 Progression to SLE (as Defined by Reaching Four or More ACR criteria) in Reported ILE Cohorts

Study	Laustrup et al., 2009 [23] ($n=37$)	Al Attia, 2006 [28] ($n=15$)	Stahl et al., 2004 [25] ($n=28$)	Swaak et al., 2001 [29] ($n=122$)	Vila et al., 2000 [26] ($n=87$)	Ganczarczyk et al., 1989 [10] ($n=22$)	Greer and Panush, 1989 [9] ($n=63$)	Combined 374
% Reaching ≥4 ACR criteria	19	0	57	21	9	32	5	13
Observation (years)	8	1.8	13	3	2.2	8	1.5	5.4
Months to ACR classification	98	—	64	14	52	60	18	43

TABLE 38.3 Median Disease Activity Scores at Various Time Points in 122 Patients Initially Diagnosed with ILE

| | Stable ILE (n=97) | | SLE progression (n=25) | |
Period	ECLAM	SLEDAI	ECLAM	SLEDAI
0–12 months	2.1	2.6	4.4	4.3
13–24 months	2.1	2	6.1	8.5
25–36 months	1.7	2	5.6	9.9

stable, or decreasing symptoms). Such data provide better insight into the active process of disease activity changes in individual patients than the small changes seen in overall SLEDAI scores (Table 38.3) and the percentage of patients with the relevant manifestation (Figure 38.5). As an example, the prevalence of arthritis changes from 15 to 19% and to 13%, but there were considerable larger variations in the number of patients remitting and experiencing new arthritic flares. The frequency of renal involvement showed a different pattern, with no new cases emerging during observation. The most frequently observed laboratory finding was leukocytopenia, which despite an overall reduction in prevalence from 36 to 20% during follow-up at the group levels, showed nearly similar proportions of new and remitting cases. Similar findings for anemia of chronic disease (ACD), where overall prevalence rose from 10 to 11% and 16%, despite a large proportion of patients with ACD remitting during observation. These data all confirm the position that ILE is not a continuous grumbling disease state, but in line with SLE in that it is also characterized by a fluctuating disease course.

Damage Development

Few data are available to determine whether the fluctuating disease activity in ILE patients will lead to permanent organ damage. In the Swedish cohort, the mean SLICC damage index did not increase significantly and only progressed from 0.1 to 0.2 during long-term follow-up of 10 years and only two out of 12 (16%) of nonprogressing ILE patients developed organ damage [25]. These data indicate that the previously mentioned persistent but low disease seems relatively benign. In contrast, ten out of 16 (63%) patients progressing to SLE developed organ damage with a mean score of 1.5 [25].

Other Outcomes

While outcome measures such as mortality, cardiovascular morbidity (given the relation between chronic inflammation and plaque formation), as well as quality of life are of clinical interest, these have not yet been reported for ILE patients. Data extractions from the referred ILE cohorts' studies describe one case of fatality in nonprogressing ILE patients. An important aspect of outcome studies is the potential influence of therapeutic interventions. Despite a lack of data, drug treatment is evidently not a rare occurrence as considerable numbers of ILE patients have been prescribed NSAIDs, antimalarials, and oral corticosteroid therapy (Table 38.4). In the ESCISIT study, 38% of patients were at inclusion already on treatment with prednisolone (none with doses higher than 10 mg/day). This figure changed to 43 and 27% in year 2 and year 3, respectively. For antimalarials, these

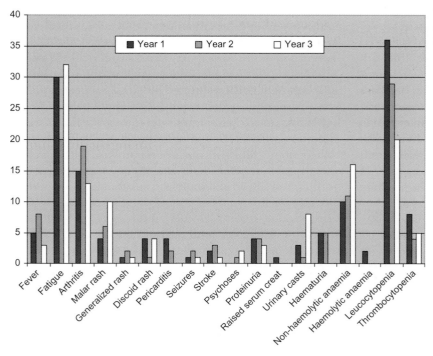

FIGURE 38.5 Frequency (% of whole cohort) of specified clinical findings in ILE cohort throughout 3 years of observation.

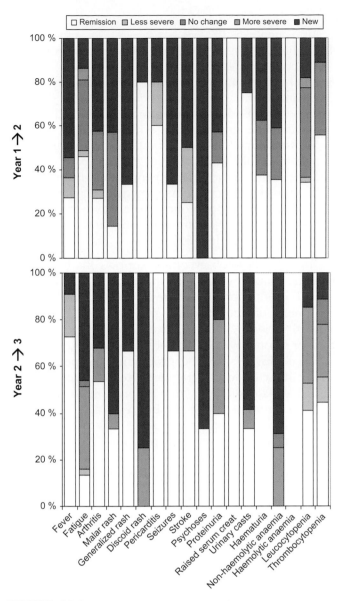

FIGURE 38.6 Dynamics of changes in disease activity in ILE patients for specified clinical findings from 1st to 2nd year and 2nd to 3rd year observation.

TABLE 38.4 Therapeutic Measures in ILE Patients[a]

	Swaak [29]	Greer [26]	Ganczarczyk [10]	Combined
NSAID	NA	47	41	44
Antimalarials	17	29	NA	23
Corticosteroids	43	32	9	28
Immunosuppressants	5	3	0	3

[a]Figures recalculated to represent percentage of patients not progressing to SLE. NA, not available.

figures were 17 and 32%, respectively, while azathioprine was prescribed in only 2, 3.5, and 4.5% of cases for years 1, 2, and 3, respectively. The influence (if any) of these treatment regimens on disease expression and the progression to SLE has not been studied in a controlled fashion. A retrospective analysis of the US military SLE study nonetheless showed that both hydroxychloroquine and corticosteroid treatment significantly increased the time from first onset of clinical symptoms to SLE progression, while OH chloroquine treatment also reduced the subsequent number of autoantibody specificities [35]. No such effect was seen for NSAID treatment. These important preliminary findings provide limited support for the early use of immunosuppressive therapies in ILE, although more detailed data on the indication and timing of therapies are clearly needed.

PATHOGENESIS

Is ILE a Subset of SLE?

Given the considerable clinical and serological similarities, it seems reasonable to presume that ILE patients share common elements of the etiopathogenesis with SLE. Based on the observations that autoantibodies can be detected in the sera of SLE patients long before they accumulate enough ACR criteria for classification [36], a hypothesis can be formulated that while ILE patients produce the same spectrum of autoantibodies as SLE patients, these antibodies fail to attain the same level of pathogenicity. In general, IgG-type autoantibodies are presumed to be more pathogenic (i.e., nephritogenic) than IgM-type antibodies in SLE as a result of affinity maturation [8, 37]. Thus the idea can be put forward that while ILE patients produce autoantibodies against the same epitope as SLE patients, they "fail" to make the switch from IgM isotype to IgG isotype autoantibody production. Comparing the autoantibody isotypes between SLE, ILE, and first-degree relatives of SLE patients, serum of ILE and SLE patients contained high levels of IgG autoantibodies to 50 autoantigens and IgM autoantibodies to 12 autoantigens. The IgG:IgM autoantibody ratio, however, showed a stepwise increase in the groups with growing disease burden, indicating that IgM autoreactivity predominates in ILE and that this may represent an early stage prior to IgG switching or developing SLE [38]. However, an alternative model can be envisioned where ILE patients form a stable group, by which unknown regulatory systems halt the affinity maturation process and the forming of IgG autoantibodies. This could form an explanation for the finding that the nonprogressing ILE patients represent a subgroup with a good prognosis and is supported by the finding that IgM autoantibodies exert protective properties against the development of autoimmune disease by enhancing the functional capabilities of immature dendritic cells for phagocytic removal of apoptotic cells [39, 40].

Abnormal Acute-Phase Reaction as a Common Pathogenetic Pathway for the Development of ILE and SLE

C-reactive protein (CRP) is an acute-phase protein that in lupus prone mice exerts therapeutic activity by protecting against the development of a glomerulonephritis [41]. The mechanism by which CRP suppresses immune complex-mediated disease may involve its binding and removal of autoantigens, which are released during apoptosis as well as the induction of suppressive macrophages through FcγReceptor activation [42]. While no such data are available in SLE patients, there are suggestions that the CRP response may represent an important aspect of the disease state.

While median CRP levels in RA are in the range of 20–40 mg/l [43], in SLE patients with an exacerbation, levels are seldom above 15 mg/l [44]. However, a marked CRP response can be found in SLE patients with serositis, synovitis, and bacterial infections [45, 46]. It is unclear whether this indicates a general abnormality of the CRP response in all SLE patients or whether this points to the presence of subsets of SLE patients with a specific alteration in the CRP response. In our studies of the acute-phase reaction, diminished CRP production was associated with increased production of ferritin during the development of nephritis [47]. This reduced CRP response may represent an isolated defect in the acute-phase reaction in SLE, as evidenced by an increase in other acute-phase proteins such as α1-acid glycoprotein and fibrogen [48–50]. The induction of the acute-phase proteins is regulated by a concerted action of cytokines in which TNF as well IL-6 play a dominant role. Overall, a correlation between IL-6 and CRP levels is claimed, but in SLE patients no deficiency in IL-6 production could be demonstrated by using whole blood cell cultures. Only in patients treated with prednisone could a diminished CRP production be demonstrated [51].

Based on a summary of data from the literature, a pathogenetic role of the acute phase response in SLE can be presumed because:

1. CRP – among other molecules – is needed for an efficient clearance of apoptotic cells.
2. SLE patients have a characteristic and clinically useful blunted CRP response, which is partly compensated by an increased production of other acute-phase proteins.
3. CRP gene polymorphism is associated with low basal CRP levels, which has been found to predispose to

lupus, production of antinuclear antibodies, and lupus nephritis [52, 53].

4. Studies in mouse models demonstrate a beneficial effect of CRP [44, 54]
5. In a number of patients with various diseases, treatment with biologicals such as anti-TNF drugs causes the production of antinuclear antibodies and clinical findings compatible with ILE [55, 56], while at the same time the CRP response is blunted in these patients by the application of an anti-TNF drug.
6. Deficiencies in the complement system including mannose binding proteins, which are also involved in the clearance of apoptotic cells, are associated with the production of antinuclear antibodies and with the existence of SLE [57–59].

Together these observations support the hypothesis that the qualitative and/or quantitative abnormalities in CRP production contribute to defective clearance of cell debris or apoptotic cell material in SLE patients. This situation where persistence of apoptotic blebs and/or secondary necrotic debris then induces an alternative pathway with the production of specific antibodies directed to cell fragments to enhance clearance of such cell debris. In the first instance, these antibodies will belong to the IgM class, but upon affinity maturation by repeated antigen exposure, it is to be expected that a later switch to IgG antibodies can take place [37]. That this pathogenic pathway may be operative is further supported by the observation that INF-α administration to patients with melanoma is followed by a decrease in CRP levels, while in contrast ferritin levels increase [60]. Gogas et al. reported an increase in autoantibody production in these patients [61].

That a similar pathogenetic mechanism is operative in both ILE and SLE is shown in Table 38.5, where both disease entities have similar Ferritin:CRP ratios, which are increased compared to RA. If ILE is a prodrome or a subset of SLE, it is expected that in ILE the profile of the autoantibodies will be of the IgM class, with antibody class switching occurring prior to or closely related with SLE development. Observations by Li et al. [38] and Fairhurst et al. [62] confirm that a stepwise increase in the

TABLE 38.5   Comparison of Ferritin:CRP Ratios in Sera of ILE, SLE, and RA Patients[a]

	RA $n=22$	ILE $n=20$	SLE $n=34$
Age (years)	45 ± 13	31 ± 12	34 ± 13
Ferritin:CRP Ratio[b]	5.6 ± 20	7.0 ± 43	8.4 ± 38

[a]*Samples were taken during their disease course after exclusion of an infection. CRP levels > 10 mg/dl and ferritin levels > 50 μg/dl were selected (A.J.G. Swaak, personal communication).*
[b]*The Ferritin/CRP ratios between ILE and SLE were not significant different, both were significantly increased compared to RA patients.*

IgG:IgM autoantibody ratio was present in normal controls, first-degree relatives of SLE, ILE, and SLE. This evidence of IgM autoreactivity in ILE and IgG autoreactivity in SLE suggests that ILE represents an early stage of SLE. It can be postulated that several relevant mechanisms may be operative in hindering progression to SLE that will need further study. While our knowledge is far from complete, important lessons can thus be learned by closely studying the immunological process in ILE patients.

SUMMARY

When a patient fulfills ACR classification for SLE, the case is considered classic and the diagnosis is readily accepted. Patients who meet less than four ACR criteria or have other representative SLE manifestations have incomplete lupus erythematosus (ILE). ILE is a relatively common distinctive disease entity and usually presents with rash, Raynaud phenomenon, arthritis, leukocytopenia, and thrombopenia together with antinuclear antibodies and up to one subspecificity. There is insufficient evidence to determine whether ILE is only a *forme fruste* of SLE that will not evolve and is more benign or whether it truly is a prodromal form of SLE where progression can be expected. Long-term follow-up of ILE patients is indicated as even patients who do not progress will have varying periods of disease activity, which may require therapy. With conservative management, 10–50% are expected to progress to SLE and this process may be impeded by early intervention with antimalarial therapy. While long-term and complete disease remission is the ultimate management goal for all patients, this remains a relatively rare occurrence. In established SLE patients, treatment is often considered efficacious, when patients no longer have acute, life-threatening complications even though many continue to have intermittent or continuous low disease activity. Actually, such patients can be considered to have returned to a *post-hoc* state of ILE, where similar doubts about the long-term prospects and optimal management are encountered as in the *pre-hoc*, nonprogressed ILE state. Closer study of ILE patients may provide important clues for our understanding of the pathophysiology and management of SLE.

References

[1] E.M. Tan, A.S. Cohen, J.F. Fries, A.T. Masi, D.J. McShane, N.F. Rothfield, et al., The 1982 revised criteria for the classification of systemic lupus erythematosus, Arthritis Rheum 25 (11) (1982) 1271–1277.

[2] M.C. Hochberg, Updating the American College of Rheumatology revised criteria for the classification of systemic lupus erythematosus, Arthritis Rheum 40 (9) (1997) 1725.

[3] E.G. Bywaters, Systemic lupus erythematosus. Classification criteria for systemic lupus erythematosus, with particular reference to lupus-like syndromes, Proc R Soc Med 60 (5) (1967 May) 463–464.

[4] I.M. Gilboe, G. Husby, Application of the 1982 revised criteria for the classification of systemic lupus erythematosus on a cohort of 346 Norwegian patients with connective tissue disease, Scand J Rheumatol 28 (2) (1999) 81–87.

[5] B. Bader-Meunier, J.B. Armengaud, E. Haddad, R. Salomon, G. Deschenes, I. Kone-Paut, et al., Initial presentation of childhood-onset systemic lupus erythematosus: a French multicenter study, J Pediatr 146 (5) (2005) 648–653.

[6] J. Nossent, N. Cikes, E. Kiss, A. Marchesoni, V. Nassonova, M. Mosca, et al., Current causes of death in systemic lupus erythematosus in Europe, 2000–2004: relation to disease activity and damage accrual, Lupus 16 (5) (2007) 309–317.

[7] K.H. Costenbader, E.W. Karlson, L.A. Mandl, Defining lupus cases for clinical studies: the Boston weighted criteria for the classification of systemic lupus erythematosus, J Rheumatol 29 (12) (2002) 2545–2550.

[8] H.C. Nossent, O.P. Rekvig, Is closer linkage between systemic lupus erythematosus and anti-double-stranded DNA antibodies a desirable and attainable goal? Arthritis Res Ther 7 (2) (2005) 85–87.

[9] J.M. Greer, R.S. Panush, Incomplete lupus erythematosus, Arch Intern Med 149 (11) (1989) 2473–2476.

[10] L. Ganczarczyk, M.B. Urowitz, D.D. Gladman, "Latent lupus", J Rheumatol 16 (4) (1989) 475–478.

[11] R.A. Asherson, R. Cervera, R.G. Lahita, Latent, incomplete or lupus at all? J Rheumatol 18 (12) (1991) 1783–1786.

[12] G.S. Alarcon, G. McGwin Jr., J.M. Roseman, A. Uribe, B.J. Fessler, H.M. Bastian, et al., Systemic lupus erythematosus in three ethnic groups. XIX. Natural history of the accrual of the American College of Rheumatology criteria prior to the occurrence of criteria diagnosis, Arthritis Rheum 51 (4) (2004) 609–615.

[13] M.R. Arbuckle, M.T. McClain, M.V. Rubertone, R.H. Scofield, G.J. Dennis, J.A. James, et al., Development of autoantibodies before the clinical onset of systemic lupus erythematosus, N Engl J Med 349 (16) (2003) 1526–1533.

[14] M.T. McClain, M.R. Arbuckle, L.D. Heinlen, G.J. Dennis, J. Roebuck, M.V. Rubertone, et al., The prevalence, onset, and clinical significance of antiphospholipid antibodies prior to diagnosis of systemic lupus erythematosus, Arthritis Rheum 50 (4) (2004) 1226–1232.

[15] L.D. Heinlen, M.T. McClain, J. Merrill, Y.W. Akbarali, C.C. Edgerton, J.B. Harley, et al., Clinical criteria for systemic lupus erythematosus precede diagnosis, and associated autoantibodies are present before clinical symptoms, Arthritis Rheum 56 (7) (2007) 2344–2351.

[16] M. Mosca, C. Tani, S. Bombardieri, Undifferentiated connective tissue diseases (UCTD): a new frontier for rheumatology, Best Pract Res Clin Rheumatol 21 (6) (2007) 1011–1023.

[17] M. Mosca, C. Baldini, S. Bombardieri, Undifferentiated connective tissue diseases in 2004, Clin Exp Rheumatol 22 (3 Suppl. 33) (2004) S14–S18.

[18] M.G. Danieli, P. Fraticelli, F. Franceschini, R. Cattaneo, A. Farsi, A. Passaleva, et al., Five-year follow-up of 165 Italian patients with undifferentiated connective tissue diseases, Clin Exp Rheumatol 17 (5) (1999) 585–591.

[19] H.J. Williams, G.S. Alarcon, R. Joks, V.D. Steen, K. Bulpitt, D.O. Clegg, et al., Early undifferentiated connective tissue disease (CTD). VI. An inception cohort after 10 years: disease remissions and changes in diagnoses in well established and undifferentiated CTD, J Rheumatol 26 (4) (1999) 816–825.

[20] P.J. Venables, Undifferentiated connective tissue diseases: mixed or muddled? Lupus 7 (2) (1998) 73—74.

[21] T.L. Rivera, P.M. Izmirly, B.K. Birnbaum, P. Byrne, J.B. Brauth, M. Katholi, et al., Disease progression in mothers of children enrolled in the Research Registry for Neonatal Lupus, Ann Rheum Dis 68 (6) (2009) 828—835.

[22] H. Laustrup, N.H. Heegaard, A. Voss, A. Green, S.T. Lillevang, P. Junker, Autoantibodies and self-reported health complaints in relatives of systemic lupus erythematosus patients: a community based approach, Lupus 13 (10) (2004) 792—799.

[23] H. Laustrup, A. Voss, A. Green, P. Junker, Occurrence of systemic lupus erythematosus in a Danish community: an 8-year prospective study, Scand J Rheumatol 38 (2) (2009) 128—132.

[24] C. Stahl-Hallengren, A. Jonsen, O. Nived, G. Sturfelt, Incidence studies of systemic lupus erythematosus in Southern Sweden: increasing age, decreasing frequency of renal manifestations and good prognosis, J Rheumatol 27 (3) (2000) 685—691.

[25] H.C. Stahl, O. Nived, G. Sturfelt, Outcome of incomplete systemic lupus erythematosus after 10 years, Lupus 13 (2) (2004) 85—88.

[26] L.M. Vila, A.M. Mayor, A.H. Valentin, M. Garcia-Soberal, S. Vila, Clinical outcome and predictors of disease evolution in patients with incomplete lupus erythematosus, Lupus 9 (2) (2000) 110—115.

[27] A.E. Wandstrat, F. Carr-Johnson, V. Branch, H. Gray, A.M. Fairhurst, A. Reimold, et al., Autoantibody profiling to identify individuals at risk for systemic lupus erythematosus, J Autoimmun 27 (3) (2006) 153—160.

[28] H.M. Al Attia, Borderline systemic lupus erythematosus (SLE): a separate entity or a forerunner to SLE? Int J Dermatol 45 (4) (2006) 366—369.

[29] A.J. Swaak, B.H. van de, R.J. Smeenk, K. Manger, J.R. Kalden, S. Tosi, et al., Incomplete lupus erythematosus: results of a multicentre study under the supervision of the EULAR Standing Committee on International Clinical Studies Including Therapeutic Trials (ESCISIT), Rheumatology (Oxford) 40 (1) (2001) 89—94.

[30] R. Jurencak, M. Fritzler, P. Tyrrell, L. Hiraki, S. Benseler, E. Silverman, Autoantibodies in pediatric systemic lupus erythematosus: ethnic grouping, cluster analysis, and clinical correlations, J Rheumatol 36 (2) (2009) 416—421.

[31] C.H. To, M. Petri, Is antibody clustering predictive of clinical subsets and damage in systemic lupus erythematosus? Arthritis Rheum 52 (12) (2005) 4003—4010.

[32] M. Mosca, R. Neri, W. Bencivelli, A. Tavoni, S. Bombardieri, Undifferentiated connective tissue disease: analysis of 83 patients with a minimum follow-up of 5 years, J Rheumatol 29 (11) (2002) 2345—2349.

[33] E. Bodolay, Z. Csiki, Z. Szekanecz, T. Ben, E. Kiss, M. Zeher, et al., Five-year follow-up of 665 Hungarian patients with undifferentiated connective tissue disease (UCTD), Clin Exp Rheumatol 21 (3) (2003) 313—320.

[34] I. Cavazzana, F. Franceschini, N. Belfiore, M. Quinzanini, R. Caporali, P. Calzavara-Pinton, et al., Undifferentiated connective tissue disease with antibodies to Ro/SSa: clinical features and follow-up of 148 patients, Clin Exp Rheumatol 19 (4) (2001) 403—409.

[35] J.A. James, X.R. Kim-Howard, B.F. Bruner, M.K. Jonsson, M.T. McClain, M.R. Arbuckle, et al., Hydroxychloroquine sulfate treatment is associated with later onset of systemic lupus erythematosus, Lupus 16 (6) (2007) 401—409.

[36] M.R. Arbuckle, M.T. McClain, M.V. Rubertone, R.H. Scofield, G.J. Dennis, J.A. James, et al., Development of autoantibodies before the clinical onset of systemic lupus erythematosus, N Engl J Med 349 (16) (2003) 1526—1533.

[37] O.P. Rekvig, J.C. Nossent, Anti-double-stranded DNA antibodies, nucleosomes, and systemic lupus erythematosus: a time for new paradigms? Arthritis Rheum 48 (2) (2003) 300—312.

[38] Q.Z. Li, J. Zhou, A.E. Wandstrat, F. Carr-Johnson, V. Branch, D.R. Karp, et al., Protein array autoantibody profiles for insights into systemic lupus erythematosus and incomplete lupus syndromes, Clin Exp Immunol 147 (1) (2007) 60—70.

[39] Y. Chen, S. Khanna, C.S. Goodyear, Y.B. Park, E. Raz, S. Thiel, et al., Regulation of dendritic cells and macrophages by an anti-apoptotic cell natural antibody that suppresses TLR responses and inhibits inflammatory arthritis, J Immunol 183 (2) (2009) 1346—1359.

[40] G.J. Silverman, C. Gronwall, J. Vas, Y. Chen, Natural autoantibodies to apoptotic cell membranes regulate fundamental innate immune functions and suppress inflammation, Discov Med 8 (42) (2009) 151—156.

[41] W. Rodriguez, C. Mold, L.L. Marnell, J. Hutt, G.J. Silverman, D. Tran, et al., Prevention and reversal of nephritis in MRL/lpr mice with a single injection of C-reactive protein, Arthritis Rheum 54 (1) (2006) 325—335.

[42] R.C. Williams Jr., M.E. Harmon, R. Burlingame, T.W. Du Clos, Studies of serum C-reactive protein in systemic lupus erythematosus, J Rheumatol 32 (3) (2005) 454—461.

[43] K.J. Evensen, T.J. Swaak, J.C. Nossent, Increased ferritin response in adult Still's disease: specificity and relationship to outcome, Scand J Rheumatol 36 (2) (2007) 107—110.

[44] G.J. Becker, M. Waldburger, G.R. Hughes, M.B. Pepys, Value of serum C-reactive protein measurement in the investigation of fever in systemic lupus erythematosus, Ann Rheum Dis 39 (1) (1980) 50—52.

[45] N. Zein, C. Ganuza, I. Kushner, Significance of serum C-reactive protein elevation in patients with systemic lupus erythematosus, Arthritis Rheum 22 (1) (1979) 7—12.

[46] H.M. Moutsopoulos, A.K. Mavridis, N.C. Acritidis, P.C. Avgerinos, High C-reactive protein response in lupus polyarthritis, Clin Exp Rheumatol 1 (1) (1983) 53—55.

[47] D.A. Hesselink, L.A. Aarden, A.J. Swaak, Profiles of the acute-phase reactants C-reactive protein and ferritin related to the disease course of patients with systemic lupus erythematosus, Scand J Rheumatol 32 (3) (2003) 151—155.

[48] K. Lanham, F.C. de Beer, G.R. Hughes, M.B. Pepys, Significance of CRP elevation in SLE, Scand J Rheumatol 12 (1) (1983) 64.

[49] C. Meijer, V. Huysen, R.T. Smeenk, A.J. Swaak, Profiles of cytokines (TNF alpha and IL-6) and acute phase proteins (CRP and alpha 1AG) related to the disease course in patients with systemic lupus erythematosus, Lupus 2 (6) (1993) 359—365.

[50] P.R. Ames, J. Alves, A.F. Pap, P. Ramos, M.A. Khamashta, G.R. Hughes, Fibrinogen in systemic lupus erythematosus: more than an acute phase reactant? J Rheumatol 27 (5) (2000) 1190—1195.

[51] A.J. Swaak, H.G. van den Brink, L.A. Aarden, Cytokine production (IL-6 and TNF alpha) in whole blood cell cultures of patients with systemic lupus erythematosus, Scand J Rheumatol 25 (4) (1996) 233—238.

[52] A.I. Russell, D.S. Cunninghame Graham, C. Shepherd, C.A. Roberton, J. Whittaker, J. Meeks, et al., Polymorphism at the C-reactive protein locus influences gene expression and predisposes to systemic lupus erythematosus, Hum Mol Genet 13 (1) (2004) 137—147.

[53] A. Jonsen, I. Gunnarsson, B. Gullstrand, E. Svenungsson, A.A. Bengtsson, O. Nived, et al., Association between SLE nephritis and polymorphic variants of the CRP and FcgammaRIIIa genes, Rheumatology (Oxford) 46 (9) (2007) 1417—1421.

[54] W. Rodriguez, C. Mold, M. Kataranovski, J. Hutt, L.L. Marnell, T.W. Du Clos, Reversal of ongoing proteinuria in autoimmune mice by treatment with C-reactive protein, Arthritis Rheum 52 (2) (2005) 642–650.

[55] P.J. Charles, R.J. Smeenk, J. De Jong, M. Feldmann, R.N. Maini, Assessment of antibodies to double-stranded DNA induced in rheumatoid arthritis patients following treatment with infliximab, a monoclonal antibody to tumor necrosis factor alpha: findings in open-label and randomized placebo-controlled trials, Arthritis Rheum 43 (11) (2000) 2383–2390.

[56] L. De Rycke, E. Kruithof, N. Van Damme, I.E. Hoffman, B.N. Van den, B.F. Van den, et al., Antinuclear antibodies following infliximab treatment in patients with rheumatoid arthritis or spondylarthropathy, Arthritis Rheum 48 (4) (2003) 1015–1023.

[57] R. Takahashi, A. Tsutsumi, K. Ohtani, Y. Muraki, D. Goto, I. Matsumoto, et al., Association of mannose binding lectin (MBL) gene polymorphism and serum MBL concentration with characteristics and progression of systemic lupus erythematosus, Ann Rheum Dis 64 (2) (2005) 311–314.

[58] R. Takahashi, A. Tsutsumi, K. Ohtani, D. Goto, I. Matsumoto, S. Ito, et al., Anti-mannose binding lectin antibodies in sera of Japanese patients with systemic lupus erythematosus, Clin Exp Immunol 136 (3) (2004) 585–590.

[59] A.P. Manderson, M. Botto, M.J. Walport, The role of complement in the development of systemic lupus erythematosus, Annu Rev Immunol 22 (2004) 431–456.

[60] T.C. Stam, A.J. Swaak, W.H. Kruit, A.M. Eggermont, Regulation of ferritin: a specific role for interferon-alpha (IFN-alpha)? The acute phase response in patients treated with IFN-alpha-2b, Eur J Clin Invest 32 (Suppl. 1) (2002) 79–83.

[61] H. Gogas, J. Ioannovich, U. Dafni, C. Stavropoulou-Giokas, K. Frangia, D. Tsoutsos, et al., Prognostic significance of autoimmunity during treatment of melanoma with interferon, N Engl J Med 354 (7) (2006) 709–718.

[62] A.M. Fairhurst, A.E. Wandstrat, E.K. Wakeland, Systemic lupus erythematosus: multiple immunological phenotypes in a complex genetic disease, Adv Immunol 92 (2006) 1–69.

II. CLINICAL ASPECTS OF DISEASE

Design of Clinical Trials for SLE

Joan T. Merrill, Kenneth C. Kalunian, Jill P. Buyon

In the past two decades more than a dozen investigational products have entered phase II/III clinical trials for lupus and none has achieved success sufficient for approval by the FDA. Only one agent, belimumab, has demonstrated superiority over placebo and this was with a small treatment effect in two very large, very expensive, international trials (including 865 and 816 patients) [1, 2]. It is obvious that an optimal trial outcome for an effective treatment should be able to be completed with an efficient sample size and, in line with that, be able to show a robust difference between the treatment and placebo groups. This chapter will address problems in trial designs for lupus that may potentially have contributed to the failure of some treatments to demonstrate efficacy and for belimumab to have shown a more pronounced treatment effect.

All of the treatments tested for lupus over the last 20 years have demonstrated adequate proof of concept in both animal models and *in vitro* human studies, so what accounts for the reason they have all fallen short of the mark in clinical trials for lupus patients? We hypothesize (and will review the evidence that underlies this argument) that clinical trials to test even the most promising investigational drugs for lupus are impeded by what at first glance might seem to be reasonable regulatory requirements, and additionally by the heterogeneity of the disease itself, the natural waxing and waning of mild to moderate lupus symptoms, and an unacceptable signal-to-noise ratio caused by disparate background treatments. Solutions to each of these problems will be proposed.

REASONABLE REGULATORY REQUIREMENTS

More than 50 years ago the FDA allowed "grandfathering" of three approved treatments for lupus: hydroxychloroquine, prednisone/prednisolone, and aspirin. Since then, requirements for government approval have been vastly improved from a scientific point of view. The following, rational sounding mandates apply across the board to any treatment currently being tested for a new indication:

1. A treatment must minimally demonstrate the ability to meet a preset primary endpoint in at least two confirmatory studies (and it is also helpful if it can additionally meet some of the preset secondary endpoints).
2. A treatment must demonstrate both efficacy (short-term improvement) and sustainability (efficacy can be shown to persist over time; usually, in the case of a lupus trial, for at least 1 year).
3. The treatment could be tested in an equivalency trial against already approved products, since these are considered to be proven efficacious.

In this third scenario it is assumed that the reason one is seeking an equivalently efficacious drug is because the new drug has less toxicity, e.g., to mitigate the infertility risk of cyclophosphamide. However, the treatment must be shown to be statistically superior in a trial using a placebo control, and all currently unapproved treatments for lupus are automatically considered unproven, and therefore equivalent to placebo.

In practice, for SLE, what this very reasonable-sounding set of rules translates into, is that cyclophosphamide, which is not approved for lupus, is presumed not to be efficacious for lupus and as such is effectively the same as placebo. By contrast, hydroxychloroquine, which is approved, can be assumed to be efficacious, even though it was grandfathered to approval and never met FDA standards of efficacy any more than cyclophosphamide did. This of course seems counterintuitive since most clinicians would acknowledge that hydroxychloroquine is not as effective a treatment for severe lupus as cyclophosphamide.

Therefore, if one were to propose a study to the FDA in which lupus nephritis were to be tested, it would not be adequate to use a current, community-accepted standard of care (such as cyclophosphamide or mycophenolate mofetil) as a comparator in an equivalency trial. In order to be considered for approval, a treatment would need to be superior to those strong immune suppressants. However, it would be perfectly acceptable (but unfortunately unethical) to use hydroxychloroquine or aspirin, approved treatments for lupus, in an equivalency trial design for nephritis.

Of course it would be likely that almost any minimally effective treatment for nephritis could be shown superior to hydroxychloroquine or aspirin, but such a trial would never pass an internal review board or be seriously entertained by any physician who treats lupus. Similarly, an equivalency trial in which prednisone alone were used as a comparator would be unacceptable in a lupus clinic for use in a year-long trial design, since both the efficacy as a single agent and the side effects would not be construed as acceptable risks. For these reasons, although it is known to be quite effective in the short-term, prednisone is only accepted by the community as an adjunctive and transitional treatment for nephritis.

Therefore the FDA rules, which sound so reasonable in theory, were evidently designed to address the highest level of clinical science possible; however, they are only appropriate for diseases in which effective and tolerable medications have already been approved. Lupus is an orphaned disease by virtue of having no strongly effective and tolerable approved medications. Thus by ignoring this disparity, and using "one size fits all" rules for good clinical science, an imponderable catch 22 exists that has effectively hindered drug development for SLE. Because of this, most clinical trials for nephritis or serious lupus have resorted to designs in which background treatments are used and the test treatment (or placebo) are added on to standard of care, a potentially seriously flawed trial design that will be considered, in all of its problematic aspects, in the section titled "Disparate Effects of Background Treatments."

The solution to this serious impediment to progress for SLE would be to find a way for cyclophosphamide or mycophenolate to be approved so that they could be used as comparators in equivalency trials. The problem with this idea is that they have only been able to be tested against each other, and the weight of the literature has found them to be roughly equal in efficacy overall [4–8], although some studies suggest that mycophenolate is safer and more tolerable [6–8].

Could not efficacy equivalency of two unapproved, community-accepted standard-of-care treatments for lupus nephritis be used as a basis for approval if one was demonstrated to be superior from the point of view of safety and tolerability? The answer from the FDA has been "no." Lupus patients and their doctors are left floundering in the wake of a governmental decision that both mycophenolate and cyclophosphamide are considered to be the same as placebo (even though the equally unproven drugs, hydroxychloroquine and aspirin, are assumed efficacious for lupus with far less basis for this belief in the literature than the strong immune suppressants can demonstrate). Therefore, proving that one is safer or more tolerable than the other (which has been found in several studies of mycophenolate) cannot form the basis for even a special case or provisional approval.

The lupus community must await the approval of one of the biologics, such as belimumab, which could never have demonstrated efficacy in a year-long trial against copious background treatments without two, hugely expensive, year-long, international clinical trials. However, it seems unlikely from the limited evidence available that using belimumab as a single-agent comparator would be immediately acceptable to the community. Staggered induction of multiple agents and medication withdrawal trials have been suggested as possible alternatives, but again, seem unlikely to generate immediate community acceptance, given the serious sequelae known from the delay of aggressive treatments for lupus nephritis [9].

So is a complicated, difficult disease with no approved medications to be left stranded with insurmountable governmental barriers to new drug approvals? It seems likely that some other, well-funded companies will follow the lead of the belimumab developers, Human Genome Sciences (HGS), and their partner Glaxo-Smith Kline (GSK), and mount large, expensive trials, leaving smaller companies in a position of needing to market their ideas to larger companies and limiting the growth potential for them of developing products for lupus. However, this chapter will review evidence that, in fact, this may not be entirely necessary.

HETEROGENEITY OF THE DISEASE ITSELF

One problem in the designs of clinical trials is the heterogeneity of the underlying biology in each patient. There has been a sophisticated precision in the preclinical study of biologics, each of which has been finely targeted to correct specific immune imbalances that are known to be present in many (but not all) lupus patients [10–12]. Still, studies continue to enter patients without regard to their variable, underlying immune states and still omit the use of logical biomarkers to guide optimal treatment regimens. No matter what the target of a biologic, the chances that it can improve all of the lupus patients all of the time at a given dose are simply not likely.

These are issues that each company must choose whether or not to address on its own, and it may be possible during the development of specific treatments to engage in productive enquiry specific to the most probable biomarker(s) for each product, in order to optimize patient selection and dosing strategies. On the other hand, pressure from marketing and the FDA has guided most developers toward a more open-ended strategy, hoping that if their treatment might be proven "good enough" in the larger lupus population, the possibility of missing important (and biologically less obvious) subsets of the population would be minimized, and the eventual approval umbrella for the product could be less restrictive.

It seems logical that the effects of background treatments on the immune disorder(s) that are specifically targeted by novel interventions should be better understood before decisions are made about which background treatments are suitable in a given clinical trial design and at what point in the protocol they should be given. This will be further discussed in the section titled "Disparate Effects of Background Treatments," which specifically addresses trial design questions.

Waxing and Waning Disease Leads to Spontaneous Improvement and Clinically Insignificant Flares of Mild to Moderate Lupus

The validated disease activity instruments that are the most favored measures for trial outcomes by regulatory agencies were originally designed for epidemiologic studies, not interventional outcome studies. It is increasingly recognized that these clinical instruments have potential strengths and weaknesses with regard to the selection of study endpoints, and that the outcomes selected in trials are not necessarily a good fit either for a physician or a patient's impressions of efficacy. Most importantly, in an attempt to capture active disease, when considering symptoms at the low (mild to moderate) end of the spectrum, these instruments have proven to be problematic in differentiating clinically inconsequential features from significant disease. By lumping this spectrum of disease into scoring systems (where minimal requirements for a given score might be far less meaningful than the maximal symptoms given the exact same score) a high signal-to-noise ratio is introduced, which narrows the potential gap that can be measured between an effective treatment and placebo. This is true both in measuring improvement (when minimal residual symptoms on an effective treatment would be scored the same as no improvement) or in measuring flares when small nuanced

increases in disease on effective treatment might not be differentiated from significant flares in the placebo group. An example of the former might be a patient who enters a trial with five active joints including the knees and with the inability to walk and this later improved to the point that there remained three mildly swollen, generally functional PIP joints. This improvement might not be captured by either the SLEDAI or BILAG. An example of difficulty differentiating clinically significant flares from minor disease fluctuations would be one patient developing a new, minimal malar rash and another a moderate, painful, weeping discoid rash, which cannot be differentiated on SLEDAI or BILAG.

The potential impact of these issues is supported by exploratory studies conducted on the phase II rituximab, abatacept, and epratuzumab trials, presented in 2008 and 2009 at the American College of Rheumatology (ACR) and European League Against Rheumatism (EULAR) meetings [13–16] and/or published [17] as well as the phase III belimumab studies [1, 2]. No firm conclusions can be drawn from these early and/or exploratory reports, some of which have not yet been published in peer-reviewed journals and most of which include some *post hoc* analyses, requiring cautious interpretation due to multiple statistical comparisons. Nevertheless, if the field is to move forward, reanalysis of failed trials and examination of the study design of a successful trial may at least generate testable hypotheses about more robust approaches to evaluating the efficacy of new treatments.

Inconsistency Between Outcome Measures and Physician and Patient Perception of Results

In many clinical trials, including several of those reviewed here, patients have not been counted as even partial responders if, during any given month in the latter part of the study, they developed clinical flares meeting the lower cutoff of moderate disease (British Isles Lupus Assessment Group [BILAG] B). However, as BILAG B ranges from moderately severe to relatively mild disease, a BILAG B flare in one patient could signify a significant change for the worse, whereas another patient who just meets the lower threshold for qualification as a flare, might be experiencing only a minor fluctuation in disease, but earn the same BILAG B score. If a minor flare reversed within a month without treatment, it is logical that physicians or patients might consider that outcome as clinically inconsequential. If there were more of these minor or transient flares in one treatment group and more of the aggressive or prolonged B flares in another, this would simply not be apparent using the lower end of BILAG B as a cutoff for nonresponse. In fact, in the abatacept study, patient-reported outcomes and physician-described flares suggested a difference between treatment and placebo, which was not seen using the lower-end BILAG B cutoff [13]. Similar trends were seen in both the abatacept and rituximab studies when only BILAG A (severe disease) flares were compared in exploratory analyses. However, it is acknowledged that using the severe definition as a cutoff eliminates important high Bs along with the potentially confounding low Bs, truncating the numbers of events available to compare [13, 14].

In the belimumab trials [1, 2], nonresponse required at least two new BILAG Bs at the final visit. With this definition, occasional B flares occurring intermittently and earlier in the trial had no impact on the primary outcome. In addition, using a multiple landmark analysis approach, this study was able to differentiate treatment and placebo at multiple points during the trial. The risk of a patient developing one minor clinically insignificant flare sometime within a year must be far higher than on any given landmark day. Thus comparing mild-to-moderate flares at multiple, separate discrete time-points seems far less likely to create background havoc than comparing the cumulative risk of meeting this endpoint. However, there are scientific issues with a multiple landmark analysis. Such an analysis does not differentiate well between clinically meaningless grumbling disease in a shifting subset of patients (which will be eliminated from high impact in this analysis) and patients who may have been extremely unwell for 1 or many more months of the year and manage to meet the primary endpoint on the single month that analysis is performed.

How instructive might it be, both for future HGS/GSK studies and for other companies, if a comparison between the multiple landmark approach and a cumulative BILAG B flare approach could be tested at minimal cost and confirmed for the entire community to be operable across trials? This study could be conducted using only the placebo groups from five or six completed phase II and III trials. The following questions could be asked, restricted to placebo-group patients who have undergone year-long trials on the most commonly used background medications in these trials: azathioprine, MMF, and methotrexate.

1. What is the actual difference in frequency between BILAG B flares ever and BILAG B flares cumulatively/per patient in placebo groups in clinical trials?
2. Since it is already known that most BILAG B flares are not treated in the course of a trial, what proportion of total BILAG B flares resolve in 1 month without treatment vs. the proportion found at monthly landmarks during a year-long study? Is this number different earlier in the year vs. later in the year in studies that do or do not include a forced steroid taper?
3. What is the actual difference in the persistence of increased disease activity for longer than 1 month when two BILAG Bs are used as a cutoff for flare in a landmark analysis vs. the first BILAG B/patient in a cumulative analysis?
4. Would the answers to questions 1–3 differentiate between analyses better if only BILAG A flares were considered separately?
5. A direct comparison of the types of flares most frequent in patients receiving azathioprine, MMF, and methotrexate would be of high value both for the community of practitioners and to treatment developers concerned about trial design. If there is an agent in development that seems particularly strong for the treatment of arthritis, then a background treatment that suppresses arthritis flares too effectively might not be optimal in that trial. If new treatments can be made available once endpoints are met, however, a patient can be allowed to complete the protocol with safety, despite nonresponse.

The unprecedented success of the belimumab phase III studies is based largely on a confident redesign of the belimumab development program based on exploratory analysis of its phase II study, which itself did not meet any of its primary or secondary outcomes [18]. The entry criteria for the phase III studies were modified to include only a defined subset of patients that, when the phase II data were restricted to this group, could have met the original predefined endpoints. Specifically, these were patients who entered with a positive ANA or anti-dsDNA. Adding a requirement for seropositivity may well have helped to ensure that the population being studied had true active disease, which begins to address the pitfalls inherent in entering patients with insignificant waxing and waning disease.

Subsequently, further outcome measure modifications were made that were, interestingly, more stringent in theory (adding lack of flare by BILAG and Physicians Global Assessment to the requirements for treatment improvement by the SLEDAI), but seemed better able to differentiate treatment from placebo when applied to the phase II data [19]. These anchors make clear clinical sense and eliminate any possibility that the four-point drop in SLEDAI could have been accompanied by a flare not captured by SLEDAI, e.g., arthritis and malar rash completely resolve but the patient develops life-threatening thrombocytopenia. Additional differences that distinguish the HGS trial from other more recent trials might account for this improved robustness in the reanalyzed responder index (SRI). Firstly, the multiple landmark analysis discussed above ensured that a given patient with a transient, minor flare at month 6 would not be eliminated as a potential responder at any other month and most importantly at the week 52 primary endpoint month. Secondly, there was also a higher cutoff for BILAG flare in the SRI definition than was used in the abatacept and rituximab trials, since a flare to B in a single organ did not count, only a flare of two Bs would automatically eliminate a patient from meeting an endpoint at a given visit. However, in case there was a one B flare that was considered clinically significant by the assessing physician, movement of their PGA score significantly to the right would capture this fact, allowing both a decrease in the spurious flares recorded and the ability to capture important flares that might otherwise be lost.

Taken together, these elements in the SRI were able to shore up the strengths and minimize the weaknesses of the combined instruments being used in the study, potentially increasing the probability that an effective treatment could be differentiated from a placebo group, while increasing the odds that any difference seen was clinically meaningful. Why, then, was the treatment effect of belimumab so modest in two huge, international trials?

Disparate Effects of Background Treatments

What remained in the revamped belimumab study design was that patients were allowed to enter the trial on a range of background medications and receive increased treatments during a significant part of the study period, continuing to risk (and indeed eliciting) a high response rate in the placebo group. In fact, there was not, in essence, a placebo group at all, but rather, like many previously failed clinical trials for lupus, it was more of a standard-of-care treatment group.

This trial succeeded in demonstrating the superiority of belimumab plus standard of care vs. standard of care alone, and in that context the treatment-effect size difference, modest as it may be, must be respected, when it is fully recognized that there must be a ceiling to the possible percent effectiveness of any targeted biologic in a disease that is biologically heterogenous, and in patients who enter the trial already impacted by heterogenous biologic interventions, who are also subjected to further modifications of these background treatments during the year. If any given treatment has the potential to impact 70% of these patients, and if some percent of their background medications might even counteract the effects of belimumab, no wonder there would be a ceiling on its efficacy. Similarly, if 30–40% of the placebo-group interventions are effective, no wonder the gap is narrowed between the placebo group and a treatment with a biologic ceiling. It is therefore not surprising that the optimal outcome achieved even in a robustly designed lupus trial involves a very small treatment-effect size.

Added to this is the fact that in the belimumab trial design, a decrease of steroids was encouraged during the latter part of the trial and data confirm that in both phase III studies there was a trend (statistically significant in BLISS 52) for more patients in the belimumab treatment group to end up on minimal steroids than those in the background treatment group. Thus, an important background treatment was removed from the treatment group more than the placebo group in both cases, which would further tend to narrow the gap between treatment and placebo on the visit being measured for the primary outcome (although it provided an additional secondary outcome to suggest belimumab efficacy). Taken together, the SRI (responder index) increased the robustness of the trial design to differentiate treatment from placebo, but the protocol for background treatments (which may have been devised on ethical and patient retention grounds) may have markedly contributed to the narrow gap between treatment and placebo.

The BLISS 52 and BLISS 76 studies of belimumab were the first pharmaceutical-sponsored, major multicenter trials in lupus to meet their primary endpoints. Exciting as these results are, these phase III studies were expensive, international trials, each requiring more than 800 patients in order to demonstrate superiority with relatively modest treatment-effect sizes. Given the study design, the small treatment-effect size should not be seen to discount in any way the effectiveness of the treatment. However, now that the ground to establish that clinical trials for lupus can succeed has been broken, might it also be possible to design smaller phase II (and maybe smaller phase III trials) with equally ethical protocols and more robust outcome measures so that new treatments for lupus can be tested more efficiently?

Can Withdrawal of Background Medications Ever be Considered?

It seems inarguable that biologic treatments must be tested on reasonably ill lupus patients, not only from the point of view of medical appropriateness, but because, as discussed above, grumbling disease interferes with measurements, and, in the best designed trials, should probably be discounted. However, it is equally inarguable that more ill patients who enter a year-long trial, often flaring through background medications, are going to require a treatment intervention and cannot be expected to have their background treatments withdrawn while receiving nothing but placebo and/or an untested study medication. In the abatacept and rituximab non-nephritis (Explorer) studies, patients who entered received mandated steroids on top of their background medications, which

[16] J. Merrill, R. Burgos-Vargas, Westhovens, et al., The efficacy and safety of abatacept in SLE: results of a 12-month exploratory study [abstract L15], Arthritis Rheum 58 (Suppl. 9) (2008).

[17] J.T. Merrill, C.M. Neuwelt, D.J. Wallace, J.C. Shanahan, K.M. Latinis, J.C. Oates, et al., Efficacy and safety of Rituximab in moderately-to-severely active SLE, Arthritis Rheum 62 (2009) 222–233.

[18] D.J. Wallace, W. Stohl, R.A. Furie, J.R. Lisse, J.D. McKay, J.T. Merrill, et al., A phase II, randomized, double-blind, placebo-controlled, dose-ranging study of belimumab in patients with active systemic lupus erythematosus, Arthritis Rheum 61 (2009) 1168–1178.

[19] R.A. Furie, M.A. Petri, D.J. Wallace, E.M. Ginzler, J.T. Merrill, W. Stohl, et al., Novel evidence-based systemic lupus erythematosus responder index, Arthritis Rheum 61 (2009) 1143–1151.

[20] T. Joan, Merrill, Molina Mhatre, Fredonna Carthen, K. Sandra, Wilson, et al., Clancy and Jill Buyon Mycophenolate Mofetil (MMF) Is Effective for Systemic Lupus (SLE) Arthritis, Final Results of An Organ-Specific, Double-Blind, Placebo-Controlled Trial ACR (2009). abstract # 264.

The Nervous System and Lupus

John G. Hanly

INTRODUCTION

Nervous system disease has been recognized in all cohorts of patients with systemic lupus erythematosus (SLE), regardless of patient demographics, characteristics, recruitment from community or academic medical centers, or geographic location. Neuropsychiatric (NP) symptoms are a frequent concern of lupus patients and a challenge for the diagnosing and treating physician. The specific manifestations vary from common entities such as mood disorders and cognitive complaints to less frequent NP events such as seizures and psychosis. Due to the lack of diagnostic specificity for SLE, the precise attribution of NP events in individual patients remains difficult despite advances in neuroimaging and other diagnostic tools for NP disease. In contrast to other organ involvement such as skin and kidney, the complexity of the nervous system and the relative lack of accessibility have slowed the speed of discovery of pathogenetic mechanisms. Despite these challenges, considerable progress has been made during the past decade in elucidating a clearer understanding of the clinical manifestations, immunopathogenic mechanisms, and treatment options for nervous system disease in SLE patients.

CLASSIFICATION AND FREQUENCY OF NPSLE

The American College of Rheumatology (ACR) classification criteria for SLE [1] include seizures and psychosis, which is an under-representation of the totality of NP disease in SLE. Central nervous system (CNS) involvement predominates over peripheral nervous system disease and may take the form of either diffuse disease (e.g., psychosis and depression) or focal disease (e.g., stroke and transverse myelitis) depending on the anatomic location of pathology. Several ad hoc classifications have been developed for NPSLE [2–4], none of which received universal acceptance. In 1999, following a multidisciplinary consensus conference, the ACR published a standard nomenclature and case definitions for 19 NP syndromes [5] that are known to occur in lupus patients, although they do not occur exclusively in SLE (Table 40.1). These include 12 CNS and 7 peripheral nervous system manifestations [5]. For each NP syndrome, potential etiologies other than SLE were identified for either exclusion or recognition as an "association," acknowledging that in some clinical presentations it is not possible to be definitive about attribution. The identification of other potential causes for NP events in SLE patients is critical and was not

TABLE 40.1 Neuropsychiatric Syndromes in SLE as Defined Using the ACR Nomenclature

Central nervous system	Peripheral nervous system
Aseptic meningitis	Guillain–Barré syndrome
Cerebrovascular disease	Autonomic neuropathy
Demyelinating syndrome	Mononeuropathy
Headache	Myasthenia gravis
Movement disorder	Cranial neuropathy
Myelopathy	Plexopathy
Seizure disorders	Polyneuropathy
Acute confusional state	
Anxiety disorder	
Cognitive dysfunction	
Mood disorder	
Psychosis	

Source: American College of Rheumatology [5].

adequately addressed in previous classifications of NPSLE. This aspect of the ACR case definitions, in conjunction with other variables, may be used to determine the attribution of NP events. Guidelines for reporting standards were also developed by the ACR research committee and specific diagnostic tests recommended for each NP syndrome.

The ACR case definitions are the current standard for describing NP events in clinical studies of SLE. Despite their utilization, there is still a wide spectrum in the reported frequency of NP disease, ranging from 37 to 95% SLE (Table 40.2) [6–10]. The most common of the 19 NP syndromes in each of these selected 5 SLE cohorts were cognitive dysfunction (55–80%), headache (24–72%), mood disorders (14–57%), cerebrovascular disease (5–18%), seizures (6–51%), polyneuropathy (3–28%), anxiety (7–24%), and psychosis (0–8%). In most of these studies, approximately half of the 19 NP events occurred in less than 1% of the patients, emphasizing the relatively low frequency of many of the individual NP syndromes.

The results of these clinical studies serve to both reinforce previous conclusions about NPSLE and introduce new concepts. Despite the improved definitions for individual NP syndromes, there continues to be substantial variability in the overall prevalence of NP disease between different populations. Whether this represents inherent differences between study cohorts, bias in data acquisition, or differences in attribution rules for NP events is unclear. None of the individual NP manifestations are unique to lupus and, indeed, some occur with comparable frequency in the general population [6] and in patients with other chronic diseases such as rheumatoid arthritis (RA) [11]. Thus, the selection of comparator groups is critical to determine whether the prevalence of certain types of NP disease in SLE patient cohorts is in excess of that found in the normal population and for other chronic diseases. Due to the fact that many of the NP syndromes are quite rare, multicenter efforts are required to assemble sufficient numbers of patients for study.

ATTRIBUTION OF NPSLE: SLICC INCEPTION COHORT STUDY

In 2000, an international, multicenter, disease inception cohort study was initiated to prospectively determine the frequency, attribution, and outcome of NP events in SLE patients [12]. In this study, which is conducted by the Systemic Lupus International Collaborating Clinics (SLICC) research network, patients are enrolled within 15 months of their diagnosis of SLE and re-evaluated on an annual basis using a standardized protocol for data capture. In the first 572 patients enrolled, with a mean disease duration of 5.2 ± 4.2 months, NP events occurred in 30% of patients, 48% of whom had more than one NP event [13]. The most common events were headache, mood disorders, cognitive dysfunction, anxiety disorder, cerebrovascular disease, seizure disorders, acute confusional states, polyneuropathy, mononeuropathy, and psychosis. The remaining eight NP syndromes occurred in less than 2% of patients for each syndrome. Decision rules were developed to determine the attribution of NP events [13]. These were based on factors such as the temporal onset of the event in relation to the diagnosis of SLE; the identification of alternative causes for the NP event as identified in the ACR case definitions [13]; and recognition that some NP events, such as headache, anxiety, mild depression, mild cognitive impairment, and polyneuropathy, lacking electrophysiological confirmation have a high frequency in the general population [6]. Two attribution models of different stringency [13] were applied and indicated that NP events directly attributable to SLE accounted for 19–38% of all events (Figure 40.1) and affected 6–12% of the study population. Of the NP events attributed to SLE, the most frequent were seizure disorders, cerebrovascular disease, mononeuropathy, acute confusional states, cranial neuropathy, myelopathy, polyneuropathy, aseptic meningitis, major mood disorders, and psychosis, with the remaining syndromes affecting no more than 2% of patients for each syndrome. Comparable findings have been reported from cross-sectional studies [8, 14]. These data suggest that non-SLE factors likely contribute

TABLE 40.2   Prevalence of NPSLE Using ACR Nomenclature

	Ainiala et al. [6]	Brey et al. [7]	Sanna et al. [9]	Hanly et al. [8]	Sibbitt et al. [10][a]
Location	Tampere, Finland	San Antonio, TX	London, UK; Cagliari, Italy	Halifax, Canada	Albuquerque, NM
No. of patients	46	128	323	111	75
Disease duration (mean ± SD years)	14 ± 8	8	11 ± 8	10 ± 1	7 ± 8
NPSLE	91%	80%	57%	37%	95%

[a]Pediatric-onset SLE population.
Source: Data from Hanly, J. G. (2004). ACR classification criteria for systemic lupus erythematosus: Limitations and revisions to neuropsychiatric variables. Lupus 13, 861–864.

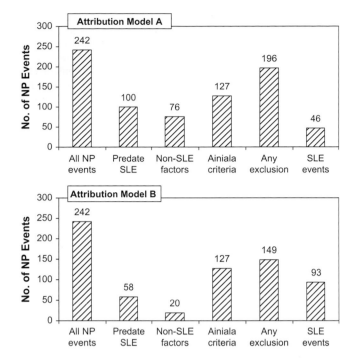

FIGURE 40.1 Attribution of NP events to SLE or other causes using models of greater (model A) or lesser stringency (model B). *Model A*, The onset of NP events more than 6 months prior to the diagnosis of SLE, the identification of non-SLE factors that contributed to or were responsible for the NP event or the occurrence of frequent NP events as defined by Ainiala *et al.* [6] indicated that the NP event was not attributed to SLE. *Model B*, The onset of NP events more than 10 years prior to the diagnosis of SLE. The identification of non-SLE factors that were responsible for the NP event ("exclusion factors") or the occurrence of frequent NP events as defined by Ainiala *et al.* [6] indicated that the NP event was not attributed to SLE. *Source: Reproduced with permission of John Wiley & Sons from Hanly et al. [13].*

to a substantial proportion of NP disease in SLE patients, particularly the "softer" NP manifestations such as headache, anxiety, and some mood disorders.

CLINICAL IMPACT AND PROGNOSIS OF NPSLE

The significance and outcome of NP events in SLE patients has been examined in cross-sectional, short-term, and longitudinal studies. Insight into the clinical significance of NP events in SLE patients has been provided through studies of self-report health-related quality of life (HRQOL). The cumulative occurrence of NP events, regardless of attribution to SLE or other causes, has been associated with a significant negative impact on patient's self-reported HRQOL, as indicated by lower scores on all eight subscales and the mental and physical component summary scores of the SF-36 (Figure 40.2) [8, 13, 14]. Jonsen et al. [15] also reported a higher frequency of disability in SLE patients with

NP disease compared to patients without NP events and the general population. These studies included all types of NP events rather than individual NP manifestations such as cognitive impairment, the occurrence of which has not been associated with lower HRQOL and work disability [16, 17].

In a study of SLE patients from a single Canadian center, the mean mental component summary (MCS) and physical component summary (PCS) scores of the SF-36 in patients with NP events were approximately 7 and 16 points below the expected Canadian population means [18]. The group mean differences over the course of 7 years (mean, 3.6 years) in MCS and PCS scores between patients with and without NP disease were 8.8 and 8.0, respectively. These values exceed the change in mean group scores in patients with scleroderma [19] and chronic obstructive pulmonary disease [20] who had concurrent clinically significant improvement in disease-specific outcomes while enrolled in clinical trials. Thus, the negative relationship between NP events and HRQOL that has been reported in cross-sectional studies of SLE [8, 13] is also apparent over time. As in the earlier studies [8, 13], there was no statistically significant difference between the effect of NP events due to SLE and non-SLE causes. In fact, the data suggest that patients with concurrent NP events attributed to both SLE and non-SLE causes have the lowest mean MCS and PCS scores over time (Figure 40.3) [14]. These findings are in contrast to those found when SLE patients with predominantly cognitive dysfunction were studied in a similar manner over 5 years [21], and they emphasize the importance of including all NP events in clinical studies of NPSLE.

The outcome of all NP events, whether attributed to SLE or non-SLE causes, has been examined in a limited number of studies. Data from clinical trials have been uncontrolled, of short duration, or focused on a single NP manifestation [22–27]. The data from observational cohorts have been inconsistent. For example, increased mortality in patients with NP events has been reported in some studies [28–30] but not in others [31–33]. In a 2-year follow-up study [34] of 32 patients hospitalized for NPSLE, the outcome was generally favorable, with either substantial improvement (69%) or stabilization (19%). In a study of SLE patients followed for up to 7 years (mean, 3.6 years), there was resolution in approximately 15% of NP events at each annual assessment, although the majority of events were persistent. Of interest, the attribution of NP events to SLE or non-SLE causes did not predict their resolution. This is in contrast to two reports with mean follow-up of 3.7 months [35] and 1.9 years [36] from a large inception cohort study of SLE in which the short-term outcome was more favorable in patients with NP events attributed to SLE (Figure 40.4). It may be that treatment of

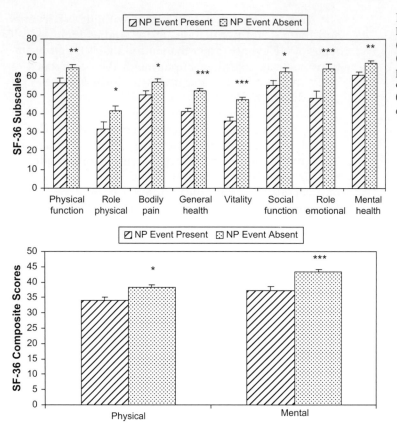

FIGURE 40.2   The difference in health-related quality of life (HRQOL) as indicated by mean (± SEM) subscale scores (*top*) and physical and mental component summary scores (*bottom*) in patients with and without NP events. Those patients with NP events, regardless of attribution, had consistently lower scores, indicating poorer HRQOL. *$p <$ 0.05, **$p < 0.01$, ***$p < 0.001$. *Source: Reproduced with permission of John Wiley & Sons from Hanly et al. [13].*

NPSLE early in the disease course may have a more favorable outcome and as such may present a therapeutic window of opportunity akin to that seen in other rheumatic diseases [37, 38].

CLINICAL MANIFESTATIONS

The 19 NP syndromes identified in the ACR case definitions can be categorized into central and peripheral nervous system manifestations [5] and into diffuse and focal manifestations [35]. Such clustering of NP events facilitates the analysis of clinical data in epidemiological studies particularly when individual NP events are a rare occurrence. In a study of 890 patients of which 271 had 407 NP events, 379 (93%) affected the CNS and 28 (7%) involved the peripheral nervous system [35]. The numbers of diffuse and focal events were 318 (78%) and 89 (22%), respectively. The outcomes as determined by physician assessment were poorer for diffuse NP events compared to focal NP events ($p < 0.001$). In subgroup analyses, this was only significant for diffuse NP events not attributed to SLE ($p = 0.019$). In addition, patients with NP events, regardless of attribution and clustered collectively into diffuse manifestations with or without concurrent focal NP events, reported poorer HRQOL as reflected by lower MCS scores compared to

patients with focal NP events only (35.71 ± 12.85 vs. 37.72 ± 13.52 vs. 47.09 ± 12.58, respectively; $p = 0.002$) (Figure 40.5) [35]. The findings were similar when the analysis was restricted to patients with NP events attributed to SLE using two attribution models of different stringency (see Figure 40.5) [35]. Similarly, patients with central NP events with or without concurrent peripheral NP events, regardless of attribution, had lower MCS scores compared to patients with peripheral NP events only (35.28 ± 12.22 vs. 38.16 ± 13.63 vs. 49.51 ± 11.54, respectively; $p = 0.020$).

Headache

The association between SLE and headache, including migraine, is controversial. The reported prevalence of headache has varied widely between 24 and 72% [6−10], but the prevalence of headache in the general population is also high, with up to 40% of individuals reporting a severe headache at least once per year [39]. Two of the more recent studies [40, 41] found no increase in the prevalence of headache in SLE, which was also the conclusion of a meta-analysis on the multiple studies on the issue [42]. Furthermore, only one study [43] reported an association between headache and other clinical features of active lupus. Aseptic meningitis is a relatively uncommon but

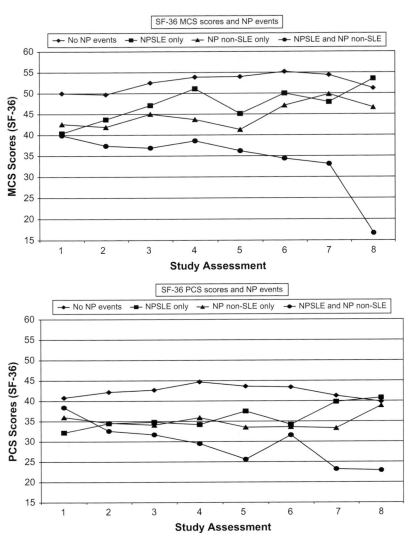

FIGURE 40.3 The mean mental component summary (MCS) scores (*top*) and the mean physical component summary (PCS) scores (*bottom*) of 209 SLE patients at enrollment and annually for up to 7 years. The group means over time for both MCS and PCS scores were significantly lower ($p < 0.001$) in patients with NP events regardless of attribution compared to those of patients without NP events. Patients with NP events were clustered into groups with events attributed only to SLE (NPSLE only), events attributed to non-SLE causes only (NP non-SLE only), and events attributed to SLE and non-SLE causes. *Source: Reproduced with permission of The Journal of Rheumatology from Hanly et al. [14].*

well-documented cause of headache in SLE [44, 45] and requires confirmation by analysis of cerebrospinal fluid (CSF). Other potential causes must be considered, including infection and idiosyncratic reactions to medications such as antibiotics and nonsteroidal anti-inflammatory drugs [46–48]. Thus, although headache may be a component of generalized active SLE in individual patients, it is likely that the majority of headaches in SLE patients, especially those that occur in isolation, are due to non-SLE causes.

Psychosis, Mood Disorders, and Anxiety

Psychosis is reported in up to 8% of SLE patients [6–10, 49] and is characterized by the presence of delusions (false belief despite evidence to the contrary) and/or hallucinations (perceptual experiences occurring in the absence of external stimuli). The latter are most frequently auditory. Psychosis is a rare but dramatic manifestation of NPSLE, and when present it must be distinguished from other causes, including high doses of corticosteroids, nonprescribed drug abuse, schizophrenia, and depression. Depression and anxiety are common symptoms in lupus and occur in 24–57% of patients [6–10, 50]. However, because there are no features of these syndromes that are unique to SLE patients, there is often uncertainty about the etiology and attribution in individual cases. The association between psychosis, depression, and antiribosomal P antibodies (anti-P) in SLE is supported by some but not all studies [50–53].

Cerebrovascular Disease

The many forms of cerebrovascular disease are reported in 5–18% of SLE patients [6–10] and are likely multifactorial in etiology. Accelerated atherosclerosis is well recognized in SLE, particularly coronary heart

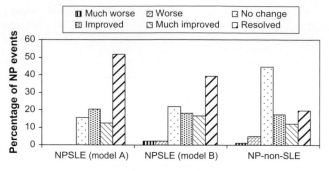

FIGURE 40.4 Physician-generated outcome scores at enrollment for NP events attributed to SLE using different attribution models A and B and for NP events attributed to non-SLE causes. *Model A,* The onset of NP events more than 6 months prior to the diagnosis of SLE, the identification of non-SLE factors that contributed to or were responsible for the NP event or the occurrence of frequent NP events as defined by Ainiala *et al.* [6] indicated that the NP event was not attributed to SLE. *Model B,* The onset of NP events more than 10 years prior to the diagnosis of SLE. The identification of non-SLE factors that were responsible for the NP event (exclusion factors) or the occurrence of frequent NP events as defined by Ainiala *et al.* [6] indicated that the NP event was not attributed to SLE. Those NP events that were attributed to SLE using either attribution model A or model B had a significantly better outcome compared to patients with NP events not attributed to lupus ($p < 0.001$). *Source: Reproduced with permission of John Wiley & Sons from Hanly et al. [35].*

disease, which is 5–10 times more frequent in SLE patients compared to control populations [54]. This also contributes to the increased rate of cerebrovascular events in SLE. An additional etiologic factor is the prothrombotic state as a consequence of antiphospholipid antibodies [55], which provides a rationale for therapeutic intervention with anticoagulants in selected cases. The objective evidence implicating vasculitis as a significant cause of cerebral infarction in SLE is limited [56–59]. The majority of cerebrovascular events are ischemic, followed by multifocal disease, intracranial hemorrhage, and sinus thrombosis.

Seizures

Generalized and focal seizures are reported in 6–51% [6–10] of SLE patients and may occur either in the setting of active generalized multisystem lupus or as isolated neurological events. The majority of patients experience a single event in the setting of generalized disease activity at approximately the time of presentation of the disease. Only a minority develop a chronic seizure disorder requiring long-term anticonvulsant therapy. Their occurrence is frequently associated with the presence of antiphospholipid antibodies [9], which co-occur with microangiopathy, arterial thrombosis, and subsequent cerebral infarction. Posterior reversible encephalopathy syndrome has been identified as another potential cause of seizures in SLE patients [60].

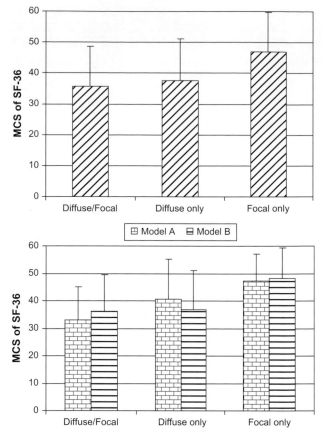

FIGURE 40.5 Mental component summary (MCS) scores (mean ± SD) at enrollment for patients with NP events assigned to groups with diffuse and focal NP disease, diffuse NP disease only, and focal NP disease only. Regardless of attribution (*top*), all patients with diffuse/focal and diffuse only events had lower MCS scores compared to those of patients with focal NP events only ($p = 0.002$). The findings were similar when the analysis was restricted to patients with NP events attributed to SLE (*bottom*) using either attribution model A ($p = 0.012$) or model B ($p = 0.001$). *Model A,* The onset of NP events more than 6 months prior to the diagnosis of SLE, the identification of non-SLE factors that contributed to or were responsible for the NP event or the occurrence of frequent NP events as defined by Ainiala *et al.* [6] indicated that the NP event was not attributed to SLE. *Model B,* The onset of NP events more than 10 years prior to the diagnosis of SLE; the identification of non-SLE factors that were responsible for the NP event (exclusion factors) or the occurrence of frequent NP event as defined by Ainiala *et al.* [6] indicated that the NP event was not attributed to SLE. *Source: Reproduced with permission of John Wiley & Sons from Hanly et al. [35].*

Demyelination, Transverse Myelopathy, and Chorea

These are rare manifestations of CNS disease in SLE and occur no more frequently than in 1–3% of patients [6–10, 61]. Clinical and neuroimaging evidence of demyelination has been described and is frequently indistinguishable from multiple sclerosis (MS) [62]. Thus, this particular syndrome may represent concordance or overlap of two autoimmune conditions.

Transverse myelopathy [63] and chorea [64] present acutely. Arterial thrombosis mediated by antiphospholipid antibodies is likely a contributory mechanism for transverse myelopathy [63, 64]. The cause of chorea is less clear, and there has been speculation that it may be a consequence of a direct interaction of antiphospholipid antibodies with neuronal structures in the basal ganglia [65]. Neuromyelitis optica (NMO), also known as Devic's syndrome, is a severe demyelinating disorder of the CNS that causes longitudinal transverse myelitis of at least three vertebral segments and recurrent optic neuritis. Devic's syndrome has been reported in patients with SLE [66] and is associated with NMO autoantibodies whose antigenic target is aquaporin-4 [67], the most abundant water channel in the CNS [68].

Neuropathy and Myasthenia Gravis

A sensorimotor neuropathy is the most common neuropathy and has been reported in up to 28% [6–10] of SLE patients. It frequently occurs independently of other disease characteristics [69]. The abnormalities are persistent, but in one study 67% of patients had no change in their neuropathy symptoms during a 7-year period [69]. Other less frequent forms of neuropathy include cranial neuropathy [8], autonomic neuropathy [70, 71], plexopathy [72], mononeuritis multiplex [73], and Guillain–Barré syndrome [74, 75]. Myasthenia gravis has been reported in SLE but is rare [76, 77].

Acute Confusional State

This term has replaced what was previously called "organic brain syndrome" and is synonymous with "encephalopathy" and "delirium." It encompasses a state of impaired consciousness or level of arousal that can progress to coma. Characteristics include reduced ability to focus, disturbed mood, and impaired cognition. It has been reported in 4–7% of SLE patients [6, 8, 9], and other causes such as metabolic abnormalities and hypertensive encephalopathy should also be considered in individual patients.

Cognition and Cognitive Dysfunction

Cognition is the sum of intellectual functions that result in thought. It includes reception of external stimuli, information processing, learning, storage, and expression. Disturbance of even one of these functions can result in disruption of normal thought production and present as cognitive dysfunction.

Assessment of Cognitive Function in SLE Patients

No simple screening test for cognitive dysfunction in SLE patients is currently available for use in the clinical setting. The Modified Mini-Mental State exam, although easily administered, is not very sensitive for detecting mild, albeit clinically significant, cognitive dysfunction, especially problems with executive function, which are common in SLE patients. Self-report of cognitive difficulties is currently the only means by which clinicians can screen patients who may have significant cognitive dysfunction. The Cognitive Symptoms Inventory is a 21-item, self-report questionnaire designed for patients with rheumatic disease to assess self-perception of one's ability to perform "everyday activities" [78]. Although it has not been validated against neuropsychological testing, a report suggests that this instrument may be useful as a bedside screening tool to identify SLE patients at risk for cognitive impairment [79]. However, formal testing of cognitive function is the only definitive way to diagnose cognitive impairment, and appropriate patients should be referred to a neuropsychologist.

The ACR has proposed a battery of neuropsychological tests for the assessment of cognitive function in SLE [5]. Although comprehensive, several factors limit the widespread use of these tests. For example, they are time-consuming, require specialized training to administer, and are subject to large practice effects. Although not a replacement for more detailed neuropsychological assessment, computerized neuropsychological testing may facilitate the rapid and efficient screening by non-experts of SLE patients for cognitive impairment.

The Automated Neuropsychological Assessment Metrics (ANAM) is one such computerized test battery, comprising subtests that are modified versions of standard neuropsychological tests [80, 81]. ANAM was designed to examine information processing efficiency in tasks ranging from basic information processing speed to attention/concentration, learning and memory, and executive functions [82]. Most cognitive functions consist of multiple independent processes with diffuse neuroanatomical substrates such that it can be difficult to determine exactly how a specific test score is achieved [83]. However, reaction time paradigms that measure response latencies in addition to performance accuracy may be able to overcome some of these limitations and better dissociate overall slowing of performance from domain-specific cognitive changes. Simple and choice reaction time tasks have been shown to provide reliable, valid, and sensitive measures of cognitive status in a variety of clinical populations, with psychometric properties comparable to those of conventional neuropsychological tests [84, 85]. Simple reaction time tasks can provide a "baseline" measure of information processing speed to account for sensorimotor processing speed, whereas choice reaction time paradigms can be used to examine efficiency within specific cognitive domains beyond this baseline slowing. One potential advantage of the ANAM is its reportedly limited dependence on proficiency in English or reading ability and its strong association with performance on the neuropsychological test battery recommended by the ACR for patients with SLE [86]. A number of studies have reported a higher frequency of cognitive impairment as detected by ANAM in SLE patients compared to healthy controls [87, 88]. However, a study that adjusted for nonspecific sensorimotor processing speed showed no significant differences among SLE patients and patients with RA and healthy controls [89]. ANAM has also been shown to be sensitive to subtle cognitive impairments in traumatic brain injury [90, 91], early dementia [92], and MS [93].

Cognitive dysfunction assessed using formal neuropsychological assessment techniques has been reported in up to 80% of SLE patients [6], although most studies have found a prevalence between 17 and 66% [94, 95]. Many individual patients have subclinical deficits. For example, a review of 14 cross-sectional studies of cognitive function in SLE revealed subclinical cognitive impairment in 11–54% of patients [94]. There are several non-SLE causes of cognitive dysfunction, many of which may also be present in lupus patients (Table 40.3).

TABLE 40.3   Non-SLE Causes of Cognitive Dysfunction

Cause	Examples
Direct CNS disease or injury	Ischemia Traumatic brain injury Cerebral hemorrhage Neurodegenerative disorders
Systemic illness	Hypertension Hyperthyroidism Hypothyroidism Fever
Medication	Beta blockers Antihistamines Antidepressants Antiepileptics
Psychological or psychiatric disturbance	Mania Depression Anxiety Psychosis
Metabolic disturbance	Hyper- or hypocalcemia Hyper- or hyponatremia Uremia Hypoxemia
Pain	Acute or chronic
Fatigue	Acute or chronic
Sleep disturbance	Fatigue/daytime somnolence Sleep apnea

Source: Modified from Hanly, J. G., and Harrison, M. J. (2005). Management of neuropsychiatric lupus. Best Pract. Res. Clin. Rheumatol. 19(5), 799–821.

Whether they contribute to or even cause cognitive dysfunction in SLE patients requires individual consideration. However, the prevalence of SLE-associated cognitive dysfunction is greater than that of both healthy controls [96–98] and several non-CNS disease populations [98–100], strongly suggesting that SLE-specific events may play a significant role in the development of cognitive dysfunction. A single pattern of SLE-associated cognitive dysfunction has not been found, but commonly identified cognitive abnormalities include overall cognitive slowing, decreased attention, impaired working memory, and executive dysfunction (e.g., difficulty with multitasking, organization, and planning). Because the majority of SLE patients with cognitive impairment have relatively mild deficits, the careful selection and assessment of cognitive performance in control groups is of critical importance in order to define expected levels of function in healthy individuals and those with other chronic diseases. Although cognitive impairment may be viewed as a distinct subset of NPSLE, it can also serve as a surrogate of overall brain health in SLE patients that may be affected by a variety of factors including other NP syndromes.

Clinical Associations with Cognitive Impairment

The association between cognitive impairment in SLE and other clinical variables has been examined in a number of studies. It is intuitive that certain clinically overt NPSLE events, such as stroke and antiphospholipid antibody associated multifocal cerebral infarction, would likely be accompanied by cognitive dysfunction. Thus, it is not surprising that the prevalence of cognitive dysfunction in patients with past or current clinically overt NPSLE is usually greater than that in those with no such history [16, 21, 96, 97].

Mood and psychological distress, even in the absence of a frank psychiatric diagnosis, are known to influence cognitive functioning as well as alter performance on neuropsychological tests. The high prevalence of psychological disturbance in SLE has led some to hypothesize that SLE-associated cognitive dysfunction is primarily due to the psychological impact of the underlying disease. Studies to date have both supported [101] and refuted [102] this hypothesis. Longitudinal data suggest that SLE patients with psychiatric involvement experience an improvement in cognition with improved psychiatric status at 1-year follow-up that is not observed in patients with persistent psychiatric disorders [103].

Most studies have detected no association between the presence of cognitive impairment and active systemic disease using several methods to measure global SLE disease activity [101, 103, 104]. Furthermore, no association has been found between cognitive dysfunction and the use [101, 103, 105, 106] or dose of corticosteroid [101, 103, 106] in patients with SLE.

The Evolution of Cognitive Impairment

The outcome of cognitive impairment in SLE patients has been examined in a number of studies. For example, in a 5-year prospective study of 70 SLE patients using a standardized panel of neuropsychological tests [21], the prevalence of overall cognitive impairment in SLE patients decreased from 21 to 13% during the period of study. Five patterns of cognitive performance were observed during the 5-year period (Figure 40.6). Eighty-three percent of patients were either never impaired or had resolution of cognitive impairment without specific therapeutic interventions. An additional 13% of patients demonstrated either an emerging or a fluctuating pattern of impairment, and only 4% (2 patients) showed persisting deficits that were stable over time. Similar benign changes in cognitive performance over time have been reported by Waterloo et al. [107] in 28 patients over 5 years, by Hay et al. [108] in a 2-year prospective study, and by Carlomagno et al. [109].

The Identification of Risk Factors for Cognitive Impairment

Predictors of cognitive decline over time have been examined. In one study [21], when patients who were cognitively impaired at the initial assessment were compared to those who were not impaired, the differences between groups in tests of recent memory and delayed free recall decreased over 5 years. A similar result was reported by Waterloo et al. [107]. However, patients who had clinically overt NPSLE at any time in their disease course had a statistically significant decline in memory performance over 5 years compared to patients without a history of clinically overt NPSLE [21]. These results suggest that the occurrence of clinically overt NP events, rather than the identification of isolated subclinical cognitive impairment per se, is a more reliable predictor of deterioration in selective aspects of memory function over time.

The association between cognitive function and anti-cardiolipin (aCL) antibodies has been examined in a number of cross-sectional and prospective studies. For example, 51 SLE patients who were studied prospectively were divided into those who were persistently aCL antibody positive and those who were persistently negative on the basis of up to seven antibody determinations during a 5-year period [110]. The relative change in performance on individual neuropsychological tests was then compared between patients who were antibody positive and patients who were negative. Those who were persistently IgG aCL antibody positive demonstrated a greater reduction in psychomotor speed compared to those who were antibody negative (Figure 40.7). In contrast, patients who were persistently IgA aCL antibody positive had significantly poorer performance in conceptual reasoning and executive ability. Similar results have been reported by Menon et al. [111] in a 2-year prospective study of 45 SLE patients. These data suggest that IgG and IgA aCL may be responsible for long-term subtle deterioration in cognitive function in SLE patients. No such association has been found between cognitive function and other lupus autoantibodies, such as antineuronal, lymphocytotoxic, anti-P, and anti-DNA antibodies [110, 112, 113].

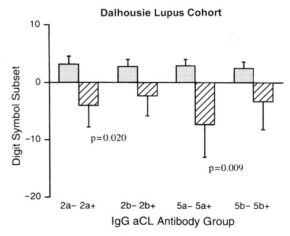

FIGURE 40.7 Change in raw scores (mean ± SEM) on the digit symbol subset of the Wechsler Adult Intelligence Scale—Revised, according to IgG anticardiolipin (aCL) status. Groups 2a+ and 5a+ represent patients who had aCL levels in excess of 2 and 5 SD, respectively, above the mean level in controls (Z score) on a minimum of four occasions during the course of 5 years. Groups 2b+ and 5b+ represent patients whose mean aCL level over the study period was in excess of 2 and 5 SD, respectively, above the mean level of controls. *Source: Reproduced with permission of John Wiley & Sons from Hanly et al. [110].*

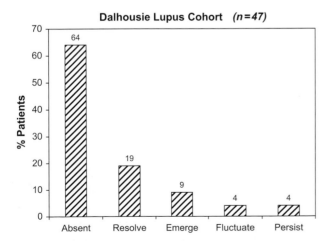

FIGURE 40.6 Change in cognitive function in 47 SLE patients assessed prospectively by formal neuropsychological evaluation on three occasions over 5 years. *Source: Reproduced with permission of John Wiley & Sons from Hanly et al. [21].*

PATHOGENESIS OF NPSLE

When considering which lupus-related immunopathogenic mechanisms contribute to NP events, it is helpful to consider whether the anatomical location of injury results in either a focal (e.g., stroke and seizure) or diffuse (e.g., acute confusion and psychosis) NP event [114]. The factors that contribute to primary NPSLE include vasculopathy, autoantibodies, and inflammatory mediators (Figure 40.8). Each plays a role in the pathogenesis of focal and diffuse NP events to a variable degree. For example, antiphospholipid antibodies are paramount in causing intravascular thrombosis leading to focal NP disease, whereas the production of intracranial inflammatory mediators is the major disease mechanism underlying diffuse NP events.

Antiphospholipid Antibodies and Vasculopathy

In addition to thrombosis of large intracranial vessels mediated by antiphospholipid antibodies, neuropathological studies of SLE patients have indicated the presence of a microvasculopathy characterized by endothelial proliferation and fibrinoid necrosis [56]. Intraluminal plugging of small vessels due to leukoagglutination has also been identified [115]. Both neuropathological abnormalities occur in close anatomical association with cerebral microinfarction, thereby implying a causal association.

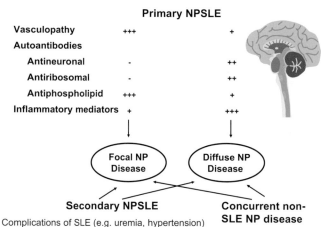

FIGURE 40.8 Factors contributing to NP events in SLE patients. Focal and diffuse nervous system events may result from direct autoimmune/inflammatory mechanisms related to SLE (primary NPSLE), may be a consequence of complications of the disease (e.g., hypertension) or its therapy (e.g., infection) (secondary NPSLE), or may be a concurrent non-SLE related NP event. *Source: Adapted with permission of Springer-Verlag from Hanly, J. G. (2001). Neuropsychiatric lupus. Curr. Rheumatol. Rep.* **3**, *205–212.*

Antineuronal Antibodies

Evidence from human studies implicating antineuronal antibodies in the pathogenesis of NPSLE is largely circumstantial and includes the temporal relationship between the presence of circulating autoantibody and the NP event, the occurrence of autoantibodies in the CSF, and, in a limited number of cases, the elution of autoantibody from brain tissue in SLE patients who have succumbed to their disease [116]. More direct evidence is available from animal studies, in which the intracranial injection of antineuronal antibodies has induced memory deficits, seizures, and neuropathological changes [116]. The source of these autoantibodies includes both the circulation, by virtue of passage through a permeabilized blood–brain barrier, and direct intrathecal production within the CNS [116]. Efforts to characterize the fine specificity of antineuronal antibodies in order to further our understanding of their immunopathogenic role have met with mixed results. Studies in murine lupus models revealed that a subset of anti-DNA antibodies react with the extracellular ligand binding domain of NR2 glutamate receptors [117], which are present throughout the forebrain and, particularly, the hippocampus, a critical location for learning and memory. In contrast to other antineuronal antibodies, the binding of anti-NR2 antibodies has functional consequences including apoptosis. In human SLE, anti-NR2 antibodies have been detected in both the circulation [118–121] and the CSF [122, 123]. Although studies of circulating anti-NR2 and NP disease have yielded conflicting results [118–121], their detection in the CSF provides a closer association with clinical NP events [122, 123].

Antiribosomal P Antibodies

Anti-P antibodies occur in up to 20% of SLE patients and were initially reported in association with psychosis and depression [52, 124]. However, these findings have been replicated in only approximately half of subsequent studies. A large meta-analysis of 1537 patients, 30% of whom had NPSLE, found that the sensitivity of anti-P antibodies for the detection of any NP event varied between 24 and 29%, with a specificity of 80% [125]. Although these findings emphasize the limited clinical utility of anti-P antibodies in the diagnosis of NPSLE, a potential new link between anti-P and the brain has been identified. Koren *et al.* [126] were the first to report an association between anti-P and antineuronal antibodies [126], and they further demonstrated that anti-P antibodies bound a 38-kDa surface protein on human neuroblastoma cells. Matus *et al.* [127] have shown that anti-P antibodies from patients with or without lupus psychosis bind a novel neuronal surface protein, which is not present on non-neuronal cells within the CNS. This antibody binding is due to the presence of amino acid sequences that are homologous to the P epitope on the surface neuronal protein, which the authors named neuronal surface P antigen (NSPA). The binding of anti-P autoantibodies to NSPA induces neuronal apoptosis mediated by deleterious intracellular calcium influx.

Inflammatory Mediators

Studies of CSF from SLE patients have implicated a number of pro-inflammatory cytokines in the pathogenesis of NPSLE, including interferon-α (IFN-α), interleukin (IL)-2, IL-6, IL-8, and IL-10 [114]. Intracranial cytokines are produced by both neuronal and glial cells [114], and they may be regulated by autoantibodies within the intrathecal space. Santer *et al.* [128] demonstrated that CSF from SLE patients is a potent inducer of IFN-α production. In fact, studies of matched CSF and serum samples from the same SLE patients indicated that CSF was more than 800-fold more potent than serum at inducing this response (Figure 40.9). The effect was also significantly greater than that seen with CSF from patients with other autoimmune inflammatory disorders and patients with MS. The induction of IFN-α, and indeed of other pro-inflammatory cytokines and chemokines, is mediated by the binding of immune complexes to FcγRII on plasmacytoid dendritic cells followed by endocytosis and activation of endosomal TLR7. The immune complexes are formed by autoantibodies and RNA—protein antigens. The plausibility of this mechanism is supported by: (1) neuronal and glial degradation products in the CSF of SLE patients [114], which provides a potential source of antigen, and (2) elevated levels of CSF matrix metalloproteinase-9 [114], which increases the permeability of the blood—brain barrier and thereby provides access of circulating autoantibodies to the intrathecal space.

Proposed Pathogenic Model for Primary NPSLE

The combined evidence from both animal and human studies suggests two separate and potentially complimentary autoimmune pathogenic mechanisms for NPSLE (Figure 40.10): (1) vascular injury involving both large- and small-caliber vessels mediated by antiphospholipid antibodies, immune complexes, and leukoagglutination, which results in focal NP events such as stroke and in diffuse NP events such as cognitive dysfunction; and (2) injury due to inflammation in which increased permeability of the blood—brain barrier, formation of immune complexes, and production of IFN-α and other inflammatory mediators lead to diffuse NP manifestations such as psychosis and acute confusional states. Some human studies indicate that these separate autoimmune pathogenic pathways correlate with clinically distinct NP events [129, 130]. Additional research involving large numbers of patients with well-characterized NP disease, in association with neuroimaging to identify structural and functional abnormalities, are required to provide further validation to this proposed pathogenic model.

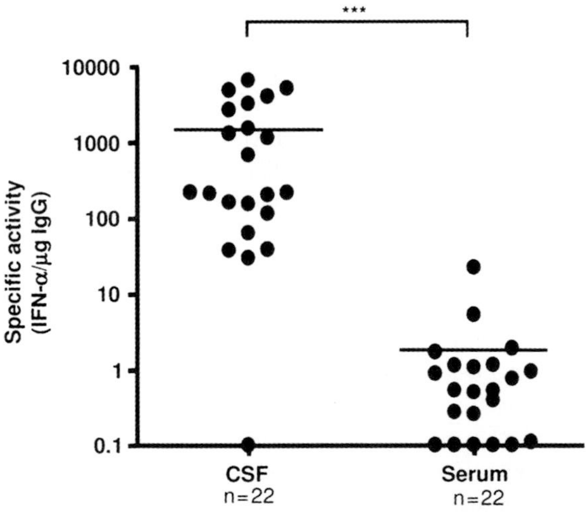

FIGURE 40.9 The specific activity (concentration of interferon-α induced per microgram of IgG of test sample) for matched serum and CSF samples from individual SLE patients. Interferon-α induction per microgram of IgG is at least 800-fold higher in CSF compared to serum (***$p < 0.001$). *Source: Modified from Santer, D. M., et al., Journal of Immunology, 2009, **182**, 1192—1202 with permission. Copyright 2009. The American Association of Immunopathologists, Inc.*

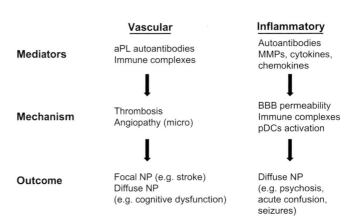

FIGURE 40.10 Autoimmune pathogenic mechanisms for NPSLE. Vascular injury involving both large- and small-caliber vessels mediated by antiphospholipid antibodies, immune complexes, and leukoagglutination results in focal NP events such as stroke and in diffuse NP events such as cognitive dysfunction. Injury due to inflammation results in increased permeability of the blood—brain barrier, formation of immune complexes, and production of IFN-α and other inflammatory mediators, leading to diffuse NP manifestations such as psychosis and acute confusional states.

DIAGNOSTIC IMAGING AND NPSLE

When considering neuroimaging in NPSLE, it is helpful to incorporate an assessment of both brain structure and brain function. Although computerized tomography (CT) scanning is the preferred technique for the diagnosis of acute intracranial hemorrhage, it has largely been replaced by magnetic resonance imaging (MRI) for the detection of other abnormalities due to its increased sensitivity [131]. Abnormalities on MRI scanning may be found in 19% [132] to 70% [133] of SLE patients. T_2-weighted MRI imaging identifies pathological processes that cause edema and is more sensitive than T_1-weighted imaging for the detection of abnormalities in patients with NPSLE. Applying the technique of fluid attenuating inversion recovery, in order to dampen the CSF signal and highlight areas of edema, further enhances the utility of T_2-weighted images [62, 134]. Focal neurological disease is associated with predominately fixed lesions in the periventricular and subcortical white matter usually in the territory of a major cerebral blood vessel [135]. However, these multiple white matter lesions are quite nonspecific and are more commonly attributed to hypertension, disease duration, and age-related small vessel disease rather than to the presence of NPSLE [136–138]. If the lesions are larger, occur in the corpus collosum, and are seen on T_1-weighted images, then the diagnosis of MS has to be considered [62]. Diffuse NP clinical presentations are associated with transient subcortical white matter lesions and patchy hyperintensities in the gray matter, which are not usually confined to the territories of major cerebral blood flow [135]. Other abnormalities detected on MRI scanning in SLE patients include cerebral infarction, venous sinus thrombosis, and increased signal in the spinal cord accompanying the clinical presentation of myelopathy [62]. MRI also provides quantitative volumetric analysis of brain atrophy.

The most objective neuroimaging study of brain function is positron emission tomography (PET) scanning, but practical considerations limit its applicability [3]. Single photon emission computed tomography (SPECT) scanning [3] provides semiquantitative analysis of regional cerebral blood flow and metabolism. It is exquisitely sensitive, and in studies of SLE patients [139–143] SPECT imaging has identified both diffuse and focal deficits that may be fixed or reversible. However, the findings are not specific for SLE [139] and do not always correlate with clinical NP manifestations [144]. In fact, it is important to remember that up to 50% of SLE patients without clinical manifestations of NP disease may have an abnormal SPECT scan [131]. The significance of these imaging abnormalities is not always clear-cut. The most common explanation is that they reflect a primary or secondary reduction in blood flow. However, in the brain there is sometimes disassociation between metabolism and blood flow. For example, changes in blood flow and metabolism can also occur in sites distant from those of the pathological lesion, a phenomenon known as diaschisis [145].

The application of several technologies to MRI scanning has provided additional opportunities to assess brain metabolism and function. Magnetic resonance angiography (MRA) permits a noninvasive visualization of cerebral blood flow, although it is probably not optimal for visualization of flow in small-caliber vessels, which are the ones primarily involved in NPSLE. Magnetic resonance spectroscopy (MRS) allows the identification and quantification of brain metabolites, thereby providing indirect evidence of cellular changes [131]. Thus, the amount of N-acetyl (NA) compounds, which reflect the quantity and integrity of neuronal cells, is reduced in lupus brains. Studies of SLE patients have found an association between reduced NA brain levels and neurocognitive dysfunction [146] and independently with elevated IgG antiphospholipid antibodies [147]. Brain lactate levels are also elevated, indicating ischemia and inflammation, whereas choline compounds are increased, reflecting damaged cell membranes and myelin destruction.

Magnetization transfer imaging (MTI) is a structural imaging modality that allows for detection of tissue damage, especially in terms of integrity of white matter tracks, that cannot be easily visualized using conventional MRI [148]. It is particularly suited to the detection and quantification of diffuse brain damage. This technique quantifies the exchange of protons between water within a macromolecule such as myelin and protons in free water. Either the loss of myelin or the accumulation of edema will alter the transfer, which is expressed as the magnetization transfer ratio (MTR) [149]. Studies to date have revealed a lower MTR in patients with NPSLE and MS, whereas there was no difference between healthy controls and SLE patients without NP disease [133, 150, 151]. The findings in SLE patients correlated with the results of cognitive assessment and psychiatric functioning [152]. Because both MRS and MTI identified abnormalities in SLE patients with normal MRI scans, these techniques provide a means of detecting and quantifying brain injury in patients with NPSLE that is not apparent with other imaging modalities.

Diffusion-weighted imaging (DWI) is very effective in the detection of hyperacute brain injury, particularly acute ischemia following stroke, when the diffusion of water is highly restricted due to the acute shift of fluid into the intracellular compartment and cytotoxic edema [62, 149]. DWI also provides additional information on white matter homogeneity and connectivity. The technique is based on the principle of isotropy (Brownian motion), which refers to unrestricted, chaotic movement of proton-containing molecules in free water. However, in the highly structured tissue of the brain, such as white matter and white matter tracks, it is easier for molecules to move in some directions than in others, thus creating preferential diffusion, or anisotropy. Pathological conditions can disturb the highly structured integrity of the white matter fibers, causing loss of anisotropy and changing the diffusion behavior of the molecules [153]. If radiofrequency pulses are applied at certain intervals, it is possible to obtain information about water molecule movement between pulses. This information is provided in the form of diffusion coefficients and contains vector information regarding the directions of diffusion. The level of fractional anisotropy (FA) can be calculated for individual pixels within a region of interest or for the whole brain, and the results are presented as a histogram with lower FA peaks indicating more pixels with higher diffusion values [153]. Lower FA peaks indicate damage or degeneration in white matter tracks [148]. DWI has been used to examine brain tissue changes in NPSLE in terms of degeneration of parenchymal structure. Although patients with NPSLE have been found to have more pixels with low FA values than healthy controls [154, 155], it is unclear whether there is a difference in diffusion pattern between patients with active disease and patients who have chronic brain damage.

An additional MRI-based technique is functional MRI (fMRI), which assesses cerebral blood flow and neuronal activity through the measurement of oxygenation status of hemoglobin. In a study [156] of 9 patients with NPSLE who were compared to the same number of patients with RA and healthy controls, blood oxygen level-dependent fMRI showed greater frontoparietal activation in SLE patients than in the other groups during a task that engaged working memory [150]. The authors interpreted the findings to indicate a compensatory adaptation of neuronal function through the recruitment of extra cortical pathways to supplement impaired function of standard pathways. An additional study [157] of 10 patients with childhood-onset SLE revealed differences in brain activation patterns compared with those observed in controls. These findings could indicate abnormalities in white matter connectivity resulting in neuronal network dysfunction.

DIAGNOSIS OF NPSLE

The first step in the management of a patient with SLE who presents with an NP event is to determine whether the event can be convincingly attributed to SLE, a complication of the disease or its therapy, or whether it reflects a coincidental disease process. This is achieved largely by a process of exclusion, given the absence of a diagnostic gold standard for most of the NP manifestations that occur in SLE. Thus, the correct diagnosis is derived from a careful analysis of the clinical, laboratory, and imaging data on a case-by-case basis, and these may be utilized to a varying extent depending on the clinical circumstances (Table 40.4). Examination of the CSF should be considered primarily to exclude infection. Measurement of CSF autoantibodies, cytokines, and biomarkers of neurological damage is still a subject of research. In considering circulating autoantibodies, those that are most likely to provide the greatest diagnostic yield are antiphospholipid antibodies. The value of measuring anti-P antibodies remains uncertain given the conflicting results to date, whereas the role of anti-NR2 antibodies in NPSLE is currently unknown. Neuroimaging should include a modality to assess brain structure and another to assess brain function. Electroencephalography and nerve conduction studies are used to investigate seizure disorders and peripheral neuropathy, respectively. Neuropsychological testing should only be done to address specific concerns about cognitive ability because the detection of isolated subclinical deficits appears to have little clinical significance.

TREATMENT OF NPSLE

Management will need to be tailored to the needs of individual patients (Table 40.5), and there remains a paucity of controlled studies to inform treatment

TABLE 40.4 Investigations in NPSLE

Autoantibodies

   Antineuronal (anti-NR2?)

   Antiribosomal P

   Antiphospholipid

Cerebrospinal fluid

   Exclude infection, assess blood—brain barrier

   Autoantibodies

   Inflammatory mediators, degradation proteins

Electrophysiological assessment

Neuropsychological assessment

Neuroimaging

   Brain structure (CT, MRI, MTI, DWI)

   Brain function (PET, SPECT, MRA, MRS, fMRI)

[24] L. Barile-Fabris, R. Ariza-Andraca, L. Olguin-Ortega, L.J. Jara, A. Fraga-Mouret, J.M. Miranda-Limon, et al., Controlled clinical trial of IV cyclophosphamide versus IV methylprednisolone in severe neurological manifestations in systemic lupus erythematosus, Ann Rheum Dis 64 (2005) 620–625.

[25] W.J. McCune, J. Golbus, W. Zeldes, P. Bohlke, R. Dunne, D.A. Fox, Clinical and immunologic effects of monthly administration of intravenous cyclophosphamide in severe systemic lupus erythematosus, N Engl J Med 318 (1988) 1423–1431.

[26] M. Tokunaga, K. Saito, D. Kawabata, Y. Imura, T. Fujii, S. Nakayamada, et al., Efficacy of rituximab (anti-CD20) for refractory systemic lupus erythematosus involving the central nervous system, Ann Rheum Dis 66 (2007) 470–475.

[27] S.D. Denburg, R.M. Carbotte, J.A. Denburg, Corticosteroids and neuropsychological functioning in patients with systemic lupus erythematosus, Arthritis Rheum 37 (1994) 1311–1320.

[28] G.M. Mody, K.B. Parag, B.C. Nathoo, D.J. Pudifin, J. Duursma, Y.K. Seedat, High mortality with systemic lupus erythematosus in hospitalized African blacks, Br J Rheumatol 33 (1994) 1151–1153.

[29] A. Samanta, J. Feehally, S. Roy, F.E. Nichol, P.J. Sheldon, J. Walls, High prevalence of systemic disease and mortality in Asian subjects with systemic lupus erythematosus, Ann Rheum Dis 50 (1991) 490–492.

[30] H. Jonsson, O. Nived, G. Sturfelt, Outcome in systemic lupus erythematosus: A prospective study of patients from a defined population, Medicine (Baltimore) 68 (1989) 141–150.

[31] J.T. Sibley, W.P. Olszynski, W.E. Decoteau, M.B. Sundaram, The incidence and prognosis of central nervous system disease in systemic lupus erythematosus, J Rheumatol 19 (1992) 47–52.

[32] E.J. Feinglass, F.C. Arnett, C.A. Dorsch, T.M. Zizic, M.B. Stevens, Neuropsychiatric manifestations of systemic lupus erythematosus: Diagnosis, clinical spectrum, and relationship to other features of the disease, Medicine (Baltimore) 55 (1976) 323–339.

[33] J.A. Kovacs, M.B. Urowitz, D.D. Gladman, Dilemmas in neuropsychiatric lupus, Rheum Dis Clin North Am 19 (1993) 795–814.

[34] F.B. Karassa, J.P. Ioannidis, K.A. Boki, G. Touloumi, M.I. Argyropoulou, K.A. Strigaris, et al., Predictors of clinical outcome and radiologic progression in patients with neuropsychiatric manifestations of systemic lupus erythematosus, Am J Med 109 (2000) 628–634.

[35] J.G. Hanly, M.B. Urowitz, L. Su, J. Sanchez-Guerrero, S.C. Bae, C. Gordon, et al., Short-term outcome of neuropsychiatric events in systemic lupus erythematosus upon enrollment into an international inception cohort study, Arthritis Rheum 59 (2008) 721–729.

[36] J.G. Hanly, M.B. Urowitz, L. Su, S.C. Bae, C. Gordon, D.J. Wallace, et al., Prospective analysis of neuropsychiatric events in an international disease inception cohort of SLE patients, Ann Rheum Dis 69 (2010) 529–535.

[37] M. Boers, Understanding the window of opportunity concept in early rheumatoid arthritis, Arthritis Rheum 48 (2003) 1771–1774.

[38] J.J. Cush, Early rheumatoid arthritis—Is there a window of opportunity? J Rheumatol Suppl 80 (2007) 1–7.

[39] N. Raskin, Headache, in: A.S. Fauci, E. Braunwald, K.J. Isselbacher, et al. (Eds.), Harrison's Principles of Internal Medicine, 14th ed., McGraw Hill, New York, 1998, pp. 68–72.

[40] A. Fernandez-Nebro, R. Palacios-Munoz, J. Gordillo, M. Abarca-Costalago, M. De Haro-Liger, J. Rodriguez-Andreu, et al., Chronic or recurrent headache in patients with systemic lupus erythematosus: A case control study, Lupus 8 (1999) 151–156.

[41] P.P. Sfikakis, D.D. Mitsikostas, M.N. Manoussakis, D. Foukaneli, H.M. Moutsopoulos, Headache in systemic lupus erythematosus: A controlled study, Br J Rheumatol 37 (1998) 300–303.

[42] D.D. Mitsikostas, P.P. Sfikakis, P.J. Goadsby, A meta-analysis for headache in systemic lupus erythematosus: The evidence and the myth, Brain 127 (2004) 1200–1209.

[43] M. Amit, Y. Molad, O. Levy, A.J. Wysenbeek, Headache in systemic lupus erythematosus and its relation to other disease manifestations, Clin Exp Rheumatol 17 (1999) 467–470.

[44] M.E. Lancman, H. Mesropian, R.J. Granillo, Chronic aseptic meningitis in a patient with systemic lupus erythematosus, Can J Neurol Sci 16 (1989) 354–356.

[45] T. Kanekura, J. Mizumoto, M. Setoyama, A case of lupus meningitis treated successfully with methylprednisolone pulse therapy, J Dermatol 20 (1993) 566–571.

[46] A. Escalante, M.M. Stimmler, Trimethoprim-sulfamethoxazole induced meningitis in systemic lupus erythematosus, J Rheumatol 19 (1992) 800–802.

[47] C. Codding, I.N. Targoff, G.A. McCarty, Aseptic meningitis in association with diclofenac treatment in a patient with systemic lupus erythematosus, Arthritis Rheum 34 (1991) 1340–1341.

[48] G.J. Gilbert, H.W. Eichenbaum, Ibuprofen-induced meningitis in an elderly patient with systemic lupus erythematosus, South Med J 82 (1989) 514–515.

[49] M. Bodani, M.D. Kopelman, A psychiatric perspective on the therapy of psychosis in systemic lupus erythematosus, Lupus 12 (2003) 947–949.

[50] R. Gerli, L. Caponi, A. Tincani, R. Scorza, M.G. Sabbadini, M.G. Danieli, et al., Clinical and serological associations of ribosomal P autoantibodies in systemic lupus erythematosus: Prospective evaluation in a large cohort of Italian patients, Rheumatology (Oxford) 41 (2002) 1357–1366.

[51] L.S. Teh, D.A. Isenberg, Antiribosomal P protein antibodies in systemic lupus erythematosus. A reappraisal, Arthritis Rheum 37 (1994) 307–315.

[52] E. Bonfa, S.J. Golombek, L.D. Kaufman, S. Skelly, H. Weissbach, N. Brot, et al., Association between lupus psychosis and antiribosomal P protein antibodies, N Engl J Med 317 (1987) 265–271.

[53] F.C. Arnett, J.D. Reveille, H.M. Moutsopoulos, L. Georgescu, K.B. Elkon, Ribosomal P autoantibodies in systemic lupus erythematosus. Frequencies in different ethnic groups and clinical and immunogenetic associations, Arthritis Rheum 39 (1996) 1833–1839.

[54] J. Wajed, Y. Ahmad, P.N. Durrington, I.N. Bruce, Prevention of cardiovascular disease in systemic lupus erythematosus— Proposed guidelines for risk factor management, Rheumatology (Oxford) 43 (2004) 7–12.

[55] J.G. Hanly, Antiphospholipid syndrome: An overview, Can Med Assoc J 168 (2003) 1675–1682.

[56] J.G. Hanly, N.M. Walsh, V. Sangalang, Brain pathology in systemic lupus erythematosus, J Rheumatol 19 (1992) 732–741.

[57] S.C. Steens, G.P. Bosma, R. ten Cate, J. Doornbos, J.M. Kros, L.A. Laan, et al., A neuroimaging follow up study of a patient with juvenile central nervous system systemic lupus erythematosus, Ann Rheum Dis 62 (2003) 583–586.

[58] S.G. Ellis, M.A. Verity, Central nervous system involvement in systemic lupus erythematosus: A review of neuropathologic findings in 57 cases, 1955–1977, Semin Arthritis Rheum 8 (1979) 212–221.

[59] R.T. Johnson, E.P. Richardson, The neurological manifestations of systemic lupus erythematosus, Medicine (Baltimore) 47 (1968) 337–369.

[60] J.K. Kur, J.M. Esdaile, Posterior reversible encephalopathy syndrome—An underrecognized manifestation of systemic lupus erythematosus, J Rheumatol 33 (2006) 2178–2183.

[61] F.G. Jennekens, L. Kater, The central nervous system in systemic lupus erythematosus. Part 1. Clinical syndromes: A literature investigation, Rheumatology (Oxford) 41 (2002) 605–618.

[62] J.W. Graham, W. Jan, MRI and the brain in systemic lupus erythematosus, Lupus 12 (2003) 891–896.

[63] B. Kovacs, T.L. Lafferty, L.H. Brent, R.J. DeHoratius, Transverse myelopathy in systemic lupus erythematosus: An analysis of 14 cases and review of the literature, Ann Rheum Dis 59 (2000) 120–124.

[64] R. Cervera, R.A. Asherson, J. Font, M. Tikly, L. Pallares, A. Chamorro, et al., Chorea in the antiphospholipid syndrome. Clinical, radiologic, and immunologic characteristics of 50 patients from our clinics and the recent literature, Medicine (Baltimore) 76 (1997) 203–212.

[65] A. Katzav, J. Chapman, Y. Shoenfeld, CNS dysfunction in the antiphospholipid syndrome, Lupus 12 (2003) 903–907.

[66] J. Birnbaum, D. Kerr, Devic's syndrome in a woman with systemic lupus erythematosus: Diagnostic and therapeutic implications of testing for the neuromyelitis optica IgG auto-antibody, Arthritis Rheum 57 (2007) 347–351.

[67] P. Waters, S. Jarius, E. Littleton, M.I. Leite, S. Jacob, B. Gray, et al., Aquaporin-4 antibodies in neuromyelitis optica and longitudinally extensive transverse myelitis, Arch Neurol 65 (2008) 913–919.

[68] V.A. Lennon, T.J. Kryzer, S.J. Pittock, A.S. Verkman, S.R. Hinson, IgG marker of optic-spinal multiple sclerosis binds to the aquaporin-4 water channel, J Exp Med 202 (2005) 473–477.

[69] R. Omdal, S. Loseth, T. Torbergsen, W. Koldingsnes, G. Husby, S.I. Mellgren, Peripheral neuropathy in systemic lupus erythematosus—A longitudinal study, Acta Neurol Scand 103 (2001) 386–391.

[70] R.H. Straub, M. Zeuner, G. Lock, H. Rath, R. Hein, J. Scholmerich, et al., Autonomic and sensorimotor neuropathy in patients with systemic lupus erythematosus and systemic sclerosis, J Rheumatol 23 (1996) 87–92.

[71] F. Liote, C.K. Osterland, Autonomic neuropathy in systemic lupus erythematosus: Cardiovascular autonomic function assessment, Ann Rheum Dis 53 (1994) 671–674.

[72] S.L. Bloch, M.P. Jarrett, M. Swerdlow, A.I. Grayzel, Brachial plexus neuropathy as the initial presentation of systemic lupus erythematosus, Neurology 29 (1979) 1633–1634.

[73] V.M. Martinez-Taboada, R.B. Alonso, J. Armona, J.L. Fernandez-Sueiro, C. Gonzalez Vela, V. Rodriguez-Valverde, Mononeuritis multiplex in systemic lupus erythematosus: Response to pulse intravenous cyclophosphamide, Lupus 5 (1996) 74–76.

[74] H.W. van Laarhoven, F.A. Rooyer, B.G. van Engelen, R. van Dalen, J.H. Berden, Guillain—Barré syndrome as presenting feature in a patient with lupus nephritis, with complete resolution after cyclophosphamide treatment, Nephrol Dial Transplant 16 (2001) 840–842.

[75] M. Lewis, T. Gibson, Systemic lupus erythematous with recurrent Guillain—Barré-like syndrome treated with intravenous immunoglobulins, Lupus 12 (2003) 857–859.

[76] E. Ben-Chetrit, A. Pollack, D. Flusser, A. Rubinow, Coexistence of systemic lupus erythematosus and myasthenia gravis: Two distinct populations of anti-DNA and anti-acetylcholine receptor antibodies, Clin Exp Rheumatol 8 (1990) 71–74.

[77] F.M. Grinlinton, N.M. Lynch, H.H. Hart, A pair of monozygotic twins who are concordant for myasthenia gravis but became discordant for systemic lupus erythematosus post-thymectomy, Arthritis Rheum 34 (1991) 916–919.

[78] T. Pincus, C. Swearingen, L.F. Callahan, A self-report cognitive symptoms inventory to assess patients with rheumatic diseases: Results in eosinophilia-myalgia syndrome (EMS), fibromyalgia, rheumatoid arthritis (RA), and other rheumatic diseases, Arthritis Rheum 39 (1996) S261.

[79] M. Sanchez, G. Alarcon, B. Fessler, H. Bastian, J. Roseman, Q. Lee, et al., Cognitive impairment (CI) in lupus is stable over time and may not be recognized by rheumatologists. Data from a multiethnic cohort, Arthritis Rheum 50 (2004) S242.

[80] D.L. Reeves, K.P. Winter, J. Bleiberg, R.L. Kane, ANAM genogram: Historical perspectives, description, and current endeavors, Arch Clin Neuropsychol 22(Suppl 1) (2007) S15–S37.

[81] D.L. Reeves, R. Kane, K. Winter, Automated neuropsychological assessment metrics (ANAM V3.11a/96) user's manual: Clinical and neurotoxicology subset (Report No. NCRF-SR-96-01), National Cognitive Foundation, San Diego, 1996.

[82] J. Bleiberg, R.L. Kane, D.L. Reeves, W.S. Garmoe, E. Halpern, Factor analysis of computerized and traditional tests used in mild brain injury research, Clin Neuropsychol 14 (2000) 287–294.

[83] D.T. Stuss, B. Levine, Adult clinical neuropsychology: Lessons from studies of the frontal lobes, Annu Rev Psychol 53 (2002) 401–433.

[84] L.I. Reicker, T.N. Tombaugh, L. Walker, M.S. Freedman, Reaction time: An alternative method for assessing the effects of multiple sclerosis on information processing speed, Arch Clin Neuropsychol 22 (2007) 655–664.

[85] T.N. Tombaugh, L. Rees, P. Stormer, A.G. Harrison, A. Smith, The effects of mild and severe traumatic brain injury on speed of information processing as measured by the computerized tests of information processing (CTIP), Arch Clin Neuropsychol 22 (2007) 25–36.

[86] T.M. Roebuck-Spencer, C. Yarboro, M. Nowak, K. Takada, G. Jacobs, L. Lapteva, et al., Use of computerized assessment to predict neuropsychological functioning and emotional distress in patients with systemic lupus erythematosus, Arthritis Rheum 55 (2006) 434–441.

[87] M. Petri, M. Naqibuddin, K.A. Carson, M. Sampedro, D.J. Wallace, M.H. Weisman, et al., Cognitive function in a systemic lupus erythematosus inception cohort, J Rheumatol 35 (2008) 1776–1781.

[88] E.Y. McLaurin, S.L. Holliday, P. Williams, R.L. Brey, Predictors of cognitive dysfunction in patients with systemic lupus erythematosus, Neurology 64 (2005) 297–303.

[89] J.G. Hanly, A. Omisade, L. Su, V. Farewell, T. Linehan, J.D. Fisk, Assessment of cognitive function in systemic lupus erythematosus, rheumatoid arthritis and multiple sclerosis using computerized neuropsychological tests, Arthritis Rheum 62 (2010) 1478–1486.

[90] J. Bleiberg, W.S. Garmoe, E.L. Halpern, D.L. Reeves, J.D. Nadler, Consistency of within-day and across-day performance after mild brain injury, Neuropsychiatry Neuropsychol Behav Neurol 10 (1997) 247–253.

[91] D.M. Levinson, D.L. Reeves, Monitoring recovery from traumatic brain injury using automated neuropsychological assessment metrics (ANAM V1.0), Arch Clin Neuropsychol 12 (1997) 155–166.

[92] D. Levinson, D. Reeves, J. Watson, M. Harrison, Automated neuropsychological assessment metrics (ANAM) measures of cognitive effects of Alzheimer's disease, Arch Clin Neuropsychol 20 (2005) 403–408.

[93] J.A. Wilken, R. Kane, C.L. Sullivan, M. Wallin, J.B. Usiskin, M.E. Quig, et al., The utility of computerized neuropsychological assessment of cognitive dysfunction in patients with relapsing-remitting multiple sclerosis, Mult Scler 9 (2003) 119–127.

[94] S.D. Denburg, J.A. Denburg, Cognitive dysfunction and antiphospholipid antibodies in systemic lupus erythematosus, Lupus 12 (2003) 883–890.

[95] J.G. Hanly, M.H. Liang, Cognitive disorders in systemic lupus erythematosus. Epidemiologic and clinical issues, Ann N Y Acad Sci 823 (1997) 60–68.

[96] R.M. Carbotte, S.D. Denburg, J.A. Denburg, Prevalence of cognitive impairment in systemic lupus erythematosus, J Nerv Ment Dis 174 (1986) 357–364.

[97] E.M. Wekking, J.C. Nossent, A.P. van Dam, A.J. Swaak, Cognitive and emotional disturbances in systemic lupus erythematosus, Psychother Psychosom 55 (1991) 126–131.

[98] J.G. Hanly, J.D. Fisk, G. Sherwood, E. Jones, J.V. Jones, B. Eastwood, Cognitive impairment in patients with systemic lupus erythematosus, J Rheumatol 19 (1992) 562–567.

[99] M.J. Prince, A.S. Bird, R.A. Blizard, A.H. Mann, Is the cognitive function of older patients affected by antihypertensive treatment? Results from 54 months of the Medical Research Council's trial of hypertension in older adults, Br Med J 312 (1996) 801–805.

[100] L.P. Lowe, D. Tranel, R.B. Wallace, T.K. Welty, Type II diabetes and cognitive function. A population-based study of Native Americans, Diabetes Care 17 (1994) 891–896.

[101] E. Kozora, L. Thompson, S. West, B. Kotzin, Analysis of cognitive and psychological deficits in systemic lupus erythematosus patients without overt central nervous system disease, Arthritis Rheum 39 (1996) 2035–2045.

[102] S. Denburg, R. Carbotte, J. Denburg, Psychological aspects of systemic lupus erythematosus: Cognitive function, mood, and self-report, J Rheumatol 24 (1997) 998–1003.

[103] E. Hay, A. Huddy, D. Black, P. Mbaya, B. Tomenson, R. Bernstein, et al., Psychiatric disorder and cognitive impairment in systemic lupus erythematosus, Arthritis Rheum 35 (1992) 411–416.

[104] R. Carbotte, S. Denburg, J. Denburg, Cognitive dysfunction in systemic lupus erythematosus is independent of active disease, J Rheumatol 22 (1995) 863–867.

[105] S. Denburg, R. Carbotte, J. Denburg, Cognitive impairment in systemic lupus erythematosus: A neuropsychological study of individual and group deficits, J Clin Exp Neuropsychol 9 (1987) 323–339.

[106] J. Fisk, B. Eastwood, G. Sherwood, J. Hanly, Patterns of cognitive impairment in patients with systemic lupus erythematosus, Br J Rheumatol 32 (1993) 458–462.

[107] K. Waterloo, R. Omdal, G. Husby, S.I. Mellgren, Neuropsychological function in systemic lupus erythematosus: A five-year longitudinal study, Rheumatology (Oxford) 41 (2002) 411–415.

[108] E.M. Hay, A. Huddy, D. Black, P. Mbaya, B. Tomenson, R.M. Bernstein, et al., A prospective study of psychiatric disorder and cognitive function in systemic lupus erythematosus, Ann Rheum Dis 53 (1994) 298–303.

[109] S. Carlomagno, S. Migliaresi, L. Ambrosone, M. Sannino, G. Sanges, G. Di Iorio, Cognitive impairment in systemic lupus erythematosus: A follow-up study, J Neurol 247 (2000) 273–279.

[110] J.G. Hanly, C. Hong, S. Smith, J.D. Fisk, A prospective analysis of cognitive function and anticardiolipin antibodies in systemic lupus erythematosus, Arthritis Rheum 42 (1999) 728–734.

[111] S. Menon, E. Jameson-Shortall, S.P. Newman, M.R. Hall-Craggs, R. Chinn, D.A. Isenberg, A longitudinal study of anticardiolipin antibody levels and cognitive functioning in systemic lupus erythematosus, Arthritis Rheum 42 (1999) 735–741.

[112] J.G. Hanly, J.D. Fisk, B. Eastwood, Brain reactive autoantibodies and cognitive impairment in systemic lupus erythematosus, Lupus 3 (1994) 193–199.

[113] J.G. Hanly, N.M. Walsh, J.D. Fisk, B. Eastwood, C. Hong, G. Sherwood, et al., Cognitive impairment and autoantibodies in systemic lupus erythematosus, Br J Rheumatol 32 (1993) 291–296.

[114] J.G. Hanly, New insights into central nervous system lupus: A clinical perspective, Curr Rheumatol Rep 9 (2007) 116–124.

[115] H.M. Belmont, S.B. Abramson, J.T. Lie, Pathology and pathogenesis of vascular injury in systemic lupus erythematosus. Interactions of inflammatory cells and activated endothelium, Arthritis Rheum 39 (1996) 9–22.

[116] J.G. Hanly, Neuropsychiatric lupus, Rheum Dis Clin North Am 31 (2005) 273–298.

[117] L.A. DeGiorgio, K.N. Konstantinov, S.C. Lee, J.A. Hardin, B.T. Volpe, B. Diamond, A subset of lupus anti-DNA antibodies cross-reacts with the NR2 glutamate receptor in systemic lupus erythematosus, Nat Med 7 (2001) 1189–1193.

[118] M.J. Harrison, L.D. Ravdin, M.D. Lockshin, Relationship between serum NR2a antibodies and cognitive dysfunction in systemic lupus erythematosus, Arthritis Rheum 54 (2006) 2515–2522.

[119] J.G. Hanly, J. Robichaud, J.D. Fisk, Anti-NR2 glutamate receptor antibodies and cognitive function in systemic lupus erythematosus, J Rheumatol 33 (2006) 1553–1558.

[120] L. Lapteva, M. Nowak, C.H. Yarboro, K. Takada, T. Roebuck-Spencer, T. Weickert, et al., Anti- N-methyl-d-aspartate receptor antibodies, cognitive dysfunction, and depression in systemic lupus erythematosus, Arthritis Rheum 54 (2006) 2505–2514.

[121] R. Omdal, K. Brokstad, K. Waterloo, W. Koldingsnes, R. Jonsson, S.I. Mellgren, Neuropsychiatric disturbances in SLE are associated with antibodies against NMDA receptors, Eur J Neurol 12 (2005) 392–398.

[122] Y. Arinuma, T. Yanagida, S. Hirohata, Association of cerebrospinal fluid anti-NR2 glutamate receptor antibodies with diffuse neuropsychiatric systemic lupus erythematosus, Arthritis Rheum 58 (2008) 1130–1135.

[123] T. Yoshio, K. Onda, H. Nara, S. Minota, Association of IgG anti-NR2 glutamate receptor antibodies in cerebrospinal fluid with neuropsychiatric systemic lupus erythematosus, Arthritis Rheum 240 (2006) 675–678.

[124] A.B. Schneebaum, J.D. Singleton, S.G. West, J.K. Blodgett, L.G. Allen, J.C. Cheronis, et al., Association of psychiatric manifestations with antibodies to ribosomal P proteins in systemic lupus erythematosus, Am J Med 90 (1991) 54–62.

[125] F.B. Karassa, A. Afeltra, A. Ambrozic, D.M. Chang, F. De Keyser, A. Doria, et al., Accuracy of anti-ribosomal P protein antibody testing for the diagnosis of neuropsychiatric systemic lupus erythematosus: An international meta-analysis, Arthritis Rheum 54 (2006) 312–324.

[126] E. Koren, M.W. Reichlin, M. Koscec, R.D. Fugate, M. Reichlin, Autoantibodies to the ribosomal P proteins react with a plasma membrane-related target on human cells, J Clin Invest 89 (1992) 1236–1241.

[127] S. Matus, P.V. Burgos, M. Bravo-Zehnder, R. Kraft, O.H. Porras, P. Farias, et al., Antiribosomal-P autoantibodies from

psychiatric lupus target a novel neuronal surface protein causing calcium influx and apoptosis, J Exp Med 204 (2007) 3221–3234.

[128] D.M. Santer, T. Yoshio, S. Minota, T. Moller, K.B. Elkon, Potent induction of IFN-alpha and chemokines by autoantibodies in the cerebrospinal fluid of patients with neuropsychiatric lupus, J Immunol 182 (2009) 1192–1201.

[129] J.G. Hanly, M.B. Urowitz, F. Siannis, V. Farewell, C. Gordon, S.C. Bae, et al., Autoantibodies and neuropsychiatric events at the time of systemic lupus erythematosus diagnosis: Results from an international inception cohort study, Arthritis Rheum 58 (2008) 843–853.

[130] C. Briani, M. Lucchetta, A. Ghirardello, E. Toffanin, S. Zampieri, S. Ruggero, et al., Neurolupus is associated with anti-ribosomal P protein antibodies: An inception cohort study, J Autoimmun 32 (2009) 79–84.

[131] W.L. Sibbitt Jr., R.R. Sibbitt, W.M. Brooks, Neuroimaging in neuropsychiatric systemic lupus erythematosus [see comments], Arthritis Rheum 42 (1999) 2026–2038.

[132] W.J. McCune, A. MacGuire, A. Aisen, S. Gebarski, Identification of brain lesions in neuropsychiatric systemic lupus erythematosus by magnetic resonance scanning, Arthritis Rheum 31 (1988) 159–166.

[133] M. Rovaris, B. Viti, G. Ciboddo, S. Gerevini, R. Capra, G. Iannucci, et al., Brain involvement in systemic immune mediated diseases: Magnetic resonance and magnetisation transfer imaging study, J Neurol Neurosurg Psychiatry 68 (2000) 170–177.

[134] W.L. Sibbitt Jr., P.J. Schmidt, B.L. Hart, W.M. Brooks, Fluid attenuated inversion recovery (FLAIR) imaging in neuropsychiatric systemic lupus erythematosus, J Rheumatol 30 (2003) 1983–1989.

[135] C.L. Bell, C. Partington, M. Robbins, F. Graziano, P. Turski, S. Kornguth, Magnetic resonance imaging of central nervous system lesions in patients with lupus erythematosus. Correlation with clinical remission and antineurofilament and anti-cardiolipin antibody titers [see comments], Arthritis Rheum 34 (1991) 432–441.

[136] M.R. Gonzalez-Crespo, F.J. Blanco, A. Ramos, E. Ciruelo, I. Mateo, M.A. Lopez Pino, et al., Magnetic resonance imaging of the brain in systemic lupus erythematosus, Br J Rheumatol 34 (1995) 1055–1060.

[137] M.M. Stimmler, P.M. Coletti, F.P. Quismorio Jr., Magnetic resonance imaging of the brain in neuropsychiatric systemic lupus erythematosus, Semin Arthritis Rheum 22 (1993) 335–349.

[138] A. Cauli, C. Montaldo, M.T. Peltz, P. Nurchis, G. Sanna, P. Garau, et al., Abnormalities of magnetic resonance imaging of the central nervous system in patients with systemic lupus erythematosus correlate with disease severity, Clin Rheumatol 13 (1994) 615–618.

[139] J.C. Nossent, A. Hovestadt, D.H. Schonfeld, A.J. Swaak, Single-photon-emission computed tomography of the brain in the evaluation of cerebral lupus, Arthritis Rheum 34 (1991) 1397–1403.

[140] K. Oku, T. Atsumi, S. Furukawa, T. Horita, Y. Sakai, S. Jodo, et al., Cerebral imaging by magnetic resonance imaging and single photon emission computed tomography in systemic lupus erythematosus with central nervous system involvement, Rheumatology (Oxford) 42 (2003) 773–777.

[141] A. Rubbert, J. Marienhagen, K. Pirner, B. Manger, J. Grebmeier, A. Engelhardt, et al., Single-photon-emission computed tomography analysis of cerebral blood flow in the evaluation of central nervous system involvement in patients with systemic lupus erythematosus [see comments], Arthritis Rheum 36 (1993) 1253–1262.

[142] M.P. Rogers, E. Waterhouse, J.S. Nagel, N.W. Roberts, S.H. Stern, P. Fraser, et al., I-123 iofetamine SPECT scan in systemic lupus erythematosus patients with cognitive and other minor neuropsychiatric symptoms: A pilot study, Lupus 1 (1992) 215–219.

[143] J.A. Kovacs, M.B. Urowitz, D.D. Gladman, R. Zeman, The use of single photon emission computerized tomography in neuropsychiatric SLE: A pilot study [see comments], J Rheumatol 22 (1995) 1247–1253.

[144] K. Waterloo, R. Omdal, H. Sjoholm, W. Koldingsnes, E.A. Jacobsen, J.A. Sundsfjord, et al., Neuropsychological dysfunction in systemic lupus erythematosus is not associated with changes in cerebral blood flow, J Neurol 248 (2001) 595–602.

[145] D.M. Feeney, J.C. Baron, Diaschisis, Stroke 17 (1986) 817–830.

[146] W.M. Brooks, R.E. Jung, C.C. Ford, E.J. Greinel, W.L. Sibbitt Jr., Relationship between neurometabolite derangement and neurocognitive dysfunction in systemic lupus erythematosus, J Rheumatol 26 (1999) 81–85.

[147] A. Sabet, W.L. Sibbitt Jr., C.A. Stidley, J. Danska, W.M. Brooks, Neurometabolite markers of cerebral injury in the anti-phospholipid antibody syndrome of systemic lupus erythematosus, Stroke 29 (1998) 2254–2260.

[148] V.K. Diwadkar, M.S. Keshavan, Newer techniques in magnetic resonance imaging and their potential for neuropsychiatric research, J Psychosom Res 53 (2002) 677–685.

[149] P.L. Peterson, F.A. Howe, C.A. Clark, J.S. Axford, Quantitative magnetic resonance imaging in neuropsychiatric systemic lupus erythematosus, Lupus 12 (2003) 897–902.

[150] G.P. Bosma, M.J. Rood, A.H. Zwinderman, T.W. Huizinga, M.A. van Buchem, Evidence of central nervous system damage in patients with neuropsychiatric systemic lupus erythematosus, demonstrated by magnetization transfer imaging, Arthritis Rheum 43 (2000) 48–54.

[151] G.P. Bosma, M.J. Rood, T.W. Huizinga, B.A. de Jong, E.L. Bollen, M.A. van Buchem, Detection of cerebral involvement in patients with active neuropsychiatric systemic lupus erythematosus by the use of volumetric magnetization transfer imaging, Arthritis Rheum 43 (2000) 2428–2436.

[152] G.P. Bosma, H.A. Middelkoop, M.J. Rood, E.L. Bollen, T.W. Huizinga, M.A. van Buchem, Association of global brain damage and clinical functioning in neuropsychiatric systemic lupus erythematosus, Arthritis Rheum 46 (2002) 2665–2672.

[153] T.W. Huizinga, S.C. Steens, M.A. van Buchem, Imaging modalities in central nervous system systemic lupus erythematosus, Curr Opin Rheumatol 13 (2001) 383–388.

[154] G.P. Bosma, T.W. Huizinga, S.P. Mooijaart, M.A. Van Buchem, Abnormal brain diffusivity in patients with neuropsychiatric systemic lupus erythematosus, AJNR Am J Neuroradiol 24 (2003) 850–854.

[155] T. Moritani, D.A. Shrier, Y. Numaguchi, C. Takahashi, T. Yano, K. Nakai, et al., Diffusion-weighted echo-planar MR imaging of CNS involvement in systemic lupus erythematosus, Acad Radiol 8 (2001) 741–753.

[156] B.M. Fitzgibbon, S.L. Fairhall, I.J. Kirk, M. Kalev-Zylinska, K. Pui, N. Dalbeth, et al., Functional MRI in NPSLE patients reveals increased parietal and frontal brain activation during a working memory task compared with controls, Rheumatology (Oxford) 47 (2008) 50–53.

[157] M.W. DiFrancesco, S.K. Holland, M.D. Ris, C.M. Adler, S. Nelson, M.P. DelBello, et al., Functional magnetic resonance imaging assessment of cognitive function in childhood-onset systemic lupus erythematosus: A pilot study, Arthritis Rheum 56 (2007) 4151–4163.

[158] L. Barile, C. Lavalle, Transverse myelitis in systemic lupus erythematosus—The effect of IV pulse methylprednisolone and cyclophosphamide, J Rheumatol 19 (1992) 370–372.

[159] S. Eyanson, M.H. Passo, M.A. Aldo-Benson, M.D. Benson, Methylprednisolone pulse therapy for nonrenal lupus erythematosus, Ann Rheum Dis 39 (1980) 377–380.

[160] V.F. Trevisani, A.A. Castro, J.F. Neves Neto, A.N. Atallah, Cyclophosphamide versus methylprednisolone for the treatment of neuropsychiatric involvement in systemic lupus erythematosus (Cochrane review). Cochrane Database Syst Rev 3 (2000).

[161] V. Baca, C. Lavalle, R. Garcia, T. Catalan, J.M. Sauceda, G. Sanchez, et al., Favorable response to intravenous methylprednisolone and cyclophosphamide in children with severe neuropsychiatric lupus, J Rheumatol 26 (1999) 432–439.

[162] F.K. Leung, P.R. Fortin, Intravenous cyclophosphamide and high dose corticosteroids improve MRI lesions in demyelinating syndrome in systemic lupus erythematosus, J Rheumatol 30 (2003) 1871–1873.

[163] D.T. Boumpas, H. Yamada, N.J. Patronas, D. Scott, J.H. Klippel, J.E. Balow, Pulse cyclophosphamide for severe neuropsychiatric lupus, Q J Med 81 (1991) 975–984.

[164] C.M. Neuwelt, S. Lacks, B.R. Kaye, J.B. Ellman, D.G. Borenstein, Role of intravenous cyclophosphamide in the treatment of severe neuropsychiatric systemic lupus erythematosus, Am J Med 98 (1995) 32–41.

[165] P.C. Ramos, M.J. Mendez, P.R. Ames, M.A. Khamashta, G.R. Hughes, Pulse cyclophosphamide in the treatment of neuropsychiatric systemic lupus erythematosus, Clin Exp Rheumatol 14 (1996) 295–299.

[166] C.C. Mok, C.S. Lau, E.Y. Chan, R.W. Wong, Acute transverse myelopathy in systemic lupus erythematosus: Clinical presentation, treatment, and outcome, J Rheumatol 25 (1998) 467–473.

[167] G. Galindo-Rodriguez, J.A. Avina-Zubieta, S. Pizarro, V. Diaz de Leon, N. Saucedo, M. Fuentes, et al., Cyclophosphamide pulse therapy in optic neuritis due to systemic lupus erythematosus: An open trial, Am J Med 106 (1999) 65–69.

[168] C. Gorman, M. Leandro, D. Isenberg, Does B cell depletion have a role to play in the treatment of systemic lupus erythematosus? Lupus 13 (2004) 312–316.

[169] K. Saito, M. Nawata, S. Nakayamada, M. Tokunaga, J. Tsukada, Y. Tanaka, Successful treatment with anti-CD20 monoclonal antibody (rituximab) of life-threatening refractory systemic lupus erythematosus with renal and central nervous system involvement, Lupus 12 (2003) 798–800.

[170] G. Sanna, M.L. Bertolaccini, M.J. Cuadrado, M.A. Khamashta, G.R. Hughes, Central nervous system involvement in the antiphospholipid (Hughes) syndrome, Rheumatology (Oxford) 42 (2003) 200–213.

[171] M.A. Khamashta, M.J. Cuadrado, F. Mujic, N.A. Taub, B.J. Hunt, G.R. Hughes, The management of thrombosis in the antiphospholipid-antibody syndrome [see comments], N Engl J Med 332 (1995) 993–997.

[172] M.A. Crowther, J.S. Ginsberg, J. Julian, J. Denburg, J. Hirsh, J. Douketis, et al., A comparison of two intensities of warfarin for the prevention of recurrent thrombosis in patients with the antiphospholipid antibody syndrome, N Engl J Med 349 (2003) 1133–1138.

[173] L.B. Krupp, C. Christodoulou, P. Melville, W.F. Scherl, W.S. MacAllister, L.E. Elkins, Donepezil improved memory in multiple sclerosis in a randomized clinical trial, Neurology 63 (2004) 1579–1585.

[174] P.M. Doraiswamy, S.M. Rao, Treating cognitive deficits in multiple sclerosis: Are we there yet? Neurology 63 (2004) 1552–1553.

[175] M.J. Harrison, K. Morris, J. Barsky, R. Horton, J. Toglia, L. Robbins, et al., Development of a unique and novel psychoeducational program for active systemic lupus erythematosus (SLE) patients with self-perceived cognitive difficulties, Arthritis Rheum 49 (2003) S1685.

[176] M.J. Harrison, K. Morris, J. Barsky, R. Horton, J. Toglia, L. Robbins, et al., Preliminary results of a novel psychoeducational program designed for systemic lupus erythematosus (SLE) patients with self-perceived cognitive difficulties, Arthritis Rheum 49 (2003) S1058.

[177] M.J. Harrison, J. Toglia, S. Chait, K. Morris, J. Barsky, R. Horton, et al., MINDFULL: Mastering the Intellectual Navigation of Daily Functioning and Undoing the Limitations of Lupus—A pilot study to improve quality of life in SLE, Arthritis Rheum 50 (2004) S242.

41

Skin

Victoria P. Werth, Cristián Vera Kellet, Jan P. Dutz

INTRODUCTION

The term "lupus" (Latin for "wolf") reflects a long history of descriptive terms in dermatology [1]. Forms of cutaneous lupus related to autoimmune skin disease are specifically called cutaneous lupus erythematosus (CLE).

The importance of thorough dermatologic evaluation for a correct diagnosis and management of LE is clear. There are a variety of clinical presentations of lupus-specific skin lesions, and recognition of these can help prognosticate the likelihood of development of systemic disease. There is a differential diagnosis for each clinical presentation of CLE, and often a confirmatory skin biopsy is helpful to exclude mimickers of cutaneous lupus and ensure appropriate diagnosis and treatment.

This chapter provides a comprehensive approach to the diagnosis, management, and treatment of patients with cutaneous manifestations of LE.

OVERVIEW

Skin disease is the second most common clinical manifestation of SLE after arthralgias, occurs in the course of disease in 70–80% of patients, and is the first manifestation of disease in 23–28% of patients [2]. Cutaneous lupus can present with varying clinical morphologies, making the clinical diagnosis sometimes challenging. Because of the implications of a diagnosis and the frequent need for systemic therapy, it is often helpful to have a confirmatory skin biopsy. In addition, there is a differential diagnosis of mimickers of cutaneous lupus for each of the different clinical presentations, again making a biopsy an important aspect for diagnosis. Often, the biopsy is diagnostic, but occasionally there is a need for clinical–pathologic correlation to arrive at the best diagnosis. In addition, dermatopathologists receive specific training in inflammatory

dermatoses, making their input often helpful. A biopsy for direct immunofluorescence (DIF) is often not necessary if the pathology is diagnostic, and there are cases of false-positive findings, particularly in photoexposed areas [3]. Thus, the DIF is most helpful if there is suspicion for cutaneous lupus and the biopsy is not diagnostic. A lupus band test, done on unaffected, non-sun-exposed skin, is mainly of historic interest. Prior to the current lupus serologic tests, the lupus band test was used to help make a diagnosis of systemic lupus erythematosus (SLE).

Skin lesions in LE are divided into lupus-specific and lupus-nonspecific findings. Those with interface dermatitis on histology are considered as lupus specific, and those with pathologic changes or clinical findings not specific to cutaneous lupus, such as vasculitis and livedo reticularis, are grouped under lupus-nonspecific findings [4].

LE-specific skin lesions include chronic cutaneous LE (CCLE), subacute cutaneous LE (SCLE), and acute cutaneous LE (ACLE). CCLE typically includes localized, generalized, and hypertrophic LE; lupus panniculitis; and LE tumidus. Patients with any of these LE-specific skin lesions can have isolated skin lesions or also SLE if they fulfill four criteria, and patients can have more than one lupus-specific skin lesion. The likelihood of meeting criteria for SLE varies with the subset of CCLE, with localized discoid LE (DLE) having an approximately 5% incidence of evolving to SLE, generalized DLE having a 20% incidence of SLE over time, and SLE seen only rarely in patients with lupus panniculitis or LE tumidus. Although patients with SCLE have an approximately 50% incidence of SLE, only 10–15% of them have severe SLE organ involvement [5–8]. Many patients are characterized as having SLE solely on the basis of skin findings, thus bringing into question the usefulness of having four skin criteria for SLE [9].

LE-nonspecific skin lesions are defined as not histopathologically distinct for LE and/or may be seen as a feature of another disease. They can include vasculitis, with a variety of presentations including leukocytoclastic, urticarial, and polyarteritis nodosa-like. Lesions due to vasculopathy include Degos-like lesions and secondary atrophie blanche. Other cutaneous vascular lesions include periungual telangiectasias, livedo reticularis, thrombophlebitis, Raynaud's phenomenon, and erythromelalgia. Nonscarring alopecia includes lupus hair, telogen effluvium, and alopecia areata. Other nonspecific LE lesions that can be associated with other diseases include sclerodactyly, rheumatoid nodules, calcinosis cutis, LE-nonspecific bullous lesions, urticaria, papulonodular mucinosis, cutis laxa/anetoderma, acanthosis nigricans, erythema multiforme, leg ulcers, and lichen planus. A number of these lupus-nonspecific skin lesions are related to antiphospholipid antibodies. LE-nonspecific skin lesions are more frequently associated with SLE than are LE-specific skin lesions [2, 10].

SUBSETS OF CUTANEOUS LUPUS ERYTHEMATOSUS

In 1981, Gilliam and Sontheimer categorized lupus-specific skin lesions of cutaneous lupus into three groups based on the amount of time that the skin symptoms are typically present: (1) CCLE, including DLE, lupus profundus, and lupus tumidus; (2) SCLE, with annular and psoriasiform variants; and (3) ACLE, including malar rash (butterfly eruption) and generalized acute cutaneous LE [4]. These different forms of cutaneous LE may occur concurrently and demonstrate overlapping features, but for didactic purposes the entities are described separately (Tables 41.1–41.3). Because different subtypes have varying associations with systemic disease, it is important to classify the skin lesions to a subtype. A regional approach to skin examination ensures that the multiple cutaneous manifestations are not overlooked (Table 41.4).

Chronic Cutaneous Lupus Erythematosus

Discoid Lupus Erythematosus

Discoid lupus erythematosus is a chronic, scarring, atrophic, photosensitive dermatosis that usually involves hair follicles. DLE is the most common form of the chronic cutaneous variants. DLE is more common in females, with a female-to-male ratio of 3:2. It has a peak incidence between the second and fourth decade of life, but it has been described in childhood. DLE has been described in all races but is more common in African-Americans than Caucasians [11].

The primary lesion is a well-defined, disk-shaped erythematous patch that varies from a few millimeters

TABLE 41.1   LE-Specific Skin Lesions

Chronic cutaneous LE

  Discoid lupus (DLE)

    Localized DLE

    Mucosal DLE

    Palmoplantar erosive LE

    Linear lupus

  Lichenoid DLE (LE/lichen planus overlap)

  Hypertrophic LE (verrucous LE)

  Lupus profundus (lupus panniculitis)

  Chilblain lupus (perniotic lupus)

  Lupus tumidus

Subacute cutaneous LE (SCLE)

  Annular-polycyclic type

  Psoriasiform/papulosquamous type

  Rowell's syndrome

  Drug-induced SCLE

  Neonatal lupus

Acute cutaneous LE

  Malar rash (butterfly eruption)

  Maculopapular lupus rash (generalized acute cutaneous LE)

  Toxic epidermal necrolysis-like acute cutaneous LE

TABLE 41.2   LE Nonspecific Skin Lesions

Photosensitivity

Cutaneous vascular disease

  Vasculitis

  Antiphospholipid syndrome

  Raynaud's phenomenon

  Degos-like

Nonscarring alopecia

  Telogen effluvium

  Lupus hair

Oral ulcers

Nail and capillary changes

Papulonodular mucinosis

Bullous lupus erythematosus

to 15 cm, with slight-to-moderate scaling. As the lesion progresses, the scale turns grayish-white and may thicken and become adherent to the skin, accumulating in dilated follicular openings with a keratinous plug

TABLE 41.3 LE-Associated Conditions

Sweet syndrome

Pyoderma gangrenosum

Palisaded neutrophilic granulomatous dermatitis

TABLE 41.4 Skin Examination, Location, and Lesions in Lupus Erythematosus

Scalp alopecia

Scarring (DLE)

Nonscarring (telogen effluvium, lupus hair)

Ear

Helix— DLE, chilblain lupus (earlobe)

Ear canal—DLE

Cheek

Butterfly (malar rash)—SLE

DLE

Lupus panniculitis

Photosensitivity (in all photodistributed areas)

Mouth

DLE

Recurrent aphthous stomatitis

Oral involvement of Rowell's syndrome or TEN-like ACLE

Trunk and proximal extremities

Psoriasiform and annular SCLE plaques

Photosensitivity

Urticarial vasculitis

Lupus panniculitis

DLE

Distal extremities

Small and medium-sized vasculitis

Palms and soles

DLE

Chilblains—Chilblain lupus

Small and medium sized vasculitis digital infarcts

Nails

Red lunula

Glomerularization of capillaries on capillaroscopy

ACLE, acute cutaneous lupus erythematosus; DLE, discoid lupus; SCLE, subacute cutaneous lupus erythematosus; SLE, systemic lupus erythematosus; TEN, toxic epidermal necrolysis.

(follicular plugging). When the scale is lifted, patients complain of pain and conical projections may extend downward, giving the "carpet-tack" sign. Lesions slowly expand centrifugally, with active inflammation and hyperpigmentation at the periphery, leaving central atrophic scarring, telangiectasia, and hypopigmentation. Scarring and/or atrophy are present in more than half of patients, and this characteristic separates these lesions from those of SCLE. The acronym "PASTE" may be used to conveniently remember the salient clinical features of DLE: *P*lugged hair follicles, *A*trophy of epidermis, *S*cale, *T*elangiectasis, and *E*rythema. DLE lesions may also demonstrate the isomorphic response (Koebner's phenomenon), meaning that skin lesions appear after local trauma [12].

DLE can be limited to the head or neck (localized DLE), or it can occur above and below the neck (generalized DLE) (Figures 41.1–41.3). The lesions typically occur on the face, scalp, ears, V area of the neck, and extensor aspects of the arms. DLE usually spares nasolabial folds, and when it affects the external ears, it involves the concha (Figure 41.4). DLE induces irreversible cicatricial alopecia (Figure 41.5) in approximately one-third of patients [13].

Although oral ulcers occur more frequently in patients with SLE, 24% of patients with DLE have mucosal involvement [14]. Oral discoid lesions present as a well-demarcated erythematous round or irregular atrophic area or ulceration, associated with radiating keratotic striae and telangiectasia. Discoid mucosal

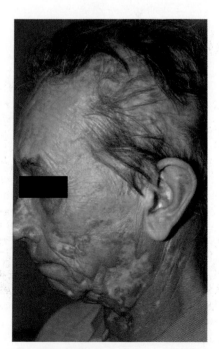

FIGURE 41.1 Generalized discoid lupus erythematosus with extensive disfiguring facial involvement.

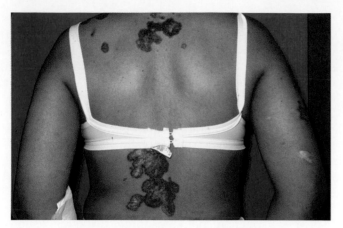

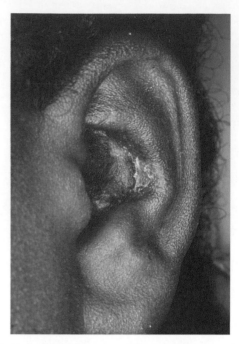

FIGURE 41.2 Generalized discoid lupus erythematosus with truncal lesions.

FIGURE 41.4 Discoid lupus erythematosus lesion of ear canal showing atrophic erythematous plaques with post-inflammatory hyperpigmentation.

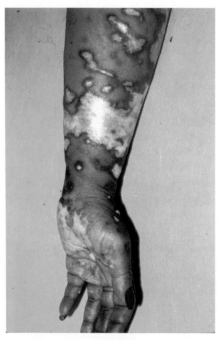

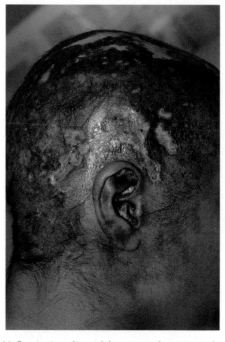

FIGURE 41.3 Generalized discoid lupus erythematosus with large lesions affecting arms and hands.

FIGURE 41.5 Active discoid lupus erythematosus lesions of the scalp showing scarring alopecia with post-inflammatory hypo- and hyperpigmentation.

lesions are often considered lupus-related oral ulcers and affect primarily nonkeratinized lining mucosa (soft palate, lips, cheeks, and alveolar process as far as the gingiva) (Figure 41.6). Oral lesions may share clinical and histological features with lichen planus [15]. DLE of the lips usually affects the vermillion border of the lip (Figure 41.7). Perioral lesions may resolve as pitted scarring.

Palmoplantar DLE lesions are uncommon and lack the specific clinical features of DLE because of the absence of hair follicles on palms and soles (Figure 41.8). Patients with palmoplantar DLE commonly have periungual involvement leading to nail dystrophy.

The natural history is variable; however, in most cases, the inflammatory component of DLE will eventually resolve without therapy. Patients tend not to develop clinical and/or serologic evidence of SLE. DLE is seen in 15—30% of SLE patients at any time

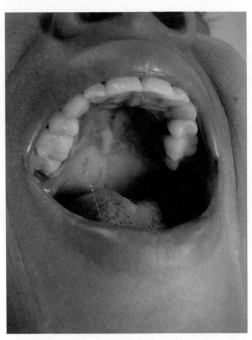

FIGURE 41.6 Atrophic lesions of discoid mucosal lupus affecting the soft palate.

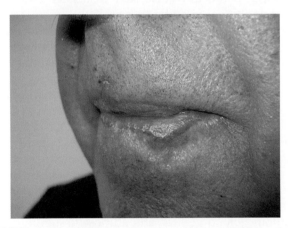

FIGURE 41.7 Discoid lupus erythematosus of the lips affecting the vermillion border.

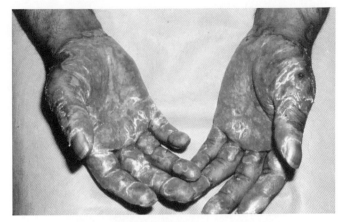

FIGURE 41.8 Palmoplantar discoid lupus erythematosus with diffuse atrophic erythematous scaling lesions affecting the palms.

during the disease course, and 5—10% of patients with clinical DLE will progress to SLE [16, 17]. Patients with localized DLE have an approximately 50% chance of remission, whereas generalized DLE becomes inactive in less than 10% of patients. Patients with generalized DLE and DLE associated with high autoantibody titers have a higher chance to develop systemic disease [16]. Patients with localized DLE, hypertrophic LE, LE panniculitis, and LE tumidus are more likely to have skin disease only [18]. Those patients with DLE that further develop systemic disease tend to have low disease systemic disease activity scores [19]. Smokers are at higher risk for DLE than nonsmokers. When comparing DLE smokers versus nonsmokers, smokers also had more extensive involvement than nonsmokers [20]. Drug-induced CCLE is rare and usually associated with fluorouracil agents [21].

HISTOLOGY

All of the clinical forms of cutaneous lupus (with the exception of LE profundus) show an extensive overlap in their histological features. Most often, the diagnosis is made on the basis of a combination of histological findings. Changes may be apparent at all levels of the skin. DLE histological findings include a lymphocytic dominant band-like infiltrate at the dermal—epithelial junction (also referred to as interface infiltrate) that disrupts the basement membrane zone and epithelial basal cell layer (vacuolar degeneration); thickening of the basement membrane; follicular plugging; dermal mucin deposition hyperkeratosis; atrophy of the epidermis; incontinence of pigment; and usually perivascular, periappendiceal, and subepidermal lymphocytic infiltrate [22].

DIF probes immunoreactants localized in patients" skin or mucous membranes such as immunoglobulins and complement on frozen sections using fluoresceinated antisera. Tissue may be examined from skin lesions (lesional) or normal skin (nonlesional). Deposition of immunoglobulin and/or complement at the basement membrane zone (BMZ) is a characteristic feature. Approximately 90% of patients with DLE manifest a positive DIF test on lesional skin. Any immunoreactant can be observed at the BMZ, but the most common immunoreactant in DLE is granular IgM deposited along the dermal—epidermal junction (Figure 41.9). IgG, IGA, and C3 are occasionally observed.

Hypertrophic (Verrucous) Lupus Erythematosus

This is a rare subtype of CCLE in which the hypertrophic changes sometimes seen in DLE are predominant.

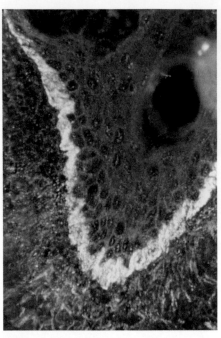

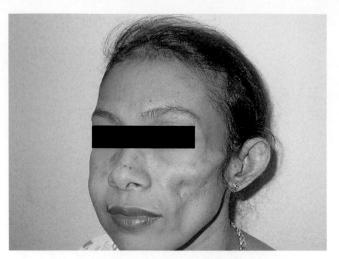

FIGURE 41.10   Lupus profundus affecting the cheeks.

FIGURE 41.9 Direct immunofluorescence examination of a discoid lupus erythematosus lesion. There is a thick granular band of IgM deposited along the dermal–epidermal junction.

It presents as oval or rounded, sharply outlined, greatly elevated tumor-like scaly and hyperkeratotic lesions, usually in sun-exposed areas. The importance of recognition of this variant lies in the fact that these lesions can be misdiagnosed as squamous cell carcinoma [23]. Given the rarity of this form, association with systemic disease is anecdotal.

Epidermal signs on histology include pseudo-epitheliomatous hyperplasia with pronounced hyperkeratosis and vacuolar degeneration of basal cells. Interface dermatitis and mucin deposition is also seen. DIF does not always reveal deposits of immunoreactants at the dermal–epidermal junction.

Lupus Panniculitis (Lupus Profundus)

Lupus panniculitis affects the subcutaneous fat. The typical presentation is of one or two deep indurated nodules or plaques involving the arms, face, buttocks, chest, and, less frequently, the abdomen, back, and neck (Figure 41.10). Generalized lupus panniculitis has been described in patients with C2 or C4 deficiency [24] and in patients with antiphospholipid syndrome [25]. Lesions may be painful. The overlying skin can be normal or show erythema, atrophy, or ulceration [26]. Lesions heal with atrophy and scarring. Patients sometimes recall a history of trauma, and there are cases in which lesions appear or exacerbate after hepatitis B vaccination [27] or treatment with interferon-β [28]. There is an uncommon clinical variant of lupus panniculitis, named "lupus mastitis," which consists of unilateral or bilateral circumscribed breast nodules that usually are misdiagnosed clinically and mammographically as carcinoma [29].

Lupus panniculitis may occur alone or before or after the onset of DLE or SLE. Seventy percent of patients with lupus panniculitis also have associated DLE, often overlying the panniculitis lesions [30]. Lesions tend to resolve spontaneously and follow a chronic course characterized by periods of remission and exacerbation with symptoms reported for an average of 6 years. Permanent disfigurement is common.

The most common histologic finding is a lobular-predominant lymphocytic panniculitis. Twenty percent of patients will have similar features as DLE, including liquefactive degeneration of the basal layer with lymphocytic infiltration of upper dermis. There can also be a focal lymphocytic panniculitis; lymphoid nodules with germinal centers in subcutaneous fat or dermis; secondary hyaline degeneration of basement membrane, blood vessels, and adipose tissue; and calcification [22].

The differential diagnosis includes other causes of panniculitis, such as neoplastic panniculitis (subcutaneous panniculitis-like T-cell lymphoma), inflammatory panniculitis (erythema nodosum, α_1-antitrypsin deficiencies, pancreatitic panniculitis, sarcoidosis, and facticial disorders), and infectious panniculitis (bacterial, mycobacterial, or fungal). Careful clinicopathologic correlation is required to rule out lymphoma [31]. Panniculitis secondary to other connective tissue diseases such as morphea, dermatomyositis, and overlap syndrome can sometimes be clinically and histologically indistinguishable from lupus panniculitis.

Lupus Tumidus

Lupus tumidus (LT) is characterized by 2- to 4-cm indurated urticaria-like single or multiple papules,

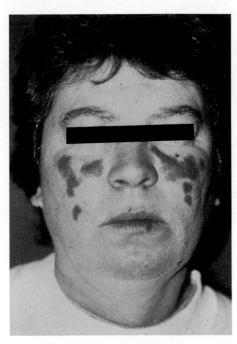

FIGURE 41.11 Face of a patient with lupus tumidus showing erythematous, edematous, infiltrative plaques on the cheeks.

nodules, or plaques with a bright reddish or violaceous color and rounded or annular configuration on sun-exposed areas such as the face, upper trunk, V area of the neck, extensor aspects of the arms, and proximal extremities (Figure 41.11). Unlike other forms of cutaneous lupus, the female-to-male ratio is usually 1:1 or there is a slight male predominance. The most important clinical feature is the absence of epidermal involvement and the swollen appearance of the lesions [32]. The main difference between LT and other forms of CCLE is the lack of scarring, hyperkeratosis, atrophy, depigmentation, and follicular plugging. In contrast to classic urticaria, lesions are nonpruritic and last longer than 24–48 h. LT may coexist with DLE, LP, or SLE. The onset of the disease is most common in summer because of sun exposure. In one series of 40 patients, 70% had photosensitivity confirmed by provocative photo testing [33].

Nonspecific LE skin lesions are uncommon in patients with LT, and few patients can be classified as SLE, reflecting that LT is a CLE variant with a more benign course [33]. When the skin activity and damage score measured by the Cutaneous Lupus Erythematosus Disease Activity and Severity Index (CLASI) was documented, LT patients displayed a significantly comparable activity to that of patients with localized DLE [34]. Permanent skin damage was significantly lower in patients with LT than in patients with DLE and SCLE [34, 35].

Histopathologic features of LT include moderate to dense superficial and deep lymphocytic inflammation in a perivascular and periadnexal distribution, with interstitial mucin deposition between collagen bundles. Involvement of the dermoepidermal junction is focal or absent [32]. DIF is usually negative.

Chilblain Lupus (Perniotic Lupus)

Chilblain lupus (CHL) is a rare form of CCLE. There is a familiar autosomal dominant form of CHL [36] that presents in children and a sporadic CHL form that usually affects middle-aged females. First symptoms typically occur during cold or damp periods and do not remit completely. CHL presents as papulo-erythematous nodules on the toes, fingers, face, and ears (Figure 41.12). At the beginning, lesions are pruritic, but then they become painful. When the lesions are located on the soles, they tend to develop necrosis rapidly.

CHL may present concurrently with DLE or other forms of CLE [37]. CHL is associated with DLE in 50% of cases and persists beyond the cold season in 34% of patients [38]. CHL lesions both clinically and pathologically overlap with DLE and classic chilblain lupus [39].

Subacute Cutaneous Lupus Erythematosus

Subacute cutaneous lupus erythematosus is a widespread, photosensitive nonscarring form of cutaneous lupus often associated with anti-Ro antibodies. SCLE can occur in patients with SLE, be induced by drugs, or may be associated with other diseases such as Sjögren's syndrome or complement deficiency. Usually, it presents in female patients in their 40s; however, the age at presentation ranges from childhood to old age. Unlike DLE, SCLE is uncommon in African-Americans.

SCLE presents as nonscarring papulosquamous or annular ill-defined noninfiltrative plaques in a characteristic photodistribution including the V area of the neck, upper trunk, extensor aspects of the shoulder,

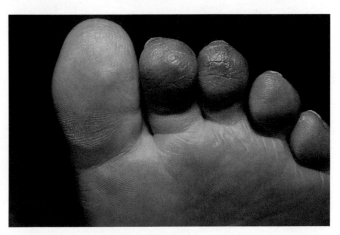

FIGURE 41.12 Papulo-erythematous nodules on the toes of a patient with chilblain lupus.

arms, forearms, and dorsal aspects of the hands and fingers, sparing the knuckles. SCLE lesions may demonstrate the Koebner phenomenon. Two classical variants of SCLE are; (1) annular, and (2) psoriasiform or papulosquamous. The annular form has more dermal involvement, a centripetal location, and sometimes an urticarial configuration (Figures 41.13 and 41.14), whereas the psoriasiform type presents with more epidermal involvement and a centrifugal distribution (Figure 41.15). Other less common morphological variants include a pityriasiform form associated with branny powdery scale [40]; an erythrodermic form with lesions associated with exfoliation affecting more than 80% of the body [41]; and a poikiloderma-like form, where the lesions are characterized by atrophy, telangiectasia, and a reticulate hyperpigmentation associated with hypopigmented lesions [42].

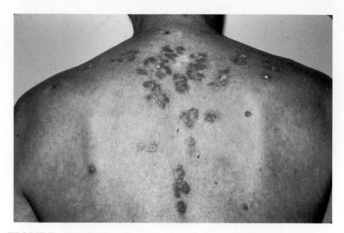

FIGURE 41.15 Papulosquamous lesions of subacute cutaneous lupus erythematosus. These lesions frequently involve the upper chest, upper back, shoulders, extensor surfaces of the arms, and dorsum of the hands and fingers and have a psoriasiform configuration.

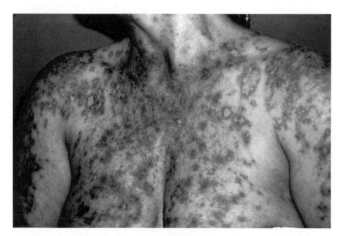

FIGURE 41.13 Chest and proximal upper extremities of a patient with annular form of subacute cutaneous lupus erythematosus.

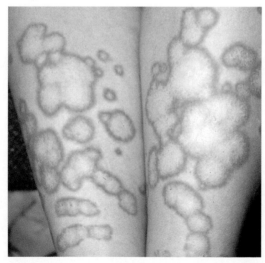

FIGURE 41.14 Arcuate erythematous plaques with discreet scaling and atrophy in a patient with annular subacute cutaneous lupus.

Early lesions of SCLE may be difficult to distinguish from polymorphous light eruption (PMLE). PMLE presents with papules, plaques, papulovesicles, or erythema multiforme0like lesions. The onset of the disease is sudden; typically the lesions occur after 30 min to several hours of exposure to sun and subside during the next 1−7 days without scarring. Interestingly, the prevalence of PMLE is increased in patients with LE, including those with SLE, SCLE, and DLE [43].

Neonatal lupus cutaneous findings are transient and resemble those of SCLE. The most common finding is "owl-eye" or "eye-mask" confluent erythema around the eyes, with erythematous and violaceous, often scaly patches [44]. The eruption is associated with maternal anti-Ro antibody, and exacerbation or initiation of lesions by exposure to sunlight has been noted (see Chapter 32).

Rowell's Syndrome

Rowell's syndrome (RS) is a rare entity characterized by the coexistence of one of the cutaneous lupus forms and erythema multiforme lesions (annular lesions with or without mucosal involvement). It is unclear if RS is a unique clinical entity or a form of SCLE. Proposed features of RS include skin changes resembling erythema multiforme, antinuclear antibody (ANA) with speckled fluorescence, anti-Ro/SSA antibodies, and a positive rheumatoid factor [45]. RS had been considered to be a type IV hypersensitivity reaction associated with certain infections, medications, and other various triggers [46].

Drug-Induced SCLE

Drug-induced cutaneous lupus is a lupus-like syndrome temporally related to continuous drug

TABLE 41.5 Drugs Associated with Cutaneous Anti-Ro LE

Antihypertensives: hydrochlorothiazides, calcium blockers, ACE inhibitors

Antifungals: terbinafine, griseofulvin

Others: statins, interferons, minocyclin, proton pump inhibitors, capacitabine

exposure (from 1 month to as long as more than a decade) that resolves after discontinuation of the drug [47]. Drug-induced SCLE has been associated with a number of drugs, including hydrochlorothiazide, calcium channel blockers, angiotensin-converting enzyme inhibitors, terbinafine, interferons, statins, proton pump inhibitors, and capacitabine [48, 49] (Table 41.5). A subset of drug-induced Ro-positive LE has been described in the literature. As opposed to patients with drug-induced SLE, these patients are antihistone negative and have striking cutaneous findings similar to SCLE lesions [49]. The mean onset of cutaneous features has been reported to be 7.27 weeks after an offending drug is introduced, and most of the symptoms improve within 6–12 weeks of discontinuation [50].

Relationship with SLE and Course of Disease

Approximately 50% of patients with SCLE fulfill at least four of the American College of Rheumatology (ACR) diagnostic criteria for SLE, often accompanied by a relatively benign form of SLE marked by musculoskeletal symptoms such as arthritis and arthralgias. Women with SCLE and anti-Ro antibodies represent the highest risk group for the development of neonatal LE in their offspring. If drug-induced SCLE is suspected, the drug must be stopped because discontinuation of the drug is commonly associated with complete resolution of the disease.

Histology

Histologic features show considerable overlap with those of DLE, and they include vacuolar alteration of the basal cell layer and a perivascular, peri-appendiceal, and subepidermal lymphocytic inflammatory cell infiltrate. In the epidermis, common findings are necrotic keratinocytes and epidermal atrophy. An abundance of mucin is often seen within the dermis. Hair follicles are generally unaffected. Identical histological changes may be found in fully developed neonatal LE or in drug-induced SCLE. A "dustlike" pattern of IgG deposition overlying epidermal basal cells just above the dermal–epidermal junction is believed to be characteristic in patients with SCLE [51], but not all studies have supported the specificity of this finding.

Acute Cutaneous Lupus Erythematosus

Acute cutaneous lupus erythematosus refers to the skin lesions in the setting of active systemic disease. ACLE may present with either a localized or a generalized distribution, or it may present in a severe acute form:

1. Localized ACLE: Known as a "butterfly" or "malar" rash, it is commonly the first manifestation of SLE, preceding by weeks or months the onset of multisystem disease. It is an erythematous rash associated with edema over the malar eminences and that extends over the bridge of the nose (Figures 41.16 and 41.17). The sharply bordered erythema is frequently mistaken by patients for sunburn. Nasolabial folds are typically spared. Smaller erythematous lesions can initially appear, which later merge and develop into papules and plaques; in addition, severe edema, scaling, erosions, and crusts may occur. This form of ACLE has a strong association with photosensitivity and always occurs in the setting of active systemic disease.

2. Generalized ACLE: Known as a "maculopapular" rash, it is an uncommon cutaneous manifestation seen in patients with SLE. This occurs in only 5–10% of SLE patients [2]. It presents as a widespread morbilliform or exanthematous eruption, consisting of multiple erythematous confluent macules and

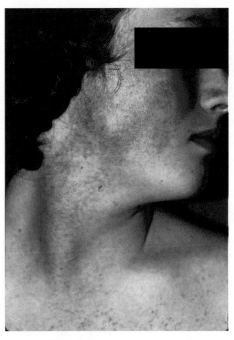

FIGURE 41.16 Localized acute cutaneous lupus erythematosus in a patient with a flare of systemic lupus erythematosus showing a photosensitive diffuse erythema of the face with a butterfly configuration.

papules that spread out symmetrically over the entire body, often also involving the palmar and plantar surfaces, as well as the backs of the hands and extensor surfaces of the fingers (Figure 41.18).

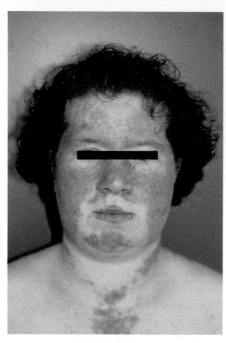

FIGURE 41.17   Acute cutaneous lupus erythematosus in a patient with a flare of systemic lupus erythematosus showing a photosensitive diffuse erythema of the face with a butterfly configuration with periorbital, perioral, and nasolabial fold sparing.

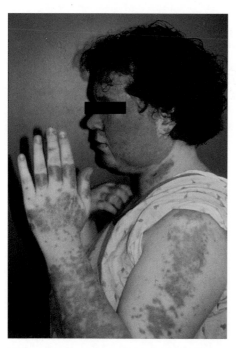

FIGURE 41.18   The same patient as shown in Figure 41.17 with a diffuse maculopapular and scaly eruption affecting the face, trunk, and upper extremities. Note the sparing of the dorsal aspects of the interphalangeal and metacarpophalangeal joints.

ACLE typically involves the interphalangeal areas and spares the knuckles. Lesions are more prominent in photoexposed areas. Generalized ACLE may present simultaneously with a malar rash. Resolution of lesions may result in post-inflammatory hyperpigmentation, especially in patients with darker skin.

3. Toxic epidermal necrolysis (TEN)-like ACLE: Lupus and other states of immune activation (connective tissue diseases, graft-versus-host disease, cancer, and HIV infection) increase the risk of developing adverse drug reactions, including TEN [52]. There are several reports of TEN-like LE with unusual gradual progression over weeks or months, absence of severe systemic involvement, and lack of any drug ingestion [53]. Usually, patients with TEN-like ACLE have a previous history of SLE and positive ANA, anti-Ro, or rheumatoid factor [53]. Many believe that this form of acute lupus is a sign of active systemic involvement. TEN-like ACLE is a result of aggressive inflammatory epidermal basal layer damage due to an increased extension of interface dermatitis.

Because most patients with ACLE have or will develop SLE, they present with significant morbidity and potential mortality secondary to their systemic manifestations. The histological changes seen in localized or generalized ACLE are often less pronounced than in other subtypes of CLE and usually show only discrete interface dermatitis with minimal vacuolization of the basement membrane. DIF is positive in almost all patients and is manifested by granular IgG and C3 along the BMZ (Figure 41.19).

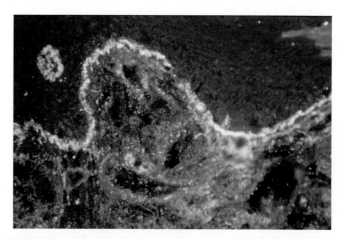

FIGURE 41.19   Direct immunofluorescence evaluation of lesional skin in a patient with acute cutaneous lupus erythematosus showing granular IgG deposition along the basement membrane zone.

LUPUS-NONSPECIFIC SKIN LESIONS

These lesions are devoid of the characteristic interface dermatitis of LE and occur in people who have been diagnosed with SLE by ACR criteria, but they can also be found in other autoimmune conditions or as an isolated finding.

Photosensitivity

The term "photosensitivity" is one of the criteria for SLE and is currently defined by the ACR as "skin rash as a result of unusual reaction to sunlight by patient history or physician observation" [7]. This is a very broad definition that may be fulfilled by other diseases, such as PMLE, photoallergic contact dermatitis, and dermatomyositis. Nevertheless, there is a clear relationship between ultraviolet (UV) radiation and the cutaneous findings in patients with CLE. Photosensitivity shows a strong association with all subtypes of cutaneous lupus. Patients with SCLE, LT, and ACLE are especially photosensitive, whereas those with DLE show less photosensitivity [33]. The presence of anti-Ro/SSA autoantibody is an additional risk factor for LE photosensitivity, thus the high rate of photosensitivity in patients with SCLE lesions. Seventy percent of SCLE patients have positive anti-Ro/SSA antibodies [54]. Both natural and artificial forms of UV radiation (unshielded fluorescent lighting and some photocopiers) may precipitate cutaneous lupus [55]. A negative history of photosensitivity does not exclude a sensitivity to sunlight because UV-induced skin lesions in CLE are characterized by a latency period of up to several weeks; therefore, the relationship between sun exposure and exacerbation of CLE may not be obvious to the patient.

Cutaneous Vasculitis

Vascular injury is an important characteristic of SLE and affects most SLE patients during their disease course. Arterioles and venules of the skin are frequently affected in SLE, although inflammation of blood vessels of any size can occur. The size of the affected blood vessels and the intensity of the inflammation determine the clinical appearance of skin vasculitis in the setting of lupus. Manifestations of cutaneous vasculitis (CV) include ulceration, gangrene, nodules, periungual infarction, splinter hemorrhages, livedo reticularis, purpura, urticaria, and bullous changes. CV is included in the Systemic Lupus Erythematosus Disease Activity Index (SLEDAI), and the score given to this item is as high as the score for neuropsychiatric manifestations [56]. CV has been associated with the development of neuropsychiatric lupus [57]. Although CV merits a high SLEDAI score, and prior reports associated CV

with active SLE and poor prognosis, isolated digital vasculitis is not invariably associated with severe systemic disease [58].

In one study of a large SLE cohort, 11% (77/670) of SLE patients had vasculitis. Cutaneous lesions were the main clinical presentation of vasculitis in 89% of the patients. A total of 86% of the patients with vasculitis had small-vessel vasculitis (associated predominantly with only cutaneous involvement) and 14% had medium-sized vessel vasculitis (associated predominantly with visceral involvement, especially of the peripheral nerves) [59]. CV typically presents as a small-vessel leukocytoclastic vasculitis, manifesting clinically as palpable purpura or occasionally as urticarial vasculitis. Small-vessel vasculitis is probably more frequent in patients with SCLE who possess anti-Ro/SS-A antibodies and have coexisting Sjögren's syndrome [60]. Urticarial vasculitis is a form of leukocytoclastic vasculitis involving the postcapillary venules commonly associated with hypocomplementemia. The lesions resemble urticaria (hives) but are painful rather than itchy, persist beyond 24 h, and leave persistent purpura or pigmentation after resolution of the acute lesions. Lesions tend to occur on the trunk and proximal extremities.

Antiphospholipid Syndrome

Patients with lupus may also have antiphospholipid syndrome (APS), a coagulation disorder resulting in a vasculopathy characterized by recurrent thrombi affecting blood vessels of all sizes. APS is associated with the presence of antiphospholipid antibodies (APAbs) (i.e., anticardiolipin antibodies, lupus anticoagulant, or both). Clinical findings correlate with arterial, venous, or capillary thrombosis. There does not appear to be a specific subset of LE associated with APAbs and APS. The prevalence of APAbs in SLE patients is increased, with 12–30% of patients testing positive for anticardiolipin antibodies and 15–34% for lupus anticoagulant [61]. A total of 50–70% of patients with SLE and APAbs will eventually develop clinical events consistent with APS. Dermatological manifestations are common, and almost 50% of patients with APS exhibit skin lesions such as livedo reticularis, ulceration, and splinter hemorrhages [62]. Skin findings are nonspecific, and they are not part of the diagnostic criteria for APS. Arterial occlusive disease presents with livedo reticularis, arterial ulcers, digital infarcts, atrophie blanche, and Degos-like lesions (Figure 41.20). Livedo reticularis is a mottling eruption accentuated by cold, partly blanchable on pressure, or warming, and it is the most common finding in patients with APS and skin manifestation, with a prevalence up to 80% [63] (Figure 41.21). Severe livedo reticularis is called livedoid vasculopathy, and

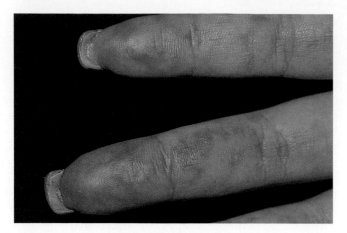

FIGURE 41.20 Mottling eruption on the palmar aspect of the digits in a patient with digital lupus vasculopathy.

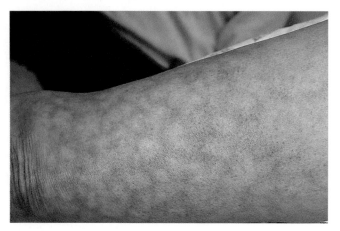

FIGURE 41.21 Livedo reticularis on the extensor upper arm clearly showing the netlike pattern of erythema.

Raynaud's Phenomenon

Raynaud's phenomenon (RP) is a functional disorder characterized by recurrent reversible vasospasm of fingers and toes often induced by exposure to cold and stress. RP occurs in 10–45% of SLE patients [66]. The classical progression consists of triphasic color changes: well-demarcated pallor of the digits leading to cyanosis, pain, and numbness, followed by a red flush upon rewarming. Among CLE patients, RP is more prevalent in those with SLE, SCLE, and generalized CCLE. The following are all predictors of LE-associated RP: the presence of periungual telangiectasia; involvement of the thumbs, toes, ears, and nose; ice-picked scars of the pulp; and the presence of high titers of ANA, anti-RNP, and nucleolar antibodies. The presence of RP in patients with SLE is associated with migraines [67] and elevation in pulmonary artery systolic pressure (pulmonary hypertension) [68].

Degos-like Lesions

Degos disease (malignant atrophic papulosis) is an occlusive arteriopathy involving small-caliber vessels characterized by the presence of typical skin lesions (white porcelain atrophic rounded macules surrounded by telangiectasia) (Figure 41.22) and visceral vascular involvement of small vessels (mainly digestive tract or central nervous system) [69]. SLE patients with Degos-like lesions follow a more benign course without the characteristic visceral involvement. Because there is a broad overlap in clinical and histological findings, Degos disease may not be a specific entity but, rather, the cutaneous lesions may represent a common endpoint to a variety of vascular insults [70].

these patients present with ulcerations that heal with stellate scaring (atrophie blanche). Venous thromboses may present as venous ulcers, lipodermatosclerosis, or superficial thrombophlebitis.

Catastrophic antiphospholipid syndrome (CAPS) is a rare and severe presentation of APLS in which patients with APLS develop disseminated intravascular coagulation-like picture with purpura fulminans [64]. CAPS occurs in less than 1% of APS cases. However, it is often associated with a high rate of mortality (i.e., 50%), usually as a consequence of multiorgan failure. The diagnosis is made by evidence of thrombosis in three or more organs, the histological finding of small-vessel occlusion in at least one organ, the rapid development of clinical manifestations, and a laboratory confirmation for the presence of APAbs. Skin involvement, such as livedo reticularis, purpura, and skin necrosis, occurs in 50.2% of patients [65].

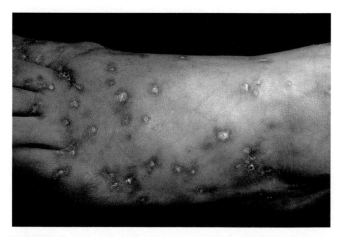

FIGURE 41.22 Degos-like lesions on the dorsal aspects of the foot of a patient with SSA antibodies. The porcelain-like central scars are clearly seen.

Nonscarring Alopecia

Patients with SLE may experience a diffuse, nonscarring alopecia associated with active disease. This type of alopecia is known as telogen effluvium, is secondary to the severe catabolic effects of the disease flare, and is the most common nonspecific skin manifestation in patients with SLE, occurring in approximately 60% of cases [71]. This alopecia can be acute, occurring 2 or 3 months after a flare of lupus, or can be chronically associated with continuous disease activity. The term "lupus hair" is used to name a nonscarring alopecia related to telogen effluvium, where a frond of miniaturized hairs are noted at the periphery of the scalp. Alopecia areata presents as a localized area of nonscarring hair loss and is more common in SLE [72].

Oral Ulcers

Oral and nasopharyngeal ulcerations are one of the cutaneous criteria proposed by the ACR for identifying patients with SLE. Oral ulcers occur more frequently in SLE than in CCLE. Activity of systemic disease is significantly increased during the time of mucosal ulceration [73]. Although these lesions may show LE-specific histopathologic changes, they are often nonspecific. Erythema, petechiae, painless ulceration of the hard palate, mucosal erosions, hemorrhage, and gingivitis may occur during acute systemic episodes. LE-related oral ulcers most commonly occur on the hard palate and are not painful. Non-LE oral ulcers, such as recurrent aphthous stomatitis (RAS), are usually quite painful. RAS is an inflammatory disease of unknown origin characterized by recurrent and painful ulcers in the oral mucosa that vary from a few millimeters (minor) to centimeters (major). RAS affects 20% of the general population and is the most common form of oral ulceration. RAS in the setting of SLE is a common feature. These ulcers have a variable degree of pain, and the size varies from a few millimeters (minor) to centimeters (major). RAS can be differentiated from LE-related ulcers because the latter are asymptomatic, occur usually in the hard palate, and improve with the treatment of other cutaneous or systemic manifestations of LE. RAS, a clinical syndrome distinct from lupus mouth ulcers, should in our opinion not be considered a criterion for the diagnosis of SLE or for assessing SLE activity.

Nail Changes

Nail-fold capillaroscopy of patients with SLE can reveal glomerularization of capillaries [74]. A red lunula is common and is characterized by replacement of the white lunula by red—pink or dusky erythematous discoloration. This feature has been associated with chilblain LE, and it occurs together with periungual erythema and painful nail [75].

Papulonodular Mucinosis

Papulonodular mucinosis (PNM) is a form of cutaneous mucinosis characterized by aberrant accumulation of glycosaminoglycans between collagen bundles. Mucin deposition is a common histological finding in patients with CLE, but it is unusual for the mucin to be present in such a quantity to produce detectable skin lesions. The lesions are asymptomatic erythematosus to skin-colored papules and nodules that present on the trunk, arm, head, and neck. It has been reported that UV light can aggravate these papulonodular lesions. These lesions may be found in patients with DLE, SCLE, or SLE. There is a higher frequency of renal disease in patients with SLE and PNM [76].

Bullous Lesions in LE

Blistering in SLE has been divided into three categories:

1. Subepidermal bullae may develop in SCLE and ACLE lesions as a reflection of hyperacute injury to the epidermal basal layer that can occur as an extension of LE-specific histopathologic injury (TEN-like ACLE and Rowell's syndrome).
2. SLE has been associated with the co-occurrence of autoimmune blistering diseases such as dermatitis herpetiformis, bullous pemphigoid, pemphigus vulgaris, pemphigus foliaceous, paraneoplastic pemphigus, epidermolysis bullosa acquisita, pseudoporphia, and IgA disease [77].
3. There is a specific subgroup defined by Gammon and Briggaman termed bullous SLE (BSLE) [78].

Bullous SLE

This form of bullous LE has distinctive clinical, histological, and immunopathological features. Because it does not share the classical histopathology of LE, it is considered an LE-nonspecific lesion. BSLE presents on the face, neck, and upper trunk, but it may also be more widespread. The bullous lesions appear on normal or erythematous skin, tend to be tense, and may rupture leaving erosions (Figure 41.23). Half of these patients may have scarring mucositis affecting the mouth and other mucosal membranes. Histologically, the most common feature is the finding of a subepidermal blister with neutrophil inflammation. In DIF studies, as opposed to the granular immunostaining seen in the LE interface dermatitis, these BSLE patients have a thick

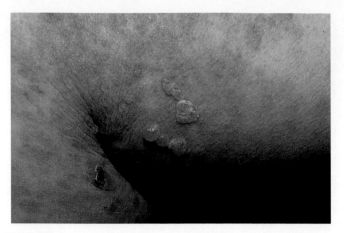

FIGURE 41.23 Lateral aspect of the knee of a patient with SLE and with bullous SLE lesions presenting as grouped tense and ruptured blisters and erosions.

linear IgG and sometimes IgA and C3 along the BMZ [79]. The circulating anti-BMZ antibodies recognize the epidermolysis bullosa acquisita autoantigen, represented by a 290-kDa protein corresponding with type VII collagen.

ASSOCIATED SKIN CONDITIONS

Many skin conditions are anecdotally reported to be associated with LE. Repeatedly reported associations include the neutrophilic dermatoses, as well as palisaded neutrophilic and granulomatous dermatitis, erythema elevatum diutinum, and Kikuchi–Fujimoto disease. We describe more fully the first two conditions because they are more common.

Neutrophilic Dermatosis

Pyoderma Gangrenosum and Sweet's Syndrome

Neutrophilic dermatoses are characterized by significant neutrophilic infiltrates of the dermis and fat without overt vasculitis; however, the presence of vasculitis does not exclude the diagnosis. These are reactive disorders in which the underlying disease is often a chronic inflammatory disease, autoimmune disease, or neoplasia. The most common associated disease is inflammatory bowel disease, although patients with connective tissue diseases and arthritis including SLE occasionally present with neutrophilic dermatosis [80].

Pyoderma gangrenosum is characterized by painful nodular, bullous, or pustular lesions that eventually ulcerate. This condition may be the presenting feature of SLE [81]. Lesions expand rapidly, developing a clean granular base and a raised dusky to violaceous border. The lesions typically affect lower extremities. The

clinical picture demonstrates a distinct absence of fever, or regional lymphadenopathy, unless the lesions become secondarily infected.

Sweet's syndrome (SS), also termed acute febrile neutrophilic dermatosis, is a reactive process characterized by the abrupt onset of tender, red-to-purple papules and also nodules that coalesce to form plaques. The plaques usually occur on the upper extremities, face, or neck and are typically accompanied by fever and peripheral neutrophilia. SS has been reported in association with or as a presenting manifestation of SLE in patients with drug-induced lupus and in neonatal lupus [82, 83].

Palisaded Neutrophilic and Granulomatous Dermatitis

Palisaded neutrophilic and granulomatous dermatitis (PNGD), also called Churg–Strauss granuloma, is an uncommon condition that has been reported with several different names in association with various disorders, most commonly with rheumatoid arthritis [84]. The clinical presentation of PNGD varies from asymptomatic papules to nodules to annular plaques. Characteristically, the lesions appear symmetrically, often on the lateral aspect of the trunk, abdomen, or extensor surfaces of the extremities, and they may occur in association with SLE.

LUPUS CRITERIA: MALAR RASH, PHOTOSENSITIVITY, ORAL ULCERS, AND DISCOID

Four of the 11 criteria for SLE are dermatologic [7]. These include malar rash, photosensitivity, oral ulcers, and discoid lesions. There is some overlap of malar rash and photosensitivity, and biopsy of oral ulcers in CLE patients frequently reveals findings consistent with discoid lupus. It is thus possible to make a diagnosis of SLE purely on the basis of skin findings, some of which may have overlap, as outlined previously. Thus, it has been argued that a more accurate diagnosis of SLE might include fewer skin criteria [9].

In addition, there are difficulties with misdiagnosis related to the lupus criteria, and thus an awareness of the potential for mimickers of CLE is critical. Rosacea may give a malar rash, polymorphous light eruption is a common cause of photosensitivity, aphthous ulcers (RAS, as described previously) may simulate ulcer of LE, cutaneous T-cell lymphoma or tinea corporis may mimic subacute cutaneous lupus, and hypertrophic LE may mimic squamous cell carcinoma of the skin. In addition, patients with CLE frequently get unrelated skin lesions, and these require differentiation from CLE. Amyopathic dermatomyositis can give all of the skin manifestations included in the lupus criteria, in addition to a positive antinuclear antibody. Skin biopsy of dermatomyositis is identical to CLE [85]. Defining the characteristic skin findings seen in dermatomyositis, such as targeting of erythematous papules over joints on the hands (Gottron's papule), is key. There are a large number of mimickers of CLE, and thus a biopsy and clinical–pathological correlation to confirm a clinical diagnosis of CLE is often optimal.

EPIDEMIOLOGY/NATURAL HISTORY

One study examined the incidence of cutaneous lupus during a 40-year period in Olmsted County, Minnesota, and found an incidence rate of CLE of 4.3 per 100,000, with 19% progressing to SLE during a 20-year period. Most cases were DLE (2.99 per 100,000), with approximately one-third of these having generalized DLE [86]. The incidence of SLE in Sweden was determined to be 4.8 per 100,000 people per year—somewhat more than that seen in Olmsted County [87, 88]. These studies, although done on relatively homogeneous populations, suggest nearly an equal incidence of SLE and CLE. In addition, skin findings are the second most frequent manifestation of SLE [89, 90].

In a study in Sweden, the incidence of Ro/SSA-positive SCLE was estimated to be 0.7 cases per 100,000 people per year [91], remarkably similar to the 0.63 cases per 100,000 observed in Olmsted County, Minnesota [86]. SCLE predominates in Caucasian females in their 40s, with 80% of SCLE cases typically occurring in females [34, 91].

Patients with generalized DLE have an approximately 20% risk of developing SLE, whereas those with localized DLE with lesions localized above the neck have an approximately 5% risk of SLE. Fifty percent of patients with SCLE meet criteria for SCLE, but often without severe systemic disease. A systematic study of anti-Ro/SSA-positive patients with skin manifestations demonstrated that these patients do develop new autoimmune diseases and are potentially at risk of drug-induced SCLE [54]. Lupus-nonspecific skin lesions are more frequently seen in patients with increased SLE activity compared to those with only LE-specific lesions and those with both kinds of lesions [10]. Overall, patients with SLE tend to be younger and SCLE patients tend to be older than patients with other forms of CLE [92].

Inflammatory skin disease is found in up to 70% of patients with SLE. The development of a web-based database for cutaneous lupus has allowed systematic characterization of the disease severity and quality of life (QOL) of cutaneous lupus. The most therapeutically resistant patients are more likely to be smokers,

confirming this risk factor for refractory disease [34, 93, 94]. Several studies have also found that smoking is a risk factor for developing CLE [20, 92]. Ultraviolet light and medications, along with a genetic predisposition, are prominent triggers of CLE. Nomenclature is confusing here because SCLE is highly UV sensitive [95–97].

Pediatric DLE is rarer but potentially more frequently associated with SLE, particularly when generalized [98].

PATHOGENESIS OF CUTANEOUS LUPUS ERYTHEMATOSUS

The phenotypic expression of CLE is influenced by environmental triggers and genetic predisposition. Skin and systemic disease do not always follow the same course because skin flares usually independently from systemic disease, a patient with SLE may or may not have any cutaneous involvement, or a particular treatment may be useful for the skin but not for systemic involvement. This suggests that there are unique features to the pathophysiology of skin disease in LE. The fact that LE, including CLE, is characterized by the production of autoantibodies and immune complexes and the observation that LE can be treated with immunosuppressive agents support an autoimmune disorder hypothesis. In the skin, antigen-presenting cells, such as immature dendritic cells (Langerhans cells) and macrophages, are thought to actively prevent the initiation of immune reactions by processing apoptotic cells and to actively promote tolerance by stimulating tolerogenic regulatory T cells [99]. Thus, the removal of Langerhans cells or of skin resident regulatory T cells (Treg) [100] may induce autoimmunity. Apoptotic cells have been detected in the skin of patients with CLE after UV irradiation and prior to lesion formation. It is suggested that impaired clearance of apoptotic cells secondary to an increase in the production of apoptotic cells, an increase in the amount of inflammatory cytokines, or a deficiency in the proteins involved in clearance may contribute to the autoimmune response [101]. As a result, apoptotic debris may be processed by antigen-presenting cells, stimulating $CD4^+$ and $CD8^+$ cells as well as antibody production [102]. A decrease of T $CD4^+CD25^+$ Treg has been observed at the site of inflammation in the skin of different subtypes of CLE and not in other inflammatory disorders, such as atopic dermatitis, psoriasis, and lichen planus. Circulating Treg cells in CLE patients were normal, suggesting an organ-specific abnormality [103].

The pathology in virtually all forms of CLE is characterized by immune cell infiltration into the dermis, culminating in keratinocyte damage. This pattern has been termed the *lichenoid tissue reaction* or *interface dermatitis* [104]. CLE lesions show a typical Th1 profile, with the predominance of a T lymphocyte infiltrate and the presence of the inflammatory cytokines interferon (IFN)-γ, interleukin (IL)-1, and tumor necrosis factor-α (TNF-α). Ultraviolet light exposure is strongly associated with lesion induction, and this may relate to modulation of antigen expression as well as induction of inflammation. Both UVA (320–400 nm) and UVB (290–320 nm) induce apoptosis in the epidermis, and it has been suggested that this promotes the expression of neoantigens that could be targeted by the immune system [105]. Autoantigens such as Ro/SSA, La/SSB, and calreticulin can be externalized from the inside of epidermal keratinocytes to the cell surface after UVB exposure, inducing autoimmunity [106]. Exposure of keratinocytes to UVB results in the synthesis of many cytokines, such as TNF-α, IL-6, IL-8, and IL-10. These cytokines cause upregulation of intercellular adhesion molecule-1 on keratinocytes and increase homing of leukocytes to the skin. TNF-α induces apoptosis and translocation of Ro/SSA and La/SSB autoantigens to the surface of keratinocytes [107]. This may allow recognition of these antigens by autoantibodies.

Genes located within the major histocompatibility complex on the short arm of human chromosome 6 regulate the immune response in CLE. Thus, there is a higher risk of developing DLE with certain gene combinations, such as HLA-Cw7, DR3, DQw1 and HLA B7, HLA-Cw7, and DR3 [108]. The human leukocyte antigen (HLA) B8, DR3, DRw52 haplotype has been associated with SCLE and anti-Ro antibody production [109].

Toll-like receptors (TLRs) bridge innate and adaptive immune responses. TLRs play an important role in the development of the autoimmune response. TLR-9 and TLR-7 ligands such as self–DNA and self–RNA, respectively, stimulate plasmacytoid dendritic cells when complexed to antimicrobial peptides, carrier proteins, and antibodies [110]. This may explain the abundant interferon-dependent proteins detected in lesional skin. IFN-α then promotes the development of lichenoid tissue reactions by increasing the expression of CXCR3 and other chemokines that attract Th1 lymphocytes [111].

Some immunodeficiency states have an increased incidence of autoimmune disorders, including CLE. Patients with C1q deficiency develop LE-like photosensitive eruptions. Almost every patient with homozygous genetic deficiency of C1q will develop SLE-like syndrome with variable skin and mucosal lesions and sometimes glomerulonephritis [112]. This paradox may be explained by altered immunoregulatory mechanisms and defects in apoptotic cell clearance in immunodeficiency states. For example, C1q binds apoptotic cells and helps in their clearance. The respective roles of apoptotic cell death, dead cell clearance, defects in

regulatory T-cell function, altered antigen-presenting function, and interferon regulation in the genesis of CLE are still a matter of ongoing investigation.

MEASUREMENT OF OUTCOMES

A number of clinical assessment instruments include skin as one of several organs evaluated in terms of LE activity. Most lack sensitivity for baseline assessment or responsiveness [19, 113]. A CLE outcome measure, the CLASI, has been developed and validated. The CLASI provides a quantitative measure of the skin-specific burden of disease, which allows for standardized assessments of disease progression. Such a standardized approach facilitates the organization of clinical trials, analysis of results, and comparisons between studies. The CLASI is a simple, single-page tool that clearly differentiates between activity and damage, providing two independent summary scores. This separation is critical because activity and damage embody two different aspects of the disease. The activity score reflects ongoing inflammation, which has the potential to decrease with treatment. The damage score represents the aftermath of inflammation, which cannot itself be treated, only prevented. As such, the activity score is most appropriate in short-term drug studies, whereas the damage score is helpful in long-term preventative studies.

In the CLASI, each part of the body is listed separately, from the scalp to the feet; in addition, there are sections focusing on mucous membrane involvement and alopecia. For the activity score, points are given for the presence of erythema, scale, mucous membrane lesions, recent hair loss, and inflammatory alopecia. For the damage score, points are given for the presence of dyspigmentation, scarring, and scarring alopecia. For both activity and damage, higher scores are awarded for more severe manifestations. Scores for each area are assigned based on the most severe lesion within the area of interest. Of note, affected body parts are weighted equally regardless of surface area and number of lesions present. Separate composite scores for activity and damage are calculated by simply summing the individual component scores. This score was developed for simple reliable and responsive measurement in the context of clinical trials, and it works well for DLE, SCLE, and tumid lupus lesions, which are typically erythematous. The CLASI is not designed to capture active lupus panniculitis, nonerythematous tumid LE, and bullous LE lesions because they are rare, not likely to be included in a controlled clinical trial, and would significantly increase the complexity of the CLASI.

Studies included inter-rater and intra-rater validation, extension of validation to rheumatology, and assessments of responsiveness during an intervention trial [114–116]. This outcome measure has been used in several trials and a prospective database of CLE [117–119].

There are several skin-specific QOL measurements, including the Skindex and the Dermatology Life Quality Index [120, 121]. Patients with CLE frequently have significant impairment of the emotion component of their QOL [122].

TREATMENT

Patients with new-onset SCLE should be screened for possible exacerbating medications. Patients should be encouraged to stop smoking because this can increase the risk of CLE and decrease the effectiveness of therapy. Patients with lupus should avoid the sun as much as possible and use sunscreens and appropriate clothing [123–125]. Sunscreens with sun protection factor numbers higher than 60 provide the best UVB block. In addition, optimal sunscreens also contain avobenzone or Mexoryl XL to block the longer UVA wavelengths.

For individual lesions, topical therapy is appropriate. Topical medications include corticosteroids and calcineurin inhibitors, which include tacrolimus, and pimecrolimus. Facial lesions should be treated with 1 or 2.5% hydrocortisone, and stronger fluorinated steroids should be avoided on the face except for very short periods of treatment of inflammatory lesions. More potent topical steroids should be reserved for use on the trunk or extremities because they can cause thinning of the skin and telangiectasias. Patients often prefer creams, although ointments may be more effective. There are corticosteroid solutions and foam preparations for the scalp. Intralesional steroids such as triamcinolone acetonide (3 mg/ml) can be helpful for individual lesions, particularly on the scalp, but care is needed to minimize the risk of atrophy and dyspigmentation. The calcineurin inhibitors can be helpful adjunctive therapy, particularly on thinner face skin [126, 127]. Although licensed and extensively used in children, patients do need to be informed of a black box warning related to toxicity in animal studies that has not been observed in humans [128]. Individual patients have been treated with liquid nitrogen [128]. Two open prospective studies suggested that the pulsed dye laser can be effective in treating DLE [119, 129].

For patients with more extensive or unresponsive disease, first-line systemic therapy is the antimalarials. The initial treatment is usually with hydroxychloroquine at a dose of <6.5 mg/kg ideal body weight/day. For therapeutically resistant patients, there is sometimes a benefit to adding quinacrine at a dose of 50–100 mg/day [130, 131]. There is no known eye toxicity with quinacrine. If patients do not respond to the combination of hydroxychloroquine and quinacrine after 2 or 3 months, then there may be additional benefit to switching from hydroxychloroquine to chloroquine at a dose of <3.5 mg/kg/day while maintaining treatment with quinacrine [132]. There are a number of potential cutaneous reactions to antimalarials, including hyperpigmentation of skin, nails, and mucous membranes; urticarial eruptions; drug exanthems; lichenoid drug reaction; hypopigmentation of the hair; and alopecia.

Other anti-inflammatory therapies have been used for CLE, but there have been very few controlled trials to demonstrate efficacy [133]. Dapsone can be useful for vasculitic lesions, urticarial vasculitis, and oral lesions, and it is usually used at a dose of 50–150 mg/day [134]. Dapsone is less effective for DLE or SCLE. The G6PD is checked and if normal or present, patients are started at 50 mg/day. The dapsone dose is titrated up by 25 mg per week, checking complete blood count and liver function tests prior to the increase, as long as the next higher dose is necessary because of continued disease activity. Oral retinoids, including isotretinoin and acitretin, have been used for CLE [135]. The response is often transient, and it can provide an approach for resistant patients with hypertrophic LE. Oral retinoids require careful monitoring to prevent pregnancy and to monitor for hepatotoxicity, hyperlipidemia, depression, and pseudotumor cerebri.

Patients who do not respond to topicals or antimalarial therapy often require treatment with immunosuppressives or thalidomide. Oral glucocorticoids may be used for brief periods of time while waiting for onset of action of more effective therapies, but their side effects relative to their efficacy make them not optimal for treatment of most lupus-specific skin lesions.

Immunosuppressives

Several immunosuppressives, including methotrexate, mycophenolate mofetil, and azathioprine, have been reported to help patients with refractory CLE [118, 136–139]. Methotrexate can be administered orally or subcutaneously for patients who do not tolerate oral administration. Folic acid is normally added to methotrexate to minimize toxicity. If cyclophosphamide is required for systemic disease, it will often improve the skin concurrently. There are no controlled trials on therapeutics of immunosuppressives.

Thalidomide is an effective therapy for refractory CLE [140, 141]. It is usually started at 50–100 mg/day. If patients respond, the maintenance dose can be as low as 50 mg/week. There are risks of teratogenicity, thrombosis, peripheral neuropathy, sleepiness, headaches, weight gain, amenorrhea that is typically reversible, and dizziness [142–144]. Patients, pharmacists, and physicians must comply with a company-administered program to prevent pregnancy, including monthly surveys and use of two forms of highly effective birth control. Due to side effects, thalidomide is a difficult drug to administer, particularly the peripheral neuropathy that occurs in up to 50% of patients [145]. There is a high incidence of sensory axonal neuropathy, the total dose of thalidomide that can cause this is relatively low, and recovery in one study occurred in 25% based on sural nerve sensory action potentials [146]. The neuropathy can persist or even progress off therapy. Patients with a history of arterial or venous thrombosis should in most cases not receive thalidomide, and patients should be maintained on antimalarials or low-dose aspirin to minimize the risk of clotting while on therapy [147].

Systemic glucocorticoids should be used only infrequently for CLE, and their use should be viewed as a bridge to onset of action of safer and more effective therapies. Low- to medium-dose systemic corticosteroids do not work well for DLE, work as an intermediate therapy for severe SCLE, and are more effective for vasculitis.

Other therapies have been tried for CLE. Efalizumab, which blocks CD11a and has effects of lymphocyte migration and antigen presentation, showed potential efficacy in CLE [148]. Unfortunately, this drug was subsequently withdrawn from the market because of increased risk of progressive multifocal encephalopathy. Chimeric anti-CD4 antibody infusion in five patients with severe, refractory CLE induced long-lasting improvement [149]. IVIG has been used for resistant CLE [150]. In general, TNF-α inhibitors have been shown to both treat and exacerbate CLE and in general cannot be recommended for these patients [151, 152]. An increase of interferon when there is therapeutic blockade of TNF-α has been demonstrated [153]. Although successful treatment with interferon or imiquimod, which induces Toll receptor 7, has been reported, there have also been numerous reports of exacerbation of systemic disease or localized CLE at sites of injection of interferon [25, 154–157]. Thus, interferon and imiquimod should normally be avoided in patients with CLE.

There may be a role for rituximab, but this has been unexplored for cutaneous LE. Because of its effect on B cells, and thus on specific subsets of CLE that are more clearly autoantibody mediated, very refractory patients with either bullous LE or SCLE may benefit from this approach [158]. One randomized study showed efficacy of R-salbutamol cream in CLE. This drug likely works by binding to and activating β_2 adrenoceptors, inhibiting pro-inflammatory cytokine release from inflammatory cells, as well as inhibiting superoxide production and peroxidase release from stimulated human granulocytes [159].

Acknowledgments

This work was supported in part by a Merit Review Grant from the Department of Veterans Health Administration, Office of Research Development, Biomedical Laboratory Research and Development and by the National Institutes of Health (NIH K24-AR 02207) to VPW. JPD is a Senior Scholar of the Michael Smith Foundation for Health Research and the Child and Family Research Institute. CVK is supported by grants from Pontificia Universidad Católica de Chile School of Medicine and the Chile Bicentennial Scholarship (Becas Chile).

References

[1] S. Fatovic-Ferencic, K. Holubar, Early history and iconography of lupus erythematosus, Clin Dermatol 22 (2004) 100–104.

[2] C. Cardinali, M. Caproni, E. Bernacchi, L. Amato, P. Fabbri, The spectrum of cutaneous manifestations in lupus erythematosus—The Italian experience, Lupus 9 (2000) 417–423.

[3] V.C. Fabre, S. Lear, M. Reichlin, S.J. Hodge, J.P. Callen, Twenty percent of biopsy specimens from sun-exposed skin of normal young adults demonstrate positive immunofluorescence, Arch Dermatol 127 (1991) 1006.

[4] J.N. Gilliam, R.D. Sontheimer, Distinctive cutaneous subsets in the spectrum of lupus erythematosus, J Am Acad Dermatol 4 (1981) 471.

[5] R.D. Sontheimer, Subacute cutaneous lupus erythematosus: A decade's perspective. Med Clin North Am 73 (1989) 1073.

[6] R.D. Sontheimer, Subacute cutaneous lupus erythematosus: 25-year evolution of a prototypic subset (subphenotype) of lupus erythematosus defined by characteristic cutaneous, pathological, immunological, and genetic findings, Autoimmun Rev 4 (2005) 253–263.

[7] E.M. Tan, A.S. Cohen, J.F. Fries, A.T. Masi, D.J. McShane, N.F. Rothfield, et al., The 1982 revised criteria for the classification of systemic lupus erythematosus, Arthritis Rheum 25 (1982) 1271.

[8] J.P. Callen, J. Klein, Subacute cutaneous lupus erythematosus. Clinical, serologic, immunogenetic, and therapeutic considerations in seventy-two patients, Arthritis Rheum 31 (1988) 1007.

[9] J.A. Albrecht, I.M. Braverman, J.P. Callen, M.K. Connolly, M. Costner, D. Fivenson, et al., Dermatology position paper on the revision of the 1982 ACR criteria for SLE, Lupus 13 (2004) 839.

[10] R.D. Zecevic, D. Vojvodic, B. Ristic, M.D. Pavlovic, D. Stefanovic, D. Karadaglic, Skin lesions—An indicator of disease activity in systemic lupus erythematosus? Lupus 10 (2001) 364–367.

[11] M. Petri, Lupus in Baltimore: Evidence-based "clinical pearls" from the Hopkins Lupus Cohort, Lupus 14 (2005) 970–973.

[12] H. Ueki, Koebner phenomenon in lupus erythematosus with special consideration of clinical findings, Autoimmunity Rev 4 (2005) 219–223.

[13] C.L. Wilson, S.M. Burge, D. Dean, R.P. Dawber, Scarring alopecia in discoid lupus erythematosus, Br J Dermatol 126 (1992) 307–314.

[14] S.M. Burge, P.A. Frith, R.P. Juniper, F. Wojnarowska, Mucosal involvement in systemic and chronic cutaneous lupus erythematosus, Br J Dermatol 121 (1989) 727–741.

[15] M. Munoz-Corcuera, G. Esparza-Gomez, M.A. Gonzalez-Moles, A. Bascones-Martinez, Oral ulcers: Clinical aspects. A tool for dermatologists. Part II. Chronic ulcers. Clin Exp Dermatol 34 (2009) 456–461.

[16] J.P. Callen, Chronic cutaneous lupus erythematosus. Clinical, laboratory, therapeutic, and prognostic examination of 62 patients, Arch Dermatol 118 (1982) 412–416.

[17] B. Tebbe, U. Mansmann, U. Wollina, P. Auer-Grumbach, A. Licht-Mbalyohere, M. Arensmeier, C.E. Orfanos, Markers in cutaneous lupus erythematosus indicating systemic involvement. A multicenter study on 296 patients. Acta Derm Venereol 77 (1997) 305–308.

[18] V.P. Werth, Clinical manifestations of cutaneous lupus erythematosus, Autoimmunity Rev 4 (2005) 296–302.

[19] A. Parodi, C. Massone, M. Cacciapuoti, M.G. Aragone, P. Bondavalli, G. Cattarini, et al., Measuring the activity of the disease in patients with cutaneous lupus erythematosus, Br J Dermatol 142 (2000) 457.

[20] H.A. Miot, L.D. Bartoli Miot, G.R. Haddad, Association between discoid lupus erythematosus and cigarette smoking, Dermatology 211 (2005) 118–122.

[21] T. Yoshimasu, A. Hiroi, K. Uede, F. Furukawa, Discoid lupus erythematosus (DLE)-like lesion induced by uracil-tegafur (UFT), Eur J Dermatol 11 (2001) 54–57.

[22] M. Baltaci, P. Fritsch, Histologic features of cutaneous lupus erythematosus, Autoimmunity Rev 8 (2009) 467–473.

[23] C. Perniciaro, H.W. Randle, H.O. Perry, Hypertrophic discoid lupus erythematosus resembling squamous cell carcinoma, Dermatol Surg 21 (1995) 255–257.

[24] H.C. Nousari, A. Kimyai-Asadi, T.T. Provost, Generalized lupus erythematosus profundus in a patient with genetic partial deficiency of C4, J Am Acad Dermatol 41 (1999) 362–364.

[25] H.C. Nousari, A. Kimyai-Asadi, F.A. Tausk, Subacute cutaneous lupus erythematosus associated with interferon beta-1a [letter], Lancet 352 (1998) 1825.

[26] J. Fraga, A. Garcia-Diez, Lupus erythematosus panniculitis, Dermatol Clin 26 (2008) 453–463.

[27] A. Choffray, L. Pinquier, H. Bachelez, Exacerbation of lupus panniculitis following anti-hepatitis-B vaccination, Dermatology (Basel) 215 (2007) 152–154.

[28] T. Gono, M. Matsuda, Y. Shimojima, K. Kaneko, H. Murata, S. Ikeda, Lupus erythematosus profundus (lupus panniculitis) induced by interferon-beta in a multiple sclerosis patient, J Clin Neurosci 14 (2007) 997–1000.

[29] R. Fernandez-Torres, F. Sacristan, J. Del Pozo, W. Martinez, L. Albaina, M. Mazaira, et al., Lupus mastitis, a mimicker of erysipelatoides breast carcinoma, J Am Acad Dermatol 60 (2009) 1074–1076.

[30] D.L. Tuffanelli, Lupus erythematosus panniculitis (profundus), Arch Dermatol 103 (1971) 231–242.

[31] L.B. Pincus, P.E. LeBoit, T.H. McCalmont, R. Ricci, C. Buzio, L.P. Fox, et al., Subcutaneous panniculitis-like T-cell lymphoma with overlapping clinicopathologic features of lupus erythematosus: Coexistence of 2 entities? Am J Dermatopathol 31 (2009) 520–526.

[32] A. Kuhn, D. Bein, G. Bonsmann, The 100th anniversary of lupus erythematosus tumidus, Autoimmunity Rev 8 (2009) 441–448.

[33] A. Kuhn, D. Richter-Hintz, C. Oslislo, T. Ruzicka, M. Megahed, P. Lehmann, Lupus erythematosus tumidus—A neglected subset of cutaneous lupus erythematosus: Report of 40 cases, Arch Dermatol 136 (2000) 1033–1041.

[34] S. Moghadam-Kia, K. Chilek, E. Gaines, M. Costner, M.E. Rose, J. Okawa, et al., Cross-sectional analysis of a collaborative Web-based database for lupus erythematosus-associated skin lesions: Prospective enrollment of 114 patients, Arch Dermatol 145 (2009) 255–260.

[35] V. Schmitt, A.M. Meuth, S. Amler, E. Kuehn, M. Haust, G. Messer, et al., Lupus erythematosus tumidus is a separate subtype of cutaneous lupus erythematosus, Br J Dermatol 162 (2010) 64–73.

[36] G. Rice, W.G. Newman, J. Dean, T. Patrick, R. Parmar, K. Flintoff, et al., Heterozygous mutations in TREX1 cause familial chilblain lupus and dominant Aicardi—Goutieres syndrome, Am J Hum Genet 80 (2007) 811–815.

[37] C.M. Hedrich, B. Fiebig, F.H. Hauck, S. Sallmann, G. Hahn, C. Pfeiffer, et al., Chilblain lupus erythematosus—A review of literature, Clin Rheumatol 27 (2008) 949–954.

[38] J.D. Bouaziz, S. Barete, F. Le Pelletier, Z. Amoura, J.C. Piette, C. Frances, Cutaneous lesions of the digits in systemic lupus erythematosus: 50 cases, Lupus 16 (2007) 163–167.

[39] B. Cribier, N. Djeridi, B. Peltre, E. Grosshans, A histologic and immunohistochemical study of chilblains, J Am Acad Dermatol 45 (2001) 924–929.

[40] M. Caproni, C. Cardinali, E. Salvatore, P. Fabbri, Subacute cutaneous lupus erythematosus with pityriasis-like cutaneous manifestations, Int J Dermatol 40 (2001) 59–62.

[41] M. Kalavala, V. Shah, S. Blackford, Subacute cutaneous lupus erythematosus presenting as erythroderma, Clin Exp Dermatol 32 (2007) 388–390.

[42] A.V. Marzano, M. Facchetti, E. Alessi, Poikilodermatous subacute cutaneous lupus erythematosus, Dermatology (Basel) 207 (2003) 285–290.

[43] F. Nyberg, T. Hasan, P. Puska, E. Stephansson, M. Hakkinen, A. Ranki, et al., Occurrence of polymorphous light eruption in lupus erythematosus, Br J Dermatol 136 (1997) 217–221.

[44] L.A. Lee, The clinical spectrum of neonatal lupus, Arch Dermatol Res 301 (2009) 107–110.

[45] N.C. Zeitouni, D. Funaro, R.A. Cloutier, E. Gagne, J. Claveau, Redefining Rowell's syndrome, Br J Dermatol 142 (2000) 343–346.

[46] A. Kacalak-Rzepka, M. Kiedrowicz, S. Bielecka-Grzela, V. Ratajczak-Stefanska, R. Maleszka, D. Mikulska, Rowell's syndrome in the course of treatment with sodium valproate: A case report and review of the literature data. Clin Exp Dermatol 34 (2009) 702–704.

[47] C.D. Vedove, M. Del Giglio, D. Schena, G. Girolomoni, Drug-induced lupus erythematosus, Arch Dermatol Res 301 (2009) 99–105.

[48] U. Floristan, R.A. Feltes, E. Sendagorta, M. Feito-Rodriguez, P. Ramirez-Marin, C. Vidaurrazaga, et al., Subacute cutaneous lupus erythematosus induced by capecitabine, Clin Exp Dermatol 34 (2009) e328–e329.

[49] A. Marzano, P. Vezzoli, C. Crosti, Drug-induced lupus: An update on its dermatologic aspects, Lupus 18 (2009) 935–940.

[50] M. Srivastava, A. Rencic, G. Diglio, H. Santana, P. Bonitz, R. Watson, et al., Drug-induced, Ro/SSA-positive cutaneous lupus erythematosus, Arch Dermatol 139 (2003) 45–49.

[51] C. Nieboer, Z. Tak-Diamand, H.E. Van Leeuwen-Wallau, Dust-like particles: A specific direct immunofluorescence pattern in sub-acute cutaneous lupus erythematosus, Br J Dermatol 118 (1988) 725–729.

[52] J.P. Kelly, A. Auquier, B. Rzany, L. Naldi, S. Bastuji-Garin, O. Correia, et al., An international collaborative case—control study of severe cutaneous adverse reactions (SCAR). Design and methods, J Clin Epidemiol 48 (1995) 1099–1108.

[53] W. Ting, M.S. Stone, D. Racila, R.H. Scofield, R.D. Sontheimer, Toxic epidermal necrolysis-like acute cutaneous lupus erythematosus and the spectrum of the acute syndrome of apoptotic pan-epidermolysis (ASAP): A case report, concept review and proposal for new classification of lupus erythematosus vesiculobullous skin lesions, Lupus 13 (2004) 941–950.

[54] K. Popovic, M. Wahren-Herlenius, F. Nyberg, Clinical follow-up of 102 anti-Ro/SSA-positive patients with dermatological manifestations, Acta Derm Venereol 88 (2008) 370–375.

[55] A. Kuhn, M. Sonntag, D. Richter-Hintz, C. Oslislo, M. Megahed, T. Ruzicka, et al., Phototesting in lupus erythematosus: A 15-year experience, J Am Acad Dermatol 45 (2001) 86–95.

[56] C. Bombardier, D.D. Gladman, M.B. Urowitz, D. Caron, C.H. Chang, Derivation of the SLEDAI. A disease activity index for lupus patients. The Committee on Prognosis Studies in SLE, Arthritis Rheum 35 (1992) 630–640.

[57] E.J. Feinglass, F.C. Arnett, C.A. Dorsch, T.M. Zizic, M.B. Stevens, Neuropsychiatric manifestations of systemic lupus erythematosus: Diagnosis, clinical spectrum, and relationship to other features of the disease, Medicine 55 (1976) 323–339.

[58] C. Gomes, J. Carvalho, E. Borba, C. Borges, M. Vendramini, C. Bueno, et al., Digital vasculitis in systemic lupus erythematosus: A minor manifestation of disease activity? Lupus 18 (2009) 990–993.

[59] M. Ramos-Casals, N. Nardi, M. Lagrutta, P. Brito-Zeron, A. Bove, G. Delgado, et al., Vasculitis in systemic lupus erythematosus: Prevalence and clinical characteristics in 670 patients, Medicine 85 (2006) 95–104.

[60] J. Sanchez-Perez, P.F. Penas, L. Rios-Buceta, J. Fernandez-Herrera, J. Fraga, A. Garcia-Diez, Leukocytoclastic vasculitis in subacute cutaneous lupus erythematosus: Clinicopathologic study of three cases and review of the literature, Dermatology (Basel) 193 (1996) 230–235.

[61] J.S. Levine, D.W. Branch, J. Rauch, The antiphospholipid syndrome. N Engl J Med 346 (2002) 752–763.

[62] R.A. Asherson, C. Frances, L. Iaccarino, M.A. Khamashta, F. Malacarne, J.C. Piette, et al., The antiphospholipid antibody syndrome: Diagnosis, skin manifestations and current therapy. Clin Exp Rheumatol 24 (2006) S46–S51.

[63] C. Frances, S. Niang, E. Laffitte, F. Pelletier, N. Costedoat, J.C. Piette, Dermatologic manifestations of the antiphospholipid syndrome: Two hundred consecutive cases, Arthritis Rheum 52 (2005) 1785–1793.

[64] O.D. Ortega-Hernandez, N. Agmon-Levin, M. Blank, R.A. Asherson, Y. Shoenfeld, The physiopathology of the catastrophic antiphospholipid (Asherson's) syndrome: Compelling evidence, J Autoimmun 32 (2009) 1–6.

[65] S. Bucciarelli, G. Espinosa, R. Cervera, The CAPS Registry: Morbidity and mortality of the catastrophic antiphospholipid syndrome, Lupus 18 (2009) 905–912.

[66] J. Dimant, E. Ginzler, M. Schlesinger, G. Sterba, H. Diamond, D. Kaplan, et al., The clinical significance of Raynaud's phenomenon in systemic lupus erythematosus, Arthritis Rheum 22 (1979) 815–819.

[67] S. Bernatsky, C.A. Pineau, J.L. Lee, A.E. Clarke, Headache, Raynaud's syndrome and serotonin receptor agonists in systemic lupus erythematosus, Lupus 15 (2006) 671–674.

[68] A. Kasparian, A. Floros, E. Gialafos, M. Kanakis, S. Tassiopoulos, N. Kafasi, et al., Raynaud's phenomenon is correlated with elevated systolic pulmonary arterial pressure in patients with systemic lupus erythematosus, Lupus 16 (2007) 505–508.

[69] N. Scheinfeld, Malignant atrophic papulosis, Clin Exp Dermatol 32 (2007) 483–487.

[70] W.A. High, J. Aranda, S.B. Patel, C.J. Cockerell, M.I. Costner, Is Degos" disease a clinical and histological end point rather than a specific disease? J Am Acad Dermatol 50 (2004) 895–899.

[71] V. Saurit, R. Campana, A. Ruiz Lascano, C. Ducasse, A. Bertoli, S. Aguero, et al., Mucocutaneous lesions in patients with systemic lupus erythematosus, Medicina 63 (2003) 283–287.

[72] V.P. Werth, W.L. White, M.R. Sanchez, A.G. Franks, Incidence of alopecia areata in lupus erythematosus, Arch Dermatol 128 (1992) 368–371.

[73] J.D. Urman, M.B. Lowenstein, M. Abeles, A. Weinstein, Oral mucosal ulceration in systemic lupus erythematosus, Arthritis Rheum 21 (1978) 58–61.

[74] N.S. Sherber, F.M. Wigley, R.K. Scher, Autoimmune disorders: Nail signs and therapeutic approaches, Dermatol Ther 20 (2007) 17–30.

[75] U. Wollina, U. Barta, C. Uhlemann, P. Oelzner, Lupus erythematosus-associated red lunula, J Am Acad Dermatol 41 (1999) 419–421.

[76] F. Rongioletti, A. Rebora, Papular and nodular mucinosis associated with systemic lupus erythematosus, Br J Dermatol 115 (1986) 631–636.

[77] A. Calebotta, A. Cirocco, E. Giansante, O. Reyes, Systemic lupus erythematosus and pemphigus vulgaris: Association or coincidence, Lupus 13 (2004) 951–953.

[78] W.R. Gammon, R.A. Briggaman, Bullous SLE: A phenotypically distinctive but immunologically heterogeneous bullous disorder, J Invest Dermatol 100 (1993) 28S–34S.

[79] H. Jedlickova, J. Bohmova, A. Sirotkova, Bullous systemic lupus erythematosus induced by the therapy for lupus nephritis, Int J Dermatol 47 (2008) 1315–1316.

[80] M.A. Waldman, J.P. Callen, Pyoderma gangrenosum preceding the diagnosis of systemic lupus erythematosus, Dermatology (Basel) 210 (2005) 64–67.

[81] S.P. Masatlioglu, F. Goktay, A.T. Mansur, A.D. Akkaya, P. Gunes, Systemic lupus erythematosus presenting as pyoderma gangrenosum in two cases, Rheumatol Int 29 (2009) 837–840.

[82] K.L. Barr, F. O'Connell, S. Wesson, V. Vincek, Nonbullous neutrophilic dermatosis: Sweet's syndrome, neonatal lupus erythematosus, or both? Modern Rheumatol 19 (2009) 212–215.

[83] E.K. Satter, W.A. High, Non-bullous neutrophilic dermatosis within neonatal lupus erythematosus, J Cutan Pathol 34 (2007) 958–960.

[84] P. Chu, M.K. Connolly, P.E. LeBoit, The histopathologic spectrum of palisaded neutrophilic and granulomatous dermatitis in patients with collagen vascular disease, Arch Dermatol 130 (1994) 1278–1283.

[85] E.S. Smith, J.R. Hallman, A.M. DeLuca, G. Goldenberg, J.L. Jorizzo, O.P. Sangueza, Dermatomyositis: A clinicopathological study of 40 patients, Am J Dermatopathol 31 (2009) 61–67.

[86] O. Durosaro, M.D. Davis, K.B. Reed, A.L. Rohlinger, Incidence of cutaneous lupus erythematosus, 1965–2005: A population-based study, Arch Dermatol 145 (2009) 249–253.

[87] O. Nived, G. Sturfelt, F. Wollheim, Systemic lupus erythematosus in an adult population in southern Sweden: Incidence, prevalence and validity of ARA revised classification criteria, Br J Rheumatol 24 (1985) 147–154.

[88] K.M. Uramoto, C.J. Michet, J. Thumboo, J. Sunku, W.M. O'Fallon, S.E. Gabriel, Trends in the incidence and mortality of systemic lupus erythematosus, 1950–1992, Arthritis Rheum 42 (1999) 46.

[89] J.N. Gilliam, R.D. Sontheimer, Skin manifestations of SLE, Clin Rheum Dis 8 (1981) 207–218.

[90] C. Vitali, W. Bencivelli, D.A. Isenberg, J.S. Smolen, M.I. Snaith, M. Sciuto, et al., Disease activity in SLE: Report of the Consensus Study Group of the European Workshop for Rheumatology Research. A descriptive analysis of 704 European lupus patients, Clin Exp Rheumatol 10 (1992) 527–539.

[91] K. Popovic, F. Nyberg, M. Wahren-Herlenius, A serology-based approach combined with clinical examination of 125 Ro/SSA-positive patients to define incidence and prevalence of subacute cutaneous lupus erythematosus, Arthritis Rheum 56 (2007) 255–264.

[92] S. Koskenmies, T.M. Jarvinen, P. Onkamo, J. Panelius, U. Tuovinen, T. Hasan, et al., Clinical and laboratory characteristics of Finnish lupus erythematosus patients with cutaneous manifestations, Lupus 17 (2008) 337—347.

[93] P. Rahman, D.D. Gladman, M.B. Urowitz, Smoking interferes with efficacy of antimalarial therapy in cutaneous lupus, J Rheumatol 25 (1998) 1716—1719.

[94] M.L. Jewell, D.P. McCauliffe, Patients with cutaneous lupus erythematosus who smoke are less responsive to antimalarial treatment, J Am Acad Dermatol 42 (2000) 983.

[95] C.J. Sanders, H. van Weelden, G.A. Kazzaz, V. Sigurdsson, J. Toonstra, C.A. Bruijnzeel-Koomen, Photosensitivity in patients with lupus erythematosus: A clinical and photobiological study of 100 patients using a prolonged phototest protocol, Br J Dermatol 149 (2003) 131.

[96] P. Lehmann, E. Holzle, P. Kind, G. Goerz, G. Plewig, Experimental reproduction of skin lesions in lupus erythematosus by UVA and UVB radiation, J Am Acad Dermatol 22 (1990) 181.

[97] R.D. Sontheimer, C.L. Henderson, R.H. Grau, Drug-induced subacute cutaneous lupus erythematosus: A paradigm for bedside-to-bench patient-oriented translational clinical investigation, Arch Dermatol Res 301 (2009) 65—70.

[98] M.C. Sampaio, Z.N. de Oliveira, M.C. Machado, V.M. dos Reis, M.A. Vilela, Discoid lupus erythematosus in children—A retrospective study of 34 patients, Pediatr Dermatol 25 (2008) 163—167.

[99] N. Romani, S. Ebner, C.H. Tripp, V. Flacher, F. Koch, P. Stoitzner, Epidermal Langerhans cells—Changing views on their function in vivo, Immunol Lett 106 (2006) 119—125.

[100] J.C. Dudda, N. Perdue, E. Bachtanian, D.J. Campbell, Foxp3+ regulatory T cells maintain immune homeostasis in the skin, J Exp Med 205 (2008) 1559—1565.

[101] V.P. Werth, Cutaneous lupus: Insights into pathogenesis and disease classification, Bull. NYU Hospital for Joint Diseases 65 (2007) 200—204.

[102] P. Rovere-Querini, I.E. Dumitriu, Corpse disposal after apoptosis, Apoptosis 8 (2003) 469—479.

[103] B. Franz, B. Fritzsching, A. Riehl, N. Oberle, C.D. Klemke, J. Sykora, et al., Low number of regulatory T cells in skin lesions of patients with cutaneous lupus erythematosus, Arthritis Rheum 56 (2007) 1910—1920.

[104] R.D. Sontheimer, Lichenoid tissue reaction/interface dermatitis: Clinical and histological perspectives, J Invest Dermatol 129 (2009) 1088—1099.

[105] J.H. Lin, J.P. Dutz, R.D. Sontheimer, V.P. Werth, Pathophysiology of cutaneous lupus erythematosus, Clin Rev Allergy Immunol 33 (2007) 85—106.

[106] L. Casciola-Rosen, A. Rosen, Ultraviolet light-induced keratinocyte apoptosis: A potential mechanism for the induction of skin lesions and autoantibody production in LE, Lupus 6 (1997) 175—180.

[107] T. Dorner, M. Hucko, W.J. Mayet, U. Trefzer, G.R. Burmester, F. Hiepe, Enhanced membrane expression of the 52 kDa Ro(SS-A) and La(SS-B) antigens by human keratinocytes induced by TNF alpha, Ann Rheum Dis 54 (1995) 904—909.

[108] J. Knop, G. Bonsmann, P. Kind, I. Doxiadis, U. Vogeler, G. Doxiadis, et al., Antigens of the major histocompatibility complex in patients with chronic discoid lupus erythematosus, Br J Dermatol 122 (1990) 723—728.

[109] R.M. Watson, P. Talwar, E. Alexander, W.B. Bias, T.T. Provost, Subacute cutaneous lupus erythematosus-immunogenetic associations, J Autoimmun 4 (1991) 73—85.

[110] M. Gilliet, W. Cao, Y.J. Liu, Plasmacytoid dendritic cells: Sensing nucleic acids in viral infection and autoimmune diseases, Nat Rev Immunol 8 (2008) 594—606.

[111] J. Wenzel, T. Tuting, An IFN-associated cytotoxic cellular immune response against viral, self-, or tumor antigens is a common pathogenetic feature in "interface dermatitis", J Invest Dermatol 128 (2008) 2392—2402.

[112] H.V. Marquart, L. Schejbel, A. Sjoholm, U. Martensson, S. Nielsen, A. Koch, A. Svejgaard, et al., C1q deficiency in an Inuit family: Identification of a new class of C1q disease-causing mutations, Clin Immunol 124 (2007) 33—40.

[113] Ad Hoc Committee Response Criteria for Cutaneous SLE, Response criteria for cutaneous SLE in clinical trials, Clin Exp Rheumatol 25 (2007) 666—671.

[114] J. Albrecht, L. Taylor, J.A. Berlin, S. Dulay, G. Ang, S. Fakhazardeh, et al., The CLASI (Cutaneous LE Disease Area and Severity Index): An outcome instrument for cutaneous lupus erythematosus, J Invest Dermatol 125 (2005) 889—894.

[115] Z. Bonilla-Martinez, J. Albrecht, L. Taylor, J. Okawa, V.P. Werth, The CLASI is a useful clinical instrument to separately follow activity and damage during therapy of cutaneous lupus erythematosus, Arch Dermatol 144 (2008) 173.

[116] M.S. Krathen, J. Dunham, E. Gaines, G. Grove, J. Junkins-Hopkins, E. Kim, et al., The Cutaneous Lupus Erythematosus Disease Activity and Severity Index (CLASI) expansion for rheumatology and dermatology, Arthritis Care Res 59 (2008) 338—344.

[117] A. Kreuter, R. Gaifullina, C. Tigges, J. Kirschke, P. Altmeyer, T. Gambichler, Lupus erythematosus tumidus: Response to antimalarial treatment in 36 patients with emphasis on smoking, Arch Dermatol 145 (2009) 244—248.

[118] A. Kreuter, N.S. Tomi, S.M. Weiner, M. Huger, P. Altmeyer, T. Gambichler, Mycophenolate sodium for subacute cutaneous lupus erythematosus resistant to standard therapy, Br J Dermatol 156 (2007) 1321—1327.

[119] A. Erceg, H.J. Bovenschen, P.C. van de Kerkhof, E.M. de Jong, M.M. Seyger, Efficacy and safety of pulsed dye laser treatment for cutaneous discoid lupus erythematosus, J Am Acad Dermatol 60 (2009) 626—632.

[120] M.M. Chren, R.J. Lasek, L.M. Quinn, et al., Skindex, a quality-of-life measure for patients with skin disease: Reliability, validity, and responsiveness, J Invest Dermatol 107 (1996) 707.

[121] A.Y. Finlay, G.K. Khan, Dermatology Life Quality Index (DLQI)—A simple practical measure for routine clinical use, Clin Exp Dermatol 19 (1994) 210—216.

[122] R.S. Klein, S. Moghadam-Kia, J. Lomonico, K. Chilek, E. Gaines, J. Okawa, et al., Quality of life in cutaneous lupus erythematosus, Paper presented at the ACR/ARHP Annual Scientific Meeting, Philadelphia, 2009. October 17—21.

[123] H. Stege, M.A. Budde, S. Grether-Beck, J. Krutmann, Evaluation of the capacity of sunscreens to photoprotect lupus erythematosus patients by employing the photoprovocation test, Photodermatol Photoimmunol Photomed 16 (2000) 256.

[124] T. Herzinger, G. Plewig, M. Rocken, Use of sunscreens to protect against ultraviolet-induced lupus erythematosus, Arthritis Rheum 50 (2004) 3045.

[125] W.L. Morison, Photoprotection by clothing, Dermatol Ther 16 (2003) 16—22.

[126] T.G. Tzellos, D. Kouvelas, Topical tacrolimus and pimecrolimus in the treatment of cutaneous lupus erythematosus: An evidence-based evaluation, Eur J Clin Pharmacol 64 (2008) 337—341.

[127] M. Sardy, T. Ruzicka, A. Kuhn, Topical calcineurin inhibitors in cutaneous lupus erythematosus, Arch Dermatol Res 301 (2009) 93—98.

[128] J. Ring, M. Mohrenschlager, V. Henkel, The US FDA "black box" warning for topical calcineurin inhibitors: An ongoing controversy, Drug Saf 31 (2008) 185—198.

[129] O. Baniandres, P. Boixeda, P. Belmar, A. Perez, Treatment of lupus erythematosus with pulsed dye laser, Lasers Surg Med 32 (2003) 327–330.

[130] I. Cavazzana, R. Sala, C. Bazzani, A. Ceribelli, C. Zane, R. Cattaneo, et al., Treatment of lupus skin involvement with quinacrine and hydroxychloroquine, Lupus 18 (2009) 735–739.

[131] E. Toubi, I. Rosner, M. Rozenbaum, A. Kessel, T.D. Golan, The benefit of combining hydroxychloroquine with quinacrine in the treatment of SLE patients, Lupus 9 (2000) 92–95.

[132] R. Feldmann, D. Salomon, J.H. Saurat, M.A. Pathak, The association of the two antimalarials chloroquine and quinacrine for treatment-resistant chronic and subacute cutaneous lupus erythematosus [see comments], Dermatology 23 (1996) 783.

[133] S. Jessop, D. Whitelaw, F. Jordaan, Drugs for discoid lupus erythematosus, Cochrane Database Syst (2001). Rev. CD002954.

[134] R.P. Hall, T.J. Lawley, H.R. Smith, S.I. Katz, Bullous eruption of systemic lupus erythematosus. Dramatic response to dapsone therapy, Ann Intern Med 97 (1982) 165.

[135] T. Ruzicka, M. Meurer, O. Braun-Falco, Treatment of cutaneous lupus erythematosus with etretinate, Acta Derm Venereol 65 (1985) 324–329.

[136] J.P. Callen, L.V. Spencer, J.B. Burruss, J. Holtman, Azathioprine. An effective, corticosteroid-sparing therapy for patients with recalcitrant cutaneous lupus erythematosus or with recalcitrant cutaneous leukocytoclastic vasculitis, Arch Dermatol 127 (1991) 515–522.

[137] I.B. Boehm, G.A. Boehm, R. Bauer, Management of cutaneous lupus erythematosus with low-dose methotrexate: Indication for modulation of inflammatory mechanisms, Rheumatol Int 18 (1998) 59.

[138] J. Wenzel, S. Brahler, R. Bauer, T. Bieber, T. Tuting, Efficacy and safety of methotrexate in recalcitrant cutaneous lupus erythematosus: Results of a retrospective study in 43 patients, Br J Dermatol 153 (2005) 157–162.

[139] S. Schanz, A. Ulmer, G. Rassner, G. Fierlbeck, Successful treatment of subacute cutaneous lupus erythematosus with mycophenolate mofetil, Br J Dermatol 147 (2002) 174–178.

[140] T.S. Housman, J.L. Jorizzo, M.A. McCarty, S.E. Grummer, A.B. Fleischer Jr., P.G. Sutej, Low-dose thalidomide therapy for refractory cutaneous lesions of lupus erythematosus, Arch Dermatol 139 (2003) 50.

[141] M.J. Cuadrado, Y. Karim, G. Sanna, E. Smith, M.A. Khamashta, G.R. Hughes, Thalidomide for the treatment of resistant cutaneous lupus: Efficacy and safety of different therapeutic regimens, Am J Med 118 (2005) 246–250.

[142] S. Bastuji-Garin, S. Ochonisky, P. Bouche, R.K. Gherardi, C. Duguet, Z. Djerradine, et al., Incidence and risk factors for thalidomide neuropathy: A prospective study of 135 dermatologic patients, J Invest Dermatol 119 (2002) 1020.

[143] B. Escudier, N. Lassau, S. Leborgne, E. Angevin, A. Laplanche, Thalidomide and venous thrombosis, Ann Intern Med 136 (2002) 711.

[144] C. Frances, S. El Khoury, A. Gompel, P.A. Becherel, O. Chosidow, J.C. Piette, Transient secondary amenorrhea in women treated by thalidomide, Eur J Dermatol 12 (2002) 63.

[145] C. Briani, G. Zara, R. Rondinone, L. Iaccarino, S. Ruggero, E. Toffanin, et al., Positive and negative effects of thalidomide on refractory cutaneous lupus erythematosus, Autoimmunity 38 (2005) 549–555.

[146] G. Zara, M. Ermani, R. Rondinone, S. Arienti, A. Doria, Thalidomide and sensory neurotoxicity: A neurophysiological study, J Neurol Neurosurg Psychiatry 79 (2008) 1258–1261.

[147] J.C. Piette, A. Sbai, C. Frances, Warning: Thalidomide-related thrombotic risk potentially concerns patients with lupus, Lupus 11 (2002) 67.

[148] T.H. Clayton, S. Ogden, M.D. Goodfield, Treatment of refractory subacute cutaneous lupus erythematosus with efalizumab, J Am Acad Dermatol 54 (2006) 892.

[149] J.C. Prinz, M. Meurer, C. Reiter, E.P. Rieber, G. Plewig, G. Riethmuller, Treatment of severe cutaneous lupus erythematosus with a chimeric CD4 monoclonal antibody, cM-T412, J Am Acad Dermatol 34 (1996) 244.

[150] M. Goodfield, K. Davison, K. Bowden, Intravenous immunoglobulin (IVIg) for therapy-resistant cutaneous lupus erythematosus (LE), J Dermatolog Treat 15 (2004) 46–50.

[151] B. Fautrel, V. Foltz, C. Frances, P. Bourgeois, S. Rozenberg, et al., Regression of subacute cutaneous lupus erythematosus in a patient with rheumatoid arthritis treated with a biologic tumor necrosis factor alpha-blocking agent: Comment on the article by Pisetsky and the letter from Aringer, Arthritis Rheum 46 (2002) 1408–1409. author reply 1409.

[152] G.S. Bleumink, E.J. ter Borg, C.G. Ramselaar, B.H. Ch Stricker, Etanercept-induced subacute cutaneous lupus erythematosus, Rheumatology (Oxford) 40 (2001) 1317–1319.

[153] A.K. Palucka, J.P. Blanck, L. Bennett, V. Pascual, J. Banchereau, Cross-regulation of TNF and IFN-alpha in autoimmune diseases, Proc Natl Acad Sci USA 102 (2005) 3372–3377.

[154] U. Gul, M. Gonul, S.K. Cakmak, A. Kilic, M. Demiriz, A case of generalized discoid lupus erythematosus: Successful treatment with imiquimod cream 5%, Adv Ther 23 (2006) 787–792.

[155] R. Gerdsen, J. Wenzel, M. Uerlich, T. Bieber, W. Petrow, Successful treatment of chronic discoid lupus erythematosus of the scalp with imiquimod, Dermatology 205 (2002) 416–418.

[156] M. Conroy, L. Sewell, O.F. Miller, T. Ferringer, Interferon-beta injection site reaction: Review of the histology and report of a lupus-like pattern, J Am Acad Dermatol 59 (2008) S48–49.

[157] I. Arrue, A. Saiz, P.L. Ortiz-Romero, J.L. Rodriguez-Peralto, Lupus-like reaction to interferon at the injection site: Report of five cases, J Cutan Pathol 34 (Suppl. 1) (2007) 18–21.

[158] V. Kieu, T. O'Brien, L.M. Yap, C. Baker, P. Foley, G. Mason, et al., Refractory subacute cutaneous lupus erythematosus successfully treated with rituximab, Australas J Dermatol 50 (2009) 202–206.

[159] G.B. Jemec, S. Ullman, M. Goodfield, A. Bygum, A.B. Olesen, J. Berth-Jones, et al., A randomized controlled trial of R-salbutamol for topical treatment of discoid lupus erythematosus, Br J Dermatol 161 (2009) 1365–1370.

Kidney

Brad H. Rovin, Isaac E. Stillman

INTRODUCTION

The kidney is frequently affected by systemic lupus erythematosus (SLE). Most often, this is due to glomerular immune complex accumulation, which leads to an inflammatory response that damages the glomeruli and the renal tubulointerstitium; if unchecked, chronic scarring of the entire kidney parenchyma may result. SLE patients with kidney involvement have worse outcomes than those with no kidney involvement [1–3]. This poor prognosis is explained only in part by the risk of chronic kidney disease (CKD) and end-stage renal disease (ESRD), suggesting that lupus nephritis (LN) is a manifestation of a more severe form of SLE.

LN is often treatable. The best outcomes occur with early recognition and prompt treatment. Early recognition requires appropriate use of screening tests for kidney involvement followed by confirmatory tests and a kidney biopsy. Current treatment regimens use intense, nonspecific immunosuppressives that confer considerable risk of severe infection and other morbidities. Efforts are underway to develop new LN therapies with greater efficacy and less toxicity, paving the way for improved outcomes in the future.

SCOPE OF LUPUS NEPHRITIS

Examination of kidney tissue obtained during research protocols from patients with SLE but without clinical signs of kidney disease showed LN was present in approximately 90% of the patients [4, 5]. Most had very mild histologic changes, but approximately 15% had moderate to severe pathology, many of whom did eventually develop proteinuria, an abnormal urine sediment, or renal insufficiency. A study comparing findings with overt LN suggested that the clinically silent period may represent the earliest stage in the natural history of LN [6].

Although the histologic changes of LN are pervasive, the reported incidence of clinically important kidney disease is approximately 38%. Of the patients destined to have clinical LN, 40–60% have overt findings of kidney disease at the time of initial diagnosis of SLE [7–9]. The incidence of kidney involvement differs among racial and ethnic groups, as seen by data compiled on 2290 lupus patients from North America, Europe, and the Middle East (Table 42.1). Caucasians (European and European Americans) are less likely to have LN than Black (African-American and Afro-Caribbean), Hispanic, or Asian (Indian and Chinese) patients. Non-White SLE patients have a 50% or greater incidence of LN.

The U.S. Renal Data Service database shows that between 1996 and 2004, the incidence of ESRD attributed to LN in adults was 4.4–4.9 cases per million in the general population [10]. Although the ESRD rate was stable during the 9 years of data collection, it is concerning that it did not decline despite the use of aggressive immunotherapy. The incidence of ESRD was greater in Blacks (17–20/million) and Hispanics (6/million) than Caucasians (2.5/million). In the United Kingdom, 19% of Caucasians and 62% of Blacks with LN progressed to ESRD [11] compared to 5.8% of patients from Saudi Arabia [9].

CKD is also an important outcome of LN. CKD is a nontraditional risk factor for cardiovascular mortality [12–14], and it is thus significant because SLE predisposes to early cardiovascular disease [15]. The prevalence of CKD in patients with SLE is difficult to estimate, but because current therapies induce complete remission in fewer than 50% of LN patients, CKD is likely to be high in the lupus population.

DIAGNOSIS OF LUPUS NEPHRITIS

The early diagnosis of lupus nephritis is critical to preserve kidney function [16–22]. This may explain the results of a retrospective study suggesting that

TABLE 42.1 Incidence of LN in Different Populations

Population	Lupus nephritis (%)
White	12–33
Black	40–69
Asian	47–53
Middle Eastern	48
Hispanic	35–61
Overall (N = 2290)	37.8

patients with initial onset of LN simultaneously with the diagnosis of SLE had higher rates of prolonged remission and a lower frequency of chronic kidney damage than those with delayed onset of LN, despite similar rates of class IV lesions [23]. The key to early diagnosis is maintaining a high index of suspicion for renal involvement in every SLE patient. Although some patients may present with overt clinical signs of renal disease, such as edema secondary to nephrotic syndrome or severe hypertension, it is more likely that the initial evidence of kidney involvement will be an abnormality of serum creatinine and/or the urinalysis. An algorithm for evaluation of the SLE patient for kidney involvement is shown in Figure 42.1.

Evaluation of Kidney Function

During the initial evaluation of SLE, or at the time of a suspected renal flare in a patient with a history of LN,

an estimate of glomerular filtration rate (GFR) as a measure of kidney function should be obtained. Although there are methods to measure GFR directly, they are generally reserved for clinical trials and research laboratories because they are difficult and expensive. Thus, for practical purposes, the serum creatinine value serves as the initial screening test of GFR. Serum creatinine values must be interpreted in the context of the individual patient. For example, a serum creatinine within the hospital laboratory normal range may actually be abnormally high for a woman with small to moderate muscle mass and thus low rates of creatinine production. Conversely, patients with high muscle mass may have normal kidney function despite a serum creatinine that is above the hospital laboratory upper limit of normal. Furthermore, in patients with severe nephrotic syndrome who become hypoalbuminemic, the tubular secretion of creatinine increases, which tends to lower serum creatinine and gives the false impression that GFR is better than it is in reality [24]. If the serum creatinine does not clarify whether the GFR is normal or abnormal, additional methods are available to estimate GFR (Table 42.2). However, all methods and formulae that use only serum creatinine as the serum component may overestimate GFR in the presence of low serum albumin.

The most common way to estimate GFR has been to calculate creatinine clearance from a serum creatinine and a 24-h urine collection. To avoid problems associated with collecting 24-h urine samples (e.g., patient compliance and over- and undercollections), estimating equations for creatinine clearance and GFR have been

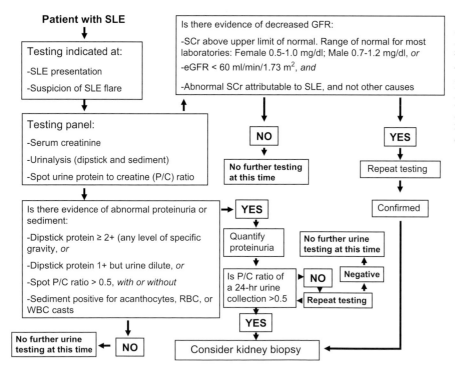

FIGURE 42.1 An algorithm for the evaluation of the kidney in patients with SLE. Note that patients with a history of LN and previous kidney biopsy may not need a repeat biopsy. Kidney biopsy should be done for new diagnoses of kidney involvement. Acanthocytes are dysmorphic RBCs specific for glomerular bleeding. GFR, glomerular filtration rate; P/C, protein to creatinine ratio in a urine sample; SCr, serum creatinine.

TABLE 42.2 Comparison of Methods for Estimating GFR

Method	Rationale	Strengths	Weaknesses
Serum creatinine (SCr)	Creatinine is usually produced at a constant rate, excreted almost entirely by the kidney, and mainly by glomerular filtration. Thus, usually changes in SCr reflect change in GFR.	Simple and relatively inexpensive	There are many conditions that can increase or decrease SCr independent of GFR. The following are the most common: • Changes in intake of cooked meat (cooking converts creatine to creatinine). SCr is approximately 25% lower on a vegetarian diet. • Drugs that decrease tubular secretion of creatinine (trimethoprim and cimetidine) raise SCr 10–30%. • Hypoalbuminemia in severely nephrotic patients is associated with an increase in tubular secretion of creatinine. Small changes in SCr (e.g., ±0.3 mg/dl) when SCr is in the normal range correspond to very large changes in GFR (approximately ±20%). Confounding this interpretation is the fact that the 95% confidence intervals of the measurement of SCr are also approximately ±0.2 mg/dl for SCr values around the normal range. Thus, often it is necessary to measure SCr two or three times to assess whether a real change in GFR has occurred versus a spontaneous laboratory variation.
Creatinine clearance (C_{Cr})	C_{Cr} refers to the renal clearance of creatinine, which is usually a close estimate of GFR. $C_{Cr} = $ (amount of creatinine in a complete 24-h urine collection)/(average SCr during the period that the urine was collected). In most patients, SCr is stable. Thus, SCr measured at the start or end of the 24-h collection reflects the average SCr during the urine collection. If SCr is changing (as in ARF), the SCr measured at the midpoint of the collection is a good estimate of the average SCr during the collection.	C_{Cr} estimate of GFR is not affected by changes in creatinine production as long as the patient is in a steady state (serum creatinine is stable).	An accurate estimate of C_{Cr} is dependent on a complete 24-h collection. Thus, the error in estimating GFR from C_{Cr} is directly proportional to the completeness of the intended 24-h collection. C_{Cr} underestimates GFR if the patient is receiving trimethoprim or cimetidine. See the weaknesses for SCr. At low GFR (e.g., <30% of normal), C_{Cr} overestimates GFR by 10–50%.

(Continued)

TABLE 42.2   Comparison of Methods for Estimating GFR—cont'd

Method	Rationale	Strengths	Weaknesses
Cockroft–Gault (C-G)	C-G estimates creatinine clearance from patients' age, sex, race, ideal body weight (IBW), and serum creatinine. C-G = $(140 - age) \times IBW$, kg $(\times 0.85$ if female$)/(72 \times SCr$, mg/dl$)$ $(\times 1.21$ if black$)$ = ml/min.	C-G takes into account the most common reasons for variation in serum creatinine: variation in body size, sex, and race.	C-G does not take into account the factors that change SCr independent of change in GFR. See weaknesses for SCr. In advanced renal failure, creatinine clearance can be 10 to 50% higher than GFR.
MDRD-4	MDRD-4 GFR (eGFR) is estimated from a complex formula based on the patient's age, sex, race, and serum creatinine. The eGFR equation was calculated from a large number of patients whose GFR was directly measured (usually by iothalamate clearance) and whose age, sex, race, and serum creatinine were known.	The MDRD-4 eGFR provides a quantitative estimate of GFR level that is not intuitively obvious from the serum creatinine level because of the hyperbolic relationship between serum creatinine and GFR.	MDRD-4 eGFR has poor precision at GFR > 60 ml/min/1.73 m². Thus, a single measure in patients with serum creatinine at or near the normal range is not a reliable measure of GFR. Multiple measures, however, improve precision. The MDRD-4 eGFR assumes everyone of the same age, race, and sex has the same rate of creatinine production. Thus, MDRD-4 underestimates GFR in large persons (or those with other mechanisms of increased creatinine production) and overestimates GFR in those with low creatinine production. MDRD-4 eGFR also suffers from confounding by all of the other factors that influence SCr, independent of GFR. See the weaknesses for SCr. The MDRD-4 eGFR is most useful in comparing GFR between large populations where the influence of body size, diet, etc. can be expected to be similar in the groups being compared.
CKD-EPI	This is the latest version of eGFR. It is calculated from a complex formula similar to that of MDRD-4 eGFR. However, CKD-EPI used a larger and more diverse data set.	CKD-EPI eGFR is somewhat more accurate than MDRD-4 eGFR in those with GFR > 60 ml/min/1.73 m².	The weaknesses are the same as for MDRD-4 eGFR.
Iothalamate clearance (C_{Iothal}) Inulin clearance (C_{In}) Iohexol clearance (C_{Iohex})	These molecules are true markers of glomerular filtration. They are freely filtered at the glomerulus, and there is little or no tubular secretion or absorption of these markers. Thus, their renal clearance reflects GFR.	C_{Iothal}, C_{In}, and C_{Iohex} are true measures of GFR.	A single measure is not a reliable estimate of the patient's prevailing GFR, particularly in those with GFR > 60 ml/min/1.73 m². It is expensive and Time-consuming. Not useful in individual patients to decide whether GFR is normal based on a single measure.

developed. These equations use readily available data such as serum creatinine, sex, race, and age. The two most popular equations are the Cockcroft–Gault (CG) formula and the Modification of Diet in Renal Disease (MDRD) equation [25, 26]. Calculators for these equations can be found at the National Kidney Foundation website (http://www.kidney.org).

The MDRD equation has received considerable attention and is often calculated by the hospital laboratory when serum creatinine is ordered. This is reported as the estimated GFR (eGFR) along with the creatinine. Caution must be used when applying these equations to SLE patients. Neither equation was developed using data from an SLE cohort or a population of mainly young women. A study compared the accuracy and precision of the CG and MDRD equations to directly measured creatinine clearance in Caucasian and African-American patients with LN [27]. For patients with impaired kidney function, the CG equation was found to be a better estimate of creatinine clearance. Importantly, these equations are not accurate in patients with GFR $\geq$ 60 ml/min. A new MDRD-like GFR estimating equation has been developed called CKD-EPI [28], which is more accurate at higher GFR than the MDRD equation. Like MDRD, CKD-EPI does not account directly for body size, and its role in LN remains to be determined. Thus, for screening purposes, CG and MDRD are not particularly useful in identifying GFR impairment of less than 30–40% [27, 29].

Verified renal insufficiency in the presence of abnormal urine findings (described later) is very suggestive of LN and should prompt consideration of a kidney biopsy. Patients with isolated renal insufficiency that cannot be explained by something other than SLE (Table 42.3) should also be considered for a kidney biopsy.

Evaluation of the Urine

A comprehensive urinalysis should be done by a qualified observer. It is usually not sufficient to rely on the clinical laboratory for microscopic evaluation of the urine sediment. The urine dipstick is a commonly used screening test, and dipstick positivity for blood and/or protein is evidence of kidney involvement in SLE. A positive test for blood is usually the result of hematuria. However, if the test for blood is strongly positive but there are only rare red blood cells (RBCs) in the urine sediment, the patient may have hemoglobinuria or myoglobinuria. If the urine is very dilute (e.g., specific gravity of 1.005 or less), red cells will lyse, mimicking hemoglobinuria.

Hematuria may be due to a number of conditions unrelated to nephritis, including menstruation. It is therefore important to examine the urine sediment to verify whether the hematuria is due to glomerular

TABLE 42.3 Etiologies of Acute Renal Insufficiency in SLE Other Than LN

Infections including bacteremia and sepsis

Volume depletion

Hypotension

Nephrotoxin exposure (e.g., radiographic contrast)

Hemolysis

Thrombosis

Cardiac failure

Commonly prescribed medications

    Nonsteroidal anti-inflammatory drugs

    Angiotensin-converting enzyme inhibitors/angiotensin receptor blockers

    Allergic interstitial nephritis (e.g., from antibiotics)

Other, non-lupus glomerular diseases

bleeding. Glomerular bleeding is suggested by the physical characteristics of urine RBCs. RBCs indicative of glomerular bleeding are called acanthocytes and are dysmorphic cells that show cell membrane blebs (Figure 42.2A,B). In addition to acanthocytes, RBCs casts, and white blood cells (WBCs) and WBC casts in the absence of infection are also indicative of glomerulonephritis (see Figure 42.2C,D). Glomerular hematuria in the absence of proteinuria and/or renal insufficiency does not require an immediate kidney biopsy for evaluation, but it needs to be followed closely for accompanying changes in kidney function and protein excretion.

Evaluation of Proteinuria

Perhaps the most important indicator of kidney involvement in lupus is proteinuria. Proteinuria magnitude has prognostic significance because heavy proteinuria may itself injure the kidney. In addition, proteinuria magnitude is a key biomarker of relapse, remission, and successful treatment. Therefore, accurate measurement of protein excretion is crucial to the ongoing management of LN. For screening purposes, the urine dipstick is rapid, easy to perform, and, although not ideal, acceptable. Indeed, in the absence of a quantitative test of proteinuria, the American College of Rheumatology (ACR) proteinuria criterion for renal lupus is a dipstick protein reading of 3+, although a requirement for a quantitative measurement is likely for future iterations of the ACR criteria.

There are several key concerns in estimating proteinuria magnitude from dipstick testing. First, very dilute urine (specific gravity < 1.010) may only be slightly dipstick positive for protein when 24-h proteinuria is

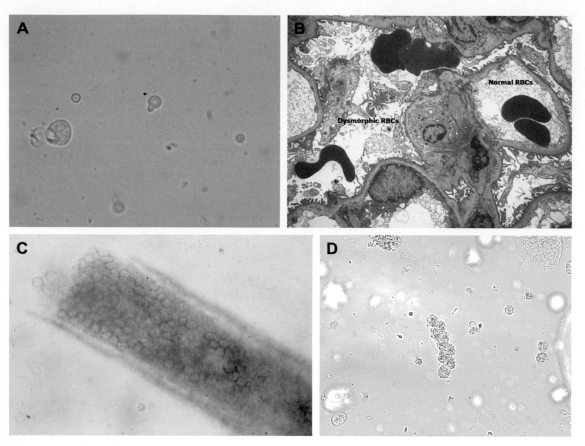

FIGURE 42.2 Urine sediment findings in LN. *A*, An acanthocyte (*) under bright field microscopy. *B*, Electron micrograph of acanthocytes forming in the glomerulus. Normal morphology RBCs are contained within the glomerular capillary lumen. Dysmorphic RBCs have traversed the glomerular basement membrane and are destined to be excreted into the urine as acanthocytes. *C*, RBC cast. *D*, WBC cast.

large (e.g., >1.0 g/day). Conversely, a small amount of protein (e.g., 24-h proteinuria < 300 mg/day) may register as a high level on the dipstick if the urine is concentrated. Second, if the urine is highly alkaline (pH > 8), the dipstick may indicate protein as high as 2+ in the absence of true abnormal proteinuria. Third, the dipstick principally detects albumin and will not detect immunoglobulin light chains or the proteins associated with tubular proteinuria such as β_2-microglobulin. A systematic study of the accuracy of the urine dipstick as a screening tool for LN found a false-negative rate in up to 30% of patients and a false-positive rate in approximately 40% of patients [30]. Thus, to improve screening, a urine protein-to-creatinine (P/C) ratio can be determined. Random spot urine P/C ratios can be used to screen but are not accurate enough to be used to make therapeutic decisions or to follow changes in proteinuria magnitude in response to therapy.

Abnormal proteinuria found on screening should be verified, and the magnitude of proteinuria should be accurately determined. The most reliable method to quantify proteinuria is to measure the P/C ratio of a 24-h urine collection or an intended 24-h collection that is at least 50% complete [31]. Measuring the P/C ratio avoids confounding the assessment of proteinuria by errors in collecting the 24-h urine. A 12-h overnight urine collection that includes the first morning void urine also provides an accurate measure of proteinuria magnitude and may be easier for patients to collect [32].

A kidney biopsy should be considered if proteinuria above 500 mg/day is confirmed, especially in the presence of hematuria or an abnormal urine sediment. Survey studies of SLE kidney biopsy collections have shown significant kidney pathology with levels of proteinuria in this range [33, 34].

KIDNEY BIOPSY

A kidney biopsy is essential to the optimal diagnosis and management of LN. This statement has historically provoked considerable debate and continues to do so. The main argument against kidney biopsy is a prevalent notion that there is only one treatment option for all the forms of LN; therefore, biopsy information would not change the approach to therapy [35]. Although this point is controversial and will be examined later, it incorrectly assumes that all renal injury in SLE patients is LN. This approach is outweighed by four arguments in favor of obtaining a kidney biopsy.

First, not all kidney disease in SLE patients is classic, immune complex-mediated glomerulonephritis (LN), so one therapy does not fit all patients. Although the findings of proteinuria and glomerular hematuria, with or without casts, are highly suggestive of LN in a patient with SLE, there are many reported cases of non-LN glomerular diseases in SLE patients with kidney and urinary abnormalities [36–41]. It is difficult to estimate the actual frequency of non-lupus glomerular disease in SLE because the literature is mostly case reports, but in a series of 252 patients, 5% were found to have pathologies such as focal segmental glomerulosclerosis, minimal change disease, thin glomerular basement membrane disease, hypertensive nephrosclerosis, and amyloidosis [37]. Focal segmental glomerulosclerosis was the most common finding in this series. The incidence of podocytopathies in LN appears to be greater than that in the general population, suggesting a causal link to the immune dysregulation of SLE [42]; however, podocyte injury does not appear to be related to the limited mesangial immune complexes present. AA amyloidosis has also been reported frequently in some series [38–41]. Differentiating these glomerulopathies from LN is important for treatment decisions and prognosis.

In addition, there are other important kidney lesions found in SLE patients. For example, patients with antiphospholipid antibodies who are hypercoagulable can develop glomerular or vascular microthrombi that lead to insidious and progressive renal insufficiency [43, 44]. Antiphospholipid syndrome (APS) nephropathy may occur in the presence or absence of LN, and it is not infrequent. In one series, it was seen in 32% of SLE patient biopsies [44]. This thrombotic microangiopathy is treated with anticoagulation as opposed to the immunosuppression needed for LN. Identifying APS nephropathy through the kidney biopsy is critical.

SLE patients can also have predominantly interstitial nephritis without glomerulonephritis and not related to medications. Although this is a rare form of kidney disease in SLE [41], it can cause renal insufficiency and activity in the urine sediment. Treatment would be different than that for LN.

Second, not all kidney disease in SLE needs to be treated with aggressive immunosuppression. The kidney biopsy assesses the degree of chronic kidney injury and therefore the risk of progressive renal failure that is related not to SLE but, rather, to natural mechanisms of kidney disease progression. Indeed, the degree of chronic injury seen on biopsy is probably the strongest histologic predictor of renal survival. The LN

patients most vulnerable to natural progression are those who have suffered the most irreversible nephron damage, specifically tubular atrophy, interstitial fibrosis, and glomerulosclerosis. Early nephron dropout cannot be inferred from estimates of GFR because as irreversible scarring occurs, surviving nephrons hypertrophy, stabilizing GFR at normal or near normal levels. Furthermore, there are currently no noninvasive biomarkers to accurately differentiate between fibrosis/scarring and active inflammation [45]. Fibrosis does not respond to immunosuppressive therapy; however, progressive kidney insufficiency can be mitigated by renal protective therapies (discussed later). These can be used with immunosuppression when concurrent inflammatory lesions need to be treated.

If extensive scarring is the dominant process found on biopsy even with some areas of active inflammation, the risk of immunosuppression may outweigh its benefits in terms of renal survival. Such patients are more appropriately treated with kidney-protective therapies alone.

Third, the kidney biopsy will be important for the proper application of future LN therapies. The most significant trend in experimental lupus therapeutics during the past several years has been trials of biologic agents for treatment. Although none of these drugs have yet earned an indication for LN, it is likely that one or more agents will be found to be effective in the near future. Because these agents are more targeted than the generalized immunosuppressants currently in use, it is possible that they will treat only specific subsets of LN. To use such novel and expensive drugs in an effective and cost-efficient manner, patients will need to be stratified into responsive and nonresponsive groups. Stratification will be based on biomarkers yet to be developed, some of which could be pathologic findings. As a hypothetical example, a specific anti-macrophage agent may be most effective in LN characterized by heavy glomerular and interstitial macrophage infiltration, independent of the biopsy class. This would require a kidney biopsy examination by techniques beyond the routine light, immunofluorescence, and electron microscopy studies currently used.

Fourth, the kidney biopsy is necessary for understanding the pathogenesis of kidney injury in human lupus. Although clinical biopsies are currently processed mainly for routine microscopy, they can and should be exploited in novel ways to better inform future drug development. For example, leukocyte subsets can be analyzed by specific staining in lupus kidneys and may yield new insights on renal inflammation [46]. Proteomic techniques can be used to search for patterns of protein expression in LN [47, 48]. Gene expression in biopsies can be analyzed with microarray techniques [48, 49]. These technologies are only beginning to be applied to kidney biopsies, but they have the potential to greatly enhance the amount of information available from the renal tissue.

The clinical utility of the kidney biopsy depends on obtaining an adequate sample of renal cortex and examination by a renal pathologist [50]. Inasmuch as every biopsy is a clinical—pathological correlation, the nephropathologist must be given all relevant clinical information to properly interpret the tissue and integrate the microscopic findings with the clinical data. Furthermore, it is essential that the clinician and pathologist review the findings together before initiation of therapy to ensure that specific clinical concerns have been addressed and that the lesions have been contextualized appropriately.

KIDNEY PATHOLOGY IN SLE

Ideally, renal pathology is evaluated using three techniques: light microscopy (LM; typically on formalin fixed tissue), immunofluorescence microscopy (IF; performed on fresh frozen tissue), and electron microscopy (EM; typically on glutaraldehyde fixed tissue). These techniques are complimentary and should be evaluated by the same nephropathologist to ensure the integration of all findings and to reduce sampling bias.

LM provides the foundation of injury pattern recognition. LM is done on serial sections stained with hematoxylin and eosin (H&E) along with other stains that complement H&E, including periodic acid—Schiff reaction (PAS), Masson trichrome (MT), and Jones (methenamine silver—periodic acid) stains. For IF, frozen tissue is routinely examined for IgG, IgA, IgM, C3, C1q, fibrin (including fibrinogen and breakdown products), kappa and lambda (light chains), and albumin (which serves as a control). EM is done for ultrastructural analysis.

Introduction to the Nephropathology of SLE

The pathogenesis of antibody formation/immune complex deposition in SLE and in LN is addressed elsewhere in this book. Although the overwhelming majority of lesions seen in LN are due to immune complex deposition primarily in glomeruli, the spectrum of lesions produced is broad. The morphologic patterns of injury reflect the varied nature of the autoantibodies/immune complexes involved and their manner of deposition. Although multiple types of autoantibodies have been eluted from LN renal tissue, anti-dsDNA antibodies are thought to predominate [51]. Autoantibodies may form *in situ* with endogenous or exogenous antigens or, alternatively, arrive from the circulation as preformed immune complexes. Factors determining the location of deposition and injury pattern include immunoglobulin class and subclass

and the ability to activate complement and cellular immunity. Other factors include avidity, charge, size, ratio of antibody to antigen, as well as specificity, rate of production, and clearance. Relatively small amounts of complexes may deposit in the mesangium, and larger amounts may result in subendothelial deposits. Low-avidity cationic deposits may be more likely to deposit across the anionic glomerular basement membrane in the subepithelial space.

Although LM is the core technique for evaluating lesions that are the consequence of immune deposition, it is not the best way to identify the deposits themselves, particularly when they are small. IF is the primary technique that defines the nature of the deposits and, to a lesser degree, their precise location. IgG is almost always the dominant immunoreactant in LN, but IgA and IgM are frequently co-deposited. The so-called "full house pattern" consisting of all three immunoglobulins (IgG, IgA, and IgM), as well as complement fractions C1q and C3, is classic for LN. When equivalent deposition is seen for IgG and IgA, the possibility of IgA nephropathy must be considered. C1q implies activation of the classic complement pathway, and when strongly present, it may suggest LN [52]. Immune complex staining in LN is typically granular (as opposed to linear) and is graded from 0 to 4+. The location of glomerular deposition can be Bowman's capsule, urinary space, mesangium, and/or capillary wall. More precise localization within the capillary wall may be possible. Deposits with a smooth outer contour are more likely to be subendothelial as opposed to the more ragged edges of subepithelial deposits. When deposits are heavy, optical resolution may be reduced, making localization more difficult. A distinctive feature of LN is the presence of deposits in the other compartments of the kidney as well, and the presence of vascular, interstitial, and tubular basement membrane staining may be of important diagnostic significance and must be noted.

EM is complimentary to IF and precisely localizes the deposits and the ultrastructural alterations that accompany them. EM is poor at characterizing their nature because most deposits appear similar [53]. Immune deposition is an extracellular process, and the typically granular electron-dense deposits appear in association with matrix or membranes but are never membrane bound or intracytoplasmic. Their location is described as mesangial, subendothelial, intramembranous, and subepithelial (Figure 42.3).The simultaneous presence of deposits in multiple sites is highly suggestive of LN. Occasionally, the deposits show foci of substructure, such as alternating bands (fingerprinting), or larger tubules. Their presence may be associated with cryoglobulinemia (typically type III, mixed) (Figure 42.4).

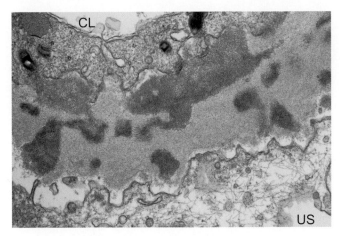

FIGURE 42.3 Granular electron-dense immune complex deposition in the glomerular capillary wall. Endothelial cells show acute injury (swelling with loss of fenestrations). Subendothelial deposits are seen to connect through a thickened basement membrane to subepithelial deposits. Scattered intramembranous deposits are also seen. Podocytes show injury with widespread foot process effacement. CL, capillary lumen; US, urinary space. EM, original magnification ×30 000.

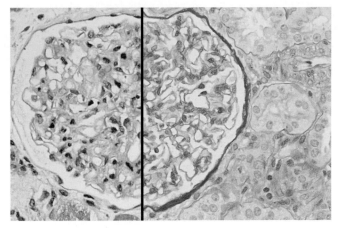

FIGURE 42.5 Normal glomerulus (class I pattern). Note the thin delicate patent capillary loops and a mesangium that is no more than mildly conspicuous. There is no hypercellularity (proliferation) within the tuft or in the urinary space. The montage demonstrates the difference between routine H&E (*left*) and PAS (*right*), which highlights mesangial matrix and basement membranes (glomerular and tubular) while only weakly staining cytoplasm. Original magnification ×400.

Lesions of Lupus Nephritis

Glomeruli

Familiarity with the normal is a prerequisite for appreciation of the pathologic. A normal glomerulus (Figure 42.5) shows thin, delicate patent capillary loops and a mesangium that is no more than mildly conspicuous. Table 42.4 lists a glossary of pathologic terminology. LM evaluation begins with an assessment of glomerular cellularity. Increased glomerular cellularity has historically, and somewhat inaccurately, been referred to as "proliferation" due to the difficulty in distinguishing infiltrating leukocytes (neutrophils

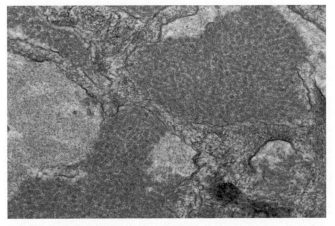

FIGURE 42.4 Electron-dense deposition within the mesangial matrix. These tubular substructures, typically measuring 25—45 nm, are typical of cryoglobulins (typically type III; mixed in LN). EM, original magnification ×40 000.

excepted) from native glomerular cells on the basis of H&E alone. Although hypercellularity may indeed be due to proliferation of cells native to the glomerulus, in many situations the increase is also due to the recruitment of exogenous inflammatory cells. Proliferation may be noted in the mesangial, endocapillary, or extracapillary (crescent) regions. The distribution of lesions is described within individual glomeruli (segmental vs. global) as well as the total number of glomeruli involved (focal vs. diffuse).

The sine qua non of LN is the deposition of immune complexes in the mesangial matrix. Glomerular anatomy helps account for the centrality of the mesangium in LN because it is in direct continuity with the subendothelial space and is separated from the capillary lumen by the endothelium alone. There is no basement membrane dividing it from the circulation. Mesangial immune complex deposition is recognized by IF (Figure 42.6) and EM (Figure 42.7). Deposition is usually associated with mesangial hypercellularity (three or more mesangial cells in the periphery of the tuft and mesangial matrix expansion, allowing it to be recognized on LM (Figure 42.8). Mesangial hypercellularity, by definition, does not involve the capillaries; thus, their lumens are patent and normocellular. The correlation between the degree of mesangial proliferation and immune deposition may be poor. Mesangial immune complex injury is typically expressed by hematuria, although the pathogenetic reason for this is unclear.

Use of the unqualified term "proliferative glomerulonephritis" in the context of LN refers to endocapillary proliferation (Figure 42.9). It typically develops on

TABLE 42.4   Glossary of Pathologic Terms[a]

Crescent (extracapillary proliferation)—Proliferation of cells in the urinary space, initially seen in association with fibrin, over time undergoes organization (fibrosis).

Diffuse—A lesion involving most (≥50%) glomeruli.

Double contours—Thickened glomerular capillary walls with reduplication of the basement membrane due to the growth of an inner neo-membrane, usually in response to mesangial interposition/subendothelial immune deposits.

Endocapillary proliferation—Endocapillary hypercellularity due to increased number of mesangial cells, endothelial cells, and infiltrating monocytes and causing narrowing of the glomerular capillary lumina.

Extracapillary proliferation or cellular crescent—Extracapillary cell proliferation of more than two cell layers occupying one-fourth or more of the glomerular capillary circumference.

Fibrinoid necrosis—Destruction of cells and matrix associated with fibrin.

Fibrocellular crescent—Proliferation that has organized, culminating in a fibrous crescent.

Focal—A lesion involving less than 50% of glomeruli.

Global—A lesion involving more than half of the glomerular tuft.

Hyaline thrombus—Intracapillary eosinophilic material of a homogeneous consistency that by immunofluorescence is shown to consist of immune deposits.

Intramembranous—Within the lamina densa of the glomerular basement membrane.

Karyorrhexis—The presence of apoptotic, pyknotic, and fragmented nuclei.

Membranoproliferative—Glomerular pattern with prominent double contour formation.

Mesangial hypercellularity (proliferation)—At least three mesangial cells per mesangial region in a 3-μm-thick section.

Mesangial interposition—Extension of cytoplasm into the subendothelial space producing double contours.

Necrosis—Characterized by fragmentation of nuclei or disruption of the glomerular basement membrane, often associated with the presence of fibrin-rich material.

Proportion of involved glomeruli—Intended to indicate the percentage of total glomeruli affected by LN, including the glomeruli that are sclerosed due to LN, but excluding ischemic glomeruli with inadequate perfusion due to vascular pathology separate from LN.

Sclerosis—Glomerular scarring by expansion of matrix and loss of normal architecture.

Segmental—A lesion involving less than half of the glomerular tuft (i.e., at least half of the glomerular tuft is spared).

Subendothelial—Space between the endothelium and the glomerular basement membrane.

Subepithelial (epimembranous)—Space between the visceral epithelial cell (podocyte) and the glomerular basement membrane.

Wire loops—Thickened hypereosinophilic segment of glomerular capillary wall due to large subendothelial immune deposits.

[a]Terms in bold as defined by the ISN/RPS classification.

a background of mesangial proliferation, but it is distinguished by hypercellular involvement of the capillary lumens. Endocapillary proliferation can be quite variable, and its glomerular distribution throughout the sample must be quantitated (Figures 42.10 and 42.11). The cells are usually a mixture of proliferating mesangial and endothelial cells, in association with infiltrating leukocytes. Heavy immune complex deposition (almost always IgG dominant), particularly in the subendothelial space, is common, and proliferation tends to parallel the distribution of subendothelial immune deposits because their location makes them most accessible to the circulation (Figure 42.12). The degree of subendothelial deposition correlates with clinical severity. Large subendothelial deposits that appear to thicken the capillary wall on LM are called "wire loops" (Figure 42.13).

Plugs of immune complexes located within the capillary lumen are descriptively known as "hyaline thrombi" despite the fact that they are neither hyaline nor thrombi (Figure 42.14). Serial examination shows many of them to be in continuity with subendothelial deposits. Intriguingly, glomerular lobules with these "thrombi" often show less proliferation then surrounding ones. True glomerular fibrin thrombi may be seen when a thrombotic microangiopathy (TMA) is superimposed. The hematoxylin body is the tissue counterpart to the LE body, and it is composed of basophilic naked nuclei that are coated by antinuclear antibodies. These are rare but specific for LN and thus diagnostically helpful when present.

Endocapillary proliferative lesions are often associated with segmental necrosis of the glomerular tuft/capillary

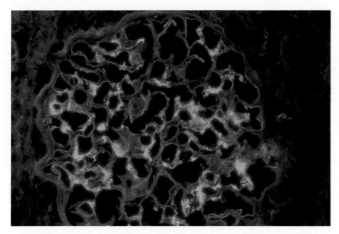

FIGURE 42.6 Mesangial immune complex deposition (class I or II pattern). Capillary walls show no positivity, and their lumens are patent. IF, anti-IgG; original magnification ×400.

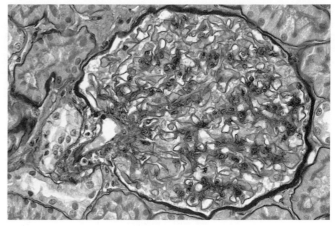

FIGURE 42.8 Mesangial proliferation (class II Pattern). Note that the capillary lumens are patent and normocellular. The vascular pole/macula densa is seen at 8 o'clock. Surrounding proximal tubules are healthy and show intact brush borders. PAS, original magnification ×400.

wall. Necrosis may be identified by the presence of endocapillary fibrin, neutrophilic exudation, apoptosis, karyorrhexis, and/or fragmentation of the basement membrane (Figure 42.15). When relatively mild, segmental inflammation may produce adhesion of capillaries to Bowman's capsule. More commonly, severe inflammation, particularly fibrinoid necrosis, may be followed by frank rupture of glomerular capillaries with hemorrhage into Bowman's space inciting formation of cellular crescents (Figure 42.16). Cellular crescents, defined as aggregates of two or more layers of cells that line one-fourth or more of the interior circumference of Bowman's capsule, are composed of a combination of proliferating epithelial cells and infiltrating macrophages and lymphocytes (Figure 42.17). Over time, persistent

crescents undergo organization into fibrocellular and ultimately fibrous crescents, often seen in association with glomerular obsolescence (Figure 42.18).

Persistence of subendothelial deposits may result in a membranoproliferative pattern, reflecting reactive basement membrane/double contour formation (Figure 42.19). In addition to the newly formed basement membrane material and subendothelial electron-dense deposits, EM often reveals the presence of cytoplasm in the subepithelial space. In some cases, these cellular segments are seen in continuity with the mesangium, leading to the term "mesangial interposition" (Figure 42.20). However, it is now clear that in many cases the cytoplasm is that of inflammatory cells.

Some degree of subepithelial immune complex deposition is common in LN and is often seen in association with reactive changes of the basement membrane. These

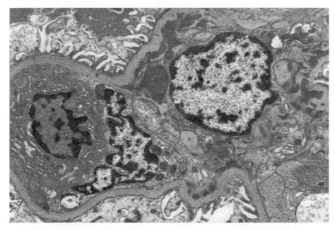

FIGURE 42.7 Extensive granular electron-dense deposits in the mesangium and paramesangium (class II pattern). The capillary lumen shows an incidental circulating inflammatory cell, but no deposits are seen there. Note that the mesangial region is in direct continuity with the subendothelial space and is separated from the capillary lumen by endothelium alone. There is no glomerular basement membrane dividing it from the circulation. Variable podocyte foot process effacement is also seen. EM, original magnification ×8000.

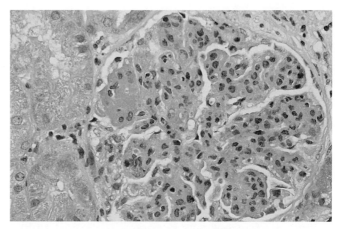

FIGURE 42.9 Global endocapillary proliferation (class III or IV pattern). Capillary lumens are occluded by the hypercellularity. The hypereosinophilic focus at 10 o'clock represents a confluent focus of immune complex deposition. Interestingly, proliferation is limited in that area. Proximal tubules are normal. H&E, original magnification ×400.

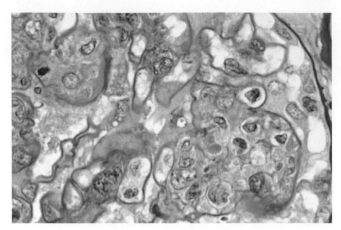

FIGURE 42.10 Segmental endocapillary proliferation (class III or IV-S Pattern). The hypercellularity occludes the capillary lumen and obscures the mesangial/capillary lumen architecture, but is bounded by the peripheral capillary wall. Some of the other loops show no proliferation. Bowman's capsule is seen in upper right. PAS, original magnification ×400.

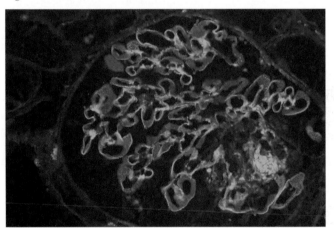

FIGURE 42.11 Segmental endocapillary immune complex deposition with global mesangial deposits (class III or IV-S pattern). Some of the capillary loops also show small granular deposits. IF, anti-IgG; original magnification ×400.

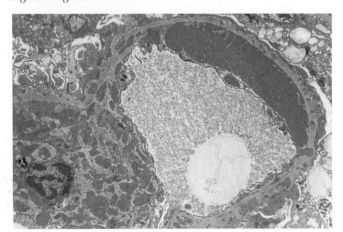

FIGURE 42.12 Heavy subendothelial and mesangial immune complex deposition (class III or IV pattern). As is common in LN, scattered subepithelial deposits, accompanied by widespread podocyte foot process effacement, are also present. EM, ×6000.

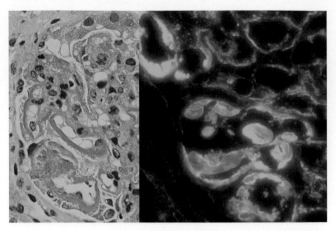

FIGURE 42.13 Subendothelial immune complex deposition by LM (*left*) and IF (*right*) (class III or IV pattern). When the deposits are large, they can be visualized on H&E, where they thicken the capillary wall ("wire loops"). Anti-IgG staining shows the large deposits to be primarily subendothelial with a smooth outer contour where they abut the glomerular basement membrane. Deposits also project in the capillary lumen. Original magnification ×400.

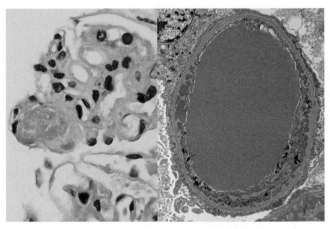

FIGURE 42.14 "Hyalin thrombus" by LM (*left*) and EM (*right*) (class III or IV pattern). A large plug of immune complexes is seen within a capillary lumen. H&E, original magnification 400×. Ultrastructural study shows the intracapillary component to be accompanied by near circumferential subendothelial deposition. EM, original magnification ×6000.

membranous changes are seen to thicken the wall on H&E, but they are best identified on Jones stain, IF, and EM (Figures 42.21−42.24). Deposits in this location, "protected" from the circulation, are not thought to lead to inflammation. However, subepithelial deposition may be associated with significant endocapillary proliferation due to deposits at other sites.

Ultrastructural study almost always discloses frequent tubuloreticular structures (or inclusions), most often in glomerular endothelial cells but occasionally elsewhere (Figure 42.25). These 24-nm interanastomosing structures, relating to the rough endoplasmic reticulum, are associated with elevated levels of circulating interferon, and

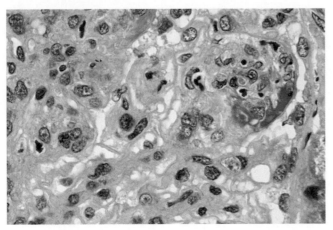

FIGURE 42.15 Segmental fibrinoid necrosis with associated karyorrhectic debris, superimposed on endocapillary proliferation (class III (A) or IV (A) pattern). The capillary wall is intact. Bowman's capsule is seen as a blue line along the right frame. MT, original magnification ×400.

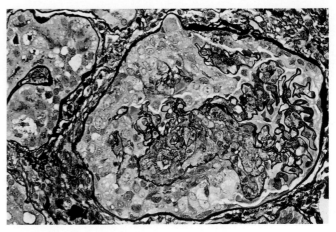

FIGURE 42.17 Cellular crescent filling urinary space and compressing the glomerular tuft, which shows endocapillary proliferation (class III or IV (A) pattern). Composed primarily of native glomerular epithelium, macrophages, and lymphocytes, cellular crescents may completely resolve in response to therapy, although the area of segmental necrosis within the tuft usually heals as segmental sclerosis. Jones, original magnification ×400.

their frequent occurrence in LN underscores the central role of interferon in this disease [54].

The active changes described so far (immune deposition, proliferation, necrosis, etc.) may, in time, progress to chronic scarring. Serial biopsies suggest that segmental necrosis, for example, often results in segmental sclerosis. More severe glomerular injury or crescent formation may ultimately result in global sclerosis. Of course, these changes, generally thought to be irreversible, may also be due to other unrelated disease processes such as hypertension and aging. Glomerulosclerosis that can be attributed to post-inflammatory scarring (as suggested by underlying proliferation) is considered involvement by LN in the current classification scheme; sclerosis due to vascular disease is not.

Tubulointerstitium

Interstitial edema and inflammation, predominantly composed of mononuclear leukocytes (lymphocytes, plasma cells, and macrophages), are seen to varying degrees in many cases of LN. The inflammatory infiltrates may be associated with tubulitis and other forms of acute tubular degenerative and regenerative changes. IF and EM often reveal immune deposits along tubular basement membranes (virtually pathognomonic for LN) and even within the interstitium (Figure 42.26). Although interstitial infiltrates may rarely be seen in the absence of significant glomerular disease, the extent of interstitial inflammation usually parallels the severity of glomerular proliferation. However, there is no

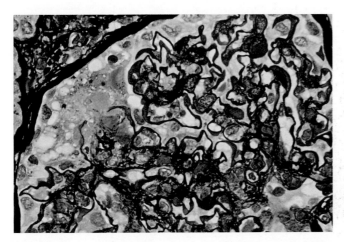

FIGURE 42.16 Segmental necrosis with rupture of the capillary wall and passage of fibrin and other inflammatory mediators into the urinary space (class III (A) or IV (A) pattern). This process initiates crescent formation. Jones, original magnification ×400.

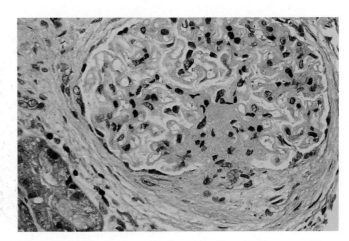

FIGURE 42.18 Segmental sclerosis with adhesion to a fibrous crescent (class III (C) or IV (C) pattern). This scar is likely the sequel of prior necrosis and crescent formation. This lesion will not respond to anti-inflammatory therapy. MT, original magnification ×400.

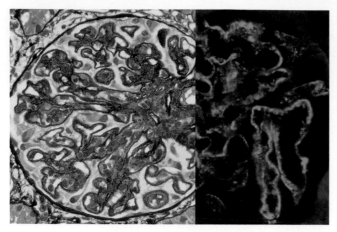

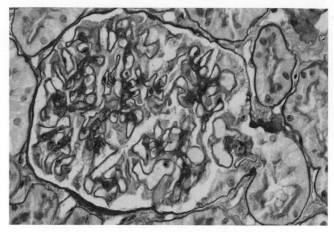

FIGURE 42.19 Membranoproliferative pattern of glomerular injury by LM (*left*) and IF (*right*) (class III-G or IV-G (A) pattern). Subendothelial deposits are seen by Jones stain to be bounded by inner neo-membrane "double contours." Anti-IgG staining shows the subendothelial deposits that have incited this reaction pattern. Original magnification ×400.

FIGURE 42.21 Membranous pattern with minimal mesangial proliferation (class V pattern). The capillary loops, while patent and normocellular, show thickened walls. The urinary pole is seen at 4 o'clock. Surrounding proximal tubules are healthy and show intact brush borders. PAS, original magnification ×400.

correlation between tubulointerstitial inflammation and tubular immune deposition [41]. The degree of interstitial inflammation has been correlated with both reduction in GFR and serologic activity [55].

Nuclei in any location may show staining for IgG on IF ("tissue ANA"). This phenomenon, likely an artifact of tissue sectioning, confirms the presence of autoantibodies but is not otherwise diagnostically significant (Figure 42.27).

The degree of cortical fibrosis is best assessed on MT stain, where expansion of the normally inconspicuous interstitium appears as varying intensities of blue, reflecting the density of collagen deposition (Figure 42.28).

Edema, signifying an acute process, appears as clear space. As in all other renal diseases, the condition of the tubulointerstitium (both acute inflammation and chronic scarring) is the histologic parameter that best correlates with renal function and is most reflective of prognosis. Morphometric evaluation has been shown to improve the assessment of interstitial injury but has not been adopted in clinical practice [56]. As tubules progress from acute injury to chronic atrophy, they undergo changes that result in diminished cytoplasm and diameter, thickening of basement membranes (best seen on PAS stain), and cast formation (Figure 42.29). Like interstitial fibrosis, these processes are typically thought to be irreversible.

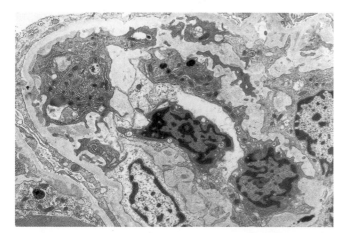

FIGURE 42.20 Membranoproliferative pattern of glomerular injury by EM (class III or IV (A) pattern). Granular electron-dense deposits are seen within a markedly thickened capillary wall. Cells are interposed between the outer basement membrane and an inner neo-membrane formed in response to the deposits (appreciated by LM as "double contours"). Endothelial cells encircle a markedly reduced capillary lumen. Original magnification ×10 000.

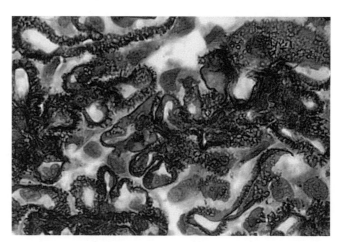

FIGURE 42.22 Membranous pattern with extensive "spikes" and "holes" in the capillary walls (class V pattern). The Jones stain, which highlights the basement membranes, does not stain immune complexes, resulting in this appearance. Original magnification ×600.

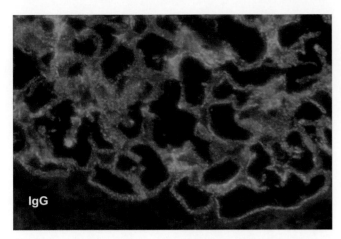

FIGURE 42.23 Membranous pattern of global subepithelial granular deposition (class V pattern). The capillary loops are patent, but mesangial regions also show deposits, seen as the zones of more intense staining. IF, anti-IgG; original magnification ×400.

Vessels

The most common changes seen on biopsy are nonspecific sclerotic lesions secondary to hypertension, aging, and atherosclerosis. Vascular lesions that are specific to SLE receive little attention, in part because they are absent from all of the classification systems. Furthermore, the terminology used to describe these lesions has been confused. Their clinical significance ranges from trivial to profound. A schema is presented in Table 42.5 [57]. In the most common lesion, consisting of uncomplicated immune deposits, vessels (predominantly small arteries and arterioles, and sometimes veins) display immune deposits usually containing IgG by IF (Figure 42.30). The presence of IgM or complement alone can reflect nonspecific injury and is therefore not diagnostic of a lupus-related process. EM localizes the deposits to the intimal

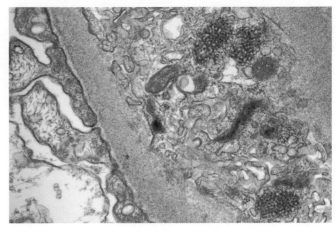

FIGURE 42.25 Tubuloreticular structures within a glomerular endothelial cell. These 24-nm structures, relating to the rough endoplasmic reticulum, are associated with elevated levels of circulating interferon, and their frequent occurrence in LN underscores the central role of interferon in this disease. Normal podocyte foot processes are seen on the other side of the membrane. EM, original magnification ×30 000.

basement membrane and media. The degree of deposition roughly correlates with glomerular proliferation and tubulointerstitial deposition. These deposits are usually clinically silent [58].

In contrast, lupus vasculopathy (non-inflammatory necrotizing vasculopathy), although far less common, is associated with a poor prognosis [59]. This is seen most commonly in the setting of severe proliferative LN. Vessels (usually arterioles) show luminal narrowing by immune deposits accompanied by endothelial and medial injury. Foci of fibrinoid necrosis may be present, but inflammation is absent (Figures 42.31 and 42.32).

True vasculitis consisting of fibrinoid necrosis of the vessel wall accompanied by inflammation is a very rare finding in LN. When present, the histologic

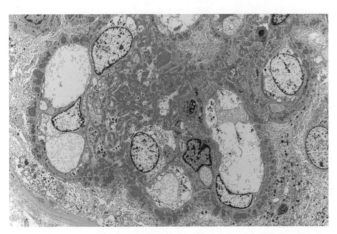

FIGURE 42.24 Membranous pattern with extensive subepithelial deposits and "spike" formation by the basement membranes (class V pattern). Intense mesangial deposition is also seen. Podocytes show diffuse injury. EM, original magnification ×2500.

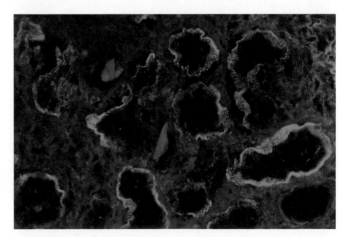

FIGURE 42.26 Granular immune complex deposition along tubular basement membranes. This finding is highly suggestive of LN. IF, anti-IgG; original magnification ×400.

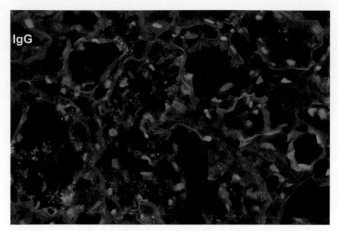

FIGURE 42.27 Nuclei of tubular epithelium showing diffuse coating by IgG ("tissue ANA"). This finding confirms the presence of autoantibodies. IF, anti-IgG; original magnification ×400.

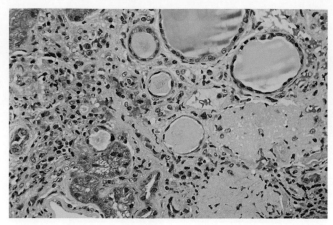

FIGURE 42.29 Cortex with end-stage renal injury (class VI pattern). There is extensive interstitial fibrosis and tubular atrophy/cystic change. Two shrunken globally sclerotic glomeruli are seen. The advanced scarring may make it impossible to determine the causative process. MT, original magnification ×400.

appearance is indistinguishable from a systemic vasculitis of the ANCA type, although immune deposits may be present. Indeed, it may represent the simultaneous occurrence of a systemic or "renal limited" vasculitis that is unrelated to LN.

Lupus-associated TMAs show no significant immune deposition (IgM and C3 may be present, secondary to vascular injury). TMAs in the setting of lupus may be seen in association with any of the general TMA clinical syndromes (e.g., malignant hypertension, systemic sclerosis, and hemolytic uremic syndrome/thrombotic thrombocytopenic purpura (TTP)) or in the setting of the antiphospholipid antibody syndrome. TMAs in SLE can also seen in the absence of any particular syndrome. Pathologically, small arteries and arterioles may show acute fibrin thrombosis, fibrinoid necrosis,

mucoid intimal hyperplasia ("onion skinning") (Figure 42.33), and chronic changes such as fibrous intimal hyperplasia and focal cortical atrophy that are the consequences of previous clotting events. Glomeruli show typical TMA changes, such as thrombosis, mesangiolysis, and double contour formation. These may be superimposed on any of the glomerular changes of LN.

Classification of Lupus Nephritis

The heterogeneity of lesions seen in LN was noted beginning with early autopsy studies [60]. The introduction of the kidney biopsy and the subsequent application of IF and EM, along with an improved understanding of the nature of SLE, led to a pioneering period of characterization and early classification in the 1950s and 1960s [61]. These early classification schemes divided glomerular changes into mild, severe proliferative, and membranous. The modern era of classification began in 1974 with the formulation by Pirani and Pollak of

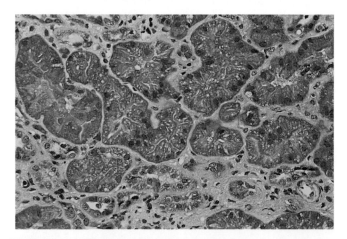

FIGURE 42.28 Cortical tubulointerstitium with signs of chronic renal injury. The normally inconspicuous interstitium shows varying degrees of fibrosis (blue). Many proximal tubules show hypertrophy, indicating nephron loss. A minority show atrophy. This compensatory process may initially maintain GFR and thus can only be recognized by biopsy. MT, original magnification ×400.

TABLE 42.5 Vascular Lesions in Lupus

Lupus specific

   Uncomplicated immune complex deposition

   Lupus vasculopathy

   Necrotizing arteritis

Lupus related

   Thrombotic microangiopathy

Non-lupus related

   Arteriosclerosis

   Arteriolar sclerosis

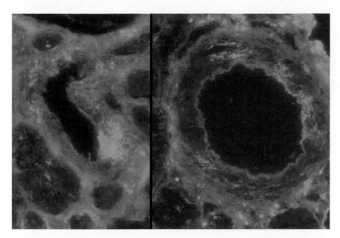

FIGURE 42.30 Uncomplicated immune deposits in the intima and media of an arteriole (*left*) and artery (*right*). The lumen is not compromised. This finding is highly suggestive of LN but does not carry an adverse prognosis. IF, anti-IgG; original magnification ×400.

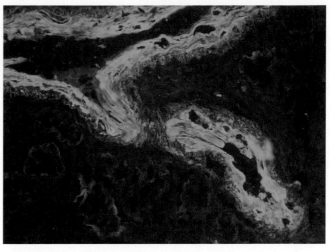

FIGURE 42.32 Lupus vasculopathy. Massive immune deposits are seen within the walls of an artery and its branch, with vascular lumenal occlusion. IF, anti-IgG; original magnification ×400.

what came to be known as the World Health Organization (WHO) classification of lupus nephritis [62, 63]. Vascular and tubulointerstitial lesions were not a part of the system. Widely adopted, it served as the framework for many subsequent and ultimately confusing adjustments. In 1982, a modification by the International Study of Kidney Diseases in Children was published [64]. It added a class VI—advanced sclerosing glomerulonephritis—but did not stipulate a percentage of glomerular involvement. Although it emphasized assessment of acute and chronic lesions, which was useful, it created numerous subclasses that made it too cumbersome for wide adoption. To further complicate matters, in 1995 WHO recommended that all cases of proliferative LN, inasmuch as they represent

a continuum, be included in class IV, and that class III be reserved for focal segmental necrotizing lesions [65]. Thus, over time a rather confusing situation developed, defeating the primary purpose of having a classification system altogether. In response, the International Society of Nephrology/Renal Pathology Society (ISN/RPS) convened a consensus conference in 2002 that produced an updated system with standard definitions and well-defined classes emphasizing "clinically relevant lesions" and encouraging "uniform and reproducible reporting between centers" [66, 67]. As in all prior systems, classification was glomerulocentric, but reporting of other lesions was encouraged. The incorporation of glomerular activity and chronicity (percentage involved) was also to be indicated. The ISN/RPS

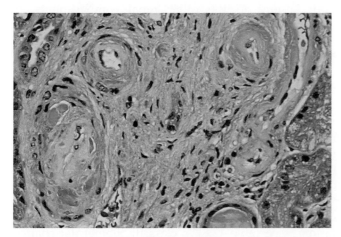

FIGURE 42.31 Lupus vasculopathy. Immune deposits are seen within the walls of these four vessels (one in longitudinal section), associated with vascular injury and lumenal occlusion. Seen most commonly in the setting of severe proliferative LN, this finding is associated with a poor prognosis. H&E, original magnification ×400.

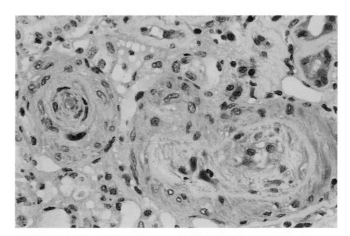

FIGURE 42.33 Thrombotic microangiopathy. Vessels show acute fibrin thrombosis, fibrinoid necrosis, and mucoid intimal hyperplasia ("onion skinning"). Immune complex deposition is not a part of this lesion. MT, original magnification ×400.

TABLE 42.6 Abbreviated ISN/RPS 2003 Classification of Lupus Nephritis

Class I Minimal mesangial lupus nephritis

Class II Mesangial proliferative lupus nephritis

Class III Focal lupus nephritis[a]

Class IV Diffuse segmental (IV-S) or global (IV-G) lupus nephritis[b]

Class V Membranous lupus nephritis[c]

Class VI Advanced sclerosing lupus nephritis

[a]Indicate the proportion of glomeruli with active and with sclerotic lesions.
[b]Indicate the proportion of glomeruli with fibrinoid necrosis and cellular crescents.
[c]Class V may occur in combination with class III or IV, in which case both will be diagnosed.
Indicate and grade (mild, moderate, severe) tubular atrophy, interstitial inflammation and fibrosis, severity of arteriosclerosis or other vascular lesions.

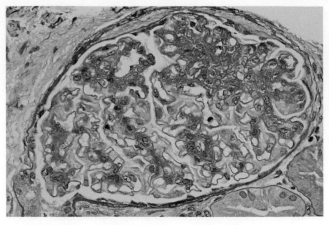

FIGURE 42.34 Segmental sclerosis. This lesion, if thought to be secondary to a prior active lesion, is incompatible with a class II designation and would be indicative of class III or IV with a chronic (C) designation. PAS, original magnification ×400.

classification has been rapidly and widely adopted, and it has essentially replaced all prior systems. Nevertheless, an understanding of the nuances of prior systems remains important to properly evaluate the older literature studies and compare them to current studies.

An overview of the ISN/RPS classification is provided in Table 42.6, with details in Table 42.7. The system, based exclusively on glomerular findings, requires the characterization of each glomerulus in the sample based on the integration of both LM and IF findings. To facilitate adoption throughout the world, the system was designed to be EM independent, but occasional cases require EM to define the glomerular lesion. The classification also requires the separate enumeration and grading of tubular atrophy, interstitial inflammation and fibrosis, and vascular lesions, together with all other pathologic processes, glomerular or otherwise, in the diagnostic section of the pathology report. The system affixes a mnemonic within each class to designate the nature of its lesions: purely active lesions (A), purely chronic lesions (C), or any combination of the two (A/C). These terms are defined in Table 42.8.

Class I (Minimal Mesangial LN)

CLINICAL Although these patients may have active extrarenal lupus, there is usually no clinical evidence of renal disease.

PATHOLOGY The glomeruli are normal by light microscopy, but mesangial immune deposits are seen by IF and EM. No deposits are seen along the capillary walls.

Class II (Mesangial Proliferative LN)

CLINICAL Proteinuria and/or hematuria may be seen in class II LN, but renal insufficiency and nephritic syndrome would be unexpected.

PATHOLOGY Glomerular changes by LM are usually limited to mesangial proliferation and mesangial matrix expansion of any degree. Mesangial deposits are rarely large enough to be see by LM but are easily identified by IF and EM. The recognition of subendothelial deposits by LM indicates class III or IV. However, small subendothelial deposits seen by IF, or more usually by EM, may be present, particularly in the region of continuity with the mesangium. Whether this finding suggests a more ominous prognosis is unclear. Of note, the presence of segmental or global sclerosis that is interpreted to represent the scars of prior proliferative, necrotizing, or crescentic lesions is incompatible with this class because they represent the chronic lesions of either class III or class IV. For example, a biopsy in which all glomeruli show only mesangial proliferation would not be class II if more than 50% of glomeruli showed scarring due to prior active lesions (Figure 42.34). Such a biopsy would be class IV with the C (chronic) designation. Lesions in the other renal compartments are unusual for class II and suggest a superimposed process.

Class III (Focal LN)

CLINICAL Hematuria and proteinuria are seen in most patients, and some may have nephrotic syndrome and renal insufficiency, unless the lesions are predominantly chronic, in which case the clinical picture may not show much activity. The clinical picture of class III is like that of class IV as the number of involved glomeruli approaches 50%.

PATHOLOGY Classes III and IV (formerly the "proliferative" classes) are defined by the percentage of glomeruli involved by "active" and/or "chronic" lesions. (see Table 42.8) In class III, glomerular changes (acute and/or chronic) are focal, which means they involve less than 50% of glomeruli in the sample. Cases in which glomerular involvement approaches 50%, particularly those with segmental necrosis, may behave

TABLE 42.7   ISN/RPS 2003 Classification of Lupus Nephritis

Class I Minimal mesangial lupus nephritis

Normal glomeruli by light microscopy, but mesangial immune deposits by immunofluorescence.

Class II Mesangial proliferative lupus nephritis

Purely mesangial hypercellularity of any degree or mesangial matrix expansion by light microscopy, with mesangial immune deposits.

May be a few isolated subepithelial or subendothelial deposits visible by immunofluorescence or electron microscopy, but not by light microscopy.

Class III Focal lupus nephritis[a]

Active or inactive focal, segmental, or global endo- or extracapillary glomerulonephritis involving <50% of all glomeruli, typically with focal subendothelial immune deposits, with or without mesangial alterations.

Class III (A) Active lesions: focal proliferative lupus nephritis

Class III (A/C) Active and chronic lesions: focal proliferative and sclerosing lupus nephritis

Class III (C) Chronic inactive lesions with glomerular scars: focal sclerosing lupus nephritis

Class IV Diffuse lupus nephritis[b]

Active or inactive diffuse, segmental, or global endo- or extracapillary glomerulonephritis involving ≥50% of all glomeruli, typically with diffuse subendothelial immune deposits, with or without mesangial alterations. This class is divided into diffuse segmental (IV-S) lupus nephritis, when ≥50% of the involved glomeruli have segmental lesions, and diffuse global (IV-G) lupus nephritis, when ≥50% of the involved glomeruli have global lesions. Segmental is defined as a glomerular lesion that involves less than half of the glomerular tuft. This class includes cases with diffuse wire loop deposits but with little or no glomerular proliferation.

Class IV-S (A) Active lesions: diffuse segmental proliferative lupus nephritis

Class IV-G (A) Active lesions: diffuse global proliferative lupus nephritis

Class IV-S (A/C) Active and chronic lesions: diffuse segmental proliferative and sclerosing lupus nephritis

Class IV-G (A/C) Active and chronic lesions: diffuse global proliferative and sclerosing lupus nephritis

Class IV-S (C) Chronic inactive lesions with scars: diffuse segmental sclerosing lupus nephritis

Class IV-G (C) Chronic inactive lesions with scars: diffuse global sclerosing lupus nephritis

Class V Membranous lupus nephritis

Global or segmental subepithelial immune deposits or their morphologic sequelae by light microscopy and by immunofluorescence or electron microscopy, with or without mesangial alterations.

Class V lupus nephritis may occur in combination with class III or IV, in which case both will be diagnosed.

Class V lupus nephritis may show advanced sclerosis.

Class VI Advanced sclerosis lupus nephritis

≥90% of glomeruli globally sclerosed without residual activity.

[a]*Indicate the proportion of glomeruli with active and with sclerotic lesions.*
[b]*Indicate the proportion of glomeruli with fibrinoid necrosis and/or cellular crescents.*
Indicate and grade (mild, moderate, severe) tubular atrophy, interstitial inflammation and fibrosis, severity of arteriosclerosis or other vascular lesions.

more like class IV. Typical active lesions include endocapillary and/or extracapillary proliferation, as well as necrosis, wire loops, hyaline thrombi, and a membranoproliferative pattern. Mesangial proliferation is typically present as well. Chronic lesions include segmental or global glomerulosclerosis attributed to LN. Very focal lesions may be missed as a result of limited sampling, leading to an inappropriate designation as class II. Glomerular necrosis can be present in either class III or class IV but is never found in other classes. Both A and C glomerular alterations may be segmental (S) and/or global (G), and they are subclassified based on the predominant pattern. In most cases, class III lesions are primarily segmental (Figure 42.35). IF and EM display significant subendothelial immune deposits in addition to mesangial deposits. Inasmuch as subendothelial deposits may be focal, and mesangial deposits are generally diffuse, limited tissue sampling (especially on EM) may create the false appearance of class II. A minor component of subepithelial deposits may be present, but when they are extensive an additional class V designation is warranted. A subset of patients with focal glomerulonephritis show extensive segmental necrotizing lesions and crescents in association with

TABLE 42.8 Active and Chronic Glomerular Lesions:
ISN/RPS 2003 Classification of Lupus Nephritis

Active lesions

    Endocapillary hypercellularity with or without leukocyte infiltration and with substantial luminal reduction

    Karyorrhexis

    Fibrinoid necrosis

    Rupture of glomerular basement membrane

    Crescents, cellular or fibrocellular

    Subendothelial deposits identifiable by light microscopy (wire loops)

    Intraluminal immune aggregates (hyaline thrombi)

Chronic lesions

    Glomerular sclerosis (segmental, global)

    Fibrous adhesions

    Fibrous crescents

limited immune deposition on IF and EM. This pattern raises the possibility of a pauci-immune glomerular disease similar to an ANCA-positive vasculitis [68].

LN, as a recurring, relapsing condition, often results in the simultaneous presence of both active and sclerosing lesions. Class III biopsies usually show some degree of acute and/or chronic tubulointerstitial injury.

Class IV (Diffuse LN)

CLINICAL Patients with class IV LN typically have the most severe renal injury. Almost all patients will have hematuria and proteinuria, and nephrotic syndrome and renal insufficiency are not uncommon. If the lesions are mainly chronic, there may not be much clinical activity other than progressive renal failure.

PATHOLOGY By definition, a majority of the glomeruli in this class are involved by lupus-related lesions. Whereas classes III and IV together often represent a spectrum of changes, it is in class IV that the histopathology of active LN typically finds its fullest expression. Although almost all cases exhibit some degree of endocapillary proliferation, the presence of extensive subendothelial deposits alone or in association with a membranoproliferative pattern is sufficient for a classification of class IV. Cases are further subclassed based on whether the lesions are predominantly segmental (S) or global (G). In contrast to class III, most cases are (G). The percentage of glomeruli with crescents and/or necrotizing lesions is recorded. As in class III, glomerulosclerosis, when thought to be due to LN, is considered evidence of prior active lesions. Lupus-related vascular disease is most commonly found

in this class, but it involves only a minority of cases. In contrast, tubulointerstitial disease is common.

Class V (Membranous LN)

CLINICAL The dominant finding in pure class V LN is proteinuria, usually nephrotic range. Hematuria may be present. Renal function is generally normal. If class III or IV is also present, the urine sediment may show more activity, and renal dysfunction is more likely.

PATHOLOGY Scattered subepithelial deposits are seen in many cases of active LN. Thus, involvement (defined as continuous subepithelial immune deposition and/or resultant basement membrane reactive changes) of the majority of the basement membranes of the majority of glomeruli is required for class V. This represents an improvement from the 1982 modified WHO classification, which included combined classes as subsets of class V membranous (i.e., what is now III + V was Vc, and IV + V was Vd) and obscured the fact that these combined classes behave clinically like "proliferative" ones. The changes typically visible by LM are primarily those of the basement membrane reaction to the deposits and not the immune deposits themselves. Thus, in early stages the subepithelial deposits may only be visible by IF and EM. By definition, the presence of significant subendothelial deposits or endocapillary proliferation warrants the addition of class III or IV (as appropriate) (Figure 42.36). As the deposits evolve, typical membranous changes such as "spikes" and vacuolation become visible, especially by PAS and Jones stains. In more chronic cases, the deposits become incorporated into the membranes, may lose their antigenicity, and become electron lucent. Segmental or global glomerulosclerosis may be part of this process. However, as in all other classes, the

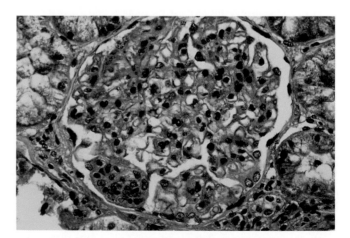

FIGURE 42.35   Segmental endocapillary proliferation may be seen in class III or IV-S. When lesions are focal and small, they may be missed due to limited tissue sampling, leading to a false class II designation. MT, original magnification ×400.

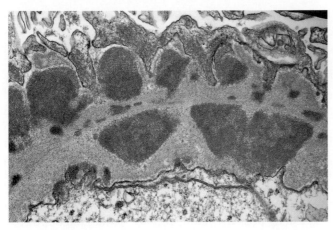

FIGURE 42.36 The simultaneous presence of extensive subepithelial and subendothelial immune deposits warrants the addition of class III or IV (focal or diffuse LN, as appropriate) to the class V (membranous) designation. EM may be necessary to identify such cases, which have a significantly worse prognosis than pure class V. EM, original magnification ×30 000.

presence of glomerular scars judged to be the sequela of prior activity warrants an additional class III or IV (as appropriate).

Distinguishing pure class V LN from idiopathic membranous glomerulonephritis is sometimes challenging. The presence of any of the following findings suggests an SLE-related membranous glomerulonephritis: mesangial hypercellularity; mesangial immune deposits; "full house" pattern of staining on IF; significant positivity for C1q; subendothelial deposits; vascular, tubular, or interstitial immune deposits; tissue ANA; and tubuloreticular structures [69].

Class V patients are also at increased risk for the development of renal vein thrombosis. Acute thrombosis may result in interstitial edema and/or hemorrhage, as well as glomerular congestion, thrombosis, and neutrophil margination. Chronic thrombosis is suggested by interstitial fibrosis and tubular atrophy that is out of proportion to glomerulosclerosis.

Class VI (Advanced Sclerosing LN)

CLINICAL The dominant clinical picture is that of progressive renal insufficiency, usually accompanied by proteinuria and sometimes hematuria.

PATHOLOGY By definition, at least 90% of the glomeruli show sclerosis, usually global, and there must be no signs of activity. This is usually accompanied by marked interstitial fibrosis and tubular atrophy, as well as chronic vascular injury. Globally sclerotic glomeruli may lose their immunoreactivity with time and thus appear negative by IF. Most of these cases represent "burned out" class III or IV LN, although class V may also be a cause.

Issues in Classification

Activity, Chronicity, Plasticity, and Prognosis

Immunosuppression is typically most effective against "active" lesions, and patients with considerable chronicity may show poor response to therapy. Thus, treatment decisions are often based on the degree of activity (reversible) and chronicity (irreversible). Furthermore, there can be a large degree of variability in the activity and chronicity present within a particular class (especially III and IV) that is not fully expressed by the A/C designations. For example, a biopsy in which 51% of the glomeruli show endocapillary proliferation without necrosis, wire loops, or crescents would be designated class IV (A), just as would one in which 95% of the glomeruli show proliferation along with crescents, necrosis, and wire loops. Because the approach to therapy may be different in these two cases, reviewing the biopsy with the nephropathologist is essential before finalizing treatment plans. Data suggest that chronicity may convey more prognostic information than class [70]. Thus, many practitioners continue to find semiquantitative scoring useful in conveying a snapshot of current disease and for comparison with subsequent biopsies. Pirani *et al.* [71] were the first to introduce the concept of "active" and "chronic" lesions. Austin *et al.* [72] devised a semiquantitative scoring system, known as the NIH scheme (Table 42.9), that has been widely adopted. Activity is scored from 0 to 24 using six histological parameters that are semiquantitatively graded from 0 to 3. The numerical weight of cellular crescents and glomerular necrosis is doubled to reflect their clinical severity. Chronicity is scored from 0 to 12 using

TABLE 42.9   NIH Activity and Chronicity Indexes

Activity index (0–24)

   Endocapillary hypercellularity: 0–3

   Glomerular neutrophils (more than two per glomerulus): 0–3

   Karyorrhexis/fibrinoid necrosis: (0–3) × 2

   Cellular crescents: (0–3) × 2

   Hyaline deposits (thrombi or wire loops): 0–3

   Interstitial inflammation: 0–3

Chronicity index (0–12)

   Glomerular sclerosis: 0–3

   Fibrous crescents: 0–3

   Tubular atrophy: 0–3

   Interstitial fibrosis: 0–3

Scale of glomerular involvement: none = 0, <25% = 1, 25–50% = 2, and >50% of glomeruli = 3.
Scale of interstitial involvement: 0 = none, mild = 1, moderate = 2, and severe = 3.

four histologic parameters that are semiquantitatively graded in a similar manner. Note that these criteria are somewhat different from those in the ISN/RPS classification (see Table 42.8), and that tubulointerstitial findings play only a rather limited role in assessing activity but a more significant role in determining the chronicity index. Vascular lesions are ignored in both indices.

As with all grading schema, the reproducibility of these scores and their prognostic ability have been questioned [73]. Correlation with outcomes improves with higher scores: activity >7 and chronicity >3 [74]. The findings on repeat biopsy, done after 6 months of treatment, are often more prognostically significant than the initial sample [46]. Although newer systems may be somewhat more prognostically powerful, they remain too cumbersome for routine use [75].

It has long been recognized that the lesions of LN are dynamic, as evidenced by the frequent clinical occurrence of relapses and spontaneous transformation from one class to another. However, recognition of these transformations is a function of how often repeat biopsies are done. The most common transformation in class is from III to IV (focal to diffuse LN) [76]. Given that those classes represent a continuum of changes, this is hardly surprising. A retrospective study addressing this issue found clinically relevant class transformation to be frequent in patients whose initial biopsy was nonproliferative. In contrast, when the reference biopsy was proliferative, transformation to nonproliferative was rare [77].

Glomerulosclerosis

Previous classification systems did not specify how sclerotic glomeruli were to be handled, leading to differences in practice when determining the percentage of "involved" glomeruli. The ISN/RPS system mandates that they be counted when they are thought to be the sequelae of prior active glomerular lesions. Although this distinction is theoretically reasonable, its application in the "real world" is difficult because there are limited histologic criteria to determine the etiology of glomerulosclerosis. Nephropathologists tend to attribute sclerosis to LN in young patients because they typically have less renal co-morbidities. Including globally sclerotic glomeruli seems to increase the percentage of class III and IV biopsies [78].

Distinction between Classes IV-G and IV-S

The distinction between classes IV-S and IV-G is undoubtedly the most controversial aspect of the ISN/RPS classification system [79]. This subclassification was introduced based on a study by the LN Collaborative Study Group [80]. Retrospectively translating their groups into the ISN/RPS classification, the study reported that despite similar initial clinical and serological parameters, IV-S had more segmental endocapillary hypercellularity and necrosis, and IV-G was typified by large immune deposits (wire loops and hyaline thrombi). The two groups were treated similarly, yet IV-S had lower remission rates and 10-year renal survival. The authors speculated that segmental lesions with a relative absence of subendothelial deposits may have a different pathogenesis than IV-G that is more like "pauci-immune" crescentic glomerulonephritis. They concluded that the two subclasses do not represent a simple continuum of disease. Subsequent retrospective studies have failed to bear this out, including a report from France that suggested possible clinical and pathogenetic differences between the IV-S and IV-G subgroups but with similar outcomes and conversion from one subgroup to the other on repeat biopsies [81–83]. Two other studies have supported these findings [70, 84]. Thus, the clinical distinction, if any, between these two subgroups remains in doubt.

ANCA, Crescentic GN, and LN

Approximately 20% of patients with SLE are ANCA positive by indirect IF (almost always p-ANCA). Some studies suggest a correlation between ANCA positivity and LN activity. MPO-ANCAs are particularly common in patients with drug-induced SLE [85]. As noted previously, some studies have suggested that class IV-S may resemble pauci-immune crescentic glomerulonephritis more than classic immune complex glomerulonephritis. Unfortunately, those studies did not report ANCA results. A 2008 series reported 10 cases of necrotizing and crescentic lupus nephritis, all with associated ANCA positivity [68]. Interestingly, only one of the cases was IV-S. The authors recommend ANCA testing in all cases of LN with "necrosis and crescent formation out of proportion to the degree of endocapillary proliferation and subendothelial immune deposition."

Reproducibility of Classification

Two studies have compared the reproducibility of the ISN/RPS and WHO classifications. A Japanese study, one of the first to examine this question, found improved consensus regarding classification and more reliable prognostication with the ISN/RPS system [82]. Furness and Taub performed the best "real-world" study of the ISN/RPS classification, comparing it with the WHO schema using the same biopsies [78]. Reproducibility of ISN/RPS class was significantly greater than that of WHO class. Reproducibility of the A/C designations was less robust but still within the range of many commonly used systems. Interestingly, the ISN/RPS classification tended to produce more class IV and less class III and V designations than did the WHO system.

The authors consider the possibility that this is due to the increased recognition of capillary immune deposition that is unrelated to proliferation.

THE TREATMENT OF LUPUS NEPHRITIS

The treatment of LN is generally tied to the pathologic class found on kidney biopsy (Figure 42.37). This is a reasonable first approximation, and treatment guidelines by class are presented here.

There is considerable variation in clinical severity (proteinuria and renal function) and histologic severity (proliferation, necrosis, crescents, and fibrosis/sclerosis) within specific LN pathologic classes. Many of these differences are reflected in the activity and chronicity index. Although there are no established guidelines for tailoring LN therapy to this level of clinical and pathologic discrimination, these variations should be taken into account for individual patients, and discretion should be used in the application of aggressive immunosuppressive therapy.

Class I and II LN

Aggressive immunosuppressive therapy is not required for class I (if diagnosed) or class II LN. These patients have minimal laboratory abnormalities and normal kidney function [86, 87]. The kidney manifestations are usually effectively treated with the regimens used to treat the patients" extrarenal SLE, which can be mild, moderate, or severe. Renal protective therapies (discussed later) are recommended if the patients have hypertension and/or proteinuria.

Class III and IV LN

"Proliferative" LN (class III or IV) is an aggressive disease that requires intense therapy. Before 1970, renal and overall survival of patients with diffuse proliferative LN was very poor, in the range of 20—25%. Since then, patient and kidney survival in class III and IV LN has dramatically improved as the rheumatology and nephrology communities have gained experience in the use of immunosuppressive regimens from randomized and nonrandomized trials. Over time, therapy has evolved: corticosteroids $\rightarrow$ corticosteroids + (long-term) cytotoxic agents $\rightarrow$ corticosteroids + (short-term) cytotoxic agents followed by antimetabolites. Therapy is generally divided into induction and maintenance phases. The rationale is to treat active disease aggressively in order to rapidly decrease inflammation and then consolidate treatment over a longer time. The basis for the addition of cytotoxic agents to corticosteroids at induction was provided by randomized trials that demonstrated by long-term follow-up that cytotoxic agents decreased the frequency of renal relapse, CKD, and ESRD [88, 89]. The beneficial effect of cytotoxic agents to preserve kidney function was only apparent after 5 years of follow-up [88—90]. Thus, early in the course of treatment the use of steroids alone or in combination with cytotoxic drugs provided similar outcomes.

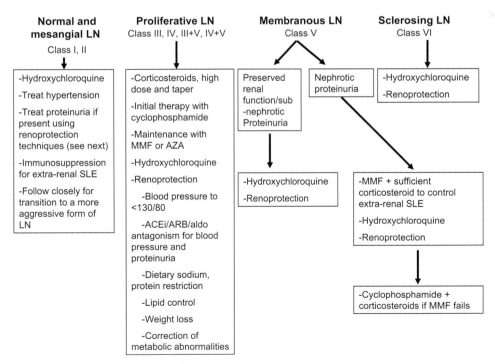

FIGURE 42.37 Approaches to the treatment of LN. See text for details of the recommended approaches.

Patients treated with short-term cytotoxic therapy (i.e., for only 6 months) alone also had an increased frequency of renal relapses [89]. The rationale for adding antimetabolites as maintenance therapy after shorter courses of cytotoxic agents is to decrease the morbidity of long-term cytotoxics while continuing to prevent relapse.

Although there are variations of the specific drugs that are used, duration of treatment, and sequence and route of administration, the commonly used therapeutics for proliferative LN remain rather limited. Newer agents such as biologics have been tested without success, but they continue to be investigated. The most common regimens are discussed next.

Protocol A: Corticosteroids in Combination with Intravenous Pulse Cyclophosphamide followed by Azathioprine or Mycophenolate Mofetil for Maintenance

Oral prednisone is given at a dose of 1 mg/kg/day and gradually tapered (Table 42.10). Oral prednisone may be preceded by up to 3 days of intravenous methylprednisolone at a dose of 0.5–1 g/day if the LN is particularly severe. Intravenous cyclophosphamide (0.5–1 g/m^2) is started with the corticosteroids and given in monthly pulses for up to seven cycles. The dose of cyclophosphamide should be decreased by 20–30% in patients with moderate to severe renal insufficiency [91] and adjusted to keep the neutrophil nadir at days 10–14 ≥2000 cells/μl. To protect fertility, women should be offered prophylaxis with leuprolide and men testosterone (Table 42.11) while cyclophosphamide is being given [92, 93]. Sperm banking and ovarian tissue cryopreservation are additional options.

Following the last cycle of intravenous cyclophosphamide, oral azathioprine (AZA; 1.5–2.5 mg/kg/day) or mycophenolate mofetil (MMF; 1000–3000 mg/day in two divided doses) is started and continued for at least 1 year after complete remission is achieved. The dose of these agents should be adjusted to keep the neutrophil count above 2000 cells/μl and, in the case of MMF, to avoid gastrointestinal toxicity. This protocol is adapted from a prospective study of class III and IV LN patients that compared short-duration intravenous cyclophosphamide (up to seven cycles) followed by an antimetabolite to what had been standard cyclophosphamide therapy of six monthly cycles followed by quarterly pulses of intravenous cyclophosphamide for 1 year after remission [94]. This study showed that over 72 months, the patients treated with maintenance AZA or MMF were significantly less likely to reach the composite endpoint of death or CKD than the cyclophosphamide group, and they experienced significantly fewer adverse side effects. Of importance, this study cohort was mainly Black and Hispanic.

TABLE 42.10 Proposed Prednisone Taper for LN

Week	Prednisone dose
Severe disease	
0–2	1 mg/kg/day ideal body weight (IBW; maximum 80 mg/day, two divided doses)
	In very severe disease, this may be preceded by 500–1000 mg/day methylprednisolone intravenously for 3 days
2–4	0.6 mg/kg/day
4–8	0.4 mg/kg/day
8–10	30 mg/day
10–11	25 mg/day
11–12	20 mg/day
12–13	17.5 mg/day
13–14	15 mg/day
14–15	12.5 mg/day
15–16	10 mg/day
16–	IBW <70 kg: 7.5 mg/day
	IBW ≥70 kg: 10 mg/day
Moderate disease	
0–2	0.4–0.6 mg/kg/day IBW (maximum 50 mg/day, two divided doses)
2–4	0.3–0.4 mg/kg/day
4–6	20 mg/day
6–7	15 mg/day
7–8	12.5 mg/day
8–9	10 mg/day
9–	IBW <70 kg: 7.5 mg/day
	IBW ≥70 kg: 10 mg/day

TABLE 42.11 Hormonal Protection of the Gonads in SLE Patients Treated with Cyclophosphamide[a]

Women: Leuprolide 3.75 mg IM every 4–6 weeks for the duration of cyclophosphamide therapy

Men: Testosterone 100 mg IM every 2 weeks for the duration of cyclophosphamide therapy

[a]*Leuprolide and testosterone should be started approximately 1 or 2 weeks before exposure to cyclophosphamide. Initially, these hormones stimulate the ovaries and testes, making them more vulnerable to the effects of the alkylating agent. In situations in which delay of cyclophosphamide is not advisable, we suggest giving the initial dose of cyclophosphamide intravenously and beginning hormonal therapy 3 days later. By this time, 15 cyclophosphamide half-lives will have elapsed. If the patient is to receive oral cyclophosphamide, it can be started 3 or 4 weeks after the initial dose of intravenous cyclophosphamide was administered.*

Protocol B: Corticosteroids in Combination with Low-Dose Intravenous Cyclophosphamide followed by AZA for Maintenance

In an effort to further reduce cyclophosphamide exposure, the low-dose Euro-Lupus regimen was compared to the higher dose NIH intravenous cyclophosphamide regimen [95, 96]. In this trial, all patients were given 3 days of methylprednisolone pulses (750 mg/day) and started on AZA (2 mg/kg/day) 2 weeks after the last cyclophosphamide infusion. Low-dose cyclophosphamide was given as six pulses of 500 mg every 2 weeks over 3 months. High-dose cyclophosphamide was given as described previously (six monthly pulses of $0.5-1$ g/m^2 followed by two quarterly pulses of the same dose range). After 10 years of follow-up, the endpoints of death, ESRD, and doubling of the serum creatinine were similar in both groups, suggesting that low-dose cyclophosphamide can be used successfully in proliferative LN. A caveat is that the patient population in this study was mostly Caucasian.

Protocol C: Corticosteroids in Combination with Oral Cyclophosphamide followed by AZA or MMF for Maintenance

Although oral cyclophosphamide is used far less frequently than intravenous cyclophosphamide for LN in the United States, it is an effective drug with outcomes comparable to those of intravenous cyclophosphamide [88]. It is easy to administer and less costly than intravenous cyclophosphamide [97], and it is used in other countries [98−102]. In addition, oral cyclophosphamide can be stopped promptly and its effects reversed more quickly if an infection or severe leucopenia develops. It appears that a major factor in the movement away from oral cyclophosphamide has been the perception of increased toxicity, especially cystitis [88]. However, many of the original studies with oral cyclophosphamide were done using doses up to 2.5 mg/kg/day with durations of 6 months or more, leading to very high cumulative cyclophosphamide exposure and considerable morbidity.

We continue to use oral cyclophosphamide for LN [103] but at a lower dosage and shorter duration, as originally described by the Lupus Nephritis Collaborative Study Group [104]. The protocol consists of corticosteroids as for intravenous cyclophosphamide, along with oral cyclophosphamide at a dose of $1.0-1.5$ mg/kg ideal body weight/day (maximum dose 150 mg/day). White blood cell counts should be monitored weekly while on cyclophosphamide, and its dose should be adjusted to keep neutrophils $\geq$2000/ul. Cyclophosphamide is continued for $2-4$ months, depending on the renal response, and then AZA or MMF is started and continued for at least 1 year beyond remission. AZA

or MMF is dosed as for the intravenous cyclophosphamide protocols. This regimen provides approximately the same cumulative amount of cyclophosphamide as 6 months of pulse intravenous dosing. Bladder toxicity is reduced by instructing patients to take cyclophosphamide in the morning and to drink extra fluid at each meal and at bedtime. Fertility considerations are addressed as in Protocol A. The regimen was found to be effective in inducing remission and was well tolerated, with fewer than 10% of patients discontinuing cyclophosphamide because of adverse effects, and no instances of hemorrhagic cystitis, malignancy, or death were observed [103]. A caveat for oral cyclophosphamide is that it should not be used in noncompliant patients. For these individuals, intravenous cyclophosphamide is preferred to ensure patients receive their medication.

Protocol D: Corticosteroids in Combination with MMF for Induction and Maintenance

Because of the toxicity of cyclophosphamide, both immediate and its predisposition to future malignancy, there has been considerable effort to find less toxic agents to treat LN. An approach that has gained widespread acceptance is corticosteroids plus oral MMF for the first 6 months of LN treatment, instead of sequential cyclophosphamide followed by MMF. The basis for this approach was three small studies of MMF conducted in Asia and a larger, highly influential study of 140 patients from the United States [101, 105−107]. The Asian studies concluded that MMF was equivalent to cyclophosphamide, but the U.S. trial demonstrated that MMF was superior to intravenous cyclophosphamide, although many did not receive the target dose of cyclophosphamide. To verify the superiority of MMF, the Aspreva Lupus Management Study (ALMS) recruited 370 patients with class III−V LN and treated them with a prednisone taper starting at 60 mg/day and either MMF at a target dose of 3 g/day or six monthly intravenous pulses of cyclophosphamide ($0.5-1$ g/m^2). Unexpectedly, the ALMS trial did not show that MMF was superior to intravenous cyclophosphamide but, rather, showed that the two drugs were approximately equivalent in inducing a response at 6 months [108]. Despite the fact that the ALMS study was not designed to be a noninferiority trial, these results, plus the perceived superior safety profile of MMF, have been used to justify MMF as the de facto standard of care for the initial therapy of LN. Interestingly, ALMS showed a similar incidence of adverse events, serious infections, and deaths for both MMF and cyclophosphamide, and although not statistically significant, withdrawals due to adverse events were almost double in the MMF arm. An unintended consequence of the MMF studies has been an increasing reluctance to perform kidney biopsies for the initial evaluation of LN, arising from the belief that MMF will be used regardless of biopsy findings.

Caution must be applied before MMF is accepted as the standard of care in LN. The benefit of cyclophosphamide in the initial treatment of LN is superior long-term preservation of kidney function compared to corticosteroids alone. However, this effect is not apparent for several years. Thus, the short-term success of MMF does not necessarily guarantee long-term preservation of kidney function to the same degree as cyclophosphamide. There has been only one long-term study of continuous MMF therapy, and this was compared to cyclophosphamide followed by AZA. There were no significant differences in renal function between the groups after a median of 64 months [102]. However, close examination of these data demonstrated a concerning trend toward more SLE relapses, prolonged proteinuria over 1 g/day, and more serum creatinine levels above 2 mg/dl in the MMF group. The combination of these clinical findings has been associated with deterioration of kidney function over time [22].

Other investigators have also tried to find an alternative to cyclophosphamide in the initial treatment of LN. Intravenous cyclophosphamide was compared to AZA plus corticosteroids, and the results showed that on repeat biopsy there was more chronic damage in the AZA group, and those treated with AZA had a higher incidence of renal relapse and doubling of the serum creatinine [109].

To address the important issue of optimal maintenance therapy, the ALMS maintenance phase will examine the durability of remission and long-term kidney function in patients treated initially with MMF or intravenous cyclophosphamide, followed by MMF or AZA.

The most provocative result from the ALMS trial was found in a post hoc analysis after stratifying response by race and ethnicity. Black or mixed-race patients who received intravenous cyclophosphamide did worse than those who received MMF, and the response rate among Hispanic patients was greater with MMF. These findings suggest that Black and Hispanic patients, generally considered to have more resistant LN [110], respond better to MMF than to intravenous cyclophosphamide. An alternative explanation is that the MMF regimen provided more immunosuppression "up front," during the first few months of treatment, than did the intravenous cyclophosphamide regimen. The notion is that immunosuppression is most effective when administered early in the course of the LN flare. Consistent with this hypothesis, we found that with daily oral cyclophosphamide the response rate for African-Americans was not different than that for European Americans [103].

Protocol E: Adjunct Therapies

All patients receiving immunosuppressives should be treated with a sulfa antibiotic or dapsone if sulfa-allergic for *Pneumocystis* prophylaxis. All patients should be treated with hydroxychloroquine unless there is a contraindication. Hydroxychloroquine may protect against vascular thrombosis [111], kidney damage [112], renal flares [113], and ESRD [114], and it has a favorable impact on lipid profiles.

In addition to neutralizing the immune injury of LN, it is essential to control blood pressure, reduce proteinuria, and take other measures to prevent chronic injury and attenuate the loss of GFR. Although successful immunosuppression will, over time, reduce proteinuria and acute damage and stabilize GFR, renoprotective measures should be initiated concomitantly to synergize with immunosuppression in realizing the goal of preservation of renal mass. Renoprotective strategies are discussed later.

Treatment Outcomes in Proliferative LN

The short-term outcomes that can be expected using these protocols for class III and IV LN are shown in Table 42.12. It is difficult to directly compare outcomes among studies because the definitions of response and complete remission are not uniform. Furthermore, pathologic classification may be somewhat different, particularly in older studies. Nevertheless, response rates appear to be similar for the different regimens. Total response rates (complete plus partial remissions) are approximately 50–70%, with fewer patients achieving complete remission. The consistent and unexplained exception to this is Chinese SLE patients, who appear to respond well to any of the regimens.

The first 6 months of LN treatment is generally considered induction [108, 115]. Although this carries an expectation of remission, the number of complete responses at 6 months is low. However, improvement is seen beyond 6 months and continues well into the maintenance phase of therapy [22, 95, 100, 103, 116, 117]. Thus, unless there is clear evidence for deterioration of renal status (rising serum creatinine, worsening proteinuria, and increased activity of the urine sediment) at 6 months, the initial treatment plan should be maintained.

The long-term outcomes documented using these protocols are shown in Table 42.13. The long-term outcome measures are less variable between studies than remissions and include death, ESRD, SLE nephritis relapse, and doubling of serum creatinine, a surrogate marker for future ESRD. The data demonstrate that mortality and ESRD associated with proliferative LN are approximately 5 and 4%, respectively, whereas doubling of the serum creatinine (8.4%) and renal relapses (29%) are more common.

TABLE 42.12 Short-Term Response of Proliferative LN to Therapy

Population	Definition of response	Treatment	Overall renal response (%)[a]	CR (%)[b]	Time (months)[c]	Reference
B, W, H	CR: nl SCr, prot ≤0.5g/day, inactive ua	MMF + P	56.2	8.6	6	[108]
	PR: Stabilize SCr, p/c <3 in nephrotic	IVC + P[d]	53	8.1	6	
	p/c ↓ by 50% in subnephrotic	IVC[e] + P	85	NA[f]	6	
B, H	Stabilize SCr, p/c <3 in nephrotic	IVC + P + AZA	84.2	NA	6	[94]
	p/c ↓ by 50% in subnephrotic	IVC + P + MMF	80	NA	6	
B, W	CR: nl SCr or ↓ to baseline p/c <0.5	POCY[g] + (MMF or AZA) + P	42	10	6	[103]
	PR: SCr ↓ at least 25%, p/c ↓ 50% and <3 in nephrotic		60.5	23	12	
B, W, H	CR: SCr, prot, ua improve to within 10% of normal	IVC[d] + P	32	7.4	6	[107]
	PR: 50% improvement in SCr, prot, ua and no single measure worse by >10	MMF + P	54	25	6	
W	Prot <1 g/day, improved ua, SCr does not double	IVC[h] (high dose)	18	NA	6	[95]
		IVC (Euro-Lupus)	20	NA	6	
		IVC (high dose)	32	NA	12	
		IVC (Euro-Lupus)	50	NA	12	
W	CR: SCr within 130% baseline, prot <0.5 g/day, improved ua	IVC[h] + P	85	32	12	[116]
	PR: SCr stable or improved, prot ↓ 50% and <3 g/day	AZA + P	85	30	12	
CH	CR: SCr stable or improved, prot ↓ <1 g/day	POCY[i] + AZA + P	89	67	12	[98]
	PR: SCr stable or improved, prot ↓ 50% and <3 g/day in nephrotic					
CH	CR: SCr stable or improved, prot ↓ <1 g/day	POCY[j] + AZA + P	90	57	24	[99]
	PR: SCr stable or improved, prot ↓ 50% and <3 g/day in nephrotic	IVC + P	73	59	24	
CH	CR: Stable Cr, prot <0.3 g/day	POCY[k] + AZA + P	91	79	12	[100]
	PR: Stable Cr, prot ↓ 50% to 0.3–3 g/day					
CH	CR: Stable Cr, prot <0.3 g/day	POCY[k] + AZA + P	90	76	12	[101]
	PR: Stable Cr, prot ↓ 50% to 0.3–3 g/day	MMF + P	95	81	12	

[a]Includes CR + PR; percentage of cohort treated.
[b]Percentage of cohort treated who achieved CR.
[c]Time at which criteria for PR and CR were assessed in months from initiation of therapy.
[d]IVC, six monthly pulses.
[e]IVC, six or seven monthly pulses in each of the cohorts; at 6 months, treatment is the same for each cohort.
[f]Not applicable; in these studies, criteria listed were only for a renal response, not for CR or PR.
[g]POCY, 1–1.5 mg/kg/day for 2–4 months.
[h]IVC, six monthly pulses followed by quarterly pulses.
[i]POCY, 1–2 mg/kg/day for 6–9 months.
[j]POCY, 50–100 mg/day for 6 months.
[k]POCY, 2.5 mg/kg/day for 6 months.
B, Black; CH, Chinese; CR, complete remission; H, Hispanic; IVC, intravenous cyclophosphamide; nl, normal; P, prednisone; p/c, 24-h urine protein-to-creatinine ratio; POCY, oral cyclophosphamide; PR, partial remission; prot, proteinuria; SCr, serum creatinine; ua, urine sediment, mainly the presence or absence of red blood cells and casts; W, White.

TABLE 42.13 Long-Term Response of Proliferative LN to Therapy

Follow-up (years)	Population	Treatment	Death[a]	Double SCr	ESRD	Relapse[b]	Composite[c]	Reference
6	W, B, H	IVC + AZA + P	0	5.3	0	31.5		[94]
6		IVC + MMF + P	5	5	5	15		
6		IVC + P	20	15	10	42		
10	W	IVC + P (Euro-Lupus)					25	[96]
10		IVC + P (high dose)					15	
6.3	W	AZA + P	8	16.2	2.7	27		[116]
5.5		IVC + P	0.04	0.04	0	0.04		
3	CH	POCY + AZA + P		6				[98]
5		POCY + AZA + P		8.4				
10		POCY + AZA + P		18.2				
8	CH	POCY + AZA + P		4.6		34.8		[100]
10	CH	PO/IV/CYC + P ± AZA	4.5	10.3	6.5			[221]

[a]Outcomes for death, doubling of SCr, ESRD, or relapse are presented as percentage of the patient population.
[b]Relapse refers to renal relapse.
[c]Composite endpoint refers to death, ESRD, or doubling of SCr.
B, Black; CH, Chinese; H, Hispanic; W, white.

Several investigations have sought to identify biomarkers present at the time of treatment initiation that are predictive of or are associated with short-term (i.e., remission) and long-term (i.e., CKD and ESRD) outcomes. Because these studies were in general small, and some were prospective whereas others were retrospective, the reliability of most of these biomarkers has not been verified. Examples of biomarkers that have been found to be significant in univariate analyses include baseline serum creatinine, baseline proteinuria, proteinuria magnitude after completion of cyclophosphamide therapy, autoantibody levels, complement levels, age at onset of treatment, race/ethnicity, comorbidities such as hypertension and other SLE manifestations, complete or partial response to therapy, biopsy class, and activity and chronicity indices. When examined by multivariate analyses, many of these markers were not independent risk factors. The independent risk factors for LN outcomes from several multivariate analyses are shown in Table 42.14. Although none of the studies examined all of the same risk factors, many were common to most of the investigations. From study to study, there is no clear consensus or pattern of predictive biomarkers for remission, relapse, or CDK/ESRD. An exception appears to be serum creatinine. The presence of renal insufficiency at the beginning of treatment diminishes the ability to achieve remission and is a risk factor for CKD and ESRD. Failure to achieve a complete remission was identified by several investigations to be a significant risk factor for relapse, CKD, ESRD, and

mortality (see Table 42.14) [22, 118]. It is possible that with a rigorous definition of complete remission, specifically requiring proteinuria to decrease below 0.5 g/day, more studies would have found achieving a complete remission to be an important factor in long-term renal preservation. For most proteinuric kidney diseases, including other immune complex GNs, resolution of proteinuria is the strongest predictor of renal survival [119–121]; thus, achieving a complete remission with disappearance of proteinuria should be protective in LN. Finally, most studies did not include socioeconomic status in their analyses, which may affect the strength of race and ethnicity as independent risk factors.

Class V LN

Class V (membranous LN) is nonproliferative and can be seen alone or with superimposed proliferative forms of LN. Pure membranous LN occurs in 8–20% of patients with LN [122–124]. This form of LN is generally regarded as less aggressive than proliferative LN, but CKD does occur in approximately 20% of the cases, and ESRD occurs in approximately 8–12% after 7–10 years [122–124]. Some report worse outcomes: 14% of patients reached the composite endpoint of death or ESRD at 5 years, and 28% reached this endpoint at 10 years [125]. Spontaneous remission of heavy proteinuria occurs in only a minority of lupus membranous patients [126, 127]. These outcomes were observed in patients with membranous LN who had

TABLE 42.14 Risk Factors for Short- and Long-Term Renal Outcomes in Proliferative LN

Outcome	Risk factor[a]	HR or RR	Study type	Reference
Ability to achieve remission	Proteinuria	0.86 (per ↑ 1 g/day)	R	[117]
	Delay therapy >3 months	0.58		
	White race	2.63	R	[110]
	SCr	0.21 (per ↑ 1 mg/dl)		
	Class IV LN	2.05		
	SCr	0.96 (per μmol/l)	P	[98]
	None[b]		P	[99]
Renal relapse	Time to remission	1.03 (per month)	R	[117]
	CNS SLE	8.41		
	Class IV LN	0.28		
	Chron index[c]	1.22 (per point)	R	[221]
	Persistently positive DS-DNA	2.94		
	Act index[d]	1.13 (per point)	P	[98]
	Failure to achieve complete remission	6.2	P	[100]
	None		P	[109]
Doubling of SCr	SCr >1 mg/dl	4.1	R	[222]
	Hispanic	3.6		
	Poverty[e]	3.5		
	Government insurance[f]	3.0		
	Nephritic flare[g]	17.7	R	[221]
	Chron index	2.1 (per point)	P	[98]
	None		P	[109]
ESRD	SCr	2.8 (per ↑ 1 mg/dl)	R	[110]
	Fail to remit	6.8		
	Class III ≥50% LN[h]	2.77		
	Anti-RO positive	2.35		
	None		R	[221]
Double SCr, ESRD, or death	SCr	1.26 (per ↑ 1 mg/dl)	R	[223]
	Chron index	1.18 (per point)		
	Mean BP	1.02 (per mmHg)		
ESRD or death	Age >50 years	3.3	R	[110]
	SCr	2.32		
	Non-White race	2.28		

[a]All clinical variables are taken at baseline before SLE LN treatment.
[b]No variables were predictive in multivariate analysis.
[c]Chronicity index on kidney biopsy.
[d]Activity index on kidney biopsy.
[e]Living in a neighborhood where >10% of residents are below federal poverty line.
[f]Medicare and Medicaid as opposed to private insurance.
[g]Nephritic flare is defined as an increase in active urinary sediment with or without an increase in proteinuria.
[h]Class III with active and/or necrotizing lesions in ≥50% of nonsclerotic glomeruli.
HR, hazard ratio; P, prospective study; R, retrospective study; RR, relative risk.

been treated, mainly with corticosteroids. Thus, the natural history of class V LN may have been modified. In addition to renal insufficiency, the heavy proteinuria characteristic of membranous LN, if chronically present, can predispose to hyperlipidemia and atherosclerosis, contributing to cardiovascular morbidity and mortality [120, 128]. Heavy proteinuria can also lead to a hypercoagulable state and arterial and venous thromboses [120, 129]. Thrombotic events occur in 13–23% of patients and have been linked to the presence of antiphospholipid antibodies and/or the nephrotic syndrome [122, 123, 130]. Membranous LN, although indolent compared to proliferative LN, is associated with significant morbidity and mortality and therefore warrants therapy.

There is no general consensus on treatment of membranous LN. Patients with mixed class V and class III or IV lesions should be treated as for the proliferative component, but they may have a less favorable prognosis [123]. Pure membranous LN with low-level proteinuria may respond to conservative treatment with renoprotective and antiproteinuria therapies as described later. Patients with nephrotic-range proteinuria and/or renal insufficiency should be considered for immunosuppression. Representative trials of immunosuppression for membranous LN are outlined in Table 42.15. These studies are in general much smaller

than the clinical trials for proliferative LN, and they are mainly open-label, retrospective, and/or uncontrolled. The single exception is the first trial listed [131], which was a prospective, randomized trial comparing the addition of cyclophosphamide or cyclosporine A to prednisone. Using the prednisone-only group as the control population, the addition of either cyclophosphamide or cyclosporine significantly increased response to therapy. However, patients treated with cyclosporine were much more likely to relapse after stopping therapy than those treated with cyclophosphamide. Within a year of finishing treatment, the cyclosporine group had a 40% probability of relapse, whereas relapses were not seen in the cyclophosphamide group until after 48 months [131]. This investigation also showed, by multivariate analysis, that the only independent predictor of failure to achieve remission was initial proteinuria over 5 g/day and that failure to achieve sustained remission was a risk factor for decline in kidney function.

In general, these trials suggest that the addition of an immunosuppressive to background prednisone will yield a complete response in 40–60% of patients within 6–12 months. Response may be more rapid with calcineurin inhibitors, but the risk of relapse is very high. Race or ethnicity does not appear to affect response. The contribution of antiproteinuric medications such as

TABLE 42.15 Short-Term Response of Membranous LN to Therapy

Population	Definition of response	Treatment	PR (%)[a]	CR (%)[b]	Time[c]	Reference
B, W, H	CR: prot ≤0.3 g/day	P[d]	13.5	13.5	12	[131]
	PR: prot <2 g/day and ↓ by 50%	IVC + P[e]	20	40	12	
B, W	CR: SCr normal, prot <0.5 g/day, ua inactive	CSA + P[f]	33	50	12	
	PR: GFR improves ≥30%, ua improved by 50%, prot ↓ by 50%	MMF + P[g]	20	40	12	[224]
B, W	CR: P/C <0.5; PR: P/C ↓ by 50% and to P/C <3	MMF + P	15	61	6	[225]
CH	CR: SCr stable or improved, prot ↓ <1 g/day	AZA + P[h]	22	67	12	[130]
	PR: SCr stable or improved, prot ↓ 50% and <3 g/day in nephrotic					
CH	CR: GFR ≤15% baseline, prot ↓<0.5 g/day, ua normal	TAC + AZA + P[i]	44	39	6	[226]
	PR: GFR stable, prot 0.5–2.9 g/day					

[a]Partial response (% of cohort).

[b]Complete response (% of cohort).

[c]Time at which criteria for PR and CR were assessed in months from initiation of therapy.

[d]Prednisone, 1 mg/kg every other day for 8 weeks, then taper.

[e]IVC, six pulses every other month, 0.5–1 g/m², plus prednisone.

[f]CSA, 5 mg/kg/d for 11 months, plus prednisone.

[g]MMF, 2–3 g/day; prednisone to control extrarenal symptoms.

[h]Prednisone, 0.8–1 mg/kg/day for 6–8 weeks, then taper; then AZA 1–2 mg/kg/day.

[i]Tacrolimus, 0.1–0.2 mg/kg/day for 6 months; pred 30 mg/day, then taper; AZA 1.5 mg/kg/day after tacrolimus.

B, Black; CH, Chinese; CR, complete remission; CSA, cyclosporine A; GFR, glomerular filtration rate; H, Hispanic; IVC, intravenous cyclophosphamide; nl, normal; P/C, 24-h urine protein-to-creatinine ratio; PR, partial remission; prot, proteinuria; SCr, serum creatinine; ua, urine sediment, mainly the presence or absence of red blood cells and casts; W, White.

angiotensin-converting enzyme (ACE) inhibitors or angiotensin receptor blockers (ARBs) to remission is difficult to assess because their use was not uniform in these studies. It may therefore be appropriate to try to induce remission with MMF plus prednisone, and if that fails, switch immunosuppression to cyclophosphamide plus corticosteroids in patients with membranous LN and heavy proteinuria. Renoprotective and antiproteinuric therapies are also recommended as adjunct strategies for controlling proteinuria and may help reduce proteinuria by as much as 30–50% in membranous LN [120, 132, 133]. These therapies are discussed later.

Class VI LN

Class VI manifests diffuse glomerulosclerosis and interstitial fibrosis/tubular atrophy, which is the final common pathway of any form of LN that has advanced to CKD. Factors contributing to the development of nephrosclerosis, in addition to the immune and inflammatory injuries associated with acute LN, include hypertension, vascular disease, and sustained proteinuria over 500 mg/day, which is itself nephrotoxic [134–138]. When LN has reached this stage, the therapeutic strategy should shift from an immunosuppression focus to a renal protection focus. The goal of renal protection is to prolong kidney function and avoid ESRD requiring renal replacement therapies for as long as possible.

RENOPROTECTIVE THERAPIES

In addition to the primary process (e.g., immune) affecting glomeruli, increased intraglomerular capillary hydrostatic pressure, systemic hypertension, proteinuria, and hyperlipidemia have all been implicated in causing kidney damage in patients with glomerular disease. Nonimmune therapies directed toward decreasing systemic and glomerular hypertension, reducing proteinuria, and controlling hyperlipidemia are kidney protective and lessen the risk of progressive loss of kidney function. These therapies are appropriate for most patients with active glomerular disease as an adjunct to specific immunosuppressive therapy, as well as patients with nonactive glomerular disease who have residual chronic kidney insufficiency and proteinuria.

Blood Pressure Control

The goal of blood pressure control in glomerular disease is to attenuate kidney damage, reduce proteinuria to below 500 mg/day, and prevent or slow further declines in glomerular filtration rate. To achieve this goal, antihypertensive therapy should target a sitting systolic blood pressure in the 120s or less, if tolerated [119, 120, 139–142]. The level of systolic blood pressure control may be more important with respect to maintenance of renal function than the diastolic.

Choice of Antihypertensives

The choice of medications used to control blood pressure for the purposes of kidney protection is important. There is ample evidence that attenuation of the renin–angiotensin system (RAS) has beneficial renal effects beyond systemic blood pressure control [120]. In particular, RAS blockade is significantly antiproteinuric and may preserve glomerular architecture [132, 133, 143–146]. RAS blockade reduces glomerular efferent arteriolar tone and thereby decreases glomerular capillary hydrostatic pressure. RAS blockade also abrogates the effects of angiotensin II on glomerular cell proliferation, growth factor expression, and pro-inflammatory cytokine expression, all of which may contribute to glomerulosclerosis. Thus, ACE inhibitors and ARBs should form the foundation of antihypertensive therapy in glomerular disease. ACE inhibitors remain the initial choice because of some theoretical advantages over ARBs. For example, ACE inhibitors lower levels of plasminogen activation inhibitor-1, and this may promote matrix remodeling [147]. Furthermore, ACE inhibitors offer better suppression of aldosterone, which has profibrotic effects [148, 149]. Nonetheless, ARBs should be tried in patients who cannot tolerate ACE inhibitors or who develop hyperkalemia on ACE inhibitors. Because RAS blockade lowers intraglomerular pressure, it may be associated with an increase in the serum creatinine level. An increase in creatinine of up to 30–50% is tolerable if it stabilizes. It is important to be aware that an increase in creatinine during RAS blockade may indicate the possibility of renal artery stenosis in the appropriate clinical setting and should be investigated.

Although kidney protection and antiproteinuric benefits have been shown with low to moderate doses of ACE inhibitors (e.g., 5 mg enalapril/day and 3 mg ramipril/day), escalation to the maximum recommended dose may be more antiproteinuric and provide better control of systemic blood pressure [150–152]. The choice of ACE inhibitor may also be relevant for kidney protection. ACE inhibitors with high affinity for tissue ACE (ramipril, benazepril, and quinapril) may be more efficacious [153].

If the blood pressure goal is not met by ACE inhibition alone, other antihypertensive medications should be added. Diuretics are a logical next choice for patients with LN in whom fluid retention may be an issue. Addition of an ARB to the ACE inhibitor should also be considered. This combination is appealing because ACE inhibition alone may not achieve maximal RAS blockade. Combination therapy appears to be more antiproteinuric than single RAS therapy [154–157]. The ACE inhibitor and ARB combination should be used cautiously in patients with left ventricular failure receiving beta-blockers.

Other antihypertensive medications that have antiproteinuric kidney protective effects can also be added to RAS blockade plus diuretics to achieve blood pressure control, including long-acting nondihydropyridine calcium channel blockers [158–160].

Certain antihypertensive medications must be used cautiously in patients with glomerular disease. The long-acting dihydropyridine calcium channel blockers are very effective antihypertensive medications; however, they decrease glomerular afferent arteriolar resistance and tend to increase glomerular capillary hydrostatic pressure. This can actually worsen proteinuria and accelerate loss of renal function [145, 161]. Dihydropyridine calcium blockers should be used only if necessary for blood pressure control. To mitigate undesirable kidney effects of these agents, it is recommended that they be used in conjunction with RAS blockade, and that systolic blood pressure be maintained at or below 110 mmHg [144, 161].

Antiproteinuric Therapy in the Absence of Hypertension

RAS blockade should also be attempted in patients with proteinuria who do not have hypertension. In this circumstance, an ACE inhibitor (or ARB) can be started at a low dose with careful monitoring to avoid hypotension. The dose can be titrated up as tolerated. In this way, many patients can achieve at least some level of RAS blockade.

Aldosterone Antagonism

Aldosterone antagonists such as spironolactone have antiproteinuric and antifibrotic effects [162, 163]. The addition of spironolactone (25 mg/day) may be considered in glomerular disease, especially in patients who do not significantly suppress aldosterone with RAS blockade. This must be used with caution because of the possibility of severe hyperkalemia, especially in patients with impaired kidney function. Serum electrolytes must be regularly monitored when aldosterone antagonists are used with RAS blockade.

Diet Control

The antiproteinuric effect of RAS blockade, as well as the antihypertensive effects of most blood pressure regimens, can be overcome by high sodium intake [164]. Unless renal sodium wasting is present, dietary sodium should be restricted to 2–2.5 g/day. This will also help control peripheral edema in the nephrotic syndrome.

A modest decrease in dietary protein may slow progression of renal failure and will reduce proteinuria [120]. For patients who are not acutely ill or catabolic, a protein reduction from the usual intake of 1.0–1.5 g/kg ideal body weight/day to 0.7 g/kg/day is well tolerated and achievable. To avoid malnutrition, these diets must have adequate calories, and protein should be of high biologic value. A dietician trained in the care of patients with kidney disease should work closely with physicians prescribing these diets. If patients are nephrotic, the amount of dietary protein allowed should be increased by 1 g/day for every gram of urinary protein exceeding 3 g/day.

Lipid Control

Because the hyperlipidemia of the nephrotic syndrome may promote nephrosclerosis and may be a risk factor for atherosclerosis (especially if present or expected to be present for a long time), it is reasonable to modify cholesterol intake. In addition, the treatment of nephrotic hyperlipidemia will often require pharmacologic intervention with statins. Statins may also be anti-inflammatory, which could contribute to the resolution of glomerular disease.

Miscellaneous

Other interventions that may be beneficial for preservation of kidney function include weight loss if obese; smoking cessation; correction of metabolic acidosis, hyperuricemia, hyperphosphatemia, and anemia; and avoidance of nonsteroidal anti-inflammatory drugs.

PATIENT FOLLOW-UP FOR LN

Duration of Therapy

The first 6 months of therapy for LN has, somewhat arbitrarily, been designated the induction period. Few patients reach complete remission by 6 months, and kidney biopsies after 6 months of induction therapies have shown that although activity indices tended to improve, complete resolution of pathologic changes was unusual [46, 105, 165, 166]. Thus, most investigators recommend continued therapy with MMF or AZA after induction. In addition, most patients continue to receive corticosteroids during this maintenance phase, albeit at small dosages (e.g., 5–10 mg/day).

Although data are limited because most investigators do not biopsy patients after complete remission, some insights into duration of therapy can be obtained from reports of kidney biopsies done several years after treatment was started. Biopsies taken 2 or more years after induction often continued to show pathologic activity. Not unexpectedly, this was particularly true of patients who still had significant proteinuria or an abnormal creatinine [76]. Of more concern, one study found that in patients with initial class IV LN biopsies, only half had reverted to class II LN on repeat biopsy after 2 years of cyclophosphamide therapy [109]. The others had converted to class III or still had class IV. The serum creatinine and level of proteinuria at the time of the second biopsy did not differentiate the group that went to class II from the group that remained with proliferative LN. Finally, most repeat biopsy studies showed an increase in the chronicity index at the second biopsy, even after successful treatment [76, 105, 109, 165–168], which suggests that each flare of LN will result in some sustained chronic damage. The strategy for LN follow-up should focus on prevention or early intervention for renal flares, as discussed later.

There are no specific guidelines for duration of maintenance therapy. In general, it is reasonable to try to taper immunosuppressive therapy after complete remission has been attained. Consensus suggests this taper begin a year after complete remission is declared. Although there is no standard definition of complete remission for LN in the literature, in terms of preservation of kidney function the most important clinical variable currently available is proteinuria. A proteinuria level less than 0.5 g/day should be the target for complete remission [132]. Serum creatinine should improve to a patient's pre-LN baseline if known. A caveat regarding serum creatinine is that it may be increased (acceptably) by renoprotective therapies. Thus, a stable serum creatinine should be the minimum requirement for complete remission. Urine sediment should not have any RBC or WBC casts, but hematuria may persist for months. Finally, at remission it would be desirable for serologic markers of lupus activity, such as complement and double-stranded DNA antibodies, to have normalized. However, there are also caveats regarding lupus serologies (discussed later).

Immunosuppression should be continued indefinitely for patients who achieve only a partial remission, and renoprotective therapies should be intensified. To affect a complete remission, immunosuppression can also be intensified, corticosteroids increased, or alternative immunosuppressive agents considered. A repeat kidney biopsy may be useful to determine the level of pathologic activity, which if severe could provide a rationale for re-induction therapy.

Renal relapse

LN often has a relapsing course. In one large study that examined renal flares in LN patients who had participated in randomized clinical trials, 40% of complete responders experienced a renal relapse within a median of 41 months after remission, and 63% of partial responders had a renal flare within a median of 11.5 months after response [169]. Putative risk factors for renal relapse are listed in Table 42.14. There is no consensus on what predisposes patients to relapse. Relapses are important to recognize and treat because it seems likely that with each episode of active LN the kidney sustains some chronic damage that may culminate in CKD or eventually ESRD [76, 105, 109, 167]. The ability to anticipate imminent renal flare, and therefore to institute therapy preemptively, could attenuate the development of chronic kidney injury. Currently, there are no biomarkers that reliably forecast future flare, but this is an area of research activity (discussed later).

Renal flare is thus diagnosed by clinical criteria based on changes in the urine sediment, rate of protein excretion, and serum creatinine from their baseline values in an individual patient. There is no consensus for renal flare criteria. Table 42.16 presents criteria used to identify renal relapses in the Ohio SLE Study cohort [170–172]. These criteria are based on the published criteria used in the Lupus Nephritis Collaborative Study of Plasmapheresis in Severe Lupus Nephritis [173] and are similar to renal flare criteria used in some clinical trials [174]. Other findings that would support

TABLE 42.16 Criteria for Diagnosis and Classification of Severity of SLE Renal Flare[a]

Mild renal flare	Moderate renal flare	Severe renal flare
↑ in glomerular hematuria from <5 to >15 RBC/hpf, with ≥2 acanthocytes/hpf *and/or*	If baseline creatinine is: <2.0 mg/dl, an ↑ of 0.20–1.0 mg/dl ≥2.0 mg/dl, an ↑ of 0.40–1.5 mg/dl *and/or*	If baseline creatinine is: <2 mg/dl, an ↑ of >1.0 mg/dl ≥2 mg/dl, an ↑ of >1.5 mg/dl *and/or*
Recurrence of ≥1 RBC cast, WBC cast (no infection), or both	If baseline Pr/Cr is: <0.5, an ↑ to ≥1.0 0.5–1.0, an ↑ to ≥2.0, but < absolute ↑ of 5.0 >1.0, an ↑ of ≥ twofold with absolute Pr/Cr <5.0	An absolute ↑ Pr/Cr >5.0

[a]*Remission of nephritis is defined as stabilization or improvement of serum creatinine to baseline or better, and a return of proteinuria to baseline or better.*

a diagnosis of flare, but are not necessarily always present, are a decrease in complement levels and an increase in anti-double-stranded DNA antibody titers. Also, flares are less likely to occur in patients who have been highly immunosuppressed by therapy. The finding of depressed serum immunoglobulin levels suggests significant immunosuppression; however, in severe nephrotic syndrome due to LN flare, serum immunoglobulins can also be low. Finally, other non-LN causes of an increase in creatinine (see Table 42.3) or an increase in proteinuria must be excluded. Increases in proteinuria can occur with pregnancy, uncontrolled hypertension, and increased sodium intake. An approach to flare therapy based on flare severity is given in Table 42.17. Caution must be exercised in using repeated courses of cyclophosphamide. It has been suggested that the cumulative lifetime dose of cyclophosphamide not exceed 36 g to decrease the chance of developing a malignancy later in life [175]. In patients approaching this limit, another agent should be used for flare.

Monitoring Patients

It is plausible that complete remission rates for LN could be improved, and less cytotoxic therapy would be required if renal flares could be anticipated and preemptive treatment started. In addition, modification of drug dose and duration of therapy based on the predicted outcome of a flare (e.g., development of chronic kidney disease) would be expected to improve treatment efficacy and reduce toxicity. Finally, because kidney biopsies are not repeated at every flare, a noninvasive predictor of renal pathology would be very useful in choosing therapy. This approach to LN treatment represents a fundamental change from a reactive to a proactive paradigm, and it will require biomarkers that accurately predict SLE nephritis activity, pathology, and prognosis to guide therapeutic decisions.

The traditional clinical biomarkers for SLE, including complement components 3 and 4 (C3 and C4) and anti-double-stranded DNA antibodies, have been used to support the diagnosis of renal flare and also to anticipate impending flare. However, these serologies have low sensitivity (49−79%) and specificity (51−74%) for concurrent renal flare, and they do not reliably predict impending flare when measured serially, with sensitivities and specificities approximately 50 and 70%, respectively [176−182]. In the Ohio SLE Study cohort, the positive predictive values for C3 and C4 to forecast impending flare were 7.4 and 5.5%, respectively [176].

Efforts are underway to identify novel biomarkers that will be more useful in forecasting LN flare. A major focus has been on developing urine markers [45] because urine reflects intrarenal events. Several urine biomarker candidates have been found (Table 42.18). None of these candidates has been validated in a large, independent, prospectively followed lupus cohort, although such studies are anticipated.

Similarly, although some studies have suggested risk factors for the development of CKD at the time treatment is started (see Table 42.14), there are no biomarkers that predict CKD in LN patients. Putative novel biomarkers of future CKD that await validation are listed in Table 42.18 [45].

Finally, a noninvasive test that accurately reflects kidney histology would be useful when planning treatment because it would provide significantly more information on what is happening within the kidney at flare than would proteinuria, urine sediment, and creatinine. Ideally, histologic biomarkers would differentiate active inflammation, necrosis, and crescents from fibrosis and chronic changes or from an etiology other than lupus, such as nephrotoxic acute tubular necrosis. Current putative biomarkers of kidney pathology (see Table 42.18) do not achieve this level of discrimination but, instead, attempt to distinguish class IV from other global pathologies [45].

THROMBOTIC MICROANGIOPATHIES THAT AFFECT THE KIDNEY IN SLE

APS is seen not infrequently in patients with SLE, and it is generally associated with antiphospholipid antibodies such as anticardiolipin and anti-β_2-glycoprotein I antibodies and also with lupus anticoagulants [183]. APS involves the kidney by causing non-inflammatory occlusions of renal blood vessels, including the intrarenal microvasculature.

In retrospective studies, renal APS occurred in approximately 30% of SLE patients and was most often found alongside LN, although it can be seen alone [44, 183]. Renal APS was found significantly more often (63 vs. 22%) in SLE patients with extrarenal APS than in SLE patients without any APS manifestations [44]. Lupus anticoagulant was present in 30−52% of cases of renal APS, whereas 72−95% of patients had anticardiolipin antibodies, and 15% had neither of these serologic markers [43, 44].

It is important to consider the diagnosis of renal APS and verify with a kidney biopsy. Renal APS is treated with chronic anticoagulation therapy plus hydroxychloroquine. Failure to treat APS can lead to CKD or ESRD, despite adequate treatment of LN with immunosuppression.

TTP may also occur in the setting of SLE and is associated with high mortality [184]. In a retrospective multivariate analysis, the presence of LN and a Systemic

TABLE 42.17 Protocols for Treatment of LN Relapses

Minor renal flare	Moderate renal flare	Severe renal flare
Minor renal flares are treated with prednisone, 20 mg/day for 2 weeks, followed by a taper back to baseline therapy over 4–8 weeks.	1. Increase prednisone dose and use the moderate disease protocol outlined in Table 42.3 for taper. 2. If patient not receiving immunosuppressive agent, begin azathioprine (1–2.5 mg/kg/day IBW) or mycophenolate (1000–1500 mg b.i.d.). If already receiving an immunosuppressive agent, increase dose or switch to cyclophosphamide. 3. Add hydroxychloroquine, 200 mg b.i.d., for anti-inflammatory/antithrombotic effects if not already receiving. 4. Add sulfamethoxazole/trimethoprim DS three times/week for PCP prophylaxis. 5. Kidney and cardiovascular protective therapies: blood pressure control, ACE inhibitors/ARBs, blood lipid control, diet—avoid high-salt, high-protein intake Maintenance therapies 1. Prednisone IBW <70 kg: 7.5 mg q.d. IBW ≥70 kg: 10 mg q.d. Maintain prednisone for at least 1 year after complete remission, then taper. Consider tapering off prednisone over next year by decreasing dose 5 mg/quarter. 2. Immunosuppressive Azathioprine: 1–2.5 mg/kg/d IBW Mycophenolate: 500–1000 mg b.i.d. Continue for at least 1 year after complete remission, then taper.	1. Increase prednisone dose and use the severe disease protocol outlined in Table 42.3 for taper. 2. Add a cytotoxic agent, usually cyclophosphamide, then switch to maintenance therapy with an immunosuppressive such as azathioprine or mycophenolate. 3. Add hydroxychloroquine, 200 mg b.i.d., for anti-inflammatory/antithrombotic effects if not already receiving. 4. Add sulfamethoxazole/trimethoprim DS three times/week for PCP prophylaxis. 5. Kidney and cardiovascular protective therapies: blood pressure control, ACE inhibitors/ARBs, blood lipid control, diet—avoid high-salt, high-protein intake Maintenance therapies 1. Prednisone IBW <70 kg: 7.5 mg q.d. IBW ≥70 kg: 10 mg q.d. Maintain prednisone for at least 1 year after complete remission, then consider tapering off prednisone over next year by decreasing dose 5 mg/quarter. 2. Immunosuppressive Azathioprine: 1–2.5 mg/kg/day IBW Mycophenolate: 500–1000 mg b.i.d. Continue for at least 1 year after remission.

TABLE 42.18   Candidate Biomarkers for Monitoring LN

Biomarker	Biomarker Source	Putative use	References
MCP-1[a]	Urine protein; urine mRNA	Predict impending LN flare	[170, 227–230]
NGAL[b]	Urine protein	Predict impending LN flare	[231–233]
Transferrin	Urine protein	Predict impending LN flare	[234, 235]
Hepcidin	Urine protein	Predict impending LN flare, remission	[236]
LFABP[c]	Urine protein	Predict CKD	[237]
mEPCR[d]	Kidney biopsy	Predict CKD	[238]
FOXP3[e]	Urine mRNA	Predict CKD	[239]
Glycoproteins[f]	Urine protein panel	Predict kidney pathology	[240, 241]
CXCL10[g]	Urine mRNA	Predict kidney pathology	[230]
CD29[h]	T cell surface marker	Predict kidney pathology	[242]

[a]MCP-1, monocyte chemoattractant protein-1, a pro-inflammatory chemokine upregulated in LN.

[b]NGAL, neutrophil gelatinase-associated lipocalin, an antibacterial protein that also transports iron and is an epithelial growth factor.

[c]LFABP, liver-type fatty acid-binding protein, produced by human proximal tubular cells.

[d]mEPCR, membrane endothelial protein C receptor, found on cortical peritubular capillaries in LN kidney biopsies.

[e]FOXP3, forkhead transcription factor, important in the development of regulatory T cells.

[f]Serum glycoproteins excreted in urine—for example, α_1 acid glycoprotein, α_1-microglobulin, and zinc α_2 glycoprotein.

[g]A Th1 chemokine upregulated in LN.

[h]A T cell β_1-integrin.

Lupus Erythematosus Disease Activity Index (SLEDAI) score greater than 10 were significant risk factors for TTP in SLE patients [184]. TTP is treated with plasma exchange in addition to high-dose steroids. Because of the high associated mortality, it is important to consider this diagnosis and treat early.

PREGNANCY AND LN

Given the patients affected by SLE, it is not surprising that questions about pregnancy frequently arise. Although pregnancy is considered in detail elsewhere in this book, a few comments about pregnancy and LN are appropriate here.

The risk of fetal loss in patients with LN has been examined in several retrospective series. In a nested case–control study of 78 pregnancies, the incidence of fetal loss was not different in patients with a history of LN compared to SLE patients with no history of LN [185]. In patients with LN in remission, fetal loss of 8–13% has been documented [186–188]. However, in patients with active LN, fetal loss was significantly higher at 35% [188]. In addition to the clinical activity of LN, hypocomplementemia appears to be a risk factor for fetal loss, whereas the use of low-dose aspirin may be protective [186].

There is also risk to the kidneys in patients with LN who become pregnant. One study noted that renal flares and progressive renal dysfunction were not different between pregnant and nonpregnant patients with LN

[185]. In other studies, renal flares were found to be higher in patients who became pregnant and had only achieved partial remission of the LN or who had more than 1 g/day proteinuria or renal insufficiency [186, 188]. Renal flare rates of 10–69% have been reported during or following pregnancy [185–188].

To protect the kidneys and the fetus, it is recommended that SLE patients with kidney involvement be advised to wait 6 months after complete renal remission before trying to become pregnant. Cytotoxic drugs such as cyclophosphamide and MMF, and antihypertensive/renoprotective agents such as ACE inhibitors and ARBs, should not be used during pregnancy. Hydroxychloroquine should be continued if patients are already taking it, and corticosteroids and AZA may be used if needed to control SLE activity.

CHILDHOOD LUPUS NEPHRITIS

Approximately 15% of all cases of SLE are diagnosed before the age of 16 years [189]. The incidence of LN in pediatric SLE patients appears to be higher than that in adults. Estimates of LN in childhood SLE range between 64 and 87%, and children tend to have more proliferative LN than do adults [190–192]. In long-term follow-up studies, the development of CKD has been reported to be as high as 45% and as low as 16–19% in children [189, 191, 192]. One study found that 19% of adolescents with LN developed ESRD compared to 2.1% of adults [191], but others found

ESRD in the 4—9% range [189, 192]. In multivariate analyses, unfavorable prognostic factors were male sex, hypertension, and inability to achieve remission, but not class IV LN [189, 192, 193]. Children with LN are currently treated with the same regimens as used for adults, but certain issues that are specific to the pediatric population demonstrate the need for therapies targeted to children. Corticosteroids can cause growth retardation, and exposure to cytotoxic agents must be carefully monitored because children have a long timeline during which they may develop cyclophosphamide-induced malignancies. In addition, compliance may be an issue with children, especially adolescents; thus, intravenous therapy as opposed to oral therapy may be necessary.

RENAL REPLACEMENT THERAPIES IN SLE

Patients who develop ESRD as a result of LN require renal replacement therapy. Both dialysis and kidney transplantation are options. There has been some uncertainty regarding whether SLE patients do worse than patients with other causes of ESRD with transplantation. This has been difficult to evaluate because of the number of confounding variables, but evidence suggests it is not a major problem.

Dialysis

Most studies that have examined dialysis in SLE patients are older, small, and observational, and they did not have some of the advantages of current artificial kidney membranes, dialysate solutions, and technologies. Nonetheless, it appears that SLE patients receiving hemodialysis have similar outcomes as patients with other causes of ESRD [194, 195]. As a group, lupus patients are generally younger than the bulk of patients on dialysis, and this may confer some survival advantage. There have been a few small studies of peritoneal dialysis and lupus. Compared to patients with other causes of ESRD, lupus patients appear to have a significantly higher mortality on peritoneal dialysis and to experience more infectious complications, including peritonitis [196, 197]. There is little information directly comparing hemodialysis to peritoneal dialysis for SLE patients, but one small study suggested 5-year survivals were similar for the two modalities [195, 198].

A persistent question regarding SLE and ESRD has been whether lupus becomes quiescent after patients reach ESRD and start on renal replacement. There does not seem to be consensus on this issue. Some investigators note significant improvement in extrarenal lupus and the ability to reduce immunosuppression [195]. Others note that activity of nonrenal SLE diminished over time after renal replacement therapy but that lupus activity was significantly worse in peritoneal dialysis patients compared to hemodialysis patients and transplant recipients [199]. Finally, there are cohorts in whom SLE activity ranged between 40 and 50%, with average durations of dialysis from 3 to more than 5 years [196, 200]. The definition of nonrenal activity varied between investigations. Often, the SLEDAI instrument was used to define flare; however, one study found SLEDAI was not accurate in hemodialysis patients [200].

Transplantation

Several small studies have examined patient and kidney allograft survival in recipients who had ESRD from LN. Some showed no difference in outcomes between SLE and non-SLE recipients [201, 202], whereas others showed worse allograft but not patient survival in the SLE group [203]. Two large studies examined this question by analyzing the U.S. Renal Data Service and United Network for Organ Sharing databases and performing multivariate analyses adjusting for comorbidities [204, 205]. These investigations had access to data from 43,000—93,000 transplant recipients, of which 2000—3000 had SLE. One study compared SLE patients to patients with diabetic nephropathy as the reference group and noted that among recipients of deceased donor kidneys, the allograft and patient survival rates were worse for SLE patients, but the hazard ratios were small at 1.14 and 1.3, respectively [204]. Worse survival was not seen in living donor recipients. The other investigation compared SLE to non-SLE recipients and did not find any differences in patient or allograft survival [205].

Several additional clinical points were made in these studies. First, SLE patients did not seem to have a higher incidence of acute rejection episodes [201, 203, 204], except in one study in which the hazard ratio for acute rejection in recipients of living (but not deceased) donor kidneys was slightly increased at 1.19 [205]. Second, the recurrence of SLE in transplant kidneys was found to be in the range of 4.3—8.6% and generally did not lead to allograft failure [201—203]. Third, lupus patients who received deceased donor kidneys had a significant decrease in allograft loss (24% decline) if MMF was used in the anti-rejection regimen [205]. Finally, SLE recipients had a higher rate of thrombotic events compared to non-SLE recipients [201, 202].

Suggested Approach to Renal Replacement Therapy for SLE Patient

Lupus patients who develop ESRD should be offered the option of a kidney transplant. Before transplantation, SLE should be quiescent. In addition, because of the higher incidence of cardiovascular disease in lupus, patients need to be carefully evaluated for this before surgery. Living donor transplants are preferred, and the use of MMF as part of the anti-rejection regimen seems appropriate. There are no data regarding prophylaxis for thrombotic events; a high index of suspicion is prudent.

Many patients will require dialysis before transplantation. Hemodialysis may be the preferred modality. While on dialysis, even though lupus can become quiescent, vigilance for extrarenal flares is appropriate, and treatment for active lupus with immunosuppression may be necessary.

EXPERIMENTAL THERAPEUTICS IN SLE: THE FUTURE OF LN TREATMENT

As discussed previously, the treatment of LN, although considerably improved since the corticosteroid era, remains inadequate and is associated with significant morbidity and mortality. To address the unmet need of more effective and less toxic therapies for LN, the recent approach has been to develop and test pharmaceuticals that target specific pathways presumed to be important in the pathogenesis of SLE and/or LN. Several of these newer drugs are termed "biologics" because they are antibodies or inhibitors of cytokines and cellular receptors. The advantage of these agents is that they affect very specific aspects of the immune system, as opposed to current therapies, and are presumably less globally immunosuppressive but theoretically more effective. Although some of these drugs have shown promise and safety in small, uncontrolled studies, larger controlled trials have been disappointing. Several problems with clinical trial design may be at fault with these studies, not the least of which is that most of these drugs are added on to regimens with corticosteroids and either cyclophosphamide or MMF. The concept of a multidrug regimen to reduce overall toxicity is rational and appealing [168], but the effects of potent immunosuppressives used in high doses may mask some of the benefits of the new agents. It is also likely that these new therapies may be effective only in certain subsets of LN, and it is currently not known which patients will receive the most benefit. Thus, in addition to developing and testing new drugs, there is an effort underway to identify biomarkers that will allow these new therapies to be used optimally and in an individualized manner [206].

B-cell Targeted Therapies

B cells appear to be critical in SLE, not only as autoantibodies but also in the role as antigen-presenting cells, cytokine-secreting cells, and regulators of T cells and dendritic cells. B cells have therefore been a major focus of new LN biologics.

Rituximab

The most widely studied anti-B-cell agent is rituximab, a monoclonal antibody against the CD20 B-cell antigen. Rituximab causes profound depletion of circulating B cells that lasts for several months. A number of small, open-label, uncontrolled trials have suggested that rituximab is effective in proliferative LN, either for refractory disease or as induction therapy [207–213]. However, in a large, prospective, double-blind controlled study of rituximab versus placebo added to MMF plus corticosteroids for proliferative LN, there was no difference in complete or partial responses between groups at 12 months [214]. The niche for rituximab in the therapy of LN thus remains unclear.

Epratuzumab

Epratuzumab is an anti-CD22 humanized monoclonal antibody. CD22 is a B-cell antigen receptor co-receptor. Epratuzumab causes partial depletion of B cells but may also interfere with the proliferation and activation of lupus B cells [207–209]. One study using this agent has been published, and only four LN patients were included [215]. Nonetheless, 75% of the LN patients showed some improvement in British Isles Lupus Assessment Group scores.

B-cell Survival Factors

B cells require cytokines such as BLyS and APRIL for survival and proliferation. Drugs that inhibit these factors, including belimumab, an anti-BLyS monoclonal antibody, and atacicept, a soluble receptor that binds to BLyS and APRIL, are being evaluated in SLE [207–209]. The atacicept trial in LN has been terminated due to infectious morbidity in combination with MMF.

B-Cell Tolerogens

B-cell tolerogens are designed to cross-link autoantibodies on the surface of lupus B cells to inactivate or deplete autoreactive cells. Abetimus sodium is a tolerogen that consists of double-stranded DNA segments attached to a carrier that cross-links anti-double-stranded DNA on B cells [207–209, 216]. This agent was studied in LN patients to determine if it could prolong time to renal flare [217]. Although it did not meet its primary endpoint, it did show trends toward improvement in some aspects of LN. Fewer abetimus sodium-treated patients had renal flares, and compared to placebo, proteinuria was decreased by at least 50% more often in the abetimus-treated patients.

Co-stimulatory Antagonists

Antigen-presenting cells communicate with T cells through a variety of co-stimulatory molecules. CD28 and CTLA4 on T cells bind to B7.1/B7.2 receptors on antigen-presenting cells. Abatacept is recombinant CTLA4 fused to IgG heavy-chain components, and it blocks the interaction between CD28 and B7.1/B7.2 [207–209]. In combination with cyclophosphamide, it was effective in reducing proteinuria in a mouse model of LN [218]. CTLA4-Ig plus cyclophosphamide is being tested in human LN.

Anticytokine Therapies

A wide array of cytokines have been implicated in the pathogenesis or tissue damage of SLE and LN. These include interleukin-6, tumor necrosis factor-α, interferon-α, and complement component C5a. Antagonists of these cytokines or their receptors have been developed and are at various stages of preclinical or clinical testing [207–209].

Leflunomide

Leflunomide is used for rheumatoid arthritis. It blocks lymphocyte proliferation and T-cell activation, and it suppresses production of cytokines such as interleukin-2. There have been a few interesting studies using leflunomide to treat LN. In one open-label, prospective trial done in China, 70 patients received leflunomide plus corticosteroids for proliferative LN and 40 patients received intravenous cyclophosphamide plus corticosteroids [219]. After 6 months, 21% of the leflunomide and 18% of the cyclophosphamide patients entered complete remission, whereas 52 and 55% achieved partial remission, respectively. Adverse effects were similar between groups. Repeat biopsies were done in 13 leflunomide patients, and whereas the activity index decreased significantly, the chronicity index increased threefold. No biopsies were done in the cyclophosphamide group for comparison. In another study of Chinese SLE patients, 31 were treated with leflunomide plus corticosteroids for 12 months [220]. The complete remission rate was 58%, and the partial remission rate was 42%. Side effects were mild. All patients had a repeat biopsy and showed a significant improvement in activity index, with a stable chronicity index. Thus, leflunomide may be an agent worth examining in a larger, randomized trial. It is worrisome that at 6 months the chronicity index increased in one trial; thus, long-term studies will be required to determine if leflunomide can preserve kidney function over time as well as cyclophosphamide.

CONCLUSION

LN is a severe manifestation of SLE and carries with it the potential of significant morbidity and mortality. Early intervention is key to improving the outcome of patients with LN. This requires the treating physician to maintain a high level of suspicion for kidney involvement, both at the time of initial diagnosis of SLE and during follow-up. When the kidney does appear to be involved, consultation with a nephrologist experienced in autoimmune diseases is appropriate, and a co-management approach for patient care should be initiated. Treatment for LN continues to be suboptimal; however, a variety of new biologic agents for LN are being tested in clinical trials or are in preclinical development and have the potential to improve results and decrease toxicities.

Acknowledgments

This chapter is dedicated to my friend and mentor, Dr. Lee Hebert. He is one of the finest clinician scientists I have known, and he motivated me to study lupus nephritis, for which I will forever be grateful. BHR. I dedicate this chapter to my mentor, Seymour Rosen—a member of the first generation of nephropathologists then and a model of experience now; he continues to inspire and challenge me. IES.

References

[1] R. Campbell Jr., G.S. Cooper, G.S. Gilkeson, Two aspects of the clinical and humanistic burden of systemic lupus erythematosus: Mortality risk and quality of life early in the course of disease, Arthritis Rheum 59 (2008) 458–464.

[2] M.I. Danila, G.J. Pons-Estel, J. Zhang, L.M. Vila, J.D. Reveille, G.S. Alarcon, Renal damage is the most important predictor of mortality within the damage index: Data from LUMINA LXIV, a multiethnic US cohort, Rheumatology (Oxford) 48 (2009) 542–545.

[3] J. Font, M. Ramos-Casals, R. Cervera, M. Garcia-Carrasco, A. Torras, A. Siso, et al., Cardiovascular risk factors and the long-term outcome of lupus nephritis, QJM 94 (2001) 19–26.

[4] M.R. Gonzalez-Crespo, J.I. Lopez-Fernandez, G. Usera, M.J. Poveda, J.J. Gomez-Reino, Outcome of silent lupus nephritis, Sem Arthritis Rheum 26 (1996) 468–476.

[5] R. Valente de Almeida, J.G. Rocha de Carvalho, V.F. de Azevedo, R.A. Mulinari, S.O. Ioshhi, S. da Rosa Utiyama, et al., Microalbuminuria and renal morphology in the evaluation of subclinical lupus nephritis, Clin Nephrol 52 (1999) 218–229.

[6] M.E. Zabaleta-Lanz, L.E. Munoz, F.J. Tapanes, R.E. Vargas-Arenas, I. Daboin, Y. Barrios, et al., Further description of early clinically silent lupus nephritis, Lupus 15 (2006) 845–851.

[7] H.M. Bastian, J.M. Roseman, G. McGwin Jr., G.S. Alarcon, A.W. Friedman, B.J. Fessler, et al., Systemic lupus erythematosus in three ethnic groups. XII. Risk factors for lupus nephritis after diagnosis, Lupus 11 (2002) 152–160.

[8] V.A. Seligman, R.F. Lum, J.L. Olson, H. Li, L.A. Criswell, Demographic differences in the development of lupus nephritis: A retrospective analysis, Am J Med 112 (2002) 726–729.

[9] A.S.a. Arfaj, N. Khalil, Clinical and immunological manifestations in 624 SLE patient in Saudi Arabia, Lupus 18 (2009) 465–473.

[10] M.M. Ward, Changes in the incidence of endstage renal disease due to lupus nephritis in the United States, 1996–2004, J Rheumatol 36 (2009) 63–67.

[11] M. Adler, S. Chambers, C. Edwards, G. Neild, D. Isenberg, An assessment of renal failure in an SLE cohort with special reference to ethnicity, over a 25-year period, Rheumatology 45 (2006) 1144–1147.

[12] A.S. Go, G.M. Chertow, D. Fan, C.E. McCulloch, C.Y. Hsu, Chronic kidney disease and the risks of death, cardiovascular events, and hospitalization, N Engl J Med 351 (2004) 1296–1305.

[13] D.E. Weiner, H. Tighiouart, M.G. Amin, P.C. Stark, B. MacLeod, J.L. Griffith, et al., Chronic kidney disease as a risk factor for cardiovascular disease and all-cause mortality: A pooled analysis of community-based studies, J Am Soc Nephrol 15 (2004) 1307–1315.

[14] M.J. Sarnak, A.S. Levey, A.C. Schoolwerth, J. Coresh, B. Culleton, L.L. Hamm, et al., Kidney disease as a risk factor for development of cardiovascular disease: A statement from the American Heart Association Councils on Kidney in Cardiovascular Disease, High Blood Pressure Research, Clinical Cardiology, and Epidemiology and Prevention, Circulation 108 (2003) 2154–2169.

[15] Y. Asanuma, A. Oeser, A.K. Shintani, E. Turner, N. Olsen, S. Fazio, et al., Premature coronary-artery atherosclerosis in systemic lupus erythematosus, N Engl J Med 349 (2003) 2407–2415.

[16] J.M. Esdaile, C. Levinton, W. Federgreen, J.P. Hayslet, M. Kashgarian, The clinical and renal biopsy predictors of long-term outcome in lupus nephritis: A study of 87 patients and review of the literature, Q J Med 269 (1989) 779–833.

[17] M. Faurschou, H. Starklint, P. Halbert, S. Jacobsen, Prognosis factors in lupus nephritis: Diagnostic and therapeutic delay increases the risk of terminal renal failure, J Rheumatol 33 (2006) 1563–1569.

[18] J.M. Esdaile, L. Joseph, T. Mackenzie, M. Kashgarian, J.P. Hayslet, The benefit of early treatment with immunosuppressive agents in lupus nephritis, J Rheumatol 21 (1994) 2046–2051.

[19] J.-H. Yen, C.-J. Chen, C.-Y. Tsai, C.-H. Lin, T.-T. Ou, C.-J. Hu, et al., Cytochrome P450 and manganese superoxide dismutase genes polymorphisms in systemic lupus erythematosus, Immunol Lett 90 (2003) 19–24.

[20] C. Fiehn, Early diagnosis and treatment in lupus nephritis: How we can influence the risk for terminal renal failure, J Rheumatol 33 (2006) 1464–1466.

[21] C. Fiehn, Y. Hajjar, K. Mueller, R. Waldherr, A.D. Ho, K. Andrassy, Improved clinical outcome of lupus nephritis during the past decade: Importance of early diagnosis and treatment, Ann Rheum Dis 62 (2003) 435–439.

[22] G. Moroni, S. Quaglini, B. Gallelli, G. Banfi, P. Messa, C. Ponticelli, The long-term outcome of 93 patients with proliferative lupus nephritis, Nephrol Dial Transplant 22 (2007) 2531–2539.

[23] Y. Takahashi, T. Mizoue, A. Suzuki, H. Yamashita, J. Kunimatsu, K. Itoh, et al., Time of initial appearance of renal symptoms in the course of systemic lupus erythematosus as a prognostic factor for lupus nephritis, Mod Rheumatol 19 (2009) 293–301.

[24] A.J.W. Branten, G. Vervoort, J.F.M. Wetzels, Serum creatinine is a poor marker of GFR in nephrotic syndrome, Nephrol Dial Transplant 20 (2005) 707–711.

[25] D.W. Cockcroft, M.H. Gault, Prediction of creatinine clearance from serum creatinine, Nephron 16 (1976) 31–41.

[26] A.S. Levey, J. Bosch, J.B. Lewis, T. Greene, N. Rogers, A more accurate method to estimate glomerular filtration rate from serum creatinine: A new prediction equation. Modification of Diet in Renal Disease Study Group, Ann Int Med 130 (1999) 461–470.

[27] N. Kasitanon, D.M. Fine, M. Haas, L.S. Magder, M. Petri, Estimating renal function in lupus neprhitis: Comparison of the modification of diet in renal disease and Cockcroft–Gault equations, Lupus 16 (2007) 887–895.

[28] A.S. Levey, L.A. Stevens, C.H. Schmid, Y.L. Zhang, A.F. Castro 3rd, H.I. Feldman, et al., A new equation to estimate glomerular filtration rate, Ann Intern Med 150 (2009) 604–612.

[29] R.J. Glassock, C. Winearls, Screening for CKD with eGFR: Doubts and dangers, Clin J Am Soc Nephrol 3 (2008) 1563–1568.

[30] M.J. Siedner, A.C. Gelber, B.H. Rovin, A.M. McKinley, L. Christopher-Stine, B. Astor, et al., Diagnostic accuracy study of urine dipstick in relation to 24-hour measurement as a screening tool for proteinuria in lupus nephritis, J Rheumatol 35 (2008) 84–90.

[31] D.J. Birmingham, B.H. Rovin, G. Shidham, H.N. Nagaraja, X. Zou, M. Bissell, et al., Spot urine protein/creatinine ratios are unreliable estimates of 24 h proteinuria in most systemic lupus erythematosus nephritis flares, Kidney Int 72 (2007) 865–870.

[32] D.M. Fine, M. Ziegenbein, M. Petri, E. Han, A. McKinley, J. Chellini, et al., A prospective study of 24-hour protein excretion in lupus nephritis: Adequacy of short-interval timed urine collections, Kidney Int 76 (2009) 1284–1288.

[33] J.P. Grande, J.E. Balow, Renal biopsy in lupus nephritis, Lupus 7 (1998) 611–617.

[34] L. Christopher-Stine, M.J. Siedner, J. Lin, M. Haas, H. Parekh, M. Petri, et al., Renal biopsy in lupus patients with low levels of proteinuria, J Rheumatol 34 (2007) 332–335.

[35] B.H. Rovin, Glomerular disease: Lupus nephritis treatment: Are we beyond cyclophosphamide? Nature Rev Nephrol 5 (2009) 492–494.

[36] L.A. Hebert, H.M. Sharma, D.D. Sedmak, W.H. Bay, Unexpected renal biopsy findings in a febrile systemic lupus erythematosus patient with worsening renal function and heavy proteinuria, Am J Kidney Dis 13 (1989) 504–507.

[37] E. Baranowska-Daca, Y.-J. Choi, R. Barrios, G. Nassar, W.N. Suki, L.D. Truong, Non-lupus nephrritides in patients with systemic lupus erythematosus: A comprehensive clinicopathologic study and review of the literature, Hum Pathol 32 (2001) 1125–1135.

[38] K.T. Ellington, L. Truong, J.J. Olivero, Renal amyloidosis in systemic lupus erythematosus, Am J Kidney Dis 21 (1993) 676–678.

[39] C. Orellana, A. Collado, M.V. Hernandez, J. Font, J.A. Del Olmo, J. Munoz-Gomez, When does amyloidosis complicate systemic lupus erythematosus? Lupus 4 (1995) 415–417.

[40] T. Pettersson, T. Tornroth, K.J. Totterman, P. Fortelius, C.P. Maury, AA amyloidosis in systemic lupus erythematosus, J Rheumatol 14 (1987) 835–838.

[41] Y. Mori, N. Kishimoto, H. Yamahara, Y. Kijima, A. Nose, Y. Uchiyama-Tanaka, et al., Predominant tubulointerstitial nephritis in a patient with systemic lupus nephritis, Clin Exp Nephrol 9 (2005) 79–84.

[42] S.W. Kraft, M.M. Schwartz, S.M. Korbet, E.J. Lewis, Glomerular podocytopathy in patients with systemic lupus erythematosus, J Am Soc Nephrol 16 (2005) 175–179.

[43] M.G. Tektonidou, F. Sotsiou, H.M. Moutsopoulos, Antiphospholipid syndrome nephropathy in catastrophic, primary, and systemic lupus erythematosus-related APS, J Rheumatol 35 (2008) 1983–1988.

[44] E. Daugas, D. Nochy, D.L. Huong, P. Duhaut, H. Beaufils, V. Caudwell, et al., Antiphospholipid syndrome nephropathy in systemic lupus erythematosus, J Am Soc Nephrol 13 (2002) 42–52.

[45] B.H. Rovin, X. Zhang, Biomarkers for lupus nephritis: The quest continues, Clin J Am Soc Nephrol 4 (2009) 1858–1865.

[46] G.S. Hill, M. Delahousse, D. Nochy, P. Remy, F. Mignon, J.-P. Mery, et al., Predictive power of the second renal biopsy in lupus nephritis: Significance of macrophages, Kidney Int 59 (2001) 304–316.

[47] J.R. Sedor, Tissue proteomics: A new investigative tool for renal biopsy analysis, Kidney Int 75 (2009) 876–879.

[48] Y. Yasuda, C.D. Cohen, A. Henger, M. Kretzler, Gene expression profiling analysis in nephrology: Towards molecular definition of renal disease, Clin Exp Nephrol 10 (2006) 91–98.

[49] K.S. Peterson, J.-F. Huang, J. Zhu, V.D. D'Agati, X. Liu, N. Miller, et al., Characterization of heterogeneity in the molecular pathogenesis of lupus nephritis from transcriptional profiles of laser-captured glomeruli, J Clin Invest 113 (2004). 1792-1733.

[50] P.D. Walker, The renal biopsy, Arch Pathol Lab Med 133 (2009) 181–188.

[51] M. Mannik, C.E. Merrill, L.D. Stamps, M.H. Wener, Multiple autoantibodies form the glomerular immune deposits in patients with systemic lupus erythematosus, J Rheumatol 30 (2003) 1495–1504.

[52] D.G. Williams, D.K. Peters, J. Fallows, A. Petrie, O. Kourilsky, L. Morel-Maroger, et al., Studies of serum complement in the hypocomplementaemic nephritides, Clin Exp Immunol 18 (1974) 391–405.

[53] G.A. Herrera, The value of electron microscopy in the diagnosis and clinical management of lupus nephritis, Ultrastruct Pathol 23 (1999) 63–77.

[54] L. Ronnblom, G.V. Alm, M.L. Eloranta, Type I interferon and lupus, Curr Opin Rheumatol 21 (2009) 471–477.

[55] M.H. Park, D. D'Agati, G.B. Appel, C.L. Pirani, Tubulointerstitial disease in lupus nephritis: Relationship to immune deposits, interstitial inflammation, glomerular changes, renal function, and prognosis, Nephron 44 (1986) 309–314.

[56] J.W. Bauer, E.C. Baechler, M. Petri, F.M. Batliwalla, D. Crawford, W.A. Ortmann, et al., Elevated serum levels of interferon-regulated chemokines are biomarkers for active human systemic lupus erythematosus, PLoS Med 3 (2006) 2274–2284.

[57] G.B. Appel, C.L. Pirani, V. D'Agati, Renal vascular complications of systemic lupus erythematosus, J Am Soc Nephrol 4 (1994) 1499–1515.

[58] E. Descombes, D. Droz, L. Drouet, J.P. Grunfeld, P. Lesavre, Renal vascular lesions in lupus nephritis, Medicine (Baltimore) 76 (1997) 355–368.

[59] G. Banfi, T. Bertani, V. Boeri, T. Faraggiana, G. Mazzucco, G. Monga, et al., Renal vascular lesions as a marker of poor prognosis in patients with lupus nephritis. Gruppo Italiano per lo Studio della Nefrite Lupica (GISNEL), Am J Kidney Dis 18 (1991) 240–248.

[60] R.C. Muehrcke, R.M. Kark, C.L. Pirani, V.E. Pollak, Lupus nephritis: A clinical and pathologic study based on renal biopsies, Medicine (Baltimore) 36 (1957) 1–145.

[61] V.E. Pollak, C.L. Pirani, F.D. Schwartz, The natural history of the renal manifestations of systemic lupus erythematosus, J Lab Clin Med 63 (1964) 537–550.

[62] S.C. Sommers, J. Bernstein, Kidney Pathology Decennial, 1966–1975, Appleton-Century-Crofts, New York, 1975.

[63] A.S. Appel, G.B. Appel, An update on the use of mycophenolate mofetil in lupus nephritis and other primary glomerular diseases, Nature Clin Practice 5 (2009) 132–142.

[64] J. Churg, L.H. Sobin, Renal Disease: Classification and Atlas of Glomerular Diseases, Igaku-Shoin, New York, 1982.

[65] J. Churg, J. Bernstein, R.J. Glassock, Renal Disease: Classification and Atlas of Glomerular Diseases, Igaku-Shoin, New York, 1995.

[66] J.J. Weening, V.D. D'Agati, M.M. Schwartz, S.V. Seshan, C.E. Alpers, G.B. Appel, et al., The classification of glomerulonephritis in systemic lupus erythematosus revisited, Kidney Int 65 (2004) 521–530.

[67] J.J. Weening, V.D. D'Agati, M.M. Schwartz, S.V. Seshan, C.E. Alpers, G.B. Appel, et al., The classification of glomerulonephritis in systemic lupus erythematosus revisited, J Am Soc Nephrol 15 (2004) 241–250.

[68] S.H. Nasr, V.D. D'Agati, H.R. Park, Necrotizing and crescentic lupus nephritis with antineutrophil cytoplasmic antibody seropositivity, Clin J Am Soc Nephrol 3 (2008) 682–690.

[69] J.C. Jennette, S.S. Iskandar, F.G. Dalldorf, Pathologic differentiation between lupus and nonlupus membranous glomerulopathy, Kidney Int 24 (1983) 377–385.

[70] N. Hiramatsu, T. Kuroiwa, H. Ikeuchi, A. Maeshima, Y. Kaneko, K. Hiromura, et al., Revised classification of lupus nephritis is valuable in predicting renal outcome with an indication of the proportion of glomeruli affected by chronic lesions, Rheumatology (Oxford) 47 (2008) 702–707.

[71] C.L. Pirani, V.E. Pollak, F.D. Schwartz, The reproducibility of semiquantitative analyses of renal histology, Nephron 29 (1964) 230–237.

[72] H.A. Austin 3rd, L.R. Muenz, K.M. Joyce, T.T. Antonovych, J.E. Balow, Diffuse proliferative lupus nephritis: Identification of specific pathologic features affecting renal outcome, Kidney Int 25 (1984) 689–695.

[73] M.M. Schwartz, S.P. Lan, J. Bernstein, G.S. Hill, K. Holley, E.J. Lewis, Irreproducibility of the activity and chronicity indices limits their utility in the management of lupus nephritis. Lupus Nephritis Collaborative Study Group, Am J Kidney Dis 21 (1993) 374–377.

[74] H.A. Austin III, D.T. Boumpas, E.M. Vaughan, J.E. Balow, Predicting renal outcomes in severe lupus nephritis: Contributions of clinical and histologic data, Kidney Int 45 (1994) 544–550.

[75] G.S. Hill, M. Delahousse, D. Nochy, E. Tomkiewicz, P. Remy, F. Mignon, et al., A new morphologic index for the evaluation of renal biopsies in lupus nephritis, Kidney Int 58 (2000) 1160–1173.

[76] G. Moroni, S. Pasquali, S. Quaglini, G. Banfi, S. Casanova, M. Maccario, et al., Clinical and prognostic value of serial renal biopsies in lupus nephritis, Am J Kidney Dis 34 (1999) 530–539.

[77] G.M.N. Daleboudt, I.M. Bajema, N.N.T. Goemaere, J.M. van Laar, J.A. Bruijn, S.P. Berger, The clinical relevance of a repeat biopsy in lupus nephritis flares, Nephrol Dial Transplant 24 (2009) 3712–3717.

[78] P.N. Furness, N. Taub, Interobserver reproducibility and application of the ISN/RPS classification of lupus nephritis—a UK-wide study, Am J Surg Pathol 30 (2006) 1030–1035.

[79] G.S. Markowitz, V.D. D'Agati, Classification of lupus nephritis, Curr Opin Nephrol Hypertens 18 (2009) 220–225.

[80] C.C. Najafi, S.M. Korbet, E.J. Lewis, M.M. Schwartz, M. Reichlin, J. Evans, Lupus Nephritis Collaborative Study Group, Significance of histologic patterns of glomerular injury

upon long-term prognosis in severe lupus glomerulonephritis, Kidney Int 59 (2001) 2156–2163.

[81] B. Mittal, S. Hurwitz, H. Rennke, A.K. Singh, New subcategories of class IV lupus nephritis: Are there clinical, histologic, and outcome differences? Am J Kidney Dis 44 (2004) 1050–1059.

[82] H. Yokoyama, T. Wada, A. Hara, J. Yamahana, I. Nakaya, M. Kobayashi, et al., The outcome and a new ISN/RPS 2003 classification of lupus nephritis in Japanese, Kidney Int 66 (2004) 2382–2388.

[83] G.S. Hill, M. Delahousse, D. Nochy, J. Bariety, Class IV-S versus class IV-G lupus nephritis: Clinical and morphologic differences suggesting different pathogenesis, Kidney Int 68 (2005) 2288–2297.

[84] Y.G. Kim, H.W. Kim, Y.M. Cho, J.S. Oh, S.S. Nah, C.K. Lee, et al., The difference between lupus nephritis class IV-G and IV-S in Koreans: Focus on the response to cyclophosphamide induction treatment, Rheumatology (Oxford) 47 (2008) 311–314.

[85] D. Sen, D.A. Isenberg, Antineutrophil cytoplasmic autoantibodies in systemic lupus erythematosus, Lupus 12 (2003) 651–658.

[86] M.A. Dooley, R.J. Falk, Human clinical trials in lupus nephritis, Sem Nephrol 27 (2007) 115–127.

[87] R.S. Flanc, M.A. Roberts, G.F.M. Strippoli, S.J. Chadban, P.G. Kerr, R.C. Atkins, Treatment for lupus nephritis. Cochrane Database Syst, Rev., CD002922 (2004).

[88] H.A. Austin, J.H. Klippel, J.E. Balow, W.G. le Riche, A.D. Steinberg, P.H. Plotz, et al., Therapy of lupus nephritis. Controlled trial of prednisone and cytotoxic drugs, N Engl J Med 314 (1986) 614–619.

[89] D.T. Boumpas, H.A. Austin, E.M. Vaughn, J.H. Klippel, A.D. Steinberg, C. Yarboro, et al., Controlled trial of pulse methylprenisolone versus two regimens of pulse cyclophosphamide in severe lupus nephritis, Lancet 340 (1992) 741–745.

[90] J.V. Donadio, K.E. Holley, R.H. Ferguson, D.M. Ilstrup, Treatment of diffuse proliferative lupus nephritis with prednisone and combined prednisone and cyclophosphamide, N Engl J Med 23 (1978) 1151–1155.

[91] M. Haubitz, F. Bohnenstengel, R. Brunkhorst, M. Schwab, U. Hofmann, D. Busse, Cyclophosphamide pharmacokinetics and dose requirements in patients with renal insufficiency, Kidney Int 61 (2002) 1495–1501.

[92] S. Pendse, E. Ginsburg, A.K. Singh, Strategies for preservation of ovarian and testicular function after immunosuppression. Am J Kidney Dis 43 (2004) 772–781.

[93] E.C. Sommers, W. Marder, G.M. Christman, V. Ognenovski, J. McCune, Use of a gonadotropin-releasing hormone analog for protection against premature ovarian failure during cyclophosphamide therapy in women with severe lupus, Arthritis Rheum 52 (2005) 2761–2767.

[94] G. Contreras, V. Pardo, B. Leclercq, O. Lenz, E. Tozman, P. O'Nan, et al., Sequential therapies for proliferative lupus nephritis, N Engl J Med 350 (2004) 971–980.

[95] F.A. Houssiau, C. Vasconcelos, D. D'Cruz, G.D. Sebastiani, R. Garrido Ed Ede, M.G. Danieli, et al., Immunosuppressive therapy in lupus nephritis: the Euro-Lupus Nephritis Trial, a randomized trial of low-dose versus high-dose intravenous cyclophosphamide, Arthritis Rheum 46 (2002) 2121–2131.

[96] F.A. Houssiau, C. Vasconcelos, D. D'Cruz, G.D. Sebastiani, E. de Ramon Garrido, M.G. Danieli, et al., The 10-year follow-up data of the Euro-Lupus Nephritis Trial comparing low-dose versus high-dose intravenous cyclophosphamide, Ann Rheum Dis 69 (2010) 61–64.

[97] E.C.F. Wilson, D.R.W. Jayne, E. Dellow, R.J. Fordham, The cost-effectiveness of mycophenolate mofetil as first-line therapy in active lupus nephritis, Rheumatology 46 (2007) 1096–1101.

[98] C.C. Mok, C.T.K. Ho, K.W. Chan, C.S. Lau, R.W.S. Wong, Outcome and prognostic indicators of diffuse proliferative lupus glomerulonephritis treated with sequential oral cyclophosphamide and azathioprine, Arthritis Rheum 46 (2002) 1003–1013.

[99] C.C. Mok, C.T.K. Ho, Y.P. Siu, K.W. Chan, T.H. Kwan, C.S. Lau, et al., Treatment of diffuse proliferative lupus glomerulonephritis: A comparison of two cyclophosphamide-containing regimens, Am J Kidney Dis 38 (2001) 256–264.

[100] T.M. Chan, K.C. Tse, C.S.O. Tang, K.N. Lai, F.K. Li, Long-term outcome of patients with diffuse proliferative lupus nephritis treated with prednisolone and oral cyclophosphamide followed by azathioprine, Lupus 14 (2005) 265–272.

[101] T.M. Chan, F.K. Li, C.S.O. Tang, R.W.S. Wong, G.X. Fang, Y.L. Ji, et al., Efficacy of mycophenolate mofetil in patients with diffuse proliferative lupus nephritis, N Engl J Med 343 (2000) 1156–1162.

[102] T.M. Chan, K.C. Tse, C.S.O. Tang, M.-Y. Mok, F.K. Li, Long-term study of mycophenolate mofetil as continuous induction and maintenance treatment for diffuse proliferative lupus nephritis, J Am Soc Nephrol 16 (2005) 1076–1084.

[103] A. McKinley, E. Park, D.N. Spetie, K. Hackshaw, L.A. Hebert, B.H. Rovin, Oral cyclophosphamide for lupus glomerulonephritis: An underused therapeutic option, Clin J Am Soc Nephrol 4 (2009) 1754–1760.

[104] E.J. Lewis, L.G. Hunsicker, S.-P. Lan, R.D. Rohde, J.M. Lachin, and the Lupus Nephritis Collaborative Study Group, A controlled trial of plasmapheresis therapy in severe lupus nephritis, N Engl J Med 326 (1992) 1373–1379.

[105] L.M. Ong, L.S. Hooi, T.O. Lim, B.L. Goh, G. Ahmad, R. Ghazalli, et al., Randomized controlled trial of pulse intravenous cyclophosphamide versus mycophenolate mofetil in the induction therapy of proliferative lupus nephritis, Nephrology 10 (2005) 504–510.

[106] W. Hu, Z. Liu, H. Chen, Z. Tang, Q. Wang, K. Shen, et al., Mycophenolate mofetil vs. cyclophosphamide therapy for patients with diffuse proliferative lupus nephritis, Chinese Med J 115 (2002) 705–709.

[107] E.M. Ginzler, M.A. Dooley, C. Aranow, M.Y. Kim, J.P. Buyon, J.T. Merrill, et al., Mycophenolate mofetil or intravenous cyclophosphamide for lupus nephritis, N Engl J Med 353 (2005) 2219–2228.

[108] G.B. Appel, G. Contreras, M.A. Dooley, E.M. Ginzler, D. Isenberg, D. Jayne, et al., Aspreva Lupus Management Study Group, Mycophenolate mofetil versus cyclophosphamide for induction treatment of lupus nephritis, J Am Soc Nephrol 20 (2009) 1103–1112.

[109] C. Grootscholten, I.M. Bajema, S. Florquin, E.J. Steenbergen, C.J. Peutz-Kootstra, R. Goldschmeding, et al., Treatment with cyclophosphamide delays the progression of chronic lesions more effectively than does treatment with azathioprine plus methylprednisolone in patients with proliferative lupus nephritis, Arthritis Rheum 56 (2007) 924–937.

[110] S.M. Korbet, M.M. Schwartz, J. Evans, E.J. Lewis, Severe lupus nephritis: Racial differences in presentation and outcome, J Am Soc Nephrol 18 (2007) 244–254.

[111] R. Kaiser, C.M. Cleveland, L.A. Criswell, Risk and protective factors for thrombosis in systemic lupus erythematosus: Results from a large, multi-ethnic cohort, Ann Rheum Dis 68 (2009) 238–241.

[112] G.J. Pons-Estel, G.S. Alarcon, G. McGwin Jr., M.I. Danila, J. Zhang, H.M. Bastian, et al., Protective effect of hydroxychloroquine on renal damage in patients with lupus nephritis: LXV. Data from a multiethnic U.S. cohort, Arthritis Rheum 61 (2009) 830–839.

[113] E. Tsakonas, L. Joseph, J.M. Esdaile, D. Choquette, J.L. Senecal, A. Cividino, et al., A long-term study of hydroxychloroquine withdrawal on exacerbations in systemic lupus erythematosus. The Canadian Hydroxychloroquine Study Group, Lupus 7 (1998) 80–85.

[114] A. Siso, M. Ramos-Casals, A. Bove, P. Brito-Zeron, N. Soria, S. Munoz, et al., Previous antimalarial therapy in patients diagnosed with lupus nephritis: Influence on outcomes and survival, Lupus 17 (2008) 281–288.

[115] F.A. Houssiau, C. Vasconcelos, D. D'Cruz, G.D. Sebastiani, E. de Ramon Garrido, M.G. Danieli, et al., Early response to immunosuppressive therapy predicts good renal outcome in lupus nephritis: Lessons from long-term follow-up of patients in the Euro-Lupus Nephritis Trial, Arthritis Rheum 50 (2004) 3934–3940.

[116] C. Grootscholten, G. Ligtenberg, E.C. Hagen, A.W.L. van den Wall Bake, J.W. de Glas-Vos, M. Biji, et al., Azathioprine/ methylprednisolone versus cyclophosphamide in proliferative lupus nephritis. A randomized, controlled trial, Kidney Int 70 (2006) 732–742.

[117] J.P.A. Ioannidis, K.A. Boki, M.E. Katsorida, A.A. Drosos, F.N. Skopouli, J.N. Boletis, et al., Remission, relapse, and re-remission of proliferative lupus nephritis treated with cyclophosphamide, Kidney Int 57 (2000) 258–264.

[118] Y.E. Chen, S.M. Korbet, R.S. Katz, M.M. Schwartz, E.J. Lewis, Value of a complete or partial remission in severe lupus nephritis, Clin J Am Soc Nephrol 3 (2008) 46–53.

[119] L.A. Hebert, W.A. Wilmer, M.E. Falkenhain, S.E. Ladson-Wofford, N.S. Nahman, B.H. Rovin, Renoprotection: One or many therapies, Kidney Int 59 (2001) 1211–1226.

[120] W.A. Wilmer, B.H. Rovin, C.J. Hebert, S.V. Rao, K. Kumor, L.A. Hebert, Management of glomerular proteinuria: A commentary, J Am Soc Nephrol 14 (2003) 3217–3232.

[121] H.N. Reich, S. Troyanov, J.W. Scholey, D.C. Cattran, Remission of proteinuria improves prognosis in IgA nephropathy, J Am Soc Nephrol 18 (2007) 3177–3183.

[122] L. Mercadal, S.T. Montcel, D. Nochy, G. Queffeulou, J.C. Piette, C. Isnard-Bagnis, et al., Factors affecting outcome and prognosis in membranous lupus nephropathy, Nephrol Dial Transplant 17 (2002) 1771–1778.

[123] S. Pasquali, G. Banfi, A. Zucchelli, G. Moroni, C. Ponticelli, P. Zucchelli, Lupus membranous nephropathy: Long-term outcome, Clin Nephrol 39 (1993) 175–182.

[124] C.C. Mok, Membranous nephropathy in systemic lupus erythematosus: A therapeutic enigma, Nature Rev Nephrol 5 (2009) 212–220.

[125] R.P. Sloan, M.M. Schwartz, S.M. Korbet, R.Z. Borok, Long-term outcome in systemic lupus erythematosus membranous glomerulonephritis, J Am Soc Nephrol 7 (1996) 299–305.

[126] J.V. Donadio, J.H. Burgess, K.E. Holley, Membranous lupus nephropathy: A clinicopathologic study, Medicine 56 (1977) 527–536.

[127] H. Gonzalez-Dettoni, F. Tron, Membranous glomerulonephropathy in systemic lupus erythematosus, Adv Nephrol 14 (1985) 347–364.

[128] J.D. Ordonez, R.A. Hiatt, E.J. Killebrew, B.H. Fireman, The increased risk of coronary heard disease associated with the nephrotic syndrome, Kidney Int 44 (1993) 638–642.

[129] J. Font, M. Ramos-Casals, R. Cervera, M. Garcia-Carrasco, A. Torras, A. Siso, et al., Cardiovascular risk factors and the long-term outcome of lupus nephritis, Q J Med 94 (2001) 19–26.

[130] C.C. Mok, K.Y. Ying, C.S. Lau, C.W. Yim, W.L. Ng, W.S. Wong, et al., Treatment of pure membranous lupus nephropathy with prednisone and azathioprine: An open-label trial, Am J Kidney Dis 43 (2004) 269–276.

[131] H.A. Austin, G.G. Illei, M.J. Braun, J.E. Balow, Randomized, controlled trial of prednisone, cyclophosphamide, and cyclosporine in lupus membranous nephropathy, J Am Soc Nephrol 20 (2009) 901–911.

[132] T.H. Jafar, C.H. Schmid, M. Landa, I. Giatras, R. Toto, G. Remuzzi, et al., Angiotensin-converting enzyme inhibitors and progression of nondiabetic renal disease. A meta-analysis of patient-level data, Ann Intern Med 135 (2001) 73–87.

[133] I. Giatras, J. Lau, A.S. Levey, Effect of angiotensin-converting enzyme inhibitors on the progression of non-diabetic renal disease: A meta-analysis of randomized trials. Angiotensin-Converting-Enzyme Inhibition and Progressive Renal Disease Study Group, Ann Intern Med 127 (1997) 337–345.

[134] M.W. Taal, B.M. Brenner, Renoprotective benefits of RAS inhibition: From ACEI to angiotensin II antagonists, Kidney Int 57 (2000) 1803–1817.

[135] C. Zoja, M. Morigi, G. Remuzzi, Proteinuria and phenotypic change of proximal tubular cells, J Am Soc Nephrol 14 (2003) S36–S41.

[136] G. Remuzzi, P. Ruggenenti, A. Benigni, Understanding the nature of renal disease progression: In proteinuric nephropathies enhanced glomerular protein traffic contributes to interstitial inflammation and renal scarring, Kidney Int 51 (1997) 2–15.

[137] M. Abbate, C. Zoja, D. Rottoli, D. Corna, N. Perico, T. Bertani, et al., Antiproteinuric therapy while preventing the abnormal protein traffic in proximal tubule abrogates protein- and complement-dependent interstitial inflammation in experimental renal disease, J Am Soc Nephrol 10 (1999) 804–813.

[138] R. Hirschberg, Bioactivity of glomerular ultrafiltrate during heavy proteinuria may contribute to renal tubulo-interstitial lesions: Evidence for an insulin-like growth factor, J Clin Invest 97 (1996) 116–124.

[139] A.S. Levey, C.D. Mulrow, An editorial update: What level of blood pressure control in chronic kidney disease? Ann Intern Med 143 (2005) 79–81.

[140] M.J. Sarnak, T. Greene, X. Wang, G. Beck, J.W. Kusek, A.J. Collins, et al., The effect of a lower target blood pressure on the progression of kidney disease: Long-term follow-up of the modification of diet in renal disease study, Ann Intern Med 142 (2005) 342–351.

[141] M.A. Pohl, S. Blumenthal, D.J. Cordonnier, F. De Alvaro, G. Deferrari, G. Eisner, et al., Independent and additive impact of blood pressure control and angiotensin II receptor blockade on renal outcomes in the irbesartan diabetic nephropathy trial: Clinical implications and limitations, J Am Soc Nephrol 16 (2005) 3027–3037.

[142] N.K. Foundation, K/DOQI clinical practice guidelines on hypertension and antihypertensive agents in chronic kidney disease, Am J Kidney Dis 47 (2004) 8–14.

[143] K.F. Hilgers, J.F. Mann, ACE inhibitors versus AT(I) receptor antagonists in patients with chronic renal disease, J Am Soc Nephrol 13 (2002) 1100–1108.

[144] H.H. Parving, G. Remuzzi, S.M. Snapinn, Z. Zhang, S. Shahinfar, Effects of losartan on renal and cardiovascular outcomes in patients with type 2 diabetes and nephropathy, N Engl J Med 345 (2001) 861–869.

[145] E.J. Lewis, L.G. Hunsicker, W.R. Clarke, T. Berl, M.A. Pohl, J.B. Lewis, et al., Renoprotective effect of the angiotensin-

receptor antagonist irbesartan in patients with nephropathy due to type 2 diabetes, N Engl J Med 345 (2001) 851–860.

[146] M. Praga, E. Gutierrez, E. Gonzalez, E. Morales, E. Hernandez, Treatment of IgA nephropathy with ACE inhibitors: A randomized and controlled trial, J Am Soc Nephrol 14 (2003) 1578–1583.

[147] A.A. Eddy, Plasminogen activator inhibitor-1 and the kidney, Am J Physiol 283 (2002) F209–F220.

[148] M. Epstein, Aldosterone as a mediator of progressive renal disease: Pathogenetic and clinical implications, Am J Kidney Dis 37 (2001) 677–688.

[149] G.L. Bakris, M. Siomos, D. Richardson, I. Janssen, W.K. Bolton, L.A. Hebert, et al., ACE inhibitor angiotensin receptor blockade: Impact on potassium in renal failure. VAL-K Study Group, Kidney Int 58 (2000) 2084–2093.

[150] L.A. Hebert, Target blood pressure for antihypertensive therapy in patients with proteinuric renal disease, Curr Hypertens Rep 1 (1999) 454–460.

[151] M. Haas, Z. Leko-Mohr, C. Erler, G. Mayer, Antiproteinuric versus antihypertensive effects of high-dose ACE inhibitor therapy, Am J Kidney Dis 30 (2002) 458–463.

[152] G. Navis, A.B. Kramer, P.E. de Jong, High-dose ACE inhibition: Can it improve renoprotection? Am J Kidney Dis 40 (2002) 664–666.

[153] J.F. Burris, The expanding role of angiotensin converting enzyme inhibitors in the management of hypertension, J Clin Pharmacol 35 (1995) 337–342.

[154] G. Wolf, E. Ritz, Combination therapy with ACE inhibitors and angiotensin II blockers to halt progression of chronic renal disease: Pathophysiology and indications, Kidney Int 67 (2005) 799–812.

[155] N. Nakao, A. Yoshimura, H. Morita, M. Takada, T. Kayano, T. Ideura, Combination treatment of angiotensin-II receptor blocker and angiotensin-converting-enzyme inhibitor in non-diabetic renal disease (COOPERATE): A randomised controlled trial, Lancet 361 (2003) 117–124.

[156] S. Yusuf, K.K. Teo, J. Pogue, L. Dyal, I. Copland, H. Schumacher, et al., Telmisartan, ramipril, or both in patients at high risk for vascular events, N Engl J Med 358 (2008) 1547–1559.

[157] T.W. Doulton, F.J. He, G.A. MacGregor, Systematic review of combined angiotensin-converting enzyme inhibition and angiotensin receptor blockade in hypertension, Hypertension 45 (2005) 880–886.

[158] N. Khosla, G.L. Bakris, Lessons learned from recent hypertension trials about kidney disease, Clin J Am Soc Nephrol 1 (2006) 229–235.

[159] K.A. Jamerson, G.L. Bakris, C.C. Wun, B. Dahlof, M. Lefkowitz, S. Manfreda, et al., Rationale and design of the Avoiding Cardiovascular Events through Combination Therapy in Patients Living with Systolic Hypertension (ACCOMPLISH) trial: The first randomized controlled trial to compare the clinical outcome effects of first-line combination therapies in hypertension, Am J Hypertens 17 (2004) 793–801.

[160] PROCOPA, Dissociation between blood pressure reduction and fall in proteinuria in primary renal disease: A randomized double-blind trial, J Hypertens 20 (2002) 729–737.

[161] P. Ruggenenti, A. Perna, R. Benini, G. Remuzzi, Effects of dihydropyridine calcium channel blockers, angiotensin-converting enzyme inhibition, and blood pressure control on chronic, nondiabetic nephropathies. Gruppo Italiano di Studi Epidemiologici in Nefrologia (GISEN), J Am Soc Nephrol 9 (1998) 2096–2101.

[162] K.J. Schjoedt, K. Rossing, T.R. Juhl, F. Boomsma, P. Rossing, L. Tarnow, et al., Beneficial impact of spironolactone in diabetic nephropathy, Kidney Int 68 (2005) 2829–2836.

[163] K.J. Schjoedt, K. Rossing, T.R. Juhl, F. Boomsma, L. Tarnow, P. Rossing, et al., Beneficial impact of spironolactone on nephrotic range albuminuria in diabetic nephropathy, Kidney Int 70 (2006) 536–542.

[164] C. Jones-Burton, S.I. Mishra, J.C. Fink, J. Brown, W. Gossa, G.L. Bakris, et al., An in-depth review of the evidence linking dietary salt intake and progression of chronic kidney disease, Am J Nephrol 26 (2006) 268–275.

[165] I. Gunnarsson, B. Sundelin, M. Heimburger, J. Forslid, R. van Vollenhoven, I. Lundberg, et al., Repeated renal biopsy in proliferative lupus nephritis—Predictive role of serum C1q and albuminuria, J Rheumatol 29 (2002) 693–699.

[166] O. Traitanon, Y. Avihingsanon, V. Kittikovit, N. Townamchai, T. Kanjanabuch, K. Praditpornsilpa, et al., Efficacy of enteric-coated mycophenolate sodium in patients with resistant-type lupus nephritis: A prospective study, Lupus 17 (2008) 744–751.

[167] J.M. Esdaile, L. Joseph, T. Mackenzie, M. Kashgarian, J.P. Hayslet, The pathogenesis and prognosis of lupus nephritis: Information from repeat renal biopsy, Sem Arthritis Rheum 23 (1993) 135–148.

[168] H. Bao, Z.-H. Liu, H.-L. Xie, W.-X. Hu, H.-T. Zhang, L.-S. Li, Successful treatment of class V+IV lupus nephritis with multitarget therapy, J Am Soc Nephrol 19 (2008) 2001–2010.

[169] G.G. Illei, K. Takada, D. Parkin, H.A. Austin, M. Crane, C.H. Yarboro, et al., Renal flares are common in patients with severe proliferative lupus nephritis treated with pulse immunosuppressive therapy: Long-term follow-up of a cohort of 145 patients participating in randomized controlled studies, Arthritis Rheum 46 (2002) 995–1002.

[170] B.H. Rovin, H. Song, D.J. Birmingham, L.A. Hebert, C.-Y. Yu, H.N. Nagaraja, Urine chemokines as biomarkers of human systemic lupus erythematosus activity, J Am Soc Nephrol 16 (2005) 467–473.

[171] B.H. Rovin, G. Nadasdy, G.J. Nuovo, H. Song, T. Nadasdy, Expression of adiponectin and its receptors in the kidney during SLE nephritis, J Am Soc Nephrol 17 (2006) 256A.

[172] D.J. Birmingham, H.N. Nagaraja, B.H. Rovin, L. Spetie, Y. Zhao, X. Li, et al., Fluctuation in self-perceived stress increases risk of flare in patients with lupus nephritis patients carrying the serotonin receptor 1A-1019G allele, Arthritis Rheum 54 (2006) 3291–3299.

[173] J.D. Clough, E.J. Lewis, J.M. Lachin, Treatment protocols of the lupus nephritis collaborative study of plasmapheresis in severe lupus nephritis. The Lupus Nephritis Collaborative Study Group, Prog Clin Biol Res 337 (1990) 301–307.

[174] M.D. Linnik, J.-Z. Hu, K.R. Heilbrunn, V. Strand, F.L. Hurley, T. Joh, Relationship between anti-double-stranded DNA antibodies and exacerbation of renal disease in patients with systemic lupus erythematosus, Arthritis Rheum 52 (2005) 1129–1137.

[175] D. Philibert, D. Cattran, Remission of proteinuria in primary glomerulonephritis: We know the goal but do we know the price? Nature Clin Pract Nephrol 4 (2008) 550–559.

[176] B.H. Rovin, D.J. Birmingham, H.N. Nagaraja, C.Y. Yu, L.A. Hebert, Biomarker discovery in human SLE nephritis, Bull NYU Hospital Joint Dis 65 (2007) 187–193.

[177] G. Moroni, A. Radice, G. Giammarresi, S. Quaglini, B. Gallelli, A. Leoni, et al., Are laboratory tests useful for monitoring the activity of lupus nephritis? A 6-year prospective study in a cohort of 228 patients with lupus nephritis, Ann Rheum Dis 68 (2009) 234–237.

[178] J.M. Esdaile, L. Joseph, M. Abrahamowicz, Y. Li, D. Danoff, A.E. Clarke, Routine immunologic tests in systemic lupus erythematosus: Is there a need for more studies? J Rheumatol 23 (1996) 1891–1896.

[179] J.M. Esdaile, M. Abrahamowicz, L. Joseph, T. Mackenzie, Y. Li, D. Danoff, Laboratory tests as predictors of disease exacerbations in systemic lupus erythematosus, Arthritis Rheum 39 (1996) 370–378.

[180] I.E.M. Coremans, P.E. Spronk, H. Bootsma, M.R. Daha, E.A.M. van der Voort, L. Kater, et al., Changes in antibodies to C1q predict renal relapse in system lupus erythematosus, Am J Kidney Dis 26 (1995) 595–601.

[181] A. Ho, S.G. Barr, L.S. Magder, M. Petri, A decrease in complement is associated with increased renal and hematologic activity in patients with systemic lupus erythematosus, Arthritis Rheum 44 (2001) 2350–2367.

[182] A. Ho, L.S. Magder, S.G. Barr, M. Petri, Decreases in anti-double-stranded DNA levels are associated with concurrent flares in patients with systemic lupus erythematosus, Arthritis Rheum 44 (2001) 2342–2349.

[183] M.G. Tektonidou, Renal involvement in the antiphospholipid syndrome (APS)-APS nephropathy, Clin Rev Allergy Immunol 36 (2009) 131–140.

[184] S.K. Kwok, J.H. Ju, C.S. Cho, H.Y. Kim, S.H. Park, Thrombotic thrombocytopenic purpura in systemic lupus erythematosus: Risk factors and clinical outcome: A single centre study, Lupus 18 (2009) 16–21.

[185] A. Tandon, D. Ibanez, D. Gladman, M. Urowitz, The effect of pregnancy on lupus nephritis, Arthritis Rheum 50 (2004) 3941–3946.

[186] E. Imbasciati, A. Tincani, G. Gregorini, A. Doria, G. Moroni, G. Cabiddu, et al., Pregnancy in women with pre-existing lupus nephritis: Predictors of fetal and maternal outcome, Nephrol Dial Transplant 24 (2009) 519–525.

[187] G. Carvalheiras, P. Vita, S. Marta, R. Trovao, F. Farinha, J. Braga, et al., Pregnancy and systemic lupus erythematosus: Review of clinical features and outcome of 51 pregnancies at a single institution, Clin Rev Allergy Immunol (2009, July 15). [online].

[188] S.J. Wagner, I. Craici, D. Reed, S. Norby, K. Bailey, H.J. Wiste, et al., Maternal and foetal outcomes in pregnant patients with active lupus nephritis, Lupus 18 (2009) 342–347.

[189] S. Hagelberg, Y. Lee, J. Bargman, G. Mah, R. Schneider, C. Laskin, et al., Long-term follow-up of childhood lupus nephritis, J Rheumatol 29 (2002) 2635–2642.

[190] H.I. Brunner, D.D. Gladman, D. Ibanez, M.D. Urowitz, E.D. Silverman, Difference in disease features between childhood-onset and adult-onset systemic lupus erythematosus, Arthritis Rheum 58 (2008) 556–562.

[191] L.B. Tucker, A.G. Uribe, M. Fernandez, L.M. Vila, G. McGwin, M. Apte, et al., Adolescent onset of lupus results in more aggressive disease and worse outcomes: Results of a nested matched case–control study within LUMINA, a multiethnic U.S. cohort (LUMINA LVII), Lupus 17 (2008) 314–322.

[192] B.S. Lee, H.Y. Cho, E.J. Kim, H.G. Kang, I.S. Ha, H.I. Cheong, et al., Clinical outcomes of childhood lupus nephritis: A single center's experience, Pediatr Nephrol 22 (2007) 222–231.

[193] P. Vachvanichsanong, P. Dissaneewate, E. McNeil, Diffuse proliferative glomerulonephritis does not determine the worst outcome in childhood-onset lupus nephritis: A 23-year experience in a single centre, Nephrol Dial Transplant 24 (2009) 2729–2734.

[194] C.F. Mojcik, J.H. Klippel, End-stage renal disease and systemic lupus erythematosus, Am J Med 101 (1996) 100–107.

[195] A. Rietveld, J.H. Berden, Renal replacement therapy in lupus nephritis, Nephrol Dial Transplant 23 (2008) 3056–3060.

[196] Y.P. Siu, K.T. Leung, M.K. Tong, T.H. Kwan, C.C. Mok, Clinical outcomes of systemic lupus erythematosus patients undergoing continuous ambulatory peritoneal dialysis, Nephrol Dial Transplant 20 (2005) 2797–2802.

[197] J.W. Huang, K.Y. Hung, C.J. Yen, K.D. Wu, T.J. Tsai, Systemic lupus erythematosus and peritoneal dialysis: Outcomes and infectious complications, Perit Dial Int 21 (2001) 143–147.

[198] H.C. Nossent, T.J. Swaak, J.H. Berden, Systemic lupus erythematosus: Analysis of disease activity in 55 patients with end-stage renal failure treated with hemodialysis or continuous ambulatory peritoneal dialysis. Dutch Working Party on SLE, Am J Med 89 (1990) 169–174.

[199] Y.S. Goo, H.C. Park, H.Y. Choi, B.S. Kim, Y.B. Park, S.K. Lee, et al., The evolution of lupus activity among patients with end-stage renal disease secondary to lupus nephritis, Yonsei Med J 45 (2004) 199–206.

[200] F.M. Ribeiro, M.A. Leite, G.C. Velarde, C.L. Fabris, R.C. Santos, J.R. Lugon, Activity of systemic lupus erythematosus in end-stage renal disease patients: Study in a Brazilian cohort, Am J Nephrol 25 (2005) 596–603.

[201] G. Moroni, F. Tantardini, B. Gallelli, S. Quaglini, G. Banfi, F. Poli, et al., The long-term prognosis of renal transplantation in patients with lupus nephritis, Am J Kidney Dis 45 (2005) 903–911.

[202] A. Ghafari, J. Etmadi, M.R. Adrdalan, Renal transplantation in patients with lupus nephritis: A single-center experience, Transplant Proc 40 (2008) 143–144.

[203] S. Lionaki, P.P. Kapitsinou, A. Iniotaki, A. Kostakis, H.M. Moutsopoulos, J.N. Boletis, Kidney transplantation in lupus patients: A case–control study from a single centre, Lupus 17 (2008) 670–675.

[204] M. Chelamcharla, B. Javaid, B.C. Baird, A.S. Goldfarb-Rumyantzev, The outcome of renal transplantation among systemic lupus erythematosus patients, Nephrol Dial Transplant 22 (2007) 3623–3630.

[205] S. Bunnapradist, P. Chung, A. Peng, A. Hong, P. Chung, B. Lee, et al., Outcomes of renal transplantation for recipients with lupus nephritis: Analysis of the Organ Procurement and Transplantation Network Database, Transplantation 82 (2006) 612–618.

[206] B.H. Rovin, A.M. Mckinley, D.J. Birmingham, Can we personalize treatment for kidney diseases? Clin J Am Soc Nephrol 4 (2009). 1970-1676.

[207] A.T. Tieng, E. Peeva, B-cell-directed therapies in systemic lupus erythematosus, Sem. Arthritis Rheum 38 (2008) 218–227.

[208] M.Y. Karim, C.N. Pisoni, M.A. Khamashta, Update on immunotherapy for systemic lupus erythematosus—What's hot and what's not!, Rheumatology (Oxford) 48 (2009) 332–341.

[209] E. Sousa, D. Isenberg, Treating lupus: From serendipity to sense, the rise of the new biologicals and other emerging therapies, Best Pract Clin Rheumatol 23 (2009) 563–574.

[210] M. Ramos-Casals, M.J. Soto, M.J. Cuadrado, M.A. Khamashta, Rituximab in systemic lupus erythematosus: A systematic review of off-label use in 188 cases, Lupus 18 (2009) 767–776.

[211] T.Y. Lu, K.P. Ng, G. Cambridge, M.J. Leandro, J.C. Edwards, M. Ehrenstein, et al., A retrospective seven-year analysis of the use of B cell depletion therapy in systemic lupus erythematosus at University College London Hospital: The first fifty patients, Arthritis Rheum 61 (2009) 482–487.

[212] E.K. Li, L.S. Tam, T.Y. Zhu, M. Li, C.L. Kwok, T.K. Li, et al., Is combination rituximab with cyclophosphamide better than rituximab alone in the treatment of lupus nephritis? Rheumatology (Oxford) 48 (2009) 892–898.

[213] I. Gunnarsson, B. Sundelin, T. Jonsdottir, S.H. Jacobson, E.W. Henriksson, R.F. van Vollenhoven, Histopathologic and clinical outcome of rituximab treatment in patients with cyclophosphamide-resistant proliferative lupus nephritis, Arthritis Rheum 56 (2007) 1263–1272.

[214] B.H. Rovin, G.B. Appel, R.A. Furie, R.J. Looney, K. Latinis, F.C. Fervenza, et al., Efficacy and safety of rituximabe in subjects with proliferative lupus nephritis: Results from the randomized, double-blind, phase III LUNAR study, J Am Soc Nephrol 20 (2009) 77A.

[215] T. Dorner, J. Kaufmann, W.A. Wegener, N. Teoh, D.M. Goldenberg, G.R. Burmester, Initial clinical trial of epratuzumab (humanized anti-CD22 antibody) for immunotherapy of systemic lupus erythematosus, Arthritis Res Ther 8 (2006) R74.

[216] F. Monneaux, S. Muller, Molecular therapies for systemic lupus erythematosus: Clinical trials and future prospects, Arthritis Res Ther 11 (2009) 1–10.

[217] M.H. Cardiel, J.A. Tumlin, R.A. Furie, D.J. Wallace, T. Joh, M.D. Linnik, Abetimus sodium for renal flare in systemic lupus erythematosus: Results of a randomized, controlled phase III trial, Arthritis Rheum 58 (2008) 2470–2480.

[218] D.I. Daikh, D. Wofsy, Cutting edge: Reversal of murine lupus nephritis with CTLA4Ig and cyclophosphamide, J Immunol 166 (2001) 2913–2916.

[219] H.Y. Wang, T.G. Cui, F.F. Hou, Z.H. Ni, X.M. Chen, F.M. Lu, et al., Induction treatment of proliferative lupus nephritis with leflunomide combined with prednisone: A prospective multicentre observational study, Lupus 17 (2008) 638–644.

[220] F.S. Zhang, Y.K. Nie, X.M. Jin, H.M. Yu, Y.N. Li, Y. Sun, The efficacy and safety of leflunomide therapy in lupus nephritis by repeat kidney biopsy, Rheumatol Int 29 (2009) 1331–1335.

[221] T. Marshall, K. Williams, Two-dimensional electrophoresis of human urinary proteins following concentration by dye precipitation, Electrophoresis 17 (1996) 1256–1272.

[222] R.G. Barr, S. Seliger, G.B. Appel, R. Zuniga, V. D'Agati, J. Salmon, et al., Prognosis in proliferative lupus nephritis: The role of socio-economic status and race/ethnicity, Nephrol Dial Transplant 18 (2003) 2039–2046.

[223] G. Contreras, V. Pardo, C. Cely, E. Borja, A. Hurtado, C. De La Cuesta, et al., Factors associated with poor outcomes in patients with lupus nephritis, Lupus 14 (2005) 890–895.

[224] N. Kasitanon, M. Petri, M. Haas, L.S. Magder, D.M. Fine, Mycophenolate mofetil as the primary treatment of membranous lupus nephritis with and without concurrent proliferative disease: A retrospective study of 29 cases, Lupus 17 (2008) 40–45.

[225] D.N. Spetie, Y. Tang, B.H. Rovin, G. Nadasdy, T. Nadasdy, T.E. Pesavento, et al., Mycophenolate therapy of SLE membranous nephropathy, Kidney Int 66 (2004) 2411–2415.

[226] C.C. Szeto, B.-C.H. Kwan, F.M.-M. Lai, L.S. Tam, E.K.-M. Li, K.-M. Chow, et al., Tacrolimus for the treatment of systemic lupus erythematosus with pure class V nephritis, Rheumatology 47 (2008) 1678–1681.

[227] R.W.-Y. Chan, F.M.-M. Lai, E.K.-M. Li, L.-S. Tam, T.Y.-H. Wong, C.Y.K. Szeto, et al., Expression of chemokine and fibrosing factor messenger RNA in the urinary sediment of patients with lupus nephritis, Arthritis Rheum 50 (2004) 2882–2890.

[228] R.W.-Y. Chan, F.M.-M. Lai, E.K.-M. Li, L.-S. Tam, K.-M. Chow, P.K.-T. Li, et al., Messenger RNA expression of RANTES in the urinary sediment of patients with lupus nephritis, Nephrology 11 (2006) 219–225.

[229] R.W.-Y. Chan, F.M.-M. Lai, E.K.-M. Li, L.-S. Tam, K.-M. Chow, P.K.-T. Li, et al., The effect of immunosuppressive therapy on the messenger RNA expression of target genes in the urinary sediment of patients with active lupus nephritis, Nephrol Dial Transplant 21 (2006) 1534–1540.

[230] Y. Avihingsanon, P. Phumesin, T. Benjachat, S. Akkasilpa, V. Kittikowit, K. Praditpornsilpa, et al., Measurement of urinary chemokine and growth factor messenger RNAs: A noninvasive monitoring in lupus nephritis, Kidney Int 69 (2006) 747–753.

[231] H.I. Brunner, M. Mueller, C. Rutherford, M.H. Passo, D.P. Witte, A. Grom, et al., Urinary neutrophil gelatinase-associated lipocalin as a biomarker of nephritis in childhood-onset systemic lupus erythematosus, Arthritis Rheum 54 (2006) 2577–2584.

[232] M. Pitashny, N. Schwartz, X. Qing, B. Hojaili, C. Aranow, M. Mackay, et al., Urinary lipocalin-2 is associated with renal disease activity in human lupus nephritis, Arthritis Rheum 56 (2007) 1894–1903.

[233] C.H. Hinze, M. Suzuki, M. Klein-Gitelman, M.H. Passo, J. Olson, N.G. Singer, et al., Neutorphil gelatinase-associated lipocalin (NGAL) anticipates the course of global and renal childhood-onset systemic lupus erythematosus disease activity, Arthritis Rheum 60 (2009) 2772–2781.

[234] M. Suzuki, G.F. Ross, K. Wiers, S. Nelson, M. Bennett, M.H. Passo, et al., Identification of a urinary proteomic signature for lupus nephritis in children, Pediatr Nephrol 22 (2007) 2047–2057.

[235] M. Suzuki, K. Wiers, E.B. Brooks, K.D. Greis, K.A. Haines, M. Klein-Gitelman, et al., Initial validation of a novel protein biomarker panel for active pediatric lupus nephritis, Pediatr Res 65 (2009) 530–536.

[236] D. Askenazi, B. Myones, A. Kamdar, R. Warren, M. Perez, M. De Guzman, et al., Outcomes of children with proliferative lupus nephritis: The role of protocol renal biopsy, Pediatr Nephrol 22 (2007) 981–986.

[237] B.H. Rovin, A. McKinley, H. Song, J. Prosek, Urine liver-fatty acid binding protein in patients with systemic lupus erythematosus, J Am Soc Nephrol 19 (2008) 776A.

[238] P.M. Izmirly, L. Barisoni, J.P. Buyon, M.Y. Kim, T.L. Rivera, J.S. Schwartzman, et al., Expression of endothelial protein C receptor in cortical peritubular capillaries associates with a poor clinical response in lupus nephritis, Rheumatology (Oxford) 48 (2009) 513–519.

[239] G. Wang, F.M. Lai, L.S. Tam, E.K. Li, B.C. Kwan, K.M. Chow, et al., Urinary FOXP3 mRNA in patients with lupus nephritis—Relation with disease activity and treatment response, Rheumatology (Oxford) 48 (2009) 755–760.

[240] S.A. Varghese, T.B. Powell, M.N. Budisavljevic, J.C. Oates, J.R. Raymond, J.S. Almeida, et al., Urine biomarkers predict the cause of glomerular disease, J Am Soc Nephrol 18 (2007) 913–922.

[241] J.C. Oates, S. Varghese, A.M. Bland, T.P. Taylor, S.E. Self, R. Stanislaus, et al., Prediction of urinary protein markers in lupus nephritis, Kidney Int 68 (2005) 2588–2592.

[242] S. Nakayamada, K. Saito, K. Nakano, Y. Tanaka, Activation signal transduction by beta1 integrin in T cells from patients with systemic lupus erythematosus, Arthritis Rheum 56 (2007) 1559–1568.

II. CLINICAL ASPECTS OF DISEASE

Heart

Kathleen Maksimowicz-McKinnon, Susan Manzi

PERICARDIAL DISEASE

Epidemiology and Pathogenesis

Pericarditis is the most commonly recognized cardiac manifestation of systemic lupus erythematosus (SLE). Depending on the criteria used for diagnosis, the reported prevalence of pericarditis in SLE ranges from 18 to 61% (Table 43.1) [1–17]. Although autopsy studies report evidence of pericardial disease in up to 61% of SLE patients, clinically symptomatic pericarditis is reported in only approximately one-fourth of patients. Patients may have a single isolated episode or recurrent episodes of pericarditis. Although rare, acute pericarditis with cardiac tamponade or congestive heart failure (CHF) has been reported as an initial presentation of SLE [18, 19].

Pericarditis and pericardial effusions in SLE result from inflammation of the pericardium. Changes of acute pericarditis may include increased pericardial vascularity, fibrin deposition with adhesion formation, and extension of inflammation to the myocardium.

Clinical Features

Asymptomatic pericardial disease in SLE is often detected on chest radiograph or with echocardiography. Signs and symptoms of acute pericarditis are detailed in Table 43.2. Other manifestations of SLE are often evident, and evidence of serositis at other sites (e.g., abdominal pain with ascites) may also be present.

Diagnosis

Diagnostic studies that may be helpful in detecting pericarditis include electrocardiography (ECG), chest radiography, and echocardiography (Figures 43.1 and 43.2). On ECG, early changes of acute pericarditis can include PR segment depression and diffuse, upwardly concave ST segment elevation. T wave inversions are a later finding, occurring several days after onset. Decreased QRS voltage and/or electrical alternans are generally associated with the presence of a large pericardial effusion. Sustained arrhythmias are not commonly seen in isolated pericarditis and suggest myocardial involvement (such as with concomitant myocarditis or ischemic heart disease). An enlarged cardiac silhouette with visible pericardial fat lines may be noted on chest radiography. Echocardiography (two-dimensional and Doppler) is noninvasive and highly sensitive for acute pericardial disease, and it is the diagnostic study of choice in this setting. Echocardiography can provide an estimate of the amount of pericardial fluid, and it can quickly identify signs suggestive of cardiac tamponade, such as dilatation and loss of collapse of the inferior vena cava during inspiration and diastolic right ventricular collapse. Right heart catheterization is sometimes needed to confirm a diagnosis of cardiac tamponade. Echocardiography may be less sensitive for the evaluation of chronic pericardial disease. Chronic manifestations such as pericardial thickening or constrictive pericarditis may be better identified by computerized tomography or magnetic resonance imaging.

In patients requiring intervention for refractory pericarditis or hemodynamically significant impairment, analysis of pericardial fluid is important, especially in patients with disease that has been refractory to treatment, mainly to eliminate other causes of pericarditis. Pericardial effusions in SLE, as in many other conditions, are frequently inflammatory with a predominance of polymorphonuclear leukocytes. The fluid is usually exudative, with elevated protein levels and normal to low glucose levels. SLE-associated effusions can also be serosanguinous or even grossly bloody. In some patients, the fluid may contain antinuclear (ANA) and double-stranded DNA antibodies, decreased complement levels, and LE cells. However, ANA may be present in pericardial fluid from other etiologies of

TABLE 43.1   Pericardial Disease in SLE

Authors	Year	No. of patients	Prevalence (%)	Diagnostic modality
Griffith and Vural [1]	1951	18	61	Autopsy
Dubois and Tuffanelli [2]	1964	520	31	
Bulkley and Roberts [3]	1975	36	53	Autopsy
Ito et al. [4]	1979	48	46	Echo
Bomalaski et al [5]	1983	47	49	Echo
Doherty and Siegel [6]	1985	50	42	Echo
Badui et al. [7]	1985	100	39	Echo
Klinkhoff et al. [8]	1985	47	21	Echo
Crozier et al. [9]	1990	50	54	Echo
Nihoyannopoulos et al. [10]	1990	93	20	Echo
Sturfelt et al. [11]	1992	75	19	Echo
Jouhikainen et al. [12]	1994	74	22	Echo
Rantapaa-Dahlqvist et al. [14]	1997	50	40	Echo
Kalke et al. [13]	1998	54	18	Echo
Houman et al. [15]	2004	100	16	
Amoroso et al. [16]	2006	34	15	Echo
Panchal et al. [17]	2006	35	26	Autopsy

TABLE 43.2   Signs and Symptoms of Acute Pericarditis

Precordial or substernal chest pain, often worsening with inspiration or when supine

Dyspnea

Fever

Tachycardia

Decreased heart sounds

Pericardial rub

Pulsus paradoxus (exaggeration of the normal pulse variation with respiration)

pericarditis, including tuberculous and malignant effusions [20, 21]. Hence, the presence of ANA in pericardial fluid should not be considered specific for SLE.

The diagnosis of pericarditis in SLE is most frequently a clinical one, established by the presence of characteristic symptoms and examination findings. It is important to remain cognizant of other disorders, some also frequently seen in SLE, which may mimic acute pericarditis. Similar presentations can be seen with pneumonia, pulmonary embolism or infarction, myocardial infarction (MI), and aortic dissection.

Treatment

Mild cases of symptomatic pericarditis often respond to therapy with nonsteroidal anti-inflammatory agents. In patients with severe or refractory symptoms, corticosteroid therapy is indicated. More invasive measures, such as pericardiocentesis or placement of a pericardial window, are indicated if patients have severe refractory symptoms associated with a pericardial effusion or in the setting of hemodynamic compromise (pericardial tamponade). In patients with recurrent pericarditis, pericardial thickening may occur. Constrictive pericarditis is a rare complication of chronic or recurrent pericarditis, and it may require pericardial stripping.

(A) **(B)**

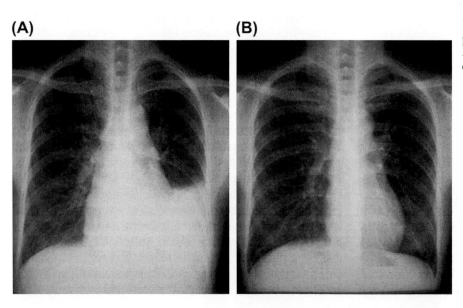

FIGURE 43.1 Pleuropericarditis in SLE pretreatment (A) and post-treatment (B). *Source: Reproduced with permission of Elsevier from Inoue et al. [19].*

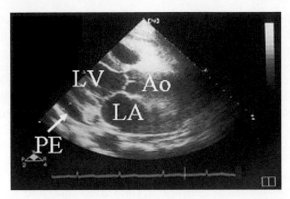

FIGURE 43.2 Two-dimensional echocardiography demonstrating a pericardial effusion (PE). Ao, aorta; LA, left atrium; LV, left ventricle. *Source: Reproduced with permission of Elsevier from Inoue et al. [19].*

MYOCARDIAL DYSFUNCTION

Epidemiology and Pathogenesis

Impaired myocardial function has been reported in SLE patients and can result from a variety of causes [1, 7, 22–34]. It is frequently associated with co-morbid conditions associated with myocardial dysfunction in the general population, such as hypertension and ischemic heart disease. SLE patients may have systolic dysfunction, diastolic dysfunction, or an abnormal response to exercise. Although this is most frequently detected by echocardiography in symptomatic patients, a study of SLE women without any clinically evident of cardiac disease who underwent coronary angiography found increased ventricular stiffness, decreased contractility, and lowered coronary artery reserve [25].

Myocarditis has been implicated as a cause of global myocardial dysfunction in SLE. However, clinical means of establishing this diagnosis are unreliable. Compared to clinical studies, autopsy studies have demonstrated higher prevalence rates of myocarditis [1, 22, 23] (Table 43.3), belying the frequent subclinical nature of this co-morbidity. In addition, since the introduction of glucocorticoid therapy, the prevalence of myocarditis by pathologic studies has decreased, with a reported prevalence as low as 7% [32]. Even with more invasive measures of myocardial function, such as angiography,

TABLE 43.3 Myocarditis and Myocardial Dysfunction in SLE

Authors	Year	No. of patients	Prevalence (%)	Diagnostic modality	Comment
Griffith and Vural [1]	1951	18	78	Autopsy	Myocarditis
Kong *et al.* [22]	1962	30	50	Autopsy	Myocarditis
Estes and Christian [23]	1971	150	8	Autopsy	Myocarditis
Maniscalco *et al.* [24]	1975	25	28	Echo	Decreased ejection fraction
Gobaira *et al.* [25]	1982	32	16	Echo	Myocarditis
Badui *et al.* [7]	1985	100	14	Echo	Depressed function
Consecutive patients					
Doherty *et al.* [26]	1988	50	8	Echo	Global hypokinesis
Leung *et al.* [27]	1990	75	9	Echo	Depressed function
Consecutive patients					
Nihoyannopoulos *et al.* [28]	1990	93	5	Echo	Regional/global dysfunction
Sasson *et al.* [29]	1992	35	64/14	Echo	Diastolic dysfunction
Active/inactive disease					
Consecutive patients					
Roldan *et al.* [30]	1996	58	13	Transesophageal echo	Congestive heart failure
Kalke *et al.* [13]	1998	54	20	Echo	Systolic/diastolic dysfunction
Wijetunga and Rockson [31]	2002	126	57	Autopsy	Combined; myocarditis
Wijetunga and Rockson [31]	2002	46	7	Autopsy	Combined; myocarditis
Apte *et al.* [33]	2008	496	11	Varied	Hemodynamic compromise/arrhythmia
Yip *et al.* [34]	2009	82	27/42	Echo	Impaired left ventricle/right ventricle long-axis function

it may be difficult to conclusively attribute myocardial dysfunction in SLE to myocarditis.

Pathologic studies suggest that SLE-associated myocarditis is a vascular rather than a primary myopathic condition. It is purported to arise from immune complex deposition, with injury occurring following pro-inflammatory cytokine and systemic and local complement activation. Interstitial and perivascular infiltrates of lymphocytes, plasma cells, and macrophages can be identified in the myocardium [35]. Immunofluorescence studies have identified immune deposits present in the walls of the myocardial vessels, with perivascular deposits of IgG and vascular deposits of C3 [36, 37]. Ultimately, myocardial depression occurs from subsequent necrosis and fibrosis of myocytes or from injury to the conduction system tissues, leading to conduction system disturbances.

Patients with SLE are at risk for other more common causes of myocardial dysfunction, which need be considered when evaluating patients with impaired ventricular function or cardiomyopathy. Chronic hypertension is frequently associated with ventricular dysfunction. SLE patients, including premenopausal women, have an increased risk of ischemic heart disease, which is a far more common etiology of myocardial dysfunction. Valvular stenosis or insufficiency can also lead to myocardial dysfunction and congestive heart failure. Pulmonary vascular disease, such as pulmonary hypertension or chronic thromboembolic disease, can result in impaired ventricular function and right-sided heart failure.

Another etiology of myocardial dysfunction in SLE is chloroquine-associated cardiomyopathy. This is a rare complication of therapy, with less than 20 cases reported in the literature [38]. In addition, less than half of these cases were substantiated by myocardial biopsy. Depressed contractility, conduction system abnormalities, and myocardial necrosis and fibrosis have been identified in animal models evaluating cardiotoxicity with chloroquine use [39–41]. Although the precise mechanism of cardiotoxicity with chloroquine therapy remains unclear, lysosomal disruption is thought to be a possible mechanism.

Clinical Features

Myocardial dysfunction in SLE may manifest with signs and symptoms of CHF as outlined in Table 43.4. In patients with mild disease, the bedside cardiac examination may be unremarkable. Nonspecific ECG findings in myocardial dysfunction may include conduction abnormalities, premature atrial or ventricular beats, or dysrhythmias. Patients with SLE-associated myocarditis may be clinically asymptomatic, with diagnostic studies such as ECG or echocardiography identifying

myocardial disease, or with a more fulminant picture with chest pain and increased cardiac enzymes, mimicking acute coronary syndrome.

SLE patients with chloroquine-associated myocardial dysfunction may manifest similarly to those with other etiologies of myocardial dysfunction. Patients may present with signs and symptoms of CHF or of cardiac dysrhythmias. There are no specific findings on imaging studies to help establish the diagnosis of chloroquine-associated cardiomyopathy. Although there are too few patients with this diagnosis to definitively identify risk factors for its development, data from Jolle and Tatum [42] suggest that older age, higher dose per unit body weight, and duration of therapy may be associated with chloroquine-associated cardiomyopathy.

Diagnosis

Echocardiographic findings in myocarditis may include global chamber dilatation, prolonged relaxation time, decreased deceleration of diastolic flow velocity, increased chamber size, and decreased E/A ratio (which reflects decreased early ventricular filling and increased late filling). Both segmental and global hypokinesis may be seen. Newer imaging modalities, such as magnetic resonance imaging and myocardial gadolinium uptake on citrate scintigraphy, are being used to evaluate myocardial dysfunction but are not able to identify specifically SLE-associated myocardial disease (Figure 43.3).

In a study of 845 patients with unexplained cardiomyopathy, myocardial biopsy was demonstrated to increase sensitivity over clinical diagnosis in establishing a diagnosis of myocarditis, but specificity was not improved [43]. Seven SLE patients were included in this study; in 6 of 7, the diagnosis was established by clinical and laboratory evaluation prior to biopsy. In lupus myocarditis, findings may be nonspecific and include fibrinoid necrosis, inflammatory cell infiltrates,

TABLE 43.4   Signs and Symptoms Associated with Myocardial Dysfunction

Fatigue
Dyspnea
Paroxysmal nocturnal dyspnea
Orthopnea
Cough
Peripheral edema
Murmur
Jugular venous distention
Rales

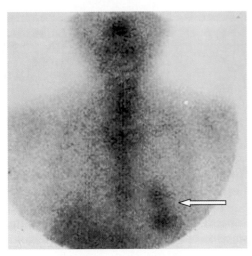

FIGURE 43.3 Diffuse myocardial uptake (*arrow*) on Ga-67 citrate scintigraphy in SLE-associated myocarditis. *Source: Reproduced with permission of Lippincott Williams & Wilkins from Jolles and Tatum [42].*

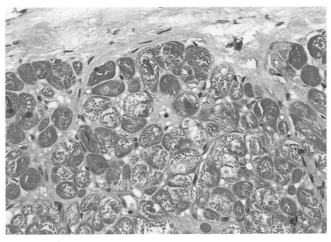

FIGURE 43.4 Vacuolated myocytes in chloroquine-associated cardiomyopathy. *Source: Reproduced with permission of BMJ Group from Jain and Halushka [35].*

immunoglobulin deposition, and fibrosis [35, 44]. However, problems with myocardial biopsy have been identified. Technical issues with myocardial biopsy include sampling error, interpretation, and limited specimen size. Given these limitations, myocardial biopsy is not routinely obtained for the diagnosis of lupus myocarditis alone. This diagnosis is usually considered in the appropriate clinical setting when other etiologies of myocardial dysfunction have been excluded.

In contrast, the relative infrequency of chloroquine-associated cardiomyopathy makes this diagnosis one of exclusion, with the only means of definitive diagnosis myocardial biopsy. Other more frequent etiologies of myocardial dysfunction in SLE, such as ischemic heart disease, infection, or pericardial disease, should be considered and investigated prior to making this diagnosis. Histopathologic findings in both humans and animal models include intracytoplasmic vacuoles, curvilinear bodies, myeloid bodies, and myocyte necrosis and fibrosis (Figure 43.4) [45–47]. However, biopsy findings may also be nonspecific and fail to definitely establish a diagnosis of chloroquine-associated cardiomyopathy. Although the presence of myocyte necrosis and fibrosis implies a more permanent state of injury, patients often demonstrate remarkable clinical improvement and improvement of myocardial function after discontinuation of antimalarial therapy.

Treatment

Myocardial dysfunction and CHF therapy in patients with SLE should be directed at the cause of the cardiomyopathy. In patients with hypertensive cardiomyopathy and associated CHF, intensive blood pressure control would be a key facet of management. Therapy

for cardiomyopathy associated with ischemic heart disease could include revascularization to improve myocardial oxygen supply. In patients in whom thorough investigations fail to reveal another etiology and cardiomyopathy is attributed to active SLE, immunosuppressive therapy would be indicated. The degree of myocardial dysfunction may also indicate the need for other interventions. Moderate to severe left ventricular dysfunction is an established risk factor for arrhythmias and sudden cardiac death; Holter monitoring and/or electrophysiologic evaluation should be considered. In patients with fixed severe left ventricular dysfunction, anticoagulation may be indicated for the prevention of thromboembolic disease.

Myocarditis in SLE is treated with high-dose glucocorticoids, often beginning with pulse intravenous therapy followed by oral therapy, based on clinical response and other disease manifestations. No randomized studies of steroid-sparing immunosuppressive agents in the treatment of myocarditis in SLE have been performed, but case reports suggest benefit from cyclophosphamide, azathioprine, and intravenous immunoglobulin in adjunct to steroid therapy [48–50]. In patients with CHF, supportive therapy with diuretics, sodium and fluid restriction, afterload reduction, and inotropic agents may help alleviate clinical symptoms. In patients with severe cardiomyopathy, anticoagulation may be considered. In patients with conduction system involvement, antiarrhythmic therapy and anticoagulation may be appropriate.

In chloroquine- and hydroxychloroquine-induced cardiomyopathy, discontinuation of antimalarial therapy is indicated. Echocardiographic and clinical improvement following discontinuation is suggestive of a diagnosis of chloroquine-induced cardiomyopathy.

CONDUCTION SYSTEM

Epidemiology and Pathogenesis

Conduction system abnormalities are present in SLE patients, with a prevalence ranging from 5 to 74%, depending on the definition of the specific abnormality and also whether co-existing cardiovascular disease is present (Table 43.5) [7, 8, 11, 48, 51–55]. Sinus tachycardia is frequently noted in patients with SLE. Low-grade atrioventricular block (first- and second-degree heart block) is more common in SLE; complete heart block is rare. Bundle branch block may also be present. Patients may manifest with transient conduction system abnormalities or may develop persistent or progressive conduction system disease. Young to middle-aged SLE women and patients with longer disease duration are at higher risk to have conduction system abnormalities.

Similar to damage occurring in other organ systems in SLE, conduction system abnormalities are purported to arise from immune-mediated injury to both the vascular supply of and the conduction system tissues. Myocarditis can result in conduction system abnormalities. One of the most widely recognized SLE-associated conduction defects is that of maternal anti-Ro antibodies and the presence of neonatal heart block [52]. However, associations between the presence of anti-U1 RNP and anti-Ro antibodies and conduction system abnormalities have been reported in adult SLE patients [53, 56]. As in other conditions and not specific to SLE, conduction system abnormalities in SLE may also be associated with ischemic heart disease, prior cardiac surgery, endocrinopathies, electrolyte disturbances, or medications.

Clinical Features

Conduction system abnormalities may be asymptomatic and identified only when cardiovascular evaluations are performed for another cause. In patients with higher grade or intraventricular blocks, manifestations can include CHF, palpitations, impaired exercise tolerance, or syncope.

Diagnosis

Findings on cardiac auscultation, such as a widely split S1 or S2, may raise concern for a conduction system disorder but are not specific findings. Electrocardiography can identify conduction system abnormalities. Investigations directed toward the cause of conduction system abnormalities other than SLE, as noted previously, are an essential part of the workup of these patients once this diagnosis is made.

Treatment

Cardiology and electrophysiologic evaluation may be helpful in risk assessment and stratification. Asymptomatic patients with a structurally normal heart, without evidence of other etiologies for conduction system abnormalities, may require only observation. Patients in whom another potentially reversible cause is found should have therapy directed at that specific abnormality. Patients with permanent high-grade block or those who are symptomatic may require medical therapy or pacemaker implantation.

TABLE 43.5   Conduction System Abnormalities in SLE

Authors	Year	No. of patients	Prevalence (%)	Diagnostic modality	Comment
Shearn [48]	1959	73	62	ECG	
Hejtmancik et al. [51]	1964	137	52	ECG	
Badui et al. [7]	1985	100	74	ECG	Any abnormality
Klinkhoff et al. [8]	1985	47	32	ECG	
Mandell [52]	1987		5	ECG	Atrioventricular conduction abnormalities
Logar et al. [53]	1990	67	10	ECG	Conduction defect
6/7 anti-Ro antibody +					
Sturfelt et al. [11]	1992	54	17	ECG	Any abnormality
O'Neill et al. [54]	1993	33	6	ECG	Conduction defect
Cardoso et al. [55]	2005	140	7	ECG	Conduction defect

VALVULAR HEART DISEASE

Epidemiology and Pathogenesis

Valvular heart disease is a common complication of SLE. Echocardiographic abnormalities are reported in up to 61% of patients; however, in the majority of patients, valvular disease is mild (Table 43.6) [3, 7–11, 17, 26, 30, 57–67]. Libman–Sacks endocarditis is the characteristic valvular pathology associated with SLE, which describes noninfectious verrucous vegetations (single or clustered) that may be identified on any or all of the valves, papillary muscles, or endocardium (Figures 43.5 and 43.6). The mitral valve is most frequently affected, followed by the aortic valve. In most patients, Libman–Sacks endocarditis manifests as thickening of the valve. Libman–Sacks-associated valvular abnormalities may result in valvular stenosis or insufficiency. Libman–Sacks lesions are often identified at autopsy, with one series reporting a prevalence of 50% [3]. However, since the advent of corticosteroid therapy, it has been reported that the prevalence of Libman–Sacks valvular disease has decreased.

Histopathologic studies of Libman–Sacks endocarditis demonstrate fibrinous exudates, hematoxylin-stained bodies, epithelial and fibroblastic proliferation, and neovascularization [35, 68]. Complement and immunoglobulin deposition in small vessels have also been identified

TABLE 43.6 Valvular Heart Disease in SLE

Authors	Year	No. of patients	Prevalence (%)	Diagnostic modality	Comment
Bulkley and Roberts [3]	1975	36	50	Autopsy	Libman–Sacks lesions
Klinkhoff et al. [8]	1985	47	21	Echo	Valvular thickening
Badui et al. [7]	1985	100	9	Echo	Valvular abnormality
Galve et al. [57]	1988	74	24	Echo	Valvular abnormality
Doherty et al. [26]	1988	50	12	Echo	Valvular thickening
Crozier et al. [9] Consecutive patients	1990	50	60	Echo	M or A regurgitation
Khamashta et al. [58] Consecutive patients	1990	132	23	Echo	Valvular abnormality
Leung et al. [27]	1990	75	25	Echo	Left-sided insufficiency
Nihoyannopoulos et al. [10]	1990	93	28	Echo	Valvular abnormality
Cervera et al. [59]	1992	70	44	Echo	Valvular abnormality
Sturfelt et al. [11]	1992	75	27	Echo	Valvular abnormality
Gleason et al. [60]	1993	20	40	Echo	Valvular abnormality
Giunta et al. [61]	1993	75	4/12	Echo	Vegetation/thickening
Gabrielli et al. [62] Consecutive patients	1995	46	39	Echo	Valvular abnormalities
Roldan et al. [30] 51% with valvular thickening 43% with vegetation	1996	69	61	TEE	Valvular abnormality
Omdal et al. [63]	2001	35	46	Echo/TEE	Mitral or aortic valvular abnormality
Jensen-Urstad et al. [64]	2002	52	27	Echo	Valvular abnormality
Leszczynski et al. [65]	2003	52	48	Echo	Valvular abnormality
Morelli et al. [66]	2003	71	43	Echo	Valvular abnormality
Panchal et al. [17]	2006	35	33	Autopsy	Valvular abnormality
Roldan et al. [67]	2007	69	55	Echo	Valvular abnormality

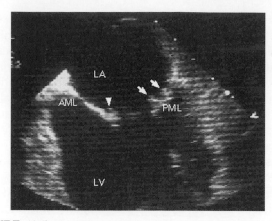

FIGURE 43.5 Two-dimensional echocardiography demonstrating mitral valvular vegetations (*arrowheads*) in Libman—Sachs endocarditis. AML, anterior mitral leaflet; LA, left atrium; LV, left ventricle; PML, posterior mitral leaflet. *Source: Reproduced with permission of Massachusetts Medical Society from Roldan et al. [30].*

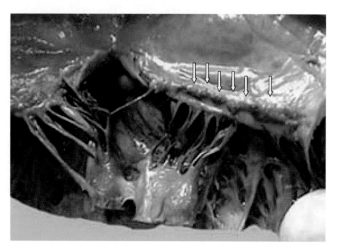

FIGURE 43.6 Multiple vegetations in a patient with Libman—Sachs endocarditis on autopsy. *Source: Reproduced with permission of Medknow Publications from Panchal et al. [17].*

by immunofluorescence, implicating a role for immune complex deposition in disease pathogenesis.

Valvular heart disease is the second most common cardiac manifestation of the antiphospholipid syndrome (APS), with valvular thickening most frequently reported [69]. The presence of anticardiolipin antibodies has also been associated with valvular heart disease in both SLE and non-SLE patients. Leszczynski et al. [65] compared echocardiographic findings in 52 SLE patients with those of 34 healthy controls. In this study, anticardiolipin antibodies were identified in 77% of patients with echocardiographic evidence of valvular heart disease. Patients may have valve thickening, thrombus formation, or vegetations, which may result in thromboembolic events. In addition, antiphospholipid antibodies (aPLs) have been associated with other cardiac abnormalities, including pericardial disease and myocardial

dysfunction. Nihoyannopoulos et al. [10] examined 50 SLE patients with aPLs and identified at least one of these findings in 78% of patients. They also examined non-SLE patients with aPLs and found similar results, where 75% of patients had at least one cardiac abnormality identified.

In contrast, Gleason et al. [60] reported that primary APS patients and SLE patients without aPLs had similar prevalences of valvular heart disease. Furthermore, a study by Gabrielli et al. [62] comparing SLE patients with primary APS patients found that 39% of SLE patients had evidence of valvular heart disease compared to no patients in the primary APS group. Although it has been suggested that aPLs could be the primary cause of valvular heart disease in patients with SLE, the data are conflicting and suggest that aPLs may be contributory but not the sole pathogenic factor in SLE-associated valvular heart disease.

Histopathologic changes seen in valvular heart disease associated with aPLs include subendothelial deposition of immunoglobulins, complement components, and antiphospholipid and anticardiolipin antibodies [35, 70].

Clinical Features

Libman—Sacks endocarditis presents similarly to any other form of endocarditis. Common manifestations include fever, tachycardia, evidence of thromboembolic disease, murmur, and anemia. Involvement of chordae tendonae, papillary muscles, or the valve ring may lead to loss of structural integrity and subsequent hemodynamic changes, including valvular insufficiency.

Data regarding outcomes of Libman—Sacks endocarditis present a conflicting picture. A prospective study of 70 SLE patients by Cervera et al. [59] suggests a more benign prognosis, with only approximately 4% of patients developing clinically significant valvular disease and only approximately 2% requiring surgical intervention. In contrast, Roldan et al. [67] found that in a study of 69 SLE patients, 22% of patients with valvular abnormalities went on to develop significant associated co-morbidities such as stroke, peripheral embolization, CHF, infective endocarditis, or the need for valve replacement. Similarly, other studies have demonstrated associations between valvular heart disease in SLE and central nervous system disease. Studies by Roldan et al. demonstrate associations between valvular heart disease in SLE and both focal brain injury and nonfocal neuropsychiatric disease [67, 71, 72].

Diagnosis

Valvular lesions may be identified by transthoracic (TTE) or transesophageal (TEE) echocardiography, but there are no features that distinguish them from lesions seen in thrombotic or infective endocarditis. A study of 81 SLE patients undergoing paired TTE and TEE evaluations demonstrated that Libman—Sacks endocarditis was significantly more likely to be identified by TEE than TTE, with poor agreement rates between these imaging studies [73]. In addition, TTE was found to have low sensitivity, specificity, and poor negative predictive value, suggesting that TEE should strongly be considered in SLE patients in whom this entity is suspected with a negative or nondiagnostic TTE.

Libman—Sacks lesions can become secondarily infected or serve as a nidus for thrombosis so that patients may have endocarditis with multiple contributing etiologies. The presence of active SLE does not preclude these other potential etiologies, so an aggressive evaluation for other potential causes of endocarditis is still warranted in this setting.

Treatment

In cases in which the underlying etiology of valvular disease is unclear, and infection is suspected, immediate initiation of broad-spectrum antibiotics should be instituted while diagnostic investigations are undertaken. Libman—Sacks endocarditis has no specific directed treatment. In patients with other manifestations of SLE disease activity, immunosuppressive therapy is indicated, but it is unclear what effect this therapy imparts on Libman—Sacks endocarditis. Data from the autopsy series reported by Bulkley and Roberts [3] suggest that the institution of glucocorticoid therapy may have impacted Libman—Sacks endocarditis in SLE patients, with post-steroid era specimens demonstrating complete or partial healing of valvular lesions.

In patients with significant hemodynamic impairment from valvular heart disease, surgical intervention would be indicated as in any other etiology of valvular heart disease. Patients may develop recurrent Libman—Sacks lesions following surgical intervention.

Anticoagulation is indicated for the management of APS, including patients with valvular heart disease. Patients with SLE and APS, or catastrophic APS, may also require immunosuppressive therapy. There is no clear indication for anticoagulation or antiplatelet therapy in patients with only aPLs with valvular heart disease without a prior history of thrombosis.

Antimicrobial prophylaxis in SLE patients with valvular heart disease is indicated in the same settings as for those with valvular heart disease without SLE. Patients considered at highest risk include those with prosthetic heart valves (including bioprosthetic and homograft) or valve repair with prosthetic material, past history of infective endocarditis, persisting or repaired cyanotic congenital heart disease with prosthetic material, or transplantation patients with valvular heart disease. The American Heart Association recommends antimicrobial prophylaxis in these patients when undergoing the following procedures that are considered high risk for bacteremia with pathogens that have been associated with endocarditis: dental procedures in which the gingival or periapical portion of teeth are manipulated or with perforation of the oral mucosa; procedures involving biopsy or incision of the respiratory tract mucosa; gastrointestinal or genitourinary invasive procedures in patients with active infection in these areas; and patients with infected skin, skin structures, or musculoskeletal tissue undergoing surgical procedures [74].

CORONARY VASCULITIS

Epidemiology and Pathogenesis

Coronary vasculitis has been reported in SLE but occurs rarely. It is unclear whether this is due to low frequency or whether this is an underreported phenomenon, given the difficulty in making this diagnosis by clinical means alone. A review of the English literature noted that of 49 SLE patients who developed acute MI before the age of 35 years, 7 had coronary arteriograms with findings suggestive of vasculitis, and 5 patients had coronary aneurysms (which is suggestive of, but not diagnostic for, a vasculitic process) [75]. An autopsy study of 35 Indian patients with SLE (mean age, 24 years) revealed 5 cases of coronary vasculitis [17]. Other studies also demonstrate this complication occurring more commonly in younger patients, most without a history of known coronary artery disease.

Clinical Features

The most common manifestation of coronary vasculitis in SLE is that of ischemic heart disease, where patients present with signs and/or symptoms of angina or with MI. There appears to be no clear association between coronary vasculitis and other manifestations of disease activity in SLE, adding to the challenge of establishing this diagnosis.

Diagnosis

Evidence of ischemic heart disease on electrocardiography or myocardial stress testing can be seen in coronary vasculitis but is not sensitive or specific for this diagnosis. Coronary aneurysms, diffuse irregularities, or isolated segments with smooth, tapering stenoses on angiography suggest, but again are not specific for, the diagnosis of coronary vasculitis (Figure 43.7) [76].

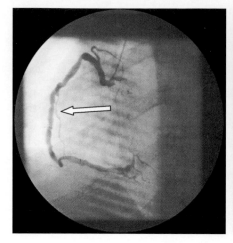
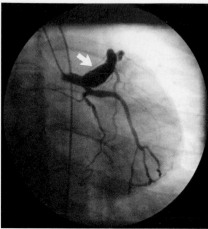

FIGURE 43.7 Diffuse irregularities and beading (*arrow*) and a large aneurysm (*arrowhead*) on coronary angiography in an SLE patient with presumed coronary vasculitis. *Source: Reproduced with permission of Lippincott Williams & Wilkins from Caracciolo et al. [76].*

The presence of rapidly developing or progressing lesions on serial angiographic studies is suggestive of vasculitis. As in other forms of autoimmune vasculitis, the gold standard for diagnosis is tissue biopsy, which is often not practical in these cases. Histopathologic findings include neutrophilic and lymphocytic infiltration, fibrinoid necrosis, cellular intimal fibrosis, and aneurismal dilatation [35, 77, 78].

Treatment

Immunosuppressive agents are indicated for the treatment of coronary vasculitis in SLE. As in other inflammatory vascular disorders, outcomes of revascularization (if necessary) are thought to be optimized if performed when vasculitis is quiescent. However, this may not be possible in patients who present with acute myocardial ischemia. Secondary preventative measures for cardiovascular disease seem reasonable in this setting, given that arterial injury frequently results in premature atherosclerotic vascular disease.

CORONARY ATHEROSCLEROSIS

Epidemiology and Pathogenesis

The contribution of coronary atherosclerotic disease (CAD) to mortality in SLE was highlighted in 1976 by Urowitz *et al.* [79]. They described a bimodal mortality pattern in SLE, in which patients dying early in the course of disease most often succumbed to complications of disease activity or therapy (e.g., infection), whereas patients who died later in the disease course often died of cardiovascular (CV) events in the setting of quiescent disease. Autopsy series have subsequently substantiated the prevalence of significant obstructive coronary artery disease in SLE, even in premenopausal

women [80, 81]. In contrast to other cardiac manifestations of SLE, which have decreased in frequency since the advent of immunosuppressive therapies, cardiovascular disease (CVD) has emerged as a more prominent source of morbidity and mortality as overall patient survival has continued to improve (Table 43.7) [11, 12, 22, 30, 66, 79, 82–89].

The LUMINA study, which evaluated 637 SLE patients, found evidence of cardiovascular damage, as defined by the presence of angina, coronary artery bypass surgery, CHF, or MI, in 6.8% of patients [89]. However, studies of asymptomatic patients with stress testing and perfusion scintigraphy suggest that subclinical CAD occurs more frequently, with abnormal studies seen in 28–40% of patients [90, 91]. Studies examining surrogate markers of coronary atherosclerotic disease and autopsy studies in SLE also support the perception of CV events in SLE patients as representing the "tip of the iceberg" of CVD in this patient population (Table 43.8) [3, 79, 89, 92–105]. Cohort studies using vascular ultrasound and electron beam tomography to evaluate CVD risk in SLE demonstrate that atherosclerotic vascular disease is present in 17–48% of patients [93, 95, 97, 99, 103–105]. Autopsy studies have demonstrated that 22–54% of SLE patients have atherosclerotic disease of the coronary arteries [3, 79, 92].

The prevalence of atherosclerotic vascular disease and associated CV events in young women with SLE is surprising, given that premenopausal women have been presumed to be at minimal risk for CVD and CV events. In contrast to the general population, in which CV events are uncommon and considered "premature" in women younger than the age of 65 years, the mean age at first CV event in SLE women is approximately 49 years [106]. Ward [107] found that young SLE women (18–44 years) were 2.27 times more likely to be hospitalized for MI and 3.8 times more likely for CHF compared

TABLE 43.7 Cardiovascular Events in SLE

Authors	Year	No. of patients	Prevalence (%)	Comment
Kong et al. [22]	1962	11	4.5	MI on autopsy
Urowitz et al. [79]	1976	81	11	MI
Gladman and Urowitz [81]	1987	507	9	Angina/MI
Jonssen et al. [82]	1989	86	20	Angina/MI
Petri et al. [83]	1992	229	8	Angina/MI
Sturfelt et al. [11]	1992	75	9	MI
Jouhikainen et al. [12]	1994	74	6	MI
Hearth-Holmes et al. [84]	1995	89	6	Angina/MI
Roldan et al. [30]	1996	58	12	Stroke
Rahman et al. [85]	2000	150	15	Angina/MI/stroke/PVD 50% hypertensive
Morelli et al. [66]	2003	50	26	TIA/stroke Consecutive patients; 23% with prior history
Bessant et al. [86]	2004	47	8/10	MI/stroke
Toloza et al. [87]	2004	546	6	MI/angina/stroke/PVD
Pons-Estel et al. [88]	2009	637	7	Angina/MI/CABG/CHF

CABG, coronary artery bypass surgery; CHF, congestive heart failure; MI, myocardial infarction; PVD, peripheral vascular disease; TIA, transient ischemic attack.

TABLE 43.8 Atherosclerotic Disease in SLE

Authors	Year	No. of patients	Prevalence (%)	Comment
Bulkley and Roberts [3]	1975	36	22	Autopsy (>50% coronary lesion)
Haider and Roberts [91]	1981	22	45	Autopsy (>75% coronary lesion)
Abu-Shakra et al. [92]	1995	40	54	Autopsy
Manzi et al. [93]	1999	175	40	Carotid ultrasound
Bruce et al. [89]	2000	130	40	Single photon emission computed tomography Dual isotope myocardial perfusion imaging Consecutive female patients
Asanuma et al. [94]	2003	65	31	Electron beam tomography (EBT) Coronary artery calcification (CAC)
Doria et al. [95]	2003	78	17	Carotid ultrasound
Manger et al. [96]	2003	75	28	EBT (CAC)
Roman et al. [97]	2003	197	37	Carotid ultrasound Consecutive patients
Sella et al. [90]	2003	82	28	Myocardial perfusion scintigraphy
Selzer et al. [98]	2004	214	32	Carotid ultrasound
Wolak et al. [99]	2004	51	28	Carotid/femoral ultrasonic biopsy Consecutive patients
Jimenez et al. [100]	2005	70	28	Carotid ultrasound
Chung et al. [101]	2006	93	19	EBT (CAC)
Maksimowicz-McKinnon et al. [102]	2006	605	14	Carotid ultrasound
Von Feldt et al. [103]	2006	152	30	EBT (CAC)
Kiani et al. [104]	2008	200	43	EBT (CAC)
Kao et al. [105]	2008	157	48	EBT (CAC)

to age-matched non-SLE women. In contrast, SLE women between the ages of 45 and 64 years were only 1.39 times more likely to be hospitalized for CHF than age-matched controls, with no difference noted for hospitalization for MI. Similarly, a study comparing SLE women with age-matched women from the Framingham offspring cohort found that SLE women between the ages of 35 and 44 had a 50-fold increased risk of MI [108]. The etiology of this marked increase in CV relative risk in young SLE women remains unclear.

Traditional CVD risk factors occur frequently in SLE. Data from the Toronto Risk Factor Study demonstrate a higher number of traditional CV risk factors per SLE patient compared with controls and an increased prevalence of hypertension, diabetes mellitus, hyperlipidemia, and hyperhomocysteinemia [109]. It is thought that disease activity in SLE may contribute to the generation of aggravation of traditional CV risk factors in these patients. Subclinical inflammation in the general population, as measured by high-sensitivity C-reactive protein levels, predicts the development of insulin resistance, diabetes mellitus, weight gain, visceral obesity, and hypertension [110–112]. It is probable that inflammation associated with disease activity in SLE imparts the same adverse effects. In addition, glucocorticoid therapy can be associated with insulin resistance, hyperglycemia, dyslipidemia, and hypertension [113–115]. However, the role of glucocorticoids in CVD in SLE remains unclear, given that the anti-inflammatory effects of therapy could also potentially decrease metabolic perturbations associated with chronic inflammation and hence decrease CVD risk by this means.

In addition, several cohort studies have demonstrated that even when modifiable CV risk factors are identified in SLE patients, they are often not treated [109, 116]. Lack of awareness of increased CV risk in young SLE patients, concerns regarding pregnancy and use of some CV medications, possible drug interactions, and lack of evidence-based guidelines addressing risk factor management in SLE likely contribute to suboptimal CV risk factor management. Complicating this issue are data from Esdaile *et al.* [117] that demonstrate that the presence of CV risk factors alone is inadequate to explain the increased incidence of CV events in SLE, implying that other factors are important in CVD pathogenesis and events in SLE. Although a "lupus factor" for CVD has been actively sought, one has not been conclusively identified. Increased disease activity, as measured by clinical scales and serologic markers, is not consistently associated with surrogate markers of CVD or CV events in patients with SLE. This could be explained by considering the more aggressive therapies that patients with severe disease receive, which might better control disease activity compared to less intensive therapies given to patients with milder disease, allowing subclinical low-grade activity and associated inflammation to persist. It may be that a longer duration of chronic, "grumbling" disease may pose more CV risk than limited episodes of more intense inflammation for these patients. In addition, patients with more severe disease could be presumed to be at higher risk and more aggressively treated, or risk factors might be more readily detected during their more frequent visits and increased surveillance.

The failure to identify a lupus factor for CVD may also reflect inadequate measures to assess CVD risk in SLE. In the general population, novel risk markers for CVD and CV risk, such as serum C-reactive protein, myeloperoxidase levels, lipoprotein subfractions, and lipoprotein-associated phospholipase A_2, have been identified as potentially beneficial in further CV risk stratification [118–124]. It may be that some of these novel risk markers will prove helpful in CV risk stratification in SLE.

aPLs are postulated to have a role in arterial endothelial injury and thus may contribute to CVD in SLE. Although several cohort studies have not identified an association between the presence of aPLs and surrogate markers of CAD, associations with hypertension and CV events in SLE have been documented [125, 126]. Whether this association occurs primarily based on thrombosis-associated processes or is also from aPL-mediated arterial injury is unclear.

Clinical Features

SLE patients with CAD manifest similarly to CAD patients in the general population. Symptoms of ischemic heart disease include fatigue; chest pressure or pain with exertion that may radiate into the jaw, shoulders, or arms; 'heartburn'; diaphoresis; upper back pain; decreased exercise tolerance; and dyspnea. It is critical to be aware of the increased risk of CVD in premenopausal SLE women, in whom these symptoms may be more likely to be attributed to serositis or other SLE- and non-SLE-associated conditions solely because of their age.

Diagnosis

The same studies utilized to diagnose CVD in the general population are appropriate for the evaluation of SLE patients. Treadmill stress testing alone has been demonstrated to have poor sensitivity and specificity for CAD in women, and it is likely to have the same limitations in SLE women [127, 128]. Stress echocardiography, nuclear stress testing, and coronary angiography are useful in the evaluation of CAD in SLE. However, it is important to be mindful of physical limitations in any individual SLE patient; those with limited exercise capacity or physical disability limiting their ability to walk at increasing speeds and inclines will not be able to successfully complete an exercise-based modality and instead should be considered for "chemical" stress testing. Vascular ultrasound (measurement of carotid plaques or vascular stiffness), cardiac magnetic resonance angiography, and electron beam tomography are currently under study in research settings for assessing CVD prevalence and risk in SLE patients, but they are not currently utilized in routine screening and assessment of SLE patients.

A key to diagnosing CVD in SLE is maintaining awareness of the increased risk in this population, especially in premenopausal women. There is ongoing discussion about whether SLE should be considered a risk factor equivalent for CVD, similar to diabetes mellitus. This designation would emphasize the need for aggressive risk factor surveillance and treatment of modifiable risk factors (Table 43.9). Currently, in patients with chest pain, ischemic heart disease should be considered as a possible etiology in addition to serositis, pulmonary embolism, or other non-SLE conditions. Any signs or symptoms suggestive of ischemic heart disease in an SLE patient, regardless of age, warrant further investigation for CAD.

Treatment

Although there are no long-term studies that conclusively demonstrate that appropriate management of modifiable CVD risk factors in SLE reduces CV events in this population, it is probable that risk factor modification will have beneficial effects on CV health and should be undertaken. Other than for family history, traditional CV risk factors are generally amenable to lifestyle and pharmacologic therapies and should be addressed.

Several factors contribute to the increased prevalence of obesity in SLE. Disease and/or treatment-associated physical disability, glucocorticoid therapy, lack of exercise, and poor dietary habits may all have a role in this risk factor in patients with SLE. Interventions directed at the prevention or treatment of obesity should be initiated at the time of diagnosis. Exercise programs and activities may be customized to the individual patient with the assistance of physical therapists and physical medicine and rehabilitation specialists. Dieticians and nutritionists can provide guidance in proper nutrition with consideration for other co-morbid conditions (e.g., diabetes mellitus). Maintenance of a healthy weight should be an ongoing part of care.

Both dietary and pharmacologic interventions may be useful in the treatment of dyslipidemia in SLE patients. Although in some patients lifestyle modifications may be adequate to attain a desirable lipid profile, in many patients this will prove inadequate for a number of reasons, including poor compliance, inherited dyslipidemias, and side effects of SLE therapies.

The beneficial effects of antihypertensive therapy in the general population derive not only from lowering blood pressure but also from the anti-inflammatory and metabolic effects of therapy. It is anticipated that SLE patients will also benefit from these effects of treatment. In SLE patients, the optimal agent may be chosen with consideration given to disease- and/or treatment-associated co-morbid conditions. For example, angiotensin-converting enzyme inhibitors are efficacious in controlling blood pressure but also help maintain renal function by the prevention of renal scarring, decreasing proteinuria, and possibly from their antioxidant effects [129–132]. In patients with CAD or CHF, β-blockers may provide additional benefit for these conditions [133]. However, Raynaud's phenomenon may be precipitated or worsened with β-blocker use. Ultimately, the focus needs to be on optimal blood pressure control, regardless of the agent chosen for treatment.

Although the prevalence of diabetes mellitus in SLE patients is relatively low in contrast to the prevalence of other traditional CVD risk factors, insulin resistance is often present and does convey CVD risk, especially when present with other metabolic syndrome features. However, the optimal management of insulin resistance in SLE has not been determined. Lifestyle modifications appear to be a reasonable starting point for management. Steroid-induced hyperglycemia occurs commonly, and treatment should be instituted when this is identified. Data from a rheumatoid arthritis cohort demonstrated that the use of hydroxychloroquine is associated with a decreased incidence of diabetes mellitus [134]. Hydroxychloroquine has also been associated with decreased glucose levels in SLE, and it may prove beneficial from a metabolic standpoint in these patients [135].

TABLE 43.9   Modifiable Cardiovascular Risk Factors

Hypertension

Hyperlipidemia

Impaired glucose metabolism

Lifestyle

   Lack of physical activity

   Tobacco use

   Diet low in fiber, fruit, and vegetables

   Psychosocial factors (e.g., depression)

Patient and provider awareness of CVD risk in SLE is essential for risk prevention and risk factor management. Although this awareness is growing, screening, counseling, and treatment remain suboptimal. In a cohort of 110 SLE patients with a mean disease duration of slightly more than 15 years, Costenbader *et al.* [136] found that only 58% of modifiable CV risk factors had been adequately addressed. Perhaps more striking, 25% of patients with a diagnosis of diabetes mellitus, which is a well-known CVD equivalent, were not receiving treatment for this co-morbidity. Patients also need to be educated about their increased risk of CVD and how they can participate in risk factor management. In addition, this knowledge may help patients commit to lifestyle changes that can decrease CVD risk, including tobacco cessation, good dietary habits, maintaining a healthy body weight, and regular exercise. CVD risk factor education and screening should be initiated at disease diagnosis and continued lifelong. The clear association between systemic inflammation and CVD and CV events in the general population suggests that maintaining good disease control may be beneficial in long-term CV risk reduction.

In patients with obstructive CAD, interventions including percutaneous coronary intervention (PCI) and coronary artery bypass surgery (CABG) may be necessary. Data suggest that SLE patients may have more complications after invasive interventions for CAD. A study of 28 SLE patients undergoing PCI found that although rates of intervention success and complications were initially not different from those of non-SLE patients, at 1 year SLE patients were more likely to experience an MI or require repeat revascularization [137]. Data regarding CABG outcomes from several small series demonstrate in-hospital mortality rates ranging from 5.7 to 17% and report complications including early graft thrombosis, bleeding, and wound complications [138].

CONCLUSION

SLE patients frequently suffer cardiovascular manifestations of disease, which can be an important source of disease-associated morbidity. Prompt identification and treatment of these manifestations can be organ- and life-saving and is facilitated by increased vigilance for signs and symptoms of cardiovascular manifestations in this vulnerable population. Patients should be educated about their increased risk of CVD, counseled on beneficial lifestyle modifications when necessary, and routinely undergo surveillance of modifiable risk factors. Regardless of age or sex, treatment and monitoring of modifiable risk factors is indicated in SLE patients.

References

[1] G.C. Griffith, I.L. Vural, Acute and subacute disseminated lupus erythematosus: A correlation of clinical and postmortem findings in eighteen cases, Circulation 3 (1951) 492–500.

[2] E.L. Dubois, D.L. Tuffanelli, Clinical manifestations of systemic lupus erythematosus. Computer analysis of 520 cases, JAMA 190 (1964) 104–111.

[3] B.H. Bulkley, W.C. Roberts, The heart in systemic lupus erythematosus and the changes induced in it by corticosteroid therapy. A study of 36 necropsy patients, Am J Med 58 (1975) 243–264.

[4] M. Ito, Y. Kagiyama, I. Omura, Y. Hiramatsu, E. Kurata, S. Kanaya, et al., Cardiovascular manifestations in systemic lupus erythematosus, Jpn Circ J 43 (1979) 985–994.

[5] J.S. Bomalaski, J.V. Talano, S. Perlman, The value of echocardiography in patients with systemic lupus erythematosus, Clin Res 31 (1983) 68.

[6] N. Doherty, R. Siegel, Cardiovascular manifestations of systemic lupus erythematosus, Am Heart J 110 (1985) 1257–1265.

[7] E. Badui, D. Garcia-Rubi, E. Robles, J. Jimenez, L. Juan, M. Deleze, et al., Cardiovascular manifestations in systemic lupus erythematosus. Prospective study of 100 patients, Angiology 36 (1985) 431–441.

[8] A. Klinkhoff, C. Thompson, G. Reid, C.D. Tomlinson, M-mode and two-dimensional echocardiographic abnormalities in systemic lupus erythematosus, JAMA 253 (1985) 3273–3277.

[9] I. Crozier, E. Li, M. Milne, M. Nicholls, Cardiac involvement in systemic lupus erythematosus detected by echocardiography, Am J Cardiol 65 (1990) 1145–1148.

[10] P. Nihoyannopoulos, P.M. Gomez, J. Joshi, S. Loizou, M.J. Walport, C.M. Oakley, Cardiac abnormalities in systemic lupus erythematosus. Association with raised anticardiolipin antibodies, Circulation 82 (1990) 369–375.

[11] G. Sturfelt, F. Eskilsson, O. Nived, L. Truedsson, S. Valind, Cardiovascular disease in systemic lupus erythematosus. A study of 75 patients from a defined population, Medicine (Baltimore) 71 (1992) 216–223.

[12] T. Jouhikainen, S. Pohjola-Sintonen, E. Stephansson, Lupus anticoagulant and cardiac manifestations in systemic lupus erythematosus, Lupus 3 (1994) 167–172.

[13] S. Kalke, C. Balakrishanan, G. Mangat, G. Mittal, N. Kumar, V.R. Joshi, Echocardiography in systemic lupus erythematosus, Lupus 7 (1998) 540–544.

[14] S. Rantapaa-Dahlqvist, G. Neumann-Andersen, C. Backman, G. Dahlen, B. Stegmayr, Echocardiographic findings, lipids, and lipoprotein (a) in systemic lupus erythematosus, Clin Rheumatol 16 (1997) 140–148.

[15] M.H. Houman, M. Smiti-Khanfir, I. Ben Ghorbell, M. Miled, Systemic lupus erythematosus in Tunisia: Demographic and clinical analysis of 100 patients, Lupus 13 (2004) 204–211.

[16] A. Amoroso, F. Cacciapaglia, S. De Castro, A. Battagliese, G. Coppolino, S. Galluzzo, et al., The adjunctive role of antiphospholipid antibodies in systemic lupus erythematosus cardiac involvement, Clin Exp Rheumatol 24 (2006) 287–294.

[17] L. Panchal, S. Divate, P. Vaideeswar, S.P. Pandit, Cardiovascular involvement in systemic lupus erythematosus: An autopsy study of 27 patients in India, J Postgrad Med 52 (2006) 5–10.

[18] T. Shimizu, M. Murata, H. Tomizawa, T. Mitsuhashi, T. Katsuki, K. Shimada, Systemic lupus erythematosus initially manifesting as acute pericarditis complicating with pericardial tamponade: A case report, J Cardiol 49 (2007) 273–276.

[19] Y. Inoue, R. Anzawa, Y. Terao, T. Tsurusaki, A. Matsuyama, M. Kunoh, et al., Pericarditis causing congestive heart failure as an initial manifestation of systemic lupus erythematosus, Int J Cardiol 119 (2007) 71—73.

[20] D.Y. Wang, P.C. Yan, W.L. Yu, D.C. Shiah, H.W. Kuo, N.Y. Hsu, Comparison of different diagnostic methods for lupus pleuritis and pericarditis: A prospective 3 year study, J Formos Med Assoc 99 (2000) 375—380.

[21] D.Y. Wang, P.C. Yang, W.L. Yu, S.H. Kuo, N.Y. Hsu, Serial antinuclear antibodies titer in pleural and pericardial fluid, Eur Respir J 15 (2000) 1106—1110.

[22] T.Q. Kong, R.E. Kellum, J.R. Haserick, Clinical diagnosis of cardiac involvement in systemic lupus erythematosus: A correlation of clinical and autopsy findings in 30 patients, Circulation 26 (1962) 7—11.

[23] D. Estes, C.L. Christian, The natural history of systemic lupus erythematosus by prospective analysis, Medicine 50 (1971) 85—95.

[24] B.S. Maniscalco, J.M. Felner, J.L. McCans, J.A. Chipella, Echocardiographic abnormalities in systemic lupus erythematosus, Circulation 52 (1975) 11—21.

[25] M. Gobaira, A. Zghaib Abad, P.A. Reyes Lopez, The heart in systemic lupus erythematosus. Study in 32 non-selected patients, Arch Inst Cardiol Mex 52 (1982) 223—228.

[26] N.E. Doherty, G. Feldman, G. Maurer, R.J. Siegel, Echocardiographic findings in systemic lupus erythematosus, Am J Cardiol 61 (1988) 1144.

[27] W.H. Leung, K.L. Wong, C.P. Lau, C.K. Wong, H.W. Liu, Association between antiphospholipid antibodies and cardiac abnormalities in patients with systemic lupus erythematosus, Am J Med 89 (1990) 411—419.

[28] P. Nihoyannopoulos, P.M. Gomez, J. Joshi, S. Loizou, M.J. Walport, C.M. Oakley, Cardiac abnormalities in systemic lupus erythematosus. Association with raised anticardiolipin antibodies, Circulation 82 (1990) 369—375.

[29] Z. Sasson, Y. Rasooly, C.W. Chow, S. Marshall, M.B. Urowitz, Impairment of left ventricular diastolic function in systemic lupus erythematosus, Am J Cardiol 69 (1992) 1629—1634.

[30] C.A. Roldan, B.K. Shively, M.H. Crawford, An echocardiographic study of valvular heart disease associated with systemic lupus erythematosus, N Engl J Med 335 (1996) 1424—1430.

[31] M. Wijetunga, S. Rockson, Myocarditis in systemic lupus erythematosus, Am J Med 113 (2002) 419—423.

[32] A. del Rio, J.J. Vazquez, J.A. Sobrino, A. Gil, J. Barbado, I. Mate, J. Ortiz-Vazques, Myocardial involvement in systemic lupus erythematosus. A noninvasive study of left ventricular function, Chest 74 (1978) 414—417.

[33] M. Apte, G. McGwin, L.M. Vila, R.A. Kaslow, G.S. Alarcon, J.D. Reveille, et al., Associated factors and impact of myocarditis in patients with SLE from LUMINA, a multiethnic U.S. cohort, Rheumatology 47 (2008) 362—367.

[34] G.W.K. Yip, Q. Shang, L.S. Tam, Q. Zhang, E.K.M. Li, J.W.H. Lung, et al., Disease chronicity and activity predict subclinical left ventricular systolic dysfunction in patients with systemic lupus erythematosus, Heart 95 (2009) 980—987.

[35] D. Jain, M.K. Halushka, Cardiac pathology of systemic lupus erythematosus, J Clin Pathol 62 (2009) 584—592.

[36] A.K. Bidani, J.L. Roberts, M.M. Schwartz, E.J. Lewis, Immunopathology of cardiac lesions in fatal systemic lupus erythematosus, Am J Med 69 (1980) 849—858.

[37]. G. Berg, J. Bodet, K. Webb, G. Williams, D. Palmer, B. Ruoff, et al., Systemic lupus erythematosus presenting as isolated congestive heart failure, J Rheumatol 12 (1985) 1182—1185.

[38] J. Baguet, F. Tremel, M. Fabre, Chloroquine cardiomyopathy with conduction disorders, Heart 81 (1999) 221—223.

[39] T. Don Michael, S. Aiwazzadeh, The effects of acute chloroquine poisoning with special reference to the heart, Am Heart J 79 (1970) 831—842.

[40] E. Lansimies, A. Lakksonen, K. Juva, L. Hirvonen, Acute cardiotoxicity of chloroquine and hydroxychloroquine in dogs, Ann Med Exp Biol Fenn 49 (1971) 45—48.

[41] A. Nelson, O. Fitzhugh, Chloroquine; pathologic changes observed in rats which for two years had been fed various proportions, Arch Pathol 45 (1948) 454—462.

[42] P. Jolles, J. Tatum, SLE myocarditis detection by Ga-67 citrate scintigraphy, Clin Nucl Med 21 (1996) 284—286.

[43] H. Ardehali, A. Qasim, T. Cappola, D. Howard, R. Hruban, J.M. Hare, et al., Endomyocardial biopsy plays a role in diagnosing patients with unexplained cardiomyopathy, Am Heart J 147 (2004) 919—923.

[44] C. August, H. Holzhausen, A. Schmoldt, R. Pompecki, S. Schroder, Histological and ultrastructural findings in chloroquine-induced cardiomyopathy, J Mol Med 73 (1995) 73—77.

[45] N. Rewcastle, J. Humphrey, Vacuolar myopathy: Clinical, histochemical, and microscopic study, Arch Neurol 12 (1965) 570—582.

[46] M.L. Estes, D. Ewing-Wilson, S. Chou, H. Mitsumoto, M. Hanson, E. Shirey, et al., Chloroquine neuromyotoxicity: clinical and pathologic perspective, Am J Med 82 (1987) 447—455.

[47] T. Pulerwitz, L. Rabbani, S. Pinney, A rationale for the use of anticoagulation in heart failure management, J Thromb Thrombolysis 17 (2004) 87—93.

[48] M.A. Shearn, The heart in systemic lupus erythematosus, Am Heart J 58 (1959) 452—466.

[49] Y. Sherer, Y. Levy, Y. Shoenfeld, Marked improvement of severe cardiac dysfunction after one course of intravenous immunoglobulin in a patient with systemic lupus erythematosus, Clin Rheumatol 18 (1999) 238—240.

[50] T. Ueda, K. Mizushige, T. Aoyama, M. Tokuda, H. Kiyomoto, H. Matsuo, Echocardiographic observation of acute myocarditis with systemic lupus erythematosus, Jpn Circ J 64 (2000) 144—146.

[51] M. Hejtmancik, J. Wright, R. Quint, F. Fennings, The cardiovascular manifestations of systemic lupus erythematosus, Am Heart J 68 (1964) 119—130.

[52] B.F. Mandell, Cardiovascular involvement in systemic lupus erythematosus, Sem Arthritis Rheum 17 (1987) 126—141.

[53] D. Logar, T. Kveder, B. Rozman, J. Dobovisek, Possible association between anti-Ro antibodies and myocarditis or cardiac conduction defects in adults with systemic lupus erythematosus, Ann Rheum Dis 49 (1990) 627—629.

[54] T.W. O'Neill, A. Mahmoud, A. Tooke, R.D. Thomas, P.J. Maddison, Is there a relationship between subclinical myocardial abnormalities, conduction defects and Ro/La antibodies in adults with systemic lupus erythematosus? Clin Exp Rheumatol 11 (1993) 409—412.

[55] C.R.L. Cardoso, M.A.O. Sales, J.A.S. Papi, G.F. Salles, QT-interval parameters are increased in systemic lupus erythematosus patients, Lupus 14 (2005) 846—852.

[56] S.D. Bilazarian, A.J. Taylor, D. Brezinski, M.C. Hochberg, T. Guarnieri, T.T. Provost, High grade atrioventricular heart block in an adult with systemic lupus erythematosus: The association of nuclear ribonucleoprotein (U1 RNP) antibodies, a case report and review of the literature, Arthritis Rheum 32 (1989) 1170—1174.

[57] E. Galve, J. Candell-Riera, G. Permanyer-Miralda, H. Garcia-Del-Castillo, J. Soler-Soler, Prevalence, morphologic types, and evolution of cardiac valvular disease in systemic lupus erythematosus, N Engl J Med 19 (1988) 817—823.

[58] M.A. Khamashta, R. Cervera, R.A. Asherson, J. Font, A. Gil, D.J. Coltart, et al., Association of antibodies against phospholipids with heart valve disease in systemic lupus erythematosus, Lancet 335 (1990) 504—505.

[59] R. Cervera, J. Font, C. Pare, M. Azqueta, F. Perez-Villa, A. Lopez-Soto, et al., Cardiac disease in systemic lupus erythematosus: prospective study of 70 patients, Ann Rheum Dis 51 (1992) 156—159.

[60] C.B. Gleason, M.F. Stoddard, S.G. Wagner, R.A. Longaker, S. Pierangeli, E.N. Harris, A comparison of cardiac valvular involvement in the primary antiphospholipid syndrome versus anticardiolipin-negative systemic lupus erythematosus, Am Heart J 125 (1993) 1123—1129.

[61] A. Giunta, U. Picillo, S. Maione, S. Migliaresi, G. Valentini, M. Arnese, et al., Spectrum of cardiac involvement in systemic lupus erythematosus: Echocardiographic, echo-Doppler observations and immunologic investigations, Acta Cardiol 48 (1993) 183—197.

[62] F. Gabrielli, E. Alcini, M.A. Di Prima, G. Mazzacurati, C. Masala, Cardiac valve involvement in systemic lupus erythematosus and primary antiphospholipid syndrome: Lack of correlation with antiphospholipid antibodies, Int J Cardiol 51 (1995) 117—126.

[63] R. Omdal, P. Lunde, K. Rasmussen, S.I. Mellgren, G. Husby, Transesophageal and transthoracic echocardiography and Doppler-examinations in systemic lupus erythematosus, Scand J Rheumatol 30 (2001) 275—281.

[64] K. Jensen-Urstad, E. Svenungsson, U. de Faire, A. Siveira, J.L. Witztum, A. Hamsten, et al., Cardiac valvular abnormalities are frequent in systemic lupus erythematosus patients with manifest arterial disease, Lupus 11 (2002) 744—752.

[65] P. Leszczynski, E. Straburzynska-Migaj, I. Korczowska, J. Lacki, S. Mackiewicz, Cardiac valvular disease in patients with systemic lupus erythematosus. Relationship with anticardiolipin antibodies, Clin Rheumatol 22 (2003) 405—408.

[66] S. Morelli, M. Bernardo, F. Viganego, A. Sgreccia, P. De Marzio, F. Conti, et al., Left sided heart valve abnormalities and risk of ischemic cerebrovascular accidents in patients with systemic lupus erythematosus, Lupus 12 (2003) 805—812.

[67] C.A. Roldan, E.A. Gelgand, C.R. Qualis, W.L. Sibbitt Jr., Valvular heart disease by transthoracic echocardiography is associated with focal brain injury and central neuropsychiatric systemic lupus erythematosus, Cardiology 108 (2007) 331—337.

[68] R.F. Shapiro, C.N. Gamble, K.B. Wiesner, J.J. Castles, A.W. Wolf, E.J. Hurley, et al., Immunopathogenesis of Libman—Sacks endocarditis: Assessment by light and immunofluorescence microscopy in two patients, Ann Rheum Dis 36 (1977) 508—516.

[69] P. Soltesz, Z. Szekanecz, E. Kilss, Y. Shoenfeld, Cardiac manifestations in antiphospholipid syndrome, Autoimmunity Rev 6 (2007) 379—386.

[70] J.J. Murphy, I.H. Leach, Findings at necropsy in the heart of a patient with anticardiolipin syndrome, Br Heart J 62 (1989) 61—64.

[71] C.A. Roldan, E.A. Glegand, C.R. Qualis, W.L. Sibbitt Jr., Valvular heart disease is associated with no focal neuropsychiatric systemic lupus erythematosus, J Clin Rheumatol 12 (2006) 3—10.

[72] C.A. Roldan, E.A. Gelgand, C. Qualis, W.L. Sibbitt Jr., Valvular heart disease as a cause of cerebrovascular disease in patients with systemic lupus erythematosus, Am J Cardiol 95 (2005) 1441—1447.

[73] C.A. Roldan, C.R. Qualis, K.S. Sopko, W.L. Sibbitt Jr., Transthoracic versus transesophageal echocardiography for detection of Libman—Sacks endocarditis: A randomized controlled study, J Rheumatol 35 (2008) 224—229.

[74] W. Wilson, K.A. Taubert, M. Gewtiz, P.B. Lockhart, L.M. Baddour, M. Levison, et al., Prevention of infective endocarditis: Guidelines from the American Heart Association: A guideline from the American Heart Association Rheumatic Fever, Endocarditis, and Kawasaki Disease Committee, Council on Cardiovascular Disease in the Young, and the Council on Clinical Cardiology, Council on Cardiovascular Surgery and Anesthesia, and the Quality of Care and Outcomes Research Interdisciplinary Working Group, J Am Dent Assoc 139 (Suppl.) (2008) 3S—24S.

[75] C. Korkmaz, D.U. Cansu, T. Kasifoglu, Myocardial infarction in young patients (≤35 years of age) with systemic lupus erythematosus: A case report and clinical analysis of the literature, Lupus 16 (2007) 289—297.

[76] E.A. Caracciolo, C.B. Marcu., A. Ghantous, T. Donohue, G. Hutchison, Coronary vasculitis with acute myocardial infarction in a young woman with systemic lupus erythematosus, J Clin Rheumatol 10 (2004) 66—68.

[77] T.A. Bonfiglio, R.E. Botti, J.W.C. Hagstrom, Coronary arteritis, occlusion, and myocardial infarction due to lupus erythematosus, Am Heart J 83 (1972) 153—158.

[78] S.M. Korbet, M.M. Schwartz, E.J. Lewis, Immune complex deposition and coronary vasculitis in systemic lupus erythematosus, Am J Med 77 (1984) 141—146.

[79] M. Urowitz, A. Bookman, B. Koehler, D. Gordon, H. Smythe, M. Ogryzlo, The bimodal mortality pattern of systemic lupus erythematosus, Am J Med 60 (1976) 221—225.

[80] M. Abu-Shakra, M.B. Urowitz, D.D. Gladman, J. Gough, Mortality studies in systemic lupus erythematosus. Results from a single center. I. Causes of death, J Rheumatol 22 (1995) 1259—1264.

[81] D.D. Gladman, M.B. Urowitz, Morbidity in systemic lupus erythematosus, J Rheumatol 14 (Suppl. 3) (1987) 223—226.

[82] H. Jonsson, O. Nived, G. Sturfelt, Outcome in systemic lupus erythematosus: A prospective study of patients from a defined population, Medicine 68 (1989) 141—150.

[83] M. Petri, S. Perez-Gutthann, D. Spence, M.C. Hochberg, Risk factors for coronary artery disease in patients with systemic lupus erythematosus, Am J Med 93 (1992) 513—519.

[84] M. Hearth-Holmes, B.A. Baethge, L. Broadwell, R.E. Wolf, Dietary treatment of hyperlipidemia in patients with systemic lupus erythematosus, J Rheumatol 22 (1995) 450—454.

[85] P. Rahman, S. Aguero, D.D. Gladman, D. Hallett, M.B. Urowitz, Vascular events in hypertensive patients with systemic lupus erythematosus, Lupus 9 (2000) 672—675.

[86] R. Bessant, A. Hignorani, L. Patel, A. MacGregor, D.A. Isenberg, A. Rahman, Risk of coronary heart disease and stroke in a large British cohort of patients with systemic lupus erythematosus, Rheumatology 43 (2004) 924—929.

[87] S.M. Toloza, A.G. Uribe, G. McGwin, G.S. Alarcon, B.J. Fessler, H.M. Bastian, et al., Systemic lupus erythematosus in a multiethnic U.S. cohort (LUMINA). Baseline predictors of vascular events, Arthritis Rheum 50 (2004) 3947—3957.

[88] G.J. Pons-Estel, L.A. Gonzalez, J. Zhang, P.I. Burgos, J.D. Reveille, L.M. Vila, et al., Predictors of cardiovascular damage in patients with systemic lupus erythematosus: Data from LUMINA (LXVIII), a multiethnic U.S. cohort, Rheumatology 48 (2009) 817—822.

[89] I.N. Bruce, R.J. Burns, D.D. Gladman, M.B. Urowitz, Single photon emission computed tomography dual isotope myocardial perfusion imaging in women with systemic lupus erythematosus. I. Prevalence and distribution of abnormalities, J Rheumatol 27 (2000) 2372–2377.

[90] E.M. Sella, E.I. Sato, W.A. Leite, J.A. Oliveira Filho, A. Barbieri, Myocardial perfusion scintigraphy and coronary disease risk factors in systemic lupus erythematosus, Ann Rheum Dis 62 (2003) 1066–1070.

[91] Y. Haider, W. Roberts, Coronary artery disease in systemic lupus erythematosus: Quantification of degrees of narrowing in 22 necropsy patients (21 women) aged 16–37 years, Am J Med 70 (1981) 775–781.

[92] M. Abu-Shakra, M.B. Urowitz, D.D. Gladman, J. Gough, Mortality studies in systemic lupus erythematosus. Results from a single center. I. Causes of death, J Rheumatol 22 (1995) 1259–1264.

[93] S. Manzi, F. Selzer, K. Sutton-Tyrrell, S. Fitzgerald, J. Rairie, R. Tracy, et al., Prevalence and risk factors of carotid plaque in women with systemic lupus erythematosus, Arthritis Rheum 42 (1999) 51–60.

[94] Y. Asanuma, A. Oeser, A.K. Shintani, E. Turner, N. Olsen, S. Fazio, et al., Premature coronary artery atherosclerosis in systemic lupus erythematosus, N Engl J Med 349 (2003) 2407–2415.

[95] A. Doria, Y. Shoenfeld, R. Wu, P.F. Gambari, M. Puato, A. Ghirardello, et al., Risk factors for subclinical atherosclerosis in a prospective cohort of patients with systemic lupus erythematosus, Ann Rheum Dis 62 (2003) 1071–1077.

[96] K. Manger, M. Kusus, C. Forster, D. Ropers, W.G. Daniel, J.R. Kalden, et al., Factors associated with coronary artery calcification in young female patients with SLE, Ann Rheum Dis 62 (2003) 846–850.

[97] M. Roman, B. Shanker, A. Davis, M. Lockshin, L. Sammaritano, R. Simantov, et al., Prevalence and correlates of accelerated atherosclerosis in systemic lupus erythematosus, N Engl J Med 349 (2003) 2399–2406.

[98] F. Selzer, K. Sutton-Tyrrell, S.G. Fitzgerald, J.E. Pratt, R.P. Tracy, L.H. Kuller, et al., Comparison of risk factors for vascular disease in the carotid artery and aorta in women with systemic lupus erythematosus, Arthritis Rheum 50 (2004) 151–159.

[99] T. Wolak, E. Todosoui, G. Szendro, A. Bolotin, B.S. Jonathan, D. Flusser, et al., Duplex study of the carotid and femoral arteries of patients with systemic lupus erythematosus: A controlled study, J Rheumatol 31 (2004) 909–914.

[100] S. Jimenez, M. Garcia-Criado, D. Tassies, J.C. Reverter, R. Cervera, M.R. Gilabert, et al., Preclinical vascular disease in systemic lupus erythematosus and primary antiphospholipid syndrome, Rheumatology 44 (2005) 756–761.

[101] C.P. Chung, A. Oeser, I. Avalos, P. Raggi, C.M. Stein, Cardiovascular risk scores and the presence of subclinical coronary artery atherosclerosis in women with systemic lupus erythematosus, Lupus 15 (2006) 562–569.

[102] K. Maksimowicz-McKinnon, L.S. Magder, M. Petri, Predictors of carotid atherosclerosis in systemic lupus erythematosus, J Rheumatol 33 (2006) 2458–2463.

[103] J.M. Von Feldt, L.V. Scalzi, A.J. Cucchiara, S. Morthala, C. Kealey, S.D. Flagg, et al., Homocysteine levels and disease duration independently correlate with coronary artery calcification in patients with systemic lupus erythematosus, Arthritis Rheum 54 (2006) 2220–2227.

[104] A.N. Kiani, L. Magder, M. Petri, Coronary calcium in systemic lupus erythematosus is associated with traditional cardiovascular risk factors, but not with disease activity, J Rheumatol 35 (2008) 300–306.

[105] A.H. Kao, M.C. Wasko, S. Krishnaswami, J. Wagner, D. Edmundowicz, P. Shaw, et al., C-reactive protein and coronary artery calcium in asymptomatic women with systemic lupus erythematosus or rheumatoid arthritis, Am J Cardiol 102 (2008) 755–760.

[106] I.N. Bruce, D.D. Gladman, M.B. Urowitz, Premature atherosclerosis in systemic lupus erythematosus, Rheum Dis Clin North Am 26 (2000) 257–278.

[107] M. Ward, Premature morbidity from cardiovascular and cerebrovascular diseases in women with systemic lupus erythematosus, Arthritis Rheum 42 (1999) 338–346.

[108] S. Manzi, E.N. Meilahn, J.E. Rairie, C.G. Conte, T.A. Medsger, L. Jansen-McWilliams, et al., Age-specific incidence rates of myocardial infarction and angina in women with systemic lupus erythematosus: Comparison with the Framingham Study, Am J Epidemiol 145 (1997) 408–415.

[109] I. Bruce, M. Urowitz, D. Gladman, D. Ibanez, G. Steiner, Risk factors for coronary heart disease in women with SLE: The Toronto Risk Factor Study, Arthritis Rheum 48 (2003) 3159–3167.

[110] T.S. Han, N. Sattar, K. Williams, C. Gonzalez-Villalpando, M.E. Lean, S.M. Haffner, Prospective study of C-reactive protein in relation to the development of diabetes and the metabolic syndrome in the Mexico City diabetes study, Diabetes Care 25 (2002) 2016–2021.

[111] Y. Saijo, N. Kiyota, Y. Kawasaki, Y. Miyazaki, J. Kashimura, M. Fukuda, et al., Relationship between C-reactive protein and visceral adipose tissue in healthy Japanese subjects, Diabetes Obesity Metab 6 (2004) 249–258.

[112] K.C. Sung, J.Y. Suh, B.S. Kim, J.H. Kang, H. Kim, M.H. Lee, et al., High sensitivity C-reactive protein as an independent risk factor for essential hypertension, Am J Hypertens 16 (2003) 429–433.

[113] L. Wei, T. MacDonald, B. Walker, Taking glucocorticoids by prescription is associated with subsequent cardiovascular disease, Ann Intern Med 141 (2004) 764–770.

[114] A. Zonana-Nacach, S. Barr, L. Magder, M. Petri, Damage in systemic lupus erythematosus and its association with corticosteroids, Arthritis Rheum 43 (2000) 1801–1808.

[115] D. Qi, T. Pulinilkunnil, D. An, S. Ghosh, A. Abrahani, J.A. Pospisilik, et al., Single-dose dexamethasone induces whole-body insulin resistance and alters both cardiac fatty acid and carbohydrate metabolism, Diabetes 53 (2004) 1790–1797.

[116] K. Costenbader, E. Wright, L. Matthew, E. Karlson, Cardiac risk factor awareness and management in patients with systemic lupus erythematosus, Arthritis Rheum 51 (2004) 983–988.

[117] J. Esdaile, M. Abrahamowicz, T. Grodzicky, Y. Li, C. Panaritis, R. du Berger, et al., Traditional Framingham risk factors fail to fully account for accelerated atherosclerosis in systemic lupus erythematosus, Arthritis Rheum 44 (2001) 2331–2337.

[118] P. Ridker, N. Cook, Clinical usefulness of very high and very low levels of C-reactive protein across the full range of Framingham risk scores, Circulation 109 (2004) 1955–1959.

[119] P.M. Ridker, M. Cushman, M.J. Stampfer, R.P. Tracy, C.H. Hennekens, Inflammation, aspirin, and the risk of cardiovascular disease in apparently healthy men, N Engl J Med 336 (1997) 973–979.

[120] R. Zhang, M. Brennan, X. Fu, R.J. Aviles, G.L. Pearce, M.S. Penn, et al., Association between myeloperoxidase levels and risk of coronary artery disease, JAMA 286 (2001) 2136–2142.

[121] M.L. Brennan, M.S. Penn, F. Van Lente., V. Nambi, M.H. Shishehbor, R.J. Aviles, et al., Prognostic value of myeloperoxidase in patients with chest pain, N Engl J Med 349 (2003) 1595–1604.

[122] J.A. Vita, M.L. Brennan, N. Gokce, S.A. Mann, M. Goormastic, M.H. Shishehbor, et al., Serum myeloperoxidase levels independently predict endothelial dysfunction in humans, Circulation 110 (2004) 1134–1139.

[123] B. Lamarche, A.C. St. Pierre, I.L. Ruel, B. Cantin, G.R. Dagenais, J.P. Despres, A prospective, population based study of low-density lipoprotein particle size as a risk factor for ischemic heart disease in men, Can J Cardiol 17 (2001) 859–865.

[124] W. Khovidhunkit, M. Kim, R. Memon, J.K. Shigenaga, A.H. Moser, K.R. Feingold, et al., Effects of infection and inflammation on lipid and lipoprotein metabolism: Mechanisms and consequences to the host, J Lipid Res 45 (2004) 1169–1196.

[125] C. Rollino, R. Boero, F. Elia, B. Montaruli, C. Massara, G. Beltrame, et al., Antiphospholipid antibodies and hypertension, Lupus 13 (2004) 769–772.

[126] T. Godfrey, M. Khamashta, G. Hughes, Antiphospholipid syndrome and renal artery stenosis, Q J Med 93 (2000) 127–129.

[127] N. Curzen, D. Patel, D. Clarke, C. Wright, D. Mulcahy, A. Sullivan, et al., Women with chest pain: Is exercise testing worthwhile? Heart 76 (1996) 156–160.

[128] A.P. Morise, G.A. Diamond, Comparison of the sensitivity and specificity of exercise electrocardiography in biased and unbiased populations of men and women, Am Heart J 130 (1995) 741–747.

[129] E.J. Henriksen, S. Jacob, Angiotensin converting enzyme inhibitors and modulation of skeletal muscle insulin resistance, Diabetes Obesity Metab 5 (2003) 214–222.

[130] A. Ward Scribner, J. Loscalzo, C. Napoli, The effect of angiotensin-converting enzyme inhibition on endothelial function and oxidant stress, Eur J Pharmacol 482 (2003) 95–99.

[131] K. Sebekova, K. Gazdikova, D. Syrova, P. Blazicek, R. Schinzel, A. Heidland, et al., Effects of ramipril in nondiabetic nephropathy: Improved parameters of oxidative stress and potential modulation of advanced glycation end products, J Hum Hypertens 17 (2003) 265–270.

[132] T. Takami, M. Shigemasa, Efficacy of various antihypertensive agents as evaluated by indices of vascular stiffness in elderly hypertensive patients, Hypertens Res 26 (2003) 609–614.

[133] S.M. Jafri, The effects of beta-blockers on morbidity and mortality in heart failure, Heart Fail Rev 9 (2004) 115–121.

[134] M.C. Wasko, H.B. Hubert, V.B. Lingala, J.R. Elliott, M.E. Luggen, J.F. Fries, et al., Hydroxychloroquine and risk of diabetes mellitus in patients with rheumatoid arthritis, JAMA 298 (2007) 187–193.

[135] M. Petri, Hydroxychloroquine use in the Baltimore Lupus Cohort: Effects on lipids, glucose, and thrombosis, Lupus 5 (Suppl. 1) (1996) S16–S22.

[136] K. Costenbader, E. Wright, L. Matthew, E. Karlson, Cardiac risk factor awareness and management in patients with systemic lupus erythematosus, Arthritis Rheum 51 (2004) 983–988.

[137] K. Maksimowicz-McKinnon, F. Selzer, S. Manzi, K.E. Kip, S.R. Mulukutla, O.C. Marroquin, et al., Poor 1-year outcomes after percutaneous coronary intervention in systemic lupus erythematosus. Report from the National Heart, Lung, and Blood Institute DYNAMIC registry, Circ Cardiovasc Intervent 1 (2008) 201–208.

[138] C.H. Lin, M.L. Lee, R.B. Hsu, Cardiac surgery in patients with systemic lupus erythematosus, Inter Cardiovas Thorac Surg 4 (2005) 618–621.

44

Vasculitis

Barri J. Fessler, Gary S. Hoffman

Vasculitis in patients with systemic lupus erythematosus (SLE) develops as a consequence of complex interactions between vascular endothelium, inflammatory cells, cytokines, autoantibodies, and immune complexes. The trigger that initiates the cascade of interactions between cells and factors that determine the location of vascular inflammation are unknown. The clinical manifestations of vasculitis may take several forms: isolated cutaneous lesions, single visceral organ involvement, or multiorgan system disease [1]. The prevalence of vasculitis in large series of SLE patients ranges from 11 to 36% [2–4], with cutaneous manifestations from small vessel involvement comprising the majority (80–90%) of cases. Medium vessel involvement is estimated to occur in less than 15%, and large vessel vasculitis is rare [4, 5]. Determination of the true incidence or prevalence of vasculitis in SLE is complicated by the fact that the diagnosis of vasculitis is often a clinical diagnosis, lacking histologic or arteriographic confirmation. In addition, noninflammatory vasculopathies are confounding variables because they may mimic or rarely co-exist with vasculitis [5–8] (see Table 44.1).

A variety of noninflammatory vasculopathies may be observed in SLE [9]. Endothelial cell injury and/or activation resulting in intimal proliferation may lead to thickening of the walls of small blood vessels. Such changes are often widely distributed and may be associated with infarcts that may or may not contribute to organ dysfunction [10, 11]. Activation of neutrophils and upregulation of endothelial cell adhesion molecules following exposure to complement split products and other chemoattractants may lead to leukoaggregation within the vasculature, causing transient or permanent end organ damage [12, 13]. Occlusion of vessels by platelet-fibrin thrombi associated with clinical features of thrombotic thrombocytopenic purpura has also been observed in a small number of SLE patients [14]. Lastly, the most common and clinically significant vasculopathy in SLE is associated with the presence of antiphospholipid antibodies. These different forms of vasculopathies are generally not inflammatory and are not regarded as true vasculitis. However, vasculitis may rarely accompany them [7, 8, 15].

THE SPECTRUM OF VASCULITIS IN SYSTEMIC LUPUS ERYTHEMATOSUS

Cutaneous Vasculitis

Cutaneous vasculitis is the most common form of vasculitis in SLE, occurring in 10–50% of patients [2–5, 16–21]. Its presence, in some instances, has correlated significantly with severity of disease and organ involvement [16, 22–24]. Cutaneous vasculitis typically occurs as an early manifestation of SLE, developing within 5 years of disease onset [25]. Non-Caucasian patients have a higher incidence of cutaneous vasculitis than do Caucasian patients [3, 20, 25]. Clinical expression of vasculitis includes petechiae, palpable purpura (Figure 44.1), urticaria, nodules, bullae, nail fold infarcts (Figure 44.2), and splinter hemorrhages, ulcerations, gangrene, and livedo reticularis [26–29]. In the early stages of evolution, cutaneous vasculitis may be flat and only later become palpable. Severe vasculitic involvement may lead to bullae formation, ulceration, or skin necrosis (Figure 44.3). Purpuric lesions, palpable or nonpalpable, are the most frequently observed manifestation of inflammatory vascular disease in SLE, occurring more often in patients with anti-SSA antibodies [30] and Sjögren's syndrome/SLE overlap [31, 32]. In addition, patients with cryoglobulins are more than twice as likely to develop cutaneous vasculitis than those without cryoglobulins [33]. Severe small vessel vasculitis may cause small microinfarcts of the tips of the digits, which heal with "ice pick" scars (Figure 44.4). Infrequently, vasculitis may manifest as urticaria (5–10%) [34–36] and requires differentiation from

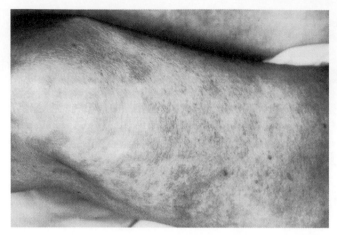

FIGURE 44.1   Palpable purpura of the lower extremity of a patient with active lupus who also had arthritis and pericarditis.

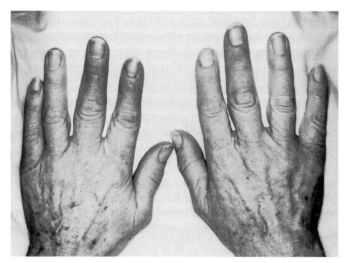

FIGURE 44.2   Nail fold infarcts. There are pinpoint lesions in the nail beds, most prominent in the second, third, and fifth fingers of the left hand. There is also Raynaud's vasospasm involving the second and third fingers of the same hand.

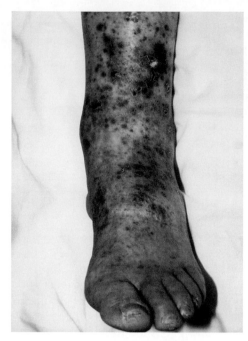

FIGURE 44.3   Severe palpable purpura with ulcerations affecting the lower extremity of a patient with active lupus manifested by glomerulonephritis, arthritis, serositis, and central nervous system disease.

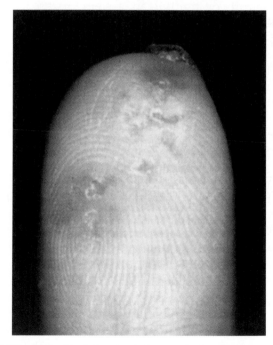

FIGURE 44.4   Healed microinfarcts of the fingertips. Often referred to as 'ice pick' scars, they appear as ischemic ulcers surrounded by areas of hyperemia.

allergic urticaria. The lesions may be hyperpathic to light touch, persist for more than 24 h, and may display stippled petechiae that heal with hyperpigmentation.

Vasospasm may be a primary event or may occur secondary to vasculitis, embolic phenomena, vasoactive drugs, vasculopathy, or toxins. Arteritis of medium-sized vessels of the skin may cause vasospasm of the ascending dermal arterioles with pooling of blood in the superficial horizontal venous plexus and give rise to mottling or a netlike rash (livedo reticularis). Alternatively, livedo reticularis may be associated with the vasculopathy of the antiphospholipid antibody syndrome. This finding may be generalized and/or accentuated about the outer arms, thighs, and over large joints. Livedo reticularis occurs in approximately 20% of SLE patients [24]. Punch biopsies of the skin are usually unrewarding because the deep dermal vessels are not accessible by this technique [28]. Anticardiolipin antibodies are found in 80% of patients with SLE and livedo reticularis [24]. Severe vasculitis of medium-sized vessels may also cause digital gangrene [17, 37], but fortunately this is rare (1—5%) [2, 17]. Cutaneous ulceration and gangrene may also develop as part of the vasculopathy associated with the antiphospholipid antibody syndrome [38—40].

Histologically, most lupus vasculitic skin lesions demonstrate leukocytoclastic vasculitis with preferential involvement of arterioles, capillaries, and venules (Figure 44.5). This is the most common form of vasculitis in SLE. In addition, small arterioles and postcapillary venules of other organs may be simultaneously affected. Infiltration and destruction of the vessel walls by polymorphonuclear cells is accompanied by polymorphonuclear fragmentation. The nuclear "debris" denotes leukocytoclasis [41]. Some cases primarily demonstrate vascular and/or perivascular infiltrate with mononuclear cells. Whether this merely represents a later stage of leukocytoclastic vasculitis is uncertain. Direct immunofluorescence of the skin lesion typically demonstrates granular IgG, complement, and fibrinogen in and/or around blood vessels [42].

Central Nervous Systems and Peripheral Nervous System Vasculitis

Neuropsychiatric manifestations may be seen in up to 75% of SLE patients [2, 17, 43—45]. Central nervous system (CNS) abnormalities are more common than peripheral neuropathy (3—18%). CNS vasculitis may rarely be the sole presenting symptom of lupus [46, 47]. In addition, CNS vasculitis may occur in the absence of other features of SLE disease activity [48—50]. Several studies have noted a correlation of cutaneous vasculitis with CNS disease [22—24, 51]. Feinglass *et al.* [23]

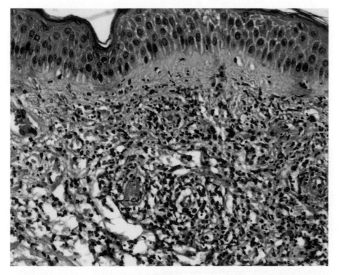

FIGURE 44.5 Skin biopsy demonstrating leukocytoclastic vasculitis involving postcapillary venules with fibrinoid necrosis within vessels and a perivascular mixed inflammatory infiltrate with nuclear dust. *Source: Photo courtesy of Kathleen J. Smith, M.D.*

observed that of a total of 140 cases, 52 had neuropsychiatric manifestations, among whom a clinical diagnosis of cutaneous vasculitis was made in 46%. In contrast, of the 88 without neuropsychiatric disease, cutaneous vasculitis was clinically recognized in only 17%.

Although CNS symptoms are usually attributed to vasculitis, demonstration of CNS vasculitis on postmortem examination of SLE patients is infrequent [44, 45, 49, 50, 52, 53]. Rather, an occlusive vasculopathy characterized by intimal proliferation of arterioles and capillaries and vascular hyalinization, associated with infarcts and hemorrhage, is observed. Johnson and Richardson [44] found vasculitic lesions in only 3 of 24 (12%) autopsy cases, and Ellis and Verity [45] recorded arteritis in 7% of 57 cases.

CNS vascular disease usually affects small vessels (diameter ≤ 2 mm). The result is multiple microinfarcts that may vary in age from region to region. This feature allows distinction from hypertensive encephalopathy and the CNS lesion of thrombotic thrombocytopenic purpura. The former has the additional feature of hypertrophy and hyalinization of the media of larger vessels, and thrombocytopenic purpura is distinguished by its microangiopathic peripheral blood smear, as well as the paucity of CNS parenchymal disease. When parenchymal vascular lesions are present in thrombotic thrombocytopenic purpura, they are usually of the same age. The CNS lesions of SLE are most common in the cerebral cortex and brain stem and correlate with generalized seizures and cranial nerve deficits, respectively. Large vessel vasculitis of the CNS is distinctly rare, with less than a dozen cases described in the literature [54]. Subarachnoid hemorrhage is an uncommon complication of SLE and may result from a ruptured intracerebral aneurysm [55—59]. Pathology specimens have revealed medium-sized vessel vasculitis.

If vasculitis affects the retina, it is usually bilateral. One eye is usually more severely affected than the other. Lesions tend to be focal and characterized by intraretinal hemorrhage and local retinal infarction with variable adjacent retinal exudation (Figures 44.6 and 44.7). Fluorescein angiography may be useful in identifying the extent of disease and response to treatment (Figure 44.8). Papillitis may result from vascular disease of nutrient vessels to the optic nerve [60]. Only a few postmortem ocular specimens have been studied. Findings have included vasculitis in the choroidal circulation [61, 62]. Retinopathy may be a marker of disease activity and CNS involvement [63, 64].

Spinal cord vascular lesions may manifest as transverse myelopathy [65]. In one series, 2 of 12 cases had histopathologic evidence of vasculitis. Perivascular round cell infiltrates, collagen proliferation, thrombotic occlusion of small arteries and arterioles, and

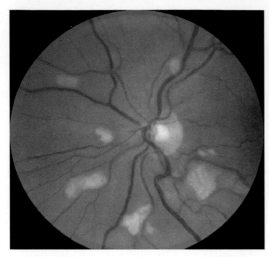

FIGURE 44.6 Fundus photograph reveals cotton wool spots or nerve fiber layer infarcts in a patient with SLE. *Source: Photo courtesy of Careen Lowder, M.D.*

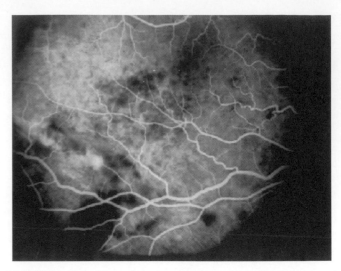

FIGURE 44.8 Fluorescein angiogram reveals focal areas of choroidal nonperfusion as seen in patients with choroidal ischemia. *Source: Photo courtesy of Careen Lowder, M.D.*

microscopic hemorrhage in the cord parenchyma were noted in all cases. Whether more extensive sampling of tissue would have demonstrated frank vasculitis in all cases is a matter of speculation.

Peripheral neuropathy in SLE may be due to vasculitis of the vasa nervorum that may result in multifocal sensorimotor neuropathy with prominent sensory symptoms [66, 67]. Mononeuritis multiplex has also been reported [68, 69]. Nerve biopsy may show evidence of necrotizing vasculitis with perivascular inflammatory infiltrates and intimal thickening [70] and tubuloreticular inclusions [71].

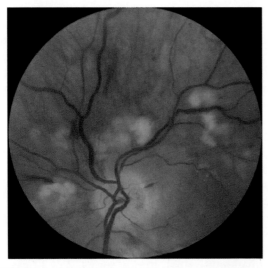

FIGURE 44.7 Fundus photograph of the right eye reveals flame-shaped hemorrhages and cotton wool spots, the two most frequently reported ocular findings in SLE. *Source: Photo courtesy of Careen Lowder, M.D.*

Pulmonary Vasculitis

The contribution of vasculitis to pulmonary pathology in SLE has been reported to range from 0 to 40% [72–76]. Fayemi [73] reported 20 autopsy cases, of which 6 had pulmonary symptoms. Four had vasculitis involving the muscular and elastic arteries, which contrasts with more commonly recognized predilection for small vessels noted in other organs. Gross *et al.* [74] reported abnormalities of the pulmonary vasculature in 8 of 43 cases, including involvement of small arteries and arterioles. In contrast, Haupt *et al.* [77] identified pulmonary vasculitis in only 2 of 120 postmortem specimens. Symptoms may include cough, dyspnea, hemoptysis, and chest pain. Pulmonary capillaritis can occur with or without diffuse alveolar hemorrhage [78]. Radiographically, lesions may appear as areas of atelectasis, infiltrates, or, rarely, nodules or cavities caused by ischemic necrosis. Despite the extensive vascular changes that can occur, cor pulmonale or right ventricular hypertrophy have been uncommon. This may reflect rapid evolution of vascular pathology.

Gastrointestinal Vasculitis

Vasculitis may affect the vessels within the mesentery (Figure 44.9) [79], pancreas [80–83], peritoneum (see Figure 44.10), and, rarely, liver [84–87] and gall bladder [88, 89], and it is associated with a high mortality rate (Table 44.2) [90].

Mesenteric vasculitis is estimated to occur in approximately 1–6% of SLE patients [3, 4, 19, 79, 90–93] and may affect the vessels supplying the stomach, small intestine, or large intestine [15, 94–96]. Clinically, gastrointestinal symptoms may range from nausea to bloating,

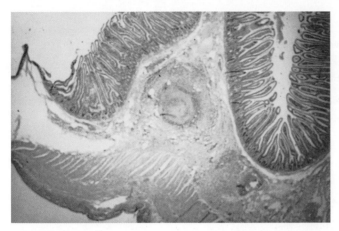

FIGURE 44.9 Gastrointestinal vasculitis. Transmural intestinal segment demonstrating a medium-sized vessel with thrombosis and whose wall is infiltrated with inflammatory cells.

vomiting, diarrhea, postprandial pain and/or fullness, frank hemorrhage, and acute abdominal pain. The less acute gastrointestinal symptoms might erroneously be attributed to side effects of commonly prescribed medications, such as corticosteroids, nonsteroidal anti-inflammatory drugs, hydroxychloroquine, and azathioprine. Physical exam findings of guarding and rebound may be absent or appear only late in the disease course in the patient taking corticosteroids or other immunosuppressive agents. Unexplained acidosis, a falling hematocrit, and hypotension are worrisome clues to a possible intra-abdominal crisis [92]. Radiographs may be normal or reveal thumbprinting, ileus, or pseudo-obstruction. Computed tomography scan may reveal focal or diffuse bowel wall thickening, abnormal bowel wall enhancement (a double halo or target sign), and prominence of mesenteric vessels (comb sign)

[97–99]. Vasculitis involving the intestinal tract is typically a small vessel vasculitis affecting arterioles and venules [100] that are beyond the resolution of arteriography. Rarely, polyarteritis nodosa-like aneurysmal lesions in medium-sized vessels may occur. When present, aneurysms may thrombose or rupture, causing hemorrhage and death [100, 101]. Intestinal vasculitis may lead to mucosal ischemia, resulting in ulceration, hemorrhage, infarction, or perforation [102, 103, 113]. Exploratory laparoscopy or laparotomy should be considered early because with delay in diagnosis of ruptured or infarcted bowel there is a very high mortality rate. Medina *et al.* [104] found that among 44 SLE patients with acute abdomen requiring surgery, none of the 33 who died were operated on within 48 h of presentation, whereas 10 of 11 who died were operated on more than 48 h after presentation. In most but not all series [105], patients with mesenteric vasculitis have evidence of lupus activity affecting other organs. Pneumatosis cystoides intestinalis, although more frequently associated with other connective tissue diseases such as scleroderma or rheumatoid arthritis, may be a benign finding [106] or may be associated with intestinal vasculitis in SLE patients [79, 107, 108]. Inflammatory vascular disease affecting venules supplying the intestines has also been implicated in the etiology of protein-losing enteropathy in SLE [109, 110].

Vasculitis may also affect other organs within the abdominal cavity. Histologic evidence of necrotizing arteritis of the liver has been reported in 18–20% of postmortem cases [87, 111]. Spontaneous hepatic rupture caused by small and medium-sized vessel vasculitis is a rare complication [86, 112]. Acute acalculous cholecystitis due to vasculitis may occur in isolation or may be associated with widespread disease activity [88, 89]. Vasculitis has been implicated as a cause of pancreatitis in a subset of SLE patients [81, 82].

Vasculitis of the Heart

Coronary artery vasculitis is exceedingly rare in SLE and has been documented primarily in case reports [114–122]. Vasculitis affecting the coronary arteries may occur in the absence of extracardiac lupus activity, thereby making the diagnosis especially challenging. Diagnosis is usually made by serial coronary angiographic studies. However, the sensitivity and specificity of angiography is unknown. Angiographic findings suggestive of vasculitis include arterial aneurysms without concomitant distal stenoses, smooth focal tapering stenoses, and/or rapidly developing arterial occlusions on repeated studies. Both the main stem and its branches including the sinus node artery have been involved [121]. Vessels that are no longer the site of active vasculitis may have persistent aneurysms or

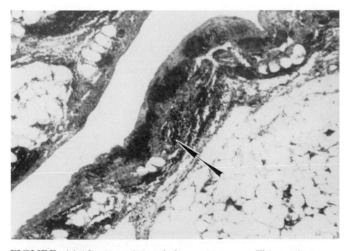

FIGURE 44.10 Vasculitis of the peritoneum. This patient presented with massive ascites. At laparotomy, no abnormal viscus was found, but a peritoneal biopsy demonstrated extensive small-vessel vasculitis.

TABLE 44.1   Incidence of Vasculitis by Organ System in SLE

Organ	Estimated frequency based on clinical diagnosis	Reported frequency based on histologic proof of vasculitis	References
Skin	10–50%	11-36%	[2–5, 16, 18, 20, 21, 25, 172]
Lung	1–2%	18–20%[a]	[73, 74, 173]
		8/44 autopsy cases	
		4/20 autopsy cases	
Gastrointestinal tract	1–6%	NA	[3, 4, 19, 80, 90–92, 111, 173–178]
		6/25 autopsy cases	
Central nervous system	Uncommon	7–12%[a]	[44, 45, 54]
		3/24 autopsy cases	
		4/57 autopsy cases	
Heart	Rare	3–37%[a]	[116, 119]
		1/30 autopsy cases	
		6/16 autopsy cases	
Genitourinary tract	Case reports	NA	[127, 132, 135–138]
Large vessels	Case reports, small series	NA	[141, 144, 147, 148]

[a]May significantly overestimate frequency due to sampling bias; that is, autopsy studies represent patients with fulminant/treatment-resistant vasculitis.
NA, not available.

TABLE 44.2   Gastrointestinal Involvement and Morbidity

Series	No. of patients	Gastrointestinal involvement (%)	Histologic proof (%)[a]	Death (%)
Matolo and Albo [176]	51	27.5	Unclear	43
Estes and Christian [2]	150	16	1.3	8
Harvey et al. [43]	106	14	58	NR
Zizic et al. [90]	140[b]	11	82	53

[a]Of all patients in whom a clinical diagnosis of gastrointestinal vasculitis was made, the percentage in whom actual histologic proof of vasculitis was documented.
[b]Patient selection emphasized acute surgical presentations.
NR, not reported.

stenoses that may thrombose and cause myocardial infarction [123]. In some cases, vasculitic changes have been accompanied by fibrosis of the conduction system [118]. Rare examples of cardiac valve dysfunction [121] and myocardial dysfunction [124] caused by small vessel vasculitis have also been documented.

Vasculitis of the Genitourinary System

True vasculitis of the kidney is the least frequent manifestation of vascular disease within the kidney, occurring in less than 5% of SLE nephritis patients [125, 126] (Figure 44.11). Vascular immune deposits, noninflammatory necrotizing vasculopathy (Figure 44.12), and thrombotic microangiopathy are other forms of vascular disease implicated in the pathogenesis of SLE-associated renal disease [10, 127, 128]. In renal vasculitis, the vessels

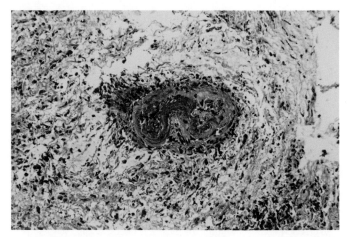

FIGURE 44.11   Acute vasculitis: necrotizing vasculitis with transmural infiltration by neutrophils and lymphocytes. Source: Photo courtesy of William J. Cook, M.D.

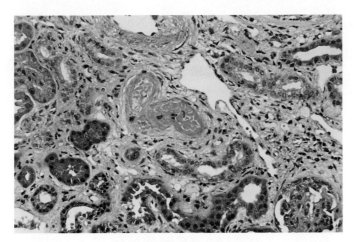

FIGURE 44.12 Lupus vasculopathy: The arteriole on this renal biopsy is narrowed by intimal deposits of eosinophilic material that likely represents a mixture of fibrin and immune deposits. Although the endothelium is necrotic, there is no inflammation. *Source: Photo courtesy of William J. Cook, M.D.*

affected are usually small and medium-sized arteries, most commonly intralobular arteries. It is most frequently associated with diffuse proliferative glomerulonephritis. The presence of vasculitis on renal biopsy is believed to be an independent indicator of adverse renal prognosis [129–131].

Reports of lower genitourinary tract involvement include examples of hydronephrosis, hydroureter, and acute renal failure [132–134]. Histologic examination of the bladder has demonstrated interstitial cystitis, hemorrhage, congestion, and vasculitis. In rare instances, ureteral vessels have been affected and have displayed necrotizing vasculitis with extensive fibrinoid changes. Such lower genitourinary tract disease should be considered in an SLE patient with deteriorating renal function in the absence of red blood cell casts and in whom calculi and extrinsic causes of compression have been excluded. In the cases reported, obstruction was relieved by corticosteroid therapy [132, 133]. In addition, rare cases of necrotizing vasculitis of the penis [135], ovaries [135–137], and uterus [138] have been reported in lupus patients.

Large Vessel Disease

Large vessel vasculitis of the aorta and its branches is exceedingly rare in SLE. Cases of Takayasu's arteritis associated with SLE have been described [139–143]. However, the co-existence of these two diseases may merely be an issue of semantics, which is reconciled when one recognizes that in fact aortitis or disease of aortic branch vessels may rarely be a part of SLE [144–147]. There is one case report documented describing SLE developing in a woman with biopsy-proven giant cell arteritis [148].

Large vessel involvement has been reported to affect the abdominal aorta, subclavian, carotid, coronary, and lower limb arteries [149–151]. Large vessel vasculitis of the extremities resulting in gangrene is very uncommon in SLE patients. In a literature review of all reported English-language cases of limb-threatening vascular occlusion, 12 cases were identified, of which 4 were due to vasculitis [152]. Pathologic specimens demonstrated full-thickness necrotizing arteritis with polymorphonuclear cell infiltration [151, 152]. Immune complexes were not identified. Sympathectomy and streptokinase have not proven to be useful in these cases. Corticosteroids have allegedly provided relief of intermittent claudication and limited the damage caused by large vessel vasculitis [149, 150].

Vasculitis of the Placenta

The contribution of placental vasculitis to fetal mortality has been subject to preliminary evaluation that has led to conflicting observations (Table 44.3). Abramowsky *et al.* [153] studied the placentas of 10 women with SLE. Five of 10 specimens demonstrated vascular fibrinoid necrosis and mononuclear and/or

TABLE 44.3 Vasculitis of the Placenta

Series	No. of patients	No. of placentas	Fetal outcome (live births)	Types of vascular lesions studied			
				Vasculitis	Thrombosis	Intimal proliferation	Tortuosity
Abramowsky *et al.* [153]	10	10	4	5	2	Reported[a]	Reported
Hanly *et al.* [154]	11	11	7	0	2	NR	NR
Benirschke and Driscoll [179]	5	5	Unknown	1	Unknown	Unknown	Unknown
Total	26	26	11	6			

[a]Not quantitated.
NR, not reported.

polymorphonuclear inflammatory infiltrates. Some vessels were tortuous and others aneurysmal. These changes were seen in the 5 multiparous patients who suffered fetal loss after the first trimester, but they were noticeably absent in the primagravids. In a prospective study of 11 cases, Hanly *et al.* [154] could not document any distinct form of placental pathology. They found no evidence of vasculitis. Demonstration of immune complexes in diseased as well as in normal placentas raises questions regarding what role immune complexes play in placental insufficiency and abortion [155]. The correlation of fetal outcome with placental thrombosis, phospholipid antibodies, vasculopathy, or vasculitis is not clear.

Autoantibodies

The role, if any, that autoantibodies play in the pathogenesis of vasculitis in lupus is unclear. Some have speculated that antiendothelial antibodies may play a role in triggering the development of vasculitis. Antiendothelial antibodies obtained from patients with lupus have been studied *in vitro* and found to upregulate endothelial cell expression of VCAM-1, ICAM-1, and E-selectin, as well as enhance leukocyte adhesion to endothelial cells [156, 157]. In addition, experimental evidence suggests the existence of cross-reactive epitopes in endothelial cells—lysophosphatidylcholine (LPC) and oxidized LDL. LPC is formed during oxidation of LDL or as a result of the action of phospholipase A_2 on phosphatidylcholine; it is an important antigen in oxidized LDL. An immune reaction to LPC leading to endothelial attack and culminating in vasculitis, thrombosis, or atherosclerosis is suggested by significantly larger amounts of antibodies to LPC in patients with SLE compared with normal controls [158]. Other proposed mechanisms by which antiendothelial antibodies may induce vascular injury include complement fixation [159], recruitment of neutrophils, and antibody-dependent cellular cytotoxicity [160]. Antineutrophil cytoplasmic antibodies (ANCA), a heterogeneous group of autoantibodies directed against a variety of antigens within the neutrophil and monocyte, have been associated with certain systemic vasculitides, such as Wegener's granulomatosis [161]. The role that ANCAs play in the pathogenesis of vasculitis is unproven [162]. ANCAs have been isolated in the sera from SLE patients with varying frequency (0–93%) [163–170]. However, the presence of ANCAs in SLE does not correlate with major organ involvement, disease activity, or the presence of vasculitis [164, 166, 168]. False-positive ANCAs in lupus patients may be due to the binding of myeloperoxidase (one of the substrate antigens in ANCA immunoassays) to DNA contained within the antigen binding site of anti-DNA antibodies [171].

DIFFERENTIAL DIAGNOSIS OF SLE VASCULITIS

Immune responses in SLE are inherently abnormal and further compromised by immunosuppressive therapy. Deterioration in clinical status must take into account the possibilities of infection, drug toxicity, disease progression, thrombosis from antiphospholipid antibodies, and, less commonly, disseminated intravascular coagulation or thrombotic thrombocytopenic purpura. Any of these multisystem processes can mimic vasculitis or, conversely, be complicated by vasculitis.

MANAGEMENT

Because the diagnosis of lupus vasculitis is often uncertain, large numbers of similarly affected patients are not available for prospective studies, and because vasculitis usually does not occur in the absence of a variety of other pathologic events, it is difficult to perform controlled therapeutic trials for lupus vasculitis. For mild to moderate manifestations, oral corticosteroids, methotrexate, azathioprine, and mycophenolate mofetil may be considered. For life-threatening major organ involvement, aggressive therapy with intravenous high-dose corticosteroids, cyclophosphamide, rituximab, intravenous immunoglobulin, and/or plasmapheresis has proven useful in selected patients. Treatment must remain a matter of clinical judgment; one must consider the patient's global illness, of which vasculitis may be only one factor and not necessarily the dominant consideration. The prognosis of vasculitis in SLE patients varies depending on extent and location of vessel involvement.

CONCLUSION

Vasculitis occurs commonly in patients with SLE and can affect any organ system. In the absence of evidence-based data, treatment is tailored according to the severity of the disease manifestations. Because there is heterogeneity in the clinical manifestations of SLE vasculitis, it is reasonable to hypothesize that different mechanisms of vascular injury may be operant in different patients, or even in the same patient in different vascular beds. For example, different factors may be responsible for the development of leukocytoclastic vasculitis in the skin compared to inflammatory aortic aneurysms. Much has yet to be learned about the determinants of vascular attack by the immune system, both from the standpoint of understanding what activates the circulating leukocyte population and from the standpoint of identifying the unique features of specific vascular beds that lead to targeting and attack by the immune system. Advances in the understanding of the pathophysiology of vasculitis will allow new targeted therapies to be developed.

References

[1] A.S. Fauci, B. Haynes, P. Katz, The spectrum of vasculitis: Clinical, pathologic, immunologic and therapeutic considerations, Ann Intern Med 89 (1978) 660—676.

[2] D. Estes, C.L. Christian, The natural history of systemic lupus erythematosus by prospective analysis, Medicine (Baltimore) 50 (1971) 85—95.

[3] C. Vitali, W. Bencivelli, D.A. Isenberg, J.S. Smolen, M.L. Snaith, M. Sciuto, et al., Disease activity in systemic lupus erythematosus: Report of the Consensus Study Group of the European Workshop for Rheumatology Research. I. A descriptive analysis of 704 European lupus patients. European Consensus Study Group for Disease Activity in SLE, Clin Exp Rheumatol 10 (1992) 527—539.

[4] M. Ramos-Casals, N. Nardi, M. Lagrutta, P. Brito-Zeron, A. Bove, G. Delgado, et al., Vasculitis in systemic lupus erythematosus: Prevalence and clinical characteristics in 670 patients, Medicine (Baltimore) 85 (2006) 95—104.

[5] C. Drenkard, A.R. Villa, E. Reyes, M. Abello, D. Alarcon-Segovia, Vasculitis in systemic lupus erythematosus, Lupus 6 (1997) 235—242.

[6] J.T. Lie, S. Kobayashi, Y. Tokano, H. Hashimoto, Systemic and cerebral vasculitis coexisting with disseminated coagulopathy in systemic lupus erythematosus associated with antiphospholipid syndrome, J Rheumatol 22 (1995) 2173—2176.

[7] D.K. Norden, B.E. Ostrov, A.B. Shafritz, J.M. Von Feldt, Vasculitis associated with antiphospholipid syndrome, Sem Arthritis Rheum 24 (1995) 273—281.

[8] P.V. Rocca, L.B. Siegel, T.R. Cupps, The concomitant expression of vasculitis and coagulopathy: Synergy for marked tissue ischemia, J Rheumatol 21 (1994) 556—560.

[9] H.M. Belmont, S.B. Abramson, J.T. Lie, Pathology and pathogenesis of vascular injury in systemic lupus erythematosus. Interactions of inflammatory cells and activated endothelium, Arthritis Rheum 39 (1996) 9—22.

[10] D.B. Bhathena, B.J. Sobel, S.D. Migdal, Noninflammatory renal microangiopathy of systemic lupus erythematosus ('lupus vasculitis'), Am J Nephrol 1 (1981) 144—159.

[11] D. Daly, Central nervous system in acute disseminated lupus erythematosus, J. Nerv. Ment. Dis 102 (1945) 461—464.

[12] S.B. Abramson, W.P. Given, H.S. Edelson, G. Weissmann, Neutrophil aggregation induced by sera from patients with active systemic lupus erythematosus, Arthritis Rheum 26 (1983) 630—636.

[13] L.W. Argenbright, R.W. Barton, Interactions of leukocyte integrins with intercellular adhesion molecule 1 in the production of inflammatory vascular injury *in vivo*. The Shwartzman reaction revisited, J Clin Invest 89 (1992) 259—272.

[14] S.K. Kwok, J.H. Ju, C.S. Cho, H.Y. Kim, S.H. Park, Thrombotic thrombocytopenic purpura in systemic lupus erythematosus: Risk factors and clinical outcome: A single centre study, Lupus 18 (2009) 16—21.

[15] E.T. Koh, M.L. Boey, P.H. Feng, Acute surgical abdomen in systemic lupus erythematosus—An analysis of 10 cases, Ann Acad Med Singapore 21 (1992) 833—837.

[16] C. Cardinali, M. Caproni, E. Bernacchi, L. Amato, P. Fabbri, The spectrum of cutaneous manifestations in lupus erythematosus—The Italian experience, Lupus 9 (2000) 417—423.

[17] E.L. Dubois, D.L. Tuffanelli, Clinical manifestations of systemic lupus erythematosus. Computer analysis of 520 cases, JAMA 190 (1964) 104—111.

[18] M.C. Hochberg, R.E. Boyd, J.M. Ahearn, F.C. Arnett, W.B. Bias, T.T. Provost, et al., Systemic lupus erythematosus: A review of clinico-laboratory features and immunogenetic markers in 150 patients with emphasis on demographic subsets, Medicine (Baltimore) 64 (1985) 285—295.

[19] M. Pistiner, D.J. Wallace, S. Nessim, A.L. Metzger, J.R. Klinenberg, Lupus erythematosus in the 1980s: A survey of 570 patients, Sem Arthritis Rheum 21 (1991) 55—64.

[20] Y. Levy, J. George, M. Hojnic, M. Ehrenfeld, M. Lorber, S. Bombardieri, et al., Comparison of clinical and laboratory parameters for systemic lupus erythematosus activity in Israelis versus Europeans, Isr J Med Sci 32 (1996) 100—104.

[21] J.A. Yell, J. Mbuagbaw, S.M. Burge, Cutaneous manifestations of systemic lupus erythematosus, Br J Dermatol 135 (1996) 355—362.

[22] J.P. Callen, J. Kingman, Cutaneous vasculitis in systemic lupus erythematosus. A poor prognostic indicator, Cutis 32 (1983) 433—436.

[23] E.J. Feinglass, F.C. Arnett, C.A. Dorsch, T.M. Zizic, M.B. Stevens, Neuropsychiatric manifestations of systemic lupus erythematosus: Diagnosis, clinical spectrum, and relationship to other features of the disease, Medicine (Baltimore) 55 (1976) 323—339.

[24] T. Yasue, Livedoid vasculitis and central nervous system involvement in systemic lupus erythematosus, Arch Dermatol 122 (1986) 66—70.

[25] B.J. Fessler, G. McGwin Jr., J.M. Roseman, H.M. Bastian, A. Friedman, J.R. Lisse, et al., Cutaneous vasculitis in systemic lupus erythematosus: Association with Hispanic ethnicity and major organ system involvement, Arthritis Rheum 43 (2000) S253.

[26] M. Tikly, S. Burgin, P. Mohanlal, A. Bellingan, J. George, Autoantibodies in black South Africans with systemic lupus erythematosus: Spectrum and clinical associations, Clin Rheumatol 15 (1996) 261—265.

[27] A.J. Wysenbeek, L. Leibovici, A. Weinberger, D. Guedj, Expression of systemic lupus erythematosus in various ethnic Jewish Israeli groups, Ann Rheum Dis 52 (1993) 268—271.

[28] R. Watson, Cutaneous lesions in systemic lupus erythematosus, Med Clin North Am. 73 (1989) 1091—1111.

[29] G. Sais, A. Vidaller, A. Jucgla, O. Servitje, E. Condom, J. Peyri, Prognostic factors in leukocytoclastic vasculitis: A clinicopathologic study of 160 patients, Arch Dermatol 134 (1998) 309—315.

[30] M.V. Fukuda, S.C. Lo, C.S. de Almeida, S.K. Shinjo, Anti-Ro antibody and cutaneous vasculitis in systemic lupus erythematosus, Clin Rheumatol 28 (2009) 301—304.

[31] T.T. Provost, N. Talal, W. Bias, J.B. Harley, M. Reichlin, E.L. Alexander, Ro(SS-A) positive Sjogren's/lupus erythematosus (SC/LE) overlap patients are associated with the HLA-DR3 and/or DRw6 phenotypes, J Invest Dermatol 91 (1988) 369—371.

[32] E. Simmons-O'Brien, S. Chen, R. Watson, C. Antoni, M. Petri, M. Hochberg, et al., One hundred anti-Ro (SS-A) antibody positive patients: A 10-year follow-up, Medicine (Baltimore) 74 (1995) 109—130.

[33] M. Garcia-Carrasco, M. Ramos-Casals, R. Cervera, O. Trejo, J. Yague, A. Siso, et al., Cryoglobulinemia in systemic lupus erythematosus: Prevalence and clinical characteristics in a series of 122 patients, Sem Arthritis Rheum 30 (2001) 366—373.

[34] R.A. Asherson, D. D'Cruz, C.J. Stephens, P.H. McKee, G.R. Hughes, Urticarial vasculitis in a connective tissue disease clinic: Patterns, presentations, and treatment, Sem Arthritis Rheum 20 (1991) 285—296.

[35] W.R. Gammon, C.E. Wheeler Jr., Urticarial vasculitis: Report of a case and review of the literature, Arch Dermatol 115 (1979) 76–80.

[36] T.T. Provost, J.J. Zone, D. Synkowski, P.J. Maddison, M. Reichlin, Unusual cutaneous manifestations of systemic lupus erythematosus: I. Urticaria-like lesions. Correlation with clinical and serological abnormalities, J Invest Dermatol 75 (1980) 495–499.

[37] E.L. Dubois, J.D. Arterberry, Gangrene as a manifestation of systemic lupus erythematosus, JAMA 181 (1962) 366–374.

[38] A.M. Barbaud, B. Gobert, S. Reichert, J.L. Schmutz, M.C. Bene, M. Weber, et al., Anticardiolipin antibodies and ulcerations of the leg, J Am Acad Dermatol 31 (1994) 670–671.

[39] D.P. Goldberg, V.L. Lewis Jr., W.J. Koenig, Antiphospholipid antibody syndrome: A new cause of nonhealing skin ulcers, Plast Reconstr Surg 95 (1995) 837–841.

[40] B.K. Jindal, M.F. Martin, A. Gayner, Gangrene developing after minor surgery in a patient with undiagnosed systemic lupus erythematosus and lupus anticoagulant, Ann Rheum Dis 42 (1983) 347–349.

[41] P.M. Zeek, C.C. Smith, J.C. Weeter, Studies on periarteritis nodosa: The differentiation between the vascular lesions of periarteritis nodosa and of hypersensitivity, Am J Pathol 24 (1948) 889–917.

[42] D.R. Synkowski, T.T. Provost, Cutaneous vasculitis in the LE patient, Clin Dermatol 3 (1985) 88–95.

[43] A.M. Harvey, L.E. Shulman, P.A. Tumulty, C.L. Conley, E.H. Schoenrich, Systemic lupus erythematosus: Review of the literature and clinical analysis of 138 cases, Medicine (Baltimore) 33 (1954) 291–437.

[44] R.T. Johnson, E.P. Richardson, The neurological manifestations of systemic lupus erythematosus, Medicine (Baltimore) 47 (1968) 337–369.

[45] S.G. Ellis, M.A. Verity, Central nervous system involvement in systemic lupus erythematosus: A review of neuropathologic findings in 57 cases, 1955–1977, Sem Arthritis Rheum 8 (1979) 212–221.

[46] A.T. Rowshani, P. Remans, A. Rozemuller, P.P. Tak, Cerebral vasculitis as a primary manifestation of systemic lupus erythematosus, Ann Rheum Dis 64 (2005) 784–786.

[47] E.A. Sanders, L.A. Hogenhuis, Cerebral vasculitis as presenting symptom of systemic lupus erythematosus, Acta Neurol Scand 74 (1986) 75–77.

[48] C.M. Everett, T.D. Graves, S. Lad, H.R. Jager, M. Thom, D.A. Isenberg, et al., Aggressive CNS lupus vasculitis in the absence of systemic disease activity, Rheumatology (Oxford) 47 (2008) 107–109.

[49] T.J. Kleinig, B. Koszyca, P.C. Blumbergs, P. Thompson, Fulminant leucocytoclastic brainstem vasculitis in a patient with otherwise indolent systemic lupus erythematosus, Lupus 18 (2009) 486–490.

[50] T. Rizos, M. Siegelin, S. Hahnel, B. Storch-Hagenlocher, A. Hug, Fulminant onset of cerebral immunocomplex vasculitis as first manifestation of neuropsychiatric systemic lupus erythematosus (NPSLE), Lupus 18 (2009) 361–363.

[51] F.B. Karassa, J.P. Ioannidis, G. Touloumi, K.A. Boki, H.M. Moutsopoulos, Risk factors for central nervous system involvement in systemic lupus erythematosus, QJM 93 (2000) 169–174.

[52] O. Devinsky, C.K. Petito, D.R. Alonso, Clinical and neuropathological findings in systemic lupus erythematosus: The role of vasculitis, heart emboli, and thrombotic thrombocytopenic purpura, Ann Neurol 23 (1988) 380–384.

[53] J.G. Hanly, N.M. Walsh, V. Sangalang, Brain pathology in systemic lupus erythematosus, J Rheumatol 19 (1992) 732–741.

[54] D.K. Weiner, N.B. Allen, Large vessel vasculitis of the central nervous system in systemic lupus erythematosus: Report and review of the literature, J Rheumatol 18 (1991) 748–751.

[55] A. Asai, M. Matsutani, T. Kohno, T. Fujimaki, K. Takakura, Multiple saccular cerebral aneurysms associated with systemic lupus erythematosus—Case report, Neurol Med Chir (Tokyo) 29 (1989) 245–247.

[56] E.P. Fody, M.G. Netsky, R.E. Mrak, Subarachnoid spinal hemorrhage in a case of systemic lupus erythematosus, Arch Neurol 37 (1980) 173–174.

[57] R.E. Kelley, N. Stokes, P. Reyes, S.I. Harik, Cerebral transmural angiitis and ruptured aneurysm: A complication of systemic lupus erythematosus, Arch Neurol 37 (1980) 526–527.

[58] Y. Nagayama, K. Kusudo, H. Imura, A case of central nervous system lupus associated with ruptured cerebral berry aneurysm, Jpn J Med 28 (1989) 530–533.

[59] T. Sakaki, T. Morimoto, S. Utsumi, Cerebral transmural angiitis and ruptured cerebral aneurysms in patients with systemic lupus erythematosus, Neurochirurgia (Stuttgart) 33 (1990) 132–135.

[60] C.D. Regan, C.S. Foster, Retinal vascular diseases: Clinical presentation and diagnosis, Int Ophthalmol Clin 26 (1986) 25–53.

[61] E.M. Graham, D.J. Spalton, R.O. Barnard, A. Garner, R.W. Russell, Cerebral and retinal vascular changes in systemic lupus erythematosus, Ophthalmology 92 (1985) 444–448.

[62] F.C. Cordes, S.D. Aiken, Ocular changes in acute disseminated lupus erythematosus: Report of a case with microscopic findings, Am J Ophthalmol 30 (1947) 1541–1555.

[63] F.J. Stafford-Brady, M.B. Urowitz, D.D. Gladman, M. Easterbrook, Lupus retinopathy: Patterns, associations, and prognosis, Arthritis Rheum 31 (1988) 1105–1110.

[64] O. Ushiyama, K. Ushiyama, S. Koarada, Y. Tada, N. Suzuki, A. Ohta, et al., Retinal disease in patients with systemic lupus erythematosus, Ann Rheum Dis 59 (2000) 705–708.

[65] A.A. Andrianakos, J. Duffy, M. Suzuki, J.T. Sharp, Transverse myelopathy in systemic lupus erythematosus. Report of three cases and review of the literature, Ann Intern Med 83 (1975) 616–624.

[66] H.M. Markusse, T.M. Vroom, A.H. Heurkens, C.J. Vecht, Polyneuropathy as initial manifestation of necrotizing vasculitis and gangrene in systemic lupus erythematosus, Neth J Med 38 (1991) 204–208.

[67] P.A. McCombe, J.G. McLeod, J.D. Pollard, Y.P. Guo, T.J. Ingall, Peripheral sensorimotor and autonomic neuropathy associated with systemic lupus erythematosus. Clinical, pathological and immunological features, Brain 110 (Pt. 2) (1987) 533–549.

[68] R.A. Hughes, J.S. Cameron, S.M. Hall, J. Heaton, J. Payan, R. Teoh, Multiple mononeuropathy as the initial presentation of systemic lupus erythematosus—Nerve biopsy and response to plasma exchange, J Neurol 228 (1982) 239–247.

[69] A.M. Bergemer, B. Fouquet, P. Goupille, J.P. Valat, Peripheral neuropathy as the initial manifestation of systemic lupus erythematosus. Report of a case, Sem Hop Paris 63 (1987) 1979–1982.

[70] A.A. Bailey, G.P. Sayre, E.C. Clark, Neuritis associated with systemic lupus erythematosus; A report of five cases, with necropsy in two, AMA Arch Neurol Psychiatry 75 (1956) 251–259.

[71] G. Midroni, S.M. Cohen, J.M. Bilbao, Endoneurial vasculitis and tubuloreticular inclusions in peripheral nerve biopsy, Clin Neuropathol 19 (2000) 70–76.

[72] J.D. Aitchison, A.W. Williams, Pulmonary changes in disseminated lupus erythematosus, Ann Rheum Dis 15 (1956) 26–32.

[73] A.O. Fayemi, Pulmonary vascular disease in systemic lupus erythematosus, Am J Clin Pathol 65 (1976) 284—290.

[74] M. Gross, J.R. Esterly, R.H. Earle, Pulmonary alterations in systemic lupus erythematosus, Am Rev Respir Dis 105 (1972) 572—577.

[75] J.L. Myers, A.A. Katzenstein, Microangiitis in lupus-induced pulmonary hemorrhage, Am J Clin Pathol 85 (1986) 552—556.

[76] W.R. Webb, G. Gamsu, Cavitary pulmonary nodules with systemic lupus erythematosus: Differential diagnosis, AJR Am J Roentgenol 136 (1981) 27—31.

[77] H.M. Haupt, G.W. Moore, G.M. Hutchins, The lung in systemic lupus erythematosus. Analysis of the pathologic changes in 120 patients, Am J Med 71 (1981) 791—798.

[78] M.R. Zamora, M.L. Warner, R. Tuder, M.I. Schwarz, Diffuse alveolar hemorrhage and systemic lupus erythematosus. Clinical presentation, histology, survival, and outcome, Medicine (Baltimore) 76 (1997) 192—202.

[79] T.J. Laing, Gastrointestinal vasculitis and pneumatosis intestinalis due to systemic lupus erythematosus: Successful treatment with pulse intravenous cyclophosphamide, Am J Med 85 (1988) 555—558.

[80] D.S. Myung, T.J. Kim, S.J. Lee, S.C. Park, J.S. Kim, J.C. Kim, et al., Lupus-associated pancreatitis complicated by pancreatic pseudocyst and central nervous system vasculitis, Lupus 18 (2009) 74—77.

[81] V.E. Pollak, W.J. Grove, R.M. Kark, R.C. Muehrcke, C.L. Pirani, I.E. Steck, Systemic lupus erythematosus simulating acute surgical condition of the abdomen, N Engl J Med 259 (1958) 258—266.

[82] J.C. Reynolds, R.D. Inman, R.P. Kimberly, J.H. Chuong, J.E. Kovacs, M.B. Walsh, Acute pancreatitis in systemic lupus erythematosus: Report of twenty cases and a review of the literature, Medicine (Baltimore) 61 (1982) 25—32.

[83] M.C. Serrano Lopez, M. Yebra Bango, E. Lopez Bonet, I. Sanchez Vegazo, F. Albarran Hernandez, L. Manzano Espinosa, et al., Acute pancreatitis and systemic lupus erythematosus: Necropsy of a case and review of the pancreatic vascular lesions, Am J Gastroenterol 86 (1991) 764—767.

[84] I. Haslock, Spontaneous rupture of the liver in systemic lupus erythematosus, Ann Rheum Dis 33 (1974) 482—484.

[85] D.F. Huang, A.H. Yang, B.C. Lin, S.R. Wang, Clinical manifestations of hepatic arteritis in systemic lupus erythematosus, Lupus 4 (1995) 152—154.

[86] P.M. Levitin, D. Sweet, C.M. Brunner, R.E. Katholi, W.K. Bolton, Spontaneous rupture of the liver. An unusual complication of SLE, Arthritis Rheum 20 (1977) 748—750.

[87] T. Matsumoto, T. Yoshimine, K. Shimouchi, H. Shiotu, N. Kuwabara, Y. Fukuda, et al., The liver in systemic lupus erythematosus: Pathologic analysis of 52 cases and review of Japanese Autopsy Registry Data, Hum Pathol 23 (1992) 1151—1158.

[88] K.M. Newbold, W.H. Allum, R. Downing, D.P. Symmons, G.D. Oates, Vasculitis of the gall bladder in rheumatoid arthritis and systemic lupus erythematosus, Clin Rheumatol 6 (1987) 287—289.

[89] C.R. Swanepoel, A. Floyd, H. Allison, G.M. Learmonth, M.J. Cassidy, M.D. Pascoe, Acute acalculous cholecystitis complicating systemic lupus erythematosus: Case report and review. Br Med J (Clin Res Ed) 286 (1983) 251—252.

[90] T.M. Zizic, J.N. Classen, M.B. Stevens, Acute abdominal complications of systemic lupus erythematosus and polyarteritis nodosa, Am J Med 73 (1982) 525—531.

[91] S.K. Kwok, S.H. Seo, J.H. Ju, K.S. Parks, C.H. Yoon, W.U. Kim, et al., Lupus enteritis: Clinical characteristics, risk factor for relapse and association with anti-endothelial antibody, Lupus 16 (2007) 803—809.

[92] S.M. Sultan, Y. Ioannou, D.A. Isenberg, A review of gastrointestinal manifestations of systemic lupus erythematosus, Rheumatology (Oxford) 38 (1999) 917—932.

[93] A.C. Buck, L.H. Serebro, R.J. Quinet, Subacute abdominal pain requiring hospitalization in a systemic lupus erythematosus patient: a retrospective analysis and review of the literature, Lupus 10 (2001) 491—495.

[94] M.S. Ho, L.B. Teh, H.S. Goh, Ischaemic colitis in systemic lupus erythematosus—Report of a case and review of the literature, Ann Acad Med Singapore 16 (1987) 501—503.

[95] P. Reissman, E.G. Weiss, T.A. Teoh, F.V. Lucas, S.D. Wexner, Gangrenous ischemic colitis of the rectum: A rare complication of systemic lupus erythematosus, Am J Gastroenterol 89 (1994) 2234—2236.

[96] L.G. Shapeero, A. Myers, P.E. Oberkircher, W.T. Miller, Acute reversible lupus vasculitis of the gastrointestinal tract, Radiology 112 (1974) 569—574.

[97] J.Y. Byun, H.K. Ha, S.Y. Yu, J.K. Min, S.H. Park, H.Y. Kim, et al., CT features of systemic lupus erythematosus in patients with acute abdominal pain: Emphasis on ischemic bowel disease, Radiology 211 (1999) 203—209.

[98] S.F. Ko, T.Y. Lee, T.T. Cheng, S.H. Ng, H.M. Lai, Y.F. Cheng, et al., CT findings at lupus mesenteric vasculitis, Acta Radiol 38 (1997) 115—120.

[99] Y. Tsushima, Y. Uozumi, S. Yano, Reversible thickening of the bowel and urinary bladder wall in systemic lupus erythematosus: A case report, Radiat Med 14 (1996) 95—97.

[100] T.R. Helliwell, D. Flook, J. Whitworth, D.W. Day, Arteritis and venulitis in systemic lupus erythematosus resulting in massive lower intestinal haemorrhage, Histopathology 9 (1985) 1103—1113.

[101] M. Yamaguchi, K. Kumada, H. Sugiyama, E. Okamoto, K. Ozawa, Hemoperitoneum due to a ruptured gastroepiploic artery aneurysm in systemic lupus erythematosus. A case report and literature review, J Clin Gastroenterol 12 (1990) 344—346.

[102] M.G. Kistin, M.M. Kaplan, J.T. Harrington, Diffuse ischemic colitis associated with systemic lupus erythematosus—Response to subtotal colectomy, Gastroenterology 75 (1978) 1147—1151.

[103] J.C. Philips, W.J. Howland, Mesenteric arteritis in systemic lupus erythematosus, JAMA 206 (1968) 1569—1570.

[104] F. Medina, A. Ayala, L.J. Jara, M. Becerra, J.M. Miranda, A. Fraga, Acute abdomen in systemic lupus erythematosus: The importance of early laparotomy, Am J Med 103 (1997) 100—105.

[105] D.D. Gladman, T. Ross, B. Richardson, S. Kulkarni, Bowel involvement in systemic lupus erythematosus: Crohn's disease or lupus vasculitis? Arthritis Rheum 28 (1985) 466—470.

[106] D. Freiman, H. Chon, L. Bilaniuk, Pneumatosis intestinalis in systemic lupus erythematosus, Radiology 116 (1975) 563—564.

[107] G.E. Cabrera, E. Scopelitis, M.L. Cuellar, L.H. Silveira, H. Mena, L.R. Espinoza, Pneumatosis cystoides intestinalis in systemic lupus erythematosus with intestinal vasculitis: Treatment with high dose prednisone, Clin Rheumatol 13 (1994) 312—316.

[108] P. Kleinman, M.A. Meyers, G. Abbott, E. Kazam, Necroqizing enterocolitis with pneumatosis intestinalis in systemic lupus erythematosus and polyarteritis, Radiology 121 (1976) 595—598.

[109] K. Kobayashi, H. Asakura, T. Shinozawa, S. Yoshida, Y. Ichikawa, M. Tsuchiya, et al., Protein-losing enteropathy in systemic lupus erythematosus. Observations by magnifying endoscopy, Dig Dis Sci 34 (1989) 1924—1928.

[110] M.M. Weiser, G.A. Andres, J.R. Brentjens, J.T. Evans, M. Reichlin, Systemic lupus erythematosus and intestinal venulitis, Gastroenterology 81 (1981) 570–579.

[111] T. Matsumoto, S. Kobayashi, H. Shimizu, M. Nakajima, S. Watanabe, N. Kitami, et al., The liver in collagen diseases: Pathologic study of 160 cases with particular reference to hepatic arteritis, primary biliary cirrhosis, autoimmune hepatitis and nodular regenerative hyperplasia of the liver, Liver 20 (2000) 366–373.

[112] J. Trambert, E. Reinitz, S. Buchbinder, Ruptured hepatic artery aneurysms in a patient with systemic lupus erythematosus: Case report, Cardiovasc Intervent Radiol 12 (1989) 32–34.

[113] B. Grimbacher, M. Huber, J. von Kempis, P. Kalden, M. Uhl, G. Kohler, et al., Successful treatment of gastrointestinal vasculitis due to systemic lupus erythematosus with intravenous pulse cyclophosphamide: A clinical case report and review of the literature, Br J Rheumatol 37 (1998) 1023–1028.

[114] T.A. Bonfiglio, R.E. Botti, J.W. Hagstrom, Coronary arteritis, occlusion, and myocardial infarction due to lupus erythematosus, Am Heart J 83 (1972) 153–158.

[115] R.H. Heibel, J.D. O'Toole, E.I. Curtiss, T.A. Medsger Jr., S.P. Reddy, J.A. Shaver, Coronary arteritis in systemic lupus erythematosus, Chest 69 (1976) 700–703.

[116] M.R. Hejtmancik, J.C. Wright, R. Quint, F.L. Jennings, The cardiovascular manifestations of systemic lupus erythematosus, Am Heart J 68 (1964) 119–130.

[117] C.J. Homcy, R.R. Liberthson, J.T. Fallon, S. Gross, L.M. Miller, Ischemic heart disease in systemic lupus erythematosus in the young patient: Report of six cases, Am J Cardiol 49 (1982) 478–484.

[118] T.N. James, C.E. Rupe, R.W. Monto, Pathology of the cardiac conduction system in systemic lupus erythematosus, Ann Intern Med 63 (1965) 402–410.

[119] T.Q. Kong, R.E. Kellum, J.R. Haserick, Clinical diagnosis of cardiac involvement in systemic lupus erythematosus. A correlation of clinical and autopsy findings in thirty patients, Circulation 26 (1962) 7–11.

[120] S.M. Korbet, M.M. Schwartz, E.J. Lewis, Immune complex deposition and coronary vasculitis in systemic lupus erythematosus. Report of two cases, Am J Med 77 (1984) 141–146.

[121] K.V. Straaton, W.W. Chatham, J.D. Reveille, W.J. Koopman, S.H. Smith, Clinically significant valvular heart disease in systemic lupus erythematosus, Am J Med 85 (1988) 645–650.

[122] W.A.C. Douglas, M.A.H. Gardner, Systemic lupus erythematosus with vasculitis confined to the coronary arteries, Aust. N.Z.J. Med 30 (1998) 283–285.

[123] T.P. Nobrega, E. Klodas, J.F. Breen, S.P. Liggett, S.T. Higano, G.S. Reeder, Giant coronary artery aneurysms and myocardial infarction in a patient with systemic lupus erythematosus, Cathet Cardiovasc Diagn 39 (1996) 75–79.

[124] M.J. Fairfax, T.G. Osborn, G.A. Williams, C.C. Tsai, T.L. Moore, Endomyocardial biopsy in patients with systemic lupus erythematosus, J Rheumatol 15 (1988) 593–596.

[125] G.B. Appel, C.L. Pirani, V. D'Agati, Renal vascular complications of systemic lupus erythematosus, J Am Soc Nephrol 4 (1994) 1499–1515.

[126] T. Tsumagari, S. Fukumoto, M. Kinjo, K. Tanaka, Incidence and significance of intrarenal vasculopathies in patients with systemic lupus erythematosus, Hum Pathol 16 (1985) 43–49.

[127] G.B. Appel, F.G. Silva, C.L. Pirani, J.I. Meltzer, D. Estes, Renal involvement in systemic lupus erythematosus (SLE): A study of 56 patients emphasizing histologic classification, Medicine (Baltimore) 57 (1978) 371–410.

[128] M.D. Hughson, T. Nadasdy, G.A. McCarty, C. Sholer, K.W. Min, F. Silva, Renal thrombotic microangiopathy in patients with systemic lupus erythematosus and the antiphospholipid syndrome, Am J Kidney Dis 20 (1992) 150–158.

[129] P. Altieri, A. Pani, P. Bolasco, P. Melis, A. Barracca, P. Todde, et al., Is renal vasculitis in patients with systemic lupus erythematosus a bad prognostic factor? Contrib Nephrol 99 (1992) 72–78.

[130] P.A. Bacon, D.M. Carruthers, Vasculitis associated with connective tissue disorders, Rheum Dis Clin North Am 21 (1995) 1077–1096.

[131] U.N. Bhuyan, A.N. Malaviya, S.C. Dash, K.K. Malhotra, Prognostic significance of renal angiitis in systemic lupus erythematosus (SLE), Clin Nephrol 20 (1983) 109–113.

[132] L. Baskin, S. Mee, M. Matthay, P.R. Carroll, Ureteral obstruction caused by vasculitis, J Urol 141 (1989) 933–935.

[133] M.H. Weisman, E.C. McDonald, C.B. Wilson, Studies of the pathogenesis of interstitial cystitis, obstructive uropathy, and intestinal malabsorption in a patient with systemic lupus erythematosus, Am J Med 70 (1981) 875–881.

[134] D. Alarcon-Segovia, C. Abud-Mendoza, E. Reyes-Gutierrez, A. Iglesias-Gamarra, E. Diaz-Jouanen, Involvement of the urinary bladder in systemic lupus erythematosus: A pathologic study, J Rheumatol 11 (1984) 208–210.

[135] B.M. Tripp, F. Chu, F. Halwani, M.M. Hassouna, Necrotizing vasculitis of the penis in systemic lupus erythematosus, J Urol 154 (1995) 528–529.

[136] D. Massasso, C. Cheruvu, F. Joshua, J. Yong, I.G. Graham, Ovarian vasculitis in an adult with fatal systemic lupus erythematosus, Lupus 18 (2009) 364–367.

[137] K.E. Meyers, S. Pfieffer, T. Lu, B.S. Kaplan, Genitourinary complications of systemic lupus erythematosus, Pediatr Nephrol 14 (2000) 416–421.

[138] S. Feriozzi, A.O. Muda, M. Amini, T. Faraggiana, E. Ancarani, Systemic lupus erythematosus with membranous glomerulonephritis and uterine vasculitis, Am J Kidney Dis 29 (1997) 277–279.

[139] T. Igarashi, S. Nagaoka, K. Matsunaga, K. Katoh, Y. Ishigatsubo, K. Tani, et al., Aortitis syndrome (Takayasu's arteritis) associated with systemic lupus erythematosus, J Rheumatol 16 (1989) 1579–1583.

[140] K. Kameyama, S. Kuramochi, T. Ueda, S. Kawada, N. Tominaga, T. Mimori, et al., Takayasu's aortitis with dissection in systemic lupus erythematosus, Scand J Rheumatol 28 (1999) 187–188.

[141] K. Kitazawa, K. Joh, T. Akizawa, A case of lupus nephritis coexisting with podocytic infolding associated with Takayasu's arteritis, Clin Exp Nephrol 12 (2008) 462–466.

[142] P.A. Saxe, R.D. Altman, Aortitis syndrome (Takayasu's arteritis) associated with SLE, J Rheumatol 17 (1990) 1251–1252.

[143] N. Washiyama, T. Kazui, M. Takinami, K. Yamashita, H. Terada, B.A. Muhammad, et al., Surgical treatment of recurrent abdominal aortic aneurysm in a patient with systemic lupus erythematosus, J Vasc Surg 32 (2000) 209–212.

[144] C. Breynaert, T. Cornelis, S. Stroobants, J. Bogaert, J. Vanhoof, D. Blockmans, Systemic lupus erythematosus complicated with aortitis, Lupus 17 (2008) 72–74.

[145] F. Rojo-Leyva, N.B. Ratliff, D.M. Cosgrove 3rd, G.S. Hoffman, Study of 52 patients with idiopathic aortitis from a cohort of 1204 surgical cases, Arthritis Rheum 43 (2000) 901–907.

[146] P.A. Saxe, R.D. Altman, Takayasu's arteritis syndrome associated with systemic lupus erythematosus, Sem Arthritis Rheum 21 (1992) 295–305.

[147] A.S. Silver, C.Y. Shao, E.M. Ginzler, Aortitis and aortic thrombus in systemic lupus erythematosus, Lupus 15 (2006) 541–543.

[148] C.B. Bunker, P.M. Dowd, Giant cell arteritis and systemic lupus erythematosus, Br J Dermatol 119 (1988) 115–120.

[149] F.C. Bakker, J.A. Rauwerda, H.J. Moens, T.A. van den Broek, Intermittent claudication and limb-threatening ischemia in systemic lupus erythematosus and in SLE-like disease: A report of two cases and review of the literature, Surgery 106 (1989) 21–25.

[150] G.S. Gladstein, R.I. Rynes, N. Parhami, L.E. Bartholomew, Gangrene of a foot secondary to systemic lupus erythematosus with large vessel vasculitis, J Rheumatol 6 (1979) 549–553.

[151] J.L. Kaufman, E. Bancilla, J. Slade, Lupus vasculitis with tibial artery thrombosis and gangrene, Arthritis Rheum 29 (1986) 1291–1292.

[152] S.M. Harmon, K.L. Oltmanns, K.W. Min, Large vessel occlusion with vasculitis in systemic lupus erythematosus, South Med J 84 (1991) 1150–1154.

[153] C.R. Abramowsky, M.E. Vegas, G. Swinehart, M.T. Gyves, Decidual vasculopathy of the placenta in lupus erythematosus, N Engl J Med 303 (1980) 668–672.

[154] J.G. Hanly, D.D. Gladman, T.H. Rose, C.A. Laskin, M.B. Urowitz, Lupus pregnancy. A prospective study of placental changes, Arthritis Rheum 31 (1988) 358–366.

[155] D.M. Grennan, J.N. McCormick, D. Wojtacha, M. Carty, W. Behan, Immunological studies of the placenta in systemic lupus erythematosus, Ann Rheum Dis 37 (1978) 129–134.

[156] D. Carvalho, C.O. Savage, D. Isenberg, J.D. Pearson, IgG anti-endothelial cell autoantibodies from patients with systemic lupus erythematosus or systemic vasculitis stimulate the release of two endothelial cell-derived mediators, which enhance adhesion molecule expression and leukocyte adhesion in an autocrine manner, Arthritis Rheum 42 (1999) 631–640.

[157] T. Springer, Traffic signals for lymphocyte recirculation and leukocyte emigration: The multi-step paradigm, Cell 76 (1994) 301–314.

[158] R. Wu, E. Svenungsson, I. Gunnarsson, C. Haegerstrand-Gillis, B. Andersson, I. Lundberg, et al., Antibodies to adult human endothelial cells cross-react with oxidized low-density lipoprotein and beta 2-glycoprotein 1 in systemic lupus erythematosus, Clin Exp Immunol 115 (1999) 561–566.

[159] D.B. Cines, A. Lyss, M. Reeber, M. Bina, R.J. Dehoratius, Presence of complement fixing anti-endothelial antibodies in systemic lupus erythematosus, J Clin Invest 73 (1984) 611–625.

[160] C. Kallenberg, Autoantibodies in vasculitis: Current perspectives, Clin Exp Rheumatol 11 (1993) 355–360.

[161] F.J. van der Woude, N. Rasmussen, S. Lobatto, A. Wiik, H. Permin, L.A. van Es, et al., Autoantibodies against neutrophils and monocytes: Tool for diagnosis and marker of disease activity in Wegener's granulomatosis, Lancet 1 (1985) 425–429.

[162] R.J. Falk, G.S. Hoffman, Controversies in small vessel vasculitis—Comparing the rheumatology and nephrology views, Curr Opin Rheumatol 19 (2007) 1–9.

[163] L. Nassberger, A.G. Sjöholm, H. Jonsson, G. Sturfelt, A. Akesson, Autoantibodies against neutrophil cytoplasm components in systemic lupus erythematosus and in hydralazine induced lupus, Clin Exp Immunol 81 (1990) 380–383.

[164] R. Puazner, M. Urowitz, D. Gladman, J. Gough, Antineutrophil cytoplasmic antibodies in systemic lupus erythematosus, J Rheumatol 21 (1994) 1670–1673.

[165] A. Schnabel, E. Csernok, D.A. Isenberg, W.L. Gross, Anti-neutrophil cytoplasmic antibodies in systemic lupus erythematosus. Prevalence, specificities and clinical significance, Arthritis Rheum 38 (1995) 633–637.

[166] P.E. Spronk, H. Bootsma, G. Horst, M.G. Huitema, P.C. Limburg, J.W. Tervaert, et al., Antineutrophil cytoplasmic antibodies in systemic lupus erythematosus, Br J Rheumatol 35 (1996) 625–631.

[167] L. Nassberger, Distribution of antineutrophil cytoplasmic autoantibodies in SLE patients with and without renal involvement, Am J Nephrol 16 (1996) 548–549.

[168] K. Nishiya, H. Chikazawa, S. Nishimura, N. Hisakawa, K. Hashimoto, Anti-neutrophil cytoplasmic antibody in patients with systemic lupus erythematosus is unrelated to clinical features, Clin Rheumatol 16 (1997) 70–75.

[169] I. Manolova, M. Dancheva, K. Halacheva, Antineutrophil cytoplasmic antibodies in patients with systemic lupus erythematosus; Prevalence, antigen specificity and clinical associations, Rheumatol Int 20 (2001) 197–204.

[170] M. Galeazzi, G. Morozzi, G.D. Sebastini, F. Bellisai, R. Marcolongo, R. Cervera, et al., Anti-neutrophil cytoplasmic antibodies in 566 European patients with systemic lupus erythematosus; Prevalence, clinical associations and correlation with other autoantibodies. European Concerted Action on the Immunogenetics of SLE, Clin Exp Rheumatol 16 (1998) 541–546.

[171] H.S. Jethwa, P. Nachman, R.J. Falk, J.C. Jennette, False-positive myeloperoxidase binding activity due to DNA/anti-DNA antibody complexes: A source for analytical error in serologic evaluation of anti-neutrophil cytoplasmic autoantibodies, Clin Exp Immunol 121 (2000) 544–550.

[172] D.L. Tuffanelli, E.L. Dubois, Cutaneous manifestations of systemic lupus erythematosus, Arch Dermatol 90 (1964) 377–386.

[173] M. Abu-Shakra, M.B. Urowitz, D.D. Gladman, J. Gough, Mortality studies in systemic lupus erythematosus. Results from a single center. I. Causes of death, J Rheumatol 22 (1995) 1259–1264.

[174] M. Abu-Shakra, M.B. Urowitz, D.D. Gladman, J. Gough, Mortality studies in systemic lupus erythematosus. Results from a single center. II. Predictor variables for mortality, J Rheumatol 22 (1995) 1265–1270.

[175] J.F. Molina, C. Drenkard, J. Molina, M.H. Cardiel, O. Uribe, J.M. Anaya, et al., Systemic lupus erythematosus in males. A study of 107 Latin American patients, Medicine (Baltimore) 75 (1996) 124–130.

[176] N.M. Matolo, D. Albo Jr., Gastrointestinal complications of collagen vascular diseases. Surgical implications, Am J Surg 122 (1971) 678–682.

[177] R.L. Nadorra, Y. Nakazato, B.H. Landing, Pathologic features of gastrointestinal tract lesions in childhood-onset systemic lupus erythematosus: Study of 26 patients, with review of the literature, Pediatr Pathol 7 (1987) 245–259.

[178] C.-K. Lee, M.S. Ahn, E.Y. Lee, J.H. Shin, Y.-S. Cho, H.K. Ha, et al., Acute abdominal pain in systemic lupus erythematosus: Focus on lupus enteritis (gastrointestinal vasculitis), Ann Rheum Dis 61 (2002) 547–550.

[179] K. Benirschke, S.G. Driscoll, The Pathology of the Human Placenta, Springer-Verlag, New York, 1967.

CHAPTER

45

Lung

Aryeh Fischer, Roland M. du Bois

INTRODUCTION

Systemic lupus erythematosus (SLE) is a multisystem autoimmune disease characterized by the presence of autoantibodies and tissue damage that may involve a myriad of organ systems. The respiratory system is commonly affected in patients with SLE, and any of its components can be impacted by the disease (Table 45.1).

The incidence of respiratory system involvement in SLE is impacted by the mode of detection and the populations studied, and it appears that respiratory involvement will occur in more than half of SLE patients at some point during the course of their disease [1–4]. The scope and the severity of involvement vary from asymptomatic abnormalities to fulminant, life-threatening disease. The presence of pulmonary disease in SLE has been associated with progression to pulmonary damage as assessed by the Systemic Lupus Collaborating Clinics/American College of Rheumatology Damage Index and an increased risk of mortality [5], but little is known about the natural history of SLE-related lung disease; we still do not know who is predisposed to develop lung involvement or how to best manage an SLE patient with lung disease.

This chapter provides an overview of the clinical assessment of SLE-associated lung disease, discusses the pulmonary manifestations associated with SLE, and concludes with thoughts on several topics of debate within the arena of SLE-associated lung disease.

CLINICAL ASSESSMENT OF SLE-ASSOCIATED LUNG DISEASE

The detection of SLE-associated lung involvement is usually by the presentation of respiratory symptoms. Less commonly, respiratory involvement is detected by an abnormal physical examination finding or by coincidental identification via testing for other medical conditions. Because there are no data to guide specific screening recommendations, detection of "early" lung involvement is left to the discretion and vigilance of the clinician. Given that lung disease is common in SLE, the clinician should maintain a high index of suspicion for its presence and have a low threshold to proceed with testing to evaluate for SLE-associated lung involvement. Ideally, the clinical assessment for SLE-associated lung disease involves effective cross-specialty collaboration among primary care providers, rheumatologists, and pulmonologists.

The clinical evaluation for lung involvement starts with a detailed assessment of respiratory symptoms such as dyspnea, cough, or pleurisy, and it includes a comprehensive cardiopulmonary physical examination. Any suspicious historical detail or physical

TABLE 45.1 Pulmonary Manifestations Associated With SLE

Infection

Pleural disease

  Pleurisy

  Pleural effusion

Parenchymal

  Acute lupus pneumonitis

  Diffuse alveolar hemorrhage

  Chronic interstitial lung disease

Shrinking lung syndrome

Pulmonary vascular disease

  Pulmonary hypertension

  Thromboembolism

Airways disease

  Upper airways (e.g., cricoarytenoid arthritis)

  Lower airways (e.g., small airways disease)

Systemic Lupus Erythematosus

examination finding should prompt further testing to define any respiratory abnormality. In addition, given that other organ involvement could be the cause of symptoms of dyspnea and cough, thorough comprehensive evaluations are often pursued to exclude nonrespiratory etiologies for these symptoms (e.g., anemia, cardiomyopathy, and gastroesophageal reflux disease).

CLINICAL ASSESSMENT

The clinical assessment of an SLE patient with respiratory symptoms is optimized by cross-specialty collaboration and often requires a combination of history and physical examination, laboratory assessment, pulmonary function testing, thoracic imaging modalities, bronchoscopy with bronchoalveolar lavage (BAL), and surgical lung biopsy (Table 45.2).

Clinical History

Given the broad array of pulmonary disorders associated with SLE, obtaining details of the clinical course of respiratory symptomatology can provide clues to diagnostic possibilities. For example, it is important to distinguish between more acute processes such as infectious pneumonia from that of a chronic insidious presentation associated with fibrotic lung disease. In addition to cataloging symptoms of dyspnea, cough, chest pain,

TABLE 45.2 Clinical Assessment of SLE-Associated Lung Disease

History

Physical examination

Laboratory studies

  Complete blood count, comprehensive metabolic panel, C-reactive protein, erythrocyte sedimentation rate, urinalysis, complement levels, anti-dsDNA antibodies, thyroid-stimulating hormone, creatinine phosphokinase

Pulmonary physiology assessment

  Pulmonary function testing

  Resting and ambulatory oximetry

Imaging modalities

  Chest radiograph

  Thoracic high-resolution computed tomography

  Transthoracic echocardiography

  CT angiography or ventilation/perfusion scan

Fiber-optic bronchoscopy and lavage

Surgical lung biopsy

pleurisy, wheeze, or hemoptysis, the comprehensive history should include an assessment for occupational and environmental exposures, smoking history, and detailed medication history. It is important to assess for any other associated symptoms because they may also provide clues to the underlying process.

Physical Examination

The clinical assessment of possible pulmonary disease in a patient with SLE mandates a thorough physical examination. The physical examination may provide important information on the overall respiratory status of the patient and help determine the acuity of presentation and potential urgency of further diagnostic evaluation. Simple assessments of respiratory rate and pulse oximetry may yield critical information. Chest auscultation may yield clues to the pulmonary problem, the presence of a pleural rub suggests pleural disease, and bi-basilar inspiratory crackles suggest fibrotic lung disease. In addition, the examination may suggest alternative processes responsible for symptoms of dyspnea (e.g., signs of cardiac decompensation or anemia) and help guide further diagnostic evaluation.

LABORATORY ASSESSMENT

A broad approach with laboratory investigation can be helpful as part of the evaluation of an SLE patient with respiratory symptomatology (see Table 45.2). The laboratory assessment may aid in the clinical assessment of the patient's SLE "status"— and may be an important component of the evaluation—but will rarely be a definitive tool in the overall evaluation. Some of the more important questions that can be addressed by laboratory evaluation and that will have direct relevance to patient management include the following: Are there serologic markers consistent with an SLE flare? Is the presentation of dyspnea a manifestation of profound anemia? Is the dyspnea a manifestation of respiratory muscle weakness related to active myositis (i.e., with significant creatine phosphokinase elevation)? Is renal function impaired? Does the white blood cell count reflect possible infection?

Pulmonary Function Assessment

Assessing the degree of any compromise in respiratory function is an important component of the overall clinical evaluation of any patient with respiratory symptoms. The tools most frequently employed include a noninvasive assessment for oxygen saturation (at rest and with walking) and complete pulmonary function testing (PFT).

Ambulatory Oximetry

The performance of ambulatory oximetry assessment is an inexpensive and reproducible modality that can be employed within most office settings and provides valuable information to the clinician about the degree of oxygen desaturation and potential need for supplemental oxygen [6, 7]. Although the presence of oxygen desaturation is not a specific tool for diagnosis and can be seen across the spectrum of respiratory illness, the recognition of hypoxemia is critical for the management of dyspnea and helps in the determination of degree of respiratory impairment. It is important to highlight that the oximetry assessment should be performed both at rest and with exertion (i.e., walking) because some patients will only demonstrate hypoxemia with increased respiratory demands. A more formal and reproducible procedure to assess exercise capacity and oxygenation is the 6-minute walk test, which measures oxygen saturation and total distance walked while the individual walks on a hard surface at his or her best pace for 6 minutes [6, 7]. Less commonly, a full cardiopulmonary exercise study may be required. This is more invasive because it requires radial artery catheter placement and will be complicated by other causes of exercise impairment, such as arthropathy and myopathy.

Pulmonary Function Testing

PFT abnormalities may not point to a specific diagnosis, but they provide reproducible and quantifiable objective assessments of which respiratory compartment is involved—airways, parenchyma, vessels, or a combination of these—and disease severity. Serial determination of the forced vital capacity (FVC) and diffusing capacity for carbon monoxide (DLCO) allows for objective quantification of ventilatory capacity and gas exchange, respectively, to be monitored. These parameters are useful in assessing the degree of respiratory impairment and are especially helpful when trying to assess for disease progression and response to therapy [8−10]. Patients who decline 10% or more of predicted FVC or 15% or more of DLCO are considered to have progressive disease by PFT. Spirometric testing provides useful information on the pattern of lung involvement. The ratio of forced expiratory volume in 1 second (FEV1)/FVC helps distinguish restrictive defects commonly seen in parenchymal lung disease from obstructive defects more commonly seen with airways disease. In obstructive lung disease, this ratio is reduced, whereas restrictive lung disease results in a normal, or sometimes elevated, ratio because the reduction in these measures is proportionate. The DLCO provides an index of gas exchange and will be reduced in a diverse spectrum of respiratory abnormalities, including vascular disease when a reduced measure is the lone abnormality.

PFT abnormalities are common in SLE patients, even among asymptomatic individuals [1, 11, 12]. A variety of studies have shown that a decreased DLCO is the most common PFT abnormality in SLE patients [11, 13, 14]. In the largest controlled study to date of PFTs in patients with SLE, Andonopoulos and colleagues [14] compared the PFT results of 70 SLE patients with those of age- and gender-matched healthy subjects. All had normal chest radiographs at study entry, and neither group included subjects with a cigarette smoking history or active pulmonary disease. They found abnormal PFTs in 63% of those with SLE compared with 17% of controls. An isolated decreased DLCO was the most common abnormality in the SLE group (31% SLE vs. 0% in controls). Small airways disease was relatively common in both groups (24% in SLE and 17% in controls) [14]. The clinical significance of PFT abnormalities in asymptomatic SLE patients has yet to be determined, particularly because the abnormalities may not be progressive in nature [15].

THORACIC IMAGING MODALITIES

Chest Radiograph

The chest radiograph is the most practical first imaging modality in any patient with symptoms of respiratory impairment. Chest radiographs are relatively inexpensive, readily available, and may provide valuable clues to the presence, type, and extent of pulmonary disease. The frequency and type of chest radiograph abnormality vary with the selection of the SLE population being studied. In a study of 70 SLE patients without smoking history or respiratory symptoms, none of the subjects had chest radiograph abnormalities [14]. In contrast, in their study of 43 SLE patients with symptoms of respiratory impairment, Silberstein and colleagues [11] found that 23% had abnormal chest radiographs. Pleural changes appear to be the most common abnormality [16, 17]. When pleural effusions are present, they tend to be small and, in 50% of patients, bilateral [4]. Peripheral parenchymal infiltrates are most commonly due to acute infectious pneumonia, although further evaluation is often needed for the evaluation of parenchymal infiltrates because these can occur due to alveolar hemorrhage and interstitial lung disease (ILD). The presence of cardiomegaly on chest radiography may provide clues to underlying pericardial disease or cardiomyopathy. Although the chest radiograph is a valuable first step in diagnostic imaging for SLE patients with respiratory symptoms, computed tomography (CT) is often much more useful for identifying and discriminating different types of specific pulmonary disease manifestations.

Thoracic High-Resolution Computed Tomography

Thoracic high-resolution computed tomography (HRCT) imaging has become standard in evaluating pulmonary disease and is particularly valuable in providing information about pattern, type, and extent of interstitial lung disease and the presence of extraparenchymal abnormalities including pleural disease (Figure 45.1) [18–20]. Serial scanning may aid in the assessment of disease progression. In many cases of ILD, a specific radiologic pattern (e.g., consistent with the histopathological pattern of usual interstitial pneumonia (UIP)) (Figures 45.2 and 45.3) can be predicted with a high degree of confidence, although nonspecific interstitial pneumonia (NSIP) is predicted with less certainty by CT and a surgical biopsy may be needed to fully elucidate the amount of fibrosis present, which impacts on treatment decisions (Figures 45.4 and 45.5) [19, 20]. This pattern recognition within specific clinical scenarios may obviate the need for surgical lung biopsy. The presence of a fibrotic radiographic pattern as evidenced by reticular opacities, traction bronchiectasis, and honeycombing are predictive of poor outcomes in both idiopathic interstitial pneumonia and rheumatoid arthritis (RA)-associated ILD [8, 19, 21]. Whether these findings can be applied to SLE-associated ILD is not known.

HRCT abnormalities (including traction bronchiectasis, ILD, lymphadenopathy, and pleuro-pericardial abnormalities) are seen in the majority of SLE patients,

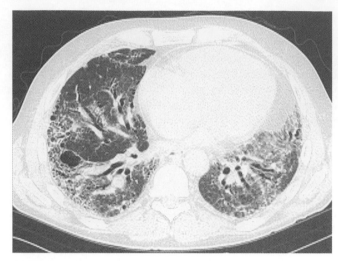

FIGURE 45.2 High-resolution computed tomography image showing extensive traction bronchiectasis and honeycombing consistent with usual interstitial pneumonia.

even in the absence of respiratory symptoms or impairment on PFTs [22]. The clinical significance of HRCT abnormalities in asymptomatic SLE patients remains to be determined.

CT Angiography and Ventilation/Perfusion Scan

The sudden onset of dyspnea in an SLE patient requires consideration of thromboembolic disease requiring imaging to assess for pulmonary embolism with either a contrast CT or ventilation/perfusion (V/Q) scanning. D-dimer assays may increase the predictive value of these imaging studies but may be unreliable in SLE patients. If the clinical scenario suggests pulmonary embolism, and the less invasive assessments prove nondiagnostic, pulmonary angiography is needed for definitive assessment.

Transthoracic Echocardiography

Transthoracic echocardiography (TTE) is useful primarily to assess the possibility of a cardiac etiology for dyspnea. However, TTE can also be useful to assess for elevated right-sided pulmonary pressures (i.e., pulmonary hypertension) that may arise secondary to parenchymal lung disease, to assess for chronic thromboembolic disease, or as a primary vasculopathy due to SLE [23]. Although providing only an approximate measure of systolic pulmonary artery pressure as derived from the speed of the recombinant jet across a leaking tricuspid valve, this measure can be helpful as a screening tool for pulmonary hypertension (PH) [23]. The definitive diagnosis of PH requires right heart catheterization [23].

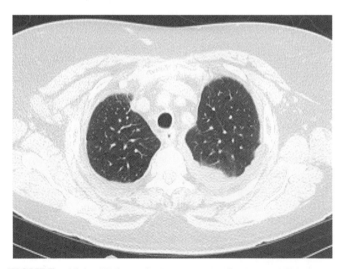

FIGURE 45.1 High-resolution computed tomography image showing extensive bilateral pleural thickening.

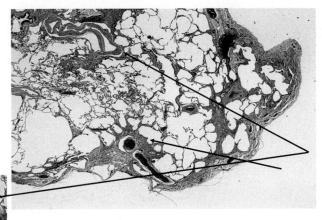

FIGURE 45.3 Photomicrograph of a surgical lung biopsy specimen demonstrating usual interstitial pneumonia. The pattern of histopathology is spatially and temporally heterogeneous with numerous accumulations of myofibroblasts—fibroblastic foci (*arrows* and *insert*).

Bronchoscopy and Bronchoalveolar Lavage

The primary role for bronchoscopic evaluation and BAL in patients with SLE is to exclude infection as the etiology of respiratory problems. Because SLE patients are immunocompromised (both by their intrinsic disease and by the medications required to manage it), infection is the most common cause of respiratory compromise [4, 24]. The SLE patient with new-onset respiratory symptoms needs to be carefully evaluated for typical and atypical infection. In this context, BAL is a valuable tool to assess particularly for *Pneumocystis* infection and to exclude fungal and mycobacterial organisms as well as more common pathogens. Once infection has been excluded, the BAL cell count differential can aid in assessing degree of cellular alveolitis related to auto-inflammation and prompting anti-inflammatory therapy accordingly. Another role for BAL is to determine if a pulmonary infiltrate is due to

alveolar hemorrhage when successive BAL returns are increasingly blood stained or if alveolar macrophages contain hemosiderin in the context of more chronic, less profuse, but recurrent hemorrhage.

The usefulness of BAL cell counts as a baseline predictor of disease progression in SLE-associated ILD is unclear. Silver and colleagues [25] have shown that BAL neutrophilia or eosinophilia in patients with systemic sclerosis (SSc)-associated ILD is useful as a predictor of progressive ILD. However, two well-designed prospective studies failed to demonstrate any prognostic significance obtained from BAL in patients with SSc-associated ILD [26, 27]. Hence, the routine use of BAL to predict the likelihood of disease progression in SSc-associated ILD, and indeed any other connective tissue disease (CTD) including SLE, is unresolved.

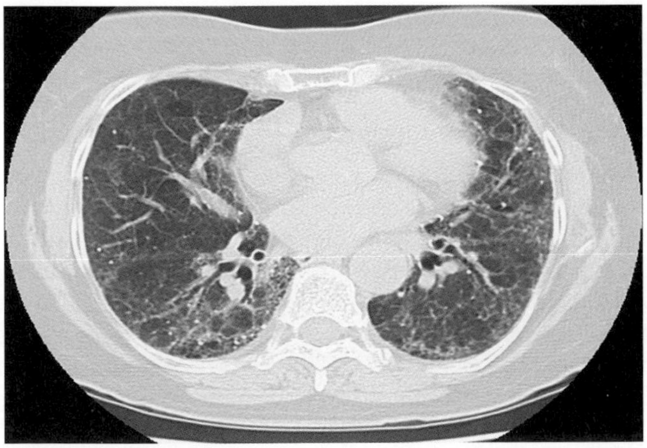

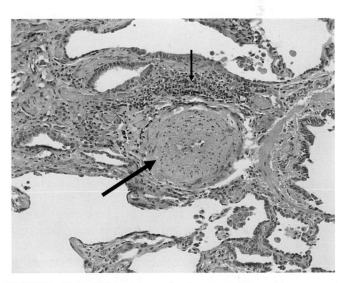

FIGURE 45.4 High-resolution computed tomography image showing diffuse ground glass opacities without honeycombing; most suggestive of nonspecific interstitial pneumonia.

FIGURE 45.5 Photomicrograph of a surgical lung biopsy specimen demonstrating fibrotic nonspecific interstitial pneumonia with lymphoid hyperplasia (*thin arrow*) and medial pulmonary arterial hyperplasia (*thick arrow*).

Surgical Lung Biopsy

Surgical lung biopsy is often needed as part of the evaluation of parenchymal abnormalities encountered in patients with SLE [24]. The specific value in determining precisely the pattern of histopathology is discussed in the following sections under each specific disease manifestation.

SLE-ASSOCIATED LUNG MANIFESTATIONS

Infection

Infection is the most common form of pulmonary involvement in patients with SLE [28] because these patients are at increased risk for infection in part due to intrinsic immune dysregulation but also as a result of the immunosuppressing medications required to manage the disease. As a result, pulmonary infections are a frequent occurrence and are associated with significant morbidity and mortality [28, 29].

SLE patients are at risk for infection with both typical and atypical pathogenic organisms, including a variety of fungal and mycobacterial species [28, 30, 31]. Due to their immunocompromised state, the clinical presentation may be subtle, particularly when infected with opportunistic organisms. The importance of a thorough evaluation that includes BAL to exclude infection in SLE patients with any pulmonary abnormality cannot be overemphasized.

For a thorough review of infectious complications associated with SLE, see Chapter 23.

Pleural Disease

Pleural abnormality is the most common respiratory manifestation of SLE [28, 32] and occurs more frequently in SLE than in any other CTD [33]. The exact prevalence of SLE-related pleural disease is not known and varies based on the population evaluated and the mode of detection. On autopsy, pleural abnormalities are quite common, with a prevalence ranging from 78 to 93% [28, 34]. The cumulative incidence of pleurisy or effusion has been evaluated in a number of studies. Cervera and colleagues [35] reported a cumulative incidence of 36% (either pleural or pericardial involvement) among a cohort of 1000 SLE patients. Dubois and Tuffanelli [36] reported pleurisy in 45% and pleural effusion in 30% of their cohort of 520 SLE patients. Several other studies have reported similar prevalence rates of pleural abnormalities, ranging from 41 to 56% [1, 37]. Pleurisy appears to be more common in men [35] and in African-Americans [38]. Pleural abnormalities may

be the presenting manifestation of SLE [39, 40], may be unilateral or bilateral [41], and may be identified in asymptomatic individuals.

Pleuritic pain can herald an SLE flare, can last for several weeks, and can be associated with fever, dyspnea, or cough. The pain is most often located at the costophrenic angle and is classically exacerbated by deep inspiration and cough. The presence of pleuritic chest pains should prompt careful evaluation to exclude thromboembolism, particularly in those with antiphospholipid (aPL) antibodies. The presence of pleural effusion should prompt investigation to exclude infection, states of volume overload, and malignancy.

SLE-associated pleural effusions are typically small to moderate in size, are often bilateral, and are exudative [41]. In contrast to RA-associated pleural effusions, the pleural glucose is usually greater than 70 mg/dl, the lactate dehydrogenase is usually less than 500 IU/l, and the pH is usually greater than 7.2 [41−43]. The role of assessing for pleural fluid ANA is not clear as there are conflicting studies regarding the specificity for ANA in pleural fluid in SLE. Leechawengwong and colleagues [44] tested pleural fluid ANA in 100 consecutive patients and found a positive ANA in all 7 patients with SLE but not in any without SLE [44]. In contrast, Khare and colleagues [45] found a positive ANA in 8 of 74 pleural effusions from non-SLE patients.

The treatment of pleural disease is not evidence-based, but in our experience pleurisy is usually rapidly responsive to nonsteroidal anti-inflammatory drugs or corticosteroids at low dose (15−20 mg/day). The presence of effusion often mandates a more protracted treatment course. In more severe or refractory cases of pleurisy or effusion, the addition of a steroid sparing agent such as hydroxychloroquine or azathioprine (AZA) is needed to treat chronic pleuritic symptoms and allow for effective corticosteroid taper. Rarely, pleural disease may require intercostal steroid injection, pleurodesis, or even pleurectomy.

Parenchymal Disease

Acute Lupus Pneumonitis

Acute lupus pneumonitis is identified in approximately 1−9% of SLE cohorts [1, 16, 35, 37, 46] and is characterized by a clinical presentation quite similar to that of infectious pneumonia [24]. Patients present with the acute onset of fever, cough, dyspnea, tachypnea, and hypoxemia. Rarely, pleurisy or hemoptysis are noted as well [47]. The chest radiograph and HRCT are abnormal and demonstrate nonspecific alveolar infiltrates, with unilateral or bilateral involvement, and typically favoring the lower lung zones [33, 48]. Acute lupus pneumonitis generally occurs in patients with pre-existing SLE, but in one series of 12 patients with acute lupus pneumonitis, 6 were diagnosed with SLE only after an episode of acute pneumonitis [47]. Given the similarity of its presentation to infectious pneumonia, and given that SLE patients are typically immunocompromised, a thorough diagnostic evaluation for an infectious etiology is indicated. Bronchoscopy is typically performed as part of this evaluation, and surgical lung biopsy may be required to help provide diagnostic certainty. Histopathologic specimens demonstrate diffuse alveolar damage (DAD) in the absence of vasculitis, capillaritis, or hemorrhage and with the presence of interstitial edema and hyaline membrane formation [32, 49].

The prognosis of acute lupus pneumonitis appears to be variable but with lethal potential. Harvey and colleagues [49] reported 3 of 38 deaths (8%) associated with lupus pneumonitis, whereas Matthay and colleagues [47] reported a mortality of 50% in their cohort of 12 patients. In a study of a multiethnic U.S. cohort of 626 SLE patients, Bertoli and colleagues [5] found that the presence of pneumonitis was an independent risk factor for the earlier development of "pulmonary damage" as defined by the Systemic Lupus Collaborating Clinics/American College of Rheumatology Damage Index. Although progression to chronic fibrotic lung disease may occur, the low incidence of chronic ILD implies that this progression rarely occurs.

The mainstay of management of acute lupus pneumonitis is the use of corticosteroids. Dosing of corticosteroids is based on the severity of the condition, with more severe cases requiring more intense dosing regimens. In addition, AZA, cyclophosphamide (CYC), and other immunosuppressive agents may be useful in an adjunctive role to help manage overall SLE disease activity, reduce the likelihood of recurrent pneumonitis, or help effect steroid taper. Despite intense immunosuppressive therapy, some patients are refractory to treatment and may require other interventions such as plasmapheresis.

Pulmonary Hemorrhage

Diffuse alveolar hemorrhage (DAH) is likely the most devastating pulmonary manifestation associated with SLE. It occurs in approximately 2% of SLE patients, and it accounts for approximately 3% of all SLE-related hospitalizations and approximately 20% of hospitalizations for SLE-related lung disease [4, 50–57]. There is a spectrum of clinical presentation ranging from mild, insidious disease to fulminant, life-threatening disease. In its most severe form, DAH is rapidly fatal, and the overall mortality associated with SLE-related DAH is reported to be 70–90% [4, 50–57].

The presentation of SLE-related DAH is not specific for the condition and typically includes the acute onset of dyspnea and cough, with hemoptysis being present in approximately 65% of patients [4, 57]. In addition, fever is not uncommon at presentation. Similar to the demographics of SLE in general, DAH most often occurs in younger women, and in patients with pre-existing SLE, but is rarely the presenting manifestation of the systemic disease [57]. The median duration of SLE at the time of DAH presentation appears to be approximately 3 years [57]. SLE patients with DAH typically already have multisystem disease, and the renal involvement appears to be the most common other visceral organ manifestation in these patients [57]. In fact, in one of the larger series of SLE-related DAH, Zamora and colleagues [57] noted that 14 of 15 (93%) of their patients had co-existent renal disease.

The clinical course is one of rapid respiratory deterioration with progressive breathlessness, tachypnea, hypoxemia, and progression to acute respiratory distress [4, 47, 50, 57, 58]. With massive hemorrhage, the hemoglobin and hematocrit will drop precipitously [57–59]. Chest radiographs typically show bilateral nonspecific peripheral infiltrates, and HRCT reveals diffuse fluffy or nodular ground glass opacification changes due to alveolar filling admixed with areas of consolidation (Figure 45.6) [57]. Among patients well enough to perform PFTs, an increase in DLCO may be noted due to the presence of blood within the alveolar compartment, but this is only true for approximately the first 24 h after the hemorrhage. Hypocomplementemia and elevated anti-dsDNA antibodies are commonly noted in these patients [57]. Co-existent acute respiratory infection with either viral or bacterial pathogens may be seen in 13–57% of cases and appears to be more common among patients receiving CYC [57, 60].

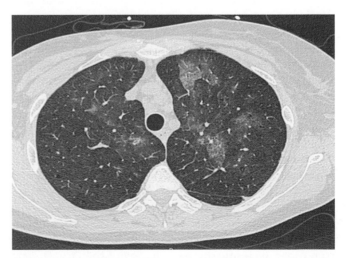

FIGURE 45.6 High-resolution computed tomography image showing patchy ground glass attenuation in a bronchocentric distribution, consistent with the intra-alveolar blood of a patient with diffuse alveolar hemorrhage.

Confirmation of the diagnosis of SLE-related DAH usually requires bronchoscopy with BAL demonstrating the presence of blood within the large airways and increasingly bloody returns from successive aliquots of BAL. The presence of hemosiderin-laden macrophages in the BAL fluid may be a valuable clue to the presence of more insidious alveolar hemorrhage. Bronchoscopic evaluation is required to exclude infection or other causes of respiratory compromise that can present acutely. Surgical lung biopsy is not usually needed because the diagnosis may be confirmed by a combination of clinical scenario and bronchoscopy and also because patients are often too ill for the surgery. When performed, histopathology reveals hemosiderin-laden macrophages; diffuse, intra-alveolar hemorrhage; co-existent capillaritis; DAD; alveolar edema and necrosis; microvascular thrombi; and vascular intimal proliferative changes [4, 55, 57, 58]. Immunofluorescence analyses of the histopathology specimens may reveal granular deposits of IgG and complement proteins within the alveolar septa and endothelium, but these immunologic features have not been demonstrated universally [58, 61–63].

The etiology of SLE-related DAH is not known. Suggestions of multifactorial causes that include pulmonary insult by infection, bleeding diathesis, combined with immune complex deposition and complement activation within the pulmonary vasculature and alveolar septa are quite possible but not yet supported by any concrete data.

SLE-related DAH often leads to respiratory failure necessitating mechanical ventilatory support in a critical care unit. Therapeutic interventions are based on anecdotal evidence and small case series and typically include high-dose corticosteroids with early addition of cytotoxic therapy in the form of intravenous CYC [64–66]. The addition of plasmapheresis may improve chances of survival but also appears to significantly increase the risk of serious infection [64, 67]. Even with such intensive regimens, the overall prognosis with SLE-related DAH remains quite poor, with approximately half of the patients dying during the course of their hospitalization [57].

Chronic Interstitial Lung Disease

Although chronic, fibrotic, parenchymal lung disease (i.e., chronic ILD) is quite common in patients with other systemic autoimmune diseases such as SSc, RA, and dermato/polymyositis, ILD rarely occurs in patients with SLE and has an estimated prevalence of approximately 3% [68, 69]. Consistent with this distinct difference from other systemic diseases, in one of the largest series of biopsy-proven CTD-associated ILD, Park and colleagues [70] described their CTD–ILD cohort of 98 patients: 35 (38%) had SSc–ILD, 28 (30%) had RA–ILD,

11 (12%) had Sjögren's–ILD, 8 (9%) had myositis–ILD, 5 (5%) had mixed connective tissue disease (MCTD)–ILD, and only 1 (1%) had SLE–ILD [70].

Chronic ILD may be the presenting manifestation of SLE [71], but it typically occurs among those with long-standing and multisystem disease [68]. The clinical presentation of chronic ILD is characterized by an insidious onset of cough, exertional dyspnea, bi-basilar crackles, and hypoxemia. PFTs typically reveal reduced lung volumes (reduced total lung capacity), a restrictive pattern on spirometry (proportionate reduction in FVC and FEV1), and impaired gas exchange (reduced DLCO). Although there are few BAL data, they may reveal an increased lymphocyte count in more cellular chronic disease but more commonly a mixed increase in neutrophils and occasional eosinophils in the more fibrotic situation [72–74]. Chest radiographs usually show bilateral pulmonary infiltrates affecting the lower lobes. Thoracic HRCT imaging provides a more detailed assessment. A prospective study by Fenlon and colleagues [22] assessed HRCT scans from 34 patients with SLE, of whom 23% had respiratory symptoms. The most common features were thickened interlobular septa, parenchymal bands, subpleural bands, pleural tags, and thickening. Only 6% had ground glass opacities, consolidation, and honeycombing. In 21% of patients, there was bronchiectasis and bronchial wall thickening. Pleural thickening was noted in 15% of patients [22].

Although the histopathologic pattern seen on surgical lung biopsy is the strongest predictor of survival in the idiopathic interstitial pneumonias, its importance in SLE-associated ILD is unclear [18, 21, 22, 70]. The full spectrum of histopathologic patterns seen in the idiopathic diffuse parenchymal lung diseases has been described in the CTDs, and the CT patterns can often be diagnostic or suggestive, including UIP (see Figures 45.2 and 45.3), NSIP (see Figures 45.4 and 45.5), organizing pneumonia (Figures 45.7 and 45.8), lymphocytic interstitial pneumonia (Figures 45.9 and 45.10), DAD (Figures 45.11 and 45.12), and DAH (see Figure 45.6), with rare cases of desquamative interstitial pneumonia and eosinophilic pneumonia also having been described [18, 70]. As with other CTDs (other than RA), NSIP is the most common histopathologic pattern seen in SLE-related ILD [70].

The precise role of surgical lung biopsy for histopathology in patients with SLE remains to be determined [75]. In the absence of otherwise convincing data, the underlying histopathologic pattern should be considered to contain important information that can help guide treatment decisions, and histopathology is also useful in evaluating concurrent pathology such as vasculitis. In this regard, although the pattern on HRCT may be known to correlate highly with a particular

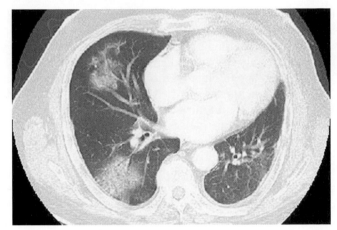

FIGURE 45.7 High-resolution computed tomography image showing focal peripheral areas of consolidation suggestive of organizing pneumonia.

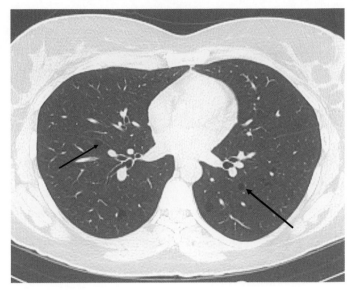

FIGURE 45.9 High-resolution computed tomography image showing numerous thin-walled cysts (i.e., cystic lung disease) most consistent with resolved lymphocytic interstitial pneumonia.

underlying histopathologic pattern that is recognized to complicate SLE and thus provide sufficient evidence in some cases to define the problem precisely, in those patients with an atypical or unclassifiable radiologic pattern, a surgical lung biopsy may assist with diagnosis, prognosis, and treatment decisions.

Most patients with SLE-related chronic ILD will have a pattern of NSIP or, less commonly, UIP [18, 70, 76]. The NSIP pattern encompasses a wide morphological spectrum with varying degrees of alveolar septa inflammation and fibrosis [18, 70, 76]. The more cellular pattern features mild to moderate alveolar mononuclear cell infiltrate, with involvement of the peribronchial region, interlobular septa, and visceral pleura. Lung architecture is preserved without associated fibrosis. Organizing pneumonia may be present as an overlapping pattern [77]. Fibrotic variants are characterized by alveolar septa fibrosis, with less cellularity and little or no honeycombing present (see Figures 45.4 and 45.5). Less commonly,

a patchy, temporally heterogeneous pattern with honeycombing is noted and denotes the presence of UIP (see Figures 45.2 and 45.3) [18].

Lymphocytic interstitial pneumonia, another pattern of chronic ILD, is rarely associated with SLE [78]. It is a relatively more benign lymphoinfiltrative disorder characterized by interstitial thickening by increased

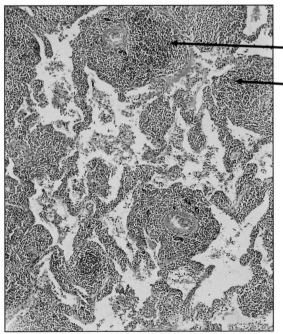

FIGURE 45.10 Photomicrograph of a surgical lung biopsy showing lymphocytic interstitial pneumonia in which the interstitium is thickened by a lymphocytic infiltrate. In places (arrows), prominent lymphoid follicles are seen.

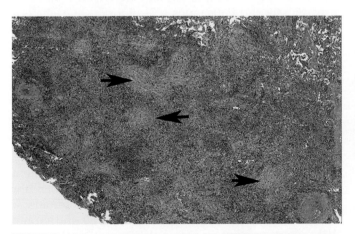

FIGURE 45.8 Photomicrograph of a surgical lung biopsy specimen demonstrating dense foci of organizing pneumonia.

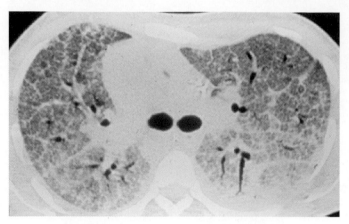

FIGURE 45.11 High-resolution computed tomography image of a patient with diffuse alveolar damage showing the widespread ground glass change becoming consolidated especially at the left base with virtually no normal lung present.

numbers of small lymphocytes and plasma cells with additional infiltration of bronchovascular bundles, interlobular septa, and pleura [18]. HRCT findings are similar to those of NSIP but may also include ill-defined centrilobular nodules (1 or 2 cm) and scattered thin-walled cysts (see Figure 45.9) [19, 20].

As reviewed by Fischer and colleagues [79], not all patients with CTD-associated ILD require pharmacologic treatment. Although radiographic abnormalities may be identified, only a subset of patients will show clinically significant, progressive disease. The decision to treat SLE—ILD often rests upon: (1) whether the patient is clinically impaired by the ILD, (2) whether the ILD is progressive, and (3) what contraindications or mitigating factors exist. Therapy for SLE—ILD is generally reserved for those patients with clinically significant, progressive disease, and this determination is based on a constellation of clinical assessment tools

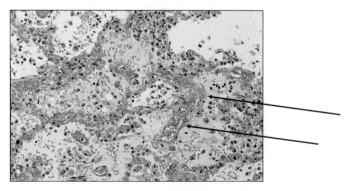

FIGURE 45.12 Photomicrograph of a surgical lung biopsy specimen demonstrating diffuse alveolar damage. The airspaces contain products of capillary leakage including red blood cells. Hyaline membranes are beginning to be formed on the inner alveolar surface (*arrows*).

that include both subjective and objective measures of respiratory impairment [79].

When approaching the choice of a disease-modifying regimen directly from an ILD perspective, there are only limited data available to guide the choice of specific pharmacologic agents for the treatment of CTD-related ILD as a group and no controlled data to guide the management of SLE-related ILD [79].

Based on a number of small, uncontrolled trials [80—86] and two controlled trials suggesting stabilization of lung function and a survival benefit to CYC for the treatment of SSc—ILD [87—89], as well as other forms of CTD—ILD, we consider CYC (either oral or intravenous) to be the preferred therapy for severe, progressive ILD associated with any of the CTDs [79].

Although commonly used in SLE in general, there are no controlled trials evaluating the efficacy of AZA for SLE—ILD. Isolated case reports, small case series, and anecdotal experience suggest that AZA has some degree of efficacy for CTD—ILD in general [90—93]. In one large trial of IPF patients, AZA plus corticosteroids (CS) plus the antioxidant N-acetylcysteine has been shown to preserve lung function as measured by FVC and DLCO compared with CS plus AZA alone [94]. It is not known whether similar efficacy can be extrapolated for SLE-related UIP.

Although data from controlled trials have supported the use of mycophenolate mofetil (MMF) in subsets of patients with lupus nephritis [95], there are no controlled trials evaluating the efficacy of MMF for SLE-related ILD. In 2006, our group published our experience with MMF for the treatment of 28 patients with CTD-associated ILD [96]. Only 1 patient in this cohort had SLE. During the course of therapy, we found that it was possible to reduce the average daily prednisone dose from 15 to 10 mg, and that the percent predicted FVC increased on average by 2.3% and the percent predicted DLCO increased by 2.6%. Moreover, the drug was well tolerated with a paucity of side effects [96]. Similarly, Liossis and colleagues [97] reported their experience with SSc—ILD in 6 patients; they found improvement in 5 of 6 patients both clinically and with objective testing. Two additional centers have published case series of 17 and 13 patients, respectively, both demonstrating that patients with SSc—ILD who were treated with MMF had stable or improved lung function, and that the drug was again well tolerated [98, 99]. It is not known whether these data can be extrapolated to SLE-related ILD.

Despite the lack of any certainty about treatment approaches in the literature, our usual regimen for SLE-related ILD includes the use of corticosteroids at a moderate to high dose (e.g., between 30 and 60 mg daily of prednisone or equivalent, or 0.25—0.75 mg/kg/day) combined typically with CYC for the most severe or progressive cases or with AZA, MMF, or other immunosuppressive agents (e.g., cyclosporine or tacrolimus) for less severe disease [79].

Shrinking Lung Syndrome

The 'shrinking lung syndrome,' first described by Hoffbrand and Beck in 1965 [100], is manifest by dyspnea and best characterized as diaphragmatic dysfunction leading to diminished lung volumes [101—104]. Radiographic evaluation usually shows diaphragmatic elevation associated with bi-basilar atelectasis—in the absence of parenchymal lung abnormalities [100—104] (Figures 45.13 and 45.14). PFT assessment typically reveals a restrictive ventilatory defect and a reduced DLCO that normalizes when corrected for alveolar volume. The hallmark of shrinking lung syndrome is the presence of these PFT features in the absence of parenchymal lung abnormalities but with disproportionately severe dyspnea (considering the absence of parenchymal radiologic changes) [103].

The precise pathogenesis of shrinking lung syndrome remains to be determined [104]. Gibson and colleagues [102] found abnormal transdiaphragmatic pressures in 5 SLE patients, suggesting diaphragmatic weakness as the underlying cause of shrinking lung syndrome. Extending these findings, Martens and colleagues [105] found that weakness of the respiratory muscles led to the restrictive defect and diaphragmatic dysfunction in their cohort of 7 patients with the syndrome. In contrast to these studies, however, Laroche and colleagues [106] found no evidence of respiratory muscle weakness or diaphragmatic abnormalities in their cohort of 12 SLE patients with the syndrome. The reasons for the discordant results are not known but

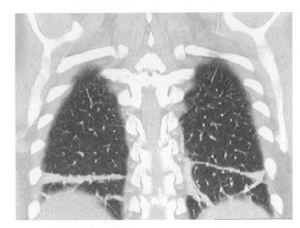

FIGURE 45.14 Coronal reconstruction of CT scan shows bilateral basal linear atelectasis in a patient with shrinking lung syndrome.

may reflect patient selection or differences in the modalities employed to assess diaphragmatic strength and function. Other investigations have failed to demonstrate any specific serologic abnormalities in these patients, global evidence of myopathy or myositis, phrenic nerve abnormalities, or diagnostic abnormalities by electromyography [103, 106—108]. Despite lack of clarity regarding its pathogenesis, assessment for diaphragmatic and other respiratory muscle dysfunction in patients suspected to have shrinking lung syndrome is often indicated, including phrenic nerve stimulation studies.

Shrinking lung syndrome is usually not progressive, but it does manifest as chronic and persistent dyspnea [104]. Small case series have shown that inhaled β-agonist or theophylline therapy may favorably impact the syndrome [109—111]. Alternatively, immunosuppression with corticosteroids or immunosuppressive agents such as AZA, methotrexate, CYC, or rituximab may improve the condition as well and could be considered on an individualized basis [103, 104, 112—114].

Pulmonary Vascular Disease

Pulmonary Hypertension

In contrast to SSc and MCTD, severe or symptomatic PH is relatively rare in SLE [4, 24, 115]. Mild, subclinical PH, however, is relatively common, with a reported prevalence range of 0.5—14% [116—118]. Data on prevalence and characteristics of SLE-associated PH are predominately based on case reports and small series with varied spectra of populations studied and modalities used to assess for PH [116]. These data do suggest that PH usually occurs within established SLE, but it may be the presenting

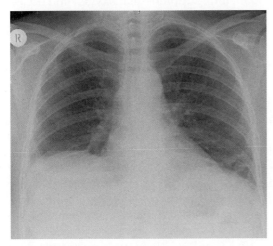

FIGURE 45.13 Chest radiograph shows bilateral diaphragmatic elevation with basal atelectasis in a patient with shrinking lung syndrome.

manifestation. SLE-associated PH typically occurs in women, but it is not clear whether PH is more common in SLE patients with Raynaud's or with aPL antibodies [116]. Co-existent parenchymal lung disease (i.e., chronic ILD) also appears to be associated with the development of PH [119]. Chronic thromboembolic disease is rarely identified as the cause for PH in SLE patients [120].

The clinical presentation of PH in patients with SLE is similar to the presentation of PH in general, with exertional fatigue and dyspnea predominating. More severe disease manifests with characteristic features of right heart failure. Clues to the presence of PH may include a disproportionate reduction in the DLCO, exertional hypoxemia, or elevated brain natriuretic peptide level [23]. Echocardiographic assessment is the best noninvasive tool for PH assessment but is not especially sensitive or specific, and confirmation of PH requires right heart catheterization [23].

The pathophysiologic mechanisms of SLE-associated PH are not known, but prevailing theories include pulmonary circulatory vasospasm, intrinsic vasculopathy, *in situ* thrombosis, and endothelial dysfunction as possible causes [116]. It is likely that the etiology is multifactorial in nature.

In patients with symptomatic SLE-associated PH, therapeutic interventions are not evidence based, but case reports and small case series support the use of vasodilator therapy with calcium channel blockers, phosphodiesterase inhibitors, endothelin receptor antagonists, and, in more severe cases, prostacyclin therapy [116, 121]. In contrast to SSc-associated PH, systemic immunosuppressive therapies (including high-dose corticosteroids and cyclophosphamide) may be useful in SLE-associated PH [122–124]. Although anticoagulation therapy is often administered in patients with idiopathic pulmonary arterial hypertension [121], its use is more controversial in SLE-associated PH due to concerns for a higher incidence of bleeding complications in this population.

Thromboembolism

Pulmonary embolism should be considered in any SLE patient with new-onset pleuritic chest pain or acute dyspnea. Thromboembolism occurs in up to 25% of SLE patients and is a major cause of morbidity and mortality [4]. SLE patients with positive aPL antibodies are at a particularly high risk for the development of thromboembolism [5, 125–129]. In their meta-analysis, Wahl and colleagues [128] found that SLE patients with aPL antibodies were more than six times more likely than SLE patients without aPL to develop thromboemboli. Similarly, Somers and colleagues [129] showed that SLE patients with positive lupus anticoagulant were five times more likely to develop

thromboembolism compared with SLE patients with negative lupus anticoagulant. Evaluation for pulmonary embolism starts with noninvasive assessment that usually includes V/Q scan or CT angiography but may require pulmonary angiography for a definitive diagnosis.

As for other patients found to have thromboembolic disease, the mainstay of therapy is anticoagulation with heparin followed by warfarin. SLE patients with thromboembolism and positive aPL antibodies usually require lifelong anticoagulation [130].

See Chapter 50 for more discussion of thromboembolism and aPL antibodies.

Acute Reversible Hypoxia Syndrome

In 1991, Abramson and colleagues [131] described a syndrome of acute reversible hypoxemia affecting 6 of 22 hospitalized SLE patients. These 6 patients all presented with acute and reversible hypoxemia in the absence of thoracic radiographic abnormalities, had symptoms of dyspnea and pleurisy, and had a wide alveolar-arterial gradient that improved with corticosteroid treatment [131]. In 1995, Martinez-Taboada and colleagues [132] described a similar clinical picture of acute reversible hypoxemia in 4 of 16 hospitalized SLE patients. The pathogenesis of this syndrome is not known. Some have suggested that the syndrome is caused by complement-mediated aggregation and activation of neutrophils within the pulmonary vasculature leading to acute reversible hypoxemia or states of active SLE that lead to increased surface expression of adhesion molecules and a resultant leuko-occlusive pulmonary vasculopathy [131–134]. Therapeutic interventions for patients with this syndrome are typically supportive in nature to manage the acute hypoxemic state and include supplemental oxygen and occasionally mandate a short-term course of corticosteroids.

Airways Disease

Airflow limitation, defined by typical PFT abnormalities, has been observed in patients with SLE, but the clinical significance of this finding is not known [13–15]. Clinically meaningful (i.e., symptomatic) airway abnormalities or obstructive lung (airways) disease are uncommon. There are isolated case reports of SLE patients having hypopharyngeal ulceration, subglottic stenosis, laryngeal inflammation, epiglottitis, vocal cord paralysis, laryngeal involvement associated with SLE vasculitis, and postintubation complications [4, 135–138]. Similarly, there are case reports describing the presence of obliterative bronchiolitis or bronchiectasis in SLE patients [22, 139, 140].

AREAS OF DEBATE

Is "Acute Lupus Pneumonitis" a Distinct Entity Specific to SLE?

There is debate regarding the definition of acute lupus pneumonitis and whether it exists as a distinct entity, specific to SLE. Swigris and colleagues [24] suggest that acute lupus pneumonitis can be considered to be DAD and suggest restriction of the use of the term "acute lupus pneumonitis" to biopsy-proven cases of DAD in the absence of hemorrhage or capillaritis. Given that the histopathologic evaluation in biopsied cases of acute lupus pneumonitis is consistent with DAD (i.e., the histopathologic correlate of the idiopathic disease, acute interstitial pneumonia AIP)—with the presence of interstitial edema and hyaline membrane formation in the absence of vasculitis, capillaritis, or hemorrhage [32, 49]—it is likely that this process reflects an acute lung injury similar to that seen in other CTDs (e.g., RA or SSc). Moreover, because this acute lung injury is known to occur among patients with CTD in general, it is not clear whether there is anything unique about the acute pneumonitis reported in patients with SLE. In addition, because there are reports of a high percentage of co-existent respiratory infection among those presenting with acute lupus pneumonitis [57, 60], the possibility that DAD develops as a consequence of an infectious insult deserves consideration as well.

What is the Intrinsic Abnormality Leading to the Shrinking Lung Syndrome and How is it Best Treated?

Of the respiratory manifestations associated with SLE, none have been as unique to SLE—and yet so poorly understood—as the shrinking lung syndrome. Since its first description in 1965 [100], theories of the pathogenesis of shrinking lung syndrome—as reviewed by Toya and Tzelepis [104]—have included those related to respiratory myopathy, phrenic nerve neuropathy, pleural adhesions, or refractory pleuritic chest pains leading to diaphragmatic dysfunction. The exact cause and pathophysiology remain unknown.

The treatment of shrinking lung syndrome is not evidence based and, in our experience, revolves around the combination of immunosuppressive therapies along with modalities to enhance diaphragmatic dysfunction. We advocate the use of moderate doses of corticosteroids initially (ranging from 0.5 to 0.75 mg/kg of prednisone per day) followed by corticosteroid tapering as a steroid-sparing agent such as AZA or MMF is added. This, in our experience, combined with β-agonist inhaled therapies and theophylline, usually leads to some degree of subjective improvement and stabilization of pulmonary function.

Should Practitioners Screen SLE Patients for Lung Disease?

An area of continued debate relates to whether screening for lung disease is indicated for CTD patients in general, including SLE. Most authorities agree that screening for lung disease in patients with SSc is indicated [141]. In this regard, although no exact screening protocols exist, clinicians are advised to screen SSc patients for both ILD and PH [141]. This recommendation is based on the fact that pulmonary disease is the leading cause of morbidity and mortality for SSc and the belief that earlier detection of SSc-associated ILD or SSc-associated PH may lead to improved survival. Data show that therapeutic interventions for SSc-associated ILD and SSc-associated PH lead to subjective and objective improvement in quality of life and pulmonary disease status [88, 89, 142]. Although not yet proven by clinical trials, it is likely that these novel therapies (particularly those for SSc-associated PH) also will lead to improved survival. Given the prevalence of lung disease in SSc, and the potentially favorable impact by therapeutic intervention, the concept of screening for the presence of lung disease in SSc seems justified.

It is of much more significant debate when the discussion of screening for lung disease is extended to the other CTDs in which the prevalence of lung disease is much lower and where there are no convincing data to suggest that specific intervention leads to improved outcomes. In our practice, we in fact do "screen" for lung disease among all of our CTD patients by carefully assessing for respiratory symptoms or physical examination signs of impairment at each clinical encounter. Any abnormality detected—even if only mild or subtle—will prompt further investigation with office spirometry, ambulatory oximetry, and chest radiography and may lead to pulmonary consultation. In other words, although we cannot advocate wide-scale screening for lung disease for all CTD (or SLE) patients, we do highlight the importance of clinical vigilance in this situation while maintaining a low threshold to proceed with pulmonary evaluation given the potentially devastating manifestations of lung disease that can occur in all CTDs, including SLE.

Forme Fruste Presentation of Lung Disease in SLE and Other CTDs

It is well-known that lung disease may be the presenting manifestation of CTD [71, 143–146]. In particular, we and others have demonstrated the importance of cross-specialty collaboration in the assessment of *de novo*

ILD and the importance of thorough evaluation to exclude forme fruste presentation of CTD [71, 143, 144, 146, 147]. An area of debate within this context relates to the accurate classification of CTD when lung disease is the only manifestation to suggest a systemic autoimmune disease. For example, some have suggested that nonspecific interstitial pneumonia may be representative of an autoimmune disease and could be considered the lung manifestation of "undifferentiated connective tissue disease" (UCTD) [148, 149]. In addition, it is not uncommon to encounter autoantibodies in patients with "idiopathic" ILD in the absence of other definitive CTD features. Does a patient with NSIP, no extrathoracic features of CTD, and positive ANA, Smith, and RNP antibodies have an underlying CTD? Is this UCTD? Given the specificity for the Smith and RNP antibodies for SLE, is this the forme fruste presentation of SLE? In our opinion, the concept of "lung-dominant CTD" or "autoimmune lung disease" exists, and further research is needed to understand its pathogenesis and to better characterize and classify these patients [150]. It is reasonable to hypothesize that some of these lung-dominant CTD patients will evolve into more defined cases of CTD and yet some may remain with autoimmune lung involvement only.

References

[1] R. Grigor, J. Edmonds, R. Lewkonia, B. Bresnihan, G.R. Hughes, Systemic lupus erythematosus. A prospective analysis, Ann Rheum Dis 37 (1978) 121–128.

[2] M. Gross, J.R. Esterly, R.H. Earle, Pulmonary alterations in systemic lupus erythematosus, Am Rev Respir Dis 105 (1972) 572–577.

[3] S.T. Holgate, D.N. Glass, P. Haslam, R.N. Maini, M. Turner-Warwick, Respiratory involvement in systemic lupus erythematosus. A clinical and immunological study, Clin Exp Immunol 24 (1976) 385–395.

[4] S. Murin, H.P. Wiedemann, R.A. Matthay, Pulmonary manifestations of systemic lupus erythematosus, Clin Chest Med 19 (1998) 641–665, viii.

[5] A.M. Bertoli, L.M. Vila, M. Apte, B.J. Fessler, H.M. Bastian, J.D. Reveille, et al., Systemic lupus erythematosus in a multiethnic U.S. cohort LUMINA XLVIII: Factors predictive of pulmonary damage, Lupus 16 (2007) 410–417.

[6] T. Eaton, P. Young, D. Milne, A.U. Wells, Six-minute walk, maximal exercise tests: Reproducibility in fibrotic interstitial pneumonia, Am J Respir Crit Care Med 171 (2005) 1150–1157.

[7] M.H. Buch, C.P. Denton, D.E. Furst, L. Guillevin, L.J. Rubin, A.U. Wells, et al., Submaximal exercise testing in the assessment of interstitial lung disease secondary to systemic sclerosis: Reproducibility and correlations of the 6-min walk test, Ann Rheum Dis 66 (2007) 169–173.

[8] K.R. Flaherty, J.A. Mumford, S. Murray, E.A. Kazerooni, B.H. Gross, T.V. Colby, et al., Prognostic implications of physiologic and radiographic changes in idiopathic interstitial pneumonia, Am J Respir Crit Care Med 168 (2003) 543–548.

[9] P.I. Latsi, R.M. du Bois, A.G. Nicholson, T.V. Colby, D. Bisirtzoglou, A. Nikolakopoulou, et al., Fibrotic idiopathic interstitial pneumonia: The prognostic value of longitudinal functional trends, Am J Respir Crit Care Med 168 (2003) 531–537.

[10] H.R. Collard, T.E. King Jr., B.B. Bartelson, J.S. Vourlekis, M.I. Schwarz, K.K. Brown, Changes in clinical and physiologic variables predict survival in idiopathic pulmonary fibrosis, Am J Respir Crit Care Med 168 (2003) 538–542.

[11] S.L. Silberstein, P. Barland, A.I. Grayzel, S.K. Koerner, Pulmonary dysfunction in systemic lupus erythematosus: Prevalence classification and correlation with other organ involvement, J Rheumatol 7 (1980) 187–195.

[12] C.T. Huang, G.R. Hennigar, H.A. Lyons, Pulmonary dysfunction in systemic lupus erythematosus, N Engl J Med 272 (1965) 288–293.

[13] D. Wohlgelernter, J. Loke, R.A. Matthay, N.J. Siegel, Systemic and discoid lupus erythematosus: Analysis of pulmonary function, Yale J Biol Med 51 (1978) 157–164.

[14] A.P. Andonopoulos, S.H. Constantopoulos, V. Galanopoulou, A.A. Drosos, N.C. Acritidis, H.M. Moutsopoulos, Pulmonary function of nonsmoking patients with systemic lupus erythematosus, Chest 94 (1988) 312–315.

[15] P.Q. Eichacker, K. Pinsker, A. Epstein, J. Schiffenbauer, A. Grayzel, Serial pulmonary function testing in patients with systemic lupus erythematosus, Chest 94 (1988) 129–132.

[16] J.G. Bulgrin, E.L. Dubois, G. Jacobson, Chest roentgenographic changes in systemic lupus erythematosus, Radiology 74 (1960) 42–49.

[17] D.M. Gould, M.L. Daves, Roentgenologic findings in systemic lupus erythematosus; An analysis of 100 cases, J Chronic Dis 2 (1955) 136–145.

[18] American Thoracic Society/European Respiratory Society, American Thoracic Society/European Respiratory Society International Multidisciplinary Consensus Classification of the Idiopathic Interstitial Pneumonias. This joint statement of the American Thoracic Society (ATS), and the European Respiratory Society (ERS) was adopted by the ATS board of directors, June 2001 and by the ERS Executive Committee, June 2001, Am J Respir Crit Care Med 165 (2002) 277–304.

[19] D.A. Lynch, Quantitative CT of fibrotic interstitial lung disease, Chest 131 (2007) 643–644.

[20] D.A. Lynch, W.D. Travis, N.L. Muller, J.R. Galvin, D.M. Hansell, P.A. Grenier, et al., Idiopathic interstitial pneumonias: CT features, Radiology 236 (2005) 10–21.

[21] S.V. Kocheril, B.E. Appleton, E.C. Somers, E.A. Kazerooni, K.R. Flaherty, F.J. Martinez, et al., Comparison of disease progression and mortality of connective tissue disease-related interstitial lung disease and idiopathic interstitial pneumonia, Arthritis Rheum 53 (2005) 549–557.

[22] H.M. Fenlon, M. Doran, S.M. Sant, E. Breatnach, High-resolution chest CT in systemic lupus erythematosus, AJR Am J Roentgenol 166 (1996) 301–307.

[23] M. McGoon, D. Gutterman, V. Steen, R. Barst, D.C. McCrory, T.A. Fortin, et al., Screening, early detection, and diagnosis of pulmonary arterial hypertension: ACCP evidence-based clinical practice guidelines, Chest 126 (2004) 14S–34S.

[24] J.J. Swigris, A. Fischer, J. Gillis, R.T. Meehan, K.K. Brown, Pulmonary and thrombotic manifestations of systemic lupus erythematosus, Chest 133 (2008) 271–280.

[25] R.M. Silver, K.S. Miller, M.B. Kinsella, E.A. Smith, S.I. Schabel, Evaluation and management of scleroderma lung disease using bronchoalveolar lavage, Am J Med 88 (1990) 470–476.

[26] C. Strange, M.B. Bolster, M.D. Roth, R.M. Silver, A. Theodore, J. Goldin, et al., Bronchoalveolar lavage and response to cyclophosphamide in scleroderma interstitial lung disease, Am J Respir Crit Care Med 177 (2008) 91–98.

[27] N.S. Goh, S. Veeraraghavan, S.R. Desai, D. Cramer, D.M. Hansell, C.P. Denton, et al., Bronchoalveolar lavage cellular profiles in patients with systemic sclerosis-associated interstitial lung disease are not predictive of disease progression, Arthritis Rheum 56 (2007) 2005–2012.

[28] S. Quadrelli, C. Alvarez, S. Arce, L. Paz, J. Sarano, E. Sobrino, et al., Pulmonary involvement of systemic lupus erythematosus: Analysis of 90 necropsies, Lupus 18 (2009) 1053–1060.

[29] H.M. Haupt, G.W. Moore, G.M. Hutchins, The lung in systemic lupus erythematosus. Analysis of the pathologic changes in 120 patients, Am J Med 71 (1981) 791–798.

[30] P.H. Feng, T.H. Tan, Tuberculosis in patients with systemic lupus erythematosus, Ann Rheum Dis 41 (1982) 11–14.

[31] D.B. Hellmann, M. Petri, Q. Whiting-O'Keefe, Fatal infections in systemic lupus erythematosus: The role of opportunistic organisms, Medicine (Baltimore) 66 (1987) 341–348.

[32] M.P. Keane, J.P. Lynch 3rd, Pleuropulmonary manifestations of systemic lupus erythematosus, Thorax 55 (2000) 159–166.

[33] H.P. Wiedemann, R.A. Matthay, Pulmonary manifestations of the collagen vascular diseases, Clin Chest Med 10 (1989) 677–722.

[34] B. Memet, E.M. Ginzler, Pulmonary manifestations of systemic lupus erythematosus, Semin Respir Crit Care Med 28 (2007) 441–450.

[35] R. Cervera, M.A. Khamashta, J. Font, G.D. Sebastiani, A. Gil, P. Lavilla, et al., Systemic lupus erythematosus: Clinical and immunologic patterns of disease expression in a cohort of 1000 patients. The European Working Party on Systemic Lupus Erythematosus, Medicine (Baltimore) 72 (1993) 113–124.

[36] E.L. Dubois, D.L. Tuffanelli, Clinical manifestations of systemic lupus erythematosus. Computer analysis of 520 cases, JAMA 190 (1964) 104–111.

[37] D. Estes, C.L. Christian, The natural history of systemic lupus erythematosus by prospective analysis, Medicine (Baltimore) 50 (1971) 85–95.

[38] M.M. Ward, S. Studenski, Clinical manifestations of systemic lupus erythematosus. Identification of racial and socioeconomic influences, Arch Intern Med 150 (1990) 849–853.

[39] W.A. Winslow, L.N. Ploss, B. Loitman, Pleuritis in systemic lupus erythematosus: Its importance as an early manifestation in diagnosis, Ann Intern Med 49 (1958) 70–88.

[40] K.S. Wan, Pleuritis and pleural effusion as the initial presentation of systemic lupus erythematous in a 23-year-old woman, Rheumatol Int 28 (2008) 1257–1260.

[41] J.T. Good Jr., T.E. King, V.B. Antony, S.A. Sahn, Lupus pleuritis. Clinical features and pleural fluid characteristics with special reference to pleural fluid antinuclear antibodies, Chest 84 (1983) 714–718.

[42] D.T. Carr, G.A. Lillington, J.G. Mayne, Pleural-fluid glucose in systemic lupus erythematosus, Mayo Clin Proc 45 (1970) 409–412.

[43] G.A. Lillington, D.T. Carr, J.G. Mayne, Rheumatoid pleurisy with effusion, Arch Intern Med 128 (1971) 764–768.

[44] M. Leechawengwong, H.W. Berger, M. Sukumaran, Diagnostic significance of antinuclear antibodies in pleural effusion, Mt Sinai J Med 46 (1979) 137–139.

[45] V. Khare, B. Baethge, S. Lang, R.E. Wolf, G.D. Campbell Jr., Antinuclear antibodies in pleural fluid, Chest 106 (1994) 866–871.

[46] J.B. Orens, F.J. Martinez, J.P. Lynch 3rd, Pleuropulmonary manifestations of systemic lupus erythematosus, Rheum Dis Clin North Am 20 (1994) 159–193.

[47] R.A. Matthay, M.I. Schwarz, T.L. Petty, R.E. Stanford, R.C. Gupta, S.A. Sahn, et al., Pulmonary manifestations of systemic lupus erythematosus: Review of twelve cases of acute lupus pneumonitis, Medicine (Baltimore) 54 (1975) 397–409.

[48] S. Carette, A.M. Macher, A. Nussbaum, P.H. Plotz, Severe, acute pulmonary disease in patients with systemic lupus erythematosus: Ten years of experience at the National Institutes of Health, Semin Arthritis Rheum 14 (1984) 52–59.

[49] A.M. Harvey, L.E. Shulman, P.A. Tumulty, C.L. Conley, E.H. Schoenrich, Systemic lupus erythematosus: Review of the literature and clinical analysis of 138 cases, Medicine (Baltimore) 33 (1954) 291–437.

[50] C. Abud-Mendoza, E. Diaz-Jouanen, D. Alarcon-Segovia, Fatal pulmonary hemorrhage in systemic lupus erythematosus. Occurrence without hemoptysis, J Rheumatol 12 (1985) 558–561.

[51] L.A. Barile, L.J. Jara, F. Medina-Rodriguez, J.L. Garcia-Figueroa, J.M. Miranda-Limon, Pulmonary hemorrhage in systemic lupus erythematosus, Lupus 6 (1997) 445–448.

[52] W.U. Kim, J.K. Min, S.H. Lee, S.H. Park, C.S. Cho, H.Y. Kim, Causes of death in Korean patients with systemic lupus erythematosus: A single center retrospective study, Clin Exp Rheumatol 17 (1999) 539–545.

[53] M.F. Liu, J.H. Lee, T.H. Weng, Y.Y. Lee, Clinical experience of 13 cases with severe pulmonary hemorrhage in systemic lupus erythematosus with active nephritis, Scand J Rheumatol 27 (1998) 291–295.

[54] C.T. Marino, L.P. Pertschuk, Pulmonary hemorrhage in systemic lupus erythematosus, Arch Intern Med 141 (1981) 201–203.

[55] G. Mintz, L.F. Galindo, J. Fernandez-Diez, F.J. Jimenez, E. Robles-Saavedra, R.D. Enriquez-Casillas, Acute massive pulmonary hemorrhage in systemic lupus erythematosus, J Rheumatol 5 (1978) 39–50.

[56] A.S. Santos-Ocampo, B.F. Mandell, B.J. Fessler, Alveolar hemorrhage in systemic lupus erythematosus: Presentation and management, Chest 118 (2000) 1083–1090.

[57] M.R. Zamora, M.L. Warner, R. Tuder, M.I. Schwarz, Diffuse alveolar hemorrhage and systemic lupus erythematosus. Clinical presentation, histology, survival, and outcome, Medicine (Baltimore) 76 (1997) 192–202.

[58] J.W. Eagen, V.A. Memoli, J.L. Roberts, G.R. Matthew, M.M. Schwartz, E.J. Lewis, Pulmonary hemorrhage in systemic lupus erythematosus, Medicine (Baltimore) 57 (1978) 545–560.

[59] E.P. Schwab, H.R. Schumacher Jr., B. Freundlich, P.E. Callegari, Pulmonary alveolar hemorrhage in systemic lupus erythematosus, Semin Arthritis Rheum 23 (1993) 8–15.

[60] J. Rojas-Serrano, J. Pedroza, J. Regalado, J. Robledo, E. Reyes, J. Sifuentes-Osornio, et al., High prevalence of infections in patients with systemic lupus erythematosus and pulmonary haemorrhage, Lupus 17 (2008) 295–299.

[61] S. Castaneda, G. Herrero-Beaumont, A. Valenzuela, J. Vidal, F. Moldenhauer, G. Renedo, et al., Massive pulmonary hemorrhage: Fatal complication of systemic lupus erythematosus, J Rheumatol 12 (1985) 186–187.

[62] A. Churg, W. Franklin, K.L. Chan, E. Kopp, C.B. Carrington, Pulmonary hemorrhage and immune-complex deposition in the lung. Complications in a patient with systemic lupus erythematosus, Arch Pathol Lab Med 104 (1980) 388–391.

[63] M.R. Desnoyers, S. Bernstein, A.G. Cooper, R.I. Kopelman, Pulmonary hemorrhage in lupus erythematosus without evidence of an immunologic cause, Arch Intern Med 144 (1984) 1398–1400.

[64] R.W. Erickson, W.A. Franklin, W. Emlen, Treatment of hemorrhagic lupus pneumonitis with plasmapheresis, Semin Arthritis Rheum 24 (1994) 114–123.

[65] D.F. Huang, S.T. Tsai, S.R. Wang, Recovery of both acute massive pulmonary hemorrhage and acute renal failure in a systemic lupus erythematosus patient with lupus

anticoagulant by the combined therapy of plasmapheresis plus cyclophosphamide, Transfus Sci 15 (1994) 283–288.

[66] R.P. Millman, T.B. Cohen, A.I. Levinson, M.A. Kelley, M.L. Sachs, Systemic lupus erythematosus complicated by acute pulmonary hemorrhage: Recovery following plasmapheresis and cytotoxic therapy, J Rheumatol 8 (1981) 1021–1023.

[67] H.H. Euler, J.O. Schroeder, P. Harten, R.A. Zeuner, H.J. Gutschmidt, Treatment-free remission in severe systemic lupus erythematosus following synchronization of plasmapheresis with subsequent pulse cyclophosphamide, Arthritis Rheum 37 (1994) 1784–1794.

[68] H. Eisenberg, E.L. Dubois, R.P. Sherwin, O.J. Balchum, Diffuse interstitial lung disease in systemic lupus erythematosus, Ann Intern Med 79 (1973) 37–45.

[69] L. Weinrib, O.P. Sharma, F.P. Quismorio Jr., A long-term study of interstitial lung disease in systemic lupus erythematosus, Semin Arthritis Rheum 20 (1990) 48–56.

[70] J.H. Park, D.S. Kim, I.N. Park, S.J. Jang, M. Kitaichi, A.G. Nicholson, et al., Prognosis of fibrotic interstitial pneumonia: Idiopathic versus collagen vascular disease-related subtypes, Am J Respir Crit Care Med 175 (2007) 705–711.

[71] V. Cottin, Interstitial lung disease: are we missing formes frustes of connective tissue disease? Eur Respir J 28 (2006) 893–896.

[72] N.B. Greene, A.M. Solinger, R.P. Baughman, Patients with collagen vascular disease and dyspnea. The value of gallium scanning and bronchoalveolar lavage in predicting response to steroid therapy and clinical outcome, Chest 91 (1987) 698–703.

[73] J.B. Martinot, B. Wallaert, P.Y. Hatron, C. Francis, C. Voisin, Y. Sibille, Clinical and subclinical alveolitis in collagen vascular diseases: Contribution of alpha 2-macroglobulin levels in BAL fluid, Eur Respir J 2 (1989) 437–443.

[74] B. Wallaert, C. Aerts, F. Bart, P.Y. Hatron, M. Dracon, A.B. Tonnel, et al., Alveolar macrophage dysfunction in systemic lupus erythematosus, Am Rev Respir Dis 136 (1987) 293–297.

[75] K.M. Antoniou, G. Margaritopoulos, F. Economidou, N.M. Siafakas, Pivotal clinical dilemmas in collagen vascular diseases associated with interstitial lung involvement, Eur Respir J 33 (2009) 882–896.

[76] A.L. Katzenstein, R.F. Fiorelli, Nonspecific interstitial pneumonia/fibrosis. Histologic features and clinical significance, Am J Surg Pathol 18 (1994) 136–147.

[77] R.B. Gammon, T.A. Bridges, H. al-Nezir, C.B. Alexander, J.I. Kennedy Jr., Bronchiolitis obliterans organizing pneumonia associated with systemic lupus erythematosus, Chest 102 (1992) 1171–1174.

[78] B. Benisch, B. Peison, The association of lymphocytic interstitial pneumonia and systemic lupus erythematosus, Mt Sinai J Med 46 (1979). 398-341.

[79] A. Fischer, K.K. Brown, S.K. Frankel, Treatment of connective tissue disease related interstitial lung disease, Clin Pulm Med 16 (2009) 74–80.

[80] A. Akesson, A. Scheja, A. Lundin, F.A. Wollheim, Improved pulmonary function in systemic sclerosis after treatment with cyclophosphamide, Arthritis Rheum 37 (1994) 729–735.

[81] R. Giacomelli, G. Valentini, F. Salsano, P. Cipriani, P. Sambo, M.L. Conforti, et al., Cyclophosphamide pulse regimen in the treatment of alveolitis in systemic sclerosis, J Rheumatol 29 (2002) 731–736.

[82] I. Pakas, J.P. Ioannidis, K. Malagari, F.N. Skopouli, H.M. Moutsopoulos, P.G. Vlachoyiannopoulos, Cyclophosphamide with low or high dose prednisolone for systemic sclerosis lung disease, J Rheumatol 29 (2002) 298–304.

[83] A. Schnabel, M. Reuter, W.L. Gross, Intravenous pulse cyclophosphamide in the treatment of interstitial lung disease due to collagen vascular diseases, Arthritis Rheum 41 (1998) 1215–1220.

[84] R.M. Silver, J.H. Warrick, M.B. Kinsella, L.S. Staudt, M.H. Baumann, C. Strange, Cyclophosphamide and low-dose prednisone therapy in patients with systemic sclerosis (scleroderma) with interstitial lung disease, J Rheumatol 20 (1993) 838–844.

[85] V.D. Steen, J.K. Lanz Jr., C. Conte, G.R. Owens, T.A. Medsger Jr., Therapy for severe interstitial lung disease in systemic sclerosis. A retrospective study, Arthritis Rheum 37 (1994) 1290–1296.

[86] B. White, W.C. Moore, F.M. Wigley, H.Q. Xiao, R.A. Wise, Cyclophosphamide is associated with pulmonary function and survival benefit in patients with scleroderma and alveolitis, Ann Intern Med 132 (2000) 947–954.

[87] R.K. Hoyles, R.W. Ellis, J. Wellsbury, B. Lees, P. Newlands, N.S. Goh, et al., A multicenter, prospective, randomized, double-blind, placebo-controlled trial of corticosteroids and intravenous cyclophosphamide followed by oral azathioprine for the treatment of pulmonary fibrosis in scleroderma, Arthritis Rheum 54 (2006) 3962–3970.

[88] D.P. Tashkin, R. Elashoff, P.J. Clements, J. Goldin, M.D. Roth, D.E. Furst, et al., Cyclophosphamide versus placebo in scleroderma lung disease, N Engl J Med 354 (2006) 2655–2666.

[89] D.P. Tashkin, R. Elashoff, P.J. Clements, M.D. Roth, D.E. Furst, R.M. Silver, et al., Effects of 1-year treatment with cyclophosphamide on outcomes at 2 years in scleroderma lung disease, Am J Respir Crit Care Med 176 (2007) 1026–1034.

[90] J.M. Cohen, A. Miller, H. Spiera, Interstitial pneumonitis complicating rheumatoid arthritis. Sustained remission with azathioprine therapy, Chest 72 (1977) 521–524.

[91] K. Dheda, U.G. Lalloo, B. Cassim, G.M. Mody, Experience with azathioprine in systemic sclerosis associated with interstitial lung disease, Clin Rheumatol 23 (2004) 306–309.

[92] O. Nadashkevich, P. Davis, M. Fritzler, W. Kovalenko, A randomized unblinded trial of cyclophosphamide versus azathioprine in the treatment of systemic sclerosis, Clin Rheumatol 25 (2006) 205–212.

[93] G. Raghu, W.J. Depaso, K. Cain, S.P. Hammar, C.E. Wetzel, D.F. Dreis, et al., Azathioprine combined with prednisone in the treatment of idiopathic pulmonary fibrosis: A prospective double-blind, randomized, placebo-controlled clinical trial, Am Rev Respir Dis 144 (1991) 291–296.

[94] M. Demedts, J. Behr, R. Buhl, U. Costabel, R. Dekhuijzen, H.M. Jansen, et al., High-dose acetylcysteine in idiopathic pulmonary fibrosis, N Engl J Med 353 (2005) 2229–2242.

[95] T.M. Chan, F.K. Li, C.S. Tang, R.W. Wong, G.X. Fang, Y.L. Ji, et al., Efficacy of mycophenolate mofetil in patients with diffuse proliferative lupus nephritis. Hong Kong-Guangzhou Nephrology Study Group, N Engl J Med 343 (2000) 1156–1162.

[96] J.J. Swigris, A.L. Olson, A. Fischer, D.A. Lynch, G.P. Cosgrove, S.K. Frankel, et al., Mycophenolate mofetil is safe, well tolerated, and preserves lung function in patients with connective tissue disease-related interstitial lung disease, Chest 130 (2006) 30–36.

[97] S.N. Liossis, A. Bounas, A.P. Andonopoulos, Mycophenolate mofetil as first-line treatment improves clinically evident early scleroderma lung disease, Rheumatology (Oxford) 45 (2006) 1005–1008.

[98] A.J. Gerbino, C.H. Goss, J.A. Molitor, Effect of mycophenolate mofetil on pulmonary function in scleroderma-associated interstitial lung disease, Chest 133 (2008) 455–460.

[99] A.C. Zamora, P.J. Wolters, H.R. Collard, M.K. Connolly, B.M. Elicker, W.R. Webb, et al., Use of mycophenolate mofetil to treat scleroderma-associated interstitial lung disease, Respir Med 102 (2008) 150–155.

[100] B.I. Hoffbrand, E.R. Beck, 'Unexplained' dyspnoea and shrinking lungs in systemic lupus erythematosus, Br Med J 1 (1965) 1273–1277.

[101] T.W. Chick, R.J. DeHoratius, B.E. Skipper, R.P. Messner, Pulmonary dysfunction in systemic lupus erythematosus without pulmonary symptoms, J Rheumatol 3 (1976) 262–268.

[102] C.J. Gibson, J.P. Edmonds, G.R. Hughes, Diaphragm function and lung involvement in systemic lupus erythematosus, Am J Med 63 (1977) 926–932.

[103] M.Y. Karim, L.C. Miranda, C.M. Tench, P.A. Gordon, P. D'Cruz D, M.A. Khamashta, et al., Presentation and prognosis of the shrinking lung syndrome in systemic lupus erythematosus, Semin Arthritis Rheum 31 (2002) 289–298.

[104] S.P. Toya, G.E. Tzelepis, Association of the shrinking lung syndrome in systemic lupus erythematosus with pleurisy: A systematic review, Semin Arthritis Rheum 39 (2009) 30–37.

[105] J. Martens, M. Demedts, M.T. Vanmeenen, J. Dequeker, Respiratory muscle dysfunction in systemic lupus erythematosus, Chest 84 (1983) 170–175.

[106] C.M. Laroche, D.A. Mulvey, P.N. Hawkins, M.J. Walport, B. Strickland, J. Moxham, et al., Diaphragm strength in the shrinking lung syndrome of systemic lupus erythematosus, Q J Med 71 (1989) 429–439.

[107] P. Hawkins, A.G. Davison, B. Dasgupta, J. Moxham, Diaphragm strength in acute systemic lupus erythematosus in a patient with paradoxical abdominal motion and reduced lung volumes, Thorax 56 (2001) 329–330.

[108] P.G. Wilcox, H.B. Stein, S.D. Clarke, P.D. Pare, R.L. Pardy, Phrenic nerve function in patients with diaphragmatic weakness and systemic lupus erythematosus, Chest 93 (1988) 352–358.

[109] S. Van Veen, A.J. Peeters, P.J. Sterk, F.C. Breedveld, The 'shrinking lung syndrome' in SLE, treatment with theophylline, Clin Rheumatol 12 (1993) 462–465.

[110] S. Branger, N. Schleinitz, S. Gayet, V. Veit, G. Kaplanski, M. Badier, et al., Shrinking lung syndrome and systemic autoimmune disease, Rev Med Interne 25 (2004) 83–90.

[111] F.J. Munoz-Rodriguez, J. Font, J.R. Badia, C. Miret, J.A. Barbera, R. Cervera, et al., Shrinking lungs syndrome in systemic lupus erythematosus: Improvement with inhaled beta-agonist therapy, Lupus 6 (1997) 412–414.

[112] P.J. Ferguson, M. Weinberger, Shrinking lung syndrome in a 14-year-old boy with systemic lupus erythematosus, Pediatr Pulmonol 41 (2006) 194–197.

[113] K.T. Oud, P. Bresser, R.J. ten Berge, R.E. Jonkers, The shrinking lung syndrome in systemic lupus erythematosus: Improvement with corticosteroid therapy, Lupus 14 (2005) 959–963.

[114] W.M. Stevens, J.G. Burdon, L.E. Clemens, J. Webb, The 'shrinking lungs syndrome'—An infrequently recognised feature of systemic lupus erythematosus, Aust N Z J Med 20 (1990) 67–70.

[115] A. Prabu, K. Patel, C.S. Yee, P. Nightingale, R.D. Situnayake, D.R. Thickett, et al., Prevalence and risk factors for pulmonary arterial hypertension in patients with lupus, Rheumatology (Oxford) 48 (2009) 1506–1511.

[116] J. Pope, An update in pulmonary hypertension in systemic lupus erythematosus—Do we need to know about it? Lupus 17 (2008) 274–277.

[117] J.Y. Shen, S.L. Chen, Y.X. Wu, R.Q. Tao, Y.Y. Gu, C.D. Bao, et al., Pulmonary hypertension in systemic lupus erythematosus, Rheumatol Int 18 (1999) 147–151.

[118] C. Haas, Pulmonary hypertension associated with systemic lupus erythematosus, Bull Acad Natl Med 188 (2004) 985–997.

[119] J.H. Ryu, M.J. Krowka, P.A. Pellikka, K.L. Swanson, M.D. McGoon, Pulmonary hypertension in patients with interstitial lung diseases, Mayo Clin Proc 82 (2007) 342–350.

[120] N.E. Anderson, M.R. Ali, The lupus anticoagulant, pulmonary thromboembolism, and fatal pulmonary hypertension, Ann Rheum Dis 43 (1984) 760–763.

[121] D.B. Badesch, S.H. Abman, G. Simonneau, L.J. Rubin, V.V. McLaughlin, Medical therapy for pulmonary arterial hypertension: Updated ACCP evidence-based clinical practice guidelines, Chest 131 (2007) 1917–1928.

[122] L. Gonzalez-Lopez, E.G. Cardona-Munoz, A. Celis, I. Garcia-de la Torre, G. Orozco-Barocio, M. Salazar-Paramo, Therapy with intermittent pulse cyclophosphamide for pulmonary hypertension associated with systemic lupus erythematosus, Lupus 13 (2004) 105–112.

[123] O. Sanchez, O. Sitbon, X. Jais, G. Simonneau, M. Humbert, Immunosuppressive therapy in connective tissue diseases-associated pulmonary arterial hypertension, Chest 130 (2006) 182–189.

[124] E. Tanaka, M. Harigai, M. Tanaka, Y. Kawaguchi, M. Hara, N. Kamatani, Pulmonary hypertension in systemic lupus erythematosus: Evaluation of clinical characteristics and response to immunosuppressive treatment, J Rheumatol 29 (2002) 282–287.

[125] M.L. Boey, C.B. Colaco, A.E. Gharavi, K.B. Elkon, S. Loizou, G.R. Hughes, Thrombosis in systemic lupus erythematosus: Striking association with the presence of circulating lupus anticoagulant, Br Med J (Clin Res Ed) 287 (1983) 1021–1023.

[126] E.N. Harris, A.E. Gharavi, M.L. Boey, B.M. Patel, C.G. Mackworth-Young, S. Loizou, et al., Anticardiolipin antibodies: Detection by radioimmunoassay and association with thrombosis in systemic lupus erythematosus, Lancet 2 (1983) 1211–1214.

[127] P. Hasselaar, R.H. Derksen, L. Blokzijl, M. Hessing, H.K. Nieuwenhuis, B.N. Bouma, et al., Risk factors for thrombosis in lupus patients, Ann Rheum Dis 48 (1989) 933–940.

[128] D.G. Wahl, F. Guillemin, E. de Maistre, C. Perret, T. Lecompte, G. Thibaut, Risk for venous thrombosis related to antiphospholipid antibodies in systemic lupus erythematosus—A meta-analysis, Lupus 6 (1997) 467–473.

[129] E. Somers, L.S. Magder, M. Petri, Antiphospholipid antibodies and incidence of venous thrombosis in a cohort of patients with systemic lupus erythematosus, J Rheumatol 29 (2002) 2531–2536.

[130] M.A. Khamashta, M.J. Cuadrado, F. Mujic, N.A. Taub, B.J. Hunt, G.R. Hughes, The management of thrombosis in the antiphospholipid-antibody syndrome, N Engl J Med 332 (1995) 993–997.

[131] S.B. Abramson, J. Dobro, M.A. Eberle, M. Benton, J. Reibman, H. Epstein, et al., Acute reversible hypoxemia in systemic lupus erythematosus, Ann Intern Med 114 (1991) 941–947.

[132] V.M. Martinez-Taboada, R. Blanco, J. Armona, J.L. Fernandez-Sueiro, V. Rodriguez-Valverde, Acute reversible hypoxemia in systemic lupus erythematosus: A new syndrome or an index of disease activity? Lupus 4 (1995) 259–262.

[133] H.M. Belmont, J. Buyon, R. Giorno, S. Abramson, Up-regulation of endothelial cell adhesion molecules characterizes disease activity in systemic lupus erythematosus. The Shwartzman phenomenon revisited, Arthritis Rheum 37 (1994) 376–383.

[134] C.M. Lloyd, J.A. Gonzalo, D.J. Salant, J. Just, J.C. Gutierrez-Ramos, Intercellular adhesion molecule-1 deficiency prolongs survival and protects against the development of pulmonary inflammation during murine lupus, J Clin Invest 100 (1997) 963–971.

[135] L. Martin, S.M. Edworthy, J.P. Ryan, M.J. Fritzler, Upper airway disease in systemic lupus erythematosus: A report of 4 cases and a review of the literature, J Rheumatol 19 (1992) 1186–1190.

[136] J.M. Toomey, G.G. Snyder 3rd, R.M. Maenza, N.F. Rothfield, Acute epiglottitis due to systemic lupus erythematosus, Laryngoscope 84 (1974) 522–527.

[137] E.D. Burgess, K.C. Render, Hypopharyngeal obstruction in lupus erythematosus, Ann Intern Med 100 (1984) 319.

[138] D.G. Scarpelli, F.W. McCoy, J.K. Scott, Acute lupus erythematosus with laryngeal involvement, N Engl J Med 261 (1959) 691–694.

[139] J. Kallenbach, S. Zwi, H.I. Goldman, Airways obstruction in a case of disseminated lupus erythematosus, Thorax 33 (1978) 814–815.

[140] F. Weber, C. Prior, E. Kowald, M. Schmuth, N. Sepp, Cyclophosphamide therapy is effective for bronchiolitis obliterans occurring as a late manifestation of lupus erythematosus, Br J Dermatol 143 (2000) 453–455.

[141] V.D. Steen, The lung in systemic sclerosis, J Clin Rheumatol 11 (2005) 40–46.

[142] C.P. Denton, J.E. Pope, H.H. Peter, A. Gabrielli, A. Boonstra, F.H. van den Hoogen, et al., Long-term effects of bosentan on quality of life, survival, safety and tolerability in pulmonary arterial hypertension related to connective tissue diseases, Ann Rheum Dis 67 (2008) 1222–1228.

[143] A. Fischer, R.T. Meehan, C.A. Feghali-Bostwick, S.G. West, K.K. Brown, Unique characteristics of systemic sclerosis sine scleroderma-associated interstitial lung disease, Chest 130 (2006) 976–981.

[144] A. Fischer, J.J. Swigris, R.M. du Bois, D.A. Lynch, G.P. Downey, G.P. Cosgrove, et al., Anti-synthetase syndrome in ANA and anti-Jo-1 negative patients presenting with idiopathic interstitial pneumonia, Respir Med 103 (2009) 1719–1724.

[145] C. Strange, K.B. Highland, Interstitial lung disease in the patient who has connective tissue disease, Clin Chest Med 25 (2004) 549–559, vii.

[146] G.E. Tzelepis, S.P. Toya, H.M. Moutsopoulos, Occult connective tissue diseases mimicking idiopathic interstitial pneumonias, Eur Respir J 31 (2008) 11–20.

[147] A. Fischer, Interstitial lung disease: A rheumatologist's perspective, J Clin Rheumatol 15 (2009) 95–99.

[148] J. Fujita, Y. Ohtsuki, T. Yoshinouchi, I. Yamadori, S. Bandoh, M. Tokuda, et al., Idiopathic non-specific interstitial pneumonia: As an "autoimmune interstitial pneumonia." Respir Med 99 (2005) 234–240.

[149] B.W. Kinder, H.R. Collard, L. Koth, D.I. Daikh, P.J. Wolters, B. Elicker, et al., Idiopathic nonspecific interstitial pneumonia: Lung manifestation of undifferentiated connective tissue disease? Am J Respir Crit Care Med 176 (2007) 691–697.

[150] A. Fischer, S.G. West, J.J. Swigris, K.K. Brown, R.M. Du Bois, CTD–ILD: A call for clarification, Chest (2010), in press.

46

Gastrointestinal: Liver

Ian R. Mackay

INTRODUCTION

Systemic Lupus Erythematosus and Liver Disease

Systemic lupus erythematosus (SLE) is a multisystem autoimmune disorder with varying major involvement of diverse organs and tissues, and it is prone to co-exist with other such disorders [1], for which the now obsolete terms "collagen vascular" or "connective tissue" disease are still occasionally used. Definition and diagnosis raise striking challenges to nosologists, clinicians, pathologists, geneticists, and epidemiologists alike. Committees of the American Rheumatism Association have twice, in 1971 [1] and 1982 [2], attempted to define SLA, and useful criteria for SLE have emanated. However, given that lupus is by definition "systemic" (i.e., "multisystem"), cases in which the disease affects only one organ or system—whether kidneys, skin, synovium, or central nervous system (CNS)—but with a positive antinuclear antibody (ANA) reaction and autoimmune-type histopathology present a nosologic dilemma. Familiar examples occurring as single entities are renal lupus and cutaneous lupus and also neuropsychiatric lupus [3]. Does the liver provide another, albeit less typical, example? Expectations have been raised by advances in autoimmune serological diagnosis since the 1980s [4], but these have so far been incompletely fulfilled. One is that serological specificities would correlate sufficiently well with clinical features to provide accurate classification of multisystem autoimmune diseases, and the other is that serological specificities would provide crucial insights into etiology and pathogenesis, including perhaps elucidation of the initial provocation, referred to as "etiology."

A rationale for this chapter is that particular cases of chronic hepatitis have in common with SLE an autoimmune background, some shared serological features, and, occasionally, an overlap of disease expressions, with such allowing for analysis of points of convergence and divergence of the two diseases.

The Logic of Naming and Diagnosis of Disease and Establishing Causality

The claim is made that too little attention is given to the precise definition of disease in clinical medicine [5]. Problems of nosology certainly arise in a disease such as SLE, in which various clinical subsets exist and many components of etiology and pathogenesis remain unclear. Commentators on the logic of diagnosis and disease terminology distinguish two schools, essentialist (reductionist) and nominalist, as providing alternative approaches [5, 6]. Essentialists have the expectation of an ideal form behind every affect or concept, expressed in medicine by the desire for a unified concept of disease that can be related to an identifiable class of agents causing illness. In other words, an underlying pathological etiology must exist, and the disease state should be defined by the essential lesion. Essentialists, in the case of autoimmune hepatitis as well as SLE, would seek to define these in terms of the immunological abnormalities that are demonstrable in the laboratory or, with advances in genetics and genomics, in terms of the promising area of genetic abnormality, as discussed later. Nominalists, on the other hand, hold that the purpose of definition is simply to state the features by which the members of the class might be individually recognized, and thus designate diseases by criteria that require no presumption of any particular underlying cause—autoimmune, infectious, or other.

Early on, chronic active hepatitis (CAH) described a miscellaneous group of persistent inflammatory liver diseases among which cases with immunological aberrations seemed prominent. After it was realized that CAH comprised separable entities including particularly chronic viral infection [7], the prototypic "CAH" of yesteryear morphed into "autoimmune hepatitis"

(AIH), which became defined by an aggregate of features, inclusive and exclusive, endorsed by an International Autoimmune Hepatitis Group (IAIHG) [8, 9]. Thus, the adjectival use here of "autoimmune" not only reflects a convenient way of emphasizing characteristics that are lacking in other types of chronic hepatitis but also allows our nosology for AIH and SLE to draw on both schools, essentialism and nominalism.

CHRONIC ACTIVE AND LUPOID HEPATITIS: EARLY DESCRIPTIONS

The historical evolution of AIH has been reviewed and referenced in detail [10]. In the mid-1940s, chronic active hepatitis was first introduced to describe a relatively benign, albeit protracted, course of epidemic (presumably viral) hepatitis affecting soldiers in World War II [11]. In the 1950s, this term (or a variant—active chronic hepatitis) became applied, more aptly, to a striking new syndrome particularly affecting young women suffering from a progressive and usually fatal hepatitis, with additional immunological and endocrine abnormalities [12–15]. Then, the occurrence in CAH of positive tests for lupus erythematosus (LE) cells, and association with features of SLE, prompted the term "lupoid hepatitis" [16]. Thereafter, the known immunologic basis of the LE cell test, together with the detection in cases of CAH of anticytoplasmic antibodies [17], led to autoimmunity being considered as a component of pathogenesis [18]. Literature from the 1960s illustrates that many authors observed a "spillover" of features of CAH into those of SLE. More recently, however, the clinical presentation of CAH has shifted from younger to older women with milder and more liver-focused disease [19, 20]. In our initial description of 7 subjects with lupoid hepatitis, features included malar rash, nonerosive arthritis, serositis, renal disorder, hematological cytopenias, and immunological expressions [16], and contemporary reports have found the same features. Examples include 26 cases among which a "surprising number" had either arthralgia or actual arthritis that, together with serositis, pneumonia, or erythema multiforme, prompted a suspicion of "collagen disease" [15], and 9 cases with liver cirrhosis in whom a diagnosis of a "lupus erythematosus-like syndrome" could be made, with 5 having Sjögren's syndrome [21]. In another series, there were 81 cases of "active juvenile cirrhosis" with accompanying diseases that included rashes, arthralgia, "lupus kidney," autoimmune thyroid disease, and ulcerative colitis but, according to the authors, "this form of chronic liver disease cannot be equated with systemic lupus erythematosus"[22]. Further reports describing "lupoid hepatitis" then followed [23, 24]. One such comprised 20 patients with CAH, most of whom had a positive test for ANA and prominent systemic features (arthritis or arthralgia, 16; pleurisy, 4; rashes, 5; ulcerative colitis, 3; convulsions, 2; and cytopenias, 11) and a high frequency of familial stigmata of autoimmunity, with such expressions deemed "remarkably similar to those encountered in systemic lupus erythematosus" [24]. However, among a group of 4 cases of "progressive hyperglobulinemic hepatitis" with extrahepatic features, the reported syndrome did not match the authors' experience with typical SLE [25]. At the Mayo Clinic, 88 cases were reported in 1975 [26] and 126 in 1983 [27]; among the latter 126 cases, there was an associated autoimmune disease in 21 and a positive test for ANA in 55%, but a relationship of the disease to SLE was discounted, as was the idea that "lupoid" features defined a subset distinguishable from others. The authors stated that "the autoimmune markers associated with CAH may reflect dysglobulinemia associated with the severity of hepatocellular inflammation and necrosis" [26].

DIAGNOSTIC CRITERIA: SLE AND AUTOIMMUNE HEPATITIS

Criteria for Systemic Lupus Erythematosus

In 1971, the American Rheumatism Association (ARA) selected a set of 14 major manifestations (criteria) that included 21 items considered to characterize SLE, and it recommended that SLE could be diagnosed if 4 or more of these criteria were present, serially or simultaneously, during a period of observation [1]. Although in 1974 these criteria were evaluated favorably [28] with high sensitivity (90%) and specificity (98%), they were deemed inadequate. In 1982, an ARA subcommittee published the following revised criteria in which defining terms were clarified, some items were eliminated, and recognized serologic markers were included [2]: (1) malar rash; (2) discoid rash; (3) photosensitivity; (4) oral ulcers; (5) nonerosive arthritis; (6) serositis; (7) renal disease; (8) neurologic disease; (9) hematological disease, either hemolytic anemia or reduced numbers of cellular elements in the blood; (10) immunoserological disease including LE cells, anti-DNA, and anti-Sm; and (11) ANA to an abnormal titer at any point in time. Diagnosis of SLE required at least 4 criteria and, when fulfilled, provided a sensitivity and specificity of 96%. Notably, liver dysfunction of any type does not appear among these criteria.

Criteria for Autoimmune Hepatitis

There have been a number of attempts to specify criteria for autoimmune hepatitis.

International Association for Study of the Liver, 1976

In 1976, criteria for "generic" CAH were presented by a criteria committee of the International Association for Study of the Liver (IASL) [29], but this definition did not accommodate the uncertainty regarding diseases to be included under that rubric or the recognized heterogeneity of CAH [7]. Indeed, "chronic active hepatitis" became seen as a spectrum of diseases [30], including those with an autoimmune or viral basis, hepatitic expressions of drug sensitivity, α_1-antitrypsin deficiency, Wilson's disease, or ethanol abuse. Next, post-transfusion non-A, non-B virus infection became identified with hepatitis C virus (HCV), later becoming the major contributor overall to the caseload of chronic hepatitis.

Recognized Attributes of AIH during the 1980s

In the 1980s, autoimmune-type CAH became better defined based on a combination of attributes characteristic of the diagnosis [31]: (1) female gender; (2) age of onset 10—30 years or older than 50 years; (3) Euro-Caucasian racial background; (4) multisystem disease expression; (5) histological features of periportal piecemeal necrosis; (6) seronegativity for HBsAg and anti-HCV and the absence of other etiologies; (7) hypergammaglobulinemia of 20—30 g/l; (8) autoantibodies to "acceptable" titers to nuclei, smooth muscle (SMA), or liver—kidney microsomes (LKM); (9) HLA-B8, DR3 by histocompatibility typing; (10) responsiveness to treatment with corticosteroids; and (11) non-occurrence of supervening hepatocellular carcinoma in contrast to other types of chronic hepatitis or cirrhosis.

International Autoimmune Hepatitis Group (Brighton) Report, 1992

In 1992, in Brighton, United Kingdom, the International Autoimmune Hepatitis Group met to review all the features of AIH and determine whether a consensus could be reached on criteria. The ensuing "Brighton Report" [8] presented diagnostic criteria for "definite" and "probable" AIH based on characteristics similar to those cited previously but validated by "gold-standard" features derived from analysis of 145 patients attending one dedicated liver unit with classic features of AIH and 250 with other liver disorders. This report was innovative in that inclusive or exclusive criteria—clinical, biochemical, histological, and serological—were given positive or negative scores or weightings according to their estimated contribution to the diagnosis. The system was then modified through three subsequent iterations based on panelists' experience with their own case material.

IASL Criteria Committee, 1994

In 1994, a criteria committee of the IASL [32] revised their 1976 manual and distinguished AIH from other types of chronic hepatitis according to the presence of significant titers of circulating tissue antibodies and the absence of other causes of chronic hepatitis, namely viruses, drugs, toxins, and metabolic abnormalities. Required titers by indirect immunofluorescence (IIF) for antibodies to nuclei, and/or to smooth muscle with specificity for actin, were 40 or greater, but laboratory conditions including test substrates and appropriate standards for performance of tests were not specified. This committee also presented indicative rather than diagnostic histologic criteria, including plasmacytosis, bridging necrosis, and pseudo-glandular rosettes of liver cells, and a quick response to corticosteroid therapy was accorded diagnostic significance.

World Congress of Gastroenterology Working Party, 1994

Also in 1994, an international working party [33] organized and funded by the World Congress of Gastroenterology reported on the terminology of chronic hepatitis. The term autoimmune hepatitis was recommended over earlier synonyms such as chronic active hepatitis, lupoid hepatitis, or plasma cell hepatitis. The definition of AIH reached was unresolving predominantly periportal hepatitis, usually with hypergammaglobulinemia and tissue autoantibodies, and responsiveness to immunosuppressive therapy. Laboratory descriptors included a fivefold selective increase in serum IgG concentration and titers by IIF of autoantibodies—ANA, SMA, or anti-LKM—of 1:40 on tissues or 1:160 for ANA on the HEp2 cell substrate. Almost all patients (Euro-Caucasoid) carry the HLA class II alleles DR3 or DR4. Evidence of virus infection should exclude AIH. Indicative histological features were bridging necrosis, plasmacytosis, and formation of liver cell "rosettes."

International Autoimmune Hepatitis Group, Revised Criteria, 1999

In 1999, IAIHG modified the "Brighton Report" of 1992 to accommodate newer diagnostic methods, consensus on histopathologic appearances on liver biopsy, and evident imprecisions in the original scoring system [8]. The revised criteria and scoring system [9] provided a score for definite or probable AIH with a sensitivity for diagnosis of 97—100%, but specificity was lower (45—92%) because diseases other than AIH fulfill criteria for a probable diagnosis.

Simplified Criteria, 2008

Simplified criteria have been propagated as an alternative to the cumbersome IAIHG scoring system [34].

The indices used, relatively minimal, were titer by IIF of relevant antibodies, level of IgG, compatible or typical liver histology, and exclusion of viral hepatitis. These criteria appeared to serve as well as those of the IAIHG [34], but there is a need for more field testing and better international consensus on tests for autoantibodies.

LUPUS AND THE LIVER: THE VISION OF THE CLINICIAN

The Liver in SLE

The liver is certainly not spared in SLE. In an exemplary 570 cases reported in 1991 [35], there was hepatomegaly in 50% attributed mainly to fatty liver or congestion, and there were increased serum aminotransferase levels or other biochemical functional abnormalities in approximately 60%. This could specify "nonspecific reactive hepatitis," which describes mild hepatic abnormalities that can accompany inflammatory lesions elsewhere in the body [36]. In some case studies, the point of inquiry has been whether the "lupus process" can involve the liver in the same way as other vulnerable tissues. In a frequently cited study published in 1980 [37], of 238 cases with a diagnosis of SLE based on fulfillment of four or more of the initial ARA criteria, results of liver function tests were available for 206. At least one test result was abnormal in 124 cases, and the authors' criteria for "liver disease" were met in 43 (21%); among these, 7 had drug-related disease. Liver biopsy was performed in 33 cases. In another study in 1981 on 81 cases of SLE, hepatomegaly was present in 42% and biochemical abnormalities of liver function in 44%, among which 14 were drug related [38]. In a prospective 1984 study on 264 cases of SLE, in which clinical and laboratory indices were derived over a 12-month period of observation, there was evidence of liver disease in 67 (25%), attributable to SLE in 29 and to causes unrelated to SLE including alcohol abuse in 13 [39]; thus, in this analysis true "lupus-associated" liver disease existed in just 25 cases (10%) among the entire cohort.

A rare SLE-like liver syndrome, recognized as a "lupus variant," has been described in which antibodies against phospholipids occur as in the antiphospholipid syndrome (aPL) syndrome together with vascular occlusive disease of the liver and/or nodular regenerative hyperplasia [40–42]. The presence of phospholipid antibody/lupus anticoagulant, whether in the primary aPL syndrome or that accompanying SLE, has been observed in several instances associated with hepatic venous thrombosis and Budd–Chiari syndrome [40] or with hepatic arterial occlusion; adverse effects of this autoantibody are not a recognized feature of AIH.

Moreover, antiphospholipid has been associated with liver infarction due to hepatic arterial occlusion, without ensuing portal hypertension [41]. Hepatic veno-occlusive disease, as seen in cases of SLE or in subjects with aPL antibody, is related to the Budd–Chiari syndrome, but the occlusive lesion affects central or sublobular veins rather than larger hepatic veins. Nodular regenerative hyperplasia is a recognized hepatic complication of SLE, particularly when there is accompanying aPL antibody. A literature survey covering 46 patients with nodular regenerative hyperplasia of the liver disclosed various accompanying multisystem disorders: SLE in 15, Felty syndrome in 12, rheumatoid arthritis in 5, systemic sclerosis in 2, Sjögren's syndrome in 1, and primary aPS in 2 [42]. On the other hand, the frequency of nodular regenerative hyperplasia of the liver among Japanese cases of SLE was only 0.3% [43].

The Liver in Autoimmune Hepatitis

Hepatic Expressions

The clinical presentation of AIH ranges from an asymptomatic disease recognized only by incidentally ascertained biochemical abnormalities of liver function to an acute (or even fulminant) hepatitis in approximately 25% of cases [44]. Usually, the disease has a slow onset with mild jaundice but, untreated, is progressively "cirrhotogenic." Among the clinical features [31, 45, 46], there is a female bias (female:male ratio, 8:1) and an age incidence that is bimodal with peaks in childhood/early adult life and after 40 years of age. Also, there is often a long presymptomatic stage, as pertains for other autoimmune disorders, that may be disclosed by expressions in young women such as amenorrhea. In symptomatic cases, hepatocellular dysfunction is revealed by nausea, anorexia, jaundice, and hepatosplenomegaly. Serum analysis shows levels of transaminase enzymes to be highly raised, alkaline phosphatase and γ-glutamyl transpeptidase modestly increased, and IgG very high (up to 100 g/l) and of polyclonal type, causing an overall increase in serum globulins, in contrast to the minimal increase seen in SLE. Among the few other autoimmune diseases with comparable increases are primary Sjögren's syndrome and Hashimoto's thyroiditis. Treatment with prednisolone induces such improvement in clinical, biochemical, and histological indices as to have diagnostic utility. As mentioned previously, CAH and lupoid hepatitis seen in the 1950s and 1960s among young women has become less prevalent than the more indolent versions that affect older females [19, 20, 47].

Serological reactants distinguish two subtypes of AIH, as described later. Autoantibodies to a reactant enriched in "microsomes" of liver and kidney, and hence

TABLE 46.1 Comparison of Type 1 and Type 2 Autoimmune Hepatitis

	Type 1	Type 2[a]
Female:male ratio	8:1	8:1
Age	Bimodal	Peak in early childhood
Course	Progressive but prednisolone responsive	Progressive but prednisolone responsive
Associated diseases[b]	Arthritis, cytopenias, colitis, thyroiditis	Thyroiditis, gastritis, diabetes, vitiligo
Genetics, MHC	B8, DR3 (DRB1 *0301), DR4, C4QA^{nul}	DR7 (DRB1 *0701)
Initiating agents	?EBV, ?hepatitis virus A, drugs (rarely)	HCV, drugs[c]
Models	No close model [103, 180]	Xeno-DNA immunization of mice with construct of CYP450 2D6/FTCD [181]
Autoantibodies to	ANA (nucleosome, etc.), atypical pANCA smooth muscle (cytoskeleton, F-actin)	LKM-1 (cytochrome P450 2D6) "thyrogastric" autoantigens

[a]Data on type 2 from Homberg et al. [50].
[b]Type 1 may present with multisystem features even resembling SLE; type 2 is more akin to organ-specific autoimmune diseases.
[c]HCV infection and certain hepatic drug reactions may be associated with anti-LKM^{+ve} variants of type 2 AIH.

called LKM [48–50], characterized a relatively infrequent type 2 AIH from the classic type 1 (Table 46.1). The clinical distinctions are that accompanying autoimmune diseases (when present) differ in being "multisystem" and SLE-like in type 1 and "organ specific" in type 2 and, in adults, there is an unexplained geographical bias in prevalence to middle and southern European countries in contrast to countries in which type 1 predominates—Sweden [51], Japan [52], the United States, Australia, and the United Kingdom.

Extrahepatic Expressions of Autoimmune Hepatitis

The earlier clinical descriptions of CAH–AIH included features resembling those seen in SLE—facial rash, arthralgia, hemolytic anemia, thrombocytopenic purpura, and mild nephropathy. A study of 108 older subjects with CAH [47] reported prominent extrahepatic features, including Sjögren's syndrome, renal tubular acidosis, fibrosing alveolitis, arthropathy, rashes, and thyroid disease. However, although there is clear overlap between AIH and other autoimmune disorders, there are clinical differences between AIH and SLE, including the usual absence in AIH of progressive renal disease (lupus nephritis) and cerebral involvement and also the presence of co-existing diseases such as ulcerative colitis [22, 24, 31, 45] that tend not to accompany SLE. The type 2 variant of AIH likewise has extrahepatic features of differing type and no associations with SLE.

Autoimmune Overlap Syndromes Involving the Liver

Multiple autoimmune diseases may aggregate in one patient or related family members, and in fact such clustering is a defining feature for an autoimmune basis for any given disease: SLE-associated and thyroiditis-associated

diseases provide familiar examples [53]. A distinction needs be drawn between "co-existence," when two distinct disease entities (albeit both autoimmune) occur concurrently in one individual, and "overlap," when there is co-expression of features of two usually distinct entities within a single organ (e.g., the liver), to such a degree that criteria for either are fulfilled [53].

AUTOIMMUNE HEPATITIS–SYSTEMIC LUPUS ERYTHEMATOSUS

Sharing of features of chronic active hepatitis and SLE was the initial prompt for the designation of lupoid hepatitis [16], although, as mentioned previously, autoimmune hepatitis and SLE are usually readily distinguished as independent diseases. The significance of shared features is examined later from the viewpoints of the clinician, histopathologist, serologist, geneticist, and epidemiologist.

AUTOIMMUNE HEPATITIS–PRIMARY BILLIARY CIRRHOSIS

Overlap between these usually distinct autoimmune liver diseases has received ongoing attention in the hepatological literature [54, 55]. It is specified by the co-existence in one individual of features specific for either disease, including clinical presentation, biochemical abnormalities of liver function, histologic lesions, and serological markers. The disease expression involves both periportal hepatocytes as in AIH and small intrahepatic bile ducts as in primary biliary cirrhosis (PBC). Perception of the overlap may depend on the perspective of the observer. Thus, among 162 cases with the primary diagnosis of autoimmune hepatitis [56], there was PBC overlap in 8 (5%), whereas among 130 cases with a primary diagnosis of PBC [57], there was an autoimmune hepatitis overlap in 12 (9%); thus a "mid-figure" for the frequency of overlap could be approximately 7%. Even so, the "basic" disease may well be PBC according to observations that the overlap is due to PBC occurring on an immunogenetic background of AIH comprising permissive HLA alleles, B8 and DR3 [58]. Given the theme of this chapter, lupus and the liver, we can inquire whether SLE is involved in any way with this overlap. It is not. In fact, PBC co-exists with SLE so infrequently as to warrant publication of a single case report [59].

What, then, are we to make of the high frequency—up to 50%—of ANAs in PBC per se and in the AIH—PBC overlap syndrome? The point is that these ANAs have very particular specificities—anticentromere, anti-Sp100 (which is directed against a transcriptional regulatory protein that aggregates in nuclear dots), and anti-gp210 and anti-p62 (which are components of the nuclear pore complex) [60]. Such ANA specificities are seldom expressed in SLE and, in fact, anti-Sp100 and antinucleoporins are virtually (and inexplicably) PBC specific. On the contrary, anti-double-stranded (ds) DNA and antichromatin are not expressed in PBC; thus, the rarity of co-existing PBC and SLE is not surprising. Regarding the presence of anticentromere reactivity in cases of PBC, clinical features of the CREST variant of systemic sclerosis (itself marked by anticentromere antibody) co-exist in approximately 10% of cases of PBC [61]. This co-existence of disease and serologic reactivity is provocative because mitochondrial autoantigens of PBC and the centromeric nuclear autoantigens of the CREST syndrome exist in separate organelles of the cell, and there is no demonstrable immunological cross-reactivity between them [62].

THE AUTOIMMUNE (IMMUNO) CHOLANGITIS VARIANT

As originally described [63], a syndrome of autoimmune cholangitis (AIC) comprised a chronic cholestatic liver disease and histological periductular lymphoid aggregates in the liver, thus simulating PBC. Tests for the antimitochondrial antibody were negative, whereas tests for ANA were positive. AIC first became aligned with autoimmune hepatitis, but the particular ANA specificities were not of the type seen in SLE or AIH but, rather, those characteristic of PBC. The current view is that AIC is either a serologic variant of PBC [64] or reflects the AIH—PBC overlap syndrome [65].

LUPUS AND THE LIVER: THE VISION OF THE HISTOPATHOLOGIST

Histopathology is the area in which AIH as usually construed moves closer to the realm of hepatology and away from the multisystem autoimmune diseases.

Histopathology of the Liver in SLE

The relevant studies are those in which there was an established diagnosis of SLE and histological data on the liver were available. A report by Mackay *et al.* in 1959 [66] included 19 cases with the following findings: normal liver or entirely nonspecific lesions including venous congestion, 6; portal fibrosis and/or minor lymphoid infiltrates interpreted as nonspecific reactive hepatitis, 11; inactive cirrhosis, 1; and hepatitis, 1 with the diagnosis revised to CAH. In the study cited previously on 238 cases [37], a histologic diagnosis was available in 33 and included cirrhosis in 4, chronic hepatitis in 4, and miscellaneous entities in 25. The authors concluded that patients who satisfy the ARA criteria for SLE and also have clinical and histologic evidence of CAH should be considered as having both diseases. A further study [38] included 81 cases, with biopsies obtained in 7, showing a portal inflammatory infiltrate in 5, and CAH in 1. Also, there is the report [39] with liver biopsy data in 14 of 264 cases, with none showing any serious or unusual lesions. The rare association of nodular regeneration hyperplasia of the liver with SLE was previously described . Thus, histology of the liver in the typical case of SLE does not reveal any "lupus-specific" pathology but, rather, normal liver, lesions attributable to drug toxicity, or "nonspecific reactive hepatitis"—the term used to indicate nondescript abnormalities in the liver that can accompany pathology elsewhere in the body [36]. Finally, it is worth noting that certain inbred strains of mice develop a lupus-like disease with immune complex-mediated glomerulonephritis, but specific liver lesions are not a reported feature of such "lupus mice."

Histopathology of the Liver in Autoimmune Hepatitis

As indicated previously, the earlier clinical and histological descriptions of CAH were more or less generic for all types of chronic hepatitis before it was realized in the 1970s that quite different liver diseases could be encompassed within the ICD9 rubric 570.71 for "chronic active hepatitis." Numerous histological descriptions of CAH became formalized by reports in 1969 and 1977 from an international group of pathologists whose deliberations have in part provided the basis for the current morphological criteria for AIH [67, 68]. A particular feature is dense inflammatory cell infiltrates in portal tracts of the liver and within lobules (Figure 46.1A), with prominence of plasma cells [16, 69, 70], succeeded in the course of the disease by fibrosis, nodule formation, and cirrhosis. Another early feature is bridging necrosis, meaning confluence of areas of hepatocellular necrosis between adjacent portal tracts and regarded

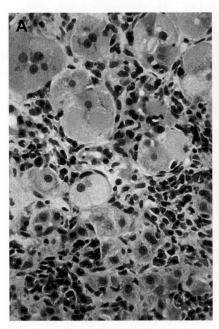
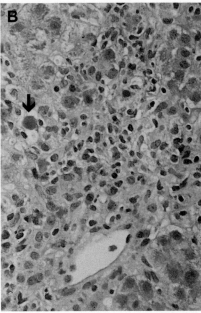

FIGURE 46.1 Histologic appearances in liver biopsy samples from autoimmune hepatitis. Intralobular lesions illustrating (*A*) cytolysis with ballooned hepatocytes with contiguous infiltrating lymphoid cells (*left*) and (*B*) apoptotic body among damaged hepatocytes (*arrow, right*). Hematoxylin & eosin; magnification ×550 and ×400, respectively. *Source: A, courtesy of Nigel Swanston.*

as a precursor to nodule formation and cirrhosis [71]. However, most characteristic is the lesion first called "piecemeal necrosis" [70] and now "interface hepatitis," representing destruction of liver cells at the junction between liver parenchyma and connective tissue in portal tracts from whence lymphoid cells invade the limiting plate of the hepatic lobule, and where hepatocellular injury is most intense. Lymphocytes can be seen in close contact with hepatocytes (peripolesis) and within them (emperipolesis), as the presumed effectors of hepatocellular destruction, resulting in hydropic swelling or apoptosis of hepatocytes. Apoptosis is evident as shrunken acidophilic remnants of hepatocytes (Figure 46.1B), equivalent to the Councilman or acidophilic bodies long familiar to hepatic histopathologists [72] and now regarded as indicative of immunologically mediated hepatocellular damage. Other morphological features include intralobular aggregates of lymphocytes contiguous with damaged hepatocytes (lobular hepatitis) and formation by hepatocytes of pseudo-glandular structures (rosettes).

There are particular aspects to the histopathology of AIH. First, many of the changes described previously are seen in all types of chronic hepatitis, whether attributable to virus infection, autoimmunity, or allergic drug-mediated reactions; that is, they are common to immune-mediated damage to hepatocytes in general. Second, histopathologists in earlier descriptions distinguished lesions with interface hepatitis as chronic aggressive (active) hepatitis from those showing only hepatic portal lymphoid infiltrates without interface hepatitis as chronic persistent hepatitis wherein

progression was deemed unlikely [68], but this distinction has since been de-emphasized [9]. Third, morphological appearances per se do not inform on the immune mechanisms of hepatocellular injury, whether cytolytic T cells, cytokines, or antibody-dependent cytotoxicity. Plasmacytosis within portal cellular infiltrates (and in the bone marrow as well), as a widely recognized feature of AIH, presumably represents a source of local antibody formation (of which the specificity is unknown). It could point to the activity of cytotoxic antibody or antibody-dependent cellular cytotoxicity (ADCC), but evidence favoring ADCC activity, such as demonstrability of a membrane-located hepatocellular-specific autoantigen, is lacking. Fourth, the types of liver parenchymal lesions in AIH described previously are not features of the recognized histopathology of "classical" SLE affecting the usual target tissues.

LUPUS AND THE LIVER: THE VISION OF THE SEROLOGIST

The recognition of positivity for antinuclear antibody reactivity in chronic hepatitis was the initial indication for an immunopathogenetic relationship between AIH and SLE. Subsequently, however, serologic profiles for the two diseases diverged to a degree because in AIH compared with SLE, the frequency/levels of autoantibodies to dsDNA and ribonucleoproteins (RNPs) are much lower, and exclusive to AIH there are reactivities such as anti-F-actin and anticytochrome P450 (anti-LKM). These are examined later in more detail.

Serological Reactants Equally Prominent in SLE and Autoimmune Hepatitis

Nuclear Constituents

ENTIRE NUCLEUS—IMMUNOFLUORESCENCE ANA

Antinuclear reactivities in SLE and, later, in AIH were first recognized by positivity on the LE cell test and then by IIF on tissue substrates (see Figure 46.2). The overall frequency of a positive ANA test in SLE approaches 100% and in AIH is approximately 70% [31, 73], varying according to test conditions. By IIF, homogeneous nuclear staining is the most frequent pattern in both SLE and active AIH (Figure 46.2A), but with remission of AIH, the frequency decreases to 34% and the homogeneous nuclear staining pattern is replaced by speckled patterns (Figure 46.2B) in 38% of cases [73, 74]; the latter specify multiple and mostly unidentified nuclear reactants. Under the aegis of the IUIS standardization committee [75], cutoffs for positivity were established in a multicenter study on normal sera [75]; percentage frequencies cited for ANA positivity on Hep-2 cells according to serum dilution at 1:40, 1:80, 1:160, and 1:320 were 32, 13, 5, and 3%, illustrating the sensitivity—specificity "trade-off." Conventional "cutoff" titers for ANA positivity by IIF are 1:40 for tissue cells and 1:160 for Hep-2 cells. Thus, interpretations of ANA should be made with the recognition that the level of reactivity will depend on tissue substrate, disease activity, and the likelihood that any given serum will have multiple ANA specificities with different levels of reactivity. Automated assays using purified or recombinant autoantigens now tend to supersede the traditional and more labor-intensive IIF procedures despite the latter providing, for some reactants, information not otherwise obtainable [76, 77].

NUCLEOSOME—CHROMATIN

Nuclear components responsible for ANA reactivities in SLE are described in detail in Chapters 13 and 34. The most typical in SLE and in AIH is the homogeneous pattern, taken to represent reactivity with nucleosome/chromatin and histones. The nucleosome comprises an octamer of histones enwrapped by two coils of DNA (146 base pairs). Histones are small conserved nuclear proteins that bind to dsDNA in chromatin, exist as five main classes—H1, H2a, H2b, H3, and H4—and have particular structural features [78]. The octamer is constituted by two H2a—H2b dimers and an H3—H4 tetramer stabilized by noncovalent bonds, and completing the structure is an H1 linker molecule that connects separate nucleosomes. A linear array of nucleosomes forms the primary chromatin fiber, 100 Å in diameter, which supercoils into increasingly larger fibers based on histone-dependent internucleosomal interactions. The condensation of chromatin into a quaternary protein structure likely contributes to the major autoimmune

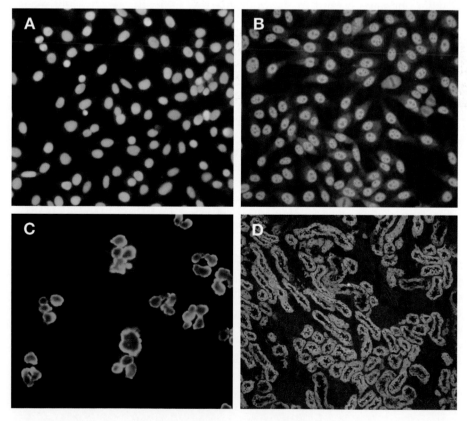

FIGURE 46.2 Staining patterns by immunofluorescence displayed by sera in AIH. (*A*) Homogeneous nuclear staining attributed to antibodies to nucleosome. (*B*) Speckled nuclear staining attributed to anti-RNP and other undefined reactants. (*C*) Antineutrophil nuclear antibody resembling pANCA. (*D*) Diffuse cytoplasmic staining of proximal renal tubules attributed to the anti-LKM characteristic of type 2 AIH. *Source: A, B, and D, courtesy of Bob Wilson; C courtesy of Daniela Zauli.*

epitope for both autoimmune T and B cells that leads to generation of homogeneously reactive ANA. The origin of an autoimmune B-cell epitope from a quaternary protein structure is not unique, also being exemplified by various other SLE-associated autoantigens, including Ku, U1-RNP, and proliferating cell nuclear antigen. If the quaternary chromatin structure were to provide the primary autoepitope for ANA, it would follow that other chromatin structures, either individual histones or other nuclear proteins, could become drawn in as reactants due to intermolecular epitope spreading.

HISTONES

Autoantibodies to histones occur in spontaneous and drug-induced SLE, as well as in AIH. All histone subtypes are recognized as autoantigens, in differing combinations. Assumptions have been made that anti-histone is the predominant antinuclear reactivity in AIH, but reported frequencies according to testing by enzyme-linked immunosorbent assay (ELISA) do not support this. Thus, in one report there were only 5 positive tests among 20 sera [79]. In another, among 65 sera previously screened by IIF for a range of ANA reactants, only 23 (35%) were positive; 12 reacted only with histones and 11 with other nuclear autoantigens as well [80]. Although in SLE specificities of ANA according to reactivity by IIF align with clinical features and/or particular subsets of the disease [4], this does not pertain in AIH.

MINOR NUCLEAR REACTANTS: CENTROMERE AND NUCLEOLI

Centromere proteins The anticentromere antibody (ACA) pattern by IIF represents reactivity to centromere proteins (Cenp) A, B, and C, also well-defined by immunoblotting and ELISA using recombinant centromere proteins. These reactivities are typically associated with the CREST variant of systemic sclerosis (SSc; scleroderma) [4] and are infrequent in SLE. General experience suggests that the frequency and serum titer of ACA in AIH are relatively low, and in line with this, reported co-existences of CREST-type SSc and AIH are scarce, contrary to what pertains for PBC [61, 62].

Nucleoli Autoantibodies to nucleolar antigens (anti-NoA) occur in SLE and related multisystem diseases, particularly SSc; the major reactant is fibrillarin [4]. In AIH, as well as in SLE and SSc, anti-NoA were detectable by IIF on monolayers of human fibroblasts at similar frequencies [81]. Another reactant for anti-NoA is a 113-kDa molecule with three functional domains, identified as poly(ADP-ribosyl)transferase (pADPRT); anti-pADRPTs were detected in 7 (14%) of 50 cases of AIH, but such antisera differed from others in not inhibiting the catalytic activity of the enzyme,

suggesting differences in the site/nature of the autoepitope [82].

Serological Reactants more Prominent in SLE than in Autoimmune Hepatitis

Nuclear Constituents

dsDNA

Anti-dsDNA is the hallmark reactivity in SLE and is seldom lacking in active disease. Earlier studies on the frequency of anti-dsDNA in AIH (designated as "CAH" at the time) gave quite divergent results attributable in part to technical issues. Some studies in the 1970s using liquid or solid phase immunoassays and commercial sources of dsDNA reactant indicated a high frequency of positivity, but equivalent frequencies were obtained in other types of chronic liver disease [83, 84]; an issue with these results is nonspecificity due to contaminant single-stranded DNA in the antigen preparation. In a more specific immunoassay for anti-dsDNA based on precipitation of immune complexes by polyethylene glycol, the positivity rate in AIH was just 10% [85]. In 1987, tests were reported for 25 cases of CAH for anti-dsDNA by IIF on *Crithidia luciliae* and also by a reliable immunoprecipitation assay; among these, 4 were positive—2 transiently and 2 persistently [86]. The variation of reactivities for anti-dsDNA in AIH prompts comment on assay formats and methodologies, noting that ELISAs can display greater variability than other types of immunoassay [87], and that disease activity influences test results. In fact, rigorous standardization of various autoimmune serologic tests in liver diseases is still an unmet need, with minimal requirements including availability of standard reference antisera, interlaboratory serum exchange to derive optimal assay formats, and the convening of international workshops to facilitate uniform performance between laboratories [77]. For the disease, precise case definition is important, and whatever reactivity is being tested for, the stage and degree of activity of disease (as is the case for SLE and AIH) influence the degree and incidence of positivity. In summary, the frequency of anti-dsDNA in AIH is low (approximately 10% at most) and positivity decreases with declining disease activity, thus differing from reactivity in SLE.

Nuclear–Cytoplasmic Proteins

RIBONUCLEOPROTEINS

RNPs, whether located in the nucleus or cytoplasm, form complexes with and thus influence the processing of low-molecular-weight ribonucleic acids. Autoantibodies particularly in SLE and Sjögren's syndrome, but in other multisystem autoimmune diseases as well, react with various RNPs and give particular speckled patterns of nuclear staining on IIF. Autoepitopes of these RNPs have been ascertained using recombinant proteins, and autoantibodies to different RNP antigens tend to align with particular clinical syndromes or disease subsets [4]. The high serum frequency in SLE of autoantibodies to characterized small nuclear RNPs—Sm, U1-RNP, Ro (SS-A), and La (SS-B)—exceeds the modest frequency in AIH [88–90]. In one study, among 30 patients with AIH there was reactivity to the 70-kDa U1-snRNP in 8 (37%) and to Ro/La (SS-A/SS-B) in 3 (10%) [89]; in another study, based on North American and Asian populations, frequencies for the former were 43% for snRNPs and 38% for Ro/La, and frequencies for the latter were even higher [90].

RIBOSOMAL P PROTEINS

Ribosomal P proteins are present in the cytoplasm of all cells. They comprise three phosphoproteins of 38 kDa P0), 19 kDa (P1), and 17 kDa (P2) that contain 22 carboxy-terminal residues, and they constitute a main immunodominant region or autoepitope [91]. Based on constituent proteins used for ELISA, autoantibodies to ribosomal P protein are demonstrable in up to 69% of patients with SLE [92]. Anti-ribosomal P is regarded as a particular marker of active SLE and, possibly and controversially, cerebral lupus, renal lupus, and hepatic lupus [92] (see Chapters 14, 28, and 40).

Autoantibodies to ribosomal P proteins have been associated with hepatic expressions of SLE and, accordingly, this is a reactivity that could be a link between AIH and SLE, although reports have come mainly from just one "school." In a patient with SLE, after 4 years of follow-up there was a loss of ANA with acquisition of reactivity for anti-P0 followed by anti-P1 and anti-P2 with coincident development of AIH [93]. This temporal relationship of anti-ribosomal P and liver dysfunction suggested a pathogenic (hepatocytotoxic) effect of such antibody. Apparently consistent with this is the presence on the surface of cultured hepatoma cells of an epitope similar to that at the carboxy terminus of ribosomal P proteins [94]. Moreover, antibody to ribosomal P (anti-P0, 38 kDa) penetrated living hepatoma cells and induced cellular dysfunction [95]. In another study, the significance of anti-ribosomal P for liver disease specifically associated with SLE was assessed in 131 patients with SLE, among whom 4 (3%) had chronic hepatitis histologically resembling classic CAH. In all 4 patients (along with 2 additional referred cases), anti-ribosomal P was demonstrable, whereas in cases of AIH without evidence of SLE, anti-ribosomal P was undetectable [96]. Moreover, in a case–control study of 20 subjects with SLE among whom anti-ribosomal P was or was not detectable, the former group had significantly greater degrees of liver dysfunction [97]. An interesting study on Japanese patients addressed relationships

between SLE, "SLE-associated hepatitis," and AIH without lupus features by ELISA using recombinant ribosomal P0 [98]. Among 61 patients with SLE, there was "liver involvement" in 16, of whom 11 had SLE-associated hepatitis (taken here to mean hepatic lupus as described previously), with anti-ribosomal P demonstrable in 69%, whereas the frequency in co-existing AIH–SLE was only 20%, and that in AIH was 0%.

Serological Reactants more Prominent in Autoimmune Hepatitis than in SLE

Nuclear Constituents

NUCLEAR ENVELOPE PROTEINS

The nuclear envelope is a complex porous assembly of proteins that separates nucleus from cytoplasm with three structural components: inner and outer nuclear membranes, nuclear pore complex as the conduit for proteins between nucleus and cytoplasm, and nuclear lamina that provide the architectural framework of the nucleus. Each has autoantigenic reactivity in liver diseases, autoantibodies to nuclear membrane, and pore complex occurring in PBC and to lamina in AIH. The lamina, depending on the cell type and differentiation state, contain one to six immunologically related polypeptides, lamins, of which there are three major members (A, B, and C in equimolar amounts) and three minor members (B2, D, and E). The lamin polypeptides have similar molecular masses (60–75 kDa) but are separable by two-dimensional gel electrophoresis, thus allowing for the identification of different reactants in disease sera. Using 51 AIH sera (16 with "active" disease) and Western blots, lamins A and C reacted at greatest frequency (12 sera (24%), all with active disease), whereas all normal sera were nonreactive, and the frequency of reactivity was low in SLE, other rheumatic diseases, and PBC [99].

Cytoplasmic Constituents

INTACT CYTOPLASM

Cytoplasmic molecules are autoantigenic in AIH, SLE, and other multisystem autoimmune diseases, as demonstrated in the 1950s by complement fixation (CF) using cell homogenates [18]. Many of these cytoplasmic reactions are now characterized at the molecular level.

NEUTROPHIL-SPECIFIC CYTOPLASMIC ANTIGENS (ANCA REACTANTS)

In the 1960s, what was then termed granulocyte-specific ANAs were identified, particularly in rheumatoid arthritis and in AIH, in which titers by IIF were considerably greater (by several logs) than those for conventional ANA [100]. A large proportion of this so-called granulocyte-specific ANA likely represented what is now defined as antineutrophil cytoplasmic antibody (ANCA), in which the cytoplasmic reactant clumps in a perinuclear distribution. The two familiar species of ANCA are cytoplasmic (c)ANCA and perinuclear (p)ANCA, which differ in their disease associations and substrate reactivities. cANCA occurs in Wegener's disease and some vasculitides, and the major antigens are located in azurophilic granules, particularly as a 29-kDa protein triplet, proteinase 3. There are different types of pANCA, with the prototype detected in systemic necrotizing vasculitis and identified as myeloperoxidase. The frequency of pANCA in type 1 AIH was 65% versus 13% in chronic hepatitis C infection with autoimmune features and 0% in type 2 AIH [101]. ANCAs of various specificities are reported in SLE, but their presence does not correspond to any particular disease subtype, nor are frequency and titer comparable with those in type 1 AIH [102]. The atypical reactant for pANCA in AIH (Figure 46.2C) and certain other gastrointestinal diseases is not the usual myeloperoxidase; proposed candidates have included lactoferrin or elastase, high mobility group proteins 1 and 2, or a 50-kDa nuclear envelope protein restricted to neutrophils and myeloid cell lines [103]. The likelihood that the reactivity is actually an antineutrophil nuclear antibody has led to the current designation as pANNA [77].

LP ANTIGEN, SLA: SLA/LP

A liver/pancreas (LP) antigen in chronic hepatitis was recognized in 1981 by CF [104] and could be co-identified with a reactant described in 1987 by radioimmunoassay and called soluble liver antigen (SLA) [105], thus becoming known as SLA/LP. The frequency of anti-SLA/LP in type 1 AIH is approximately 25%. SLA/LP is particularly associated with type 1 AIH, and it can occur as the single serological marker. The liver-specific reactant for SLA/LP, after being attributed to various liver proteins, was definitively characterized as the UGA—serine transfer RNA—protein complex or tRNA-associated protein (tRAP) [106, 107], with structural considerations placing this protein in the family of serine hydroxymethyl transferases [108]. SLA/LP is an enzyme and functions as selenocysteinyl tRNA synthase to generate selenocysteinyl tRNA^{Sec} from a phosphoseryl precursor; the relationship between the function of SLA/LP and its autoantigenicity raises interesting questions for liver autoimmunity [109]. Notably, serological reactivity with SLA/LP does not occur in SLE.

"MICROSOMAL" AUTOANTIGENS: LKM ISOFORMS AND OTHERS

In CAH, autoantibodies reactive by IIF with cytoplasm of liver cells and the P3 segment of proximal renal tubular cells [45] were identified with "microsomes," hence LKM (Figure 46.2D). "Microsomes" in fact refers to particles identifiable *in vitro* as derivatives of endoplasmic reticulum that contains functionally important degradative enzymes. An apparent serological dichotomy between the "lupoid" and "anti-LKM" types of chronic hepatitis was substantiate in 1987 in an analysis of 87 cases [46]. The clinical features, histopathology, and outcome for types 1 and 2 are mostly similar, but there are demographic and genetic distinctions (Table 46.1).

Subsequently, a degree of heterogeneity of anti-LKM was recognized according to the disease background [110]: The LKM autoantigen associated with spontaneous type 2 AIH became LKM-1, and in other settings were called anti-LKM-2 and -3. Notably, it is only type 1 AIH, and not type 2, that exhibits overlap with SLE.

LKM-1, CYTOCHROME (CYP)450 2D6 Molecular identification of the LKM antigens, achieved first by immunoblotting [111] and then by use of reactive sera to screen gene expression libraries [112], revealed LKM-1 as an isoform 2D6 of the cytochrome oxidase P450 enzyme family (P450). Three linear epitope sequences were identified; the most reactive was sequenced as DPAQPPRD within residues 257—269 [113]. Interestingly, molecular identification of the specific autoimmune reactant for the infrequent type 2 AIH has long preceded identification of any reactant(s) for the more prevalent type 1.

LKM-2, CYP450 2C9 A microsomal reactant for serum was recognized in cases of hepatitis induced occasionally by exposure to medicinal drugs, particularly tienilic acid [114]. Called LKM-2, it was identified as another CYP450 isoform, 2C9 [115].

LM, CYP450 1A2 In cases of drug-induced hepatitis provoked by dihydralazine, sera react by IIF with microsomes only of liver cells. This reactant, called liver microsomal (LM) antigen, proved to be yet another isoform, CYP450 1A2 [116].

OTHER AUTOANTIGENIC CYP450 ISOFORMS

LKM-3 (UDP GLUCURONOSYL TRANSFERASE)

An anti-LKM reactivity was recognized in cases of hepatitis associated with delta hepatitis virus infection and called anti-LKM-3 [117]. This microsomal autoantigen was identified not as a CYP450 isoform but instead as the bilirubin conjugating enzyme uridine 5'-diphosphate glucuronosyltransferase (UGT) 1.6 [118]. In addition, a rare, spontaneously occurring type 2 AIH was associated with anti-UGT (LKM-3) reactivity of serum [119]; these autoantibodies reacted with various UGT isoforms, most strongly against UGT1A1, which is the main isoform required for the glucuronidation and disposal of bilirubin [120].

ANTI-LKM ASSOCIATED WITH DISEASES OTHER THAN AIH

AUTOIMMUNE POLYENDOCRINE SYNDROME TYPE 1 (APS-1) APS-1, also known by the acronym APECED (autoimmune polyendocrine, candidiasis, ectodermal dysplasia) [121], is of wide interest because of insights it has provided into natural thymic (central) tolerance. APS-1 has mostly a childhood onset characterized by T-cell-dependent immune deficiency and "organ-specific" endocrine diseases and their corresponding autoantibody expressions, including reactivity to various CYP450 isoforms [122]. In 10–20% of APS-1 cases, there is an associated hepatitis accompanied by anti-LKM reactivity to CYP450 isoforms 1A2 and 2A6 [122]. The APS-1/APECED syndrome results from defective function of a nuclear transcriptional autoimmune regulator (AIRE) protein encoded by the *AIRE* gene (21q22.3). Homozygous inheritance of mutant AIRE, expressed particularly in stromal cells in the thymic medulla, curtails the establishment of central immune tolerogenesis because the AIRE protein facilitates intrathymic expression of normally organ-specific "cryptic" autoantigens [123]. Thus, autoimmune expressions of AIRE deficiency in human APS-1 and the AIRE-deficient gene knockout mouse are organ specific rather than multisystem, and the associated hepatitis underscores intriguing aspects to the categorization of AIH into types 1 and 2. In other words, establishment of central T-cell tolerance to the autoantigens of type 1 AIH, ubiquitous nuclear constituents and cellular F-actin, would not require AIRE, whereas (presumably) tolerance to the LKM CYP450 autoantigens of type 2 does: Accordingly, AIRE-deficient AIH aligns with the organ-specific autoimmune diseases (see Table 46.1).

HEPATITIS C VIRUS INFECTION HCV infection resulting in chronic hepatitis C (CHC) is a suspected instigator of autoimmunity [124], as discussed subsequently. In one study [125], HCV infection provoked autoantibodies associated with type 1 AIH including SMA (actin-specificity not stated), ANA at low frequency, and rheumatoid factor, but AMA only very rarely. Mostly, however, anti-LKM are reported in CHC [126–129].

LIVER CYTOSOL TYPE 1 (LC-1)

Autoantibodies were recognized in 1988 [130] and further characterized [131] to a soluble liver cytosolic antigen in cases of type 2 AIH. Notably, anti-LC-1 reactivity is organ (liver) specific and also relatively disease specific for type 2 AIH. Unlike anti-LKM-1, it is seldom demonstrable in CHC and only occasionally (in children) in type 1 AIH [132]. It is not autoantigenic in SLE.

The LC-1 molecule has been eluted from liver cytosol as a protein of 240–290 kDa, and because the signal by immunoblot under reducing conditions is at 62 kDa, it likely exists as a tetramer [133]. Anti-LC1 shows by IIF a characteristically decreasingly homogeneous staining of hepatocytes toward the central vein, although this is usually obscured by accompanying anti-LKM; reactivity is also demonstrable by immunodiffusion, immunoblot, or counterimmunoelectrophoresis [134]. Anti-LC-1 has been identified as formiminotransferase cyclodeaminase (FTCA) [134], which has two domains—a globular FT domain joined by a short linker to the CA domain. Epitopes for autoantibodies have been mapped, and according to one study [135], there is a conformational epitope spanning much of the molecule and two linear epitopes located in the C-terminal domain—amino acids 428–434 and 440–447.

Cytoskeletal Autoantigens: the "Smooth Muscle" Antigen and F-actin

The cytoskeleton comprises structural proteins of the cell that support subcellular organization, cellular contractility, and locomotion. The three major types of cytoskeletal filaments or fibers are actin microfilaments (6 nm); intermediate-sized filaments (15 nm) vimentin, desmin and others; and microtubules (30 nm). Autoantibodies are revealed by IIF on substrates such as smooth muscle that are enriched in these components [136]. Smooth muscle in rodent gastric mucosal preparations remains the convenient substrate to demonstrate these diagnostically (Figure 46.3A).

SMOOTH MUSCLE ANTIBODY

Since the 1960s, SMA has provided a reliable serological marker for AIH, with numerous studies supporting the initially cited frequency of reactivity by IIF of approximately 80% [137], although tempered by a degree of positivity at lower titers in some other destructive liver diseases, PBC, chronic viral hepatitis, and others. As mentioned previously, SMA is not detected in type 2 AIH. Importantly, because SMA, like anti-LKM, segregates AIH from SLE, this reactivity must in some way be associated with liver cell destruction of a type that does not occur in SLE. Technical procedures for testing for SMA are important [77, 138, 139], but most critical is ascertainment of the filament subspecificity of the SMA reaction. Thus, generic SMA represents reactivity with various filament types, but it is reactivity with polymerized F-actin, which has a wider cellular distribution than intestinal smooth muscle and hence characteristic expressions on different tissue types, that characterizes type 1 AIH.

F-ACTIN

Monomeric G-actin is a 46-kDa globular protein present in all cells, wherein it exists in a dynamic state undergoing polymerization in four stages [140] to form F-actin microfilaments. This, together with the six similar isoforms of G-actin, and a structural model of the F-actin microfilament, are reviewed by Chhabra and dos Remedios [140]. Reactivity of SMA-positive sera with renal glomerular mesangium was the first indication of the wider cellular distribution of the SMA reactant [141], subsequently demonstrated as actin in the 1970s [142]. The autoantigenic reactivity specifying type 1 AIH depends on a presumably conformational epitope of the native molecule because reactivity is lost under the conventional denaturing conditions of immunoblotting. Demonstration of anti-F-actin by IIF requires unfixed or lightly fixed tissue substrates [76], and with potent autoimmune antisera, intracellular anti-F actin is demonstrable in most cell types. Initially, vinblastin-treated cultured cell monolayers were used [136], but there are characteristic features by IIF staining that clearly specify anti-F-actin reactivity [76, 77, 143–145]. Thus, on kidney, anti-F-actin autoantibodies give characteristic staining by IIF of the mesangium of renal glomeruli (SMA-G) and the outer border of tubular cells (SMA-T) (Figure 46.3B), in contrast to the more polyspecific staining of smooth muscle in vessel walls (SMA-V). On liver, anti-F-actin gives a typical "polygonal" pattern of IIF staining outlining margins of hepatocytes (Figure 46.3C) due to the abundant submembranous actomyosin [146] that functionally enables hepatocytes to contract around biliary duct ductules to propel bile flow [147]. On cultured cell lines, there are linear staining "actin cables" (Figure 46.3D).

Antibody to F-actin to high titer is reported in 70–80% of cases of AIH in acute or active phases [77, 148], but frequencies and titers may decrease in remission phases. In 99 patients with type 1 AIH, there was a positivity rate (74%) correlating with younger age, progressive disease, and HLA alleles B8 and DR3 [146]. Regarding specificity, autoantibody reactivity to F-actin is far superior to that to generic SMA, for which positivity occurs in various viral infections and miscellaneous liver diseases, being attributable to antibodies to monomeric G-actin or non-actin cytoskeletal elements, often intermediate filaments (vimentin and desmin), or tubulin [136].

Whether antibody to F-actin has any immunopathogenic relevance in AIH and how it might be provoked are undecided. If liver cell destruction of any cause were to result in spillage of the abundant intracellular actin, this could act as a perpetuating autoantigen in individuals at genetic risk for autoimmunity. However, just how actin is distributed in apoptosis fragments after cell death is not evident. Alternatively, if the permeability of the liver cell membrane were to be impaired in the course of hepatocyte injury, the copious submembranous actin would become exposed to autoimmune attack, but this remains to be demonstrated. For type 1

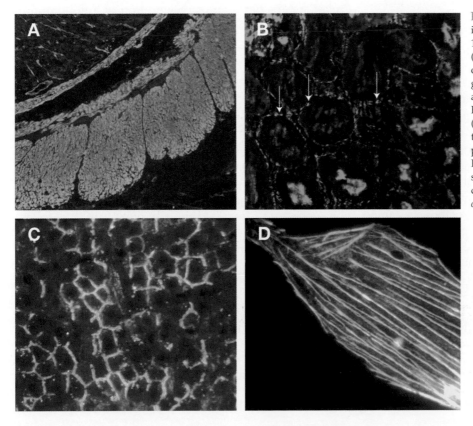

FIGURE 46.3 Staining patterns by immunofluorescence displayed by sera in type 1 AIH reactive for smooth muscle antibody (SMA) that differentiates AIH from SLE and other diseases. (A) Classic SMA on mouse gastric mucosa showing staining of muscularis and muscularis mucosae. (B) Kidney showing F-actin-specific "picket fence" staining (arrows) of peripheral border of proximal renal tubules. (C) Liver showing polygonal staining pattern due to reactivity with submembranous F-actin. (D) Cultured fibroblasts showing staining of F-actin microfilaments (actin cables). *Source: D, reproduced by kind permission of Taylor & Francis from Mackay and Toh [103].*

AIH, other insufficiently addressed issues relating to F-actin include the potential role of anti-F-actin in cytopathogenesis, analysis of antibody epitopes, and T-cell responsiveness.

Liver Membrane Antigens

ASIALOGLYCOPROTEIN RECEPTOR

The liver-specific asialoglycoprotein receptor (ASGPR) was identified in 1984 as a component of liver membrane protein, long suspected of harboring an auto-antigenic reactant with sera in AIH [149, 150]. The ASGPR, itself a glycoprotein, affects the binding and endocytosis of serum glycoproteins for disposal by the liver. It is a well-conserved protein except for differences in the number of subunits according to species: There are two in humans—H1 and H2. The properties and isolation of ASGPR, detection of autoantibodies by radioimmuno-assay, and frequencies are fully described [148]. However, studies have been limited because of difficulties in biochemical purification and expression to high yield as recombinant protein from a cDNA. A multinational survey of the positivity rate for anti-ASGPR in AIH reported it to be 76%, but the rate ranged from 7 to 19% for various other categories of liver disease, indicative of low specificity [151]. Interest in anti-ASGPR has now declined, perhaps because it is seen as consequential to, rather than causal of, liver damage, although levels are claimed to be a good marker of activity of AIH. The successful expression in a mammalian cell line of recombinant subunit H1 of human ASGPR yielded a protein with all the reactivity of the complete receptor [152], and this could reactivate interest. Notably, ASGPR is not demonstrable as an autoantigen in SLE.

Multiple Liver Membrane Autoantigens?

Identification of further autoantigenic reactants located in liver cell membrane has led to several studies on electrophoretically separated constituents using Western immunoblotting. This consistently reveals multiple reactants at differing molecular weights, some possibly cytoplasmic contaminants of incompletely purified membrane preparation. Few, if any, of the ascertained components have emerged as promising liver membrane-specific autoimmune reactants because for most there is a marked decrease in intensity of signal with increasing serum dilution beyond 1:1000, and after therapeutically induced remissions [153], and normal sera show equivalent (albeit much weaker) reactivities, further suggesting a consequential rather than causal relationship to hepatocellular damage. In the latest and most detailed of the studies [154], several of the reactive components were molecularly identified as potential constituents of the liver membrane (liver arginase, cytokeratins, heat shock proteins, and valosin-containing protein), but as in previous studies, these membrane components reacted weakly with normal sera (i.e., with natural antibodies), becoming "amplified" in the course of hepatocellular destruction. Thus, the explanation for their appearance could be trivial or, alternatively, "destruction-dependent" amplification of reactivity could be integral to pathogenetic processes in autoimmunity.

T-Cell Reactivity in Autoimmune Hepatitis

T-cells are important in the induction and effector phases of adaptive autoimmune responses, but understanding the role of autoimmune T-cells requires knowledge of the antigenic peptide epitopes to which such T-cells are responsive. Functional assays are based on exposure of T-cells in blood or tissues to autoantigen and the measurement of either cell proliferation or lymphokine release and also on numeric assays on binding of T-cells to antigenic peptides on major histocompatibility complex (MHC) tetramers to allow for enumeration in blood or tissues of autoantigen-specific CD8 or CD4 T-cells. More data are required regarding the moieties of autoantigens that are relevant to induction, and effector activities, of autoimmune T-cells in type 1 AIH, as well as in SLE, before their role can be adequately addressed. Such data are accumulating for type 2 AIH from studies on autoantigenic peptides of CYP450 2D6 for CD8[+] and CD4[+] T-cells [155, 156]. However, such data do not pertain to our enquiry into relationships between or provocation of autoimmunity in chronic hepatitis and SLE.

LUPUS AND THE LIVER: THE VISION OF THE IMMUNOGENETICIST

Genetic Predisposition to SLE

In Chapter 5 of the third edition of this book, there was an optimistic forecast that "by the time the next edition of this book is published the genetic contribution to SLE will be almost completely defined." Indeed, rapid progress was described in Chapter 4 of the fourth edition and in a benchmark review published in 2006 [157], and contemporary data appear in Chapters 1–4 of this edition.

The long-known genetic contributions to SLE are those associated with female gender, alleles of MHC (HLA) including molecules DR2 and DR3, null alleles at the C4QA complement locus, and Fcγ receptors. In addition to these are polymorphisms of the common immune response genes identified by a number of genome-wide association (GWA) studies in SLE cohorts as low-level risk alleles, including some that are also implicated in other prevalent autoimmune diseases, such as thyroiditis, type 1 diabetes, and rheumatoid arthritis [53]. Long suspected are immunomodulatory genes on the X chromosome, and one such gene has been identified at locus Xq28, specified as IRAK1, of which an allele contributes risk for SLE [158].

There are many other chromosomal regions of interest, and some of these susceptibility loci in humans are syntenic with genome segments identified in one or another of the various lupus-susceptible mouse strains. In particular, in mice, there is within a chromosome one "hot spot," the *sle1* region (1q 41–42), that includes a gene locus or loci enhancing the likelihood of ANA production [159]; activation of this is seen as the initial step in the cascade of events that result in lupus. Numerous other loci are implicated in the pathogenesis of SLE via diverse functional pathways, including disposal of antigen–immune complexes, lymphoid signaling, apoptosis, and epitope modification.

Genetic Predisposition to Autoimmune Hepatitis

Female Gender

As for most autoimmune diseases, including SLE, there is a strong gender bias to females (~6:1 female:male ratio) in the occurrence of AIH of either type 1 or type 2. Reasons for this sexual dimorphism in immune function, and in autoimmunity in general, have been examined [160]. Hormonal (estrogenic) influences certainly contribute through effects on immune responsiveness, which for B cells, as judged by antibody production, is demonstrably higher for females than for males [161].

MHC Alleles

TYPE 1 AIH

HLA In 1971, an association was recognized in cases of CAH (type 1 AIH) between the class I HLA alleles A1 and B8 [162]. Later, an association with DR3 was defined, and family studies established a close association with the ancestral HLA A1–B8–DR3 haplotype [163]; DR3 was seen as the likely primary risk allele, with linkage disequilibrium accounting for the earlier A1–B8 association. The association with HLA DR3 was widely confirmed, and the estimated relative risk for AIH was cited as six- or sevenfold [164]. The DR3 allele was shown to be part of an extended haplotype that included C4A gene deletions, thus providing a genetic basis for known low levels of the C4 component of complement in AIH [165]. A subsequently recognized association was with HLA DR4 [166]. Perhaps there are two nonexclusive immunogenetic subsets of type 1 AIH—the HLA DR3 subset, with a younger age of onset, more progressive disease, a greater tendency

to relapse despite corticosteroid therapy, and a higher requirement for liver transplantation, and an HLA DR4 subset, with an older age of onset and more benign disease [167]. Molecular genotyping revealed the HLA-related immunogenetic background of type 1 AIH as HLA DRB1*0301—DRB3*0101—DQA1*0501—DQB1*0201 genotypes, 52 vs. 19% of controls, and the strong secondary association with the DR4 allele DRB1*0401, 54% of DRB3*0101-negative patients compared with 23% of DRB3*0101-negative controls [168]. A lysine residue at position 71 of the HLA DRβ1 polypeptide (lysine-71) was considered to define a disease susceptibility motif with affinity for the as yet unknown disease-initiating autoantigenic moiety in type 1 AIH [169, 170], so far unverified. Like other HLA-dependent autoimmune diseases, there are different risk alleles for type 1 AIH among ethnically different populations [171], although most of the ascertained risk alleles express the DRβ1 lysine-71 disease susceptibility motif. Thus, comparing type 1 AIH with SLE, HLA DR3 is a clear risk allele albeit with a lower level of risk for SLE than for AIH (two- or threefold vs. six- or sevenfold), whereas DR2 confers risk only for SLE and DR4 only for AIH.

COMPLEMENT: C4QA Genes encoding complement components are located within the MHC. Homozygosity for null alleles for *C4QA* is recognized in type 1 AIH [166] as well as in SLE, and it is phenotypically expressed as low serum levels of C4. This results in impaired disposal of infectious agents, and tissue breakdown products, thus potentially contributing to the pathogenesis of either disease.

TYPE 2 AIH

Type 2 AIH is immunogenetically distinct from type 1 AIH, but sufficient numbers of reports/cases have not been assembled for susceptibility alleles to be confidently specified. Carriers of DR7 (*DRBI*0701*), among whom hepatitis is more aggressive and the outcome worse, are believed to have an increased risk [171].

Non-MHC Susceptibility Alleles

TYPE 1 AIH

TUMOR NECROSIS FACTOR-α The gene for tumor necrosis factor-α (TNF-α) contains alleles in promoter region 1. The *TNF-A* allele induces high constitutive and induced levels of TNF-α and thereby could confer risk for type 1 AIH via heightened T-cell effector activity [170].

CTLA-4 Cytotoxic T lymphocyte antigen 4 (CTLA-4) is a molecule that normally serves to downmodulate T-cell reactivity after a T-cell immune response and thus is an important participant in regulatory T-cell activity. The *CTLA-4 G* allele represents a polymorphism in exon 1 of the gene (position 49 A/G), and the ensuing allele is associated with reduced activity of regulatory T-cells. This allele is reported at higher frequency in type 1 AIH, as well as in SLE and other autoimmune diseases [170]. A study of families and children of patients with AIH [172] implicated a different CTLA-4 allele, with the interesting finding that only type 1 cases and not type 2 cases showed this association.

sle1 **GENE REGION** Studies on the numerous murine models have disclosed that chromosome 1 carries a gene region, *sle1*, comprising several loci conferring risk for lupus-like disease, and GWA studies in human SLE suggest the existence of an ortholog on the 1q 22−24 region. In mice at least, this region dictates ANA production and then disease [158]. It would be relevant to the theme of this chapter if this could be shown to pertain in type 1 AIH as well.

TYPE 2 AIH

There is a paucity of genetic data for type 2 AIH other than for the variant associated with AIRE mutations.

Comment on Genetic Susceptibility to Autoimmunity

Ascertaining the degree of sharing of susceptibility genes for AIH and SLE is a crucial question. In SLE, evolving molecular genetic technologies for specifying loci of interest and identifying candidate gene products conferring susceptibility are rapidly expediting a deeper understanding of causes. This could likewise apply to AIH, although GWA studies have not been developed because the prevalence and hence the "disease profile" are relatively lower than those for SLE, and funding sources are less accessible; the lack of valid disease models in animals of type AIH is another handicap.

The influences of the multiple genetic determinants of autoimmunity, and the many stages of the immune response at which these can operate, raise questions beyond the scope of this chapter. However, in brief, there are at least three groupings of gene variants.

The first group, common to many autoimmune diseases [53], comprises genes whose molecular products are required in one way or another for either orderly progression of normal responses to "foreign" antigens or for maintenance of tolerance to "self" antigens. Examples include genes encoding molecules that function as cell signaling or transcriptional factors, mediators of apoptosis, participants in immunoregulatory functions, chemoattractants, or inflammatory cytokines (tumor necrosis factor and interferons). Such genes would account for the characteristic clustering of various autoimmune expressions within affected individuals or related family members [53]. Notwithstanding, among autoimmune hepatitis families, reported co-existence of cases of AIH, or SLE, seem less frequent than might be expected.

The second group comprises gene products responsible for immunogenetic targeting of immune responses to specific autoantigens, resulting in either disease affecting particular tissues (organ-specific immunopathies) or disease that can be systemically distributed. The main example is the MHC (HLA) that encodes alleles having (or lacking) high binding affinity for particular autoantigens, noting that among AIH and SLE there are risk alleles either shared or disease specific, as described previously. This is, as stated previously, a disease-susceptibility motif on DRβ1 in type 1 AIH, but how this could direct the autoimmune process to hepatocytes is unknown. Another example could be inheritance of risk genes that influence the assembly/structure of T- or B-cell antigen receptors that, although basically randomly assembled, could be for certain reactivities "hardwired" in the primary B-cell repertoire, as suggested for antinuclear reactivity in lupus-prone mice [173].

The third group, on which more information is needed, comprises gene variants whose products influence vulnerability of a particular target tissue to immune attack, exemplified by the genetic defect in thyroid cells that in obese chickens predisposes to thyroiditis [174]. Such "end organ" genes conceivably could confer predisposition to autoimmune damage to the kidney in SLE or hepatocytes in AIH.

LUPUS AND THE LIVER: THE VISION OF THE EPIDEMIOLOGIST

The Elusive Environmental Agent

Certain susceptibility genes require interaction with an environmental agent for the expression of an autoimmune disease, but such agents are seldom evident in either human or spontaneous autoimmunity in animals. Perhaps the environmental agent acts merely by causing cellular injury or apoptosis, allowing exposure to intracellular (cryptic) autoantigens or apoptotic fragments (apotopes) to which tolerance is not fully established, thus resulting in induction of autoimmunity. In addition, the putative infectious "trigger" may contain a molecular sequence that sufficiently resembles that of a host component so that autoimmune induction occurs by cross-reactivity and epitope spreading. This molecular mimicry hypothesis is validated by some indisputable examples (streptococcal- or virally induced myocarditis and *Campylobacter*-induced polyneuropathy), but more evidence is needed for its general applicability to autoimmunity [175].

Of more direct concern here is whether we can implicate a provocative initiating agent common to SLE and AIH. Among the suspected environmental infectious determinants of SLE is the Epstein−Barr virus (EBV), as discussed elsewhere in this book, and in AIH, suspected initiating viruses likewise include EBV [176].

For AIH (but not for SLE), hepatitis viruses and particularly A and C (HAV and HCV) have come under scrutiny. Type 1 AIH is occasionally reported to supervene on infection with HAV [46], and the frequently reported cases of lupoid hepatitis in young women in the 1960s coincided with a concurrent endemic of HAV infection. HCV is associated more with type 2 than type 1 AIH; however, type 1 AIH is predominant in areas of the world where HCV is least endemic, and there is the usual absence of markers of HCV infection in either type of spontaneous AIH. Anti-LKM is expressed in CHC at varying frequencies according to geographical location; among 83 French patients, the frequency of anti-LKM (anti-CYP450 2D6) was as high as 77% [126], but as with other studies on HCV-associated anti-LKM-1, titers were relatively low and males were over-represented. The P450 isoform usually engaged in CHC is 2D6, but autoantibodies are directed to epitopes other than those targeted in "spontaneous" type 2 AIH [127]. Also provoked in CHC are autoantibodies to other CYP450 reactants, including 2A6 [110] and 2E1 [128], and to other liver autoantigens as well—anti-LC-1 in 12% of cases [177] and anti-SLA/LP in 10.4% [178]—although levels by ELISA were lower than in spontaneous AIH. These associations raise questions regarding etiopathogenesis and therapy, including mechanisms whereby HCV might initiate serological if not actual autoimmunity. However, the dissociation of HCV infection in type 1 AIH overall, and type 2 AIH in children, argues strongly against HCV being an antecedent to AIH.

Therapeutic drugs are infrequent instigators of either SLE or AIH but nonetheless raise interesting considerations. One is that such conditions, whether expressed as AIH-like or SLE-like syndromes, are uniquely drug dependent such that the syndrome recedes when the offending drug is withdrawn [179]. Medications implicated in drug-induced SLE include hydralazine and procaine amide. In drug-induced AIH with an "autoimmune overlay," the features can simulate either type of AIH—type 1 with seropositivity or ANA and SMA or type 2 with seropositivity for anti-LKM/CYP450. For the type 1 instances that are instigated by the drugs minocycline and α-methyl dopa, there is a commonality with drug-induced SLE, but the exact immunopathogenesis is unknown. For the type 2 instances, disposal of the drug by the relevant CYP450 isoform creates a drug isoform-specific epitope with immunogenic potential [179], without any commonality with SLE.

CONCLUSIONS

This chapter has considered issues raised by the nosologic problems of autoimmune hepatitis with features of SLE and liver lesions that occur in classical SLE. Accordingly, the chapter was introduced by a brief discussion of problems of disease nomenclature, particularly with those diseases in which autoimmunity is prominent among multifactorial components. Three propositions provided the infrastructure to the chapter. First, there is a disease known as autoimmune hepatitis (previously called CAH) that is intrinsic to the liver and in the domain of hepatology but, by reason of its autoimmune causation, can exhibit sufficient multisystem disease criteria for it to become co-identified with SLE, the former lupoid hepatitis. Second, classical SLE is itself a disease with widespread inflammatory expressions, partly attributable to "toxic" effects resulting from tissue breakdown, such that pathology in the liver equivalent to that seen in toxemias or severe infections can occur but with no resemblance to the pathology of AIH. Third, the immune effector mechanisms, particularly the generation of cytopathogenic immune complexes (the "lupus" process), that elicit the characteristic inflammatory damage to the tissues usually affected in SLE, kidney, brain, skin, lungs, etc. are apparently not operative within the liver.

Serologically, there are commonalities in SLE and AIH, namely the presence of antibodies to chromatin/histones as the major ANA reactant. However, there are also unique differences, including a strong expression of anti-dsDNA in SLE but only weak, or no, expression in AIH and the absence in SLE of reactivities that are characteristic of one or another type of AIH, namely to F-actin, soluble liver antigen (tRNA-associated protein), various microsomal (CYP450) enzymes, and the liver-specific molecule FTCA (LC-1). Another commonality is a degree of sharing of immunogenetic determinants including HLA DR3, and critical here will be the results of GWA studies in AIH for comparison or contrast with the detailed data accrued for SLE.

For AIH and SLE, the differences seem to outweigh the similarities. A final point to consider is that when there does appear to be convergence of SLE and AIH, this is usually attributable to multisystemic features that accumulate during the course of an initially occurring AIH, like that seen with other SLE overlaps such as "rhupus," rather than the reverse wherein a case presenting initially as SLE traverses to AIH. We wonder how many editions of this classic volume will be printed before there is a complete understanding of the relationship between SLE and chronic autoimmune hepatitis.

Acknowledgments

Appreciation for illustrative material is expressed to Nigel Swanston (Figure 46.1A) Bob Wilson (Figures 46.2A, B, D, 3B and Daniela Zauli Figure 46.2C). We thank Elaine Pearson for assistance with the preparation of the manuscript.

References

[1] A.S. Cohen, W.E. Reynolds, E.C. Franklin, J.P. Kulka, M.W. Ropes, L.E. Shulman, et al., Preliminary criteria for the classification of systemic lupus erythematosus, Bull Rheum Dis 21 (1971) 643–648.

[2] E.M. Tan, A.S. Cohen, J.F. Fries, A.T. Masi, D.J. McShane, N.F. Rothfield, et al., The 1982 revised criteria for the classification of systemic lupus erythematosus, Arthritis Rheum 25 (1982) 1271–1277.

[3] Anonymous, The American College of Rheumatology nomenclature and case definitions for neuropsychiatric lupus syndromes, Arthritis Rheum 42 (1999) 599–608.

[4] E.M. Tan, Antinuclear antibodies: Diagnostic markers for autoimmune diseases and probes for cell biology, Adv Immunol 44 (1989) 93–151.

[5] L.K.F. Temple, R.S. McLeod, S. Gallinger, V.G. Wright, Defining disease in the genomics era, Science 293 (2001) 807–808.

[6] J.G. Scadding, Essentialism and nominalism in medicine: Logic of diagnosis in disease terminology, Lancet 348 (1996) 594–596.

[7] I.R. Mackay, The prognoses of chronic hepatitis, Ann Intern Med 77 (1972) 649–651.

[8] P.J. Johnson, I.G. McFarlane, Meeting report: International Autoimmune Hepatitis Group, Hepatology 18 (1993) 998–1005.

[9] F. Alvarez, P.A. Berg, F.B. Bianchi, et al., International Autoimmune Hepatitis Group report: Review of criteria for diagnosis of autoimmune hepatitis, J Hepatol 31 (1999) 929–938.

[10] I.R. Mackay, Historical reflections on autoimmune hepatitis, World J Gastroenterol 14 (21) (2008) 3292–3300.

[11] M.H. Barker, R.B. Capps, F.W. Allen, Chronic hepatitis in the Mediterranean theatre, JAMA 129 (1945) 653–659.

[12] I.J. Wood, W.E. King, P.J. Parsons, J.W. Perry, M. Freeman, L. Limbrick, Non-suppurative hepatitis: A study of acute and chronic forms with special reference to biochemical and histological changes, Med J Aust 11 (1948) 249–261.

[13] J. Waldenstrom, Leber, Blutproteine und Nahrungseiweiss. Dtsch, Gesellschaft Verdauungs-und Stoffwechselkrankheiten 15 (1950) 113–119.

[14] H.G. Kunkel, J.R. Ahrens, W.J. Eisenmenger, A.M. Bongiovanni, R.J. Slater, Extreme hypergammaglobulinemia in young women with liver disease of unknown etiology [abstract], J Clin Invest 30 (1951) 654.

[15] A.G. Bearn, H.G. Kunkel, R.J. Slater, The problem of chronic liver disease in young women, Am J Med 21 (1956) 3–15.

[16] I.R. Mackay, L.I. Taft, D.C. Cowling, Lupoid hepatitis, Lancet 2 (1956) 1323–1326.

[17] I.R. Mackay, D.C. Gajdusek, An "autoimmune" reaction against human tissue antigens in certain acute and chronic diseases: II. Clinical correlations, Arch Intern Med 101 (1958) 30–46.

[18] I.R. Mackay, S. Weiden., J. Hasker, Autoimmune hepatitis, Ann N Y Acad Sci 124 (1965) 767–780.

[19] R.A. Bradbear, W.N. Robinson, W.G.E. Cooksley, J.W. Halliday, O.D. Harris, L.W. Powell, Are the causes and presentation of chronic hepatitis changing? An analysis of 104 cases over 15 years, Q J Med 53 (1984) 279–288.

[20] D.R. Parker, I.G.C. Kingham, Type 1 autoimmune hepatitis is primarily a disease of later life, Q J Med 90 (1997) 289–296.

[21] H. Krook, Liver cirrhosis in patients with a lupus erythematosus-like syndrome, Acta Med Scand 169 (1961) 713–726.

[22] A.E. Read, S. Sherlock, C.V. Harrison, Active "juvenile" cirrhosis as part of a systemic disease and the effect of corticosteroid therapy, Gut 4 (1963) 378–393.

[23] T.B. Reynolds, H.A. Edmondson, R.L. Peters, A. Redeker, Lupoid hepatitis, Ann Intern Med 61 (1964) 650–666.

[24] M.J. MacLachlan, G.P. Rodnan, W.M. Cooper, R.H. Fennell Jr., Chronic active ("lupoid") hepatitis: A clinical, serological and pathological study of 20 patients, Ann Intern Med 62 (1965) 425–462.

[25] P.A. Miescher, A. Braverman, E. Amorosi, Progressive hypergammaglobulinaemic hepatitis, Dtsch Med Wschr 91 (1966) 1525–1532.

[26] R.D. Soloway, D.M. Summerskill, A.H. Baggenstoss, L.J. Shoenfield, "Lupoid" hepatitis, a nonentity in the spectrum of chronic active liver disease, Gastroenterology 63 (1975) 458–465.

[27] A.J. Czaja, G.L. Davis, J. Ludwig, A.H. Baggenstoss, H.F. Taswell, Autoimmune features as determinants of prognosis in steroid-treated chronic active hepatitis of uncertain etiology, Gastroenterology 85 (1983) 713–717.

[28] R.B. Trimble, A.S. Townes, H. Robinson, S.B. Kaplan, R.W. Chandler, A.S. Hanissian, et al., Preliminary criteria for the classification of systemic lupus erythematosus (SLE). Evaluation in early diagnosed SLE and rheumatoid arthritis, Arthritis Rheum 17 (1974) 184–188.

[29] C.M. Leevy, H. Popper, S. Sherlock, Diseases of the liver and biliary tract. Standardization of nomenclature, diagnostic criteria and diagnostic methodology, in: "Fogarty International Centre Proceedings No. 22," U.S. Government Printing Office, Washington, DC, 1976, pp. 76–725. DHEW Publication No. (NIH).

[30] J.R. Hodges, G.H. Millward-Sadler, R. Wright, Chronic active hepatitis: The spectrum of disease, Lancet 1 (1982) 550–552.

[31] I.R. Mackay, Autoimmune diseases of the liver, in: N.R. Rose, I.R. Mackay (Eds.), The Autoimmune Diseases, Academic Press, Orlando, FL, 1985, pp. 291–337.

[32] International Hepatology Informatics Group, Diseases of the liver and biliary tract, in: C.M. Leevy, S. Sherlock, N. Tygstrup, R. Zetterman (Eds.), Standardisation of Nomenclature, Diagnostic Criteria and Prognosis, Raven, New York, 1994.

[33] International Working Party, Terminology of chronic hepatitis, Am J Gastroenterol 90 (1995) 181–189.

[34] E.M. Hennes, M. Zeniya, A.J. Czaja, A. Parés, G.N. Dalekos, E.L. Krawitt, et al., International Autoimmune Hepatitis Group. Simplified criteria for the diagnosis of autoimmune hepatitis, Hepatology 48 (1) (2008) 169–176.

[35] M. Pistina, D.J. Wallace, S. Nessim, A.L. Metzger, J.R. Klinenberg, Lupus erythematosus in the 1980s: A survey of 570 patients, Semin Arthritis Rheum 21 (1991) 55–64.

[36] F. Schaffner, H. Popper, Non-specific reactive hepatitis in aged and infirm people, Am J Dig Dis 4 (1959) 399.

[37] B.A. Runyon, D.R. LaBreque, S. Anuras, The spectrum of liver disease in systemic lupus erythematosus, Am J Med 69 (1980) 187–194.

[38] T. Gibson, A.R. Myers, Subclinical liver disease in systemic lupus erythematosus, J Rheumatol 8 (1981) 752–759.

[39] M.H. Miller, M.B. Urouitz, D.D. Gladman, L.M. Blendis, The liver in systemic lupus erythematosus, Q J Med 53 (1984) 401–409.

[40] W. Van Steenbergen, J. Beyls, J. Vermylen, J. Fevery, G. Marchal, V. Desmet, et al., Lupus anticoagulant and thrombosis of the hepatic veins (Budd–Chiari syndrome). Report of three patients and review of the literature, J Hepatol 3 (1986) 87–94.

[41] B.A. Leggett, The liver in systemic lupus erythematosus, J Gastroenterol Hepatol 8 (1993) 84–88.

[42] F.P. Ruiz, J.O. Martinez, A.C.Z. Mendoza, L.R. Del Arbol, A.M. Caparros, Nodular regenerative hyperplasia of the liver in rheumatic diseases: Report of seven cases and review of the literature, Semin Arthritis Rheum 21 (1991) 47–54.

[43] M. Sekiya, I. Sekigawa, T. Hishikawa, N. Iida, H. Hashimoto, S. Hirose, Nodular regenerative hyperplasia of the liver in systemic lupus erythematosus, Scand J Rheumatol 26 (1997) 215–217.

[44] R.M. Crapper, P.S. Bhathal, I.R. Mackay., I.H. Frazer, "Acute" autoimmune hepatitis, Digestion 34 (1986) 216–225.

[45] A.J. Czaja, Autoimmune chronic active hepatitis, in: A.J. Czaja, E.R. Dickson (Eds.), Chronic Active Hepatitis. The Mayo Clinic Experience, Dekker, New York, 1986, pp. 105–126.

[46] E.L. Krawitt, Autoimmune hepatitis, N Engl J Med 334 (1996) 897–903.

[47] P.L. Golding, M. Smith, R. Williams, Multisystem involvement in chronic liver disease. Studies on the incidence and pathogenesis, Am J Med 55 (1973) 772–782.

[48] M. Rizzeto, G. Swana, D. Doniach, Microsomal antibodies in active chronic hepatitis and other disorders, Clin Exp Immunol 15 (1973) 331–344.

[49] M. Rizzeto, F.B. Bianchi, D. Doniach, Characterization of the microsomal antigen related to a subclass of active chronic hepatitis, Immunology 26 (1974) 589–601.

[50] J.C. Homberg, N. Abuaf, O. Bernard, et al., Chronic active hepatitis with anti-liver/kidney microsome antibody type 1: A second type of autoimmune hepatitis, Hepatology 7 (1987) 1333–1339.

[51] S. Lindgren, H.-B. Braun, G. Michel, A. Nemeth, S. Nilsson, B. Thome-Kromer, et al., Absence of LKM antibody reactivity in autoimmune and hepatitis C-related chronic liver disease in Sweden, Scand J Gastroenterol 32 (1997) 175–178.

[52] E. Tanaka, K. Kiyosawa, T. Seki, A. Matsumoto, T. Sodeyama, S. Furuta, et al., Low prevalence of hepatitis C infection in patients with autoimmune hepatitis type 1, J Gastroenterol Hepatol 8 (1993) 442–447.

[53] I.R. Mackay, Clustering and commonalities among autoimmune diseases, J Autoimmun 2009 33 (3-4) (2009) 170–177.

[54] J. Woodward, J. Neuberger, Autoimmune overlap syndromes, Hepatology 33 (2001) 994–1002.

[55] C. Rust, U. Beuers, Overlap syndromes among autoimmune liver diseases, World J Gastroenterol 14 (21) (2008) 3368–3373.

[56] R.P. Kenny, A.J. Czaja, J. Ludwig, E.R. Dickson, Frequency and significance of antimitochondrial antibodies in severe chronic hepatitis, Dig Dis Sci 31 (1986) 705–711.

[57] O. Chazouillères, D. Wendum, L. Serfaty, et al., Primary biliary cirrhosis-autoimmune hepatitis overlap syndrome: clinical features and response to therapy, Hepatology 28 (1998) 296–301.

[58] A.W. Lohse, K.-H. Meyer zum Büschenfelde, B. Franz., et al., Characterization of the overlap syndrome of primary biliary cirrhosis and autoimmune hepatitis: Evidence for it being a hepatitic form of PBC in genetically susceptible individuals, Hepatology 29 (1999) 1078–1084.

[59] S. Hall, P.H. Axelsen, D.E. Larson, T.W. Bunch, Systemic lupus erythematosus developing in patients with primary biliary cirrhosis, Ann Intern Med 100 (3) (1984) 388–389.

[60] I.R. Mackay, S. Whittingham, S. Fida, M. Myers, N. Ikuno, M.E. Gershwin, et al., The peculiar immunity of primary biliary cirrhosis, Immunol Rev 174 (2000) 226–237.

[61] F.C. Powell, A.L. Schroeter, E.R. Dickson, Primary biliary cirrhosis and the CREST syndrome: A report of 22 cases, Q J Med 62 (1987) 75–82.

[62] I.R. Mackay, M.J. Rowley, S.F. Whittingham, Nuclear autoantibodies in primary biliary cirrhosis, in: K.-H. Meyer zum Buschenfelde, J. Hoofnagle, M. Manns (Eds.), Immunology and Liver: Falk Symposium Number 70, Immunology and Liver, Kluwer, Dordrecht, 1993, pp. 397–409.

[63] G. Brünner, O. Klinge, A cholangitis with antinuclear antibodies (immuno-cholangitis) resembling chronic destructive non-suppurative cholangitis, Dtsch Med Wschr 112 (1987) 1454–1458.

[64] Z.D. Goodman, P.R. McNally, D.R. Davis, K.G. Ishak, Autoimmune cholangitis: A variant of primary biliary cirrhosis. Clinicopathologic and serologic correlation in 200 cases, Dig Dis Sci 40 (1995) 1232–1242.

[65] H. Kinoshita, K. Omagari, I. Matsuo, et al., Autoimmune cholangitis: Serological features in 21 Japanese cases compared with those in primary biliary cirrhosis and autoimmune hepatitis, Liver 14 (1999) 122–128.

[66] I.R. Mackay, L.I. Taft, D.C. Cowling, Lupoid hepatitis and the hepatic lesions of systemic lupus erythematosus, Lancet 1 (1959) 65–69.

[67] J. de Groote, V.J. Desmet, P. Gedigk, et al., A classification of chronic hepatitis, Lancet 2 (1968) 626–628.

[68] International group of pathologists, Acute and chronic hepatitis revisited, Lancet 2 (1977) 914–919.

[69] A.R. Page, R.A. Good, Plasma cell hepatitis with special attention to steroid therapy, J Dis Child 99 (1960) 288–314.

[70] H. Popper, F. Paronetto, F. Schaffner, Immune processes in the pathogenesis of liver disease, Ann N Y Acad Sci 124 (1965) 781–799.

[71] W.C.E. Cooksley, R.A. Bradbear, W. Robinson, H. Harrison, J.W. Halliday, L.W. Powell, et al., The prognosis of chronic active hepatitis without cirrhosis in relation to bridging necrosis, Hepatology 6 (1986) 345–348.

[72] J. Searle, B.V. Harmon, C.S. Bishop, J.F.R. Kerr, The significance of cell death by apoptosis in hepatobiliary disease, J Gastroenterol Hepatol 2 (1987) 77–96.

[73] A.J. Czaja, M. Nishioka, S. Morshed, T. Hachiya, Patterns of nuclear immunofluorescence and reactivities to recombinant nuclear antigens in autoimmune hepatitis, Gastroenterology 107 (1994) 200–207.

[74] A.J. Czaja, H.A. Homburger, Autoantibodies in liver disease, Gastroenterology 120 (1) (2001) 239–249.

[75] E.M. Tan, T.E.W. Feltkamp, J.S. Smolen, et al., Range of antinuclear antibodies in "healthy" individuals, Arthritis Rheum 40 (1997) 1601–1611.

[76] D. Vergani, F. Alvarez, F.B. Bianchi, E.L. Cançado, I.R. Mackay, M.P. Manns, et al., Liver autoimmune serology: A consensus statement from the Committee for Autoimmune Serology of the International Autoimmune Hepatitis Group, J Hepatol 41 (4) (2004) 677–683.

[77] D.P. Bogdanos, P. Invernizzi, I.R. Mackay, D. Vergani, Autoimmune liver serology: Current diagnostic and clinical challenges, World J Gastroenterol 14 (21) (2008) 3374–3387.

[78] I.R. Mackay, Antinuclear (chromatin) autoantibodies in autoimmune hepatitis, J Gastroenterol Hepatol 16 (2001) 245–247.

[79] F. Konikoff, M. Swissa, Y. Schoenfeld, Autoantibodies to histones and their subfractions in chronic liver diseases, Clin Immunol Immunopathol 51 (1989) 77–82.

[80] A.J. Czaja, C. Ming, M. Shirai, M. Nishioka, Frequency and significance of antibodies to histones in autoimmune hepatitis, J Hepatol 23 (1995) 32–38.

[81] D. Kenneally, I.R. Mackay, B.H. Toh, Antinucleolar autoantibodies demonstrated by monolayers of human fibroblasts in sera from patients with systemic lupus erythematosus, progressive systemic sclerosis and chronic active hepatitis, J Clin Lab Immunol 14 (1984) 13–16.

[82] J. Wesierska-Gadek, E. Penner, Nuclear antigens, in: I.G. McFarlane, R. Williams (Eds.), Molecular Basis of Autoimmune Hepatitis, Landes, Austin, TX, 1996, pp. 23–44.

[83] S. Jain, T. Markham, H.C. Thomas, Double stranded DNA-binding capacity of serum in acute and chronic liver disease, Clin Exp Immunol 26 (1976) 35—41.

[84] A. Peretz, F. Mascart-Lemone, G. Nuttin, J.P. Famaey, Chronic active hepatitis with a high level of anti-ds-DNA detected by a solid phase radioimmunoassay, Clin Rheumatol 1 (1982) 208—211.

[85] R. Smeenk, G. van der Lelij, T. Swaak, Specificity in systemic lupus erythematosus of antibodies to double-stranded DNA measured with the polyethylene glycol precipitation assay, Arthritis Rheum 25 (1982) 631—638.

[86] B.A. Leggett, R.V. Collins, W.G.E. Cooksley, D.I. Prentice, L.W. Powell, Evaluation of the Crithidia assay to distinguish between chronic active hepatitis and systemic lupus erythematosus, J Gastroenterol Hepatol 2 (1987) 202—211.

[87] W. Emlen, L. O'Neill, Clinical significance of antinuclear antibodies. Comparison of detection with immunofluorescence and enzyme-linked immunosorbent assays, Arthritis Rheum 40 (1997) 1612—1618.

[88] F. Konikoff, Y. Schoenfeld, D.A. Isenberg, I. Barrison, T. Sobe, E. Theodor, et al., Anti-Rnp antibodies in chronic liver diseases, Clin Exp Rheumatol 5 (1987) 359—361.

[89] E. Penner, I. Kindas-Mugge, E. Hitchman, G. Sauermann, Nuclear antigens recognized by autoantibodies present in liver disease sera, Clin Exp Immunol 63 (1986) 428—433.

[90] M. Nishioka, S.A. Morshed, S. Parveen, C. Ming, Heterogeneity of antinuclear antibodies in autoimmune liver diseases, in: E.L. Krawitt, R.H. Wiesner, M. Nishioka (Eds.), Autoimmune Liver Diseases, second ed., Elsevier, Amsterdam, 1998, pp. 179—216.

[91] K. Elkon, S. Skelly, A. Parnassa, A.P. Moller, W. Danho, H. Weissbach, et al., Identification and chemical synthesis of a ribosomal protein antigenic determinant in systemic lupus erythematosus, Proc Natl Acad Sci USA 83 (1986) 7419—7423.

[92] J.L. Lin, V. Dubljevic, M.J. Fritzler, B.H. Toh, Major immunoreactive domains of human ribosomal P proteins lie N-terminal to a homologous C-22 sequence: Application to a novel ELISA for systemic lupus erythematosus, Clin Exp Immunol 141 (1) (2005) 155—164.

[93] E. Koren, W. Schnitz, M. Reichlin, Concomitant development of chronic active hepatitis and antibodies to ribosomal P proteins in a patient with systemic lupus erythematosus, Arthritis Rheum 36 (1993) 1325—1328.

[94] E. Koren, M.W. Reichlin, M. Koscec, R.D. Fugate, M. Reichlin, Autoantibodies to the ribosomal P proteins react with a plasma membrane related target on human cells, J Clin Invest 89 (1992) 1236—1241.

[95] M. Koscec, E. Koren, M. Wolfson-Reichlin, R.D. Fugate, E. Trieu, I.N. Targoff, et al., Autoantibodies to ribosomal P proteins penetrate into live hepatocytes and cause cellular dysfunction in culture, J Immunol 159 (1997) 2033—2041.

[96] F.C. Arnett, M. Reichlin, Lupus hepatitis: An under-recognized feature associated with autoantibodies to ribosomal P, Am J Med 99 (1995) 465—472.

[97] M. Hulsey, R. Coldstein, L. Scully, W. Surbeck, M. Reichlin, Anti-ribosomal P antibodies in systemic lupus erythematosus: A case control study correlating hepatic and renal disease, Clin Immunol Immunopathol 74 (1995) 252—256.

[98] H. Ohira, J. Takiguchi, T. Rai, K. Abe, J. Yokokawa, Y. Sato, et al., High frequency of anti-ribosomal P antibody in patients with systemic lupus erythematosus-associated hepatitis, Hepatol Res 28 (3) (2004) 137—139.

[99] J. Wesierska-Gadek, E. Penner, E. Hitchman, G. Sauermann, Antibodies to nuclear lamins in autoimmune liver disease, Clin Immunol Immunopathol 49 (1988) 107—115.

[100] M.J. Smalley, I.R. Mackay, S. Whittingham, Antinuclear factors and human leucocytes: Reaction with granulocytes and lymphocytes, Aust Ann Med 17 (1968) 28—32.

[101] D. Zauli, S. Ghetti, A. Grassi, C. Descovich, F. Cassam, G. Ballardini, et al., Neutrophil cytoplasmic antibodies in type 1 and type 2 autoimmune hepatitis, Hepatology 25 (1997) 1005—1007.

[102] P.E. Spronk, H. Bootsma, G. Horst, M.G. Huitema, P.C. Limburg, J.W. Cohen Tervaert, et al., Antineutrophil cytoplasmic antibodies in systemic lupus erythematosus, Br J Rheumatol 35 (1996) 625—631.

[103] I.R. Mackay, B.-H. Toh, Autoimmune hepatitis. The way we were, the way we are today, and the way we hope to be, Autoimmunity 35 (2002) 293—305.

[104] P.A. Berg, E. Stechemesser, J. Strienz, et al., Hypergammaglobulinämische chronisch aktive Hepatitis mit Nachweis von leber-pankreas-spezifischen komplemtbindenden Antikörpern, Verh Dtsch Ges Inn Med 87 (1981) 921—927.

[105] M. Manns, G. Gerken, A. Kyriatsoulis, et al., Characterization of a new subgroup of autoimmune chronic active hepatitis by autoantibodies against a soluble liver antigen, Lancet 1 (1987) 292—294.

[106] I. Wies, S. Brunner, J. Henninger, et al., Identification of target antigen for SLA/LP autoantibodies in autoimmune hepatitis, Lancet 355 (2000) 1510—1515.

[107] M. Costa, J.L. Rodriguez-Sanchez, A.J. Czaja, C. Gelpi, Isolation and characterization of cDNA encoding the antigenic protein of the human tRNP (ser) sec complex recognized by autoantibodies from patients with type-1 autoimmune hepatitis, Clin Exp Immunol 121 (2000) 364—374.

[108] T. Kevenbeck, A.W. Lohse, J.A. Grötzinger, A conformational approach suggests the function of the autoimmune hepatitis target antigen soluble liver antigen/liver pancreas, Hepatology 34 (2001) 230—233.

[109] J. Herkel, M.P. Manns, A.W. Lohse, Selenocystine, soluble liver antigen/liver-pancreas, and autoimmune hepatitis, Hepatology 46 (2007) 275—277.

[110] M.P. Manns, P. Obermayer-Straub, Cytochromes P450 and UDP-glucuronyl transferases: Model antigens to study drug-induced, virus-induced and autoimmune liver disease, Hepatology 26 (1997) 1054—1066.

[111] D.J. Waxman, D.P. Lapenson, M. Krishnan, O. Bernard, G. Kreibich, F. Alvarez, Antibodies to liver/kidney microsomes in chronic active hepatitis recognize specific forms of hepatic cytochrome P450, Gastroenterology 95 (1988) 1326—1331.

[112] M.P. Manns, E.F. Johnson, K.J. Griffin, E.M. Tan, K.F. Sullivan, Major antigen of liver kidney microsomal autoantibodies in idiopathic autoimmune hepatitis is cytochrome P450db, J Clin Invest 83 (1989) 1066—1072.

[113] M.P. Manns, K.J. Griffin, K.F. Sullivan, E.F. Johnson, LKM-1 autoantibodies recognize a short linear sequence in P450IID6, a cytochrome P450 monooxygenase, J Clin Invest 88 (1991) 1370—1378.

[114] J.-C. Homberg, C. André, N. Abuaf, A new anti-liver/kidney microsome antibody (anti-LKM2) in tienilic induced hepatitis, Clin Exp Immunol 55 (1984) 561—570.

[115] P.H. Beaune, P.M. Dansette, D. Mansuy, L. Kiffel, M. Finck, C. Amar, et al., Human anti-endoplasmic reticulum autoantibodies appearing in a drug-induced hepatitis directed against a human liver cytochrome P450 that hydroxylates the drug, Proc Natl Acad Sci USA 84 (1987) 551—555.

[116] M. Bourdi, D. Larrey, J. Nataf, J. Bernuau, D. Pessayre, M. Iwasaki, et al., Anti-liver endoplasmatic reticulum autoantibodies are directed against human cytochrome P450 IA2: A

specific marker of dihydralazine-induced hepatitis, J Clin Invest 85 (1990) 1967—1973.

[117] O. Crivelli, C. Lavarini, E.A. Chiaberge, A. Amoroso, P. Farci, F. Negro, et al., Microsomal auto-antibodies in chronic infection with the HBsAg associated delta (δ) agent, Clin Exp Immunol 54 (1983) 232—238.

[118] M. Durazzo, T. Philipp, F.N.A.M. van Pelt, B. Lüttig, E. Borghesio, G. Michel, et al., Heterogeneity of microsomal auto-antibodies (LKM) in chronic hepatitis C and D virus infection, Gastroenterology 108 (1995) 455—462.

[119] C.P. Strassburg, P. Obermayer-Straub, B. Alex, M. Durazzo, M. Rizzetto, R.H. Tukey, et al., Autoantibodies against glucuronosyltransferases differ between viral hepatitis and autoimmune hepatitis, Gastroenterology 111 (1996) 1582—1592.

[120] T. Bachrich, T. Thalhammer, W. Jäger, P. Haslmayer, B. Alihodzic, S. Bakos, et al., Characterization of autoantibodies against uridine-diphosphate glucuronosyltransferase in patients with inflammatory liver diseases, Hepatology 33 (2001) 1053—1059.

[121] J. Perheentupa, Autoimmune polyendocrinopathy—candidiasis—ectodermal dystrophy, J Clin Endocrinol Metab 91 (2006) 2843—2850.

[122] P. Peterson, R. Uibo, O. Kampe, Polyendocrine syndromes, in: N.R. Rose, I.R. Mackay (Eds.), The Autoimmune Diseases, fourth ed., Academic Press, San Diego, 2006, pp. 515—526.

[123] J.J. De Voss, A.K. Shum, K.P. Johannes, W. Lu, A.K. Krawisz, P. Wang, et al., Effector mechanisms of the autoimmune syndrome in the murine model of autoimmune polyglandular syndrome type 1, J Immunol 181 (6) (2008) 4072—4079.

[124] R.W. McMurray, K. Elbourne, Hepatitis C virus infection and autoimmunity, Semin. Arthritis Rheum 26 (1997) 689—701.

[125] B.D. Clifford, D. Donahue, L. Sonith, E. Cable, B. Lüttig, M. Manns, et al., Prevalence of serological markers of autoimmunity in patients with hepatitis C, Hepatology 21 (1995) 613—619.

[126] F. Lunel, N. Abauf, L. Frangeul, et al., Liver/kidney microsome antibody type 1 and hepatitis C virus infection, Hepatology 16 (1992) 630—636.

[127] K. Choudhuri, G. Miele-Vergani, D. Vergani, Cytochrome P4502D6: Understanding an autoantigen, Clin Exp Immunol 108 (1997) 381—383.

[128] M. Vidali, E. Hildestrand, E. Eliasson, E. Motteran, R. Reale, G. Rolla, et al., Use of molecular stimulation for mapping conformational CYP2E1 epitopes, J Biol Chem 279 (2004) 50949—50955.

[129] A.R. Kammer, S.H. Van der Burg, B. Grabscheid, I.P. Hunziger, K.M. Kwappenberg, J. Reichen, et al., Molecular mimicry of human cytochrome P450 by hepatitis C virus at the level of cytotoxic T cell recognition, J Exp Med 190 (1999) 169—176.

[130] E. Martin, N. Abauf, F. Cavalli, V. Durand, C. Johanet, J-C. Homberg, Antibody to liver cytosol (anti-LC1) in patients with autoimmune chronic active hepatitis type 2, Hepatology 8 (1988) 1662—1666.

[131] N. Abauf, C. Johanet, P. Chretien, E. Martini, E. Soulier, S. Laperche, et al., Characterization of the liver cytosol antigen type 1 reacting with autoantibodies in chronic active hepatitis, Hepatology 16 (1992) 892—898.

[132] S. Han, M. Tredger, G.V. Gregorio, G. Mieli-Vergani, D. Vergani, Anti-liver cytosolic antigen type 1 (LC1) antibodies in childhood autoimmune liver disease, Hepatology 21 (1995) 58—62.

[133] L. Muratori, M. Cataleta, P. Muratori, P. Manotti, M. Lenzi, F. Cassani, et al., Detection of anti-liver cytosol antibody type 1 (anti-LC1) by immunodiffusion, counterimmunoelectrophoresis and immunoblotting: Comparison of different techniques, J Immunol Methods 187 (1995) 259—264.

[134] P. Lapierre, O. Hajoui, J.-C. Homberg, F. Alvarez, Form-aminotransferase cyclodeaminase is an organ-specific auto-antigen recognised by sera of patients with autoimmune hepatitis, Gastroenterology 116 (1999) 643—649.

[135] P. Lapierre, C. Johanet, F. Alvarez, Characterization of the B cell response of patients with anti-liver cytosol autoantibodies in type 2 autoimmune hepatitis, Eur J Immunol 33 (7) (2003) 1869—1878.

[136] B.-H. Toh, Smooth muscle autoantibodies and autoantigens, Clin Exp Immunol 38 (1979) 621—628.

[137] S. Whittingham, J. Irwin, I.R. Mackay, M. Smalley, Smooth muscle autoantibody in "autoimmune" hepatitis, Gastroenterology 51 (1966) 499—505.

[138] M. Fusconi, F. Cassani, M. Lenzi, G. Ballardini, U. Volta, F.B. Bianchi, Anti-actin antibodies: A new test for an old problem, J Immunol Methods 130 (1990) 1—8.

[139] E.L.R. Cancado, C.P. Abrantes-Lemos, L.S. Vilas-Boas, N.F. Novo, F.J. Carrilho, A.A. Laudanna, Thermolabile and calcium-dependent serum factor interferes with polymerized actin, and impairs anti-actin antibody detection, J Autoimmun 17 (2001) 223—228.

[140] D. Chhabra, C.G. dos Remedios, Actin: An overview of its structure and function, in: C. dos Remedios, D. Chhabra (Eds.), Actin, Actin-Binding Proteins and Disease, Springer, New York, 2008, pp. 50—64.

[141] S. Whittingham, I.R. Mackay, J. Irwin, Autoimmune hepatitis. Immunofluorescence reactions with cytoplasm of smooth muscle and renal glomerular cells, Lancet 1 (7451) (1966) 1333—1335.

[142] K. Lidman, G. Biberfeld, A. Fagraeus, R. Norberg, R. Torstensson, G. Utter, et al., Anti-actin specificity of human smooth muscle antibodies in chronic active hepatitis, Clin Exp Immunol 24 (2) (1976) 266—272.

[143] R.A. Silvestrini, E.M. Benson, Whither smooth muscle antibodies in the third millennium, J Clin Pathol 54 (2001) 677—678.

[144] I.R. Mackay, A.J. Czaja, I.G. McFarlane, M.P. Manns, Chronic hepatitis, in: N.R. Rose, I.R. Mackay (Eds.), The Autoimmune Diseases, fourth ed., Academic Press, San Diego, 2006, pp. 729—747.

[145] I.R. Mackay, R. Martinez-Neira, S. Whittingham, D. Nicolay, B.-H. Toh, Autoantigenicity of actin, in: C. dos Remedios, D. Chhabra (Eds.), Actin, Actin-Binding Proteins and Disease, Springer, New York, 2008, pp. 50—64.

[146] S. Yasuura, T. Veno, S. Watanabe, M. Hirose, T. Namahisa, Immunocytochemical localization of myosin in normal and phalloidin-treated rat hepatocytes, Gastroenterology 97 (1989) 982—989.

[147] S. Watanabe, A. Miyazaki, M. Hirose, M. Takeuchi, H. Ohide, T. Kitamura, et al., Myosin in hepatocytes is essential for bile canalicular contraction, Liver 11 (3) (1991) 185—189.

[148] A.J. Czaja, F. Cassani, M. Cataleta, P. Valentine, F.B. Bianchi, Frequency and significance of antibodies to actin in type 1 autoimmune hepatitis, Hepatology 24 (1996) 1068—1073.

[149] I.G. McFarlane, B.M. McFarlane, G.N. Major, P. Tolley, R. Williams, Identification of the hepatic asialoglycoprotein receptor (hepatic lectin) as a component of liver specific membrane lipoprotein (LSP), Clin Exp Immunol 55 (1984) 347—354.

[150] B. McFarlane, Hepatocellular membrane antigens, in: I.G. McFarlane, R. Williams (Eds.), Molecular Basis of Autoimmune Hepatitis, Landes, Austin, TX, 1996, pp. 75—104.

[151] U. Treichel, B. McFarlane, T. Seki, et al., Demographics of anti-asialoglycoprotein receptor autoantibodies in autoimmune hepatitis, Gastroenterology 107 (1994) 799—804.

II. CLINICAL ASPECTS OF DISEASE

[152] T. Schreiter, C. Liu, G. Gerken, U. Treichel, Detection of circulating autoantibodies directed against the asialoglycoprotein receptor using recombinant receptor subunit H1, J Immunol Methods 301 (1-2) (2005) 1−10.

[153] I. Matsuo, N. Ikuno, K. Omagari, H. Kinoshita, M. Oka, H. Yamaguchi, et al., Autoimmune reactivity of sera to hepatocyte plasma membrane in type 1 autoimmune hepatitis, J Gastroenterol 35 (2000) 226−234.

[154] F. Tahiri, F. Le Naour, S. Huguet, R. Lai-Kuen, D. Samuel, C. Johanet, et al., Identification of plasma membrane autoantigens in autoimmune hepatitis type 1 using a proteomics tool, Hepatology 47 (3) (2008) 937−948.

[155] Y. Ma, D.P. Bogdanos, M.J. Hussain, J. Underhill, S. Bansal, M.S. Longhi, et al., Polyclonal T-cell responses to cytochrome P450IID6 are associated with disease activity in autoimmune hepatitis type 2, Gastroenterology 130 (3) (2006) 868−882.

[156] M.S. Longhi, M.J. Hussain, D.P. Bogdanos, A. Quaglia, G. Mieli-Vergani, Y. Ma, et al., Cytochrome P450IID6-specific CD8 T cell immune responses mirror disease activity in autoimmune hepatitis type 2, Hepatology 46 (2) (2007) 472−484.

[157] A.M. Fairhurst, A.E. Wandstrat, E.K. Wakeland, Systemic lupus erythematosus: Multiple immunological phenotypes in a complex genetic disease, Adv Immunol 92 (2006) 1−69.

[158] E.K. Wakeland, K. Liu, R.R. Graham, T.W. Behrens, Delineating the genetic basis of systemic lupus erythematosus, Immunity 15 (2001) 397−408.

[159] C.O. Jacob, J. Zhu, D.L. Armstrong, M. Yan, J. Han, X.J. Zhou, et al., Identification of IRAK1 as a risk gene with critical role in the pathogenesis of systemic lupus erythematosus, Proc Natl Acad Sci USA 106 (15) (2009) 6256−6261.

[160] P.A. McCombe, J.M. Greer, I.R. Mackay, Sexual dimorphism in autoimmune disease, Curr Mol Med 9 (2009) 1058−1079.

[161] M.J. Rowley, I.R. Mackay, Measurement of antibody-producing capacity in man, I. The normal response to flagellin from Salmonella adelaide. Clin Exp Immunol 5 (1969) 407−418.

[162] I.R. Mackay, P.J. Morris, Association of autoimmune chronic hepatitis with HLA-A1-B8, Lancet 2 (1972) 793−795.

[163] I.R. Mackay, B.D. Tait, HLA associations with autoimmune-type chronic active hepatitis: Identification of B8-DRw3 haplotype by family studies, Gastroenterology 79 (1980) 95−98.

[164] I.R. Mackay, R.M. O'Brien., S. Whittingham, B.D. Tait, Immunogenetic aspects, in: N. Farid (Ed.), The Immunogenetics of Autoimmune Disease, Autoimmune hepatitis and other diseases of the liver, Vol. 2, CRC Press, Boca Raton, FL, 1991, pp. 119−213.

[165] B. Tait, I.R. Mackay, P.H. Board, M. Coggan, P. Emery, G. Eckardt, HLA A1, B8, DR3 extended haplotypes in autoimmune chronic hepatitis, Gastroenterology 97 (1989) 479−481.

[166] D.G. Doherty, J.A. Underhill, P.T. Donaldson, K. Manabe, G. Mieli-Vergani, A.L.F.W. Eddleston, et al., Polymorphism in the human complement C4 genes and genetic susceptibility to autoimmune hepatitis, Autoimmunity 18 (1994) 243−249.

[167] P.T. Donaldson, D.G. Doherty, K.M. Hayllar, I.G. McFarlane, P.J. Johnson, R. Williams, Susceptibility to autoimmune chronic active hepatitis: Human leukocyte antigens DR4 and A1-B8-DR3 are independent risk factors, Hepatology 13 (1991) 701−706.

[168] A.J. Czaja, M.D. Strettell, L.J. Thomson, P.J. Santrachi, S.B. Moore, P.T. Donaldson, et al., Associations between alleles of the major histocompatibility complex and type 1 autoimmune hepatitis, Hepatology 25 (1997) 317−323.

[169] D.G. Doherty, P.T. Donaldson, J.A. Underhill, et al., Allelic sequence variation in the HLA class II genes and proteins in patients with autoimmune hepatitis, Hepatology 19 (1994) 609−615.

[170] A.J. Czaja, P.T. Donaldson, Genetic susceptibilities for immune expression and liver cell injury in autoimmune hepatitis, Immunol. Rev 174 (2000) 250−259.

[171] D. Vergani, G. Mieli-Vergani, Aetiopathogenesis of autoimmune hepatitis, World J Gastroenterol 14 (2008) 3306−3312.

[172] I. Djilali-Saiah, P. Ouellette, S. Caillat-Zucman, D. Debray, J.I. Kohn, F. Alvarez, CTLA-4/CD 28 region polymorphisms in children from families with autoimmune hepatitis, Hum Immunol 62 (12) (2001) 1356−1362.

[173] S. Chang, L. Yang, Y.M. Moon, Y.G. Cho, S.Y. Min, T.J. Kim, et al., Anti-nuclear antibody reactivity in lupus may be partly hard-wired into the primary B-cell repertoire, Mol Immunol 46 (16) (2009) 3420−3426.

[174] G. Wick, K. Hála, H. Wolf, A. Ziemiecki, R.S. Sundick, M. Stöffler-Meilicke, et al., The role of genetically-determined primary alterations of the target organ in the development of spontaneous autoimmune thyroiditis in obese strain (OS) chickens, Immunol Rev 94 (1986) 113−136.

[175] Multiauthor Review (J.M. Davies, Coordinator), Molecular mimicry, Cell Mol Life Sci 57 (2000) 523−578.

[176] S. Vento, L. Guella, F. Mirandola, F. Cainelli, G. Di Perri, M. Solbiati, et al., Epstein−Barr virus as a trigger for autoimmune hepatitis in susceptible individuals, Lancet 346 (1995) 608−609.

[177] K. Béland, P. Lapierre, G. Marceau, F. Alvarez, Anti-LC1 autoantibodies in patients with chronic hepatitis C virus infection, J Autoimmun 22 (2) (2004) 159−166.

[178] S. Vitozzi, P. Lapierre, I. Djilali-Saiah, G. Marceau, K. Beland, F. Alvarez, Anti-soluble liver antigen (SLA) antibodies in chronic HCV infection, Autoimmunity 37 (3) (2004) 217−222.

[179] I.R. Mackay, Immune mechanisms and liver toxicity, in: R.G. Cameron, G. Feuer, F.A. de la Inglesia (Eds.), Handbook of Experimental Pharmacology Series: Drug-Induced Hepatotoxicity, Springer, Berlin, 1996, pp. 221−247.

[180] M. Biburger, G. Tiegs, Animal models of autoimmune liver disease, in: M.E. Gershwin, J.M. Vierling, M.P. Manns (Eds.), Liver Immunology. Principles and Practice, second ed., Humana Press, Totowa, NJ, 2007, pp. 293−308.

[181] P. Lapierre, I. Djilali-Saiah, S. Vitozzi, F. Alvarez, A murine model of type 2 autoimmune hepatitis: Xenoimmunization with human antigens, Hepatology 39 (4) (2004) 1066−1074.

47

Gastrointestinal: Nonhepatic

Iris Dotan, Lloyd Mayer

When considering the "systemic" nature of lupus erythematosus, the gastrointestinal (GI) tract tends to be overlooked, not by the disease itself but by the primary physician. William Osler's [1] description in 1895 of significant gastrointestinal crises in "erythema exudativum multiforme" illustrates that the GI tract can overshadow many of the other aspects of systemic lupus erythematosus (SLE). In general, the major GI complications of SLE are less frequently seen than the skin rash, arthritis, and nephritis that commonly bring patients to clinical attention. However, if we look at various series of patients with SLE [2–15], we find that there are quite a number of listed signs and symptoms relating to the GI tract (Table 47.1). For example, in the largest such series, by Dubois and Tuffanelli [5], of 520 patients studied, 53% had nausea or vomiting, 49% complained of anorexia, and 19% had abdominal pain. Although many of these symptoms are nonspecific and may relate to other systemic manifestations of SLE (e.g., nausea and vomiting secondary to uremia) or may be side effects of drug therapy, some may represent inapparent GI tract disease.

As just suggested, a major problem in recognizing GI tract disease as a complication of SLE is that much of the same GI pathology can be induced by the various medications used in SLE treatment. For example, there are controversies about lupus pancreatitis and about liver disease caused directly by SLE versus aspirin-induced focal necrosis or steroid-induced fatty liver or pancreatitis. Another problem is that many of the symptoms relating to the GI tract are not as severe as the more pressing arthritis or nephritis and thus tend to be glossed over. In this chapter, a review of the diverse GI disorders seen as sequelae of SLE is presented, beginning with the oral cavity. GI ailments relating to therapy for SLE and SLE induced by drugs prescribed for GI disorders are considered separately.

ORAL CAVITY

Clinical Features

The GI tract extends from the mouth to the anus, so involvement of the oral cavity in SLE should be considered to be a GI manifestation. Oral lesions are such a central and common sign in SLE that the presence of oral ulceration is one of the criteria in the diagnosis of SLE [16]. They tend to precede severe systemic disease flares [17]. Involvement of the oral cavity is reported to occur in 7–57% of SLE patients [2, 6, 8–12, 14, 18]. Oral lesions are in general of three types: erythematous, discoid, and ulcerative. According to one study [12], erythematous lesions occur most often (35%), followed by discoid (16%) and ulcerative (6%) lesions. Another study [9] found 9% of patients to have more than one lesion at a time. The lesions may merge into one another and may be associated with edema and petechiae [11].

Discoid Lesions

Discoid lesions may be painful. They are characterized by a central area of erythema with white dots surrounded by rays of white striae and telangiectasia, are firm on palpation, and have a well-defined elevated border [11–19]. They are found in decreasing frequency in the buccal mucosa, gingiva, labial mucosa, and vermilion border [9, 11, 19], and they are generally seen in patients with active disease [11]. They may be ulcerated and are infected by yeast in approximately 50% of cases [19].

Ulcers

Ulcers are usually shallow, 1 or 2 cm in diameter, tend to occur in crops, and are most commonly found on the hard palate (89% of cases) [9]. They may also be found on the buccal mucosa and extend into the pharynx in approximately one-third of patients [9]. Involvement of

887

TABLE 47.1 GI-Related Symptoms Reported (as Percentages) in SLE Patients

Reference	No. of patients	Anorexia	Vomiting/ nausea	Diarrhea	Abdominal pain	Hemorrhage	Oral lesions	Esophageal dysphagia	Serositis/ ascites	Pancreatitis	Gastroesophageal ulcers	Small intestine ulcers/large intestine ulcers	Splenomegaly
Jessar et al. [2]	44	—	18	18	22	—	18 ulcers	—	—	—	—	—	27
	168	—	13	13	17	—	15 ulcers	—	—	—	—	—	17
Harvey et al. [3]	138	—	11/14	8	10	5	—	5/6	—	—	0	2/4	—
Brown et al. [4]	87	—	8	3	2	—	—	—	—	—	—	4/2	—
Dubois and Tuffanelli [5]	520	49	53	6	19	6	—	—/15	—/11	—	1.5	0/<1	9
Estes and Christian [6]	150	—	—	—	—	<1	?	—	—	<1	<1	2/0	18
Fries and Holman [7]	—	36	27/36	25	34	10	—	—/5	—	—	—	—	—
Ropes [8]	99	—	—	—	—	—	41	63	9	—	—	—	67
Urman et al. [9]	182	—	—	—	62	40	Ulcers	3/22	Autopsy/8	Autopsy	22	1/0	—
Al Rawi et al. [10]	67	—	—	—	—	—	52	—	—	—	—	—	18
Jonsson et al. [11]	51	—	—	—	—	—	45	—	—	—	—	—	—
Tsianos et al. [12]	25	—	—	—	—	—	57	—	—	—	—	—	—
Nadorra et al. [13] Autopsy	26	8	27	19	65	46	—	—/4	42/90	23	33	23	—
Burge et al. [14]	53	—	—	—	—	36	—	—	—	—	—	—	—
Castrucci et al. [15]	18	—	—	—	—	Ulcers	72	Manometric abnormalities	—	—	—	—	—

the nasal and laryngeal mucosa has also been described [20–22]. Oral ulcers often present with pain, a burning sensation, soreness, dry mouth, and, when extending into the pharynx, odynophagia and even dysphagia [7]. These lesions may persist for years [11, 23, 24] or may be intermittent, displaying cyclical remissions and exacerbations [7]. As with discoid lesions, oral ulcers tend to occur with disease flares [11, 17, 24]. The presence of oral ulcers has been considered by some to be a poor prognostic sign. One study reported that patients with oral ulcers had a higher mortality than did patients without oral ulcers [7]. This has not been corroborated by a second study [9].

Erythematous Lesions

Erythematous lesions are typically painless, flat, red lesions with poorly defined borders. Occasionally, they are accompanied by edema and petechiae. They most commonly occur on the hard palate [11].

Pathology/Pathophysiology

The histopathology and immunopathology of the discoid lesions in SLE are similar to those of discoid lupus erythematosus except that the inflammatory infiltrate found deep in the connective tissue is more diffuse and less intense in SLE [11, 19, 24, 25]. Keratotic plugging, pseudoepitheliomatous hyperplasia, and epithelial cell islands within the connective tissue are also found. The inflammatory infiltrate found in the connective tissue is typically lymphohistiocytic with an occasional plasma cell, and the inflammatory infiltrate is typically separated from the epithelium by an eosinophilic-free zone. The infiltrating cells are predominantly $CD4^+$ and $CD8^+$ T cells distributed in the lamina propria, submucosa, and, occasionally, the epithelium. Less than 5% of the infiltrating inflammatory cells express interleukin-2 receptors or transferrin receptors [26]. Epithelial cell changes are seen and include the presence of colloid bodies, multinucleated cells, occasional nuclear hyperchromatism, and, rarely, dysplasia. Some epithelial cells express major histocompatibility class II antigens, with HLA-DR being more commonly present than HLA-DP. No HLA-DQ is seen. Liquefaction degeneration of the basal epithelial cell layers is also noted [26]. Hyalinization of the subepithelial connective tissue may occur, as can edema of the lamina propria [11, 19, 24, 25]. Immunoglobulin and complement as well as fibrinogen are deposited in the basement membrane zone and are found in both discoid lesions and clinically normal mucosa of SLE patients [11, 19, 27–32].

Ultrastructurally, the lesions show cytomorphologic features peculiar to keratinizing epithelium. Dyskeratotic cells and filamentous bodies, which probably correspond to the colloid bodies seen on light microscopy, are present [19, 33–37].

As suggested by the occasional presence of dysplastic epithelial cells in discoid lesions, it is important to be aware that cancer can develop in long-standing discoid lesions of the lip at the vermilion border [11].

Erythematous and ulcerative lesions show nonspecific changes, including liquefaction degeneration of the basal layer and a diffuse subepithelial inflammatory infiltrate [11, 19, 24]. Erythematous lesions also show subepithelial and epithelial inflammation, disturbed epithelial maturation, and acanthosis [11]. On immunopathology, IgM deposits are noted in the basement membrane [11].

Diagnosis

Because a significant proportion of oral lesions may be asymptomatic, careful examination of the oral cavity is mandatory in all lupus patients. Besides the difficulty in distinguishing discoid lesions from lichen planus and leukoplakia, there may be difficulty in distinguishing SLE oral lesions from monilia, erythema multiforme, pemphigus, and herpetic lesions [38–40]. Bacterial, fungal, and viral cultures, as well as detailed drug history, can help in these cases. In addition, the finding of immunoglobulin and complement deposition in the basement membrane in discoid lesions may be useful as a distinguishing feature because immunoglobulin deposition is rare in lichen planus and leukoplakia [19]. Schiodt [25] described five features that may distinguish discoid lesions from the similar-appearing lesions of lichen planus and leukoplakia: (1) hyperkeratosis with keratotic plugs; (2) atrophy of the rete processes; (3) edema of the lamina propria; (4) deep inflammatory infiltrate; and (5) thick, patchy, or continuous periodic acid–Schiff-positive deposits in the basement membrane zone. The presence of two or more of these criteria was specific and sensitive for the diagnosis of SLE discoid lesions.

Therapy

There are no well-controlled trials demonstrating efficacy of therapy of any of these lesions [19]. Treatments that have been recommended for oral ulcerations include oral and topical steroids [19]. For discoid lesions, antimalarials [9, 21, 39, 41] have been tried, as have oral and topical steroids [41, 42] and dapsone [43, 44]. Treatment directed at controlling the SLE flare may often prove helpful in ameliorating the oral lesions. Antimicrobial, antiviral, or antifungal therapy should be used where infection is suspected or proven. Drug withdrawal, such as the withdrawal of hydralazine (shown to resolve oral ulcerations in at least one patient with SLE) [45], may occasionally prove useful.

SLE and Sjögren's Syndrome

Sjögren's syndrome has been associated with SLE in 13–20% of patients [46–48]. A higher percentage of sicca symptoms associated with SLE has been reported, but because the criteria for Sjögren's syndrome classification were not met, this is probably an overestimation [49, 50]. Although Sjögren's syndrome usually appears in later stages of the disease, it may precede the diagnosis of SLE [51]. Compared to patients with SLE alone, those with secondary Sjögren's syndrome tend to exhibit fewer systemic manifestations, especially renal involvement [52].

Other Oral Lesions

Other oral lesions have been seen in lupus patients. Acute necrotizing ulcerative gingivitis has been associated with SLE [24, 53]. It is manifested by ulceration, bleeding, facial swelling, dysphagia, halitosis, and even sloughing of the gingival tissue with loss of teeth. Steroids may contribute to the severity of the disease. Treatment includes: (1) reduction in steroid dose; (2) oral, topical, and, if severe, intravenous antibiotics against usual mouth flora or against a specific organism if one is identified; and (3) topical gentian violet. Jaworski *et al.* [53] suggested that altered host immune mechanisms, such as decreased chemotaxis of polymorphonuclear cells, in addition to further suppression of the immune system by immunosuppressive agents used in the treatment of SLE, probably contribute to tip the host–organism balance in favor of the symbiotic organism. All SLE patients, especially those taking steroids or other immunosuppressive drugs, should be instructed to maintain excellent oral hygiene, and dentists should be alerted to the greater risks of postoperative bleeding and infection in this group [54].

Other rare lesions include Raynaud's phenomenon of the tip of the tongue [55] and tense bullae of the oral cavity [56]. The bullae resolve with steroid therapy. A case of oral hairy leukoplakia has also been documented [57]. SLE patients may also get yeast infections or oral herpes simplex virus, especially during exacerbations or steroid therapy [7, 11, 21]. In addition, drugs used in the treatment of SLE may cause oral lesions. For example, antimalarials have been reported to produce brown or blue-gray pigmentation of the oral mucosa [58].

ESOPHAGUS

The incidence of esophageal disease is difficult to ascertain. Reports on the incidence of symptomatic esophageal disease vary widely (1.5–22%) [3, 5, 7, 8, 13]. One problem is the fact that no correlation between symptomatology and pathology exists [15].

Compounding this problem is the possibility that some patients with mixed connective tissue disease, with known esophageal pathology, may have been misclassified as having SLE in the earlier literature.

Clinical Features

Patients with SLE may present with dysphagia, heartburn typical of acid reflux, as well as atypical chest pain. Esophagitis with ulcerations may be seen in approximately 3.5% of patients [3, 8], rarely with perforation [13]. Mucosal bridging, a finding occasionally seen in patients with a history of peptic esophagitis, has also been reported [59]. In addition, corticosteroids or immunosuppressive therapy can predispose patients to *Candida* esophagitis, and patients may present with oral candidiasis and odynophagia.

Pathology/Pathophysiology

Controversy exists regarding the etiology of the esophagitis and chest pain seen in lupus patients. Some consider these findings to be peptic in origin. Evidence for this comes from the findings of abnormal esophageal motility, specifically hypoperistalsis and aperistalsis in as many as 72% of lupus patients [15, 60–65], that might result in decreased clearance of refluxed acid. Others have found a high incidence of gastroesophageal reflux [15]. The type of dysmotility seen varies. In one survey of unselected SLE patients studied by manometry, X-ray, and history, 14% had aperistalsis in the lower two-thirds of the esophagus, lower esophageal sphincter, or both (i.e., the smooth muscle portion of the esophagus) consistent with a progressive systemic sclerosis picture; 14% had a motility disorder of the upper one-third of the esophagus (i.e., the skeletal muscle portion of the esophagus) consistent with a dermatomyositis picture; and two had involvement of the entire esophagus [61].

This has led other authors to suggest that the pathologic findings in the lupus esophagus are vasculitic rather than peptic in origin. Castrucci *et al.* [15] hypothesized that the hypoperistalsis or aperistalsis found in the esophagus of SLE patients may be due to an inflammatory reaction in the esophageal muscle or to ischemic vasculitic damage of the Auerbach plexus. Harvey *et al.* [3] noted arteritis in the esophagus of four SLE patients in association with esophageal ulcers. A large proportion of patients with esophageal ulcers in these studies had been taking steroids, which suggests that these may have been stress ulcers.

A lesser incidence of involvement of the lower esophageal sphincter (LES) in SLE has been reported [66]. In contrast to systemic sclerosis patients, 81.8% of whom had LES abnormalities, 30% classified as severe, none of the SLE patients evaluated had severe involvement of the LES. This may explain the interesting observation that the complications of acid reflux so pronounced in other collagen vascular diseases such as progressive systemic sclerosis are not noted in SLE.

Despite manometric and histologic evidence of esophageal disease, the lack of correlation of abnormalities to symptoms makes one wonder about the significance of these findings. With the exception of the occasional esophageal ulcer [5, 6, 8, 13], which only rarely has been reported to perforate, esophageal symptoms are generally not severe.

Diagnosis

Esophagitis and esophageal ulcer may be diagnosed by esophagogastroduodenoscopy or by barium esophagram. Esophageal dysmotility may be detected by manometric studies or, occasionally, on barium esophagram.

Therapy

Because it is not clear whether the esophagitis and ulcerations found are vasculitic in origin, peptic in origin, or a complication of steroid therapy, it is difficult to determine the appropriate mode of therapy. Empiric proton pump inhibitors, H_2 blockers, or other antacid therapy in patients taking steroids is controversial. A controlled trial of antireflux therapy in patients with SLE would greatly help determine whether such esophageal lesions are peptic in nature and help determine whether antacid and antireflux therapy should be routinely recommended to SLE patients. Treatment of the lupus may help if in fact the esophageal lesions are vasculitic in origin. A trial of calcium channel blockers for esophageal dysmotility may be of benefit. Patients with *Candida* esophagitis will require appropriate antifungal therapy, including nystatin or clotrimazole troches for mild infections and fluconazole for more severe cases.

STOMACH AND DUODENUM

Clinical features

Lupus patients with intestinal complications referable to the stomach or duodenum may have gastric or duodenal ulcer disease. Abdominal pain is the most common symptom, often described as burning, boring, or nagging. Pain is usually epigastric and relieved by food or antacids. Nausea and vomiting may occur with gastric ulcer or duodenal ulcer complicated by gastric outlet obstruction. A small proportion of patients may present with bleeding or perforation. On physical examination, abdominal tenderness or fecal occult blood may be found, but the exam is often nonspecific. Corticosteroid use may mask the signs and symptoms of disease.

Pathology/Pathophysiology

The actual incidence of ulcer disease in SLE has never been directly addressed. In Brown *et al.*'s review [4], it is mentioned that 5% (4 of 85) of patients had ulcer disease unrelated to therapy, whereas Ropes [8] quotes a figure of 20% (7 of 35). Dubois and Tuffanelli [5] cite a 1.5% incidence of ulcer perforation, which resulted in death in one patient taking steroids. Although the role of *Helicobacter pylori* and nonsteroidal anti-inflammatory agents in the etiology of gastroduodenal ulcers is now well-established [67, 68], in SLE patients the possibility still exists that such ulcers are a direct consequence of lupus. One group reported a case of giant gastric ulceration associated with the antiphospholipid antibody syndrome, which resolved on treatment with anticoagulants and corticosteroids [69]. In a review of the digestive tract findings in connective tissue disorders, Siurala *et al.* [70] performed gastric biopsies on 17 patients with SLE. Four showed superficial gastritis, and an additional 8 showed varying degrees of atrophic gastritis. There appeared to be no correlation to therapy, and of note was the presence of similar histologic findings in 8 of 20 control patients with celiac disease. No data were available regarding the presence of antiparietal cell antibodies, which have been seen in these patients [71] and may relate to the atrophic gastritis. Nadorra *et al.* [13] found heterotopic calcification of the superficial gastric mucosa in 3 of 36 autopsies of patients with childhood-onset SLE, albeit 2 of the 3 had been on hemodialysis. Thus, there does not appear to be sufficient data to determine whether SLE is causative in the development of peptic ulcer disease.

Okayasu *et al.* [72] performed an autopsy study in Japan examining the risk of cancer in patients with connective tissue disorders. Although it was found that dermatomyositis was strongly associated with the development of gastric carcinoma, no such association was found for SLE either in the stomach or elsewhere. Thus, there appears to be no link between SLE and the development of gastric malignancy.

Diagnosis

The diagnosis of gastric or duodenal ulcer may be made on clinical grounds or by esophagogastroduodenoscopy or upper GI series. *Helicobacter pylori* infection may be detected through serology for IgG and IgA anti-*Helicobacter* antibodies, by [13C]-urea breath test following an oral urea load, or with endoscopic biopsy using urease or histologic examination.

Therapy

It is reasonable to treat all documented benign gastric and duodenal ulcers with proton pump inhibitors, H₂ blockers, or other antacid therapy, and it has been argued that it is reasonable to give prophylactic therapy to those patients taking steroids, aspirin, or nonsteroidal anti-inflammatory drugs (NSAIDs). Misoprostol may also be considered for prophylaxis, although the complication of misoprostol-induced diarrhea (up to 40% of patients) may be worse than the symptoms of peptic ulcer disease. Special consideration should be given to bleeding peptic ulcers in SLE patients. Hiraishi *et al.* [73] reported a patient with SLE who presented with severe upper gastrointestinal bleeding as a result of gastric ulcer. The bleeding did not stop with conventional anti-ulcer therapy or by surgery in which an antral ulcer penetrating into the ileum with histologic stigmata of vasculitis was found. The bleeding was controlled only after pulse methylprednisolone therapy was initiated, stressing the issue of etiology-oriented therapy.

Other Gastric Manifestations

Watermelon Stomach

A case of iron deficiency anemia due to chronic blood loss attributed to watermelon stomach in an SLE patient with anticardiolipin antibodies was reported by Archimandritis *et al.* [74]. This condition of gastric antral vascular ectasia is associated with connective tissue disorders in 62% of cases. Treatment with methylprednisolone and subsequently hydroxychloroquine in addition to iron supplements was effective in controlling the anemia.

Eosinophilic Gastroenteritis

This disease, once perceived to be uncommon with an obscure etiology, is increasingly reported, specifically in young populations. Eosinophilic gastroenteritis (EG) has been reported in association with food allergies as well as several connective tissue diseases. The first description of EG in an SLE patient was reported in 2004 by Barbie *et al.* [75] in a 37-year-old woman with idiopathic thrombocytopenic purpura presenting with nausea, vomiting, abdominal pain, gastric and intestinal thickening, as well as ascites on computed tomography (CT) scan. The presence of antinuclear antibodies (ANAs) and dsDNA Abs and renal findings established the diagnosis of SLE. Gastric and intestinal biopsies were consistent with EG, and treatment with steroids alleviated her symptoms. Similar cases in which enteric expression of EG was more prominent have been reported [76, 77]. A case report by Jaimes-Hernandez *et al.* [76] presented a 36-year-old woman with

abdominal pain, nausea, vomiting, melena, and intestinal thickening as well as ascites on CT. She underwent an exploratory laparotomy and the finding of eosinophilic infiltration to the lamina propria established the diagnosis of EG. The simultaneous diagnosis of SLE was established based on the findings of hemolytic anemia, thrombocytopenia, positive ANAs and anticardiolipin antibodies, as well as a pleural effusion. Steroid-treatment was beneficial, and a complication of intestinal pseudo-obstruction after cessation resolved spontaneously. The association of intestinal pseudo-obstruction, EG, and SLE was reported by other authors in adult [78] as well as pediatric patients [79]. In children, intestinal pseudo-obstruction as a result of EG was reported as the first symptom in two cases. The presumed immunopathology, for both adult and pediatric patients, is that activated lymphocytes in the intestinal tract attract eosinophils that then secrete inflammatory cytokines, promoting a cycle of more migration and more inflammation.

SMALL AND LARGE INTESTINE

Some of the more relevant and potentially dangerous GI complications of SLE occur in the small and large intestines. Although some complications described previously (e.g., aspirin-induced gastritis or ulcer) may be related to therapy, complications seen in the small and large intestines (e.g., vasculitis of the bowel with concomitant bleeding perforation) are clearly lupus related.

Vasculitis

Clinical Features

Patients with vasculitis may present with any of a variety of abdominal pain syndromes. They may present with nondescript bloating, nausea, and diarrhea or may have abrupt and massive GI hemorrhage or acute abdominal pain. Lupus-related symptoms referable to the intestine may be insidious or acute, localized or diffuse, benign or catastrophic. Steroid use may cloak symptoms. To avert disaster, one must maintain a high index of suspicion when the lupus patient reports abdominal complaints.

Vasculitis of the small or large intestine is the most serious GI complication of SLE [21]. It may present as GI hemorrhage, occurring in 46% of cases in the Nadorra *et al.* [13] study and resulting in death in half of these cases, thus accounting for approximately 25% of the deaths in their series. It may also present as an acute abdomen secondary to intestinal ischemia or infarction with perforation [80–85]. In the Zizic *et al.* study [82], for example, 5 of 107 SLE patients seen in their clinic died from intestinal perforation, making it the major cause of death in their series. In general, the outcome with perforation is dismal, resulting in death in more than two-thirds of cases [86] when the outcome was known.

Patients with intestinal vasculitis may present with less severe symptoms such as nausea, vomiting, bloating, diarrhea, and fever on the basis of intestinal ischemia with or without ileus, infection, obstruction, or malabsorption [80, 84, 87, 88]. These symptoms are often insidious, may be masked by corticosteroids, often precede the more serious complications of perforation and hemorrhage, and should alert the clinician to the possibility of an approaching disaster. Unexplained acidosis, hypotension, abdominal distention, or bowel dilation on X-ray should alert the clinician to the possibility of a perforated viscus. The absence of bowel sounds and the presence of guarding are not reliably found [89].

In addition, SLE patients with peripheral vasculitis, circulating rheumatoid factor, central nervous system involvement, and thrombocytopenia appear to be more at risk of developing an acute abdominal event according to one study [82], although others dispute this [89].

Pathology/Pathophysiology

Grossly, the appearance of the bowel ranges from segmental edema [4, 21, 87, 90] to discrete ulceration [91–93], gangrene [93], and perforation [80–85]. Histologically, both small vessel arteritis [13, 81, 84, 85, 94] and venulitis [84, 93, 95] have been seen, although larger vessels may be involved as in polyarteritis nodosa [21] (Figure 47.1). Although lesions of the larger vessels, such as medium-sized vessel vasculitis, are rare, they

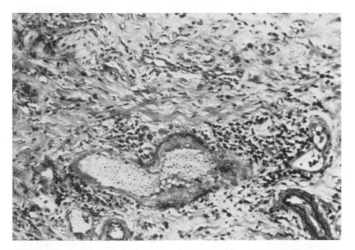

FIGURE 47.1   A cut section of a sigmoid biopsy from the patient whose X-ray is depicted in Figure 47.2. Note the perivascular lymphocytic infiltrate and destruction of a large artery wall. *Source: Courtesy of Dr. David Sachar.*

tend to produce more necrosis and gangrene [96, 97]. Associated histologic findings may include atrophy and degeneration of the media of small and medium arteries, fibrinoid necrosis of vessel walls, old thrombosis, phlebitis, and monocyte infiltrate in the lamina propria. Acute and chronic inflammatory infiltrate is seen, often confined to the mucosa but occasionally transmural [13].

Pneumatosis cystoides intestinalis, a usually benign entity that consists of intramural air in the intestine, has multiple etiologies and may be seen in SLE [98] as a possible result of vasculitis. In one case, this finding was associated with a cecal perforation [83], whereas it was benign in others [98, 99].

Diagnosis

In patients with the more insidious onset of symptoms, appropriate studies can be obtained, such as barium studies (or gastrograffin when perforation is suspected) or endoscopy searching for signs of ischemia secondary to vasculitis (e.g., thumbprinting, ulcerations, and pallor) (Figure 47.2). CT may reveal nonspecific findings such as marked bowel wall thickening and enhancement of the mucosa and serosa creating a "target sign" [100–102]. In some cases, ascites is seen. Arteriography may play a role in the diagnosis of ischemia secondary to vasculitis, as reported in one case [94], whereas a second report [103] describes the use of an indium-111 granulocyte scan to identify four of five lupus patients with GI symptoms who were then determined to have vasculitis. These latter tests have not been generally used. In patients presenting with acute abdominal pain, CT findings may assist in the diagnosis of mesenteric ischemia. abdominal pain and report that 31 (79%) had signs of ischemic bowel disease, defined by at least three of the following: bowel wall thickening, target sign (thickened bowel wall with peripheral rim enhancement or an enhancing inner and outer rim with hypoattenuation in the center), dilatation of intestinal segments, engorgement of mesenteric vessels, and increased attenuation of mesenteric fat [104]. This suggests that CT may be useful for detecting the primary cause of GI symptoms, particularly in the patient presenting with acute abdominal pain. In patients presenting with an acute abdomen, emergency exploratory laparotomy is in order.

Colitis as a result of vasculitis is unusual, occurring in less than 1% of GI vasculitis cases in SLE (suggested by the presence of isolated colonic lesions) [105, 106]. If ischemic colitis is suspected, colonoscopy and biopsies are warranted. The endoscopic findings may include erythematous friable mucosa and diffuse ulceration with exudates, and although there is no pathognomonic histologic finding, thickening of the small vessel wall

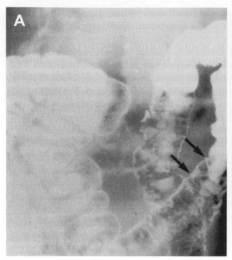
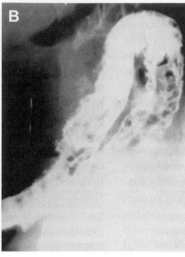

Figure 47.2   Barium enema X-ray presentation of a patient with acute lupus vasculitis of the bowel. There are discrete ulcers (collar button ulcers—*arrows*; *A*) seen as well as increased secretions ("graying" of the barium; *B*). *Source: Courtesy of Dr. David Sachar.*

and lymphocyte infiltration around vessels support the diagnosis.

Therapy

Intestinal perforation or infarction, as well as small bowel ischemia, requires emergency surgery for resection of the involved area because mortality is high in untreated cases. Early laparotomy has been shown to improve survival [107]. High-dose intravenous steroids are commonly used, and in patients who do not rapidly improve, cyclophosphamide can be added [108]. Large bowel ischemia may be treated expectantly with blood products, fluids, and antibiotics when appropriate. As with small bowel ischemia, high-dose prednisone is the most common and recommended treatment, and it was reported to resolve one case of pneumatosis cystoides intestinalis in a lupus patient with vasculitis [109]. In a case in which this failed, high-dose cyclophosphamide was successful in the treatment of a patient with peritonitis and colitis due to intestinal vasculitis [110].

Other SLE-related Intestinal Diseases

Intestinal Thrombosis

In addition to vasculitis-associated intestinal ischemia and infarction, infarction and ischemia have also been seen in the lupus intestine in association with thrombosis. Lupus patients with the antiphospholipid antibody syndrome seem to be especially at risk. Asherson *et al.* [111] describe a case of bowel infarction necessitating colectomy. Histologic examination of the specimen revealed the presence of thrombus in branches of the inferior mesenteric artery with the total absence of any vasculitis. The patient's serum demonstrated antibodies to phospholipid (the lupus anticoagulant) and

to cardiolipin, leading the authors to theorize that a thrombotic diathesis, on the basis of the presence of these antibodies, precipitated the bowel infarction. Another group has reported a case of repeated intestinal ulcerations secondary to thrombosis in a patient with lupus and high serum antiphospholipid antibody levels [112]. Diagnosis and therapy are similar to those for intestinal vasculitis and include appropriate anticoagulation or antithrombotic medication. Plasmapheresis for patients with antiphospholipid antibody syndrome may be of some benefit [108].

Inflammatory Bowel Disease

An association between SLE and both ulcerative colitis [91, 113, 114] and Crohn's disease [115, 116] has been suggested. Although this may imply that some lupus patients may have intestinal findings secondary to inflammatory bowel disease (IBD) rather than vasculitis, there is no convincing proof. In most disputed cases, the pathology seemed most likely to be vasculitic and the association with IBD, when seen, to be coincidental [114]. Diagnosis is made by clinical history in addition to typical barium X-ray findings or colonoscopic or histologic appearance or both. Treatment includes sulfasalazine, 5-aminosalicylate derivatives, steroids, and azathioprine/6-mercaptopurine.

Drug-Related Disease

As for other parts of the GI tract, drug therapy may contribute to intestinal disease in SLE. For example, Khoury [118] describes the case of an SLE patient who developed ulcerative proctitis twice, each time after starting ibuprofen, which remitted when the ibuprofen was stopped. NSAIDs have been shown to trigger the onset of ulcerative colitis in predisposed patients, so this may be the unmasking of an underlying disease. One must always be aware that patients with inflammatory bowel disease may develop drug-induced SLE from drug therapy such as anti-tumor necrosis factor-α (TNF-α) therapy and sulfasalazine [119]. A review of the literature regarding reported cases of drug-induced SLE (DILE) in patients treated with anti-TNF-α therapy revealed that all agents could precipitate DILE, although the majority were associated with infliximab treatment [120]. Whereas this review included patients treated with anti-TNF-α therapies for different diagnoses, Colombel *et al.* [121] specifically assessed IBD patients treated with infliximab, reporting that only 3 of 500 treated patients developed DILE [121]. It is important to differentiate a positive ANA resulting from infliximab or adalimumab therapy (these agents induce apoptosis of activated T cells) from a true DILE syndrome. Often, patients will develop a serum sickness-like reaction related to Abs to infliximab that can be confused with true SLE or DILE in the setting of a positive ANA. The former requires manipulation of dose and alterations in the interval between therapy. The latter may require cessation of this therapy and mitigate against using other agents in this class.

Diagnosis may be suggested by history. Treatment includes withdrawal of the offending drug. Interestingly, TNF-α blocker-induced DILE cases have a higher prevalence of antibodies to dsDNA, rash, and hypocomplementemia than do DILE cases due to other drugs. Fever is common in both types of DILE. Renal disease, which is rare in classic DILE, has been reported in cases of anti-TNF-α-induced DILE. Given the issues described previously, it may be advisable to consider whether all patients who are begun on anti-TNF therapy should have pretreatment serologic evaluation for SLE.

Protein-Losing Enteropathy

There are 24 reported cases of protein-losing enteropathy in association with SLE [122–128]. It is defined as excessive transintestinal mucosal loss of plasma proteins. It typically occurs in young women. Clinically, one may see diarrhea and typical manifestations of hypoproteinemia such as peripheral edema. It is characterized by coarse, thickened folds in the stomach and small bowel with or without ulcerations. The diagnosis is suspected in patients with hypoproteinemia and no other obvious source of protein loss and may be confirmed by radioisotopic studies or 24-h stool α_1-antitrypsin clearance [127]. In some cases [122, 128–131], the enteropathy preceded the diagnosis of SLE. The reported cases responded to low-dose steroid therapy, and it was postulated that immune complex deposition led to increased permeability of intestinal capillaries and subsequently of the epithelial lining cells [123], analogous to glomerular basement membrane deposition in lupus nephritis.

Fat Malabsorption

Malabsorption of fat, with or without carbohydrate malabsorption, has also been described [69, 88, 132–136]. Clinically, it is manifest by the presence of voluminous oily, foul-smelling stools and weight loss in a patient with adequate oral intake. Diagnosis is suggested by a positive staining of stool sample with Sudan black and may be confirmed by elevated levels of fat in a 72-h stool collection. Pathologically, Weisman *et al.* [133] reported widening and blunting of the villi and immune complex deposition in their case of fat malabsorption. This lesion responded dramatically to steroids, with reversal of the histologic picture.

In another report [136], two cases of fat malabsorption in SLE patients are described that are attributed to lupus-induced gut hypomotility with subsequent bacterial overgrowth. One would expect antibiotics to improve fat malabsorption in these cases.

It is also theoretically possible, though unproven, that serositis leading to adhesions and partial bowel obstruction as described by Miller *et al.* [137] might lead to stasis and bacterial overgrowth with subsequent fat malabsorption, and that lupus- or drug-induced pancreatitis [21, 88, 138–143] might lead to pancreatic insufficiency with concomitant fat malabsorption. Treatment of antibiotic- and steroid-unresponsive cases would include low-fat diet, lipid-soluble vitamin (A, E, D, and K) supplements, cholestyramine, and judicious use of antidiarrheal agents.

Gluten-Sensitive Enteropathy (Celiac Sprue)

The co-existence of lupus and celiac sprue has been rarely reported [144]. Siurala *et al.* [70] noted nonspecific inflammatory changes in villus blunting on random small bowel biopsy specimens and likened these to celiac sprue. Davies and Marks [145] reported the case of a patient with SLE who developed dermatitis herpetiformis, a dermatologic disease with a known association to celiac sprue. This suggests the possibility of an association between SLE and celiac disease in these patients. In fact, several patients with collagen vascular disease and protein malabsorption with diarrhea, including one patient with SLE, have reportedly responded to a gluten-free diet [69, 134]. Classically, the diagnosis of celiac sprue is made by the presence of flattened small intestinal villi on biopsy that normalize on gluten-free diet and recur with reintroduction of gluten into the diet. However, many lupus patients with celiac-like changes on intestinal biopsy do not respond to a gluten-free diet. Immunohistologic studies reveal deposition of C3 in the capillary walls of the small intestine [146]. Steroids may prove effective in resistant cases.

Chronic Intestinal Pseudo-Obstruction

Several small case series reported the association of chronic intestinal pseudo-obstruction (CIPO) with lupus [147–150]. This complication, characterized by ineffective intestinal propulsion and manifested by abdominal pain and distention, bloating, vomiting, and constipation, reflects a dysfunction of the visceral smooth muscle or the enteric nervous system. Out of 19 cases reported, 9 had CIPO as the initial presentation of SLE. Interestingly, in 12 of 19 there was bilateral ureterohydronephrosis [150], suggesting a central smooth muscle motility problem of neuropathic or myogenic origin. The diagnosis is made by radiographic studies showing fluid levels and dilated bowel, without an organic cause for obstruction, and by manometric studies. Treatment is with high-dose steroids, parenteral nutrition, promotility agents, and maintenance immunosuppression. Perlemuter *et al.* [151] used octreotide, 50 µg twice daily, when the previous measures failed, with no recurrence of CIPO. Even with aggressive therapy, CIPO remains a serious complication of SLE, associated with a mortality of 26% (5/19 patients).

Amyloidosis

Amyloidosis in association with lupus has very rarely been reported [152, 153]. Patients may present with recurrent diarrhea and GI bleeding.

PERITONEAL CAVITY

Clinical Features

The presence of peritoneal inflammation has been well documented in autopsy and surgical studies as well as in several case reports [7, 8, 88, 137, 143, 154–159] in as many as 63% of patients [7]. Peritoneal inflammation may present in a number of ways, ranging from an acute abdomen to painless ascites. Overt signs may not be present due to the masking effect of steroids.

Pathology/Pathophysiology

The inflammatory change in the peritoneum is presumably due to deposition of immune complexes in mesenteric vessels [160–162] and can result in abdominal pain compatible with an acute abdomen [88, 163] or can be painless [141, 154, 155]. Patchy serosal and peritoneal plaques have been observed [156]. Peritoneal inflammation can also result in ascites, which in various series is present in 8–88% of cases [5, 8, 13]. Paracentesis usually reveals an exudative effusion with white blood cells [164] that is sterile when cultured. Lupus erythematosus cells were found in approximately one-third of the patients in one study [165]. Interestingly, the ascitic fluid has been shown to have high DNA-binding activity and low total complement, supporting a role for immune complexes in the development of ascites. Ascites can also be due to the nephrotic syndrome or constrictive pericarditis, both of which should be considered first before ascribing the ascites to peritoneal inflammation. Lastly, Budd–Chiari syndrome (thrombosis of the portal vein and potentially the inferior vena cava) as a consequence of clotting diathesis in SLE patients with lupus anticoagulant has also been reported to produce ascites [166] and should be considered in the differential diagnosis.

The presence of ascites increases the patient's susceptibility to spontaneous bacterial peritonitis. This complication has been reported in SLE [167, 168]. Conn [169] suggests that it is relatively more common in patients with "disseminated" SLE than in patients with alcoholic cirrhosis. The great majority of cases of spontaneous bacterial peritonitis develop in the hospital, usually more than 1 week after admission [169]. In one of the SLE patients reported to have developed spontaneous bacterial peritonitis, *Streptococcus pneumoniae* was the causative organism [163].

Lupus patients with renal failure on continuous ambulatory peritoneal dialysis are obviously at risk for bacterial peritonitis. For example, Allais *et al.* [154] reported a case of *Listeria monocytogenes* in a lupus patient on continuous ambulatory peritoneal dialysis. Abdominal abscesses have also been found [13, 98]. Perforation of the intestine in a patient with SLE can lead to peritonitis [83].

Hemoperitoneum has also occurred in SLE [170]. In this case, the patient had a ruptured gastroepiploic aneurysm, which showed evidence of vasculitis (fibrinoid necrosis with destruction of muscle and elastic fibers in the vasa vasorum as well as disintegration of the connective tissue layer of the tunica media). A fall prior to the development of the hemoperitoneum was probably the precipitating event. McCollum *et al.* [171] also reported on the development of a ruptured aneurysm in a patient with SLE—in this case, the hepatic artery.

Diagnosis

The value of paracentesis is obvious in the lupus patient with ascites or signs of peritonitis. Zizic *et al.* [82] emphasized the urgency in evaluation and diagnosis of such patients, maintaining a high index of suspicion, because many of the gross signs of peritonitis, such as rebound tenderness and guarding, may be masked by the high steroid doses that these patients are often taking.

Therapy

Once a diagnosis is made, appropriate therapy should be initiated. In the case of SLE-induced peritonitis, increasing the dose of steroids has been shown to be effective [164]. In the presence of ascites, steroids alone appear to be less effective, and some have suggested that the addition of cyclophosphamide improves outcome [161]. For bacterial peritonitis, antibiotics should be instituted, and for bowel perforation or aneurysm rupture, emergency surgery is imperative.

Serositis may also lead to other complications. For example, as noted previously, Miller *et al.* [137] described a case of small bowel obstruction secondary to extensive small bowel adhesions in a patient with SLE serositis.

PANCREAS

Clinical Features

A number of lupus patients develop episodes of acute "idiopathic" pancreatitis, unrelated to the known causes of mechanical obstruction of the pancreatic duct or toxic-metabolic etiologies [72, 88, 139, 142, 143, 170, 172, 173]. The incidence is variable, ranging from 1 to 23% [6, 8, 13], and in most reported cases it occurs in the setting of SLE with multiorgan involvement. In one study, 77 cases of lupus-associated pancreatitis were reviewed [174]. As expected, most of them were young females. Interestingly, in most cases there was no association with steroids or azathioprine therapy. Clinically, the presenting symptoms of acute pancreatitis in SLE are no different than those found in non-SLE patients: severe abdominal pain, often radiating to the back; nausea; and vomiting. X-rays frequently reveal a localized ileus of either the small bowel or the colon (sentinel loop). The activity of the underlying SLE appears to be greater in patients with pancreatitis than in those without [138, 175].

Complications of pancreatitis are not common. Fistulae and abscess have not been noted. Pseudocyst formation [98, 175] and resolution [176] have been reported. However, the mortality rate was higher than in non-SLE-associated pancreatitis both in the Breuer *et al.* [174] and in the Pollak *et al.* [89] series, in which three-fourths of the patients with pancreatitis died. Active lupus and different biochemical abnormalities, but not treatment with steroids or azathioprine, were significantly associated with increased mortality.

The signs and symptoms of chronic pancreatitis are more subtle and may include evidence of diabetes and fat malabsorption.

Pathology/Pathophysiology

Because steroids as well as other drugs used in SLE, such as azathioprine, have been implicated in pancreatitis [143, 177–180], there have been some questions regarding whether SLE in and of itself can cause pancreatitis. The weight of evidence points toward an actual vasculitic lesion as etiologic in some cases of lupus pancreatitis [139]. Several cases occurred in the pre-steroid era [88, 142] and in patients not taking steroids or immunosuppressives [138, 175]. Furthermore, in some patients the pancreatitis resolved while either maintaining or increasing their steroid dose [175]. Four of 20 patients in the Reynolds *et al.* [175] review had no predisposing factor for their pancreatitis other than SLE. Histologically, steroid-induced pancreatitis differs from the vasculitic picture of pancreatitis seen in the available animal models [181, 182]. A role for antiphospholipid and anticardiolipin antibodies, seen in some patients with lupus, has been suggested [183, 184].

Ropes [8] described the pancreas in three lupus patients as containing, respectively: (1) moderate fibrosis and lymphocytic infiltration, (2) ectasia of the acini, and (3) acute and chronic inflammatory changes in the vessels with fibrinoid necrosis. The absence of clinically evident pancreatitis in Ropes' study, concurrent with the finding of pancreatitis in 6% of his autopsied cases, suggests the possibility of subclinical pancreatitis in lupus patients. Further evidence for this comes from a study by Tsianos *et al.* [12], in which 30.5% (11 of 36) of SLE patients who had no abdominal pain or other gastrointestinal symptoms at the time of blood drawing had hyperamylasemia. The authors suggest that asymptomatic pancreatic damage in SLE may occur frequently, and that the hyperamylasemia in these patients probably reflects a slow, subclinical, inflammatory process of the pancreas. Another possibility is hyperamylasemia from small bowel perforation.

Diagnosis

Acute pancreatitis may be diagnosed by typical symptoms and blood work. The serum amylase level is elevated, but amylase isoenzymes, serum lipase, and amylase/creatinine clearance ratio (>5.5) are suggested to differentiate pancreatic from salivary hyperamylasemia and the hyperamylasemia seen in renal disease [185]. In addition, an inflamed, edematous pancreas on sonogram or CT scan can confirm the diagnosis.

Therapy

If steroids or azathioprine are suspected to be the inciting agent, they should be withdrawn. Treatment of uncomplicated acute pancreatitis should include fluid

[122] W.N. Pachas, W.G. Linscheer, R.S. Pinals, Protein-losing enteropathy in systemic lupus erythematosus, Am J Gastroenterol 55 (1971) 162–167.

[123] D.E. Trentham, A.T. Masi, Systemic lupus erythematosus with a protein-losing enteropathy, JAMA 236 (1976) 287–288.

[124] M. Tsukahara, K. Matsuo, H. Kojima, Protein-losing enteropathy in a boy with systemic lupus erythematosus, J Pediatr 97 (1980) 778–780.

[125] T.A. Waldaman, R.D. Wochner, W. Strober, The role of the gastrointestinal tract in plasma protein metabolism, Am J Gastroenterol 46 (1969) 275.

[126] S. Murao, Y. Taooka, Y. Yamanishi, H. Mukuzono, K. Aoi, Y. Isibe, et al., Protein-losing enteropathy and cerebral infarction associated with systemic lupus erythematosus, Ryumachi 34 (1994) 59–63.

[127] D.A. Perednia, N.A. Curosh, Lupus-associated protein-losing enteropathy, Arch Intern Med 150 (1990) 1806–1810.

[128] K.A. Northcott, E.M. Yoshida, U.P. Steinbrecher, Primary protein losing enteropathy in anti-double-stranded DNA disease: The initial and sole clinical manifestation of occult systemic lupus erythematosus? J Clin Gastroenterol 33 (2001) 340–341.

[129] U. Chung, M. Oka, Y. Nakagawa, T. Nishishita, N. Sekine, Y. Tanaka, et al., A patient with protein-losing enteropathy associated with systemic lupus erythematosus, Intern Med 31 (1992) 521–524.

[130] J.F. Molina, R.F. Brown, A. Gedalia, L.R. Espinoza, Protein losing enteropathy as the initial manifestation of childhood systemic lupus erythematosus, J Rheumatol 23 (1996) 1269–1271.

[131] R.L. Sunheimer, C. Finck, S. Mortazavi, C. McMahon, M.R. Pincus, Primary lupus-associated protein-losing enteropathy, Ann Clin Lab Sci 24 (1994) 239–242.

[132] P. Bazinet, G.A. Marin, Malabsorption in systemic lupus erythematosus, Am J Dig Dis 16 (1971) 460–466.

[133] M.H. Weisman, E.C. McDanald, C.B. Wilson, Studies of the pathogenesis of interstitial cystitis, obstructive uropathy, and intestinal malabsorption in a patient with systemic lupus erythematosus, Am J Med 70 (1981) 875–881.

[134] A.K. Rustai, M.A. Peppercorn, Gluten-sensitive enteropathy and systemic lupus erythematosus, Arch Intern Med 148 (1988) 1583–1584.

[135] D.J. Kurlander, J.B. Kirsner, The association of chronic "nonspecific" inflammatory bowel disease with lupus erythematosus, Ann Intern Med 60 (1964) 799.

[136] Anonymous, Severe acquired hypocholesterolemia: Two case reports, Nutr Rev 47 (1989) 202.

[137] M.H. Miller, M.B. Urowitz, D.D. Gladman, E.C. Tozman, Chronic adhesive lupus serositis as a complication of systemic lupus erythematosus. Refractory chest pain and small-bowel obstruction, Arch Intern Med 144 (1984) 1863–1864.

[138] Y.A. Mekori, A. Yaretzky, M. Schneider, A. Klajman, Pancreatitis in systemic lupus erythematosus—A case report and review of the literature, Postgrad Med J 56 (1980) 145–147.

[139] M. Baron., M.L. Brisson, Pancreatitis in systemic lupus erythematosus, Arthritis Rheum 25 (1982) 1006–1009.

[140] L.J. Herskowitz, S. Olansky, P.G. Lang, Acute pancreatitis associated with long-term azathioprine therapy. Occurrence in a patient with systemic lupus erythematosus, Arch Dermatol 115 (1979) 179.

[141] H. Kawanishi, E. Rudolph, F.E. Bull, Azathioprine induced acute pancreatitis, N Engl J Med 289 (1973) 357.

[142] E.C. Reifenstein, E.C. Reifenstein Jr., G.H. Reifenstein, Variable symptom complex of undetermined etiology with total termination including conditions described as visceral erythema group (Osler) disseminated lupus erythematosus, atypical verrucous endocarditis (Libman–Sacks), fever of unknown origin (Christian) and a diffuse peripheral vascular disease (Baehr and others), Arch Intern Med 63 (1939) 553.

[143] A. Paulino-Netto, D.A. Dreiling, Pancreatitis in disseminated lupus erythematosus. A case report, J Mt Sinai Hosp 27 (1960) 291.

[144] G.R. Komatireddy, J.B. Marshall, R. Aqel, L.E. Spollen, G.C. Sharp, Association of systemic lupus erythematosus and gluten enteropathy, South Med J 88 (1995) 673–676.

[145] M.G. Davies, R. Marks, Simultaneous systemic lupus erythematosus and dermatitis herpetiformis, Arch Dermatol 112 (1976) 1292.

[146] A. Tsutsumi, T. Sugiyama, R. Matsumura, M. Sueishi, K. Takabayashi, T. Koike, et al., Protein losing enteropathy associated with collagen diseases, Ann Rheum Dis 50 (1991) 178–181.

[147] P. Cacoub, Y. Benhamou, P. Barbet, J.C. Piette, A. Le Cae, S. Chaussade, et al., Systemic lupus erythematosus and chronic intestinal pseudoobstruction, J Rheumatol 20 (1993) 377–381.

[148] G. Perlemuter, S. Chaussade, B. Wechsler, P. Cacoub, M. Dapoigny, A. Kahan, et al., Chronic intestinal pseudoobstruction in systemic lupus erythematosus, Gut 43 (1998) 117–122.

[149] P. Munyard, M. Jaswon, Systemic lupus erythematosus presenting as intestinal pseudo-obstruction, JR Soc Med 39 (1997) 877–879.

[150] M.Y. Mok, R.W.S. Wong, C.S. Lau, Intestinal pseudo-obstruction in systemic lupus erythematosus, Lupus 9 (2000) 11–18.

[151] G. Perlemuter, P. Cacoub, S. Chaussade, B. Wechsler, D. Couturier, J.C. Piette, Octreotide treatment of chronic intestinal pseudoobstruction secondary to connective tissue diseases, Arthritis Rheum 42 (1999) 1545–1549.

[152] I. Al Hoqail, H. Naddaf, A. Al Rikabi, H. Al Arfaj, A. Al Arfaj., Systemic lupus erythematosus and amyloidosis, Clin Rheumatol 16 (1997) 422–424.

[153] T. Betsuyaku, T. Adachi, H. Haneda, J. Suzuki, M. Nishimura, S. Abe, et al., A secondary amyloidosis associated with systemic lupus erythematosus, Intern Med 32 (1993) 391–394.

[154] J.M. Allais, S.J. Cavalieri, M.H. Bierman, R.B. Clark, Listeria monocytogenes peritonitis in a patient on continuous ambulatory peritoneal dialysis, Nebr Med J 74 (1989) 303–305.

[155] T.M. Zizic, J.N. Classen, M.B. Stevens, Acute abdominal complications of systemic lupus erythematosus and polyarteritis nodosa, Am J Med 73 (1982) 525–531.

[156] V.H. Low, P.D. Robins, D.J. Sweeney, Systemic lupus erythematosus serositis, Australas Radiol 39 (1995) 300.

[157] P.M. Houtman, S.S. Hofstra, Lupus peritonitis presented as vague abdominal complaints in a SLE patient, Neth J Med 40 (1992) 232–235.

[158] S. Wakiyama, K. Yoshimura, M. Shimada, K. Sugimachi, Lupus peritonitis mimicking acute surgical abdomen in a patient with systemic lupus erythematosus: Report of a case, Surg Today 26 (1996) 715–718.

[159] M. Hammoudeh, A.R. Siam, Recurrent peritonitis with ascites as the predominant manifestation of systemic lupus erythematosus, Clin Rheumatol 14 (1995) 352–354.

[160] A.L. Schocket, D. Lain, P.F. Kohler, J. Steigerwald, Immune complex vasculitis as a cause of ascites and pleural effusions in systemic lupus erythematosus, J Rheumatol 5 (1978) 33–38.

[161] J. Biran, D. McShane, M.H. Ellman, Ascites as the major manifestation of systemic lupus erythematosus, Arthritis Rheum 19 (1976) 782.

[162] P.E. Jones, P. Rawcliffe, N. White, A.W. Segal, Painless ascites in systemic lupus erythematosus, Br Med J 1 (1977) 1513.

[163] N.M. Matolo, D. Albo Jr., Gastrointestinal complications of collagen vascular diseases: Surgical implications, Am J Surg 122 (1971) 678.

[164] D.R. Mushner, Systemic lupus erythematosus: A cause of medical peritonitis, Am J Surg 124 (1972) 368.

[165] B. Naylor, Cytological aspects of pleural, peritoneal and pericardial fluids from patients with systemic lupus erythematosus, Cytopathology 3 (1992) 1—8.

[166] R.A. Asherson, R.P. Thompson, N. MacLachlan, E. Baguley, P. Hicks, G.R. Hughes, Budd Chiari syndrome, visceral arterial occlusions, recurrent fetal loss and the "lupus anticoagulant" in systemic lupus erythematosus, J Rheumatol 16 (1989) 219—224.

[167] B.F. Shesol, F.E. Rosato, Concomitant acute lupus erythematosus and primary pneumococcal peritonitis, Am J Gastroenterol 63 (1975) 324.

[168] P.E. Lipsky, J.A. Hardin, L. Shour, et al., Spontaneous peritonitis and systemic lupus erythematosus: Importance of accurate diagnosis of gram positive bacterial infections, JAMA 232 (1975) 929.

[169] H.O. Conn, Spontaneous bacterial peritonitis: Variant syndromes, South Med J 80 (1987) 1343—1346.

[170] M. Yamaguchi, K. Kumada, H. Sugiyama, E. Okamoto, K. Ozawa, Hemoperitoneum due to a ruptured gastroepiploic artery aneurysm in systemic lupus erythematosus. A case report and literature review, J Clin Gastroenterol 12 (1990) 344—346.

[171] C.N. McCollum, M.E. Sloan, A.M. Davison, G.R. Giles, Ruptured hepatic aneurysm in systemic lupus erythematosus, Ann Rheum Dis 38 (1979) 396—398.

[172] K.P. Leong, M.L. Boey, Systemic lupus erythematosus (SLE) presenting as acute pancreatitis—A case report, Singapore Med J 37 (1996) 323—324.

[173] M. Takasaki, Y. Yorimitsu, I. Takahashi, S. Miyake, T. Horimi, Systemic lupus erythematosus presenting with drug-unrelated acute pancreatitis as an initial manifestation, Am J Gastroenterol 90 (1995) 1172—1173.

[174] G.S. Breuer, A. Baer, D. Dahan, G. Nesher, Lupus-associated pancreatitis, Autoimmunity Rev 5 (2006) 314—318.

[175] J.C. Reynolds, R.D. Inman, R.P. Kimberly, J.H. Chuong, J.E. Kovacs, M.B. Walsh, Acute pancreatitis in systemic lupus erythematosus: Report of twenty cases and a review of the literature, Medicine 61 (1982) 25—32.

[176] M. Borum, W. Steinberg, M. Steer, S. Freedman, P. White, Chronic pancreatitis: A complication of systemic lupus erythematosus, Gastroenterology 104 (1993) 613—615.

[177] C.A. Dujoune, D.L. Azarnoff, Clinical complications of corticosteroid therapy, Med Clin North Am 57 (1973) 1331.

[178] J.F. Patterson, S.J. Wierzbinski, Digestive system manifestations of collagen disease, Med Clin North Am 46 (1962) 1387.

[179] W.B. Nelp, Acute pancreatitis associated with systemic lupus erythematosus, Arch Intern Med 108 (1961) 102.

[180] M. Sparberg, Recurrent acute pancreatitis associated with systemic lupus erythematosus. Report of a case, Am J Dig Dis 12 (1967) 522.

[181] S.S. Lazarus, S.A. Bencosme, Development and regression of cortisone induced lesions in rabbit pancreas, Am J Clin Pathol 26 (1956) 1146.

[182] H.H. Stumpf, S.L. Wilens, C. Somoza, Pancreatic lesions and peripancreatic fat necrosis in cortisone-treated rabbits, Lab Invest 5 (1956) 224.

[183] T.S. Yeh, C.R. Wang, Y.T. Lee, C.Y. Chuang, C.Y. Chen, Acute pancreatitis related to anticardiolipin antibodies in lupus patients visiting an emergency department, Am J Emerg Med 11 (1993) 230—232.

[184] C.R. Wang., H.C. Hsieh, G.L. Lee, C.Y. Chuang, C.Y. Chen, Pancreatitis related to antiphospholipid antibody syndrome in a patient with systemic lupus erythematosus, J Rheumatol 19 (1992) 1123—1125.

[185] A.L. Warshaw, A.F. Fuller Jr., Specificity of increased renal clearance of amylase in diagnosis of acute pancreatitis, N Engl J Med 292 (1975) 325—328.

[186] C.R. Swanepoel, A. Floyd, H. Allison, G.M. Learmonth, M.J. Cassidy, M.D. Pascoe, Acute acalculous cholecystitis complicating systemic lupus erythematosus: Case report and review, Br Med J (Clin Res Ed) 286 (1983) 251—252.

[187] A. Suwa, N. Hama, S. Kawai, K. Ishiyama, M. Tanabe, T. Yamada, et al., A case of Sjogren's syndrome and systemic lupus erythematosus complicated with necrotizing angiitis of the gallbladder, Ryumachi 35 (1995) 904—909.

[188] A.J. Rhoton, J.H. Gilliam, K.R. Geisinger, Hemobilia in systemic lupus erythematosus, South Med J 86 (1993) 1049—1051.

[189] C.C. Papaioannou, G.G. Hunder, J.T. Lie, Vasculitis of the gallbladder in a 70-year-old man with giant cell (temporal) arteritis, J Rheumatol 6 (1979) 71—76.

[190] A. Tolaymat, F. Al Mousily, A.B. Haafiz., N. Lammert, S. Afshari, Spontaneous rupture of the spleen in a patient with systemic lupus erythematosus, J Rheumatol 22 (1995) 2344—2345.

[191] F. Liote, J. Angle, N. Gilmore, C.K. Osterland, Asplenism and systemic lupus erythematosus, Clin Rheumatol 14 (1995) 220—223.

[192] M. Ostensen, P.M. Villiger, Nonsteroidal anti-inflammatory drugs in systemic lupus erythematosus, Lupus 10 (2001) 135—139.

[193] J.M. Falk, F.B. Thomas, A cute pancreatitis due to procaineamide induced lupus erythematosus, Ann Intern Med 83 (1975) 832.

[194] M.A. Stratton, Drug-induced systemic lupus erythematosus, Clin Pharm 4 (1985) 657—663.

[195] D.J. Veale, M. Ho, K.D. Morley, Sulphasalazine-induced lupus in psoriatic arthritis, Br J Rheumatol 34 (1995) 383—384.

[196] V.J. Bray, S.G. West, K.T. Schultz, D.T. Boumpas, R.L. Rubin, Antihistone antibody profile in sulfasalazine induced lupus, J Rheumatol 21 (1994) 2157—2158.

[197] I. Gunnarsson, E. Pettersson, S. Lindblad, B. Ringertz, Olsalazine-induced lupus syndrome, Scand J Rheumatol 26 (1997) 65—66.

[198] M.S. Al-Hakeem, M.A. McMillen, Evaluation of abdominal pain in systemic lupus erythematosus, Am J Surg 176 (1998) 291—294.

48

Cellular Hematology

Seetha U. Monrad, Mariana J. Kaplan

INTRODUCTION

Immune dysregulation characteristic of lupus frequently targets the hematologic system. As such, hematologic abnormalities are considered as part of the American College of Rheumatology (ACR) criteria for the diagnosis of systemic lupus erythematosus (SLE) [1, 2]. The four characteristics emphasized in these criteria (hemolytic anemia, leukopenia, lymphopenia, and thrombocytopenia) are the most common immune-mediated hematologic abnormalities described in patients with lupus. Because nearly every patient will present some disturbance in number or function of blood cellular elements at some point during their disease course, hematologic abnormalities represent important prognostic variables that predict the course of lupus and are considered to be markers of disease severity [3, 4]. Autoimmune hemolytic anemia (AHA) and immune thrombocytopenic purpura (ITP) have been reported as a complication of SLE in approximately 7—13% and 16—27% of cases, respectively [5, 6], with a significant percentage representing serious manifestations of the disease. Similarly, leukopenia (including more commonly lymphopenia but also neutropenia) has been reported in 20—50% of patients [5]. Thrombocytopenia and AHA have been associated with increased mortality in various SLE populations [7—9]. As such, early diagnosis and treatment of hematologic complications in SLE is crucial and may significantly impact prognosis and survival. This chapter discusses the mechanisms leading to the various hematologic abnormalities described in lupus, and it reviews the clinical presentation and epidemiology of these abnormalities as well as the current and potential future treatments for SLE-associated blood disorders. Coagulation disorders other than those related to platelets and thrombotic thrombocytopenic purpura (TTP) are discussed elsewhere in this book.

PATHOGENIC MECHANISMS OF CYTOPENIAS IN SLE

Autoimmune Hemolytic Anemia

Studies on animal and human AHA suggest that loss of immunological tolerance toward red blood cell self-antigens may be due to various non-mutually exclusive mechanisms, including molecular mimicry, lymphocyte activation, errors in central or peripheral tolerance, and aberrant cytokine production.

Antibody-induced damage of blood cells by complement-dependent or -independent mechanisms is a common pathogenic mechanism for cytopenias in SLE. Erythrocyte autoantibodies are classified into two main groups based on their thermal requirements: (1) "warm" antibodies of IgG or IgA isotype that associate with red cells optimally at 37°C, and (2) IgM "cold" antibodies, which characteristically bind at 20—25°C and dissociate when temperatures increase to 35°C or higher but coat the cell with complement. The most common form of AHA is warm antibody AHA, mediated by the binding of IgG and/or complement to antigens on the erythrocyte. Most cases of SLE are caused by warm or "mixed" (warm and cold) autoantibodies. It is currently thought that these autoantibodies probably arise due to an underlying defect in immune regulation in patients with AHA secondary to SLE. The autoantibodies are frequently directed toward an antigen in the Rhesus (Rh) system. Destruction of the autoantibody-coated red blood cell is mediated primarily by sequestration and phagocytosis by macrophages of the Billroth cords in the spleen and, to some extent, by Kupffer cells of the liver [10]. After binding to the erythrocyte, the Fc portion of the antibody promotes binding to Fc receptors on splenic macrophages. This phenomenon leads to the destruction of the erythrocyte and its clearance from the circulation. A trapped red blood cell may be partially or completely ingested by a macrophage. Partial phagocytosis is most common and leads to formation of spherocytes. After a portion of the adherent red blood cell membrane is internalized by the macrophage, the remainder of the cell escapes back into the circulation and assumes a spherical shape. Spherical red blood cells are more rigid and less deformable than normal erythrocytes, which makes them more susceptible to further fragmentation and destruction in future passages through the spleen [11].

The precise specificity of the anti-erythrocyte antibodies in most SLE patients with AHA is not well-defined. In primary AHA, the non−Rh-specific IgG autoantibody reacts with either the band 3 anion transporter protein of membrane erythrocytes or an epitope formed by band 3 protein and glycophorin A [12]. The lupus-prone mouse model New Zealand Black (NZB) spontaneously develops AHA and also synthesizes anti-erythrocyte autoantibodies exhibiting anti-anion channel protein band 3 specificity [13], and $CD4^+$ T cells from these mice respond to this antigen. Although specific autoantibodies generated from NZB mice have been useful in characterizing red blood cell epitopes recognized by autoantibodies in AHA [14, 15], considerable heterogeneity of autoantigens implicated in anti−red blood cell autoimmune responses has been proposed [16]. Indeed, there is an ability for autoaggressive responses to red blood cells in NZB mice to switch to targets other than the dominant autoantigen [17]. This observation suggests that, even where a single dominant target for pathogenic responses can be clearly identified, reinstatement of tolerance to that, autoantigen alone may not be an effective treatment. Two major loci on NZB chromosome 7 and chromosome 1 have been linked to anti-erythrocyte autoantibody production [18]. Furthermore, cytokines appear to regulate the development of hemolytic anemia in this model [19, 20].

A series of events during aging of circulating red blood cells occurs that leads to cell deterioration and subsequent removal from the circulation. To what extent changes occurring during red blood cell aging are involved in AHA is not clear. Interestingly, anti-band 3 IgG antibodies are also formed in healthy subjects, likely functioning to eliminate senescent erythrocytes, which, upon aging, express band 3 protein-derived neo-antigens [21]. However, the link between naturally occurring anti-band 3 autoantibodies and pathological antibodies to the same epitope remains unknown. It has been proposed that these antigenic neo-epitopes, when exposed on senescent red blood cells that are not appropriately cleared, may drive autoantibody responses and trigger a hemolytic process [22]. Indeed, in murine lupus, the binding of red blood cell autoantibodies to specific antigens involved in AHA is progressively increased with red blood cell senescence in the circulation [16]. It has been speculated that in predisposed individuals, pathogenic antigens may be progressively exposed on circulating red blood cells with aging, partly as a consequence of contact with proteolytic enzymes present in plasma and released locally in the spleen [16]. The presence of these multiple autoantigenic epitopes on red blood cells may facilitate interaction with low-affinity autoreactive B cells. On the other hand, in another murine model, the anti−red blood cell Ig H + L (HL) mice demonstrate hemolytic anemia despite clonal deletion of autoreactive B cells in the bone marrow (BM) and peripheral lymph nodes. In these mice, peritoneal cavity lymphocytes are sequestered without exposure to "self" red blood cell antigens and expand, subsequently undergoing activation by environmental antigens resulting in AHA [23]. The importance of T-helper cells in the expansion and activation of autoreactive B-cells in the development of AHA has been supported by studies of $RAG-2^{-/-}$ mice (which lack mature B- and T-cells) crossed with HL mice. The $RAG-2^{-/-}$ × HL mice do not develop AHA, indicating that the presence of T-helper cells is required for the manifestation of the disease [24, 25].

Various studies have shown a significant association between the presence of hemolytic anemia and antiphospholipid antibodies (aPLs) These observations indicate that aPL generation could have pathogenic significance in certain cases of AHA by acting as anti-erythrocyte autoantibodies [25, 26]. Some patients with SLE and associated AHA show an acquired deficiency of CD55 and/or CD59 on red blood cells, whereas SLE patients without associated AHA do not display these abnormalities [27]. CD55 (decay accelerating factor) and CD59 (protectin) are glycosylphosphatidylinositol-anchored proteins with complement inhibitory properties. CD55 regulates C3/C5 convertases, whereas CD59 regulates assembly of the terminal components of the membrane attack complex. It has been proposed that downregulation of these molecules may play a facilitator role in the development of hemolysis. The loss of another membrane regulatory protein of complement, CR1, has also been described in SLE [28]. Interestingly, CD55 and CD59 are deficient on red blood cells from patients with paroxysmal nocturnal hemoglobinuria, a condition characterized by enhanced predisposition to intravascular hemolysis [29].

Immune-Mediated Thrombocytopenia and Qualitative Platelet Defects in SLE

The pathogenesis of immune-mediated thrombocytopenia in SLE appears to be multifactorial and includes antiplatelet glycoprotein antibodies, immune complexes, aPLs, thrombotic microangiopathy, bone marrow stromal alterations, vasculitis, hemophagocytosis, and autoantibodies to thrombopoietin (TPO) [5, 30−32].

Antiplatelet autoantibodies are found in approximately 40% of SLE patients with and without thrombocytopenia but are not typically seen in patients who have recovered from thrombocytopenia, supporting a potential pathogenic role. Autoantibodies to two platelet-specific antigens, glycoprotein IIb/IIIa (GPIIb/IIIa) and TPO, are seen more frequently in SLE patients with thrombocytopenia than in those without it, and the frequencies of these antibodies are comparable between SLE patients with thrombocytopenia and patients with idiopathic thrombocytopenia. Anti-TPO antibodies have been associated with decreased plasma TPO concentrations and lower platelet values in long-term follow-up. In SLE patients with thrombocytopenia, the anti-TPO-positive patients had significantly higher frequencies of megakaryocytic hypoplasia and poorer therapeutic responses to corticosteroids and intravenous immunoglobulin than did the anti-TPO-negative patients, most of whom had the anti-GPIIb/IIIa Ab alone [30, 33]. Antibodies to other specific target antigens on platelets of patients with SLE and thrombocytopenia have been described [34, 35]. Attempts have been made to differentiate immune-mediated thrombocytopenia in SLE patients from "classic" ITP. Similar levels of platelet-associated IgG have been observed, but differences in T-cell responses have been noted [36].

The antibody-coated platelets are ingested by macrophages of the spleen, liver, lymph nodes, and bone marrow, which carry receptors for the Fc region of immunoglobulin. Phagocytosis is more efficient when the antiplatelet antibodies belong to the IgG_1 and IgG_2 subtypes [37]. The mechanism of platelet destruction then involves the coating of platelets with IgG autoantibodies and clearance from the circulation by the phagocytic cells of the spleen and other organs, similar to that described previously with regard to red blood cell destruction.

aPLs are also associated with the development of mild thrombocytopenia. Patients with a circulating lupus anticoagulant or aPL antibody are three times more likely to have a history of a moderate to severe thrombocytopenia than antibody-negative patients. In most cases, the thrombocytopenia is mild and without consequences. Three main possible mechanisms have been proposed to explain the role of aPL in thrombocytopenia: direct binding of the antibody to platelet phospholipids, immune mediated through antiplatelet antibodies similar to ITP, and platelet activation and/or damage [38].

SLE patients also manifest autoimmune-mediated qualitative platelet defects. Antibodies directed against GPIIb/IIIa can induce a thrombasthenia-like syndrome if they interact with or block functional sites on the GPIIb/IIIa complex [39]. Acquired von Willebrand's disease is a rare complication of SLE [40]. The pathogenic mechanisms responsible include binding of anti-von Willebrand factor (VWF) IgG antibodies to VWF on endothelial cells and platelets leading to early removal of the VWF and a diminished level of large multimeric VWF, as well as the inactivation of VWF function by circulating anti-VWF [41]. Autoantibodies against VWF can be directed against either functional or nonfunctional domains of the VWF. Consequently, VWF autoantibodies may induce a functional impairment in plasma VWF but usually lead to reduced factor VIII and VWF levels due to the formation of antigen—antibody complexes that are rapidly cleared from the circulation [41].

Thrombotic Thrombocytopenic Purpura in SLE

TTP is a thrombotic microangiopathy that presents with hemolysis and thrombocytopenia, and it is the primary form of microangiopathic hemolytic anemia (MAHA) described in SLE. The development of TTP appears to be tightly linked to VWF. In circulation, binding of VWF to injured vasculature results in the unfolding or activation of VWF. This conformational form of VWF promotes platelet binding (prothrombotic) and proteolysis by ADAMTS13 (anticoagulant). ADAMTS13 is a circulating zinc metalloprotease whose only known substrate is VWF; it is constitutively active in the circulation with a narrow range of activity (79—127%) in healthy individuals. Inhibition of the ADAMTS13 function leads to loss of normal VWF proteolysis and the accumulation of thrombogenic forms of VWF termed ultra-large VWFs. Although the specific mechanism remains unclear, ultra-large VWFs have an increased propensity to bind circulating platelets and may lead to the development of platelet-rich microthrombi. The embolization of these microthrombi into the circulation of the brain, heart, and kidney are the pathogenic hallmarks of TTP and may be responsible for the clinical symptoms of TTP [42]. Hemolytic anemia is thought to occur because of turbulent flow in occluded microcirculatory beds, with resultant shear stress on and fragmentation of erythrocytes.

Two types of ADAMTS13 deficiency have been recognized: autoimmune IgG inhibitors of ADAMTS13 (by far the most common) and mutations of the ADAMTS13 gene. Plasma mixing studies have detected the presence of inhibitory antibodies of ADAMTS13 in 50—90% of the acquired TTP cases. In patients whose inhibitory antibodies are too low to be detectable by plasma mixing studies, IgG isolated from the plasma samples may yield positive inhibition [42]. Anti-ADAMTS13 antibody is detectable by enzyme-linked immunosorbent assay in 97—100% of patients with acquired ADAMTS13 deficiency. The etiologies of ADAMTS13 inhibitory antibodies are unknown in most cases.

ADAMTS13 activity has been found to be decreased in patients with thrombotic microangiopathies associated with various autoimmune diseases including SLE. There are reports of SLE patients with severe deficiency of ADAMTS13 activity, and this has been closely associated with anti-ADAMTS13 IgG antibodies. The role of anti-IgM antibodies and TTP development in SLE is still a matter of debate [43]. However, there are reports in SLE that, despite low but detectable ADAMTS13 levels, no autoantibodies against the activity of this protease were detected [44]. It is also possible that low levels of ADAMTS13 synergize with other prothrombotic factors in SLE to promote overt TTP development.

Cases of SLE-associated thrombotic microangiopathy in the absence of acquired TTP are also described, especially in the setting of active lupus nephritis [45, 46]. It is postulated that these forms of thrombotic microangiopathy are due to other mechanisms of endothelial cell injury, with resultant shear stress-induced damage to erythrocytes and subsequent features of intravascular hemolytic anemia.

Mechanisms of Immune-Mediated Leukopenia in SLE

Accelerated apoptosis, a phenomenon described in human SLE and in murine lupus, has been proposed to play an important pathogenic role in SLE and may also contribute to the leukopenia observed in this disease [47].

Aberrant/accelerated apoptosis may lead to presentation of modified autoantigens to the immune system and promotion of autoimmune responses. Enhanced lymphocyte, monocyte/macrophage, and neutrophil apoptosis have been described in SLE, and this phenomenon may contribute to the development of leukopenia [48–51]. Autoreactive T cells are well described in this disease in both murine and human systems. Lupus autoreactive CD4$^+$ T cells acquire potent cytolytic capabilities in SLE, and various target cells have been described, including monocytes/macrophages and neutrophils [49, 52]. Circulating lymphocytotoxic antibodies have been demonstrated for decades in a large number of SLE patients [53]. The levels of these antibodies correlate with the degree of lymphopenia [54] and with lupus disease activity [55]. These antibodies appear to be predominantly cold-reactive IgM, and they may act by opsonization and promotion of phagocytosis rather than by being directly cytotoxic [56]. Interestingly, these autoantibodies have also been observed in relatives of SLE patients [57], react with T cells, and disrupt T-cell function [58]. Another suggested

mechanism of antileukocyte antibody-induced lymphopenia is the induction of cytokines, which may in turn lead to reduced lymphocyte number. Antilymphocyte antibodies have also been implicated in the pathogenesis of neuropsychiatric manifestations of SLE [59].

The mechanisms of neutropenia in SLE include neutrophil-reactive autoantibodies, bone marrow suppression, and increased neutrophil apoptosis [60, 61]. Enhanced apoptosis in neutrophils as well as abnormal clearance of apoptotic neutrophils have been proposed to be secondary to various mechanisms, including decreased CD44 expression [61], enhanced Fas expression [62], anti-extractable nuclear antigen antibodies [63], and a potential role of soluble or membrane-bound tumor necrosis factor (TNF)-related apoptosis-inducing ligand [52]. Anti-neutrophil antibodies (ANCAs) have been described in patients with SLE in several series [64]. Although antigen specificity has been described, its significance is not known. Indeed, ANCAs from a series of pediatric patients with SLE were directed toward a number of neutrophil antigens, including myeloperoxidase, cathepsin G, and elastase. These specificities, however, did not correlate with disease activity or with target organ involvement [65]. Specific autoantibodies to ribonucleoprotein particles (anti-Ro) have been shown to bind to a cross-reactive neutrophil membrane protein [66].

An association between neutropenia and aPL has also been described [67]. Circulating neutrophils in SLE commonly bear increased amounts of membrane immunoglobulin, related to both immune complex deposition and antineutrophil antibody. Evidence for antibody-mediated peripheral destruction in the autoimmune neutropenias is well documented [68].

Qualitative abnormalities in various granulocyte functions have also been reported in SLE, but the precise etiology of these abnormalities remains to be determined. SLE serum induces increased neutrophil aggregation [69] and interferes with phagocytosis and lysosomal release of normal neutrophils [70]. A subset of neutrophils that contaminate mononuclear cell preparations in SLE has been described, but the functional and pathogenic roles of these cells remain to be determined [71, 72]. Increased expression of early neutrophil genes has been reported in the bone marrow of SLE patients. These genes encode products of immature granulocytes, and their expression is regulated during myeloid differentiation [73]. Neutrophil abnormalities in SLE are discussed more extensively elsewhere in this book.

Other abnormalities in immune cell subsets, including dysfunction of lymphocytes, monocytes, macrophages, and natural killer cells, are discussed at length in other chapters of this book.

Bone Marrow Abnormalities in SLE

Hypocellularity, dysplasia, increased fibrosis, and BM necrosis have been described in SLE patients with hemocytopenias. These findings suggest that BM may be an important target of immune-mediated damage in this disease. However, the exact mechanisms of BM damage are poorly characterized in SLE and may involve a variety of factors, including autoantibodies, immune complexes, T-cells, and cytokines.

BM CD34$^+$ hematopoetic progenitor cells are decreased in individuals with active SLE compared to healthy controls, and this is associated with an increase in their cell surface expression of the proapopotic molecule Fas [74, 75] and enhanced apoptosis [76]. Upregulation and induction of CD40 on BM CD34$^+$ cells from SLE patients has been reported to contribute to the amplification of Fas-mediated apoptosis of progenitor cells [77]. Stromal cells from SLE patients fail to support allogeneic progenitor cell growth [74]. Furthermore, due to a diminished activity of lupus monocytes (and possibly enhanced apoptosis of these cells), the production of hematopoietic growth factors by BM fibroblasts appears to be insufficient [78].

Dyserythropoiesis and megakaryocytic atypias are commonly seen in SLE patients with cytopenias, along with disruptions in normal BM architecture and hypocellularity. BM necrotic alterations in SLE are associated with vascular changes including dilatation of sinuses. Hemoglobin levels in SLE have correlated with abnormal localization of immature precursors in their BMs [79]. The number of myeloid colony-forming units-C is lower in SLE patients with severe neutropenia, suggesting an inhibition of myeloid development in patients with this disease. Whether this inhibition is mediated by cellular immunity, autoantibody effects, or specific changes in BM microenvironment is not known [60].

There are many case reports of pure red cell aplasia (PRCA) and myelofibrosis in SLE [80, 81]. In SLE patients who develop aplastic anemia, complement-dependent and/or -independent autoantibodies have been found to suppress erythroid and granulocytic colony formation of BM progenitor cells [82]. IgG fractions of patients with active SLE and hemocytopenias suppress BM progenitor growth *in vitro* by directly binding CD34$^+$ hematopoetic cells but not more differentiated cells. The nature of the antigen on these early progenitors is unknown [83]. Inhibitory autoantibodies against erythroid progenitor cells and erythropoietin have been described in these patients [84].

A role for T cells in BM damage has also been proposed in a model in which autoreactive T cells that home to the BM may affect hematopoetic stem cells through cytotoxic destruction. T-cell removal from SLE BM samples increases the progenitor cell clonogenic potential [85], and T-cells from anemic SLE patients inhibit autologous or allogeneic BM erythroid clone formation *in vitro* [86].

Type I interferons (IFNs) are also potential culprits in BM damage in SLE. IFN-α is elevated in lupus serum, and type I IFN signatures in lupus PBMCs have been described [71]. Furthermore, type I IFNs are increasingly being proposed as major players in lupus pathogenesis. Type I IFNs have long been described to have a suppressive effect in the BM in both human and murine models [87]. Heterogeneity of progenitor cell sensitivity to growth suppression by IFN-α has been described and appears to be influenced by hematopoietic lineage and degree of differentiation of the progenitor cell [88]. A clear effect of IFN-α on erythroid progenitors has been reported and may potentially play a role in the development of anemia in SLE [89]. Increased levels of IFN-α have been found in BMs from lupus-prone mice [90]. A suppressive effect on megakaryocytes has also been described [91]. The inhibitory signal on hematopoietic progenitor cells mediated by IFN-α appears to be transduced by two signaling pathways, one regulated by Tyk2 and the other dependent on Stat1 [92]. Serum from leukopenic SLE patients has been shown to induce apoptosis of CD34$^+$ BM progenitors. Although this was not explained by autoantibodies, TNF-α, or IFN-α, it is possible that other type I IFNs could play a role [93]. Also, various groups have reported a striking decrease in the levels of BM-derived endothelial progenitor cells (EPCs), a subset of progenitors involved in vascular repair and prevention of vascular damage. Interestingly, IFN-α appears to be largely responsible for driving aberrant EPC phenotype and function in SLE [94, 95].

Defects in megakariopoiesis have also been described in SLE patients with thrombocytopenia. It has been proposed that peripheral destruction due to platelet autoantibodies, anti-TPO antibodies, lower effective circulating levels of TPO, and impaired compensatory responses from BM damage may interact in SLE to elicit significant thrombocytopenia [31].

Mechanisms of Other Hematological Abnormalities Seen in SLE

Patients with chronic inflammatory disorders, including SLE, commonly have features of anemia of chronic disease (ACD). Although ACD is not classically viewed as an immune-mediated anemia, the abnormal cytokine milieu present in chronic inflammatory conditions can result in dysregulation of iron homeostasis, impaired proliferation of erythroid progenitor cells, and a blunted erythropoietin (EPO) response—all implicated in the pathogenesis of ACD [96]. In various autoimmune diseases, an insufficient supply of hematopoietic cells with EPO, and their resistance to its proliferative action, has been proposed [97]. Anti-EPO Abs have been described in SLE patients and associated with red cell aplasia and AHA [98]. EPO levels are significantly decreased in SLE when correlated with hemoglobin levels [99]. Although impaired EPO production has been attributed to type I IFNs, TNF-α and TGF-β, when examining *in vitro* cultures [100], the role of these molecules in decreased EPO levels in SLE remains to be determined.

Mechanisms of Malignancies in SLE

Immune perturbations, including autoimmune diseases, have been associated with an increased risk of hematological malignancies including non-Hodgkin's and Hodgkin's lymphoma, myeloid malignancies, and myelodysplastic syndromes [101]. Although immunosuppressive drugs may play an important role in this increased malignancy risk in SLE, the exact oncogenic mechanisms remain unclear. Lupus and various malignancies may share a common pathogenic background, including genetic influence, mutual provoking factors such as viral infections, hormonal factors, and lifestyle-related risks. Disproportional humoral autoimmune responses, cell cycle regulation abnormalities affecting chronic antigenic stimulation, and aberrant function of NK cells and CD8$^+$ T cells may play an additional role [102].

CLINICAL MANIFESTATIONS AND TREATMENT OF CYTOPENIAS IN SLE

General Considerations

Hematologic abnormalities in lupus generally present in a similar manner as they do in non-lupus disorders. For instance, patients with anemia are usually asymptomatic unless the anemia is significant or acute, in which case patients may exhibit typical signs and symptoms, including fatigue, pallor, tachycardia, shortness of breath, or chest pain. Patients with massive hemolysis can be jaundiced. Thrombocytopenia is usually asymptomatic until counts drop below 20,000, at which point petechiae, ecchymoses, and mucosal bleeding can develop. Lupus patients may, however, experience earlier or more prominent symptoms and complications from their cytopenias, depending on their other attendant lupus manifestations and co-morbidities. A number of serologic and biochemical abnormalities can be present, depending on the etiology of the hematologic abnormality, and these are discussed in more detail later. However, because SLE patients often have multiple co-existing causes of cytopenia, interpretation of laboratory studies must take this into account and be performed cautiously.

An important initial step in evaluation is determining whether the primary underlying process is immune or nonimmune in origin. Although hematologic abnormalities can develop in isolation, there is often clinical and serological evidence of other lupus activity; this can help identify the cytopenia as being secondary to ongoing autoimmunity. Again, multiple mechanisms can be at play, and thus management usually involves several approaches, including screening for and treating infections, medication side effects, nutritional deficiencies, occult blood loss, etc.

Similar to clinical manifestations, management of hematologic abnormalities in lupus generally follows the principles of treatment in non-lupus patients. In part, this is because there are few or no trials assessing the use of specific therapeutics in lupus patients; the literature is predominantly limited to case reports and series. However, lupus patients often receive corticosteroids and other immunomodulatory agents earlier or to a greater degree than are used in idiopathic cases. Sometimes this is because the abnormality is immune mediated and responds directly to immunosuppression. In other cases, the rationale is to manage the underlying lupus disease activity thought to be driving the processes responsible for the cytopenias. In addition, patients often have other clinical manifestations requiring immunosuppression, providing additional incentive for their use.

With a few exceptions, the mechanisms by which immunomodulatory medications improve the hematologic perturbances seen in lupus are poorly understood and likely due to multiple mechanisms discussed previously and elsewhere in this book. However, rituximab is being used with increasing frequency in the management of many immune-mediated hematologic processes, and it warrants further discussion here. Rituximab is an anti-CD20 monoclonal antibody that selectively depletes B cells. This can result in a number of downstream immunologic effects potentially relevant in the management of autoimmune conditions, including diminished antibody responses, altered antigen presentation and co-stimulation, modification of the cytokine milieu, and normalization of aberrant autoreactive T-cell responses [103]. In light of the rapid effects that rituximab can exhibit, it has also been postulated that rituximab-opsonized B cells could block monocyte and macrophage Fc receptors, thereby reducing peripheral sequestration and destruction of antibody-coated cells [104]. There is a growing body of literature documenting the efficacy of rituximab in AHA, ITP, and

TTP [105]. Although the role of rituximab in the overall treatment of lupus is not fully clarified, it appears to be a promising tool in the armamentarium of therapeutics for managing autoimmune hematologic manifestations.

ANEMIA

Descriptions in the literature of the prevalence of anemia in SLE vary somewhat, depending on the cohort of patients studied and how anemias are classified. Historical estimates have ranged between 57–78% [106–108], although more recent studies have reported slightly lower prevalences [98, 109].

Nonimmune Causes of Anemia

ACD is the most common type of anemia seen in SLE patients [98, 107]. Patients have a mild to moderate normochromic, normocytic anemia; low reticulocyte counts; reduced iron and transferrin saturation concentrations; and normal or increased ferritin levels. Management is predominantly aimed at controlling systemic inflammation and active lupus manifestations. ACD needs to be distinguished from iron deficiency anemia, which is also frequently present in SLE patients [98] and usually due to menstrual or gastrointestinal blood loss and, more rarely, from pulmonary hemorrhage. Although iron deficiency will result in similarly reduced iron and transferrin saturation levels, the ferritin level will be low, and the anemia is usually hypochromic and microcytic. Treatment involves oral or parenteral iron supplementation. Patients with chronic renal insufficiency have diminished EPO production contributing to their anemia; this can be treated with recombinant human EPO. Although EPO has been shown to be beneficial in chronic anemia associated with rheumatoid arthritis [97], it is unclear which subsets of SLE patients would benefit the most from this drug and whether this depends, at least in part, on the prevalence of anti-EPO antibodies described previously [98, 99]. Nevertheless, EPO is often used in lupus anemia, particularly in the setting of renal insufficiency.

Numerous medications used in the management of SLE can contribute to anemia and need to be considered as possible etiologic agents. Aspirin, nonsteroidal anti-inflammatory drugs, steroids, and oral anticoagulants can precipitate/exacerbate occult gastrointestinal blood loss. Immunosuppressive agents, including azathioprine, methotrexate, mycophenolate mofetil, and cyclophosphamide, can suppress marrow production of erythrocytes. Patients on these medications require regular monitoring of their blood counts.

Autoimmune Hemolytic Anemia

The prevalence of AHA ranges from 7 to 13% [5, 110, 111], with a significant percentage of patients manifesting early in the course of their disease [112]. Patients with AHA will have laboratory abnormalities indicative of hemolysis, including decreased haptoglobin, elevated indirect bilirubin, elevated lactate dehydrogenase, and spherocytes on peripheral smear (Figure 48.1). In the absence of concomitant BM abnormalities, there will be evidence of increased erythropoeisis, including reticulocytosis and circulating nucleated red blood cells. Patients not infrequently have concomitant immune-mediated thrombocytopenia, known as Evan's syndrome. A positive direct antiglobulin test (DAT; or direct Coomb's test) helps distinguish immune-mediated hemolysis from other types of hemolytic anemia. This agglutination assay detects IgG or C3b directly bound to the red blood cell membrane. It is important to emphasize that the presence of a positive DAT is not diagnostic of active autoimmune hemolysis in the absence of other biochemical markers. The indirect antiglobulin test, which detects circulating autoantibodies to erythrocytes, can also be positive. If the DAT is positive, the bound antibodies are eluted off and then incubated with reagent red blood cells to identify the autoantibody specificity. In recently transfused patients, alloantibodies are also screened for. These types of testing are particularly important to perform in the actively hemolyzing patient requiring blood transfusions to prevent further hemolysis.

Treatment generally follows the same principles as those of idiopathic AHA. The briskly hemolyzing patient may require blood transfusions; patients with chronic hemolysis will require supplementation with iron and folate to support marrow production of erythrocytes. Initial immune-targeted therapy involves

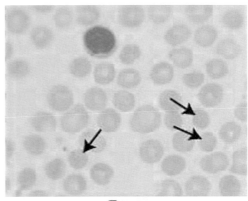

Evans

FIGURE 48.1 Representative smear from a patient with Evan's syndrome (autoimmune hemolytic anemia and thrombocytopenia) demonstrating spherocytes (*arrows*).

corticosteroids, usually at doses of 1—1.5 mg/kg of prednisone or equivalent doses for other corticosteroids. There are numerous putative mechanisms for how corticosteroids diminish hemolysis, including diminishment of antierythrocyte antibody production by B cells. However, the most important early effect is likely decreased Fc receptor expression on macrophages, thereby inhibiting clearance of antibody-coated red blood cells in the reticuloendothelial system [113, 114]. Responses are usually rapid and seen within 1—3 weeks, with a rise in hematocrit and decrease in reticulocyte count, after which time steroids are gradually tapered. A retrospective analysis of 26 female lupus patients with severe AHA found a 96% response rate to corticosteroids, with 77% maintaining their response upon steroid taper [115].

Patients who do not respond to steroids, who cannot be tapered down, or who relapse can be treated with other modalities. Splenectomy has historically been used as a second-line treatment, and it can be quite effective in creating remission in refractory idiopathic warm-antibody AHA. However, it has associated surgical and postsurgical morbidity; in addition, SLE patients with AHA often require immunosuppressive treatment for other lupus manifestations. Thus, splenectomy is often reserved for when other modalities have failed. Immunosuppressive agents used in the management of AHA have included azathioprine, cyclosporine, and cyclophosphamide, and these have been successfully used in SLE-associated AHA [116, 117]. Successful treatment with mycophenolate mofetil, often used for other manifestations of SLE, has also been described [115]. Danazol, an attenuated androgen, has been used in refractory SLE-associated cases [118], as have intravenous immunoglobulin (IVIG) infusions [115]. Rituximab is being successfully used for isolated AHA and Evan's syndrome associated with SLE [103, 115], as well as in the rarer cold agglutinin disease [119].

Thrombocytopenic Purpura

TTP is one of the true emergencies associated with SLE, and early treatment is crucial and differs significantly from conventional treatment of lupus flares. Clinical manifestations can overlap significantly with lupus-associated signs and symptoms and result in diagnostic confusion and/or delay [120, 121], in part because TTP usually presents in patients with active lupus [122, 123]. Thus, there should be a strong suspicion for TTP in any lupus patient presenting with profound thrombocytopenia and MAHA.

Despite numerous case reports and series in the literature, the incidence and prevalence of TTP in SLE is unknown. One group reported that 4 of 103 patients with TTP (3.8%) had a diagnosis of SLE [124]. Although

most lupus patients have been diagnosed prior to developing microangiopathy, TTP can occasionally be a presenting or even preceding manifestation [125]. The primary manifestations of TTP include MAHA, thrombocytopenia, fever, neurologic features, and renal abnormalities, although it is now appreciated that patients do not always present with this classic pentad. Patients can also have nonspecific prodromal systemic symptoms, including fatigue and malaise. MAHA presents with serologic evidence of hemolysis, including decreased haptoglobin, increased indirect bilirubin, and reticulocytosis. In particular, the lactate dehydrogenase may be elevated, reflecting both hemolysis and tissue ischemia. However, TTP is not associated with a positive DAT, reflecting the nonimmune mechanism thought to underlie hemolysis. On peripheral smear, the hallmark finding is red blood cell fragmentation with resultant schistocyte forms and markedly diminished platelets (Figure 48.2); there can also be nucleated red blood cells.

Patients should be assessed for other potential causes of thrombotic microangiopathy, including sepsis, disseminated intravascular coagulation, predisposing medications, or catastrophic aPL syndrome. Patients with SLE can have a positive DAT as part of their autoantibody milieu even in the absence of autoimmune hemolysis, and thus its presence should not exclude the diagnosis of TTP in the appropriate setting [126]. ADAMTS-13 activity and inhibitor levels can be ordered; however, these are send-out assays for most institutions, and initiation of treatment should not be delayed in the appropriate clinical setting while awaiting results. In addition, although extremely low levels of ADAMTS-13 are seen in TTP, they are not diagnostic; similarly, normal or mildly decreased levels do not rule the disorder out.

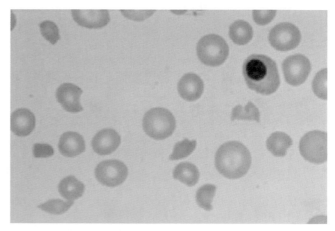

FIGURE 48.2   Representative smear from a patient with thrombotic thrombocytopenic purpura demonstrating schistocytes, a nucleated red blood cell, and absent platelets.

Plasma exchange is the first line of treatment because this is thought to clear anti-ADAMTS-13 antibodies in addition to replacing the deficient protein; it should be initiated as soon as possible to prevent further organ damage. Although plasma exchange has significantly improved survival, lupus patients still have higher mortality rates than patients with idiopathic TTP [122, 125, 127]. High-dose steroids and immunosuppressives including cyclosporine and cyclophosphamide have been given in conjunction to presumptively treat the underlying lupus activity driving the TTP. Vincristine has also been used. Similarly to AHA, there are increasing reports of the beneficial use of rituximab for refractory or recurrent TTP [128, 129]. An important issue is the timing of rituximab administration in conjunction with plasma exchange to prevent clearance of the monoclonal antibody. Although there is still debate, current literature suggests that plasma exchange can be administered 24—36 h after rituximab infusion [105].

THROMBOCYTOPENIA

Thrombocytopenia, defined as a platelet count less than $100,000/mm^3$, is found in 10—25% of patients with SLE [130]. Patients can develop acute severe thrombocytopenia, usually in the presence of significant disease activity or occasionally as a medication side effect. More commonly, thrombocytopenia is chronic and moderate in level, and it rarely causes major symptoms.

Numerous medications have been associated with thrombocytopenia, via both peripheral destructive and myelosuppressive effects. An important consideration in the hospitalized lupus patient is heparin effects on platelet counts. Heparin can cause a nonimmune-mediated mild drop in platelet count usually within the first few days of heparin exposure, known as heparin-associated thrombocytopenia; this usually corrects itself within a few days without intervention or drug withdrawal. In contrast, heparin-induced thrombocytopenia (HIT) is a severe prothrombotic immune-mediated drug reaction resulting from the development of heparin—antiplatelet factor 4 complex antibodies. There can be serious venous or arterial thrombosis that can mimic catastrophic aPL syndrome or systemic vasculitis. Platelet counts drop 4 or more days after heparin exposure. HIT antibodies can be ordered and, if negative, exclude the presence of HIT; however, nonpathogenic antibodies can also generate a positive test. This is particularly relevant in lupus patients, in which a high false-positive rate of HIT antibody testing has been reported, especially in patients with aPLs [131]. The gold standard for diagnosis is the serotonin release assay, which detects the release of labeled serotonin from platelets in the presence of low concentrations of heparin. Treatment involves discontinuation of all heparin exposure and immediate initiation of nonheparin anticoagulants such as argatroban, lepirudin, or factor X inhibitors [132].

Immune-Mediated Thrombocytopenia

In the absence of other hematologic abnormalities, the peripheral smear in immune-mediated thrombocytopenia will show decreased platelet numbers and occasional large platelet forms (Figure 48.3), with normal red blood cell and leukocyte numbers and morphology. Bone marrow evaluation reveals increased megakaryocytes, reflecting peripheral platelet destruction. Patients should be assessed for potentially contributing medications or co-morbid conditions, including aPLs, chronic hepatitis, or HIV if clinically indicated, and evidence of hemolysis. There is no clinical utility to ordering anti-platelet antibodies, even if a potential pathogenic role has been described.

Treatment follows that of idiopathic ITP [133]. Consensus statements have favored withholding treatment unless platelet counts are below 30,000 or there is symptomatic bleeding. Initial therapy is corticosteroids at doses of 1 or 2 mg/kg/day of prednisone or equivalent dose for other corticosteroids, although occasionally high-dose pulses of methylprednisolone are used. Steroids result in a short-term response in 60% of patients; patients often require more chronic treatment. IVIG 1 or 2 g/kg over 2—5 days is used in cases of severe thrombocytopenia accompanied by bleeding, and it has an 80% response rate, which is usually nonsustained. Anti-D, a polyclonal antibody against the Rho(D) blood antigen, is used successfully in Rh(+) patients and is

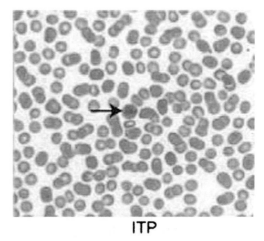

ITP

FIGURE 48.3 Representative ITP smear demonstrating low platelet counts and a megathrombocyte (*arrow*).

cheaper and easier to administer than IVIG, although there have been no head-to-head comparison studies. Anti-D is generally ineffective in splenectomized patients, suggesting that its primary mode of action involves blockade of peripheral clearance. Relapsing cases are often treated with additional steroids and/or immunosuppressive agents, vinca alkaloids, and, increasingly, rituximab, which can result in a lasting response in 30% of patients [134]. There have been numerous reports of the use of danazol for refractory thrombocytopenia, with good response and tolerance of the medication; successful use of dapsone has also been reported. The role of splenectomy is controversial. Although it can be safely performed and results in durable responses in more than 50% of patients in some reports [4, 135, 136], others have found poorer response rates with frequent relapses [137]. There are a number of investigational agents being used for idiopathic ITP; however, their role in the management of immune thrombocytopenia in lupus is unclear.

LEUKOPENIA

Leukopenia is fairly common, with reported prevalences of approximately 50% [106].

Lymphopenia

Reports of the prevalence of lymphopenia vary from as high as approximately 75% in several series [138, 139] to much lower in others [5]. Some of this variability may reflect the confounding leukopenic effects of immunosuppressive treatment. Lymphopenia also occurs more frequently with increasing age in SLE [140]. In addition, it is worth considering that defining lymphopenia as counts below $1500/mm^3$ (as established in the ACR criteria for SLE) may be a misnomer because many laboratories nowadays only consider lymphocyte counts below $800/mm^3$ as abnormal.

Although different groups have found variable clinical features associated with the presence of lymphopenia, a more consistent association with anti-dsDNA and anti-Ro antibodies has been found [139]. Significant lymphopenia is also associated with disease activity and damage accrual [141]. Whether lymphopenia is associated with an increased rate of infection is unclear. One study of 91 diagnosed Chinese patients found a lymphocyte count less than $1000/mm^3$ to be a risk factor for major infection [142], whereas another group found no increased infections with lymphocyte counts less than $1500/mm^3$ [5]. Literature suggests that severe lymphoablative conditioning in the absence of HIV infection, such as seen in patients with autoimmune diseases undergoing autologous stem cell transplantation, is relatively safe with regard to infection risk [143]. In general, levels of lymphocytes are not used as a criteria to establish antimicrobial prophylaxis in patients without HIV infection.

Neutropenia

Significant neutropenia ($<1000/mm^3$) is much less common than lymphopenia in SLE. A study of 126 lupus patients found neutrophil counts of less than $2000/mm^3$ in 47% but moderate to severe neutropenia ($<1000/mm^3$) in less than 5% [5]. Neutropenia occasionally presents concomitantly with AHA and thrombocytopenia. One retrospective review of 33 SLE neutropenic patients found drug toxicity to be the most common cause [144]. Regardless of cause, neutropenia is a major risk factor for infection, with frequency and severity of infections proportional to the degree of neutropenia. Management should include discontinuation of implicated medications and treatment of underlying lupus activity. Granulocyte colony-stimulating factor (GCSF) can be administered to treat selected cases of severe refractory neutropenia in the setting of infection. One series of 9 patients found GCSF administration associated with lupus flares in one-third [145]. Therefore, the authors recommend that patients who require GCSF should be carefully monitored for increased disease activity.

BONE MARROW DISEASE

BM involvement in lupus can manifest as low cell counts in single or multiple lines; lupus patients often require BM biopsy to distinguish between peripheral versus marrow-based causes of cytopenia. The major nonimmune-mediated cause of BM suppression is drug effect/toxicity. Antimetabolite agents used in the treatment of lupus have variable degrees of myelotoxicity and can contribute to low cell counts. Azathioprine-induced myelosuppression in large part occurs from abnormal metabolism of the drug due to genetic polymorphisms of thiopurine methyltransferase (TPMT); 1 in 300 persons have severe deficiency of the enzyme. Thus, patients being initiated on azathioprine should have their TPMT phenotype determined because low enzyme activity is associated with an increased risk of myelosuppression. This is done by measuring TMPT activity in red blood cell lysates using radiochemical or high-performance liquid chromatography assays. Although the prevalence of clinically significant TPMT deficiency is relatively low, several models have suggested that TPMT phenotyping prior to azathioprine initiation is cost-effective in a variety of situations [146]. TPMT genotyping can also be performed but is currently of less clinical utility [147]. Bone marrow suppression is the main dose-limiting toxicity of cyclophosphamide and other alkylating agents; in addition, there is a cumulative dose effect, and patients who have received large amounts of cyclophosphamide in the past have poor BM reserve.

Pure Red Cell Aplasia

Lupus is a fairly rare cause of PRCA. Anemia usually develops gradually and therefore can be quite severe. Patients have a normochromic, normocytic anemia and a significantly reduced reticulocyte count ($<1\%$); other cell counts are normal, and other lupus disease manifestations are often quiescent [148]. Patients should be screened for the use of potential causative medications as well as serologies suggestive of acute parvovirus B_{19} infection. Nutritional deficiencies should be excluded by assessing iron, B_{12}, and folate levels. Diagnosis is made by examination of the BM, with absence of the erythroid lineage and preservation of the others.

Treatment requires transfusion if the anemia is symptomatic, with iron chelation if required. Prednisone 1 mg/kg is the initial treatment, with response expected within 4–6 weeks, followed by a slow taper over several months. Cytotoxic agents including azathioprine, methotrexate, and cyclophosphamide have been used successfully in refractory cases, as have IVIG, danazol, and rituximab.

Myelofibrosis

SLE can rarely be associated with autoimmune myelofibrosis, occasionally as a presenting feature of the disease. Patients usually develop pancytopenia and have a leukoerythroblastic peripheral smear; BM biopsy reveals hypocellularity and variable amounts of marrow fibrosis. Steroid treatment can reverse marrow abnormalities and result in improved counts [149].

LYMPHOPROLIFERATIVE DISEASE

Lymph node (LN) enlargement is common, present in approximately 25–50% of lupus patients at some course in their disease, and can be associated with increased disease activity [106, 150]. Cervical LN enlargement is the most common; LNs are usually soft, nontender, and smaller than 2 cm. Histologically, lymph nodes in SLE are characterized by varying degrees of coagulative necrosis with nonspecific reactive follicular hyperplasia; more rarely, but unique to lupus, hematoxylin bodies can be present in necrotic loci. The pathologic findings can resemble those seen in Kikuchi–Fujimoto's disease (histiocytic necrotizing lymphadenitis), a benign, self-limited condition associated with cervical lymphadenopathy, fever, and leukopenia; there are also rare reports of the co-existence of both conditions. However, an increasing amount of literature has documented the presence of atypical lymphoproliferative features in SLE lymph node lesions [151]. These abnormalities may represent intermediary stages in the spectrum between benign LN enlargement and frank lymphoma. Furthermore, SLE is associated with an increased risk of non-Hodgkin's lymphoma (NHL) development. A multisite international cohort of more than 9500 SLE patients was linked to regional tumor registries to determine standardized incidence ratios for assorted malignancies. The standardized incidence ratio for developing NHL was found to be 3.64 [101]. A pooled analysis of nearly 30,000 participants in 12 case–control studies examining the risk of NHL associated with self-reported autoimmune conditions found that patients with SLE had a 2.7-fold increased risk of developing NHL [152]. Risk factors for the development of NHL include the presence of hematologic disease or sicca symptoms, but not the use of cytotoxic therapy. Thus, patients with lymphadenopathy should be monitored for development of possible malignant transformation, including focal LN enlargement, systemic symptoms, and serologic evidence of monoclonality.

References

[1] E.M. Tan, A.S. Cohen, J.F. Fries, A.T. Masi, D.J. McShane, N.F. Rothfield, et al., The 1982 revised criteria for the classification of systemic lupus erythematosus, Arthritis Rheum 25 (1982) 1271–1277.

[2] M.C. Hochberg, Updating the American College of Rheumatology revised criteria for the classification of systemic lupus erythematosus, Arthritis Rheum 40 (1997) 1725.

[3] M. Abu-Shakra, M.B. Urowitz, D.D. Gladman, J. Gough, Mortality studies in systemic lupus erythematosus. Results from a single center. II. Predictor variables for mortality, J Rheumatol 22 (1995) 1265–1270.

[4] C. Arnal, J.C. Piette, J. Leone, B. Taillan, E. Hachulla, F. Roudot-Thoraval, et al., Treatment of severe immune thrombocytopenia associated with systemic lupus erythematosus: 59 cases, J Rheumatol 29 (2002) 75–83.

[5] J.C. Nossent, A.J. Swaak, Prevalence and significance of haematological abnormalities in patients with systemic lupus erythematosus, Q J Med 80 (1991) 605–612.

[6] M. Pistiner, D.J. Wallace, S. Nessim, A.L. Metzger, J.R. Klinenberg, Lupus erythematosus in the 1980s: A survey of 570 patients, Semin Arthritis Rheum 21 (1991) 55–64.

[7] M.M. Ward, E. Pyun, S. Studenski, Mortality risks associated with specific clinical manifestations of systemic lupus erythematosus, Arch Intern Med 156 (1996) 1337–1344.

[8] C.C. Mok, K.W. Lee, C.T. Ho, C.S. Lau, R.W. Wong, A prospective study of survival and prognostic indicators of systemic lupus erythematosus in a southern Chinese population, Rheumatology (Oxford) 39 (2000) 399–406.

[9] N. Kasitanon, L.S. Magder, M. Petri, Predictors of survival in systemic lupus erythematosus, Medicine (Baltimore) 85 (2006) 147–156.

[10] A.D. Schreiber, M.M. Frank, Role of antibody and complement in the immune clearance and destruction of erythrocytes. I. *In vivo* effects of IgG and IgM complement-fixing sites, J Clin Invest 51 (1972) 575–582.

[11] A.F. LoBuglio, R.S. Cotran, J.H. Jandl, Red cells coated with immunoglobulin G: binding and sphering by mononuclear cells in man, Science 158 (1967) 1582–1585.

[12] E.J. Victoria, S.W. Pierce, M.J. Branks, S.P. Masouredis, IgG red blood cell autoantibodies in autoimmune hemolytic anemia bind to epitopes on red blood cell membrane band 3 glycoprotein, J Lab Clin Med 115 (1990) 74–88.

[13] R.N. Barker, G.G. de Sa Oliveira, C.J. Elson, P.M. Lydyard, Pathogenic autoantibodies in the NZB mouse are specific for erythrocyte band 3 protein, Eur J Immunol 23 (1993) 1723–1726.

[14] J.P. Leddy, J.L. Falany, G.E. Kissel, S.T. Passador, S.I. Rosenfeld, Erythrocyte membrane proteins reactive with human (warm-reacting) anti-red cell autoantibodies, J Clin Invest 91 (1993) 1672–1680.

[15] G.G. Oliveira, P.R. Hutchings, P.M. Lydyard, Anti-CD4 treatment of NZB mice prevents the development of erythrocyte autoantibodies but hastens the appearance of anaemia, Immunol Lett 39 (1994) 153–156.

[16] L. Fossati-Jimack, S. Azeredo da Silveira, T. Moll, T. Kina, F.A. Kuypers, P.A. Oldenborg, et al., Selective increase of autoimmune epitope expression on aged erythrocytes in mice: Implications in anti-erythrocyte autoimmune responses, J Autoimmun 18 (2002) 17–25.

[17] F. Burling, J. Ng, H. Thein, J. Ly, M.R. Marshall, P. Gow, Ethnic, clinical and immunological factors in systemic lupus erythematosus and the development of lupus nephritis: Results from a multi-ethnic New Zealand cohort, Lupus 16 (2007) 830–837.

[18] S. Kikuchi, H. Amano, E. Amano, L. Fossati-Jimack, M.L. Santiago-Raber, T. Moll, et al., Identification of 2 major loci linked to autoimmune hemolytic anemia in NZB mice, Blood 106 (2005) 1323–1329.

[19] A.R. Youssef, C.R. Shen, C.L. Lin, R.N. Barker, C.J. Elson, IL-4 and IL-10 modulate autoimmune haemolytic anaemia in NZB mice, Clin Exp Immunol 139 (2005) 84–89.

[20] M.L. Santiago-Raber, R. Baccala, K.M. Haraldsson, D. Choubey, T.A. Stewart, D.H. Kono, et al., Type-I interferon receptor deficiency reduces lupus-like disease in NZB mice, J Exp Med 197 (2003) 777–788.

[21] M.M. Kay, J.J. Marchalonis, J. Hughes, K. Watanabe, S.F. Schluter, Definition of a physiologic aging autoantigen by using synthetic peptides of membrane protein band 3: Localization of the active antigenic sites, Proc Natl Acad Sci USA 87 (1990) 5734–5738.

[22] S. Giannouli, M. Voulgarelis, P.D. Ziakas, A.G. Tzioufas, Anaemia in systemic lupus erythematosus: From pathophysiology to clinical assessment, Ann Rheum Dis 65 (2006) 144–148.

[23] M. Murakami, T. Tsubata, R. Shinkura, S. Nisitani, M. Okamoto, H. Yoshioka, et al., Oral administration of lipopolysaccharides activates B-1 cells in the peritoneal cavity and lamina propria of the gut and induces autoimmune symptoms in an autoantibody transgenic mouse, J Exp Med 180 (1994) 111–121.

[24] S. Nisitani, M. Murakami, T. Honjo, Anti-red blood cell immunoglobulin transgenic mice. An experimental model of autoimmune hemolytic anemia, Ann N Y Acad Sci 815 (1997) 246–252.

[25] D. Alarcon-Segovia, M. Deleze, C.V. Oria, J. Sanchez-Guerrero, L. Gomez-Pacheco, J. Cabiedes, et al., Antiphospholipid antibodies and the antiphospholipid syndrome in systemic lupus erythematosus. A prospective analysis of 500 consecutive patients, Medicine (Baltimore) 68 (1989) 353–365.

[26] J.L. Chen, X.M. Huang, X.J. Zeng, Y. Wang, M.X. Zhou, Y.H. Ma, et al., Hematological abnormalities in systemic lupus erythematosus and clinical significance thereof: Comparative analysis of 236 cases, Zhonghua Yi Xue Za Zhi 87 (2007) 1330–1333.

[27] Y. Richaud-Patin, B. Perez-Romano, E. Carrillo-Maravilla, A.B. Rodriguez, A.J. Simon, J. Cabiedes, et al., Deficiency of red cell bound CD55 and CD59 in patients with systemic lupus erythematosus, Immunol Lett 88 (2003) 95–99.

[28] G.D. Ross, W.J. Yount, M.J. Walport, J.B. Winfield, C.J. Parker, C.R. Fuller, et al., Disease-associated loss of erythrocyte complement receptors (CR1, C3b receptors) in patients with systemic lupus erythematosus and other diseases involving autoantibodies and/or complement activation, J Immunol 135 (1985) 2005–2014.

[29] M.E. Medof, A. Gottlieb, T. Kinoshita, S. Hall, R. Silber, V. Nussenzweig, et al., Relationship between decay accelerating factor deficiency, diminished acetylcholinesterase activity, and defective terminal complement pathway restriction in paroxysmal nocturnal hemoglobinuria erythrocytes, J Clin Invest 80 (1987) 165–174.

[30] M. Kuwana, Y. Okazaki, M. Kajihara, J. Kaburaki, H. Miyazaki, Y. Kawakami, et al., Autoantibody to c-Mpl (thrombopoietin receptor) in systemic lupus erythematosus: Relationship to thrombocytopenia with megakaryocytic hypoplasia, Arthritis Rheum 46 (2002) 2148–2159.

[31] P.D. Ziakas, J.G. Routsias, S. Giannouli, A. Tasidou, A.G. Tzioufas, M. Voulgarelis, Suspects in the tale of lupus-associated thrombocytopenia, Clin Exp Immunol 145 (2006) 71–80.

[32] D.B. Cines, H. Liebman, R. Stasi, Pathobiology of secondary immune thrombocytopenia, Semin Hematol 46 (2009) S2–S14.

[33] M. Kuwana, J. Kaburaki, Y. Okazaki, H. Miyazaki, Y. Ikeda, Two types of autoantibody-mediated thrombocytopenia in patients with systemic lupus erythematosus, Rheumatology (Oxford) 45 (2006) 851–854.

[34] T. Jouhikainen, R. Kekomaki, M. Leirisalo-Repo, T. Backlund, G. Myllyla, Platelet autoantibodies detected by immunoblotting in systemic lupus erythematosus: Association with the lupus anticoagulant, and with history of thrombosis and thrombocytopenia, Eur J Haematol 44 (1990) 234–239.

[35] D.T. Boumpas, H.A. Austin 3rd, B.J. Fessler, J.E. Balow, J.H. Klippel, M.D. Lockshin, Systemic lupus erythematosus: Emerging concepts. Part 1: Renal, neuropsychiatric, cardiovascular, pulmonary, and hematologic disease, Ann Intern Med 122 (1995) 940–950.

[36] M. Kuwana, J. Kaburaki, Y. Ikeda, Autoreactive T cells to platelet GPIIb-IIIa in immune thrombocytopenic purpura. Role in production of anti-platelet autoantibody, J Clin Invest 102 (1998) 1393–1402.

[37] R.S. Schwartz, Treating chronic idiopathic thrombocytopenic purpura—A new application of an old treatment, N Engl J Med 330 (1994) 1609–1610.

[38] I. Uthman, B. Godeau, A. Taher, M. Khamashta, The hematologic manifestations of the antiphospholipid syndrome, Blood Rev 22 (2008) 187–194.

[39] S.L. Lee, A.B. Miotti, Disorders of hemostatic function in patients with systemic lupus erythematosus, Semin Arthritis Rheum 4 (1975) 241–252.

[40] J.V. Simone, J.A. Cornet, C.F. Abildgaard, Acquired von Willebrand's syndrome in systemic lupus erythematosus, Blood 31 (1968) 806–812.

[41] C. Sucker, J.J. Michiels, R.B. Zotz, Causes, etiology and diagnosis of acquired von Willebrand disease: A prospective

diagnostic workup to establish the most effective therapeutic strategies, Acta Haematol 121 (2009) 177–182.

[42] H.M. Tsai, The molecular biology of thrombotic microangiopathy, Kidney Int 70 (2006) 16–23.

[43] M. Rieger, P.M. Mannucci, J.A. Kremer Hovinga, A. Herzog, G. Gerstenbauer, C. Konetschny, et al., ADAMTS13 autoantibodies in patients with thrombotic microangiopathies and other immunomediated diseases, Blood 106 (2005) 1262–1267.

[44] P.M. Mannucci, M. Vanoli, I. Forza, M.T. Canciani, R. Scorza, Von Willebrand factor cleaving protease (ADAMTS-13) in 123 patients with connective tissue diseases (systemic lupus erythematosus and systemic sclerosis), Haematologica 88 (2003) 914–918.

[45] S. Dold, R. Singh, H. Sarwar, Y. Menon, L. Candia, L.R. Espinoza, Frequency of microangiopathic hemolytic anemia in patients with systemic lupus erythematosus exacerbation: Distinction from thrombotic thrombocytopenic purpura, prognosis, and outcome, Arthritis Rheum 53 (2005) 982–985.

[46] B.J. Hunt, S. Tueger, J. Pattison, J. Cavenagh, D.P. D'Cruz, Microangiopathic haemolytic anaemia secondary to lupus nephritis: An important differential diagnosis of thrombotic thrombocytopenic purpura, Lupus 16 (2007) 358–362.

[47] M.J. Kaplan, Apoptosis in systemic lupus erythematosus, Clin Immunol 112 (2004) 210–218.

[48] M.F. Denny, P. Chandaroy, P.D. Killen, R. Caricchio, E.E. Lewis, B.C. Richardson, et al., Accelerated macrophage apoptosis induces autoantibody formation and organ damage in systemic lupus erythematosus, J Immunol 176 (2006) 2095–2104.

[49] M.J. Kaplan, E.E. Lewis, E.A. Shelden, E. Somers, R. Pavlic, W.J. McCune, et al., The apoptotic ligands TRAIL, TWEAK, and Fas ligand mediate monocyte death induced by autologous lupus T cells, J Immunol 169 (2002) 6020–6029.

[50] D.J. Armstrong, A.D. Crockard, B.G. Wisdom, E.M. Whitehead, A.L. Bell, Accelerated apoptosis in SLE neutrophils cultured with anti-dsDNA antibody isolated from SLE patient serum: A pilot study, Rheumatol Int 27 (2006) 153–156.

[51] W. Emlen, J. Niebur, R. Kadera, Accelerated *in vitro* apoptosis of lymphocytes from patients with systemic lupus erythematosus, J Immunol 152 (1994) 3685–3692.

[52] W. Matsuyama, W. Yamamoto, I. Higashimoto, K. Oonakahara, M. Watanabe, K. Machida, et al., TNF-related apoptosis-inducing ligand is involved in neutropenia of systemic lupus erythematosus, Blood 104 (2004) 184–191.

[53] R.J. Winchester, J.B. Winfield, F. Siegal, P. Wernet, Z. Bentwich, H.G. Kunkel, Analyses of lymphocytes from patients with rheumatoid arthritis and systemic lupus erythematosus. Occurrence of interfering cold-reactive antilymphocyte antibodies, J Clin Invest 54 (1974) 1082–1092.

[54] W.T. Butler, J.T. Sharp, R.D. Rossen, M.D. Lidsky, K.K. Mittal, D.A. Gard, Relationship of the clinical course of systemic lupus erythematosus to the presence of circulating lymphocytotoxic antibodies, Arthritis Rheum 15 (1972) 251–258.

[55] M.B. Magalhaes, L.M. da Silva, J.C. Voltarelli, E.A. Donadi, P. Louzada-Junior, Lymphocytotoxic antibodies in systemic lupus erythematosus are associated with disease activity irrespective of the presence of neuropsychiatric manifestations, Scand J Rheumatol 36 (2007) 442–447.

[56] J.B. Winfield, R.J. Winchester, H.G. Kunkel, Association of cold-reactive antilymphocyte antibodies with lymphopenia in systemic lupus erythematosus, Arthritis Rheum 18 (1975) 587–594.

[57] R.J. DeHoratius, R. Pillarisetty, R.P. Messner, N. Talal, Antinucleic acid antibodies in systemic lupus erythematosus patients and their families. Incidence and correlation with lymphocytotoxic antibodies, J Clin Invest 56 (1975) 1149–1154.

[58] C. Osman, A.J. Swaak, Lymphocytotoxic antibodies in SLE: A review of the literature, Clin Rheumatol 13 (1994) 21–27.

[59] S.D. Denburg, S.A. Behmann, R.M. Carbotte, J.A. Denburg, Lymphocyte antigens in neuropsychiatric systemic lupus erythematosus. Relationship of lymphocyte antibody specificities to clinical disease, Arthritis Rheum 37 (1994) 369–375.

[60] M. Arenas, A. Abad, V. Valverde, P. Ferriz, R. Pascual, Selective inhibition of granulopoiesis with severe neutropenia in systemic lupus erythematosus, Arthritis Rheum 35 (1992) 979–980.

[61] A.P. Cairns, A.D. Crockard, J.R. McConnell, P.A. Courtney, A.L. Bell, Reduced expression of CD44 on monocytes and neutrophils in systemic lupus erythematosus: Relations with apoptotic neutrophils and disease activity, Ann Rheum Dis 60 (2001) 950–955.

[62] P.A. Courtney, A.D. Crockard, K. Williamson, A.E. Irvine, R.J. Kennedy, A.L. Bell, Increased apoptotic peripheral blood neutrophils in systemic lupus erythematosus: relations with disease activity, antibodies to double stranded DNA, and neutropenia, Ann Rheum Dis 58 (1999) 309–314.

[63] S.C. Hsieh, H.S. Yu, W.W. Lin, K.H. Sun, C.Y. Tsai, D.F. Huang, et al., Anti-SSB/La is one of the antineutrophil autoantibodies responsible for neutropenia and functional impairment of polymorphonuclear neutrophils in patients with systemic lupus erythematosus, Clin Exp Immunol 131 (2003) 506–516.

[64] H.F. Pan, X.H. Fang, G.C. Wu, W.X. Li, X.F. Zhao, X.P. Li, et al., Anti-neutrophil cytoplasmic antibodies in new-onset systemic lupus erythematosus and lupus nephritis, Inflammation 31 (2008) 260–265.

[65] A. Bakkaloglu, R. Topaloglu, U. Saatci, S. Ozdemir, S. Ozen, Y. Bassoy, et al., Antineutrophil cytoplasmic antibodies in childhood systemic lupus erythematosus, Clin Rheumatol 17 (1998) 265–267.

[66] B.T. Kurien, J. Newland, C. Paczkowski, K.L. Moore, R.H. Scofield, Association of neutropenia in systemic lupus erythematosus (SLE) with anti-Ro and binding of an immunologically cross-reactive neutrophil membrane antigen, Clin Exp Immunol 120 (2000) 209–217.

[67] M. Deleze, D. Alarcon-Segovia, C.V. Oria, J. Sanchez-Guerrero, L. Fernandez-Dominguez, L. Gomez-Pacheco, et al., Hemocytopenia in systemic lupus erythematosus. Relationship to antiphospholipid antibodies, J Rheumatol 16 (1989) 926–930.

[68] M. Akhtari, B. Curtis, E.K. Waller, Autoimmune neutropenia in adults, Autoimmun Rev. 9 (2009) 62–66.

[69] Y. Hashimoto, M. Ziff, E.R. Hurd, Increased endothelial cell adherence, aggregation, and superoxide generation by neutrophils incubated in systemic lupus erythematosus and Felty's syndrome sera, Arthritis Rheum 25 (1982) 1409–1418.

[70] R.B. Zurier, Reduction of phagocytosis and lysosomal enzyme release from human leukocytes by serum from patients with systemic lupus erythematosus, Arthritis Rheum 19 (1976) 73–78.

[71] L. Bennett, A.K. Palucka, E. Arce, V. Cantrell, J. Borvak, J. Banchereau, et al., Interferon and granulopoiesis signatures in systemic lupus erythematosus blood, J Exp Med 197 (2003) 711–723.

[72] E. Hacbarth, A. Kajdacsy-Balla, Low density neutrophils in patients with systemic lupus erythematosus, rheumatoid arthritis, and acute rheumatic fever, Arthritis Rheum 29 (1986) 1334–1342.

[73] M. Nakou, N. Knowlton, M.B. Frank, G. Bertsias, J. Osban, C.E. Sandel, et al., Gene expression in systemic lupus erythematosus: Bone marrow analysis differentiates active from inactive disease and reveals apoptosis and granulopoiesis signatures, Arthritis Rheum 58 (2008) 3541–3549.

[74] H.A. Papadaki, D.T. Boumpas, F.M. Gibson, D.R. Jayne, J.S. Axford, E.C. Gordon-Smith, et al., Increased apoptosis of bone marrow CD34(+) cells and impaired function of bone marrow stromal cells in patients with systemic lupus erythematosus, Br J Haematol 115 (2001) 167–174.

[75] L.Y. Sun, K.X. Zhou, X.B. Feng, H.Y. Zhang, X.Q. Ding, O. Jin, et al., Abnormal surface markers expression on bone marrow CD34$^+$ cells and correlation with disease activity in patients with systemic lupus erythematosus, Clin Rheumatol 26 (2007) 2073–2079.

[76] A.L. Hepburn, I.A. Lampert, J.J. Boyle, D. Horncastle, W.F. Ng, M. Layton, et al., In vivo evidence for apoptosis in the bone marrow in systemic lupus erythematosus, Ann Rheum Dis 66 (2007) 1106–1109.

[77] K. Pyrovolaki, I. Mavroudi, P. Sidiropoulos, A.G. Eliopoulos, D.T. Boumpas, H.A. Papadaki, Increased expression of CD40 on bone marrow CD34$^+$ hematopoietic progenitor cells in patients with systemic lupus erythematosus: Contribution to Fas-mediated apoptosis, Arthritis Rheum 60 (2009) 543–552.

[78] T. Otsuka, K. Nagasawa, M. Harada, Y. Niho, Bone marrow microenvironment of patients with systemic lupus erythematosus, J Rheumatol 20 (1993) 967–971.

[79] M. Voulgarelis, S. Giannouli, A. Tasidou, D. Anagnostou, P.D. Ziakas, A.G. Tzioufas, Bone marrow histological findings in systemic lupus erythematosus with hematologic abnormalities: A clinicopathological study, Am J Hematol 81 (2006) 590–597.

[80] M.O. Arcasoy, N.J. Chao, T-cell-mediated pure red-cell aplasia in systemic lupus erythematosus: Response to cyclosporin A and mycophenolate mofetil, Am J Hematol 78 (2005) 161–163.

[81] A.Y. Chan, E.K. Li, L.S. Tam, G. Cheng, P.C. Choi, Successful treatment of pure red cell aplasia associated with systemic lupus erythematosus with oral danazol and steroid, Rheumatol Int 25 (2005) 388–390.

[82] J. Rossert, Erythropoietin-induced, antibody-mediated pure red cell aplasia, Eur J Clin Invest 35 (Suppl. 3) (2005) 95–99.

[83] H. Liu, K. Ozaki, Y. Matsuzaki, M. Abe, M. Kosaka, S. Saito, Suppression of haematopoiesis by IgG autoantibodies from patients with systemic lupus erythematosus (SLE), Clin Exp Immunol 100 (1995) 480–485.

[84] K.R. Hartman, V.F. LaRussa, S.W. Rothwell, T.O. Atolagbe, F.T. Ward, G. Klipple, Antibodies to myeloid precursor cells in autoimmune neutropenia, Blood 84 (1994) 625–631.

[85] P.D. Kiely, C.P. McGuckin, D.A. Collins, D.H. Bevan, J.C. Marsh, Erythrocyte aplasia and systemic lupus erythematosus, Lupus 4 (1995) 407–411.

[86] K. Yamasaki, Y. Niho, T. Yanase, Erythroid colony forming cells in systemic lupus erythematosus, J Rheumatol 11 (1984) 167–171.

[87] H.E. Broxmeyer, S. Cooper, B.Y. Rubin, M.W. Taylor, The synergistic influence of human interferon-gamma and interferon-alpha on suppression of hematopoietic progenitor cells is additive with the enhanced sensitivity of these cells to inhibition by interferons at low oxygen tension in vitro, J Immunol 135 (1985) 2502–2506.

[88] E.M. Mazur, W.J. Richtsmeier, K. South, Alpha-interferon: Differential suppression of colony growth from human erythroid, myeloid, and megakaryocytic hematopoietic progenitor cells, J Interferon Res 6 (1986) 199–206.

[89] S.W. Mamus, M.M. Oken, E.D. Zanjani, Suppression of normal human erythropoiesis by human recombinant DNA-produced alpha-2-interferon in vitro, Exp Hematol 14 (1986) 1015–1022.

[90] Z.X. Lian, K. Kikuchi, G.X. Yang, A.A. Ansari, S. Ikehara, M.E. Gershwin, Expansion of bone marrow IFN-alpha-producing dendritic cells in New Zealand Black (NZB) mice: High level expression of TLR9 and secretion of IFN-alpha in NZB bone marrow, J Immunol 173 (2004) 5283–5289.

[91] Z.C. Han, J. Briere, J.F. Abgrall, G. Le Gall, D. Parent, L. Sencebe, Effects of recombinant human interferon alpha on human megakaryocyte and fibroblast colony formation, J Biol Regul Homeost Agents 1 (1987) 195–200.

[92] K. Kato, K. Kamezaki, K. Shimoda, A. Numata, T. Haro, K. Aoki, et al., Intracellular signal transduction of interferon on the suppression of haematopoietic progenitor cell growth, Br J Haematol 123 (2003) 528–535.

[93] M. Tiefenthaler, N. Bacher, H. Linert, O. Muhlmann, S. Hofer, N. Sepp, et al., Apoptosis of CD34$^+$ cells after incubation with sera of leukopenic patients with systemic lupus erythematosus, Lupus 12 (2003) 471–478.

[94] M.F. Denny, S. Thacker, H. Mehta, E.C. Somers, T. Dodick, F.J. Barrat, et al., Interferon-alpha promotes abnormal vasculogenesis in lupus: A potential pathway for premature atherosclerosis, Blood 110 (2007) 2907–2915.

[95] P.Y. Lee, Y. Li, H.B. Richards, F.S. Chan, H. Zhuang, S. Narain, et al., Type I interferon as a novel risk factor for endothelial progenitor cell depletion and endothelial dysfunction in systemic lupus erythematosus, Arthritis Rheum 56 (2007) 3759–3769.

[96] G. Weiss, L.T. Goodnough, Anemia of chronic disease, N Engl J Med 352 (2005) 1011–1023.

[97] H.R. Peeters, M. Jongen-Lavrencic, C.H. Bakker, G. Vreugdenhil, F.C. Breedveld, A.J. Swaak, Recombinant human erythropoietin improves health-related quality of life in patients with rheumatoid arthritis and anaemia of chronic disease; utility measures correlate strongly with disease activity measures, Rheumatol Int 18 (1999) 201–206.

[98] M. Voulgarelis, S.I. Kokori, J.P. Ioannidis, A.G. Tzioufas, D. Kyriaki, H.M. Moutsopoulos, Anaemia in systemic lupus erythematosus: Aetiological profile and the role of erythropoietin, Ann Rheum Dis 59 (2000) 217–222.

[99] G. Schett, U. Firbas, W. Fureder, H. Hiesberger, S. Winkler, D. Wachauer, et al., Decreased serum erythropoietin and its relation to anti-erythropoietin antibodies in anaemia of systemic lupus erythematosus, Rheumatology (Oxford) 40 (2001) 424–431.

[100] R.T. Means Jr., S.B. Krantz, Inhibition of human erythroid colony-forming units by interferons alpha and beta: Differing mechanisms despite shared receptor, Exp Hematol 24 (1996) 204–208.

[101] S. Bernatsky, J.F. Boivin, L. Joseph, R. Rajan, A. Zoma, S. Manzi, et al., An international cohort study of cancer in systemic lupus erythematosus, Arthritis Rheum 52 (2005) 1481–1490.

[102] S. Bernatsky, R. Ramsey-Goldman, A. Clarke, Malignancy and autoimmunity, Curr Opin Rheumatol 18 (2006) 129–134.

[103] M. Tokunaga, K. Fujii, K. Saito, S. Nakayamada, S. Tsujimura, M. Nawata, et al., Down-regulation of CD40 and CD80 on B cells in patients with life-threatening systemic lupus erythematosus after successful treatment with rituximab, Rheumatology (Oxford) 44 (2005) 176–182.

[104] R.P. Taylor, M.A. Lindorfer, Drug insight: The mechanism of action of rituximab in autoimmune disease—The immune complex decoy hypothesis, Nat Clin Pract Rheumatol 3 (2007) 86–95.

[105] B. Garvey, Rituximab in the treatment of autoimmune haematological disorders, Br J Haematol 141 (2008) 149–169.

[106] S.R. Michael, I.L. Vural, F.A. Bassen, L. Schaefer, The hematologic aspects of disseminated (systemic) lupus erythematosus, Blood 6 (1951) 1059–1072.

[107] D.R. Budman, A.D. Steinberg, Hematologic aspects of systemic lupus erythematosus. Current concepts, Ann Intern Med 86 (1977) 220–229.

[108] L.M. Vila, A.M. Mayor, A.H. Valentin, M. Garcia-Soberal, S. Vila, Clinical and immunological manifestations in 134 Puerto Rican patients with systemic lupus erythematosus, Lupus 8 (1999) 279–286.

[109] R. Segal, Y. Baumoehl, O. Elkayam, D. Levartovsky, I. Litinsky, D. Paran, et al., Anemia, serum vitamin B_{12}, and folic acid in patients with rheumatoid arthritis, psoriatic arthritis, and systemic lupus erythematosus, Rheumatol Int 24 (2004) 14–19.

[110] S.I. Kokori, J.P. Ioannidis, M. Voulgarelis, A.G. Tzioufas, H.M. Moutsopoulos, Autoimmune hemolytic anemia in patients with systemic lupus erythematosus, Am J Med 108 (2000) 198–204.

[111] M. Jeffries, F. Hamadeh, T. Aberle, S. Glenn, D.L. Kamen, J.A. Kelly, et al., Haemolytic anaemia in a multi-ethnic cohort of lupus patients: A clinical and serological perspective, Lupus 17 (2008) 739–743.

[112] S.M. Sultan, S. Begum, D.A. Isenberg, Prevalence, patterns of disease and outcome in patients with systemic lupus erythematosus who develop severe haematological problems, Rheumatology (Oxford) 42 (2003) 230–234.

[113] J.P. Atkinson, A.D. Schreiber, M.M. Frank, Effects of corticosteroids and splenectomy on the immune clearance and destruction of erythrocytes, J Clin Invest 52 (1973) 1509–1517.

[114] L.F. Fries, C.M. Brickman, M.M. Frank, Monocyte receptors for the Fc portion of IgG increase in number in autoimmune hemolytic anemia and other hemolytic states and are decreased by glucocorticoid therapy, J Immunol 131 (1983) 1240–1245.

[115] E. Gomard-Mennesson, M. Ruivard, M. Koenig, A. Woods, N. Magy, J. Ninet, et al., Treatment of isolated severe immune hemolytic anaemia associated with systemic lupus erythematosus: 26 cases, Lupus 15 (2006) 223–231.

[116] M. Tokunaga, K. Saito, K. Nakatsuka, S. Nakayamada, K. Nakano, S. Tsujimura, et al., Successful treatment of intravenous cyclophosphamide pulse therapy for systemic lupus erythematosus complicated with steroid-resistant hemolytic anemia, Nihon Rinsho Meneki Gakkai Kaishi 26 (2003) 304–309.

[117] S.W. Wang, T.T. Cheng, Systemic lupus erythematosus with refractory hemolytic anemia effectively treated with cyclosporin A: A case report, Lupus 14 (2005) 483–485.

[118] A.C. Chan, K. Sack, Danazol therapy in autoimmune hemolytic anemia associated with systemic lupus erythematosus, J Rheumatol 18 (1991) 280–282.

[119] T. Kotani, T. Takeuchi, Y. Kawasaki, S. Hirano, Y. Tabushi, M. Kagitani, et al., Successful treatment of cold agglutinin disease with anti-CD20 antibody (rituximab) in a patient with systemic lupus erythematosus, Lupus 15 (2006) 683–685.

[120] F. Musio, E.M. Bohen, C.M. Yuan, P.G. Welch, Review of thrombotic thrombocytopenic purpura in the setting of systemic lupus erythematosus, Semin Arthritis Rheum 28 (1998) 1–19.

[121] J.N. George, S.K. Vesely, J.A. James, Overlapping features of thrombotic thrombocytopenic purpura and systemic lupus erythematosus, South Med J 100 (2007) 512–514.

[122] P. Letchumanan, H.J. Ng, L.H. Lee, J. Thumboo, A comparison of thrombotic thrombocytopenic purpura in an inception cohort of patients with and without systemic lupus erythematosus, Rheumatology (Oxford) 48 (2009) 399–403.

[123] S.K. Kwok, J.H. Ju, C.S. Cho, H.Y. Kim, S.H. Park, Thrombotic thrombocytopenic purpura in systemic lupus erythematosus: Risk factors and clinical outcome: A single centre study, Lupus 18 (2009) 16–21.

[124] C. Porta, E. Bobbio-Pallavicini, R. Centurioni, R. Caporali, C.M. Montecucco, Thrombotic thrombocytopenic purpura in systemic lupus erythematosus. Italian Cooperative Group for TTP, J Rheumatol 20 (1993) 1625–1626.

[125] K. Hamasaki, T. Mimura, H. Kanda, K. Kubo, K. Setoguchi, T. Satoh, et al., Systemic lupus erythematosus and thrombotic thrombocytopenic purpura: A case report and literature review, Clin Rheumatol 22 (2003) 355–358.

[126] A. Aleem, S. Al-Sugair, Thrombotic thrombocytopenic purpura associated with systemic lupus erythematosus, Acta Haematol 115 (2006) 68–73.

[127] W.R. Bell, H.G. Braine, P.M. Ness, T.S. Kickler, Improved survival in thrombotic thrombocytopenic purpura—hemolytic uremic syndrome. Clinical experience in 108 patients, N Engl J Med 325 (1991) 398–403.

[128] A. Hundae, S. Peskoe, E. Grimsley, S. Patel, Rituximab therapy for refractory thrombotic thrombocytopenic purpura and autoimmune-mediated thrombocytopenia in systemic lupus erythematosus, South Med J 101 (2008) 943–944.

[129] N. Limal, P. Cacoub, D. Sene, I. Guichard, J.C. Piette, Rituximab for the treatment of thrombotic thrombocytopenic purpura in systemic lupus erythematosus, Lupus 17 (2008) 69–71.

[130] D.M. Keeling, D.A. Isenberg, Haematological manifestations of systemic lupus erythematosus, Blood Rev 7 (1993) 199–207.

[131] R. Pauzner, A. Greinacher, K. Selleng, K. Althaus, B. Shenkman, U. Seligsohn, False-positive tests for heparin-induced thrombocytopenia in patients with antiphospholipid syndrome and systemic lupus erythematosus, J Thromb Haemost 7 (2009) 1070–1074.

[132] E. Shantsila, G.Y. Lip, B.H. Chong, Heparin-induced thrombocytopenia. A contemporary clinical approach to diagnosis and management, Chest 135 (2009) 1651–1664.

[133] B. Godeau, D. Provan, J. Bussel, Immune thrombocytopenic purpura in adults, Curr Opin Hematol 14 (2007) 535–556.

[134] C. Lindholm, K. Borjesson-Asp, K. Zendjanchi, A.C. Sundqvist, A. Tarkowski, M. Bokarewa, Long-term clinical and immunological effects of anti-CD20 treatment in patients with refractory systemic lupus erythematosus, J Rheumatol 35 (2008) 826–833.

[135] W.W. Coon, Splenectomy for cytopenias associated with systemic lupus erythematosus, Am J Surg 155 (1988) 391–394.

[136] Y.N. You, A. Tefferi, D.M. Nagorney, Outcome of splenectomy for thrombocytopenia associated with systemic lupus erythematosus, Ann Surg 240 (2004) 286–292.

[137] S. Hall, J.L. McCormick Jr., P.R. Greipp, C.J. Michet Jr., C.H. McKenna, Splenectomy does not cure the thrombocytopenia of systemic lupus erythematosus, Ann Intern Med 102 (1985) 325–328.

[138] S.J. Rivero, E. Diaz-Jouanen, D. Alarcon-Segovia, Lymphopenia in systemic lupus erythematosus. Clinical, diagnostic, and prognostic significance, Arthritis Rheum 21 (1978) 295–305.

[139] L.M. Vila, G.S. Alarcon, G. McGwin Jr., H.M. Bastian, B.J. Fessler, J.D. Reveille, Systemic lupus erythematosus in a multiethnic U.S. cohort, XXXVII: Association of lymphopenia with clinical manifestations, serologic abnormalities, disease activity, and damage accrual, Arthritis Rheum 55 (2006) 799–806.

[140] M.M. Ward, S. Studenski, Age associated clinical manifestations of systemic lupus erythematosus: A multivariate regression analysis, J Rheumatol 17 (1990) 476–481.

[141] M.J. Mirzayan, R.E. Schmidt, T. Witte, Prognostic parameters for flare in systemic lupus erythematosus, Rheumatology (Oxford) 39 (2000) 1316–1319.

[142] W.L. Ng, C.M. Chu, A.K. Wu, V.C. Cheng, K.Y. Yuen, Lymphopenia at presentation is associated with increased risk of infections in patients with systemic lupus erythematosus, QJM 99 (2006) 37–47.

[143] J. Storek, Z. Zhao, E. Lin, T. Berger, P.A. McSweeney, R.A. Nash, et al., Recovery from and consequences of severe iatrogenic lymphopenia (induced to treat autoimmune diseases), Clin Immunol 113 (2004) 285–298.

[144] D. Martinez-Banos, J.C. Crispin, A. Lazo-Langner, J. Sanchez-Guerrero, Moderate and severe neutropenia in patients with systemic lupus erythematosus, Rheumatology (Oxford) 45 (2006) 994–998.

[145] H.H. Euler, P. Harten, R.A. Zeuner, U.M. Schwab, Recombinant human granulocyte colony stimulating factor in patients with systemic lupus erythematosus associated neutropenia and refractory infections, J Rheumatol 24 (1997) 2153–2157.

[146] G.P. Clunie, L. Lennard, Relevance of thiopurine methyltransferase status in rheumatology patients receiving azathioprine, Rheumatology (Oxford) 43 (2004) 13–18.

[147] M.A. Naughton, E. Battaglia, S. O'Brien, M.J. Walport, M. Botto, Identification of thiopurine methyltransferase (TPMT) polymorphisms cannot predict myelosuppression in systemic lupus erythematosus patients taking azathioprine, Rheumatology (Oxford) 38 (1999) 640–644.

[148] G.S. Habib, W.R. Saliba, P. Froom, Pure red cell aplasia and lupus, Semin Arthritis Rheum 31 (2002) 279–283.

[149] E. Kiss, I. Gal, E. Simkovics, A. Kiss, A. Banyai, S. Szakall, et al., Myelofibrosis in systemic lupus erythematosus, Leuk Lymphoma 39 (2000) 661–665.

[150] Y. Shapira, A. Weinberger, A.J. Wysenbeek, Lymphadenopathy in systemic lupus erythematosus. Prevalence and relation to disease manifestations, Clin Rheumatol 15 (1996) 335–338.

[151] M. Kojima, T. Motoori, S. Asano, S. Nakamura, Histological diversity of reactive and atypical proliferative lymph node lesions in systemic lupus erythematosus patients, Pathol Res Pract 203 (2007) 423–431.

[152] K. Ekstrom Smedby, C.M. Vajdic, M. Falster, E.A. Engels, O. Martinez-Maza, J. Turner, et al., Autoimmune disorders and risk of non-Hodgkin lymphoma subtypes: A pooled analysis within the InterLymph Consortium, Blood 111 (2008) 4029–4038.

Musculoskeletal System: Articular Disease, Bone Metabolism

Diane Horowitz, Galina Marder, Richard Furie

Although typically neither life-threatening nor even generally deserving of aggressive interventions, articular manifestations are nonetheless quite common and can be quite debilitating. Many of the musculoskeletal complications, such as myositis and arthritis, are a reflection of disease activity, whereas others, such as osteonecrosis and osteoporosis, are generally unfortunate side effects of treatment.

ARTHRITIS

Arthralgia and arthritis are extremely common manifestations, affecting the clear majority of patients with lupus. For most patients, the symptoms are limited to pain and swelling because the arthritis is typically nonerosive. However, for a restricted number of patients, there can be significant destruction and functional loss. Fernandez *et al.* [1] provide a very comprehensive historical review of lupus arthropathy. It has become apparent that there are different patterns of arthritis, and these are outlined here.

Nonerosive Arthritis

Generally insidious in onset and usually an early disease manifestation, nonerosive arthritis is the most common type of arthritis that occurs in patients with lupus. It is generally symmetric and has a predilection for the hands, wrists, and knees. Morning stiffness, a striking symptom in rheumatoid arthritis (RA), may be a feature of lupus. Joint erythema, tenderness, warmth, swelling, or effusion may be present on physical examination. Except in cases of Jaccoud's arthropathy (described later), alignment of the metacarpals and phalanges is normal, and deformities are notably

absent. Table 49.1 provides a summary of clinical, laboratory, and imaging findings in lupus arthritis.

Although most patients with lupus arthritis do not develop large synovial effusions, occasionally joint fluid is obtained for analysis by arthrocentesis. The fluid generally ranges in appearance from clear to mildly opaque. Viscosity may be normal or slightly abnormal. The white blood cell count of lupus synovial fluid is generally low, ranging from the hundreds to thousands of cells per milliliter [2]. Occasionally, the white blood count may exceed 10,000 cells/ml. Differentials performed on synovial fluid have demonstrated a predominance of mononuclear cells, although neutrophils have been found in abundance in those fluids with marked elevations of white cells. Red cells are scant in lupus synovial fluid. Although investigators have demonstrated lupus erythematosus cells, antinuclear antibodies, and low complement concentrations in synovial fluid, these tests are of no diagnostic or therapeutic value. Radiographs may demonstrate soft tissue swelling, but erosive changes and joint space narrowing are typically absent [3]. With more sensitive imaging techniques, such as high-resolution ultrasound with power Doppler or magnetic resonance imaging (MRI), one is more apt to detect inflammatory changes in some patients [4].

In contradistinction to rheumatoid arthritis, synovial specimens from patients with lupus do not demonstrate exuberant inflammatory changes. In a study of 30 percutaneous synovial knee biopsies taken from lupus patients, Natour *et al.* [5] noted the following in these tissues: (1) synoviocyte hyperplasia, (2) scarce inflammatory infiltrates, (3) vascular proliferation, (4) edema and congestion, (5) fibrinoid necrosis and intimal fibrous hyperplasia of blood vessels, and (6) fibrin on the synovial surface. The mechanism by which lupus patients with arthritis are generally spared from erosive changes

TABLE 49.1   Clinical, Laboratory, and Imaging Characteristics of Lupus Arthritis

Manifestation	Systemic lupus erythematosus
SYMPTOMS	
Joint pain	Present
Joint swelling	Mild to moderate
Morning stiffness	Mild to moderate
SIGNS	
Distribution	More restricted on average (hands, wrists, knees)
Pattern	Symmetric
Joint swelling	Mild to moderate
Deformity	
Laxity	Common
Reducible deformity	Common
Fixed deformity	Rare
Rheumatoid nodules	Rare
LABORATORY VALUES	
ESR; CRP	Often elevated
RF	Positive in 20%
Anti-SSA/SSB	May be positive
Anti-CCP	Positive in 15%
ANA	Positive
Anti-DNA	Positive
Anti-Sm/RNP	Positive
Synovial fluid	
White count	Mild elevation
Differential	Generally mononuclear cell predominance
IMAGING TESTS	
Radiographs	
Juxta-articular osteopenia	Uncommon
Malalignment	Rare (except in Jaccoud's)
Joint space narrowing	Rare
Marginal erosions	Absent
MRI	
Joint space narrowing	Rare
Erosions	Rare
Malalignment	Rare (except in Jaccoud's)
Synovial proliferation	Rare

was the subject of investigation by Mensah *et al.* [6]. They postulated that an interferon-rich milieu, as is seen in many lupus patients, supports the differentiation of myelomonocytic precursors to myeloid dendritic cells and away from osteoclasts. Inhibition of osteoclastogenesis results in diminished bone erosion.

There are no specific clinical features of lupus arthritis that distinguish it from other types of inflammatory arthritis early in the course of disease. The diagnosis of lupus arthritis is thus dependent on the presence of other clinical manifestations as well as serologic evidence of disease.

Erosive Arthritis ('Rhupus')

A syndrome marked by the articular features of RA, namely erosive arthritis, and clinical as well as serologic features of lupus has been referred to as "rhupus" [7, 8]. Physical findings in patients with rhupus may include synovitis, synovial proliferation, rheumatoid nodules, and malalignment. Autoantibodies present in this subset of patients may include rheumatoid factor and anticyclic citrullinated antibodies (CCPs) [9]. Arthritis of this type clearly has destructive potential; therefore, a more aggressive treatment approach may be warranted.

Jaccoud's Arthropathy

Sigismond Jaccoud, a Swiss physician who lived in Paris in the 1800s, described an arthritis associated with rheumatic fever. In the modern era, Jaccoud's arthropathy is a term that is used to describe an arthropathy characterized by the presence of reducible deformities and the absence of erosions [10]. Jaccoud's arthropathy has been observed in other joints of the body, such as the feet and knees. Although typically associated with lupus, it is seen in other conditions.

Bleifeld and Inglis [11] published their observations of the hands of 50 patients with lupus. Abnormalities included laxity (50%) and reducible deformities (38%), but radiographs of patients in this series failed to demonstrate erosive changes. Figure 49.1 demonstrates the typical reducible hand deformities of Jaccoud's arthropathy, and the patient's radiographs are shown in Figure 49.2. To the trainee, Jaccoud's arthropathy may be mistaken for RA. Features that distinguish Jaccoud's arthropathy from RA include reducibility of the deformities and the lack of articular damage on X-ray. Although hand function is preserved early on, patients with advanced Jaccoud's may develop compromised hand function as a result of contractures or severe ulnar deviation. The pathogenesis of joint laxity in Jaccoud's arthropathy is not understood. Although rarely obtained, biopsy specimens show normal synovium.

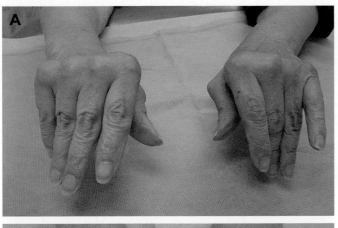

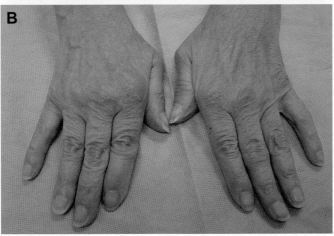

FIGURE 49.1 Female patient with lupus and Jaccoud's arthropathy. *A*, Flexion deformities, ulnar deviation, and mild swan neck deformities. *B*, The deformities are reduced with minimal pressure against the underlying table.

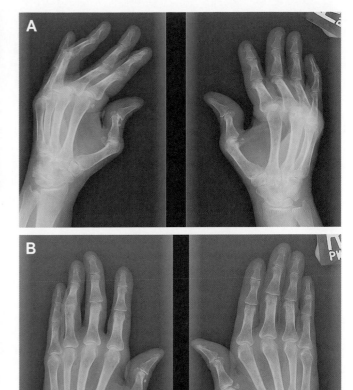

FIGURE 49.2 Radiographs of the patient in Figure 49.1. *A*, Flexion deformities, ulnar deviation, and swan neck deformities but no erosive changes. *B*, The deformities are reduced with minimal pressure against the X-ray table.

Treatment

Treatment regimens need to be customized according to the patient's disease activity. Nonsteroidal anti-inflammatory drugs and low doses of corticosteroids are often quite effective in managing the arthritic symptoms. For those refractory to mild interventions, hydroxychloroquine or immunosuppressives may be warranted. Wong and Esdaile [12] reviewed the experience with methotrexate in controlled and uncontrolled studies. Although benefit was demonstrated in some studies, the effects were not always consistent. Fortin *et al.* [13] published the experience of the Canadian Network for Improved Outcomes in Systemic Lupus Erythematosus. In this randomized, double-blind, placebo-controlled study of 86 patients, the steroid-sparing effects of methotrexate were demonstrated. Although the study did not focus solely on articular manifestations, more than 90% of participants had musculoskeletal involvement at baseline. Azathioprine, approved for the treatment of rheumatoid arthritis,

represents another treatment option. A multicenter trial comparing the efficacy of mycophenolate to cyclophosphamide in patients with lupus nephritis yielded data on extrarenal responses to these interventions [14]. Of those with BILAG A or B musculoskeletal domain scores at baseline, more than 85% of the patients in both arms had improvements in their domain scores at 24 weeks. In a study of 27 patients with lupus arthritis, mycophenolate, administered to a target dose of 3 g/day, outperformed placebo in the achievement of a musculoskeletal domain BILAG C response [15]. Although most rheumatologists do not employ the off-label use of tumor necrosis factor (TNF) inhibitors in lupus because of fear of disease exacerbations, anecdotal experience suggests that some patients with lupus arthritis benefit from their use. Biologics in development, such as belimumab and tocilizumab, have been reported to favorably affect arthritis [16, 17].

Splints will correct the maligned digits in a patient with Jaccoud's arthropathy. However, it is the authors' experience that the digits will deform once the splint is removed. Similarly, pharmacologic interventions, although capable of reducing pain and swelling in the more typical subset of lupus arthritis, do not affect the laxity and deformity associated with the Jaccoud's subset. Occasionally, surgical intervention of the hand is required. Reported results have been variable [18].

TENDON RUPTURE

Spontaneous tendon rupture, albeit rare, has been reported in lupus [19]. The most commonly affected tendons have been the patellar and Achilles tendons. Ruptures may be acute and are sometime bilateral. Predisposing factors include trauma and steroids. Whether lupus patients more commonly develop quinolone-induced tendon rupture than the general population is unknown. Histology, reported in a limited number of cases, has shown variable degrees of inflammatory changes, ranging from no inflammation to exuberant synovial proliferation.

MYOSITIS

Clinical Features

Clinical evidence of skeletal muscle involvement in lupus varies from diffuse muscle tenderness and myalgia to frank myositis. Muscle tenderness and generalized myalgia are relatively common in lupus and occur in nearly half of patients [20, 21]. However, myositis is rare, being reported in 3–16% of lupus patients by different authors [22–24]. Inflammatory myopathy in lupus has features identical to idiopathic inflammatory myositis (IIM). It is characterized by: (1) proximal muscle weakness with or without myalgia, (2) elevated creatine phosphokinase (CPK) and/or aldolase, (3) electromyographic abnormalities, and (4) characteristic histologic changes on biopsy.

Although the development of a second distinct autoimmune disease, polymyositis, in patients with lupus is well-established, the prevalence of patients with a polymyositis/systemic lupus erythematosus (SLE) overlap is quite low compared with scleroderma or RA overlaps with lupus. This is an interesting phenomenon considering similarities in pathogenic mechanisms (e.g., interferon signature) between SLE and inflammatory myositis [25, 26].

Garton and Isenberg [21] compared the clinical and laboratory features of lupus myositis to primary IIM. They found that a relapsing and remitting course was the most common pattern of disease in both groups;

monophasic disease was the least common pattern. No significant differences were found in muscle strength and CPK levels between both groups of patients. The frequencies of ENA, Ro, and La antibodies were almost equally distributed in the lupus myositis group, whereas they were not present in the IIM group. According to a follow-up study from the same center, lupus patients with myositis were more likely to have alopecia, oral ulcers, erosive joint disease, or pulmonary involvement but less likely to have Sjögren's syndrome or nephritis [27]. Interestingly, dermatomyositis was seen in the majority of patients in a subsequent lupus cohort, and a high association with RNP antibodies was again noted [27].

Earlier studies suggested that myositis of lupus follows a much milder course than primary IIM. Foote and others [23] reviewed the records of patients with documented lupus who developed an inflammatory myopathy. Excluding RNP-positive patients, they reported mortality rates that were significantly lower than those reported for polymyositis complicating scleroderma or RA (18 vs. 47%, respectively). However, in a subsequent study, Garton and Isenberg [21] retrospectively evaluated the clinical courses during a mean follow-up period of 7.4 years and found that patients with lupus myositis had equally severe courses. The overall mortality of 10% (1/11) was directly attributable to myositis and similar to the mortality rate (2/19) of patients with primary IIM.

Moreover, the overall prognosis of lupus patients with myositis depends on other manifestations of the disease. Although according to Dayal and Isenberg [27], mortality in lupus patients with myositis was not statistically different from that of lupus patients without myositis, the mean age of lupus patients with myositis at the time of their deaths was significantly less (24.7 vs. 51 years).

Histologic Features

A wide spectrum of histological findings are reported from multiple biopsy and autopsy series of lupus myositis. These include: (1) inflammatory myopathy with lymphocytic and plasma cell infiltrates found in perivascular, perimysial, and endomysial locations; (2) vasculitis with mononuclear inflammation of vessel walls and rare vessel wall necrosis; (3) perifasicular atrophy; and (4) vacuolar myopathy in the absence of corticosteroid or chloroquine exposure [28]. Oxenhandler and others [28] demonstrated immunoglobulin and complement deposition in the skeletal muscle of patients with SLE, underscoring similar mechanisms of injury in lupus myositis as are described for cutaneous and renal involvement in lupus. Moreover, histologic findings characteristic of inflammatory myopathy were found even in the clinically uninvolved muscles of lupus patients [28].

Lim and others [29] documented a significantly higher frequency (87 vs. 58%) of abnormalities in muscle biopsy specimens in patients with lupus compared with a control group that included patients with myalgia and arthralgia who were clinically thought to have fibromyalgia. They noted a statistically significant higher frequency of type II fiber atrophy and lymphocytic vasculitis. Although only small numbers of patients in this series had inflammatory myositis on histological evaluation, an elevated serum CK correlated with proximal weakness and with myositis on muscle biopsy. An increased incidence of RNP antibodies was again found in those lupus patients with myositis [29].

Using electron microscopy, Pallis and others [30] measured the thickness of unilaminar capillary basement membranes and quantified basement membrane laminae and pericyte layers in the skeletal muscle biopsies of 31 patients with lupus, including 14 with mixed connective tissue disease, and 11 controls. Capillary basement membrane thickness was significantly increased in SLE compared with controls ($p < 0.01$) and was significantly associated with the level of C3d-g ($p < 0.01$). Further histopathological evidence of a microvascular pathogenesis was provided by Bronner and others [31]. They demonstrated endothelial microtubular inclusions on muscle biopsies of different types of myositis patients, including lupus-related myositis.

Treatment

Corticosteroids are the mainstay of therapy for lupus myositis. They are used at doses that vary according to the severity of the myositis. The dose may range from 0.5—1.5 mg/kg to intravenous "pulse" methylprednisolone. In Tymm and Webb's [32] analysis of 105 cases, 23% of patients improved and were maintained in remission with only low-dose corticosteroids; outcomes were no better in the group that received high-dose steroids. Initiation of therapy within 4 months of the onset of symptoms was associated with a favorable outcome, whereas a poor response after 6 months of therapy was associated with a poor long-term outcome. There is no consensus regarding an immunosuppressive regimen for induction or maintenance. However, according to most observations, the majority of patients will not require cytotoxic agent [21, 26, 32]. As when designing treatment plans for other manifestations of lupus, customization of the treatment regimen for myositis should take into account other factors, such as concomitant manifestations and drug toxicities.

OSTEONECROSIS

Introduction

Simply stated, osteonecrosis is the death of bone marrow and trabecular bone resulting in compromise to the structural integrity of the bone architecture. Although there is a myriad of etiologies and varied clinical presentations, cell death in osteonecrosis due to the interruption of local blood supply to periarticular bone is the common pathogenetic mechanism. Osteonecrosis is synonymous with ischemic necrosis of the bone, avascular necrosis, osteochondritis dissecans, and aseptic necrosis of bone [33].

Osteonecrosis can be nontraumatic or traumatic. As seen in Figure 49.3, the nontraumatic subset can be associated with autoimmune, hematologic, vascular, or orthopedic diseases as well as iatrogenic conditions and environmental factors. Other than SLE, rheumatic conditions associated with osteonecrosis include polymyositis, polymyalgia rheumatica, rheumatoid arthritis, ankylosing spondylitis, Sjögren's syndrome, temporal arteritis, Ehler—Danlos syndrome, and primary antiphospholipid syndrome. Less common associations include hyperparathyroidism, Cushing's disease, chronic renal failure, Fabry's disease, human immunodeficiency virus, meningococcemia, Caisson disease, alcoholism, radiation exposure, organ transplantation, Legg—Calve—Perthes disease, and slipped capital femoral epiphysis [33].

Clinical Impact

Silent osteonecrosis refers to asymptomatic disease that is diagnosed based on imaging findings. Asymptomatic osteonecrosis is usually associated with less

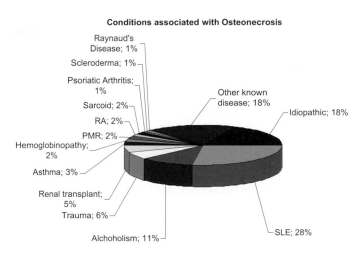

Conditions associated with Osteonecrosis

FIGURE 49.3 Conditions associated with osteonecrosis. Analysis of 169 patients with avascular necrosis. *Source: Adapted with permission of Wiley Interscience from Zizic et al. [51].*

severe changes on imaging studies than symptomatic osteonecrosis. Nevertheless, patients with asymptomatic disease have the potential to progress to symptomatic disease. A 1-year study of 8 patients with asymptomatic osteonecrosis of the hip demonstrated stable lesions that did not progress during the observed year. Furthermore, the patients did not develop symptoms during the study period [34]. A 3-year prospective study of 23 SLE patients free of hip pain but on glucocorticoid therapy for active SLE found contrasting results. In this study, baseline testing revealed osteonecrosis by MRI in 35% of subjects and by bone scan in 26% of subjects. At the conclusion of the study, 2 of the 8 patients with MRI changes at the onset had worsening of osteonecrosis seen on MRI. One of these patients developed hip pain and radiographic changes. The remaining 6 patients had no changes in their MRI findings. A patient with bone scan abnormalities at the onset of the study developed MRI and radiographic changes as well as hip pain by the end of the study [35]. The disparity between these studies can be explained by the small study group size and the difference in duration of follow-up. Asymptomatic and symptomatic osteonecrosis can be seen as a spectrum of disease ranging from the clinically unaffected patient with mild changes on MRI to the severely disabled patient with advanced cortical destruction seen on radiograph. In this spectrum, some patients will progress, whereas other patients' disease activity arrests. It is unclear why some patients have a more fulminant course, but more severe disease may be linked to prolonged exposure to risk factors and/or a genetic predisposition.

Symptomatic osteonecrosis is a source of disability and negatively affects quality of life. In SLE patients, especially those who develop osteonecrosis at a relatively young age, the disability associated with osteonecrosis significantly impacts quality of life. A case–control study of 95 SLE patients concluded that the patients with osteonecrosis had higher HAQ scores and lower scores on the physical functioning domain of the SF-20. In this group, the disability was not attributed to pain alone because the pain domain of the SF-20 and the HAQ pain scores were similar in the osteonecrosis and nonosteonecrosis groups [36].

Epidemiology

Although there is a disparity in the reported incidence of osteonecrosis in patients with SLE, these patients have an increased risk of osteonecrosis compared to the normal population. Among patients with SLE, symptomatic osteonecrosis has a reported prevalence of 5–15%, whereas the prevalence of asymptomatic osteonecrosis has been reported to be more than 50% (Table 49.2) [33, 36–52]. A few years after corticosteroids were introduced into clinical practice, Dubois and Cozen [40] first reported an association between SLE and osteonecrosis. In 1960, they reported that 11 of 400 SLE patients developed symptomatic osteonecrosis. In the Lupus in Minorities: Nature v. Nurture (LUMINA) cohort of 571 patients with systemic lupus nephritis, 33 (5.8%) subjects developed symptomatic osteonecrosis [39]. A long-term study demonstrated that the duration of SLE affects the prevalence of osteonecrosis. In this study, 1241 SLE patients divided into four entry groups (1970–1978, 1979–1987, 1988–1996, and 1997–2005) were followed for consecutive 9-year periods. The patients in the earliest entry group had a 12.1% prevalence of osteonecrosis in the first period (1970–1978), 17.6% in the second period (1979–1987), 27.5% in the third period (1988–1996), and 25% in the fourth period (1997–2005). The second entry group also had an increase in the prevalence of osteonecrosis as time progressed, with a prevalence of 4.9% in the second period, 13.9% in the third period, and 20.5% in the fourth period [49]. The prevalence of asymptomatic osteonecrosis has also been shown to vary by the duration of disease. In a study of 66 patients with SLE on corticosteroids and no history of hip pain who were evaluated by MRI, 11 hips in 8 patients (12%) were found to have osteonecrosis on MRI [34].

Osteonecrosis is also a source of morbidity in patients with juvenile SLE. The prevalence of osteonecrosis in juvenile SLE ranges from 5.4 to 22.7% in symptomatic patients and up to 50% when including asymptomatic patients [53–55].

The long bones, such as the femur and humerus, are the most susceptible to osteonecrosis [56]. In a cohort of 95 patients with osteonecrosis followed over 8.5 years, there were 217 sites of osteonecrosis. Whereas 29.5% of patients had involvement at a single site, 45.3% had involvement at two sites, and 25.2% of patients had osteonecrosis at three or more sites. In this group, the hip was the most commonly involved joint and was most likely to have bilateral involvement. Overall, sites of involvement included hip (72%), knee (36%), shoulder (15%), ankle (11%), elbow (3%), and wrist (3%) [36]. In a study of 38 SLE patients with osteonecrosis, 95% of patients had osteonecrosis of the hip, and 72% of patients had bilateral hip osteonecrosis. Furthermore, 13% had involvement in the femoral condyle of the knee, 3% had involvement of the humerus, and 3% had carpal bone involvement [44]. Case reports have described osteonecrosis in the talus, vertebral body, and small bones of the feet [57, 58]. Although these data are derived from relatively small groups of patients, they demonstrate that multijoint osteonecrosis, especially of the hips, is common in adults with SLE.

TABLE 49.2   Selected Studies of Prevalence of Osteonecrosis

Authors	Publication year	Study years	No. of patients	Prevalence (%)
Abeles *et al.* [37]	1978	1957–1968	365	4.7
Zizic *et al.* [52]	1985	1978–1981	54	51.9
Petri [48]	1995	1989–1995	407	14.5
Mok *et al.* [44]	1998	1971–1997	320	12.0
Zonana-Nacach *et al.* [62]	2000	1987–1998	539	8.7
Gladman *et al.* [36]	2001	1970–1995	744	12.8
Nagasawa *et al.* [46]	2005	1994–1997	45	44.4
		1994–1997	45	11.1
		1994–1997	45	33.3
Calvo-Alen *et al.* [39]	2006	1995–1998	571	0.1
Prasad *et al.* [142]	2007	1970–2004	570	11.4
Fialho *et al.* [63]	2007	2004–2005	46	21.7
Urowitz *et al.* [49]	2008	1970–2005	1241	—
		Cohort 1: 1970–1978	228	12.1
		Cohort 1: 1970–1987	228	17.6
		Cohort 1: 1970–1996	228	27.5
		Cohort 1: 1970–2005	228	25
		Cohort 2: 1979–1987	364	4.9
		Cohort 2: 1979–1996	364	13.9
		Cohort 2: 1979–2005	364	20.5
		Cohort 3: 1988–1996	260	10.4
		Cohort 3: 1988–2005	260	15.7
		Cohort 4: 1997–2005	389	9.4

Pathogenesis

Osteonecrosis results from ischemia, but the pathogenesis of the ischemia is unclear, especially in patients with SLE. Commonly proposed mechanisms for the ischemia include intravascular occlusion, diminished blood flow from increased extravascular pressure, or inhibition of angiogenesis [56, 59, 60]. An emerging school of thought has proposed that osteonecrosis may be secondary to primary osteocyte death, but this has not been studied in SLE patients or a comparable animal model [33]. An elegant mechanism that incorporates some of the aforementioned hypotheses was proposed in 2006. Systemic inflammation with a resultant increase in the amount of circulating oxidized low-density lipoprotein induces the production of adipocytes and suppresses the production of osteoblasts. The increased adipocyte mass increases the intracortical pressure. The inflammatory response also increases levels of TNF, which shifts the balance in favor of osteoclast formation over osteoblast formation. The increased TNF levels also raise levels of homocysteine, a known prothombotic amino acid. Furthermore, the pro-inflammatory cytokines can reduce angiogenic factors leading to osteocyte necrosis [56]. In a rat model of osteonecrosis in which rats injected with lipopolysaccharide and methylprednisolone develop osteonecrosis of the femoral head, there was an increase in toll-like receptor (TLR) 4-inducible cytokines, such as interleukin (IL)-1β, IL-2, IL-4, IL-6, IL-10, granulocyte macrophage colony-stimulating factor, interferon-γ, and TNF-α, within the first week [61]. It is unclear which mechanism, if any, is responsible for the development of osteonecrosis, but analysis of risk factors may elucidate the pathogenesis of osteonecrosis (Figure 49.4).

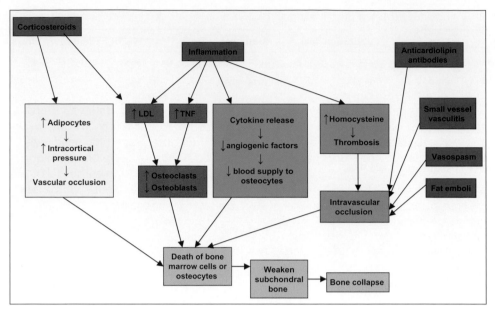

FIGURE 49.4　Proposed pathogenic mechanisms of osteonecrosis in SLE. *Source: Adapted with permission of Nature Publishing Group from Lane [56].*

Risk Factors

Given the increased prevalence of osteonecrosis in patients with SLE, investigation has focused on this group of patients to identify specific risk factors. Although corticosteroids have been strongly correlated with osteonecrosis, factors such as disease activity, use of cytotoxic agents, Raynaud's phenomenon, vasculitis, and antiphospholipid antibodies have also been implicated as potential risk factors for osteonecrosis. Increasingly, studies have shown that the pattern of steroid use is an important factor in the development of osteonecrosis. For example, Mok *et al.* [44] observed that the dose of steroid used early in the course of treatment increased the risk of osteonecrosis. Nagasawa *et al.* [46] showed that the use of pulse steroids was also associated with an increased risk of osteonecrosis. A study of 539 patients in the Hopkins lupus cohort revealed that each 2-month exposure to high-dose prednisone conferred a 1.2-fold increase in the risk of developing avascular necrosis [62]. Increasing evidence links the temporal proximity of corticosteroid use to the development of osteonecrosis. Fialho *et al.* [63] demonstrated that the cumulative steroid dose in the year preceding osteonecrosis, but not in the 13–24 months preceding diagnosis, was associated with osteonecrosis. In a study of 60 newly diagnosed patients with SLE who required high-dose steroids (prednisolone 40 mg/day or more), 89% of the patients who developed silent osteonecrosis had evidence of osteonecrosis within 3 months of initiation of treatment [64]. Furthermore, a study of 72 newly diagnosed SLE patients treated with high-dose corticosteroids who underwent MRI of the knees and hips demonstrated osteonecrosis in 32/72 (44%) within the first 5 months of the initiation of high-dose corticosteroids. In this study, exposure to pulse corticosteroids did not correlate with the development of osteonecrosis [47]. Clearly, there is no explanation for the differential susceptibility to corticosteroid-induced osteonecrosis, but it is well accepted that high-dose corticosteroids do pose a substantial risk.

Cytotoxic agents have also been shown to be an independent risk factor for the development of osteonecrosis in SLE patients. Further evidence supporting the causative role of cytotoxic agents is the increased risk of osteonecrosis in cancer patients receiving chemotherapy regimens consisting of high doses of cytotoxic agents with corticosteroids [39, 65]. The mechanism by which cytotoxic agents cause osteonecrosis is unclear.

Disease activity independent of corticosteroid use has been implicated by some investigators. Although mechanisms are not understood, occlusive vascular disease mediated by vasospasm or thrombosis may be operable. A prospective study of 28 SLE patients with osteonecrosis demonstrated a signal (albeit not statistically significant) that patients with Raynaud's phenomenon had higher rates of osteonecrosis. Although steroid use was associated with osteonecrosis in this study, of the patients with ischemic necrosis of bone in this cohort, those with Raynaud's phenomenon received less prednisone than those without Raynaud's [52]. Although anticardiolipin antibodies were only significantly correlated in one study, additional evidence to support the role of anticardiolipin antibodies in osteonecrosis derives from the reports of osteonecrosis in the primary antiphospholipid syndrome [66, 67]. In a cohort of 30 subjects with primary antiphospholipid syndrome, 20% of the patients were found to have asymptomatic osteonecrosis on a screening MRI of the femoral head. Three subjects had intermediate bilateral osteonecrosis, 2 subjects had early unilateral osteonecrosis, and 1 subject had bilateral early osteonecrosis [66].

Although steroid use is strongly associated with osteonecrosis, there are many patients who are treated with steroids who do not develop osteonecrosis. As with many conditions, it is likely that there are genetic factors that contribute to susceptibility to corticosteroid-induced osteonecrosis. Polymorphisms in MDR1 (ABCB1) have been shown to be associated with steroid-induced osteonecrosis in patients with SLE. In one study, 127 SLE patients who were initially treated with 40 mg/day or more of prednisolone were followed with MRI for 5 years after steroid treatment. In this cohort, 21 patients developed osteonecrosis. Patients who had the MDR1 3435 TT and MDR1 2677 TT had a lower rate of osteonecrosis (adjusted odds ratio of 0.14 with 95% confidence interval (CI) of 0.017–1.153 and adjusted odds ratio of 0.21 with 95% CI of 0.018–1.301, respectively) [68]. MDR1 encodes P-glycoprotein, a transport protein involved in the metabolism of steroids and many other therapeutic agents. The decreased rate of osteonecrosis in patients carrying these single nucleotide polymorphisms was also seen in Asian patients from a renal transplant center in Japan who had received corticosteroids [69]. Further research needs to be done to confirm this association, evaluate other ethnic groups for this association, and delineate the pathway by which P-glycoprotein affects drug metabolism and its role in the development of osteonecrosis.

Diagnosis

Typical clinical features of osteonecrosis include the insidious onset of deep throbbing pain localized to the bone with radiation to adjacent joints. Pain is exacerbated with weight bearing. Imaging modalities used for osteonecrosis include computed tomography (CT), MRI, radionucleotide bone scan, and radiographs (Table 49.3). Radiographs are the least sensitive modality, and MRI is the most sensitive [35]. Although MRI is the gold standard for staging, conventional radiographs can be used for initial screening. In most cases, MRI will be required for follow-up either to stage disease or to search for smaller lesions not visible on radiograph. Due to the high prevalence of osteonecrosis in the hips, it is recommended that all SLE patients with osteonecrosis at sites other than the hip undergo imaging of both hips [60] (Figures 49.5–49.8).

TABLE 49.3 Radiologic Findings in Osteonecrosis

CT	Reactive sclerosis and subchondral collapse
MRI	Decreased signal in a segmental pattern
Radionucleotide bone scan	Decreased early uptake and increased late uptake
Radiograph	Early: Ill-defined mottling in trabecular pattern Late: Medullary space involvement with dense serpiginous calcification and subchondral fracture

Source: Assouline-Dayan et al. [33] and Jones and Hungerford [59].

Four staging systems for osteonecrosis of the femoral head are commonly used: the Steinberg system, the Marcus system, the Ficat and Arlet system, and the ARCO (International Classification of Osteonecrosis of the Femoral Head). The main features highlighted in these systems are the degree of involvement and severity of bone destruction (Table 49.4). Lesions are considered small if they encompass less than 15% of the femoral head, medium if they encompass 15–30% of the femoral head, and large if they encompass greater than 30% of the femoral head [33, 60].

Treatment

Treatment of osteonecrosis can be divided into two categories: conservative and surgical. Data supporting the effectiveness of conservative therapeutic approaches

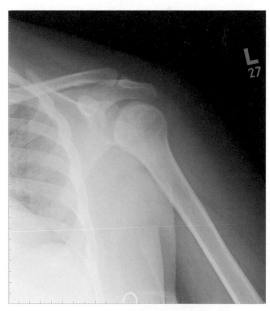

FIGURE 49.5 Shoulder Radiograph: Internally rotated radiograph of the left shoulder demonstrates a crescent-shaped area of subchondral lucency with early subarticular collapse ("crescent sign") and surrounding serpentine-shaped sclerosis consistent with avascular necrosis of the left humeral head.

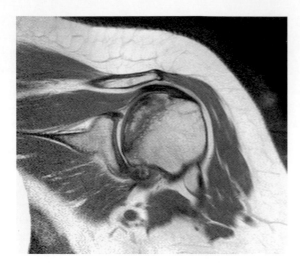

FIGURE 49.6 Shoulder MRI: Oblique coronal proton density weighted MR image of the left shoulder demonstrates a serpentine area of low signal within the subarticular humeral head with some maintained normal marrow fat signal centrally consistent with avascular necrosis of the humeral head.

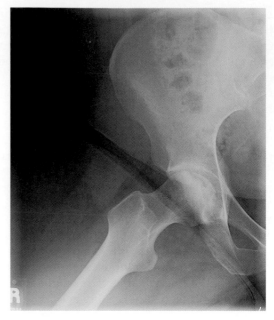

FIGURE 49.8   Digital Radiograph of hip: Lateral view/frog leg: Patchy areas of lucency and subchondral fracture. Ficat stage III changes in the hip.

are quite limited. Therefore, conservative treatments are best for precollapse sites with small lesions. Such approaches include modification of activity with restricted weight bearing, stimulation of repair with ultrasound or electrical signals, and possible pharmacologic treatment. Examples of modification of activity include limiting overhead reaching in patients with shoulder disease or orthotic use in patients with disease of the talus [38, 56, 60]. Ultrasound and electric signals have been shown to stimulate osteogenesis and neovascularization, although there are very limited data for osteonecrosis [60]. The role of pharmaceutical treatments in osteonecrosis is also unclear. Because of the expansion of interosseous adipocytes with glucocorticoid treatment, statin therapy was proposed for the prevention of osteonecrosis. In a cohort of 43 patients with SLE undergoing treatment with high-dose glucocorticoids, atorvastatin failed to prevent the development of osteonecrosis [70]. In a study of 60 newly diagnosed SLE patients who required therapy with high-dose glucocorticoids (40 mg/day or more of prednisolone) and in which the intervention was warfarin (1—5 mg/day with goal INR of 1.5—2), there was a higher rate of symptomatic osteonecrosis in the control group.

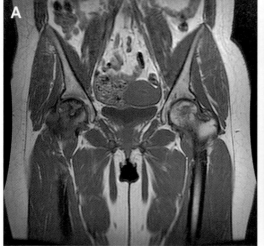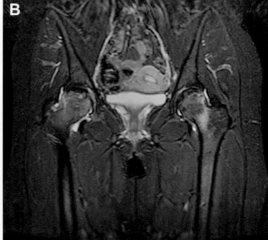

FIGURE 49.7   HIP MRI: A, Coronal T1-weighted image of the pelvis demonstrates serpentine-shaped low signal abnormalities within the bilateral femoral heads (right greater than left). Normal fatty marrow signal is seen centrally within this region. B, On the corresponding coronal fat-suppressed T2-weighted image, bright signals are seen in the remainder of the head and neck regions bilaterally consistent with bone marrow edema (right greater than left). There is articular collapse noted of the right femoral head.

TABLE 49.4   The Ficat and Arlat System for Osteonecrosis of Femoral Head

Stage	Clinical	Radiograph	MRI
0	Asymptomatic	Normal	Normal or edema pattern
1	Pain	Normal	Same as stage 0
2	Pain	Cysts and/or sclerosis	Geographic zone of demarcation
3	Pain	Femoral head collapse Crescent sign and/or step-off in contour of subchondral bone	Same as stage 2 + cortical collapse
4	Pain	Flattening of the femoral head Progressive joint space loss Osteophytes Subchondral cyst	Stage III + joint space narrowing

Source: Adapted with permission of Lippincott Williams and Wilkins from Berquist [143].

During the 5 years of follow-up in this study, 14% of the control subjects and 4.8% of the intervention group had avascular necrosis of the hip. Although there was a statistically significant difference for symptomatic osteonecrosis, there was not a statistically significant difference in rates of asymptomatic osteonecrosis [64]. An open-label trial of alendronate (10 mg/day or 70 mg week), calcium (500—1000 mg/day), and vitamin D supplementation in 60 patients with known osteonecrosis demonstrated a decrease in visual analog scores for pain and disability and an increase in standing and walking time in the first 2 years of treatment. Furthermore, there was a decrease in bone marrow edema seen on MRI; radiographs remained stable or progressed one grade only. The lack of a comparator group makes it impossible to draw conclusions about the value of the pharmacologic interventions used in this study [71].

Surgical treatment of osteonecrosis is reserved for patients with uncontrollable pain or immobility. Patients who fit these criteria usually have advanced changes on imaging studies. Small or medium-sized lesions can be treated with joint-preserving procedures such as support of subchondral bone, vascularized and nonvascularized bone grafting, cementation, implantation of trabecular metal rods, core decompression and osteotomy. The goal of osteotomy is to move the osteonecrotic lesion away from the weight-bearing area of the joint. Osteotomy has fallen out of favor because of its relatively low success rates and significant complication rates. Large lesions as well as small or medium-sized lesions that have failed other treatments are eligible for total joint replacement or limited resurfacing. These surgical techniques have been best studied in the hip joint, which is the most common site of osteonecrosis [59, 60]. A retrospective study of 47 consecutive arthroplasties in 36 SLE patients with osteonecrosis showed that patients who underwent total hip arthroplasty had significantly higher scores in the SF-36 physical functioning and bodily pain domains compared with those who underwent bipolar hemiarthroplasty [72]. The following concerns need to be addressed when discussing total joint replacement in SLE patients: (1) the age of the patient at the time of joint replacement, (2) postsurgical expectations, and (3) co-morbidities that might increase the surgical risk. In general, total joint replacement greatly improves the quality of life in patients with osteonecrosis of the hip.

OSTEOPOROSIS

Introduction

As long-term survival rates in lupus increase and the diagnostic modalities to assess bone mineral density have become more refined, the importance of bone health in lupus patients has shifted into focus. Although lupus patients are subject to the traditional risk factors for reduced bone mass, factors associated with disease pathogenesis, disease activity, and treatment have also been associated with reduced bone mass. Bone density evaluations and appropriate treatments are the key to minimizing reductions in bone mineral density in patients with lupus [73, 74].

Epidemiology of Osteopenia and Osteoporosis

According to the Centers for Disease Control and Prevention, data collected from 1988 to 1994 in men and women older than the age of 50 years revealed a prevalence of osteopenia of 16% in men and 40% in women and a prevalence of osteoporosis of 2% in men and 16% in women [75]. Whether patients with lupus are at increased risk of osteoporosis as a specific result of the disease has been the subject of several studies. A study of bone mass in otherwise healthy women in Vietnam revealed the prevalence of osteoporosis to be 2.24, 8.4, and 30.5% in women 20—49, 50—59, and 60—69 years old, respectively [76]. In a study of 100 females with lupus and 100 controls who were matched for age, race, and menopausal status, the women with lupus had significantly lower lumbar spine bone mineral density (−0.221 vs. −0.356; $p = 0.001$) and more new fractures (13 vs.4; $p = 0.035$) during a 2-year period [77]. In a study of 702 women with lupus followed for 5951 person-years, there was an almost fivefold increase in the risk of fracture seen in the subjects with lupus compared to the general female population [78].

Numerous studies of the prevalence of osteopenia and osteoporosis in patients with lupus have been performed. These studies have mainly been cross-sectional and focused on female patients but mixed in terms of menopausal status. Studies of patients with lupus yielded prevalence figures for osteopenia as high as 50.8% and for osteoporosis as high as 23% [79, 80]. In these studies, the average age was younger than 50 years, although the majority of patients were postmenopausal. A cohort of 163 Swedish female patients with lupus (median age of 47 years, 55% postmenopausal, and a median disease duration of 11 years) was analyzed for bone mineral density at several sites [80]. Dual-energy X-ray absorptiometry (DXA) at the lumbar spine demonstrated a 30% prevalence rate of osteopenia and a 9% rate of osteoporosis, and DXA of the femoral neck demonstrated a 32% rate of osteopenia and a 9% rate of osteoporosis. In this study, 23% of patients had osteoporosis of at least one site, and 4% had osteoporosis at three or more sites. Of these patients, 52% were taking glucocorticoids at the time of the study, and 85% had taken glucocorticoids at some point during their disease [80]. In a cross-sectional study of 107 lupus patients (93% female, 72% premenopausal, and mean age of 41.1 years), the prevalence of osteopenia was 39%, and the prevalence of osteoporosis was 4% [81]. A study of 100 premenopausal subjects with lupus (mean age of 32.8 years and mean disease duration of 73.2 months) yielded prevalence rates of 40 and 5% for osteopenia and osteoporosis, respectively [82]. Osteoporosis is not a health issue that is restricted to women. A 2008 case–control study of 40 men with lupus (mean age of 42.6 years, disease duration of 84.7 months, and 85% on long-term glucocorticoids) demonstrated a significant decrease in bone mineral density at the lumbar spine and femoral neck in the male patients [83].

The morbidity associated with low bone mineral density stems from fractures. Although bone mineral density scores are the commonly used surrogate to assess bone strength, they account for only 60–80% of bone strength. The remaining bone strength is attributed to bone quality, which is determined by bone microarchitecture, underlying microfractures, and bone turnover [74, 84]. A 2007 cross-sectional study of 304 women with lupus demonstrated a self-reported fracture rate of 12.5% during a 6-year period. One particularly interesting finding in this study was that more than 60% of the patients with fractures had normal bone mineral density scores [85]. This observed discordance in some patients between bone mineral density and fracture risk supports other risk factors for fracture beyond abnormal bone mineral density. In contrast to these findings, Yee et al. [79] noted in a study of 242 patients with lupus that 9.1% had a fragility fracture, and 90.9% with fractures had diminished bone mineral density.

Assessment of Bone Mineral Density and Fracture Risk

The most common method to determine bone mineral density is the DXA test. The central DXA is the gold standard test and measures bone density at the hip and the spine, whereas the peripheral DXA test uses the same technology but measures bone density in the extremities, such as the forearm, wrist, finger, or heel. Other methods include quantitative ultrasound, quantitative computed tomography, and peripheral quantitative computed tomography. These methods are used to calculate the T- and z-scores. A T-score is the number of standard deviations above or below the mean bone density score of 30-year-old healthy adults of the same sex and ethnicity. A z-score is the number of standard deviations above or below the mean score of adults matched for age, sex, and ethnicity [86, 87].

Osteopenia is defined as a T-score between 1 and 2.5 standard deviations below that of the young healthy adult population, whereas osteoporosis is defined as a T-score worse than −2.5 standard deviations. According to the World Health Organization (WHO) criteria, the definitions of osteopenia and osteoporosis are the same for lupus patients as for the general population [88, 89]. The clinical diagnosis of osteoporosis is made by either a bone density result meeting the previously mentioned WHO criteria or the presence of a fragility fracture [90].

The main concern with low bone mass is the increased fracture risk and the morbidity associated with fracture. Bone fracture occurs when a force applied to the bone exceeds the load-bearing capacity of the bone. As the structural integrity of the bone is compromised by reduced mineral density and/or declining bone quality, the minimal force needed to cause fracture decreases and the risk of fracture increases [91]. A fragility fracture is a fracture resulting from a fall from standing height or lower [90].

WHO's FRAX algorithm is a commonly accepted method to help asses fracture risk. It takes into account bone mineral density measurements at the femoral neck, risk factors, age (must be older than 40 years), and ethnicity, and it returns a 10-year fracture probability [88, 89].

Pathophysiology of Bone Loss in Lupus

Peak bone mass is achieved between 20 and 30 years of age. Thereafter, bone is remodeled through a balance of osteoclastic and osteoblastic activity. In the general population, an imbalance that leads to osteoporosis is thought to be caused by hormonal changes associated with menopause and aging [92–94]. Inflammation, metabolic changes, and corticosteroid use are more detrimental to trabecular bone than cortical bone due to the higher turnover rate of trabecular bone. Furthermore, the ratio of trabecular bone to cortical bone is greater in the lumbar spine compared to the hip [93—95]. The variability of bone mineral density by site of measurement was demonstrated in a study of 32 corticosteroid-treated lupus patients and 16 corticosteroid-naïve patients. There were statistically significant site-to-site differences in bone mineral density scores when the lateral spine was compared to the total hip or forearm. Greater trabecular bone loss occurred in the corticosteroid-treated patients [96].

Glucocorticoids are thought to adversely affect bone formation, bone resorption, and calcium metabolism (both vitamin D-dependent and vitamin D-independent) [97, 98]. Glucocorticoids decrease the number of functional osteoblasts [99, 100]. Other factors, such as vitamin D deficiency, in patients with lupus can be associated with heightened osteoclastic activity.

Risk Factors for Bone Loss in Lupus

Traditional risk factors that are more prevalent in the lupus population include female sex, premature menopause, low vitamin D levels, and sedentary lifestyle. Independent traditional risk factors include weight less than 58 kg, age 65 years or older, personal or family history of fractures as an adult, inadequate calcium intake, excessive alcohol use, and smoking. Lupus-related risk factors include inflammatory-mediated bone loss, glucocorticoid use, nonglucocorticoid medications, reduced mobility and corresponding decline in muscle mass, myopathy, renal disease, endocrine factors, amenorrhea, low plasma androgen levels, hyperprolactinemia, and chronic induction of bone-resorbing cytokines (Table 49.5 and Figure 49.9) [84].

The roles of age, body mass index, and menopause as risk factors for decreased bone mass have been confirmed in many studies of patients with lupus. In a study of 163 women with lupus, a multiple stepwise regression showed that age was associated with low bone mineral density at the radius, total hip, femoral neck, and lumbar spine ($p < 0.001$ at all sites) [80]. In addition, body mass index correlated with bone density of the total hip ($p < 0.001$) [80]. A study of 242 lupus patients (average age, 39.9 years; age range, 18—80 years; 63.2% Caucasian; average disease duration, 7 years) revealed that menopause was a predictor of osteoporosis (OR, 13.3; CI, 1.6—111.1) [79]. In a cohort of 107 lupus patients described previously, low bone mineral density determinations at the lumbar spine or hip were associated with postmenopausal status (spine, $p < 0.01$; hip, $p < 0.037$) and low body mass index (spine, $p = 0.025$; hip, $p < 0.0001$) [81]. In a study of body composition in 52 premenopausal and 30

TABLE 49.5   Risk Factors for Osteoporosis in Patients with Lupus

TRADITIONAL RISK FACTORS

Modifiable

  Low body weight (<127 lbs)

  Recurrent falls

  Low physical activity

  Smoking

  Excessive alcohol (>2 drinks/day)

  Low dietary calcium or vitamin D

Nonmodifiable

  Age

  Female sex

  Personal or maternal history of fragility fractures

  Menopause before age 40 years

LUPUS-RELATED RISK FACTORS

Lupus activity/Inflammation

Hypogonadism (early menopause secondary to disease activity, renal failure, or therapy)

Suboptimal peak bone mass secondary to lupus activity

Sun avoidance

Medications

  Glucocorticoids

  Anticonvulsants

  Low-molecular-weight heparin

  Immunosuppressive therapy (cyclophosphamide, azathioprine, cyclosporine)

Source: Adapted with permission of Elsevier from Lee and Ramsey-Goldman [84].

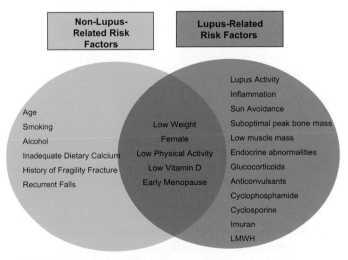

FIGURE 49.9   Risk factors for osteoporosis in lupus. LMWH, low-molecular-weight heparin.

postmenopausal females with lupus, total body bone mineral density was independently and indirectly related to corticosteroid exposure and disease severity. Whereas fat mass was significantly associated with a higher lumbar spine bone mineral density ($p = 0.008$), fat-free mass was associated with a higher bone mineral density at the lumbar spine ($p = 0.025$), femoral neck ($p = 0.003$), and total body bone mineral density measurements ($p = 0.007$) [101].

The prevalence rates of osteopenia and osteoporosis in male lupus patients are less than those of female lupus patients but higher than those of healthy male controls [79, 80, 83]. A study of 40 men with lupus and 40 controls demonstrated that increasing age ($p < 0.001$ for DXA at hip and spine), habitual drinking ($p = 0.038$ for DXA at spine), and decreased body mass index ($p = 0.007$ for DXA at spine) were significantly associated with lower bone mineral density [83].

In the general population, patients of Caucasian and Asian ethnicities are at increased risk of osteoporosis compared to African-Americans. However, the protection seen in the African-American population is questionable in the lupus population. In a study of 298 lupus patients conducted between 1996 and 2002 with 77 African-Americans and 221 Caucasians, a multivariate analysis corrected for disease activity, glucocorticoid use, renal disease, body mass index, and caffeine use. The authors concluded that the African-American race was associated with lower bone mineral density at the lumbar spine (adjusted OR = 4.43; 95% CI, 2.19–8.91) but not at the hip (adjusted OR, 1.54; 95% CI, 0.69–3.46) [102]. These findings are particularly interesting due to the fact that the lumbar spine is richer in trabecular bone than the hip, and trabecular bone is more sensitive to the bone loss attributed to inflammation and metabolic and age-related factors. A study of premenopausal Chinese females with lupus who had exposure to corticosteroids found a prevalence of osteoporosis of 4–6%, which is lower than the reported prevalence in Caucasian premenopausal females with lupus [103].

Glucocorticoids

Patients with lupus are exposed to glucocorticoids at a much higher rate than the general population. Bone loss associated with corticosteroid use is most dramatic in the first year of treatment but does continue throughout treatment [104, 105]. Fortunately, some of the bone loss associated with steroids is reversible after withdrawal of steroid treatment, but this has not been studied in the lupus population [106].

Both the maximum daily dose and the cumulative dose of steroids have been associated with bone loss in lupus patients. A study of 242 lupus patients (mean age, 39.9 years (range, 18–80); 63.2% Caucasian; mean disease duration, 7 years (range, 0–42 years)) demonstrated that exposure to more than 10 mg/day of prednisone (OR, 2.1; 95% CI, 1.1–4.2) was a predictor of reduced bone mineral density [79]. A 2000 longitudinal study evaluated bone mineral density at two time points (mean interval, 21 months; standard deviation, 11 months) in 35 premenopausal females. This study demonstrated significant bone loss at the lumbar spine (-1.22% per year) but not at the hip. In a subgroup analysis, the 15 patients with a mean prednisolone daily dose greater than or equal to 7.5 mg/day had lumbar spine bone loss of -2.12% per year. The steroid naïve patients and those patients who received a mean daily dose of less than 7.5 mg of prednisolone had no significant bone loss. Lumbar bone loss was seen only in patients who had received a cumulative steroid dose of less than 5 g before entering the study. Patients who had received more than 5 g of cumulative prednisolone at the study inception did not demonstrate a loss in bone mass over the duration of the study; this supports the theory that glucocorticoid-induced bone loss occurs early in the treatment with corticosteroids [105]. Analysis of 539 patients in the Hopkins Lupus Cohort demonstrated that when corrected for age, sex, race, and exposure to pulse steroids, the cumulative prednisone dose was associated with the development of osteoporotic fractures (relative risk, 2.5; 95% CI, 1.7–3.7). Note that in this study, 93% of the patients were female, and 52% were Caucasian. Although this study did correct for age, it did not correct for menopausal status or disease activity [107].

Disease Activity and Damage

The increased prevalence of low bone mineral density in lupus patients is not solely due to the use of corticosteroids. There are many other risk factors that contribute to the increased prevalence of osteopenia and osteoporosis in this population. A small study that demonstrated the role of factors other than glucocorticoids in the development of low bone mineral density involved 11 steroid-naïve female lupus patients. When compared to healthy controls, these patients had significantly lower bone mineral density at the hip, which has a higher cortical bone composition than the spine [108]. Inflammation is thought to affect bone mass via a complex network of cytokines. In addition to the direct effect of cytokines, there are other elements of disease activity, such as nephritis, immobility secondary to disease complications (arthritis, osteonecrosis, and myopathy), and sun avoidance (secondary to sun-sensitive rashes, physician advice, or inability to spend time outdoors secondary to disease), that can all increase the risk of osteoporosis. In a Swedish study by the Systemic Lupus International Collaborative Clinical/American College of Rheumatology (SLICC/ACR), elevated erythrocyte sedimentation rate (ESR) and serum creatinine were found to be independent factors associated with low bone mineral density [80].

A cross-sectional analysis of 307 women stratified into four groups by corticosteroid use and disease damage as measured by the SLICC/ACR Damage Index suggested that disease damage correlates with low bone mineral density in patients with lupus. Of those enrolled, 35% were postmenopausal, and the mean age was 41.7 years (±11.1 years). A total of 161 patients had disease damage, 23 of whom were corticosteroid naïve; 146 patients had no disease damage, 46 of whom were steroid naïve. Patients with disease damage and corticosteroid use served as the reference group. Mean bone mineral density in this group was not significantly different from the mean value of patients with disease damage and no corticosteroid use. However, when the reference group was compared to patients with no damage either with or without a history of corticosteroid use, significant differences were observed (both $p < 0.05$) [109]. These findings support the theory that disease damage is a contributor to bone loss in lupus. In another study of 64 patients with lupus and a mean disease duration of 7.7 ± 5.7 years, disease duration and disease damage index were both predictors of low bone mineral density at the lumbar spine and femoral neck [110].

Inflammation and the cytokines associated with inflammation are postulated to contribute to the increased bone loss in lupus. IL-1, IL-6, TNF-α, vitamin D_3, and IL-11 are thought to affect the RANKL-OPG (receptor activator of nuclear factor—κB ligand—osteoprotegerin) pathway and stimulate bone loss. Activated T cells can produce RANKL and TNF. The role of TNF in bone turnover is not limited to the RANKL-OPG pathway; TNF can also stimulate osteoclast maturation. Furthermore, parathyroid hormone, elevated in renal disease, and glucocorticoids can affect the RANKL-OPG pathway [73, 74, 84].

Hormonal Factors

Amenorrhea, low androgen levels, premature menopause (related to disease activity, renal failure, or from cyclophosphamide), and hyperprolactinemia have been associated with low bone mineral density in lupus [73, 111–113]. In a study of 94 lupus patients (40% on high-dose steroids and 68% treated with glucocorticoids and cyclophosphamide), 54% had decreased progesterone, 25% had reduced estradiol concentrations, 13% had increased luteinizing hormone, 9% had increased follicle-stimulating hormone, and 10% had increased prolactin [114].

Medications Other Than Glucocorticoids

Chronic anticoagulation, cyclophosphamide, mycophenalate mofetil, and cyclosporine have been associated with reduced bone density. Cyclophosphamide is known to cause ovarian failure and premature menopause, which can have a deleterious effect on bone mineral density [73]. Cyclophosphamide may contribute to increased fracture risk in lupus patients, and this is supported by the correlation of cyclophosphamide with decreased bone mineral density measured at the total hip in a Swedish cohort [80]. Warfarin is postulated to interfere with osteocalcin synthesis and therefore could have a detrimental effect on bone density, although there is no literature quantifying these effects. Low-molecular-weight heparin is assumed to decrease bone mineral density via effects on bone cells, vitamin D levels, and ionized calcium concentrations [73]. Cyclosporine has been implicated in the increased fracture risk in transplant patients. It is difficult to correlate this observation with lupus patients because transplant patients are on higher doses of cyclosporine than are lupus patients. In addition, transplant patients are on other immunosuppressive agents and steroids [73]. Azathioprine has been mentioned as a possible cause of bone loss, but the evidence is not convincing [115]. In a Swedish cohort, anti-malarials were shown to be independently correlated with low bone mineral density, but a multiple regression analysis failed to demonstrate statistical significance. In contrast, anti-malarial therapy was independently associated with higher bone mineral density in a 2001 study. A study evaluated 92 patients with lupus for risk factors associated with low bone mineral density. In this cohort, 98% of patients had received prednisone, 51% were postmenopausal, and 68% were on hydroxychloroquine; 15% were osteoporotic. Treatment with hydroxychloroquine was associated with a higher bone mineral density in the hip and spine in both univariate and multivariate analysis [116]. Although immunosuppressive medications might have negative effects on bone mineral density, it is important to take into consideration the steroid-sparing effect of some of these medications.

Vitamin D

A 2005 study of 107 lupus patients (93% female; mean age, 41 years) showed that vitamin D deficiency ($p = 0.047$) was associated with low bone mineral density [81]. In a study of 36 lupus patients (12 with high activity and 24 with minimal activity) compared to normal controls, the subgroup of patients with high activity had significantly lower vitamin D levels compared to the controls or the lupus patients with minimal disease activity [117]. The etiology of hypovitaminosis D in patients with lupus is multifactorial and can be due to poor dietary intake, sun avoidance (secondary to history of photosensitive rash, medication side effects, or physical limitations secondary to disease), renal disease, or chronic steroid therapy [98, 118].

Despite the increased fracture risk in patients with familial and nonfamilial hyperhomocysteinemia [119], the association of homocysteine levels and fracture risk was not replicated in a lupus cohort. In a case–control study reported in 2008 in which 100 female lupus patients were matched with 100 controls for age, race, and menopausal status, homocysteine levels were higher in the lupus cohort ($p < 0.001$), but the homocysteine levels were not significantly associated with bone mineral density. Furthermore, the homocysteine levels were not predictive of fracture in either cohort [77].

Specific Risk Factors for Fracture

Unfortunately, there are limited numbers of studies that focus on fracture risk in the lupus population, and the results from the available data are conflicting. Fracture risk has been associated with male sex and use of intravenous methylprednisolone in a cohort of 107 lupus patients (93% female) who were followed between 2001 and 2003 [81]. In a study of 242 patients with lupus (95.5% female; mean age, 39.9 years), 22 patients sustained a fracture. Multivariate analysis demonstrated that reduced bone mineral density (OR, 8.1; 95% CI, 1.7–40.0) and age (OR, 1.2; 95% CI, 1.1–1.3) were predictors of fractures, but corticosteroid use, sex, race, and disease damage were not. In this study, it is pertinent to note that 60% of the patients with fractures had a normal bone mineral density [79]. Furthermore, in a cross-sectional study of 304 females with lupus, self-reported fractures were associated with longer disease duration [85]. In summary, risk factors specifically related to fracture risk include low bone mineral density, disease duration, age, and possibly corticosteroid use.

Evaluation of Bone Health

The clinician needs to be cognizant of the bone health of his or her lupus patients. Individualized risk factor assessments should be done for all patients to help determine appropriate screening and treatment. It should be noted that the current practice guidelines for osteoporosis screening apply in the lupus population. The U.S. Preventive Services Task Force recommendation is to screen all women older than 65 years for osteoporosis. The recommendation extends to start at age 60 years for women at increased risk for osteoporotic fracture [120]. The National Osteoporosis Foundation provides the following indications for screening: women 65 years or older, men 70 years or older, postmenopausal or perimenopausal women younger than 65 years with risk factors, men between 50 and 69 years with risk

factors, all adults with fracture after age 50 years, and exposure to more than 5 mg of prednisone per day for 3 or more months. These guidelines also call for screening adults with a medical condition that increases the risk of osteoporosis [88, 89]. The National Osteoporosis Foundation's guidelines would thus pertain to all patients with lupus. A 2005 report issued by Lee and Ramsey-Goldman [84] recommended bone mineral density determinations in patients with lupus who are classified into one of the following groups: (1) significant glucocorticoid use, (2) premature menopause or hypogonadism, (3) 50 years or older and with risk factors, and (4) postmenopausal. Patients with a T-score of less than 1 should have an appropriate laboratory evaluation, including complete blood count, liver function tests, metabolic panel, thyroid function tests, parathyroid hormone. Furthermore, 25-hydroxy vitamin D, testosterone in men, luteinizing and follicle-stimulating hormones in women, and N-telopeptides may be helpful in some patients [74]. A quality indicator set for lupus published in 2009 recommended bone mineral density testing for patients with lupus if they received prednisone at a dose of 7.5 mg or more daily for 3 or more months (or glucocorticoid equivalent). Bone density screening should be documented within the 12 months prior to glucocorticoid initiation or within the first 6 months of glucocorticoid treatment. This recommendation does not take into account risk factors other than steroid use, and it does not apply to patients already on antiresorptive or anabolic treatments for osteoporosis or osteopenia [121]. The need to evaluate bone health in premenopausal patients on corticosteroids is supported by a study of 98 Malaysian patients with lupus. The mean age was 30.05 ± 7.54 years, the mean prednisolone dose at bone mineral density measurement was 18.38 ± 10.85, and the mean duration of corticosteroid use was 2.5 years. Of note, the mean SLEDAI score in this cohort was 8.5. Analysis of bone mineral density by DXA with T-scores referenced to a database of Japanese women revealed that the cumulative steroid dose and duration of corticosteroid use were inversely correlated with bone mineral density. Cumulative steroid dose was associated with lumbar spine T-scores ($p = 0.019$), and duration of steroid use was associated with femoral neck T-scores and trochanter T-scores ($p = 0.04$ and $p = 0.008$, respectively). In this study, traditional risk factors such as race, smoking, self-reported calcium intake, and exercise were not associated with low bone mineral density. Also, the lupus disease activity index was not correlated with bone mineral density. This study shows the importance of evaluating and educating premenopausal women on corticosteroids [122].

The diagnosis of osteopenia and osteoporosis in this patient population is not as straightforward as that in the general population. The current WHO definitions of osteoporosis were derived by comparing bone density measurements to those of young Caucasian controls, and the currently accepted definitions were designed for the evaluation of postmenopausal Caucasian females. The effects of extrapolating these definitions to males, non-Caucasians, and premenopausal females are not clear. Although the most commonly used T-scores represent the comparison of bone mineral density to those of sex-matched young healthy controls at their peak bone mass, it may be better to use the z-score, which compares bone mineral density to those of sex- and age-matched controls. In the pediatric population, z-scores are the standard [123]. The best way to determine fracture risk is to examine both the absolute bone density and the patient's specific risk factors for fracture.

Treatment

Nonpharmacological Interventions

Treatment begins with modifications of risk factors. Smoking and excessive alcohol intake can increase fracture risk and should be avoided by patients. Exercise, as deemed safe for the patient, should be encouraged. In many cases, a supervised exercise routine might be helpful, at least when initiating this modification.

Pharmacologic Interventions

A 2009 quality indicator set for systemic lupus erythematosus provides recommendations for both the evaluation and the treatment of osteoporosis in patients with lupus [121]. These recommendations delineate the minimal acceptable standard of care based on available evidence. In these recommendations, patients are stratified by their prednisone exposure (or glucocorticoid equivalent) and not any other risk factors, such as sex, age, or menopausal status. Patients taking prednisone 7.5 mg or greater per day for 3 or more months should be treated with at least supplemental calcium and vitamin D. Patients taking prednisone 7.5 mg or greater per day for 1 or more months and with a central T-score indicative of osteoporosis should be treated with an antiresorptive agent or anabolic agent [121].

In 2001, the American College of Rheumatology published recommendations for treatment of glucocorticoid-induced osteoporosis. For patients starting glucocorticoid therapy for a presumed duration of 3 or more months with prednisone 5 mg or more per day, the recommendations are: (1) lifestyle modifications such as smoking cessation and alcohol consumption reduction, (2) instruction in weight-bearing exercise, (3) calcium supplementation and vitamin D supplementation, and (4) bisphosphonates in postmenopausal women and some premenopausal women at the physician's

discretion. The ACR further recommended bone mineral density testing for all patients receiving long-term glucocorticoids at 5 mg or more of prednisone per day. If the bone mineral density is normal (i.e., T-score > -1), annual or biannual bone mineral density testing might be the sole recommendation. If bone mineral density testing is abnormal (i.e., T score < -1), bisphosphonate therapy (with caution in premenopausal women) is recommended. For those intolerant of bisphosphonates, calcitonin may be an alternative therapeutic intervention [124].

Although the association of homocysteine with fracture risk in this population is unclear, some recommend folic acid 1 mg daily if patients have elevated homocysteine levels [93].

Although supplemental calcium and vitamin D therapy is common practice, there is minimal direct evidence for the role of supplemental calcium and vitamin D for primary or secondary prevention of osteopenia in patients with lupus. A 2007 study measuring the dietary intake of calcium in 60 premenopausal women with lupus on corticosteroid treatment demonstrated that bone mineral density was not correlated with calcium intake [125]. Nevertheless, supplemental vitamin D and calcium are recommended in this patient population because this intervention is recommended for the general population and has minimal side effects [91]. Furthermore, calcium supplementation has been shown to improve bone mineral density in one or more sites in perimenopausal and postmenopausal women without lupus [126]. In 2002, a meta-analysis of calcium supplementation in postmenopausal women demonstrated that calcium had a minor positive effect on bone mineral density but did not affect the rates of vertebral and nonvertebral fractures [127]. In 2002, a meta-analysis of vitamin D effects on bone mineral density showed that supplemental vitamin D was associated with a reduction in nonvertebral fractures [128].

In a study of 98 premenopausal lupus patients of Asian or Southeast Asian descent followed for 2 years, the subjects were divided into three groups based on their bone health regimen. All subjects were treated with calcium carbonate 500 mg orally twice per day. Group 1 was only treated with calcium, group 2 was treated with calcium and calcitriol (0.25 µg twice a day), and group 3 was treated with calcium and alendronate (70 mg orally weekly). All of the patients were taking corticosteroids; the mean duration of corticosteroid use was 2.5 years at study entry. The study showed a preservation in lumbar spine bone mineral density at 2 years in subjects in groups 1 and 2, whereas subjects in group 3 had an increase in bone mineral density at 2 years ($p < 0.001$) [129].

Although steroid use is a contributor to bone loss in lupus patients, systemic inflammation is also a risk factor for bone loss. The proper use of steroids and the prudent use of steroid-sparing agents might prevent bone loss secondary to inflammation. Minimizing inflammation will limit the inflammation-mediated activation of osteoclasts and increase functional osteoblasts [93, 94]. Future therapies targeted at inflammatory cytokines may be helpful to mitigate bone loss, but these therapies are not currently available.

Bisphosphonates, including etidronate, risedronate, and alendronate, have been shown to mitigate bone loss in both postmenopausal and glucocorticoid-induced osteoporosis [74, 84, 91]. Studies of bisphosphonate use in lupus patients are scant. Alendronate and calcium were shown to be superior to calcium alone or calcium and calcitriol in nonsteroid-naïve lupus patients [129]. Bisphosphonates, which bind bone hydroxyapatite and inhibit osteoclastic activity, have a long half-life. This raises concerns about their use in premenopausal women desirous of future child-bearing [93, 94, 130].

Teriparatide has been approved by the Food and Drug Administration for use in glucocorticoid-induced osteoporosis and may be a treatment option for lupus patients who are unable to tolerate bisphosphonates. A 36-month, randomized, double-blind controlled trial of teriparatide versus alendronate for glucocorticoid-induced osteoporosis in 428 subjects who received 5 mg or more of daily prednisone (or equivalent) for 3 or more months demonstrated that teriparatide was superior to alendronate in terms of increasing bone mineral density and preventing new vertebral fractures [131]. There have been no studies of teriparatide in lupus patients.

Although estrogen replacement therapy has been shown to be effective in mitigating bone loss in postmenopausal women, long-term toxicity associated with hormone replacement therapy has greatly reduced enthusiasm for this therapeutic approach. Furthermore, exogenous estrogen is contraindicated in patients with known hypercoagulability, such as those with antiphospholipid antibodies [74, 84]. Given the unwanted risks associated with estrogen replacement therapy, it is not the first-line treatment for primary or secondary prevention of osteopenia or osteoporosis in lupus patients. In addition, there is the frequently discussed association of lupus activity with supplemental estrogen. In a study of 32 postmenopausal lupus patients, treatment with a transdermal estrogen replacement patch (estradiol 50 µg/day) was associated with an increase in lumbar spine bone mineral density at 6 months in the treated group. It is important to note that both the treatment group and the placebo group in this trial received 5 mg of medroxyprogesterone, 500 mg of calcium, and 400 IU of vitamin D_3 daily. Lupus activity was not affected by the use of estradiol in this study [132].

Raloxifine (60 mg daily), a selective estrogen receptor modulator, plus calcium 1200 mg daily was shown in 33 postmenopausal lupus patients to maintain bone mineral density in the lumbar spine and femoral hip compared to calcium treatment alone. The patients were excluded if they had a history of antiphospholipid antibodies or required 10 mg or more of daily prednisone. There was no difference in lupus activity between the study and control groups, but there was a more favorable lipid profile in the treated group [133].

Synthetic dehydroepiandrosterone (DHEA) has been of interest in treating bone loss in lupus because DHEA levels have been shown to be low in lupus patients. Low DHEA levels in lupus patients may be due to suppression by steroids or by other disease-associated factors. A study of 155 lupus patients on chronic glucocorticoids who were treated with prasterone (synthetic DHEA) found a dose-dependent increase in bone mineral density after 18 months of treatment (100 vs. 200 mg; $p = 0.021$). In the open-label phase of the study, there was a gain in bone mineral density at the lumbar spine ($p = 0.042$), but no effect at the hip for the 200-mg/day dosing group was seen. The results of this study, although not conclusive, provided a signal that prasterone may have a small protective effect against bone loss in women with lupus [134].

Special Populations: Juvenile Systemic Lupus Erythematosus

Patients with juvenile lupus are particularly susceptible to decreased bone mass. The reported prevalence of osteopenia in juvenile lupus is approximately 40% in two studies [135, 136]. In a study of 70 cases of juvenile-onset lupus, 76% of the subjects were female, the mean age was 26.4 years, the mean disease duration was 10.8 years, and 7% of the subjects were corticosteroid naïve. Forty-one percent of the subjects had osteopenia at one or more sites, 6% had a history of vertebral collapse, and there was a statistically significant difference in the lower spine bone mineral density at the lumbar spine compared to that of 70 control subjects [136]. A 2007 study of 64 patients with juvenile lupus revealed lumbar spine osteopenia in 37.5% of patients and osteoporosis in 20.3% of patients. In this study, 76.6% of the patients were female, the mean age was 14.3 years, and mean disease duration was 2.9 years [137].

In healthy adults, peak bone mass is achieved in the third decade [92, 94], and by definition, patients with juvenile lupus develop disease before achieving peak bone mass. Proposed risk factors are the same for juvenile lupus as for adult lupus, but the effects of these risk factors may be greater given that these young patients were in the process of developing their peak bone mass when they were affected by the disease.

In the aforementioned study of 64 patients with juvenile lupus, multivariate analysis revealed an association of cumulative corticosteroid dose ($p = 0.0026$) with osteopenia at the lumbar spine, the association of disease duration ($p = 0.0028$) with osteoporosis at the lumbar spine, and the association of abnormal hip bone mineral density with disease duration ($p = 0.0249$). Factors associated with low bone mineral density in univariate analysis that were not confirmed in the multivariate analysis included duration of corticosteroid use, noncorticosteroid medications (azathioprine, cyclophosphamide, and mycophenolate mofetil), and nephritis (class III and IV) [137]. Other studies have shown that the cumulative corticosteroid dose is associated with low bone mineral density in pediatric lupus patients [136, 138].

Because steroids have been shown to be an independent risk factor for bone loss in patients with juvenile lupus, it is important to use corticosteroids judiciously in this population. Whereas calcium and vitamin D supplementation are recommended in the adult population, there are no formal guidelines for the pediatric population. Although it has not been specifically studied, it is reasonable to supplement children with calcium and vitamin D so that they meet the recommended vitamin D and calcium requirements [123]. Adjunctive therapies, such as bisphosphonates, are controversial in children because of concerns of long-term persistence of bisphosphonates in bone. Three studies of alendronate in children (82 children, 17 reported with lupus) demonstrated an increase in bone mineral density during a 1-year period of therapy, and there were no reported effects on longitudinal bone growth [139–141]. Nevertheless, larger studies need to be performed to evaluate the efficacy and long-term safety of these medications in children [123].

References

[1] A. Fernandez, G. Quintana, E.L. Matteson, J.F. Restrepo, F. Rondon, A. Sanchez, et al., Lupus arthropathy: Historical evolution from deforming arthritis to rhupus, Clin Rheumatol 23 (2004) 523–526.

[2] T.K. Pekin, N.J. Zvaifler, Synovial fluid findings in systemic lupus erythematosus (SLE), Arthritis Rheum 13 (6) (1970) 777–785.

[3] B.N. Weissman, A.S. Rappoport, J.L. Sosman, P.H. Schur, Radiographic findings in the hands in patients with systemic lupus erythematosus, Radiology 126 (2) (1978) 313–317.

[4] S. Wright, E. Filippucci, W. Grassi, A. Grey, A. Bell, Hand arthritis in systemic lupus erythematosus: An ultrasound pictorial essay, Lupus 15 (8) (2006) 501–506.

[5] J. Natour, L.C. Montezzo, L.A. Moura, E. Atra, A study of synovial membrane of patients with systemic lupus erythematosus (SLE), Clin Exp Rheumatol 9 (3) (1991) 221–225.

[6] K.A. Mensah, A. Mathian, L. Ma, L. Xing, C.T. Ritchlin, E.M. Schwarz, Nonerosive arthritis in lupus is mediated by IFN-α-stimulated monocyte differentiation that is nonpermissive of osteoclastogenesis, Arthritis Rheum 62 (2010) 1127–1137.

[7] R.S. Panush, N.L. Edwards, S. Longley, E. Webster, Rhupus syndrome, Arch Intern Med 148 (7) (1988) 1633–1636.

[8] A. Fernández, G. Quintana, F. Rondón, J.F. Restrepo, A. Sánchez, E.L. Matteson, et al., Lupus arthropathy: A case series of patients with rhupus, Clin Rheumatol 25 (2) (2006) 164–167.

[9] L.M. Amezcua-Guerra, R. Springall, R. Marquez-Velasco, L. Gomez-Garcia, A. Vargas, R. Bojalil, Presence of antibodies against cyclic citrullinated peptides in patients with 'rhupus': A cross-sectional study, Arthritis Res Ther 8 (2006) R144.

[10] M.B. Santiago, V. Galvão, Jaccoud arthropathy in systemic lupus erythematosus: Analysis of clinical characteristics and review of the literature, Medicine 87 (1) (2008) 37–44.

[11] C.J. Bleifeld, A.E. Inglis, The hand in systemic lupus erythematosus, J Bone Joint Surg Am 56 (6) (1974) 1207–1215.

[12] J.M. Wong, J.M. Esdaile, Methotrexate in systemic lupus erythematosus, Lupus 14 (2) (2005) 101–105.

[13] P.R. Fortin, M. Abrahamowicz, D. Ferland, D. Lacaille, C.D. Smith, M. Zummer, Steroid-sparing effects of methotrexate in systemic lupus erythematosus: A double-blind, randomized, placebo-controlled trial, Arthritis Care Res 59 (12) (2008) 1796–1804.

[14] E.M. Ginzler, D. Wofsy, D. Isenberg, C. Gordon, L. Lisk, M. Dooley, Nonrenal disease activity following mycophenolate mofetil or intravenous cyclophosphamide as induction treatment for lupus nephritis: Findings in a multicenter, prospective, randomized, open-label, parallel-group clinical trial, Arthritis Rheum 62 (1) (2010) 211–221.

[15] J.T. Merrill, F.C. Carthen, S.K. Wilson, S. Kamp, J. Hutcheson, J. Rawdon, et al., Mycophenolate mofetil (MMF) for treatment of arthritis in patients with systemic lupus erythematosus (SLE) [abstract No. 1069], Arthritis Rheum 58 (2008).

[16] D.J. Wallace, W. Stohl, R.A. Furie, J.R. Lisse, J.D. McKay, J.T. Merrill, et al., A phase II, randomized, double-blind, placebo-controlled, dose-ranging study of belimumab in patients with active systemic lupus erythematosus, Arthritis Care Res 61 (9) (2009) 1168–1178.

[17] G.G. Illei, Y. Shirota, C.H. Yarboro, J. Daruwalla, E. Tackey, K. Takada, et al., Tocilizumab in systemic lupus erythematosus: Data on safety, preliminary efficacy, and impact on circulating plasma cells from an open-label phase I dosage-escalation study, Arthritis Rheum 62 (2) (2010) 542–552.

[18] E.A. Nalebuff, Surgery of systemic lupus erythematosus arthritis of the hand, Hand Clin 12 (3) (1996) 591–602.

[19] R.A. Furie, E.K. Chartash, Tendon rupture in systemic lupus erythematosus, Semin Arthritis Rheum 18 (1988) 127–133.

[20] D. Estes, C.L. Christian, The natural history of systemic lupus erythematosus by prospective analysis, Medicine (Baltimore) 50 (1971) 85–95.

[21] M.J. Garton, D.A. Isenberg, Clinical features of lupus myositis versus idiopathic myositis: A review of 30 cases, Br J Rheum 36 (1997) 1067–1074.

[22] G.C. Tsokos, H.M. Moutsopoulos, A.D. Steinberg, Muscle involvement in systemic lupus erythematosus, J Am Med Assoc 246 (1981) 766–768.

[23] R.A. Foote, S.M. Kimbrough, J.C. Stevens, Lupus myositis, Muscle Nerve 5 (1982) 65–68.

[24] J.E. McDonagh, D.A. Isenberg, Development of additional autoimmune diseases in a population of patients with systemic lupus erythematosus, Ann Rheum Dis 59 (2000) 230–232.

[25] R.J. Walsh, S.W. Kong, Y. Yao, B. Jallal, P.A. Kiener, J.L. Pinkus, et al., Type I interferon-inducible gene expression in blood is present and reflects disease activity in dermatomyositis and polymyositis, Arthritis Rheum 56 (11) (2007) 3784–3792.

[26] E.C. Baechler, J.W. Bauer, C.A. Slattery, W.A. Ortmann, K.J. Espe, J. Novitzke, et al., An interferon signature in the peripheral blood of dermatomyositis patients is associated with disease activity, Mol Med 13 (1-2) (2007) 59–68.

[27] N.A. Dayal, D.A. Isenberg, SLE/myositis overlap: Are the manifestations of SLE different in overlap disease? Lupus 11 (5) (2002) 293–298.

[28] R. Oxenhandler, M.N. Hart, J. Bickel, D. Scearce, J. Durham, W. Irvin, Pathologic features of muscle in systemic lupus erythematosus, Hum Pathol 13 (1982) 745–757.

[29] K.L. Lim, R. Abdul-Wahab, J. Lowe, R.J. Powell, Muscle biopsy abnormalities in systemic lupus erythematosus: Correlation with clinical and laboratory parameters, Ann Rheum Dis 53 (1994) 178–182.

[30] M. Pallis, J. Lowe, R. Powell, An electron microscopic study of muscle capillary wall thickening in systemic lupus erythematosus, Lupus 3 (5) (1994) 401–407.

[31] I.M. Bronner, J.E. Hoogendijk, H. Veldman, M. Ramkema, Weerman MA van den Bergh, et al., Tubuloreticular structures in different types of myositis: Implications for pathogenesis, Ultrastruct Pathol 32 (4) (2008) 123–126.

[32] K.E. Tymms, J. Webb, Dermatopolymyositis and other connective tissue diseases: A review of 105 cases, J Rheumatol 12 (1985) 1140–1148.

[33] Y. Assouline-Dayan, C. Chang, A. Greenspan, Y. Shoenfeld, M.E. Gershwin, Pathogenesis and natural history of osteonecrosis, Semin Arthritis Rheum 32 (2) (2002) 94–124.

[34] C. Aranow, S. Zelicof, D. Leslie, S. Solomon, P. Barland, A. Norman, et al., Clinically occult avascular necrosis of this hip in SLE, J Rheumatol 24 (12) (1997) 2318–2322.

[35] K. Nagasawa, H. Tsukamoto, Y. Tada, T. Mayumi, H. Satoh, H. Onitsuka, et al., Imaging study on the mode of development and changes in avascular necrosis of the femoral head in SLE: Long-term observations, Br J Rheumatol 33 (1994) 343–347.

[36] D.D. Gladman, V. Chaudhry-Ahluwalia, D. Ibañez, E. Bogoch, M.B. Urowitz, Outcomes of symptomatic osteonecrosis in 95 patients with SLE, J Rheumatol 28 (2001) 2226–2229.

[37] M. Abeles, J.D. Urman, N.F. Rothfield, Aseptic necrosis of bone in SLE: Relationship to corticosteroid therapy, Arch Intern Med 138 (1978) 750–754.

[38] M. Abu-Shakra, D. Buskila, Y. Shoenfeld, Osteonecrosis in patients with SLE, Clin Rev Allergy Immunol 25 (2003) 13–23.

[39] J. Calvo-Alen, G. McGwin, S. Toloza, M. Fernandez, J.M. Roseman, H.M. Bastian, et al., SLE in a multiethnic U.S. cohort (LUMINA): XXIV. Cytotoxic treatment is an additional risk factor for the development of symptomatic osteonecrosis in lupus patients: Results of a nested matched case–control study, Ann Rheum Dis 65 (2006) 785–790.

[40] E. Dubois, L. Cozen, Avascular necrosis associated with lupus erythematosus, JAMA 174 (1960) 966–971.

[41] D.D. Gladman, V. Chaudhry-Ahluwalia, D. Ibanez, E. Bogoch, M.B. Urowitz, Predictive factors for symptomatic osteonecrosis in patients with SLE, J Rheumatol 28 (4) (2001) 761–765.

[42] F.A. Houssiau, A. N'Zeusseu Toukap, G. Depresseux, B.E. Maldague, J. Malghem, J.P. Devogelaer, et al., Magnetic resonance imaging-detected avascular necrosis in SLE: Lack of correlation with antiphospholipid antibodies, Br J Rheumatol 37 (1998) 448–453.

[43] S. Migliaresi, U. Picillo, L. Ambrosone, G. Di Palma, M. Mallozzi, E.R. Tesone, et al., Avascular osteonecrosis in patients with SLE: Relation to corticosteroid therapy and anticardiolipin antibodies, Lupus 3 (1) (1994) 37–41.

[44] C.C. Mok, C.S. Lau, R.W. Wong, Risk factors for avascular necrosis in SLE, Br J Rheumatol 37 (8) (1998) 895–900.

[45] M.Y. Mok, V.Y. Farerelee, D.A. Isnberg, Risk factors for avascular necrosis of bone in patients with SLE: Is there a role for antiphospholipid antibodies? Ann Rheum Dis 59 (6) (2000) 462–467.

[46] K. Nagasawa, Y. Tada, S. Koarada, T. Horiuchi, H. Tsukamoto, K. Murai, et al., Early development of steroid-associated osteonecrosis of femoral head in SLE: prospective study by MRI, Lupus 14 (2005) 385–390.

[47] K. Oinuma, Y. Haradaa, Y. Nawatab, K. Takabayashib, I. Abea, K. Kamikawaa, et al., Osteonecrosis in patients with SLE develops very early after starting high dose corticosteroid treatment, Ann Rheum Dis 60 (2001) 1145–1148.

[48] M. Petri, Musculoskeletal complications of SLE in the Hopkins Lupus Cohort: An update, Arthritis Rheum 8 (3) (1995) 137–145.

[49] M.B. Urowitz, D.D. Gladman, B.D.M. Tom, D. Ibañez, V.T. Farewell, Changing patterns in mortality and disease outcomes for patients with SLE, J Rheum 35 (11) (2008) 2152–2158.

[50] T. Watanabe, T. Tsuchida, N. Kanda, K. Tamaki, Avascular necrosis of bone in SLE, Scand J Rheumatol 26 (1997) 184–187.

[51] T.M. Zizic, C. Marcoux, D.S. Hungerford, M.B. Stevens, The early diagnosis of ischemic necrosis of bone, Arthritis Rheum 29 (10) (1986) 1177–1186.

[52] T.M. Zizic, C. Marcoux, D.S. Hungerford, J.V. Dansereau, M.B. Stevens, Corticosteroid therapy associated with ischemic necrosis of bone in SLE, Am J Med 79 (1985) 596–604.

[53] H. Brunner, E. Silverman, T. To, C. Bombardier, B. Feldman, Risk factors for damage in childhood-onset SLE, Arthritis Rheum 46 (2003) 436–444.

[54] L.T. Hiraki, J. Hamilton, E.D. Silverman, Review: Measuring permanent damage in pediatric SLE, Lupus 16 (2007) 657–662.

[55] A. Ravelli, C. Duarte-Salazar, S. Buratti, et al., Assessment of damage in juvenile-onset SLE: A multicenter cohort study, Arthritis Rheum 49 (2003) 501–507.

[56] N.E. Lane, Therapy insight: Osteoporosis and osteonecrosis in SLE. Nat Clin Pract Rheumatol 2 (10) (2006) 562–569.

[57] M.Y. Mok, D.A. Isenberg, Avascular necrosis of a single vertebral body, an atypical site of disease in a secondary APLS, Ann Rheum Dis 59 (6) (2000) 494–495.

[58] D. Resnick, C. Pineda, D. Trudell, Widespread osteonecrosis of the foot in systemic lupus erythematosus: Radiographic and gross pathologic correlation, Skeletal Radiol 13 (1) (1985) 33–38.

[59] L.C. Jones, D.S. Hungerford, Osteonecrosis: Etiology, diagnosis, and treatment, Curr Opin Rheumatol 16 (2004) 443–449.

[60] M.A. Mont, L.C. Jones, Management of osteonecrosis in SLE, Rheum Dis Clin North Am 26 (2) (2000) 279–309.

[61] S. Okazaki, Y. Nishitani, S. Nagoya, M. Kaya, T. Yamashita, H. Matsumoto, Femoral head osteonecrosis can be caused by disruption of the systemic immune response via the toll-like receptor 4 signalling pathway, Rheumatology 48 (3) (2009) 227–232.

[62] A. Zonana-Nacach, S.G. Barr, L.S. Magder, M. Petri, Damage in SLE and its association with corticosteroids, Arthritis Rheum 43 (8) (2000) 1801–1808.

[63] S.C.M.S. Fialho, E. Bonfá, L.F. Vitule, et al., Disease activity as a major risk factor for osteonecrosis in early SLE, Lupus 16 (2007) 239–244.

[64] K. Nagasawa, Y. Tada, S. Koarada, et al., Prevention of steroid-induced osteonecrosis of femoral head in SLE by anti-coagulant, Lupus 15 (2006) 354–357.

[65] D.L. Sweet, D.G. Roth, R.K. Desser, et al., Avascular necrosis of the femoral head with combination therapy, Ann Intern Med 85 (1976) 67–68.

[66] M.G. Tektonidiou, K. Malagari, P.G. Vlachoyiannopoulos, et al., Asymptomatic avascular necrosis in patients with primary antiphospholipid syndrome in the absence of corticosteroid use, Arthritis Rheum 48 (3) (2003) 732–736.

[67] J.C. Rueda, M.A. Duque, R.D. Mantilla, A. Iglesias-Gamarra, Osteonecrosis and antiphospholipid syndrome, J Clin Rheumatol 15 (3) (2009) 130–132.

[68] X.Y. Yang, D.H. Xu, MDR1(ABCB1) gene polymorphisms associated with steroid-induced osteonecrosis of femoral head in SLE, Pharmazie 62 (12) (2007) 930–932.

[69] T. Asano, K.A. Takahashi, M. Fujioka, et al., ABCB1 C3435T and G2677T/A polymorphism decreased the risk for steroid induced osteonecrosis of the femoral head after kidney transplantation, Pharmacogenetics 13 (11) (2003) 675–682.

[70] E.J. Lydon, M. Schweitzer, J.D. Godberg, H.M. Belmont, Atorvastatin to prevent avascular nectosis of bone in systemic lupus erythematosus [abstract], Arthritis Rheum 54 (2008) S432.

[71] S. Agarwala, D. Jain, R. Joshi, A. Sule, Efficacy of alendronate, a bisphosphonate, in the treatment of AVN of the hip. A prospective open-label study, Rheumatology 44 (2005) 352–359.

[72] H. Ito, T. Matsuno, T. Hirayama, H. Tanino, A. Minami, Health-related quality of life in patients with systemic lupus erythematosus after medium to long-term follow-up of hip arthroplasty, Lupus 16 (5) (2007) 318–323.

[73] O. Di Munno, M. Mazzantini, A. Delle Sedie, et al., Risk factors for osteoporosis in female patients with systemic lupus erythematosus, Lupus 13 (2004) 724–730.

[74] C. Lee, R. Ramsey-Goldman, Bone health and systemic lupus erythematosus, Curr Rheumatol Rep 7 (2005) 482–489.

[75] National Health and Nutrition Survey, CDC/NHANES (2001). http://www.cdc (accessed December 22, 2009).

[76] V.T.T. Hien, N.C. Khan, N.T. Lam, et al., Determining the prevalence of osteoporosis and related factors using quantitative ultrasound in Vietnamese adult women, Am J Epidemiol 161 (9) (2005) 824–830.

[77] E.Y. Rhew, C. Lee, P. Eksarko, et al., Homocysteine, bone mineral density and fracture risk over 2 years of follow-up in women with and without systemic lupus erythematosus, J Rheumatol 35 (2) (2008) 230–236.

[78] R. Ramsey-Goldman, J.E. Dunn, C.E. Huang, et al., Frequency of fractures in women with systemic lupus erythematosus: Comparison with United States population data, Arthritis Rheum 42 (5) (1999) 882–890.

[79] C.S. Yee, N. Crabtree, J. Skan, et al., Prevalence and predictors of fragility fractures in systemic lupus erythematosus, Ann Rheum Dis 64 (2005) 111–113.

[80] K. Almehed, H. Forsblad d'Elia, G. Kvist, et al., Prevalence and risk factors for osteoporosis in female SLE patients—Extended report, Rheumatology 46 (2007) 1185–1190.

[81] I.E.M. Bultink, W.F. Lems, P.J. Kostense, et al., Prevalence of and risk factors for low bone mineral density and vertebral fractures in patients with systemic lupus erythematosus, Arthritis Rheum 54 (7) (2005) 2044–2050.

[82] C. Mendoza-Pinto, M. Garcia-Carrasco, H. Sandoval-Cruz, et al., Risks factors for low bone mineral density in premenopausal Mexican women with systemic lupus erythematosus, Clin Rheumatol 28 (2009) 65–70.

[83] C.C. Mok, S.K. Yee, Y. Ying, et al., Bone mineral density and body composition in men with systemic lupus erythematosus: A case control study, Bone 43 (2008) 327–331.

[84] C. Lee, R. Ramsey-Goldman, Osteoporosis in systemic lupus erythematosus mechanisms, Rheum Dis Clin North Am 31 (2005) 363–385.

[85] C. Lee, O. Almagor, D. Dunlop, et al., Self-reported fractures and associated factors in women with systemic lupus erythematosus, J Rheumatol 34 (2007) 2018–2023.

[86] K.G. Saag, Osteoporosis: A. An epidemiology and clinical assessment, in: J.H. Klippel, J.H. Stone, L.J. Crofford, P.H. White (Eds.), Primer on the Rheumatic Diseases, 13th ed)., Springer, New York, (2008), pp. 576–583.

[87] P. Sambrook, Osteoporosis: B. Pathology and pathophysiology, in: J.H. Klippel, J.H. Stone, L.J. Crofford, P.H. White (Eds.), Primer on the Rheumatic Diseases, 13th ed)., Springer, New York, (2008), pp. 584–591.

[88] National Osteoporosis Foundation, America's Bone Health: The state of osteoporosis and low bone mass (2009). http://www.nof.org/advocacy/prevalence (accessed December 2, 2009).

[89] National Osteoporosis Foundation, Clinician's Guide to Preventions and Treatment of Osteoporosis (2008). http://www.nof.org/professionals/NOF_Clinicians_Guide.pdf (accessed December 22, 2009).

[90] A. Qaseem, V. Snow, P. Skekelle, et al., Screening for osteoporosis in men: A clinical practice guideline from the American College of Physicians, Ann Intern Med 148 (9) (2008) 680–684.

[91] P. Panopalis, J. Yazdany, Bone health in systemic lupus erythematosus, Curr Rheumatol Rep 11 (2009) 177–184.

[92] J.D. Alele, D.L. Kamen, The importance of inflammation and vitamin D status in SLE-associated osteoporosis, Autoimmun Rev 9 (3) (2009) 137–139.

[93] N. Lane, Osteoporosis: Is there a rational approach to fracture prevention, Bull Hosp Joint Dis 64 (2006) 67–71.

[94] N.E. Lane, Therapy insight: Osteoporosis and osteonecrosis in systemic lupus erythematosus, Nat Clin Pract Rheumatol 2 (10) (2006) 562–569.

[95] N.E. Lane, Q. Rehman, Osteoporosis in the rheumatic disease patient, Lupus 11 (2002) 675–679.

[96] M. Boyanov, R. Robeva, P. Popivanov, Bone mineral density changes in women with systemic lupus erythematosus, Clin Rheumatol 22 (2003) 318–323.

[97] I.R. Reid, Glucocorticoid osteoporosis—Mechanisms and management, Eur J Endocrinol 137 (1997) 209–217.

[98] R.G. Klein, S.B. Arnaud, J.C. Gallagher, et al., Intestinal calcium absorption in exogenous hypercortisolism. Role of 25-hydroxyvitamin D and corticosteroid dose, J Clin Invest 60 (1977) 253–259.

[99] S.C. Manolagas, R.S. Weinstein, New developments in pathogenesis and treatment of steroid-induced osteoporosis, J Bone Miner Res 14 (1999) 1061–1066.

[100] R.S. Weinstein, R.L. Jilka, A.M. Parfitt, et al., Inhibition of osteoblastogenesis and promotion of apoptosis of osteoblasts and osteocytes by glucocorticoids, J Clin Invest 102 (1998) 274–282.

[101] Y. Kipen, B.J.G. Strauss, E.F. Morand, Body composition in systemic lupus erythematosus, Br J Rheumatol 37 (1998) 514–519.

[102] C. Lee, O. Almagor, D. Dunlop, et al., Association between African American race/ethnicity and low bone mineral density in women with systemic lupus erythematosus, Arthritis Rheum 57 (4) (2007) 585–592.

[103] E.K. Li, L.S. Tam, R.P. Young, Loss of bone mineral density in Chinese pre-menopausal women with systemic lupus erythematosus treated with corticosteroids, Br J Rheumatol 37 (4) (1998) 405–410.

[104] V. LoCascio, E. Bonucci, B. Imbimbo, et al., Bone loss in response to long-term glucocorticoid therapy, Bone Miner 8 (1990) 39–51.

[105] D. Jardinet, C. Lefebvre, G. Depresseux, Longitudinal analysis of bone mineral density in pre-menopausal female systemic lupus erythematosus patients: Deleterious role of glucocorticoid therapy at the lumbar spine, Rheumatology 39 (2000) 389–392.

[106] G. Rizzato, L. Montemurro, Reversibility of exogenous corticosteroid-induced bone loss, Eur Respir J 6 (1993) 116–119.

[107] A. Zonana-Nacach, S.G. Barr, L.S. Magder, M. Petri, Damage in systemic lupus erythematosus and its association with corticosteroids, Arthritis Rheum 43 (8) (2000) 1801–1808.

[108] F.A. Houssiau, C. Lefebvre, G. Depresseux, et al., Trabecular and cortical bone loss in systemic lupus erythematosus, Br J Rheumatol 35 (1996) 244–247.

[109] C. Lee, O. Almagor, D.D. Dunlop, et al., Disease damage and low bone mineral density: An analysis of women with systemic lupus erythematosus ever and never receiving corticosteroids, Rheumatology 45 (2006) 53–60.

[110] A. Becker, R. Fisher, W.A. Scherbaum, M. Schneider, Osteoporosis screening in systemic lupus erythematosus: Impact of disease duration and organ damage, Lupus 10 (2001) 809–814.

[111] L. Sinigaglia, M. Varenna, L. Binelli, et al., Bone mass in systemic lupus erythematosus, Clin Exp Rheumatol 18 (2) (2000) S27–S34.

[112] R.G. Lahita, H.L. Bradlow, E. Ginzler, et al., Low plasma androgens in women with systemic lupus erythematosus, Arthritis Rheum 30 (1987) 241–248.

[113] R.G. Lahita, Sex hormones and systemic lupus erythematosus, Rheum Dis Clin North Am 26 (4) (2000) 951–968.

[114] S.S. Shabanova, L.P. Ananieva, Z.S. Alekberova, Guzov II, Ovarian function and disease activity in patients with systemic lupus erythematosus, Clin Exp Rheumatol 26 (3) (2008) 436–441.

[115] L. Sinigaglia, M. Varenna, G. Girasole, G. Bianchi, Epidemiology of osteoporosis in rheumatic diseases, Rheum Dis Clin North Am 32 (2006) 631–658.

[116] S. Lakshminarayanan, S. Walsh, M. Mohanraj, N. Rothfeld, Factors associated with low bone mineral density in female patients with systemic lupus erythematosus, J Rheumatol 28 (2001) 102–108.

[117] V.C.Z. Borba, J.G.H. Vieira, T. Kasamatsu, Vitamin D deficiency in patient with active systemic lupus erythematosus, Osteoporosis Int 20 (2009) 427–433.

[118] N. Shoenfeld, H. Amital, Y. Shoenfled, The effect of melanism and vitamin D synthesis on the incidence of autoimmune disease, Nat Clin Pract Rheumatol 5 (2) (2009) 99–105.

[119] R.R. McLean, P.F. Jacques, J. Selhub, et al., Homocysteine as a predictive factor for hip fracture in older persons, N Engl J Med 350 (2004) 2042–2049.

[120] U.S. Preventive Services Task Force, Osteoporosis Screening, Agency for Healthcare Research and Quality, Rockville, MD, 2002, September. http://www.ahrq.gov/clinic/3rduspstf/osteoporosis (accessed December 22, 2009).

[121] J. Yazdany, P. Panopalis, J. Zell, et al., A quality indicator set for systemic lupus erythematosus, Arthritis Rheum 61 (3) (2009) 370–377.

[122] S.S. Yeap, A.R. Fauzi, N.C.T. Kong, et al., Influences on bone mineral density in Malaysian premenopausal systemic lupus erythematosus patients on corticosteroids, Lupus 18 (2009) 178–181.

[123] V. Lilleby, Bone status in juvenile systemic lupus erythematosus, Lupus 16 (2007) 580–586.

[124] American College of Rheumatology Ad Hoc Committee on Glucocorticoid-Induced Osteoporosis, Recommendation for the prevention and treatment of glucocorticoid-induced osteoporosis, Arthritis Rheum 44 (7) (2001) 1496–1503.

[125] H.C. Chong, S.S. Chee, E.M. Goh, et al., Dietary calcium and bone mineral density in premenopausal women with systemic lupus erythematosus, Clin Rheumatol 26 (2007) 182–185.

[126] B.E. Nordin, Calcium and osteoporosis, Nutrition 13 (7–8) (1997) 664–686.

[127] B. Shea, G. Wells, A. Cranney, et al., Osteoporosis Methodology Group and the Osteoporosis Research Advisory Group. Meta-analysis of therapies for postmenopausal osteoporosis VII: Meta-analysis of calcium supplementation for the prevention of postmenopausal osteoporosis, Endocr Rev 23 (2002) 552–559.

[128] E. Papadimitropoulos, G. Wells, B. Shea, et al., Osteoporosis Methodology Group and the Osteoporosis Research Advisory Group. Meta-analysis of therapies for postmenopausal osteoporosis VIII: Meta-analysis of the efficacy of vitamin D treatment in preventing osteoporosis in postmenopausal women, Endocr Rev 23 (2002) 560–569.

[129] S.S. Yeap, A.R. Fauzi, N.C.T. Kong, et al., A comparison of calcium, calcitriol, and aledronate in corticosteroid-treated premenopausal patients with systemic lupus erythematosus, J Rheumatol 35 (12) (2008) 2344–2347.

[130] A. Kelman, N.E. Lane, The management of secondary osteoporosis, Best Pract Res Clin Rheumatol 19 (6) (2005) 1021–1037.

[131] K.G. Saag, J.R. Zanchetta, J.P. Devogelaer, et al., Effects of teriparatide versus alendronate for treating glucocorticoid-induced osteoporosis, Arthritis Rheum 60 (11) (2009) 3346–3355.

[132] H.P. Bhaattoa, P. Bettembuk, A. Balogh, G. Szegedi, The effect of 1-year transdermal estrogen replacement therapy on bone mineral density and biochemical markers of bone turnover in osteopenic postmenopausal systemic lupus erythematosus patients: A randomized, double-blind, placebo-controlled trial, Osteoporosis Int 15 (2004) 396–404.

[133] C.C. Mok, C.H. To, A. Mak, K.M. Ma, Raloxifene for postmenopausal women with systemic lupus erythematosus: A pilot randomized controlled study, Arthritis Rheum 52 (12) (2005) 3997–4002.

[134] J. Sanchez-Guerrero, H.E. Fragoso-Loyo, C.M. Nuewelt, Effects of prasterone on bone mineral density in women with active systemic lupus erythematosus receiving glucocorticoid therapy, J Rheumatol 35 (2008) 1567–1575.

[135] K.A. Alsufyani, O. Ortiz-Alvarez, D.A. Cabral, Bone mineral density in children and adolescents with systemic lupus erythematosus, juvenile dermatomyositis, and systemic vasculitis: Relationship to disease duration, cumulative corticosteroid dose, calcium intake and exercise, J Rheumatol 32 (2005) 729–733.

[136] V. Lilleby, G. Lien, K. Frey Froslie, et al., Frequency of osteopenia in children and young adults with childhood-onset systemic lupus erythematosus, Arthritis Rheum 52 (2005) 2051–2059.

[137] S. Compeyrot-Lacassagne, P.N. Tyrrell, E. Atenafu, et al., Prevalence and etiology of low bone mineral density in juvenile systemic lupus erythematosus, Arthritis Rheum 56 (6) (2007) 1966–1973.

[138] S. Trapani, R. Civnini, M. Ermini, et al., Osteoporosis in juvenile systemic lupus erythematosus: A longitudinal study on the effect of steroids on bone mineral density, Rheumatol Int 18 (1998) 45–49.

[139] M.L. Bianchi, R. Cimaz, M. Bardare, et al., Efficacy and safety of alendronate for the treatment of osteoporosis in diffuse connective tissue diseases in children: A prospective multicenter study, Arthritis Rheum 43 (2000) 1960–1966.

[140] S. Rudge, S. Hailwood, A. Horne, Effects of once-weekly oral alendronate on bone in children on glucocorticoid treatment, Rheumatology (Oxford) 44 (2005) 813–818.

[141] E. Unal, A. Abaci, E. Bober, A. Büyükgebiz, Efficacy and safety of oral alendronate treatment in children and adolescents with osteoporosis, J. Pediatr. Endocrinol. Metab. 19 (4) (2006) 523–528.

[142] R. Prasad, D. Ibanez, D. Gladman, M. Urowitz, The role of non-corticosteroid related factors in osteonecrosis (ON) in SLE: A nested case–control study of inception patients, Lupus 16 (2007) 157–161.

[143] T.H. Berquist, Pelvis, hips and thighs, in: T.H. Berquist (Ed.), Musculoskeletal Imaging Companion, Lippincott Williams & Wilkins, Philadelphia, 2007.

PART III

ANTI-PHOSPHOLIPID SYNDROME

A. Pathogenesis
(Chapters 50–51)

B. Clinical
(Chapters 52–55)

Antiphospholipid Syndrome: Pathogenesis

Tatsuya Atsumi, Olga Amengual, Takao Koike

HISTORICAL BACKGROUND

Antiphospholipid antibodies (aPLs) include a heterogeneous group of circulating immunoglobulins present in a wide range of infectious and autoimmune diseases. In particular, anticardiolipin antibodies (aCL) and lupus anticoagulants (LA) are associated with the antiphospholipid syndrome (APS). The term APS is used to link thrombosis and/or pregnancy morbidity to the persistence of aPLs [1].

Since the first description of APS in 1983 [2], the range of manifestations associated with aPLs has considerably broadened with recognition of other manifestations such as livedo reticularis, nonthrombotic neurological syndromes, psychiatric manifestations, skin ulcers, hemolytic anemia, thrombocytopenia, pulmonary hypertension, nephropathy, heart valve abnormalities, and atherosclerosis [3]. Nowadays, APS is recognized as a condition that is a frequent cause of acquired thrombophilia.

Antiphospholipid antibodies were first discovered in 1906, when Wassermann *et al.* [4] described a method to detect syphilis and noticed that sera from syphilitic patients could agglutinate a lipoid tissue extract. In the early 1940s, Pangborn [5] found that the relevant antigenic component of the tissue extracts used in these tests was a phospholipid named cardiolipin. The antibodies reactive to cardiolipin were termed "reagins." Reagins in serum samples can be detected with the antigen mixture of cardiolipin, cholesterol, and phosphatidylcholine (Venereal Disease Research Laboratory or VDRL antigen), resulting in a visible flocculation in *in vitro* assays. Although reagins are found predominantly in association with syphilitic infection, they are not specific to antigens of *Treponema pallidum*. Tests for treponema-specific antigens (*T. pallidum* hemagglutination assay) developed in the 1950s revealed that individuals with chronically biological false-positive serological test for syphilis (BFP-STS), when followed up for many years, often developed systemic lupus erythematosus (SLE) [6–8]. Some patients with SLE and chronically BFP-STS had recurrent spontaneous fetal loss, thrombocytopenia, and thromboembolic problems compared with age-matched SLE patients with a negative STS [9].

In addition, BFP-STS results were noted in early reports of SLE patients with acquired inhibitors of *in vitro* phospholipid-dependent coagulation tests, particularly the activation of prothrombin [10]. This association was investigated by Laurell and Nilsson [11], who suggested that the anticoagulant effect and the positive reaction with phospholipid antigen were ascribable to the same causal factor.

The name lupus anticoagulant, coined by Feinstein and Rapaport [12], represents the first of the major misnomers in this field. These antibodies, designated LA because they were originally detected in the plasma of patients with SLE, are more frequently present in people who do not have SLE. Although LA prolong coagulation reactions *in vitro*, clinically they can cause thrombotic diseases, and individuals with LA do not have a bleeding tendency, unless a second coagulation defect such as factor VIII deficiency or thrombocytopenia is present. Soon thereafter, the 'LA phenomenon' was known to be caused by an antibody reacting to a lipid antigen. Against this background, evidence supports that these antibodies are specific for phospholipid complexed with plasma proteins, such as β_2-glycoprotein I (β_2GPI).

In the early 1980s, radioimmunoassay and enzyme-linked immunosorbent assay (ELISA), which directly detect circulating aCL, were devised by Harris *et al.* [13] and Koike *et al.* [14], respectively. Investigators noticed that aCL cross-reacted with negatively charged phospholipids, such as phosphatidylserine and phosphatidylglycerol [15]. Thus, the name aCL was expanded to aPLs. Studies on the aCL assay demonstrated the requirement of additive bovine serum in the blocking buffer, diluent, or both to enhance aCL

945

binding to the targeted phospholipid. In 1990, three groups independently reported the necessity of a cofactor for the binding of autoimmune aCL to the solid-phase phospholipids [16–18]. β_2GPI was identified as this cofactor. Subsequently, it was shown that the epitope for aCL develops when β_2GPI is adsorbed on polyoxygenated polystyrene plates, and that β_2GPI interacts with a lipid membrane composed of negatively charged phospholipids [19].

Since then, a number of "cofactors" have been described, and the antigenic specificities of aPLs include other proteins or phospholipid–protein complexes such as prothrombin, annexin V, protein S, protein C, and high- and/or low-molecular-weight kininogen [20–22]. In particular, prothrombin has been recognized as the "second" major antigenic target of autoimmune aPLs, and antiprothrombin antibodies were shown to be responsible for LA activity in many APS patients [20]. The expression of epitopes by at least some of these phospholipid-binding proteins does not depend on the presence of phospholipids. The spectrum of potential immunogenic targets facilitates subclassification of these heterogeneous antibodies.

In 1998, an international consensus on classification criteria for definite APS was stated in Sapporo; the criteria were thus called Sapporo criteria [23], and they were revised in 2006 [1]. In the revised criteria, IgG/M aCL, IgG/M anti-β_2GPI antibodies, and LA are listed to categorize APS. Other aPLs, as markers of APS, are still under discussion.

CELL MEMBRANE PHOSPHOLIPIDS

Phospholipids are compounds composed of 1,2-diacylglycerol and a phosphodiester bridge linking the glycerol backbone to the nitrogenous bases of choline, ethanolamine, the inositol derivatives, and glycerol. Cardiolipin is made up of two phospholipid molecules joined by a central glycerol molecule attached to the phosphodiester group. Phospholipids are abundant in the brain, spinal cord, and body fluids such as plasma, but they occur primarily in various organelle and cellular membranes, including mitochondria, endothelial cells, and aggregated and/or activated platelets.

The distribution of phospholipids over the two halves of the cellular membrane bilayer is extremely asymmetric for both phosphatidylserine and sphingomyelin, the first being almost exclusively present in the inner monolayer, whereas sphingomyelin is confined to the outer monolayer. Phosphatidylcholine and phosphatidylethanolamine are distributed more evenly, although phosphatidylcholine seems to have a preference for the outer monolayer and phosphatidylethanolamine for the inner leaflet. The result of this distribution is that the outer surface of cells is approximately 80% composed of phosphatidylcholine. The inner monolayer of the cellular membrane almost exclusively comprises negatively charged phospholipids [24].

Many studies led to the concept that membrane phospholipid asymmetry is ubiquitous. Plasma membrane phospholipid asymmetry is assembled and maintained by specific mechanisms that control transbilayer lipid movement. An adenosine triphosphate (ATP)-dependent aminophospholipid translocase or "flipase," another ATP-dependent translocating activity referred to "flopase," and a calcium-dependent phospholipid scramblase-1 may regulate the differential transbilayer orientation of phospholipids in cell membranes [25, 26].

Loss of asymmetry, especially the appearance of phosphatidylserine at the cell surface, is associated with many physiological and pathological phenomena, including hemostasis and thrombosis [27]. Exposure of phosphatidylserine on the cell surface plays a central role in promoting blood coagulation because phosphatidylserine serves as a catalytic surface for the assembly of the coagulation factors, including the prothrombinase and tenase complex [28, 29]. The externalization of phosphatidylserine is also essential for the binding of aPLs to procoagulant cells.

ANTIPHOSPHOLIPID ANTIBODIES

The specificities of autoantibodies associated with the APS have been largely investigated, and it is clear that despite their name, aPLs do not direct to anionic phospholipids. They react with a complex of phospholipid and plasma proteins, including β_2GPI, prothrombin, annexin V, high- and/or low-molecular-weight kininogen, protein S, and protein C, previously known as 'cofactors.'

Anticardiolipin Antibodies and β_2-Glycoprotein I

Antigenic Target: $\beta2$-Glycoprotein I

β_2GPI is a 50-kDa plasma protein with a carbohydrate content of 17%, and it is present in normal human plasma at approximately 200 µg/ml [30]. β_2GPI was first described by Schultze et al. [31] in 1961 as a perchloric acid-soluble human plasma protein with unknown functions. However, it was not until the early 1990s that clear evidence emerged of its major role in the binding of aCL to phospholipids in patients with APS.

Northern blots revealed that β_2GPI is mainly synthesized in the liver but also in various cells. After eliminating sialic acid from the molecule, multiple isoelectric subspecies were identified. β_2GPI has been identified as a component of several human plasma proteins, such as chylomicrons and low-density, very low-density, and high-density lipoproteins; therefore, it is termed apolipoprotein H.

The complete amino acid sequence of human β_2GPI, as determined by peptide sequencing, reveals a single polypeptide chain composed of 326 amino acid residues with five oligosaccharide attachment sites [30]. The complete nucleotide sequence and the deduced amino acid sequence were defined by cDNA cloning from human liver-originated cells (e.g., a hepatoma cell line [HepG2]) and sequencing [32, 33]. β_2GPI belongs to a superfamily of proteins characterized by repeating stretches of approximately 60 amino acid residues, each with a set of 16 conserved residues and two fully conserved disulfide bonds. These repeating units are designated as short consensus repeat (SCR), complement control protein repeats, or sushi domains. β_2GPI consists of five SCR domains. The first four are regular SCR domains with respect to their amino acid sequences and are characterized by a framework of four conserved half-cysteine residues related to the formation of two internal disulfide bridges [34]. The fifth C-terminal domain deviates significantly from the common SCR folds. Domain V contains 82 amino acid residues and six half-cysteines, and it is stabilized by three internal disulfide bonds. This region, carrying a lysine-rich domain, has a definite positive charge character and is responsible for adhesion to anionic phospholipids [35–37] (Figures 50.1).

In 1999, two individual groups crystallized human β_2GPI and characterized its tertiary structure [38, 39]. The X-ray crystallographic structures of this heavily glycosylated protein reveal an elongated fishhook-like arrangement of the five globular SCR domains (Figure 50.2). Both of these crystal analyses consistently indicated that the C-terminal fifth domain deviates strongly from the standard fold, as observed in domains I–IV.

β_2GPI binds to solid-phase phospholipids through a specific region in the fifth domain CKNKEKKC (281–288), and the phospholipid binding property is significantly reduced by cleavage of one particular site (K317–T318) in domain V. Crystallographic studies of β_2GPI indicate that the aberrant non-SCR-like half of the fifth domain forms a specific phospholipid binding site. A large patch of 14 positively charged amino acid residues provides electrostatic interactions with anionic phospholipid head groups and an exposed membrane insertion loop yields specificity for lipid layers. Whitin domain V, the exposed C-terminal loop (S311–K317) is considered to be highly mobile, and it was not visible

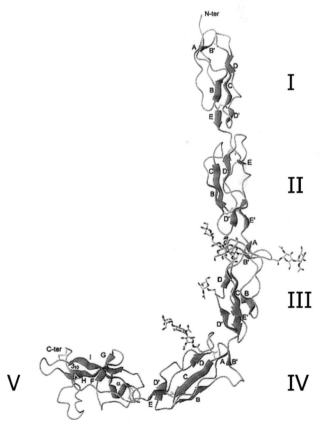

FIGURE 50.2 Crystal structure of human β_2-glycoprotein I: structure representation of human blood plasma β_2GPI shown as a ribbon drawing. The structure reveals the extended chain of the five short consensus domains. The β_2 strands are shown in blue, and disulfide bonds are displayed in yellow. *Source: Adapted with permission of Macmillan Publishers Ltd from Schwarzenbacher et al. [39].*

FIGURE 50.1 Amino acid sequence of human β_2-glycoprotein I.

TABLE 50.1   Functions of β₂-Glycoprotein I

ANTICOAGULANT

Inhibition of prothrombinase activity (thrombin generation)

Inhibition of factor X activation

Inhibition of factor XI activation

Inhibition of factor XII activation

Inhibition of platelet aggregation

Inhibition of tissue factor activity (extrinsic coagulation reaction)

PROCOAGULANT

Inhibition of protein C activation

Inhibition of inactivation of factor Va by activated protein C

Inhibition of protein S

Inhibition of protein Z anticoagulant pathway

Inhibition of tPA activity

tPA, tissue plasminogen activator.

in the final electron density map in one of the two previously mentioned studies on the crystallized β₂GPI [38]. Furthermore, Hoshino et al. [40] determined the three-dimensional solution structure of domain V by heteronuclear multidimensional magnetic resonance and showed that the molecule is composed of four well-defined anti-parallel β-strands and two short α-helices, as well as a long, highly flexible loop. There is a sequence of hydrophobic residues (L313−W316) on the tip of the long C-terminal loop. This cluster of hydrophobic residues, together with the surrounding positively charged residues, seems to construct an ideal binding site on β₂GPI for negatively charged phospholipids.

β₂GPI binds to negatively charged substances such as heparin, dextran sulfate, lipoproteins, DNA, and anionic phospholipids such as cardiolipin. β₂GPI adheres to activated platelets, inhibits the contact activation of blood coagulation, inhibits prothrombinase activity, activates lipoprotein lipase [41], lowers triglyceride levels [42], binds to oxidized low-density lipoprotein [43], and binds to non-self-particles or apoptotic bodies to allow their clearance [44, 45].

The in vitro properties of β₂GPI as a natural regulator of coagulation have been proposed in many reports, but the roles of β₂GPI in coagulation are not completely elucidated [46−49]. Table 50.1 summarizes the in vitro effects of β₂GPI in the coagulation pathway.

Polymorphism of β2-Glycoprotein I

The human β₂GPI gene is localized on chromosome 17q23-qter. When the molecular basis of β₂GPI polymorphisms was defined, four major polymorphisms (Ser/Asn88, Leu/Val247, Cys/Gly306, and Trp/Ser316) became evident [50−52].

Two siblings with a complete β₂GPI deficiency, one apparently healthy 36-year-old woman and her brother, were identified [53, 54]. Analysis of their β₂GPI genes revealed that a thymine corresponding to position 379 of the β₂GPI cDNA was deleted (β₂GPI-Sapporo); hence, a frame shift would occur, thus making the gene code for an amino acid sequence unrelated to β₂GPI beyond this position. The siblings of homozygous β₂GPI-Sapporo subjects had no thrombotic episodes. In contrast, there is a case report of a carrier with both Gly306 and Ser316 β₂GPI mutation, a low phospholipid binding variant of β₂GPI, who had recurrence of deep vein thrombosis and stroke until the age of 35 years [55].

β₂GPI knockout mice showed decreased in vitro ability for thrombin generation [56]. Furthermore, although mice lacking β₂GPI are fertile, the success of early pregnancy is moderately compromised, suggesting that β₂GPI is necessary for optimal implantation and placental morphogenesis.

Nicked β2-Glycoprotein I

The nicked (cleaved) form of β₂GPI, produce by plasmin and activated factor X (Xa) through cleaving at specific residues (K317−T318) in the fifth domain [57, 58], lacks the ability to interact with anionic phospholipids [35, 36, 59, 60].

Matsuura et al. [61] attempted to predict tertiary structures of intact and nicked domain V of β₂GPI through protein modeling and epitope mapping of anti-β₂GPI monoclonal antibody, which is specific to the intact β₂GPI, using phage library technology. Their data suggested that the two flanking cysteines were important for peptide conformational presentation of the fifth domain when binding to phospholipids. In contrast, the pathophysiological implications of plasma/serum levels of the nicked β₂GPI have been discussed. Horbach et al. [62] reported that plasma β₂GPI levels were lower in patients with disseminated intravascular coagulation syndrome compared with controls, and nicked β₂GPI levels were contrarily high. They suggested that the generation of the nicked β₂GPI occurred due to absorption to apoptotic cells and activation of fibrinolysis, respectively. Furthermore, Itoh et al. [63] reported highly increased plasma levels of the nicked β₂GPI in patients with leukemia and with LA, and we found that nicked β₂GPI plasma levels correlated with cerebral infarction lesions [64].

The properties of nicked β₂GPI were investigated. Instead of losing phospholipid binding activity, nicked β₂GPI gains binding capacity to plasminogen and suppresses the plasmin generation in the presence of tissue plasminogen activator (tPA) and fibrin. Therefore, nicked β₂GPI is a natural regulator of extrinsic fibrinolysis by negative feedback pathway.

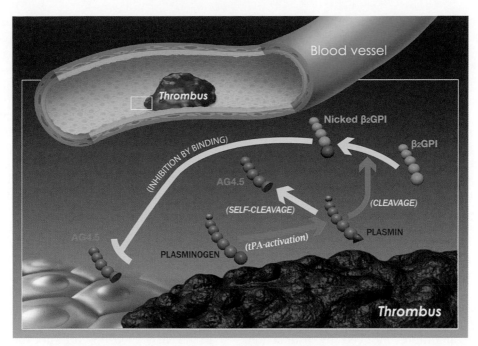

FIGURE 50.3 Effect of nicked β_2-glycoprotein I on thrombotic ischemia. When ischemic thrombus is formed in the artery, upregulation of fibrinolysis occurs and plasmin is generated from its precursor plasminogen. Plasmin cleaves β_2GPI into nicked β_2GPI, whereas angiostatin is generated via autoproteolysis of plasmin. Nicked β_2GPI binds angiostatin and attenuates its antiangiogenic property, resulting in promoted angiogenesis. AG4.5, angiostatin4.5; tPA, tissue plasminogen activator.

Angiostatin is a plasminogen fragment with an anti-angiogenic property. Among the isoforms of angiostatin, angiostatin4.5, also known as plasminogen kringle 1−5 (K1−5), which consists of plasminogen kringle 1−4 and most of the kringle 5, is the product of plasmin auto-proteolysis. Nakagawa *et al.* [65] showed that plasmin-nicked β_2GPI promotes angiogenesis by interacting with plasmin-generated angiostatin4.5, suggesting that the nicked form of β_2GPI plays a role in the regulation of angiogenesis at the site of thrombosis (Figure 50.3).

Antibodies against β2-Glycoprotein I

Anticardiolipin antibodies present in patients with APS can be detected by solid-phase immunoassays, typically ELISAs, with the putative antigen, cardiolipin, dried onto microtiter plates [13, 14]. However, in addition to detecting antibodies to cardiolipin, this assay also detects antibodies against phospholipid-binding plasma or serum proteins (i.e., cofactors), specifically β_2GPI bound to the cardiolipin. These proteins are present in the serum or plasma samples and/or can be components of sample diluent and blocking buffers. Most, if not all, aCL associated with APS are directed against epitopes expressed on β_2GPI, not on cardiolipin or other negatively charged phospholipids such as phosphatidylserine, phosphatidic acid, or phosphatidylinositol [16, 17, 19, 59, 66−68]. In contrast, aPLs associated with infectious diseases bind directly to these negatively charged phospholipids [69]. Anti-β_2GPI autoantibodies recognize cryptic epitope(s) on the β_2GPI molecule that is exposed when β_2GPI interacts with lipid membranes composed of negatively charged phospholipids or when β_2GPI is adsorbed on a polyoxygenated polystyrene plate treated with γ-irradiation or electrons [19, 70].

The mechanisms involved in the binding of anti-β_2GPI antibodies to β_2GPI are still a matter of discussion. Several groups have reported that anti-β_2GPI autoantibodies in APS patients have a relatively low affinity. The location of the epitopic sites on β_2GPI for anti-β_2GPI autoantibodies from APS patients is also controversial. Anti-β_2GPI antibodies have been shown to recognize different epitopes located in all five domains of β_2GPI. In 1996, Igarashi *et al.* [70] first reported that domains IV or I are candidates for epitopic location in the β_2GPI molecule by using a series of deletion mutant proteins of β_2GPI. Later, George *et al.* [71] demonstrated that domain IV of β_2GPI is one of the major epitopic locations for monoclonal aCL raised in APS patients. In contrast, Iverson *et al.* [72] and McNeeley *et al.* [73] showed that anti-β_2GPI antibodies in the major population of APS patients bind to a particular structure in domain I of β_2GPI, and antibody binding was diminished by replacement of a related amino acid located in the domain. However, some specific mutations made in domain IV were reported to affect antibody binding to β_2GPI in the anti-β_2GPI antibody ELISA [74].

Moreover, de Laat *et al.* [75] showed that pathogenic aCL bind to a cryptic epitope (G40−R43) on domain I of β_2GPI, which is accessible for aCL only after conformational change, and it is induced by the binding of β_2GPI to a negatively charged surface via a positive-charge patch in domain V. Ioannou et al. [76] reported that the binding of aPLs to β_2GPI domain I is complex and likely involves discontinuous epitopes that include R39 in addition to G40—R43.

Kasahara and colleagues [77] demonstrated that epitopic structures recognized by anti-β_2GPI antibodies are cryptic and that three electrostatic interactions between domains IV and V (D193—K246, D222 —K317, and E228—K308) participate in their exposure. Binding of anti-β_2GPI antibodies to solid-phase β_2GPI was significantly diminished by L replacement of W235, an amino acid residue partly located in an inner region on domain IV and commonly found in the epitopic structures. Thus, cryptic epitopes in domain IV might by recognized by anti-β_2GPI antibodies found in patients with APS, and epitope spreading may occur in the region surrounding W235.

Amino acid differences of β_2GPI can affect the nature of conformational alterations induced by interaction with phospholipids. Therefore, polymorphism on or near the phospholipid binding site of the antigenic site can affect aCL production and the development of APS. Among the polymorphisms, β_2GPI Leu/Val247 polymorphism locates in domain V of β_2GPI, between the phospholipid binding sites [59] and the potential site of the epitopes for anti-β_2GPI antibodies [70, 78]. Atsumi and colleagues [79] and Hirose et al. [80] analyzed the genetic polymorphism of β_2GPI and reported that the allele containing Val247 may be important for β_2GPI antigenicity. In addition, Yasuda and colleagues [81] found that autoimmune anti-β_2GPI antibodies showed stronger binding to Val247—β_2GPI than to Leu247—β_2GPI. Conformational optimization indicated that the replacement of Leu247 with Val247 led to a significant alteration in the tertial structure of domain V and/or domain IV—V interaction.

Animal Models of Antiphospholipid Syndrome and β2-Glycoprotein I Knockout Mice

In NZW × BXSB F1 (WB F1) mice with systemic lupus-like disease, several autoantibodies, circulating immune complexes, and lupus nephritis are present. Myocardial infarction is the hallmark in WB F1 male mice. Yoshida et al. [82] found that the incidence of myocardial infarction in these mice increased with age and that more than 80% of male WB F1 mice with myocardial infarction had small multiple infarctions in the right ventricular free wall and anterior, lateral, posterior, and septal ventricular walls. In the affected coronary arteries, the intimate thickening consisted of cellular components with frequent recanalization or an eosinophil thrombus-like substance. Thrombocytopenia is a frequent finding in patients with aPLs. WB F1 mice also develop thrombocytopenia with age, and both platelet-associated antibodies and circulating antiplatelet antibodies have been detected in this animal. The mechanism by which thrombocytopenia occurs is unknown.

Hashimoto et al. [83] were the first to report that WB F1 male mice produced autoantibodies against cardiolipin and that the titer of aCL increases with age. These antibodies were detected by conventional aCL ELISA [14], and alternatively could be detected by ELISA for anti-β_2GPI antibody detection, using β_2GPI complexed with solid-phase cardiolipin [68] or using β_2GPI coated on a polyoxygenated polystyrene plate as an antigen [19].

As another approach to establish the animal model of human APS, active immunization with aCL, β_2GPI, or a combination of anti-β_2GPI antibodies and β_2GPI was considered. Immunization of animals with human β_2GPI induces antibodies that resemble aPLs from APS patients with regard to binding characteristics [84, 85] and pathogenic properties [86].

In the early 2000s, β_2GPI knockout mice were generated by using a homologous recombination approach [56]. When β_2GPI heterozygotes on a mixed genetic background were intercrossed, only 9% of the resulting 336 offspring possessed both disrupted alleles. They had an impaired in vitro ability to generate thrombin relative to wild-type mice. Thus, β_2GPI appears to play an important role in thrombin-mediated coagulation.

Antiphospholipid Antibodies and T-Cells

Interesting research investigating T-cells responsible for the development of APS has been performed. Hattori et al. [87] identified HLA class II-restricted β_2GPI-reactive CD4$^+$ T-cells in the peripheral blood of APS patients. These β_2GPI-reactive T-cells possess helper activity that induces the production of antibodies directed to β_2GPI immobilized on cardiolipin. The same group established T-cell clones specific to β_2GPI from patients with APS and investigated their antigen recognition profiles and helper activity involved in promoting anti-β_2GPI antibody. They showed that β_2GPI—specific CD4$^+$ T-cells in patients with APS preferentially recognized a cryptic peptide encompassing amino acid residues 276—290 (p276—290) of β_2GPI, which contains the major phospholipid binding site. This cryptic peptide has the capacity to stimulate B cells to produce anti-β_2GPI antibodies through interleukin (IL)-6 expression and CD40—CD40 ligand engagement [88].

Kuwana et al. [89] demonstrated that the binding of β_2GPI to anionic phospholipid facilitates processing and presentation of a cryptic epitope that activates pathogenic autoreactive CD4$^+$ T cells. Furthermore, it was reported that β_2GPI-reactive CD4$^+$ T-cell proliferation depends on monocytes as antigen-presenting cells [90]. The reaction was observed only in the presence of anti-β_2GPI antibodies and was mediated by Fcγ receptor I. These results suggest that the opsonization of the β_2GPI/phosphatidylserine complex by the IgG anti-β_2GPI antibody is essential for efficient antigen presentation from monocytes to T-cells and eventually for maintaining the pathogenic anti-β_2GPI antibody response in patients with APS.

Lupus Anticoagulant

History and Definition of Lupus Anticoagulant

Circulating anticoagulants are defined as endogenously produced substances that may interfere with coagulation either *in vivo* or *in vitro*. Most of the inhibitors are immunoglobulins. Other inhibitors, such as elevated levels of fibrin split products, are exceptional. LA are those immunoglobulins (IgG, IgM, IgA, or their combination) that interfere with the *in vitro* phospholipid-dependent tests of coagulation (prothrombin time (PT), activated partial thromboplastin time (aPTT), kaolin clotting time (KCT), and dilute Russell' s viper venom time). LA was first described in 1952 in lupus patients with bleeding tendency [10] and was thus called lupus anticoagulant, but this term is a misnomer because the majority of patients with LA do not have SLE. The paradoxical relationship between LA and a susceptibility to thrombosis was first identified in 1963 [91], and further studies clearly identified LA as a risk factor for both venous and arterial thromboembolic events. In 1980, Thiagarajan *et al.* [92] reported that monoclonal IgM antibody with LA activity cross-reacted with anionic phospholipids, and they proposed that LA bind to anionic but not dipolar ionic phospholipids, thereby limiting the exposed surface required for binding the prothrombinase complex. The *in vitro* action of LA appears to inhibit, albeit at a slow rate, the prothrombinase complex, which consists of factor Xa, activated factor V (Va; Xa/Va complex), calcium, and phospholipids [93].

β2-Glycoprotein I-Dependent Lupus Anticoagulant and Prothrombin-Dependent Lupus Anticoagulant

There are some subtypes of LA classified according to their clotting inhibitory behavior. In 1978, Exner *et al.* [94] described three types of LA mixing pattern by KCT. In 1990, β_2GPI was identified as a cofactor of solid-phase aCL assay, as described previously. Since this finding, cofactors of LA have been also investigated.

Oosting *et al.* [95] and Roubey *et al.* [96] demonstrated that many LA-positive plasmas depended on the presence of β_2GPI in their anticoagulant activity; thus, β_2GPI was identified as one of the major cofactors of LA. β_2GPI-dependent LA likely comprises antibodies directed against β_2GPI (anti-β_2GPI antibodies), but

many patients with anti-β_2GPI antibodies did not have LA. Takeya *et al.* [97] showed the heterogeneity of LA activity in a series of human/murine monoclonal anti-β_2GPI antibodies. Therefore, anti-β_2GPI antibodies can be either antibodies with LA activity or those without, and anti-β_2GPI antibodies with LA activity were assumed to be more pathogenic than anti-β_2GPI antibodies without LA. On the other hand, murine monoclonal anti-human β_2GPI antibodies behaved as LA [98], suggesting that not only autoantibodies but also immunized antibodies can have LA activity, and that antibody binding to β_2GPI is not necessarily related to the presentation of cryptic or neo-epitopes on the molecule.

β_2GPI binding to membranes with physiological anionic phospholipid content is relatively weak in comparison to that of plasma coagulation proteins [99]. As LA and/or its cofactors compete with coagulation factors for the available catalytic phospholipid surface, it is assumed that anti-β_2GPI antibodies with LA activity enhance the affinity of β_2GPI for phospholipids. Using ellipsometry, Willems *et al.* [100] demonstrated that high-affinity binding of aCL−β_2GPI complexes (purified from a single patient's serum) to phospholipids was due to bivalent interactions between them and lipid-bound β_2GPI. By solid-phase ELISA technique, it was shown that murine monoclonal anti-β_2GPI antibodies enhanced the β_2GPI binding to fixed phospholipids on ELISA plates [97]. These authors demonstrated that the F(ab')$_2$ (bivalent) fragment of the monoclonal anti-β_2GPI antibody had also promoted the enhancement, whereas the Fab' (monovalent) fragment did not, confirming the significance of the bivalency. Using real-time biospecific interaction analysis based on plasmon surface resonance technology, Arnout *et al.* [101] found a high-affinity interaction between β_2GPI and phospholipids in the presence of murine monoclonal anti-β_2GPI antibody with LA activity but not in the presence of monoclonal anti-β_2GPI antibodies without LA activity. Fab' fragments inhibited the formation of stable high-affinity complexes on phospholipid surfaces. These authors noted that bivalency may be essential to cross-link two β_2GPI molecules and induce a correct spatial orientation of the phospholipid binding domains of both β_2GPI molecules, thus markedly increasing their affinity for phospholipid surface.

Prothrombin was first reported by Loeliger in 1959 to be a probable cofactor for LA [102]. One year later, Rapaport *et al.* [103] described the case of a child with LA who had recurrent episodes of bleeding. Further investigations showed a severe prothrombin deficiency and prolonged coagulation times (PT and aPTT). In 1991, Bevers *et al.* [20] emphasized the importance of prothrombin in causing LA activity. Furthermore, Oosting *et al.* [21] reported the inhibitory effect of LA on endothelial cell-mediated prothrombinase activity and that the IgG fraction containing LA activity bound to the phospholipid—prothrombin complex. Therefore, prothrombin was recognized as another target for auto-antibodies with LA activity. Accordingly, it is widely accepted that anti-β$_2$GPI antibodies and antiprothrombin antibodies are the two major autoantibodies responsible for LA activity: anti-β$_2$GPI antibodies for β$_2$GPI-dependent LA and antiprothrombin antibodies for prothrombin-dependent LA. Unlike β$_2$GPI, the cofactor effect of prothrombin was highly species specific and bovine prothrombin failed to support the inhibitory activity of LA [104].

In addition to β$_2$GPI and prothrombin, evidence indicates that a variety of plasma proteins may also act as cofactors of LA. These proteins are high- and/or low-molecular-weight kininogen, annexin V, protein C, and protein S, but their role as potential cofactors in LA activity is not yet clarified.

Antiprothrombin Antibodies

ANTIGENIC TARGET: PROTHROMBIN

Prothrombin (factor II) is a vitamin K-dependent single-chain glycoprotein of 579 amino acid residues with a molecular weight of 72 kDa, and it is present at a concentration of approximately 100 μg/ml in normal plasma [105]. Prothrombin undergoes γ-carboxylation during liver biosynthesis. These γ-carboxyglutamic residues, known as the Gla domain, are located on fragment 1 of the prothrombin molecule. The Gla domain is essential for the calcium-dependent binding of phospholipids to prothrombin, which is necessary for the conversion of prothrombin to biologically active α-thrombin. A kringle domain containing two kringle structures and a C-terminal serine protease follows the Gla domain.

Prothrombin is physiologically activated by the prothrombinase complex (factor Xa, factor Va, phospholipids, and calcium). Once negatively charged phospholipids bind prothrombin, prothrombinase complex converts prothrombin to thrombin, which triggers fibrinogen polymerization into fibrin [28]. Moreover, thrombin binds thrombomodulin on the surface of endothelial cells and activates protein C, which then exerts its anticoagulant activity by digesting factor Va and depriving the prothrombinase complex of its most important cofactor. Because of this negative feedback pathway, prothrombin/thrombin behaves as "indirect" anticoagulant.

PROTHROMBIN—ANTIPROTHROMBIN ANTIBODY REACTION

In 1983, Bajaj et al. [106] reported the presence of prothrombin-binding antibodies in patients with LA and severe hypoprothrombinemia. They postulated that hypoprothrombinemia results from the rapid clearance of prothrombin—antiprothrombin antibody complexes from the circulation. Then, prothrombin—antiprothrombin antibody complexes were detected by counterimmunoelectrophoresis in the plasma of patients with LA but without severe hypoprothrombinemia [107]. Fleck et al. [108] subsequently confirmed that antiprothrombin antibodies are responsible for the LA activity.

Antiprothrombin antibodies bind to prothrombin coated on γ-irradiated or activated polyvinyl chloride ELISA plates, antibodies directed against prothrombin alone (APT-A) [109, 110], or to prothrombin exposed to immobilized phosphatidylserine, phosphatidylserine-dependent antiprothrombin antibodies (anti-PS/PT) [110, 111].

Unlike β$_2$GPI, prothrombin requires calcium ions for its binding to anionic phospholipids. Antiprothrombin antibodies may be directed against cryptic or neo-epitopes exposed when prothrombin binds to anionic phospholipids. Upon binding to phosphatidylserine-containing surface in the presence of calcium ions, human prothrombin undergoes a conformational change that results in the exposure of a hydrophobic patch thought to be crucial for functional phospholipid binding [112]. It was proposed that a second conformational change creates a surface-exposed hydrophilic cleft that may be complementary in shape and charge to that of the polar group [113].

On the other hand, antiprothrombin antibodies may be low-affinity antibodies recognized more efficiently when the prothrombin is bound to phosphatidylserine coated on ELISA plates [110], or they may bind bivalently to immobilized prothrombin [114]. Thus, prothrombin complexed with phosphatidylserine could allow clustering and better orientation of the antigen, offering optimal conditions for antibody recognition.

The epitope(s) recognized by antiprothrombin antibodies remains to be defined. Rao et al. [115] demonstrated binding of antiprothrombin antibodies to prethrombin 1 and fragment 1, as well as to the whole prothrombin molecule. However, none of the antibodies reacted with immobilized thrombin. These findings suggest that dominant epitopes are likely to be located near the phospholipid binding site of the prothrombin, although they may have heterogeneous distribution. Moreover, Akimoto et al. [116] found that 62% of antiprothrombin antibodies were dominantly directed to the fragment 1 of prothrombin and 38% to the prethrombin site (fragment 2 plus α-thrombin).

The mechanisms by which antiprothrombin antibodies cause LA activity are not completely elucidated. Pierangeli et al. [117] suggested that antiprothrombin antibodies cause prolongation of in vitro clotting by inhibiting the conversion of prothrombin to thrombin.

However, it seems unlikely that antiprothrombin antibodies prolong the clotting times by hampering the activation of prothrombin through binding near the activation sites in the molecule because such a mechanism would not explain the neutralizing effects of high phospholipid concentrations. Simmelink et al. [118] showed that affinity-purified antiprothrombin antibodies from LA-positive plasma inhibited both prothrombinase and tenase complex. Thus, in the procoagulant pathway, antiprothrombin antibodies might increase the affinity of prothrombin for negatively charged phospholipids, thereby competing with clotting factors for the available catalytic phospholipid surface, a mechanism similar to that of anti-β_2GPI antibodies. The model, based on increased affinity of protein—antibody complexes (β_2GPI or prothrombin) for negatively charged phospholipids, can explain why LA activity caused by both anti-β_2GPI antibodies and antiprothrombin antibodies can be neutralized by the addition of extra phospholipids. Furthermore, in their investigation of the anticoagulant activity of the protein C system, Galli et al. [119] demonstrated the dominant inhibitory effect on activated protein C activity of anti-β_2GPI antibodies compared with the effect of antiprothrombin antibodies using isolated total IgG fractions. Thus, the phenomena may depend on the specificity of the antibodies used in the experiments.

Sakai et al. [120] established a phosphatidylserine-dependent monoclonal antiprothrombin antibody, 231D. The 231D spiked plasma showed strong LA activity, and in vitro thrombin generation was significantly reduced in the presence of high factor Va/factor Xa ratio. The anticoagulant activity of 231D may depend on its interaction with the factor Va binding site of the prothrombin molecule.

Other Autoantibodies Related to the Antiphospholipid Syndrome

Autoantibodies against phosphatidylethanolamine have been associated with thrombosis and pregnancy morbidity [121], but its utility in clinical practice is still under discussion [122]. Phosphatidylethanolamine is a zwitterionic phospholipid with an important role in phospholipid-dependent reactions of the protein C system. Sugi and McIntyre [123] demonstrated that the phospholipid-binding proteins responsible for antiphosphatidylethanolamine antibodies were high- and/or low-molecular-weight kininogen. Therefore, kininogen was recognized as one of the cofactors of aPLs. High-molecular-weight kininogen is present in low concentrations at birth. To avoid false-negative kininogen-dependent antiphosphatidylethanolamine antibody findings, fetal and newborn bovine serum should not be used as a blocking agent or in serum diluents [124]. Yamada

and colleagues showed the predictive value of antiphosphatidylethanolamine antibody for the adverse pregnancy outcome in a prospective study (Sapporo study) of the general population [125].

Oosting et al. [21] demonstrated that antibodies responsible for the inactivation of factor Va mediated by activated protein C are directed against phospholipid-bound protein C or protein S. Antibodies against factor XII, which plays a role in initiating intrinsic coagulation and/or intrinsic fibrinolysis, have also been related to the APS [126, 127].

PATHOGENESIS OF ANTIPHOSPHOLIPID ANTIBODIES

Because of their strong relationship with clinical symptoms, aPLs have been considered pathogenic antibodies. Studies on the pathogenicity of aPLs have been carried out mainly on the corresponding antigens, especially on the function of β_2GPI and their modifications by the antibodies. However, the recent trend is to favor the hypothesis that the action of the autoantibodies on prothrombotic cells via β_2GPI is more important than the function of β_2GPI. A number of molecules that take part in the activation of prothrombotic cells have been identified.

It is recognized that the inhibition of the natural anticoagulant systems, the impairment of fibrinolytic activity, and the direct effect of aPLs on cell functions are some of the mechanisms of aPL-mediated thrombosis. Evidence suggests that complement activation is also required for aPL-mediated tissue injury, and findings have revealed that aPLs play a role in the T-cell responses (Table 50.2).

Interference with the Coagulation Pathway

Protein C System

The protein C system is one of the important antithrombotic pathways mediated by the vessel wall, and its impairment may result in blood clot. Thrombin—thrombomodulin complex activates protein C. The activated protein C functions as anticoagulant, as related to proteolytically inactivation of factor Va and activated factor VIII (VIIIa) in the presence of protein S. Because both protein C and its cofactor, protein S, are phospholipid-binding plasma proteins, this system is one of the most likely to be involved in the pathogenesis of thrombosis in patients with APS.

The phospholipid-dependent reaction of the protein C system may be inhibited by aPLs in different ways. First, aPLs can hamper the activation of protein C by the thrombin—thrombomodulin complex [128]. Second, aPLs may inhibit the proteolytic effect of activated

TABLE 50.2   Proposed Mechanisms of Antiphospholipid Antibody Pathogenicity

Interference with the coagulation pathway

  Protein C pathway

    aPLs hamper protein C activation

    aPLs impair the catalytic function of activated protein C

    aPLs interfere with protein S

  Protein Z

  Contact activation pathway

  β_2GPI–thrombin interaction

  Antithrombin pathway

Impairment of fibrinolysis

  Upregulation of plasminogen activator inhibitor-1

  Interference with activated factor XII

  Effects of lipoprotein(a)

Cell interaction

  Induction of pro-coagulant activity on cells (endothelial cells and monocytes)

  Release of membrane-bound microparticles by endothelial cells and platelets

  Stimulation of platelet function

  Disruption of the annexin V shield

Complement activation

aPLs, antiphospholipid antibodies.

protein C on factors Va and VIIIa. A reduction of factor Va degradation was found in plasma of patients with LA [129]. The inhibitory effect of IgG purified from patients with aPLs on degradation of factor Va mediated by activated protein C was subsequently confirmed [130]. In addition, purified IgG/M from aPL-positive patients disturbed the anticoagulant activity of activated protein C on human endothelial cells [131]. Ieko *et al.* [49] demonstrated that human monoclonal anti-β_2GPI antibodies inhibit activated protein C function. Furthermore, aCL bound to protein C in the presence of both phospholipids and β_2GPI, and binding activities strongly correlated with anti-β_2GPI antibody titers, indicating that protein C could be a target of aCL leading to protein C dysfunction [132]. Galli *et al.* [133] reported that most prothrombin–antiprothrombin antibody immune complexes may predispose to thrombosis by interfering with the inactivation of factor Va by the activated protein C in the presence of protein S. This inhibitory effect of antiprothrombin antibodies on activity of activated protein C was also demonstrated in the absence of protein S [134].

Finally, aPLs may alter the effect of protein S in the protein C pathway. Reduced levels of protein S have been found in plasma from APS patients [135, 136]. Some of the IgG that inhibited factor Va degradation was directed not only to phospholipid-bound protein C but also to phospholipid-bound protein S [21].

Protein Z

The inhibition of protein Z, a natural anticoagulant system, has been proposed as an additional thrombotic mechanism in APS. However, the thrombotic risk related with protein Z deficiency is unclear. Protein Z is a vitamin K-dependent protein that serves as a cofactor for the plasma protein Z-dependent protease inhibitor for the inactivation of factor Xa. Plasma protein Z deficiency was detected in patients with aPLs [137] and reported to be associated with thrombosis [138]. In the presence of β_2GPI, aPLs greatly impair the inhibition of factor Xa by protein Z/protein Z-dependent protease inhibitor [139].

Contact Activation Pathway

The contact pathway of coagulation is initiated with activation of factor XII by contact with negatively charges surfaces. Activated factor XII (XIIa) cleaves factor XI to activated factor XI (XIa) in the presence of high-molecular kininogen and prekalikrein. β_2GPI inhibited the phospholipid-mediated autoactivation of factor XII and the contact activation pathway of coagulation [47]. Shi *et al.* [140] demonstrated that β_2GPI binds directly to factor XI and inhibits activation of factor XI by thrombin and factor XIIa; this inhibition attenuates thrombin generation. Dysregulation of the contact activation pathway by aPLs may be an important mechanism of thrombosis in patients with APS.

Monoclonal anti-β_2GPI antibodies enhanced the inhibition of factor XI activation by β_2GPI and thrombin complex [141]. The direct effect of anti-β_2GPI antibodies on thrombin-dependent factor XI activation may, in part, explain why anti-β_2GPI antibodies have LA activity.

Thrombin

Thrombin is one of the most potent enzymes involved in the regulation of a number of important biological functions *in vivo*. Thrombin is a member of the serine protease family, and it is generated from its inactive precursor prothrombin by factor Xa, as part of the prothrombinase complex, on the surface of activated cells. The hemostatic reaction is the best known function of thrombin. Thrombin acts as procoagulant not only by cleaving fibrinogen to fibrin but also by interacting with protease-activated receptors to activate various procoagulant cells, and it also interacts with glycoprotein (GP) Ib–IX–V complex on the surface of platelets to promote platelet aggregation and activation. On the other hand,

955

thrombin behaves as anticoagulant upon binding to thrombomodulin to favor activation of protein C.

Thrombin takes part in many other reactions, including the establishment of inflammation, neoplastic transformation, angiogenesis, arteriosclerosis, and tissue repair.

β_2GPI is involved in thrombin generation as demonstrated by the significant reduction of in vitro ability to generate thrombin observed in plasma from β_2GPI null mice [56]. Rahgozar et al. [141] showed, for the first time, that domain V of β_2GPI directly binds exosites I and II on thrombin. This finding revealed that interaction between β_2GPI and thrombin may interfere not only with the coagulation system but also with many of the biologic functions in which thrombin participates, thus providing new insight into the pathophysiology of APS.

Antithrombin Pathway

Antithrombin III, referred to as antithrombin, is a member of the serine protease inhibitor family and one of the most important natural proteins responsible for the prevention of spontaneous intravascular clot formation. As a physiologic inhibitor of clot formation, antithrombin inhibits thrombin formation, as well as serine protease activated factor IX (IXa), Xa, Xia, and XIIa. The antithrombotic activity of antithrombin is markedly accelerated by the presence of heparin. Deficiency of antithrombin results in an increased risk of venous thromboembolism. It has been reported that IgG antiheparin antibodies, detected in patients with APS, inhibit the acceleration of antithrombin activity [142]. These authors speculated that subpopulations of aPLs with high affinity for heparin, or antiheparin antibodies, may be effective in promoting a pathologic procoagulant state.

Impairment of Fibrinolysis

Disruption of fibrinolysis is one of the proposed pathogenic mechanisms for APS. Fibrinolytic reactions involve the formation of plasmin from plasminogen by tPA and the hydrolytic cleavage of fibrin to fibrin degradation products by plasmin. The regulation of plasmin generation and activity is very important for maintaining the homeostatic balance in vivo. Endothelial cells, when activated, secrete the plasminogen activator inhibitor-1 (PAI-1), as well as release tPA, to depress fibrinolysis by blocking tPA activity.

The effect of aPLs in the fibrinolytic system has been examined, with controversial results apparently due to the heterogeneity and the small size of the cohorts. Patients with connective tissue diseases, including APS, may have a hypofibrinolytic condition related to high PAI-1 levels [143]. Ames et al. [135] reported upregulation of PAI-I levels in female patients with primary

APS. They further showed a reduction in tPA release by endothelial cell stimulation, suggesting that tPA/PAI-1 balance was crucial to the development of thrombosis in some APS patients. Several other reports pointed toward a hypofibrinolytic state in APS characterized by elevated PAI-1, indicating a perturbation of endothelial cells with consequent fibrinolytic impairment, but no direct evidence was reported for the induction of PAI-1 by aPLs [144, 145]. Antibodies specifically interacting with the catalytic domain of tPA have been detected in patients with APS, representing a possible cause of hypofibrinolysis [146].

Deficiencies in contact proteins were found in patients with unexplained recurrent abortions [147]. Factor XII, a protein of the contact activation system, has a key role in intrinsic fibrinolysis, as well as in intrinsic coagulation. Intrinsic fibrinolytic activities were suppressed by β_2GPI and monoclonal anti-β_2GPI antibody [148]. This inhibition was attributed to a reduced contact activation reaction started by factor XIIa. Impaired factor XIIa-dependent activation of fibrinolysis has been reported in pregnant women with APS who developed late-pregnancy complications [149].

Lipoprotein (a) (Lp(a)) is an apoprotein that shares some sequence homology with plasminogen. Lp(a) inhibits fibrinolytic activity by acting as an uncompetitive inhibitor of tPA but also by increasing PAI-1 expression in endothelial cells [150]. This behavior confers Lp(a) a prothrombotic potential. APS patients had high plasma levels of Lp(a) [151, 152]. Furthermore, elevated levels of Lp(a) in conjunction with low levels of D-dimer and high levels of PAI-1 were found in plasma from APS patients, indicating that deranged fibrinolysis was related to Lp(a) in APS [151]. Plasma Lp(a) is genetically determined, and it is possible that high Lp(a) levels in APS patients might account for the impairment of fibrinolysis independent of the presence of aPLs.

Antiphospholipid Antibody and Cell Interaction

Antiphospholipid Antibody—Endothelial Cell/Monocyte Interactions

Injured and/or activated endothelial cells or monocytes are a predominant target of aPLs. Cultured endothelial cells incubated with aPLs expressed increased levels of adhesion molecules, such as intercellular cell adhesion molecule-1, vascular cell adhesion molecule-1, and P-selectin [153, 154]. This effect is mediated by β_2GPI and cell surface receptors and may promote inflammation and thrombosis [155].

The fact that aPLs induce tissue factor (TF) activity, antigens, or mRNA in endothelial cells and monocytes has been confirmed by many studies [156–158]. In addition, elevated levels of endothelin-1, a potent endothelium-derived contracting factor, were found in patients with APS and arterial thrombosis. Human monoclonal aCL induced prepro-endothelin-1 mRNA, confirming the direct effect of aPLs on endothelium [159]. These results suggest that the interaction between antibody and endothelial cells or monocytes plays a significant role in the onset of thrombosis.

Prothrombin binds to endothelial cells, and this binding was enhanced by a human monoclonal IgG anti-prothrombin antibody, IS6 [160]. IS6 upregulates expression of TF and E-selectin on endothelial cells [161].

Phosphatidylserine is the main phospholipid for β_2GPI or prothrombin binding, and it is not found on the cell surface under normal conditions. Externalization of phosphatidylserine occurs in activated cells and is crucial for the binding of aPLs to procoagulant cells and for the production of procoagulant substances. Therefore, although the antigens for aPLs are present in large amounts in the blood of APS patients, they normally do not develop thrombosis, unless an additional trigger is present.

Microparticle production is a hallmark of cell activation. Antiphospholipid antibodies stimulated the release of microparticles from endothelial cells [162]. However, the role of circulating microparticles in the pathophysiology of thrombosis has not been elucidated.

Antiphospholipid Antibody—Platelet Interaction

Membranes of activated platelets with negatively charged phospholipids are an important source of catalytic surface for blood coagulation. Factor Xa and thrombin are generated on activated platelets, and procoagulant microparticles shed by platelet activation.

Platelets are prone to agglutinate and aggregate after exposure to aPLs [163], and circulating activated platelets have been detected in patients with APS [164]. β_2GPI binds to surface membranes of activated platelets and inhibits the generation of factor Xa. Anti-β_2GPI antibodies interfere with this inhibition [165]. Thus, activated platelets may be a predominant immune target of anti-β_2GPI antibodies.

Antiphospholipid Antibodies and Disruption of the Annexin V Anticoagulant Shield

Annexin V (previously known as placental anticoagulant protein I and vascular anticoagulant α) is a glycoprotein found in a variety of tissues, including the placenta and the vascular endothelium. Annexin V binds anionic phospholipids with high affinity and may form a protective anticoagulant shield on vascular cells. The potent anticoagulant properties of the protein are based on its capacity to displace coagulation factors from phospholipid surfaces. Annexin V is abundant on

the apical surfaces in placental syncytiotrophoblasts, and annexin V expression is decreased on trophoblasts of placentas from patients with preeclampsia. The degree of the reduction correlates with elevation of the markers of activation of blood coagulation [166].

Both aPLs and annexin V have affinity for anionic phospholipids; thus, aPLs may interfere with the formation of the antithrombotic annexin V over phospholipids on apical cytoplasmic membranes. In fact, levels of this protein are markedly reduced on placental villi in patients with APS [167] and on apical membranes of placental villi when cultured with IgG fractions from APS patients [168]. The aPL-mediated reduction on annexin V occurs via displacement by aPLs in the presence of β_2GPI, resulting in accelerated coagulation [169].

Anti-annexin V autoantibodies have been detected in patients with SLE and APS [170, 171] and presumably play some role in the pathogenesis of APS.

CELL RECEPTORS FOR ANTIPHOSPHOLIPID ANTIBODY INTERACTIONS

The cell activation-mediated by aPLs might require an interaction between phospholipid-binding proteins and a specific cell receptor(s).

Annexin II, also known as annexin A2, is a receptor for tPA and plasminogen. Annexin II is found on the surface of endothelial cells and monocytes and also on the brush-border membrane of placental syncytiotrophoblasts [172, 173]. Annexin II interacts with the β_2GPI/anti-β_2GPI antibody complex on the endothelial cell and monocyte surfaces, mediating cell activation [174, 175]. The involvement of annexin II in aPL-mediated pathogenic effects has been reported *in vitro* and *in vivo* [176]. However, it is not clear whether such a receptor is actually involved in cell activation because annexin II is not a transmembrane protein. Furthermore, it has been proposed that the activation of signaling responses requires the presence of another transmembrane adaptor protein(s) that associates with annexin II on the endothelial cell surface [177].

The Toll-like receptor (TLR) family, particularly TLR-2 and TLR-4 [178–180], may also play a role in the interaction of the β_2GPI/anti-β_2GPI antibody complex [177]. Adhered β_2GPI interacts with TLR-4, and anti-β_2GPI antibodies cross-link β_2GPI and TLR-4 complex, eventually triggering signaling cascade activation. Moreover, TLR-4 acts as the putative adaptor protein for annexin II [175].

Megalin/gp330 is an endocytic receptor that internalizes multiple ligands, including apolipoprotein E and B100. Megalin is the main antigenic target in passive

Heymann nephritis, in which it binds circulating auto-antibodies, leading to the formation of subepithelium immune deposits. Megalin was reported to be a receptor of the β_2GPI molecule and the β_2GPI—phospholipid complex [181]. Pennings *et al.* [182] demonstrated that dimeric β_2GPI can interact with different low-density lipoprotein receptor (LDL-R) family members, including megalin.

Apolipoprotein E receptor 2 (ApoER2'), the only member of the LDL-R family present on the platelet membranes, has been suggested as a receptor of β_2GPI on platelets. ApoER2' is found in many other cell types thus, it is possible that ApoER2' may also be a receptor for β_2GPI in endothelial cells and monocytes [183]. Blockage of ApoER2' on platelets using a receptor-associated protein leads to a loss of the increased adhesion of platelets to collagen induced by β_2GPI/anti-β_2GPI antibody complex [184]. Furthermore, dimeric β_2GPI interacts with ApoER2', and the β_2GPI/anti-β_2GPI antibody complex has to form on exposed phosphatidylserine on the cell surface before interacting with ApoER2' [184]. This procedure requires an initial independent priming stimulus to lead to the exposure of phosphatidylserine on the surface of platelets. The binding site on β_2GPI for ApoER2' is located in domain V but does not overlap with the phospholipid binding site [185].

β_2GPI directly binds to the GPIbα subunit of the platelet adhesion receptor GPIb/IX/V *in vitro* [186, 187]. The platelet GPIbα subunit, having the von Willebrand factor as the most important ligand, also serves to localize factor XI and thrombin on the platelet surface [188]. Binding of β_2GPI to GPIbα enables anti-β_2GPI antibodies, directed against domain I, to activate platelets, resulting in thromboxane production, and also to activate the phosphoinositol-3 kinase/Akt pathway [186], contributing to platelet adhesion and aggregation.

The involvement of Fcγ receptor on cellular activation has been investigated *in vivo* [189] and in *in vitro* studies on platelets [184], monocytes [190], and endothelial cells [175]. Results suggest that this receptor is not strictly necessary for cellular activation.

SIGNALING PATHWAYS OF CELL ACTIVATION

The mechanisms of aPL-mediated cell activation have been elucidated at the molecular level. The adapter molecule myeloid differentiation protein (MyD88)-dependent signaling pathway and the nuclear factor-κB (NF-κB) have been involved in endothelial cell activation by aPLs [177, 191, 192]. IgG purified from patients with APS induced the nuclear translocation of NF-κB, leading to the transcription of a large number of genes that have an NF-κB-responsive element in their promoter. This nuclear translocation of NF-κB mediates, at least in part, the increased expression of TF and adhesion molecules by endothelial cells.

The decisive role of p38 mitogen-activated protein kinase (MAPK) pathway in aPL-mediated cell activation has been reported by several groups. Bohgaki and colleagues [193] demonstrated that monocytes stimulated by monoclonal anti-β_2GPI antibodies from APS patients induce phosphorylation of p38 MAPK, a locational shift of NF-κB into the nucleus, and upregulation of TF expression. Such activation was not seen in the absence of β_2GPI, indicating that the disturbance of monocyte by anti-β_2GPI antibodies is started by interaction between the cell and the autoantibody-bound β_2GPI. The importance of p38 MAPK in cell activation has also been demonstrated in platelets [194] and endothelial cells [195]. Pretreatment of platelets with a p38 MAPK-specific inhibitor, SB203580, completely abrogated aPL-mediated platelet aggregation.

Activation of the p38 MAPK pathway increases activities of cytokines such as tumor necrosis factor-α, IL-1β, and macrophage inflammatory cytokine 3β [180, 193, 196]. The p38 MAPK pathway has also been involved in the regulation of TF expression in monocytes, endothelial cells, and smooth muscle cells. Furthermore, the induction of TF expression was reported through the simultaneous activation of NF-κB via the MAPK pathway and of the MEK-1/ERK pathway [197]. However, inhibitor of the MEK-1/ERK pathway could not suppress the expression of TF, suggesting that the p38 MAPK system plays the main role in those reactions. Figure 50.4 shows some of the mechanisms of procoagulant cell activation mediated by aPLs.

ANTIPHOSPHOLIPID ANTIBODIES AND COMPLEMENT ACTIVATION

Complement activation has received much attention since it was determined to be relevant to the pathophysiology of APS, especially with regard to pregnancy morbidity [198]. Placenta trophoblast cells are targeted by phospholipid-binding protein—aPLs complexes, leading to activation of the complement cascade through the classical pathway. C3 and subsequently C5 are activated. Generated C5a recruits and activates polymorphonuclear leukocytes and monocytes and stimulates the release of mediators of inflammation, which ultimately results in thrombosis and fetal injury.

Elevated levels of C3 and C4 predicted subsequent miscarriages in patients with unexplained recurrent pregnancy loss [199]. Moreover, higher levels of complement activation products were detected in the plasma of APS patients with cerebral ischemic events

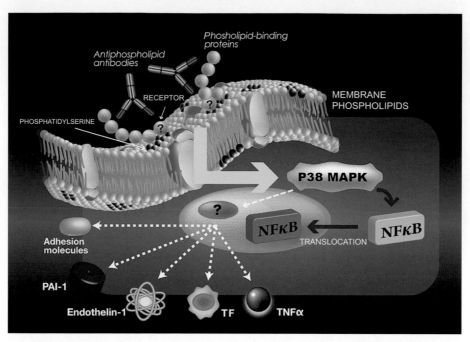

FIGURE 50.4   Proposed mechanisms of procoagulant cell activation mediated by antiphospholipid antibodies. Antiphospholipid antibodies recognize phospholipid-binding protein (β_2-glycoprotein I or prothrombin) bound to endothelial cells and induce p38 MAPK phosphorylation, leading to the transcription of procoagulant substances, adhesion molecules, and, subsequently, thrombus formation. NFκB, nuclear factor-κB; p38 MAPK, p38 mitogen activated protein kinase; PAI-1, plasminogen activator inhibitor-1; TF, tissue factor; TNFα, tumor necrosis factor-α.

compared to patients with non-APS-related cerebral ischemia [200].

The involvement of complement activation in the pathogenesis of thrombosis is supported by data from animal models. Pierangeli *et al.* [201], using an *in vivo* murine model of thrombosis and microcirculation, showed that activation of complement C3 and C5 by aPLs mediates the induction of thrombosis and the activation of endothelial cells. Fischetti *et al.* [202] demonstrated the implication of complement in the development of thrombosis in a rat animal model. In this *in vivo* animal model, polyclonal IgG aPLs purified from the serum of APS patients were transferred to rats pretreated with lipopolysaccharide. Rats receiving IgG aPLs developed thrombus, whereas those receiving polyclonal IgG from healthy subjects did not.

Oku and colleagues [203] have clarified the profile of complement activation in patients with APS. The serum complement levels were clearly lower in patients with primary APS than in healthy persons or controls with non-SLE rheumatic diseases. The hypocomplementemia was related to consumption of complement as a result of complement activation. In primary APS patients, the serum complement levels correlated with LA activity and plasma levels of tumor necrosis factor-α, implying that complement activation induced by aPLs may be one of the responsible mechanisms of the thrombophilic state in APS.

The IgG isotype of aPLs is the most frequently found isotype in patients with APS, and the IgG2 subclass is the most prevalent [204, 205]. IgG2 and IgG4 subclasses have a relatively weak ability to fix complement via the classical pathways; thus, other additional mechanisms may be involved in the enhancement of complement activation in patients with aPLs. The inflammatory process accompanied by complement activation is an important phenomenon that mediates the crossroads between the immune response and thrombosis.

References

[1]   S. Miyakis, M.D. Lockshin, T. Atsumi, D.W. Branch, R.L. Brey, R. Cervera, et al., International consensus statement on an update of the classification criteria for definite antiphospholipid syndrome (APS), J Thromb Haemost 4 (2006) 295−306.

[2]   G.R. Hughes, Thrombosis, abortion, cerebral disease, and the lupus anticoagulant, Br Med J (Clin Res Ed) 287 (1983) 1088−1089.

[3]   R. Cervera, J.C. Piette, J. Font, M.A. Khamashta, Y. Shoenfeld, M.T. Camps, et al., Antiphospholipid syndrome: Clinical and immunologic manifestations and patterns of disease expression in a cohort of 1000 patients, Arthritis Rheum 46 (2002) 1019−1027.

[4]   A. Wassermann, A. Neisser, C. Bruck, Eine serodiagnostishe reaction bei syphilis, Dtsch Med Wochenschr 32 (1906) 745−746.

[5]   M.C. Pangborn, A new serologically active phospholipid from beef heart, Proc Soc Exp Biol Med 48 (1941) 484−486.

[6] J.F. Mueller, O. Ratnoff, R.W. Henile, Observations on the characteristics of an unusual circulating anticoagulant, J Lab Clin Med 38 (1951) 254–261.

[7] J.R. Haserick, R. Long, Systemic lupus erythematosus preceded by false positive test for syphilis: Presentation of five cases, Ann Intern Med 37 (1951) 559–565.

[8] J.E. Moore, W.B. Lutz, The natural history of systemic lupus erythematosus: An approach to its study through chronic biological false positive reactors, J Chronic Dis 1 (1955) 297–316.

[9] J.E. Moore, C.F. Mohr, Biologically false positive serologic test for syphilis: Type, incidence, and cause, JAMA 150 (1952) 467–473.

[10] C.L. Conley, R.C.A. Hartman, A hemorrhagic disorder caused by circulating anticoagulants in patients with disseminated lupus erythematosus, J Clin Invest 31 (1952) 621–622.

[11] A.B. Laurell, I.M. Nilsson, Hypergammaglobulinemia, circulating anticoagulant, and biologic false positive Wassermann reaction: A study in two cases, J Lab Clin Med 49 (1957) 694–707.

[12] D.I. Feinstein, S.I. Rapaport, Acquired inhibitors of blood coagulation, Prog Hemost Thromb 1 (1972) 75–95.

[13] E.N. Harris, A.E. Gharavi, M.L. Boey, B.M. Patel, C.G. Mackworth-Young, S. Loizou, et al., Anti-cardiolipin antibodies: Detection by radioimmunoassay and association with thrombosis in systemic lupus erythematosus, Lancet 2 (1983) 1211–1214.

[14] T. Koike, M. Sueishi, H. Funaki, H. Tomioka, S. Yoshida, Antiphospholipid antibodies and biological false positive serological test for syphilis in patients with systemic lupus erythematosus, Clin Exp Immunol 56 (1984) 193–199.

[15] E.N. Harris, A.E. Gharavi, S. Loizou, G. Derue, J.K. Chan, B.M. Patel, et al., Cross-reactivity of antiphospholipid antibodies, J Clin Lab Immunol 16 (1985) 1–6.

[16] H.P. McNeil, R.J. Simpson, C.N. Chesterman, S.A. Krilis, Anti-phospholipid antibodies are directed against a complex antigen that induces a lipid-binding inhibitor of coagulation: β_2-glycoprotein I (apolipoprotein H), Proc Natl Acad Sci USA 87 (1990) 4120–4124.

[17] M. Galli, P. Comfurius, C. Maassen, H.C. Hemker, M.H. de Baets, P.J. van Breda-Vriesman, et al., Anticardiolipin antibodies (ACA) directed not to cardiolipin but to a plasma protein cofactor, Lancet 335 (1990) 952–953.

[18] E. Matsuura, Y. Igarashi, M. Fujimoto, K. Ichikawa, T. Koike, Anticardiolipin cofactor(s) and differential diagnosis of autoimmune disease, Lancet 336 (1990) 177–178.

[19] E. Matsuura, Y. Igarashi, T. Yasuda, D.A. Triplett, T. Koike, Anticardiolipin antibodies recognize β_2-glycoprotein I structure altered by interacting with an oxygen modified solid phase surface, J Exp Med 179 (1994) 457–462.

[20] E.M. Bevers, M. Galli, T. Barbui, P. Comfurius, R.F. Zwaal, Lupus anticoagulant IgG's (LA) are not directed to phospholipids only, but to a complex of lipid-bound human prothrombin, Thromb Haemost 66 (1991) 629–632.

[21] J.D. Oosting, R.H.W.M. Derksen, I.W.G. Bobbink, T.M. Hackeng, B.N. Bouma, P.G. de Groot, Antiphospholipid antibodies directed against a combination of phospholipids with prothrombin, protein C, or protein S: An explanation for their pathogenic mechanism? Blood 81 (1993) 2618–2625.

[22] T. Sugi, J.A. McIntyre, Autoantibodies to phosphatidylethanolamine (PE) recognize a kininogen–PE complex, Blood 86 (1995) 3083–3089.

[23] W.A. Wilson, A.E. Gharavi, T. Koike, M.D. Lockshin, D.W. Branch, J.C. Piette, et al., International consensus statement on preliminary classification criteria for definite antiphospholipid syndrome: Report of an international workshop, Arthritis Rheum 42 (1999) 1309–1311.

[24] P.F. Devaux, Static and dynamic lipid asymmetry in cell membranes, Biochemistry 30 (1991) 1163–1173.

[25] J. Connor, C.H. Pak, R.F. Zwaal, A.J. Schroit, Bidirectional transbilayer movement of phospholipid analogs in human red blood cells. Evidence for an ATP-dependent and protein-mediated process, J Biol Chem 267 (1992) 19412–19417.

[26] Q. Zhou, J. Zhao, J.G. Stout, R.A. Luhm, T. Wiedmer, P.J. Sims, Molecular cloning of human plasma membrane phospholipid scramblase. A protein mediating transbilayer movement of plasma membrane phospholipids, J Biol Chem 272 (1997) 18240–18244.

[27] R.F. Zwaal, A.J. Schroit, Pathophysiologic implications of membrane phospholipid asymmetry in blood cells, Blood 89 (1997) 1121–1132.

[28] K.G. Mann, M.E. Nesheim, W.R. Church, P. Haley, S. Krishnaswamy, Surface-dependent reactions of the vitamin K-dependent enzyme complex, Blood 76 (1990) 1–16.

[29] T. Wiedmer, Q. Zhou, D.Y. Kwoh, P.J. Sims, Identification of three new members of the phospholipid scramblase gene family, Biochim Biophys Acta 1467 (2000) 244–253.

[30] J. Lozier, N. Takahashi, F.W. Putnam, Complete amino acid sequence of human plasma β_2-glycoprotein I, Proc Natl Acad Sci USA 81 (1984) 3640–3644.

[31] H.E. Schultze, K. Heide, H. Haput, Uber ein bisher ubekanntes niedermolekulars β2-Globulin des Humanserums, Naturwissenschaften 48 (1961) 719.

[32] E. Matsuura, M. Igarashi, Y. Igarashi, H. Nagae, K. Ichikawa, T. Yasuda, et al., Molecular definition of human β_2-glycoprotein I (β_2-GPI) by cDNA cloning and interspecies differences of β_2-GPI in alternation anticardiolipin binding, Int Immunol 3 (1991) 1217–1221.

[33] A. Steinkasserer, C. Estaller, E.H. Weiss, R.B. Sim, A.J. Day, Complete nucleotide and deduced amino acid sequence of human beta 2-glycoprotein I, Biochem J 277 (1991) 387–391.

[34] P. Bork, A.K. Downing, B. Kieffer, I.D. Campbell, Structure and distribution of modules in extracellular proteins, Q Rev Biophys 29 (1996) 119–167.

[35] A. Steinkasserer, P.N. Barlow, A.C. Willis, Z. Kertesz, I.D. Campbell, R.B. Sim, et al., Activity, disulphide mapping and structural modelling of the fifth domain of human β_2-glycoprotein I, FEBS Lett 313 (1992) 193–197.

[36] Y. Sheng, A. Sali, H. Herzog, J. Lahnstein, S.A. Krilis, Site-directed mutagenesis of recombinant human beta 2-glycoprotein I identifies a cluster of lysine residues that are critical for phospholipid binding and anti-cardiolipin antibody activity, J Immunol 157 (1996) 3744–3751.

[37] D.K. Sanghera, D.R. Wagenknecht, J.A. McIntyre, M.I. Kamboh, Identification of structural mutations in the fifth domain of apolipoprotein H (beta 2-glycoprotein I) which affect phospholipid binding, Hum Mol Genet 6 (1997) 311–316.

[38] B. Bouma, P.G. de Groot, J.M. van den Elsen, R.B. Ravelli, A. Schouten, M.J. Simmelink, et al., Adhesion mechanism of human $\beta(2)$-glycoprotein I to phospholipids based on its crystal structure, EMBO J 18 (1999) 5166–5174.

[39] R. Schwarzenbacher, K. Zeth, K. Diederichs, A. Gries, G.M. Kostner, P. Laggner, et al., Crystal structure of human beta$_2$-glycoprotein I: Implications for phospholipid binding and the antiphospholipid syndrome, EMBO J 18 (1999) 6228–6239.

[40] M. Hoshino, Y. Hagihara, I. Nishii, T. Yamazaki, H. Kato, Y. Goto, Identification of the phospholipid-binding site of human beta (2)-glycoprotein I domain V by heteronuclear magnetic resonance, J Mol Biol 304 (2000) 927–939.

[41] Y. Nakaya, E.J. Schaefer, H.B. Brewer Jr., Activation of human post heparin lipoprotein lipase by apolipoprotein H (beta 2-glycoprotein I), Biochem Biophys Res Commun 95 (1980) 1168–1172.

[42] H. Wurm, E. Beubler, E. Polz, A. Holasek, G. Kostner, Studies on the possible function of beta 2-glycoprotein-I: Influence in the triglyceride metabolism in the rat, Metabolism 31 (1982) 484–486.

[43] K. Kobayashi, E. Matsuura, Q. Liu, J. Furukawa, K. Kaihara, J. Inagaki, et al., A specific ligand for beta(2)-glycoprotein I mediates autoantibody-dependent uptake of oxidized low density lipoprotein by macrophages, J Lipid Res 42 (2001) 697–709.

[44] A.A. Manfredi, P. Rovere, G. Galati, S. Heltai, E. Bozzolo, L. Soldini, et al., Apoptotic cell clearance in systemic lupus erythematosus. I. Opsonization by antiphospholipid antibodies, Arthritis Rheum 41 (1998) 205–214.

[45] A.A. Manfredi, P. Rovere, S. Heltai, G. Galati, G. Nebbia, A. Tincani, et al., Apoptotic cell clearance in systemic lupus erythematosus. II. Role of beta$_2$-glycoprotein I, Arthritis Rheum 41 (1998) 215–223.

[46] W. Shi, B. Chong, P. Hogg, C. Chesterman, Anticardiolipin antibodies block the inhibition by β$_2$-glycoprotein I of the factor Xa generating activity of platelets, Thromb Haemost 70 (1993) 342–345.

[47] I. Schousboe, M.S. Rasmussen, Synchronized inhibition of the phospholipid mediated autoactivation of factor XII in plasma by beta 2-glycoprotein I and anti-beta 2-glycoprotein I, Thromb Haemost 73 (1995) 798–804.

[48] J. Nimpf, H. Wurm, G.M. Kostner, Beta(2)-glycoprotein-I (apo-H) inhibits the release reaction of human platelets during ADP-induced aggregation, Atherosclerosis 63 (1987) 109–114.

[49] M. Ieko, K. Ichikawa, D.A. Triplett, E. Matsuura, T. Atsumi, K. Sawada, et al., Beta$_2$-glycoprotein I is necessary to inhibit protein C activity by monoclonal anticardiolipin antibodies, Arthritis Rheum 42 (1999) 167–174.

[50] A. Steinkasserer, C. Dorner, R. Wurzner, R.B. Sim, Human beta 2-glycoprotein I: Molecular analysis of DNA and amino acid polymorphism, Hum Genet 91 (1993) 401–402.

[51] D.K. Sanghera, T. Kristensen, R.F. Hamman, M.I. Kamboh, Molecular basis of the apolipoprotein H (β$_2$-glycoprotein I) protein polymorphism, Hum Genet 100 (1997) 57–62.

[52] F.C. Gushiken, F.C. Arnett, C. Ahn, P. Thiagarajan, Polymorphism of beta$_2$-glycoprotein I at codons 306 and 316 in patients with systemic lupus erythematosus and antiphospholipid syndrome, Arthritis Rheum 42 (1999) 1189–1193.

[53] S. Yasuda, A. Tustsumi, H. Chiba, H. Yanai, Y. Miyoshi, R. Takeuchi, et al., Beta(2)-glycoprotein I deficiency: Prevalence, genetic background and effects on plasma lipoprotein metabolism and hemostasis, Atherosclerosis 152 (2000) 337–346.

[54] R. Takeuchi, T. Atsumi, M. Ieko, H. Takeya, S. Yasuda, K. Ichikawa, et al., Coagulation and fibrinolytic activities in 2 siblings with beta(2)-glycoprotein I deficiency, Blood 96 (2000) 1594–1595.

[55] F.C. Gushiken, F.C. Arnett, P. Thiagarajan, Primary antiphospholipid antibody syndrome with mutations in the phospholipid binding domain of beta(2)-glycoprotein I, Am J Hematol 65 (2000) 160–165.

[56] Y. Sheng, S.W. Reddel, H. Herzog, Y.X. Wang, T. Brighton, M.P. France, et al., Impaired thrombin generation in β$_2$-glycoprotein I null mice, J Biol Chem 276 (2001) 13817–13821.

[57] Y. Hagihara, K. Enjyoji, T. Omasa, Y. Katakura, K. Suga, M. Igarashi, et al., Structure and function of the recombinant fifth domain of human β$_2$-glycoprotein I: Effects of specific cleavage between Lys77 and Thr78, J Biochem 121 (1997) 128–137.

[58] N. Ohkura, Y. Hagihara, T. Yoshimura, Y. Goto, H. Kato, Plasmin can reduce the function of human beta$_2$ glycoprotein I by cleaving domain V into a nicked form, Blood 91 (1998) 4173–4179.

[59] J. Hunt, S. Krilis, The fifth domain of β$_2$-glycoprotein I contains a phospholipid binding site (Cys281–Cys288) and a region recognized by anticardiolipin antibodies, J Immunol 152 (1994) 653–659.

[60] J.E. Hunt, R.J. Simpson, S.A. Krilis, Identification of a region of β$_2$-glycoprotein I critical for lipid binding and anti-cardiolipin antibody cofactor activity, Proc Natl Acad Sci USA 90 (1993) 2141–2145.

[61] E. Matsuura, J. Inagaki, H. Kasahara, D. Yamamoto, T. Atsumi, K. Kobayashi, et al., Proteolytic cleavage of beta(2)-glycoprotein I: Reduction of antigenicity and the structural relationship, Int Immunol 12 (2000) 1183–1192.

[62] D.A. Horbach, E. van Oort, T. Lisman, J.C. Meijers, R.H. Derksen, P.G. de Groot, Beta$_2$-glycoprotein I is proteolytically cleaved *in vivo* upon activation of fibrinolysis, Thromb Haemost 81 (1999) 87–95.

[63] Y. Itoh, K. Inuzuka, I. Kohno, H. Wada, H. Shiku, N. Ohkura, et al., Highly increased plasma concentrations of the nicked form of beta(2) glycoprotein I in patients with leukemia and with lupus anticoagulant: Measurement with a monoclonal antibody specific for a nicked form of domain V, J Biochem 128 (2000) 1017–1024.

[64] S. Yasuda, T. Atsumi, M. Ieko, E. Matsuura, K. Kobayashi, J. Inagaki, et al., Nicked beta$_2$-glycoprotein I: A marker of cerebral infarct and a novel role in the negative feedback pathway of extrinsic fibrinolysis, Blood 103 (2004) 3766–3772.

[65] H. Nakagawa, S. Yasuda, E. Matsuura, K. Kobayashi, M. Ieko, H. Kataoka, et al., Nicked β$_2$-glycoprotein I binds angiostatin 4.5 (plasminogen kringle 1-5) and attenuates its antiangiogenic property, Blood 114 (2009) 2553–2559.

[66] A. Tincani, L. Spatola, E. Prati, F. Allegri, P. Ferremi, R. Cattaneo, et al., The anti-β$_2$-glycoprotein I activity in human anti-phospholipid syndrome sera is due to monoreactive low-affinity autoantibodies directed to epitopes located on native β$_2$-glycoprotein I and preserved during species' evolution, J Immunol 157 (1996) 5732–5738.

[67] T. Koike, E. Matsuura, What is the "true" antigen for anticardiolipin antibodies? Lancet 337 (1991) 671–672.

[68] E. Matsuura, Y. Igarashi, M. Fujimoto, K. Ichikawa, T. Suzuki, S. Yoshida, et al., Heterogeneity of anticardiolipin antibodies defined by the anticardiolipin cofactor, J Immunol 148 (1992) 3885–3891.

[69] J.E. Hunt, H.P. McNeil, G.J. Morgan, R.M. Crameri, S.A. Krilis, A phospholipid–beta$_2$-glycoprotein I complex is an antigen for anticardiolipin antibodies occurring in autoimmune disease but not with infection, Lupus 1 (1992) 75–81.

[70] M. Igarashi, E. Matsuura, Y. Igarashi, H. Nagae, K. Ichikawa, D.A. Triplett, et al., Human β$_2$-glycoprotein I as an anticardiolipin cofactor determined using deleted mutants expressed by a Baculovirus system, Blood 87 (1996) 3262–3270.

[71] J. George, B. Gilburd, M. Hojnik, Y. Levy, P. Langevitz, E. Matsuura, et al., Target recognition of beta$_2$-glycoprotein I (beta$_2$GPI)-dependent anticardiolipin antibodies: Evidence for involvement of the fourth domain of beta$_2$GPI in antibody binding, J Immunol 150 (1998) 3917–3923.

[72] G.M. Iverson, S. Reddel, E.J. Victoria, K.A. Cockerill, Y.X. Wang, M.A. Marti-Renom, et al., Use of single point mutations in domain I of beta 2-glycoprotein I to determine fine antigenic specificity of antiphospholipid autoantibodies, J Immunol 169 (2002) 7097–7103.

[73] P.A. McNeeley, J.S. Dlott, R.A. Furie, R.M. Jack, T.L. Ortel, D.A. Triplett, et al., Beta$_2$-glycoprotein I-dependent anti-cardiolipin antibodies preferentially bind the amino terminal domain of beta$_2$-glycoprotein I, Thromb Haemost 86 (2001) 590—595.

[74] T. Koike, K. Ichikawa, T. Atsumi, H. Kasahara, E. Matsuura, β$_2$-Glycoprotein I—anti-β$_2$-glycoprotein I interaction, J Autoimmun 15 (2000) 97—100.

[75] B. de Laat, R.H. Derksen, R.T. Urbanus, P.G. de Groot, IgG antibodies that recognize epitope Gly40—Arg43 in domain I of beta 2-glycoprotein I cause LAC, and their presence correlates strongly with thrombosis, Blood 105 (2005) 1540—1545.

[76] Y. Ioannou, C. Pericleous, I. Giles, D.S. Latchman, D.A. Isenberg, A. Rahman, Binding of antiphospholipid antibodies to discontinuous epitopes on domain I of human beta (2)-glycoprotein I: Mutation studies including residues R39 to R43, Arthritis Rheum 56 (2007) 280—290.

[77] H. Kasahara, E. Matsuura, K. Kaihara, D. Yamamoto, K. Kobayashi, J. Inagaki, et al., Antigenic structures recognized by anti-beta$_2$-glycoprotein I auto-antibodies, Int Immunol 17 (2005) 1533—1542.

[78] K. Ichikawa, M. Khamashta, T. Koike, E. Matsuura, G.R.V. Hughes, Reactivity of monoclonal anticardiolipin antibodies from patients with the antiphospholipid syndrome to β$_2$-glycoprotein I, Arthritis Rheum 37 (1994) 1453—1461.

[79] T. Atsumi, A. Tsutsumi, O. Amengual, M.A. Khamashta, G.R. Hughes, Y. Miyoshi, et al., Correlation between beta$_2$-glycoprotein I valine/leucine247 polymorphism and anti-beta$_2$-glycoprotein I antibodies in patients with primary antiphospholipid syndrome, Rheumatology 38 (1999) 721—723.

[80] N. Hirose, R. Williams, A.R. Alberts, R.A. Furie, E.K. Chartash, R.I. Jain, et al., A role for the polymorphism at position 247 of the beta$_2$-glycoprotein I gene in the generation of anti-beta$_2$-glycoprotein I antibodies in the antiphospholipid syndrome, Arthritis Rheum 42 (1999) 1655—1661.

[81] S. Yasuda, T. Atsumi, E. Matsuura, K. Kaihara, D. Yamamoto, K. Ichikawa, et al., Significance of valine/leucine247 polymorphism of beta$_2$-glycoprotein I in antiphospholipid syndrome: Increased reactivity of anti-beta$_2$-glycoprotein I autoantibodies to the valine247 beta$_2$-glycoprotein I variant, Arthritis Rheum 52 (2005) 212—218.

[82] H. Yoshida, H. Fujiwara, T. Fujiwara, S. Ikehara, Y. Hamashima, Quantitative analysis of myocardial infarction in (NZW × BXSB) F1 hybrid mice with systemic lupus erythematosus and small coronary artery disease, Am J Pathol 129 (1987) 477—485.

[83] Y. Hashimoto, M. Kawamura, K. Ichikawa, T. Suzuki, T. Sumida, S. Yoshida, et al., Anticardiolipin antibodies in NZW × BXSB F1 mice. A model of antiphospholipid syndrome, J Immunol 149 (1992) 1063—1068.

[84] M. Blank, A. Faden, A. Tincani, J. Kopolovic, I. Goldberg, B. Gilburd, et al., Immunization with anticardiolipin cofactor (β-2-glycoprotein I) induces experimental antiphospholipid syndrome in naive mice, J Autoimmun 7 (1994) 441—455.

[85] S. Kouts, M.X. Wang, S. Adelstein, S.A. Krilis, Immunization of a rabbit with beta 2-glycoprotein I induces charge-dependent cross-reactive antibodies that bind anionic phospholipids and have similar reactivity as autoimmune anti-phospholipid antibodies, J Immonol 155 (1995) 958—966.

[86] S.S. Pierangeli, X.W. Liu, G. Anderson, J.H. Barker, E.N. Harris, Thrombogenic properties of murine anti-cardiolipin antibodies induced by β$_2$ glycoprotein I and human immunoglobulin G antiphospholipid antibodies, Circulation 94 (1996) 1746—1751.

[87] N. Hattori, M. Kuwana, J. Kaburaki, T. Mimori, Y. Ikeda, Y. Kawakami, T cells that are autoreactive to beta$_2$-glycoprotein I in patients with antiphospholipid syndrome and healthy individuals, Arthritis Rheum 43 (2000) 65—75.

[88] T. Arai, K. Yoshida, J. Kaburaki, H. Inoko, Y. Ikeda, Y. Kawakami, et al., Autoreactive CD4(+) T-cell clones to beta$_2$-glycoprotein I in patients with antiphospholipid syndrome: Preferential recognition of the major phospholipid-binding site, Blood 98 (2001) 1889—1896.

[89] M. Kuwana, E. Matsuura, K. Kobayashi, Y. Okazaki, J. Kaburaki, Y. Ikeda, et al., Binding of beta 2-glycoprotein I to anionic phospholipids facilitates processing and presentation of a cryptic epitope that activates pathogenic autoreactive T cells, Blood 105 (2005) 1552—1557.

[90] Y. Yamaguchi, N. Seta, J. Kaburaki, K. Kobayashi, E. Matsuura, M. Kuwana, Excessive exposure to anionic surfaces maintains autoantibody response to beta(2)-glycoprotein I in patients with antiphospholipid syndrome, Blood 110 (2007) 4312—4318.

[91] E.J.W. Bowie, J.H. Thompson, C.A. Pascuzzi, Thrombosis in systemic lupus erythematosus despite circulating anticoagulants, J Lab Clin Med 62 (1963) 416.

[92] P. Thiagarajan, S.S. Shapiro, L. De Marco, Monoclonal immunoglobulin M lambda coagulation inhibitor with phospholipid specificity. Mechanism of a lupus anticoagulant, J Clin Invest 66 (1980) 397—405.

[93] E.T. Yin, L.W. Gaston, Purification and kinetic studies on a circulating anticoagulant in a suspected case of lupus erythematosus, Thromb Haemost 14 (1965) 88—115.

[94] T. Exner, K.A. Rickard, H. Kronenberg, A sensitive test demonstrating lupus anticoagulant and its behavioural patterns, Br J Haematol 40 (1978) 143—151.

[95] J.D. Oosting, R.H. Derksen, H.T. Entjes, B.N. Bouma, P.G. de Groot, Lupus anticoagulant activity is frequently dependent on the presence of beta 2-glycoprotein I, Thromb Haemost 67 (1992) 499—502.

[96] R.A.S. Roubey, C.W. Pratt, J.P. Buyon, J.B. Winfield, Lupus anticoagulant activity of autoimmune antiphospholipid antibodies is dependent upon β$_2$-glycoprotein I, J Clin Invest 90 (1992) 1100—1104.

[97] H. Takeya, T. Mori, E.C. Gabazza, K. Kuroda, H. Deguchi, E. Matsuura, et al., Anti-beta$_2$-glycoprotein I (beta$_2$GPI) monoclonal antibodies with lupus anticoagulant-like activity enhance the beta$_2$GPI binding to phospholipids, J Clin Invest 99 (1997) 2260—2268.

[98] J. Arnout, M. Vanrusselt, C. Wittevrongel, J. Vermylen, Monoclonal antibodies against beta-2-glycoprotein I: Use as reference material for lupus anticoagulant tests, Thromb Haemost 79 (1998) 955—958.

[99] M.F. Harper, P.M. Hayes, B.R. Lentz, R.A.S. Roubey, Characterization of β$_2$-glycoprotein I binding to phospholipid membranes, Thromb Haemost 80 (1998) 610—614.

[100] G.M. Willems, M.P. Janssen, M.M. Pelsers, P. Comfurius, M. Galli, R.F. Zwaal, et al., Role of divalency in the high-affinity binding of anticardiolipin antibody—β$_2$-glycoprotein I complexes to lipid membranes, Biochemistry 35 (1996) 13833—13842.

[101] J. Arnout, C. Wittevrongel, M. Vanrusselt, M. Hoylaerts, J. Vermylen, Beta-2-glycoprotein I dependent lupus anticoagulants form stable bivalent antibody beta-2-glycoprotein I complexes on phospholipid surfaces, Thromb Haemost 79 (1998) 79—86.

[102] A. Loeliger, Prothrombin as a co-factor of the circulating anticoagulant in systemic lupus erythematosus? Thromb Diath Haemorrh 3 (1959) 237—256.

III. ANTI-PHOSPHOLIPID SYNDROME

[103] S.I. Rapaport, S.B. Ames, B.J. Duvall, A plasma coagulation defect in systemic lupus erythematosus arising from hypoprothrombinemia combined with antiprothrombinase activity, Blood 15 (1960) 212–217.

[104] L.V.M. Rao, A.D. Hoang, S.I. Rapaport, Differences in the interactions of lupus anticoagulant IgG with human prothrombin and bovine prothrombin, Thromb Haemost 73 (1995) 668–674.

[105] B.K. Chow, V. Ting, F. Tufaro, R.T. MacGillivray, Characterization of a novel liver-specific enhancer in the human prothrombin gene, J Biol Chem 1991 (1991) 18927–18933.

[106] S.P. Bajaj, S.I. Rapaport, D.S. Fierer, K.D. Herbst, D.B. Schwartz, A mechanism for the hypoprothrombinemia of the acquired hypoprothrombinemia–lupus anticoagulant syndrome, Blood 61 (1983) 684–692.

[107] J.R. Edson, J.M. Vogt, D.K. Hasegawa, Abnormal prothrombin crossed-immunoelectrophoresis in patients with lupus inhibitors, Blood 1984 (64) (1984) 807–816.

[108] R.A. Fleck, S.I. Rapaport, L.V.M. Rao, Anti-prothrombin antibodies and the lupus anticoagulant, Blood 72 (1988) 512–519.

[109] V. Pengo, A. Biasiolo, T. Brocco, S. Tonetto, A. Ruffatti, Autoantibodies to phospholipid-binding plasma proteins in patients with thrombosis and phospholipid-reactive antibodies, Thromb Haemost 75 (1996). 721–714.

[110] M. Galli, G. Beretta, M. Daldossi, E.M. Bevers, T. Barbui, Different anticoagulant and immunological properties of antiprothrombin antibodies in patients with antiphospholipid antibodies, Thromb Haemost 77 (1997) 486–491.

[111] T. Atsumi, M. Ieko, M.L. Bertolaccini, K. Ichikawa, A. Tsutsumi, E. Matsuura, et al., Association of autoantibodies against the phosphatidylserine–prothrombin complex with manifestations of the antiphospholipid syndrome and with the presence of lupus anticoagulant, Arthritis Rheum 43 (2000) 1982–1993.

[112] J.R. Wu, B.R. Lentz, Phospholipid-specific conformational changes in human prothrombin upon binding to procoagulant acidic lipid membranes, Thromb Haemost 71 (1994) 596–604.

[113] J.F. McDonald, A.M. Shah, R.A. Schwalbe, W. Kisiel, B. Dahlback, G.L. Nelsestuen, Comparison of naturally occurring vitamin K-dependent proteins: Correlation of amino acid sequences and membrane binding properties suggests a membrane contact site, Biochemistry 36 (1997) 5120–5127.

[114] E.M. Bevers, R.F. Zwaal, G.M. Willems, The effect of phospholipids on the formation of immune complexes between autoantibodies and beta2-glycoprotein I or prothrombin, Clin Immunol 112 (2004) 150–160.

[115] L.V. Rao, A.D. Hoang, S.I. Rapaport, Mechanism and effects of the binding of lupus anticoagulant IgG and prothrombin to surface phospholipid, Blood 88 (1996) 4173–4182.

[116] T. Akimoto, T. Akama, I. Kono, T. Sumida, Relationship between clinical features and binding domains of antiprothrombin autoantibodies in patients with systemic lupus erythematosus and antiphospholipid syndrome, Lupus 8 (1999) 761–766.

[117] S.S. Pierangeli, G.H. Goldsmith, D.W. Branch, E.N. Harris, Antiphospholipid antibody: Functional specificity for inhibition of prothrombin activation by the prothrombinase complex, Br J Haematol 97 (1997) 768–774.

[118] M.J. Simmelink, D.A. Horbach, R.H. Derksen, J.C. Meijers, E.M. Bevers, G.M. Willems, et al., Complexes of antiprothrombin antibodies and prothrombin cause lupus anticoagulant activity by competing with the binding of clotting factors for catalytic phospholipid surfaces, Br J Haematol 113 (2001) 621–629.

[119] M. Galli, L. Ruggeri, T. Barbui, Differential effects of antibeta2-glycoprotein I and antiprothrombin antibodies on the anticoagulant activity of activated protein C, Blood 91 (1998) 1999–2004.

[120] Y. Sakai, T. Atsumi, M. Ieko, O. Amengual, S. Furukawa, A. Furusaki, et al., The effects of phosphatidylserine-dependent antiprothrombin antibody on thrombin generation, Arthritis Rheum 60 (2009) 2457–2467.

[121] M. Sanmarco, S. Gayet, M.C. Alessi, M. Audrain, E. de Maistre, J.C. Gris, et al., Antiphosphatidylethanolamine antibodies are associated with an increased odds ratio for thrombosis. A multicenter study with the participation of the European Forum on Antiphospholipid Antibodies, Thromb Haemost 97 (2007) 949–954.

[122] M.L. Bertolaccini, B. Roch, O. Amengual, T. Atsumi, M.A. Khamashta, G.R.V. Hughes, Multiple antiphospholipid tests do not increase the diagnostic yield in antiphospholipid syndrome, Br J Rheumatol 37 (1998) 1229–1232.

[123] T. Sugi, J.A. McIntyre, Phosphatidylethanolamine induces specific conformational changes in the kininogens recognizable by antiphosphatidylethanolamine antibodies, Thromb Haemost 76 (1996) 354–360.

[124] J.A. McIntyre, D.R. Wagenknecht, Anti-phosphatidylethanolamine (aPE) antibodies: A survey, J Autoimmun 15 (2000) 185–193.

[125] H. Yamada, T. Atsumi, G. Kobashi, C. Ota, E.H. Kato, N. Tsuruga, et al., Antiphospholipid antibodies increase the risk of pregnancy-induced hypertension and adverse pregnancy outcomes, J Reprod Immunol 79 (2009) 188–195.

[126] D.W. Jone, P.J. Nicholls, S. Donohoe, M.J. Gallimore, M. Winter, Antibodies to factor XII are distinct from antibodies to prothrombin in patients with the anti-phospholipid syndrome, Thromb Haemost 87 (2002) 426–430.

[127] M.L. Bertolaccini, K. Mepani, G. Sanna, G.R. Hughes, M.A. Khamashta, Factor XII autoantibodies as a novel marker for thrombosis and adverse obstetric history in patients with systemic lupus erythematosus, Ann Rheum Dis 66 (2007) 533–536.

[128] R. Cariou, G. Tobelem, S. Bellucci, J. Soria, C. Soria, J. Maclouf, et al., Effect of lupus anticoagulant on antithrombotic properties of endothelial cells—Inhibition of thrombomodulin-dependent protein C activation, Thromb Haemost 60 (1988) 54–58.

[129] E. Marciniak, E.H. Romond, Impaired catalytic function of activated protein C: a new in vitro manifestation of lupus anticoagulant, Blood 71 (1989) 2426–2432.

[130] R.G. Malia, S. Kitchen, M. Greaves, F.E. Preston, Inhibition of activated protein C and its cofactor protein S by antiphospholipid antibodies, Br J Haematol 76 (1990) 101–107.

[131] M. Borrell, N. Sala, C. de Castellarnau, S. Lopez, M. Gari, J. Fontcuberta, Immunoglobulin fractions isolated from patients with antiphospholipid antibodies prevent the inactivation of factor Va by activated protein C on human endothelial cells, Thromb Haemost 68 (1992) 101–107.

[132] T. Atsumi, M.A. Khamashta, O. Amengual, S. Donohoe, I. Mackie, K. Ichikawa, et al., Binding of anticardiolipin antibodies to protein C via β_2-glycoprotein I (β_2-GPI): A possible mechanism in the inhibitory effect of antiphospholipid antibodies on the protein C system, Clin Exp Immunol 112 (1998) 325–333.

[133] M. Galli, G.M. Willems, J. Rosing, R.M. Janssen, J.W. Govers-Riemslag, P. Comfurius, et al., Anti-prothrombin IgG from patients with anti-phospholipid antibodies inhibits the inactivation of factor Va by activated protein C, Br J Haematol 129 (2005) 240–247.

[134] S.L. Field, C.N. Chesterman, P.J. Hogg, Dependence on prothrombin for inhibition of activated protein C activity by lupus antibodies, Thromb Haemost 84 (2000) 1132—1133.

[135] P.R. Ames, C. Tommasino, L. Iannaccone, M. Brillante, R. Cimino, V. Brancaccio, Coagulation activation and fibrinolytic imbalance in subjects with idiopathic antiphospholipid antibodies—A crucial role for acquired free protein S deficiency, Thromb Haemost 76 (1996) 190—194.

[136] J. Ginsberg, C. Demers, P. Brill-Edwards, R. Bona, M. Johnston, A. Wong, et al., Acquired free protein S deficiency is associated with antiphospholipid antibodies and increased thrombin generation in patients with systemic lupus erythematosus, Am J Med 98 (1995) 379—383.

[137] M.D. McColl, A. Deans, P. Maclean, R.C. Tait, I.A. Greer, I.D. Walker, Plasma protein Z deficiency is common in women with antiphospholipid antibodies, Br J Haematol 20 (2003) 913—914.

[138] J. Pardos-Gea, J. Ordi-Ros, S. Serrano, E. Balada, I. Nicolau, M. Vilardell, Protein Z levels and anti-protein Z antibodies in patients with arterial and venous thrombosis, Thromb Res 121 (2008) 727—734.

[139] R.R. Forastiero, M.E. Martinuzzo, L. Lu, G.J. Broze, Autoimmune antiphospholipid antibodies impair the inhibition of activated factor X by protein Z/protein Z-dependent protease inhibitor, J Thromb Haemost 1 (2003) 1764—1770.

[140] T. Shi, G.M. Iverson, J.C. Qi, K.A. Cockerill, M.D. Linnik, P. Konecny, et al., Beta 2-glycoprotein I binds factor XI and inhibits its activation by thrombin and factor XIIa: Loss of inhibition by clipped beta 2-glycoprotein I, Proc Natl Acad Sci USA 101 (2004) 3939—3944.

[141] S. Rahgozar, Q. Yang, B. Giannakopoulos, X. Yan, S. Miyakis, S.A. Krilis, Beta$_2$-glycoprotein I binds thrombin via exosite I and exosite II: Anti-beta$_2$-glycoprotein I antibodies potentiate the inhibitory effect of beta$_2$-glycoprotein I on thrombin-mediated factor XIa generation, Arthritis Rheum 56 (2007) 605—613.

[142] S. Shibata, P.C. Harpel, A. Gharavi, J. Rand, H. Fillit, Autoantibodies to heparin from patients with antiphospholipid antibody syndrome inhibit formation of antithrombin III—thrombin complexes, Blood 83 (1994) 2532—2540.

[143] M. Jurado, J.A. Paramo, M. Gutierrez-Pimentel, E. Rocha, Fibrinolytic potential and antiphospholipid antibodies in systemic lupus erythematosus and other connective tissue disorders, Thromb Haemost 68 (1992) 516—520.

[144] R.B. Francis, J.A. McGehee, D.I. Feinstein, Endothelial-dependent fibrinolysis in subjects with the lupus anticoagulant and thrombosis, Thromb Haemost 59 (1988) 412—414.

[145] D.M. Keeling, S.J. Campbell, I.J. Mackie, S.J. Machin, D.A. Isenberg, The fibrinolytic response to venous occlusion and the natural anticoagulant in patients with antiphospholipid syndrome both with and without systemic lupus erythematosus, Br J Haematol 77 (1991) 354—359.

[146] M. Cugno, M. Cabibbe, M. Galli, P.L. Meroni, S. Caccia, R. Russo, et al., Antibodies to tissue-type plasminogen activator (tPA) in patients with antiphospholipid syndrome: Evidence of interaction between the antibodies and the catalytic domain of tPA in 2 patients, Blood 103 (2004) 2121—2126.

[147] T. Sugi, T. Makino, Autoantibodies to contact proteins in patients with recurrent pregnancy losses, J Reprod Immunol 53 (2002) 269—277.

[148] R. Takeuchi, T. Atsumi, M. Ieko, Y. Amasaki, K. Ichikawa, T. Koike, Suppressed intrinsic fibrinolytic activity by monoclonal anti-beta-$_2$ glycoprotein I autoantibodies: Possible mechanism for thrombosis in patients with antiphospholipid syndrome, Br J Haematol 119 (2002) 781—788.

[149] F. Carmona, I. Lazaro, J.C. Reverter, D. Tassies, J. Font, R. Cervera, et al., Impaired factor XIIa-dependent activation of fibrinolysis in treated antiphospholipid syndrome gestations developing late-pregnancy complications, Am J Obstet Gynecol 194 (2006) 457—465.

[150] O.R. Etingin, D.P. Hajjar, K.A. Hajjar, P.C. Harpel, R.L. Nachman, Lipoprotein (a) regulates plasminogen activator inhibitor-1 expression in endothelial cells, J Biol Chem 266 (1991) 2459—2465.

[151] T. Atsumi, M.A. Khamashta, C. Andujar, M.J. Leandro, O. Amengual, P.R. Ames, et al., Elevated plasma lipoprotein(a) level and its association with impaired fibrinolysis in patients with antiphospholipid syndrome, J Rheumatol 25 (1998) 69—73.

[152] M. Yamazaki, H. Asakura, H. Jokaji, M. Saito, C. Uotani, I. Kumabashiri, et al., Plasma levels of lipoprotein(a) are elevated in patients with the antiphospholipid antibody syndrome, Thromb Haemost 71 (1994) 424—427.

[153] R. Simantov, J.M. LaSala, S.K. Lo, A.E. Gharavi, L.R. Sammaritano, J.E. Salmon, et al., Activation of cultured vascular endothelial cells by antiphospholipid antibodies, J Clin Invest 96 (1995) 2211—2219.

[154] S.S. Pierangeli, R.G. Espinola, X. Liu, E.N. Harris, Thrombogenic effects of antiphospholipid antibodies are mediated by intercellular cell adhesion molecule-1, vascular cell adhesion molecule-1, and P-selectin, Circ Res 88 (2001) 245—250.

[155] N. Del Papa, Y.H. Sheng, E. Raschi, D.A. Kandiah, A. Tincani, M.A. Khamashta, et al., Human beta 2-glycoprotein I binds to endothelial cells through a cluster of lysine residues that are critical for anionic phospholipid binding and offers epitopes for anti-beta 2-glycoprotein I antibodies, J Immunol 160 (1998) 5572—5578.

[156] A. Kornberg, M. Blank, S. Kaufman, Y. Shoenfeld, Induction of tissue factor-like activity in monocytes by anti-cardiolipin antibodies, J Immunol 153 (1994) 1328—1332.

[157] O. Amengual, T. Atsumi, M.A. Khamashta, G.R.V. Hughes, The role of the tissue factor pathway in the hypercoagulable state in patients with the antiphospholipid syndrome, Thromb Haemost 79 (1998) 276—281.

[158] D.W. Branch, G.M. Rodgers, Induction of endothelial cell tissue factor activity by sera from patients with antiphospholipid syndrome: A possible mechanism of thrombosis, Am J Obstet Gynecol 168 (1993) 206—210.

[159] T. Atsumi, M.A. Khamashta, R.S. Haworth, G. Brooks, O. Amengual, K. Ichikawa, et al., Arterial disease and thrombosis in the antiphospholipid syndrome: A pathogenic role for endothelin 1, Arthritis Rheum 41 (1998) 800—807.

[160] Y. Zhao, R. Rumold, M. Zhu, D. Zhou, A.E. Ahmed, D.T. Le, et al., An IgG antiprothrombin antibody enhances prothrombin binding to damaged endothelial cells and shortens plasma coagulation times, Arthritis Rheum 42 (1999) 2132—2138.

[161] M. Vega-Ostertag, X. Liu, H. Kwan-Ki, P. Chen, S. Pierangeli, A human monoclonal antiprothrombin antibody is thrombogenic *in vivo* and upregulates expression of tissue factor and E-selectin on endothelial cells, Br J Haematol 135 (2006) 214—219.

[162] F. Dignat-George, L. Camoin-Jau, F. Sabatier, D. Arnoux, F. Anfosso, N. Bardin, et al., Endothelial microparticles: A potential contribution to the thrombotic complications of the antiphospholipid syndrome, Thromb Haemost 91 (2004) 667—673.

[163] M.H. Wiener, M. Burke, M. Fried, I. Yust, Thromboagglutination by anticardiolipin antibody complex in the antiphospholipid syndrome: A possible mechanism of immune-mediated thrombosis, Thromb Res 103 (2001) 193—199.

III. ANTI-PHOSPHOLIPID SYNDROME

[164] L. Emmi, C. Bergamini, A. Spinelli, F. Liotta, T. Marchione, A. Caldini, et al., Possible pathogenetic role of activated platelets in the primary antiphospholipid syndrome involving the central nervous system, Ann N Y Acad Sci 823 (1997) 188—200.

[165] W. Shi, B.H. Chong, C.N. Chesterman, β_2-Glycoprotein I is a requirement for anticardiolipin antibodies binding to activated platelets: Differences with lupus anticoagulants, Blood 81 (1993) 1255—1262.

[166] F. Shu, M. Sugimura, N. Kanayama, H. Kobayashi, T. Kobayashi, T. Terao, Immunohistochemical study of annexin V expression in placentae of preeclampsia, Gynecol Obstet Invest 49 (2000) 17—23.

[167] J.H. Rand, X.X. Wu, S. Guller, J. Gil, A. Guha, J. Scher, et al., Reduction of annexin-V (placental anticoagulant protein-I) on placental villi of women with antiphospholipid antibodies and recurrent spontaneous abortion, Am J Obstet Gynecol 171 (1994) 1566—1572.

[168] J.H. Rand, X.X. Wu, S. Guller, J. Scher, H.A. Andree, C.J. Lockwood, Antiphospholipid immunoglobulin G antibodies reduce annexin-V levels on syncytiotrophoblast apical membranes and in culture media of placental villi, Am J Obstet Gynecol 177 (1997) 918—923.

[169] J.H. Rand, X.X. Wu, R. Lapinski, W.L. van Heerde, C.P. Reutelingsperger, P.P. Chen, et al., Detection of antibody-mediated reduction of annexin A5 anticoagulant activity in plasmas of patients with the antiphospholipid syndrome, Blood 104 (2004) 2783—2790.

[170] J. Matsuda, N. Saitoh, K. Gohchi, M. Gotoh, M. Tsukamoto, Anti-annexin V antibody in systemic lupus erythematosus patients with lupus anticoagulant and/or anticardiolipin antibody, Am J Hematol 47 (1994) 56—58.

[171] N. Nakamura, T. Ban, K. Yamaji, Y. Yoneda, Y. Wada, Localization of the apoptosis-inducing activity of lupus anticoagulant in an annexin V-binding antibody subset, J Clin Invest 101 (1998) 1951—1959.

[172] K.A. Hajjar, A.T. Jacovina, J. Chacko, An endothelial cell receptor for plasminogen/tissue plasminogen activator. I. Identity with annexin II, J Biol Chem 269 (1994) 21191—21197.

[173] D. Kaczan-Bourgois, J.P. Salles, F. Hullin, J. Fauvel, A. Moisand, I. Duga-Neulat, et al., Increased content of annexin II (p36) and p11 in human placenta brush-border membrane vesicles during syncytiotrophoblast maturation and differentiation, Placenta 17 (1996) 669—676.

[174] K. Ma, R. Simantov, J.C. Zhang, R. Silverstein, K.A. Hajjar, K.R. McCrae, High affinity binding of beta 2-glycoprotein I to human endothelial cells is mediated by annexin II, J Biol Chem 275 (2000) 15541—15548.

[175] J. Zhang, K.R. McCrae, Annexin A2 mediates endothelial cell activation by antiphospholipid/anti-beta$_2$ glycoprotein I antibodies, Blood 105 (2005) 1964—1969.

[176] Z. Romay-Penabad, M.G. Montiel-Manzano, T. Shilagard, E. Papalardo, G. Vargas, A.B. Deora, et al., Annexin A2 is involved in antiphospholipid antibody-mediated pathogenic effects *in vitro* and *in vivo*, Blood 114 (2009) 3074—3083.

[177] E. Raschi, C. Testoni, D. Bosisio, M.O. Borghi, T. Koike, A. Mantovani, et al., Role of the MyD88 transduction signaling pathway in endothelial activation by antiphospholipid antibodies, Blood 101 (2003) 3495—3500.

[178] N. Satta, S. Dunoyer-Geindre, G. Reber, R.J. Fish, F. Boehlen, E.K. Kruithof, et al., The role of TLR2 in the inflammatory activation of mouse fibroblasts by human antiphospholipid antibodies, Blood 109 (2007) 1507—1514.

[179] S.S. Pierangeli, M.E. Vega-Ostertag, E. Raschi, X. Liu, Z. Romay-Penabad, V. De Micheli, et al., Toll-like receptor and antiphospholipid mediated thrombosis: *In vivo* studies, Ann Rheum Dis 66 (2007) 1327—1333.

[180] M. Sorice, A. Longo, A. Capozzi, T. Garofalo, R. Misasi, C. Alessandri, et al., Anti-beta$_2$-glycoprotein I antibodies induce monocyte release of tumor necrosis factor alpha and tissue factor by signal transduction pathways involving lipid rafts, Arthritis Rheum 56 (2007) 2687—2697.

[181] S.K. Moestrup, I. Schousboe, C. Jacobsen, J.R. Leheste, E.I. Christensen, T.E. Willnow, Beta$_2$-glycoprotein-I (apolipoprotein H) and beta$_2$-glycoprotein-I—phospholipid complex harbor a recognition site for the endocytic receptor megalin, J Clin Invest 102 (1998) 902—909.

[182] M.T. Pennings, M. van Lummel, R.H. Derksen, R.T. Urbanus, R.A. Romijn, P.J. Lenting, et al., Interaction of beta$_2$-glycoprotein I with members of the low density lipoprotein receptor family, J Thromb Haemost 4 (2006) 1680—1690.

[183] O.M. Andersen, D. Benhayon, T. Curran, T.E. Willnow, Differential binding of ligands to the apolipoprotein E receptor 2, Biochemistry 42 (2004) 9355—9364.

[184] B.C. Lutters, R.H. Derksen, W.L. Tekelenburg, P.J. Lenting, J. Arnout, P.G. de Groot, Dimers of beta 2-glycoprotein I increase platelet deposition to collagen via interaction with phospholipids and the apolipoprotein E receptor 2', J Biol Chem 278 (2003) 33831—33838.

[185] M. van Lummel, M.T. Pennings, R.H. Derksen, R.T. Urbanus, B.C. Lutters, N. Kaldenhoven, et al., The binding site in β_2-glycoprotein I for ApoER2' on platelets is located in domain V, J Biol Chem 280 (2005) 36729—36736.

[186] T. Shi, B. Giannakopoulos, X. Yan, P. Yu, M.C. Berndt, R.K. Andrews, et al., Anti-beta$_2$-glycoprotein I antibodies in complex with beta$_2$-glycoprotein I can activate platelets in a dysregulated manner via glycoprotein Ib-IX-V, Arthritis Rheum 54 (2006) 2558—2567.

[187] M.T. Pennings, R.H. Derksen, M. van Lummel, J. Adelmeijer, K. VanHoorelbeke, R.T. Urbanus, et al., Platelet adhesion to dimeric beta-glycoprotein I under conditions of flow is mediated by at least two receptors: Glycoprotein Ibalpha and apolipoprotein E receptor 2', J Thromb Haemost 5 (2007) 369—377.

[188] F.A. Baglia, C.N. Shrimpton, J. Emsley, K. Kitagawa, Z.M. Ruggeri, J.A. Lopez, et al., Factor XI interacts with the leucine-rich repeats of glycoprotein Ibalpha on the activated platelet, J Biol Chem 279 (2004) 49323—49329.

[189] M. Jankowski, I. Vreys, C. Wittevrongel, D. Boon, J. Vermylen, M.F. Hoylaerts, et al., Thrombogenicity of beta 2-glycoprotein I-dependent antiphospholipid antibodies in a photochemically induced thrombosis model in the hamster, Blood 101 (2003) 157—162.

[190] H. Zhou, A.S. Wolberg, R.A. Roubey, Characterization of monocyte tissue factor activity induced by IgG antiphospholipid antibodies and inhibition by dilazep, Blood 104 (2004) 2353—2358.

[191] P.L. Meroni, E. Raschi, C. Testoni, A. Tincani, G. Balestrieri, R. Molteni, et al., Statins prevent endothelial cell activation induced by antiphospholipid (anti-beta$_2$-glycoprotein I) antibodies: Effect on the proadhesive and proinflammatory phenotype, Arthritis Rheum 44 (2001) 2870—2878.

[192] S. Dunoyer-Geindre, P. De Moerloose, B. Galve-De Rochemonteix, G. Reber, E. Kruithof, NFkappaB is an essential intermediate in the activation of endothelial cells by anti-beta(2) glycoprotein 1 antibodies, Thromb Haemost 88 (2002) 851—857.

[193] M. Bohgaki, T. Atsumi, Y. Yamashita, S. Yasuda, Y. Sakai, A. Furusaki, et al., The p38 mitogen-activated protein kinase (MAPK) pathway mediates induction of the tissue factor gene in monocytes stimulated with human monoclonal anti-beta$_2$ glycoprotein I antibodies, Int Immunol 16 (2004) 1633—1641.

[194] M. Vega-Ostertag, E.N. Harris, S.S. Pierangeli, Intracellular events in platelet activation induced by antiphospholipid antibodies in the presence of low doses of thrombin, Arthritis Rheum 50 (2004) 2911—2919.

[195] M. Vega-Ostertag, K. Casper, R. Swerlick, D. Ferrara, E.N. Harris, S.S. Pierangeli, Involvement of p38 MAPK in the up-regulation of tissue factor on endothelial cells by anti-phospholipid antibodies, Arthritis Rheum 52 (2005) 1545—1554.

[196] K. Nakajima, Y. Tohyama, S. Kohsaka, T. Kurihara, Protein kinase C alpha requirement in the activation of p38 mitogen-activated protein kinase, which is linked to the induction of tumor necrosis factor alpha in lipopolysaccharide-stimulated microglia, Neurochem Int 44 (2004) 205—214.

[197] C. Lopez-Pedrera, P. Buendia, M.J. Cuadrado, E. Siendones, M.A. Aguirre, N. Barbarroja, et al., Antiphospholipid anti-bodies from patients with the antiphospholipid syndrome induce monocyte tissue factor expression through the simul-taneous activation of NF-kappaB/Rel proteins via the p38 mitogen-activated protein kinase pathway, and of the MEK-1/ ERK pathway, Arthritis Rheum 54 (2006) 301—311.

[198] G. Girardi, P. Redecha, J.E. Salmon, Heparin prevents anti-phospholipid antibody-induced fetal loss by inhibiting complement activation, Nat Med 10 (2004) 1222—1226.

[199] M. Sugiura-Ogasawara, K. Nozawa, T. Nakanishi, Y. Hattori, Y. Ozaki, Complement as a predictor of further miscarriage in couples with recurrent miscarriages, Hum Reprod 21 (2006) 2711—2714.

[200] W.D. Davis, R.L. Brey, Antiphospholipid antibodies and complement activation in patients with cerebral ischemia, Clin Exp Rheumatol 10 (1992) 455—460.

[201] S.S. Pierangeli, G. Girardi, M. Vega-Ostertag, X. Liu, R.G. Espinola, J. Salmon, Requirement of activation of complement C3 and C5 for antiphospholipid antibody-mediated thrombophilia, Arthritis Rheum 52 (2005) 2120—2124.

[202] F. Fischetti, P. Durigutto, V. Pellis, A. Debeus, P. Macor, R. Bulla, et al., Thrombus formation induced by antibodies to beta$_2$-glycoprotein I is complement dependent and requires a priming factor, Blood 106 (2005) 2340—2346.

[203] K. Oku, T. Atsumi, M. Bohgaki, O. Amengual, H. Kataoka, T. Horita, et al., Complement activation in patients with primary antiphospholipid syndrome, Ann Rheum Dis 68 (2009) 1030—1035.

[204] L.R. Sammaritano, S. Ng, R. Sobel, S.K. Lo, R. Simantov, R. Furie, et al., Anticardiolipin IgG subclasses: Association of IgG2 with arterial and/or venous thrombosis, Arthritis Rheum 40 (1997) 1998—2006.

[205] O. Amengual, T. Atsumi, M.A. Khamashta, M.L. Bertolaccini, G.R. Hughes, IgG2 restriction of anti-beta$_2$-glycoprotein I as the basis for the association between IgG2 anticardiolipin anti-bodies and thrombosis in the antiphospholipid syndrome, Arthritis Rheum 41 (1998) 1513—1515.

Endothelial Cell Damage and Atherosclerosis

Joan T. Merrill

INTRODUCTION

Endothelial damage is implicated in the development of atherosclerosis regardless of whether the disease is incited by hypertension, diabetes, smoking, or poorly regulated lipid metabolism [1] and regardless of the specific alignment of these various risk factors in individual patients. Infections and autoimmune diseases, including systemic lupus erythematosus (SLE) and the primary antiphospholipid syndrome (APS), have become strongly identified with accelerated atherosclerosis, and an extensive literature of *in vitro* and *in vivo* studies supports a model in which chronic immune imbalance provides a critical and ubiquitous underpinning to the development of arterial plaque. Furthermore, it is increasingly evident that the progressive events that lead to atherosclerosis are promulgated to a great extent by disordered endothelial cells interacting with inflammatory mediators.

Coronary vascular disease has come to be regarded in and of itself as a chronic, autoimmune, inflammatory disease [2], and the processes that are characteristically played out along the blood vessel walls may not be fundamentally different in overt inflammatory diseases from what is considered the more "usual" wear and tear on blood vessels, even if this drama is relatively subtle in the larger aging population. For this reason, the study of aberrant endothelial changes in SLE may be of value not only for the protection of our patients but also potentially for the invention of targeted blood vessel-stabilizing agents for a wider population at long-term risk.

The specific risk for premature atherosclerosis in autoimmune patients such as those with lupus is known to include traditional risk factors [3, 4], chronic generalized inflammation, and target-specific autoimmune mechanisms [4–7]. The effects of all of these elements might be exacerbated by chronic use of medications such as corticosteroids that add an additional disease burden by contributing to obesity, hyperglycemia, and hyperlipedemia [3–5]. Although both traditional and autoimmune mechanisms must certainly combine to contribute broadly to atherosclerotic plaque in lupus patients over time, a full analysis of endothelial risks for atherosclerosis in lupus, which includes a complete review of traditional risk factors, is beyond the scope of this chapter. Therefore, with a brief disclaimer that traditional, reversible risk factors remain of the highest priority in lupus patients and should be aggressively addressed by clinicians, the focus of this review is primarily the impact on endothelial cells from prototypic pathologic features of lupus and the question of whether the spectrum of autoantibodies associated with APS may be directly or indirectly involved in this process.

EVIDENCE FOR AN AUTOIMMUNE ETIOLOGY FOR ATHEROSCLEROSIS AND VASCULAR ENDOTHELIAL DYSFUNCTION IN SLE

Although many physiologic measurements for plaque and endothelial dysfunction are now available, most of the reports on lupus populations have employed ultrasound measurements of carotid artery intimal media thickness or endothelium-dependent flow-mediated dilatation (FMD) of the upper extremities, which is an indicator of endothelial responsiveness [8–10].

Both measurements have been repeatedly associated with other relevant measures for cardiovascular disease in different populations [8–10] and are widely accepted surrogates for plaque and endothelial dysfunction, respectively. High-resolution B-mode ultrasound is a noninvasive technique that is useful for measuring atherosclerosis in vessels that are accessible to the surface of the body, such as the carotid arteries. It provides the ability to specifically gauge the distance between the easily visualized blood—intima and media—adventitia interfaces in the carotid wall, which results in a fairly precise measurement of carotid intima-media thickness (IMT). Although there has been some controversy in the past about whether measurement of the carotid arteries is a good surrogate for what might be going on in coronary vessels, work by Amato *et al.* suggests that if ultrasound techniques are applied to the deep coronary vessels, the carotid measurements are found to be quite correlative [8].

FMD measures the increase in blood vessel diameter (usually measuring the brachial artery) as a response to an increased shear in blood flow. Temporary ischemia is induced by occluding blood flow in the forearm. This generates endothelial release of nitric oxide, which causes smooth muscle relaxation and vessel dilation. The magnitude of change in the diameter of the vessel as visualized on an ultrasound image is measured 60 s after the release of the mechanical occlusion. This number is generally thought to be decreased in patients with endothelial dysfunction [9, 10], and it may become apparent earlier in the development of atherosclerosis than increases in IMT.

IMT has been found to be increased in multiple studies of SLE patients compared to control populations [4, 5, 11–18]. There is no doubt that increased IMT in SLE is associated with traditional risk factors for atherosclerosis [13]; however, biochemical evidence for an increase in nontraditional risk factors, such as inflammatory endothelial activation, impaired vascular remodeling and other inflammatory features have been suggested to play an additional role [19, 20]. The degree to which cardiovascular risk in SLE may be due to direct vascular effects of the disease or indirect promotion of traditional risk factors by the disease or its treatments remains controversial.

In at least one study, it has been reported that patients with SLE presented a higher mean IMT of the common carotid artery than healthy subjects despite equivalent exposure to traditional cardiovascular risk factors, supporting the hypothesis that SLE pathology may contribute directly to atherosclerosis. However, this has not been confirmed in every series [21]. Interestingly, disrupted endothelial function and various proatherogenic and prothrombotic vascular disease risk factors have been reported to be prevalent even in clinically quiescent SLE, supporting the notion that at least some of the effects of the disease might be indirect and perhaps, by inciting traditional types of endothelial impairment that are self-perpetuating, permanently damaging [22]. This model is supported by the finding that compared with controls, SLE patients have a significantly higher rate of classical risk factors for atherosclerosis, including hypertension hypercholesterolemia and hypertriglyceridemia [13].

Another perspective on contradictory findings regarding cause and effect across various lupus study populations is provided by Dr. Manzi's group. Their evidence suggests that conventional risk factors for atherosclerosis may predominate in older patients, and that lupus-specific associations may be more apparent in younger patients [23] and have relevance to earlier signals for vascular pathology such as arterial stiffness as opposed to plaque detection [5]. However, the tendency for lupus disease activity to improve with age might explain these observations to some extent, and it does not rule out the long-term risks that might have been contributed from earlier lupus activity in older patients. In line with this, many other studies correlate age with IMT even within a lupus population so that although age is certainly already known to be a true risk factor for atherosclerosis, it may also confound the analysis of lupus-induced acceleration of the disease.

FMD, then, may be a better surrogate for premature atherosclerosis, which is relevant to lupus pathology, than IMT measurements. FMD has been reported to be impaired in a number of controlled studies of lupus populations [11, 13, 15, 21, 23–27]. In one report, patients with SLE had endothelial dysfunction that remained significant even after adjustment for other classic cardiac risk factors [26]. FMD abnormalities have been described in SLE patients who have no prior ischemic events and with the same IMT findings as controls [21], confirming that this likely represents an earlier stage of pathology. Importantly, these signs of endothelial dysfunction have been directly associated with the degree of disease activity in patients [15, 21]. In another study, when only SLE patients were considered in the absence of controls, endothelial dysfunction correlated negatively with IMT [27], supporting the hypothesis that the accumulated burden of plaque in older people who no longer have active lupus disease might impair the assessment of early versus later disease findings.

FMD, however, has also been found to significantly differ between SLE patients with ($5.54 \pm 4.36\%$) and those without ($8.81 \pm 5.28\%$) clinical cardiovascular complications ($p = 0.01$) [25]. Therefore, a case can be made that FMD can distinguish SLE patients from healthy subjects, as well as from lupus without atherosclerosis. To some extent, if the goal were to define specific pathologic risks associated with lupus, this technique could be preferable over IMT, which might be more likely to pick up accumulated damage in the absence of current lupus-specific risk factors. However, these surrogate markers are, at best, approximations of disease, and neither has been demonstrated to be completely consistent from study to study.

INFLAMMATION AND ATHEROSCLEROSIS: RELEVANCE TO ENDOTHELIAL CELLS AND SLE

It is now appreciated that inflammation drives the entire spectrum of events that lead to atherosclerotic disease, from initial injury to endothelial cells to the trapping and accumulation of foam cells in the arterial wall, plaque buildup, and the critical end-stage events that trigger rupture of the fibrous cap, which close down arteries and initiate life-threatening clinical sequelae [28]. Specific lupus-driven pathologies have been implicated in the dysregulated endothelium that becomes so vulnerable to these progressive events in patients with accelerated atherosclerosis. Before considering the potential role of vascular autoantibodies that are associated with APS, a review of these lupus-relevant inflammatory mechanisms and their associations with endothelial cell dysregulation is in order.

Oxidized Phospholipids

Phospholipids are known to become oxidized in chronic inflammatory states such as hyperlipidemia and atherosclerosis, and it seems logical that this may be accentuated in acute inflammatory disorders such as lupus. Oxidized phospholipids may have both pro-inflammatory and anti-inflammatory effects on endothelium [24], so their roles as mediators of endothelial function or dysfunction may be complex.

Oxidation of low-density lipoprotein (LDL) particles is considered to be a key step in the progression of atherosclerosis. Oxidatized LDL is extremely pro-inflammatory, with chemotactic, cytotoxic, and immunogenic properties [29]. Autoantibodies to this modified moiety can be detected within atherosclerotic lesions beneath the endothelium or circulating as immune complexes in many diseases, including lupus [30], and have been implicated as risk factors for both subclinical atherosclerosis [31] and acute arterial thrombosis [30]. Antibodies to oxidized high-density lipoprotein (HDL) have also been found to be circumstantially correlated to anti-endothelial cell antibodies, and binding inhibition studies have suggested significant cross-reactivity between epitopes on endothelial cells and oxidized HDL [32].

Lysophosphatidylcholine (LPC) is a phospholipid that can be detected both in oxidized LDL and in damaged endothelium, and evidence suggests that it might contribute to the antigenicity of oxidized LDL

[32]. LPC is formed by oxidation of phospholipids and/or pro-inflammatory factors leading to activation of phospholipase A2, which can activate cells and contribute to further inflammation.

HDL is also known to be abnormal in the inflammatory milieu of lupus, losing what is normally a significant protective, antioxidant role in the stabilization of the bloodstream [33]. In one of the most definitive studies of its type, McMahon et al. [33] performed IMT measurements by carotid artery ultrasound in 276 women with SLE and evaluated the antioxidant function of HDL as the change in oxidation of LDL that it could produce. Two important anti-inflammatory proteins of HDL, paraoxonase 1 and apolipoprotein A1, were also low. A total of 86.7% of patients with plaque on carotid ultrasound were found to have abnormal HDL, but only 40.7% of patients without plaque had abnormal HDL ($p < 0.001$). An association between the abnormal, pro-inflammatory HDL and plaque held up in a multivariate analysis, adjusting for other, more conventional, risk factors that also had some impact [33]. Both paroxynase and apolipoprotein A1 have been independently studied in lupus in the context of specific autoantibodies, including those that appear to bind to HDL and specific autoantibodies to apolipoprotein A1 that can be detected when it is surrounded by a negative charge but separated from the other components of HDL [34–37].

Dr. Hahn's group had previously performed a study in which 154 women with SLE were compared to 48 women with rheumatoid arthritis (RA) and 72 healthy controls [38]. Pro-inflammatory HDL was identified in 44.7% of SLE patients, 4.1% of controls, and 20.1% of RA patients. In addition, they found that SLE patients with cardiovascular arterial disease had significantly higher pro-inflammatory HDL scores than patients without evidence for this complication ($p < 0.001$) [38]. In this study, the levels of pro-inflammatory HDL also correlated with the levels of oxidized LDL ($r = 0.37$; $p < 0.001$), suggesting that the pro-inflammatory HDL, by failing in its usual task of protecting other lipoproteins, may have a role in creating a permissive environment for the accumulation of oxidized LDL.

Hahn and McMahon [39] have theorized that in the context of an inflammatory, oxidative microenvironment, HDLs are converted to pro-inflammatory HDLs, which have lower levels of the major cholesterol transport protein apolipoprotein A1 as well as impaired function of the major HDL antioxidant enzyme, paroxonase. They suggest that this development may be a critical factor in the breakdown of regulation of LDL oxidation. A decrease in apolipoprotein A1 may also impair efficient transport of cholesterol out of subendothelial deposits, leading to the buildup of toxic elements in the blood vessel wall in inflammatory, proatherogenic conditions such as SLE.

Circulating Endothelial Cells and Endothelial Progenitor Cells

Circulating endothelial cells are rare in the healthy bloodstream [40] but are known to appear during inflammation and can also be incited by other mechanisms that promote vascular damage. Circulating endothelial cells have been specifically observed in active flares of lupus, correlating with markers of complement activation such as C3a [15]. This supports a model in which autoantibody-mediated vascular injury could be promoting the presence of these cells in the circulation.

A related and somewhat more controversial population of cells, which can be detected in peripheral blood when there is endothelial damage, have been designated as endothelial progenitor cells [40] because they have surface markers that suggest features of both immature cells and endothelial cells. In addition, they can be made to form colonies on plates in the laboratory. The question of whether these cells, when detected in increased numbers in the circulation, are damaged or dying remains unanswered, but they are thought to have pro-inflammatory or pro-coagulant properties [40], as would be expected with wounded cells.

Circulating endothelial progenitor cells are being widely studied as potential biomarkers for vascular injury and/or atherosclerosis [41] or as elements that can contribute to vascular regeneration and repair. Indeed, evidence suggests that they can reverse endothelial dysfunction [5] or the vascular damage that predisposes to atherosclerosis. In a study of 44 inactive SLE patients and 35 age-matched female controls, the numbers of circulating endothelial progenitor cells were found to be reduced in the SLE patients and some of their *in vitro* functionality was partially inhibited compared to those in the controls [42]. This suggests that in lupus, these cells may either be chronically depleted or damaged, impeding or delaying the normal endothelial repair processes.

In a study comparing circulating endothelial progenitor cells in 35 SLE patients and controls, the number of endothelial progenitor cells in blood samples from SLE patients was not lower than that in controls, but these cells were found to have different properties than control cells, with some functions decreased and some functional and/or inflammatory functions increased, including the expression of inducible nitric oxide synthase, interleukin-6, and the endothelial adhesion molecule intracellular adhesion molecule-1 (ICAM-1) [41]. This finding provides evidence of activation of endothelial progenitor cells in the lupus bloodstream and possible pro-inflammatory potential. It is not known, however, whether these findings would be relevant to a role for these cells in endothelial repair. In fact, another report suggests that the abnormal phenotype of circulating endothelial progenitor cells in SLE includes impairments in the synthesis of pro-angiogenic factors and poor performance at differentiating into mature endothelial cells [43].

Because endothelial progenitor cells may be derived from CD34+ hematopoietic stem cells, Westerweel *et al.* [44] compared the numbers of these stem cells and endothelial progenitor cells in patients with lupus and stained the cells with annexin V to identify how many of each type might be apoptotic cells, which are cells undergoing programmed cell death. Patients with SLE had lower levels of both types of cells, and this seemed to be associated with increased apoptosis of the earlier lineage of stem cells. This might signify a chronic depletion of materials necessary to promote normal endothelial repair and remodeling.

Because endothelial progenitor cells have been reported to synthesize high levels of interferon-α (IFN-α), it has been hypothesized that elevated IFN-I levels could lead to endothelial dysfunction by decreasing the number and function of these cells [45]. A study of 70 lupus patients and 31 healthy controls demonstrated markedly reduced levels of colony-forming units of endothelial progenitor cells compared with controls ($p < 0.0001$), and it found that the decrease in these cells was more striking in the subset of patients with high levels of type I IFNs [45]. Elevations in type I IFN levels were also associated with impaired endothelial function in these lupus patients [45].

Modulators of Endothelial Function and Their Relevance to Lupus

Endothelial cells secrete the potent mediators of inflammation and vascular self-regulation that help to control the selective adhesion, emigration and activation of leukocytes, the oxidation of the local environment, the responsiveness of endothelium to physiologic mediators of vasoconstriction and vasodilation, and the pro-coagulant potential of the blood vessel surface as well of the surrounding particles that interact on the endothelial membrane [46]. A healthy artery is required to keep a number of complex, sometimes conflicting signals in balance as they come from a myriad of mediators.

At the same time, the endothelium must contend with the direct effect of uninvited bloodstream stressors arising from toxins, infections, diet, or the burdens of a chronic inflammatory disease. This suggests a critical need for both elasticity and balance in the daily performance of the endothelial cell. Many modulators of endothelium have been described as abnormally increased or decreased in lupus patients, but many of these reports are conflicting, and all are operative only in subsets of the lupus patients. It seems reasonable that for generalizable purposes, issues of balance and adaptability must take precedence over the presence, absence, or direction of an abnormality in any given regulatory element. As with the blood vessel, the physician must have adaptability when evaluating individual patient data to determine the risk for accelerated atherosclerosis or the rate at which an artery is likely to become diseased.

Endothelial cells are a critical source of circulating von Willebrand factor [46]. In a prospective study of 182 SLE patients followed for a mean of 8.3 years, cardiovascular events were predicted by several markers of endothelial activation (which included elevated von Willebrand factor, soluble vascular cellular adhesion molecule-1 (s-VCAM-1), and fibrinogen) [47]. In a study of 74 SLE patients and 74 age- and sex-matched controls, the von Willebrand factor was also found to be higher in lupus patients [12].

SLE patients have also been reported to have significant elevations in plasma vascular endothelial growth factor levels [16], which were correlated with IMT measurements of plaque. The endothelial-specific Ang-Tie ligand receptor system has been identified as an important regulator of vascular responsiveness to inflammatory stimuli, and it has been found to be disrupted in SLE. Also in this case, the abnormality was correlated to active disease [48]. Taken together, these data suggest a number of potential targets for treatments that might prevent endothelial activation in SLE, but again, only subsets of patients will demonstrate these markers at a given point in time, and the significance of each one (as opposed to the others) may never be elucidated.

In vitro studies also support the contention that there may be many factors in lupus plasma that could be damaging to endothelial cells. Annexin V binding to human umbilical vein endothelial cells was decreased when plasma from SLE cases was added, and this effect was greater with samples from lupus patients with a history of cardiovascular disease than from those without this history [49]. There was also a positive association between annexin V binding in SLE plasma and IMT results in the lupus patients who had donated the plasma ($r = 0.73$; $p < 0.001$) [49].

In a study that measured endothelial regulators at multiple visits, soluble E-selectin and adiponectin were significantly higher in the SLE patients with plaque compared to those without plaque, extending across multiple visits, and this was not associated with treatments or disease activity [18]. The elevation of E-selectin would have been anticipated because this biomarker reflects activation of the endothelium and has been previously associated with atherosclerosis and cardiovascular risk in both SLE and non-SLE cohorts [50–52]. However, the association of adiponectin and coronary heart disease in other populations has tended to go the other way (high adiponectin seemingly decreasing the risk for atherosclerosis). For example, this was found to be the case in studies of otherwise healthy aging populations [53, 54], and it has been thought that adiponectin might generally have a protective effect in coronary arterial disease [55].

Furthermore, high adiponectin levels in SLE have not been confirmed in every lupus study [56, 57], although they have been found in children with diabetes [58]. One possible explanation is that increased adiponectin represents, in some subsets of patients, an increase in a protective feedback loop, responsive to progressive damage as atherosclerosis progresses. Alternatively, it should be considered that either an increase or a decrease in measurements of endothelial regulators might be pathologic to the viability of arterial walls, potentially increasing risk for damage and eventual atherosclerosis.

Levels of adiponectin (or any of the mediators that are elevated in large subsets of the lupus population but for which this finding does not apply across the board in any cross-sectional study) could differ at different points in time during the development of diseased blood vessels, reactive to different inciting factors and different stages of pathology. For this reason, a more sophisticated understanding of the balance of factors involved in endothelial function, injury, and repair may be needed to sort out the meaning of these various findings in different populations at risk for atherosclerosis.

The endothelial protein C receptor (EPCR) may also promote abnormal vascular functions in SLE. In a study performed by Dr. Robert Clancy's team at New York University [64], membrane expression of this endothelial protein was found in lupus kidney biopsies in the medulla, arterial endothelium, and cortical peritubular capillaries. Where significant staining was found in the peritubular capillaries (evident in 16/59 lupus nephritis biopsies but not in healthy control kidneys), this predicted a poor prognosis for the patient's response to therapy. This finding was also associated, but only in part, with renal damage [59]. These observations suggest a role for EPCR as a biomarker for impairment of arterial circulation in lupus.

When EPCR is shed, it can promote an acute risk for thrombosis. Dr. Clancy's group also found that mean soluble levels of this endothelial receptor correlated positively with serum creatinine ($r = 0.3429$; $p < 0.0001$) in patients with nephritis [60], suggesting the likelihood of impaired circulation in this subset of patients as well as increased tissue damage. The prevalence of a polymorphism of EPCR that is associated with an enhanced propensity for endothelial shedding (polymorphism A6936G) was also higher in SLE patients (41%; $n = 27$) than controls (7%; $n = 29$) ($p = 0.0039$). Although a direct link between EPCR and a clinical risk for atherosclerosis has not been established in patients with lupus, such a connection seems intuitive. For example, in a report of an *in vitro* experiment, the inflammatory cytokine IFN-γ, which is a known atherogenic mediator, also seemed to contribute to membrane shedding of EPCR [60].

Endothelial cells are a major source of nitric oxide, which is synthesized by the oxidation of arginine. The enzymes that promote these reactions can be either constitutively active or inducible, and they are thought to play an important role in immunity and inflammation [61], with variable inflammatory and anti-inflammatory effects [62]. Nitric oxide is likely to help mediate important vascular events relevant to both SLE and atherosclerosis because it functions as a signaling molecule that regulates many physiologic and immunologic functions by permeating cell membranes and modifying proteins, lipids, and DNA [63]. In the course of these reactions, nitric oxide is known to be capable of altering the function or the immunogenicity of a large number of molecules.

Nitric oxide produced by inducible nitric oxide synthase is thought to frequently promote vasodilation, vascular dysfunction, and cell damage. The production of nitric oxide by constitutive endothelial cell enzymes, however, may have a protective, or even anti-inflammatory, function for the blood vessel wall by preventing the adhesion and damage caused by myeloid cells, helping to maintain the ultimate integrity of endothelial cells and promoting healthy, well-regulated blood vessel functions. However, the extent to which these generalizations hold in a disordered vasculature remains uncertain.

Nitric oxide can be induced *in vitro* by treating endothelial cells with acetylcholine. Clancy's group found that smooth muscle relaxation (a biologic surrogate for vasodilation) was prevented by the inhibition of nitric oxide production [64]. Further experiments by this group suggested the possibility that when the endothelial adhesion protein ICAM-1 (CD54) is bound by invading neutrophils or other myeloid cells, this induces changes in actin leading to activation of the endothelial cells. However, natural nitric oxide production by healthy endothelial cells may be able to disrupt this process and protect vascular endothelium from leukocyte-mediated injury [64].

Nitric oxide may also inhibit neutrophil superoxide production, which may be important in inflammatory conditions that lead to endothelial injury. Dr. Clancy's group performed an elegant series of experiments that evaluated the effect of nitric oxide on superoxide anion production stimulated either by activating intact neutrophils or via a biochemical trigger using the xanthine oxidase/hypoxanthine pathway. Nitric oxide inhibited superoxide anion production by stimulated, intact neutrophils when it was added well in advance of the cell stimulant, but it inhibited the xanthine oxidase-induced reaction only briefly, which would be expected based on the short half-life of the molecule [65]. This suggests the possibility that there may be formation of a stable intermediate within living cells that prolongs the ultimate effects of the transiently viable nitric oxide moiety.

The effect of nitric oxide on cell-free, NADPH oxidase superoxide generation was also studied [65], and it was found that nitric oxide was not inhibitory unless it was added before the activation of arachidonate in this reaction (i.e., early in the process). However, treatment of only the membrane component of the system was sufficient to cause the inhibition. Taken together, these data support a model in which endothelial nitric oxide could be protective to endothelium against the toxic production of reactive oxygen species by invading myeloid cells, but once these inflammatory cells assemble the activated machinery and start provoking endothelial damage, nitric oxide might not be as helpful.

Two genetic polymorphisms in the promoter region of the inducible nitric oxide synthase enzyme (G-954C and CCTTT) are common in people of African descent, possibly due to a protective effect that this genetic inheritance may have on malaria. Both of these nucleotide polymorphism variants were found to be more frequent in African-American SLE patients than in controls [66].

Overall serum levels of nitrate/nitrite have been reported to be significantly higher in SLE patients than in controls, with significant correlation to general disease activity scores and renal activity [67]. Oates *et al.* [68] demonstrated that various measurements reflecting nitric oxide or its modified proteins correlated with disease activity in different subgroups of lupus patients. This group also performed a longitudinal study of 83 SLE patients and 40 control subjects examining serum nitrate—nitrite levels at intervals, with each sample drawn from patients after they had eaten a low-nitrate diet, which may be a critical variable in assessing the accuracy of global nitrate—nitrite levels as a biomarker for vascular activation [69]. Levels were higher in lupus patients than in controls and were associated, over time, with increased disease activity, proliferative nephritis, and poor response to therapy, suggesting the possibility of a role for nitric oxide imbalance in lupus vascular damage.

In another study by the same group, blood samples were obtained from 67 SLE patients and 31 healthy controls. Apoptosis (programmed cell death) of cells exposed to SLE plasma in the presence of nitric oxide [70] was increased compared to that in controls—another potential, if indirect, indicator of long-term risk to endothelial viability in these patients. Together with the evidence discussed previously that lupus patients may harbor defects in endothelial repair, it could be hypothesized that the risk to blood vessel competency may be coming from both directions. Inhibition of nitric oxide synthesis was shown in a murine model to reduce urinary 8-isoprostane F2α and nitric oxide systemic markers of oxidant stress [71], confirming the possibility that modulating this system in SLE might have benefits.

Not surprisingly, there is a large, controversial literature with many disparate reports on how nitric oxide works and whether it is pro-inflammatory or anti-inflammatory or both [63]. As stressed in a sophisticated discussion of this topic by Dr. James Oates, both the concentration of nitric oxide in a given process and the physiologic environment in which it is acting may affect its ultimate effects on the vasculature [63]. Thus, any generalizations about the role that it may play in SLE or APS must be formulated cautiously. Philosophically, however, it seems likely that a disordered endothelium might lead to decreased anti-inflammatory and/or increased pro-inflammatory effects of this agent under the influence of various different types of stimuli. Regardless of the fine details, it may be that "too much" or "too little" of any given function of nitric oxide may not explain the net atherosclerosis risk in lupus or APS quite as usefully as the more generally applicable concept of "imbalanced" functioning.

Thus, there are many examples observed of abnormalities of endothelial regulators in SLE, and it is evident that the chances of developing one cohesive model to explain all of these observations are slim. Many of these studies are not in agreement regarding which elements are increased in their relatively small samples of SLE patients, and there is often a poor consistency between abnormalities that might be considered pro-atherogenic and those that might be considered protective within one study. Different pathways may be involved in endothelial damage in different patients or in the same patients at different milestones in their illness as the disease flares and is treated with various inflammatory modulators. Keeping this in mind, the general issue for future treatment development would be to promote vascular stability and flexibility of endothelial function in response to stressors or to define and decrease common early promulgators of these complex stressors rather than to follow the favored model for twenty-first century drug development—that of attempting to identify one or a simple cocktail of finely targeted treatments—because it is becoming increasingly apparent that each rational microtarget is the consequence of disparate and fluid abnormalities,

each of which might not be relevant in most of the patients most of the time.

ANTIPHOSPHOLIPID SYNDROME, ENDOTHELIAL DAMAGE, AND ATHEROSCLEROSIS

A series of consensus projects undertaken since the late 1990s have been attended by most of the world's experts in APS, who have sought to redefine this entity based on rational clinical and biologic grounds [72–75]. This entirely laudable wish to construct classification criteria that might facilitate clinical research has resulted in a great deal of diagnostic confusion in the clinic, as more science emerges to describe relevant pathologic antibodies and even clinical features that are not included in the previous years' consensus definitions. It has also become increasingly apparent that major overlaps exist between this disease, even in its primary form, and both clinical and laboratory features that are found in, but not formally classified as, systemic lupus.

A range of vascular abnormalities can arise associated with the novel autoantibodies and clinical events that most frequently overlap in those patients who are designated as having systemic lupus and those designated as having primary APS. However, in fact, APS may rarely be purely a primary disease because its manifestations are found not only in lupus patients but also in the context of numerous other autoimmune, infectious, or neoplastic diseases [76–79], many of which may not have been recognized in patients who receive the diagnosis of primary APS.

Although it may not be generally acceptable to refer to the spectrum of intravascular autoantibodies that these patients develop as part of APS, for the purposes of reviewing the many papers that confirm their association with both lupus and primary APS (as currently defined), they are discussed with the assumption of relevance in this review.

Because a wide spectrum of vascular pathology is well documented in primary antiphospholipid patients, this spectrum extends far beyond the strictly defined clinical classification criteria [80–86]. In particular, because atherosclerosis is elusive and multifactorial [87], it is apparent that the study of each condition contributes its own level of complexity. It seems logical that it could be very difficult to provide definitive associations between one quixotic disorder and the other, and this has proven to be the case, as evidenced by the variable results among some clinical studies [88–93].

In sorting through the contradictory literature that has supported or dismissed a causality relationship between antiphospholipid antibodies (aPLs) and atherosclerosis, the best conclusion that can be drawn is that all of these studies are underpowered for adjusting the multitudinal variables that are likely to be influencing the outcomes. Indeed, the modeling done with these factors has been as different from study to study as the prevalence of each variable in the variegated populations that were examined. Furthermore, many of these studies have examined different subsets of aPLs, tested under varying conditions in different laboratories. Furthermore, like the inflammation of SLE, aPLs may instigate or promote some of the more traditional risk factors for atherosclerosis, creating even more obscurity in the study of their impact.

As in lupus studies, the variable of age has confounded these analyses because atherosclerosis undoubtedly increases with age and detectable aPLs may have a tendency to decrease in some of these patients as the years go by. Even adjusting for age alone in small populations may not be useful because there simply may not be enough raw data to interpret the observations. Therefore, for the purposes of the current focus on evidence for an association between aPLs and endothelial damage, a critical analysis of the quality of the clinical associations between atherosclerosis and aPLs (or lack thereof) that have been reported in these many underpowered study designs will not be undertaken here.

A number of studies have correlated aPLs to measurements of plaque or FMD abnormalities. For example, 28 patients with newly diagnosed primary APS, 26 patients with stable coronary disease, and 38 healthy individuals underwent carotid duplex ultrasound measurements of IMT and FMD studies. The FMD measures in the patients with primary APS were significantly lower than those of the controls, whereas IMT was higher [94]. In another small, controlled study, FMD of the brachial artery and IMT of carotid arteries were measured in 16 APS patients and 16 healthy controls matched for age, gender, and other cardiovascular risk factors. FMD was significantly lower in patients than in controls (6.3 ± 5.2 vs. $18.2 \pm 2.7\%$; $p < 0.005$), whereas IMT did not differ [95]. Charakida et al. [96] studied 77 patients with positive aPLs who were matched for age and cardiovascular risk factors to controls, and they found that carotid IMT was increased in those with aPLs, and these patients also had lower FMD and paraoxonase activity.

In a study of 182 SLE patients, subjects were followed prospectively for a mean of 8.3 years. Cardiovascular events were predicted by positive aPL, as well as by markers of endothelial activation (which included elevated von Willebrand factor, s-VCAM-1, and fibrinogen) [47]. Within the subpopulation of patients who had aPLs, IMT measurements seemed to be greater in those with thrombotic history than in those who did not yet have pathologic consequences associated with

these antibodies [97]. Furthermore, using stepwise multiple regression analysis, the titer of IgG anticardiolipin independently predicted carotid IMT measurements ($p < 0.005$) [97]. This suggests that clinically pathogenic prothrombotic subsets of aPLs might be predictive of risk for advancing atherosclerosis. If these subsets could be better defined, perhaps more consistency in the literature could be obtained.

Carotid and femoral IMT and stiffness were measured in 58 patients with APS and 58 controls by Belizna et al. [98]. A significant difference between patients and controls was also found in this study, with increased IMT, arterial stiffness, and the presence of plaque in the APS patients. In another study of patients with primary APS, flow-mediated vasodilation was lower, IMT was larger, and von Willebrand factor was higher in patients compared to controls [99].

In a different approach, brachial artery FMD was performed in 40 patients with APS, none of whom had any history of coronary complications. Results from these patients were compared to those of 40 age- and sex-matched controls. Circulating endothelial cells were quantified, and soluble endothelial mediators were also studied, including plasma levels of adhesion molecules (s-ICAM-1, s-VCAM-1, s-E-selectin), thrombomodulin, von Willebrand factor, and tissue plasminogen activator (t-PA) [100]. There was a marked decrease in FMD responses by patients with APS compared to controls ($p = 0.0001$). Although no differences were observed in levels of circulating thrombomodulin, E-selectin, or VCAM-1, significant increases in s-ICAM-1 and von Willebrand factor were confirmed ($p = 0.003$ and $p = 0.002$, respectively). Mature circulating endothelial cells were also higher in APS patients. These data confirm that both a mechanical measurement of arterial endothelial dysfunction and plasma markers of stressed endothelium may be associated with APS in the absence of clinical evidence for arterial disease.

In an additional small controlled study of 25 patients with primary APS and 25 healthy people who had been matched for age, sex, and atherosclerosis risk factors, adhesion molecules, t-PA, and plasminogen activator inhibitor-1 (PAI-1) were measured [101]. FMD was significantly lower in the patients with primary APS than in controls (8 ± 5 vs. $15 \pm 6\%$; $p < 0.001$). Concentrations of VCAM-1 ($p < 0.001$), ICAM-1 ($p < 0.001$), and fibrinogen ($p < 0.05$) were higher in patients than in controls, and the first two correlated directly with FMD. No differences were found between groups for D-dimer, t-PA, and PAI-1 levels or activities [101].

However, in another small, but very judiciously controlled study in which FMD was compared to markers of endothelial function, 20 patients with primary APS were specifically included only if there were no known risk factors for cardiovascular diseases, use of exogenous estrogens, pregnancy, or intake of treatments likely to impact on endothelial function, including vitamins or antioxidants. A total of 30 age- and sex-matched healthy controls were also subjected to the same restrictive entry criteria. FMD of the brachial artery and some plasma markers of endothelial and platelet activation were measured. There was an insignificant trend toward lowered FMD in the patients and no differences were observed in plasma von Willebrand factor, soluble P-selectin, or soluble CD40L. Circulating progenitor and endothelial cells were also comparable between a subset of each group [102].

A Canadian publication reported results from a prospective cohort of 415 people who were followed for arterial and venous events for a median of 7.4 years [103]. In this nested case–control study, the predictive value of endothelial and inflammatory markers was examined. Patients with aPLs were more likely to suffer new arterial events [103], but otherwise conventional risk factors seemed to have the most impact on short-term risk for arterial events in this somewhat larger, prospective cohort.

Margarita *et al.* [104] compared IMT and other cardio-vascular risk factors in 44 patients with primary APS, 25 patients with inherited thrombophilia, and 34 healthy controls. In this study, it was found that traditional risk factors (e.g., smoking habits, hypertension, and dyslipidemia) and IMT measurements were similar between these groups at an age younger than 40 years, but IMT was greater in the primary APS patients than controls at an age older than 40 years, including measurements of the common carotid ($p = 0.01$), carotid bifurcation ($p = 0.003$), and the internal carotid ($p = 0.005$) [104]. These data confirm that the handling of variables in small studies may have a major impact on the interpretation of the data.

Risk of endothelial dysfunction may not be equivalent in all patients with primary APS, as was found in a study of 31 patients with primary APS and 27 age- and sex-matched, healthy controls [105]. FMD was significantly lower overall in patients with primary APS than in the controls (6.9 ± 4.9 vs. $14.8 \pm 4.1\%$; $p < 0.0001$). However, when further examining the data from those patients in the group who had primary APS, comparing those who had a history of arterial thrombosis and those who had a history of only venous events, those with arterial involvement ($n = 17$) had significantly lower FMD than those with a history of venous thrombosis (12 patients) (4.6 ± 3.9 vs. $7.4 \pm 4.1\%$; $p = 0.02$) [105]. Vascular location is another potentially important variable that might be considered when interpreting these studies. Interestingly, this study also compared FMD to a nitroglycerin-induced protocol, which is considered to be an endothelium-independent dilatation. Although the endothelium-dependent FMD

was impaired in the patients with APS, the endothelium-independent measurement was not different between the groups [105].

Endothelial dysfunction in patients with primary APS has also been suggested using a measurement of myocardial blood flow in a three-phase protocol—rest, cold pressor test, and adenosine positron emission tomography [106]. Patients with primary APS seemed to have lower myocardial flow reserve and endothelial-dependent vasodilation and less change in myocardial blood flow in response to a cold pressor test, suggesting impaired endothelial dysfunction using these sophisticated methods [106].

Many *in vitro* and animal studies have demonstrated that specific novel autoantibodies, found frequently in patients with APS, could play regulatory or dysregulatory roles in endothelial-dependent interactions that might lead to atherosclerosis, even though no causality for these mechanisms can be definitively confirmed by human studies of long-term outcome. This does not detract from the potential relevance of the many papers that support the hypothesis of a strong, subterranean influence of these antibodies on the vasculature.

HDL from women with aPLs has been reported to inhibit nitric oxide production in human aortic endothelial cells, in contrast to HDL from controls [96]. Furthermore, the expected beneficial effects of HDL on VCAM-1 expression by and superoxide production in monocytes and their adhesion to activated human aortic endothelial cells were decreased in women with APS [96]. Reiss *et al.* [107] found that both IFN-γ and immune complexes decrease expression of cholesterol 27-hydroxylase in human aortic endothelial cells. Downregulation of cholesterol 27-hydroxylase by immune complexes was also dependent on complement fixation [107], again demonstrating the co-dependence of many identified variables contributing to accelerated atherosclerosis mediated by intravascular autoimmunity.

Elevated lipoprotein (a) (Lp(a)) has also been implicated in premature atherosclerosis. High levels of Lp(a) have been described in patients designated as having primary APS as well as in those with SLE, and antibodies to oxidized Lp(a) have also been described in both of these populations [108]. Vascular dysfunction in APS may also involve upregulation of various cell surface and intracellular signaling molecules, as well as pro-inflammatory cytokine release from activated endothelial cells. Endothelial microparticles have been found to be elevated in patients with circulating aPLs, and evidence suggests they may be prothrombotic [109].

Murine studies have provided further proof of concept for the potential pathologic role of aPLs in inciting or perpetuating the progression of atherosclerosis. For example, immunization of LDL-RKO mice with anticardiolipin antibodies derived from a patient with APS enhanced primary atherogenesis (as measured by arterial fatty streak formation) [110]. CD1 mice were given an injection of monoclonal aPLs derived from male (BXSB × NZW) F1 mice. (This murine model develops lupuslike disease associated with APS and coronary artery disease.) Five of 8 antiphospholipid monoclonal antibodies induced vascular functional changes suggestive of endothelial dysfunction in these healthy CD1 mice [111].

Annexin A2 is a receptor for t-PA that is produced by endothelial cells. In one study, human aPLs, which induce thrombosis in normal mice, caused comparatively less thrombus when injected into annexin A2-deficient mice. Tissue factor activity was also decreased in these mice [112]. An anti-annexin A2 monoclonal antibody also interfered with the effects of aPLs on cultured endothelial cells, inhibiting their expression of ICAM-1 and E-selectin and induction of tissue factor activity. This suggests that annexin A2 may mediate the effects of some aPLs on endothelial cells and also the possibility of a protective effect of some antiphospholipid-related autoantibodies on endothelial cells (discussed later) [112].

An array of human and murine monoclonal anticardiolipin and anti-β_2-glycoprotein I antibodies or monoclonal isotype controls were injected into female BALB/c severe combined immunodeficiency (SCID) mice. Several surrogate measures for cardiovascular risk activation were measured in the plasma of these mice, including paraoxonase activity, peroxynitrite, superoxide, nitric oxide, and nitrotyrosine. The results were variable between subsets of antibodies. Paraoxonase activity and nitric oxide (nitrate plus nitrite) were decreased, whereas peroxynitrite, nitrotyrosine, superoxide, and expression of total antioxidant capacity of plasma were increased in the mice receiving human anticardiolipin IgG. Both paraoxonase activity and nitric oxide were decreased in mice that received murine anti-β_2-glycoprotein IgG and anticardiolipin IgM. Reduction of inducible nitric oxide synthase was demonstrated in the hearts of animals receiving IgG anticardiolipin by Western blotting and immunostaining [113], suggesting that aPLs may have profound, if unpredictable, effects on cardiovascular risks relevant to endothelial cell survival [113]. That all aPLs do not have identical effects seems clear.

In another experiment, mice were injected with IgG from a patient with APS in the presence or absence of an inhibitor of mitogen-activated protein kinase, and the inhibitor reduced the thrombus that was induced in femoral veins by the aPLs [114]. In additional *in vitro* experiments, the inhibitor reduced the number of myeloid cells (THP-1 cell line) that adhered to endothelial cells and tissue factor activity in carotid arteries, which suggests that aPLs may, at least in part, induce thrombus by direct signaling through endothelial cells [114].

Prothrombin is a pivotal regulator of clotting that associates with the assemblage of coagulation factors on endothelial cells and other intravascular structures. Prothrombin is known to be one of the antigens targeted by some of the spectrum of autoantibodies associated with APS. The effects of a monoclonal anti-prothrombin antibody derived from a patient with APS have been studied in several models. This antibody increased the thrombus size in an induced thrombosis mouse model, increased the expression of tissue factor and E-selectin on human endothelial cells when compared with controls, and induced increased mRNA expression of tissue factor in these cells [115]. Therefore, various subsets of autoantibodies can affect the endothelium and contribute to vascular dysfunction to an extent that could be hypothesized to be quite relevant to long-term cardiovascular risk.

From a different perspective, aPLs have been reported to be more frequent in patients with cardiovascular abnormalities. Marai *et al.* [116] compared 45 non-autoimmune patients with coronary arterial or cerebrovascular disease to 62 healthy controls. FMD and nitroglycerin-mediated vasodilatation were measured along with a spectrum of autoantibodies associated with APS, which included anticardiolipin IgG, IgM, and IgA; antinuclear antibody; anti-β_2-glycoprotein I (IgG, IgM, and IgA); and oxidized LDL. FMD was significantly lower in the cardiovascular patients, as expected, and there was a trend to decreased nitroglycerin-mediated vasodilatation as well, but this had only borderline significance ($p = 0.084$). Oxidized LDL and most of the autoantibody measurements did not differ between patients and controls, but mean IgG anticardiolipin was significantly higher in patients with arterial history [116].

Non-autoimmune men with borderline hypertension have been reported to have significantly higher anti-endothelial cell antibodies and anti-β_2-glycoprotein I IgG antibodies than normotensive control subjects ($p = 0.029$ and $p = 0.0001$ respectively) as well as higher IgM anti-endothelial antibodies ($p = 0.012$) [117]. Individuals in this study who were also found to have atherosclerotic plaque had significantly higher anti-endothelial cell antibodies of both IgG and IgM type ($p = 0.042$ and $p = 0.018$, respectively) than those without plaque [117].

What is the significance of anti-endothelial cell antibodies in APS? Certainly, various types of antiphospholipid autoantibodies can bind to vascular endothelial cells. One study examined the effect of serum and IgG from patients with APS who were positive for anti-endothelial antibodies on the clearance of apoptotic cells from endothelial cell monolayers. The antiphospholipid serum and IgG inhibited the clearance of endothelial cells undergoing programmed cell death and also opsonized these apoptotic endothelial cells, causing an enhanced Fc-dependent ingestion by phagocytic cells [118]. In both cases where the antibodies affected a normal clearance function (albeit in seemingly different directions), this was accompanied by increased thrombin generation.

Furthermore, these effects of patient serum and IgG on apoptotic endothelial cells were reproduced using a monoclonal antibody derived from a patient with APS. This antibody is known to specifically bind to endothelial cells and to be thrombogenic in experimental models. These findings suggest that anti-endothelial antibodies can be prothrombotic in patients with APS by inhibiting the clearance of pro-coagulant dying endothelial cells and possibly by inducing inflammatory changes in phagocytes [118].

A cell-based enzyme-linked immunosorbent assay (ELISA) was used to confirm the binding of aPLs to endothelial cells, and then the effects of these antibodies on fibrin formation and lysis were studied on cultured endothelial cell monolayers, with a measurement of PAI-1 in the supernatants [119]. Four of 14 aPL-positive sera prolonged the fibrin clot lysis time, and this correlated with the secretion of PAI-1 by the cultured endothelial cells [119]. A direct effect on annexin V was not noticed in this study. Importantly, this illustrates that some prothrombotic mechanisms that can be confirmed using aPLs are only elicited by some, but not all, of these antibodies, again underscoring the probable variability in effects of these antibodies in individual patients, which might be impacted by the antibody structures, the availability of their targets, and the microenvironment of the vasculature in individual patients.

Simoncini et al. [120] found that incubation of human endothelial cells with IgG from 12 APS patients increased reactive oxygen species, which was then inhibited by antioxidants such as vitamin C and N-acetyl-L-cysteine. Further study of these endothelial cells suggested that the activation of reactive oxygen species led to the stimulation of signals through the p38 mitogen-activated protein kinase, leading to upregulation of VCAM-1 expression and increased adhesion by a myeloid cell line (THP-1) [120]. This suggests that aPLs can induce oxidative stress in endothelial cells and alter their adhesiveness in an inflammatory milieu.

In vitro and in vivo models have confirmed the ability of aPLs to upregulate tissue factor on endothelial cells. Ferrara et al. [121] administered IgG from four patients with APS or control sera to cultured human umbilical vein endothelial cells. All four aPLs as well as the cell-activating stimulant, phorbol myristate acetate, upregulated tissue factor on the endothelial cells, and fluvastatin inhibited these effects in all cases in a dose-dependent manner—an effect that was reversed by mevalonate. These data suggest the possibility that, at least in some cases, statins might have some benefit in stabilizing the vasculature in APS patients.

Because it is well-known that in vitro testing for antibodies to β_2 glycoprotein I requires the antigen to be surrounded by a negative charge (e.g., a bed of phosphatidylserine or a γ-irradiated solid-phase plastic surface), the question of how the endothelium may be uniquely altered at time points when β_2-glycoprotein I is able to attract these antibodies may be relevant. A cell-based ELISA was used to examine the binding between endothelial cells and human sera known to be positive for aPLs as well as monoclonal aPLs [122]. Neither type of aPL could stick to resting endothelial cells, but one of the monoclonal aPLs bound well to activated and apoptotic endothelial cells [122]. This suggests that inflammation and/or activation of the vasculature may be required to induce pathologic binding of aPLs and again points out that not all antibodies will behave the same way, even in identical experimental circumstances. Thus, again, APS may have heterogeneous pathologies in different microenvironments, and the contributions of aPLs of individual specificities to the risk for atherosclerosis could be variable, and even disparate, over time.

IgG from 14 patients with APS was used to induce the expression of VCAM-1 on human umbilical vein endothelial cells [123]. Eight of the samples produced a more than 50% increase in VCAM-1, and this was significantly decreased by aspirin [123], suggesting a direct effect of aspirin on stabilizing an overreactive endothelium that might be triggered by some (but perhaps not all) aPLs. It is also possible that aPLs might cause direct damage to endothelial cells. Patients with APS have been reported to have elevated plasma levels of endothelial microparticles compared to healthy controls, and in a comparative study these particles were found in patients with SLE and aPLs and in patients with APS but not in SLE patients without aPLs or patients with other types of thrombotic risk [124]. Furthermore, plasma from patients with APS, but not plasma from any of the other groups, seemed to induce the release of endothelial microparticles that had pro-coagulant activity [124]. This suggests the possibility of direct interactions between antiphospholipid-type autoantibodies and the endothelium that might lead to the release of damaged, pro-coagulant endothelial particles into the blood.

Espinola et al. [125] found that surface expression of E-selectin on endothelial cells can be dramatically increased by aPLs in roughly the same degree as was elicited by treatment with tumor necrosis factor-α. In this carefully designed series of experiments, aPLs also induced activation of the central signaling moiety, nuclear factor-κB [125], and caused an increase in the number of leukocytes adhering to the endothelium of mice compared to the effects of isotype control antibodies in the same model system. This effect was not operable, however, in E-selectin-deficient mice. In an established protocol for measuring antiphospholipid-induced thrombosis, the thrombus size was also decreased when E-selectin-deficient mice were compared to control mice [125].

Endothelial cells have been found to share antigenic epitopes with β2-glycoprotein I and with oxidized LDL, especially the type with LPC [32]. Oxidized HDL may also share some antigenic epitopes with endothelial cells. A study of 184 patients with SLE and 85 healthy controls found a close correlation between measurements of anti-oxidized LDL, anticardiolipin, anti-LPC, anti-β2-glycoprotein I, and anti-endothelial cell antibodies [32]. Furthermore, binding of patient sera to endothelial cells could be competitively inhibited by β2-glycoprotein I, LPC, and oxidized LDL, supporting the possibility of either a specific shared epitope or shared charge-dependent interactions. In either case, it might explain why β2-glycoprotein I antibodies may be detectable for years without thrombotic events. It is possible that a critical mass of inflammatory components need to assemble together on a membrane (or the endothelial cells need to be activated to expose negatively charged or activated phospholipids) before β2-glycoprotein I becomes a pathologic antigen.

Oxidized LDL can also bind to β2-glycoprotein I, and the complexes of oxidized LDL and β2-glycoprotein-I seem to be immunogenic for their own subset of autoantibodies, which have been associated with arterial thrombosis and evidence of atherosclerosis by IMT measurements in patients with lupus and APS [30, 31]. A convenient surrogate for oxidized LDL that can be used to complex with β2-glycoprotein I in order to measure these antibodies is 7-ketocholesteryl-9-carboxy-nonanoate (oxLig-1).

Ames et al. [126] studied 29 patients with primary APS, 10 with aPLs but no clinical sequelae, 17 with thrombophilia but no aPLs, and 23 healthy controls. The subjects underwent measurement of carotid IMT and were examined for paraoxonase activity and autoantibodies. The complex of β2-glycoprotein I and oxidized LDL was found to be highest in the healthy group, but IgG antibodies to that structure as well as to the surrogate β2-glycoprotein I/oxLig1 were highest in the patients with primary APS ($p < 0.0001$). Autoantibodies to β2-glycoprotein I/oxLig1 independently predicted lower paraoxonase activity and greater carotid IMT in the population. The results of this study suggest an additional mechanism by which a subset from the spectrum of autoantibodies associated with APS might contribute to dysfunctional, highly oxidized endothelium and the progression of atherosclerosis.

Hahn and McMahon [39] reviewed much of these data and suggested that there is strong evidence for a model supporting inflammatory-related risks for atherosclerosis in lupus and/or APS. Based on in vitro studies, from animal models and from clinical series, it is apparent that aPLs can directly interact with endothelium and induce signaling changes and a pro-inflammatory, pro-coagulant phenotype. Their own evidence supports a model in which inflammatory stimuli might convert the normal protective HDL to what they designated as a pro-inflammatory HDL that carries less of the important cholesterol transport protein apolipoprotein A1 and also is associated with decreased function of the antioxidant enzyme paroxonase. As discussed previously, this could lead to increased oxidation of LDL in patients with lupus.

However, in those patients with autoimmunity that extends to the potential for making aPLs, increased oxidized LDL could stimulate immunogenicity of otherwise benign antigens such as β2-glycoprotein I, which may lead to a spectrum of cross-reacting autoantibodies that can further perturb oxidized lipoproteins and endothelial cells. This further promotes an inflammatory and pro-coagulant vascular surface, as well as generalized impairment of the endothelium in its normal vasoprotective and vasoreactive functions, egress of inflammatory cells into subendothelial space, dysfunctional lipid transport, and the progression or destabilization of atheromatous plaque.

NATURAL AUTOANTIBODIES: CAN THEY PROTECT AGAINST ENDOTHELIAL DAMAGE, ARE THEY RELATED TO ANTIPHOSPHOLIPID ANTIBODIES, AND CAN THEY EXPLAIN WHY SOME STUDIES DO NOT CONFIRM A PATHOGENIC ROLE FOR ANTIPHOSPHOLIPID ANTIBODIES IN ATHEROSCLEROSIS?

There is supposed to be an array of "natural" autoantibodies in healthy people, some of which target protein sequences that are highly redundant in many species and that have been conserved during evolution [127]. Evidence suggests that these particular autoantibodies might be considered not only benign but also perhaps even protective to homeostasis of the vasculature [127—130]. Natural autoantibodies have been described in blood samples from healthy people, but they can only be detected by special methods because they are usually masked by policing antibodies that recognize their active sites and interfere with their exposure [131, 132].

Anticardiolipin antibodies are apparent in 12% of the healthy elderly population and 2% of younger people when using standard ELISA [133, 134], but they are able to be detected in a much wider population when chemical treatments are used to unstrip the masking autoantibodies that are apparently clinging to the active sites of these autoantibodies in most healthy people. It is now recognized that there are large regulatory networks of these policing "anti-idiotype" autoantibodies, which recognize and mask the active sites of natural autoantibodies and keep them in check. Anti-idiotypes are thought to be another group of natural autoantibodies that serve the purpose of specifically targeting the active sites of potentially misbehaving antibodies, presumably rendering them inactive until needed for a particular purpose [131, 132, 135, 136].

Both natural autoantibodies and aPLs have been implicated in a beneficial role that might contribute to the maintenance of a stable endothelium, in that they can bind to the membranes of dying (apoptotic) cells, opsonizing them to facilitate their clearance [127]. It has been reported that the cell membranes on apoptotic cells develop oxidatized epitopes that attract both natural antibodies and aPLs [127, 129, 130, 137], reminiscent of the model described previously in which dysregulated oxidation and inflamed vasculature may promote the appearance of this spectrum of autoantibodies. This is consistent with the model that if the control over this system is dysregulated, endothelial damage might be exacerbated in a highly oxidized, inflamed vasculature, but also supports a potential helpful role for some subset of aPLs in this situation.

Not only can aPLs be unmasked in healthy people by chemical treatments of their interlaced network of circulating proteins but also autoantibodies from patients with autoimmune diseases can become masked after similar chemical treatments [135, 136]. In fact, it is possible to simultaneously mask and unmask different autoantibodies in a sample of blood, suggesting that disorderly oxidation or other inflammatory changes in blood vessels might cause small modifications to how many "benign" or "nefarious" antibodies interact with each other and with the endothelium, potentially disturbing the homeostasis of controlled release of natural autoantibodies or their effects on the vasculature [135, 138, 139].

By an extension of this logic, whether or not autoantibodies with a given specificity contribute to atherosclerosis might depend on how and when they bind to intravascular structures, what other variables may be disturbing the intravascular environment, and how the specific structural–functional relationships of these interactions are affected in the process. It may be that not only are some subsets of autoantibodies supposed to have beneficial functions but also this designation might be turned on its head in a disturbed vascular microenvironment. Nevertheless, the possibility that some aPLs might be protective against atherosclerosis cannot be ruled out.

References

[1] A. Doria, L. Iaccarino, P. Sarzi-Puttini, F. Atzeni, M. Turriel, M. Petri, Cardiac involvement in systemic lupus erythematosus, Lupus 14 (2005) 683–686.

[2] R.B. Singh, S.A. Mengi, Y.J. Xu, A.S. Arneja, N.S. Dhalla, Pathogenesis of atherosclerosis: A multifactorial process, Exp Clin Cardiol 7 (2002) 40–53.

[3] S. Manzi, E.N. Meilahn, J.E. Rairie, C.G. Conte, T.A. Medsger Jr., L. Jansen-McWilliams, et al., Age-specific incidence rates of myocardial infarction and angina in women with systemic lupus erythematosus: Comparison with the Framingham Study, Am J Epidemiol 145 (1997) 408–415.

[4] S. Manzi, F. Selzer, K. Sutton-Tyrrell, S.G. Fitzgerald, J.E. Rairie, R.P. Tracy, et al., Prevalence and risk factors of carotid plaque in women with systemic lupus erythematosus, Arthritis Rheum 42 (1999) 51–60.

[5] F. Selzer, K. Sutton-Tyrrell, S.G. Fitzgerald, J.E. Pratt, R.P. Tracy, L.H. Kuller, et al., Comparison of risk factors for vascular disease in the carotid artery and aorta in women with systemic lupus erythematosus, Arthritis Rheum 50 (2004) 151–159.

[6] G. Foteinos, Q. Xu, Immune-mediated mechanisms of endothelial damage in atherosclerosis, Autoimmunity 42 (2009) 627–633.

[7] S. Zampieri, L. Iaccarino, A. Ghirardello, E. Tarricone, S. Arienti, P. Sarzi-Puttini, et al., Systemic lupus erythematosus, atherosclerosis, and autoantibodies, Ann N Y Acad Sci 1051 (2005) 351–361.

[8] M. Amato, P. Montorsi, A. Ravani, E. Oldani, S. Galli, P.M. Ravagnani, et al., Carotid intima-media thickness by B-mode ultrasound as surrogate of coronary atherosclerosis: Correlation with quantitative coronary angiography and coronary intravascular ultrasound findings, Eur Heart J 28 (2007) 2094–2101.

[9] S. Laurent, P. Lacolley, P. Brunel, B. Laloux, B. Pannier, M. Safar, Flow-dependent vasodilation of brachial artery in essential hypertension, Am J Physiol 258 (1990) H1004–H1011.

[10] D.S. Celermajer, K.E. Sorensen y, V.M. Gooch, Non-invasive detection of endothelial dysfunction in children and adults at risk of atherosclerosis, Lancet 340 (1992) 1111–1115.

[11] P. Ghosh, A. Kumar, S. Kumar, A. Aggarwal, N. Sinha, R. Misra, Subclinical atherosclerosis and endothelial dysfunction in young South-Asian patients with systemic lupus erythematosus, Clin Rheumatol 28 (2009) 1259–1265.

[12] K. de Leeuw, A.J. Smit, E. de Groot, A.M. van Roon, C.G. Kallenberg, M. Bijl, Longitudinal study on premature atherosclerosis in patients with systemic lupus erythematosus, Atherosclerosis 206 (2009) 546–550.

[13] C.Y. Zhang, L.J. Lu, F.H. Li, H.L. Li, Y.Y. Gu, S.L. Chen, et al., Evaluation of risk factors that contribute to high prevalence of premature atherosclerosis in Chinese premenopausal systemic lupus erythematosus patients, J Clin Rheumatol 15 (2009) 111–116.

[14] F. Cacciapaglia, E.M. Zardi, G. Coppolino, F. Buzzulini, D. Margiotta, L. Arcarese, et al., Stiffness parameters, intima-media thickness and early atherosclerosis in systemic lupus erythematosus patients, Lupus 18 (2009) 249–256.

[15] Q. Shang, L.S. Tam, E.K. Li, G.W. Yip, C.M. Yu, Increased arterial stiffness correlated with disease activity in systemic lupus erythematosus, Lupus 17 (2008) 1096–1102.

[16] B.M. Colombo, F. Cacciapaglia, M. Puntoni, G. Murdaca, E. Rossi, G. Rodriguez, et al., Traditional and non traditional risk factors in accelerated atherosclerosis in systemic lupus erythematosus: Role of vascular endothelial growth factor (VEGATS Study), Autoimmun Rev 8 (2009) 309–315.

[17] B.M. Colombo, G. Murdaca, M. Caiti, G. Rodriguez, L. Grassia, E. Rossi, et al., Intima-media thickness: A marker of accelerated atherosclerosis in women with systemic lupus erythematosus, Ann N Y Acad Sci 1108 (2007) 121–126.

[18] H.R. Reynolds, J. Buyon, M. Kim, T. Rivera, P. Izmirly, P. Tunicka, et al., Association of plasma soluble E-selectin and adiponectin with carotid plaque in patients with systemic lupus erythematosus atherosclerosis, Artherosclerosis (2009). [Epub ahead of print].

[19] K. de Leeuw, B. Freire, A.J. Smit, H. Bootsma, C.G. Kallenberg, M. Bijl, Traditional and non-traditional risk factors contribute to the development of accelerated atherosclerosis in patients with systemic lupus erythematosus, Lupus 15 (2006) 675–682.

[20] A. Doria, Y. Shoenfeld, R. Wu, P.F. Gambari, M. Puato, A. Ghirardello, et al., Risk factors for subclinical atherosclerosis in a prospective cohort of patients with systemic lupus erythematosus, Ann Rheum Dis 62 (2003) 1071–1077.

[21] P. Valdivielso, J.J. Gómez-Doblas, M. Macias, M. Haro-Liger, A. Fernández-Nebro, M.A. Sánchez-Chaparro, et al., Lupus-associated endothelial dysfunction, disease activity and arteriosclerosis, Clin Exp Rheumatol 26 (2008) 827–833.

[22] A.B. Lee, T. Godfrey, K.G. Rowley, C.S. Karschimkus, G. Dragicevic, E. Romas, et al., Traditional risk factor assessment does not capture the extent of cardiovascular risk in systemic lupus erythematosus, Intern Med J 36 (2006) 237–243.

[23] F. Selzer, K. Sutton-Tyrrell, S. Fitzgerald, R. Tracy, L. Kuller, S. Manzi, Vascular stiffness in women with systemic lupus erythematosus, Hypertension 37 (2001) 1075–1082.

[24] P. Fu, K.G. Birukov, Oxidized phospholipids in control of inflammation and endothelial barrier, Transl Res 153 (2009) 166–176.

[25] E. Kiss, P. Soltesz, H. Der, Z. Kocsis, T. Tarr, H. Bhattoa, et al., Reduced flow-mediated vasodilation as a marker for cardiovascular complications in lupus patients, J Autoimmun 27 (2006) 211–217.

[26] A. Cederholm, E. Svenungsson, D. Stengel, G.Z. Fei, A.G. Pockley, E. Ninio, et al., Platelet-activating factor–acetylhydrolase and other novel risk and protective factors for cardiovascular disease in systemic lupus erythematosus, Arthritis Rheum 50 (2004) 2869–2876.

[27] M. El-Magadmi, H. Bodill, Y. Ahmad, P.N. Durrington, M. Mackness, M. Walker, et al., Systemic lupus erythematosus: An independent risk factor for endothelial dysfunction in women, Circulation 110 (2004) 399–404.

[28] A.H. Kao, J.M. Sabatine, S. Manzi, Update on vascular disease in systemic lupus erythematosus, Curr Opin Rheumatol 15 (2003) 519–527.

[29] A. Steinerová, J. Racek, F. Stozický, T. Zima, L. Fialová, A. Lapin, Antibodies against oxidized LDL—Theory and clinical use, Physiol Res 50 (2001) 131–141.

[30] L.R. Lopez, K.J. Dier, D. Lopez, J.T. Merrill, C.A. Fink, Anti-beta 2-glycoprotein I and antiphosphatidylserine antibodies are predictors of arterial thrombosis in patients with antiphospholipid syndrome, Am J Clin Pathol 121 (2004) 142–149.

[31] L.R. Lopez, M. Salazar-Paramo, C. Palafox-Sanchez, B.L. Hurley, E. Matsuura, I. Garcia-De La Torre, Oxidized low-density lipoprotein and beta2-glycoprotein I in patients with

systemic lupus erythematosus and increased carotid intima-media thickness: Implications in autoimmune-mediated atherosclerosis, Lupus 15 (2006) 80–86.

[32] R. Wu, E. Svenungsson, I. Gunnarsson, C. Haegerstrand-Gillis, B. Andersson, I. Lundberg, et al., Antibodies to adult human endothelial cells cross-react with oxidized low-density lipoprotein and beta 2-glycoprotein I (beta 2-GPI) in systemic lupus erythematosus, Clin Exp Immunol 115 (1999) 561–566.

[33] M. McMahon, J. Grossman, B. Skaggs, J. Fitzgerald, L. Sahakian, N. Ragavendra, et al., Dysfunctional proinflammatory high-density lipoproteins confer increased risk of atherosclerosis in women with systemic lupus erythematosus, Arthritis Rheum 60 (2009) 2428–2437.

[34] J. Delgado Alves, P.R. Ames, S. Donohue, L. Stanyer, J. Nourooz-Zadeh, C. Ravirajan, et al., Antibodies to high-density lipoprotein and beta2-glycoprotein I are inversely correlated with paraoxonase activity in systemic lupus erythematosus and primary antiphospholipid syndrome, Arthritis Rheum 46 (2002) 2686–2694.

[35] J.R. Batuca, P.R. Ames, D.A. Isenberg, J.D. Alves, Antibodies toward high-density lipoprotein components inhibit paraoxonase activity in patients with systemic lupus erythematosus, Ann N Y Acad Sci 1108 (2007) 137–146.

[36] A.R. Dinu, J.T. Merrill, C. Shen, I.V. Antonov, B.L. Myones, R.G. Lahita, Frequency of antibodies to the cholesterol transport protein apolipoprotein A1 in patients with SLE, Lupus 7 (1998) 355–360.

[37] J.T. Merrill, E. Rivkin, C. Shen, R.G. Lahita, Selection of a gene for apolipoprotein A1 using autoantibodies from a patient with systemic lupus erythematosus, Arthritis Rheum 38 (1995) 1655–1659.

[38] M. McMahon, J. Grossman, J. FitzGerald, E. Dahlin-Lee, D.J. Wallace, B.Y. Thong, et al., Proinflammatory high-density lipoprotein as a biomarker for atherosclerosis in patients with systemic lupus erythematosus and rheumatoid arthritis, Arthritis Rheum 54 (2006) 2541–2549.

[39] B.H. Hahn, M. McMahon, Atherosclerosis and systemic lupus erythematosus: The role of altered lipids and of autoantibodies, Lupus 17 (2008) 368–370.

[40] A.D. Blann, A. Woywodt, F. Bertolini, T.M. Bull, J.P. Buyon, R.M. Clancy, et al., Circulating endothelial cells. Biomarker of vascular disease, Thromb Haemost 93 (2005) 228–235.

[41] X.L. Deng, X.X. Li, X.Y. Liu, L. Sun, R. Liu, Comparative study on circulating endothelial progenitor cells in systemic lupus erythematosus patients at active stage, Rheumatol Int 22 (2009).

[42] J.R. Moonen, K. de Leeuw, X.J. van Seijen, C.G. Kallenbergm, M. van Luyn, M. Bijl, et al., Reduced number and impaired function of circulating progenitor cells in patients with systemic lupus erythematosus, Arthritis Res Ther 9 (2007) 4.

[43] M.F. Denny, Mehta H. Thacker, E.C. Somers, T. Dodick, F.J. Barrat, W.J. McCune, et al., Interferon-alpha promotes abnormal vasculogenesis in lupus: A potential pathway for premature atherosclerosis, Blood 110 (2007) 2907–2915.

[44] P.E. Westerweel, R.K. Luijten, I.E. Hoefer, H.A. Koomans, R.H. Derksen, M.C. Verhaar, Haematopoietic and endothelial progenitor cells are deficient in quiescent systemic lupus erythematosus, Ann Rheum Dis 66 (2007) 865–870.

[45] P.Y. Lee, Y. Li, H.B. Richards, F.S. Chan, H. Zhuang, S. Narain, et al., Type I interferon as a novel risk factor for endothelial progenitor cell depletion and endothelial dysfunction in systemic lupus erythematosus, Arthritis Rheum 56 (2007) 3759–3769.

[46] J.D. Pearson, Normal endothelial cell function, Lupus 9 (2000) 183–188.

[47] J. Gustafsson, I. Gunnarsson, O. Borjesson, S. Pettersson, S. Moller, G. Fei, et al., Predictors of the first cardiovascular event in patients with systemic lupus erythematosus—A prospective cohort study, Arthritis Res Ther 11 (2009) R186.

[48] P. Kümpers, S. David, M. Haubitz, J. Hellpap, R. Horn, V. Bröcker, et al., The Tie2 receptor antagonist angiopoietin 2 facilitates vascular inflammation in systemic lupus erythematosus, Ann Rheum Dis 68 (2009) 1638–1643.

[49] A. Cederholm, E. Svenungsson, K. Jensen-Urstad, C. Trollmo, A.K. Ulfgren, J. Swedenborg, et al., Decreased binding of annexin v to endothelial cells: A potential mechanism in atherothrombosis of patients with systemic lupus erythematosus, Arterioscler Thromb Vasc Biol 25 (2005) 198–203.

[50] Y.H. Rho, C.P. Chung, A. Oeser, J. Solus, P. Raggi, T. Gebretsadik, et al., Novel cardiovascular risk factors in premature coronary atherosclerosis associated with systemic lupus erythematosus, J Rheumatol 35 (2008) 1789–1794.

[51] S.J. Hwang, C.M. Ballantyne, A.R. Sharrett, L.C. Smith, C.E. Davis, A.M. Gotto Jr., et al., Circulating adhesion molecules VCAM-1, ICAM-1, and E-selectin in carotid atherosclerosis and incident coronary heart disease cases: The Atherosclerosis Risk in Communities (ARIC) study, Circulation 96 (1997) 4219–4225.

[52] L.E. Rohde, R.T. Lee, J. Rivero, M. Jamacochian, L.H. Arroyo, W. Briggs, et al., Circulating cell adhesion molecules are correlated with ultrasound-based assessment of carotid atherosclerosis, Arterioscler Thromb Vasc Biol 18 (1998) 1765–1770.

[53] T.A. Hopkins, N. Ouchi, R. Shibata, K. Walsh, Adiponectin actions in the cardiovascular system, Cardiovasc Res 74 (2007) 11–18.

[54] A.M. Kanaya, C. Wassel Fyr, E. Vittinghoff, P.J. Havel, M. Cesari, B. Nicklas, et al., Health ABC Study, Serum adiponectin and coronary heart disease risk in older Black and White Americans, J Clin Endocrinol Metab 91 (2006) 5044–5050.

[55] J.R. Kizer, J.I. Barzilay, L.H. Kuller, J.S. Gottdiener, Adiponectin and risk of coronary heart disease in older men and women, J Clin Endocrinol Metab 93 (2008) 3357–3364.

[56] C.P. Chung, A.G. Long, J.F. Solus, Y.H. Rho, A. Oeser, P. Raggi, et al., Adipocytokines in systemic lupus erythematosus: Relationship to inflammation, insulin resistance and coronary atherosclerosis, Lupus 18 (2009) 799–806.

[57] Y. Asanuma, A. Oeser, A.K. Shintani, E. Turner, N. Olsen, S. Fazio, et al., Premature coronary-artery atherosclerosis in systemic lupus erythematosus, N Engl J Med 349 (2003) 2407–2415.

[58] K. Heilman, M. Zilmer, K. Zilmer, P. Kool, V. Tillmann, Elevated plasma adiponectin and decreased plasma homocysteine and asymmetric dimethylarginine in children with type 1 diabetes, Scand J Clin Lab Invest 69 (2009) 85–91.

[59] P.M. Izmirly, L. Barisoni, J.P. Buyon, M.Y. Kim, T.L. Rivera, J.S. Schwartzman, et al., Expression of endothelial protein C receptor in cortical peritubular capillaries associates with a poor clinical response in lupus nephritis, Rheumatology (Oxford) 48 (2009) 513–519.

[60] C.A. Sesin, X. Yin, C.T. Esmon, J.P. Buyon, R.M. Clancy, Shedding of endothelial protein C receptor contributes to vasculopathy and renal injury in lupus: In vivo and in vitro evidence, Kidney Int 68 (2005) 110–120.

[61] S.B. Abramson, A.R. Amin, R.M. Clancy, M. Attur, The role of nitric oxide in tissue destruction, Best Pract Res Clin Rheumatol 15 (2001) 831–845.

[62] M.C. Levesque, J.B. Weinberg, The dichotomous role of nitric oxide in the pathogenesis of accelerated atherosclerosis associated with systemic lupus erythematosus, Curr Mol Med 4 (7) (2004) 777–786.

[63] J.C. Oates, The biology of reactive intermediates in systemic lupus erythematosus, Autoimmunity 43 (2010) 56–63.

[64] R.M. Clancy, S.B. Abramson, Acetylcholine prevents intercellular adhesion molecule 1 (CD54)-induced focal adhesion complex assembly in endothelial cells via a nitric oxide–cGMP-dependent pathway, Arthritis Rheum 43 (2000) 2260–2264.

[65] R.M. Clancy, J. Leszczynska-Piziak, S.B. Abramson, Nitric oxide, an endothelial cell relaxation factor, inhibits neutrophil superoxide anion production via a direct action on the NADPH oxidase, J Clin Invest 90 (1992) 1116–1121.

[66] J.C. Oates, M.C. Levesque, M.R. Hobbs, E.G. Smith, I.D. Molano, G.P. Page, et al., Nitric oxide synthase 2 promoter polymorphisms and systemic lupus erythematosus in African-Americans, J Rheumatol 30 (2003) 60–67.

[67] G. Gilkeson, C. Cannon, J. Oates, C. Reilly, D. Goldman, M. Petri, Correlation of serum measures of nitric oxide production with lupus disease activity, J Rheumatol 26 (1999) 318–324.

[68] J.C. Oates, E.F. Christensen, C.M. Reilly, S.E. Self, G.S. Gilkeson, Prospective measure of serum 3-nitrotyrosine levels in systemic lupus erythematosus: Correlation with disease activity, Proc Assoc Am Physicians 111 (1999) 611–621.

[69] J.C. Oates, S.R. Shaftman, S.E. Self, G.S. Gilkeson, Association of serum nitrate and nitrite levels with longitudinal assessments of disease activity and damage in systemic lupus erythematosus and lupus nephritis, Arthritis Rheum 58 (2008) 263–272.

[70] J.C. Oates, L.W. Farrelly, A.F. Hofbauer, W. Wang, G.S. Gilkeson, Association of reactive oxygen and nitrogen intermediate and complement levels with apoptosis of peripheral blood mononuclear cells in lupus patients, Arthritis Rheum 56 (2007) 3738–3747.

[71] C.J. Njoku, K.S. Patrick, P. Ruiz Jr., J.C. Oates, Inducible nitric oxide synthase inhibitors reduce urinary markers of systemic oxidant stress in murine proliferative lupus nephritis, J Invest Med 53 (2005) 347–352.

[72] S. Miyakis, M.D. Lockshin, T. Atsumi, D.W. Branch, R.L. Brey, R. Cervera, et al., International consensus statement on an update of the classification criteria for definite antiphospholipid syndrome (APS), J Thromb Haemost 4 (2006) 295–306.

[73] W.A. Wilson, A.E. Gharavi, J.C. Piette, International classification criteria for antiphospholipid syndrome: synopsis of a postconference workshop held at the Ninth International (Tours) aPL symposium, Lupus 10 (2001) 457–460.

[74] M.D. Lockshin, L.R. Sammaritano, S. Schwartzman, Validation of the Sapporo criteria for antiphospholipid syndrome, Arthritis Rheum 43 (2000) 440–443.

[75] W. Wilson, A.E. Gharavi, T. Koike, M.D. Lockshin, D.W. Branch, J.C. Piette, et al., International consensus statement on preliminary classification criteria for definite antiphospholipid syndrome: Report of an international workshop, Arthritis Rheum 42 (1999) 1309–1311.

[76] R.A. Ostrowski, J.A. Robinson, Antiphospholipid antibody syndrome and autoimmune diseases, Hematol Oncol Clin North Am 22 (2008) 53–65.

[77] L. Pugliese, I. Bernardini, E. Pacifico, M. Viola-Magni, E. Albi, Antiphospholipid antibodies in patients with cancer, Int J Immunopathol Pharmacol 19 (2006) 879–888.

[78] J.M. Grossman, Primary versus secondary antiphospholipid syndrome: Is this lupus or not? Curr Rheumatol Rep 6 (2004) 445–450.

III. ANTI-PHOSPHOLIPID SYNDROME

[79] I. Marai, G. Zandman-Goddard, Y. Shoenfeld, The systemic nature of the antiphospholipid syndrome, Scand J Rheumatol 33 (2004) 365–372.

[80] J. Nojima, H. Kuratsune, E. Suehisa, Y. Futsukaichi, H. Yamanishi, T. Machii, et al., Association between the prevalence of antibodies to beta(2)-glycoprotein I, prothrombin, protein C, protein S, and annexin V in patients with systemic lupus erythematosus and thrombotic and thrombocytopenic complications, Clin Chem 47 (2001) 1008–1015.

[81] M.L. Bertolaccini, G. Sanna, S. Ralhan, L.C. Gennari, J.T. Merrill, M.A. Khamashta, et al., Antibodies directed to protein S in patients with systemic lupus erythematosus: Prevalence and clinical significance, Thromb Haemost 90 (2003) 636–641.

[82] R.R. Forastiero, M.E. Martinuzzo, G.J. Broze, High titers of autoantibodies to tissue factor pathway inhibitor are associated with the antiphospholipid syndrome, J Thromb Haemost 1 (2003) 718–724.

[83] G.M. Iverson, C.A. von Mühlen, H.L. Staub, A.J. Lassen, W. Binder, G.L. Norman, Patients with atherosclerotic syndrome, negative in anti-cardiolipin assays, make IgA autoantibodies that preferentially target domain 4 of beta2-GPI, J Autoimmun 27 (2006) 266–271.

[84] M. Reichlin, J. Fesmire, A. Quintero-Del-Rio, M. Wolfson-Reichlin, Autoantibodies to lipoprotein lipase and dyslipidemia in systemic lupus erythematosus, Arthritis Rheum 46 (2002) 2957–2963.

[85] I. Palomo, F. Segovia, C. Ortega, S. Pierangeli, Antiphospholipid syndrome: A comprehensive review of a complex and multisystemic disease, Clin Exp Rheumatol 27 (2009) 668–677.

[86] T.P. Greco, A.M. Conti-Kelly, T. Greco Jr., R. Doyle, E. Matsuura, J.R. Anthony, et al., Newer antiphospholipid antibodies predict adverse outcomes in patients with acute coronary syndrome, Am J Clin Pathol 132 (2009) 613–620.

[87] G.J. Blake, P.M. Ridker, Inflammatory mechanisms in atherosclerosis: From laboratory evidence to clinical application, Ital Heart J 2 (2001) 796–800.

[88] G. Medina, D. Casaos, L.J. Jara, O. Vera-Lastra, M. Fuentes, L. Barile, et al., Increased carotid artery intima-media thickness may be associated with stroke in primary antiphospholipid syndrome, Ann Rheum Dis 62 (2003) 607–610.

[89] M. Galli, Antiphospholipid antibodies and thrombosis: Do test patterns identify the patients' risk? Thromb Res 114 (2004) 597–601.

[90] V. Pengo, E. Bison, A. Ruffatti, S. Iliceto, Antibodies to oxidized LDL/beta2-glycoprotein I in antiphospholipid syndrome patients with venous and arterial thromboembolism, Thromb Res 122 (4) (2008) 556–559.

[91] C.K. Shortell, K. Ouriel, R.M. Green, J.J. Condemi, J.A. DeWeese, Vascular disease in the antiphospholipid syndrome: A comparison with the patient population with atherosclerosis, J Vasc Surg 15 (1992) 158–165.

[92] A. Farzaneh-Far, M.J. Roman, M.D. Lockshin, R.B. Devereux, S.A. Paget, M.K. Crow, et al., Relationship of antiphospholipid antibodies to cardiovascular manifestations of systemic lupus erythematosus, Arthritis Rheum 54 (2006) 3918–3925.

[93] M.J. Roman, M.K. Crow, M.D. Lockshin, R.B. Devereux, S.A. Paget, L. Sammaritano, et al., Rate and determinants of progression of atherosclerosis in systemic lupus erythematosus, Arthritis Rheum 56 (2007) 3412–3419.

[94] P. Soltesz, H. Der, K. Veres, R. Laczik, S. Sipka, G. Szegedi, et al., Immunological features of primary anti-phospholipid syndrome in connection with endothelial dysfunction, Rheumatology (Oxford) 47 (2008) 1628–1634.

[95] F. Bilora, M.T. Sartori, E. Zanon, E. Campagnolo, M. Arzenton, A. Rossato, Flow-mediated arterial dilation in primary antiphospholipid syndrome, Angiology 60 (2009) 104–107.

[96] M. Charakida, C. Besler, J.R. Batuca, S. Sangle, S. Marques, M. Sousa, et al., Vascular abnormalities, paraoxonase activity, and dysfunctional HDL in primary antiphospholipid syndrome, JAMA 16 (2009) 1210–1217.

[97] P.R. Ames, A. Margarita, J. Delgado Alves, C. Tommasino, L. Iannaccone, V. Brancaccio, Anticardiolipin antibody titre and plasma homocysteine level independently predict intima media thickness of carotid arteries in subjects with idiopathic antiphospholipid antibodies, Lupus 11 (2002) 208–214.

[98] C.C. Belizna, V. Richard, E. Primard, J.M. Kerleau, N. Cailleux, J.P. Louvel, et al., Early atheroma in primary and secondary antiphospholipid syndrome: an intrinsic finding, Semin Arthritis Rheum 37 (2008) 373–380.

[99] H. Der, G. Kerekes, K. Veres, P. Szodoray, J. Toth, G. Lakos, et al., Impaired endothelial function and increased carotid intima-media thickness in association with elevated von Willebrand antigen level in primary antiphospholipid syndrome, Lupus 16 (2007) 497–503.

[100] M. Cugno, M.O. Borghi, L.M. Lonati, L. Ghiadoni, M. Gerosa, C. Grossi, et al., Patients with antiphospholipid syndrome display endothelial perturbation, J Autoimmun 34 (2) (2009) 105–110.

[101] M. Stalc, P. Poredos, P. Peternel, M. Tomsic, M. Sebestjen, T. Kveder, Endothelial function is impaired in patients with primary antiphospholipid syndrome, Thromb Res 118 (2006) 455–461.

[102] P. Gresele, R. Migliacci, M.C. Vedovati, A. Ruffatti, C. Becattini, M. Facco, et al., Patients with primary antiphospholipid antibody syndrome and without associated vascular risk factors present a normal endothelial function, Thromb Res 123 (2009) 444–451.

[103] C. Neville, J. Rauch, J. Kassis, S. Solymoss, L. Joseph, P. Belisle, et al., Antiphospholipid antibodies predict imminent vascular events independently from other risk factors in a prospective cohort, Thromb Haemost 101 (2009) 100–107.

[104] A. Margarita, J. Batuca, G. Scenna, J.D. Alves, L. Lopez, L. Iannaccone, et al., Subclinical atherosclerosis in primary antiphospholipid syndrome, Ann NY Acad Sci 1108 (2007) 475–480.

[105] F. Mercanoglu, D. Erdogan, H. Oflaz, R. Kücükkaya, F. Selcukbiricik, A. Gül, et al., Impaired brachial endothelial function in patients with primary anti-phospholipid syndrome, Int J Clin Pract 58 (2004) 1003–1007.

[106] E. Alexanderson, P. Cruz, A. Vargas, A. Meave, A. Ricalde, J.A. Talayero, et al., Endothelial dysfunction in patients with antiphospholipid syndrome assessed with positron emission tomography, J Nucl Cardiol 14 (2007) 566–572.

[107] A.B. Reiss, N.W. Awadallah, S. Malhotra, M.C. Montesinos, E.S. Chan, N.B. Javitt, et al., Immune complexes and IFN-gamma decrease cholesterol 27-hydroxylase in human arterial endothelium and macrophages, J Lipid Res 42 (2001) 1913–1922.

[108] F.I. Romero, M.A. Khamashta, G.R. Hughes, Lipoprotein(a) oxidation and autoantibodies: A new path in athero-thrombosis, Lupus 9 (2000) 206–209.

[109] C. Pericleous, I. Giles, A. Rahman, Are endothelial microparticles potential markers of vascular dysfunction in the antiphospholipid syndrome? Lupus 18 (2009) 671–675.

[110] J. George, A. Afek, A. Gilburd, Y. Levy, M. Blank, J. Kopolovic, et al., Atherosclerosis in LDL-receptor knockout mice is accelerated by immunization with anticardiolipin antibodies, Lupus 6 (1997) 723–729.

[111] C. Belizna, A. Lartigue, J. Favre, D. Gilbert, F. Tron, H. Lévesque, et al., Antiphospholipid antibodies induce vascular functional changes in mice: a mechanism of vascular lesions in antiphospholipid syndrome? Lupus 17 (2008) 185—194.

[112] Z. Romay-Penabad, M.G. Montiel-Manzano, T. Shilagard, E. Papalardo, G. Vargas, A.B. Deora, et al., Annexin A2 is involved in antiphospholipid antibody-mediated pathogenic effects *in vitro* and *in vivo*, Blood 114 (2009) 3074—3083.

[113] J. Delgado Alves, L.J. Mason, P.R. Ames, P.P. Chen, J. Rauch, J.S. Levine, et al., Antiphospholipid antibodies are associated with enhanced oxidative stress, decreased plasma nitric oxide and paraoxonase activity in an experimental mouse model, Rheumatology (Oxford) 44 (2005) 1238—1244.

[114] M.E. Vega-Ostertag, D.E. Ferrara, Z. Romay-Penabad, X. Liu, W.R. Taylor, M. Colden-Stanfield, et al., Role of p38 mitogen-activated protein kinase in antiphospholipid antibody-mediated thrombosis and endothelial cell activation, J Thromb Haemost 5 (2007) 1828—1834.

[115] M. Vega-Ostertag, X. Liu, H. Kwan-Ki, P. Chen, S. Pierangeli, A human monoclonal antiprothrombin antibody is thrombogenic in vivo and upregulates expression of tissue factor and E-selectin on endothelial cells, Br J Haematol 135 (2006) 214—219.

[116] I. Marai, M. Shechter, P. Langevitz, B. Gilburd, A. Rubenstein, E. Matssura, et al., Anti-cardiolipin antibodies and endothelial function in patients with coronary artery disease, Am J Cardiol 101 (2008) 1094—1097.

[117] J. Frostegård, R. Wu, C. Gillis-Haegerstrand, C. Lemne, U. de Faire, Antibodies to endothelial cells in borderline hypertension, Circulation 98 (1998) 1092—1098.

[118] A. Graham, I. Ford, R. Morrison, R.N. Barker, M. Greaves, L.P. Erwig, Anti-endothelial antibodies interfere in apoptotic cell clearance and promote thrombosis in patients with antiphospholipid syndrome, J Immunol 182 (2009) 1756—1762.

[119] A.M. Patterson, I. Ford, A. Graham, N.A. Booth, M. Greaves, The influence of anti-endothelial/antiphospholipid antibodies on fibrin formation and lysis on endothelial cells, Br J Haematol 133 (2006) 323—330.

[120] S. Simoncini, C. Sapet, L. Camoin-Jau, N. Bardin, J.R. Harlé, J. Sampol, et al., Role of reactive oxygen species and p38 MAPK in the induction of the pro-adhesive endothelial state mediated by IgG from patients with anti-phospholipid syndrome, Int Immunol 17 (2005) 489—500.

[121] D.E. Ferrara, R. Swerlick, K. Casper, P.L. Meroni, M.E. Vega-Ostertag, E.N. Harris, et al., Fluvastatin inhibits upregulation of tissue factor expression by antiphospholipid antibodies on endothelial cells, J Thromb Haemost 2 (2004) 1558—1563.

[122] Q. Chen, P.R. Stone, S.T. Woon, L.M. Ching, S. Hung, L.M. McCowan, et al., Antiphospholipid antibodies bind to activated but not resting endothelial cells: Is an independent triggering event required to induce antiphospholipid antibody-mediated disease? Thromb Res 114 (2004) 101—111.

[123] S. Dunoyer-Geindre, E.K. Kruithof, F. Boehlen, N. Satta-Poschung, G. Reber, P. de Moerloose, Aspirin inhibits endothelial cell activation induced by antiphospholipid antibodies, J Thromb Haemost 2 (2004) 1176—1181.

[124] F. Dignat-George, L. Camoin-Jau, F. Sabatier, D. Arnoux, F. Anfosso, N. Bardin, et al., Endothelial microparticles: A potential contribution to the thrombotic complications of the antiphospholipid syndrome, Thromb Haemost 1 (2004) 667—673.

[125] R.G. Espinola, X. Liu, M. Colden-Stanfield, J. Hall, E.N. Harris, S.S. Pierangeli, E-selectin mediates pathogenic effects of antiphospholipid antibodies, J Thromb Haemost 1 (2003) 843—848.

[126] P.R. Ames, J. Delgado Alves, L.R. Lopez, F. Gentile, A. Margarita, L. Pizzella, et al., Antibodies against beta2-glycoprotein I complexed with an oxidised lipoprotein relate to intima thickening of carotid arteries in primary antiphospholipid syndrome, Clin Dev Immunol 13 (2006) 1—9.

[127] T. Czompoly, K. Olasz, D. Simon, Z. Nyarady, L. Palinkas, L. Czirjak, et al., A possible new bridge between innate and adaptive immunity: Are the anti-mitochondrial citrate synthase autoantibodies components of the natural antibody network? Mol Immunol 43 (2006) 1761—1768.

[128] N. Baumgarth, J.W. Tung, L.A. Herzenberg, Inherent specificities in natural antibodies: A key to immune defense against pathogen invasion, Springer Semin Immunopathol 26 (2005) 347—362.

[129] M.K. Chang, A. Boullier, K. Hartvigsen, S. Horkko, Y.I. Miller, D.A. Woelkers, et al., The role of natural antibodies in atherogenesis, J Lipid Res 46 (2005) 1353—1363.

[130] C.J. Binder, G.J. Silverman, Natural antibodies and the autoimmunity of atherosclerosis, Springer Semin Immunopathol 26 (2005) 385—404.

[131] J.A. McIntyre, The appearance and disappearance of antiphospholipid autoantibodies subsequent to oxidation—reduction reactions, Thromb Res 114 (2004) 579—587.

[132] M. Blank, R.A. Asherson, R. Cervera, Y. Shoenfeld, Antiphospholipid syndrome infectious origin, J Clin Immunol 24 (2004) 12—23.

[133] R.A. Fields, H. Toubbeh, R.P. Searles, A.D. Bankhurst, The prevalence of anticardiolipin antibodies in a healthy elderly population and its association with antinuclear antibodies, J Rheumatol 66 (1989) 623—625.

[134] W. Shi, S.A. Krilis, B.H. Chong, S. Gordon, C.N. Chesterman, Prevalence of lupus anticoagulant and anticardiolipin antibodies in a healthy population, Aust N Z J Med 20 (1990) 231—236.

[135] J.A. McIntyre, D.R. Wagenknecht, W.P. Faulk, Redox-reactive autoantibodies: Detection and physiological relevance, Autoimmun Rev 5 (2006) 76—83.

[136] J.A. McIntyre, D.R. Wagenknecht, W.P. Faulk, Autoantibodies unmasked by redox reactions, J Autoimmun 24 (2005) 311—317.

[137] D. Stahl, M. Hoemberg, U. Cassens, U. Pachmann, W. Sibrowski, Influence of isotypes of disease-associated autoantibodies on the expression of natural autoantibody repertoires in humans, Immunol Lett 102 (2006) 50—59.

[138] J.R. Yuste, J. Prieto, Anticardiolipin antibodies in chronic viral hepatitis. Do they have clinical consequences? Eur J Gastroenterol Hepatol 15 (2003) 717—726.

[139] R.A. Asherson, R. Cervera, Antiphospholipid antibodies and infections, Ann Rheum Dis 62 (2003) 388—393.

III. ANTI-PHOSPHOLIPID SYNDROME

Laboratory Testing for Antiphospholipid Syndrome

Yiannis Ioannou, Steven A. Krilis

INTRODUCTION

The antiphospholipid syndrome (APS) is an autoimmune condition characterized by arterial and venous thrombosis, recurrent fetal loss, thrombocytopenia, and livedo reticularis. Laboratory criteria for the diagnosis of APS include autoantibodies directed against cardiolipin (aCL), lupus anticoagulant (LA), and β_2-glycoprotein I (β_2GPI) [1]. This group of autoantibodies is collectively termed antiphospholipid antibodies (aPLs), although this term is misleading given that this also includes anti-β_2GPI antibodies and most likely autoantibodies to other proteins such as prothrombin [2].

The diagnosis is made through a combination of clinical features and persistent positivity ($\geq$12 weeks) for aPLs, in keeping with the updated consensus Sydney classification criteria proposed for APS (Figure 52.1) [3]. Although there is no one laboratory test diagnostic for APS, these assays form the cornerstone for establishing the diagnosis in conjunction with the clinical classification criteria. This chapter describes the nature of the laboratory tests in detail, their historical development, methodology including standardization, specificity, sensitivity, association with clinical phenotypes, and clinical utility.

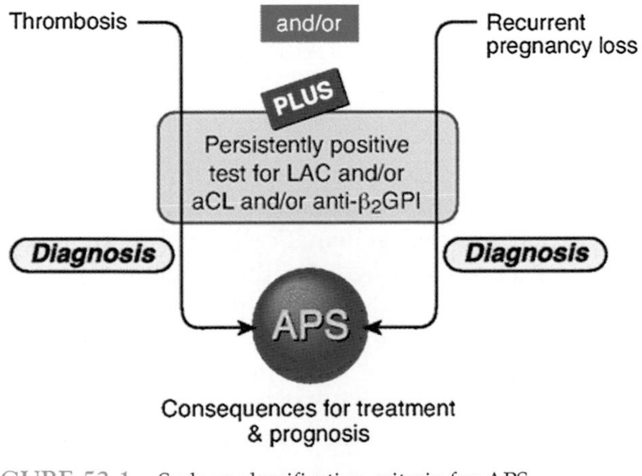

HISTORICAL PERSPECTIVE

Serological False-Positive Test for Syphilis

Initial studies leading up to the discovery of aPLs were centered around the development of assays for syphilis. In 1906, a reagin test developed by Wasserman and coined the Wasserman reaction identified sera from patients infected with syphilis that reacted with syphilitic tissues [4]. It was initially thought that infected serum was reacting with antigens derived from *Treponema pallidum* present in the syphilitic tissues. However, it was realized that infected sera must be reacting with a native antigen because the same effect was seen when noninfected normal human or mammalian tissue was used [5]. It was another 35 years before Pangborn in 1941 identified the antigenic component of the reagin test as being CL, sourced from bovine hearts [6, 7].

Subsequently, the combination of CL, lecithin, and cholesterol formed the basis of the test for syphilis known as the Venereal Disease Research Laboratory (VDRL) test. However, with the development of more specific tests such as the *T. pallidum* immobilization test it was realized that the Wasserman reaction and VDRL test could produce false-positive results for syphilis in noninfected individuals. In 1952, Moore and Mohr [8] identified two circumstances in which false-positive results for syphilis using the reagin test could be observed—an "acute" reaction characterized

FIGURE 52.1 Sydney classification criteria for APS.

by transient false positives observed in patients following an acute viral infection or postvaccination and a "chronic" persistent false-positive test (more than 6 months). In 1955, Moore and Lutz [9] subsequently linked the latter "persistent" false-positive group to the presence of autoimmunity, notably systemic lupus erythematosus (SLE).

Lupus Anticoagulant

In 1952, Conley and Hartman [10] were the first to describe a "circulating anticoagulant" that caused prolongation of the prothrombin time in the serum of two patients with SLE who also had a false-positive serological test for syphilis. Subsequent studies in the mid- to late 1950s demonstrated that the circulating anticoagulant could also be attributed to the biological false-positive test for syphilis in patients with SLE [11, 12], although it was not until 1972 that the term "lupus anticoagulant" was suggested by Feinstein and Rapaport [13]. At the time, it was puzzling that an anticoagulant activity *in vitro*, associated with SLE patients, was not accompanied by a tendency to bleed [14]. Bowie and colleagues from the Mayo clinic in 1963 were the first to describe the paradoxical association of an increased tendency to thrombose rather than bleed with the presence of this plasma anticoagulant activity [15]. However, it was Laurell and Nilsson [12] in 1957 who first suggested the possible association of the anticoagulant phenomenon with a syndrome of recurrent miscarriages and also suggested that the circulating anticoagulant *in vitro* [16] might be associated with inhibitory effects against PL. In 1980, Thiagarajan *et al.* [17] described the properties of a purified IgM antibody from serum with marked LA activity. This purified IgM was found to inhibit PL-dependent coagulation tests, but only in the presence of anionic PL, and not when neutral phosphatidylcholine or phosphatidylethanolamine was used as the lipid source. The first description of a method for measuring LA was published in 1986, and this method used a variation on the "Stypven time," a test making use of Russell's viper venom and hence termed the dilute Russell's viper venom test (dRVVT) [18].

The possible association of LA activity with recurrent fetal loss was first noted in 1957, but it was not until the early 1980s that multiple studies described the association of LA with recurrent fetal loss and prothrombotic tendencies [19–21]. Hughes at St. Thomas's Hospital (London) first described the association of cerebral thrombosis with recurrent miscarriage and LA in an editorial in the *British Medical Journal* in 1983 and suggested the presence of a distinct syndrome [22].

Anticardiolipin Antibody Test

Although CL was discovered in 1941 to be the main antigen in bovine hearts accounting for the false-positive syphilis serology test [6], it was another 42 years before the first solid-phase radioimmunoassay was developed for the detection of autoantibodies reacting against CL [23]. Subsequently, in 1985, an enzyme-linked immunosorbent assay (ELISA) for the detection of aCL was described [24]. A year later, the term anticardiolipin syndrome was introduced by Hughes *et al.* [25] and was subsequently renamed APS by the same group in 1987 [26].

Anti-β_2GPI Antibodies and Co-Factor Activity

It was apparent in this early work that the presence of bovine or human serum was necessary for optimal binding of aCL to CL-coated ELISA wells. In 1989, Krilis and colleagues (St. George Hospital, Sydney, Australia) performed experiments demonstrating that purified aCL IgG derived from patients with APS did not bind CL in a modified CL ELISA in which the blocking agent did not contain serum components [1]. However, when the purified aCL IgG was supplemented with serum, recapitulation of aCL activity was observed. This was confirmed by two other independent groups [27, 28]. In the following year, the purification and sequencing of the co-factor as β_2GPI was reported by McNeil *et al.* [1]. In 1992, it was reported that the majority of autoantibodies responsible for plasma LA activity were also dependent on β_2GPI [29, 30]. Given that these antibodies bind β_2GPI alone coated on an irradiated plate in the absence of anionic PL [31, 32], rather than refer to β_2GPI as a "co-factor," it is more accurate to label these autoantibodies anti-β_2GPI. Although other protein "cofactors," such as protein C [33], protein S [33], and prothrombin [34], have been described for aCL, β_2GPI is the clinically most relevant and best studied.

LUPUS ANTICOAGULANT TEST

Overview

The coagulation pathway consists of an intrinsic (or "contact activation") and extrinsic (or tissue factor) pathway as depicted schematically in Figure 52.2. In clinical practice, testing for the intrinsic pathway involves the activated partial thromboplastin time (aPTT), and that for the extrinsic pathway involves the prothrombin time (PT). The principles underlying the LA test involve three essential and sequential steps as outlined in the Second International Workshop

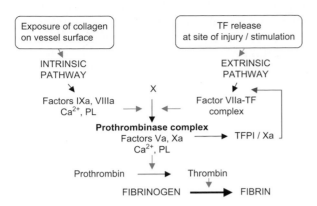

FIGURE 52.2 Summary of coagulation pathway. Extrinsic and intrinsic coagulation pathways. *Blue arrows* indicate promotion and *red arrows* inhibition of pathway. Ca^{2+}, calcium ions; PL, phospholipids; TF, tissue factor.

convened under the auspices of the International Society of Thrombosis and Hemostasis (ISTH) [35, 36]:

1. A screening test to demonstrate the presence of prolongation of a PL-dependent clotting time.
2. A mixing test: This is achieved by using a mix of 1:1 of pooled normal and patient plasma. If prolongation of the clotting time as observed in step 1 is due to a coagulation factor deficiency, this would be supplemented by the normal plasma and the prolongation corrected to within 5 s of the aPTT of the pooled normal plasma sample.
3. A confirmatory test to ensure that the prolongation is PL dependent: This is termed the platelet-neutralization procedure when utilizing activated platelets as the source of anionic PL. The methods most commonly employed to activate platelets are with calcium ionophore or by exposing the platelets to repeated freeze–thaw cycles resulting in lysis and exposure of anionic PL [37]. This is the reason why it is imperative to employ platelet-free plasma: A significant amount of residual platelets or platelet fragments may lead to a false-negative result.

The three tests most frequently employed in clinical practice to determine the presence of a prolonged clotting time and then subsequently the presence of LA activity are the aPTT, the dRVVT, and the kaolin clotting time (KCT). Less commonly employed tests not routinely used in clinical practice are the Textran time test and the thromboplastin inhibition test. Immunoglobulins within the patients' plasma are responsible for exhibiting the LA phenomenon. Given that these immunoglobulins with LA activity are heterogeneous in nature, it is perhaps not surprising that no single LA test is 100% sensitive or indeed specific to detect all LA. Early studies revealed that a single test may only identify up to 70% of LA-positive samples [38]. This is why it is recommended that at least two screening tests are performed [39]. Ideally, one of the screening tests should be based on a clotting time dependent on the intrinsic pathway (e.g., aPTT) and the other on direct activation of factor X (e.g., dRVVT) [40].

aPTT and KCT Tests

These are tests of the intrinsic coagulation system (see Figure 52.2). In order to activate this pathway, PL, an activator (e.g., silica, celite, or kaolin), and calcium (to reverse the effect of citrate or oxalate) are added to plasma and the time taken for a fibrin clot to form is measured. The aPTT is termed partial in view of the absence of tissue factor from this mixture. Prolongation of the aPTT that does not correct after mixing with normal plasma indicates the possible presence of LA; however, a false-positive result may occur in the presence of heparin or coagulation factor specific inhibitors.

The sensitivity of the aPTT will vary according to the nature of the commercial reagents, including the source, physical nature, and composition of the PL and the nature of the activator employed. Increasing dilution of the PL or the aPTT reagent (daPTT) has been employed in an effort to increase the sensitivity of this method. It was postulated that such a method could also be used to detect LA in heparinized samples [41]. However, rising factor VIII levels have been found to limit the use of aPTT as an LA test in pregnancy, with the standard aPTT shown to be more useful in pregnancy than daPTT [42].

Rauch *et al.* [43] showed that LAs from patients with SLE specifically recognize hexagonal (II) phase phosphatidylethanolamine (central aqueous channels surrounded by cylinders packed in a hexagonal manner) but not bilayer phosphatidylethanolamine [43]. Although in principle the specificity of this assay should be good, some variability has been observed, particularly in the inability to differentiate between factor VIII inhibitor and LA [44]. The Staclot reagent has incorporated a heparin blocker and hexagonal II lipids in the incubation mixture for enhanced specificity.

Kaolin is a clay mineral, and because of its low content of PLs, it was originally regarded as a sensitive test for LA. However, this test has a number of technical problems in that it has a tendency to form sediment within coagulometer dispensing systems affecting reproducibility. Micronized silica (SCT) has been shown to be a better alternative activator to employ compared to kaolin because SCT can be used with optical coagulometers without the formation of sediment [45]. Consequently, the KCT has been excluded as a recommended test for estimating LA in updated guidelines [36].

dRVVT Test

This test was first described in 1986 [18] and is increasingly being employed worldwide as the test of choice for both screening and confirmation of the presence for LA [40]. Activation of factor X occurs when a fraction derived from the venom of the Russell viper combines with PL. Hence, this bypasses factors VIII and IX and contact factors (see Figure 52.2). Because this test is independent of these coagulation factors, it is the ideal LA test during pregnancy given the physiological rise that occurs with these factors, and it is also much less likely to yield false-positive results with contact factor deficiencies.

Variation, Standardization, and Quantification of the LA Test

A considerable degree of variation may exist in the detection of LA. In a large French survey of 4500 laboratories, less than half (41%) reported detection of LA in LA-positive patient plasma [46]. However, such surveys were undertaken pre-ISTH guidelines, which were published in 1995 [39]. Surveys conducted after these guidelines were issued reveal that error rates are generally low, most often less than 5%, suggesting that adoption of these guidelines has improved reliability [47, 48].

There are also considerable problems when trying to quantify the LA test in a manner that can be reproducible between different centers. LA activity in single patient plasma is caused by a heterogeneous population of autoantibodies with varying specificities to multiple antigens. This is the reason why at least two separate LA tests need to be done to confirm LA activity and why standardizing the amount of LA activity poses such challenges. LA activity may be due to anti-β_2GPI antibodies or anti-prothrombin [49]. Although both of these populations may occur together, only anti-β_2GPI antibodies with LA activity have convincingly been shown to harbor a strong association with thrombogenic pathogenicity [50]. This has led to the development of assays that can distinguish between anti-prothrombin- and anti-β_2GPI-dependent LAs. One group employed monoclonal antibodies against both β_2GPI and prothrombin to spike normal plasma in an effort to create a calibrating standard of LA that could then allow for quantification [51]. It has been postulated that this method, when combined with results of the anti-β_2GPI assay, may be a good predictor of thrombosis [52]. However, the degree of heterogeneity in LA activity is such that semiquantitative measuring against local values is most often used, and no universally accepted standard has been adopted to allow for accurate quantification.

Integrated tests exist whereby screening, mixing, and confirmation occur in one single procedure. These tests performed as a single procedure involve screening the plasma for LA activity first at low PL concentrations, representing the screening test, and then at high PL concentrations, representing the confirmatory test. Often, the plasma samples are tested by means of the dRVVT and aPTT at low and high PL concentrations run in parallel. The results may then be expressed according to prespecified cutoff values by calculating the percentage correction (screen confirm/screen × 100) or LA ratio (screen/confirm), allowing for semi-quantitation [53]. Both such results (percentage correction and the LA ratio) may be normalized against a pooled normal plasma sample run in parallel: (screen/confirm of patient)/(screen/confirm of pooled normal plasma) [36].

Effect of Anticoagulation on the LA Test

Detecting LA in patients on vitamin K antagonists presents challenges, particularly in those who are weakly LA positive. Interpretation of screening, mixing, and confirmatory tests is difficult because prolonged coagulation times may be superimposed by such treatments. Given the potential for problems, the revised Sydney criteria recommended that, if possible, laboratory investigation be postponed until discontinuation of treatment. Alternatively, the LA diagnostic procedure could be performed on equal mixtures of normal and patient plasma [3]. Such recommendations have also been supported by the latest consensus guideline update for LA detection [36]. Other alternatives that have been suggested are to perform integrated tests such as the aPTT, SCT, or dRVVT at both low and high PL concentrations [54, 55], although the validity of such approaches is yet to be confirmed in appropriate larger cohorts. No screening LA test is valid in patients on heparin, and the sample must be taken prior to commencing such treatment.

ANTICARDIOLIPIN ANTIBODY TEST

This is the most commonly employed aPL antibody assay, performed by coating an ELISA plate with CL. Other anionic PLs that have also been employed include phosphatidylserine [56]. The plate is then incubated with bovine-derived serum, which serves to both block nonspecific binding sites and provide a source of β_2GPI. Diluted patient sera is then added, and the presence of IgG or IgM aCL is detected by adding anti-human IgG or IgM, respectively, conjugated with an enzyme, followed by the addition of the relevant substrate for detection. The presence of bovine serum here is essential in capturing relevant "aCL" antibodies because it provides the source of β_2GPI, which acts as a co-factor. In essence, the aCL ELISA detects both direct aCL and other antibodies that bind CL-bound antigens present in bovine serum, such as β₂GPI (Figure 52.3). Hence, in essence, aCL positivity may be either β₂GPI dependent or β₂GPI independent. There is also the potential to miss antibodies that bind human and not bovine β₂GPI [57]. This factor has been the driving force in the development of commercial assays precoated with CL and human β₂GPI, which also incorporate human β₂GPI in the incubation buffer.

Standardization of aCL Assays

The association of aPL with thrombotic events or pregnancy morbidity underpins the diagnosis of APS. However, epidemiological studies designed to demonstrate these associations show seemingly heterogeneous results and consequently draw conclusions that differ from one another. These discrepancies may be due to a variety of factors specifically relating to patient inclusion/exclusion criteria, diagnostic criteria for a clinical event, clinical endpoints, assay method employed to detect aPL, and predefined cutoff points for determining a positive result. The final points are likely to be a major contributing factor toward the heterogeneity between studies observed.

IgG and IgM aCL are expressed in international standardized units as GPL and MPL units, respectively.

According to the revised classification criteria, a result is deemed to be positive if the titer is at a medium or high level (≥40 GPL or MPL units) [3]. These standardized units are derived using standardized anti-IgG aCL and anti-IgM aCL calibrators distributed from the Antiphospholipid Standardization Laboratory [58]. The first international aCL workshop in 1986 defined GPL and MPL units as binding observed with 1 μg of affinity-purified polyclonal IgG and IgM sample that was distributed to participating laboratories [59]. Secondary standards were ultimately calibrated against the primary standards. Hence, different batches of calibrators will contain heterogeneous polyclonal aPL from different batches of patients. This has led to suggestions that these calibrators may not necessarily behave in a homogeneous manner when assayed at different dilutions or using different kits [60]. This is underlined by the observation of considerable interlaboratory variability [58] even when using the same batch of calibrators. In one interlaboratory study, 12 serum samples were tested for both IgG and IgM aCL using the same assay in 56 laboratories. General consensus defined as an interlaboratory agreement of more than 90% was achieved in only 10 of the 24 cases (42%) [61]. There is also a high degree of variability between different commercial kits for aCL detection when assessed within the same laboratory, with lower variability seen with different commercial anti-β₂GPI kits [62]. For the

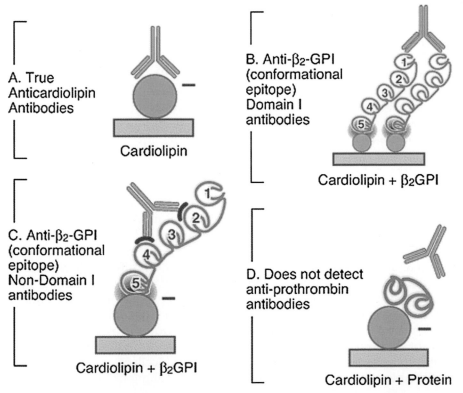

FIGURE 52.3 Schematic representation of the cardiolipin ELISA.

A. True Anticardiolipin Antibodies
Cardiolipin

B. Anti-β₂-GPI (conformational epitope) Domain I antibodies
Cardiolipin + β₂GPI

C. Anti-β₂-GPI (conformational epitope) Non-Domain I antibodies
Cardiolipin + β₂GPI

D. Does not detect anti-prothrombin antibodies
Cardiolipin + Protein

reasons stated, the European Antiphospholipid Forum has validated the efficacy of two other calibrators, HCAL (IgG) and EY2C9 (IgM), and claims less inter-laboratory variation than when using the standard aCL calibrators [63]. These calibrators are derived from a monoclonal anti-β$_2$GPI antibody developed for use as a standard in assays to detect β$_2$GPI-dependent aCL and anti-β$_2$GPI antibodies [64, 65]. It is unlikely that these standards will displace the currently accepted international calibrators in use for detecting aCL; however, the European Antiphospholipid Forum did suggest that they be used as external controls two or three times a year in laboratories that undertake routing aCL titer estimations [66]. Given the current absence of an internationally accepted anti-β$_2$GPI calibrator, the same European group proposed the use of HCAL and EY2C9 humanized monoclonals as calibrators for the anti-β$_2$GPI to validate comparisons between different assay systems [67]. Internationally accepted criteria for testing LA were proposed in 1995 by the International Society for Thrombosis and Haemostasis [39] and were updated in 2009 [36].

ANTI-β$_2$-GLYCOPROTEIN I ASSAY

β$_2$-Glycoprotein I

The discovery in 1990 that β$_2$GPI is an essential co-factor of aCL activity in patient APS serum samples has directly led to an exponential increase in research on this protein. β$_2$GPI was first described in 1961 as a component of the β-globulin fraction of human serum [68]. The gene encoding β$_2$GPI maps to chromosome 17q23-24 [69], and the promoter has also been characterized [70]. Moreover this protein has been termed apolipoprotein H because approximately 40% was initially described as being bound to lipoprotein [71]. However, β$_2$GPI bears no structural similarity to other lipoproteins, and studies have been unable to detect this protein in lipoprotein fractions, leading to proposals to abandon the term apolipoprotein H and simply use the term β$_2$GPI [72].

The major source of production of this glycoprotein is the liver, although mRNA coding for β$_2$GPI has also been found in endothelial cells, intestinal epithelial cells, trophoblasts, neurons, and astrocytes [73–75]. It is synthesized as a single polypeptide chain and consists of 326 amino acids with a calculated molecular weight of 37.1 kDa. However, the protein has four N-linked carbohydrate side chains that account for approximately 20% (w/w) of the total molecular mass. Hence, the actual molecular weight of reduced β$_2$GPI as assessed using protein gel electrophoresis is approximately 50kDa [76]. This noncomplement

protein belongs to the complement control protein (CCP) superfamily [77], which also includes molecules such as complement receptor 1 and 2 [78], IL-2 receptor, or DAF [79]. It consists of five short CCP repeating domains, which are also termed "sushi" domains. Of the five CCP domains (DI–DV) comprising β$_2$GPI, four consist of approximately 60 amino acids each. Within each of these four domains, there are four cysteine residues forming two disulfide bonds that contribute to a "loop-back" structure, with sequence homology between the domains ranging from 20 to 40%. DIII and DIV are the only domains to have three and one carbohydrate side chains, respectively (see Figure 52.5). The fifth C-terminal domain (DV) is aberrant, consisting of 82 amino acids due to a six-residue insertion and a 19-residue C-terminal extension cross-linked by an additional disulfide bond. Two independent groups have studied the crystal structure of β$_2$GPI, describing it as having a fishhook-like appearance (which is shown schematically in Figure 52.4) [80, 81]. Studies from multiple independent groups support the theory that the majority of pathogenic aPLs target a conformational epitope on the N-terminal (DI) of β$_2$GPI [82–85].

β$_2$GPI exhibits a high degree of conservation among mammals. Human, bovine, canine, and mouse proteins all have five domains with 60–80% homology in the amino acid sequence, with DV being the most conserved domain [86–88]. Chimpanzee β$_2$GPI has a 99.4% amino acid sequence homology with human β$_2$GPI and even shares common polymorphisms [89]. β$_2$GPI is one of the most abundant proteins in human serum, with a mean level of 200 µg/ml, second only to fibrinogen among proteins involved in clotting. Given these observations, it is perhaps reasonable to presume that β$_2$GPI possesses important homeostatic functions, although its physiological function has not been fully characterized and seems likely to be complex. Data from early *in vitro* studies demonstrate the ability of β$_2$GPI to bind structures or surfaces coated with negatively charged macromolecules, such as anionic PL, heparin, DNA, and mitochondria [90–93]. Specifically relating to the coagulation pathways, β$_2$GPI has been shown to inhibit activation of the intrinsic coagulation cascade [94], inhibit activation of factor XII [95], inhibit ADP-mediated platelet aggregation [96], bind to factor XI and inhibit its activation by factor XIIa and thrombin [97], bind to receptors on the surface of platelets such as glycoprotein Iba [98] and ApoER2 [99], and aid clearance of oxidized low-density lipoprotein [100]. This pleiotropic protein has also been linked to the regulation of processes involved in angiogenesis [101, 102], apoptotic cell clearance [103, 104], oxidative stress [105], and malignancy [106].

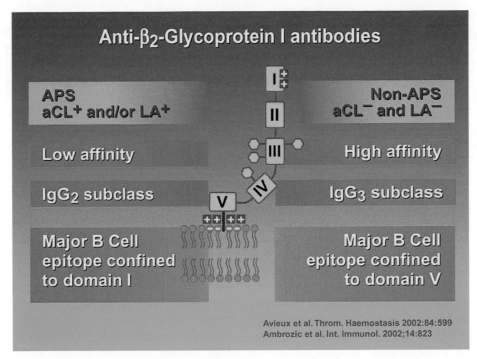

FIGURE 52.4 Schematic of β₂-glycoprotein I and summary of APS- and non-APS-related properties of anti-β₂-glycoprotein I antibodies. aCL, anticardiolipin antibodies; LA, lupus anticoagulant. *Source: Based on Arvieux et al. [117] and Ambrozic et al. [173].*

Anti-β₂GPI assay

In the revised Sydney criteria, a result is defined as positive if the titer is greater than the 99th percentile [3]. This assay involves coating purified human β₂GPI on the surface of an irradiated plate. The patient serum is then added and the presence of IgG or IgM anti-β₂GPI antibodies detected by the addition of a human anti-IgG (or anti-IgM) antibody conjugated with an enzyme that reacts with a substrate used to quantify the reaction. In theory, this should detect a greater number of pathogenic samples than the aCL assay. However, observations demonstrating that there exists nonpathogenic antibodies that react in this assay [84] suggest that "neo-epitopes" may be exposed when β₂GPI associates with an artificial plastic surface [57].

There remains a lack of a formally accepted method for performing this ELISA, with commercial kits employing tailor-made calibrators. Two potential calibrators that have been proposed for universal use in this assay to aid standardization are the humanized monoclonal antibodies HCAL (IgG) and EY2C9 (IgM) [64, 107]. Despite the lack of a universally standardized calibrator for this assay, there generally seems to be superior interlaboratory consensus with this assay compared to the aCL ELISA [57, 108, 109].

ASSOCIATION OF LABORATORY TESTS WITH APS CLINICAL PHENOTYPE

According to the revised classification, no one test, even if persistently positive, is enough to make the diagnosis of APS: The presence of a clinical event (e.g., a vascular thrombosis or pregnancy morbidity) is an absolute prerequisite (see Figure 52.1) [3]. Hence, laboratory tests cannot be described in isolation and must be discussed in the context of their epidemiological significance and clinical associations.

Epidemiology of aPL and APS

There are some limitations in comparing different epidemiological studies due to lack of uniformity in assay methods and, until 1999, lack of agreed classification criteria for APS [110]. Furthermore, many studies investigating the prevalence of aPL in patient populations presenting with thromboembolic events or pregnancy morbidity cannot be extended to form conclusions regarding the prevalence of APS in these groups. This is because the majority of studies have estimated aCL or LA only once and not serially as required by the original Sapporo and updated Sydney classification criteria [3, 110].

It is recognized that there is a relatively high prevalence of aPL and even LA in the general population, although the majority of people do not have APS [111]. Estimates for aCL vary from 1% [112] to 5.6% [113] and for LA from 1.2% [112] to 3.8% [114]. However, these are estimates derived from the general adult population. For those older than 48 years, the frequency of aCL (both β_2GPI dependent and independent) increases and approximates up to 26% [115]. The frequency of anti-β_2GPI antibodies in the healthy population has been little studied. The largest case—control study investigating the frequency of anti-β_2GPI antibodies in patients presenting with stroke or myocardial infarction tested the serum of 1360 age-matched (between ages 48 and 70 years), healthy controls. The frequency of β_2GPI-dependent IgG aCL was 12.1% and that of anti-β_2GPI positivity 1.9% in this normal group [115].

Clearly, APS is not as prevalent in the population, and this is because transient aPL positivity is seen with certain environmental triggers. Numerous infections have been linked to the development of aPL, as reviewed by Cervera and Asherson [116]. Examples include viruses (e.g., hepatitis C, HIV, Epstein—Barr virus, or cytomegalovirus), bacterial infection (e.g., tuberculosis, salmonella, and *Coxiella burnetti*), spirochaetes (*T. pallidum*), and parasitic infections (malaria and toxoplasmosis). However, in the vast majority of cases, the induction of aPL in response to infection is transient and does not result in the development of APS. If the infection is chronic and persistent, such as in leprosy and HIV, then these nonpathogenic aPLs may also be persistent. For example, one study found evidence of persistent anti-β_2GPI and anti-prothrombin antibodies in patients with leprosy. However, the properties of these "nonpathogenic" anti-β_2GPI antibodies differed from those seen in APS patients in terms of subclass, avidity, and epitope specificity within the target autoantigen [117] (see Figure 52.4). Drugs such as phenytoin, propranolol, hydralazine, and cocaine have also been linked to the development of aPL but not APS [118].

The prevalence of APS in the general population is unknown. One editorial suggested that 2—5% of the population has experienced a deep venous thrombosis [119] and that primary APS may account for 15—20% of all deep venous thrombotic episodes [120]; thus, the prevalence of APS in the population is likely to be as high as 0.3—1% [121]. However, caution should be exercised when drawing conclusions from different studies investigating different population groups. Without formal epidemiological studies designed to address this question, the true figure of APS within the population remains unknown.

In the largest survey of APS patients to date, compiled by the Euro-Phospholipid Project Group, the clinical features of 1000 patients with APS were analyzed [122]. Of these APS patients, 53.1% had APS alone, 36% also had SLE, 5% had a lupus-like disease, 5.9% had other autoimmune diseases (e.g., Sjögren's syndrome (2.2%) and rheumatoid arthritis (1.8%)), and 1% had the catastrophic form of APS known as CAPS. The ratio of females to males is high at 5:1 in all APS patients, with a ratio of 7:1 in those with APS and SLE compared to 3.5:1 in those with APS alone. The presence of aPL as defined by the Sapporo criteria [110] was as follows: aCL in 87.9%, LA in 53.6%, and antinuclear antibodies in 59.7% [122]. An update of this cohort confirms that morbidity and mortality remain high in APS even despite anticoagulation, and it is interesting to note that no clinical or immunological characteristic could be identified that may predict recurrence of a clinical event relating to vascular thrombosis or pregnancy morbidity [123]. However, one smaller study comparing 30 male and 38 female patients with APS and no other autoimmune disease found strokes to be significantly more prevalent in females [124]. Interestingly, in children, the phenotype of APS is probably different. Studies in children presenting with idiopathic ischemic stroke have found a high prevalence of persistent aCL [125], whereas almost all patients in one cohort with children and APS were positive for LA [126]. Also in children, as observed in the Ped-APS Registry Cohort, of the 133 patients with APS, 30% of those with SLE presented with APS as the "primary" autoimmune disorder, eventually progressing to full-blown SLE [127, 128], which is more than four times the rate of that seen in adults with SLE [129].

In patients with SLE, the frequency of LA has been estimated to be between 11% [130] and 30% [131]. The frequency of aCL has been estimated to be approximately 20—35% [132—134]. The frequency of APS in patients with SLE has been reported to be approximately 30% [111, 135]. However, these epidemiological data predate the original Sapporo classification criteria [110]; hence, it is difficult to be certain whether such patients, originally classified as having APS, would fulfill the current criteria. In one large cohort of patients with SLE, the frequency of APS according to current classification criteria approximates to 10% (University College Hospital London cohort of 550 patients; D. A. Isenberg, personal communication). In a cohort of 380 patients (Hopkins Lupus Cohort) followed up for a mean of 12.3 years, dual positivity for LA and aCL was predictive of both venous and arterial thrombosis, whereas LA only was predictive of myocardial infarction and neither were predictive of atherosclerosis [136]. Anti-β_2GPI antibodies in patients with SLE or other autoimmune diseases have also been shown to predict thrombosis, especially if patients are also LA positive [50], and may help stratify risk in SLE patients with LA and no history of a thrombotic event [137]. In women with SLE, the presence of aCL and LA is associated with increased risk of pregnancy loss [133].

aPL Positivity and Risk of Thrombosis

Multiple prospective studies have demonstrated a statistically significant association between the presence of aCL and LA and the occurrence of venous thrombosis [138–140]. Furthermore, this risk is enhanced if patients have additional risk factors for the development of venous thrombosis [141, 142]. Ginsburg et al. [120] performed a case–control substudy as part of the Physicians Health Study and found a statistical association between aCL and venous thrombosis. Unfortunately, LA was not measured, and methods for diagnosing thrombosis were not stated. In a case–control study of 256 patients with venous thromboembolism, there was a statistically significant association with the presence of LA but not of aCL compared to the nonthrombotic arm [143]. However, a weakness of this study was that the mean age of patients enrolled was 55 years, thus accounting for the high prevalence of aCL in the nonthrombotic arm. Given this, the significance of aCL positivity in those older than age 50 years is uncertain. The presence of anti-β_2GPI antibody positivity and the risk of developing venous thrombosis were extensively reviewed in a meta-analysis by Galli et al. [144]. Of the 21 studies identified, 12 showed a statistical association with venous thrombosis. The most likely reason for this heterogeneity is that these studies employed "home-made" anti-β_2GPI detection assays with varying protocols and varying cutoff points. Most of the studies were performed between 1993 and 1998, prior to the publication of the 1999 Sapporo classification criteria.

If a patient is aCL or LA positive, what is the risk of developing a second thrombotic episode after cessation of anticoagulation? A study by Schulman et al. [139] addressed this question. During a 4-year follow-up of 412 patients post 6 months anticoagulation after a first venous thrombotic episode, 29% with aCL developed a further episode compared to 14% without aCL ($p = 0.0012$). This positive association was seen despite a number of shortcomings in this study. Abnormal aCL was defined as any value above 5 GPL units, and only 8 patients had significantly positive aCL (i.e., more than 40 GPL units), of whom 3 developed a recurrence. The study did not estimate LA or attempt to define anti-β_2GPI reactivity.

Differentiating between β_2GPI-dependent and -independent aCL reactivity has been employed by Brey et al. [115] in identifying risk factors for developing arterial thrombosis. In a large case–control study of 374 men presenting with myocardial infarction and 259 with stroke, compared to an age-matched control group of 1360 men, there was a significantly greater incidence of β_2GPI-dependent IgG aCL in the group with arterial thromboses. However, a relatively low cutoff value of 23 GPL units was used, and LA activity was not estimated. Other studies have also demonstrated an association between aCL and stroke [145, 146] and myocardial infarction [147], whereas a few have failed to demonstrate an association with stroke [120, 148] or ischemic heart disease [149]. Part of the reason for this may be due to small cohort numbers included in these studies and the fact that these conditions occur with high frequency in elderly populations that generally have a higher incidence of aCL. This may account for the stronger association of stroke with aCL and LA seen in patients younger than age 50 years [150–152]. A large multicenter case–control study in Holland unequivocally showed a strong association between LA positivity and stroke in young patients. The serology of 175 patients presenting with ischemic stroke and 203 patients with myocardial infarction (all ≤50 years of age) was compared to that of 628 healthy age- and sex-matched controls. LA was found in 17% of patients with ischemic stroke and 3% of patients with myocardial infarction versus 0.7% in the control group, giving an odds ratio of 43.1 (95% confidence interval (CI), 12.2–152.0) for stroke and 5.3 (95% CI, 1.4–20.8) for myocardial infarctions, as shown in Figure 52.5. Interestingly, in women who were positive for LA and who also smoked, the odds ratio for having an ischemic stroke increased to 87.0 (95% CI, 14.5–523.0), and it increased to 201.0 (95% CI, 22.1–1828.0) in those women who were LA positive and on the oral contraceptive pill. There was also an association, albeit weaker, for anti-β_2GPI positivity and stroke, but there was no association with aCL or anti-prothrombin positivity [153], although others have demonstrated that aCL in excess of 100 GPL units constitutes an independent risk factor for stroke [154]. de Laat et al. [50] identified 58 blood samples from 198 patients with autoimmune disease that all had LA activity but only half the samples exhibited β_2GPI dependency. The presence of LA activity associated with β_2GPI dependency had a very strong association with thrombosis (odds ratio, 42.3; 95% CI, 194.3–9.9), whereas the β_2GPI-independent samples with LA activity had no association.

aPL Positivity and Risk of Pregnancy Morbidity

This has been extensively reviewed by Giannakopoulos et al. [57]. Before discussing associated risks, it is worth emphasizing the importance of simple terminology. Early criteria proposed for the diagnosis of APS used terms such as recurrent "fetal loss" [155] or recurrent "spontaneous abortions" [156] to describe the main obstetric condition. However, this nomenclature was confusing because at that time pregnancy loss was defined as an abortion if the event occurred prior to 20 weeks of gestation and termed stillbirth or fetal loss after this period [157]. Subsequently, it was

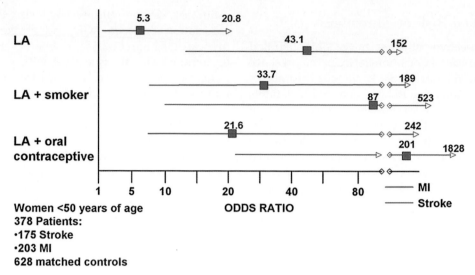

FIGURE 52.5 Lupus anticoagulant activity in young women carries a strong association with ischemic stroke. Positivity for anticardiolipin antibodies and anti-prothrombin antibodies had no correlation to onset of ischemic stroke in this case–control study [151]. LA, lupus anti-coagulant; MI, myocardial infarction.

recognized that the embryonic period exists prior to 10 weeks of gestation and the fetal period extends beyond 10 weeks [157].

One meta-analysis assessed the strength of association between recurrent miscarriage and LA, anti-β_2GPI, or aCL positivity (women without SLE) [158]. This study found a strong association with LA positivity and late gestation miscarriage (between 13 and 24 weeks, as defined by the authors). Anti-β_2GPI positivity associated with early gestation miscarriage (late gestation was not assessed), and aCL associated with both early and late gestation miscarriage [158].

There is evidence that anti-β_2GPI positivity with LA activity associates with pregnancy morbidity, in line with that seen in thrombosis [50]. Two retrospective studies have shown that recurrent fetal loss ($\geq$10 weeks of gestation) and thromboembolic events have a strong association in women who test positive to all three standard aPL tests (LA, aCL, and anti-β_2GPI) compared to those who are positive to two or less of these tests [159, 160]. Other studies have also shown that anti-β_2GPI antibodies associate with pregnancy morbidity in women who are also LA positive [161].

Triple Positivity

A number of retrospective analyses have shown a strong correlation between triple positivity (LA, anti-β_2GPI, and aCL) and thrombosis and miscarriage, compared to patients who test positive to either two or only one assay [159, 160, 162–164]. Pengo et al. [165] analyzed the clinical profile of 160 patients with persistent triple positivity. Despite the majority of patients being on oral anticoagulation, the cumulative incidence

of thromboembolic events was 12.2% (95% CI, 9.6–14.8) after 1 year, 26.1% (95% CI, 22.3–29.9) after 5 years, and 44.2% (95% CI, 38.6–49.8) after 10 years. Hence, not only does triple positivity associate with pathogenicity but also recurrence rates are frequent, suggesting a more severe and resistant clinical phenotype in this serological setting [165]. It is interesting, however, that in a study examining the risk of stroke in women younger than age 50 years, triple positivity associated with the occurrence of a stroke no more than LA positivity alone [153]. The evidence suggests that triple positivity may help in predicting recurrence in a patient with a history of one or more clinical events, but uncertainty remains whether this immunological phenotype is associated with a greater risk of a clinical event over LA positivity alone in patients with no history of a clinical event. Prospective trials designed to answer this question are yet to be performed.

Other Antibody Tests

Of the antibody tests that are not standard and not currently included in the classification criteria, anti-prothrombin antibodies have been most extensively studied. A systematic review by Galli et al. [144] concluded that the prothrombin ELISA did not consistently associate with thromboembolic events, a finding also supported by other studies that have used this assay in conjunction with standard tests [166]. However, other groups have developed an ELISA that detects the pro-thrombin–phosphatidylserine complex and have shown a consistent association with pathogenicity [167]. Other antibody assays have also been described that target phosphatidylethanolamine, whereby associations have been found with both thrombosis [168] and miscarriage [169]. However, other reports have questioned whether such commercially available PL assays can improve the yield of APS diagnosis [16], necessitating the need for future validation of such assays. Some studies have explored the potential utility of investigating other subtypes of aCL or anti-β_2GPI, such as IgA. Although some studies have reported a poor association with pathogenicity [170], others have reported an increased association with IgA aCL and pregnancy morbidity [171]. Finally, as has been described previously, the pathogenic subpopulation of anti-β_2GPI antibodies are thought to be those directed to domain DI of β_2GPI. Hence, recent attempts have focused on the development of an ELISA that specifically detects the presence of anti-DI antibodies rather than detecting nonpathogenic anti-β_2GPI antibodies that may recognize other domains within β_2GPI. Initial screening studies with such an anti-DI antibody ELISA have revealed a good association with pathogenicity [172].

CONCLUSION

Persistent positivity for aPL in conjunction with clinical events form the cornerstone for the diagnosis of APS. The three laboratory tests that are in common and universal use are the aCL, LA, and anti-β_2GPI antibody tests. Standardization and interlaboratory variability continue to be issues, despite universal calibrators being employed for aCL testing, for example. Despite this, persistent positivity for these tests, particularly when triple positive, has a high association with APS disease. Of the three tests, persistent LA positivity probably harbors the greatest association with pathogenicity, particularly when anti-β_2GPI positive. However, given the high percentage of aPL positivity in the normal population, the risk of developing a clinical event in the setting of persistent aPL positivity remains largely unknown due to the lack of prospective studies designed to address this question.

References

[1] H.P. McNeil, R.J. Simpson, C.N. Chesterman, S.A. Krilis, Antiphospholipid antibodies are directed against a complex antigen that includes a lipid-binding inhibitor of coagulation: beta 2-glycoprotein I (apolipoprotein H), Proc Natl Acad Sci USA 87 (1990) 4120—4124.

[2] T. Atsumi, M. Ieko, M.L. Bertolaccini, K. Ichikawa, A. Tsutsumi, E. Matsuura, et al., Association of autoantibodies against the phosphatidylserine—prothrombin complex with manifestations of the antiphospholipid syndrome and with the presence of lupus anticoagulant, Arthritis Rheum 43 (2000) 1982—1993.

[3] S. Miyakis, M.D. Lockshin, T. Atsumi, D.W. Branch, R.L. Brey, R. Cervera, et al., International consensus statement on an update of the classification criteria for definite antiphospholipid syndrome (APS), J Thromb Haemost 4 (2006) 295—306.

[4] A. Wasserman, A. Neisser, C. Bruck, Eine serodiagnostiche reaktion bei syphilis, Deutsche Medizinische Wochenschrift 32 (1906) 745—746.

[5] K. Landsteiner, R. Muller, D. Potzl, Zur Frage der Komplementbingdunsreaktion bei Syphilis, Wein Klin Wschr 20 (1907) 1565—1567.

[6] M. Pangborn, A new serologically active phospholipid from beef heart, Proc Soc Exp Biol Med 48 (1941) 484—486.

[7] M. Pangborn, Isolation and purification of a serologically active phospholipid from beef heart, J Biol Chem 143 (1942) 247—256.

[8] J.E. Moore, C.F. Mohr, Biologically false positive serologic tests for syphilis, JAMA 150 (1952) 467—473.

[9] J.E. Moore, W.B. Lutz, The natural history of systemic lupus erythematosus: An approach to its study through chronic biologic false positive reactors, J Chronic Dis 1 (1955) 297—316.

[10] C.L. Conley, R.C. Hartman, A hemorrhagic disorder caused by circulating anticoagulant in patients with disseminated lupus erythematosus, J Clin Invest 31 (1952) 621.

[11] S.L. Lee, M. Sanders, A disorder of blood coagulation in systemic lupus erythematosus, J Clin Invest 34 (1955) 1814—1822.

[12] A.B. Laurell, I.M. Nilsson, Hypergammaglobulinemia, circulating anticoagulant, and biologic false positive Wassermann reaction: A study in two cases, J Lab Clin Med 49 (1957) 694—707.

[13] D.I. Feinstein, S.I. Rapaport, Acquired inhibitors of blood coagulation, Prog Hemost Thromb 1 (1972) 75—95.

[14] S.S. Shapiro, Lupus anticoagulants and anticardiolipin antibodies: Personal reminiscences, a little history, and some random thoughts, J Thromb Haemost 3 (2005) 831—833.

[15] E.J. Bowie, J.H. Thompson Jr., C.A. Pascuzzi, C.A. Owen Jr., Thrombosis in systemic lupus erythematosus despite circulating anticoagulants, J Lab Clin Med 62 (1963) 416—430.

[16] A.E. Tebo, T.D. Jaskowski, A.R. Phansalkar, C.M. Litwin, D.W. Branch, H.R. Hill, Diagnostic performance of phospholipid-specific assays for the evaluation of antiphospholipid syndrome, Am J Clin Pathol 129 (2008) 870—875.

[17] P. Thiagarajan, S.S. Shapiro, L. De Marco, Monoclonal immunoglobulin M lambda coagulation inhibitor with phospholipid specificity. Mechanism of a lupus anticoagulant, J Clin Invest 66 (1980) 397—405.

[18] P. Thiagarajan, V. Pengo, S.S. Shapiro, The use of the dilute Russell viper venom time for the diagnosis of lupus anticoagulants, Blood 68 (1986) 869—874.

[19] J.P. Soulier, M.C. Boffa, Repetitive abortions with thromboembolic accidents and anti-thromboplastine anticoagulant. Three cases, Nouv Presse Med 9 (1980) 859—864.

[20] L.O. Carreras, G. Defreyn, S.J. Machin, J. Vermylen, R. Deman, B. Spitz, et al., Arterial thrombosis, intrauterine death and "lupus" anticoagulant: Detection of immunoglobulin interfering with prostacyclin formation, Lancet 1 (1981) 244—246.

[21] L.O. Carreras, J. Vermylen, B. Spitz, A. Van Assche, "Lupus" anticoagulant and inhibition of prostacyclin formation in patients with repeated abortion, intrauterine growth retardation and intrauterine death, Br J Obstet Gynaecol 88 (1981) 890—894.

[22] G.R. Hughes, Thrombosis, abortion, cerebral disease, and the lupus anticoagulant, Br Med J (Clin Res Ed) 287 (1983) 1088—1089.

[23] E.N. Harris, A.E. Gharavi, M.L. Boey, B.M. Patel, C.G. Mackworth-Young, S. Loizou, et al., Anticardiolipin antibodies: Detection by radioimmunoassay and association with thrombosis in systemic lupus erythematosus, Lancet 2 (1983) 1211–1214.

[24] S. Loizou, J.D. McCrea, A.C. Rudge, R. Reynolds, C.C. Boyle, E.N. Harris, Measurement of anti-cardiolipin antibodies by an enzyme-linked immunosorbent assay (ELISA): Standardization and quantitation of results, Clin Exp Immunol 62 (1985) 738–745.

[25] G.R. Hughes, N.N. Harris, A.E. Gharavi, The anticardiolipin syndrome, J Rheumatol 13 (1986) 486–489.

[26] E.N. Harris, E. Baguley, R.A. Asherson, G.R.V. Hughes, Clinical and serological features of the "antiphospholipid syndrome" (APS) [abstract], Br J Rheumatol 26 (1987) 19.

[27] E. Matsuura, Y. Igarashi, M. Fujimoto, K. Ichikawa, T. Koike, Anticardiolipin cofactor(s) and differential diagnosis of auto-immune disease, Lancet 336 (1990) 177–178.

[28] M. Galli, P. Comfurius, C. Maassen, H.C. Hemker, M.H. de Baets, P.J. van Breda-Vriesman, et al., Anticardiolipin antibodies (ACA) directed not to cardiolipin but to a plasma protein cofactor, Lancet 335 (1990) 1544–1547.

[29] J.D. Oosting, R.H. Derksen, H.T. Entjes, B.N. Bouma, P.G. de Groot, Lupus anticoagulant activity is frequently dependent on the presence of beta 2-glycoprotein I, Thromb Haemost 67 (1992) 499–502.

[30] R.A. Roubey, C.W. Pratt, J.P. Buyon, J.B. Winfield, Lupus anticoagulant activity of autoimmune antiphospholipid antibodies is dependent upon beta 2-glycoprotein I, J Clin Invest 90 (1992) 1100–1104.

[31] R.A. Roubey, R.A. Eisenberg, M.F. Harper, J.B. Winfield, "Anticardiolipin" autoantibodies recognize beta 2-glycoprotein I in the absence of phospholipid: Importance of Ag density and bivalent binding, J Immunol 154 (1995) 954–960.

[32] E. Matsuura, Y. Igarashi, T. Yasuda, D.A. Triplett, T. Koike, Anticardiolipin antibodies recognize beta 2-glycoprotein I structure altered by interacting with an oxygen modified solid phase surface, J Exp Med 179 (1994) 457–462.

[33] J.D. Oosting, R.H. Derksen, I.W. Bobbink, T.M. Hackeng, B.N. Bouma, P.G. de Groot, Antiphospholipid antibodies directed against a combination of phospholipids with prothrombin, protein C, or protein S: An explanation for their pathogenic mechanism? Blood 81 (1993) 2618–2625.

[34] E.M. Bevers, M. Galli, T. Barbui, P. Comfurius, R.F. Zwaal, Lupus anticoagulant IgG's (LA) are not directed to phospholipids only, but to a complex of lipid-bound human prothrombin, Thromb Haemost 66 (1991) 629–632.

[35] J.T. Brandt, L.K. Barna, D.A. Triplett, Laboratory identification of lupus anticoagulants: Results of the Second International Workshop for Identification of Lupus Anticoagulants. On behalf of the Subcommittee on Lupus Anticoagulants/Antiphospholipid Antibodies of the ISTH, Thromb Haemost 74 (1995) 1597–1603.

[36] V. Pengo, A. Tripodi, G. Reber, J.H. Rand, T.L. Ortel, M. Galli, et al., Update of the guidelines for lupus anticoagulant detection, J Thromb Haemost 7 (2009) 1737–1740.

[37] M. Greaves, H. Cohen, S.J. MacHin, I. Mackie, Guidelines on the investigation and management of the antiphospholipid syndrome, Br J Haematol 109 (2000) 704–715.

[38] J.T. Brandt, D.A. Triplett, K. Musgrave, C. Orr, The sensitivity of different coagulation reagents to the presence of lupus anticoagulants, Arch Pathol Lab Med 111 (1987) 120–124.

[39] J.T. Brandt, D.A. Triplett, B. Alving, I. Scharrer, Criteria for the diagnosis of lupus anticoagulants: An update. On behalf of the Subcommittee on Lupus Anticoagulant/Antiphospholipid Antibody of the Scientific and Standardisation Committee of the ISTH, Thromb Haemost 74 (1995) 1185–1190.

[40] A. Tripodi, Laboratory testing for lupus anticoagulants: A review of issues affecting results, Clin Chem 53 (2007) 1629–1635.

[41] B.M. Alving, P.E. Baldwin, R.L. Richards, B.J. Jackson, The dilute phospholipid APTT: A sensitive assay for verification of lupus anticoagulants, Thromb Haemost 54 (1985) 709–712.

[42] A.N. Blanco, B.E. Grand, G. Pieroni, L.B. Penalva, L.S. Voto, M.A. Lazzari, Behavior of diluted activated partial thrombo-plastin time in pregnant women with a lupus anticoagulant, Am J Clin Pathol 100 (1993) 99–102.

[43] J. Rauch, M. Tannenbaum, A.S. Janoff, Distinguishing plasma lupus anticoagulants from anti-factor antibodies using hexagonal (II) phase phospholipids, Thromb Haemost 62 (1989) 892–896.

[44] A. Tripodi, M.E. Mancuso, V. Chantarangkul, M. Clerici, R. Bader, P.L. Meroni, et al., Lupus anticoagulants and their relationship with the inhibitors against coagulation factor VIII: Considerations on the differentiation between the 2 circulating anticoagulants, Clin Chem 51 (2005) 1883–1885.

[45] V. Chantarangkul, A. Tripodi, A. Arbini, P.M. Mannucci, Silica clotting time (SCT) as a screening and confirmatory test for detection of the lupus anticoagulants, Thromb Res 67 (1992) 355–365.

[46] J. Roussi, J.P. Roisin, A. Goguel, Lupus anticoagulants: First French interlaboratory Etalonorme survey, Am J Clin Pathol 105 (1996) 788–793.

[47] E.J. Favaloro, R. Bonar, J. Sioufi, M. Wheeler, J. Low, M. Aboud, et al., Multilaboratory testing of thrombophilia: Current and past practice in Australasia as assessed through the Royal College of Pathologists of Australasia Quality Assurance Program for Hematology, Semin Thromb Hemost 31 (2005) 49–58.

[48] I. Jennings, M. Greaves, I.J. Mackie, S. Kitchen, T.A. Woods, F.E. Preston, Lupus anticoagulant testing: Improvements in performance in a UK NEQAS proficiency testing exercise after dissemination of national guidelines on laboratory methods, Br J Haematol 119 (2002) 364–369.

[49] J. Nojima, H. Kuratsune, E. Suehisa, Y. Futsukaichi, H. Yamanishi, T. Machii, et al., Anti-prothrombin antibodies combined with lupus anti-coagulant activity is an essential risk factor for venous thromboembolism in patients with systemic lupus erythematosus, Br J Haematol 114 (2001) 647–654.

[50] H.B. de Laat, R.H. Derksen, R.T. Urbanus, M. Roest, P.G. de Groot, Beta2-glycoprotein I-dependent lupus anticoagulant highly correlates with thrombosis in the antiphospholipid syndrome, Blood 104 (2004) 3598–3602.

[51] A. Le Querrec, J. Arnout, D. Arnoux, J.Y. Borg, C. Caron, L. Darnige, et al., Quantification of lupus anticoagulants in clinical samples using anti-beta2GP1 and anti-prothrombin monoclonal antibodies, Thromb Haemost 86 (2001) 584–589.

[52] K. Devreese, K. Peerlinck, M.F. Hoylaerts, Thrombotic risk assessment in the antiphospholipid syndrome requires more than the quantification of lupus anticoagulants, Blood 115 (2010) 870–887.

[53] E.M. Jacobsen, L. Barna-Cler, J.M. Taylor, D.A. Triplett, F. Wisloff, The lupus ratio test—An interlaboratory study on the detection of lupus anticoagulants by an APTT-based, integrated, and semi-quantitative test. Fifth International Survey of Lupus Anticoagulants—ISLA 5, Thromb Haemost 83 (2000) 704–708.

[54] A. Tripodi, V. Chantarangkul, M. Clerici, P.M. Mannucci, Laboratory diagnosis of lupus anticoagulants for patients on oral anticoagulant treatment. Performance of dilute Russell

viper venom test and silica clotting time in comparison with Staclot LA, Thromb Haemost 88 (2002) 583–586.

[55] D.A. Triplett, L.K. Barna, G.A. Unger, A hexagonal (II) phase phospholipid neutralization assay for lupus anticoagulant identification, Thromb Haemost 70 (1993) 787–793.

[56] N.S. Rote, D. Dostal-Johnson, D.W. Branch, Antiphospholipid antibodies and recurrent pregnancy loss: Correlation between the activated partial thromboplastin time and antibodies against phosphatidylserine and cardiolipin, Am J Obstet Gynecol 163 (1990) 575–584.

[57] B. Giannakopoulos, F. Passam, Y. Ioannou, S.A. Krilis, How we diagnose the antiphospholipid syndrome, Blood 113 (2009) 985–994.

[58] E.N. Harris, S.S. Pierangeli, Revisiting the anticardiolipin test and its standardization, Lupus 11 (2002) 269–275.

[59] E.N. Harris, A.E. Gharavi, S.P. Patel, G.R. Hughes, Evaluation of the anti-cardiolipin antibody test: Report of an international workshop held 4 April 1986, Clin Exp Immunol 68 (1987) 215–222.

[60] R.C. Wong, Consensus guidelines for anticardiolipin antibody testing, Thromb Res 114 (2004) 559–571.

[61] E.J. Favaloro, R. Silvestrini, Assessing the usefulness of anti-cardiolipin antibody assays: A cautious approach is suggested by high variation and limited consensus in multilaboratory testing, Am J Clin Pathol 118 (2002) 548–557.

[62] M.A. Audrain, F. Colonna, F. Morio, M.A. Hamidou, J.Y. Muller, Comparison of different kits in the detection of autoantibodies to cardiolipin and beta2glycoprotein 1, Rheumatology (Oxford) 43 (2004) 181–185.

[63] A. Tincani, F. Allegri, M. Sanmarco, M. Cinquini, M. Taglietti, G. Balestrieri, et al., Anticardiolipin antibody assay: A methodological analysis for a better consensus in routine determinations—A cooperative project of the European Anti-phospholipid Forum, Thromb Haemost 86 (2001) 575–583.

[64] K. Ichikawa, A. Tsutsumi, T. Atsumi, E. Matsuura, S. Kobayashi, G.R. Hughes, et al., A chimeric antibody with the human gamma1 constant region as a putative standard for assays to detect IgG beta2-glycoprotein I-dependent anti-cardiolipin and anti-beta2-glycoprotein I antibodies, Arthritis Rheum 42 (1999) 2461–2470.

[65] K. Ichikawa, M.A. Khamashta, T. Koike, E. Matsuura, G.R. Hughes, Beta 2-glycoprotein I reactivity of monoclonal anticardiolipin antibodies from patients with the anti-phospholipid syndrome, Arthritis Rheum 37 (1994) 1453–1461.

[66] A. Tincani, F. Allegri, G. Balestrieri, G. Reber, M. Sanmarco, P. Meroni, et al., Minimal requirements for antiphospholipid antibodies ELISAs proposed by the European Forum on Antiphospholipid Antibodies, Thromb Res 114 (2004) 553–558.

[67] G. Reber, A. Tincani, M. Sanmarco, P. de Moerloose, M.C. Boffa, Proposals for the measurement of anti-beta2-glycoprotein I antibodies. Standardization group of the European Forum on Antiphospholipid Antibodies, J Thromb Haemost 2 (2004) 1860–1862.

[68] S. Schultze, K. Heide, H. Haupt, Uber ein bisher unbekanntes niedermolekulares Beta2-Globulin des Humanserums, Naturwissenschaften 48 (1961) 719.

[69] A. Steinkasserer, D.J. Cockburn, D.M. Black, Y. Boyd, E. Solomon, R.B. Sim, Assignment of apolipoprotein H (APOH: beta-2-glycoprotein I) to human chromosome 17q23-qter; Determination of the major expression site, Cytogenet Cell Genet 60 (1992) 31–33.

[70] H.H. Wang, A.N. Chiang, Cloning and characterization of the human beta2-glycoprotein I (beta2-GPI) gene promoter: Roles of the atypical TATA box and hepatic nuclear factor-1alpha in regulating beta2-GPI promoter activity, Biochem J 380 (2004) 455–463.

[71] N.S. Lee, H.B. Brewer Jr., J.C. Osborne Jr., Beta 2-glycoprotein I. Molecular properties of an unusual apolipoprotein, apolipoprotein H, J Biol Chem 258 (1983) 4765–4770.

[72] C. Agar, P.G. de Groot, J.H. Levels, J.A. Marquart, J.C. Meijers, Beta2-glycoprotein I is incorrectly named apolipoprotein H, J Thromb Haemost 7 (2009) 235–236.

[73] B. Caronti, C. Calderaro, C. Alessandri, F. Conti, R. Tinghino, G. Palladini, et al., Beta2-glycoprotein I (beta2-GPI) mRNA is expressed by several cell types involved in anti-phospholipid syndrome-related tissue damage, Clin Exp Immunol 115 (1999) 214–219.

[74] M. Averna, G. Paravizzini, G. Marino, E. Lanteri, G. Cavera, C.M.. Barbagallo, et al., Liver is not the unique site of synthesis of beta 2-glycoprotein I (apolipoprotein H): Evidence for an intestinal localization, Int J Clin Lab Res 27 (1997) 207–212.

[75] L.W. Chamley, J.L. Allen, P.M. Johnson, Synthesis of beta2 glycoprotein 1 by the human placenta, Placenta 18 (1997) 403–410.

[76] J. Lozier, N. Takahashi, F.W. Putnam, Complete amino acid sequence of human plasma beta 2-glycoprotein I, Proc Natl Acad Sci USA 81 (1984) 3640–3644.

[77] K.B. Reid, A.J. Day, Structure–function relationships of the complement components, Immunol Today 10 (1989) 177–180.

[78] D.T. Fearon, Regulation of the amplification C3 convertase of human complement by an inhibitory protein isolated from human erythrocyte membrane, Proc Natl Acad Sci USA 76 (1979) 5867–5871.

[79] A. Nicholson-Weller, J. Burge, D.T. Fearon, P.F. Weller, K.F. Austen, Isolation of a human erythrocyte membrane glycoprotein with decay-accelerating activity for C3 convertases of the complement system, J Immunol 129 (1982) 184–189.

[80] R. Schwarzenbacher, K. Zeth, K. Diederichs, A. Gries, G.M. Kostner, P. Laggner, et al., Crystal structure of human beta2-glycoprotein I: Implications for phospholipid binding and the antiphospholipid syndrome, EMBO J 18 (1999) 6228–6239.

[81] B. Bouma, P.G. de Groot, J.M. van den Elsen, R.B. Ravelli, A. Schouten, M.J. Simmelink, et al., Adhesion mechanism of human beta(2)-glycoprotein I to phospholipids based on its crystal structure, EMBO J 18 (1999) 5166–5174.

[82] S.W. Reddel, Y.X. Wang, Y.H. Sheng, S.A. Krilis, Epitope studies with anti-beta 2-glycoprotein I antibodies from auto-antibody and immunized sources, J Autoimmun 15 (2000) 91–96.

[83] G.M. Iverson, E.J. Victoria, D.M. Marquis, Anti-beta2 glyco-protein I (beta2GPI) autoantibodies recognize an epitope on the first domain of beta2GPI, Proc Natl Acad Sci USA 95 (1998) 15542–15546.

[84] B. de Laat, R.H. Derksen, M. van Lummel, M.T. Pennings, P.G. de Groot, Pathogenic anti-β2-glycoprotein I antibodies recognize domain I of β2-glycoprotein I only after a conformational change, Blood 107 (2006) 1916–1924.

[85] Y. Ioannou, C. Pericleous, I. Giles, D.S. Latchman, D.A. Isenberg, A. Rahman, Binding of antiphospholipid antibodies to discontinuous epitopes on domain I of human beta (2)-glycoprotein I: Mutation studies including residues R39 to R43, Arthritis Rheum 56 (2007) 280–290.

[86] A. Steinkasserer, C. Estaller, E.H. Weiss, R.B. Sim, A.J. Day, Complete nucleotide and deduced amino acid sequence of human beta 2- glycoprotein I, Biochem J 277 (1991) 387–391.

[87] Y. Sheng, H. Herzog, S.A. Krilis, Cloning and characterization of the gene encoding the mouse beta 2-glycoprotein I, Genomics 41 (1997) 128–130.

III. ANTI-PHOSPHOLIPID SYNDROME

[88] B. Gao, M. Virmani, E. Romm, E. Lazar-Wesley, K. Sakaguchi, E. Appella, et al., Sequence of a cDNA encoding bovine apolipoprotein H, Gene 126 (1993) 287–288.

[89] D.K. Sanghera, C.S. Nestlerode, R.E. Ferrell, M.I. Kamboh, Chimpanzee apolipoprotein H (beta2-glycoprotein I): Report on the gene structure, a common polymorphism, and a high prevalence of antiphospholipid antibodies, Hum Genet 109 (2001) 63–72.

[90] I. Schousboe, Purification, characterization and identification of an agglutinin in human serum, Biochim Biophys Acta 579 (1979) 396–408.

[91] H. Wurm, Beta 2-glycoprotein-I (apolipoprotein H) interactions with phospholipid vesicles, Int J Biochem 16 (1984) 511–515.

[92] E. Polz, H. Wurm, G.M. Kostner, Investigations on beta 2-glycoprotein-I in the rat: Isolation from serum and demonstration in lipoprotein density fractions, Int J Biochem 11 (1980) 265–270.

[93] J. Kroll, J.K. Larsen, H. Loft, M. Ezban, K. Wallevik, M. Faber, DNA-binding proteins in Yoshida ascites tumor fluid, Biochim Biophys Acta 434 (1976) 490–501.

[94] I. Schousboe, Beta 2-glycoprotein I: A plasma inhibitor of the contact activation of the intrinsic blood coagulation pathway, Blood 66 (1985) 1086–1091.

[95] I. Schousboe, M.S. Rasmussen, Synchronized inhibition of the phospholipid mediated autoactivation of factor XII in plasma by beta 2-glycoprotein I and anti-beta 2-glycoprotein I, Thromb Haemost 73 (1995) 798–804.

[96] J. Nimpf, H. Wurm, G.M. Kostner, Beta 2-glycoprotein-I (apo-H) inhibits the release reaction of human platelets during ADP-induced aggregation, Atherosclerosis 63 (1987) 109–114.

[97] T. Shi, G.M. Iverson, J.C. Qi, K.A. Cockerill, M.D. Linnik, P. Konecny, et al., Beta 2-glycoprotein I binds factor XI and inhibits its activation by thrombin and factor XIIa: Loss of inhibition by clipped beta 2-glycoprotein I, Proc Natl Acad Sci USA 101 (2004) 3939–3944.

[98] T. Shi, B. Giannakopoulos, X. Yan, P. Yu, M.C. Berndt, R. Andrews, et al., Anti-beta2-glycoprotein I antibodies in complex with beta2-glycoprotein I can activate platelets in a dysregulated manner via glycoprotein Ib-IX-V, Arthritis Rheum 54 (2006) 2558–2567.

[99] B.C. Lutters, R.H. Derksen, W.L. Tekelenburg, P.J. Lenting, J. Arnout, P.G. de Groot, Dimers of beta 2-glycoprotein I increase platelet deposition to collagen via interaction with phospholipids and the apolipoprotein E receptor 2′, J Biol Chem 278 (2003) 33831–33838.

[100] Y. Hasunuma, E. Matsuura, Z. Makita, T. Katahira, S. Nishi, T. Koike, Involvement of beta 2-glycoprotein I and anti-cardiolipin antibodies in oxidatively modified low-density lipoprotein uptake by macrophages, Clin Exp Immunol 107 (1997) 569–573.

[101] P. Yu, F.H. Passam, D.M. Yu, G. Denyer, S. Krilis, β2-Glyco-protein I inhibits vascular endothelial growth factor and basic fibroblast growth factor induced angiogenesis through its amino terminal domain, J Thromb Haemost 6 (2008) 1215–1223.

[102] T. Sakai, K. Balasubramanian, S. Maiti, J.B. Halder, A.J. Schroit, Plasmin-cleaved beta-2-glycoprotein 1 is an inhibitor of angiogenesis, Am J Pathol 171 (2007) 1659–1669.

[103] S.N. Maiti, K. Balasubramanian, J.A. Ramoth, A.J. Schroit, Beta-2-glycoprotein 1-dependent macrophage uptake of apoptotic cells. Binding to lipoprotein receptor-related protein receptor family members, J Biol Chem 283 (2008) 3761–3766.

[104] K. Balasubramanian, S.N. Maiti, A.J. Schroit, Recruitment of beta-2-glycoprotein 1 to cell surfaces in extrinsic and intrinsic apoptosis, Apoptosis 10 (2005) 439–446.

[105] Y. Ioannou, J.Y. Zhang, F.H. Passam, S. Rahgozar, J.C. Qi, B. Giannakopoulos, et al., Naturally occurring free thiols within β2-glycoprotein I in vivo: Nitrosylation, redox modification by endothelial cells and regulation of oxidative stress induced cell injury, Blood (2010), in press.

[106] T.A. Luster, J. He, X. Huang, S.N. Maiti, A.J. Schroit, P.G. de Groot, et al., Plasma protein beta-2-glycoprotein 1 mediates interaction between the anti-tumor monoclonal antibody 3G4 and anionic phospholipids on endothelial cells, J Biol Chem 281 (2006) 29863–29871.

[107] G. Reber, P. de Moerloose, Anti-beta2-glycoprotein I anti-bodies—When and how should they be measured? Thromb Res 114 (2004) 527–531.

[108] E.J. Favaloro, R.C. Wong, S. Jovanovich, P. Roberts-Thomson, A review of beta2-glycoprotein-l antibody testing results from a peer-driven multilaboratory quality assurance program, Am J Clin Pathol 127 (2007) 441–448.

[109] G. Reber, I. Schousboe, A. Tincani, T. Sanmarco, T. Kveder, P. de Moerloose, et al., Inter-laboratory variability of anti-beta2-glycoprotein I measurement. A collaborative study in the frame of the European Forum on Antiphospholipid Antibodies Standardization Group, Thromb Haemost 88 (2002) 66–73.

[110] W.A. Wilson, A.E. Gharavi, T. Koike, M.D. Lockshin, D.W. Branch, J.C. Piette, et al., International consensus statement on preliminary classification criteria for definite anti-phospholipid syndrome: Report of an international workshop, Arthritis Rheum 42 (1999) 1309–1311.

[111] M. Petri, Epidemiology of the antiphospholipid antibody syndrome, J Autoimmun 15 (2000) 145–151.

[112] N.S. Pattison, L.W. Chamley, E.J. McKay, G.C. Liggins, W.S. Butler, Antiphospholipid antibodies in pregnancy: Prevalence and clinical associations, Br J Obstet Gynaecol 100 (1993) 909–913.

[113] W. Shi, S.A. Krilis, B.H. Chong, S. Gordon, C.N. Chesterman, Prevalence of lupus anticoagulant and anticardiolipin anti-bodies in a healthy population, Aust N Z J Med 20 (1990) 231–236.

[114] C. Infante-Rivard, M. David, R. Gauthier, G.E. Rivard, Lupus anticoagulants, anticardiolipin antibodies, and fetal loss. A case–control study, N Engl J Med 325 (1991) 1063–1066.

[115] R.L. Brey, R.D. Abbott, J.D. Curb, D.S. Sharp, G.W. Ross, C.L. Stallworth, et al., Beta(2)-glycoprotein 1-dependent anti-cardiolipin antibodies and risk of ischemic stroke and myocardial infarction: The Honolulu Heart Program, Stroke 32 (2001) 1701–1706.

[116] R. Cervera, R.A. Asherson, Antiphospholipid syndrome associated with infections: Clinical and microbiological characteristics, Immunobiology 210 (2005) 735–741.

[117] J. Arvieux, Y. Renaudineau, I. Mane, R. Perraut, S.A. Krilis, P. Youinou, Distinguishing features of anti-beta2 glycoprotein I antibodies between patients with leprosy and the antiphospholipid syndrome, Thromb Haemost 87 (2002) 599–605.

[118] B. Robertson, M. Greaves, Antiphospholipid syndrome: An evolving story, Blood Rev 20 (2006) 201–212.

[119] W.W. Coon, P.W. Willis 3rd, J.B. Keller, Venous thromboembolism and other venous disease in the Tecumseh community health study, Circulation 48 (1973) 839–846.

[120] K.S. Ginsburg, M.H. Liang, L. Newcomer, S.Z. Goldhaber, P.H. Schur, C.H. Hennekens, et al., Anticardiolipin antibodies and the risk for ischemic stroke and venous thrombosis, Ann Intern Med 117 (1992) 997–1002.

[121] R.A. Roubey, Antiphospholipid antibodies: Immunological aspects, Clin Immunol 112 (2004) 127–128.

[122] R. Cervera, J.C. Piette, J. Font, M.A. Khamashta, Y. Shoenfeld, M.T. Camps, et al., Antiphospholipid syndrome: Clinical and

immunologic manifestations and patterns of disease expression in a cohort of 1000 patients, Arthritis Rheum 46 (2002) 1019–1027.

[123] R. Cervera, M.A. Khamashta, Y. Shoenfeld, M.T. Camps, S. Jacobsen, E. Kiss, et al., Morbidity and mortality in the antiphospholipid syndrome during a 5-year period: A multi-centre prospective study of 1000 patients, Ann Rheum Dis 68 (2009) 1428–1432.

[124] L.J. Jara, G. Medina, O. Vera-Lastra, L. Barile, The impact of gender on clinical manifestations of primary antiphospholipid syndrome, Lupus 14 (2005) 607–612.

[125] G. Kenet, S. Sadetzki, H. Murad, U. Martinowitz, N. Rosenberg, S. Gitel, et al., Factor V Leiden and anti-phospholipid antibodies are significant risk factors for ischemic stroke in children, Stroke 31 (2000) 1283–1288.

[126] Y. Berkun, S. Padeh, J. Barash, Y. Uziel, L. Harel, M. Mukamel, et al., Antiphospholipid syndrome and recurrent thrombosis in children, Arthritis Rheum 55 (2006) 850–855.

[127] T. Avcin, R. Cimaz, E.D. Silverman, R. Cervera, M. Gattorno, S. Garay, et al., Pediatric antiphospholipid syndrome: Clinical and immunologic features of 121 patients in an international registry, Pediatrics 122 (2008) e1100–e1107.

[128] T. Avcin, R. Cimaz, B. Rozman, The Ped-APS Registry: The antiphospholipid syndrome in childhood, Lupus 18 (2009) 894–899.

[129] T. Tarr, G. Lakos, H.P. Bhattoa, G. Szegedi, Y. Shoenfeld, E. Kiss, Primary antiphospholipid syndrome as the forerunner of systemic lupus erythematosus, Lupus 16 (2007) 324–328.

[130] K.L. Wong, F.Y. Chan, C.P. Lee, Outcome of pregnancy in patients with systemic lupus erythematosus. A prospective study, Arch Intern Med 151 (1991) 269–273.

[131] R. Cervera, J. Font, A. Lopez-Soto, F. Casals, L. Pallares, A. Bove, et al., Isotype distribution of anticardiolipin antibodies in systemic lupus erythematosus: Prospective analysis of a series of 100 patients, Ann Rheum Dis 49 (1990) 109–113.

[132] E.L. Radway-Bright, C.T. Ravirajan, D.A. Isenberg, The prevalence of antibodies to anionic phospholipids in patients with the primary antiphospholipid syndrome, systemic lupus erythematosus and their relatives and spouses, Rheumatology (Oxford) 39 (2000) 427–431.

[133] W.H. Kutteh, E.C. Lyda, S.M. Abraham, M.C. Wacholtz, Association of anticardiolipin antibodies and pregnancy loss in women with systemic lupus erythematosus, Fertil Steril 60 (1993) 449–455.

[134] R. Cervera, M.A. Khamashta, J. Font, G.D. Sebastiani, A. Gil, P. Lavilla, et al., Systemic lupus erythematosus: Clinical and immunologic patterns of disease expression in a cohort of 1000 patients. The European Working Party on Systemic Lupus Erythematosus, Medicine (Baltimore) 72 (1993) 113–124.

[135] P.E. Love, S.A. Santoro, Antiphospholipid antibodies: Anti-cardiolipin and the lupus anticoagulant in systemic lupus erythematosus (SLE) and in non-SLE disorders. Prevalence and clinical significance, Ann Intern Med 112 (1990) 682–698.

[136] M. Petri, The lupus anticoagulant is a risk factor for myocardial infarction (but not atherosclerosis): Hopkins Lupus Cohort, Thromb Res 114 (2004) 593–595.

[137] C. Zoghlami-Rintelen, R. Vormittag, T. Sailer, S. Lehr, P. Quehenberger, H. Rumpold, et al., The presence of IgG antibodies against beta2-glycoprotein I predicts the risk of thrombosis in patients with the lupus anticoagulant, J Thromb Haemost 3 (2005) 1160–1165.

[138] G. Finazzi, V. Brancaccio, M. Moia, N. Ciaverella, M.G. Mazzucconi, P.C. Schinco, et al., Natural history and risk factors for thrombosis in 360 patients with antiphospholipid antibodies: A four-year prospective study from the Italian Registry, Am J Med 100 (1996) 530–536.

[139] S. Schulman, E. Svenungsson, S. Granqvist, Anticardiolipin antibodies predict early recurrence of thromboembolism and death among patients with venous thromboembolism following anticoagulant therapy. Duration of Anticoagulation Study Group, Am J Med 104 (1998) 332–338.

[140] M. Galli, Antiphospholipid antibodies and thrombosis: Do test patterns identify the patients' risk? Thromb Res 114 (2004) 597–601.

[141] J. Kassis, C. Neville, J. Rauch, L. Busque, E.R. Chang, L. Joseph, et al., Antiphospholipid antibodies and thrombosis: Association with acquired activated protein C resistance in venous thrombosis and with hyperhomocysteinemia in arterial thrombosis, Thromb Haemost 92 (2004) 1312–1319.

[142] M. Hudson, A.L. Herr, J. Rauch, C. Neville, E. Chang, R. Ibrahim, et al., The presence of multiple prothrombotic risk factors is associated with a higher risk of thrombosis in individuals with anticardiolipin antibodies, J Rheumatol 30 (2003) 2385–2391.

[143] J.S. Ginsberg, P.S. Wells, P. Brill-Edwards, D. Donovan, K. Moffatt, M. Johnston, et al., Antiphospholipid antibodies and venous thromboembolism, Blood 86 (1995) 3685–3691.

[144] M. Galli, D. Luciani, G. Bertolini, T. Barbui, Anti-beta 2-glycoprotein I, antiprothrombin antibodies, and the risk of thrombosis in the antiphospholipid syndrome, Blood 102 (2003) 2717–2723.

[145] Antiphospholipid Antibodies in Stroke Study, Anticardiolipin antibodies are an independent risk factor for first ischemic stroke. The Antiphospholipid Antibodies in Stroke Study (APASS) group, Neurology 43 (1993) 2069–2073.

[146] S. Tuhrim, J.H. Rand, X.X. Wu, J. Weinberger, D.R. Horowitz, M.E. Goldman, et al., Elevated anticardiolipin antibody titer is a stroke risk factor in a multiethnic population independent of isotype or degree of positivity, Stroke 30 (1999) 1561–1565.

[147] O. Vaarala, M. Manttari, V. Manninen, L. Tenkanen, M. Puurunen, K. Aho, et al., Anti-cardiolipin antibodies and risk of myocardial infarction in a prospective cohort of middle-aged men, Circulation 91 (1995) 23–27.

[148] E. Ahmed, B. Stegmayr, J. Trifunovic, L. Weinehall, G. Hallmans, A.K. Lefvert, Anticardiolipin antibodies are not an independent risk factor for stroke: An incident case-referent study nested within the MONICA and Vasterbotten cohort project, Stroke 31 (2000) 1289–1293.

[149] K.V. Phadke, R.A. Phillips, D.T. Clarke, M. Jones, P. Naish, P. Carson, Anticardiolipin antibodies in ischaemic heart disease: Marker or myth? Br Heart J 69 (1993) 391–394.

[150] A.W. de Jong, W. Hart, M. Terburg, J.L. Molenaar, P. Herbrink, W.C. Hop, Cardiolipin antibodies and lupus anticoagulant in young patients with a cerebrovascular accident in the past, Neth J Med 42 (1993) 93–98.

[151] R.L. Brey, R.G. Hart, D.G. Sherman, C.H. Tegeler, Antiphospholipid antibodies and cerebral ischemia in young people, Neurology 40 (1990) 1190–1196.

[152] G.E. Tietjen, S.R. Levine, E. Brown, E. Mascha, K.M. Welch, Factors that predict antiphospholipid immunoreactivity in young people with transient focal neurological events, Arch Neurol 50 (1993) 833–836.

[153] R.T. Urbanus, B. Siegerink, M. Roest, F.R. Rosendaal, P.G. de Groot, A. Algra, Antiphospholipid antibodies and risk of myocardial infarction and ischaemic stroke in young women in the RATIO study: A case–control study, Lancet Neurol 8 (2009) 998–1005.

[154] P. Verro, S.R. Levine, G.E. Tietjen, Cerebrovascular ischemic events with high positive anticardiolipin antibodies, Stroke 29 (1998) 2245−2253.

[155] E.N. Harris, Syndrome of the black swan, Br J Rheumatol 26 (1987) 324−326.

[156] M.A. Khamashta, G.R. Hughes, ACP Broadsheet No 136: February 1993. Detection and importance of anticardiolipin antibodies, J Clin Pathol 46 (1993) 104−107.

[157] R.H. Derksen, M.A. Khamashta, D.W. Branch, Management of the obstetric antiphospholipid syndrome, Arthritis Rheum 50 (2004) 1028−1039.

[158] L. Opatrny, M. David, S.R. Kahn, I. Shrier, E. Rey, Association between antiphospholipid antibodies and recurrent fetal loss in women without autoimmune disease: A meta-analysis J Rheumatol 33 (2006) 2214−2221.

[159] A. Ruffatti, M. Tonello, T. Del Ross, A. Cavazzana, C. Grava, F. Noventa, et al., Antibody profile and clinical course in primary antiphospholipid syndrome with pregnancy morbidity, Thromb Haemost 96 (2006) 337−341.

[160] A. Ruffatti, M. Tonello, A. Cavazzana, P. Bagatella, V. Pengo, Laboratory classification categories and pregnancy outcome in patients with primary antiphospholipid syndrome prescribed antithrombotic therapy, Thromb Res 123 (2009) 482−487.

[161] T. Sailer, C. Zoghlami, C. Kurz, H. Rumpold, P. Quehenberger, S. Panzer, et al., Anti-beta2-glycoprotein I antibodies are associated with pregnancy loss in women with the lupus anticoagulant, Thromb Haemost 95 (2006) 796−801.

[162] E.Y. Lee, C.K. Lee, T.H. Lee, S.M. Chung, S.H. Kim, Y.S. Cho, et al., Does the anti-beta2-glycoprotein I antibody provide additional information in patients with thrombosis? Thromb Res 111 (2003) 29−32.

[163] A. Detkov, Gil-Aguado, P. Lavilla, M.V. Cuesta, G. Fontan, D. Pascual-Salcedo, Do antibodies to beta2-glycoprotein 1 contribute to the better characterization of the antiphospholipid syndrome? Lupus 8 (1999) 430−438.

[164] V. Pengo, A. Biasiolo, T. Brocco, S. Tonetto, A. Ruffatti, Autoantibodies to phospholipid-binding plasma proteins in patients with thrombosis and phospholipid-reactive antibodies, Thromb Haemost 75 (1996) 721−724.

[165] V. Pengo, A. Ruffatti, C. Legnani, P. Gresele, D. Barcellona, N. Erba, et al., Clinical course of high-risk patients diagnosed

with antiphospholipid syndrome, J Thromb Haemost 8 (2010) 237−242.

[166] A. Tincani, G. Morozzi, A. Afeltra, C. Alessandri, F. Allegri, O. Bistoni, et al., Antiprothrombin antibodies: A comparative analysis of homemade and commercial methods. A collaborative study by the Forum Interdisciplinare per la Ricerca nelle Malattie Autoimmuni (FIRMA), Clin Exp Rheumatol 25 (2007) 268−274.

[167] T. Atsumi, O. Amengual, S. Yasuda, T. Koike, Antiprothrombin antibodies—Are they worth assaying? Thromb Res 114 (2004) 533−538.

[168] M. Sanmarco, S. Gayet, M.C. Alessi, M. Audrain, E. de Maistre, J.C. Gris, et al., Antiphosphatidylethanolamine antibodies are associated with an increased odds ratio for thrombosis. A multicenter study with the participation of the European Forum on antiphospholipid antibodies, Thromb Haemost 97 (2007) 949−954.

[169] J.C. Gris, I. Quere, M. Sanmarco, B. Boutiere, E. Mercier, J. Amiral, et al., Antiphospholipid and antiprotein syndromes in non-thrombotic, non-autoimmune women with unexplained recurrent primary early foetal loss. The Nimes Obstetricians and Haematologists Study—NOHA, Thromb Haemost 84 (2000) 228−236.

[170] A. Selva-O'Callaghan, J. Ordi-Ros, F. Monegal-Ferran, N. Martinez, F. Cortes-Hernandez, M. Vilardell-Tarres, IgA anticardiolipin antibodies—Relation with other antiphospholipid antibodies and clinical significance, Thromb Haemost 79 (1998) 282−285.

[171] S. Carmo-Pereira, M.L. Bertolaccini, A. Escudero-Contreras, M.A. Khamashta, G.R. Hughes, Value of IgA anticardiolipin and anti-beta2-glycoprotein I antibody testing in patients with pregnancy morbidity, Ann Rheum Dis 62 (2003) 540−543.

[172] B. de Laat, V. Pengo, I. Pabinger, J. Musial, A.E. Voskuyl, I.E. Bultink, et al., The association between circulating antibodies against domain I of beta2-glycoprotein I and thrombosis: An international multicenter study, J Thromb Haemost 7 (2009) 1767−1773.

[173] A. Ambrozic, T. Avicin, K. Ichikawa, T. Kveder, E. Matsuura, M. Hojnik, et al., Anit-beta 2-glycoprotein I antibodies in children with atopic dermatitis, Int Immunol 14 (2002) 823−830.

Clinical Presentation of Antiphospholipid Syndrome

Munther A. Khamashta

Antiphospholipid syndrome (APS) is a noninflammatory autoimmune disease. The most critical pathologic process is thrombosis, which results in most of the clinical features suffered by these patients. Recurrent thrombosis together with an adverse pregnancy history and the presence of persistently elevated levels of antiphospholipid antibodies (aPLs) defines the syndrome [1].

There has been a rapid increase in worldwide interest related to APS because of the biological peculiarity of the disease in which autoantibodies cause thrombosis and because APS is recognized as one of the most frequent causes of acquired thrombophilia [2].

DIAGNOSTIC APPROACH TO ANTIPHOSPHOLIPID SYNDROME

The diagnosis of APS is first and foremost clinical (the patient must have one or more thrombotic or obstetric features of the condition). Laboratory testing for aPLs is used to confirm or refute the diagnosis. The 1999 "International Consensus Statement on Preliminary Classification Criteria for Definite Antiphospholipid Syndrome" provides simplified criteria for the classification of APS [3], and these criteria have been updated (Table 53.1) [4]. A patient with APS must manifest at least one of two clinical criteria (vascular thrombosis or pregnancy morbidity) and at least one of three laboratory criteria: positive lupus anticoagulant (LA), or medium- to high-titer IgG or IgM anticardiolipin antibody (aCL), or medium- to high-titer β_2-glycoprotein I IgG or IgM antibody confirmed on two separate occasions, at least 12 weeks apart. Many of the patients reported to have the syndrome have systemic lupus erythematosus (SLE) and can be regarded as having

secondary APS. Some patients do not have any underlying systemic disease. These patients may be regarded as having primary APS [5]. For research and classification purposes, the term "primary" is useful, although there appear to be few differences in complications related to aPL antibody or in antibody specificity in the presence or absence of SLE [6, 7]. Although some patients with primary APS progress to systemic lupus, most do not show such progression [8].

The aCL is the more sensitive test for APS, whereas LA is more specific. However, the specificity of aCL antibodies for APS increases with an increasing antibody titer [9]. Low positive aCL results should be viewed with suspicion; they may be found in up to 5% of normal individuals and should not be used to make the diagnosis of APS.

The demonstration that autoimmune aCLs are directed against a β_2-glycoprotein I or an epitope formed by the interaction of phospholipids and β_2-glycoprotein I led to the development of assays for anti-β_2-glycoprotein I antibodies and other phospholipid-binding plasma proteins [10]. A number of studies have highlighted the significance of testing for antibodies directed to β_2-glycoprotein I and prothrombin—phosphatidylserine complex as an alternative enzyme-linked immunosorbent assay (ELISA) with higher specificity than the conventional aCL assay [11, 12].

Clinicians should recognize that the international consensus criteria were developed primarily for research purposes to ensure more uniform characterization, as well as subcategorization, of patients included in studies. We view this objective as crucial for credible investigative efforts and for appreciation of subtleties of treatment. As with other autoimmune conditions, such as SLE, there are individuals who present with one or more clinical or laboratory features suggestive

1001

TABLE 53.1   Revised Classification Criteria for Antiphospholipid Syndrome

CLINICAL CRITERIA

1. Vascular thrombosis

   One or more clinical episodes of arterial, venous, or small vessel thrombosis, in any tissue or organ.

   Thrombosis must be confirmed by objective validated criteria (i.e., unequivocal findings of appropriate imaging studies or histopathology). For histopathological confirmation, thrombosis should be present without significant evidence of inflammation in the vessel wall.

2. Pregnancy morbidity

   (a) One or more unexplained deaths of a morphologically normal fetus at or beyond the 10th week of gestation, with normal fetal morphology documented by ultrasound or by direct examination of the fetus, *or*

   (b) One or more premature births of a morphologically normal neonate before the 34th week of gestation because of: (i) eclampsia or severe preeclampsia defined according to standard definitions, or (ii) recognized features of placental failure, *or*

   (c) Three or more unexplained consecutive spontaneous abortions before the 10th week of gestation, with maternal anatomic or hormonal abnormalities and paternal and maternal chromosomal causes excluded.

   In studies of populations of patients who have more than one type of pregnancy morbidity, investigators are strongly encouraged to stratify groups of subjects according to a, b, or c above.

LABORATORY CRITERIA

1. LA present in plasma, on two or more occasions at least 12 weeks apart, detected according to the guidelines of the International Society on Thrombosis and Haemostasis (Scientific Subcommittee on LA/Phospholipid-Dependent Antibodies).

2. aCLs of IgG and/or IgM isotype in serum or plasma, present in medium or high titers (i.e., >40 GPL or MPL, or >99th percentile), on two or more occasions, at least 12 weeks apart, measured by a standardized ELISA.

3. Anti-β_2-glycoprotein I antibody of IgG and/or IgM isotype in serum or plasma (in titers >99th percentile), present on two or more occasions, at least 12 weeks apart, measured by a standardized ELISA, according to recommended procedures.

aCLs, anticardiolipin antibodies; LA, lupus anticoagulant.
Source: Miyakis et al. [4].

of APS but in whom the diagnosis cannot be made by the relatively strict international consensus criteria. In such cases, experienced clinical judgment is required for best care.

EPIDEMIOLOGY

The epidemiology of aPLs is still being investigated worldwide. Efforts are being made in clinics throughout the world to assess the importance of this factor in recurrent abortion, stroke, myocardial infarction, epilepsy, and so on. Prospective studies have shown an association between aPLs and the first episode of venous thrombosis [13], the first myocardial infarction [14], and the first ischemic stroke [15]. A critical issue, therefore, is the identification of patients with aPLs who are at increased risk for a thrombotic event.

Antiphospholipid antibodies, using standardized techniques, are detected in less than 1% of apparently normal individuals and in up to 3% of the elderly population without clinical manifestations of APS. Among patients with SLE, the prevalence of aPLs is much higher, ranging from 30 to 40% [16]. For otherwise healthy control subjects, there are insufficient data to determine what percentage of those with aPL antibodies will eventually have a thrombotic event or a complication of pregnancy consistent with APS. In contrast,

APS may develop in 50–70% of patients with both SLE and aPLs after 10–20 years of follow-up [17, 18].

Recurrent miscarriage occurs in approximately 1% of the general, reproductive-aged population attempting to have children. Approximately 10–15% of women with recurrent miscarriage are diagnosed as having APS. Fetal death in the second or third trimester of pregnancy occurs in up to 5% of pregnancies, depending on the span of gestational age studied, but is less likely as pregnancy advances. Although fetal death is linked to APS, the overall contribution of APS to the problem of fetal death is uncertain [19].

The effect of race has been adequately studied only in African-American and White populations [18, 20, 21]. LA and high-titer aCL are significantly less common in African-Americans than in Whites, and APS is more frequent in Arabs than in the Indian population.

CLINICAL FEATURES

Thrombosis

Arterial and venous thrombosis can be present in APS, distinguishing this from other prothrombotic states such as protein C, S, or antithrombin deficiency, in which only venous thrombosis occurs. Any organ and any size of vessel (small, medium, or large) can be affected; thus, the range of clinical features is extremely wide.

In the venous circulation, thrombosis of the deep veins of the lower extremities has been reported most frequently (occasionally after the use of oral contraceptive pills containing estrogen). It is often recurrent and may be accompanied by pulmonary embolism. It has been estimated that up to 19% of patients with deep vein thrombosis and/or pulmonary thromboembolism are suffering from aPL coagulopathy and may demonstrate a positive LA test, aCL, or both. Other reported sites of thrombosis include the axillary, ocular, renal, hepatic, and sagittal veins and the inferior vena cava. The APS is now considered one of the most frequent causes of Budd–Chiari syndrome [22]. aPLs have been implicated in the development of adrenal vein thrombosis leading to Addison's disease [23]. Venous events usually occur at single sites, and these can recur at the same or different sites months or years apart.

Unlike other known clotting disorders, arterial thromboses are a major feature of APS. Occlusions of the intracranial arteries have been reported most frequently, with the majority of patients presenting with stroke and transient ischemic attacks (TIAs). Other arterial thromboses have involved the retina, coronary, mesenteric, and peripheral arteries. The clinical presentation depends on the anatomic site occluded. Malignant hypertension with renal insufficiency secondary to thrombosis of the renal glomeruli and renal thrombotic microangiopathy (without classic lupus nephritis) has also been associated with the presence of aPLs [24]. As with venous thrombosis, arterial events usually occur at single sites and can recur months or years later.

Central Nervous System Manifestations

Central nervous system (CNS) involvement is common in patients with APS. The original description of the syndrome in 1983 stressed the importance of cerebral features in these patients [25] and highlighted the frequency of intractable headache or migraine, epilepsy, chorea, and cerebrovascular accidents (TIAs or visual field defects or progressive cerebral ischemia) [26]. Although the mechanism of neurological involvement in patients with APS is thought to be thrombotic in origin, a number of other neuropsychiatric manifestations cannot be explained solely by hypercoagulability. Table 53.2 summarizes CNS manifestations reported to be associated with the presence of aPLs.

Stroke and TIAs are the most common neurological complications of APS. These are also the most frequently reported presentation of arterial thrombosis in APS [27]. The association between cerebrovascular disease and aPLs was described in early studies [28, 29], and these antibodies are now internationally recognized as an

TABLE 53.2 Central Nervous System Manifestations Associated with the Presence of Antiphospholipid Antibodies

Cerebrovascular disease
   Transient ischemic attacks
   Ischemic strokes
   Acute ischemic encephalopathy
   Cerebral venous thrombosis
Epilepsy
Headache
Chorea
Multiple sclerosis
Transverse myelitis
Idiopathic intracranial hypertension
Other neurological syndromes
   Sensorineural hearing loss
   Guillain–Barré syndrome
   Transient global amnesia
   Ocular syndromes
   Dystonia–Parkinsonism
Cognitive dysfunction
Dementia
Other psychiatric disorders
   Depression
   Psychosis

important etiological factor and may be present in up to 7% of all patients who have suffered a stroke [30]. They should be sought especially in young patients with strokes, in whom they may account for up to 18% [31]. A growing body of evidence supports an association between aPLs and ischemic stroke not only in SLE and/or APS but also in unselected populations [15, 32–34].

An ischemic stroke can be isolated or multiple and recurrent. The risk of recurrent stroke appears to be increased in APS, and patients occasionally present with multi-infarct dementia [35]. It is unclear whether the presence of aCL at the time of initial stroke increases the risk of recurrence in an unselected population. Levine *et al.* [36] examined in a prospective study 81 consecutive patients with aPLs who developed focal cerebral ischemia. The mean age of this cohort was younger than the average atherothromboembolic stroke victim, with women being more commonly affected than men. Significantly, the frequency of conventional stroke risk factors was lowest in the group of stroke patients with the highest levels of IgG aCL reactivity. Moreover, patients with highest IgG aCL had the shortest times to subsequent thrombo-occlusive events, mainly represented by cerebral infarction often occurring within the first year of follow-up, supporting that IgG aCL represented a risk factor for recurrent stroke.

Angiograms typically demonstrate thrombotic occlusion of intracranial branch or trunk vessels, but in one-third of cases, these studies may be normal. Magnetic resonance imaging (MRI) is consistently more sensitive than computed tomography in detecting infarcts in patients with APS [37]. Brain MRI in aPL patients with ischemic stroke shows cortical abnormalities consistent with large vessel occlusion. aPL patients often present small foci of high signal in brain white matter, which are often defined as consistent with the presence of small vessel disease. In some cases, emboli from cardiac valvular vegetations may account for stroke or TIAs, and echocardiograms are advisable, particularly where there is no evidence of large vessel occlusion [38–41].

A less common form of cerebral thrombotic disease associated with aPL is sagittal venous sinus thrombosis [42]. As with stroke, venous sinus thrombosis in patients with APS occurs at a younger age than in individuals without aPL and shows a higher rate of post-thrombosis migraine and more cerebral infarctions in imaging studies. However, other causes, such as factor V Leiden mutation resulting in activated protein C resistance, should be excluded.

Acute ischemia of the eye is a major feature of APS, with retinal and choroidal vessel involvement being the most common finding [43–45]. Patients usually complain of transient blurred vision or amaurosis fugax, transient diplopia, decreased vision, transient field loss associated with headache, or photopsy. It is important to distinguish this aPL-induced thrombotic retinopathy from that seen in SLE patients as a result of vasculitis or atherosclerosis. Optic neuropathy also occurs in SLE patients.

Sudden sensorineural hearing loss, often presenting as sudden deafness, can occur in some patients with APS. Toubi et al. [46] reported a 27% prevalence of positivity for aCL in 30 unselected patients with sudden or progressive sensorineural hearing loss. Naarendorp and Spiera [47] reported on 6 SLE patients with sudden sensorineural hearing loss; all 6 patients were positive for aCL or LA. They concluded that acute onset of sensorineural hearing loss in the presence of aPLs may be a manifestation of APS, and that anticoagulation treatment was to be recommended for these patients.

Clinical syndromes mimicking multiple sclerosis (MS), mainly in its relapsing–remitting pattern, are reported to occur in association with aPLs [48–50]. It can be difficult to distinguish MS from APS clinically and on MRI. Certainly the ischemic changes produced by APS in the white matter may be indistinguishable on MRI from those of MS (Figure 53.1). In our experience, a careful medical history, a previous history of thrombosis or fetal loss, an abnormal localization of the lesions in brain MRI, and the response to anticoagulant therapy might be helpful in the differential diagnosis. Although controversial, we believe that testing for aPLs should become routine in all patients with MS, especially if there are atypical features [48, 51, 52]. For the patient, the importance of a treatable differential diagnosis of APS cannot be understated [53].

Many other neurological abnormalities have been reported in APS, but these are not clearly linked to thrombosis. These include transverse myelopathy, chorea, seizures, Guillain–Barré syndrome, psychosis, and migraine headaches.

One of the most prominent features in patients with APS is the headache. This symptom, a common complaint of APS patients in clinical practice, can vary from classic migraine to an almost continuous incapacitating headache. Frequently, the migraine history goes back to childhood, and, interestingly, there is often a family history. Although important in clinical practice, the association between migraine and aPL in the published literature is still controversial. Tietjen et al. [54] studied a large population and failed to demonstrate an association between the presence of aCL and migraine in subjects younger than 60 years. Similar results have been reported previously [55]. Our most recent experience in applying the 1999

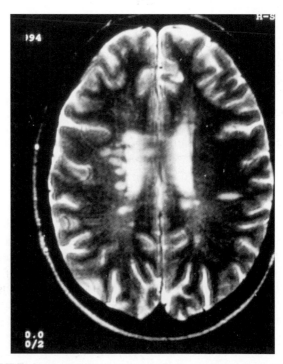

FIGURE 53.1   Cerebral Ischemic Lesions associated with APS

American College of Rheumatology criteria for neuropsychiatric syndromes to a large cohort of SLE patients (323 patients) showed a prevalence of headache similar to that of the general population, but aPLs were significantly more prevalent in the group of patients with headache compared to those without headache [56]. We did not find a higher prevalence of aPLs in migraine than in other types of headache. Interestingly, in some of our patients with severe headache resistant to conventional treatment, complete or partial remission of the headache was noticed when treatment with aspirin, heparin, or warfarin was started for thrombotic events [57, 58]. Furthermore, the relationship between migraine and stroke has been widely studied [59–62]. The predictive value of variables such as gender, the presence or absence of a family history of migraine, use of oral contraceptives, high blood pressure, diabetes, heart disease, and smoking has been analyzed and related to the development of "migrainous stroke." None of these studies analyzed the presence of aPLs as an independent risk factor for stroke in their patients. In our series, 8 patients, all young women, developed stroke after some years of migraine [57]. This highlights the importance of considering the presence of aPLs in patients with migrainous stroke or even with severe persistent headache, especially in young women.

Neuropsychiatric manifestations in APS range from mild cognitive dysfunction to severe dementia [63–66]. Patients affected by mild cognitive dysfunction often complain of poor concentration or forgetfulness. Verbal memory deficits, decreased psychomotor speed, and decreased overall productivity have been significantly correlated with elevated aPL levels. Some of these patients had never had a history of stroke, and gross neurological examinations were normal. Computed tomography scans in these patients are usually normal, and MRI studies frequently show scattered areas of nonspecific increased signal intensity. These areas may increase or decrease in both size and number over time and may not represent permanent injury.

The recognition of subtle forms of cognitive dysfunction has been greatly facilitated by the application of formal neuropsychological assessment mainly in patients with SLE. Hanly *et al.* [65] and Menon *et al.* [66] showed that SLE patients persistently positive for aCL (medium high titer) had significantly lower scores on a variety of neuropsychological tests than did SLE patients negative for aPLs. Denburg *et al.* [64] also found that LA-positive patients performed worse than LA-negative patients in measures of verbal memory, cognitive flexibility, and psychomotor speed. Some of these patients had no previous history of any neuropsychiatric involvement, suggesting a direct relationship between the presence of aPLs and cognitive impairment. Whether these cognitive deficits result from recurrent cerebral ischemia or whether there are other underlying mechanisms remains unknown. Anecdotal reports of improvement of cognitive disorders after initiation of anticoagulation therapy for other reasons in APS patients may provide some support for the theory that arterial thrombosis and/or ischemia represent the primary cause of this type of CNS dysfunction. Although MRI is a highly sensitive method for detecting cerebral lesions, new imaging techniques such as positron emission tomography or single photon emission computed tomography may provide objective information about cerebral damage and the possible underlying mechanisms in APS patients with cognitive dysfunction [67].

Mosek *et al.* [68] examined the relationship of aPLs and dementia in the elderly. They found that 5 of the 87 (6%) demented patients, but none of the 69 controls, had significantly elevated IgG aCL levels. None of the patients had features of an immune-mediated disease. The role of aPL antibodies in these patients, with apparently diffuse brain disease, is unknown.

Although depression and psychosis have been associated with aPLs, it had been postulated that autoantibodies, and specifically aPLs, may represent an adverse response to neuroleptic treatment. Schwartz *et al.* [69] studied 34 unmedicated patients without known autoimmune disorders admitted with acute psychosis. aCL and LA were determined before and after neuroleptic treatment. They found that 32% of the unmedicated psychotic patients had positive aPLs (24% IgG aCL and 9% LA). Of the 22 patients followed up after medication, 32% showed moderate titers of IgG aCL and 18% LA positivity. This study showed an increased incidence of aPLs in psychotic patients and that the presence of these antibodies cannot be simply assumed to be a result of the treatment.

Seizures have been consistently associated with the presence of aPLs in SLE patients and are well described in patients with primary APS [56, 70–73]. The etiology of seizures in APS patients may be associated with cerebral infarction, but some authors have suggested that aPLs interact directly with neuronal tissue [37]. Intriguingly, aPLs can also reduce a γ-aminobutyric acid receptor-mediated chloride current in snail neurons [74]. This effect might lower seizure threshold.

Other syndromes that are rarer but no less important have been associated with the presence of aPLs; for example, chorea was reported in early descriptions of the syndrome [25, 75]. In an extensive review of 50 APS patients with chorea, the role of estrogens in this disorder was highlighted (6 patients were receiving the oral contraceptive pill and 4 developed chorea during pregnancy or the puerperium) [76]. Although striatal ischemia can cause this symptom, some studies using positron emission tomography have shown increases

TABLE 53.3   Skin Manifestations Associated with Antiphospholipid Syndrome

Livedo reticularis

Cutaneous ulceration

Cutaneous gangrene and necrosis

Livedoid vasculitis

Skin nodules

Erythematous macules

Subungual splinter hemorrhages

Thrombophlebitis

Degos-like lesions

Anetoderma

in lentiform and caudate nucleus metabolism, which returns to normal when the patient recovers [77].

Skin Manifestations

A wide variety of cutaneous manifestations have been reported in patients with APS (Table 53.3), and these very often constitute early signs of the disease. The majority of these lesions can be explained by vascular occlusion, frequently demonstrable by histopathological examination [78].

Livedo reticularis (racemosa) is the most frequently associated cutaneous manifestation in patients with APS (Figure 53.2) [79]. Livedo reticularis is a dusky, violaceous, vascular discoloration of the skin with a network pattern, which usually appears over the upper and lower limbs. The association between livedo reticularis and aPLs was first reported by Hughes in 1983 [26] and was confirmed in subsequent studies [80–82]. Many cases of Sneddon's syndrome, defined

FIGURE 53.2   Livedo Reticularis

as the clinical triad of stroke, livedo reticularis, and hypertension, may represent undiagnosed APS [83]. The relationship between livedo reticularis and other clinical manifestations and its value as a predictor for the risk of thrombosis need further investigation.

Skin ulcers are also common in APS. Histopathological analysis of the skin specimens often demonstrates thrombosis of the small vessels with vascular proliferation and minimal inflammatory changes [84]. Skin ulcers normally appear in the extremities, although they may be more widespread and sometimes leave white atrophic scars. Treatment with steroids or immunosuppression is usually ineffective, and some successfully treated cases can be explained by the coexistence of an underlying connective tissue disease with a vasculitic process [85]. Treatment with antiplatelet agents, oral anticoagulation, or even antifibrinolytic therapy can be effective [86, 87].

The prevalence of subungual splinter hemorrhages is less than 5% in patients with APS [88]. Although the causes are many, including infectious processes and vasculitis, multiple lesions in different fingers in a patient with primary or secondary APS should alert the physician to the occurrence of other thrombotic events [89].

Digital gangrene has been described in primary and secondary APS, and widespread superficial skin necrosis appears in approximately 3% of APS patients. Its onset is often acute; lesions are purpuric and painful, and they are localized to the limbs, head, and buttocks. There are several case reports of extensive skin necrosis as the first manifestation of APS [90]. Some of them occurred without any obvious trigger factor [91], but in others, skin necrosis followed the onset of infection or surgery [92].

A variety of other cutaneous lesions resembling vasculitis, including purpura, tender nodules, papules, and palmar–plantar erythema, have been described in APS [78]. Histologically, the absence of vasculitis and the finding of noninflammatory thrombosis of small arteries or veins of the dermis and hypodermis are characteristic of skin lesions in APS.

Heart Manifestations

APS has a variety of cardiac manifestations. The most important are valve disease, coronary artery disease, and, less commonly, cardiomyopathy and intracardiac thrombus. These lesions appear in primary and secondary APS [93].

A number of studies have documented the prevalence of valve disease in SLE and primary APS patients [41, 94, 95]. Valve disease is present in approximately 48% of patients with SLE and aPLs and in only 21% of SLE patients negative for aPLs. Studies involving only patients with APS showed that 35–75% had valvulopathy [42, 43, 96]. In addition, aPLs were significantly higher (21%) in a group of unselected patients with heart valve disease compared with control subjects (9%) [97]. There are also some data supporting the relationship between valve lesions and the presence of aPLs, with evidence suggesting that markers of endothelial cell activation are upregulated in valves from APS patients [98]. Deposition of immunoglobulins in the valves suggested a possible association between this deposition and the activation of the endothelial cells in APS.

Two morphologic echocardiographic patterns can be discerned: valve masses (vegetations) and valvular thickening. The predominant functional abnormality is regurgitation, whereas stenosis is rarely seen. The mitral valve is the most commonly affected site, followed by the aortic valve. Involvement of the tricuspid or pulmonary valve has also been described [93]. Mitral or aortic regurgitation is rarely severe enough to cause symptoms, and requirement for valve replacement is distinctly uncommon. Valvular abnormalities of APS are different from those seen in rheumatic heart disease. In APS, valvular thickening is diffuse, whereas in rheumatic heart disease thickening is more localized, present at leaflet tips, and often associated with thickening, fusion, and calcification of the chordae tendinae.

A study described the echocardiographic lesions of 29 patients with primary APS [96]. Transesophageal echocardiography was performed in all these patients at the onset of the study and in 13 of 19 patients 1 year later. This second echocardiogram showed no changes in the lesions in 6 patients and new lesions in the remaining 7 patients. All received anticoagulant or antiaggregant treatment during the year of the study. The authors concluded that anticoagulant or antiaggregant treatment does not diminish the size of the valve lesions, although there are anecdotal reports of vegetations resolving with high-intensity anticoagulation [99].

Coronary artery disease has been documented in APS patients. Early descriptions by Hamsten et al. in 1986 [100] have been confirmed by prospective studies showing that elevated levels of aPLs in a non-SLE population imply an increased risk for the development of myocardial infarction [14, 15, 32, 101, 102]. The prevalence of aCL in patients with myocardial infarction is between 5 and 15% [103, 104]. Although this prevalence may not justify routine screening in patients with myocardial infarction, its measurement may be important in patients younger than age 45 years if they have previous venous or arterial thrombosis or an adverse pregnancy history (e.g., recurrent miscarriages and fetal death). In patients with SLE, particularly women between 35 and 44 years of age, the risk of cardiovascular events is more than 10 times higher than that in healthy women of similar age [105]. Approximately

35% of SLE patients are positive for aPLs, and the contribution of aPLs to the development of myocardial infarction in this population may indeed be substantial.

Different antibodies have been implicated in the development of coronary artery disease in APS patients. aCL is directed against different antigenic structures. Some of them recognize phospholipids, some recognize phospholipid binding proteins such as β_2-glycoprotein I, and some may be directed against cross-reactive epitopes common with oxidized low-density lipoprotein (LDL) [106]. The presence of antioxidized LDL is considered to be a marker of atherosclerosis. Two prospective studies have shown that these antibodies are predictive for myocardial infarction in the general population [104, 107]. aCL levels were associated with antioxidized LDL antibodies, and the joint effect of these two different populations of antibodies was additive for the risk of myocardial infarction. In another study, including a large population of patients with different manifestations of coronary artery disease (109 patients with coronary artery bypass surgery, 106 patients with balloon angioplasty, 101 patients with acute myocardial infarction, and 99 patients with acute myocardial ischemia), it was shown that antioxidized LDL antibody titers were significantly higher in the group with acute myocardial infarction than in other groups in men but not in women. The titers of aCL did not differ among the patient groups. Neither of the autoantibodies was associated with recurrent coronary events. aCLs were inversely correlated with dietary intake of vitamin E and polyunsaturated fat in men with coronary heart disease [108].

Kidney Manifestations

Renal involvement is a prominent feature in APS. Initially underestimated, it is currently the subject of numerous studies [109].

Hypertension is the most common clinical manifestation, present in more than 40% of patients. It can be present as mild, severe, or even malignant hypertension. Pathogenic mechanisms leading to hypertension include thrombosis in a trunk of the renal artery (renovascular hypertension) and intrarenal vascular lesions [24]. These lesions can cause hypertension via stimulation of the renin–angiotensin–aldosterone system. All patients with APS and hypertension should be investigated for renal involvement (especially renal artery stenosis) because hypertension is often the only early clinical manifestation [110]. Although renal involvement, when it occurs, is usually chronic, with renal insufficiency slowly progressing, acute renal failure can also occur. In the study by Nochy et al. [111], which included 16 patients with primary APS, 90% of patients had hypertension, which was severe in 31% and malignant in

12%. Renal insufficiency was present in 87%, and 2 patients had acute renal failure.

Thrombosis of renal arteries and veins has been described in APS patients [102]. This occlusion can be caused by different mechanisms such as in situ thrombosis, or in the case of arterial thrombosis, it can also be the consequence of an embolic event resulting from a damaged cardiac valve. Treatment with anticoagulant drugs or transluminal angioplasty has been successful in cases in which the problem is detected at an early stage [112].

Antiphospholipid antibodies have been anecdotally detected in some patients with end-stage renal disease (ESRD), although the data are conflicting [113, 114]. Whereas Brunet et al. [113] found a prevalence of aPLs of 31% (aCL, 15.5%; LA, 16.5%) and a clear association between the presence of LA and vascular access thrombosis, Fabrizi et al. [114] reported a lower prevalence of aPL (8.8%). The presence of aPLs was not associated with factors such as age, cause of ESRD, time on hemodialysis, type of hemodialysis membrane, or thrombotic events. Only one patient positive for LA had recurrent thrombosis of the access graft and native veins.

Although the outcome of renal transplant in patients with SLE is not different from that of other populations, it is possible that the presence of aPLs may modify the prognosis. A number of studies have reported a poor outcome, with high rates of graft loss as a result of thrombotic events in patients with SLE positive for aPLs [115–117]. Most of the patients included in these studies were not fully anticoagulated. In patients with primary APS, the observed outcome has been even worse, with recurrence of the disease as a thrombotic microangiopathy in the graft despite intensive anticoagulant therapy [118]. Vaidya et al. [117] undertook a multi-center study, including 502 ESRD patients waiting for renal transplant. They found that 93 (19%) of these patients were positive for aCL. Only 23 of 93 cases fulfilled the criteria for APS (previous thrombotic episodes, including documented thrombotic microangiopathy or adverse pregnancy history). Eleven of these 23 patients received kidney transplants. Four were treated with oral anticoagulation, and 7 did not receive any anticoagulant drug. All 7 patients without anticoagulant treatment lost their allografts within 1 week as a result of thrombosis. Only 1 of the 4 patients treated lost the graft because of thrombosis. Interestingly, of the remaining 70 aCL-positive patients without any other features of APS, 37 received a renal transplant, and none of them lost the allografts as a result of thrombosis. These observations suggest that all patients with APS should be fully anticoagulated as part of the post-transplant therapeutic regimen [119].

The histologic findings of the renal lesions in APS have been extensively reviewed [111, 120, 121]. Daugas et al.

[121] delineated the clinical and histologic manifestations found in patients with APS and renal involvement. APS nephropathy is clinically manifest by a syndrome of vascular nephropathy, associating hypertension, acute and/or chronic renal insufficiency, and low-grade proteinuria. Histologically, APS nephropathy is a vaso-occlusive process associating, side-by-side, acute thromboses (thrombotic microangiopathy) and chronic vascular lesions (arterial fibrous intimal hyperplasia, arteriosclerosis, and organized thromboses, with or without recanalization). The authors emphasized the importance of recognizing this characteristic histologic picture for diagnostic and therapeutic purposes. The presence of APS nephropathy in conjunction with lupus nephropathy probably augments the risk of evolution toward ESRD because it is associated with higher creatinine levels at the time of diagnosis, increased interstitial fibrosis, and systemic hypertension.

Lung Manifestations

Patients with APS may develop a broad spectrum of lung disease. Pulmonary thromboembolism and pulmonary hypertension are the most common complications, but microvascular pulmonary thrombosis, pulmonary capillaritis, and alveolar hemorrhage have also been reported [122].

Pulmonary embolism is a frequent complication of deep venous thrombosis and occurs in approximately 30% of APS patients. It may be the first manifestation of the disease. Recurrent pulmonary embolism may lead to thromboembolic pulmonary hypertension.

The prevalence of pulmonary hypertension in APS associated with SLE and primary APS has been estimated to be 1.8 and 3.5%, respectively [6]. In most cases, the mechanism has been pulmonary emboli, although in situ thrombosis remains a possibility [123].

Although quite a rare manifestation, some APS patients may present with diffuse alveolar pulmonary hemorrhage. The common symptoms are fever, cough, dyspnea with or without hemoptysis, hypoxemia, and diffuse pulmonary infiltrates. The diagnosis is usually made by open lung biopsy, which shows microvascular thrombosis and secondary alveolar hemorrhage with or without pulmonary capillaritis. Treatment with corticosteroids usually leads to dramatic improvement [124].

Hematologic Manifestations

The most frequent hematologic manifestation of APS is thrombocytopenia. It is present in 25% of patients, although it is rarely severe enough to cause hemorrhage [125]. The platelet count often remains stable for many years; then, for reasons that are often obscure, the count drops, sometimes catastrophically. Occasionally, patients with APS may present only with severe thrombocytopenia but later develop pregnancy loss or thrombosis. The exact role of aPLs in thrombocytopenia is unclear [126]. Data have shown that thrombocytopenia probably results from specific antiplatelet glycoprotein antibodies [127−129]. Galli *et al.* [127] found antibodies to the specific platelet membrane glycoproteins IIb/IIIa and Ib/IX in 40% of patients positive for aPLs; there was a significant correlation between these antibodies and thrombocytopenia.

Hemolytic anemia may be present in some patients with APS and is sometimes associated with the presence of thrombocytopenia, the so-called "Evans syndrome." Although a positive direct Coombs' test is not rare in APS patients (10−20%), hemolytic anemia is uncommon. A significant correlation has been found between aPLs and hemolytic anemia, although the association with different isotypes of aCL is still controversial. Some authors have found an association with IgM isotype [81, 130]; however, this has not been confirmed in other studies [5, 6, 131]. The pathogenic mechanisms for aPL-related hemolytic anemia are not clear. aPLs may bind directly to the red blood cell membrane, but factors such as antierythrocyte antibodies or immune complexes fixed to red blood cells may also play a role.

Obstetric Manifestations

Recurrent pregnancy loss, typically in the second trimester, is one of the most consistent features of APS [19, 132]. Many cases of APS are diagnosed after investigation for recurrent miscarriage. The risk of fetal loss is directly related to the antibody titer, particularly the IgG aCL [9, 133], although many women with recurrent miscarriage have IgM aCL antibodies only. It is impossible to predict which women are likely to develop complications in pregnancy, and some women with persistently elevated aPL titers and a history of thromboses with or without thrombocytopenia have no fetal complications. Previous poor obstetric outcome remains the most important predictor of future risk [134].

In pregnancies that do not end in miscarriage or fetal loss, there is a high incidence of preeclampsia, intrauterine growth restriction (IUGR), placental abruption, and premature delivery [134]. It is important to note that the most recent classification criteria for APS have been amended to highlight the fact that not only fetal loss but also premature birth before 34 weeks as a result of preeclampsia, placental abruption, or IUGR, and positive aPLs or LA may allow the patient to be labeled as having APS [4].

Varying high rates of preeclampsia have been reported in women with APS, which contributes to the high rate of preterm delivery in this condition. Raised aCL levels have been reported in association with severe early onset preeclampsia [135]. The weight of evidence supports aPL testing in women with early onset (<34 weeks of gestation) severe preeclampsia [136].

Units that manage patients with severe manifestations of APS, such as thrombosis and previous stillbirth [134, 137], have a higher incidence of complications in pregnancy than those that recruit women predominantly from recurrent miscarriage clinics [138, 139]. In our unit, we run a multidisciplinary team service, in which most APS cases have been identified by either rheumatology (association with SLE) or hematology (previous thrombosis) colleagues. Many patients are referred for specialist management after previous poor obstetric outcome. Our live birth rate is greater than 80%, and in a previous study from our unit, the incidence of preeclampsia was 18%, the percentage of babies born with birth weights less than the 10th percentile for gestational age was 31%, and the percentage of infants delivered prematurely (<37 weeks) with a mean gestation of approximately 34 weeks was approximately 43% [140]. Approximately 70% of these women were delivered by cesarean section, and 7% of babies died in the neonatal period as a result of problems related to prematurity. Regular ultrasound scanning for fetal growth is recommended in these patients and, in specialist units (including our own), assessment of uterine artery Doppler waveforms is performed in the mid-trimester [134, 141, 142]. The presence of mid-trimester uterine artery notches in high-risk pregnancies is associated with preeclampsia, IUGR, and intrapartum asphyxia with a sensitivity of 90% and a positive predictive value of 60% [143]. In high-risk pregnancies, abnormal uterine artery Doppler velocimetry is also of some value in predicting placental abruption, a common feature of APS pregnancies, but gives no indication regarding the timing of this event [144].

Catastrophic Antiphospholipid Syndrome

A minority of patients with APS present with an acute and devastating syndrome characterized by multiple simultaneous vascular occlusions throughout the body, often resulting in death. This syndrome, termed catastrophic APS, is defined by the clinical involvement of at least three different organ systems over a period of days or few weeks with histopathological evidence of multiple occlusions of large or small vessels [145]. Although the same clinical manifestations seen with primary and secondary APS occur as part of catastrophic APS, there are important differences in prevalence and in the caliber of the vessels predominantly affected. Ischemia of the kidneys, bowels, lungs, heart, and/or brain is most frequent, but rarely adrenal, testicular, splenic, pancreatic, or skin involvement have been described. Occlusion of small vessels (thrombotic microangiopathy) is characteristic, resulting in symptoms related to dysfunction of the affected organs. Depending on the organs involved, patients may present with hypertension and renal impairment, acute respiratory distress syndrome, alveolar hemorrhage and capillaritis, confusion and disorientation, or abdominal pain and distension secondary to bowel infarction. Precipitating factors of catastrophic APS include infections, surgical procedures, withdrawal of anticoagulant therapy, and the use of drugs such as oral contraceptives.

Other Manifestations

Avascular necrosis of bone is an uncommon complication in lupus patients and clearly associated with high steroid dosage. We have noted an increased risk of avascular necrosis in individuals positive for aPLs, possibly as a result of small arterial occlusions, notably of the head of the femur [146, 147]. Rare cases of bone marrow necrosis have been reported secondary to APS [148, 149].

Many patients with APS seem to develop widespread arteriopathy. The systemic narrowing of major arteries is similar in many respects to the widespread endarterial disease seen in some patients after heart–lung transplantation. Thus, aPLs might be associated with accelerated vascular disease, including atherosclerosis [150]. More recent work has suggested that aPLs are directed against oxidized phospholipids, following the observation that oxidation of cardiolipin is required, even in the presence of β_2-glycoprotein I, for it to be recognized by purified IgG fractions and monoclonal antibodies from APS patients [151]. These findings raise the possibility that oxidized LDL, as oxidized cardiolipin, may act as an immunizing antigen leading to the generation of aPLs, providing an additional link between atherogenesis and APS.

Although there have been suggestions that aPLs may cause failure of *in vitro* fertilization, a literature review of 16 studies found no association [152].

References

[1] G.R.V. Hughes, The antiphospholipid syndrome: Ten years on, Lancet 342 (1993) 341–344.

[2] M.A. Khamashta, Hughes Syndrome—Antiphospholipid Syndrome, Springer-Verlag, London, 2000.

[3] W.A. Wilson, A.E. Gharavi, T. Koike, M.D. Lockshin, D.W. Branch, J.C. Piette, et al., International consensus statement on preliminary classification criteria for definite antiphospholipid syndrome: Report of an international workshop, Arthritis Rheum 42 (1999) 1309–1311.

[4] S. Miyakis, M.D. Lockshin, T. Atsumi, D.W. Branch, R.L. Brey, R. Cervera, et al., International consensus statement on an update of the classification criteria for definite antiphospholipid syndrome (APS), J Thromb Haemost 4 (2006) 295–306.

[5] R.A. Asherson, M.A. Khamashta, J. Ordi-Ros, R.H. Derksen, S.J. Machin, J. Barquinero, et al., The "primary" antiphospholipid syndrome: Major clinical and serological features, Medicine 68 (1989) 366–374.

[6] J.L. Vianna, M.A. Khamashta, J. Ordi-Ros, J. Font, R. Cervera, A. Lopez-Soto, et al., Comparison of the primary and secondary antiphospholipid syndrome: A European multicenter study of 114 patients, Am J Med 96 (1994) 3–9.

[7] R. Cervera, J.C. Piette, J. Font, M.A. Khamashta, Y. Shoenfeld, M.T. Camps, et al., Antiphospholipid syndrome: Clinical and immunologic manifestations and patterns of disease expression in a cohort of 1000 patients, Arthritis Rheum 46 (2002) 1019–1027.

[8] J.A. Gómez-Puerta, H. Martín, M.C. Amigo, M.A. Aguirre, M.T. Camps, M.J. Cuadrado, et al., Long-term follow-up in 128 patients with primary antiphospholipid syndrome: Do they develop lupus? Medicine (Baltimore) 84 (4) (2005) 225–230.

[9] N.E. Harris, J.K.H. Chan, R.A. Asherson, V.R. Aber, A.E. Gharavi, G.R.V. Hughes, Thrombosis, recurrent fetal loss and thrombocytopenia, Arch Intern Med 146 (1986) 2153–2156.

[10] R.A. Roubey, Autoantibodies to phospholipid-binding plasma proteins: A new view of lupus anticoagulants and other antiphospholipid autoantibodies, Blood 84 (1994) 2854–2867.

[11] O. Amengual, T. Atsumi, M.A. Khamashta, T. Koike, G.R. Hughes, Specificity of ELISA for antibody to beta 2-glycoprotein I in patients with antiphospholipid syndrome, Br J Rheumatol 35 (1996) 1239–1243.

[12] T. Atsumi, M. Ieko, M.L. Bertolaccini, K. Ichikawa, A. Tsutsumi, E. Matsuura, et al., Association of autoantibodies against the phosphatidylserine–prothrombin complex with manifestations of the antiphospholipid syndrome and with the presence of lupus anticoagulant, Arthritis Rheum 43 (2000) 1982–1993.

[13] K.S. Ginsburg, M.H. Liang, L. Newcomer, S.Z. Goldhaber, P.H. Schur, C.H. Hennekens, et al., Anticardiolipin antibodies and the risk for ischemic stroke and venous thrombosis, Ann Intern Med 117 (1992) 997–1002.

[14] O. Vaarala, M. Manttari, V. Manninen, L. Tenkanen, M. Puurunen, K. Aho, et al., Anticardiolipin antibodies and risk of myocardial infarction in a prospective cohort of middle-aged men, Circulation 91 (1995) 23–27.

[15] R.L. Brey, R.D. Abbott, J.D. Curb, D.S. Sharp, G.W. Ross, C.L. Stallworth, et al., Beta(2)-glycoprotein 1-dependent anticardiolipin antibodies and risk of ischemic stroke and myocardial infarction: The Honolulu Heart Program, Stroke 32 (2001) 1701–1706.

[16] R. Cervera, M.A. Khamashta, J. Font, G.D. Sebastiani, A. Gil, P. Lavilla, et al., Systemic lupus erythematosus: Clinical and immunologic patterns of disease expression in a cohort of 1000 patients. The European Working Party on Systemic Lupus Erythematosus, Medicine 72 (1993) 113–124.

[17] N.M. Shah, M.A. Khamashta, T. Atsumi, G.R. Hughes, Outcome of patients with anticardiolipin antibodies: A 10 year follow-up of 52 patients, Lupus 7 (1998) 3–6.

[18] M. Petri, Epidemiology of the antiphospholipid antibody syndrome, J Autoimmun 15 (2000) 145–151.

[19] R.H. Derksen, M.A. Khamashta, D.W. Branch, Management of the obstetric antiphospholipid syndrome, Arthritis Rheum 50 (4) (2004) 1028–1039.

[20] A.N. Malaviya, R. Marouf, K. Al-Jarrallah, A. Al-Awadi, K. Al-Saied, S. Al-Gaurer, et al., Hughes syndrome: A common problem in Kuwait hospitals, Br J Rheumatol 35 (1996) 1132—1136.

[21] J.F. Molina, S. Gutierrez-Urena, J. Molina, O. Uribe, S. Richards, C. De Ceulaer, et al., Variability of anticardiolipin antibody isotype distribution in 3 geographic populations of patients with systemic lupus erythematosus, J Rheumatol 24 (1997) 291—296.

[22] G. Espinosa, J. Font, J.C. Garcia-Pagan, D. Tassies, J.C. Reverter, et al., Budd—Chiari syndrome secondary to antiphospholipid syndrome. Clinical and immunologic characteristics of 43 patients, Medicine 80 (2001) 345—354.

[23] R.A. Asherson, G.R. Hughes, Hypoadrenalism, Addison's disease and antiphospholipid antibodies, J Rheumatol 18 (1991) 1—3.

[24] M.C. Amigo, R. Garcia-Torres, M. Robles, T. Bochicchio, P.A. Reyes, Renal involvement in the primary antiphospholipid syndrome, J Rheumatol 18 (1992) 1181—1185.

[25] G.R.V. Hughes, Thrombosis, abortion, cerebral disease, and the lupus anticoagulant, Br Med J 287 (1983) 1088—1089.

[26] G.R.V. Hughes, The Prosser—White oration 1983. Connective tissue disease and the skin, Clin Exp Dermatol 9 (1984) 535—544.

[27] E. Toubi, M.A. Khamashta, A. Panarra, G.R. Hughes, Association of antiphospholipid antibodies with central nervous system disease in systemic lupus erythematosus, Am J Med 99 (1995) 397—401.

[28] E.N. Harris, A.E. Gharavi, R.A. Asherson, M.L. Boey, G.R.V. Hughes, Cerebral infarction in systemic lupus erythematosus: Association with anticardiolipin antibodies, Clin Exp Rheumatol 2 (1984) 47.

[29] R.A. Asherson, M.A. Khamashta, A. Gil, J.J. Vazquez, O. Chan, E. Baguley, et al., Cerebrovascular disease and antiphospholipid antibodies in systemic lupus erythematosus, lupus-like disease, and the primary antiphospholipid syndrome, Am J Med 86 (1989) 391—399.

[30] J. Montalban, A. Codina, J. Ordi, M. Vilardell, M.A. Khamashta, G.R.V. Hughes, Antiphospholipid antibodies in cerebral ischemia, Stroke 22 (1991) 750—753.

[31] P. Nencini, M.C. Baruffi, R. Abbate, G. Massai, L. Amaducci, D. Inzitari, Lupus anticoagulant and anticardiolipin antibodies in young adults with cerebral ischemia, Stroke 23 (1992) 189—193.

[32] R.T. Urbanus, B. Siegerink, M. Roest, F.R. Rosendaal, P.G. de Groot, A. Algra, Antiphospholipid antibodies and risk of myocardial infarction and ischaemic stroke in young women in the RATIO study: A case—control study, Lancet Neurol 8 (2009) 998—1005.

[33] A. Danowski, M.N. de Azevedo, J.A. de Souza Papi, M. Petri, Determinants of risk for venous and arterial thrombosis in primary antiphospholipid syndrome and in antiphospholipid syndrome with systemic lupus erythematosus, J Rheumatol 36 (2009) 1195—1199.

[34] S.R. Levine, L. Salowich-Palm, K.L. Sawaya, M. Perry, H.J. Spencer, H.J. Winkler, et al., IgG anticardiolipin antibody titer >40 GPL and the risk of subsequent thrombo-occlusive events and death. A prospective cohort study, Stroke 28 (1997) 1660—1665.

[35] R.A. Asherson, D. Mercey, G. Phillips, N. Sheehan, A.E. Gharavi, E.N. Harris, et al., Recurrent stroke and multi-infarct dementia in systemic lupus erythematosus: Association with antiphospholipid antibodies, Ann Rheum Dis 46 (1987) 605—611.

[36] S.R. Levine, R.L. Brey, K.L. Sawaya, L. Salowich-Palm, J. Kokkinos, B. Kostrzema, et al., Recurrent stroke and thrombo-occlusive events in the antiphospholipid syndrome, Ann Neurol 38 (1995) 119—124.

[37] S.R. Levine, M.J. Deegan, N. Futrell, K.M. Welch, Cerebrovascular and neurologic disease associated with antiphospholipid antibodies: 48 cases, Neurology 40 (1990) 1181—1189.

[38] M.A. Khamashta, R. Cervera, R.A. Asherson, J. Font, A. Gil, D.J. Coltart, et al., Association of antibodies against phospholipids with heart valve disease in systemic lupus erythematosus, Lancet 335 (1990) 1541—1544.

[39] R. Cervera, M.A. Khamashta, J. Font, P.A. Reyes, J.L. Vianna, A. Lopez-Soto, et al., High prevalence of significant heart valve lesions in patients with the "primary" antiphospholipid syndrome, Lupus 1 (1991) 43—47.

[40] E. Galve, J. Ordi, J. Barquinero, A. Evangelista, M. Vilardell, J. Soler-Soler, Valvular heart disease in the primary antiphospholipid syndrome, Ann Intern Med 116 (1992) 293—298.

[41] R. Garcia-Torres, M.C. Amigo, A. de la Rosa, A. Moron, P.A. Reyes, Valvular heart disease in primary antiphospholipid syndrome (PAPS): Clinical and morphological findings, Lupus 5 (1996) 56—61.

[42] M.A. Deschiens, J. Conard, M.H. Horellou, A. Ameri, M. Preter, F. Chedru, et al., Coagulation studies, factor V Leiden, and anticardiolipin antibodies in 40 cases of cerebral venous thrombosis, Stroke 27 (1996) 1724—1730.

[43] C. Castanon, M.C. Amigo, J.L. Banales, A. Nava, P.A. Reyes, Ocular vaso-occlusive disease in primary antiphospholipid syndrome, Ophthalmology 102 (1995) 256—262.

[44] J.P. Dunn, S.W. Noorily, M. Petri, D. Finkelstein, J.T. Rosenbaum, D.A. Jabs, Antiphospholipid antibodies and retinal vascular disease, Lupus 5 (1996) 313—322.

[45] B. Wiechens, J.O. Schroder, B. Potzsch, R. Rochels, Primary antiphospholipid antibody syndrome and retinal occlusive vasculopathy, Am J Ophthalmol 123 (1997) 848—850.

[46] E. Toubi, J. Ben-David, A. Kessel, L. Podoshin, T.D. Golan, Autoimmune aberration in sudden sensorineural hearing loss: Association with anticardiolipin antibodies, Lupus 6 (1997) 540—542.

[47] M. Naarendorp, H. Spiera, Sudden sensorineural hearing loss in patients with systemic lupus erythematosus or lupus-like syndromes and antiphospholipid antibodies, J Rheumatol 25 (1998) 589—592.

[48] M.J. Cuadrado, M.A. Khamashta, A. Ballesteros, T. Godfrey, M.J. Simon, G.R. Hughes, Can neurologic manifestations of Hughes (antiphospholipid) syndrome be distinguished from multiple sclerosis? Analysis of 27 patients and review of the literature, Medicine 79 (2000) 57—68.

[49] J.W. Ijdo, A.M. Conti-Kelly, P. Greco, M. Abedi, M. Amos, J.M. Provenzale, et al., Antiphospholipid antibodies in patients with multiple sclerosis and MS-like illnesses: MS or APS? Lupus 8 (1999) 109—115.

[50] T.F. Scott, D. Hess, J. Brillman, Antiphospholipid antibody syndrome mimicking multiple sclerosis clinically and by magnetic resonance imaging, Arch Intern Med 154 (1994) 917—920.

[51] D. Karussis, R.R. Leker, A. Ashkenazi, O. Abramsky, A subgroup of multiple sclerosis patients with anticardiolipin antibodies and unusual clinical manifestations: Do they represent a new nosological entity? Ann Neurol 44 (1998) 629—634.

[52] J. Sastre-Garriga, J.C. Reverter, J. Font, M. Tintore, G. Espinosa, X. Montalban, Anticardiolipin antibodies are not a useful screening tool in a nonselected large group of patients with multiple sclerosis, Ann Neurol 49 (2001) 408—411.

[53] G. Ruiz-Irastorza, M.A. Khamashta, Warfarin for multiple sclerosis? QJ Med 93 (2000) 497—499.

[54] G.E. Tietjen, M. Day, L. Norris, S. Aurora, A. Halvorsen, L.R. Schultz, et al., Role of anti-cardiolipin antibodies in young persons with migraine and transient focal neurologic events: A prospective study, Neurology 50 (1998) 1433–1440.

[55] J. Montalban, R. Cervera, J. Font, J. Ordi, J. Vianna, H.J. Haga, et al., Lack of association between anticardiolipin antibodies and migraine in systemic lupus erythematosus, Neurology 42 (1992) 681–682.

[56] G. Sanna, M.L. Bertolaccini, M.J. Cuadrado, H. Laing, M.A. Khamashta, A. Mathieu, et al., Neuropsychiatric manifestations in systemic lupus erythematosus: Prevalence and association with antiphospholipid antibodies, J Rheumatol 30 (2003) 985–992.

[57] M.J. Cuadrado, M.A. Khamashta, G.R. Hughes, Migraine and stroke in young women, QJ Med 93 (2000) 317–318.

[58] M.J. Cuadrado, M.A. Khamashta, D. D'Cruz, G.R. Hughes, Migraine in Hughes syndrome—Heparin as a therapeutic trial? QJ Med 94 (2001) 114–115.

[59] C. Tzourio, A. Tehindrazanarivelo, S. Iglesias, A. Alperovitch, F. Chedru, J. d'Anglejan-Chatillon, et al., Case—control study of migraine and risk of ischaemic stroke in young women, Br Med J 310 (1995) 830–833.

[60] A. Carolei, C. Marini, G. De Matteis, History of migraine and risk of cerebral ischaemia in young adults. The Italian National Research Council Study Group on Stroke in the Young, Lancet 347 (1996) 1503–1506.

[61] K.R. Merikangas, B.T. Fenton, S.H. Cheng, M.J. Stolar, N. Risch, Association between migraine and stroke in a large-scale epidemiological study of the United States, Arch Neurol 54 (1997) 362–368.

[62] C.L. Chang, M. Donaghy, N. Poulter, Migraine and stroke in young women: Case—control study. World Health Organization Collaborative Study of Cardiovascular Disease and Steroid Hormone Contraception, Br Med J 318 (1999) 13–18.

[63] B.M. Coull, S.H. Goodnight, Antiphospholipid antibodies, prethrombotic states, and stroke, Stroke 21 (1990) 1370–1374.

[64] S.D. Denburg, R.M. Carbotte, J.S. Ginsberg, J.A. Denburg, The relationship of antiphospholipid antibodies to cognitive function in patients with systemic lupus erythematosus, J Int Neuropsychol Soc 3 (1997) 377–386.

[65] J.G. Hanly, C. Hong, S. Smith, J.D. Fisk, A prospective analysis of cognitive function and anticardiolipin antibodies in systemic lupus erythematosus, Arthritis Rheum 42 (1999) 728–734.

[66] S. Menon, E. Jameson-Shortall, S.P. Newman, M.R. Hall-Craggs, R. Chinn, D.A. Isenberg, A longitudinal study of anticardiolipin antibody levels and cognitive functioning in systemic lupus erythematosus, Arthritis Rheum 42 (1999) 735–741.

[67] M.G. Tektonidou, N. Varsou, G. Kotoulas, A. Antoniou, H.M. Moutsopoulos, Cognitive deficits in patients with antiphospholipid syndrome: Association with clinical, laboratory, and brain magnetic resonance imaging findings, Arch Intern Med 166 (2006) 2278–2284.

[68] A. Mosek, I. Yust, T.A. Treves, N. Vardinon, A.D. Korczyn, J. Chapman, Dementia and antiphospholipid antibodies, Dement Geriatr Cogn Disord 11 (2000) 36–38.

[69] M. Schwartz, M. Rochas, B. Weller, A. Sheinkman, I. Tal, D. Golan, et al., High association of anticardiolipin antibodies with psychosis, J Clin Psychiatry 59 (1998) 20–23.

[70] M.T. Herranz, G. Rivier, M.A. Khamashta, K.U. Blaser, G.R. Hughes, Association between antiphospholipid antibodies and epilepsy in patients with systemic lupus erythematosus, Arthritis Rheum 37 (1994) 568–571.

[71] H.H. Liou, C.R. Wang, C.J. Chen, R.C. Chen, C.Y. Chuang, I.P. Chiang, et al., Elevated levels of anticardiolipin antibodies and epilepsy in lupus patients, Lupus 5 (1996) 307–312.

[72] Y. Shoenfeld, S. Lev, I. Blatt, M. Blank, J. Font, P. von Landenberg, et al., Features associated with epilepsy in the antiphospholipid syndrome, J Rheumatol 31 (2004) 1344–1348.

[73] J.T. Peltola, A. Haapala, J.I. Isojarvi, A. Auvinen, J. Palmio, K. Latvala, et al., Antiphospholipid and antinuclear antibodies in patients with epilepsy or new-onset seizure disorders, Am J Med 109 (2000) 712–717.

[74] H.H. Liou, C.R. Wang, H.C. Chou, V.L. Arvanov, R.C. Chen, Y.C. Chang, et al., Anticardiolipin antisera from lupus patients with seizures reduce a GABA receptor-mediated chloride current in snail neurons, Life Sci 54 (1994) 1119–1125.

[75] R.A. Asherson, G.R.V. Hughes, Antiphospholipid antibodies and chorea, J Rheumatol 15 (1988) 377–379.

[76] R. Cervera, R.A. Asherson, J. Font, M. Tikly, L. Pallares, A. Chamorro, et al., Chorea in the antiphospholipid syndrome: Clinical, radiologic, and immunologic characteristics of 50 patients from our clinics and recent literature, Medicine 76 (1997) 203–212.

[77] J. Sunden-Cullberg, J. Tedroff, S.M. Aquilonius, Reversible chorea in primary antiphospholipid syndrome, Mov Disord 13 (1998) 147–149.

[78] G.E. Gibson, W.P. Su, M.R. Pittelkow, Antiphospholipid syndrome and the skin, J Am Acad Dermatol 36 (1997) 970–982.

[79] I.W. Uthman, M.A. Khamashta, Livedo racemosa: A striking dermatological sign for the antiphospholipid syndrome, J. Rheumatol 33 (2006) 2379–2382.

[80] R.A. Asherson, S.C. Mayou, P. Merry, M.M. Black, G.R. Hughes, The spectrum of livedo reticularis and anti-cardiolipin antibodies, Br J Dermatol 120 (1989) 215–221.

[81] D. Alarcon-Segovia, M. Delezé, C.V. Oria, J. Sanchez-Guerrero, L. Gomez-Pacheco, J. Cabiedes, et al., Antiphospholipid antibodies and the antiphospholipid syndrome in systemic lupus erythematosus: A prospective analysis of 500 consecutive patients, Medicine 68 (1989) 353.

[82] H.J. Englert, S. Loizou, G.G. Derue, M.J. Walport, G.R. Hughes, Clinical and immunologic features of livedo reticularis in lupus: A case—control study, Am J Med 87 (1989) 408–410.

[83] C. Frances, J.C. Piette, The mystery of Sneddon syndrome: Relationship with antiphospholipid syndrome and systemic lupus erythematosus, J Autoimmun 15 (2000) 139–143.

[84] D.P. Goldberg, V.L. Lewis Jr., W.J. Koenig, Antiphospholipid antibody syndrome: A new cause of nonhealing skin ulcers, Plast Reconstr Surg 95 (1995) 837–841.

[85] P.V. Rocca, L.B. Siegel, T.R. Cupps, The concomitant expression of vasculitis and coagulopathy: Synergy for marked tissue ischemia, J Rheumatol 21 (1994) 556–560.

[86] M.A. Aguirre, A. Jurado, F. Mujic, M.J. Cuadrado, Oral anticoagulation therapy of chronic skin ulcers in a patient with primary antiphospholipid syndrome, Clin Exp Rheumatol 16 (1998) 628–629.

[87] E. Gertner, J.T. Lie, Systemic therapy with fibrinolytic agents and heparin for recalcitrant nonhealing cutaneous ulcer in the antiphospholipid syndrome, J Rheumatol 21 (1994) 2159–2161.

[88] F. Mujic, M. Lloyd, M.J. Cuadrado, M.A. Khamashta, G.R. Hughes, Prevalence and clinical significance of subungual splinter haemorrhages in patients with the antiphospholipid syndrome, Clin Exp Rheumatol 13 (1995) 327–331.

[89] C. Frances, J.C. Piette, V. Saada, T. Papo, B. Wechsler, O. Chosidow, et al., Multiple subungual splinter hemorrhages in the antiphospholipid syndrome: A report of five cases and review of the literature, Lupus 3 (1994) 123–128.

[90] D. Creamer, B.J. Hunt, M.M. Black, Widespread cutaneous necrosis occurring in association with the antiphospholipid syndrome: A report of two cases, Br J Dermatol 142 (2000) 1199–1203.

[91] S. Paira, S. Roverano, A. Zunino, M.E. Oliva, M.L. Bertolaccini, Extensive cutaneous necrosis associated with anticardiolipin antibodies, J Rheumatol 26 (1999) 1197–1200.

[92] L.F. Del Castillo, C. Soria, C. Schoendorff, C. Garcia Garcia, N. Diez-Caballero, A. Rodriguez Alen, et al., Widespread cutaneous necrosis and antiphospholipid antibodies: Two episodes related to surgical manipulation and urinary tract infection, J Am Acad Dermatol 36 (1997) 872–875.

[93] M. Lockshin, F. Tenedios, M. Petri, G. McCarty, R. Forastiero, S. Krilis, et al., Cardiac disease in the antiphospholipid syndrome: Recommendations for treatment. Committee consensus report, Lupus 12 (2003) 518–523.

[94] C.A. Roldan, B.K. Shively, C.C. Lau, F.T. Gurule, E.A. Smith, M.H. Crawford, Systemic lupus erythematosus valve disease by transesophageal echocardiography and the role of antiphospholipid antibodies, J Am Coll Cardiol 20 (1992) 1127–1134.

[95] R. Cervera, Recent advances in antiphospholipid antibody-related valvulopathies, J Autoimmun 15 (2000) 123–125.

[96] N. Espinola-Zavaleta, J. Vargas-Barron, T. Colmenares-Galvis, F. Cruz-Cruz, A. Romero-Cardenas, C. Keirns, et al., Echocardiographic evaluation of patients with primary antiphospholipid syndrome, Am Heart J 137 (1999) 973–978.

[97] O. Bouillanne, A. Millaire, P. de Groote, F. Puisieux, J.Y. Cesbron, B. Jude, et al., Prevalence and clinical significance of antiphospholipid antibodies in heart valve disease: A case–control study, Am Heart J 132 (1996) 790–795.

[98] A. Afek, Y. Shoenfeld, R. Manor, I. Goldberg, L. Ziporen, J. George, et al., Increased endothelial cell expression of alpha3beta1 integrin in cardiac valvulopathy in the primary (Hughes) and secondary antiphospholipid syndrome, Lupus 8 (1999) 502–507.

[99] M.A. Agirbasli, D.E. Hansen, B.F. Byrd III, Resolution of vegetations with anticoagulation after myocardial infarction in primary antiphospholipid syndrome, J Am Soc Echocardiogr 10 (1997) 877–880.

[100] A. Hamsten, R. Norberg, M. Bjorkholm, U. de Faire, G. Holm, Antibodies to cardiolipin in young survivors of myocardial infarction: An association with recurrent cardiovascular events, Lancet 1 (1986) 113–116.

[101] E. Zuckerman, E. Toubi, A. Shiran, E. Sabo, Z. Shmuel, T.D. Golan, et al., Anticardiolipin antibodies and acute myocardial infarction in non-systemic lupus erythematosus patients: A controlled prospective study, Am J Med 101 (1996) 381–386.

[102] T.P. Greco, A.M. Conti-Kelly, T. Greco Jr., R. Doyle, E. Matsuura, J.R. Anthony, et al., Newer antiphospholipid antibodies predict adverse outcomes in patients with acute coronary syndrome, Am J Clin Pathol 132 (2009) 613–620.

[103] Y. Adler, Y. Finkelstein, G. Zandeman-Goddard, M. Blank, M. Lorber, A. Lorber, et al., The presence of antiphospholipid antibodies in acute myocardial infarction, Lupus 4 (1995) 309–313.

[104] R. Wu, S. Nityanand, L. Berglund, H. Lithell, G. Holm, A.K. Lefvert, Antibodies against cardiolipin and oxidatively modified LDL in 50-year-old men predict myocardial infarction, Arterioscler Thromb Vasc Biol 17 (1997) 3159–3163.

[105] S. Manzi, E.N. Meilahn, J.E. Rairie, C.G. Conte, T.A. Medsger Jr., L. Jansen-McWilliams, et al., Age-specific incidence rates of myocardial infarction and angina in women with systemic lupus erythematosus: Comparison with the Framingham study, Am J Epidemiol 145 (1997) 408–415.

[106] O. Vaarala, M. Puurunen, M. Lukka, G. Alfthan, M. Leirisalo-Repo, K. Aho, et al., Affinity-purified cardiolipin-binding antibodies show heterogeneity in their binding to oxidized low-density lipoprotein, Clin Exp Immunol 104 (1996) 269–274.

[107] M. Puurunen, M. Manttari, V. Manninen, L. Tenkanen, G. Alfthan, C. Ehnholm, et al., Antibody against oxidized low-density lipoprotein predicting myocardial infarction, Arch Intern Med 154 (1994) 2605–2609.

[108] A.T. Erkkila, O. Narvanen, S. Lehto, M.I. Uusitupa, S. Yla-Herttuala, Autoantibodies against oxidized low-density lipoprotein and cardiolipin in patients with coronary heart disease, Arterioscler Thromb Vasc Biol 20 (2000) 204–209.

[109] C.M. Nzerue, K. Hewan-Lowe, S. Pierangeli, E.N. Harris, "Black swan in the kidney": Renal involvement in the antiphospholipid antibody syndrome, Kidney Int 62 (2002) 733–744.

[110] S.R. Sangle, D.P. D'Cruz, W. Jan, M.Y. Karim, M.A. Khamashta, I.C. Abbs, et al., Renal artery stenosis in the antiphospholipid (Hughes) syndrome and hypertension, Ann Rheum Dis 62 (2003) 999–1002.

[111] D. Nochy, E. Daugas, D. Droz, H. Beaufils, J.P. Grunfeld, J.C. Piette, et al., The intrarenal vascular lesions associated with primary antiphospholipid syndrome, J Am Soc Nephrol 10 (1999) 507–518.

[112] G.I. Remondino, E. Mysler, M.N. Pissano, M.C. Furattini, M.C. Basta, J.L. Presas, et al., A reversible bilateral renal artery stenosis in association with antiphospholipid syndrome, Lupus 9 (2000) 65–67.

[113] P. Brunet, M.F. Aillaud, M. San Marco, C. Philip-Joet, B. Dussol, D. Bernard, et al., Antiphospholipids in hemodialysis patients: Relationship between lupus anticoagulant and thrombosis, Kidney Int 48 (1995) 794–800.

[114] F. Fabrizi, R. Sangiorgio, G. Pontoriero, M. Corti, F. Tentori, E. Troina, et al., Antiphospholipid (aPL) antibodies in end-stage renal disease, J Nephrol 12 (1999) 89–94.

[115] J. Radhakrishnan, G.S. Williams, G.B. Appel, D.J. Cohen, Renal transplantation in anticardiolipin antibody-positive lupus erythematosus patients, Am J Kidney Dis 23 (1994) 286–289.

[116] J.H. Stone, W.J. Amend, L.A. Criswell, Antiphospholipid antibody syndrome in renal transplantation: Occurrence of clinical events in 96 consecutive patients with systemic lupus erythematosus, Am J Kidney Dis 34 (1999) 1040–1047.

[117] S. Vaidya, R. Sellers, P. Kimball, T. Shanahan, J. Gitomer, K. Gugliuzza, et al., Frequency, potential risk and therapeutic intervention in end-stage renal disease patients with antiphospholipid antibody syndrome: A multicenter study, Transplantation 69 (2000) 1348–1352.

[118] G. Mondragon-Ramirez, T. Bochicchio, R. Garcia-Torres, M.C. Amigo, M. Martinez-Lavin, P. Reyes, et al., Recurrent renal thrombotic angiopathy after kidney transplantation in two patients with primary antiphospholipid syndrome (PAPS), Clin Transplant 8 (1994) 93–96.

[119] J.A. McIntyre, D.R. Wagenknecht, Antiphospholipid antibodies. Risk assessments for solid organ, bone marrow, and tissue transplantation, Rheum Dis Clin North Am 27 (2001) 611–631.

[120] M.C. Amigo, R. Garcia-Torres, Kidney disease in antiphospholipid syndrome, in: M.A. Khamashta (Ed.), Hughes Syndrome. Antiphospholipid Syndrome, Springer-Verlag, London, 2000, pp. 70–81.

[121] E. Daugas, D. Nochy, L.T. Huong du, P. Duhaut, H. Beaufils, V. Caudwell, et al., Antiphospholipid syndrome nephropathy

in systemic lupus erythematosus, J Am Soc Nephrol 13 (2002) 42−52.

[122] G. Espinosa, R. Cervera, J. Font, R.A. Asherson, The lung in the antiphospholipid syndrome, Ann Rheum Dis 61 (2002) 195−198.

[123] R.A. Asherson, T.W. Higenbottam, A.T. Dinh Xuan, M.A. Khamashta, G.R. Hughes, Pulmonary hypertension in a lupus clinic: Experience with twenty-four patients, J Rheumatol 17 (1990) 1292−1298.

[124] E. Gertner, Diffuse alveolar hemorrhage in the antiphospholipid syndrome: Spectrum of disease and treatment, J Rheumatol 26 (1999) 805−807.

[125] M.J. Cuadrado, F. Mujic, E. Munoz, M.A. Khamashta, G.R. Hughes, Thrombocytopenia in the antiphospholipid syndrome, Ann Rheum Dis 56 (1997) 194−196.

[126] I. Uthman, B. Godeau, A. Taher, M. Khamashta, The hematologic manifestations of the antiphospholipid syndrome, Blood Rev 22 (2008) 187−194.

[127] M. Galli, M. Daldossi, T. Barbui, Anti-glycoprotein Ib/IX and IIb/IIIa antibodies in patients with antiphospholipid antibodies, Thromb Haemost 71 (1994) 571−575.

[128] B. Godeau, J.C. Piette, P. Fromont, L. Intrator, A. Schaeffer, P. Bierling, Specific antiplatelet glycoprotein autoantibodies are associated with the thrombocytopenia of primary antiphospholipid syndrome, Br J Haematol 98 (1997) 873−879.

[129] S. Panzer, M.E. Gschwandtner, D. Hutter, S. Spitzauer, I. Pabinger, Specificities of platelet autoantibodies in patients with lupus anticoagulants in primary antiphospholipid syndrome, Ann Hematol 74 (1997) 239−242.

[130] K.Y. Fong, S. Loizou, M.L. Boey, M.J. Walport, Anticardiolipin antibodies, haemolytic anaemia and thrombocytopenia in systemic lupus erythematosus, Br J Rheumatol 31 (1992) 453−455.

[131] S.I. Kokori, J.P. Ioannidis, M. Voulgarelis, A.G. Tzioufas, H.M. Moutsopoulos, Autoimmune hemolytic anemia in patients with systemic lupus erythematosus, Am J Med 108 (2000) 198−204.

[132] G. Ruiz-Irastorza, M.A. Khamashta, Antiphospholipid syndrome in pregnancy, Rheum Dis Clin North Am 33 (2007) 287−297.

[133] A. Lynch, R. Marlar, J. Murphy, G. Davila, M. Santos, J. Rutledge, et al., Antiphospholipid antibodies in predicting adverse pregnancy outcome. A prospective study, Ann Intern Med 120 (1994) 470−475.

[134] K. Bramham, B.J. Hunt, S. Germain, I. Calatayud, M. Khamashta, S. Bewley, et al., Pregnancy outcome in different clinical phenotypes of antiphospholipid syndrome, Lupus 19 (2010) 58−64.

[135] G.A. Dekker, J.I. de Vries, P.M. Doelitzsch, P.C. Huijgens, B.M. von Blomberg, C. Jakobs, et al., Underlying disorders associated with severe early-onset preeclampsia, Am J Obstet Gynecol 173 (1995) 1042−1048.

[136] K. Duckitt, D. Harrington, Risk factors for pre-eclampsia at antenatal booking: Systematic review of controlled studies, Br Med J 330 (2005) 565−567.

[137] D.W. Branch, R.M. Silver, J.L. Blackwell, J.C. Reading, J.R. Scott, Outcome of treated pregnancies in women with antiphospholipid syndrome: An update of the Utah experience, Obstet Gynecol 80 (1992) 614−620.

[138] M. Backos, R. Rai, N. Baxter, I.T. Chilcott, H. Cohen, L. Regan, Pregnancy complications in women with recurrent miscarriage associated with antiphospholipid antibodies treated with low dose aspirin and heparin, Br J Obstet Gynaecol 106 (1999) 102−107.

[139] R.G. Farquharson, S. Quenby, M. Greaves, Antiphospholipid syndrome in pregnancy: A randomized, controlled trial of treatment, Obstet Gynecol 100 (2002) 408−413.

[140] F. Lima, M.A. Khamashta, N.M. Buchanan, S. Kerslake, B.J. Hunt, G.R. Hughes, A study of sixty pregnancies in patients with the antiphospholipid syndrome, Clin Exp Rheumatol 14 (1996) 131−136.

[141] A. Caruso, S. De Carolis, S. Ferrazzani, G. Valesini, L. Caforio, S. Mancuso, Pregnancy outcome in relation to uterine artery flow velocity waveforms and clinical characteristics in women with antiphospholipid syndrome, Obstet Gynecol 82 (1993) 970−977.

[142] D. Le Thi Huong, B. Wechsler, D. Vauthier-Brouzes, P. Duhaut, N. Costedoat, M.R. Andreu, et al., The second trimester Doppler ultrasound examination is the best predictor of late pregnancy outcome in systemic lupus erythematosus and/or the antiphospholipid syndrome, Rheumatology 45 (2006) 332−338.

[143] M.A. Coleman, L.M. McCowan, R.A. North, Mid-trimester uterine artery Doppler screening as a predictor of adverse pregnancy outcome in high-risk women, Ultrasound Obstet Gynecol 15 (2000) 7−12.

[144] K. Harrington, D. Cooper, C. Lees, K. Hecher, S. Campbell, Doppler ultrasound of the uterine arteries: The importance of bilateral notching in the prediction of pre-eclampsia, placental abruption or delivery of a small-for-gestational-age baby, Ultrasound Obstet Gynecol 7 (1996) 182−188.

[145] R.A. Asherson, R. Cervera, P.G. de Groot, D. Erkan, M.C. Boffa, J.C. Piette, et al., Catastrophic antiphospholipid syndrome: International consensus statement on classification criteria and treatment guidelines, Lupus 12 (2003) 530−534.

[146] R.A. Asherson, F. Liote, B. Page, O. Meyer, N. Buchanan, M.A. Khamashta, et al., Avascular necrosis of bone and antiphospholipid antibodies in systemic lupus erythematosus, J Rheumatol 20 (1993) 284−288.

[147] S. Sangle, D.P. D'Cruz, M.A. Khamashta, G.R. Hughes, Antiphospholipid antibodies, systemic lupus erythematosus, and non-traumatic metatarsal fractures, Ann Rheum Dis 63 (2004) 1241−1243.

[148] J. Moore, D.D. Ma, A. Concannon, Nonmalignant bone marrow necrosis: A report of two cases, Pathology 30 (1998) 318−320.

[149] S. Paydas, R. Kocak, S. Zorludemir, F. Baslamisli, Bone marrow necrosis in antiphospholipid syndrome, J Clin Pathol 50 (1997) 261−262.

[150] J. George, D. Haratz, Y. Shoenfeld, Accelerated atheroma, antiphospholipid antibodies, and the antiphospholipid syndrome, Rheum Dis Clin North Am 27 (2001) 603−610.

[151] S. Horkko, T. Olee, L. Mo, D.W. Branch, V.L.J. Woods, W. Palinski, et al., Anticardiolipin antibodies from patients with the antiphospholipid antibody syndrome recognize epitopes in both beta(2)-glycoprotein 1 and oxidized low-density lipoprotein, Circulation 103 (2001) 941−946.

[152] Practice Committee of American Society for Reproductive Medicine, Anti-phospholipid antibodies do not affect IVF success, Fertil Steril 90 (5 Suppl.) (2008) S172−S173.

Pregnancy and Antiphospholipid Syndrome

Kristina E. Milan, D. Ware Branch

INTRODUCTION

Clinicians first recognized that thrombosis is associated with antiphospholipid antibodies (aPLs) more than 50 years ago. In the 1970s, pregnancy loss was also noted to be associated with lupus anticoagulant. By the mid-1980s, there was growing excitement about the possibility that fetal loss attributable to aPLs was treatable [1]. Some experts noted that fetal loss associated with aPLs was often seen in the setting of preeclampsia or severe placental insufficiency. Others found aPLs similarly linked to the problem of recurrent early miscarriage [2]. By the late 1990s, international experts formulated consensus criteria for the diagnosis of definite antiphospholipid syndrome (APS) that included explicit obstetric criteria to include the spectrum of recurrent preembryonic and embryonic pregnancy loss as well as severe preeclampsia or placental insufficiency manifested as fetal death or intrauterine growth restriction requiring delivery prior to 34 weeks of gestation [3]. The classification criteria have since been further refined [4]. They include both the clinical and the laboratory criteria necessary for the diagnosis of APS (Table 54.1).

The association between aPLs and thrombosis is firmly established, and the high recurrence risk for thrombosis warrants consideration of long-term anticoagulation. However, the mechanism of obstetric complications, the security of the various obstetric clinical associations, and best treatments are subjects of ongoing study. In addition, controversies regarding clinical associations and treatments persist. This chapter provides an overview of our current understanding of obstetric APS and concepts regarding treatment.

MECHANISM OF ANTIPHOSPHOLIPID SYNDROME-RELATED OBSTETRIC DISEASE

An aPL-mediated cause of pregnancy loss has been strongly suggested in murine models using passive immunization with human aPL [5]. In addition, circulating human and mouse aPLs are associated with larger and more persistent experimentally induced thrombi than those in mice treated with control antibodies [6]. Evidence suggests that antiphospholipids do not interact with phospholipids directly but, rather, indirectly through phospholipid-binding plasma proteins. The plasma proteins thought to be most relevant to APS are the β_2-glycoprotein I proteins. Specifically, some investigators have found that autoantibodies that bind to a specific epitope on domain 1 of β_2-gylcoprotein I correlate best with thrombosis, a primary feature of APS [7]. It should be noted, however, that this is an area of active research. In a meta-analysis of women with aPLs and fetal loss, lupus anticoagulant was found to best correlate with recurrent fetal loss, and a similar correlation was not found for β_2-glycoprotein I antibodies [8]. Therefore, it remains unclear whether domain I-specific autoantibodies correlate specifically with fetal loss.

Research on the pathogenesis of APS describes the importance of monocytes, platelets, endothelial cells, and complement in inducing thrombosis and adverse pregnancy outcomes. Monocytes and endothelial cells incubated with aPLs and β_2-gylcoprotein I may be activated through toll-like receptor (TLR) signaling, specifically TLR-4 [9]. Activation of monocytes further induces expression of intercellular cell adhesion modecule-1, vascular cell adhesion molecule-1, and E-selectin, resulting in further activation of monocytes [10]. Importantly,

TABLE 54.1   Criteria for the Diagnosis of Antiphospholipid Antibody Syndrome

APS is present if at least one of the clinical criteria and one of the laboratory criteria are met.

CLINICAL CRITERIA

Vascular Thrombosis

One or more clinical episodes of arterial, venous, or small vessel thrombosis, occurring in any tissue or organ. Thrombosis must be confirmed by validated criteria such as imaging findings or histopathology. For histopathologic confirmation, thrombosis should be present without significant inflammation in the vessel wall.

Pregnancy Morbidity[a]

a. One or more unexplained deaths of a morphologically normal fetus at or beyond the 10th week of gestation. Normal morphology may be confirmed by normal ultrasonographic examination and/or gross examination of the tissue, *or*

b. One of more premature births of a morphologically normal neonate before the 34th week of gestation secondary to:

   i. Severe preeclampsia or eclampsia
   ii. Recognized features of placental insufficiency,[b] *or*

c. Three or more unexplained consecutive spontaneous abortions before the 10th week of gestation, with maternal anatomic or hormonal etiologies as well as parental chromosomal causes excluded.

LABORATORY CRITERIA

Lupus Anticoagulant

Lupus anticoagulant (LA) present in plasma on two or more occasions, at least 12 weeks apart, detected according to the guidelines of the International Society on Thrombosis and Hemostasis in the following steps.

a. Prolonged phospholipids-dependent coagulation demonstrated on a screening test such as activated partial thromboplastin time, kaolin clotting time, dilute Russell's viper venom time, dilute prothrombin time, or textarin time

b. Failure to correct the prolonged coagulation time on the screening test by mixing with normal, platelet-poor plasma

c. Shortening or correction of prolonged coagulation time on the screening test by the addition of excess phospholipid

d. Exclusion of other coagulopathies

Anticardiolipin Antibody

Must be of IgG and/or IgM isotype in serum or plasma, present in medium or high titer[c] (>40 GPL or MPL, or >99th percentile), on two or more occasions, at least 12 weeks apart, measured with standard ELISA.

Anti-β_2-Glycoprotein I Antibody

Must be of IgG and/or IgM isotype in serum or plasma, with titer >99th percentile, present on two or more occasions, at least 12 weeks apart, measured by standard ELISA.

[a]*In patients with more than one type of pregnancy morbidity, investigators are strongly encouraged to stratify groups based on a, b, and c.*
[b]*Generally accepted features of placental insufficiency include oligohydramnios, elevated umbilical artery Doppler flow velicimetry waveform analysis for gestational age, non-reassuring fetal surveillance testing including nonreactive non-stress test suggestive of fetal hypoxemia, or postnatal birth weight <10th percentile.*
[c]*The threshold used to distinguish medium- or high-titer from low-titer aCL antibodies has not been standardized. Many laboratories use 15 or 20 international phospholipid units as the threshold separating low from medium and high titers. Others define the threshold as 2—2.5 times the median titer of aCL antibodies or as the 99th percentile of aCL antibody titers within a normal population. The current international consensus suggests more that 40 GPL or MPL units or greater than the 99th percentile.*
Source: Miyakis et al. [4].

the activation of monocytes and endothelial cells upregulates the production of tissue factor (TF), which is the most potent *in vivo* inductor of the coagulation cascade [11]. Platelets are also activated by aPLs, resulting in increased synthesis of thromboxane A$_2$, a potent proaggregant and mediator of vasoconstriction.

Despite the seemingly direct correlation with thrombosis, contemporary work emphasizes inflammation, and specifically the complement system, as playing a major role in fetal loss associated with aPLs. C3 activation is a primary event leading to fetal loss in a mouse model of APS [12, 13]. Not only does blocking the complement cascade prevent the aPL-induced pregnancy loss but also mice deficient in either C3 or C5 are resistant

to aPL-induced pregnancy loss and thrombosis after passive administration of aPLs [14]. Placental studies in patients with aPLs have demonstrated placental deposition of C4d and C3d fractions, thereby suggesting local complement activation [15]. Activated complement has also been shown to further potentiate the activation of TF; in addition, mice deficient in the downstream receptor target of TF, protease-activated receptor 2 (PAR-2), did not demonstrate pregnancy failure despite immunization with aPLs [16].

The beneficial effect of heparin in preventing fetal loss in animal models seems likely to be related, at least in part, to the capacity of heparin to block activation of complement rather than via its anticoagulant action.

Girardi *et al.* [17] have shown that antithrombotic doses of hirudin and fondaparinux, both known to have no effect on complement activation, are ineffective in preventing aPL-induced fetal loss in the same murine model.

Thus, a current model for aPL-mediated clinical sequelae involves aPLs binding β_2-glycoprotein I on relevant cell membranes and the subsequent activation of endothelial cells, monocytes, and platelets. This in turn leads to the upregulation of cellular adhesion and prothrombotic molecules. Local complement activation occurs, increasing levels of C3a/C5a/C5b membrane attack complex and other products of the complement cascade to further potentiate endothelial, monocyte, and platelet activation. In addition, interaction of aPLs with proteins implicated in clotting regulation may hinder inactivation of procoagulant activated factors and may also impede fibrinolysis, further promoting a procoagulant state [18]. The proposed result of this local inflammation, cellular activation, and thrombosis is clinically expressed obstetrically by poor placentation, with inadequate trophoblast invasion and subsequent spiral arteriolar vasculopathy (Figure 54.1).

DIAGNOSTIC APPROACH TO ANTIPHOSPHOLIPID SYNDROME

The diagnosis of APS is first and foremost clinical. The patient must manifest at least one of two clinical criteria—vascular thrombosis or pregnancy morbidity—and at least one of the following three laboratory criteria: a positive lupus anticoagulant, medium- to high-titer immunoglobulin IgG or IgM isotype anticardiolipin antibodies, or medium- to high-titer anti-β_2-glycoprotein I IgG, or IgM isotype antibodies. The international consensus statements on preliminary classification criteria for definite APS provide simplified criteria (see Table 54.1) [3]. In 2006, the classification criteria were updated with several important changes. The pertinent differences between the original and revised criteria are as follows:

- The inclusion of anti-β_2-glycoprotein I IgG and IgM antibodies as diagnostic of APS. Lupus anticoagulant and anticardiolipin antibodies were retained as pertinent aPLs.
- Recognition that the threshold used to distinguish medium levels of anticardiolipin and anti-β_2-glycoprotein I antibodies from low levels has no standard; thus, a positive test for either of these antibodies should be >40 GPL or MPL units or >99th percentile.
- Repeat testing for an initially positive test for aPLs, which is required for the diagnosis of the syndrome, should be performed ≥12 weeks after the initial clinical manifestation and positive test(s).
- The distinction between "primary" and "secondary" was abandoned because of the lack of differences in clinical consequences between the two categories.

Clinical Criteria

Thrombosis as a manifestation of APS is discussed elsewhere in this book. The obstetric features of APS include either: (1) death of the conceptus manifest as

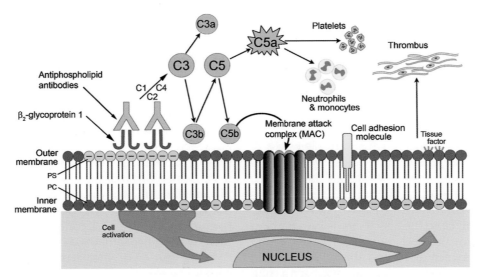

FIGURE 54.1 Mechanistic approach to antiphospholipid antibodies (aPLs). aPLs binds to a specific epitope of β_2-gylcoprotein I on relevant cell membranes. This subsequently activates endothelial cells, monocytes, and platelets and leads to upregulation of cellular adhesion and prothrombotic molecules such as tissue factor. Local complement activation potentiates further activation of endothelial cells, monocytes, and platelets. This is expressed clinically by poor placentation and inadequate trophoblast invasion. *Source: Adapted with permission of Lippincott Williams & Wilkins from Branch and Khamashta [25].*

recurrent early pregnancy loss (three or more consecutive preembryonic or embryonic miscarriages), (2) one or more fetal deaths (death *in utero* >10 weeks of gestation), or (3) severe preeclampsia or evidence of placental insufficiency requiring delivery prior to 34 weeks of gestation. Placental insufficiency is diagnosed by the presence of intrauterine growth restriction, oligohydramnios, or nonreassuring fetal surveillance.

Laboratory Criteria

The most commonly detected aPLs (and the only three currently recognized by the 2006 international consensus statement) include lupus anticoagulant, anticardiolipin antibodies, and anti-β_2-glycoprotein I antibodies.

Lupus anticoagulant (LA) correlates best with thrombosis when compared with anticardiolipin antibodies [19]. It is detected in platelet-poor plasma and is so named for its prolongation of phospholipid-dependent clotting tests such as the activated partial thromboplastin time (aPTT). Screening is usually performed using a sensitive aPTT and the dilute Russell's viper venom time. Prolongation of these tests may be due to plasma factor deficiencies, anticoagulant medications, or factor-specific inhibitors; thus, when a screening test is positive, subsequent confirmatory tests are done to exclude other possible etiologies. These typically include: (1) mixing the patient's plasma with normal (factor-sufficient) plasma to test for factor deficiencies as a cause of prolonged clotting times, and (2) phospholipid readdition tests such as the platelet neutralization or hexagonal phase phospholipid neutralization procedures. The final report should indicate that LA is either "present" or "absent." Rarely, assays that discriminate between LA and anticoagulant activity caused by other, rare antibodies are required [20]. In patients with a positive LA who also test positive for anti-β_2-glycoprotein I antibody, correlation with thrombosis is dramatically increased compared with the two separate assays [21]. Because transient positive results can occur, a positive LA (or other positive aPL) must be repeated after at least 12 weeks.

Anticardiolipin antibodies (aCLs) are detected in serum using enzyme-linked immunosorbent assays (ELISAs). Both IgG and IgM isotypes are included in the international diagnostic criteria [4]. Most experts believe that the sensitivity of aCLs for APS is greater than that of LA. However, the specificity is low, particularly for IgM-only results, and some studies have found poor correlation between aCLs and thrombosis [19]. In addition, standardization for aCL assays has been a challenge [22], resulting in poor interlaboratory concordance. Standard reference aCL reagents are available and should be used. Results are then expressed in units of GPL for IgG and MPL for IgM. Because of the

quantitative variation, final results are best interpreted is semiquantitative terms, such as negative, low, medium, or high titer. In general, results >40 GPL or MPL units (>99th percentile) are considered medium or high positive and potentially clinically relevant.

Anti-β_2-glycoprotein I antibodies are also detected in serum via ELISA, and both IgG and IgM isotypes are included in the international diagnostic criteria [4]. As with aCLs, results are determined using standard reference sera, with potentially relevant results being >99th percentile for the laboratory or >40 standard units for IgG or IgM. Initial excitement regarding possible high sensitivity and specificity of this assay for APS has been tempered by studies suggesting a poor correlation with thrombosis [23]. In our experience, medium-positive IgM results are particularly problematic. Others have also observed that not all antibodies against β_2-glycoprotein I are pathologic [7].

For both aCLs and anti-β_2-glycoprotein I antibodies, the specificity for APS increases with increasing antibody titer. Low-positive antibody results can be found in up to 5% of normal individuals and up to 35% of patients with systemic lupus erythematosus (SLE) [24]; therefore, results less than 40 international units should be viewed with caution. The specificity is also increased when the titer for aCLs or anti-β_2-glycoprotein I IgG is elevated versus the IgM isotype.

With the strict clinical and laboratory requirements devised with the purpose of research standardization, there are certainly some patients who will meet one or more of the clinical or laboratory features without meeting full diagnostic criteria for APS. In such cases, sound clinical judgment will be required to adjudicate the risk—benefit of potential treatment.

POTENTIAL COMPLICATIONS OF ANTIPHOSPHOLIPID SYNDROME IN PREGNANCY

The potential complications in women with APS accepted as attributable or strongly associated with APS include fetal death at or beyond 10 weeks of gestation, recurrent preembryonic or embryonic pregnancy loss, maternal and fetal risks of preeclampsia or placental insufficiency potentially necessitating preterm delivery, as well as the risk of maternal thrombosis (including stroke), and complications related to treatment with anticoagulants.

Fetal loss (>10 menstrual weeks of gestation) is certainly one hallmark of APS; at least 40% of pregnancy losses reported by women with LA or medium- to high-positive IgG aCL occur in the fetal period [25]. APS-related pregnancy loss has also been extended to include women with early recurrent pregnancy loss, including those occurring in the preembryonic (<6 menstrual weeks of gestation) and embryonic periods (6–9 menstrual weeks of gestation) [3]. In serologic evaluation of women with recurrent pregnancy loss, 10–20% have detectable aPLs [26–29].

The rate of gestational hypertension/preeclampsia and/or evidence of uteroplacental insufficiency in pregnancies complicated by APS may be as high as 50% in women with SLE, prior thrombosis, and other concurrent medical conditions [24, 27, 30–32], and preeclampsia may develop as early as 15–17 weeks of gestation. In contrast, women presenting with recurrent preembryonic or embryonic miscarriage and without significant medical histories have relatively low rates of adverse second- or third-trimester outcomes [33–37].

In contrast to the increased rates of preeclampsia observed in some case series of women previously diagnosed with APS, aPLs are not found in a statistically significant proportion of a general obstetric population presenting with preeclampsia [38]. Similarly, aPLs are not significantly increased in women at moderate risk (previous history of preeclampsia or underlying hypertension) for developing preeclampsia [39]. However, two groups of investigators have reported that women with early onset, severe preeclampsia are more likely to test positive for aPLs than are healthy controls [40, 41].

Placental insufficiency may be manifest clinically as oligohydramnios, fetal growth restriction, and/or abnormal antepartum fetal surveillance. Placental insufficiency is often seen concomitantly with gestational hypertension/preeclampsia, particularly in women with SLE, prior thrombosis, and other medical conditions [27, 32]. Some investigators have found relatively high rates of fetal growth restriction in association with aPLs [42] and in women with thrombotic APS [43]. Even with currently used treatment protocols, the rate of fetal growth restriction approaches 30% in some series [1, 32].

As part of the clinical spectrum of placental insufficiency, pregnancies complicated by APS are also more likely to exhibit nonreassuring fetal heart rate patterns during intrapartum monitoring and antenatal testing of fetal well-being, such as biophysical profiles and external fetal monitoring [1, 32]. Not surprisingly, the rate of preterm birth in these series ranges from 32 to 65% [1, 31, 32].

Venous thrombotic events that have been associated with aPLs include deep venous thrombosis and pulmonary embolus. The most common arterial events include transient ischemic attacks and cerebral vascular accidents. A meta-analysis of studies in the general population found an association with both arterial and venous thromboses in patients with LA (odds ratio, 8.6–10.8 and 4.1–16.2, respectively) [19]. Seventy percent of documented thrombotic events in patients with APS occur within the venous system.

The potential complications from heparin therapy during pregnancy include hemorrhage, osteoporosis, and heparin-induced thrombocytopenia. Clinically significant hemorrhage is rare, and it is usually the result of an underlying obstetric etiology such as abruption of the placenta. Despite relatively low rates of osteoporosis associated with heparin therapy (likely <1%), the rate may be higher in women with underlying autoimmune disorders who have required corticosteroid treatment. In addition to prenatal vitamins, it is reasonable to recommend calcium supplementation on the order of 1500–2000 mg of calcium carbonate daily. Heparin-induced thrombocytopenia (HIT) is a rare but serious and potentially lethal complication of heparin therapy. Rates of HIT outside of pregnancy are approximately 3%; however, there are insufficient data to provide reasonable estimates of the incidence of HIT during pregnancy [44]. Because of the potential risk, it is recommended for obstetric patients upon taking any dose of unfractionated heparin (UFH) or full anticoagulation doses of low-molecular-weight heparin (LMWH) to be assessed for thrombocytopenia just prior to the initiation of treatment and then every 2 or 3 days from Days 4 to 14 [45].

THERAPEUTIC APPROACH TO ANTIPHOSPHOLIPID SYNDROME IN PREGNANCY

The goals of therapeutic treatment of APS during pregnancy should maximize maternal–fetal outcomes by minimizing the possibility of adverse pregnancy outcomes, such as pregnancy loss, preeclampsia, placental insufficiency, and iatrogenic preterm birth. In addition, therapeutic goals should reduce or eliminate the risk of maternal thrombosis during pregnancy and in the initial postpartum period.

One of the first treatments used for APS in pregnancy was oral corticosteroids. Despite its relatively widespread use in APS in the 1980s, use of this treatment in pregnancy waned after a randomized trial showed heparin to be an equivalent with regard to overall outcomes and less likely than prednisone to be associated with early preterm birth and hypertension [33]. Likewise, intravenous immune globulin (IVIG) during pregnancy has been the subject of three randomized trials [46–48]. These have shown that the addition of IVIG to heparin or IVIG alone offer no better outcomes than heparin and low-dose aspirin or prednisone and low-dose aspirin. A subsequent Cochrane analysis concluded that IVIG was associated with an increased risk of pregnancy loss or premature birth compared to heparin and low-dose aspirin [49]. Thus, IVIG is not recommended in the treatment of APS in pregnancy.

Heparin and low-dose aspirin are now established as the mainstays of obstetric APS therapy, and currently recommended regimens are based on patient history and are reviewed here as such.

Antiphospholipid Syndrome without Prior Thrombosis

APS patients without a prior thrombotic event may be categorized into one of two groups for the purpose of treatment: (1) those with recurrent early (preembryonic or embryonic) miscarriage and no other features of APS or (2) those with one or more prior fetal deaths (>10 weeks of gestation) or prior early delivery (<34 weeks of gestation) due to severe preeclampsia or placental insufficiency. A third group of patients is composed of those with a history of thrombosis, irrespective of pregnancy history.

Recurrent Early (Preembryonic or Embryonic) Miscarriage

Several studies have reported 79—100% pregnancy success rates with low-dose aspirin alone [31, 32, 36, 50—52]. In contrast, two studies have reported no significant reduction in miscarriage with aspirin alone compared with placebo [36, 53]. The combination of UFH and low-dose aspirin has been shown to decrease rates of miscarriage in patients with APS with predominantly recurrent early miscarriage [35, 37, 49, 54]. Two other randomized trials comparing LMWH and aspirin with aspirin alone demonstrated this combination to be of little benefit [34, 55]. Conversely, two pilot studies suggest that LMWH with aspirin may be just as effective in preventing recurrent pregnancy loss as UFH and aspirin [56, 57].

Despite the potentially contradictory evidence, a 2005 Cochrane systematic review concluded that management of women with recurrent miscarriage and APS should consist of a combination of heparin 5000 U subcutaneously b.i.d. and low-dose aspirin [49], which is currently in line with expert opinion [58].

In pregnancies that progress beyond 20 weeks of gestation, the risk of fetal death, preeclampsia, severe placental insufficiency, and iatrogenic preterm birth is not thought to be significantly different from that of the general obstetric population. There is evidence, however, that women at high risk of preeclampsia treated with antiaggregant medications such as aspirin have a significant reduction in pregnancy complications [59].

Fetal Death (>10 Weeks of Gestation) or Prior Early Delivery (<34 Weeks of Gestation) Due to Severe Preeclampsia or Placental Insufficiency

The optimal treatment for women in this category is not defined by randomized trials. Case series suggest that these women are at risk of thrombosis [60]. Favorable pregnancy outcomes were seen when therapy was initiated in the early first trimester after documentation of a live intrauterine pregnancy. Most experts recommend low-dose aspirin and either prophylactic or intermediate-dose heparin (Table 54.2) [25, 49].

Antiphospholipid Syndrome with Thrombosis

As previously reviewed, pregnant women with APS and previous history of thrombosis are at increased risk for both adverse maternal and neonatal events. Because of this, most experts recommend low-dose aspirin and full anticoagulation with heparin or LMWH (see Table 54.2) [58]. For patients on vitamin K antagonists prior to conception, two options exist: (1) transition to UFH preconceptionally, or (2) maintenance on therapeutic vitamin K antagonist until 6 weeks of gestation with subsequent transition to UFH [58]. Vitamin K antagonists are known to cross the placenta and have been associated with embryopathy from 6 through 12 weeks of gestation and a possible fetopathy thereafter [61]. As noted in Table 54.2, anticoagulant coverage of the postpartum period is recommended in women with APS, with or without prior thrombosis [58].

For women with a particularly serious thrombotic history, such as recurrent thrombotic events or cerebral thrombotic events, the risk of thrombosis is generally accepted to be higher. Some experts have recommended that such patients be treated with warfarin in the second and early third trimesters, using heparin in the early first trimester to avoid the risk of embryopathy and also prior to the expected time of delivery (34—36 weeks of gestation).

If treated with therapeutic LMWH, changes in heparin metabolism related to pregnancy necessitate checking peak and/or trough anti-factor Xa levels at regular intervals, perhaps monthly, with dosing adjustment as needed. In women on intermediate UFH, peak and trough levels should be monitored somewhat more frequently. If women are negative for LA, aPTTs may be used to monitor UFH.

Compared with UFH, LMWH appears to have an increased risk of neuraxial hematoma associated with regional anesthesia such as an epidural or spinal [62]. As such, some experts advocate switching from LMWH to UFH at 36 weeks of gestation to perhaps increase the likelihood of the patient being a candidate for regional anesthesia in the case of preterm labor or the need for urgent delivery.

Treatment of Antiphospholipid Syndrome in "Refractory" Cases

Despite treatment with heparin, recurrent pregnancy losses occur in 20—30% of cases in most APS case series

TABLE 54.2 Subcutaneous Heparin Regimens Used in the Treatment of APS During Pregnancy and Postpartum

PROPHYLACTIC REGIMENS WITH CONCURRENT LOW-DOSE ASPIRIN

Recurrent Preembryonic and Embryonic Loss, No History of VTE

Standard unfractionated heparin (UFH):

Prophylactic dose subcutaneous UFH with 5000[a] to 7500[b] units every 12 h

or

Intermediate dose subcutaneous UFH every 12 h for goal anti-Xa level of 0.1—0.3 units/ml.

Low-molecular-weight heparin (LMWH):

Prophylactic dose LMWH with enoxaparin 40 mg once daily or dalteparin 5000 units once daily

or

Intermediate dose LMWH with enoxaparin 40 mg every 12 h or dalteparin 5000 units every 12 h

Prior Fetal Death or Early Delivery Secondary to Severe Preeclampsia or Severe Placental Insufficiency, No History of VTEs

UFH

Prophylactic dose subcutaneous UFH with 5000[a] to 7500[b] units every 12 h

or

Intermediate dose subcutaneous UFH every 12 h for goal anti-Xa level of 0.1—0.3 units/ml.

LMWH

Prophylactic dose LMWH with enoxaparin 40 mg once daily or dalteparin 5000 units once daily

or

Intermediate dose LMWH with enoxaparin 40 mg every 12 h or dalteparin 5000 units every 12 h

ANTICOAGULATION REGIMENS WITH CONCURRENT LOW-DOSE ASPIRIN FOR WOMEN WITH HISTORY OF THROMBOTIC EVENTS AND RECEIVING LONG-TERM ANTICOAGULATION

UFH

Adjusted dose subcutaneous UFH every 8—12 h to target aPTT in therapeutic range

LMWH

Adjusted dose subcutaneous enoxaparin 1 mg/kg every 12 h (or 1.5 mg/kg as a single dose every 24 h) or dalteparin 100 units/kg every 12 h (or 200 units/kg as a single dose every 24 h)

POSTPARTUM THROMBOPROPHYLAXIS

Vitamin K antagonists for 4—6 weeks with a target INR of 2.0—3.0, with initial UFH or LMWH overlap until the INR is ≥2.0, or prophylactic LMWH for 4—6 weeks

[a]*Recommended in Bates et al. [58].*
[b]*Recommended in Branch and Khamashta [64].*
VTEs, venous thrombotic events.

and trials. The best approach to such cases is unknown because there are currently no properly designed, published trials. Thus, treatment recommendations are little more than speculative in this scenario. However, most investigators would likely agree that for women on prophylactic regimens, increasing to full anticoagulation in subsequent pregnancies is rational. If the pregnancy failed while the patient was taking full anticoagulation, some investigators have considered the addition of immunomodulatory agents such as glucocorticoids or hydroxychloroquine.

Future Therapeutic Considerations

Contemporary work highlights the increasing importance of inflammation in APS-related pregnancy loss. The upregulation of TF on blood cells and vascular endothelium is thought to further potentiate the thrombotic and inflammatory effects of aPLs [63]. There is evidence in murine models that treatment with pravastatin and simvastatin decreases TF and PAR-2 expression on neutrophils, thus preventing pregnancy loss [16]. However, statins are Food and

Drug Administration pregnancy risk category X (contraindicated in pregnancy) and thus would not seem to be a good candidate for human pregnancy treatment until adequate studies have been done to show safety during pregnancy.

OBSTETRIC MANAGEMENT IN THE PATIENT WITH ANTIPHOSPHOLIPID SYNDROME

Prepregnancy

Ideally, women with APS will have sought preconception counseling. Clinically significant levels of aPLs should be confirmed and potential adverse maternal and obstetric complications discussed, including thrombosis, pregnancy loss, gestational hypertension/preeclampsia, uteroplacental insufficiency, and the potential need for preterm birth. It is also reasonable to assess APS patients for baseline evidence of anemia, thrombocytopenia, and underlying renal disease (urinalysis, serum creatinine, 24-h urine for creatinine clearance, and total protein) in addition to routine preconception screening.

Potential treatment regimens should be discussed with the patient, along with the potential risks including HIT and osteoporosis and the potential for embryopathy if maintained on a vitamin K antagonist. Many authorities recommend instituting daily low-dose aspirin (81 mg) in addition to prenatal vitamins prior to conception.

Prenatal

Women with APS who suspect they are pregnant should be evaluated immediately. An early transvaginal ultrasound is useful to confirm an intrauterine pregnancy as well as provide accurate dating of the pregnancy. One of the anticoagulation prophylaxis regimens discussed previously (see Table 54.2) should be instituted, and when appropriate, a strategy to monitor for HIT should also be instituted. Calcium supplementation is encouraged, as well as the performance of daily weight-bearing exercise for the prevention of osteoporosis.

Prenatal visits should occur every 2–4 weeks until 20–24 weeks of gestation and every 1 or 2 weeks thereafter. Patient visits should be specifically designed to monitor for the development of preeclampsia and thrombosis as well as to assess fetal well-being. These visits should consist of regular blood pressure checks as well as screening for symptoms of preeclampsia. Given the risk of fetal growth restriction and oligohydramnios (secondary to uteroplacental insufficiency), serial ultrasound examinations should be performed every 3 or 4 weeks after 18–20 weeks of gestation. Antenatal surveillance should consist of daily maternal fetal kick counts and at least once-weekly non-stress tests with amniotic fluid volume measurements starting at 30–32 weeks of gestation. If there is evidence of placental insufficiency or preeclampsia, more frequent ultrasound examinations and/or fetal testing may be warranted.

Labor and Delivery

Labor and delivery in women with APS should be managed in the same way as in any patient who is considered at high risk for preeclampsia and uteroplacental insufficiency. Most authorities recommend continuous electronic fetal monitoring throughout labor, given the increased risk of nonreassuring fetal heart rate tracings noted in women with APS.

The most common management dilemma in women with APS involves the timing of anticoagulation regimens near term to minimize thrombotic events peridelivery without placing the patient at increased risk of bleeding. Approaches vary, and there is no evidence that one method is better than another. Patients receiving prophylactic anticoagulation with heparin can be instructed to withhold their injections at the onset of labor. Alternatively, injections can be discontinued 12 h before a planned induction. The most common practice in women with APS on full-dose anticoagulation (UFH or LMWH) is to hold the last injection 24 h prior to a planned induction of labor or cesarean delivery. As an alternative for women deemed at extremely high risk for thromboembolism, including those with an event within 2 weeks of delivery, intravenous heparin can be started in labor and discontinued 2–4 h prior to anticipated delivery. Intravenous heparin can be resumed 4–6 h after vaginal delivery and 12 h after cesarean.

Spontaneous labor is problematic for women who are fully anticoagulated, particularly those receiving LMWH preparations. Anti-factor Xa levels are no longer recommended prior to placement of regional anesthesia because they are a poor predictor of the risk for spinal hematoma [62]. Protamine sulfate may be necessary in the event of surgical intervention.

Given the concern for neuraxial hematoma formation in women undergoing regional anesthesia while on anticoagulants, the American Society of Regional Anesthesia (ASRA) has established guidelines for the care of patients during labor or prior to cesarean delivery [62]:

• Neuraxial blockade should be withheld until 24 h after the last injection in women on full anticoagulation with LMWH or on adjusted dose anticoagulation with UFH.

- For women receiving low-dose thromboprophylaxis (low-dose twice-daily UFH or once-daily LMWH), ASRA recommends that needle placement be delayed until 10–12 h after the last dose.

Postpartum

Anticoagulant coverage of the postpartum period in women with APS and prior thrombosis is critical. We prefer initiating vitamin K antagonist thromboprophylaxis as soon as possible after delivery, with doses adjusted to achieve an international normalized ratio of 2.0–3.0. There is no consensus regarding the postpartum management of APS patients without prior thrombosis. The current recommendation in the United States is to treat with anticoagulant therapy for 4–6 weeks after delivery [58]. The need for postpartum anticoagulation in women with APS diagnosed solely on the basis of recurrent preembryonic or embryonic losses is uncertain. In addition, both heparin and warfarin are safe for breast-feeding mothers. Finally, oral contraceptives containing estrogen are contraindicated.

References

[1] D.W. Branch, J.R. Scott, N.K. Kochenour, E. Hershgold, Obstetric complications associated with the lupus anticoagulant, N Engl J Med 313 (1985) 1322–1326.

[2] R.S. Rai, K. Clifford, H. Cohen, L. Regan, High prospective fetal loss rate in untreated pregnancies of women with recurrent miscarriage and antiphospholipid antibodies, Hum Reprod 10 (1995) 3301–3304.

[3] W.A. Wilson, A.E. Gharavi, T. Koike, M.D. Lockshin, D.W. Branch, J.C. Piette, et al., International consensus statement on preliminary classification criteria for definite antiphospholipid syndrome: Report of an international workshop, Arthritis Rheum 42 (1999) 1309–1311.

[4] S. Miyakis, M.D. Lockshin, T. Atsumi, D.W. Branch, R.L. Brey, R. Cervera, et al., International consensus statement on an update of the classification criteria for definite antiphospholipid syndrome (APS), J Thromb Haemost 4 (2006) 295–306.

[5] J.E. Salmon, P.G. de Groot, Pathogenic role of antiphospholipid antibodies, Lupus 17 (2008) 405–411.

[6] S.S. Pierangeli, A.E. Gharavi, E.N. Harris, Experimental thrombosis and antiphospholipid antibodies: New insights, J Autoimmun 15 (2000) 241–247.

[7] B. de Laat, R.H. Derksen, R.T. Urbanus, P.G. de Groot, IgG antibodies that recognize epitope Gly40-Arg43 in domain I of beta 2-glycoprotein I cause LAC, and their presence correlates strongly with thrombosis, Blood 105 (2005) 1540–1545.

[8] L. Opatrny, M. David, S.R. Kahn, I. Shrier, E. Rey, Association between antiphospholipid antibodies and recurrent fetal loss in women without autoimmune disease: A meta-analysis, J Rheumatol 33 (2006) 2214–2221.

[9] E. Raschi, M.O. Borghi, C. Grossi, V. Broggini, S. Pierangeli, P.L. Meroni, Toll-like receptors: Another player in the pathogenesis of the anti-phospholipid syndrome, Lupus 17 (2008) 937–942.

[10] S.S. Pierangeli, P.P. Chen, E.B. Gonzalez, Antiphospholipid antibodies and the antiphospholipid syndrome: An update on treatment and pathogenic mechanisms, Curr Opin Hematol 13 (2006) 366–375.

[11] C. Lopez-Pedrera, M.J. Cuadrado, V. Herandez, P. Buendia, M.A. Aguirre, N. Barbarroja, et al., Proteomic analysis in monocytes of antiphospholipid syndrome patients: Deregulation of proteins related to the development of thrombosis, Arthritis Rheum 58 (2008) 2835–2844.

[12] V.M. Holers, G. Girardi, L. Mo, J.M. Guthridge, H. Molina, S.S. Pierangeli, et al., Complement C3 activation is required for antiphospholipid antibody-induced fetal loss, J Exp Med 195 (2002) 211–220.

[13] D. Ware Branch, A.G. Eller, Antiphospholipid syndrome and thrombosis, Clin Obstet Gynecol 49 (2006) 861–874.

[14] S.S. Pierangeli, G. Girardi, M. Vega-Ostertag, X. Liu, R.G. Espinola, J. Salmon, Requirement of activation of complement C3 and C5 for antiphospholipid antibody-mediated thrombophilia, Arthritis Rheum 52 (2005) 2120–2124.

[15] J.M. Shamonki, J.E. Salmon, E. Hyjek, R.N. Baergen, Excessive complement activation is associated with placental injury in patients with antiphospholipid antibodies, Am J Obstet Gynecol 196 (2007). 167.e1-167.e5.

[16] P. Redecha, C.W. Franzke, W. Ruf, N. Mackman, G. Girardi, Neutrophil activation by the tissue factor/factor VIIa/PAR2 axis mediates fetal death in a mouse model of antiphospholipid syndrome, J Clin Invest 118 (2008) 3453–3461.

[17] G. Girardi, P. Redecha, J.E. Salmon, Heparin prevents antiphospholipid antibody-induced fetal loss by inhibiting complement activation, Nat Med 10 (2004) 1222–1226.

[18] S.S. Pierangeli, P.P. Chen, E. Raschi, S. Scurati, C. Grossi, M.O. Borghi, et al., Antiphospholipid antibodies and the antiphospholipid syndrome: Pathogenic mechanisms, Semin Thromb Hemost 34 (2008) 236–250.

[19] M. Galli, D. Luciani, G. Bertolini, T. Barbui, Lupus anticoagulants are stronger risk factors for thrombosis than anticardiolipin antibodies in the antiphospholipid syndrome: A systematic review of the literature, Blood 101 (2003) 1827–1832.

[20] H.B. de Laat, R.H. Derksen, R.T. Urbanus, M. Roest, P.G. de Groot, Beta2-glycoprotein I-dependent lupus anticoagulant highly correlates with thrombosis in the antiphospholipid syndrome, Blood 104 (2004) 3598–3602.

[21] V. Pengo, A. Biasiolo, C. Pegoraro, U. Cucchini, F. Noventa, S. Iliceto, Antibody profiles for the diagnosis of antiphospholipid syndrome, Thromb Haemost 93 (2005) 1147–1152.

[22] D.A. Triplett, Antiphospholipid antibodies, Arch Pathol Lab Med 126 (2002) 1424–1429.

[23] M. Galli, D. Luciani, G. Bertolini, T. Barbui, Anti-beta 2-glycoprotein I, antiprothrombin antibodies, and the risk of thrombosis in the antiphospholipid syndrome, Blood 102 (2003) 2717–2723.

[24] J.S. Levine, D.W. Branch, J. Rauch, The antiphospholipid syndrome, N Engl J Med 346 (2002) 752–763.

[25] D.W. Branch, M.A. Khamashta, Antiphospholipid syndrome: Obstetric diagnosis, management, and controversies, Obstet Gynecol 101 (2003) 1333–1344.

[26] K. Aoki, A.B. Dudkiewicz, E. Matsuura, M. Novotny, G. Kaberlein, N. Gleicher, Clinical significance of beta 2-glycoprotein I-dependent anticardiolipin antibodies in the reproductive autoimmune failure syndrome: Correlation with conventional antiphospholipid antibody detection systems, Am J Obstet Gynecol 172 (1995) 926–931.

[27] D.W. Branch, Antiphospholipid antibodies and reproductive outcome: The current state of affairs, J Reprod Immunol 38 (1998) 75–87.

[28] D.W. Branch, R. Silver, S. Pierangeli, I. van Leeuwen, E.N. Harris, Antiphospholipid antibodies other than lupus

anticoagulant and anticardiolipin antibodies in women with recurrent pregnancy loss, fertile controls, and antiphospholipid syndrome, Obstet Gynecol 89 (1997) 549–555.

[29] D.L. Yetman, W.H. Kutteh, Antiphospholipid antibody panels and recurrent pregnancy loss: Prevalence of anticardiolipin antibodies compared with other antiphospholipid antibodies, Fertil Steril 66 (1996) 540–546.

[30] D.W. Branch, Antiphospholipid antibodies and pregnancy: Maternal implications, Semin Perinatol 14 (1990) 139–146.

[31] D.L. Huong, B. Wechsler, O. Bletry, D. Vauthier-Brouzes, G. Lefebvre, J.C. Piette, A study of 75 pregnancies in patients with antiphospholipid syndrome, J Rheumatol 28 (2001) 2025–2030.

[32] F. Lima, M.A. Khamashta, N.M. Buchanan, S. Kerslake, B.J. Hunt, G.R. Hughes, A study of sixty pregnancies in patients with the antiphospholipid syndrome, Clin Exp Rheumatol 14 (1996) 131–136.

[33] F.S. Cowchock, E.A. Reece, D. Balaban, D.W. Branch, L. Plouffe, Repeated fetal losses associated with antiphospholipid antibodies: A collaborative randomized trial comparing prednisone with low-dose heparin treatment, Am J Obstet Gynecol 166 (1992) 1318–1323.

[34] R.G. Farquharson, S. Quenby, M. Greaves, Antiphospholipid syndrome in pregnancy: A randomized, controlled trial of treatment, Obstet Gynecol 100 (2002) 408–413.

[35] W.H. Kutteh, Antiphospholipid antibody-associated recurrent pregnancy loss: Treatment with heparin and low-dose aspirin is superior to low-dose aspirin alone, Am J Obstet Gynecol 174 (1996) 1584–1589.

[36] N.S. Pattison, L.W. Chamley, M. Birdsall, A.M. Zanderigo, H.S. Liddell, J. McDougall, Does aspirin have a role in improving pregnancy outcome for women with the antiphospholipid syndrome? A randomized controlled trial, Am J Obstet Gynecol 183 (2000) 1008–1012.

[37] R. Rai, H. Cohen, M. Dave, L. Regan, Randomised controlled trial of aspirin and aspirin plus heparin in pregnant women with recurrent miscarriage associated with phospholipid antibodies (or antiphospholipid antibodies), BMJ 314 (1997) 253–257.

[38] M. Dreyfus, G. Hedelin, R. Kutnahorsky, M. Lehmann, B. Viville, B. Langer, et al., Antiphospholipid antibodies and preeclampsia: A case–control study, Obstet Gynecol 97 (2001) 29–34.

[39] D.W. Branch, T.F. Porter, L. Rittenhouse, S. Caritis, B. Sibai, B. Hogg, et al., Antiphospholipid antibodies in women at risk for preeclampsia, Am J Obstet Gynecol 184 (2001) 825–834.

[40] D.W. Branch, R. Andres, K.B. Digre, N.S. Rote, J.R. Scott, The association of antiphospholipid antibodies with severe preeclampsia, Obstet Gynecol 73 (1989) 541–545.

[41] J. Moodley, V. Bhoola, J. Duursma, D. Pudifin, S. Byrne, D.G. Kenoyer, The association of antiphospholipid antibodies with severe early-onset pre-eclampsia, S Afr Med J 85 (1995) 105–107.

[42] W.J. Polzin, J.N. Kopelman, R.D. Robinson, J.A. Read, K. Brady, The association of antiphospholipid antibodies with pregnancies complicated by fetal growth restriction, Obstet Gynecol 78 (1991) 1108–1111.

[43] K. Bramham, B.J. Hunt, S. Germain, I. Calatayud, M. Khamashta, S. Bewley, et al., Pregnancy outcome in different clinical phenotypes of antiphospholipid syndrome, Lupus 19 (2010) 58–64.

[44] T.E. Warkentin, Heparin-induced thrombocytopenia: Yet another treatment paradox? Thromb Haemost 85 (2001) 947–949.

[45] T.E. Warkentin, A. Greinacher, A. Koster, A.M. Lincoff, Treatment and prevention of heparin-induced thrombocytopenia:

American College of Chest Physicians evidence-based clinical practice guidelines (8th edition, Chest 133 (2008) 340S–380S.

[46] D.W. Branch, A.M. Peaceman, M. Druzin, R.K. Silver, Y. El-Sayed, R.M. Silver, et al., A multicenter, placebo-controlled pilot study of intravenous immune globulin treatment of antiphospholipid syndrome during pregnancy. The Pregnancy Loss Study Group, Am J Obstet Gynecol 182 (2000) 122–127.

[47] G. Triolo, A. Ferrante, F. Ciccia, A. Accardo-Palumbo, A. Perino, A. Castelli, et al., Randomized study of subcutaneous low molecular weight heparin plus aspirin versus intravenous immunoglobulin in the treatment of recurrent fetal loss associated with antiphospholipid antibodies, Arthritis Rheum 48 (2003) 728–731.

[48] E. Vaquero, N. Lazzarin, H. Valensise, S. Menghini, G. Di Pierro, F. Cesa, et al., Pregnancy outcome in recurrent spontaneous abortion associated with antiphospholipid antibodies: A comparative study of intravenous immunoglobulin versus prednisone plus low-dose aspirin, Am J Reprod Immunol 45 (2001) 174–179.

[49] M. Empson, M. Lassere, J. Craig, J. Scott, Prevention of recurrent miscarriage for women with antiphospholipid antibody or lupus anticoagulant, Cochrane Database Syst Rev, CD002859 (2005).

[50] K.A. Granger, R.G. Farquharson, Obstetric outcome in antiphospholipid syndrome, Lupus 6 (1997) 509–513.

[51] R.K. Silver, S.N. MacGregor, J.S. Sholl, J.M. Hobart, M.G. Neerhof, A. Ragin, Comparative trial of prednisone plus aspirin versus aspirin alone in the treatment of anticardiolipin antibody-positive obstetric patients, Am J Obstet Gynecol 169 (1993) 1411–1417.

[52] F.J. Munoz-Rodriguez, J. Font, R. Cervera, J.C. Reverter, D. Tassies, G. Espinosa, et al., Clinical study and follow-up of 100 patients with the antiphospholipid syndrome, Semin Arthritis Rheum 29 (1999) 182–190.

[53] M. Tulppala, M. Marttunen, V. Soderstrom-Anttila, T. Foudila, K. Ailus, T. Palosuo, et al., Low-dose aspirin in prevention of miscarriage in women with unexplained or autoimmune related recurrent miscarriage: Effect on prostacyclin and thromboxane A2 production, Hum Reprod 12 (1997) 1567–1572.

[54] R.D. Franklin, W.H. Kutteh, Antiphospholipid antibodies (APA) and recurrent pregnancy loss: Treating a unique APA positive population, Hum Reprod 17 (2002) 2981–2985.

[55] C.A. Laskin, K.A. Spitzer, C.A. Clark, M.R. Crowther, J.S. Ginsberg, G.A. Hawker, et al., Low molecular weight heparin and aspirin for recurrent pregnancy loss: Results from the randomized, controlled HepASA trial, J Rheumatol 36 (2009) 279–287.

[56] M.D. Stephenson, P.J. Ballem, P. Tsang, S. Purkiss, S. Ensworth, E. Houlihan, et al., Treatment of antiphospholipid antibody syndrome (APS) in pregnancy: A randomized pilot trial comparing low molecular weight heparin to unfractionated heparin, J Obstet Gynaecol Can 26 (2004) 729–734.

[57] L.S. Noble, W.H. Kutteh, N. Lashey, R.D. Franklin, J. Herrada, Antiphospholipid antibodies associated with recurrent pregnancy loss: Prospective, multicenter, controlled pilot study comparing treatment with low-molecular-weight heparin versus unfractionated heparin, Fertil Steril 83 (2005) 684–690.

[58] S.M. Bates, I.A. Greer, I. Pabinger, S. Sofaer, J. Hirsh, Venous thromboembolism, thrombophilia, antithrombotic therapy, and pregnancy: American College of Chest Physicians evidence-based clinical practice guidelines (8th edition, Chest 133 (2008) 844S–886S.

[59] L.M. Askie, L. Duley, D.J. Henderson-Smart, L.A. Stewart, Antiplatelet agents for prevention of pre-eclampsia: A meta-analysis of individual patient data, Lancet 369 (2007) 1791–1798.

[60] D. Erkan, J.T. Merrill, Y. Yazici, L. Sammaritano, J.P. Buyon, M.D. Lockshin, High thrombosis rate after fetal loss in anti-phospholipid syndrome: Effective prophylaxis with aspirin, Arthritis Rheum 44 (2001) 1466—1467.

[61] W.S. Chan, S. Anand, J.S. Ginsberg, Anticoagulation of pregnant women with mechanical heart valves: A systematic review of the literature, Arch Intern Med 160 (2000) 191—196.

[62] T.T. Horlocker, D.J. Wedel, H. Benzon, D.L. Brown, F.K. Enneking, J.A. Heit, et al., Regional anesthesia in the anticoagulated patient: Defining the risks (the second ASRA Consensus Conference on Neuraxial Anesthesia and Anti-coagulation), Reg Anesth Pain Med 28 (2003) 172—197.

[63] A.V. Kinev, R.A. Roubey, Tissue factor in the antiphospholipid syndrome, Lupus 17 (2008) 952—958.

[64] D.W. Branch, M.A. Khamashta, Antiphospholipid syndrome, in: J.T. Queenan (Ed.), High-Risk Pregnancy, American College of Obstetricians and Gynecologists, Washington, DC, 2007, pp. 60—72.

III. ANTI-PHOSPHOLIPID SYNDROME

Treatment of Antiphospholipid Syndrome

Gerard Espinosa, Ricard Cervera

INTRODUCTION

Antiphospholipid syndrome (APS) is characterized by the development of venous or arterial thromboses, fetal losses, and thrombocytopenia in the presence of antiphospholipid antibodies (aPLs), namely lupus anticoagulant (LA), anticardiolipin antibodies (aCLs), or antibodies directed to various proteins, mainly β_2-glycoprotein I (β_2GPI), or all three [1–3]. Regarding the thrombotic clinical spectrum, APS may have various presentations. Any combination of vascular occlusive events may occur in the same individual, and the time interval between them also varies considerably from weeks to months or even years. Other clinical manifestations, not directly associated with the presence of underlying thrombotic lesions, have been less frequently described in patients with APS (Table 55.1) [4–6].

In 1992, Asherson [7] described the "catastrophic" variant of APS as a condition characterized by multiple

TABLE 55.1 Clinical Features and Laboratory Characteristics Not Directly Associated with Underlying Thrombosis But Probably Related to Antiphospholipid Antibodies

Clinical feature	Laboratory characteristic
Thrombocytopenia	IgA isotype of anticardiolipin and anti-β_2-GPI antibodies
Hemolytic anemia	Antiprothrombin antibodies
Livedo reticularis	Antiphosphatidyletanolamine antibodies
Heart valve disease	
Nephropathy	
Myelopathy	
Chorea	
Multiple sclerosis-like	
Pulmonary hypertension	

vascular occlusive events, usually affecting small vessels, presenting over a short period of time, with laboratory confirmation of the presence of aPLs. The hallmark of this disorder is the diffuse thrombotic microvasculopathy, with microthrombosis the main finding in necropsy studies [8]. Catastrophic APS represents less than 1% of APS cases [3]. However, patients with catastrophic APS usually end up in a life-threatening situation, with a mortality rate of approximately 30% in the largest published series with the best available therapy [8–10].

This chapter focuses on the treatment of thrombotic and obstetric manifestations of APS. In addition, the evidence-based information about management of catastrophic APS is summarized. Finally, the management of some difficult cases (e.g., "atypical" features of APS) is discussed.

PRIMARY THROMBOPROPHYLAXIS IN PATIENTS WITH ANTIPHOSPHOLIPID ANTIBODIES

This section discusses the therapeutic approach in carriers of aPLs but without previous thrombotic events—that is, patients with systemic lupus erythematosus (SLE) and aPLs, patients with only obstetrical manifestations of APS, and healthy individuals who carry aPLs.

The key questions for these groups of patients are whether they need any treatment and, if so, what is the most effective prophylactic treatment. Current evidence-based knowledge does not support the widespread use of aspirin in all these aPL-positive patients. First, the annual thrombosis risk in asymptomatic aPL patients ranges from 0 to 3.8% [11–15], equivalent to that of major bleeding associated with the use of aspirin. Second, the only randomized clinical trial (the APLASA study), in which 98 asymptomatic, persistently aPL-positive individuals were randomized to receive a daily dose of 81 mg of aspirin (48 patients) or placebo (50 patients), showed that these patients have a low overall annual incidence rate of acute thrombosis and develop vascular events when additional thrombosis risk factors are present [16]. Specifically, the overall thrombosis incidence rate was 1.33/100 patient-years; it was 2.75/100 patient-years for aspirin-treated subjects and 0/100 patient-years for placebo-treated subjects (hazard ratio, 1.04; 95% confidence interval (CI), 0.69—1.56). All but 1 patient with thrombosis in either study had concomitant thrombosis risk factors or systemic autoimmune disease at the time of thrombosis. Therefore, asymptomatic, persistently aPL-positive individuals seem not to benefit from low-dose aspirin for primary thromboprophylaxis.

However, some points need to be addressed. Studies on annual thrombotic risk of asymptomatic aPL-positive patients present some limitations, such as the lack of control groups and the unknown clinical characteristics (non-aPL thrombosis risk factors or aPL profile) of patients. The randomized controlled trial also had some limitations, including the small sample size (98 patients) and the short follow-up period (mean ± SD, 2.3 ± 0.95 years) [16]. In addition, included subjects presented with a titer of aCL ≥ 20 GPL or MPL units, which is lower than the threshold established to define the positivity of aCL (>40 GPL or MPL units). These points might lead to an insufficient power to conclude that in all aPL-positive patients, aspirin is not useful as a primary thromboprophylaxis.

A more realistic approach with a lower degree of evidence would be to stratify these individuals according to clinical features such as the presence of traditional congenital or acquired procoagulant risk factors, the profile of aPLs (persistently positive aCL and/or anti-β_2GPI antibodies at moderate/high titers and/or unequivocally LA), and the co-existence of an underlying autoimmune disease (SLE in particular) to consider a primary prophylactic therapy with low-dose (75—100 mg daily) aspirin (LDA) [17]. In this sense, it is known that SLE represents a prothrombotic condition *per se* [18]. The disease itself acts as a strong thrombophilic risk factor, primarily related to the chronic systemic inflammation and renal involvement. Furthermore, one study suggested that prophylactic aspirin should be given to all patients with SLE to prevent both arterial and venous thrombotic manifestations, especially in patients with aPLs [19]. In the same study, the authors suggested that in selected patients with LA and a low bleeding risk, prophylactic oral anticoagulant therapy may provide higher utility. There is consensus for primary prophylaxis in these patients, mainly with LDA.

An alternative to aspirin in SLE patients may be hydroxychloroquine. There is a significant amount of evidence supporting the protective role of this drug against the development of both venous and arterial thrombosis [20—22]. This drug has an immunomodulatory effect based on interference with antigen processing, inhibition of T-cell receptor- and B-cell antigen receptor-induced calcium signaling, and inhibition of toll-like receptor signaling. In addition, hydroxychloroquine inhibits interleukin (IL) 1β-induced inducible nitric oxide (NO) synthase expression and NO production, suppressing IL-1β-induced activation of nuclear factor-κB. These results may also open the door for the use of hydroxychloroquine in non-SLE aPL-positive patients in the future.

Another interesting group of nonthrombotic patients with aPLs is represented by women with obstetrical morbidity, the so-called obstetrical APS. Two retrospective studies showed that there was a low incidence of subsequent thrombosis after delivery in the group of APS patients who were treated with LDA compared to the untreated group [23, 24]. These data suggest the beneficial role of aspirin in this subgroup of APS patients.

An Italian collaborative study group confirmed some of these recommendations [25]. The group assessed risk factors for a first thrombotic event in aPL-positive carriers and evaluated the efficacy of prophylactic treatments, including LDA and long-term warfarin or LDA/ heparin administered during high-risk periods (pregnancy/puerperium, immobilization, and surgery). Long-term LDA was initiated arbitrarily for at least one of the following reasons: SLE, other autoimmune diseases, obstetric APS, or LA positivity. Warfarin was administered for thrombophylic conditions such as pulmonary hypertension, atrial fibrillation, or posttraumatic tetraplegia. Upper normal values of aPLs were 8.3—10 GPL and 9.6—10 MPL, respectively. The cutoff value for medium/high titers was ≥ 20 GPL (52.1% of patients) and ≥ 20 MPL (21.7% of patients). Prophylaxis was administered for a total of 906.1 patient-years, whereas 921.6 patient-years were without prophylactic treatment. No study participants developed serious adverse events, including major bleedings, during prophylaxis. The thrombosis incidence rate was 1.64 per 100 patients per year. Specifically, seven events (23.3%) developed during high-risk periods (puerperium, oral contraceptives, or surgery) and eight (26.6%) during prophylactic treatment (LDA or heparin). Multivariate logistic regression analysis showed that only hypertension and medium/high titers of IgG aCL were independent risk factors for a first thrombotic event in asymptomatic aPL carriers and, more interestingly, that primary prophylaxis is protective [25].

It is important to note that all patients or subjects who present with persistently positive aPLs should control, modify, or discontinue additional vascular risk factors such as hypertension, hypercholesterolemia, or tobacco use. In addition, estrogen-containing oral contraceptives

TABLE 55.2 Proposed Therapeutic Strategies for Primary Thromboprophylaxis of Patients with Antiphospholipid Antibodies

GENERAL MEASURES

To avoid vascular risk factors such as hypertension, hypercholesterolemia, and tobacco

Estrogen-containing contraceptives contraindicated

Prophylaxis with LMWH in higher risk situations (surgery, immobilization)

PRIMARY THROMBOPROPHYLAXIS

Individuals without systemic autoimmune disease	
With high thrombotic risk aPL profile[a]	LDA
Without high thrombotic risk aPL profile[a]	General measures
Patients with SLE	
With high thrombotic risk aPL profile[a]	HDX + LDA
Without high thrombotic risk aPL profile[a]	HDX
Women with obstetric APS but without previous thrombosis	LDA

[a]Persistently positive LA, aCL at moderate to high titers, or anti-β-GPI + LA or aCL. Abbreviations: aPL, antiphospholipid antibodies; APS, antiphospholipid syndrome; HDX, hydroxychloroquine; LDA, low-dose aspirin; SLE, systemic lupus erythematosus.

should be avoided in women with aPLs, and prophylaxis with low-molecular-weight heparin should be given in higher risk situations, such as surgery (Table 55.2).

Regarding drug therapy, long-term LDA (75—100 mg daily) is advised in individuals without systemic autoimmune disease with a high thrombotic risk aPL profile (i.e., with persistently positive LA, aCL at moderate to high titers, or anti-β_2GPI plus LA or aCL). In the remaining asymptomatic carriers of aPLs—that is, subjects without systemic autoimmune disease with transiently positive LA or aCL at low titers—general measures can be sufficient. In SLE patients with aPLs, hydroxychloroquine may be useful to prevent the development of thrombosis. If SLE patients present with any feature of the high thrombotic risk aPL profile, long-term LDA is also advisable. In women with obstetric APS but without previous thrombosis, long-term LDA is recommended (see Table 55.2).

SECONDARY THROMBOPROPHYLAXIS IN PATIENTS WITH ANTIPHOSPHOLIPID SYNDROME

Two meta-analyses performed in patients with SLE [26] and primary APS [27] demonstrated that the presence of aPLs is related to an increased risk of thrombotic events. Furthermore, a systematic review of published articles on APS showed that LA was a clear risk factor for thrombosis, irrespective of the site and type of thrombosis, the presence of SLE, and the methods used to detect them [28]. Moreover, a high level of aCL [12, 29] and concomitant positivity for anti-β2GPI and LA or aCL have been recognized to increase the risk of thrombosis in patients with aPLs [15].

There are evidence-based data on the therapeutic management of patients with aPLs and a previous thrombotic event [30]. The best secondary thromboprophylaxis in patients with definite APS—that is, those who have suffered from thrombosis and at least two positive determinations of aPLs [1]—is long-term anticoagulation [31]. This point is very important considering that some studies have included patients with only a single positive determination of aPLs [32]. In other words, most patients included in some studies on secondary thromboprophylaxis did not have APS.

In accordance with an excellent systematic review [31], patients with definite APS with first venous thrombosis have to be treated with prolonged oral anticoagulation at a target INR of 2.0—3.0, and a target INR of 3.0—4.0 should be used for those with arterial or recurrent thrombotic events. These conclusions are based on the analysis of nine cohort studies [13, 33—40], five subgroup analyses [41—44], and two randomized controlled studies [45, 46]. The main limitation of this review is the low quality of some of the included studies (observational, nonrandomized, and retrospective cohorts). However, it is important to note that despite the potential risks of missing information and reporting bias, these studies offer a more realistic view of these patients. One of the most important findings regarding the efficacy/safety of high- and moderate-intensity anticoagulation comes from the randomized, double-blind trial of Crowther et al. [45]. They showed that high-intensity (INR, 3.0—4.0) was not superior to moderate-intensity (INR, 2.0—3.0) warfarin for thromboprophylaxis in patients with aPLs and previous thrombosis. The guidelines for the treatment of venous thromboembolic disease also recommend an INR of 2.0—3.0 as the preferred intensity of long-term anticoagulant treatment with vitamin K antagonists in all patients with venous thromboembolism, including patients with APS [47]. One of the main conclusions of this review was that the general population and patients with only a single positive aPL determination seemed to have a similar recurrent thrombotic rate [31]. Otherwise, among patients fulfilling laboratory criteria for definite APS [1], the risk of recurrent events was lower in patients with predominant first venous thromboses than in patients presenting with arterial and/or recurrent events. Furthermore, standard-intensity oral anticoagulation (target INR, 2.0—3.0) protected the former from further thrombosis, whereas in the latter, a better outcome was demonstrated with higher intensity anticoagulation (INR > 3.0) [31].

The Antiphospholipid Antibody in Stroke Study (APASS) group, in collaboration with the Warfarin-Aspirin Recurrent Stroke Study (WARSS) group, designed a prospective study of the role of aPLs in recurrent ischemic stroke [48]. They compared the risk of recurrent stroke and other thromboembolic disease during a 2-year follow-up period in patients with ischemic stroke who were randomized to either aspirin therapy (325 mg/day) or warfarin therapy at a dose to maintain the INR between 1.4 and 2.8. Conversely to the previous review [31], in both warfarin and aspirin arms, the recurrent thrombosis rate was not different between patients who were positive for both aCL and LA and those who were LA^+/aCL^-, LA^-/aCL^+, or LA^-/aCL^-. However, the WARSS/APASS study included patients with only a single aCL value of 10 GPL or higher at the time of an initial ischemic stroke. Therefore, this trial was not conducted in patients with definite APS, and its conclusions should not be applied to these patients.

One of the problems with high-intensity anticoagulation may be the higher risk of secondary bleeding—a point that clinicians must consider when deciding the best treatment for these patients. In a study of 66 patients with definite APS with previous thrombosis treated with oral anticoagulation to a target INR of 3.5, the risk of intracranial and fatal bleeding was similar to that for groups of patients treated to lower target ratios [38]. As a whole, the rate of major bleeding was 6 cases per 100 patient-years (95% CI, 1.6—15.0). The rate of intracranial bleeding was 1.5 per 100 patient-years (95% CI, 0.04—8.4), and the rate of thrombotic recurrence was 9.1 cases per 100 patient-years (95% CI, 3.3—19.6). Nevertheless, in the systematic review, repeated thromboses were more frequent and associated with a higher mortality than hemorrhagic complications in patients taking warfarin [31]. On the other hand, 31 of 420 (7.4%) patients in the prospective study of the "Euro-Phospholipid" cohort receiving oral anticoagulants presented with hemorrhages: In 13 cases, hemorrhage occurred in internal organs (cerebral in 7, gastrointestinal in 4, and intra-abdominal in 2), and in 6 cases hemorrhage was the main cause of death [49].

In conclusion, in addition to anticoagulant therapy, it is important to take into account that in this group of patients, the identification and avoidance of vascular risk factors, such as hypertension, hypercholesterolemia, or tobacco use, is very important. Estrogen-containing oral contraceptives are forbidden in women with aPLs. Otherwise, patients with definite APS and a first venous event should be treated with long-term oral anticoagulation to an INR of 2.0—3.0. In patients

TABLE 55.3   Proposed Therapeutic Strategies for Secondary Thromboprophylaxis of Patients with Antiphospholipid Antibodies

Patients with a single positive result of aPLs after several determinations and either venous or arterial thrombotic event	Same plan as that for the general population
Patients with definite APS	
And a first venous thrombotic event	OA (INR 2.0—3.0)
And an arterial thrombotic event	OA (INR 3.0—4.0)
And recurrent thrombotic events while on OA	1. Warrant the therapeutic range of INR
	2. INR 3.0—4.0
And recurrent thrombotic events while on oral anticoagulation achieving a target INR > 3.0	INR 3.0—4.0 + LDA Consider hydroxychloroquine (mainly in patients with SLE)

Abbreviations: aPLs, antiphospholipid antibodies; APS, antiphospholipid syndrome; HDX, hydroxichloroquine; INR, international normalized ratio; IVIG, intravenous immunoglobulins; LDA, low-dose aspirin; LMWH, low molecular weight heparin; OA, oral anticoagulation; SLE, systemic lupus erythematosus.

with definite APS and arterial thrombosis, oral anticoagulation to an INR greater than 3.0 may be advisable (Table 55.3).

Recurrent Thrombosis Despite Optimal Anticoagulation

An important therapeutic challenge is recurrent thrombosis despite optimal anticoagulation therapy. In the 5-year follow-up of the Euro-Phospholipid cohort of 1000 APS patients, recurrent thrombotic events occurred in 166 (16.6%) of them, the most common of which were strokes (2.4%), transient ischemic attacks (2.3%), deep vein thromboses (2.1%), and pulmonary embolism (2.1%) [49]. The best support for making any therapeutic recommendation for this group of patients comes from the previously mentioned systematic review of the literature [31]. Of the 180 recurrent thrombotic events reported, 49 (27%) occurred in patients treated with warfarin. Within this group, the actual INR at the time of the event was less than 3.0 in 42 cases (86%). In the Euro-Phospholipid cohort study [49], the INR at the time of the recurrent thrombotic event was difficult to determine in most patients, and unfortunately these data were not consistently obtained in this study. Interestingly, strokes and transient ischemic attacks were the most common recurrent thrombotic events. Because most of these patients were receiving oral anticoagulants at a target INR between 2 and 3, this might indicate that this treatment mainly protects against venous thrombosis but may be not sufficiently protective against arterial thrombosis. However, a subtherapeutic INR at the time of thrombosis may only represent inadequate anticoagulation and not treatment failure. Recurrences were infrequent among patients effectively receiving oral anticoagulation at an INR of 3.0–4.0. Therefore, patients with APS with recurrent venous events should be treated with warfarin at an INR greater than 3.0. This recommendation is based on cohort studies because randomized controlled trials have included few patients with this profile. In addition, there are no evidence-based data to recommend additional antithrombotic treatment such as aspirin for patients who experience recurrent events while receiving oral anticoagulants at an INR greater than 3.0. However, it may be a reasonable option to add LDA to oral anticoagulation in these cases.

Another therapeutic option, mainly in patients with SLE, is the addition of hydroxychloroquine. First, it has an excellent safety profile. Second, there is evidence from several studies that hydroxychloroquine has a protective effect on the development of both venous and arterial thrombosis in SLE patients with aPLs [20–22]. Specific studies in patients with primary APS are still lacking. According to these data, antimalarial treatment may be a possible complement of the anticoagulant therapy in patients with APS.

In summary, in patients with definite APS and recurrent thrombotic events while on oral anticoagulants, it is mandatory for the INR to be in the therapeutic range. In this case, the best option is oral anticoagulation to an INR greater than 3.0. In patients who have recurrent events while on oral anticoagulants achieving a target INR greater than 3.0, an option is to add LDA. In patients with SLE and those with a high risk of hemorrhage, antimalarial treatment may be a possible complement to the anticoagulant therapy (see Table 55.3).

IMPROVING SECONDARY THROMBOPROPHYLAXIS IN PATIENTS WITH ANTIPHOSPHOLIPID SYNDROME

A novel aspect of APS is that patients should be stratified and treated according to some clinical and immunologic characteristics in addition to aPL positivity [1]. One of these immunologic characteristics is the profile of aPLs. The latest revised classification criteria recommend to classify APS patients into different categories according their aPL profile [1]. In this way, patients with positivity for multiple aPLs in any combination belong to category I and those with positivity for a single aPL to category II (further divided based on the type of antibody). This subclassification may be especially important for patients' enrolment in clinical studies.

Using this classification, patients with LA, aCL IgG at high titers, or anti-β2GPI antibodies plus LA or aCL had the highest thrombotic risk in one study [50]. It is well-known that the combination of positive assays is a better predictor for thrombotic risk than when a single test is positive. In a retrospective study by Pengo et al. [51] on 100 patients with aPLs, positivity for LA, aCL, and anit-β2GPI conferred the highest risk for thrombosis (odds ration (OR) exceeding 33). Interestingly, no other assay combination was associated with significant risk for thrombosis. The same results were found by Forastiero et al. [15]. In this study, the triple positivity for LA, IgG, and anti-β2GPI and positivity for IgG anti-prothrombin antibodies (aPTs) gave the highest annual rate of thrombosis (8.4%). However, from a therapeutic standpoint, there is no evidence on the effectiveness of more intensive therapy in these patients. Common sense dictates the need for closer clinical and therapeutic monitoring (to ensure a correct INR), and this is advisable in patients with thrombosis and any of these immunological profiles.

In addition to the profile, another point to consider is the persistence of aPL positivity over time. Currently, there is no evidence on the usefulness of repeat aPL testing on patients who meet criteria for APS. However, a prospective study of patients with SLE demonstrated that the risk of thrombosis for LA-positive patients was significantly increased at both the arterial and the venous level. Interestingly, LA-negative patients who had persistently positive aCL (defined as positive in more than two-thirds of the determinations) had an increased risk of thrombosis at the expense of arterial events, whereas in LA-negative and transiently aCL-positive patients (defined as positive on at least two occasions but on less than two-thirds of the determinations), the risk of thrombosis—both arterial and venous—was no different from that in aPL-negative SLE patients [52]. Our group obtained similar results in patients with APS [53]. Adjusted risk for recurrent thrombosis during follow-up was increased in persistently positive aPL patients (defined as more than 75% of the aPL determinations positive during follow-up) compared with transiently positive aPL patients. The profile of persistently positive aPL related with the appearance of thrombosis during follow-up was the combination of aCL IgG plus LA. The role of high aCL titers (≥40 GPL or MPL), which are a laboratory criterion for APS diagnosis, in the recurrent thrombosis risk was not determined in these two studies.

The final clinical feature is the co-existence of an inherited thrombophilia, mainly the Leiden mutation of factor V and the G20210A mutation of prothrombin. However, their role in the thrombotic risk of patients with APS is contradictory. In general, the prevalence of factor V Leiden and prothrombin gene mutation is similar in patients with APS and healthy individuals, and the presence of these does not increase the risk of a thrombotic event [54–61]. Conversely, the presence of factor V Leiden has been related with an increased risk of thrombosis in patients with APS [62–64]. Galli *et al.* [65] suggested that factor V Leiden was associated with the thrombotic risk of patients with LA. Regarding prothrombin mutation, its presence has not been related with increased thrombotic risk in patients with APS [58, 60, 65].

No studies have evaluated whether thrombophilic defects are risk factors for recurrent venous thrombosis in patients with APS under anticoagulant therapy. Unpublished data from our cohort of patients indicated that patients with inherited thrombophilic defects, such as factor V Leiden or prothrombin mutation, did not show increased risk of recurrent thrombosis under anticoagulant therapy. Further studies are necessary to establish the exact role of these genetic thrombophilic defects in patients with APS and whether a modified treatment is advisable.

MANAGEMENT OF "DIFFICULT" CASES

The topic of this section is the management of difficult cases, such as patients for whom their aPL test turns negative or those who do not display formal classification criteria for APS. In this context, there is not sufficient evidence to make recommendations, and in most cases common sense dictates the therapeutic approach.

Patients for Whom Their Antiphospholipid Antibodies Test Turns Negative

In a few patients with APS and previously positive aPLs, their tests may become negative over time [66–71]. Erkan *et al.* [69] demonstrated that aPLs remained stable for at least three-fourths of subsequent tests, regardless of which laboratory performed the tests. The factors related to the "disappearance" of aPLs are unknown. One important question is whether these patients really do suffer from APS. In a series of 10 patients with primary APS according to the Sapporo criteria [70] who became negative over time, 4 presented with another known precipitating factor for venous thrombosis [71]. In addition, some of these patients had low titers (<40 GPL/MPL units) of aCLs. Therefore, it is difficult to confirm the exact pathogenic role of aCLs in these patients. There is no evidence regarding the increased thrombotic risk and the role of prophylactic treatment in this group of patients. Also, the question of whether treatment should be stopped after (spontaneous) disappearance of aPLs needs further study. Although anticoagulation withdrawal may be safe in APS patients when aPLs become negative [71], further evidence on the clinical importance of a disappearance of aPLs is needed to recommend this approach. A

therapeutic option may be to switch anticoagulation by antiaggregation and hydroxychloroquine [72–74] and, in the case in which aPLs are persistently negative over time, to stop any treatment with a strict control of classic thrombotic risk factors.

Patients Who Do Not Display Classification Criteria for Antiphospholipid Syndrome

Other difficult cases are represented by patients with aPLs but who do not display formal laboratory or clinical classification criteria for APS [1]. Examples of this group are patients with thrombosis and repeated low titers of aCLs (<40 GPL/MPL units or 99th percentile) or anti-β_2GPI antibodies (<99th percentile) and negative LA or patients with aPLs and clinical features not included as clinical criteria, such as nonbacterial thrombotic endocarditis, seizures, or nephropathy [1]. In the first example, the diagnostic problem is due to the absence of data to establish the threshold between moderate/high levels and low levels. Unfortunately, gold standards for aPL enzyme-linked immunosorbent assays (ELISAs) are lacking, which makes standardization very difficult. In addition, the interlaboratory reproducibility of aCL and anti-β_2GPI measurement is unacceptably poor. Specifically, for β_2GPI assays, no accepted common calibrator is available; therefore, comparison of numerical results obtained with different assays is not possible. As a consequence, interpretation of the degree of positivity shows important discrepancies between laboratories [75]. Thus, the status of medium and high positive samples as defined in the laboratory criteria of APS currently has no relevance for anti-β_2GPI results. However, it is imperative to remember that these criteria have only a classificatory purpose, and their usefulness is directed to include patients in clinical or therapeutic studies. Therefore, in daily clinical practice, the therapeutic approach is similar: long-term anticoagulation in a similar manner as for patients who fulfill the laboratory criteria of APS.

In the second case, the updated classification criteria for APS indicate that these "atypical" clinical characteristics are frequently related to aPLs and attempt to define the characteristics of these clinical pictures. However, their inclusion as classification criteria for definite APS may decrease diagnostic specificity [1].

Cardiac valve involvement is frequently found in APS patients. Valvular involvement including thickening and vegetations is the most common cardiac manifestation in APS, ranging from 30 to 50% in different series [76]. A committee consensus report recommended anticoagulation for patients with valvulopathy who have had any evidence of thromboembolic disease. In addition, prophylactic antiplatelet therapy may be appropriate for asymptomatic patients. There is no evidence regarding the beneficial effect of immunosuppressive treatment [77].

A particular case is represented by patients with hematological abnormalities such as autoimmune thrombocytopenia or autoimmune hemolytic anemia related to aPLs [78]. Thrombocytopenia is frequently found in APS patients, ranging from 22 to 42% in different series. It is usually moderate ($>50 \times 10^9/l$) and rarely requires treatment. However, when thrombocytopenia is less than $30-50 \times 10^9/l$ or symptomatic with bleeding, the same treatments that are used in patients with SLE, such as steroids with or without intravenous immunoglobulins, should be considered in view of the similar pathophysiology [78]. Rituximab, the monoclonal anti-CD20 antibody, has been increasingly used to treat SLE patients with hemocytopenias. The effectiveness of this novel treatment is yet to be established for the treatment of thrombocytopenia in patients with APS, and only few conflicting results have been reported [79]. In case of failure of all these treatments, splenectomy should be considered. Of note, patients remain at risk of thrombosis despite thrombocytopenia, and anticoagulation should be continued unless the platelet count falls below $50 \times 10^9/l$, but the intensity of anticoagulation may need to be reduced.

In the largest cohort of APS patients, Cervera et al. [3] reported that autoimmune hemolytic anemia was seen in 6.6% of patients. In general, the therapeutic approach is similar to that for patients with SLE—that is, glucocorticoids as the cornerstone of treatment. If the glucocorticoids are ineffective, intravenous immunoglobulins and rituximab may be useful [78].

Other "atypical" cases are patients with classical clinical features related to aPLs other than LA, aCL, or anti-β₂GPI. The family of aPLs includes a number of antibodies with specificities other than that for β₂GPI. Among them, the antibodies directed against prothrombin (aPT) have been extensively studied [80]. However, their clinical significance is still under debate. On the one hand, no consensus exists regarding the best test to perform aPT measurement. On the other hand, a systematic review of the literature did not show any consistent association between aPT and thrombosis [28]. Studies suggest that IgG against human PT or the PT/phosphatidylserine complex may represent a risk factor for venous thrombosis [50]. Protein S (PS) is another target for aPLs. Antibodies to PS are reported in children during or after infections (mainly viral) [81], but they are commonly transient and their routine detection could be of little use in adult patients suspected of suffering from APS.

Antibodies directed against phosphatidylethanolamine (aPE), a zwitterionic phospholipid, deserve particular attention because of their relationship with the thrombotic events of APS as reported in several studies [82]. Specifically, aPE and the conventional aPLs were measured in a large cohort of thrombotic patients with or without the main known clinical and biological risk factors for thrombosis. Most of the subjects were suffering from venous thrombosis. Interestingly, aPE was present in 15% of the thrombotic patients versus 3% of the controls ($p < 0.001$) and mostly alone (62%). Moreover, aPE was found to be an independent risk factor for venous thrombosis with an odds ratio of 6 [82]. From the clinical practice standpoint, in some patients with a high clinical degree of suspicion of APS but without aPLs (the so-called "seronegative APS"), the detection of these antibodies with specificities other than that for β₂GPI should not be neglected. In case of positive results, long-term anticoagulation should be assessed.

IgA Isotype of Anticardiolipin Antibodies and/or Anti-β₂-glycoprotein I

The exact role of the IgA isotype of aCLs and anti-β₂GPI in the pathogenesis of APS is unknown because there is a paucity of data and contradictory results concerning their clinical relevance. One reason for the variability in the data may be the lack of standardized assays and cutoff values. In addition, in some studies, the prevalence of both aCL IgA and anti-β₂GPI IgA may possibly depend on the ethnicity of the study population [83–85]. On the contrary, several reports support the clinical utility of the IgA isotype, especially anti-β₂GPI IgA [86]. In a prospective study of patients with known aCL-associated illnesses, IgA was the most common isotype in both aCL and anti-β₂GPI, and anti-β₂GPI was superior to aCL for the diagnosis of APS [87]. In a retrospective study, a strong relationship between elevated titers of anti-β₂GPI IgA and a history of venous thrombosis, thrombocytopenia, valvular heart disease, livedo reticularis, and epilepsy was found [88]. Shen et al. [89] found that elevated titers of the IgA isotype of any ELISA-based aPLs appeared to be an independent risk factor for thromboses even in the absence of LA in 472 patients with thrombosis. Given these data, in patients with a high degree of suspicion of APS and who are aPL negative, testing for IgA aCLs and anti-β₂GPI may be performed. In case of positivity, long-term anticoagulation may be assessed (see Table 55.2).

MANAGEMENT OF OBSTETRIC ANTIPHOSPHOLIPID SYNDROME

In addition to recurrent pregnancy loss, either early miscarriage or fetal death, other reproductive processes such as unexplained infertility and mainly implantation failure after *in vitro* fertilization and embryo transfer may be affected by aPLs [90]. Because of the uncertain etiology and pathogenesis of obstetric APS, treatment has remained empirical and speculative. On the one hand, steroids and intravenous immunoglobulin (IVIG) may reduce the production of aPLs. On the other hand, the use of antiplatelet/anticoagulant agents, mainly aspirin and heparin, prevents thrombosis in the uteroplacental circulation. Interventions such as these drug therapies and monitored pregnancy have increased fetal survival, but no gold standard has been determined [90].

The first treatment proposed for prevention of fetal loss associated with aPLs was the combination of high doses of prednisone (40 mg/day or more) and LDA [91, 92]. Steroids were initially prescribed in the hope of lowering antibody levels. However, there was no significant association between the disappearance of aPLs and live birth [93]. Results from three randomized trials that tested corticosteroids are relevant in this respect. One compared heparin (20,000 U/day) versus prednisone (40 mg/day) (both plus LDA) [94]. Another compared prednisone at a starting dose of 20 mg/day plus LDA versus aspirin alone; a similar live birth rate was found for the two treatments, but there was a higher incidence of premature rupture of the membranes, preeclampsia, abruptio placenta, and preterm delivery in the prednisone-treated groups [95]. The live birth rate was also similar in a randomized trial in women with a history of at least two fetal losses associated with various autoimmune anomalies comparing prednisone (0.5–0.8 mg/kg daily) plus LDA to placebo [96]. More infants were born prematurely and a higher frequency of major side effects of therapy in the mother (hypertension and diabetes mellitus) was reported in the treated group.

In addition, it soon became apparent that corticosteroid treatment for pregnant women was associated with considerable perinatal risks, mainly preterm delivery and premature rupture of the membranes. Systemic steroids administration is also associated with considerable maternal morbidity, including hypertension, cushingoid features, acne, infection, osteopenia and vertebral collapse, and diabetes. Therefore, based on the previously discussed evidence, it is clear that prednisone treatment in pregnancy is ineffective but associated with maternal and fetal morbidity previously thought to be related to APS and should thus be avoided [97].

The effectiveness of IVIG therapy can be related to several immunological mechanisms, such as blockade of antibody binding to receptors on macrophages, an increase in T-suppressor cells, or a decrease in antibody synthesis [98]. The first multicenter, placebo-controlled pilot study of IVIG treatment of APS during pregnancy was designed to assess the impact on obstetric and neonatal outcomes of the addition of IVIG (1 g/kg body weight for 2 consecutive days each month) to a heparin and LDA regimen [99]. Immunoglobulin therapy did not improve obstetric and neonatal outcomes beyond those achieved with a heparin and LDA regimen.

A prospective, nonrandomized, two-center trial including 82 recurrent aborters (two or more first-trimester spontaneous abortions) showed similar live birth rates but higher pregnancy complication rates (pregnancy-induced hypertension, gestational diabetes, and preterm delivery) in patients treated with prednisone plus LDA than in IVIG-treated women, which may be explained by prednisone-dependent side effects [100].

In a third prospective and randomized trial, Triolo et al. [101] compared IVIG and anticoagulation with low-molecular-weight heparin plus LDA in 40 women with recurrent pregnancy loss (at least three occurrences) associated with aPLs. Both therapies were started when the women were pregnant. The women treated with low-molecular-weight heparin plus LDA had a higher rate of live births (84%) than those treated with IVIG (57%). Therefore, on the basis of these results, prednisone and IVIG cannot be recommended for the initial treatment of APS patients, and its use in such patients remains anecdotal. In some cases, however, corticosteroids, usually at low doses, are needed as part of antepartum care for nonobstetrical indications, such as flares of SLE or coincident immune thrombocytopenic purpura. It remains possible (and untested) that women for whom previous anticoagulant therapy during pregnancy has failed may derive sufficient benefit from the addition of IVIG for a successful pregnancy [102].

Good results with LDA alone, with success rates higher than 70%, have been achieved in APS patients with two or more pregnancy losses [95, 103–107]. Aspirin daily doses used in these studies ranged between 75 and 100 mg. The optimal antiplatelet dose for aspirin is still uncertain. Although doses as high as 325 mg three times a day have been used in the past, there is no evidence that doses higher than 75 mg/day are more effective in preventing thrombotic events, whereas toxicity is probably dose related [108].

The use of heparin was a logical approach to treatment for a disorder resulting from thrombosis. In the earliest published case series in 1990 [109], it was observed that under heparin (mean dose, 24 700 U/day), 14 of 15 pregnancies ended in live births in 14 women with aPLs and history of 28 miscarriages out of 29 previous pregnancies. There is accumulating experience with the use of low-molecular-weight heparins both in pregnant and in nonpregnant patients for the prevention of complications associated with aPLs. There is also evidence that low-molecular-weight heparins do not cross the placenta and that they are safe and effective in pregnancy [110]. Low-molecular-weight heparins have potential advantages over unfractionated heparin during pregnancy because they cause less heparin-induced thrombocytopenia, have the potential for once-daily administration because of better bioavailability and longer half-life, and may result in a lower risk for heparin-induced osteoporosis [110].

No considerable difference in terms of fetal outcome between pregnancies treated with only aspirin and those treated with heparin (alone or in combination with aspirin), whether associated with prednisone or not, has been observed. However, this may be explained on the basis of selection criteria for heparin use, which represents higher risk groups of patients because it is administered in cases of vascular history and/or when aspirin alone has failed in a previous pregnancy.

The most recommended treatment for women with recurrent pregnancy wastage and aPLs is heparin and LDA, starting therapy when pregnancy is confirmed [90, 97, 111–115]. This recommendation is essentially based on three clinical trials that reported better obstetric outcomes using aspirin plus heparin rather than aspirin alone [116–118]. Two of them were randomized trials [116, 118], whereas Kutteh [117] assigned treatment in a consecutive way, which limits the validity of his results. Results from two studies [116, 117] were quite similar. The live birth rate in the heparin-treated groups was higher compared to that in women treated with aspirin alone. On the contrary, Farquharson et al. [118] found similar results with LDA alone or in combination with dalteparin. Moreover, no differences were found between treatment groups with respect to obstetrical complications. However, potential limitations of these studies [117, 118] need to be addressed. The study by Kutteh [117] excluded women with LA, whereas in the study of Rai et al. [116], 80% of the patients had LA in the absence of aCLs. This is a very infrequent profile because aCLs are usually present in more than 80% of patients with APS [3]. Importantly, in both reports there is no mention of whether patients had been treated with aspirin alone in a previous pregnancy ending in miscarriage. Most patients recruited for both studies had early pregnancy losses—a condition usually seen and treated by specialists dealing with fertility difficulties. It is thus plausible that APS referral centers are receiving patients who have failed in their first treatment attempt with LDA, and in these cases therapy with heparin plus LDA is recommended [90, 119]. In the study by Rai et al. [116], most miscarriages occurred before 13 weeks of gestation, and there was no difference in outcome between the two treatments in pregnancies that advanced beyond this time. Undoubtedly, a daily 100-mg aspirin

can be started when pregnancy is contemplated, but beginning heparin or prednisone prior to conception exposes the patient to an unknown duration of these drugs, with potential risks. In the study by Rai et al. [116], aspirin treatment was started after diagnosis of pregnancy. In the other two studies [117, 118], the outcomes of the aspirin-only group were worse than those of other series with the same schedule [95, 104–107, 119]. Two studies—a double-blinded, randomized, placebo-controlled trial [120] and a subgroup analysis of one randomized, controlled trial performed by Cowchock and Reece [121]—have not shown any benefit of adding aspirin to intensive obstetric care and placebo treatment. The prognosis in both the aspirin group and the control group was remarkably good, with success rates higher than 80%. However, most patients included in these studies had only low-titer aCLs [120] or lacked significant adverse obstetric histories [121]. In addition, treatment was started when pregnancy was diagnosed or on discovery of aPLs during pregnancy but not before conception. Thus, these two studies emphasize a very important aspect of the management of these patients, and the only one for which there is general agreement: These patients should undergo close fetal and maternal surveillance by a well-coordinated multidisciplinary team that includes obstetricians, internists/rheumatologists, and hematologists.

The American College of Chest Physicians' evidence-based clinical practice guidelines for the management of antithrombotic therapy in pregnancy recommend that women with recurrent early pregnancy loss or unexplained late pregnancy loss who test positive for aPLs and have no history of venous or arterial thrombosis should be treated with antepartum administration of prophylactic or intermediate-dose unfractioned heparin or prophylactic low-molecular-weight heparin combined with aspirin (grade 1B) [110].

In summary, due to the lack of controlled clinical trials, the treatment should be discussed with the patient. In general practice, preconceptional counseling should be carried out to explain the risks of pregnancy. Patients must understand that they have an increased risk of potentially severe complications, such as preeclampsia, prematurity, or thrombosis, during pregnancy but also the puerperium. Therefore, close fetal and maternal surveillance during pregnancy by a coordinated multidisciplinary team is mandatory. Thus, a first Doppler study of the uterine arteries should be performed in all pregnant women with APS at weeks 20 and 24 of gestation, with serial examinations of umbilical arteries afterwards. Moreover, blood pressure and the presence of protein in the urine must be closely monitored at each visit, mainly during the second and third trimesters (Table 55.4).

TABLE 55.4   Proposed Therapeutic Strategies for Patients with Obstetric Manifestations and Antiphospholipid Antibodies

GENERAL CARE

Close fetal and maternal surveillance by a coordinated multidisciplinary team
Doppler study of uterine arteries at weeks 20 and 24 of gestation and afterwards
Monitoring blood pressure and proteinuria
Peripartum thromboprophylaxis mandatory

Women with aPLs and a first pregnancy or with previous normal pregnancies	General care or LDA
Women with obstetric APS in form of early miscarriages	LDA[a] If fail: LDA + LMWH (prophylactic dose)
Women with obstetric APS in form of fetal losses or previous preeclampsia or prematurity	LDA + LMWH (prophylactic dose) If fail: LDA + LMWH (full dose) If fail: to add IVIG
Women with APS and previous thrombosis	LDA + LMWH (full dose) Cumarins may be used after 12th week of gestation If fail: to add IVIG

[a]*Whatever the chosen option, it is advisable to start aspirin before conception.*
aPLs, antiphospholipid antibodies; APS, antiphospholipid syndrome; IVIG, intravenous immunoglobulin; LDA, low-dose aspirin; LMWH, low-molecular-weight heparin.

For women with aPLs and a first pregnancy or with previous normal pregnancies, close fetal and maternal control may be adequate, without specific treatment. However, depending on the patients' characteristics (e.g., age), LDA may be required. In women with early miscarriages without previous thrombosis, aspirin in monotherapy is an option, especially in younger patients with several fertile years remaining. In women with previous fetal losses without previous thrombosis and those with previous preeclampsia or prematurity, combination treatment with LDA and low-molecular-weight heparin at prophylactic doses could be the first option. In any case and whatever the chosen option, it is advisable to start aspirin before conception. If this therapeutic plan fails, the following step is to combine aspirin with low-molecular-weight heparin at full doses. The last option in the case of lack of efficiency is to add monthly IVIG during pregnancy (see Table 55.4).

In APS patients with previous thrombosis, a full therapeutic dose of low-molecular-weight heparin should be prescribed. It is important to understand that cumarins are associated with fetal malformations, especially during organogenesis (between 6 and 12 weeks of pregnancy). Therefore, patients should be switched to the full dose of low-molecular-weight heparin before this period. In patients with a high risk of thrombosis, such those with previous arterial events or recurrent events, oral anticoagulation may be an option during pregnancy, but only after week 12. These patients should be placed back on oral anticoagulation as soon as possible after delivery.

CATASTROPHIC ANTIPHOSPHOLIPID SYNDROME

Catastrophic APS is a rare form of APS, representing less than 1% of reported APS cases [122]. In the earliest published series, the mortality rate was approximately 50% [9, 10]. However, our group found that the mortality rate had declined by approximately 20% [123]. This is clearly due to the use, as first-line therapies, of full anticoagulation (AC), corticosteroids (CS), plasma exchanges (PE), and IVIG.

The mechanisms of causation and pathogenesis of catastrophic APS are not completely understood. It is unclear why some patients develop recurrent thromboses, mainly affecting large vessels, whereas others develop rapidly recurring vascular occlusions, predominantly affecting small vessels. A possible mechanism of catastrophic APS is systemic inflammatory response syndrome, which is presumed to be due to excessive cytokine release from affected and necrotic tissues [124].

As mentioned previously, the higher recovery rate was achieved by the combination of AC + CS + PE (77.8%), followed by AC + CS + PE and/or IVIG (69%). In contrast, concomitant treatment with cyclophosphamide did not demonstrate additional benefit [123]. However, Bayraktar *et al.* [125] demonstrated that cyclophosphamide use improved survival in SLE-associated patients.

When patients were divided according to year of diagnosis and treatment, the mortality rate decreased from 53% in those diagnosed before 2000 to 33.3% in

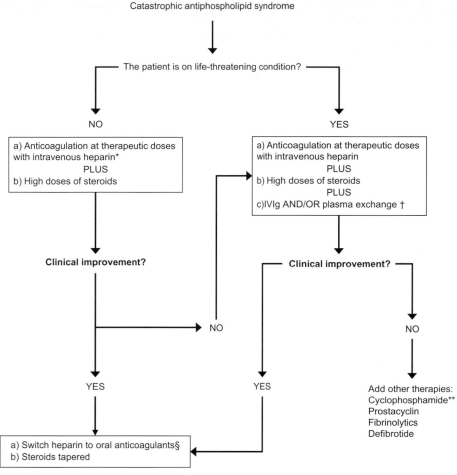

FIGURE 55.1   Treatment algorithm of catastrophic antiphospholipid syndrome.

those diagnosed from 2001 to February 2005 ($p = 0.005$; OR, 2.25; 95% CI, 1.27—3.99) [126]. Patients in the second period were younger than those in the first (34.4 ± 11.8 and 39.4 ± 14.8 years, respectively; $p = 0.016$), and a higher number of precipitating factors for catastrophic APS episodes were identified in the second period. In addition, treatments including AC + CS + PE and/or IVIG were more frequently administered in the episodes of catastrophic APS diagnosed after 2001 compared with the previous period (28.6 vs. 13.3%). We consider that the difference in the mean age at the time of catastrophic APS between patients in the first and the second period, although statistically significant, was not significant enough to explain the decrease in mortality rate in the second period. The higher number of identifiable precipitating factors in the second period may indicate that physicians are increasingly recognizing catastrophic APS and, therefore, earlier and more specific therapies for both precipitating factors and catastrophic

event are being prescribed. However, we consider that the main explanation for this significant reduction of mortality was the more frequent use of treatment with AC + CS + PE and/or IVIG. According to the results of this study, we strongly advocate the use of a combined treatment of AC + CS + PE as first-line therapy for patients with catastrophic APS [126]. This is in accordance with the international consensus on guidelines for the management of catastrophic APS (Figure 55.1).

FUTURE TRENDS IN THE MANAGEMENT OF THE ANTIPHOSPHOLIPID SYNDROME

Although there is consensus regarding the treatment of APS patients with thrombosis with long-term oral anticoagulation and regarding the use of aspirin plus heparin to prevent obstetric manifestations, the

TABLE 55.5   Main Current Pathogenetic Views of Antiphospholipid Antibodies and Potential Future Therapeutic Strategies

Pathogenic Mechanisms	Potential Future Therapeutic Strategies
EFFECT ON HEMOSTATIC REACTIONS	
Coagulation	New thrombin inhibitors
Acquired protein C resistance	
Inhibition of protein S cofactor activity	
Inhibition of antithrombin	
Inhibition of tissue factor pathway inhibitor	
Deposition of immune complexes	
Induction of microparticle formation	
Fibrinolyis	
Inhibition of tPA activity	
Inhibition of fibrinolysis via interaction with antiplasmin	
Activation of factor XI	
ACTIVATION OF CELLULAR ELEMENTS	
Endothelial cells	Statins
Proinflammatory and prothombotic phenotype	P38MAPK inhibitors
TLR4 and annexin 2	Hydroxychloroquine
Monocytes	Anticytokine therapies
Induction of tissue factor	
TLR4	Inhibition of TF upregulation
Platelets	
Activation and induction of aggregation	
LRP-8, GPIba	Platelet GP IIb/IIIa antagonists
Activation of complement	Complement inhibitors
Production of aPL	Rituximab

Abbreviations: aPLs: antiphospholipid antibodies, LRP-8: LDL receptor related-protein 8, GP: glycoprotein, TF: tissue factor, TLR: toll-like receptor, tPA: tissue plasminogen activator.

discovery of new pathogenic mechanisms might result in the identification of novel therapeutic targets and therefore may improve the management of these patients. Table 55.5 highlights the main current pathogenetic views of aPLs and potential future therapeutic strategies.

References

[1] S. Miyakis, M.D. Lockshin, T. Atsumi, D.W. Branch, R.L. Brey, R. Cervera, et al., International consensus statement on an update of the classification criteria for definite antiphospholipid syndrome (APS), J Thromb Haemost 4 (2006) 295–306.

[2] R.A. Asherson, M.A. Khamashta, J. Ordi-Ros, R.H. Derksen, S.J. Machin, J. Barquinero, et al., The "primary" antiphospholipid syndrome: Major clinical and serological features, Medicine (Baltimore) 68 (1989) 366–374.

[3] R. Cervera, J.C. Piette, J. Font, M.A. Khamashta, Y. Shoenfeld, M.T. Camps, et al., Antiphospholipid syndrome: Clinical and immunologic manifestations and patterns of disease expression in a cohort of 1000 patients, Arthritis Rheum 46 (2002) 1019–1027.

[4] R.A. Asherson, R. Cervera, Unusual manifestations of the antiphospholipid syndrome, Clin Rev Allergy Immunol 25 (2003) 61–78.

[5] K.D. Deane, S.G. West, Antiphospholipid antibodies as a cause of pulmonary capillaritis and diffuse alveolar hemorrhage: A case series and literature review, Semin Arthritis Rheum 35 (2005) 154–165.

[6] M. Sanmarco, M.C. Boffa, Antiphosphatidylethanolamine antibodies and the antiphospholipid syndrome, Lupus 18 (2009) 920–923.

[7] R.A. Asherson, The catastrophic antiphospholipid syndrome, J Rheumatol 19 (1992) 508–512.

[8] S. Bucciarelli, G. Espinosa, R. Cervera, D. Erkan, J.A. Gómez-Puerta, Ramos-Casals, et al., Mortality in the catastrophic

antiphospholipid syndrome: Causes of death and prognostic factors in a series of 250 patients, Arthritis Rheum 54 (2006) 2568–2576.

[9] R.A. Asherson, R. Cervera, J.C. Piette, J. Font, J.T. Lie, A. Burcoglu, et al., Catastrophic antiphospholipid syndrome. Clinical and laboratory features of 50 patients, Medicine (Baltimore) 77 (1998) 195–207.

[10] R.A. Asherson, R. Cervera, J.C. Piette, Y. Shoenfeld, G. Espinosa, M.A. Petri, et al., Catastrophic antiphospholipid syndrome: Clues to the pathogenesis from a series of 80 patients, Medicine (Baltimore) 80 (2001) 355–377.

[11] N.M. Shah, M.A. Khamashta, T. Atsumi, G.R. Hughes, Outcome of patients with anticardiolipin antibodies: A 10-year follow-up of 52 patients, Lupus 7 (1998) 3–6.

[12] G. Finazzi, V. Brancaccio, M. Moia, N. Ciaverella, M.G. Mazzucconi, P.C. Schinco, et al., Natural history and risk factors for thrombosis in 360 patients with antiphospholipid antibodies: A four-year prospective study from the Italian Registry, Am J Med 100 (1996) 530–536.

[13] J.A. Giron-Gonzalez, E. Garcia del Rio, C. Rodriguez, J. Rodríguez-Martorell, A. Serrano, Antiphospholipid syndrome and asymptomatic carriers of antiphospholipid antibody: Prospective analysis of 404 individuals, J Rheumatol 31 (2004) 1560–1567.

[14] E. Somers, L.S. Magder, M. Petri, Antiphospholipid antibodies and incidence of venous thrombosis in a cohort of patients with systemic lupus erythematosus, J Rheumatol 29 (2002) 2531–2536.

[15] R. Forastiero, M. Martinuzzo, G. Pombo, D. Puente, A. Rossi, L. Celebrin, et al., A prospective study of antibodies to beta2-glycoprotein I and prothrombin, and risk of thrombosis, J Thromb Haemost 3 (2005) 1231–1238.

[16] D. Erkan, M.J. Harrison, R. Levy, M. Peterson, M. Petri, L. Sammaritano, et al., Aspirin for primary thrombosis prevention in the antiphospholipid syndrome: A randomized, double-blind, placebo-controlled trial in asymptomatic antiphospholipid antibody-positive individuals, Arthritis Rheum 56 (2007) 2382–2391.

[17] M. Gerosa, C. Chighizola, P.L. Meroni, Aspirin in asymptomatic patients with confirmed positivity of antiphospholipid antibodies? Yes (in some cases), Intern Emerg Med 3 (2008) 201–203.

[18] D. Erkan, Lupus and thrombosis, J Rheumatol 33 (2006) 1715–1717.

[19] D.G. Wahl, H. Bounameaux, P. de Moerloose, F.P. Sarasin, Prophylactic antithrombotic therapy for patients with systemic lupus erythematosus with or without antiphospholipid antibodies: Do the benefits outweigh the risks? A decision analysis, Arch Intern Med 160 (2000) 2042–2048.

[20] M. Petri, Thrombosis and systemic lupus erythematosus: The Hopkins Lupus Cohort perspective, Scand J Rheumatol 25 (1996) 191–193.

[21] G. Ruiz-Irastorza, M.V. Egurbide, J.I. Pijoan, M. Garmendia, I. Villar, A. Martinez-Berriotxoa, et al., Effect of antimalarials on thrombosis and survival in patients with systemic lupus erythematosus, Lupus 15 (2006) 577–583.

[22] D. Erkan, Y. Yazici, M.G. Peterson, L. Sammaritano, M.D. Lockshin, A cross-sectional study of clinical thrombotic risk factors and preventive treatments in antiphospholipid syndrome, Rheumatology (Oxford) 41 (2002) 924–929.

[23] D. Erkan, J.T. Merrill, Y. Yazici, L. Sammaritano, J.P. Buyon, M.D. Lockshin, High thrombosis rate after fetal loss in antiphospholipid syndrome: Effective prophylaxis with aspirin, Arthritis Rheum 44 (2001) 1466–1467.

[24] A. Tincani, M. Taglietti, C. Biasini, M. Frassi, R. Gorla, G. Balestrieri, Thromboembolic events after fetal loss in patients with antiphospholipid syndrome: Comment on the article by Erkan et al, Arthritis Rheum 46 (2002) 1126–1127.

[25] A. Ruffatti, T. Del Ross, M. Ciprian, M. Nuzzo, M. Rampudda, M.T. Bertero, et al., Risk factors for a first thrombotic event in antiphospholipid antibody carriers. A multicenter retrospective follow-up study, Ann Rheum Dis 68 (2009) 397–399.

[26] D.G. Wahl, F. Guillemin, E. de Maistre, C. Perret, T. Lecompte, G. Thibaut, Risk for venous thrombosis related to antiphospholipid antibodies in systemic lupus erythematosus—A meta-analysis, Lupus 6 (1997) 467–473.

[27] D.G. Wahl, F. Guillemin, E. de Maistre, C. Perret-Guillaume, T. Lecompte, G. Thibaut, Meta-analysis of the risk of venous thrombosis in individuals with antiphospholipid antibodies without underlying autoimmune disease or previous thrombosis, Lupus 7 (1998) 15–22.

[28] M. Galli, D. Luciani, G. Bertolini, T. Barbui, Lupus anticoagulants are stronger risk factors for thrombosis than anticardiolipin antibodies in the antiphospholipid syndrome: A systematic review of the literature, Blood 101 (2003) 1827–1832.

[29] M. Turiel, P. Sarzi-Puttini, R. Peretti, E. Rossi, F. Atzeni, W. Parsons, et al., Thrombotic risk factors in primary antiphospholipid syndrome: A 5-year prospective study, Stroke 36 (2005) 1490–1494.

[30] G. Ruiz-Irastorza, M.A. Khamashta, The treatment of antiphospholipid syndrome: A harmonic contrast, Best Pract Res Clin Rheumatol 21 (2007) 1079–1092.

[31] G. Ruiz-Irastorza, B.J. Hunt, M.A. Khamashta, A systematic review of secondary thromboprophylaxis in patients with antiphospholipid antibodies, Arthritis Rheum 57 (2007) 1487–1495.

[32] W. Lim, M.A. Crowther, J.W. Eikelboom, Management of antiphospholipid antibody syndrome: A systematic review, JAMA 295 (2006) 1050–1057.

[33] M.H. Rosove, P.M. Brewer, Antiphospholipid thrombosis: Clinical course after the first thrombotic event in 70 patients, Ann Intern Med 117 (1992) 303–308.

[34] R.H. Derksen, P.G. de Groot, L. Kater, H.K. Nieuwenhuis, Patients with antiphospholipid antibodies and venous thrombosis should receive long term anticoagulant treatment, Ann Rheum Dis 52 (1993) 689–692.

[35] M.A. Khamashta, M.J. Cuadrado, F. Mujic, N.A. Taub, B.J. Hunt, G.R. Hughes, The management of thrombosis in the antiphospholipid-antibody syndrome, N Engl J Med 332 (1995) 993–997.

[36] S. Krnic-Barrie, C.R. O'Connor, S.W. Looney, S.S. Pierangeli, E.N. Harris, A retrospective review of 61 patients with antiphospholipid syndrome. Analysis of factors influencing recurrent thrombosis, Arch Intern Med 157 (1997) 2101–2108.

[37] F.J. Muñoz-Rodriguez, J. Font, R. Cervera, J.C. Reverter, D. Tàssies, G. Espinosa, et al., Clinical study and follow-up of 100 patients with the antiphospholipid syndrome, Semin Arthritis Rheum 29 (1999) 182–190.

[38] G. Ruiz-Irastorza, M.A. Khamashta, B.J. Hunt, A. Escudero, M.J. Cuadrado, G.R. Hughes, Bleeding and recurrent thrombosis in definite antiphospholipid syndrome: Analysis of a series of 66 patients treated with oral anticoagulation to a target international normalized ratio of 3.5, Arch Intern Med 162 (2002) 1164–1169.

[39] P.R. Ames, A. Ciampa, M. Margaglione, G. Scenna, L. Iannaccone, V. Brancaccio, Bleeding and re-thrombosis in primary antiphospholipid syndrome on oral anticoagulation: An 8-year longitudinal comparison with mitral valve

replacement and inherited thrombophilia, Thromb Haemost 93 (2005) 694—699.

[40] A.K. Wittkowsky, J. Downing, J. Blackburn, E. Nutescu, Warfarin-related outcomes in patients with antiphospholipid antibody syndrome managed in an anticoagulation clinic, Thromb Haemost 96 (2006) 137—141.

[41] J.S. Ginsberg, P.S. Wells, P. Brill-Edwards, D. Donovan, K. Moffatt, M. Johnston, et al., Antiphospholipid antibodies and venous thromboembolism, Blood 86 (1995) 3685—3691.

[42] P. Prandoni, P. Simioni, A. Girolami, Antiphospholipid antibodies, recurrent thromboembolism, and intensity of warfarin anticoagulation, Thromb Haemost 75 (1996) 859.

[43] A. Rance, J. Emmerich, J.N. Fiessinger, Anticardiolipin antibodies and recurrent thromboembolism, Thromb Haemost 77 (1997) 221—222.

[44] S. Schulman, E. Svenungsson, S. Granqvist, Anticardiolipin antibodies predict early recurrence of thromboembolism and death among patients with venous thromboembolism following anticoagulant therapy. Duration of Anticoagulation Study Group, Am J Med 104 (1998) 332—338.

[45] M.A. Crowther, J.S. Ginsberg, J. Julian, J. Denburg, J. Hirsh, J. Douketis, et al., A comparison of two intensities of warfarin for the prevention of recurrent thrombosis in patients with the antiphospholipid antibody syndrome, N Engl J Med 349 (2003) 1133—1138.

[46] G. Finazzi, R. Marchioli, V. Brancaccio, P. Schinco, F. Wisloff, J. Musial, et al., A randomized clinical trial of high-intensity warfarin vs. conventional antithrombotic therapy for the prevention of recurrent thrombosis in patients with the antiphospholipid syndrome (WAPS), J Thromb Haemost 3 (2005) 848—853.

[47] C. Kearon, S.R. Kahn, G. Agnelli, S. Goldhaber, G.E. Raskob, A.J. Comerota, Antithrombotic therapy for venous thromboembolic disease, Chest 133 (2008) 454S—545S.

[48] R.L. Sacco, S. Prabhakaran, J.L. Thompson, A. Murphy, R.R. Sciacca, B. Levin, et al., Comparison of warfarin versus aspirin for the prevention of recurrent stroke or death: Subgroup analyses from the Warfarin-Aspirin Recurrent Stroke Study, Cerebrovasc Dis 22 (2006) 4—12.

[49] R. Cervera, M.A. Khamashta, Y. Shoenfeld, et al., Euro-Phospholipid Project Group (European Forum on Antiphospholipid Antibodies), Morbidity and mortality in the antiphospholipid syndrome during a 5-year period: A multicenter prospective study of 1000 patients, Ann Rheum Dis 68 (2009) 1428—1432.

[50] M. Galli, G. Borrelli, E.M. Jacobsen, R.M. Marfisi, G. Finazzi, R. Marchioli, et al., Clinical significance of different antiphospholipid antibodies in the WAPS (warfarin in the antiphospholipid syndrome) study, Blood 110 (2007) 1178—1183.

[51] V. Pengo, A. Biasiolo, C. Pegoraro, U. Cucchini, F. Noventa, Antibody profiles for the diagnosis of antiphospholipid syndrome, Thromb Haemost 93 (2005) 1147—1152.

[52] A. Martinez-Berriotxoa, G. Ruiz-Irastorza, M.V. Egurbide, M. Garmendia, J. Gabriel Erdozain, I. Villar, et al., Transiently positive anticardiolipin antibodies and risk of thrombosis in patients with systemic lupus erythematosus, Lupus 16 (2007) 810—816.

[53] G. Espinosa, S. Bucciarelli, D. Tàssies, J.C. Reverter, R. Cervera, Persistently positive antiphospholipid antibodies are related with the appearance of thrombosis during follow-up of patients with antiphospholipid syndrome [abstract], Arthritis Rheum 56 (Suppl.) (2007) S554.

[54] D. Dizon-Townson, C. Hutchison, R. Silver, D.W. Branch, K. Ward, The factor V Leiden mutation which predisposes to thrombosis is not common in patients with antiphospholipid syndrome, Thromb Haemost 74 (1995) 1029—1031.

[55] M. Torresan, T.F. Machado, L.H. Siqueira, M.C. Ozelo, V.R. Arruda, J.M. Annichino-Bizzacchi, The impact of the search for thrombophilia risk factors among antiphospholipid syndrome patients with thrombosis, Blood Coagul Fibrinolysis 11 (2000) 679—682.

[56] N. Chopra, S. Koren, W.L. Greer, P.R. Fortin, J. Rauch, I. Fortin, et al., Factor V Leiden, prothrombin gene mutation, and thrombosis risk in patients with antiphospholipid syndrome, J Rheumatol 29 (2002). 1683-1638.

[57] R. Forastiero, M. Martinuzzo, Y. Adamczuk, M.L. Varela, G. Pombo, L.O. Carreras, The combination of thrombophilic genotypes is associated with definite antiphospholipid syndrome, Haematologica 86 (2001) 735—741.

[58] S. Bentolila, L. Ripoll, L. Drouet, I. Crassard, E. Tournier-Lasserve, J.C. Piette, Lack of association between thrombosis in primary antiphospholipid syndrome and the recently described thrombophilic 3′-untranslated prothrombin gene polymorphism, Thromb Haemost 78 (1997) 1415.

[59] G.J. Ruiz-Arguelles, J. Garces-Eisele, G.J. Ruiz-Delgado, D. Alarcon-Segovia, The G20210A polymorphism in the 3′-untranslated region of the prothrombin gene in Mexican mestizo patients with primary antiphospholipid syndrome, Clin Appl Thromb Hemost 5 (1999) 158—160.

[60] M.L. Bertolaccini, T. Atsumi, B.J. Hunt, O. Amengual, M.A. Khamashta, G.R. Hughes, Prothrombin mutation is not associated with thrombosis in patients with antiphospholipid syndrome, Thromb Haemost 80 (1998) 202—203.

[61] J.L. Pablos, R.A. Caliz, P.E. Carreira, T. Atsumi, L. Serrano, O. Amengual, et al., Risk of thrombosis in patients with antiphospholipid antibodies and factor V Leiden mutation, J Rheumatol 26 (1999) 588—590.

[62] R. Simantov, S.K. Lo, J.E. Salmon, L.R. Sammaritano, R.L. Silverstein, Factor V Leiden increases the risk of thrombosis in patients with antiphospholipid antibodies, Thromb Res 84 (1996) 361—365.

[63] B. Montaruli, A. Borchiellini, G. Tamponi, L. Giorda, P. Bessone, J.A. van Mourik, et al., Factor V Arg506→Gln mutation in patients with antiphospholipid antibodies, Lupus 5 (1996) 303—306.

[64] P.R. Ames, C. Tommasino, G. D'Andrea, L. Iannaccone, V. Brancaccio, M. Margaglione, Thrombophilic genotypes in subjects with idiopathic antiphospholipid antibodies—Prevalence and significance, Thromb Haemost 79 (1998) 46—49.

[65] M. Galli, G. Finazzi, F. Duca, F. Norbis, M. Moia, The G1691→A mutation of factor V, but not the G20210→A mutation of factor II or the C677→T mutation of methylenetetrahydrofolate reductase genes, is associated with venous thrombosis in patients with lupus anticoagulants, Br J Haematol 108 (2000) 865—870.

[66] G.R. Hughes, M.A. Khamashta, Seronegative antiphospholipid syndrome, Ann Rheum Dis 62 (2003) 1127.

[67] M. Sanmarco, Clinical significance of antiphosphatidylethanolamine antibodies in the so-called "seronegative antiphospholipid syndrome.", Autoimmun Rev 9 (2) (2009) 90—92.

[68] M. Galli, G.M. Willems, J. Rosing, R.M. Janssen, J.W. Govers-Riemslag, P. Comfurius, et al., Anti-prothrombin IgG from patients with anti-phospholipid antibodies inhibits the inactivation of factor Va by activated protein C, Br J Haematol 129 (2005) 240—247.

[69] D. Erkan, W.J. Derksen, V. Kaplan, L. Sammaritano, S.S. Pierangeli, R. Roubey, et al., Real world experience with antiphospholipid antibody tests: How stable are results over time? Ann Rheum Dis 64 (2005) 1321—1325.

[70] W.A. Wilson, A.E. Gharavi, T. Koike, M.D. Lockshin, D.W. Branch, J.C. Piette, et al., International consensus statement on preliminary classification criteria for definite antiphospholipid syndrome. Report of an international workshop, Arthritis Rheum 42 (1999) 1309–1311.

[71] J. Criado-García, R.A. Fernández-Puebla, L. López Jiménez, F. Velasco, M. Santamaría, A. Blanco-Molina, Anticoagulation treatment withdrawal in primary antiphospholipid syndrome when anticardiolipin antibodies become negative, Rev Clin Esp 208 (2008) 135–137.

[72] D.J. Wallace, Does hydroxychloroquine sulphate prevent clot formation in systemic lupus erythematosus? Arthritis Rheum 30 (1987) 1435–1436.

[73] D.J. Wallace, M. Linker-Israeli, A.L. Metzger, V.J. Stecher, The relevance of antimalarial therapy with regard to thrombosis, hypercholesterolemia and cytokines in SLE, Lupus 2 (Suppl. 1) (1993) S13–S15.

[74] M. Petri, Hydroxychloroquine use in the Baltimore Lupus Cohort: Effects on lipids, glucose and thrombosis, Lupus 5 (Suppl. 1) (1996) S16–S22.

[75] E.J. Favaloro, R.C. Wong, S. Jovanovich, P. Roberts-Thomson, A review of beta2-glycoprotein I antibody testing results from a peer-driven multilaboratory quality assurance program, Am J Clin Pathol 127 (2007) 441–448.

[76] P. Soltész, Z. Szekanecz, E. Kiss, Y. Shoenfeld, Cardiac manifestations in antiphospholipid syndrome, Autoimmun Rev 6 (2007) 379–386.

[77] M. Lockshin, F. Tenedios, M. Petri, G. McCarty, R. Forastiero, S. Krilis, et al., Cardiac disease in the antiphospholipid syndrome: Recommendations for treatment. Committee consensus report, Lupus 12 (2003) 518–523.

[78] I. Uthman, B. Godeau, A. Taher, M. Khamashta, The hematologic manifestations of the antiphospholipid syndrome, Blood Rev 22 (2008) 187–194.

[79] P.R. Ames, C. Tommasino, G. Fossati, G. Scenna, V. Brancaccio, F. Ferrara, Limited effect of rituximab on thrombocytopaenia and anticardiolipin antibodies in a patient with primary antiphospholipid syndrome, Ann Hematol 86 (2007) 227–228.

[80] K. Oku, T. Atsumi, O. Amengual, T. Koike, Antiprothrombin antibody testing: Detection and clinical utility, Semin Thromb Hemost 34 (2008) 335–339.

[81] C.H. van Ommen, M. van Wijnen, P.G. de Groot, C.M. van der Horst, J. Peters, Postvaricella purpura fulminans caused by acquired protein S deficiency resulting from antiprotein S antibodies: Search for the epitopes, J Pediatr Hematol Oncol 24 (2002) 413–416.

[82] M. Sanmarco, S. Gayet, M.C. Alessi, M. Audrain, E. de Maistre, J.C. Gris, et al., Antiphosphatidylethanolamine antibodies are associated with an increased odds ratio for thrombosis. A multicenter study with the participation of the European Forum on antiphospholipid antibodies, Thromb Haemost 97 (2007) 949–954.

[83] J.F. Molina, S. Gutierrez-Urena, J. Molina, O. Uribe, S. Richards, C. De Ceulaer, et al., Variability of anticardiolipin antibody isotype distribution in 3 geographic populations of patients with systemic lupus erythematosus, J Rheumatol 24 (1997) 291–296.

[84] K.L. Wong, H.W. Liu, K. Ho, K. Chan, R. Wong, Anticardiolipin antibodies and lupus anticoagulant in Chinese patients with systemic lupus erythematosus, J Rheumatol 18 (1991) 1187–1192.

[85] E. Cucurull, A.E. Gharavi, E. Diri, E. Mendez, D. Kapoor, L.R. Espinoza, IgA anticardiolipin and anti-beta2-glycoprotein I are the most prevalent isotypes in African American patients with systemic lupus erythematosus, Am J Med Sci 318 (1999) 55–60.

[86] M. Samarkos, K.A. Davies, C. Gordon, S. Loizou, Clinical significance of IgA anticardiolipin and anti-beta2-GP1 antibodies in patients with systemic lupus erythematosus and primary antiphospholipid syndrome, Clin Rheumatol 25 (2006) 199–204.

[87] T.P. Greco, M.D. Amos, A.M. Conti-Kelly, J.D. Naranjo, J.W. Ijdo, Testing for the antiphospholipid syndrome: Importance of IgA antibeta 2-glycoprotein I, Lupus 9 (2000) 33–41.

[88] G. Lakos, E. Kiss, N. Regeczy, P. Tarján, P. Soltész, M. Zeher, et al., Isotype distribution and clinical relevance of anti-beta2-glycoprotein I (beta2-GPI) antibodies: Importance of IgA isotype, Clin Exp Immunol 117 (1999) 574–579.

[89] Y.M. Shen, R. Lee, R. Frenkel, R. Sarode, IgA antiphospholipid antibodies are an independent risk factor for thromboses, Lupus 17 (2008) 996–1003.

[90] M. Petri, U. Qazi, Management of antiphospholipid syndrome in pregnancy, Rheum Dis Clin North Am 32 (2006) 591–607.

[91] W.F. Lubbe, W.S. Butler, S.J. Palmer, G.C. Liggins, Fetal survival after prednisone suppression of maternal lupus-anticoagulant, Lancet 1 (1983) 1361–1363.

[92] W.F. Lubbe, G.C. Liggins, Lupus anticoagulant and pregnancy, Am J Obstet Gynecol 153 (1985) 322–327.

[93] H.J. Out, H.W. Bruinse, G.C. Christiaens, M. van Vliet, P.G. de Groot, H.K. Nieuwenhuis, et al., A prospective, controlled multicenter study on the obstetric risks of pregnant women with antiphospholipid antibodies, Am J Obstet Gynecol 167 (1992) 26–32.

[94] F.S. Cowchock, E.A. Reece, D. Balaban, D.W. Branch, L. Plouffe, Repeated fetal losses associated with antiphospholipid antibodies: A collaborative randomized trial comparing prednisone with low-dose heparin treatment, Am J Obstet Gynecol 166 (1992) 1318–1323.

[95] R.K. Silver, S.N. MacGregor, J.S. Sholl, J.S. Sholl, J.M. Hobart, M.G. Neerhof, et al., Comparative trial of prednisone plus aspirin versus aspirin alone in the treatment of anticardiolipin antibody-positive obstetric patients, Am J Obstet Gynecol 169 (1993) 1411–1417.

[96] C.A. Laskin, C. Bombardier, M.E. Hannah, F.P. Mandel, J.W. Ritchie, V. Farewell, et al., Prednisone and aspirin in women with autoantibodies and unexplained recurrent fetal loss, N Engl J Med 337 (1997) 148–153.

[97] M. Greaves, H. Cohen, S.J. MacHin, I. Mackie, Guidelines on the investigation and management of the antiphospholipid syndrome, Br J Haematol 109 (2000) 704–715.

[98] J.F. Seite, Y. Shoenfeld, P. Youinou, S. Hillion, What is the content of the magic draft IVIg? Autoimmun Rev 7 (2008) 435–439.

[99] D.W. Branch, A.M. Peaceman, M. Druzin, R.K. Silver, Y. El-Sayed, R.M. Silver, et al., A multicenter, placebo-controlled pilot study of intravenous immune globulin treatment of antiphospholipid syndrome during pregnancy. The Pregnancy Loss Study Group, Am J Obstet Gynecol 182 (2000) 122–127.

[100] E. Vaquero, N. Lazzarin, H. Valensise, S. Menghini, G. Di Pierro, F. Cesa, et al., Pregnancy outcome in recurrent spontaneous abortion associated with antiphospholipid antibodies: A comparative study of intravenous immunoglobulin versus prednisone plus low-dose aspirin, Am J Reprod Immunol 45 (2001) 174–179.

[101] G. Triolo, A. Ferrante, F. Ciccia, A. Accardo-Palumbo, A. Perino, A. Castelli, et al., Randomized study of subcutaneous low molecular weight heparin plus aspirin versus intravenous immunoglobulin in the treatment of recurrent fetal loss associated with antiphospholipid antibodies, Arthritis Rheum 48 (2003) 728–731.

III. ANTI-PHOSPHOLIPID SYNDROME

[102] M.A. Khamashta, Systemic lupus erythematosus and pregnancy, Best Pract Res Clin Rheumatol 20 (2006) 685—694.

[103] F. Carmona, J. Font, M. Azulay, M. Creus, F. Fábregues, R. Cervera, et al., Risk factors associated with fetal losses in treated antiphospholipid syndrome pregnancies: A multivariate analysis, Am J Reprod Immunol 46 (2001) 274—279.

[104] F. Lima, M.A. Khamashta, N.M. Buchanan, S. Kerslake, B.J. Hunt, G.R. Hughes, A study of sixty pregnancies in patients with the antiphospholipid syndrome, Clin Exp Rheumatol 14 (1996) 131—136.

[105] K.A. Granger, R.G. Farquharson, Obstetric outcome in antiphospholipid syndrome, Lupus 6 (1997) 509—513.

[106] D.L. Huong, B. Wechsler, O. Bletry, D. Vauthier-Brouzes, G. Lefebvre, J.C. Piette, A study of 75 pregnancies in patients with antiphospholipid syndrome, J Rheumatol 28 (2001) 2025—2030.

[107] L.H. Silveira, C.L. Hubble, L.J. Jara, S. Saway, P. Martínez-Osuna, M.J. Seleznick, et al., Prevention of anticardiolipin antibody-related pregnancy losses with prednisone and aspirin, Am J Med 93 (1992) 403—411.

[108] G. Ruiz-Irastorza, M.A. Khamashta, G.R. Hughes, Anti-aggregant and anticoagulant therapy in systemic lupus erythematosus and Hughes' syndrome, Lupus 10 (2001) 241—245.

[109] M.H. Rosove, K. Tabsh, N. Wasserstrum, P. Howard, B.H. Hahn, K.C. Kalunian, Heparin therapy for pregnant women with lupus anticoagulant or anticardiolipin antibodies, Obstet Gynecol 75 (1990) 630—634.

[110] S.M. Bates, I.A. Greer, I. Pabinger, S. Sofaer, J. Hirsh, et al., Venous thromboembolism, thrombophilia, antithrombotic therapy, and pregnancy: American College of Chest Physicians evidence-based clinical practice guidelines (8th edition), Chest 133 (Suppl. 6) (2008) 844S—886S.

[111] W.H. Kutteh, N.S. Rote, R. Silver, Antiphospholipid antibodies and reproduction: The antiphospholipid antibody syndrome, Am J Reprod Immunol 41 (1999) 133—152.

[112] ACOG Practice Bulletin No. 68, Antiphospholipid syndrome, Obstet Gynecol 106 (2005) 1113—1121.

[113] R.H. Derksen, P.G. De Groot, H.K. Nieuwenhuis, G.C. Christiaens, How to treat women with antiphospholipid antibodies in pregnancy? Ann Rheum Dis 60 (2001) 1—3.

[114] B.L. Myones, D. McCurdy, The antiphospholipid syndrome: Immunologic and clinical aspects. Clinical spectrum and treatment, J Rheumatol Suppl 58 (2000) 20—28.

[115] M. Empson, M. Lassere, J. Craig, J. Scott, Prevention of recurrent miscarriage for women with antiphospholipid antibody or lupus anticoagulant, Cochrane Database Syst Rev (2005) CD002859.

[116] R. Rai, H. Cohen, M. Dave, L. Regan, Randomised controlled trial of aspirin and aspirin plus heparin in pregnant women with recurrent miscarriage associated with phospholipid antibodies (or antiphospholipid antibodies), Br Med J 314 (1997) 253—257.

[117] W.H. Kutteh, Antiphospholipid antibody-associated recurrent pregnancy loss: Treatment with heparin and low-dose aspirin is superior to low-dose aspirin alone, Am J Obstet Gynecol 174 (1996) 1584—1589.

[118] R.G. Farquharson, S. Quenby, M. Greaves, Antiphospholipid syndrome in pregnancy: A randomized, controlled trial of treatment, Obstet Gynecol 100 (2002) 408—413.

[119] J. Balasch, J. Font, A. Lopez-Soto, R. Cervera, I. Jové, F.J. Casals, et al., Antiphospholipid antibodies in unselected patients with repeated abortion, Hum Reprod 5 (1990) 43—46.

[120] N.S. Pattison, L.W. Chamley, M. Birdsall, A.M. Zanderigo, H.S. Liddell, J. McDougall, Does aspirin have a role in improving pregnancy outcome for women with the antiphospholipid syndrome? A randomized controlled trial, Am J Obstet Gynecol 183 (2000) 1008—1012.

[121] S. Cowchock, E.A. Reece, Do low-risk pregnant women with antiphospholipid antibodies need to be treated? Organizing Group of the Antiphospholipid Antibody Treatment Trial, Am J Obstet Gynecol 176 (1997) 1099—1100.

[122] R. Cervera, Lessons from the "Euro-Phospholipid" project, Autoimmun Rev 7 (2008) 174—178.

[123] S. Bucciarelli, R. Cervera, G. Espinosa, J.A. Gómez-Puerta, M. Ramos-Casals, J. Font, Mortality in the catastrophic antiphospholipid syndrome: Causes of death and prognostic factors, Autoimmun Rev 6 (2006) 72—75.

[124] G. Espinosa, R. Cervera, R.A. Asherson, Catastrophic antiphospholipid syndrome and sepsis. A common link? J Rheumatol 34 (2007) 923—926.

[125] U.D. Bayraktar, D. Erkan, S. Bucciarelli, G. Espinosa, R. Asherson, et al., The clinical spectrum of catastrophic antiphospholipid syndrome in the absence and presence of lupus, J Rheumatol 34 (2007) 346—352.

[126] R.A. Asherson, R. Cervera, P.G. de Groot, D. Erkan, M.C. Boffa, J.C. Piette, et al., Catastrophic antiphospholipid syndrome: International consensus statement on classification criteria and treatment guidelines, Lupus 12 (2003) 530—534.

PART IV

TREATMENT

Corticosteroid and Nonsteroidal Anti-Inflammatory Drug use in Systemic Lupus Erythematosus

Olga Dvorkina, Ellen M. Ginzler

BIOLOGIC EFFECTS OF GLUCOCORTICOIDS

A hallmark of systemic lupus erythematosus (SLE) activity is acute and chronic inflammation, resulting from activation of inflammatory genes which encode cytokines, chemokines, adhesion molecules, inflammatory proteins, and receptors [1]. A loss of tolerance to self antigens with polyclonal activation of B lymphocytes resulting in increased general antibody and autoantibody production is characteristic of the immunologic function in SLE. Glucocorticoids (GCs) exert their biologic effect through specific GC receptors on multiple cell types, with resulting anti-inflammatory and immunosuppressive actions, which correlate with the dose and duration of GC therapy (Table 56.1). After GC administration, the biologic activities of monocytes, macrophages, and endothelial cells are rapidly affected, in turn mediating leucocyte trafficking and adhesion to vascular walls. Vascular dilatation and permeability are reduced by downregulation of nitric oxide production, secretion of anti-inflammatory cytokines such as IL10 and TGF-β are increased, and pro-inflammatory cytokines including TNF, IL-1, and GMC-SF are inhibited. The presence and activity of both T and B lymphocytes are affected by steroid administration, ultimately resulting in decreased autoantibody formation [2]. The balance of Th1/Th2 T lymphocytes is shifted, with a decrease in circulating CD4+/Th1 cells and activation of Th2 cells. In 2008, studies suggested that circulating CD4+CD25+ T regulatory cells (T_{REG}) are increased compared to control subjects in SLE patients receiving chronic GC therapy [3]. This finding may reflect increased activity of GC-induced TNF-receptor in T_{REGs} or a general increase in the expression of cytokine receptors secondary to GC therapy. Direct effects on B cells are less prominent, predominantly driven by the changes in T-cell regulatory functions. Serum immunoglobulin levels may decrease as a response to short-term GC administration. The effects of corticosteroids on B cells *in vitro* are better understood, and differ at different phases of B-cell activation. The proliferation of activated B cells is markedly depressed by steroids; however, the differentiation into immunoglobulin-secreting cells is not suppressed *in vitro* by steroids [4].

THE EFFECT OF CORTICOSTEROIDS UPON SURVIVAL IN SLE

One of the earliest reports of a series of SLE patients receiving steroid therapy was published in 1952, describing the course of 18 patients treated with 200–300 mg/day of cortisone; steroids were tapered as clinical manifestations improved. In a follow-up ranging from 3–20 months, six of the 18 patients died [5]. As steroids became a more common therapy for SLE, each new study based on the date of publication reported generally higher survival rates over time. Kaplan reviewed these findings, and had an interesting and unusual interpretation of the data [6]. He made the argument that through the early 1980s the reported studies did not substantiate an improvement in survival due to steroid therapy. Instead, he suggested that differences in outcome were related to disease severity. He also pointed out that the pre-steroid era corresponded temporally to the pre-antibiotic, pre-biopsy era, and prior to the development of more effective antihypertensive medications.

TABLE 56.1   Anti-inflammatory and immunosuppressive mechanisms of corticosteroid action

Anti-inflammatory effects	Mechanism of action	Clinical outcome
Antagonism of transcription factors (NF-κB, AP-1, NF-AT)	1. Up-regulates synthesis of transcription factor inhibitors (I-κB, GILZ)	1. Reduction in prodution of TNF-α, IL-1, GM-CSF, IL-6, IFN-γ, IL-12
	2. Direct interaction between GR and transcription factors, inhibiting protranscriptional activity	
	3. Competition for essential transcriptional cofactors (CREB-binding protein, SRC-1)	2. Reduced expression of IL-8, ICAM-1, ELAM-1, VCAM-1, E-selectin, RANTES, MCP-1
Reduction of cytokine and chemokine expression	1. Synthesis of destabilizing protein reduces mRNA half-life	3. Decreased substance P and inducible nitric oxide synthetase expression
	2. Transcriptional inhibition via direct GR or GRE antagonism and reduction of transcription factor activity (see earlier)	
Inhibition of arachidonic acid release by enchanced lipocortin activity	1. Rapid tanslocation of cytosolic lipocortin to cell surface	Reduced production of prostaglandin mediators of inflammation
	2. Increased transcription of lipocortin mRNA	

Immunosuppression	Mechanism of action	Clinical outcome
Inhibition of IL-2/IL-2R activity	1. Direct suppression of IL-2R mRNA levels (possibly through transrepression)	Modulation of T-cell activation
	2. Up-regulation of GILZ and other mechanisms of NF-κB inhibition)	
	3. Inhibition of IL-2 signaling by blockade of STAT 5 activation (possibly by inhibition of tyrosine phosphorylation)	
Inhibition of dendritic cell (DC) function	1. Impaired antigen-presenting function	
	2. Reduced expression of costimulatory molecules	
	3. Suppression of IL-12 production	
	4. Apoptosis of dermal/interstitial DC precursors	
Preferential enhancement of Th2-type cytokine profile	Down-regulation of IL-12 activity	1. Enhanced TGF-β and IL-10 production
	1. Inhibition of STAT 4 phosphorylation (rapid onset but mechanism unknown)	2. Reduction of inflammatory cytokines
	3. Down-regulation of IL-12 receptor	

Urman and Rothfield presented an interesting comparison of two lupus cohorts using life table analysis as support for the concept that steroid therapy was associated with improved survival [7]. The two cohorts consisted of 209 patients treated in New York City (NYC) from 1957–1968, and the other included 156 patients treated in Connecticut from 1968–1976. A significantly greater number of nonwhite (black and Puerto Rican) patients were treated in the NYC group (85 vs. 55%), whereas the serum creatinine and BUN levels were similar at baseline in both groups. There was a higher incidence of death from infection in group 1 and from lupus nephritis in group 2, but the differences were not statistically significant. Differences in survival were significant at 1 year (89 vs. 99%, $p < 0.001$), 5 years (70 vs. 93%, $p < 0.001$), and 10 years (63 vs. 84%, $p < 0.003$). The authors point out a difference in treatment philosophy among the two cohorts; in group 2 patients the serum C3 and anti-DNA titers were evaluated in all patients at every visit, and a decision to reduce or increase the steroid dose was influenced by these studies. No data was provided on the mean or median daily doses, cumulative doses, or duration of steroids in either group.

Albert, Hadler and Ropes also examined the hypothesis that corticosteroid therapy was responsible for the dramatic improvement in survival among lupus patients [8]. In a meta-analysis of 52 publications and a separate assessment of outcome among 142 SLE patients treated at the Massachusetts General Hospital from 1922–1966, they concluded that steroids had no discernible effect on overall survival in low- and medium-risk groups. However, the use of steroids was associated with improved survival among high-risk patients. Subsequent to these assessments of the contribution of steroids to overall survival in SLE, there have been no further systematic studies in relatively large cohorts over an extended period of time. Controlled studies of corticosteroids alone vs. other therapies would not be warranted, and are probably unethical. Attention has turned to steroid efficacy as acute management of specific clinical and serologic manifestations, and prevention of disease exacerbations and progression of features of damage.

INDICATIONS FOR CORTICOSTEROIDS BASED ON DISEASE MANIFESTATIONS

From the time of the FDA approval of cortisone for the treatment of lupus, little has changed with regard to recommendations for acute therapy of severe or life-threatening manifestations. Presentation with high fever, obtundation, rapidly progressive nephritis with renal insufficiency, severe hemolytic anemia, and/or thrombocytopenia requires rapid and aggressive therapy. Initiation of high-dose steroids, whether oral or intravenous, should be considered immediately; over time the steroid formulations have changed and the definition of high-dose therapy has escalated (Table 56.2) [2, 9–16]. Current oral therapy for severe lupus activity most often begins with a prednisone dose of 1 mg/kg/day. Divided doses may be necessary initially, especially with features such as fever and changes in mental status. A single daily dose may be sufficient in the absence of these clinical manifestations. Failure to respond quickly may necessitate increases in dose, frequently with doubling of the initial dose. Intravenous steroids, most often methylprednisolone, should be considered in patients who cannot tolerate oral steroids or in whom absorption of medications is impaired, usually secondary to underlying gastrointestinal malfunction. Extremely high-dose steroid therapy (i.e. intravenous pulse therapy) will be discussed below.

Mild to moderate presentations or exacerbations of lupus activity may not require initiation of steroid therapy, or treatment may range from low to moderate (e.g. 5–40 mg/day) doses of prednisone or its equivalent. Any time that steroid therapy is initiated, or its dose is increased as a result of a flare of disease activity, consideration should be made of the need for a change in the maintenance regimen that will allow for steroid tapering and prevention of further flares.

Lupus nephritis

Lupus nephritis (LN) has generally been recognized as the clinical manifestation which confers the worst

TABLE 56.2 Therapeutic glucocorticoid compounds

Systemic preparations	Potency equivalent
Short-acting	
Hydrocortisone	1
Prednisone	4
Prednisolone	4
Methylprednisolone	5
Intermediate-acting	
Triamcinolone	4–5
Long-acting	
Betamethasone	30
Dexamethasone	25–30
Topical preparations	**Brand names**
Lowest potency	
Hydrocortisone, 1%, 2.5%	
Low potency	
Fluocinolone acetonide, 0.01%	Synalar
Triamcinolone acetonide, 0.025%	Aristocort
Intermediate potency	
Betamethasone valerate, 0.1%	Beta-Val
Fluocinolone acetonide, 0.025% Cordran	Synalar, Synemol flurandrenolide, 0.05%
Triamcinolone acetonide, 0.1%	Aristocort, Triacet
High potency	
Betamethasone dipropionate, 0.05%	Diprolene, Lotrisone
Fluocinonide, 0.05%	Lidex
Halcinonide, 0.1%	Halog
Triamcinolone acetonide, 0.5%	Aristocort
Very high potency	
Betamethasone dipropionate (with propylene glycol), 0.05%	Diprolene
Clobetasol, 0.05%	Temovate, Cormax
Halobetasol, 0.05%	Ultravate

prognosis for both overall and organ-based survival, contributing to the worst short-time mortality among lupus cohorts. In an uncontrolled retrospective study comparing a 6-month regimen of 15–20 mg/day of prednisone vs. 60–100 mg/day, Pollak and colleagues pointed out more than 35 years ago that prednisone dose correlated with short-term survival, preservation of renal function, and histologic improvement in patients with lupus nephritis [17, 18].

Reviewing studies from the early 1970s, it has become difficult to isolate the benefits of either short- or long-term steroid therapy, as a result of the common use of immunosuppressive agents in combination with corticosteroids. Early uncontrolled studies reported that azathioprine (AZA) was associated with improvement in azotemia and proteinuria in 50–100% of patients with severe lupus nephritis who were previously unresponsive to high-dose corticosteroids alone [19–21]. Shortly thereafter, publications followed reporting the efficacy of various cyclophosphamide (CTX) regimens for the treatment of LN. In a study by Donadio and colleagues, 50 patients with nephritis were randomized, 26 to prednisone alone and 24 to prednisone plus oral CTX [22]. The prednisone alone group were treated with a dose starting at 60 mg per day orally for 1–3 months, followed by gradual tapering to approximately 20 mg per day by 6 months. In the combination therapy group, prednisone dose was continued for 6 months in patients receiving less than 60 mg per day at the time they entered the study; for those on 60 mg per day at entry, the dose was tapered in the same manner as for the patients treated with prednisone alone. Twenty-one patients in each group improved, but 10 had a subsequent renal flare over a mean 41.2 (range 11–70) month follow-up. A substantially higher rate of later renal flares was observed in the patients treated with prednisone alone, and 4 patients in this group progressed to end-stage renal disease [22]. A number of studies followed comparing prednisone alone to combination regimens for LN, allowing one to draw conclusions about the efficacy of steroids as a stand-alone regimen. Most of these originated at the National Institute for Health (NIH) in the 1980s, with reports over 5 and 10 years of progressive follow-up [23, 24]. One hundred and eleven LN patients at the NIH were randomized to one of five regimens; those randomized to any of the regimens containing CTX (IV, oral, or oral AZA + CTX) had significantly better preservation of renal function compared to those randomized to prednisone (1 mg/kg/day for 4–8 weeks, then tapered slowly based on clinical response) alone. Flares were treated with increases in prednisone up to 1 mg/kg for 4–6 weeks in the prednisone only group; in the four groups with immunosuppressive drugs, increase from the 10–20 mg/day maintenance dose was allowed of no more than 0.5 mg/kg/day for 1 month. Patient survival was somewhat reduced in the prednisone only group, but no comparisons reached statistical significance, even when combining all of the immunosuppressive groups. Progression to renal failure was significantly reduced in the CTX groups compared to prednisone alone.

Since the late twentieth-century, many other trials, both randomized and retrospectively reviewed, have examined the outcome of LN based on the combination of steroids plus immunosuppressive or biologic therapies. None have compared prednisone to placebo, or prednisone to a newer immunosuppressive regimen without steroids. For example, Levey and colleagues assessed the response to therapy within 48 weeks of 86 patients with LN followed for a mean of 136 weeks, based on initial clinical and pathologic features [25]. As part of a protocol including randomization to plasmapheresis or not, all patients were initially treated with prednisone 60–80 mg/day for 4 weeks, tapered to 20–25 mg every other day over 32 weeks, plus oral CTX 1–2 mg/kg for first 8 weeks. While the addition of plasmapheresis did not improve outcome for patients with severe LN, there was an inverse correlation of response to therapy over the initial 48 weeks with subsequent development of end-stage renal disease. Favorable outcomes, on the other hand, were considered to be remission or improvement in abnormal baseline parameters: of 55 patients with initial elevated serum creatinine, 20 (36%) had resolution during follow-up, while only 4 (7%) had resolution of both azotemia and proteinuria. Based on these data, the authors propose treatment for LN with an initial 8-week course of high-dose prednisone and low-dose oral CTX, followed by long-term low-dose alternate-day prednisone without another immunosuppressive agent, especially in patients with features that confer a low risk of developing renal failure.

More recently published clinical trials evaluating the efficacy of other immunosuppressive regimens compared to placebo have all included some dose of steroids as part of the protocol. For example, in Chan et al.'s study of mycophenolate mofetil compared to oral CTX for proliferative nephritis, all patients were started on prednisolone 0.8 mg/kg/day, which was subsequently reduced by 5 mg/day every 2 weeks until the dose was 20 mg/day, then by 2.5 mg per day every 2 weeks for 4 weeks, with a maintenance dose of 10 mg/day at approximately 6 months [26]. It is not clear from the article when the tapering regimen was started. The conclusion to be drawn from experiences to date is that at least high-dose oral steroid remains a component of the standard of care for active LN; data to establish the dose and duration of therapy are not clear.

Neuropsychiatric lupus

Central and peripheral nervous system activity in SLE encompasses a wide spectrum of manifestations, ranging from focal to diffuse in anatomic location and mild to severe or life-threatening. The American College of Rheumatology (ACR) has classified neuropsychiatric (NP) features into 19 syndromes with a standard nomenclature, which aids in the diagnosis and treatment of individual patients [27]. In some cases, high-dose aggressive corticosteroid therapy, often in combination with immunosuppressive or psychotropic agents, is indicated. For other features, the use of steroids is controversial, and may even be contraindicated. The two most common NP features are represented in the ACR Classification Criteria for SLE—psychosis and seizures. Both may present in association with other NP manifestations or other severe clinical features of active lupus, warranting high-dose steroid therapy as part of an overall regimen. In other cases, these manifestations may occur alone; the appropriate therapy is then not so clear. In a retrospective analysis of 28 patients with 52 episodes (37 organic central nervous system disease and 15 psychiatric) of NP lupus, Sergent and colleagues assessed outcome in relationship to steroid treatment [28]. Based on their findings, they challenged the previously held notion that such patients should be treated with high-dose steroids. Twelve of the 28 patients had major complications of therapy after receiving a mean prednisone dose of 170 mg/day for more than a month (range 100–400 mg/day, with or without concomitant immunosuppressive agents), most often serious infections, five of which were fatal. Functional psychosis usually occurred in the setting of prior steroid therapy and responded to a decrease in steroid dose and the addition of psychotropic drugs. They noted that all of the surviving patients with seizures and functional psychoses recovered completely without sequelae.

In a later report, Sibley and colleagues reinforced the conclusion that therapy with steroids did not offer substantial benefit to patients with CNS lupus manifestations [29]. Among the 48 patients in their cohort of 266, CNS disease occurred in the absence of other features of active lupus in 81%. Most CNS episodes were self-limited, and eventual discontinuation of anticonvulsants was successful without further seizure activity. Recurrence of psychosis was rare. In contrast, Pego-Reigosa and Isenberg describe a subset of 11 patients from their cohort of 485 who presented with psychosis at or shortly after SLE diagnosis, all of whom had other features of active lupus [30]. Aggressive immunosuppression with high-dose steroids in all plus monthly intravenous CTX in 40% led to eventual clinical response, although chronic mild psychotic symptoms were persistent in 30% of patients. In general, we agree with the recommendations offered by Sanna and colleagues in their review of CNS lupus stressing the avoidance of steroids when other agents are likely to be effective, such as symptomatic therapy for anxiety and depression, analgesics for severe headache, and anticoagulation for features of the antiphospholipid syndrome [31].

In lupus patients who present with obtundation or coma, rapid and thorough assessment is essential to rule out infectious and metabolic causes. It may be difficult initially to differentiate infection from aseptic meningitis in the presence of an abnormal cerebrospinal fluid examination. While awaiting culture evidence, we believe it is appropriate to treat aggressively with corticosteroids (1–2 mg/kg/day in divided doses) in combination with empirical use of antibiotics. In the face of confirmed sepsis in a patient with other severe or life-threatening manifestations of active lupus, steroids are indicated along with antibiotic therapy.

Transverse myelopathy represents a rare but devastating feature of NP lupus, which must be recognized and treated without delay, to avoid irreversible neurologic damage, including a significant risk of fatality. In addition to identifying an association with antiphospholipid antibodies and treatment with anticoagulation, it is universally accepted that therapy with high-dose steroids, at least 1–2 mg/kg/day in divided doses or even in intravenous pulse doses, usually along with intravenous cyclophosphamide, is warranted [32–34].

Memory problems and other cognitive deficits are a frequent complaint among lupus patients. It is first essential to attempt to distinguish causes other than active lupus itself, such as fatigue, depression, and anxiety. Antiphospholipid antibodies must also be considered. When cognitive problems are thought to be a primary manifestation of lupus, steroid therapy may offer a benefit. In a study using patients as their own controls, Denburg et al. found that, in patients with mild NP symptoms but otherwise inactive lupus, treatment for periods of up to 6 months with prednisone up to 0.5 mg/kg/day can result in improvement in cognition and mood, with the effect lasting beyond the drug exposure [35].

Appropriate therapy for focal nervous system manifestations may be difficult to determine, especially as the pathology of lesions may not be clear in the absence of a tissue diagnosis, although imaging modalities are constantly being improved. It is often assumed that such lesions are secondary to vasculitis or antiphospholipid antibody-related thrombosis. Unusual cases have been reported, however, which are associated with a good therapeutic response to high-dose steroids. For example, a case report describes a lupus patient with isolated bilateral cranial nerve involvement of the vagus nerves resulting in progressive difficulty swallowing along with transient respiratory failure [36]. MRI showed a small brainstem lesion at the ponto-medullary junction near the location of the vagus nerve nuclei. Intravenous methylprednisolone for 3 weeks resulted in complete recovery.

Other visceral manifestations

Cardiopulmonary and gastrointestinal features of lupus range from mild and often intermittent to acute, severe and potentially life-threatening. Corticosteroids in varying doses and duration are the most common initial therapy. Acute lupus pneumonitis (ALP) is associated with significant morbidity and appreciable mortality, while diffuse alveolar hemorrhage (DAH), if untreated, is often fatal. As with CNS manifestations, a search for infectious causes is essential, but rapid and aggressive anti-inflammatory therapy must be instituted. High-dose IV steroids are standard of care, with the addition of IV CTX or plasmapheresis in most patients with DAH [37, 38]. More common but less serious cardiopulmonary manifestations include pleural and pericardial effusions, which may accompany disease activity in other systems, or there may be isolated findings that tend to recur for brief periods with symptoms of pleuritic chest pain and/or shortness of breath. Low to moderate doses of steroids are generally effective; however, nonsteroidal anti-inflammatory drugs may be effective for mild disease or as a steroid-sparing regimen [39].

Several groups have identified an interesting phenomenon of an acute leuco-occlusive vasculopathy in several organ systems that are responsive to steroid therapy. Abramson and colleagues reported the course of acutely ill, hospitalized patients who developed acute reversible hypoxemia with widened A-a O_2 gradients and diffusion abnormalities but with normal imaging studies [40]. They hypothesized that excessive complement activation in the presence of endothelial cells activated by elevated plasma complement split products (C3a, C5a) can result in neutrophil—endothelial cell adhesion and subsequent leuco-occlusive vasculopathy in various organs during a flare of lupus. In this case hypoxemia would lead to a pulmonary leak. The syndrome was responsive to short-term high-dose steroid therapy.

A similar syndrome has been described in two lupus patients with mesenteric vasculopathy, who presented with repeated acute episodes of nausea, vomiting, and abdominal pain, associated with significant hypocomplementemia in the absence of antiphospholipid antibodies [41]. CT scans demonstrated severe intestinal wall edema. The acute flares responded rapidly to bowel rest and intravenous methylprednisolone, suggesting an acute gastrointestinal distress syndrome (AGDS) with bowel wall ischemia caused by leukoaggregation and a gut capillary leak syndrome. Despite excellent short-term responses to corticosteroids, repeated episodes and an inability to taper steroids led to a requirement for maintenance immunosuppressive therapy.

Although uncommon among patients with lupus, the development of pancreatitis causes a dilemma in terms of identifying the underlying cause and deciding on a therapeutic plan. It has been seen during generalized acute flares of lupus disease activity, in association with antiphospholipid antibodies, isolated from other disease manifestations, and during recovery after treatment for a flare with steroids and immunosuppressive drugs. Despite a number of case series publications and literature reviews, the etiologic explanation, and thus the therapeutic plan, remains controversial [42—44]. Reports in 2004 and 2007 appear to suggest an increased occurrence of lupus-associated pancreatitis, and recommendations for treating with steroids outweigh those for avoiding or tapering steroids [45, 46]. We agree with this approach, keeping in mind that care should include bowel rest, generally necessitating intravenous steroid therapy, even in low to moderate doses.

Hematologic manifestations

Thrombocytopenia is the most common hematologic manifestation of lupus requiring therapy. Depending upon the level of platelets (generally less than 30,000/mm^3) and its association with other organ system manifestations, prednisone may be instituted at as low a dose as 20 mg/day, especially in the absence of symptoms of bleeding or severe bruising. Profound thrombocytopenia is generally treated with 1—2 mg/kg/day of prednisone. The response to initial therapy will dictate the need for higher steroid doses or the addition of immunosuppressive agents or intravenous immunoglobulin. It has been suggested that thrombocytopenia in the setting of SLE is less steroid-responsive than idiopathic thrombocytopenia. A study by Abbasi and colleagues supports this concept. They performed a retrospective review of 41 patients aged at least 15 who presented with platelets <30,000/mm^3 to assess the clinical significance of a positive ANA on their presentation and response to steroids [47]. After 2 weeks of therapy with prednisone in a dose of 1—2 mg/kg/day, the mean platelet count in the 10 patients with a positive ANA was 32,800/mm^3, compared to 99,323 in those with a negative ANA. Patients with a positive ANA were 6.25 less likely to achieve a complete response (platelets >100,000 per mm^3 for at least 3 months after discontinuation of therapy). There was no relationship between positive ANA and age, gender, presence of autoimmune disease, family history of autoimmune disease, platelet count, hemoglobin, or ESR prior to steroid treatment. The authors did not comment on subsequent follow-up with regard to whether ANA-positive patients developed manifestations of SLE.

Thrombotic thrombocytopenic purpura, in the setting of established SLE, a much rarer phenomenon, should be considered life-threatening, and treatment with high-dose steroids and plasmapheresis is indicated as soon as the diagnosis is confirmed [48].

Hemolytic anemia may be low-grade, generally responsive to low-to-moderate-dose prednisone. Severe hemolytic anemia with rapid red blood cell destruction should be treated initially with high-dose steroids, with attention also paid to secondary complications such as high-output congestive heart failure, necessitating blood transfusion. Modest degrees of neutropenia and lymphopenia may not require any specific therapy.

Constitutional, mucocutaneous, and musculoskeletal manifestations

Fatigue, low-grade fever, skin rashes, nasopharyngeal ulcers, arthralgias, and synovitis are common features of active lupus; they may also be chronic or intermittent. Patients presenting with an initial diagnosis of SLE or with an acute systemic flare may require a low-to-moderate dose of prednisone (10−40 mg/day) to control the initial symptoms; others may require only non-steroidal anti-inflammatory drugs (see later discussion of NSAIDs). Systemic steroids may be avoided for many cutaneous manifestations with the use of topical preparations (see Table 56.2). Early initiation of antimalarials and/or disease-modifying antirheumatic drugs (DMARDs) is important to consider in order to minimize the ongoing need for steroids.

PATTERNS OF CORTICOSTEROID THERAPY

The dose and duration of steroid therapy are important considerations in treating active lupus, based on the organ systems involved and the degree of functional impairment and clinical symptoms. Other considerations include the route of administration (oral, parenteral), the timing of doses, tapering regimens, and protocols designed to prevent new flares of disease activity.

Intravenous high-dose pulse therapy

Cathcart *et al.* first proposed the use of very high-dose intravenous methylprednisolone (IVMP) for severe lupus nephritis in 1976 [49]. The rationale for this aggressive therapy was the similarity in renal interstitial histologic findings (interstitial edema, round-cell infiltration) between a LN patient with rapid deterioration in renal function and the histology of acute rejection in renal transplants and their good response to IV pulse steroids. The aim of an intense pulse of steroids with subsequent return to a much lower dose sufficient to control other active lupus manifestations was to avoid high-dose long-term steroids and their complications. Initially 7 LN patients were reported, all with evidence of immune complex disease (low C3, + anti-DNA titers). Only one had received low-dose prednisone and none was receiving immunosuppressive therapy. The regimen consisted of IVMP, 1 gm/day for 3 consecutive days; low-dose prednisone (40 mg/day) was withheld for as long as possible (14 days later in 6 patients) to assess immediate effects of the IV pulse. All 7 patients had a return of serum creatinine to pre-flare levels and remained free of renal manifestations 6−30 months after pulse, although 1 developed hemiplegia and psychotic behavior 3 days after last bolus, thought to be due to SLE activity, as it finally resolved with prednisone 120 mg/day. The authors considered why pulse therapy may work, offering the hypothesis that it may be due to reversal of immune complex deposition secondary to interference with anti-DNA antibody synthesis. It may also block the deleterious effects of circulating lymphotoxins, as it results in profound B- and T-cell lymphopenia within 4−6 hours in normal subjects.

The long-term beneficial effects of pulse steroids observed in this first case series were not duplicated in a larger series reported by Isenberg and colleagues several years later [50]. They treated 20 lupus patients with IVMP, assessing outcome at 4, 12, and 24 weeks, based on the daily oral prednisone requirement. Four patients had a complete sustained improvement, as defined by a marked reduction in daily prednisone requirement. Ten patients had a partial sustained improvement, and another 2 had some improvement which was not sustained over 12−24 weeks. No improvement was achieved in 6 patients, while 1 had deterioration in renal status at 4 weeks and another at 12 weeks. No major side effects were noted.

Identification of patients likely to respond to IV pulse steroid therapy was sought in a series of 34 patients, 12 of whom had an improvement in serum creatinine of at least 20% within 2 weeks post-treatment [51]. Distinguishing features of the responder group, 60% of whom maintained their response for at least 6 months, included a recent deterioration in renal function compared to a more stable level of renal dysfunction prior to therapy in the nonresponder group. Responders also had more diffuse lesions on renal biopsy and more abnormal serologic features, including higher anti-DNA

agents only for a clinical flare. The flare rate was higher in the conventional treatment group (20 vs. 2, $p < 0.001$). Immunosuppressive therapy was required for 7 flares in the conventional group and only 2 in the pre-emptive group.

Tseng and colleagues carried out a study based on a similar hypothesis that increases in anti-dsDNA level of at least 25% and $\geq$50% increase in the level of the complement activation product C3a together predict a flare of SLE activity in patients with stable or inactive disease, and that short-term steroid therapy will prevent an overt flare [76]. In a prospective randomized double-blind control trial, 154 patients were followed monthly for up to 19 months to detect these serologic changes; C3, C4, and Ch50 levels were also followed monthly. Forty-one patients underwent a serologic flare, of whom 21 were randomized to prednisone (30 mg/day × 2 weeks, 20 mg/day × 1 week, 10 mg/day × 1 week) and 20 to placebo. Mild, moderate, and severe flares were defined using the SELENA SLEDAI instrument. During a 90-day follow-up after randomization, 11 placebo patients had 13 flares (six mild/moderate, seven severe), whereas none in the prednisone-treated group had a severe flare ($p = 0.0086$). Eight of the total of 21 flares were renal flares. Significant improvement in outcome measures in the steroid group at 1 month included SLEDAI score, anti-DNA antibody level, and C4, with a trend toward improved C3a. The mean daily dose of prednisone was calculated for each group at randomization and at 3 monthly follow-up visits. As per the protocol, the prednisone dose increased in the first month for the treatment group, but in the 6 patients who had a disease flare the prednisone dose increased in the second month and remained higher for the duration of follow-up. The median cumulative dose overall was 350 mg and 135 mg for the placebo and treatment groups. It should be noted that this study had very rigorous enrollment criteria, with all patients having a prior history of positive anti-DNA titers, and monthly follow-up in stable patients may not be practical in clinical practice.

Treatment of lupus flares, once clinically symptomatic disease is recognized, depends upon the particular organ system manifestations and degree of dysfunction. Rapid reversal of disease activity has been attempted in the FLOAT trial, which enrolled 50 patients with mild/moderate flare (musculoskeletal, mucocutaneous, fatigue, and pleurisy) as defined by SELENA SLEDAI [77]. Patients were randomized to either triamcinolone 100 mg IM ($n = 24$) on the clinic visit day of flare or a Medrol dose-pack ($n = 26$) started on the day of the clinic visit. Assessment of efficacy was based on the Medical Outcomes Study Short Form 36 health status questionnaire (SF-36) and a 5-point Likert scale of activity administered on randomization day, day 1,

day 2, week 1, week 2, week 3 and week 4. Some improvement occurred in 69.5, 79.1, 78.2, 91.3, 75, and 79.1% of the triamcinolone group over successive time evaluations, compared to 41.6, 87.5, 79.1, 58.3, 83.3, and 75% with the Medrol dose-pack. For complete improvement, assessments were 4.3, 8.6, 12.5, 30.4, and 34.7% for triamcinolone and 0, 0, 8.3, 20.8, and 25% for oral MP. None of the differences were statistically significant. Although no side effects were reported, compliance with oral MP was inconsistent, in that some patients started the medication late and two failed to take the medication at all. This study points out the virtue of parenterally administered medication in patients who have a history of noncompliance or previous failure to respond to a medical regimen.

STEROID TAPERING

There is no single appropriate recommendation for the reduction in steroid dose after successful control of active disease manifestations, nor are there any controlled studies of different regimens. As discussed above, the addition of antimalarials or immunosuppressive agents may be important not only as a part of the initial induction regimen, but also to facilitate steroid tapering. Although it is usually prudent to first consolidate the total daily dose of steroids from divided to a single morning dose in order to coincide with the normal peak cortisone level, this may not always be possible, especially in patients with features such as fever and obtundation. They may require steroid administration on an 8- or 12-hour basis in order to suppress symptoms. Al-Maini and Urowitz offer logical guidelines for reducing the steroid dose, generally beginning with no more than 25% decrements, later decreasing the dose by smaller amounts at each taper [78]. Close follow-up of clinical symptoms and laboratory evidence is essential in deciding the interval between dose decrements. For patients who are clinically steroid-dependent or reluctant to reduce their dose, we again embrace the philosophy that "more is less." Biweekly, or even monthly, reduction in prednisone dose of 1 mg/day may result in a steady decline in the dose, rather than a fluctuating dose with a more rapid taper and subsequent dose increases. We rarely prescribe alternate-day tapering regimens, not only because patients may develop constitutional symptoms late in the second day, but also because compliance in terms of remembering to take the dose may become an issue.

Reduction and withdrawal of corticosteroid therapy can lead to an array of symptoms ranging from frank adrenal insufficiency to pseudo-rheumatic complaints to flares of the underlying disease [79, 80]. The primary challenge to the clinician is to determine the nature and etiology of the symptoms so that an appropriate adjustment in medication can be made [79−83]. Patients usually respond to reinstitution of the lowest previous dose of steroids.

SIDE EFFECTS OF STEROIDS

Systemic steroids

Use of corticosteroids is associated with significant adverse effects, which generally depend on the dose and duration of treatment, although individual patients may tolerate widely varying steroid regimens without clinical symptoms [84]. Early side effects such as cushingoid features, hirsutism, acne vulgaris, and dyspepsia improve or resolve with dose tapering. Some features may be difficult to distinguish from ongoing active lupus. Behavioural changes such as irritability, difficulty concentrating, depression, or even frank psychosis [85−87] may lag beyond a reduction in steroid dose. Muscle weakness secondary to steroids versus that resulting from partially treated lupus myositis may require muscle biopsy [88, 89].

Both immunosuppression from lupus itself as well as the combined effects of steroids and immunosuppressive agents are well known to increase the risk for infection with common and opportunistic pathogens [90−94].

Hypertension, hyperlipidemia, and diabetes mellitus also may improve with lowering of the steroid dose, but carry the risk for renal insufficiency and premature atherosclerosis. While there is no question that steroids as well as other traditional risk factors contribute to the accelerated development of atherosclerosis in SLE [95−101], a study in 2001 emphasized the pathophysiologic contribution of immune dysregulation [102]. Endothelial dysfunction leading to intimal damage may occur as a consequence of inflammatory vascular processes related to active lupus. Turner and colleagues investigated endothelial function with the technique of flow-mediated dilatation (FMD), but failed to find a significant relationship between FMD and markers of disease activity, duration of disease, or traditional cardiovascular risk factors; nevertheless they concluded that endothelial dysfunction is associated with long-term steroid exposure [103]. El-Magadmi *et al.* found significant impairment of FMD in lupus patients compared to control subjects, but failed to find an association of antiphospholipid antibody status, Raynaud phenomenon, or current steroid use with this measure of endothelial function [104]. Wajed and colleagues propose guidelines for aggressive recognition and management of risk factors to minimize the development and progression of atherosclerosis in SLE patients [105].

Premenopausal women with SLE have reduced bone mineral density at the spine and hip, which can be attributed to a number of factors, including disease duration and activity, lifestyle (i.e. sedentary versus physically active, smoking), and exposure to steroids [106−108]. Osteopenia and early osteroporosis without fractures may be reversible with appropriate treatment regimens [109]; of concern is the finding that alternate-day steroid therapy does not change the risk for osteroporosis [110]. Inclusion of supplemental dietary calcium and vitamin D should be part of a prophylactic or therapeutic regimen for virtually all SLE patients receiving corticosteroids [111−13]. The American College of Rheumatology guidelines for prevention or reversal of bone loss include the use of a bisphosphonate for all postmenopausal women and men being started on long-term glucocorticoid therapy ($\geq$5 mg/day prednisone or equivalent) [113]. Bisphosphonate treatment in premenopausal women is controversial, and should include consideration of future plans for child-bearing.

Avascular necrosis (AVN) occurs as an irreversible complication of exogenous glucocorticoid therapy in lupus patients [114−116]. Although cumulative steroid dose is strongly associated with the incidence of AVN [117], other lupus-related factors such as the presence of antiphospholipid antibodies have been implicated [118]. AVN can occur with minimal steroid use, but high-dose intravenous methylprednisolone does not seem to increase the risk of its occurrence [119, 120].

Progressive cognitive deficit is a troubling feature in some patients with lupus. Although it is a well-known complication of antiphospholipid antibodies, even in the absence of other manifestations of active disease, corticosteroids have also been implicated, primarily in the middle-age group [121]. Zanardi and colleagues compared the frequency and severity of cerebral atrophy using brain CT in 107 lupus patients, 39 non-lupus subjects, and 50 normal individuals. The first two groups were receiving prednisone 1 mg/kg/day for at least 3 months [122]. Cerebral atrophy was more frequent in both groups of patients on steroids compared to controls. Severity of brain atrophy, however, was higher in lupus patients. Other neuropsychiatric manifestations were similar in patients with or without cerebral atrophy, with the exception of seizures, which were more frequent in lupus patients with cerebral atrophy.

Ophthalmologic side effects of steroids may be both acute and chronic. The risk of glaucoma results from increased intraocular pressure, which may be secondary to steroids administered topically, periocularly, or systemically [123]. The prevalence of posterior subcapsular cataracts has been observed to be as high as 30−40% in individuals receiving prednisone in a dose of 10 mg/day or more over a 2-year period

[124]. The cause of cataract formation is generally thought to be due to binding of glucocorticoids to lens proteins, resulting in destabilization of the protein structure [125].

Topical steroids

Topical preparations may cause side effects which are related to both the potency and the absorption of the given preparation. Local atrophy, telangiectasia, purpura, and striae can occur especially with extended use of occlusive dressings or with natural occlusion in the groin or axilla. These effects are attributed to local catabolic effects leading to degeneration of collagen. Atrophy may occur on the face even without occlusion [126]. An eruption similar to rosacea, as well as steroid acne characterized by papules and comedones, may also occur on the face with the use of potent topical preparations.

NONSTEROIDAL ANTI-INFLAMMATORY DRUGS IN SLE: USES AND SIDE EFFECTS

Nonsteroidal anti-inflammatory drugs (NSAIDs) have been prescribed for nearly 80% of SLE patients, both as symptomatic treatment before the diagnosis is made as well as subsequently for features such as fever, headaches, arthralgias, myalgias, mild synovitis, and serositis [127]. Prior to the availability of newer immunosuppressive agents, indometacin was reported to be effective in a small number of patients with LN and refractory nephrotic syndrome [128].

NSAID-induced side effects involve almost all organ systems. Dyspepsia has been reported in 10–50% of NSAID-treated lupus patients [129, 130], although the actual incidence of peptic ulcer disease in SLE has not been reported; in a British cohort of 226 SLE patients, however, no deaths secondary to peptic ulcer disease were found [131].

Renal toxicity from NSAIDs may be the most clinically significant, as there is substantial risk of irreversible damage. Because the mechanism of action of NSAIDs involves inhibition of the regulation of total renal blood flow, distribution of renal blood flow, sodium and water balance, and renin release, side effects include sodium and water retention, and a possible decrease in glomerular filtration rate or even acute renal failure [130]. Patients with active LN may be at increased risk of these NSIAD-induced side effects [131]. Interstitial nephritis, renal papillary necrosis, and nephrotic syndrome have also been described as idiosyncratic reactions [130]. Little evidence exists to support the possibility that selective COX-2 inhibitors have less renal toxicity than other NSAIDS in patients with lupus.

It may be difficult to distinguish the CNS side effects of NSAIDs (aseptic meningitis, psychosis, cognitive dysfunction) from neuropsychiatric features of active lupus. Although ibuprofen is most commonly reported to cause aseptic meningitis, other NSAIDs have also been implicated [132]. Indometacin in particular has been associated with acute psychosis, especially in the elderly [132].

The COX-2 inhibitors, of which only celecoxib is currently available in the United States, have a potential for toxicity not described with nonselective NSAIDs. One such side effect is the increased pro-thrombotic profile, due to unopposed action of thromboxane A-2, a potent stimulator of platelet aggregation [133]. While not specific to SLE, there is also the possibility of allergic responses to the sulfa component of celecoxib [133].

References

[1] P.J. Barnes, How corticosteroids control inflammation: Quintiles Prize Lecture 2005, Br J Pharmacol 148 (2006) 245–254.

[2] W.W. Chatham, R.P. Kimberly, Treatment of lupus with corticosteroids, Lupus 10 (2001) 140–147.

[3] N.A. Azab, I.H. Bassyouni, Y. Emad, G.A. Abd El-Wahab, G. Hamdy, M.A. Mashahit, CD4+CD25+ regulatory T cells (T_{REG}) in systemic lupus erythematosus (SLE) patients: The possible influence of treatment with corticosteroids, Clin Immunol 127 (2008) 151–157.

[4] T.R. Cupps, T.L. Gerrard, J.M. Falkoff, G. Whalen, A.S. Fauci, Effects of in vitro corticosteroids on B cell activation, proliferation, and differentiation, J Clin Invest 75 (1985) 754–761.

[5] L. Sofer, R. Bader, Corticotropin and cortisone in acute disseminated lupus erythematosus, J Am Med Assoc 149 (1952) 1002–1008.

[6] D. Kaplan, Systemic lupus erythematosus — Corticosteroids, Clin Rheum Dis 9 (1983) 601–616.

[7] J.D. Urman, N.F. Rothfield, Corticosteroid treatment in systemic lupus erythematosus. Survival studies, J Am Med Assoc 238 (1977) 2272–2278.

[8] D.A. Albert, N.M. Hadler, M.W. Ropes, Does corticosteroid therapy affect the survival of patients with systemic lupus erythematosus? Arthritis Rheum 22 (1979) 945–953.

[9] E.L. Dubois, Prednisone and prednisolone in the treatment of systemic lupus erythematosus, J Am Med Assoc 161 (1956) 427–433.

[10] E.L. Dubois, Triamcinolone in the treatment of systemic lupus erythematosus, J Am Med Assoc 167 (1958) 1590–1599.

[11] E.L. Dubois, Methylprednisolone (Medrol) in the treatment of systemic lupus erythematosus, J Am Med Assoc 170 (1959) 1537–1542.

[12] E.L. Dubois, Current therapy of systemic lupus erythematosus. A comparative evaluation of corticosteroids and their side-effects with emphasis on fifty patients treated with dexamethasone, J Am Med Assoc 173 (1960) 1833–1840.

[13] E.L. Dubois, Paramethasone in the treatment of systemic lupus erythematosus. Analysis of results in 51 patients with emphasis on single daily oral doses, J Am Med Assoc 184 (1963) 463–469.

[14] M.B. Stevens, B.H. Hahn, Management of systemic lupus erythematosus, Bull Rheum Dis 32 (1982) 35–42.

[15] K.A. Kirou, D.T. Boumpas, Systemic glucocorticoid therapy in systemic lupus erythematosus in "Dubois" Lupus Erythematosus, in: D.J. Wallace, B.H. Hahn (Eds.), seventh ed., Lippincott, Williams & Wilkins, Philadelphia, 2007, pp. 1175–1198.

[16] F. Buttgereit, J.A.P. da Silva, M. Boers, G.-R. Burmester, M. Cutolo, J. Jacobs, et al., Standardised nomenclature for glucocorticoid dosages and glucocorticoid treatment regimens: current questions and tentative answers in rheumatology, Ann. Rheum Dis 61 (2002) 718–722.

[17] V.E. Pollak, C.L. Pirani, R.M. Kark, Effect of large doses of prednisone on the renal lesions and life span of patients with lupus glomerulonephritis, J Lab Clin Med 57 (1961) 495–511.

[18] V.E. Pollak, C.L. Pirani, F.D. Schwartz, The natural history of the renal manifestations of systemic lupus erythematosus, J Lab Clin Med 63 (1964) 537–550.

[19] J.P. Hayslett, M. Kashgarian, C.D. Cook, B.H. Spargo, The effect of azathioprine on lupus glomerulonephritis, Medicine 51 (1972) 393–412.

[20] M. Sztejnbok, A. Stewart, H. Diamond, D. Kaplan, Azathioprine in the treatment of systemic lupus erythematosus. A controlled study, Arthritis Rheum 14 (1971) 639–645.

[21] E. Ginzler, E. Sharon, H. Diamond, D. Kaplan, Long-term maintenance therapy with azathioprine in systemic lupus erythematosus, Arthritis Rheum 18 (1975) 27–35.

[22] J.V. Donadio Jr., K.E. Holley, R.H. Ferguson, D.M. Ilstrup, Treatment of diffuse proliferative lupus nephritis with prednisone and combined prednisone and cyclophosphamide, N Engl J Med 299 (1978) 1151–1155.

[23] H.A. Austin, J.H. Klippel, J.E. Balow, N.G. le Riche, A.D. Steinberg, P.H. Plotz, et al., Therapy of lupus nephritis. Controlled trial of prednisone and cytotoxic drugs, N Engl J Med 314,614-619

[24] A.D. Steinberg, S.C. Steinberg, Long-term preservation of renal function in patients with lupus nephritis receiving treatment that includes cyclophosphamide versus those treated with prednisone only, Arthritis Rheum 34 (1991) 945–950.

[25] A.S. Levey, S.-.P. Lan, H.L. Corwin, B.S. Kasinath, J. Lachin, E.G. Neilson, et al., Progression and remission of renal disease in the Lupus Nephritis Collaborative Study. Results of treatment with prednisone and short-term oral cyclophosphamide, Ann Intern Med 116 (1992) 114–123.

[26] T.K. Chan, F.K. Li, C.S.O. Tang, R.W.S. Wong, G.X. Fang, Y.L. Ji, et al., Efficacy of mycophenolate mofetil in patients with diffuse proliferative nephritis, New Engl J Med 343 (2000) 1156–1162.

[27] The American College of Rheumatology nomenclature and case definitions for neuropsychiatric lupus syndromes, Arthritis Rheum 42 (1999) 599–608.

[28] J.S. Sergent, M.D. Lockshin, M.S. Klempner, B.A. Lipsky, Central nervous system disease in systemic lupus erythematosus, Am J Med 58 (1975) 644–654.

[29] J.T. Sibley, W.P. Olszynski, W.E. Decoteau, M.B. Sundaram, The incidence and prognosis of central nervous system disease in systemic lupus erythematosus, J Rheumatol 19 (1992) 47–52.

[30] J.M. Pego-Reigosa, D.A. Isenberg, Psychosis due to systemic lupus erythematosus: characteristics and long-term outcome of this rare manifestation of the disease, Rheumatology 47 (2008) 1498–1502.

[31] G. Sanna, M.L. Bertolaccini, A. Mathieu, Central nervous system lupus: a clinical approach to therapy, Lupus 12 (2003) 935–942.

[32] V. Harisdangkul, D. Doorenbos, S.H. Subramony, Lupus transverse myelopathy: better outcome with early recognition and aggressive, high-dose intravenous corticosteroid pulse treatment, J Neurol 242 (1995) 326–331.

[33] M Lopez Dupla, M.A. Khamashta, A.D. Sanchez, F.P. Ingles, P.L. Uriol, A.G. Aguado, Transverse myelitis as a first manifestation of systemic lupus erythematosus: a case report, Lupus 4 (1995) 239–242.

[34] K.-F. Chan, M.-L. Boey, Transverse myelopathy in SLE: clinical features and functional outcomes, Lupus 5 (1996) 294–299.

[35] S.D. Denburg, R.M. Carbotte, J.A. Denburg, Corticosteroid and neuropsychological functioning in patients with systemic lupus erythematosus, Arthritis Rheum 37 (1994) 1311–1320.

[36] K.-H. Yu, C.-H. Yang, C.-C. Chu, Swallowing disturbance due to isolated vagus nerve involvement in systemic lupus erythematosus, Lupus 16 (2007) 746–749.

[37] B. Memet, E.M. Ginzler, Pulmonary manifestations of systemic lupus erythematosus, Semin Respir Crit Care Med 28 (2007) 441–450.

[38] A.S. Santos-Ocampo, B.F. Mandell, B.J. Fessler, Alveolar hemorrhage in systemic lupus erythematous. Presentation and management, Chest 118 (2000) 1083–1090.

[39] D.-Y. Wang, Diagnosis and management of lupus pleuritis, Curr Opin Pulm Med 8 (2002) 312–316.

[40] S.B. Abramson, J. Dobro, M.A. Eberle, M. Benton, J. Reibman, H. Epstein, et al., Acute reversible hypoxemia in systemic lupus erythematosus, Ann Intern Med 114 (1991) 941–947.

[41] M. Kishimoto, A. Nasir, A. Mor, H.A. Belmont, Acute gastrointestinal distress syndrome in patients with systemic lupus erythematosus, Lupus 16 (2007) 137–141.

[42] J.C. Reynolds, R.D. Inman, R.P. Kimberly, J.H. Chuong, J.E. Kovacs, M.B. Walsh, Acute pancreatitis in systemic lupus erythematosus: Report of twenty cases and a review of the literature, Medicine 61 (1982) 25–32.

[43] M.C. Serrano Lopez, M. Yebra Bango, E. Lopez Bonet, I. Sanchez Vegazo, F. Albarran Hernandez, L. Manzano Espinosa, et al., Acute pancreatitis and systemic lupus erythematosus: Necropsy of a case and review of the pancreatic vascular lesions, Am J Gastroenterol 86 (1991) 764–767.

[44] S. Saab, M.P. Corr, M.H. Weisman, Corticosteroids and systemic lupus erythematosus pancreatitis: A case series, J Rheumatol 25 (1998) 801–806.

[45] V. Pascual-Ramos, A. Duarte-Rojo, A.R. Villa, B. Hernandez-Cruz, D. Alarcon-Segovia, Alcocer-Varela, et al., Systemic lupus erythematosus as a cause and prognostic factor of acute pancreatitis, J Rheumatol 31 (2004) 707–712.

[46] S. Kobayashi, M. Yoshida, T. Kitahara, Y. Abe, A. Tsuchida, Y. Nojima, Autoimmune pancreatitis as the initial presentation of systemic lupus erythematosus, Lupus 16 (2007) 133–136.

[47] S.Y. Abbasi, M. Milhem, L. Zaru, A positive antinuclear antibody test predicts for a poor response to initial steroid therapy in adults with idiopathic thrombocytopenic purpura, Ann Hematol 87 (2008) 459–462.

[48] D.C. Hess, K. Sethi, E. Awad, Thrombotic thrombocytopenic purpura in systemic lupus erythematosus and antiphospholipid antibodies: effective treatment with plasma exchange and immunosuppression, J Rheumatol 19 (1992) 1474–1478.

[49] E.S. Cathcart, B.A. Idelson, M.A. Scheinberg, W.G. Couser, Beneficial effects of methylprednisolone "pulse" therapy in diffuse proliferative lupus nephritis, Lancet 1 (1976) 163–166.

[50] D.A. Isenberg, W.J.W. Morrow, M.L. Snaith, Methylprednisolone pulse therapy in the treatment of systemic lupus erythematosus, Ann Rheum Dis 41 (1982) 347–351.

[51] R.P. Kimberly, M.D. Lockshin, R.L. Sherman, J.S. McDougal, R.D. Inman, C.L. Christian, High-dose intravenous methyl prednisolone pulse therapy in systemic lupus erythematosus, Am J Med 70 (1981) 817–824.

[52] A.J. Wysenbeek, L. Leibovici, J. Zoldan, Acute central nervous system complications after pulse steroid therapy in patients with systemic lupus erythematosus, J Rheumatol 17 (1990) 1695–1696.

[53] K.S. Barron, D.A. Person, E.J. Brewer Jr., M.G. Beale, A.M. Robson, Pulse methylprednisolone therapy in diffuse proliferative lupus nephritis, J Pediat 101 (1982) 37–141.

[54] F.A. Houssiau, C. Vasconcelos, D. D'Cruz, G.D. Sebastiani, E. de Ramon Garrido, M.G. Danieli, et al., Immunosuppressive therapy in lupus nephritis. The Euro-Lupus Nephritis Trial, a randomized trial of low-dose versus high-dose intravenous cyclophosphamide, Arthritis Rheum 46 (2002) 2121–2131.

[55] M.R. Liebling, K. McLaughlin, S. Boonsue, J. Kasdin, E.V. Barnett, Monthly pulses of methylprednisolone in SLE nephritis, J Rheumatol 9 (1982) 543–548.

[56] J.C. Edwards, M.L. Snaith, D.A. Isenberg, A double blind controlled trial of methylprednisolone infusions in systemic lupus erythematosus using individualised outcome assessment, Ann Rheum Dis 46 (1987) 773–776.

[57] D.T. Boumpas, H.A. Austin III, E.M. Vaughn, J.H. Klippel, A.D. Steinberg, C.H. Yarboro, et al., Controlled trial of pulse methylprednisolone versus two regimens of pulse cyclophsphamide in severe lupus nephritis, Lancet 340 (1992) 741–745.

[58] R. Sesso, M. Monteiro, E. Sato, G. Kirsztan, L. Silva, H. Aizen, A controlled trial of pulse cyclophosphamide versus pulse methylprednisolone in severe lupus nephritis, Lupus 3 (1994) 107–112.

[59] M.F. Gourley, H.A. Austin III, D. Scott, C.H. Yarboro, E.M. Vaughn, J. Muir, et al., Methylprednisolone and cyclophosphamide, alone or in combination, in patients with lupus nephritis. A randomized, controlled trial, Ann Intern Med 125 (1996) 549–557.

[60] S. Eyanson, M.H. Passo, M.A. Aldo-Benson, M.D. Benson, Methylprednisolone pulse therapy for nonrenal lupus erythematosus, Ann Rheum Dis 39 (1980) 377–380.

[61] C.G. Mackworth-Young, J. David, S.H. Morgan, G.R.V. Hughes, A double blind, placebo controlled trial of intravenous methylprednisolone in systemic lupus erythematosus, Ann Rheum Dis 47 (1988) 496–502.

[62] L. Barile, C. Lavalle, Transverse myelitis in systemic lupus erythematosus – the effect of IV pulse methylprednisolone and cyclophosphamide, J Rheumatol 19 (1992) 370–372.

[63] L. Barile-Fabris, R. Ariza-Andraca, L. Olguin-Ortega, L.J. Jara, A. Fraga-Mauret, J.M. Miranda-Limon, et al., Controlled clinical trial of IV cyclophosphamide versus IV methylprednisolone in severe neurological manifestations in systemic lupus erythematosus, Ann Rheum Dis 64 (2005) 620–625.

[64] D. Lurie, M. Basher Kahaleh, Pulse corticosteroid therapy for refractory thrombocytopenia in systemic lupus erythematosus, J Rheumatol 9 (1882) 311–313.

[65] J. Goldberg, M. Lidskey, Pulse methylprednisolone therapy for persistent subacute cutaneous lupus, Arthritis Rheum 27 (1984) 837–838.

[66] B.J. Parker, I.N. Bruce, High dose methylprednisolone therapy for the treatment of severe systemic lupus erythematosus, Lupus 16 (2007) 387–393.

[67] H. Badsha, K.O. Kong, T.Y. Lian, S.P. Chan, C.-J. Edwards, H.H. Chng, Low-dose pulse methylprednisolone for systemic lupus erythematosus flares is efficacious and has a decreased risk of infectious complications, Lupus 11 (2002) 508–513.

[68] G. Franchin, B. Diamond, Pulse steroids: How much is enough? Autoimmunity Reviews 5 (2006) 111–113.

[69] P.C. Avgerinos, G.B. Cutler Jr., G.C. Tsokos, P.W. Gold, P. Feuillan, W.T. Gallucci, et al., Dissociation between cortisol and adrenal androgen secretion in patients receiving alternate day prednisone therapy, J Clin Endocrinol Metab 65 (1987) 24–29.

[70] A.S. Fauci, D.C. Dale, Alternate-day prednisone therapy and human lymphocyte subpopulations, J Clin Invest 55 (1975) 22–32.

[71] G.L. Ackerman, Alternate-day steroid therapy in lupus nephritis, Ann Intern Med 72 (1970) 511–519.

[72] M.J. Bell, L.W. Martin, L.L. Gonzales, P.T. McEnery, C.D. West, Alternate-day single-dose prednisone therapy: a method of reducing steroid toxicity, J Pediat Surg 7 (1965) 223–229.

[73] S.P. Ballou, M.A. Khan, I. Kushner, Intravenous pulse methylprednisolone followed by alternate day corticosteroid therapy in lupus erythematosus: a prospective evaluation, J Rheumatol 12 (1985) 944–948.

[74] F.J. Frey, M.K. Ruegsegger, B.M. Frey, The dose-dependent system availability of prednisone: one reason for the reduced biologic effect of alternate-day prednisone, Br J Pharma 21 (1986) 183–189.

[75] H. Bootsma, P. Sprunk, R. Derksen, G. de Boer, H. Wolters-Dicke, J. Hermans, et al., Prevention of relapses in systemic lupus erythematosus, Lancet 345 (1995) 1595–1599.

[76] C.-E. Tseng, J.P. Buyon, M. Kim, H.M. Belmont, M. Mackay, B. Diamond, et al., The effect of moderate-dose corticosteroids in preventing severe flares in patients with serologically active, but clinically stable, systemic lupus erythematosus, Arthritis Rheum 54 (2006) 3623–3632.

[77] A. Danowski, L. Magder, M. Petri, Flares in lupus: Outcome assessment trial (FLOAT), a comparison between oral methylprednisolone and intramuscular triamcinolone, J Rheumatol 33 (2006) 57–60.

[78] M. Al-Maini, M. Urowitz, Systemic steroids, in "Systemic Lupus Erythematosus. A Companion to Rheumatology", in: G.C. Tsokos, C. Gordon, J.S. Smolen (Eds.), Mosby Elsevier, Philadelphia, 2007, p. 493.

[79] R.B. Dickson, N.P. Christy, On the various forms of corticosteroid withdrawal syndrome, Am J Med 68 (1980) 224–230.

[80] K.J. Newmark, S. Mitra, L.B. Berman, Acute arthralgia following high-dose intravenous methylprednisolone therapy, Lancet 2 (1974) 229.

[81] H.G. Morgan, J. Boulnois, C. Burns-Cox, Addiction to prednisone, Br Med J 2 (1973) 93–94.

[82] D.K. Flavin, P.A. Frederickson, J.W. Richardson, T.C. Merritt, Corticosteroid abuse—an unusual manifestation of drug dependence, Mayo Clin Proc 58 (1983) 764–766.

[83] J.D. Semel, Fever on a drug-free day of alternate day steroid therapy, Am J Med 76 (1984) 315–317.

[84] A. Zonana-Nacach, S.G. Barr, L.S. Magder, M. Petri, Damage in systemic lupus erythematosus and its association with corticosteroids, Arthritis Rheum 43 (2000) 1801–1808.

[85] The Boston Collaborative Drug Surveillance Program, Acute adverse reactions to prednisone in relation to dosage, Clin Pharmacol Ther 13 (1972) 694–698.

[86] R.W.C. Hall, M.K. Popkin, S.K. Stickney, E.R. Gardner, Presentation of the steroid psychoses, J Nervous Mental Dis 167 (1979) 229–236.

[87] M.H.M. Ling, P.J. Perry, M.T. Tsuang, Side effects of corticosteroid therapy. Psychiatric aspects, Arch Gen Psychiatr 38 (1981) 471–477.

[88] A.K. Afifi, K.A. Bergman, J.C. Harvey, Steroid myopathy: Clinical, histologic and cytologic observations, Johns Hopkins Med J 123 (1968) 158–173.

[89] A. Askari, P.J. Vignos Jr., R.W. Moskowitz, Steroid myopathy in connective tissue disease, Am J Med 61 (1976) 485–492.

[90] E. Ginzler, H.S. Diamond, D. Kaplan, M. Weiner, M. Schlesinger, M. Seleznik, Computer analysis of factors

influencing frequency of infection in systemic lupus erythematosus, Arthritis Rheum 21 (1978) 37−44.

[91] P.J. Staples, D.N. Gerding, J.L. Decker, R.S. Gordon Jr., Incidence of infection in systemic lupus erythematosus, Arthritis Rheum 17 (1974) 1−10.

[92] V. Noel, O. Lortholary, P. Casassus, T. Genereau, M.-H. Andre, L. Guillevin, Risk factors and prognostic influence of infection in a single cohort of 87 adults with systemic lupus erythematosus, Ann Rheum Dis 60 (2001) 1141−1144.

[93] I. Kang, S.H. Park, Infectious complications in SLE after immunosuppressive therapies, Curr Opin Rheumatol 15 (2003) 528−534.

[94] H.-J. Kim, Y.-J. Park, W.-U. Kim, S.-H. Park, C.-S. Cho, Invasive fungal infections in patients with systemic lupus erythematosus: experience from affiliated hospitals of Catholic University of Korea, Lupus 18 (2009) 661−666.

[95] B.H. Bulkley, W.C. Roberts, The heart in SLE and the changes induced in it by corticosteroid therapy, Am J Med 58 (1975) 243−264.

[96] M.B. Urowitz, A.A.M. Bookman, B.E. Koehler, D.A. Gordon, H.A. Smythe, M.A. Ogryzlo, The bimodal pattern of mortality in systemic lupus erythematosus, Am J Med 60 (1976) 221−225.

[97] M. Petri, C. Lakatta, L. Magder, D. Goldman, Effect of prednisone and hydroxychloroquine on coronary artery disease risk factors in systemic lupus erythematosus: A longitudinal data analysis, Am J Med 96 (1994) 254−259.

[98] S. Manzi, E.N. Meilahn, J.E. Rairie, C.G. Conte, T.A. Medsger Jr., L. Jansen-McWilliams, et al., Age-specific incidence rates of myocardial infarction and angina in women with systemic lupus erythematosus: Comparison with the Framingham Study, Am J Epidemiol 145 (1997) 408−415.

[99] M. Petri, Detection of coronary artery disease and the role of traditional risk factors in the Hopkins Lupus Cohort, Lupus 9 (2000) 170−175.

[100] E. Svenungsson, K. Jensen-Urstad, M. Heimburger, A. Silveira, A. Hamsten, U. de Faire, et al., Risk factors for cardiovascular disease in systemic lupus erythematosus, Circulation 104 (2001) 1887−1893.

[101] K. Manger, M. Kusus, C. Forster, D. Ropers, W.G. Daniel, J.R. Kalden, et al., Factors associated with coronary artery calcification in young female patients with SLE, Ann Rheum Dis 62 (2003) 846−850.

[102] J.M. Esdaile, M. Abrahamowicz, T. Grodzicky, Y. Li, C. Panaritis, R. Du Berger, et al., Tradional Framingham risk factors fail to fully account for accelerated atherosclerosis in systemic lupus erythematosus, Arthritis Rheum 44 (2001) 2331−2337.

[103] E. Turner, V. Dishy, C.P. Chung, P. Harris, R. Pierce, Y. Asanuma, et al., Endothelial function in systemic lupus erythematosus: relationship to disease activity, cardiovascular risk factors, corticosteroid therapy, and coronary calcification, Vascular Health and Rick Management 1 (2005) 357−360.

[104] M. El-Magadmi, H. Bodill, Y. Ahmad, P.N. Durrington, M. Mackness, M. Walker, et al., Systemic lupus erythematosus. An independent risk factor for endothelial dysfunction in women, Circulation 110 (2004) 399−404.

[105] J. Wajed, Y. Ahmad, P.N. Durrington, I.N. Bruce, Prevention of cardiovascular disease in systemic lupus erythematosus − proposed guidelines for risk factor management, Rheumatology 43 (2004) 7−12.

[106] A.A. Kalla, A.B. Fataar, S.J. Jessop, L. Bewerunge, Loss of trabecular bone mineral density in systemic lupus erythemtosus, Arthritis Rheum 36 (1993) 1726−1734.

[107] L. Sinigaglia, M. Varenna, L. Binelli, F. Zucchi, D. Ghiringhelli, M. Gallazzi, et al., Determinants of bone mass in systemic lupus erythematosus: a cross sectional study of premenopausal women, J Rheumatol 26 (1999) 1280−1284.

[108] C. Lee, R. Ramsey-Goldman, Osteoporosis in systemic lupus erythematosus mechanisms, Rheum Dis Clin N Am 31 (2005) 363−385.

[109] E.G. Lufkin, H.W. Wahner, E.J. Bergstralh, Reversibility of steroid-induced osteoporosis, Am J Med 85 (1988) 887−888.

[110] O.S. Gluck, W.A. Murphy, T.J. Hahn, B. Hahn, Bone loss in adults receiving alternate day glucocorticoid therapy: A comparison with daily therapy, Arthritis Rheum 24 (1981) 892−898.

[111] J.D. Adachi, W.G. Bensen, F. Bianchi, A. Cividino, S. Pillersdorf, R.J. Sebalt, et al., Vitamin D and calcium in the prevention of corticosteroid induced osteoporosis, J Rheumatol 23 (1996) 995−1000.

[112] J.D. Adachi, G. Ioannidis, Calcium and vitamin D therapy in corticosteroidinduced bone loss: what is the evidence? Calcif Tissue Int 65 (1999) 332−336.

[113] Recommendations for the prevention and treatment of glucocorticoid-induced osteoporosis, 2001 update. American College of Rheumatology Ad Hoc Committee on Glucocorticoid-Induced Osteoporosis, Arthritis Rheum 44 (2001) 1496−1503.

[114] A.R. Klipper, M.R. Stevens, T. Zizic, D.S. Hungerford, Ischemic necrosis of bone in systemic lupus erythematosus, Medicine 55 (1976) 251−257.

[115] E.S. Weiner, M. Abeles, Aseptic necrosis and glucocorticoids in systemic lupus erythematosus: A reevaluation, J Rheumatol 16 (1989) 604−608.

[116] T.M. Zizic, C. Marcoux, D.S. Hungerford, J.-V. Dansereau, M.B. Stevens, Corticosteroid therapy associated with ischemic bone necrosis in systemic lupus erythematosis, Am J Med 79 (1985) 596−604.

[117] E.S. Weiner, M. Abeles, More on aseptic necrosis and glucocorticoids in systemic lupus erythematosus (letter), J Rheumatol 17 (1990) 119.

[118] M.J. Seleznick, L.H. Silveira, L.R. Espinoza, Avascular necrosis associated with anticardiolipin antibodies, J Rheumatol 18 (1991) 1416−1417.

[119] M.E. Shipley, P.A. Bacon, H. Berry, B.L. Hazleman, R.D. Sturrock, D.R. Swinson, et al., Pulsed methylprednisolone in active early rheumatoid disease: A dose-ranging study, Br J Rheum 27 (1988) 211−214.

[120] I.A. Williams, A.D. Mitchell, W. Rothman, P. Tallett, K. Williams, P. Pitt, Survey of the long term incidence of osteonecrosis of the hip and adverse medical events in rheumatoid arthritis after high dose intravenous methylprednisolone, Ann Rheum Dis 47 (1988) 930−933.

[121] E.Y. McLaurin, S.L. Holliday, P. Williams, R.L. Brey, Predictors of cognitive dysfunction in patients with systemic lupus erythematosus, Neurology 64 (2005) 297−303.

[122] V.A. Zanardi, L.A. Magna, L.T.L. Costallat, Cerebral atrophy related to corticotherapy in systemic lupus erythematosus (SLE), Clin Rheumatol 20 (2001) 245−250.

[123] R.C. Tripathi, S.K. Parapuram, B.J. Tripathi, Y. Zhong, K.V. Chalam, Corticosteroids and glaucoma risk, Drugs & Aging 15 (1999) 439−450.

[124] B. Becker, The side effects of corticosteroids, Invest Ophthalmol 3 (1964) 492−497.

[125] J.E. Dickerson, E. Dotzel, A.F. Clark, Steroid-induced cataract: New perspectives from *in vitro* and lens culture studies, Experimental Eye Research 65 (1997) 507−515.

[126] T.O. McMeekin, S.L. Moschella, Iatrogenic complications of dermatologic therapy, Med Clin North Am 63 (1979) 441−452.

[127] D.J. Wallace, A.L. Metzger, J.R. Klinenberg, NSAID usage patterns by rheumatologists in the treatment of SLE, J Rheumatol 16 (1989) 557−560.

IV. TREATMENT

[128] L.R. Espinoza, L.J. Jara, P. Martinz-Osuna, L.H. Silveira, M.L. Cuellar, M. Seleznick, Refractory nephrotic syndrome in lupus nephritis: favorable response to indomethacin therapy, Lupus 2 (1993) 9—14.

[129] A.A. Horizon, D.J. Wallace, Risk:benefit ratio of nonsteroidal anti-inflammatory drugs in systemic lupus erythematosus, Expert Opin Drug Saf 3 (2004) 273—278.

[130] M. Ostensen, P.M. Villiger, Nonsteroidal anti-inflammatory drugs in systemic lupus erythematosus, Lupus 10 (2001) 135—139.

[131] S.M. Sultan, Y. Ioannou, D.A. Isenberg, A review of gastrointestinal manifestations of systemic lupus erythematosus, Rheumatology 38 (1999) 917—932.

[132] R.A. Hoppmann, J.G. Peden, S.K. Ober, Central nervous system side effects of nonsteroidal anti-inflammatory drugs. Aseptic meningitis, psychosis, and cognitive dysfunction, Arch Intern Med 151 (1991) 1309—1313.

[133] S.A. Lander, D.J. Wallace, M.H. Weisman, Celecoxib for systemic lupus erythematosus: case series and literature review of the use of NSAIDs in SLE, Lupus 11 (2002) 340—347.

Antimalarials and SLE

Nathalie Costedoat-Chalumeau,
Gaëlle Leroux, Jean-Charles Piette, Zahir Amoura

INTRODUCTION

Numerous new drugs, mainly therapeutic mono-clonal antibodies, have now been developed for patients with systemic lupus erythematosus (SLE). Meantime, interest for a drug used for more than 50 years in the treatment of SLE has grown. Indeed, increasing evidence suggests that hydroxychloroquine (HCQ) is an essential medication for SLE. First, HCQ prevents and alleviates SLE manifestations [1, 2]. Second, HCQ is the only treatment that reduces the risk of damage accrual in SLE patients [3]. Third, apart from its direct effects on SLE activity, HCQ appears to protect against thrombotic events and has beneficial effect on lipid profiles and glycemia which may contribute to reduce the high cardiovascular risk of SLE patients. Additionally, HCQ is a long-acting drug with a high efficacy/toxicity ratio. Finally, while most of the new drugs developed for the treatment of SLE are expensive and consequently are available in rich countries, HCQ is cheap and available even in developing countries.

This chapter mainly focuses on HCQ and chloroquine (CQ), which are the best-studied antimalarials used in SLE. Quinacrine is an older drug, progressively replaced by HCQ that is safer. Quinacrine is no longer commercially available in many countries including the U.S. and Canada. However, given that quinacrine has no retinal toxicity, it has been advocated as a second-line substitution for HCQ in patients with pre-existing macular degeneration or those unable to tolerate HCQ [4]. It may also be used in combination with HCQ, in patients with lupus skin lesions unresponsive to HCQ alone [5], and, as suggested by a small study, in patients with refractory SLE with major organ involvement [6]. When available, data on quinacrine will be discussed.

ANTIMALARIALS AND SLE

Historical

It is believed that the use of antimalarials in the treatment of inflammatory rheumatism goes back to the seventeenth century. The natives of South America had indeed shown the Jesuit priests a "potion" to treat malarial fevers that contained Cinchona bark. The priests noticed that this potion also alleviated rheumatism. Its action on cutaneous lesions of SLE was reported around 1890. CQ was the first widely prescribed antimalarial agent. Since side effects including retinal toxicity, neuromyotoxicity, and cardiotoxicity have been described, a second generation of antimalarial drugs, namely HCQ, has been developed.

HCQ is a cationic, amphilic drug that differs in structure to CQ only by the presence of a hydroxyl group at the end of the side chain [7]. This difference is thought to account for the reduced ocular [8] and cardiac toxicity of HCQ [9].

Efficacy of Antimalarials on SLE Activity

HCQ is effective in SLE, as demonstrated by a randomized, double-blinded, placebo-controlled study from the Canadian Hydroxychloroquine Study Group [1, 10]. This study included 47 SLE patients on stable doses of HCQ. Patients were randomized to continue drug therapy ($n = 25$) or to be changed to a placebo ($n = 22$). During the 6-month period after HCQ discontinuation, the risk of clinical SLE flares increased by 2.5 (95% CI: 1.08–5.58; 9 flares out of 25 patients versus 16 out of 22). The relative risk of severe SLE exacerbations including vasculitis, transverse myelitis, and lupus nephritis was 6.1 times higher (95% CI: 0.72–52.44) in the group of patients who discontinued HCQ (1 flare out of 25 patients versus 5 out of 22) [1]. These results have been confirmed in pregnant patients [11]. A 12-month, double-blind, placebo-controlled trial of CQ (250 mg daily) similarly showed that CQ improved non-major organ manifestations, reduced steroid requirements, and prevented disease exacerbations [12].

Other studies showed that HCQ is useful in patients with SLE glomerulonephritis. A study of 450 SLE patients found HCQ use protective against renal insufficiency [13]. HCQ is also an independent predictor of complete renal remission in SLE patients treated with mycophenolate mofetil for membranous lupus nephritis [2]. In that study, patients treated with mycophenolate mofetil plus HCQ had a remission rate 5.2 times higher than those treated with mycophenolate mofetil alone (95% CI: 1.2–22.2; $p = 0.026$).

Early HCQ use was associated with delayed SLE onset in a retrospective study of a cohort of 130 US military personnel who later met ACR SLE criteria [14]. The authors studied patients who had initially less than four criteria of the ACR SLE classification. Those treated with HCQ at the onset of symptoms ($n = 26$) had a longer time between the first clinical symptom and the "diagnosis" of SLE according to the ACR SLE classification (i.e. presence of four or more items; median: 1.08 versus 0.29 years, $p = 0.018$). Patients treated with prednisone before diagnosis also more slowly fulfilled the four classification criteria ($p = 0.011$). No difference was found according to nonsteroidal anti-inflammatory drug (NSAID) use. Patients treated with HCQ had a lower rate of autoantibodies and a decreased number of autoantibody specificities at and after diagnosis of SLE [14].

Finally, in 2008, using the GRADE system to analyze the quality of the evidence, Ruiz-Irastorza *et al.* extensively reviewed the clinical efficacy and side effects of antimalarials in SLE [15]. They found 11 studies that included data on the effect of antimalarials on SLE activity: four randomized controlled trials (double-blind in two); four prospective and two retrospective cohort studies; and one retrospective analysis of data from the extension of a randomized controlled trial. All the studies consistently found that antimalarials significantly reduced SLE activity (over 50% in most studies). The reduction of severe flares was closed to statistical significance. In two retrospective studies of patients with lupus nephritis, HCQ was identified as an adjuvant treatment for achieving remission. A steroid-sparing effect of antimalarials was found in three studies [15].

Other Beneficial Effects

The beneficial effects of antimalarials largely exceed their direct effect on SLE activity, especially in this setting of patients exposed to side effects of treatments and to atherothrombosis. Indeed, HCQ and CQ appear to protect against the occurrence of thrombotic events [16–21], diabetes [22], low bone mineral density related to steroids [23, 24], and even perhaps cancer [25]. Antimalarials have a beneficial effect on lipid profiles of patients with rheumatic diseases, especially those treated with steroids [26–30].

Effects on Thrombosis

Antimalarials, especially HCQ, reduce platelet aggregation, and this effect is pronounced enough for HCQ to have been proposed, a few decades ago, as an agent for deep vein thrombosis prophylaxis in patients undergoing hip replacements.

The antithrombotic properties of HCQ have been demonstrated in an animal model: after a standardized thrombogenic injury, the mice injected with immunoglobulin G purified from a patient with the antiphospholipid syndrome showed significantly smaller thrombi that persisted for a shorter period of time if they were fed with HCQ compared with placebo [16].

Petri reported in 1996 that HCQ use was protective against future thrombosis in the Johns Hopkins Lupus Cohort [17]. These results have been confirmed in Chinese patients with an odds ratio of thrombotic complications of 0.17 (95% CI: 0.07–0.44; $p < 0.0001$) when taking HCQ [18]. Ruiz-Irastorza *et al.* found similar results with an OR of 0.28 (95% CI: 0.08–0.90) [19]. In 442 SLE patients from the multiethnic LUMINA (Lupus in Minorities: Nature vs. Nurture) cohort, HCQ was protective against thrombotic events only in the univariate analyses [31]. In Greek SLE patients, the duration of HCQ use was protective for thrombosis both in SLE patients with ($n = 144$) and without ($n = 144$) antiphospholipid antibodies [21].

Finally, in 2009, Kaiser *et al.* analyzed risk factors for thrombosis in a large ($n = 1930$), multiethnic SLE cohort [20]. After adjusting for disease severity and incorporating propensity scores, HCQ use was significantly protective for thrombosis (OR 0.62, $p < 0.001$) [20].

The observed reduction of thrombosis in human and experimental antiphospholipid syndrome by antimalarials may be explained by a reversible reduction in the formation of antiphospholipid-beta2–glycoprotein complexes to phospholipid surfaces and monocytes, as demonstrated by Rand *et al.* [32]. Interestingly, using the techniques of ellipsometry and atomic force microscopy, the authors showed that this effect was observed for HCQ concentrations of 1000 ng/ml and greater [32], which is consistent with the target concentration of blood HCQ in SLE patients that we have defined [33].

Effects on Glycemia

Hypoglycemia is a well-recognized adverse effect of treatment with antimalarials [34]. In *in vitro* and in animal studies, antimalarials improved insulin secretion and peripheral insulin sensitivity. During an intensive outpatient intervention including 38 decompensated patients with type 2 diabetes resistant to treatment, those treated with HCQ 200 mg, 3 times per day during 6 months had an absolute reduction in glycated hemoglobin A1c level of 3.3% (95% CI: −3.9 to −2.7; $p =$ 0.001), while no significant changes were seen in patients on placebo. The daily insulin dose in patients treated with HCQ had to be reduced to an average of 30% [35].

In a large prospective, multicenter observational study of 4905 adults with rheumatoid arthritis and no diabetes, Wasko et al. [22] showed that use of HCQ was associated with a reduced risk of diabetes during the 21.5 years of follow-up. The hazard ratio for incident diabetes among patients who had taken HCQ ($n = 1808$) was 0.62 (95% CI: 0.42−0.92) compared with those who had not taken HCQ ($n = 3097$) [22]. Risk reduction increased with duration of HCQ exposure: the adjusted relative risk of developing diabetes among patients who had taken HCQ for more than 4 years ($n = 384$) was 0.23 (95% CI: 0.11−0.50; $p < 0.001$), compared to those who had not taken HCQ.

Effects on Lipid Profile

Antimalarials have a beneficial effect on lipid profile of patients with rheumatoid arthritis and SLE, especially those treated with steroids [26−30].

Concordant studies showed a reduced total cholesterol levels in patients taking antimalarials. Petri et al. [27] studied 264 SLE patients from the Baltimore Lupus Cohort. In the longitudinal regression analysis, HCQ was associated with lower serum cholesterol whatever the HCQ dose (200 or 400 mg/day). The authors calculated that HCQ was able to "balance" the adverse effect of 10 mg of prednisone on cholesterol. Rahman et al. (26) studied 815 SLE patients. Initiation of antimalarials without steroids reduced the baseline total cholesterol by 4.1% at 3 months ($p = 0.020$) and by 0.6% at 6 months ($p = $ NS), whereas initiation of antimalarials on a stable dose of steroids reduced the total cholesterol by 11.3% at 3 months ($p = 0.0002$) and 9.4% at 6 months ($p = 0.004$). The cessation of antimalarials increased the total cholesterol by 3.6% at 3 months ($p = $ NS) and 5.4% at 6 months ($p = $ NS). In 181 patients taking steroids and antimalarials, the mean total cholesterol level was 11% less than for 201 patients receiving a comparable dose of steroids alone ($p = 0.0023$).

Others studies, in addition to confirming reduction in total cholesterol level, have focused on lipid profile. Results are concordant except for the effect on high-density lipoprotein (HDL) level, which was found unchanged [29], decreased [28], or increased [30, 36]. Indeed, Wallace et al. (29) studied serum levels of total cholesterol, triglycerides, HDL and low-density lipoprotein (LDL) in 155 women with rheumatoid arthritis or SLE (treated or untreated with HCQ and/or steroids). HCQ use was associated with lower cholesterol levels ($p = 0.0006$), triglycerides ($p = 0.0459$), and LDL ($p = 0.0004$), irrespective of concomitant steroid administration. No HDL differences were observed. Hodis et al. [28] found in 18 SLE patients that those taking HCQ had lower total triglyceride, LDL, HDL, VLDL-cholesterol, and apolipoprotein CIII levels (with a reduction ranging from 35 to 54%; $p < 0.03$). By contrast, Munro et al. [30] performed a 1-year prospective randomized trial in 100 patients with rheumatoid arthritis. The HCQ group ($n = 51$, including 35 who completed the study) had a rise in the HDL values compared with a decline in the gold group ($p = 0.034$ at 6 months, and $p = 0.040$ at 12 months). Borba and Bonfa [36] studied 60 SLE female patients treated or untreated with CQ. Patients receiving CQ had significantly higher levels of HDL compared to the no therapy group ($p < 0.05$). In addition, among patients receiving steroids, those treated with CQ had higher levels of HDL and lower levels of triglycerides and VLDL ($p < 0.05$).

In 65 consecutive SLE Chinese patients, Tam et al. [37] only found a trend toward a lower triglyceride level in patients on HCQ (1.0 [range: 0.7−1.4] mmol/L vs. 1.2 [0.8−1.6] mmol/L), whereas other parameters of the serum lipid profile were similar between patients whether receiving or not receiving HCQ. A possible explanation for these discrepancies was the low doses of steroids used and a lower background lipid level in this study.

The mechanism of cholesterol lowering by antimalarials is not clear but may involve an overall reduction in hepatic cholesterol synthesis, as demonstrated in animal studies [38]. This could be explained by the inhibition of lysosomal function by antimalarials (see section on Mechanism of Action), that leads to an accumulation of LDL in the lysosome. Indeed, CQ is known to block cholesterol transport out of the lysosome and is used for these properties in in vitro models studying cholesterol metabolism [39]. Moreover, using radioactive cholesterol-rich nanoemulsion that mimics the LDL structure, it has been demonstrated in vivo that plasma LDL removal by LDL receptor was increased in SLE patients taking CQ with a consequent beneficial decrease in LDL levels [40].

In conclusion, antimalarials, alone or added to steroids, appear to have a beneficial effect in dyslipidemia. This suggests that antimalarials, especially HCQ, may have a role in reducing the key risk factors for atherosclerosis in SLE, namely platelet aggregation, diabetes, and dyslipidemia. Accordingly, despite high rates of use of HCQ in their study, Roman et al. found a significant negative relation between the use of HCQ and the presence of atherosclerosis (carotid plaques) in 197 SLE patients [41].

Effects on Bone Mineral Density

Two articles have analyzed the relationship between the bone mineral density and antimalarial use in SLE patients. The first [24] studied dual X-ray absorptiometry in 34 postmenopausal SLE patients who had received long-term steroids. By multivariate analysis the use of HCQ, either current or past, was associated with a higher spinal bone mineral density. The second [23] studied dual X-ray absorptiometry in 92 consecutive SLE patients (98% had received prednisone, 68% had received HCQ and 51% were postmenopausal). In the multivariate analysis, the use of HCQ was the only factor associated with higher bone mineral density of the hip and spine. This protective role of HCQ against low bone mineral density has to be confirmed with further studies.

Effects on Cancer

There are data showing that CQ may play a role in the treatment of some cancers, especially glioblastoma. Indeed, in a small randomized, double-blind, placebo-controlled study of 30 patients, CQ, added to conventional treatment, improved survival in patients with glioblastoma multiforme [42]. If confirmed, this effect could be due to protective properties of antimalarials against DNA damage. As seen above, antimalarials are weak bases that concentrate in lysosomes and are strong DNA-intercalating agents that prevent mutations in cells [42]. CQ has shown an inhibitory action on telomerase, which is involved in the unlimited replication of tumorous cells. CQ improves cellular mechanisms of DNA repair after damage caused by alkylating therapy [42]. Rahim and Strobl [43] found that the growth of human breast cancer cell lines in $vitro$ was inhibited by CQ and HCQ via a regulation of protein acetylation events.

Regarding SLE patients, the data are limited to a study by Ruiz-Irastorza et $al.$ who found in an observational prospective cohort study of 235 SLE patients that antimalarials may have a protective action against cancer [25]. Adjusted hazard ratio for cancer among patients who received antimalarials at any time compared with patients never treated with these drugs was 0.15 (95% CI: 0.02–0.99; p = 0.049). As emphasized by the authors, these results should be confirmed in larger multicenter studies.

Anti-infectious Effects

In addition to their antimalarial properties, CQ and HCQ have antibacterial, antifungal, and antiviral actions [44, 45]. Doxycycline in association with HCQ is already the reference treatment for chronic Q fever [45], and preliminary in $vivo$ clinical trials suggest that CQ alone or in combination with antiretroviral drugs might represent an interesting way to treat human immunodeficiency virus (HIV) infection [45].

The clinical consequences of these effects in SLE patients are unknown.

Of note, HCQ is not effective against CQ-resistant strains of $Plasmodium$ $falciparum$ and is not active against the exo-erythrocytic forms of $Plasmodium$ $vivax$, $Plasmodium$ $ovale$ and $Plasmodium$ $malariae$. Therefore, HCQ will neither prevent infection due to these organisms when given prophylactically, nor prevent relapse of infection due to these organisms. This is important in the counseling of SLE patients who are traveling in malaria endemic area.

Effects on Damage Accrual

We know of two studies that have assessed the relationship between HCQ use and the risk of overall damage accrual in SLE patients [3, 46]. The first included 151 SLE patients from Israel [46]. After a mean follow-up of 45 months, treatment with HCQ was associated with a higher damage-free survival ($p < 0.0001$).

The other study was produced by the LUMINA study group [3]. A total of 518 SLE patients with less than 5 years' disease duration at inclusion were followed up annually. At study entry, 291 (56%) were treated with HCQ. Untreated patients on enrollment had higher SLAM (Systemic Lupus Activity Measure) and SDI (Systemic Lupus International Collaborating Clinics Damage Index) scores and were significantly more likely to have major organ involvement such as renal disease ($p < 0.0001$) or central nervous system disease ($p < 0.0025$), and to accrue damage (HR = 0.68; 95% CI: 0.53–0.93; $p < 0.014$). After adjustment for the propensity scores for differences in treatment assignment, HCQ use was still associated with a reduced risk of developing new damage (HR = 0.73; 95% CI: 0.52–1.00; $p = 0.05$). This benefit was demonstrated only in patients receiving HCQ who had no damage at study entry (HR = 0.55; 95% CI: 0.34–0.87; $p = 0.0111$). The authors stated that HCQ usage is independently associated with a reduced risk of damage accrual in SLE patients who had not yet accrued damage at the time of treatment initiation.

In 2009, the LUMINA study group specifically assessed whether HCQ can delay renal damage occurrence in lupus nephritis patients with no renal damage before inclusion ($n = 203$) [47]. Sixty-three (31%) patients developed renal damage (mainly proteinuria "damage") over a mean disease duration of 5.2 ± 3.5 years. Patients treated with HCQ (79.3%) had a lower frequency of WHO class IV glomerulonephritis, a lower disease activity, and received lower steroid doses than those who did not take HCQ. After adjusting for possible confounders factors, HCQ was protective in retarding renal damage development in full (HR = 0.12; 95%

CI: 0.02–0.97; $p = 0.0464$) and reduced (HR = 0.29; 95% CI: 0.13–0.68; $p = 0.0043$) models [47].

Effects on Survival

In light of the data above, antimalarials appear to have a protective effect on survival in SLE patients, as demonstrated by two studies [19, 48].

The first one, published by Ruiz-Irastorza *et al.* [19] was an observational prospective cohort study of 232 SLE patients, including 150 patients (64%) who received antimalarials. Among the 23 patients who died, 19 (83%) had never received antimalarials. No patient treated with antimalarials died of cardiovascular complications. Cumulative 15-year survival rates were 0.68 for never versus 0.95 for ever HCQ-treated patients ($p < 0.001$). Antimalarial use remained protective on survival even after adjusting for patient characteristics [19].

Alarcon *et al.* [48] performed a case-control study among 608 SLE patients from the multiethnic LUMINA cohort. Sixty-one deceased patients (cases) were matched for disease duration with alive patients (controls) in a 3 to 1 proportion. Propensity scores were derived by logistic regression to adjust for confounding by indication (since SLE patients with milder disease manifestations are more likely to be prescribed HCQ). The authors found that HCQ had a protective effect on survival (OR = 0.128; 95% CI: 0.054–0.301 for HCQ alone, and OR = 0.319; 95% CI: 0.118–0.864 after adding the propensity score) [48].

Synthesis

Finally, in their analysis using the GRADE system of the clinical efficacy and side effects of antimalarials in SLE, Ruiz-Irastorza *et al.* [15] found:

- High evidence that antimalarials prevent SLE flares (including during pregnancy), and increase long-term survival of SLE patients.
- Moderate evidence that antimalarials protect against irreversible organ damage, thrombosis and bone mass loss.
- Low evidence that antimalarials have an effect on severe lupus activity, lipid levels and subclinical atherosclerosis.

MECHANISM OF ACTION

Classical Mechanism of Action

Antimalarials have been used for many years in inflammatory rheumatism, but their mechanism of action remains controversial.

It is generally accepted that these molecules interfere with the function of phagocytosis through an increase of pH in intracellular compartments, leading to a disruption in the process of antigen presentation. Indeed, CQ and HCQ are weak bases and are unchanged at neutral pH of serum. Lipophilic, they can pass freely through cell membranes and enter cells and then acidic cytoplasmic vesicles. In acidic conditions, they acquire an ion proton and become positively charged. They become fat-insoluble and cannot pass membranes. This fixing of hydrogen ions by HCQ or CQ results in a higher pH, as measured by immunochemical methods [49]. This process can occur throughout the body, but predominate in the cells of the macrophage lineage in which a substantial portion of ATP is transferred to the internalization of cell membranes (or phagocytosis) and the acidification of lysosomes for digestion of proteins. CQ and HCQ are then highly concentrated within mononuclear cells. This elevation of pH in lysosomal storage vesicles induced by antimalarials inhibits phagocytosis, migratory properties, and the metabolism of membrane phospholipids. This is supposed to reduce the selective presentation of self antigens (antigens which are of low affinity), while respecting that of exogenous antigens and blocking the interaction between antigen-presenting cells and T lymphocytes [50].

HCQ has been shown to induce apoptosis in peripheral blood lymphocytes [51–56]. However, some experimental results have found that HCQ exhibits an anti-apoptotic action and that this anti-apoptotic effect is dependent on the presence of cells of the macrophage lineage in SLE patients [57].

Besides altering the process of antigen presentation, CQ and HCQ may also block the proliferative responses of T cells after stimulation by mitogens or alloantigens [58–62]. According to experimental results, the production of cytokines (interleukin 1, interleukin 2, interleukin 6, interferon gamma and/or TNF alpha) is inhibited [58, 61, 63–67] or on the contrary increased (for interleukin-1) [51].

Other mechanisms of action involve the inhibition of DNA polymerases [68] or the inhibition of the activity of phospholipases A2 [69].

Toll-Like Receptor

Data strongly support the notion that the key activity of antimalarial drugs in SLE could be an inhibition of the Toll-like receptor (TLR) activation. TLRs are receptors involved in innate immunity that seem to have a particularly important role in autoimmune diseases including SLE [70]. TLRs were first known for their ability to discriminate microbial macromolecules from host tissue and thereby rapidly activate the innate immune system. However, it has become increasingly apparent that select host molecules, especially nucleic acids, can serve as endogenous TLR ligands, perhaps promoting responses to damaged tissue [70]. The acidic lysosomal environment is favorable to binding of nucleic acids to intracellular TLRs.

As previously seen, HCQ inhibits endosomal acidification, on which signaling of the intracellularly located TLRs, such as TLRs 3, 7, 8, and 9, depend [70]. The inhibitory action of CQ and HCQ on TLRs interaction with nucleic acid ligands has been confirmed [70–72]. Leadbetter *et al.* have demonstrated, in a transgenic mouse model, that the production of rheumatoid factors by activated B lymphocytes requires activation of TLR 9 [73]. In this study, CQ was able to block the production of rheumatoid factors in inhibiting TLR 9 in the endosomes. Brentano *et al.* [71] showed that HCQ, used *in vitro* as an anti-TLR, inhibits the production of cytokines and chemokines by fibroblasts stimulated by cells from necrotic synovial fluid of patients with rheumatoid arthritis.

Finally, as emphasized by Lafyatis *et al.* [70], it is notable that the *in vivo* concentrations (above 1000 ng/ml) associated with decreased frequency of subsequent SLE flare in our study [33] were in the same range as those shown to block intracellular TLRs *in vitro* [73].

TOXICITY

In general, antimalarials are well tolerated and rarely need to be discontinued because of an adverse reaction. Two types of side-effects may be encountered. The first includes gastrointestinal intolerance, aquagenic pruritus [74], and other cutaneous manifestations [75]. These manifestations are not rare, usually disappear with dose reduction, and rarely require withdrawal of the treatment. The second type of toxicity is rare but potentially severe and involves various combinations of retinal, neuromuscular, cardiac, and hematological impairments. Discontinuation of the treatment is usually required and is generally associated with a slow but not necessarily complete resolution of the symptoms (see Table 57.1).

In their review of clinical efficacy and side effects of antimalarials in SLE using the GRADE system, Ruiz-Irastorza *et al.* found high evidence supporting the global safety of antimalarials, both HCQ and CQ, and a moderate grade of evidence that HCQ offers a safer profile than CQ [15]. In their review, the frequency of adverse effects reported was low, mainly gastrointestinal and cutaneous, and usually mild. They found only one study comparing the toxicity of HCQ and CQ [76]. In this large series of 940 patients (including 178 with SLE) there was a higher frequency of adverse effects in patients treated with CQ than in those receiving HCQ (28.4 vs. 14.7%, $p = 0.001$). Overall, 15% of patients discontinued antimalarials due to toxicity, patients receiving HCQ being less likely to discontinue the drug due to side effects than those taking CQ (OR 0.62; 95% CI: 0.40–0.96) [76].

Ophthalmological Toxicity

Antimalarial Ophthalmologic Toxicity

Much of the concern regarding antimalarials has focused on potential ocular toxicity.

Deposition of the drug in the cornea may be associated with complaints of blurred vision, photophobia, focusing difficulties, and visual halos. These side effects are most commonly seen within the first several weeks of treatment and typically resolve despite continuation of therapy. To our knowledge, this is not associated with further retinal toxicity.

In the retina, antimalarials bind to the melanin of the pigmented epithelial layer and may damage rods and cones. Early retinal changes (so-called premaculopathy) are typically first detected in the macula with findings of macular edema, increased pigmentation and granularity, and loss of the foveal reflex. Although patients with premaculopathy generally have no visual complaints, on testing, a paracentral scotomata to a red, but not white, test object may be detected. These types of retinal changes are considered reversible upon discontinuation of the antimalarial drug.

Advance macular disease (maculopathy) is characterized by a central area of patchy depigmentation of the macula surrounded by a concentric ring of pigmentation ("bull's eye" lesion). If drug exposure continues, the

TABLE 57.1 Potential toxicities of antimalarials in SLE

Agent	Brand name(s)
Hydroxychloroquine	Plaquenil
Chloroquine	Aralen/Avloclor
Quinacrine	Atabrine/Mepacrine
Potential toxicities	
Constitutional	Malaise, irritability, weight loss
Gastrointestinal	Anorexia, nausea, vomiting, diarrhea, abdominal pain, bloating flatulence
Dermatologic	Pruritis, urticaria, multiple forms of rash, skin discoloration, alopecia, bleaching of hair
Opthalmologic	Blurred vision, preretinopathy, retinopathy
Neuromuscular	Headache, dizziness, insomnia, tinnitus, psychosis, seizures, peripheral neuropathy, myopathy of proximal muscles, cardiomyopathy
Hematologic	Leukopenia, agranulocytosis, aplastic anemia

pigment epithelial atrophy and functional disturbance may gradually spread over the entire fundus. Advanced cases show widespread pigment epithelial and retinal atrophy with loss of visual acuity, peripheral vision and night vision. Vascular narrowing may appear, so that advanced antimalarial toxicity affects visual function much like retinitis pigmentosa [77]. Patients may be initially entirely asymptomatic or may complain of nyctalopia and scotomatous vision with field defects of paracentral, pericentral ring types, and typical temporal scotomas. When a maculopathy is present, even after cessation of the drug, there is little if any visual recovery, and sometimes a progression of visual loss [77].

Of note, symptoms or fundus changes that are unilateral are generally not considered sufficient to implicate drug toxicity [77].

Incidence of retinopathy in clinical practice is very small and is lower with HCQ than CQ [78]. Studies of several large series of patients with rheumatic disease report little or no toxicity among thousands of subjects [77, 79]. In a retrospective study of 1207 patients who had been treated with HCQ, one patient had definite toxicity (0.08%) and five other patients had indeterminate but probable toxicity (0.4%) [79].

In their review of clinical efficacy and side effects of antimalarials in SLE using the GRADE system, Ruiz-Irastorza et al. compared the ocular toxicity of CQ and HCQ in the literature [15]. In four studies including 647 patients treated with CQ for a mean of > 10 years, they found that 16 (2.5%) patients were diagnosed with definite retinal toxicity, in comparison with only two (0.1%) of 2043 patients taking HCQ for a similar period included in six studies (OR 25.88; 95% CI: 6.05–232.28; $p < 0.001$). When patients classified as having probable retinal toxicity were added, there were 17/647 (2.6%) CQ users and 6/2043 (0.3%) HCQ users with toxicity (OR 9.16; 95% CI: 3.42–28.47; $p < 0.001$) [15].

Toxic retinopathy is associated with higher doses and longer duration of use, but it is not clear whether the critical factor is daily dose, duration of use, cumulative dose, or genetic susceptibility [78].

Screening Recommendations

Even if retinal toxicity from HCQ is rare, its potential permanence and severity make it imperative that physicians take measures to minimize its occurrence and effects. The goal is to detect premaculopathy, in order to stop the treatment before the occurrence of irreversible damage. However, there is no universally accepted method for screening or for judging risks [78].

The American Academy of Ophthalmology [80] first updated their recommendations on screening for HCQ retinopathy in 2002. It was recommended that all individuals starting HCQ or CQ have a complete baseline ophthalmologic screening within the first year of treatment. This should include both a complete ophthalmologic examination (visual acuity and dilated examination of the cornea and retina), and a field testing (with Amsler grid or Humphrey 10-2 fields). Optional investigations include color testing, fundus photography, and/or specialized tests (especially multifocal electroretinogram).

Thereafter, high-risk patients (those with more than 5 years of treatment or those with HCQ doses > 6.5 mg/kg/day, CQ doses > 3 mg/kg/day or with other risk factors such as high body fat level, concomitant kidney or liver disease, concomitant retinal disease, or age > 60 years) should undergo annual screening. Similarly, this should include both a complete ophthalmologic examination (visual acuity and dilated examination of the cornea and retina), and a field testing (with Amsler grid or Humphrey 10-2 fields). Optional investigations can also be done.

Those with no risk factors should have a retinal examination performed and Amsler grid or Humphrey 10-2 fields every 1–4 years.

If toxicity is suspected, more elaborate tests can be performed, including, for example, multifocal electroretinography, although it requires further evaluation [78].

In 2008, an American study assessed current screening practices and knowledge of patient risk factors for HCQ toxicity [81]. Among 67 ophthalmologists, 90% used either central automated threshold perimetry or Amsler grid as recommended by the published 2002 American Academy of Ophthalmology recommendations. However, most physicians could not correctly identify the evidence-based risk factors and screened more frequently than recommended. The authors calculate that if all patients were screened using recommendations, financial savings could exceed $150 million every 10 years in the USA [81].

In 2004, French experts proposed to associate clinical examination with two tests of macular function among color vision, Humphrey central visual field testing, and electroretinography (pattern or multifocal) [82]. These experts have recommended carrying out such monitoring on an annual basis among SLE patients. Of note, the clinical examination, color vision, and Humphrey central visual field testing can be performed by nearly all ophthalmologists.

In 2009, the new American Academy of Ophthalmology guidelines were being prepared. New recommendations should simplify the screening program, and discuss newer diagnostic technology (especially multifocal electroretinogram). Because some patients may have idiosyncratic toxicity after limited exposure, guidelines will now recommend that all patients on HCQ or CQ should begin annual screening within the first year of usage. Screening will mainly consist of an ocular exam with Humphrey central visual field testing.

Multifocal electroretinogram will be associated when available. Color vision and Amsler grid will no longer be recommended. At the time of publication, no reference is available.

Cardiac Toxicity

Cardiac toxicity related to antimalarials has been reviewed by us with a total of 54 cases analyzed [9]. Since this review of 54 cases, few additional cases have been reported [83, 84].

Cardiotoxicity includes both heart conduction disturbances and congestive heart failure. These cardiac toxic effects described both singly or in combination, have been reported with CQ and less frequently with HCQ use alone.

In our review, congestive heart failure was found in 25 patients leading to death in 11 cases (46%) and to transplantation in two cases including ours. Mean age of the patients was 52 years (range: 27—81). Antimalarials were mainly given for lupus, either systemic or discoid, rheumatoid arthritis, or malaria. Heart conduction disorders were associated in 16 cases (64%), and were diagnosed prior to the development of cardiac failure in four cases. Patients were treated with CQ ($n = 16$; 64%), HCQ ($n = 7$; 28%) or both ($n = 2$; 8%). Duration of antimalarial use varied widely, ranging from 3 months to 27 years (mean: 10 years) with a similarly wide range of cumulative dose of antimalarial drugs (0.270—9.125 kg). Associated toxicity was found in 15 patients and included myopathy ($n = 12$; 48%), retinopathy ($n = 6$; 24%), neuropathy ($n = 5$; 20%), and/or skin coloration ($n = 3$; 12%). It should be noted that these manifestations might have been underdiagnosed since they were not systematically assessed. Clinical and echocardiographic presentations often included a restrictive pattern and biventricular hypertrophy that can mimic amyloidosis.

Heart conduction disturbances secondary to long-term antimalarial treatment were found in 45 patients [9] and included bundle-branch block and atrioventricular block. Complete atrioventricular block often required the insertion of a pacemaker. There was little information on electrocardiographic evolution over time. However, when available, it seems that there was a slow progression from bundle-branch block (including left anterior hemiblock) or first and second atrioventricular block, to complete atrioventricular block. Just two of these 45 patients had been treated with HCQ only, whereas most of them had received CQ. Withdrawal of antimalarials was associated with regression of heart conduction disorders in only three cases, whereas 12 patients (80%) did not show any improvement.

Histological findings are essential to confirm the diagnosis, and to exclude differential diagnosis, such as lupus-related myocarditis, viral myocarditis in patients with immunosuppressive agents, and ischemic heart disease. Light microscopy discloses enlarged and vacuolated cardiocytes, whereas electron microscopy shows the presence of curvilinear bodies in cardiac myocytes. These typical myocardial changes are very similar to muscles changes encountered in antimalarial myopathy. Differential histological diagnosis includes toxic medications (e.g., amiodarone) and many of the lysosomal storage diseases, like Fabry's disease but curvilinear bodies in cardiac myocytes have never been observed in these conditions and appear to be the most distinctive and reliable morphologic indicator of antimalarial-associated myocyte damage. When cardiac biopsy is contraindicated due to the severity of the cardiomyopathy, the muscle biopsy may help to establish the diagnosis, since lesions are usually very similar [9].

After withdrawal of antimalarials, improvement of myocardial involvement has been more frequently reported than regression of heart conduction disorders. Antimalarials have been stopped in 12 cases of congestive heart failure, leading to partial or complete improvement in eight cases after a period ranging from 3 months to 5 years. Cardiac symptoms persisted despite cessation of CQ therapy in five cases, leading to death ($n = 3$) or to heart transplantation ($n = 2$) within a period lasting from 1 week to 3 months. Given that toxic cardiomyopathy needs months or years to resolve and that histological findings may persist as long as 9 years, the lack of improvement in these cases might be explained by the short delay since withdrawal. This emphasizes the importance of recognizing early signs of toxicity in patients treated with antimalarials in order to withdraw antimalarials before the occurrence of life-threatening complications.

Questions regarding potential risk factors for cardiotoxicity remain largely unanswered. Similarly to retinal toxicity, it has been suggested that cardiotoxicity may be enhanced by renal insufficiency, older age, pre-existing cardiac disease, elevated per-kilogram daily dose of antimalarials, longer duration of treatment, and use of CQ rather than HCQ. However, cardiotoxicity has been described in the absence of these factors.

In contrast to what is proposed for ophthalmologic assessment, there is no recommendation concerning electrocardiographic screening and survey of patients with prolonged treatment with antimalarials [75]. Cardiac assessment with baseline and annual ECG should be discussed. This seems particularly important for patients treated with CQ, for those with other manifestations of toxicity, and in cases of high daily doses and prolonged duration of treatment.

Finally, where a high incidence of heart conduction disorders, including bundle-branch block and incomplete or complete atrioventricular block, has been observed among patients treated with CQ [85], a series of electrocardiograms (ECGs) in 85 unselected patients with connective tissue diseases treated with HCQ as the sole antimalarial was reassuring since PR interval, QTc interval, and heart rate were not different from normal values. The rate of heart conduction disorders was also similar to what was expected in the general population [86].

Neuromyotoxicity

Antimalarials can cause a marked neuromyopathy, characterized by slowly progressive weakness of insidious onset. This weakness usually first affects the proximal muscles of the legs and may be associated with peripheral neuropathy. Myasthenia-like syndrome including ptosis related to CQ [87], and respiratory failure due to paresis of respiratory muscles related to HCQ have also been reported [89]. Reflexes are usually diminished symmetrically. Creatinine kinase levels are often normal. Electromyogram shows reduction in nerve conduction time and both neuropathic and myopathic changes. Muscle biopsy reveals a vacuolar myopathy including curvilinear bodies and type 1 and type 2 muscle fiber atrophy. These histological findings are very similar to those found in cardiac toxicity. Nerve biopsy may show axonal degeneration and rarely demyelination [88]. Other manifestations of antimalarial toxicity may be associated. Slow partial to complete recovery has been reported on cessation of antimalarials [88]. Few patients have been rechallenged with antimalarials at a lower dose without ill effect whereas others have shown recurrence of symptoms [88].

This adverse effect is rare. A review published in 2000 [88] found 12 reported cases of neuromyotoxicity related to HCQ, including three occurring in SLE patients. Possible predisposing factors include Caucasian race, concomitant renal failure, elevated per-kilogram daily dose of antimalarials, longer duration of treatment, and use of CQ rather than HCQ [88, 89]. The exact mechanism of neuromyotoxicity remains unclear, but lysosomal damage is likely an important step (see the section on Mechanism of Action). Indeed, vacuolar changes and curvilinear bodies found in biopsy are generally thought to correspond to affected lysosomes [88].

Patients treated with antimalarials in whom proximal myopathy, neuropathy, or cardiomyopathy develop should be evaluated for possible treatment toxicity. Clinicians should be aware of this rare complication of antimalarials, as discontinuation of the agent may result in clinical improvement.

Gastrointestinal Toxicity

Common gastrointestinal symptoms include nausea, vomiting, and diarrhea. Their frequency has been evaluated at around 30% for quinacrine, 20% for CQ, and 10% for HCQ [90]. Others symptoms, including anorexia, heartburn, abdominal distension, and elevated transaminases are rare.

Gastrointestinal manifestations are generally seen during the first weeks of treatment. They are associated with higher dosages of antimalarials and generally resolve with time, with a decreased dosage or with an administration twice a day, during meals. They are probably related to blood antimalarial levels. Munster *et al.* [91], studied blood HCQ levels in 123 patients with rheumatoid arthritis who began a 6-week, double-blind trial comparing three different doses of HCQ at 400, 800, or 1200 mg/day, followed by 18 weeks of open-label HCQ treatment at 400 mg/day. The relationship between adverse event data and drug levels was evaluated using logistic regression analysis. In a pooled univariate analysis of data, gastrointestinal adverse events were correlated with higher blood HCQ concentrations in the first 3 weeks ($p < 0.001$). The proportion of patients with these manifestations was 10% for patients with blood HCQ concentrations below 750 ng/ml, 23% for those with concentrations between 750 and 1500 ng/ml, and superior to 30% for higher blood HCQ concentration.

Cutaneous Toxicity

Cutaneous Pigmentation

Yellow-brown to slate-gray or black hyperpigmentation is observed in 10–25% of patients who receive long-term antimalarial therapy [92]. Quinacrine is associated with darker pigmentation compared to HCQ or CQ [93]. This molecule can also induce a lemon-yellow discoloration, which may involve all of the skin, thereby mimicking jaundice [92].

The majority of cases reported in the literature relate to CQ [92], and to our knowledge, only nine cases of hyperpigmentation to HCQ have been reported (Costedoat-Chalumeau, submitted). This pigmentation can begin after a few months of receiving the antimalarials. In our experience, it predominates on the anterior side of the shins (pretibial), and can also be seen in the face, hard palate, forearms, and nail bed (with transversal bands or diffuse pigmentation) (Figure 57.1). Pigmentary changes have also been reported in deeper structures such as joint tissue, trachea, and cartilage in the nose and ears. Lesions primarily occur as isolated, oval macules and spread progressively in large lesions. If a skin biopsy is performed, melanin granules and hemosiderin deposits are generally observed within

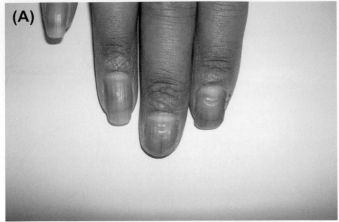

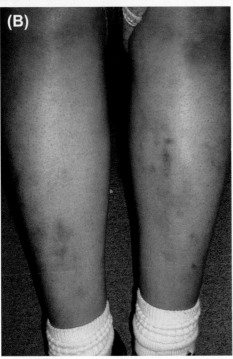

FIGURE 57.1 Hyperpigmentation secondary to HCQ: in this patient, hyperpigmentation was seen in the nail bed with transversal bands on the third finger and diffuse pigmentation that predominates on the second finger (A), and on the anterior side of the shins (B).

the hair can become bleached or blond (so-called "acquired poliosis") [94]. This hypochromia generally disappears within a few months after interruption of the drug.

Pruritus

Pruritus may occur in patients taking antimalarials [95]. Jimenez-Alonso *et al.* studied 105 SLE patients and 31 patients with cutaneous lupus, of whom 104 had taken antimalarial drugs, namely HCQ or CQ [74]. They observed that 44.2% of the patients in the antimalarial group versus 5.6% of those not receiving these treatments had pruritus ($p < 0.01$). Six patients were classified as having probable or definite pruritus related to antimalarials. They all had an aquagenic or post-wetness type of generalized pruritus which started 1–3 weeks after initiation of antimalarials. This pruritus mainly developed a few minutes after a hot shower, lasting at a high intensity for around 10 minutes, and remaining at a low intensity for several hours. No skin changes were visible. The withdrawal of antimalarials was necessary in four cases, and a reintroduction was possible in two cases [74]. In our experience, improvement may be obtained with a reduction of the dosage and/or with use of antihistaminics.

Psoriasis Exacerbation

The coexistence of psoriasis with SLE is rare (around 1.1%) [96], and data concerning the use of antimalarials in such patients are conflicting [96, 97].

Antimalarials have been used in patients with psoriatic arthritis, as it reduces inflamed joints. Such patients have been reported to present flare up of their psoriasis, pustular eruption, and erythroderma [97]. By contrast, in a large study, no exacerbation of psoriatic skin lesions was observed in 50 patients treated with HCQ, whereas the arthropathy responded to HCQ in 40 patients [98]. In their review of the literature in 1999, Wolf and Schiavo estimated that up to 18% of patients with psoriasis would develop an exacerbation of their disease following antimalarial therapy [99]. Interestingly, in contrast to other treatments such as lithium and beta-blockers, antimalarials do not induce *de novo* psoriasis, but they only trigger already existing psoriasis, probably via an alteration of the activity of enzymes involved in the epidermal proliferation process [99]. The authors further demonstrated that HCQ induces hyperproliferation and irregular keratinization in cultured skin of psoriatic patients [100].

Finally, in 2006, a systematic review analyzed 31 case series and case reports (since no randomized trial was found), and concluded that there is no strong evidence to either refute or support a role of antimalarials in the exacerbation of psoriasis [97].

the dermis. The Perl's iron stain is negative [92]. This coloration may partially or, less frequently, completely slowly resolve after cessation of therapy. Antimalarial drug-induced hyperpigmentation has also been suggested as a marker for patients at risk of ocular side effects.

Apart from pigmentation, antimalarials, mainly CQ, may also give rise to hypopigmentation, and are sometimes associated with hyperpigmented macules on other parts of the body. Vitiligo has been reported to occur in African patients with dark skin. Similarly, the roots of

Other Cutaneous Manifestations, Including Hypersensitivity Skin Eruptions

Quinacrine is associated with a lichen pianos or eczema-like eruption that can be the first sign in a chain of events that ultimately leads to aplastic anemia.

Other cutaneous reported manifestations are rare, and include morbilliform eruption, exfoliative dermatitis, urticaria, eczematous, photosensitivity, erythroderma, acute generalized exanthematous pustulosis, and erythema annulare centrifugum [93].

Some of these manifestations appear in the first days or weeks of treatment, and are indicative of a hypersensitivity rash, that is considered as a contraindication to these treatments. The frequency of this side-effect is difficult to evaluate and is, in our experience, very rare in patients treated with HCQ. Mates *et al.* have described their experience of slow oral desensitization in four patients with hypersensitivity skin eruption due to HCQ [101]. The outpatients were started on a daily dose of 1 ml of a suspension of 0.1 mg/ml of HCQ. This suspension was prepared by the hospital pharmacy from mashed HCQ tablets. The dose was then gradually increased to 400 mg/day in 36 days. All four patients completed the procedure successfully without significant difficulty. All patients were being treated with HCQ at time of publication [101].

Hematological Toxicity

Agranulocytosis and aplastic anemia have been exceptionally reported in patients treated with CQ or quinacrine. To our knowledge, this side effect has been observed in only two patients treated with HCQ, both before 1969 [102, 103]. However, in the Physician Desk Reference it is recommended to monitor patients with a periodic complete blood cell count [75]. The hematologic adverse effects caused by antimalarials are reversed with cessation of therapy.

There is some concern that antimalarials could cause hemolysis in patients with glucose-6-phosphate dehydrogenase (G6PD) deficiency. Indeed, methemoglobin formation and haemolytic anemia have been reported in G6PD-deficient patients treated with primaquine for malaria. The severity varied considerably among affected individuals. Only a few cases have been reported in malaria patients treated with CQ. To our knowledge, no cases have been reported in SLE patients or in patients treated with HCQ.

Others

Patients may rarely present nightmares and headaches. Irritability, nervousness, psychosis, seizures, and hyperexcitability have also been reported. This is supposed to be explained by a stimulation of the cerebral cortex related to antimalarials. These neuropsychiatric side effects have been reported with quinacrine and to a lesser effect with CQ. They seem very rare with HCQ [104, 105].

Tinnitus and hearing impairment or loss have been rarely reported following treatment with CQ [75].

Hypoglycemia is a rare adverse effect of antimalarials, with few cases reported with CQ and only two with HCQ (one in a patient with rheumatoid arthritis and diabetes mellitus, and one in a patient with rheumatoid arthritis without other predisposing disorder) [106].

Finally, non-specific symptoms like dizziness and constitutional complaints can be reported by the patients.

PREGNANCIES AND BREAST-FEEDING

Pregnancies

Data on Pregnancies

It has long been recommended that HCQ should be discontinued during pregnancy. This arose from the observation of cases of retinal toxicity and hearing in animals [107] and humans [107−109], when exposed *in utero* to CQ. Moreover, even if no specific abnormality has been reported in humans with CQ or HCQ, CQ inhibits the DNA−RNA and DNA−DNA polymerases [107] and is teratogenic at high doses in animals [108].

We have demonstrated that HCQ passes across the placenta, with cord blood concentrations nearly identical to those found in maternal blood, suggesting that during pregnancy the level of exposure to HCQ in mother and fetus is similar [110]. These results raised the issue of potential teratogenic and toxic effects of HCQ. In the other hand, discontinuation of HCQ doubles the risk of SLE flare during the next 6 months [1], and it is generally agreed that pregnancy *per se* increases disease activity in SLE patients. Therefore, withdrawal of HCQ at the onset of pregnancy may result in exacerbation of SLE while discontinuation of the drug might not prevent side effects in fetuses because of its very long terminal elimination half-life.

Parke [111−113] and other investigators [11, 114−116] have first proposed continuation of this treatment throughout gestation and a total of 101 pregnancies resulting in live births were then reported [11, 111−116]. One of these studies [11] was a randomized, controlled study to assess efficacy of HCQ in 20 pregnant patients with SLE ($n = 17$) or discoid lupus erythematosus ($n = 3$). The HCQ group had no flare-ups (compared to three flares in the placebo group), lower SLEPDAI (SLE in pregnancy activity index), and a decrease in prednisone dose. The efficacy of HCQ in pregnant SLE patients was later confirmed by Clowse et al. [117].

In 2003, we reported our experience concerning HCQ treatment of women with SLE, at a single center, throughout 133 pregnancies that resulted in 117 live births [118]. The children of mothers enrolled in the study were followed up for a mean of 26 months. Eighty-eight percent of pregnancies in the HCQ group and 84% of those in the control group ended successfully with a live birth. The outcomes of pregnancies were not statistically different between groups. One child in each group died of causes related to prematurity. Three malformations were observed in the HCQ group (one hypospadias, one craniostenosis, and one cardiac malformation) versus four in the control group. No visual, hearing, growth, or developmental abnormalities were reported in any of the children at the last follow-up (ages 12—108 months; mean age: 26 months) [118].

Clowse et al. are in agreement with these results [117]. Finally, a systematic review of HCQ use in pregnant patients with autoimmune diseases showed that HCQ is not associated with any increased risk of congenital defects, spontaneous abortions, fetal death, prematurity, and decreased numbers of live births in patients [119].

Data on Children Follow-up

As previously stated, the theoretical concerns regarding possible neonate toxicity of antimalarials, (especially CQ) are primarily visual and to a lesser extent hearing [107—109]. To our knowledge, such side effects have never been reported with HCQ. In our cohort, no visual or auditory deficit was clinically observed at an average age of 26 months [118], this being consistent with the literature [120, 121].

Studies including ophthalmic examinations are scarce, given the difficulties to accurately assess the vision or hearing of young children. Klinger et al. [116] performed ophthalmic examinations in 21 children born to women who took CQ ($n = 7$) or HCQ ($n = 14$) during pregnancy. Ophthalmologic examinations and tests included slitlamp biomicroscopy of the anterior segment, dilated retinal examination using indirect ophthalmoscopy, cyclopegic refraction, visual acuity testing, visual field assessment, and color vision assessment. The mean duration of gestational exposure was 7.2 months. No ophthalmic abnormality was detected in these children at an average age of 2.8 years. These results were confirmed in five studies (using dilated retinal examination with indirect ophthalmoscopy and/or flash electroretinography) of 44 additional observations [11, 115, 122—124]. Only one study found electroretinogram abnormalities in some infants exposed to HCQ in utero [125]. However, the methodology used, especially tests in mesopic conditions, and

the absence of an adequate control group, was highly subject to caution [Ingster-Moati, in press].

Breast-Feeding

We measured the concentration of HCQ in the blood and milk of two breast-feeding patients [110]. The HCQ concentrations in the milk were 344 and 1424 ng/ml, respectively. We later confirmed these results in two more mothers, whose breast milk HCQ concentrations were 1131 ng/ml and 1392 ng/ml [123]. It can thus be calculated that the HCQ ingestion by the infants was no more than 0.2 mg/kg/day. These levels are concordant with the daily ingestion of 0.11 mg/kg in one infant reported by Nation et al. [126]. The amount of HCQ received by children throughout lactation is then very small compared with the daily therapeutic dosage (6.5 mg/kg in adults). Additionally, given that HCQ concentrations in breast milk are low compared with those found in cord blood (transplacental passage), it does not seem logical to advise against breast-feeding if HCQ therapy has been maintained throughout pregnancy.

Conclusion

In 2004, during the 4th International Conference "Sex Hormones, Pregnancy and the Rheumatic Diseases", international experts have concluded that: (1) when indicated, antimalarials should be continued during pregnancy and lactation (evidence level II, according to the classification by Miyakis et al.), (2) HCQ is the antimalarial of choice in fertile women in need of treatment (evidence level IV), and (3) both CQ and HCQ are compatible with breastfeeding (evidence level IV) [127].

CONTRAINDICATIONS

Absolute Contraindications

The absolute contraindications of antimalarials are rare and involve: (1) retinal or visual field changes attributable to any 4-aminoquinoline compound, (2) known hypersensitivity to any 4-aminoquinoline compound, and (3) long-term use in children [75]. Since desensitization may be effective for patients with hypersensitivity [101] (see the section on Toxicity, Other Cutaneous Manifestations), retinal toxicity remains the only absolute contraindication in adult SLE patients.

Once again, in view of the absence of documented HCQ-related adverse effects during pregnancy [118] and because the likely risk of increased SLE flares due to HCQ discontinuation [1] is then added to the risk of flares secondary to pregnancy itself, pregnancy and breast-feeding are no longer considered as a contraindication for this treatment [127].

Relative Contraindications

Caution should be used in patients with documented psychotic disorders, liver disease, alcoholism, psoriasis, G6PD deficiency, porphyria cutanea tarda (PCT), and neuromuscular disorders, including myasthenia gravis [75]. Additionally, CQ should be administered with caution in patients with pre-existing auditory damage, in patients with history of epilepsy and in cases of concurrent administration with known hepatotoxic drugs [75].

Psoriasis and G6PD deficiency have been discussed in the section on Toxicity: data are reassuring and we believe that HCQ can be cautiously used in SLE patients with these co-existing diseases.

To our knowledge, no data are available regarding the effects of antimalarials in patients with known myasthenia gravis. Concerns regarding use of these treatments in such patients are due to reversible myasthenia-like symptoms that have been reported with CQ use (see the section on Toxicity). Indeed, in asymptomatic individuals who have a pre-existing subclinical abnormality of the neuromuscular junction, CQ acts on the neuromuscular junction to provoke clinically apparent myasthenia [128]. However, in our experience, HCQ is well tolerated by patients with SLE and myasthenia gravis.

The co-existence of SLE and PCT is considered rare, but lupus (either cutaneous lupus or SLE) was found in 15 of 676 patients with PCT (2.2%) seen at the Mayo Clinic over a 20-year period [129]. This association poses therapeutic challenges, since the relationships between antimalarials and PCT are complex. First, cases of antimalarial drug-induced PCT (or porphyria unmasked by antimalarials) have been reported. Second, some patients with PCT who received HCQ or CQ may present an acute exacerbation of hepatic disease followed by a long-term clinical remission after recovery from the acute exacerbation [93]. Interestingly, this observation has led to the use of antimalarials (in very low dosage) as treatment for PCT [130]. It is then needed to recognize the co-existence of PCT and SLE, since standard or high-dose antimalarial therapy may be dangerous. In patients with co-existent disease, it is usually recommended to cautiously initiate antimalarials either in very low dosage to prevent the acute exacerbation (i.e. 100 mg of HCQ twice weekly) or, more rarely, in higher dosage in hospitalized patients with close hepatic monitoring. In some cases, patients may also be pretreated with phlebotomies [93].

Finally, as both G6PD deficiency and PCT are rare, routine testing for these conditions is not recommended before initiating CQ or HCQ treatment.

PHARMACOLOGY, PHARMACOKINETIC PROPERTIES AND BLOOD HCQ DOSAGE

Pharmacology and Pharmacokinetic Properties

Chloroquine

CQ is a 4-aminoquinoline (7-chloro-4-[[4-(diethyla-mino)-1-methylbutyl]-amino] quinoline phosphate) (Figure 57.2). The isoquinoline nucleus has double benzene like structure with a chloride atom at position 7 and an alkyl side chain at the 4-amino site. It is a chiral molecule with D and L forms. The drug is manufactured

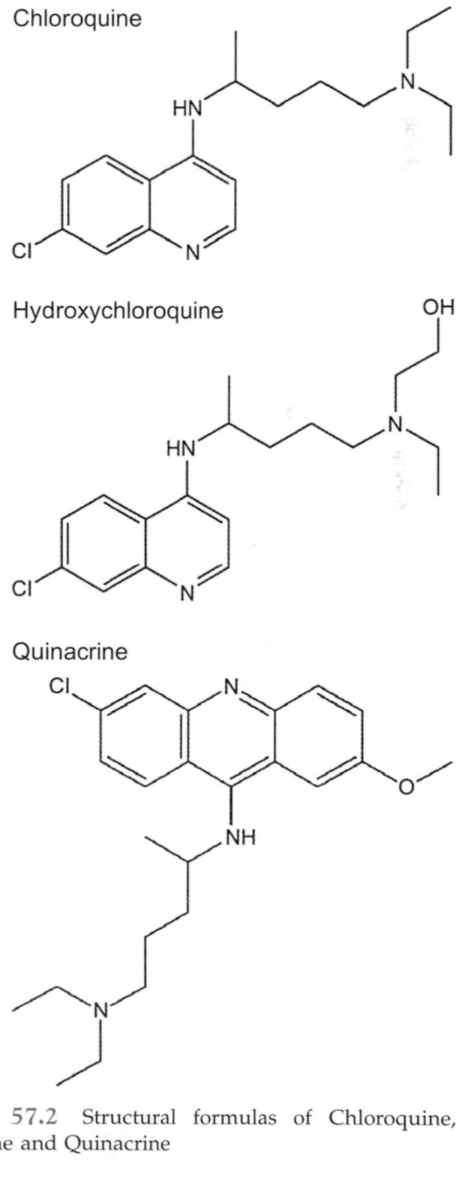

FIGURE 57.2 Structural formulas of Chloroquine, Hydroxychloroquine and Quinacrine

as white, bitter-tasting tablets, containing 100, 250 or 500 mg of CQ phosphate or sulfate (Aralen, Avloclor, Nivaquine, Resochin, Dawaquin). CQ base 150 mg is equivalent to CQ sulfate 200 mg and to CQ phosphate 250 mg.

CQ is readily and almost completely absorbed from the gastrointestinal tract, and only a small portion of the administered dose is found in the stools [75]. After oral administration, peak plasma concentrations are reached within 8–12 hours. Approximately 55% of the drug in the plasma is bound to protein [75]. Renal clearance accounts for half of the total systemic clearance and is increased by acidification of the urine. CQ undergoes hepatic biotransformation via the cytochrome P450 enzymes into two active metabolites: desethylchloroquine and bisdesethylchloroquine [93].

CQ is deposited in the tissues in considerable amounts. Highest concentrations of CQ are found in melanin-containing cells in the retina and skin. In animals, from 200–700 times the plasma concentration may be found in the liver, spleen, kidney, and lung [75]. Leukocytes also concentrate the drug. As a result of this widespread tissue deposition, CQ has a large volume or distribution (over 100 L/kg).

Doses of 250 mg daily lead to stable plasma concentrations of 100–500 ng/ml [93]. The pharmacokinetics is complex with plasma levels determined by the rate of distribution rather than by the rate of elimination. When the drug is discontinued after daily dosage for 2 weeks, plasma half-life is initially 6–7 days with a gradual increase to 17 days after 4 weeks and a terminal half-life of 30–60 days [93]. CQ remains in the skin for more than 6 months after cessation of therapy, at a time when the drug is no longer detectable in the plasma [131].

Hydroxychloroquine

HCQ is a 4-aminoquinoline that differs from CQ by a hydroxyl group at the end of a side chain (2-[[4-[(7-chloro-4-quinolyl)-amino] pentyl] ethylamino] ethanol sulfate) (Figure 57.2). The drug is manufactured as HCQ sulfate (Plaquenil). Plaquenil is available as white to off-white, film-coated, peanut-shaped tablets, containing 200 mg HCQ sulfate (equivalent to 155 mg base).

Pharmacokinetics of HCQ has been less studied with data coming mainly from the group of Suzan Tett. HCQ has similar pharmacokinetic properties to CQ with rapid gastrointestinal absorption and a very large volume of distribution (with values estimated at 605 liters [132] or 5522 liters [133]). Very little data are available on the mechanism of elimination of HCQ. The renal elimination of the unchanged drug seems low, around 21% of the dose [134]. HCQ is also metabolized by cytochrome P450 enzymes (cytochrome 2D6, 2C8, 3A4 and 3A5) into three metabolites: desethylhydroxychloroquine,

desethylchloroquine, and bidesethylchloroquine. To our knowledge, no studies are available regarding the link between genotype differences in cytochrome P450 enzymes and the efficacy or toxicity of HCQ.

The interindividual variability of blood HCQ concentrations is important with more than a 10-fold range of drug concentrations found after similar doses. This has been observed in apparently compliant patients with rheumatoid arthritis [91, 135, 136] and, moreover, in healthy volunteers [137, 138]. One explanation is the significant variability in bioavailability between individuals, as shown by Tett *et al.* (139). In this study, six healthy volunteers received a tablet of 200 mg of HCQ, on three occasions. Maximum blood HCQ concentration for the three tablet doses showed significant differences between subjects (range 135–422 ng/ml; $p < 0.009$) but not within subjects ($p = 0.32$), with a coefficient of variation of 34% and 11%, respectively. Truncated areas under the blood HCQ concentration–time curve of the three tablets were different between ($p = 0.0001$) but not within subjects ($p = 0.13$). These results were confirmed in nine patients treated with 200 mg of HCQ for rheumatoid arthritis [140]. Additionally to the variability in bioavailability between individuals, it is possible that the wide range of blood HCQ concentrations arises partly from genetic differences in the capacity to metabolize HCQ via cytochrome, as has been shown for many other drugs.

Using a population pharmacokinetic model, Carmichael *et al.* [132] studied a total of 780 blood HCQ concentrations obtained from 123 patients. With this model, the authors estimated the values of bioavailability (0.75), clearance (9.9 liter/hour) and volume of distribution (605 liters). The half-life of terminal elimination was 43 days. The terminal half-life of elimination of HCQ is indeed very long. It was estimated at 1 or 2 months, and at 123 hours in a more recent study [91]. This long half-life led to the conclusion that the blood HCQ concentration is stable over time in a given patient. Accordingly, Miller *et al.* found, in a pharmacokinetic study of five patients treated with HCQ for rheumatoid arthritis, that the concentrations obtained at steady state showed limited fluctuations in the day, to overcome the scheduling problems in clinical studies [141].

As a consequence, maximal clinical efficacy may take up to 3–6 months to be achieved. Dose loading, with administration of higher doses (up to 1200 mg/day) for the first 6 weeks, has been shown to accelerate clinical response in rheumatoid arthritis [91].

Quinacrine

Quinacrine is a chloroquine 9-aminocridine which differs from CQ only in having an acridine nucleus (i.e., an extra benzene ring) instead of a quinoline

(6-chloro-9-[[4-(diethylamino)-l-methylbutyl]amino]-2-methoxyacridine dihydrochloride) (Figure 57.2). The drug is manufactured as Mepacrine and Atabrine.

The pharmacokinetics of this drug is similar to the 4-aminoquinolones.

Blood HCQ Assay

HCQ and its metabolite levels can be quantified by high-performance liquid chromatography (HPLC). For reasons of sensitivity and reproducibility, HCQ concentrations should be measured in whole blood [91]. Of note, since HCQ accumulates in leucocytes, HCQ concentrations are clearly higher when determined in whole blood than in plasma or serum.

As previously stated, there is great interindividual variability in blood HCQ concentrations, leading to the question of whether or not there is a relationship between concentrations and efficacy, and whether or not it is necessary to individualize dosing to target HCQ concentrations associated with optimal outcomes.

A relationship between whole-blood concentrations of HCQ and clinical efficacy has been reported in patients with rheumatoid arthritis [91, 135, 136]. Similar results have been demonstrated in SLE patients [33]. We have measured whole-blood HCQ concentrations in 143 unselected SLE patients who were all receiving 400 mg/day of HCQ. We first studied the relationship between HCQ concentrations and SLE activity on the day of HCQ assay (day 0), and then examined whether these baseline HCQ concentrations were predictive of flares during the subsequent 6 months. The mean blood HCQ concentration on day 0 was 1017 ± 532 ng/ml with large interindividual variations similar to that observed in healthy volunteers [137–139] and in patients with rheumatoid arthritis [91, 135, 136]. Low blood HCQ concentrations were strongly associated with ongoing disease activity. At baseline, 23 patients had active disease (mean SLEDAI 12.4 ± 7.5). Their mean blood HCQ concentration was significantly lower than that in the 120 patients with inactive disease (694 ± 448 vs. 1079 ± 526 ng/ml; $p = 0.001$). We also found that low baseline blood HCQ concentrations in patients with inactive SLE were strongly associated with the risk of developing SLE flares during the subsequent 6 months: among the 120 patients who had inactive disease at baseline, the mean HCQ concentration at baseline in the 14 (12%) who had disease exacerbations during follow-up was significantly lower than that in the patients whose disease remained inactive (703 ± 534 vs. 1128 ± 507 ng/ml; $p = 0.006$). We used receiver operating characteristic (ROC) curve analysis to determine the blood HCQ concentration associated with the lowest risk of SLE flare in the subsequent 6 months. A threshold value of 1000 ng/ml provided the best trade-off between sensitivity and specificity, as well as a high negative predictive value for SLE flares (96%). We therefore proposed 1000 ng/ml as the target whole-blood HCQ concentration in patients with SLE.

A French multicenter randomized prospective study is currently ongoing (PLUS study with more than 500 patients included in September 2009) to determine the potential benefits of individualized HCQ dosing schedules aimed at maintaining the blood HCQ concentration above 1000 ng/ml (ClinicalTrials.gov number, NCT00413361).

Additionally, blood HCQ assay may be a simple, objective and reliable marker of non-compliance with medications in patients with SLE [142]. Indeed, as HCQ has a long terminal elimination half-life, the patients who have undetectable blood HCQ concentrations necessarily have not taken HCQ for a long time. In our experience, regular drug assays (available within a few days in our center), by showing the patients with very low blood HCQ concentrations, help physicians to detect noncompliance and could serve as the basis for counseling and supporting patients with poor adherence to therapeutic regimens. Additionally, we strongly believe that a SLE flare should lead to an assessment of blood HCQ level in order to prevent incorrect interpretation of consequences of poor compliance as a lack of response, and to avoid unnecessary or even dangerous regimen escalation.

DRUG INTERACTIONS AND CIGARETTE SMOKING

Pharmacokinetic interaction studies are limited. However, some data indicate that antimalarials may exhibit drug interactions.

Drug Interactions

Levels of some drugs, including penicillamine, digoxin, metoprolol, and cyclosporine may be increased with HCQ or CQ usage probably via a cytochrome P450 inhibition [143]. Indeed, CQ, and probably HCQ, are potent inhibitors of cytochrome P450 2D6-catalysed pathways *in vitro* [144]. It has been demonstrated in a randomized, double-blind crossover study with seven healthy volunteers, that concurrent use of metoprolol, a cytochrome P450 2D6 enzyme substrate, and HCQ increases metoprolol exposure. The area under the plasma concentration–time curve increased by $65 \pm 4.6\%$, whereas the maximal plasma concentrations of metoprolol increased by $72 \pm 6.9\%$ [144]. Consequently, this association requires close monitoring for metoprolol adverse effects such as bradycardia [143].

In vitro, cytochrome P450 inhibitors, troleandomycine, cimetidine and quinidine, increase the CQ concentration and reduce the concentration of its metabolites [145]. To our knowledge, no data are available on the clinical consequences of such interactions in SLE patients.

Classically, kaolin and antacids may decrease absorption of CQ.

Finally, an antagonizing effect of the combination of omeprazole with CQ has been demonstrated on antimalarial activity, suggesting that proton pump inhibitors may have an antagonizing effect on the pharmacologic actions of CQ and HCQ [146].

Cigarette Smoking

Cigarette smoking has been suspected to increase the risk and the activity of SLE and cutaneous lupus. Some data indicate that cigarette smoking might interfere with the effectiveness of HCQ and CQ in cutaneous lupus. As these antimalarial agents are metabolized via the cytochrome P450 enzyme system and as the constituents of cigarette smoke are known potent inducers of cytochrome P450, it has been hypothesized that the resistance of cutaneous lupus might be explained by a modification of the metabolism of these drugs [147].

However, we did not find any significant relationship between cigarette smoking and blood HCQ concentrations in 223 patients treated with 400 mg/day of HCQ for at least 6 months [147]. This is a strong argument against a direct effect of smoking on HCQ metabolism. We concluded that another mechanism of interaction (as a modification of the lysosomal accumulation of antimalarial agents) or a direct deleterious effect of smoking on cutaneous lesions seems more likely.

DOSAGE AND FOLLOW-UP

When initiating a treatment with antimalarials, it is important to keep in mind that the use of antimalarials is characterized by a long delay in the onset of action because of their long half-life. As a result, patients may incorrectly interpret this delay as a lack of efficacy of the treatment, and may have poor adherence [142]. Then, when prescribing these drugs, the physician should cautiously explain to the patient that the efficacy may take 2—8 weeks to be achieved.

Dosage

Hydroxychloroquine

Few data have been published on the pharmacologic management of HCQ therapy, and the optimal daily dosage in SLE is controversial. The initial dose recommended for SLE patients in the Physicians Desk Reference is 400—800 mg daily (310—620 mg base), continued for several weeks or months until remission [75]. The average maintenance dosage is 200—400 mg daily (155—310 milligrams base) [75].

Ophthalmologists have suggested that the daily dosage should not exceed 6.5 mg/kg, in order to avoid retinal toxicity. If dosage is superior, ophthalmologic follow-up must be reinforced (see the section on Toxicity).

Considering the wide interindividual variations in blood HCQ concentration, the question is whether the daily HCQ dosage should be still adapted to the body weight or whether it should be individualized and adapted to its blood concentration, similarly to what is done for other drugs. The French multicenter randomized prospective study PLUS (ClinicalTrials.gov number, NCT00413361) should allow us to answer this question in the next few years.

Chloroquine

In SLE, it is generally believed that the daily dosage should not exceed 2.5 mg/kg, in order to avoid retinal toxicity. No recommendations were found in the Physicians Desk Reference.

Quinacrine

No recommendations were found in the Physicians Desk Reference.

Finally, the combination of HCQ with CQ should not be used due to potential retinotoxicity. By contrast, the association of Quinacrine with CQ or with HCQ may have some indications [4].

Follow-up

When initiating HCQ, ophthalmologic examinations (e.g., visual acuity, funduscopic, visual field tests, and/or multifocal electroretinogram) are recommended at baseline, periodically thereafter, and if the patient experiences any visual disturbances [75]. Knee and ankle reflexes should be regularly monitored [75]. We think that an electrocardiogram should be performed at baseline and probably repeated thereafter. Finally, it is recommended in the Physician's Desk Reference to make periodic blood cell count [75] despite the extreme rarity of hematological toxicity.

Quinacrine is not retinotoxic but blood counts need to be monitored, because it can cause aplastic anemia. Any new rashes should be immediately reported to a physician and the drug stopped until it is evaluated [4]. Indeed, some rashes are associated with a high risk of aplastic anemia.

Since children are especially sensitive to the 4-aminoquinoline compounds, with a number of fatalities reported following accidental ingestion of only a few tablets of CQ, patients should be strongly warned to keep these drugs out of the reach of children [75]. There are limited data on pediatric HCQ overdoses, but given its similarity to CQ, HCQ is considered potentially toxic at small doses [148].

CONCLUSION

In conclusion, whereas HCQ has a high efficacy/toxicity ratio, the percentage of SLE patients receiving HCQ remains rather low, even in important studies including, *per nature*, selected patients and experienced clinicians [149, 150], and, more strikingly this percentage is not mentioned in major trials [151–153]. Then, we and others believe that the management of SLE patients could probably be easily improved with a more systematic use of this old and inexpensive treatment. Unless toxicity supervenes, it can be administered for the duration of the disease [48].

The interest of blood HCQ concentration monitoring is important in assessing adherence with treatment. However, the potential benefits of individualized HCQ dosing schedules aimed at maintaining appropriate blood HCQ concentration remains to be demonstrated.

References

[1] A randomized study of the effect of withdrawing hydroxychloroquine sulfate in systemic lupus erythematosus, The Canadian Hydroxychloroquine Study Group, N Engl J Med 324 (3) (1991) 150–154.

[2] N. Kasitanon, D.M. Fine, M. Haas, L.S. Magder, M. Petri, Hydroxychloroquine use predicts complete renal remission within 12 months among patients treated with mycophenolate mofetil therapy for membranous lupus nephritis, Lupus 15 (6) (2006) 366–370.

[3] B.J. Fessler, G.S. Alarcon, G. McGwin Jr., J. Roseman, H.M. Bastian, A.W. Friedman, et al., Systemic lupus erythematosus in three ethnic groups: XVI. Association of hydroxychloroquine use with reduced risk of damage accrual, Arthritis Rheum 52 (5) (2005) 1473–1480.

[4] D.J. Wallace, Is there a role for quinacrine (Atabrine) in the new millennium? Lupus 9 (2) (2000) 81–82.

[5] I. Cavazzana, R. Sala, C. Bazzani, A. Ceribelli, C. Zane, R. Cattaneo, et al., Treatment of lupus skin involvement with quinacrine and hydroxychloroquine, Lupus 18 (8) (2009) 735–739.

[6] E. Toubi, I. Rosner, M. Rozenbaum, A. Kessel, T.D. Golan, The benefit of combining hydroxychloroquine with quinacrine in the treatment of SLE patients, Lupus 9 (2) (2000) 92–95.

[7] M.L. Estes, D. Ewing-Wilson, S.M. Chou, H. Mitsumoto, M. Hanson, E. Shirey, et al., Chloroquine neuromyotoxicity. Clinical and pathologic perspective, Am J Med 82 (3) (1987) 447–455.

[8] M. Easterbrook, An ophthalmological view on the efficacy and safety of chloroquine versus hydroxychloroquine, J Rheumatol 26 (9) (1999) 1866–1868.

[9] N. Costedoat-Chalumeau, J.S. Hulot, Z. Amoura, A. Delcourt, T. Maisonobe, R. Dorent, et al., Cardiomyopathy related to

antimalarial therapy with illustrative case report, Cardiology 107 (2) (2006) 73–80.

[10] E. Tsakonas, L. Joseph, J.M. Esdaile, D. Choquette, J.L. Senecal, A. Cividino, et al., A long-term study of hydroxychloroquine withdrawal on exacerbations in systemic lupus erythematosus. The Canadian Hydroxychloroquine Study Group, Lupus 7 (2) (1998) 80–85.

[11] R.A. Levy, V.S. Vilela, M.J. Cataldo, R.C. Ramos, J.L. Duarte, B.R. Tura, et al., Hydroxychloroquine (HCQ) in lupus pregnancy: double-blind and placebo-controlled study, Lupus 10 (6) (2001) 401–404.

[12] I.M. Meinao, E.I. Sato, L.E. Andrade, M.B. Ferraz, E. Atra, Controlled trial with chloroquine diphosphate in systemic lupus erythematosus, Lupus 5 (3) (1996) 237–241.

[13] M. Petri, G.S. Alarcon, R.P. Kimberly, J.D. Reveille, Predictors of renal insufficiency in systemic lupus erythematosus (abstract), Arthritis Rheum (2005).

[14] J.A. James, X.R. Kim-Howard, B.F. Bruner, M.K. Jonsson, M.T. McClain, M.R. Arbuckle, et al., Hydroxychloroquine sulfate treatment is associated with later onset of systemic lupus erythematosus, Lupus 16 (6) (2007) 401–409.

[15] G. Ruiz-Irastorza, M. Ramos-Casals, P. Brito-Zeron, M.A. Khamashta, Clinical efficacy and side effects of antimalarials in systemic lupus erythematosus: a systematic review, Ann Rheum Dis (2009).

[16] M.H. Edwards, S. Pierangeli, X. Liu, J.H. Barker, G. Anderson, E.N. Harris, Hydroxychloroquine reverses thrombogenic properties of antiphospholipid antibodies in mice, Circulation 96 (12) (1997) 4380–4384.

[17] M. Petri, Thrombosis and systemic lupus erythematosus: the Hopkins Lupus Cohort perspective, Scand J Rheumatol 25 (4) (1996) 191–193.

[18] M.Y. Mok, E.Y. Chan, D.Y. Fong, K.F. Leung, W.S. Wong, C.S. Lau, Antiphospholipid antibody profiles and their clinical associations in Chinese patients with systemic lupus erythematosus, J Rheumatol 32 (4) (2005) 622–628.

[19] G. Ruiz-Irastorza, M.V. Egurbide, J.I. Pijoan, M. Garmendia, I. Villar, A. Martinez-Berriotxoa, et al., Effect of antimalarials on thrombosis and survival in patients with systemic lupus erythematosus, Lupus 15 (9) (2006) 577–583.

[20] R. Kaiser, C.M. Cleveland, L.A. Criswell, Risk and protective factors for thrombosis in systemic lupus erythematosus: results from a large, multi-ethnic cohort, Ann Rheum Dis 68 (2) (2009) 238–241.

[21] M.G. Tektonidou, K. Laskari, D.B. Panagiotakos, H.M. Moutsopoulos, Risk factors for thrombosis and primary thrombosis prevention in patients with systemic lupus erythematosus with or without antiphospholipid antibodies, Arthritis Rheum 61 (1) (2009) 29–36.

[22] M.C. Wasko, H.B. Hubert, V.B. Lingala, J.R. Elliott, M.E. Luggen, J.F. Fries, et al., Hydroxychloroquine and risk of diabetes in patients with rheumatoid arthritis, Jama 298 (2) (2007) 187–193.

[23] S. Lakshminarayanan, S. Walsh, M. Mohanraj, N. Rothfield, Factors associated with low bone mineral density in female patients with systemic lupus erythematosus, J Rheumatol 28 (1) (2001) 102–108.

[24] C.C. Mok, A. Mak, K.M. Ma, Bone mineral density in postmenopausal Chinese patients with systemic lupus erythematosus, Lupus 14 (2) (2005) 106–112.

[25] G. Ruiz-Irastorza, A. Ugarte, M.V. Egurbide, M. Garmendia, J.I. Pijoan, A. Martinez-Berriotxoa, et al., Antimalarials may influence the risk of malignancy in systemic lupus erythematosus, Ann Rheum Dis 66 (6) (2007) 815–817.

[26] P. Rahman, D.D. Gladman, M.B. Urowitz, K. Yuen, D. Hallett, I.N. Bruce, The cholesterol lowering effect of antimalarial drugs is enhanced in patients with lupus taking corticosteroid drugs, J Rheumatol 26 (2) (1999) 325–330.

[27] M. Petri, C. Lakatta, L. Magder, D. Goldman, Effect of prednisone and hydroxychloroquine on coronary artery disease risk factors in systemic lupus erythematosus: a longitudinal data analysis, Am J Med 96 (3) (1994) 254–259.

[28] H.N. Hodis, F.P. Quismorio Jr., E. Wickham, D.H. Blankenhorn, The lipid, lipoprotein, and apolipoprotein effects of hydroxychloroquine in patients with systemic lupus erythematosus, J Rheumatol 20 (4) (1993) 661–665.

[29] D.J. Wallace, A.L. Metzger, V.J. Stecher, B.A. Turnbull, P.A. Kern, Cholesterol-lowering effect of hydroxychloroquine in patients with rheumatic disease: reversal of deleterious effects of steroids on lipids, Am J Med 89 (3) (1990) 322–326.

[30] R. Munro, E. Morrison, A.G. McDonald, J.A. Hunter, R. Madhok, H.A. Capell, Effect of disease modifying agents on the lipid profiles of patients with rheumatoid arthritis, Ann Rheum Dis 56 (6) (1997) 374–377.

[31] K.T. Ho, C.W. Ahn, G.S. Alarcon, B.A. Baethge, F.K. Tan, J. Roseman, et al., Systemic lupus erythematosus in a multi-ethnic cohort (LUMINA): XXVIII. Factors predictive of thrombotic events, Rheumatology (Oxford) 44 (10) (2005) 1303–1307.

[32] J.H. Rand, X.X. Wu, A.S. Quinn, P.P. Chen, J.J. Hathcock, D.J. Taatjes, Hydroxychloroquine directly reduces the binding of antiphospholipid antibody-beta2-glycoprotein I complexes to phospholipid bilayers, Blood 112 (5) (2008) 1687–1695.

[33] N. Costedoat-Chalumeau, Z. Amoura, J.S. Hulot, H.A. Hammoud, G. Aymard, P. Cacoub, et al., Low blood concentration of hydroxychloroquine is a marker for and predictor of disease exacerbations in patients with systemic lupus erythematosus, Arthritis Rheum 54 (10) (2006) 3284–3290.

[34] K. Shojania, B.E. Koehler, T. Elliott, Hypoglycemia induced by hydroxychloroquine in a type II diabetic treated for poly-arthritis, J Rheumatol 26 (1) (1999) 195–196.

[35] A. Quatraro, G. Consoli, M. Magno, F. Caretta, A. Nardozza, A. Ceriello, et al., Hydroxychloroquine in decompensated, treatment-refractory noninsulin-dependent diabetes mellitus. A new job for an old drug? Ann Intern Med 112 (9) (1990) 678–681.

[36] E.F. Borba, E. Bonfa, Longterm beneficial effect of chloroquine diphosphate on lipoprotein profile in lupus patients with and without steroid therapy, J Rheumatol 28 (4) (2001) 780–785.

[37] L.S. Tam, E.K. Li, C.W. Lam, B. Tomlinson, Hydroxychloroquine has no significant effect on lipids and apolipoproteins in Chinese systemic lupus erythematosus patients with mild or inactive disease, Lupus 9 (6) (2000) 413–416.

[38] A.C. Beynen, A.J. van der Molen, M.J. Geelen, Inhibition of hepatic cholesterol biosynthesis by chloroquine, Lipids 16 (6) (1981) 472–474.

[39] T.A. Pagler, A. Neuhofer, H. Laggner, W. Strobl, H. Stangl, Cholesterol efflux via HDL resecretion occurs when cholesterol transport out of the lysosome is impaired, J Lipid Res 48 (10) (2007) 2141–2150.

[40] J.C. Sachet, E.F. Borba, E. Bonfa, C.G. Vinagre, V.M. Silva, R.C. Maranhao, Chloroquine increases low-density lipoprotein removal from plasma in systemic lupus patients, Lupus 16 (4) (2007) 273–278.

[41] M.J. Roman, B.A. Shanker, A. Davis, M.D. Lockshin, L. Sammaritano, R. Simantov, et al., Prevalence and correlates of accelerated atherosclerosis in systemic lupus erythematosus, N Engl J Med 349 (25) (2003) 2399–2406.

[42] J. Sotelo, E. Briceno, M.A. Lopez-Gonzalez, Adding chloroquine to conventional treatment for glioblastoma multiforme: a randomized, double-blind, placebo-controlled trial, Ann Intern Med 144 (5) (2006) 337–343.

[43] R. Rahim, J.S. Strobl, Hydroxychloroquine, chloroquine, and all-trans retinoic acid regulate growth, survival, and histone acetylation in breast cancer cells, Anticancer Drugs (2009).

[44] D.S. Fedson, Confronting an influenza pandemic with inexpensive generic agents: can it be done? Lancet Infect Dis 8 (9) (2008) 571–576.

[45] J.M. Rolain, P. Colson, D. Raoult, Recycling of chloroquine and its hydroxyl analogue to face bacterial, fungal and viral infections in the 21st century, Int J Antimicrob Agents 30 (4) (2007) 297–308.

[46] Y. Molad, A. Gorshtein, A.J. Wysenbeek, D. Guedj, R. Majadla, A. Weinberger, et al., Protective effect of hydroxychloroquine in systemic lupus erythematosus. Prospective long-term study of an Israeli cohort, Lupus 11 (6) (2002) 356–361.

[47] G.J. Pons-Estel, G.S. Alarcon, G. McGwin Jr., M.I. Danila, J. Zhang, H.M. Bastian, et al., Protective effect of hydroxychloroquine on renal damage in patients with lupus nephritis: LXV, data from a multiethnic US cohort, Arthritis Rheum 61 (6) (2009) 830–839.

[48] G.S. Alarcon, G. McGwin Jr., A.M. Bertoli, B.J. Fessler, J. Calvo-Alen, H.M. Bastian, et al., Effect of hydroxychloroquine in the survival of patients with systemic lupus erythematosus. data from lumina, a multiethnic us cohort (LUMINA L), Ann Rheum Dis 66 (9) (2007) 1168–1172.

[49] S. Ohkuma, B. Poole, Fluorescence probe measurement of the intralysosomal pH in living cells and the perturbation of pH by various agents, Proc Natl Acad Sci USA 75 (7) (1978) 3327–3331.

[50] H.K. Ziegler, E.R. Unanue, Decrease in macrophage antigen catabolism caused by ammonia and chloroquine is associated with inhibition of antigen presentation to T cells, Proc Natl Acad Sci USA 79 (1) (1982) 175–178.

[51] F. Potvin, E. Petitclerc, F. Marceau, P.E. Poubelle, Mechanisms of action of antimalarials in inflammation: induction of apoptosis in human endothelial cells, J Immunol 158 (4) (1997) 1872–1879.

[52] P. Boya, R.A. Gonzalez-Polo, D. Poncet, K. Andreau, H.L. Vieira, T. Roumier, et al., Mitochondrial membrane permeabilization is a critical step of lysosome-initiated apoptosis induced by hydroxychloroquine, Oncogene 22 (25) (2003) 3927–3936.

[53] L. Lagneaux, A. Delforge, M. Dejeneffe, M. Massy, M. Bernier, D. Bron, Hydroxychloroquine-induced apoptosis of chronic lymphocytic leukemia involves activation of caspase-3 and modulation of Bcl-2/bax/ratio, Leuk Lymphoma 43 (5) (2002) 1087–1095.

[54] J.H. Lai, L.J. Ho, K.C. Lu, D.M. Chang, M.F. Shaio, S.H. Han, Western and Chinese antirheumatic drug-induced T cell apoptotic DNA damage uses different caspase cascades and is independent of Fas/Fas ligand interaction, J Immunol 166 (11) (2001) 6914–6924.

[55] X.W. Meng, J.M. Feller, J.B. Ziegler, S.M. Pittman, C.M. Ireland, Induction of apoptosis in peripheral blood lymphocytes following treatment in vitro with hydroxychloroquine, Arthritis Rheum 40 (5) (1997) 927–935.

[56] W.U. Kim, S.A. Yoo, S.Y. Min, S.H. Park, H.S. Koh, S.W. Song, et al., Hydroxychloroquine potentiates Fas-mediated apoptosis of rheumatoid synoviocytes, Clin Exp Immunol 144 (3) (2006) 503–511.

[57] S.T. Liu, C.R. Wang, G.D. Yin, M.F. Liu, G.L. Lee, M.Y. Chen, et al., Hydroxychloroquine sulphate inhibits in vitro apoptosis of

circulating lymphocytes in patients with systemic lupus erythematosus, Asian Pac J Allergy Immunol 19 (1) (2001) 29–35.

[58] K. Sperber, H. Quraishi, T.H. Kalb, A. Panja, V. Stecher, L. Mayer, Selective regulation of cytokine secretion by hydroxychloroquine: inhibition of interleukin 1 alpha (IL-1-alpha) and IL-6 in human monocytes and T cells, J Rheumatol 20 (5) (1993) 803–808.

[59] A.L. Gilman, F. Beams, M. Tefft, A. Mazumder, The effect of hydroxychloroquine on alloreactivity and its potential use for graft-versus-host disease, Bone Marrow Transplant 17 (6) (1996) 1069–1075.

[60] F.D. Goldman, A.L. Gilman, C. Hollenback, R.M. Kato, B.A. Premack, D.J. Rawlings, Hydroxychloroquine inhibits calcium signals in T cells: a new mechanism to explain its immunomodulatory properties, Blood 95 (11) (2000) 3460–3466.

[61] R.B. Landewe, A.M. Miltenburg, M.J. Verdonk, C.L. Verweij, F.C. Breedveld, M.R. Daha, et al., Chloroquine inhibits T cell proliferation by interfering with IL-2 production and responsiveness, Clin Exp Immunol 102 (1) (1995) 144–151.

[62] K.R. Schultz, S. Bader, D. Nelson, M.D. Wang, K.T. HayGlass, Immune suppression by lysosomotropic amines and cyclosporine on T-cell responses to minor and major histocompatibility antigens: does synergy exist? Transplantation 64 (7) (1997) 1055–1065.

[63] S. Picot, F. Peyron, J.P. Vuillez, B. Polack, P. Ambroise-Thomas, Chloroquine inhibits tumor necrosis factor production by human macrophages in vitro, J Infect Dis 164 (4) (1991) 830.

[64] S. Namiuchi, S. Kumagai, H. Imura, T. Suginoshita, T. Hattori, F. Hirata, Quinacrine inhibits the primary but not secondary proliferative response of human cytotoxic T cells to allogeneic non-T cell antigens, J Immunol 132 (3) (1984) 1456–1461.

[65] B.E. van den Borne, B.A. Dijkmans, H.H. de Rooij, S. le Cessie, C.L. Verweij, Chloroquine and hydroxychloroquine equally affect tumor necrosis factor-alpha, interleukin 6, and interferon-gamma production by peripheral blood mononuclear cells, J Rheumatol 24 (1) (1997) 55–60.

[66] M. Tishler, I. Yaron, I. Shirazi, M. Yaron, Hydroxychloroquine treatment for primary Sjögren's syndrome: its effect on salivary and serum inflammatory markers, Ann Rheum Dis 58 (4) (1999) 253–256.

[67] J. Bondeson, R. Sundler, Antimalarial drugs inhibit phospholipase A2 activation and induction of interleukin 1beta and tumor necrosis factor alpha in macrophages: implications for their mode of action in rheumatoid arthritis, Gen Pharmacol 30 (3) (1998) 357–366.

[68] S.N. Cohen, K.L. Yielding, Inhibition of DNA and RNA polymerase reactions by chloroquine, Proc Natl Acad Sci USA 54 (2) (1965) 521–527.

[69] A. Filippov, G. Skatova, V. Porotikov, E. Kobrinsky, M. Saxon, Ca2+-antagonistic properties of phospholipase A2 inhibitors, mepacrine and chloroquine, Gen Physiol Biophys 8 (2) (1989) 113–118.

[70] R. Lafyatis, M. York, A. Marshak-Rothstein, Antimalarial agents: Closing the gate on toll-like receptors? Arthritis Rheum 54 (10) (2006) 3068–3070.

[71] F. Brentano, O. Schorr, R.E. Gay, S. Gay, D. Kyburz, RNA released from necrotic synovial fluid cells activates rheumatoid arthritis synovial fibroblasts via Toll-like receptor 3, Arthritis Rheum 52 (9) (2005) 2656–2665.

[72] D. Kyburz, F. Brentano, S. Gay, Mode of action of hydroxychloroquine in RA-evidence of an inhibitory effect on toll-like receptor signaling, Nat Clin Pract Rheumatol 2 (9) (2006) 458–459.

[73] E.A. Leadbetter, I.R. Rifkin, A.M. Hohlbaum, B.C. Beaudette, M.J. Shlomchik, A. Marshak-Rothstein, Chromatin-IgG complexes activate B cells by dual engagement of IgM and Toll-like receptors, Nature 416 (6881) (2002) 603–607.

[74] J. Jimenez-Alonso, J. Tercedor, L. Jaimez, E. Garcia-Lora, Antimalarial drug-induced aquagenic-type pruritus in patients with lupus, Arthritis Rheum 41 (4) (1998) 744–745.

[75] Physician's Desk Reference. 2005 [cited 2005; Available from:

[76] J.A. Avina-Zubieta, G. Galindo-Rodriguez, S. Newman, M.E. Suarez-Almazor, A.S. Russell, Long-term effectiveness of antimalarial drugs in rheumatic diseases, Ann Rheum Dis 57 (10) (1998) 582–587.

[77] M.F. Marmor, New American Academy of Ophthalmology recommendations on screening for hydroxychloroquine retinopathy, Arthritis Rheum 48 (6) (2003) 1764.

[78] M.F. Marmor, The dilemma of hydroxychloroquine screening: new information from the multifocal ERG, Am J Ophthalmol 140 (5) (2005) 894–895.

[79] G.D. Levy, S.J. Munz, J. Paschal, H.B. Cohen, K.J. Pince, T. Peterson, Incidence of hydroxychloroquine retinopathy in 1,207 patients in a large multicenter outpatient practice, Arthritis Rheum 40 (8) (1997) 1482–1486.

[80] M.F. Marmor, R.E. Carr, M. Easterbrook, A.A. Farjo, W.F. Mieler, Recommendations on screening for chloroquine and hydroxychloroquine retinopathy: a report by the American Academy of Ophthalmology, Ophthalmology 109 (7) (2002) 1377–1382.

[81] A.E. Semmer, M.S. Lee, A.R. Harrison, T.W. Olsen, Hydroxychloroquine retinopathy screening, Br J Ophthalmol 92 (12) (2008) 1653–1655.

[82] F. Rigaudiere, I. Ingster-Moati, J.C. Hache, J. Leid, R. Verdet, P. Haymann, et al., [Up-dated ophthalmological screening and follow-up management for long-term antimalarial treatment], J Fr Ophtalmol 27 (2) (2004) 191–199.

[83] V.A. Manohar, K.G. Moder, W.D. Edwards, K. Klarich, Restrictive cardiomyopathy secondary to hydroxychloroquine therapy, J Rheumatol 36 (2) (2009) 440–441.

[84] T.R. Soong, L.A. Barouch, H.C. Champion, F.M. Wigley, M.K. Halushka, New clinical and ultrastructural findings in hydroxychloroquine-induced cardiomyopathy—a report of 2 cases, Hum Pathol 38 (12) (2007) 1858–1863.

[85] P. Godeau, L. Guillevin, J. Fechner, O. Bletry, G. Herreman, [Disorders of conduction in lupus erythematosus: frequency and incidence in a group of 112 patients], Ann Med Interne (Paris) 132 (4) (1981) 234–240.

[86] N. Costedoat-Chalumeau, J.S. Hulot, Z. Amoura, G. Leroux, P. Lechat, C. Funck-Brentano, et al., Heart conduction disorders related to antimalarials toxicity: an analysis of electrocardiograms in 85 patients treated with hydroxychloroquine for connective tissue diseases, Rheumatology (Oxford) 46 (5) (2007) 808–810.

[87] A. Sghirlanzoni, R. Mantegazza, M. Mora, D. Pareyson, F. Cornelio, Chloroquine myopathy and myasthenia-like syndrome, Muscle Nerve 11 (2) (1988) 114–119.

[88] M. Stein, M.J. Bell, L.C. Ang, Hydroxychloroquine neuromyotoxicity, J Rheumatol 27 (12) (2000) 2927–2931.

[89] A.K. Siddiqui, S.I. Huberfeld, K.M. Weidenheim, K.R. Einberg, L.S. Efferen, Hydroxychloroquine-induced toxic myopathy causing respiratory failure, Chest 131 (2) (2007) 588–590.

[90] M.J. Van Beek, W.W. Piette, Antimalarials. Dermatol Clin 19 (1) (2001) 147–160. ix.

[91] T. Munster, J.P. Gibbs, D. Shen, B.A. Baethge, G.R. Botstein, J. Caldwell, et al., Hydroxychloroquine concentration-response relationships in patients with rheumatoid arthritis, Arthritis Rheum 46 (6) (2002) 1460–1469.

[92] P.K. Puri, N.I. Lountzis, W. Tyler, T. Ferringer, Hydroxy-chloroquine-induced hyperpigmentation: the staining pattern, J Cutan Pathol 35 (12) (2008) 1134–1137.

[93] S. Kalia, J.P. Dutz, New concepts in antimalarial use and mode of action in dermatology, Dermatol Ther 20 (4) (2007) 160–174.

[94] S. Meller, P.A. Gerber, B. Homey, Clinical image: blonde by prescription, Arthritis Rheum 58 (8) (2008) 2286.

[95] U. Gul, S.K. Cakmak, A. Kilic, M. Gonul, S. Bilgili, A case of hydroxychloroquine induced pruritus, Eur J Dermatol 16 (5) (2006) 586–587.

[96] S. Sorbara, E. Cozzani, A. Rebora, A. Parodi, Hydroxy-chloroquine in psoriasis: is it really harmful? Acta Derm Venereol 86 (5) (2006) 450–451.

[97] S.M. Herman, M.H. Shin, A. Holbrook, D. Rosenthal, The role of antimalarials in the exacerbation of psoriasis: a systematic review, Am J Clin Dermatol 7 (4) (2006) 249–257.

[98] G.M. Kammer, N.A. Soter, D.J. Gibson, P.H. Schur, Psoriatic arthritis: a clinical, immunologic and HLA study of 100 patients, Semin Arthritis Rheum 9 (2) (1979) 75–97.

[99] R. Wolf, V. Ruocco, Triggered psoriasis, Adv Exp Med Biol 455 (1999) 221–225.

[100] R. Wolf, A.L. Schiavo, M.L. Lombardi, F. de Angelis, V. Ruocco, The in vitro effect of hydroxychloroquine on skin morphology in psoriasis, Int J Dermatol 38 (2) (1999) 154–157.

[101] M. Mates, S. Zevin, G.S. Breuer, P. Navon, G. Nesher, Desen-sitization to hydroxychloroquine—experience of 4 patients, J Rheumatol 33 (4) (2006) 814–816.

[102] D. Chernof, K.S. Taylor, Hydroxychloroquine-induced agran-ulocytosis, Arch Dermatol 97 (2) (1968) 163–164.

[103] R.P. Propp, J.S. Stillman, Agranulocytosis and hydroxy-chloroquine, N Engl J Med 277 (9) (1967) 492–493.

[104] V. Ferraro, F. Mantoux, K. Denis, M.A. Lay-Macagno, J.P. Ortonne, J.P. Lacour, [Hallucinations during treatment with hydrochloroquine], Ann Dermatol Venereol 131 (5) (2004) 471–473.

[105] W.Q. Ward, W.G. Walter-Ryan, G.M. Shehi, Toxic psychosis: a complication of antimalarial therapy, J Am Acad Dermatol 12 (5 Pt 1) (1985) 863–865.

[106] D.U. Cansu, C. Korkmaz, Hypoglycaemia induced by hydroxychloroquine in a non-diabetic patient treated for RA, Rheumatology (Oxford) 47 (3) (2008) 378–379.

[107] R. Roubenoff, J. Hoyt, M. Petri, M.C. Hochberg, D.B. Hellmann, Effects of antiinflammatory and immunosuppressive drugs on pregnancy and fertility, Semin Arthritis Rheum 18 (2) (1988) 88–110.

[108] P.A. Phillips-Howard, D. Wood, The safety of antimalarial drugs in pregnancy, Drug Saf 14 (3) (1996) 131–145.

[109] G.J. Matz, R.F. Naunton, Ototoxicity of chloroquine, Arch Otolaryngol 88 (4) (1968) 370–372.

[110] N. Costedoat-Chalumeau, Z. Amoura, G. Aymard, T.H. Le, B. Wechsler, D. Vauthier, et al., Evidence of transplacental passage of hydroxychloroquine in humans, Arthritis Rheum 46 (4) (2002) 1123–1124.

[111] A.L. Parke, Antimalarial drugs, systemic lupus erythematosus and pregnancy, J Rheumatol 15 (4) (1988) 607–610.

[112] A. Parke, Antimalarial drugs and pregnancy, Am J Med 85 (4A) (1988) 30–33.

[113] A. Parke, B. West, Hydroxychloroquine in pregnant patients with systemic lupus erythematosus, J Rheumatol 23 (10) (1996) 1715–1718.

[114] N.M. Buchanan, E. Toubi, M.A. Khamashta, F. Lima, S. Kerslake, G.R. Hughes, Hydroxychloroquine and lupus pregnancy: review of a series of 36 cases, Ann Rheum Dis 55 (7) (1996) 486–488.

[115] M. Motta, A. Tincani, D. Faden, E. Zinzini, G. Chirico, Anti-malarial agents in pregnancy, Lancet 359 (9305) (2002) 524–525.

[116] G. Klinger, Y. Morad, C.A. Westall, C. Laskin, K.A. Spitzer, G. Koren, et al., Ocular toxicity and antenatal exposure to chloroquine or hydroxychloroquine for rheumatic diseases, Lancet 358 (9284) (2001) 813–814.

[117] M.E. Clowse, L. Magder, F. Witter, M. Petri, Hydroxy-chloroquine in lupus pregnancy, Arthritis Rheum 54 (11) (2006) 3640–3647.

[118] N. Costedoat-Chalumeau, Z. Amoura, P. Duhaut, L.T. Huong du, D. Sebbough, B. Wechsler, et al., Safety of hydroxy-chloroquine in pregnant patients with connective tissue diseases: a study of one hundred thirty-three cases compared with a control group, Arthritis Rheum 48 (11) (2003) 3207–3211.

[119] K. Sperber, C. Hom, C.P. Chao, D. Shapiro, J. Ash, Systematic review of hydroxychloroquine use in pregnant patients with autoimmune diseases, Pediatr Rheumatol Online J 7 (2009) 9.

[120] M. Frassi, C. Biasini, M. Taglietti, E. Danieli, D. Faden, A. Lojacono, et al., Hydroxychloroquine in pregnant patients with rheumatic disease: a case control observation of 76 treated pregnancies (abstract), Lupus 13 (9) (2004) 755.

[121] M.E. Clowse, L.S. Magder, M. Petri, The Effect of Hydroxy-chloroquine on Pregnancy Outcomes and Disease Activity in Lupus Patients (abstract), Arthritis Rheum 50 (2004) 689–690.

[122] R. Cimaz, A. Brucato, E. Meregalli, M. Muscara, P. Sergi, Electroretinograms of children born to mothers treated with hydroxychloroquine during pregnancy and breast-feeding: comment on the article by Costedoat-Chalumeau et al., Arthritis Rheum 50 (9) (2004) 3056–3057. author reply 7–8.

[123] N. Costedoat-Chalumeau, Z. Amoura, D. Sebbough, J.C. Piette, Reply. Arthritis Rheum 50 (9) (2004) 3057.

[124] M. Motta, A. Tincani, D. Faden, E. Zinzini, A. Lojacono, A. Marchesi, et al., Follow-up of infants exposed to hydroxy-chloroquine given to mothers during pregnancy and lactation, J Perinatol 14 (2004) 14.

[125] F. Renault, R. Flores-Guevara, C. Renaud, P. Richard, A.I. Vermersch, F. Gold, Visual neurophysiological dysfunction in infants exposed to hydroxychloroquine in utero, Acta Pae-diatr 98 (9) (2009) 1500–1503.

[126] R.L. Nation, L.P. Hackett, L.J. Dusci, K.F. Ilett, Excretion of hydroxychloroquine in human milk, Br J Clin Pharmacol 17 (3) (1984) 368–369.

[127] M. Ostensen, M. Khamashta, M. Lockshin, A. Parke, A. Brucato, H. Carp, et al., Anti-inflammatory and immuno-suppressive drugs and reproduction, Arthritis Res Ther 8 (3) (2006) 209.

[128] P.R. Fischer, E. Walker, Myasthenia and malaria medicines, J Travel Med 9 (5) (2002) 267–268.

[129] G.E. Gibson, M.T. McEvoy, Coexistence of lupus eryth-ematosus and porphyria cutanea tarda in fifteen patients, J Am Acad Dermatol 38 (4) (1998) 569–573.

[130] P. Harper, S. Wahlin, Treatment options in acute porphyria, porphyria cutanea tarda, and erythropoietic protoporphyria, Curr Treat Options Gastroenterol 10 (6) (2007) 444–455.

[131] G. Sjolin-Forsberg, B. Berne, C. Blixt, M. Johansson, B. Lindstrom, Chloroquine phosphate: a long-term follow-up of drug concentrations in skin suction blister fluid and plasma, Acta Derm Venereol 73 (6) (1993) 426–429.

[132] S.J. Carmichael, B. Charles, S.E. Tett, Population pharmacoki-netics of hydroxychloroquine in patients with rheumatoid arthritis, Ther Drug Monit 25 (6) (2003) 671–681.

[133] S.E. Tett, D.J. Cutler, R.O. Day, K.F. Brown, A dose-ranging study of the pharmacokinetics of hydroxy-chloroquine following intravenous administration to healthy volunteers, Br J Clin Pharmacol 26 (3) (1988) 303–313.

[134] D.J. Cutler, A.C. MacIntyre, S.E. Tett, Pharmacokinetics and cellular uptake of 4-aminoquinoline antimalarials, Agents Actions Suppl 24 (1988) 142–157.

[135] S.E. Tett, D.J. Cutler, C. Beck, R.O. Day, Concentration-effect relationship of hydroxychloroquine in patients with rheumatoid arthritis—a prospective, dose ranging study, J Rheumatol 27 (7) (2000) 1656–1660.

[136] S.E. Tett, R.O. Day, D.J. Cutler, Concentration-effect relationship of hydroxychloroquine in rheumatoid arthritis—a cross sectional study, J Rheumatol 20 (11) (1993) 1874–1879.

[137] S.E. Tett, D.J. Cutler, R.O. Day, Bioavailability of hydroxychloroquine tablets assessed with deconvolution techniques, J Pharm Sci 81 (2) (1992) 155–159.

[138] S.E. Tett, D.J. Cutler, R.O. Day, K.F. Brown, Bioavailability of hydroxychloroquine tablets in healthy volunteers, Br J Clin Pharmacol 27 (6) (1989) 771–779.

[139] S. Tett, R. Day, D. Cutler, Hydroxychloroquine relative bioavailability: within subject reproducibility, Br J Clin Pharmacol 41 (3) (1996) 244–246.

[140] A.J. McLachlan, S.E. Tett, D.J. Cutler, R.O. Day, Disposition and absorption of hydroxychloroquine enantiomers following a single dose of the racemate, Chirality 6 (4) (1994) 360–364.

[141] D.R. Miller, S.K. Khalil, G.A. Nygard, Steady-state pharmacokinetics of hydroxychloroquine in rheumatoid arthritis patients, Dicp 25 (12) (1991) 1302–1305.

[142] N. Costedoat-Chalumeau, Z. Amoura, J.S. Hulot, G. Aymard, G. Leroux, H.A. Hammoud, et al., Very low blood hydroxychloroquine concentrations as an objective marker of poor adherence to treatment in systemic lupus erythematosus, Ann Rheum Dis 66 (6) (2007) 821–824.

[143] Micromedex® Healthcare Series; http://www.thomsonhc.com/hcs/librarian. Access 2009.

[144] M. Somer, J. Kallio, U. Pesonen, K. Pyykko, R. Huupponen, M. Scheinin, Influence of hydroxychloroquine on the bioavailability of oral metoprolol, Br J Clin Pharmacol 49 (6) (2000) 549–554.

[145] A. Jamshidzadeh, H. Niknahad, H. Kashafi, Cytotoxicity of chloroquine in isolated rat hepatocytes, J Appl Toxicol 27 (4) (2007) 322–326.

[146] M.R. Namazi, The potential negative impact of proton pump inhibitors on the immunopharmacologic effects of chloroquine and hydroxychloroquine, Lupus 18 (2) (2009) 104–105.

[147] G. Leroux, N. Costedoat-Chalumeau, J.S. Hulot, Z. Amoura, C. Frances, G. Aymard, et al., Relationship between blood hydroxychloroquine and desethylchloroquine concentrations and cigarette smoking in treated patients with connective tissue diseases, Ann Rheum Dis 66 (11) (2007) 1547–1548.

[148] E.R. Smith, W. Klein-Schwartz, Are 1-2 dangerous? Chloroquine and hydroxychloroquine exposure in toddlers, J Emerg Med 28 (4) (2005) 437–443.

[149] R.K. Burt, A. Traynor, L. Statkute, W.G. Barr, R. Rosa, J. Schroeder, et al., Nonmyeloablative hematopoietic stem cell transplantation for systemic lupus erythematosus, Jama 295 (5) (2006) 527–535.

[150] J. Sanchez-Guerrero, A.G. Uribe, L. Jimenez-Santana, M. Mestanza-Peralta, P. Lara-Reyes, A.H. Seuc, et al., A trial of contraceptive methods in women with systemic lupus erythematosus, N Engl J Med 353 (24) (2005) 2539–2549.

[151] M. Petri, M.Y. Kim, K.C. Kalunian, J. Grossman, B.H. Hahn, L.R. Sammaritano, et al., Combined oral contraceptives in women with systemic lupus erythematosus, N Engl J Med 353 (24) (2005) 2550–2558.

[152] E.M. Ginzler, M.A. Dooley, C. Aranow, M.Y. Kim, J. Buyon, J.T. Merrill, et al., Mycophenolate mofetil or intravenous cyclophosphamide for lupus nephritis, N Engl J Med 353 (21) (2005) 2219–2228.

[153] M.Y. Karim, P. Alba, M.J. Cuadrado, I.C. Abbs, D.P. D'Cruz, M.A. Khamashta, et al., Mycophenolate mofetil for systemic lupus erythematosus refractory to other immunosuppressive agents, Rheumatology (Oxford) 41 (8) (2002) 876–882.

IV. TREATMENT

58

Systemic Lupus Erythematosus: Cytotoxic Agents

Eva D. Papadimitraki, George Bertsias, George Chamilos, Dimitrios T. Boumpas

INTRODUCTION

Cytotoxic agents were initially introduced in medicine for their ability to interrupt nucleic acid and protein synthesis in malignant cells. Subsequently, and on the basis of their immune-suppressing capacities, they were used for the prevention of rejection of allogenic transplanted organs (azathioprine, cyclosporine, and mycophenolate mofetil). The realization that these agents exhibit immunomodulatory capacities led to their use in autoimmune diseases including systemic lupus erythematosus (SLE). Ever since their introduction, the management and prognosis have greatly improved. Because of their significant side effects, cytotoxic therapy in lupus is reserved only for patients with moderate to severe disease (Table 58.1).

To date, most experts agree that the treatment of moderate-to-severe lupus requires a period of intensive immunosuppressive therapy (induction therapy) to control aberrant immunological activity, recover function and halt tissue injury, followed by a longer period of less intense and less toxic therapy (maintenance therapy). The latter is based on agents that carry a lower risk for complications and are more convenient to the patient, to consolidate remission and prevent flares [1].

In this chapter, we discuss the use of cytotoxic/immunosuppressive drug therapy in SLE. For each drug under discussion, we provide background information on the mode of action, the pharmacokinetics, clinical use in lupus, and side effects. We also discuss both the monitoring of side effects and strategies to minimize them. Because azathioprine, cyclophosphamide, and mycophenolate mofetil are the most widely used cytotoxic drugs, we review these drugs in more detail.

AZATHIOPRINE

Azathioprine (AZA, Imuran) is a cycle-specific antimetabolite that is commonly included in maintenance regimens for lupus nephritis and in regimens against mild-to-moderate SLE. Both AZA and its metabolite, 6-mercaptopurine (6-MP), mediate their effects by affecting

TABLE 58.1 General indications for cytotoxic drug use in systemic lupus erythematosus

GENERAL

- Involvement of major organs and/or extensive involvement of non-major organs (skin) refractory to other agents
- Failure to respond to or inability to taper corticosteroids to acceptable doses for long-term use

SPECIFIC ORGAN INVOLVEMENT

- **Renal**
- Proliferative and/or membranous nephritis (nephritic or nephritic syndrome)
- **Hematological**
- Severe thrombocytopenia (platelets less than 20,000 K)
- Thrombotic thrombocytopenic purpura (TTP)-like syndrome
- Severe hemolytic or aplastic anemia, or immune neutropenia not responding to corticosteroids
- **Pulmonary**
- Lupus pneumonitis and/or alveolar hemorrhage
- **Cardiac**
- Myocarditis with depressed left ventricular function, pericarditis with impeding tamponade
- **Gastrointestinal**
- Abdominal vasculitis
- **Nervous system**
- Transverse myelitis, cerebritis, psychosis refractory to corticosteroids, mononeuritis multiplex, severe peripheral neuropathy

cell-mediated and humoral immune responses. This includes decreasing circulating lymphocytes, inhibiting lymphocyte proliferation, reducing antibody production and suppressing natural killer (NK) cell activity [2].

Pharmacokinetics

AZA is rapidly converted to 6-MP through enzymatic and non-enzymatic mechanisms; 6-MP is then converted to: (a) active analogs such as thiopurine via hypoxanthine-guanine-phosphoribosyl-transferase (HGPRT), and (b) inactive analogs such as 6-methylmercaptopurine (via thiopurine methyltransferase) and 6-methylthiouric acid (via xanthine oxidase) metabolites [3].

One daily dose is sufficient for therapeutic purposes. Measurement of neither AZA nor 6-MP plasma levels is helpful in predicting a drug's therapeutic or toxic effects. Approximately 1% of 6-MP is excreted in the urine; dose modification helps in reducing toxicity in the case of renal impairment (25% dose reduction if creatinine clearance [CrCl] is 10−30 ml/min; 50% reduction if CrCl is < 10 ml/min). The drug is slightly dialyzable (5−20%) and is administered post-hemodialysis.

Drug Interactions

Concurrent administration of allopurinol should be avoided since the combination of these two drugs may dramatically increase the toxicity of AZA [4]. Resistance to warfarin has been associated with administration of AZA [5].

Adverse Effects

Gastrointestinal complaints such as nausea, vomiting, and diarrhea are the most common side effects of AZA leading approximately 15−30% of patients to the discontinuation of the drug within 6 months [6]. Mild increases in liver-associated enzymes are not uncommon but severe liver injury is rare. Reversible, dose-related myelosuppression is also not uncommon; leucopenia occurs in approximately 4.5% and thrombocytopenia in 2% of patients receiving low-dose AZA [7].

Notably, AZA toxicity is highly idiosyncratic and has been associated with genetic polymorphisms leading to decreased thiopurine methyltransferase (TPMT) activity and thus, to impaired ability to detoxify 6-MP. Individuals with low or absent TPMT activity are at increased risk of developing severe AZA-induced myelotoxicity, which is of delayed (4−10 weeks after initiation of the treatment) but abrupt onset. Genetic testing for TPMT polymorphisms represents an option to identify patients with impaired TPMT activity. Alternatively, TPMT activity can be measured in red cell membranes using commercially available kits. When neither of the above-mentioned options are available, a low initial dose and careful dose titration of AZA with frequent monitoring of the white blood cell (WBC) count is warranted (every 1−2 weeks for the first 3 months of treatment and every 1−3 months thereafter). Less common side effects include acute hypersensitivity syndromes — usually within the first 2 weeks of treatment — and predisposition to infection. The implication of AZA in the pathogenesis of lymphoproliferative malignancies has not been confirmed in large controlled studies [6, 8].

Clinical Use

The starting dose is 1 mg/kg/day with the usual dose at 2−3 mg/kg/day in 1−3 doses taken with food. Monitoring includes complete blood count (CBC) with platelets, creatinine, aspartate transaminase (AST) or alanine transaminase (ALT). Liver enzyme activity should be measured every 2−3 months (Table 58.2). The drug should be used cautiously in patients with renal or liver disease or patients who use allopurinol. In moderate-to-severe lupus, AZA has been used as a maintenance therapy at doses ranging from 1−3 mg/kg/day, and it is preferred in women of reproductive age due to its acceptable safety profile during pregnancy (see below). Although discontinuation of the drug due to side effects is not uncommon, more often the drug is discontinued for lack of efficacy.

Grootscholten and colleagues evaluated the efficacy of AZA as induction/maintenance therapy for proliferative lupus nephritis (see Table 58.4) [9]. A total of 87 European patients with proliferative lupus nephritis were randomized to receive intravenous cyclophosphamide (IVCY) (6 monthly pulses of 750 mg/m^2 followed by seven pulses every 3 months) plus prednisone or intravenous methylprednisolone (IVMP) (three daily pulses of 1 g on 0, 2, and 6 weeks) plus AZA (2 mg/kg/day) plus prednisone (20 mg/day for the first 5 months then tapered to 10 mg/day). After 2 years, both groups were receiving maintenance therapy with AZA (2 mg/kg/day) plus prednisone (10 mg/day). Mean follow-up was 5.7 years and the two groups were similar in terms of renal function and SLE activity. The two groups did not differ on the basis of induction of remission, mean serum creatinine and proteinuria. However, the AZA group exhibited a higher relapse rate and an incremented frequency of infection (mostly herpes zoster virus, perhaps attributed to the higher cumulative dose of steroids in this arm) than the cyclophosphamide (CY) group. Repeat biopsy after 2 years of therapy in a representative sample of participants revealed a higher mean chronicity index in the AZA group, which indicates a better efficacy of CY to halt progression of chronic lesions [9, 10].

TABLE 58.2 Recommended monitoring of cytotoxic therapy in SLE

Drug	Dosage	Dose adjustment	Toxicities requiring monitoring	Baseline evaluation	Laboratory Monitoring
Azathioprine	50–100 mg/d in 1–3 doses with food	25% reduction if CrCl 10–30 ml/min; 50% reduction if CrCl <10 ml/min	Myelosuppression, hepatotoxicity, lymphoproliferative diseases	CBC, platelets, creatinine, AST or ALT	CBC and platelets every 2 weeks, with changes in dosage; baseline tests every 1–3 months
Mycophenolate mofetil	1–3 g/d in 2 divided doses with food	Maximum 1 g/d if CrCl <25 ml/min	Myelosuppression, hematotoxicity, infection	CBC, platelet, creatinine, AST or ALT	CBCs and platelets every 1–2 weeks with changes in dosage; baseline tests every 1–3 months
Cyclophosphamide	50–150 mg/d in a single dose with breakfast. Lots of fluids, empty bladder before bedtime	25% reduction if CrCl 25–50 ml/min; 30–50% reduction if CrCl <25 ml/min. 25% reduction if serum bilrubin 3.1–5 mg/dl or transaminases >3 times ULN	Myelosuppression, hemorrhagic cystitis, myeloproliferative disease, malignancies	CBC, platelet, creatinine, AST or ALT, urinalysis	CBC with differential every 1–2 weeks, with changes in dosage and then every 1–3 months. Keep WBC above 4000/mm^3 with dose adjustment. Urinalysis, AST or ALT every 3 months. Urinalysis every 6–12 months following cessation
Methotrexate	7.5–15 mg/wk in 1–3 doses with food or milk/water	50% reduction if CrCl 10–50 ml/min; avoid use if CrCl <10 ml/min. Avoid use in hepatic dysfunction (serum bilrubin 3.1–5 mg/dl or transaminases >3 times ULN)	Myelosuppression, hepatic fibrosis, pneumonitis	CXR, Hepatitis B and C, serology in high risk patients, AST or ALT, albumin, alkaline phosphatase and creatinine	CBC with platelet, AST, albumin, creatinine every 1–3 months
Cyclosporine A	100–400 mg/d in 2 doses at the same time every day with meal or between meals	Avoid in impaired renal function	Renal insufficiency, anemia, hypertension	CBC, creatinine, uric acid, LFTs, blood pressure	Creatinine every 2 wks until dose is stable, then monthly; CBC, potassium and LFTs every 1–3 months. Cyclosporine levels only with high doses
Leflunomide	Loading dose: 100 mg/d in a single dose × 3 ds. Then 10–20 mg/d	Avoid in hepatic dysfunction (serum bilrubin 3.1–5 mg/dl or transaminases >3 times ULN)	Myelosuppression, hepatotoxicity, fetal toxicity	CBC, creatinine, LFTs, albumin	CBC and LFTs monthly × 6 months, every 1–3 months thereafter. Continue monthly monitoring if MTX co-administered

CBC, complete blood count; AST, aspartate transaminase; ALT, alanine transaminase; LFTs, liver function tests; ULN, upper limit of normal.
*Modified from ACR Ad Hoc committee on clinical guidelines for monitoring drug therapy in rheumatoid arthritis. Arthritis Rheum 1996; 39:723–731.

AZA has long been considered as a reasonable and efficacious option for maintenance therapy in SLE. Moroni *et al.* compared AZA (1.5–2 mg/kg/day for 2–4 years) to CsA (4 mg/kg/day for one month then tapered to 2.5–3 mg/kg/day) as maintenance regimen in a RCT of 69 diffuse proliferative lupus nephritis patients with preserved renal function. The rate of flare ups, the blood pressure level, and the level of protein-uria did not differ between the two groups and the results of this trial, although limited by the inclusion of only mild-to-moderate nephritis cases, the moderate doses of the immunosuppressants, and the short dura-tion of induction therapy (3 month scheme of steroids and oral CY), underscore the efficacy of AZA as a main-tenance regimen in patients with proliferative lupus nephritis and preserved renal function [11].

AZA can be considered as first-line treatment agent for pure lupus membranous nephritis (LMN). Mok *et al.* studied the long-term effect (mean follow-up duration, 12 years) of AZA (2 mg/kg/d) combined with prednisolone (0.8–1 mg/kg/d for 6–8 weeks, then tapered) in a cohort of 38 SLE patients with class Va/Vb LMN. Twenty percent of these patients had CrCl <60 ml/min at baseline and 58% had nephritic-range proteinuria. At 12 months, the rates of complete, partial and no response were 67, 22 and 11%, respec-tively, whereas the cumulative incidence of renal flares at 5 and 10 years was calculated at 19 and 32%, respec-tively. Of note, 79% of flares were proteinuric and responded to another immunosuppressive agent. Doubling of serum creatinine or decline of CrCl by $\geq$30% occurred at 18% of patients at the end of follow-up [12].

CYCLOPHOSPHAMIDE

Cyclophosphamide (Cytoxan, Endoxan) is an alkylat-ing agent whose administration results in cell death which can occur at any stage during the cell cycle. CY depletes both T and B cells, and reduces the production of pathogenic autoantibodies [13].

Pharmacokinetics

CY is well absorbed with peak plasma levels occur-ring one hour following oral administration. Oral and intravenous administrations of CY result in similar plasma concentrations and the serum half life is approx-imately 6 hours. CY, an inactive merchlorethamine derivative, is rapidly metabolized to a variety of active metabolites such as 4-hydroxycyclophosphamide, phos-phoramide mustard and acrolein, by cytochrome P-450 in the liver, or other tissues such as transitional epithelial

cells of the bladder or lymphocytes [14]. Approximately 20% of the drug is excreted by the kidney whereas 80% is processed by the liver. A dose modification is necessary for patients with renal impairment (25% reduction in patients with CrCl 25–50 ml/min; 30–50% reduction if CrCl is below 25 ml/min). CY may be administered in patients with end-stage renal disease on dialysis with dialysis performed 8–12 hours later. No dose modification is required in liver disease. Concurrent administration of other drugs such as allopurinol, carba-mazepine, phenytoin, and succinylcholine, may increase the toxicity of CY. However, plasma concentrations of CY are not used as clinical predictors of either efficacy or toxicity.

Modes of Administration and Monitoring

Pulse CY Therapy

The NIH trials demonstrated equivalent efficacy yet lower toxicity with monthly intravenous (0.5–1 g/m^2) versus oral regimens, which led to the predominant use of intravenous regimens (IVCY) in current clinical practice [15]. Table 58.3 summarizes the NIH protocol for IVCY administration and monitoring. Reversible myelotoxicity is a common, dose-related, adverse effect. After pulse therapy, the nadir of lymphocyte count occurs on approximately days 7 to 10 and that of granu-locyte count on approximately days 10 to 14. The WBC nadir is about 3000 cells/mm^3 after a dose of 1 g/m^2 and 1,500 cells/mm^3 after a dose of 1.5 g/m^2 [16]. Because the risk of infection increases substantially with WBC below 3000 cells/mm^3 the dose is adjusted to keep it above this level. A prompt recovery from gran-ulocytopenia usually occurs after 21–28 days. On the other hand, thrombocytopenia is extremely rare in monotherapy with CY.

Oral CY

Monitoring for daily oral administration includes: CBC with differential and platelet count every 1–2 weeks initially, and every 1–3 months thereafter along with urinalysis, serum creatinine, AST, or ALT. WBC less than 3500/mm^3 should lead to dose reductions, because lower counts are associated with a markedly increased risk of opportunistic infection. Urinalysis is performed every 6–12 months after cessation to monitor for bladder carcinoma (see below).

Adverse Effects

Reversible alopecia and nausea are the most commonly observed side effects of CY. Myelotoxicity, gonadal toxicity, and malignancy represent less frequent though much more serious adverse effects.

TABLE 58.3 NIH Protocol for administration and monitoring of pulse cyclophosphamide therapy

Estimate creatinine clearance by standard methods

Calculate body surface area (m^2): BSA = $\sqrt{\text{Height (cm)} \times \text{Weight (kg)}/3600}$

Cyclophosphamide (Cytoxan) (CY) dosing and administration:

Initial dose CY is 0.75 grams/m^2 (0.5 grams/m^2 of CY if creatinine clearance rate is less than one-third of expected normal)

Administer CY in 150 ml normal saline IV over 30–60 min (alternative: equivalent CY dose is taken orally in compliant patients)

WBC at days 10 & 14 after each CY treatment (patient should delay prednisone until after blood tests drawn to avoid transient corticosteroid-induced leukocytosis)

Adjust subsequent doses of CY to maximum 1.0 gram/m^2 to keep nadir WBC >1,500/μL. If WBC nadir falls <1,500/μL decrease next dose by 25%

Repeat CY doses monthly (every 3 weeks in patients with extremely aggressive disease) for 6 months (7 pulses), then quarterly for one year after remission is achieved [inactive urine sediment, proteinuria <1 gram/day, normalization of complement (and ideally anti-DNA), and minimal or no extra-renal lupus activity]. Alternative maintenance therapy: AZA or MMF for 1–2 years

Protection bladder against CY-induced hemorrhagic cystitis:

Diuresis with 5% dextrose and 0.45% saline (2 liters at 250 ml/h). Frequent voiding; continue high-dose oral fluids for 24 hours. Patients return to clinic if they cannot sustain inadequate fluid intake

Consider Mesna (each dose 20% of total CY dose) intravenously or orally at 0, 2, 4 and 6 hours after CY dosing. Mesna is important to use if sustained diuresis may be difficult to achieve, or if pulse CY is administered in outpatient setting

If anticipated difficulty with sustaining diuresis (e.g. severe nephrotic syndrome) or with voiding (e.g. neurogenic bladder) insert a 3-way urinary catheter with continuous bladder flushing with standard antibiotic irrigating solution (e.g. 3 liters) or normal saline for 24 hours to minimize risk of hemorrhagic cystitis

Anti-emetics (usually administered orally):

Dexamethasone 10 mg single dose plus:

Serotonin receptor antagonists: granisetron 1 mg with CY dose (usually repeat dose in 12 hours); ondasetrone 8 mg tid for 1–2 days

Monitor fluid balance during hydration. Use diuresis if patient develops progressive fluid accumulation

Complications of pulse CY:

Nausea and vomiting (central effect) mostly controlled by serotonin receptor antagonists; transient hair thinning (rarely severe at CY doses ≤1 g/m^2).

Common: significant infection diathesis only if leukopenia not carefully controlled; modest increase in herpes zoster (very low risk of dissemination); infertility (male and female); amenorrhea proportional to age of the patient during treatment and to the cumulative dose of CY. In females at high risk for persistent amenorrhea may consider using leuprolide 3.75 mg subcutaneously 2 weeks prior to each dose of cyclophosphamide. In males may use testosterone 100 mg intramuscularly every 2 weeks.

Infections

A variety of different infections can occur including bacterial infections, opportunistic infections (*Pneumocystis jiroveci*, fungal infections, and nocardia) and reactivation of latent herpes zoster, mycobacterium tuberculosis, and human papilloma virus. An increased rate of herpes zoster and bacterial infections has been documented in patients receiving CY and has been associated with higher doses of corticosteroids and a nadir of WBC <3000 cells/mm^3 at some point during treatment [17]. Oral CY regimens also may pose a greater risk of infection compared with intravenous pulse regimens [18]. CY may be an independent risk factor for death due to acute viral infection [19]. Opportunistic infections such as candidiasis or *Pneumonocystis* pneumonia may be seen in patients on concomitant high-dose corticosteroid therapy. Studies have shown that the dosage of corticosteroids is the overriding independent determinant of the risk of infection among SLE patients receiving CY with concomitant high doses of corticosteroids [20].

Gonadal Toxicity

Premature ovarian failure (POF) is a well-documented side effect of CY. Several mechanisms including marked acceleration in follicular maturation, depletion and eventually exhaustion, and direct toxicity of the drug and its metabolites to gonadal cells, have been implicated in its pathogenesis. The risk of developing POF depends on the age of the patient at the initiation of treatment and the cumulative dose of CY as we first reported in 1993 [21]. In this study, the rates of sustained amenorrhea after a short course (≤ 7 pulses) of CY were 0% for patients aged <25 years, 12% for those aged 26–30 years, and

25% for patients aged ≥30 years. On the other hand, a long course (≥15 pulses) of CY induced sustained amenorrhea in 17% of patients aged <25 years, 43% of patients aged 26–30 years, and 100% of patients aged ≥30 years. Male gonadal toxicity may be observed with as little as 7 g cumulative dose corresponding to an approximately 2 month daily oral therapy [22].

A number of strategies to preserve fertility in patients with SLE taking CY have been tried with encouraging initial results. Some authors have suggested that co-administration of GnRH antagonists confers protection against premature ovarian failure, and therefore recommend a GnRH antagonist-based protocol in CY-treated female patients (Table 58.3) [23–25]. A non-randomized Chinese study by Liang *et al.* showed a reduction of premature ovarian failure in women 30–39 years old with lupus who received GnRH analogs concomitantly to IVCY treatment [26]. The results of the PREGO study (prospective randomized study on protection against gonadal toxicity) which randomized young female lupus patients on CY to receive monthly injections of GnRH analogs or placebo are awaited at present [27]. Other strategies for preserving fertility, such as cryopreservation of unfertilized ova and ovarian tissue germ cell transplantation are currently under investigation and should be considered experimental − at best − at the present time. In male patients receiving CY for malignancies, the frequency of azoospermia ranges from 30 to 90% [28, 29]. The administration of testosterone and sperm banking represent valid strategies for the preservation of testicular function and fertility (Table 58.3) [30].

Malignancy

Patients with SLE are at increased risk of developing lymphoma independent of treatment. An increased risk for gynecological, lung, and other types of cancer has also been suggested [31]. Immunosuppressive drug exposure may enhance this risk and the precise role of alkylating agents and other immunosuppressive medications to lymphoma/carcinogenesis in these patients is subject to ongoing evaluation. Mechanisms of alkylating agent-induced malignancy include direct chromosomal damage and impaired immune surveillance. The duration of therapy is an important risk factor; the incidence is greatest in patients treated for more than 2–3 years or with cumulative doses >100 g [32]. Patients with previous exposure to CY are at increased risk for hematologic malignancies including myelodysplastic syndrome, myeloproliferative disease including acute leukaemia, and multiple myeloma. Importantly, an increased risk of cervical dysplasia has been associated with CY exposure, and thus following established guidelines for regular cervical cancer screening is of paramount importance in women with a history of CY exposure (see below) [33].

Bladder Toxicity Including Bladder Carcinoma

Hemorrhagic cystitis and bladder carcinoma have been well documented in patients receiving long-term oral CY [34]. The role of the BK virus, present in the majority of adults in latent form in the urogenital tract, and its reactivation following CY therapy is currently been explored. Non-glomerular microscopic or gross hematuria represent the most common manifestation of CY-induced hemorrhagic cystitis; the value of urine cytology has been questioned; in our opinion, this does not represent a useful test to monitor for bladder cancer [35, 36]. The risk for bladder malignancy is life-long after CY therapy and patients with non-glomerular hematuria should undergo cystoscopy regardless of the time of appearance of hematuria. An up to 30-fold increase in the risk of developing bladder cancer has been documented in large trials, although this might be an overestimation. Among patients receiving CY, a cumulative dose of above 100 g of CY and smoking are independent risk factors for bladder cancer [35].

The use of intermittent pulse CY and adequate hydration has practically eliminated the cases of bladder carcinoma, although hemorrhagic cystitis may be seen in cases with an inability to empty the bladder (for instance, neurogenic bladder in patients with transverse myelitis) or when the practice of adequate hydration and frequent emptying of the bladder are not followed meticulously (Table 58.3). We routinely use sodium 2-mercaptoethane sulfonate (mesna) − an agent that has been advocated to reduce the concentration of acrolein and probably other toxic metabolites in the bladder − although this practice has not been supported by controlled studies in lupus.

Clinical Use

Although some centers still employ daily oral CY regimens for short periods of time (2 mg/kg/day every morning in a single dose for 3–12 months until remission), the NIH protocol has become the protocol of choice for most physicians (Table 58.3). CY may retard progressive renal scarring, preserve renal function, and reduce the risk for the development of end-stage renal disease (ESRD) requiring dialysis or renal transplantation [36–39]. Studies suggest a similar important role in children with lupus nephritis (see below).

Following induction therapy, a maintenance regimen is essential to decrease the risk of flares [40]. Studies from the NIH have demonstrated that combination pulse therapy with CY and IVMP improves renal outcomes without increasing toxicity [40, 41]. Based on these studies the NIH group proposed 7 monthly pulses of IVCY (0.5–1 g/m^2) followed by quarterly pulses for at least 1 year beyond remission. For patients with moderate-to-severe disease, the addition of monthly pulses of IVMP is combined during the induction period. Ovarian toxicity (found to be both age- and dose-related), infections (especially with herpes zoster), flares (observed in approximately one third of patients), incomplete response and in rare cases, refractoriness to treatment have emerged from these studies as significant limitations of current cytotoxic therapy.

Because of concerns about toxicity, together with the appreciation that the disease may be less severe in Caucasians, European investigators sought alternative protocols to administer the drug (Euro-Lupus Nephritis Trial — ELNT) [42, 43]. In studies involving for the most part patients with milder forms of disease (mean serum creatinine 1.2 mg/dl, mean proteinuria 3.0 g/day), less intensive regimens of CY (6 fortnightly pulses at a fixed dose of 500 mg each in combination with 3 daily doses of 750 mg of IVMP) followed by AZA as maintenance had comparable efficacy but less toxicity than a short-course of high-dose IVCY (8 pulses). By multivariate analysis, early response to therapy at 6 months (defined as a decrease in serum creatinine level and proteinuria <1 g/24 hours) was the best predictor of good long-term renal outcome. This study demonstrated that low-dose IVCY could serve as an alternative therapeutic protocol for non-high-risk Caucasian SLE patients with moderate-to-severe nephritis [44].

The experience from the use of CY for the treatment of LMN is limited, coming only from uncontrolled and retrospective studies in populations with both pure and mixed membranous/proliferative lesions. In the single RCT, 41 LMN patients were randomized to receive alternate day oral prednisone only, or oral prednisone plus either alternate month IVCY (5 doses) or cyclosporine (CsA) for 11 months. Although IVCY and CsA were each more effective than prednisone alone in inducing remissions (complete remission rate 46% for prednisone plus CY or CsA vs. 13% for prednisone alone) relapse of nephrotic syndrome was more frequent in the CsA group [45].

In addition to proliferative and membranous lupus nephritis, case reports, case series, and uncontrolled clinical studies support the efficacy of IVCY in severe thrombocytopenia, neurological disease (myelitis, encephalitis, psychosis, momoneuritis multiplex, and polyneuropathy), abdominal vasculitis, acute pneumonitis/alveolar haemorrhage, extensive skin disease, and other severe manifestations of lupus [46, 47]. A single randomized controlled trial in neuropsychiatric lupus has confirmed its efficacy in severe neuropsychiatric lupus [48]. In this trial, 32 patients with non-thrombotic neurological syndromes were randomized to receive three pulses of 1 g IVMP followed by either pulses of 1 g of MP — monthly for 4 months, then bimonthly for 6 months and subsequently every 3 months for 1 year — or IVCY (0.75 g/m^2 monthly for 1 year, then every 3 months for another year). Seizure

was the commonest syndrome ($n = 11$); other manifestations included peripheral neuropathy ($n = 7$), optic neuritis ($n = 5$), transverse myelitis ($n = 4$), brainstem disease ($n = 2$), coma ($n = 2$) and intranuclear ophthalmoplegia ($n = 1$). Clinical response, defined as $\geq 20\%$ improvement of clinical, laboratory or specific neurological testing variables, was observed in 18 out of 19 patients who received IVCY compared with 7 out of 13 who received IVMP. Accordingly, the combination of IVMP pulses with IVCY is considered the treatment of choice for several inflammatory neurologic manifestations of SLE [49].

MYCOPHENOLATE MOFETIL

Mycophenolate mofetil (MMF) is a morpholinoethyl ester and a prodrug of mycophenolic acid (MPA), a potent inhibitor of inosine monophosphate dehydrogenase (IMP-DH) which is indispensible for the *de novo* synthesis of guanosine nucleotides [50]. It is derived from the fungus *Penicillium stoloniferum* and is converted to MMA by the liver. First approved by the US Food and Drug administration for use in renal transplantation in 1995, MMF has now been used for over 10 years as an immunosuppressive agent. Its antilymphocytic activity has been targeted for use in lupus and RCTs have documented its value as both induction and maintenance therapy of lupus nephritis in several different geographic and ethnic populations. The latter in association with the drug's relatively benign adverse effect profile have rendered MMF the standard therapy for lupus nephritis in some centers. In our opinion, longer-term and further validation of data are needed before its widespread use for severe disease is established (see below).

MMF suppresses DNA synthesis and proliferation of T and B lymphoctytes (preferential inhibition of type II isoform IMP-DH). The lack of any salvage nucleotide synthesis pathway within T and B cells renders them a selective target for MMA. Conversely, other tissues with high proliferative activity (skin, intestine, neutrophils) which possess an intrinsic salvage guanosine synthesis pathway can escape the antiproliferative effects of the antimetabolite, which explains its more favorable toxicity profile compared with CY [51, 52]. *In vitro* actions of MMF include inhibition of both T-cell and B-cell proliferation upon mitogenic stimulation, suppression of antibody production, abrogation of the antigen-presenting and migration capacities of myeloid dendritic cells and reduction of adhesion molecules which are necessary for the migration of lymphocytes into sites of inflammation [51, 52]. The drug decreases cytokine-induced nitric oxide production in mouse and rat endothelial cells, a finding that if extrapolated to human disease, may explain its activity against tissue damage of inflamed tissue. Moreover, MMF may inhibit vascular smooth-muscle proliferation, and retard the development of atherosclerosis associated with organ transplantation, a feature highly desirable in lupus where premature atherosclerosis represents a major issue [52].

Pharmacokinetics

MMF is converted to an inactive metabolite in the liver and is then excreted into the gastrointestinal tract, where after undergoing deglucuronidation, enters the enterohepatic circulation and is excreted by the kidney. The drug has excellent oral bioavailability and peak levels occur within 1–2 hours after administration. Its half-life is 17 hours. MPA area under the curve (AUC) measurements show marked interindividual variation with fixed doses. Although therapeutic drug monitoring to guide MMF dosing has been proposed in renal transplantation, validation of therapeutic MPA monitoring in lupus nephritis is still required. Serum levels of MMF are increased in patients with impaired renal function and the dose should be adjusted in such patients (1 g/d maximum if CrCl is <25 ml/min). Anti-acids and chlolestyramine decrease the bioavailability of MMF and the co-administration with AZA should be avoided. MMF may be teratogenic and should not be administered during pregnancy [53].

Adverse Effects

Gastrointestinal toxicity including nausea, vomiting, and diarrhea – all alleviated by reducing or splitting the daily dose – are the most common side effects of MMF. Although results of different studies regarding infectious complications have been inconclusive, respiratory tract infections, herpes zoster, and cellulitis are the most frequent but overall they are uncommon [54, 55]. Most data implicating MMF in the pathogenesis of severe respiratory, generalized varicella, CMV and BK virus infections, come from cohorts of renal transplant patients where higher doses of MMF and concomitant treatment with other immunosuppressive medications were employed [56–58].

Clinical Use

Recommended mode of administration and monitoring are summarized in Table 58.2. Initially exhibiting efficacy in the treatment of refractory-to-corticosteroids SLE in small case series, MMF was subsequently compared with other cytotoxic agents as induction and/or maintenance regimen in large RCTs (Table 58.4) [59–63].

Induction Therapy

To date, several RCTs have compared the efficacy of MMF versus CY as induction therapy in proliferative lupus nephritis. Two of these trials [59, 63] were conducted among Asian patients with moderate-to-severe lupus nephritis and demonstrated equal efficacy of MMF (2 g/day for 6 months) and CY (monthly IV pulses of 0.75–1 g/m^2 in one study and 2.6 g/kg/day orally in the other) with comparable rates of remission. The results of these studies were limited by the small numbers of participants, short follow-up, and inclusion of only patients with an Asian background [59, 63].

Ginzler *et al.* reported that MMF was superior to CY as induction therapy in a RCT of 140 patients (56% African-Americans). Patients were assigned to receive MMF (3 g/day) or IVCY (0.5–1 g/m^2) for 6 months combined with a tapering dose of steroids (starting from 1 mg/kg/day) and the study allowed cross-over at 3 months for reasons of failure or toxicity. Complete and partial remission rates were 23 and 30%, respectively in the MMF group, as compared to 6 and 35%, in the CY group, and treatment with CY was associated with higher failure (69 vs. 48% in the MMF group) and cross-over rates (20 vs. 8% in the MMF group). The results of this trial should be interpreted with caution, in view of the fact that MMF was compared to CY alone, not combined with monthly IVMP pulses as routinely recommended for severe lupus nephritis. Moreover, patients with severe renal impairment (serum creatinine >3 mg/dl) were not included. In conclusion, this study should rather be considered as a non-superiority trial because of the inadequate statistical power for a superiority trial [61].

In agreement with the results of the aforementioned trials, four meta-analyses of RCTs have claimed superiority of MMF against CY as an induction regimen for proliferative lupus nephritis but the results of these meta-analyses may suffer from the same flaws as the original studies [64–66]. This issue was addressed by the Aspreva Lupus Management Study (ALMS), one of the largest and most racially diverse studies in lupus nephritis, which included a total of 370 patients, 27% of which had CrCl <60 ml/min/1.73 m^2. This study showed similar response rates (56% for MMF and 53% for IVCY) with no differences in the adverse events between the two groups [67–69]. The most recent meta-analysis of 10 RCTs involving a total of 847 SLE patients has also concluded that MMF and CY offer similar efficacy in terms of renal remission (RR for complete or partial remission 1.05; 95% CI 0.95–1.17) but MMF exhibits a more favorable safety profile (RR for amenorrhea 0.21; 95% CI 0.09–0.48; RR for leukopenia 0.47; 95% CI 0.27–0.83), and may be more efficacious in certain populations [70]. The RR of all-cause

TABLE 58.4 Randomized Controlled Trials (RCTs) for the use of cytotoxic drugs in lupus nephritis (2000–2009)

Trial	No.	Baseline characteristics		Randomized treatments	Follow-up	Results
		Race	Renal			
Chan et al. (2000) [59] (induction)	42	100% Chinese	• Class IV nephritis • Mean SCr 1.2 mg/dl (29% with elevated Cr) • Mean serum albumin 2.8 mg/dl	• MMF 1 g/d for 6 months, then halved • CY 2.5 mg/kg/d, replaced by AZA 1.5 mg/k/d at 6 mo • Both arms: prednisone 0.8 mg/kg/d tapered to 10 mg/d at 6 mo, then AZA 1 mg/kg/d at 12 mo	12 mo	• CR 81% in MMF, 76% in CY • Improvements in proteinuria, serum albumin and Cr similar in year 1 • Relapse rates in year 1: 15% in MMF, 11% in CY
Chan et al. (2005) [60] (maintenance)	64	100% Chinese	• Class IV nephritis • Elevated SCr: 27% • Proteinuria: >3g/d: 63% • Newly diagnosed disease: 50/64	• MMF 2g/d for 6 mo, 1.5 g/d for 6 mo, 1 g/d for 1 yr (treatment duration 12 mos in 20 pts, ≥24 mos in 12 pts) • CY 2.5 mg/kg/d, replaced by AZA 1.5 mg/k/d at 6 mo	63 mo	• CR/PR 90% in both arms • MMF arm had fewer infections • Composite end-point ESRD/death: 4 pts in CY vs 0 in MMF (p = 0.06)
Ginzler et al. (2003) [61] (induction)	140	56% African-American	• Class IV nephritis • Nephrotic: 42% of MMF, 46% of CY • SCr >1.3 mg/dl: 18% in MMF, 27% in CY	• MMF as Evaluation induction (1–at 3 g/d) • IVCY (NIH protocol) • Patients with no improvement >30% in ≥1 renal parameter that was abnormal on entry were crossed to other regimen at 12 wks	24 wks	• More CR in MMF than CY (16 vs 4, p < 0.05) • More PR in MMF than in CY (21 vs 17) • Mean SCr, proteinuria, urinary sediment at 24 wks: NS (intention to treat analysis) • 3 patients (all randomized to CY) died with severe SLE
Contreras et al. (2004) [62] (maintenance)	59	5% white, 95% Hispanic or African-American	• Class III-IV nephritis • Nephrotic: 42% of MMF, 60% of CY/MMF • Mean SCr: 1.7 mg/dl (AZA) 1.5 mg/dl (CY); 1.6 mg/dl (MMF) • Chronicity index: 3.2 (AZA); 1.9 (CY) (p < 0.01 vs MMF); 3.8 (MMF)	• Induction: mo pulses IVCY (0.5–1 g/m²) & prednisone (mean dose 0.6 mg/kg/d) ± IVMP pulses (3–7) until remission • Maintenance therapy: (a) quarterly IVCY (0.5–1 mg/kg/d), (b) AZA (1–3 mg/kg), (c) MMF (0.5–3 g/d). All oral prednisone (≤0.5 mg/k/d) for 1–3 yrs	34 mo	• 72-month, event-free survival rate for composite end point ESRD/death: higher in MMF (p = 0.05) and AZA (p < 0.01) than CY • Rate of relapse free survival: higher in MMF than CY (p < 0.05) • Lower rates of hospitalization, infections and gastrointestinal side effects in MMF and AZA arms
Ong et al. (2005) [63] (induction)	54	90% Malaysian and Chinese	• Class III, IV, IV+V • Proteinuria (MMF 1.8 g/d; CY 1.2 g/d) • Activity index: 9.0 (MMF), 8.6 (CY)	• Induction: prednisone (60 mg/d 4–6 wks, then tapered to 5–10 mg/d) plus: • IVCY 0.75–1 g/m² adjusted to WBCs monthly × 6 months or MMF 1 g BD adjusted to WBCs and adverse effects	6 mo	• PR+CR 52% in CY, 58% in MMF; CR 12% in CY, 26% in MMF (p = 0.22) • Increase in chronicity index in CY arm (2.4→4.4, p = 0.003 vs 2.9→3.6 in MMF, p = 0.08) • Similar rates of infections and GI effects

(Continued)

TABLE 58.4   Randomized Controlled Trials (RCTs) for the use of cytotoxic drugs in lupus nephritis (2000–2009)—cont'd

Trial	No.	Baseline characteristics		Randomized treatments	Follow-up	Results
		Race	Renal			
Grootscholten et al. (2006) [9] (induction)	87	76% Caucasians	• Class IV/Vd (91%), III/Vc (9%) • eGFR <70 ml/min: 56% • Proteinuria >3.5 g/24 h: 53% • Activity index: 9.3 (AZA), 9.5 (CY) • Chronicity index: 2.3 (AZA); 2.7 (CY)	• CY: 6 monthly pulses IV CY (750 mg/m²) followed by 7 pulses every 12 weeks. Oral prednisone 1 mg/kg/d, tapered to 10 mg/d • AZA: 2 mg/kg/d at day 1, combined with 3×3 pulses IV MP 1000 mg. Oral prednisone 20 mg/d, then tapered to 10 mg/d	5.7 years	• Doubling of SCr more frequent in AZA vs CY (RR 4.1, 95% CI 0.8–20, p = NS) • Relapses more frequent in AZA vs CY (RR 8.8, 95% CI 1.5–32). • SCr, proteinuria similar in two groups • Increased infections in AZA vs CY (37 vs 18 per 100 patient-years), mostly HZV
Moroni et al. (2006) [11] (maintenance)	69	100% Caucasians	• Class IV (87%), Vc/Vd (13%) • Mean Scr 0.9 mg/dl • Mean proteinuria 2.8 g/d (CsA), 2.2 g/d (AZA) • Mean activity index: 7 • Mean chronicity index: 2.5 (CsA), 2.8 (AZA)	• Induction: pulses IV MP 1g ×3d → prednisone (1mg/kg, then tapered to 0.5 mg/kg) and oral CY (1–2 mg/kg) ×3 mos • CsA 4 mg/kg/d ×1 mo → tapered to 2.5–3 mg/kg/d • AZA 1.5–2 mg/kg/d	4 years	• Flare-ups: n = 7 in the CsA group vs n = 8 in the AZA group • Similar improvements in proteinuria, SCr, blood pressure • Both agents well tolerated
Bao et al. (2008) [96] (induction)	40	100% Asian	• Class V+IV • SCr >1.2 mg/dl: 8% • Mean proteinuria 4.4 g/24 h • Activity index: 8.0 (CY); 9.8 (multitarget)	• CY: monthly pulses IV CY 1 g/m² ×6–9 • Multi-target: tacrolimus 4 mg/d + MMF 2 g/day • All received IV MP (0.5 g/d) ×3 days, then oral prednisone	6 mos	• CR: 50% in multitarget vs 5% in CY • PR: 40% in both groups • No differences in mean SCr, CrCl between the groups • No significant adverse events
Austin et al. (2009) [45] (induction)	42	White 20% (prednisone), 53% (CY), 8% (CsA)	• Class V • Median GFR 80–89 ml/min/1.73 m² • Median proteinuria 5.0–5.8 g/24 h • Median SAlb 2.5–3.0 g/dl	• All received oral prednisone (1 mg/kg qod ×8 weeks, then tapered) • IV CY: 6 doses every other month, 0.5–1 g/m², adjusted to leukocyte nadir) • CsA: 5 mg/kg/d, adjusted to SCr	12 mos	• Cumulative probability of remission: 27% with prednisone, 60% with IV CY (p = 0.04), 83% with CsA (p = 0.002) • Nephrotic relapses per 100 patient-months: 2.0 with CsA vs 0.2 with IV CY (p = 0.02)
Appel et al. (2009) [69] (induction)	370	Genetically diverse	• Class III–V • 20/185 in MMF arm, 12/185 in CY arm with CrCl <30 ml/min/1.73 m² • 16–17% in both arms pure Class V nephritis	• Induction (24 wks): prednisone (60 mg/d progressively tapered) plus: • IVCY 0.75–1 g/m² × 24 wks or • MMF (targeted dose 3 g/d) × 24 wks • Re-randomization at 24 wks for maintenance scheme	6 mo	• Similar response: 56% in MMF, 53% in CY • Similar rates of CR, systemic disease activity, damage • Similar rates of adverse effects (28% MMF, 23% CY) and infections (69% MMF, 62% CY) • Better response rates to MMF in non-white, non-Asian participants

Abbreviations: AZA, azathioprine; Cr, Creatinine; CT, combination therapy; CY, cyclophosphamide; gp, group, IV, intravenous; IVMP, intravenous methylprednizolone; MMF, mycophenolate mofetil; mo, months; MP, methylprednizolone, n, number of participants; NS, non-significant; pts, patients; SLE, systemic lupus erythematosus; BD, twice daily; CR, complete remission, PR, partial remission.

mortality was comparable between MMF and CY (RR 0.71; 95% CI 0.37—1.4) as well as the risk for ESRD (RR 0.45; 95% CI 0.18—1.12).

Maintenance Therapy

Contreras *et al.* studied MMF as maintenance therapy in a RCT of 69 patients (predominantly African-Americans and Hispanics) with severe lupus nephritis and evidence of major renal involvement (mean protein/creatinine ratio > 5) [62]. After receiving 4—7 monthly pulses of IVCY with steroids, patients were assigned to quarterly IVCY pulses or oral AZA (1—3 mg/kg/day) or MMF (0.5—3 g/d). The primary endpoints of death and chronic renal failure were reached in significantly fewer patients in the AZA and MMF compared with the IVCY group (80 and 87 vs. 45%, respectively). AZA and MMF also performed better than IVCY in terms of relapse-free survival (58 and 78 vs. 43%, respectively) with a concomitant higher incidence of adverse event, hospitalization day, and mortality rate in the IVCY group [62]. MMF and AZA were also compared by Chan *et al.* as maintenance therapy in an RCT of 64 Chinese patients following induction with either oral CY or MMF (2 g/day) [59]. Steroids were also administered and the MMF dose was reduced by 25% after 6 months and to 1 g/d after 1 year in stable patients. Although rates of chronic renal failure, relapse and mortality were comparable in both groups, the risks of amenorrhea and infection were significantly lower in MMF-treated patients. The results of these two trials are summarized in the meta-analysis by Zhu *et al.* who have concluded that maintenance therapy with MMF does not reduce the rate of death, end-stage renal disease, renal relapse or doubling of serum creatinine values compared to AZA [66]. At present and until the results of two other ongoing multi-center RCTs (Mycophenolate Mofetil Versus Azathioprine for Maintenance Therapy for Lupus Nephritis: [MAINTAIN] trial and ALMS: maintenance part) are available, MMF and AZA should be considered as comparable maintenance therapy options for moderate-to-moderately severe lupus nephritis.

Membranous Nephritis

The optimal regimen and duration of immunosuppressive treatment in LMN remains unclear because of the lack of large RCTs and the fact that the results of open-label and retrospective studies on the use of MMF in pure LMN have been conflicting [71—74]. A pooled analysis of 84 patients with pure LMN who participated in the ALMS trial and in the study by Ginzler *et al.*, has demonstrated comparable remission rates (percentage change of proteinuria and serum creatinine were the primary end points) in MMF- and IVCY-treated patients [75]. However, because of the paucity of well-designed RCTs and long-term data, the use of MMF as a first-line agent in pure LMN is encouraged only by a minority of authorities at present.

Extra-renal Lupus Manifestations

Limited evidence from case series or open-label trials suggests that MMF may be effective in refractory hematological or skin manifestations of SLE, whereas its efficacy for neuropsychiatric lupus remains to be determined [76—78]. The results of a prospective, randomized, open-label trial in patients with active lupus nephritis (classes III—V) revealed that MMF was equally effective with IVCY in inducing extrarenal remission, notably in the mucocutaneous, musculoskeletal, cardiovascular/respiratory, and vasculitis systems, and flares were rare. Levels of C3, C4, and CH50 and titers of anti-dsDNA antibodies were also normalized after treatment with either MMF or IVCY [79].

Of note, the drug showed no efficacy in preventing extra-renal flares in 75 patients who were followed at a single center for 5 years [80].

Collectively in these studies the drug demonstrated comparable efficacy and fewer side effects than CY. In the absence of long-term data and hard primary outcomes (such as doubling of serum creatinine or ESRD) claims of superiority in terms of efficacy over CY-based regimens cannot be adequately substantiated at present. This is especially true for the most severe cases, where CY has a track record of efficacy, something that hopefully will also be demonstrated for MMF. RCTs are needed for the establishment of MMF for extra-renal disease manifestations although preliminary evidence has provided encouraging results.

CALCINEURIN INHIBITORS

Cyclosporine

Cyclosporine A (CsA) is a fungus-derived lipophilic endecapeptide that inhibits calcineurin and T-cell activation. The therapeutic effects of CsA in lupus are linked to its ability to reduce antigen presentation and T-cell-mediated autoantibody production by B lymphocytes.

Pharmacokinetics

Peak serum concentration occurs within 1—8 hours after administration. Drug concentration is measured in whole blood but this is rarely needed in autoimmune diseases, unless CsA is used in doses ≥3 mg/kg/day. Clinical response to CsA is relatively slow and occurs within 1—2 months after treatment initiation. CsA is metabolized to more than 20 metabolites and excreted mainly in the bile. Although its elimination is not altered in renal insufficiency, CsA should be avoided in patients with impaired renal function, due to its nephrotoxic effects (see below) [81].

Drug Interactions

A variety of drugs interact with CsA leading to reduced (rifampin, phenyntoin, phenobarbital, nafcillin) or increased drug concentrations (erythromycin, clarithromycin, azoles, calcium channel blockers, amiodarone, allopurinol, colchicine) or augmenting its nephrotoxic effects (NSAIDs, aminoglycosides, quinolones, ACE inhibitors, amphotericin B) [81].

Adverse Effects

Transient gastrointestinal complaints, hypertrichosis, gingival hyperplasia, and a clinically insignificant rise in serum alkaline phosphatase levels, are the most commonly observed side effects. Tremor, paresthesias, electrolyte disturbances (hyperkalemia, hypomagnesemia), and hyperuricemia may also occur. Hypertension occurs in approximately 20% of patients receiving CsA and is controlled either by reduction of the dose, or by antihypertensive treatment [82]. A major side effect of CsA is nephrotoxicity, which is reversible after adjustment of the dose or discontinuation of the drug. Risk factors are high-dose CsA (>5 mg/kg/day) and an increase in serum creatinine >50% of the baseline value [82]. Monitoring guidelines to decrease toxicity are shown in Table 58.2. The drug has been associated with an increased risk of skin cancer and lymphoma in transplant recipients; such an association, however, has not been confirmed in patients with autoimmune disorders.

Clinical Use

Cyclosporine is most commonly used for LMN at doses of 3–5 mg/kg/d (Table 58.2). Radhakrishnan *et al.* treated 10 LMN patients with CsA alone or in combination with low-dose steroids for 23–43 months and reported a reduction of proteinuria to <1 g/day in six of them [83]. The efficacy of combined CsA (2–6 mg/kg/day) and prednisone was further demonstrated by Hallegua *et al.* who reported a reduction of mean proteinuria from 5.6 g/d to 1.4 g/d after 2 years in 10 patients with LMN. Common side effects included reversible nephrotoxicity and hypertension [84]. Hu *et al.* showed complete renal remission (defined as proteinuria <0.4 g/day with normal albumin and normal serum creatinine) in 12 out of 24 Chinese patients who received CsA (5 mg/kg/day, then tapered) with various doses of steroids. The relapse rate was 33% [85].

The relative efficacy of CsA, IVCY, and corticosteroids was studied in a RCT of 41 patients with pure LMN (median GFR 83 ml/min/1.73m², median proteinuria 5.4 g/d) [45]. All patients received alternate day oral prednisone (1 mg/kg for 8 weeks, then tapered to 0.25 mg/kg). The three groups received corticosteroids alone, IVCY every alternate month (0.5–1 g/m² adjusted to leukocyte nadir), or CsA (started at 5 mg/kg/d subsequently adjusted to serum creatinine) for 11 months, respectively. At 1 year, the cumulative probability of remission was 27% with prednisone, 60% with IVCY, and 83% with CsA. Although both IVCY and CsA were more effective than prednisone in inducing remissions of proteinuria, relapse of nephrotic syndrome occurred significantly more often after completion of CsA than after IVCY therapy (rates of relapse per 100 patient-months was 2.0 for CsA vs. 0.2 for IVCY, $p = 0.02$). The results of this study are limited by the small number of participants and the short follow-up but indicate that although CsA is effective as induction therapy in LMN, it may require maintenance therapy (with smaller doses of CsA, IVCY or MMF) to prevent flares [45]. Some authorities advise against the long-term use of CsA due to potential risks of nephrotoxicity, hyperlipidemia, and hypertension, which could superimpose additional thrombotic risk factors to LMN patients. Of note, the use of CsA did not confer additional thrombotic risk in 162 Chinese patients with lupus glomerulonephritis who were studied retrospectively for 8 years. However, only 18% of these patients had pure LMN, and thus the lack of long-term prethrombotic effects of CsA in LMN cannot be documented upon these data only [86]. Small case series suggest a clinical benefit of CsA in other manifestations of SLE such as skin rashes and thrombocytopenia and aplastic anemia [87].

CsA has also been used in refractory-to-conventional-treatment proliferative nephritis either in combination with corticosteroids or in between quarterly doses of IVCY, demonstrating good efficacy (complete remission rates up to 90% in some uncontrolled studies). Relapse of proteinuria is common after discontinuation of the treatment. Moroni *et al.* compared AZA with CsA as maintenance therapies in a RCT of 69 patients with diffuse proliferative lupus nephritis and preserved renal function [11]. Induction therapy patients received three consecutive daily pulses of 1 g IVMP followed by prednisone and oral CY for 3 months. They were then assigned to either CsA (4 mg/kg/d for 1 month and then tapered to 2.5–3mg/kg/d) or AZA (1.5–2 mg/kg/day) for 2–4 years. The rate of flare-ups was comparable in the two groups and there was no difference in proteinuria and blood pressure levels. Both agents were well tolerated. The results of this trial highlighted the potential use of CsA as maintenance treatment in lupus nephritis patients with preserved renal function.

Tacrolimus

Tacrolimus is another calcineurin inhibitor that is 10–100 times more potent than CsA. Similarly to CsA, tacrolimus inhibits T-cell activation, whereas further actions of the drug-relevant mostly to its use in allergic skin reactions include down-regulation of histamine release and leukotriene production.

Pharmakokinetics and Adverse Effects

After a single oral dose, tacrolimus reaches peak blood levels within 30–180 minutes, with a mean bioavailability 16–22% and a half-life of 40 hours. Tacrolimus is strongly bound to plasma proteins and red blood cells and is eliminated mainly through hepatic circulation [88]. Similarly to CsA, systemic tacrolimus administration has been associated with a dose-dependent reversible nephrotoxicity. Tacrolimus has been shown to lead to a lower incidence of hypertension than CsA in liver transplant patients [89]. Less common side effects include cardiomyopathy in children, anxiety, seizures, delerium and tremor, diabetes, and hyperlipidemia.

Clinical Use

Intraperitoneal administration of tacrolimus resulted in remission of skin lesions in the MRL-*lpr* lupus mouse model, and the efficacy of topical tacrolimus application for skin disease has been shown in several case series of lupus patients [90, 91]. Two uncontrolled trials demonstrated efficacy of tacrolimus in proliferative and membranous lupus nephropathy [92, 93]. The combination of a 6-month regimen comprising tacrolimus (0.1–0.2 mg/kg/day) and oral corticosteroids followed by maintenance treatment with AZA and corticosteroids achieved faster remission and lower flares than AZA or CY plus corticosteroids had shown retrospectively in LMN patients [94]. A multitarget aiming regimen of MMF (2 g/d), tacrolimus (4 mg/d) and corticosteroids (three pulses of IVMP at 0.5 g/d followed by oral prednisone) was compared with 6 monthly pulses of IVCY (1 g/m^2) plus steroids in a prospective study of 30 Chinese patients with mixed class IV+V nephritis, mean proteinuria 4.4 g/day and preserved renal function. At 6 months, remission rates were superior in the multitherapy group compared with the IVCY group (50 vs. 5%), with a similar rate of adverse events. However, the lack of long-term data does not allow for the assessment of mortality and ESRD outcomes. Despite the limitations of short study duration, small number of patients and the exclusion of patients with a high chronicity index score, these results indicate that multitarget therapy may represent a valuable option in mixed diffuse proliferative/membranous nephritis which is refractory to conventional immunosuppressive treatment [95, 96]. Meanwhile, the results of an ongoing RCT directly comparing the efficacy of tacrolimus versus MMF in proliferative or membranous lupus nephritis in 100 Chinese patients are still awaited [97].

Experimental therapies

Sirolimus (Rapamycin, Rapamune) is a hydrophobic lipophilic macrocyclic lactone derived from the actinomycyte *Streptomyces hydroscopicus*. It bears structural resemblance to erythromycin and tacrolimus, and together with everolimus, its synthetic derivative, they belong to a new class of immunosuppressive drugs known as proliferation signaling inhibitors (PSIs). Sirolimus inhibits m-TOR, an enzyme that catalyzes distinctive reactions mediating the co-stimulatory cascade and the G1 build-up following T-cell or B-cell activation. Furthermore, through m-TOR inhibition, it controls the expression of Fox-P3 and contributes to the expansion of regulatory T cells. Finally, it reduces dendritic cell activation, and all of these actions are likely to contribute to the efficacy of the drug in murine and perhaps human disease.

Unlike the calcineurin inhibitors, sirolimus is not associated with nephrotoxicity. Hypertension, hyperlipidemia, new-onset diabetes, an increased prevalence of urinary tract infections and lower limb edema are the most commonly reported side effects. Interestingly, data coming from transplantation trials have shown that PSIs attenuate neo-intimal thickening and transplant atherosclerosis from cardiac allograft vasculopathy. Accordingly, PSI-eluting stents have demonstrated lower rates of re-stenosis and late lumen loss. Whether these effects will translate to cardioprotective effect in lupus patients remains to be shown.

Sirolimus has been shown to reduce proteinuria in animal models of lupus nephritis and a prospective, open-label, phase II study is currently underway to assess efficacy in SLE [98, 99].

OTHER AGENTS

Methotrexate

Methotrexate (MTX), an antimetabolite, is a structural analog of folic acid that inhibits the *de novo* synthesis of purine metabolites through inhibition of dihydrofolate reductase (DHFR) and other folate-dependent enzymes. It exerts both anti-inflammatory and immunosuppressive effects.

Pharmacokinetics and Adverse Effects

Similar to rheumatoid arthritis (RA), MTX can be administered weekly either orally or parenterally in SLE [100]. Hepatic, hematologic, and pulmonary toxicity are the most important side effects (Table 58.2). Concomitant administration of folic acid (2.5–5 mg/week, not until 24 hours after the intake of MTX) is recommended and has been proven to limit MTX-related toxicity.

Clinical Use in Lupus

Although there has been substantial evidence supporting the effectiveness of MTX in RA, there are scarce data on its use in lupus. Current recommendations suggest its use as a steroid-sparing agent for articular and cutaneous manifestations of the disease. A single randomized double-blind placebo-controlled trial of MTX in lupus patients showed that a weekly dose of 15–20 mg for 6 months effectively controlled disease activity and allowed for corticosteroid reduction [100].

Leflunomide

Leflunomide is an isoxazole derivative currently employed as a disease-modifying drug for the treatment of RA. It is rapidly absorbed after oral administration and is converted into the active metabolite A77-1726, which has a half-life of approximately 15 days. Leflunomide blocks pyrimidine synthesis through inhibition of the enzyme dihydroorotate dehydrogenase leading to decreased production of r-UMP and reduced production of inflammatory cytokines by lymphocytes [101, 102].

Because of its long half-life, leflunomide requires a loading dose of 100 mg/day for 3 days followed by 20 mg/day thereafter (Table 58.2). The drug is eliminated through the feces and urine. Leflunomide is teratogenic and is contraindicated in patients who are trying to become or are pregnant.

In a prospective uncontrolled multicenter observational study oral leflunomide (30 mg/d after a 3-day loading dose of 1 mg/kg/d) was compared with IVCY (0.5 g/m^2) in 110 patients with biopsy-proven proliferative nephritis. All patients received steroids (starting dose of 0.8 mg/kg/day progressively tapered to 10 mg/d). The study showed comparable complete (2–58%) and partial (41–52%) response rates with similar rates of adverse events in the two arms. However, the small duration of the study (6–12 months), the concomitant use of steroids, and the small dose of IVCY preclude any definite conclusions [103, 104]. Better-designed RCTs are needed to establish the use of leflunomide in proliferative lupus nephritis.

Chlorambucil

Chlorambucil (CAB, Leukeran) is an aromatic bifunctional alkylating agent with substitution of the N-methyl group of mechlorethamine with phenylbutyric acid. The drug is given orally (0.1–0.20 mg/kg/d) with good absorption. Serum half-life is approximately 90 minutes with drug metabolism involving both methylation and beta-oxidation. Drug effects on immune functions are comparable to those described for CY. Side effects are also similar to those of CY, with the exception of bladder toxicity. A more prolonged and less predictable bone marrow suppression compared to that of CY can be observed. Furthermore, a markedly increased risk of leukemia, particularly acute myeloblastic leukemia (AML) has been associated with CAB [105, 106].

There are few clinical studies of CAB in SLE, but favorable outcomes in nephritis as well as extrarenal manifestations such as neuropsychiatric, vasculitis, and multisystem involvement have been reported. However, the drug is rarely used for these indications. In a retrospective study of 19 patients with predominantly class Va or Vb lupus nephritis, Moroni *et al.* showed that CAB combined with alternate month cycles of IVMP was more effective than MP alone in inducing remission of nephrotic syndrome (64 vs. 38%) and preserving renal function over the observation period of 83 months [106]. The use of CAB for LMN or other lupus manifestations is limited nowadays.

Abetimus sodium (LJP394)

Abetimus sodium (LJP394), a tetrameric oligonucleotide analog has been used to prevent renal flares in lupus. The effect of weekly intravenous administration of 100 mg of abetimus on preventing flares was evaluated in a placebo-controlled study of 317 patients who were followed for 22 months. Although the primary end point of the study was not reached, the authors reported a 25% reduction in renal flares in abetimus compared with placebo-treated patients with a parallel reduction in the levels of anti-dsDNA titers and proteinuria [107].

Clofazimine

Clofazimine (CFZ) is a riminophenazine dye that has been extensively used for the treatment of leprosy and other skin diseases. Clofazimine (100 mg/d) was compared to chloroquine (CDP, 250 mg/d) in a RCT of 33 active cutaneous SLE patients who were followed for 6 months. At the end of the study, 75% of CFZ- and 82% of CDP-treated patients had complete or near-complete remission of skin lesions and the authors concluded that the two drugs were equally effective in controlling cutneous lesions in SLE. However, five CFZ- and one CDP-treated patients dropped out of the treatment protocol after experiencing a severe lupus flare. Since the possibility that CFZ treatment may predispose to systemic lupus exacerbation cannot be ruled out, at present its use should be reserved for patients with exclusively cutaneous manifestations [108].

INTRAVENOUS GAMMA GLOBULIN

Intravenous immunoglobulin (IVIG) represents an immunoregulatory agent and has been used for the treatment of a variety of severe manifestations of lupus. Mechanisms of action include Fc-gamma receptor blockade as well as immunomodulation of complement and T-cells, all of which have been thought to contribute to its therapeutic effect. In 2008 and 2009, it was suggested that suppression of autoreactive B lymphocytes through signaling on the Fcγ-RIIB, idiotype-mediated inhibition of B-cell receptors, and neutralization of B-cell survival factors such as BAFF and APRIL, are important determinants of the drug's effects in autoimmune disease [109, 110].

Intravenous immunoglobulin is administered at doses of 400 mg/kg/day for five consecutive days, and is usually reserved for the treatment of severe, refractory thrombocytopenia achieving a rapid rise in the number of platelets within hours of the administration. Nephritis, arthritis, fever, rashes, and immunologic parameters improve with IVIG [111]. Despite the lack of evidence from randomized trials, IVIG can serve as first-line agent for certain manifestations such as neurological involvement, when CY is not an option, or in the context of concomitant infection. Side effects of IVIG include fever, neutropenia, myalgias, headache, arthralgias, and rarely, aseptic meningitis. It is contraindicated in IgA deficiency and should not be given to patients with thrombophilia, renal failure, or substantial cardiovascular disease.

Intravenous gamma globulin demonstrated good efficacy and safety profile in preventing spontaneous abortions and controlling disease manifestations in pregnant patients with SLE with or without antiphospholipid syndrome and a history of recurrent spontaneous abortions (RSA) [150]. In the study by Pericone et al., 12 SLE-RSA patients treated with 0.5 mg/kg IVIG every 3 weeks until the 33rd week of gestation were compared with 12 SLE-RSA controls treated with prednisone (0.25 mg/kg) and aspirin (100 mg/d) until the 34th gestational week. Both groups had received NSAIDs plus steroids prior to the onset of pregnancy. At the end of the study, 100% of IVIG-treated patients versus 75% of controls had a successful pregnancy and IVIG treatment was associated with better LAI-P scores with no major adverse effects after completion of treatment [150]. Given the considerable cost of IVIG and the lack of better controlled data, further studies are warranted to clarify the potential role of the drug as a substitute/adjunct to conventional treatments in the context of SLE-RSAs in pregnancy.

CYTOTOXIC THERAPY IN SLE: THE NEED FOR A STRATEGY THAT BEST FITS THE PATIENT

Treatment of moderate and severe SLE entails a period of intensive induction therapy followed by a longer period of less intensive maintenance therapy.

Irrespective of the agent used, early — within 6 months — effective cytotoxic therapy is of paramount importance, as first demonstrated in lupus nephritis by Esdaile et al. [111]. Pulses of IVCY in combination with steroids are considered the "gold standard" for induction therapy in severe lupus. The 10-year follow up of the Euro-Lupus nephritis trial demonstrated comparable efficacy of low- and high-dose CY induction regimens; thus the former may be favored as a less toxic therapeutic approach for selected patients. There is no doubt that MMF is less toxic than CY, does not cause ovarian failure, and is more acceptable to patients. Data from RCTs suggest that MMF as an induction regimen may be at least as good as IVCY in terms of efficacy and safety for selected cases and that it may be superior to IVCY in black populations. In light of the limitations of the MMF trials and of the paucity of long-term data, MMF should be used with close observation as induction therapy for moderately severe SLE. For severe cases or the cases whereby the disease does not remit after the first 6 months of therapy with MMF (or does not improve substantially after the first 3 months of therapy), the combination of pulses of IVCY and IVMP until remission remains the treatment of choice. Preliminary evidence suggests that MMF with tacrolimus may have added benefit over CY in such cases. In resistant-to-conventional-treatment SLE, the use of biological agents (rituximab) or establishment of multitarget therapy (MMF plus calcineurin inhibitors plus steroids) are valuable options. MMF is at least as good as AZA as a maintenance therapy — for the initial 1—2 years of remission — and CsA is an alternative in patients with stable renal function.

In the absence of RCTs, the optimal regimen and duration of therapy for LMN remains elusive. It is generally agreed that mixed proliferative and membranous lesions should be treated in the same way as pure proliferative lupus nephritis. If LMN is associated with nephrotic range proteinuria, renal impairment, or inadequate response to supportive/non-immunosuppressive treatment, immunosuppressive agents should be employed even in the absence of proliferative lesions. In this case, treatment options include the combination of steroids with AZA, calcineurin inhibitors (CsA or tacrolimus), or alkylating agents (CY or CAB). Whether MMF is effective in membranous lupus nephritis lesions remains controversial. In the absence of comparative trials, first-line treatment of LMN with MMF is advocated only by a minority of authorities at present. Further controlled studies are needed to compare the efficacy of AZA, calcineurin inhibitors, MMF, and alkylating agents in LMN. The majority of available RCTs mainly involve patients with lupus nephritis, with recommendations of non-renal SLE manifestations mainly being based on the extrapolation of these data.

GENERAL ISSUES

Cytotoxic Drugs in Severe, Life-Threatening or Refractory Disease

Controlled trials and clinical experience suggest that IVCY in combination with IVMP is the treatment of choice for most patients with severe, life-threatening lupus [1, 112]. For refractory patients, based on initial experience, availability and potential side-effects, combinations of IVCY with rituximab may be an acceptable strategy, although the EXPLORER [113] and LUNAR trials which evaluate the safety and efficacy of rituximab in generalized lupus and lupus nephritis, respectively, have failed to meet their primary clinical end-points. Limitations in trial design and follow-up have been invoked to explain these results. Moreover, rituximab has demonstrated good long-term efficacy as a rescue agent when combined with CY for refractory disease. In keeping with this, Lu *et al.* reported a significant decrease in BILAG score from 12 to 5 in 50 refractory lupus patients who were treated with 1 g rituximab, 750 mg of CY, and 100–250 mg of MP administered 2 weeks apart, and who were followed retrospectively for 7 years [114]. Until the issue of rituximab-associated risk for potentially fatal progressive multifocal encephalopathy (PML) is dispassionately examined and elucidated, it is advocated that the use of B-cell depletion (BCD) agents should be restricted to physicians and patients enrolled in specific programs where vigilance for infection is ensured and case-finding efforts are officially encouraged [115]. MMF may rescue a few refractory patients and its combination with IVCY is also an option for patients with severe refractory disease. However, its efficacy in critically ill patients requires further documentation. For selected patients (patients with neurological disease, antiphosholipid syndrome, thrombocytopenia) IVIG may be considered as an adjunct therapy.

CY may be administered either orally or as pulse therapy in the intensive care unit. The major concern is bladder protection and infections from the respirator and the intravenous lines. In such cases, bladder irrigation through a three-way catheter in case of urine output lower than 100 ml/h is essential, along with diligent care of intravenous lines and tapering of corticosteroids. An aggressive search for infection prior to and after therapy is essential. Until infection is ruled out, we usually use high-dose corticosteroids. An aggressive tapering of corticosteroids – once the patient improves – is essential to minimize the risk of infectious and other complications. IVIG may be considered in such cases. The role of the multidisciplinary approach involving several medical subspecialties cannot be overemphasized.

Prevention and Management of Infection: Immunizations

Infections attributed predominantly to corticosteroids and cytotoxic drugs are an important cause for hospital admissions and death accounting for approximately one quarter of all deaths in SLE patients [116]. Judicious use of corticosteroids may decrease the frequency of infections. Other strategies to decrease the impact of infections include: (a) simple hygiene measures and education aimed at both patients and doctors; (b) antimicrobial prophylaxis in cohorts of patients with increased prevalence of certain infections, patients who receive heavy doses of immunosuppressive agents, or undergo procedures associated with transient bacteremia; and (c) immunizations (Table 58.5).

The initial evaluation of a patient with lupus who receives cytotoxic drugs and presents with symptoms or signs suggestive of infection is of paramount importance and possesses significant diagnostic and therapeutic challenges. In particular, because of the broad list of potential pathogens, the subtle clinical manifestations of infection especially in neutropenic patients and those receiving high doses of corticosteroids, and the frequent overlap in symptoms of infection with those of the underlying disease, the diagnostic approach and management need to follow a rational plan which is largely determined by the following three factors: (a) the dominant clinical syndrome; (b) history of epidemiologic exposures; and (c) the net-state of immunosuppression of the patient. In the first case, important questions include: is the clinical presentation suggestive of a particular pathogen or syndrome; what is the timetable and the pace of infection (acute or subacute) and if the patient is clinically stable or acutely ill (septic); is the infection breaking through an existing outpatient or inpatient antibiotic treatment regiment and why? The epidemiologic history entails the exposure history of the host to potential pathogens including opportunistic, endemic, or multidrug-resistant organisms. Significant epidemiologic exposures to be considered include: (a) community exposures (particularly those associated with exposure to transmissible source patient such as influenza (H1N1), mycobacterium tuberculosis (MTB), varicella zoster virus (VZV); travel; occupation; pets or other animals); (b) nosocomial exposures such as recent construction or epidemic outbreaks; and (c) latent or dormant pathogens such as a history of positive PPD or serology for chronic viral infections. Finally, the net state of immunosuppression is a complex entity that reflects the effect of various qualitative and qualitative defects in the immune status of an individual, and is best determined by the composite assessment of: (a) type of drugs, dose, intensity and duration of immunosuppressive treatment; (b) presence of native

TABLE 58.5 Malignancy screening, immunizations and infection control for SLE patients on cytotoxic treatment [123]

Indication	Recommendation
Malignancy screening	American Cancer Society Guidelines for the early detection of cancer relative to breast cancer, cervical cancer and colon and rectal cancer Cervical cytology once or twice in the first year, then annually Consider HPV DNA testing in initial smears. Repeat cytology at 6 mo if positive for HPV DNA
Vaccinations	Annual influenza (including H1N1) vaccination and pneumococcal vaccinations advised. Preferably administer when disease is inactive Live attenuated vaccines are contraindicated during immunosuppressive and /or high dose steroid (prednisone >20 mg/d) treatment Tetanus toxoid immunization Hepatitis B vaccine may be safe HPV vaccine: efficacy and target groups not yet determined
Infection control	HIV screening according to patient's risk factors HCV and HBV screening according to patient's risk factors or prior to high dose steroids or immunosuppressants Baseline serology (IgG/IgM) for CMV. Assess CMV antigemia in selected cases (high intensity steroid/immunosuppressive schedule with clinical picture consistent with CMV disease) PPD routinely in endemic areas only. PPD should be offered prior to starting high dose steroids (>15 mg/d) or immunosuppressants Consider PCP prophylaxis (one double strength tablet of trimethoprine-sulfomethoxazole three times a week or dapsone 100 mg/day if allergic to sulfamethoxazole) if long-term high dose steroids +/− immunosuppressants and CD4 <300 cells/μl * Consider endocarditis prophylaxis * Monitor for severe neutropenia (<500 PMN/μl) or lymphopenia (<500/μl) in patients taking immunosuppressants

Not supported by 2009 EULAR recommendations for monitoring of SLE patients.

Abbreviations: HPV, human papilloma virus; CIN, cervical intraepithelial neoplasia; HIV, human immunodeficiency virus; HCV, hepatitis C virus; HBV, hepatitis B virus; CMV, cytomegalovirus; PPD, purified protein derivative PCP, pneumonocystis carinii.

immunosuppression such as asplenia (anatomic or functional) or hypogammaglobulinemia; (c) co-morbidities such as diabetes, renal or hepatic failure, malnutrition; (d) disruption of anatomic barriers; and (e) miscellaneous conditions (pregnancy, total parenteral nutrition). The careful assessment of all these variables should allow the clinician to formulate a plan in the management of the patient and decide for the need of hospital admission and initiation of empirical antimicrobial therapy. Although guidelines or algorithms cannot replace good clinical judgment, which should always include a thorough history and physical examination of the patient and the need for individualized therapy, our preferred approach is depicted in Figure 58.1.

Bacterial Endocarditis

Clinical or subclinical valvular abnormalities are common in patients with moderate to severe lupus and may be predisposing to bacterial endocarditis. Moyssakis *et al.* reported 11% prevalence of Libman-Sacks endocarditis (causing valvular dysfunction, mostly mitral and secondarily aortic regurgitation) in 342 SLE patients who were studied with transthoracic echocardiography [117]. The updated recommendations of the European Society of Cardiology (ESC 2009) and the latest guidelines from the American Heart Association (AHA 2007) have proposed limitation of antibiotic prophylaxis only to patients with

previous endocarditis, prosthetic valves or prosthetic valve materials, and several congenital defects. There are no controlled data to support the cost-effectiveness of prophylactic antibiotics in patients with autoimmune conditions under cytotoxic treatment [118, 119]. In this respect, strict adherence to the above-mentioned guidelines would exclude the majority of SLE patients from endocarditis prophylaxis. However, since these guidelines reflect expert consensus opinions rather than being based on strong evidence, and given the paucity of controlled data on the real prevalence of bacteremia and endocarditis in lupus patients (especially those under immunosuppression), the decision on endocarditis prophylaxis should be individualized and ideally discussed with the patient on an informed consent basis. In our opinion, the employment of antibiotic prophylaxis in lupus patients receiving high-intensity immunosuppressive therapy is a reasonable suggestion although firm data to support this are lacking at present [120]. Of interest, current ESC guidelines no longer recommend antibiotic prophylaxis for gastrointestinal, genitourinary, or respiratory procedures in the absence of infection, even for high-risk patients.

Tuberculosis and Pneumocystis Pneumonia

There is no prospective study assessing the risk of tuberculosis (TB) infection in patients receiving corticosteroids but population-based studies indicate that the

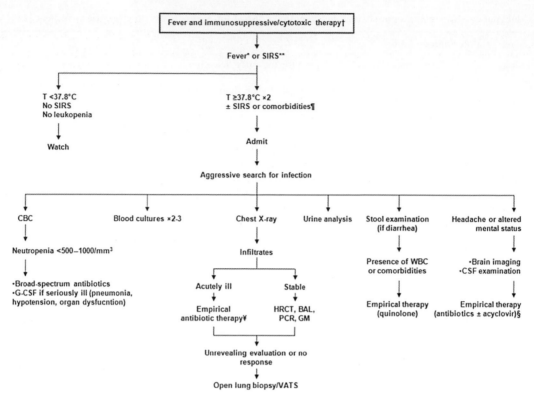

FIGURE 58.1   Initial assessment and management of SLE patients who receive immunosuppressive/cytotoxic therapy and present with fever, systemic inflammatory response syndrome, or other symptoms and signs suggestive of infection (dyspnea, abdominal pain, diarrhea, confusion). Management is determined by net state of immunosuppression, epidemiologic exposure and clinical presentation (see text). *Abbreviations*: HRCT, high-resolution chest CT, BAL, broncho-alveolar lavage, PCR, polymerase chain reaction, GM, galactomannaneVATS, video-assisted thoracoscopy, MRSA, methicillin resistant *Staphylococcus aureus*; H1A1, influenza H1A1.

relative risk for developing active TB is 4—6-fold that of healthy individuals. In general, prevalence of TB in SLE patients appears to be higher compared to the general population (6—60-fold higher rates varying with the study) [121, 122]. Tuberculin skin testing (TST) is recommended for patients who are candidates for subsequent treatment with long-term prednisone equal or greater than 15 mg/d or immunosuppressive drugs, according to the CDC guidelines. Because of the isoniazid (INH)-age-related hepatotoxicity (4.6% in individuals aged >65 years versus 0.2—0.3% in 20—30-year-old individuals), an age cut-off of 65 is indicated, so that potential risks of INH toxicity in older patients do not outweigh potential benefit. Therefore, it is acceptable that in moderate-risk patients, TST is advocated only for those under the age of 65 [123].

The incidence of *Pneumocystis* pneumonia (PCP) in SLE patients on CY approximates 0.15%, whereas the overall prevalence of PCP in lupus patients is largely unknown [124]. Of note, MMF confers *in vitro* protection against PCP but whether this translates into clinical benefit has not been determined [50]. In contrast to systemic vasculitis, we do not routinely employ PCP prophylaxis in lupus. However, some authors recommend it for patients on high-dose corticosteroids alone

or in combination with cytotoxic drugs especially if CD4 count is below 300 cells/μl (one double strength tablet of trimethoprine-sulfomethoxazole three times a week or dapsone 100 mg/day if allergic to sulfamethoxazole). Clearly there is a need for consensus guidelines addressing prophylactic antibiotics in these patients.

Viral Infections

Lupus patients do not display an increased risk for HIV infection and therefore screening should be based on individualized risk factors [123]. Similarly, HBV and HCV infections do not occur more frequently in SLE patients. Due to risks of occurrence and reactivation of the infection following immunosuppressive therapy, particularly when steroids are administered, screening for HCV and HBV is prudent prior to starting high-dose steroids or other immunosuppressive medications.

Parvovirus B19 mimicking a lupus flare and cytomegalovirus (CMV) mostly associated with intense immunosuppression, may be observed in SLE. Varicella-zoster virus (VZV) reactivation is an important issue in SLE patients and risk factors include history of CY or AZA exposure, the presence of concurrent or previous malignancy and lupus nephritis [125]. Shingles dissemination and bacterial superinfection are mostly linked to high-dose corticosteroid treatment.

Immunizations

Although vaccination may theoretically induce polyclonal activation in lupus exacerbating a flare, it is felt to be safe. In view of the morbidity and mortality associated with infections, SLE patients would greatly benefit from vaccination programs. However, it is recommended that vaccines should not be given to patients with very active SLE or those with active nephritis [123]. Inactivated live vaccines (measles, mumps, rubella, polio, varicella, VZV (zoster vaccine) and vaccinia (smallpox)) are contraindicated in patients taking immunosuppressive drugs and/or prednisone at a dose >20 mg/d, due to the risk of viral multiplication. Influenza vaccine has been shown to be safe and effective while pneumococcal vaccine is also safe but the resultant antibody titers may be decreased in patients with SLE as compared to controls [127]. Use of corticosteroids may contribute to the blunted antibody response. Protective immune response can be achieved safely in SLE patients with both tetanus toxoid and *Haemophilus influenzae* type B in addition to pneumococcus. According to the Advisory Committee on Immunization Practices (ACIP) guidelines, individuals aged >60 years should be vaccinated with a single dose of VZV vaccine. Zoster vaccine should be administered at least 14 days before initiation or should be deferred for at least 1 month after discontinuation of high-dose corticosteroids ($\geq$20 mg/day prednisone or equivalent) or immunosuppressive therapy. Therapy with low doses of methotrexate (<0.4 mg/kg/week) or AZA (<3.0 mg/kg/day) are also not considered sufficiently immunosuppressive to create vaccine safety concerns.

In the past there has been a lot of concern over the development of rheumatic-related syndromes temporally related to vaccinations, particularly for hepatitis B; however, a causative relationship has not been established. In 2007 Kuruma *et al.* reported on hepatitis B vaccination of 28 SLE patients with inactive SLE, treated with low-dose glucocorticoids and not receiving immunosuppressive medications, negative anti-dsDNA and anti-cardiolipin antibodies. Patients were administered a recombinant vaccine. All patients developed protective antibodies (26/28 after the third dose and 2 after the fourth dose); no increase in SLE flares was observed. The authors concluded that hepatitis B vaccination might be safe in SLE patients [128].

Monitoring of Immune Function

Few studies have assessed whether leucopenia and low immunoglobulin levels are predictive of infections in SLE patients. In a study of 82 SLE patients, Mosca *et al.* showed that a lymphocyte count $\leq 1 \times 10^9$/L was associated with an increased infection risk (odds ratio (OR) 4.7). Low levels of IgG3 ($\leq$ 60 μg/ml) or IgG4 ($\leq$ 20 μg/ml) have also been associated with an increased risk of infections [123]. The infection risk of SLE patients could be assessed, taking into consideration neutropenia, lymphopenia, and IgG levels. In particular, total IgG and subclass levels should be assessed at the first patient assessment and during follow-up visits in patients taking immunosuppressive drugs. Severe neutropenia (<500/μl), severe lymphopenia (<400/μl), and low IgG levels (<500 mg/dl) increase the risk for major infections [123].

Pulmonary Infiltrates

Pulmonary complications remain a major cause of morbidity and mortality in lupus patients receiving cytotoxic drugs. The differential diagnosis in this setting is broad and includes both infectious and non-infectious processes. Rarely are the radiographic findings specific for one disease, and most potential etiologies have overlapping clinical and radiographic appearances. An aggressive work-up to identify a specific cause is of paramount importance; this includes early use of CT scanning (usually high-resolution) and a low threshold for bronchoscopy [129].

Screening for Malignancy

Certain cancers occur more frequently in patients with SLE as compared with the general population. Hematological malignancies (particularly non-Hodgkin lymphoma), cervical, breast, and lung cancer have a higher prevalence among lupus sufferers [130, 131]. In addition, an abnormal cervico-vaginal cytology is reported in 16% of examined patients with an odds ratio (OR) that ranges between 1.8 and 28 compared to the general population [132]. Importantly, existing studies have failed to demonstrate an increased incidence of bladder cancer, not even in association with CY treatment, irrespective of the protocol used (NIH protocol, Eurolupus protocol, or oral CY). The association between the administration of CY and bladder cancer reported among patients with vasculitis or hematological malignancies seems to be related to cumulative doses of CY above 100 g [133, 134].

Immunomodulating/immunosuppressive treatment can contribute to lymphoma/carcinogenesis either by directly causing mutagenesis of by impairing the "immune surveillance" which can lead to uncontrolled inappropriate cell proliferation. According to the study by Bernatsky *et al.*, the largest multicenter study conducted up-to-date including more than 9500 SLE patients, the increased risk of cancer in SLE compared with the general population manifests early during the disease course — implying that risk factors other than the cumulative effects of immunosuppressive drugs may contribute to the increased cancer risk [135]. In a nested case cohort study that investigated the relationship between medication exposure and cancer in SLE, there was suggestion of an increased risk of hematological cancer for all immunosuppressive agents combined, which was observed 5 or more years after the initial drug exposure [136]. However, this study did not optimally control for disease activity. At present, with the best available data, it is considered that disease-related factors may be at least as important as medication exposures to at least certain cancers, such as NHL. With the exception of cervical intraepithelial neoplasia (CIN), existing studies have failed to identify a clear association between cancer and exposure to immunosuppressive drugs [137].

Based on these data, vigilance in cancer-preventive strategies at least similar to that of the general population is essential and is currently recommended (Table 58.5). This is particularly important in view of some data suggesting that appropriate cancer screening may be overlooked in patients with SLE. Considering the unique epidemiological and clinical characteristics of SLE patients, development of specific SLE guidelines for malignancy screening may be considered in the future [123, 138].

Lymphoma and Hematological Malignancy

The standardized incidence ratio (SIR) of NHL ranged from 3.7 to 11.6 in lupus patients, involving mostly aggressive-type NHL (diffuse large cell lymphoma). Bernatsky et al. have suggested an increased risk of hematological malignancy after immunosuppressive drug exposure, particularly when these were lagged by a period of 5 years, in their multisite international SLE cohort (adjusted HR 2.29) [135]. Since the clinical appearance of NHL and active lupus may be similar (lymphadenopathy, splenomegaly, cytopenias, constitutional symptoms, and positive ANA), watchful surveillance and thorough investigations including lymph node biopsies are sometimes warranted. In patients with prominent lymphadenopathy, massive splenomegaly, or expansion of the $CD19^+$ $CD22^+$ peripheral blood population, a search for NHL is warranted. Chronic antigen stimulation and polyclonal B-cell activation may also contribute to the development of Hodgkin lymphoma which displays an increased prevalence in SLE (SIR ranging 2.4—17.8) [135].

Gynecological Malignancies

An increased frequency — up to 6-fold — of abnormal cervical Papanicolaou (Pap) smears in women with SLE has been reported by several groups, and both disease and/or drug (particularly CY and secondarily AZA) mediated immunosuppression have been implicated in its incremented prevalence [132, 139]. Interestingly, Tam et al. reported an increased prevalence of high-risk HPV infection and multiple HPV infections among 85 SLE patients who were compared with age- and sex-matched controls [139]. Most experts agree that SLE should be regarded as a risk factor for cervical malignancy and high-risk HPV infection. Extrapolating screening guidelines for cervical cancer among HIV patients and taking into consideration the latest EULAR guidelines for monitoring SLE patients, we would recommend cervical cytology for cancer screening once (EULAR guidelines) or twice in the first year (United States Preventive Services Task Force) and then annually, adding HPV testing to the first year obtained cervical smears and then modifying subsequent screening based on these results (cervical cytology screening every 6 months for women with detectable HPV DNA and annually for all the others) [123]. Although CDC-approved for immunocompromised individuals, the issue of the efficacy/antibody response and target age group of HPV vaccine has not been addressed in SLE.

An increased risk of breast cancer has been suggested by some early cohort data (SIR 0.9—2.1) and further studies have been carried out to investigate the role of hormonal and reproductive risk factors in women with SLE. Available data suggest that risk of breast cancer may be more pronounced in white women, rather than being uniform across all lupus populations [140]. At present, screening guidelines are similar for those in the general population.

Lung Cancer

An increased incidence of lung cancer, particularly adenocarcinoma and secondarily small-cell carcinoma, has been reported on several cohorts. There is substantial evidence suggesting that smoking is an important predictor of lung cancer in SLE patients. In one of the largest cohorts studying 9500 SLE patients from 23 multinational centers, 71% of cancer cases were smokers, whereas only 20% were previously exposed to immunosuppressive drugs [141].

Use of Cytotoxic Drugs in Pregnancy

Pregnant patients with active lupus and mild disease are generally managed with corticosteroids. In moderate-to-severe disease, corticosteroids, AZA, CsA, and IVIG are considered acceptable [142]. In life-threatening disease, CY may be used only if there are no alternative therapies. CY and MTX are abortifaciant and teratogenic and are contraindicated in pregnancy (FDA Category D-positive evidence for risk). MMF has been assigned as Category D since November 2007 and thus should not be administered. Leflunomide is also contraindicated. Inadvertent exposure to potentially fetotoxic drugs is not an absolute indication for pregnancy termination and such a decision should be based on careful assessment of any risk to the fetus. The American Academy of Pediatrics states that nursing is permissible for women receiving corticosteroids but the interval between dose and nursing should be at least 4 hours if the prednisone dose is more than 20 mg/d. Because AZA may be excreted in the breast milk, breast feeding is not recommended. Table 58.6 summarizes data on the safety of cytotoxic drugs during pregnancy and lactation.

Use of Cytotoxic Agents in Childhood

Up to 15–20% of lupus cases are diagnosed in children younger than 16 years (childhood-onset lupus,

c-SLE). A higher prevalence of nephritis and central nervous system (CNS) involvement implying a more sustained need for immunosuppressive treatment and steroids, and a higher prevalence of progression to ESRD are distinguishing features of c-SLE [143].

RCTs comparing the efficacy and safety of the previously analyzed adult therapeutic regimens in pediatric lupus are generally lacking. Lau *et al.* retrospectively studied the effect of IVCY versus MMF in 13 children with class III nephritis [144]. After a follow-up of 6 months, remission rates were better in the MMF compared with the IVCY group (complete remission 66 vs. 0%; partial remission 57 vs. 17%, respectively). The small number of participants, lack of randomization, limited follow-up, and the non-inclusion of patients with

TABLE 58.6   Safety of immunosuppressive drugs during pregnancy and lactation [149, 150]

Drug	Safe in pregnancy (FDA category)	Comments
Azathioprine	Yes* (C)	Generally safe although fetal malformation reported Do not exceed 2 mg/kg for risk of defective hematopoiesis in infants Not recommended for breast-feeding but case reports testify to possible safety in nursing mothers
Cyclosporine	Yes* (C)	Increased risk of obstetric cholestatis Not teratogenic, can be given at a dose 2.5-5.0 mg/kg throughout pregnancy Contraindicated during breast-feeding
Corticosteroids	Yes* (C)	Increased risk of gestational diabetes and premature rupture of membranes. Slight increase in cleft palate, fetal growth restriction and intrauterine infection particularly in doses of prednisone >15 mg/d Administer the lowest effective dose during first trimester If long-term steroid treatment, peripartum stress doses of steroids should be administered. Pulse, bolus and fluorinated steroids should be avoided. Safe during breast-feeding but interval between lactation and steroid intake should be >4 hrs if prednisone dose >20 mg/d
Cyclophosphamide	No (D)	High rate of fetal loss and fetal abnormalities Use only in life threatening disease if no other alternatives
Leflunomide	No (X)	Potentially mutagenic and teratogenic Contraindicated in pregnancy and breast-feeding
Methotrexate	No (X)	Pregnancy should be avoided if either partner is receiving methotrexate Avoid pregnancy for at least 3 months in males and at least one ovulatory cycle in females (optimum 3 months) Contraindicated during breast-feeding
Mycophenolate mofetil	No (D)	Increases risk of spontaneous abortion and congenital malformations Category D since November 2007 Stop at least 6 weeks prior to pregnancy Contraindicated during breast-feeding
Tacrolimus	Yes* (C)	May increase risk of gestational diabetes May induce neonatal hyperkalemia and renal dysfunction, but no controlled data on human pregnancies Should be avoided during breast-feeding
IVIG	Yes* (C)	Reduces RSA and halts disease activity in SLE +/− APS patients when administered during pregnancy Unknown effects when used in nursing mothers

*No drug is felt to be absolutely safe in pregnancy. The authors assign drugs as "safe" when classified at least as FDA category C (animal studies have shown an adverse effect and there are no adequate and well controlled studies in pregnant women OR no animal studies have been conducted and there are no adequate and well controlled studies in women) pointing to case report/series based bibliographic evidence testifying to safe administration during pregnancies in cases where benefits clearly outweigh risks.
Abbreviations: FDA, food and drug administration; RSA, recurrent spontaneous abortions; SLE, systemic lupus erythematosus; APS, antiphospholipid syndrome.

severe renal disease, raise questions about the superiority of MMF over IVCY in c-SLE nephritis [144]. In keeping with this, despite showing efficacy in membranous nephropathy, MMF failed to attenuate disease progression in pediatric patients with proliferative lesions [145]. In a study published online in 2009, Baskin *et al.* evaluated the efficacy of short-term IVCY treatment as induction treatment followed by AZA or MMF as maintenance, in 20 c-SLE patients with class III–IV nephritis, who were followed for 49 months. All patients received three IVMP pulses, followed by oral prednisone 0.5–1 mg/kg per day and one IV pulse of CY per month for 6 months. AZA was started as a remission-maintaining treatment. In 10 of 20 patients, treatment was switched to MMF. Complete, partial, and no remission rates were 70, 15, and 5%, respectively and 10% of patients progressed to ESRD. Of note, switching of AZA to MMF increased complete remission rates by 25% [146]. Although these results suggest that MMF may be useful in refractory c-SLE cases, RCTs are needed to better establish its efficacy and safety for this indication.

Significant CNS involvement in children usually warrants a combination of high-dose steroids with an immunosuppressive agent, such as CY or AZA, although further studies and a general consensus are needed to establish the optimal therapeutic approach for neuropsychiatric involvement in c-SLE [147]. Importantly, preliminary evidence suggests that BCD with rituximab is safe and efficacious for severe hematological or cardiorespiratory manifestations of c-SLE, or for refractory nephritis [148]. To date, there is no study documenting an increased rate of serious adverse effects such as life-threatening infections or secondary neoplasia in children with SLE receiving cytotoxic therapy compared with adults. In summary, the overall treatment strategies are similar between c-SLE and adult SLE patients. Although the role of MMF seems to be well documented in adult moderately severe to severe lupus nephritis, more evidence is needed to assess its long-term safety and efficacy in c-SLE.

References

[1] D.T. Boumpas, P. Sidiropoulos, G. Bertsias, Optimum therapeutic approaches for lupus nephritis; what therapy and for whom? Nat Clin Pract Rheumatol 1 (2005) 22–30.

[2] B.K. Pedersen, J.M. Beyer, A. Rasmussen, et al., Azathioprine as a single drug in the treatment of rheumatoid arthritis induces complete suppression of natural killer cell activity, APMIS 92 (1984) 221–225.

[3] S.K. Van, C.A. Johnson, W.R. Porter, The pharmacology and metabolism of the thiopurine drugs 6-mercaptopurine and azathioprine, Drug Metabol Rev 16 (1985) 157–174.

[4] D. Cummins, M. Sekar, O. Halil, N. Banner, Myelosuppression associated with azathioprine-allopurinol interaction after heart and lung transplantation, Transplantation 61 (1996) 1661–1662.

[5] J. Walker, H. Mendelson, A. McClure, M.D. Smith, Warfarin and azathioprine: clinically significant drug interaction, J Rheumatol 29 (2002) 398–399.

[6] G. Singh, J.F. Fries, P. Spitz, C.A. Williams, Toxic effects of azathioprine in rheumatoid arthritis: a national port-marketing perspective, Arthritis Rheum 32 (1989) 837–843.

[7] G. Leopold, E. Schutz, J.P. Haas, M. Oellerich, Azathioprine-induced severe pancytopenia due to a homozygous two-point mutation of the thiopurine methyltransferase gene in a patient with juvenile HLA-B-27-associated spondylarthritis, Arthritis Rheum 40 (1997) 1896–1898.

[8] B. Lofstrom, C. Backlin, C. Sundstrom, et al., A closer look at non-Hodgkin's lymphoma cases in a national Swedish systemic lupus erythematosus cohort: a nested case-control study, Ann Rheum Dis 66 (2007) 1627–1632.

[9] C. Grootscholten, C. Ligtenberg, E.C. Hagen, et al., Azathioprine/methylprednisolone versus cyclophosphamide in proliferative lupus nephritis. A randomized controlled trial, Kidn Int 70 (2006) 732–742.

[10] C. Grootshalten, I.M. Bajema, S. Florquin, et al., Treatment with cyslophosphamide delays the progression of chronic lesions more effectively than does treatment with azathioprine and methylprednisolone in patients with proliferative lupus nephritis, Arthritis Rheum 56 (3) (2007) 924.

[11] G. Moroni, A. Doria, M. Mosca, et al., A randomized pilot trial comparing cyclosporine and azathioprine as maintenance therapy in diffuse proliferative lupus nephritis over four years, Clin J Am Soc Nephrol 1 (5) (2006) 925–932.

[12] C.C. Mok, K.Y. Yin, C.W. Yim, et al., Very long-term outcome of pure nephropathy treated with glucocorticoid and azathioprine, Lupus 18 (2009) 1091–1095.

[13] T.R. Cupps, L.C. Edgar, A.S. Fauci, Suppression of human B lymphocyte function by cyclophosphamide, J Clin Invest 128 (6) (1982) 2453–2457.

[14] B.D. Pryor, S.G. Bologna, L.E. Kahl, Risk factors for serious infection during treatment with cyclophosphamide and high dose corticosteroids for systemic lupus erythematosus, Arthritis Rheum 39 (1996) 1475–1482.

[15] R. Martinelli, L.J.C. Pereira, E.S.C. Santos, H. Rocha, Clinical effects of intermittent, intravenous cyclophosphamide in severe systemic lupus erythematosus, Nephron 74 (1996) 313.

[16] L.B. Grochow, M. Colvin, Clinical pharmacokinetics of cyclophosphamide, Clin Pharmacokinet 4 (1979) 380–394.

[17] B.D. Pryor, S.G. Bologna, L.E. Kahl, Risk factors for serious infection during treatment with cyclophosphamide and high dose corticosteroids for systemic lupus erythematosus, Arthritis Rheum 39 (1996) 1475–1482.

[18] L. Guillevin, J.F. Cordier, F. Lhote, et al., A prospective, multicenter, randomized trial comparing steroids and oral cyclophosphamide in the treatment of generalized Wegener's granulomatosis, Arthritis Rheum 40 (1997) 2099–2104.

[19] M. Ramos-Casals, M.J. Cuadrado, P. Alba, G. Sanna, et al., Acute viral infections in patients with systemic lupus erythematosus. Description of 23 cases and review of the literature, Medicine (Baltimore) 87 (6) (2008) 311–318.

[20] D.D. Gladman, F. Hussain, D. Ibanez, M.B. Urowitz, The nature and outcome of infection in systemic lupus erythematosus, Lupus 11 (2002) 234–239.

[21] D.T. Boumpas, Austin HA 3d, E.M. Vaughan, et al., Risk for sustained amenorrhea in patients with systemic lupus erythematosus receiving intermittent pulse cyclophosphamide therapy, Ann Intern Med 119 (1993) 366–369.

[22] S.A. Rivkees, J.D. Crawford, The relationship of gonadal activity and chemotherapy-induced gonadal damage, JAMA 259 (1988) 2123.

[23] C.A. Slater, M.H. Liang, J.W. McCune, G.M. Christman, M.R. Laufer, Preserving ovarian function in patients receiving cyclophosphamide, Lupus 8 (1999) 3—10.

[24] M.A. Dooley, C.C. Patterson, L. Susan, et al., Preservation of ovarian function using depot leuprolide acetate during cyclophosphamide therapy for severe lupus nephritis, Arthritis Rheum 43 (2000) 2858.

[25] E.C. Somers, W. Marder, G.M. Christman, V. Ognenovski, W.J. McCune, Use of a gonadotropin-releasing hormone analog for protection against premature ovarian failure during cyclophosphamide therapy in women with severe lupus, Arthritis Rheum 52 (2005) 2761—2767.

[26] L.Q. Liang, Q. Qiu, X.Y. Yang, H.S. Xu, Y.J. Ye, Z.P. Zhan, et al., Role of gonadotropin releasing hormone analogues for ovarian protection in systemic lupus erythematosus patients treated with cyclophosphamide, Zhonghua Yi Xue Za Zhi 88 (15) (2008) 1009—1011.

[27] K. Manger, L. Wildt, J.R. Kalden, B. Manger, Prevention of gonadal toxicity and preservation of gonadal function and fertility in young women with systemic lupus erythematosus treated by cyclophosphamide: the PREGO study, Autoimmun Rev 5 (4) (2006) 269—272.

[28] H.S. Nicholson, J. Byrne, Fertility and pregnancy after treatment for cancer during childhood or adolescence, Cancer 71 (1993) 3392—3399.

[29] A. Masala, R. Faedda, S. Aigna, et al., Use of testosterone to prevent cyclophosphamide-induced azoospermia, Ann Intern Med 126 (1997) 292—295.

[30] A. Raptopoulou, P. Sidiropoulos, D. Boumpas, Ovarian failure and strategies for fertility preservation in patients with systemic lupus erythematosus, Lupus 13 (2004) 887—890. Review.

[31] E. Kiss, L. Kovacs, P. Szodoray, Malignancies in systemic lupus erythematosus, Autoimmun Rev 9 (4) (2010) 195—199. Epub ahead of print. Review.

[32] T.J. Stillwell, R.C.J. Benson, R.A. DeRemee, et al., Cyclophosphamide induced bladder toxicity in Wegener's granulomatosis, Arthritis Rheum 31 (1988) 465—470.

[33] M. Gayed, S. Bernatsky, R. Ramsey-Goldman, et al., Lupus and cancer, Lupus 18 (6) (2009) 479—485. Review.

[34] C.D. Radis, L.E. Kahl, G.L. Baker, et al., Effects of cyclophosphamide on the development of malignancy and on long-term survival on patients with rheumatoid arthritis. A 20-year followup study, Arthritis Rheum 38 (1995) 1120.

[35] C. Talar-Williams, Y.M. Hijazi, M.M. Walther, et al., Cyclophosphamide induced cystitis and bladder cancer in patients with Wegener's granulomatosis, Ann Int Med 124 (1996) 477—484.

[36] H.A. Austin III, J.H. Klippel, J.E. Balow, et al., Therapy of lupus nephritis. Controlled trial of prednizone and cytotoxic drugs, N Engl J Med 314 (1986) 614—619.

[37] D.T. Boumpas., H.A. Austin, E.M. Vaughan, et al., Severe lupus nephritis: Controlled trial of pulse methylprednisolone versus two different regimens of pulse cyclophosphamide, Lancet 340 (1992) 41.

[38] M.F. Gourley, H.A. Austin, D. Scott, et al., Methylprednisolone and cyclophosphamide, alone or in combination, in patients with lupus nephritis. A randomized, controlled trial, Ann Intern Med 125 (1996) 549.

[39] R. Faedda, D. Palomba, A. Satta, et al., Immunosuppressive treatment of the glomerulonephritis of systemic lupus, Clin Nephrol 44 (1995) 3678.

[40] G.G. Illei, H.A. Austin, M. Crane, et al., Combination therapy with pulse cyclophosphamide plus pulse methylprednisolone improves long-term renal outcome without adding toxicity in patients with lupus nephritis, Ann Intern Med 135 (2001) 248—257.

[41] M.A. Dooley, S. Hogan, C. Jennette, et al., Cyclophosphamide therapy for lupus nephritis: poor renal survival for black Americans. Glomerular Disease Collaborative Network, Kidney Int 51 (1997) 1188—1195.

[42] F.A. Houssiau, C. Vasconcelos, D. D'Cruz, et al., Immunosuppressive therapy in lupus nephritis: the Euro-Lupus Nephritis Trial, a randomised trial of low dose versus high-dose intravenous cyclophosphamide, Arthritis Rheum 46 (2002) 2121—2131.

[43] F.A. Houssiau, C. Vasconcelos, D.P. CruzD', et al., Early response to immunosuppressive therapy predicts good renal outcome in lupus nephritis: lessons from long-term follow-up of patients in Euro-Lupus Nephritis Trial, Arthritis Rheum 50 (2004) 3934—3940.

[44] D.P.D. Cruz, F.A. Houssiau, The Euro-Lupus Nehpritis Trial: the development of the sequential treatment protocol, Lupus 118 (2009) 175—177.

[45] H.A. Austin, G.G. Illei, M.J. Braun, J.E. Balow, Randomized, controlled trial of prednisone, cyclophosphamide and cyclosporine in lupus membranous nephropathy, J Am Soc Nephrol 20 (4) (2009) 901—911.

[46] D.T. Boumpas, H. Yamada, N.J. Patronas, et al., Pulse cyclophosphamide for severe neuropsychiatric lupus, Q J Med 81 (296) (1991) 975—984.

[47] K. Takada, G.G. Illei, D.T. Boumpas, Cyclophosphamide for the treatment of systemic lupus erythematosus, Lupus 10 (3) (2001) 154—161.

[48] L. Barile-Fabris, R. Ariza-Andraca, L. Olguin-Ortega, et al., Controlled clinical trial of IV cyclophosphamide versus IV methylprednisolone in severe neurological manifestations in systemic lupus erythematosus, Ann Rheum Dis 64 (4) (2005) 620—625.

[49] G. Bertsias, J.P. Ioannidis, J. Boletis, et al., EULAR recommendations for the management of systemic lupus erythematosus. Report of a Task Force of the EULAR Standing Committee for the International Clinical Studies Including Therapeutics, Ann Rheum Dis 67 (2008) 195—205.

[50] M.L. Ritter, L. Pirofski, Mycophenolate mofetil: effects on cellular immune subsets, infectious complications and antimicrobial activity, Transpl Inf Dis 11 (2009) 290—297. Review.

[51] A.S. Appel, G.S. Appel, An update on the use of mycophenolate mofetil in lupus nephritis and other primary glomerular diseases, Nature Clin Pract Nephrol 5 (3) (2009) 132—142.

[52] W.T. Gibson, M. Reuben Heyden, Mycophenolate modfetil. Results of animal and human studies, Ann NY Acad Sci 1101 (2007) 209—221. Review.

[53] M.T. Anderka, A.E. Lin, D.N. Abuelo, A.A. Mitchell, S.A. Rassmussen, Reviewing the evidence of mycophenolate mofetil as a new teratogen: case report and review of the literature, Am J Med Gen 149 (A) (2009) 1241—1248.

[54] C.N. Pisoni, F.J. Sanchez, Y. Karim, et al., Mycophenolate mofetil treatment in systemic lupus erythemstosus, Arthritis Rheum 50 (2004) S415 (abstract).

[55] M.M. Riskalla, E.C. Somer, R.A. Fatica, W.J. McCune, Tolerability of mycophenolate mofetil in patients with systemic lupus erythematosus, J Rheumatol 30 (2003) 1508—1512.

[56] H.H. Hirsch, J. Steiger, B.K. Polyomavirus, Lancet Infect Dis 3 (2003) 611—633.

[57] R. San Juan, J.M. Aguado, C. Lambreras, et al., Impact of current transplantation management on the development of cytomegalovirus disease after renal transplantation, Clin Infect Dis 37 (2008) 875—882.

[58] M.H. Schiff, B. Leishman, Long term safety of CellCept (mycophenolate mofetil), a new therapy for rheumatoid arthritis patients, Br J Clin Pharmacol 41 (1996) 513—516.

[59] T.M. Chan, F.K. Li, C.S. Tang, et al., Efficacy of mycophenolate mofetil in patients with diffuse proliferative lupus nephritis.

Hong Kong–Guangzhou Nephrology Study Group, N Engl J Med 343 (2000) 1156–1162.

[60] T.M. Chan, W.S. Wong, C.S. Lau, et al., Long term study of mycophenolate mofetil as continuous induction and maintenance treatment for proliferative lupus nephritis, J Am Soc Nephrol 16 (2005) 1076–1084.

[61] E.M. Ginzler, C. Aranow, J. Buyon, et al., A multicenter study of mycophenolate mofetil (MMF) vs. intravenous cyclophosphamide (IVC) as induction therapy for severe lupus nephritis (LN): preliminary results, Arthritis Rheum 48 (9) (2003) 1690.

[62] G. Contreras, V. Pardo, B. Leclercq, et al., Sequential therapies for proliferative lupus nephritis, N Engl J Med 350 (10) (2004) 971–980.

[63] L.M. Ong, L.S. Hooi, T.O. Lim, et al., Randomized control trial of pulse intravenous cyclophosphamide versus mycophenolate mofetil in the induction therapy of proliferative lupus nephritis, Nephrol (Carlton) 10 (2005) 504–510.

[64] R.A. Moore, S. Derry, Systematic review and metaanalysis of randomized trials and control studies of mycophenolate mofetil in lupus nephritis, Arthritis Res Therapy 8 (2006) R 182.

[65] M. Walsh, M. James, D. Jayne, et al., Mycophenolate mofetil for induction therapy of lupus nephritis: a systematic review and metaanalysis, Clin J Am Soc Nephrol 2 (2007) 968–975.

[66] B. Zhu, N. Chen, Y. Lin, et al., Mycophenolate mofetil induction and maintenance therapy of lupus nephritis; a meta-analysis of randomized control trials, Nephrol Dial Transplant 22 (2007) 1933–1942.

[67] A. Sinclair, G. Appel, M.A. Dooley, et al., Mycophenolate mofetil as induction and maintenance therapy for lupus nephritis. Rationale and protocol for the randomized, controlled Asperva Lupus Management Study (ALMS), Lupus 16 (2007) 972–980.

[68] Ginzler E.M., Appel G.B., Dooley M.A., et al. Mycophenolate mofetil and intravenous cyclophosphamide in the Asperva Lupus Management Study (ALMS) (2007). Efficacy by racial group. American College of Rheumatology National Scientific Meeting. http://acr.confex.com.acr/2007/webprogram/Paper8393.html (website).

[69] G.B. Appel, G. Contreras, M.A. Dooley, et al., Mycophenolate mofetil versus cyclophosphamide for induction treatment of lupus nephritis, J Am Soc Nephrol 20 (5) (2009) 1103–1112.

[70] A. Mak, A.A. Cheak, J.Y. Tan, et al., Mycophenolate mofetil is as efficacious as but safer than, cyclophosphamide in the treatment of proliferative lupus nephritis: a meta-analysis and meta-regression, Rheumatology (Oxf) 48 (8) (2009) 944–952.

[71] P.P. Kapitsinou, N. Boletis, F.N. Skopouli, et al., Lupus nephritis: treatment with mycophenolate mofetil, Rheumatology (Oxf) 65 (2004) 2411–2415.

[72] D.N. Spetie, Y. Tang, B.H. Rovin, et al., Mycophenolate mofetil atherapy of SLE membranous nephropathy, Kindy Int 66 (2004) 2411–2415.

[73] M.Y. Karim, S.N. Pisoni, L. Ferro, et al., Reduction of proteinuria with mycophenolate mofetil in predominantly membranous lupus nephropathy, Rheumatology 44 (2005) 1317–1321.

[74] N. Kasitanon, M. Petri, M. Haas, et al., Mycophenolate mofetil as the primary treatment of membranous lupus nephritis with and without concurrent proliferative disease a retrospective study of 29 cases, Lupus 17 (2008) 40–45.

[75] J. Radhakrishnan, D.A. Moutzouris, E.M. Ginzler, et al., Mycophenolate mofetil and intravenous cyclophosphamide are similar as induction therapy for class V lupus nephritis, Kindy Int 77 (2) (2010) 152–160. Nov 4 [Epub ahead of print].

[76] M. Gaubitz, A. Schorat, H. Schotte, P. Kern, W. Domschke, Mycophenolate mofetil for treatment of systemic lupus erythematosus: an open pilot trial, Lupus 8 (1999) 731–737.

[77] S. Vasoo, J. Thumboo, K.Y. Fong, Refractory immune thrombocytopenia in systemic lupus erythematosus.: response to mycophenolate mofetil, Lupus 12 (2003) 630–632.

[78] M.A. Dooley, F.G. Cosio, P.H. Nachman, et al., Mycophenolate mofetil therapy in lupus nephritis: clinical observations, J Am Soc Nephrol 10 (4) (1999) 833–839.

[79] E.M. Ginzler, D. Wofsy, D. Isenberg, C. Gordon, L. Lisk, A. Dooley M-, Nonrenal disease activity following mycophenolate mofetil or intravenous cyclophosphamide as induction treatment for lupus nephritis. Findings in a multicenter, prospective, randomized, open-label, parallel-group clinical trial, Arthritis Rheum 62 (1) (2010) 211–221.

[80] J.D. Posalski, M. Ishimori, D.J. Wallace, et al., Does mycophenolate mofetil prevent extra-renal flares in systemic lupus erythamatosus? Results from an observational study of patients in a single practice treated for up to five years, Lupus 18 (6) (2009) 516–521.

[81] C. Campana, M.B. Regazzi, I. Buggia, M. Molinaro, Clinically significant drug interactions with cyclosporin: an update, Clin Pharmacokinet 30 (1996) 141–179.

[82] F. Rodriguez, J.C. Krayenbuhl, W.B. Harrison, et al., Renal biopsy findings and follow up of renal function in rheumatoid arthritis patients treated with cyclosporine A: an update from the international Kidney Biopsy Registry, Arthritis Rheum 39 (1996) 1491–1498.

[83] J. Radhakrishnan, C.L. Kunis, V. D'Agati, et al., Ciclosporine treatment of lupus membranous nephropathy, Clin Nephrol 42 (3) (1994) 147–154.

[84] D. Hallegua, D.J. Wllace, A.L. Metjer, et al., Cyclosporine for membranous lupus nephritis: experience with ten patients and review of the literature, Lupus 9 (4) (2000) 241–251.

[85] W. Hu, Z. Liu, S. Shen, et al., Cyclosporine A in the treatment of membranous lupus nephropathy, Chin Med (Engl) 116 (12) (2003) 1827–1830.

[86] C.C. Mok, K.H. Tong, C.H. To, et al., Risk and predictors of arterial thrombosis in lupus and non-lupus primary glomerulonephritis. A comparative study, Medicine (Baltimore). 86 (4) (2007) 203–209.

[87] B. Griffiths, P. Emery, The treatment of lupus with cyclosporine A, Lupus 10 (3) (2001) 165–170.

[88] I.S. Nasr, Topical tacrolimus in dermatology, Clin Exp Dermatol 25 (3) (2000) 250–254.

[89] C.H. Smith, New approaches to topical therapy, Clin Exp Dermatol 25 (7) (2000) 567–574.

[90] T. Assmann, B. Homey, T. Ruzicka, Topical tacrolimus for the treatment of inflammatory skin diseases, Expert Opin Pharmacother 2 (7) (2001) 1167–1175.

[91] F. Furukawa, S. Imamura, M. Takigawa, FK506: therapeutic effects on lupus dermatoses in autoimmune-prone MRL/Mp-lpr/lpr mice, Arch Dermatol Res 287 (6) (1995) 558–563.

[92] T. Yoshimasu, T. Ohtani, T. Sakamoto, et al., Topical FK506 (tacrolimus) therapy for facial erythematous lesions of cutaneous lupus erythematosus and dermatomyositis, Eur J Dermatol 12 (1) (2002) 50–52.

[93] C.C. Mok, K.H. Tong, C.H. To, et al., Tacrolimus for induction therapy for diffuse proliferative lupus nephritis. An open label pilot study, Kidn Int 68 (2005) 813–817.

[94] K.C. Tse, M.F. Lam, S.C. Tang, et al., A pilot study on tacrolimus treatment in membranous or quiscient lupus nephritis with proteinuria resistant to angiotensin inhibition or blockade, Lupus 16 (2007) 46–51.

[95] C.C. Szeto, B.C. Kwan, F.M. Lai, et al., Tacrolimus for the treatment of systemic lupus erythematosus with pure class V nephritis, Rheumatol (Oxf) 47 (11) (2008) 1678–1681.

[96] H. Bao, Z.H. Liu, H.L. Xie, et al., Succesful treatment of class V+IV lupus nephritis with multitarget therapy, J Am Soc Nephrol 19 (2008) 2001–2010.

[97] Comparing the Efficacy of Tacrolimus and Mycophenolate Mofetil for the Initial Therapy of Active Lupus Nephritis. www.clinicaltrials.gov/ct2/results?term=NCT00371319 (website).

[98] F. Medina, J. Fuentes, I. Carranza, et al., Sirolimus: a potential treatment of diffuse proliferative lupus nephritis, Ann Rheum Dis 65 (Suppl 2) (2006) s351.

[99] B. Sayin, H. Krakayali, T. Kolak, et al., Conversion to sirolimus for chronic allograft nephropathy and calcineurin inhibitor toxicity and the adverse effects of sirolimus after conversion, Transplant Proc 41 (7) (2009) 2789–2793.

[100] J.R. Carneiro, E.I. Sato, Double blind, randomized, placebo control trial of methotrexate in systemic lupus erythematosus, J Rheumatol 26 (1999) 1275–1279.

[101] G.H. Thoenes, T. Sitter, K.H. Langer, et al., Leflunomide (HWA 486) inhibits experimental autoimmune tubulointerstitial nephritis in rats, Int J Immunopharmacol 11 (8) (1989) 921–929.

[102] R.R. Bartlett, S. Popovic, R.X. Raiss, Development of autoimmunity in MRL/lpr mice and the effects of drugs on this murine disease, Scand J Rheumatol Suppl. 75 (1988) 290–299.

[103] H.Y. Wang, T.G. Cui, F.F. Hou, et al., Induction treatment of proliferative lupus nephritis with leflunamide combined with prednisone: a prospective multicentre observational study, Lupus 17 (2008) 638.

[104] F.S. Zhang, Y.K. Nie, X.M. Jin, et al., The efficacy and safety of leflunamide therapy in lupus nephritis by repeat kidney biopsy, Rheumatol 29 (11) (2009) 1331–1335.

[105] J.B. Ellman, Q.E. Whiting-O'Keefe, W.V. Epstein, Chlorambucil (CAB) treatment in systemic lupus erythematosus (SLE), Arthritis Rheum 23 (1980) S70.

[106] G. Moroni, M. Maccario, G. Banfi, et al., Treatment of membranous lupus nephritis, Am J Kidn Dis 31 (1998) 681–686.

[107] M.H. Cardiel, J.A. Tumlin, R.A. Furie, et al., Abetimus sodium for renal flare in systemic lupus erythematosus: results of a randomized controlled phase III trial, Arthritis Rheum 58 (2008) 2470–2480.

[108] E.L. Bezera, M.J. Villa, P.B. da Trindade Neto, et al., Double blind, randomized controlled clinical trial of clofazimine compared with chloroquine in patients with systemic lupus erythematosus, Arthritis Rheum 52 (10) (2005) 3078–3083.

[109] S.V. Kaveri, S. Lacroix-Desmazes, J. Bayry, The anti-inflammatory IgG, N Engl J Med 359 (2008) 307–309.

[110] G. Zandman-Goddard, M. Blank, E. Shoenfeld, Intravenous immunoglobulins in systemic lupus erythematosus: from the bench to the bedside, Lupus 18 (2009) 884–888. Review.

[111] J.M. Esdaile, L. Joseph, T. MacKenzie, M. Kashgarian, J.P. Hayslett, The benefit of early treatment with immunosuppressive drugs in lupus nephritis, J Rheumatol 22 (1995). 1211-1211.

[112] W.J. McCune, Mycophenolate mofetil for lupus nephritis, N Engl J Med 353 (21) (2005) 2282–2284.

[113] J.T. Merrill, C.M. Neuwelt, D.J. Wallace, et al., Efficacy and safety of rituximab in moderately-to-severely active systemic lupus erythematosus. The randomized, double-blind, phase II/III systemic lupus erythematosus evaluation of rituximab trial, Arthritis Rheum 62 (1) (2010) 222–233.

[114] T.I. Lu, K.P. Ng, G. Cambridge, et al., A retrospective seven year analysis of the use of B cell depletion therapy in systemic lupus erythematosus at university college London Hospital, Arthritis Rheum 61 (2009) 482–487.

[115] K.R. Carson, A.M. Evens, E.A. Richey, et al., Progressive multifocal encephalopathy in HIV-negative patients: a report of 57 cases from the research on adverse drug events and reports project, Blood 113 (2009) 4834–4840.

[116] R. Cervera, M.A. Khamashta, J. Font, et al., Morbidity and mortality in systemic lupus erythematosus during a 10-year period: a comparison of early and late manifestations in a cohort of 1,000 patients, Medicine (Baltimore) 82 (5) (2003) 299–308.

[117] I. Moyssakis, M.G. Tektonidou, V.A. Vassiliou, et al., Libman-Sacks endocarditis in systemic lupus erythematosus: prevalence, associations and evolution, Am J Med 120 (7) (2007) 636–642.

[118] G. Habib, B. Hoen, P. Tornos, et al., The Task Force on the prevention, diagnosis and treatment of infective endocarditis of the European Society of Cardiology (ESC). Guidelines on the prevention, diagnosis and treatment of infective endocarditis, Eur Heart J 30 (19) (2009) 2369–2413.

[119] W. Wilson, K.A. Taubert, M. Gewitz, et al., Prevention of infective endocarditis. Guidelines from the American Heart Association. A guideline from the American Heart Association, rheumatic fever, endocarditis and Kawasaki disease Committee, Council of cardiovascular disease in the young, and the Council of clinical cardiology, Council on cardiovascular surgery and anesthesia, and the Quality of care and outcomes research interdisciplinary work group, Circulation 116 (2007) 1736–1754.

[120] W.R. Gilliland, G.C. Tsokos, Prophylactic use of antibiotics and immunisations in patients with SLE, Ann Rheum Dis 61 (2002) 191–192.

[121] J.G. Erdozain, G. Ruiz-Irastorza, M.V. Egurbide, et al., High risk of tuberculosis in systemic lupus erythematosus, Lupus 15 (2006) 232–235.

[122] A.D. Chu, A.H. Polesky, G. Bhatia, et al., Active and latent TB infection in patients with systemic lupus erythematosus living in the United States, J Clin Rheumatol 15 (5) (2009) 226–229.

[123] M. Mosca, C. Tani, M. Aringer, et al., European League Against Rheumatism recommendations for monitoring systemic lupus erythematosus patients in clinical practice and in observational studies, Ann Rheum Dis 69 (7) (2010) 1269–1274. Epub ahead of print.

[124] D. Gupta, A. Zachariah, H. Roppelt, et al., Prophylactic antibiotic usage for Pneumocystis jirovecii pneumonia in patients with systemic lupus erythematosus on cyclophosphamide. A survey of US rheumatologists and the review of the literature, J Clin Rheumatol 14 (2008) 267–272.

[125] M. Ramos-Casals, M.J. Cuadrado, P. Alba, G. Sanna, P. Brito-Zerón, L. Bertolaccini, et al., Acute viral infections in patients with systemic lupus erythematosus: description of 23 cases and review of the literature, Medicine 87 (2008) 311–318.

[126] T. Glück, B. Kiefmann, M. Grohmann, W. Falk, R.H. Straub, J. Schölmerich, Immune status and risk for infection in patients receiving chronic immunosuppressive therapy, J Rheumatol 32 (2005) 1473–1480.

[127] M. Abu-Shakra, Safety of vaccination of patients with systemic lupus erythematosus, Lupus 18 (2009) 1205–1208.

[128] K.A.M. Kuruma, E.F. Borba, M.H. Lopes, J.F. De Carvalho, E. Bonfa, Safety and efficacy of hepatitis B vaccine in systemic lupus erythematosus, Lupus 16 (2007) 350–354.

[129] A.F. Shorr, G.M. Susla, N.P. O'Grady, Pulmonary infiltrates in the non-HIV-infected immunocompromised patient: Etiologies, diagnostic strategies, and outcomes Chest 125 (2004) 260–271.

[130] S. Bernatsky, J.F. Boivin, L. Joseph, R. Rajan, A. Zoma, S. Manzi, et al., An international cohort study of cancer in systemic lupus erythematosus, Arthritis Rheum 52 (2005) 1481–1490.

[131] S. Bernatsky, A. Clarke, R. Ramsey-Goldman, L. Joseph, J.F. Boivin, R. Rajan, et al., Hormonal exposures and breast cancer in a sample of women with systemic lupus erythematosus, Rheumatology 43 (2004) 1178–1181.

IV. TREATMENT

TABLE 59.1   Biologics in the treatment of SLE

Target	Treatment	Mode of action	Clinical Trials
BLyS APRIL	Belimumab (monoclonal Ab) Atacicept (TACI-Ig)BR3-Ig	Blocks BLyS (and/or APRIL) effect on B cells	Efficacious in animal models.Positive large trial, albeit with moderate effects in patients with SLE (phase III)
Interleukin-6 receptor (IL-6R)	Tocilizumab (Monoclonal Ab)	Blocks IL-6 effect on lymphocytes such as immunoglobulin production	Succesful in mice. Phase I in patients
Interferon (IFN)	Rontalizumab (monoclonal Ab) Sifalizumab (monoclonal Ab)	Reverse interferon signature	Phase I trial demonstrated safety. Phase II dose-escalating trials in moderate to severe SLE are underway
TNF-α	Etanercept (Soluble receptor) Infliximab (Chimeric Ab)	Block TNF-α mediated signaling	Trials in nephritis underway. Mixed results in case series
C5	Eculizumab (Monoclonal Ab)	Blocks MAC formation	Successful in mice. Effective and safe in patients with PNH
CD20	Rituximab (Chimeric Ab) Ocrelizumab (Humanized Ab) TRU-015 (small polypeptide)	B cell depletion	Phase II/III trial in patients with SLE is ongoing
CD22	Epratuzumab (Humanized Ab)	Modulation of B cell signaling	Safe in phase I in SLE patients. Phase II is ongoing
CD11a	Efalizumab	Blocks LFA-1:ICAM interaction	Case reports showed usefulness in cutaneous lupus. Withdrawn from the market
dsDNA B cell receptor	Abetimus	Blocks the production of anti-dsDNA antibodies	Some effect on quality of life; better effect on patients with high levels of anti-dsDNA antibodies
CD28:CD80/86	Abatacept (CTLA-4-Ig)	Blocks T cell: Antigen-presenting cell interaction	Proven effectiveness in mice (with and without cyclophosphamide). Phase II/III in SLE patients
CD40L	Monoclonal antibodies	Blocks T−B-cell cross-talk	Effective in mice and patients with SLE. Significant side-effects lead to discontinuation of the trials
Syk	Fostamatinib (small molecule)	Blocks Fc receptor and T cell receptor mediated signal transduction	Successful in mice. Proven efficacious in rheumatoid arthritis

Indeed, mice transgenic for BLyS [13] developed B-cell hyperplasia resulting in hypergammaglobulinemia, anti-dsDNA antibody production and immunoglobulin deposition in the kidneys. Furthermore, lupus-prone mice (both NZB/NZW F1 and MRL-*lpr/lpr*) were found to have elevated BLyS levels at the onset of disease [14]. These observations led to the design of a fusion molecule of the BLyS receptor TACI with the Fc portion of the immunoglobulin molecule (TACI-Ig) that could bind and neutralize circulating BLyS. Consistent with previous observations, TACI-Ig was successful in ameliorating disease in the NZB/NZW F1 lupus model [14] as well as in a rheumatoid arthritis murine model [15]. Studies using MRL-*lpr/lpr* mice transfected with an adenovirus encoding TACI-Ig showed similar results with decreased proteinuria and improved survival of the TACI-Ig expressing lupus-prone mice when compared to control-treated mice [16].

Patients with SLE were also found to have elevated levels of BLyS; 25% of SLE had persistent increase and another 25% had intermittent increase of this cytokine in their peripheral blood [17, 18]. BLyS levels correlated with anti-dsDNA levels [19, 20] further suggesting a role of BLyS in the generation or the maintenance of autoreactive B-cell clones in SLE.

Inhibition of BLyS using a monoclonal antibody (belimumab) is currently tested in the largest phase III clinical trial ever done with SLE patients. A total of 865 patients with mild to moderate SLE have been randomized to receive monthly infusions of belimumab 10 mg/kg, belimumab 1 mg/kg or placebo with the primary end point being reduction in SLEDAI by at least 4 points at 52 weeks and no worsening of the BILAG and the physician's global assessment indices. Preliminary analysis showed that at 52 weeks 57.6% of patients with SLE receiving belimumab at the higher dose reached the primary study endpoint as compared to 43.6% of SLE patients receiving placebo [21].

At the time of writing, atacicept, a TACI-Ig fusion molecule, is being tried in phase II/III trials in patients with SLE and other autoimmune diseases. The advantage of atacicept is that theoretically it can block the effect of both BLyS and APRIL thus influencing early as well as mature B cells. An initial phase Ib trial in SLE showed favorable tolerability profile and led to a decrease in immunoglobulin, especially IgM production [22]. Following that, a phase II/III trial in lupus nephritis of atacicept in combination with mycophenolate mofetil and corticosteroids was halted because of high infection rates possibly due to concomitant use of multiple immunosuppressive drugs. Another phase II/III trial in SLE without severe renal involvement is still underway. This 1-year trial will assess the efficacy of biweekly (for 4 weeks) and then weekly atacicept 150 mg vs. atacicept 75 mg vs. placebo in the prevention of SLE flares. A fusion molecule of BR3 (BAFFR) and Ig is also currently at the early stages of development for use in patients. It has to be stated though that current data from clinical trials do not support the use of BLyS inhibitors as sole treatment especially for severe SLE manifestations such as nephritis or CNS involvement. Nevertheless, BLyS inhibitors may prove instrumental in the maintenance of long-term remission by blocking the maturation of autoreactive B cells especially at the early stages of their development.

Interleukin-6 (IL-6)

Monocytes secrete IL-6 [23] that subsequently binds to the IL-6 receptor (IL-6R) either on the surface of the cells or to its soluble form. The IL-6:IL-6R complex can directly affect both B- and T-cells by promoting their differentiation and maturation. IL-6 synergizes with other cytokines such as IL-1 or TNF-α to promote inflammatory responses [24]. It can, together with TGF-β, promote the differentiation of the newly recognized Th17 pro-inflammatory T-cell subset, which may be involved in the pathophysiology of lupus nephritis [25, 26].

IL-6 was shown to be elevated in murine lupus models and the inhibition of its action by blocking either the cytokine itself or its receptor resulted in clinical improvement of the disease (decreased anti-dsDNA antibodies and proteinuria, and improved survival) [27, 28]. Patients with active SLE have increased IL-6 levels in the serum and urine [29]. In addition, IL-6 is expressed in the glomeruli of patients with lupus nephritis [30]. *In vitro* experiments showed that IL-6 promotes immunoglobulin production by SLE B-cells and IL-6 neutralization decreases it [31].

A preliminary open-label phase I trial of the humanized monoclonal anti-IL6 receptor antibody tocilizumab, already proven efficacious for rheumatoid arthritis, showed a favorable profile in SLE patients [32]. Tocilizumab led to a decrease in acute-phase reactants and a small decrease in immunoglobulin and anti-dsDNA levels; two patients though developed severe neutropenia. Further studies are needed to assess its clinical effectiveness and address the issue of toxicity.

Interferon (IFN)

Once activated, T-cells secrete IFN-γ, a cytokine regarded as the signature Th1 cytokine. IFN-γ plays an important role in the pathophysiology of murine lupus [33] as exhibited by alleviation of the disease once it is blocked. Blocking IFN-γ either before disease onset or even after the establishment of lupus nephritis led to a significant decrease in autoantibody production and improved survival of MRL-*lpr/lpr* mice [34, 35].

Similarly, the interferon pathway has been shown to be highly active in patients with SLE. Peripheral blood mononuclear cells (PBMC) from patients with SLE show a characteristic genetic pattern with several genes in the interferon pathway being dysregulated [36]. This interferon signature was most prominent among patients with severe disease [36, 37]. Additionally, plasma from patients with SLE can induce these interferon-responsive genes in normal PBMCs. However, contrary to animals, the type I interferons (alpha and beta) are the instigators of this characteristic interferon genetic signature in SLE. IFN-α may be produced by plasmacytoid dendritic cells in response to immune complexes and apoptotic material [38]. IFN-α in turn, further activates dendritic cells [39] as well as T- and B-cells, thus leading to worsening immune dysfunction in SLE [40]. Rontalizumab (rhuMab IFNAlpha), a humanized monoclonal IgG1 antibody, was shown to be safe and successful in turning off interferon-inducible genes in a phase I trial in SLE patients. A phase II trial of rontalizumab in SLE patients is underway. Sifalimumab, a fully human monoclonal antibody is also going to be tested in a phase II study of moderately to severely active SLE patients who are not responding to conventional treatment.

Tumor Necrosis Factor (TNF)-α

A mainstay for the treatment of various forms of autoimmune arthritis [41], TNF-alpha inhibitors have not proven helpful in the treatment of SLE. Indeed, several patients who did not have SLE and were treated with these medications developed antinuclear antibodies and anti-dsDNA antibodies; clinical lupus was observed in a minority of these patients. TNF-alpha inhibitors though may be helpful in some cases of SLE. For example, UVB light which plays a role in the development of cutaneous disease, turns on TNF-α production in the skin [42]. Interestingly, TNF-α is produced in the skin of patients with refractory subacute cutaneous lupus [43]. There have been some case reports of severe lupus skin manifestations responding to TNF-α inhibition and a phase II trial of Etanercept in discoid lupus is under way. On the other hand, the NZB/NZW F1 lupus model developed worsening nephritis with the use of TNF-α inhibitor. Additionally, TNF receptor-deficient NZM2328 mice had worse disease with increased numbers of Th17 cells in their circulation [44]. It seems therefore that abrogation of TNF-α-mediated signaling results in worsening kidney disease in the mice with lupus. Nevertheless, case reports and case series showed somewhat positive effect of TNF-α inhibition in lupus nephritis while concomitant use of other immunosuppressives may abrogate the appearance of increased titers of autoantibodies [45, 46]. A trial of etanercept in lupus nephritis is underway.

COMPLEMENT SYSTEM

SLE nephritis flares are characterized by activation of the complement, eventually leading to the production of potent pro-inflammatory C5 split products C5a and C5b and the assembly of the membrane attack complex C5b-9 [47, 48], that attract inflammatory cells and directly damage the tissues. Lupus-prone mice were treated with a monoclonal antibody that prevents the cleavage of C5 [49]. The treated mice had prolonged survival and decreased proteinuria as compared to control-treated mice. In humans, eculizumab, an anti-C5 antibody has been shown to be efficacious in paroxysmal nocturnal hemoglobinuria [50] and is currently evaluated for a variety of immunologic conditions such as hemolytic uremic syndrome. Eculizumab may prove efficacious in SLE and especially lupus nephritis as it may prevent end-organ damage by blocking activation of complement.

CELL SURFACE MOLECULES

CD20

SLE B cells produce an array of auto-antibodies and pro-inflammatory cytokines while at the same time playing the role of antigen-presenting cells [51]. In 2002, a chimeric anti-human CD20 antibody (Rituximab) has been used successfully for the treatment of lymphoma [52] and rheumatoid arthritis [53]. This molecule depletes B cells, although its effect on immunoglobulin production may be limited, as it does not affect plasma cells. It has shown promise in small studies and case series in SLE, with as many as 90% of patients achieving partial or total remission [54—61]. These studies also showed that a rather high percentage of patients with SLE developed human anti-chimeric antibodies (HACA) [59] as compared to lymphoma and rheumatoid arthritis patients; this phenomenon may affect the efficacy of the drug or lead to infusion reactions. A trial of rituximab with mycophenolate mofetil in SLE nephritis failed to reach its primary endpoint at remission at 52 weeks of treatment. Another phase II/III trial of rituximab in moderate to severe non-renal SLE is currently underway.

In addition to rituximab, a trial of ocrelizumab, another anti-CD20 antibody, is underway. Ocrelizumab is a humanized (90%) anti-CD20 antibody that is evaluated for efficacy in phase III trials in both renal (BELONG) and non-renal (BEGIN) SLE. Finally, TRU-015, a small modular immunopharmaceutical (SMIP) compound was tried in a phase I trial in patients with lupus nephritis. Already showing promise in rheumatoid arthritis [62], this polypeptide can bind to CD20 and deplete B cells similarly to the larger anti-CD20 antibodies.

Yet, despite the design and marketing of several new compounds, the exact role of these B-cell-depleting molecules in SLE remains to be determined.

CD22

CD22, similarly to CD20, is expressed by mature B-cells, and not by antigen-forming cells or memory B-cells. It binds to its ligand, CD22L, and influences the transduction of signal through the B-cell receptor. Interestingly, CD22L expression on circulating B- and T-cells from lupus-prone mice increases as the mice age and develop worsening disease [63].

A humanized anti-CD22 antibody (epratuzumab) is currently undergoing a phase II trial in patients with moderately active SLE after an initial phase I trial that showed a favorable safety profile with minimal immunogenicity of the compound and only moderate decrease in the number of circulating B cells. Furthermore, the fact that epratuzumab is fully humanized and causes moderate B-cell depletion may make this a medication of choice for patients that are intolerant to rituximab.

CD11a

Efalizumab, a monoclonal humanized IgG1 antibody that specifically targets the alpha subunit (CD11a) of Lymphocyte Function associated Antigen-1 (LFA-1) has been approved for use in psoriasis [64]. LFA-1 expressed primarily on T cells, binds to ICAM-1 on antigen-presenting cells stabilizing the immune synapse. Efalizumab prevents this interaction and has been found to be helpful in cases of recalcitrant cutaneous lupus [65]. However, reports of progressive multifocal leucoencephalopathy resulted in the withdrawal of this medicine [66].

B-cell Receptor

SLE patients, especially patients with active nephritis have anti-dsDNA antibodies in their sera [67]. These antibodies may play a role in the pathophysiology of SLE by binding to the basement membrane of the glomeruli.

In an attempt to block the production of anti-dsDNA antibodies, the small artificial molecule LJP-394 (abetimus sodium) was designed. LJP-394 is made of four deoxynucleotide-like molecules in a tetrameric form, mimicking the dsDNA-containing immune complexes. When injected in mice with lupus, LJP-394 resulted in decreased anti-dsDNA antibody levels. Several trials in patients with SLE showed mixed results [68—70]. LJP-394 led to the decrease of anti-dsDNA antibody titers as predicted. SLE patients who took LJP-394 had improved quality of life but there were no significant differences between and LJP-394 and placebo in disease outcome. These trials have shown that LJP-394 may be somewhat useful in a subset of patients in whom anti-dsDNA antibodies are elevated.

CO-STIMULATORY MOLECULES

SLE is characterized by an aberrant function of all components of both the innate immune and the adaptive immune system. Importantly, T-cells seem to be in a state of pre-activation with aggregated lipid rafts and rapid mobilization of calcium and tyrosine phosphorylation once their receptor engages its ligand [71—73]. These T-cells provide help to B-cells and, by acquiring CD44 [74], they invade tissues and locally produce pro-inflammatory cytokines.

Blocking T-cell—antigen-presenting cell interaction may result in an increased threshold for T-cell activation. This can be achieved by inhibiting the binding of the CD28 co-stimulatory molecule on T-cells to CD80/86 on the antigen-presenting cells. To this end, a fusion molecule of CTLA4 with immunoglobulin, abatacept (CTLA-4-Ig), was constructed and proven to be effective in clinical trials of patients with rheumatoid arthritis [75]. CTLA-4 is a molecule that is expressed on the surface of T-cells following stimulation and is 100 times more potent as a ligand for CD80/86 than CD28. Its physiologic role is to end the stimulation of T-cells by out-competing the CD28 for binding to CD80/86. The fusion molecule CTLA-4-Ig when injected in lupus-prone mice resulted in clinical improvement alone [76] or especially when combined with cyclophosphamide [77]. Currently, abatacept with background treatment with mycophenolate mofetil is in phase II/III clinical trial in SLE patients with nephritis. Another phase II trial aims at answering whether adding abatacept on a cyclophosphamide/azathioprine regimen in patients with lupus nephritis would result in better renal outcomes.

Another co-stimulatory pair of molecules closely related to CD28 is the inducible co-stimulator (ICOS). ICOS is expressed on activated T-cells and interacts with its ligand B7-related peptide-1 (B7RP-1); this interaction is of particular importance for the T-cell-instructed B-cell class switching and IgG production. Blocking of B7RP-1 resulted in amelioration of the disease manifestations in animal models of lupus nephritis and inflammatory arthritis [78—80]. This effect seemed to result from inhibition of the development of T follicular helper cells that provide guidance to B-cells in the germinal centers. A fully human anti-B7RP-1 antibody (AMG557) is currently in a phase I clinical trial in SLE patients.

Finally, another co-stimulatory molecule that has been targeted in SLE is the CD40-CD40 ligand (designated CD154 or CD40L) pair. CD40L is expressed shortly after activation on the surface of T-cells and through engagement with the B-cell residing CD40 molecule, leads to a more robust activation of the B-cell. T-cells from patients with SLE show persistent upregulation of CD40L upon stimulation, a phenomenon that partly explains B-cell overactivity and production of auto-antibodies. Mice with lupus nephritis that were treated with anti-CD40L antibody had decreased severity of disease and increased survival when compared to placebo-treated mice [81–83].

Clinical trials were conducted with two different [84] anti-CD40L antibodies (BG9588 and IDEC-1). BG9588 [85] but not IDEC-1 [86] showed positive effect with serological improvement of disease activity (decrease in anti-dsDNA titers, increase in C3 concentration, decrease in hematuria). The trials with both antibodies resulted in thromboembolic events [84], deemed to be caused by engagement of the CD40L molecule expressed on platelets by these antibodies. Currently the clinical trials with these antibodies are suspended.

INTRACELLULAR TARGETS

Syk

Syk kinase associates with FcεRγ receptor found on monocyte lineage cells. In 2003, it was shown that FcεRγ substitutes the TCR ζ chain in SLE T-cells and associates with the TCR/CD3 complex [87]. This substitution results in the recruitment of Syk in the TCR/CD3 complex in SLE T-cells. Since Syk is a more potent signal transducer than the normally found Zap70, its recruitment to the TCR/CD3 complex explains the very robust intracellular calcium mobilization observed once the T cells get activated.

A new anti-Syk molecule, R406 and its orally available pro-drug R788 was developed and was shown to be effective in treating established nephritis in NZB/NZW F1 mice [88]. Additionally *in vitro* work showed that R406 treatment of SLE T-cells resulted in reduction of the calcium influx [89]. Although clinical trials in SLE are not currently under way, R788 (fostamatinib) has been successful in the treatment of patients with rheumatoid arthritis [90]. Based on current preclinical data, fostamatinib may prove to be useful in preventing the abnormal activation of T-cells and monocyte lineage cells in SLE patients.

CONCLUSION

Presently, non-specific immunosuppression remains the treatment of choice for the induction of remission in SLE while antimalarials and low-dose immunosuppressives are used to maintain remission. New medications are already in the pipeline that will give options for the treatment of SLE even at its most severe forms. One can speculate that targeting B-cells, T–B-cell interaction and/or co-stimulatory pathways will prove efficacious in the induction of remission in SLE. At the same time targeting BLyS-mediated B-cell maturation or increasing the threshold for T-cell and monocyte

activation may help maintain remission. Finally, certain manifestations of SLE such as refractory skin disease and nephritis may be successfully treated with medications such as TNA-α antagonists and anti-IL-6R antibody respectively.

In conclusion, it is apparent that although no new drugs have been approved for SLE for over three decades, uncovering pathophysiologic pathways and designing novel approaches to correct them will eventually lead to the development of safe and effective medications that will revolutionize the treatment of SLE.

References

[1] K. Takada, G.G. Illei, D.T. Boumpas, Cyclophosphamide for the treatment of systemic lupus erythematosus, Lupus 10 (2001) 154–161.

[2] M.F. Gourley, H.A. Austin 3rd, D. Scott, C.H. Yarboro, E.M. Vaughan, J. Muir, et al., Methylprednisolone and cyclophosphamide, alone or in combination, in patients with lupus nephritis. A randomized, controlled trial, Ann Intern Med 125 (1996) 549–557.

[3] J.E. Balow, H.A. Austin 3rd, Progress in the treatment of proliferative lupus nephritis, Curr Opin Nephrol Hypertens 9 (2000) 107–115.

[4] A. Ippolito, M. Petri, An update on mortality in systemic lupus erythematosus, Clin Exp Rheumatol 26 (2008) S72–S79.

[5] P.E. Lipsky, D.M. van der Heijde, E.W. St Clair, D.E. Furst, F.C. Breedveld, J.R. Kalden, et al., Infliximab and methotrexate in the treatment of rheumatoid arthritis. Anti-Tumor Necrosis Factor Trial in Rheumatoid Arthritis with Concomitant Therapy Study Group, N Engl J Med 343 (2000) 1594–1602.

[6] W.J. Sandborn, J.F. Colombel, R. Enns, B.G. Feagan, S.B. Hanauer, I.C. Lawrance, et al., Natalizumab induction and maintenance therapy for Crohn's disease, N Engl J Med 353 (2005) 1912–1925.

[7] J.D. Gorman, K.E. Sack, J.C. Davis Jr., Treatment of ankylosing spondylitis by inhibition of tumor necrosis factor alpha, N Engl J Med 346 (2002) 1349–1356.

[8] P.A. Moore, O. Belvedere, A. Orr, K. Pieri, D.W. LaFleur, P. Feng, et al., BLyS: member of the tumor necrosis factor family and B lymphocyte stimulator, Science 285 (1999) 260–263.

[9] M.P. Cancro, D.P. D'Cruz, M.A. Khamashta, The role of B lymphocyte stimulator (BLyS) in systemic lupus erythematosus, J Clin Invest 119 (2009) 1066–1073.

[10] Y. Liu, X. Hong, J. Kappler, L. Jiang, R. Zhang, L. Xu, et al., Ligand-receptor binding revealed by the TNF family member TALL-1, Nature 423 (2003) 49–56.

[11] R.K. Do, E. Hatada, H. Lee, M.R. Tourigny, D. Hilbert, S. Chen-Kiang, Attenuation of apoptosis underlies B lymphocyte stimulator enhancement of humoral immune response, J Exp Med 192 (2000) 953–964.

[12] B.L. Hsu, S.M. Harless, R.C. Lindsley, D.M. Hilbert, M.P. Cancro, Cutting edge: BLyS enables survival of transitional and mature B cells through distinct mediators, J Immunol 168 (2002) 5993–5996.

[13] S.D. Khare, I. Sarosi, X.-Z. Xia, S. McCabe, K. Miner, I. Solovyev, et al., Severe B cell hyperplasia and autoimmune disease in TALL-1 transgenic mice, PNAS 97 (2000) 3370–3375.

[14] J.A. Gross, J. Johnston, S. Mudri, R. Enselman, S.R. Dillon, K. Madden, et al., TACI and BCMA are receptors for a TNF homologue implicated in B-cell autoimmune disease, Nature 404 (2000) 995–999.

[15] J.A. Gross, S.R. Dillon, S. Mudri, J. Johnston, A. Littau, R. Roque, et al., TACI-Ig neutralizes molecules critical for B cell development and autoimmune disease. impaired B cell maturation in mice lacking BLyS, Immunity 15 (2001) 289–302.

[16] W. Liu, A. Szalai, L. Zhao, D. Liu, F. Martin, R.P. Kimberly, et al., Control of spontaneous B lymphocyte autoimmunity with adenovirus-encoded soluble TACI, Arthritis Rheum 50 (2004) 1884–1896.

[17] W. Stohl, Targeting B lymphocyte stimulator in systemic lupus erythematosus and other autoimmune rheumatic disorders, Expert Opin Ther Targets 8 (2004) 177–189.

[18] W. Stohl, S. Metyas, S.M. Tan, G.S. Cheema, B. Oamar, D. Xu, et al., B lymphocyte stimulator overexpression in patients with systemic lupus erythematosus: longitudinal observations, Arthritis Rheum 48 (2003) 3475–3486.

[19] G.S. Cheema, V. Roschke, D.M. Hilbert, W. Stohl, Elevated serum B lymphocyte stimulator levels in patients with systemic immune-based rheumatic diseases, Arthritis Rheum 44 (2001) 1313–1319.

[20] J. Zhang, V. Roschke, K.P. Baker, Z. Wang, G.S. Alarcon, B.J. Fessler, et al., Cutting edge: a role for B lymphocyte stimulator in systemic lupus erythematosus, J Immunol 166 (2001) 6–10.

[21] S. Navarra, R. Guzman, A. Gallacher, R.A. Levy, E.K. Li, M. Thomas, et al. (October 2009). Belimumab, a BLyS-specific inhibitor, reduced disease activity, flares and prednisone use in patients with active SLE: efficacy and safety results from the Phase 3 BLISS-52 study. In 73rd Annual Scientific Meeting of the American College of Rheumatology, Philadelphia.

[22] I. Nestorov, O. Papasouliotis, C. Pena Rossi, A. Munafo, Pharmacokinetics and immunoglobulin response of subcutaneous and intravenous atacicept in patients with systemic lupus erythematosus, J Pharm Sci (2009).

[23] T. Naka, N. Nishimoto, T. Kishimoto, The paradigm of IL-6: from basic science to medicine. Arthritis Res, 4 Suppl 3 (2002) S233–S242.

[24] E. Tackey, P.E. Lipsky, G.G. Illei, Rationale for interleukin-6 blockade in systemic lupus erythematosus, Lupus 13 (2004) 339–343.

[25] J.C. Crispin, M. Oukka, G. Bayliss, R.A. Cohen, C.A. Van Beek, I.E. Stillman, et al., Expanded double negative T cells in patients with systemic lupus erythematosus produce IL-17 and infiltrate the kidneys, J Immunol 181 (2008) 8761–8766.

[26] Z. Zhang, V.C. Kyttaris, G.C. Tsokos, The role of IL-23/IL-17 axis in lupus nephritis, J Immunol 183 (2009) 3160–3169.

[27] B.K. Finck, B. Chan, D. Wofsy, Interleukin 6 promotes murine lupus in NZB/NZW F1 mice, J Clin Invest 94 (1994) 585–591.

[28] M. Mihara, N. Takagi, Y. Takeda, Y. Ohsugi, IL-6 receptor blockage inhibits the onset of autoimmune kidney disease in NZB/W F1 mice, Clin Exp Immunol 112 (1998) 397–402.

[29] C.Y. Tsai, T.H. Wu, C.L. Yu, J.Y. Lu, Y.Y. Tsai, Increased excretions of beta2-microglobulin, IL-6, and IL-8 and decreased excretion of Tamm-Horsfall glycoprotein in urine of patients with active lupus nephritis, Nephron 85 (2000) 207–214.

[30] A. Fukatsu, S. Matsuo, H. Tamai, N. Sakamoto, T. Matsuda, T. Hirano, Distribution of interleukin-6 in normal and diseased human kidney, Lab Invest 65 (1991) 61–66.

[31] M. Linker-Israeli, R.J. Deans, D.J. Wallace, J. Prehn, T. Ozeri-Chen, J.R. Klinenberg, Elevated levels of endogenous IL-6 in systemic lupus erythematosus. A putative role in pathogenesis, J Immunol 147 (1991) 117–123.

[32] G.G. Illei, C. Yarboro, Y. Shirota, E. Tackey, L. Lapteva, T. Fleisher, et al., Tocilizumab (Humanized Anti IL-6 Receptor Monoclonal Antibody) In Patients With Systemic Lupus Erythematosus (SLE): Safety, Tolerability And Preliminary Efficacy. In 70th Americal College of Rheumatology Annual Scientific Meeting, Tocilizumab, Washington, DC, 2006.

[33] G.J. Prud'homme, D.H. Kono, A.N. Theofilopoulos, Quantitative polymerase chain reaction analysis reveals marked overexpression of interleukin-1 beta, interleukin-1 and interferon-gamma mRNA in the lymph nodes of lupus-prone mice, Mol Immunol 32 (1995) 495–503.

[34] G.J. Prud'homme, Y. Chang, Prevention of autoimmune diabetes by intramuscular gene therapy with a nonviral vector encoding an interferon-gamma receptor/IgG1 fusion protein, Gene Ther 6 (1999) 771–777.

[35] Y. Chang, G.J. Prud'homme, Intramuscular administration of expression plasmids encoding interferon-gamma receptor/IgG1 or IL-4/IgG1 chimeric proteins protects from autoimmunity, J Gene Med 1 (1999) 415–423.

[36] E.C. Baechler, F.M. Batliwalla, G. Karypis, P.M. Gaffney, W.A. Ortmann, K.J. Espe, et al., Interferon-inducible gene expression signature in peripheral blood cells of patients with severe lupus, Proc Natl Acad Sci USA 100 (2003) 2610–2615.

[37] K.A. Kirou, C. Lee, S. George, K. Louca, M.G. Peterson, M.K. Crow, Activation of the interferon-alpha pathway identifies a subgroup of systemic lupus erythematosus patients with distinct serologic features and active disease, Arthritis Rheum 52 (2005) 1491–1503.

[38] T. Lovgren, M.L. Eloranta, U. Bave, G.V. Alm, L. Ronnblom, Induction of interferon-alpha production in plasmacytoid dendritic cells by immune complexes containing nucleic acid released by necrotic or late apoptotic cells and lupus IgG, Arthritis Rheum 50 (2004) 1861–1872.

[39] P. Blanco, A.K. Palucka, M. Gill, V. Pascual, J. Banchereau, Induction of dendritic cell differentiation by IFN-alpha in systemic lupus erythematosus, Science 294 (2001) 1540–1543.

[40] K.N. Schmidt, W. Ouyang, Targeting interferon-alpha: a promising approach for systemic lupus erythematosus therapy, Lupus 13 (2004) 348–352.

[41] N.J. Olsen, C.M. Stein, New Drugs for Rheumatoid Arthritis, N Engl J Med 350 (2004) 2167–2179.

[42] V.N. Foltyn, T.D. Golan, In vitro ultraviolet irradiation induces pro-inflammatory responses in cells from premorbid SLE mice, Lupus 10 (2001) 272–283.

[43] S. Zampieri, M. Alaibac, L. Iaccarino, R. Rondinone, A. Ghirardello, P. Sarzi-Puttini, et al., TNF-{alpha} is expressed in refractory skin lesions from subacute cutaneous lupus erythematosus patients, Ann Rheum Dis 65 (2006) 545–548. doi:10.1136/ard.2005.039362.

[44] N. Jacob, H. Yang, L. Pricop, Y. Liu, X. Gao, S.G. Zheng, et al., Accelerated pathological and clinical nephritis in systemic lupus erythematosus-prone New Zealand Mixed 2328 mice doubly deficient in TNF receptor 1 and TNF receptor 2 via a Th17-associated pathway, J Immunol 182 (2009) 2532–2541.

[45] M. Aringer, F. Houssiau, C. Gordon, W.B. Graninger, R.E. Voll, E. Rath, et al., Adverse events and efficacy of TNF-alpha blockade with infliximab in patients with systemic lupus erythematosus: long-term follow-up of 13 patients, Rheumatology (Oxford) 48 (2009) 1451–1454.

[46] M. Aringer, J.S. Smolen, Efficacy and safety of TNF-blocker therapy in systemic lupus erythematosus, Expert Opin Drug Saf 7 (2008) 411–419.

[47] R.P. Rother, C.F. Mojcik, E.W. McCroskery, Inhibition of terminal complement: a novel therapeutic approach for the treatment of systemic lupus erythematosus, Lupus 13 (2004) 328–334.

[48] D.T. Boumpas, H.A. Austin 3rd, B.J. Fessler, J.E. Balow, J.H. Klippel, M.D. Lockshin, Systemic lupus erythematosus: emerging concepts. Part 1: Renal, neuropsychiatric,

cardiovascular, pulmonary, and hematologic disease, Ann Intern Med 122 (1995) 940—950.

[49] Y. Wang, Q. Hu, J.A. Madri, S.A. Rollins, A. Chodera, L.A. Matis, Amelioration of lupus-like autoimmune disease in NZB/WF1 mice after treatment with a blocking monoclonal antibody specific for complement component C5, Proc Natl Acad Sci USA 93 (1996) 8563—8568.

[50] P. Hillmen, C. Hall, J.C.W. Marsh, M. Elebute, M.P. Bombara, B.E. Petro, et al., Effect of Eculizumab on Hemolysis and Transfusion Requirements in Patients with Paroxysmal Nocturnal Hemoglobinuria, N Engl J Med 350 (2004) 552—559.

[51] J. Anolik, I. Sanz, B cells in human and murine systemic lupus erythematosus, Curr Opin Rheumatol 16 (2004) 505—512.

[52] S. Akhtar, I. Maghfoor, Rituximab plus CHOP for diffuse large-B-cell lymphoma, N Engl J Med 346 (2002) 1830—1831. author reply 1830—1.

[53] J.C. Edwards, L. Szczepanski, J. Szechinski, A. Filipowicz-Sosnowska, P. Emery, D.R. Close, et al., Efficacy of B-cell-targeted therapy with rituximab in patients with rheumatoid arthritis, N Engl J Med 350 (2004) 2572—2581.

[54] P.P. Sfikakis, J.N. Boletis, G.C. Tsokos, Rituximab anti-B-cell therapy in systemic lupus erythematosus: pointing to the future, Curr Opin Rheumatol 17 (2005) 550—557.

[55] P.P. Sfikakis, J.N. Boletis, S. Lionaki, V. Vigklis, K.G. Fragiadaki, A. Iniotaki, et al., Remission of proliferative lupus nephritis following B cell depletion therapy is preceded by down-regulation of the T cell costimulatory molecule CD40 ligand: an open-label trial, Arthritis Rheum 52 (2005) 501—513.

[56] M. Tokunaga, K. Fujii, K. Saito, S. Nakayamada, S. Tsujimura, M. Nawata, et al., Down-regulation of CD40 and CD80 on B cells in patients with life-threatening systemic lupus erythematosus after successful treatment with rituximab, Rheumatology (Oxford) 44 (2005) 176—182.

[57] K. Saito, M. Nawata, S. Nakayamada, M. Tokunaga, J. Tsukada, Y. Tanaka, Successful treatment with anti-CD20 monoclonal antibody (rituximab) of life-threatening refractory systemic lupus erythematosus with renal and central nervous system involvement, Lupus 12 (2003) 798—800.

[58] J.E. Gottenberg, L. Guillevin, O. Lambotte, B. Combe, Y. Allanore, A. Cantagrel, et al., Tolerance and short term efficacy of rituximab in 43 patients with systemic autoimmune diseases, Ann Rheum Dis 64 (2005) 913—920.

[59] R.J. Looney, J.H. Anolik, D. Campbell, R.E. Felgar, F. Young, L.J. Arend, et al., B cell depletion as a novel treatment for systemic lupus erythematosus: a phase I/II dose-escalation trial of rituximab, Arthritis Rheum 50 (2004) 2580—2589.

[60] M. Ramos-Casals, M.J. Soto, M.J. Cuadrado, M.A. Khamashta, Rituximab in systemic lupus erythematosus: A systematic review of off-label use in 188 cases, Lupus 18 (2009) 767—776.

[61] T.Y. Lu, K.P. Ng, G. Cambridge, M.J. Leandro, J.C. Edwards, M. Ehrenstein, et al., A retrospective seven-year analysis of the use of B cell depletion therapy in systemic lupus erythematosus at University College London Hospital: the first fifty patients, Arthritis Rheum 61 (2009) 482—487.

[62] D.J. Burge, S.A. Bookbinder, A.J. Kivitz, R.M. Fleischmann, C. Shu, J. Bannink, Pharmacokinetic and pharmacodynamic properties of TRU-015, a CD20-directed small modular immunopharmaceutical protein therapeutic, in patients with rheumatoid arthritis: a Phase I, open-label, dose-escalation clinical study, Clin Ther 30 (2008) 1806—1816.

[63] F. Lajaunias, A. Ida, S. Kikuchi, L. Fossati-Jimack, E. Martinez-Soria, T. Moll, et al., Differential control of CD22 ligand expression on B and T lymphocytes, and enhanced expression in murine systemic lupus, Arthritis Rheum 48 (2003) 1612—1621.

[64] M. Lebwohl, S.K. Tyring, T.K. Hamilton, D. Toth, S. Glazer, N.H. Tawfik, et al., A novel targeted T-cell modulator, efalizumab, for plaque psoriasis, N Engl J Med 349 (2003) 2004—2013.

[65] N. Usmani, M. Goodfield, Efalizumab in the treatment of discoid lupus erythematosus, Arch Dermatol 143 (2007) 873—877.

[66] K.R. Carson, D. Focosi, E.O. Major, M. Petrini, E.A. Richey, D.P. West, C.L. Bennett, Monoclonal antibody-associated progressive multifocal leucoencephalopathy in patients treated with rituximab, natalizumab, and efalizumab: a Review from the Research on Adverse Drug Events and Reports (RADAR) Project. Lancet Oncol 10 (2009) 816—824.

[67] D.T. Boumpas, H.A. Austin III, B.J. Fessler, J.E. Balow, J.H. Klippel, M.D. Lockshin, Systemic lupus erythematosus: emerging concepts. Part 1: Renal, neuropsychiatric, cardiovascular, pulmonary, and hematologic disease, Annals of Internal Medicine 122 (1995) 940—950.

[68] D.J. Wallace, J.A. Tumlin, LJP 394 (abetimus sodium, Riquent) in the management of systemic lupus erythematosus, Lupus 13 (2004) 323—327.

[69] V. Strand, C. Aranow, M.H. Cardiel, D. Alarcon-Segovia, R. Furie, Y. Sherrer, et al., Improvement in health-related quality of life in systemic lupus erythematosus patients enrolled in a randomized clinical trial comparing LJP 394 treatment with placebo, Lupus 12 (2003) 677—686.

[70] D. Alarcon-Segovia, J.A. Tumlin, R.A. Furie, J.D. McKay, M.H. Cardiel, V. Strand, et al., LJP 394 for the prevention of renal flare in patients with systemic lupus erythematosus: results from a randomized, double-blind, placebo-controlled study, Arthritis Rheum 48 (2003) 442—454.

[71] S.N. Liossis, X.Z. Ding, G.J. Dennis, G.C. Tsokos, Altered pattern of TCR/CD3-mediated protein-tyrosyl phosphorylation in T cells from patients with systemic lupus erythematosus. Deficient expression of the T cell receptor zeta chain, J Clin Invest 101 (1998) 1448—1457.

[72] D. Vassilopoulos, B. Kovacs, G.C. Tsokos, TCR/CD3 complex-mediated signal transduction pathway in T cells and T cell lines from patients with systemic lupus erythematosus, J Immunol 155 (1995) 2269—2281.

[73] S. Krishnan, M.P. Nambiar, V.G. Warke, C.U. Fisher, J. Mitchell, N. Delaney, et al., Alterations in lipid raft composition and dynamics contribute to abnormal T cell responses in systemic lupus erythematosus, J Immunol 172 (2004) 7821—7831.

[74] Y. Li, T. Harada, Y.T. Juang, V.C. Kyttaris, Y. Wang, M. Zidanic, et al., Phosphorylated ERM Is Responsible for Increased T Cell Polarization, Adhesion, and Migration in Patients with Systemic Lupus Erythematosus, J Immunol 178 (2007) 1938—1947.

[75] M.C. Genovese, J.C. Becker, M. Schiff, M. Luggen, Y. Sherrer, J. Kremer, et al., Abatacept for rheumatoid arthritis refractory to tumor necrosis factor alpha inhibition, N Engl J Med 353 (2005) 1114—1123.

[76] B.K. Finck, P.S. Linsley, D. Wofsy, Treatment of murine lupus with CTLA4Ig, Science 265 (1994) 1225—1227.

[77] D.I. Daikh, D. Wofsy, Cutting edge: reversal of murine lupus nephritis with CTLA4Ig and cyclophosphamide, J Immunol 166 (2001) 2913—2916.

[78] H. Iwai, M. Abe, S. Hirose, F. Tsushima, K. Tezuka, H. Akiba, et al., Involvement of inducible costimulator-B7 homologous protein costimulatory pathway in murine lupus nephritis, J Immunol 171 (2003) 2848—2854.

[79] H. Iwai, Y. Kozono, S. Hirose, H. Akiba, H. Yagita, K. Okumura, et al., Amelioration of collagen-induced arthritis by blockade of inducible costimulator-B7 homologous protein costimulation, J Immunol 169 (2002) 4332—4339.

[80] Y.L. Hu, D.P. Metz, J. Chung, G. Siu, M. Zhang, B7RP-1 blockade ameliorates autoimmunity through regulation of follicular helper T cells, J Immunol 182 (2009) 1421–1428.

[81] S.L. Kalled, A.H. Cutler, S.K. Datta, D.W. Thomas, Anti-CD40 ligand antibody treatment of SNF1 mice with established nephritis: preservation of kidney function, J Immunol 160 (1998) 2158–2165.

[82] D.I. Daikh, B.K. Finck, P.S. Linsley, D. Hollenbaugh, D. Wofsy, Long-term inhibition of murine lupus by brief simultaneous blockade of the B7/CD28 and CD40/gp39 costimulation pathways, J Immunol 159 (1997) 3104–3108.

[83] G.S. Early, W. Zhao, C.M. Burns, Anti-CD40 ligand antibody treatment prevents the development of lupus-like nephritis in a subset of New Zealand black x New Zealand white mice. Response correlates with the absence of an anti-antibody response, J Immunol 157 (1996) 3159–3164.

[84] P.I. Sidiropoulos, D.T. Boumpas, Lessons learned from anti-CD40L treatment in systemic lupus erythematosus patients, Lupus 13 (2004) 391–397.

[85] D.T. Boumpas, R. Furie, S. Manzi, G.G. Illei, D.J. Wallace, J.E. Balow, et al., A short course of BG9588 (anti-CD40 ligand antibody) improves serologic activity and decreases hematuria in patients with proliferative lupus glomerulonephritis, Arthritis Rheum 48 (2003) 719–727.

[86] K.C. Kalunian, J.C. Davis Jr., J.T. Merrill, M.C. Totoritis, D. Wofsy, Treatment of systemic lupus erythematosus by inhibition of T cell costimulation with anti-CD154: a randomized, double-blind, placebo-controlled trial, Arthritis Rheum 46 (2002) 3251–3258.

[87] S. Krishnan, V.G. Warke, M.P. Nambiar, G.C. Tsokos, D.L. Farber, The FcR gamma subunit and Syk kinase replace the CD3 zeta-chain and ZAP-70 kinase in the TCR signaling complex of human effector CD4 T cells, J Immunol 170 (2003) 4189–4195.

[88] F.R. Bahjat, P.R. Pine, A. Reitsma, G. Cassafer, M. Baluom, S. Grillo, et al., An orally bioavailable spleen tyrosine kinase inhibitor delays disease progression and prolongs survival in murine lupus, Arthritis Rheum 58 (2008) 1433–1444.

[89] S. Krishnan, Y.T. Juang, B. Chowdhury, A. Magilavy, C.U. Fisher, H. Nguyen, et al., Differential expression and molecular associations of Syk in systemic lupus erythematosus T cells, J Immunol 181 (2008) 8145–8152.

[90] M.E. Weinblatt, A. Kavanaugh, R. Burgos-Vargas, A.H. Dikranian, G. Medrano-Ramirez, J.L. Morales-Torres, et al., Treatment of rheumatoid arthritis with a Syk kinase inhibitor: a twelve-week, randomized, placebo-controlled trial, Arthritis Rheum 58 (2008) 3309–3318.

Index